THE UBIQUITIN PROTEASOME SYSTEM IN THE CENTRAL NERVOUS SYSTEM: FROM PHYSIOLOGY TO PATHOLOGY

THE UBIQUITIN PROTEASOME SYSTEM IN THE CENTRAL NERVOUS SYSTEM: FROM PHYSIOLOGY TO PATHOLOGY

MARIO DI NAPOLI
AND
CEZARY WOJCIK
EDITORS

Nova Biomedical Books
New York

For permission to use material from this book please contact us:
Telephone 631-231-7269; Fax 631-231-8175
Web Site: http://www.novapublishers.com

LIBRARY OF CONGRESS CATALOGING-IN-PUBLICATION DATA

The ubiquitin proteasome system in the central nervous system : from physiology to pathology / Mario Di Napoli and Cezary Wojcik, (editors).

p. ; cm.

Includes bibliographical references and index.

ISBN-13: 978-1-60021-749-4 (hardcover)

ISBN-10: 1-60021-749-4 (hardcover)

1. Ubiquitin. 2. Central nervous system--Physiology. 3. Central nervous system--Pathophysiology. I. Napoli, Mario Di. II. Wójcik, Cezary, 1968-

[DNLM: 1. Central Nervous System--metabolism. 2. Central Nervous System Diseases--metabolism. 3. Proteasome Endopeptidase Complex--physiology. 4. Proteins--metabolism. 5. Ubiquitin-Protein Ligase Complexes--physiology. WL 300 U15 2007]

QP552.U24U2577 2007

612.8'2--dc22

2007016546

Published by Nova Science Publishers, Inc. ✢ New York

CONTENTS

PREFACE

The book focused on the role of ubiquitin protesome system (UPS) in central nervous system. Proteasomes are large multicatalytic proteinase complexes that are found in the cytosol and in the nucleus of eukaryotic cells with a central role in cellular protein turnover. The UPS has a central role in the selective degradation of intracellular proteins. In addition to serving as a means to rapidly eliminate short-lived regulatory proteins involved in cell cycle, cell growth, and differentiation, in periods of stress rapid elimination of denatured, misfolded and damaged proteins by the proteasome becomes a critical determinant of cell fate. These aspects are analyzed in central nervous system physiology and pathology.

Chapter 1 - Multiple critical cellular processes are regulated by maintaining the appropriate levels of proteins. Whereas *de novo* protein synthesis is a comparatively slow process, proteins are rapidly degraded at a rate compatible with the control of cell cycle transitions, signaling events and induction of cell death. The ubiquitin-proteasome system (UPS) plays a pivotal role in the degradation of short-lived and regulatory proteins important in a variety of basic cellular processes, including regulation of the cell cycle, modulation of cell surface receptors and ion channels, and antigen presentation. On the other hand the UPS also displays an important quality control function, removing abnormal proteins from the cytosol, the nucleus and the endoplasmic reticulum. The pathway involves an enzymatic cascade through which multiple 76–amino acid ubiquitin monomers are covalently attached via a three-step process to the protein substrate, which is then degraded by the 26S proteasome complex, a cylindrical organelle that recognizes ubiquitinated proteins, degrades the proteins, and recycles ubiquitin. It is now clear that regulated protein degradation by the UPS affects practically every cellular process. In the nervous system, ubiquitination plays a role, among others, in neuronal signaling, synapse formation and function, as well as, in various diseases. It is becoming increasingly evident that altered activities of the UPS are crucially involved in the pathophysiology of Parkinson's disease, Alzheimer's disease, Huntington's disease, prion diseases and in spinocerebellar ataxia, just to name a few. Protein degradation pathways are also targets for therapy as shown by the successful results obtained with the inhibitors of the 26S proteasome. Further work in this area holds great promise toward our understanding and treatment of a wide range of neurological disorders.

Chapter 2 - Between the 1950s and 1980s, scientists were focusing mostly on how the genetic code is transcribed to RNA and translated to proteins, but how proteins are degraded has remained a neglected research area. With the discovery of the lysosome by Christian de Duve, it was assumed that cellular proteins are degraded within this organelle. Yet, several independent lines of experimental evidence strongly suggested that intracellular proteolysis is largely nonlysosomal, but the mechanisms involved remained obscure. The discovery of the ubiquitin–proteasome system (UPS) resolved the enigma. The authors now recognize that degradation of intracellular proteins is involved in regulation of a broad array of cellular processes, such as cell cycle and division, regulation of transcription factors, and assurance of the cellular quality control. Not surprisingly, aberrations in the system have been implicated in the pathogenesis of human disease, such as malignancies and neurodegenerative disorders, which led subsequently to an increasing effort to develop mechanism based drugs.

Chapter 3 - Cells contain many different kinds of proteins, each fulfilling structural, functional, or regulatory roles. The presence of either damaged or mutated proteins, or of altered levels of normal proteins could cause pathological conditions and even cell death. Therefore, monitoring the state of all these proteins, as well as continuously adjusting their levels to suit demands, is paramount to survival. To exercise such quality control, cells are continuously spending energy both to synthesize new proteins, and to simultaneously degrade them, even though many may still be functional. An important characteristic of regulatory degradation is that it is specific; only the correct proteins are removed in a time-coordinated manner. Such extraordinary specificity is achieved by a modular system that identifies proteins that are to be degraded, marks them by covalently attaching ubiquitin to an amino residue, and finally proteolyses them into amino acids. This sequence of events is executed by the following components. Recognition of target proteins is carried out by a specific ubiquitin-protein ligase, called an E3. This protein recognizes the substrate and usually directs a ubiquitin-conjugating enzyme, an E2, to attach ubiquitin, a small 76 amino acid protein, onto an amino group on the substrate. Ubiquitin molecules are often added one to another, resulting in chains of ubiquitin extending from the protein targeted for degradation. These polyubiquitin conjugates are then shuttled to the 26S proteasome, a large ATP-dependent proteolytic complex, where they are degraded. Interestingly, ubiquitination is a reversible process, with deubiquitinating enzymes able to remove ubiquitin from the target before it can be recognized by the proteasome. Hence, transfer of the polyubiquitinated conjugate to the proteasome must happen swiftly or be shielded from these enzymes. The cumulative balance of these processes allows the ubiquitin-proteasome system to control the cellular levels and half lives of thousands of proteins making it a key player in basic biological pathways such as cell division, differentiation, signal transduction, trafficking, and quality control. Not surprisingly, aberrations in the system have been implicated in the pathogenesis of many diseases, certain malignancies, neurodegenerative disorders, inflammation and immune response. Understanding of the underlying mechanisms involved is important for the development of novel, mechanism-based drugs.

Chapter 4 - Covalent modification of proteins by ubiquitin is a key mechanism for the control of cellular processes as diverse as cell proliferation, differentiation, apoptosis. Deubiquitination, reversal of this modification, is catalyzed by deubiquitinating enzymes that belong to the superfamily of proteases. Deubiquitinating enzymes occupy the second largest family of enzymes in the ubiquitin system, implying their functions in the control of diverse cellular processes by regulating the fate, function of ubiquitinated proteins. Cellular functions of deubiquitinating enzymes include the regulation of proteasome activity, protein stability, signal transduction, DNA repair, chromatin dynamics, transcription, endocytosis. Deubiquitinating enzymes also play important roles in the processing of inactive ubiquitin precursors for the generation of matured ubiquitin monomers, the removal of ubiquitin from 'distal' end of poly-ubiquitination chains conjugated to proteins for controlling the fidelity of the ubiquitination process, the cleavage of ubiquitin adducts for the release of free ubiquitin. Deubiquitinating enzymes consist of five families that have distinct catalytic domain structures: the ubiquitin-specific protease (USP) family, the ubiquitin C-terminal hydrolase (UCH) family, the ovarian tumor protease (OTU) family, the Machado-Joseph disease protein (MJD) family, the Jab1/MPN/Mov34-domain protease (JAMM) family. While the JAMM family members are metalloproteases, the other family members are cysteine proteases. As the names of certain families imply, deubiquitinating enzymes play critical regulatory roles in a multitude of processes from cancer to neurodegenerative diseases. In this chapter, the authors summarize the catalytic properties of deubiquitinating enzymes so far been identified, the recent findings on their functions as cellular regulators. The authors also describe the specific features of deubiquitinating enzymes that are related with neuronal diseases.

Chapter 5 - Ubiquitin-like (UBL) domain proteins (UDPs) constitute a family of proteins with a modular architecture, which is characterized by an integral UBL-domain. Although members of the UDP family display a variety of different functions, many of them are on some level connected with the ubiquitin-proteasome system, a central pathway, which accommodates intracellular protein degradation in eukaryotic cells. While some UDPs are involved in substrate recruitment for the 26S proteasome, also a ubiquitin-specific hydrolase, an ER-membrane resident protein, a co-chaperone, and a ubiquitin ligase belong to this family. Several of these proteins have been implicated in the development of neurodegenerative diseases. Of the initially studied UDPs, most bound the proteasome in a UBL-dependent manner. Therefore it appeared that proteasome binding was a general feature of this protein family. However, evidence is accumulating that a number of UDPs also bind to other components of the ubiquitin pathway, while some appear not to bind the proteasome at all. Hence UDPs appear functionally more diverse than one would expect from their structural appearance. Here the authors provide insight into the UDP family and attempt to summarize what is known about their physiological role, especially with respect to neurodegenerative diseases. The authors come to the conclusion that, despite their striking structural similarity, UDPs display rather diverse binding features, and appear to be part of a sophisticated protein network within the ubiquitin system.

Chapter 6 - The 20S proteasome, a 700 kDa multicatalytic proteinase complex, is responsible for the extralysosomal protein degradation that occurs in the cytosol and nucleus of eukaryotic cells. It represents the proteolytic core of the 26S proteasome, a 2000 KDa elongated structure formed by the 20S capped, at each side, by the 19S regulatory complex (also called PA700). The 26S complex is involved in the ATP, ubiquitin-dependent and ubiquitin-independent proteolytic pathways. The proteasome constitutes up to 1% of protein in the cells and the free 20S proteasomes are the major portion of the total amount of proteasomes. Its molecular architecture is extremely conserved from archaebacteria to higher eukaryotes and is organized in four stacked 7-membered rings of α and β subunits, in a cylinder-like shape. The two inner rings are composed of β subunits, harbouring the active sites, flanked by the two outer rings made up of non-catalytic α subunits which regulate the substrate access through the opening of the outer ring and the binding of regulators. The 20S proteasome is a member of the N-terminal nucleophile (Ntn)-hydrolases family. Its N-terminal threonine residues are exposed as the nucleophile in peptide bond hydrolysis. The three β subunits, β1, β5 and β2 (also called Y/delta, X and Z, respectively) express the three catalytic activities, designated peptidyl-glutamyl peptide hydrolyzing, chymotrypsin-like and trypsin-like, based on the ability to cleave peptide bonds on the carboxyl side of hydrophobic, basic and acidic aminoacids, respectively. Furthemore, two additional activities cleaving bonds after branched chain and small neutral amino acids have been described and called branched chain amino acid preferring and small neutral amino acid preferring. They enable the 20S proteasome to degrade alone a wide variety of protein substrates: poorly folded or unfolded proteins and oxidized proteins characterized by an increased surface hydrophobicity. Under the influence of γ-interferon, a major immunomodulatory cytokine, vertebrate proteasomes assemble catalytically-active subunits, named β5i, β1i and β2i (also called LMP7, LMP2 and MECL1, respectively) which replace their constitutive homologues, β5, β1 and β2, respectively, and associate to a regulatory particle, PA28, (or 11S regulatory complex) also induced by γ-interferon. Such a complex has been demonstrated to be specialized in generating MHC class I antigenic peptides. This review focuses on recent progress concerning the structure, including the assembly pathway, and the enzymatic activities that are involved in physiological/pathological functions exerted by the eukaryotic 20S proteasomes in the cell.

Chapter 7 - 26S proteasomes are ~2.4 MDa supramolecular assemblies that function as protein degrading complexes in neuronal as well as other cell types. They constitute the final, common destination of the proteins degraded by the ubiquitin-proteasome pathway, and perhaps by some non-ubiquitin-dependent pathways as well. 26S proteasomes are formed by association of the core 20S proteasomes with one or two PA700 activators (19S caps). While the core 20S proteasomes harbor the proteolytic activities, the remaining features of 26S proteasomes are conferred by components of the PA700. Mammalian PA700 is composed of 18 subunits, including 6 AAA ATPases (Rpt1-6) and several non-ATPase subunits (Rpn1-3, Rpn5-12 and Uch37). PA700 is physically divided into the lid and base subcomplexes. PA700 allows the recognition of polyubiquitinated proteins, their attachment, unfolding, opening of the closed proteasomal 'gates' and translocation of the unfolded polypeptide chain of the substrate towards the central catalytic cavity of the proteasome. At the same time PA700 allows the release of free ubiquitin through at least two different deubiquitinating activities. All of these functions are coupled to the ATP-ase activity of the complex, making them highly susceptible to ATP depletion such as during episodes of limited hypoxia or ischemia. Moreover, under those conditions 26S proteasomes tend to separate into free 20S proteasomes and PA700 complexes. Besides the canonical 26S proteasome subunits, several proteins associate loosely with the 26S proteasome, including additional deubiquitinating enzymes, ubiquitin ligases and polyubiquitin binding and delivery factors.

Chapter 8 - The 26S proteasome has long been viewed as a major therapeutic target. However, in the past 20 years only inhibitors of the proteolytic sites have been developed. Such a focus was primarily the result of the limited availability of assays for monitoring activity of the 26S proteasome. Due to the difficulties in preparation of naturally polyubiquitinated proteins, these assays were based on artificial model substrates, typically monomeric proteins that could be either polyubiquitinated *in vitro* without a specific E3 ubiquitin ligase (lysozyme, DHFR, Ub-Pro-β-gal) or degraded without polyubiquitination (fluorogenic peptides, loosely structured caseine, denatured ovalbumin). Although these reagents proved invaluable in uncovering the basic principles of proteasomal function, it becomes increasingly clear that they did not allow one to address the puzzling complexity of the 19S cap composition, indispensable in the highly controlled and rapid ($T_{1/2}$ <5 min) degradation of naturally unstable regulatory proteins and signalling molecules. Recent development of *in vitro* assays for specific ubiquitination and degradation of natural substrates from yeast, pioneered by several research groups including ours, has turned the tide. As the authors will discuss these studies, although in their early stages, have already revealed an unanticipated complexity of the substrate recruitment mechanism and the catalytic cycle itself.

Chapter 9 - The ubiquitin-proteasome system is responsible for the majority of regulated intracellular protein degradation. The importance of this system is reflected in its involvement in a large number of biological processes such as cell cycle traverse, apoptosis, antigen presentation, circadian rhythms, protein quality control, etc., as well as many aspects of neuronal development and function. The proteolytic component of the system is the 26S proteasome, a large ATP-dependent enzyme that degrades proteins marked for destruction by polyubiquitin chains. The 26S proteasome is composed of two subunits, the 20S proteasome and the 19S regulatory complex. The 20S proteasome is a cylindrical particle composed of α and β subunits arranged as four heptameric rings housing the proteolytic activity. Indiscriminate protein degradation by the 20S proteasome is prevented by the fact that the active sites are sequestered within the central chamber of the cylinder, and substrate access is blocked by N-terminal extensions of the α subunits constituting the outer rings. For substrates to gain entry into the 20S proteasome, the α N-terminal extensions must be reoriented in order to open the axial channel leading into the proteasome catalytic chamber. This is accomplished by binding of the 19S regulatory complex and/or other proteasome activator complexes to one or both ends of the 20S cylinder. The 19S regulatory complex is a multiprotein structure that recognizes, unfolds and pumps polyubiquitylated substrates into the 20S catalytic core. Other identified proteasome activators include PA28α and PA28αβ that are involved in MHC class I antigen presentation, PA28γ, thought to be involved in apoptosis, and PA200 recently linked to DNA repair. Inhibitors of the 20S proteasome include Hsp90 and PI31 — which presumably compete with proteasome activators for binding the 20S proteasome α rings — PR39, a noncompetitive reversible inhibitor of proteasomes and PA28αβ-proteasome complexes, and many viral proteins that inhibit 20S and 26S proteasomes as well as other components of the ubiquitin-proteasome system. In addition to activators and inhibitors of 20S proteasomes, there are numerous proteins that bind to the 26S holoenzyme and modulate its activity. These include a host of ubiquitin-conjugating enzymes (E2s), ubiquitin-ligases (E3s), ubiquitin-chain elongation factors (E4s), isopeptidases, and an increasing number of polyubiquitin chain receptors thought to deliver polyubiquitylated substrates to the 26S proteasome. Finally, the HEAT repeat-containing protein Ecm29 has been proposed to function as an adaptor that links 26S proteasomes to protein quality control and endoplasmic reticulum-associated protein degradation pathways, endocytosis and vesicular trafficking, and transport processes through interaction of Ecm29 with molecular motors. In this review, the authors will discuss recent developments concerning the aforementioned proteasome-interacting proteins with emphasis on those proteins which likely function within the normal physiology and pathophysiology of the nervous system.

Chapter 10 - The accumulation of misfolded or damaged proteins causes the failure of normal cell structure and functions needed for growth and viability. The toxicity of misfolded species has been linked to human diseases, neurodegenerative disorders in particular, including Alzheimer's, Parkinson's and Huntington's diseases, characterized by the accumulation of intracellular aggregates or inclusion bodies. To interrupt this adverse development, defective proteins must be rapidly repaired by molecular chaperones or destroyed by energy-dependent cytoplasmic proteases. A balance among these processes ultimately maintains the cellular homeostasis. In eukaryotes, the 26S proteasome, a protease/chaperone complex that generally acts as an ubiquitination system, is a central component in the protein triage decision process, though it also selectively degrades structurally abnormal proteins in a ubiquitin-independent manner. In either case, all substrate proteins must undergo the structural changes and stabilization necessary for a rapid degradation. It has, therefore, often been suggested that several chaperone functions are closely related to the stimulation of proteasomal degradation. This chapter summarizes recent discoveries pertaining to chaperone activities in the ubiquitin-proteasome system, and to their regulation of protein breakdown mediated by the proteasome.

Chapter 11 - Proteasomes are present in the cytoplasm and in the nuclei of all cell types including neurons and glial cells. Their relative abundance within subcellular compartments is highly variable depending on the cell type as well as depending on physiological and/or pathological stimuli. Cytoplasmic proteasomes are not uniformly distributed; instead they are enriched at the centrosomes at the cytoskeletal networks and at the outer surface of membranaceous organelles such as the endoplasmic reticulum and the endosome system. In the nuclei proteasomes are present throughout the nucleoplasm but are usually not found within the nucleoli. Nuclear proteasomes often associate with structures of the nuclear matrix in particular with discrete subnuclear domains called the PML nuclear bodies (POD domains). PML bodies in the nucleus and the pericentrosomal area of the cytoplasm may function as proteolytic centers of the cell since they are enriched in components of the proteasome system. Under conditions of impaired proteolysis proteasomes and ubiquitinated proteins further accumulate at those locations. The knowledge about intracellular distribution of proteasomes is important for our understanding of the dynamic organization of the ubiquitin- and proteasome-dependent proteolysis within cells. The present chapter will review available information about subcellular localization of proteasomes with emphasis on neuronal and glial cells; however since most data were obtained in other cell types they will be reviewed as well whenever this is relevant.

Chapter 12 - The regulated degradation of a majority of cellular proteins is catalyzed by the ubiquitin-proteasome system (UPS). The catalytic engine at the center of the UPS is the proteasome, a large multi-subunit self-compartmentalized protease. The UPS is engaged in a variety of functions critical to protein quality control, including surveillance and elimination of misfolded proteins that are irreparably damaged and potentially toxic. Unfortunately, when the capacity of the UPS is exceeded, misfolded protein substrates accumulate and tend to self associate, a characteristic hallmark of a growing class of aggregation diseases such as Alzheimer's and Parkinson's. These aggregates are subsequently assembled through an active and regulated process to form aggresomes. Aggresomes are dynamic structures, formed in response to an overload of improperly folded proteins. Assembly of aggresomes occurs at the centrosome, a juxtanuclear structure with roles historically limited to microtubule organization as it relates to cellular division and directional motor protein-dependent transport within the cell. Moreover, studies characterizing aggresome formation have demonstrated that the centrosome serves as a site for the recruitment and concentration of UPS components, including the proteasome, its regulators and a host of other proteins involved in protein quality control. Therefore, in addition to other cellular activities, the centrosome may play a central role in protein quality control, spatially regulating critical processes in protein folding, degradation, and aggregation.

Chapter 13 - Proteins co-translationally inserted into the endoplasmic reticulum (ER) of neuronal cells must undergo proper folding, association and posttranslational modifications in order to assure their function in the different segments of the secretory pathway, plasma membrane and extracellualar space. A quality control (QC) system assures that only properly folded and assembled proteins exit the ER, while misfolded and aberrant proteins are degraded by a process known as ER-associated degradation (ERAD). The most studied form of ERAD relies on the activity of the cytosolic ubiquitin-proteasome system (UPS) and requires recognition and extraction (a.k.a. retrotranslocation or dislocation) of misfolded proteins prior to their degradation. Nevertheless, ample evidence exists for UPS-independent cytosolic and lumenal ERAD as well. ER and cytoplasmic chaperones assist in the retrotranslocation process, which is mediated through the Sec61 translocone or derlin-associated channels. Emerging substrates are ubiquitinated on the cytosolic face of the ER by a limited number of specialized E2s and E3s. Studies in yeast have implied in ERAD a cytosolic AAA ATP-ase called VCP (valosin-containing protein; Cdc48, p97, TER94) in association with the Ufd1-Npl4 dimer. Inefficient ERAD and/or increased load with misfolded proteins induce ER stress, which in turn activates a transcriptional response known as UPR (unfolded protein response). UPR involves transient attenuation of translation followed by increased synthesis of ER chaperones, components of ERAD and other proteins involved in relieving ER stress. However, persistent ER stress leads to the initiation of apoptosis. Inefficient ERAD contributes to the formation of cytoplasmic inclusions characteristic of neurodegenerative disorders, while acute ER stress is often associated with neuronal cell loss during cerebral ischemia. VCP is mutated in a hereditary dominant disorder known as inclusion body myopathy, Paget's disease of the bone and frontotemporal dementia (IBMPFD).

Chapter 14 - The proteasome is an intracellular multisubunit protease. It plays an important role in a myriad of intracellular processes and in removing misfolded proteins by degradation. Protein misfolding and aggregation are common to most neurodegenerative diseases, suggesting that abnormalities of protein homeostasis contribute to pathogenesis. Recent cell-based and genetic studies suggest that perturbations in the proteasome degradation mechanisms contribute to neurodegenerative disease processes Although, significant progress has been made in the understanding of substrate recognition and proteolysis, very little is known on the intermediate steps of transportation of misfolded protein into the proteolytic proteasome chamber and their effects on proteasome proteolytic activity. Mathematical models can help to understand proteasome dysfunction in neurodegenerative diseases. Using the in silico informatics work done on the specificity and kinetics of the proteasome molecules, different non-exclusive models of proteasome dysfunction can be predicted. If the influx and cleavage rate are impaired a low degradation rate is predicted, while if the cleavage rate is limiting, such as for substrates containing polyglutamine tracks, the degradation rate increases at the decrease of the influx rate. From these standpoints, misfolded protein accumulation in neurodegenerative disease could be due to a low cleavage capacity of catalytic subunits toward specific substrates, such as polyglutamine rich sequence proteins, with a consequent direct 'clogging' of proteasomes by undegraded longer fragments. An alternative hypothesis suggests an impaired transport into the proteolytic chamber, mainly due to a reduced capacity of proteasome 19S gating subunits to attach substrate recruitment fragments. ATP deficiency and steric modification of the substrates are envisaged as potential mechanisms generating such impairment. These theoretical studies, although in their early stages, have already revealed an unanticipated complexity of the dysfunction of proteolytic activity in neurodegenerative diseases at the level of substrate recruitment mechanism as well as at the level of the catalytic cycle itself.

Chapter 15 - The brain is a very complex structure. The complexity of the brain is necessary for it to carry out its varied functions, such as receiving and processing stimuli, learning and memory, effecting motor output, is critical to the survival of the organism. Gaining a complete understanding of the structure and function of the brain requires an understanding of its embryonic development. Many billions of neurons must differentiate into neurons and subsequently be wired up correctly and form functional synapses. These processes occur during neural development, which has a number of distinct stages: neurogenesis, axon guidance, arborisation and synaptogenesis that are multifaceted and tightly regulated. One mechanism for the regulation of intricate cellular processes, such as the cell cycle, synaptic plasticity and neural development, is through the control of protein levels via their synthesis and or degradation. The ubiquitin-proteasome system (UPS) is able to regulate protein levels by targeting specific proteins for proteasome-mediated proteolysis or for their endocytosis and subsequent lysosomal-mediated proteolysis. Evidence is now accumulating that these mechanisms are critical for the correct development of the nervous system during its distinct stages. A number of E3 ubiquitin ligases such as *Neuralized, ligand of num-protein X*, *Sel-10* and *mind bomb* have been identified as being required for regulating lateral inhibition and neurogenesis via the ubiquitin-mediated internalisation of the Notch and Delta transmembrane proteins critical for the differentiation of neurons. Genetic screens in *Drosophila* have identified components of the UPS such as *bendless*, a ubiquitin-conjugating enzyme (E2), *non-stop*, a ubiquitin-specific protease and *ariadne-1* which interacts with a novel ubiquitin-conjugating enzyme in the navigation of axons to their targets. Recently the UPS has been identified as mediating the chemotropic responses of *Xenopus* retinal growth cones to specific guidance cues such as netrin-1 and brain-derived neurotrophic factor, since proteasome inhibition blocks chemotropic responses to these cues and leads to increases in ubiquitin-protein complexes in growth cones. The UPS is also able to regulate axon guidance at the level of guidance cue receptors at the cell surface which may enable a growth cone to change its responsiveness to particular cues during it journey. For example, as growth cones cross the *Drosophila* midline the internalisation of the guidance cue receptor, *roundabout,* is controlled by *commissureless* interacting with DNedd4, an E3 ubiquitin ligase, resulting in the removal of Robo from the growth cone surface. Once they have reached their targets growth cones undergo morphological changes such as branch formation during arborisation and subsequently synaptogenesis. Recent studies suggest that the UPS is involved in branch retraction in mushroom body neurons during metamorphosis. Moreover, two E3 ligases, *highwire*, and the anaphase promoting complex (APC), a de-ubiquitinating enzyme *fat facets*, regulate synaptogenesis via a balance of ubiquitination and de-ubiquitination at the neuromuscular junction in *Drosophila.* These studies establish a fundamental role for the UPS in regulating the development of the nervous system via both ubiquitin-mediaded endocytosis and ubiquitin-mediated proteolysis.

Chapter 16 - The central nervous system (CNS) is the most complex structure known, and understanding its function is considered the 'final frontier' of biology. Neurons and their supporting glial cells form the cellular building blocks of the CNS. Individually these cells express a vast array of proteins that receive and transmit information; collectively integrating and processing this information allowing us to make sense of our environment and modify our behaviors. The process of protein degradation and turnover is essential for physiology of neuronal cells. The proteasome-ubiquitin system (UPS) plays central role in cytosolic proteolysis. It regulates the distribution of cell cycle phases, gene expression, transcription and antigen processing. In the CNS the proteasome pathway has additional functions which are due to physiology of the CNS. Many neuronal specific proteins interact with the 20S proteasome, moreover, brain proteasomes are thought to have a different proteolytic profile than proteasomes from other tissues. Proteasomes are present ubiquitously, but not homogenously, in CNS cells (both in neurons and in glia). The subcellular localization of proteasomes differs in various parts of the CNS. Both neurons and glia are able to induce immunosubunits of proteasome under inflammatory conditions. UPS also shapes the function, development, and plasticity of synaptic connections and UPS elements are recruited in post-synaptic densities (PDS). The UPS is also essential for Long-Term Facilitation and is required for the establishment of long-term memory, moreover the blockade of UPS produces full retrograde amnesia in rat. The knowledge of the UPS in particular areas of the CNS and of the function of this system in every neuronal cell, enables to draft a more complex vision on protein turnover in the CNS, but details of all the processes remain to be discovered.

Chapter 17 - Compared to the central nervous system (CNS), the involvement of the ubiquitin-proteasome system (UPS) in peripheral nerve biology has received less attention. Nonetheless, there are several emerging areas of peripheral nerve biology, as well as pathobiology, in which the ubiquitin-proteasome pathway appears to play critical roles. A major contribution to understanding the role of the UPS in nervous system function came from studies of the *Drosophila* neuromuscular junction. Overexpression of the deubiquitinating protease *fat facets* in motor neurons led to a profound disruption of synaptic growth control, indicating that a balance between ubiquitination and deubiquitination is key for synaptic development. Recent studies in a variety of neuronal model systems revealed that ubiquitination is a critical factor in basic presynaptic and postsynaptic mechanisms, as well as synaptic plasticity throughout the nervous system. In addition to the role in synaptic communication, the UPS is involved in controlling the survival and proliferation of peripheral glial cells, called Schwann cells. Similarly, the survival of cultured sympathetic neurons is also influenced by the UPS. In response to injury, peripheral nerves degenerate by an apoptosis independent mechanism known as Wallerian degeneration. This mechanism involves the fragmentation of axonal microtubules and the destruction of the axolemma. The application of proteasome inhibitors slows this degenerative process and profoundly delays Wallerian degeneration. Therefore, it is possible that proteasome inhibitors can help to maintain axonal integrity in neurodegenerative events where axonal degeneration is involved. The UPS has also been linked to at least two distinct hereditary disorders of the PNS, namely neurofibromatosis and demyelinating neuropathies. In neurofibromatosis, Schwann cell proliferation is deregulated leading to multiple tumors, mostly schwannomas. Mutated forms of the neurofibrinin 2 tumor suppressor gene are rapidly degraded by the proteasome and thought to contribute to a loss of function phenotype of the wild type allele. Peripheral myelin protein 22 (PMP22), also known as growth arrest specific gene 3, is a short-lived Schwann cell protein that is also degraded by the proteasome. Mutations within and duplication of the PMP22 gene are linked to demyelinating neuropathies, including Charcot-Marie-Tooth disease type IA. When PMP22 is mutated or overexpressed, the degradation of the protein is slowed leading to its accumulation in cytosolic aggregates, termed aggresomes. The formation of PMP22 aggresomes is associated with an impairment of UPS activity and the recruitment of proteasome substrates and chaperones to the aggregates. These events can disrupt the cell cycle of the Schwann cells and enhance neural dysfunction. The findings described here indicate that there has been a tremendous progress in our understanding of the role of the UPS in PNS biology. Significantly, findings from the PNS have provided ground-breaking insights into basic neural mechanisms, such as synaptic communication and axonal degeneration. The involvement of the UPS in peripheral nerve disease reveals another commonality between the CNS and PNS, and will help to speed the development of novel therapeutic strategies aimed at influencing protein turnover in the nervous system.

Chapter 18 - The ubiquitin-proteasome pathway (UPS) is understood to have a major role in the regulation of the cell cycle of dividing cells. More recently, it has become clear that even non-dividing cells, like neurons, rely on the specific regulated protein degradation pathways of the UPS. Through genetic studies as well as elegant biochemical and cell biological experiments it is now clear that the UPS plays an important role in neuron specific functions, such as pathfinding during development, neurodegeneration and synaptic plasticity. The regulation of synapse formation and re-formation is central to neuronal function. Recent work has demonstrated that the regulated destruction of specific components of the synapse is a crucial determinant of the long-term status of a neuronal connection. Moreover, it is clear that neurotransmitter stimulation of receptors can begin a cascade resulting in the targeted removal of specific synapse scaffold proteins. The permanent removal of these proteins through degradation allows for the rearrangement of synapse components in a manner specific to the neurotransmitter-mediated signal.

Chapter 19 - Organisms respond to acute environmental change by orchestrating a stress response to prevent further damage. The key aspect of the stress response is a reaction to any form of macromolecular damage that exceeds a set threshold, independent of the underlying cause. It is becoming clear that higher organisms developed a complex, sensitive and maybe equally important network of regulatory pathways, relying largely on protein interactions, post-translational modifications and proteolysis. Molecular chaperones help hundreds of signalling molecules to keep their activation-competent state, and regulate various signalling processes ranging from signalling at the plasma membrane to transcription. Furthermore, molecular chaperones recognize proteins of non-native structure, prevent them from irreversible intracellular aggregation, and then act with regulatory co-chaperones in the conversion of proteins to be properly folded and in a functional state, stabilizing the phenotypes of various cells and organisms. This may be related to their low affinity for the proteins they interact with, which means that they represent weak links in protein networks. However, not every non-native protein is folded successfully. Those proteins that are not accurately folded/refolded are then directed to the ubiquitin-proteasome system (UPS) for destruction. The UPS is the predominant nonlysosomal protein degradation pathway which insures the viability, proliferation and signaling of eukaryotic organisms. In periods of stress, rapid elimination of denatured, misfolded and damaged proteins by proteasomes also becomes a critical determinant of cell fate. Stress response including its associated oxidative, nitrosylative and energetic stresses underlie neurodegeneration and evoke a discreet set of transcriptional events which have a complex and interdependent relationship with proteasomal function. Both chaperones and proteasomes act jointly together for selective removal of proteins with aberrant structure so as to keep protein homeostasis in cells. Though the precise nature of the cooperative linkage between chaperone and UPS pathways remains largely elusive so far, how slowly folding or misfolded polypeptides are targeted for proteasomal degradation, accumulating evidence from *in vivo* and *in vitro* studies shed some light on the molecular mechanisms that link proteasomes and molecular chaperones. Generally, selection of proteins for degradation is mediated by E3 ubiquitin ligases of the mechanistically distinct HECT and RING domain sub-types. Recent studies suggest that the U-box protein family represents a third class of E3 enzymes. CHIP, a U-box-containing protein, is a degradatory co-chaperone of heat-shock protein 70 (Hsp70) and Hsp90 that facilitates polyubiquitination of chaperone substrates. This mechanism affords time for a separate set of stressor-specific adaptations, designed to re-establish cellular homeostasis, to take action. Finally, the disruption of this protein folding quality control results in the accumulation of non-native protein species that can form oligomers, aggregates, and inclusions indicative of neurodegenerative disease. Many neurodegenerative diseases such as amyotrophic lateral sclerosis, Alzheimer's disease, Parkinson's disease and polyglutamine diseases, to cite the well studied, are characterized by conformational changes in proteins that result in misfolding, aggregation and intra- or extra-neuronal accumulation of amyloid fibrils. A common feature may be the formation of off-pathway folding intermediates that are unstable, self-associate, and with time lead to a chronic imbalance in protein homeostasis acting on the correct functioning of molecular chaperones and the UPS with deleterious consequences on cellular function. This has led to a hypothesis that enhancement of components of the cellular quality control machinery, specifically the levels and activities of molecular chaperones and the UPS suppress aggregation and toxicity phenotypes to allow cellular function to be restored. A detailed understanding of the molecular basis of chaperone-UPS mediated protection against neurodegeneration might lead to the development of therapies for neurodegenerative disorders that are associated with protein misfolding and aggregation.

Chapter 20 - Glucocorticoid hormones exert a variety of effects on the brain and impact memory, anxiety, and CNS responses to stress. The action of these hormones is mediated primarily by soluble receptors, the corticosteroid or glucocorticoid receptors, which primarily act directly in the nucleus to regulate select networks of target genes. Multiple mechanisms account for the diversity of glucocorticoid action including cell and tissue-specific response of target genes, the differential action of receptor variants as well as complex interactions between a plethora of accessory factors that directly or indirectly modulate glucocorticoid receptor activity. Furthermore, many *in vitro* and *in vivo* model studies have revealed a relationship between expression levels of glucocorticoid receptors and cellular responsiveness to glucocorticoid hormone. Various intrinsic and extrinsic factors influence the expression of the glucocorticoid receptors and thereby impact cellular output to glucocorticoid hormone exposure. This review will focus on the current state of knowledge regarding the regulation of glucocorticoid receptor protein expression and highlight a number of recent studies that illustrate the critical importance of cellular maintenance of appropriate receptor levels in complex neuronal responses.

Chapter 21 - The apoptotic pathway and the ubiquitin- and proteasome system (UPS) are mutually dependent and interconnected. Active caspases are able to cleave proteasomal subunits leading to a decrease in proteasome activity, while proteasomes are able to degrade active caspases. Proteasome inhibitors easily induce apoptosis in rapidly cycling cells and in particular in malignant cells, however they are relatively well tolerated by differentiated, postmitotic cells such as neurons. Moreover, they confer neuroprotection likely through the induction of heat shock proteins and inhibit apoptotic and excitotoxic neuronal death triggered by different mechanisms. Nevertheless, at higher doses they eventually induce apoptosis. Some populations of neurons are more susceptible to the proapoptotic effects of proteasome inhibitors than other, in particular the dopaminergic neurons in the striatum. Proteasome dependent steps are located both in the induction and the execution stages of apoptosis, upstream and downstream from caspase activation. Multiple mechanisms of action may be involved such as inhibition of the cell cycle, oxidative stress, protein aggregation and inhibition of NFκB activation, just to name a few. The clinical use of proteasome inhibitors raises the question of their possible neurotoxic effect, which may surface when this class of drugs will be used in the treatment of chronic disorders such as rheumatoid arthritis. The relationship between the UPS and apoptosis in neuronal cells is therefore highly complex and far from being fully understood.

Chapter 22 - Unrepaired protein damage leads to the formation of lethal protein aggregates in cells and ultimately causes cell death. Protein damage accumulates in cells due to oxidative stress, transduction with prion particles with dominant conformations or due to genetic alterations in proteins that lead to formation of insoluble aggregates. All prokaryotic and eukaryotic cells possess two main strategies to counteract these changes and avoid the accumulation of protein aggregates. These pathways for protein quality control include: (1) the protein chaperone and refolding systems and (2) targeted proteolysis of the malfolding protein. In mammalian cell the molecular chaperones heat shock proteins 70 and 90 (Hsp70 and 90) appear to play key regulatory roles in protein triage after damage. These molecular chaperones can bind to malfolded proteins, deter the aggregation cascade and then target the protein substrates towards either: (1) the pathways of refolding by chaperonin- containing folding structures or (2) can promote ubiquination of its target through mechanisms involving the ubiquitin ligase CHIP and deliver the ubiquinated protein to the proteasome for degradation. Dysregulation of this system occurs during aging and is amplified during a range of degenerative disease states. Failure of this defense system may occur at many levels and decreased expression of proteins that mediate pathways 1 and 2 appears to be involved in aging, particularly of neuronal cells.

Chapter 23 - The causes of various neurodegenerative diseases, particularly sporadic cases, remain unknown, but increasing evidence suggests that these diseases may share similar molecular and cellular mechanisms of pathogenesis. One prominent feature common to most neurodegenerative diseases is the accumulation of misfolded proteins in the form of insoluble protein aggregates or inclusion bodies. Although these aggregates have different protein compositions, they all contain ubiquitin and proteasome subunits, implying a failure of the ubiquitin-proteasome system (UPS) in the removal of misfolded proteins. A direct link between UPS dysfunction and neurodegeneration has been provided by recent findings that genetic mutations in UPS components cause several rare, familial forms of neurodegenerative diseases. Furthermore, it is becoming increasingly clear that oxidative stress, which results from aging or exposure to environmental toxins, can directly damage UPS components, thereby contributing to the pathogenesis of sporadic forms of neurodegenerative diseases. Aberrations in the UPS often result in defective proteasome-mediated protein degradation, leading to accumulation of toxic proteins and eventually to neuronal cell death. Interestingly, emerging evidence has begun to suggest that impairment in substrate-specific components of the UPS, such as E3 ubiquitin-protein ligases, may cause aberrant ubiquitination and neurodegeneration in a proteasome-independent manner. This chapter provides an overview of the molecular components of the UPS and their impairment in familial and sporadic forms of neurodegenerative diseases, and summarizes present knowledge about the pathogenic mechanisms of UPS dysfunction in neurodegeneration.

Chapter 24 - The accumulation of unfolded, misfolded or damaged proteins in cells is a threat to cell survival. The ubiquitin-proteasome system (UPS) is responsible for the degradation of these abnormal proteins. UPS dysfunction has been postulated to play a key role in the pathogenesis of neurodegenerative diseases such as Alzheimer's and Parkinson's diseases. Both normal and misfolded proteins can undergo highly specific degradation by the UPS. Selective degradation of correctly folded proteins underlies many cellular regulations. Examples include degradation of cyclins or their inhibitors in the regulation of the cell cycle and the degradation of IκB in the activation of immunity responses. The endoplasmic reticulum (ER) is the site of synthesis of membrane proteins and secretory proteins. In the ER, defective or unfolded proteins are degraded [a process known as ER-associated proteins degradation (ERAD)], whereas correctly folded proteins are spared. In the familial form of Alzheimer's disease, transcriptional misreading of the stress-induced polyubiquitin gene produces ubiquitin with aberrant C-terminal extensions that competitively inhibit proteasomal function. This inhibition of UPS may impair ERAD, thereby causing the accumulation of misfolded proteins in the ER, resulting in ER stress and induction of cell death through the activation of calpain and caspase-3. Proteasome inhibitors such as lactacystin have been reported to activate the pro-apoptotic transcription factor C/EBP-homologous protein (CHOP) and to cause cell death in cultured cortical neurons. Although the inhibition of proteasomes has been linked to cell death, recent studies have shown that, below a threshold level, proteasome inhibition can activate neuroprotective responses — proteasome inhibition has been shown to induce various molecular chaperones such as heat shock proteins (HSPs) that increase cell tolerance to the accumulation of unfolded and damaged proteins, stimulate the expression of UPS components through a feedback mechanism, and suppress inflammatory responses by inhibiting IκB degradation. Although the inhibition of proteasomes may stimulate neuroprotective responses, prolonged ER stress ultimately leads to apoptosis. Further studies to elucidate the impact of proteasomal inhibition on other cellular signaling pathways may provide insights on the interplay between the UPS and cell physiology. A better understanding of the function and activation of the neuroprotective or pro-apoptotic responses would provide a means to manipulate this pathway in order to cure diseases associated with unfolded proteins. Much remains to be discovered about the inducibility and functioning of chaperones and neuroprotective ubiquitin-proteasome pathways in neurons. Such studies would be useful, since genetic polymorphism in these protective systems and changes in their expression with ageing may play critical roles in the accumulation of unfolded or damaged proteins, and in the pathogenesis of disease. Moreover, pharmacological induction or activation of these protein repair-and-degradative systems could in future be developed into innovative therapies for neurodegenerative diseases.

Chapter 25 - The ubiquitin-proteasome system (UPS) has a central role in the selective degradation of many intracellular proteins. Functional failure of the UPS may result in an abnormal accumulation of ubiquitinated, misfolded, aggregated, or oxidated proteins that should be removed from cells, finally resulting in cell death. Recent advances in genetic studies in familiar Parkinson's disease (PD) have provided important insight into the molecular pathways involved in disease pathogenesis. Proteins coded by the causal genes of familial parkinsonism, such as α-synuclein, parkin, ubiquitin carboxy-terminal hydrolase L1 (UCH-L1), and DJ-1, are possibly related to the ubiquitin-proteasome protein degradation system. Mutations in these proteins may interfere with normal protein degradation by UPS caused by 'loss of function' or 'gain of function'. A major constituent of Lewy bodies is aggregated α-synuclein. Mutant α-synuclein aggregates and resists degradation by the UPS, eventually disturbing normal cellular functions. Both parkin and UCH-L1 are components of the UPS that contribute to normal ubiquitination and de-ubiquitination mechanisms, respectively. Loss-of-function of parkin or UCH-L1 can interrupt normal protein degradation by the UPS. In addition, DJ-1 may also function to alleviate protein misfolding. The accumulated findings suggest the hypothesis that UPS failure and the subsequent proteolytic stress contribute to the etiopathogenesis that underlies dopaminergic neurodegeneration in both hereditary and sporadic PD. Experimental studies have revealed that dopaminergic neurons may be particularly vulnerable to proteasome inhibition *in vitro* and *in vivo*. Local administration of a proteasome inhibitor into the nigrostriatal system (the substantia nigra, striatum, or medial forebrain bundle) in rodents is sufficiently to induce dopaminergic neuron degeneration in the substantia nigra and the formation of α-synuclein-immunopositive intracytoplasmic inclusions similar to Lewy bodies. In addition, it has recently been reported that dopaminergic neurons progressively degenerate with Lewy-body-like inclusion formation following the systemic administration of proteasome inhibitors in rats. These experimental findings suggest that the inhibition of the UPS may be a common pathway for dopaminergic neuron death in PD, although further studies are required to establish the proteasome-inhibitor-induced PD model. If UPS failure is a key mechanism underlying dopaminergic neuron death, the next question, which is probably a more essential one, is why UPS failure occurs in dopaminergic neurons in PD. There must be primary, upstream events that affect UPS function. To date, several factors have been stressed in the pathogenetic mechanisms underlying dopaminergic neuron degeneration in PD, such as deficits in mitochondrial function, oxidative stress, neuroinflammation, and the accumulation of aberrant or misfolded proteins. Determining the principal molecular pathways that exaggerate UPS dysfunction will provide relevant clues to the understanding of the pathogenesis of sporadic and familial forms of PD. This chapter reviews the most recent advances in the authors knowledge on the relationships between UPS failure and dopaminergic neuron degeneration.

Chapter 26 - The ubiquitin-proteasome system (UPS) has multiple roles in axon degeneration. An efficiently functioning UPS is essential for maintenance of healthy axons, but the UPS is also required to activate pathways of programmed axon death designed to remove axons in injury, disease and development. Thus, as in the cell body (Chapter 24) the UPS is a double-edged sword for axons. Genetic defects in the UPS often cause progressive degeneration of synapses and distal axons. These structures appear more vulnerable than neuronal cell bodies to failure of normal protein turnover. Similarly, pharmacological blockade with proteasome inhibitors causes neurite death *in vitro* and peripheral neuropathy *in vivo*, suggesting long axons are critically dependent on a fully functioning UPS. Thus, axon degeneration could make an important contribution to neurodegenerative disorders where UPS defects are reported, such as Alzheimer's disease, Parkinson's disease and Huntington's disease. Prominent and early axon and synapse pathology has been reported in each of these disorders. The UPS also controls axon survival by regulating nuclear and axonal events. A chimeric nuclear protein containing an N-terminal region of multiubiquitination factor Ube4b delays the degeneration of injured axons in the slow Wallerian degeneration mutant mouse (*Wld^S*) for several weeks. The Ube4b domain of Wld^S protein is required for its neuroprotective effect *in vivo* and interacts with valosin containing protein (VCP/p97) within the nucleus, an event that may influence downstream axonal mediators of this phenotype. Proteasome inhibition in axons also delays Wallerian degeneration, possibly by preventing downregulation of the MEK/ERK pathway. Rapid Wallerian degeneration in wild-type axons seems to be a proactive, regulated process, similar in principle to apoptosis, albeit different in molecular details. Physical axon injury is not the only way to trigger this process because Wld^S delays also axon degeneration caused by several genetic and toxic insults. The common factor may be a blockade of anterograde axonal transport. Transport from the cell body may deliver an inhibitor of Wallerian degeneration that stops it from being triggered in healthy axons. Axonal pruning, the large-scale elimination of excess axon branches formed during development, also requires cell-autonomous action of the UPS. Cell specific deletion of genes encoding E1 and proteasome subunits in *Drosophila* blocks this process suggesting that the UPS degrades key regulators of the pruning process, or participates in the execution phase. There are interesting parallels with Wallerian degeneration, which also requires UPS activity in *Drosophila*. Both are proactive, cell-autonomous axon death programmes regulated in part by the UPS. However, at least some of the molecular details are distinct. Thus, several specific actions of the UPS participate in programmed degeneration of axons, whereas non-specific failure of the UPS can cause axon pathology. The apparent contradiction reflects the many roles of the UPS in the normal biochemistry of axons, roles that will be important to consider for effective targeting of therapeutic strategies.

Chapter 27 - Chronic pain states involve long-term biochemical and anatomical changes including plasticity at the first synapse in the spinal cord, which is crucial to the development of hyperalgesia (increased sensitivity to noxious stimuli) and allodynia (perception of innocuous stimuli as painful). Spinal dorsal horn neurons become hyperexcitable in the process of "central sensitisation", which shows partial similarity to other forms of synaptic plasticity, such as hippocampal long-term potentiation (LTP). Both processes involve pre- and post-synaptic changes, and rely on the NMDA receptor and associated proteins. The ubiquitin-proteasome system (UPS) has been implicated in central sensitisation and the development of neuropathic pain. In an animal model of neuropathic pain, proteasome inhibitors have been shown to rapidly attenuate behavioural hyperalgesia and allodynia, inhibited firing of dorsal horn neurons evoked by noxious and innocuous stimuli in neuropathic animals, or by mustard oil in normal animals. Expression of the enzyme UCH-L1 (Ubiquitin C-terminal hydrolase) was further increased in the spinal cord dorsal horn ipsilateral to neuropathy, supporting a central role for the UPS in neuropathic pain. Studies of other CNS areas have emphasised the importance of the UPS in regulation of synapse structure and neurotransmitter release and its role of such changes in plasticity. Postsynaptically, the UPS mediates changes in composition of the postsynaptic density (PSD) since activity-dependent ubiquitination regulates PSD composition and several key scaffolding molecules which are involved in pain sensitisation, undergo activity-dependent ubiquitination. PSD-95 in particular plays a key role in neuropathic pain. PSD-95 links to the NMDA receptor which is essential for central sensitisation and may regulate AMPA receptor synaptic insertion. NMDA receptor activation causes PSD-95 ubiquitination and degradation and blockade of this process prevents NMDA receptor induced AMPA-receptor recycling and long-term depression. Since alteration in levels of surface glutamate receptor expression is a key means by which synaptic strength is altered, these observations further support the idea of acute regulation of synapse function by the ubiquitin-proteasome pathway. In addition, proteasome inhibitors reduce NMDA receptor-dependent activation of the CREB and ERK/MAPK signalling cascades, indicating a further mechanism by which the UPS may influence long-term plasticity during chronic pain. The UPS is also involved in synapse development and in morphological changes in dendritic spines. Thus, a wide array of changes in protein: protein interactions, signalling events and cytostructural changes depend on UPS function during the synaptic plasticity that underlies chronic pain states. These processes may represent promising targets for the development of novel analgesic strategies.

Chapter 28 - Protein misfolding and aggregation are common to most neurodegenerative diseases, suggesting that abnormalities of protein homeostasis contribute to pathogenesis. Protein folding inside cells is assisted by various chaperones and folding factors, and misfolded proteins are eliminated by the ubiquitin-proteasome system (UPS) and macroautophagy to ensure high fidelity of protein expression. Under certain circumstances, misfolded proteins escape the degradation process, yielding to deposit of protein aggregates such as loop-sheet polymer and amyloid fibril. Dysfunction of the UPS or macroautophagy pathways might contribute in a wide variety of neurodegenerative diseases. Some proteins, when not properly degraded through the UPS, tend to form aggregates by binding to one another to form an insoluble structure that is very difficult to disassemble. Many of the components of neurodegenerative disease aggregates have been studied for their ability to form independent aggregates *in vitro* and *in vivo* and their biological activity described. Consistent with this view, protein aggregates have been regarded in a pathogenic connotation, with most aspects of neurologic pathogenesis being largely attributed to their presence in nerve tissues. However, the neurotoxicity of protein aggregates remains ambiguous as direct evidence substantiating it have long remained elusive. Primary UPS involvement in neurodegenerative diseases seems even more probable when the UPS is viewed not simply as an isolated degradation machine but rather as a complex cascade linked both to other degradation processes and to chaperone systems. In neurodegenerative diseases, perturbations of proteasome function may occur through its recruitment to or sequestration into protein aggregates, through chronic overloading of proteasome capacity by misfolded protein, or through still undetermined effects on other activities of the UPS. Collectively, these dysfunctions in proteasome activity could also be called *proteasomepathies* and differentiated from other specific protein degradation system disturbance. The identification of several degradation system disturbances will assist in choosing therapies when protein-specific disease-modifying treatments are available.

Chapter 29 - The ubiquitin-proteasome system (UPS) is a highly regulated and fundamental pathway for protein degradation. It controls many key cellular mechanisms critical for cell viability and function and also removes abnormal and toxic proteins generated by a lifetime of environmental damage. Most proteins are tagged by ubiquitin prior to degradation by the UPS. Notably, abnormal protein deposits containing ubiquitinated proteins are detected in a variety of neurodegenerative diseases. Such disorders include Alzheimer's disease, Parkinson's disease, amyotrophic lateral sclerosis and Huntington's disease, to name a few. Whether these protein deposits are pathogenic or represent a coping mechanism to prolong survival of the affected cells is a hotly debated issue. Lately, tremendous strides have been made to elucidate the mechanisms regulating the accumulation and aggregation of ubiquitinated proteins associated with neurodegeneration. This chapter provides a critical overview of the latest studies addressing these mechanisms. First, the authors will focus on oxidative stress. The brain is considered to be unusually sensitive to oxidative damage. Moreover, many age-related neurodegenerative disorders exhibit abnormal accumulation of oxidatively damaged proteins. Although there are many tantalizing clues indicating that the proteasome is crucial for the degradation of oxidatively modified proteins, the authors still lack a clear understanding of how these proteins are targeted for proteasomal degradation. The controversy is over the requirement of ubiquitination for the degradation of oxidatively modified proteins. Secondly, the authors will address the relationship between inflammation and UPS impairment. The brain was long considered to be an immunologically privileged site, particularly because of the blood brain barrier and the lack of a lymphatic system. However, more recently it has been shown that the brain mounts an inflammatory response, as noted from the occurrence of edema, microglia and astrocyte activation, local invasion of circulating immune cells and production of cytokines and other immune factors. There is abundant evidence supporting that an inflammatory reaction is mounted within the CNS following trauma, stroke, infection and seizures, all of which can augment brain damage. The authors will discuss how products of inflammation induce oxidative stress and the accumulation and aggregation of ubiquitinated proteins. A better understanding of the mechanisms that regulate the aggregation of ubiquitinated proteins is of clinical importance for developing therapeutic strategies to prevent and treat neurodegenerative diseases. One of the major challenges that we are faced with is to single out the UPS as a therapeutic target for preventing neurodegeneration. This challenge rests on developing therapeutic strategies that will enhance degradation of abnormal and toxic proteins without compromising the normal function of the UPS.

Chapter 30 - Tau proteins belong to the family of Microtubule-Associated Proteins (MAPs). They are mostly expressed in neurons where they act on the assembly and the stability of tubulin. Depending on their phosphorylation state, tau proteins will modulate the polymerisation and the stability of microtubules within the axon. The primary function of tau and its differential phosphorylation enable tau to be involved in neurite outgrowth and in axonal transport. Hence, these proteins appear important for the physiology and the normal function of a neuron. However, tau proteins are also the major constituents of intraneuronal and glial inclusions described in Alzheimer's disease (AD) and many other related disorders called tauopathies. They are thought to be directly linked to the progression of neurodegeneration. For instance, the gravity of the symptoms observed in AD was shown to be closely related to the progression of the '*tau pathology*'. The recent discovery of mutations within the *tau* gene has strengthened the role of tau in the neurodegenerative processes observed in all these disorders. Indeed, the presence of certain mutations within tau can lead to its intraneuronal and intraglial aggregation and the death of the cells affected. In all these diseases tau proteins are abnormally modified when aggregated. In particular, molecular analysis revealed that this protein is hyperphosphorylated in its filamentous state. This implies that some kinases or phosphatases are involved in the abnormal processing of tau and may be responsible for its aggregation and the subsequent neurodegeneration. The intracellular accumulation of tau may also induce its hyperphosphorylation and fibrillogenesis. Inhibition of tau protein proteolysis may then be one of the mechanisms involved in its intracytoplasmic accumulation. In AD or Parkinson's disease (PD), it was shown that the activity of proteasome was inhibited. In its aggregated form, tau is also ubiquitinated suggesting that the proteasome can be involved in the degradation of tau protein. A major question concerning the proteasome impairment in AD and its involvement in the degradation of tau is its place in the cascade of events leading to tau aggregation and neurodegeneration. On one hand, a proteasome defect could contribute to the failure of the clearance of tau inclusions noticed in the disease as observed by the presence of ubiquitinated tau in the neurofibrillary tangles. On the other hand, the deficiency of this proteolytic system can also lead to tau protein accumulation, hyperphosphorylation, ubiquitination and finally to its intraneuronal aggregation.

This review focuses on the recent advances made in the understanding of the relationships between tau protein malfunction, its hyperphosphorylation and the ubiquitin-dependent proteasomal degradation of tau in AD and other tauopathies. The challenge is to pinpoint the role of the proteasome in the cascade of events leading to neurodegeneration.

Chapter 31 - Parkinson's disease (PD) is a movement disorder characterized by the selective loss of dopaminergic (DA) neurons in substantia nigra. It is generally believed that a combination of environmental and genetic factors underlie the selective death of these DA neurons and ensuing locomotor symptoms. Significant breakthroughs in human genetic studies have recently led to the identification of several genes linked to PD. Among these genes, α-synuclein, parkin, UCH-L1 and DJ-1 play diverse roles in ubiquitin-dependent proteolysis by the 26 S proteasome, while the functions of PINK1 and LRRK2 are still largely unknown. Pathogenic mutations of α-synuclein enhance its propensity to misfold and aggregate. Parkin has a protein-ubiquitin E3 ligase activity towards a variety of substrates. When parkin is mutated, accumulation of its substrates may significantly contribute to the demise of dopaminergic neurons. Ubiquitin Carboxyl-terminal Hydroxylase L1 (UCH-L1) is a brain-specific deubiquitinating enzyme, whose catalytic activity is significantly reduced by its mutations found in a few PD cases. Among the many functions of DJ-1, its ability to counteract reactive oxygen species appears to be critically involved in PD. Mutations of DJ-1 greatly affect its stability and dimerization, as well as interactions with a variety of proteins including parkin. The disparate functions of these PD-linked proteins all fall within the framework of how the cell handles misfolded and aggregated proteins. An emerging common theme is that unfolded or misfolded proteins, if not promptly removed through ubiquitin-dependent proteolysis by the 26 S proteasome, may induce Unfolded Protein Response (UPR). Cells activate the UPR program to increase the production of proteins that help to handle misfolded or unfolded proteins. Another cellular program that is operating in parallel and perhaps connected with UPR is the ability of the cell to accumulate misfolded proteins in the form of aggresomes. The formation of a single large inclusion at the centrosome area greatly minimizes the impact of dispersed aggregates of misfolded proteins. Increasing evidence suggests that the cell may activate autophagy to degrade proteins in the aggresome. Thus, a unifying theme may connect the diverse functions of several PD-linked genes to the three sets of overlapping, interdependent cellular programs: UPR, Aggresome Formation and Autophagy. Cell death could be triggered when these protective mechanisms fail. The greatest challenge is to understand the selectivity of cell death. The general functions of the ubiquitin-dependent proteolysis system (UPS), as well as those of the three protective cellular programs, must be considered within the unique cellular and physiological context of dopaminergic neurons to ultimately answer the question. Superimposed on this complexity is the impact of various toxins implicated in PD. Although this multilayered view may be an oversimplification of the involvement of UPS in PD, it allows us to summarize many pieces of seemly unrelated information into a coherent model that can be tested experimentally. Future explorations under this framework would provide us a more comprehensive view on the molecular and cellular basis of PD.

Chapter 32 - Huntington's disease (HD) belongs to a group of nine polyglutamine (polyQ) tract disorders, which also includes spinocerebellar ataxias (SCA's) types -1, 2, 3, 6, 7, and 17, spinobulbar muscular atrophy (SMBA) and dentatorubral-pallidoluysian atrophy (DRPLA). The proteins involved in each of these disorders show no homology to one another except for an expanded polyQ tract. Although each protein is ubiquitously expressed throughout the central nervous system (and most non-neural tissues), only a distinct subset of neurons is affected in each disease, with only partial overlap between each. A common feature of these diseases is the formation of polyQ-containing intraneuronal inclusions, which are typically also immunoreactive for ubiquitin. However, the pathogenesis of these diseases is unknown, and there is much debate as to whether it is the inclusions themselves that are pathogenic or whether they are merely markers of disease. One suggestion, on the basis of numerous studies showing co-localisation of various other proteins with the inclusions, has been that the inclusions contribute to pathogenesis, interfering with normal cellular functioning by trapping components such as transcription factors, molecular chaperones, and components of the ubiquitin-proteasome system (UPS), and thus preventing them from carrying out their normal functions. However, this theory is disputed, with other studies suggesting that the inclusions may in fact be a form of cellular defense. Other evidence against a toxic role for aggregates includes the short-stop HD animal model, where increased inclusion formation accompanies decreased neuronal death, and SCA-1 models in which the protein is mutated so as not to form aggregates but toxicity is still seen. As the inclusions in these polyQ diseases are ubiquitinated, the role of the UPS in their pathogenesis has come under scrutiny. While some studies show that the function of the UPS is impaired in these disorders, other studies report no loss of function. Any impairment of the UPS may relate to difficulty with degradation of expanded or perhaps just aggregated polyQ proteins, although the evidence for such difficulty is also conflicting. It has been reported that UPS components are sequestered irreversibly into aggregates of polyQ-containing huntingtin (Htt) fragments. These Htt fragments were incompletely degraded, and a stable interaction between the polyQ-bearing proteins and the proteasome was seen. Such a stable interaction with non-degradable, aggregated polyQ proteins might result in depletion of proteasomal activity. In support of this suggestion, other studies have reported that UPS impairment is seen in the presence of aggregated polyQ-bearing proteins, and that this is evident in both the cytoplasmic and the nuclear compartments, even when the aggregated protein sequences are targeted to either the nucleus or cytoplasm alone. However, this impairment is also seen in cells where there are no detectable aggregates or toxicity, suggesting that UPS overload may not be a factor in neurotoxicity. Furthermore, an animal model of SCA-7 has shown that while there is neuronal damage in susceptible cells, the UPS remains functional in these neurons. Here we review the conflicting evidence from previous studies of the UPS in polyQ disorders, and discuss both the possible roles of the UPS in the pathogenesis of these diseases and the effect of inclusion formation on the UPS.

Chapter 33 - Prion diseases are a group of neurodegenerative diseases that affect humans and animals. They are distinct from other neurodegenerative disorders in that they can be infectious as well as familial or sporadic. Prion diseases are characterized by long incubation periods prior to onset of symptoms, and the pathology is limited to the central nervous system consisting mainly of vacuolation in neuronal cell bodies, neuronal cell death, deposition of protein aggregates, and astrocytosis. Prion diseases were originally classified as slow viral infections; however, there is mounting evidence to support the claim that the infectious unit is a protein. A small endogenous protein, the prion protein (PrP^C; c: cellular), is a key factor in these diseases. The expression of PrP^C is highest in neurons and its precise physiological role is not clear. It is processed through the secretory system to the plasma membrane where it is predominantly an extracellular glycosyl-phosphatidyl-inositol anchored protein that contains one disulfide bond and it is di-glycosylated. The protein aggregates detected in diseased individuals contain a structurally altered protease resistant form of PrP^C, called PrP^{Sc} (Sc: scrapie). PrP^{Sc} is though to be the major part of the infectious unit. The neurotoxic mechanisms behind neuronal death in prion diseases are not clear. Loss of functional PrP^C and/or PrP^{Sc} toxicity have been suggested; however, PrP^C knockout mice are apparently normal, suggesting that loss of PrP^C is not the major cause, and toxic effects of PrP^{Sc} are limited to PrP^C expressing tissue. Therefore, alternate pathways of neurotoxicity have been proposed, e.g., transmembrane forms of PrP^C, and an interplay between the ubiquitin-proteasome system (UPS) and cytosolic PrP^C or cytosolic PrP^{Sc}. As with other neurodegenerative diseases the UPS has been linked with prion diseases. There are, for example, reports of ubiquitinated PrP^{Sc}, increased polyubiquitin expression, and impaired proteasome activity in prion disease and disease models. The majority of PrP^C is topologically located in the secretory system and the extracellular space. However, there are reports describing a small subset of PrP^C in the cytosol, cytosolic PrP^C, where it is subject to efficient ubiquitin-proteasome degradation. This subset of PrP^C can either arise from inefficient translocation into the endoplasmic reticulum (ER) or retrotranslocation from the ER via the ER-associated-degradation (ERAD) pathway. Cytosolic PrP^C has a tendency to aggregate if proteasome activity is inhibited. Initial reports of toxic effects of cytosolic PrP^C on cells prompted speculation whether impairment of cytosolic PrP^C degradation by the UPS due to, e.g., PrP^C mutations or perturbed ubiquitin-proteasome activity, with a resulting rise in cytosolic PrP^C concentration, could explain some of the neurotoxicity in prion diseases. However, the effects of cytosolic PrP^C between studies are in conflict, with some studies reporting toxic effects and others reporting neuroprotective effects. A recent study of the effect of mild proteasome inhibition on viability in scrapie cell-culture models has shown that cytosolic aggresome formation of PrP^{Sc}, rather than PrP^C, caused apoptosis, suggesting that accumulation of cytosolic PrP^{Sc} due to impairment of the UPS could be an important factor in the neurotoxic mechanisms at work in prion diseases.

Chapter 34 - The proteasome plays a pivotal role during proteolytic processing of cellular proteins required for the generation of antigenic peptides presented to cytotoxic T cells by major histocompatibility complex class I molecules. The process of peptide generation is greatly improved by formation of immunoproteasomes through replacement of three β subunits with β1i (also called LMP2), β5i (LMP7) and β2i (MECL-1) and expression of PA28, a heptameric activator complex. The assembly of immunoproteasomes is stimulated by interferon-γ, a cytokine that is produced shortly after viral infection and is one of the mediators that link innate and adaptive immune responses. Numerous infectious microorganisms developed sophisticated strategies to avoid presentation of their antigenic peptides including production of proteasome-modulating molecules. Recent studies indicate a unique mechanism of epitope generation by proteasomes referred to as peptide splicing. In the brain, microglial cells are the major antigen presenting cells. However, during inflammation virtually all cells in the central nervous system can be induced to express immunoproteasomes and to present antigens in association with MHC class I molecules. The proteasome-mediated generation of peptide epitopes evolutionarily serves to present antigens derived from intracellular infectious microorganisms. However, proteolytic processing of intracellular proteins is by no means selective and includes processing of all proteins including self molecules that can become targets for cytotoxic T cells during several inflammatory and degenerative central nervous system diseases, such as multiple sclerosis and paraneoplastic neurological disorders. The role of ubiquitin-proteasome pathway in neuroinflammatory disorders extends beyond antigen processing for MHC class I presentation. Activation of NF-κB a key modulator of inflammatory reaction results from proteasomal degradation of its inhibitor – IκB. Proteasomes are also involved in the regulation of the activity of other transcription factors involved in the inflammatory responses including STAT proteins and Egr-1. Understanding of the underlying mechanisms involved in proteasome-mediated inflammatory processes is important for the development of novel, mechanism-based drugs.

Chapter 35 - The proteasome is involved in a number of critical intracellular processes. A major function of the proteasome is non-lysosomal degradation of intracellular proteins, in particular defect or damaged proteins. Targeted proteins are attached to ubiquitin, a process which is catalysed by three enzymes, E1 – E3. In human brain, the activity of proteasome is varying in different regions and with age. The normal function of proteasome in the nervous system is as essential as in other tissues, and perhaps even more, due to the limited capability of renewal of neurons and glial cells. Indeed, inhibition of proteasome alone has been shown to induce neuron death in vitro. There is considerable interest concerning the role of the ubiquitin-proteasome pathway in pathological neurodegeneration. This is partly due to the observation that proteasome activity decreases with normal aging of the brain. The main reason, however, is the accumulation of disease-related proteins in aggregates within neurons or glial cells that is a major feature of many neurodegenerative diseases, as Parkinson's disease, Huntington's disease, amyotrophic lateral sclerosis and Alzheimer's disease. Besides proteolysis, another vital task of the proteasome is the processing of intracellular proteins to be presented by the MHC class I molecules to cytotoxic T lymphocytes on the surface of the cell. Antibodies to proteasome have been identified in patients with autoimmune diseases as systemic lupus erythematosus and Sjögren syndrome. Proteasome antibodies have recently also been identified in serum from patients with immune-mediated neurological diseases, as multiple sclerosis (MS) and paraneoplastic cerebellar degeneration (PCD). In addition, autoreactive T cells to proteasome have been identified in MS patients. The presence of circulating proteasome antibodies suggests a more widespread affection of the immune system than indicated by the organ-specific nature of MS and PCD. A humoral response to proteasome can be triggered by the elevated proteasome levels that are found in some autoimmune diseases secondary to tissue damage. Primary damage of cells is a plausible explanation for an immune response targeted to an intracellular organelle. PCD is associated with a systemic tumour, and apoptosis of dying tumour cells could result in cross-presentation of intracellular antigens to the immune system and evoke immune responses, whereas the mechanism of initiation of systemic immune responses to proteasome in MS is unknown. The functional role of antibodies to proteasome in chronic inflammatory neurological disease remains to be determined. If the antibodies are indeed of pathogenic importance, as has been shown for antibodies to intracellular targets in SLE, the action of these antibodies might mimic the action of synthetic proteasome inhibitors. Such inhibitors exert their action through disturbance of protein breakdown, inhibition of antigen presentation and inhibition of proliferation, as well as the induction of apoptosis. In animal models, proteasome inhibitors also have potent anti-inflammatory effects. Thus, immune responses to proteasome in chronic inflammatory disease can potentially be both harmful, through interference with normal proteasome function, and beneficial, by suppressing inflammation. This review article aims to evaluate the current literature on antibodies to proteasome in neurological diseases, and to discuss the potential importance of these responses.

Chapter 36 - In higher eukaryotic cells, the 26S proteasome is the central component of the ubiquitin-proteasome system (UPS), in which it provides for the degradation of cytoplasmic and nuclear proteins, usually tagged with ubiquitin oligomers, and their resolution into short peptides. This pathway is involved in the control of a large array of cellular processes including protein turnover, digestion of damaged, mutant and viral proteins, cell cycle regulation, cell division, differentiation and development. Furthermore, it is also implicated in DNA repair, stress, immune and inflammatory responses, apoptosis, cell surface receptor modulation, transcription factor processing and activation, etc. Proteins belonging to different molecular pathways playing an important role in glioma progression or regression may undergo degradation or processing via the UPS, and consequently be inactivated, or conversely activated following proteasome inhibition. In GBM there is a striking shift of the balance constitutive/immunoproteasome towards the latter; paralleled by depression of the chymotrypsin-like activity. This is in opposition to its expected enhancement, being this activity higher in the immunoproteasome with respect to the standard proteasome. A better understanding of this discrepancy as well as of the enhanced apoptosis associated with proteasome inhibition, as observed in GBM, could be helpful in designing novel therapeutic strategies.

Chapter 37 - Pituitary tumors are usually benign lesions, but their tumorigenic process may constitute a model of the initial stages of carcinogenesis. Two major theories have been subject to most investigation: hormonal (usually hypothalamic factors) and/or growth factor over-stimulation, or a molecular defect within the pituitary itself. Oncogenes and tumor suppressor genes involved in other types of tumor do not appear to play a major role in the pathogenesis of pituitary tumors. In addition, germline genetic disorders, in which pituitary tumors are a common feature, have not shed much light on the pathogenesis of the more common sporadic tumors. An increasing number of reports point to deregulation of the cell cycle in these tumors, while transgenic disruption of the cell cycle machinery frequently leads to pituitary tumors in animal models. Cell cycle progression during G1, S and G2 phases is normally regulated by the fluctuation in the concentration of cyclins, cyclin-dependent kinases (CDKs) and their inhibitors, while securin, separin and cohesin regulate progression through M phase. This is mainly achieved through the programmed degradation of these proteins within the ubiquitin-proteasome system (UPS), but also by transcriptional regulation and subcellular compartmentalization. Alterations of these processes result in uncontrolled proliferation, aneuploidy and tumorigenesis. Aberrations of one or more components of the pRb/p16/cyclinD1/CDK4 pathway have been shown in 80% of pituitary tumors. The authors have shown that low levels of nuclear p27 in human pituitary tumors associate with increased degradation of the protein through the UPS. Human securin, identified as the product of pituitary tumor transforming gene (PTTG), is over-expressed in human pituitary tumors. This can cause aneuploidy and inhibition of p53 actions towards cell cycle arrest, DNA repair and apoptosis. PTTG also contributes to pituitary tumorigenesis by modulation of angiogenesis. Degradation of PTTG is ubiquitin-dependent and promotes the initiation of anaphase and exit from mitosis. Incomplete PTTG degradation through the anaphase-promoting complex/cyclosome (APC/C) secondary to PTTG over-expression results in doubling of chromosome numbers. Whether the cell cycle changes reported in pituitary tumors are truly causal remains uncertain and it is more likely that alteration in signaling pathways feed into the cell cycle which then executes an aberrant set of instructions that result in cell proliferation. Excessive regulatory hormone stimulation can lead to an increased number of cells in the pituitary in various physiological or pathological states. Animal models also provide data that in the presence of excessive hypothalamic hormone stimulation, adenoma formation can occur. Hormonal (usually hypothalamic factors) and/or growth factor over-stimulation of the pituitary is dependent on signaling through membrane and/or nuclear receptors. A number of these receptors such as protein G- coupled receptors, tyrosine-kinase receptors, growth hormone, glucocorticoid and estrogen receptors are down-regulated via degradation through the ubiquitin proteasome system. Various anomalies of receptor expression observed in pituitary tumors may be explained through excessive or incomplete degradation, which may then cause aberrant signaling in different proliferative pathways to result in tumor formation. Increasing research in the field of ubiquitin-proteasome degradation of various proteins involved in pituitary proliferation is likely to provide new insights into pituitary tumorigenesis.

Chapter 38 - One of the most common neurological disorders today is epilepsy. Epilepsy is a chronic brain disease characterized by recurrent, spontaneous seizures resulting from abnormal synchronization of neurons in the central nervous system. Seizures can stem from a variety of brain insults including head trauma, fever, illness, and electroconvulsive shock. However, one of the most important factors governing seizure susceptibility appears to be genetic predisposition. Epilepsy often results from inheritance of one or a combination of several predisposing genetic factors that disturb the balance of excitatory and inhibitory neural networks in the brain. More than 70 genes have been linked to epilepsy from work done on inherited disorders in humans, mice, and fruit flies. These genes encode a wide variety of products ranging from ion channel proteins to tRNAs. Recently, a relationship has begun to emerge between epilepsy and genes of the ubiquitin-proteasome system (UPS). The UPS is the molecular machinery responsible for the degradation of cytoplasmic proteins in the cell. A protein is marked for proteolytic processing in the UPS by the addition of ubiquitin molecules that target the protein to the proteasome, a multisubunit complex that reduces it to small peptides and amino acids. Defects in UPS genes have been linked to epilepsy and altered seizure susceptibility in humans, mice, dogs, and most recently in fruit flies. The two human UPS genes linked to epilepsy are *UBE3A* and *EPM2B*. These genes both encode E3 ubiquitin ligase proteins, the enzymes directly responsible for mediating the transfer of ubiquitin to substrates to mark them for proteasomal degradation. Defects in the human *UBE3A* gene lead to Angelman syndrome, a complex genetic disease marked by epilepsy in conjunction with other neurological manifestations such as mental retardation and ataxia. Mutation of the human *EPM2B* gene causes a severe and ultimately fatal form of progressive myoclonus epilepsy known as Lafora disease, which is characterized by the occurrence of starchy inclusion bodies within cells. Similar to humans, disruption of the homologs of *UBE3A* and *EPM2B* in animal models leads to altered seizure susceptibility and epilepsy. Finally, the most recent addition to the list of epilepsy-related UPS genes is *mei-P26* in the fruit fly *Drosophila melanogaster*. Mutation of the *mei-P26* gene has recently been shown to drastically decrease seizure susceptibility, essentially curing epilepsy in *Drosophila* models of the disease. The protein encoded by *mei-P26* resembles E3 ubiquitin ligases but this function has not yet been tested biochemically. Although the mechanisms by which these genes regulate seizure susceptibility are still under investigation, it is becoming increasingly clear that UPS-related genes play a critical role in the etiology of epilepsy and human seizure disorders.

Chapter 39 - Ischemic stroke is caused by obstruction of blood flow to the brain, resulting in energy failure that initiates a complex series of metabolic events, ultimately causing neuronal death. Cell death occurs by a necrotic pathway characterized by either ischemic/homogenizing cell change or edematous cell change. Death also occurs via an apoptotic-like pathway that is characterized, minimally, by DNA laddering and a dependence on caspase activity and, optimally, by those properties, additional characteristic protein and phospholipid changes, and morphological attributes of apotosis. Death may also occur by autophagocytosis. This review is directed at understanding how the ubiquitin-proteasome system (UPS) participates in global and focal cerebral ischemia. These are the two principal rodent models for human disease. Proteasomes are large multicatalytic protease complexes that are found in the cytosol and in the nucleus of eukaryotic cells with a central role in cellular protein turnover. The UPS is the predominant nonlysosomal protein degradation pathway which insures the viability, proliferation and signaling of eukaryotic organisms. Overwhelming data exists implicating a critical role of the UPS in cerebral ischemic injury. Ischemic and hypoxic trauma and their associated oxidative, nitrosylative and energetic stress underlie neurodegeneration following stroke and evoke a discreet set of transcriptional events which have a complex and interdependent relationship with proteasomal function. Rapid elimination of denatured, misfolded and damaged proteins by the proteasome becomes a critical determinant of cell fate. Proof of principle has been obtained from animal models of cerebral ischemia in which proteasome inhibitors reduce neuronal and astrocytic degeneration, cortical infarct volume, infarct neutrophil infiltration, and nuclear factor-κB (NF-κB) immunoreactivity. This neuroprotective efficacy has been observed when proteasome inhibitors have been used 6 hours after ischemic insult. Strategies aimed at effecting long lasting changes in proteasomal function are not recommended given the growing body of evidence implicated long term proteasomal dysfunction in chronic neurodegenerative disease. These effects are likely due to the fact that the UPS is also essential for cellular growth, metabolism and repair. These effects of proteasomal inhibition make development of short lived proteasome inhibitors or compounds which can spatially and temporally regulated the UPS desirable clinical targets. Preclinical studies in animal models indicate that the use of specific proteasome inhibitors may be valuable in treating a host of acute neurological disorders including ischemic stroke. Proteasome inhibition could be a potential treatment option for stroke.

Chapter 40 - The ubiquitin-proteasome system (UPS) is of key importance in the degradation of misfolded/abnormal proteins, viral proteins and many short-lived proteins that play vital roles in cell proliferation, differentiation, apoptosis and inflammatory processes. Therefore, both the ubiquitin - protein conjugation system and the 26S proteasomes constitute important target for pharmacological intervention. A number of small molecule inhibitors that target the 20S proteasome are described in the literature as possible anti-cancer and anti-inflammatory agents. Among them, a dipeptidyl boronic acid bortezomib (Velcade®, PS-341) is currently used in clinical practice for the treatment of relapsed or refractory myeloma, while MLN-519, a synthetic analog of microbial lactacystin, is under clinical evaluation for the treatment of ischemic cerebral stroke. Novel highly selective inhibitors of the 20S proteasome that bind noncovalently to the substrate binding-sites only and that exhibit less cytotoxic effects against normal cells, have recently been generated by combinatorial chemistry or identified by high-throughput screening of the pharmaceutical company's compound archives. Furthermore, a number of agents used in conventional therapies, dietary chemopreventive compounds and toxins block 26S proteasome-dependent protein degradation, either by inhibiting the 20S proteolytically active subunits or by regulating the expression or function of non-proteolytic subunits of this complex. Moreover, selected small molecules function as 20S proteasome activators and may be considered as potential therapeutics in pathological states resulting from the loss of proteasome activity, such as neurodegenerative disorders. Finally, more recently particular attention has been focused on small molecules that could block protein degradation at the level of their ubiquitination or could direct disease- promoting proteins for ubiquitination and degradation. These groups of modulators are currently being tested in preclinical settings for their therapeutic potential in cancer and inflammatory diseases.

The chapter summarizes the current knowledge on the chemistry of synthetic and natural inhibitors and modulators of the UPS, their mode of action and potential therapeutic relevance in the therapy of various human diseases with special respect to the central nervous system (CNS) pathology.

Chapter 41 - The ubiquitin-proteasome system (UPS), aside from a major degradation pathway of intracellular proteins, involves the modulation of key proteins that control cellular physiology through cell cycle regulation, immune response, and activation of gene expression. The core enzymatic molecule of the UPS is the 20S proteasome. Alterations in the proteasome proteolytic pathway have been contributed to protein alterations associated with aging and, in fact, dysregulation of the UPS has been linked to several disease states including neurodegenerative diseases (i.e. Parkinson's, Alzheimer's, and Huntington's disease), malignancies, and inflammatory-related diseases. As such, strong preclinical data now exist supporting the use of reversible proteasome inhibitors to treat a variety of disease states including cancer, rheumatoid arthritis, asthma, psoriasis, autoimmune encephalomyelitis, myocardial infarction, and ischemic brain injury. Currently, the proteasome inhibitor Velcade® is approved for treatment of multiple myeloma. Phase I safety trials have also been completed with the proteasome inhibitor MLN-519, at doses capable of reducing blood proteasome activity by 80%. Experimental studies with MLN-519 have indicated significant neuroprotective treatment effects in animal models of ischemia/reperfusion injury at doses that reduce blood 20S proteasome activity by 40-80%. Following focal ischemic brain injury in rats, treatment with MLN-519 has been associated with a significant reduction of brain infarction along with improved neurological outcome and electrophysiological brain activity as evaluated up to two weeks post-injury. Importantly, MLN-519 exhibited a wide therapeutic treatment window with a delayed initial treatment of up to 6-10 h post-injury. The therapeutic efficacy has been linked to an attenuation of aberrant gene expression; in particular, studies with MLN-519 have indicated that treatment of ischemic brain injury in rats is associated with a reduction of the nuclear factor (NF)-κB mediated neuro-inflammatory response, where, following injury, MLN-519 treatment has been shown to reduce activated NF-κB immunoreactivity and attenuate the increase in both cytokine (TNF-α, IL-1β, and IL-6) and cellular adhesion molecule (ICAM-1 and E-selectin) expression. MLN-519 also provided dramatic reductions of both neutrophil and macrophage infiltration into the injured rat brain. Similar anti-inflammatory effects of proteasome inhibition have been observed in other experimental inflammatory disease models as well. The aim of this chapter is to review the experimental and clinical data relating to the role of the proteasome in CNS disorders and to evaluate the potential use of proteasome inhibitors to treat CNS disease.

Chapter 42 - The proteasome is an enzyme, which is present within all cells, from yeast to humans. It has a central role in the proteolytic degradation of the vast majority of intracellular proteins. Among the key proteins modulated by the proteasome are those involved in controlling inflammatory processes, cell cycle regulation, and gene expression. Agents that inhibit the proteasome have been shown to be active in numerous animal models of inflammation and cancer. Two proteasome inhibitors are under clinical evaluation. MLN-519 is being studied for the treatment of reperfusion injury that occurs following cerebral ischemia and myocardial infarction. The other, Bortezomib (Velcade®), has recently been licensed for the clinical treatment of multiple myeloma. It is also undergoing further evaluation for the treatment of chronic lymphocytic leukemia and a variety of solid tumors. The proteasome may also have an important role in the evolution of HIV-related disorders including AIDS and inflammatory disorders. Therapeutic strategies using proteasome inhibitors for the treatment of these conditions have now entered preclinical development. MLN-519 is a small-molecular-weight lactacystin analogue developed by Millenium (LeukoSite) for the potential treatment of inflammatory disease and stroke using a novel ubiquitin proteasome enzyme inhibitor approach. The reperfusion that follows an ischemic event provides both positive and negative factors that affect the overall outcome of the cerebral tissue. The ischemic endothelium upregulates the expression of cell adhesion molecules, which then attract the circulating leukocytes. Once bound to the endothelium, these cells diapedese into the tissue and are responsible for the destruction and much of the subsequent tissue damage. MLN-519 attenuates the expression of these cellular proteins, reduces the invasion of leukocytes and hence limits tissue damage. MLN-519 has demonstrated a neuroprotective effect in rat models of middle cerebral artery temporary occlusion. MLN-519 reduces infarct volume, brain edema and increases neurological recovery with a reported therapeutic window of at least 6-hours. These effects are associated with a temporary reduction of circulanting 20S proteasome activity (70-80%), with a reduced leukocyte infiltration and decreased nuclear factor-κB activation. Similar protective results have also been reported in experimental myocardial infarction models in rats and pigs: MLN-519 protects cardiac tissue from ischemia and maintains its functionality as demonstrated by preserved left ventricular developed pressure and contractile function. These data demonstrate substantial clinical value, since many patients are admitted to the hospital hours after the stroke or heart attack has occurred and reperfusion has begun. That inhibition of the proteasome can be of benefit under these clinically-relevant conditions demonstrates their potential in these common life-threatening diseases. An explorative phase I trial has demonstrated that MLN-519 is well tolerated by healthy subjects at levels that are maximally neuroprotective in experimental conditions. It is currently undergoing further evaluation for clinical trials in acute stroke and myocardial infarction.

Chapter 43 - Bortezomib (Velcade®) is a dipeptide boronic acid proteasome inhibitor that specifically targets the chymotryptic-like proteolytic activity of the 20S proteasome. It has shown great potential as a novel anti-cancer agent and has been approved for the treatment of multiple myeloma. Therapeutic development as a single agent or in combination with other agents is ongoing in hematological malignancies as well as in various solid tumor types. Furthermore, application of proteasome inhibition therapy in other areas of disease is being explored, such as prevention of reperfusion injury following acute ischemic stroke and management of chronic inflammatory diseases. Bortezomib is generally well tolerated. However, one of the most serious overall as well as dose limiting toxicities has been peripheral neuropathy. Bortezomib-induced peripheral neuropathy constitutes a length-dependent, sensory rather than motor, axonal, small rather than large fiber, polyneuropathy. In agreement with its small fiber neuropathy characteristics, neuropathic pain and symptoms of autonomic dysfunction have also been frequently reported upon bortezomib treatment. Risk factors for bortezomib neurotoxicity include pre-existent neuropathy and prior treatment with neurotoxic (anti-cancer) agents. Additionally, individual susceptibility, and not only cumulative dose, is of great importance. After discontinuing bortezomib therapy, neuropathy resolves in approximately half of the patients. Nevertheless, in severe cases, pharmacologic management of (autonomic) neuropathy and neuropathic pain is required up to several months after discontinuation of bortezomib. The exact biological mechanism of peripheral neuropathy induced by systemic proteasome inhibition therapy still has to be elucidated. Detrimental effects of proteasome inhibition on nerve terminal protein homeostasis as well as myelin production by Schwann cells might explain the high incidence of neurotoxicity in bortezomib-treated patients.

In: The Ubiquitin Proteasome System... ISBN 978-1-60021-749-4
Eds: Mario Di Napoli and Cezary Wojcik, pp. 1-14

Chapter 1

FOCUSING ON THE UBIQUITIN PROTEASOME SYSTEM IN NERVOUS SYSTEM

***Mario Di Napoli*[1,2,*] *and Cezary Wójcik*[3,¥]**

[1]Neurological Section, SMDN – Center for Cardiovascular Medicine and Cerebrovascular Disease Prevention, Sulmona (L'Aquila) I-67039, Italy;
[2]Neurological Service, *San Camillo de'Lellis* General Hospital, Rieti I-02100, Italy;
[3]Department of Anatomy and Cell Biology, Indiana University School of Medicine, Evansville, IN 47712, USA.

ABSTRACT

Multiple critical cellular processes are regulated by maintaining the appropriate levels of proteins. Whereas *de novo* protein synthesis is a comparatively slow process, proteins are rapidly degraded at a rate compatible with the control of cell cycle transitions, signaling events and induction of cell death. The ubiquitin-proteasome system (UPS) plays a pivotal role in the degradation of short-lived and regulatory proteins important in a variety of basic cellular processes, including regulation of the cell cycle, modulation of cell surface receptors and ion channels, and antigen presentation. On the other hand the UPS also displays an important quality control function, removing abnormal proteins from the cytosol, the nucleus and the endoplasmic reticulum. The pathway involves an enzymatic cascade through which multiple 76–amino acid ubiquitin monomers are covalently attached via a three-step process to the protein substrate, which is then degraded by the 26S proteasome complex, a cylindrical organelle that recognizes

* Correspondence concerning this article should be addressed to Dr. Mario Di Napoli, MD; SMDN – Center for Cardiovascular Medicine and Cerebrovascular Disease Prevention, Neurological Section, Via Trento, 41 I-67039 – Sulmona (AQ) Italy. Phone: +39 (0)864 52716; Fax: +39 (0)864 52716; Email: mariodinapoli@katamail.com.

¥ Correspondence concerning this article should be addressed to Dr. Cezary Wójcik, MD, PhD; Department of Anatomy and Cell Biology, Indiana University School of Medicine, 8600 University Blvd., Evansville, IN 47712, USA. Phone: +1 (812) 461-5437; Fax: +1 (812) 465-1184; E-mail: cwojcik@iupui.edu.

ubiquitinated proteins, degrades the proteins, and recycles ubiquitin. It is now clear that regulated protein degradation by the UPS affects practically every cellular process. In the nervous system, ubiquitination plays a role, among others, in neuronal signaling, synapse formation and function, as well as, in various diseases. It is becoming increasingly evident that altered activities of the UPS are crucially involved in the pathophysiology of Parkinson's disease, Alzheimer's disease, Huntington's disease, prion diseases and in spinocerebellar ataxia, just to name a few. Protein degradation pathways are also targets for therapy as shown by the successful results obtained with the inhibitors of the 26S proteasome. Further work in this area holds great promise toward our understanding and treatment of a wide range of neurological disorders.

Keywords: protein degradation, ubiquitin, proteasome, ubiquitin-proteasome system, protein misfolding disease.

Abbreviations

ALS, amyotrophic lateral sclerosis; AMP, adenosine monophosphate; AP2, amyloid ß-protein; ATP, adenosine triphosphate; CNS, central nervous system; DUB, deubiquitinating enzyme;E1, ubiquitin activating enzyme; E2, ubiquitin conjugating enzyme; E3, ubiquitin ligating enzyme; ER, endoplasmic reticulum; ERAD, endoplasmic reticulum associated degradation; HD, Huntington's disease; HSPs, heat shock proteins; MHC, major histocompatibility complex; MPC; multicatalytic proteinase complex; PD, Parkinson's disease; PKC; protein kinase-C; PolyQ, polyglutamine diseases; PPi, inorganic pyrophosphate; PS2, presenilin-2 membrane protein; SUMO, small ubiquitin-relayed modifier; UCH-L1, ubiquitin C-terminal hydrolases-L1; UPS, ubiquitin-proteasome system; VCP, Valosin Containing Protein.

Introduction

Protein degradation was for a long time disregarded as a topic of little importance for medicine and physiology in general and for neuroscience in particular. Proteolysis took a back seat while gene expression and signal transduction held center stage as the dominant regulatory mechanisms. However, cellular proteins exist in a state of a dynamic equilibrium, being constantly degraded and renewed in order to keep high quality necessary for the proper performance of their functions. When we look each day in a mirror and see our face, in reality we see every day a different person, since the proteins making up our cells undergo a constant turnover. Proteins differ in their stability and longevity within cells: some have a half-life of seconds, while others have a half-life of days. Moreover, external stimuli and internal clocks can affect the stability of many proteins. Completely blocking protein degradation inevitably leads to cell death, demonstrating that regulated intracellular degradation of proteins is essential for cell survival.

It was generally accepted for a long time that cell proteins end up in the lysosomes, regarded as a cellular *trash can*, where they are degraded in exoergic reactions by an array of acid hydrolases. Protein degradation was considered an unimportant and boring area of study, where nothing new can be expected. Although lysosomal enzymes were implicated in protein degradation, the notion was abandoned when it was discovered that lysosomal inhibitors like leupeptin, antipain, and chymostatin were inactive as inhibitors of basal protein breakdown. Moreover, protein turnover occurred in cells that lacked lysosomes such as reticulocytes. It was first shown more than half a century ago that degradation of some enzymes requires adenosine triphosphate (ATP), implying that an active, energy-requiring process is responsible for protein turnover. In the 1970s Avram Hershko and Aaron Ciechanover have discovered an ATP-dependent proteolytic system in reticulocyte lysates. It required conjugation of a small polypeptide, that was essential for degradation but lacked any proteolytic activity on its own. It was subsequently shown that this 7-kd factor, later identified as ubiquitin, was ligated to larger proteins and targeted them for ATP-dependent degradation, a process known as ubiquitination (or ubiquitylation). The search for the protease degrading polyubiquitinated proteins ended when it was discovered that it is a huge 26S complex, which at its core has a 20S particle, previously characterized by Sherwin Wilk and Marian Orlowski as the multicatalytic proteinase complex (MPC) and described earlier by Klaus Scherrer as the prosome, a ribonucleoprotein particle. In 1988 the identity of both particles was confirmed and the name '*proteasome*' was proposed to replace former nomenclature (MPC, prosome, macropain, etc.).

During the decade of 1980s and early 1990s the research on this field continued at a slow pace: the crowd gathered at the few scientific meetings dedicated to ubiquitin and proteasome seldom exceeded a hundred. Finally, by mid-1990s, the scientific community realized the importance of the ubiquitin-proteasome system (UPS) in various aspects of the function and structure of cells and tissues, including its role in the nervous system. Over the last decade the number of reports on the role of UPS in nervous system found annually in Medline has steadily grown (Figure 1). The field has received its ultimate recognition when Avram Hershko, Aaron Ciechanower and Irwin Rose have been awarded the 2004 Nobel Prize in Chemistry. Gossips are in the air that a possibility of a second Nobel Prize in the field does exist, this time in Medicine or Physiology. It is now clear that regulated protein degradation by the UPS is a critical component of numerous cellular processes in all eukaryotes, including the cell cycle, cell growth and differentiation, embryogenesis, apoptosis, signal transduction, DNA repair, regulation of transcription and DNA replication, transmembrane transport, endocytosis, stress responses, antigen presentation and other aspects of the immune response. The well-known physiological functions of UPS in the nervous system include circadian rhythm regulation, axon guidance and acquisition of memory. Therefore, if you are a neuroscientist working in basic research or a clinician working in clinical neurology the chances are very high that at some stage of your practice you will encounter some aspect of the UPS.

For many years it has been thought that activity of the UPS is limited to the cytosol and the nucleus, where its components are located. However, recent experimental evidence has demonstrated that membrane-anchored and even secretory pathway-compartmentalized proteins are also targeted by the UPS. The fact that ubiquitin- and proteasome-dependent proteolysis directs so many cellular processes may not seem all that surprising. Proteolysis leads to a rapid and irreversible destruction of proteins; it can thereby dismantle existing cellular programmes and allow new ones to take root: it is not surprising that the UPS is involved in the regulation of many basic cellular processes. Protein ubiquitylation /deubiquitylation can now be seen to be as important as protein phosphorylation /dephosphorylation in the control of macromolecular mechanisms in the cell.

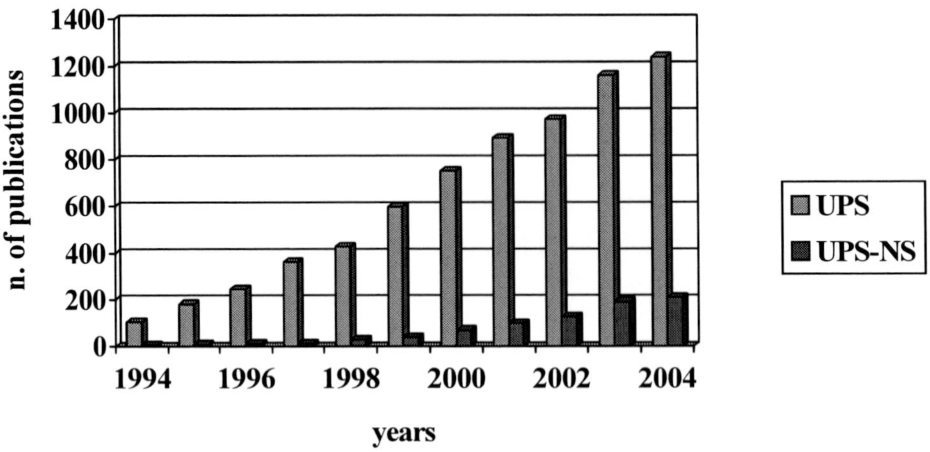

Figure 1. The number of publications examining the role of ubiquitin-proteasome system (UPS) in nervous system (NS). Number of reports published on Medline between 1994 to 2004 using the search strategies UPS and the combination UPS and NS (UPS-NS).

Components of the UPS

At the very core of the UPS is the E1-E2-E3 ubiquitination cascade (Figure 2), which eventually delivers the polyubiquitinated protein to the 26S proteasome for degradation into short peptides and free ubiquitin. The latter is recycled within the cell. While at first glance simple, the ubiquitination cascade is in fact very complex, taking into account that there are over 700 different E3s in the human genome, grouped into several families. Many of the E3 are assembled from modular subunits and are regulated by a different posttranslational modification with the ubiquitin-like protein called Nedd8 (neddylation). Selection of the proteins to be degraded by the different E3s depends on multiple criteria, including their primary amino acid sequence, partial cleavage by other proteases, phosphorylation status, modification by yet another ubiquitin-like protein called SUMO (sumoylation), association with other proteins, etc. Moreover, sometimes the E1-E2-E3 cascade does not suffice for an efficient ubiquitination of a substrate and an additional factor called E4 or polyubiquitin chain elongation factor is required for efficient protein degradation. According to the model depicted on Figure 2, 26S proteasomes bind polyubiquitinated substrates. This is true, however for the efficient degradation of many proteins additional '*delivery*' or '*shuttle*' factors such as valosin-containing protein (VCP, p97) or Rad23 are required for their delivery or handling to the proteasome. The ubiquitination cascade is constantly counteracted by multiple deubiquitinating enzymes (DUBs), however they are required for the expression of the ubiquitin genes, recycling of ubiquitin and prevention of the build-up of harmful polyubiquitin chains. Yet another layer of complexity is added by the fact that while Lys48 is used for the formation of common polyubiquitin chains targeting proteins for degradation, the remaining five Lys residues of the ubiquitin molecule may be also used for the formation of polyubiquitin chains with many different functions. Moreover, monoubiquitination and multiubiquitination (monoubiquitination at multiple lysines of one substrate) have important roles in protein trafficking and vesicle internalization. Furthermore, there are many ubiquitin-like proteins, which can be covalently attached to proteins performing different functions. Nedd8 and SUMO are just examples of this group of covalent modifiers. Ubiquitin-like motifs are also found within larger and more complex proteins.

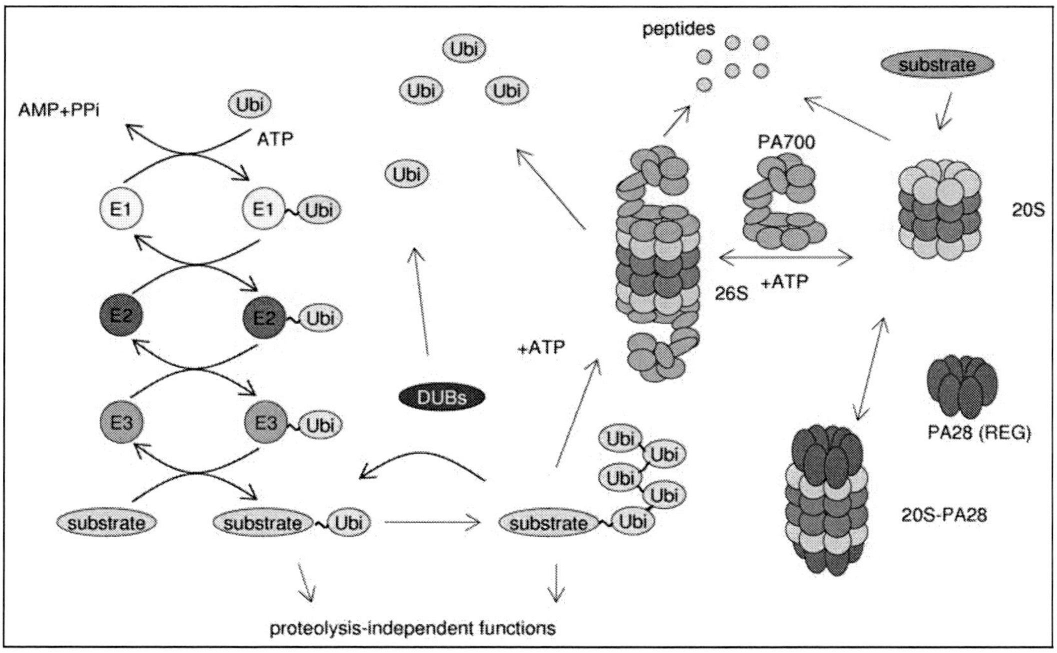

Figure 2. An overview of the ubiquitin- and proteasome-dependent system of protein degradation. The ubiquitin activating enzyme (E1) forms a highly energetic thiolester intermediate with ubiquitin (Ubi), transferring it to one of several Ubi conjugating enzymes (E2). E2 interacts with one of hundreds Ubi ligases (E3), which recognize different substrates to be ubiquinated. Ubi moieties are then transferred to the substrates forming poly-Ubi chains. Lys-48 linked poly-Ubi chains are recognized and bound by the 26S proteasomes. Once bound, the substrates are degraded into peptides, while free Ubi is recycled. Monoubiquitinated proteins and chains with linkages other than Lys-48 serve non-proteolytical functions. Ubiquitinated substrates can be de-ubiquitinated by the action of one of several deubiquitinating enzymes (DUBs). 26S proteasome is composed of the 20S proteasome and two PA700 caps. 20S proteasome by itself is able to degrade unfolded and oxidized proteins. It can also associate with different activators, such as PA28. ATP indicates adenosine triphosphate, AMP adenosine monophosphate, PPi inorganic pyrophosphate.

While the structure and function of the core 20S proteasome is well known, the 26S proteasome guards its secrets very well, avoiding crystallization. The function of many PA700 subunits is poorly understood. They undergo different posttranslational modifications which may change their activity and influence the distribution of the complex within cells. There are contradicting theories explaining the workings of the PA700 cap. A static view has been recently challenged by a model involving the breakdown of PA700 into its components during every catalytic cycle. There also exists a possibility that different forms of PA700 may be present in the cells, some involving alternative components, such as the signalosome or the VCP hexamer. Moreover, the function of other proteasome activators and associated proteins, including PA28 and PA200, remains obscure. However, mathematical models and computer simulation of proteasome functions can help to understand the 26S proteasome. Recent advances in this topic have already revealed an unanticipated complexity of proteolytic activity acting on the substrate recruitment mechanism and the catalytic cycle itself. The knowledge of the detailed structure and function of the 26S proteasome allows us to predict its dysfunction in different pathological situations, leading to the definition of a possible group of disorders characterize by compromised proteasome function, which can be collectively designated as the '*proteasomepathies*'.

Involvement of the UPS in Pathogenesis of Multiple CNS Disorders

In the nervous system, the physiological significance of the UPS is just beginning to be explored. Dissecting the role of the UPS during neuronal differentiation and neuronal apoptosis in cell culture models has laid the groundwork for our current understanding of ubiquitination and proteasomal protein degradation in the brain. As in all other mammalian tissues, the UPS is fundamental to normal brain function and during its development in several aspects of neuronal physiology such as neuronal signaling, synapse formation and function, memory processing, etc. Degradation of proteins by the UPS seems to be spatially organized within cells. Protein which fail to be degraded tend to aggregate in specific cellular locations corresponding to specific '*proteolysis centers*'. Such aggregates, inclusion bodies or '*aggresomes*' are found in many neurodegenerative pathologies of the central nervous system (CNS) both inherited and acquired, such as Alzheimer's (AD), Huntington's (HD), Parkinson's (PD) and Lewy body diseases, amyotrophic lateral sclerosis (ALS) and Creutzfeld–Jakob disease (CJD). A consistent feature of major human neurodegenerative disorders is the accumulation of disease-related proteins, in non-native conformations, as protein aggregates within neurons or glial cells. Often the proteins in these aggregates are post-translationally conjugated with ubiquitin, suggesting a link between pathological protein-aggregation events in the nervous system and dysfunction of the UPS. The ubiquitinated protein aggregates are believed to result from malfunction or overload of the UPS or from structural changes in the protein substrates which prevent their recognition and degradation by the UPS. Impaired proteolysis might also contribute to the synaptic dysfunction seen early in neurodegenerative diseases because the UPS is known to play a role in normal functioning of synapses. Further, the switch from short to long-term facilitation in synapses is mediated by proteasomal degradation of inhibitors of the '*long term*' pathway. Another interesting aspects is the participation of the UPS in the cleavage of components involved in AD: the amyloid ß-protein (AP2) and the presenilin-2 membrane protein (PS2). Several neurodegenerative diseases are associated with mutations of UPS components or of proteins that may resist proteasomal degradation if mutated. Mutations in *parkin*, encoding an ubiquitin ligase (E3) result in juvenile recessive PD. α-Synuclein, which is mutated in some familiar forms of PD, is highly enriched in presynaptic terminals and Lewy-bodies. Recently, the higher than normal level of wild-type α-synuclein was found in a family with early onset PD, and a contributing mechanism could be an insufficient clearance by the UPS. Genetic evidence clearly demonstrates that disruption of ubiquitin-mediated processes can lead to neurodegeneration; however, thc relationship between the UPS and idiopathic neurodegenerative disorders is less clear. In the latter cases, although a number of different mechanisms could potentially contribute to dysfunction of the UPS and promote the neurodegenerative process, whether UPS dysfunction is causally related to disease pathogenesis, or alternatively arises as a result of the pathological state, and indeed whether ubiquitinated inclusions are harmful or beneficial to cells, remains to be clarified. Recent studies on HD and other polyglutamine diseases (PolyQ) show that the respective mutated proteins can interfere with axonal transport. The huntingtin protein involved in HD also interacts with synaptic vesicles and proteins involved in neurotransmission. In spinocerebellar ataxia 1, there is an altered trafficking of glutamate receptor subunits and protein kinase-C (PKC) in Purkinje cells. Given the role of the UPS in disease pathophysiology, it is important to study the key proteins in the axons and in the synapses that may be altered in the different disorders.

The involvement of aberrant protein folding combined with their inefficient clearance from neurons by the UPS has led to a proposal of calling collectively the various neurodegenerative disorders as '*protein misfolding diseases*' or '*conformational diseases*'. For most of proteins to be active, they need well-defined three-dimensional structures alone or in complex. Folding is a process through which newly synthesized proteins get to the native state. Protein folding inside cells is assisted by various chaperones and folding factors, and misfolded proteins are eliminated by the UPS to ensure high fidelity of protein expression. Under certain circumstances, specific peptides or proteins misfold altering the folding pathway or final conformation of a protein, often as a result of mutations, give rise to protein aggregates; misfolded proteins escape the degradation process, yielding to deposit of protein aggregates such as loop-sheet polymer and AP2 fibril. Protein misfolding gives rise to the malfunctioning of living systems. Many neurological '*conformational diseases*' have the accumulation of misfolded proteins and this combined accumulation results in the promotion of insoluble protein deposits and neuronal cell death (Table 1).

All of these diseases are characterized by the failure of the cell's '*quality control system*' to remove toxic, misfolded proteins—they then accumulate and cause disease, via various pathways (in some cases, as in AD, involving triggering apoptosis)—that are only partially understood. The first step of the cell's '*quality control*' system is the binding of molecular chaperones to nascent proteins as they emerge from the ribosomes. Molecular chaperones are themselves proteins, and they promote correct folding of nascent proteins and prevent harmful protein-protein interactions or aggregation. However, a fraction of newly-translated proteins fail to fold properly. These defective proteins are then degraded by the final step in the cellular '*quality control system*', which is the UPS. Although mutation is involved in producing defective, aggregation-prone toxic proteins (or, in the case of certain types of PD, defective components of the '*quality control system*') in the case of familial forms of AD and PD, it is not exactly known how pathways for removal of toxic proteins may break down in the more common, late-onset sporadic forms of these diseases. Most likely there is a gradual decrease with age of the UPS capability to cope with misfolded proteins, which in some individuals is faster than in another, and which can be enhanced by complex multiple genetic and environmental factors. While recruitment of misfolded proteins in the cytosol to proteolytic centers and formation of '*aggresomes*' may have a beneficial, protective function, accumulation of misfolded proteins within the ER triggers the so called unfolded protein response (UPR) in an attempt to overcome the burden of misfolded proteins. Persistent UPR leads to apoptosis in both acute and chronic pathologies of the CNS, such as stroke and PD, respectively. The UPS counteracts the ER stress by degrading proteins retrotranslocated from the ER to cytosol in a process known as ERAD (ER-associated degradation). Mutations of VCP, an ATP-ase with a supposed role in ERAD of many proteins, are associated with a newly discovered form of frontotemporal dementia, providing yet another proof of the role of the UPS in CNS pathology.

Table 1. 'Conformational diseases' - neurological diseases caused by defects in protein folding, stability and aggregation.

Disease	Aggregate type	UPS components	Location	Characteristic pathology
Alzheimer disease	Plaques	Ubiquitin Proteasome	Extracellular	Extracellular neuritic plaques
	Tangles	Ubiquitin Proteasome	Cytoplasmic	Neurofibrillary tangles of hyperphosphorylated tau
Parkinson disease	Lewy bodies	Proteasome Ubiquitin HSPs E3–Parkin DUB– UCH-L1	Cytoplasmic	Intracellular Lewy bodies, Lewy neurites, fibrillar, α-synuclein
Multiple system atrophy	Glial/neuronal inclusions	Ubiquitin	Cytoplasmic	Oligodendroglial inclusions immunostained with tau and ubiquitin
Polyglutamine disease	Inclusions	Proteasome Ubiquitin HSPs	Cytoplasmic	Aggregates and fibrillar, huntingtin fragments
			Nuclear	neuronal inclusions
Prion diseases	Aggresome-like	Ubiquitin HSPs	Cytoplasmic	Intracellular deposits, and occasional synaptic and axonal deposits
			Extracellular	Amyloid plaques
Familial encephalopathy with neuroserpin inclusion bodies	Collins bodies neuronal inclusions	Neuroserpin	Cytoplasmic	Eosinophilic neuronal inclusions of neuroserpin (Collins' bodies) in the deeper layers of the cerebral cortex and the substantia nigra.
Stroke	Aggresome-like	Ubiquitin HSPs	Cytoplasmic/ nuclear	Protein aggregates surrounding nuclei and along dendrites in postischemic neurons in the neuronal soma, dendrites, and axons. Ubiquinated proteins are associated with intracellular membranous structures in neuronal lysosomal vesicles and in late endosome-like organelles in the ischemic area

DUB indicates deubiquitinating enzyme, UCH-L1, ubiquitin C-terminal hydrolases-L1, HSPs, heat shock proteins.

Today, only a handful of diseases, particularly neurodegenerative ones, are known to be caused by malfunction of the UPS. With perhaps as many as 1000 human genes encoding components of ubiquitin and ubiquitin-related modification pathways, it is almost certain that many more diseases will be found to arise from genetic errors in the UPS or by pathogen subversion of the system. The high sensitivity of postmitotic, fully differentiated cells to oxidized or misfolded proteins, which need to be eliminated rapidly in order to maintain neuronal metabolism, makes neurons particularly vulnerable to defects in UPS-mediated protein degradation. At the moment, for few disorders, we can temporarily relieve symptoms by pharmacological or surgical manipulation of the neurotransmitters emitted by the degenerating neurons, but the discovery during the past decade of the role of UPS in several neurodegenerative diseases could permit us to interrupt the process of neuronal loss itself by repair and possibly augmentation of the neuronal UPS, which will impose a major challenge. Various aspects of the role played by the UPS in the different neurodegenerative disorders are discussed extensively in several chapters of this book. This opens several avenues for the development of new therapies because the components of the UPS could be potential targets for therapy of CNS diseases and disorders. Introducing the concept of UPS dysfunction into our thinking will also allow us to transcend the classic category of clinical and anatomical pathology. An excellent example is the obsolescence of the term olivopontocerebellar atrophy. The sporadic form of olivopontocerebellar atrophy has been found to harbour cytoplasmic inclusions in oligodendrocytes consisting chiefly of α-synuclein. The same is true for Shy-Drager syndrome and striatonigral degeneration. These three entities have now been combined into a pathogenetically based category called multiple system atrophy. The familial form of olivopontocerebellar atrophy has been subsumed into the various forms of spinocerebellar ataxia, which are differentiated by their genetic defects and by the nature of their protein aggregates. The term olivopontocerebellar atrophy has therefore proved useless and has virtually disappeared from the literature. As we consider the pathogenesis and classification of neurodegenerative disease, we must consider the identity of the abnormally aggregating protein, the cause of its misfolding, causes of protein aggregation other than misfolding, the causes of failure of the UPS to dispose of the abnormally folded or aggregated protein, and the mechanism by which abnormally aggregated protein causes cellular damage. This framework will bring a more rational classification of diseases and a very high probability of specific treatments or prevention. It was proposed the concept that neurodegenerative diseases such as AD and PD, as well as, other conditions collectively represent '*conformational diseases*' in which altered function of the UPS can cause or directly contribute to disease pathogenesis. Many disease causing phenotypes are common to multiple neurodegenerative disorders and caused by a gain of function associated with protein misfolding and aggregation, while individual variations can be attributed to a particular group of neurons affected and a loss of function phenotypes, both dependent on the particular protein affected by the disease. However, it has become increasingly apparent that there are a variety of conditions *in vivo* where, even with chaperones and the proteolytic machinery present in the same compartment as a misfolding protein, these mechanisms of quality control fail and the misfolded proteins proceed to form aggregates. A protective action of aggregate formation, as opposed to an immediately pathogenic role, has been increasingly supported. For example, in the case of HD, serial examination of neuronal cells in culture overproducing a polyQ-expanded huntingtin-GFP fusion revealed that those cells containing morphologically visible fluorescent aggregates exhibited better viability than those bearing diffusely fluorescent material. This seems consistent with the concept that it is small assemblies of misfolded proteins, not morphologically visible inclusions, that exert toxic effects on cells. Correspondingly, a recent comparison of size versus toxicity for aggregates of the prion protein PrP suggested that small aggregates containing one to two dozen molecules were the most toxic to cells. Apparently, the cells do not have mechanisms for clearing these small aggregates while have a better possibility in controlling intracellular levels of bigger aggregates. Earlier studies suggested that the UPS might be the mainstay of removal of aggregation-prone species. Indeed, aggregates detected in neurodegenerative disease are usually reactive with anti-ubiquitin antibodies, implying that the misfolding species have been recognized by the ubiquitin conjugation system and further experimental evidences indicate that these modified proteins present a particular challenge to proteasomes, possibly leading to their inhibition. From this standpoint, we probably should consider several '*conformational diseases*' as '*proteasomopathies*' where the final pathogenetic determinant is the inability of proteasomes to degrade misfolded proteins. However, recent work with autophagy-deficient mice demonstrates that downstream of the UPS, neurons may still clear the aggregated proteins by autophagic mechanisms, which constitute the last line of defense. When those fail, massive neurodegeneration takes place. The mechanism of this inactivation remains unknown but two possible models could account for the impairment of UPS function by protein aggregation and to explain the accumulation of misfolded ubiquinated substrates in neurodegenerative diseases (Figure 3). One possibility is that the aggregated or aggregation-prone proteins directly inhibit or '*choke*' the 26S proteasome—a situation that might result from their engagement by degradation-resistant or hard-to-unfold proteins: a reduced cleavage strength towards specific amino acid sequences keep the proteasome sequestrated with undegraded substrate and intermediate long fragments. Perturbations of proteasome function may also occur through its recruitment to or sequestration into protein aggregates, through chronic overloading of proteasomal capacity by misfolded protein, or through still undetermined effects on other activities of the UPS. In this manner protein aggregates directly inhibit or sequester 26S proteasomes (Figure3; Panel A). A second possibility, not mutually exclusive with the first, is that protein aggregates indirectly interfere with UPS function by sequestering or directly '*clogging*' proteasomes, it is possible that they could impair UPS function by influencing the proteasome activity or distribution of UPS modulators, inactivating or depleting a UPS activator (Figure 3; Panel B) where aggregation would result due low influx of substrate into the 20S core particle. Proteasome substrates with a low affinity binding to 19S sites which regulates the influx (and unfolding) mechanisms will have strongly reduced transportation rates. However, no true '*proteasomopathy*' has been identified, i.e. no mutations in the genes coding for subunits of the 26S proteasome have been identified to be associated with disease susceptibility. Obviously, mutations leading to the loss of proteasome activity would have been fatal, however it is easy to envision alleles of proteasome genes, which induce very discrete changes into such a complex organelle as the 26S proteasome. Such changes may gradually over the years tilt the delicate balance of protein degradation and aggregation within cells towards aggregation, contributing to the pathogenetic cascade. Genetic abnormalities within the ubiquitin pathway, either in ubiquitin-ligase (E3) enzymes or in deubiquitinating enzymes have been however identified. They cause disease because of problems associated with substrate recognition or supply of free ubiquitin, respectively. In some cases, mutations in protein substrates of the UPS may directly contribute to disease progression because of inefficient substrate recognition by elements of the ubiquitination cascade, delivery factors or the proteasome itself. Mutations in transcripts of ubiquitin (as a result of '*molecular misreading*') also affect ubiquitin-dependent proteolysis with catastrophic consequences. This has been shown in AD and tauopathies and could apply to other age-associated neurodegenerative conditions.

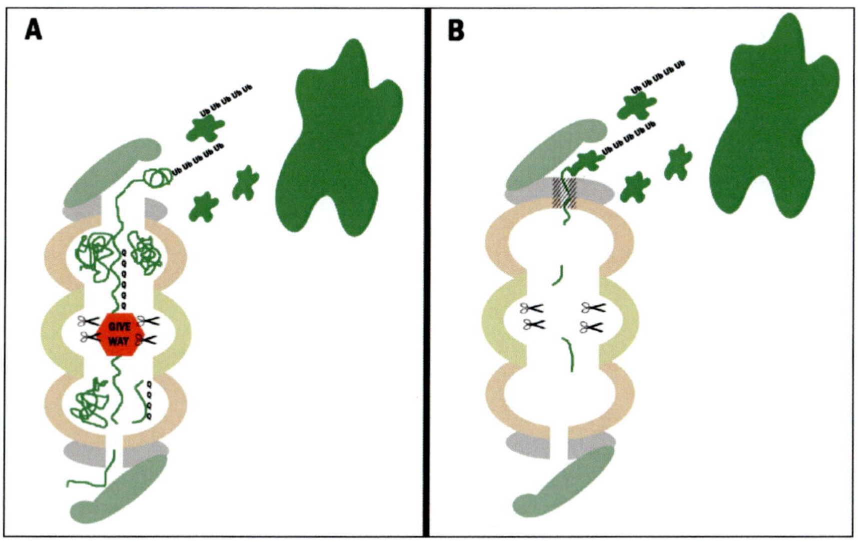

Figure 3. Mechanisms of proteasome block: simple sketch of impaired 26S proteasome degradation of ubiquitinated substrate. Two possible scenarios are conjectured. *Panel A*. Inhibition of the 26S proteasome activity due to filling up with substrates and intermediates that have a reduced number of cleavage sites. Substrates are ubiquitinated and transported into the proteasome core via the binding with the 19S particle. Substrates are then cleaved by the proteasome according to the cleavage specificities. These are defined by the interaction between the amino acid sequence and the proteasome catalytic sites. PolyQ repeats strongly impair the cleavage activity with a concomitant reduction of the degradation rate. The internal volume will be progressively filled with long intermediates which have a low efflux rate and undegraded substrate. Opening the gate favors the efflux of intermediate subsrates, which then exit in a length dependent manner. Therefore, long intermediates have a low efflux rate. According to the model, the filling of the proteasome stops the influx of new substrate molecules, causing an accumulation of substrate outside the proteasome. *Panel B*: Inhibition of 26S proteasome activity due to impaired transport of substrates caused by amino acid sequence changes on the substrate that effect 19S binding. The interaction between substrate and proteasome regulating the influx of substrate is largely unknown. Energetic barrier between the amino acid sequence constituting the substrate and the structure of the 19S can be hypothesized. Here, for simplicity we depict an impaired transport of substrate. A limited influx of substrates results in an accumulation of non-degraded proteins outside the proteasome (see *Chapter 14*). Courtesy of Fabio Luciani, School of Biotechnology and Biochemical Sciences, UNSW, Sydney, Australia.

It appears now that ubiquitin modification may in fact recruit aggregated species for clearance via an independent mechanism, the '*autophagy*' pathway. Autophagy involves the recognition and packaging/engulfment of targeted proteins or organelles into autophagosome vesicles that become fused with lysosomes, wherein both vesicles and their contents are broken down. Thus, it seems that ubiquitin modification of defective proteins may provide an entryway to either the proteasomal system or, in contexts where aggregation is occurring, the autophagy system.

Within the nervous system, accumulation of misfolded proteins as a result of defective ubiquitin-dependent proteolysis may contribute to aggregation events, which underlie the pathogenesis of several major human neurodegenerative diseases. The role played by the UPS is also evident in autoimmune diseases and inflammatory responses of the CNS, in the pathogenesis of different CNS tumors, in cerebral ischemia but is also involved in epilepsy and seizure susceptibility.

Drugs Interfering with the UPS Function

Because of the central role the UPS plays in such a broad array of basic cellular processes, development of drugs that modulate the activity of the system may provide a highly specific, target-oriented therapeutic approach to many neurological diseases. Proteasome inhibitors have been originally developed in the early 1990s as a tool to study the UPS, however they quickly made they way to the clinic. Velcade® (Bortezomib) is a potent proteasome inhibitor, which has been approved in 2003 by the FDA for the treatment of chemotherapy-resistant multiple myeloma. Although the current focus of potential proteasomal drugs is on cancer, the prevalence of this pathway in neuronal function makes it a tantalizing target for future CNS therapeutics. However, inhibition of enzymes common to the entire pathway, such as the proteasome, affects many processes non specifically although a narrow window between beneficial effects and toxicity can be identified for a short-term treatment. Surprisingly, many cell types, including neurons, but not cancer cells, resist short intervals of inhibition of proteasome activity. Recent experimental evidence strongly suggests that such inhibitors may indeed be beneficial in brain infarct and autoimmune encephalomyelitis. Because specificity of the ubiquitin proteasome mediated proteolysis is determined by specific ubiquitin ligases (E3s), identification of specific E3s and their allosteric modulators and inhibitors are likely to provide effective therapeutic targets for the treatment of several CNS disorders. New proteasome inhibitors and drugs inhibiting ubiquitin ligases are under development using different methods of the high-throughput screening of compounds obtained by combinatorial chemistry. A better approach may be the development of small molecules that are substrate specific and bind, preferably, to specific substrates or to their ancillary proteins rather then to an E3. Finally, drugs may be envisioned which will interact specifically with delivery/shuttle factors, such as inhibitors of VCP. Although many proteasome inhibitors are available, no effective drugs exist that can stimulate the proteasome. Since abnormal protein aggregation is a common feature of different neurodegenerative diseases, enhancement of proteasome activity might be an efficient way to remove the aggregates that accumulate in the brain.

Proteasome inhibitors have been tested in different experimental systems for the treatment of various disorders of the nervous system in particular those associated with ischemia-reperfusion injury. MLN-519 is a proteasome inhibitor with pharmacokinetic properties different from those of Velcade®, which entered phase I clinical trials for the treatment of stroke. While the clinical experience with proteasome inhibitors is limited, they seem to be quite safe to use. Nevertheless, they have the potential of producing detrimental side effects including the induction of some kind of neurodegenerative changes.

Concluding Remarks

The Editors hope that the information gathered in this single volume by a pleiade of authors from divergent areas of expertise will provide the readers with the basic, necessary information in this emerging field. Of course, this collection of articles is not meant to provide a comprehensive coverage of the field; instead, we have attempted to *zoom in* covering some topics of particular interest and emerging issues. We hope that this volume serves to bring both the expert and the novice up to date. We have attempted to draw together contributions from experts in the field to illustrate the comprehensive manner in which the UPS regulates cell physiology and pathology in nervous system. There is no doubt that when the full implications of protein modification by ubiquitin and ubiquitin-like molecules are going to be fully understood, then we will gain fundamental new insights into crucial life processes. We will also have come to an understanding of those pathological processes resulting from UPS malfunction. The medical implications of the UPS research should have considerable impact on the pharmaceutical industry and should open new avenues for therapeutic intervention in human diseases. However, everyone also needs to keep in mind that due to the accelerating pace of biomedical research in general and the study of the UPS in particular we are likely to see within the months and years to come new advances which will require the updating of this book. We will certainly welcome any suggestions for changes and amendments of this book which we may include in the future editions. Despite its all limitations, we believe that this book will provide the readers with an important source of information.

The basic science chapters of this book are going to present an overview of the different aspects of the UPS, while the following chapters will address more specifically the UPS function in the context of neuronal tissue and the nervous system. We tried to select the most important topics covering this vast area, however many topics must have been omitted or covered in less detail due to spatial constrains. Even though the field is young, it has now reached the point at which the scientific community at large needs reference works in which contributing authors indicate the fundamental roles of the UPS in the physiology and pathology of the nervous system, therefore the appearance of this book on the UPS in nervous system is extremely timely. There are no doubts that the involvement of the UPS in the pathogenesis of multiple diseases, the continuous use of proteasome inhibitors in the clinic, as well as, the foreseen introduction of a new generation of drugs inhibiting or activating selected ubiquitin ligases creates a need for a concise source of information discussing the role of the UPS in the nervous system. In this volume, we are pleased to present a collection of articles that discuss recent developments and emerging themes discussing the role of UPS in the physiology and pathology of nervous system. This collection of articles highlights the story of UPS as we currently know it: from the regulation of basic cellular processes to quality control and the pathogenetic mechanisms of neurological diseases, from X-ray crystallography of the 20S proteasome to the interaction between substrates and their ligases, to the development of mechanism-based drugs, and to target-specific aberrant processes. On the pages of this volume, the basic knowledge about the UPS will be dissected in terms of the intracellular localization, structures and functions of proteasomes and related particles, the emerging roles of enzymes of ubiquitylation and deubiquitylation, the roles of ubiquitin-like proteins and the formation of aggresomes as the consequence of proteasome inhibition.

Finally, we would like to dedicate this book to our loving families in acknowledgement of their understanding and support and to extend our appreciation and thanks to the Authors for their work in reviewing this extraordinary progress and for providing a basis for understanding future advances of the UPS in the physiology and pathology of nervous system.

Good reading!

Mario Di Napoli, M.D. Sulmona, L'Aquila

Cezary Wójcik, M.D. Ph.D., Italy Evansville, IN, United States

In: The Ubiquitin Proteasome System… ISBN 978-1-60021-749-4
Eds: Mario Di Napoli and Cezary Wojcik, pp. 15-40

Chapter 2

INTRACELLULAR PROTEIN DEGRADATION: FROM A VAGUE IDEA THRU THE LYSOSOME AND THE UBIQUITIN-PROTEASOME SYSTEM AND ONTO HUMAN DISEASES AND DRUG TARGETING†

*Aaron Ciechanover**

Cancer and Vascular Biology Research Center, The Rappaport Faculty of Medicine and Research Institute, Technion-Israel Institute of Technology, Haifa 31096, Israel.

ABSTRACT

Between the 1950s and 1980s, scientists were focusing mostly on how the genetic code is transcribed to RNA and translated to proteins, but how proteins are degraded has remained a neglected research area. With the discovery of the lysosome by Christian de Duve, it was assumed that cellular proteins are degraded within this organelle. Yet, several independent lines of experimental evidence strongly suggested that intracellular proteolysis is largely nonlysosomal, but the mechanisms involved remained obscure. The discovery of the ubiquitin–proteasome system (UPS) resolved the enigma. We now recognize that degradation of intracellular proteins is involved in regulation of a broad array of cellular processes, such as cell cycle and division, regulation of transcription factors, and assurance of the cellular quality control. Not surprisingly, aberrations in the system have been implicated in the pathogenesis of human disease, such as malignancies and neurodegenerative disorders, which led subsequently to an increasing effort to develop mechanism based drugs.

† Nobel Lecture 2004, © The Nobel Foundation 2004.

* Correspondence concerning this article should be addressed to Address for correspondence: Dr. Aaron Ciechanover MD, DSc; Faculty of Medicine, Technion-Israel Institute of Technology, Efron Street, Bat Galim, PO Box 9649, Haifa 31096, Israel. Phone: +972-4-829-5379, 5356; Fax: +972-4-852-3947, 3922; E-mail: c_tzachy@netvision.net.il.

Keywords: ubiquitin, proteasome, protein degradation, lysosome.

Abbreviations

APC, Adenomatous Polyposis Coli; APF-1, ATP-dependent Proteolysis Factor 1 (ubiquitin); BSA, bovine serum albumin; CP, 20S core particle (of the proteasome); G6PD, glucose-6-phosphate dehydrogenase; MCP, multicatalytic proteinase complex (26S proteasome); MVBs, multivesicular bodies; ODC, ornitihine decarboxylase; PEPCK, phosphoenol-pyruvate carboxykinase; RP, 19S regulatory particle (of the proteasome); TAT, tyrosine aminotransferase; Ub, ubiquitin; UBIP, ubiquitous immunopoietic polypeptide (ubiquitin); UBLs, ubiquitin-like proteins; UPS, ubiquitin–proteasome system.

Introduction

The concept of protein turnover is hardly 60 years old. Beforehand, body proteins were viewed as essentially stable constituents that were subject to only minor 'wear and tear': dietary proteins were believed to function primarily as energy providing fuel, which were independent of the structural and functional proteins of the body. The problem was hard to approach experimentally, as research tools were not available. An important research tool that was lacking at that time were stable isotopes. While radioactive isotopes were developed earlier by George de Hevesy (de Hevsey G, Chemistry 1943. In Nobel Lectures in Chemistry 1942–1962. World Scientific 1999, pp. 5–41), they were mostly unstable and could not be used to follow metabolic pathways). The concept that body structural proteins are static and the dietary proteins are used only as a fuel was challenged by Rudolf Scheonheimer in Columbia University in New York city. Schoenheimer escaped from Germany and joined the Department of Biochemistry in Columbia University founded by Hans T Clarke [1–3]. There he met Harold Urey, who was working in the Department of Chemistry and who discovered deuterium, the heavy isotope of hydrogen, a discovery that enabled him to prepare heavy water, D_2O. David Rittenberg who had recently received his Ph.D. in Urey's laboratory, joined Schoenheimer, and together they entertained the idea of '*employing a stable isotope as a label in organic compounds, destined for experiments in intermediary metabolism, which should be biochemically indistinguishable from their natural analog*' [1] Urey later succeeded in enriching nitrogen with ^{15}N, which provided Schoenheimer and Rittenberg with a 'tag' for amino acids and as a result for the study of protein dynamics. They discovered that following administration of ^{15}N-lablled tyrosine to rat, only ~50% was recovered in the urine, '*while most of the remainder is deposited in tissue proteins. An equivalent of protein nitrogen is excreted*' [4]. They further discovered that from the half that was incorporated into body proteins, '*only a fraction was attached to the original carbon chain, namely to tyrosine, while the bulk was distributed over other nitrogenous groups of the proteins*' [4] mostly as an αNH_2 group in other amino acids. These experiments demonstrated unequivocally that the body structural proteins are in a dynamic state of synthesis and degradation, and that even individual amino acids are in a state of dynamic interconversion. Similar results were obtained using ^{15}N-lablled leucine [5]. This series of findings shattered the paradigm in the field at that time that: (1) ingested proteins are completely metabolized and the products are excreted, and (2) body structural proteins are stable and static. Schoenheimer was invited to deliver the prestigious Edward K Dunham lecture at Harvard University, where he presented his revolutionary findings. After his untimely tragic death in 1941, his lecture notes were edited by Hans Clarke, David Rittenberg and Sarah Ratner, and were published in a small book by Harvard University Press. The editors called the book 'The Dynamic State of Body Constituents' [6], adopting the title of Schoenheimer's presentation. In the book, the new hypothesis is clearly presented: '*The simile of the combustion engine pictured the steady-state flow of fuel into a fixed system, and the conversion of this fuel into waste products. The new results imply that not only the fuel, but the structural materials are in a steady state of flux. The classical picture must thus be replaced by one which takes account of the dynamic state of body structure*'. However, the idea that proteins are turning over was not accepted easily and was challenged as late as the mid-1950s. For example, Hogness *et al.* [7] studied the kinetics of β-galactosidase in *Escherichia coli* and summarized their findings: '*To sum up: there seems to be no conclusive evidence that the protein molecules within the cells of mammalian tissues are in a dynamic state. Moreover, our experiments have shown that the proteins of growing E. coli are static. Therefore, it seems necessary to conclude that the synthesis and maintenance of proteins within growing cells is not necessarily or inherently associated with a 'dynamic state*'. While the experimental study involved the bacterial β-galactosidase, the conclusions were broader, including also the authors' hypothesis on mammalian proteins.

The use of the term 'dynamic state' was not incidental, as they challenged directly Schoenheimer's studies. Now, after more then six decades of research in the field and with the discovery of the lysosome and later the complex ubiquitin–proteasome system (UPS) with its numerous tributaries, it is clear that the area has been revolutionized. We now realize that intracellular proteins are turning over extensively, that this process is specific, and that the stability of many proteins is regulated individually and can vary under different conditions. From a scavenger, unregulated and nonspecific end process, it has become clear that proteolysis of cellular proteins is a highly complex, temporally controlled and tightly regulated process that plays major roles in a broad array of basic pathways. Among these processes are cell cycle, development, differentiation, regulation of transcription, antigen presentation, signal transduction, receptor-mediated endocytosis, quality control and modulation of diverse metabolic pathways. Subsequently, it has changed the paradigm that regulation of cellular processes occurs mostly at the transcriptional and translational levels, and has set regulated protein degradation in an equally important position. With the multitude of substrates targeted and processes involved, it is not surprising that aberrations in the pathway have been implicated in the pathogenesis of many diseases, among them certain malignancies, neurodegeneration, and disorders of the immune and inflammatory system. As a result, the system has become a platform for drug targeting, and mechanism-based drugs are currently developed, one of them is already on the market.

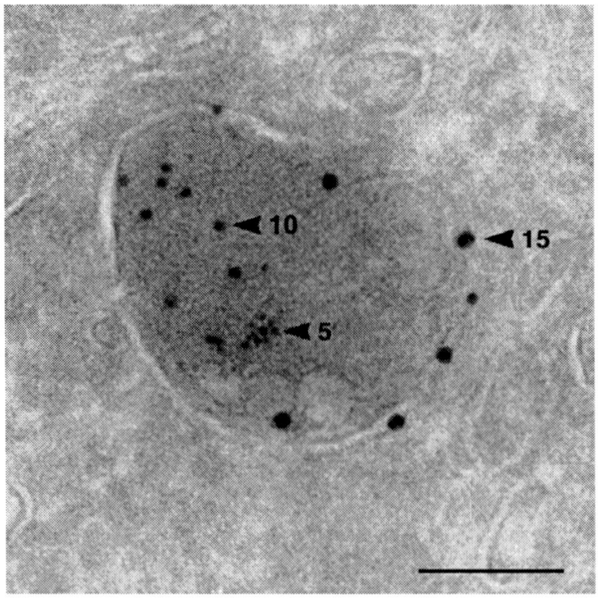

Figure 1. The lysosome. Ultrathin cryosection of a rat PC12 cell that had been loaded for 1 h with bovine serum albumin (BSA)-gold (5 nm particles) and immunolabelled for the lysosomal enzyme cathepsin B (10-nm particles) and the lysosomal membrane protein LAMP1 (15 nm particles). Lysosomes are recognized also by their typical dense content and multiple internal membranes. Bar, 100 nm. Courtesy of Viola Oorschot and Judith Klumperman, Department of Cell Biology, University Medical Centre Utrecht, The Netherlands.

The Lysosome and Intracellular Protein Degradation

In the mid-1950s, Christian de Duve discovered the lysosome (see, e.g., de Duve *et al.*, [8] and Gianetto and de Duve [9] and Figure 1). The lysosome was first recognized biochemically in rat liver as a vacuolar structure that contains various hydrolytic enzymes, which function optimally at an acidic pH. It is surrounded by a membrane that endows the contained enzymes latency that is required to protect the cellular contents from their action (see below). The definition of the lysosome has been broadened over the years. This is because it has been recognized that the digestive process is dynamic and involves numerous stages of lysosomal maturation together with the digestion of both exogenous proteins (which are targeted to the lysosome through receptor mediated endocytosis and pinocytosis) and exogenous particles (which are targeted via phagocytosis; the two processes are known as heterophagy), as well as digestion of endogenous proteins and cellular organelles (which are targeted by micro- and macroautophagy; see Figure 2). The lysosomal/vacuolar system as we currently recognize is a discontinuous and heterogeneous digestive system that also includes structures that are devoid of hydrolases – for example, early endosomes which contain endocytosed receptor–ligand complexes and pinocytosed/phagocytosed extracellular contents. On the other extreme, it includes the residual bodies – the end products of the completed digestive processes of heterophagy and autophagy. In between these extremes one can observe primary/nascent lysosomes that have not yet been engaged yet in any proteolytic process; early autophagic vacuoles that might contain intracellular organelles; intermediate/late endosomes and phagocytic vacuoles (heterophagic vacuoles) that contain extracellular/intracellular protein degradation contents/particles; and multivesicular bodies (MVBs), which are the transition vacuoles between endosomes/phagocytic vacuoles and the digestive lysosomes.

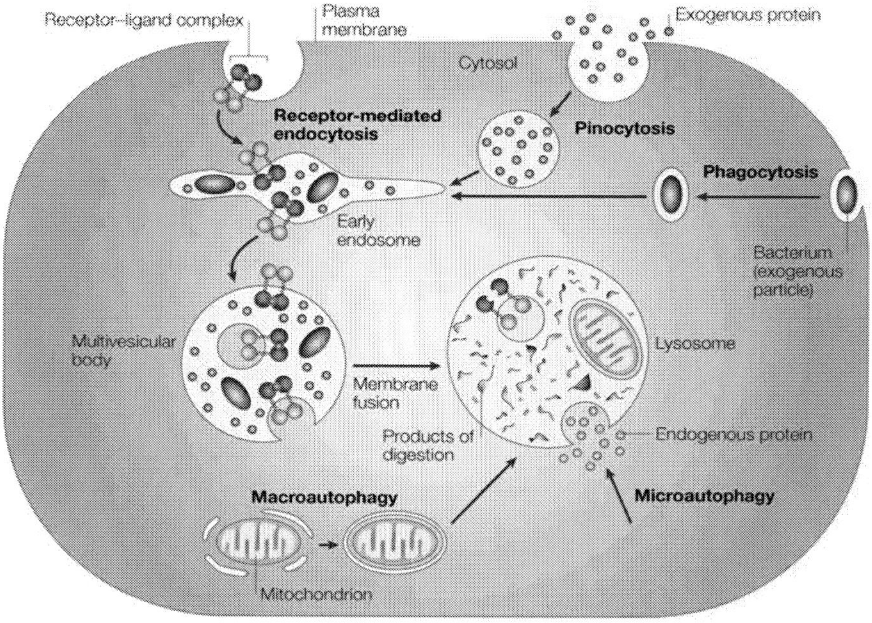

Figure 2. The four digestive processes mediated by the lysosome. (*i*) specific receptor-mediated endocytosis, (*ii*) pinocytosis (nonspecific engulfment of cytosolic droplets containing extracellular fluid), (*iii*) phagocytosis (of extracellular particles), and (*iv*) autophagy (micro- and macro-; of intracellular proteins and organelles) (with permission from Nature Publishing Group, copyright MacMillan Magazines Ltd. Published originally in Ciechanover [83]).

The discovery of the lysosome along with independent experiments that were carried out at the same time and that have further strengthened the notion that cellular proteins are indeed in a constant state of synthesis and degradation (see, e.g., Simpson [10]), led scientists to feel, for the first time, that they have at hand an organelle that can potentially mediate degradation of intracellular proteins. The fact that the proteases were separated from their substrates by a membrane provided an explanation for controlled degradation, and the only problem left to be explained was how the substrates are translocated into the lysosomal lumen, exposed to the activity of the lysosomal proteases and degraded. An important discovery in this respect was the unravelling of the basic mechanism of action of the lysosome – autophagy (reviewed in Mortimore and Poso [11]). Under basal metabolic conditions, portions of the cytoplasm, which contain the entire cohort of cellular proteins, are segregated within a membrane-bound compartment, and are then fused to a primary nascent lysosome and their contents digested. This process was denoted microautophagy. Under more extreme conditions, for example starvation, mitochondria, endoplasmic reticulum membranes, glycogen bodies and other cytoplasmic entities can also be engulfed by a process called macroautophagy (see, e.g., Ashford and Porter [12]; the different modes of action of the lysosome in digesting extra- and intracellular proteins are shown in Figure 2). However, over a period of more than two decades, between the mid-1950s and the late 1970s, it has become gradually more and more difficult to explain several aspects of intracellular protein degradation based on the known mechanisms of lysosomal activity: accumulating lines of independent experimental evidence indicated that the degradation of at least certain classes of cellular proteins must be nonlysosomal.

Yet, in the absence of any 'alternative', researchers came up with different explanations, some more substantiated and others less, to defend the 'lysosomal' hypothesis. First was the gradual discovery, coming from different laboratories, that different proteins vary in their stability and their half-life times can span three orders of magnitude, from a few minutes to many days. Thus, the $t_{1/2}$ of ornitihine decarboxylase (ODC) is ~10 min, while that of glucose-6-phosphate dehydrogenase (G6PD) is 15 h (for review articles, see, e.g., Schimke and Doyle [13] and Goldberg and St John [14]). Also, the rates of degradation of many proteins were shown to change with changing physiological conditions, such as availability of nutrients or hormones. It was conceptually difficult to reconcile the findings of distinct and changing half-lives of different proteins with the mechanism of action of the lysosome, where the microautophagic vesicle contains the entire cohort of cellular (cytosolic) proteins that are therefore expected to degrade at the same rate. Similarly, changing pathophysiological conditions, such as starvation or resupplementation of nutrients, were expected to affect the stability of all cellular proteins to the same extent. Clearly, this was not the case.

Another source of concern about the lysosome as the organelle in which intracellular proteins are degraded was the finding that specific and general inhibitors of lysosomal proteases have different effects on different populations of proteins, making it clear that distinct classes of proteins are targeted by different proteolytic machineries. Thus, the degradation of endocytosed/pinocytosed extracellular proteins was significantly inhibited, a partial effect was observed on the degradation of long-lived cellular proteins, and almost no effect was observed on the degradation of short-lived and abnormal/mutated proteins.

Finally, the thermodynamically paradoxical observation that the degradation of cellular proteins requires metabolic energy, and more importantly, the emerging evidence that the proteolytic machinery uses the energy directly, were in contrast with the known mode of action of lysosomal proteases that under the appropriate acidic conditions, and similar to all known proteases, degrade proteins in an exergonic manner.

The assumption that the degradation of intracellular proteins is mediated by the lysosome was nevertheless logical. Proteolysis results from direct interaction between the target substrates and proteases, and therefore it was clear that active proteases cannot be free in the cytosol, which would have resulted in destruction of the cell. Thus, it was recognized that any suggested proteolytic machinery that mediates degradation of intracellular proteins must also be equipped with a mechanism that separates – physically or virtually – the proteases and their substrates, and enables them to associate only when needed. The lysosomal membrane provided this fencing mechanism. Obviously, nobody could have predicted that a new mode of posttranslational modification – ubiquitination – could function as a proteolysis signal, and that untagged proteins will remain protected. Thus, while the structure of the lysosome could explain the separation necessary between the proteases and their substrates, and autophagy could explain the mechanism of entry of cytosolic proteins into the lysosomal lumen, major problems have remained unsolved. Important among them were: (*i*) the varying half-lives, (*ii*) the energy requirement, and (*iii*) the distinct response of different populations of proteins to lysosomal inhibitors. Thus, according to one model, it was proposed that different proteins have different sensitivities to lysosomal proteases, and their half-lives *in vivo* correlate with their sensitivity to the action of lysosomal proteases *in vitro* [15].

To explain an extremely long half-life for a protein that is nevertheless sensitive to lysosomal proteases, or alterations in the stability of a single protein under various physiological states, it was suggested that although all cellular proteins are engulfed into the lysosome, only the short-lived proteins are degraded, whereas the long-lived proteins exit back into the cytosol: '*To account for differences in half-life among cell components or of a single component in various physiological states, it was necessary to include in the model the possibility of an exit of native components back to the extralysosomal compartment*' [16]. According to a different model, selectivity is determined by the binding affinity of the different proteins for the lysosomal membrane, which controls their entry rates into the lysosome, and subsequently their degradation rates [17]. For a selected group of proteins, such as the gluconeogenetic enzymes phosphoenol-pyruvate carboxykinase (PEPCK) and fructose-1,6-biphosphatase, it was suggested, though not firmly substantiated, that their degradation in the yeast vacuole is regulated by glucose via a mechanism called 'catabolite inactivation' that possibly involves their phosphorylation. However this regulated mechanism for vacuolar degradation is limited only to a small and specific group of proteins (see, e.g., Müller *et al.* [18]; reviewed in Holzer [19]). More recent studies have shown that at least for stress-induced macroautophagy, a general sequence of amino acids, KFFERQ, directs, via binding to a specific 'receptor' and along with cytosolic and lysosomal chaperones, the regulated entry of many cytosolic proteins into the lysosomal lumen. While further corroboration of this hypothesis is still required, it explains the mass entry of a large population of proteins that contain a homologous sequence, but not the targeting for degradation of a specific protein under defined conditions (reviewed in Majeski and Dice [20], and Cuervo and Dice [21]). The energy requirement for protein degradation was described as indirect, and necessary, for example, for protein transport across the lysosomal membrane [22] and/or for the activity of the H^+ pump and the maintenance of the low acidic intralysosomal pH that is necessary for optimal activity of the proteases [23] We now know that both mechanisms require energy. In the absence of any alternative, and with lysosomal degradation as the most logical explanation for targeting all known classes of proteins at the time, Christian de Duve summarized his view on the subject in a review article published in the mid-1960s, saying: '*Just as extracellular digestion is successfully carried out by the concerted action of enzymes with limited individual capacities, so, we believe, is intracellular digestion*' [24]. The problem of different sensitivities of distinct protein groups to lysosomal inhibitors has remained unsolved, and may have served as an important trigger in future quest for a nonlysosomal proteolytic system. Progress in identifying the elusive, nonlysosomal proteolytic system(s) was hampered by the lack of a cell-free preparation that could faithfully replicate the cellular proteolytic events – degrading proteins in a specific and energy requiring mode. An important breakthrough was made by Rabinovitz and Fisher, who found that rabbit reticulocytes degrade abnormal, amino-acid analogue-containing hemoglobin [25]. Their experiments modelled known disease states, the hemoglobinopathies. In these diseases, abnormal mutated hemoglobin chains (such as sickle cell hemoglobin) or excess of unassembled normal hemoglobin chains (which are synthesized normally, but also excessively in thalassemias, diseases in which the pairing chain is not synthesized at all or is mutated and rapidly degraded, and consequently the bi-heterodimeric hemoglobin complex is not assembled) are rapidly degraded in the reticulocyte [26,27]. Reticulocytes are terminally differentiating red blood cells that do not contain lysosomes. Therefore, it was postulated that the degradation of hemoglobin in these cells is mediated by a nonlysosomal machinery. Etlinger and Goldberg [28] were the first to isolate and characterize a cell-free proteolytic preparation from reticulocytes. The crude extract selectively degraded abnormal hemoglobin, required ATP hydrolysis, and acted optimally at a neutral pH, which further corroborated the assumption that the proteolytic activity was of a nonlysosomal origin.

A similar system was isolated and characterized later by Hershko, Ciechanover, and their co-workers [29]. Additional studies by this group led subsequently to resolution, characterization and purification of the major enzymatic components from this extracts and to the discovery of the ubiquitin (Ub) signalling system (see below).

The Lysosome Hypothesis is Challeged

As mentioned above, the unravelled mechanism(s) of action of the lysosome could explain only partially, and at times not satisfactorily, several key emerging characteristics of intracellular protein degradation. Among them were the heterogeneous stability of individual proteins, the effect of nutrients and hormones on their degradation, and the dependence of intracellular proteolysis on metabolic energy. The differential effect of selective inhibitors on the degradation of different classes of cellular proteins (see above but mostly below) could not be explained at all. The evolution of methods to monitor protein kinetics in cells together with the development of specific and general lysosomal inhibitors has resulted in the identification of different classes of cellular proteins, long- and short-lived, and the discovery of the differential effects of the inhibitors on these groups (see, e.g., Knowles and Ballard [30] and Neff *et al.* [31]). An elegant experiment in this respect was carried out by Brian Poole and his co-workers in the Rockefeller University. Poole was studying the effect of lysosomotropic agents, weak bases such as ammonium chloride and chloroquine, that accumulate in the lysosome and dissipate its low acidic pH. It was assumed that this mechanism underlies also the antimalarial activity of chloroquine and similar drugs where they inhibit the activity of parasite's lysosome, 'paralyzing' its ability to digest the host's hemoglobin during the intraerythrocytic stage of its life cycle. Poole and his co-workers metabolically labelled endogenous proteins in living macrophages with ^{3}H-labelled leucine and 'fed' them with dead macrophages that had been previously labelled with ^{14}C-leucine. They assumed, apparently correctly, that the dead macrophages debris and proteins will be phagocytosed by live macrophages and targeted to the lysosome for degradation. They monitored the effect of lysosomotropic agents on the degradation of these two protein populations. In particular, they studied the effect of the weak bases chloroquine and ammonium chloride (which enter the lysosome and neutralize the H^+ ions), and the acid ionophore X537A, which dissipates the H^+ gradient across the lysosomal membrane. They found that these drugs specifically inhibited the degradation of extracellular proteins, but not that of intracellular proteins [32]. Poole summarized these experiments and explicitly predicted the existence of a nonlysosomal proteolytic system that degrades intracellular proteins: '*Some of the macrophages labelled with tritium were permitted to endocytise the dead macrophages labelled with ^{14}C. The cells were then washed and replaced in fresh medium. In this way, we were able to measure in the same cells the digestion of macrophage proteins from two sources. The exogenous proteins will be broken down in the lysosomes, while the endogenous proteins will be broken down wherever it is that endogenous proteins are broken down during protein turnover*' [33].

The requirement for metabolic energy for the degradation of both prokaryotic [34] and eukaryotic [10,35] proteins was difficult to understand. Proteolysis is an exergonic process and the thermodynamically paradoxical energy requirement for intracellular proteolysis made researchers believe that energy cannot be consumed directly by proteases or the proteolytic process per se, and is used indirectly. As Simpson summarized his findings [10] '*The data can also be interpreted by postulating that the release of amino acids from protein is itself directly dependent on energy supply. A somewhat similar hypothesis, based on studies on autolysis in tissue minces, has recently been advanced, but the supporting data are very difficult to interpret. However, the fact that protein hydrolysis as catalyzed by the familiar proteases and peptidases occurs exergonically, together with the consideration that autolysis in excised organs or tissue minces continues for weeks, long after phosphorylation or oxidation ceased, renders improbable the hypothesis of the direct energy dependence of the reactions leading to protein breakdown*'. Being cautious, however, and probably unsure about this unequivocal conclusion, Simpson still left a narrow orifice opened for a proteolytic process that requires energy in a direct manner: '*However, the results do not exclude the existence of two (or more) mechanisms of protein breakdown, one hydrolytic, the other energy-requiring*'. Since any proteolytic process must be at one point or another hydrolytic, the statement that makes a distinction between a hydrolytic process and an energy-requiring, yet nonhydrolytic one, is not clear. Judging the statement from a historical point of view and knowing the mechanism of action of the Ub system, where energy is required also in the prehydrolytic step (Ub conjugation), Simpson may have thought of a two-step mechanism, but did not give it a clear description. At the end of this clearly understandable and apparently difficult deliberation, he left us with a vague explanation linking protein degradation to protein synthesis, a process that was known to require metabolic energy: '*The fact that a supply of energy seems to be necessary for both the incorporation and the release of amino acids from protein might well mean that the two processes are interrelated. Additional data suggestive of such a view are available from other types of experiments. Early investigations on nitrogen balance by Benedict, Folin, Gamble, Smith, and others point to the fact that the rate of protein catabolism varies with the dietary protein level. Since the protein level of the diet would be expected to exert a direct influence on synthesis rather than breakdown, the altered catabolic rate could well be caused by a change in the rate of synthesis*' [10]. With the discovery of lysosomes in eukaryotic cells it could be argued that energy is required for the transport of substrates into the lysosome or for maintenance of the low intralysosomal pH (see above), for example. The observation by Hershko and Tomkins that the activity of tyrosine aminotransferase (TAT) was stabilized following depletion of ATP [35,36] indicated that energy may be required at an early stage of the proteolytic process, most probably before proteolysis occurs. Yet, it did not provide a clue as for the mechanism involved: energy could be used, for example, for specific modification of TAT, for example phosphorylation, that would sensitize it to degradation by the lysosome or by a yet unknown proteolytic mechanism, or for a modification that activates its putative protease. It could also be used for a more general lysosomal mechanism, one that involves transport of TAT into the lysosome, for example. The energy inhibitors inhibited almost completely degradation of the entire population of cell proteins, confirming previous studies (e.g. Simpson [10]) and suggesting a general role for energy in protein catabolism. Yet, an interesting finding was that energy inhibitors had an effect that was distinct form that of protein synthesis inhibitors which affected only enhanced degradation (induced by steroid hormone depletion), but not basal degradation. This finding ruled out, at least partially, a tight linkage between protein synthesis and degradation. In bacteria, which lack lysosomes, an argument involving energy requirement for lysosomal degradation could not have been proposed, but other indirect effects of ATP hydrolysis could have affected proteolysis in *E. coli*, such as phosphorylation of substrates and/or proteolytic enzymes, or maintenance of the 'energized membrane state'. According to this model, proteins could become susceptible to proteolysis by changing their conformation, for example, following association with the cell membrane that maintains a local, energy-dependent gradient of a certain ion. While such an effect was ruled out [37], and since there was no evidence for a phosphorylation mechanism (although the proteolytic machinery in prokaryotes had not been identified at that time), it seemed that at least in bacteria, energy is required directly for the proteolytic process. In any event, the requirement for metabolic energy for protein degradation in both prokaryotes and eukaryotes, a process that is exergonic thermodynamically, strongly indicated that in cells proteolysis is highly regulated, and that a similar principle/mechanism has been preserved along evolution of the two kingdoms. Implying from the possible direct requirement for ATP in degradation of proteins in bacteria, it was not too unlikely to assume a similar direct mechanism in the degradation of cellular proteins in eukaryotes. Supporting this notion was the description of the cell-free proteolytic system in reticulocytes [28,29], a cell that lacks lysosomes, which indicated that energy is probably required directly for the proteolytic process, although here too, the underlying mechanisms had remained enigmatic at the time.

Yet, the description of the cell-free system paved the road for detailed dissection of the underlying mechanisms involved.

The Ubiquitin-Proteasome System

The cell-free proteolytic system from reticulocytes [28,29] turned out to be an important and rich source for the purification and characterization of the enzymes that are involved in the UPS. Initial fractionation of the crude reticulocyte cell extract on the anion-exchange resin diethylaminoethyl cellulose yielded two fractions, which were both required to reconstitute the energy-dependent proteolytic activity that is found in the crude extract. The unadsorbed flow through material was denoted fraction I, and the high salt eluate of the adsorbed proteins which was denoted fraction II (Table 1) [38]. This was an important observation and a lesson for the future dissection of the system. For one it suggested that the system is not composed of a single 'classical' protease that has evolved evolutionarily to acquire energy dependence (although such energy-dependent proteases, the mammalian 26S proteasome (see below) and the prokaryotic *Lon* gene product have been described later), but that it is made of at least two components. This finding of a two-component, energy-dependent protease, left the researchers with no paradigm to follow, and in attempts to explain the finding, they suggested, for example, that the two fractions could represent an inhibited protease and its activator. Second, learning from this reconstitution experiment and the essential dependence between the two active components, we continued to reconstitute activity from resolved fractions whenever we encountered a loss of activity along further purification steps. This biochemical 'complementation' approach resulted in the discovery of additional enzymes of the system, all required to be present in the reaction mixture in order to catalyze the multistep proteolysis of the target substrate. We chose first to purify the active component from fraction I. It was found to be a small, ~8.5 kDa heat stable protein that was designated ATP-dependent Proteolysis Factor 1, APF-1. APF-1 was later identified as Ub (see below; I am using the term APF-1 to the point in which it was identified as Ub and then change terminology accordingly). In retrospect, the decision to start the purification efforts with fraction I turned out to be important, as fraction I contained only one single protein – APF-1 – that was necessary to stimulate proteolysis of the model substrate we used at the time, while fraction II turned out to contain many more. Later studies showed that fraction I contains other components necessary for the degradation of other substrates, but these were not necessary for the reconstitution of the system at that time. This enabled us not only to purify APF-1 but also to quickly decipher its mode of action. If we would have started our purification efforts with fraction II, we would have encountered a significantly bumpier road. A critically important finding that paved the way for future developments in the field was that multiple moieties of APF-1 are covalently conjugated to the target substrate when incubated in the presence of fraction II, and the modification requires ATP (Ciechanover *et al.* [39] and Hershko *et al.* [40]; Figures 3 and 4). It was also found that the modification is reversible, and APF-1 can be removed from the substrate or its degradation products [40]. The discovery that APF-1 is covalently conjugated to protein substrates and stimulates their proteolysis in the presence of ATP and crude fraction II, led in 1980 to the proposal of a model according to which protein substrate modification by multiple moieties of APF-1 targets it for degradation by a downstream, at that time unidentified, protease that cannot recognize the unmodified substrate; following degradation, reusable APF-1 is released [40].

Table 1. Resolution of the ATP-dependent proteolytic activity from crude reticulocyte extract into two essentially required complementing activities (adapted from Ciechanover *et al.* [38]; with permission from Elsevier/*Biochem Biophys Res Commun*)

Fraction	Degradation of [^{3}H]globin (%)	
	-ATP	+ATP
Lysate	1.5	10
Fraction I	0.0	0.0
Fraction II	1.5	2.7
Fraction I and Fraction II	1.6	10.6

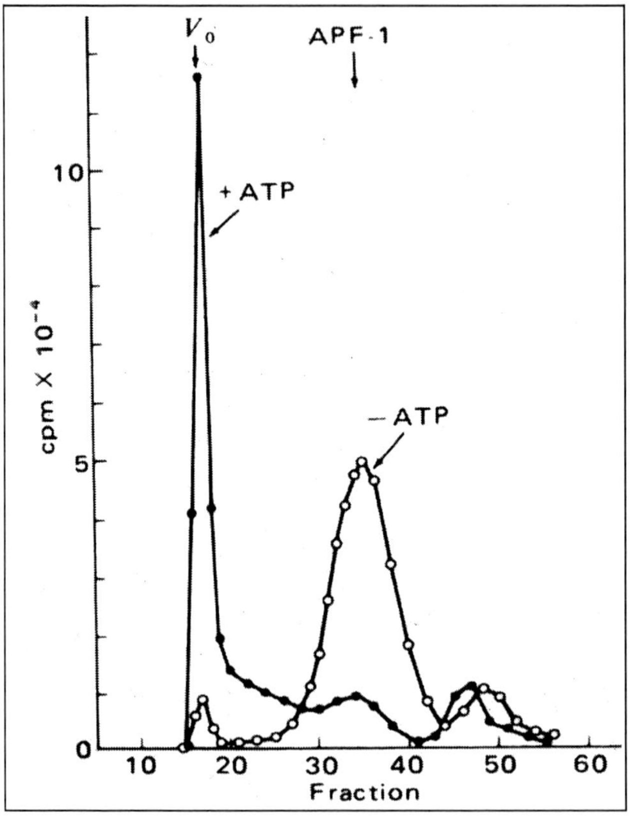

Figure 3. APF-1/ubiquitin is shifted to high molecular mass compound(s) following incubation in ATP-containing crude cell extract. ^{125}I-labelled APF-1/ubiquitin was incubated with reticulocyte crude Fraction II in the absence (open circles) or presence (closed circles) of ATP, and the reaction mixtures were resolved via gel filtration chromatography. The radioactivity measured in each fraction is shown. As can be seen, following addition of ATP, APF-1/ubiquitin becomes covalently attached to some component(s) in fraction II, which could be another enzyme of the system or its substrate(s) (with permission from Proceedings of the National Academy of the USA; published originally in Ciechanover *et al.* [39]).

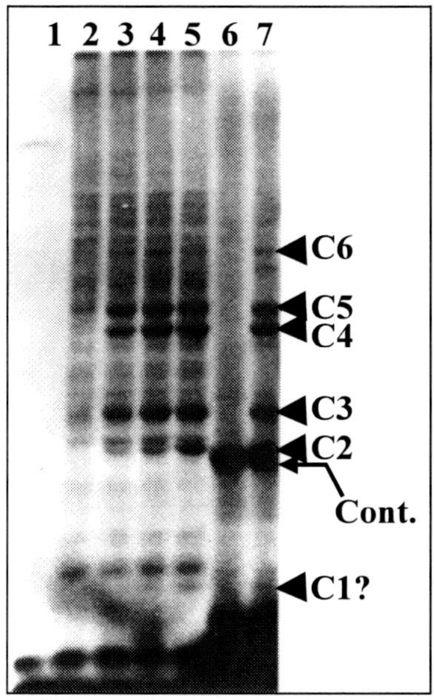

Figure 4. Multiple molecules of APF-1/ubiquitin are conjugated to the proteolytic substrate, probably signalling it for degradation. To interpret the data described in the experiment depicted in Figure 2 and to test the hypothesis that APF-1 is conjugated to the target proteolytic substrate, ^{125}I-APF-1/ubiquitin was incubated along with crude Fraction II (*Figure 3* and text) in the absence (lane 1) or presence (lanes 2–5) of ATP and in the absence (lanes 1,2) or presence (lanes 3–5) of increasing concentrations of unlabelled lysozyme. Reaction mixtures resolved in lanes 6 and 7 were incubated in the absence (lane 6) or presence (lane 7) of ATP, and included unlabelled APF-1/ubiquitin and ^{125}I-labelled lysozyme. C1–C6 denote specific APF-1/ubiquitin–lysozyme adducts in which the number of APF-1/ubiquitin moieties bound to the lysozyme moiety of the adduct is increasing, probably from 1 to 6. Reactions mixtures were resolved via sodium dodecyl sulfate-polyacrylamide gel electrophoresis (SDS-PAGE) and visualized following exposure to an X-ray film (autoradiography) (with permission from Proceedings of the National Academy of the USA; published originally in Hershko *et al.* [40]).

Aminoacid analysis of APF-1, along with its known molecular mass and other general characteristics raised the suspicion that APF-1 is Ub [41], a known protein of previously unknown function. Indeed, Wilkinson *et al.* [42] confirmed unequivocally that APF-1 is indeed Ub. Ub is a small, heat-stable and highly evolutionarily conserved protein of 76 residues. It was first purified during the isolation of thymopoietin [43] and was subsequently found to be ubiquitously expressed in all kingdoms of living cells, including prokaryotes [44]. Interestingly, it was initially found to have lymphocyte-differentiating properties, a characteristic that was attributed to the stimulation of adenylate cyclase [44,45]. Accordingly, it was named UBIP for ubiquitous immunopoietic polypeptide [44]. However, later studies showed that Ub is not involved in the immune response [46], and that it was a contaminating endotoxin in the preparation that generated the adenylate cyclase and the T-cell differentiating activities. Furthermore, the sequence of several eubacteria and archaebacteria genomes as well as biochemical analyses in these organisms (unpublished) showed that Ub is restricted only to eukaryotes. The finding of Ub in bacteria [44] was probably due to contamination of the bacterial extract with yeast Ub derived from the yeast extract in which the bacteria were grown. While in retrospect the name Ub is a misnomer, as it is restricted to eukaryotes and is not ubiquitous as was previously thought, from historical reasons it has still remained the name of the protein. Accordingly, and in order to avoid confusion, I suggest that the names of other novel enzymes and components of the Ub system, but of other systems as well, should remain as were first coined by their discoverers.

An important development in the Ub research field was the discovery that a single Ub moiety can be covalently conjugated to histones, particularly to histones H2A and H2B. While the function of these adducts has remained elusive until recently, their structure was unravelled in the mid-1970s. The structure of the Ub conjugate of H2A (uH2A; was also designated protein A24) was deciphered by Goldknopf and Busch [47,48] and by Hunt and Dayhoff [49], who found that the two proteins are linked through a fork-like, branched isopeptide bond between the carboxy-terminal glycine of Ub (Gly^{76}) and the ε-NH_2 group of an internal lysine (Lys^{119}) of the histone molecule. The isopeptide bond found in the histone–Ub adduct was suggested to be identical to the bond that was found between Ub and the target proteolytic substrate [50] and between the Ub moieties in the polyubiquitin chain [51,52] that is synthesized on the substrate and that functions as a proteolysis recognition signal for the downstream 26S proteasome. In this particular polyubiquitin chain, the linkage is between Gly^{76} of one Ub moiety and internal Lys48 of the previously conjugated moiety. Only Lys^{48}-based Ub chains are recognized by the 26S proteasome and serve as proteolytic signals. In recent years, it has been shown that the first Ub moiety can also be attached in a linear mode to the N-terminal residue of the proteolytic target substrate [53]. However, the subsequent Ub moieties are generating Lys^{48}-based polyubiquitin chain on the first linearly fused moiety. N-terminal ubiquitination is clearly required for targeting naturally occurring lysine-less proteins for degradation. Yet, several lysine-containing proteins have also been described that traverse this pathway, for example the muscle specific transcription factor MyoD. In these proteins the internal lysine residues are probably not accessible to the cognate ligases. Other types of polyubiquitin chains have also been described that are not involved in targeting the conjugated substrates for proteolysis. Thus, a Lys^{63}-based polyubiquitin chain has been described that is probably necessary to activate transcription factors (reviewed recently in Muratani and Tansey [54]). Interestingly, the role of monoubiquitination of histones has also been identified recently and this modification is also involved in regulation of transcription, probably via modulation of the structure of the nucleosomes (for recent reviews, see, e.g., Zhang [55] and Osley [56]).

The identification of APF-1 as Ub, and the discovery that a high-energy isopeptide bond, similar to the one that links Ub to histone H2A, links it also to the target proteolytic substrate, resolved at that time the enigma of the energy requirement for intracellular proteolysis (see however below) and paved the road to the untangling of the complex mechanism of isopeptide bond formation. This process turned out to be similar to that of peptide bond formation that is catalyzed by tRNA synthetase following amino-acid activation during protein synthesis or during the nonribosomal synthesis of short peptides [57]. Using the unravelled mechanism of Ub activation and immobilized Ub as a 'covalent' affinity bait, the three enzymes that are involved in the cascade reaction of Ub conjugation were purified by Ciechanover, Hershko, and their co-workers. These enzymes are: (*i*) E1, the Ub-activating enzyme; (*ii*) E2, the Ub-carrier protein; and (*iii*) E3, the Ub-protein ligase [58,59]. The discovery of an E3, which is a specific substrate-binding component, indicated a possible solution to the problem of the varying stabilities of different proteins – they might be specifically recognized and targeted by different ligases.

In a short period, the Ub tagging hypothesis received substantial support. For example, Chin and co-workers injected into HeLa cells labelled Ub and hemoglobin and denatured the injected hemoglobin by oxidizing it with phenylhydrazine. They found that Ub conjugation to globin is markedly enhanced by denaturation of hemoglobin and the concentration of globin–Ub conjugates was proportional to the rate of hemoglobin degradation [60] Hershko *et al.* [61] observed a similar correlation for abnormal, aminoacid analogue-containing short-lived proteins. A previously isolated cell cycle arrest mutant that loses the Ub–histone H2A adduct at the permissive temperature [62] was found by Finley, Ciechanover and Varshavsky to harbor a thermolabile E1 [63]. Following heat inactivation, the cells fail to degrade normal short-lived proteins [64]. Although the cells did not provide direct evidence for substrate ubiquitination as a destruction signal, they still provided the strongest direct linkage between Ub conjugation and degradation.

At this point, the only missing link was the identification of the downstream protease that would specifically recognize ubiquitinated substrates. Tanaka *et al.* [65] identified a second ATP-requiring step in the reticulocyte proteolytic system, which occurred after Ub conjugation, and Hershko *et al.* [66] demonstrated that the energy is required for conjugate degradation. An important advance in the field was a discovery by Hough and co-workers, who partially purified and characterized a high-molecular mass alkaline protease that degraded Ub adducts of lysozyme but not untagged lysozyme, in an ATP-dependent mode [67]. This protease which was later called the 26S proteasome (see below) provided all the necessary criteria for being the specific proteolytic arm of the Ub system. This finding was confirmed, and the protease was further characterized by Waxman *et al.* [68], who found that it is an unusually large, is ~1.5MDa enzyme, unlike any other known protease. A further advance in the field was the discovery [69] that a smaller neutral multisubunit 20S protease complex that was discovered together with the larger 26S complex, is similar to a 'multicatalytic proteinase complex' (MCP) that was described earlier in bovine pituitary gland by Wilk and Orlowski [70]. This 20S protease is ATP-independent and has different catalytic activities, cleaving on the carboxy-terminal side of hydrophobic, basic and acidic residues. Hough *et al.* [69] raised the possibility – although they did not show it experimentally – that this 20S protease can be a part of the larger 26S protease that degrades the Ub adducts. Later studies showed that indeed, the 20S complex is the core catalytic particle of the larger 26S complex [71,72]

However, a strong evidence that the active 'mushroom'-shaped 26S protease is generated through the assembly of two distinct subcomplexes – the catalytic 20S cylinder-like MCP and an additional 19S ball-shaped subcomplex (that was predicted to have a regulatory role) – was provided only in the early 1990s by Hoffman *et al.* [73], who mixed the two purified particles and generated the active 26S enzyme. The proteasome is a large, 26S, multicatalytic protease that degrades polyubiquitinated proteins to small peptides. It is composed of two subcomplexes: a 20S core particle (CP) that carries the catalytic activity, and a regulatory 19S regulatory particle (RP). The 20S CP is a barrel-shaped structure composed of four stacked rings, two identical outer α rings and two identical inner β rings. The eukaryotic α and β rings are composed each of seven distinct subunits, giving the 20S complex the general structure of $\alpha_{1-7}\beta_{1-7}\beta_{1-7}\alpha_{1-7}$. The catalytic sites are localized to some of the β subunits. Each extremity of the 20S barrel can be capped by a 19S RP each composed of 17 distinct subunits, nine in a 'base' subcomplex, and eight in a 'lid' subcomplex. One important function of the 19S RP is to recognize ubiquitinated proteins and other potential substrates of the proteasome. Several Ub binding subunits of the 19S RP have been identified, although their biological roles and mode of action have not been discerned. A second function of the 19S RP is to open an orifice in the a ring that will allow entry of the substrate into the proteolytic chamber. Also, since a folded protein would not be able to fit through the narrow proteasomal channel, it is assumed that the 19S particle unfolds substrates and inserts them into the 20S CP. Both the channel opening function and the unfolding of the substrate require metabolic energy, and indeed, the 19S RP 'base' contains six different ATPase subunits. Following degradation of the substrate, short peptides derived from the substrate are released, as well as reusable Ub (for a scheme describing the Ub system, see Figure 5; for the structure of the 26S proteasome, see Figure 6).

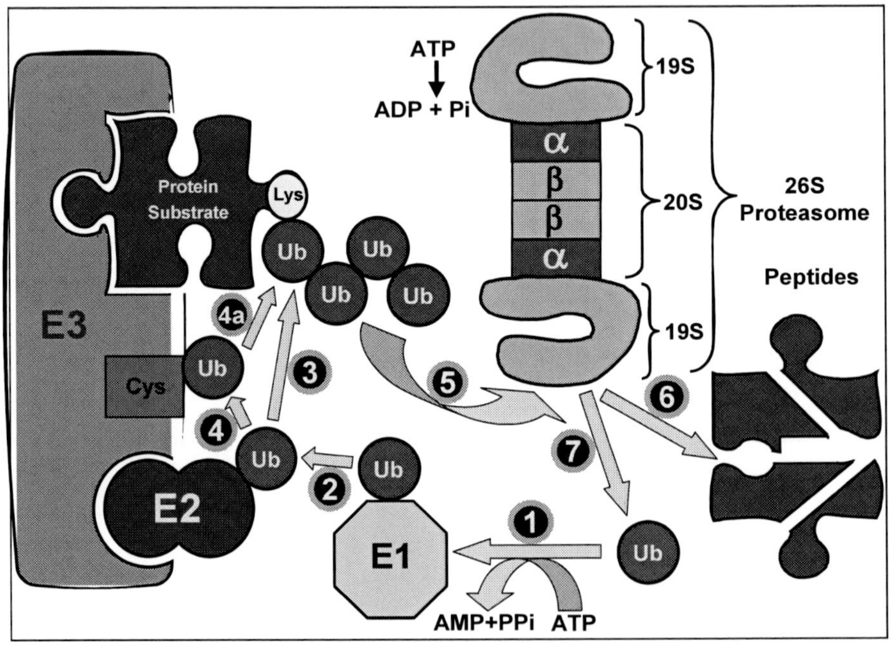

Figure 5. The ubiquitin–proteasome proteolytic system. Ubiquitin is activated by the ubiquitin-activating enzyme, E1 (1) followed by its transfer to a ubiquitin carrier protein (ubiquitin-conjugating enzyme, UBC), E2 (2). E2 transfers the activated ubiquitin moieties to the protein substrate that is bound specifically to a unique ubiquitin ligase E3. The transfer is either direct [(3) in the case of RING finger ligases] or via an additional thiol-ester intermediate on the ligase [(4, 4a) in case of HECT domain ligases]. Successive conjugation of ubiquitin moieties to one another generates a polyubiquitin chain that serves as the binding (5) and degradation signal for the downstream 26S proteasome. The substrate is degraded to short peptides (6), and free and reusable ubiquitin is released by de-ubiquitinating enzymes (DUBs) (7).

Concluding Remarks

The evolution of proteolysis as a centrally important regulatory mechanism is a remarkable example for the evolution of a novel biological concept and the accompanying battles to change paradigms. The five decade journey between the early 1940s and early 1990s began with fierce discussions on whether cellular proteins are static as has been thought for a long time, or are turning over. The discovery of the dynamic state of proteins was followed by the discovery of the lysosome, which was believed – between the mid-1950s and mid-1970s – to be the organelle within which intracellular proteins are destroyed. Independent lines of experimental evidence gradually eroded the lysosomal hypothesis and resulted in a new idea that the bulk of intracellular proteins are degraded – under basal metabolic conditions – via a nonlysosomal machinery. This resulted in the discovery of the Ub system in the late 1970s and early 1980s.

With the identification of the reactions and enzymes that are involved in the Ub–proteasome cascade, a new era in the protein degradation field began at the late 1980s and early 1990s. Studies that showed that the system is involved in targeting of key regulatory proteins – such as light-regulated proteins in plants, transcriptional factors, cell cycle regulators and tumor suppressors and promoters – started to emerge (see, e.g., Shanklin *et al.* [74], Hochstrasser and Varshavsky [75], Scheffner *et al.* [76], Glotzer *et al.* [77] and Ciechanover *et al.* [78]). They were followed by numerous studies on the underlying mechanisms involved in the degradation of specific proteins, each with its own unique mode of recognition and regulation. The unravelling of the human genome revealed the existence of hundreds of distinct E3s, attesting to the complexity and the high specificity and selectivity of the system. Two important advances in the field were the discovery of the nonproteolytic functions of Ub such as activation of transcription and routing of proteins to the vacuole, and the discovery of modification by Ub-like proteins (UBLs), which are also involved in numerous nonproteolytic functions such as directing proteins to their subcellular destination, protecting proteins from ubiquitination, or controlling entire processes such as autophagy (see, e.g., Mizushima *et al.* [79]) (for the different roles of modifications by Ub and UBLs, see Figure 7).

All these studies have led to the emerging realization that this novel mode of covalent conjugation plays a key role in regulating a broad array of cellular process – among them cell cycle and division, growth and differentiation, activation and silencing of transcription, apoptosis, the immune and inflammatory response, signal transduction, receptor mediated endocytosis, various metabolic pathways, and the cell quality control – through proteolytic and nonproteolytic mechanisms. The discovery that Ub modification plays a role in routing proteins to the lysosome/ vacuole and that modification by specific and unique UBLs and modification system controls autophagy closed an exciting historical cycle, since it demonstrated that the two apparently distinct systems communicate with one another.

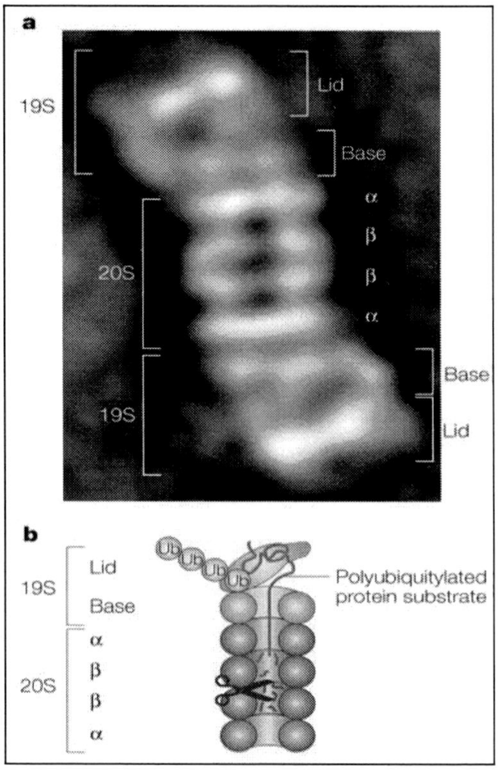

Figure 6. The Proteasome. The proteasome is a large, 26S, multicatalytic protease that degrades polyubiquitinated proteins to small peptides. It is composed of two subcomplexes: a 20S core particle (CP) that carries the catalytic activity, and a regulatory 19S regulatory particle (RP). The 20S CP is a barrel-shaped structure composed of four stacked rings, two identical outer a rings and two identical inner b rings. The eukaryotic α and β rings are composed each of seven distinct subunits, giving the 20S complex the general structure of $\alpha_{1\text{-}7}\beta_{1\text{-}7}\beta_{1\text{-}7}\alpha_{1\text{-}7}$. The catalytic sites are localized to some of the β subunits. Each extremity of the 20S barrel can be capped by a 19S RP each composed of 17 distinct subunits, 9 in a 'base' subcomplex, and 8 in a 'lid' subcomplex. One important function of the 19S RP is to recognize ubiquitinated proteins and other potential substrates of the proteasome. Several ubiquitin-binding subunits of the 19S RP have been identified, however, their biological roles mode of action have not been discerned. A second function of the 19S RP is to open an orifice in the α ring that will allow entry of the substrate into the proteolytic chamber. Also, since a folded protein would not be able to fit through the narrow proteasomal channel, it is assumed that the 19S particle unfolds substrates and inserts them into the 20S CP. Both the channel opening function and the unfolding of the substrate require metabolic energy, and indeed, the 19S RP 'base' contains six different ATPase subunits. Following degradation of the substrate, short peptides derived from the substrate are released, as well as reusable ubiquitin (with permission from Nature Publishing Group copyright MacMillan Magazines Ltd. Published originally in Ciechanover [83]). (a) Electron microscopy image of the 26S proteasome from the yeast *S. cerevisiae*. (b) Schematic representation of the structure and function of the 26SA proteasome.

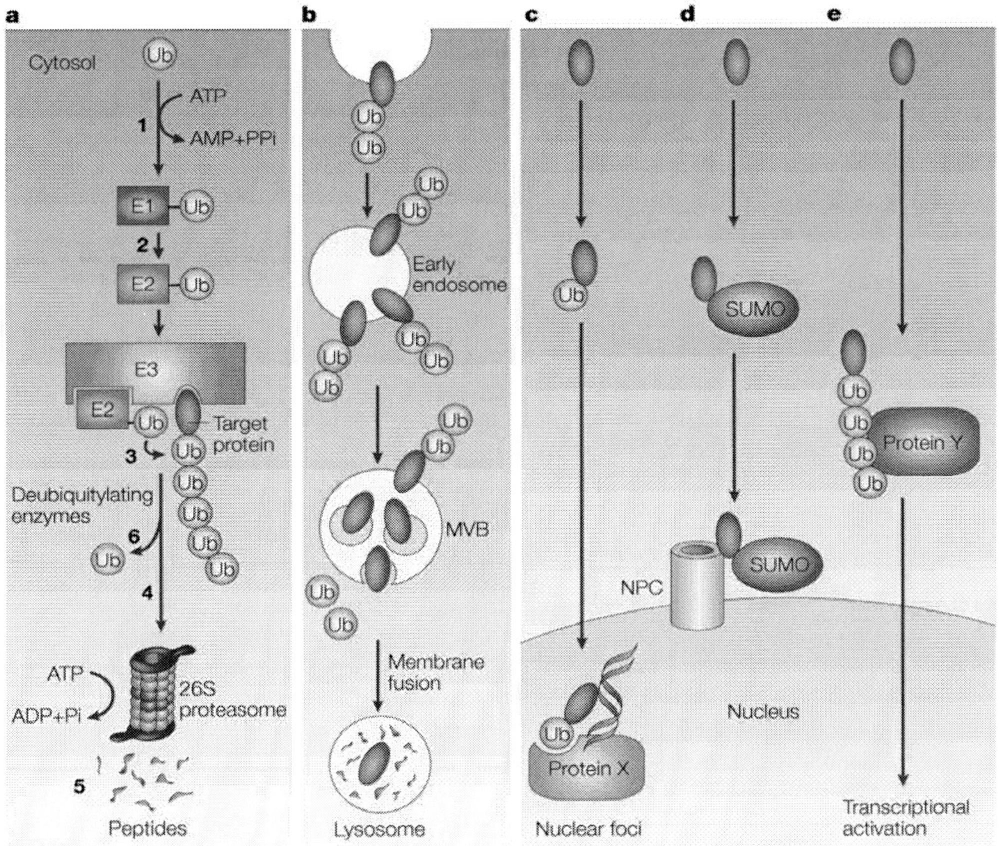

Figure 7. Some of the different functions of modification by ubiquitin and ubiquitin-like proteins. (a) Proteasomal-dependent degradation of cellular proteins (see Figure 4). (b) Mono or oligoubiquitination targets membrane proteins to degradation in the lysosome/vacuole. (c) Monoubiquitination, or (d) a single modification by a ubiquitin-like (UBL) protein, for example SUMO, can target proteins to different subcellular destinations such as nuclear foci or the nuclear pore complex (NPC). Modification by UBLs can serve other, nonproteolytic, functions, such as protecting proteins from ubiqutination or activation of E3 complexes. (e) Generation of a Lys63-based polyubiquitin chain can activate transcriptional regulators, directly or indirectly (via recruitment of other proteins (Protein Y; shown), or activation of upstream components such as kinases). Ub denotes ubiquitin, K denotes Lys, and S denotes Cys. (with permission from Nature Publishing Group, copyright MacMillan Magazines Ltd. Published originally in Ciechanover [83]).

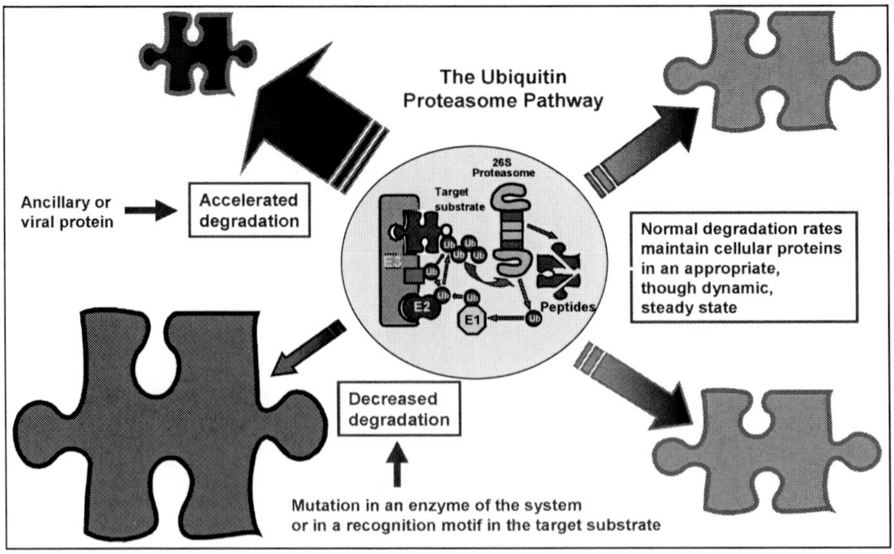

Figure 8 Aberrations in the ubiquitin–proteasome system and pathogenesis of human diseases. Normal degradation of cellular proteins maintains them in a steady-state level, though this level may change under various pathophysiological conditions (upper and lower right side). When degradation is accelerated due an increase in the level of an E3 (Skp2 in the case of p27, for example), or overexpression of an ancillary protein that generates a complex with the protein substrate and targets it for degradation (the Human Papillomavirus E6 oncoprotein that associates with p53 and targets it for degradation by the E6-AP ligase, or the cytomegalovirus-encoded ER proteins US2 and US11 that target MHC class I molecules for ERAD), the steady-state level of the protein decreases (upper left side). A mutation in a ubiquitin ligase [such as occurs in Adenomatous Polyposis Coli – APC, or in E6-AP (Angelmans' Syndrome)] or in the substrate's recognition motif (such as occurs in β-catenin or in ENaC) will result in decreased degradation and accumulation of the target substrate intracellular protein degradation.

With the many processes and substrates targeted by the Ub pathway, it is not surprising to find that aberrations in the system underlie, directly or indirectly, the pathogenesis of many diseases. While inactivation of a major enzyme such as E1 is obviously lethal, mutations in enzymes or in recognition motifs in substrates that do not affect vital pathways or that affect the involved process only partially, may result in a broad array of phenotypes. Likewise, acquired changes in the activity of the system can also evolve into certain pathologies. The pathological states associated with the Ub system can be classified into two groups: (a) those that result from loss of function – mutation in a Ub system enzyme or in the recognition motif in the target substrate that results in stabilization of certain proteins; and (b) those that result from gain of function – abnormal or accelerated degradation of the protein target (for aberrations in the Ub system that result in disease states, see Figure 8). Studies that employ targeted inactivation of genes coding for specific Ub system enzymes and substrates in animals can provide a more systematic view into the broad spectrum of pathologies that may result from aberrations in Ub-mediated proteolysis. Better understanding of the processes and identification of the components involved in the degradation of key regulatory proteins will lead to the development of mechanism-based drugs that will target specifically only the involved proteins. While the first drug, a specific proteasome inhibitor is already on the market [80] it appears that one important hallmark of the new era we are entering now will be the discovery of novel drugs based on targeting of specific processes such as inhibiting aberrant Mdm2- or E6-AP mediated accelerated targeting of the tumor suppressor p53 which will lead to regain of its lost function. Many reviews have been published on different aspects of the Ub system. The purpose of this article was to bring to the reader several milestones along the historical pathway along which the Ub system has been evolved. For additional reading on the Ub system the reader is referred to the many reviews written on the system, among them for example are Glickman and Ciechanover [81] Pickart and Cohen [82]. Some parts of this review, including several figures, are based on another recently published review article [83].

Acknowledgements

Research in the laboratory of Aaron Ciechanover has been supported along the years by grants from the US-Israel Binational Science Foundation (BSF), the Israel Science Foundation (ISF) founded by the Israeli National Academy of Humanities, Arts and Sciences, the German-Israeli Foundation (GIF) for Scientific Research and Development, the Israel Cancer Research Fund (ICRF) USA, the Deutsche-Israeli Cooperation Program (DIP), the European Union (EU), the Israel Cancer Society (ICS), the Prostate Cancer Foundation (PCF) – Israel, the Foundation for Promotion of Research in the Technion and various research grants administered by the Vice President of the Technion for Research. Infrastructural equipment for the laboratory of AC and for the Cancer and Vascular Center has been purchased with the support of the Wolfson Charitable Fund – Center of Excellence for Studies on Turnover of Cellular Proteins and its Implications to Human Diseases.

The review was obtained thanks to the kind help of the Nobel Foundation in Stockholm, http://nobelprize.org (© The Nobel Foundation 2004). Printed with permission from the Nobel Foundation.

References

[1] Clarke HT: Impressions of an organic chemist in biochemistry. *Annu Rev Biochem* 1958, 27:1–14.

[2] Kennedy EP: Hitler's gift and the era of biosynthesis. *J Biol Chem* 2001, 276:42619–42631.

[3] Simoni RD, Hill RL, Vaughan M The use of isotope tracers to study intermediary metabolism: Rudolf Schoenheimer: *J Biol Chem* 2002, 277:e1–e3 (available on-line at: http:/www.jbc.org)

[4] Schoenheimer R, Ratner S, Rittenberg D: Studies in protein metabolism: VII. The metabolism of tyrosine. *J Biol Chem* 1939, 127:333–344

[5] Ratner S, Rittenberg D, Keston AS, Schoenheimer R: Studies in protein metabolism: XIV. The chemical interaction of dietary Glycine and body proteins in rats. *J Biol Chem* 1940, 134:665–676

[6] Schoenheimer R: The Dynamic State of Body Constituents (Cambridge, MA, USA: Harvard University Press, 1942.

[7] Hogness DS, Cohn M, Monod J: Studies on the induced synthesis of β-galactosidase in *Escherichia coli*: The kinetics and mechanism of sulfur incorporation. *Biochim Biophys Acta* 1955, 16:99–116

[8] de Duve C, Gianetto R, Appelmans F Wattiaux R: Enzymic content of the mitochondria fraction. *Nature (London)* 1953, 172:1143–1144

[9] Gianetto R, de Duve C: Tissue fractionatin studies 4. Comparative study of the binding of acid phosphatase, β-glucoronidase and cathepsin by rat liver particles. *Biochem J* 1955, 59:433–438

[10] Simpson MV: The release of labeled amino acids from proteins in liver slices. *J Biol Chem* 1953, 201:143–154

[11] Mortimore GE, Poso AR: Intracellular protein catabolism and its control during nutrient deprivation and supply. *Annu Rev Nutr* 1987, 7:539–564

[12] Ashford TP, Porter KR: Cytoplasmic components in hepatic cell lysosomes. *J Cell Biol* 1962, 12:198–202

[13] Schimke RT, Doyle D: Control of enzyme levels in animal tissues. *Annu Rev Biochem* 1970, 39:929–976

[14] Goldberg AL, St John AC: Intracellular protein degradation in mammalian and bacterial cells: Part 2. *Annu Rev Biochem* 1976, 45:747–803

[15] Segal HL, Winkler JR, Miyagi MP: Relationship between degradation rates of proteins in vivo and their susceptibility to lysosomal proteases. *J Biol Chem* 1974, 249:6364–6365

[16] Haider M, Segal HL: Some characteristics of the alanineaminotransferase and arginase-inactivating system of lysosomes. *Arch Biochem Biophys* 1972, 148:228–237

[17] Dean RT: Lysosomes, protein degradation. *Acta Biol Med Ger* 1977, 36:1815–1820

[18] Müller M, Müller H, Holzer H: Immunochemical studies on catabolite inactivation of phosphoenolpyruvate carboxykinase in *Saccharomyces cerevisiae*. *J Biol Chem* 1981, 256:723–727

[19] Holzer H: Proteolytic catabolite inactivation in *Saccharomyces cerevisiae*. *Revis Biol Celular 1989*, 21:305–319

[20] Majeski AE, Dice JF: Mechanisms of chaperone-mediated autophagy. *Intl J Biochem Cell Biol* 2004, 36:2435–2444

[21] Cuervo AM, Dice JF: Lysosomes, a meeting point of proteins, chaperones, and proteases. *J Mol Med* 1998, 76:6–12

[22] Hayashi M, Hiroi Y, Natori Y: Effect of ATP on protein degradation in rat liver lysosomes. *Nat New Biol* 1973, 242:163–166

[23] Schneider DL: ATP-dependent acidification of intact and disrupted lysosomes: evidence for an ATP-driven proton pump. *J Biol Chem* 1981, 256:3858–3864

[24] de Duve C, Wattiaux R: Functions of lysosomes. *Annu Rev Physiol* 1966, 28:435–492

[25] Rabinovitz M, Fisher JM: Characteristics of the inhibition of hemoglobin synthesis in rabbit reticulocytes by threo-α-amino-β-chlorobutyric acid. *Biochim Biophys Acta* 1964, 91:313–322

[26] Carrell RW, Lehmann H: The unstable haemoglobin haemolytic anaemias. *Semin Hematol* 1969, 6:116–132

[27] Huehns ER, Bellingham AJ: Diseases of function and stability of haemoglobin. *Br J Haematol* 1969, 17:1–10

[28] Etlinger JD, Goldberg AL: A soluble ATP-dependent proteolytic system responsible for the degradation of abnormal proteins in reticulocytes. *Proc Natl Acad Sci USA* 1977, 74:54–58

[29] Hershko A, Heller H, Ganoth D, Ciechanover A: Mode of degradation of abnormal globin chains in rabbit reticulocytes In Protein Turnover and Lysosome Function. Edited by Segal HL, Doyle DJ. New York: Academic Press; 1978:149–169.

[30] Knowles SE, Ballard FJ: Selective control of the degradation of normal and aberrant proteins in Reuber H35 hepatoma cells. *Biochem J* 1976, 156:609–617

[31] Neff NT, DeMartino GN, Goldberg AL: The effect of protease inhibitors and decreased temperature on the degradation of different classes of proteins in cultured hepatocytes. *J Cell Physiol* 1979, 101:439–457

[32] Poole B, Ohkuma S, Warburton MJ: The accumulation of weakly basic substances in lysosomes and the inhibition of intracellular protein degradation. *Acta Biol Med Germ* 1977, 36:1777–1788

[33] Poole B, Ohkuma S, Warburton MJ: Some aspects of the intracellular breakdown of exogenous and endogenous proteins In Protein Turnover and Lysosome Function. Edited by Segal HL, Doyle DJ. New York: Academic Press; 1978:43–58.

[34] Mandelstam J: Turnover of protein in growing and non-growing populations of *Escherichia coli. Biochem J* 1958, 69:110–119

[35] Steinberg D, Vaughan M: Observations on intracellular protein catabolism studied in vitro. *Arch Biochem Biophys* 1956, 65:93–105

[36] Hershko A, Tomkins GM: Studies on the degradation of tyrosine aminotransferase in hepatoma cells in culture: influence of the composition of the medium and adenosine triphosphate dependence. *J Biol Chem* 1971, 246:710–714

[37] Goldberg AL, Kowit JD, Etlinger JD: Studies on the selectivity and mechanisms of intracellular protein degradation In Proteolysis and Physiological Regulation. Edited by Ribbons DW, Brew K. New York: Academic Press; 1976:313–337.

[38] Ciechanover A, Hod Y, Hershko A: A heat-stable polypeptide component of an ATP-dependent proteolytic system from reticulocytes. *Biochem Biophys Res Common* 1978, 81:1100–1105

[39] Ciechanover A, Heller H, Elias S, Haas AL, Hershko A: ATPdependent conjugation of reticulocyte proteins with the polypeptide required for protein degradation. *Proc Natl Acad Sci USA* 1980, 77:1365–1368

[40] Hershko A, Ciechanover A, Heller H, Haas AL, Rose IA: Proposed role of ATP in protein breakdown: conjugation of proteins with multiple chains of the polypeptide of ATP-dependent proteolysis. *Proc Natl Acad Sci USA* 1980, 77:1783–1786

[41] Ciechanover A, Elias S, Heller H, Ferber S, Hershko A: Characterization of the heat-stable polypeptide of the ATP-dependent proteolytic system from reticulocytes. *J Biol Chem* 1980, 255:7525–7528

[42] Wilkinson KD, Urban MK, Haas AL: Ubiquitin is the ATP-dependent proteolysis factor I of rabbit reticulocytes. *J Biol Chem* 1980, 255:7529–7532

[43] Goldstein G: Isolation of bovine thymin, a polypeptide hormone of the thymus. *Nature (London)* 1974, 247:11–14

[44] Goldstein G, Scheid M, Hammerling U, Schlesinger DH, Niall HD, Boyse EA: Isolation of a polypeptide that has lymphocyte-differentiating properties and is probably represented universally in living cells. *Proc Natl Acad Sci USA* 1975, 72:11–15

[45] Schlessinger DH, Goldstein G, Niall HD: The complete amino acid sequence of ubiquitin, an adenylate cyclase stimulating polypeptide probably universal in living cells. *Biochemistry* 1975, 14:2214–2218

[46] Low TLK, Goldstein AL: The chemistry and biology of thymosin:amino acid analysis of thymosin a1 and polypeptide β1. *J Biol Chem* 1979, 254:987–995

[47] Goldknopf IL, Busch H: Remarkable similarities of peptide fingerprints of histone 2A and non-histone chromosomal protein A24. *Biochem Biophys Res Commun* 1975, 65:951–955

[48] Goldknopf IL, Busch H: Isopeptide linkage between non-histone and histone 2A polypeptides of chromosome conjugate-protein A24. *Proc Natl Acad Sci USA* 1977, 74:864–868

[49] Hunt LT, Dayhoff MO: Amino-terminal sequence identity of ubiquitin and the non-histone component of nuclear protein A24. *Biochim Biophys Res Commun* 1977, 74:650–655

[50] Hershko A, Ciechanover A, Rose IA: Identification of the active amino acid residue of the polypeptide of ATP-dependent protein breakdown. *J Biol Chem* 1981, 256:1525–1528

[51] Hershko A, Heller H: Occurrence of a polyubiquitin structure in ubiquitin-protein conjugates. *Biochem Biophys Res Common* 1985, 128:1079–1086

[52] Chau V, Tobias JW, Bachmair A, Mariott D, Ecker D, Gonda DK, Varshavsky A: A multiubiquitin chain is confined to specific lysine in a targeted short-lived protein. *Science* 1989, 243:1576–1583

[53] Ciechanover A, Ben-Saadon R: N-terminal ubiquitination: more protein substrates join in. *Trends Cell Biol* 2004, 14:103–106

[54] Muratani M, Tansey WP: How the ubiquitin–proteasome system controls transcription. *Nat Rev Mol Cell Biol* 2003, 4:192–201

[55] Zhang Y: Transcriptional regulation by histone ubiquitination and deubiquitination. *Genes Dev* 2003, 17:2733–2740

[56] Osley MA: H2B ubiquitylation: the end is in sight. *Biochim Biophys Acta* 2004, 1677:74–78

[57] Lipman F: Attempts to map a process evolution of peptide biosynthesis. *Science* 1971, 173:875–884

[58] Ciechanover A, Elias S, Heller H, Hershko A: 'Covalent affinity' purification of ubiquitin-activating enzyme. *J Biol Chem* 1982, 257:2537–2542

[59] Hershko A, Heller H, Elias S, Ciechanover A: Components of ubiquitin-protein ligase system: resolution, affinity purification and role in protein breakdown. *J Biol Chem* 1983, 258:8206–8214

[60] Chin DT, Kuehl L, Rechsteiner M: Conjugation of ubiquitin to denatured hemoglobin is proportional to the rate of hemoglobin degradation in HeLa cells. *Proc Natl Acad Sci* USA 1982, 79:5857–5861

[61] Hershko A, Eytan E, Ciechanover A, Haas AL: Immunochemical analysis of the turnover of ubiquitin-protein conjugates in intact cells: relationship to the breakdown of abnormal proteins. *J Biol Chem* 1982, 257:13964–13970

[62] Matsumoto Y, Yasuda H, Marunouchi T, Yamada M: Decrease in uH2A (protein A24) of a mouse temperature-sensitive mutant. *FEBS Lett* 1983, 151:139–142

[63] Finley D, Ciechanover A Varshavsky A: Thermolability of ubiquitinactivating enzyme from the mammalian cell cycle mutant ts85. *Cell* 1984, 37:43–55

[64] Ciechanover A, Finley D, Varshavsky A: Ubiquitin dependence of selective protein degradation demonstrated in the mammalian cell cycle mutant ts85. *Cell* 1984, 37:57–66

[65] Tanaka K, Waxman L, Goldberg AL: ATP serves two distinct roles in protein degradation in reticulocytes, one requiring and one independent of ATP. *J Cell Biol* 1983, 96:1580–1585

[66] Hershko A, Leshinsky E, Ganoth D, Heller H: ATP-dependent degradation of ubiquitin-protein conjugates. *Proc Natl Acad Sci USA* 1984, 81:1619–1623

[67] Hough R, Pratt G, Rechsteiner M: Ubiquitin–lysozyme conjugates. Identification and characterization of an ATP-dependent protease from rabbit reticulocyte lysates. *J Biol Chem* 1986, 261:2400–2408

[68] Waxman L, Fagan J, Goldberg AL: Demonstration of two distinct high molecular weight proteases in rabbit reticulocytes, one of which degrades ubiquitin conjugates. *J Biol Chem* 1987, 262:2451–2457

[69] Hough R, Pratt G, Rechsteiner M: Purification of two high molecular weight proteases from rabbit reticulocyte lysate. *J Biol Chem* 1987, 262:8303–8313

[70] Wilk S, Orlowski M: Cation-sensitive neutral endopeptidase: isolation and specificity of the bovine pituitary enzyme. *J Neurochem* 1980, 35:1172–1182

[71] Eytan E, Ganoth D, Armon T, Hershko A: ATP-dependent incorporation of 20S protease into the 26S complex that degrades proteins conjugated to ubiquitin. *Proc Natl Acad Sci USA* 1989, 86:7751–7755

[72] Driscoll J, Goldberg AL: The proteasome (multicatalytic protease) is a component of the 1500-kDa proteolytic complex which degrades ubiquitinconjugated proteins. *J Biol Chem* 1990, 265:4789–4792

[73] Hoffman L, Pratt G, Rechsteiner M: Multiple forms of the 20S multicatalytic and the 26S ubiquitin/ATP-dependent proteases from rabbit reticulocyte lysate. *J Biol Chem* 1992, 267:22362–22368

[74] Shanklin J, Jaben M, Vierstra RD: Red light-induced formation of ubiquitin-phytochrome conjugates: identification of possible intermediates of phytochrome degradation. *Proc Natl Acad Sci USA* 1987, 84:359–363

[75] Hochstrasser M, Varshavsky A: In vivo degradation of a transcriptional regulator: the yeast a2 repressor. *Cell* 1990, 61:697–708

[76] Scheffner M, Werness BA, Huibregtse JM, Levine AJ, Howley PM: The E6 oncoprotein encoded by human papillomavirus types 16 and 18 promotes the degradation of p53. *Cell* 1990, 63:1129–1136

[77] Glotzer M, Murray AW, Kirschner MW: Cyclin is degraded by the ubiquitin pathway. *Nature* 1991, 349:132–138

[78] Ciechanover A, DiGiuseppe JA, Bercovich B, Orian A, Richter JD, Schwartz AL, Brodeur GM: Degradation of nuclear oncoproteins by the ubiquitin system in vitro. *Proc Natl Acad Sci USA* 1991, 88:139–143

[79] Mizushima N, Noda T, Yoshimori T, Tanaka Y, Ishii T, George MD, Klionsky DJ, Ohsumi M Ohsumi Y: A protein conjugation system essential for autophagy. *Nature* 1998, 395:395–398

[80] Adams J: Potential for proteasome inhibition in the treatment of cancer. *Drug Discov Today* 2003, 8:307–315

[81] Glickman MH, Ciechanover A: The ubiquitin–proteasome pathway: destruction for the sake of construction. *Physiol Rev* 2002, 82:373–428

[82] Pickart CM, Cohen RE: Proteasomes and their kin: proteases in the machine age. *Nat Rev Mol Cell Biol* 2004, 5:177–187

[83] Ciechanover A: From the lysosome to ubiquitin and the proteasome. *Nat Rev Mol Cell Biol* 2005, 6:79–86.

In: The Ubiquitin Proteasome System… ISBN 978-1-60021-749-4
Eds: Mario Di Napoli and Cezary Wojcik, pp. 41-70

Chapter 3

UBIQUITIN AND UBIQUITINATION: AN OVERVIEW OF THE UBIQUITIN-PROTEASOME SYSTEM FOR PROTEIN DEGRADATION

Yulia Matiuhin and Michael H. Glickman*

Department of Biology and the Institute for Catalysis Science and Technology, Technion – Israel Institute of Technology, 32000 Haifa, Israel.

ABSTRACT

Cells contain many different kinds of proteins, each fulfilling structural, functional, or regulatory roles. The presence of either damaged or mutated proteins, or of altered levels of normal proteins could cause pathological conditions and even cell death. Therefore, monitoring the state of all these proteins, as well as continuously adjusting their levels to suit demands, is paramount to survival. To exercise such quality control, cells are continuously spending energy both to synthesize new proteins, and to simultaneously degrade them, even though many may still be functional. An important characteristic of regulatory degradation is that it is specific; only the correct proteins are removed in a time-coordinated manner. Such extraordinary specificity is achieved by a modular system that identifies proteins that are to be degraded, marks them by covalently attaching ubiquitin to an amino residue, and finally proteolyses them into amino acids. This sequence of events is executed by the following components. Recognition of target proteins is carried out by a specific ubiquitin-protein ligase, called an E3. This protein recognizes the substrate and usually directs a ubiquitin-conjugating enzyme, an E2, to attach ubiquitin, a small 76 amino acid protein, onto an amino group on the substrate. Ubiquitin molecules are often added one to another, resulting in chains of ubiquitin

* Correspondence concerning this article should be addressed to Dr. Michael H. Glickman, PhD; Department of Biology and the Institute for Catalysis Science and Technology, Technion – Israel Institute of Technology, 32000 Haifa, Israel. Phone: +972-4-8294552; Fax: +972-4-8225153; E-mail: glickman@tx.technion.ac.il.

extending from the protein targeted for degradation. These polyubiquitin conjugates are then shuttled to the 26S proteasome, a large ATP-dependent proteolytic complex, where they are degraded. Interestingly, ubiquitination is a reversible process, with deubiquitinating enzymes able to remove ubiquitin from the target before it can be recognized by the proteasome. Hence, transfer of the polyubiquitinated conjugate to the proteasome must happen swiftly or be shielded from these enzymes. The cumulative balance of these processes allows the ubiquitin-proteasome system to control the cellular levels and half lives of thousands of proteins making it a key player in basic biological pathways such as cell division, differentiation, signal transduction, trafficking, and quality control. Not surprisingly, aberrations in the system have been implicated in the pathogenesis of many diseases, certain malignancies, neurodegenerative disorders, inflammation and immune response. Understanding of the underlying mechanisms involved is important for the development of novel, mechanism-based drugs.

Keywords: ubiquitin, proteasome, protein degradation, ubiquitin-like proteins UPS.

Abbreviations

AD, Alzheimer's disease; APG, autophagy; AR-JP, autosomal recessive juvenile parkinsonism; ATP, adenosine triphosphate; CF, cystic fibrosis; CFTR, cystic fibrosis transmembrane regulator; CP, core particle; CUE, coupling of ubiquitin conjugation to endoplasmic reticulum degradation domain; D-box, destruction box; DNA, deoxyribonucleic acid; DUB, deubiquitinating enzyme; ER, endoplasmic reticulum; GTP, guanidine triphosphate; HECT, homologous to the E6-AP C-terminus; LBs, Lewy bodies; MHC, major histocompatibility complex; MVB, multivesicular body; ORF, open reading frame ubiquitin; PD, Parkinson's disease; RNA, ribonucleic acid; RP, regulatory particle; SCF, Skp1; Cullin; F-box; SP, stationary phase; TGN, trans golgi network; Ub, ubiquitin; UBC, ubiquitin-conjugating enzyme; UIM, ubiquitin interacting motif; UBL, ubiquitin-like;UBR, ubiquitin ligase; UPS, ubiquitin-proteasome system.

Introduction

Until recently, the scientific world concentrated on aspects of the flow of genetic information from DNA to RNA to proteins. Regulation was attributed to key steps in transcription and translation, whereas proteolysis was considered an unregulated dead-end process. The approach changed completely with the remarkable discovery of the ubiquitin-proteasome proteolytic pathway about 30 years ago. Then, in the late 1970s, the pioneering biochemical studies of Hershko, Rose, Ciechanover and coworkers revealed that energy, in the form of ATP, is needed to break down proteins, elevating the importance of this process from 'uncontrolled' to 'highly regulated' (see Chapter 2 and [1-6]). Later, through cumulative work of many colleagues, it transpired that specific and timely degradation of cellular proteins in this manner is tightly coordinated and used to regulate many – if not most –

biological processes (reviewed in [7-10]). The discovery of this pathway is described in the preceding chapter.

Enzymes that hydrolyze peptide bonds are termed proteases. Proteases are numerous, varying in size, architecture, catalysis mechanism, substrate specificity and, most important, mode of regulation. For example, trypsin – one of the gastrointestinal enzymes - degrades dietary proteins 'outside' the body, in the lumen of the gastrointestinal tract in an uncontrolled fashion. Self ('body') proteins are degraded via two distinct mechanisms. Extracellular proteins (and some intracellular proteins) are taken up by cells via pinocytosis or receptor-mediated endocytosis and degraded in lysosomes by a mixture of non-specific trypsin-like proteases. Most intracellular proteins are hydrolyzed by a nonlysosomal proteolytic system that is endowed with a high degree of specificity toward its substrates - the ubiquitin-proteasome proteolytic pathway (reviewed in [7]).

In the ubiquitin-proteasome system (UPS), a small protein – ubiquitin (Ub) – serves to target proteins for degradation by the 26S proteasome – a large ATP-dependent protease – once it is covalently attached to its target. Recognition of target proteins is carried out by a specific Ub-protein ligase, called an E3. This protein recognizes the substrate and usually directs a Ub-conjugating enzyme, an E2, to attach Ub, onto an amino group on the substrate. Ub molecules are often added one to another, resulting in chains of Ub extending from the protein targeted for degradation. Linkage of successive Ub molecules is through an isopeptide bond between the C-terminus of the distal molecule and the amine side chain on Lys-48 of the previous Ub in the chain generating Lys-48-linked chains (other linkages may occur; to be elaborated on later). A similar bond anchors the polyubiquitin (polyUb) chain to an amino-group on the substrate, usually a lysine side chain, occasionally the free amino-terminus itself.

An ensemble of enzymes orchestrates the creation of Ub chains: first, each Ub molecule is activated via thiol-ester formation with an E1 enzyme that uses energy from ATP hydrolysis for these purposes. Next, the activated molecule is transferred to one of a number of E2 (ubiquitin-conjugating; UBC) enzymes, and finally, Ub is attached to a substrate mediated by an E3 (ubiquitin ligase; UBR) enzyme. The process of ubiquitination is reversible: several families of deubiquitinating (DUB) enzymes exist, which catalyze hydrolysis of an isopeptide linkage.

The outcome of Ub conjugation is not limited to proteasome targeting and proteolysis. For instance, fusion of monoubiquitin (monoUb) does not target a protein to the proteasome but rather participates in lysosomal trafficking and sorting, in histone acetylation, virus budding and in DNA damage repair (reviewed in [11-15]). Modification of target proteins with polyUb chains that are linked via lysine residues other than the typical lysine48 (modification of lysine 6, 11, 27, 29, 33, or 63 on each Ub moiety within a chain) have been reported to function in DNA damage repair, intracellular localization, and possibly for proteolytic functions too [16-18]. The possibility that 'mixed' or 'branched' chains exist (containing heterogeneous linkages, or multiple modifications on a single Ub within the chain), adds yet another layer of complexity and further broadens the potential outcomes of what until recently was simply termed 'ubiquitination'. The ubiquitous reach of the Ub system extends however even beyond these boundaries, as multiple homologues of Ub are also present in cells. Ub-like molecules are generally divided into two groups. Proteins of the first type, also known as ubiquitin-like modifiers (UBL), are, like Ub, capable of covalently modifying protein substrates. This family includes Rub1 (Nedd8 in mammals), Smt3 (SUMO-1, -2, -3), Hub1, Urm1, APG8 and APG12 and several more [19-21]. Most members of this family possess high degree of homology to Ub in their primary structure and overall architecture. Like Ub, these molecules undergo reversible attachment to their substrates mediated by E1-like, E2-like set of enzymes via isopeptide bond and can be cleaved off by UBL-specific isopeptidases. Nonetheless, modification by UBLs are apparently not recognized by the proteasome, and thus do not target substrates for proteolysis, highlighting a substantial regulatory difference between these two similar modes of modification. On the contrary, occasionally, modification of a protein by a UBL (such as SUMO, or Nedd8) antagonizes that of Ub as both modifications can occur on identical lysine residues within a given protein [22]. The second group of Ub-like molecules constitutes a family of proteins that bear a Ub-like domain fused to the rest of the protein in a non-cleavable manner. Multiple members of this family have been reported, among them Rad23, Dsk2, Ddi1 and Ubp6 [19-21]. In summary, Ub, together with Ub-like proteins constitute a large and diverse family of protein-modifiers-of-other-proteins, whose structural and functional characteristics are the main subject of this chapter.

Before we go into detail, first some numbers: there are at least one and a half thousand genes encoding components of the Ub pathway in the human genome, taking into account both proteolytic and non- proteolytic functions. Among them, there are hundreds of genes containing RING-finger domains which serve as ubiquitinating enzymes, and hundreds more of substrate-recognition components of multisubunit RING-finger E3 complexes, approximately fifty E2s, more than seventy deubiquitinating enzyme genes, dozens of Ub-like genes and 32 independent genes encoding the subunits of the 26S proteasome.

Ubiquitin

Ub is the prototype of a family of proteins that display remarkably similar structures, but variable sequences. Ub itself presents the most striking case of evolutionary conservation, differing at only 3 of 76 positions between yeast and humans (and even fewer differences are found between most other eukaryotes), underlying its conserved functional importance.

Ub terminates with a flexible tail ending in two glycine residues. It is the C-terminal carboxyl group that is attached to one or more amino groups on a substrate protein. This linkage usually occurs through the ε-amino group of a lysine side chain, creating what is referred to as an isopeptide bond. In all known cases, Ub is coded by a fusion gene and synthesized in a precursor form, either fused to certain ribosomal proteins or made as a tandem repeat of head-to-tail Ub units that always terminates with at least one additional amino acid [16]. Therefore, for Ub to be active it must first be processed to its mature form, a task performed by members of the deubiquitinating enzyme family (see Chapter 4).

Ub can be linked to substrates as a monomer, or in the form of isopeptide-linked polymers called polyubiquitin chains, whose diverse structure can influence the substrate's fate. The polymerization state of Ub is extremely important, though none of the various modes of modification by Ub are understood in full yet. For example, polyubiquitin chains linked through Lys-48 of Ub (Lys-48) target substrates to proteasomes, resulting in the essential function for this side chain [16]. Other kinds of linkages are known to exist, some of them performing non-proteolytic functions. All known roles of monoubiquitination are proteasome-independent. Modification by monoubiquitin plays an important role in lysosomal trafficking and sorting, in histone acetylation, virus budding and in DNA damage repair.

FUNCTIONS OF POLYUBIQUITIN

Ub is a unique protein as it can be covalently attached to other Ub molecules. This property means that posttranslational modification of a target protein by Ub is not a simple on/off switch (as – for comparison's sake – is phosphorylation/dephosphorylation), but can result in a heterogeneous population depending on the extent of Ub polymerization. The length of polyubiquitin modification varies from a few molecules to well over ten. Compounding the situation, Ub contains multiple lysines, any of which can serve as a receptor for a subsequent molecule of Ub, leading to differently-linked chains each with unique spatial properties. For these chains to serve regulatory functions, mechanism a mechanism should be in place to carefully define chain linkage and length. Once assembled, receptors and binding proteins should be able to distinguish between the various options, sending each to its appropriate outcome.

THE UBIQUITIN-PROTEASOME PROTEOLYTIC SYSTEM

As mentioned above, attachment of lysine48-linked polyubiquitin chains on a substrate leads to its degradation by the 26S proteasome. The degradation process involves three discrete and successive steps (see Figure 1): (a) recognition of the substrate by a specialized E3 ubiquitinating enzyme and tagging it by covalent attachment of multiple Ub molecules, (b) delivery of the tagged protein to the site of its degradation – the 26S proteasome, and (c) irreversible degradation of the protein into short peptides by the proteolytic complex accompanied by release of free and reusable Ub. This last process is mediated by Ub recycling isopeptidases (deubiquitinating enzymes) (reviewed in Chapter 4 and [7,23,24]).

Ubiquitin Conjugation (Ubiquitination)

Initially, the Ub-activating enzyme, E1, activates Ub in an ATP-requiring reaction to generate a high-energy thiol ester intermediate, E1-S~Ub (see Figure 1). One of several E2 enzymes (Ub-carrier proteins or Ubiquitin-Conjugating enzymes, UBCs) transfers the activated Ub from E1, via an additional high-energy thiol ester intermediate, E2-S~Ub, to the substrate that is specifically bound to a member of the Ub-protein ligase family, E3. There are two main classes of E3 enzymes: RING fingers and HECT domains. For the HECT (Homologous to the E6-AP C-Terminus) domain E3s, the Ub is transferred once again from

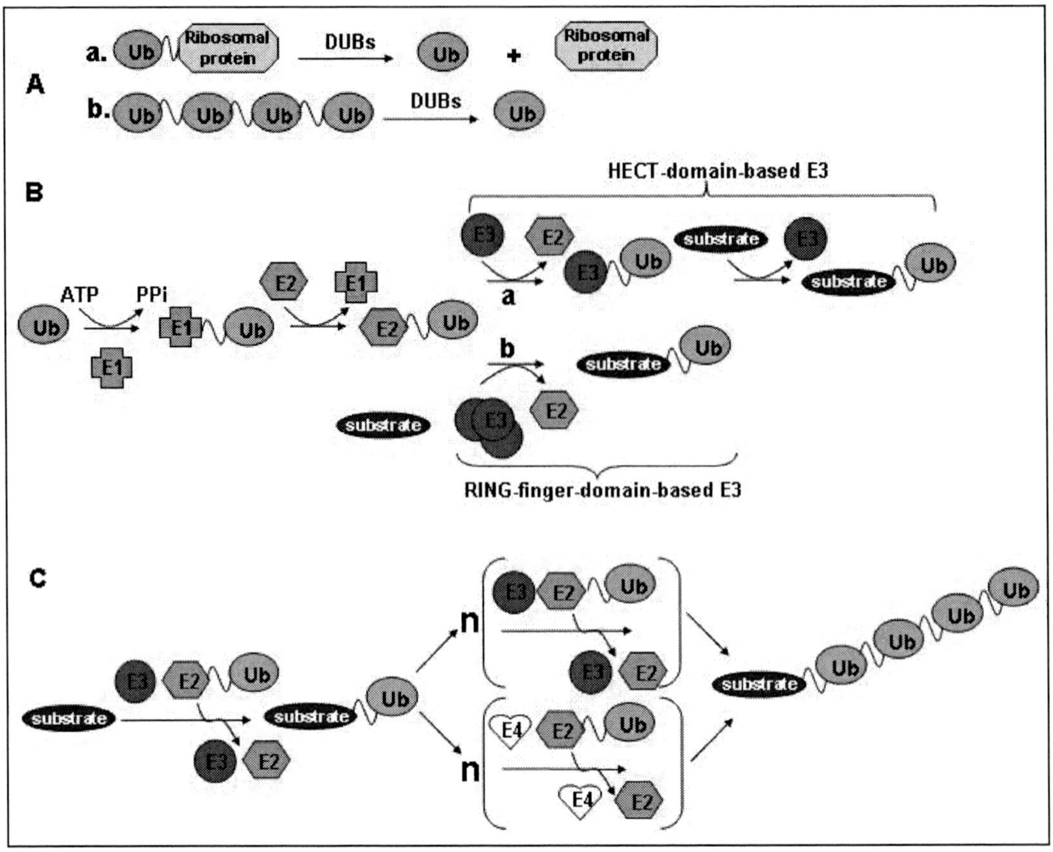

Figure 1. Ubiquitin biosynthesis and conjugation. A. Ubiquitin (Ub) biosynthesis. Ub is translated as a fusion to ribosomal proteins or as a linear repeat of multiple copies. Free Ub is generated when deubiquitinating enzymes (DUBs) cleave the peptide bond between the Ub moiety and the ribosomal protein (a) or in between two adjacent Ub molecules in a chimeric repeat (b). B. Ub congugation (monoubiquitination). Ub conjugation necessitates a series of steps. First, a Ub molecule is activated by the Ub-activating enzyme (E1) in the presence of ATP forming a thiol-ester bond between the carboxyl terminus of Ub and a cysteine on E1. Next, the Ub is transferred to one of the Ub-conjugating enzymes (E2). From this point in the pathway there are two options for Ub conjugation to the substrate. 1) Occurs in HECT-domain-based Ub ligases (E3). The activated Ub moiety is transferred to a cysteine on the respective E3 and then to the internal lysine of the intended protein in a two-step manner. 2) Occurs in RING-finger-domain-based Ub ligases. In this class of enzymes, the role of the E3 is to coordinate the interaction between the substrate and E2-Ub conjugate. The Ub molecule is transferred to the substrate in a concerted manner. C. Extension of the Ub chain (polyubiquitination). After the first Ub molecule is attached to a substrate, the chain elongation can occur if necessary. The elongation occurs in the presence of the same set of the ubiquitination enzymes that perform the attachment of the first Ub (upper) or in the presence of a so-called chain elongation factor - E4 (lower). Multiple cycles of ubiquitination in either case result in long polyubiquitin chains.

the E2 enzyme to an active site Cys residue on the E3, to generate a third high-energy thiol ester intermediate, Ub-S~E3, prior to its transfer to the ligase-bound substrate. RING finger containing E3s catalyze direct transfer of the activated Ub moiety to the E3-bound substrate. Some RING finger ligases contain an additional domain for interacting with their substrates, thus can function as stand alone ligases by simultaneously binding both the substrate and the

E2 conjugating enzyme that transfers the activated Ub directly to the substrate. However, other RING finger proteins are small polypeptides that fit into a modular complex that also contains substrate-recruiting or attachment subunits; the entire complex is required for ubiquitination to be carried out.

Overall, E3s play a key role in the Ub-mediated proteolytic cascade since they serve as the specific recognition factors of the system. In certain cases the first Ub moiety is conjugated to the substrate by one E3, while chain elongation is catalyzed by a different ligase often termed E4. E4 binds to the Ub moieties of preformed short conjugates and catalyzes Ub chain elongation in conjunction with E1, E2, and E3. It thus renders them preferred substrates for proteasomal degradation. The current model is that E4 enzymes define a novel protein family that works with the upstream components of the pathway to ensure commitment of a substrate for proteolysis. A deviant RING finger version, known as U box is found in a number of proteins that were shown to elongate short chains dependent on E1 and E2, but independent on E3. Whether other types of chain-elongation factors or E4s exist waits to be addressed [25].

Seeing as there are different modes of Ub conjugation, resulting in attachment of monoubiquitin to a single site, multiple modifications by mono Ub (to different sites on a single protein) termed monoubiquitination or attachment of polyubiquitin chains which themselves could be elongated through different lysines on each Ub molecule within the chain, it is of primary interest to understand what determines the end design. Are E3 Ub-ligases specialized to carry out certain types of ubiquitination? Are substrates preprogrammed to preferentially receive ubiquitination of a given type? Is the outcome determined by auxiliary factors that attach to the E3-substrate complex and divert ubiquitination to a certain length or linkage type? Are chain length and linkages determined primarily by E4 elongation factors that take over from the E3 enzyme that places the proximal Ub in a chain, or is the outcome determined by the delicate balance of ubiquitination versus deubiquitination? DUBs, for instance, could snip and trim budding chains to 'carve' them into a preferred configuration? These are all fundamental questions which await detailed answers.

Substrate Recognition

A major question is how the system achieves its high specificity and selectivity. Why are certain proteins extremely stable in the cell, while others are extremely short-lived? Why are certain proteins degraded only at a particular time point during the cell cycle or only following specific extracellular stimuli, yet they are stable under most other conditions? It appears that specificity of the Ub system is determined by two distinct and unrelated groups of proteins: (*i*) E3s and (*ii*) ancillary proteins [7]. First, within the Ub system, substrates must be specifically recognized by an appropriate E3 as a prerequisite to their ubiquitination. In most cases however, substrates are not recognized in a constitutive manner, and in some cases they are not recognized directly by the E3. In some instances, the E3 must 'be switched on' by undergoing post-translational modification in order to yield an active form that recognizes the substrate. In many other cases, it is the substrate that must undergo a certain change that renders it susceptible for recognition such as phosphorylation or dephosphorylation. The stability of additional proteins depends on association with ancillary proteins such as molecular chaperones that act as recognition elements in *trans* and serve as a link to the appropriate ligase. Others, such as certain transcription factors, have to dissociate from the specific DNA sequence to which they bind in order to be recognized by the system. Stability of yet other proteins depends on oligomerization. Thus, in addition to the E3s themselves, modifying enzymes (such as kinases), ancillary proteins, chaperones, partners in complexes or quaternary structures, or DNA sequences to which substrates bind also play an important role in the recognition process.

A variety of *degradation signals* or *degrons* have been described for both natural and engineered substrates of the UPS [7]. These signals are structural features or sequences within the substrate that are sufficient to cause its rapid proteolysis. Elements of degrons are recognized and bound directly by the ubiquitination apparatus. In several well-described examples, very short sequences within a substrate (as few as 7 amino acids), or well defined motifs such as 'destruction boxes' (D-Box) or PEST sequences (rich in proline, glutamate, serine and threonine) are largely responsible for the binding of the E3. Sometimes even a single residue can serve as a degron, for example, basic (arginine, lysine) or aromatic (tyrosine, phenylalanine, tryptophan) N-terminal residues instead of the standard methionine can be recognized by a designated E3 Ub ligase (known in yeast as Ubr1) and confer onto any substrate the status of a 'short-lived protein'. This property has been widely used in biological and medical research for the study of the importance pf cellular levels of numerous substrates of interest by designing gene products that are post-translationally processed to reveal this 'N-end rule degron'. The site at which the protein is modified with Ub is not always within the segment to which the E3 binds. In some cases, a protein segment must be post-translationally modified, for example, by a phosphate group or hydroxylation of a specific proline - before the E3 can bind tightly to it. This allows the rate of degradation of the substrate to be regulated by signal transduction pathways or by changes in the cellular environment.

A very different type of degradation determinant is a hydrophobic surface exposed on the exterior of a protein. This feature is an effective degron because hydrophobic amino acids are normally buried within the interiors of most properly folded proteins, or occur at the interface between the subunits of a protein complex and are covered when they have assembled correctly. An exposed hydrophobic surface is therefore a likely indicator of an improperly folded or assembled protein. Such proteins have the potential to form toxic aggregates within the cell and must be eliminated before they can do so. This logic suggests that hydrophobic determinants may be utilized primarily for protein quality control eliminating misfolded/damaged proteins.

Some degrons are very complex. The hydrophobic regions exposed on improperly folded proteins, for example, are unlikely to be a specific sequence of amino acids. Several hydrophobic surfaces, possibly contributed by different regions within a protein, may have to be recognized simultaneously by the ubiquitination system. Other proteins contain complex degrons in order to help ensure that they are degraded only under the appropriate conditions.

Delivery

Following ubiquitination, the protein substrates are directed to the 26S complex for degradation. The mechanism of delivery of polyubiquitinated substrates to the proteasome is currently somewhat ambiguous (reviewed in [26,27]). In vitro - using purified components – polyubiquitinated substrates can diffuse to the proteasome, be recognized and then degraded. It should be emphasized that often, the rates and efficiency of such reactions are markedly less than proteolysis rates measured in vivo for the same substrate. It is clear, that in order to maintain the continuous flow of substrates to the proteasome in vivo, all steps should be tightly regulated. In principal, a delivery mechanism could aid substrates to reach the proteasome without 'wandering off along the way' or protecting them from being disassembled by deubiquitinating enzymes that are abundant in the cell (see Figure 1). Thus, it has been proposed that polyubiquinated substrates reach the proteasome in two parallel and complementary pathways: direct, and delivered by shuttle proteins; the main distinction being the site of the key polyubiquitin-binding protein. In the former, the polyubiquitin-receptor is an integral component of the proteasome; in the latter scenario, a polyubiquitin-binding protein attaches to designated substrate and tows it to the proteasome, forming a ternary complex proteasome-shuttle-substrate.

The 26S complex binds polyubiquitin [28,29]. Therefore tagged substrates do bind to it directly. The delivery of the ubiquitinated substrates could be alternatively mediated by so-called shuttles - proteins or protein complexes capable of simultaneously binding both proteasome and polyUb [26,27]. Finally, in some cases, ubiquitination and proteasome-dependent proteolysis may not be independent processes at all. There might be physical contact between components of the ubiquitination and destruction machineries ensuring direct transfer of substrates upon attachment of the polyUb signal. All three mechanisms of targeting substrates to the proteasome seem to coexist in cells.

There are increasing reports that there are direct contacts between the Ub system for tagging myriad proteins with Ub and the proteasome, suggesting that ubiquitination and degradation might not be completely independent processes. E2s such as Ubc1, Ubc2, and Ubc4 coimmunoprecipitate with the proteasome [30]. In addition, interactions of several other E3s, Ubr1, Ufd4, Hul5, and the mammalian HECT-E3, KIAA10, have been mapped to components of the proteasome, suggesting that at least some substrates are ubiquitinated at (or in close proximity to) the proteasome [7,31].

Other substrates (the majority, most probably) are selected and ubiquitinated in other cellular locals (nucleus, cytosol, ER) away from concentrations of the proteasome, after which they are transferred to the proteasome. A slew of shuttle proteins – Rad23/hHR23, Dsk2/hPLIC, Ddi1 to name a few – was proposed to escort these tagged proteins to the site of destruction as all members of the family associate both with the proteasome via the Ub-like domain (UBL) and with polyUb via the Ub-associated domain (UBA). There may be additional shuttles, for example, a protein complex containing VCP/Cdc48 was found to co-purify with the proteasome and with polyUb [32,33].

Finally, one subunit of the 26S proteasome itself, Rpn10/S5a, might be an ideal candidate for shuttling polyubiquitinated substrates to the proteasome and anchoring them once recognized. These two functions correspond to the two pools in which Rpn10 is found:

a proteasome-unassociated pool that is capable of binding polyubiquitin, and a proteasome-incorporated pool that serves as an integral proteasomal subunit [34-37].

Indirect interactions between the ubiquitination machinery and the proteasome are also implied from studying the behavior of Dsk2/hPLICs. HPLIC-1 and hPLIC-2, two human homologs of yeast Dsk2, have been shown to interact both with the proteasome as well as with two different E3s (E6-AP and β-TrCP), and they can all colocalize as part of a multiprotein complex in vivo. An additional protein that has UBL domain and can bind to the proteasome is the chaperone cofactor BAG-1 [38]. BAG-1, in addition to promoting the association of Hsc70/Hsp70 with the proteasome, also interacts with CHIP, a U box-containing ubiquitinating enzyme. This assembly seems to target misfolded proteins to degradation (see Chapter 19).

Degradation

As mentioned above, degradation of polyubiquitinated substrates is carried out by the 26S proteasome, a large protease complex. While proteolysis of proteins tagged by the polyubiquitin signal is the best studied function of the proteasome, it should be noted that in some instances proteins that are modified by other signals or even unmodified proteins can be efficiently degraded by the proteasome [7].

The proteasome is a multicatalytic protease, meaning that it has the capacity to endoproteolyticaly cleave proteins at multiple –and different- sites within the polypeptide chain (see Chapter 6). The outcome is a repertoire of short peptides generated from each substrate. The proteasome is composed of two sub-complexes: a 20S core particle (CP) that carries the proteolytic activity and a regulatory 19S regulatory particle (RP) that binds substrates and prepares them for proteolysis by the 20S CP (see Figure 2 and reviewed in [39-42]). As determined by crystallography, 20S is a hollow cylindrical structure made of four stacked heptagonal rings. Each of the two outer rings contains seven genetically related α-type subunits (α1 – α7). Likewise, seven homological β-type subunits (β1 – β7) form each of the two inner rings. In eukaryotes, three of the seven β subunits - β1, β2 and β5 – have been shown to harbor functional protease active sites. Their protease specificities are distinct: each cleaves specifically after different amino acids. Overall, each proteasome has six (three different) proteolytic sites such that together they can cleave almost all peptide bonds within a given substrate. Nevertheless, due to a complex set of restrictions imposed by flanking regions of the substrate, some peptide bonds are cleaved preferentially over others, leading to a typical repertoire of peptides generated from each substrate. Such tight proteolytic specificity is a direct consequence of the proteolytic active sites β subunits facing innerwards to 20S CP barrel cavity. Substrates can access this cavity only through a narrow channel leading from the surface of the α-rings. Recently it was found that the N-termini of the α-subunits block the entrance to the proteolytic cavity suggesting that the proteasome channel is gated, probably as a means to regulate substrate traffic through the proteasome.

Each extremity of the 20S barrel can be capped by a 19S RP (see Chapter 7). One important function of the 19S RP is to recognize ubiquitinated proteins and other potential substrates of the proteasome. So far, two subunits of the 19S RP capable of binding polyubiquitin have indeed been identified (Rpn10/S5a and the ATPase Rpt5), however, their importance and mode of action have not been discerned [28,36,37,43]. A second function of the 19S RP is to open an orifice in the α ring that will allow entry of the substrate into the proteolytic chamber [44,45]. Also, since a folded protein would not be able to fit through the narrow proteasomal channel, it is assumed that the 19S particle unfolds substrates and inserts them into the 20S CP. Both the channel opening function and the unfolding of the substrate require metabolic energy, and indeed, the 19S RP contains six different ATPase subunits (termed Rpt1-6).

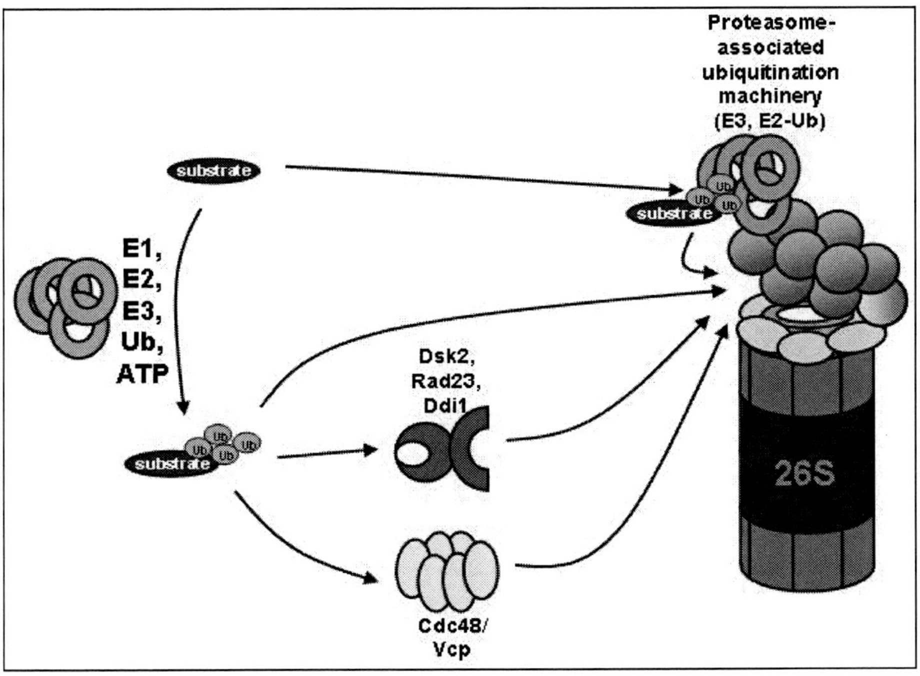

Figure 2. Substrates delivery to the proteasome. Substrate can be recruited to the proteasome in a number of manners. Some substrates are polyubiquitinated on the proteasome by proteasome-associated ubiquitination machinery. Most substrates are recognized and polyubiquitinated independently of the proteasome, and are then targeted either directly binding to the ubiquitin-binding sites within the proteasome, or shuttled by delivery proteins capable of binding both to polyubiquitin chains and to the proteasome (examples of shuttles: Dsk2, Rad23, Ddi1, Cdc48/Vcp).

In budding yeast, the RP is formed by at least 18 different subunits of two types: 1) ATPases of the AAA family designated as Rpt1-6 (for regulatory particle triple-A protein) and 2) non-ATPases designated as Rpn1-13 (for regulatory particle non-ATPase). Additional proteins (among them Rpn14, Sem1/DSS1 and Nas6/Gankyrin) are found associated to a subset of proteasomes, however genes encoding these putative subunits are nonessential and the contribution of the encoded proteins to proteasome function is unclear [30,46,47]. The 19S particle can be dissected into two subcomplexes – lid and base. The interaction of the two is stabilized by (though does not strictly necessitate) the Rpn10 subunit of the proteasome [45,48].

The disk-shaped Lid formation consists of eight subunits – Rpn3, Rpn5, Rpn6, Rpn7, Rpn8, Rpn9, Rpn11 and Rpn12 (and probably Rpn13 and Sem1) - can detach from the 19S RP and reattach reversibly [35,48]. Degradation of ubiquitinated substrates requires the presence of this Lid. This requirement suggests interaction with the polyubiquitin chain through a subunit that has Ub-binding ability in the Lid. This subunit has not been identified yet, but deubiquitinating activity has been reported in the lid and it is attributed to Rpn11, a metalo-protease capable for trimming linked Ub molecules [49,50]. Apparently, removal of Ub from the target substrate is an essential and integral aspect of preparing substrates for translocation into the 20S CP where they are hydrolyzed into peptides (as discussed above).

All the ATPase subunits of the RP – Rpt1-6 – are present in the other RP subcomplex called base [7]. In addition to these six ATPases, the base contains three non-ATPase subunits – Rpn1, Rpn2 and Rpn10. It is believed that the Rpt subunits form a six-member ATPase ring and at least some of them can contact directly the 20S α-ring. Such an arrangement is essential in order to open the central channel and translocate substrates into the destruction chamber of the 20S. Akin to chaperonins, these ATPases probably unfold proteins, a process that serves as a prerequisite for their translocation into the narrow channel leading into the chamber of the 20S CP. Another subunit of the base is Rpn10, though it attaches the lid as well. Based on the documented phenotypes, the major role of Rpn10 as a proteasomal subunit is to tether the lid and the base, though Rpn10 alone is neither sufficient nor essential for this task [35,45]. There are some other functions of Rpn10 that are mapped to its C-terminal Ub-interacting motif (UIM). For instance, Rpn10 was thought to serve as a polyubiquitin receptor in the proteasome. Rpn10 is the only subunit that exists also outside the proteasome in sufficient amounts suggesting it may recruit polyubiquitinated proteins and escort them to the 19S RP [36,44]. Interestingly, another subunit at the Base – Rpt5 – shows ability to bind polyubiquitin too [43]. As if that were not enough, Rpn1 situated in the Base alongside Rpn10 and the RPT subunits shows high affinity for Ub-like molecules [31,51]. Rpn1 was proposed as a principal recognition site of UBL domain-containing proteins. The location of Rpn1 and Rpn2 within the base is therefore a subject of hot debate in the field. In summary, there is increasing evidence that Ub binding, whether directly or indirectly, occurs in the base.

Following degradation of the substrate, short peptides derived from the substrate are released, as is reusable Ub. Peptides products of varying length are rapidly trimmed by downstream peptidases such that they do not accumulate in vivo. Nevertheless, some peptides may serve regulatory roles. The best studied case is the presentation of proteasome end-products as antigenic peptides to the immune system by somatic cells as a marker of the proteome within each cell. Infected or damaged cells that produce large amounts of abnormal peptides will be recognized in this manner as foreign to the organism and attacked by its own immune system. Proteasomal degradation is not always complete. In some cases, the proteasome, rather than completely destroying its target, processes the ubiquitinated substrate precisely, releasing a truncated product. In the case of the NF-κB transcriptional regulator, an active subunit (p50 or p52) is thus released from a longer inactive precursor (p105 or p100) [7,52,53].

Regulation

The UPS can be regulated at the level of ubiquitination, at the level of delivery to the proteasome, or at the level of proteolysis. Since conjugation and proteasomal degradation is required for multitude of cellular functions, regulation must be delicately and specifically tuned. Furthermore, Ub levels, levels of the ubiquitination machinery, and the cellular levels of the proteasome are dynamic and rapidly responsive to external signals or cellular needs.

In a few cases general rather then specific components of the pathway can be modulated by physiological signals. For example, upregulation of the pathway is observed during massive degradation of skeletal muscle proteins that occurs under normal fasting, but also under pathological conditions such as cancer cachexia, severe sepsis, metabolic acidosis, or following denervation [3,54,55]. Likewise, mild exposure to denaturing conditions such as heat, oxidation, alkylating or modifying reagents also elevates components of the Ub-proteasome pathway [56-58]. In contrast, prolonged starvation, entry into stationary phase (SP), or severe heat shock, promote proteasome dissociation into 20S CP and 19S RP subcomplexes resulting in a rapid decline in intracellular proteolysis [59]. These conditions may require repressed proteolysis for survival as proteasome mutants with persistent unregulated proteolysis are extremely vulnerable. Severe oxidative stress conditions results in a similar decline in proteasome function, with spiraling deleterious effects on cell survival.

In most cases however, regulation is specific and the target substrates are recognized by specific ligases that bind to defined motifs. As mentioned earlier, the targeting motif can be a single amino acid residue (e.g. the N-terminal residue) or a sequence (the Destruction box in cyclins) or a domain (such as an hydrophobic patch) that is not normally exposed. In other cases the motif is a post-translational modification such as phosphorylation that is generated in response to cell needs or external signals. Phosphorylation can occur either on the substrate or the ligase [7].

Another notable mode of regulation is different subcellular localization of the substrates and the components of ubiquitination/degradation machinery. One of the Cdc28 cyclin-dependent kinase inhibitors – Far1 - bifunctional protein that can function either in the nucleus, where it is required to arrest the cell cycle, or in the cytosol, where it establishes cell polarity during yeast mating. The nuclear pool of Far1 undergoes rapid degradation when a cell exits cell-cycle arrest. However, the cytosolic pool of Far1 is stable due to the residing of the substrate-recognition subunit of the E3 complex responsible for its ubiquitination exclusively in the nucleus [60]. In another example, degradation of misfolded proteins in the ER requires their retrotranslocation to the cytoplasm where they are exposed to the ubiquitination machinery and to the proteasome.

The ubiquitination state is a dynamic and transient condition; the levels of ubiquitinated proteins depend on the rate at which they are tagged, but also on the rate at which they are either degraded or deubiquitinated. There are two types of deubiquitinating activity: proteasome-associated and extraproteasomal. The former is responsible for Ub salvage and recycling during the process of degradation while the latter is required for Ub biosynthesis but may also result in rescuing the substrate from degradation by dismantling the polyUb chains attached to it. Thus, for efficient proteolysis, once a substrate is polyubiquitinated it must be quickly transferred to the proteasome or protected from the counteracting DUB activity. Delivery proteins may double up as 'DUB-protectors' in their capacity to bind and shuttle polyubiquitinated targets.

The abundance of the delivery proteins is also tightly regulated. Upregulation as well as downregulation of shuttle proteins lead to stabilization of myriad substrates and accumulation of polyubiquitinated species. In an extreme case (such as observed for Dsk2), high levels of a shuttle can be extremely toxic [61,62].

UPS Substrates and Associated Diseases

Ub-mediated proteolysis of a variety of cellular proteins plays an important role in many basic cellular processes. Among these are regulation of cell cycle and division, differentiation and development, involvement in the cellular response to stress and extracellular effectors, morphogenesis of neuronal networks, modulation of cell surface receptors, ion channels and the secretory pathway, DNA repair, transcriptional regulation, transcriptional silencing, long-term memory, circadian rhythms, regulation of the immune and inflammatory responses, and biogenesis of organelles. The list of cellular proteins that are targeted by Ub is growing rapidly. Among them are cell cycle regulators such as cyclins, cyclin-dependent kinase inhibitors, and proteins involved in sister chromatid separation, tumor suppressors, transcriptional activators and their inhibitors. Cell surface receptors, and endoplasmic reticulum (ER) proteins are also targeted by the system. Finally, mutated and denatured/misfolded proteins are recognized specifically, and are removed efficiently. In this capacity, the system is a key player in the cellular quality control and defense mechanisms. The products of the proteasome can play an important role in the immune response. In the case of degradation of foreign proteins - such as those of viral origin – the resulting short peptides are presented by MHC class I molecules to the cytotoxic T cell that lyse the presenting cell (see Chapter 34).

While inactivation of a major enzyme such as E1 is obviously lethal, mutations or acquired changes in enzymes or in recognition motifs in substrates that do not affect vital pathways or that affect the involved process only partially may result in a broad array of diseases.

The pathological states associated with the Ub system can be classified into two groups: (a) those that result from loss of function - mutation in a Ub system enzyme or target substrate that result in stabilization of certain proteins, and (b) those that result from gain of function - abnormal or accelerated degradation of the protein target.

Alterations in ubiquitination and deubiquitination reactions have been directly implicated in the etiology of many malignancies [7]. In general, cancers can result from stabilization of oncoproteins, or destabilization of tumor suppressor genes. Some of the natural substrates for degradation by the proteasome are oncoproteins that if not properly removed from the cell, can promote cancer. For instance, Ub targets N-myc, c-myc, c-fos, c-jun, Src, and the adenovirus E1A proteins [22,63,64]. Destabilization of tumor suppressor proteins such as p53 and p27 has also been implicated in the pathogenesis of malignancies.

The Cystic Fibrosis gene encodes the CF transmembrane regulator (CFTR) that is a chloride channel. Only a small fraction of the protein matures to the cell surface, whereas most of it is degraded from the Endoplasmic Reticulum (ER) prior to its maturation by the Ub system. One frequent mutation in the channel is ΔF508. The mutation leads to an autosomal recessive inherited multisystem disorder characterized by chronic obstruction of airways and severe maldigestion due to exocrine pancreatic dysfunction. Despite normal ion channel function, CFTRΔF508 does not reach the cell surface at all, and is retained in the ER from which it is degraded (see Chapter 13). It is possible that the rapid and efficient degradation results in complete lack of cell surface expression of the F508 protein, and therefore contributes to the pathogenesis of the disease [7,65,66].

In neurons proteasomes play a role in neuronal connectivity, synaptic transmission, synaptic plasticity, and protection against neurodegenerative diseases [67]. It is therefore not surprising that accumulation of Ub conjugates and/or inclusion bodies associated with Ub, proteasome, and certain disease-characteristic proteins, have been reported in a broad array of chronic neurodegenerative diseases, such as the neurofibrillary tangles of Alzheimer's disease (AD), brainstem Lewy bodies (LBs) - the neuropathological hallmark in Parkinson's disease (PD), and nuclear inclusions in CAG repeat expansion (poly-glutamine extension) disorders such as occurring Huntington's disease [68-72]. However, in all these cases, a direct pathogenetic linkage to aberrations in the Ub system has not been established. Accumulation of Ub conjugates in Lewy inclusion bodies in many of these cases may be secondary, and reflects unsuccessful attempts by the Ub and proteasomal machineries to remove damaged/abnormal proteins. While the initial hypothesis was that inclusion bodies are generated because of the inherent tendency of the abnormal proteins to associate with one another and aggregate, it is now thought that the process maybe more complex and involves active cellular machineries, including inhibition of the Ub system by the aggregated proteins. This aggregation of brain proteins into defined lesions is emerging as a common but poorly understood mechanistic theme in sporadic and hereditary neurodegenerative disorders (see Chapter 28).

The case of Parkinson's disease highlights the complexity of the involvement of the Ub system in the pathogenesis of neurodegeneration [68,69]. Aberrations in several proteins such as mutations in α-synuclein, an important neuronal protein, or in the deubiquitinating enzyme UCH-L1, have been described that link the Ub system to the pathogenesis of the disease. One important player in the pathogenesis of Parkinson's disease is Parkin which is a RING-finger E3. Mutations in the gene appear to be responsible for the pathogenesis of autosomal recessive juvenile parkinsonism (AR-JP), one of the most common familial forms of Parkinson's disease. Parkin ubiquitinates and promotes the degradation of several substrates. It is possible that aberration in the degradation of one of these substrates that leads to its accumulation, is neurotoxic and underlies the pathogenesis of AR-JP (see Chapters 25 and 31).

NON-PROTEOLYTIC FUNCTIONS OF POLYUB

Recent MS-MS analysis of polyUb chains occurring *in vivo* indicates that Ub chains in the cell can be formed through all seven lysines within Ub (K6, K11, K27, K29, K33, Lys-48, and K63) as well as through its N-terminus in a head-to-tail fashion [73]. Many of these chains have been synthesized in vitro, suggesting that the forms of polyubiquitin that could regulate cellular functions are numerous [17,74]. As of yet, there is no evidence for mixed chains (made up of different Ub- linkages successively within a single chain), nor for branched chains (in which multiple ubiquitins can anchor to different lysines on a single preceding Ub within the chain) yet these are possibilities that should be investigated.

Lys48-linked polyubiquitin chains have been studied extensively, and they regulate nuclear, cytosolic, and endoplasmic reticulum membrane proteins by targeting these proteins for degradation by the 26S proteasome. The quaternary structure of Ub polymers and the exact spatial relationship between each Ub molecule is extremely critical for their ability to target substrates for degradation by the proteasome as the other types of polyUb do not necessarily serve as a proteolytic signal.

Lys63-linked chains are known to regulate DNA repair, signal transduction, and endocytosis and are likely to control other basic cellular processes as well but probably do not target proteins for degradation by the proteasome. It appears that conjugation in position 63 requires distinct conjugation enzymes. Another mode of conjugation involves linking Ub molecules via Lys-29 of Ub. At least in one case it was shown that short Lys-29 chains are involved in recruitment of the chain-elongating factor E4 [25]. Enzymes that can catalyze formation of multiubiquitin chains linked via Lys-6, or Lys-11, are also known but it is not clear what is the purpose of these 'alternative' chains and whether they can even target substrates to the 26S proteasome [13,14,17,75].

It is important to note, however, that multiubiquitin chains linked via Lys-6, Lys-11, or Lys-48, all bind to the proteasomal subunit Rpn10/S5a with similar affinities [75]. Likewise, a number of delivery proteins (Rad23 for example) bind to polyubiquitin of different linkages with no apparent ability to distinguish between them [74,75]. Furthermore, linear Ub fusions are competitive inhibitors of Lys-48-linked polyubiquitinated substrates for proteasomal binding, indicating they probably bind to the same site on the proteasome. These observations indicate that there is likely to be a separation between binding to the proteasome and correctly preparing a substrate for degradation. In light of such observations, the *reason-d'etre* of nonlysine48-linked chains awaits illumination. The possibility of targeting substrates to the degradation by the proteasome by non-Lys-48 polyUb chains is probable but under normal conditions Lys-48 chains seems to carry a function of high priority proteolytic signal. In any case, binding of non-Lys-48 polyUb to the proteasome interferes with its degradation activity by means of competition for common polyUb binding site. Hence, chain assembly should be properly regulated. It is obvious that if such a regulation mode exists it happens upstream to the binding to the proteasome, probably aided by DUBs that snip or trim rogue chains, and by delivery proteins with unique affinities for different types of remaining polyUb chains. So, apart from aiding the polyubiquitinated substrates to the proteasome and protecting them from deubiquitination, the shuttle proteins may sift low-priority or undesirable degradation substrates on the way to the proteasome. Chimeric chains enabling linking to different lysine residues within one chain though can exist theoretically have not been documented in vivo.

FUNCTIONS OF MONOUBIQUITIN

So far this introductory chapter in this book has dealt with the aspects of polyubiquitination, which refers to attachment of a polymeric chain consisting of Ub moieties. Another type of modification by Ub includes fusion of a single Ub molecule to a lysine residue of the target protein – the process called monoubiquitination. If multiple internal lysine residues on a substrate are modified with monoubiquitin the substrate is said to be multimonoubiquitinated or multiubiquitinated, in distinction from polyubiquitinated substrates (see Figure 3).

Monoubiquitination may serve as a cellular targeting or localization signal, but it does not seem to target proteins to the proteasome. Monoubiquitination plays important roles in virus budding, intracellular trafficking and transcription regulation [13].

Researchers studying Ub in the mid-1980s sincerely believed that the Ub-proteasome pathway had no point of intersection with lysosomal proteolysis.This assumption reflected the fact that agents which disrupt lysosomal functioning, have no effect on the ATP-dependent turnover of short-lived and abnormal proteins. In direct contradiction of this formerly strict rule, we now know that ubiquitination is sometimes required for lysosomal proteolysis. This could not be detected in the 1984 study because only a small fraction of short-lived proteins is targeted to lysosomes. Had the researchers of the Ub pathway been less lucky they could eventually have ended up investigating the aspects of lysosomal degradation [76]. In fact, endocytosis is just one of many protein trafficking steps that depend on Ub conjugation.

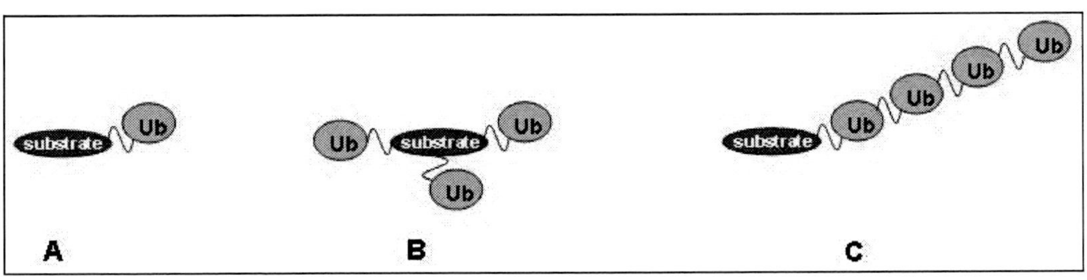

Figure 3. Diversity of modification by Ubiquitin. (A) Substrate modified by one ubiquitin molecule on a single lysine residue are said to be monoubiquitinated. (B) When monoubiquitin molecules are attached to multiple lysine residues, the substrate is said to be multi-monoubiquitinated or multiubiquitinated; with a ubiquitin chain, the substrate is said to be polyubiquitinated (C). Polyubiquitin. Chains of varying length (even upwards of 20 molecules) are found. Additional complexity is found in the nature of the linkage between neighboring ubiquitin moieties in the chain. Chains can elongate by conjugating to any of the lysines in the preceding ubiquitin, though lysine48-linked chains are the most studied and best understood.

Plasma membrane proteins destined for degradation are collected by endocytosis. This pathway is sometimes triggered by the binding of a ligand to its receptor, and serves to remove the receptor from the surface of the cell so that the cell becomes less sensitive to the ligand [14]. Here, Ub often plays a crucial role because ubiquitination of the cytoplasmic portion of both the target protein and the components of the endocytic machinery serves as an essential step in the recruitment of the protein into clathrin-coated pits. A single Ub is usually sufficient, unlike the ubiquitination that directs proteins to the proteasome. Although endocytosed proteins that are not modified with Ub can be recycled to the cell surface, those that are ubiquitinated are generally routed to the lysosome for degradation.

Proteins comprising the components of the sorting and budding machinery such as epsins, Eps15, Hrs bear several remarkable features [13,14]. For example, they carry a Ub-binding domain (usually, the Ubiquitn-interacting motif (UIM) or the Coupling of Ub conjugation to endoplasmic reticulum degradation domain (CUE) and are themselves ubiquitinated. This observation hints that the pairing of Ub binding and ubiquitination may be an important regulatory mechanism. Ub binding by and ubiquitination of the same protein could provide one type of signal for the sequential assembly of a protein network. Moreover, an intramolecular interaction is possible between the appended monoubiquitin and the Ub-binding domain. This interaction may protect a monoubiquitin-binding domain from binding to free Ub in the cytosol rather than its intended ubiquitinated partner, or from deubiquitinating enzymes, or could result in a change of protein conformation to affect protein activity. In addition, a Ub-UIM/CUE interaction might regulate monoubiquitination versus polyubiquitin chain formation. In support of this idea, the three dimensional structure of a CUE-Ub complex reveals that binding of the CUE domain to Ub blocks access to Ub Lys-48, thus limiting chain formation at this residue. Monoubiquitination of the components of the endocytic machinery is not restricted for those possessing Ub-binding properties as CIN85 and Numb – two other endocytic proteins are monoubiquitinated though they do not carry known Ub-binding domains.

Many plasma membrane proteins can be endocytosed without ubiquitination and still degraded in the lysosome when necessary [12]. Degradation of these proteins often depends on ubiquitination after they have arrived in the endosome. Unmodified proteins are recycled to the plasma membrane, while those that have been ubiquitinated are sent on to the lysosome. Ub attached to cargo proteins allows such sorting by directing them to sites of involution in the membrane of the endosome for selective incorporation into internalized vesicles. Ubiquitinated proteins congregate at these sites because of the assembly there of large complexes of proteins that include Ub binding proteins as well as others that orchestrate the budding of vesicles into the lumen of the endosome. Once the formation of vesicles has begun, the role of Ub on the target proteins is apparently complete because it is removed before the vesicles pinch off into the interior of the organelle. The endosome matures into a multivesicular body (MVB), and its subsequent fusion with the lysosome releases the internal vesicles into the lysosomal interior.

In both yeast and mammalian cells, monoubiquitylation is sufficient to trigger internalization into primary endocytic vesicles. Internalization information is carried in the Ub polypeptide itself because Ub can stimulate internalization when it is fused in-frame to receptors that lack lysine ubiquitylation sites or other internalization signals, or to heterologous proteins that are not normally internalized.

Apart from endocytosis, monoubiquitination is required to activate/deactivate certain transcription events. For instance, ubiquitination of histone H2B promotes site-specific methylation of histone H3, with an ultimate readout of transcriptional silencing [15]. Monoubiquitin regulates the activity of transcription factors in the nucleus as well. Deletion of a yeast E3 responsible for the degradation of a model transcription factor blocks transcriptional activation. Furthermore, fusion of a single Ub moiety to the transcription factor restores activity but not degradation, suggesting that monoubiquitination is required for normal transactivation activity. Yet another mode of monoubiquitination is to change subcellular localization of the target protein. For example, FANCD2 is monoubiquitinated in S phase of the cell cycle, a modification that stimulates its translocation into discrete nuclear foci [77].

The ability of Ub to enhance protein-protein interactions has also been exploited by enveloped viruses to capture the cellular machinery that mediates MVB vesicle budding [13]. Enveloped viruses bud from the plasma membrane of infected cells in the opposite direction to that of vesicles during endocytosis, and evidence is accumulating that monoubiquitin is an essential part of this process. During the process of budding, Gag – the viral protein essential for budding - is monoubiquitylated. Ubiquitination of Gag is important for its function in virus budding because the depletion of intracellular Ub levels inhibits budding.

Two fascinating facts should be mentioned with regard to monoubiquitination and its role in endocytosis. First, is that the cell clearly discriminates between monoubiquitination and polyubiquitination: while the former triggers early endocytic events as well as sorting into MVB, the latter is accountable for targeting of proteins from TGN to the lysosome for degradation. Second, is that the degree of ubiquitination is so regulated that proteins that acquire a monoubiquitin signal are not inappropriately multi-ubiquitinated and degraded. How does the cell achieve such a specificity is not known because some combinations of the standard ubiquitylation enzymes — E1s, E2s and E3s — seem to participate in both monoubiquitylation and the formation of multi-Ub chains. So it seems that the outcome of ubiquitination is determined by both spatial and temporal arrangement of both ubiquitination, deubiquitination enzymes and a variety of Ub-binding proteins.

Ubiquitin-Like Proteins (UBLs)

From structurally distinct Ub modifications, it is only a small step to a remarkable recent development — structurally distinct ubiquitins [17,21,78]. We now know that Ub defines a family of structurally related signaling proteins which share a common biochemical mechanism of isopeptide tagging. The interferon-induced ISG15 protein was the first such protein to be discovered; other examples followed in short order. The functional range of individual family members varies widely. Nedd8/Rub1, for example, seems to function only as an activator of cullin-based RING fingers E3s, whereas SUMO modifies numerous cellular proteins and may signal several different fates for its substrates. Cellular processes, in which other Ub-like molecules are involved, extend from autophagy to immune response.

Ub-like proteins fall into two separate classes [19-21]. Proteins of the first class function as modifiers in a manner analogous to that of Ub. They exist either in a free form or, when catalysed by specific enzymes, attached covalently to other proteins by their C-termini; hence 'Ub-like modifiers', or UBLs. Proteins of the second class bear protein domains that are related to Ub but are otherwise unrelated in sequence to each other. In contrast to UBLs, these proteins are not conjugated to other proteins. This chapter deals with the UBL proteins.

With the notable exception of APG12, APG8 and URM1, all currently known UBLs are related in sequence to Ub. APG8, APG12, URM1 and FAT10 are the only ones that are expressed as mature proteins, whereas all other UBLs (including Ub) are expressed as inactive precursors; they are made initially as fusions with C-terminal extensions, which prevent conjugation. These tails can be either single amino acids or polypeptides. These proteins also share a common biochemical mechanism: an isopeptide bond is formed between the modifier's terminal diglycine and an amino group of the target protein (Hub1, which uniquely terminates in a dityrosine motif, is an exception). Ub-like domains which occur as stable elements within other proteins cannot be processed or conjugated because they lack the terminal diglycine motif.

Rub1/NEDD8

Rub1 from yeast (related to ubiquitin1) displays 53% identity with Ub and 59% identity with its mammalian homologue Nedd8. Among important structural features conserved between Rub1/Nedd8 and Ub are (*i*) 5 internal Lys residues (Lys-6, 11,27,33,48); (*ii*) two glycine residues at the C-terminus that are important for recognition by isopeptidases; (*iii*) a cluster of surface hydrophobic residues (Leu-8, Ile-44, Val-70), important for recognition by the 26S proteasome. Comparison of the crystal structures of Ub and Nedd8 in human revealed very similar spatial architecture [78]. Two remarkable differences between Ub and Rub1 must be admitted: (*i*) while Ub is usually translated as a fusion to other proteins or to Ub, Rub1 is translated in a precursor form containing an asparagine residue as a C-terminal extension; (*ii*) residue 72 in Rub1 sequence which is encoded by Ala (Arg in Ub) was shown to be responsible for the inability of Rub1 to be activated by Uba1 (E1).

As previously mentioned, Rub1 is synthesized in an inactive precursor form. In order to be efficiently conjugated to its targets, the last asparagine residue should be clipped off. Lately, a piece of evidence was presented that yeast Yuh1 (one of deubiquitinating enzymes) is also responsible for Rub1 processing [79]. Following maturation, Rub1/Nedd8 can be conjugated to its targets in the presence of a set of Rub1-specific activating enzymes – Ula1 and Uba3, which resemble the N-terminal and C-terminal domains of the Uba1 E1 enzyme respectively, a Ub-conjugating enzyme - Ubc12, and core components of a specific RING finger E3 complex (SCF) [78]. The process of Rub1/Nedd8 attachment, termed rubylation (or neddylation in higher eukaryotes), was shown to occur in all eukaryotes from budding yeast to human. The type of linkage between Rub1/Nedd8 and its substrates is a classic isopeptide bond between a C-terminal carboxyl of Rub1 and an internal Lys side chain of a substrate. So far, only monorubylated species have been discovered, however existence of polyRub chains cannot be excluded. A cycle of rubylation is usually followed by deconjugation of Rub1 from its substrate. Derubylation is mediated by the Csn5 subunit of the COP9 signalosome in yeast. In humans, three additional DUBs were found to remove Nedd8.

The only identified targets for rubylation are members of the Cullin family [80,81]. Cdc53/Cul1 is the most intensively studied cullin in yeast. It was first recognized as a part of a multisubunit E3 complex termed SCF (for ***S***kp2/***C***ullin/***F***-box) (reviewed in [82]). SCF consists of at least 4 subunits: Cul1, which serves as a scaffold for assembly of the rest of subunits; Rbx1 – a RING finger protein, which recruits an E2 to the complex; Skp2 – an adaptor which brings an F-box protein whose function is in substrate binding. The interaction between Rbx1 and Cul1 is mediated by the C-terminal cullin repeats in Cul1 and recently it was shown that the rubylation site is mapped close to the contact surface between the two subunits. Rubylation of Cul1 facilitates recruitment of an E2 to the Cul1/Rbx1 module without affecting the affinity of the SCF for its substrates. Recently, in agreement with predictions, two other cullins were found to be a target for rubylation. Subsequently, all cullins were shown to be neddylated *in vitro* [80].

The systematic deletion of Rub1 in *S. cerevisiae* is viable. Moreover, the null mutants grow at wild type rates and exhibit no phenotypes sensitive to a series of stress conditions such as starvation, heat shock, cadmium and canavanine that are common to other mutations in the UPP [81]. In *Arabidopsis thaliana*, Nedd8 deletants exhibit severe phenotypes in embryo development and render insensitive to phytohormone auxin. In *S. pombe*, *C. elegans*, *D. melanogaster*, mice, hamster and higher organisms the deletion of Nedd8 is lethal.

The only known biochemical function of Rub1/Nedd8 is to modify members of Cullin family. In many organisms, rubylation is important for cell cycle progression, development, embriogenesis, photomorphogenesis and cytoskeletal regulation. Collectively, these phenotypes point to the importance of Rub1 in aiding the assembly or modifying the ubiquitination activity of SCF. Recent results suggest that Rub1/Nedd8 is required for fine-tuning of the ubiquitination activity of SCF by increasing its affinity towards E2s.

Smt3/SUMO

Smt3, a yeast version of a mammalian protein SUMO-1 (small Ub-related modifier) also known as sentrin, displays 18% identity with Ub (reviewed in [21]). In mammals, SUMO-1 has two additional relatives, SUMO-2 and SUMO-3. Smt3 is encoded by a 101-residue polypeptide. In domain architecture as well as in primary sequence Smt3/SUMO is significantly less similar to Ub than Rub1. In addition, Smt3 differs from Ub and Rub1 by possessing short N-terminal extension known as a non-Ub-like domain.

The Smt3 precursor, possessing a 3-residue C-terminal extension after the conserved diglycine, is processed by the Smt3-specific isopeptidase Ulp1 [78]. Conjugation of Smt3 (sumoylation) requires the E1 heterodimer Aos1/Uba2 resembling N-terminal and C-terminal portions of Uba1 respectively, a single E2 enzyme Ubc9 and three different SUMO ligases (E3s). *S. cerevisiae* contains two known Smt3 E3s – Siz1 and Siz2 that mediate conjugation to the target. Again, an isopeptide bond is formed between Smt3 and an internal lysine. In contrast to Rub1, first, SUMO-1and SUMO-2/3 and subsequently Smt3 were shown to form polySUMO conjugates [83,84]. The major branch sites are lysines in the N-terminal non-Ub-like domain. Sumoylation is a reversible modification, and a family of Smt3-specific isopeptidases, including the yeast proteins Ulp1 and Ulp2 removes Smt3 from its substrates. Ulp2 also catalyses dismantling of polySUMO chains [84].

Unlike Rub1, Smt3 has numerous substrates and it seems that much more remain to be discovered. The first identified substrate of SUMO was RanGAP1, a GTPase-activating protein of the Ras-related GTPase Ran. Since then a list of Smt3 targets is continuously being extended; notable Smt3 substrates are presented in Table 1 (reviewed in [21,78]).

Table 1. Targets for sumoylation.

Substrate	Role of sumoylation
RanGAP1	Targeting to nuclear pore
PML	Transcription activation
Sp3	Transcription repression
Top2	Chromosome structure
Dictiostelium MAP kinase	Signal transduction activation
IκBα Mdm2	Stabilization
p53	Stabilization and transcription enhancement
GLUT1 and GLUT4	Stabilization/destabilization
PCNA	Enzymatic efficiency

In budding yeast as well as in higher organisms, the deletion of Smt3/SUMO is lethal. Moreover, mutations in components of SUMO machinery are either lethal or exhibit severe phenotypes in cell cycle progression. In contrast to *S.cerevisiae*, *RUB1* deletants in *S.pombe* are viable but grow poorly.

With so many substrates to modify and severe associated phenotypes, it is not surprising that functions of Smt3 are so overwhelming. No typical role for sumoylation can be elucidated in contrast to ubiquitination and rubylation. In some cases, sumoylation serves as an antagonist of ubiquitination [7]. For instance, in IκBα,Mdm2 and p53 SUMO targets the same lysine residues that constitute ubiquitination site thereby blocking the degradation of the proteins by UPP (Table 1). In some cases, for example in p53, the same effect is achieved upon attachment of acetyl group to the described ubiquitination site. Therefore, the Ub machinery clearly discriminates between Ub and Ub-like proteins. In some cases, sumoylation most likely alters the conformation of a target protein thereby promoting its association/dissociation with other cellular factors as demonstrated for various transcription factors. It must be noticed that sumoylation can convert transcription factors to activators as well as to repressors of transcription [85]. Yet in other cases sumoylation is required for reducing cellular levels of target proteins as was demonstrated for GLUT4 glucose transporter [21]. Finally, sumoylation may alter intracellular localization of its targets as was shown for RanGAP1 protein [21]. Apparently, sumoylation represents a very important mode of regulation but its actual meaning is still to be discovered.

APG Proteins

Remarkably, two other Ub-like proteins, called APG8 (Aut7) and APG12 (for 'autophagy'), function in the starvation response known as autophagy (reviewed in [19-21]). APG8 and APG12, which are significantly larger than Ub and have no sequence relationship to it, exhibit sequence similarity to one another, and one (APG8) has a three-dimensional structure very similar to Ub. Like Ub, the C-termini of both proteins are activated with ATP by an enzyme related to E1 before being covalently attached to targets. As with other UBLs, APG8 and APG12 utilize an E1–E2 enzyme couple for their covalent linkage to substrates. Interestingly, the two modifiers share a single E1-like enzyme but have different E2-like proteins. This is the only known example in which a single E1 can activate two different UBLs.

The detailed roles of these Ub-like proteins in autophagy have not yet been worked out. It is clear, however, that they do not direct the destruction of their targets by the proteasome.

APG8 is necessary for formation of intermediate membrane structures that arise during formation of the autophagosome. Remarkably, the target of APG8 ligation is not a protein, but a phospholipid – the triglyceride phosphatidylethanolamine. APG12 conjugation is required for either autophagosome precursor elongation or completion of autophagosome formation. There is a single known target for APG12, the APG5 protein. This conjugation appears to be constitutive and there is no known protease for cleaving APG12 from its substrate.

Other UBLs

The catalogue of UBL proteins possesses further members, but, for most of them, only preliminary accounts are available. The list is still growing, and hence this survey is restricted to only a few, better characterized examples [19-21,78].

Two mammalian proteins, UCRP (Ub cross-reacting protein; also known as ISG15) and FAT10, resemble two Ub moieties fused in tandem. The functions of Fat10 remain mysterious; this Ubl is encoded in the MHC class I locus and may play a role in cytokine-induced apoptosis. ISG15 conjugation plays an important role in normal development and in interferon a/h-mediated responses to viral infection, although the specific purposes served by ISG15 modification are not yet known. A recently discovered modifier - HUB (homologous to Ub) is highly conserved from yeast to mammals and displays only 22% sequence identity with Ub. HUB possesses an invariant C-terminal double-tyrosine motif, followed by a single variable residue. Whether UCRP or HUB pathways are connected to the Ub system is not yet known [21].

Another recently identified modifier is URM1 from yeast, which is unrelated to both Ub and APG8/12 but is possibly related remotely to SUMO. URM1 is similar and its activating enzyme UBA4 are related to prokaryotic proteins involved in molybdopterin and thiamin biosynthesis, suggesting that prokaryotes might use similar enzymatic reactions in these pathways. The URM1 system from yeast does not function in these biosynthetic pathways, but mutant cells lacking the URM1 system are sensitive to heat. These recent findings are particularly intriguing with respect to the evolution of UBL pathways and suggest that UBLs might be ancient inventions.

Can Ub family members engage in cross-talk? Increasing evidence suggests an answer in the affirmative. The modification of the same protein by more than one UBL, sometimes at the same residue, is possible. For instance, both SUMO and Ub can modify the same residues of IκBα, MEK1 and PCNA [19-21,78]. It is possible that the ISG15 protein can also protect certain proteins from proteasomal degradation by an analogous mechanism. Not surprisingly, there are other post-translational modifications that can also directly modulate Ub or UBL ligation. For instance, the balance between acetylation and ubiquitination of the same pair of lysine residues in the Smad7 signal-transduction protein determines the intracellular levels of this inhibitory Smad protein.

Some of the UBL conjugation systems retain any functional overlap or cross-regulation. The modification of Ub E3s by RUB is one obvious example of such cross-regulation. In a recent unexpected development, UbcH8—a bona fide conjugating enzyme for Ub—was found to have a second role as a conjugating enzyme for ISG15. In another example, Rub1 is only poorly activated by the Ub E1, but a single change in Rub1 – at residue 72 – to the amino acid found in Ub, makes RUB virtually indistinguishable in its ability to bind the Ub E1 [19-21,78]. In all likelihood, additional Ub family members remain to be discovered since Ub-like ORFs generally fall below the length cutoff for annotation in genome sequencing.

Conclusions

Our understanding of intracellular protein degradation mechanisms has undergone enormous changes over the past decade and a half. We now know that Ub ligation is involved in the majority of regulated protein degradation in the cell through both proteasomal and non-proteasomal routes. Ub-like proteins have also been found to be regulators of certain Ub ligation enzymes and of specialized lysosomal targeting and degradation pathways such as autophagy.

Much remains to be learned. We have a very limited knowledge of the nature of naturally occurring degrons or the degrons that define certain proteins as abnormal or misfolded. What is the cytosolic unfolded protein response, for example (as distinct from the elaborate detection machinery for such proteins in the ER)? Structural analysis of Ub ligase-degron complexes will be required for a full understanding of this problem. How Ub ligase complexes assemble the polyubiquitin chains that serve as signals for proteasome binding is poorly understood as well. The mechanics of substrate protein binding, unfolding, and translocation within the 26S proteasome are just beginning to be studied as is the assembly of this fascinating 2,500 kDa proteolytic machine.

Similarly, in pathways such as endocytosis and autophagy, Ub or Ub-like protein attachments function as sorting or regulatory signals, but studies of the mechanisms of action of these various signals are in their infancy. Many cell biological issues remain to be addressed: How does protein localization modulate the ability of a protein to be ubiquitinated or modified by another Ubl? Conversely, how do such modifications affect protein localization and protein-protein interactions? Where do the many different enzymes of the Ub system localize? Are there still more Ubls awaiting discovery? What additional physiological pathways are regulated by the Ub system and how?

The UPS is an extremely complex array of events and reactions. The attempts made so far in order to classify and simplify the steps of the pathway allowed much better understanding of the system. But this oversimplification may also lead to overlooking of other important aspects of ubiquitination. The field of the Ub and proteasome is very young and rapidly developing and yesterday's dogmas may fall at once. One thing is evident: despite coming along way in the three decades since the discovery of Ub as a small protein modifier that targets proteins for elimination, the complete understanding of the Ub system is still far off, and its complexity is still greater than one can imagine.

ACKNOWLEDGEMENTS AND COMMENTS

We would like to dedicate this chapter to the memory of the late Prof. Cecile Pickart, a pioneer of the Ub field. Always original in your thoughts, critical in your comments, and humble in your ways - you will remain a role model and be sorely missed.

Work in our lab is supported by grants from the Israel Academy of Sciences, The Taub Family Foundation and the Wolfson Foundation for the Study of Ubiquitin at the Technion.

REFERENCES

[1] Hershko A, Ciechanover A, Heller H, Haas AL, Rose IA: Proposed role of ATP in protein breakdown: conjugation of protein with multiple chains of the polypeptide of ATP-dependent proteolysis. *Proc Natl Acad Sci USA* 1980, 77:1783-1786.

[2] Ciechanover A, Finley D, Varshavsky A: Ubiquitin dependence of selective protein degradation demonstrated in the mammalian cell cycle mutant ts85. *Cell* 1984, 37:57-66.

[3] Ciechanover A: Intracellular protein degradation: from a vague idea thru the lysosome and the ubiquitin-proteasome system and onto human diseases and drug targeting. *Cell Death Differ* 2005, 12: 1178-1190.

[4] Finley D, Ciechanover A, Varshavsky A: Thermolability of ubiquitin-activating enzyme from the mammalian cell cycle mutant ts85. *Cell* 1984, 37:43-55.

[5] Hershko A: The ubiquitin system for protein degradation and some of its roles in the control of the cell division cycle. *Cell Death Differ* 2005, 12:1191-1197.

[6] Rose IA: Ubiquitin at Fox Chase. *PNAS* 2005, 102:11575-11577.

[7] Glickman MH, Ciechanover A: The Ubiquitin-proteasome Proteolytic Pathway: Destruction for the sake of construction. *Physiol Rev* 2002, 82:373-428.

[8] Goldberg AL: Nobel Committee Tags Ubiquitin for Distinction. *Neuron* 2005, 45:339-344.

[9] Pickart CM: Mechanisms underlying ubiquitination. *Annu Rev Biochem* 2001, 70:503-533.

[10] Varshavsky A: The early history of the ubiquitin field. *Protein Sci* 2006, 15:647-654.

[11] Hicke L: Gettin' down with ubiquitin: turning off cell-surface receptors, transporters and channels. *Trends Cell Biol* 1999, 9:107-112.

[12] Hicke L: Protein regulation by monoubiquitin. *Nat Rev Mol Cell Biol* 2001, 2:195-201.

[13] Hicke L: A New Ticket for Entry into Budding Vesicles-Ubiquitin. *Cell* 2001, 106:527-530.

[14] Hicke L, Dunn R: Regulation of membrane protein transport by ubiquitin and ubiquitin-binding proteins *Annu Rev Cell Dev Biol* 2003, 19:141-172.

[15] Schnell JD, Hicke L: Non-traditional Functions of Ubiquitin and Ubiquitin-binding Proteins. *J Biol Chem* 2003, 278:35857-35860.

[16] Pickart CM: Ubiquitin in chains. *Trends Biochem Sci* 2000, 25:544-548.

[17] Pickart CM, Eddins MJ: Ubiquitin: structures, functions, mechanisms. Biochimica et Biophysica Acta (BBA) - *Molecular Cell Research* 2004, 1695:55-72.

[18] You J, Pickart CM: A HECT domain E3 enzyme assembles novel polyubiquitin chains. *J Biol Chem* 2001, 276:19871-19878.

[19] Hochstrasser M: Evolution and function of ubiquitin-like protein-conjugating systems. *Nature Cell Biol* 2000, 2:E153-E157.

[20] Hochstrasser M: Molecular biology: New Proteases in a Ubiquitin Stew. *Science* 2002, 298:549-552.

[21] Jentsch S, Pyrowolakis G: Ubiquitin and its kin: how close are family ties? *Trends Cell Biol* 2000, 10:335-342.

[22] Bode AM, Dong Z: Post-translational modification of p53 in tumorigenesis. *Nature Reviews Cancer* 2004, 4:793-805.

[23] Amerik AY, Hochstrasser M: Mechanism and function of deubiquitinating enzymes *Biochim Biophys Acta*. 2004 1695:189-207.

[24] Guterman A, Glickman MH: Deubiquitinating enzymes are *IN*(trinsic to proteasome function). *Curr Prot Pep Sci* 2004, 5:201-210.

[25] Koegl M, Hoppe T, Schlenker S, Ulrich HD, Mayer TU, Jentsch S: A novel ubiquitination factor, E4, is involved in multiubiquitin chain assembly. *Cell* 1999, 96:635-644.

[26] Hartmann-Petersen R, Seeger M, Gordon C: Transferring substrates to the 26S proteasome. *Trends Biochem Sci* 2003, 28:26-31.

[27] Hartmann-Petersen R, Gordon C: Protein Degradation: Recognition of Ubiquitinylated Substrates. *Current Biology* 2004, 14:R754-R756.

[28] Deveraux Q, Ustrell V, Pickart C, Rechsteiner M: A 26S subunit that binds ubiquitin conjugates. *J Biol Chem* 1994, 269:7059-7061.

[29] Elsasser S, Chandler-Militello D, Muller B, Hanna J, Finley D: Rad23 and Rpn10 Serve as Alternative Ubiquitin Receptors for the Proteasome. *J Biol Chem* 2004, 279:26817-26822.

[30] Glickman MH, Raveh D: Proteasome plasticity. *FEBS Letters* 2005, 579:3214-3223

[31] Legget DS, Hanna J, Borodovsky A, Crossas B, Schmidt M, Baker RT, Walz T, Ploegh HL, Finley D: Multiple associated proteins regulate proteasome structure and function. Mol *Cell* 2002, 10:495-507.

[32] Richly H, Rape M, Braun S, Rumpf S, Hoege C, Jentsch S: A series of ubiquitin binding factors connects cdc48/p97 to substrate multiubiquitylation and proteasomal targeting. *Cell* 2005, 120:73-84.

[33] Bazirgan OA, Hampton RY: Cdc48-Ufd2-Rad23: the road less ubiquitinated? *Nat Cell Biol* 2005, 7:207.

[34] Fu H, Sadis S, Rubin DM, Glickman MH, van Nocker S, Finley D, Vierstra RD: Multiubiquitin chain binding and protein degradation are mediated by distinct domains within the 26S proteasome subunit Mcb1. *J Biol Chem* 1998, 273:1970-1989.

[35] Fu HY, Reis N, Lee Y, Glickman MH, Vierstra R: Subunit interaction maps for the regulatory particle of the 26s proteasome and the cop9 signalosome reveal a conserved core structure. *EMBO J* 2001, 20:7096-7107.

[36] van Nocker S, Sadis S, Rubin DM, Glickman MH, Fu H, Coux O, Wefes I, Finley D, Vierstra RD: The multiubiquitin chain binding protein Mcb1 is a component of the 26S proteasome in *S. cerevisiae* and plays a nonessential, substrate-specific role in protein turnover. *Mol Cell Biol* 1996, 11:6020-6028.

[37] van Nocker S, Deveraux Q, Rechsteiner M, Vierstra RD: *Arabidopsis MBP1* gene encodes a conserved ubiquitin recognition component of the 26S proteasome. *Proc Natl Acad Sci USA* 1996, 93:856-860.

[38] Lueders J, Demand J, Hoehfeld J: The ubiquitin-related BAG-1 provides a link between the molecular chaperones Hsc70/Hsp70 and the proteasome. *J Biol Chem* 2000, 275:4613-4617.

[39] Baumeister W, Walz J, Zuhl F, Seemuller E: The proteasome: paradigm of a self-compartmentalizing protease. *Cell* 1998, 92:367-380.

[40] Glickman MH, Maytal V: Regulating the 26S proteasome, in *Curr Top Microbiol and Immunol*, Zwickl P, Baumeister W, Editors. 2002, Springer-Verlag: Germany. p. 43-72.

[41] Pickart CM, Cohen RE: Proteasomes and their kin: proteases in the machine age. *Nat Rev Mol Cell Biol* 2004, 5:177-187.

[42] Zwickl P, Seemueller E: 20S proteasomes, in *Curr Topics Microbiol Immunol*, Zwickl P, Baumeister W, Editors. 2002, Springer-Verlag: Germany. p. 23-41.

[43] Lam YA, Lawson TG, Velayutham M, Zweier JL, Pickart CM: A proteasomal ATPase subunit recognizes the polyubiquitin degradation signal. *Nature* 2002, 416:763-767.

[44] Glickman MH, Rubin DM, Fried VA, Finley D: The regulatory particle of the *S. cerevisiae* proteasome. *Mol Cell Biol* 1998, 18:3149-3162.

[45] Glickman MH, Rubin DM, Larsen CN, Finley D: The regulatory particle of the yeast proteasome., in *Proteasomes: The world of regultory proteolysis*, Hilt W, Wolf DH, Editors. 2000, Eurekah.com / LANDES BIOSCIENCE Publishing Company: Georgetown; Texas; USA. p. 71-90.

[46] Verma R, Chen S, Feldman R, Schieltz D, Yates J, Dohmen J, Deshaies RJ: Proteasomal Proteomics: Identification of Nucleotide-sensitive proteasome-interacting proteins by mass spectrometric analysis of affinity-purified proteasomes. *Mol Biol Cell* 2000, 11:3425-3439.

[47] Guerrero C, Tagwerker C, Kaiser P, Huang L: An Integrated Mass Spectrometry-based Proteomic Approach: Quantitative Analysis of Tandem Affinity-purified in vivo Cross-linked Protein Complexes (qtax) to Decipher the 26 s Proteasome-interacting Network. *Mol Cell Proteomics* 2006, 5:366-378.

[48] Glickman MH, Rubin DM, Coux O, Wefes I, Pfeifer G, Cjeka Z, Baumeister W, Fried VA, Finley D: A subcomplex of the proteasome regulatory particle required for ubiquitin-conjugate degradation and related to the COP9/Signalosome and eIF3. *Cell* 1998, 94:615-623.

[49] Verma R, Aravind L, Oania R, McDonald WH, Yates III JR, Koonin EV, Deshaies RJ: Role of Rpn11 Metalloprotease in Deubiquitination and Degradation by the 26S Proteasome. *Science* 2002, 298:611-615.

[50] Guterman A, Glickman MH: Complementary roles for Rpn11 and Ubp6 in deubiquitination and proteolysis by the proteasome. *J Biol Chem* 2004, 279:1729-1738.

[51] Elsasser S, Gali RR, Schwickart M, Larsen CN, Leggett DS, Muller B, Feng MT, Tubing F, Dittmar GA, Finley D: Proteasome subunit Rpn1 binds ubiquitin-like protein domains. *Nat Cell Biol* 2002, 4:725-730.

[52] Rape M, S Jentsch: Productive RUPture: activation of transcription factors by proteasomal processing. *Biochim Biophys Acta* 2004, 1695:209-213.

[53] Orian A, Schwartz AL, Israel A, Whiteside S, Kahana C, Ciechanover A: Structural motifs involved in ubiquitin-mediated processing of the NF-kappaB precursor p105: roles of the glycine-rich region and a downstream ubiquitination domain. *Mol Cell Biol* 1999, 19:3664-3673.

[54] Kornitzer D, Ciechanover A: Modes of regulation of ubiquitin-mediated protein degradation. *J Cell Physiol* 2000, 182:1-11.

[55] Lecker SH, Solomon V, Mitch WE, Goldberg AL: Muscle protein breakdown and the critical role of the ubiquitin-proteasome pathway in normal and disease states. *J Nutr* 1999, 129:227S-237S.

[56] Finley D, Ozkaynak E, Varshavsky A: The yeast polyubiquitin gene is essential for resistance to high temperatures, starvation, and other stresses. *Cell* 1987, 48:1035-1046.

[57] Shang F, Gong X, Taylor A: Activity of ubiquitin-dependent pathway in response to oxidative stress. *J Biol Chem* 1997, 272:23086-23093.

[58] Cheng L, Watt R, Piper PW: Polyubiquitin gene expression contributes to oxidative stress resistance in respiratory yeast. *Mol Gen Genet* 1994, 243:358-362.

[59] Bajorek M, Finley D, Glickman MH: Proteasome Disassembly and Downregulation Is Correlated with Viability during Stationary Phase. *Current Biology* 2003, 13:1140-1144.

[60] Blondel M, Galan J-M, Chi Y, Lafourcade C, Longaretti C, Deshaies RJ, Peter M: Nuclear -specific degradation of Far1 is controlled by localization of the F-box protein Cdc4. *EMBO J* 2000, 19:6085-6097.

[61] Funakoshi M, Sasaki T, Nishimoto T, Kobayashi H: Budding yeast Dsk2p is a polyubiquitin-binding protein that can interact with the proteasome. *Proc Natl Acad Sci USA* 2002, 99:745-750.

[62] Biggins S, Ivanovska I, Rose MD: Yeast ubiquitin-like genes are involved in duplication of the microtubule organizing center. *J Cell Biol* 1996, 133:1331-1346.

[63] Wilkinson KD: Signal transduction: aspirin, ubiquitin and cancer. *Nature* 2003, 424:738-739.

[64] Zhang HG, Wang J, Yang X, Hsu HC, Mountz JD: Regulation of apoptosis proteins in cancer cells by ubiquitin. *Oncogene* 2004 23:2009-2015.

[65] Ward CL, Omura S, Kopito RR: Degradation of CFTR by the ubiquitin proteasome pathway. *Cell* 1995, 83:121-127.

[66] Kopito RR: Biosynthesis and degradation of CFTR. *Physiol Rev* 1999, 79:S167-S173.

[67] DiAntonio A, Hicke L: Ubiquitin-dependent regulation of the synapse. *Ann Rev Neurosci* 2004, 27:223-246.

[68] Alves-Rodrigues A, Gregori L, Figueiredo-Pereira ME: Ubiquitin, cellular inclusions, and their role in neurodegeneration. *Trends Neurosci* 1998, 21:516-520.

[69] Sherman MY, Goldberg AL: Cellular defense mechanisms against unfolded proteins: A cell biologist thinks about neurodegenerative diseases. *Neuron* 2001, 29:15-32.

[70] Kalchman MA, Graham RK, Xia G, Koide HB, Hodgson JG, Graham KC, xGoldberg YP, Gietz RD, Pickart CM, Hayden MR: Huntingtin is ubiquitinated and interacts with a specific ubiquitin-conjugating enzyme. *J Biol Chem* 1996, 271:19385-19394.

[71] Ross CA, Wood JD, Schilling G, Peters MF, Nucifora FC Jr, Cooper JK, Sharp AH, Margolis RL, Borchelt DR: Polyglutamine pathogenesis. *Philos Trans R Soc Lond B Biol Sci* 1999, 354:1005-1011.

[72] Cummings CJ, Reinstein E, Sun Y, Antalffy B, Jiang Y, Ciechanover A, Orr HT, Beaudet AL, Zoghbi HY: Mutation of the E6-AP ubiquitin ligase reduces nuclear inclusion frequency while accelerating polyglutamine-induced pathology in SCA mice. *Neuron* 1999, 4:879-892.

[73] Peng J, Schwartz D, Elias JE, Thoreen CC, Cheng D, Marsischky G, Roelofs J, Finley D, Gygi SP: A proteomics approach to understanding protein ubiquitination. *Nat Biotechnol* 2003, 21:921-926.

[74] Thrower JS, Hoffman L, Rechsteiner M, Pickart C: Recognition of the polyubiquitin proteolytic signal. *EMBO J* 2000, 19:94-102.

[75] Raasi S, Varadan R, Fushman D, Pickart CM: Diverse polyubiquitin interaction properties of ubiquitin-associated domains. *Nat Struct Mol Biol* 2005, 12:708-714.

[76] Pickart CM: Back to the Future with Ubiquitin. *Cell* 2004, 116:181-190.

[77] Garcia-Higuera I, Taniguchi T, Ganesan S, Meyn MS, Timmers C, Hejna J, Grompe M, D'Andrea AD: Interaction of the Fanconi anemia proteins and BRCA1 in a common pathway. *Mol Cell* 2001, 7:249-262.

[78] Schwartz DC, Hochstrasser M: A superfamily of protein tags: ubiquitin, SUMO and related modifiers. *Trends Bioch Sci* 2003, 28:321-328.

[79] Linghu B, Callis J, Goebl MG: Rub1p Processing by Yuh1p Is Required for Wild-Type Levels of Rub1p Conjugation to Cdc53p. *Eukaryotic Cell* 2002, 1:491-494.

[80] Ohh M, Kim WY, Moslehi JJ, Chen Y, Chau V, Read MA, Kaelin WGJ: An intact NEDD8 pathway is required for Cullin-dependent ubiquitylation in mammalian cells. *EMBO Rep* 2002 177-182.

[81] Liakopoulos D, Doenges G, Matuschewski K, Jentsch S: A novel protein modification pathway related to the ubiquitin system. *EMBO J* 1998, 17:2208-2214.

[82] Joazeiro CAP, Weissman AM: RING Finger proteins: Mediators of ubiquitin ligase activity. *Cell* 2000, 102:549-552.

[83] Li S-J, Hochstrasser M: The Ulp1 SUMO isopeptidase: distinct domains required for viability, nuclear envelope localization, and substrate specificity. *J Cell Biol* 2003, 160:1069-1082.

[84] Bylebyl GR, Belichenko I, Johnson ES: The SUMO isopeptidase Ulp2 prevents accumulation of SUMO chains in yeast. *J Biol Chem* 2003, 278:835720-835728.

[85] Aggarwal BB, Kumar A, Bharti AC: Anticancer potential of curcumin: preclinical and clinical studies. *Anticancer Res* 2003, 23:363-398.

In: The Ubiquitin Proteasome System... ISBN 978-1-60021-749-4
Eds: Mario Di Napoli and Cezary Wojcik, pp. 71-92

Chapter 4

DIVERSITY AND CELLULAR FUNCTIONS OF DEUBIQUITINATING ENZYMES

Kyuhee Oh, Ok Sun Bang and Chin Ha Chung*

NRL of Protein Biochemistry, School of Biological Sciences, Seoul National University, Seoul 151-742, Korea.

ABSTRACT

Covalent modification of proteins by ubiquitin is a key mechanism for the control of cellular processes as diverse as cell proliferation, differentiation, apoptosis. Deubiquitination, reversal of this modification, is catalyzed by deubiquitinating enzymes that belong to the superfamily of proteases. Deubiquitinating enzymes occupy the second largest family of enzymes in the ubiquitin system, implying their functions in the control of diverse cellular processes by regulating the fate, function of ubiquitinated proteins. Cellular functions of deubiquitinating enzymes include the regulation of proteasome activity, protein stability, signal transduction, DNA repair, chromatin dynamics, transcription, endocytosis. Deubiquitinating enzymes also play important roles in the processing of inactive ubiquitin precursors for the generation of matured ubiquitin monomers, the removal of ubiquitin from 'distal' end of poly-ubiquitination chains conjugated to proteins for controlling the fidelity of the ubiquitination process, the cleavage of ubiquitin adducts for the release of free ubiquitin. Deubiquitinating enzymes consist of five families that have distinct catalytic domain structures: the ubiquitin-specific protease (USP) family, the ubiquitin C-terminal hydrolase (UCH) family, the ovarian tumor protease (OTU) family, the Machado-Joseph disease protein (MJD) family, the Jab1/MPN/Mov34-domain protease (JAMM) family. While the JAMM family members are metalloproteases, the other family members are cysteine proteases. As the names of certain families imply, deubiquitinating enzymes play critical regulatory roles in a multitude of processes from cancer to neurodegenerative diseases. In this

* Correspondence concerning this article should be addressed to Prof. Chin Ha Chung, PhD; School of Biological Science, Seoul National University, Seoul, 151-742, Korea. Phone: +82 2 880 6693; Fax: +82 2 871 9193; E-mail: chchung@snu.ac.kr.

chapter, we summarize the catalytic properties of deubiquitinating enzymes so far been identified, the recent findings on their functions as cellular regulators. We also describe the specific features of deubiquitinating enzymes that are related with neuronal diseases.

Keywords: deubiquitinating enzyme, JAMM metalloprotease, MJD protease, OTU protease, ubiquitin C-terminal hydrolase, ubiquitin-specific protease.

Introduction

Ubiquitin (Ub) is a well-conserved 76 amino acid polypeptide that is present in all eukaryotic cells [1]. It is conjugated to a variety of cellular proteins on ε-amino group of lysine residue or rarely on N-terminal amino group [2-4]. Covalent modification of Ub to target proteins is mediated by the enzymatic cascade system, consisting of Ub-activating enzymes (E1), Ub-conjugating enzymes (E2), Ub ligases (E3). This post-translational modification changes the fate of target proteins, such as the alterations in their stability, protein-protein interaction, subcellular localization, so on [5,6] (see Chapter 3).

Ub itself is also a target for ubiquitination, thereby forming poly-Ub chains that are also linked by isopeptide bonds. Ub has 7 conserved lysine residues facing outward, all of which are known to be ubiquitinated at least in yeast [7]. Therefore, there could be at least 7 types of Ub modification, discriminated by which lysine residue is used to make a poly-Ub chain. Whether different lysine residues are used in a poly-Ub chain *in vivo* remains to be elucidated, since no diagnostic reagent, such as linkage-specific antibodies, is available at present [8].

Modification of proteins by Ub plays a key role in the control of diverse important cellular processes [5,6]. Chromosome structure regulation, DNA repair, signal transduction, antigen presentation, viral pathogenesis, stress response, protein trafficking are the examples. Many of these processes, but not all, are mediated by the ubiquitin-proteasome system (UPS) that leads to the degradation of poly-ubiquitinated proteins by the 26S proteasome that consists of two major components: the 19S regulatory particle (also called PA700) harboring the ATPase activity, the 20S proteasome having the proteolytic core in its chamber [6] (see Chapters 6, 7). Therefore, defects in the UPS are known to have critical relations to various diseases, including neurodegenerative diseases like Alzheimer's disease (AD) (see Chapter 30), Parkinson's disease (PD) (see Chapter 31), Huntington's disease (HD) [8] (see Chapter 32).

Analogous to protein phosphorylation, dephosphorylation, protein modification by Ub is a reversible process. Deubiquitinating enzymes (DUBs) catalyze the removal of Ub from Ub-conjugated proteins [9]. Since deubiquitination is antagonistic to ubiquitination, it is well expected that deubiquitination is also involved in the control of diverse cellular functions that are mediated by ubiquitination of target proteins. However, the study on deubiquitination is in the beginning stage in comparison to that on ubiquitination. So what DUBs do in cells, how they are regulated largely remain to be investigated.

In addition to the reversal of protein ubiquitination, DUBs display other catalytic functions. They generate Ub monomer from its precursors, since in the cell Ub is not synthesized as a free form but as Ub-ribosomal peptide fusion proteins (i.e., Ub C-terminal extension peptides of 52, 80 amino acids), head-to-tail linked Ub multimers with additional amino acids following the last Ub molecule [10]. DUBs also disassemble free poly-Ub chains by releasing Ub monomer from either 'distal' or 'proximal' end of the chains. Some DUBs appear to perform 'editing' function by removing Ub from 'distal' end of poly-Ub chains that are conjugated to target proteins, thus controlling the fidelity of the conjugation process. DUBs must also cleave off small nucleophiles from Ub adducts, which are generated by side reactions with cellular thiols, amines, for maintaining the intracellular pool of free Ub [11]. These catalytic functions of DUBs are summarized in Figure 1.

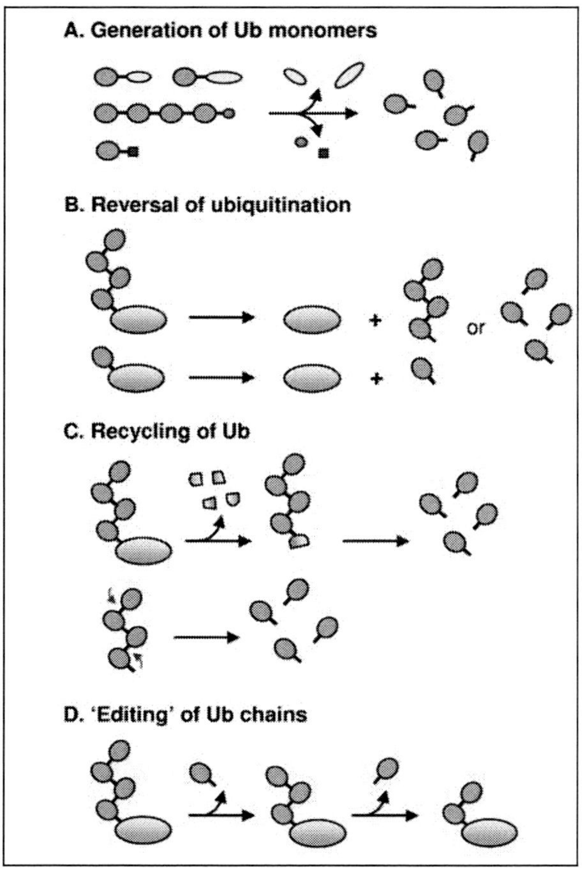

Figure 1. Catalytic function of DUBs. (A) Ub adducts, Ub-ribosomal peptide fusions are preferentially processed by UCHs for releasing free Ub monomers, ribosomal peptides. USPs generate Ub monomers from head-to-tail linked Ub polymers. Ubs are colored in orange, ribosomal peptides in yellow, a single amino acid in green, small nucleophiles in red. The black tail of Ub monomer represents the C-terminal di-glycine. (B) DUBs may cleave off poly-Ub chain from target proteins or successively remove Ub monomer from 'distal' end of poly-Ub chain. Reversal of mono-ubiquitination (e.g., Ubp8, USP8) is involved in the control of cellular pathways other than UPS. (C) Poly-Ub chain bound to proteolytic remnants can be disassembled by DUBs (e.g., Doa4) for recycling Ub monomers. DUBs also disassemble free poly-Ub chains synthesized *de novo* by releasing Ub monomer from either 'distal' (red arrow) or 'proximal' end (blue arrow) of the chains. (D) DUBs (e.g., UCH37) may perform 'editing' function by removing Ub from distal ends of poly-Ub chain conjugated to proteins for controlling the fidelity of ubiquitination process.

Deubiquitinating Enzyme Families

Since the deubiquitinating activity toward histone H2A (A24) was first reported in 1981 [12], a large number of DUBs have been identified by biochemical, genomic screening approaches from various organisms. Numerous methods for assaying DUBs *in vitro* have also been developed. The most widely used method is SDS-polyacrylamide gel electrophoresis, resolving the cleavage products after incubation of DUBs with linear Ub fusions (e.g., Ub-β-galactosidase or Ub-peptide) or branched Ub oligomers (e.g., Lys48- or Lys63-linked tetra-Ub). Perhaps the easiest way to assay DUBs is the use of Ub-7-amido-4-methylcoumarin (AMC). After incubation, DUBs can be assayed by simple measurement of the fluorescence of released AMC [13]. A new method for assaying DUBs has recently been developed [14]. This method employs the use of a linear fusion of poly-His-glutathione-S-transferase-Ub-ecotin (His-GST-Ub-ecotin) as a substrate. Since ecotin, a trypsin inhibitor protein from *E. coli*, is heat stable, the activity of DUBs can be indirectly assayed by determining the ability of ecotin to inhibit trypsin after incubation of any DUB with His-GST-Ub-ecotin followed by heating at 100°C. This method can also be used for assaying the proteases that process Ub-like proteins (Ubls) using the substrate, in which Ub is replaced by any Ubl.

Two classical families of DUBs were found by screening for the cleavage of Ub-fusion peptides: the UCH (Ub C-terminal hydrolase), USP (Ub-specific protease) families. DUBs of these families are cysteine proteases. A new family of DUBs with different catalytic mechanism was later identified. The JAMM (Jab1/MPN/Mov34-domain metalloprotease) family including the Rpn11/POH1 subunit in the 19S regulatory particle of the 26S proteasome has the metal-binding JAMM motif that is essential for deubiquitinating activity [15]. OTU (ovarian tumor protease), MJD (Machado-Joseph disease proteins) are two families of cysteine proteases that were most recently added to DUBs. These families were found by bio-informatical approaches [16,17]. As OTUs, MJDs have no homolog in yeast genome, consist of relatively small members compared to other DUB families, they are regarded as DUBs that were introduced lately in evolution.

UCH (Ubiquitin C-Terminal Hydrolase) Family

For the first time in 1989, two DUBs, called Yuh1, UCH-L1, were purified from yeast, human brain tissues, respectively, their genes were cloned [18,19]. Based on their sequence information, researchers could find other homologous DUBs that were grouped as the UCH family. Examples of UCH family members include Yuh1, the only UCH in budding yeasts, UCH-L1, UCH-L3, UCH37 (also called UCH-L5), BAP1 (BRCA-associated protein-1) in mammals. UCHs have a core catalytic domain of about 230 amino acids that is structurally defined by the presence of a catalytic triad consisting of positionally conserved cysteine, histidine, aspartic acid residues. They have relatively small size of 20-30 kDa with a few exceptions (e.g., 80-kDa BAP1 [20]).

UCHs remove peptides, small molecules from the C-terminus of Ub, thus functioning in generation of free Ub from Ub adducts as well as from Ub-ribosomal peptide fusion proteins. However, most of them cannot disassemble poly-Ub chains, due to the steric hindrance by a loop that occludes the active sites of UCHs [21]. However, UCH-L3 has recently been shown to efficiently cleave a 13-residue peptide in isopeptide linkage with Ub [22]. In addition, UCH37 is capable of releasing Ub from the 'distal' end of poly-Ub chains [23]. Thus, some UCH family members seem to show considerable flexibility to the substrate size.

USP (Ubiquitin-Specific Protease) Family

USPs also belong to the classical DUB families. The USP family members were initially called as UBPs (Ub-specific processing proteases). The human genome sequencing leads to the finding of as many as 58 putative UBP members in humans. Therefore, the Human Genome Organization (HUGO) Nomenclature Committee has proposed a systematic nomenclature for the enzymes using the abbreviation USP for Ub-specific protease [24]. However, the yeast enzymes are still called as UBPs, because they show low sequence homology to the human enzymes.

USPs have a core catalytic domain of about 350 amino acids. Unlike the highly conserved UCHs, the USP family exhibits strong homology only in two regions that surround the catalytic cysteine, histidine residues, so called the Cys box (about 20 amino acids), the His box (60-90 amino acids) [11]. They vary in size from 40 to 300 kDa with a variety of N-terminal extensions, occasional C-terminal extensions, and some insertions in the catalytic domains. The functions of these divergent sequences remain poorly understood, although the extensions, insertions have been suggested to function in substrate recognition, subcellular localization of the enzymes, protein-protein interactions. In addition to processing of Ub precursors, salvage of trapped ubiquitins, USPs are responsible for removing Ub from poly-ubiquitinated proteins, for disassembly of free poly-Ub chains.

OTU (Ovarian Tumor Protease) Family

Bio-informatical analysis by Makarova, coworker [16] have initially revealed that OTU proteins in *Drosophila* have homology to viral cysteine proteases, that OTU proteins form a new eukaryotic cysteine protease family. Two otubains (OTU domain Ub-aldehyde-binding proteins), otubain-1, otubain-2, that belong to this family are the first two mammalian proteases that exhibit deubiquitinating activity [25]. Cezanne (cellular zinc finger anti-NFκB) that also belongs to the OTU family has later been shown to cleave the isopeptide bond in poly-Ub chains [26]. Interestingly, A20 having the OTU domain shows not only the deubiquitinating activity but also the Ub E3 ligase activity [27]. The deubiquitinating activity of A20 is specific to Lys63-linkage in poly-Ub chain conjugated to RIP (TNF receptor-interacting protein), whereas the E3 ligase activity is responsible for Lys48-specific ubiquitination of the same protein. Since Lys48-linked, but not Lys63-linked, poly-Ub chains are recognized by the 26S proteasome, these two opposing activities of A20 lead to the degradation of RIP (i.e., first by deubiquitination from Lys63-linked chains, then by ubiquitination to form Lys48-linked chains).

MJD (Machado-Joseph Disease Protein) Family

MJD is the most recently found family of DUBs, its representative member is ataxin-3. Because the poly-glutamine expansion in ataxin-3 is a causative of Machado-Joseph disease, this family is named after it. Five MJD family members have been found in the human genome: ataxin-3, an ataxin-like protein, three Josephin domain-containing proteins. However, ataxin-3 is the only one whose deubiquitinating activity was confirmed [17].

Ataxin-3 is composed of a Josephin domain followed by two Ub interacting motifs (UIMs), a C-terminal poly-glutamine tract. The UIM domain binds to poly-Ub, the Josephin domain has the typical properties of deubiquitinating activities: disassembling Ub-lysozyme conjugates, cleaving the artificial substrate, Ub-AMC, binding to Ub-aldehyde [28]. Thus, ataxin-3 was proposed to function as a poly-Ub chain 'editing' enzyme.

JAMM (Jab1/MPN/Mov34-Domain Metalloprotease) Family

Hershko, co-workers initially identified an ATP-dependent, Ub-aldehyde insensitive deubiquitinating activity associated with the 26S proteasome from reticulocyte lysates [29]. Subsequent studies have shown that a motif within the MPN (Mpr1, Pad1 N-terminal) domain of Rpn11 in the 19S regulatory particle is responsible for the deubiquitinating activity [15,30]. This $EX_nHXHX_{10}D$ motif, termed JAMM or MPN+, is frequently found in the active site of metallo-enzymes. Moreover, the deubiquitinating activity of Rpn11 is insensitive to classical DUB inhibitors, such as Ub-aldehyde, but is destroyed by metal chelators. Thus, Rpn11 with other eukaryotic proteins that contain the metal-binding JAMM motif has been grouped as a new DUB family. In addition, Jab1 (also called Csn5) in COP9 signalosome has the JAMM motif, shows deneddylating activity that removes the Ub-like protein Nedd8 from its target cullin proteins [31].

AMSH (associated molecule with the SH3-domain of STAM) that regulates receptor sorting at the endosomes was also identified as a JAMM domain-containing DUB [32]. Unlike Rpn11, Jab1 that are embedded in macromolecular complexes for their deubiquitinating activity, the purified AMSH protein itself displays deubiquitinating activity *in vitro*.

Structure, Reaction Mechanism of Deubiquitinating Enzymes

Structures of DUBs

The three-dimensional structures of several UCH family members have been determined by X-ray crystallography or NMR (Figure 2). Human UCH-L3 [33], yeast Yuh1 [21] show high structural similarity with papain-like cysteine proteases, especially in the region including the active site catalytic triad. A distinctive feature of these enzymes from papain is the presence of a loop that lies directly over the active site. The active site cross-over loop appears to function to restrict the size, shape of substrates, thus limiting the entry of bulky substrates into the active site of UCHs. Recently, the crystal structure of UCH-L1 has also been solved [34]. The overall architecture of UCH-L1 closely resembles that of UCH-L3, Yuh1. However, the geometry of the catalytic residues in the active site of UCH-L1 is distorted in a way that the distance between the cysteine, histidine residues is too long to form a productive active site conformation. Therefore, it was proposed that the binding of substrates might induce a conformational change that brings the histidine residue in close proximity to the cysteine residue for UCH-L1 to be active.

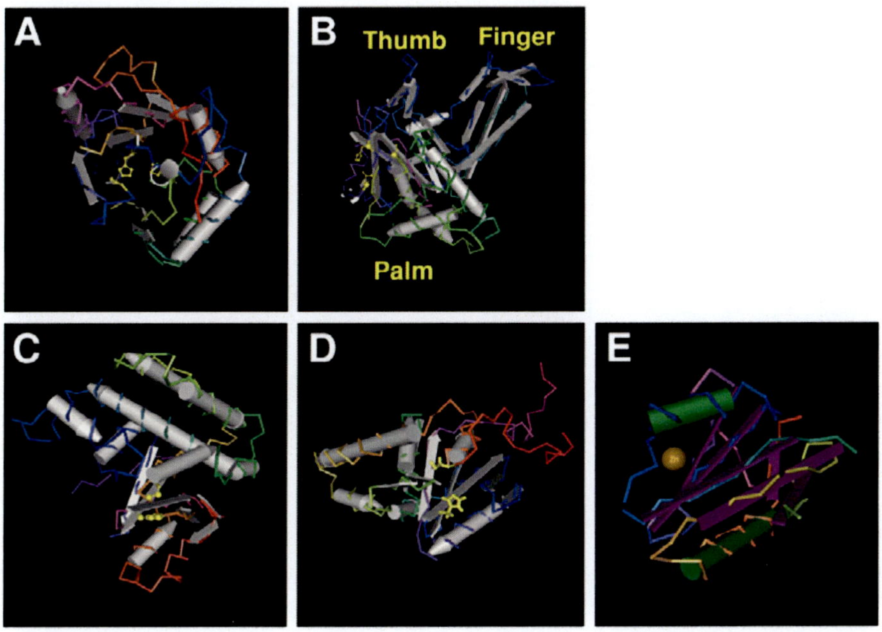

Figure 2. Three-dimensional structures of the catalytic domains of 5 DUB families. The catalytic domain structures of yeast Yuh1 (A), human USP7 (B), human otubain-2 (C), human ataxin-3 (D), *Archeoglobus* AfJAMM (E) are shown as the representatives of their corresponding family members. Information for the structures was from the Protein Databank (PDB): Yuh1, 1CMX; USP7, 1NBF; otubain-2, 1TFF; ataxin-3, 2AGA; JAMM, 1R5X. The photographic images were made by using the CN3D program. The catalytic histidine, cysteine residues are shown in yellow (A-D). The zinc ion is shown as the brown ball (E).

Among the USP family members, the crystal structures of the catalytic domains of USP7, USP14 were solved in isolation, in complex with Ub-aldehyde [35,36]. The catalytic domain of USP7 (also called HAUSP; herpes-virus associated Ub-specific protease) is composed of three domains: Fingers, Palm, Thumb. This three-domain architecture of USP7 is conserved among other USP family members, including USP14. The highly conserved Cys, His boxes are positioned on the opposite sides of the catalytic cleft created by the Palm, Thumb scaffold. This open-cleft structure of the USP7 catalytic domain, the three-domain architecture are suitable for accommodating large substrates, such as poly-ubiquitinated proteins, free poly-Ub chains, which are cleaved by USPs but not by most UCHs. However, the interaction of USP7 with its substrates, such as p53, is mediated by the N-terminal TRAF-like domain that is located apart from the catalytic domain. Thus, it appears that the highly divergent N-, /or C-terminal extensions of USPs contribute to the recruitment of specific substrates to the catalytic domains of the enzymes.

Similar to the geometry of the active site residues in UCH-L1, the distance between the catalytic cysteine, histidine residues of USP7 is too long to form a productive active site conformation [35]. However, the binding of Ub induces dramatic conformational changes around the catalytic cleft, which realign the active site residues in close proximity for productive catalysis. In contrast to the deformed active site conformation of USP7, the active site of free USP14 already adopts a productive conformation [36]. However, the active site is covered by two surface loops, preventing the access of the C-terminus of Ub to the catalytic residues of USP14. The interaction of Ub to the loops induces considerable conformational changes that widen the binding groove for the C-terminus of Ub, thereby allowing its access to the catalytic cysteine residue. Thus, USP7, USP14 appear to exhibit quite distinct active site conformations, different activation mechanisms, although they share the conserved three-domain architecture. Both mechanisms, however, serve to activate the deubiquitinating activity, appear to ensure appropriate substrate specificity.

Otubain-2 is the only protein whose crystal structure was solved among the OTU family members [37]. Although otubain-2 is a cysteine protease, its overall structure is distinct from that of previously characterized DUBs, other cysteine proteases, with no similarity to any known structures. The geometry of the active site residues in otubain-2 is in catalytically productive conformation in the absence of bound Ub, similar to that of USP14. However, the active site cross-over loop is positioned closer to the active site residues, thus making a novel oxyanion hole organization. This distinct active site topology may provide restricted substrate specificity of otubain-2.

The solution structure of the Josephin domain of ataxin-3 has recently been solved by NMR [38,39]. Despite the divergent primary sequence, the Josephin domain has structural similarity with the conserved active site region of papain-like cysteine proteases, particularly with that of UCH-L3. A unique structural feature of the Josephin domain is the presence of a flexible helical hairpin formed by an insertion sequence. Because of its proximity to the active site, the hairpin has been proposed to function in determining the substrate specificity.

The structure of the eukaryotic JAMM family members has not yet been determined. However, the structure of the JAMM domain protein AfJAMM from the thermophilic prokaryote *Archaeoglobus fulidus* shows no similarity to any known DUB structures. Rather, the arrangement of a set of amino acids that binds a zinc ion, forms the proposed active site of AfJAMM resembles that of a well-known metalloprotease, thermolysin [40]. The mutagenesis of expected active site amino acid residues in another JAMM protein called Jab1 abrogated the isopeptidase activity [31]. Therefore, the JAMM proteins have been suggested to represent a new family of metalloproteases.

Reaction Mechanism

The reaction mechanism for cysteine proteases has been characterized in detail [41]. Since the core catalytic sites of DUBs that belong to the family of cysteine proteases have high structural similarity with papain, it is well expected that the catalytic reaction is mediated by a similar mechanism. The sulfur atom located in the side chain of active site cysteine functions as a nucleophile that attacks the carbonyl group of the C-terminal glycine of Ub. Deprotonation from the thiol group is facilitated by adjacent histidine. Positively charged surroundings called oxyanion hole then pull the oxygen in the carbonyl group to promote the nucleophilic attack, increase the stability of oxyanion-containing tetrahedral intermediate. This intermediate is made of covalent thioester linkage between the cysteine thiol group of the enzyme, the C-terminus of Ub moiety [42]. The covalently bound Ub-enzyme intermediates react with a water molecule to finish the whole reaction. The specific DUB inhibitors, such as Ub-aldehyde or Ub-vinyl sulfone, also covalently bind to the enzymes with same processes but cannot escape from the enzyme.

The JAMM metalloprotease family, like thermolysin, appears to use zinc ion to make the water molecule as a nucleophile attacker [43]. The highly conserved two histidine, aspartic acid residues coordinate the zinc ion at proper location. Mutagenesis of these residues in Rpn11 makes the enzyme inert [15]. Thus, the DUBs of the JAMM family are insensitive to classical cysteine protease inhibitors, such as iodoacetamide, N-ethylmaleimide, Ub-aldehyde. Rather, they are inactivated by the chelating agents, such as EDTA, *o*-penanthroline, TPEN [N,N,N',N'-tetrakis(2-pyridylmethyl)-ethylenediamine] that is a specific chelator of zinc ion.

Deubiquitinating Enzymes as Cellular Regulators

Regulation of Proteasome Functions

DUBs are involved in the control of Ub-dependent proteolysis by direct, indirect association with the 26S proteasome. The failure of removing the Ub moiety from protein substrate leads to inappropriate degradation of Ub molecules that are conjugated to the substrate [44] or results in interruption of the proteasomal degradation of the substrate itself [15], such as by choking the narrow entry into the proteolytic core of the 20S proteasome by the unprocessed poly-ubiquitinated protein substrate. Thus, deubiquitinating activity is indispensable for normal proteolytic function of the 26S proteasome.

Rpn11 [15] in yeast, its human homolog POH1 [30] are the DUBs that are built-in as the subunit of the 19S regulatory particle of the 26S proteasome. Mutation of the predicted active site histidine residue to alanine in the JAMM domain of Rpn11 is lethal, leads to accumulation of poly-ubiquitinated proteins. In addition, the 26S proteasome carrying the mutant Rpn11 is unable to deubiquitinate or degrade ubiquitinated protein substrate *in vitro*. Thus, the deubiquitinating activity of Rpn11 appears critical for the proteasomal degradation of target proteins. Interestingly, the deubiquitinating activity of Rpn11 in the isolated 26S proteasome is fully ATP-dependent. Since the Rpn11 isopeptidase activity releases poly-Ub chains, since the ATPase activity of the 19S regulatory particle functions in unfolding of protein substrates, the coupling of these two activities is likely to facilitate the translocation of the deubiquitinated, unfolded substrate into the proteolytic core of the 20S proteasome.

UCH37 has also been identified as an integral subunit of the 19S regulatory particle of the 26S proteasome in mammals [23,45]. UCH37 is involved in 'editing' of ubiquitinated substrates according to the length of poly-Ub chains rather than the structure of the target protein themselves. By successive release of Ub from the 'distal' end of Lys48-linked poly-Ub chains, UCH37 could selectively rescue poorly ubiquitinated or slowly degraded Ub-conjugated proteins from proteolysis. The fission yeast has an ortholog of human UCH37, called Uch2, while the budding yeast does not. The 26S proteasome isolated from *uch2* null mutant cells shows much lower deubiquitinating activity toward Ub-AMC, Lys48-linked tetra-Ub than that from wild-type cells [46]. However, the mutant cells are viable without significant phenotypic abnormalities, suggesting that other DUBs may compensate for the lack of the Uch2 subunit in the 26S proteasome of fission yeasts.

Doa4 (also called Ubp4) in yeasts is required for the rapid degradation of protein substrates by the 26S proteasome [47]. In *doa4* cells, small poly-ubiquitinated peptide remnants accumulate with reduction of free Ub level. This enzyme interacts physically with the 26S proteasome. Thus, Doa4 appears to promote proteolysis through removal of Ub from proteolytic intermediates on the 26S proteasome before or after initiation of substrate breakdown.

Ubp6 in yeast also physically interacts with the purified proteasomes, its deubiquitinating activity is dramatically stimulated by binding to the 26S proteasome [44]. This association is mediated by the interaction between the N-terminal Ub-like domain of Ubp6, the Rpn1 subunit of the 19S regulatory particle. In *ubp6* deletion mutants, the level of free Ub is markedly reduced due to its abnormally rapid degradation by the 26S proteasome. Thus, Ubp6 seems to function in the recycling of Ub by preventing the translocation of Ub moieties that are conjugated to substrate proteins into the proteolytic core of the 20S proteasome [48,49]. USP14, a mammalian homolog of yeast Ubp6, also associates with the 26S proteasome [50].

The mammalian USP5 (also called isopeptidase-T), the yeast Ubp14 are functionally associated with the 26S proteasome [51]. USP5 acts only on unanchored poly-Ub chains that are generated *de novo* by Ub E3 ligases or released from poly-ubiquitinated proteins [52]. Since USP5, Ubp14 cannot act on poly-Ub chains that are conjugated to proteins, the Ub chains must first be released from protein substrates by other DUBs. In direct contrast to UCH37, USP5 successively releases Ub from the free 'proximal' end (i.e., from the end with an unattached Gly76 carboxyl group) of poly-Ub chain. In *ubp14* mutant cells, free poly-Ub chain accumulates, proteasome-dependent proteolysis is severely attenuated [53]. Thus, USP5, Ubp14 appear to prevent the accumulation of excessive free poly-Ub chains, which could compete with poly-Ub-conjugated protein substrates for binding to the 26S proteasome, thereby inhibit the proteasomal degradation [51].

Regulation of Protein Stability

Since the Lys48-linked poly-ubiquitination leads the target proteins to the 26S proteasomes for degradation, it is expected that the deubiquitinating activity of DUBs specific to the Lys48 linkage antagonizes the ubiquitination, stabilizes the protein. USP7 is a representative example of DUBs that regulate the protein stability. The p53 protein is a tumor suppressor that induces cell growth arrest, apoptosis in response to various stresses, such as

DNA damage [54,55]. Therefore, p53 level is kept low in normal cells, mainly by the action of Mdm2 Ub ligase that governs the ubiquitination, degradation of p53. USP7 was shown to deubiquitinate, stabilize p53 [56]. Moreover, overexpression of USP7 prevents the Mdm2-mediated degradation of p53, leading to p53-dependent cell growth inhibition, apoptosis. However, USP7 has also been shown to deubiquitinate, stabilize Mdm2 [57]. Furthermore, complete ablation of endogenous USP7 level by RNAi leads to rapid degradation of self-ubiquitinated Mdm2 by the 26S proteasome, resulting in the stabilization of p53. It has also been shown that targeted disruption of the *USP7* gene causes p53 stabilization [58]. In addition, USP7 interacts more strongly with Mdm2 than with p53 even in the presence of excess p53 [59]. Thus, it appears that Mdm2 is a preferential target of USP7. Taken together, the balance between self-ubiquitination, USP7-mediated deubiquitination of Mdm2 seems to play an important role in the control of p53 stability in cells.

Another example for DUBs that regulate protein stability is USP2a (also called UBP69 [60]). USP2a deubiquitinates, stabilizes fatty acid synthase, which is often overexpressed in prostate cancer, protects the cancer cells from apoptosis [61]. Overexpression of USP2a in prostate cancer cells results in deubiquitination of fatty acid synthase, thus in protection from its degradation by the 26S proteasome, whereas knockdown of USP2a by RNAi leads to an increase in the ubiquitination of fatty acid synthase. Moreover, apoptosis can be induced by RNAi-mediated knockdown of USP2a as can be seen by that of fatty acid synthase. Thus, USP2a appears to play a critical role in prostate cancer cell survival through deubiquitination, stabilization of fatty acid synthase.

Regulation of Signal Transduction: NF-κB Signaling

A representative signaling transduction pathway that is regulated by deubiquitination as well as by ubiquitination is the NF-κB pathway. The transcription factor NF-κB is normally sequestered in the cytoplasm as a complex with IκB. Binding of a ligand, such as tumor necrosis factor-α (TNF-α), to its receptor leads to recruitment of several signal adaptor proteins, including TNF-α receptor associating factor-2 (TRAF2), the receptor-interacting protein (RIP) [62,63]. IκB kinase (IKK) that is recruited by RIP to the receptor complex is activated, leading to phosphorylation of IκB. The phosphorylated IκB is then poly-ubiquitinated through Lys48 linkages, degraded by the 26S proteasome, resulting in the release of NF-κB as a free form that can translocate into the nucleus, activate transcription of its target genes.

CYLD is a DUB that is linked to a human genetic disorder, called familial cylindromatosis, the loss of its deubiquitinating activity is correlated with the turban tumor syndrome [64]. CYLD removes Ub from Lys63-linked poly-Ub chain that is conjugated to TRAF2. TRAF2 is a Lys63-specific Ub ligase that targets RIP, itself. Upon binding to the activated TNF receptor, TRAF2 is self-ubiquitinated, recruits IKK kinase that phosphorylates, activates IKK. Therefore, deubiquitination of TRAF2 by CYLD down-regulates the NF-κB pathway by preventing the activation of IKK [65].

Another DUB that is involved in the control of the NF-κB pathway is A20, a member of the OTU family [66]. The Lys63-linked poly-ubiquitination of RIP by TRAF2 leads to the recruitment of IKK to the activated TNF receptor. IKK has two catalytic subunits, IKKα, IKKβ, a regulatory subunit, NEMO (also called IKKγ). NEMO binds to the Lys63-linked poly-Ub chains of RIP. Therefore, the mutations in NEMO that block binding to the Ub chain abrogate the binding of NEMO to RIP, the recruitment of RIP to the TNF receptor, the activation of IKK, NF-κB [67]. Remarkably, A20 has two distinct catalytic domains, both of which cooperate to down-regulate NF-κB signaling: the N-terminal OTU-domain removes Lys63-linked poly-Ub chains from active RIP, which then permits the C-terminal zinc finger-domain to add Lys48-linked poly-Ub chains to RIP for the proteasomal degradation. Thus, A20 that has both deubiquitination, ubiquitination functions in a single polypeptide appears to play a critical role in the down-regulation of NF-κB signaling.

Regulation of DNA Repair

Proliferating cell nuclear antigen (PCNA) is a trimeric ring-shaped complex that binds DNA, acts as a sliding clamp, processivity factor for DNA polymerases [68]. The post-replication DNA repair activity of PCNA is differentially modulated by two distinct types of DNA damage-induced Ub modification. Mono-ubiquitination of PCNA initiates error-prone DNA repair, subsequent Lys63-linked poly-ubiquitination promotes error-free DNA repair [69]. Recently, PCNA was shown to be a target substrate of USP1 [70]. In the absence of DNA damage, USP1 removes Ub from mono-ubiquitinated PCNA. However, DNA damage induces the degradation of USP1, causes an increase in mono-ubiquitinated PCNA level, error-prone trans-lesion DNA synthesis. Interestingly, USP1 degradation is mediated by the auto-cleavage of its own internal di-glycine motif that is normally found at the C-terminus of free Ub. Moreover, depletion of USP1 by RNAi causes an increase in mutation frequency in cells. Thus, USP1 appears to play a critical role in the regulation of post-replication DNA damage repair.

Regulation of Chromatin Dynamics, Transcription

Covalent modification of histones plays a critical role in the regulation of chromatin dynamics, transcription. One example of such modification is the mono-ubiquitination, deubiquitination of histone H2B. In yeast, H2B is ubiquitinated by the Bre1 Ub ligase, deubiquitinated by two DUBs, Ubp10 (also called Dot4), Ubp8 that is a component of the histone acetylation SAGA (Spt-Ada-Gcn5-acetyltransferase) complex. Deubiquitination of H2B by Ubp10 leads to recruitment of the histone deacetylase Sir2 to telomere for gene silencing [71]. In contrast to Ubp10, deubiquitination of H2B by Ubp8 correlates with transcriptional activation of certain genes that are regulated by the SAGA complex [72]. Since sequential ubiquitination, deubiquitination of H2B occurs in normal gene induction, since disruption of either ubiquitination or Ubp8-mediated deubiquitination of H2B results in a decreased transcription, the dynamic H2B modification by Ub appears to be required for optimal transcription. However, Ubp8 is not required for telomeric silencing in contrast to Ubp10, Ubp10 is not involved in the regulation of the SAGA-dependent genes. Thus, although both Ubp8, Ubp10 act on H2B as a common substrate, they function in distinct physiological pathways [71].

USP7 in *Drosophila* contributes to epigenic silencing of homeotic genes by Polycomb [73]. USP7 forms a stable heteromeric complex with GMP synthetase, deubiquitinates H2B, but not H2A, under *in vitro* conditions. Interestingly, the binding of USP7 to GMP synthetase is required for its deubiquitinating activity. In addition, GMP synthetase markedly stimulates the deubiquitination of p53 by USP7. Moreover, homeotic transformation by Polycomb is strongly enhanced by the mutations that abrogate the catalytic activity of USP7 or the binding of GMP synthetase to USP7. Thus, USP7 appears to play an important role in chromatin silencing in collaboration with GMP synthetase.

In mammals, Mdm2, a major Ub ligase for p53, interacts with, ubiquitinates H2B *in vivo* as well as *in vitro* [74]. This H2B ubiquitination requires the RING finger domain of Mdm2, which is essential for its Ub ligase activity. Moreover, Mdm2 associates with chromatin, promotes p53-independent transcriptional repression in a RING domain-dependent fashion, suggesting that H2B ubiquitination by Mdm2 may facilitate repression, gene silencing. Since USP7 deubiquitinates, stabilizes Mdm2, mammalian USP7 may also take an important part in chromatin silencing by regulating the dynamics of H2B Ub modification.

Regulation of Endocytosis

Mono-ubiquitination plays an important role in endocytosis of receptors, sorting of proteins [75,76]. For example, mono-ubiquitination of ligand-activated receptor tyrosine kinases serves as an endocytosis signal that directs their internalization from the plasma membrane, as a sorting signal for their trafficking from endosomes to lysosomes. Endosomal sorting of ubiquitinated receptor tyrosine kinases is initiated by a complex of two Ub-binding proteins, hepatocyte growth factor-regulated substrate (Hrs), signal-transducing adaptor molecule (STAM). USP8 (also called UBPY) binds to the Hrs-STAM endosomal sorting complex, deubiquitinates the activated EGF-receptor, resulting in the inhibition of receptor sorting [77]. Whereas overexpression of USP8 reduces the level of ubiquitinated EGF-receptor, delays its degradation in EGF-stimulated cells, knockdown of USP8 by RNAi results in elevated ubiquitination, accelerated degradation of the activated receptor. Thus, USP8 appears to plays a role in EGF-receptor down-regulation.

Another DUB that also binds to the Hrs-STAM endosomal sorting complex is AMSH that belongs to the JAMM family [32]. Expression of catalytically inactive AMSH causes an accumulation of Ub on endosomes, knockdown of AMSH by RNAi results in acceleration of the activated EGF-receptor degradation. Thus, AMSH appears to act negatively on the Ub-dependent sorting of EGF-receptor to lysosomes.

In yeast, deubiquitination itself is not mandatory for receptor sorting to the vacuole, as Ub can confer sorting after being fused in frame to a cargo protein [78], but is required for maintenance of the free Ub pool, upon which receptor trafficking depends. Inactivation of Doa4 leads to an arrest of receptor sorting, which can be restored by overexpression of free Ub [79]. Thus, Doa4 appears to play a role in recycling of Ub at the endosomes by rescuing it from degradation.

In *Drosophila*, Fat facets (Faf) is a DUB that is required during eye development to limit the number of photoreceptors in each facet to eight [80]. Liquid facets (Lqf) is a homolog of the vertebrate protein epsin that acts as an adaptor molecule associated with the clathrin endocytotic complex [81]. Faf deubiquitinates Lqf, prevents its proteasomal degradation. Thus, the stabilized Lqf then facilitates the endocytosis of the receptor Delta that is implicated in cell patterning [82].

Neuronal Diseases, Deubiquitinating Enzymes

Parkinson's Disease

Aberrations in the UPS have been implicated, either as a primary cause or secondary consequences, in the pathogenesis of both inherited, acquired neurodegenerative diseases [83]. Several cases of genetic abnormalities within DUBs have also been reported to contribute to neuronal disease pathogenesis. In two siblings with a strong family history of PD, an autosomal dominant point mutation (I93M) in UCH-L1 was identified [84]. The I93M mutation reduces the *in vitro* deubiquitinating activity of UCH-L1 by about 50%. Moreover, UCH-L1 co-localized with α-synuclein in synaptic vesicles, its overexpression leads to an accumulation of α-synuclein. Therefore, it was proposed that UCH-L1 might modulate the turnover of α-synuclein, a major fibrillar component of the Lewy bodies that are the hallmark of sporadic PD. However, it has later been reported that the I93M variant is a rare cause of PD or a harmless substitution [85]. Moreover, the gracile axonal dystrophy (GAD) mice that have a deletion in the *UCH-L1* gene does not show the progressive death of dopaminergic neurons in the *substantia nigra* [86]. Since the mutation is not 100% penetrant in the family, since GAD mice do not develop a Parkinsonian syndrome, it was unclear whether the loss of deubiquitinating activity underlies the pathology observed in patients with the I93M mutation. One possible explanation can be that UCH-L1, UCH-L3 may have redundant, overlapping functions in the maintenance of neurons.

Another polymorphic variant of UCH-L1 (S18Y) that reduces the risk of PD was later found [87]. Surprisingly, the dimeric form of UCH-L1 displays an Ub ligase activity, in contrast to the monomeric enzyme that catalyzes deubiquitination. The UCH-L1 dimer adds ubiquitins to mono-, di-ubiquitinated α-synuclein through the Lys63 linkage. Since α-synuclein conjugated with Lys63-linked poly-Ub chains is not degraded by the 26S proteasome, the ligase function of dimeric UCH-L1 would lead to an elevation of cytoplasmic concentration of α-synuclein that could form pathogenic aggregates. Remarkably, the S18Y mutant shows a reduced tendency of dimer formation, thereby having reduced ligase activity, but has comparable deubiquitinating activity as wild-type UCH-L1. These findings are consistent with the observations that protective effect of the S18Y mutation is dependent on the allele dosage, that Ser18 is not involved in the normal hydrolytic activity of UCH-L1. Thus, it is also possible that the ligase activity of the I93M mutant plays also a pathogenic role, as a decreased activity of the mutant enzyme may result in a net increase in the ligase activity.

Poly-Glutamine Related Diseases

Machado-Joseph disease (MJD), also known as spinocerebellar ataxia type-3 (SCA3), is the most common dominantly inherited ataxia, is among the frequent neurodegenerative diseases caused by poly-glutamine (polyQ) encoding CAG repeat expansion [88,89]. By using the *Drosophila* model system, ataxin-3 that is a MJD family member of DUBs has recently been shown to suppress polyQ-induced retinal degeneration. Ataxin-3 itself is a polyQ protein that also has a Josephin domain for deubiquitinating activity, two UIM domains for Ub binding. Co-expression of normal ataxin-3 with its pathogenic form carrying expanded polyQ leads to a reduction in the formation of nuclear inclusions, thus in the polyQ toxicity. Nuclear inclusions are one of the hallmarks of the polyQ diseases as well as of other neurodegenerative diseases. However, introduction of mutations to the UIM domains results in a decrease in the suppressive effect of normal ataxin-3. Furthermore, inactivation of the deubiquitinating activity by substituting the catalytic cysteine residue with alanine in the Josephin domain causes a complete loss of the protective effect of normal ataxin-3 over its pathogenic form. Thus, the Ub-associated activities, especially the deubiquitination function, of ataxin-3 contribute to the suppression of polyQ-induced neurodegeneration.

Using the similar fly model system, overexpression of the truncated huntingtin protein with expanded polyQ was shown to induce a severe retinal degeneration [90]. Interestingly, ataxin-3 is found in nuclear inclusions of other polyQ diseases, such as HD, spinocerebellar ataxia type-1 (SCA1). Co-expression of normal ataxin-3 with the pathogenic huntingtin fully suppresses the polyQ-induced eye disruption. Likewise, co-expression of ataxin-3 with the SCA1 disease protein ataxin-1, which is not related in sequence with ataxin-3 outside of the polyQ domain [91], reduces the polyQ toxicity. Thus, ataxin-3 is capable of suppressing neurodegeneration caused by unrelated pathogenic polyQ proteins.

Other Neurological Diseases

The ataxia mutation (ax^J) in mice is a spontaneously arising mutation that results in reduced expression of USP14 [92,93]. Mice homozygous for ax^J are retarded for growth, suffer from behavioral disorders, including a resting tremor, hindlimb paralysis. Unlike other neurodegenerative disorders, such as PD, SCA1 in human, neither Ub-positive protein aggregates nor neuronal cell death is detectable in the central nervous system of ax^J mice. Instead, ax^J mice have defects in synaptic transmission in both the central, peripheral nervous systems, due to abnormalities associated with neurotransmitter release at the neuromuscular junctions. In addition, the levels of free Ub in most of the ax^J tissues are reduced about 35% in comparison with that in normal tissues, suggesting the role of USP14 in maintaining the cellular level of free Ub in mammalian cells [94]. Ubp6, is a yeast homolog of mammalian USP14, associates with the 26S proteasome, functions in the recycling of Ub by preventing the proteasomal degradation [48,49]. Thus, the alteration in the level of free Ub due to reduced expression of USP14 in axJ mice may contribute to the neurological disease.

CONCLUSIONS

Although the importance of DUBs as cellular regulators has recently been uncovered in numerous cases, the studies on DUBs are still in the beginning stage, considering their diversity. Genomic analyses have identified 95 potential genes encoding DUBs from human, 62 from *Arabidopsis thaliana*, 34 from *Drosophila melanogaster*, 20 from *Saccharomyces cerevisiae.* Moreover, many of these genes generate multiple transcripts, which would further increase the diversity of DUBs. However, only a handful of them have been assigned for their physiological functions. Thus, a major challenge in the research of DUBs is to elucidate specific physiological roles of the DUB family members through identification of specific substrates for each DUB.

All ubiquitinated proteins are potential substrates of DUBs. DUBs can act as negative regulators for the proteasomal degradation of ubiquitinated proteins, leading to stabilization of the substrates. DUBs can also regulate nonproteolytic Ub-mediated processes by acting on mono-ubiquitinated proteins or poly-ubiquitinated proteins that are conjugated through the Lys63 linkages. However, these activities of DUBs should be timely, /or spatially regulated to prevent unnecessary action of the enzymes on ubiquitinated target substrates. Thus, another major challenge in this field is the determination of the mechanisms for the control of DUB activities. Examples include covalent modifications (e.g., inhibition of CYLD by phosphorylation [95]), requirement of cofactors (e.g., GMP synthetase, Bre5 for optimal activities of USP7, Ubp3, respectively [73,96]), transcriptional control (e.g., USP1 [97]) for activation, inhibition of DUBs under specific circumstances. In summary, what DUBs do in cells, how they are regulated are to be further investigated in the coming years.

ACKNOWLEDGEMENTS

This work was supported by the grants from Korea Science, Engineering Foundation (M10533010001-05N3301-00100, R11-2000-083-00000-0).

REEFERENCES

[1] Schlesinger DH, Goldstein G, Niall HD: The complete amino acid sequence of ubiquitin, an adenylate cyclase stimulating polypeptide probably universal in living cells. *Biochemistry* 1975, 14:2214-2218.

[2] Ciechanover A, Heller H, Elias S, Haas AL, Hershko A: ATP-dependent conjugation of reticulocyte proteins with the polypeptide required for protein degradation. *Proc Natl Acad Sci USA* 1980, 77:1365-1368.

[3] Hershko A, Heller H, Elias S, Ciechanover A: Components of ubiquitin-protein ligase system. Resolution, affinity purification, role in protein breakdown. *J Biol Chem* 1983, 258:8206-8214.

[4] Breitschopf K, Bengal E, Ziv T, Admon A, Ciechanover A: A novel site for ubiquitination: the N-terminal residue, not internal lysines of MyoD, is essential for conjugation, degradation of the protein. *EMBO J* 1998, 17:5964-5973.

[5] Hershko A, Ciechanover A: The ubiquitin system. *Annu Rev Biochem* 1998, 67:425-479.

[6] Pickart CM, Eddins MJ: Ubiquitin: structures, functions, mechanisms. *Biochim Biophys Acta* 2004, 1695:55-72.

[7] Peng J, Schwartz D, Elias JE, Thoreen CC, Cheng D, Marsischky G, Roelofs J, Finley D, Gygi SP: A proteomics approach to understanding protein ubiquitination. *Nat Biotechnol* 2003, 21:921-926.

[8] Ross CA, Pickart CM: The ubiquitin-proteasome pathway in Parkinson's disease, other neurodegenerative diseases. *Trends Cell Biol* 2004, 14:703-711.

[9] Nijman SM, Luna-Vargas MP, Velds A, Brummelkamp TR, Dirac AM, Sixma TK, Bernards R: A genomic, functional inventory of deubiquitinating enzymes. *Cell* 2005, 123:773-786.

[10] Finley D, Chau V: Ubiquitination. *Annu Rev Cell Biol* 1991, 7:25-69.

[11] Kim JH, Park KC, Chung SS, Bang O, Chung CH: Deubiquitinating enzymes as cellular regulators. *J Biochem (Tokyo)* 2003, 134:9-18.

[12] Andersen MW, Goldknopf IL, Busch H: Protein A24 lyase is an isopeptidase. *FEBS Lett* 1981, 132:210-214.

[13] Dang LC, Melandri FD, Stein RL: Kinetic, mechanistic studies on the hydrolysis of ubiquitin C-terminal 7-amido-4-methylcoumarin by deubiquitinating enzymes. *Biochemistry* 1998, 37:1868-1879.

[14] Kang SH, Park JJ, Chung SS, Bang OS, Chung CH: Strategies for assaying deubiquitinating enzymes. *Methods Enzymol* 2005, 398:500-508.

[15] Verma R, Aravind L, Oania R, McDonald WH, Yates JR, 3rd, Koonin EV, Deshaies RJ: Role of Rpn11 metalloprotease in deubiquitination, degradation by the 26S proteasome. *Science* 2002, 298:611-615.

[16] Makarova KS, Aravind L, Koonin EV: A novel superfamily of predicted cysteine proteases from eukaryotes, viruses, Chlamydia pneumoniae. *Trends Biochem Sci* 2000, 25:50-52.

[17] Scheel H, Tomiuk S, Hofmann K: Elucidation of ataxin-3, ataxin-7 function by integrative bioinformatics. *Hum Mol Genet* 2003, 12:2845-2852.

[18] Miller HI, Henzel WJ, Ridgeway JB, Kuang W, Chrisholm V, Liu, C.: Cloning, expression of a yeast ubiquitin-protein cleaving activity in Escherichia coli. *Biotechnology* 1989, 7:698-704.

[19] Wilkinson KD, Lee KM, Deshpande S, Duerksen-Hughes P, Boss JM, Pohl J: The neuron-specific protein PGP 9.5 is a ubiquitin carboxyl-terminal hydrolase. *Science* 1989, 246:670-673.

[20] Jensen DE, Proctor M, Marquis ST, Gardner HP, Ha SI, Chodosh LA, Ishov AM, Tommerup N, Vissing H, Sekido Y, Minna J, Borodovsky A, Schultz DC, Wilkinson KD, Maul GG, Barlev N, Berger SL, Prendergast GC, Rauscher FJ, 3rd: BAP1: a novel ubiquitin hydrolase which binds to the BRCA1 RING finger, enhances BRCA1-mediated cell growth suppression. *Oncogene* 1998, 16:1097-1112.

[21] Johnston SC, Riddle SM, Cohen RE, Hill CP: Structural basis for the specificity of ubiquitin C-terminal hydrolases. *EMBO J* 1999, 18:3877-3887.

[22] Misaghi S, Galardy PJ, Meester WJ, Ovaa H, Ploegh HL, Gaudet R: Structure of the ubiquitin hydrolase UCH-L3 complexed with a suicide substrate. *J Biol Chem* 2005, 280:1512-1520.

[23] Lam YA, DeMartino GN, Pickart CM, Cohen RE: Specificity of the ubiquitin isopeptidase in the PA700 regulatory complex of 26 S proteasomes. *J Biol Chem* 1997, 272:28438-28446.

[24] Baker RT, Wang XW, Woollatt E, White JA, Sutherland GR: Identification, functional characterization, chromosomal localization of USP15, a novel human ubiquitin-specific protease related to the UNP oncoprotein, a systematic nomenclature for human ubiquitin-specific proteases. *Genomics* 1999, 59:264-274.

[25] Balakirev MY, Tcherniuk SO, Jaquinod M, Chroboczek J: Otubains: a new family of cysteine proteases in the ubiquitin pathway. *EMBO Rep* 2003, 4:517-522.

[26] Evans PC, Smith TS, Lai MJ, Williams MG, Burke DF, Heyninck K, Kreike MM, Beyaert R, Blundell TL, Kilshaw PJ: A novel type of deubiquitinating enzyme. *J Biol Chem* 2003, 278:23180-23186.

[27] Wertz IE, O'Rourke KM, Zhou H, Eby M, Aravind L, Seshagiri S, Wu P, Wiesmann C, Baker R, Boone DL, Ma A, Koonin EV, Dixit VM: De-ubiquitination, ubiquitin ligase domains of A20 downregulate NF-kappaB signalling. *Nature* 2004, 430:694-699.

[28] Burnett B, Li F, Pittman RN: The polyglutamine neurodegenerative protein ataxin-3 binds polyubiquitylated proteins, has ubiquitin protease activity. *Hum Mol Genet* 2003, 12:3195-3205.

[29] Eytan E, Armon T, Heller H, Beck S, Hershko A: Ubiquitin C-terminal hydrolase activity associated with the 26 S protease complex. *J Biol Chem* 1993, 268:4668-4674.

[30] Yao T, Cohen RE: A cryptic protease couples deubiquitination, degradation by the proteasome. *Nature* 2002, 419:403-407.

[31] Cope GA, Suh GS, Aravind L, Schwarz SE, Zipursky SL, Koonin EV, Deshaies RJ: Role of predicted metalloprotease motif of Jab1/Csn5 in cleavage of Nedd8 from Cul1. *Science* 2002, 298:608-611.

[32] McCullough J, Clague MJ, Urbe S: AMSH is an endosome-associated ubiquitin isopeptidase. *J Cell Biol* 2004, 166:487-492.

[33] Johnston SC, Larsen CN, Cook WJ, Wilkinson KD, Hill CP: Crystal structure of a deubiquitinating enzyme (human UCH-L3) at 1.8 Å resolution. *EMBO J* 1997, 16:3787-3796.

[34] Das C, Hoang QQ, Kreinbring CA, Luchansky SJ, Meray RK, Ray SS, Lansbury PT, Ringe D, Petsko GA: Structural basis for conformational plasticity of the Parkinson's disease-associated ubiquitin hydrolase UCH-L1. *Proc Natl Acad Sci USA* 2006, 103:4675-4680.

[35] Hu M, Li P, Li M, Li W, Yao T, Wu JW, Gu W, Cohen RE, Shi Y: Crystal structure of a UBP-family deubiquitinating enzyme in isolation, in complex with ubiquitin aldehyde. *Cell* 2002, 111:1041-1054.

[36] Hu M, Li P, Song L, Jeffrey PD, Chenova TA, Wilkinson KD, Cohen RE, Shi Y: Structure, mechanisms of the proteasome-associated deubiquitinating enzyme USP14. *EMBO J* 2005, 24:3747-3756.

[37] Nanao MH, Tcherniuk SO, Chroboczek J, Dideberg O, Dessen A, Balakirev MY: Crystal structure of human otubain 2. *EMBO Rep* 2004, 5:783-788.

[38] Nicastro G, Menon RP, Masino L, Knowles PP, McDonald NQ, Pastore A: The solution structure of the Josephin domain of ataxin-3: structural determinants for molecular recognition. *Proc Natl Acad Sci USA* 2005, 102:10493-10498.

[39] Mao Y, Senic-Matuglia F, Di Fiore PP, Polo S, Hodsdon ME, De Camilli P: Deubiquitinating function of ataxin-3: insights from the solution structure of the Josephin domain. *Proc Natl Acad Sci USA* 2005, 102:12700-12705.

[40] [Anonymous]: Structure, Implications of JAMM, a Novel Metalloprotease. *PLoS Biol* 2004, 2:E17.

[41] Otto H-H, Schirmeister T: Cysteine Proteases, Their Inhibitors. *Chem Rev* 1997, 97:133-171.

[42] Harrison MJ, Burton NA, Hillier IH: Catalytic mechanism of the enzyme papain: predictions with a hybrid quantum mechanical/molecular mechanical potential. *J Am Chem Soc* 1997, 119:12285-12291.

[43] Matthews BW: Structural basis of the action of thermolysin, related zinc peptidases. *Acc Chem Res* 1988, 21:333-340.

[44] Leggett DS, Hanna J, Borodovsky A, Crosas B, Schmidt M, Baker RT, Walz T, Ploegh H, Finley D: Multiple associated proteins regulate proteasome structure, function. *Mol Cell* 2002, 10:495-507.

[45] Lam YA, Xu W, DeMartino GN, Cohen RE: Editing of ubiquitin conjugates by an isopeptidase in the 26S proteasome. *Nature* 1997, 385:737-740.

[46] Stone M, Hartmann-Petersen R, Seeger M, Bech-Otschir D, Wallace M, Gordon C: Uch2/Uch37 is the major deubiquitinating enzyme associated with the 26S proteasome in fission yeast. *J Mol Biol* 2004, 344:697-706.

[47] Papa FR, Amerik AY, Hochstrasser M: Interaction of the Doa4 deubiquitinating enzyme with the yeast 26S proteasome. *Mol Biol Cell* 1999, 10:741-756.

[48] Chernova TA, Allen KD, Wesoloski LM, Shanks JR, Chernoff YO, Wilkinson KD: Pleiotropic effects of Ubp6 loss on drug sensitivities, yeast prion are due to depletion of the free ubiquitin pool. *J Biol Chem* 2003, 278:52102-52115.

[49] Hanna J, Leggett DS, Finley D: Ubiquitin depletion as a key mediator of toxicity by translational inhibitors. *Mol Cell Biol* 2003, 23:9251-9261.

[50] Borodovsky A, Kessler BM, Casagrande R, Overkleeft HS, Wilkinson KD, Ploegh HL: A novel active site-directed probe specific for deubiquitylating enzymes reveals proteasome association of USP14. *EMBO J* 2001, 20:5187-5196.

[51] Amerik AY, Hochstrasser M: Mechanism, function of deubiquitinating enzymes. *Biochim Biophys Acta* 2004, 1695:189-207.

[52] Wilkinson KD, Tashayev VL, O'Connor LB, Larsen CN, Kasperek E, Pickart CM: Metabolism of the polyubiquitin degradation signal: structure, mechanism, role of isopeptidase T. *Biochemistry* 1995, 34:14535-14546.

[53] Amerik A, Swaminathan S, Krantz BA, Wilkinson KD, Hochstrasser M: *In vivo* disassembly of free polyubiquitin chains by yeast Ubp14 modulates rates of protein degradation by the proteasome. *EMBO J* 1997, 16:4826-4838.

[54] Levine AJ: p53, the cellular gatekeeper for growth, division. *Cell* 1997, 88:323.

[55] Vousden KH, Lu X: Live or let die: The cell's response to p53. *Nat Rev Cancer* 2002, 2:594.

[56] Li M, Chen D, Shiloh A, Luo J, Nikolaev AY, Qin J, Gu W: Deubiquitination of p53 by HAUSP is an important pathway for p53 stabilization. *Nature* 2002, 416:648-653.

[57] Li M, Brooks CL, Kon N, Gu W: A dynamic role of HAUSP in the p53-Mdm2 pathway. *Mol Cell* 2004, 13:879-886.

[58] Cummins JM, Rago C, Kohli M, Kinzler KW, Lengauer C, Vogelstein B: Tumour suppression: disruption of HAUSP gene stabilizes p53. *Nature* 2004, 428:1 p following 486.

[59] Hu M, Gu L, Li M, Jeffrey PD, Gu W, Shi Y: Structural basis of competitive recognition of p53, MDM2 by HAUSP/USP7: implications for the regulation of the p53-MDM2 pathway. *PLoS Biol* 2006, 4:e27.

[60] Park KC, Kim JH, Choi EJ, Min SW, Rhee S, Baek SH, Chung SS, Bang O, Park D, Chiba T, Tanaka K, Chung CH: Antagonistic regulation of myogenesis by two deubiquitinating enzymes, UBP45, UBP69. *Proc Natl Acad Sci USA* 2002, 99:9733-9738.

[61] Graner E, Tang D, Rossi S, Baron A, Migita T, Weinstein LJ, Lechpammer M, Huesken D, Zimmermann J, Signoretti S, Loda M: The isopeptidase USP2a regulates the stability of fatty acid synthase in prostate cancer. *Cancer Cell* 2004, 5:253-261.

[62] Zhang SQ, Kovalenko A, Cantarella G, Wallach D: Recruitment of the IKK signalosome to the p55 TNF receptor: RIP, A20 bind to NEMO (IKKgamma) upon receptor stimulation. *Immunity* 2000, 12:301-311.

[63] Dixit V, Mak TW: NF-kappaB signaling. Many roads lead to madrid. *Cell* 2002, 111:615-619.

[64] Bignell GR, Warren W, Seal S, Takahashi M, Rapley E, Barfoot R, Green H, Brown C, Biggs PJ, Lakhani SR, Jones C, Hansen J, Blair E, Hofmann B, Siebert R, Turner G, Evans DG, Schrander-Stumpel C, Beemer FA, van Den Ouweland A, Halley D, Delpech B, Cleveland MG, Leigh I, Leisti J, Rasmussen S: Identification of the familial cylindromatosis tumour-suppressor gene. *Nat Genet* 2000, 25:160-165.

[65] Trompouki E, Hatzivassiliou E, Tsichritzis T, Farmer H, Ashworth A, Mosialos G: CYLD is a deubiquitinating enzyme that negatively regulates NF-kappaB activation by TNFR family members. *Nature* 2003, 424:793-796.

[66] Song HY, Rothe M, Goeddel DV: The tumor necrosis factor-inducible zinc finger protein A20 interacts with TRAF1/TRAF2, inhibits NF-kappaB activation. *Proc Natl Acad Sci USA* 1996, 93:6721-6725.

[67] Wu CJ, Conze DB, Li T, Srinivasula SM, Ashwell JD: NEMO is a sensor of Lys 63-linked polyubiquitination, functions in NF-kappaB activation. *Nat Cell Biol* 2006, 8:398-406.

[68] Paunesku T, Mittal S, Protic M, Oryhon J, Korolev SV, Joachimiak A, Woloschak GE: Proliferating cell nuclear antigen (PCNA): ringmaster of the genome. *Int J Radiat Biol* 2001, 77:1007-1021.
[69] Hoege C, Pfander B, Moldovan GL, Pyrowolakis G, Jentsch S: RAD6-dependent DNA repair is linked to modification of PCNA by ubiquitin, SUMO. *Nature* 2002, 419:135-141.
[70] Huang TT, Nijman SM, Mirchandani KD, Galardy PJ, Cohn MA, Haas W, Gygi SP, Ploegh HL, Bernards R, D'Andrea AD: Regulation of monoubiquitinated PCNA by DUB autocleavage. *Nat Cell Biol* 2006, 8:339-347.
[71] Emre NC, Ingvarsdottir K, Wyce A, Wood A, Krogan NJ, Henry KW, Li K, Marmorstein R, Greenblatt JF, Shilatifard A, Berger SL: Maintenance of low histone ubiquitylation by Ubp10 correlates with telomere-proximal Sir2 association, gene silencing. *Mol Cell* 2005, 17:585-594.
[72] Henry KW, Wyce A, Lo WS, Duggan LJ, Emre NC, Kao CF, Pillus L, Shilatifard A, Osley MA, Berger SL: Transcriptional activation via sequential histone H2B ubiquitylation, deubiquitylation, mediated by SAGA-associated Ubp8. *Genes Dev* 2003, 17:2648-2663.
[73] van der Knaap JA, Kumar BR, Moshkin YM, Langenberg K, Krijgsveld J, Heck AJ, Karch F, Verrijzer CP: GMP synthetase stimulates histone H2B deubiquitylation by the epigenetic silencer USP7. *Mol Cell* 2005, 17:695-707.
[74] Minsky N, Oren M: The RING domain of Mdm2 mediates histone ubiquitylation, transcriptional repression. *Mol Cell* 2004, 16:631-639.
[75] Hicke L: Protein regulation by monoubiquitin. *Nat Rev Mol Cell Biol* 2001, 2:195-201.
[76] Dupre S, Volland C, Haguenauer-Tsapis R: Membrane transport: ubiquitylation in endosomal sorting. *Curr Biol* 2001, 11:R932-934.
[77] Mizuno E, Iura T, Mukai A, Yoshimori T, Kitamura N, Komada M: Regulation of epidermal growth factor receptor down-regulation by UBPY-mediated deubiquitination at endosomes. *Mol Biol Cell* 2005, 16:5163-5174.
[78] Reggiori F, Pelham HR: Sorting of proteins into multivesicular bodies: ubiquitin-dependent, -independent targeting. *EMBO J* 2001, 20:5176-5186.
[79] Swaminathan S, Amerik AY, Hochstrasser M: The Doa4 deubiquitinating enzyme is required for ubiquitin homeostasis in yeast. *Mol Biol Cell* 1999, 10:2583-2594.
[80] Fischer-Vize JA, Rubin GM, Lehmann R: The fat facets gene is required for Drosophila eye, embryo development. *Development* 1992, 116:985-1000.
[81] Chen H, Fre S, Slepnev VI, Capua MR, Takei K, Butler MH, Di Fiore PP, De Camilli P: Epsin is an EH-domain-binding protein implicated in clathrin-mediated endocytosis. *Nature* 1998, 394:793-797.
[82] Overstreet E, Fitch E, Fischer JA: Fat facets, Liquid facets promote Delta endocytosis, Delta signaling in the signaling cells. *Development* 2004, 131:5355-5366.
[83] Ciechanover A, Brundin P: The ubiquitin proteasome system in neurodegenerative diseases: sometimes the chicken, sometimes the egg. *Neuron* 2003, 40:427-446.
[84] Leroy E, Boyer R, Auburger G, Leube B, Ulm G, Mezey E, Harta G, Brownstein MJ, Jonnalagada S, Chernova T, Dehejia A, Lavedan C, Gasser T, Steinbach PJ, Wilkinson

KD, Polymeropoulos MH: The ubiquitin pathway in Parkinson's disease. *Nature* 1998, 395:451-452.

[85] Lincoln S, Vaughan J, Wood N, Baker M, Adamson J, Gwinn-Hardy K, Lynch T, Hardy J, Farrer M: Low frequency of pathogenic mutations in the ubiquitin carboxy-terminal hydrolase gene in familial Parkinson's disease. *Neuroreport* 1999, 10:427-429.

[86] Kurihara LJ, Kikuchi T, Wada K, Tilghman SM: Loss of Uch-L1, Uch-L3 leads to neurodegeneration, posterior paralysis, dysphagia. *Hum Mol Genet* 2001, 10:1963-1970.

[87] Maraganore DM, Lesnick TG, Elbaz A, Chartier-Harlin MC, Gasser T, Kruger R, Hattori N, Mellick GD, Quattrone A, Satoh J, Toda T, Wang J, Ioannidis JP, de, rade M, Rocca WA: UCHL1 is a Parkinson's disease susceptibility gene. *Ann Neurol* 2004, 55:512-521.

[88] Gusella JF, MacDonald ME: Molecular genetics: unmasking polyglutamine triggers in neurodegenerative disease. *Nat Rev Neurosci* 2000, 1:109-115.

[89] Zoghbi HY, Orr HT: Glutamine repeats, neurodegeneration. *Annu Rev Neurosci* 2000, 23:217-247.

[90] Jackson GR, Salecker I, Dong X, Yao X, Arnheim N, Faber PW, MacDonald ME, Zipursky SL: Polyglutamine-expanded human huntingtin transgenes induce degeneration of Drosophila photoreceptor neurons. *Neuron* 1998, 21:633-642.

[91] Fernandez-Funez P, Nino-Rosales ML, de Gouyon B, She WC, Luchak JM, Martinez P, Turiegano E, Benito J, Capovilla M, Skinner PJ, McCall A, Canal I, Orr HT, Zoghbi HY, Botas J: Identification of genes that modify ataxin-1-induced neurodegeneration. *Nature* 2000, 408:101-106.

[92] D'Amato CJ, Hicks SP: Neuropathologic alterations in the ataxia (paralytic) mouse. *Arch Pathol* 1965, 80:604-612.

[93] Wilson SM, Bhattacharyya B, Rachel RA, Coppola V, Tessarollo L, Householder DB, Fletcher CF, Miller RJ, Copeland NG, Jenkins NA: Synaptic defects in ataxia mice result from a mutation in Usp14, encoding a ubiquitin-specific protease. *Nat Genet* 2002, 32:420-425.

[94] Anderson C, Crimmins S, Wilson JA, Korbel GA, Ploegh HL, Wilson SM: Loss of Usp14 results in reduced levels of ubiquitin in ataxia mice. *J Neurochem* 2005, 95:724-731.

[95] Reiley W, Zhang M, Wu X, Granger E, Sun SC: Regulation of the deubiquitinating enzyme CYLD by IkappaB kinase gamma-dependent phosphorylation. *Mol Cell Biol* 2005, 25:3886-3895.

[96] Cohen M, Stutz F, Belgareh N, Haguenauer-Tsapis R, Dargemont C: Ubp3 requires a cofactor, Bre5, to specifically de-ubiquitinate the COPII protein, Sec23. *Nat Cell Biol* 2003, 5:661-667.

[97] Nijman SM, Huang TT, Dirac AM, Brummelkamp TR, Kerkhoven RM, D'Andrea AD, Bernards R: The deubiquitinating enzyme USP1 regulates the Fanconi anemia pathway. *Mol Cell* 2005, 17:331-39.

In: The Ubiquitin Proteasome System... ISBN 978-1-60021-749-4
Eds: Mario Di Napoli and Cezary Wojcik, pp. 93-115

Chapter 5

UBIQUITIN DOMAIN PROTEINS FUNCTIONAL VARIATIONS OF A COMMON STRUCTURE

***Andrea Schulze*[1], *Michael Seeger*[1] *and Rasmus Hartmann-Petersen*[2,*]**
[1]Charité - Universitaetsmedizin Berlin, Campus Charité Mitte, Institut für Biochemie, Monbijoustr. 2, 10117 Berlin, Germany;
[2]Institute of Molecular Biology and Physiology, University of Copenhagen, Universitetsparken 13, DK-2100 Copenhagen, Denmark.

ABSTRACT

Ubiquitin-like (UBL) domain proteins (UDPs) constitute a family of proteins with a modular architecture, which is characterized by an integral UBL-domain. Although members of the UDP family display a variety of different functions, many of them are on some level connected with the ubiquitin-proteasome system, a central pathway, which accommodates intracellular protein degradation in eukaryotic cells. While some UDPs are involved in substrate recruitment for the 26S proteasome, also a ubiquitin-specific hydrolase, an ER-membrane resident protein, a co-chaperone, and a ubiquitin ligase belong to this family. Several of these proteins have been implicated in the development of neurodegenerative diseases. Of the initially studied UDPs, most bound the proteasome in a UBL-dependent manner. Therefore it appeared that proteasome binding was a general feature of this protein family. However, evidence is accumulating that a number of UDPs also bind to other components of the ubiquitin pathway, while some appear not to bind the proteasome at all. Hence UDPs appear functionally more diverse than one would expect from their structural appearance. Here we provide insight into the UDP family and attempt to summarize what is known about their physiological role, especially with respect to neurodegenerative diseases. We come to the conclusion that, despite their

* Correspondence concerning this article should be addressed to Dr. Rasmus Hartmann-Petersen, PhD; Institute of Molecular Biology and Physiology, University of Copenhagen, Universitetsparken 13, DK-2100 Copenhagen, Denmark. Phone : +45 3532 1502; Fax: +45 3532 1567 E-mail: RHPetersen@aki.ku.dk.

striking structural similarity, UDPs display rather diverse binding features, and appear to be part of a sophisticated protein network within the ubiquitin system.

Keywords: chaperone, neurodegenerative disease, protease, proteasome, ubiquitin, UBL.

INTRODUCTION

The modification of proteins with the conserved polypeptide, ubiquitin, has a crucial role in a wide range of cellular processes within the eukaryotic cell. It is a prerequisite for most proteasome-dependent protein degradation, and therefore essential for the disposal of damaged and misfolded polypeptides, but also for the control of cellular processes by specific and timed breakdown of regulatory proteins. Apart from its role in protein degradation, ubiquitylation is also involved in the regulation of many non-proteolytic cellular processes such as DNA repair, transcription and translation [1].

The attachment of ubiquitin to a substrate protein requires a series of catalyzed steps. First ubiquitin is activated and bound to a ubiquitin-activating enzyme (E1) in an ATP-consuming process. Next the ubiquitin molecule is transferred to a ubiquitin-conjugating enzyme (E2), which associates with a ubiquitin-protein ligase (E3) and the substrate (see Chapter 3). This results in the formation of an isopeptide bond between the C-terminus of ubiquitin and the ε-amino-group of a lysine side chain within the substrate [2,3]. Several rounds of this conjugation process lead to substrate proteins, modified with a chain of ubiquitin molecules. In some cases also the activity of a ubiquitin-chain elongation factor (E4) is required for efficient generation of ubiquitin chains [4]. The process of ubiquitylation is highly dynamic and reversible and several deubiquitylating enzymes (DUBs) play important roles in trimming or cleaving the ubiquitin chains, prior to degradation (see Chapter 4).

Different lysine residues in ubiquitin are utilized in chain formation and whereas the modification with polyubiquitin linked via Lys48 generally targets proteins for degradation by the 26S proteasome, Lys63-linked chains have been described to play a role in NFκB signaling and the formation of Lewy bodies in Parkinson's disease [6-8]. Monoubiquitylation, on the other hand, plays a role in processes such as receptor internalization and transcriptional regulation [9].

Once a substrate protein has been modified with a Lys48-linked chain, it can be degraded by the 26S proteasome. The proteolytic component of this abundant intracellular protease is the 20S core complex, a cylindrical structure enclosing a central chamber harboring the catalytic sites [10]. Access to the lumen is provided by the 19S regulatory complex attached to one or both ends of the 20S core [11]. The 19S regulatory complex can dissociate into two subcomplexes called base and lid [12]. The base, which binds the 20S core, contains six ATPase subunits as well as two non-ATPase subunits, S1/Rpn2 and S2/Rpn1, and mediates binding and unfolding of the substrate [13,14]. The lid covers the base and is involved in deubiquitylation of substrate proteins prior to their degradation. The ubiquitin-binding S5a/Rpn10 subunit, which is most likely localized in the hinge area between the base and lid, is involved in the recognition of ubiquitylated substrate proteins [12,15].

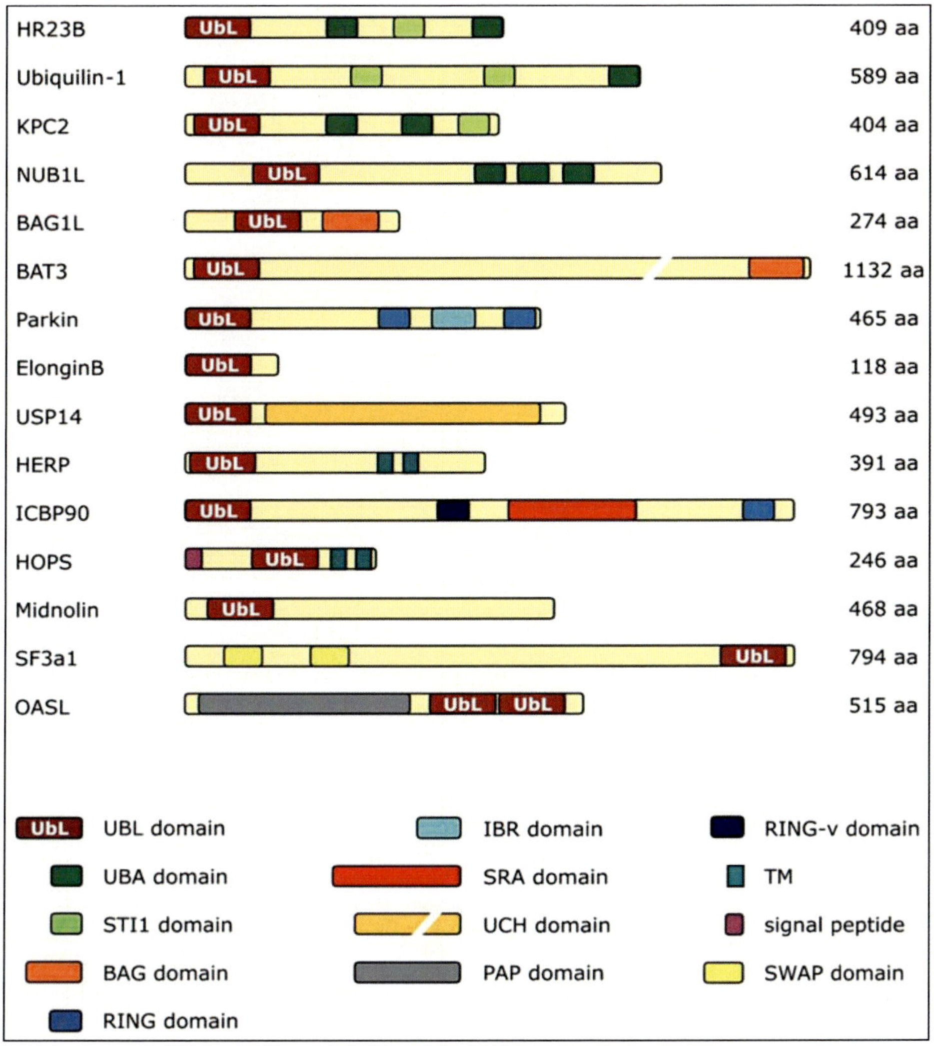

Figure 1. Domain architecture of selected human UDPs. Schematic representation of the domain architecture of selected human UDPs. The various domains are color-coded according to the key below. Details are described in the text.

Ubiquitin has a highly stable and compact structure consisting of two α-helices and five β-sheets arranged in the order ββαββαβ, which form the ubiquitin superfold [16]. Recently a number of ubiquitin-like proteins have been identified, which despite displaying rather limited sequence homology, all contain the ubiquitin superfold (Figure 1). These proteins can be divided into two groups: ubiquitin-like modifiers (UBLs) and ubiquitin domain proteins (UDPs) [17]. The UBL group includes ubiquitin itself, NEDD8, SUMO, FAT10, ATG8 and others, which can be covalently attached to substrate proteins in a process similar to ubiquitylation. Despite their structural similarity, different UBLs are involved in diverse cellular functions [17]. However, we shall not discuss these proteins further here, but instead focus on the group of UDPs.

The UDPs display a modular architecture (Figure 1) and therefore contain, apart from their UBL-domain, an individual set of specific structural features. The first UDPs, which were characterized, were responsible for the recruitment of proteasome substrates, and bound the proteasome in a UBL-domain dependent manner [18]. Thus the interaction of integral UBL domains with the proteasome was considered to be a common property of all UDPs. However, it turned out that different UDPs associate with different sites within the 26S proteasome. Moreover, some UDPs appear not to bind the proteasome at all. Hence, despite their obvious structural similarities the UDPs display a remarkable diversity in terms of their binding properties.

Here, we give an overview on UDPs and describe the binding features of their UBL-domains with respect to their cellular function, focusing especially on the nervous system and neurodegenerative diseases.

UBL/UBA Domain Proteins Like RAD23 and DSK2 Recruit Substrates for the Proteasome

A subgroup of UDPs, known as the UBL/UBA domain proteins, contain apart from their UBL-domain, one or more ubiquitin-associated (UBA)-domains (Figure 1). Rad23 as well as Dsk2 and their homologues are among the most extensively studied proteins of this group. *S. cerevisiae* Rad23 was first identified as a factor involved in nucleotide excision repair (NER) of UV-damaged DNA [19]. Rad23 null mutants displayed sensitivity to UV-radiation and the human homologue HHR23B was shown to associate with other nucleotide excision repair proteins such as the *Xeroderma pigmentosum* group C protein (XPC) [20].

In 1998 Schauber and co-workers demonstrated that the UBL-domain of Rad23 interacts with the proteasome, which suggested an involvement of the proteasome system in DNA repair [21]. However, the molecular function of the UBA-domain remained enigmatic for some time, until it was reported that it preferentially binds polyubiquitin and that the fission yeast orthologues of Rad23 and Dsk2 recruit ubiquitylated substrates to the proteasome by binding substrate proteins via their UBA domains and the proteasome via their UBL-domain [18]. This model was in line with earlier data on the human Dsk2 homologues ubiquilin-1 and ubiquilin-2 (also called hPLIC-1 and hPLIC-2) and was shortly afterwards confirmed by work in budding yeast [22,23]. Later independent groups have exploited cell free systems to investigate the function of UBL/UBA proteins as recruitment factors for proteasomal substrates [24,25]. Verma and co-workers isolated 26S proteasomes from *rpn10*-null and *rad23*-null yeast strains, which both displayed a deficiency with respect to the deubiquitylation and degradation of the proteasome substrate Sic1; effects which were rescued by the addition of recombinant Rpn10 or Rad23 [25]. Interestingly, only low concentrations of the recombinant proteins led to a rescue of the Sic1 phenotype, whereas the effect decreased at higher concentrations, indicating that an optimal balance between the proteasome and its substrate recruitment factors is essential. Although Rad23 and Dsk2 as well as the HHR23-proteins and the ubiquilins were shown to bind the proteasome, significant differences in terms of the proteasomal subunits recognized by the individual UBL-domains have been reported. The human ubiquitin-binding proteasome subunit S5a has two ubiquitin-interacting motifs (UIMs), called UIM1 and UIM2, with UIM2 positioned closer to the C-terminus. Hiyama and colleagues clearly demonstrated that the human Rad23 homologue HHR23A binds UIM2 [26]. However, in yeast Rad23 interacts with the proteasome base subunit Rpn1 and not with the S5a orthologue Rpn10 [27,28]. This observation was explained by the fact that UIM2 is not present in Rpn10. Remarkably, fission yeast Dph1, a Dsk2 orthologue can bind the UIM of the Rpn10 orthologue Pus1 (our unpublished observations). Whether HHR23A is also able to bind the human orthologue of Rpn1 has not yet been determined. However, it seems that the diversity between different UIMs and other UBL-domain binding structures is matched by variability among different UBL-domains.

Interestingly, Dsk2 and Rad23 compete for Rpn1-binding and data from cross-linking experiments suggest that also Rpn2, the second non-ATPase subunit of the proteasome base complex, is involved in Rad23 and Dsk2 docking at the proteasome [27,28]. Although there is no controversy on the fact that the proteasome subunit Rpn1 interacts with Rad23, different binding sites within Rpn1 were identified in budding and fission yeast. While, in *S. cerevisiae*, a leucine-rich repeat-like (LRR-like) domain was found to be the minimum binding site for the Rad23 UBL-domain, in fission yeast a region closer to the N-terminus of the protein was identified to be critical for this interaction [28,29]. Overall, the data suggest that there are significant structural differences between the UBL-domains of Rad23 and HHR23A, which result in the recognition of different targets. Individual UBL/UBA proteins therefore possess UBL-domains that all seem to bind the proteasome, but appear to be different in terms of their target at the proteasome.

A new aspect of UDP function was revealed by the discovery that, apart from the proteasome, also the E4 Ufd2 can associate with the Rad23 UBL-domain [30]. Moreover, Ufd2 was demonstrated to compete with the proteasome subunit Rpn1 for binding Rad23. Ufd2 is also known to bind to the Cdc48 ATPase complex. A recently published study suggests a model explaining the functional consequences of the interaction between Ufd2 and the Cdc48 complex [31]. One possibility is that the E4 activity of Ufd2 elongates the short ubiquitin chain on the substrate and the now polyubiquitylated substrate protein is then recruited to the proteasome via S5a/Rpn10. Another option would be that Rad23 or Dsk2 associate with Ufd2 in a UBL-domain dependent manner, possibly inhibiting its E4 activity. This enables the UDP to bind the oligo-ubiquitylated substrate with its UBA-domain(s) and dissociate from the Ufd2-Cdc48-complex, enabling UBL-domain mediated docking on the proteasome and delivery of the ubiquitylated cargo [31].

Genetic data demonstrating that Rad23 and Dsk2 are essential for the degradation of proteins derived from the endoplasmic reticulum (ER), were the first evidence that these UDPs also have, like Cdc48, a role in ER-associated protein degradation (ERAD) [32]. Accordingly, an interaction between HHR23B and PNGase, which is responsible for deglycosylation of ERAD substrates, has been characterized [33,34]. Thus, in the course of ERAD, ER-derived protein substrates are probably extracted to the cytosol with the help of Cdc48, deglycosylated by PNGase and then transferred to the proteasome by the UDPs.

In another study, HHR23A and HHR23B were shown to interact with ataxin-3, a protein responsible for the neurodegenerative Machado-Joseph disease [35]. As ataxin-3 contains UIMs, it was obvious to assume that it interacts with the HHR23B-UBL-domain in a UIM-dependent manner. However, detailed study of the interaction demonstrated that the UBL-domain of HHR23B interacts with the Ataxin-3 Josephin-domain, a papain-like cysteine protease similar to other deubiquitylating enzymes [35,36]. Since ataxin-3 was shown to associate with the proteasome, p97 (the human homologue of Cdc48), and ubiquitin-chains, it is believed to participate in recruiting ubiquitylated substrates to the proteasome [37]. However, ataxin-3 was also shown to be a substrate of the E4 Ufd2 [38]. Considering that Ufd2 and Rad23 act in a common pathway, it is feasible that Ataxin-3 is a substrate of HHR23B [30,31]. Thus, further studies are required to understand the physiological background of the interaction between ataxin-3 and HHR23B.

Ubiquilins are human homologues of yeast Dsk2. There are four ubiquilins, 1 through 4, which all contain, besides their N-terminal UBL-domains, C-terminal UBA-domains and Sti1-like repeats, that can interact with the molecular chaperone Hsp70 [39].

Ubiquilin-1 and ubiquilin-2, which were both shown to bind the proteasome, are also known as PLIC1 (*p*rotein *l*inking *I*AP with the *c*ytoskeleton) and PLIC2, as they were found to mediate the interaction between the plasma membrane receptor IAP and vimentin-containing intermediate filaments [22,40,41]. Interestingly, both ubiquilins were linked to Alzheimer's disease, as they were proposed to bind presenilins and localize to Lewy bodies and neurofibrillar tangles [42]. Presenilins are required for the generation of the C-terminus of amyloid β-protein (Aβ) and mutations in the *presenilin-1* and the *presenilin-2* genes lead to an increased ratio of Aβ42/Aβ40 generation from β-amyloid precursor protein (APP). These Aβ-peptides accumulate outside the cell resulting in amyloid plaques, one mayor lesion in Alzheimer's disease [43]. The interaction with the presenilins was mapped to the C-terminal UBA-domain of ubiquilin-1 [42]. However, in these studies, in which GST-ubiquilin fusions were used to precipitate *in vitro* translated presenilins, primarily a slow migrating smear was co-precipitated with the ubiquilin UBA-domain. Therefore it is likely that only ubiquitylated presenilins can interact with ubiquilin-1. The same might be true for the interaction shown between ubiquilin-1 and the γ-aminobutyric acid A ($GABA_A$) receptors [44]. These receptors are the mayor sites of fast synaptic inhibition in the brain. For efficient inhibitory synaptic transmission, the number of $GABA_A$-receptors at inhibitory synapses has to be tightly regulated. This is sustained by the internalization at clathrin-coated pits. Ubiquilin-1 has been described to bind $GABA_A$-receptors and to facilitate their membrane insertion by increasing the stability of their intracellular pool [44]. However, based on the data shown it is feasible that the UDP interacts only with ubiquitylated $GABA_A$ receptors. As proteasomal degradation of the receptors could not be observed, ubiquilin-1 appears also to have a function in endocytic processes [44]. Accordingly, the interaction of ubiquilin-1 with Eps15, an essential component of the clathrin-mediated endocytic pathway, has been reported [45]. Eps15 contains two UIMs and ubiquilin-1 was shown to bind the N-terminal one. Apart from Eps15, other UIM-containing proteins linked to the endocytic pathway, were also shown to bind the UBL-domain of ubiquilin-1. However, although these proteins co-localized with ubiquilin-1, they were not localized in endocytic compartments. Instead ubiquilin-1 as well as Eps15 were found in cytoplasmic aggregates and aggresomes. Thus ubiquilin-1 was proposed to be involved in the sequestration of certain UIM-containing endocytic proteins to ubiquitin-rich aggregates [45]. This is consistent with the finding that ubiquilin-1 and ubiquilin-2 localize to ubiquitin-positive structures and are both present in aggresomes [46]. Perhaps ubiquilin-1 recruits aggregated, ubiquitylated proteins as well as UIM-containing endocytic proteins to aggresomes. Based on the fact that it can bind the proteasome as well as

Eps15 it is possible, that, in addition to its role as a substrate recruitment factor of the proteasome, ubiquilin-1 might also promote the proteasome independent disposal of aggregated ubiquitin conjugates.

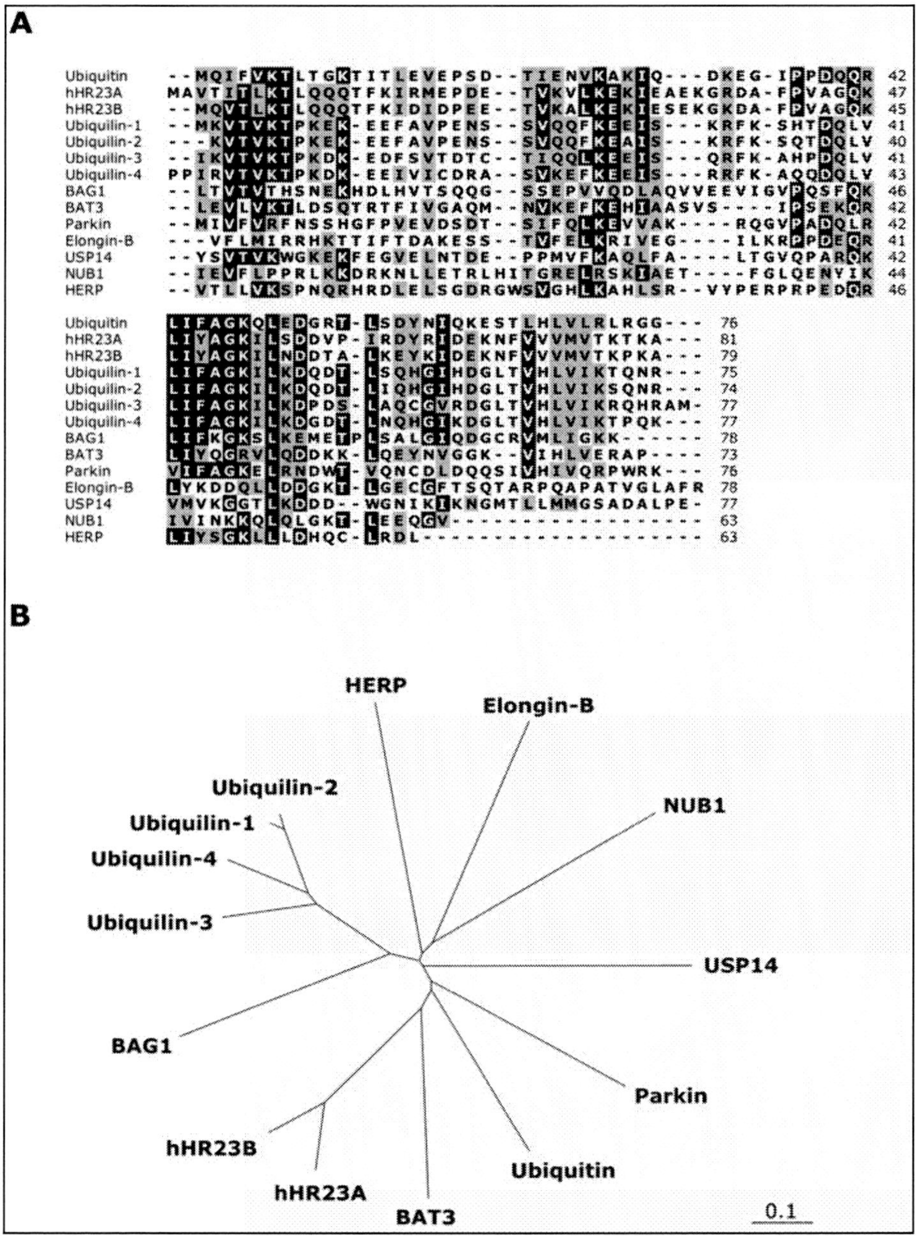

Figure 2 Primary structures of selected UBL domains. (A) Alignment of amino acid sequences from selected human UBL-domains. Identical amino acids are highlighted in black, similar amino acids in grey. (B) Phylogenetic tree of the UBL-domains. The scale bar is a measure for amino acid substitutions per site.

Remarkably, HHR23B does, in contrast to ubiquilin-1, not bind to Eps15, although both of them interact with the proteasome (M.S., unpublished data). Moreover, the primary structures of the UBL-domains from of ubiquilin-1 and HHR23B appear as similar to each other as to UDPs like HERP, which do not bind the proteasome (Figure 2). Therefore, it seems fair to speculate that the ubiquitin superfold might provide some sort of basic homologous interaction scaffold, which then has developed in different directions to bind various targets. Proteasome binding by different UBL-domains might therefore be rather based on analogous than homologues structures.

Ubiquilin-4 is linked to the neurodegenerative disease, spinocerebellar ataxia type 1 (SCA1) [47]. This inherited progressive disease primarily affects the brainstem, spinocerebellar tracts and cerebellar Purkinje cells. The disease is caused by the expansion of a polyglutamine stretch within the SCA1 protein, ataxin-1. Based on its interaction with ataxin-1, ubiquilin-4 is also called 'ataxin-1-interacting ubiquitin-like protein' or A1Up. Similar to other UBL/UBA proteins it binds the proteasome subunit S5a via its UBL-domain and polyubiquitin via its UBA-domain [47,48].

Other UBL/UBA Proteins

Ddi1 is a UBL/UBA protein from *S. cerevisiae*, which like Dsk2 contains an N-terminal UBL-domain and a C-terminal UBA-domain. It also interacts with the proteasome as well as with ubiquitylated substrates and was shown to mediate the degradation of an artificial model substrate [23,30]. The transcription of Ddi1 is induced in response to DNA damage [49,50]. However, its function in the DNA damage response pathway was unknown until Kaplun *et al.* demonstrated that the endonuclease Ho is stabilized and accumulates in the cytoplasm of *ddi1*-deletion strains [51]. Ho is known to introduce site-specific DNA double-strand breaks and has been shown to be rapidly degraded by the 26S proteasome during DNA-damage response. Interestingly, Ho was not found to interact with other UBL/UBA proteins such as Rad23 or Dsk2, revealing a specificity of Ho for Ddi1. This raises the question whether the UBA-domain of Ddi1 binds a certain type of ubiquitin chains, which are only generated in the course of Ho ubiquitylation, or whether specificity is introduced by other means.

Another UBL/UBA protein is KPC2, which was first identified in a complex with KPC1, an E3 ligase responsible for p27 ubiquitylation [52]. KPC2 binding to KPC1 results in the stabilization of the E3, which is dependent on the KPC2 Sti1-domain [53]. siRNA-mediated knock down of KPC2 results in a stabilization of p27, which is consistent with the observation that KPC2 binds ubiquitylated p27 and the proteasome via its C-terminal UBA-domain and its UBL-domain, respectively. Interestingly, the KPC2 UBL-domain was essential for the interaction with KPC1. Moreover, KPC1 competes with the proteasome for binding to the UBL-domain of KPC2 [52,53]. It therefore seems likely that, in analogy to the E4 Ufd2 and Rad23, p27 is ubiquitylated by KPC1, which then binds KPC2 that can shuttle modified p27 to the proteasome.

NUB1 (NEDD8 ultimate buster-1L) is an interferon-inducible UBL/UBA protein, which is predominantly localized to the nucleus. It was initially described as a down-regulator of the neddylation system, which recruits the NEDD8 monomer as well as neddylated proteins to the proteasome for degradation [54,55]. Later on it was revealed that NUB1 also interacts with the ubiquitin-like modifier FAT10 and promotes proteasome dependent degradation of

FAT10-modified proteins [56,57]. Moreover, when compared to its interaction with NEDD8, NUB1-binding of FAT10 was four-fold more efficient. As NUB-1 was also shown to interact with the proteasome subunit S5a, it seemed likely that its UBL-domain is crucial to recruit FAT10, its conjugates and perhaps NEDD8 to the proteasome [55]. However, detailed analyses of the interaction sites revealed that S5a binds to a C-terminal region of NUB1 and not to its UBL-domain [58].

Parkin – an E3 Ubiquitin Ligase Linked to Juvenile Parkinsonism

In patients suffering from autosomal-recessive juvenile parkinsonism (AR-JP), an early onset and slow progression of the disease, as well as classical symptoms associated with sporadic Parkinsonism, are observed. The first gene found to be involved in AR-JP was the *parkin* gene [59] and subsequent studies suggested that mutations in this gene are the most common cause for recessive, early-onset parkinsonism [60,61]. Parkin contains a UBL-domain and two E3-type RING-finger motifs separated by an in-between-RING (IBR) domain (Figure 1) indicating a function in the ubiquitin-proteasome system. In fact, parkin was shown to be a functional E3-ubiquitin protein ligase and pathological mutations in *parkin* impair the catalytic activity of the gene product [62-64]. Several targets of the parkin E3 activity have been identified. These include the synaptic vesicle associated GTPase CD-Crel-1, Cyclin-E, the α-synuclein interacting protein synphilin, the parkin-associated endothelin receptor-like receptor-1 (PaelRl) and the glycosylated form of α-synuclein [63,65-68]. By implication, loss of parkin function may result in the accumulation of its substrates, which then leads to the death of dopaminergic neurons. Support for this idea comes from experiments in which over-expression of the parkin substrate PaelRl leads to cell death in neuroblastoma cells. This effect can be rescued by parkin, but not by parkin mutants, which are deficient in their ubiquitylation activity [67]. Moreover, the glycosylated form of α-synuclein was found to accumulate in the brain of parkin-deficient AR-JP patients [68].

Although the majority of missense mutations cluster to the C-terminus of parkin, harboring the RING-domain, a few mutations were also identified in the UBL-domain-encoding region. NMR studies have revealed an interaction between the parkin UBL-domain and the proteasome subunit S5a/Rpn10. Furthermore, it was demonstrated that in the pathogenic parkin mutant R42P, this interaction is disrupted. However, in an *in vitro* pull-down assay a physical interaction between the parkin UBL-domain and S5a/Rpn10 could not be verified although HHR23A, which was used as a positive control, clearly interacted with S5a/Rpn10 under the same experimental conditions [69]. This indicates that the interaction is very weak and therefore its physiological relevance may be questioned. These doubts are supported by data showing that parkin as well as other proteins, which are linked to Parkinson's disease, promote substrate modification with ubiquitin chains linked via Lys63 [70]. Therefore it appears likely that parkin has a molecular function, which is not linked to proteasome-dependent protein degradation.

The parkin UBL-domain was also shown to regulate the stability of the parkin protein [71,72]. However, the results obtained from different studies are somewhat controversial.

Pathogenic mutations within the UBL-domain of parkin were shown to destabilize the parkin protein [72]. Over-expression demonstrated that the steady-state levels of a truncated parkin-version, which did not contain the UBL-domain, were dramatically enhanced in comparison to full-length parkin. These data imply that the UBL-domain might harbor the lysine residue essential for parkin ubiquitylation. However, ubiquitylated forms of the truncated parkin version were also detectable and the ubiquitylation site of parkin was localized to a region C-terminal of the UBL-domain. By deletion analysis of the parkin UBL-domain, the first six residues were found to be responsible for destabilizing parkin [71]. Thus the parkin UBL-domain is not required for parkin ubiquitylation, but instead harbors a destabilization signal responsible for its rapid turnover. Interestingly, parkin contains an integral start-codon at position 80, just C-terminally of the UBL-domain and there is experimental evidence that this start-codon is indeed used [72]. Thus, human cells contain a truncated parkin version lacking the UBL-domain. The functional significance of this parkin version and its stability compared to the UBL-containing version is unclear so far.

Elongin B is a UDP Present in a Multisubunit E3 Ubiquitin Ligase

Von Hippel-Lindau (VHL) disease is an autosomal dominant cancer syndrome, which manifests as angiomas of the retina, hemangioblastomas of the central nervous system, renal clear cell carcinomas and pheochromocytoma [73]. Patients with the VHL disease harbor a germ line mutation in one allele of the *VHL* gene. This tumor suppressor gene encodes the VHL protein, which is a component of the multisubunit E3 ubiquitin-protein ligase CBC^{VHL}(*C*ullin-Elongin BC-VHL). Apart from the VHL protein, which is responsible for substrate binding, the E3 complex contains the structural component cullin-2, the catalytic subunit RBX1(ROC1) and elongin B and C, which link the VHL protein and the cullin [74].

The CBC^{VHL} complex mediates ubiquitylation of hypoxia-inducible transcription factor α (HIFα and a number of other substrates, in order to target them for proteasome dependent degradation [75,76]. Under constitutive conditions, HIFα is ubiquitylated by the CBC^{VHL} complex, leading to low levels of HIFα and moderate expression of HIFα target genes. In patients suffering from the VHL disease the VHL gene is mutated, resulting in a protein that is incapable of binding and thus ubiquitylating the HIFα protein. Therefore HIFα accumulates, as it cannot be degraded by the proteasome. This leads to increased levels of HIFα target genes, which are mostly angiogenic factors, such as the vascular endothelial growth factor. Accordingly, these angiogenic factors cause extensive proliferation of capillaries, which is crucial for tumor development in VHL disease [74].

The adapter proteins elongin B and elongin C were first described as subunits of the transcription elongation regulatory factor SIII (or elongin ABC) [77]. Apart from CBC^{VHL} elongin B and elongin C were also found in other CBCs, where they either link other VHL-box proteins and cullin-2 or SOCS-box proteins and cullin-5 [78]. Whether elongin B has further functions besides its role as adapter component of CBCs is not yet known.

BAG-1 COORDINATES CHAPERONES WITH THE DEGRADATION PATHWAY

The co-chaperone BAG-1 (Bcl2-associated athanogene) was first identified as a binding partner of the anti-cell death protein Bcl-2, which is involved in the regulation of apoptosis [79]. It has been linked to a wide variety of cellular functions ranging from transcriptional regulation to control of cell migration [80]. As a result of alternative translation initiation, there are multiple isoforms of mammalian BAG-1 called BAG-1L, BAG-1M, BAG-1 and BAG-1S. Due to their different N-termini, these isoforms are characterized by the presence or absence of certain structural and functional elements such as nuclear localization signals or DNA binding motifs [81]. However, the more C-terminal parts of the protein, which contain the UBL-domain and the BAG-domain, is identical in all isoforms. The BAG-domain was shown to interact with the ATPase domain of the molecular chaperones Hsc70 and Hsp70. Binding of BAG-1 leads to conformational changes in Hsc70/Hsp70, which induce nucleotide exchange as well as substrate release from the peptide binding pocket of the chaperone. Thus BAG-1 acts as nucleotide exchange factor and modulator, which negatively regulates the refolding events of Hsc70/Hsp70 [82-85].

As the BAG-1 UBL-domain also binds the proteasome, it is likely that its function is to trigger the release of unfolded proteins from Hsc70/Hsp70, subjecting them to proteasome-dependent degradation [86]. This idea is supported by the finding that BAG-1 and Hsp70/Hsc70 form a ternary complex with the U-box ubiquitin-protein ligase CHIP.

CHIP in a complex with parkin and Hsp70 was shown to support the degradation of unfolded Pael receptor [87]. In addition, it mediates ubiquitylation of the glucocorticoid hormone receptor and the microtubule-binding protein tau, which forms intracellular protein aggregates in the brain of Alzheimer patients [88,89]. Although BAG-1 promotes CHIP-dependent degradation of the glucocorticoid hormone receptor, it has not yet been shown whether it is also involved in the degradation of CHIP substrates such as tau or Pael receptor, which play a role in neurodegenerative diseases [90]. Remarkably, BAG-1 itself is ubiquitylated by CHIP, which instead of leading to its degradation rather stimulates its association with the proteasome [91].

BAG-1 has six relatives in humans, all of which contain BAG-domains. However, only BAG-6, better known as scythe or Bat3, also contains a UBL-domain [80]. The *Xenopus* homologue of BAG-6 called scythe was originally identified as an interaction partner of the potent apoptotic inducer called reaper in *Drosophila* [92]. Scythe does, like BAG-1, interact with the ATPase domain of Hsp70 and inhibits Hsp70 mediated protein refolding. Binding of reaper leads to the dissociation of BAG-6 from Hsp70, and the liberation of a yet unidentified pro-apoptotic molecule, which in turn triggers cytochrome c release from mitochondria, resulting in the induction of apoptosis [93,94]. It seems therefore fair to speculate that in the absence of reaper, scythe promotes the proteasome-dependent degradation of the anti-apoptotic factor. However, whether the UBL-domain of scythe is indeed capable of binding the proteasome has not yet been studied.

Neurological Symptoms in Ataxia Mice are Caused by Recessive Mutations in the Gene Encoding USP14

Ubiquitin-protein conjugates are highly dynamic structures. In addition to many different sets of ubiquitylating enzymes, cells also possess a number of deubiquitylating enzymes (DUBs), which hydrolyze ubiquitin chains. These deubiquitylating activities are crucial for the cell, as the degradation of ubiquitylated proteins by the proteasome requires the removal of the ubiquitin signal, before the substrate protein is unfolded and enters the proteolytic core of the enzyme complex. Moreover, this recycling process is required to maintain the cellular level of free ubiquitin [95]. Apart from the proteasome subunits Uch37 and Rpn11, which also display deubiquitylating activity, there are several DUBs that can associate with the proteasome [96-100]. USP14 and its yeast homologue Ubp6 are DUBs, which use their N-terminal UBL-domain (Figure 1) to interact with the proteasome. Ubp6 was shown to bind specifically to the base subunit Rpn1 and proteasome binding increases its deubiquitylating activity [98,100,101].

Interestingly, a mutation in the *USP14* gene was linked to neurological symptoms observed in ataxia mice [102]. These animals develop severe tremors within two to three weeks followed by hindlimb paralysis and die at six to ten weeks. In contrast to many other neurodegenerative disorders like Parkinson's disease, Alzheimer's disease or SCA neither ubiquitin-positive aggregates nor cell death or loss of nervous tissue are detectable in the central nervous system of ataxia mice. Due to an insertion in intron 5 of *USP14*, the expression of *USP14* in these mice is reduced [102]. This reduction of USP14 leads to low ubiquitin levels in ataxia mice, which is consistent with the finding that loss of the Ubp6 in yeast also results in reduced ubiquitin levels [103,103]. Addition of proteasome inhibitors leads to a partial recovery of the cellular ubiquitin supply, suggesting that in the absence of Ubp6 ubiquitin might be degraded by the proteasome. These results accentuate the requirement of deubiquitylating enzymes in ubiquitin recycling to prevent proteasomal degradation of the modifier and to maintain the pool of free ubiquitin in the cell. However, as Usp14 was shown to be specific for monoubiquitylated proteins *in vitro*, it might also have a role in processes important for synaptic transmission such as protein trafficking [104]. Indeed, synaptic transmission in both the central and the peripheral nervous system is deficient in ataxia mice. Furthermore the finding that USP14 is present in the synaptoneurosome fraction, which is enriched in synaptic vesicles, is consistent with this role [102].

The UDP Herp is an Essential Component of a High Molecular Mass Erad Complex

The accumulation of misfolded proteins in the ER leads to the induction of the unfolded protein response pathway (UPR), which helps the cell to cope with this situation. The membrane-resident UDP HERP (Homocysteine-responsive endoplasmic reticulum-resident ubiquitin-like domain member 1 protein) is strongly up-regulated by this pathway and was shown to protect cells from ER stress-induced apoptosis in a UBL-domain dependent manner. Upon prolonged exposure to ER stress, HERP is cleaved by a Caspase3/7 activity. [105-107]. Interestingly and similar to ubiquilin, HERP was also shown to interact with presenilins and over-expression of HERP led to an enhancement of amyloid beta-protein generation. This interaction though, was independent of the HERP UBL-domain [106,108].

The function of HERP in the UPR pathway was poorly understood until it was recently shown to be essential for the efficient degradation of ERAD substrates [107,109]. Moreover, it was revealed that HERP is part of a high molecular weight complex, which also contains other components of the ERAD pathway, such as the membrane resident ubiquitin ligase HRD1 and the ATPase p97/Cdc48 [109]. Therefore HERP protects the cell from ER-stress by promoting the disposal of misfolded protein via ERAD. However, although the HERP UBL domain was required for the efficient degradation of substrate proteins, it appears not to mediate the observed interaction with the ERAD complex nor does it bind the proteasome [109]. Thus the UBL-domain of HERP might recruit other factors to the ERAD complex, which are also required for the degradation of misfolded substrates. To better understand the molecular function of HERP it will be essential to identify the target of its UBL-domain.

OTHER UDPS

Although some of the UDPs described above do not directly interact with the proteasome they all have a role in the ubiquitin system. However, there are a number of UDPs, which do not display features connecting them to the ubiquitin pathway.

One of them is HOPS (hematode odd protein shuttling), a UDP, which has recently been described to be involved in liver regeneration after partial hepatectomy [110]. HOPS was shown to be exported from the nucleus to the cytoplasm during liver regeneration and to interfere with protein synthesis by binding to EF-1A. It contains a UBL-domain which, in contrast to those in most other UDPs, is localized in the center of the protein sequence. Its targets have not been elucidated yet.

ICBP90 (inverted CCAAT binding protein of 90 kDa) is a nuclear phosphoprotein, containing an N-terminal UBL-domain, a leucine zipper, a RING variant domain, a SRA-domain and a RING domain. The protein was first described to bind to the topoisomerase II promotor [111]. ICBP90 is highly expressed in proliferating tissue and accordingly, cancer cell lines display increased expression levels of the protein [111,112]. In addition, ICBP90 was shown to interact with methylated CpGs through its SRA domain. As it targets methylated promotors of various tumor suppressors, it is discussed to play a crucial role in carcinogenesis [113]. ICBP90 has also been linked to the ubiquitin-proteasome pathway, since its murine homologue is able to ubiquitylate histone H3 in a RING domain-dependent manner [114]. The ubiquitylation of histones does usually not lead to their degradation, but rather results in transcriptional activation of associated genes. Whether the UBL-domain of ICBP90 is involved in transcriptional activation by recruiting factors, which are necessary for this process, remains to be seen.

The splicosome-associated protein SF3a1 also known as SAP114 is a component of the SF3a splicing factor, which is part of the splicosome, a macromolecular complex responsible for removing introns from nuclear pre-mRNA. The UBL-domain of SF3a1 is located to the C-terminus of the protein. Its role in the splicing process is not understood yet. Upstream of the UBL-domain SAP114 contains two SWAP-domains (Figure 1), responsible for the interaction with another component of the SF3a splicing factor [115].

The murine UDP midnolin (midbrain nucleolar protein) is exclusively found in the nucleus and nucleolus. Its expression was shown to be regulated during development, as in mouse embryos a strong expression in the midbrain was observed at day 12.5. Thus the protein might be involved in the regulation of genes responsible for neurogenesis [116].

IKKβ is part of the IκB kinase complex which regulates the activity of NFκB. It contains a centrally positioned degenerate UBL-domain also designed as UBL-like domain. This protein domain was shown to be essential for the catalytic activity IKKβ, as a mutant version in which a leucine within the UBL-like domain was replaced by an alanine was incapable of dissociating the NFκ-B p65 subunit from the IκB-kinase complex [117]. Thus the UBL-like domain of IKKβ plays a role in the detachment of p65 from IKKβ.

OASL (OAS-like) is a member of the 2′-5′-oligoadenylate synthetases (OAS). This protein family is responsible for the elimination of double stranded RNA after viral infection. In response to interferons, the OAS proteins are synthesized as latent enzymes, which are activated by the interaction with double-stranded RNA. Active OAS proteins synthesize 2′-5′-oligoadenylates, which in turn activate RNaseL, leading to the degradation of RNA and therefore to the inhibition of viral protein synthesis [118]. The N-terminus of OASL contains an OAS core domain (also designated as PAP domain), that is highly homologous to other OAS family members and was shown to be responsible for poly-A-polymerase-activity. However, in contrast to other OAS proteins, OASL has no oligoadenylate synthetase activity [119,120]. As OASL is the only family member that contains two UBL-domains at the C-terminus, it may have distinct biological functions. In fact, OASL was shown to interact with the methyl CpG-binding protein 1 (MBD1) [121]. Similar to OASL, MBD1 is induced by interferon and there is evidence that genetically modified mice, which lack a functional MBD1 gene, exhibit increased expression of endogenous provirus [122]. Hence, MBD1 and OASL are co-induced by interferon and might cooperate in inhibiting viral transcription. Interestingly, the interaction of OASL with MBD1 was dependent on its UBL domains, which is further evidence, that the UBL-domain can interact with a wide variety of proteins not necessarily linked to the ubiquitin-proteasome pathway [121]. Apart from the mentioned proteins an increasing number of UDPs, whose functions are mainly unclear, are emerging [123] indicating that members of this protein family are linked to an even broader spectrum of cellular functions.

CONCLUSIONS

As evident from our literature review of the various UDPs, these proteins display a remarkable variety of functions encompassing many facets of cell biology. However, clearly most UDPs are somehow connected with the ubiquitin system either directly by virtue of a UBL-proteasome interaction or more indirectly.

Though, at least the ubiquilins, USP14 and parkin appear to be directly involved in the pathology of certain neurodegenerative diseases. This is in no way a general trait of the UDPs, but goes to show the importance of the ubiquitin-system in maintaining an appropriate protein milieu in neuronal cells.

Interestingly, many generally less characterized UDPs are apparently not connected with the ubiquitin-system at all. Hence, UDPs appear functionally more diverse than one would expect from a cursory inspection of their primary structure. Surely, future studies will reveal even more exciting functional aspects of the UDPs.

Acknowledgements

The authors apologize to those whose work was not cited due to space constraints. M.S. is supported by a grant from the Deutsche Forschungsgemeinschaft (SE 938/3-1) and R.H.-P. by grants from the Danish Natural Science Research Council and the Lundbeck Foundation.

References

[1] Glickman MH, Ciechanover A: The ubiquitin-proteasome proteolytic pathway: destruction for the sake of construction. *Physiol Rev* 2002, 82:373-428.

[2] Pickart CM: Mechanisms underlying ubiquitination. *Annu Rev Biochem* 2001, 70:503-533.

[3] Weissman AM: Themes and variations on ubiquitylation. *Nat Rev Mol Cell Biol* 2001, 2:169-178.

[4] Koegl M, Hoppe T, Schlenker S, Ulrich HD, Mayer TU, Jentsch S: A novel ubiquitination factor, E4, is involved in multiubiquitin chain assembly. *Cell* 1999, 96:635-644.

[5] Wilkinson KD: Ubiquitination and deubiquitination: targeting of proteins for degradation by the proteasome. *Semin Cell Dev Biol* 2000, 11:141-148.

[6] Thrower JS, Hoffman L, Rechsteiner M, Pickart CM: Recognition of the polyubiquitin proteolytic signal. *EMBO J* 2000, 19:94-102.

[7] Deng L, Wang C, Spencer E, Yang L, Braun A, You J, Slaughter C, Pickart C, Chen ZJ: Activation of the IkappaB kinase complex by TRAF6 requires a dimeric ubiquitin-conjugating enzyme complex and a unique polyubiquitin chain. *Cell* 2000, 103:351-361.

[8] Lim KL, Chew KC, Tan JM, Wang C, Chung KK, Zhang Y, Tanaka Y, Smith W, Engelender S, Ross CA, Dawson VL, Dawson TM: Parkin mediates nonclassical, proteasomal-independent ubiquitination of synphilin-1: implications for Lewy body formation. *J Neurosci* 2005, 25:2002-2009.

[9] Hicke L: Protein regulation by monoubiquitin. *Nat Rev Mol Cell Biol* 2001, 2:195-201.

[10] Groll M, Ditzel L, Lowe J, Stock D, Bochtler M, Bartunik HD, Huber R: Structure of 20S proteasome from yeast at 2.4 A resolution. *Nature* 1997, 386:463-471.

[11] Groll M, Bajorek M, Kohler A, Moroder L, Rubin DM, Huber R, Glickman MH, Finley D: A gated channel into the proteasome core particle. *Nat Struct Biol* 2000, 7:1062-1067.

[12] Glickman MH, Rubin DM, Coux O, Wefes I, Pfeifer G, Cjeka Z, Baumeister W, Fried VA, Finley D: A subcomplex of the proteasome regulatory particle required for ubiquitin-conjugate degradation and related to the COP9-signalosome and eIF3. *Cell* 1998, 94:615-623.

[13] Braun BC, Glickman M, Kraft R, Dahlmann B, Kloetzel PM, Finley D, Schmidt M: The base of the proteasome regulatory particle exhibits chaperone-like activity. *Nat Cell Biol* 1999, 1:221-226.

[14] Lam YA, Lawson TG, Velayutham M, Zweier JL, Pickart CM: A proteasomal ATPase subunit recognizes the polyubiquitin degradation signal. *Nature* 2002, 416:763-767.

[15] Deveraux Q, Ustrell V, Pickart C, Rechsteiner M: A 26 S protease subunit that binds ubiquitin conjugates. *J Biol Chem* 1994, 269:7059-7061.

[16] Vijay-Kumar S, Bugg CE, Cook WJ: Structure of ubiquitin refined at 1.8 A resolution. *J Mol Biol* 1987, 194:531-544.

[17] Jentsch S, Pyrowolakis G: Ubiquitin and its kin: how close are the family ties? *Trends Cell Biol* 2000, 10:335-342.

[18] Wilkinson CR, Seeger M, Hartmann-Petersen R, Stone M, Wallace M, Semple C, Gordon C: Proteins containing the UBA domain are able to bind to multi-ubiquitin chains. *Nat Cell Biol* 2001, 3:939-943.

[19] Miller RD, Prakash L, Prakash S: Defective excision of pyrimidine dimers and interstrand DNA crosslinks in rad7 and rad23 mutants of Saccharomyces cerevisiae. *Mol Gen Genet* 1982, 188:235-239.

[20] Masutani C, Sugasawa K, Yanagisawa J, Sonoyama T, Ui M, Enomoto T, Takio K, Tanaka K, van der Spek PJ, Bootsma D, Hoeijmakers JH, Hanaoka F: Purification and cloning of a nucleotide excision repair complex involving the xeroderma pigmentosum group C protein and a human homologue of yeast RAD23. *EMBO J* 1994, 13:1831-1843.

[21] Schauber C, Chen L, Tongaonkar P, Vega I, Lambertson D, Potts W, Madura K: Rad23 links DNA repair to the ubiquitin/proteasome pathway. *Nature* 1998, 391:715-718.

[22] Kleijnen MF, Shih AH, Zhou P, Kumar S, Soccio RE, Kedersha NL, Gill G, Howley PM: The hPLIC proteins may provide a link between the ubiquitination machinery and the proteasome. *Mol Cell* 2000, 6:409-419.

[23] Saeki Y, Saitoh A, Toh-e A, Yokosawa H: Ubiquitin-like proteins and Rpn10 play cooperative roles in ubiquitin-dependent proteolysis. *Biochem Biophys Res Commun* 2002, 293:986-992.

[24] Raasi S, Pickart CM: Rad23 ubiquitin-associated domains (UBA) inhibit 26 S proteasome-catalyzed proteolysis by sequestering lysine 48-linked polyubiquitin chains. *J Biol Chem* 2003, 278:8951-8959.

[25] Verma R, Oania R, Graumann J, Deshaies RJ: Multiubiquitin chain receptors define a layer of substrate selectivity in the ubiquitin-proteasome system. *Cell* 2004, 118:99-110.

[26] Hiyama H, Yokoi M, Masutani C, Sugasawa K, Maekawa T, Tanaka K, Hoeijmakers JH, Hanaoka F: Interaction of hHR23 with S5a. The ubiquitin-like domain of hHR23 mediates interaction with S5a subunit of 26 S proteasome. *J Biol Chem* 1999, 274:28019-28025.

[27] Saeki Y, Sone T, Toh-e A, Yokosawa H: Identification of ubiquitin-like protein-binding subunits of the 26S proteasome. *Biochem Biophys Res Commun* 2002, 296:813-819.

[28] Elsasser S, Gali RR, Schwickart M, Larsen CN, Leggett DS, Muller B, Feng MT, Tubing F, Dittmar GA, Finley D: Proteasome subunit Rpn1 binds ubiquitin-like protein domains. *Nat Cell Biol* 2002, 4:725-730.

[29] Seeger M, Hartmann-Petersen R, Wilkinson CR, Wallace M, Samejima I, Taylor MS, Gordon C: Interaction of the anaphase-promoting complex/cyclosome and proteasome protein complexes with multiubiquitin chain-binding proteins. *J Biol Chem* 2003, 278:16791-16796.

[30] Kim I, Mi K, Rao H: Multiple interactions of Rad23 suggest a mechanism for ubiquitylated substrate delivery important in proteolysis. *Mol Biol Cell* 2004, 15:3357-3365.

[31] Richly H, Rape M, Braun S, Rumpf S, Hoege C, Jentsch S: A series of ubiquitin binding factors connects CDC48/p97 to substrate multiubiquitylation and proteasomal targeting. *Cell* 2005, 120:73-84.

[32] Medicherla B, Kostova Z, Schaefer A, Wolf DH: A genomic screen identifies Dsk2p and Rad23p as essential components of ER-associated degradation. *EMBO Rep* 2004, 5:692-697.

[33] Biswas S, Katiyar S, Li G, Zhou X, Lennarz WJ, Schindelin H: The N-terminus of yeast peptide: N-glycanase interacts with the DNA repair protein Rad23. *Biochem Biophys Res Commun* 2004, 323:149-155.

[34] Katiyar S, Li G, Lennarz WJ: A complex between peptide:N-glycanase and two proteasome-linked proteins suggests a mechanism for the degradation of misfolded glycoproteins. *Proc Natl Acad Sci U S A* 2004, 101:13774-13779.

[35] Wang G, Sawai N, Kotliarova S, Kanazawa I, Nukina N: Ataxin-3, the MJD1 gene product, interacts with the two human homologs of yeast DNA repair protein RAD23, HHR23A and HHR23B. *Hum Mol Genet* 2000, 9:1795-1803.

[36] Nicastro G, Menon RP, Masino L, Knowles PP, McDonald NQ, Pastore A: The solution structure of the Josephin domain of ataxin-3: structural determinants for molecular recognition. *Proc Natl Acad Sci U S A* 2005, 102:10493-10498.

[37] Doss-Pepe EW, Stenroos ES, Johnson WG, Madura K: Ataxin-3 interactions with rad23 and valosin-containing protein and its associations with ubiquitin chains and the proteasome are consistent with a role in ubiquitin-mediated proteolysis. *Mol Cell Biol* 2003, 23:6469-6483.

[38] Matsumoto M, Yada M, Hatakeyama S, Ishimoto H, Tanimura T, Tsuji S, Kakizuka A, Kitagawa M, Nakayama KI: Molecular clearance of ataxin-3 is regulated by a mammalian E4. *EMBO J* 2004, 23:659-669.

[39] Kaye FJ, Modi S, Ivanovska I, Koonin EV, Thress K, Kubo A, Kornbluth S, Rose MD: A family of ubiquitin-like proteins binds the ATPase domain of Hsp70-like Stch. *FEBS Lett* 2000, 467:348-355.

[40] Ko HS, Uehara T, Tsuruma K, Nomura Y: Ubiquilin interacts with ubiquitylated proteins and proteasome through its ubiquitin-associated and ubiquitin-like domains. *FEBS Lett* 2004, 566:110-114.

[41] Wu AL, Wang J, Zheleznyak A, Brown EJ: Ubiquitin-related proteins regulate interaction of vimentin intermediate filaments with the plasma membrane. *Mol Cell* 1999, 4:619-625.

[42] Mah AL, Perry G, Smith MA, Monteiro MJ: Identification of ubiquilin, a novel presenilin interactor that increases presenilin protein accumulation. *J Cell Biol* 2000, 151:847-862.

[43] Selkoe DJ: Cell biology of protein misfolding: the examples of Alzheimer's and Parkinson's diseases. *Nat Cell Biol* 2004, 6:1054-1061.

[44] Bedford FK, Kittler JT, Muller E, Thomas P, Uren JM, Merlo D, Wisden W, Triller A, Smart TG, Moss SJ: GABA(A) receptor cell surface number and subunit stability are regulated by the ubiquitin-like protein Plic-1. *Nat Neurosci* 2001, 4:908-916.

[45] Regan-Klapisz E, Sorokina I, Voortman J, de Keizer P, Roovers RC, Verheesen P, Urbe S, Fallon L, Fon EA, Verkleij A, Benmerah A, van Bergen en Henegouwen PM: Ubiquilin recruits Eps15 into ubiquitin-rich cytoplasmic aggregates via a UIM-UBL interaction. *J Cell Sci* 2005, 118:4437-4450.

[46] Massey LK, Mah AL, Ford DL, Miller J, Liang J, Doong H, Monteiro MJ: Overexpression of ubiquilin decreases ubiquitination and degradation of presenilin proteins. *J Alzheimers Dis* 2004, 6:79-92.

[47] Davidson JD, Riley B, Burright EN, Duvick LA, Zoghbi HY, Orr HT: Identification and characterization of an ataxin-1-interacting protein: A1Up, a ubiquitin-like nuclear protein. *Hum Mol Genet* 2000, 9:2305-2312.

[48] Riley BE, Xu Y, Zoghbi HY, Orr HT: The effects of the polyglutamine repeat protein ataxin-1 on the UbL-UBA protein A1Up. *J Biol Chem* 2004, 279:42290-42301.

[49] Zhu Y, Xiao W: Two alternative cell cycle checkpoint pathways differentially control DNA damage-dependent induction of MAG1 and DDI1 expression in yeast. *Mol Genet Genomics* 2001, 266:436-444.

[50] Jelinsky SA, Samson LD: Global response of Saccharomyces cerevisiae to an alkylating agent. *Proc Natl Acad Sci U S A* 1999, 96:1486-1491.

[51] Kaplun L, Ivantsiv Y, Kornitzer D, Raveh D: Functions of the DNA damage response pathway target Ho endonuclease of yeast for degradation via the ubiquitin-26S proteasome system. *Proc Natl Acad Sci U S A* 2000, 97:10077-10082.

[52] Kamura T, Hara T, Matsumoto M, Ishida N, Okumura F, Hatakeyama S, Yoshida M, Nakayama K, Nakayama KI: Cytoplasmic ubiquitin ligase KPC regulates proteolysis of p27(Kip1) at G1 phase. *Nat Cell Biol* 2004, 6:1229-1235.

[53] Hara T, Kamura T, Kotoshiba S, Takahashi H, Fujiwara K, Onoyama I, Shirakawa M, Mizushima N, Nakayama KI: Role of the UBL-UBA protein KPC2 in degradation of p27 at G1 phase of the cell cycle. *Mol Cell Biol* 2005, 25:9292-9303.

[54] Kito K, Yeh ET, Kamitani T: NUB1, a NEDD8-interacting protein, is induced by interferon and down-regulates the NEDD8 expression. *J Biol Chem* 2001, 276:20603-20609.

[55] Kamitani T, Kito K, Fukuda-Kamitani T, Yeh ET: Targeting of NEDD8 and its conjugates for proteasomal degradation by NUB1. *J Biol Chem* 2001, 276:46655-46660.

[56] Hipp MS, Raasi S, Groettrup M, Schmidtke G: NEDD8 ultimate buster-1L interacts with the ubiquitin-like protein FAT10 and accelerates its degradation. *J Biol Chem* 2004, 279:16503-16510.

[57] Hipp MS, Kalveram B, Raasi S, Groettrup M, Schmidtke G: FAT10, a ubiquitin-independent signal for proteasomal degradation. *Mol Cell Biol* 2005, 25:3483-3491.

[58] Tanji K, Tanaka T, Kamitani T: Interaction of NUB1 with the proteasome subunit S5a. *Biochem Biophys Res Commun* 2005, 337:116-120.

[59] Kitada T, Asakawa S, Hattori N, Matsumine H, Yamamura Y, Minoshima S, Yokochi M, Mizuno Y, Shimizu N: Mutations in the parkin gene cause autosomal recessive juvenile parkinsonism. *Nature* 1998, 392:605-608.

[60] Farrer M, Chan P, Chen R, Tan L, Lincoln S, Hernandez D, Forno L, Gwinn-Hardy K, Petrucelli L, Hussey J, Singleton A, Tanner C, Hardy J, Langston JW: Lewy bodies and parkinsonism in families with parkin mutations. *Ann Neurol* 2001, 50:293-300.

[61] West AB, Maidment NT: Genetics of parkin-linked disease. *Hum Genet* 2004, 114:327-336.

[62] Shimura H, Hattori N, Kubo S, Mizuno Y, Asakawa S, Minoshima S, Shimizu N, Iwai K, Chiba T, Tanaka K, Suzuki T: Familial Parkinson disease gene product, parkin, is a ubiquitin-protein ligase. *Nat Genet* 2000, 25:302-305.

[63] Zhang Y, Gao J, Chung KK, Huang H, Dawson VL, Dawson TM: Parkin functions as an E2-dependent ubiquitin- protein ligase and promotes the degradation of the synaptic vesicle-associated protein, CDCrel-1. *Proc Natl Acad Sci U S A* 2000, 97:13354-13359.

[64] Imai Y, Soda M, Takahashi R: Parkin suppresses unfolded protein stress-induced cell death through its E3 ubiquitin-protein ligase activity. *J Biol Chem* 2000, 275:35661-35664.

[65] Staropoli JF, McDermott C, Martinat C, Schulman B, Demireva E, Abeliovich A: Parkin is a component of an SCF-like ubiquitin ligase complex and protects postmitotic neurons from kainate excitotoxicity. *Neuron* 2003, 37:735-749.

[66] Chung KK, Zhang Y, Lim KL, Tanaka Y, Huang H, Gao J, Ross CA, Dawson VL, Dawson TM: Parkin ubiquitinates the alpha-synuclein-interacting protein, synphilin-1: implications for Lewy-body formation in Parkinson disease. *Nat Med* 2001, 7:1144-1150.

[67] Imai Y, Soda M, Inoue H, Hattori N, Mizuno Y, Takahashi R: An unfolded putative transmembrane polypeptide, which can lead to endoplasmic reticulum stress, is a substrate of Parkin. *Cell* 2001, 105:891-902.

[68] Shimura H, Schlossmacher MG, Hattori N, Frosch MP, Trockenbacher A, Schneider R, Mizuno Y, Kosik KS, Selkoe DJ: Ubiquitination of a new form of alpha-synuclein by parkin from human brain: implications for Parkinson's disease. *Science* 2001, 293:263-269.

[69] Sakata E, Yamaguchi Y, Kurimoto E, Kikuchi J, Yokoyama S, Yamada S, Kawahara H, Yokosawa H, Hattori N, Mizuno Y, Tanaka K, Kato K: Parkin binds the Rpn10 subunit of 26S proteasomes through its ubiquitin-like domain. *EMBO Rep* 2003, 4:301-306.

[70] Lim KL, Dawson VL, Dawson TM: Parkin-mediated lysine 63-linked polyubiquitination: A link to protein inclusions formation in Parkinson's and other conformational diseases? *Neurobiol Aging* 2006, 27:524-529.

[71] Finney N, Walther F, Mantel PY, Stauffer D, Rovelli G, Dev KK: The cellular protein level of parkin is regulated by its ubiquitin-like domain. *J Biol Chem* 2003, 278:16054-16058.

[72] Henn IH, Gostner JM, Lackner P, Tatzelt J, Winklhofer KF: Pathogenic mutations inactivate parkin by distinct mechanisms. *J Neurochem* 2005, 92:114-122.

[73] Lonser RR, Glenn GM, Walther M, Chew EY, Libutti SK, Linehan WM, Oldfield EH: von Hippel-Lindau disease. *Lancet* 2003, 361:2059-2067.

[74] Kaelin WG, Jr.: Molecular basis of the VHL hereditary cancer syndrome. *Nat Rev Cancer* 2002, 2:673-682.

[75] Maxwell PH, Wiesener MS, Chang GW, Clifford SC, Vaux EC, Cockman ME, Wykoff CC, Pugh CW, Maher ER, Ratcliffe PJ: The tumour suppressor protein VHL targets hypoxia-inducible factors for oxygen-dependent proteolysis. *Nature* 1999, 399:271-275.

[76] Cockman ME, Masson N, Mole DR, Jaakkola P, Chang GW, Clifford SC, Maher ER, Pugh CW, Ratcliffe PJ, Maxwell PH: Hypoxia inducible factor-alpha binding and ubiquitylation by the von Hippel-Lindau tumor suppressor protein. *J Biol Chem* 2000, 275:25733-25741.

[77] Duan DR, Pause A, Burgess WH, Aso T, Chen DY, Garrett KP, Conaway RC, Conaway JW, Linehan WM, Klausner RD: Inhibition of transcription elongation by the VHL tumor suppressor protein. *Science* 1995, 269:1402-1406.

[78] Kamura T, Burian D, Yan Q, Schmidt SL, Lane WS, Querido E, Branton PE, Shilatifard A, Conaway RC, Conaway JW: Muf1, a novel Elongin BC-interacting leucine-rich repeat protein that can assemble with Cul5 and Rbx1 to reconstitute a ubiquitin ligase. *J Biol Chem* 2001, 276:29748-29753.

[79] Takayama S, Sato T, Krajewski S, Kochel K, Irie S, Millan JA, Reed JC: Cloning and functional analysis of BAG-1: a novel Bcl-2-binding protein with anti-cell death activity. *Cell* 1995, 80:279-284.

[80] Takayama S, Reed JC: Molecular chaperone targeting and regulation by BAG family proteins. *Nat Cell Biol* 2001, 3:E237-241.

[81] Takayama S, Krajewski S, Krajewska M, Kitada S, Zapata JM, Kochel K, Knee D, Scudiero D, Tudor G, Miller GJ, Miyashita T, Yamada M, Reed JC: Expression and location of Hsp70/Hsc-binding anti-apoptotic protein BAG-1 and its variants in normal tissues and tumor cell lines. *Cancer Res* 1998, 58:3116-3131.

[82] Hohfeld J, Jentsch S: GrpE-like regulation of the hsc70 chaperone by the anti-apoptotic protein BAG-1. *EMBO J* 1997, 16:6209-6216.

[83] Zeiner M, Gebauer M, Gehring U: Mammalian protein RAP46: an interaction partner and modulator of 70 kDa heat shock proteins. *EMBO J* 1997, 16:5483-5490.

[84] Takayama S, Bimston DN, Matsuzawa S, Freeman BC, Aime-Sempe C, Xie Z, Morimoto RI, Reed JC: BAG-1 modulates the chaperone activity of Hsp70/Hsc70. *EMBO J* 1997, 16:4887-4896.

[85] Sondermann H, Scheufler C, Schneider C, Hohfeld J, Hartl FU, Moarefi I: Structure of a Bag/Hsc70 complex: convergent functional evolution of Hsp70 nucleotide exchange factors. *Science* 2001, 291:1553-1557.

[86] Luders J, Demand J, Hohfeld J: The ubiquitin-related BAG-1 provides a link between the molecular chaperones Hsc70/Hsp70 and the proteasome. *J Biol Chem* 2000, 275:4613-4617.

[87] Imai Y, Soda M, Hatakeyama S, Akagi T, Hashikawa T, Nakayama KI, Takahashi R: CHIP is associated with Parkin, a gene responsible for familial Parkinson's disease, and enhances its ubiquitin ligase activity. *Mol Cell* 2002, 10:55-67.

[88] Connell P, Ballinger CA, Jiang J, Wu Y, Thompson LJ, Hohfeld J, Patterson C: The co-chaperone CHIP regulates protein triage decisions mediated by heat-shock proteins. *Nat Cell Biol* 2001, 3:93-96.

[89] Shimura H, Schwartz D, Gygi SP, Kosik KS: CHIP-Hsc70 complex ubiquitinates phosphorylated tau and enhances cell survival. *J Biol Chem* 2004, 279:4869-4876.

[90] Demand J, Alberti S, Patterson C, Hohfeld J: Cooperation of a ubiquitin domain protein and an E3 ubiquitin ligase during chaperone/proteasome coupling. *Curr Biol* 2001, 11:1569-1577.

[91] Alberti S, Demand J, Esser C, Emmerich N, Schild H, Hohfeld J: Ubiquitylation of BAG-1 suggests a novel regulatory mechanism during the sorting of chaperone substrates to the proteasome. *J Biol Chem* 2002, 277:45920-45927.

[92] Thress K, Henzel W, Shillinglaw W, Kornbluth S: Scythe: a novel reaper-binding apoptotic regulator. *EMBO J* 1998, 17:6135-6143.

[93] Thress K, Evans EK, Kornbluth S: Reaper-induced dissociation of a Scythe-sequestered cytochrome c-releasing activity. *EMBO J* 1999, 18:5486-5493.

[94] Thress K, Song J, Morimoto RI, Kornbluth S: Reversible inhibition of Hsp70 chaperone function by Scythe and Reaper. *EMBO J* 2001, 20:1033-1041.

[95] Amerik AY, Hochstrasser M: Mechanism and function of deubiquitinating enzymes. *Biochim Biophys Acta* 2004, 1695:189-207.

[96] Verma R, Aravind L, Oania R, McDonald WH, Yates JR, 3rd, Koonin EV, Deshaies RJ: Role of Rpn11 metalloprotease in deubiquitination and degradation by the 26S proteasome. *Science* 2002, 298:611-615.

[97] Papa FR, Amerik AY, Hochstrasser M: Interaction of the Doa4 deubiquitinating enzyme with the yeast 26S proteasome. *Mol Biol Cell* 1999, 10:741-756.

[98] Leggett DS, Hanna J, Borodovsky A, Crosas B, Schmidt M, Baker RT, Walz T, Ploegh H, Finley D: Multiple associated proteins regulate proteasome structure and function. *Mol Cell* 2002, 10:495-507.

[99] Lam YA, Xu W, DeMartino GN, Cohen RE: Editing of ubiquitin conjugates by an isopeptidase in the 26S proteasome. *Nature* 1997, 385:737-740.

[100] Borodovsky A, Kessler BM, Casagrande R, Overkleeft HS, Wilkinson KD, Ploegh HL: A novel active site-directed probe specific for deubiquitylating enzymes reveals proteasome association of USP14. *EMBO J* 2001, 20:5187-5196.

[101] Chernova TA, Allen KD, Wesoloski LM, Shanks JR, Chernoff YO, Wilkinson KD: Pleiotropic effects of Ubp6 loss on drug sensitivities and yeast prion are due to depletion of the free ubiquitin pool. *J Biol Chem* 2003, 278:52102-52115.

[102] Wilson SM, Bhattacharyya B, Rachel RA, Coppola V, Tessarollo L, Householder DB, Fletcher CF, Miller RJ, Copeland NG, Jenkins NA: Synaptic defects in ataxia mice result from a mutation in Usp14, encoding a ubiquitin-specific protease. *Nat Genet* 2002, 32:420-425.

[103] Anderson C, Crimmins S, Wilson JA, Korbel GA, Ploegh HL, Wilson SM: Loss of Usp14 results in reduced levels of ubiquitin in ataxia mice. *J Neurochem* 2005, 95:724-731.

[104] Yin L, Krantz B, Russell NS, Deshpande S, Wilkinson KD: Nonhydrolyzable diubiquitin analogues are inhibitors of ubiquitin conjugation and deconjugation. *Biochemistry* 2000, 39:10001-10010.

[105] Kokame K, Agarwala KL, Kato H, Miyata T: Herp, a new ubiquitin-like membrane protein induced by endoplasmic reticulum stress. *J Biol Chem* 2000, 275:32846-32853.

[106] Sai X, Kawamura Y, Kokame K, Yamaguchi H, Shiraishi H, Suzuki R, Suzuki T, Kawaichi M, Miyata T, Kitamura T, De Strooper B, Yanagisawa K, Komano H: Endoplasmic reticulum stress-inducible protein, Herp, enhances presenilin-mediated generation of amyloid beta-protein. *J Biol Chem* 2002, 277:12915-12920.

[107] Hori O, Ichinoda F, Yamaguchi A, Tamatani T, Taniguchi M, Koyama Y, Katayama T, Tohyama M, Stern DM, Ozawa K, Kitao Y, Ogawa S: Role of Herp in the endoplasmic reticulum stress response. *Genes Cells* 2004, 9:457-469.

[108] Sai X, Kokame K, Shiraishi H, Kawamura Y, Miyata T, Yanagisawa K, Komano H: The ubiquitin-like domain of Herp is involved in Herp degradation, but not necessary for its enhancement of amyloid beta-protein generation. *FEBS Lett* 2003, 553:151-156.

[109] Schulze A, Standera S, Buerger E, Kikkert M, van Voorden S, Wiertz E, Koning F, Kloetzel PM, Seeger M: The Ubiquitin-domain Protein HERP forms a Complex with Components of the Endoplasmic Reticulum Associated Degradation Pathway. *J Mol Biol* 2005, 354:1021-1027.

[110] Della Fazia MA, Castelli M, Bartoli D, Pieroni S, Pettirossi V, Piobbico D, Viola-Magni M, Servillo G: HOPS: a novel cAMP-dependent shuttling protein involved in protein synthesis regulation. *J Cell Sci* 2005, 118:3185-3194.

[111] Hopfner R, Mousli M, Jeltsch JM, Voulgaris A, Lutz Y, Marin C, Bellocq JP, Oudet P, Bronner C: ICBP90, a novel human CCAAT binding protein, involved in the regulation of topoisomerase IIalpha expression. *Cancer Res* 2000, 60:121-128.

[112] Mousli M, Hopfner R, Abbady AQ, Monte D, Jeanblanc M, Oudet P, Louis B, Bronner C: ICBP90 belongs to a new family of proteins with an expression that is deregulated in cancer cells. *Br J Cancer* 2003, 89:120-127.

[113] Unoki M, Nishidate T, Nakamura Y: ICBP90, an E2F-1 target, recruits HDAC1 and binds to methyl-CpG through its SRA domain. *Oncogene* 2004, 23:7601-7610.

[114] Citterio E, Papait R, Nicassio F, Vecchi M, Gomiero P, Mantovani R, Di Fiore PP, Bonapace IM: Np95 is a histone-binding protein endowed with ubiquitin ligase activity. *Mol Cell Biol* 2004, 24:2526-2535.

[115] Kramer A, Mulhauser F, Wersig C, Groning K, Bilbe G: Mammalian splicing factor SF3a120 represents a new member of the SURP family of proteins and is homologous to the essential splicing factor PRP21p of Saccharomyces cerevisiae. *Rna* 1995, 1:260-272.

[116] Tsukahara M, Suemori H, Noguchi S, Ji ZS, Tsunoo H: Novel nucleolar protein, midnolin, is expressed in the mesencephalon during mouse development. *Gene* 2000, 254:45-55.

[117] May MJ, Larsen SE, Shim JH, Madge LA, Ghosh S: A novel ubiquitin-like domain in IkappaB kinase beta is required for functional activity of the kinase. *J Biol Chem* 2004, 279:45528-45539.

[118] Justesen J, Hartmann R, Kjeldgaard NO: Gene structure and function of the 2'-5'-oligoadenylate synthetase family. *Cell Mol Life Sci* 2000, 57:1593-1612.

[119] Hartmann R, Olsen HS, Widder S, Jorgensen R, Justesen J: p59OASL, a 2'-5' oligoadenylate synthetase like protein: a novel human gene related to the 2'-5' oligoadenylate synthetase family. *Nucleic Acids Res* 1998, 26:4121-4128.

[120] Rebouillat D, Marie I, Hovanessian AG: Molecular cloning and characterization of two related and interferon-induced 56-kDa and 30-kDa proteins highly similar to 2'-5' oligoadenylate synthetase. *Eur J Biochem* 1998, 257:319-330.

[121] Andersen JB, Strandbygard DJ, Hartmann R, Justesen J: Interaction between the 2'-5' oligoadenylate synthetase-like protein p59 OASL and the transcriptional repressor methyl CpG-binding protein 1. *Eur J Biochem* 2004, 271:628-636.

[122] Zhao X, Ueba T, Christie BR, Barkho B, McConnell MJ, Nakashima K, Lein ES, Eadie BD, Willhoite AR, Muotri AR, Summers RG, Chun J, Lee KF, Gage FH: Mice lacking methyl-CpG binding protein 1 have deficits in adult neurogenesis and hippocampal function. *Proc Natl Acad Sci U S A* 2003, 100:6777-6782.

[123] Hartmann-Petersen R, Gordon C: Integral UBL domain proteins: a family of proteasome interacting proteins. *Semin Cell Dev Biol* 2004, 15:247-259.

In: The Ubiquitin Proteasome System… ISBN 978-1-60021-749-4
Eds: Mario Di Napoli and Cezary Wojcik, pp. 117-136

Chapter 6

STRUCTURE AND FUNCTION OF THE 20S PROTEASOMES

Manila Amici and Anna Maria Eleuteri*

Post-Graduate School in Clinical Biochemistry, Department of Molecular, Cellular and Animal Biology, University of Camerino, 62032 Camerino (MC), Italy.

ABSTRACT

The 20S proteasome, a 700 kDa multicatalytic proteinase complex, is responsible for the extralysosomal protein degradation that occurs in the cytosol and nucleus of eukaryotic cells. It represents the proteolytic core of the 26S proteasome, a 2000 KDa elongated structure formed by the 20S capped, at each side, by the 19S regulatory complex (also called PA700). The 26S complex is involved in the ATP, ubiquitin-dependent and ubiquitin-independent proteolytic pathways. The proteasome constitutes up to 1% of protein in the cells and the free 20S proteasomes are the major portion of the total amount of proteasomes. Its molecular architecture is extremely conserved from archaebacteria to higher eukaryotes and is organized in four stacked 7-membered rings of α and β subunits, in a cylinder-like shape. The two inner rings are composed of β subunits, harbouring the active sites, flanked by the two outer rings made up of non-catalytic α subunits which regulate the substrate access through the opening of the outer ring and the binding of regulators. The 20S proteasome is a member of the N-terminal nucleophile (Ntn)-hydrolases family. Its N-terminal threonine residues are exposed as the nucleophile in peptide bond hydrolysis. The three β subunits, β1, β5 and β2 (also called Y/delta, X and Z, respectively) express the three catalytic activities, designated peptidyl-glutamyl peptide hydrolyzing, chymotrypsin-like and trypsin-like, based on the ability to cleave peptide bonds on the carboxyl side of hydrophobic, basic and acidic aminoacids, respectively. Furthemore, two additional activities cleaving bonds after branched chain and small neutral amino acids have been described and called branched chain amino acid

* Correspondence concerning this article should be addressed to Dr. Anna Maria Eleuteri, PhD; Post-Graduate School in Clinical Biochemistry, University of Camerino, Via Scalzino, 7; 62032 Camerino (MC) – Italy. Phone: +39 0737 403267; Fax: +39 0737 403350; E-mail: annamaria.eleuteri@unicam.it.

preferring and small neutral amino acid preferring. They enable the 20S proteasome to degrade alone a wide variety of protein substrates: poorly folded or unfolded proteins and oxidized proteins characterized by an increased surface hydrophobicity. Under the influence of γ-interferon, a major immunomodulatory cytokine, vertebrate proteasomes assemble catalytically-active subunits, named β5i, β1i and β2i (also called LMP7, LMP2 and MECL1, respectively) which replace their constitutive homologues, β5, β1 and β2, respectively, and associate to a regulatory particle, PA28, (or 11S regulatory complex) also induced by γ-interferon. Such a complex has been demonstrated to be specialized in generating MHC class I antigenic peptides. This review focuses on recent progress concerning the structure, including the assembly pathway, and the enzymatic activities that are involved in physiological/pathological functions exerted by the eukaryotic 20S proteasomes in the cell.

Keywords: 20S proteasome, structure, assembly, catalytic activities, physiological functions, role in pathologies.

Abbreviations:

AD, Alzheimer's disease; APP, amyloid protein precursor; BrAAP, branched chain amino acid preferring; ChT-L, chymotrypsin-like; CNS, central nervous system; DCI, 3,4-dichloroisocoumarin; DFP, diisopropylfluorophosphate; HNE, 4-hydroxy-2-nonenal; Hsp, heat shock protein; IFNγ, interferon-γ; IRI, Ischemia-Reperfusion Injury; LMP, low molecular weight protein; MHC, major histocompatibility complex; Ntn, N-terminal nucleophile; Nob1p, Nin One Binding Protein; PARP, poly-ADP-ribose polymerase; PGPH, peptidyl-glutamyl peptide hydrolyzing; SDS, sodium dodecylsolphate; SNAAP, small neutral amino acids preferring; TAP, transporter associated with antigen presentation; T-L, trypsin-like; Ump1, underpin the maturation of the proteasome; Z, benzyloxycarbonyl.

Introduction

Over the past few years intensive research has focused on the 20S proteasome and its molecular structure and function. In eukaryotes the proteasomes are ubiquitous and essential for cellular viability; they represent the major proteolytic activity in mammalian cells and constitute up to 1% of the cell protein [1-5].

The 20S proteasome was originally discovered by several laboratories, working in different areas, each of these giving different names to the particle. In 1988 the name 'proteasome' was proposed to indicate the proteolytic and particulate nature of the complex and the definitions 20S proteasome and 26S proteasome are currently used to distinguish the two forms of the molecule [6-8].

The 20S proteasome is present in the cytoplasm and nucleus of the eukaryotic cells and, associated with regulatory particles, it forms the 26S proteasome of which represents the catalytic core (for details, see Chapter 7). The 26S proteasome is an essential component of the ubiquitin-dependent and ubiquitin-indipendent proteolysis. The degradation of most proteins is generally associated to the ubiquitin-dependent pathway, however the possibility that proteins can be degraded without ubiquitination either by the 26S proteasome in the presence of ATP or, directly, by the free 20S proteasome exists. The free 20S proteasome constitutes a major portion of the total amount of proteasomes in the cells evidencing an independent involvement of this molecule in protein degradation [9]. Among its substrates there are natural unfolded proteins, some short-lived regulatory proteins, oxidized proteins, misfolded, mutated or damaged proteins. The free 20S proteasome is present in the cells in a latent form, in fact it can be artificially activated in various ways such as low concentration of sodium dodecylsulfate (SDS), heat exposure, low ionic strength of solution, small hydrophobic peptides [10-12].

Furthermore, interferon-γ (IFNγ) induces the replacement of three constitutively expressed catalytic subunits with three inducible subunits, resulting in the so-called 'immunoproteasomes', responsible for efficient immunological processing of intracellular antigens [13-14].

The present article provides a review of the eukaryotic 20S proteasomes' structure, assembly, proteolytic activities and physiological/pathological functions.

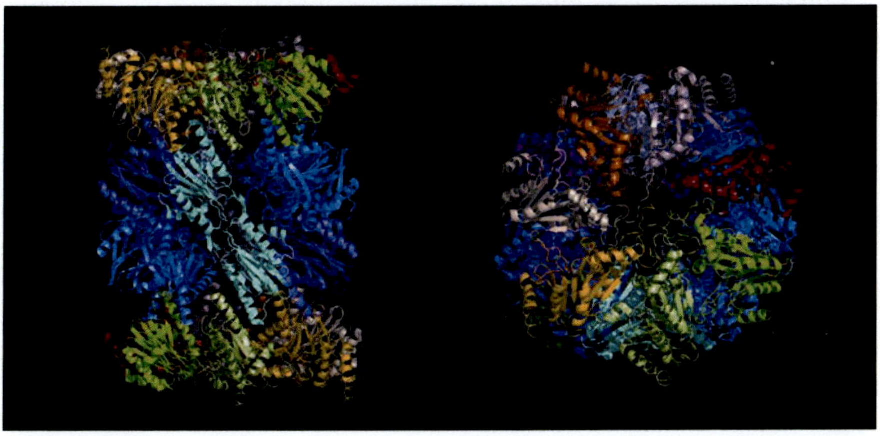

Figure 1. Structure of the bovine 20S proteasome. Ribbon drawings of the mammalian 20S proteasome complex indicating the $\alpha_{1\text{-}7}\beta_{1\text{-}7}\beta_{1\text{-}7}\alpha_{1\text{-}7}$ structure (left: side view; right: top view). Drawings made with PyMOL Molecular Graphics System, Version 0.97 © 2004 by DeLano Scientific LLC. The atomic parameters for the 20S proteasome from bovine liver derive from the Brookhaven Protein Data Bank (code 1IRU).

STRUCTURE

In 1995 the X-ray crystal structure of the 20S proteasome from the *Thermoplasma acidophilum* showed for the first time the architecture of the complex at atomic resolution [15]. The molecule has a cylindrical shape composed of four stacked heptameric rings. In archea two different genes code for subunits termed α and β [16] assembled into a $\alpha_7\beta_7\beta_7\alpha_7$ structure. This general architecture is also found in eukaryotes but the α and β subunits have diverged into seven different subunits present in unique locations in two copies per particle leading to a $\alpha_{1-7}\beta_{1-7}\beta_{1-7}\alpha_{1-7}$ composition. Thus the multiple axes of symmetry of the archea proteasome are reduced to a C2 symmetry in the eukaryotic proteasome (Figure 1).

The α and β subunits have a common fold characterized by a sandwich of two β-sheets each consisting of five strands, surrounded by two α-helixes on each side (Figure 2). The H1 and H2 helixes mediate the interaction of α- and β-rings; H3 and H4 provide contacts between the β-rings. The α subunit has, at the N-terminus, the HO helix which in the β subunit precursor is substituted by a prosequence that will be removed during maturation [15].

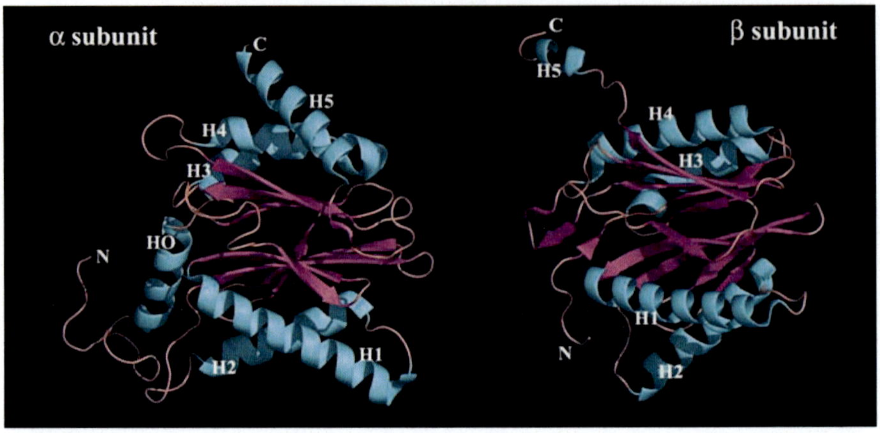

Figure 2. Ribbon diagram of an individual α and an individual β subunits of the mammalian proteasome. The two subunits are shown in similar orientation. The major difference resides in the N-terminal HO helix of the α subunit. The N-terminus of the β subunit is formed by the catalytic threonine. Drawings made with PyMOL Molecular Graphics System, Version 0.97 © 2004 by DeLano Scientific LLC. The atomic parameters for the 20S proteasome from bovine liver derive from the Brookhaven Protein Data Bank (code 1IRU).

As shown in Figure 3, a systematic nomenclature for the subunits location within the rings has been proposed: subunits are numbered α1 to α7 and β1 to β7, and those related by C2 symmetry are distinguished by the prime symbol (α1' to α7' and β1' to β7'). In mammals, the complexity of proteasomes is further enhanced by the fact that, after IFNγ stimulation, three of the constitutive subunits (β1, β2 and β5) can be replaced to form immunoproteasomes by closely related subunits termed β1i, β2i and β5i [17] (Figure 3).

Crystal structures of the yeast [17] and the bovine 20S proteasomes [18] clarify that the complex topology is conserved from yeast to mammals, however some structural features, such as C-terminal extension and internal loops necessary to determine the fixed subunit arrangement, are different in the yeast 20S particle compared to the bovine one. This evidence seems to be related to the ability of mammals to assemble either the constitutive or the inducible subunits in a given location.

The overall size of the four-layered proteasome cylinder was determined to be 14.8 nm in lenght and 11.3 nm in diameter. A tight packing of the subunits towards the outer solvent seems to make the cylinder impermeable to substrates from the side, leaving entrance only through the central channel [19] which, in eukaryotes, is plugged by the interdigitating N-terminal tails of α subunits. This structure corresponds to the latent state of the 20S proteasome obtained by certain purification procedures [20]. Association of the 20S particle with activator complexes, such as PA28 and PA700, seems to trigger channel opening.

The central channel of the cylinder is characterized by three large cavities: a central chamber, where the active sites are sheltered, flanked by two slightly smaller antechambers. The catalytic cavity is described by the two β-rings whereas the outer chambers are formed jointly by one α and one β ring.

The hydrophobic character and the small size of the gates forming these cavities explain the finding that only poorly folded or unfolded proteins are substrates of the 20S proteasome [21].

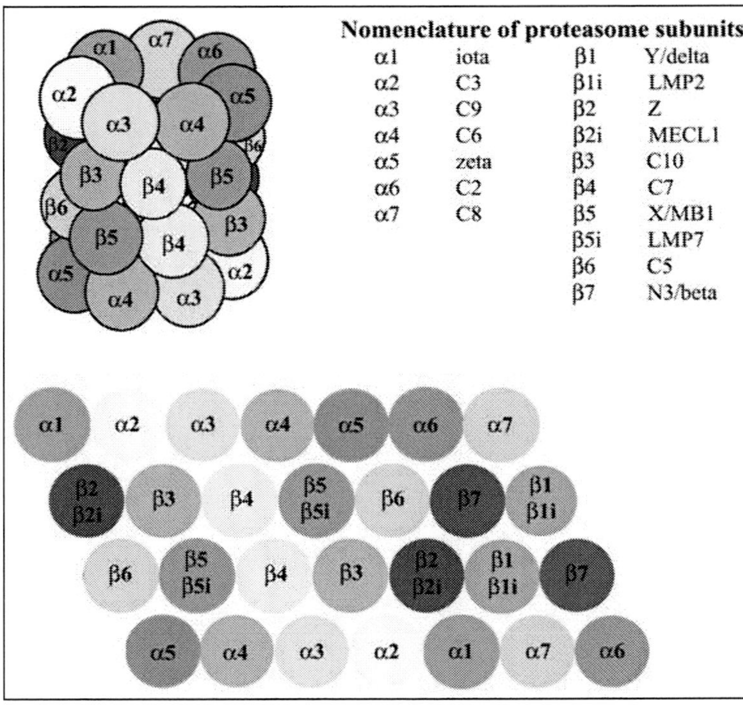

Nomenclature of proteasome subunits

α1	iota	β1	Y/delta
α2	C3	β1i	LMP2
α3	C9	β2	Z
α4	C6	β2i	MECL1
α5	zeta	β3	C10
α6	C2	β4	C7
α7	C8	β5	X/MB1
		β5i	LMP7
		β6	C5
		β7	N3/beta

Figure 3. Model of the 20S human proteasome. Schematic representation of the eukaryotic 20S proteasome showing the seven α and the seven β subunits location into a barrel shaped structure and a roll-out vision. The systematic nomenclature of the subunits is reported in the table: IFN-γ inducible β subunits are indicated as βi.

Assembly

In eukaryotes the early steps of the assembly pathway are not fully understood. It seems that α subunits have the tendency to form ring structures. The assembly of complexes, constituted by one to four stacked heptameric α-rings, has been documented by electron microscopy for the *Trypanosoma brucei* 20S proteasome expressed in *E. coli*. [22-23].

Attempts to identify the early intermediates in the eukaryotic proteasome assembly have not succedeed. It has been recently published that two chaperones, named PAC1 and PAC2, mediate the formation of α-rings. The PAC1-PAC2 complex associates with the α-subunits before α-rings are complete and functions as a scaffold for α-ring assembly [24]. Works on the assembly process of yeast and mammalian proteasomes have characterized more complex intermediates [25-28]. An incomplete precursor complex is made by a seven-membered α ring and β2, β3 and β4 subunits in which the C-terminal extension of β2 wraps around β3 stabilizing the intermediate. Such a complex is not able to dimerize making it a more long-lived precursor [29]. Its association with the other β subunits leads to the ability to dimerize into preholoproteasome in which the C-terminal extension of the β7 subunit inserting between β1 and β2 subunits in the opposing ring has a key role in the efficiency of the process [30]. Experiment of RNAi of β5 demonstrated that the presence of this subunit is necessary for a correct assembly of the 20S proteasome [31]. In 1998 a short-lived protein called Ump1 was identified [28] as a chaperone necessary for a correct proteasome assembly and maturation. Another interacting protein Nob1p (Nin One Binding Protein) has been indicated as crucial for the maturation of the 20S proteasome by Ump1 [32]. It had been shown that Ump1 and the propeptide of β5 are both indispensable for dimerization of halfproteasome precursor complexes and the subsequent maturation steps leading to the 20S proteasome functionality. In fact the β5 propeptide and the Ump1 mutually induce conformational or positional changes of each other during the dimerization process. So far Ump1 proteins have been identified only in eukaryotes most likely related to the presence of seven distinct β subunits, five of which expressed with propeptides (β1, β2, β5, β6, β7). Studies using the ump1-Δ mutant showed that Ump1 partecipates in the dimerization pathway keeping the halfproteasomes in the correct conformation and it triggers the maturation of inactive β subunit precursors. In the catalytic β subunits the N-terminal propeptides are autocatalitically removed, whereas they are processed by neighboring active subunits in the β6 and β7 subunits following the 20S proteasome assembly. Propeptides seem to play a key role in protecting the N-terminal threonin residues from acetylation until the dimerization process has been completed [33-34]. Furthermore they are not important for the β subunits positioning within the β-rings, however, as discussed above, they cooperate with Ump1 to obtain a correct assembly of the 20S complex. They are indispensable for the incorporation of γ-interferon inducible β subunits into the immunoproteasomes. It looks like that this assembly process proceeds in a defined, tightly regulated, cooperative pathway which is different from the constitutive β subunits assembly. First the β1i subunit is incorporated, a necessary event for the next correct assembly of the β2i subunit. In order to have β5i inserted in the immunoproteasome structure, the presence of β5/X is also important. The above described sequence of incorporations is driven by the propeptides. It is now clear that the immunoproteasome or the constitutive proteasome biogenesis relays on several mixed proteasome precursors containing both inducible and housekeeping β subunits [35]. The interactions of Ump1 with the propeptides of constitutive and IFNγ-inducible subunits, being mutually exclusive, affect the subunit addition and assembly specificity.

CATALYTIC MECHANISM

The 20S proteasome belongs to the N-terminal nucleophile hydrolases (Ntn-hydrolases), a class of enzymes which perform their catalytic activities relaying on the N-terminal aminoacid residue as nucleophile [15]. Mutation analyses of the *Thermoplasma acidophilum* involving all the serines, two histidines, the only cystein and two aspartate residues of the β subunit led to the conclusion that the 20S proteasome did not belong to the four classical proteases classes. The only mutation which produced inactivation was the deletion of the N-terminal threonine or its mutation to an alanine [36], making the enzyme also deficient in autolysis. The N-terminal threonine/serine mutant conserved intact its activity but was more sensitive to the DCI inactivation [36]. Crystallographic data obtained for the 20S archeaproteasome in complex with the inhibitor N-acetyl-Leu-Leu-nLeu-CHO showed that the aldehyde group of the inhibitor was in proximity of the OH group of the N-terminal threonine, demostrating that the hydroxyl group functions as the nucleophile [15]. Another aminoacid residue necessary to the catalytic activity is the Lys33: its exchange with either an alanine or an arginine residue deactivates the enzyme. At the beginning it was thought the Lys33 acted as proton acceptor but, at neutral pH, it is charged, therefore unable to accept protons. It is expected that it lowers the pKa of the amino group of the Thr1 by its electrostatical potential, facilitating the proton acceptance [17,33]. The catalytic mechanism involves also the residues Glu/Asp17, Ser129, Asp166 and Ser169, with the glutamate/aspartate and lysine aminoacids playing the roles of histidine and aspartate in serine proteases [15,17,36-37].

The hydroxyl group of Thr1 attacks the carbonyl carbon of the peptide bond to be cleaved, while its amino group acts as the proton acceptor which enhances the nucleophilicity by stripping the proton from the side chain hydroxyl; a water molecule is likely to mediate the proton transfer.

During the autocatalytic cleavage of the β subunits prosequences the N-terminal amino group is not available as proton acceptor: the Thr1Oγ of the β subunit precursor adds to the carbonyl carbon of Gly-1 and a water molecule is supposed to play the base function [38].

Studies with specific inhibitors and mutant forms of archeal, bacterial, yeast and mammalian β subunits have shown a common catalytic mechanism in all proteasomes [27,37,39-43]. Nevertheless, in eukaryotic proteasomes only three β subunits (β1, β2 and β5), after the autocatalytic removal of the N-terminal peptide extensions, undergo activation, while it is not completely known the role of the other four β type subunits.

ACTIVITIES OF THE 20S PROTEASOMES

Several studies using short synthetic substrates defined the primary hydrolytic activities of the proteasome based on the nature of the P1 residue directly adjacent to the scissible bond [5,7,44].

A 'chymotrypsin-like' (ChT-L, cleaving after hydrophobic residues), a 'trypsin-like' (T-L, cleaving after basic residues) and a 'peptidylglutamyl-peptide hydrolysing' (PGPH, cleaving after acidic residues) activities were initially proposed [20,45-46]. Each of these classical activities has been assigned to a specific β subunit according to a number of mutagenesis, inhibition and X-ray diffraction studies [2,27,41-42,47-49]. The β1 subunit has been linked to the PGPH activity, the β2 subunit to the T-L activity and the β5 subunit to the ChT-L activity.

The main and one of the most active catalytic components of the proteasome is the ChT-L activity that can be fully expressed in the presence of certain activators like as low concentrations of SDS, fatty acids, some phospholipids [50] and the protein activator PA28. It is also sensitive to inactivation by inhibitors that act directly to the active site, acylating the hydroxyl group of Thr1 [51]. Examples of them are: DCI [52], lactacystin [39], epoxomicin [53], peptidyl vinyl sulfones [54], peptidyl aldehydes [55-56] and peptide boronates [57]; moreover it is the only catalytic component inhibited by the serine protease inhibitor DFP [58] (Table 1).

Table 1. Proteasome inhibitors

Class	Compound	ChT-L (K_i)	T-L (K_i)	PGPH (K_i)	BrAAP Ki
Peptide aldehydes	Z-LLL-CHO (MG132)	4[a] nM 6.9[b] μM	2.76[c] μM	900[c] μM	6.8[d] μM
	Z-LGF-CHO	1.1[b] μM			NI[e]
	Z-LAF-CHO	1.5[b] μM			NI[e]
	Z-LLF-CHO	1.6[f] μM			110[g] μM
	Z-LGL-CHO	1.76[h] μM			76.4[h] μM
	Z-FF-CHO	4.0 μM (spleen) 56 μM (pituitary)[i]			1.46 μM (spleen) NI (pituitary)[i]
	Z-GPFF-CHO	5.6 μM (spleen) 58 μM (pituitary)[i]			1.2 μM (spleen) 310 μM (pituitary)[i]
	Z-GPFL-CHO	16.2 μM (spleen) 52.7 μM (pituitary)			1.04 μM (spleen) 4.66 μM (pituitary)[i]
	Z-IG(OtBu)AL-CHO (PSI)	IC50 = 250[j] nM			
Peptide boronates	Z-LLL-B(OH)2 (MG262)	0.03[a] nM			
	Pyrazylcarbonyl-FL-B(OH)2 (PS341)	0.62[a] nM			
		Kass ($M^{-1}s^{-1}$)	Kass ($M^{-1}s^{-1}$)	Kass ($M^{-1}s^{-1}$)	
3,4-dichloro-isocumarin (DCI)		1201			NI
Lactacystin and derivates	Lactacystin	194[m]	10[m]	4.2[m]	3.7[g]
	Clasto-lactacystin-β-lactone	8500[n]	253[n]	37[n]	
Peptide vinyl sulfones	Nip-LLL-CH-SO2-CH3 (NLVS)	13400[m]	422[m]	100[m]	
	YLLL-CH-SO2-CH3 (YLVS)	1500[m]	560[m]	20[m]	
Peptide epoxyketones	Dihydroeponemycin	65[o]	4.4[o]	61[o]	
	Ac(CH3)-IIWL-EX (Epoxomicin)	37000[p]	79[p]	37[p]	
	Ac-hFLFL-EX (YU101)	166000[p]	7.1[p]	21[p]	

a: [56]; b: [37]; c: [49]; d: [7]; e: [61]; f: [62]; g: [63]; h: [43]; i: [64]; j: [53]; l: [65]; m: [66]; n: [67]; o: [51]; p: [68]; NI: no inhibition.

The T-L activity prefers to cleave after Arg residues with respect to Lys. It shows more resistance to inhibition by lactacystin, DCI, peptidyl vinyl sulfones and epoxomicin than the ChT-L component [39,52-54]. It is specifically inhibited by leupeptin and shows a more sensitive inactivation by thiol-blocking agents than the other catalytic components [7,44,59] (Table 1).

The PGPH activity, also known as caspase-like or post-acidic activity, cleaves bonds after glutamate and aspartyl residues and it is more resistant to inactivation by active site-directed inhibitors than the ChT-L. In the constitutive proteasome it is in a latent state and can be activated by low concentration of SDS, fatty acids and magnesium ions [60-62].

Moreover, through the use of peptide and selected protein substrates, several works investigated the cleavage specificity of the 20S proteasomes leading to the evidence that residues beyond P1 (P2, P3, P4) may determine the absolute substrate specificity of the multiple active sites of the proteasome [20,68,71-72]. It has been demonstrated that the residues around the active site affect the substrate selectivity for the P1 residue influencing degradation [68,73-74] which, in addition, undergoes the interferences of neighboring subunits of the catalytic ones [34,37,42].

Mammalian proteasomes show additional endopeptidase activities: one cleaving preferentially after branched-chain amino acids (BrAAP activity) and the other after small neutral amino acids (SNAAP activity) [75]. At the beginning, these two activities were considered an expression of overlapping specificities of the classical three catalytic components and the result of cooperation between various subunits [20].

The discovery that, in eukaryotes, under the influence of IFNγ, the replacement of the housekeeping catalytic subunits with the inducible counterparts occurs, has partially disclosed the site responsible for the BrAAP activity. In fact it has been demonstrated that the assembly of β5i/LMP7 enhances the activity cleaving bonds after branched chain and aromatic amino acids [66,76-78].

This catalytic component is susceptible to inactivation by some peptide aldehyde inhibitors such as Z-LLF-CHO, Z-GPFL-CHO, Z-GPFF-CHO and Z-FF-CHO, whereas it is activated by DCI and PA28 [79-80].

Furthermore, the immunoproteasome is characterized by different specificities compared to the constitutive complex. In fact, an almost complete loss of the PGPH activity, due to the replacement of β1/Y subunit with β1i/LMP2 which expresses a ChT-L activity, and a decreased ChT-L component, caused both by the replacement of β5/X with β5i/LMP7 and by the β1i/LMP2 incorporation, are evident [20,66,77].

Besides the peptidase activities measured with short synthetic peptides, the 20S proteasome functionality has been investigated using macromolecular substrates such as β-casein, lysozyme, histone H3, insulin β-chain, and many others. From these studies it cames out that the rate-limiting step in the 20S proteasome protein processing is the substrate entrance to the catalytic chamber. In eukaryotic 20S proteasomes, the gate in correspondence of the α-rings closes the access to the interior cavity. Several reports have demonstrated that unfolded, partially folded proteins and macromolecular substrates characterized by an increased surface hydrophobicity are degraded by the 20S complex. Furthermore, substrates with flexible regions can enter the 20S in a hairpin conformation: a paper of Liu *et al.* showed that p21 and α-synuclein were proteolized when they were made as covalently closed circular constructs with no free termini referring to the ability of the 20S proteasome to cleave substrates in an internal region even when the ends are unable to enter the channel [81].

The mechanism of how the 20S proteasome degrades protein substrates is not fully understood. It is now established that it doesn't degrade proteins randomly, on the contrary the degradation process is specific and allosterically regulated [82]. In fact, the binding of hydrophobic peptides to not catalytic sites promotes peptides hydrolysis by all the active sites [12]. This observation is consistent with a two-state model, proposed by Osmulski and Gaczynska [83], which involves two allosteric states: a closed-gate barrel-like conformation called R and an open-gate cylinder-like conformation called T, stabilized by hydrophobic substrates. Binding of several proteins to the 20S particle affect selectively distinct catalytic components, either activating or inhibiting them, through a mechanism based on allosteric effects. Examples of that are: PA28γ [84] PA200 [85] and histone H3 [86] as activators and Hsp90 [87] and Ritonavir [88] as inhibitors.

The size of degradation products believed to range from 7-9 amino acids [15] has been later found to be comprised within 3-30 residues [89]. Based on the observation that regulatory components binding the 20S proteasome influence the product size modifying the axial gate [90], it is reasonable to suppose that the dimensions of the axial gates formed by the α-rings determine the products size.

It has been demonstrates that immunoproteasomes gain properties which facilitate the generation of 7-9 amino acid-containing class I antigenic peptides [91-92]. Most class I antigenic peptides have aromatic, branched chain or basic residues at the C-terminus [93]. These features fit perfectly with the specificities acquired with the incorporation of the IFNγ inducible subunits in the 20S proteasome [66,94-96].

This evidence is supported by experiments in which, following exposure to proteasome inhibitors, cells failed to assemble, at normal rates, MHC class I stable complexes because of the lack of antigenic peptides [3].

Interestingly, IFNγ induces the synthesis of most MHC class I antigen presentation pathway components, such as the βi subunits, PA28 subunits, TAP proteins and the MHC class I heavy chain. PA28 caps on both ends the 20S immunoproteasome inducing its activation through a change in the complex conformation to a T state [97].

Functions of the 20S Proteasomes

Protein degradation by the 20S proteasome represents an uncompletely explored field in fact the number of proteins recognized to be hydrolyzed by the 20S proteasome, independently from ubiquitin and ATP, is increasing.

It is now accepted that the 20S proteasome plays a central role in the overall intracellular proteolysis. Poorly folded, unfolded and oxidized proteins [98-101], as well as proteins with the reduction of disulfide bonds are susceptible to degradation [86]. Oxidation of proteins causes an exposure of hydrophobic moieties from the inner part of the protein to the surface, rendering them prone to the selective 20S proteasome removal and preventing the accumulation of highly oxidized and cross-linked proteins, which are no longer attackable.

The recognition process preceding the degradation involves the binding of the hydrophobic patches of the partially denaturated substrate to the α-subunits promoting the opening and the activation of the 20S particle [101].

The primary role of the 20S proteasome in the selective degradation of oxidized proteins is of great physiological and pathological relevance. In ageing, a decline of the 20S proteasome activity parallels an increase of oxidized proteins and protein aggregates, which, in turn, inhibit the proteasome functionality leading to cell damage and death (see Chapter 22). Moreover, in aging spinal cord a 20S proteasome inhibition together with a loss of proteasome expression have been documented [102].

It has been reported that the 20S proteasome is altered in structure and function in some areas of the CNS of patients affected by Parkinson's disease [103-104; Chapters 25 and 31].

Specifically, a loss of α subunits and a reduction of the ChT-L, T-L and PGPH activities were shown. Taking into account that the clearance of oxidized proteins and α-synuclein is relied on the 20S proteasome [105], its impairment generates an accumulation of both α-synuclein and oxidized proteins which aggregate in neurons contributing to an overall dysfunction and a further inhibition of the 20S complex. This mechanism could participate to the pathogenesis of Parkinson's disease.

Several evidences have demonstrated the involvement of the 20S proteasome in another neurodegenerative disorder, the Alzheimer's disease (AD) (see Chapters 28 and 30). It is now clear that the 20S proteasome is responsible for the processing of amyloid protein precursor (APP) [106] and, *in vitro*, is able to hydrolyze tau proteins [107-108]. A loss of ChT-L and PGPH activities, not associated with a decrease in subunit expression, in the AD brain has been reported [109]; in particular, a significant decrease was observed in specific regions of the brain, such as the hippocampus and parahippocampal gyrus, superior and middle temporal gyri and inferior parietal lobule. In addition it has been published that β-amyloid peptides inhibit the proteasome functionality [110] and that an increase in oxidative stress occurs in AD brain [111] leading to an accumulation of oxidized and aggregated proteins, among them tau proteins [112], having detrimental effects on both the 20S proteasome and cell viability.

A role for oxidative stress and proteasomal inhibition are reported also during Ischemia-Reperfusion Injury (IRI): a time-dependent decrease in proteasome activity, which is not associated with decreased expression of proteasome subunits, has been demonstrated after cerebral ischemia-reperfusion injury [113; and Chapter 19]. In particular, the IRI-induced proteasome inhibition has been correlated both to the lipid peroxidation and the lipid peroxidation product HNE, and to the increasing nitric oxide and nitrosylated glutathione [114-115].

On the contrary, proteasome complexes isolated from erythrocytes and bovine organs, upon exposure to various oxidative conditions are stimulated, reinforcing the theory of the proteasome key role in the cellular antioxidant defense [116-118]. It is evident that the constitutive 20S proteasome has a different susceptibility to oxidants compared the immunoproteasome, with the former being activated and the latter being inhibited.

As described above, the 20S proteasome is also present in the nucleus, where it is responsible for the removal of oxidized proteins and among them histones [119-120]; it seems to be highly regulated by nuclear proteins like histone H3 [86] and the poly-ADP-ribose polymerase (PARP) [121]. In fact, PARP binds to the DNA strand breaks, initiating the poly-ADP ribosylation and activating the proteasome: this event could contribute to the specific removal of damaged histones, playing an important role in the secondary antioxidant defenses.

The turnover of $p21^{WAF1/CIP1}$, a cyclin-dependent kinase inhibitor, which is a regulator of the cell-cycle progression, is mediated by either the 26S or the 20S proteasome; in detail, when the latter pathway is undertaken p21 binds to the α7 subunit, triggering the opening of the α-ring gate and promoting a rapid degradation of itself.

As already mentioned, the 20S proteasome plays a key role in immunosurveillance against viruses and cancer [122] being responsible for the generation of most antigenic peptides. In particular, several studies, using specific proteasome inhibitors, have demonstrated the involvement of the IFNγ inducible 20S proteasome in this process. Peptidyl aldehydes and lactacystin have been shown to block the generation of class I antigenic peptides and their presentation on the cell surface to cytotoxic lymphocytes [3].

Conclusions

In this review we have tried to give a picture of the 20S proteasome structure and functionality and the way they influence each other. It is evident that the 20S proteasome plays a pivotal role in the overall protein turnover, independently from ATP and ubiquitin. Some proteins may have a dual fate: either the 26S proteasome degradation or the 20S degradation, whereas poorly folded, unfolded and oxidized proteins are selectively removed by the 20S complex. Therefore, in these cases, substrates can bind to α-subunits and promote the opening of the gate and subsequent activation of the particle. In addition the 20S complex can be stimulated also through the interaction with other proteins, such as histones and hydrophobic peptides, bypassing the action of the binding of regulators to the 20S proteasome. In the future the number of proteins degraded by the 20S proteasome will increase as well as the mechanisms which modulate differently its functionality depending on the various cellular compartments.

Further studies will be needed to better understand how the 20S proteasome is able to perform the highly controlled process of protein breakdown, also with the aim to design specific inhibitors affecting a single catalytic component.

A wider knowledge both of the 20S proteasome regulation and of its involvement in the genesis and outcomes of several pathologies will allow to consider the 20S particle as a target for the prevention and treatment of diseases.

Acknowledgements

To Professor Marian Orlowski whose pioneering work on the proteasome inspired most of the research cited in this review. AM Eleuteri is personally grateful to him for his guidance and scientific support. The authors wish to thank Dr. Matteo Mozzicafreddo for technical assistance.

References

[1] Ciechanover A: The ubiquitin-proteasome proteolytic pathway. *Cell* 1994, 79:13-21.

[2] Rock KL, Gramm C, Rothstein L, Clark K, Stein R, Dick L, Hwang D, Goldberg AL: Inhibitors of the proteasome block the degradation of most cell proteins and the generation of peptides presented on MHC class I molecules. *Cell* 1994, 78:761-771.

[3] Hilt W, Heinemeyer W, Wolf DH: Studies on the yeast proteasome uncover its basic structural features and multiple in vivo functions. *Enzym Protein* 1993, 47:189-201.

[4] Tanaka K: Molecular biology of proteasomes. *Mol Biol Rep* 1995, 21:21-26.

[5] Tanaka K, Li K, Ichihara A, Waxman L, Goldberg AL: A high molecular weight protease in the cytosol of rat liver. I. Purification, enzymological properties, and tissue distribution. *J Biol Chem* 1986, 261:15197-15203.

[6] DeMartino GN, Goldberg AL: Identification and partial purification of an ATP-stimulated alkaline protease in rat liver. *J Biol Chem* 1979, 254:3712-3715.

[7] Wilk S, Orlowski M: Evidence of pituitary cation-sensitive neutral endopeptidase is a multicatalytic protease complex. *J Neurochem.* 1983, 40:842-849.

[8] Arrigo AP, Tanaka K, Goldberg AL, Welch WJ: Identity of the 19S 'prosome' particle with the large multifunctional protease complex of mammalian cells (the proteasome). *Nature* 1988, 331:190-192.

[9] Orlowski M, Wilk S: Ubiquitin-independent proteolytic functions of the proteasome. *Arch Biochem Bioph* 2003, 415:1-5.

[10] Groll M, Bajorek M, Koehler A, Moroder L, Rubin D, Huber R, Glickman MH, Finley D: A gated channel into the proteasome core particle. *Nat Struct Biol* 2000, 7:1062-1067.

[11] Bajorek M, Finley D, Glickman MH: Proteasome disassembly and down regulation is correlated with viability during stationary phase. *Curr Biol* 2003, 13:1140-1144.

[12] Kisselev AF, Kaganovich D, Goldberg AL: Binding of hydrophobic peptides to several non-catalytic sites promotes peptide hydrolysis by all active sites of 20S proteasomes. Evidence for peptide-induced channel opening in the alpha-rings. *J Biol Chem* 2002, 277:22260-22270.

[13] Tanaka K, Kasahara M: The MHC class I ligand-generating system: roles of immunoproteasomes and the interferon-gamma-inducible proteasome activator PA28. *Immunol Rev* 1998, 163:161-176.

[14] Fruh K, Yang Y: Antigen presentation by MHC class I and its regulation by interferon gamma. *Curr Opin Immunol* 1999, 11:76-81.

[15] Lowe J, Stock D, Jap B, Zwickl P, Baumeister W, Huber R: Crystal structure of the 20S proteasome from the archeon T. acidophilum at 3.4 A resolution. *Science* 1995, 28:533-539.

[16] Dahalmann B, Kopp F, Kuehn L, Niedel B, Pfeifer G, Hegerl R, Baumeister W: The multicatalytic proteinase (prosome) is ubiquitous from eukaryotes to archeabacteria. *FEBS Lett* 1989, 17:125-131.

[17] Groll M, Ditzel L, Lowe J, Stock D, Bochtler M, Bartunik HD, Huber R: Structure of 20S proteasome from yeast at 2.4 A resolution. *Nature* 1997, 386:463-471.

[18] Unno M, Mizushima T, Morimoto Y, Tomisugi Y, Tanaka K, Yasuaka N, Tsukihara T: The structure of the mammalian 20S proteasome at 2.75 A resolution. *Structure* 2002, 10:609-618.

[19] Stock D, Ditzel L, Baumeister W, Huber R, Lowe J: Catalytic mechanism of the 20S proteasome of Thermoplasma acidophilum revealed by X-ray crystallography. *Cold Spring Harb Symp Quant Biol* 1995, 60:525-532.

[20] Orlowski M, Wilk S: Catalytic activities of the 20S proteasome, a multicatalytic proteinase complex. *Arch Biochem Biophys* 2000, 383:1-16.

[21] Wenzel T, Baumeister W: Conformational constraints in protein degradation by the 20S proteasome. *Nat Struct Biol* 1995, 2:199-204.

[22] Yao Y, Toth CR,Huang L,Wong ML, Dias P, Burlingame AL, Coffino P, Wang CC: alpha5 subunit in Trypanosoma brucei prpteasome can self-assembleto form a cylinder of four stacked heptameric rings. *Biochem J* 1999, 344:349-358.

[23] Huang L, Shen M, Chernushevich I, Burlingame AL, Wang CC, Robertson CD: Identification and isolation of three proteasome subunits and their encoding genes from Trypanosoma brucei. *Mol Biochem Parasitol* 1999, 102:211-223.

[24] Hirano Y, Hendil KB, Yashiroda H, Iemura S, Nagane R, Hioki Y, Natsume T, Tanaka K, Murata S: A heterodimeric complex that promotes the assembly of mammalian 20S proteasomes. *Nature* 2005, 437: 1381-1385.

[25] Frentzel S, Pesold-Hurt B, Seeling A, Kloetzel PM: 20S proteasomes are assembledvia distinct precursor complexes. Processing of LMP2 and LMP7 proproteins takes place in 13-16S preproteasome complexes. *J Mol Biol* 1994, 236:975-981.

[26] Nandi D, Woodward E, Ginsburg DB, Monaco JJ: Intermediates in the formation of mouse 20S proteasomes: implications for the assembly of precursor beta subunits. *EMBO J* 1997, 16:5363-5375.

[27] Chen P, Hochstrasser M: Autocatalytic subunit processing couples active site formation in the 20S proteasome to completion of the assembly. *Cell* 1996, 86:961-972.

[28] Ramos PC, Höckendorff J, Johnson ES, Varshavsky A, Dohmen RJ: Ump1p is required for proper maturation of the 20S proteasome and becomes its substrate upon completion of the assembly. *Cell* 1998, 92:489-499.

[29] Heinemeyer W, Ramos PC, Dohmen RJ: The ultimate nanoscale mincer: assembly, structure and active site of the 20S proteasome core. *Cell Mol Life Sci* 2004, 61:1562-1578.

[30] Ramos PC, Marques AJ, London MK, Dohmen RJ: Role of C-terminal extensions of subunits beta2 and beta7 in assembly and activity of eukatyiotic proteasomes. *J Biol Chem* 2004, 279:14323-14330.

[31] Wojcik C, DeMartino GN: Analysis of Drosophila 26S proteasome using RNA interference. *J Biol Chem* 2002, 277:6188-6197.

[32] Tone Y, Toh-e A: Nob1p is required for biogenesis of the 26S proteasome and degraded upon its maturation in Saccharomyces cerevisiae. *Genes Dev* 2002, 16:3142-3157.

[33] Arendt CS, Hochstrasser M: Eukaryotic 20S proteasome catalytic subunit propeptides prevent active site inactivation by N-terminal acetylation and promote particle assembly. *EMBO J* 1999, 18:3575-3585.

[34] Jäger S, Groll M, Huber R, Wolf DH, Heinemeyer W: Proteasome β-subunits: unequal roles of propeptides in core particle maturation and a hierarchy of active site function. *J Mol Biol* 1999, 291:997-1013.

[35] Griffin TA, Nandi D, Cruz M, Fehling HJ, Kaer LV, Monaco JJ, Colbert RA: Immunoproteasome assembly: cooperative incorporation of interferon gamma (IFN-gamma)-inducible subunits. *J Exp Med* 1998, 187:97-104.

[36] Seemuller E, Lupas A, Stock D, Lowe J, Huber R, Baumeister W: Proteasome from Thermoplasma acidophilum: a threonine protease. *Science* 1995, 268:579-582.

[37] Heinemeyer W, Fischer M, Krimmer T, Stachon U, Wolf DH: The active sites of the eukaryotic 20 S proteasome and their involvement in subunit precursor processing. *J Biol Chem* 1997, 272:25200-25209.

[38] Ditzel L, Huber R, Mann K, Heinemeyer W, Wolf DH, Groll M: Conformational constraints for protein self-cleavage in the proteasome. *J Mol Biol* 1998, 279:1187-1191.

[39] Fenteany G, Standaert RF, Lane WS, Choi S, Corey EJ, Schreiber SL: Inhibition of proteasome activities and subunit-specific amino-terminal threonine modification by lactacystin. *Science* 1995, 268:726-731.

[40] Mayr J, Seemuller E, Muller SA, Engel A, Baumeister W: Late events in the assembly of 20S proteasomes. *J Struct Biol* 1998, 124:179-188.

[41] Arendt CS, Hochstrasser M: Identification of the yeast 20S proteasome catalytic centers and subunit interactions required for active-site formation. *Proc Natl Acad Sci USA* 1997, 94:7156-7161.

[42] Groll M, Heinemeyer W, Jager S, Ullrich T, Bochtler M, Wolf DH, Huber R: The catalytic sites of 20S proteasomes and their role in subunit maturation: a mutational and crystallographic study. *Proc Natl Acad Sci USA* 1999, 96:10976-10983.

[43] Salzmann U, Kral S, Braun B, Standera S, Schmidt M, Kloetzel PM, Sijts A: Mutational analysis of subunit i beta2 (MECL-1) demonstrates conservation of cleavage specificity between yeast and mammalian proteasomes. *FEBS Lett* 1999, 454:11-15.

[44] Orlowski M, Wilk S: A multicatalytic protease complex from pituitary that forms enkephalin and enkephalin containing peptides. *Biochem Biophys Res Commun* 1981, 101:814-822.

[45] Orlowski M: The multicatalytic proteinase complex, a major extralysosomal proteolytic system. *Biochemistry* 1990, 29:10289-10297.

[46] Cardozo C: Catalytic components of the bovine pituitary multicatalytic proteinase complex (proteasome). *Enzyme Protein* 1993, 47:296-305.

[47] Enenkel C, Lehmann H, Kipper J, Guckel R, Hilt W, Wolf DH: PRE3, highly homologous to the human major histocompatibility complex-linked LMP2 (RING12) gene, codes for a yeast proteasome subunit necessary for the peptidylglutamyl-peptide hydrolyzing activity. *FEBS Lett* 1994, 341:193-196.

[48] Dick TP, Nussbaum AK, Deeg M, Heinemeyer W, Groll M, Schirle M, Keilholz W, Stevanovic S, Wolf DH, Huber R, Rammensee HG, Schild H: Contribution of proteasomal beta-subunits to the cleavage of peptide substrates analyzed with yeast mutants. *J Biol Chem* 1998, 273:25637-25646.

[49] McCormack TA, Cruikshank AA, Grenier L, Melandri FD, Nunes SL, Plamondon L, Stein RL, Dick LR: Kinetic studies of the branched chain amino acid preferring peptidase activity of the 20S proteasome: development of a continuous assay and inhibition by tripeptide aldehydes and clasto-lactacystin beta-lactone. *Biochemistry* 1998, 37:7792-7800.

[50] Wilk S, Chen WE: Synthetic peptide-based activators of the proteasome. Mol Biol Rep 1997, 24:119-124.

[51] Kisselev AF, Goldberg AL: Proteasome inhibitors: from research tools to drug candidates. *Chem Biol* 2001, 8:739-758.

[52] Orlowski M, Michaud C: Pituitary multicatalytic proteinase complex. Specificity of components and aspects of proteolytic activity. *Biochemistry* 1989, 28:9270-9278.

[53] Meng L, Mohan R, Kwok BH, Elofsson M, Sin N, Crews CM: Epoxomicin, a potent and selective proteasome inhibitor, exhibits in vivo antiinflammatory activity. *Proc Natl Acad Sci USA* 1999, 96:10403-10408.

[54] Bogyo M, McMaster JS, Gaczynska M, Tortorella D, Goldberg AL, Ploegh H: Covalent modification of the active site threonine of proteasomal beta subunits and the Escherichia coli homolog HslV by a new class of inhibitors. *Proc Natl Acad Sci USA* 1997, 94:6629-6634.

[55] Figueiredo-Pereira ME, Berg KA, Wilk S: A new inhibitor of the chymotrypsin-like activity of the multicatalytic proteinase complex (20S proteasome) induces accumulation of ubiquitin-protein conjugates in a neuronal cell. *J Neurochem* 1994, 63:1578-1581.

[56] Vinitsky A, Michaud C, Powers JC, Orlowski M: Inhibition of the chymotrypsin-like activity of the pituitary multicatalytic proteinase complex. *Biochemistry* 1992, 31:9421-9428.

[57] Adams J, Behnke M, Chen S, Cruickshank AA, Dick LR, Grenier L, Klunder JM, Ma YT, Plamondon L, Stein RL: Potent and selective inhibitors of the proteasome: dipeptidyl boronic acids. *Bioorg Med Chem Lett* 1998, 8:333-338.

[58] Wagner BJ, Margolis JW, Abramovitz AS, Fu SC: Differential inhibition of two proteolytic activities in bovine lens neutral-proteinase preparations. *Biochem J* 1985, 228:517-519.

[59] Loidl G, Groll M, Musiol HJ, Ditzel L, Huber R, Moroder L: Bifunctional inhibitors of the trypsin-like activity of eukaryotic proteasomes. *Chem Biol* 1999, 6:197-204.

[60] Dahlmann B, Becher B, Sobek A, Ehlers C, Kopp F, Kuehn L: In vitro activation of the 20S proteasome. Enzyme Protein 1993, 47:274-284.

[61] Orlowski M, Cardozo C, Hidalgo MC, Michaud C: Regulation of the peptidylglutamyl-peptide hydrolyzing activity of the pituitary multicatalytic proteinase complex. *Biochemistry* 1991, 30:5999-6005.

[62] Pereira ME, Yu B, Wilk S: Enzymatic changes of the bovine pituitary multicatalytic proteinase complex, induced by magnesium ions. *Arch Biochem Biophys* 1992, 294:1-8.

[63] Cardozo C, Michaud C, Orlowski M: Components of the bovine pituitary multicatalytic proteinase complex (proteasome) cleaving bonds after hydrophobic residues. *Biochemistry* 1999, 38:9768-9777.

[64] Martinez CK, Monaco JJ: Homology of proteasome subunits to a major histocompatibility complex-linked LMP gene. *Nature* 1991, 353:664-667.

[65] Kopp F, Dahlmann B, Hendil KB: Evidence indicating that the human proteasome is a complex dimer. *J Mol Biol* 1993, 229:14-19.

[66] Eleuteri AM, Kohanski RA, Cardozo C, Orlowski M: Bovine spleen multicatalytic proteinase complex (proteasome). Replacement of X, Y, and Z subunits by LMP7, LMP2, and MECL1 and changes in properties and specificity. *J Biol Chem* 1997, 272:11824-11831.

[67] Harper JW, Powers JC: Reaction of serine proteases with substituted 3-alkoxy-4-chloroisocoumarins and 3-alkoxy-7-amino-4-chloroisocoumarins: new reactive mechanism-based inhibitors. *Biochemistry* 1985, 24:7200-7213.

[68] Bogyo M, Shin S, McMaster JS, Ploegh HL: Substrate binding and sequence preference of the proteasome revealed by active-site-directed affinity probes. *Chem Biol* 1998, 5:307-320.

[69] Meng L, Kwok BH, Sin N, Crews CM: Eponemycin exerts its antitumor effect through the inhibition of proteasome function. *Cancer Res* 1999, 59:2798-2801.

[70] Elofsson M, Splittgerber U, Myung J, Mohan R, Crews CM: Towards subunit-specific proteasome inhibitors: synthesis and evaluation of peptide alpha',beta'-epoxyketones. *Chem Biol* 1999, 6:811-822.

[71] Harris JL, Alper PB, Li J, Rechsteiner M, Backes BJ: Substrate specificity of the human proteasome. *Chem Biol* 2001, 8:1131-1141.

[72] Nazif T, Bogyo M: Global analysis of proteasomal substrate specificity using positional-scanning libraries of covalent inhibitors. *Proc Natl Acad Sci USA* 2001, 98:2967-2972.

[73] Cardozo C, Vinitsky A, Michaud C, Orlowski M: Evidence that the nature of amino acid residues in the P3 position directs substrates to distinct catalytic sites of the pituitary multicatalytic proteinase complex (proteasome). *Biochemistry* 1994, 33:6483-6489.

[74] Groll M, Nazif T, Huber R, Bogyo M: Probing structural determinants distal to the site of hydrolysis that control substrate specificity of the 20S proteasome. *Chem Biol* 2002, 9:655-662.

[75] Orlowski M, Cardozo C, Michaud C: Evidence for the presence of five distinct proteolytic components in the pituitary multicatalytic proteinase complex. Properties of two components cleaving bonds on the carboxyl side of branched chain and small neutral amino acids. *Biochemistry* 1993, 32:1563-1572.

[76] Cardozo C, Eleuteri AM, Orlowski M: Differences in catalytic activities and subunit pattern of multicatalytic proteinase complexes (proteasomes) isolated from bovine pituitary, lung, and liver. Changes in LMP7 and the component necessary for expression of the chymotrypsin-like activity. *J Biol Chem* 1995, 270:22645-22651.

[77] Eleuteri AM, Angeletti M, Lupidi G, Tacconi R, Bini L, Fioretti E: Isolation and characterization of bovine thymus multicatalytic proteinase complex. *Protein Expr Purif* 2000, 18:160-168.

[78] Orlowski M, Cardozo C, Eleuteri AM, Kohanski R, Kam CM, Powers JC: Reactions of [14C]-3,4-dichloroisocoumarin with subunits of pituitary and spleen multicatalytic proteinase complexes (proteasomes). *Biochemistry* 1997, 36:13946-13953.

[79] Mykles DL: Differential effects of bovine PA28 on six peptidase activities of the lobster muscle proteasome (multicatalytic proteinase). *Arch Biochem Biophys* 1996, 325:77-81.

[80] Ustrell V, Realini C, Pratt G, Rechsteiner M: Human lymphoblast and erythrocyte multicatalytic proteases: differential peptidase activities and responses to the 11S regulator. *FEBS Lett* 1995, 376:155-158.

[81] Liu CW, Corboy MJ, DeMartino GN, Thomas PJ: Endoproteolytic activity of the proteasome. *Science* 2003, 299:408-411.

[82] Maupin-Furlow JA, Kaczowka SJ, Reuter CJ, Zuobi-Hasona K, Gil MA: Archaeal proteasomes: potential in metabolic engineering. *Metab Eng* 2003, 5:151-163.

[83] Osmulski PA, Gaczynska M: Nanoenzymology of the 20S proteasome: proteasomal actions are controlled by the allosteric transition. *Biochemistry* 2002, 41:7047-7053.

[84] Wilk S, Chen WE, Magnusson RP: Properties of the nuclear proteasome activator PA28gamma (REGgamma). *Arch Biochem Biophys* 2000, 383:265-271.

[85] Ustrell V, Hoffman L, Pratt G, Rechsteiner M: PA200, a nuclear proteasome activator involved in DNA repair. *EMBO J* 2002, 21:3516-3525.

[86] Orlowski M: Selective activation of the 20 S proteasome (multicatalytic proteinase complex) by histone h3. *Biochemistry* 2001, 40:15318-15326.

[87] Eleuteri AM, Cuccioloni M, Bellesi J, Lupidi G, Fioretti E, Angeletti M: Interaction of Hsp90 With 20S Proteasome: Thermodynamic and Kinetic Characterization. *PROTEINS: Structure, Function, and Genetics* 2002, 48:169–177.

[88] Schmidtke G, Holzhutter HG, Bogyo M, Kairies N, Groll M, de Giuli R, Emch S, Groettrup M: How an inhibitor of the HIV-I protease modulates proteasome activity. *J Biol Chem* 1999, 274:35734-35740.

[89] Kisselev AF, Akopian TN, Goldberg AL: Range of sizes of peptide products generated during degradation of different proteins by archaeal proteasomes. *J Biol Chem* 1998, 273:1982-1989.

[90] Emmerich NP, Nussbaum AK, Stevanovic S, Priemer M, Toes RE, Rammensee HG, Schild H: The human 26 S and 20 S proteasomes generate overlapping but different sets of peptide fragments from a model protein substrate. *J Biol Chem* 2000, 275:21140-21148.

[91] Tanaka K: Role of proteasomes modified by interferon-gamma in antigen processing. *J Leukoc Biol* 1994, 56:571-575.

[92] Goldberg AL, Cascio P, Saric T, Rock KL: The importance of the proteasome and subsequent proteolytic steps in the generation of antigenic peptides. *Mol Immunol* 2002, 39:147-164.

[93] Elliott T, Smith M, Driscoll P, McMichael A: Peptide selection by class I molecules of the major histocompatibility complex. *Curr Biol* 1993, 3:854-866.

[94] Aki M, Shimbara N, Takashina M, Akiyama K, Kagawa S, Tamura T, Tanahashi N, Yoshimura T, Tanaka K, Ichihara A: Interferon-gamma induces different subunit

organizations and functional diversity of proteasomes. *J Biochem (Tokyo)* 1994, 115:257-269.

[95] Driscoll J, Brown MG, Finley D, Monaco JJ: MHC-linked LMP gene products specifically alter peptidase activities of the proteasome. *Nature* 1993, 365:262-264.

[96] Gaczynska M, Rock KL, Goldberg AL: Gamma-interferon and expression of MHC genes regulate peptide hydrolysis by proteasomes. *Nature* 1993, 365:264-267.

[97] Gray CW, Slaughter CA, DeMartino GN: PA28 activator protein forms regulatory caps on proteasome stacked rings. *J Mol Biol* 1994, 236:7-15.

[98] Amici M, Sagratini D, Pettinari A, Pucciarelli S, Angeletti M, Eleuteri AM: 20S proteasome mediated degradation of DHFR: implications in neurodegenerative disorders. *Arch Biochem Biophys* 2004, 422: 168–174.

[99] Rivett AJ: Preferential degradation of the oxidatively modified form of glutamine synthetase by intracellular mammalian proteases. *J Biol Chem* 1985, 260:300-305.

[100] Grune T: Oxidative stress, aging and the proteasomal system. *Biogerontology* 2000, 1:31-40.

[101] Grune T, Merker K, Sandig G, Davies KJ: Selective degradation of oxidatively modified protein substrates by the proteasome. *Biochem Biophys Res Commu* 200, 305:709-718.

[102] Keller JN, Huang FF, Markesbery WR: Decreased levels of proteasome activity and proteasome expression in aging spinal cord. *Neuroscience* 2000, 98:149-156.

[103] McNaught KS, Belizaire R, Isacson O, Jenner P, Olanow CW: Altered proteasomal function in sporadic Parkinson's disease. *Exp Neurol* 2003, 179:38-46.

[104] McNaught KS, Belizaire R, Jenner P, Olanow CW, Isacson O: Selective loss of 20S proteasome alpha-subunits in the substantia nigra pars compacta in Parkinson's disease. *Neurosci Lett* 2002, 326:155-158.

[105] Tofaris GK, Layfield R, Spillantini MG: alpha-synuclein metabolism and aggregation is linked to ubiquitin-independent degradation by the proteasome. *FEBS Lett* 2001, 509:22-26.

[106] Nunan J, Shearman MS, Checler F, Cappai R, Evin G, Beyreuther K, Masters CL, Small DH: The C-terminal fragment of the Alzheimer's disease amyloid protein precursor is degraded by a proteasome-dependent mechanism distinct from gamma-secretase. *Eur J Biochem* 2001, 268:5329-5336.

[107] Cardozo C, Michaud C: Proteasome-mediated degradation of tau proteins occurs independently of the chymotrypsin-like activity by a nonprocessive pathway. *Arch Biochem Biophys* 2002, 408:103-110.

[108] David DC, Layfield R, Serpell L, Narain Y, Goedert M, Spillantini MG: Proteasomal degradation of tau protein. *J Neurochem* 2002, 83:176-185.

[109] Keller JN, Hanni KB, Markesbery WR: Impaired proteasome function in Alzheimer's disease. *J Neurochem* 2000, 75:436-439.

[110] Gregori L, Fuchs C, Figueiredo-Pereira ME, Van Nostrand WE, Goldgaber D: Amyloid beta-protein inhibits ubiquitin-dependent protein degradation in vitro. *J Biol Chem* 1995, 270:19702-19708.

[111] Smith MA, Rottkamp CA, Nunomura A, Raina AK, Perry G: Oxidative stress in Alzheimer's disease. *Biochim Biophys Acta* 2000, 1502:139-144.

[112] Keck S, Nitsch R, Grune T, Ullrich O: Proteasome inhibition by paired helical filament-tau in brains of patients with Alzheimer's disease. *J Neurochem* 2003, 85:115-122.

[113] Keller JN, Huang FF, Zhu H, Yu J, Ho YS, Kindy TS: Oxidative stress-associated impairment of proteasome activity during ischemia-reperfusion injury. *J Cereb Blood Flow Metab* 2000, 20:1467-1473.

[114] Chan PH: Role of oxidants in ischemic brain damage. *Stroke* 1996, 27:1124-1129.

[115] Keller JN, Kindy MS, Holtsberg FW, St Clair DK, Yen HC, Germeyer A, Steiner SM, Bruce-Keller AJ, Hutchins JB, Mattson MP: Mitochondrial manganese superoxide dismutase prevents neural apoptosis and reduces ischemic brain injury: suppression of peroxynitrite production, lipid peroxidation, and mitochondrial dysfunction. *J Neurosci* 1998, 18:687-697.

[116] Strack PR, Waxman L, Fagan JM: Activation of the multicatalytic endopeptidase by oxidants. Effects on enzyme structure. *Biochemistry* 1996, 35:7142-7149.

[117] Amici M, Lupidi G, Angeletti M, Fioretti E, Eleuteri AM: Peroxynitrite-induced oxidation and its effects on isolated proteasomal systems. *Free Radic Biol Med* 2003, 34:987-996.

[118] Amici M, Forti K, Nobili C, Lupidi G, Angeletti M, Fioretti E, Eleuteri AM: Effect of neurotoxic metal ions on the proteolytic activities of the 20S proteasome from bovine brain. *J Biol Inorg Chem* 2002, 7:750-756.

[119] Ullrich O, Grune T: Proteasomal degradation of oxidatively damaged endogenous histones in K562 human leukemic cells. *Free Radic Biol Med* 2001, 31:887-893.

[120] Ullrich O, Sitte N, Sommerburg O, Sandig V, Davies KJ, Grune T: Influence of DNA binding on the degradation of oxidized histones by the 20S proteasome. *Arch Biochem Biophys* 1999, 362:211-216.

[121] Ullrich O, Reinheckel T, Sitte N, Hass R, Grune T, Davies KJ: Poly-ADP ribose polymerase activates nuclear proteasome to degrade oxidatively damaged histones. *Proc Natl Acad Sci USA* 1999, 96:6223-6228.

[122] Rock KL, Goldberg AL: Degradation of cell proteins and generation of MHC class I-presented peptides. *Annu Rev Immunol* 1999, 17:739-779.

In: The Ubiquitin Proteasome System... ISBN 978-1-60021-749-4
Eds: Mario Di Napoli and Cezary Wojcik, pp. 137-157 © 2007 Nova Science Publishers, Inc.

Chapter 7

STRUCTURE AND FUNCTION OF THE 26S PROTEASOMES

***Cezary Wójcik*[1,*] *and George N. DeMartino*[2,¥]**

[1]Department of Anatomy and Cell Biology, Indiana University School of Medicine, Evansville, Indiana, USA;

[2]Department of Physiology, University of Texas Southwestern Medical Center, Dallas, Texas, USA.

ABSTRACT

26S proteasomes are ~2.4 MDa supramolecular assemblies that function as protein degrading complexes in neuronal as well as other cell types. They constitute the final, common destination of the proteins degraded by the ubiquitin-proteasome pathway, and perhaps by some non-ubiquitin-dependent pathways as well. 26S proteasomes are formed by association of the core 20S proteasomes with one or two PA700 activators (19S caps). While the core 20S proteasomes harbor the proteolytic activities, the remaining features of 26S proteasomes are conferred by components of the PA700. Mammalian PA700 is composed of 18 subunits, including 6 AAA ATPases (Rpt1-6) and several non-ATPase subunits (Rpn1-3, Rpn5-12 and Uch37). PA700 is physically divided into the lid and base subcomplexes. PA700 allows the recognition of polyubiquitinated proteins, their attachment, unfolding, opening of the closed proteasomal 'gates' and translocation of the unfolded polypeptide chain of the substrate towards the central catalytic cavity of the proteasome. At the same time PA700 allows the release of free ubiquitin through at least two different deubiquitinating activities. All of these functions are coupled to the ATP-ase activity of the complex, making them highly susceptible to ATP depletion such as

* Correspondence concerning this article should be addressed to Dr. Cezary Wójcik, MD, PhD; Department of Anatomy and Cell Biology, Indiana University School of Medicine, 8600 University Blvd., Evansville, IN, USA, Phone: (812) 461 5437; Fax: (812) 465 1184; E-mail: cwojcik@iupui.edu.

¥ Correspondence concerning this article should be addressed to Dr. George N. DeMartino, PhD; Department of Physiology, University of Texas Southwestern Medical Center, 5323 Harry Hines Blvd., Dallas, TX, USA, Phone: (214) 648-3770; Fax (214) 648 4771; E-mail: George.DeMartino@UTSouthwestern.edu.

during episodes of limited hypoxia or ischemia. Moreover, under those conditions 26S proteasomes tend to separate into free 20S proteasomes and PA700 complexes. Besides the canonical 26S proteasome subunits, several proteins associate loosely with the 26S proteasome, including additional deubiquitinating enzymes, ubiquitin ligases and polyubiquitin binding and delivery factors.

Keywords: proteasome, ubiquitin, PA700, proteasome activator, isopeptidase, chaperone, protein degradation.

Abbreviations

AAA, ATPases with multiple cellular activities; CSN, COP9 signalosome; eIF3, eukaryotic initiation factor; ODC, Ornithine decarboxylase; PA, proteasome activator; PC, proteasome/cyclosome; PCI, Proteasome, COP9, Initiation factor 3; PINT, Proteasome, Int-6, Nip-1 and TRIP-15; RP, Regulatory Particle; Rpn, regulatory particle non-ATPases; Rpt, Regulatory particle triple-A protein; UIM, ubiquitin-interacting motif; UPS, ubiquitin and proteasome dependent proteolytic system, VCP, Valosin Containg Protein; VWA, von Willebrand factor, type A.

Introduction

The ubiquitin- and proteasome dependent proteolytic system (UPS) is responsible for the degradation of most intracellular proteins in eukaryotic cells [1,2]. Several hundreds of ubiquitin ligases in concert with several ubiquitin conjugating enzymes mediate the specific ubiquitination of individual substrates, as described in Chapter 3. Ubiquitinated substrates are targeted to the 26S proteasome, which functions as the final common step in the UPS pathway. The 26S proteasome is composed of the core 20S proteasome (described in detail in Chapter 6) associated with a specialized complex of proteins forming the PA700 (Proteasome Activator of 700, 000 daltons), also known as the 19S RP (19S Regulatory Particle). The general function of PA700 is to impart specific catalytic and regulatory features to the resulting 26S proteasome. PA700 is a defined complex equipped with multiple enzymatic activities that binds to one or both ends of the cylinder-shaped 20S proteasome [3-9]. The resulting 'singly capped' or 'doubly capped' complexes are both called 26S proteasomes [10,11]. It is not known, whether there are physiological differences between the singly capped versus the double capped 26S proteasomes. In addition, hybrid proteasomes have also been described, which are formed when PA700 attaches to one end of the 20S proteasome while another regulatory complex known as PA28 attaches to the other [12]. Since the functions of the 20S proteasome have been described in the preceding chapter, our presentation will focus on the structure and function of the PA700 component of the 26S proteasome and the mechanisms by which this regulatory complex mediates selective degradation of ubiquitinated proteins, i.e. the main function of the 26S proteasomes.

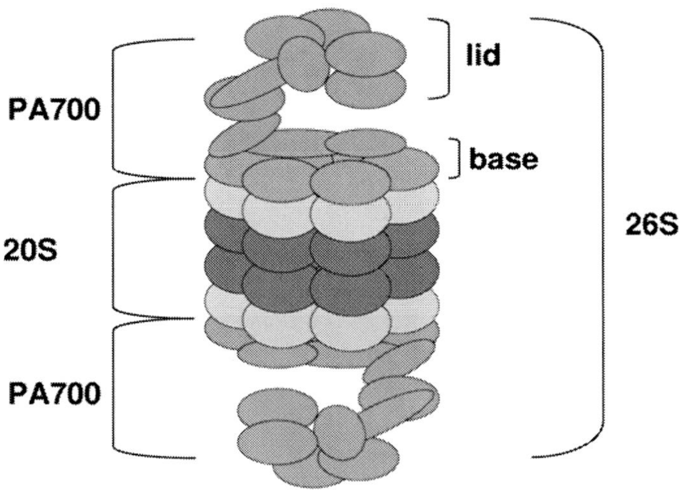

Figure 1. Schematic representation of the 26S proteasome. The picture indicates its subdivision into the core 20S proteasome and two PA700/19S RP complexes. The latter are further subdivided into the base and lid subcomplexes.

AN OVERVIEW OF THE SUBUNITS OF PA700/19S RP

Double-capped 26S proteasomes have a molecular mass of ~2400 kDa; 700 kDa is contributed by the 20S core and the remaining 1400 kDa by two PA700 complexes. Figure 1 shows a schematic representation of the structure of 26S proteasomes. The C2 symmetry of the 20S proteasome dictates the overall symmetry of the entire complex. While 20S proteasome is a regular structure composed of similar subunits, PA700 is composed of approximately 18 different gene products forming an asymmetrical mass whose crystal structure has not been yet solved. Cryo-electron microscopy combined with image averaging gave an insight into the shape and structure of this PA700 mass. It has an appearance described sometimes as an opened 'alligator mouth'. Part of PA700 seems to be attached tightly to the α ring of the 20S proteasome, while the other portion is attached to it by an angle of about 45 degrees [10,11]. While some subunits of PA700 bear structural similarity to each other, others are completely different. Their sizes vary from 28 to 112 kDa in contrast to the much more uniform population of 20S proteasome subunits which are in the 20-30 kDa range. Each PA700 gene product is present in a single copy per complex [3,5]. The overall structure and function of PA700 is highly conserved in eukaryotes, and only minor differences in subunit composition appear to exist among species (see below). Additional proteins have been shown to associate more or less strictly with PA700 in different stoichiometric proportions. While many of these interactions are of unknown or questionable biological significance, many others are likely physiologically meaningful. The functions of some of these interacting proteins raise questions as to whether they should be classified as *bona fide* subunits of the PA700.

Nomenclature of proteasome subunits is extremely confusing since the proteasome has been studied in many species by different groups of investigators prior to the realization that they all were components of the same complex. A systematic Rpn/Rpt nomenclature has been suggested for PA700 subunits, to complement the α/β nomenclature of the 20S proteasome subunits [13]. While this nomenclature was originally derived from yeast, it has been adapted and used in other organisms as well. Recently it has been implemented by *Nature-Alliance for Cellular Signaling* as its official nomenclature of human PA700 subunits used on its 'Molecule Pages' [14]. Therefore, we use this systematic nomenclature throughout the text. However, since literature data are full of references to alternative names of proteasome subunits Appendix 1 is provided, presenting a complete list of synonyms for PA700 subunits. Rpn/Rpt nomenclature distinguishes between the AAA ATPase subunits (Regulatory particle triple-A protein) and the non-AAA ATPase subunits (Regulatory particle non-ATPase). Even the systematic nomenclature is not free from confusion, since one yeast gene product originally classified as PA700 subunit called Rpn4 is actually a transcriptional regulator of the expression of proteasomal genes. Moreover, Rpn4 is short-lived itself being a proteasome substrate [15]. On the other hand, mammalian PA700 subunits which do not have yeast homologs (such as S15 or Uch37) lack Rpn/Rpt names.

The six homologous Rpts form a well defined group within the PA700 subunits (Figure 2). The proteasomal ATPases are members of the AAA family (ATPases Associated with various cellular Activities) [16-19] . They all have similar molecular weights (42-56 kDa) and contain a 200-amino acid domain characteristic of the AAA protein family [17], which has Walker A and B nucleotide-binding motifs [19]. Despite high homology of their AAA domains their functions are non-redundant [93], perhaps because of the high divergence of the remaining portions of the ATPases. The relative contributions of individual ATPase subunits to the overall ATPase activity and thus to the ATP-dependent functions of the 26S proteasome are ill defined [20,21].

The 'Rpns' or non-ATPase subunits of PA700 (Rpn1 to 3 and 5 to 12) represent a much more diverse group of proteins, with different protein domains (Figure 2). Rpn1 and Rpn2 share a low degree of sequence similarity and contain multiple PC (proteasome/cyclosome) repeats. PC repeats are weakly conserved modules of unknown function from the Armadillo superfamily, which beside Rpn1 and 2 have been detected in one subunit of the APC/cyclosome E3 ligase complex [22]. Both Rpn1 and Rpn2 also contain the KEKE motifs likely involved in protein-protein interactions [20,23-26]. The remaining subunits have little similarity to one another, and their primary structures generally provide little specific information about their functions. Lid subunits share close homology with subunits of two other eukaryotic protein coplexes, namely the COP9/signalosome and eIF3 [33]. Therefore four lid subunits, Rpn3, Rpn7, Rpn5 and Rpn9 share with components of those complexes the C-terminal PINT motif (Proteasome, Int-6, Nip-1 and TRIP-15), also known as the PCI domain (Proteasome, COP9, Initiation factor 3). Since all proteins with PINT domains are part of larger multi-protein complexes, it may mediate protein-protein interactions but its exact role is unknown [27,28]. Rpn10 binds poly-Ub chains through two UIFs (ubiquitin-interacting motifs) [29,30] but it also contains the VWA (von Willebrand factor, type A) domain, which is involved in the formation of multi-protein complexes [31]. Rpn8 and Rpn11 contain the JAB/JAMM domain also known as the MPN domain or PAD-1-like domain. This domain is found also in homologous subunits of the COP9/signalosome and eIF3 as well as various members of the MEROPS peptidase family M67 (clan M-) [3]. Rpn11 has been shown to display deubiquitinating activity [32,33]. Uch37, a subunit of PA700 present in mammalian 26S proteasomes but not in yeast, also functions as a deubuiqutinating enzyme and contains a conserved cysteine residue characteristic of the active site family enzymes [34-36].

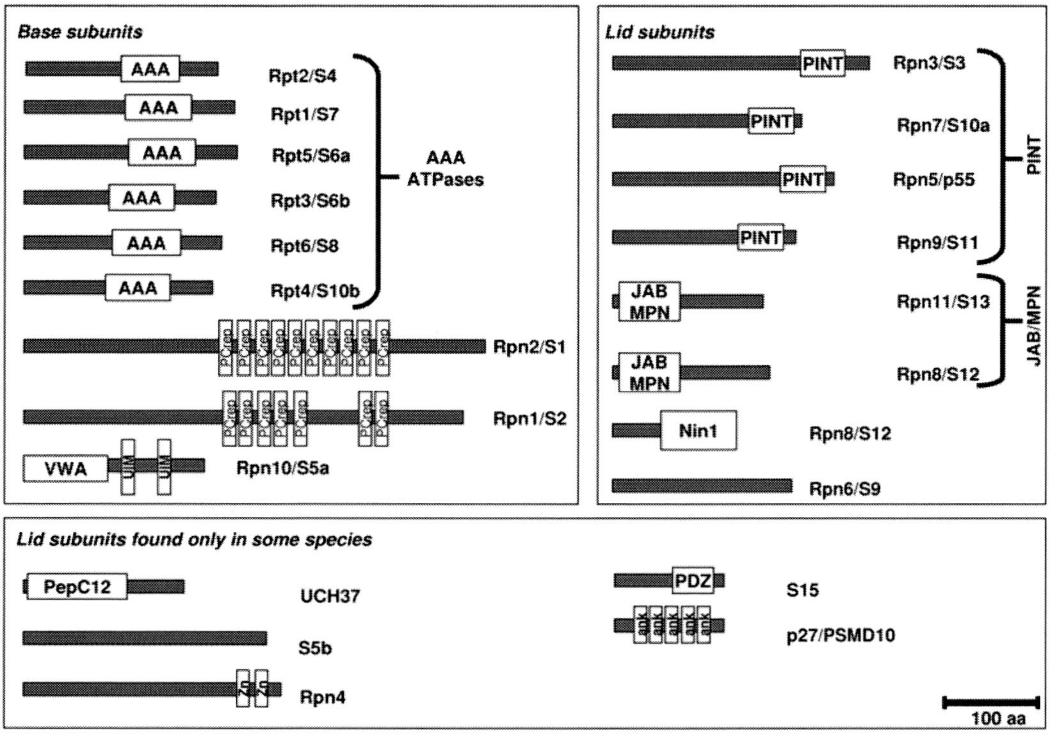

Figure 2. Schematic representation of the domain architecture of the different components of the PA700. PA700 subunits are grouped into the base subunits and lid subunits. In addition, the lid subunits which are not universal components of PA700 have been arranged in a third group. Known protein domains are indicated. See text for abbreviations of the protein domains. The size of each protein representation corresponds to the size of each subunit. The bar in the right lower corner corresponds to 100 aminoacids in length.

Several subunits of PA700 have been identified only in preparations from some species or groups of organisms (Figure 2). While in some cases these differences reflect authentic distinctions among species, in other cases they may reflect differences in experimental procedures. Special consideration must be given to Rpn4, which was identified as a subunit of yeast 26S proteasome, but has no ortholog in other species [37]. In contrast to other proteasome subunits, Rpn4 is a constitutively short-lived proteasome substrate. Inhibition of the proteasome activity results in accumulation of Rpn4, which functions as a positive transcriptional factor for global expression of proteasome subunits [15,38]. An unidentified functional counterpart of Rpn4 may exist in higher eukaryotes because inhibition of proteasome function can upregulate expression of proteasome subunits [39,40]. Uch37 has been identified in mammals and in fission yeast as an bona fide subunit of the 26S proteasome, however it is absent from the genome of budding yeast, where its function is performed by other deubiquitinating enzymes, loosely associated with the 26S proteasome [35]. The lid of mammalian 26S proteasomes are also characterized by the presence of unique and poorly characterized additional subunits, such as S15/PSMD9, S5b/PSMD5 and p27/PSMD10, which therefore lack Rpt/Rpn names derived from the yeast nomenclature. While S5b has no know structural features, the other two subunits have common domains involved in protein-protein interactions, PDZ (S15) and ankyrin repeats (p27).

Subdivision of PA700 into the Base and the Lid

Despite the lack of a crystal structure, the general architecture of PA700 has been established, including most subunit-subunit interactions, which are depicted on Figure 3 [10,41-43]. A major advance in understanding the general architecture of PA700 has been the identification and characterization of two component subcomplexes, termed the 'base' and the 'lid' [44]. The base subcomplex contains eight subunits - six AAA ATPases (numbered from Rpt1 to Rpt6) and two non-ATPase subunits, Rpn1 and Rpn2, the two largest subunits of the 26S proteasome. The proteasomal ATPases form a heterologous six-membered ring that directly abuts the terminal α-ring of the 20S proteasome. The center of the ATPase ring is likely coaxial with the opening of the α ring of the 20S proteasome forming a tunnel through which substrates must pass to enter the central cavity of the proteasome, where the active sites are located (see Chapter 6). The exact orientation of Rpn1 and Rpn2 relative to the ATPase ring is uncertain. A molecular modeling study has predicted that these subunits form an α-helical toroid with a central pore that extends the axial channel of the proteasome and ATPase ring [45], however a direct experimental evidence to support this model is lacking. The base probably serves multiple roles in degradation of polyubiquitinated proteins, some of them mediated through common functions of all ATPases while some of them mediated by non-redundant functions, specific and unique to each. The overall ATP dependence of the 26S proteasome function depends on this ring of ATPases. Two best examples of non-redundant functions of the proteasomal ATPases are provided by Rpt2, which appears to gate the central access pore to the proteasome [14], and by Rpt5 which binds poly-Ub chains [50]. In addition, the non-ATPase base subunit Rpn2 binds Ub-like domains (UBLs) of proteins such as Rad23 or Dsk2, which are likely to deliver substrates to the proteasome (see Chapter 5). Four ATPases are phosphorylated [46], a modification that may modulate the interaction of PA700 with the 20S proteasome [47]. Moreover, Rpt2 has been shown to be reversibly modified by O-linked N-acetylglucosamine. This modification modulates ATP-ase activity and decreases peptidase activity of 26S proteasome both *in vitro* and *in vivo* [48]. This modification may link proteasomal activity to the nutrition status of the cells. Another study has claimed that additional subunits of the proteasome are modified by addition of O-linked N-acetylglucosamine as well [49]. The base likely evolved from a ring of identical ATP-ases associating with the proteasome, such as the PAN ATPase found in archaeabacteria [50]. PAN has the basic properties attributed to the ring of proteasomal ATPases in eukaryotes, since it stimulates protein refolding and promotes protein degradation by associated 20S proteasomes.

The lid subcomplex seems to be added later in evolution and it does not have a homologous structure in archeabacteria. It is linked to the base through the 'hinge' Rpn10 subunit [51], however it is likely that this interaction is stabilized by additional proteins loosely associated with the proteasome, in particular Ecm29 [52]. The lid contains the remaining Rpn subunits as well as additional species-specific subunits such as UCH37 in mammalian proteasomes. The precise function of most lid subunits is not understood. Rpn10 binds poly-Ub chains, however curiously, this property is dispensable for most normal proteasome functions [53]. Moreover, since appreciable amounts of Rpn10 are present in the cells as a free protein, unbound to the PA700/26S complexes, free Rpn10 may recruit, bind and deliver polyubiquitinated substrates to the proteasome.

Two lid subunits, Rpn11 and UCH37 mediate deubiquitination (see Chapter 4). Rpn11 is a Zn^{2+} metalloproteinase able to cleave poly-Ub chains proximally, i.e. from their attachment points on substrate proteins [86]. In contrast, Uch37 is a cysteine protease that cleaves poly-Ub distally, cleaving off ubiquitin monomers from the ends of polyubiquitin chains, thereby progressively decreasing the length of the chain [51]. The entire lid subcomplex bears remarkable resemblance to two different the ubiquitous protein complexes present within cells, the COP9 signalosome (CSN) and the eukaryotic initiation factor eIF3, both with subunit-for-subunit homology to the proteasome lid [44]. All three complexes probably evolved from one common ancestor complex.

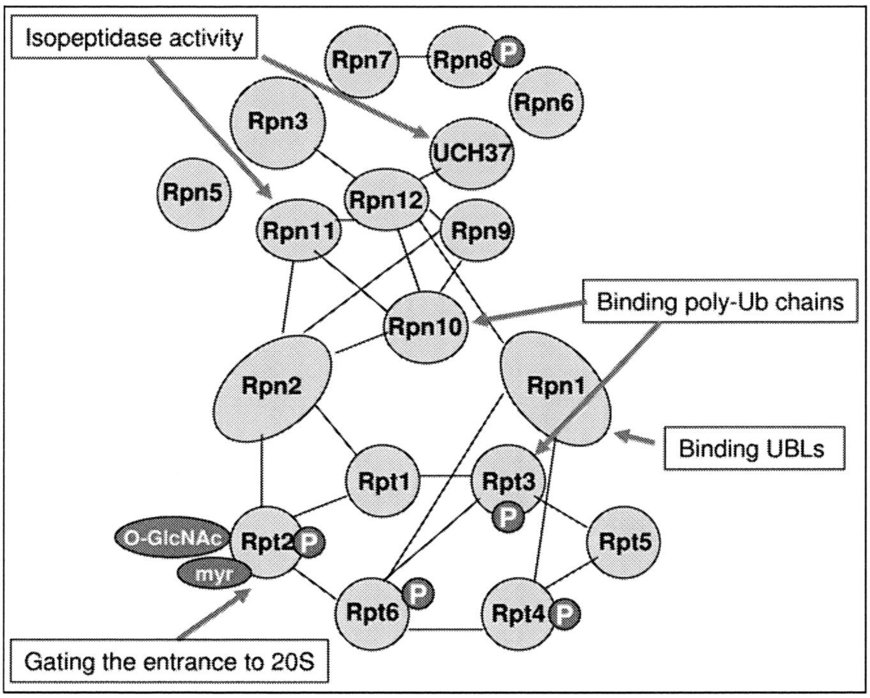

Figure 3. Protein interactiosn within PA700. Schematic representation of the known interactions between the different components of the PA700, known functions attributed to individual subunits and known post-translational modificatiosn likely to modify the function of the 26S proteasomes.

Assembly of the 26S Proteasome

The exact cellular process by which the 26S proteasome is assembled remains unknown, but the best evidence suggests that it results from binding of independently assembled 20S proteasome and PA700 [10,54]. There is a considerable, but incomplete information about the assembly of the 20S proteasome as discussed in the previous chapter [55,56], whereas very little is known about the assembly of PA700. Formation of 26S proteasome from purified 20S proteasome and PA700 can be achieved *in vitro* by an ATP-dependent process [7]. This suggests that PA700 and 20S proteasome are sufficient for assembly of 26S proteasome. This process, however, is inefficient *in vitro*. Several studies have identified protein complexes that improve 26S proteasome assembly, but little is known about the molecular basis of these effects [10,57]. Another study has indicated that Hsp90 mediates 26S proteasome assembly [58]. Recently, a protein termed Ecm29 was identified as a stoichiometric component of the 26S proteasome from *Saccharomyces cerevisiae* purified by affinity chromatography without exposure to high salt concentrations [53]. Ecm29 is a ~200 kDa protein that binds to both the 20S proteasome and PA700 probably tethering the two subcomplexes. Electron microscopy reveals Ecm29 as a V-shaped protein that may act as 'clip' between the α rings of the 20S proteasome and the base of PA700, but little is known about the molecular basis of such binding. Ecm29 stabilizes 26S proteasome in the absence of ATP, further supporting a role for it in physically linking 20S proteasome to PA700. Orthologs of Ecm29 are distributed widely among species, and evidence in mammalian cells suggests that it can function as an adaptor to localize the proteasome to membranes ([59] see Chapter 9). Thus, further study of the protein in yeast and other organisms will be required to establish its precise function in 26S proteasome structure and function.

Functions of the 26S Proteasome Mediated through the PA700 Complex

The 20S core proteasome dictates the overall architecture of the 26S proteasome complex. It imparts the structural attachment to the PA700 complexes. 20S proteasome has the catalytic sites which perform the cleavage of digested polypeptides. The structure of the 20S core and the nature and regulation of its proteolytic activities are described in detail in Chapter 6, therefore we will concentrate our discussion on the unique features imparted upon the 26S proteasome by its PA700 component. PA700 serves multiple roles in mediating proteasomal degradation of polyubiquitinated proteins. It relieves the structurally-imposed inhibition of proteolytic activity by opening the gates to the central channel of the 20S proteasome, which is normally obstructed by N-terminal extensions of the α-subunits. PA700 also serves as the recognition and binding element for the poly-Ub chain, whose only role is to bring substrate sin close proximity of the degradation machinery. Moreover, PA700 prepares the substrate for degradation through its 'unfoldase' or 'reverse chaperone' activity followed by translocation of the unfolded polypeptide chain through the central proteasome channel towards the central 'chamber of doom', where proteolysis takes place. Finally,

PA700 recycles ubiquitin by removing it from the substrate through its isopeptidase activity. If ubiquitin is not removed, proteolysis still can proceed but at a very slow rate, due to the high stability of the ubiquitin fold. In addition, depletion of cellular levels of free ubiquitin inhibits ubiquitination of new substrates and therefore prevents their degradation. ATP hydrolysis catalyzed by the ATP-ases of the PA700 is obligatory for overall proteolysis. It is not known, whether all or only several of the PA700-mediated processes listed above are coupled to ATPase activity. In order to allow a detailed discussion, individual functions of 26S proteasome are described in separately, however it is likely that these functions are intimately associated.

Roles of Atrase Activity in PA700 Function

A fundamental feature of the UPS pathway is the requirement for metabolic energy in form of ATP hydrolysis for the degradation of proteins. ATP is consumed at the two extremes of the UPS pathway – at the beginning, when ATP is needed to activate ubiquitin by an E1 (Chapter 3) and at the end, when ATP is required for the attachment of PA700 to the 20S proteasomes and subsequently for the processing and degradation of substrates. It is unknown how the proteolysis is mechanistically linked to ATPase activity. ATP hydrolysis probably mediates multiple elements of 26S proteasome function, including assembly of the complex from 20S proteasome and PA700 subcomplexes, proteasome activation, poly-Ub chain binding, and substrate unfolding, translocation, and deubiquitination. The ATPase activity of PA700 is catalyzed by all ATPase subunits of the base, however the relative roles and contributions of the individual ATPases to various functions remain unclear. Nevertheless, individual ATPases play distinct and non-redundant roles [60], however there are insufficient data to conclude with certainty whether there is a complete division of labor for these ATPases among different ATP-dependent processes. Whatever is the case, the assembly of the ATP-ases into a hexameric ring and the attachment of this ring onto the α-ring of the 20S proteasome are most likely critical to their functions, since it is conserved in the VAT ATP-ase binding to proteasomes in archaea [7]. Recently, it has been reported that ATP hydrolysis triggers rapid dissociation of PA700 from immunopurified 26S proteasomes in a manner coincident with release of the bulk of proteasome-interacting proteins (Chapter 8). Moreover, the PA700 further disassembled into lids, bases and free Rpn10. Interaction with a purified poly-Ub substrate reconstituted the holoenzyme from its components [61].

Activation of the Core 20S Proteasome by PA700

20S proteasomes have a low catalytic activity since the catalytic sites are self-compartmentalized within their central chamber and substrates are excluded from entering it through occlusion of the proteasomal gates [62]. Even in an open, active conformation, substrates reach the sequestered catalytic sites only after passing through a narrow 13 Å passage formed by the terminal α rings of the proteasome [62-64]. This first structural constraint prevents the entry of substrates with appreciable tertiary structure, but even short or unfolded polypeptides must overcome a second structural impediment posed by the proteasome. Activation of the proteasome is equivalent to the opening of proteasomal gates through a rearrangement of the N-terminal extensions of the α subunits. While such activation can be obtained by physical means or interaction with low levels of detergents, the physiological mechanism relies on the interaction with specific regulatory proteins. A detailed molecular explanation for PA700-induced proteasome activation is lacking in the absence of a crystal structure of the 26S proteasome, however it is likely similar in principle to proteasome activation by a different class of proteasome activators, such as PA28/PA26. A co-crystal structure has revealed that binding of such activator to the α ring of 20S proteasome promotes an rearrangement of the occluding N-terminal extensions of the α-subunits from a position roughly perpendicular to the central proteasome channel, to one roughly parallel to the channel, thereby opening a pore through which substrates may pass [65]. In contrast to PA700, this class of activators (see Chapter 9) does not promote the degradation of ubiquitinated proteins, presumably because PA28 lacks other essential features present in PA700 which are necessary for processing such substrates. PA700 most likely also activates the proteasome by relieving occlusion of the proteasome pore, through a physical interaction between the heterohexameric AAA ATPase ring of the base and the heteroheptameric α ring of the proteasome. Binding of PA700 to the proteasome greatly enhances the hydrolysis of short peptide substrates, probably suggesting that PA700 binding increases access of these substrates to the catalytic sites [7]. Moreover, this activation can be accomplished entirely by the base subcomplex of the PA700, indicating that the interaction of the ATPase ring is sufficient for activation [44]. Deletion of the pore-occluding peptide of the α3 subunit of the 20S proteasome results in a constitutively active 20S proteasome, whose activity is not stimulated further by binding to PA700 [66]. Complementation experiments using the α3 deletion mutant and Rpt2 ATPase mutant have indicated that the Rpt2 subunit of PA700 activates the proteasome by a mechanism involving gating of the α pore [67].

Binding of Poly-UB Chains

While the base itself stimulates the activity of proteasome towards small peptides, it is unable to promote the recognition and degradation of ubiquitinated proteins without the association of the lid subcomplex. The entire PA700 links therefore the proteasome to the ubiquitin pathway bringing upon the recognition and recruitment of poly-Ub proteins. 26S proteasome binds efficiently K48-G76 linked poly-Ub chains composed of four or more ubiquitin moieties, however the exact molecular basis for this interaction remains poorly understood [68]. Rpn10 and Rpt6 have been identified by different methods as poly-Ub chain binding proteins. Moreover, cells contain other poly-Ub binding proteins that interact with the 26S proteasome and may function to deliver substrates to it for degradation. Historically, Rpn10 was the first PA700 subunit identified as a receptor for K48-G76 linked poly-Ub chains [69-72]. It binds poly-Ub through a short hydrophobic sequence termed the ubiquitin-interacting motif (UIM), which has been identified in many other proteins, some of them involved in various aspects of ubiquitin metabolism ([73] see Chapter 5). The interaction between poly-Ub and Rpn10 involves direct binding of complementary hydrophobic patches

[74,75]. Rpn10 from *Saccharomyces cerevisiae* and *Arabidopsis* has only one C-terminal UIM [75], while human and *Drosophila* Rpn10 has two UIMs [76]. Each of the two UIMs binds poly-Ub with different affinities and may have some degree of cooperativity in intact Rpn10. The initial data about the relevance of Rpn10 in poly-Ub binding were contradicted by later findings indicating that disruption of this gene produces only a weak phenotype. Deletion of the Rpn10 gene in yeast is not lethal and inhibits the degradation of only a subclass of ubiquitinated proteins [85]. Likewise, downregulation of Rpn10 by RNA interference in *Drosophila* cells does not inhibit growth or overall ubiquitin-dependent protein degradation [39]. Moreover, deletion of the single conserved UIM in yeast has no effect on the degradation of Ub-Pro-β-galactosidase, a substrate whose cellular degradation otherwise requires expression of Rpn10, indicating that most of the Rpn10 role may be limited to its scaffolding function as a hinge between the lid and the base [75].

The 'far-western' methodology originally used to identify Rpn10 as a poly-Ub binding protein failed to identify other poly-Ub receptors in the 26S proteasome. Chemical crosslinking has been used to identify Rpt5 as a different poly-Ub receptor [77]. Electron paramagnetic resonance, a direct indicator of binding, has shown an ATP-dependent physical interaction between a poly-Ub chain and Rpt5. Strikingly, the same cross-linking methodology failed to detect an interaction between Rpn10 and poly-Ub in an intact 26S proteasome, while it detected an interaction between free Rpn10 and poly-Ub. The molecular basis for the interaction between Rpt5 and poly-Ub is unknown. Analysis of Rpt5 sequence does not show any similarities to known Ub-binding motifs. Unlike Rpt10, Rpt5 is an essential protein in yeast, and RNAi of Rpt5 significantly reduced growth of *Drosophila* S2 cells, however similar results have been obtained for other proteasomal ATPases as well [39,60]. Despite the fact that Rpt5 seems to bind poly-Ub, 'lidless' proteasomes from yeast lacking Rpn10 are defective in the degradation of model ubiquitinated proteins, suggesting that other features of the lid are important for manifestation of normal degradation of ubiquitinated proteins [44]. Altogether, the molecular basis of poly-Ub chain binding to 26S proteasomes remains poorly understood. It likely depends on the combination of binding and interaction with different components of this supramolecular assembly, which can not be reproduced as a mere sum of the binding affinities towards its individual components.

Unfolding of Substrates

Generally, ubiquitination does not change most or all of the native tertiary structure of the numerous proteins degraded by 26S proteasomes. Due to the structural constraints of the 26S proteasomes described above, the tertiary structures of these proteins must be destabilized prior to their proteolysis. Some proteins are ubiquitinated even while they are components of multimeric complexes or when they are embedded in a lipid bilayer, implicating that also the quaternary structure of such complexes must be destabilized, in order to allow the ubiquitinated protein to be selectively dislodged and degraded. It is generally assumed that in most of the cases the PA700 portion of the 26S proteasome harbors the necessary 'unfoldase' or 'reverse chaperone' activity which is necessary and sufficient to unfold the tertiary protein structure and even extract proteins from quaternary complexes. An excellent example of the ability of PA700 to accomplish this function directly is the selective degradation of ubiquitinated Sic1 from a Sic1/Cdk/cyclin complex by purified 26S proteasome [78]. However, in the case of other quaternary structures, additional proteins are required. Such proteins are often called 'shuttles' or 'delivery factors'. A prime example is given by the VCP (Valosin Containg Protein) ATPase, which forms a hexameric ring similar to the ring of proteasomal ATPasaes. VCP has been dubbed the 'segregase' since it is able to segregate and extract ubiquitinated proteins from multimeric complexes found in the cytosol (such as IkBα associated with NFκB) or at the surface of the ER (substrates associated with the retrotranslocation complex, see Chapter 13) [79,80]. Numerous additional proteins have been shown to interact physically with isolated PA700 subunits, intact PA700, or intact 26S proteasome [34,41,81]. Most of those interacting proteins support or modulate the functions of the 26S proteasome, however their detailed description is beyond the scope of the present chapter.

The exact mechanisms by which the 26S proteasome carries out protein unfolding is unclear. Isolated PA700 has chaperone-like properties, since it inhibits the aggregation of misfolded proteins and catalyze the refolding of certain heat- and chemically-denatured proteins [82-84]. These properties are inherent to the base subcomplex, which has an overall architecture similar to other known molecular chaperones such as Hsp90. Similar to Hsp90, it has been shown that ATPase activity is required for the chaperone function of the proteasomal ring of ATPases [82]. Those results indicate that PA700 recognizes and interacts with certain structural features of nonnative proteins, likely to occur transiently or in limited regions of proteins with otherwise high global stability. This type of secondary interaction probably follows the primary targeting of most proteins through a poly-Ub chain. Once recognized by the 26S proteasomes, such features would promote further destabilization and unfolding of the protein, perhaps linked to cycles of ATP hydrolysis and/or processive proteolysis. Support for this model has been obtained by examining UPS-dependent degradation of stable model proteins in reticulocyte extracts [85]. These elegant experiments indicate that 26S proteasome unfolds and degrades proteins processively from a point near the polyubiquitination site. Despite considerable progress, a detailed molecular mechanism for protein unfolding by the 26S proteasomes remains poorly understood. Some outstanding issues to be resolved include, the exact nature of the interaction between PA700 and the substrate, the role of ATP hydrolysis in substrate unfolding, and details of the possible mechanistic linkages among substrate unfolding, translocation, and proteolysis.

Translocation of Substrates from the PA700 Towards the Core 20S Proteasome

Most substrates of the 26S proteasome are degraded completely to short peptides and amino acids once they are engaged by the proteasome, suggesting that proteolysis is processive as shown for several model substrates [85-87]. Since folded protein domains are unable to penetrate through the limiting annulus in the α ring of the proteasome, processive proteolysis may be initiated at a free termini or at a loop of an unstructured region of a protein [88,89]. Single or double polypeptide chains can all easily pass through the α ring of the proteasome. Processive proteolysis could then proceed by a mechanism linked to successive unfolding and translocation of the remaining substrate. Such model also is compatible with the few known examples of limited proteasomal proteolysis, because degradation could start at one terminus of a protein and proceed processively until reaching a 'stop translocation/degradation' site dictated by a structural feature of the substrate. Stalled substrates could be released from the proteasome, thereby generating the mature processed protein. The p105 subunit of NFκB is the best example of a protein processed by this mechanism. In this instance, the C-terminal half of p105 is degraded by the proteasome to yield the mature p50 subunit [90]. Strong evidence for co-translational processing by endoproteolysis has also been presented for p105 and both mechanisms may operate in cells [91]. Little is known about the roles of individual subunits of the PA700 in this translocation process and how it is coupled to the ATP hydrolysis.

Deubiquitination of Substrates

Release of free ubiquitin linked to the degradation of substrates tagged by poly-Ub chains is essential for the function of the UPS pathway (see Chapter 4). Once recognized through specific poly-Ub chain receptors and engaged through secondary interactions, the poly-Ub chain is no longer needed for the degradation of the substrate to occur. Deubiquitination of the substrate is obligatory for and coupled to substrate degradation. Degradation of artificial substrates with uncleavable ubiquitin moiety is much slower than of those with a cleavable ubiquitin moiety, indicating that unfolding and degradation of a stable ubiquitin fold requires considerable effort slowing down the degradation process. Two subunits of the lid, Rpn11 and Uch37, as well as several proteins loosely associated with the 26S proteasomes, have the deubiquitinating or isopeptidase activity. Rpn11 is present in 26S proteasomes isolated from all sources and is an essential protein [32,33,92]. It cleaves the proximal isopeptide bond linking the poly-Ub chain to the substrate. This reaction is catalyzed by both free PA700 and 26S proteasome, but curiously depends on ATP hydrolysis by only 26S proteasome. This feature highlights another likely role of ATPase activity in proteolysis by the 26S proteasome. Because deubiquitination *per se* is unlikely to require ATP hydrolysis, the energy dependence of Rpn11-catalyzed deubiquitination may be linked to translocation or unfolding of the substrate such that the isopeptide bond is spatially positioned for cleavage. Removal of the poly-Ub chain probably is important for overall substrate degradation due to steric considerations because the bulky chain would impede translocation of the attached polypeptide substrate through the opened pore of the proteasome. In fact, inhibition of Rpn11 severely reduces rates of proteolysis by the 26S proteasome [32,33]. Uch37 is the second deubiquitinating subunit of PA700 [33,34]. This protein does not exist in budding yeast, but is found in *Schizosaccharomyces pombe*, *Drosophila*, and all mammals [93]. Like Rpn11 it is found in the lid, and immunoelectron microscopy has localized it to a peripheral site of this structure [94]. Uch37 cleaves ubiquitin from the distal end of poly-Ub chains [34,35]. The exact significance of this type of activity is unclear. However, it is conceivable that it represents an 'editing' function whereby tagged proteins that do not become engaged in degradation within a reasonable time are deubiqutinated and released from the proteasome. Unlike Rpn11, decreased expression of Uch37 has little effect on cell viability, proteasome function, or global ubiquitin-dependent protein degradation [39,93].

Ubiquitin-Independent Degradation of Proteins by the 26S Proteasome

26S proteasomes are the only known enzymes that selectively degrade polyubiquitinated proteins. Nevertheless, evidence suggests that they also can degrade certain non-ubiquitinated proteins as well. To accomplish this task, 26S proteasomes must recognize and interact with features other than poly-Ub chains for selection of certain protein substrates. Indeed, recent work has directly demonstrated that the only function of the poly-Ub chain in the proteasome-dependent degradation is to deliver the substrate directly to this protease. Ornithine decarboxylase (ODC) is a well-studied example of a natural non-ubiqutinated protein to be degraded by the 26S proteasome [95-97]. ODC is targeted to the 26S proteasome through its binding to antizyme, an endogenous protein inhibitor [98]. Surprisingly, antizyme does not interact directly with the 26S proteasome, but it probably induces a conformational state of ODC that permits interaction of its carboxyterminus with a the 26S proteasome [99]. The same element recognizes both poly-Ub chains and the C-terminus of ODC, since poly-Ub chains competitively inhibit antizyme-induced ODC binding and degradation [100]. Once the C-terminus of ODC is engaged by PA700, ODC is inactivated and unfolded prior to degradation, and either or both of these processes require PA700-catalyzed ATP hydrolysis in mechanism that probably is related to the chaperone-like properties of PA700 [101]. Interestingly, the carboxylterminus of ODC is probably disordered, a structural feature that may dictate initial interaction with PA700. Other unstructured non-ubiquitinated proteins also interact with PA700 and are degraded by the 26S proteasome *in vitro* [88]. Thus, the ability of PA700 to interact with features of unstructured proteins, as might be required for the unfolding or translocation processes, also could dictate initial targeting of certain proteins to PA700. It is unclear to what extent this type of targeting occurs in intact cells.

Summary and Perspective

26S proteasomes are the main proteolytic organelles in neuronal cells as in other cell types, degrading most cytosolic, nuclear and ER-derived proteins with or without previous ubiquitination. The structure and function of the 26S proteasomes depends on the supply of ATP, therefore in conditions when ATP productions is impaired, such as during hypoxia or ischaemia, 26S proteasome function is generally inhibited. Moreover, they tend to separate into the core 20S proteasomes and the PA700/19S cap complexes. Those two biochemical changes have likely deep consequences on the metabolism of the affected neurnal cells, which have not been studied in detail. To better understand the pathophysiology of those events, a detailed knowledge of the structure and function of the 26S proteasomes is required. While their overall architecture and properties are quite well known, many aspects remain unsolved. The 20S core proteasome has been crystallized and extensively studied, while the detailed structure of the PA700 component of 26S proteasome remains elusive. Despite numerous attempts a crystal structure of PA700 or of the entire 26S proteasome has not been solved. This may reflect a high variability of the positioning of individual subunits within this subcomplex and/or their multiple posttranslational modifications, as well as substoichiometric association with other proteins, not considered *bona fide* components of the 26S proteasome.

REFERENCES

[1] Craiu A, Gaczynska M, Akapian T, Gramm CF, Fenteany G, Goldberg AL, Rock KL: Lactacystin and *clasto*-lactacystin b-lactone modify multiple proteasome b-subunits and inhibit intracellular protein degradation and major histocompatibility complex class I antigen presentation. *J Biol Chem* 1997, 272:13437-13445.

[2] Rock KL, Gramm C, Rothstein L, Clark K, Stein R, Dick L, Hwang D, Goldberg AL: Inhibitors of the proteasome block the degradation of most cell proteins and the generation of peptides presented on MHC class I molecules. *Cell* 1994, 78:761-771.

[3] Voges D, Zwickl P, Baumeister W: The 26S proteasome: a molecular machine designed for controlled proteolysis. *Ann Rev Biochem* 1999, 68:1015-1068.

[4] DeMartino GN, Slaughter CA: The proteasome, a novel protease regulated by multiple mechanisms. *J Biol Chem* 1999, 274:22123-22126.

[5] Glickman MH, Rubin DM, Fried VA, Fischer JE, Finley D: The regulatory particle of the *Saccharomyces cerevisiae* proteasome. *Mol Cell Biol* 1998, 18:3149-3162.

[6] Hoffman L, Pratt G, Rechsteiner M: Multiple forms of the 20S multicatalytic and the 26S ubiquitin-ATP-dependent proteases from rabbit reticulocyte lysate. *J Biol Chem* 1992, 267:22362-22368.

[7] Ma C-P, Vu JH, Proske RJ, Slaughter CA, DeMartino GN: Identification, purification, and characterization of a high-molecular weight, ATP-dependent activator (PA700) of the 20S proteasome. *J Biol Chem* 1994, 269:3539-3547.

[8] Udvardy A: Purification and characterization of a multiprotein component of the Drosophila 26S (1500 kDa) proteolytic complex. *J Biol Chem* 1993, 268:9055-9062.

[9] Peters J-M, Walsh MJ, Franke WW: An abundant and ubiquitous homo-oligomeric ring-shaped ATPase particle related to the putative vesicle fusion proteins Sec 18p and NSF. *EMB* 1990, 9:1757-1767.

[10] Adams GM, Falke S, Goldberg AL, Slaughter CA, DeMartino GN, Gogol EP: Structural and functional effects of PA700 and modulator protein on proteasomes. *J Mol Biol* 1997, 273:646-657.

[11] Coux O, Tanaka K, Goldberg AL: Structure and functions of the 20S and 26S proteasomes. *Ann Rev Biochem* 1996, 65:801-847.

[12] Tanahashi N, Murakami Y, Minami Y, Shimbara N, Hendil KB, Tanaka K: Hybrid proteasomes. Induction by interferon-gamma and contribution to ATP-dependent proteolysis. *J Biol Chem* 2000, 275:14336-14345.

[13] Finley D, Tanaka K, Mann C, Feldmann H, Hochstrasser M, Vierstra R, Johnston S, Hampton R, Haber J, Mccusker J, Silver P, Frontali L, Thorsness P, Varshavsky A, Byers B, Madura K, Reed SI, Wolf D, Jentsch S, Sommer T, Baumeister W, Goldberg A, Fried V, Rubin DM, Toh-e A: Unified nomenclature for subunits of the Saccharomyces cerevisiae proteasome regulatory particle. *TIBS* 1998, 23:244-245.

[14] Wojcik C: Rpt2. *AfCS-Nature Molecule Pages* 2005, doi:10.1038/mp.a003572.01.

[15] Xie Y, Varshavsky A: RPN4 is a ligand, substrate, and transcriptional regulator of the 26S proteasome: a negative reedback circuit. *Proc Natl Acad Sci (USA)* 2001, 98:3056-3061.

[16] Beyer A: Sequence analysis of the AAA protein family. *Protein Science* 1997, 6:2043-2058.

[17] Confalonieri F, Duguet M: A 200-amino acid ATPase module in search of a basic function. *BioEss* 1995, 17:639-650.

[18] Vale RD: AAA proteins: lords of the ring. *JCB* 2000, 150:F13-F19.

[19] Ogura T, Wilkinson AJ: AAA + superfamily ATPases: common structure-diverse function. *Genes to Cells* 2001, 6:575-597.

[20] DeMartino GN, Moomaw CR, Zagnitko OP, Proske RJ, Ma C-P, Afendis SJ, Swaffield JC, Slaughter CA: PA700, an ATP-dependent activator of the 20S proteasome, is an ATPase containing multiple members of a nucleotide-binding protein family. *J Biol Chem* 1994, 269:20878-20884.

[21] Hoffman L, Rechsteiner M: Nucleotidase activities of the 26S proteasome and its regulatory complex. *J Biol Chem* 1996, 271:32538-32545.

[22] Lupas A, Baumeister W, Hofmann K: A repetitive sequence in subunits of the 26S proteasome and 20S cyclosome (anaphase-promoting complex). *TIBS* 1997, 22:195-196.

[23] DeMarini DJ, Papa FR, Swaminathan S, Ursic D, Rasmussen TP, Culbertson MR, Hochstrasser M: The yeast SEN3 gene encodes a regulatory subunit of the 26S proteasome complex required for ubiquitin-dependent protein degradation in vivo. *MCB* 1995, 15:6311-6321.

[24] Hampton RY, Gardner RG, Rine J: Role of the 26S proteasome and *HRD* genes in the degradation of 3-hydroxy-3methylglutaryl-CoA reductase, an integral endoplasmic reticulum membrane protein. *Molecular Biology of the Cell* 1996, 7:2029-2044.

[25] Yokota K, Kagawa S, Shimizu Y, Akioka H, Tsurumi C, Noda C, Fujimura M, Yokosawa H, Fujiwara T, Takahashi E, Ohba M, Yamasaki M, DeMartino GN, Slaughter CA, Toh-e A, Tanaka K: cDNA cloning of p112, the largest regulatory subunit of the human 26S proteasome, and functional analysis of its yeast homogue, Sen3p. *Molecular Biology of the Cell* 1996, 7:853-870.

[26] Realini C, Rogers SW, Rechsteiner M: KEKE motifs. Proposed roles in protein-protein association and presentation of peptides by MHC class I receptors. *FEBS Lett* 1994, 348:109-113.

[27] Aravind L, Ponting CP: Homologues of 26S proteasome subunits are regulators of transcription and translation. *Protein Sci* 1998, 7:1250-1254.

[28] Hofmann K, Bucher P: The PCI domain: a common theme in three multiprotein complexes. *TIBS* 1998, 23:204-205.

[29] Deveraux Q, Ustrell V, Pickart C, Rechsteiner M: A 26S protease subunit that binds ubiquitin conjugates. *J Biol Chem* 1994, 269:7059-7061.

[30] Hiyama H, Yokoi M, Masutani C, Sugasawa K, Maekawa T, Tanaka K, Hoeijamakers JHJ, Hanaoka F: Interaction of hHR23 with S5a: the ubiquitin-like domain of hHR23 mediates interaction with S5a subunit of 26S proteasome. *J Biol Chem* 1999, 274:28019-28025.

[31] Bork P, Rohde K: More von Willebrand factor type A domains? Sequence similarities with malaria thrombospondin-related anonymous protein, dihydropyridine-sensitive calcium channel and inter-alpha-trypsin inhibitor. *Biochem J* 1991, 279 (Pt 3):908-910.

[32] Verma R, Aravind L, Oania R, McDonald WH, Yates JR, Koonin EV, Deshaies RJ: Role of Rpn11 metalloprotease in deubiquitination and degradation by the 26S proteasome. *Science* 2002, 298:611-615.

[33] Yao T, Cohen RE: a cryptic protease couples deubiquitination and degradation by the proteasome. *Natur* 2002, 419:403-407.

[34] Lam YA, Xu W, DeMartino GN, Cohen RE: Editing of ubiquitin conjugates by an isopeptidase in the 26S proteasome. *Nature* 1997, 385:737-740.

[35] Lam YA, DeMartino GN, Pickart CM, Cohen RE: Specificity of the ubiquitin isopeptidase in the PA700 regulatory complex of 26S proteasomes. *J Biol Chem* 1997, 272:28483-28446.

[36] Wing SS: Deubiquitinating enzymes-the importance of driving in reverse along the ubiquitin-proteasome pathway. *Inter J Biochem Cell Biol* 2003, 35:590-605.

[37] Fujimuro M, Tanaka K, Yokasawa H, Toh-e A: Son1p is a component of the 26S proteasome of the yeast *Saccharomyces cerevisiae*. *FEBS Lett* 1998, 423:149-154.

[38] Mannhaupt G, Schnall R, Karpov V, Vetter I, Feldmann H: Rpn4 acts as a transcriptional factor by binding to PACE, a nonamer box found upstream of 26S proteasomal and other genes in yeast. *FEBS Lett* 1999, 450:27-34.

[39] Wojcik C, DeMartino GN. Analysis of *Drosophila* 26S proteasome using RNA interference. J Biol Chem 2002, 277:6188-6197.

[40] Meiners S, Heyken D, Weller A, Ludwig A, Stangl K, Kloetzel P-M: Inhibition of proteasome activity induces concerted expression of proteasome genes and *de novo* formation of mammalian proteasomes. *J Biol Chem* 2003, 278:21517-21525.

[41] Ferrell K, Wilkinson CRM, Dubiel W, Gordon C: Regulatory subunit interactions of the 26S proteasome, a complex problem. *TIBS* 2000, 25:83-88.

[42] Kopp F, Kuehn L: Orientation of the 19S regulator relative to the 20S core proteasome: an immunoelectron microscopic study. *JMB* 2003, 329:9-14.

[43] Yoshimura T, Kameyama K, Takagi T, Ikai A, Tokunaga F, Koide T, Tanahashi N, Tamura T, Cejka Z, Baumeister W, Tanaka K, Ichihara A: Molecular characterization of the "26S" proteasome complex from rat liver. *JSB* 1993, 111:200-211.

[44] Glickman MH, Rubin DM, Coux O, Wefes I, Pfeifer G, Cjeka Z, Baumeister W, Fried V, Finley D: A subcomplex of the proteasome regulatory particle required for

ubiquitin-conjugate degradation and related to the COP9-signalosome and eIF3. *Cell* 1998, 94:615-623.

[45] Kajava AV: What curves alpha-solenoids? Evidence for an alpha-helical toroid structure of Rpn1 and Rpn2 proteins of hte 26S proteasome. *J Biol Chem* 2002, 277:49791-49798.

[46] Mason GGF, Murray RZ, Pappin D, Rivett AJ: Phosphorylation of ATPase subunits of the 26S proteasome. *FEBS Lett* 1998, 430:269-274.

[47] Bose S, Brooks P, Mason GGF, Rivett AJ: g-Interferon decreases the level of 26S proteasomes and changes the pattern of phosphorylation. *Biochem J* 2001, 353:291-297.

[48] Zhang F, Su K, Yang X, Bowe DB, Paterson AJ, Kudlow JE: O-GlcNAc modification is an endogenous inhibitor of the proteasome. *Cell* 2003, 115:715-725.

[49] Sümegi M, Hunyadi-Gulyás E, Medzihradszky KF, Udvardy A: 26S proteasome subunits are O-linked *N*-acetylglucosamine-modified in *Drosophila meanogaster*. *Biochem Biophys Res Commun* 2003, 312:1284-1289.

[50] Benaroudj N, Goldberg AL: PAN, the proteasome-activating nucleotidase from archaebacteria, is a protein-unfolding molecular chaperone. *Nat Cell Biol* 2000, 2:833-839.

[51] Elsasser S, Gali RR, Schwickart M, Larsen CN, Leggett DS, Muller B, Feng MT, Tubing F, Dittmar GA, Finley D: Proteasome subunit Rpn1 binds ubiquitin-like protein domains. *Nat Cell Biol* 2002, 4:725-730.

[52] Leggett DS, Hanna J, Borodovsky A, Crosas B, Schmidt M, Baker RT, Waltz T, Ploeugh H, Finley D: Multiple associated proteins regulate proteasome structure and function. *Mol Cell* 2002, 10:498-507.

[53] van Nocker S, Sadis S, Rubin DM, Glickman M, Fu H, Coux O, Wefes I, Finley D, Vierstra RD: The multiubiquitin-chain-binding protein Mcb1 is a component of the 26S proteasome in *Saccharomyces cerevisiae* and plays a nonessential, substrate-specific role in protein turnover. *Mol Cell Biol* 1996, 16:6020-6028.

[54] Yang Y, Früh K, Ahn K, Peterson PA: *In vivo* assembly of the proteasomal complexes, implications for antigen processing. *J Biol Chem* 1995, 270:27687-27694.

[55] Baumeister W, Walz J, Zühl F, Seemüller E: The proteasome: paradigm of a self-compartmentalizing protease. *Cell* 1998, 92:367-380.

[56] Schmidtke G, Kraft R, Kostka S, Henklein P, Frömmel C, Löwe J, Huber R, Kloetzel P-M, Schmidt M: Analysis of mammalian 20S proteasome biogenesis: the maturation of b-subunits is an ordered two-step mechanism involving autocatalysis. *EMB* 1996, 15:6887-6898.

[57] DeMartino GN, Proske RJ, Moomaw CR, Strong AA, Song X, Hisamatsu H, Tanaka K, Slaughter CA: Identification, purification, and characterization of a PA700-dependent activator of the proteasome. *J Biol Chem* 1996, 271:3112-3118.

[58] Imai J, Maruya M, Yashiroda H, Yahara I, Tanaka K: The molecular chaperone Hsp90 plays a role in the assembly and maintenance of the 26S proteasome. *EMB* 2003, 22:3557-3567.

[59] Gorbea C, Goellner GM, Teter K, Holmes RK, Rechsteiner M: Characterization of mammalian Ecm29, a 26 S proteasome-associated protein that localizes to the nucleus and membrane vesicles. *J Biol Chem* 2004, 279:54849-54861.

[60] Rubin DM, Glickman MH, Larsen CN, Druvakumar S, Finley D: Active site mutants in the six regulatory particle ATPases reveal multiple roles for ATP in the proteasome. *EMBO J* 1998, 17:4909-4919.

[61] Babbitt SE, Kiss A, Deffenbaugh AE, Chang YH, Bailly E, Erdjument-Bromage H, Tempst P, Buranda T, Sklar LA, Baumler J, Gogol E, Skowyra D: ATP hydrolysis-dependent disassembly of the 26S proteasome is part of the catalytic cycle. *Cell* 2005, 121:553-565.

[62] Groll M, Ditzel L, Lowe J, Stock D, Bochtler M, Bartunik HD, Huber R: Structure of the 20S proteasome from yeast at 2.4A resolution. *Nature* 1997, 386:463-471.

[63] Bochtler M, Ditzel L, Groll M, Hartmann C, Huber R: The proteasome. *Ann Rev Biophys Biomol Struct* 1999, 28:295-317.

[64] Löwe J, Stock D, Jap B, Zwickl P, Baumeister W, Huber R: Crystal structure of the 20S proteasome from the archaeon T. acidophilum at 3.4 Ä resolution. *Sci* 1995, 268:533-539.

[65] Whitby FG, Masters EI, Kramer L, Knowlton JR, Yao Y, Wang CC, Hill CP: Structural basis for the activation of 20S proteasomes by 11S regulators. *Natur* 2001, 408:115-120.

[66] Groll M, Bajorek M, Kohler A, Moroder L, Rubin DM, Huber R, Glickman MN, Finley D: A gated channel into the proteasome core particle. *Nat Struct Biol* 2001, 11:1062-1067.

[67] Köhler A, Cascio P, Leggett DS, Woo KM, Goldberg AL, Finley D: The axial channel of the proteasome core particle is gated by the Rpt2 ATPase and controls both substrate entry and product release. *Mol Cell* 2001, 7:1143-1152.

[68] Thrower JS, Hoffman L, Rechsteiner M, Pickart CM: Recognition of the polyubiquitin proteolytic signal. *EMBO* 2001, 19:94-102.

[69] Ferrell K, Deveraux Q, van Nocker S, Rechsteiner M: Molecular cloning and expression of a multiubiquitin chain binding subunit of the human 26S protease. *FEBS Lett* 1996, 381:143-148.

[70] van Nocker S, Deveraux Q, Rechsteiner M, Vierstra RD: *Arabidopsis MBP1* gene encodes a conserved ubiquitin recognition component of the 26S proteasome. *PNAS* 1996, 93:856-860.

[71] Beal R, Deveraux Q, Xia G, Rechsteiner M, Pickart C: Surface hydrophobic residues of multiubiquitin chains essential for proteolytic targeting. *PNAS* 1996, 93:861-866.

[72] Kominami K, Okura N, Kawamura M, DeMartino GN, Slaughter CA, Simbara N, Chung CH, Fujimuro M, Yokosawa H, Shimizu Y, Tanahashi N, Tanaka K, Toh-e A: Yeast counterparts of subunits S5a and p58 (S3) of the human 26S proteasome are encoded by two multicopy suppressors of *nin1-1*. *Molecular Biology of the Cell* 1997, 8:171-187.

[73] Hofmann K, Falquet L: A ubiquitin-interacting motif conserved in components of the proteasomal and lysosomal protein degradation systems. *Trends Biochem Sci* 2001, 26:347-350.

[74] Mueller TD, Feigon J: Structural determinants for the binding of ubiquitin-like domains to the proteasome. *EMBO J* 2003, 22:4634-4645.

[75] Fu H, Sadis S, Rubin DM, Glickman M, van Nocker S, Finley D, et al. Multiubiquitin chain binding and protein degradation are mediated by distinct domains within the 26S proteasome subunit Mcb1. J Biol Chem 273, 1970-1981. 1998.

[76] Young P, Deveraux Q, Beal RE, Pickart CM, Rechsteiner M. Characterization of two polyubiquitin binding sites in the 26S protease subunit 5a. Journal of Biological Chemistry 273, 5461-5467. 1998.

[77] Lam YA, Lawson TG, Velayultham M, Zweierm J.L., Pickart CM. A proteasomal ATPase subunit recognizes the polyubiquitin degradation signal. Nature (London) 416, 763-767. 2002.

[78] Verma R, McDonald H, Yates JR, Deshaies RJ: Selective degradation of ubiquitinated Sic1 by purified 26S proteasome yields active S phase cyclin-Cdk. *Mol Cell* 2001, 8:439-448.

[79] Braun S, Matuschewski K, Rape M, Thoms S, Jentsch S: Role of the ubiquitin-selective CDC48(UFD1/NPL4)chaperone (segregase) in ERAD of OLE1 and other substrates. *EMB* 2002, 21:615-621.

[80] Dai RM, Chen E, Longo DL, Gorbea CM, Li C-CH: Involvement of valsoin-containing protein, an ATPase co-purified with IkBa and 26S proteasome, in ubiquitin-proteasome-mediated degradation of IkBa. *J Biol Chem* 1998, 273:3562-3573.

[81] Verma R, Chen S, Feldman R, Schieltz D, Yates J, Dohmen J, Deshaies RJ: Proteasomal proteomics: identification of nucleotide-sensitive proteasomal-interacting proteins by mass spectrometric analysis of affinity-purified proteasomes. *Mol Biol Cell* 2000, 11:3425-3439.

[82] Braun BC, Glickman M, Kraft R, Dahlmann B, Kloetzel P-M, Finley D, Schmidt M: The base of the proteasome regulatory particle exhibits chaperone-like activity. *Nature Cell Biol* 1999, 1:221-226.

[83] Strickland E, Hakala K, Thomas PJ, DeMartino GN: Recognition of misfolding proteins by PA700, the regulatory subcomplex of the 26S proteasome. *J Biol Chem* 2000, 275:5565-5572.

[84] Liu C, Millen L, Roman TB, Xiong H, Gilbert HF, Novia R, DeMartino GN, Thomas PJ: Conformational remodeling of proteasomal substrates by PA700, the 19S regulatory complex of the 26S proteasome. *J Biol Chem* 2002, 277:26815-26820.

[85] Lee C, Schwartz MP, Prakash S, Iwakura M, Matouschek A: ATP-dependent proteases degrade their substrates by processively unraveling them from the degradation signal. *Mol Cell* 2001, 7:627-737.

[86] Akopian TN, Kisselev AF, Goldberg AL: Processive degradation of proteins and other catatlytic properties of the proteasome from *Thermoplasma acidophilum*. *J Biol Chem* 1997, 272:1791-1798.

[87] Dick LR, Moomaw CR, DeMartino GN, Slaughter CA: Degradation of oxidized insulin B chain by the multiproteinase complex macropain (proteasome). *Biochemistry* 1991, 30:2725-2734.

[88] Liu C-W, Corboy MJ, DeMartino GN, Thomas PJ: Endoproteolytic activity of the proteasome. *Science* 2004, 299:408-411.

[89] Rape M, Jentsch S. Taking a bite: proteasomal protein processing. Nature Cell Biol. 4, E113-E116. 2002.

[90] Palombella VJ, Rando OJ, Goldberg AL, Maniatis T: The ubiquitin-proteasome pathway is required for processing th NF-kB1 precursor protein and the activation of NF-kB. *Cell* 1994, 78:773-785.

[91] Lin L, DeMartino GN, Greene WC: Cotranslational biogenesis of NF-kappaB p50 by the 26S proteasome. *Cell* 1998, 92:819-828.

[92] Lundgren J, Masson P, Realini C, Young P: Use of RNA interference and complementation to study the function of the *Drosophila* and human 26S proteasome subunits. *Mol Cell Biol* 2004, 23:5320-5330.

[93] Li T, Naqvi NI, Yanf H, Teo TS: Identification of a 26S proteasome-associated UCH in fission yeast. *Biochem Biophys Res Commun* 2000, 272:270-275.

[94] Hölzl H, Kapelari B, Kellermann J, Seemüller E, Sümegi M, Udvardy A, Medalia O, Sperling J, Müller SA, Engel A, Baumeister W: The regulatory complex of *Drosophila melanogaster* 26S proteasomes: subunit composition and localization of a deubiquitylating enzyme. *JCB* 2000, 150:119-129.

[95] Murakami Y, Matsufuji S, Kameji T, Hayashi S, Igarashi K, Tamura T, Tanaka K, Ichihara A: Ornithine decarboxylase is degraded by the 26S proteasome without ubiquitination. *Natur* 1992, 360:597-599.

[96] Tokunaga F, Goto T, Koide T, Murakami Y, Hayashi S, Tamura T, Tanaka K, Ichihara A: ATP- and antizyme-dependent endoproteolysis of ornithine decarboxylase to oligopeptides by the 26S proteasome. *J Biol Chem* 1994, 269:17382-17385.

[97] Elias S, Bercovich B, Kahana C, Coffino P, Fiischer M, Hilt W, Wolf DH, Ciechanover A: Degradation of ornithine decarboxylase by the mammalian and yeast 26S proteasome complexes requires all the components of the protease. *EJB* 1995, 229:276-283.

[98] Coffino P: Regulation of polyamines by antizyme. *Nat Rev Mol Cell Biol* 2001, 2:188-194.

[99] Murakami Y, Matsufuji S, Hayashi S, Tanahashi N, Tanaka K: Degradation of ornithine decarboxylase by the 26S proteasome. *BBRC* 2000, 267:1-6.

[100] Zhang M, Pickart CM, Coffino P: Determinants of proteasome recognition of ornithine decarboxylase, a ubiquitin-independent substrate. *EMB* 2003, 22:1488-1496.

[101] Murakami Y, Matsufuji S, Hayashi S, Tanahashi N, Tanaka K: ATP-dependent inactivation and sequestration of ornithine decarboxylase by the 26S proteasome are prerequisites for degradation. *Mol Cell Biol* 1999, 19:7216-7227.

In: The Ubiquitin Proteasome System… ISBN 978-1-60021-749-4
Eds: Mario Di Napoli and Cezary Wojcik, pp. 159-168 © 2007 Nova Science Publishers, Inc.

Chapter 8

Assays with Natural Substrates: Nivel Tools to Address the Complexity of Protein Degradation by 26S Proteasomes

K. Matthew Scaglione and Dorota Skowyra*

Edward A. Doisy Department of Biochemistry and Molecular Biology, Saint Louis University School of Medicine, St. Louis, MO, USA.

Abstract

The 26S proteasome has long been viewed as a major therapeutic target. However, in the past 20 years only inhibitors of the proteolytic sites have been developed. Such a focus was primarily the result of the limited availability of assays for monitoring activity of the 26S proteasome. Due to the difficulties in preparation of naturally polyubiquitinated proteins, these assays were based on artificial model substrates, typically monomeric proteins that could be either polyubiquitinated *in vitro* without a specific E3 ubiquitin ligase (lysozyme, DHFR, Ub-Pro-β-gal) or degraded without polyubiquitination (fluorogenic peptides, loosely structured caseine, denatured ovalbumin). Although these reagents proved invaluable in uncovering the basic principles of proteasomal function, it becomes increasingly clear that they did not allow one to address the puzzling complexity of the 19S cap composition, indispensable in the highly controlled and rapid ($T_{1/2}$ <5 min) degradation of naturally unstable regulatory proteins and signalling molecules. Recent development of *in vitro* assays for specific ubiquitination and degradation of natural substrates from yeast, pioneered by several research groups including ours, has turned the tide. As we discuss below, these studies,

* Correspondence concerning this article should be addressed to Dr. Dorota Skowyra, PhD; Edward A. Doisy Department of Biochemistry and Molecular Biology, Saint Louis University School of Medicine, St. Louis, MO, USA. Phone: (314) 977 9280; Fax: (314) 977 9205 ; E-mail: skowyrad@slu.edu.

although in their early stages, have already revealed an unanticipated complexity of the substrate recruitment mechanism and the catalytic cycle itself.

Keywords: 26S proteasome, PIPs (proteasome-interacting proteins), SCF ubiquitin ligase.

Abbreviations

AAA-type, ATPases associated with cellular activities; ADP adenosine diphosphate; APC, anaphase promoting complex; ATP, adenosine triphosphate; Cdc4, cell division cycle 4; CDK, cyclin dependent kinase; Ddi1, DNA damage inducible 1; DHFR, dihydrofolate reductase; E2, ubiquitin-conjugating enzyme; E3, ubiquitin ligase; Hul5, HECT [homologous to E6-associated protein (E6AP) C-terminus] ubiquitin ligase 5; Hs1U, heat shock locus U; I□B, inhibitor of NF (nuclear factor)–□B; K48-type, lysine 48-type; Met4, methionine requiring 4; NF□B, nuclear factor-□B; ODC, ornithine decarboxylase; PIPs, proteasome-interacting proteins; Rad23, radiation sensitive 23; Rpn3ts, regulatory particle non-ATPase 3 temperature sensitive; S-CDK, S-phase cyclin-dependent kinase; SCF, Skp1, Cullin, F-Box protein; Sic1, substrate/subunit inhibitor of cyclin-dependent protein kinase; Skp1, suppressor of kinetochore protein mutant; Ub-Pro-□-gal, ubiquitin-proline-beta-gal; UBA, ubiquitin-associated; UbL, ubiquitin-like; Ufd1, ubiquitin fusion degradation protein 1.

The Emerging Complexity of Substrate Recruitment

The 26S proteasome recruits most of its natural substrates via a post-translational modification with a K48-type of polyubiquitin [1,2] (see Chapter 3). The currently dominating view is that the timing and specificity of ubiquitin-mediated proteolysis are controlled mainly at the step of substrate recognition by a specific ubiquitin ligase. The variety of substrate recognition mechanisms employed by ubiquitin ligases well reflects the large number of proteins that are targeted for proteolysis. However, in addition to the polyubiquitin chain, at least three other factors have recently been implicated in substrate recruitment. First, many ubiquitin ligases interact with the 26S proteasome, including Ubr1 and Ufd4 [3], SCF [4,5], APC [4], and Hul5 [6]. At least in the case of Ufd4 the interaction is direct and essential for substrate instability [7]. Second, various adapter proteins can bind the 26S proteasome and participate in recognition of the polyubiquitin chain [8]. These include the UbL-UBA domain proteins Rad23, Dsk2, and Ddi1, and the AAA-type ATPase Ufd1/Cdc48/p97. Since different adapter proteins facilitate degradation of different naturally unstable substrates, the adapter proteins may contribute to the specificity of proteolysis by a yet unknown mechanism [9]. This possibility extends the conclusion first derived from genetic analysis of *rpn3ts* alleles, which provided an elegant demonstration that the degradation of several naturally unstable cell cycle regulatory proteins is differentially regulated on the level of the 26S proteasome [10]. Finally, on the example of the Met4 transcriptional activator of the methionine biosynthetic network in yeast we have recently shown that specific sulfur amino acids promote degradation of polyubiquitinated Met4 by destabilizing its interaction with SCF^{Met30} [11], revealing yet another layer of complexity in the mechanism of substrate recruitment.

Dissecting Degradation of Natural Substrates as an Attractive Experimental Approach

Several features make the naturally unstable substrates ideal for dissecting the complexity of proteasomal function. The natural substrates are usually degraded more rapidly than artificial model substrates both *in vitro* and *in vivo* ($T_{1/2}$~ 2-5 min vs ~30 min or longer in the case of artificial model substrates), carry all the evolutionarily selected features ensuring their instability, and are linked to the network of cellular signaling. In fact, since the role of degradation of many naturally unstable regulatory molecules is to facilitate signal transduction via promoting release and/or remodeling of the associating proteins, many natural substrates are recruited to the 26S proteasome as part of multiprotein complexes. This aspect of proteasomal function is well illustrated by the classic examples of NFκB, which is activated *in vivo* by selective degradation of its inhibitor IκB [12], and S-phase cyclin-dependent kinase (S-CDK), which is activated at the G1/S phase transition upon degradation of its inhibitor Sic1 *in vivo* [13], and *in vitro* [14]. In the latter case, polyubiquitinated Sic1 remains in a tight complex with the SCF^{Cdc4} ubiquitin ligase [15] and with S-CDK [14] both of which interact with the 26S proteasome [4,5,16]. This phenomenon is not limited to SCF substrates, as a large number of proteins can interact with the 26S proteasome, either via binding to a substrate or via direct interaction with the 19S. Among the proteasome-interacting proteins (PIPs) that can be trapped in a complex with the 26S in the presence of ATPγS are ubiquitination machineries, deubiquitinating enzymes, UbL-UBA adapters, molecular chaperones and many other proteins whose role in proteolysis is yet unknown [4,5].

Proteasome-Interacting Proteins (PIPS) Could Play a Role in the Catalytic Cycle of the 26S Proteasome – The 'Chew and Spew' Model

The possibility that PIPs play a role in the catalytic cycle of the 26S proteasome is supported by the observation that release of the bulk of PIPs from immunopurified 26S particles is accompanied by ATP hydrolysis-dependent dissociation and disassembly of the 19S [5]. 30-60% of total 26S particles from yeast extracts undergoes the disassembly in a manner linked to both ATP hydrolysis and dissociation of the PIPs, suggesting that a large number of endogenous substrates are degraded via this mechanism. Indeed, using a biochemical reconstruction approach with purified Sic1, the prototype substrate of SCF^{Cdc4} ubiquitin ligase, we have shown that the ATP hydrolysis-mediated dissociation and disassembly of the 19S cap is strictly dependent on the degradation of a substrate and coincides with release of both the product peptides and the PIPs, including SCF^{Cdc4} [5]. We

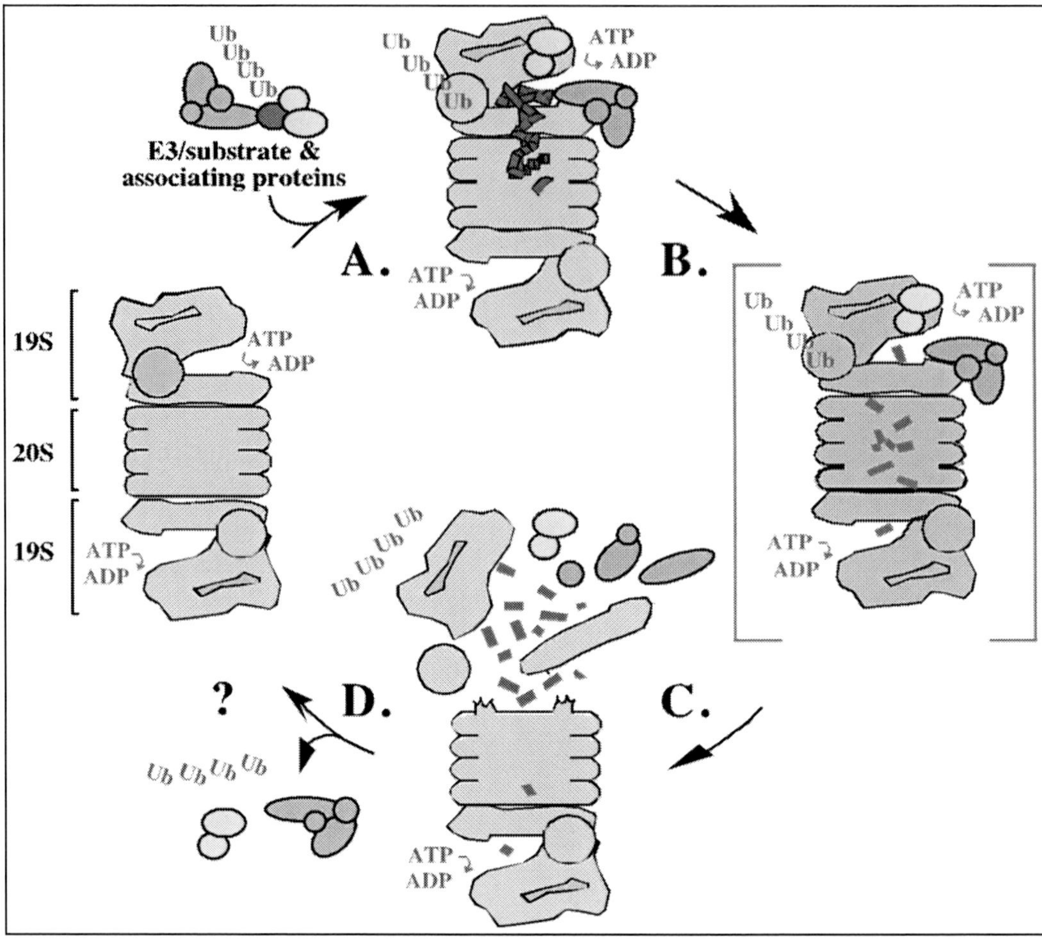

Figure 1. The 'chew and spew' model for the catalytic cycle of the 26S proteasome with a naturally unstable substrate. For simplicity, the model focuses on the doubly capped proteasomes. (A). Recruitment of a polyubiquitinated substrate (red) in a complex with the E3 ubiquitin ligase (orange) and associating proteins (yellow), at least some of which directly interact the 26S proteasome and remain bound to the 26S after substrate unfolding. (B). Degradation of a substrate triggers coupling between ATP hydrolysis and a conformational change in the ATPases via a mechanism possibly linked to the appearance of product peptides (red brackets symbolically mark the coupling event with red ATP->ADP transition indicating the critical hydrolysis reaction). (C). The coupling mechanism generates a powerful mechanical force, which triggers the disassembly of the 19S into subunits and/or sub-complexes, the release and disassembly of the E3 and other PIPs, and a burst-type 'spew-like' product peptides release. (D). The 19S components re-associate with the 20S, resetting the machinery for a new round of the catalytic cycle. Note that the model shows only one possible scheme for the disassembly: dissociation of the E3-bound 19S cap, with a singly capped proteasome as a product. In this scheme, the 19S-free end of the singly capped proteasome could facilitate accelerated release of product peptides, essentially as proposed by Kisselev *et al.* [21,22], based on model derived from 11S-20S-19S hybrid particles. While disassembly of two 19S caps is also possible, as a result of positive cooperativity between the 20S ends [19,25], disassembly of the E3-free 19S is unlikely because it could not explain release of the E3.

obtained similar results with several other substrates, including the Met4 substrate of SCF^{Met30}, and the Cdc34 E2 ubiquitin-conjugating enzyme autoubiquitinated in SCF-dependent manner. Based on these observations, we proposed that the ATP hydrolysis-dependent disassembly of the 26S proteasome is part of the catalytic cycle (Figure 1).

The proposed 'chew and spew' model establishes a new view on the catalytic cycle of the 26S proteasome, in which degradation of a substrate triggers coupling between ATP hydrolysis and disassembly of the 19S. What could be the molecular basis of this mechanism? One possibility is that product peptides induce a conformational change in the adjacent ATPases by inducing an allosteric change within the 20S. Indeed, analysis of the proteasome-associated ATPase HslU has demonstrated the potential of AAA-type ATPases to change conformation in a nucleotide-dependent manner [17]. In addition, several studies have suggested that allosteric transitions play a role in function of the proteasome [18,19,20,21,22]. Especially interesting are the effects associated with presence of hydrophobic peptides, which have been proposed to bind to several noncatalytic sites in the 20S and to accelerate proteolysis via two mechanisms. First is to allosterically activate the catalytic sites, leading to a more effective 'bite and chew' degradation [20,21]. Second is to accelerate the release of product peptides by promoting an opening in the α rings [21]. The role of this mechanism is thought to be functionally similar to the role of the 11S activator, which acts by opening of the 19S cap-free end of the proteasome [23]. A similar role could be assigned to the ATP hydrolysis-dependent dissociation of the 19S cap. This mechanism not only generates the 19S cap-free end of the proteasome, but it was also linked to the quickest type of degradation typical of naturally unstable proteins, and to the release of the product peptides [5].

A second model predicts that product peptides are required, but not sufficient, and that the coupling mechanism depends on a change induced directly in the 19S cap. Such a change could be induced in the ATPases in response to interaction with some 'coupling' factor recruited with the substrate. S-CDK and the SCF^{Cdc4} are good candidates for such a factor, as both remain in a complex with polyubiquitinated Sic1 [14,15] and interact with the 26S proteasome [4,5,16]. Interestingly, endogenous SCF interacts with the 26S proteasome even when thermal inactivation of the *cdc34-1* mutant protein blocks ubiquitination [4]. If substrate ubiquitination plays no role in SCF recruitment to the proteasome, its degradation may be insufficient for SCF release. In such a case, direct interaction with the E3 may play a role in coordinating release of the degradation products with disassembly of the 26S.

Regardless of the precise molecular mechanism, the ultimate proof that the disassembly is part of the catalytic cycle will come with testing the prediction that the 19S subcomplexes and/or subunits formed as result of the ATP hydrolysis-dependent disassembly can rapidly reassemble with the 20S, either spontaneously or with help of cofactors. An interesting question is whether this type of reassembly differs from the *de novo* assembly of the 19S caps and from the reassembly of an intact 19S cap generated from the 26S via nucleotide depletion (see Chapter 7). Although the 19S cap generated by this mechanism can reassemble with the 20S proteasome upon addition of ATP, purified 19S cap generated by nucleotide depletion binds the 20S proteasome very inefficiently, with ~50-fold excess of the 19S cap required for quantitative reassembly with the 20S proteasome even in the presence of a modulator [24,25]. Thus, it is possible that the ATP hydrolysis-dependent disassembly of the 19S cap is necessary for its quick and efficient reassembly with the 20S proteasome. Whether and how the reassembly process is affected by the PIPs remains to be determined.

The 'Chew and Spew' Mechanism Could Play a Role in Degradation of the most rapidly Turned-Over but not Necessarily all Substrates

Two observations suggest that the ATP hydrolysis-dependent disassembly of the 19S cap from the 26S particles plays a role in degradation of the most rapidly turned over class of substrates, characterized by a half-life in the range of a few minutes. First, this mechanism has been discovered in assays with the rapidly degraded natural substrates ($T_{1/2}$ of Sic1 and Met4 ~ 2-5 min) and linked to a similarly rapid release of the bulk of endogenous PIPs, including the E3 components of ubiquitination machineries [5]. Since many E3s directly interact with the proteasome, it appears that they would have to be removed prior to recruitment of the next E3-substrate complex. In this view, release of the E3 might be an obligatory step in the degradation of this class of substrates. Second, we have linked the disassembly to a burst-type, spew-like release of product peptides [5]. Although product peptides could leak out via a standard opening of the gated channel in assembled 26S particles and such mechanism could be sufficient to support degradation of the slowly turned-over substrates, the model derived from the 11S-20S-19S hybrid proteasomes suggests that the accelerated release of product peptides could promote more rapid degradation (see Chapter 9).

These possibilities suggest that what is essential for function of the 26S proteasome needs to be evaluated from the perspective of how the lack of this function affects the precise cellular activity and/or the transition it evolved to facilitate rather than the mechanistic capacity of the machinery to do the job out of biological context. Indeed, different features are essential for different activities of the machinery. Even dependence on a polyubiquitin tag, the most established hallmark of ubiquitin-mediated proteolysis, is not absolutely necessary for degradation of proteasomal substrates. Not only the proteasome can degrade many model substrates without polyubiquitination (casein, calmoduline, lysozyme, ODC, p21, are all established substrates of the proteasome), but also blocking ubiquitination of many naturally unstable proteins does not make them 'rock' stable [26]. For example, Sic1 is stabilized five fold in *cdc34, skp1, cdc53, and cdc4 ts* mutants of SCF, from a half-life of 5 to 25 minutes. Under these conditions, Sic1 is still a very unstable protein (at least as unstable as casein). However, the key observation here is not that the unubiquitinated Sic1 is degraded, but that the change in its half-life is lethal to cells simply because the degradation is too slow in context of the precise signaling role it evolved to facilitate.

In the case of the ATP hydrolysis-dependent disassembly of the 26S particles, an essential role could be convincingly demonstrated only by isolating a mutant in the ATPase or ATPases with a defect in the dissociation of the 19S cap, but not in substrate unfolding and translocation to the 20S core. For this reason, such mutants would need to be first carefully characterized *in vitro*, prior to *in vivo* analysis. Two lines of evidence suggest that an identification of ATPase mutants with defects in individualized functions is a feasible task. First, Hershko and colleagues have observed that while various nucleotides can support degradation of a model substrate and undergo hydrolysis by the 26S proteasome, they cannot replace ATP in the formation of 26S [27]. If the same mechanism controls the assembly and disassembly of the 26S particle, the difference in nucleotide requirement may allow a separation of function. Second, a growing number of reports suggest individualized roles for the ATPases. Genetic analysis in yeast first suggested individualized roles of proteasomal ATPases by demonstrating distinct phenotypes associated with the inactivation of the conserved ATP-binding motif of each ATPase [28]. Since then, inactivation of the Rpt2 ATPase alone has been shown to inhibit the opening of the gating channel [28,29]. In contrast, a cross-linking approach has shown that the Rpt5 ATPase interacts with a substrate-attached polyubiquitin chain, suggesting a role in substrate recruitment [30]. Finally, the Rpt1 and Rpt6 ATPases were shown to bind Ubr1 and Ufd4 E3s, implying a role in recruitment of ubiquitination machineries [3,7]. Although it is unknown which ATPases are involved in substrate unfolding and translocation to the proteolytic core, the hypothesis that the individual ATPases have specialized functions becomes increasingly attractive. In this view, the coupling between ATP hydrolysis and disassembly of the 19S cap could depend on a single ATPase. On the other hand, it is hard to imagine that the machinery operates without coordinating the ATPase activities, especially if they have the potential to induce disassembly of the 19S and, thus, interrupt proteasomal function if not controlled. An interesting possibility is that such coordination exists, and depends on the function a single 'master' ATPase that controls the assembly and disassembly of the 26S.

The availability of *in vitro* systems that faithfully recapitulate the degradation of naturally unstable SCF substrates [4,5,9,11,14,31,32], and allows monitoring several distinct steps in the catalytic cycle, including recruitment of the SCF-bound substrate to the proteasome, substrate deubiquitination, unfolding, degradation, and disassembly of the 26S, puts us and other investigators in a position to initiate the dissection of the contributions of each ATPase. Full dissection is necessary because identification of the ATPase or ATPases that control the disassembly of the 19S during the catalytic cycle can only be achieved via separation of the individual roles of the ATPases. Only with ATPase mutants that cannot perform a specific task, could we get sufficient evidence to conclude whether this function is or is not essential for proteasomal activity and in what context.

Concluding Remarks

In the past few years, the combination of genetic analysis *in vivo* and biochemical reconstruction *in vitro* have led to the development of assays for specific and rapid ubiquitination and degradation of several naturally unstable regulatory proteins from yeast. The unanticipated complexity of the substrate recruitment mechanism and the catalytic cycle itself uncovered in these studies suggests that the reconstruction of protein degradation with natural substrates *in vitro* is the key to understanding the complexity of proteasomal function. It is also appealing to think that these assays will provide a basis for the development of a variety of small molecule regulators that can affect the catalytic cycle of the 26S proteasome via mechanisms distinct from interfering with its enzymatic activities *per se*. Conformational transitions of the 26S machinery, the dynamics of its assembly and disassembly, and the interaction between the 26S and PIPs are only a few of the emerging targets for a possible pharmacological regulation.

REFERENCES

[1] Chau V, Tobias JW, Bachmair A, Marriott D, Ecker DJ, Gonda DK, and Varshavsky A: A multiubiquitin chain is confined to specific lysine in a targeted short-lived protein. *Science* 1989, 243:1576-1583.

[2] Thrower JS, Hoffman L, Rechsteiner M, Pickart CM: Recognition of the polyubiquitin signal. *EMBO J* 2000, 19:94-102.

[3] Xie Y, Varshavsky A: Physical association of ubiquitin ligases and the 26S proteasome. *Proc Natl Acad Sci USA* 2000, 97:2497-2502.

[4] Verma R, Chen S, Feldman R, Schieltz D, Yates J, Dohmen RJ, Deshaies RJ: Proteasomal proteomics: identification of nucleotide-sensitive proteasome-interacting proteins by mass spectrometric analysis of affinity-purified proteasomes. *Mol Biol Cell* 2000, 10:3425-3439.

[5] Babbitt SE, Kiss A, Deffenbaugh AE, Chang Y-H, Bailly E, Erdjument-Bromage H, Tempts P, Buranda T, Sklar L, Baumler J, Gogol E, Skowyra D: ATP hydrolysis-dependent disassembly of the 26S proteasome is part of the catalytic cycle. *Cell* 2005, 121:553-565.

[6] Leggett DS, Hanna J, Borodovsky A, Crosas B, Schmidt M, Baker RT, Walz T, Ploegh H, Finley D: Multiple associated proteins regulate proteasome structure and function. *Molecular Cell* 2002, 10:495-507.

[7] Xie Y, Varshavsky A: UFD4 lacking the proteasome-binding region catalyzes ubiquitination but is impaired in proteolysis. *Nature Cell Biology* 2002, 4:1003-1007.

[8] Elsasser S, Finley D: Delivery of ubiquitinated substrates to protein-unfolding machines *Nature Cell Biology* 2005, 7:16-23.

[9] Verma R, Oania R, Graumann J, Deshaies R: Multiubiquitin chain receptors define a layer of substrate selectivity in the ubiquitin-proteasome system. *Cell* 2004, 118:99-110

[10] Bailly E, and Reed S: Functional Characterization of Rpn3 Uncovers a Distinct 19S Proteasomal Subunit Requirement for Ubiquitin-Dependent Proteolysis of Cell Cycle Regulatory Proteins in Budding Yeast. *Mol Cell Biol* 1999, 19:6872-6890.

[11] Chandrasekaran S, Deffenbaugh AE, Ford D, Bailly E, Mathias N, Skowyra D: Destabilization of binding to SCF^{Met30} is a rate-limiting regulatory step in degradation of polyubiquitinated Met4 by the 26S proteasome.*Molecular Cell*. In press.

[12] Maniatis T: A ubiquitin ligase complex essential for NFkappaB, Wnt/Wingless, and Hedgehog signaling pathways. *Genes Dev* 1999, 13:505-510.

[13] Schwob E, Bohm T, Mendenhall M, Nasmyth K: The B-type cyclin kinase inhibitor p40SIC1 controls the G1/S transition in S. cerevisiae. *Cell* 1994, 79:233-244.

[14] Verma R, McDonald H, Yates III JR, Deshaies R: Selective degradation of ubiquitinated Sic1 by purified 26S proteasome yields active S-phase Cyclin-Cdk. *Molecular Cell* 2001, 8: 439-448.

[15] Deffenbaugh AE, Scaglione KM, Zhang L, Moore JM, Buranda T, Sklar LA, Skowyra D: Release of ubiquitin-charged Cdc34 from the RING domain is essential for ubiquitination of the SCF^{Cdc4}-bound substrate Sic1. *Cell* 2003, 114:611-22.

[16] Kaiser P, Mancollin V, Clarke DJ, Watson MH, Bartolaet BI, Reed SI, Bailly E: Cyclin-dependent kinase and Cks/Suc1 interact with the proteasome in yeast to control proteolysis of M-phase targets. *Genes Dev* 1999, 13:1190-1202.

[17] Wang J, Song JJ, Seong IS, Franklin MC, Kamtekar S, Eom SH, Chung CH: Nucleotide-dependent conformational changes in a protease-associated ATPase HslU. *Structure* 2001, 9: 1107-1116.

[18] Osmulski PA, Gaczynska M: Nanoenzymology of the 20S proteasome: proteasomal actions are controlled by the allosteric transition. *Biochemistry* 2002, 41:7047-7053.

[19] Hutschenreiter S, Tinazli A, Model K, Tampe R: Two-substrate association with the 20S proteasome at single-molecule level. *EMBO J* 2004, 23:2488-2497.

[20] Kisselev AF, Akopian TN, Castillo V, Goldberg AL: Proteasome active sites allosterically regulate each other, suggesting a cyclical bite-chew mechanism for protein breakdown. *Molecular Cell* 1999, 4:395-402.

[21] Kisselev AF, Kaganovich D, Goldberg AL: Binding of hydrophobic peptides to several non-catalytic sites promotes peptide hydrolysis by all active sites of the 20S proteasomes. *J Biol Chem* 2002, 277:22260-22270.

[22] Kisselev AF, Garcia-Calvo M, Overkleeft HS, Peterson E, Pennington MW, Ploegh HL, Thornberry NA, Godberg AL: The caspase-like sites of proteasomes, their substrate specificity, new inhibitors and substrates, and allosteric interactions with the trypsin-like sites. *J Biol Chem* 2003, 278:35869-35877.

[23] Whitby FG, Masters EI, Kramer L, Knowlton JR, Yao Y, Wang CC, Hill CP: Structural basis for the activation of 20S proteasomes by 11S regulators. *Nature* 2000, 408:115-120.

[24] Adams GA, Falke S, Goldberg AL, Slaughter CA, DeMartino GN, Gogol EP: Structural and functional effects of PA700 and modulator protein on proteasomes. *J Mol Biol* 1997, 273: 646-657.

[25] Adams GA, Crotchett B, Slaughter CA, DeMartino GN, Gogol EP: Formation of proteasome-PA700 complexes directly correlates with activation of peptidase activity. *Biochemistry* 1998, 37:12927-12932.

[26] Verma R, Deshaies RJ: A proteasome howdunit: the case of a missing signal. *Cell* 2000, 101: 341-344.

[27] Armon T, Ganoth D, Hershko A: Assembly of the 26S complex that degrades proteins ligated to ubiquitin is accompanied by the formation of ATPase activity. *J Biol Chem* 1990, 265:20723-20726.

[28] Rubin D M, Glickman MH, Larsen C N, Dhruvakumar S, Finley D: Active site mutants in the six regulatory particle ATPases reveal multiple roles for ATP in the proteasome. *EMBO J* 1998, 17:4909-4919.

[29] Kohler A, Cascio P, Leggett DS, Woo KM, Goldberg AL, Finley D: The Axial channel of the proteasome core particle is gated by the Rpt2 ATPases and control both substrate entry and product release. *Molecular Cell* 2001, 7:1143-1152.

[30] Lam YA, Lawson TG, Velayutham M, Zweier JL, Pickart C: A proteasomal ATPase subunits recognizes the polyubiquitin degradation signal. *Nature* 2002, 416:763-767.

[31] Verma R, Aravind L, Oania R, McDonald WH, Yates JR 3rd, Koonin EV, Deshaies RJ: Role of Rpn11 metalloprotease in deubiquitination and degradation by the 26S proteasome. *Science* 2002, 298:611-615.

[32] Verma R, Deshaies RJ: Assaying degradation and deubiquitination of a ubiquitinated substrate by purified 26S proteasomes. *Methods Enzymol* 2005, 398:391-399.

In: The Ubiquitin Proteasome System… ISBN 978-1-60021-749-4
Eds: Mario Di Napoli and Cezary Wojcik, pp. 169-205

Chapter 9

PROTEASOME ACTIVATORS, INHIBITORS AND ASSOCIATED PROTEINS

Geoffrey M. Goellner and Carlos Gorbea*
Department of Biochemistry, University of Utah School of Medicine, Salt Lake City, UT 84112, USA.

ABSTRACT

The ubiquitin-proteasome system is responsible for the majority of regulated intracellular protein degradation. The importance of this system is reflected in its involvement in a large number of biological processes such as cell cycle traverse, apoptosis, antigen presentation, circadian rhythms, protein quality control, etc., as well as many aspects of neuronal development and function. The proteolytic component of the system is the 26S proteasome, a large ATP-dependent enzyme that degrades proteins marked for destruction by polyubiquitin chains. The 26S proteasome is composed of two subunits, the 20S proteasome and the 19S regulatory complex. The 20S proteasome is a cylindrical particle composed of α and β subunits arranged as four heptameric rings housing the proteolytic activity. Indiscriminate protein degradation by the 20S proteasome is prevented by the fact that the active sites are sequestered within the central chamber of the cylinder, and substrate access is blocked by N-terminal extensions of the α subunits constituting the outer rings. For substrates to gain entry into the 20S proteasome, the α N-terminal extensions must be reoriented in order to open the axial channel leading into the proteasome catalytic chamber. This is accomplished by binding of the 19S regulatory complex and/or other proteasome activator complexes to one or both ends of the 20S cylinder. The 19S regulatory complex is a multiprotein structure that recognizes, unfolds and pumps polyubiquitylated substrates into the 20S catalytic core. Other identified proteasome activators include PA28α and PA28αβ that are involved in MHC class I antigen presentation, PA28γ, thought to be involved in

* Correspondence concerning this article should be addressed to Dr. Carlos Gorbea, PhD; Department of Biochemistry, University of Utah School of Medicine, 15 N Medical Dr. East Rm. 4100, Salt Lake City, UT 84112-5650, USA.. Phone: +1-801-581-8890; Fax: +1-801-581-7959; E-mail: gorbea@biochem.utah.edu.

apoptosis, and PA200 recently linked to DNA repair. Inhibitors of the 20S proteasome include Hsp90 and PI31 — which presumably compete with proteasome activators for binding the 20S proteasome α rings — PR39, a noncompetitive reversible inhibitor of proteasomes and PA28αβ-proteasome complexes, and many viral proteins that inhibit 20S and 26S proteasomes as well as other components of the ubiquitin-proteasome system. In addition to activators and inhibitors of 20S proteasomes, there are numerous proteins that bind to the 26S holoenzyme and modulate its activity. These include a host of ubiquitin-conjugating enzymes (E2s), ubiquitin-ligases (E3s), ubiquitin-chain elongation factors (E4s), isopeptidases, and an increasing number of polyubiquitin chain receptors thought to deliver polyubiquitylated substrates to the 26S proteasome. Finally, the HEAT repeat-containing protein Ecm29 has been proposed to function as an adaptor that links 26S proteasomes to protein quality control and endoplasmic reticulum-associated protein degradation pathways, endocytosis and vesicular trafficking, and transport processes through interaction of Ecm29 with molecular motors. In this review, we will discuss recent developments concerning the aforementioned proteasome-interacting proteins with emphasis on those proteins which likely function within the normal physiology and pathophysiology of the nervous system.

Keywords: proteasome activators, proteasome inhibitors, proteasome-binding proteins, ubiquitin receptors, deubiquitylating enzymes, Ecm29.

Abbreviations:

AAA, ATPases associated with a variety of cellular activities; AdE1A, adenovirus early region 1; AR-JP, Autosomal Recessive Juvenile Parkinsonism; CFTR, cystic fibrosis transmembrane conductance regulator; CHO, Chinese Hamster Ovary cells; ChT-L, chymotrypsin-like; CTL, MHC class I restricted T-lymphocytes; DSBs, DNA double strand breaks; DSS1, Deleted-in-Split-Hand/Split-Foot-1; DUBs, deubiquitylating enzymes; EBNA1, Epstein-Barr virus nuclear antigen 1; EBV, Epstein-Barr virus; ECM29, Extracellular Mutant 29 gene; eEF1A, eukaryotic elongation factor 1A; ERAD, endoplasmic reticulum-associated degradation pathway; ERGIC, ER-Golgi intermediate compartment; GFP, green fluorescent protein; Gar, glycine and alanine repeat; HBX, hepatitis B virus X protein; HCV, hepatitis C virus; HEAT, Huntingtin, Elongation factor 3, HIV-1, human immunodeficiency virus-1; HR, homologous recombination; Hsp90, heat shock protein 90; IFN-γ, interferon γ; MMS, methyl methane sulfonate; MVB, multivesicular bodies; NHEJ, non-homologous end joining; PAAF1, proteasomal ATPase-associated factor 1; PCM1, pericentrin; PGPH, post-glutamyl-peptidyl hydrolyzing; PNGase, peptide:N-glycanase; polyQ, polyglutamine; polyUb, polyubiquitin; RC, 19S regulatory complex or PA700; SCA, Spinocerebellar Ataxia; SCF, skp1-cullin-F-box complex; TAP, tandem affinity purification; T-L, trypsin-like; Ub, ubiquitin; UBA, ubiquitin-associated domain; UBL, ubiquitin-like domain; UIM, ubiquitin-interacting motif; UPS, ubiquitin-proteasome system; VCP, valosin-containing protein or p97.

INTRODUCTION

It is increasingly clear that cellular homeostasis is the result of maintaining appropriate levels of key enzymes and other regulatory proteins that control essential life processes. Although transcription and translation play a major role in determining the intracellular concentration of proteins, their timely destruction by proteolysis is also critical for normal cell function. Covalent attachment of the small 76-residue protein ubiquitin (Ub) to eukaryotic proteins is a post-translational modification of remarkable scope, complexity and importance. Although Ub conjugation can serve non-destructive purposes [1-5], regulated extralysosomal degradation of intracellular proteins is the principal function of this modification [6]. The Ub system regulates a large number of biological processes including cell cycle traverse [7,8], apoptosis [9], antigen presentation [10], circadian rhythms [11,12], protein quality control [13,14], transcription [1], as well as many aspects of neuronal cell biology [15-26]. To target proteins for destruction, the C-terminus of ubiquitin is first activated by an ATP-consuming, ubiquitin-activating enzyme or E1, and the activated ubiquitin is then transferred in thiol-ester linkage to a number of carrier proteins known as ubiquitin-conjugating enzymes or E2s [6]. The ultimate transfer of Ub to substrate proteins is in most cases mediated by hundreds of ubiquitin ligases or E3s, which constitute the substrate-recognition components of the system [27-32], and E2s and E3s generally collaborate to generate polyubiquitin (polyUb) chains attached to lysine ε-amino groups in the protein substrates. In many instances, elongation of the polyUb chain to optimal degradation-competent lengths requires the further action of ubiquitin elongation factors or E4s [33]. The polyUb protein is ultimately targeted to the 26S proteasome, a large (2.5 MDa) ATP-dependent protease present in both the cytosol and nucleus of eukaryotic cells, for degradation [34].

The Ubiquitin-Proteasome System in the Central Nervous System

The ubiquitin-proteasome system (UPS) plays many important roles in the physiology of normal and diseased neuronal cells. Components of the UPS have been implicated in axonal degeneration [17,18], development of synapses [35-39], and neurotransmission [16,40,41]. It also contributes to synaptic plasticity, particularly with respect to the association of glutamate receptors with postsynaptic densities and the establishment of long-term facilitation [42-46]. In addition, it is becoming increasingly evident that the UPS contributes to the development of several neurodegenerative diseases including Alzheimer's, Parkinson's, and those that result from polyglutamine expansions, such as Huntington's and a number of Spinocerebellar Ataxias (SCAs) [47,48]. Although parkin and SCA-3 are themselves components of the UPS, mutated proteins in the majority of neurodegenerative diseases have no obvious functional or evolutionary relationship. The abnormal accumulation of ubiquitylated proteins or peptides in the nervous system, however, suggests that insufficient rates of protein degradation by the UPS could underlie the pathogenesis of these diseases.

The 26S Proteasome

The 26S proteasome is the only enzyme known to degrade polyubiquitylated proteins in eukaryotic cells. The proteolytic core of the 26S proteasome is the 20S proteasome, a cylindrical particle composed of four-stacked heptameric rings of α and β subunits arranged in an αββα configuration that houses 3 proteolytic activities termed chymotrypsin-like (ChT-L), trypsin-like (T-L), and post-glutamyl-peptidyl hydrolyzing (PGPH) activities [49,50] (see Chapter 6) Indiscriminate proteolysis of protein substrates is prevented by the sequestration of the 20S proteasome's active sites within a central chamber formed by the inner β rings [49,50]. Although, the 20S proteasome can itself degrade a limited number of substrates [51], mechanisms are required to facilitate the transfer of protein substrates into the central chamber. This is accomplished when one or two 19S regulatory complexes (RCs or PA700) bind to the end α rings of the 20S proteasome [52] forming the 26S proteasome (see Chapter 7). The RC — which in turn can be divided into two subcomplexes termed the base and the lid [53] — provides subunits that recognize, unfold, and translocate protein substrates into the 20S cylinder where they are degraded [54]. The 19S complex also contains a subunit that disassembles polyUb chains attached to the protein substrates [55,56] and associates with other cellular isopeptidases to ensure complete release and recycling of the Ub tagging molecules [57,58].

PROTEASOME ACTIVATORS

In addition to the 19S RC, which utilizes ATP to stimulate the proteasomal degradation of full-length proteins, a number of ATP-independent activators of the 20S proteasome that stimulate the hydrolysis of short peptides *in vitro* have been identified [59-62]. However, the *in vivo* roles of these activators have yet to be firmly established. Activators of the 20S proteasome, which include three homologous proteins denominated PA28 α, β, γ (or 11S REG α, β, γ) and the recently identified protein PA200, bind to and 'cap' the ends of the 20S core particle in an analogous manner as the 19S regulatory complex [63-65]. Moreover, it has become increasingly evident that 'hybrid' proteasomes, composed of a single 20S catalytic core particle capped on one end by a 19S RC and on the other end by an ATP-independent activator, are relatively prevalent —comprising perhaps 25% of the overall cellular proteasomal content [66]. Thus, it appears that cells are quite dynamic in terms of their proteasomal content, and the ability to change proteasomal 'caps' endows them with multiple levels of control over protein degradation.

PA28 α, β, γ

Identified in the 1990's, PA28 proteins (α, β, γ) are small 28-30 kDa proteins that share significant sequence homology with one another, and appear to function as either homo- (α/α,γ/γ) or hetero- (α/β) heptamers that cap one or both ends of the 20S proteasome [59,60]. Important differences exist among the three PA28 proteins. Although PA28α can function independently, complexes containing both PA28α and PA28β have the highest affinity for the proteasome, whereas PA28γ homoheptamers have an intermediate affinity towards the 20S enzyme [67]. Furthermore, PA28α and PA28αβ stimulate proteasomal cleavage after acidic, basic, and hydrophobic amino acids in substrate peptides, while PA28γ only stimulates the β2 active site responsible for cleavage after basic residues [67,68]. Intracellular localizations also vary among the PA28 homologs in that PA28α and PA28β are predominantly cytoplasmic proteins whereas PA28γ localizes to the nucleus of cells [59,60]. Finally, while all three proteins display a widespread tissue distribution, PA28γ is notably elevated in the brain and PA28αβ levels are highest in immune organs [59].

PA28 proteins accelerate the digestion of peptides by the 20S proteasome *in vitro*, however it is unlikely that their sole purpose for the cell is to perform an equivalent function *in vivo* since short peptides are rapidly removed by a host of cellular peptidases [69]. Although redundancy may be part of their purpose, PA28 activators almost certainly perform other useful functions for the cell. A high-resolution crystal structure of the yeast 20S proteasome in complex with PA26 (*Trypanosoma brucei* homolog of PA28α) has been solved which depicts at least one consequence of 20S-PA28 complex formation. Binding of PA26 to the 20S proteasome rearranges the N-terminal sequences of the α2, α3, α4 and α5 proteasomal subunits, thereby opening the axial channel into the central chamber of the 20S cylinder (see Figure 1) [63]. Thus, one primary function of PA28 proteins is likely 'gating' peptide entry into and/or egress from the proteasome core particle. As mentioned above, the cell appears to have little need to promote peptide entry into the proteasome (especially into 20S core particles already capped with a 19S RC). Consequently, control of peptide release seems a particularly attractive hypothesis. In this way, either larger or shorter peptides might be released from a 26S proteasome depending on the presence or absence of an activator capping the downstream aspect of the 20S channel, respectively. That being said, experimental evidence to date suggests that PA28αβ has little effect on length of peptides released from the proteasome [70].

Although PA28 molecules may play a role in gating peptides into and out of the 20S core, they do not by themselves promote proteasomal degradation of full-length native proteins. Interestingly, however, recent experiments have linked PA28αβ with the chaperone system [71]. It is thus theoretically possible that chaperones provide the environment to unfold proteins while the PA28 cap opens a channel into the 20S catalytic chamber thereby reconstituting an alternative proteasomal system capable of degrading full-length proteins. Alternatively, PA28 proteins could act in a triage or sorting role in conjunction with chaperones and, similar to the CHIP E3 ligase [72], 'hand off' improperly folded proteins incapable of regaining their native state to the 26S complex for degradation.

A number of other PA28 functions have been hypothesized as well. For example, PA28 activators might serve as adaptors to localize the proteasome to specific subcellular sites of protein degradation [59,60]. In addition, by capping 26S proteasomes with diverse PA28 proteins, differential activation of specific proteasomal catalytic sites might be achieved *in vivo* resulting in altered cleavage patterns of substrate proteins. Thus, by modulating peptide release length via opening of an egress channel and/or altering cleavage patterns by differentially activating specific proteasomal catalytic sites, PA28 could help cells establish some degree of control over which types of peptides are released from the 20S core particle.

Binding of different PA28 molecules could thereby have significant downstream physiological consequences for cells and for organisms as a whole.

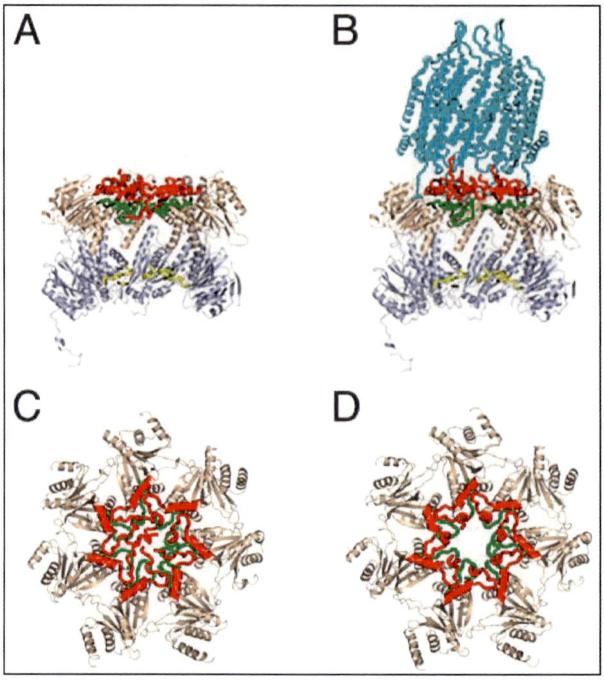

Figure 1. Gating of the 20S proteasome by PA26. Ribbon representation of portions of the 20S proteasome alone (A, C) or in complex with trypanosomal PA26 (B, D). A) Sideview of uncomplexed 20S proteasome showing N-terminal tails (red) of 20S α subunits extending into and filling the annulus (green) that separates the α (wheat) and β (purple) subunits of the 20S complex. Note that the image has been cut away to reveal internal features, and only one-half of a full 20S molecule (single α and β rings) is depicted for simplicity. B) Sideview of the 20S proteasome in complex with PA26 (cyan) displaying a significant reorientation of the 20S α subunits (red) up and out of the annulus (green). C) Top view of uncomplexed 20S proteasome showing obstruction of the annulus (green) by the α-subunit N-terminal tails (red). D) Top view of 20S proteasome complexed with PA26 and showing displacement of the N-terminal tails (red) out of the 20S annulus (green) allowing a pore to be formed through which substrate proteins can be pumped into or out of the 20S catalytic chamber.

Considerable evidence exists linking PA28αβ and proteasomes to class I antigen presentation (see also Chapter 34). Three proteasomal subunits, PA28αβ (but not PA28γ) and several components of the MHC class I pathway are all induced by the cytokine interferon (IFN)-γ [60]. Furthermore, lactacystin, a specific inhibitor of the proteasome, has been shown to significantly diminish class I antigen presentation [73,74] and recent investigations in mice have substantiated the notion that PA28α and PA28β are intimately involved in cellular immunity [75,76]. However, precisely how PA28αβ contributes to presentation of class I epitopes is still unknown. Several hypotheses have been purported, all of which we have alluded to above. First, since 20S proteasomes most often release peptides of 6-8 residues in length [77] and MHC class I molecules typically present peptides between 8-11 amino acids long [78], there is a disparity in optimal length which could be made up for if PA28αβ opened the exit gate of the 20S particle allowing slightly longer than normal peptides to be released. Alternatively, binding of PA28αβ might affect the relative *in vivo* catalytic activity of the three-proteasomal active sites thereby altering the cleavage sites within an antigenic peptide and the ultimate epitope presented. A third potential mechanism has PA28αβ serving as an adaptor directly coupling hybrid proteasomes and their released peptides to the MHC class I peptide loading complex located on the endoplasmic reticulum membrane [59,60]. In support of this hypothesis, immunoproteasomes (20S proteasomes containing the three proteasomal β subunits induced by IFN-γ) have been shown to be enriched at the ER [79] where TAP complexes transport peptides for association with newly synthesized MHC class I molecules. It should be noted that these hypotheses regarding PA28αβ function are not mutually exclusive and any combination of them could in fact serve to enhance the production of particular MHC class I epitopes.

Although a conclusive physiological role for PA28γ remains to be established, PA28γ function has been linked to cell cycle traverse and apoptosis [59]. RNA interference (RNAi) knockdown of PA28γ expression in *Drosophila melanogaster* cells results in an increased number of cells found in the G1 phase of the cell cycle [80]. PA28γ knockout mice confirm this result and, in addition, fibroblasts cultured from PA28γ deficient animals display increased rates of apoptosis [81]. The hypothesis that PA28γ has an anti-apoptotic function is strengthened by yeast two-hybrid screens for PA28γ-interacting proteins, which have pulled out a number of apoptosis-related proteins including MEKK3, FLASH, Daxx, RanBPM, and PIAS1 ([59] and Gao and Rechsteiner, manuscript in preparation). However, the specific role that PA28γ plays in cell cycle control and apoptosis remains to be elucidated.

PA28γ and the proteasome have also been implicated in the pathogenesis of a group of severe neurodegenerative disorders called polyglutamine (polyQ) diseases [82]. Huntington's Disease, Spinal Bulbar Muscular Atrophy, Dentatorubral Pallidoluysian Atrophy, and a number of SCAs (1,2,3,6,7,17) all share the common structural feature of an expanded glutamine tract in their respective gene products. Expansion of the polyQ tract causes these proteins to inappropriately self-aggregate [82]. Indeed, one of the primary hallmarks of polyQ disease is the presence of intracellular, often nuclear, polyQ inclusion bodies deposited within the diseased brains of polyQ patients— although it is not entirely clear what role these inclusions play in the mechanism of polyQ pathogenesis [83,84]. Interestingly, it has been noted that chaperones and components of the 26S proteasomal degradation machinery are present at polyQ inclusions, strongly suggesting that the cell is at least attempting to degrade the aggregates [82]. Other studies have reported that proteasomes have significant difficulty degrading polyQ proteins *in vitro* [85,86]. Furthermore, PA28γ has been directly implicated in the putative difficulty of proteasomes to clear polyQ aggregates because of its nuclear localization, high expression in the brain, and inability to stimulate the 20S proteasomal active sites responsible for cleavage after glutamine residues [82]. These observations have led to the hypothesis that expanded polyQ tracts may actually 'clog' the central chamber of the 20S proteasome —particularly in a PA28γ hybrid context, and that polyQ diseases might be thought of as proteasomal storage diseases [82]. Recent studies in mice, however, do not support the notion that significant clogging of 26S proteasomes is central to the mechanism underlying polyQ neurodegeneration [87,88; and Chapter 32].

PA200

This activator of the 20S proteasome was first identified in 1992 as a bound component of a slow-migrating, activated form of the 20S proteasome in rabbit reticulocyte lysates [52]. A decade later, large-scale proteomic screens in yeast identified the product of the *BLM3* gene as a proteasome-interacting component [89,90]. Purification and initial characterization of the mammalian homolog of blm3p, PA200, was reported the same year by Ustrell *et al.*, who presented evidence that PA200 is a proteasome activator involved in DNA repair [91]. PA200 is a HEAT (for Huntingtin, Elongation factor 3, PR65/A subunit of protein phosphatase 2A and Tor)-repeat containing protein that forms a flexible, α-solenoid toroidal structure that binds either one or both ends of the 20S proteasome and activates peptide hydrolysis by the 20S particle *in vitro* (see Figure 2 and [64,65]). The mechanism of proteasome activation by PA200 appears to be analogous to the activation of 20S proteasomes by PA28 [64]. Binding of PA200 to the 20S proteasome's ends presumably displaces α-subunit N-terminal sequences that open the axial pore to allow substrate entry into and/or product egress out of the 20S proteasome's catalytic chamber [63]. PA200 is a nuclear protein that is highly expressed in testis and spleen [91], two tissues that generate a high number of DNA double strand breaks (DSBs) through meiotic and V(D)J recombination mechanisms, respectively. Below, we discuss recent evidence regarding the involvement of 26S proteasomes in DNA repair and the role(s) PA200 may play in the context of this proteasomal function.

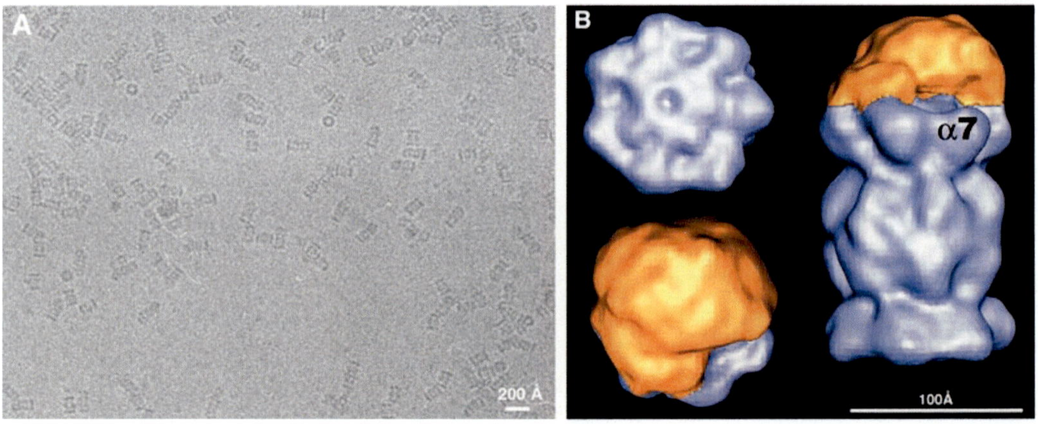

Figure 2. PA200 binds to the α rings of the 20S proteasome. (A) Cryo-electron microscopy of the PA200-20S proteasome complex. (B) Three-dimensional reconstruction of a singly bound PA200-20S proteasome complex at 23 Å resolution. Left, bottom and top views of the complex. PA200 (gold cap) is an asymmetrical dome structure that contacts six of the seven proteasomal α subunits, excluding the α7 subunit. Right, side view of the PA200-20S complex featuring an entrance to the cavity of PA200 adjacent to the α7 subunit.

The 26S Proteasome and DNA Repair

Four recent reports link the 19S RC lid subunit sem1p (rpn15p) to DNA repair [92-95]. Sem1p in *Saccharomyces cerevisiae* and its DSS1 (Deleted-in-Split-Hand/Split-Foot-1) orthologs in fission yeast and humans are small (~10 kDa), acidic proteins that interact with the product of the breast cancer susceptibility gene BRCA2 [96]. BRCA2 is a component of a large protein complex, termed BRCC that displays E3 ubiquitin ligase activity and participates in the DNA repair pathway by homologous recombination (HR). Deletion and/or mutations in the *SEM1* gene cause accumulation of polyUb proteins, show impaired proteasomal degradation and decreased 26S proteasome stability, and exhibit synthetic growth defects when grown in the presence of a number of genotoxins [92-94]. In addition, deletion of *SEM1* causes synthetic growth defects when combined with deletions of genes involved in both HR and non-homologous end joining (NHEJ) [94]. Thus, it has become clear that the 26S proteasome is implicated in the repair of DNA double-strand breaks.

Possible Role of PA200 in DNA Repair

Immunoprecipitation experiments by Krogan *et al.* have shown that 26S proteasomes are indeed recruited to sites of DSBs [94]. However, sem1p does not seem to be involved in recruiting 26S proteasomes to such sites, thus raising the question as to how 26S proteasomes are targeted to sites of DNA double-strand breaks. *BLM3* (now called *BLM10* [65]) was initially identified as an extragenic suppressor of the *blm3-1* mutation in a genetic screen to detect yeast genes controlling sensitivity to bleomycin, a drug that induces DNA DSBs [97]. This suggested that a function for blm3p (or blm10p) and PA200 might be to link proteasomes to DNA repair mechanisms. In fact, the message for blm10p is increased ~5-fold in yeast cells exposed to the DNA damaging agent methyl methane sulfonate (MMS) [98], the blm10p protein is present in complexes containing sir4p, a component that binds DSBs [90], PA200 localizes to nuclear foci generated upon γ-irradiation of mammalian cells [91], and although *blm10* deletion mutants exhibit modest sensitivity to bleomycin, truncation of the C-terminal 339 amino acid residues produces bleomycin hypersensitive yeast cells [65]. Moreover, Schmidt *et al.* have inferred that in *Saccharomyces cerevisiae* all of the blm10p protein resides in hybrid 19S-20S-blm10p complexes [65], implying that in budding yeast, any DNA-repair functions of blm10p are linked to the 26S proteasome. Thus, significant evidence has accrued to support a role for PA200 in DNA repair. A question remains though —what role does PA200 play in DNA repair? The fact that *blm3Δ* mutants are not hypersensitive to bleomycin indicates that PA200's involvement in DNA repair is non-essential. A synthetic phenotype of defective growth at 30°C and strong sensitivity to the arginine analog L-canavanine is, however, observed in *BLM10* and *ECM29* (which encodes another proteasome binding protein) double mutants [65], which further suggests that the functions of blm10p and ecm29p overlap. In this case, we would suspect the *blm10Δ ecm29Δ* double mutation to exhibit hypersensitivity to bleomycin. As discussed below, we have proposed that Ecm29 functions as an adaptor to recruit 26S proteasomes to different cellular locations. Thus, the possibility of PA200 serving a similar function —its being an adaptor to recruit 26S proteasomes to DSBs— must be considered.

Proteasome Inhibitors

There are an increasing number of reports describing chemical inhibitors of the proteasome (see Chapter 40), some of which are undergoing clinical trials to treat a diverse number of human pathologies. In this review, however, we briefly delve into natural protein inhibitors of the proteasome that might affect its activity *in vivo*. Protein inhibitors of the 20S and 26S proteasomes include the chaperone heat shock protein 90 (Hsp90) [99,100], PI31 [101-104], the proteasomal ATPase-associated factor 1 (PAAF1) [105], and the angiogenic peptide PR39 [106-108]. In addition, recent studies have also linked protein aggregation and inclusion body formation to inhibition of the UPS [109-112]. Finally, a handful of viral proteins associate with subunits of both the 20S particle and the 19S RC to inhibit protease activity [113-118], and the 26S proteasome is subject to chemical modifications such as O-glycosylation [119] and proteolytic cleavage [120,121] that suppress its activity *in vivo*.

Using fluorogenic peptides as substrates, a 2001 study measured the effect of uncomplexed Hsp90 on purified 20S proteasomes and reported that while this chaperone inhibited two of the three canonical catalytic sites of constitutive 20S particles, immunoproteasomes were refractory to inhibition by Hsp90 [100]. This *in vitro* inhibitory effect of the heat shock protein on the pituitary 20S enzyme was reversed by addition of small quantities of the activator PA28αβ. However, since all organs but the brain express PA28αβ, it could call into question the *in vivo* relevance of the observed inhibition. Furthermore, results from two recent studies show that Hsp90 actually promotes proteasomal degradation of proteins damaged by oxidation [122] or for processing in the MHC class I antigen presentation pathway [123]. Thus, Hsp90's effect on *in vivo* proteasome activity remains to be firmly established although the inhibition of constitutive proteasomes by Hsp90 might be contingent upon the chaperone's lack of association with a protein substrate.

Although the effects of Hsp90 on proteasome activity in living organisms are controversial, the chaperone does have two clear functions within the UPS (see Chapter 19). First, Hsp90 participates in the ATP-dependent assembly of the 26S proteasome [124]. Functional loss of Hsp90 in the *Hsp82-4 Δhsc82* yeast temperature-sensitive mutant causes dissociation of the 26S proteasome into 20S and 19S particles, and reassembly of the 26S enzyme is inhibited in the presence of geldanamycin, an Hsp90 inhibitor. In addition, there is genetic interaction between Hsp90 and several 19S RC subunits in the yeast mutants. Thus, the Hsp90-dependent assembly-disassembly cycle of the 26S proteasome appears to be a cell stress response caused by heat shock [124]. Second, in a more general role, Hsp90 'antagonizes' the UPS by binding and stabilizing client proteins, many of which are oncogenic signaling components such as protein kinases and tumor suppressors [125]. Inhibition of Hsp90, e.g., by geldanamycin, causes dissociation of the Hsp90-bound molecules and induces the proteasomal degradation of Hsp90 client proteins.

PI31 is a proline-rich, 30-kDa protein that binds the α rings of 20S proteasomes and inhibits the hydrolysis of proteins and fluorogenic peptide substrates [101-103]. PI31 inhibits both constitutive and immunoproteasomes alike *in vitro* [102], and competes with PA28αβ and the 19S RC for binding to the 20S particle [102,103]. The affinity of PI31 for 20S proteasomes has been estimated to be at least 50 times higher than that of PA28αβ [102], meaning that inhibition of the 20S proteasome by PI31 can only be reversed at high concentrations of the proteasomal activator. Yet, inhibition of 20S proteasome-PA28αβ complexes or 26S proteasomes reaches a maximum of 50% even at very high concentrations of the inhibitor [103], the physiological meaning of which is still unclear. It is possible that in binding one end of the 20S cylinder, PI31 induces long-range conformational changes that lock in place PA28αβ or 19S RC bound at the opposite end of the 20S enzyme, resulting in 'hybrid' complexes whose activity is suppressed by one-half. At any rate, overexpression of PI31 in mouse embryonic fibroblasts has shown that PI31 does not function as a general inhibitor of proteasome activity *in vivo* [104]. Rather by localizing to the nuclear envelope/ER membrane, it interferes with the maturation of immunoproteasome precursor complexes, thus modulating the surface expression of specific MHC class I-antigenic peptide complexes [104]. Rechsteiner and Hill [59] have pointed out, however, that modulation of proteasome-mediated antigen processing cannot be the only biological function of PI31 since the protein is expressed in plants and in invertebrates that lack immunoproteasomes or MHC class I immune responses.

A 43-kDa protein designated proteasomal ATPase-associated factor 1 or PAAF1 was recently found associated with affinity-purified 26S proteasomes [105]. The protein contains 7 WD40 repeats and homologs are present in yeast and higher eukaryotes. PAAF1 message levels vary among human organs and are highest in kidney, brain and testis and lowest in lung, colon, thymus, small intestine and leukocytes. PAAF1 inhibits the UPS by associating with the base ATPase subunits of the 19S regulatory complex (preferentially S8) and preventing reconstitution of the 26S proteasome. In this regard, PAAF1 serves a function opposite to that of Hsp90 [124] in the assembly/disassembly cycle of 26S proteasomes.

PR39 is a 39-amino acid, proline- and arginine-rich peptide antibiotic secreted by macrophages [106]. Like PI31 and Hsp90, PR39 inhibits preferentially the ChT-L and PGPH activities of proteasomes [108]. Inhibition is mediated by direct binding of PR39 to the α7 subunits of the 20S particle [107]. PR39 reportedly inhibits 26S proteasome-mediated degradation of hypoxia-inducible factor-1α [106] and IκBα [107], thus promoting vascularization and suppressing inflammation, respectively. Furthermore, PR39 plays an antiapoptotic role by increasing expression of IAP-2, the inhibitor of apoptosis-2 protein, thus reducing caspase-3 activity [126]. Recent biochemical studies have shown that PR39 inhibits proteasomes noncompetitively through an allosteric mechanism that allows for selective inhibition of certain protein substrates without exercising a global effect over the UPS. Atomic force microscopy studies have revealed structural perturbations in 20S and 26S proteasomes treated with PR39 or PR39 truncations, leading Gaczynska *et al.* to propose that binding of PR39 to the α7 subunit imposes structural constrains that prevents switching between open and closed proteasomal conformations [108].

Protein Aggregates

Recent reports suggest that protein aggregates (see Chapter 12) inhibit proteasomes *in vitro* and in cultured mammalian cells. For instance, β-amyloid and α-synuclein oligomers inhibit the chymotrypsin-like (ChT-L) activity of the 20S proteasome *in vitro* [109,110], and aggregates of huntingtin, cystic fibrosis transmembrane conductance regulator (CFTR), rhodopsin, and ataxin-1 have all been shown to inhibit the UPS within cells [111,112]. Thus, it would appear that any cellular protein aggregate might be expected to inhibit the UPS. However, to date, studies in mice do not support the hypothesis that protein inclusions characteristic of many neurodegenerative disorders compromise the overall activity of the UPS [87,88].

Viral Proteins

Many viruses have evolved mechanisms to suppress proteasome disposal of viral proteins, and consequently suppress cellular class I immune responses. An assortment of viral proteins binds subunits of both the 20S and 26S proteasomes and inhibits their enzymatic activity. These include the hepatitis C viral protein NS3 [113], the Epstein-Barr virus nuclear antigen 1 (EBNA1) [114], the adenovirus early region 1A protein (AdE1A) [115,116], the human immunodeficiency virus-1 (HIV-1) Tat protein [117], and the X protein of hepatitis B virus [118]. Targets in the 20S proteasomes include both α- and β-type subunits, but within the 26S enzyme, there appears to be preference for subunits within the base complex of the 19S RC.

Chronic infection with the plus-strand enveloped RNA hepatitis C virus (HCV) often leads to serious liver disease, such as cirrhosis, steatosis, and liver cancer. NS3 is a non-structural protein of HCV with helicase and protease functions that plays an essential role in viral RNA replication, translation, and processing of the viral polyprotein. Yeast two-hybrid screens identified NS3 as an LMP-7 (β5 subunit of immunoproteasomes)-interacting protein that specifically binds the LMP-7 pro-sequence region [113]. Although binding to the pro-sequence of LMP-7 does not appear to hinder LMP-7 processing, HCV infection specifically inhibits the activity of immunoproteasomes after treatment with IFN-γ. The mechanism of inhibition is unclear but it is possible that NS3 thwarts immunoproteasome complex maturation and assembly.

The Epstein-Barr (EBV) viral protein EBNA1 contains an internal repeat composed exclusively of glycines and alanines (GAr) that inhibits the presentation of endogenous antigens to MHC class I restricted T-lymphocytes (CTL) [127]. Resistance to CTL recognition is thought to play an important role in maintaining EBV latency. Interestingly, unstructured repeat sequences are also present in some transcription factors that are processed — but not fully degraded— by proteasomes [128], as well as in other γ-herpesviral proteins related to EBNA1, and in the latency-associated nuclear antigen of Kaposi sarcoma herpes virus [129]. The exact mechanism of inhibition is unknown, but neither ubiquitylation nor interaction of ubiquitylated GAr-containing substrates with the 26S proteasome is prevented.

Thus, a mechanism whereby GArs destabilize the interaction of polyUb substrates with proteasomes and promote premature substrate release has been proposed [130].

Three subunits of the 19 RC base complex, namely S2, S4 and S8, bind to the N-terminal region of the adenovirus 12S E1A gene product [115,116], a viral region known to target several proteins involved in cellular transformation, induction of S phase, transcriptional repression, and transcriptional activation [131]. Binding of the AdE1A protein to S4 and S8 reduces ATPase activity, and thus correlates with the ability of AdE1A to inhibit ubiquitin-mediated proteolysis [115]. Proteasomal inhibition by AdE1A stabilizes, among many substrates, p53 [115,116]. Also, AdE1A expression down-regulates LMP2, LMP7, MHC class 1 molecules and TAPs, thereby impairing class I immune responses in general [131].

Another viral protein that binds ATPase subunits within the base complex of the 26S proteasome is the HIV-1 Tat protein [132], an 86-residue protein that functions as a transcriptional activator. Binding to the 26S proteasome is also mediated through the 20S catalytic core via interaction with the α4 and α7 subunits, six constitutive β subunits, or with the LMP7 and MECL1 subunits of immunoproteasomes [133]. Hence, Tat inhibits the peptidase activity of the 20S proteasome, competes with PA28αβ for binding to the 20S enzyme [117], and may suppress class I PA28αβ-mediated epitope presentation [134]. Like HIV-1 Tat, the hepatitis B virus X protein (HBX) (residues 116-138) binds to the proteasomal α4 subunit and competes with PA28αβ for binding to the 20S cylinder *in vitro* [135,136]. In addition, HBX binds two distinct ATPase subunits within the 19S RC, S4 and S6' [137,138]. Expression of HBX in cells modestly inhibits the ChT-L and T-L sites of the 20S proteasome [118], and decreases activation of the p50 subunit of NF-κB [139] and the cellular turnover of c-Jun and Ub-Arg-β-galactosidase [137]. The functional consequences of HBX's interaction with proteasomes are not well understood. But, in principle, proteasome inhibition could help stabilize viral gene products and suppress antigen presentation.

26S Proteasome-Associated Proteins

Since the 26S proteasome is the only enzyme that degrades polyUb substrates, it is a target for regulation by a host of cellular proteins. Some of these proteins escort polyUb substrates to the 26S proteasome, thus controlling substrate access to the proteasome [140]. Others catalyze the disassembly of the polyUb chain either to rescue substrates from destruction [141], or to allow translocation of the substrate into the 20S catalytic core [55,56], or to replenish the cellular pools of ubiquitin molecules [57,58]. And yet others are E2s, E3s and E4s that bind to the 26S enzyme and directly deliver Ub-modified substrates for destruction [33,142-144].

S5a was the first polyUb-binding protein identified [145]. It exists either as a subunit of the 26S proteasome [145,146] or as a free protein outside the proteasome [147]. However, it is not known what role free S5a plays in Ub-mediated proteolysis or whether it exchanges with 26S proteasome-bound S5a. S5a exhibits specificity towards polyUb chains and shows little affinity towards monoubiquitin [145]. Binding of S5a to polyUb trees is mediated by UIM (Ubiquitin Interacting Motif) domains, ~20 amino acid sequences that are present in a number of different proteins including some that function along the endocytic/lysosomal pathway [148,149]. S5a interacts with S1/rpn2 and S2/rpn1, non-ATPase subunits within the base subcomplex of the 19S RC, through leucine-rich repeats designated PC domains found within these subunits [150]. A PC domain that may interact with S5a is also found in the Apc1 subunit of the anaphase-promoting complex (APC)/cyclosome, a multisubunit Ub ligase that degrades substrates controlling G1 phase of the cell cycle, DNA replication, mitotic progression, and cellular signaling [150,151]. Consistent with this possibility, the APC has indeed been found to copurify with affinity-purified 19S regulatory complexes [152].

Ubiquitin Receptors

A second group of ubiquitin receptors is involved in targeting polyUb substrates to the 26S proteasome, and is characterized by the presence of an N-terminal ubiquitin-like (UBL) domain and one or more ubiquitin-associated (UBA) domains [137,153-159] (see Chapter 5). This group includes the yeast proteins rad23p, dsk2p and ddi1p, and their homologs in higher eukaryotes. The co-chaperone Bag1 also contains an UBL domain and might serve a similar adaptor/targeting function linking the proteasome to chaperone/E2/CHIP complexes [72] (see Chapter 9). UBL domains are recognized by the 26S proteasome, where they presumably dock on the S1 and S2 subunits of 19S complex [150,160,161]. By contrast, UBA domains have a relatively high affinity for Lys-48-linked polyUb chains [162,163] and are substrate specific [164], thus implying that UBAs recognize binding determinants on both the polyUb chains and the conjugated substrates. Furthermore, mammalian Rad23 docks onto S5a as well, suggesting that Rad23 and S5a might function cooperatively [154]. In yeast, *rad23 dsk2 ddi1* and *rad23 dsk2 rpn10* triple mutants are viable, indicating that the flow of polyUb substrates to the 26S enzyme is only partially blocked [157,165,166]. This suggests that other yet uncharacterized polyUb receptors must exist in yeast. For instance, elongation factor 1A (eEF1A), which promotes binding of aminoacyl-tRNA to ribosomes, also binds both polyUb targets and proteasomes even though it lacks obvious UBL and UBA domains [140,167].

Rad23p contains another functional domain termed the XPC-binding domain which interacts with cytosolic peptide:N-glycanase (PNGase) [168-170]. PNGase is responsible for deglycosylation of unfolded N-linked glycoproteins dislocated from the ER into the cytosol [171]. When complexed with Rad23, PNGase associates with the 26S proteasome, thus facilitating the proteasomal degradation of N-linked glycoproteins by removing the attached glycan moieties [169]. Therefore, Rad23 plays a vital adaptor role beyond escorting polyUb substrates to the 26S proteasome by also linking PNGase, an important component of the ER-associated degradation pathway (ERAD), to the 26S enzyme.

There is a recently identified family of Ub receptors that bears similarity to the UBL and UBA proteins. Members of this family are characterized by the presence of a UBX domain within their sequence [172]. UBX domains are 80-residue C-terminal modules evolutionarily related to ubiquitin and present in a variety of proteins including p47, a cofactor of the chaperone p97 or valosin-containing protein (VCP) [172,173]. VCP is a hexameric ATPase of the AAA (ATPases associated with a variety of cellular activities) superfamily that functions in several processes, such as homotypic vesicle fusion, ERAD, and mitotic spindle disassembly [174,175]. VCP carries out these seemingly disparate functions by virtue of its ability to bind a large number of adaptors [140]. All seven UBX proteins encoded by the budding yeast genome have been shown to bind cdc48p, the yeast homolog of VCP, suggesting that the UBX domain is a general VCP/cdc48p-binding module [176]. Interestingly, one of the best characterized VCP adaptors —the Ufd1/Npl4 heterodimer critical for ERAD— does not contain UBX or UBA domains, but rather binds polyUb proteins through an UT3 domain in Ufd1 and a NZB domain in Npl4 [177]. Another well-characterized VCP cofactor, the p47 protein, recruits cdc48p during post-mitotic reassembly of the nuclear envelope, ER and Golgi apparatus [178,179]. In addition to the UBX domain in its sequence, p47 also contains an N-terminal UBA domain that binds polyUb [176,180]. In fact, all three UBA/UBX proteins in budding yeast bind ubiquitylated proteins, and at least two, p47 and Ubx2, are involved in protein degradation [176]. Whether involvement in protein degradation and membrane fusion are related or totally independent functions of p47 remains to be elucidated.

The functions of VCP in protein degradation are complex. VCP not only participates in the degradation of cytosolic proteins [164,174,180-183], but also is fundamental for the degradation of ER membrane proteins, as it seems to extricate these proteins from the membrane after their ubiquitylation, and prior to their delivery to the 26S proteasome [184-186]. The N-terminal domain of VCP also binds polyUb chains directly [187], and cdc48p binds the E4 Ufd2 inhibiting its polyUb-chain elongation activity [188]. Proteins that require cdc48p to be degraded also have a requirement for ufd2p and rad23p or dsk2p, which has allowed Richly *et al.* to determine an ordered sequence of factors that escort polyUb substrates from cdc48p to the proteasome [188]. Thus, in ERAD and during activation of a membrane-bound transcription factor, VCP coordinates the recruitment of multiple factors that allow appropriate targeting of the substrates from the site of polyubiquitylation to the site of degradation. Whether this coordinated transfer of polyUb substrates to the proteasome is limited to proteins exhibiting physical constraints, such as being present in the ER lumen or bound to the ER membrane, is an open question. However, there are many examples of E2s and E3s that associate with the 26S proteasomes (see below), and VCP itself copurifies with proteasomes [152,183] suggesting that some proteins might be polyubiquitylated on the proteasome and that VCP may directly deliver some substrates to the 26S enzyme.

Since 26S proteasomes already have intrinsic affinity towards polyUb substrates, then why does the cell need ubiquitin receptors in the first place? As explained above, in certain scenarios, an ordered sequence of protein-protein interactions would be required to ensure delivery of a substrate from the site of polyubiquitylation, i.e., the ER, to the proteasome. Another possibility is that ubiquitin receptors are in fact adaptors to link 26S proteasomes to components of the ubiquitin conjugation system, such as E3s or PNGase, to constitute a 'degradasome' or coupled system to efficiently funnel target proteins to the proteasome. Coupling within the UPS could therefore explain why ubiquitin receptors protect polyUb-conjugated proteins from the action of deubiquitylating enzymes [189] and why overexpression of S5a and Rad23 inhibits rather than promotes proteolysis [190,191]. This apparent paradox can be resolved by considering the possibility that high concentrations of S5a and Rad23 titrate out components of the degradation machinery that assemble around 26S proteasomes.

It seems that some degradasomes do not need adaptor proteins such as Rad23 and Bag1. It has been shown that certain E2s and E3s not only catalyze the transfer of Ub to substrate proteins, but also bind to 26S proteasome themselves. The E2s— Ubc1, Ubc2, Ubc4, and Ubc5 have all been shown to co-immunoprecipitate with the 26S complex [192], and a growing list of E3s including KIAA10, Ubr1, Ufd4, parkin, skp1-cullin-F-box complexes (SCF) and SNEV have also been found to associate with either the 20S or 19S subcomplexes [144,193]. For example, parkin, the E3 ligase associated with Autosomal Recessive Juvenile Parkinsonism (AR-JP), contains a C-terminal RING finger motif that serves as a recruiting module for E2s and an N-terminal UBL domain that binds to the proteasomal subunit S5a [194]. Therefore, parkin not only facilitates substrate tagging with ubiquitin by its partner E2s, but it also serves to appropriately localize the Ub-tagged substrates to the proteasome. Indeed, a single point mutation at position 42 of parkin's UBL domain has been identified in AR-JP patients [194], suggesting that the ability of parkin to adequately bind the proteasome is critical for the timely degradation of parkin substrates. In theory then, the ability of E2s and E3s to bind the proteasome may allow nascent polyUb substrates to be shuttled directly to the 26S complex for their immediate destruction thereby coupling these two processes into a single efficient pathway.

Deubiquitylating Enzymes (DUBs)

Deubiquitylating enzymes (DUBs) are a large group of proteases (561 predicted in the human genome) that catalyze the removal of ubiquitin from Ub-conjugated proteins by specifically cleaving after the C-terminal glycine residue of Ub [195,196; and Chapter 4]. Liberation of ubiquitin from polyUb substrates serves a number of primary functions for the cell [196]. For example, because efficient docking of polyUb substrates to the proteasome requires at least 4 Ub moieties [196], DUBs can negatively affect rates of substrate degradation by trimming the polyUb tree length. In this way, DUBs can perform a 'proofreading' function for the cell to remove short Ub chains from poorly ubiquitylated or slowly degraded 26S proteasome substrates. Furthermore, DUBs are required to activate newly synthesized Ub, and they serve to keep the 26S proteasome clear of unanchored 'free' polyUb chains that might compete with *bona fide* polyUb-tagged substrates. Finally, by removing polyUb chains before substrates are actively engaged and pumped into the proteasomal catalytic core, DUBs function to recycle ubiquitin monomers and replenish the cellular pool of free Ub available for subsequent degradation reactions.

Since DUBs play fundamental roles in Ub-dependent protein degradation, it is not surprising that a number of them intimately interact with the proteasome [195,196]. Indeed, some of them, such as S13/rpn11, are actually components of the 26S complex. Rpn11 is a metalloprotease that resides in the 19S RC and likely couples deubiquitylation with substrate degradation by releasing attached polyUb chains from substrates thus allowing their full translocation into the proteolytic chamber of the 20S proteasome [55,56]. Proper functioning of rpn11 is critical for the timely degradation of proteasomal substrates in general, and its function is required for viability in yeast [196]. The Ub hydrolase UCH37 is an intrinsic subunit of the 19S RC of fission yeast and higher eukaryotes (not present in the

Saccharomyces cerevisiae genome) that has been shown to disassemble polyUb degradation signals from their distal to proximal ends [141]. Thus, UCH37 could serve a proofreading function within the UPS by editing or 'trimming' the lengths of attached polyUb chains. Furthermore, by modestly shortening the length of the polyUb signal on a substrate bound to the 26S proteasome, UCH37 could facilitate repositioning of the bound substrate for optimal access into the proteasome's central chamber.

Several other DUBs have been shown to transiently interact with the 26S proteasome such as UchL5, Doa4, and Ubp6/USP14 [195,196]. For example, Ubp6 utilizes its N-terminal UBL domain to bind the S2/rpn1 subunit of the 19S complex [58]. Interestingly, Ubp6's enzymatic activity is dramatically stimulated (~300 fold) upon association with the proteasome [58], and genetic analyses in yeast suggest that Ubp6 plays a critical role in maintaining normal cellular levels of free Ub [58,196]. Genetic reduction of Ubp6 also results in decreased efficiency of the proteasome to degrade certain substrates without compromising overall 26S activity [196]. Thus, it appears that Ubp6 specifically functions on only certain polyUb substrates, and does not play a general 'housekeeping' role like S13/rpn11. Furthermore, these results suggest that other DUBs can also deubiquitylate the same substrates as Ubp6, and that there is at least some level of redundancy in the system.

Another DUB that transiently associates with the proteasome is ataxin-3. Expansion of the polyQ tract within ataxin-3 beyond its normal polymorphic range causes the severe neurodegenerative disorder Spinocerebellar Ataxia 3 (also known as Machado Joseph Disease), and like other polyQ disorders, brains of SCA3 patients display the pathological hallmarks of neuronal inclusion bodies composed of aggregated polyQ protein and components of the UPS [197]. Normal cellular ataxin-3 has been shown to interact with both VCP/p97 and Rad23 [198], co-immunoprecipitate with the proteasome [198], possess Ub-hydrolase activity that trims Lys-48-linked Ub chains [199,200], and to protect against polyQ neurotoxicity *in vivo* [201]. The normal cellular function of ataxin-3 therefore appears to be that of a *bona fide* DUB with neuroprotective properties. Ataxin-3 itself is polyubiquitylated and degraded by the 26S proteasome [202]. Polyubiquitylation of ataxin-3 requires E4B (UFD2a), an E4 that suppresses neurodegeneration induced by mutant ataxin-3 when expressed in flies [203]. Expanded ataxin-3 also binds the co-chaperone CHIP, which increases the rates of ubiquitylation and degradation of expanded ataxin-3 [204]. Thus, it appears that E4B and CHIP are critical limiting factors in the degradation of expanded ataxin-3 and could potentially become targets for therapeutic intervention. Precisely how expansion of ataxin-3's polyQ tract results in neuronal cell death remains to be fully elucidated. However, a recent report shows that polyQ expanded ataxin-3 induces apoptosis in neuronal cultures through a mechanism involving activation of caspases 3 and 9, and upregulation of the proapoptotic protein Bax [205].

Ecm29

The product of the Extracellular Mutant 29 gene (ECM29) was originally isolated in a screen for yeast proteins defective in cell wall biosynthesis [206]. Large-scale proteomic screens in yeast later identified ecm29p as a 26S proteasome-binding component [89,90].

Using conventional biochemical fractionation procedures, D. Finley and co-workers subsequently demonstrated the association of ecm29p with the 26S proteasome, and suggested that Ecm29 is a stoichiometric component of yeast 26S proteasomes [58]. Since 26S proteasomes isolated from ecm29 deletion strains readily dissociate into 19S and 20S complexes in the absence of nucleotide, Leggett *et al.* have also proposed that ecm29p tethers the 20S proteasome to the 19S RC [58].

Unlike many 26S proteasome-associated proteins, the amino acid sequence of Ecm29 does not suggest an obvious connection to the UPS. Ecm29 is a highly conserved, HEAT-repeat-containing protein [207] expressed in most cells and organisms as a ~200-kDa polypeptide. It also contains a putative VHS-like domain (~150 amino acid residue sequence found in Vps27, Hrs and STAM proteins [208]), two helical regions with homology to the N-terminus of β-adaptins, and multiple clathrin-interacting motifs throughout its sequence [209]. Below, we discuss the possible significance of the presence of HEAT repeats and a VHS-like domain in the structure and function of the Ecm29 protein.

We recently produced anti-peptide antibodies to human Ecm29 and used them to examine its distribution among mouse organs, in mammalian cells in culture, and within regions of the mouse brain [209,210]. We found that the levels of Ecm29 vary markedly among mouse organs, being highly expressed in testis, brain and pancreas, low in liver and almost absent in heart and kidney [209]. This distribution pattern does not match the distribution of several 26S proteasome subunits ([209] and unpublished data), which argues against mammalian Ecm29 being a stoichiometric component of 26S proteasomes. In fact, there appears to be in yeast a 5- to 10-fold excess of 26S proteasomes over the number of Ecm29 molecules based on the expression of *Saccharomyces cerevisiae* protein libraries fused to either green fluorescent protein (GFP) or the tandem affinity purification (TAP) tag [211,212]. Ecm29 is exclusively bound to the 26S proteasome in HeLa, HEK293 and COS7 cells ([209] and unpublished data), and overexpression of epitope-tagged Ecm29 in HEK293 cells results in complete association of the tagged Ecm29 molecules with 26S proteasomes [209]. This indicates that in HEK293 cells, 26S proteasomes are also present in excess relative to the number of Ecm29 molecules. In addition, we have not observed Ecm29 bound to the 20S proteasome either after reconstitution attempts or during purification from bovine brain or mammalian cells in culture ([209] and unpublished data). Thus, if Ecm29 tethers the 19S regulatory complex to the 20S proteasome, it does so only on a limited basis, and other protein(s) must serve a similar stabilizing function in some cells and tissues. However, as we discuss below, rather than functioning to stabilize the 26S proteasome, we favor the hypothesis that Ecm29 functions as an adaptor to localize the 26S proteasome to specific sites within the cell.

Confocal immunofluorescence microscopy using anti-peptide antibodies to Ecm29 have localized the protein to the centrosome, the nucleus, and to a subset of secretory compartments (including the ER, the ER-Golgi intermediate compartment (ERGIC) and endosomes) [209]. These cellular organelles all constitute locations where enhanced ubiquitin-dependent proteolysis occurs, and several lines of evidence support our hypothesis that Ecm29 functions as an adaptor to recruit 26S proteasomes to these sites:

1. Computer analyses predict that the protein is composed almost exclusively of HEAT repeats [207]. These protein modules form flexible, curved solenoids [213] or globular interfaces [214] that bind multiple protein ligands. Also, HEAT repeats are found in several adaptor proteins [213].
2. Ecm29 contains a region with homology to the VHS domains of Hrs, STAMs, and GGAs, proteins that serve an adaptor function in endosomal sorting and vesicular trafficking, and two regions that are homologous to the β2 subunit of AP2 adaptor complexes [209].
3. Chinese Hamster Ovary (CHO) mutant cells with increased ERAD express 2-3-fold higher levels of Ecm29 than wild type CHO cells [209]. The fact that these mutants exhibit simultaneous resistance to toxins that are endocytosed by different pathways and modify different cellular targets suggest that multitoxin resistance arises from changes in a common toxin processing mechanism. Indeed, Teter *et al.* have found that toxin resistance in these mutants is caused by increased coupling efficiency between toxin retrotranslocation from the ER lumen into the cytosol and proteasomal toxin degradation in the cytosol [215].
4. Yeast two-hybrid screens using human Ecm29 constructs (Prolexys, Inc., Salt Lake City) have revealed E2s, E3s, endocytic components, and molecular motors as Ecm29-interacting proteins (Gorbea and Rechsteiner, manuscript in preparation). The centrosomal protein pericentrin (PCM1) has also been identified as an Ecm29-interacting protein in a stringent, high-throughput yeast two-hybrid system that tested pairwise interactions among the available Gateway-cloned open reading frames [216].
5. Affinity capture of epitope-tagged, Ecm29-binding proteins present in cell extracts has identified 26S proteasome subunits, E3s, molecular motors, centrosomal proteins, chaperones, and proteins associated with the endoplasmic reticulum (Gorbea and Rechsteiner, manuscript in preparation).

Thus, the sequence of Ecm29, the presence of the Ecm29 protein and its interacting partners in multiple cellular locations, and the increased coupling efficiency between retrotranslocation and proteasomal degradation of bacterial toxins in ERAD mutants expressing higher levels of Ecm29 all suggest an adaptor function for this 26S proteasome-associated protein, as summarized schematically in Figure 3.

The recruitment of the 26S proteasome to centrosomes or to the cytoplasmic face of the ER serves clear biological purposes, e.g., degradation of abnormal proteins that concentrate in aggresomes and ERAD [217,218]. However, the role of the proteasome in endocytosis remains a mystery. Yet, our work [209] and others' [219] have shown that both 20S and 26S proteasomes are present on endocytic vesicles. Although it is clear that monoubiquitylation plays a key role in the internalization and sorting of protein cargo along the endocytic pathway, the attached Ub moiety does not target the modified cargo to the 26S proteasome; rather it functions as a routing signal to late endosomes, multivesicular bodies (MVBs) and lysosomes. There are, however, an increasing number of reports implicating the 26S proteasome in several steps along the endocytic pathway. When proteasome activity is inhibited, some membrane receptors are not internalized [43,220]; are preferentially recycled to the plasma membrane [221]; or their trafficking is blocked at a late endosome-to-lysosome stage [222,223]. In neurons, it is now clear that proteasomal degradation of the scaffolding protein PSD-95 at postsynaptic densities is required for signal-induced internalization of glutamate receptors [42,43] even though the receptors themselves are not degraded. Thus, localization of Ecm29 and 26S proteasomes to endosomes [209,219] raises the possibility that the enzyme degrades components involved in the endocytic process or plays a remodeling role on endocytic vesicles. Accordingly, it has been reported that STAMs, which bind both ubiquitin and Hrs during entry into the MVB pathway, are themselves polyubiquitylated and rapidly degraded [224].

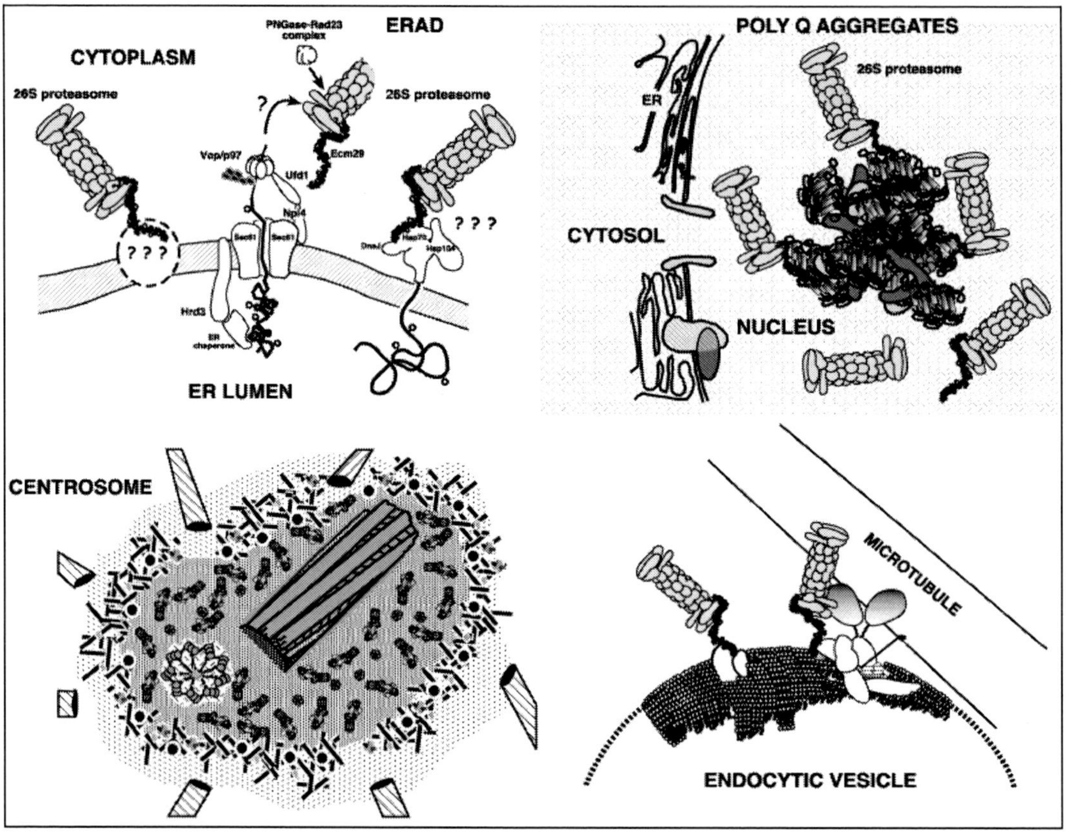

Figure 3. Working model that Ecm29 functions as an adaptor for localizing the 26S proteasome within cells. Current evidence suggest that Ecm29 may bind distinct cellular proteins to link the 26S enzyme to ER-associated protein degradation pathways (*top, left*), to intranuclear polyglutamine aggregates (*top, right*), to the centrosome (*bottom, left*), or to endocytic components and molecular motors (*bottom, right*).

Ecm29 in the Central Nervous System

Western blots using 3 specific anti-Ecm29 peptide antibodies, e.g., ECM1, ECM2, and ECM3, have revealed that mouse brain expresses substantial amounts of what appear to be at least eleven specific Ecm29 immunoreactive species ranging in molecular weight from 55 to greater than 400 kDa [209,210]. The largest Ecm29 species reacts only with ECM1 and ECM3, raised to peptide sequences close to the N- or at the C-termini of the protein, respectively. Hence, it is likely that this and perhaps other Ecm29 variants are generated by an alternative splicing mechanism. Pulse-chase analysis of newly synthesized Ecm29 in brain or cultured neurons should answer whether any smaller isoforms arise by post-translational proteolytic processing.

Sedimentation of brain homogenates on glycerol gradients have demonstrated that only the ~200-kDa Ecm29 species is clearly associated with the 26S proteasome; with one exception the other variants sediment slower than 20S [210]. These results suggest that the different ECM immunoreactive species carry out distinct functions in nervous tissue. Western blots of dissected mouse brain regions with the three Ecm29 antibodies have revealed that Ecm29 isoforms are present throughout the central nervous system, although the distribution patterns vary among different parts of the brain. Immunohistochemical staining of brain sections using the three ECM antibodies have revealed marked differences in staining patterns as well. For instance, strong ECM2 and ECM3 staining occurs throughout all regions of the hippocampal formation, and staining of the hippocampal subregions CA1 and CA3 is both axonal and dendritic [210]. Cell-type specificity of staining is particularly noteworthy in the CA1 subregion since ECM staining is clearly not restricted to neurons. Astrocytes are labeled with ECM1 and ECM3, whereas ECM2 staining is virtually absent. Examination of other brain regions further reveals the complexity in Ecm29 staining. For example, in the hypothalamus all three ECM antibodies stain the arcuate hypothalamic nucleus, but in the dorsal-medial hypothalamic region, staining of neuronal cell bodies is only seen with ECM1 and ECM2 [210]. It should also be noted that expression of Ecm29 isoforms in the mouse brain does not conform to the general trend of expression of 20S proteasomes and 19S regulatory complexes.

Ecm29 in Cultured Cortical Neurons

The complex pattern of expression of Ecm29 variants seems to be, to a great extent, neuronal in origin. For instance, various Ecm29 isoforms were observed in fractionated synaptosomes (Gorbea, Rogers and Rechsteiner, unpublished data) and in extracts from primary cortical neurons (Gorbea, Rogers and Rechsteiner, manuscript in preparation). By contrast, only the ~200-kDa form of Ecm29 was present in extracts from a microglial cell line. Immunofluorescense microscopic analyses of cultured cortical neurons with the three ECM antibodies also produced differential patterns of staining. For instance, ECM1 stained cell bodies and dendrites; staining of dendrites with ECM1 was consistent with the presence of Ecm29 in spines and/or filopodia. By contrast, ECM3 staining produced a vesicular pattern in axons (Gorbea, Rogers and Rechsteiner, manuscript in preparation). As mentioned above, our two-hybrid screen data have identified molecular motors and endocytic components as 26S proteasome-interacting proteins, which may explain the significant enrichment of Ecm29 in axons and synaptosomes. Since 26S proteasomes are present at synapses [23,24], there must be mechanisms whereby the enzyme reaches regions that can be quite distant from the neuronal cell body. If Ecm29 serves to couple the 26S proteasome to motors or vesicles, this could explain the need for such high levels of the protein in brain. Sorting out the intracellular distribution and molecular nature of the isoforms will aid in ascertaining the biological roles of Ecm29 isoforms in the brain.

Conclusions

The past few years have witnessed an explosive increase in our understanding concerning the breadth of importance the UPS has in all essential aspects of life. The identification of increasing numbers of cellular proteins that bind to proteasomes demonstrates that these enzymes are highly dynamic protein degradation complexes subject to many levels of regulation. Among the host of cellular proteins that interact with the proteasome we now count activators, inhibitors, components of the ubiquitylation machinery, ubiquitin hydrolases, polyUb-chain receptors, and the HEAT-repeat-containing protein Ecm29. These interacting proteins are thought to not only affect 26S catalytic activity, but also to localize the proteasome to specific sites within the cell, and assist delivery of polyUb substrates to the proteasome for degradation. Still, many questions remain regarding the structure as well as the biological significance of proteasome-activator/inhibitor complexes. Furthermore, many novel interactions between proteasomes and other cellular proteins are likely to be discovered in the future, expanding our knowledge about the range of biological processes influenced by the UPS.

A growing appreciation for the role of proteasomes in neurons is evident from this volume, and it seems likely that crucial proteasomal functions will be confirmed in the normal and pathophysiology of the nervous system. To date, however, what we know about the impact of the UPS in neuronal cell biology is relatively scant. Anything we can learn from the distribution of UPS components in the brain, and from the identification of neuronal proteasome-interacting proteins will undoubtedly assist us in elucidating the roles that the UPS plays in neuronal cell function, and should be a high priority for future research.

Acknowledgements

This review was written with support from grant NS042892 from the National Institutes of Health to Martin Rechsteiner. We thank Frank Whitby (University of Utah School of Medicine, Salt Lake City, UT) for assistance with Figure 1, and Joaquín Ortega (McMaster University, Hamilton, ON, Canada) for providing the images in Figure 2.

References

[1] Dhananjayan SC, Ismail A, Nawaz Z: Ubiquitin and control of transcription. *Essays Biochem* 2005, 41:69-80.

[2] Osley MA: H2B ubiquitylation: the end is in sight. *Biochim Biophys Acta* 2004, 1677:74-78.

[3] Hicke L, Dunn R: Regulation of membrane protein transport by ubiquitin and ubiquitin-binding proteins. *Annu Rev Cell Dev Biol* 2003, 19:141-172.

[4] Urbe S: Ubiquitin and endocytic protein sorting. *Essays Biochem* 2005:81-98.

[5] Morita E, Sunquist WI: Retrovirus budding. *Annu Rev Cell Dev Biol* 2004, 20:395-425.

[6] Hershko A, Ciechanover A: The ubiquitin system. *Annu Rev Biochem* 1998, 67:425-479.

[7] Hershko A: The ubiquitin system for protein degradation and some of its roles in the control of the cell division cycle. *Cell Death Differ* 2005, 12:1191-1197.

[8] DeSalle LM, Pagano M: Regulation of the G1 to S transition by the ubiquitin pathway. *FEBS Lett* 2001, 490:179-189.

[9] Jesenberger V, Jentsch S: Deadly encounter: ubiquitin meets apoptosis. *Nat Rev Mol Cell Biol* 2002, 3:112-121.

[10] Kloetzel PM: The proteasome and MHC class I antigen processing. *Biochim Biophys Acta* 2004, 1695:217-225.

[11] Ko HW, Jiang J, Edery I: Role for Slim in the degradation of Drosophila Period protein phosphorylated by Doubletime. *Nature* 2002, 420:673-678.

[12] Eide EJ, Woolf MF, Kang H, Woolf P, Hurst W, Camacho F, Vielhaber EL, Giovanni A, Virshup DM: Control of mammalian circadian rhythm by CKIε-regulated proteasome-mediated PER2 degradation. *Mol Cell Biol* 2005, 25:2795-2807.

[13] Ellgaard L, Helenius A: Quality control in the endoplasmic reticulum. *Nat Rev Mol Cell Biol* 2003, 4:181-191.

[14] Meusser B, Hirsch C, Jarosch E, Sommer T: ERAD: the long road to destruction. *Nat Cell Biol* 2005, 7:766-772.

[15] Cooper HM: Axon guidance receptors direct growth cone pathfinding: rivalry at the leading edge. *Int J Dev Biol* 2002, 46:621-631.

[16] Speese SD, Trotta N, Rodesch CK, Aravamudan B, Broadie K: The ubiquitin proteasome system acutely regulates presynaptic protein turnover and synaptic efficacy. *Curr Biol* 2003, 13:899-910.

[17] Ehlers MD: Deconstructing the axon: Wallerian degeneration and the ubiquitin-proteasome system. *Trends Neurosci* 2004, 27:3-6.

[18] Korhonen L, Lindholm D: The ubiquitin proteasome system in synaptic and axonal degeneration: a new twist to an old cycle. *J Cell Biol* 2004, 165:27-30.

[19] van Roessel P, Elliott DA, Robinson IM, Prokop A, Brand AH: Independent regulation of synaptic size and activity by the anaphase-promoting complex. *Cell* 2004, 119:707-718.

[20] DiAntonio A, Hicke L: Ubiquitin-dependent regulation of the synapse. *Annu Rev Neurosci* 2004, 27:223-246.

[21] Bryan B, Cai Y, Wrighton K, Wu G, Feng XH, Liu M: Ubiquitination of RhoA by Smurf1 promotes neurite outgrowth. *FEBS* Lett 2005, 579:1015-1019.

[22] Kato A, Rouach N, Nicoll RA, Bredt DS: Activity-dependent NMDA receptor degradation mediated by retrotranslocation and ubiquitination. *Proc Natl Acad Sci* USA 2005, 102:5600-5605.

[23] Yi JJ, Ehlers MD: Ubiquitin and protein turnover in synapse function. *Cell* 2005, 47:629-632.

[24] Bingol B, Schuman EM: Synaptic protein degradation by the ubiquitin proteasome system. *Curr Opin Neurobiol* 2005, 15:536-541.

[25] Chisholm AD, Jin Y: Neuronal differentiation in C. elegans. *Curr Opin Cell Biol* 2005, 17:682-689.

[26] Teng FY, Tang BL: APC/C regulation of axonal growth and synaptic functions in postmitotic neurons: the Liprin-α connection. *Cell Mol Life Sci* 2005, 62:1571-1578.

[27] Pintard L, Willems A, Peter M: Cullin-based ubiquitin ligases: Cul3-BTB complexes join the family. *EMBO J* 2004, 23:1681-1687.

[28] Willems AR, Schwab M, Tyers M: A hitchhiker's guide to cullin ubiquitin ligases: SCF and its kin. *Biochim Biophys Acta* 2004, 1695:133-170.

[29] Ang XL, Wade Harper J: SCF-mediated protein degradation and cell cycle control. *Oncogene* 2005, 24:2860-2870.

[30] Castro A, Bernis C, Vigneron S, Labbé JC, Lorca T: The anaphase-promoting complex: a key factor in the regulation of cell cycle. *Oncogene* 2005, 24:314-325.

[31] d'Azzo A, Bongiovanni A, Nastasi T: E3 ubiquitin ligases as regulators of membrane protein trafficking and degradation. *Traffic* 2005, 6:429-441.

[32] Ardley HC, Robinson PA: E3 ubiquitin ligases. *Essays Biochem* 2005, 41:15-30.

[33] Hoppe T: Multiubiquitylation by E4 enzymes: 'one size' doesn't fit all. *Trends Biochem Sci* 2005, 30:183-187.

[34] Wolf DH, Hilt W: The proteasome: a proteolytic nanomachine of cell regulation and waste disposal. *Biochim Biophys Acta* 2004, 1695:19-31.

[35] Fischer JA, Overstreet E: Fat facets does a Highwire act at the synapse. *Bioessays* 2002, 24:13-16.

[36] McCabe BD, Hom S, Aberle H, Fetter RD, Marques G, Haerry TE, Wan H, O'Connor MB, Goodman CS, Haghighi AP: Highwire regulates presynaptic BMP signaling essential for synaptic growth. *Neuron* 2004, 41:891-905.

[37] Nakata K, Abrams B, Grill B, Goncharov A, Huang X, Chisholm AD, Jin Y: Regulation of a DLK-1 and p38 MAP kinase pathway by the ubiquitin ligase RPM-1 is required for presynaptic development. *Cell* 2005, 120:407-420.

[38] Wu C, Wairkar YP, Collins CA, DiAntonio A: Highwire function at the Drosophila neuromuscular junction: spatial, structural, and temporal requirements. J *Neurosci* 2005, 25:9557-9566.

[39] Kuo CT, Jan LY, Jan YN: Dendrite-specific remodeling of Drosophila sensory neurons requires matrix metalloproteases, ubiquitin-proteasome, and ecdysone signaling. *Proc Natl Acad Sci USA* 2005, 102:15230-15235.

[40] Bedford FK, Kittler JT, Muller E, Thomas P, Uren JM, Merlo D, Wisden W, Triller A, Smart TG, Moss SJ: GABA(A) receptor cell surface number and subunit stability are regulated by the ubiquitin-like protein Plic-1. *Nat Neurosci* 2001, 4:908-916.

[41] Araki T, Milbrandt J: ZNRF proteins constitute a family of presynaptic E3 ubiquitin ligases. *J Neurosci* 2003, 23:9385-9394.

[42] Colledge M, Snyder EM, Crozier RA, Soderling JA, Jin Y, Langeberg LK, Lu H, Bear MF, Scott JD: Ubiquitination regulates PSD-95 degradation and AMPA receptor surface expression. *Neuron* 2003, 40:595-607.

[43] Patrick GN, Bingol B, Weld HA, Schuman EM: Ubiquitin-mediated proteasome activity is required for agonist-induced endocytosis of GluRs. *Curr Biol* 2003, 13:2073-2081.

[44] Chain DG, Schwartz JH, Hegde AN: Ubiquitin-mediated proteolysis in learning and memory. *Mol Neurobiol* 1999, 20:125-142.

[45] Zhao Y, Hegde AN, Martin KC: The ubiquitin proteasome system functions as an inhibitory constraint on synaptic strengthening. *Curr Biol* 2003, 13:887-898.

[46] Upadhya SC, Smith TK, Hegde AN: Ubiquitin-proteasome-mediated CREB repressor degradation during induction of long-term facilitation. *J Neurochem* 2004, 91:210-219.

[47] Ciechanover A, Brundin P: The ubiquitin proteasome system in neurodegenerative diseases: sometimes the chicken, sometimes the egg. *Neuron* 2003, 40:427-446.

[48] Ross CA, Pickart CM: The ubiquitin-proteasome pathway in Parkinson's disease and other neurodegenerative diseases. *Trends Cell Biol* 2004, 14:703-711.

[49] Groll M, Ditzel L, Löwe J, Stock D, Bochtler M, Bartunik HD, Huber R: Structure of 20S proteasome from yeast at 2.4 Å resolution. *Nature* 1997, 386:463-471.

[50] Unno M, Mizushima T, Morimoto Y, Tomisugi Y, Tanaka K, Yasuoka N, Tsukihara T: The structure of the mammalian 20S proteasome at 2.75 Å resolution. *Structure* 2002, 10:609-618.

[51] Orlowski M, Wilk S: Ubiquitin-independent proteolytic functions of the proteasome. *Arch Biochem Biophys* 2003, 415:1-5.

[52] Hoffman L, Pratt G, Rechsteiner M: Multiple forms of the 20 S multicatalytic and the 26 S ubiquitin/ATP-dependent proteases from rabbit reticulocyte lysate. *J Biol Chem* 1992, 267:22362-22368.

[53] Glickman MH, Rubin DM, Coux O, Wefes I, Pfeifer G, Cjeka Z, Baumeister W, Fried VA, Finley D: A subcomplex of the proteasome regulatory particle required for ubiquitin-conjugate degradation and related to the COP9-signalosome and eIF3. *Cell* 1998, 94:615-623.

[54] Rechsteiner M: The 26S proteasome. In Protein Degradation. Ubiquitin and the Chemistry of Life. Edited by Mayer RF, Ciechanover A, Rechsteiner M. Weinheim: WILEY-VCH Verlag GmbH; 2005:220-247.

[55] Yao T, Cohen RE: A cryptic protease couples deubiquitination and degradation by the proteasome. *Nature* 2002, 419:403-407.

[56] Verma R, Aravind L, Oania R, McDonald WH, Yates III JR, Koonin EV, Deshaies RJ: Role of rpn11 metalloprotease in deubiquitination and degradation by the 26S proteasome. *Science* 2002, 298:611-615.

[57] Swaminathan S, Amerik AY, Hochstrasser M: The Doa4 deubiquitinating enzyme is required for ubiquitin homeostasis in yeast. *Mol Biol Cell* 1999, 10:2583-2594.

[58] Leggett DS, Hanna J, Borodovsky A, Crosas B, Schmidt M, Baker RT, Walz T, Ploegh H, Finley D: Multiple associated proteins regulate proteasome structure and function. Mol *Cell* 2002, 10:495-507.

[59] Rechsteiner M, Hill CP: Mobilizing the proteolytic machine: cell biological roles of proteasome activators and inhibitors. *Trends Cell Biol* 2005, 15:27-33.
[60] Rechsteiner M, Realini C, Ustrell V: The proteasome activator 11S REG (PA28) and class I antigen presentation. *Biochem J* 2000, 345:1-15.
[61] Ma C-P, Slaughter CA, DeMartino GN: Identification, purification, and characterization of a protein activator (PA28) of the 20S proteasome (macropain). *J Biol Chem* 1992, 267:10515-10523.
[62] Dubiel W, Pratt G, Ferrell K, Rechsteiner M: Purification of an 11S regulator of the multicatalytic protease. *J Biol Chem* 1992, 267:22369-22377.
[63] Whitby FG, Masters EI, Kramer L, Knowlton JR, Yao Y, Wang CC, Hill CP: Structural basis for the activation of 20S proteasomes by 11S regulators. *Nature* 2000, 408:115-120.
[64] Ortega J, Heymann JB, Kajava AV, Ustrell V, Rechsteiner M, Steven AC: The axial channel of the 20S proteasome opens upon binding of the PA200 activator. *J Mol Biol* 2005, 346:1221-1227.
[65] Schmidt M, Haas W, Crosas B, Santamaria PG, Gygi SP, Walz T, Finley D: The HEAT repeat protein Blm10 regulates the yeast proteasome by capping the core particle. *Nat Struct Mol Biol* 2005, 12:294-303.
[66] Tanahashi N, Murakami Y, Minami Y, Shimbara N, Hendil KB, Tanaka K: Hybrid proteasomes. Induction by interferon-γ and contribution to ATP-dependent proteolysis. *J Biol Chem* 2000, 275:14336-14345.
[67] Realini C, Jensen CC, Zhang Z, Johnston SC, Knowlton JR, Hill CP, Rechsteiner M: Characterization of recombinant REGα, REGβ, and REGγ proteasome activators. *J Biol Chem* 1997, 272:25483-25492.
[68] Gao X, Li J, Pratt G, Wilk S, Rechsteiner M: Purification procedures determine the proteasome activation properties of REGγ (PA28γ). *Arch Biochem Biophys* 2004, 425:158-164.
[69] Saric T, Graef CI, Goldberg AL: Pathway for degradation of peptides generated by proteasomes: a key role for thimet oligopeptidase and other metallopeptidases. *J Biol Chem* 2004, 279:46723-46732.
[70] Cascio P, Call M, Petre BM, Walz T, Goldberg AL: Properties of the hybrid form of the 26S proteasome containing both 19S and PA28 complexes. *EMBO J* 2002, 21:2636-2645.
[71] Minami Y, Kawasaki H, Minami M, Tanahashi N, Tanaka K, Yahara I: A critical role for the proteasome activator PA28 in the HSP90-dependent protein refolding. *J Biol Chem* 2000, 275:9055-9061.
[72] Esser C, Alberti S, Höhfeld J: Cooperation of molecular chaperones with the ubiquitin/proteasome system. *Biochim Biophys Acta* 2004, 1695:171-178.
[73] Rock KL, Gramm C, Rothstein L, Clark K, Stein R, Dick L, Hwang D, Goldberg AL: Inhibitors of the proteasome block the degradation of most cell proteins and the generation of peptides presented on MHC class I molecules. *Cell* 1994, 78:761-771.
[74] Cerundolo V, Benham A, Braud V, Mukherjee S, Gould K, Macino B, Neefjes J, Townsend A: The proteasome-specific inhibitor lactacystin blocks presentation of

cytotoxic T lymphocyte epitopes in human and murine cells. *Eur J Immunol* 1997, 27:336-341.

[75] Preckel T, Fung-Leung WP, Cai Z, Vitiello A, Salter-Cid L, Winqvist O, Wolfe TG, Von Herrath M, Angulo A, Ghazal P, Lee JD, Fourie AM, Wu Y, Pang J, Ngo K, Peterson PA, Früh K, Yang Y: Impaired immunoproteasome assembly and immune response in PA28-/- mice. *Science* 1999, 286:2162-2165.

[76] Murata S, Udono H, Tanahashi N, Hamada N, Watanabe K, Adachi K, Yamano T, Yui K, Kobayashi N, Kasahara M, Tanaka K, Chiba T: Immunoproteasome assembly and antigen presentation in mice lacking both PA28α and PA28β. *EMBO J* 2001, 20:5898-5907.

[77] Kisselev AF, Akopian TN, Woo KM, Goldberg AL: The sizes of peptides generated from protein by mammalian 26 and 20S proteasomes. Implications for understanding the degradative mechanism and antigen presentation. *J Biol Chem* 1999, 274:3363-3371.

[78] Engelhard VH: Structure of peptides associated with class I and class II MHC molecules. *Annu Rev Immunol* 1994, 12:181-207.

[79] Rivett AJ, Bose S, Brooks P, Broadfoot KI: Regulation of proteasome complexes by γ-interferon and phosphorylation. *Biochimie* 2001, 83:363-366.

[80] Masson P, Lundgren J, Young P: Drosophila proteasome regulator REGγ: transcriptional activation by DNA replication-related factor DREF and evidence for a role in cell cycle progression. *J Mol Biol* 2003, 327:1001-1012.

[81] Barton LF, Runnels HA, Schell TD, Cho Y, Gibbons R, Tevethia SS, Deepe GS Jr, Monaco JJ: Immune defects in 28-kDa proteasome activator γ-deficient mice. *J Immunol* 2004, 172:3948-3954.

[82] Goellner GM, Rechsteiner M: Are Huntington's and polyglutamine-based ataxias proteasome storage diseases? *Int J Biochem Cell Biol* 2003, 35:562-571.

[83] Bates G: Huntingtin aggregation and toxicity in Huntington's disease. *Lancet* 2003, 361:1642-1644.

[84] Arrasate M, Mitra S, Schweitzer ES, Segal MR, Finkbeiner S: Inclusion body formation reduces levels of mutant huntingtin and the risk of neuronal death. *Nature* 2004, 431:805-810.

[85] Holmberg CI, Staniszewski KE, Mensah KN, Matouschek A, Morimoto RI: Innefficient degradation of truncated polyglutamine proteins by the proteasome. *EMBO J* 2004, 23:4307-4318.

[86] Venkatraman P, Wetzel R, Tanaka M, Nukina N, Goldberg AL: Eukaryotic proteasomes cannot digest polyglutamine sequences and release them during degradation of polyglutamine-containing proteins. *Mol Cell* 2004, 14:95-104.

[87] Bett JS, Goellner GM, Woodman B, Pratt G, Rechsteiner M, Bates GP: Proteasome impairment does not contribute to pathogenesis in R6/2 Huntington's disease mice: exclusion of proteasome activator REGγ as a therapeutic target. *Hum Mol Genet* 2006, 15:33-44.

[88] Bowman AB, Yoo SY, Dantuma NP, Zoghbi HY: Neuronal dysfunction in a polyglutamine disease model occurs in the absence of ubiquitin-proteasome system

impairment and inversely correlates with the degree of inclusion formation. *Hum Mol Genet* 2005, 14:679-691.

[89] Gavin AC, Bösche M, Krause R, Grandi P, Marzioch M, Bauer A, Schultz J, Rick JM, Michon AM, Cruciat CM, Remor M, Höfert C, Schelder M, Brajenovic M, Ruffner H, Merino A, Klein K, Hudak M, Dickson D, Rudi T, Gnau V, Bauch A, Bastuck S, Huhse B, Leutwein C, Heurtier MA, Copley RR, Edelmann A, Querfurth E, Rybin V, Drewes G, Raida M, Bouwmeester T, Bork P, Seraphin B, Kuster B, Neubauer G, Superti-Furga G: Functional organization of the yeast proteome by systematic analysis of protein complexes. *Nature* 2002, 415:141-147.

[90] Ho Y, Gruhler A, Heilbut A, Bader GD, Moore L, Adams SL, Millar A, Taylor P, Bennett K, Boutilier K, Yang L, Wolting C, Donaldson I, Schandorff S, Shewnarane J, Vo M, Taggart J, Goudreault M, Muskat B, Alfarano C, Dewar D, Lin Z, Michalickova K, Willems AR, Sassi H, Nielsen PA, Rasmussen KJ, Andersen JR, Johansen LE, Hansen LH, Jespersen H, Podtelejnikov A, Nielsen E, Crawford J, Poulsen V, Sørensen BD, Matthiesen J, Hendrickson RC, Gleeson F, Pawson T, Moran MF, Durocher D, Mann M, Hogue CW, Figeys D, Tyers M: Systematic identification of protein complexes in Saccharomyces cerevisiae by mass spectrometry. *Nature* 2002, 415:180-183.

[91] Ustrell V, Hoffman L, Pratt G, Rechsteiner M: PA200, a nuclear proteasome activator involved in DNA repair. *EMBO J* 2002, 21:3516-3525.

[92] Sone T, Saeki Y, Toh-e A, Yokosawa H: Sem1p is a novel subunit of the 26 S proteasome from Saccharomyces cerevisiae. *J Biol Chem* 2004, 279:28807-28816.

[93] Funakoshi M, Li X, Velichutina I, Hochstrasser M, Kobayashi H: Sem1, the yeast ortholog of a human BRCA2-binding protein, is a component of the proteasome regulatory particle that enhances proteasome stability. *J Cell Sci* 2004, 117:6447-6457.

[94] Krogan NJ, Lam MHY, Fillingham J, Keogh MC, Gebbia M, Li J, Datta N, Cagney G, Buratowski S, Emili A, Greenblatt JF: Proteasome involvement in the repair of DNA double-strand breaks. *Mol Cell* 2004, 16:1027-1034.

[95] Jossé L, Harley ME, Pires IMS, Hughes DA: Fission yeast Dss1 associates with the proteasome and is required for efficient ubiquitin-dependent proteolysis. *Biochem J* 2006, 393:303-309.

[96] Yang H, Jeffrey PD, Miller J, Kinnucan E, Sun Y, Thomä NH, Zheng N, Chen PL, Lee WH, Pavletich NP: BRCA2 function in DNA binding and recombination from a BRCA2-DSS1-sDNA structure. *Science* 2002, 297:1837-1848.

[97] Moore CW: Further characterizations of bleomycin-sensitive (blm) mutants of Saccharomyces cerevisiae with implications for a radiomimetic model. *J Bacteriol* 1991, 173:3605-3608.

[98] Jelinsky SC, Estep P, Church GM, Samson LD: Regulatory networks revealed by transcriptional profiling of damaged Saccharomyces cerevisiae cells: Rpn4 links base excision repair with proteasomes. *Mol Cell Biol* 2000, 20:8157-8167.

[99] Tsubuki S, Saito Y, Kawashima S: Purification and characterization of an endogenous inhibitor specific to the Z-Leu-Leu-Leu-MCA degrading activity in proteasome and its identification as heat-shock protein 90. *FEBS Lett* 1994, 344:229-233.

[100] Lu X, Michaud C, Orlowski M: Heat shock protein-90 and the catalytic activities of the 20 S proteasome (multicatalytic proteinase complex). *Arch Biochem Biophys* 2001, 387:163-171.

[101] Chu-Ping M, Slaughter CA, DeMartino GN: Purification and characterization of a protein inhibitor of the 20S proteasome (macropain). *Biochim Biophys Acta* 1992, 1119:303-311.

[102] Zaiss DMW, Standera S, Holzhütter H, Kloetzel PM, Sijts AJAM: The proteasome inhibitor PI31 competes with PA28 for binding to 20S proteasomes. *FEBS Lett* 1999, 457:333-338.

[103] McCutchen-Maloney SL, Matsuda K, Shimbara N, Binns DD, Tanaka K, Slaughter CA, DeMartino GN: cDNA cloning, expression, and functional characterization of PI31, a proline-rich inhibitor of the proteasome. *J Biol Chem* 2000, 275:18557-18565.

[104] Zaiss DMW, Standera S, Kloetzel PM, Sijts AJAM: PI31 is a modulator of proteasome formation and antigen processing. *Proc Natl Acad Sci USA* 2002, 99:14344-14349.

[105] Park Y, Hwang YP, Lee JS, Seo SH, Yoon SK, Yoon JB: Proteasomal ATPase-associated factor 1 negatively regulates proteasome activity by interacting with proteasomal ATPases. *Mol Cell Biol* 2005, 25:3842-3853.

[106] Li J, Post M, Volk R, Gao Y, Li M, Metais C, Sato K, Tsai J, Aird W, Rosenberg RD, Hampton TG, Sellke F, Carmeliet P, Simons M: PR39, a peptide regulator of angiogenesis. *Nat Med* 2000, 6:49-55.

[107] Gao Y, Lecker S, Post MJ, Hietaranta AJ, Li J, Volk R, Li M, Sato K, Saluja AK, Steer ML, Goldberg AL, Simons M: Inhibition of ubiquitin-proteasome pathway-mediated IκBα degradation by a naturally occurring antibacterial peptide. *J Clin Invest* 2000, 106:439-448.

[108] Gaczynska M, Osmulski PA, Gao Y, Post MJ, Simons M: Proline- and arginine-rich peptides constitute a novel class of allosteric inhibitors of proteasome activity. *Biochemistry* 2003, 42:8663-8670.

[109] Gregori L, Fuchs C, Figueiredo-Pereira ME, Van Nostrand WE, Goldgaber D: Amyloid β protein inhibits ubiquitin-dependent protein degradation in vitro. *J Biol Chem* 1995, 270:19702-19708.

[110] Lindersson E, Beedholm R, Højrup P, Moos T, Gai W, Hendil KB, Jensen PH: Proteasomal inhibition by α-synuclein filaments and oligomers. *J Biol Chem* 2004 279:12924-12934.

[111] Bence NF, Sampat RM, Kopito RR: Impairment of the ubiquitin-proteasome system by protein aggregation. *Science* 2001, 292:1552-1555.

[112] Bennett EJ, Bence NF, Jayakumar R, Kopito RR: Global impairment of the ubiquitin-proteasome system by nuclear or cytoplasmic protein aggregates precedes inclusion body formation. *Mol Cell* 2005, 17:351-365.

[113] Khu YL, Tan YJ, Lim SG, Hong W, Goh PY: Hepatitis C virus non-structural proteins NS3 interacts with LMP7, a component of the immunoproteasome, and affects its proteasome activity. *Biochem J* 2004, 384:401-409.

[114] Levitskaya J, Sharipo A, Leonchiks, Ciechanover A, Masucci M: Inhibition of ubiquitin/proteasome-dependent protein degradation by the Gly-Ala repeat domain of

the Epstein-Barr-virus nuclear antigen 1. *Proc Natl Acad Sci USA* 1997, 94:12616-12621.

[115] Turnell AS, Grand RJ, Gorbea C, Zhang X, Wang W, Mymryk JS, Gallimore PH: Regulation of the 26S proteasome by adenovirus E1A. *EMBO J* 2000, 19:4759-4773.

[116] Zhang X, Turnell AS, Gorbea C, Mymryk JS, Gallimore PH, Grand RJ: The targeting of the proteasomal regulatory subunit S2 by adenovirus E1A causes inhibition of proteasomal activity and increased p53 expression. *J Biol Chem* 2004, 279:25122-25133.

[117] Seeger M, Ferrel K, Frank R, Dubiel W: HIV-1 Tat inhibits the 20 S proteasome and its 11 S regulator-mediated activation. *J Biol Chem* 1997, 272:8145-8148.

[118] Hu Z, Zhang Z, Doo E, Coux O, Goldberg AL, Liang TJ: Hepatitis B virus X protein is both a substrate and a potential inhibitor of the proteasome complex. *J Virol* 1999, 73:7231-7240.

[119] Zhang F, Su K, Yang X, Bowe DB, Paterson AJ, Kudlow JE: O-GlcNAc modification is an endogenous inhibitor of the proteasome. *Cell* 2003, 115:715-725.

[120] Sun XM, Butterworth M, MacFarlene M, Dubiel W, Ciechanover A, Cohen GM: Caspase activation inhibits proteasome function during apoptosis. *Mol Cell* 2004, 14:81-93.

[121] Adrain C, Creagh EM, Cullen SP, Martin SJ: Caspase-dependent inactivation of proteasome function during programmed cell death in Drosophila and man. *J Biol Chem* 2004, 279:36923-36930.

[122] Whittier JE, Xiong Y, Rechsteiner MC, Squier TC: Hsp90 enhances degradation of oxidized calmodulin by the 20 S proteasome. *J Biol Chem* 2004, 279:46135-46142.

[123] Yamano T, Murata S, Shimbara N, Tanaka N, Chiba T, Tanaka K, Yui K, Udono H: Two distinct pathways mediated by PA28 and hsp90 in major histocompatibility complex class I antigen processing. *J Exp Med* 2002, 196:185-196.

[124] Imai J, Maruya M, Yashiroda H, Yahara I, Tanaka K: The molecular chaperone Hsp90 plays a role in the assembly and maintenance of the 26S proteasome. *EMBO J* 2003, 22:3557-3567.

[125] Zhang H, Burrows F: Targeting multiple signal transduction pathways through inhibition of Hsp90. *J Mol Med* 2004, 82:488-499.

[126] Wu J, Parungo C, Wu G, Kang PM, Laham RJ, Sellke FW, Simons M, Li J: PR39 inhibits apoptosis in hypoxic endothelial cells: role of inhibitor apoptosis protein-2. *Circulation* 2004, 109:1660-1667.

[127] Levitskaya J, Coram M, Levitsky V, Imreh S, Steigerwald-Mullen PM, Klein G, Kurilla MG, Masucci MG: Inhibition of antigen processing by the internal repeat region of the Epstein-Barr Virus nuclear antigen-1. *Nature* 1995, 375:685-688.

[128] Rape M, Jentsch S: Taking a bite: proteasomal protein processing. *Nat Cell Biol* 2002, 4:E113-E116.

[129] Chang Y, Cesarman E, Pessin MS, Lee F, Culpepper J, Knowles DM, Moore PS: Identification of herpesvirus-like DNA sequences in AIDS-associated Kaposi's sarcoma. *Science* 1994, 266:1865-1869.

[130] Leonchiks A, Stavropoulou V, Sharipo A, Masucci MG: Inhibition of ubiquitin-dependent proteolysis by a synthetic glycine-alanine repeat peptide that mimics an inhibitory viral sequence. *FEBS Lett* 2002, 522:93-98.

[131] Gallimore PH, Turnell AS: Adenovirus E1A: remodelling the host cell, a life or death experience. *Oncogene* 2001, 20:7824-7835.

[132] Nelbock P, Dillon PJ, Perkins A, Rosen CA: A cDNA for a protein that interacts with the human immunodeficiency virus Tat transactivator. *Science* 1990, 248:1650-1653.

[133] Apcher GS, Heink S, Zantopf D, Kloetzel PM, Schmid HP, Mayer RJ, Krüger E: Human immunodeficiency virus-1 Tat protein interacts with distinct protasomal α and β subunits. *FEBS Lett* 2003, 553:200-204.

[134] Huang X, Seifert U, Salzmann U, Henklein P, Preissner R, Henke W, Sijts AJ, Kloetzel PM, Dubiel W: The RTP site shared by the HIV-1 Tat protein and the 11 S regulator a is crucial for their effects on proteasome function including antigen processing. *J Mol Biol* 2002, 323:771-782.

[135] Huang J, Kwong J, Sun EC, Liang TJ: Proteasome complex as a potential cellular target of hepatitis B virus X protein. *J Virol* 1996, 70:5582-5591.

[136] Stohwasser R, Holzhutter HG, Lehmann U, Henklein P, Kloetzel PM: Hepatitis B virus HBX peptide 116-138 and proteasome activator PA28 compete for binding to the proteasome α4/MC6 subunit. *Biol Chem* 2003, 384:39-49.

[137] Zhang Z, Torii N, Furusaka A, Malayaman N, Hu Z, Liang TJ: Structural and functional characterization of interaction between hepatitis B virus X protein and the proteasome complex. *J Biol Chem* 2000, 275:15157-15165.

[138] Barak O, Aronheim A, Shaul Y: HBV X protein target HIV Tat-binding protein 1. *Virology* 2001, 283:110-120.

[139] Sirma H, Weil R, Rosmorduc O, Urban S, Israel A, Kremsdorf D, Brechot C: Cytosol is the prime compartment of hepatitis B virus X protein where it colocalizes with the proteasome. *Oncogene* 1998, 16:2051-2063.

[140] Elsasser S, Finley D: Delivery of ubiquitinated substrates to protein-unfolding machines. *Nat Cell Biol* 2005, 7:742-749.

[141] Lam YA, Xu W, DeMartino GN, Cohen RE: Editing of ubiquitin conjugates by an isopeptidase in the 26S proteasome. *Nature* 1997, 385:737-740.

[142] Tongaonkar P, Chen L, Lambertson D, Ko B, Madura K: Evidence for an interaction between ubiquitin-conjugating enzymes and the 26S proteasome. *Mol Cell Biol* 2000, 20:4691-4698.

[143] Hendil KB, Hartmann-Petersen R: Proteasomes: a complex story. *Curr Protein Pept Sci* 2004, 5:135-151.

[144] Löscher M, Fortschegger K, Ritter G, Wostry M, Voglauer R, Schmid JA, Watters S, Rivett JA, Ajuh P, Lamond AI, Katinger H, Grillari J: Interaction of U-box E3 ligase SNEV with PSMB4, the α7 subunit of the 20S proteasome. *Biochem J* 2005, 388:593-603.

[145] Deveraux Q, Ustrell V, Pickart C, Rechsteiner M: A 26 S protease subunit that binds ubiquitin conjugates. *J Biol Chem* 1994, 269:7059-7061.

[146] van Nocker S, Sadis S, Rubin DM, Glickman M, Fu H, Coux O, Wefes I, Finley D, Vierstra RD: The multiubiquitin-chain-binding protein Mcb1 is a component of the 26S

proteasome in Saccharomyces cerevisiae and plays a nonessential, substrate-specific role in protein turnover. *Mol Cell Biol* 1996, 16:6020-6028.
[147] [147] Hartmann-Petersen R, Gordon C: Proteins interacting with the 26S proteasome. *Cell Mol Life Sci* 2004, 61:1589-1595.
[148] Young P, Deveraux Q, Beal RE, Pickart CM, Rechsteiner M: Characterization of two polyubiquitin binding sites in the 26 S protease subunit 5a. *J Biol Chem* 1998, 273:5461-5467.
[149] Hofmann K, Falquet L: A ubiquitin-interacting motif conserved in components of the proteasomal and lysosomal protein degradation systems. *Trends Biochem Sci* 2001, 26:347-350.
[150] Seeger M, Hartmann-Petersen R, Wilkinson CRM, Wallace M, Samejima I, Taylor MS, Gordon C: Interaction of the anaphase-promoting complex/cyclosome and proteasome protein complexes with multiubiquitin chain-binding proteins. *J Biol Chem* 2003, 278:16791-16796.
[151] Castro A, Bernis C, Vigneron S, Labbé JC, Lorca T: The anaphase promoting complex: a key factor in the regulation of cell cycle. *Oncogene* 2005, 24:314-325.
[152] Verma R, Chen S, Feldman R, Schieltz D, Yates J, Dohmen J, Deshaies RJ: Proteasomal proteomics: identification of nucleotide-sensitive proteasome-interacting proteins by mass spectrometric analysis of affinity-purified proteasomes. *Mol Biol Cell* 2000, 11:3425-3439.
[153] Schauber C, Chen L, Tongaonkar P, Vega I, Lambertson D, Potts W, Madura K: Rad23 links DNA repair to the ubiquitin/proteasome pathway. *Nature* 1998, 391:715-718.
[154] Hiyama H, Yokoi M, Masutani C, Sugasawa K, Maekawa T, Tanaka K, Hoeijmakers JH, Hanaoka F: Interaction of hHR23 with S5a. The ubiquitin-like domain of hHR23 mediates interaction with S5a subunit of 26 S proteasome. *J Biol Chem* 1999, 274:28019-28025.
[155] Wilkinson CR, Seeger M, Hartmann-Petersen R, Stone M, Wallace M, Semple C, Gordon C: Proteins containing the UBA domain are able to bind to multi-ubiquitin chains. *Nat Cell Biol* 2001, 3:939-943.
[156] Funakoshi M, Sasaki T, Nishimoto T, Kobayashi H: Budding yeast Dsk2p is a polyubiquitin-binding protein that can interact with the proteasome. *Proc Natl Acad Sci USA* 2002, 99:745-750.
[157] Elsasser S, Chandler-Militello D, Muller B, Hanna J, Finley D: Rad23 and Rpn10 serve as alternative ubiquitin receptors for the proteasome. *J Biol Chem* 2004, 279:26817-26822.
[158] Chen L, Madura K: Rad23 promotes the targeting of proteolytic substrates to the proteasome. *Mol Cell Biol* 2002, 22:4902-4913.
[159] Hofmann K, Bucher P: The UBA domain: a sequence motif present in multiple enzyme classes of the ubiquitination pathway. *Trends Biochem Sci* 1996, 21:172-173.
[160] Elsasser S, Gali RR, Schwickart M, Larsen CN, Leggett DS, Muller B, Feng MT, Tubing F, Dittmar GA, Finley D: Proteasome subunit Rpn1 binds ubiquitin-like protein domains. *Nat Cell Biol* 2002, 4:725-730.

[161] Saeki Y, Sone T, Toh-e A, Yokosawa H: Identification of ubiquitin-like protein-binding subunits of the 26S proteasome. *Biochem Biophys Res Commun* 2002, 296:813-819.

[162] Hicke L, Schubert HL, Hill CP: Ubiquitin-binding domains. *Nat Rev Mol Cell Biol* 2005, 6:610-621.

[163] Raasi S, Orlov I, Fleming KG, Pickart CM: Binding of polyubiquitin chains to ubiquitin-associated (UBA) domains of HHR23A. *J Mol Biol* 2004, 341:1367-1379.

[164] Verma R, Oania R, Graumann J, Deshaies RJ: Multiubiquitin chain receptors define a layer of substrate selectivity in the ubiquitin-proteasome system. *Cell* 2004, 118:99-110.

[165] Saeki Y, Saitoh A, Toh-e A, Yokosawa H: Ubiquitin-like proteins and Rpn10 play cooperative roles in ubiquitin-dependent proteolysis. *Biochem Biophys Res Commun* 2002, 293:986-992.

[166] Kim I, Mi K, Rao H: Multiple interactions of rad23 suggest a mechanism for ubiquitylated substrate delivery important in proteolysis. *Mol Biol Cell* 2004, 15:3357-3365.

[167] Chuang SM, Chen L, Lambertson D, Anand M, Kinzy TG, Madura K: Proteasome-mediated degradation of cotranslationally damaged proteins involves translation elongation factor 1A. *Mol Cell Biol* 2005, 25:403-413.

[168] Park H, Suzuki T, Lennarz WJ: Identification of proteins that interact with mammalian peptide:N-glycanase and implicate this hydrolase in the proteasome-dependent pathway for protein degradation. *Proc Natl Acad Sci USA* 2001, 98:11163-11168.

[169] Suzuki T, Park H, Kwofie MA, Lennarz WJ: Rad23 provides a link between the Png1 deglycosylating enzyme and the 26 S proteasome in yeast. *J Biol Chem* 2001, 276:21601-21607.

[170] Lee JH, Choi JM, Lee C, Yi KJ, Cho Y: Structure of a peptide:N-glycanase-Rad23 complex: insight into the deglycosylation for denatured glycoproteins. *Proc Natl Acad Sci USA* 2005, 102:9144-9149.

[171] Suzuki T, Park H, Kitajima K, Lennarz WJ: Peptides glycosylated in the endoplasmic reticulum of yeast are subsequently deglycosylated by a soluble peptide: N-glycanase activity. *J Biol Chem* 273:21526-21530.

[172] Buchberger A, Howard MJ, Proctor M, Bycroft M: The UBX domain: a widespread ubiquitin-like module. *J Mol Biol* 2001, 307:17-24.

[173] Dreveny I, Kondo H, Uchiyama K, Shaw A, Zhang X, Freemont PS: Structural basis of the interaction between the AAA ATPase p97/VCP and its adaptor protein p47. *EMBO J* 2004, 23:1030-1039.

[174] Cao K, Nakajima R, Meyer HH, Zhang Y: The AAA ATPase Cdc48/p97 regulates spindle disassembly at the end of mitosis. *Cell* 2003, 115:355-367.

[175] Woodman PG: p97, a protein coping with multiple identities. *J Cell Sci* 2003, 116:4283-4290.

[176] Schuberth C, Richly H, Rumpf S, Buchberger A: Shp1 and Ubx2 are adaptors of Cdc48 involved in ubiquitin-dependent protein degradation. *EMBO Rep* 2004, 5:818-824.

[177] Meyer HH, Wang Y, Warren G: Direct binding of ubiquitin conjugates by the mammalian p97 adaptor complexes, p47 and Ufd1-Npl4. *EMBO J* 2002, 21:5645-5652.

[178] Meyer HH: Golgi reassembly after mitosis: the AAA family meets the ubiquitin family. *Biochim Biophys Acta* 2005, 1744:481-492.

[179] Hetzer M, Meyer HH, Walther TC, Bilbao-Cortés D, Warren G, Mattaj IW: Distinct AAA-ATPase p97 complexes function in discretesteps of nuclear assembly. *Nat Cell Biol* 2001, 3:1086-1091.

[180] Hartmann-Petersen R, Wallace M, Hofmann K, Koch G, Johnsen AH, Hendil KB, Gordon C: The Ubx2 and Ubx3 cofactors direct Cdc48 activity to proteolytic and nonproteolytic ubiquitin-dependent processes. *Curr Biol* 2004, 14:824-828.

[181] Ghislain M, Dohmen RJ, Levy F, Varshavsky A: Cdc48p interacts with Ufd3p, a WD repeat protein required for ubiquitin-mediated proteolysis in Saccharomyces cerevisiae. *EMBO J* 1996, 15L 4884-4899.

[182] Fu X, Ng C, Feng D, Liang C: Cdc48p is required for the cell cycle commitment point at Start via degradation of the G1-CDK inhibitor Far1p. *J Cell Biol* 2003, 163:21-26.

[183] Dai RM, Chen E, Longo DL, Gorbea CM, Li CC: Involvement of valosin-containing protein, an ATPase co-purified with IκBα and 26 S proteasome, in ubiquitin-proteasome-mediated degradation of IκBα. *J Biol Chem* 1998, 273:3562-3573.

[184] Ye Y, Meyer HH, Rapoport TA: The AAA ATPase Cdc48/p97 and its partners transport proteins from the ER into the cytosol. *Nature* 2001, 414:652-656.

[185] [185] Braun S, Matuschewski K, Rape M, Thoms S, Jentsch S: Role of the ubiquitin-selective CDC48(UFD1/NPL4) chaperone (segregase) in ERAD of OLE1 and other substrates. *EMBO J* 2002, 21:615-621.

[186] Jarosch E, Taxis C, Volkwein C, Bordallo J, Finley D, Wolf DH, Sommer T: Protein dislocation from the ER requires polyubiquitination and the AAA-ATPase Cdc48. *Nat Cell Biol* 4:134-139.

[187] Dai RM, Li CC: Valosin-containing protein is a multi-ubiquitin chain-targeting factor required in ubiquitin-proteasome degradation. *Nat Cell Biol* 2001, 3:740-744.

[188] Richly H, Rape M, Braun S, Rumpf S, Hoege C, Jentsch S: A series of ubiquitin binding factors connects CDC48/p97 to substrate multiubiquitylation and proteasomal targeting. *Cell* 120:73-84.

[189] [189] Hartmann-Petersen R, Hendil KB, Gordon C: Ubiquitin binding proteins protect ubiquitin conjugates from disassembly. *FEBS Lett* 2003, 535:77-81.

[190] Deveraux Q, van Nocker S, Mahaffey D, Vierstra R, Rechsteiner M: Inhibition of ubiquitin-mediated proteolysis by the Arabidopsis 26 S protease subunit S5a. *J Biol Chem* 1995, 270:29660-29663.

[191] Raasi S, Pickart CM: Rad23 ubiquitin-associated domains (UBA) inhibit 26 S proteasome-catalyzed proteolysis by sequestering lysine 48-linked polyubiquitin chains. *J Biol Chem* 2003, 278:8951-8959.

[192] Tongaonkar P, Chen L, Lambertson D, Ko B, Madura K: Evidence for an interaction between ubiquitin-conjugating enzymes and the 26S proteasome. *Mol Cell Biol* 2000, 20:4691-4698.

[193] Hendil KB, Hartmann-Petersen R: Proteasomes: a complex story. *Curr Protein Pept Sci* 2004, 5:135-151.

[194] Sakata E, Yamaguchi Y, Kurimoto E, Kikuchi J, Yokoyama S, Yamada S, Kawahara H, Yokosawa H, Hattori N, Mizuno Y, Tanaka K, Kato K: Parkin binds the Rpn10

subunit of 26S proteasomes through its ubiquitin-like domain. *EMBO J* 2003, 4:301-306.

[195] Nijman SMB, Luna-Vargas MPA, Velds A, Brummelkamp TR, Dirac AMG, Sixma TA, Bernards R: A genomic and functional inventory of deubiquitinating enzymes. *Cell* 2005, 123:773-786.

[196] Amerik AY, Hochstrasser M: Mechanism and function of deubiquitinating enzymes. *Biochim Biophys Acta* 2004, 1695:189-207.

[197] Ross CA, Poirier MA: Protein aggregation and neurodegenerative disease. *Nat Med* 2004, 10:S10-S17.

[198] Doss-Pepe EW, Stenroos ES, Johnson WG, Madura K: Ataxin-3 interactions with Rad23 and Valosin-containing protein and its associations with ubiquitin chains and the proteasome are consistent with a role in ubiquitin-mediated proteolysis. *Mol Cell Biol* 2003, 23:6469-6483.

[199] Burnett B, Li F, Pittman RN: The polyglutamine neurodegenerative protein ataxin-3 binds polyubiquitylated proteins and has ubiquitin protease activity. *Hum Mol Genet* 2003, 12:3195-3205.

[200] Chai Y, Berke SS, Cohen RE, Paulson HL: Poly-ubiquitin binding by the polyglutamine disease protein ataxin-3 links its normal function to protein surveillance pathways. *J Biol Chem* 2004, 279:3605-3611.

[201] Warrick JM, Morabito LM, Bilen J, Gordesky-Gold B, Faust LZ, Paulson HL, Bonini NM: Ataxin-3 suppresses polyglutamine neurodegeneration in Drosophila by a ubiquitin-associated mechanism. *Mol Cell* 2005, 18:37-48.

[202] Shoesmith Berke SJ, Chai Y, Marrs GL, Wen H, Paulson HL: Defining the role of ubiquitin-interacting motifs in the polyglutamine disease protein, ataxin-3. *J Biol Chem* 2005, 280:32026-32034.

[203] Matsumoto M, Yada M, Hatakeyama S, Ishimoto H, Tanimura T, Tsuji S, Kakizuka A, Kitagawa M, Nakayama KI: Molecular clearance of ataxin-3 is regulated by a mammalian E4. *EMBO J* 2004, 23:659-669.

[204] Jana NR, Dikshit P, Goswami A, Kotliarova S, Murata S, Tanaka K, Nukina N: Co-chaperone CHIP associates with expanded polyglutamine protein and promotes their degradation by proteasomes. *J Biol Chem* 2005, 280:11635-11640.

[205] Chou A-H, Yeh T-H, Kuo Y-L, Kao Y-C, Jou M-J, Hsu C-Y, Tsai S-R, Kakizuka A, Wang H-L: Polyglutamine-expanded ataxin-3 activates mitochondrial apoptotic pathway by upregulating Bax and downregulating Bcl-xL. *Neurobiol Dis* 2006, 21:333-45

[206] Lussier M, White AM, Sheraton J, di Paolo T, Treadwell J, Southard SB, Horenstein CI, Chen-Weiner J, Ram AF, Kapteyn JC, Roemer TW, Vo DH, Bondoc DC, Hall J, Zhong WW, Sdicu AM, Davies J, Klis FM, Robbins PW, Bussey H: Large scale identification of genes involved in cell surface biosynthesis and architecture in Saccharomyces cerevisiae. *Genetics* 1997, 147:435-450.

[207] Kajava AV, Gorbea C, Ortega J, Rechsteiner M, Steven AC: New HEAT-like repeat motifs in proteins regulating proteasome structure and function. *J Struct Biol* 2004, 146:425-430.

[208] Lohi O, Poussu A, Mao Y, Quiocho F, Lehto VP: VHS domain — a longshoreman of vesicle lines. *FEBS Lett* 2002, 513:19-23.

[209] Gorbea C, Goellner GM, Teter K, Holmes RK, Rechsteiner M: Characterization of mammalian Ecm29, a 26 S proteasome-associated protein that localizes to the nucleus and membrane vesicles. *J Biol Chem* 2004, 279:54849-54861.

[210] Gorbea C, Kaufmann AG, Pratt G, Rechsteiner M, Rogers SW: Multiple forms of the 26S proteasome-associated protein Ecm29 in mouse brain. *J Chem* 2006. In press.

[211] Huh WK, Falvo JV, Gerke LC, Carroll AS, Howson RW, Weissman JS, O'Shea EK: Global analysis of protein localization in budding yeast. *Nature* 2003, 425:686-691.

[212] Ghaemmaghami S, Huh WK, Bower K, Howson RW, Bell A, Dephoure N, O'Shea EK, Weissman JS: Global analysis of protein expression in yeast. *Nature* 2003, 425:737-741.

[213] Andrade MA, Petosa C, O'Donoghue SI, Müller CW, Bork P: Comparison of ARM and HEAT protein repeats. *J Mol Biol* 2001, 309:1-18.

[214] Yamamoto Y, Singh CR, Marintchev A, Hall NS, Hannig EM, Wagner G, Asano K: The eukaryotic initiation factor (eIF) 5 HEAT domain mediates multifactor assembly and scanning with distinct interfaces to eIF1, eIF2, eIF3, and eIF4G. *Proc Natl Acad Sci USA* 102:16164-16169.

[215] Teter K, Jobling MG, Holmes RK: A class of mutant CHO cells resistant to cholera toxin rapidly degrades the catalytic polypeptide of cholera toxin and exhibits increased endoplasmic reticulum-associated degradation. *Traffic* 4:232-242.

[216] Rual JF, Venkatesan K, Hao T, Hirozane-Kishikawa T, Dricot A, Li N, Berriz GF, Gibbons FD, Dreze M, Ayivi-Guedehoussou N, Klitgord N, Simon C, Boxem M, Milstein S, Rosenberg J, Goldberg DS, Zhang LV, Wong SL, Franklin G, Li S, Albala JS, Lim J, Fraughton C, Llamosas E, Cevik S, Bex C, Lamesch P, Sikorski RS, Vandenhaute J, Zoghbi HY, Smolyar A, Bosak S, Sequerra R, Doucette-Stamm L, Cusick ME, Hill DE, Roth FP, Vidal M: Towards a proteome-scale map of the human protein-protein interaction network. *Nature* 2005, 437:1173-1178.

[217] Kopito RR: Aggresomes, inclusion bodies and protein aggregation. *Trends Cell Biol* 2000, 10:524-530.

[218] Meusser B, Hirsch C, Jarosch E, Sommer T: ERAD: the long road to destruction. *Nature* 2005, 7:766-772.

[219] Dong J, Chen W, Welford A, Wandinger-Ness A: The proteasome α-subunit XAPC7 interacts specifically with Rab7 and late endosomes. *J Biol Chem* 2004, 279:21334-21342.

[220] van Kerkhof P, Govers R, Alves dos Santos CM, Strous GJ: Endocytosis and degradation of the growth hormone receptor are proteasome-dependent. *J Biol Chem* 2000, 275:1575-1580.

[221] Melman L, Geuze HJ, Li Y, McCormick LM, Van Kerkhof P, Strous GJ, Schwartz AL, Bu G: Proteasome regulates the delivery of LDL receptor-related protein into the degradation pathway. *Mol Biol Cell* 2002, 13:3325-3335.

[222] Longva KE, Blystad FD, Stang E, Larsen AM, Johannessen LE, Madshus IH: Ubiquitination and proteasomal activity are required for transport of the EGF receptor to inner membranes of multivesicular bodies. *J Cell Biol* 2002, 156:843-854.

[223] van Kerkhof P, Alves dos Santos CM, Sachse M, Klumperman J, Bu G, Strous GJ: Proteasome inhibitors block a late step in lysosomal transport of selected membrane but not soluble proteins. *Mol Biol Cell* 2001, 12:2556-2566.

[224] Kobayashi H, Tanaka N, Asao H, Miura S, Kyuuma M, Semura K, Ishii N, Sugamura K: Hrs, a mammalian master molecule in vesicular transport and protein sorting, suppresses the degradation of ESCRT proteins signal transducing adaptor molecule 1 and 2. *J Biol Chem* 2005, 280:10468-10477.

In: The Ubiquitin Proteasome System... ISBN 978-1-60021-749-4
Eds: Mario Di Napoli and Cezary Wojcik, pp. 207-225

Chapter 10

The Molecular Chaperones in the Ubiquitin-Proteasome System

Mihiro Yano* and Hiroshi Kido

Division of Enzyme Chemistry, Institute for Enzyme Research, The University of Tokushima, Tokushima 770-8503, Japan.

Abstract

The accumulation of misfolded or damaged proteins causes the failure of normal cell structure and functions needed for growth and viability. The toxicity of misfolded species has been linked to human diseases, neurodegenerative disorders in particular, including Alzheimer's, Parkinson's and Huntington's diseases, characterized by the accumulation of intracellular aggregates or inclusion bodies. To interrupt this adverse development, defective proteins must be rapidly repaired by molecular chaperones or destroyed by energy-dependent cytoplasmic proteases. A balance among these processes ultimately maintains the cellular homeostasis. In eukaryotes, the 26S proteasome, a protease/chaperone complex that generally acts as an ubiquitination system, is a central component in the protein triage decision process, though it also selectively degrades structurally abnormal proteins in a ubiquitin-independent manner. In either case, all substrate proteins must undergo the structural changes and stabilization necessary for a rapid degradation. It has, therefore, often been suggested that several chaperone functions are closely related to the stimulation of proteasomal degradation. This chapter summarizes recent discoveries pertaining to chaperone activities in the ubiquitin-proteasome system, and to their regulation of protein breakdown mediated by the proteasome.

Keywords: Molecular chaperone, proteolysis, proteasome, ubiquitin.

* Correspondence concerning this article should be addressed to Dr. Mihiro Yano, MD, PhD; Division of Enzyme Chemistry, Institute for Enzyme Research, The University of Tokushima, Tokushima 770-8503, Japan. Phone: 81-88-633-7424, Fax: 81-88-633-7425, E-mail: yano@ier.tokushima-u.ac.jp.

Abbreviations

AAA, ATPases associated with various cellular activities; ADP, adenosine diphosphate; ATP, adenosine triphosphate; CaM, calmodulin; CFTR, cystic fibrosis transmembrane regulator; CHIP, C-terminus of Hsp70-interacting protein; ER, endoplasmic reticulum; GFP, green fluorescent protein; HOP, Hsp70-Hsp90-organizing protein; NDP, nucleoside diphosphate; ODC, ornithine decarboxylase; PA700, proteasome activator, 700 kDa; Pael-R, pael receptor; PAN, proteasome-activating nucleotidase; UPS, ubiquitin-proteasome system; VHL, von Hippel-Lindau; VCP, valosin-containing protein.

Introduction

Functional proteins are continuously regulated by a post-translational quality control machinery that preserves their proper intracellular structure and function. The fate of many cellular proteins depends on a life cycle encompassed by their folding and degradation [1]. Once proteins have been damaged, molecular chaperones and proteases must decide whether to repair or destroy them. The failure of the quality control system to repair or remove misfolded proteins allows diseases to progress, such as neurodegenerative disorders associated with protein inclusions [2]. There is growing evidence indicating that the potentially toxic aggregates formed in several areas of the brain often contain molecular chaperones that are involved in the refolding of misfolded proteins and the components of the ubiquitin-proteasome system (UPS) that eliminate them [3]. Recent studies have found independently that the large chaperone-protease complex (ATP-dependent protease) recognizes the state of a protein, and that their combined functions play a key role in the protein quality control self-triage [4].

In prokaryotes, the ClpAP and ClpXP proteases are energy-dependent proteolytic complexes, in which a molecular chaperone ClpA (or ClpX) combines with a self-compartmentalizing core protease (ClpP), capable of both functions together [5,6]. Protein substrates must be unfolded before their translocation to the proteolytic chamber, since ClpP has an axial pore, precluding the access of larger proteins. ClpA (ClpX), composed of ATPase particles capping a protease (ClpP), mediates the unfolding of substrates and facilitates their access to catalytic sites within the chamber prior to their degradation [7]. In an eukaryotic cytosol, a machinery similar to that of ClpA(X)P has been described most precisely for the 26S proteasome. It is composed of the catalytic 20S core and a terminal regulatory component containing ATPase particles, known as 19S cap or PA700, which docks at both ends of the entry-exit channel of the 20S proteasome [8,9].

In the degradation process by the 26S proteasome, it is commonly assumed, with few exceptions, that the 19S cap recognizes substrate proteins that have been modified covalently with a polyubiquitin chain as a prerequisite for their proteolysis, then unfolds and translocates them to the catalytic sites of the 20S proteasome through a narrow gate [10]. In addition, 19S cap acts as a defining molecular chaperone that preferentially binds to non-native, non-ubiquitinated proteins, inhibiting their aggregation and promoting their reactivation [11,12]. These characteristics predict that, besides exercising its general chaperone functions, 19S cap may also indirectly promote proteolysis by maintaining misfolded proteins in a non-aggregated state. While ATP, or its hydrolytic energy, is critical in modulating the rate of substrate turnover by molecular chaperones as they inhibit aggregation and promote folding [13,14], the regulatory subcomplexes of ATP-dependent proteases also utilize ATP in events responsible for the breakdown of proteins [15-17]. In the case of the 26S proteasome, the six ATPases reside within the regulatory particles comprising the outer ring, the so-called base, covering the entry ports into the 20S core [18]. The putative functions of the base are to unfold and translocate substrate proteins, and mediate the opening of the pore in the 20S proteasome, to promote ATP-dependent degradation. Thus, binding of the 19S cap regulatory components to the 20S proteasome is essential to activate the degradation of ubiquitinated proteins. While this cooperation is the most unique characteristic of the proteasomal degradation pathway, the 20S proteasome alone is capable of directly degrading some non-ubiquitinated proteins such as $p21^{WAF/CIP1}$, scRNAase and oxidized proteins [19-21]. This observation suggests that the 20S proteasome itself binds to substrates and leads them into the catalytic cavity by regulating the gate of the channel, since the ends of the 20S core are found in a closed state.

To clarify the direct entry process of the substrates into the 20S core, we have examined the chaperone functions of the 20S proteasome, which assists the recruitment of substrates directly into it, with noteworthy results [22]. First, when the human 20S proteasome was incubated in presence of ATP and ADP, simulating physiological conditions, it catalyzed the weak, though linear ATP-ADP exchange reaction, with enzymatic properties similar to those of the nucleoside diphosphate (NDP) kinase-like activity of molecular chaperones, such as Hsp70 and 14-3-3 [23,24]. While this might indicate that the 20S proteasome functions as a general chaperone, the role of the nucleotide exchange activity remains unclear. Second, the 20S proteasome is capable of recognizing non-native conformations of model proteins directly, and of preventing their aggregation [25]. This process is not followed by the refolding of denatured substrates to their native state, while the 19S cap or 26S proteasome promotes their reactivation, suggesting that the 20S proteasome, which exhibits proteolytic activity, plays an active role of molecular chaperone in the degradation of proteins.

In addition to these intrinsic chaperone activities of the proteasome, the ubiquitin proteasome pathway requires the assistance of molecular chaperones for substrate degradation. The targeting of misfolded proteins to the proteasome by chaperones has been established in various ways. Previous studies have shown that Hsp70 and Hsp90 are required in the degradation of some misfolded proteins, though they do not clarify how these chaperones participate in the process that targets polypeptides to the proteasome [26,27]. C terminus of Hsp70-interacting protein (CHIP) and BAG-1, two Hsp70 cofactors with chaperone interaction domains, have recently been identified as linkers between chaperones and proteasome (see Chapter 19). CHIP, an Hsp70-associated ubiquitin ligase, ubiquitylates misfolded proteins associated with cytoplasmic chaperones, promoting their delivery for degradation by the proteasome [28-30]. In this process, BAG-1, the nucleotide exchange factor of Hsp70, seems to cooperate with CHIP by facilitating the transfer of ubiquitylated proteins to the proteasome [31,32]. A similar function seems to be provided by cdc48, also known as p97 and VCP. Cdc48 is a hexameric ATPase that acts as a multiubiquitin chain-targeting factor, recruiting ubiquitinated substrates to the proteasome [33,34]. Most strikingly, Cdc48 can also unfold bulky ubiquitinated proteins, and convert them into a state that is suitable for degradation by the proteasome [35]. This chapter summarizes the state-of-the-art, focusing on the chaperone activities linked to protein breakdown by the proteasome.

The 19S Cap is a Molecular Chaperone that Regulates Proteasome-Mediated Proteolysis

The 19S cap, a regulatory component of the 26S proteasome originally isolated as an ATP-dependent activator of the 20S proteasome [15,36], participates in ATP-dependent proteolysis, along with a proteolytic component of the 20S proteasome. It can be dissociated into 'base' and 'lid' subcomplexes [18]. The base complex contains six AAA ATPases, creating a ring that binds directly at both ends of the 20S core particle, and two non-ATPase subunits. The lid complex, which consists of eight non-ATPase subunits, is attached to the base complex.

When substrates are efficiently identified by the proteasome-dependent degradation system, the 19S cap plays three prominent roles in the proteolytic process. *First*, a role of substrate recognition, divided between ubiquitin-dependent and ubiquitin-independent degradation pathways. Most cellular proteins are targeted for the 19S cap after having been covalently modified by a polyubiquitin chain as a degradation signal [37,38]. In such case, the 19S cap recognizes and interacts directly with a polyubiquitin chain, without requiring the elongated and unfolded conformation of an ubiquitinated substrate [39]. In some cases, however, the conjugation of ubiquitin may be activated by the unfolding of a protein [40]. A more intriguing observation was recently made, showing that S6a/Rpt5, an ATPase subunit of the 19S cap, serves as a ubiquitin receptor for the proteasome in an ATP-dependent manner [41], whereas Rpn10, a non-ATPase subunit, was considered its best candidate. In contrast, few examples of ubiquitin-independent protein degradation have been described. In these cases, a conformational change in the substrate, which facilitates its binding to the 19S cap, may participate in the substrate recognition. Ornithine decarboxylase (ODC), a rate-limiting enzyme that catalyzes the first step of polyamine biosynthesis, is a prototypical, ubiquitin-independent substrate of the 26S proteasome [42]. However, the degradation of ODC depends on antizyme, an inducible protein inhibitor that binds to, and induces conformational changes in ODC, exposing the C-terminal region to target ODC to the 26S proteasome [43,44]. The cyclin-dependent kinase inhibitor, $p21^{WAF1/CIP1}$, ovalbumin and scrambled RNase A in the non-native state are other known ubiquitin-independent substrates [19,20,45]. Furthermore, proteasome-activating nucleotidase (PAN), an archaeal ATPase complex analogous to the 19S cap directly recognizes and binds substrates without ubiquitin [46]. Therefore, the 19S cap reportedly interacts directly with non-ubiquitinated substrates, a function similar to that of molecular chaperones for non-native substrates. The *second role* played by the 19S cap is unfolding of the substrates and their translocation into the catalytic sites of the 20S proteasome. The substrates need to be unfolded before their translocation to the catalytic sites because of the usually quite narrow gate of the 20S proteasome. An unfoldase activity of the 19S cap has been suspected, however not confirmed. Recent observations suggest that the 19S cap is responsible for the unfolding of ODC and for its translocation into the proteolytic compartment. In the process of ODC degradation, the 26S proteasome irreversibly inactivates ODC, which is equivalent to unfolding, before its degradation within the 20S proteasome [47]. Surprisingly, the 19S cap is incapable, by itself, to inactivate ODC, and its association with the 20S proteasome seems indispensable for this inactivation. Although there is no direct evidence, ATPase subunits in the 19S cap may mediate the conformational changes in ODC that allow its entry into the proteolytic cavity. It is noteworthy that PAN catalyzes the unfolding and translocation of a global protein [46]. PAN shows rapid loss of the green fluorescent protein (GFP) ssrA, by which its unfolding can be monitored during ATP-dependent proteolysis. Furthermore, PAN appears to translocate GFPssrA into the 20S proteasome in a C- to N-terminal direction during the degradation process [48]. In this case, the PAN-mediated unfolding of GFPssrA occurs even when its translocation is inhibited, suggesting that translocation follows ATP-dependent unfolding, and that both steps can be dissociated. On the other hand, the ubiquitinated substrates must be detached from the multi-ubiquitin chain, since it is difficult to thread a bulky ubiquitin chain through the narrow axial channel of the proteasome. It has recently been proposed that Rpn11/POH1, a subunit of lid, is involved in the deubiquitination of substrate polypeptides during proteasomal degradation [49]. These observations indicate that the functions that change the conformation of the substrates, and transfer them to the active sites of the 20S proteasome as a prerequisite for degradation, are attributable to a regulatory component of the 26S proteasome. *The third role* played by the 19S cap is that of channel gatekeeper. Opening of the channel for entry of the substrate into the 20S proteasome is apparently coupled to the ATP-dependent reaction of the 19S cap. From the crystal structure of the yeast 20S proteasome, the gates into the chamber at terminal rings are sealed, with the N-terminal tails of the α3-subunit comprising the outer ring of the 20S proteasome preventing access by the substrate [50]. It has recently become apparent that the ATPase domain of Rpt2 in yeast proteasome, one of six ATPases in the regulatory component, causes movement of the N-terminal tails of the α3-subunit, leading to opening of the gate. This indicates that the ATPase activity of Rpt2 may be involved in a gating mechanism, although it has not yet been verified experimentally that gate opening requires the hydrolysis of ATP [51]. In the case of PAN, the N-terminal deletion of the 20S α-subunit, which achieves the opening of the gate artificially, facilitates the degradation of substrates without PAN and ATP, whereas the degradation by the wild-type 20S proteasome requires PAN and ATP, indicating that PAN opens the gate in an ATP-dependent manner [48]. These multiple roles confer the 19S cap unique functions with respect to the various processes of protein degradation by the 26S proteasome as a molecular chaperone.

The 19S Cap Interacts with, and Repairs Misfolded Proteins

With regard to the entry of substrates into the 20S particle, the protein-unfolding activity of the regulatory particle seems linked to the stimulation of proteasomal degradation. On the other hand, most cellular proteins that escape degradation are probably repaired by general molecular chaperones. Recent observations pertaining to the 'chaperone-like activities' of the

19S cap, which mediate the anti-aggregation and refolding of several model proteins, suggest that it plays a direct role in the repair of a given target as a result of protein triage [13,14]. Further evidence that these functions are defining characteristics of molecular chaperones was obtained from a study of misfolded proteasomal substrates of the 26S proteasome, notably scrambled ribonuclease A (scRNase A) and pentaubiquitinated dihydrofolate reductase reactivated by the 19S cap [20]. In prokaryotes, PAN is similarly capable of preventing the aggregation of proteins and of promoting the refolding of misfolded proteins.

What is the biological importance of the 19S cap's chaperone-like activity? It is noteworthy that the refolding of misfolded scRNase A by free 19S cap dissociated from the 20S proteasome results in a proteasome-resistant form, whereas the degradation of misfolded scRNase A is stimulated during the formation of the 26S proteasome complex. Furthermore, the observation that PAN can act independently and exist predominantly as a structure separate from the 20S proteasome in cells, except during the substrate degradation process, suggests that it prevents aggregation and promotes refolding of misfolded proteins *in vivo*. This anti-aggregating effect on misfolded proteins may assist the proteasomal degradation in the trapping of unfolded substrates before their entry into the proteasome, since aggregated proteins are relatively resistant to proteolysis. In contrast, the refolding of misfolded proteins into proteasome-resistant forms does not appear directly coupled with the degradation process, although it may be operative in its early phase by stabilizing the structures adjacent to the degradation signal. It is noteworthy that the nucleotide dependence of chaperone activities varies widely among regulatory particles. Neither the 19S cap nor PAN require the hydrolysis of ATP to limit the aggregation of denatured proteins and stimulate their refolding. On the other hand, the unfolding of substrates by PAN, as well as the regulation of proteolysis by a complex between PAN and the 20S proteasome requires ATP hydrolysis. In this respect, energy-dependent unfolding is likely to be a more powerful reaction in proteasomal degradation. However, it is of interest that these chaperone-like properties operate in parallel to partition the pathway for repair versus degradation of a given target, therefore serve as a triage system for damaged proteins.

Ubiquitin- and 19S Cap-Independent Proteolysis by the 20S Proteasome

The 20S proteasome, the catalytic core of the 26S proteasome, is a 700 kDa, cylinder-shaped protease arranged in a stack of four heptameric rings composed of two outer α and two inner β rings, each containing seven distinct subunits [52; and Chapter 6]. It has a hollow central cavity with multiple proteolytically active sites, and α-rings occluded at both ends [50]. This restriction of access to the interior, probably sequestrating the active sites from unwanted substrates, represents the function of the 19S cap in the regulation of substrate entry, mediating the recognition, unfolding, translocation and opening of channels into the catalytic chamber in preparation for degradation [10]. Although the assistance by the 19S cap is a general characteristic of proteasome-dependent turnover, the 20S proteasome is capable of degrading proteins directly as an alternative mechanism, bypassing the ubiquitination and the 19S cap.

Various oxidized proteins [21,53-55], $p21^{WAF/CIP1}$ [19], and scRNase A are degraded by the 20S proteasome in ATP- and ubiquitin-independent manners *in vitro*. The molecular mechanisms of recognition and degradation of oxidized proteins are poorly understood, though they may involve the exposure of hydrophobic surfaces after oxidative modification, enhancing the proteolysis [56-58]. However, a recent study of the degradation of oxidized calmodulin (CaM) showed that the signal for its degradation by the 20S proteasome was a secondary structural loss caused by oxidation or calcium binding, rather than a hydrophobic change on the surface. It is noteworthy that oxidized CaM was degraded into multiple large fragments by the 20S proteasome, suggesting that the latter recognizes and degrades oxidized CaM in a state of incomplete unfolding, and that the global unfolding of CaM is not necessary for its targeting [21]. It has recently been shown that Hsp90, in association with the 20S proteasome, recognizes oxidized CaM selectively, and promotes its degradation by the proteasome [59]. Hsp90 cannot facilitate the degradation of fluorogenic peptides by the 20S proteasome. It might, therefore, play a crucial role in modulating oxidized CaM to access it into the catalytic core, but does not induce the conformational change in the 20S proteasome allowing its entry. The cdk inhibitor p21Cip is a non-ubiquitinated substrate of the 20S proteasome, since its mutant form, which lacks all the target lysine residues for ubiquitylation, is degraded by the 20S proteasome [60]. As for the turnover of $p21^{WAF/CIP1}$, a recent study has found that the 20S proteasome recognizes p21Cip directly through the C8 α-subunit in the outer ring, and degrades it without the 19S cap. These observations suggest that ubiquitin-independent degradation by the 20S proteasome is more common than generally believed. Whether the 20S proteasome functions *in vivo* by itself, without binding to regulatory proteins, remains uncertain. Although free $p21^{WAF/CIP1}$ has a loosely folded structure similar to that of oxidized CaM [61], these proteins appear to be too large to clear the narrow entry port. If the 20S proteasome does degrade these proteins intracellularly, it implies that either the 20S proteasome or the substrate itself unfolds selectively the N-terminal portions of the α subunit which seals the gates, resulting in opening of the channel.

The 19S cap-independent-proteolysis of the 20S proteasome predicts some intrinsic function of the enzyme, which targets substrates to interior catalytic sites. The eukaryotic 20S proteasome contains seven α-type and seven β-type subunits. Subunits X, Y and Z in the β-subunit ring provide the active-site nucleophiles for the three distinct proteolytic activities [53,62-64], while the functions of the remaining β- and of the seven α-subunits remain to be determined. A study of the functions of these unknown subunits to clarify the mechanism of degradation by the 20S proteasome would be valuable. We have recently found that the purified human 20S proteasome exerts activities of ATP hydrolysis and ATP-ADP exchange, similar to those of the molecular chaperone Hsp70 [22]. Since, the ATPase activity of the 20S proteasome in absence of ADP is generally difficult to detect, it has, thus far, not been reported. However, the addition of ADP to the assay system resulted in a prominently enhanced rate of ATP hydrolysis (Figure 1A). The 20S proteasome also catalyses the reverse reaction, i.e. ATP synthesis (Figure 1B). The *K*m value of ATP for its hydrolysis in presence of 0.5 mM ADP, and *K*m value of ADP for ATP synthesis are 2.86 and 0.14 mM, respectively. Consequently, changes in the concentrations of these nucleotides in the cytosol are expected to affect the enzyme activities *in vivo*, since both *K*m values are slightly lower than the physiological cytosolic concentrations of ATP (5 mM) and ADP (0.5 mM). These enzymatic properties are similar to those of NDP kinase, which catalyzes the transfer of γ-phosphate between NDPs and nucleoside triphosphates [65-67], whereas the 20S proteasome expresses a preference for ATP as a phosphate donor, distinct from NDP kinase. During the transfer of γ-phosphate, the 20S proteasome forms acid-labile phosphohistidine as an autophosphorylated intermediate, followed by NDP-dependent dephosphorylation of the latter. These properties are strikingly similar to those of molecular chaperones, such as Hsp70 and 14-3-3 proteins. These observations suggest that the 20S proteasome has an ATP-binding site other than the catalytic site.

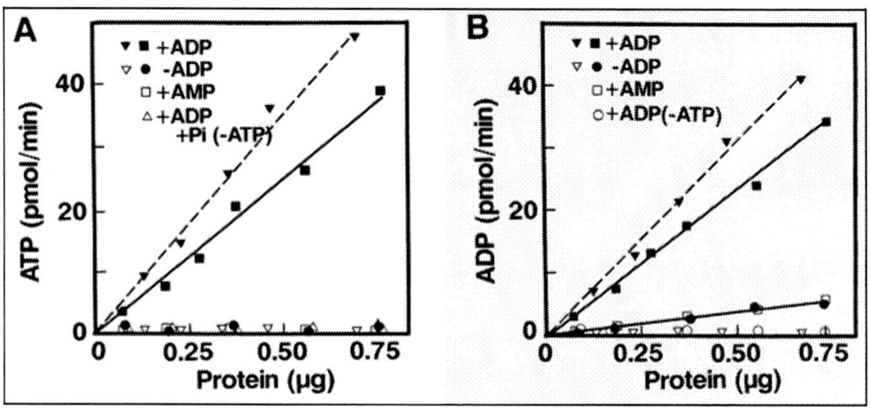

Figure 1. ATP hydrolysis and ATP synthesis of human 20S proteasome. Measurements of ATP hydrolysis (A) and ATP synthesis (B) of human 20S proteasome was made at 37 °C, for 1 h, in presence of 5 mM ATP and 0.5 mM ADP (■), and 0.05 □Ci of [14C]ATP for assays of ATP hydrolysis, or 0.02 μCi of [14C]ADP for assays of ATP synthesis. The nucleotide was removed or replaced in the reaction mixture as follows: activities in absence of ADP in the reaction mixture (●), in presence of AMP (0.5 mM) instead of ADP (□), in presence of Pi (5 mM) instead of ATP for assays of ATP synthesis (Δ), and in absence of ATP for assays of ATP hydrolysis (○). The activities of ATP hydrolysis and ATP synthesis of yeast proteasome in presence (▼) versus absence (▽) of 0.5 mM ADP in the reaction mixture were also assayed.

We then pursued the identification of the subunit(s) responsible for this newly discovered activity. The separation of the autophosphorylated and ATP-photoaffinity-labeled 20S proteasome into individual subunits by reverse-phase high-performance liquid chromatography revealed that the subunits C5 and C8, are involved in the phosphate transfer and binding of ATP, although the 20S proteasome loses its ATP-ADP exchange activity upon separation of the subunits. These observations might sharpen the understanding of previously unknown mechanisms, including the targeting of substrates and regulation of gate opening.

Besides the recognition of partial unfolded substrates, such as those of oxidized CaM and p21Cip, the ATP-ADP exchange activity predicts that the 20S proteasome merely functions as a general chaperone for non-native proteins. In pursuit of this hypothesis, we have shown that the 20S proteasome recognizes preferentially several heat- or chemically-denatured proteins, and protects them against irreversible aggregation *in vitro* [25]. The 20S proteasome prevented the aggregation of misfolded proteins, such as citrate synthase, malate dehydrogenase and glutaraldehyde dehydrogenase in a dose-dependent manner. This was not followed by refolding of these denatured substrates into their native states, whereas the 26S proteasome or the 19S cap generally promote their reactivation. Thus, unlike other known chaperones, the prevention of aggregation may simply reflect the ability of the 20S proteasome to interact with an unfolded protein. However, partially folded proteins tend to aggregate, including those that are misfolded [14], and the degradation of these species should be hampered without an anti-aggregation factor [1]. Therefore, this could be a critical step in the proteasomal degradation without the 19S cap, by which unfolded substrates are bound to inhibit competing aggregation, facilitating their entry into the catalytic site of the 20S proteasome. In some cases, it may be an intermediate step during refolding, in conjunction with other chaperones, such as the 19S cap.

The Role of Non-Proteasomal Molecular Chaperones in the Ubiquitin-Proteasome System

In the general pathway of a triage system, molecular chaperones have the first opportunity to convert damaged proteins to a functional native conformation. However, the failure to rescue non-native proteins allows molecular chaperones to passively facilitate partitioning to the UPS [68]. A recent analysis of misfolded von Hippel-Lindau (VHL) tumor suppressor provided new insights into this transition from chaperone-mediated folding to chaperone-mediated degradation [69]. Two distinct pathways regulated by the molecular chaperones mediate VHL folding and degradation. Newly synthesized VHL trapped by Hsp70 is initially recruited to the folding process by the chaperonin TRiC/CCT. On the other hand, folding-defective VHL with Hsp70 interacts with the Hsp90 complex, leading to its degradation by the UPS. VHL degradation requires the Hsp70 cofactor *STI* 1/HOP that has the unique ability to bridge a complex between Hsp70 and Hsp90. It is noteworthy that TRiC is dispensable for the degradation of VHL, whereas Hsp90, which is not required for the folding of VHL, is essential for its degradation. The folding and degradation of VHL both require Hsp70, suggesting that the latter's role is to maintain the solubility of unfolded VHL, to recruit it for the process of folding or degradation. On the other hand, Hsp90 might facilitate the delivery of misfolded VHL to the proteasome, for example by generating a specific conformation that can subsequently be recognized by an ubiquitin ligase, although the precise mechanisms remain unknown. Like the degradation of oxidized CaM, these results indicate that Hsp90 can function as a physiological regulator of proteasomal protein degradation, since expression levels of Hsp90 are induced during environmental stress, including oxidation, which cause the pool of intracellular misfolded proteins [59]. CHIP and BAG-1 have recently been established as the candidate proteins that link molecular chaperones to the UPS. The E3 ubiquitin ligase, CHIP, was initially identified as the cofactor of Hsp70 that inhibits its chaperone functions, such as protein folding and ATPase activity [70]. CHIP also has a U-box at its C terminus, which is structurally related to RING-finger domains found in many ubiquitin ligases, indicating that CHIP may participate in ubiquitin conjugation but not in productive folding of non-native proteins [71]. Some studies have, indeed, shown that CHIP ubiquitylates various chaperone substrates, such as CFTR and glucocorticoid hormone receptor, promoting their degradation by the proteasome [28,29]. As a consequence, the recruitment of CHIP to Hsp70/Hsp90 shifts the chaperones from a protein folding to a degradation mode by ubiquitylating associated substrates. It is noteworthy that an efficient ubiquitination of unfolded proteins is accomplished by the CHIP conjugation machinery associated with Hsp70 or Hsp90, in which the chaperones act as substrate recognition factors to select substrates for CHIP-mediated ubiquitination. It has been suggested that BAG-1 plays a role, in cooperation with CHIP, in the degradation of protein in the UPS. Since BAG-1 has an ubiquitin-like domain utilized for proteasome binding, and a BAG domain that mediates Hsp70 binding, it might act as a coupling factor that facilitates the delivery of Hsp70-bound substrates to the proteasome [32,68]. In fact, an increase in BAG-1 concentrations induces the association of Hsp70 with the proteasome, and promotes the CHIP-mediated degradation of glucocorticoid hormone receptor [31]. The unique function of CHIP in the UPS, along with Hsp70, makes it also a particularly desirable target in the pathogenesis of various neurodegenerative diseases. CHIP influences the turnover of disease-related misfolded proteins via the UPS. For example, tau, which forms protein aggregates in patients suffering from Alzheimer's disease, is ubiquitinated by the CHIP-Hsc70 complex for degradation by the proteasome [72]. Further participation of CHIP in neurodegenerative diseases has become apparent from studies that link it to Parkinson's disease. Parkin, an E3 ubiquitin ligase responsible for familial Parkinson's disease [73], ubiquitylates three pathologically relevant proteins, α-synuclein, synphilin and plasma membrane protein Pael-R, which are then degraded by the proteasome [74]. It was recently reported that CHIP is involved in the regulation of Parkin-mediated Pael-R ubiquitination. Accumulation of unfolded Pael-R in the endoplasmic reticulum (ER) results in ER stress-induced degeneration of dopaminergic neurons, a characteristic of Parkinson's disease [75]. Of further interest, CHIP appears to facilitate the Parkin-dependent ubiquitination of unfolded Pael-R, which is retrotranslocated from the ER, through the release of Hsp70 from Parkin-Pael-R complexes, while Hsp70 transiently binds to unfolded Pael-1 and inhibits the ubiquitin ligase activity of Parkin [76]. The pathological characteristic of polyglutamine diseases is the neuronal accumulation of expanded-polyglutamine proteins, causing the formation of intracellular protein inclusions and aggregates [77,78; and Chapter 32]. CHIP, which has recently been shown to associate with expanded polyglutamine proteins, is likely to promote their ubiquitation and degradation and, ultimately, suppress their aggregation and toxicity [79]. On the other hand, a discordant observation was reported that the suppression of aggregates by CHIP is independent of its enhancement of ubiquitination and degradation, and depends on its ability to keep polyglutamine proteins in a soluble state through its interaction with Hsp70/Hsc70 [80]. Further studies will clarify the mechanism by CHIP inhibits the aggregation of polyglutamine proteins. Besides Hsp70/Hsp90 and their cofactors cooperating with the UPS, other chaperones have been implicated in the recognition and delivery of multi-ubiquitinated substrates to the proteasome. The 97-kDa valosin-containing protein (p97 or VCP), the mammalian homolog of yeast cdc48 and a member of the AAA proteins (ATPases associated with various cellular activities), is a hexameric, barrel-shaped structure reminiscent of a molecular chaperone [81,82]. VCP is implicated in various cellular processes, such as cell cycle progression [83], membrane fusion [84,85], apoptosis [86], ER-associated degradation [87,88; and Chapter 13] and protein degradation. Notably, nearly all these activities are directly or indirectly regulated by the UPS. However, there is evidence indicating that VCP binds polyubiquitinated proteins directly, and targets them to the proteasome for degradation. Indeed, the loss of VCP function results in inhibition of the degradation of substrate proteins, such as IκBα and cyclinB1 and, consequently, in the accumulation of ubiquitinated proteins, whereas re-adding purified VCP restores their degradation [33,89,90]. This establishes VCP as a preferential targeting factor playing an important role in proteasome-mediated degradation, though the molecular mechanisms behind the targeting of a substrate to the proteasome remain poorly understood. Based on the various cellular functions through the UPS and structural feature of VCP, it is likely that VCP acts as a chaperone in the ubiquitin-proteasome degradation pathway. All AAA proteins possess weak ATPase activity, and their chaperone-like activity seems generally linked to the conserved ATPase domains. A relationship between ATPase of VCP and ubiquitin-proteasome-mediated degradation has been demonstrated recently [91]. VCP binds directly to multiubiquitin chains attached to substrates through its N-terminal, the deletion of which causes a defect of proteasome-mediated degradation. Though it is necessary, the N-terminal is not sufficient to mediate the degradation by the ubiquitin-proteasome, and the D2 domain of VCP responsible for the major ATPase activity is also required. In the ubiquitin-proteasome degradation pathway, the 19S cap very likely plays a role in regulating the accessibility of substrate by mediating conformational changes, for example unfolding, that facilitate entry of the substrate into the 20S proteasome. However, the close space between the base and lid of the 19S cap might prevent the access of larger proteins to the unfolding site within the base. In addition, VCP differs from the proteasome in that (*i*) it does not interact with a proteolytic core and (*ii*) its central channel leads into the cytoplasma, suggesting that VCP could unfold large substrate proteins and convert them into a state that is suitable for degradation upstream from the proteasome [35]. The *in vitro* unfoldase activity of a model protein of VAT, the archaeal homologue of VCP, might support this hypothesis [92]. These combined considerations suggest that a major function of VCP in the UPS is to unfold the substrate before its delivery to the proteasome. Besides acting as an unfoldase, VCP functions as a chaperone, disassembling the substrate/cofactors complex for subsequent degradation, and using the energy from the hydrolysis of ATP [93,94]. The functional importance of VCP as a molecular chaperone has been implicated in ER-associated degradation [87,88]. VCP, with its partner proteins, Ufd1 and Npl4, binds the substrates onto the ER membrane, and releases them into the cytosol as polyubiquitinated species in an ATP-dependent manner, recruiting substrates to the proteasome. The ER-associated degradation protein substrates accumulate in the ER when Cdc48 fails to function, suggesting that VCP acts as a molecular chaperone, which disassembles proteins complexes, or unfolds targeted substrates, or both. Thus, these activities, together with its ability to bind multiubiquitinated proteins, position the VCP as a preproteasomal chaperone that targets substrate proteins to the proteasome. In addition, VCP represents a more general chaperone activity that recognizes non-native proteins and prevents their aggregation [95], making VCP a key molecule in charge of triage decisions toward the repair versus degradation of targets in eukaryotic cells.

Conclusions

The functions of molecular chaperones in the UPS we have discussed can be classified into proteasomal versus non-proteasomal (Figure 2). Following their delivery to the 26S proteasome, the latter unfolds and translocates the ubiquitinated substrates from its 19S cap-binding site to the 20S core in an ATP-dependent manner. If substrate proteins encounter the free 19S cap first, they are unfolded and the substrate-19S complexes may associate with the 20S proteasome for subsequent degradation. However, an initial association of misfolded proteins with the free 19S cap might favor refolding over degradation. In such case, the regulatory component interacts directly with non-native structures of nonubiquitinated proteins and enhances their refolding. On the other hand, the association with 20S core, forming the 26S proteasome, allows the 19S cap to facilitate passively the partitioning to degradation pathway. Consequently, the 19S cap can open the way to either repair or degradation of a given target. Misfolded or partially folded proteins eventually aggregate because their exposed hydrophobic surfaces cause inappropriate intra- and intermolecular interactions. While the 20S proteasome or the 19S cap can prevent the aggregation of misfolded proteins as a molecular chaperone, it is noteworthy that this function results in promoting proteolysis indirectly by maintaining unfolded proteins in an non-aggregated state, rendering them, in turn, more accessible to the 20S proteasome or 19S cap. Growing evidence also points to the involvement of non-proteasomal chaperones in the UPS. General chaperones, such as Hsp70 and Hsp90, attempt to rescue the misfolded proteins, first by refolding them to a functional native state. However, failure to repair may lead the molecular chaperones to the UPS to eliminate the chaperone-bound substrates. In this process, substrates ubiquitination and sorting to the proteasome are directed by the interaction of the cofactors, such as CHIP and BAG-1, with molecular chaperones. VCP (Cdc48) is the molecular chaperone that has received most attention for its function mediating the delivery of ubiquitinated proteins to the proteasome. Substrates modified with a multi-ubiquitin chain are recruited to VCP via its cofactors, where they might be unfolded to access the recognition site within the 19S cap. Further studies of the chaperone activities in the UPS could provide important clues toward the understanding of the roles of UPS in cellular regulation.

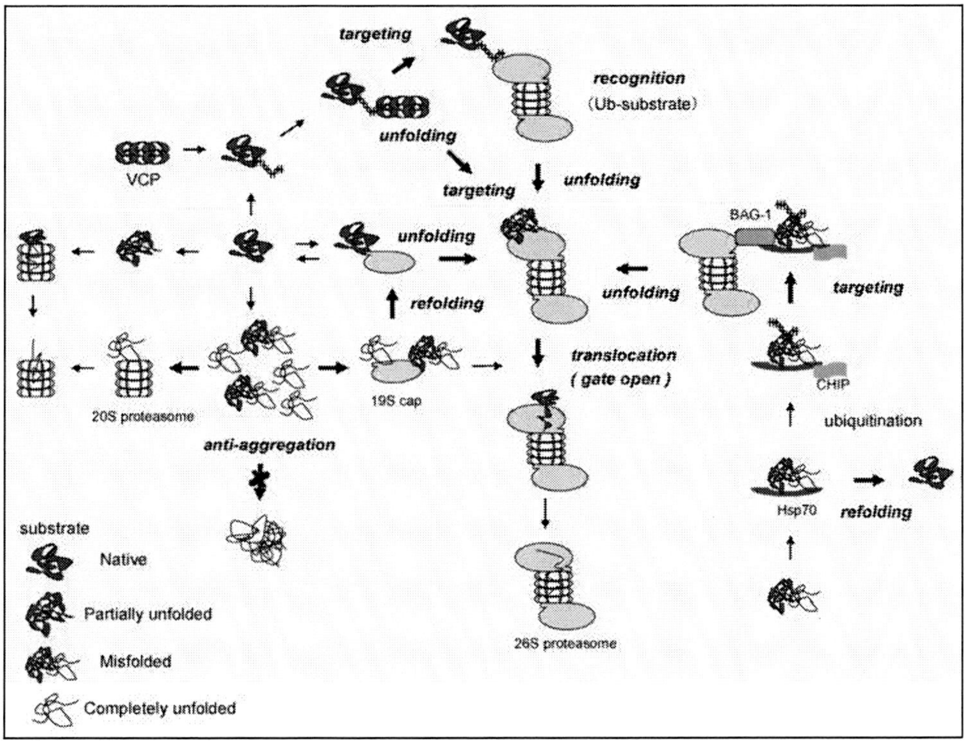

Figure 2. Model of chaperone functions in the ubiquitin-proteasome degradation pathway. The steps regulated by the molecular chaperone are indicated by thick arrows.

References

[1] Wickner S, Maurizi MR, Gottesman S: Posttranslational Quality Control: Folding, Refolding, and Degrading Proteins. *Science* 1999, 286:1888-1893.

[2] Muchowski PJ, Wacker JL: Modulation of neurodegeneration by molecular chaperones. *Nat Rev Neurosci* 2005, 6:11-22.

[3] Ross CA, Pickart CM: The ubiquitin-proteasome pathway in Parkinson's diseases and other neurodegenerative diseases. *Trends Cell Biol* 2004, 14:703-711.

[4] Gottesman S, Wickner S, Maurizi MR: Protein quality control triage by chaperones: and proteases. *Genes Dev* 1997, 11:815-823.

[5] Schirmer EC, Glover JR, Singer MA, Lindquist S: HSP100/Clp proteins: a common mechanisms explains diverse functions. *Trends Biochem Sci* 1996, 21:289-296.

[6] Gottesman S, Maurizi MR, Wickner S: Regulatory Subunits of Energy-Dependent Proteases. *Cell* 1997, 91:435-438.

[7] Singh SK, Grimaud R, Hoskins JR, Wickner S, Maurizi MR: Unfolding and internalization of proteins by the ATP-dependent proteases ClpXP and ClpAP. *Proc Natl Acad Sci USA* 2000, 97:8898-8903.

[8] Coux O, Tanaka K, Goldberg AL: Structure and functions of the 20S and 26S proteasomes. *Annu Rev Biochem* 1996, 65:801-847.

[9] DeMartino G.N, Slaughter CA: The Proteasome, a Novel Protease Regulated by Multiple Mechanisms. *J Biol Chem* 1999, 274:22123-22126.

[10] Baumeister W, Walz J, Zühl F, Seemüller E: The proteasome: paradigm of a self-compartmentalizing protease. *Cell* 1998, 92:367-380.

[11] Strickland E, Hakala K, Thomas PJ, DeMartino G.N: Recognition of misfolding proteins by PA700, the regulatory subcomplex of the 26S proteasome. *J Biol Chem* 2000, 275:5565-5572.

[12] Braun BC, Glickman M, Kraft R, Dahlmann B, Kloetzel PM, Finley D, Schmidt M: The base of the proteasome regulatory particle exhibits chaperone-like activity. *Nat Cell Biol* 1999, 1:221-226.

[13] Bukau B, Horwich AL: The Hsp70 and Hsp60 chaperone machines. *Cell* 1998, 92:351-366.

[14] Hartl FU, Hayer-Hartl M: Molecular Chaperones in the Cytosol: from Nascent Chain to Folded Protein. *Science* 2000, 295:1852-1858.

[15] Hoffman L, Pratt G., Rechsteiner M: Multiple forms of the 20S multicatalytic and the 26S ubiquitin/ATP-dependent proteases from rabbit reticulocyte lysate. *J Biol Chem* 1992, 267:22362-22368.

[16] Larsen CN, Finley D: Protein translocation channels in the proteasome and other Proteases *Cell* 1997, 91:431-434.

[17] DeMartino GN, Moomaw CR, Zagnitko OP, Proske RJ, Ma C-P, Afendis SJ, Swaffield JC, Slaughter CA: PA700, an ATP-dependent activator of the 20S proteasome, is an ATPase containing multiple members of a nucleotide-binding protein family. *J Biol Chem* 1994, 269:20878-20884.

[18] Glickman MH, Rubin DM, Coux O, Wefes I, Pfeifer G, Cjeka Z, Baumeister W, Fried V, Finley D: A subcomplex of the proteasome regulatory particle required for ubiquitin-conjugate degradation and related to the COP9-signalosome and eIF3. *Cell* 1998, 94:615-623.

[19] Touitou R, Richardson J, Bose S, Nakanishi M, Rivett J, J Allday M: A degradation signal located in the C-terminus of p21 WAF1/CIP1 is a binding site for the C8 alpha-subunit of the 20S proteasome. *EMBO J* 2001, 20:2367-2375.

[20] Liu C-W, Millen L, Roman TB, Xiong H, Gilbert HF, Noiva R, DeMartino GN, Thomas PJ: Conformational Remodeling of Proteasomal Substrates by PA700, the 19S Regulatory Complex of the 26S proteasome. *J Biol Chem* 2002, 277:26815-26820.

[21] Ferrington DA, Sun H, Murray KK, Costa J, Williams TD, Bigelow DJ, Squier TC: Selective degradation of oxidized calmodulin by the 20S proteasome. *J Biol Chem* 2001, 276:937-943.

[22] Yano M, Mori S, Kido H: Intrinsic nucleoside diphosphate kinase-like activity is a novel function of the 20S proteasome. *J Biol Chem* 1999, 274:34375-34382.

[23] Hiromura M, Yano M, Mori H, Inoue M, Kido H: Intrinsic ADP-ATP exchange activity is a novel function of the molecular chaperone, Hsp70. *J Biol Chem* 1998, 273:5435-5438.

[24] Yano M, Mori S, Niwa Y, Inoue M, Kido H: Intrinsic nucleoside diphosphate kinase-like activity as a novel function of 14-3-3 proteins. *FEBS Lett* 1997, 419:244-248.

[25] Yano M, Koumoto Y, Kanesaki Y, Wu X, Kido H: 20S Proteasome prevents aggregation of heat-denatured proteins without PA700 regulatory subcomplex like a molecular chaperone. *Biomacromolecules* 2004, 5:1465-1469.

[26] Bercovich B, Stancovski I, Mayer A, Blumenfeld N, Laszlo A, Schwartz AL, Ciechanover A: Ubiquitin-dependent degradation of certain protein substrates in vitro requires the molecular chaperone Hsc70. *J Biol Chem* 1997, 272:9002-9010.

[27] Youker RT, Walsh P, Beilharz T, Lithgow T, Brodsky JL: Distinct roles for the Hsp40 and Hsp90 molecular chaperones during cystic fibrosis transmembrane conductance regulator degradation in yeast. *Mol Biol Cell* 2004, 15: 4787-4797.

[28] Meacham GC, Patterson C, Zhang W, Younger JM, Cyr DM: The Hsc70 co-chaperone CHIP targets immature CFTR for personal degradation. *Nat Cell Biol* 2001, 3: 100-105

[29] Connell P, Ballinger CA, Jiang J, Wu Y, Thompson LJ, Hohfeld J, Patterson C: The co-chaperone CHIP regulates protein triage decisions mediated by heat-shock proteins. *Nat Cell Biol* 2001, 3:93-96.

[30] Murata S, Minami Y, Minami M, Chiba T, Tanaka K: CHIP is a chaperone-dependent E3 ligase that ubiquitylates unfolded protein. *EMBO Rep* 2001, 2:1133-1138.

[31] Demand J, Alberti S, Patterson C, Hohfeld J: Cooperation of a ubiquitin domain protein and an E3 ubiquitin ligase during chaperone/proteasome coupling. *Curr Biol* 2001, 11:1569-1577.

[32] Esser C, Alberti S, Hohfeld J: Cooperation of molecular chaperones with the ubiquitin/proteasome system. *Biochim Biophys Acta* 2004, 1695:171-188.

[33] Dai RM, Li CC: Valosin-containing protein is a multiubiquitin chain-targeting factor required in ubiquitin-proteasome degradation. *Nat Cell Biol* 2001, 3:740-744.

[34] Hartmann-Peterson R, Seeger M, Gordon C. Transferring substrates to the 26S proteasome. *Trends Biochem Sci* 2003, 28:26-31.

[35] Elsasser S, Finley D: Delivery of ubiquitinated substrates to protein-unfolding machines. *Nat Cell Biol* 2005, 7:742-749.

[36] Chu-Ping M, Vu JH, Proske RJ, Slaughter CA, DeMartino GN: Identification, purification, and characterization of a high molecular weight, ATP-dependent activator (PA700) of the 20S proteasome. *J Biol Chem* 1994, 269:3539-3547.

[37] Hershko A, Ciechanover A: The ubiquitin system. *Annu Rev Biochem* 1998, 67:425-479.

[38] Hochstrasser M: Ubiquitin-dependent protein degradation. *Annu Rev Genet* 1996, 30:405-439.

[39] Horwich AL, Weber-Ban EU, Finley D: Chaperone rings in protein folding and degradation. *Proc Natl Acad Sci USA* 1999, 96:11033-11040.

[40] Varshavsky A: The ubiquitin system. *Trends Biochem Sci* 1997, 22:383-387.

[41] Lam YA, Lawson TG, Velayutham M, Zweier JL, Pickart CM: A proteasomal ATPase subunit recognizes the polyubiquitin degradation signal. *Nature* 2002, 416:763-767.

[42] Murakami Y, Matsufuji S, Kameji T, Hayashi S, Igarashi K, Tamuta T, Tanaka K, Ichihara A: Ornithine decarboxylase is degraded by the 26S proteasome without ubiquitination. *Nature* 1992, 360:597-599.

[43] Li X, Coffino P: Degradation of ornithine decarboxylase : exposure of the C-terminal target by a polyamine-indelible inhibitory protein. *Mol Cell Biol* 1993, 13:2377-2383.

[44] Li X, Stubbiness B, Hoffman L, Pratt G, Rechsteiner M, Coffino P: The N-terminus of antizyme promotes degradation of heterologous proteins. *J Biol Chem* 1996, 271:4441-4446.

[45] Benaroudj N, Tarcsa E, Cascio P, Goldberg AL: The unfolding of substrates and ubiquitin-independent protein degradation by proteasome. *Biochimie* 2001, 83:311-318.

[46] Benaroudj N, Goldberg AL: PAN, the proteasome-activating nucleotidase from archaebacteria, is a protein-unfolding molecular chaperone. *Nat Cell Biol* 2000, 2:833-839.

[47] Murakami Y, Matsufuji S, Hayashi S, Tanahashi N, Tanaka K: ATP-dependent inactivation and sequestration of ornithine decarboxylase by the 26S proteasome are prerequisites for degradation. *Mol Cell Biol* 1999, 19:7216-7227.

[48] Navon A, Goldberg AL: Proteins are unfolded on the surface of the ATPase ring before transport into the proteasome. *Mol Cell* 2001, 8:1339-1349.

[49] Guterman A, Glickman MH: Complementary roles for Rpn11 and Ubp6 in deubiquitination and proteolysis by the proteasome. *J Biol Chem* 2004, 279:1729-1738.

[50] Groll M, Bajorek M, Kohler A, Moroder L, Rubin DM, Huber R, Glickman MH, Finley D: A gate channel into the proteasome core particle. *Nat Struct Biol* 2000, 7:1062-1067.

[51] Köhler A, Cascio P, Leggett DS, Woo KM, Goldberg AL, Finley D: The Axial Channel of the proteasome core particle is gated by the Rpt2 ATPase and controls both substrate entry and product release. *Mol Cell* 2001, 7:1143-1152.

[52] Groll M, Ditzel L, Löwe J, Stock D, Bochtler M, Bartunik HD, Huber R: Structure of the 20S proteasome from yeast at 2.4 A resolution. *Nature* 1997, 386:463-471.

[53] Merker K, Grune T: Proteolysis of oxidized proteins and cellular senescence. *Exp Gerontol* 2000, 35:779-786.

[54] Wang R, Chait BT, Wolf I, Kohanski RA, Cardozo C: Lysozyme degradation by the bovine multicatalytic proteinase complex (proteasome): evidence for a nonprocessive mode of degradation. *Biochemistry* 1999, 38:14573-14581.

[55] Lasch P, Petras T, Ullrich O, Backmann J, Naumann D, Grune T: Hydrogen peroxide-induced structural alterations. *J Biol Chem* 2001, 276:9492-9502.

[56] Levine RL, Oliver CN, Fulks RM, Stadtman ER: Turnover of bacterial glutamine synthetase: oxidative inactivation precedes proteolysis. *Proc Natl Acad Sci USA* 1981, 4:2120-2124.

[57] Rivett AJ: Preferential degradation of the oxidatively modified form of glutamine synthetase by intracellular mammalian proteases. *J Biol Chem* 1985, 260:300-305.

[58] Grune T, Reinheckel T, Davies KJ: Degradation of oxidized proteins in mammalian cells. *FASEB J* 1997, 11:526-534.

[59] Whittier JE, Xiong Y, Rechsteiner MC, Squier TC: Hsp90 Enhances Degradation of Oxidized Calmodulin by the 20S proteasome. *J Biol Chem* 2004, 279:46135-46142.

[60] Sheaff RJ, Singer JD, Swanger J, Smitherman M, Roberts JM, Clurman BE: Proteasomal turnover of p21 Cip1 does not require p21 Cip1 ubiquitination. *Mol Cell* 2000, 5:403-410.

[61] Kriwacki RW, Hengst L, Tennant L, Reed SI, Wright PE: Structural studies of p21 Waf/Cip1/sdi1 in the free and cdk2-bound state: Conformational disorder mediates binding diversity. *Proc Natl Acad Sci USA* 1996, 93:11504-11509.

[62] Heinemeyer W, Kleinschmidt LA, Saidowsky J, Escher C, Wolf DH: Proteinase yscE, the yeast proteasome/multicatalytic-multifunctional proteinase: mutants unravel its function in stress induced proteolysis and uncover its necessity for cell survival. *EMBO J* 1991, 10:555-562.

[63] Dick LR, Moomaw CR, Pramanik BC, DeMartino G.N, Slaughter CA: Identification and localization of a cysteinyl residue critical for the trypsin-like catalytic activity of the proteasome. *Biochemistry* 1992, 31:7347-7355.

[64] Heinemeyer W, Gruhler A, Mohrle V, Mahe Y, Wolf DH: PRE2, highly homologous to the human major histocompatibility complex-linked RING10 gene, codes for a yeast proteasome subunit necessary for chymotryptic activity and degradation of ubiquitinated proteins. *J Biol Chem* 1993, 268:5115-5120.

[65] Sanger F, Nicklen S, Coulson AR: DNA sequencing with chain-terminating inhibitors. *Proc Natl Acad Sci USA* 1977, 74:5463-5467.

[66] Deville-Bonne D, Sellam O, Merola F, Lascu I, Desmadril M, Verson M: Phosphorylation of nucleoside diphosphate kinase at the active site studied by steady-state and time-resolved fluorescence. *Biochemistry* 1996, 35:14643-14650.

[67] Freije JMP, Blay P, MacDonald NJ, Manrow RE, Steeg PS: Site-directed mutation of Nm23-H1. Mutations lacking motility suppressive capacity upon transfection are deficient in histidine-dependent protein phosphotransferase pathways in vitro. *J Biol Chem* 1997, 272:5525-5532.

[68] McClellan AJ, Tam S, Kaganovich D, Frydman J: Protein quality control: chaperone culling corrupt conformations. *Nat Cell Biol* 2005, 7:36-741.

[69] McClellan AJ, Scott MD, Frydman J: Folding and quality control of the VHL tumor suppressor proceed through distinct chaperone pathways. *Cell* 2005, 121:739-748.

[70] Ballinger CA, Connell P, Wu Y, Hu Z, Thompson LJ, Yin LY, Patterson C: Identification of CHIP, a novel tetratricopeptide repeat-containing protein that interacts with heat shock proteins and negatively regulates chaperone functions. *Mol Cell Biol* 1999, 19:4535-4545.

[71] Cyr DM, Hohfeld J, Patterson C: Protein quality control: U-box-containing E3 ubiquitin ligases join the fold. *Trends Biochem Sci* 2002, 27:368-375.

[72] Petrucelli L, Dickson D, Kehoe K, Taylor J, Snyder H et al: CHIP and Hsp70 regulate tau ubiquitination, degradation and aggregation. *Hum Mol Genet* 2004, 13:703-714.

[73] Kitada T, Asakawa S, Hattori N, Matsumine H, Yamamura Y, Minoshima S, Yokochi M, Mizuno Y, Shimizu N: Mutations in the parkin gene cause autosomal recessive juvenile parkinsonism. *Nature* 1998, 392:605-608.

[74] Tanaka K, Suzuki T, Hattori N, Mizuno Y: Ubiquitin, proteasome and Parkin. *Biochim Biophys Acta* 2004, 1695:235-247.

[75] Hoppe, T: Multiubiquitylation by E4 enzymes: 'one size' doesn't fit all. *Trends Biochem Sci* 2005, 30:183-187.

[76] Imai Y, Soda M, Hatakeyama S, Akagi T, Hashikawa T, Nakayama KI, Takahashi R: CHIP Is Associated with Parkin, a Gene Responsible for Familial Parkinson's Disease, and Enhances Its Ubiquitin Ligase activity: *Mol Cell* 2002, 10:55-67.

[77] Perutz MF: Glutamine repeats and neurodegenerative diseases: molecular aspects. *Trends Biochem Sci* 1999, 24:58-63.

[78] Zoghbi HY, Orr HT: Glutamine repeats and neurodegeneration. *Annu Rev Neurosci* 2000, 23:217-247.

[79] Jana NR, Dikshit P, Goswami A, Kotliarova S, Murata S, Tanaka K, Nukina N. Co-chaperone CHIP associates with expanded polyglutamine protein and promotes their degradation by proteasomes. *J Biol Chem* 2005, 280:11635-11640.

[80] Miller VM, Nelson RF, Gouvion CM, Williams A, Rodriguez-Lebron E, Harper SQ, Davison BL, Rebagliati MR, Paulson HL: CHIP suppresses polyglutamine aggregation and toxicity *in vitro* and *in vivo*. *J Neurosci* 2005, 25:9152-9161.

[81] Patel S, Latterich M: The AAA team: related ATPases with diverse functions. *Trends Biochem Sci* 1998, 8:65-71.

[82] Maurizi MR, Li CC: AAA proteins: in search of a common molecular basis. *EMBO Rep* 2001, 21:980-985.

[83] Fröhlich KU, Fries HW, Rudiger M, Erdmann R, Botstein D, Mecke D: Yeast cell cycle protein CDC48p shows fulllength homology to the mammalian protein VCP and a member of a protein family involved in secretion, peroxisome formation, and gene expression. *J Cell Biol* 1991, 114:443-453.

[84] Latterich M, Frohlich KU, Schekman R: Membrane fusion and the cell cycle: Cdc48p participates in the fusion of ER membranes. *Cell* 1995, 82:885-893.

[85] Rabouille C, Levine TP, Peter J-M, Warren, G.: An NSF-like ATPase, p97, and NSF mediate cisternal regrowth from mitotic Golgi fragments. *Cell* 1995, 82:905-914.

[86] Madeo F, Fröhlich E, Fröhlich KU: A yeast mutant showing diagnostic markers of early and late apoptosis. *J Cell Biol* 1997, 137:729-734.

[87] Ye Y, Meyer HH, Rapoport TA: The AAA ATPase Cdc48/p97 and its partners transport proteins from the ER into the cytosol. *Nature* 2001, 414:652-656.

[88] Rabinovich E, Kerem A, Frohlich K-U, Diamant N, Bar-Nun S: AAA-ATPase p97/Cdc48, a cytosolic chaperone required for endoplasmic reticulum-associated protein degradation. *Mol Cell Biol* 2002, 22:626-634.

[89] Dai RM, Chen E, Longo DL, Gorbea CM, Li CC: Involvement of valosin-containing protein, an ATPase co-purified with IκBα and 26S proteasome, in Ubiquitin-Proteasome-mediated Degradation of IκBα. *J Biol Chem* 1998, 273:3562-3573.

[90] Wojcik C, Yano M, DeMartino GN: RNA interference of valosin-containing protein (VCP/p97) reveals multiple cellular roles linked to ubiquitin/proteasome-dependent proteolysis. *J Cell Sci* 2004, 117:281-292.

[91] Song C, Wang Q, Li CC: ATPase Activity of p97-Valosin-containing Protein (VCP). *J Biol Chem* 2002, 278:3648-3655.

[92] Golbik R, Lupas AN, Koretke KK, Baumeister W, Peters J: The Janus face of the archaeal Cdc48/p97 homologue VAT: protein folding versus unfolding. *Biol Chem* 1999, 380:1049-1062.

[93] Braun S, Matuschewski K, Rape M, Thomas S, Jentsch S: Role of the ubiquitin-selective CDC48$^{UFD1/NPL4}$ chaperone (segregase) in ERAD of OLE1 and other substrates. *EMBO J* 2002, 21:615-621.

[94] Ye Y, Meyer HH, Rapoport TA: Function of the p97-Ufd1-Npl4 complex in retrotranslocation from the ER to the cytosol: dual recognition of nonubiquitinated polypeptide segments and polyubiquitin chains. *J Cell Biol* 2003, 162:71-84.

[95] Thomas S: Cdc48 can distinguish between native and non-native proteins in the absence of cofactors. *FEBS Lett* 2002, 520:107-110.

In: The Ubiquitin Proteasome System... ISBN 978-1-60021-749-4
Eds: Mario Di Napoli and Cezary Wojcik, pp. 227-240 © 2007 Nova Science Publishers, Inc.

Chapter 11

INTERCELLULAR LOCALIZATION OF PROTEASOMES

Cezary Wójcik[1,*] *and George N. DeMartino*[2,¥]
[1]Department of Anatomy and Cell Biology Indiana University School of Medicine Evansville Indiana USA;
[2]Department of Physiology University of Texas Southwestern Medical Center Dallas Texas USA.

ABSTRACT

Proteasomes are present in the cytoplasm and in the nuclei of all cell types including neurons and glial cells. Their relative abundance within subcellular compartments is highly variable depending on the cell type as well as depending on physiological and/or pathological stimuli. Cytoplasmic proteasomes are not uniformly distributed; instead they are enriched at the centrosomes at the cytoskeletal networks and at the outer surface of membranaceous organelles such as the endoplasmic reticulum and the endosome system. In the nuclei proteasomes are present throughout the nucleoplasm but are usually not found within the nucleoli. Nuclear proteasomes often associate with structures of the nuclear matrix in particular with discrete subnuclear domains called the PML nuclear bodies (POD domains). PML bodies in the nucleus and the pericentrosomal area of the cytoplasm may function as proteolytic centers of the cell since they are enriched in components of the proteasome system. Under conditions of impaired proteolysis proteasomes and ubiquitinated proteins further accumulate at those locations. The knowledge about intracellular distribution of proteasomes is important for our understanding of the dynamic organization of the ubiquitin- and proteasome-dependent

* Correspondence concerning this article should be addressed to Dr. Cezary Wójcik, MD, PhD; Department of Anatomy and Cell Biology, Indiana University School of Medicine, 8600 University Blvd. Evansville IN USA. Phone: (812) 461 5437; Fax: (812) 465 1184; E-mail: cwojcik@iupui.edu.

¥ Correspondence concerning this article should be addressed to Dr. George N. DeMartino, PhD; Department of Physiology, University of Texas Southwestern Medical Center, 5323 Harry Hines Blvd. Dallas TX USA. Phone: (214) 648-3770; Fax (214) 648 4771; E-mail: George.DeMartino@UTSouthwestern.edu.

proteolysis within cells. The present chapter will review available information about subcellular localization of proteasomes with emphasis on neuronal and glial cells; however since most data were obtained in other cell types they will be reviewed as well whenever this is relevant.

Keywords: proteasome, microsocopy, cytoplasm, nucleus, centrosome, aggresome, proteolytic center.

Abbreviations

CNS central nervous system; ER endoplasmic reticulum; PML promyelocytic leukemia; UPS ubiquitin and proteasome dependent proteolytic system.

Introduction

Previous chapters have discussed the structure and function of different components of the ubiquitin- and proteasome system of protein degradation (UPS). 20S proteasomes are abundant (approximately 0.6% of total cell protein in HeLa cells [1]) and ubiquitously present within eukaryotic cells both in the nucleus and in the cytoplasm [2]. Highly proliferating and transformed cell lines usually have higher proteasome levels and activity than quiescent and non-transformed cells [3]. Proteasomes are abundant in the central nervous system (CNS) from which they were successfully purified [4-8]. Their distribution in different subcellular compartments is variable depending of the cell type. In rat CNS most neurons display mainly nuclear immunolabeling with anti-proteasome antibodies whereas the pyramidal cortical neurons of layer 5 and the motor neurons of the ventral horn in the spinal cord show an intense nuclear and cytoplasmic labeling as well as labeling in the processes. Additionally some populations of large neurons in the mesencephalon and brainstem also display a moderate immunoreactivity in their perikarya [9]. This pattern changes in some neurodegenerative disorders which are characterized by the formation of proteasome immunoreactive intracytoplasmic inclusions such as Lewy bodies in Parkinson's disease and neurofibrillary tangles in Alzheimer's disease [10,11].

UPS is a complex and hierarchical system whose different components are not randomly dispersed within the cells but rather are distributed in a very complex pattern allowing sequential interaction. The UPS has been studied mostly by biochemists often approaching the cells as bags full of proteins which are purified and studied *in vitro* in order to be fully understood. Discoveries achieved with purified proteasomes and other components of the UPS set the base for their understanding however once those elements have been analyzed in detail it is necessary to turn back and to consider their function in intact cells. In contrast to the beauty of purified biochemical assays the study of proteins in intact cells is often obscure since it involves a multitude of factors which may directly or indirectly affect the outcome of the experiments. Proteasome localization has been reviewed in the recent years by several authors (e.g. [12-16]). Much less is known about the intracellular distribution of the different enzymes forming the ubiquitination cascade albeit some components have been characterized in detail. For example, the cytosolic E3 enzyme parkin preferentially localizes to the microtubules [17] while another E3 called gp78 specifically resides within endoplasmic reticulum [18].

Data describing the intracellular localization of proteasomes are often contradictory and are biased by the use of different cell types or different methods such as immunocytochemical labeling at the level of light, fluorescent and electron microscopy observation of GFP-proteasome fusions as well as subcellular fractionation. Immunodetection of proteasomes depends on the antibodies used and their crossreactivity with other cellular components as well as on the type of fixation used [19]. Some epitopes may be masked at certain locations by interactions with compartment-specific proteins. Detection of proteasome subunits fused with GFP raises the question about the functionality of such hybrid proteins in the context of a multisubunit complex which is poised to interact with multiple proteins within the cells. Subcellular fractionation experiments are subject to crosscontamination and/or loss of some components from a given fraction. There is also a considerable variability in the intracellular distribution of proteasomes depending on the source of material used for the fractionation. Altogether no single method is perfect to study the intracellular localization of proteasomes however the data obtained with the use of different methods complement each other delivering a synthetic approach to the problem.

The efforts to describe localization of proteasomes in intact cells are additionally obstructed by the uncertainty which molecular entity we describe as ‘the proteasome’ since proteasomes actually may exist in multiple forms: free subunits, proteasome precursors, free 20S proteasomes, 26S proteasomes (20S proteasome + one or two PA700 caps), 20S proteasomes associated with PA28 and hybrid proteasomes (20S proteasomes with one PA700 and one PA28 cap), just to name a few [20,21]. The different forms exist in the cells in a dynamic equilibrium and the relative proportions of each molecular species is highly variable depending on cell type, cell cycle phase and metabolic conditions [22-24]. In addition to that proteasomes can be present either as constitutive proteasomes or as immunoproteasomes. The latter contain three interferon-γ inducible subunits of the β-type (β5i β1i and β2i) replacing their constitutive counterparts (β5 β1 and β2; see Chapter 34). It has been suggested that intermediate forms with mixed composition may also exist [25]. Proteasome subunits can be posttranslationally modified, e.g. by phosphorylation, what further contributes to the diversity of different proteasome subpopulations [26-28]. Immunoproteasomes in some instances have different localization from constitutive 20S proteasomes as detected with antisera labeling their specific immunosubunits [24,29]. Inflammation in the central nervous system dramatically changes the content of immunoproteasomes in the microglia, however it is not known what is the impact of the inflammatory process on their intracellular distribution [30].

CYTOPLASMIC PROTEASOMES

The relative proportion of nuclear versus cytoplasmic proteasomes varies greatly and probably depends on the cell type, growth conditions, cell density, fixation and the assay method used for detection. In rat CNS most neurons display mainly nuclear immunolabeling with anti-proteasome antibodies whereas some big motor neurons from the cortex and the ventral horn of the spinal cord show also an intense nuclear labeling as well as labeling in the processes. Additionally some populations of large neurons in the mesencephalon and brainstem also display a moderate immunoreactivity in their perikarya [47]. While in rat hepatocytes only 17% of 20S proteasomes are nuclear more than 50% of proteasomes are nuclear in L-123 cells [31].

There are several examples evidencing dynamic changes in the intracellular distribution of proteasomes. In a lung cancer cell line poor growth conditions and/or high cell density result in a switch from a nuclear labeling pattern to a predominantly cytoplasmic pattern [45] while in nonstimulated A-431 cells proteasomes are both in the cytoplasm and nuclei; however after a 15 min treatment with EGF most proteasomes relocate to the nuclei. After another hour proteasomes are released from the nuclei and return to the cytoplasm [19]. Studies with the aid of specific inhibitors demonstrated that the nuclear export of proteasomes depend on the signaling through EGF-mediated tyrosine phosphorylation and phospholipase C activity [32]. Moreover during apoptosis proteasomes are transported from the nucleus into the cytoplasm where they associate with cytoplasmic blebs [33]. The demonstrated mobility of proteasome pools between cytosol and nuclei stays in clear conflict with the claim that proteasomes can not exit the nucleus unless the nuclear envelope breaks down during mitosis [34]. If the latter claim is indeed true it would mean that in postmitotic cells such as neurons proteasomes trapped in the nuclei are trapped there for the entire lifespan of the cell. Despite the fact that proteasomes in the neurons seem to be mostly nuclear such scenario seems unrealistic.

Using a variety of antisera detecting proteasome subunits several authors have observed a uniform labeling of the cytoplasm with negative shadows corresponding to vacuolar structures and a slightly stronger labeling of the nuclei with negative shadows corresponding to the nucleoli [35-38]. This pattern of distribution of the 20S proteasomes was later confirmed with the use of a GFP-LMP2 fusion proteasomal subunit which incorporated into active 20S particles in living cells [56].

The use of different fixation and extraction techniques has shown that a fraction of cytoplasmic proteasomes is tightly associated with cytoskeletal elements especially with intermediate filaments [39]. This association is regulated in a cell cycle dependent manner e.g. in HeLa and PtK2 cells more proteasomes are associated with cytokeratin filaments during the G2 phase [51]. Proteasomes also have been observed in association with actin filaments [40] and actin-myosin complexes [41]. It is not clear how much o this early results may be due to fixation artifacts and how much they actually reflect real proteasome characteristics.

A fraction of proteasomes in mammalian cells is associated with cellular membranes in particular with the the outer surface of the endoplasmic reticulum (ER). Electron microscopy determined that 14% of cytoplasmic proteasomes in rat hepatocytes is associated with the ER [42] while cellular subfractionation methods estimate that only 1% of proteasomes is bound to the ER in crustacean muscles [43]). Most of 20S proteasomes and 26S proteasomes in yeast are associated with the ER and the nuclear envelope as shown by the localization of functional GFP-hybrids of 26S proteasome subunits [44]. In contrast the use of three different antibodies detecting 26S proteasome subunits in yeast has shown that they are predominantly nuclear [60]. The reason for the discrepancy between these results is unclear. Proteasomes associated with the ER may be preferentially involved in the process known as ERAD (ER-associated degradation) which consists in the selective retrograde export of structurally abnormal or misfolded proteins from the ER into the cytosol as well as constitutive turnover of several proteins associated with their ubiquitination and proteasome-dependent degradation [45-47] (see Chapter 13).

Detailed fractionation of the ER has shown that proteasomes associate mostly with the smooth ER and with the cis-Golgi while they do not associate with the rough ER [52]. ER-associated proteasomes are enriched with the interferon-γ-inducible subunits [48]. It is at the ER that immunoproteasomes come in close contact with the TAP transporter which transports peptides generated by proteasomes into the ER lumen where they associate with class I MHC molecules [24]. PI31 a specific protein initially purified as a proteasome inhibitor also localizes to the outer surface of the ER where it interferes with the assembly of active immunoproteasomes [24].

Proteasomes associate only with the cytoplasmic face of ER and other cellular membranes and are not found within the lumen of membranaceous organelles. However after starvation or after inhibition of lysosomal enzymes they are found within the lumen of the lysosomes probably as a consequence of nonselective autophagy or chaperone-mediated transport [49]. Since proteasomes are inactive in the low pH found in lysosomes it is likely that proteasomes themselves are degraded within lysosomes. Recently a subpopulation of proteasomes have been identified which is associated with the outer surface of the endosomes. This association may be mediated by the proteasome-interacting protein Ecm29 ([50] see Chapter 9).

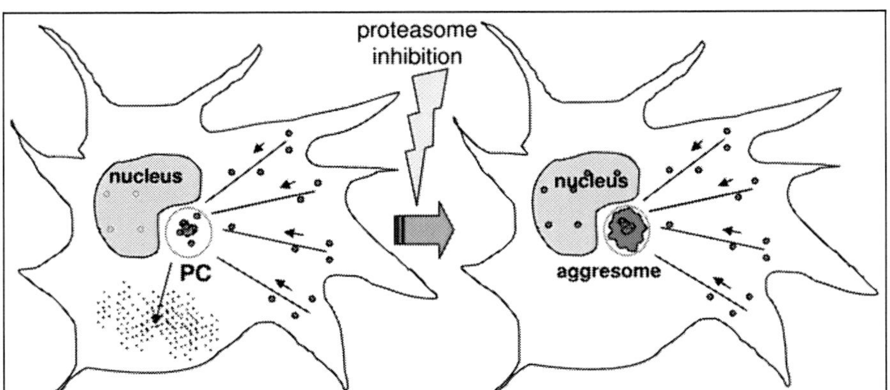

Figure 1. Localization of UPS-dependent proteolysis at the proteolytic center. Despite the ubiquitous presence of UPS components throughout the cell at least some proteolysis is localized to a perinuclear area around the centrosome called the proteolytic center (PC). PC is enriched in proteasomes and some other components of the UPS. Substrates arrive to the PC via a microtubule dependent transport. Inhibition of the proteasome or its overwhelming with an excess of misfolded proteins leads to a situation where more substrates arrive at the PC than are degraded or otherwise removed from the PC inducing their aggregation around the centrosome in form of a structure often referred to as the 'aggresome'.

There are indications that UPS-dependent proteolysis is particularly enhanced in the perinuclear region around the centrosome called therefore the proteolytic center of the cell which is enriched in proteasomes and some components of the UPS. Proteasome substrates are targeted to this proteolytic center via microtubule-assisted transport from the cell periphery. When cells are treated with proteasome inhibitors or when there is an excessive protein misfolding more substrates arrive to the proteolytic center than are degraded or otherwise cleared from that area leading to aggregation of the substrate protein at the proteolytic center in form of organized structures called 'aggresomes' (Figure 1). The process of aggresome formation and its consequences are described in detail in Chapter 12 [51-57].

Nuclear Proteasomes

Albeit in neuronal cells nuclear proteasomes constitute probably the largest pool of this multicatalytic complex it is unlikely that it remains permanently trapped within the nucleus as suggested by studies with a GFP-proteasome subunit fusion protein [56]. A hypothetical situation of excessive proteasome recruitment within neuronal nuclei is depicted on Figure 2. It is more likely that in neurons proteasomes recirculate continuously between the nucleus and the cytosol. Moreover the model of nuclear trapping of proteasomes is contradicted by data from yeast where the nucleus is the birthplace of mature proteasomes since they are imported there in form of precursor complexes from the cytosol [58]. Once in the nucleus they assemble with the aid of the nuclear Blm3 chaperone [59] which is homologous to the PA200 activator from mammalian cells [60]. Several subunits of 20S proteasomes have nuclear localization signals associated with clusters of acidic amino acids that control their nuclear import. Combination of clusters with positive and negative charges may serve to regulate the translocation of proteasomes from the cytoplasm to the nucleus and tyrosine phosphorylation of same subunits has been suggested to play additional role in the nucleo-cytoplasmic transfer of proteasomes [61,62].

Changes in proteasome localization during the cell cycle are not relevant for neuronal cells; however they are relevant for the glial cells [63]. In HeLa and PtK2 cells there was an increased nuclear localization of proteasomes as the cells progressed through the cell cycle approaching mitosis [64]. In the nucleus the proteasomes are localized in the euchromatin regions as well as at the periphery of the heterochromatin and nucleoli [65]. A subpopulation of nuclear proteasomes is associated with discrete nuclear subdomains called PML (promyelocytic leukemia) bodies (a.k.a. ND10 bodies or POD domains nuclear bodies) [66,67]. One group of researchers has characterized a nuclear body rich in proteasomes and other elements of the UPS called the clastosome. Despite the fact that clastosomes do contain the PML protein, their characteristics are different from classical PML bodies [42]. For example while upon proteasome inhibition ubiquitinated proteins accumulate at PML bodies they are actually dispersed from the clastosomes. In some instances proteasomes have been detected also within nucleoli [68]. However, a detailed proteomic analysis of nucleoli has not supported this conclusion [69]. Recently it has been reported that a variant of survivin is specifically degraded in the nucleoli by the UPS [70]. Therefore the possible nucleolar localization of proteasomes remains open for discussion. The association of proteasomes with different elements of the nuclear matrix is generally loose since they can be easily extracted [71] however during the progression of the cell cycle in HeLa cells increased levels of proteasomes remain bound much more tightly [72].

When the cells enter mitosis 20S proteasomes accumulate around the condensing chromosomes which remain devoid of labeling throughout cell division. In late anaphase 20S proteasomes colocalize with α-tubulin of the spindle fibers. In telophase and early interphase of the daughter cells intensive staining of proteasomes persists in the nuclei [73]. Detergent extraction of most soluble proteasomes from mitotic cells reveal tight association of proteasomes with the spindle poles especially during prometaphase and metaphase and with the midbody during telophase [79]. The observed spatial and temporal distribution pattern of proteasomes during mitosis is highly reminiscent of the behavior of their important mitotic substrates cyclins [74].

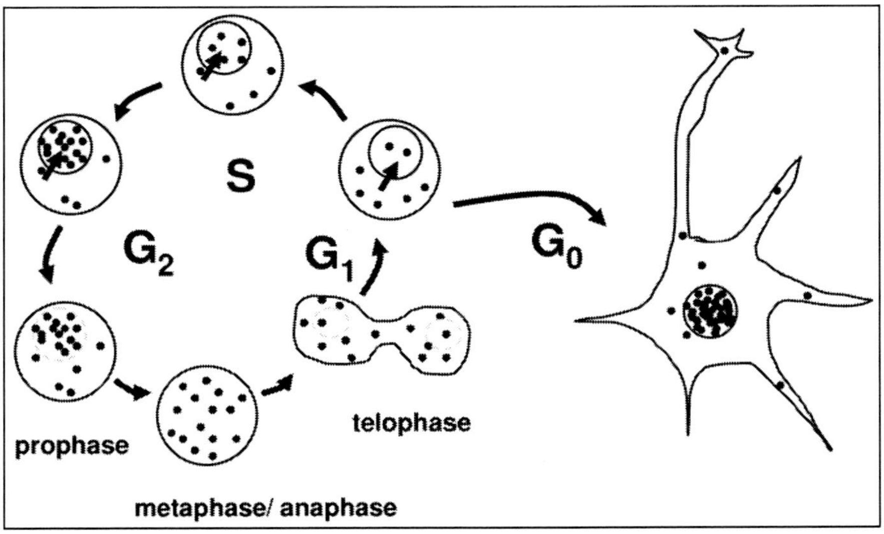

Figure 2. Model of nuclear trapping of proteasomes. A study with GFP-labeled proteasome subunit integraded into functional proteasomes has suggested that proteasomes are synthesized in the cytosol and then translocated to the nucleus where they are trapped until the nuclear envelope breakdown during mitosis which allows their release to the cytoplasm. While such scenario is possible in rapidly dividing cells grown *in vitro* it is unlikely in postmitotic neurons. It would have lead to a depletion of cytoplasmic proteasomes and an excessive accumulation of proteasomes in the nuclei. Studies in yeast suggest that the final steps of proteasome assembly take place in the nucleus before mature proteasomes are exported to the cytoplasm. Nevertheless, in many neurons a nuclear localization of proteasomes is prevailing.

ARE THERE EXTRACELLULAR PROTEASOMES?

Proteasomes and the entire ubiquitination pathway function inside of the cells in the nucleus and cytosol. Even when they degrade or interact with proteins on the ER, other compartments of the secretory pathway and on the plasma membrane, they do it exclusively on the cytosolic side. Nevertheless, there are several reports on proteasomes being found in the extracellular space. It is likely that proteasomes found in the extracellular space are released there as the result of cellular lysis and breakdown. Indeed plasma levels of proteasome antigens increase abruptly in necrotizing cancers when they are massively released to the extracellular space from dying cells. Levels of circulating proteasomes may be even used in such conditions for diagnostic and prognostic purposes. It is likely that in ischemic brain proteasomes are released from damaged neurons leaking to the neuropilus however no studies were performed to clarify this issue. Free proteasomes in body fluids may contribute to the pathogenesis of certain autoimmune disorders characterized by the presence of antiproteasome antibodies such as systemic lupus erythematosus or rheumatoid arthritis (see Chapter 35). Levels of circulating proteasomes are markedly elevated in patients with systemic autoimmune diseases apparently correlating with disease activity and reflecting the magnitude of cellular damage [75].

It is more difficult to explain other findings, such as association of proteasomal antigens with the surface of T, B and NK cells producing a characteristic pattern of antigens depending on the CD classification [76] or the EGF-dependent association of proteasome with plasma membrane in A-431 cells [77]. Moreover there are reports that specific proteasome subpopulations are secreted into the media of cultured cells [78]. In case of the ascidian sperm not only proteasome is secreted during fertilization but there is also evidence for an extracellular ubiquitination event followed by a proteasome-dependent degradation of a component of the vitelline coat [79]. Secretion of proteasomes or proteasome-like particles may be also achieved by some tumor cells such as the C6 astrocytoma. It has been reported that this secreted proteasome is crucial to the destruction of the type IV collagen found in the basement membrane, being therefore important for the initiation of tumor-associated angiogenesis [80,81]. On the other hand it has been reported that NK cells accumulate proteasomes in form of mucoid masses in the cytosol which may be secreted during their cytotoxic action against tumor cells [82]. It is difficult to envision how proteasomes enter in that case the secretory pathway since normally they are absent from the lumen of the ER, Golgi and the endosome system.

Conclusions

Current evidence suggests that cellular proteasomes form several intracellular pools which are in a state of dynamic equilibrium. Generally in neuronal cells most proteasomes are nuclear; however abundant cytosolic and membrane-associated proteasome pools exist as well. While the general architecture of proteasomes within each pool is the same they can often associate with different regulators and interactors specific to the individual compartments. Upon several stimuli proteasome pools can be rapidly relocated between subcellular compartments what is likely an adaptive mechanism. There are no experimental data available regarding changes in intracellular localization of proteasomes in neuronal cells nevertheless there are no reasons to believe that in neurons proteasomal populations are more static than in other cell types. Relocation of proteasomes upon different insults (e.g. hypoxia, oxidative damage, proteasome inhibition) may form part of an important adaptive mechanism in neuronal cells. Mobilization of subcellular pools of proteasomes or – on the contary – blockage of such mobilization may be future targets for the development of new drugs.

References

[1] Hendil KB: The 19 S multicatalytic "prosome" proteinase is a constitutive enzyme in HeLa cells. *Biochem Int* 1988, 17:471-477.

[2] Tanaka K, Kumatori A, Ii K, Ichihara A: Direct evidence for nuclear and cytoplasmic colocalization of proteasomes (multiprotease complexes) in liver. *J Cell Physiol* 1989, 139:34-41.

[3] Kumatori A, Tanaka K, Inamura N, Sone S, Ogura T, Matsumoto T, Tachikawa T, Shin S, Ichihara A: Abnormally high expression of proteasomes in human leukemic cells. *Proc Natl Acad Sci U S A* 1990, 87:7071-7075.

[4] Piccinini M, Tazartes O, Mostert M, Musso A, DeMarchi M, Rinaudo MT: Structural and functional characterization of 20S and 26S proteasomes from bovine brain. *Brain Res Mol Brain Res* 2000, 76:103-114.

[5] Akaishi T, Shiomi T, Sawada H, Yokosawa H: Purification and properties of the 26S proteasome from the rat brain: evidence for its degradation of myelin basic protein in a ubiquitin-dependent manner. *Brain Res* 1996, 722:139-144.

[6] Azaryan A, Banay-Schwartz M, Lajtha A: The presence of ATP + ubiquitin-dependent proteinase and multicatalytic proteinase complex in bovine brain. *Neurochem Res* 1989, 14:995-1001.

[7] Dresdner K, Barker LA, Orlowski M, Wilk S: Subcellular distribution of prolyl endopeptidase and cation-sensitive neutral endopeptidase in rabbit brain. *J Neurochem* 1982, 38:1151-1154.

[8] Wilk S, Orlowski M: Cation-sensitive neutral endopeptidase: isolation and specificity of the bovine pituitary enzyme. *J Neurochem* 1980, 35:1172-1182.

[9] Mengual E, Arizti P, Rodrigo J, Gimenez-Amaya JM, Castano JG: Immunohistochemical distribution and electron microscopic subcellular localization of the proteasome in the rat CNS. *J Neurosci* 1996, 16:6331-6341.

[10] Kwak S, Masaki T, Ishiura S, Sugita H: Multicatalytic proteinase is present in Lewy bodies and neurofibrillary tangles in diffuse Lewy body disease brains. *Neurosci Lett* 1991, 128:21-24.

[11] Fergusson J, Landon M, Lowe J, Dawson SP, Layfield R, Hanger DP, Mayer RJ: Pathological lesions of Alzheimer's disease and dementia with Lewy bodies brains exhibit immunoreactivity to an ATPase that is a regulatory subunit of the 26S proteasome. *Neurosci Lett* 1996, 219:167-170.

[12] Rivett AJ: Intracellular distribution of proteasomes. *Curr Opin Immunol* 1998, 10:110-114.

[13] Rivett AJ, Knecht E: Proteasome location. *Curr Biol* 1993, 3:127-129.

[14] Knecht E, Rivett AJ. Intracellular localization of proteasomes. In: Hilt W, Wolf DH, editors. Proteasomes: The World of Regulatory Proteolysis.Austin, TX: Landes Bioscience; 2000. p. 181-93.

[15] Gordon C: The intracellular localization of the proteasome. *Curr Top Microbiol Immunol* 2002, 268:175-184.

[16] Wojcik C, DeMartino GN: Intracellular localization of proteasomes. *Int J Biochem Cell Biol* 2003, 35:579-589.

[17] Yang F, Jiang Q, Zhao J, Ren Y, Sutton MD, Feng J: Parkin stabilizes microtubules through strong binding mediated by three independent domains. *J Biol Chem* 2005.

[18] Fang S, Ferrone M, Yang C, Jensen JP, Tiwari S, Weissman AM: The tumor autocrine motility factor receptor, gp78, is a ubiquitin protein ligase implicated in degradation from the endoplasmic reticulum. *Proc Natl Acad Sci U S A* 2001, 98:14422-14427.

[19] Machiels BM, Henfling ME, Broers JL, Hendil KB, Ramekers FC: Changes in immunocytochemical detectability of proteasome epitopes depending on cell growth and fixation conditions of lung cancer cell lines. *Eur J Cell Biol* 1995, 66:282-292.

[20] DeMartino GN, Slaughter CA: The proteasome, a novel protease regulated by multiple mechanisms. *J Biol Chem* 1999, 274:22123-22126.

[21] Glickman MH, Ciechanover A: The ubiquitin-proteasome proteolytic pathway: destruction for the sake of construction. *Physiol Rev* 2002, 82:373-428.

[22] Tanahashi N, Murakami Y, Minami Y, Shimbara N, Hendil KB, Tanaka K: Hybrid proteasomes. Induction by interferon-gamma and contribution to ATP-dependent proteolysis. *J Biol Chem* 2000, 275:14336-14345.

[23] Russell SJ, Steger KA, Johnston SA: Subcellular localization, stoichiometry, and protein levels of 26 S proteasome subunits in yeast. *J Biol Chem* 1999, 274:21943-21952.

[24] Brooks P, Fuertes G, Murray RZ, Bose S, Knecht E, Rechsteiner MC, Hendil KB, Tanaka K, Dyson J, Rivett J: Subcellular localization of proteasomes and their regulatory complexes in mammalian cells. *Biochem J* 2000, 346 Pt 1:155-161.

[25] Kloetzel PM: Antigen processing by the proteasome. *Nat Rev Mol Cell Biol* 2001, 2:179-187.

[26] Haass C, Kloetzel PM: The Drosophila proteasome undergoes changes in its subunit pattern during development. *Exp Cell Res* 1989, 180:243-252.

[27] Pal JK, Martins dS, Scherrer K: Differential synthesis and cytolocalization of prosomes in chick embryos during development. *Int J Dev Biol* 1994, 38:525-534.

[28] Briane D, Olink-Coux M, Vassy J, Oudar O, Huesca M, Scherrer K, Foucrier J: Immunolocalization of a specific type of prosome close to the bile canaliculi in fetal and adult rat liver. *Eur J Cell Biol* 1992, 57:30-39.

[29] Fabunmi RP, Wigley WC, Thomas PJ, DeMartino GN: Interferon gamma regulates accumulation of the proteasome activator PA28 and immunoproteasomes at nuclear PML bodies. *J Cell Sci* 2001, 114:29-36.

[30] Stohwasser R, Giesebrecht J, Kraft R, Muller EC, Hausler KG, Kettenmann H, Hanisch UK, Kloetzel PM: Biochemical analysis of proteasomes from mouse microglia: induction of immunoproteasomes by interferon-gamma and lipopolysaccharide. *Glia* 2000, 29:355-365.

[31] Rivett AJ, Palmer A, Knecht E: Electron microscopic localization of the multicatalytic proteinase complex in rat liver and in cultured cells. *J Histochem Cytochem* 1992, 40:1165-1172.

[32] Evdonin AL, Tsupkina NV, Konstantinova IM, Medvedeva ND: [Effect of EGF on nuclear-cytoplasmic distribution of proteasomes in A-431 cells]. *Tsitologiia* 2001, 43:142-147.

[33] Pitzer F, Dantes A, Fuchs T, Baumeister W, Amsterdam A: Removal of proteasomes from the nucleus and their accumulation in apoptotic blebs during programmed cell death. *FEBS Lett* 1996, 394:47-50.

[34] Reits EA, Benham AM, Plougastel B, Neefjes J, Trowsdale J: Dynamics of proteasome distribution in living cells. *EMBO J* 1997, 16:6087-6094.

[35] Kloetzel PM, Falkenburg PE, Hossl P, Glatzer KH: The 19S ring-type particles of Drosophila. Cytological and biochemical analysis of their intracellular association and distribution. *Exp Cell Res* 1987, 170:204-213.

[36] Kamakura K, Ishiura S, Nonaka I, Sugita H: Localization of ingensin in rat central nervous system and skeletal muscle. *J Neurosci Res* 1988, 20:473-478.

[37] Grossi de Sa MF, Martins dS, Harper F, Coux O, Akhayat O, Pal JK, Florentin Y, Scherrer K: Cytolocalization of prosomes as a function of differentiation. *J Cell Sci* 1988, 89 (Pt 2):151-165.

[38] Knecht E, Palmer A, Sweeney ST, Rivett AJ: Immunocytochemical localization of the multicatalytic proteinase in rat liver and in L-132 cells. *Biochem Soc Trans* 1991, 19:293S.

[39] Olink-Coux M, Arcangeletti C, Pinardi F, Minisini R, Huesca M, Chezzi C, Scherrer K: Cytolocation of prosome antigens on intermediate filament subnetworks of cytokeratin, vimentin and desmin type. *J Cell Sci* 1994, 107 (Pt 3):353-366.

[40] Arcangeletti C, Sutterlin R, Aebi U, De Conto F, Missorini S, Chezzi C, Scherrer K: Visualization of prosomes (MCP-proteasomes), intermediate filament and actin networks by "instantaneous fixation" preserving the cytoskeleton. *J Struct Biol* 1997, 119:35-58.

[41] Ryabova LV, Virtanen I, Olink-Coux M, Scherrer K, Vassetzky SG: Distribution of prosome proteins and their relationship with the cytoskeleton in oogenesis of Xenopus laevis. *Mol Reprod Dev* 1994, 37:195-203.

[42] Palmer A, Rivett AJ, Thomson S, Hendil KB, Butcher GW, Fuertes G, Knecht E: Subpopulations of proteasomes in rat liver nuclei, microsomes and cytosol. *Biochem J* 1996, 316 (Pt 2):401-407.

[43] Beyette JR, Mykles DL: Immunocytochemical localization of the multicatalytic proteinase (proteasome) in crustacean striated muscles. *Muscle Nerve* 1992, 15:1023-1035.

[44] Enenkel C, Lehmann A, Kloetzel PM: GFP-labelling of 26S proteasomes in living yeast: insight into proteasomal functions at the nuclear envelope/rough ER. *Mol Biol Rep* 1999, 26:131-135.

[45] Werner ED, Brodsky JL, McCracken AA: Proteasome-dependent endoplasmic reticulum-associated protein degradation: an unconventional route to a familiar fate. *Proc Natl Acad Sci U S A* 1996, 93:13797-13801.

[46] Brodsky JL, McCracken AA: ER protein quality control and proteasome-mediated protein degradation. *Semin Cell Dev Biol* 1999, 10:507-513.

[47] Tsai B, Ye Y, Rapoport TA: Retro-translocation of proteins from the endoplasmic reticulum into the cytosol. *Nat Rev Mol Cell Biol* 2002, 3:246-255.

[48] Brooks P, Fuertes G, Murray RZ, Bose S, Knecht E, Rechsteiner MC, Hendil KB, Tanaka K, Dyson J, Rivett J: Subcellular localization of proteasomes and their regulatory complexes in mammalian cells. *Biochem J* 2000, 346 Pt 1:155-161.

[49] Cuervo AM, Palmer A, Rivett AJ, Knecht E: Degradation of proteasomes by lysosomes in rat liver. *Eur J Biochem* 1995, 227:792-800.

[50] Gorbea C, Goellner GM, Teter K, Holmes RK, Rechsteiner M: Characterization of mammalian Ecm29, a 26 S proteasome-associated protein that localizes to the nucleus and membrane vesicles. *J Biol Chem* 2004, 279:54849-54861.

[51] Wojcik C, Schroeter D, Wilk S, Lamprecht J, Paweletz N: Ubiquitin-mediated proteolysis centers in HeLa cells: indication from studies of an inhibitor of the chymotrypsin-like activity of the proteasome. *Eur J Cell Biol* 1996, 71:311-318.

[52] Wojcik C: On the spatial organization of ubiquitin-dependent proteolysis in HeLa cells. *Folia Histochem Cytobiol* 1997, 35:117-118.

[53] Wojcik C: An inhibitor of the chymotrypsin-like activity of the proteasome (PSI) induces similar morphological changes in various cell lines. *Folia Histochem Cytobiol* 1997, 35:211-214.

[54] Fabunmi RP, Wigley WC, Thomas PJ, DeMartino GN: Activity and regulation of the centrosome-associated proteasome. *J Biol Chem* 2000, 275:409-413.

[55] Wigley WC, Fabunmi RP, Lee MG, Marino CR, Muallem S, DeMartino GN, Thomas PJ: Dynamic association of proteasomal machinery with the centrosome. *J Cell Biol* 1999, 145:481-490.

[56] Johnston JA, Ward CL, Kopito RR: Aggresomes: a cellular response to misfolded proteins. *J Cell Biol* 1998, 143:1883-1898.

[57] Kopito RR: Aggresomes, inclusion bodies and protein aggregation. *Trends Cell Biol* 2000, 10:524-530.

[58] Lehmann A, Janek K, Braun B, Kloetzel PM, Enenkel C: 20 S proteasomes are imported as precursor complexes into the nucleus of yeast. *J Mol Biol* 2002, 317:401-413.

[59] Fehlker M, Wendler P, Lehmann A, Enenkel C: Blm3 is part of nascent proteasomes and is involved in a late stage of nuclear proteasome assembly. *EMBO Rep* 2003, 4:959-963.

[60] Ustrell V, Hoffman L, Pratt G, Rechsteiner M: PA200, a nuclear proteasome activator involved in DNA repair. *EMBO J* 2002, 21:3516-3525.

[61] Tanaka K, Yoshimura T, Tamura T, Fujiwara T, Kumatori A, Ichihara A: Possible mechanism of nuclear translocation of proteasomes. *FEBS Lett* 1990, 271:41-46.

[62] Wang HR, Kania M, Baumeister W, Nederlof PM: Import of human and Thermoplasma 20S proteasomes into nuclei of HeLa cells requires functional NLS sequences. *Eur J Cell Biol* 1997, 73:105-113.

[63] Lafarga M, Berciano MT, Pena E, Mayo I, Castano JG, Bohmann D, Rodrigues JP, Tavanez JP, Carmo-Fonseca M: Clastosome: a subtype of nuclear body enriched in 19S and 20S proteasomes, ubiquitin, and protein substrates of proteasome. *Mol Biol Cell* 2002, 13:2771-2782.

[64] Palmer A, Mason GG, Paramio JM, Knecht E, Rivett AJ: Changes in proteasome localization during the cell cycle. *Eur J Cell Biol* 1994, 64:163-175.

[65] De Conto F, Pilotti E, Razin SV, Ferraglia F, Geraud G, Arcangeletti C, Scherrer K: In mouse myoblasts nuclear prosomes are associated with the nuclear matrix and accumulate preferentially in the perinucleolar areas. *J Cell Sci* 2000, 113 (Pt 13):2399-2407.

[66] Mattsson K, Pokrovskaja K, Kiss C, Klein G, Szekely L: Proteins associated with the promyelocytic leukemia gene product (PML)- containing nuclear body move to the nucleolus upon inhibition of proteasome-dependent protein degradation. *Proc Natl Acad Sci U S A* 2001, 98:1012-1017.

[67] Lallemand-Breitenbach V, Zhu J, Puvion F, Koken M, Honore N, Doubeikovsky A, Duprez E, Pandolfi PP, Puvion E, Freemont P, de The H: Role of promyelocytic leukemia (PML) sumolation in nuclear body formation, 11S proteasome recruitment, and As2O3-induced PML or PML/retinoic acid receptor alpha degradation. *J Exp Med* 2001, 193:1361-1371.

[68] Arabi A, Rustum C, Hallberg E, Wright AP: Accumulation of c-Myc and proteasomes at the nucleoli of cells containing elevated c-Myc protein levels. *J Cell Sci* 2003, 116:1707-1717.

[69] Andersen JS, Lyon CE, Fox AH, Leung AK, Lam YW, Steen H, Mann M, Lamond AI: Directed proteomic analysis of the human nucleolus. *Curr Biol* 2002, 12:1-11.

[70] Song Z, Wu M: Identification of a novel nucleolar localization signal and a degradation signal in Survivin-deltaEx3: a potential link between nucleolus and protein degradation. *Oncogene* 2005.

[71] Wojcik C: Proteasome activator subunit PA28 alpha and related Ki antigen (PA28 gamma) are absent from the nuclear fraction purified by sucrose gradient centrifugation. *Int J Biochem Cell Biol* 1999, 31:273-276.

[72] Wojcik C, Paweletz N, Schroeter D: Localization of proteasomal antigens during different phases of the cell cycle in HeLa cells. *Eur J Cell Biol* 1995, 68:191-198.

[73] Amsterdam A, Pitzer F, Baumeister W: Changes in intracellular localization of proteasomes in immortalized ovarian granulosa cells during mitosis associated with a role in cell cycle control. *Proc Natl Acad Sci U S A* 1993, 90:99-103.

[74] Pines J, Hunter T: Human cyclins A and B1 are differentially located in the cell and undergo cell cycle-dependent nuclear transport. *J Cell Biol* 1991, 115:1-17.

[75] Egerer K, Kuckelkorn U, Rudolph PE, Ruckert JC, Dorner T, Burmester GR, Kloetzel PM, Feist E: Circulating proteasomes are markers of cell damage and immunologic activity in autoimmune diseases. *J Rheumatol* 2002, 29:2045-2052.

[76] Bureau JP, Olink-Coux M, Brouard N, Bayle-Julien S, Huesca M, Herzberg M, Scherrer K: Characterization of prosomes in human lymphocyte subpopulations and their presence as surface antigens. *Exp Cell Res* 1997, 231:50-60.

[77] Evdonin AL, Tsupkina NV, Konstantinova IM, Medvedeva ND: [Intracellular distribution of proteasomes in A-431 cells]. *Tsitologiia* 1999, 41:895-899.

[78] Kulichkova VA, Mittenberg AG, Ermolaeva YB, Tsimokha AS, Volkova IV, Evteeva IN, Kozyukharova IV, Gauze LN, Konstantinova IM: Specificity of the proteasome

population secreted from cells into the culture medium. *Dokl Biol Sci* 2004, 399:503-506.

[79] Sawada H, Takahashi Y, Fujino J, Flores SY, Yokosawa H: Localization and roles in fertilization of sperm proteasomes in the ascidian Halocynthia roretzi. *Mol Reprod Dev* 2002, 62:271-276.

[80] Del Maestro RF, Vaithilingam IS, McDonald W: Degradation of collagen type IV by C6 astrocytoma cells. *J Neurooncol* 1995, 24:75-81.

[81] Vaithilingam IS, McDonald W, Malott DW, Del Maestro RF: An extracellular proteasome-like structure from C6 astrocytoma cells with serine collagenase IV activity and metallo-dependent activity on alpha-casein and beta-insulin. *J Biol Chem* 1995, 270:4588-4593.

[82] Nannmark U, Kitson RP, Johansson BR, Rivett AJ, Goldfarb RH: Immunocytochemical localization of multicatalytic protease complex (proteasome) during generation of murine IL-2-activated natural killer (A-NK) cells. *Eur J Cell Biol* 1996, 71:402-408.

In: The Ubiquitin Proteasome System... ISBN 978-1-60021-749-4
Eds: Mario Di Napoli and Cezary Wojcik, pp. 241-270

Chapter 12

Aggresome Formation: A Failure of the Ubiquitin-Proteasome System

Patrick H. Thibodeau[2], Michael J. Corboy[2] and W. Christian Wigley[1,*]
[1]Reata Pharmaceuticals, Inc. Dallas, TX 75207, USA;
[2]UT Southwestern Medical Center, Dallas, TX 75390, USA.

Abstract

The regulated degradation of a majority of cellular proteins is catalyzed by the ubiquitin-proteasome system (UPS). The catalytic engine at the center of the UPS is the proteasome, a large multi-subunit self-compartmentalized protease. The UPS is engaged in a variety of functions critical to protein quality control, including surveillance and elimination of misfolded proteins that are irreparably damaged and potentially toxic. Unfortunately, when the capacity of the UPS is exceeded, misfolded protein substrates accumulate and tend to self associate, a characteristic hallmark of a growing class of aggregation diseases such as Alzheimer's and Parkinson's. These aggregates are subsequently assembled through an active and regulated process to form aggresomes. Aggresomes are dynamic structures, formed in response to an overload of improperly folded proteins. Assembly of aggresomes occurs at the centrosome, a juxtanuclear structure with roles historically limited to microtubule organization as it relates to cellular division and directional motor protein-dependent transport within the cell. Moreover, studies characterizing aggresome formation have demonstrated that the centrosome serves as a site for the recruitment and concentration of UPS components, including the proteasome, its regulators and a host of other proteins involved in protein quality control. Therefore, in addition to other cellular activities, the centrosome may play a central role in protein quality control, spatially regulating critical processes in protein folding, degradation, and aggregation.

* Correspondence concerning this article should be addressed to Dr. W. Christian Wigley, PhD; Reata Pharmaceuticals, Inc., 1950 Stemmons Freeway, Suite 5001, Dallas, TX, USA, 75207, Phone: 972-865-2209, Fax: 214-550-2714, E-mail: chris.wigley@reatapharma.com.

Keywords: aggregation, aggresome, centrosome, proteasome, protein misfolding, quality control, UPS.

Abbreviations

ABC, ATP-binding cassette; ADRP, autosomal dominant retinitis pigmentosa; AR, androgen receptor; CDK, cyclin-dependent kinase; CFTR, cystic fibrosis transmembrane conductance regulator; ERAD, ER-associated degradation; FAD, familial Alzheimer's disease; FALS, familial amyotrophic lateral sclerosis; Htn, huntingtin; IF, intermediate filaments; IFN-γ, interferon gamma; MHC, major histocompatibility complex; MTOC, microtubule organizing center; NFκB, nuclear factor κB; POD, promyelocytic leukemia oncogenic domain; PS1, presenilin 1, SBMA, spinobulbar muscular atrophy; SOD-1, superoxide dismutase 1; TAP, transporter for antigen presentation; UPS, ubiquitin-proteasome system.

Introduction

The ubiquitin-proteasome system (UPS) is the primary mechanism for the regulated elimination and recycling of cellular proteins in eukaryotes. For example, a critical role of the proteasome is the removal of potentially toxic misfolded and damaged proteins. When the degradation capacity of the UPS is surpassed, the misfolded substrates accumulate in the cell as distinct centrosomal inclusions of aggregated protein [1,2] that have been termed aggresomes [3]. Aggresome formation is not an arbitrary, indiscriminate event, but is instead part of a highly organized and regulated process, which both delivers the inclusions and concentrates the degradation machinery at a specific site. Aggresome formation was first described for the cystic fibrosis transmembrane conductance regulator (CFTR), a complex multidomain membrane protein whose mutation results in development of cystic fibrosis. Since, aggresomes have been shown to form from misfolding and aggregation of an increasingly diverse spectrum of cytoplasmic, transmembrane and secretory proteins, suggesting a general feature of the cell's attempts to deal with these potentially toxic species. In this regard, a number of studies indicate that the formation of aggresomes may in some instances serve a protective role, confining toxic conformers when degradation lags. Additionally, a growing body of evidence suggests that a number of viruses may have evolved to take advantage of the process of aggresome formation, thereby facilitating the assembly of new viral particles in infected cells. This chapter gives an overview of the process of protein folding and quality control (with particular emphasis on membrane protein folding), the proteasome's role in the degradation of misfolded proteins, the formation of aggresomes when degradation fails, and potential implications of aggresome formation for other cellular processes.

PROTEIN FOLDING

The central dogma of modern molecular biology, formalized first by Crick in the late 1950s, is that the information necessary for biological activities via proteins is encoded by DNA in units known as genes which are transcribed into mRNA molecules and subsequently translated into their respective functional protein components [4]. The properties by which the coding information is transmitted from DNA to RNA and then to protein are well understood. DNA is transcribed by RNA polymerases, into RNA and modified by a series of enzymes to make mRNA [5-7]. These mRNA molecules are then translated into protein sequence by the ribosome, a large RNA-protein complex. This transfer of information form DNA to RNA to polypeptide chain is known as the first genetic code. The nascent polypeptide chain is the first non-nucleotide product. The individual amino acids in this polymer are linked *via* peptide bonds and the specific combination and order of this sequence dictates both the final structure and function of the fully folded protein [8].

However, this linear polypeptide chain does not generally have intrinsic protein function. Protein function is usually attained only after the proper formation of the specific three-dimensional protein structure, provided by the linear amino acid sequence [9]. This process has been termed, at least anecdotally, the second-half of genetic code, is not well understood, and is largely determined by the physical and chemical composition and order of the polypeptide chain.

Several principles have emerged with growing number of available folding studies, structures and sequences. First, the folding process is regulated. Sampling all possible conformations available to a small polypeptide chain of amino acids is not possible in the timescale that most proteins fold. This paradox, describing the differences between the time needed to sample all possible physical states and the timeframe within which most proteins fold, has been termed the 'Levinthal Paradox' after Cyrus Levinthal who first formalized this discrepancy [10]. Second, protein structure and function is provided by the sequence of the linear polypeptide chain. Studies by Anfinsen, utilizing purified RNase demonstrated that proteins could spontaneously refold away from all other cellular components, demonstrating that the information for the structure and activity of a protein is contained in the polypeptide chain [11]. Third, the vast majority of protein sequence does not contribute directly to protein function. Rather, the majority of amino acids in a protein contribute indirectly to function through the folding and stabilization of complex three-dimensional structures. Finally, it is the interaction between amino acids that drives the formation of appropriate three-dimensional structure. Viewed very simply, a protein structure is a three-dimensional jigsaw puzzle whose final assembly is dependent on the shapes of all of the individual pieces as they relate to one another.

PHYSICAL BASIS OF PROTEIN FOLDING AND STRUCTURE

A variety of parameters drive the formation of these compact, folded structures. Though a conceptual description of these processes exists, a first principle understanding of the physical processes associated with protein folding is not understood (*i.e.* prediction of high-resolution protein structure, based solely on this first principle understanding of protein folding determinants and sequence is not possible across all families of proteins). However, the basic physical processes underlying protein folding have been described and include interactions within the polypeptide chain and between the polypeptide chain and its environment.

The most significant physical process that drives protein folding is the desolvation of hydrophobic amino acids by the hydrophobic effect. Amino acid hydrophobicity generally correlates with decreased surface accessibility and exclusion from the solvent environment [12,13]. Hydrophobic amino acids form the core of most globular proteins, which are normally tightly packed and solvent excluded, and it is the three-dimensional assembly of these sequences that largely drive the formation of the appropriate protein structure. Thermodynamic studies have demonstrated that the solvation of these core hydrophobic residues by protein denaturation and amino acid exposure correlates well with solvent partitioning studies of free amino acids, suggesting that the partitioning away from aqueous solvent is a major determinant in protein folding [14,15].

Electrostatic interactions also contribute to protein folding and stabilization of the native state. Formal charge-charge interactions between residues, such as salt bridging interactions, can stabilize specific protein structures and conformations. These interactions generally involve surface exposed residues and may contribute to native state stability. It is unlikely, however, that these electrostatic interactions are the predominant force driving for folding [16,17].

Hydrogen bonding is also critical to protein folding and structure. Formation and stabilization of secondary structure elements is highly dependant upon neighboring-residue hydrogen bonds. While these interactions are weak individually, the contribution of hydrogen bonding in a protein structure is not insignificant due to the large quantity of hydrogen bonds in a given structure [14].

van der Waals interactions and dipole interactions contribute to protein folding and stability, although individual interactions contribute only marginally energetically. Dipoles may be generated by small groups of amino acids or generated by the formation of secondary structure. Both have been shown to impact protein structure and function. Similarly, van der Waals interactions, while small individually, contribute significantly due to the large number of these interactions in any given protein [14].

Thermodynamic and Kinetic Regulation of Protein Folding

The physical and chemical forces acting on the polypeptide chain may exert their effects by regulating the thermodynamic properties of the polypeptide chain or by controlling the kinetic properties of the folding process or some combination of both [18]. The simplest folding pathways are two-state with the free energy of the native state being 5-10 kcal/mol more stable than that of the denatured state (Figure 1) [19]. However, deviations from a simple, two-state model have been reported for a number of proteins [18,20].

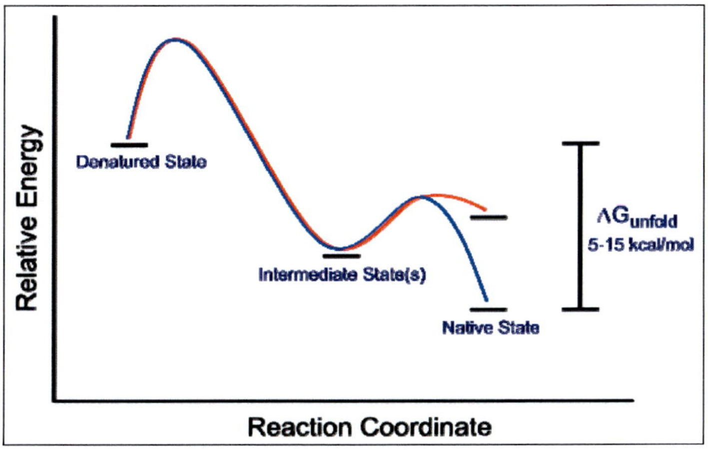

Figure 1. Thermodynamic regulation of protein folding. The protein folding process is generally spontaneous, with the native, or folded state, being between 5-10 kcal/mol more stable than the denatured state. A hypothetical reaction coordinate is shown with a spontaneous three-state reaction shown in black with a hypothetical mutation that destabilizes the native state shown in red. Such a mutation would be predicted to cause the reaction to arrest in the intermediate state.

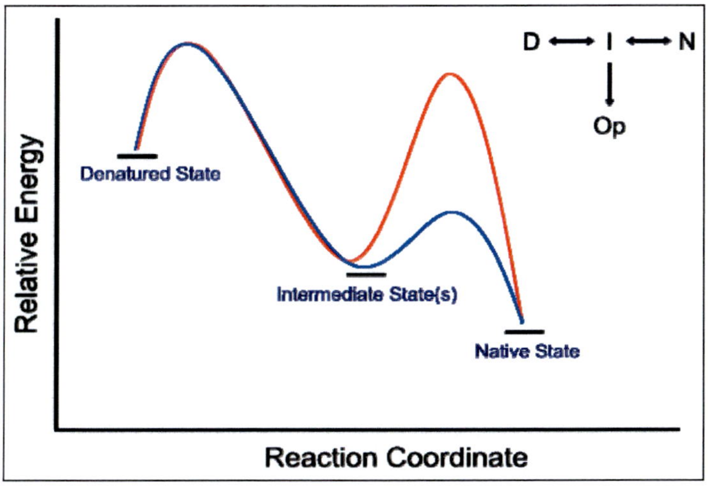

Figure 2. Kinetic regulation of protein folding. The kinetic properties of the folding process are also critical to the proper folding of a nascent polypeptide. A hypothetical reaction coordinate is shown. The thermodynamic properties of the wild-type, black, and mutant, red, proteins are identical. The kinetic barrier between the intermediate and native states is increased by the hypothetical mutation. Such a mutation may alter the folding pathway by slowing the formation of the native state structure from the intermediate thereby facilitating the formation of off-pathway structures that are often most easily accessible from these intermediate states. A simple kinetic scheme is inset describing this relationship between on- and off-pathway reactions. D, I and N, represent the denatured, intermediate, and native states, respectively; Op represents non-productive, off-pathway reactions.

The main resistance to folding is an entropic penalty associated with the constraint of the polypeptide chain [21]. The combination of physical forces, specified by the polypeptide sequence, drives the folding process and stabilizes the native state. Stability altering mutations, are often found in, but not exclusive to, the protein core and can be associated with changes in hydrophobic or electrostatic interactions [17,22].

The effects of the chemical and physical forces may be exerted on the kinetics of these reactions as well. While many mutations directly impact the relative stabilities of one or more of the structural states of a protein, others appear to have little effect on the stability of the structural states, but instead directly impact the kinetics of the folding process (Figure 2). As protein folding reactions are often a series of complex, competing reactions, whereby the rate of a productive reaction has evolved to out-compete the rate of a non-productive reaction, alterations in the kinetics of these processes can have adverse effects on the ability of the polypeptide to attain the native state. Mutations may have little impact on the final structure of the native state, but instead impact the fraction of protein that is able to reach this state by 'kinetically trapping' these proteins in intermediate, non-native states. Previous studies have demonstrated that these partially folded protein structures are prone to aggregation *via* off-pathway reactions [23]. Kinetic changes in the interactions between the folding polypeptide and the quality control machinery have also been implicated as a mechanism by which the cell recognizes improperly or partially folded protein structures, facilitating their degradation [24].

Protein Conformation and Physiology

Independent of the mechanism by which mutations affect the protein folding process, the loss of properly folded protein *via* alterations to the folding pathways is increasingly being recognized as a major source in disease. A growing number of human diseases are now recognized as being caused by either the improper folding of a protein or a change in the conformational state of a folded protein [25-28].

Table 1. Protein Misfolding Diseases – Loss of Function

Disease	Protein involved
Cystic fibrosis	CFTR
Cancer	p53, BRCA
Cataracts	crystallins
Fabry's disease	α-galactosidase
Hypercholesterolemia	ABCG5
Leprechaunism	insulin receptor
Long QT	HERG
Maple syrup urine disease	α-ketoacid dehydrogenase complex
Marfan syndrome	fibrillin
Retinitis pigmentosa	rhodopsin
Scurvy	collagen
Tay-Sachs disease	β-hexosaminidase

In many cases, the simple loss of functional protein is associated with the disease state (Table 1) [25]. The loss of functional protein may be the result of gross misfolding or a decreased steady state level of protein due to perturbation of protein stability. Such mutations are usually genetically recessive, although dominant negative phenotypes are possible if the associated proteins form oligomers.

In addition to classic loss-of-function mutations, gain-of-function mutations associated with altered protein conformation are being recognized as associated both with disease (Table 2) [26,27,29-32]. A large number of neurodegenerative and systemic diseases are associated with altered protein conformation, although the exact physiological role of these protein conformations are not known. One of the classic hallmarks of this class of gain-of-function diseases is the presence of amyloid protein structure. Amyloid structures are highly stable, β-rich structures which are capable of nucleating the conversion of native protein structure into amyloid structure. This nucleating and self-propagating capability underlies the gain-of-function like phenotype associated with the dominant genetic nature of these diseases.

Table 2. Protein Misfolding Diseases – Gain of Toxic Function

Disease	Protein involved
Alzheimer's disease	β-amyloid
Amyotrophic lateral sclerosis	superoxide dismutase 1
Creutzfeldt-Jacob 'mad cow'	prion protein
Diabetes (type II)	islet amyloid polypeptide (amylin)
Familial amyloid polyneuropathy	transthyretin
Finnish hereditary systemic amyloidosis	gelsolin
Hemodialysis-related amyloidosis	lysozyme
Huntington's disease	huntingtin
Parkinson's disease	α-synuclein

Another class of proteins, the prion proteins, is involved in both disease and non-disease related cellular function and undergoes similar nucleated structural conversion. This conversion of globular proteins from a native state conformation to a prion conformation has been directly implicated in several human and animal diseases, most notably variant-Creutzfeldt-Jakob (mad cow) disease and scrapie, as well as a variety of normal physiological processes. Similar structural conversion has been shown to be biologically important in a variety of organisms, providing mechanisms for evolutionary adaptation to changing environmental conditions [33,34]. In yeast and fungi, conversion of normal protein structure to the prion state induces changes in physiological traits allowing for non-Mendelian, heritable adaptation under stressful environmental conditions which are propagated after cellular division.

Modulating Folding and Misfolding as a Therapy for Disease

A variety of therapies has been devised for diseases associated with protein folding and conformation, although few have been developed to the point of clinical use. Therapeutic strategies developed to overcome the loss-of-function type mutations are simple: facilitate the proper folding and maturation of the disease associated polypeptide. One such example is a therapy for patients of the hereditary disease, Fabry's disease [35,36]. Mutation of the α-gal gene results in destabilization of the protein which causes a lysosomal storage and metabolic disorder that significantly reduces the patients' quality of life. The destabilized α-gal protein fails to leave the ER and, as a result, cannot participate in lysosomal metabolic processes. Administration of a high-affinity inhibitor of α-gal restores this folding and trafficking by stabilizing the native state structure in the ER and promoting protein trafficking. Protein half-life is significantly longer than the bioavailability of the drug, allowing for the dis-inhibition of the native, lysosomally resident α-gal protein. Similar strategies, utilizing high affinity inhibitors, are being developed for a number of other protein folding diseases [37-40].

Treatment of the gain-of-function neurodegenerative diseases is more difficult. While aberrant protein structures are correlated with disease, it is not known whether these structures are causative of, a simple by-product of, or a protection from the disease state [41-43]. As such, interference with these processes may not be directly beneficial and additional work to identify the role these protein conformations play in disease is necessary before effective therapeutic strategies can be safely developed. Recent studies have shown that these amyloid structures are dynamic and that protein deposits are cleared, often in a proteasome-dependent manner from tissues, suggesting that a therapeutic approach to modulating these processes may be possible [44,45].

The Ubiquitin-Proteasome System

In eukaryotic organisms, the vast majority of cellular proteins are degraded by the ubiquitin-proteasome pathway (see Chapter 3). This system is functionally coupled to a wide variety of cellular processes, including the generation of antigenic peptides for immune surveillance, regulation of protein levels of various short-lived regulatory proteins and transcription factors, and prevention of accumulation of misfolded mutant and damaged proteins [46-49]. Degradation by the 26S particle [50] (Chapter 7 and 8), generally requires binding and hydrolysis of ATP in a tightly regulated and coupled series of concerted steps. In most cases, targeting of substrate requires the covalent conjugation of ubiquitin to the target [47,48]. Ubiquitination is accomplished through the sequential activity of three classes of enzymes, which catalyze, in turn, activation (E1), conjugation (E2) of ubiquitin, followed by its ligation (E3) to substrate proteins [51]. Additional ubiquitin molecules are added by a similar set of sequential processive conjugation reactions to ubiquitin itself. The resulting polyubiquitin chain serves as a marker for degradation by the 26S proteasome. The polyubiquitinated substrates are then recognized by the 26S proteasome, de-ubiquitinated, unfolded and degraded [47,48].

The 26S proteasome is a 2.5-mDa multi-subunit molecular machine comprised by two major subcomplexes, the 20S proteasome and PA700 (or 19S). The cylindrical 20S proteasome, constituting the catalytic core of the protease, represents a paradigm of self-compartmentalization (Chapter 6) [52,53]. The 20S particle is a 700-kDa cylinder composed of four stacked heptameric rings, two outer α rings and two inner β rings, totaling 28 subunits [54,55]. The eukaryotic proteasome consists of two copies each of seven different α subunits and seven different β subunits that form heterooligomeric rings [(α1-α7)(β1-β7)(β1-β7)(α1-α7)]. The proteolytic activities of the 20S core are catalyzed by three of the β subunits (β1, β2, and β5). Each catalyzes a distinct proteolytic reaction. These catalytic activities have been termed chymotrypsin-like, trypsin-like, and peptidylglutamyl peptide hydrolytic activities based on the side chain specificities they target for hydrolysis [56,57]. The overall 20S architecture sequesters the catalytic sites within the hollow central cavity of the protease core. Access to the interior of the cylinder is restricted by narrow openings at either end of the cylinder, limiting substrates to short peptides and unfolded proteins [55,58-60]. As such, in most cases, the proteasome is necessarily coupled to other accessory complexes that mediate recognition, binding, and unfolding of target proteins. In mammalian cells, these roles are fulfilled by PA700 [47,48] and another regulatory complex, PA28 [61].

PA700 (19S) is a 700-kDa complex composed of 18-20 subunits that binds to one or both ends of the 20S core particle, forming the 26S complex [52,62,63]. Binding of PA700 to 20S activates the proteasome [50,63]. PA700 affords specificity for ubiquitinated substrates to the proteasome, mediated by its polyubiquitin binding [64,65] and cleavage [66,67] activities. PA700 also possesses multiple ATPase activities [68], likely enhancing degradation of ubiquitinated proteins by actively unfolding substrates and facilitating their translocation into the central cavity of the protease [50,52,69,70]. Interestingly, isolated PA700 can function *in vitro* in a manner similar to molecular chaperones [71-73]. This activity possibly reflects the complex's ability to recognize misfolded proteins as part of the 26S particle *in vivo*. Unlike the 20S particle, for which crystallographic structures have been solved from several organisms [55,74,75], little is known about the detailed three-dimensional structure of PA700. However, the relative arrangement of subunits and some of their functions have been identified [47], including the 'base', which is composed of six separate AAA ATPase subunits that may function to unfold substrate proteins [47,76].

A second regulatory complex, PA28, also associates with the 20S proteasome *in vivo* [61,77] (see Chapter 9). PA28 (11S) is a 180-kDa multisubunit heterocomplex containing two polypeptides arranged in a single hexameric/heptameric ring [78-81]. PA28 binds to and activates the proteasome by opening the access pore of the core particle [82]. However, rather than increasing the processing of ubiquitinated substrates [61,77], PA28 functions to increase proteasomal processing of small peptides [83,84]. Moreover, PA28α, PA28β, as well as subunits of the transporter for antigen presentation (TAP) are upregulated by the immunostimulatory cytokine interferon-γ (IFN-γ) [46,85,86]. Together, these observations suggest a potential role for PA28 in modulating the processing of substrates presentation as antigens (Chapter 34).

Proteasome Function

Proteasome-mediated protein degradation is essential for a wide array of critical cellular processes, including immune surveillance, regulation, and quality control [47,48]. The continual turnover of proteins by the proteasome provides a rich source of antigenic peptides for presentation on major histocompatibility complex (MHC) class I molecules. Upon cleavage by the proteasome, the resulting 8-10 residue oligopeptides are subsequently translocated into the ER by TAP, where they are loaded onto newly synthesized MHC-1 molecules, and delivered to the cell surface [46]. Under normal conditions, all the antigens presented by MHC-I molecules are derived from cellular proteins and are viewed as self by the immune surveillance system. However, in virally infected cells, the display of antigens derived from the degradation of proteins of viral origin marks the cell for destruction by cytotoxic T cells [46]. As mentioned earlier, IFN-γ stimulation enhances antigen presentation by increasing expression of PA28 subunits and TAP [46,47,85,86]. Furthermore, IFN-γ promotes formation of a proteasome variant called the immunoproteasome. This is the result of IFNγ stimulated expression of an alternate set of catalytic β-subunits (LMP2, LMP7, and MECL-1) that are preferentially incorporated into newly assembled 20S particles [46,47]. Although the overall degradation activity remains unchanged, immunoproteasomes exhibit altered cleavage site specificities that lead to the generation of a different ensemble of oligopeptides, perhaps with more favorable for antigenic profiles (Chapter 14 and 34).

In addition to its role in immune surveillance, the proteasome also controls the levels of short-lived cell-cycle regulatory proteins. Progression through the cell cycle in eukaryotes is governed by variations in the activity of cyclin-dependent kinases (Cdks) [87]. Because Cdks themselves are quite stable, their periodic oscillations in activity are instead achieved by cyclical changes in the levels of positive (cyclins) and negative (Cdk inhibitors) regulatory factors. Specific cyclins accumulate at different times to activate Cdks, which then mediate passage through various cell-cycle checkpoints. Upon checkpoint passage, the cyclins are selectively ubiquitinated and rapidly degraded by the proteasome, thus inactivating the particular Cdk [87]. By contrast, Cdk inhibitors bind to and maintain the kinase in an inactive conformation, even in the presence of the activator cyclin [87]. In this case, targeted ubiquitination and degradation of the inhibitor by the proteasome relieves inhibition, allowing cyclin-mediated activation and cell-cycle progression to occur.

The UPS also regulates the levels of numerous transcription factors, tumor suppressors, and oncoproteins. Perhaps the most well characterized example is nuclear factor κB (NFκB), which is regulated by the proteasome at multiple levels. NFκB denotes a family of inducible transcription factors with relevant roles in immunity, inflammation, stress response, and development [88]. The best-known form is NFκB-1, a heterodimer of p50 and p65 (RelA). In resting cells, the p50/p65 dimer is maintained in an inactive complex in the cytosol through interaction with the inhibitory protein IκBα. Upon stimulation by any of a number of intra- and extracellular signals, IκBα becomes phosphorylated, ubiquitinated, and consequently degraded by the proteasome [88]. IκBα degradation liberates active p50/p65 dimers, which then translocate to the nucleus and activate expression of target genes. In another level of proteasome-mediated regulation, the p50 subunit itself is produced by the proteasome. *Via* limited co-translational proteolysis of the p50 precursor p105, proteasomal activity releases the N-terminal p50 moiety and the C-terminal remainder is degraded [89-91]. p50 generation from p105 is the first instance where a protein was shown to be discretely, rather than processively processed from either terminus by the ubiquitin-proteasome pathway without being completely destroyed. This processing related degradation event is believed to be mediated by a previously unknown endoproteolytic activity, which was only recently formally demonstrated for the proteasome [92]. The yeast NFκB relatives SPT23 and MGA2 are similarly processed by the proteasome *via* limited proteolysis [93,94]. Thus, the proteasome regulates NFκB-1 activity at two levels, both by controlling the amounts of the p50 subunit and the levels of the IκBα inhibitor. The proteasome also regulates (through more traditional degradation) the levels of several transcription factors (*e.g.,* the E2F family) [95], cellular (but not viral) forms of oncoproteins (Myc, Fos and Jun) [96-99], and tumor suppressors (p53 and PTEN) [100,101].

Finally, the proteasome is the primary protease involved in protein quality control, and, in this regard, is responsible for the degradation of misfolded and damaged proteins [52]. The role of UPS in the selective removal of damaged cytosolic and nuclear proteins has long been appreciated. In addition, it has recently become apparent that the cytosolic proteasome is also responsible for the majority of quality control degradation in the secretory pathway [102,103]. Given that the proteasome's role in the surveillance of protein folding and misfolding is at the heart of aggresome formation, this function will be discussed in greater detail below. In addition, while the aggresome formation process is essentially the same regardless of the substrate, initial studies characterizing this structure focused primarily on a polytopic integral membrane protein the cystic fibrosis transmembrane conductance regulator (CFTR) [2], and, as such, the remaining discussion will focus primarily on the fate of misfolded membrane proteins, with particular emphasis on CFTR [50].

Quality Control of Membrane Protein Folding

The cystic fibrosis transmembrane conductance regulator, a member of the ATP-Binding Cassette (ABC) transporter supergene family, is one of the best-characterized systems for studying integral membrane protein folding and quality control. CFTR is a complex multi-domain integral membrane protein composed of two cytosolic nucleotide-binding domains, two transmembrane domains, and a PKA-sensitive regulatory domain [104]. In addition to its role in chloride conductance [105-107], CFTR regulates the activity of several other critical transport systems in the apical membrane of epithelial cells [108,109,110], including bicarbonate secretion mediated by members of the SLC26 family of anion exchangers [111]. Disease-causing mutations in CFTR lead to loss of one or more of these activities attributable to a variety of molecular mechanisms, the most common of which is defective folding [112]. Several mutations, including the common ΔF508 mutation, result in misfolded CFTR proteins that are unable to properly traffic to their appropriate location in the apical membrane. Even for wild-type CFTR, folding and trafficking are inefficient, with only an estimated 30% of the wild type molecules ever achieving mature form in cultured cells [113,114,115] . Thus, most of the immature wild type (~70%) and nearly all of the immature

ΔF508 CFTR never reach the membrane but are instead detained by the ER quality control system and degraded.

Initial events delineating the folding and maturation of integral membrane proteins like CFTR occur in the ER. A particularly sensitive feature of ER quality control is its ability to ensure that prior to their being allowed to progress through the secretory pathway to their ultimate sub-cellular destination, nascent proteins adopt their specific native conformation. In general, correctly folded, processed and assembled proteins are transported, whereas terminally misfolded proteins and persistently unassembled subunits are retained and after retrograde translocation into the cytosol subsequently degraded by the proteasome [116,117]. ER-associated degradation (ERAD) describes the process by which cytosolic proteasomes gain access to and degrade misfolded membrane proteins [102,118,119].

ERAD is a multi-step process that is initiated by the recognition and retention of misfolded proteins in the ER (Chapter 13). Proteins recognized as improperly folded are targeted, in turn, for ubiquitination, dislocation from the membrane, deglycosylation, and finally de-ubiquitination and degradation by the 26S proteasome. Ultimate degradation of ER polypeptides destined for elimination was historically believed to occur within the ER itself. However, placing these critical and often inefficient folding reactions in such close proximity to (sequestered within the same organelle) and in competition with the proteolytic machinery presented a conceptual dilemma. Early clues from studies of the yeast ubiquitin conjugating enzyme Ubc6p strongly suggested the cytosolic ubiquitin proteasome system was involved in the elimination of mutant sec61-2, an integral membrane component of the ER protein translocation machinery [111]. Although the molecular mechanisms of several individual steps are still not well understood, CFTR has served as an informative model system for the elucidation of ERAD [120-123]. Consistent with earlier observations and evidence from several studies of CFTR degradation [122,124,125], the current model for dislocation of misfolded proteins from the ER involves retrograde transport of ubiquitinated substrates through the Sec61 translocon channel [126-128], [122]. However, the channel itself and the substrate ubiquitination machinery (e.g. Ubc6p and Ubc7p) are not sufficient for translocation. Additional components of the system include the lumenal chaperone BiP and Sec63p, a BiP interacting protein [111]. The extent to which extraction and degradation are linked is unclear. For example, in yeast, lumenal ERAD substrates such as carboxypeptidase Y appear to be completely extracted to the cytosol prior to initiation of degradation [129]. By contrast, studies suggest that for transmembrane ERAD substrates like CFTR, dislocation and proteolysis, while not explicitly coupled, are more tightly coordinated [130]. Synchronized extraction and degradation would be logically expected to inhibit the cytosolic accumulation of solvent exposed aggregation-prone hydrophobic transmembrane regions.

The undesired exposure of normally buried hydrophobic regions is typically the initial manifestation of a misfolded protein, whether soluble or transmembrane [131,132]. Left unattended, these exposed hydrophobic patches tend to self-associate, thus driving aggregation of the nonnative proteins resulting in the deposition of insoluble inclusions. Because such improperly folded structures are potentially toxic [25], cellular compartments have evolved quality control mechanisms to inhibit their misfolding and either repair or eliminate terminally misfolded proteins once formed. Molecular chaperones (*e.g.,* Hsp70, Hsp90, BiP; see also Chapter 10 and 19) as well as regulatory components of the proteasome (*e.g.*, PA700) rely on binding to the same or similar determinants on misfolded proteins, in most cases, to the surface-exposed hydrophobic regions [133,134]. Cytosolic Hsp70 is a nucleotide-regulated chaperone with specificity for short hydrophobic sequences of un- or partially-folded substrates. Hsp70 impacts the folding of nascent polypeptides as it goes through co-chaperone (HDJ1, BAG1) mediated substrate binding and release cycles. Hsp90 functions in concert with Hsp70 and, in an ATP-hydrolysis dependent manner, recognizes and interacts with a broad spectrum of substrates [121]. More recently, it has been suggested that later folding intermediates might represent the common feature of Hsp90 substrate recognition.

The ultimate fate of misfolded proteins (repair *versus* elimination) has been proposed to be dependant on the outcome of a kinetic competition between binding by molecular chaperones, leading to refolding, and binding by regulatory components of the proteasome, leading to degradation. Moreover, and in an added level of complexity, there appears to be significant cross-talk between the two systems. In addition to its role in preventing aggregation and promoting folding, the cytosolic molecular chaperone Hsp70 is also required for the ubiquitination and degradation of several proteins [135]. Likewise, as mentioned above, the proteasome regulatory complex PA700 exhibits chaperone-like activity *in vitro* [136]. In any case, when neither the salvage nor disposal systems are capable of dealing with the misfolded protein, the typical result is aggregation.

Evidence from both *in vivo* and *in vitro* biochemical studies have demonstrated that misfolded CFTR is prone to aggregation. Pharmacologic inhibition of the proteasome in cultured cells expressing ΔF508 CFTR leads to the formation of high molecular weight, detergent-insoluble complexes of ubiquitinated CFTR molecules [3,120]. In the absence of degradation machinery, CFTR forms similar complexes when expressed in a cell-free model of ERAD [125]. As visualized in cultured cells by immunocytochemistry, the misfolded CFTR that cannot be degraded accumulates as a large juxtanuclear inclusion localized to the centrosome and colocalized with proteasomal components of the UPS and relevant molecular chaperones [2,3]. The formation of these inclusions is consistent with an ER extraction process in which withdrawl from the membrane and degradation are not explicitly coupled [125,137]. Unless exquisitely coordinated with proteasomal activity, inhibition of the proteasome by pharmacological inhibitors such as lactacystin would not necessarily be expected to impair dislocation of the ubiquitinated protein. When extraction of CFTR is not quickly followed by or linked to proteolysis, exposure of the large hydrophobic transmembrane domains apparently drives the protein into a nonnative conformation that is recognized by packaging machinery and/or rapidly forms insoluble aggregates. These characteristic centrosomal inclusions of misfolded protein, enriched with quality control machinery, constitute the aggresome [3].

THE AGGRESOME

The two studies credited with originally describing aggresomes of misfolded CFTR [2,3] were not the first reported indications of the structure. Earlier, Vidair *et al.* observed an increase in the aggregation of insoluble, misfolded protein at or near the centrosome in cultured cells subjected to heat shock [138]. Also, Wojcik and coworkers found that in HeLa cells treated with proteasome inhibitors, pronounced juxtanuclear aggregates formed that were rich in ubiquitin and proteasomes [1,139]. Furthermore, treatment of cells with the microtubule-disrupting drug nocodazole blocked the juxtanuclear localization, resulting in cytoplasmic dispersal of the aggregates [1]. Similar observations were subsequently made in several additional cell types [140]. Wojcik concluded that the bulk of ubiquitin-dependent proteolysis occurs at a site termed the 'proteolysis center', and suggested that they represent an active cellular mechanism to concentrate the proteasome and its substrates at the same intracellular location [1,139].

Founded upon these early observations, a series of subsequent reports have painted a clearer picture of aggresome location, composition, structure, and possible function(s). The juxtanuclear aggresome is surrounded by the ER and adjacent to the Golgi, yet distinct from both organelles. Indeed, aggresome-like inclusions likely have been seen for years upon overexpression of heterologous proteins but were perhaps mistakenly classified as Golgi localization due to the limitations of light microscopy and the proximity of the structure to that organelle. Aggresomes are in fact located at the microtubule organizing center (MTOC) surrounding the centrioles, as evidenced by colocalization with the centrosomal marker γ-tubulin and ultrastructural analysis [2,3]. At the core of aggresomes is a concentrated mass of aggregated protein that is usually (but not always) ubiquitinated, surrounded by intermediate filaments, and enriched in quality control and degradation components. In addition to components of the UPS [141], including 20S proteasomes, the proteasomal activators PA700 and PA28 [2,139], and ubiquitin [1-3,139], a growing list of molecular chaperones have been described as components of the aggresome. These include Hsp70 [2], Hsp90 [2], the Hsp40 homologues Hdj1 and Hdj2, and the TCP-1 chaperonin [142]. The aggresome is also surrounded by the intermediate filament (IF) vimentin [3]. Normally, type III IFs like vimentin form extended networks throughout the cytoplasm, but coincident with aggresome formation, vimentin collapses to form a cage as the aggregates accumulate at the MTOC.

Mechanism of Aggresome Formation

The single large juxtanuclear aggresome (Figure 3) is actually formed by microtubule-dependent trafficking and assembly of multiple smaller, peripheral aggregates. Garcia-Mata *et al.* studied the dynamics of aggresome formation using an aggregation-prone 250-amino acid fragment of p115 fused to GFP (GFP-250). Time-lapse analysis in living cells showed that small aggregates of GFP-250 first form at the cell periphery and then travel to the MTOC, where they merge to form a single large inclusion [142]. Formation of the initial peripheral aggregates is highly specific, as different misfolded proteins coexpressed in the same cell aggregate into discrete homogeneous foci that are only later united by coincident trafficking to the MTOC [143]. As mentioned above, initial studies demonstrated that disruption of microtubules with nocodazole blocked aggresome formation, resulting in the persistent dispersal of small punctate aggregates [1-3]. More specifically, trafficking of aggregates to the MTOC exploits minus-end directed transport mediated by the dynein/dynactin motor complex, as overexpression of the dynactin inhibitor p50/dynamitin specifically inhibits aggregate trafficking to the MTOC [142]. Consistent with such a role in aggresome formation, cytoplasmic dynein has been shown to redistribute to aggresomes of misfolded ΔF508 CFTR [144]. The precise manner by which proteins destined for aggresome inclusion are recognized by the transport complex remains unknown. However, recent evidence suggests that specific 'aggresome determinants' exist in proteins that form aggresomes [145], distinguishing these aggregation-prone polypeptides from closely related sequences that lack this property. It is possible that these same determinants represent a transport signature that is recognized by the microtubule motor machinery or adapter proteins associated with retrograde trafficking of aggresomal cargo.

Beyond UPS and microtubule machinery, a number of other proteins have been demonstrated to mediate the formation of aggresomes, further expanding our understanding of the key players in this general response to inadequate proteasomal activity. For example, Ataxin 3 (AT3) is a polyglutamine containing protein with ubiquitin binding and hydrolytic activity, which has been implicated in polyglutamine expansion-mediated neurodegeneration. RNAi knockdown studies have demonstrated a critical role for AT3 in the formation and maintenance of ΔF508 CFTR aggresomes. Moreover, this function is dependent upon the ability of AT3 to bind and catalyze the removal of polyubiquitin [146]. Similarly, RNAi has been used to reveal a role for valosin-containing protein (VCP) in aggresome formation, perhaps mediated by its interaction with proteasome substrates prior to their degradation [147].

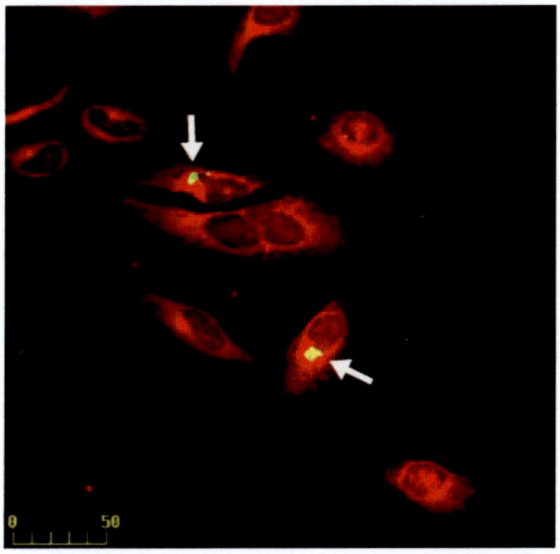

Figure 3. Aggresome formation at the centrosome. Transiently transfected Hela cells expressing GFP fused to human huntingtin exon 1 containing polyglutamine expansion (Htn 103Q), (gift of Dr. E. Signer) were stained with rabbit polyclonal anti pericentrin, a centrosomal marker protein (gift of Dr. S. Doxsey) followed by rhodamine-labeled goat anti-rabbit IgG and imaged by fluorescence microscopy. Arrows indicate juxtanuclear aggresomes formed at the centrosome in expressing cells. The scale bar indicates size in μm.

In addition to ΔF508 CFTR, an increasing spectrum of proteins have been found to form aggresomes when their degradation is impaired. As with ΔF508 CFTR, in a number of medically relevant cases disease-causing mutations increase aggresome formation. In particular, many proteins implicated in a variety of neurodegenerative conditions have been shown to form centrosomal aggregates under various conditions [148-154]. For example, the membrane protein presenilin 1 (PS1) forms aggresomes in cultured cells in the absence of proteasome activity, and formation is amplified by the early-onset familial Alzheimer's disease (FAD) mutation A246E [3]. Similar results were seen with the P23H mutant form of rhodopsin, which is associated with autosomal dominant retinitis pigmentosa (ADRP) [155,156], with peripheral myelin protein 22 (PMP22), an inefficiently folded Schwann cell glycoprotein associated with a number of heritable peripheral neuropathies [157], and with a polyglutamine expansion in the androgen receptor (AR) associated with spinobulbar muscular atrophy (SBMA) [158]. These structures have also been shown to form from the aggregation of misfolded soluble proteins, as well as from extracted membrane proteins. In addition to the 'synthetic' GFP-250 substrate discussed above [142] aggresome formation has also been shown for secreted (surfactant protein C) [159], cytosolic (huntingtin exon I, Htn) [160], mitochondrial (superoxide dismutase 1, SOD-1) [161], and cytoskeletal/filamentous (cytokeratin) [162] proteins. Interestingly, Htn proteins with pathological polyglutamine repeats (Htn-51Q or Htn-83Q) form aggresomes (Figure 3), but the non-pathological repeat (Htn-25Q) does not [163]. Similarly, mutations in SOD-1 associated with familial amyotrophic lateral sclerosis (FALS) have been reported to give rise to aggresomes, while wild type SOD-1 does not [164].

Functional Significance of Aggresomes

Centrosomal association of quality control components such as the UPS and chaperones is not strictly a function of aggresome formation, but also a general feature of resting cells [2,165-168]. 20S proteasomes, PA700, PA28, ubiquitin, Hsp70, and Hsp90 have all been shown by immunocytochemistry and biochemistry to co-localize and co-purify with centrosomes in several different cultured cell lines under basal conditions. In addition, purified centrosome-associated proteasomes are active in degrading ubiquitinated proteins and proteasome-specific peptide substrates, and demonstrate the same ATP-dependence and inhibitor profile as soluble proteasomes [165]. When combined with overexpression of mutant CFTR, pharmacological inhibition of the proteasome results in striking recruitment of cytosolic proteasome components to the centrosomal inclusions [2]. This recruitment is accompanied by appreciable expansion of the centrosome and redistribution of proteasomes, γ-tubulin, and chaperones to an insoluble fraction. Therefore, it appears the centrosomal proteolysis centers normally function to concentrate and recruit components of the ubiquitin-proteasome and chaperone systems to balance protein folding and degradation [1,2,139,165] (Figure 4). Thus, aggresomes likely represent the endpoint of a normal cellular function gone awry resulting from UPS failure when the burden of misfolded protein cannot be adequately handled.

Such centrosomal proteolysis centers also play a crucial role in antigen processing and, ironically, productive viral assembly. The MTOC is one of the major sites of proteasomal degradation of viral proteins for antigen presentation, along with nuclear promyelocytic leukemia oncogenic domains (PML bodies or PODs) [169-171]. Consistent with this, PML bodies are also enriched with components of the ubiquitin-proteasome system, especially

PA28 and immunoproteasomes [169,172]. Increased delivery of viral proteins to the MTOC enhances their processing into MHC-I linked antigens, as seen with a human pappiloma virus 16 (HPV-16) E7-γ-tubulin fusion protein [173]. In addition, inflammatory stimulation increases transient aggregation of ubiquitinated proteins at the MTOC [174], suggesting a regulated delivery system to concentrate viral proteins at the site of their processing. However, this delivery system appears to have been corrupted in some cases for production of viruses rather than degradation. So-called viral factories are high concentrations of viral proteins centered at the MTOC to facilitate viral assembly, which require minus-end directed micotubular trafficking [175-177]. Consistent with such a function for the centrosomal region, Mre11, a component of the Mre11/Rad50/Nbs1 DNA repair complex, is sequestered as a centrosomal aggregate in response to expression of the adenovirus 5 (Ad5) E4orf3 gene product. This is thought to directly result in inactivation of the cellular repair complex during Ad5 infection [178], further emphasizing the importance of this intracellular region to the replication cycle of viral particles. The similarity between these juxtanuclear assembly centers and aggresomes suggests viral factories represent the subversion of a normal cellular process designed to deliver viral antigens to the MTOC for antigen processing.

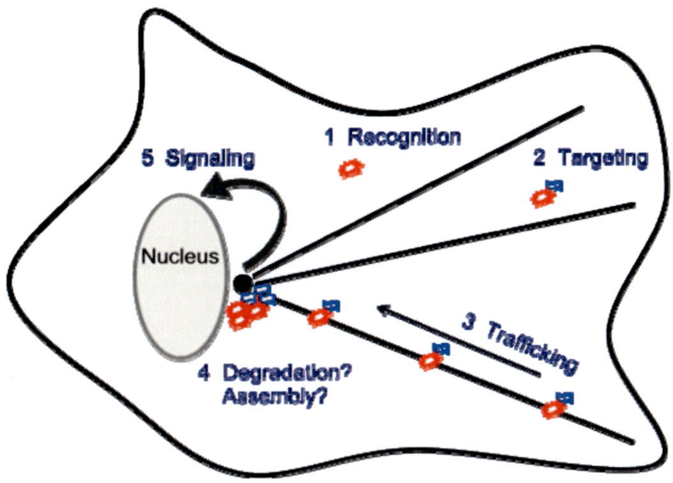

Figure 4. Schematic of aggresome formation in a eukaryotic cell. The nucleus (gray) and centrosome (black circle) are as indicated. Black lines represent microtubules radiating from the microtubule organizing center to the cell periphery. The schematic is drawn to highlight several important questions regarding the aggresome described herein and elsewhere: 1. How is the substrate (depicted in red) recognized and what proteins are involved? 2. How is the substrate targeted for centrosomal localization? 3. Does trafficking to the centrosome serve to simply deposit aggregated substrates, or also to concentrate them, thereby increasing their exposure/susceptibility to the centrosome-associated proteasomes? 4. Is assembly into vimentin-surrounded juxtanuclear centrosomal inclusions an active, protein-mediated process, or simply a cellular response to aggregation? 5. What signals arise from the processes occurring at this location, how are they interpreted, and what are their targets?

Conclusions

Deposition of misfolded protein into aggresome-like cytoplasmic inclusions is a common cytopathological feature in a number of neurodegenerative diseases. The majority of cases are idiopathic, but a number of familial cases can be linked to toxic gain of function mutations [179-181], which have also been shown to increase aggresome formation in cultured cells, *e.g.,* Htn and SOD-1 [182,183]. Protein aggregation in general, and aggresome formation in particular, impairs function of the UPS, as has been demonstrated with ΔF508 CFTR, a pathological-repeat length Htn variant [184], and paired helical filaments of the microtubule binding protein Tau [185]. This inhibition suggests a possible mechanism directly linking aggregation to cell death due to the central role of the ubiquitin-proteasome pathway in cell regulation. However, current research has not reached a consensus on whether these inclusions are a cause or a consequence of the pathology. In fact, several lines of recent evidence suggest that aggresome formation may even be a protective response. In this regard, blocking aggresome formation exacerbates the toxicity of aggregated AR [186]. Also, aggresomes formed by α-synuclein and synphilin-1 have been reported to not directly result in induction of apoptosis [187]. Moreover, in another culture model relevant to Parkinson's disease, dopamine was shown to enhance aggregate formation without decreasing cell viability, providing further evidence for aggresome-mediated cytoprotection [188], although this point remains the focus of ongoing debate [189]. The aggresome may even play an unexpected physiologic role in regulating the function of certain proteins such as iNOS [190]. Thus, conditions used to describe and unravel mechanistic and functional aspects of aggresome formation, where overt centrosomal aggregation is observed may mask more subtle and yet more relevant roles for this unique structure. Clearly, additional work will be required to convincingly address these and other significant questions regarding aggresome biology.

Acknowledgements

The authors thank Drs. Philip J. Thomas and George N. DeMartino for helpful discussions.

References

[1] Wójcik C, Schroeter D, Wilk S, Lamprecht J, Paweleta N: Ubiquitin-mediated proteolysis centers in HeLa cells: indication from studies of an inhibitor of the chymotrypsin-like activity of the proteasome. *Eur J Cell Biol* 1996, 71:311-318.

[2] Wigley WC, Fabunmi RP, Lee MG, Marino CR, Muallem S, DeMartino GN, Thomas PJ: Dynamic association of proteasomal machinery with the centrosome. *J Cell Biol* 1999, 145:481-490.

[3] Johnston JA, Ward CL, Kopito RR: Aggresomes: A Cellular Response to Misfolded Proteins. *J Cell Biol* 1998, 143:1883-1898.

[4] Crick FH: On protein synthesis. *Symp Soc Exp Biol* 1958, 12:138-163.

[5] Gilmartin GM: Eukaryotic mRNA 3' processing: a common means to different ends. *Genes Dev* 2005, 19:2517-2521.

[6] Grishok A: RNAi mechanisms in Caenorhabditis elegans. *FEBS Lett* 2005, 579:5932-5939.

[7] Moore MJ: From birth to death: the complex lives of eukaryotic mRNAs. *Science* 2005, 309:1514-1518.

[8] Anfinsen CB, Redfield RR: Protein structure in relation to function and biosynthesis. *Adv Protein Chem* 1956, 48:1-100.

[9] Sela M, Anfinsen CB, Harrington WF: The correlation of ribonuclease activity with specific aspects of tertiary structure. *Biochim Biophys Acta* 1957, 26:502-512.

[10] Levinthal C: Are there pathways for protein folding? *J Chim Phys* 1968, 65:44-45.

[11] Cuatrecasas P, Taniuchi H, Anfinsen CB: The structural basis of the catalytic function of staphylococcal nuclease. *Brookhaven Symp Biol* 1968, 21:172-200.

[12] Nozaki Y, Tanford C: The solubility of amino acids and two glycine peptides in aqueous ethanol and dioxane solutions. Establishment of a hydrophobicity scale. *J Biol Chem* 1971, 246:2211-2217.

[13] Moelbert S, Emberly E, Tang C: Correlation between sequence hydrophobicity and surface-exposure pattern of database proteins. *Protein Sci* 2004, 13:752-762.

[14] Dill KA: Dominant Forces in Protein Folding. *Biochemistry* 1990, 29:7133.

[15] Matsumura M, Becktel WJ, Matthews BW: Hydrophobic stabilization in T4 lysozyme determined directly by multiple substitutions of Ile 3. *Nature* 1988, 334:406-410.

[16] Fersht AR: Conformational equilibria in -and -chymotrypsin. The energetics and importance of the salt bridge. *J Mol Biol* 1972, 64:497-509.

[17] Anderson DE, Becktel WJ, Dahlquist FW: pH-induced denaturation of proteins: a single salt bridge contributes 3-5 kcal/mol to the free energy of folding of T4 lysozyme. *Biochemistry* 1990, 29:2403-2408.

[18] Dinner AR, Sali A, Smith LJ, Dobson CM, Karplus M: Understanding protein folding via free-energy surfaces from theory and experiment. *Trends Biochem Sci* 2000, 25:331-339.

[19] Pace CN, Shirley BA, McNutt M, Gajiwala K: Forces contributing to the conformational stability of proteins. *Faseb J* 1996, 10:75-83.

[20] Religa TL, Markson JS, Mayor U, Freund SM, Fersht AR: Solution structure of a protein denatured state and folding intermediate. *Nature* 2005, 437:1053-1056.

[21] Brady GP, Sharp KA: Entropy in protein folding and in protein-protein interactions. *Curr Opin Struct Biol* 1997, 7:215-221.

[22] Karpusas M, Baase WA, Matsumura M, Matthews BW: Hydrophobic packing in T4 lysozyme probed by cavity-filling mutants. *Proc Natl Acad Sci U S A* 1989, 86:8237-8241.

[23] King J, Haase-Pettingell C, Robinson AS, Speed M, Mitraki A: Thermolabile folding intermediates: inclusion body precursors and chaperonin substrates. [Review]. *FASEB Journal* 1996, 10:57-66.

[24] Meacham GC, Lu Z, King S, Sorscher E, Tousson A, Cyr DM. The Hdj-2Hsc70 chaperone pair facilitates early steps in CFTR biogenesis. EMBO J. 1999, 18:1492-1505.

[25] Thomas PJ, Qu B-H, Pedersen PL: Defective protein folding as a basis of human disease. *TIBS* 1995, 20:456-459.

[26] Forman MS, Trojanowski JQ, Lee VM: Neurodegenerative diseases: a decade of discoveries paves the way for therapeutic breakthroughs. *Nat Med* 2004, 10:1055-1063.

[27] Ross CA, Poirier MA: Protein aggregation and neurodegenerative disease. *Nat Med* 2004, 10 Suppl:S10-S17.

[28] Gregersen N, Bolund L, Bross P: Protein misfolding, aggregation, and degradation in disease. *Methods Mol Biol* 2003, 232:3-16.

[29] True HL, Berlin I, Lindquist SL: Epigenetic regulation of translation reveals hidden genetic variation to produce complex traits. *Nature* 2004, 431:184-187.

[30] Si K, Lindquist S, Kandel ER: A neuronal isoform of the aplysia CPEB has prion-like properties. *Cell* 2003, 115:879-891.

[31] Shorter J, Lindquist S: Prions as adaptive conduits of memory and inheritance. *Nat Rev Genet* 2005, 6:435-450.

[32] Serio TR, Lindquist SL: [PSI+], SUP35, and chaperones. *Adv Protein Chem* 2001, 57:335-366.

[33] Wickner RB, Edskes HK, Roberts BT, Baxa U, Pierce MM, Ross ED, Brachmann A: Prions: proteins as genes and infectious entities. *Genes Dev* 2004, 18:470-485.

[34] Ross ED, Minton A, Wickner RB: Prion domains: sequences, structures and interactions. *Nat Cell Biol* 2005, 7:1039-1044.

[35] Asano N, Ishii S, Kizu H, Ikeda K, Yasuda K, Kato A, Martin OR, Fan JQ: In vitro inhibition and intracellular enhancement of lysosomal alpha-galactosidase A activity in Fabry lymphoblasts by 1-deoxygalactonojirimycin and its derivatives. *Eur J Biochem* 2000, 267:4179-4186.

[36] Fan JQ, Ishii S, Asano N, Suzuki Y: Accelerated transport and maturation of lysosomal alpha-galactosidase A in Fabry lymphoblasts by an enzyme inhibitor. *Nat Med* 1999, 5:112-115.

[37] Gong Q, Jones MA, Zhou Z: Mechanisms of pharmacological rescue of trafficking defective hERG mutant channels in human long QT syndrome. *J Biol Chem* 2006, 281:4069-4074.

[38] Delisle BP, Anderson CL, Balijepalli RC, Anson BD, Kamp TJ, January CT: Thapsigargin selectively rescues the trafficking defective LQT2 channels G601S and F805C. *J Biol Chem* 2003, 278:35749-35754.

[39] Rajamani S, Anderson CL, Anson BD, January CT: Pharmacological rescue of human K(+) channel long-QT2 mutations: human ether-a-go-go-related gene rescue without block. *Circulation* 2002, 105:2830-2835.

[40] Ficker E, Obejero-Paz CA, Zhao S, Brown AM: The binding site for channel blockers that rescue misprocessed human long QT syndrome type 2 ether-a-gogo-related gene (HERG) mutations. *J Biol Chem* 2002, 277:4989-4998.

[41] Morgan C, Colombres M, Nunez MT, Inestrosa NC: Structure and function of amyloid in Alzheimer's disease. *Prog Neurobiol* 2004, 74:323-349.

[42] Volles MJ, Lansbury PT, Jr: Zeroing in on the pathogenic form of alpha-synuclein and its mechanism of neurotoxicity in Parkinson's disease. *Biochemistry* 2003, 42:7871-7878.

[43] Lansbury PT, Jr: Structural neurology: are seeds at the root of neuronal degeneration? *Neuron* 1997, 19:1151-1154.

[44] Lansbury PT, Jr.: Inhibition of amyloid formation: a strategy to delay the onset of Alzheimer's disease. *Curr Opin Chem Biol* 1997, 1:260-267.

[45] Johnson SM, Wiseman RL, Sekijima Y, Green NS, Adamski-Werner SL, Kelly JW: Native state kinetic stabilization as a strategy to ameliorate protein misfolding diseases: a focus on the transthyretin amyloidoses. *Acc Chem Res* 2005, 38:911-921.

[46] Rock KL, Goldberg AL: Degradation of cell proteins and the generation of MHC class I- presented peptides. *Annu Rev Immunol* 1999, 17:739-779.

[47] Voges D, Zwickl P, Baumeister W: The 26S proteasome: a molecular machine designed for controlled proteolysis. *Annu Rev Biochem* 1999, 68:1015-1068.

[48] Hershko A, Ciechanover A: The ubiquitin system. *Annu Rev Biochem* 1998, 67:425-479.

[49] Schwartz AL, Ciechanover A: The ubiquitin-proteasome pathway and pathogenesis of human diseases. *Annu Rev Med* 1999, 50:57-74.

[50] Ma C-P, Vu JH, Proske RJ, Slaughter CA, DeMartino GN: Identification, purification, and characterization of a high molecular weight ATP-dependent activator (PA700) of the 20S proteasome. *J Biol Chem* 1994, 269:3539-3547.

[51] Weissman AM: Themes and variations on ubiquitylation. *Nat Rev Mol Cell Biol* 2001, 2:169-178.

[52] Coux O, Tanaka K, Goldberg AL: Structure and functions of the 20S and 26S proteasomes. *Annu Rev Biochem* 1996, 65:801-847.

[53] Baumeister W, Walz J, Zuhl F, Seemuller E: The proteasome: paradigm of a self-compartmentalizing protease. *Cell* 1998, 92:367-380.

[54] Lowe J, Stock D, Jap B, Zwickl P, Baumeister W, Huber R: Crystal structure of the 20S proteasome from the archaeon T. acidophilum at 3.4 A resolution [see comments]. *Science* 1995, 268:533-539.

[55] Groll M, Ditzel L, Lowe J, Stock D, Bochtler M, Bartunik HD, Huber R: Structure of 20S proteasome from yeast at 2.4 A resolution. *Nature* 1997, 386:463-471.

[56] Heinemeyer W, Fischer M, Krimmer T, Stachon U, Wolf DH: The active sites of the eukaryotic 20 S proteasome and their involvement in subunit precursor processing. *J Biol Chem* 1997, 272:25200-25209.

[57] Dick TP, Nussbaum AK, Deeg M, Heinemeyer W, Groll M, Schirle M, Keilholz W, Stevanovic S, Wolf DH, Huber R, Rammensee HG, Schild H: Contribution of proteasomal beta-subunits to the cleavage of peptide substrates analyzed with yeast mutants. *J Biol Chem* 1998, 273:25637-25646.

[58] Wenzel T, Baumeister W: Conformational constraints in protein degradation by the 20S proteasome. *Nat Struct Biol* 1995, 2:199-204.

[59] Lowe J, Stock D, Jap B, Zwickl P, Baumeister W, Huber R: Crystal structure of the 20S proteasome from the archaeon T. acidophilum at 3.4 A resolution [see comments]. *Science* 1995, 268:533-539.

[60] Lowe J, Stock D, Jap B, Zwickl P, Baumeister W, Huber R: Crystal structure of the 20S proteasome from the archaeon T. acidophilum at 3.4 A resolution [see comments]. *Science* 1995, 268:533-539.
[61] Ma C-P, Slaughter CA, DeMartino GN: Identification, purification, and characterization of a protein activator (PA28) of the 20 S proteasome (macropain). *J Biol Chem* 1992, 267:10515-10523.
[62] Yoshimura T, Kameyama K, Takagi T, Ikai A, Tokunaga F, Koide T, Tanahashi N, Tamura T, Cejka Z, Baumeister W, Tanaka K, Ichihara A: Molecular characterization of the "26S" proteasome complex from rat liver. *J Struct Biol* 1993, 111:200-211.
[63] Adams GM, Falke S, Goldberg AL, Slaughter CA, DeMartino GN, Gogol EP: Structural and functional effects of PA700 and modulator protein on proteasomes. *J Mol Biol* 1997, 273:646-657.
[64] Deveraux Q, Ustrell V, Pickart C, Rechsteiner M: A 26 S protease subunit that binds ubiquitin conjugates. *J Biol Chem* 1994, 269:7059-7061.
[65] Lam YA, Lawson TG, Velayutham M, Zweier JL, Pickart CM: A proteasomal ATPase subunit recognizes the polyubiquitin degradation signal. *Nature* 2002, 416:763-767.
[66] Yao T, Cohen RE: A cryptic protease couples deubiquitination and degradation by the proteasome. *Nature* 2002, 419:403-407.
[67] Verma R, Aravind L, Oania R, McDonald WH, Yates JR, III, Koonin EV, Deshaies RJ: Role of Rpn11 metalloprotease in deubiquitination and degradation by the 26S proteasome. *Science* 2002, 298:611-615.
[68] DeMartino GN, Moomaw CR, Zagnitko OP, Proske RJ, Chu-Ping M, Afendis, SJ, Swaffield JC, Slaughter CA: PA700, an ATP-dependent activator of the 20 S proteasome, is an ATPase containing multiple members of a nucleotide-binding protein family. *J Biol Chem* 1994, 269:20878-20884.
[69] Strickland E, Hakala K, Thomas PJ, DeMartino GN: Recognition of misfolding proteins by PA700, the regulatory subcomplex of the 26 S proteasome. *J Biol Chem* 2000, 275:5565-5572..
[70] Liu CW, Millen L, Roman TB, Xiong H, Gilbert HF, Noiva R, DeMartino GN, Thomas PJ: Conformational remodeling of proteasomal substrates by PA700, the 19 S regulatory complex of the 26 S proteasome. *J Biol Chem* 2002, 277:26815-26820.
[71] Strickland E, Hakala K, Thomas PJ, DeMartino GN: Recognition of misfolding proteins by PA700, the regulatory subcomplex of the 26 S proteasome. *J Biol Chem* 2000, 275:5565-5572.
[72] Braun BC, Glickman M, Kraft R, Dahlmann B, Kloetzel PM, Finley D, Schmidt M: The base of the proteasome regulatory particle exhibits chaperone-like activity. *Nat Cell Biol* 1999, 1:221-226.
[73] Liu CW, Millen L, Roman TB, Xiong H, Gilbert HF, Noiva R, DeMartino GN, Thomas PJ: Conformational remodeling of proteasomal substrates by PA700, the 19 S regulatory complex of the 26 S proteasome. *J Biol Chem* 2002, 277:26815-26820.
[74] Lowe J, Stock D, Jap B, Zwickl P, Baumeister W, Huber R: Crystal structure of the 20S proteasome from the archaeon T. acidophilum at 3.4 A resolution [see comments]. *Science* 1995, 268:533-539.

[75] Unno M, Mizushima T, Morimoto Y, Tomisugi Y, Tanaka K, Yasuoka N, Tsukihara T: The structure of the mammalian 20S proteasome at 2.75 A resolution. *Structure (Camb)* 2002, 10:609-618.

[76] Liu CW, Millen L, Roman TB, Xiong H, Gilbert HF, Noiva R, DeMartino GN, Thomas PJ: Conformational remodeling of proteasomal substrates by PA700, the 19 S regulatory complex of the 26 S proteasome. *J Biol Chem* 2002, 277:26815-26820.

[77] Dubiel W, Pratt G, Ferrell K, Rechsteiner M: Purification of an 11S regulator of the multicatalytic protease. *J Biol Chem* 1992, 267:22369-22377.

[78] Song X, Mott JD, von Kampen J, Pramanik B, Tanaka K, Slaughter CA, DeMartino GN: A Model for the Quaternary Structure of the Proteasome Activator PA28. *J Biol Chem* 1996, 271:26410-26417.

[79] Ahn K, Erlander M, Leturcq D, Peterson PA, Früh K, Yang Y: In *Vivo* Characterization of the Proteasome Regulator PA28. *J Biol Chem* 1996, 271:18237-18242.

[80] Whitby FG, Masters EI, Kramer L, Knowlton JR, Yao Y, Wang CC, Hill CP: Structural basis for the activation of 20S proteasomes by 11S regulators. *Nature* 2000, 408:115-120.

[81] Rechsteiner M, Realini C, Ustrell V: The proteasome activator 11 S REG (PA28) and class I antigen presentation. *Biochem J* 2000, 345 Pt 1:1-15.

[82] Whitby FG, Masters EI, Kramer L, Knowlton JR, Yao Y, Wang CC, Hill CP: Structural basis for the activation of 20S proteasomes by 11S regulators. *Nature* 2000, 408:115-120.

[83] Dick T, Ruppert T, Groettrup M, Kloetzel P-M, Kuehn L, Koszinowski UH, Stevanovic S, Schild H, Rammensee H-G: Coordinated dual cleavage induced by the proteasome regulator PA28 lead to dominant MHC ligands. *Cell* 1996, 86:253-262.

[84] Groettrup M, Soza A, Eggers.M., Kuehn L, Dick TP, Schild H, Rammensee H-G, Koszinowski UH, Kloetzel P-M: A role for the proteasome regulator PA28a in antigen presentation. *Nature* 1996, 381:166-168.

[85] Realini C, Dubiel W, Pratt G, Ferrell K, Rechsteiner M: Molecular cloning and expression of a g-interferon-inducible activator of the multicatalytic protease. *J Biol Chem* 1994, 269:20727-20732.

[86] Ahn JY, Tanahashi N, Akiyama K, Hisamatsu H, Noda C, Tanaka K, Chung CH, Shibmara N, Willy PJ, Mott JD, Slaughter CA, DeMartino GN: Primary structures of two homologous subunits of PA28, a g-inrerferon-inducible protein activator of the 20S proteasome. *FEBS Lett* 1995, 366:37-42.

[87] Johnson DG, Walker CL: Cyclins and cell cycle checkpoints. *Annu Rev Pharmacol Toxicol* 1999, 39:295-312.

[88] Baldwin AS, Jr.: The NF-kappa B and I kappa B proteins: new discoveries and insights. *Annu Rev Immunol* 1996, 14:649-683.

[89] Palombella VJ, Rando OJ, Goldberg AL, Maniatis T: The ubiquitin-proteasome pathway is required for processing the NF-kappa B1 precursor protein and the activation of NF-kappa B. *Cell* 1994, 78:773-785.

[90] Orian A, Whiteside S, Israel A, Stancovski I, Schwartz AL, Ciechanover A: Ubiquitin-mediated processing of NF-kappa B transcriptional activator precursor p105. Reconstitution of a cell-free system and identification of the ubiquitin-carrier protein,

E2, and a novel ubiquitin-protein ligase, E3, involved in conjugation. *J Biol Chem* 1995, 270:21707-21714.

[91] Lin L, DeMartino GN, Greene WC: Cotranslational biogenesis of NF-kappaB p50 by the 26S proteasome. *Cell* 1998, 92:819-828.

[92] Liu CW, Corboy MJ, DeMartino GN, Thomas PJ: Endoproteolytic activity of the proteasome. *Science* 2003, 299:408-411.

[93] Hoppe T, Matuschewski K, Rape M, Schlenker S, Ulrich HD, Jentsch S: Activation of a membrane-bound transcription factor by regulated ubiquitin/proteasome-dependent processing. *Cell* 2000, 102:577-586.

[94] Rape M, Hoppe T, Gorr I, Kalocay M, Richly H, Jentsch S: Mobilization of processed, membrane-tethered SPT23 transcription factor by CDC48(UFD1/NPL4), a ubiquitin-selective chaperone. *Cell* 2001, 107:667-677.

[95] Campanero MR, Flemington EK: Regulation of E2F through ubiquitin-proteasome-dependent degradation: stabilization by the pRB tumor suppressor protein. *Proc Natl Acad Sci U S A* 1997, 94:2221-2226.

[96] Bonvini P, Nguyen P, Trepel J, Neckers LM: In vivo degradation of N-myc in neuroblastoma cells is mediated by the 26S proteasome. *Oncogene* 1998, 16:1131-1139.

[97] Tsurumi C, Ishida N, Tamura T, Kakizuka A, Nishida E, Okumura E, Kishimoto T, Inagaki M, Okazaki K, Sagata N, .: Degradation of c-Fos by the 26S proteasome is accelerated by c-Jun and multiple protein kinases. *Mol Cell Biol* 1995, 15:5682-5687.

[98] Stancovski I, Gonen H, Orian A, Schwartz AL, Ciechanover A: Degradation of the proto-oncogene product c-Fos by the ubiquitin proteolytic system in vivo and in vitro: identification and characterization of the conjugating enzymes. *Mol Cell Biol* 1995, 15:7106-7116.

[99] Jariel-Encontre I, Pariat M, Martin F, Carillo S, Salvat C, Piechaczyk M: Ubiquitinylation is not an absolute requirement for degradation of c-Jun protein by the 26 S proteasome. *J Biol Chem* 1995, 270:11623-11627.

[100] Maki CG, Huibregtse JM, Howley PM: In vivo ubiquitination and proteasome-mediated degradation of p53(1). *Cancer Res* 1996, 56:2649-2654.

[101] Torres J, Pulido R: The tumor suppressor PTEN is phosphorylated by the protein kinase CK2 at its C terminus. Implications for PTEN stability to proteasome-mediated degradation. *J Biol Chem* 2001, 276:993-998.

[102] Werner ED, Brodsky JL, McCracken AA: Proteasome-dependent endoplasmic reticulum-associated protein degradation: an unconventional route to a familiar fate. *Proc Natl Acad Sci USA* 1996, 13797-137801.

[103] Plemper RK, Wolf DH: Endoplasmic reticulum degradation. Reverse protein transport and its end in the proteasome. *Mol Biol Rep* 1999, 26:125-130.

[104] Riordan JR, Rommens JM, Kerem B-S, Alon N, Rozmahel R, Grzelczak Z, Zielenski J, Lok S, Plavsic N, Chou J-L, Drumm ML, Iannuzzi MC, Collins FS, Tsui L-C: Identification of the cystic fibrosis gene: cloning and characterization of complementary DNA. *Science* 1989, 245:1066-1073.

[105] Anderson MP, Gregory RJ, Thompson S, Souza DW, Paul S, Mulligan RC, Smith AE, Welsh MJ: Demonstration that CFTR is a chloride channel by alteration of its anion selectivity. *Science* 1991, 253:202-205.

[106] Bear CE, Duguay F, Naismith AL, Kartner N, Hanrahan JW, Riordan JR: Cl- channel activity in Xenopus oocytes expressing the cystic fibrosis gene. *J Biol Chem* 1991, 266:19142-19145.

[107] Sheppard DN, Welsh MJ: Structure and function of the CFTR chloride channel. *Physiol Rev* 1999, 79:S23-S45.

[108] Lee MG, Wigley WC, Zeng W, Noel LE, Marino CR, Thomas PJ, Muallem S: Regulation of Cl^-/HCO_3^- exchange by cystic fribrosis transmembrane conductance regulator expressed in NIH 3T3 and HEK 293 cells. *J BiolChem* 1999, 274:3414-3421.

[109] Schwiebert EM, Benos DJ, Egan ME, Stutts MJ, Guggino WB. CFTR Is a Conductance Regulator as well as a Chloride Channel. *Physiol Rev* 1999, 79:S145-S166.

[110] Choi JY, Muallem D, Kiselyov K, Lee MG, Thomas PJ, Muallem S: Aberrant CFTR-dependent. *Nature* 2001, 410:94-97.

[111] Ko SB, Shcheynikov N, Choi JY, Luo X, Ishibashi K, Thomas PJ, Kim JY, Kim KH, Lee MG, Naruse S, Muallem S: A molecular mechanism for aberrant CFTR-dependent HCO(3)(-) transport in cystic fibrosis. *EMBO J* 2002, 21:5662-5672.

[112] Thomas PJ, Ko YH, Pedersen PL: Altered protein folding may be the molecular basis of most cases of cystic fibrosis. *FEBS Lett* 1992, 312:7-9.

[113] Cheng SH, Gregory RJ, Marshall J, Paul S, Souza DW, White GA, O'Riordan CR, Smith AE: Defective intracellular transport and processing of CFTR is the molecular basis of most cystic fibrosis. *Cell* 1990, 63:827-834.

[114] Lukacs GL, Mohamed A, Kartner N, Chang X-B, Riordan JR, Grinstein S: Conformational maturation of CFTR but not its mutant counterpart (F508) occurs in the endoplasmic reticulum and requires ATP. *EMBO J* 1994, 13:6076-6086.

[115] Ward CL, Kopito RR: Intracellular turnover of cystic fibrosis transmembrane conductance regulator. *J Biol Chem* 1994, 269:25710-25718.

[116] Ellgaard L, Molinari M, Helenius A: Setting the standards: quality control in the secretory pathway. *Science* 1999, 286:1882-1888.

[117] Plemper RK, Wolf DH: Endoplasmic reticulum degradation. Reverse protein transport and its end in the proteasome. *Mol Biol Rep* 1999, 26:125-130.

[118] Plemper RK, Wolf DH: Retrograde protein translocation: ERADication of secretory proteins in health and disease. *Trends Biochem Sci* 1999, 24:266-270.

[119] Sommer T, Wolf DH: Endoplasmic reticulum degradation: reverse protein flow of no return. *FASEB J* 1997, 11:1227-1233.

[120] Ward CL, Omura S, Kopito RR: Degragation of CFTR by the ubiquitin-proteasome pathway. *Cell* 1995, 83:121-127.

[121] Jensen TJ, Loo MA, Pind S, Williams DB, Goldberg AL, Riordan JR: Multiple proteolytic systems, including the proteasome, contribute to CFTR processing. *Cell* 1995, 83:129-135.

[122] Bebok Z, Mazzochi C, King SA, Hong JS, Sorscher EJ. The mechanism underlying cystic fibrosis transmembrane conductance regulator transport from the endoplasmic

reticulum to the proteasome includes Sec61beta and a cytosolic deglycosylated intermediary. *J Biol Chem* 1998, 273:29873-29878.

[123] Xiong X, Bragin A, Widdicombe JH, Cohn J, Skach WR: Structural cues involved in endoplasmic reticulum degradation of G85E and G91R mutant cystic fibrosis transmembrane conductance regulator. *J Clin Invest* 1997, 100:1079-1088.

[124] Xiong X, Bragin A, Widdicombe JH, Cohn J, Skach WR: Structural cues involved in endoplasmic reticulum degradation of G85E and G91R mutant cystic fibrosis transmembrane conductance regulator. *J Clin Invest* 1997, 100:1079-1088.

[125] Xiong X, Chong E, Skach WR. Evidence that endoplasmic reticulum (ER)-associated degradation of cystic fibrosis transmembrane conductance regulator is linked to retrograde translocation from the ER membrane. *J Biol Chem* 1999, 274:2616-2624.

[126] Plemper RK, Wolf DH: Retrograde protein translocation: ERADication of secretory proteins in health and disease. *Trends Biochem Sci* 1999, 24:266-270.

[127] Plemper RK, Wolf DH: Endoplasmic reticulum degradation. Reverse protein transport and its end in the proteasome. *Mol Biol Rep* 1999, 26:125-130.

[128] Sommer T, Wolf DH: Endoplasmic reticulum degradation: reverse protein flow of no return. *FASEB J* 1997, 11:1227-1233.

[129] Plemper RK, Deak PM, Otto RT, Wolf DH: Re-entering the translocon from the lumenal side of the endoplasmic reticulum. Studies on mutated carboxypeptidase yscY species. *FEBS Lett* 1999, 443:241-245.

[130] Plemper RK, Egner R, Kuchler K, Wolf DH: Endoplasmic reticulum degradation of a mutated ATP-binding cassette transporter Pdr5 proceeds in a concerted action of Sec61 and the proteasome. *J Biol Chem* 1998, 273:32848-32856.

[131] Wickner S, Maurizi MR, Gottesman S: Posttranslational quality control: folding, refolding, and degrading proteins. *Science* 1999, 286:1888-1893.

[132] Agashe VR, Hartl FU: Roles of molecular chaperones in cytoplasmic protein folding. *Semin Cell Dev Biol* 2000, 11:15-25.

[133] Wickner S, Maurizi MR, Gottesman S: Posttranslational quality control: folding, refolding, and degrading proteins. *Science* 1999, 286:1888-1893.

[134] Gottesman S, Wickner S, Maurizi MR: Protein quality control: triage by chaperones and proteases. *Genes Dev* 1997, 11:815-823.

[135] Bercovich B, Stancovski I, Mayer A, Blumenfeld N, Laszlo A, Schwartz A, Ciechanover A: Ubiquitin-dependent Degradation of Certain Protein Substrates *in Vitro* Requires the Molecular Chaperone Hsc70. *J Biol Chem* 1997, 272:9002-9010.

[136] Strickland E, Hakala K, Thomas PJ, DeMartino GN: Recognition of misfolding proteins by PA700, the regulatory subcomplex of the 26 S proteasome. *J Biol Chem* 2000, 275:5565-5572.

[137] Plemper RK, Egner R, Kuchler K, Wolf DH: Endoplasmic reticulum degradation of a mutated ATP-binding cassette transporter Pdr5 proceeds in a concerted action of Sec61 and the proteasome. *J Biol Chem* 1998, 273:32848-32856.

[138] Vidair CA, Huang RN, Doxsey SJ: Heat shock causes protein aggregation and reduced protein solubility at the centrosome and other cytoplasmic locations. *Int J Hyperther* 1996, 12:681-695.

[139] Wojcik C: On the spatial organization of ubiquitin-dependent proteolysis in HeLa cells. *Folia Histochem Cytobiol* 1997, 35:117-118.

[140] Wojcik C: An inhibitor of the chymotrypsin-like activity of the proteasome (PSI) induces similar morphological changes in various cell lines. *Folia Histochem Cytobiol* 1997, 35:211-214.

[141] Wojcik C, DeMartino GN: Intracellular localization of proteasomes. *Int J Biochem Cell Biol* 2003, 35:579-589.

[142] Garcia-Mata R, Bebök Z, Sorcher EJ, Sztul ES: Characterization and dynamics of aggresome formation by a cytosolic GFP-chimera. *J Cell Biol* 1999, 146:1239-1254.

[143] Rajan RS, Illing ME, Bence NF, Kopito RR: Specificity in intracellular protein aggregation and inclusion body formation. *Proc Natl Acad Sci U S A* 2001, 98:13060-13065.

[144] Johnston JA, Illing ME, Kopito RR: Cytoplasmic dynein/dynactin mediates the assembly of aggresomes. *Cell Motil Cytoskeleton* 2002, 53:26-38.

[145] Gautreau a, Fieve BT, Brault E, Antony C, Houdusse A, Louvard D, Arpin M: Isolation and characterization of an aggresome determinant in the NF2 tumor suppressor. *J Biol Chem* 2003, 278:6235-6242.

[146] Burnett BG, Pittman RN: The polyglutamine neurodegenerative protein ataxin 3 regulates aggresome formation. *Proc Natl Acad Sci U S A* 2005, 102:4330-4335.

[147] Wojcik C, Yano M, DeMartino GN: RNA interference of valosin-containing protein (VCP/p97) reveals multiple cellular roles linked to ubiquitin/proteasome-dependent proteolysis. *J Cell Sci* 2004, 117:281-292.

[148] Meriin AB, Sherman MY: Role of molecular chaperones in neurodegenerative disorders. *Int J Hyperthermia* 2005, 21:403-419.

[149] Kristiansen M, Messenger MJ, Klohn PC, Brandner S, Wadsworth JD, Collinge J, Tabrizi SJ: Disease-related prion protein forms aggresomes in neuronal cells leading to caspase activation and apoptosis. *J Biol Chem* 2005, 280:38851-38861.

[150] Keck S, Nitsch R, Grune T, Ullrich O: Proteasome inhibition by paired helical filament-tau in brains of patients with Alzheimer's disease. *J Neurochem* 2003, 85:115-122.

[151] Dimcheff DE, Portis JL, Caughey B: Prion proteins meet protein quality control. *Trends Cell Biol* 2003, 13:337-340.

[152] Biasini E, Fioriti L, Ceglia I, Invernizzi R, Bertoli A, Chiesa R, Forloni G: Proteasome inhibition and aggregation in Parkinson's disease: a comparative study in untransfected and transfected cells. *J Neurochem* 2004, 88:545-553.

[153] Wang C, Tan JM, Ho MW, Zaiden N, Wong SH, Chew CL, Eng PW, Lim TM, Dawson TM, Lim KL: Alterations in the solubility and intracellular localization of parkin by several familial Parkinson's disease-linked point mutations. *J Neurochem* 2005, 93:422-431.

[154] Olanow CW, Perl DP, DeMartino GN, McNaught KS: Lewy-body formation is an aggresome-related process: a hypothesis. *Lancet Neurol* 2004, 3:496-503.

[155] Illing ME, Rajan RS, Bence NF, Kopito RR: A rhodopsin mutant linked to autosomal dominant retinitis pigmentosa is prone to aggregate and interacts with the ubiquitin proteasome system. *J Biol Chem* 2002, 277:34150-34160.

[156] Rajan RS, Illing ME, Bence NF, Kopito RR: Specificity in intracellular protein aggregation and inclusion body formation. *Proc Natl Acad Sci U S A* 2001, 98:13060-13065.

[157] Notterpek L, Ryan MC, Tobler AR, Shooter EM: PMP22 accumulation in aggresomes: implications for CMT1A pathology. *Neurobiol Dis* 1999, 6:450-460.

[158] Taylor JP, Tanaka F, Robitschek J, Sandoval CM, Taye A, Markovic-Plese S, Fischbeck KH: Aggresomes protect cells by enhancing the degradation of toxic polyglutamine-containing protein. *Hum Mol Genet* 2003, 12:749-757.

[159] Kabore AF, Wang WJ, Russo SJ, Beers MF: Biosynthesis of surfactant protein C: characterization of aggresome formation by EGFP chimeras containing propeptide mutants lacking conserved cysteine residues. *J Cell Sci* 2001, 114:293-302.

[160] Waelter S, Boeddrich A, Lurz R, Scherzinger E, Lueder G, Lehrach H, Wanker EE: Accumulation of mutant huntingtin fragments in aggresome-like inclusion bodies as a result of insufficient protein degradation. *Mol Biol Cell* 2001, 12:1393-1407.

[161] Johnston JA, Dalton MJ, Gurney ME, Kopito RR: Formation of high molecular weight complexes of mutant Cu, Zn- superoxide dismutase in a mouse model for familial amyotrophic lateral sclerosis. *Proc Natl Acad Sci U S A* 2000, 97:12571-12576.

[162] Bardag-Gorce F, Riley NE, Nan L, Montgomery RO, Li J, French BA, Lue YH, French SW: The proteasome inhibitor, PS-341, causes cytokeratin aggresome formation. *Exp Mol Pathol* 2004, 76:9-16.

[163] Waelter S, Boeddrich A, Lurz R, Scherzinger E, Lueder G, Lehrach H, Wanker EE: Accumulation of mutant huntingtin fragments in aggresome-like inclusion bodies as a result of insufficient protein degradation. *Mol Biol Cell* 2001, 12:1393-1407.

[164] Johnston JA, Dalton MJ, Gurney ME, Kopito RR: Formation of high molecular weight complexes of mutant Cu, Zn- superoxide dismutase in a mouse model for familial amyotrophic lateral sclerosis. *Proc Natl Acad Sci U S A* 2000, 97:12571-12576.

[165] Fabunmi RP, Wigley WC, Thomas PJ, DeMartino GN: Activity and regulation of the centrosome-associated proteasome. *J Biol Chem* 2000, 275:409-413.

[166] Brown CR, Doxsey SJ, Hong-Brown LQ, Martin RL, Welch WJ: Molecular chaperones and the centrosome. A role for TCP-1 in microtubule nucleation. *J Biol Chem* 1996, 271:824-832.

[167] Brown CR, Hong-Brown LQ, Doxsey SJ, Welch WJ: Molecular chaperones and the centrosome. A role for HSP 73 in centrosomal repair following heat shock treatment. *J Biol Chem* 1996, 271:833-840.

[168] Lange BM, Bachi A, Wilm M, Gonzalez C: Hsp90 is a core centrosomal component and is required at different stages of the centrosome cycle in Drosophila and vertebrates. *EMBO J* 2000, 19:1252-1262.

[169] Anton LC, Schubert U, Bacik I, Princiotta MF, Wearsch PA, Gibbs J, Day PM, Realini C, Rechsteiner MC, Bennink JR, Yewdell JW: Intracellular Localization of Proteasomal Degradation of a Viral Antigen. *J Cell Biol* 1999, 146:113-124.

[170] Lacaille VG, Androlewicz MJ: Targeting of HIV-1 Nef to the centrosome: implications for antigen processing. *Traffic* 2000, 1:884-891.

[171] Lelouard H, Gatti E, Cappello F, Gresser O, Camosseto V, Pierre P: Transient aggregation of ubiquitinated proteins during dendritic cell maturation. *Nature* 2002, 417:177-182.

[172] Fabunmi RP, Wigley WC, Thomas PJ, DeMartino GN: Interferon gamma regulates accumulation of the proteasome activator PA28 and immunoproteasomes at nuclear PML bodies. *J Cell Sci* 2001, 114:29-36.

[173] Hung CF, Cheng WF, He L, Ling M, Juang J, Lin CT, Wu TC: Enhancing major histocompatibility complex class I antigen presentation by targeting antigen to centrosomes. *Cancer Res* 2003, 63:2393-2398.

[174] Lelouard H, Gatti E, Cappello F, Gresser O, Camosseto V, Pierre P: Transient aggregation of ubiquitinated proteins during dendritic cell maturation. *Nature* 2002, 417:177-182.

[175] Sanchez V, Greis KD, Sztul E, Britt WJ: Accumulation of virion tegument and envelope proteins in a stable cytoplasmic compartment during human cytomegalovirus replication: characterization of a potential site of virus assembly. *J Virol* 2000, 74:975-986.

[176] Sanchez V, Sztul E, Britt WJ: Human cytomegalovirus pp28 (UL99) localizes to a cytoplasmic compartment which overlaps the endoplasmic reticulum-golgi-intermediate compartment. *J Virol* 2000, 74:3842-3851.

[177] Heath CM, Windsor M, Wileman T: Aggresomes resemble sites specialized for virus assembly. *J Cell Biol* 2001, 153:449-455.

[178] Araujo FD, Stracker TH, Carson CT, Lee DV, Weitzman MD: Adenovirus type 5 E4orf3 protein targets the Mre11 complex to cytoplasmic aggresomes. *J Virol* 2005, 79:11382-11391.

[179] Skipper L, Farrer M: Parkinson's genetics: molecular insights for the new millennium. *Neurotoxicology* 2002, 23:503-514.

[180] Valentine JS, Hart PJ: Misfolded CuZnSOD and amyotrophic lateral sclerosis. *Proc Natl Acad Sci U S A* 2003, 100:3617-3622.

[181] Bates G: Huntingtin aggregation and toxicity in Huntington's disease. *Lancet* 2003, 361:1642-1644.

[182] Waelter S, Boeddrich A, Lurz R, Scherzinger E, Lueder G, Lehrach H, Wanker EE: Accumulation of mutant huntingtin fragments in aggresome-like inclusion bodies as a result of insufficient protein degradation. *Mol Biol Cell* 2001, 12:1393-1407.

[183] Johnston JA, Dalton MJ, Gurney ME, Kopito RR: Formation of high molecular weight complexes of mutant Cu, Zn- superoxide dismutase in a mouse model for familial amyotrophic lateral sclerosis. *Proc Natl Acad Sci U S A* 2000, 97:12571-12576.

[184] Bence NF, Sampat RM, Kopito RR: Impairment of the ubiquitin-proteasome system by protein aggregation. *Science* 2001, 292:1552-1555.

[185] Keck S, Nitsch R, Grune T, Ullrich O: Proteasome inhibition by paired helical filament-tau in brains of patients with Alzheimer's disease. *J Neurochem* 2003, 85:115-122.

[186] Taylor JP, Tanaka F, Robitschek J, Sandoval CM, Taye A, Markovic-Plese S, Fischbeck KH: Aggresomes protect cells by enhancing the degradation of toxic polyglutamine-containing protein. *Hum Mol Genet* 2003, 12:749-757.

[187] Tanaka M, Kim YM, Lee G, Junn E, Iwatsubo T, Mouradian MM: Aggresomes formed by alpha-synuclein and synphilin-1 are cytoprotective. *J Biol Chem* 2004, 279:4625-4631.

[188] Yoshimoto Y, Nakaso K, Nakashima K: L-dopa and dopamine enhance the formation of aggregates under proteasome inhibition in PC12 cells. *FEBS Lett* 2005, 579:1197-1202.

[189] Webb JL, Ravikumar B, Rubinsztein DC: Microtubule disruption inhibits autophagosome-lysosome fusion: implications for studying the roles of aggresomes in polyglutamine diseases. *Int J Biochem Cell Biol* 2004, 36:2541-2550.

[190] Kolodziejska KE, Burns AR, Moore RH, Stenoien DL, Eissa NT: Regulation of inducible nitric oxide synthase by aggresome formation. *Proc Natl Acad Sci U S A* 2005, 102:4854-4859.

In: The Ubiquitin Proteasome System... ISBN 978-1-60021-749-4
Eds: Mario Di Napoli and Cezary Wojcik, pp. 271-305 © 2007 Nova Science Publishers, Inc.

Chapter 13

Endoplasmic Reticulum (ER) Stress, ER-Associated Degradation and Unfolded Protein Response

Cezary Wójcik[1,*] and Mario Di Napoli[2,3,¥]

[1]Department of Anatomy and Cell Biology, Indiana University School of Medicine, Evansville, IN 47712, USA;

[2]Neurological Section, SMDN – Center for Cardiovascular Medicine and Cerebrovascular Disease Prevention, Sulmona (L'Aquila) I-67039, Italy;

[3]Neurological Service, San Camillo de'Lellis General Hospital, Rieti I-02100, Italy.

Abstract

Proteins co-translationally inserted into the endoplasmic reticulum (ER) of neuronal cells must undergo proper folding, association and posttranslational modifications in order to assure their function in the different segments of the secretory pathway, plasma membrane and extracellualar space. A quality control (QC) system assures that only properly folded and assembled proteins exit the ER, while misfolded and aberrant proteins are degraded by a process known as ER-associated degradation (ERAD). The most studied form of ERAD relies on the activity of the cytosolic ubiquitin-proteasome system (UPS) and requires recognition and extraction (a.k.a. retrotranslocation or dislocation) of misfolded proteins prior to their degradation. Nevertheless, ample evidence exists for UPS-independent cytosolic and lumenal ERAD as well. ER and cytoplasmic chaperones assist in the retrotranslocation process, which is mediated through the Sec61 translocone or derlin-associated channels. Emerging substrates are

* Correspondence concerning this article should be addressed to Dr. Cezary Wójcik, MD, PhD; Department of Anatomy and Cell Biology, Indiana University School of Medicine, 8600 University Blvd., Evansville, IN 47712, USA. Phone: +1 (812) 461-5437; Fax: +1 (812) 465-1184; E-mail: cwojcik@iupui.edu.

¥ Correspondence concerning this article should be addressed to Dr. Mario Di Napoli, MD; SMDN – Center for Cardiovascular Medicine and Cerebrovascular Disease Prevention, Neurological Section, Via Trento, 41 I-67039 – Sulmona (AQ) Italy. Phone: +39 (0)864 52716; Fax: +39 (0)864 52716; Email: mariodinapoli@katamail.com.

ubiquitinated on the cytosolic face of the ER by a limited number of specialized E2s and E3s. Studies in yeast have implied in ERAD a cytosolic AAA ATP-ase called VCP (valosin-containing protein; Cdc48, p97, TER94) in association with the Ufd1-Npl4 dimer. Inefficient ERAD and/or increased load with misfolded proteins induce ER stress, which in turn activates a transcriptional response known as UPR (unfolded protein response). UPR involves transient attenuation of translation followed by increased synthesis of ER chaperones, components of ERAD and other proteins involved in relieving ER stress. However, persistent ER stress leads to the initiation of apoptosis. Inefficient ERAD contributes to the formation of cytoplasmic inclusions characteristic of neurodegenerative disorders, while acute ER stress is often associated with neuronal cell loss during cerebral ischemia. VCP is mutated in a hereditary dominant disorder known as inclusion body myopathy, Paget's disease of the bone and frontotemporal dementia (IBMPFD).

Keywords: ERAD (ER-associated degradation), ER (endoplasmic reticulum) stress, unfolded protein response, ubiquitin proteasome system, protein misfolding disease, VCP (valosin containing protein).

ABBREVIATIONS

ALS, amyotrophic lateral sclerosis; AMFR, autocrine motility factor receptor; AR-JP, autosomal recessive juvenile parkinsonism; ATF3, activating transcription factor 3; ATF4, activating transcription factor 4 ATF6, activating transcription factor 6; BiP, binding Ig protein; CBF, CCAAT-binding factor; CFTR, cystic fibrosis transmembrane regulator; CHOP,C/EBP homologous transcription factor; cLD , core lumenal domain; CMV, cytomegalovirus; CNS, central nervous system; CPY, carboxypeptidase Y; EDEM, ER-degradation enhancing alpha-mannosidase-like protein; eIF2, eukaryotic translation initiation factor 2; ERAD, endoplasmic reticulum associated proteins degradation; ERSE, ER stress element; FENIB, Familial Encephalopathy with Neuroserpin Inclusion Bodies; GRP78, glucose regulated protein 78 (or BiP); HMG-CoA, 3-Hydroxy-3-Methylglutaryl Coenzyme A; Hsp70, heat shock protein 70; IBMPFD, inclusion body myopathy, Paget's disease of the bone and frontotemporal dementia; IP3, inositol-3,4,5-trisphosphate; Ire1, inositol requiring enzyme 1; JNK, cJun NH2-terminal kinase pathway; Met-tRNAi, initiator methionyl-transfer RNA; MHC, major histocompatibility complex; Pael, parkin-associated endothelin receptor-like; PD, Parkinson's disease; PDI, protein disulfide isomerase; PERK, double stranded RNA-activated protein kinase-like ER kinase; PNGase, peptide N-glycanase; PolyQ, polyglutamine; QC, quality control; S1P (or S2P), site 1 (or 2) protease; UPR, unfolded protein response; UPRE, UPR element; UPS, ubiquitin-proteasome system; VCP, valosin-containing protein (p97, Cdc48, TER94); XBP1, X-box-binding protein-1.

INTRODUCTION

The endoplasmic reticulum (ER) is an extensive intracellular membrane system. It is important for a number of cellular functions including translocation of secretory proteins across the membrane, insertion of membrane proteins, lipid synthesis, calcium storage and signaling, and separation of nucleoplasm from cytoplasm (reviewed in [1-3]). The ER is responsible for the structural maturation of the roughly one-quarter of the proteome that traverses the secretory pathway [4,5]. Its structure varies depending on cell type. The ER lumen is similar to the extracellular space: it has high calcium concentrations, is more oxidizing than the cytosol [6,7], and contains a specialized set of chaperones and enzymes. The structural maturation of many proteins synthesized in the ER is slow and inefficient [6], probably because they require several post-translational modifications: folding is often accompanied by and dependent on signal sequence cleavage, N-linked glycosylation, disulphide-bond formation and reshuffling, addition of glycosylphosphatidylinositol anchors, and insertion of membrane proteins in the lipid bilayer. These events are occurring more slowly by orders of magnitude than the typical conformational changes that accompany folding. Coordinating these covalent modifications is a challenging task for the folding machinery in the ER. Correspondingly, the ER provides an environment optimized to face these challenges, including high concentrations of general chaperones as well as a range of strategies specifically tailored to aid folding of secretory proteins.

Besides providing a unique folding environment, the ER has a crucial quality-control (QC) role [6,7]. When folding or assembly intermediates expose hydrophobic surfaces, unpaired cysteines or immature glycans, ER resident chaperones or oxidoreductases interact with them, and as a consequence they are retained in the ER or retrieved from the Golgi complex [6,8]. By forming multimolecular complexes [9], folding factors in the ER may provide matrices that couple retention to folding and assembly. Immature proteins may also form aggregates that are excluded from vesicles exiting from the ER [1,6]. All proteins are subjected to a QC that monitors their architectural design through ubiquitous folding sensors and facilitate export of individual proteins or classes of proteins [1,6,7]. Although some proteins can be rerouted to and degraded in the endolysosomal compartment, the ER is the main test bench where molecules destined for the extracellular space are scrutinized for their potential toxicity. The reasons for having a quality-control system in the ER are easy to understand where protein folding and function are concerned, especially in multicellular organisms where development relies on the fidelity of protein secretion. QC can also regulate the transport or the activity of certain proteins during differentiation or in response to stress or metabolic requirements [10].

ER is widely distributed within neurons, being present in the perikarya, dendrites and dendritic spines, axons and presynaptic nerve terminals, and in growth cones. ER regulates functional and structural changes in nerve cell circuits in both the developing and adult nervous systems [11-16]. ER is continuous with the outer nuclear membrane and is often associated intimately with plasma membrane and mitochondria, which suggests functional coupling between these structures [17]

Two domains, rough ER, which contains ribosomes and is responsible for protein synthesis, and smooth ER, which has a particularly important role in Ca^{2+} signaling, can be distinguished [18]. Rough ER is very abundant in neurons where it forms the characteristic tigroid (Nissl bodies) in the perikarya. Several millions of proteins are synthesized and co-translationally inserted into the ER of each neuron per minute, where they need to fold, assemble and undergo modification by trimming and addition of oligosaccharides in a crowded, oxidative environment where protein concentrations reach 100 mg/mL [19]. Homeostasis within the ER is important for proper protein folding and stresses including perturbations in Ca^{2+} homeostasis, elevated secretory protein synthesis, expression of misfolded or mutant proteins, nutrient or glucose reduction and overload of cellular cholesterol can lead to an accumulation of unfolded proteins or protein aggregates [20].

Both N-glycosylation and disulphide-bond formation play a crucial role in the folding of secretory and membrane proteins in the ER. The glycan moieties bind the lectin chaperones calnexin and calreticulin, whose role in glycoprotein folding has been extensively reviewed [21]. Both calnexin and calreticulin form a complex with ERp57, an ER oxidoreductase, coupling folding assistance to disulphide-bond formation. The impressive number of oxidoreductases in the ER suggests that catalysis and regulation of disulphide-bond formation is crucial for folding. Energywise, in most cases, the contribution of a disulphide bond is hardly more than that of a single hydrogen bond, yet, without disulphide bonds, native conformations are not obtained. Disulphide bonds cannot force a folding protein into a given conformation: in the sampling of conformations during folding in the ER, native and non-native disulphide cross-links are transiently formed. Continuous activity of oxidoreductases probably ensures that these covalent links remain flexible until folding is completed. Secreted proteins enter the ER with reduced cysteines and leave it with oxidized cystines. The requirement for oxidative equivalents is fulfilled by Ero1 proteins, which transfer electrons from protein disulphide isomerase (PDI) to oxygen through a series of specific interchange reactions [22,23]. Additional electron transport pathways exist, and involve proteins such as Erv2p from yeast, although their role under normal conditions remains to be determined [23]. The redox gradient between the ER and the cytosol seems to be important for intercompartmental signalling, particularly in the integrated response to oxidative stress, in which adaptive responses emanating from different compartments are coordinated [24]. Redox reactions with opposite electron fluxes must take place in the ER to mediate formation, isomerization and reduction of disulphides [22,23]. The wealth of redox assistants allows these fluxes to be separate, and channels electron transport through specific protein–protein interactions. The main role of glutathione in redox homeostasis in the ER seems to be that of buffering the oxidative power of Ero1 [22,23]. Because disulphide bonds are so important for folding, we may conclude from the number of ER-resident oxidoreductases that secretory proteins need more help, possibly because they are often larger than cytosolic ones [25]. Perhaps secretory proteins, which are designed to act extracellularly and often must carry their biological messages over long distances, need more protection from denaturing forces outside the cell (see Chapter 19). This is an issue that glycans are likely to contribute to as well.

Additionally, a sophisticated quality-control system exists in the ER, which retains and retrieve proteins that have not yet reached their native state [6]. Protein misfolding is a common occurrence and an amazing QC mechanism prevents misfolded or otherwise abnormal proteins from leaving the ER and wreaking havoc in further elements of the secretory pathway, the cell membrane or in the extracellular space [6,26]. ER-associated degradation (ERAD) [27] and unfolded protein response (UPR) [28] are two distinct mechanisms used by the ER in response to the presence of misfolded proteins. Misfolded proteins are recognized by multiple resident ER chaperones: if the misfolded proteins fail to refold properly, they are degraded, usually after retrotranslocation to the cytosol where they undergo degradation mediated by the ubiquitin-proteasome system (UPS) [6,10,29-32]. Besides its QC function, ERAD is also exploited in the regulated degradation of properly folded proteins, such as 3-Hydroxy-3-Methylglutaryl Coenzyme A (HMG-CoA) reductase and inositol-3,4,5-trisphosphate (IP3) receptors [33,34]. Studies in yeast have identified Cdc48p, whose mammalian homolog is valosin-containing protein (VCP), as a crucial component of ERAD, involved in retrotranslocation of emerging substrates and their delivery to the proteasome [35,36].

Failure to dispose misfolded proteins or an increased load with misfolded proteins within the ER (such as occurring during oxidative damage associated with ischemia and reperfusion injury) leads to an integrated cellular response, which involves translational attenuation, decreasing the input of new proteins, followed by a transcriptional reaction known together as UPR [37-40]. UPR allows for an increase in the capacity for protein folding, reduction in newly translated proteins entering into the ER, increase in the degradation of misfolded proteins leading to a concerted upregulation of multiple proteins, including components of ERAD, which counteract at different levels the ER dysfunction caused by protein misfolding. Thus, UPR is an adaptive mechanism which promotes survival, but if it else has failed, the UPR signals for cell apoptosis to preserve the surrounding eukaryotic cells [19,41]. These two systems are intimately linked: UPR induction increases ERAD capacity, loss of ERAD leads to constitutive UPR induction, and simultaneous loss of ERAD and the UPR greatly decreases cell viability.

ER-Associated Degradation (ERAD)

It was originally assumed, that ERAD is carried on by protease(s) present in the ER or cis-Golgi compartments. Since components of the UPS are absent from the lumen of the ER [42] the discovery that UPS actually mediates ERAD was surprising [43-45]. It implicated that ERAD substrates must be recognized, retrotranslocated to the cytosol and then ubiquitinated before they can be degraded by the proteasomes. However, the exact features that are being monitored and how substrates are delivered to the retrotranslocation machinery remain important open questions for most substrates. Skyrocketing popularity of the UPS research has created the false and widespread impression that all ERAD in mammalian cells is UPS-dependent [32], what is far from being true. As might be expected by the diversity of proteins that fold in the ER, recent studies argue that ERAD encompasses a number of different systems, each responsible for the degradation of subsets of proteins that share common physical properties. This is perhaps most clearly shown in yeast, where there are at least two distinct surveillance mechanisms for identifying terminally misfolded ER proteins. The first, designated ERAD-L, inspects for proteins that contain misfolded lumenal (soluble or membrane-tethered) domains such as CPY*, a mutant form of the endogenous CPY protein that is incapable of folding. The second, termed ERAD-C, detects misfolded cytosolic domains of transmembrane proteins [46]. Although both of these pathways ultimately converge on the UPS, they depend on different sets of ER-associated components to detect and deliver misfolded species to the cytosol. In the case of ERAD-C (but not ERAD-L), degradation is typically dependent on a specific subset of cytosolic chaperones including Hsp70 and Hsp40 members [47]. Numerous proteins undergo nonproteasomal degradation in which both deglycosylation and proteolysis take place in the ER lumen [26,31,48-52]. Moreover, proteasome inhibition does not completely block the degradation of even the best characterized ERAD substrates such as CPY*, CFTR or α1-antitrypsin [44,48,51,53,54].

A second common erroneous generalization is that the activity of VCP/Cdc48 is indispensable for ERAD [32], while in the cell multiple mechanistically different ERAD pathways can exist as indicated on Figure 1. Nevertheless, the UPS-mediated ERAD is the best studied and has drawn the most attention, since it is involved in the degradation of several important substrates. The exact character of the non-proteasomal pathways remain elusive. A fascinating problem is how molecules that have not been given the time to fold (and therefore are unfolded) are discriminated from those that have failed to fold after many attempts (misfolded), and must therefore be disposed of. One way of timing glycoprotein QC involves the sequential processing of N-glycans and in particular mannose trimming in the ER [30]. It remains to be seen how substrates are eventually targeted to the retrotranslocation channels, how these are opened, and to what extent proteins must be unfolded to negotiate dislocation [22,30].

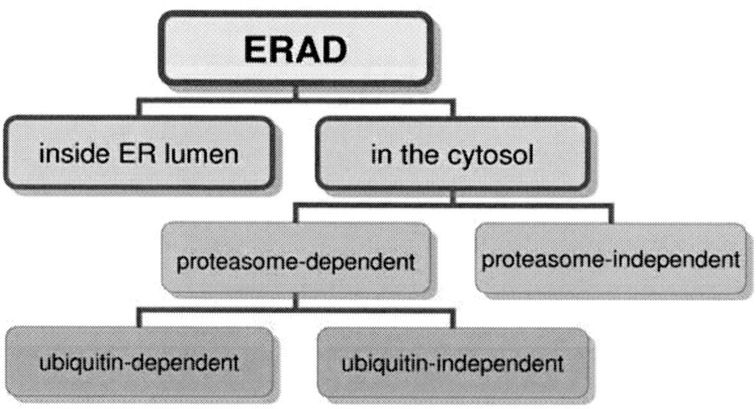

Figure 1. Multiple pathways of ERAD. While most emphasis has been recently granted to the UPS-dependent form of ER-associated degradation (ERAD), it actually encompasses multiple mechanistically different pathways, many of them poorly understood.

The Model of UPS-Dependent ERAD

The key players of the UPS-dependent ERAD have been usually first identified and best characterized in yeast and include ER chaperones (such as BiP, calnexin, calreticulin, protein-disulphide isomerase), cytosolic chaperones (such as Hsp70 and Hsp90), ER-associated E3s (gp78, HRD1 and TEB4) and E2s (Ubc1, 6 and 7), 26S proteasome and the complex of Cdc48p (VCP in mammals) with Npl4p and Ufd1p [6,10,29-31]. Their subcellular localization and interplay in the degradation of a prototypical ERAD substrate are depicted in Figure 2.

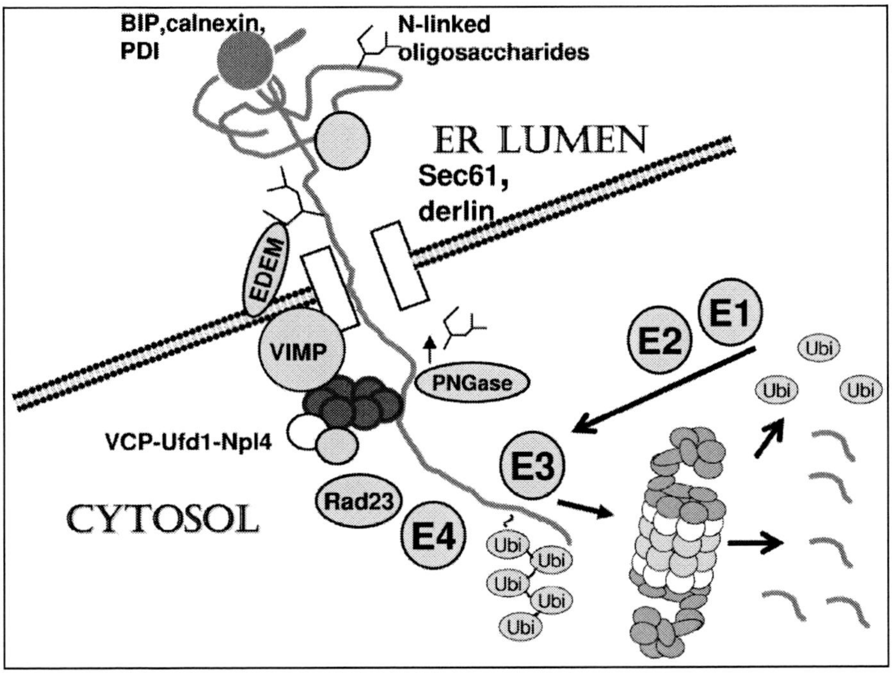

Figure 2. The current model of UPS-dependent ERAD. Misfolded proteins are bound in the ER lumen by chaperones such as BiP, calnexin, calreticulin or protein disulfide isomerase (PDI) in numerous refolding attempts; when those fail, misfolded proteins are handled over to the retrotranslocation channel with the assistance of the lectin EDEM. The retrotranslocation channel made by Sec61 and/or derlin-1 allows the passage of the unfolded polypeptide chain to the cytosolic face of the ER, where N-linked oligosaccharides are removed by the action of peptide N-glycanase (PNGase) and ubiquitination takes place, mediated by the E1-E2-E3 cascade of enzymes. VCP-Ufd1-Npl4 binds the emerging polypeptide chain both prior and after ubiquitination, brings additional factors such as Rad23 and associated E4 enzyme (Ufd2) and finally deliver the substrate to 26S proteasomes for degradation.

Most ERAD substrates are glycoproteins, which interact with two lectin-like chaperones, calnexin and calreticulin, calreticulin is soluble in the ER lumen, while calnexin is attached to the inner leaflet of the ER membrane. Misfolded glycoproteins interact with those lectins only when they are monoglucosylated. Removal of glucose by glucosidase II is followed by their release from these chaperones, while subsequent glucosylation mediated by glycoprotein glucosyltransferase allows their binding. Alternating rounds of glucosylation and deglucosylation and therefore of binding and release constitutes the 'calnexin/calreticulin cycle' which allows many attempts of refolding. Properly folded proteins leave the cycle with an oligosaccharide harboring nine mannose residues, exit from the ER and undergo mannose trimming and attachement of complex oligosaccharides mediated by Golgi enzymes.

Failure to fold properly in a certain amount of time leads to an irreversible cleavage of mannose residues by ER mannosidases creating truncated monoglycosylated glycans resistant to the glucosidase II action [6,26] Such proteins interact with a lectin called EDEM (ER-degradation enhancing alpha-mannosidase-like protein), which has mannosidase fold without the enzymatic activity [55,56,56]. It is not clear how EDEM handles over the doomed proteins to the retrotranslocation channel, but its role in ERAD is well established since overexpression of EDEM accelerates ERAD, while downregulation by RNAi prevents it [57,58]. It is thought that ERAD substrates which are not glycoproteins are retained and targeted for degradation by a less studied mechanism involving other ER chaperones, such as PDI. Retrotranslocation of most misfolded proteins is mediated by the Sec61 translocone, as evidenced by co-immunoprecipitation experiments and by yeast Sec61 mutants, which are defective in retro-translocation [59,60]. However, tail-anchored ER proteins such as Ubc6 do not require Sec61 or another known ERAD components for their degradation by the UPS [61]. A different protein called derlin1 was postulated to form an alternative retrotranslocation channel in mammalian cells [62,63]. Finally, such channel may be formed by the Doa10 ubiquitin ligase in yeast and its mammalian homolog TEB4 [64,65].

The fate of ERAD substrates depends on the location of bulk of the polypeptide chain with respect to the ER membrane. Distinct machinery is required for the degradation in yeast of a multispanning membrane protein versus a soluble luminal protein [64]. Soluble luminal and membrane attached ERAD substrates require ER chaperones and derlin-1 for efficient degradation, while ERAD substrates with bulky cytosolic domains do not require them, requiring cytosolic chaperones instead [66]. The location of a misfolded domain is often more important that the location of the bulk of the protein. Misfolding within a cytosolic domain triggers the cytosolic variant of ERAD, while misfolding within the luminal domain recruits the luminal ERAD machinery [46]. All ERAD pathways seem to converge once the polypeptides emerge through the retrotranslocation channel, since in both cases $Cdc48^{Npl4/Ufd1}$ is required. Numerous evidences exist to support the notion that $Cdc48^{Npl4/Ufd1}$ extracts and delivers the substrates to the proteasome prior and/or after ubiquitination [67]. Yeast mutants in Ccd48, Ufd1 or Npl4 are defective in ERAD and accumulate poly-Ub substrates in association with the ER [68-72]. However, even in yeast some ERAD substrates are efficiently extracted and degraded without Cdc48 [73]. Moreover, proteasomes are capable to directly extract proteins from the ER membrane, since yeast mutants expressing functionally attenuated proteasomes are defective in the degradation of transmembrane segments of ER proteins whose N-terminal cytosolic domains are degraded normally [74]. In yeast most proteasomes are associated with the ER [75] where they directly associated with Sec61 translocones [76], and even more proteasomes are recruited to ER membranes after induction of ER stress [77]. In addition to that, specialized receptors exist on the ER membrane which are able to recruit proteasomes, such as HERP/Mif1 [77,78] and VIMP [62,63].

The emerging ER proteins destined for proteasomal degradation undergo ubiquitination on the cytosolic face of the ER. While ubiquitin activation is mediated by the same E1 required for the ubiquitination of all substrates, several E2s and E3s are specialized in ubiquitinating ERAD substrates. First ERAD-specific E2s identified in yeast were Ubc6p and Ubc7p: Ubc6p is directly anchored to the ER membrane, while Ubc7p is recruited by an adaptor membrane protein Cue1p [79,80]. Both have mammalian homologs, which are involved in ERAD as well [81]. More recently other E2s have also been involved in ERAD of different substrates, including Ubc1p and Ubc4p/Ubc5p [69,82,83]. Several E3s have been reported to participate in ERAD, including HRD1, gp78, TEB4, $SCF^{Fbx2/Fbs1}$, parkin and

CHIP. HRD1 is an ER-resident RING-finger protein implicated in the degradation of unfolded ERAD substrates and HMG-CoA reductase [84]. HRD1 is induced after cerebral ischemia and its offer expression protects neuronal cells from apoptosis induced by hypoxia [84,85]. Another ER-resident RING finger E3 is gp78, which was initially characterized as an autocrine tumor motility factor receptor [86,87]. TEB4 is a multispanning membrane protein similar to yeast Doa10, harboring a RING finger, which is involved in degradation of transmembrane proteins apparently not requiring the participation of the Sec61 translocone [65,88]. Finally, an E3 involved in ERAD with great importance for neuroscientists is parkin, which is mutated in autosomal recessive juvenile parkinsonism (AR-JP). Parkin is characterized by an ubiquitin-like domain at its N-terminus and two RING finger motifs and an IBR motif (in between RING fingers) at its C-terminus. It is upregulated by conditions causing ER stress, while its overexpression protects cells from ER stress, demonstrating its involvement in ERAD [89]. A putative G protein-coupled transmembrane polypeptide, named Pael (parkin-associated endothelin receptor-like) receptor is ubiquitinated by parkin. Failure of its ubiquitination leads to ER stress, UPR and cell death. Pael-R is highly expressed in the dopaminergic neurons in the substantia nigra and it accumulates in the brain of AR-JP patients [90,91]. Parkin-mediated degradation of Pael-R is assisted by CHIP, another E3 interacting with cytososlic chaperones, which can also degrade ERAD substrates dislocated to the cytosol [92,93]. N-linked oligosaccharides attached to retrotranslocated substrates are removed by peptide N-glucanase either co-retrotranslocationally or after retrotranslocation [26]. While it is possible that PNGase activity may also reside in the ER lumen, the best characterized PNGase is cytosolic and it associates with other proteins involved in ERAD [94]. The complex of PNGase with the ubiquitin-binding protein Rad23 has been recently postulated to regulate glycoprotein turnover [95]. Nevertheless, many proteins appear to escape PNGase activity since there are E3s which are specialized in the recognition of cytosolic proteins bearing N-linked glycans. In contrast to those four monomeric E3s, the E3s recognizing N-lnked glycans are of the multisubunit SCF family, with two different F-box proteins called Fbs1 and Fbs2 [96-98]. A single substrate can be ubiquitinated by different E3s, as is the case of the prototypical ERAD substrate αTCR, which may be ubiquitinated by HRD1, gp78 or the $SCF^{Fbx2/Fbs1}$ complex [86,97-99]. It has been recently demonstrated that Pael-R, which is a substrate of parkin, and whose accumulation is thought to promote apoptosis in parkin-deficient neurons, can be ubiquitinated and targeted for degradation by HRD1 [100].

Many emerging ERAD substrates may be only oligoubiquitinated and therefore their full ubiquitination requires the activity of a ubiquitin chain elongation factor or E4 which is encoded by the *Ufd2* gene in yeast and has two homologs in the human genome, UFD2A and UFD2B [101]. UFD2A is expressed predominantly in neuronal tissues where it interacts with VCP [102]. It is cleaved by caspases during apoptosis [103], while both *Ufd2* homologs are often deleted in neuroblastoma and are subject to mutations in other brain tumours [104]. It is widely accepted that polypeptide chains emerging from the retrotranslocone or otherwise destined for degradation interact with VCP associated either with the Ufd1-Npl4 complex or with one of the Ubx family of proteins [32]. VCP associates with the emerging substrate by a dual recognition mechanism, involving binding to the polyubiquitin chain as well as binding to misfolded, hydrophobic regions of the emerging polypeptide [67]. VCP also binds several other proteins involved in the retrotranslocation process including PNGase, Rad23 and the 26S proteasome itself [94,102,105]. In the widely accepted model, VCP is proposed to provide a pulling force for the emerging substrates through hydrolysis of ATP, then a framework for sequential interaction with multiple proteins, which ends up with handling of the substrate over to the 26S proteasome [32,68,106]. One of the proteins interacting with VCP is Rad23, which is itself a putative adaptor molecule interacting with ubiquitin chains through its ubiquitin-associated motifs (UBA) and with the *Ufd2* and the proteasome through a ubiquitin-like element (UBL) [107,108]. Whatever factors are involved in the handling of the substrate over to the proteasome, it is the final degradation station for many ERAD substrates. When proteasome is pharmacologically inhibited, ERAD substrates accumulate, some of them within the ER lumen, while others in the cytosol in form of aggresomes. It is not clear, what determines where an ERAD substrate will accumulate upon proteasome inhibition, since for example two closely related components of the TCR complex accumulate in different compartments upon proteasome inhibition: δCD3 accumulates in the ER, while αTCR accumulates in the cytosol [109].

Is VCP Involved in Mammalian ERAD?

The awesome power of yeast genetics unveiled a lot about ERAD mechanisms in *S. cerevisiae* [30]. It has been accepted without major objections that mammalian ERAD works in a very similar way. Results obtained with Cdc48 are often extrapolated to VCP, despite the fact that mammalian VCP fails to rescue yeast *Cdc48* mutations [110]. Other important difference between mammalian and yeast ERAD is that yeast lack glycoprotein glucosylatransferase, suggesting that the deglucosylation-glucosylation cycle may not exist in this organism [26]. The only direct evidence about a role of $VCP^{Ufd1/Npl4}$ in mammalian ERAD arises from an *in vitro* model of permeabilized cells developed by the laboratory of Tom Rapoport, where the US11 protein of the cytomegalovirus (CMV) induces retrotranslocation of MHC class I heavy chains to the cytosol [62,67,68]. Contrary to the yeast model, impairment of the VCP ATP-ase activity reduces the size of polyubiquitin chains synthesized on the MHC class I heavy chains [68] indicating that the main role of VCP may consist in the recruitment of the E4 to elongate short polyubiquitin chains [108]. The direct role of VCP in mammalian ERAD may be thus limited to this function. A recent paper has unveiled that US11 induces UPR, which in turn hyperactivates ERAD [111], therefore Rapoport's model system represents a pathological condition rather than physiological ERAD in mammalian cells. Moreover, Römisch has recently suggested that Rapoport's observations do not correspond to retrotranslocation of heavy chain but to its disaggregation on the cytoplasmic face of the ER, reflecting the well known chaperone/segregase activity of VCP [27]. It has been known for some time that proteasomes associate with the ER [75,112] and that such association is increased by ER stress [77], however it has been recently shown that 26S proteasomes bind directly to Sec61 complexes competing with ribosomes [76]. In the case of such direct binding, which *in vitro* does not seem to require VCP, it is difficult to envision a model which will accommodate this bulky molecule as well as why such association would be actually required. In the case of most

ERAD substrates, the ER-associated E3s alone can build up enough polyubiquitin that no further E4 activity would be required similar to most cytosolic substrates. If this hypothesis is correct, most ERAD substrates would not require VCP function, while the US11-mediated MHC class I degradation is an example of a rare condition which requires the E4 activity for efficient degradation [68]. While yeast Ccd48, Ufd1, and Npl4 mutants are defective in ERAD and accumulate polyubiquitinated substrates in association with the ER [68-72], other substrates are extracted from the ER and degraded without those factors [73,74]. Cholera toxin is known to exploit mammalian retrotranslocation machinery and as such has been used to study ERAD [29]. While one group reported indirect evidence that VCP is required for retrotranslocation of cholera toxin [113], it has been subsequently demonstrated that it reaches the cytosol without the participation of VCP [114]. Finally, it needs to be kept in mind that VCP itself has the property to bind polyubiquitin chains, which is increased by its association with Ufd1-Npl4, which also bind polyubiquitin [67,68,115]. Since those proteins are abundant, they are likely to be pulled down in any assays where polyubiquitinated proteins are present. VCP interacts with numerous proteins [35,36], however it is hard to judge which interactions are physiological and which ones are artificial. Moreover, since VCP displays chaperone activity it may bind basically any protein in a misfolded state, whenever it exposes hydrophobic patches [70].

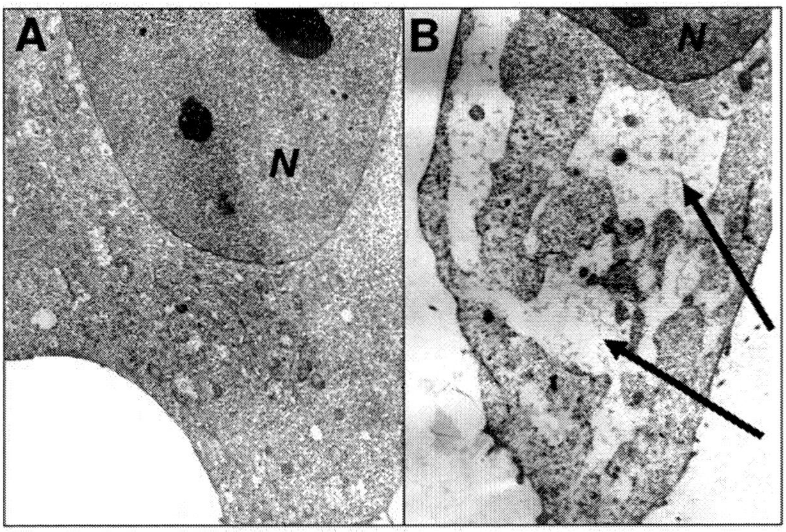

Figure 3. VCP is required for proper ER structure and function. Electron micrographs showing prominent ultrastructural changes in cells submitted to RNAi of VCP (B) compared to control cells (A). Depletion of VCP is associated with cell vacuolation evidenced by distension of ER cisternae (arrows), which is associated with the triggering of unfolded protein response. This effect of the depletion of VCP does not seem to be caused by a massive accumulation of ERAD substrates, since the lumen of the distended cisternae is electron-lucent and does not show aggregates of misfolded proteins.

A recent argument in favor of the Rapoport's model consisted on an RNAi experiment, where a 60% decrease in VCP levels prevents the downregulation of ER-localized IP3 receptors upon binding IP3. However, in contrast to the QC substrates those receptors are properly folded proteins degraded in a regulated fashion and therefore do not represent typical ERAD [34]. RNAi of VCP leading to a 90% decrease in its levels induces swelling of ER cisternae (Figure 3) associated with induction of ER stress and accumulation of cytosolic UPS substrates, however it fails to inhibit the degradation of some lumenal and transmembrane ERAD substrates (Wójcik *et al.*, unpublished observations). Another argument in favor of the involvement of VCP in ERAD comes from a study where overexpression of the IBMPFD mutant VCP leads to accumulation of bulk polyubiquitinated proteins as well as of a co-transfected ERAD substrate ΔF508CFTR [116]. Unfortunately, it is not clear whether this is an effect specific to ERAD substrates or whether it reflects a global impairment of the UPS, which could be measured by a co-transfection of IBMPFD mutant VCP with a cytosolic UPS substrate. It is therefore clear that much work must be still done in order to establish what is the exact role of VCP in mammalian ERAD.

THE STRUCTURE AND FUNCTION OF VCP

VCP is an abundant and highly conserved hexameric protein of the AAA family of ATP-ases, also known as p97 in *Xenopus* [117-119], Cdc48p in yeast [120,121] and TER94 in *Drosophila* [122]. The distruption of the VCP gene in yeast, *Drosophila* or mouse is lethal [120,122,123], while depletion of VCP by RNAi in human cells leads to apoptosis [124]. VCP has been involved in the UPS-mediated degradation of different proteins, probably as a 'shuttle' factor which delivers polyubiquitinated proteins to the proteasome [32,105,108,125,126]. It has two adjacent AAA domains D1 and D2, which form a hexameric ring, while the N-terminal domains are free to interact with numerous proteins. Crystal structure of VCP has been solved leading to the proposal of alternative hypotheses on how ATP hydrolysis drives the conformational changes [127-129]. D2 domain contribute most ATP-ase activity, thus mutants within this domain have a dominant negative phenotype [130,131]. VCP interacts with numerous proteins within the cytosol and the nuclei [35,36]. It is a factor mediating NSF-independent fusion between membranes of the ER [132], mitotic Golgi fragments [119], low density microsomes [133] and fragments of the nuclear envelope [134]. VCP releases the t-SNARE syntaxin 5 from a complex with p47 and VCIP135 [135]. Surprisingly, overexpression of a dominant-negative VCP mutant does not affect any membrane fusion events *in vivo* [131]. VCP has been also implicated in mitotic spindle disassembly [124,136], replication and nucleotide excision repair [137,138], UPS-dependent degradation of cytosolic proteins [105,124,125,139], ERAD [126] and regulated ubiquitin-dependent processing [140,141]. It has the ability to separate substrate proteins from complexes with other proteins ('segregase' activity) [72]. Not only VCP itself binds polyubiquitin chains but it also associates with the Ufd1-Npl4 dimer, whose components bind poly-Ub, therefore the $VCP^{Ufd1/Npl4}$ complex binds polyubiquitin with higher affinity [67,115,125,142]. According to some reports after proteasome inhibition VCP is localized to aggresomes [124,143-145]. RNAi of VCP induces the formation of multiple dispersed cytosolic ubiquitin-positive aggregates rather than single aggresomes [124,145,146]. When cells are submitted to RNAi of VCP and then treated with proteasome inhibitors the formation of aggresomes is prevented [124], indicating that VCP participates in the microtubule-dependent transport of dispersed aggregates towards the cell center paralleling its involvement in chromosome congression [136]. This finding was recently confirmed by a dominant-negative approach [147]. VCP interacts with histone deacetylase-6 [148], which also has been implicated in aggresome formation [149,150]. It has anti-apoptotic and prosurvival properties, since RNAi of VCP in HeLa cells induces apoptosis [124]. Moreover, the VCP homolog in *C. elegans* called MAC-1 also prevents apoptosis in the nematode. It interacts with the caspase CED3, the apoptotic activator CED4 and the anti-apoptotic molecule CED9, a homolog of Bcl-2 [151]. Surprisingly RNAi of VCP inhibits cell death induced by ER stressors such as thapsigargin and BFA [152]. Overexpression of Ter94 in *Drosophila* cells enhanced cell death induced by aggregates of polyQ proteins [153]. Contradictory results about the role of VCP in apoptosis can be reconciled if VCP is considered a central signaling molecule regulating ER structure and function. It is frequently forgotten that the function of VCP in protein degradation is often independent from Ufd1-Npl4 and relies on the interaction with other proteins such as p47 and UBX [124,154,155]. While the role of yeast Cdc48 in ERAD is well established, implication of mammalian VCP in ERAD is based mostly upon reconstitution experiments, with dominant negative VCP mutants inhibiting ERAD of a model substrate in a permeabilized cell [29,62,67,68]. Several different protein kinases may phosphorylate VCP, what probably has a regulatory function [156-158]. It has been demonstrated recently that VCP is a downstream target of Akt signaling necessary for its antiapoptotic action, however it has not been shown how the phosphorylation event affects if at all VCP activity and/or interactions with other proteins [159]. VCP is also regulated by oxidative stress, where oxidative modification of one of the cystein residues strengthens the interaction with Ufd1-Npl4 dimer and decreases its ATP-ase activity [160].

Unfolded Protein Response (UPR)

To maintain the efficiency of quality-control mechanisms in diverse physiological conditions, living cells have evolved regulatory circuits that monitor the levels of available chaperones. This is true for both the cytosol and the ER, and compartment-specific responses clearly exist that selectively restore optimal levels of the desired folding factors. ER stress can be triggered by increased protein misfolding within the ER caused for example by increased protein synthesis, heavy metals, oxidative damage or mutation of a substrate protein, or by decreased clearance of misfolded proteins from the ER, caused by impaired degradation or mutation of proteins involved in ERAD. UPR is often induced by different drugs, including thapsigargin, which induces calcium efflux from the ER, reducing agents, which disrupt disulfide bridges in the ER proteins, tunicamycin, which prevents protein N-linked glycosylation and brefeldin A, which induces the influx of Golgi enzyme sinto the ER and prevents the clearance of newly synthesized proteins by disrupting the secretory pathway thus selectively targeting the ER folding machinery [20,37]. The accumulation of aberrant proteins in the cytosol triggers the heat-shock response, resulting in *de novo* synthesis of Hsp70 and other cytosolic chaperones [161]; when the load of misfolded proteins in the ER exceeds the buffering capacity of ER chaperones UPR is triggered through the depletion of free BiP [19,37-41] (Figure 4).

BiP, also know as glucose regulated protein 78 (GRP78), is a member of the Hsp70 family residing within the ER lumen that binds transiently to newly synthesized proteins. BiP has an *N*-terminal ATPase and a *C*-terminal substrate binding domain. In the ADP-bound form BiP has high affinity for protein substrates. Substrates bound to BiP are locked in their conformation and stimulate the ATPase activity of BiP [162,163]. Exchange of ADP with ATP releases the substrate from BiP [162], which then progresses on its folding pathway. Subsequent ATP-hydrolysis returns BiP into the ADP, high affinity state. Thus, by cycling through the BiP ADP–ATP cycle a folding polypeptide chain consumes ATP. ATP is imported into the ER via antiport with ADP and AMP [164].

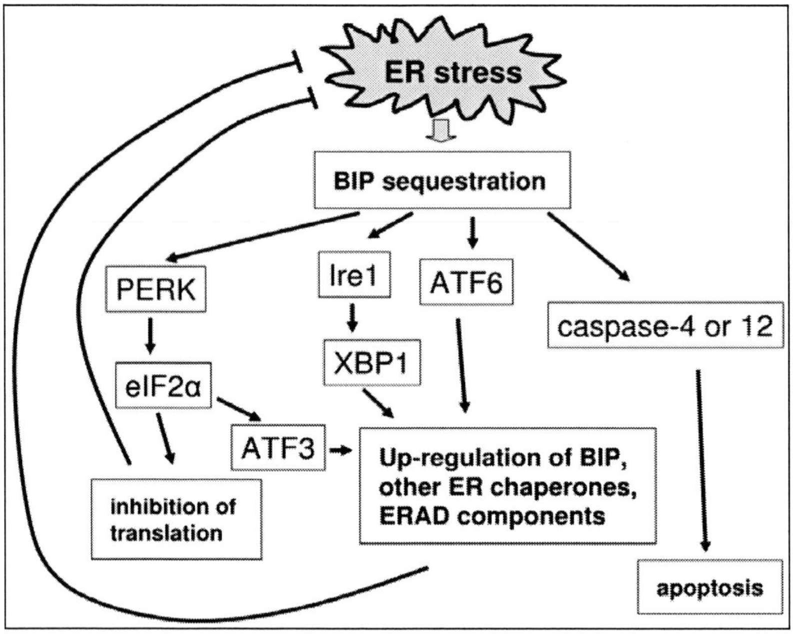

Figure 4. The current model of unfolded protein response. Misfolded proteins within the ER lumen sequester the BiP chaperone leading to the activation of PEK/PERK, Ire1 and ATF6. PERK phosphorylates eIF2α, which leads to transcriptional attenuation and induction of the ATF3 transcription factor. Ire1 induces cytosolic splicing of the XBP1 mRNA, leading to the production of the active XBP1 transcription factor. ATF6 migrates to the Golgi, where a cytosolic transcriptional activation domain is cut off and released. ATF6, ATF3 and XBP1 all induce transcription of ER chaperones, ERAD components and other factors counteracting ER stress. However, if ER stress persists, ultimately caspase-4 (caspase-12 in mice) is activated leading to ER-stress-induced apoptosis.

Whatever is the mechanism inducing ER stress, misfolded proteins recruit ER chaperones, including BiP, which under normal conditions is bound to luminal domains of three different transmembrane ER proteins: Ire1 (inositol requiring enzyme 1), ATF6 (Activating transcription factor 6) , and PEK/PERK (double stranded RNA-activated protein kinase-like ER kinase). When BiP is titrated by misfolded proteins, the intralumenal portions of those three proteins form homooligomers, what leads to their activation. BiP specifically binds Ire1 and that this interaction disappears under conditions of ER stress. Additionally, overexpression of BiP can suppress the induction of the UPR. This had suggested a titration model, in which BiP acts as a negative regulator of Ire1 and the accumulation of misfolded forms leads indirectly to Ire1 activation by the sequestration of BiP by misfolded proteins. Upon the release of BiP, Ire1 dimerizes to activate its cytosolic ribonuclease domain, which cleaves a 26-nucleotide intron sequence from X-box DNA-binding protein (XBP1) mRNA [40]. The resulting XBP1 is a potent activator of UPR genes after it migrates to the nucleus and binds the upstream DNA UPR element (UPRE) [165]. Translation of the spliced XBP1 message creates an active transcription factor that directly mediates transcription of UPR targets including ER chaperones, the ERAD machinery, and a range of other secretory proteins. The UPR genes regulating XBP1 are essential for protein folding, maturation and degradation.

The elegant model of BiP titration has been recently challenged, when mutational analysis found that deletion of the region of Ire1 responsible for BiP binding did not impair the regulation of Ire1 in the response to unfolded protein [166]. The crystal structure of the conserved core lumenal domain (cLD) of yeast Ire1 reveals a deep hydrophobic groove reminiscent of the binding pocket in the major histocompatibility complexes (MHCs) that is responsible for peptide recognition. That Ire1 may directly bind misfolded polypeptides is an appealing idea [167]. By directly recognizing misfolded forms, the initiation of Ire1 induction could occur prior to the full titration of BiP, which, given BiP's extremely high abundance in the ER, might occur only after a catastrophic accumulation of misfolded proteins. Additionally, direct recognition of misfolded forms by Ire1 (and PERK in higher eukaryotes) could allow for a more shaded set of responses in which different misfolded forms could be preferentially recognized by the different sensors (BiP, Ire1, or PERK). Thus, in principle, the nature and timing of the UPR could be tailored to the specific class of misfolded forms that are prevalent in the ER.

While Ire is the only ER stress sensor in yeast, higher eukaryotes have two Ire isoforms, α and β. Moreover, they utilize two other sensors, the ER transmembrane transcription factor ATF6 and the ER transmembrane kinase PERK. ATF6 was found in mammalian cells using yeast two-hybrid screens as another UPR transducer that promotes UPR responsive genes. Two forms of ATF6, ATF6α and ATF6β, both reside in the ER and can each travel to the Golgi upon release of BiP [20]. The 50 kDa cytosolic region of ATF6 α/β is cleaved by site-1 protease (S1P) and site-2 protease (S2P) in the Golgi compartment [168]. This active cytosolic ATF6 domain migrates to the nucleus and binds to the ER stress response element (ERSE) with CCAAT-binding factor (CBF) to promote the transcription of ER-resident molecular chaperones and other assistant folding enzymes [1,165]. ATF6 may work in parallel with XBP1 to promote proteins needed to assist in the folding of unfolded proteins in the ER [169]. Since the timing of activation for either the XBP1 pathway or the ATF6 pathway is unclear, it is unknown if these pathways are redundant or are used for a UPR gradient response. Further study of these pathways is required to better understand their role in mammalian cells.

PEK/PERK contains a lumenal sensor that is highly related to that of Ire1, suggesting that either can be readily activated by BiP release [37]. Oligomerization of PEK/PERK, leads to its autophosphorylation and activation of its eIF2α kinase activity, providing the first line of defense against overload with unfolded proteins. PERK activity transiently attenuates protein synthesis through phosphorylation of cytosolic α-subunit of eukaryotic translation initiation factor 2 (eIF2α), leading to a generalized repression of translation [170,171] associated with a selective translation of mRNAs bearing upstream open reading frames (ORFs). When eIF2α is phosphorylated, it can no longer exchange GDP for GTP in the eIF2-GDP-eIF2α complex, which prevents initiation events on almost all cellular mRNAs within the cell [20,37]. Since eIFα is required to bring the initiator methionyl-transfer RNA (Met-tRNAi) to the 40S ribosome, the absence of free eIF2 will prevent the 40S and 60S ribosomal subunits from forming the 80S initiation complex for translation [20,37]. However, selective mRNAs can be preferentially translated with eIF2α phosphorylated, provided that there are upstream ORFs within the 5' untranslated region of the mRNA [20]. Activation of the kinase by the presence of misfolded proteins results in a generalized inhibition of translation as well as the upregulation of a specific transcription factor, activating transcription factor 4 (ATF4), which has this additional upstream ORF. ATF4 activates genes involved in amino acid metabolism and transport, oxidative-reduction reactions and ER stress-induced apoptosis [40]. ATF4 has been shown to bind to a C/EBP-ATF composite site and regulate CHOP (C/EBP homologous transcription factor) transcription [38]. CHOP is implicated in both growth arrest and cellular apoptosis; with prolonged UPR activation ATF4 can induce expression of CHOP leading to caspase-3 activation and cell death [38].

Oligomerization of Ire1 and its autophosphorylation activates its endonuclease activity, mediating a unique cytosolic splicing of the mRNA coding for the XBP1 transcription factor [165]. Spliced XBP1 mRNA is translated into the active XBP1. Release of BiP from ATF6 allows its transfer to the Golgi, where it is cleaved by S1P and S2P. This leads to the release of a free cytosolic portion of ATF6, which is a transcription factor, relocating to the nucleus [172,173]. XBP1, ATF6, and ATF4 all induce a concerted transcription of multiple genes coding various ER chaperones, components of ERAD, and elements of the secretory pathway [174,175].

ATF6 drives the transcriptional upregulation of many ER-resident proteins and folding assistants. Ire1 activates XBP1, which in turn induces transcription of factors that facilitate ERAD. The two-step activation of XBP1 (transcriptionally induced by ATF6 and post-transcriptionally regulated by Ire1) guarantees the proper timing of the UPR; attempts to fold proteins precede the decision to degrade them25. If the response fails to clear the ER, apoptosis is induced through several pathways [1,20,37]. This link has important consequences for the pathogenesis of degenerative disorders.

Thus, UPR is an adaptative response allowing the cell to cope with ER stress: however, when UPR is persistent, it leads to apoptosis via transcriptional induction of CHOP/GADD153, TRAF-mediated activation of cJUN N-terminal kinase (JNK), and/or the activation of caspase-4 (caspase 12 in mice) [19,41]. Under prolonged UPR, ER calcium homeostasis is significantly disrupted, leading to activation of caspase-12. Caspase-4/Caspase-12 is an ER-associated effector of caspase activity that leas to cellular apoptosis though the activation of caspases 9 and 3 [40]. For example, in G_{M1}-gangliosidosis, G_{M1}-ganglioside accumulates in the ER and leads to loss of ER calcium. This depletion signals activation of UPR in these cells resulting, in prolonged activation of UPR genes and UPR signalled apoptosis [176].

ERAD constitutively counteracts ER stress, eliminating misfolded proteins from the ER. Nevertheless, when the load of misfolded proteins is increased, UPR is required to increase ERAD activity [82,177]. Inhibition of proteasomes induce UPR, as measured by XBP1 splicing and BiP induction (Wojcik *et al.*, manuscript in preparation). Pharmacological inhibition of the UPS upregulates multiple genes involved in UPR [178], either by accumulation of ERAD substrates and/or through stabilization of the active nuclear ATF6 [179]. However, in some systems, it also suppresses XBP1 splicing [180], inhibits PERK, and stabilizes the GADD34 phosphatase, thus preventing eIF2α phosphorylation [181].

Changes in ERAD and UPR Associated with Human Disease

Because ERAD is a central element of the secretory pathway, it is not surprising that it has major implications for the generation of human diseases. Failure of ERAD, excessive ERAD and induction of UPR all play an important role in human pathology, including diseases such as cancer, stroke, cystic fibrosis or pulmonary emphysema [29,31,40,182,183]. In the case of many diseases, the roles of individual components of the ERAD pathway – including VCP – have began to be elucidated.

The underlying disorders can be classified into two groups. The first group results from loss-of-function mutations in ERAD components that stabilize aberrant proteins, which in turn accumulate and damage the cell. Insufficient ERAD is also associated with various pathological conditions, which involve an accumulation of undegraded substrates. ERAD substrates encounter at least two bottlenecks: dislocation across the ER membrane and degradation by the proteasome [10]. A prominent example of this is Parkinson's disease (PD). Many cases of familial PD are associated with Parkin mutations linking ERAD to the generation of this neurodegenerative disorder [184]. Parkin functions in a complex with CHIP as a ubiquitin ligase [93]. A number of putative substrates for this complex have been suggested. One of them is wild-type Pael-R, a membrane-spanning receptor protein that tends to aggregate in a misfolded form [185]. Overexpression of Pael-R leads to cell death, which can be prevented by coexpressing Parkin. The structure, interactions and topology determine which of those events is time-limiting. ERAD substrates which are able to exit the ER under conditions of impaired UPS function accumulate in the cytosol forming aggresomes, while ERAD substrates which are unable to exit the ER form aggregates within its lumen, producing Russell bodies [186,187]. While it is still a matter of dispute, whether aggresomes are deleterious or protective [187,188], deposits of aggregated proteins within the lumen of the ER may impair ER function, induce UPR and ultimately lead to cell death, as in the case of mutated uromodulin associated with polycystic kidney disease [189]. Some misfolded proteins which should be degraded by ERAD fail to be recognized by the QC system and are secreted, what leads to their extracellular accumulation, a situation known as amyloidosis. For example, only the most highly destabilized variants of transthyretin are degraded by ERAD while other misfolded variants are secreted with wild type efficiency, leading to amyloid deposits. Inducing and enhancing ERAD of such proteins would provide means of treating various amyloidosis [190].

The second group of disorders is caused by premature degradation of secretory or membrane proteins, preventing their deployment to distal compartments, Various genetic disorders are characterized by an excessive ERAD of an otherwise functional protein. Cystic fibrosis (CF) is the most common genetic disorder in Caucasians, caused by the lack of expression of CFTR (CF conductance regulator) on the surface of epithelial cells [183]. CFTR is normally synthesized in the ER and transported to the cell surface, however a deletion of a Phe residue at position 508 induces retention of ΔF508CFTR in the ER and its degradation by the UPS [44,45]. If the mutant protein is allowed to reach the cell surface it is able to perform its physiological function. Proteasomal inhibitors that are effective in treating other diseases [191] lead to the accumulation of aggregated polyubiquitinated ΔF508CFTR in the cytoplasm and cannot rescue its surface expression [192]. When ubiquitination of ΔF508CFTR is blocked by expressing an inactive version of the involved E2/E3 complex (UbcH5a/CHIP), ΔF508CFTR is not dislocated but retained within the ER [193]. However, ΔF508CFTR is delivered to the cell surface upon exposure to chemical chaperones, which increase the levels of properly folded ΔF508CFTR and allow it to evade the QC system [54]. Patients with CF can thus benefit from an inhibition of ERAD at the level of retrotranslocation, but not at the level of degradation by proteasomes.

A similar mechanism operates in the X-linked form of the Charcot-Marie-Tooth disease which is caused by mutations of connexin 32 [194], the leukodystrophy Pelizaeus-Merzbacher disease [195], which is caused by mutations of the proteolipid protein by splice site mutations, duplications of the PLP1 gene or null alleles; in hereditary emphysema of the lungs, caused by mutations of α1-antitrypsin [196]; and in the recent identified autosomal dominant inclusion body (Collins' bodies) dementia, FENIB (*F*amilial *E*ncephalopathy with *N*euroserpin *I*nclusion *B*odies) [197-199], characterized by eosinophilic neuronal inclusions of neuroserpin in the deeper layers of the cerebral cortex and the substantia nigra and is associated with a mutation, Ser_{49}→Pro, in the neuroserpin gene [118].

Perfectly normal proteins can be aberrantly targeted for ERAD when they are bound to viral proteins. For example, the US11 protein of the CMV induces the degradation of MHC class I heavy chains thus preventing an efficient presentation of viral antigens to immunocompetent cells [200]. Instead of a viral protein it is an endogenous sterol-sensing protein called Insig-1 which induces a quick degradation of HMG-CoA reductase whenever cholesterol is abundant [201-203]. Inefficient degradation of HMG-CoA and thus increased production of endogenous cholesterol despite abundant external supply relates ERAD to heart disease. In addition to that homocysteine, a known risk factor in cardiovascular disease, induces the retention in the ER and consequent ERAD of several important factors which are normally secreted by endothelial cells such as the von Willebrand factor [204]. Increased blood levels of homocysteine lead to atherosclerosis and thrombosis, by a pathogenetic mechanisms which involve ER stress and UPR activation in endothelial cells [205-207]. Moreover, continuous ER stress leads to apoptosis of endothelial cells mediated by UPR activation [169]. Homocysteine induces apoptosis in endothelial cells by activating Ire1 [169]. Homocysteine-induced cell death was suppressed with the overexpression of point mutants that inactivated the RNase domain of Ire1, suggesting that excessive homocysteine uncontrollably activates UPR though the false activation of XBP1.

In acute pathological states of the brain, including stroke, neurotrauma and epileptic seizures, as well as in degenerative diseases, ER function is often affected in multiple ways [208]. These include oxidative stress, nitric oxide-induced inactivation of the ER calcium pump resulting in disturbances of ER calcium homeostasis and impairment of UPR and

ERAD. Furthermore, proteasomal function is impaired which causes secondary ER dysfunction. The only way to escape this potentially lethal cycle is to induce UPR and thus to activate new synthesis of the ER chaperone BiP to levels sufficient to refold unfolded proteins. ER dysfunction may induce a state of tolerance, impair cellular functions, or induce apoptosis, depending on the severity and duration and the cell type affected. In different neurodegenerative conditions associated with protein misfolding [including PD, Huntington's disease (HD), Alzheimer's disease, Prion-related disorders, amyotrophic lateral sclerosis (ALS) and others], irreversible alteration of ER homeostasis has been proposed to be a critical mediator of neuronal dysfunction.

An increase of ERAD associated with UPR is found in PD [209] and in neurodegeneration resulting from accumulation of polyglutamine (polyQ) tract proteins [210]. Two different neurotoxins known to induce parkinsonism (6-hydroxydopamine and 1-methyl- 4-phenylpyridinium) induce features of UPR in a dopaminergic cell line, including XBP1 splicing, eIF2α phosphorylation and induction of ER chaperones, the transcription factor CHOP/Gadd153 and elements of the UPS [209]. SOD1 proteins genetically linked to ALS may induce motoneuron dysfunction by alteration in the function of ER and Golgi [211]. Experimental brain hypoxia-ischemia triggers ER dysfunction and subsequent UPR accompanied by induction of the HRD1 Ub ligase [85]. Blocking the proapoptotic branch of UPR while enhancing the prosurvival response, including an enhancement in ERAD may be exploited as a new therapeutic strategy in stroke. Proteasome inhibitors are considered in the treatment of stroke to limit the damage caused by secondary inflammation. Proteasome inhibitors normally do not penetrate the blood-brain barrier, however they may penetrate the nervous tissue in the ischemic core and the penumbra leading to a potentially deleterious inhibition of ERAD and induction of UPR in neuronal cells [212].

Involvement of VCP in Human Disease

VCP plays an important role in the pathogenesis of neurodegenerative diseases. A mutation in the N-terminal domain of VCP creates a dominant trait referred to as inclusion body myopathy Paget's disease of the bone and frontotemporal dementia (IBMPFD) [213-215]. VCP has been observed in ubiquitin-positive intraneuronal inclusions found in common neurodegenerative diseases such as PD, HD, and ALS [130,143,146]. Such inclusions are often regarded as signs of 'intracellular indigestion' that is associated with impaired UPS function and can be modeled by protein aggregates induced with proteasome inhibitors or 'aggresomes' [188,192]. RNA interference experiments have demonstrated that VCP is required for aggresome formation [124], a finding which was recently supported by other authors who used a dominant negative approach [147]. Furthermore, VCP interacts with proteins containing polyQ expansions [130]. VCP also functionally regulates the activity of dorfin, an E3 ubiquitinating superoxide dismutase 1 (SOD1), which is mutated in familial forms of ALS [146]. Overexpression of VCP protects the cells against the formation of discrete cytoplasmic aggregates of misfolded proteins [216], while in another system it is associated with a worsening of the neurodegenerative phenotype [153].

While impaired VCP activity seems to be associated with neurodegeneration, overexpression and increased activity of VCP is associated with poor prognosis in several types of human cancer, including esophageal squamous cell carcinoma [217], non-small-cell lung carcinoma [218], pancreatic ductal adenocarcinoma [219], colorectal adenocarcinoma [220], gastric carcinoma [221], and hepatocellular carcinoma [222]. Poor prognosis in cancer is also associated with increased expression of the gp78 ER-anchored ubiquitin ligase [223,224], which directly interacts with VCP [225]. Gp78, originally identified as the autocrine motility factor receptor (AMFR), is a surface glycoprotein which induces transformation of NIH 3T3 fibroblasts [226]. It is likely that upregulation of gp78 and/or VCP increases the resistance of cancer cells to ER stress [227]. Therefore, development of potential inhibitors of VCP is of great interest for cancer therapy [228].

Conclusions and Perspectives

ER stress, UPR and ERAD are interconnected phenomena, which are extremely important in the physiology of neuronal cells and are directly connected to the ethipathology of multiple acute and chronic disorders of the CNS, ranging from stroke to PD. UPS is involved in ERAD and controls the UPR, however not all ERAD is UPS-mediated, while the effects of UPS on UPR are ambiguous: proteasome inhibition sometimes trigger UPR while sometimes prevents it. Therefore while it is tempting to use simplified models, UPR and ERAD are extremely complex processes. The signaling network associated with UPR involves numerous signaling nodes and it has begun to be clarified only recently. Misfolded ER proteins follow different pathways of ERAD requiring different sets of proteins. The use of yeast as a model organism to study ERAD has provided many insights on how ERAD functions, however the indiscriminate extrapolation of results obtained in yeast to mammalian systems is dangerous, since many differences do exist. Studies in yeast have established the Cdc48p ATPase in complex with Ufd1-Npl4 as a crucial molecule involved in ERAD, however the evidence for the involvement of its mammalian homolog, VCP, in ERAD is far less convincing. Mammalian ERAD and UPR seems to be much more complex than in yeast, since new pathways have evolved (eg. ATF6 is absent from yeast), old pathways have been duplicated (e.g. Ire1, Rad23 and Ufd2 have each two homologs in mammals) while other pathways have acquired new and different functions (eg. this may be the fate of VCP). However, there is no doubt that VCP participates in many aspects of UPS function, as well as, regulates ER structure and function, both aspects not necessarily connected with ERAD. More importantly, VCP seems to be directly involved in the pathogenesis of several neurological disorders. There is no doubt that the next few years will bring a burst of new discoveries about the different aspects of ER stress, UPR and ERAD and in particular about their role in human disease. These data pretend to reinforce the notion that irreversible alteration of ER function has deleterious effects to the cell and the organism. But, as initially described in yeast models, the UPR is a prosurvival pathway and its beneficial effect can be also translated into pathological conditions. Pharmacological targeting of different components of the UPR/ER stress pathway may have therapeutic application for the treatment of many neurological disorders.

References

[1] Ma Y, Hendershot LM: The unfolding tale of the unfolded protein response. *Cell* 2001, 107:827-830.

[2] Matlack KE, Mothes W, Rapoport TA: Protein translocation: tunnel vision. *Cell* 1998, 92:381-390.

[3] McMaster CR: Lipid metabolism and vesicle trafficking: more than just greasing the transport machinery. *Biochem Cell Biol* 2001, 79:681-692.

[4] van Anken E, Braakman I: Endoplasmic reticulum stress and the making of a professional secretory cell. *Crit Rev Biochem Mol Biol* 2005, 40:269-283.

[5] van Anken E, Braakman I: Versatility of the endoplasmic reticulum protein folding factory. *Crit Rev Biochem Mol Biol* 2005, 40:191-228.

[6] Ellgaard L, Helenius A: Quality control in the endoplasmic reticulum. *Nat Rev Mol Cell Biol* 2003, 4:181-191.

[7] Fewell SW, Travers KJ, Weissman JS, Brodsky JL: The action of molecular chaperones in the early secretory pathway. *Annu Rev Genet* 2001, 35:149-191.

[8] Ellgaard L, Molinari M, Helenius A: Setting the standards: quality control in the secretory pathway. *Science* 1999, 286:1882-1888.

[9] Meunier L, Usherwood YK, Chung KT, Hendershot LM: A subset of chaperones and folding enzymes form multiprotein complexes in endoplasmic reticulum to bind nascent proteins. *Mol Biol Cell* 2002, 13:4456-4469.

[10] Sitia R, Braakman I: Quality control in the endoplasmic reticulum protein factory. *Nature* 2003, 426:891-894.

[11] Banno T, Kohno K: Conformational changes of the smooth endoplasmic reticulum are facilitated by L-glutamate and its receptors in rat Purkinje cells. *J Comp Neurol* 1998, 402:252-263.

[12] Cheng TP, Reese TS: Polarized compartmentalization of organelles in growth cones from developing optic tectum. *J Cell Biol* 1985, 101:1473-1480.

[13] Dailey ME, Bridgman PC: Dynamics of the endoplasmic reticulum and other membranous organelles in growth cones of cultured neurons. *J Neurosci* 1989, 9:1897-1909.

[14] Deitch JS, Banker GA: An electron microscopic analysis of hippocampal neurons developing in culture: early stages in the emergence of polarity. *J Neurosci* 1993, 13:4301-4315.

[15] Kanaseki T, Ikeuchi Y, Tashiro Y: Rough surfaced smooth endoplasmic reticulum in rat and mouse cerebellar Purkinje cells visualized by quick-freezing techniques. *Cell Struct Funct* 1998, 23:373-387.

[16] Levesque L, Annaert W, Craessaerts K, Mathews PM, Seeger M, Nixon RA, Van Leuven F, Gandy S, Westaway D, George-Hyslop P, De Strooper B, Fraser PE: Developmental expression of wild-type and mutant presenilin-1 in hippocampal neurons from transgenic mice: evidence for novel species-specific properties of human presenilin-1. *Mol Med* 1999, 5:542-554.

[17] Rizzuto R, Pinton P, Carrington W, Fay FS, Fogarty KE, Lifshitz LM, Tuft RA, Pozzan T: Close contacts with the endoplasmic reticulum as determinants of mitochondrial Ca2+ responses. *Science* 1998, 280:1763-1766.

[18] Voeltz GK, Rolls MM, Rapoport TA: Structural organization of the endoplasmic reticulum. *EMBO Rep* 2002, 3:944-950.

[19] Wu J, Kaufman RJ: From acute ER stress to physiological roles of the Unfolded Protein Response. *Cell Death Differ* 2006, 124:587-599.

[20] Kaufman RJ, Scheuner D, Schroder M, Shen X, Lee K, Liu CY, Arnold SM: The unfolded protein response in nutrient sensing and differentiation. *Nat Rev Mol Cell Biol* 2002, 3:411-421.

[21] Schrag JD, Procopio DO, Cygler M, Thomas DY, Bergeron JJ: Lectin control of protein folding and sorting in the secretory pathway. *Trends Biochem Sci* 2003, 28:49-57.

[22] Fassio A, Sitia R: Formation, isomerisation and reduction of disulphide bonds during protein quality control in the endoplasmic reticulum. *Histochem Cell Biol* 2002, 117:151-157.

[23] Sevier CS, Kaiser CA: Formation and transfer of disulphide bonds in living cells. *Nat Rev Mol Cell Biol* 2002, 3:836-847.

[24] Harding HP, Zhang Y, Zeng H, Novoa I, Lu PD, Calfon M, Sadri N, Yun C, Popko B, Paules R, Stojdl DF, Bell JC, Hettmann T, Leiden JM, Ron D: An integrated stress response regulates amino acid metabolism and resistance to oxidative stress. *Mol Cell* 2003, 11:619-633.

[25] van Anken E, Romijn EP, Maggioni C, Mezghrani A, Sitia R, Braakman I, Heck AJ: Sequential waves of functionally related proteins are expressed when B cells prepare for antibody secretion. *Immunity* 2003, 18:243-253.

[26] Spiro RG: Role of N-linked polymannose oligosaccharides in targeting glycoproteins for endoplasmic reticulum-associated degradation. *Cell Mol Life Sci* 2004, 61:1025-1041.

[27] Romisch K: Endoplasmic Reticulum-Associated Degradation. *Annu Rev Cell Dev Biol* 2005.

[28] Schroder M, Kaufman RJ: ER stress and the unfolded protein response. *Mutat Res* 2005, 569:29-63.

[29] Tsai B, Ye Y, Rapoport TA: Retro-translocation of proteins from the endoplasmic reticulum into the cytosol. *Nat Rev Mol Cell Biol* 2002, 3:246-255.

[30] Kostova Z, Wolf DH: For whom the bell tolls: protein quality control of the endoplasmic reticulum and the ubiquitin-proteasome connection. *EMBO J* 2003, 22:2309-2317.

[31] Ahner A, Brodsky JL: Checkpoints in ER-associated degradation: excuse me, which way to the proteasome? *Trends Cell Biol* 2004, 14:474-478.

[32] Bar-Nun S: The role of p97/Cdc48p in endoplasmic reticulum-associated degradation: from the immune system to yeast. *Curr Top Microbiol Immunol* 2005, 300:95-125.

[33] Song BL, Sever N, bose-Boyd RA: Gp78, a Membrane-Anchored Ubiquitin Ligase, Associates with Insig-1 and Couples Sterol-Regulated Ubiquitination to Degradation of HMG CoA Reductase. *Mol Cell* 2005, 19:829-840.

[34] Alzayady KJ, Panning MM, Kelley GG, Wojcikiewicz RJ: Involvement of the p97-Ufd1-Npl4 Complex in the Regulated Endoplasmic Reticulum-associated Degradation of Inositol 1,4,5-Trisphosphate Receptors. *J Biol Chem* 2005, 280:34530-34537.

[35] Wang Q, Song C, Li CC: Molecular perspectives on p97-VCP: progress in understanding its structure and diverse biological functions. *J Struct Biol* 2004, 146:44-57.

[36] Dreveny I, Pye VE, Beuron F, Briggs LC, Isaacson RL, Matthews SJ, McKeown C, Yuan X, Zhang X, Freemont PS: p97 and close encounters of every kind: a brief review. *Biochem Soc Trans* 2004, 32:715-720.

[37] Harding HP, Calfon M, Urano F, Novoa I, Ron D: Transcriptional and translational control in the Mammalian unfolded protein response. *Annu Rev Cell Dev Biol* 2002, 18:575-599.

[38] Ma Y, Hendershot LM: The mammalian endoplasmic reticulum as a sensor for cellular stress. *Cell Stress Chaperones* 2002, 7:222-229.

[39] Hampton RY: IRE1: a role in UPREgulation of ER degradation. *Dev Cell* 2003, 4:144-146.

[40] Shen X, Zhang K, Kaufman RJ: The unfolded protein response-a stress signaling pathway of the endoplasmic reticulum. *J Chem Neuroanat* 2004, 28:79-92.

[41] Boyce M, Yuan J: Cellular response to endoplasmic reticulum stress: a matter of life or death. *Cell Death Differ* 2006, 13:363-373.

[42] Wojcik C, DeMartino GN: Intracellular localization of proteasomes. *Int J Biochem Cell Biol* 2003, 35:579-589.

[43] Sommer T, Jentsch S: A protein translocation defect linked to ubiquitin conjugation at the endoplasmic reticulum. *Nature* 1993, 365:176-179.

[44] Jensen TJ, Loo MA, Pind S, Williams DB, Goldberg AL, Riordan JR: Multiple proteolytic systems, including the proteasome, contribute to CFTR processing. *Cell* 1995, 83:129-135.

[45] Ward CL, Omura S, Kopito RR: Degradation of CFTR by the ubiquitin-proteasome pathway. *Cell* 1995, 83:121-127.

[46] Vashist S, Ng DT: Misfolded proteins are sorted by a sequential checkpoint mechanism of ER quality control. *J Cell Biol* 2004, 165:41-52.

[47] Nishikawa S, Brodsky JL, Nakatsukasa K: Roles of molecular chaperones in endoplasmic reticulum (ER) quality control and ER-associated degradation (ERAD). *J Biochem (Tokyo)* 2005, 137:551-555.

[48] Mancini R, Aebi M, Helenius A: Multiple endoplasmic reticulum-associated pathways degrade mutant yeast carboxypeptidase Y in mammalian cells. *J Biol Chem* 2003, 278:46895-46905.

[49] Schmitz A, Herzog V: Endoplasmic reticulum-associated degradation: exceptions to the rule. *Eur J Cell Biol* 2004, 83:501-509.

[50] Heinemann FS, Mziaut H, Korza G, Ozols J: A microsomal endopeptidase from liver that preferentially degrades stearoyl-CoA desaturase. *Biochemistry* 2003, 42:6929-6937.

[51] Cabral CM, Choudhury P, Liu Y, Sifers RN: Processing by endoplasmic reticulum mannosidases partitions a secretion-impaired glycoprotein into distinct disposal pathways. *J Biol Chem* 2000, 275:25015-25022.

[52] Fayadat L, Siffroi-Fernandez S, Lanet J, Franc JL: Degradation of human thyroperoxidase in the endoplasmic reticulum involves two different pathways depending on the folding state of the protein. *J Biol Chem* 2000, 275:15948-15954.

[53] Hiller MM, Finger A, Schweiger M, Wolf DH: ER degradation of a misfolded luminal protein by the cytosolic ubiquitin-proteasome pathway. *Science* 1996, 273:1725-1728.

[54] Gelman MS, Kannegaard ES, Kopito RR: A principal role for the proteasome in endoplasmic reticulum-associated degradation of misfolded intracellular cystic fibrosis transmembrane conductance regulator. *J Biol Chem* 2002, 277:11709-11714.

[55] Hosokawa N, Wada I, Hasegawa K, Yorihuzi T, Tremblay LO, Herscovics A, Nagata K: A novel ER alpha-mannosidase-like protein accelerates ER-associated degradation. *EMBO Rep* 2001, 2:415-422.

[56] Mast SW, Diekman K, Karaveg K, Davis A, Sifers RN, Moremen KW: Human EDEM2, a novel homolog of family 47 glycosidases, is involved in ER-associated degradation of glycoproteins. *Glycobiology* 2005, 15:421-436.

[57] Olivari S, Galli C, Alanen H, Ruddock L, Molinari M: A novel stress-induced EDEM variant regulating endoplasmic reticulum-associated glycoprotein degradation. *J Biol Chem* 2005, 280:2424-2428.

[58] Molinari M, Calanca V, Galli C, Lucca P, Paganetti P: Role of EDEM in the release of misfolded glycoproteins from the calnexin cycle. *Science* 2003, 299:1397-1400.

[59] Wiertz EJ, Tortorella D, Bogyo M, Yu J, Mothes W, Jones TR, Rapoport TA, Ploegh HL: Sec61-mediated transfer of a membrane protein from the endoplasmic reticulum to the proteasome for destruction. *Nature* 1996, 384:432-438.

[60] Zhou M, Schekman R: The engagement of Sec61p in the ER dislocation process. *Mol Cell* 1999, 4:925-934.

[61] Walter J, Urban J, Volkwein C, Sommer T: Sec61p-independent degradation of the tail-anchored ER membrane protein Ubc6p. *EMBO J* 2001, 20:3124-3131.

[62] Ye Y, Shibata Y, Yun C, Ron D, Rapoport TA: A membrane protein complex mediates retro-translocation from the ER lumen into the cytosol. *Nature* 2004, 429:841-847.

[63] Lilley BN, Ploegh HL: A membrane protein required for dislocation of misfolded proteins from the ER. *Nature* 2004, 429:834-840.

[64] Huyer G, Piluek WF, Fansler Z, Kreft SG, Hochstrasser M, Brodsky JL, Michaelis S: Distinct machinery is required in Saccharomyces cerevisiae for the endoplasmic reticulum-associated degradation of a multispanning membrane protein and a soluble luminal protein. *J Biol Chem* 2004, 279:38369-38378.

[65] Kreft SG, Wang L, Hochstrasser M: Membrane topology of the yeast endoplasmic reticulum-localized ubiquitin ligase Doa10 and comparison with its human ortholog TEB4 (MARCH-VI). *J Biol Chem* 2006, 281:4646-4653.

[66] Taxis C, Hitt R, Park SH, Deak PM, Kostova Z, Wolf DH: Use of modular substrates demonstrates mechanistic diversity and reveals differences in chaperone requirement of ERAD. *J Biol Chem* 2003, 278:35903-35913.

[67] Ye Y, Meyer HH, Rapoport TA: Function of the p97-Ufd1-Npl4 complex in retrotranslocation from the ER to the cytosol: dual recognition of nonubiquitinated polypeptide segments and polyubiquitin chains. *J Cell Biol* 2003, 162:71-84.

[68] Ye Y, Meyer HH, Rapoport TA: The AAA ATPase Cdc48/p97 and its partners transport proteins from the ER into the cytosol. *Nature* 2001, 414:652-656.

[69] Bays NW, Wilhovsky SK, Goradia A, Hodgkiss-Harlow K, Hampton RY: HRD4/NPL4 is required for the proteasomal processing of ubiquitinated ER proteins. *Mol Biol Cell* 2001, 12:4114-4128.

[70] Rabinovich E, Kerem A, Frohlich KU, Diamant N, Bar-Nun S: AAA-ATPase p97/Cdc48p, a cytosolic chaperone required for endoplasmic reticulum-associated protein degradation. *Mol Cell Biol* 2002, 22:626-634.

[71] Jarosch E, Taxis C, Volkwein C, Bordallo J, Finley D, Wolf DH, Sommer T: Protein dislocation from the ER requires polyubiquitination and the AAA-ATPase Cdc48. *Nat Cell Biol* 2002, 4:134-139.

[72] Braun S, Matuschewski K, Rape M, Thoms S, Jentsch S: Role of the ubiquitin-selective CDC48(UFD1/NPL4)chaperone (segregase) in ERAD of OLE1 and other substrates. *EMBO J* 2002, 21:615-621.

[73] Lee RJ, Liu CW, Harty C, McCracken AA, Latterich M, Romisch K, DeMartino GN, Thomas PJ, Brodsky JL: Uncoupling retro-translocation and degradation in the ER-associated degradation of a soluble protein. *EMBO J* 2004, 23:2206-2215.

[74] Mayer TU, Braun T, Jentsch S: Role of the proteasome in membrane extraction of a short-lived ER-transmembrane protein. *EMBO J* 1998, 17:3251-3257.

[75] Enenkel C, Lehmann A, Kloetzel PM: Subcellular distribution of proteasomes implicates a major location of protein degradation in the nuclear envelope-ER network in yeast. *EMBO J* 1998, 17:6144-6154.

[76] Kalies KU, Allan S, Sergeyenko T, Kroger H, Romisch K: The protein translocation channel binds proteasomes to the endoplasmic reticulum membrane. *EMBO J* 2005, 24:2284-2293.

[77] van Laar T, van der Eb AJ, Terleth C: Mif1: a missing link between the unfolded protein response pathway and ER-associated protein degradation? *Curr Protein Pept Sci* 2001, 2:169-190.

[78] Hori O, Ichinoda F, Yamaguchi A, Tamatani T, Taniguchi M, Koyama Y, Katayama T, Tohyama M, Stern DM, Ozawa K, Kitao Y, Ogawa S: Role of Herp in the endoplasmic reticulum stress response. *Genes Cells* 2004, 9:457-469.

[79] Biederer T, Volkwein C, Sommer T: Role of Cue1p in ubiquitination and degradation at the ER surface. *Science* 1997, 278:1806-1809.

[80] Galan JM, Cantegrit B, Garnier C, Namy O, Haguenauer-Tsapis R: 'ER degradation' of a mutant yeast plasma membrane protein by the ubiquitin-proteasome pathway. *FASEB J* 1998, 12:315-323.

[81] Tiwari S, Weissman AM: Endoplasmic reticulum (ER)-associated degradation of T cell receptor subunits. Involvement of ER-associated ubiquitin-conjugating enzymes (E2s). *J Biol Chem* 2001, 276:16193-16200.

[82] Friedlander R, Jarosch E, Urban J, Volkwein C, Sommer T: A regulatory link between ER-associated protein degradation and the unfolded-protein response. *Nat Cell Biol* 2000, 2:379-384.

[83] Gitan RS, Eide DJ: Zinc-regulated ubiquitin conjugation signals endocytosis of the yeast ZRT1 zinc transporter. *Biochem J* 2000, 346 Pt 2:329-336.

[84] Kaneko M, Ishiguro M, Niinuma Y, Uesugi M, Nomura Y: Human HRD1 protects against ER stress-induced apoptosis through ER-associated degradation. *FEBS Lett* 2002, 532:147-152.

[85] Qi X, Okuma Y, Hosoi T, Kaneko M, Nomura Y: Induction of murine HRD1 in experimental cerebral ischemia. *Brain Res Mol Brain Res* 2004, 130:30-38.

[86] Fang S, Ferrone M, Yang C, Jensen JP, Tiwari S, Weissman AM: The tumor autocrine motility factor receptor, gp78, is a ubiquitin protein ligase implicated in degradation from the endoplasmic reticulum. *Proc Natl Acad Sci U S A* 2001, 98:14422-14427.

[87] Nabi IR, Watanabe H, Silletti S, Raz A: Tumor cell autocrine motility factor receptor. *EXS* 1991, 59:163-177.

[88] Hassink GC, Kikkert M, van VS, Lee SJ, Spaapen R, van Laar T, Coleman CS, Bartee E, Fruh K, Chau V, Wiertz EJ: TEB4 is a C4HC3 RING finger-containing ubiquitin ligase of the endoplasmic reticulum. *Biochem J* 2005.

[89] Imai Y, Soda M, Takahashi R: Parkin suppresses unfolded protein stress-induced cell death through its E3 ubiquitin-protein ligase activity. *J Biol Chem* 2000, 275:35661-35664.

[90] Takahashi R, Imai Y: Pael receptor, endoplasmic reticulum stress, and Parkinson's disease. *J Neurol* 2003, 250 Suppl 3:III25-III29.

[91] Takahashi R, Imai Y, Hattori N, Mizuno Y: Parkin and endoplasmic reticulum stress. *Ann N Y Acad Sci* 2003, 991:101-106.

[92] Imai J, Hasegawa H, Maruya M, Koyasu S, Yahara I: Exogenous antigens are processed through the endoplasmic reticulum-associated degradation (ERAD) in cross-presentation by dendritic cells. *Int Immunol* 2005, 17:45-53.

[93] Imai Y, Soda M, Hatakeyama S, Akagi T, Hashikawa T, Nakayama KI, Takahashi R: CHIP is associated with Parkin, a gene responsible for familial Parkinson's disease, and enhances its ubiquitin ligase activity. *Mol Cell* 2002, 10:55-67.

[94] Suzuki T, Park H, Lennarz WJ: Cytoplasmic peptide:N-glycanase (PNGase) in eukaryotic cells: occurrence, primary structure, and potential functions. *FASEB J* 2002, 16:635-641.

[95] Kim I, Ahn J, Liu C, Tanabe K, Apodaca J, Suzuki T, Rao H: The Png1-Rad23 complex regulates glycoprotein turnover. *J Cell Biol* 2006, 172:211-219.

[96] Yoshida Y, Adachi E, Fukiya K, Iwai K, Tanaka K: Glycoprotein-specific ubiquitin ligases recognize N-glycans in unfolded substrates. *EMBO Rep* 2005, 6:239-244.

[97] Yoshida Y, Tokunaga F, Chiba T, Iwai K, Tanaka K, Tai T: Fbs2 is a new member of the E3 ubiquitin ligase family that recognizes sugar chains. *J Biol Chem* 2003, 278:43877-43884.

[98] Yoshida Y, Chiba T, Tokunaga F, Kawasaki H, Iwai K, Suzuki T, Ito Y, Matsuoka K, Yoshida M, Tanaka K, Tai T: E3 ubiquitin ligase that recognizes sugar chains. *Nature* 2002, 418:438-442.

[99] Kikkert M, Doolman R, Dai M, Avner R, Hassink G, van VS, Thanedar S, Roitelman J, Chau V, Wiertz E: Human HRD1 is an E3 ubiquitin ligase involved in degradation of proteins from the endoplasmic reticulum. *J Biol Chem* 2004, 279:3525-3534.

[100] Kaneko M, Nomura Y: [Protective effects of HRD1 and 4-phenylbutyric acid against neuronal cell death]. *Nippon Yakurigaku Zasshi* 2004, 124:391-398.

[101] Hoppe T: Multiubiquitylation by E4 enzymes: 'one size' doesn't fit all. *Trends Biochem Sci* 2005, 30:183-187.

[102] Kaneko C, Hatakeyama S, Matsumoto M, Yada M, Nakayama K, Nakayama KI: Characterization of the mouse gene for the U-box-type ubiquitin ligase UFD2a. *Biochem Biophys Res Commun* 2003, 300:297-304.

[103] Mahoney JA, Odin JA, White SM, Shaffer D, Koff A, Casciola-Rosen L, Rosen A: The human homologue of the yeast polyubiquitination factor Ufd2p is cleaved by caspase 6 and granzyme B during apoptosis. *Biochem J* 2002, 361:587-595.

[104] Caren H, Holmstrand A, Sjoberg RM, Martinsson T: The two human homologues of yeast UFD2 ubiquitination factor, UBE4A and UBE4B, are located in common neuroblastoma deletion regions and are subject to mutations in tumours. *Eur J Cancer* 2006, 42:381-387.

[105] Dai RM, Chen E, Longo DL, Gorbea CM, Li CC: Involvement of valosin-containing protein, an ATPase Co-purified with IkappaBalpha and 26 S proteasome, in ubiquitin-proteasome-mediated degradation of IkappaBalpha. *J Biol Chem* 1998, 273:3562-3573.

[106] Ye Y, Shibata Y, Kikkert M, van VS, Wiertz E, Rapoport TA: Inaugural Article: Recruitment of the p97 ATPase and ubiquitin ligases to the site of retrotranslocation at the endoplasmic reticulum membrane. *Proc Natl Acad Sci U S A* 2005, 102:14132-14138.

[107] Kim I, Mi K, Rao H: Multiple interactions of rad23 suggest a mechanism for ubiquitylated substrate delivery important in proteolysis. *Mol Biol Cell* 2004, 15:3357-3365.

[108] Richly H, Rape M, Braun S, Rumpf S, Hoege C, Jentsch S: A series of ubiquitin binding factors connects CDC48/p97 to substrate multiubiquitylation and proteasomal targeting. *Cell* 2005, 120:73-84.

[109] Yang M, Omura S, Bonifacino JS, Weissman AM: Novel aspects of degradation of T cell receptor subunits from the endoplasmic reticulum (ER) in T cells: importance of oligosaccharide processing, ubiquitination, and proteasome-dependent removal from ER membranes. *J Exp Med* 1998, 187:835-846.

[110] Madeo F, Schlauer J, Frohlich KU: Identification of the regions of porcine VCP preventing its function in Saccharomyces cerevisiae. *Gene* 1997, 204:145-151.

[111] Tirosh B, Iwakoshi NN, Lilley BN, Lee AH, Glimcher LH, Ploegh HL: Human cytomegalovirus protein US11 provokes an unfolded protein response that may facilitate the degradation of class I major histocompatibility complex products. *J Virol* 2005, 79:2768-2779.

[112] Brooks P, Fuertes G, Murray RZ, Bose S, Knecht E, Rechsteiner MC, Hendil KB, Tanaka K, Dyson J, Rivett J: Subcellular localization of proteasomes and their regulatory complexes in mammalian cells. *Biochem J* 2000, 346 Pt 1:155-161.

[113] Abujarour RJ, Dalal S, Hanson PI, Draper RK: p97 is in a complex with cholera toxin and influences the transport of cholera toxin and related toxins to the cytoplasm. *J Biol Chem* 2005.

[114] Kothe M, Ye Y, Wagner JS, De Luca HE, Kern E, Rapoport TA, Lencer WI: Role of p97 AAA-ATPase in the retrotranslocation of the cholera toxin A1 chain, a non-ubiquitinated substrate. *J Biol Chem* 2005, 280:28127-28132.

[115] Meyer HH, Wang Y, Warren G: Direct binding of ubiquitin conjugates by the mammalian p97 adaptor complexes, p47 and Ufd1-Npl4. *EMBO J* 2002, 21:5645-5652.

[116] Weihl CC, Dalal S, Pestronk A, Hanson PI: Inclusion body myopathy associated mutations in p97/VCP impair endoplasmic reticulum associated degradation. *Hum Mol Genet* 2005.

[117] Peters JM, Harris JR, Lustig A, Muller S, Engel A, Volker S, Franke WW: Ubiquitous soluble Mg(2+)-ATPase complex. A structural study. *J Mol Biol* 1992, 223:557-571.

[118] Lomas DA, Finch JT, Seyama K, Nukiwa T, Carrell RW: Alpha 1-antitrypsin Siiyama (Ser53-->Phe). Further evidence for intracellular loop-sheet polymerization. *J Biol Chem* 1993, 268:15333-15335.

[119] Rabouille C, Levine TP, Peters JM, Warren G: An NSF-like ATPase, p97, and NSF mediate cisternal regrowth from mitotic Golgi fragments. *Cell* 1995, 82:905-914.

[120] Moir D, Stewart SE, Osmond BC, Botstein D: Cold-sensitive cell-division-cycle mutants of yeast: isolation, properties, and pseudoreversion studies. *Genetics* 1982, 100:547-563.

[121] Frohlich KU, Fries HW, Rudiger M, Erdmann R, Botstein D, Mecke D: Yeast cell cycle protein CDC48p shows full-length homology to the mammalian protein VCP and is a member of a protein family involved in secretion, peroxisome formation, and gene expression. *J Cell Biol* 1991, 114:443-453.

[122] Leon A, McKearin D: Identification of TER94, an AAA ATPase protein, as a Bam-dependent component of the Drosophila fusome. *Mol Biol Cell* 1999, 10:3825-3834.

[123] Mouse Genome Informatics (MGI). Lexicon Genetics Data Submission to MGI via the NIH Knockout Project MGI:3609573 J:103485. http://www informatics jax org/external/ko/lexicon/3468 html 2005. Available from: URL: http://www.informatics.jax.org/external/ko/lexicon/3468.html

[124] protein (VCP/p97) reveals multiple cellular roles linked to ubiquitin/proteasome-dependent proteolysis. *J Cell Sci* 2004, 117:281-292.

[125] Dai RM, Li CC: Valosin-containing protein is a multi-ubiquitin chain-targeting factor required in ubiquitin-proteasome degradation. *Nat Cell Biol* 2001, 3:740-744.

[126] Bays NW, Hampton RY: Cdc48-Ufd1-Npl4: stuck in the middle with Ub. *Curr Biol* 2002, 12:R366-R371.

[127] Zhang X, Shaw A, Bates PA, Newman RH, Gowen B, Orlova E, Gorman MA, Kondo H, Dokurno P, Lally J, Leonard G, Meyer H, van Heel M, Freemont PS: Structure of the AAA ATPase p97. *Mol Cell* 2000, 6:1473-1484.

[128] DeLaBarre B, Brunger AT: Complete structure of p97/valosin-containing protein reveals communication between nucleotide domains. *Nat Struct Biol* 2003, 10:856-863.

[129] DeLaBarre B, Brunger AT: Nucleotide dependent motion and mechanism of action of p97/VCP. *J Mol Biol* 2005, 347:437-452.

[130] Hirabayashi M, Inoue K, Tanaka K, Nakadate K, Ohsawa Y, Kamei Y, Popiel AH, Sinohara A, Iwamatsu A, Kimura Y, Uchiyama Y, Hori S, Kakizuka A: VCP/p97 in abnormal protein aggregates, cytoplasmic vacuoles, and cell death, phenotypes relevant to neurodegeneration. *Cell Death Differ* 2001, 8:977-984.

[131] Dalal S, Rosser MF, Cyr DM, Hanson PI: Distinct roles for the AAA ATPases NSF and p97 in the secretory pathway. *Mol Biol Cell* 2003.

[132] Latterich M, Frohlich KU, Schekman R: Membrane fusion and the cell cycle: Cdc48p participates in the fusion of ER membranes. *Cell* 1995, 82:885-893.

[133] Roy L, Bergeron JJ, Lavoie C, Hendriks R, Gushue J, Fazel A, Pelletier A, Morre DJ, Subramaniam VN, Hong W, Paiement J: Role of p97 and syntaxin 5 in the assembly of transitional endoplasmic reticulum. *Mol Biol Cell* 2000, 11:2529-2542.

[134] Hetzer M, Meyer HH, Walther TC, Bilbao-Cortes D, Warren G, Mattaj IW: Distinct AAA-ATPase p97 complexes function in discrete steps of nuclear assembly. *Nat Cell Biol* 2001, 3:1086-1091.

[135] Uchiyama K, Jokitalo E, Lindman M, Jackman M, Kano F, Murata M, Zhang X, Kondo H: The localization and phosphorylation of p47 are important for Golgi disassembly-assembly during the cell cycle. *J Cell Biol* 2003, 161:1067-1079.

[136] Cao K, Nakajima R, Meyer HH, Zheng Y: The AAA-ATPase Cdc48/p97 regulates spindle disassembly at the end of mitosis. *Cell* 2003, 115:355-367.

[137] Zhang H, Wang Q, Kajino K, Greene MI: VCP, a weak ATPase involved in multiple cellular events, interacts physically with BRCA1 in the nucleus of living cells. *DNA Cell Biol* 2000, 19:253-263.

[138] Yamada T, Okuhara K, Iwamatsu A, Seo H, Ohta K, Shibata T, Murofushi H: p97 ATPase, an ATPase involved in membrane fusion, interacts with DNA unwinding factor (DUF) that functions in DNA replication. *FEBS Lett* 2000, 466:287-291.

[139] Ghislain M, Dohmen RJ, Levy F, Varshavsky A: Cdc48p interacts with Ufd3p, a WD repeat protein required for ubiquitin-mediated proteolysis in Saccharomyces cerevisiae. *EMBO J* 1996, 15:4884-4899.

[140] Rape M, Hoppe T, Gorr I, Kalocay M, Richly H, Jentsch S: Mobilization of processed, membrane-tethered SPT23 transcription factor by CDC48(UFD1/NPL4), a ubiquitin-selective chaperone. *Cell* 2001, 107:667-677.

[141] Hitchcock AL, Krebber H, Frietze S, Lin A, Latterich M, Silver PA: The conserved npl4 protein complex mediates proteasome-dependent membrane-bound transcription factor activation. *Mol Biol Cell* 2001, 12:3226-3241.

[142] Meyer HH, Shorter JG, Seemann J, Pappin D, Warren G: A complex of mammalian ufd1 and npl4 links the AAA-ATPase, p97, to ubiquitin and nuclear transport pathways. *EMBO J* 2000, 19:2181-2192.

[143] Mizuno Y, Hori S, Kakizuka A, Okamoto K: Vacuole-creating protein in neurodegenerative diseases in humans. *Neurosci Lett* 2003, 343:77-80.

[144] Schroder R, Watts GD, Mehta SG, Evert BO, Broich P, Fliessbach K, Pauls K, Hans VH, Kimonis V, Thal DR: Mutant valosin-containing protein causes a novel type of frontotemporal dementia. *Ann Neurol* 2005, 57:457-461.

[145] Nan L, Wu Y, Bardag-Gorce F, Li J, French BA, Wilson LT, Khanh NS, French SW: RNA interference of VCP/p97 increases Mallory body formation. *Exp Mol Pathol* 2005, 78:1-9.

[146] Ishigaki S, Hishikawa N, Niwa J, Iemura S, Natsume T, Hori S, Kakizuka A, Tanaka K, Sobue G: Physical and functional interaction between Dorfin and Valosin-containing protein that are colocalized in ubiquitylated inclusions in neurodegenerative disorders. *J Biol Chem* 2004, 279:51376-51385.

[147] Kitami MI, Kitami T, Nagahama M, Tagaya M, Hori S, Kakizuka A, Mizuno Y, Hattori N: Dominant-negative effect of mutant valosin-containing protein in aggresome formation. *FEBS Lett* 2006, 580:474-478.

[148] Seigneurin-Berny D, Verdel A, Curtet S, Lemercier C, Garin J, Rousseaux S, Khochbin S: Identification of components of the murine histone deacetylase 6 complex: link between acetylation and ubiquitination signaling pathways. *Mol Cell Biol* 2001, 21:8035-8044.

[149] Kawaguchi Y, Kovacs JJ, McLaurin A, Vance JM, Ito A, Yao TP: The deacetylase HDAC6 regulates aggresome formation and cell viability in response to misfolded protein stress. *Cell* 2003, 115:727-738.

[150] Corcoran LJ, Mitchison TJ, Liu Q: A novel action of histone deacetylase inhibitors in a protein aggresome disease model. *Curr Biol* 2004, 14:488-492.

[151] Wu D, Chen PJ, Chen S, Hu Y, Nunez G, Ellis RE: C. elegans MAC-1, an essential member of the AAA family of ATPases, can bind CED-4 and prevent cell death. *Development* 1999, 126:2021-2031.

[152] Rao RV, Poksay KS, Castro-Obregon S, Schilling B, Row RH, del RG, Gibson BW, Ellerby HM, Bredesen DE: Molecular components of a cell death pathway activated by endoplasmic reticulum stress. *J Biol Chem* 2004, 279:177-187.

[153] Higashiyama H, Hirose F, Yamaguchi M, Inoue YH, Fujikake N, Matsukage A, Kakizuka A: Identification of ter94, Drosophila VCP, as a modulator of polyglutamine-induced neurodegeneration. *Cell Death Differ* 2002, 9:264-273.

[154] Neuber O, Jarosch E, Volkwein C, Walter J, Sommer T: Ubx2 links the Cdc48 complex to ER-associated protein degradation. *Nat Cell Biol* 2005, 7:993-998.

[155] Decottignies A, Evain A, Ghislain M: Binding of Cdc48p to a ubiquitin-related UBX domain from novel yeast proteins involved in intracellular proteolysis and sporulation. *Yeast* 2004, 21:127-139.

[156] Egerton M, Ashe OR, Chen D, Druker BJ, Burgess WH, Samelson LE: VCP, the mammalian homolog of cdc48, is tyrosine phosphorylated in response to T cell antigen receptor activation. *EMBO J* 1992, 11:3533-3540.

[157] Mayr PS, Allan VJ, Woodman PG: Phosphorylation of p97(VCP) and p47 in vitro by p34cdc2 kinase. *Eur J Cell Biol* 1999, 78:224-232.

[158] Geussova G, Kalab P, Peknicova J: Valosine containing protein is a substrate of cAMP-activated boar sperm tyrosine kinase. *Mol Reprod Dev* 2002, 63:366-375.

[159] Vandermoere F, El Yazidi-Belkoura I, Slomianny C, Demont Y, Bidaux G, Adriaenssens E, Lemoine J, Hondermarck H: The valosin-containing protein (VCP) is a target of AKT signaling required for cell survival. *J Biol Chem* 2006.

[160] Noguchi M, Takata T, Kimura Y, Manno A, Murakami K, Koike M, Ohizumi H, Hori S, Kakizuka A: ATPase activity of p97/Valosin-containing protein is regulated by oxidative modification of the evolutionally conserved 522nd cysteine residue in Walker A motif. *J Biol Chem* 2005.

[161] Morimoto RI: Regulation of the heat shock transcriptional response: cross talk between a family of heat shock factors, molecular chaperones, and negative regulators. *Genes Dev* 1998, 12:3788-3796.

[162] Flynn GC, Chappell TG, Rothman JE: Peptide binding and release by proteins implicated as catalysts of protein assembly. *Science* 1989, 245:385-390.

[163] Flynn GC, Pohl J, Flocco MT, Rothman JE: Peptide-binding specificity of the molecular chaperone BiP. *Nature* 1991, 353:726-730.

[164] Hirschberg CB, Robbins PW, Abeijon C: Transporters of nucleotide sugars, ATP, and nucleotide sulfate in the endoplasmic reticulum and Golgi apparatus. *Annu Rev Biochem* 1998, 67:49-69.

[165] Yoshida H, Matsui T, Yamamoto A, Okada T, Mori K: XBP1 mRNA is induced by ATF6 and spliced by IRE1 in response to ER stress to produce a highly active transcription factor. *Cell* 2001, 107:881-891.

[166] Kimata Y, Oikawa D, Shimizu Y, Ishiwata-Kimata Y, Kohno K: A role for BiP as an adjustor for the endoplasmic reticulum stress-sensing protein Ire1. *J Cell Biol* 2004, 167:445-456.

[167] Credle JJ, Finer-Moore JS, Papa FR, Stroud RM, Walter P: On the mechanism of sensing unfolded protein in the endoplasmic reticulum. *Proc Natl Acad Sci U S A* 2005, 102:18773-18784.

[168] Ye J, Rawson RB, Komuro R, Chen X, Dave UP, Prywes R, Brown MS, Goldstein JL: ER stress induces cleavage of membrane-bound ATF6 by the same proteases that process SREBPs. *Mol Cell* 2000, 6:1355-1364.

[169] Zhang C, Cai Y, Adachi MT, Oshiro S, Aso T, Kaufman RJ, Kitajima S: Homocysteine induces programmed cell death in human vascular endothelial cells through activation of the unfolded protein response. *J Biol Chem* 2001, 276:35867-35874.

[170] Shi Y, Vattem KM, Sood R, An J, Liang J, Stramm L, Wek RC: Identification and characterization of pancreatic eukaryotic initiation factor 2 alpha-subunit kinase, PEK, involved in translational control. *Mol Cell Biol* 1998, 18:7499-7509.

[171] Harding HP, Zhang Y, Ron D: Protein translation and folding are coupled by an endoplasmic-reticulum-resident kinase. *Nature* 1999, 397:271-274.

[172] Yoshida H, Haze K, Yanagi H, Yura T, Mori K: Identification of the cis-acting endoplasmic reticulum stress response element responsible for transcriptional induction of mammalian glucose-regulated proteins. Involvement of basic leucine zipper transcription factors. *J Biol Chem* 1998, 273:33741-33749.

[173] Haze K, Yoshida H, Yanagi H, Yura T, Mori K: Mammalian transcription factor ATF6 is synthesized as a transmembrane protein and activated by proteolysis in response to endoplasmic reticulum stress. *Mol Biol Cell* 1999, 10:3787-3799.

[174] Travers KJ, Patil CK, Wodicka L, Lockhart DJ, Weissman JS, Walter P: Functional and genomic analyses reveal an essential coordination between the unfolded protein response and ER-associated degradation. *Cell* 2000, 101:249-258.

[175] Ng DT, Spear ED, Walter P: The unfolded protein response regulates multiple aspects of secretory and membrane protein biogenesis and endoplasmic reticulum quality control. *J Cell Biol* 2000, 150:77-88.

[176] Tessitore A, del PM, Sano R, Ma Y, Mann L, Ingrassia A, Laywell ED, Steindler DA, Hendershot LM, d'Azzo A: GM1-ganglioside-mediated activation of the unfolded protein response causes neuronal death in a neurodegenerative gangliosidosis. *Mol Cell* 2004, 15:753-766.

[177] Casagrande R, Stern P, Diehn M, Shamu C, Osario M, Zuniga M, Brown PO, Ploegh H: Degradation of proteins from the ER of S. cerevisiae requires an intact unfolded protein response pathway. *Mol Cell* 2000, 5:729-735.

[178] Anderson KC: Moving disease biology from the lab to the clinic. *Cancer* 2003, 97:796-801.

[179] Hong M, Li M, Mao C, Lee AS: Endoplasmic reticulum stress triggers an acute proteasome-dependent degradation of ATF6. *J Cell Biochem* 2004, 92:723-732.

[180] Lee AH, Iwakoshi NN, Anderson KC, Glimcher LH: Proteasome inhibitors disrupt the unfolded protein response in myeloma cells. *Proc Natl Acad Sci U S A* 2003, 100:9946-9951.

[181] Nawrocki ST, Carew JS, Dunner K, Jr., Boise LH, Chiao PJ, Huang P, Abbruzzese JL, McConkey DJ: Bortezomib inhibits PKR-like endoplasmic reticulum (ER) kinase and induces apoptosis via ER stress in human pancreatic cancer cells. *Cancer Res* 2005, 65:11510-11519.

[182] Szegezdi E, Fitzgerald U, Samali A: Caspase-12 and ER-stress-mediated apoptosis: the story so far. *Ann N Y Acad Sci* 2003, 1010:186-194.

[183] Rowe SM, Miller S, Sorscher EJ: Cystic Fibrosis. *N Engl J Med* 2005, 352:1992-2001.

[184] Tanaka K, Suzuki T, Hattori N, Mizuno Y: Ubiquitin, proteasome and parkin. *Biochim Biophys Acta* 2004, 1695:235-247.

[185] Imai Y, Soda M, Inoue H, Hattori N, Mizuno Y, Takahashi R: An unfolded putative transmembrane polypeptide, which can lead to endoplasmic reticulum stress, is a substrate of Parkin. *Cell* 2001, 105:891-902.

[186] Valetti C, Grossi CE, Milstein C, Sitia R: Russell bodies: a general response of secretory cells to synthesis of a mutant immunoglobulin which can neither exit from, nor be degraded in, the endoplasmic reticulum. *J Cell Biol* 1991, 115:983-994.

[187] Kopito RR, Sitia R: Aggresomes and Russell bodies. Symptoms of cellular indigestion? *EMBO Rep* 2000, 1:225-231.

[188] Sherman MY, Goldberg AL: Cellular defenses against unfolded proteins: a cell biologist thinks about neurodegenerative diseases. *Neuron* 2001, 29:15-32.

[189] Scolari F, Caridi G, Rampoldi L, Tardanico R, Izzi C, Pirulli D, Amoroso A, Casari G, Ghiggeri GM: Uromodulin storage diseases: clinical aspects and mechanisms. *Am J Kidney Dis* 2004, 44:987-999.

[190] Sekijima Y, Wiseman RL, Matteson J, Hammarstrom P, Miller SR, Sawkar AR, Balch WE, Kelly JW: The biological and chemical basis for tissue-selective amyloid disease. *Cell* 2005, 121:73-85.

[191] Rajkumar SV, Richardson PG, Hideshima T, Anderson KC: Proteasome inhibition as a novel therapeutic target in human cancer. *J Clin Oncol* 2005, 23:630-639.

[192] Johnston JA, Ward CL, Kopito RR: Aggresomes: a cellular response to misfolded proteins. *J Cell Biol* 1998, 143:1883-1898.

[193] Younger JM, Ren HY, Chen L, Fan CY, Fields A, Patterson C, Cyr DM: A foldable CFTR{Delta}F508 biogenic intermediate accumulates upon inhibition of the Hsc70-CHIP E3 ubiquitin ligase. *J Cell Biol* 2004, 167:1075-1085.

[194] Kleopa KA, Yum SW, Scherer SS: Cellular mechanisms of connexin32 mutations associated with CNS manifestations. *J Neurosci Res* 2002, 68:522-534.

[195] Gow A, Southwood CM, Lazzarini RA: Disrupted proteolipid protein trafficking results in oligodendrocyte apoptosis in an animal model of Pelizaeus-Merzbacher disease. *J Cell Biol* 1998, 140:925-934.

[196] Sandhaus RA: alpha1-Antitrypsin deficiency . 6: new and emerging treatments for alpha1-antitrypsin deficiency. *Thorax* 2004, 59:904-909.

[197] Bradshaw CB, Davis RL, Shrimpton AE, Holohan PD, Rea CB, Fieglin D, Kent P, Collins GH: Cognitive deficits associated with a recently reported familial neurodegenerative disease: familial encephalopathy with neuroserpin inclusion bodies. *Arch Neurol* 2001, 58:1429-1434.

[198] Davis RL, Shrimpton AE, Holohan PD, Bradshaw C, Feiglin D, Collins GH, Sonderegger P, Kinter J, Becker LM, Lacbawan F, Krasnewich D, Muenke M, Lawrence DA, Yerby MS, Shaw CM, Gooptu B, Elliott PR, Finch JT, Carrell RW, Lomas DA: Familial dementia caused by polymerization of mutant neuroserpin. *Nature* 1999, 401:376-379.

[199] Davis RL, Holohan PD, Shrimpton AE, Tatum AH, Daucher J, Collins GH, Todd R, Bradshaw C, Kent P, Feiglin D, Rosenbaum A, Yerby MS, Shaw CM, Lacbawan F, Lawrence DA: Familial encephalopathy with neuroserpin inclusion bodies. *Am J Pathol* 1999, 155:1901-1913.

[200] Wiertz EJ, Jones TR, Sun L, Bogyo M, Geuze HJ, Ploegh HL: The human cytomegalovirus US11 gene product dislocates MHC class I heavy chains from the endoplasmic reticulum to the cytosol. *Cell* 1996, 84:769-779.

[201] Anderson RG, Orci L, Brown MS, Garcia-Segura LM, Goldstein JL: Ultrastructural analysis of crystalloid endoplasmic reticulum in UT-1 cells and its disappearance in response to cholesterol. *J Cell Sci* 1983, 63:1-20.

[202] Hampton RY, Gardner RG, Rine J: Role of 26S proteasome and HRD genes in the degradation of 3-hydroxy-3-methylglutaryl-CoA reductase, an integral endoplasmic reticulum membrane protein. *Mol Biol Cell* 1996, 7:2029-2044.

[203] Sever N, Yang T, Brown MS, Goldstein JL, bose-Boyd RA: Accelerated degradation of HMG CoA reductase mediated by binding of insig-1 to its sterol-sensing domain. *Mol Cell* 2003, 11:25-33.

[204] Lentz SR, Sadler JE: Homocysteine inhibits von Willebrand factor processing and secretion by preventing transport from the endoplasmic reticulum. *Blood* 1993, 81:683-689.

[205] Kokame K, Kato H, Miyata T: Homocysteine-respondent genes in vascular endothelial cells identified by differential display analysis. GRP78/BiP and novel genes. *J Biol Chem* 1996, 271:29659-29665.

[206] Outinen PA, Sood SK, Liaw PC, Sarge KD, Maeda N, Hirsh J, Ribau J, Podor TJ, Weitz JI, Austin RC: Characterization of the stress-inducing effects of homocysteine. *Biochem J* 1998, 332 (Pt 1):213-221.

[207] Ron D: Hyperhomocysteinemia and function of the endoplasmic reticulum. *J Clin Invest* 2001, 107:1221-1222.

[208] Paschen W: Endoplasmic reticulum: a primary target in various acute disorders and degenerative diseases of the brain. *Cell Calcium* 2003, 34:365-383.

[209] Holtz WA, O'Malley KL: Parkinsonian mimetics induce aspects of unfolded protein response in death of dopaminergic neurons. *J Biol Chem* 2003, 278:19367-19377.

[210] Thomas M, Yu Z, Dadgar N, Varambally S, Yu J, Chinnaiyan AM, Lieberman AP: The unfolded protein response modulates toxicity of the expanded glutamine androgen receptor. *J Biol Chem* 2005.

[211] Vlug AS, Teuling E, Haasdijk ED, French P, Hoogenraad CC, Jaarsma D: ATF3 expression precedes death of spinal motoneurons in amyotrophic lateral sclerosis-SOD1 transgenic mice and correlates with c-Jun phosphorylation, CHOP expression, somato-dendritic ubiquitination and Golgi fragmentation. *Eur J Neurosci* 2005, 22:1881-1894.

[212] Wojcik C, Di Napoli M: Ubiquitin-proteasome system and proteasome inhibition: new strategies in stroke therapy. *Stroke* 2004, 35:1506-1518.

[213] Watts GD, Wymer J, Kovach MJ, Mehta SG, Mumm S, Darvish D, Pestronk A, Whyte MP, Kimonis VE: Inclusion body myopathy associated with Paget disease of bone and frontotemporal dementia is caused by mutant valosin-containing protein. *Nat Genet* 2004, 36:377-381.

[214] Haubenberger D, Bittner RE, Rauch-Shorny S, Zimprich F, Mannhalter C, Wagner L, Mineva I, Vass K, Auff E, Zimprich A: Inclusion body myopathy and Paget disease is linked to a novel mutation in the VCP gene. *Neurology* 2005, 65:1304-1305.

[215] Kimonis VE, Watts GD: Autosomal dominant inclusion body myopathy, Paget disease of bone, and frontotemporal dementia. *Alzheimer Dis Assoc Disord* 2005, 19 Suppl 1:S44-S47.

[216] Yamanaka K, Okubo Y, Suzaki T, Ogura T: Analysis of the two p97/VCP/Cdc48p proteins of Caenorhabditis elegans and their suppression of polyglutamine-induced protein aggregation. *J Struct Biol* 2004, 146:242-250.

[217] Yamamoto S, Tomita Y, Hoshida Y, Iizuka N, Kidogami S, Miyata H, Takiguchi S, Fujiwara Y, Yasuda T, Yano M, Nakamori S, Sakon M, Monden M, Aozasa K: Expression level of valosin-containing protein (p97) is associated with prognosis of esophageal carcinoma. *Clin Cancer Res* 2004, 10:5558-5565.

[218] Yamamoto S, Tomita Y, Hoshida Y, Iizuka N, Monden M, Yamamoto S, Iuchi K, Aozasa K: Expression level of valosin-containing protein (p97) is correlated with progression and prognosis of non-small-cell lung carcinoma. *Ann Surg Oncol* 2004, 11:697-704.

[219] Yamamoto S, Tomita Y, Hoshida Y, Nagano H, Dono K, Umeshita K, Sakon M, Ishikawa O, Ohigashi H, Nakamori S, Monden M, Aozasa K: Increased expression of valosin-containing protein (p97) is associated with lymph node metastasis and prognosis of pancreatic ductal adenocarcinoma. *Ann Surg Oncol* 2004, 11:165-172.

[220] Yamamoto S, Tomita Y, Hoshida Y, Sakon M, Kameyama M, Imaoka S, Sekimoto M, Nakamori S, Monden M, Aozasa K: Expression of valosin-containing protein in colorectal carcinomas as a predictor for disease recurrence and prognosis. *Clin Cancer Res* 2004, 10:651-657.

[221] Yamamoto S, Tomita Y, Hoshida Y, Takiguchi S, Fujiwara Y, Yasuda T, Yano M, Nakamori S, Sakon M, Monden M, Aozasa K: Expression level of valosin-containing protein is strongly associated with progression and prognosis of gastric carcinoma. *J Clin Oncol* 2003, 21:2537-2544.

[222] Yamamoto S, Tomita Y, Nakamori S, Hoshida Y, Nagano H, Dono K, Umeshita K, Sakon M, Monden M, Aozasa K: Elevated expression of valosin-containing protein (p97) in hepatocellular carcinoma is correlated with increased incidence of tumor recurrence. *J Clin Oncol* 2003, 21:447-452.

[223] Hirono Y, Fushida S, Yonemura Y, Yamamoto H, Watanabe H, Raz A: Expression of autocrine motility factor receptor correlates with disease progression in human gastric cancer. *Br J Cancer* 1996, 74:2003-2007.

[224] Timar J, Raso E, Dome B, Ladanyi A, Banfalvi T, Gilde K, Raz A: Expression and function of the AMF receptor by human melanoma in experimental and clinical systems. *Clin Exp Metastasis* 2002, 19:225-232.

[225] Zhong X, Shen Y, Ballar P, Apostolou A, Agami R, Fang S: AAA ATPase p97/valosin-containing protein interacts with gp78, a ubiquitin ligase for endoplasmic reticulum-associated degradation. *J Biol Chem* 2004, 279:45676-45684.

[226] Onishi Y, Tsukada K, Yokota J, Raz A: Overexpression of autocrine motility factor receptor (AMFR) in NIH3T3 fibroblasts induces cell transformation. *Clin Exp Metastasis* 2003, 20:51-58.

[227] Ma Y, Hendershot LM: The role of the unfolded protein response in tumour development: friend or foe? *Nat Rev Cancer* 2004, 4:966-977.

[228] Adams J, Kauffman M: Development of the proteasome inhibitor Velcade (Bortezomib). *Cancer Invest* 2004, 22:304-311.

In: The Ubiquitin Proteasome System… ISBN 978-1-60021-749-4
Eds: Mario Di Napoli and Cezary Wojcik, pp. 307-341

Chapter 14

THEORETICAL MODELS FOR PROTEASOME FUNCTION: PREDICTIVE METHODS TO UNDERSTAND UBIQUITIN PROTEIN SYSTEM IN NEURODEGENERATIVE DISEASES

Fabio Luciani[1,*], Morten Nielsen[2], Can Kesmir[2] and Mario Di Napoli[3,4]

[1]School of Biotechnology and Biomolecular Sciences, UNSW, Sydney, Australia;
[2]Center for Biological Sequence Analysis, BioCentrum, DTU, Denmark;
[3]Neurological Section, SMDN – Center for Cardiovascular Medicine and Cerebrovascular Disease Prevention, Sulmona, Italy;
[4]Neurological Service, San Camillo de'Lellis General Hospital, Rieti, Italy.

ABSTRACT

The proteasome is an intracellular multisubunit protease. It plays an important role in a myriad of intracellular processes and in removing misfolded proteins by degradation. Protein misfolding and aggregation are common to most neurodegenerative diseases, suggesting that abnormalities of protein homeostasis contribute to pathogenesis. Recent cell-based and genetic studies suggest that perturbations in the proteasome degradation mechanisms contribute to neurodegenerative disease processes Although, significant progress has been made in the understanding of substrate recognition and proteolysis, very little is known on the intermediate steps of transportation of misfolded protein into the proteolytic proteasome chamber and their effects on proteasome proteolytic activity. Mathematical models can help to understand proteasome dysfunction in neurodegenerative diseases. Using the in silico informatics work done on the specificity

* Correspondence concerning this article should be addressed to Dr. Fabio Luciani, PhD; School of Biotechnology and Biomolecular Sciences, UNSW, Sydney 2052, Australia. Phone: +61-293-853701, Fax: +61-293-851483, E-mail luciani@unsw.edu.au.

and kinetics of the proteasome molecules, different non-exclusive models of proteasome dysfunction can be predicted. If the influx and cleavage rate are impaired a low degradation rate is predicted, while if the cleavage rate is limiting, such as for substrates containing polyglutamine tracks, the degradation rate increases at the decrease of the influx rate. From these standpoints, misfolded protein accumulation in neurodegenerative disease could be due to a low cleavage capacity of catalytic subunits toward specific substrates, such as polyglutamine rich sequence proteins, with a consequent direct 'clogging' of proteasomes by undegraded longer fragments. An alternative hypothesis suggests an impaired transport into the proteolytic chamber, mainly due to a reduced capacity of proteasome 19S gating subunits to attach substrate recruitment fragments. ATP deficiency and steric modification of the substrates are envisaged as potential mechanisms generating such impairment. These theoretical studies, although in their early stages, have already revealed an unanticipated complexity of the dysfunction of proteolytic activity in neurodegenerative diseases at the level of substrate recruitment mechanism as well as at the level of the catalytic cycle itself.

Keywords: proteasome, protein degradation, protein folding, mathematical mode, neurodegeneration.

ABBREVIATIONS

A, alanine; AAA, ATPases Associated with diverse cellular Activities; α-Syn, α-synuclein; Aβ, β-amyloid AD, Alzheimer disease; D, aspartic acid; degrons, destabilizing amino acid sequences; E, glutamic acid; G, Glycine ; GFP, green fluorescent protein; HD, Huntington's disease; Htt, huntingtin; Hsp, heat shock protein; IBs, cytoplasmatic inclusion bodies; IFN-γ, interferon γ; I, isoleucin; L, leucin; L-model, Luciani's model; MHC, major histocompatibility complex; P, proline; PA700, proteasome activator MW 700; Pael-R, parkin associated endothelin-receptor-like receptor, PD, Parkinson disease; PolyQ, polyglutamine; PrP, prion protein; PS1, presenilin 1; RC, regulatory complex; RING, real interesting new gene; SCA, spinocerebellar ataxia; SMM, stabilized matrix method; TAP, transporter associated with antigen presentation; TNF-α, tumor necrosis factor α; Ub, ubiquitin; Ub–UBB^{+1}, Ubiquitinated UBB^{+1}; UBB^{+1}, frame-shifted ubiquitin B protein; UPS, ubiquitin-proteasome system.

THE PROTEASOME

The proteasome is an intracellular multisubunit neutral protease that catalyzes selective proteolytic protein degradation (Figure 1). It constitutes the effector arm of the ubiquitin–proteasome system (UPS), which first tags proteins, usually by ubiquitin (Ub) chain conjugation, and then delivers them to the proteasome, where they are cleaved to peptides. Proteasomes are ATP-dependent proteases. The 26S proteasome complex consists of a 20S proteolytic core and typically a 19S cap. The 19S complex is a 17-subunit machine, itself composed of two subcomplexes: the base, thought to be responsible for the denaturation of

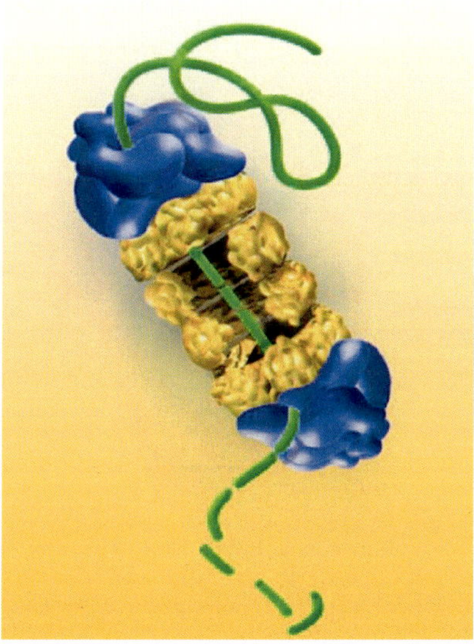

Figure 1. The 26S proteasome. Proteasomes are large complexes that carry out crucial roles in many cellular pathways by degrading proteins in the cytosol and nucleus of eukaryotic cells to enforce quality control and to regulate many cellular processes. The catalytic heart of these complexes, the 20S proteasome (yellow), has been highly conserved from yeast to humans, with simpler versions also found in some archaea and prokaryotes. The 20S proteasome is a barrel-shaped assembly of 28 protein subunits that possesses three distinct proteolytic active sites with different specificities. Together, the three active sites, present in the two central rings of β subunits, hydrolyze almost all peptide bonds, having trouble only with those bonds that follow glycine (G) and proline (P). The potentially catastrophic elimination of inappropriate substrates is prevented by sequestration of active sites within the hollow structure of the 20S proteasome. Substrates access the central catalytic chamber through axial ports in the end rings of subunits, although in the absence of activators these channels are closed and proteasome activity is repressed. Proteasomes are activated by protein complexes that bind to the end rings of subunits (blue). The best known activator is PA700 [proteasome activator MW 700, also known as 19S or regulatory complex (RC)], which has been highly conserved from yeast to humans and binds to the 20S proteasome to form the 26S proteasome. PA700 is the only proteasome activator that is known to stimulate degradation of protein substrates, which it generally recognizes by a poly-Ub modification and which it processes by an ATP-dependent mechanism. Thus, PA700 is thought to mediate most of the biological effects of the proteasome by facilitating substrate degradation. This biological role is well established, and PA700 and 26S proteasomes have been reviewed extensively elsewhere see *Chapter 9*. Image credit: U.S. Department of Energy Genomics: GTL Program, http://doegenomestolife.org

the substrate via the action of six ATPase subunits, and the lid, which sits over the base and is believed to recognize polyubiquitinated proteasome substrates and remove the poly-Ub tag via the action of an inherent isopeptidases. The cylindrical 20S proteolytic complex comprises four rings, each containing seven subunits. The two inner rings possess three catalytic sites each, which are characterized by their individual preference for cleavage after specific amino acid residues, and degrade the substrate into short peptides. This inner proteolytic chamber is accessible only through two openings at each end of the 20S cylinder, each of which is gated by one of the 19S complexes. These complexes include at least 18 distinct proteins, of which six are ATPases [1]. In eukaryotes each of the six is distinct and in yeast all are essential [2]. Hexameric ATPase rings, members of the very diverse AAA

(ATPases Associated with diverse cellular Activities) ATPase family, have the general capacity to change the conformation of their substrates [3], fuelled by a cycle of ATP binding, hydrolysis and release. These general considerations imply that the ATPases of the proteasome are the key mediators of substrate unfolding and insertion, an inference with experimental support [4]. In addition to 19S and 20S core constituents, many proteins associate transiently with the 26S proteasome, further increasing the complexity of this macromolecular system. More details on proteasome structure and functions are reviewed in Chapters 6 and 7 and the reader is urged to consult these chapters. Hydrolytic cleavage of peptide bonds takes place in the 20S core. Within that closed space, destruction is relatively independent of substrate sequence. Access is through a narrow pore (<15Å) positioned axially at either end of the barrel-shaped chamber [5]. Entry thus demands tag recognition, unfolding and insertion. Unfolding is essential, as folded proteins are too bulky to traverse the entryway.

Recognition and unfolding of protein substrates and their translocation into the 20S core complex is mediated by regulatory protein complexes such as 11S and 19S, which, *in vivo*, may associate with one or both ends of the 20S core [1] (see Chapter 8). The function of the eukaryotic proteasome to serve as supplier of epitopes presented by the MHC class I has greatly intensified experimental work aimed at elucidating the structural and kinetic basis for the high selectivity with which the proteasome cuts out antigenic peptides from precursor proteins [6].

It has long been known that small peptides can serve as substrates for the purified 20S proteasome but that, in general *in vivo*, full-length proteins must be polyubiquitinated, recognized by the 19S regulatory complex, deubiquitinated and unfolded in an ATP-dependent manner before translocation into the proteolytic 20S chamber [7]. Significant progress has been made in understanding substrate recognition and proteolysis but the intermediate processes of unfolding and insertion have substantially eluded investigation. Experiments have shown that access to the 20S cylinder is gated and thus regulated by a key protein within the 19S complex, namely the Rpt2 subunit [8]. X-ray crystallographic analysis of yeast 20S proteasome revealed that the amino termini of the 20S α subunits are juxtaposed to block entry into the 20S cylinder. Deletion of several amino-terminal residues of the α3 subunit results in a disordered axial gate, which allows the free entry of protein substrates into the proteolytic chamber [9]. Because substrate proteins must enter the 20S chamber through either end of the cylinder via these gates, these studies suggest that one role of the 19S regulatory complex is to regulate the opening of the proteolytic chamber. In fact, a co-crystalisation study of wild-type yeast 20S with the 19S regulatory complex from *Trypanosome* showed that the amino termini of three α subunits were dislocated outwardly, thus opening a channel into the 20S complex [9]. These constitutively open channel 20S mutants have proven useful in the investigation of substrate entry/product release. Crystallographic study of the yeast 20S proteasome revealed small openings in the side walls of the proteolytic chamber through which peptide products were hypothesized to pass [9]. After entry into the proteolytic chamber, protein substrates are cleaved to yield oligopeptides that range from 3 to 24 residues [10,11]. The length of the peptides produced has been hypothesized to reflect the length of time that the substrate is present in the proteolytic chamber (i.e. the longer the retention of substrate, the shorter the peptide produced) [11].

Support for this model came from an analysis of the peptide length generated by the open channel 20S mutants. Substrate access to the channel is greatly increased relative to wild-type 20S, and conceivably peptide product release is increased. It is plausible that peptide length of products generated by the open channel 20S mutant would be on average longer due to the shorter time spent within the proteolytic chamber [11]. This indeed proved to be the case, in that the median length of peptides produced by the open channel 20S mutant was 40% longer than those generated by wild-type 20S. These findings also support the idea that product release proceeds through the gated axial channel rather than the side wall openings observed in the crystallographic analysis.

Two forms of proteasome exist: the 'immunoproteasome', which is expressed in cells stimulated by interferon γ (IFN-γ) or tumor necrosis factor α (TNF-α), and is commonly found in primary and secondary lymphoid organs, and the 'constitutive proteasome', which is expressed in healthy normal tissues and in particular in immune-privileged organs such as the brain [12-14]. The activity of the proteasome during inflammation is altered via induction of the regulatory units and replacement of the constitutive active subunits (β-1[δ,Y], β-2[MC14, LMP9, Z] and β-5[MB1, X]) by their immuno (β-1i[LMP2], β–2i[MECL-1] and β-5i[LMP7]) counterparts in newly synthesized proteasomes [15,16]. During an antiviral or antibacterial immune response, immunoproteasomes largely replace constitutive proteasomes [17]. This replacement has a positive effect on major histocompatibility complex (MHC) class I restricted antigen presentation, as has been demonstrated in several systems [6,14,17-22]. This argument is reviewed in Chapter 34. Finally, intermediate, hybrid chimeric forms of the proteasomes have also been detected [23,24] where alternate caps in addition to the 19S can be form in a small subpopulation of proteasomes. The physiological role of these alternative caps is not completely clear. They may have a special function during the immune response and facilitating the exit of peptide-products from the 20S. The reader can consult Chapter 9 for an updated review on this topic.

The Specificity of the Proteasome

The specificity of the proteasome is often studied by *in vitro* proteasome digest experiments, where whole proteins are digested by the proteasome. Toes *et al.* [25] have performed such proteasome digest experiments to characterize the differences between immunoproteasome and constitutive proteasome specificity. It has been suggested that the amino acid on the N-terminal of the cleaved bond (P1 position according to nomenclature suggested by Berger and Schechter, [26]) is the most important one determining cleavage [27], although the flanking region may also be important [28-30]. The data from Toes *et al.* [25] permit to estimate which amino acids are differentially preferred by the two forms of the proteasome, estimating the frequency by which the different amino acids are used for cleavage by the two proteasomes. It is found that the immunoproteasome cleaves after leucine (L) and isoleucine (I) residues with a significantly higher frequency than the constitutive proteasome. Fifteen and 5 of the 56 cleavages sites used only by the immunoproteasome are L and I, respectively. For the constitutive proteasome the corresponding values are 2, and 1 sites, respectively, out of the total of 73 sites. The immunoproteasome, on the other hand, cleaves less frequent after aspartic acid (D) and glutamic acid (E) residues compared to the constitutive proteasome. Out of 56 sites used solely by the immunoproteasome only one site is E (D is never used). For the constitutive proteasome these values are 9 and 7 for D and E, respectively. Comparing the amino acid usage of the two proteasomes to the average amino acid distribution in the enolase protein, only the immunoproteasome has an amino acid preference significantly different from the background [31]. This suggests that the constitutive proteasome uses semi-specific and degenerate sequence signals to cleave a protein. The cleavage by the immunoproteasome is restricted to fewer amino acids, and thus, the immunoproteasome is more specific than the constitutive proteasome [31]. Moreover, the cleavage sites preferred by the immunoproteasome fit much better with the human MHC class I binding motifs than the sites preferred by the constitutive proteasome [31].

Predicting Proteasome Specificity

Degradation of intracellular proteins by the proteasome is crucial for many cellular processes. Full inhibition of the proteasome activity results in cell death within 48 hours, although the resistance to proteasome inhibition differs in cellular subtypes [32]. Deficiencies in the proteasomal activity can cause pathology (see e.g., below the section on the role of proteasome in neurodegenerative diseases). Moreover, the proteasomes shape the immune response of higher vertebrates as reviewed in Chapter 34. Successful prediction of the proteasome cleavage site specificity is necessary to understand the pathology associated with protein degradation and immunity. For example, predictions could help to find which regions of the pathogen genomes are most immunogenic and these regions can be used in designing optimal vaccines. Similarly, predictions can be used to identify cytotoxic peptides in neurodegenerative diseases. However, the enzymatic specificity of the proteasome is very complex and thus the prediction of proteasomal specificity is a difficult task.

Several proteasomal cleavage prediction methods have been published. The first method, FragPredict, was developed by Holzhutter *et al.* [33] and is publicly available as a part of MAPPP service (www.mpiib-berlin.mpg.de/MAPPP/). It combines proteasomal cleavage prediction with MHC and TAP binding prediction. FragPredict consists of two algorithms. The first algorithm uses a statistical analysis of cleavage-enhancing and -inhibiting amino acid motifs to predict potential proteasomal cleavage sites [33]. The second algorithm, which uses the results of the first algorithm as an input, predicts which fragments are most likely to be generated. This model takes the time-dependent degradation into account based on a kinetic model of the 20S proteasome [34]. At the moment, FragPredict is the only method that can predict fragments, rather than only possible cleavage sites.

PAProC (www.paproc.de) is a prediction method for cleavages by human as well as wild-type and mutant yeast proteasomes. The influences of different amino acids at different positions are determined by using a stochastic hill-climbing algorithm [35] based on experimentally *in vitro* verified cleavage and non-cleavage sites [36].

Recently Tenzer *et al.* [37] have published a method for predicting which peptides can be presented by the MHC class I pathway. In this work they characterize the cleavage specificity of the proteasome in terms of a stabilized matrix method (SMM) defining the specificity of the constitutive and immunoproteasome separately. This method is available at www.mhc-pathway.net.

All these methods make use of limited *in vitro* data for characterizing the specificity of the proteasome. Moreover, both FragPredict and the matrix based methods by Tenzer *et al.* [37] are linear methods, and may not capture the nonlinear features of the specificity of the proteasome. Another prediction method called NetChop [38,39] has two important extensions: first, the prediction system is trained on multilayered artificial neural networks. This allows the method to incorporate higher order sequence correlations in the prediction scheme, making it potentially more powerful than both PAProC and the matrix based methods, which use a linear method to predict proteasome cleavage. Second, the method exists in two versions. One version is trained to predict proteasomal cleavage on *in vitro* digest data, similarly to the other previous methods, and the other is using naturally processed MHC class I ligands. The latter version should predict the combined specificity of both forms of proteasomes because some of trained ligands are generated by immunoproteasomes and some by the constitutive proteasome. Details on NetChop method are given elsewhere [38,39].

In large scale benchmark calculations the predictive performance of the different methods are compared [37,39]. According to these tests, at the moment, NetChop and SMM methods provide most reliable predictions of proteasomal cleavage.

Table 1. Cleavage characteristics of human constitutive proteasomes

Position	Positive effect on cleavage	Negative effect on cleavage
P1	Phenyalanine, Leucine, Tyrosine	Proline, Glycine, Threonine, Asparagine, Lysine
P2	Glutamine, Tyrosine, Valine	Proline, Aspartic acid
P3	Valine	Glycine, Glutamine
P4	Proline, Threonine	Lysine, Aspartic acid
P2’	Histidine	Lysine, Serine, Arginine, Glutamic acid, Proline

The position names are according to the nomenclature suggested by Berger and Schechter, 1970 [26]. The cleavage occurs between the P1 and the P1′ position. The residues on the N-terminal flanking region of the cleavage cite are called P1, P2, P3, P4, ..., while the C-terminal side is referred to as P1′, P2′, ...

Besides providing useful predictions of the proteasomal cleavage, a prediction method can also help us in learning more about the biology of the proteasome. For example, sequence features used by a prediction method to discriminate possible cleavage sites from unlikely ones can be extracted. In an attempt to do this a very small neural network trained on the human constitutive proteasome data was analyzed. Cleavage and inhibiting amino acids found in this analysis are summarized in Table 1. In the P1 position large hydrophobic residues promote cleavage prediction by the network. Proline at P1 and P2 is strictly cleavage-inhibiting, whereas at P4 it is cleavage-promoting, as suggested earlier [40,41].

Glycine seems to be cleavage-inhibiting when present at positions P1 and P3. The P2′ position may have as much influence as P2; charged residues at P2′ are cleavage-inhibiting. In the P1′ position both experimental results and theoretical studies suggest a preference for small, β-turn promoting amino acids for cleavage [27,35]; however, this could not be detected in the weight analysis. For rare amino acids like tryptophan and cysteine, it was not possible to draw any conclusions since these amino acids have a very low frequency in the data used. Interestingly, these characteristics are very similar to the ones suggested earlier for the yeast proteasome [35].

MODELS OF PROTEASOME KINETICS

Although significant progress has been made in understanding substrate recognition and proteolysis by proteasomes (see above), very little is known about the intermediate steps of transportation into the proteolytic chamber of the proteasome. Few models have been proposed addressing the question of the kinetics of fragments during protein degradation. Many of them investigate the degradation of short fluorogenic peptides and do not consider the role of substrate length and of the gate functioning, as well as the effect of interaction with other residues present in the proteasome chamber [42-45]. These issues are very important in studying the role of proteasome in neurodegenerative diseases. To our knowledge only two theoretical models have addressed the effect of these quantities. Holzhuetter and colleagues suggested a linear kinetics model to investigate the effect of substrate orientation and length [34], where the transport rate within the 20S particle is an exponential function of the substrate length. This model showed that the degradation follows a C- to N- term rule. Recent data have however shown that degradation can proceed also in the N to C direction, as well as, endoproteolytic cleavages are possible therefore undermining the reliability of this model [46,47]. Moreover, the degradation process is described as a length dependent process: the longer substrates are degraded at a lower rate. A quantitative comparison between degradation of substrates by constitutive and immunoproteasome suggested that constitutive and immunoproteasome have the same cleavage specificities, but the immunoproteasome is able to cleave more efficiently the substrate molecules [34,42]. Recently, experimental data provided clear evidence that the gate functioning, and most importantly, the finite volume within the chamber can strongly affect the degradation kinetics [8,48,49]. These issues have been investigated by Luciani *et al.* [50]. This general model (referred to as L-model from now on) describes the degradation of proteins without a particular sequence. Influx, efflux and cleavage are the major processes included in this model and depending on the length of the fragment. The model is based on several assumptions that are summarized below (see also Table 2): (*i*) the protein enters into the proteasome with an influx rate a_0 as a long string of amino acids; (*ii*) the influx is limited by the maximum number of amino acids that proteasome chamber has space for [5,46,51] and by the opening of the gate [8]; (*iii*) the amino acid composition of the protein does not influence the influx rate in the model; (*iv*) the cleavage can occur, with rate c, at both N- and C-termini: the probability of the cleavage is modelled as a Gaussian function, where on average a cleavage would occur at every 3-4 amino acids; (*v*) the efflux of the fragments from the proteasome is modelled as a declining function with the fragment length. Very short fragments leave the proteasome fast, and for intermediate length fragments (~25 amino acids) the efflux rate starts decreasing. Long fragments have a negligible rate of efflux, and thus they remain in the proteasome for further degradation. L-model obeys to Michaelis-Menten kinetics, as suggested by *in vitro* proteasomal degradation experiments [10,52-59]. Further details and mathematical computations are given in Luciani *et al* [50].

Table 2. Parameters of the Luciani model (L-model)

Parameter	Definition	Description
a_0	Influx rate	Substrates are transported into the 20S with a given rate
c	Cleavage rate	Substrate are cleaved with a given rate c on average each 4 amino acids.
e_0	Max Efflux rate	Fragments exit the proteasome with a length dependent mechanism.
L	Substrate length	Long fragments residue longer in the core particle.
V	Volume 20S	The proteasome 20S core particle can accommodate until 1000 aa

aa indicates aminoacids.

Mosfolded Proteins as Key Effectors in Neurodegenerative Diseases

A broad array of human neurodegenerative diseases are characterized by the accumulation of extracellular and/or intracellular protein aggregates sharing strikingly similar histopathological features that may hold the key to their molecular pathogenesis [60,61]. This suggests an underlying role in the biophysical properties of protein stability, aggregation and degradation in the pathology of nervous system disease (Table 3) [62-64] (see also Chapter 12). The pathological hallmark is the presence of insoluble intra- or extracellular inclusion bodies in affected regions of the brain. The major constituents of these inclusions are misfolded proteins. Their aggregation into proteinaceous inclusions is perhaps surprising given that there are cellular processes present designed to prevent such events. Moreover, what is also puzzling is the observation that these inclusions are found almost exclusively within certain subgroups of neurons in specific regions of the brain regardless of the widespread tissue distribution of the core proteins [65]: aggresome like structures in prionopathies [66,67], neurofibrillary tangles and plaques in Alzheimer disease (AD) and tauopathies, Lewy bodies in Parkinson disease (PD) and related synucleinopathies [68-70], cytoplasmic and nuclear inclusion bodies in polyglutamine (PoliQ) expansion diseases such as Huntington's disease (HD) and spinocerebellar ataxia (SCA) [71-74]. Aggresome like structures that clumped protein aggregates surrounding nuclei, along dendrites and Ub-immunoreactive nuclear inclusions are also formed after acute neurological lesions such as cerebral ischemia [75-79].

The presence of insoluble proteinaceous deposits differ in their protein content but invariably contains components of the UPS [60]. Although the UPS involvement in the clearance of misfolded proteins is not completely known, it has led to the suggestion that a chronic imbalance between generation of misfolded proteins and processing may be the primary cause for the formation of protein deposits [61,80]. The classic causes for protein misfolding, which lead to a loss of function, are missense mutations, protein modifications or post-translational damage, or expansion of amino acid repeats as is observed in PolyQ disorders.

Table 3. Neurological diseases caused by defects in protein folding, stability and aggregation

Disease	Aggregate type	UPS components	Location	Characteristic pathology
Alzheimer disease	Plaques	Ubiquitin Proteasome	Extracellular	Extracellular neuritic plaques
	Tangles	Ubiquitin Proteasome	Cytoplasmic	Neurofibrillary tangles of hyperphosphorylated tau
Parkinson disease	Lewy bodies	Proteasome Ubiquitin Hsps E3–Parkin DUB– UCH-L1	Cytoplasmic	Intracellular Lewy bodies, Lewy neurites, fibrillar, α-synuclein
Multiple system atrophy	Glial/ neuronal inclusions	Ubiquitin	Cytoplasmic	Oligodendroglial inclusions immunostained with tau and ubiquitin
Polyglutamine disease	Inclusions	Proteasome Ubiquitin Hsps	Cytoplasmic	Aggregates and fibrillar, huntingtin fragments
			Nuclear	neuronal inclusions
Prion diseases	Aggresome-like	Ubiquitin Hsps	Cytoplasmic	Intracellular deposits, and occasional synaptic and axonal deposits
			Extracellular	Amyloid plaques
Stroke	Aggresome-like	Ubiquitin Hsps	Cytoplasmic/ nuclear	Protein aggregates surrounding nuclei and along dendrites in postischemic neurons in the neuronal soma, dendrites, and axons. Ubiquinated proteins are associated with intracellular membranous structures in neuronal lysosomal vesicles and in late endosome-like organelles in the ischemic area

DUB indicates deubiquitinating enzyme, UCH-L1, ubiquitin C-terminal hydrolases-L1, Hsp, heat shock protein.

These pathological hallmarks share certain biochemical features. First, they are characterized by the presence of aggregates of one major 'core' protein aberrantly polymerized or aggregated disease protein, either full-length protein or proteolytic fragments including cytoskeletal elements such as tau, tubulin-associated proteins and neurofilaments [81-84]. Second, they display amyloid-like properties, suggesting a similar fibrilar structure. Third, they harbour components of the UPS machinery: Ub chains, parkin, proteasome subunits [85-88], and heat shock protein (Hsp) chaperones such as Hsp70 and Hsp27 [89,90]. Finally, they often contain other proteins that, although not themselves intrinsically aggregation-prone, become recruited, concentrated, or trapped within the inclusions. The molecular forces driving their co-localization to inclusions and the potential role this plays in pathogenesis are still largely unknown.

As protein misfolding poses a major threat to cell function and viability, molecular mechanisms must have evolved to prevent the accumulation of misfolded proteins and thus aggregate formation. Two protective strategies appear to be followed (see Chapter 12): (*i*) molecular chaperones are employed to stabilize nonnative protein conformations and to promote folding to the native state whenever possible [91]; (*ii*) misfolded proteins are removed by degradation, involving, for example, the UPS [65] or through basal autophagy [92-94]. Protein fate thus appears to be determined by a tight interplay of cellular protein-folding and protein-degradation systems.

Neurodegenerative diseases are typically manifested at an advanced age: it is suspected that the loss of efficiency in molecular and/or cellular clearance mechanisms increases with rising age, probably due to impaired antioxidative systems [95]. In this case, the UPS regulating the cellular breakdown of cellular misfolded proteins could play a role. These diseases are characterized by the accumulation of aberrant proteins either intracellularly or extracellularly in specific groups of cells that subsequently undergo death [96], although the precise association between protein accumulation and cell death remains incompletely understood with several differences between different disease processes. In some cases, misfolded protein accumulations may themselves be toxic or exert spatial constraints on cells that affect their ability to function normally. In other cases, the sequestering of proteins in aggregates may itself be a protective mechanism, and it is the overwhelming of pathways that consolidate aberrant proteins that is the toxic event. Taken together, the pathophysiology of neurodegenerative diseases provides a compelling demonstration of the importance of the regulated metabolism of misfolded proteins and provides direct evidence of the role of both molecular chaperones and the UPS in guarding against protein misfolding and its consequent toxicity [65,97,98] (see Chapter 19).

PROTEIN AGGREGATION IN NEURODEGENERATION

Recent studies suggest that up to 30% of all newly synthesized proteins never reach their native state [99]. As protein misfolding poses a major threat to cell function and viability, molecular mechanisms must have evolved to prevent the accumulation of misfolded proteins and thus aggregate formation. Accumulation of protein aggregates in cytoplasmatic inclusion bodies (IBs) is a nearly universal feature of sporadic and hereditary neurodegenerative disease [100] (Table 3). However, the cellular mechanisms by which these abnormal protein structures are linked to underlying cellular pathology have remained elusive. Although misfolded proteins are also considered to accumulate in neurons and glia as a result of physiological processes associated with ageing, protein overexpression and conformational rigidity has been shown to be particularly toxic to neurons under disease conditions. Many of the mutations that cause dominantly inherited neurodegenerative diseases have been demonstrated to dramatically increase the propensity of the mutant gene products to self-associate into protein aggregates *in vitro* and *in vivo* [101,102], supporting the widely considered hypothesis that aggregation underlies the molecular pathogenesis of many neurodegenerative disorders [103]. Such mutations affect proteins, irrespective of their native function, and cause them to misfold. An example of this one is HD where the protein encoded by the huntingtin (Htt) gene contains a stretch of glutamine residues (or PolyQ repeat) subject to misreading and expansion. When the length of the PolyQ repeat in Htt reaches a critical threshold of approximately 35 residues, the protein becomes prone to misfolding and aggregation [73] and it probably obstructs the proteasomes as a partially degraded substrate, which could explain its deleterious effects on cellular function [104]. PolyQ proteins are incompletely degraded, even when directly targeted for proteasomal degradation, *in vitro* and *in vivo* [104]. The presence of the expanded PolyQ stretch associated with the proteasome could generate additional surfaces for molecular interaction with aggregation-prone proteins, thus promoting recruitment to the aggregates. Moreover, one would predict that kinetic trapping of PolyQ-containing proteins within the proteasome might interfere with degradation of other cellular substrates of the proteasome.

In other diseases, protein misfolding occurs due to other mutations that induce misfolding and aggregation; for example, mutations in superoxide dismutase-1 lead to aggregation and neurotoxicity in amyotrophic lateral sclerosis [105-107]. Other mutations that result in neurodegenerative diseases are instructive in that they directly implicate the UPS in the pathogenesis of these diseases. For example, mutations in the gene encoding the protein parkin are associated with juvenile-onset PD [108; and Chapter 31]. Parkin is a RING finger-containing Ub ligase, and mutations in this Ub ligase cause accumulation of target proteins that ultimately result in the neurotoxicity and motor dysfunction associated with PD [108]. Repressor screens of neurodegeneration phenotypes in animal models have also linked the molecular chaperone machinery to neurodegeneration [65,97,98].

In cells, aggregated proteins are gathered into IB that often harbor components of the protein surveillance machinery: Ub chains, proteasome subunits, Hsp chaperones, and other proteins that, although not themselves intrinsically aggregation-prone, become recruited, concentrated, or trapped within the inclusions such as intermediate filaments, protein kinases, and transcription factors [96,103]. Misfolded and aggregated peptides appear to owe their toxicity to protein regions that become exposed on their surfaces while being buried in the interior of correctly folded native state structures. Surface exposure of large hydrophobic groups favors interactions of misfolded proteins with cell membranes with a subsequent loss of the regulation of intracellular ion balance and redox status [69,109,110].

Other proposed deleterious effects of misfolded proteins also include the pathological interplay with intracellular substrates [86], leading to the generation of reactive oxygen species with subsequent oxidative stress and cellular demise through classical apoptotic pathways transcriptional dysregulation [111], disruption of microtubule-dependent axonal transport [112], perturbation of membrane permeability [113], and impaired function of the UPS [114,115].

The UPS protects cells against the potentially toxic effects of protein aggregation by recognizing and selectively degrading misfolded and damaged proteins. Many studies have shown that proteasome inhibition, pharmacologically or genetically induced, increases aggregation or inclusion formation in cells and invertebrate models [73,115-123]. If during the course of disease the UPS becomes compromised for any reason, this would further reduce the global rate of protein degradation and foster a cellular environment that increasingly favors aggregation and inclusion formation. Because a build-up of aggregated protein in the cell may itself directly compromise the UPS [114], a vicious cycle of protein aggregation and proteasomal perturbation might ensue. In addition, at least some disease proteins appear to be degraded less efficiently than their normal counterparts — a property that also should increase their steady-state levels and favor aggregation [115,124-126]. An inducible mouse model of HD shows that inclusions may be cleared by the UPS once the Htt transgene was repressed [127], however usually aggregation is irreversible and proteasomes do not have access to the proteins embedded in IBs. In that case, autophagy provides the mechanism of clearance of aggregated proteins as first shown by Wójcik *et al.* [128] and recently confirmed in several models of neurodegenerative disorders [129]. The significance of this latter role is underscored by the discovery that loss-of-function mutations in genes encoding UPS components can cause neurodegenerative diseases in humans [130] and rodents [131,132] and enhance the cytotoxicity of aggregation-prone proteins linked to dominantly inherited neurodegenerative diseases [125,133].

This strongly implicates the UPS in these disease-associated inclusions, either due to malfunction (of specific UPS components) or overload of the system (due to aggregation of unfolded/mutant proteins), resulting in subsequent cellular toxicity. Although defective UPS function is a robust and reproducible response to protein aggregation, the mechanism by which protein aggregation is linked to UPS impairment remains an open and compelling question. Inclusion formation is also, at least in part, a cellular response that serves to concentrate misfolded proteins and perhaps facilitate their degradation — in other words, a cell-driven process. The best example is the aggresome: an inclusion body described in many cell models that overexpress mutant proteins [96,120,134,135]. Aggresomes are cytoplasmic deposits of aggregated protein that form in a microtubule-dependent manner and localize to the microtubule organizing center. Aggresomes arise when the rate of abnormal protein production exceeds the cell's capacity to handle it, including when the proteasome is compromised [96,120,134,135]. An unresolved issue is whether the cytoplasmic inclusions seen in human diseases — such as Lewy bodies, glial cytoplasmic inclusions and perinuclear PolyQ inclusions — are in fact aggresomes.

MODELS OF PROTESOME DYSFUNCTION

The demonstration that components of the UPS often are involved in neurodegeneration prompted to examine whether a general impairment of the proteolytic machinery may contribute to the pathology. There is no doubt that proteasome inhibition enhances aggregation of mutant proteins in cells. However, it is less clear whether aggregates directly inhibit proteolytic activities of the proteasome in the disease state. The observation of severe UPS impairment in compartments lacking detectable aggregates or aggregation-prone protein, together with the lack of interference of protein aggregates on 26S proteasome function *in vitro*, suggests that UPS impairment is unlikely to be a consequence of direct clogging or choking of proteasomes by protein aggregates. The UPS controls the stability of most nuclear and cytoplasmic proteins and is therefore essential for virtually all aspects of cellular function. The UPS is impaired in the presence of aggregated proteins that become deposited into cytoplasmic IBs. However, production of protein aggregates specifically targeted to either the nucleus or cytosol leads to global impairment of UPS function in both cellular compartments and is independent of sequestration of aggregates into IBs. These data suggest a common proteotoxic mechanism for nuclear and cytoplasmic protein aggregates in the pathogenesis of neurodegenerative disease. Two possible models could account for the impairment of UPS function by protein aggregation and therefore explain the accumulation of misfolded ubiquinated substrates in neurodegenerative diseases (Figure 2). In Figure 2 the two kinetic schemes explaining the two possible models for the UPS impairing are shown. The first picture (Figure 2, Panel A) depicts the accumulation of proteasome substrate as the result of a filling up of the internal proteasome chamber due to slow degradation. The aggregated or aggregation-prone proteins directly inhibit or 'choke' the 26S proteasome—a situation that might result from their engagement by degradation-resistant [104,136] or hard-to-unfold proteins: a reduced cleavage strength towards specific amino acid sequences keep the proteasome sequestrated with undegraded substrate and intermediate long fragments. A consequence of this dynamics is an impaired influx rate and a slower degradation process. Since proteasomal proteolysis is highly processive [137], aggregates could be the inhibitors of proteasomal degradation, as they are undegradable and slowly released. This 'proteasome choking' model predicts that in cells exhibiting near complete loss of UPS function, a substantial fraction of total cellular proteasomes should be associated with protein aggregates. Thus, aggregates could simply sequester proteasomes away from cellular sites where they are required, as suggested from the observation of proteasome subunits in IBs in brains from human [138] and animal models [80] of neurodegenerative disease. Recent studies suggest that even short, non-pathogenic (Q19) PolyQ tracts are inefficiently degraded [104] or completely indigestible [136] by proteasomes *in vitro* and *in vivo*, despite the presence of endogenous ubiquitylation. Kopito and co-workers observed failed UPS activity as suggested by increasing in GFP fluorescence when mutant proteins became aggregated in the cell [114] Similar results have been observed in cellular models of PD [122,123,139,140], prion diseases [120,141,142], and PolyQ diseases [114,115,117].

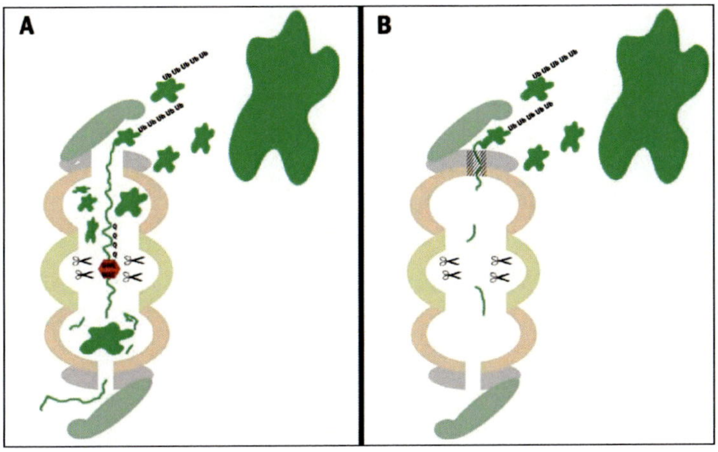

Figure 2. Mechanisms of proteasome block: simple sketch of impaired 26S proteasome degradation of ubiquitinated substrate. Two possible scenarios are conjectured. *Panel A*. Inhibition of the 26S proteasome activity due to filling up with substrates and intermediates that have a reduced number of cleavage sites. Substrates are ubiquitinated and transported into the proteasome core via the binding with the 19S particle. Substrates are then cleaved by the proteasome according to the cleavage specificities. These are defined by the interaction between the amino acid sequence and the proteasome catalytic sites. PolyQ repeats strongly impair the cleavage activity with a concomitant reduction of the degradation rate. The internal volume will be progressively filled with long intermediates, which have a low efflux rate and undegraded substrate. Opening the gate favors the efflux of intermediate substrates, which then exit in a length dependent manner. Therefore, long intermediates have a low efflux rate. According to the model, the filling of the proteasome stops the influx of new substrate molecules, causing an accumulation of substrate outside the proteasome. *Panel B*: Inhibition of 26S proteasome activity due to impaired transport of substrates caused by amino acid sequence changes on the substrate that effect 19S binding. The interaction between substrate and proteasome regulating the influx of substrate is largely unknown. Energetic barrier between the amino acid sequence constituting the substrate and the structure of the 19S can be hypothesized. Here, for simplicity we depict an impaired transport of substrate. A limited influx of substrates results in an accumulation of non-degraded proteins outside the proteasome.

These findings of proteasome inhibition have, so far, been observed in cell-culture models where the mutant protein is overexpressed. Whether this accurately mirrors the *in vivo* state, where expression levels of mutant protein are lower and the pathophysiological insult occurs over years rather than days, is unknown. Indeed, despite the existence of many mouse models with intraneuronal inclusions, no one has yet reported inhibition of proteasome activity by inclusions *in vivo*. There is even some negative evidence in animal models suggesting that the proteasome is not significantly impaired by inclusions [143]. Further studies of proteasome function in animal models, including inducible models, are required to answer whether direct compromise of proteasomal activity plays any role in pathogenesis.

A second model, not mutually exclusive with the first, is that protein aggregates indirectly interfere with UPS function by sequestering or directly clogging proteasomes. It is possible that they could impair UPS function by influencing the proteasome activity or distribution of UPS modulators, inactivating or depleting a UPS activator (Figure, 2 Panel B), where aggregation would result due low influx of substrate into the 20S core particle. Proteasome substrates with a low affinity binding to 19S sites, which is responsible for the influx and unfolding mechanisms have a strongly reduced transportation rates. Several indirect experimental studies suggest this point of view: (*i*) molecular misreading of a Ub gene leads to accumulation of an aberrant, frame-shifted Ub, (UBB^{+1}) [144]. UBB^{+1} accumulates in degenerating neurons of tauopathies, such as AD, progressive supranuclear palsy and Pick's disease, and of PolyQ diseases (see Chapter 28). UBB^{+1} is the result of a GU dinucleotide deletion adjacent to the first GAGAG repeat in *UBB* mRNA. The frame shifted protein can be ubiquitinated at lysine 29 and 48. UBB^{+1} has lost its carboxy-terminal G and therefore the ability to form isopeptide bonds with lysine residues in target molecules. Ubiquitinated UBB^{+1} (Ub–UBB^{+1}) is normally degraded by the UPS. However, a decrease in the activity of the proteasome results in accumulation of Ub–UBB^{+1} and, subsequently, a further inhibition of the proteasome. Strong proteasome inhibition induces apoptosis in neuronal cells. In short, UBB^{+1} is unable to ubiquitinate, is a target for the UPS and can inhibit the proteasome [145]. Furthermore, under oxidative stress, 26S proteasome tend to be partially unstable, while 20S increases its stability and amount in cells allowing it to degrade partially unfolded proteins without ubiquitination [146]. (*ii*) Proteins containing expanded PolyQ have been shown to interact with and inactivate PolyQ-containing transcription factors [147,148]; recent studies have suggested that this interaction can occur with an early, oligomeric form of PolyQ repeat and may depend more on the PolyQ conformation than on its aggregation state [149]. Smaller intermediate forms of protein aggregates are more toxic to cells than fibrillar forms [109,150]. Consistent with this, it has been argued that the formation of higher order aggregates and their subsequent coalescence into IBs may be cytoprotective [151,152]. Most disease-linked aggregated proteins accumulate in IBs that are characteristically restricted to either the nucleus or cytoplasm, and the effective toxicity associated with the presence of these IBs appears to be strongly influenced by the cellular compartment in which the aggregates accumulate [153,154]. Finally, it should be noted that nonnative pathogenic conformers of polyQ proteins could interact with and activate UPS inhibitors. For example, the presence of nonnative undegraded protein aggregates could initiate caspase activation; one recent study reported that caspase activation results in irreversible inhibition of proteasomes via cleavage of 19S regulatory particle subunits S5a, S6', and S1 [155] (see Chapter 21). It is important to recognize that perturbations of the UPS could be subtle and still compromise neurons by rendering them more susceptible to other cellular stresses. This was evident in a cell model in which expanded PolyQ protein caused a decrease in proteasome activity only when the cells were treated with heat shock [156].

Models of Protesome Kinetics in Neurodegenerative Disorders

Mathematical models of proteasome degradation provide a qualitative and quantitative analysis of the mechanism underlying the degradation process. Hypothesis suggested by experimental data can be tested and quantitative predictions can be achieved. In this chapter two possible scenarios explaining protein agglomerates have been proposed. A quantitative and complete mathematical analysis to test such scenarios requires very detailed modeling of the mechanisms behind the interactions between proteasome subunits and substrate molecules. For instance, modeling the ATP-dependent activity of the 19S base and the mechanical transport of substrate molecules within the 20S core are suggestive and intriguing phenomena that can help the understanding of UPS blockage and the impaired activity observed in neurodegenerative associated protein degradation. Unfortunately to date, no such model has been proposed, the main reason being the very little information about these mechanisms, although experimental works in this field is constantly growing [157,158].

The L-model [50] offers a simple and appropriate mathematical design to test the two suggested hypothesis. Proteasome inhibition can be observed with a block of the catalytic subunits by specific molecules that prevent the cleavage activity (for a review see Chapters 40). Nevertheless, the L-model shows that proteasome inhibition can be also achieved with impairment of the kinetics mechanisms underlying the degradation process. These mechanisms are not mutually exclusive, and can concertedly induce severe blockage of the proteasome activity.

To test the first scenario explained previously and in Figure 2A, the degradation of an extended PolyQ substrate is considered. The most important determinant of the degradation kinetics of a given protein by 26S is its ubiquitination, however once ubiquitinated, the number of predicted cleavage sites within a protein can be used as a measure of resistance to degradation. NetChop method predicts that a Poly-A or PolyQ tract will be almost fully resistant to cleavage by the proteasome (results not shown, but can easily be obtained from NetChop server at www.cbs.dtu.dk/services/NetChop).

After a preliminary phase in which the model proteasome fills up, a quasi steady state is reached, where few substrates may be simultaneously present within the proteasome chamber, and fragments of different length are released to the outside. Recently, by electron microscopy and tandem mass spectrometry it has been shown that the proteasome can store up to three GFP and two Cyt-C molecules simultaneously. Some of these molecules reside in a partially folded state within the antechambers of the 20S proteasome [48]. The degradation rate in the model is a function of the cleavage rate, *c*, the efflux rate, e_0, and finally of the influx rate, a_0. In a simplified toy model it is possible to analytically express the steady state degradation rate in terms of the influx efflux and cleavage rate parameters (See Luciani *et al.* [50] for more details)

$$\frac{dS}{dt} = \frac{ca_0e_0S}{a_0vcLS + e_0c} \quad (1)$$

where S is the substrate concentration, L is the substrate length and v is the fraction of volume filled within the core particle. The maximum degradation rate $V_{\max}$ is

$$V_{\max} = \frac{e_0c}{vL(c + e_0)} \quad (2)$$

which is approached for very high substrate concentrations. These parameters regulate the fraction of proteasome volume filled and the number of fragments that are unleashed by the proteasome at any given time.

If the substrate is transported into the proteasome at a given rate a_0, this rate is limited by the filling of the internal chambers with substrates and fragments produced, which are not yet removed. Figure 3 shows the fraction of proteasome core particle filled at a given time during the degradation of a 100 amino acid long substrate. The fraction of volume filled increases with the increase of the influx rate, and decreases with the increase of the cleavage rate c. Furthermore, the filling of the proteasome influences the degradation rate. The number of substrate molecules degraded per unit time should intuitively increase with increasing cleavage rate, c. The degradation rate is increased if the cleavage rate increases as well as predicted by the model (Figure 3, Panel B, and Equation 1). However, the model shows an additional complex kinetic. If the influx rate is very low, the proteasome is filled less than 70% (Figure 3, Panel A). In this parameter regime the increase of the cleavage rate does not influence the degradation rate, which remains constant around 0.1 substrate molecules per unit time. The amount of substrates degraded is determined by the influx rate a_0 and the substrate concentration outside. Figure 3 clearly shows that for low values of the influx rate the degradation is fairly independent on the cleavage rate c. Interestingly, increasing the influx rate does not necessarily means that the degradation rate increases. For limited cleavage rate (see Figure 3, Panel B), the number of substrate molecules degraded decreases with the increasing influx rate. The reason for this more complex behavior is probably the internal dynamics of undegraded substrates and intermediate long fragments. If the influx and cleavage rate of substrate is very low, the proteasome is almost empty and the substrates entering the proteasome are immediately degraded and fragments are ejected. Increasing the influx of substrate, for instance increasing the concentration of ubiquitinated substrate outside the core particle, the proteasome fills up much faster and if the cleavage activity is reduced (e.g. due to PolyQ repeats) many long intermediate fragments will be generated. The result is an almost full proteasome, and a low number of substrate molecules entering the proteasome chamber. A quasi steady state regime will be reached where proteasomes filled with long intermediate fragments strongly impair the degradation. This unexpected result might be experimentally explored: experiments with different substrate concentrations in a low cleavage regime could prove this theoretical expectation.

The L-model can also be used for the investigation of the second scenario (see above, and Figure 2, Panel B) that considers a substrate with a limited influx such as the UBB^{+1}, which is a target of proteasome degradation [144]. At low substrate concentration the proteasome degrades the UBB^{+1} and these molecules are removed from the cell. At high concentrations of UBB^{+1} in the system, the proteasome activity is blocked, and an accumulation of ubiquitinated UBB^{+1} molecules is observed [159]. The strong reduction of proteasome activity at high concentration can be explained by a weakened or impaired transport of UBB^{+1} within the proteasome particle, which can eventually reduce the gate opening effect. In our model an impaired influx can result in a low number of substrate molecules degraded per time unit (see Figure 3).

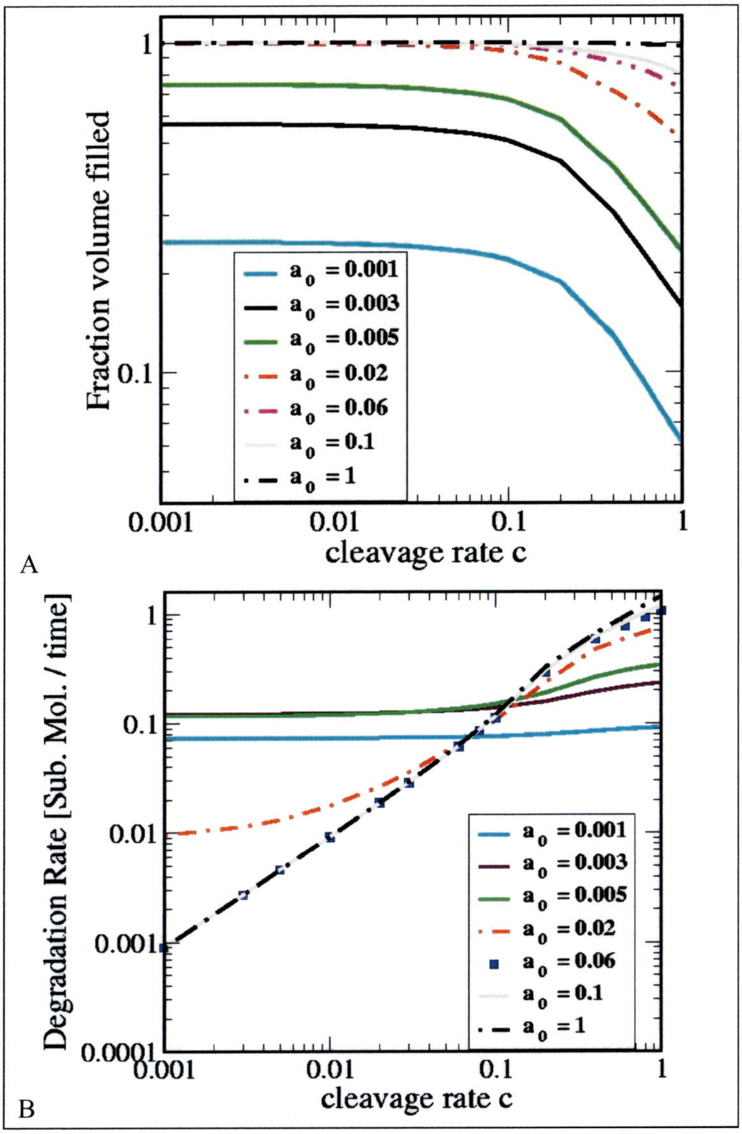

Figure 3. Simulations of proteasome degradation with a Substrate of L=100 amino acids and substrate/proteasome ratio of 100. A: Fraction of the proteasome core particle (20S) filled vs. the cleavage rate c. Each curve represents a different value of the influx rate a_0 (See also *Table 2*). The fraction of volume filled with substrate and intermediate fragments is a function of both the cleavage and the influx rate. Increasing the influx rate results in a higher fraction of volume filled. On the other hand, increasing the cleavage rate, for instance if the substrate contains a region having high density of cleavage sites, decrease the filling up of the internal volume. B: Degradation rate as a function of the cleavage rate *c*. Each curve represents a simulation for a different influx rate a_0. The number of substrate molecules degraded per unit time is influenced by the cleavage rate *c* and the influx rate. Obviously, the degradation rate increases with the increasing cleavage rate *c*. However, when both the cleavage and the influx rate are very small, the behavior is non-intuitive: the degradation rate decreases with the increase of the influx rate a_0. In this condition the volume is filled less than 60-70% (see A panel). Increasing the influx rate fills up the proteasome very fast augmenting the sequestration time of proteasomes by undegraded and partially processed fragments. Therefore, high influx and low cleavage rate result with impaired degradation kinetics.

As already pointed out, the internal dynamics of the proteasome chamber influences the degradation process, and for low cleavage rate the degradation process is of a non-trivial fashion. The net degradation rate is dependent on the internal kinetics as well as the gate opening mechanism.

Model Results

Recent experimental data support the first scenario, i.e. an impaired UPS caused by the filling of the proteasome chamber by non degraded substrates. For instance, Homberg *et al.* [104] reported kinetically trapped PolyQ-containing substrates within the proteasome. Venkatraman *et al.* [136] showed that purified proteasomes in an open gate conformation were unable to cleave long substrates containing long glutamine repeats. These last results clearly evidence the fact that substrates may enter into the proteasome particle but then are poorly processed by the catalytic sites. Also proteasome cleavage prediction programs (e.g., NetChop, see results given in Table 4) suggest that PolyQ, polyG and polyA tracks to be resistant to the proteasomal cleavage.

Table 4. Predicted cleavages sites on start of huntingtin (Htt) protein by Net Chop (www.cbs.dtu.dk/servers/NetChop)

```
Htt(Positions 1-80)

MATLEKLMKAFESLKSFQQQQQQQQQQQQQQQQQQQQQQQPPPPPPPPPPPQLPQPPPQAQPLLPQPQPPPPPPPPPPGP

SS.SS..SS.S..S..S.SS.................................S......SS..................
```

An S indicates a predicted cleavage sites, and "." denotes sites that are unlikely to be cleaved, ie. these sites generate regions that are resistant to proteasomal cleavage.

Many aggregation prone proteins, such as modified Htt proteins or PolyG repeats, are characterised by an elongated version of the native protein. According to the L-model the transport of elongated molecules may require a longer time, and thus, these sequences reside longer in the core particle. Such a condition would lower the degradation rate further, and therefore, can contribute to protein accumulation and UPS sequestration (in fact polyG sequence greater than 35 amino acids unfold and are degraded by the proteasome at a very low rate [136]).

Regarding the second scenario, it has been shown that stabilizing or destabilizing a folded domain within substrate proteins by altering the amino acid sequence, for example introducing a sequence containing only G and A residues, impairs proteasome activity due to an high energy requirement in the translocation of this substrate from the ATP ring of the 19S into the 20S [160].

A major prediction of this model is the regulation of the degradation rate by the gate opening mechanism. The number of substrate molecules degraded per time unit increases when the proteasome gate is open. Evidence for such behaviour can be found *in vitro* and *in vivo* degradation with archeal 20S proteasome and the α3N mutant eukaryotic proteasome, which are characterized by an open gate channel [8,158]. The archeal proteasome degrades PolyQ repeats *in vivo* and *in vitro*. This proteasome is characterized by an open gate conformation, and by the presence of 14 identical catalytic subunits [161]. According to the L-model, such a proteasome is kinetically equivalent to the model condition in which substrates are degraded at a high rate because of high cleavage rate c and high efflux rate, e_0. Eukaryotic proteasomes, which are characterized by only six (three pairs of different specificity) catalytic subunits and often by a regulated gate with 19S or PA28 caps, may have a reduced degradation activity compared to the archeal forms.

The accumulation of undegraded substrate in proximity of the UPS can be explained according to our kinetics model in which the degradation rate is influenced by the filling up of the proteolytic chamber within the 20S particle. A non-trivial scenario arises from the L-model of proteasome degradation. The degradation rate can be substantially reduced with the increase of the influx of substrate within the proteasome. This unexpected result is a direct consequence of the filling of the proteasome, which is strongly dependent on both the influx and the efflux of substrates and intermediate fragments. In a scenario where cleavage is rate limiting, such as PolyQ rich substrates, this result should be experimentally investigated.

In vitro experiments show that the rate of substrate transport and degradation is a length dependent process [10,50,162]. This can be tested also for PolyQ substrates, for example by monitoring fluorescence signals with different length of GFP-PolyQ reporters. Such experiments show conflicting results. Jana *et al.* found that the degradation rate is inversely related to the PolyQ length, i.e., if PolyQ stretch is longer, the degradation rate is slower [115]. On the other hand, Verhoef *et al.* found only a slight dependence of the degradation rate on the PolyQ length, and a fast degradation rate for substrates carrying an N-end rule degradation signal [163]. These conflicting results arise because these experiments do not address specifically the transport into the proteasome, but they monitor the complete process of protein agglomeration and proteasome sequestration and degradation.

It has been shown that the accumulation of aggregation prone substrates such as Htt, UBB^{+1}, and polyQ rich sequences, depends on the substrate length [163-166]. Aggregated substrates might not enter completely the proteasome causing impairment in the transport mechanism within the core particle. Models describing the biophysical length dependent mechanism of polyQ agglomeration have been also proposed [167,168], but a clear description of the process remains largely unknown.

The kinetics model and the prediction algorithms described in this chapter allow the prediction of the specificity of the proteasomal cleavage and its kinetics. So, leaving the agglomeration process out of the picture for a moment, proteasomal cleavage prediction algorithms (e.g., NetChop) suggest that PolyQ degradation is not length dependent because five or more repeating glutamine residues produce a cleavage resistant region. However, combined with the proteasomal kinetics, i.e. the L-model, we predict that longer PolyQ sequences should be degraded slowly, because longer PolyQ sequences fill up the proteasome much faster and thus can easily block degradation. Other substrate length-dependent mechanisms might affect the UPS activity. For example, the lysine number at the ubiquitination site is known to affect the sequestration rate of UPS [169]. This is yet another issue that we cannot address using our model, because our model is a simple description of the transport mechanism within the proteasome. More experimental work is needed to identify the explicit mechanisms underlying the interaction between 19S subunits and ubiquitinated substrates. Recent experiments [170] with post-translational modified Htt filaments extracted from IBs cause impairment of 26S but not of 20S proteasome degradation. Moreover, these mutant Htt filaments cause proteasome impairment only when extracted from IB. IBs do not modify degradation rates in both 20S and 26S proteasome systems. These data support the hypothesis that the interaction between 19S and modified substrates is a major limiting factor in the degradation of some neurodegenerative-associated substrates. According to the L-model these data might be well explained by a more 'close-gate configuration' of the 26S with respect to the 20S proteasome as expected by a decreased V_{max} in 26S activity [170] (see for more details [50]).

In vitro experiments quantifying the intermediate products generated by proteasomal degradation could help to understand the specific mechanisms underlying the transport of aggregation prone substrate, and help to rule out many hypotheses, clarifying the future experimental directions. This information can allow us to discriminate between the effect of cleavage rate, transport rate, and agglomeration on the degradation of proteins playing a role in neurodegenerative diseases.

Conclusions and Future Perspectives

In this chapter, some published models for the kinetics of proteasome degradation and for the prediction of fragments generated are reviewed. In particular the model proposed by Luciani is discussed, and its applications in the context of neurodegenerative diseases are proposed. Cleavage prediction models provide a likelihood of the cleavage sites within a specific substrate, although they lack a precise and quantitative description of the kinetics of degradation. On the other hand, kinetics models provide a quantitative and qualitative study of the kinetics of the degradation despite the lack of a detailed description of the specificity of the cleavage mechanism. To understand better the complex behavior of the proteasome models integrating both aspects (specificity and kinetics) are required. Such models will be more conclusive in studying the effect of the proteasome in pathological cases.

Here, using a kinetics model, we proposed at least two possible scenarios to explain proteasome dysfunction in neurodegenerative diseases. In the first scenario, the reduced substrate degradation is explained by a reduced catalytic activity towards the substrate molecule, and by the filling of the proteolytic chamber as demonstrated by this model. The second scenario explains the accumulation of the substrate by a very low transport to the cavity of the 20S, mostly because of an impaired interaction between the ubiquitinated substrate and the 19S cap. In summary we have demonstrated that the accumulation of a ubiquitinated substrate can be explained without an inhibition of proteasome, because the interaction between specific substrates and the proteasome structure can also slow down the degradation rate of proteasomes.

Predictions with mathematical models can be tested by manipulating the activity of the proteasome *in vivo*. This is a challenging task but will undoubtedly provide more insight into the pathogenesis of neurodegenerative diseases and in new therapeutic options. Currently, a more complex and realistic model is being investigated. A sequence specific cleavage mechanism has been introduced in the L-model. The cleavage specificities predicted by software, such as NetChop, are embedded in the kinetic model as rate of cleavage per site. As a result, cleavage kinetics is strongly related to the particular substrate considered. The degradation rate is a function of the substrate specificities and of the gate functioning. Preliminary results of this extended model show that a reduced cleavage activity, such as a PolyQ repeat substrates, strongly reduces the degradation rate. This new model seems to support the predicted scenario that a reduced cleavage rate into the proteolytic chamber results in an impaired degradation rate.

A more detailed model describing the interaction between 19S subunit and substrate would be useful to address more difficult issue, e.g., quantification of the impairment of the UPS system when the ubiquitination is defective or when substrate-19S interaction is of low affinity. Moreover, such a model might be useful for a more quantitative and qualitative description of the transport of long substrates characterized by repetitive mutant amino acid sequences causing a block in the transport mechanism. One candidate is the modified ratchet-kinetic model, which describes the function of the ATPase ring of the 19S. This ring provides the energy to attach to the substrate and pull it inside, resulting in a net force towards the proteasome core particle [160]. A ratchet is a mechanical device that restricts movement in one direction and allows movement in the opposite direction. Currently a quantitative description of the proteasome transportation mechanism is unavailable and further studies are required to elucidate the correlation between the transportation rate of substrates and the degradation rate (see Chapter 8).

Therapeutic application of these future results can be envisaged. For instance, the deficiency of ATP as major cause of deficient transport for elongated and/or modified molecules can be tested quantitatively. Providing an ATP source to this impaired UPS could result in a reduction of substrate accumulation, and a potential therapeutic option in the treatment of neurodegenerative disease.

Acknowledgement

FL is supported by the Australian Research Council through the Discovery Grant scheme (DP0556732).

References

[1] Glickman MH, Rubin DM, Coux O, Wefes I, Pfeifer G, Cjeka Z, Baumeister W, Fried VA, Finley D: A subcomplex of the proteasome regulatory particle required for ubiquitin-conjugate degradation and related to the COP9-signalosome and eIF3. *Cell* 1998, 94:615-23.

[2] Rubin DM, Glickman MH, Larsen CN, Dhruvakumar S, Finley D: Active site mutants in the six regulatory particle ATPases reveal multiple roles for ATP in the proteasome. *EMBO J* 1998, 17:4909-4919.
[3] Vale RD: AAA proteins. Lords of the ring. *J Cell Biol* 2000, 150:F13-F19.
[4] Liu CW, Millen L, Roman TB, Xiong H, Gilbert HF, Noiva R, DeMartino GN, Thomas PJ: Conformational remodeling of proteasomal substrates by PA700, the 19 S regulatory complex of the 26 S proteasome. *J Biol Chem* 2002, 277:26815-26820.
[5] Groll M, Ditzel L, Lowe J, Stock D, Bochtler M, Bartunik HD, Huber R: Structure of 20S proteasome from yeast at 2.4 A resolution. *Nature* 1997, 386:463-471.
[6] Ehring B, Meyer TH, Eckerskorn C, Lottspeich F, Tampe R: Effects of major-histocompatibility-complex-encoded subunits on the peptidase and proteolytic activities of human 20S proteasomes. Cleavage of proteins and antigenic peptides. *Eur J Biochem* 1996, 235:404-415.
[7] Glickman MH, Ciechanover A: The ubiquitin-proteasome proteolytic pathway: destruction for the sake of construction. *Physiol Rev* 2002, 82:373-428.
[8] Kohler A, Cascio P, Leggett DS, Woo KM, Goldberg AL, Finley D: The axial channel of the proteasome core particle is gated by the Rpt2 ATPase and controls both substrate entry and product release. *Mol Cell* 2001, 7:1143-1152.
[9] Groll M, Bajorek M, Kohler A, Moroder L, Rubin DM, Huber R, Glickman MH, Finley D: A gated channel into the proteasome core particle. *Nat Struct Biol* 2000, 7:1062-1067.
[10] Akopian TN, Kisselev AF, Goldberg AL: Processive degradation of proteins and other catalytic properties of the proteasome from Thermoplasma acidophilum. *J Biol Chem* 1997, 272:1791-1798.
[11] Kisselev AF, Akopian TN, Woo KM, Goldberg AL: The sizes of peptides generated from protein by mammalian 26 and 20 S proteasomes. Implications for understanding the degradative mechanism and antigen presentation. *J Biol Chem* 1999, 274:3363-3371.
[12] Noda C, Tanahashi N, Shimbara N, Hendil KB, Tanaka K: Tissue distribution of constitutive proteasomes, immunoproteasomes, and PA28 in rats. *Biochem Biophys Res Commun* 2000, 277:348-354.
[13] Dahlmann B, Ruppert T, Kuehn L, Merforth S, Kloetzel PM: Different proteasome subtypes in a single tissue exhibit different enzymatic properties. *J Mol Biol* 2000, 303:643-653.
[14] Kuckelkorn U, Ruppert T, Strehl B, Jungblut PR, Zimny-Arndt U, Lamer S, Prinz I, Drung I, Kloetzel PM, Kaufmann SH, Steinhoff U: Link between organ-specific antigen processing by 20S proteasomes and CD8(+) T cell-mediated autoimmunity. *J Exp Med* 2002, 195:983-990.
[15] Tanaka K, Kasahara M: The MHC class I ligand-generating system: roles of immunoproteasomes and the interferon-gamma-inducible proteasome activator PA28. *Immunol Rev* 1998, 163:161-176.
[16] Groettrup M, Soza A, Kuckelkorn U, Kloetzel PM: Peptide antigen production by the proteasome: complexity provides efficiency. *Immunol Today* 1996, 17:429-435.

[17] Khan S, van den BM, Schwarz K, de Giuli R, Diener PA, Groettrup M: Immunoproteasomes largely replace constitutive proteasomes during an antiviral and antibacterial immune response in the liver. *J Immunol* 2001, 167:6859-6868.

[18] van Hall T, Sijts A, Camps M, Offringa R, Melief C, Kloetzel PM, Ossendorp F: Differential influence on cytotoxic T lymphocyte epitope presentation by controlled expression of either proteasome immunosubunits or PA28. *J Exp Med* 2000, 192:483-494.

[19] Morel S, Levy F, Burlet-Schiltz O, Brasseur F, Probst-Kepper M, Peitrequin AL, Monsarrat B, Van Velthoven R, Cerottini JC, Boon T, Gairin JE, Van den Eynde BJ: Processing of some antigens by the standard proteasome but not by the immunoproteasome results in poor presentation by dendritic cells. *Immunity* 2000, 12:107-117.

[20] Chen W, Norbury CC, Cho Y, Yewdell JW, Bennink JR: Immunoproteasomes shape immunodominance hierarchies of antiviral CD8(+) T cells at the levels of T cell repertoire and presentation of viral antigens. *J Exp Med* 2001, 193:1319-1326.

[21] Schultz ES, Chapiro J, Lurquin C, Claverol S, Burlet-Schiltz O, Warnier G, Russo V, Morel S, Levy F, Boon T, Van den Eynde BJ, Van Der BP: The production of a new MAGE-3 peptide presented to cytolytic T lymphocytes by HLA-B40 requires the immunoproteasome. *J Exp Med* 2002, 195:391-399.

[22] Sijts AJ, Standera S, Toes RE, Ruppert T, Beekman NJ, van Veelen PA, Ossendorp FA, Melief CJ, Kloetzel PM: MHC class I antigen processing of an adenovirus CTL epitope is linked to the levels of immunoproteasomes in infected cells. *J Immunol* 2000, 164:4500-4506.

[23] Tanahashi N, Murakami Y, Minami Y, Shimbara N, Hendil KB, Tanaka K: Hybrid proteasomes. Induction by interferon-gamma and contribution to ATP-dependent proteolysis. *J Biol Chem* 2000, 275:14336-14345.

[24] Dahlmann B: Proteasomes. *Essays Biochem* 2005, 41:31-48.

[25] Toes RE, Nussbaum AK, Degermann S, Schirle M, Emmerich NP, Kraft M, Laplace C, Zwinderman A, Dick TP, Muller J, Schonfisch B, Schmid C, Fehling HJ, Stevanovic S, Rammensee HG, Schild H: Discrete cleavage motifs of constitutive and immunoproteasomes revealed by quantitative analysis of cleavage products. *J Exp Med* 2001, 194:1-12.

[26] Berger A, Schechter I: Mapping the active site of papain with the aid of peptide substrates and inhibitors. *Philos Trans R Soc Lond B Biol Sci* 1970, 257:249-264.

[27] Altuvia Y, Margalit H: Sequence signals for generation of antigenic peptides by the proteasome: implications for proteasomal cleavage mechanism. *J Mol Biol* 2000, 295:879-890.

[28] Mo AX, van Lelyveld SF, Craiu A, Rock KL: Sequences that flank subdominant and cryptic epitopes influence the proteolytic generation of MHC class I-presented peptides. *J Immunol* 2000, 164:4003-4010.

[29] Beekman NJ, van Veelen PA, van Hall T, Neisig A, Sijts A, Camps M, Kloetzel PM, Neefjes JJ, Melief CJ, Ossendorp F: Abrogation of CTL epitope processing by single amino acid substitution flanking the C-terminal proteasome cleavage site. *J Immunol* 2000, 164:1898-1905.

[30] Ossendorp F, Eggers M, Neisig A, Ruppert T, Groettrup M, Sijts A, Mengede E, Kloetzel PM, Neefjes J, Koszinowski U, Melief C: A single residue exchange within a viral CTL epitope alters proteasome-mediated degradation resulting in lack of antigen presentation. *Immunity* 1996, 5:115-124.

[31] Kesmir C, van N, V, de Boer RJ, Hogeweg P: Bioinformatic analysis of functional differences between the immunoproteasome and the constitutive proteasome. *Immunogenetics* 2003, 55:437-449.

[32] Di Napoli M, McLaughlin B: The ubiquitin-proteasome system as a drug target in cerebrovascular disease: therapeutic potential of proteasome inhibitors. *Curr Opin Investig Drugs* 2005, 6:686-699.

[33] Holzhutter HG, Frommel C, Kloetzel PM: A theoretical approach towards the identification of cleavage-determining amino acid motifs of the 20 S proteasome. *J Mol Biol* 1999, 286:1251-1265.

[34] Holzhutter HG, Kloetzel PM: A kinetic model of vertebrate 20S proteasome accounting for the generation of major proteolytic fragments from oligomeric peptide substrates. *Biophys J* 2000, 79:1196-1205.

[35] Kuttler C, Nussbaum AK, Dick TP, Rammensee HG, Schild H, Hadeler KP: An algorithm for the prediction of proteasomal cleavages. *J Mol Biol* 2000, 298:417-429.

[36] Nussbaum AK, Kuttler C, Hadeler KP, Rammensee HG, Schild H: PAProC: a prediction algorithm for proteasomal cleavages available on the WWW. *Immunogenetics* 2001, 53:87-94.

[37] Tenzer S, Peters B, Bulik S, Schoor O, Lemmel C, Schatz MM, Kloetzel PM, Rammensee HG, Schild H, Holzhutter HG: Modeling the MHC class I pathway by combining predictions of proteasomal cleavage, TAP transport and MHC class I binding. *Cell Mol Life Sci* 2005, 62:1025-1037.

[38] Kesmir C, Nussbaum AK, Schild H, Detours V, Brunak S: Prediction of proteasome cleavage motifs by neural networks. *Protein Eng* 2002, 15:287-296.

[39] Nielsen M, Lundegaard C, Lund O, Kesmir C: The role of the proteasome in generating cytotoxic T-cell epitopes: insights obtained from improved predictions of proteasomal cleavage. *Immunogenetics* 2005, 57:33-41.

[40] Nussbaum AK, Dick TP, Keilholz W, Schirle M, Stevanovic S, Dietz K, Heinemeyer W, Groll M, Wolf DH, Huber R, Rammensee HG, Schild H: Cleavage motifs of the yeast 20S proteasome beta subunits deduced from digests of enolase 1. *Proc Natl Acad Sci U S A* 1998, 95:12504-12509.

[41] Shimbara N, Ogawa K, Hidaka Y, Nakajima H, Yamasaki N, Niwa S, Tanahashi N, Tanaka K: Contribution of proline residue for efficient production of MHC class I ligands by proteasomes. *J Biol Chem* 1998, 273:23062-23071.

[42] Peters B, Janek K, Kuckelkorn U, Holzhutter HG: Assessment of proteasomal cleavage probabilities from kinetic analysis of time-dependent product formation. *J Mol Biol* 2002, 318:847-862.

[43] Stein RL, Melandri F, Dick L: Kinetic characterization of the chymotryptic activity of the 20S proteasome. *Biochemistry* 1996, 35:3899-3908.

[44] Stohwasser R, Salzmann U, Giesebrecht J, Kloetzel PM, Holzhutter HG: Kinetic evidences for facilitation of peptide channelling by the proteasome activator PA28. *Eur J Biochem* 2000, 267:6221-6230.

[45] Schmidtke G, Emch S, Groettrup M, Holzhutter HG: Evidence for the existence of a non-catalytic modifier site of peptide hydrolysis by the 20 S proteasome. *J Biol Chem* 2000, 275:22056-22063.

[46] Forster A, Hill CP: Proteasome degradation: enter the substrate. *Trends Cell Biol* 2003, 13:550-553.

[47] Liu CW, Corboy MJ, DeMartino GN, Thomas PJ: Endoproteolytic activity of the proteasome. *Science* 2003, 299:408-411.

[48] Sharon M, Witt S, Felderer K, Rockel B, Baumeister W, Robinson CV: 20S proteasomes have the potential to keep substrates in store for continual degradation. *J Biol Chem* 2006, 281:9569-9575.

[49] Hutschenreiter S, Tinazli A, Model K, Tampe R: Two-substrate association with the 20S proteasome at single-molecule level. *EMBO J* 2004, 23:2488-2497.

[50] Luciani F, Kesmir C, Mishto M, Or-Guil M, de Boer RJ: A mathematical model of protein degradation by the proteasome. *Biophys J* 2005, 88:2422-2432.

[51] Whitby FG, Masters EI, Kramer L, Knowlton JR, Yao Y, Wang CC, Hill CP: Structural basis for the activation of 20S proteasomes by 11S regulators. *Nature* 2000, 408:115-120.

[52] Cardozo C, Vinitsky A, Michaud C, Orlowski M: Evidence that the nature of amino acid residues in the P3 position directs substrates to distinct catalytic sites of the pituitary multicatalytic proteinase complex (proteasome). *Biochemistry* 1994, 33:6483-6489.

[53] Cardozo C, Michaud C, Orlowski M: Components of the bovine pituitary multicatalytic proteinase complex (proteasome) cleaving bonds after hydrophobic residues. *Biochemistry* 1999, 38:9768-9777.

[54] Djaballah H, Rivett AJ: Peptidylglutamyl-peptide hydrolase activity of the multicatalytic proteinase complex: evidence for a new high-affinity site, analysis of cooperative kinetics, and the effect of manganese ions. *Biochemistry* 1992, 31:4133-4141.

[55] Hortin GL, Murthy J: Substrate size selectivity of 20S proteasomes: analysis with variable-sized synthetic substrates. *J Protein Chem* 2002, 21:333-337.

[56] Kisselev AF, Kaganovich D, Goldberg AL: Binding of hydrophobic peptides to several non-catalytic sites promotes peptide hydrolysis by all active sites of 20 S proteasomes. Evidence for peptide-induced channel opening in the alpha-rings. *J Biol Chem* 2002, 277:22260-22270.

[57] Orlowski M, Cardozo C, Hidalgo MC, Michaud C: Regulation of the peptidylglutamyl-peptide hydrolyzing activity of the pituitary multicatalytic proteinase complex. *Biochemistry* 1991, 30:5999-6005.

[58] Realini C, Jensen CC, Zhang Z, Johnston SC, Knowlton JR, Hill CP, Rechsteiner M: Characterization of recombinant REGalpha, REGbeta, and REGgamma proteasome activators. *J Biol Chem* 1997, 272:25483-25492.

[59] Reidlinger J, Pike AM, Savory PJ, Murray RZ, Rivett AJ: Catalytic properties of 26 S and 20 S proteasomes and radiolabeling of MB1, LMP7, and C7 subunits associated with trypsin-like and chymotrypsin-like activities. *J Biol Chem* 1997, 272:24899-24905.

[60] Ciechanover A, Orian A, Schwartz AL: Ubiquitin-mediated proteolysis: biological regulation via destruction. *Bioessays* 2000, 22:442-451.

[61] Sherman MY, Goldberg AL: Cellular defenses against unfolded proteins: a cell biologist thinks about neurodegenerative diseases. *Neuron* 2001, 29:15-32.

[62] Trzesniewska K, Brzyska M, Elbaum D: Neurodegenerative aspects of protein aggregation. *Acta Neurobiol Exp (Wars)* 2004, 64:41-52.

[63] Ross CA, Poirier MA: Opinion: What is the role of protein aggregation in neurodegeneration? *Nat Rev Mol Cell Biol* 2005, 6:891-898.

[64] Dimakopoulos AC: Protein aggregation in Alzheimer's disease and other neoropathological disorders. *Curr Alzheimer Res* 2005, 2:19-28.

[65] Berke SJ, Paulson HL: Protein aggregation and the ubiquitin proteasome pathway: gaining the UPPer hand on neurodegeneration. *Curr Opin Genet Dev* 2003, 13:253-261.

[66] Dimcheff DE, Portis JL, Caughey B: Prion proteins meet protein quality control. *Trends Cell Biol* 2003, 13:337-340.

[67] Chiesa R, Harris DA: Prion diseases: what is the neurotoxic molecule? *Neurobiol Dis* 2001, 8:743-763.

[68] Thompson AJ, Barrow CJ: Protein conformational misfolding and amyloid formation: characteristics of a new class of disorders that include Alzheimer's and Prion diseases. *Curr Med Chem* 2002, 9:1751-1762.

[69] Hashimoto M, Rockenstein E, Crews L, Masliah E: Role of protein aggregation in mitochondrial dysfunction and neurodegeneration in Alzheimer's and Parkinson's diseases. *Neuromolecular Med* 2003, 4:21-36.

[70] Driscoll M, Gerstbrein B: Dying for a cause: invertebrate genetics takes on human neurodegeneration. *Nat Rev Genet* 2003, 4:181-194.

[71] Wyttenbach A: Role of heat shock proteins during polyglutamine neurodegeneration: mechanisms and hypothesis. *J Mol Neurosci* 2004, 23:69-96.

[72] Jana NR, Nukina N: Recent advances in understanding the pathogenesis of polyglutamine diseases: involvement of molecular chaperones and ubiquitin-proteasome pathway. *J Chem Neuroanat* 2003, 26:95-101.

[73] Chan HY, Warrick JM, Gray-Board, Paulson HL, Bonini NM: Mechanisms of chaperone suppression of polyglutamine disease: selectivity, synergy and modulation of protein solubility in Drosophila. *Hum Mol Genet* 2000, 9:2811-2820.

[74] Chai Y, Berke SS, Cohen RE, Paulson HL: Poly-ubiquitin binding by the polyglutamine disease protein ataxin-3 links its normal function to protein surveillance pathways. *J Biol Chem* 2004, 279:3605-3611.

[75] Hu BR, Janelidze S, Ginsberg MD, Busto R, Perez-Pinzon M, Sick TJ, Siesjo BK, Liu CL: Protein aggregation after focal brain ischemia and reperfusion. *J Cereb Blood Flow Metab* 2001, 21:865-875.

[76] Liu CL, Martone ME, Hu BR: Protein ubiquitination in postsynaptic densities after transient cerebral ischemia. *J Cereb Blood Flow Metab* 2004, 24:1219-1225.

[77] Wojcik C, Di Napoli M: Ubiquitin-Proteasome System and Proteasome Inhibition: New Strategies in Stroke Therapy. *Stroke* 2004, 35:1506-1518.

[78] Di Napoli M, Papa F: The proteasome system and proteasome inhibitors in stroke: controlling the inflammatory response. *Curr Opin Investig Drugs* 2003, 4:1333-1342.

[79] Gubellini P, Bisso GM, Ciofi-Luzzatto A, Fortuna S, Lorenzini P, Michalek H, Scarsella G: Ubiquitin-mediated stress response in a rat model of brain transient ischemia/hypoxia. *Neurochem Res* 1997, 22:93-100.

[80] Cummings CJ, Mancini MA, Antalffy B, DeFranco DB, Orr HT, Zoghbi HY: Chaperone suppression of aggregation and altered subcellular proteasome localization imply protein misfolding in SCA1. *Nat Genet* 1998, 19:148-154.

[81] Lee DH, Wang HY: Differential physiologic responses of alpha7 nicotinic acetylcholine receptors to beta-amyloid1-40 and beta-amyloid1-42. *J Neurobiol* 2003, 55:25-30.

[82] Lee FJ, Liu F, Pristupa ZB, Niznik HB: Direct binding and functional coupling of alpha-synuclein to the dopamine transporters accelerate dopamine-induced apoptosis. *FASEB J* 2001, 15:916-926.

[83] Lee KS, Linden R, Prado MA, Brentani RR, Martins VR: Towards cellular receptors for prions. *Rev Med Virol* 2003, 13:399-408.

[84] Lee SJ, Liyanage U, Bickel PE, Xia W, Lansbury PT, Jr., Kosik KS: A detergent-insoluble membrane compartment contains A beta in vivo. *Nat Med* 1998, 4:730-734.

[85] Abramov AY, Canevari L, Duchen MR: Calcium signals induced by amyloid beta peptide and their consequences in neurons and astrocytes in culture. *Biochim Biophys Acta* 2004, 1742:81-87.

[86] Lacor PN, Buniel MC, Chang L, Fernandez SJ, Gong Y, Viola KL, Lambert MP, Velasco PT, Bigio EH, Finch CE, Krafft GA, Klein WL: Synaptic targeting by Alzheimer's-related amyloid beta oligomers. *J Neurosci* 2004, 24:10191-10200.

[87] Lain E, Penke B, Delacourte A, Gundisch D, Schroder H, Witter B: Effects of Abeta1-42 fibrils and of the tetrapeptide Pr-IIGL on the phosphorylation state of the tau-protein and on the alpha7 nicotinic acetylcholine receptor in vitro. *Eur J Neurosci* 2005, 21:879-888.

[88] Le Y, Gong W, Tiffany HL, Tumanov A, Nedospasov S, Shen W, Dunlop NM, Gao JL, Murphy PM, Oppenheim JJ, Wang JM: Amyloid (beta)42 activates a G-protein-coupled chemoattractant receptor, FPR-like-1. *J Neurosci* 2001, 21:RC123.

[89] Koenigsknecht J, Landreth G: Microglial phagocytosis of fibrillar beta-amyloid through a beta1 integrin-dependent mechanism. *J Neurosci* 2004, 24:9838-9846.

[90] Lambert MP, Viola KL, Chromy BA, Chang L, Morgan TE, Yu J, Venton DL, Krafft GA, Finch CE, Klein WL: Vaccination with soluble Abeta oligomers generates toxicity-neutralizing antibodies. *J Neurochem* 2001, 79:595-605.

[91] Muchowski PJ: Protein misfolding, amyloid formation, and neurodegeneration: a critical role for molecular chaperones? *Neuron* 2002, 35:9-12.

[92] Hara T, Nakamura K, Matsui M, Yamamoto A, Nakahara Y, Suzuki-Migishima R, Yokoyama M, Mishima K, Saito I, Okano H, Mizushima N: Suppression of basal

autophagy in neural cells causes neurodegenerative disease in mice. *Nature* 2006, 441:885-889.

[93] Komatsu M, Kominami E, Tanaka K: Autophagy and Neurodegeneration. *Autophagy* 2006, 2.

[94] Komatsu M, Waguri S, Chiba T, Murata S, Iwata J, Tanida I, Ueno T, Koike M, Uchiyama Y, Kominami E, Tanaka K: Loss of autophagy in the central nervous system causes neurodegeneration in mice. *Nature* 2006, 441:880-884.

[95] Tabner BJ, Turnbull S, Fullwood NJ, German M, Allsop D: The production of hydrogen peroxide during early-stage protein aggregation: a common pathological mechanism in different neurodegenerative diseases? *Biochem Soc Trans* 2005, 33:548-550.

[96] Kopito RR: Aggresomes, inclusion bodies and protein aggregation. *Trends Cell Biol* 2000, 10:524-530.

[97] Rao RV, Bredesen DE: Misfolded proteins, endoplasmic reticulum stress and neurodegeneration. *Curr Opin Cell Biol* 2004, 16:653-662.

[98] Soti C, Csermely P: Chaperones and aging: role in neurodegeneration and in other civilizational diseases. *Neurochem Int* 2002, 41:383-389.

[99] Schubert U, Anton LC, Gibbs J, Norbury CC, Yewdell JW, Bennink JR: Rapid degradation of a large fraction of newly synthesized proteins by proteasomes. *Nature* 2000, 404:770-774.

[100] Tran PB, Miller RJ: Aggregates in neurodegenerative disease: crowds and power? *Trends Neurosci* 1999, 22:194-197.

[101] Fink AL: Protein aggregation: folding aggregates, inclusion bodies and amyloid. *Fold Des* 1998, 3:R9-23.

[102] Wanker EE: Protein aggregation and pathogenesis of Huntington's disease: mechanisms and correlations. *Biol Chem* 2000, 381:937-942.

[103] Taylor JP, Hardy J, Fischbeck KH: Toxic proteins in neurodegenerative disease. *Science* 2002, 296:1991-1995.

[104] Holmberg CI, Staniszewski KE, Mensah KN, Matouschek A, Morimoto RI: Inefficient degradation of truncated polyglutamine proteins by the proteasome. *EMBO J* 2004, 23:4307-4318.

[105] Cluskey S, Ramsden DB: Mechanisms of neurodegeneration in amyotrophic lateral sclerosis. *Mol Pathol* 2001, 54:386-392.

[106] Liu J, Lillo C, Jonsson PA, Vande VC, Ward CM, Miller TM, Subramaniam JR, Rothstein JD, Marklund S, Andersen PM, Brannstrom T, Gredal O, Wong PC, Williams DS, Cleveland DW: Toxicity of familial ALS-linked SOD1 mutants from selective recruitment to spinal mitochondria. *Neuron* 2004, 43:5-17.

[107] Wood JD, Beaujeux TP, Shaw PJ: Protein aggregation in motor neurone disorders. *Neuropathol Appl Neurobiol* 2003, 29:529-545.

[108] Imai Y, Soda M, Hatakeyama S, Akagi T, Hashikawa T, Nakayama KI, Takahashi R: CHIP is associated with Parkin, a gene responsible for familial Parkinson's disease, and enhances its ubiquitin ligase activity. *Mol Cell* 2002, 10:55-67.

[109] Bucciantini M, Giannoni E, Chiti F, Baroni F, Formigli L, Zurdo J, Taddei N, Ramponi G, Dobson CM, Stefani M: Inherent toxicity of aggregates implies a common mechanism for protein misfolding diseases. *Nature* 2002, 416:507-511.

[110] Caughey B, Lansbury PT: Protofibrils, pores, fibrils, and neurodegeneration: separating the responsible protein aggregates from the innocent bystanders. *Annu Rev Neurosci* 2003, 26:267-298.

[111] Fu L, Gao YS, Sztul E: Transcriptional repression and cell death induced by nuclear aggregates of non-polyglutamine protein. *Neurobiol Dis* 2005, 20:656-665.

[112] Gunawardena S, Her LS, Brusch RG, Laymon RA, Niesman IR, Gordesky-Gold B, Sintasath L, Bonini NM, Goldstein LS: Disruption of axonal transport by loss of huntingtin or expression of pathogenic polyQ proteins in Drosophila. *Neuron* 2003, 40:25-40.

[113] Lashuel HA, Hartley D, Petre BM, Walz T, Lansbury PT, Jr.: Neurodegenerative disease: amyloid pores from pathogenic mutations. *Nature* 2002, 418:291.

[114] Bence NF, Sampat RM, Kopito RR: Impairment of the ubiquitin-proteasome system by protein aggregation. *Science* 2001, 292:1552-1555.

[115] Jana NR, Zemskov EA, Wang G, Nukina N: Altered proteasomal function due to the expression of polyglutamine-expanded truncated N-terminal huntingtin induces apoptosis by caspase activation through mitochondrial cytochrome c release. *Hum Mol Genet* 2001, 10:1049-1059.

[116] Chai Y, Koppenhafer SL, Shoesmith SJ, Perez MK, Paulson HL: Evidence for proteasome involvement in polyglutamine disease: localization to nuclear inclusions in SCA3/MJD and suppression of polyglutamine aggregation in vitro. *Hum Mol Genet* 1999, 8:673-682.

[117] Chan HY, Warrick JM, Andriola I, Merry D, Bonini NM: Genetic modulation of polyglutamine toxicity by protein conjugation pathways in Drosophila. *Hum Mol Genet* 2002, 11:2895-2904.

[118] Junn E, Lee SS, Suhr UT, Mouradian MM: Parkin accumulation in aggresomes due to proteasome impairment. *J Biol Chem* 2002, 277:47870-47877.

[119] Kim S, Nollen EA, Kitagawa K, Bindokas VP, Morimoto RI: Polyglutamine protein aggregates are dynamic. *Nat Cell Biol* 2002, 4:826-831.

[120] Ma J, Lindquist S: Wild-type PrP and a mutant associated with prion disease are subject to retrograde transport and proteasome degradation. *Proc Natl Acad Sci U S A* 2001, 98:14955-14960.

[121] Peters PJ, Ning K, Palacios F, Boshans RL, Kazantsev A, Thompson LM, Woodman B, Bates GP, D'Souza-Schorey C: Arfaptin 2 regulates the aggregation of mutant huntingtin protein. *Nat Cell Biol* 2002, 4:240-245.

[122] Rideout HJ, Larsen KE, Sulzer D, Stefanis L: Proteasomal inhibition leads to formation of ubiquitin/alpha-synuclein-immunoreactive inclusions in PC12 cells. *J Neurochem* 2001, 78:899-908.

[123] Tofaris GK, Layfield R, Spillantini MG: alpha-synuclein metabolism and aggregation is linked to ubiquitin-independent degradation by the proteasome. *FEBS Lett* 2001, 509:22-26.

[124] Bennett MC, Bishop JF, Leng Y, Chock PB, Chase TN, Mouradian MM: Degradation of alpha-synuclein by proteasome. *J Biol Chem* 1999, 274:33855-33858.

[125] Cummings CJ, Reinstein E, Sun Y, Antalffy B, Jiang Y, Ciechanover A, Orr HT, Beaudet AL, Zoghbi HY: Mutation of the E6-AP ubiquitin ligase reduces nuclear inclusion frequency while accelerating polyglutamine-induced pathology in SCA1 mice. *Neuron* 1999, 24:879-892.

[126] Yoo SY, Pennesi ME, Weeber EJ, Xu B, Atkinson R, Chen S, Armstrong DL, Wu SM, Sweatt JD, Zoghbi HY: SCA7 knockin mice model human SCA7 and reveal gradual accumulation of mutant ataxin-7 in neurons and abnormalities in short-term plasticity. *Neuron* 2003, 37:383-401.

[127] Yamamoto A, Lucas JJ, Hen R: Reversal of neuropathology and motor dysfunction in a conditional model of Huntington's disease. *Cell* 2000, 101:57-66.

[128] Wojcik C, Schroeter D, Wilk S, Lamprecht J, Paweletz N: Ubiquitin-mediated proteolysis centers in HeLa cells: indication from studies of an inhibitor of the chymotrypsin-like activity of the proteasome. *Eur J Cell Biol* 1996, 71:311-318.

[129] Nixon RA: Autophagy in neurodegenerative disease: friend, foe or turncoat? *Trends Neurosci* 2006.

[130] Mizuno Y, Hattori N, Mori H, Suzuki T, Tanaka K: Parkin and Parkinson's disease. *Curr Opin Neurol* 2001, 14:477-482.

[131] He L, Lu XY, Jolly AF, Eldridge AG, Watson SJ, Jackson PK, Barsh GS, Gunn TM: Spongiform degeneration in mahoganoid mutant mice. *Science* 2003, 299:710-712.

[132] Saigoh K, Wang YL, Suh JG, Yamanishi T, Sakai Y, Kiyosawa H, Harada T, Ichihara N, Wakana S, Kikuchi T, Wada K: Intragenic deletion in the gene encoding ubiquitin carboxy-terminal hydrolase in gad mice. *Nat Genet* 1999, 23:47-51.

[133] Fernandez-Funez P, Nino-Rosales ML, de Gouyon B, She WC, Luchak JM, Martinez P, Turiegano E, Benito J, Capovilla M, Skinner PJ, McCall A, Canal I, Orr HT, Zoghbi HY, Botas J: Identification of genes that modify ataxin-1-induced neurodegeneration. *Nature* 2000, 408:101-106.

[134] Johnston JA, Ward CL, Kopito RR: Aggresomes: a cellular response to misfolded proteins. *J Cell Biol* 1998, 143:1883-1898.

[135] Waelter S, Boeddrich A, Lurz R, Scherzinger E, Lueder G, Lehrach H, Wanker EE: Accumulation of mutant huntingtin fragments in aggresome-like inclusion bodies as a result of insufficient protein degradation. *Mol Biol Cell* 2001, 12:1393-1407.

[136] Venkatraman P, Wetzel R, Tanaka M, Nukina N, Goldberg AL: Eukaryotic proteasomes cannot digest polyglutamine sequences and release them during degradation of polyglutamine-containing proteins. *Mol Cell* 2004, 14:95-104.

[137] Goldberg AL, Akopian TN, Kisselev AF, Lee DH: Protein degradation by the proteasome and dissection of its in vivo importance with synthetic inhibitors. *Mol Biol Rep* 1997, 24:69-75.

[138] Schmidt T, Lindenberg KS, Krebs A, Schols L, Laccone F, Herms J, Rechsteiner M, Riess O, Landwehrmeyer GB: Protein surveillance machinery in brains with spinocerebellar ataxia type 3: redistribution and differential recruitment of 26S proteasome subunits and chaperones to neuronal intranuclear inclusions. *Ann Neurol* 2002, 51:302-310.

[139] Liu Y, Fallon L, Lashuel HA, Liu Z, Lansbury PT, Jr.: The UCH-L1 gene encodes two opposing enzymatic activities that affect alpha-synuclein degradation and Parkinson's disease susceptibility. *Cell* 2002, 111:209-218.

[140] Petrucelli L, O'Farrell C, Lockhart PJ, Baptista M, Kehoe K, Vink L, Choi P, Wolozin B, Farrer M, Hardy J, Cookson MR: Parkin protects against the toxicity associated with mutant alpha-synuclein: proteasome dysfunction selectively affects catecholaminergic neurons. *Neuron* 2002, 36:1007-1019.

[141] Ma J, Wollmann R, Lindquist S: Neurotoxicity and neurodegeneration when PrP accumulates in the cytosol. *Science* 2002, 298:1781-1785.

[142] Yedidia Y, Horonchik L, Tzaban S, Yanai A, Taraboulos A: Proteasomes and ubiquitin are involved in the turnover of the wild-type prion protein. *EMBO J* 2001, 20:5383-5391.

[143] Bowman AB, Yoo SY, Dantuma NP, Zoghbi HY: Neuronal dysfunction in a polyglutamine disease model occurs in the absence of ubiquitin-proteasome system impairment and inversely correlates with the degree of nuclear inclusion formation. *Hum Mol Genet* 2005, 14:679-691.

[144] van Leeuwen FW, de Kleijn DP, van den Hurk HH, Neubauer A, Sonnemans MA, Sluijs JA, Koycu S, Ramdjielal RD, Salehi A, Martens GJ, Grosveld FG, Peter J, Burbach H, Hol EM: Frameshift mutants of beta amyloid precursor protein and ubiquitin-B in Alzheimer's and Down patients. *Science* 1998, 279:242-247.

[145] De Vrij FM, Sluijs JA, Gregori L, Fischer DF, Hermens WT, Goldgaber D, Verhaagen J, van Leeuwen FW, Hol EM: Mutant ubiquitin expressed in Alzheimer's disease causes neuronal death. *FASEB J* 2001, 15:2680-2688.

[146] Bose S, Stratford FL, Broadfoot KI, Mason GG, Rivett AJ: Phosphorylation of 20S proteasome alpha subunit C8 (alpha7) stabilizes the 26S proteasome and plays a role in the regulation of proteasome complexes by gamma-interferon. *Biochem J* 2004, 378:177-184.

[147] Kazantsev A, Preisinger E, Dranovsky A, Goldgaber D, Housman D: Insoluble detergent-resistant aggregates form between pathological and nonpathological lengths of polyglutamine in mammalian cells. *Proc Natl Acad Sci U S A* 1999, 96:11404-11409.

[148] Nucifora FC, Jr., Sasaki M, Peters MF, Huang H, Cooper JK, Yamada M, Takahashi H, Tsuji S, Troncoso J, Dawson VL, Dawson TM, Ross CA: Interference by huntingtin and atrophin-1 with cbp-mediated transcription leading to cellular toxicity. *Science* 2001, 291:2423-2428.

[149] Schaffar G, Breuer P, Boteva R, Behrends C, Tzvetkov N, Strippel N, Sakahira H, Siegers K, Hayer-Hartl M, Hartl FU: Cellular toxicity of polyglutamine expansion proteins: mechanism of transcription factor deactivation. *Mol Cell* 2004, 15:95-105.

[150] Kayed R, Head E, Thompson JL, McIntire TM, Milton SC, Cotman CW, Glabe CG: Common structure of soluble amyloid oligomers implies common mechanism of pathogenesis. *Science* 2003, 300:486-489.

[151] Arrasate M, Mitra S, Schweitzer ES, Segal MR, Finkbeiner S: Inclusion body formation reduces levels of mutant huntingtin and the risk of neuronal death. *Nature* 2004, 431:805-810.

[152] Taylor JP, Tanaka F, Robitschek J, Sandoval CM, Taye A, Markovic-Plese S, Fischbeck KH: Aggresomes protect cells by enhancing the degradation of toxic polyglutamine-containing protein. *Hum Mol Genet* 2003, 12:749-757.
[153] Klement IA, Skinner PJ, Kaytor MD, Yi H, Hersch SM, Clark HB, Zoghbi HY, Orr HT: Ataxin-1 nuclear localization and aggregation: role in polyglutamine-induced disease in SCA1 transgenic mice. *Cell* 1998, 95:41-53.
[154] Saudou F, Finkbeiner S, Devys D, Greenberg ME: Huntingtin acts in the nucleus to induce apoptosis but death does not correlate with the formation of intranuclear inclusions. *Cell* 1998, 95:55-66.
[155] Sun XM, Butterworth M, MacFarlane M, Dubiel W, Ciechanover A, Cohen GM: Caspase activation inhibits proteasome function during apoptosis. *Mol Cell* 2004, 14:81-93.
[156] Ding Q, Lewis JJ, Strum KM, Dimayuga E, Bruce-Keller AJ, Dunn JC, Keller JN: Polyglutamine expansion, protein aggregation, proteasome activity, and neural survival. *J Biol Chem* 2002, 277:13935-13942.
[157] Elsasser S, Finley D: Delivery of ubiquitinated substrates to protein-unfolding machines. *Nat Cell Biol* 2005, 7:742-749.
[158] Smith DM, Kafri G, Cheng Y, Ng D, Walz T, Goldberg AL: ATP binding to PAN or the 26S ATPases causes association with the 20S proteasome, gate opening, and translocation of unfolded proteins. *Mol Cell* 2005, 20:687-698.
[159] Hol EM, van Leeuwen FW, Fischer DF: The proteasome in Alzheimer's disease and Parkinson's disease: lessons from ubiquitin B+1. *Trends Mol Med* 2005, 11:488-495.
[160] Hoyt MA, Zich J, Takeuchi J, Zhang M, Govaerts C, Coffino P: Glycine-alanine repeats impair proper substrate unfolding by the proteasome. *EMBO J* 2006, 25:1720-1729.
[161] Groll M, Huber R: Substrate access and processing by the 20S proteasome core particle. *Int J Biochem Cell Biol* 2003, 35:606-616.
[162] Kisselev AF, Goldberg AL: Proteasome inhibitors: from research tools to drug candidates. *Chem Biol* 2001, 8:739-758.
[163] Verhoef LG, Lindsten K, Masucci MG, Dantuma NP: Aggregate formation inhibits proteasomal degradation of polyglutamine proteins. *Hum Mol Genet* 2002, 11:2689-2700.
[164] Kaytor MD, Wilkinson KD, Warren ST: Modulating huntingtin half-life alters polyglutamine-dependent aggregate formation and cell toxicity. *J Neurochem* 2004, 89:962-973.
[165] Krobitsch S, Lindquist S: Aggregation of huntingtin in yeast varies with the length of the polyglutamine expansion and the expression of chaperone proteins. *Proc Natl Acad Sci U S A* 2000, 97:1589-1594.
[166] Lindsten K, De Vrij FM, Verhoef LG, Fischer DF, van Leeuwen FW, Hol EM, Masucci MG, Dantuma NP: Mutant ubiquitin found in neurodegenerative disorders is a ubiquitin fusion degradation substrate that blocks proteasomal degradation. *J Cell Biol* 2002, 157:417-427.

[167] Chen S, Ferrone FA, Wetzel R: Huntington's disease age-of-onset linked to polyglutamine aggregation nucleation. *Proc Natl Acad Sci U S A* 2002, 99:11884-11889.

[168] Stork M, Giese A, Kretzschmar HA, Tavan P: Molecular dynamics simulations indicate a possible role of parallel beta-helices in seeded aggregation of poly-Gln. *Biophys J* 2005, 88:2442-2451.

[169] Pickart CM: Back to the future with ubiquitin. *Cell* 2004, 116:181-190.

[170] Diaz-Hernandez M, Valera AG, Moran MA, Gomez-Ramos P, Alvarez-Castelao B, Castano JG, Hernandez F, Lucas JJ: Inhibition of 26S proteasome activity by huntingtin filaments but not inclusion bodies isolated from mouse and human brain. *J Neurochem* 2006, 98:1585-1596.

In: The Ubiquitin Proteasome System… ISBN 978-1-60021-749-4
Eds: Mario Di Napoli and Cezary Wojcik, pp. 343-371

Chapter 15

THE UBIQUITIN-PROTEASOME SYSTEM AND THE DEVELOPMENT OF THE NERVOUS SYSTEM

Douglas S. Campbell*
Department of Neurobiology and Anatomy, University of Utah Medical Center, Salt Lake City, Utah, USA.

ABSTRACT

The brain is a very complex structure. The complexity of the brain is necessary for it to carry out its varied functions, such as receiving and processing stimuli, learning and memory, effecting motor output, is critical to the survival of the organism. Gaining a complete understanding of the structure and function of the brain requires an understanding of its embryonic development. Many billions of neurons must differentiate into neurons and subsequently be wired up correctly and form functional synapses. These processes occur during neural development, which has a number of distinct stages: neurogenesis, axon guidance, arborisation and synaptogenesis that are multifaceted and tightly regulated. One mechanism for the regulation of intricate cellular processes, such as the cell cycle, synaptic plasticity and neural development, is through the control of protein levels via their synthesis and or degradation. The ubiquitin-proteasome system (UPS) is able to regulate protein levels by targeting specific proteins for proteasome-mediated proteolysis or for their endocytosis and subsequent lysosomal-mediated proteolysis. Evidence is now accumulating that these mechanisms are critical for the correct development of the nervous system during its distinct stages. A number of E3 ubiquitin ligases such as *Neuralized, ligand of num-protein X*, *Sel-10* and *mind bomb* have been identified as being required for regulating lateral inhibition and neurogenesis via the ubiquitin-mediated internalisation of the Notch and Delta transmembrane proteins critical for the differentiation of neurons. Genetic screens in *Drosophila* have identified

* Correspondence concerning this article should be addressed to Dr. Douglas S. Campbell, PhD; Laboratory for Developmental Gene Regulation, Neuronal Growth and Regeneration Research Group, RIKEN Brain Science Institute, 2-1 Hirosawa, Wako, Saitama, 351-0198, Japan. Phone: +1-81-48-467-9712; Fax: +1-81-48-467-9714; Email: douglas.campbell@cantab.net.

components of the UPS such as *bendless*, a ubiquitin-conjugating enzyme (E2), *non-stop*, a ubiquitin-specific protease and *ariadne-1* which interacts with a novel ubiquitin-conjugating enzyme in the navigation of axons to their targets. Recently the UPS has been identified as mediating the chemotropic responses of *Xenopus* retinal growth cones to specific guidance cues such as netrin-1 and brain-derived neurotrophic factor, since proteasome inhibition blocks chemotropic responses to these cues and leads to increases in ubiquitin-protein complexes in growth cones. The UPS is also able to regulate axon guidance at the level of guidance cue receptors at the cell surface which may enable a growth cone to change its responsiveness to particular cues during it journey. For example, as growth cones cross the *Drosophila* midline the internalisation of the guidance cue receptor, *roundabout,* is controlled by *commissureless* interacting with DNedd4, an E3 ubiquitin ligase, resulting in the removal of Robo from the growth cone surface. Once they have reached their targets growth cones undergo morphological changes such as branch formation during arborisation and subsequently synaptogenesis. Recent studies suggest that the UPS is involved in branch retraction in mushroom body neurons during metamorphosis. Moreover, two E3 ligases, *highwire*, and the anaphase promoting complex (APC), a de-ubiquitinating enzyme *fat facets*, regulate synaptogenesis via a balance of ubiquitination and de-ubiquitination at the neuromuscular junction in *Drosophila*. These studies establish a fundamental role for the UPS in regulating the development of the nervous system via both ubiquitin-mediaded endocytosis and ubiquitin-mediated proteolysis.

Keywords: Axons, Growth and development, Growth cones, Proteasome, endopeptidase complex, Synapses, Ubiquitins.

ABBREVIATIONS

Aei, After eight; APC, Anaphase promoting complex; Ari-1, Ariane; BDNF, Brain derived neurotrophic factor, Ben, Bendless; bHLH, Basic helix-loop-helix; BMP, Bone morphogenetic protein; Comm, Commissureless; CSL, CBF1, Su(H), Lag-1; DCC, Deleted in colorectal carcinoma; Des, Deadly seven; CBF1, C promoter binding factor DLK-1 (DAP (Death Associated Protein kinase) Like Kinase; DUB, Deubiquitinating enzyme; Dx, Deltex; E1, Ubiquitin activating enzyme; E2, ubiquitin-conjugating enzyme; E3, Ubiquitin ligase; Eno, Ectopic neurite outgrowth; Eph, Erythropoietin producing hepatocellular; Faf, Fat facets, Fraz, Frazzled; Gbb, Glass bottomed boat; Gmc, Ganglion mother cell; HECT, Homologous to E6-Associated Protein (E6AP) C-Terminus; Hiw, Highwire; IAP, Inhibitor of apoptosis proteins; Lac, Lactacystin; Lag1, lin-12- and glp-1; LnLL, N-Acetyl-Leu-Leu-NorLeu-Al; LNX, Ligand of numb-protein X; LPA L-α-lysophosphatidic acid; Lqf, Liquid facets; MAPK, Mitogen activated protein kinase; MB, Mushroom body; Mib, Mind bomb; Med, Medea; Nedd4, Neuronally expressed developmentally down regulated gene 4; Neur, Neuralized; NHR, Neuralized homology repeat; NMJ, Neuromuscular junction; PAM, Protein associated with myc; PHR-1 PAM, highwire and rpm-1; PIC, Protease inhibitor cocktail; PNS, Peripheral nervous system; RBP-JK, Recombination signal sequence binding protein for Jk genes; RING, Really interesting new gene; Robo, Roundabout; SCF, Skp1 Cullin, F-Box protein; Sema1a, Semaphorin 1a; Sema3A, Semaphorin 3A; SFRP1, Secreted

Frizzled-related protein 1; SMAD, Small mothers against decapentaplegic; Smurf1, Smad ubiquitin regulatory factor 1; Su(dx), Suppressor of deltex; Su(H), Suppressor of hairless; TGF-α, Transforming growth factor α; Ubc, Ubiquitin-conjugating enzyme; UBP, Ubiquitin-specific protease; UPS, Ubiquitin-proteasome system; Wit, Wishful thinking; Wlds, Wallerian degeneration slow; WWE, Tryptophan- tryptophan-glutamate.

Introduction

The ubiquitin-proteasome system (UPS) came to light from studies on intracellular proteolysis requiring metabolic energy and was subsequently shown to be the major pathway of protein degradation in eukaryotic cells [1,2]. Since the importance of the UPS was identified almost twenty five years ago many key details regarding the mechanism of proteolysis have been resolved, along with an appreciation for the cellular contexts in which ubiquitin-mediated proteolysis plays a role reviewed in [2-4] and Chapter 3. At the same time Thomas and colleagues reported the *Drosophila* mutant *bendless* with defects in the giant fibre neuron and escape-jump response [5]. Cloning of the affected gene identified *bendless* as a ubiquitin-conjugating enzyme providing the first evidence that the UPS plays a role during the development of the nervous system [6,7]. However, given the unknown nature of potential ubiquitinated substrates of the *bendless* gene it was difficult to explain how such a gene could specifically affect axon growth and synaptogenesis. It is only recently that the UPS has received more attention by developmental neurobiologists via its functions in both ubiquitin-mediated proteolysis and ubiquitin-mediated regulation of cell surface receptors via endocytosis, both of which are necessary for the development of the nervous system, a very tightly regulated process. This chapter will introduce the different stages during the development of the nervous system from neurogenesis to synaptogenesis where roles for the UPS have been identified (summarized in Table 1).

Neurogenesis

During the process of neural development a large number of different types of neurons are formed. Neurogenesis is the stage of development when neuronal precursor cells proliferate to produce neurons. They derive from progenitor cells known as neuroblasts, which delaminate from neurogenic regions of the neurectoderm (reviewed in [8]). In insects each neuroblast proliferates and divides into a large cell (neuroblast) which continues to divide and a small cell known as a ganglion mother cell (gmc) that is able to divide one more time, the daughters of which differentiate as neurons. In *Drosophila,* for example, basic helix-loop-helix (bHLH) transcription factors such as the *achaete scute* complex are involved in the segregation of neuroblasts from other epidermal cells. The expression of these types of genes, known as proneural genes, results in the competence of cells to become neurons. A single cell within a cell cluster will delaminate as a neuroblast and continues to express the *achaete scute* complex, whereas the *achaete scute* complex will become down-regulated in the other cells of the cluster. Laser ablation of the neuroblast leads to another cell from the cell cluster turning it into a neuroblast, suggesting that the other cells are no longer inhibited from forming neuroblasts. This process is called lateral inhibition [9]. Lateral inhibition has since been found to be conserved between invertebrates and vertebrates with the neuroblast able to inhibit other cells in the cluster from becoming neurons. Genetic screens in *Drosophila* led to the identification of the neurogenic genes which when mutated lead to more neurons [10]. Lateral inhibition is mediated largely by the Notch, Delta and Serrate families of transmembrane receptors. Cells which become neuroblasts have high levels of the cell surface Notch receptor, with neighbouring cells expressing the Notch ligand Delta, and thus inactivating Notch signalling and inhibiting the Notch expressing cell from becoming a neuroblast. A model has been proposed whereby the stochastic change in expression of Notch-Delta signalling in one cell of the proneural cluster signals it out to become the neuroblast. Notch controls gene expression via the enhancer of split complex.

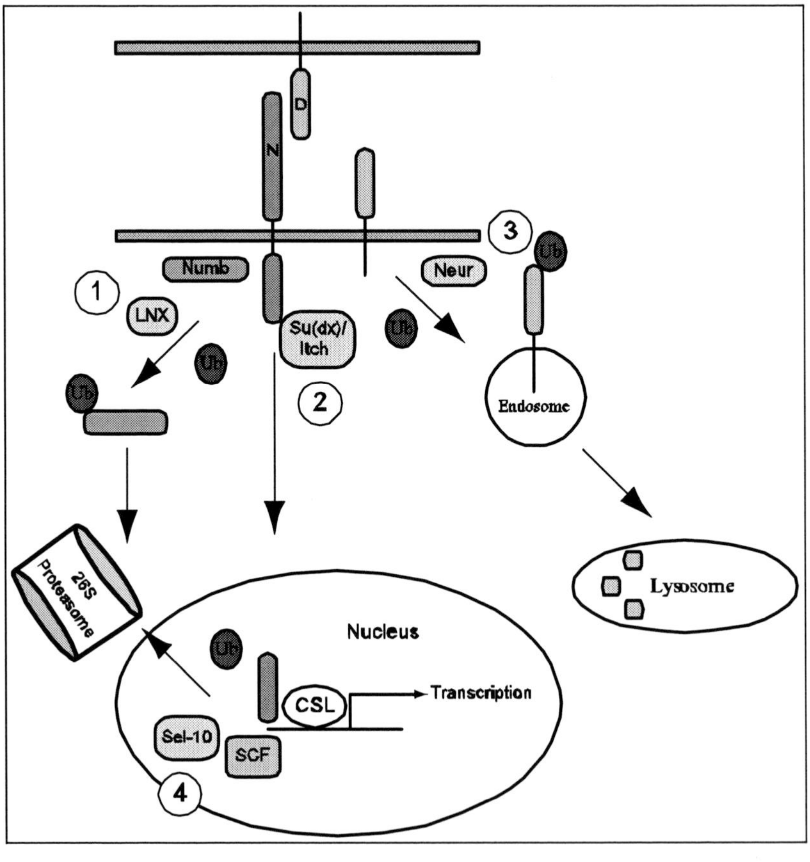

Figure 1. Regulation of Notch-Delta signalling by the ubiquitin-proteasome system. The Notch signalling pathway is regulated by at least four ubiquitin ligases and other components of the UPS. (1). *Ligand of numb-protein X* (*LNX*) ubiquitinates (Ub) and targets *Numb* for proteasome-mediated proteolysis. (2). *Su(dx)/ itch* has been proposed to ubiquitinate and regulate the cell surface levels of *notch* (N). (3). *Neuralize*d (Neur) ubiquitinates *Delta* (D) promoting its endocytosis and subsequent degradation presumably in the lysosome compartment (4). *Sel-10* in conjunction with an SCF (Skp1, Cullin and F box protein) complex ubiquitinates and targets the intracellular domain of Notch for proteasome-mediated proteolysis. CSL (CBF1/ RBK-Jk Su(H), Lag-1).

Table 1. Summary of ubiquitin-proteasome system components with roles in neural development

Stage of neural development	UPS component or gene	Species	Enzyme and target	References
Neurogenesis				
	LNX1	Drosophila	E3, Ubiquitin ligase, Numb	[11]
	Su(dx)/ itch	Drosophila	E3, Ubiquitin ligase, Notch	[12]
	Neuralised	Drosophila	E3, Ubiquitin ligase, Delta	[13-15]
	Mind bomb	Zebrafish	E3, Ubiquitin ligase, Delta	[16, 17]
	Mind bomb 1	Mouse	E3, Ubiquitin ligase, Delta	[18]
	Sel-10	Drosophila	E3, Ubiquitin ligase, Notch	[19, 20]
	Cullin1	Xenopus	E3, Ubiquitin ligase	[21]
	UCH-L1	Mouse	Ubiquitin C-terminal Hydrolase L1	[22]
Neurite outgrowth				
	Lin-23	C. elegans	E3, Ubiquitin ligase	[23]
	HR6B	Rat	N-end rule Ubiquitin-conjugating enzyme	[24]
	Smurf1	Mouse	RhoA	[25]
Axon guidance				
	Ariadne	Drosophila	E2, Ubc D10	[26]
	Bendless	Drosophila	E2, Ubc	[5-7]
	Non-stop	Drosophila	UBP	[27, 28]
	Rnf6	Mouse	E3, Ubiquitin ligase Regulates levels of LIM1 kinase in growth cones	[29]
	DNedd4	Drosophila	Comm/ Robo	[30]
Chemotropic responses of growth cones			Guidance cue	
	Proteasome	Xenopus	Netrin-1	[31]
	Proteasome	Mouse	Netrin-1-induced down regulation of DCC	[32]
	Proteasome	Xenopus	BDNF	[31]
	Proteasome	Xenopus	LPA	[31]
	Proteasome	Xenopus	EphB2	[33]
	Proteasome	Xenopus	SFRP1	[34]
Arborisation and synaptogenesis			Effect on arborisation and synaptogenesis	
	Fat facets	Drosophila	Promotes NMJ arborisation and synaptogenesis	[35]
	Usp9x	Mouse	Ubiquitin-specific protease	[36]
	Highwire	Drosophila	Inhibitor of NMJ arborisation and synaptogenesis	[35, 37]
	Esrom	Zebrafish	Retinal axon topography; prevents arborisation in inappropriate regions of the tectum	[38-40]
	RPM-1	C. elegans	Promotes synaptogenesis by inhibiting the p38 MAP kinase pathway	[41]
	PHR1	Mouse	Inhibits synaptogenesis	[42]
	Anaphase promoting complex	Drosophila	Inhibits NMJ arborisation and synaptogenesis	[43]
	Cullin3	Drosophila	Induces axonal and dendritic arborisation in mushroom body neurons	[44]
Branch retraction	Proteasome	Drosophila	Mushroom body neurons	[45]

The earliest evidence that the UPS may regulate the notch pathway came from a study using conditional dominant-negative mutations in the β2 and β6 proteasome subunit genes [46]. Since the mutations are temperature-sensitive the onset of proteasome inhibition can be controlled. When grown at the restrictive temperature a proportion of external sense organs grew abnormally resulting in two sockets and no shaft. The neuron to sheath and shaft to socket cell fate transformations are similar to mutations in the notch pathway and the 'double socket' phenotype was enhanced when notch signalling was up-regulated, suggesting that a reduction in proteasome activity increased notch signalling. Furthermore, the intracellular form of Notch was stabilised in cells with mutated proteasome components suggesting that notch itself is a target of proteasome-mediated degradation [46]. Recent studies have confirmed this hypothesis by the identification of E3 ubiquitin ligases which regulate the Notch signalling pathway: *suppressor of deltex* (*Su*(*dx*)) / *itchy* [12], and Sel-10 [19,20] (which regulate notch), *neuralized* (*Neur*) which regulates Delta [13-15] and *ligand of numb-protein X* (*LNX*) which regulates numb, a Notch antagonist [11] (Figure 1).

Su(dx) was identified in a genetic screen in *Drosophila* as a suppressor of *deltex (dx),* mutation of which results in a thickening of wing veins. This has been shown to genetically interact with the notch pathway [47]. Another mutation, *itchy,* resulted in persistent itching behaviour and pleiotropic immunological abnormalities in mice and appeared to be a totally different phenotype to *dx* [48]. However, cloning of the affected genes identified that both and *Su*(*dx*) and *itchy* contain an N-terminal Ca^{2+}-dependent lipid/ protein binding domain, a (C2) domain, two WW domains (a protein-interaction module composed of 35-40 amino acids with two conserved tryptophan residues) and a C-terminal catalytic domain homologous to the E6AP C terminus (HECT)-domain characteristic of the HECT family of E3 ubiquitin ligases [12,49,50]. The domain composition of *Su(dx),* and *itchy* is identical to the Nedd4 /Rsp5p family involved in ubiquitin-dependent endocytosis of plasma membrane proteins [51], suggesting a role for ubiquitin-mediated endocytosis in regulating Notch signalling.

Su(*dx*) is a negative regulator of notch signaling with mutations able to suppress notch and delta haploinsuficient phenotypes and enhance that of hairless, a negative regulator of Notch signaling [49]. Over expression enhanced *notch* and *delta* mutations and is dependent on the really interesting new gene (RING) domain [49]. RING domains are involved in transferring ubiquitin to heterologous substrates, or themselves, promoting degradation by the proteasome [52] domain. Complementary studies with *itchy* in mouse suggested it may promote the ubiquitination of the intracellular domain of notch via its HECT domain [53], further suggesting Notch is regulated via ubiquitination and endocytosis. More recently, analysis of the expression of genes regulated by notch suggest that *Su*(*dx*) acts as a negative regulator of notch *in vivo* [54]. Additionally, *deltex* itself was found to encode a RING-H2 domain at its C-terminus and two copies of a WWE (tryptophan-tryptohpan-glutamate) protein-protein interaction domain (which are necessary and sufficient for *deltex* to interact with the ankyrin repeats of Notch) [55-57] indicating that it may be another ubiquitin ligase regulating the Notch pathway.

Sel-10 was first identified as a suppressor mutation of an egglaying defect caused by a hypomorph in the lin-12c gene, the *C. elegans* homologue of Notch [58]. Isolation of the affected gene revealed a cdc42-related protein containing an F box (containing approximately

50 amino acids involved in protein-protein interactions) and 7 WD40 repeats (approximately 40 amino acids with conserved tryptophan and aspartate residues) [59]. The F box interacts with the SCF- (Skp1, Cullin and an F-Box protein) ubiquitin ligases which ubiquitinate the intracellular domain of Notch, removal of which mimics the effect of proteasome inhibitors [20,60]. The interaction of Sel-10 makes the intraceullular domain of Notch very labile and enables very tight control over the sub-cellular localisation of Notch.

Notch levels and signalling are also regulated by Numb. Numb undergoes asymmetric localisation during the process by which sibling cells are selected by Notch signaling and has been shown to be a Notch antagonist through its interaction with the intracellular domain of Notch. LNX-1 is a RING finger containing E3 ubiquitin ligase that targets numb for ubiqiuitin-mediated degradation [11] and has been proposed to regulate Notch signaling by decreasing protein levels of Numb. LNX-1 may be a vertebrate specific regulator of Notch signaling, since homologues have not been found in invertebrates such as *Drosophila* and *C. elegans*.

Neur was first identified as a *Drosophila* mutation resulting a in a neurogenic phenotype, i.e. the over production of neurons at the expense of epidermal cells, similar to mutations in the notch pathway [61,62]. Cloning of the affected gene revealed that *Neur* contained two copies of a neuralized homology repeat (NHR) and a c-terminal RING finger [62,63]. Subsequent clonal analysis of *neur* mutants and misexpression of wild-type and mutant forms identified that *neur* is required cell autonomously in Notch receiving cells, i.e. those which express Delta and that the RING domain is critical for this function [14,64]. Recent studies in *Drosophila* and *Xenopus* have identified that *Neur* functions as an E3 ubiquitin ligase *in vitro* and is able to promote the ubiquitination of Delta *in vitro* [13,14,64]. Furthermore, in *neur* mutants the level of Delta protein on the cell surface is increased and Neur promotes endocytosis and subsequent degradation of Delta in a RING finger-dependent way *in vivo* [13-15]. Neur undergoes mono-ubiquitination [13] suggestive of targeting for endocytosis and subsequent lysosome-mediated proteolysis [65]. How Neur regulates the down regulation of Delta is still unclear and both cell autonomous and non-cell-autonomous roles have been proposed. Consistent with a cell autonomous role Neur may trigger the down regulation of Delta within notch-expressing cells. Neur can also act non cell-autonomously in cells that send the Notch signal rather than receive it [66] where Delta internalisation occurs concomitantly with the transendocytosis of the extracellular domain of notch in the cell receiving the signal [67].

The E3 ubiquitin ligase Neur is not essential for lateral inhibition in vertebrates such as mouse [68] (in contrast to its role in *Xenopus* [13]) suggesting that other E3 ubiquitin ligases may play a role in vertebrate CNS development. In the zebrafish embryo one of the roles of Notch signalling as in *Drosophila* is to select a subset of neuronal progenitors to become neurons. Mutations in notch1a (*deadly seven*, *des*) or *deltaD* (*after eight*, *aei*) have relatively mild neurogenic phenotypes, possibly due redundancy of *notch* and *delta* genes in zebrafish [69-71]. In contrast, zebrafish *mind bomb* (*mib*) mutants have a very severe neurogenic phenotype with additional phenotypes in the somites, neural crest and vasculature [72]. Positional cloning of *mib* revealed that it encoded a RING E3 ubiquitin ligase [16]. Additionally, Itoh and colleagues have demonstrated that *mib* acts as an E3 ubiquitin ligase and that it interacts with and is able to ubiquitinate Delta *in vitro* promoting its internalisation via endocytosis [16]. Transplantation experiments suggested that *mib* is required in the signalling cell for efficient activation of Notch and the authors proposed a model where once Delta interacts with Notch, Delta undergoes endocytosis promoted by *mib* along with transendocytosis of the extracellular domain of Notch by the signalling cell. Transendocytosis promotes the cleavage of Notch generating the intracellular fragment of Notch required for activation of target genes [16,17].

AXON GUIDANCE

Subsequent to neurogenesis the neuron undergoes a morphological polarisation leading to one long axon and multiple dendrites projecting from the cell body through the processes of axon and dendritogenesis. The axons and dendrites must navigate to their correct targets to establish connections between neurons, necessary for the correct functioning of the mature nervous system. Axons navigate using the growth cone at the axon tip. The growth cone guides the growing axon through the embryo to locate the cell with which the neuron will form a synapse. Over the last decade dramatic progress has been made in identifying the ligands and receptors necessary for growth cone guidance reviewed in [73,74] and it is only relatively recently that we are beginning to understand more about the intracellular signalling pathways downstream of guidance cue receptors. Studies on the UPS in axon guidance have relied largely on *Drosophila* as a model system, owing to its suitability for genetic screens, although more recently the visual system of the frog *Xenopus laevis* has been employed owing to its ease of visualisation and manipulation [75,76].

MUTATIONS IN COMPONENTS OF THE INVERTEBRATE UPS HAVE DEFECTS IN AXON GUIDANCE

The first indication that the UPS may function in the process of axon guidance came from genetic screens in *Drosophila* looking for mutants with axon pathfinding defects. The *Drosophila bendless* (*ben*) gene, homologous to ubiquitin-conjugating enzymes (Ubc; E2), was originally isolated as a mutation affecting the escape jump response [5-7]. This behavioural defect was ascribed to a single lesion affecting the connectivity between the giant fiber and the tergotrochanter motor neurons. Subsequently, pathfinding errors were observed in the visual system with disruption of the highly ordered arrays of photoreceptor and lamina axons. For example, R7 and R8 axons terminated in inappropriate superficial or deep positions in the medulla [6], suggesting that *ben* may function in growth cone guidance through targeting specific proteins for degradation.

ariadne (ari-1), another *Drosophila* mutant, identified by Ferrus and colleagues, is a novel RING finger containing protein [26]. A yeast two-hybrid screen showed that *ari-1* interacts via its N-terminal RING-finger motif with a novel Ubc, UbcD10 [26]. The recent biochemical characterisation of the Skp1/ cullin/ F-box protein (SCF) and anaphase promoting complex (APC) E3 ubiquitin ligase complexes suggests that RING fingers play a central role in the combinatorial set of protein interactions that determine substrate specificity in the ubiquitin pathway [77]. *ari-1* is expressed in all tissues during development. Null mutants of *ari-1* are lethal, exhibiting a wide range of abnormalities, including those in axon pathfinding. *ari-1* shares the RING-B-box-RING (RBR) motif with the Parkinson disease related protein, parkin [78], suggesting a similar mechanism of action of the two proteins and suggesting a possible link between neural development and neurodegenerative diseases such as Parkinson disease in which abnormal protein aggregation and turnover may play a role [79].

Mutations in other components of the ubiquitin-proteasome pathway, such as the ubiquitin-specific proteases (UBPs) that are involved in releasing ubiquitin from tagged proteins (see Chapter 5), can also result in pathfinding errors. The *Drosophila* mutant *non-stop*, a UBP, exhibits abnormal pathfinding and target recognition of photoreceptor neurons. *non-stop* expression is required for a subset of glial cells to migrate into their appropriate position in the lamina target region and emit the necessary targeting signals [27,28]. In contrast, the mechanisms underlying the axon pathfinding defects observed in *ben* and *ari-1* mutants are less well understood. It is possible that specificity in what would appear to be a general system for degrading a large proportion of cellular proteins could be imparted by the combinatorial association of E3 ligases or Ubcs with other proteins such as F-box proteins [77]. These interactions may determine the specificity of the proteins degraded and the phenotypes observed when particular components of the multiprotein complex are mutated.

Recently, a genetic screen in the nematode worm *C. elegans* has implicated the UPS system in the termination of axon growth. Hobert and colleagues [23] identified a new allele[(*eno*) ectopic / erratic neurite outgrowth] of the LIN-23 gene that in addition to having axon termination defects, also had defects in other aspects of axon pathfinding such as fasciculation and branching. LIN-23 had previously been identified as an F-box containing protein with WD40 repeats. The latter is a component of the SCF-type of E3 ubiquitin ligases, which negatively regulates post-embryonic cell divisions [80]. The mutation in the *eno* allele of LIN-23 is a missense mutation, hypothesised to disrupt particular protein-protein interactions mediated by LIN-23. It has axon guidance defects but does not affect its role in regulating the cell cycle, potentially because it regulates the ubiquitin-dependent degradation of different substrates for these two processes.

The above studies have implicated a role for the UPS in guiding invertebrate axons to their targets via effects on neurite outgrowth and in axon pathfinding. However, how the mutation leads to a particular phenotype, as discussed above for the *Drosophila* mutants, remains unclear. Complementary *in vitro* studies on vertebrate growth cones [31,33] have identified a role for the UPS in vertebrate axon guidance and in particular in mediating growth cone response to chemotropic cues and may help to suggest a framework for understanding the *Drosophila* mutant phenotypes.

Involvement of the UPS in Mediating Chemotropic Responses of Vertebrate Growth Cones

The growth cone is instrumental in the process of axon guidance through its ability to recognise and respond to cues in its environment, enabling it and its cognate axon to navigate to their target. Growth cones are able to navigate long distances in the developing embryo and can be operating at a significant distance away from the cell body. Using fluorescent lipophilic dye labelling of *Xenopus* retinal axons *in vivo* in the living embryo Harris and colleagues found that retinal axons are able to navigate to their targets in the optic tectum correctly where they underwent normal changes in morphology and arborisation even when the retinal axons had been severed from their cell bodies suggesting that axon guidance can be under local control [81]. Growth cones need to respond rapidly to the changing environment. Since transport along axons can be relatively slow and the speed at which changes in growth cone motility take place are rapid, the transport of components from the cell body unlikely. Synthesis and degradation of proteins could be one way of overcoming this. To investigate the possibility that the UPS may be involved in mediating growth cone responses to chemotropic cues Campbell and Holt identified ubiquitin proteasome components and the E1 ubiquitin activating enzyme in *Xenopus* retinal growth cones [31] (Figure 2), demonstrating that vertebrate growth cones contain components of the UPS.

A number of *in vitro* assays have been developed for determining the responses and intracellular signalling pathways of growth cones to particular cues, in particular, the growth cone turning and collapse assays developed in the Poo and Raper laboratories respectively [82,83] (Figure 2). To test the role of the proteasome in mediating chemotropic responses Campbell and Holt, applied proteasome inhibitors to the culture medium immediately prior to performing growth cone collapse and turning assays. Proteasome inhibitors blocked growth cone collapse and turning responses to the chemotropic cues, netrin-1, L-α-lysophosphatidic acid (LPA) and brain derived neurotrophic factor (BDNF) (Figure 2).

Growth cones can respond rapidly to guidance cues, for example, *Xenopus* retinal growth cones begin to collapse in response to bath application of semaphorin 3A (Sema3A) in two minutes and maximum collapse is reached in ten minutes [84]. Therefore, the effects of the proteasome inhibitors suggest that rapid proteasome-mediated proteolysis is required to mediate the chemotropic responses to netrin-1, LPA and BDNF. Furthermore, Campbell and Holt were able to demonstrate, using an antibody which specifically recognises ubiquitin attached to proteins (FK2, [85,86]), that netrin-1 and LPA stimulated a doubling of ubiquitin-protein conjugates in growth cones within five minutes. This suggests that the cues stimulate rapid production of ubiquitinylated proteins destined for proteasome-mediated proteolysis.

In the same study, Campbell and Holt demonstrated that rapid and local protein synthesis is also required to mediate chemotropic responses of growth cones to cues such as netrin-1 and Sema3A and that growth cones, like other subcellular compartments, contain protein synthetic and degradative machinery in close proximity. These observations, and those of Ming and colleagues and Brittis and colleagues who also identified a role for local protein synthesis in growth cone guidance [87,88], suggest that the processes of protein synthesis and degradation may be tightly regulated. The authors proposed a model for growth cone turning involving the rapid and local synthesis and/ or degradation of proteins which are necessary to mediate the cytoskeletal changes necessary for growth cone turning [31,76]. Different chemotropic cues can elicit chemotropic responses through particular combinations of protein synthesis and / or degradation.

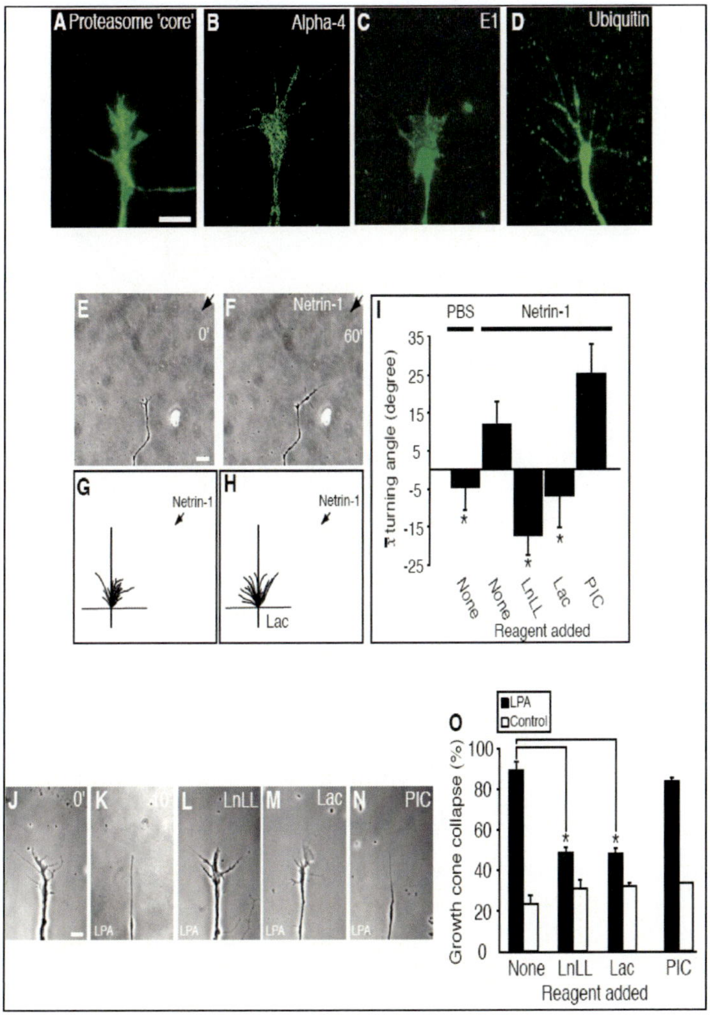

Figure 2. Chemotropic responses of growth cones mediated by the ubiquitin-proteasome system. Stage 24 Xenopus retinal growth cones showing localisation of proteasome (A; 20S proteasome 'core' and B; anti α-4 subunit antibodies), and ubiquitin system (C; anti-ubiquitin D; anti-ubiquitin activating enzyme E1 antibodies). Stage 24 retinal growth cones on 1μg/ml fibronectin are attracted by a gradient of netrin-1 ejected from a micropipette over 1 hour (E-G and I). Netrin-1-induced attractive turning is inhibited by the proteasome inhibitors N-Acetyl-Leu-Leu-NorLeu-Al (50μM LnLL) and lactacystin (10μM Lac) but not by a protease inhibitor cocktail of cell permeable non-proteasome protease inhibitors (10μg/mlPIC; H and I). Superimposed neurite trajectories with netrin-1 in the pipette and no reagent added to bathing medium (G) and in the presence of lactacystin (H). Proteasome inhibitor-treated conditions are significantly different from netrin-1 (* = $P<0.05$) but not from PBS control (I; Kolmogorov-Smirnov test). Arrows indicate position of pipette (E and F). LPA (1μM) induces retinal growth cone collapse in 10 minutes (K and O). LPA-induced collapse is prevented in the presence of the proteasome inhibitors LnLL and lactacystin but not by the protease inhibitor cocktail (L-O). * = $P<0.05$; Mann-Whitney U test (O). Scale bars =10μm. Reprinted from Neuron, Volume 32, Campbell D. S. and Holt C. E., Chemotropic responses of retinal growth cones mediated by rapid local protein synthesis and degradation, Pages 1013-1026 Copyright (2001), with permission from Elsevier.

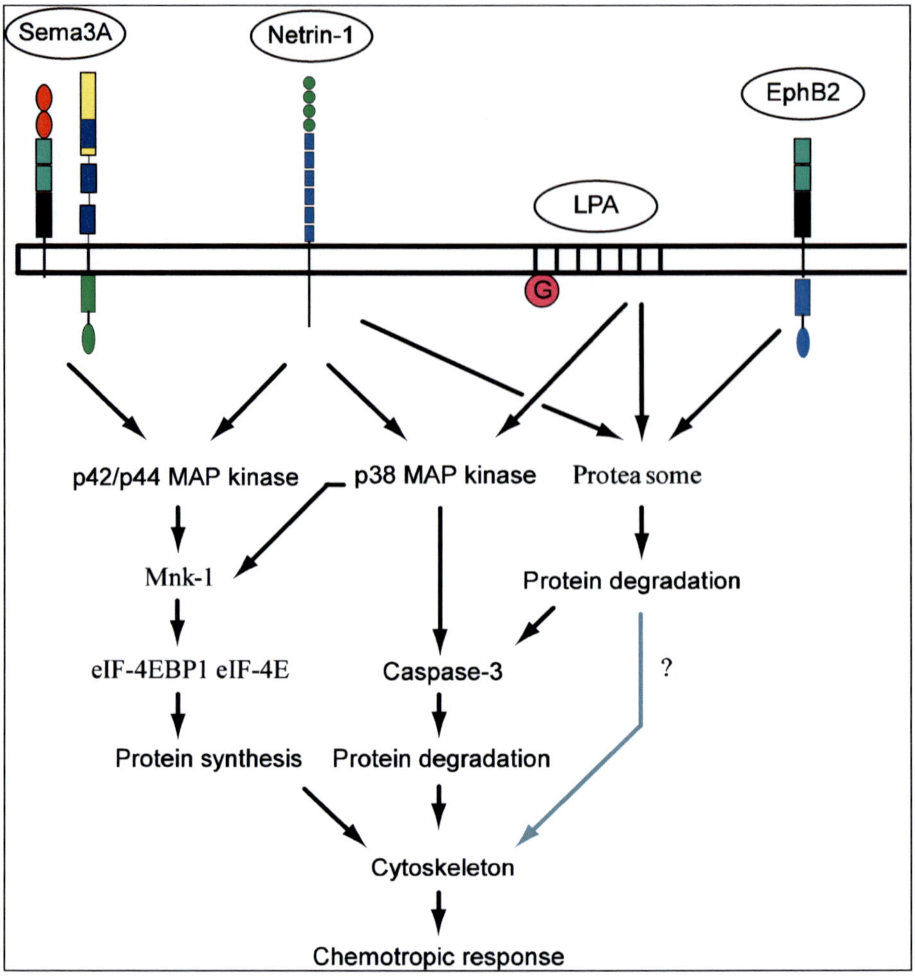

Figure 3. Intracellular signalling pathways regulating local protein levels in growth cones in response to guidance cues. Netrin-1-induced growth cone responses require proteasome-mediated proteolysis, protein synthesis regulated by p42/p44 and p38 MAPKs and caspase-3 activation downstream of p38 MAPK and proteasome-mediated proteolysis. Sema3A-induced chemotropic responses require only p42/p44 MAPK-dependent protein synthesis and are independent of p38 MAPK, proteasome-mediated proteolysis and caspase-3. LPA-induced growth cone collapse requires proteasome- and p38 MAPK-mediated activation of caspase-3. EphB2-induced growth cone collapse requires proteasome-mediated proteolysis and is independent of protein synthesis. The position of proteasome function with respect to the p42/p44 MAPK and p38 MAPK pathways has not been determined and a role for proteasome-mediated proteolysis functioning independently of caspase-3 activation cannot be excluded (grey arrow). Reprinted and modified from Neuron, Volume 37, Campbell D. S. and Holt C. E., Apoptotic pathway and MAPKs differentially regulate chemotropic responses of retinal growth cones, Pages 939-952 Copyright (2003), with permission from Elsevier.

However, there is not an absolute requirement for either on its own or in combination since LPA requires proteasome-mediated proteolysis but not protein synthesis [31].

Furthermore, the ability of unclustured ephrinB1 ectodomains to induce retinal growth cone collapse is not blocked by protein synthesis or proteasome inhibitors [33]. The ability of different guidance cues to utilise different pathways (Figure 3) may increase the flexibility and control of a growth cone's response to a particular cue as it navigates through its envrironment.

A further study from the Holt laboratory identified the rapid activation of caspase-3, another protease involved in mediating chemotropic responses of growth cones, to netrin-1 and LPA [89] (Figure 3). Interestingly, caspase-3 activation in response to these cues, was blocked by proteasome inhibitors suggesting that proteasome-mediated degradation of proteins is necessary for caspase-3 activation [89]. Candidate proteins to be degraded are the inhibitor of apoptosis proteins (IAPs), proteasome-mediated proteolysis of which is required for caspase activation in other systems [90] (see Chapter 21). Presumably caspase-3 activation in growth cones would be tightly regulated and would not lead to apoptosis of the cell. This could be achieved by its transient and local activation, or by its subsequent degradation, since IAPs are able to target caspase-3 for proteasome-mediated proteolysis [91]. Additionally, the netrin receptor deleted in colorectal carcinoma (DCC) can undergo netrin-1-induced endocytosis [32,92] and ubiquitin-mediated proteolysis [32,93] suggesting a further level of complexity for the UPS in regulating growth cone chemotropic responses of growth cones. The identity of the proteins synthesised and degraded by chemotropic cues is currently an active area of investigation, as well as determining the *in vivo* significance of these observations.

The UPS and *Roundabout*-Dependent Midline Crossing of Axons in *Drosophila*

During the process of growth cone pathfinding, the growth cone makes decisions in order to navigate towards its correct target. Since the path taken by a growth cone can be a relatively long one the route taken may be divided up into a number of smaller segments. The decisions taken by growth cones can take place at intermediate targets known as 'choice points' which subdivide the route taken to reach a particular target. One such intermediate target is the midline of the central nervous system. In vertebrates, co-culture experiments of commissural neuronal explants cultured with ectopic floor plate tissue (a source of the chemoattractant netrin-1) demonstrated that axons which have experienced netrin-1 at the floor plate are no longer attracted to the floor plate or to cells secreting netrin-1 (which would prevent axons re-crossing) [94]. The mechanism of the change in responsiveness of commissural axons remains unclear. Crossing of axons at the *Drosophila* midline has been used to study the mechanisms of how the growth cone changes its responsiveness to cues at an intermediate target. Recently the UPS has been identified as one such mechanism [30]. At the *Drosophila* midline glial cells secrete the attractive cue, netrin, and the repulsive cue, slit. Growth cones that respond to these cues at the midline express the netrin receptor, DCC/ Frazzled (Fraz), and the slit receptor, roundabout (*Robo*). Growth cones which cross the midline are firstly attracted toward it by netrin, while Robo, the repulsive receptor is down-regulated [95], allowing growth cones to cross. The expression of Robo at the cell surface is regulated by the product of the *commissureless* gene (Comm) [96]. Comm is a type I transmembrane protein and facilitates the removal of Robo from the growth cone surface [95-98]. After crossing the midline growth cones change their responsiveness to slit by increasing the cell surface receptor expression of Robo, that prevents them from re-crossing the midline [97,98]. Two models have been proposed for the regulation of Robo by Comm in axons crossing the midline [99]. In the 'sorting' model Robo would be trafficked to the endosomal / lysosomal pathway from the trans-golgi network in the presence of Comm rendering the growth cone insensitive to slit. The clearance model proposes that Robo is removed rapidly from the growth cones surface via endocytosis mediated by Comm.

To gain a better understanding of how growth cones adapt their responses to environmental cues on reaching an intermediate target, a milestone which enables them to continue along the pathway, Tear and colleagues [30] performed a yeast two-hybrid screen to identify proteins interacting with Comm, which might regulate cell surface levels of Robo. Using the cytoplasmic domain of Comm as bait they identified the *Drosophila* homologue of Nedd4 (*n*euronally *e*xpressed *d*evelopmentally *d*own regulated 4; DNedd4; [30]). DNedd4 is an E3 ubiquitin ligase, the HECT domain of which is implicated in the direct ubiquitination of substrates [100]. Comm was shown to be ubiquitinated by DNedd4. When Comm was coexpressed with Robo in *Drosophila* S2 cells Robo was internalised to endosomes. This could be blocked by mutating the putative DNedd4 ubiquitination sites on Comm leading to the proposal that ubiquitination is important in regulating growth cone's behaviour at the midline. In this case Comm interacts with Robo and the complex is internalised by the action of DNedd4 in accordance with the clearance model [30]. Embryos lacking DNedd4 function exhibit defects at sites where axons would normally cross the midline. However, a reduction in DNedd4 expression does not lead to a *comm- or robo*-like mutant phenotype, as would be expected if it was acting purely in the Robo-Comm pathway, suggesting that DNedd4 may be able to regulate the expression at of other cell surface axon guidance receptors. Intriguingly, Nedd4 is also present in *Xenopus* retinal growth cones (D.S.C. and C.E. Holt unpublished observations), though its function awaits further investigation.

Using a heterologous expression system (293 T Cells), cotransfection of Comm and active DNedd4 resulted in a reduction in Comm levels relative to the amount coexpressed with a catalytically inactive mutant of DNedd4, consistent with the idea that DNedd4 targets Comm for proteasome-mediated degradation [30]. However, in addition to its role in targeting proteins for degradation, Nedd4, and Rsp5p, the yeast orthologue, may also be involved in targeting cell surface proteins for endocytosis, perhaps leading to intracellular signalling events or lysosomal-mediated degradation [51,101]. A role for Comm in initiating synaptogenesis via endocytosis at the neuromuscular junction in *Drosophila* has been identified ([102]; discussed below in the section 'The role of the UPS in synaptogenesis'). A more recent study from Dickson and colleagues has brought into question the for of DNedd4 in midline crossing and they were unable to detect the ubiquitination of Comm *in vitro* and *in vivo* [103]. It is unclear why identical experiments yielded opposite results. Further experiments examining the trafficking of Robo in the presence and absence of Comm in neurons using time-lapse imaging of *Drosophila* PNS neurons supported the sorting model [103].

The studies on *Xenopus* retinal growth cones [31,33,89,92] and at the *Drosophila* midline [30] illustrate a number of roles in which the UPS may function in axon guidance: in mediating the growth cone's response to a particular cue via intracellular signalling pathways and possibly in the regulation of a growth cone's response to a particular cue by regulating its responsiveness at the level of receptor expression on the growth cone surface.

INVOLVEMENT OF THE UPS IN BRANCH RETRACTION

Once growth cones have reached their targets in the nervous system, morphological changes take place which result in the formation of the pre-synaptic nerve ending and the formation of synapses critical for the functioning of the nervous system. A recurring theme in establishing connections between the growing axon and post-synaptic dendrites in both vertebrates and invertebrates is the development of very complex and elaborate axonal and dendritic processes and branches when the connections are first established. This is followed by a refinement phase of selective pruning resulting in the stabilisation of a sub-set of processes. The extensive axonal and dendritic remodelling which occurs during metamorphosis in the *Drosophila* mushroom bodies (MB), sites of olfactory learning in insects, has been used as a model system for the study of the molecular and cellular mechanisms of axon pruning, of which very little is currently known. Recently, Luo and colleagues have identified a role for the UPS in mediating local degeneration resulting in axon pruning [45].

During the development of the *Drosophila* MB neurons, the MB neuroblast divides successively to generate three different types of neurons (γ, α'/β' and α/β) with different axonal projections in the adult fly [104,105]. The γ neurons initially establish a dendritic arbor and an axonal projection that bifurcates into a medial and dorsal branch.

Time-course analysis of axon pruning in MB γ neurons expressing membrane targeted GFP as a marker revealed that axon pruning of the dorsal and medial branches occurred via local degeneration [45]. The apparent degeneration of axons rather than retraction of branches led the authors to suspect that protein turnover may be involved in these processes leading to an examination of the role of the UPS. To test the role of the UPS a yeast ubiquitin protease (UBP2) known to function in *Drosophila* was overexpressed in MB γ neurons, resulting in a failure of axon pruning, and suggesting that protein ubiquitination is important in axon pruning. Taking advantage of the power of *Drosophila* genetics and existing collections of mutants, Watts and colleagues found that MB γ neuron expression of a putative loss of function mutation encoding a predicted ubiquitin activating enzyme (E1) also blocked axon pruning. Axon pruning was also reduced by the expression of mutated forms of *mov64* and *Rpn6*, two subunits of the 19S particle of the proteasome. However, blocking endocytosis, a function of mono-ubiquitination, did not lead to aberrant axon pruning, suggesting that the role in regulating axon pruning is via ubiquitin-dependent proteasome-mediated proteolysis. Mutations in a number of E2 and E3 ubiquitin ligases did not block axon pruning though. Since there are many E2 and E3 ligases identified from the *Drosophila* genome this observation may suggest that the specificity of proteins degraded by the UPS is critical for axon pruning and is determined by the combinations of E2 and E3 enzymes expressed in MB γ neurons. The axon pruning observed in MB neurons is specific to the MB γ neuron, leaving the α'/β' neuron unaffected. Previously, Luo and colleagues have demonstrated a cell autonomous requirement for the ecdysone receptor MB γ neuron to mediate axon pruning [106]. Components of the ecdysone signalling pathway are absent in α'/β' neurons, suggesting a possible mechanism for mediating the specificity of pruning potentially resulting in differential activation of the UPS in the different MB cell types. It remains to be established at which stage of pruning the UPS is required and which proteins undergo proteasome-mediated proteolysis during pruning.

The local degeneration observed in axon pruning during metamorphosis is reminiscent of the fragmentation and degeneration which occurs to the distal portion of an axon when it is cut leading to a process known as Wallerian degeneration [107]. The *Wallerian degeneration slow* (*Wld*^{s}) mouse, in which Wallerian degeneration occurs more slowly, results in the upregulation of a fusion protein comprising a region of UFD2 / E4 involved in poly-ubiquitination, suggesting that the UPS may be involved in delaying the degeneration. Indeed, a recently study by He and colleagues [108] demonstrated that blocking proteasome function pharmacologically or genetically delayed axon degeneration *in vitro* and *in vivo,* further demonstrating a requirement for the UPS in degeneration and indicating that the mechanisms involved in axon pruning and in degeneration are similar in vertebrates and invertebrates. Additionally, a study by Fawcett and colleagues identified a role for the UPS in growth cone reformation after axotomy indicating similar processes are involved in axonal degeneration and regeneration [109]. Chapter 26 reviews the UPS and axon degeneration.

The Role of the UPS in Synaptogenesis

Synapses are critical to the functioning of the nervous system since they form connections between neurons. The neuromuscular junction (NMJ) of the *Drosophila* peripheral nervous system (PNS) has served as a very powerful system for studying the mechanisms of synaptogenesis and synaptic plasticity [110-112]. Synapses of the NMJ, i.e. those formed by motor neurons on to body wall muscle fibres are uniquely specified, can be reproducibly identified between different flies and are known as 'identified synapses'. NMJ synapses are relatively large and superficial making them much more accessible for study than those synapses in the central nervous system, particularly of vertebrates. Yet the NMJ shares important features with central excitatory synapses in the vertebrate brain. For example, the NMJ synapses are glutaminergic and contain similar ionotropic glutaminergic receptors [113] making their study of general importance. Additionally, the suitability of *Drosophila* for performing mutagenesis screens has led to the identification of some of the components mediating synapse formation and plasticity.

Using a behavioural screen performed in *Drosophila* for identifying walking mutants Wan and colleagues performed a secondary screen for anatomical phenotypes of the larval NMJ [37]. Synapses from *highwire* (*hiw*) mutants have more pre-synaptic boutons, increased branches and are electrophysiologically abnormal (despite being ultrastructurally normal when viewed with electron microscopy) compared to wild-type synapses, consistent with a synaptic overgrowth phenotype, indicating it is a negative regulator of synaptogenesis [37]. However, the specificity of connections is not altered and the initial axon pathfinding and target selection appear normal, suggesting the phenotype is specific to synaptogenesis. Cloning of the *hiw* locus revealed the gene encoded a large protein (5233 amino acids) containing a RING-H2 domain characteristic of a class of E3 ubiquitin ligases [52] suggesting that it may be involved in mediating protein-protein interactions for determining substrate specificity for ubiquitin-mediated proteolysis. It is found at pre-synaptic terminals and synapses. Recently DiAntonio and colleagues have demonstrates that the presynaptic function of *highwire* is necessary and sufficient to regulate synaptic development and additionally that it may play a general role throughout larval development to regulate synaptic structure and physiological function [114]. RPM-1, the *C. elegans* homologue of *hiw,* is similarly expressed pre-synaptically and has been shown to act cell autonomously to negatively regulate synaptogenesis of GABAergic motor neurons [41,115]. Furthermore, Jin and colleagues have recently shown that RPM-1 regulates presynaptic development via negatively regulating the p38 MAP kinase pathway probably via the ubiquitination of DLK-1 (DAP (*d*eath associated protein kinase) *l*ike *k*inase) a MAP kinase kinase kinase [116]. A further screen searching for proteins interacting with RPM-1 identified FSN-1, a novel F-box protein that interacts with the *C. elegans* homologues of cullin and SKP1 to form a new type of SCF complex required for the localised regulation of presynaptic differentiation [117]. Additional evidence from mice suggests that similar ubiquitin-mediated pathways may be involved in vertebrate synaptogenesis as mutations in the mouse homologue of *hiw*, *PHR-1* (PAM (protein associated with myc), *highwire* and rpm-1) result in a similar disruption of nerve terminal morphology to that found in *hiw* and *rpm-1* mutants [42]. *Esrom*, the zebrafish homologue of *hiw,* functions as an E3 ubiqutin ligase and plays roles in the topographic mapping of RGC axons and in preventing arborisation in inappropriate regions of the tectum [38,40] A role of *Esrom* in synaptogenesis remains to be investigated.

A role for the ubiquitination in synaptogenesis at the NMJ was further established by another genetic screen performed by DiAntonio and colleagues in which they set out to identify genes which, when overexpressed in neurons, alter synaptic growth [35]. Two *Drosophila* lines were identified that had a synaptic overgrowth, i.e. increased synaptic bouton number and enhanced branching pattern phenotype. Both encoded *fat facets (faf)*, a member of a family of deubiquitinating enzymes (DUBs), which remove ubiquitin from target proteins and hence antagonise ubiqutinin-dependent mechanisms. *faf* inhibits proteasome-mediated proteolysis [118,119] in the developing eye where *faf* has been shown to interact physically with and deubiquitinate *liquid facets* (*lqf*), the homologue of the vertebrate endocytic protein *epsin* (adaptors in endocytosis; [120]) preventing its degradation [121] and possibly promoting its endocytosis via mono-ubiquitination. Over-expression of the yeast deubiquitinating protease UBP2 in the nervous system of *Drosophila,* which has overlapping substrate specificity with *faf,* also leads to synaptic overgrowth similar to *faf* over-expression, suggesting that blocking ubiquitin-dependent processes alters synaptic development [35]. This most likely occurs through the prolonged presence of specific synaptic proteins.

The *faf* over-expression phenotype in the NMJ is very similar to the loss of function mutations in *hiw* discussed above and a further genetic screen performed by DiAntonio and colleagues to identify molecular components regulated by *faf* identified mutant alleles of *hiw,* suggesting that *faf* function can antagonise *hiw* activity as a negative regulator of synaptic growth. Furthermore, loss of function mutations in *faf* partially suppress the physiological abnormalities in *hiw* mutants, revealing a role for endogenous *faf* in synapse formation [35].

The genetic interaction between *faf,* a DUB and *hiw*, a potential E3 ubiquitin ligase, suggest that synapse formation requires a balance between ubiquitination and deubiquitination of target proteins.

Recently, a yeast two-hybrid screen to identify proteins interacting with *hiw* identified the Smad protein *medea* (*Med*), a component of the bone morphogenetic protein (BMP) signalling pathway [122]. BMPs are members of the transforming growth factor β TGF- β family which have been implicated as retrograde signalling molecules at *Drosophila* synapses [123-125] (Figure 4). Mutations in the BMP pathways, such as the BMP type II receptor *wishful thinking* (*wit*), its ligand *glass bottomed boat* (*gbb*) and *med,* all have reduced synaptic growth [122-125], i.e. the opposite phenotype to *hiw* mutants. BMP mutants are able to suppress the synaptic overgrowth phenotype of *hiw* mutants, while activation of BMP signalling in *hiw* mutants (which in wild-type embryos does not lead to an additional increase in synaptic growth) leads to further expansion of synapses [122]. The authors proposed that the growth of synapses at the NMJ requires a balance between positive signalling mediated by BMPs and negative regulation by *hiw*. Since *hiw* encodes a putative E3 ubiquitin ligase the possibility exists that *hiw* may be able to target *med* for proteasome-mediated proteolysis, which could reduce the amount of *med* available for the regulation of transcription of genes necessary for synapse formation [122,126] (Figure 4). In other systems, such as in embryonic pattern formation, BMP signalling is regulated by a ubiquitin ligase (Smurf1; Smad ubiquitin regulatory factor) which targets receptor-regulated SMADs (small mothers against decapentaplegic) for proteasome-mediated degradation [127]. A critical test of this hypothesis would be the demonstration that *hiw* encodes a functionally active E3 ubiquitin ligase and that *med* is one of its substrates.

Whilst identifying a role for ubiquitination these studies, such as that of DiAntonio and colleagues have not elucidated the nature of the target proteins to be ubiquitinated and whether proteasome-mediated proteolysis and / or ubiquitin-dependent endocytosis are required for synapse formation. Endocytosis has previously been shown to be important for the initiation of synapse formation [102]. In the absence of Comm, motor neuron growth cones fail to initiate synaptogenesis. This can be rescued by the expression of Comm in the muscles (Figure 4). Expression of a mutant form of Comm lacking the cytoplasmic domain required for endocytosis phenocopies the defects in synaptogenesis, suggesting that Comm is important in the cell surface dynamics and remodelling which occurs during synaptogenesis [102]. It is currently unknown whether the endocytosis of Comm during the initiation of synaptogenesis is dependent on DNedd4 or ubiquitination, as has previously been suggested in the regulation of Robo levels to prevent axons from re-crossing the *Drosophila* midline [30]. Additionally, over-expression of semaphorin1a (Sema1a) and Robo at the *Drosophila* NMJ results in a '*bendless*-like' phenotype, due to decreased synaptic function, suggesting that Sem1a and Robo are inhibitory to synaptogenesis [128,129] (Figure 4). Since the over-expression phenotypes are similar to those of *bendless*, a Ubc E2, it may be hypothesised that internalisation of Sema1a and Robo, regulated by the UPS, is necessary for synaptogenesis to proceed. Furthermore, the identification of *lqd*, a homologue of the vertebrate epsin involved

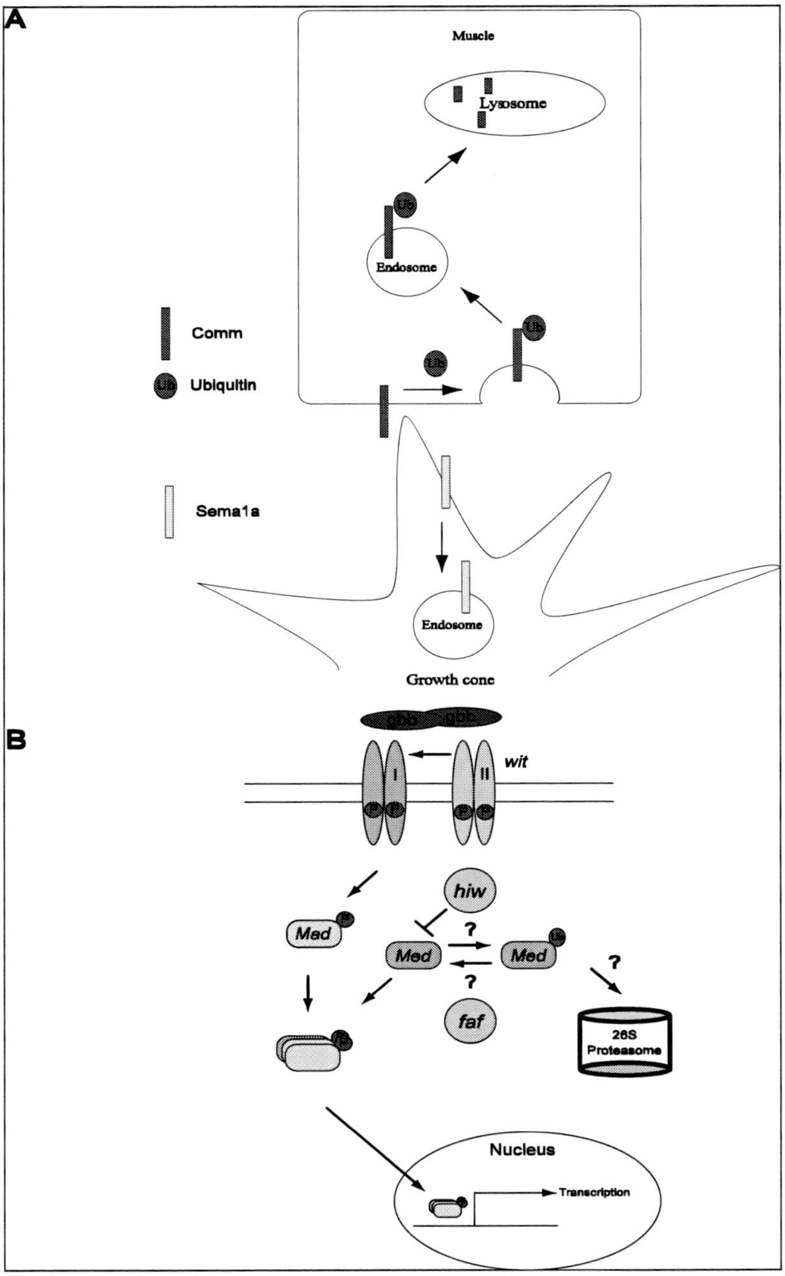

Figure 4. Regulation of Drosophila NMJ synaptogenesis by the ubiquitin-proteasome system and bone morphogenetic protein signalling pathways. Synapse initiation at the Drosophila NMJ requires the endocytosis of *commissureless* (*comm*) from the cell surface of the muscles and the endocytosis of Sema1a from the surface of the growth cone presumably via a ubiquitin-dependent mechanism. Comm may regulate the removal of a hypothesised synapse inhibitory factor to allow synaptogenesis to proceed (A). Bone morphogenetic protein (BMP) signalling involves the interaction of BMP or the Drosophila homologue *glass bottomed boat* (*gbb*) with type-1 and type-2 serine/ threonine kinase receptors *thick vein* (*tkv*), *saxophone* (*sax*) and *wishful thinking* (*wit*) which are able to phosphorylate R-Smad (mad). Phosphorylated mad interacts with a co-Smad (*Medea*) to regulate transcription (B). *highwire* (*hiw*), a presumptive E3 ubiquitin ligase inhibits *Medea* function to regulate synaptogenesis. *hiw* may regulate *Medea* via its ubiquitination and subsequent proteasome-mediated proteolysis. *fat facets* (*faf*), a

deubiquitinating enzyme is able to anatagonise the effects of hiw on synaptogenesis, possibly by the removal of ubiquitin from a common substrate such as *Medea* (B). Hypothetical pathways are indicated by a '?'.

in endocytosis (possibly facilitating neurotransmitter release at the synapse), as a substrate of *faf* opens up the possibility that ubiquitin-mediated endocytosis may function in synaptogenesis. The involvement of endocytosis in determining a growth cone's response to chemotropic cues (reviewed in [76]) and in regulating a growth cone's responsiveness to Robo at the midline [30] demonstrates the general importance of endocytosis in neuronal development. It will be interesting to determine the involvement of ubiquitination-mediated endocytosis in these processes. Recently, studies in *Aplysia* [130] *Drosophila* [131] and mammals [132,133] have implicated the UPS in the regulation of synaptic strength of mature synapses, suggesting that the regulation of protein levels via ubiquitin-dependent proteolysis, in addition to its role in synaptogenesis, is also important in the plasticity of mature synapses (reviewed in [134] and Chapters 16 and 18).

The anaphase promoting complex (APC), an E3 ubiquitin ligase, is well known for its role in mediating cell cycle transitions via the ubiquitin-dependent degradation of target proteins such as cyclins. Recent evidence from Brand and colleagues indicates a role for the APC in synaptogenesis at the *Drosophila* NMJ [43]. Loss of function of APC components results in synaptic bouton overgrowth requiring liprin-α a potential target of the APC, indicating the APC acts as a negative regulator of synaptic size [43]. The phenotypes of APC2 / *morula* and *hiw* mutants are different, suggesting that the role played by the UPS in synaptogenesis requires at least two E3 ubiquitin ligase activities.

CONCLUSIONS

Since the discovery of the *bendless* mutation affecting axon pathfinding in *Drosophila* twenty years ago as a mutation in an E2 ubiquitin conjugating enzyme, studies have identified key roles for the UPS in regulating many aspects of the development of the nervous system, from neurogenesis to synaptogenesis. There appear to be two main types and roles of ubiquitination: mono-ubiquitination that results in the endocytosis of the target proteins and poly-ubiquitination which leads to degradation of the target proteins by proteasome-mediated proteolysis. Both types of ubiquitination occur during the development of the nervous system and serve as mechanisms for the regulation of cellular processes. The promotion of endocytosis by ubiquitination may be of particular importance since endocytosis is fundamental to the functioning of the neuron, for example, its capacity to regulate responsiveness of cells to signals such as the Notch-Delta signalling pathway during neurogensis, or the growth cone's ability to respond to chemotropic cues, or to the synaptic vesicle cycling which occurs during neurotransmitter release at the synapse.

The identification of the UPS in particular biological processes raises many questions from the nature of the proteins degraded to the exact role of the UPS in the process and why mutations in components of the UPS, which would be expected to be required for the degradation of many proteins result in specific phenotypes such as *bendless* [5]. Genetic approaches, in particular in *Drosophila*, mouse and zebrafish, have been responsible for the identification of many genes involved in developmental processes *in vivo*. The identification of the roles of the UPS in neural development has come about largely through the use of genetics. Recently, however, particularly in the study of growth cone guidance, an *in vitro* system has enabled the examination of the behaviours of individual growth cones [31,33,76,89]. In addition to identifying the proteins that undergo ubiquitination and proteasome-mediated proteolysis, the role of the UPS in axon pathfinding *in vivo* remains to be determined.

One of the most intriguing findings regarding the involvement in the UPS in the development of the nervous system is that mutations in components of this system have much more highly specific phenotypes than one would have initially predicted given the general usage of the UPS in many cellular processes. These observations suggest that individual components of the UPS play very specific roles in specific cellular processes. The specificity may lie in the combination of E2 and E3 enzymes expressed in particular cell types and their specific involvement in particular processes, implied from the study by Watts and colleagues [45], who found that mutations in a number of components of the UPS did not affect axon pruning, whereas mutations in others did. In *Drosophila* there are more than thirty E2s and fifty E3s and humans have more than forty E2s, more than five hundred E3s and more than 80 DUBs, which would allow components of the UPS to act in many different combinations, imparting specificity and controllability of the proteins targeted by the UPS. Identifying which E2, E3 and Dubs are in involved in which processes and their substrates will result in a better understanding of the UPS in the development of the nervous system.

ACKNOWLEDGEMENTS

I am extremely grateful to Professor Christine E. Holt for her constant support and encouragement and for introducing me to the ubiquitin-proteasome system. I am also very grateful to Drs Paul Goldsmith and Susan Chapman and Prof. Christine Holt for critical reading of the manuscript and Professor S. Jack for helpful insights. Funding provided by the BBSRC (UK) and an EMBO Long-Term Fellowship. I apologize to those authors whose work I have been unable to cite or discuss in detail due to space constraints.

REFERENCES

[1] Pickart CM: Back to the future with ubiquitin. *Cell* 2004, 116:181-190.

[2] Glickman MH, Ciechanover A: The ubiquitin-proteasome proteolytic pathway: destruction for the sake of construction. *Physiol Rev* 2002, 82:373-428.

[3] Finley D, Ciechanover A, Varshavsky A: Thermolability of ubiquitin-activating enzyme from the mammalian cell cycle mutant ts85. *Cell* 1984, 37:43-55.

[4] Ciechanover A, Finley D, Varshavsky A: Ubiquitin dependence of selective protein degradation demonstrated in the mammalian cell cycle mutant ts85. *Cell* 1984, 37:57-66.

[5] Thomas JB, Wyman RJ: Mutations altering synaptic connectivity between identified neurons in Drosophila. *J Neurosci* 1984, 4:530-538.

[6] Muralidhar MG, Thomas JB: The Drosophila bendless gene encodes a neural protein related to ubiquitin-conjugating enzymes. *Neuron* 1993, 11:253-266.

[7] Oh CE, McMahon R, Benzer S, Tanouye MA: bendless, a Drosophila gene affecting neuronal connectivity, encodes a ubiquitin-conjugating enzyme homolog. *J Neurosci* 1994, 14:3166-3179.

[8] Harris WA, Hartenstein, V. Cellular Determination. In *Fundamental Neuroscience*. Edited by Zigmond MJ, Bloom FE, Landis SC, Roberts JL, Squire LR. San Diego: Academic Press; 1999: 481-517.

[9] Taghert PH, Doe CQ, Goodman CS: Cell determination and regulation during development of neuroblasts and neurones in grasshopper embryo. *Nature* 1984, 307:163-165.

[10] Campos-Ortega JA: Genetic mechanisms of early neurogenesis in Drosophila melanogaster. *Mol Neurobiol* 1995, 10:75-89.

[11] Nie J, McGill MA, Dermer M, Dho SE, Wolting CD, McGlade CJ: LNX functions as a RING type E3 ubiquitin ligase that targets the cell fate determinant Numb for ubiquitin-dependent degradation. *Embo J* 2002, 21:93-102.

[12] Cornell M, Evans DA, Mann R, Fostier M, Flasza M, Monthatong M, Artavanis-Tsakonas S, Baron M: The Drosophila melanogaster Suppressor of deltex gene, a regulator of the Notch receptor signaling pathway, is an E3 class ubiquitin ligase. *Genetics* 1999, 152:567-576.

[13] Deblandre GA, Lai EC, Kintner C: Xenopus neuralized is a ubiquitin ligase that interacts with XDelta1 and regulates Notch signaling. *Dev Cell* 2001, 1:795-806.

[14] Lai EC, Deblandre GA, Kintner C, Rubin GM: Drosophila neuralized is a ubiquitin ligase that promotes the internalization and degradation of delta. *Dev Cell* 2001, 1:783-794.

[15] Yeh E, Dermer M, Commisso C, Zhou L, McGlade CJ, Boulianne GL: Neuralized functions as an E3 ubiquitin ligase during Drosophila development. *Curr Biol* 2001, 11:1675-1679.

[16] Itoh M, Kim CH, Palardy G, Oda T, Jiang YJ, Maust D, Yeo SY, Lorick K, Wright GJ, Ariza-McNaughton L, Weissman AM, Lewis J, Chandrasekharappa SC, Chitnis AB: Mind bomb is a ubiquitin ligase that is essential for efficient activation of Notch signaling by Delta. *Dev Cell* 2003, 4:67-82.

[17] Chen W, Casey Corliss D: Three modules of zebrafish Mind bomb work cooperatively to promote Delta ubiquitination and endocytosis. *Dev Biol* 2004, 267:361-373.

[18] Barsi JC, Rajendra R, Wu JI, Artzt K: Mind bomb1 is a ubiquitin ligase essential for mouse embryonic development and Notch signaling. *Mech Dev* 2005, 122:1106-1117.

[19] Wu G, Lyapina S, Das I, Li J, Gurney M, Pauley A, Chui I, Deshaies RJ, Kitajewski J: SEL-10 is an inhibitor of notch signaling that targets notch for ubiquitin-mediated protein degradation. *Mol Cell Biol* 2001, 21:7403-7415.

[20] Oberg C, Li J, Pauley A, Wolf E, Gurney M, Lendahl U: The Notch intracellular domain is ubiquitinated and negatively regulated by the mammalian Sel-10 homolog. *J Biol Chem* 2001, 276:35847-35853.

[21] Voigt J, Papalopulu N: A dominant-negative form of the E3 ubiquitin ligase Cullin-1 disrupts the correct allocation of cell fate in the neural crest lineage. *Development* 2006, 133:559-68.

[22] Sakurai M, Ayukawa K, Setsuie R, Nishikawa K, Hara Y, Ohashi H, Nishimoto M, Abe T, Kudo Y, Sekiguchi M, Sato Y, Aoki S, Noda M, Wada K: Ubiquitin C-terminal hydrolase L1 regulates the morphology of neural progenitor cells and modulates their differentiation. *J Cell Sci* 2006, 119:162-171.

[23] Mehta N, Loria PM, Hobert O: A genetic screen for neurite outgrowth mutants in Caenorhabditis elegans reveals a new function for the F-box ubiquitin ligase component LIN-23. *Genetics* 2004, 166:1253-1267.

[24] Kavakebi P, Hausott B, Tomasino A, Ingorokva S, Klimaschewski L: The N-end rule ubiquitin-conjugating enzyme, HR6B, is up-regulated by nerve growth factor and required for neurite outgrowth. *Mol Cell Neurosci* 2005, 29:559-568.

[25] Bryan B, Cai Y, Wrighton K, Wu G, Feng XH, Liu M: Ubiquitination of RhoA by Smurf1 promotes neurite outgrowth. *FEBS Lett* 2005, 579:1015-1019.

[26] Aguilera M, Oliveros M, Martinez-Padron M, Barbas JA, Ferrus A: Ariadne-1: a vital Drosophila gene is required in development and defines a new conserved family of ring-finger proteins. *Genetics* 2000, 155:1231-1244.

[27] Martin KA, Poeck B, Roth H, Ebens AJ, Ballard LC, Zipursky SL: Mutations disrupting neuronal connectivity in the Drosophila visual system. *Neuron* 1995, 14:229-240.

[28] Poeck B, Fischer S, Gunning D, Zipursky SL, Salecker I: Glial cells mediate target layer selection of retinal axons in the developing visual system of Drosophila. *Neuron* 2001, 29:99-113.

[29] Tursun B, Schluter A, Peters MA, Viehweger B, Ostendorff HP, Soosairajah J, Drung A, Bossenz M, Johnsen SA, Schweizer M, Bernard O, Bach I: The ubiquitin ligase Rnf6 regulates local LIM kinase 1 levels in axonal growth cones. *Genes Dev* 2005, 19:2307-2319.

[30] Myat A, Henry P, McCabe V, Flintoft L, Rotin D, Tear G: Drosophila Nedd4, a ubiquitin ligase, is recruited by Commissureless to control cell surface levels of the roundabout receptor. *Neuron* 2002, 35:447-459.

[31] Campbell DS, Holt CE: Chemotropic responses of retinal growth cones mediated by rapid local protein synthesis and degradation. *Neuron* 2001, 32:1013-1026.

[32] Kim TH, Lee HK, Seo IA, Bae HR, Suh DJ, Wu J, Rao Y, Hwang KG, Park HT: Netrin induces down-regulation of its receptor, Deleted in Colorectal Cancer, through the ubiquitin-proteasome pathway in the embryonic cortical neuron. *J Neurochem* 2005, 95:1-8.

[33] Mann F, Miranda E, Weinl C, Harmer E, Holt CE: B-type Eph receptors and ephrins induce growth cone collapse through distinct intracellular pathways. *J Neurobiol* 2003, 57:323-336.

[34] Rodriguez J, Esteve P, Weinl C, Ruiz JM, Fermin Y, Trousse F, Dwivedy A, Holt C, Bovolenta P: SFRP1 regulates the growth of retinal ganglion cell axons through the Fz2 receptor. *Nat Neurosci* 2005, 8:1301-1309.

[35] DiAntonio A, Haghighi AP, Portman SL, Lee JD, Amaranto AM, Goodman CS: Ubiquitination-dependent mechanisms regulate synaptic growth and function. *Nature* 2001, 412:449-452.

[36] Xu J, Taya S, Kaibuchi K, Arnold AP: Spatially and temporally specific expression in mouse hippocampus of Usp9x, a ubiquitin-specific protease involved in synaptic development. *J Neurosci Res* 2005, 80:47-55.

[37] Wan HI, DiAntonio A, Fetter RD, Bergstrom K, Strauss R, Goodman CS: Highwire regulates synaptic growth in Drosophila. *Neuron* 2000, 26:313-329.

[38] Karlstrom RO, Trowe T, Klostermann S, Baier H, Brand M, Crawford AD, Grunewald B, Haffter P, Hoffmann H, Meyer SU, Muller BK, Richter S, van Eeden FJ, Nusslein-Volhard C, Bonhoeffer F: Zebrafish mutations affecting retinotectal axon pathfinding. *Development* 1996, 123:427-438.

[39] Le Guyader S, Maier J, Jesuthasan S: Esrom, an ortholog of PAM (protein associated with c-myc), regulates pteridine synthesis in the zebrafish. *Dev Biol* 2005, 277:378-386.

[40] D'Souza J, Hendricks M, Le Guyader S, Subburaju S, Grunewald B, Scholich K, Jesuthasan S: Formation of the retinotectal projection requires Esrom, an ortholog of PAM (protein associated with Myc). *Development* 2005, 132:247-256.

[41] Schaefer AM, Hadwiger GD, Nonet ML: rpm-1, a conserved neuronal gene that regulates targeting and synaptogenesis in C. elegans. *Neuron* 2000, 26:345-356.

[42] Burgess RW, Peterson KA, Johnson MJ, Roix JJ, Welsh IC, O'Brien TP: Evidence for a conserved function in synapse formation reveals Phr1 as a candidate gene for respiratory failure in newborn mice. *Mol Cell Biol* 2004, 24:1096-1105.

[43] van Roessel P, Elliott DA, Robinson IM, Prokop A, Brand AH: Independent regulation of synaptic size and activity by the anaphase-promoting complex. *Cell* 2004, 119:707-718.

[44] Zhu S, Perez R, Pan M, Lee T: Requirement of Cul3 for axonal arborization and dendritic elaboration in Drosophila mushroom body neurons. *J Neurosci* 2005, 25:4189-4197.

[45] Watts RJ, Hoopfer ED, Luo L: Axon pruning during Drosophila metamorphosis: evidence for local degeneration and requirement of the ubiquitin-proteasome system. *Neuron* 2003, 38:871-885.

[46] Schweisguth F: Dominant-negative mutation in the beta2 and beta6 proteasome subunit genes affect alternative cell fate decisions in the Drosophila sense organ lineage. *Proc Natl Acad Sci U S A* 1999, 96:11382-11386.

[47] Xu T, Artavanis-Tsakonas S: deltex, a locus interacting with the neurogenic genes, Notch, Delta and mastermind in Drosophila melanogaster. *Genetics* 1990, 126:665-677.

[48] Hustad CM, Perry WL, Siracusa LD, Rasberry C, Cobb L, Cattanach BM, Kovatch R, Copeland NG, Jenkins NA: Molecular genetic characterization of six recessive viable alleles of the mouse agouti locus. *Genetics* 1995, 140:255-265.

[49] Fostier M, Evans DA, Artavanis-Tsakonas S, Baron M: Genetic characterization of the Drosophila melanogaster Suppressor of deltex gene: A regulator of notch signaling. *Genetics* 1998, 150:1477-1485.

[50] Perry WL, Hustad CM, Swing DA, O'Sullivan TN, Jenkins NA, Copeland NG: The itchy locus encodes a novel ubiquitin protein ligase that is disrupted in a18H mice. *Nat Genet* 1998, 18:143-146.

[51] Rotin D, Staub O, Haguenauer-Tsapis R: Ubiquitination and endocytosis of plasma membrane proteins: role of Nedd4/Rsp5p family of ubiquitin-protein ligases. *J Membr Biol* 2000, 176:1-17.

[52] Joazeiro CA, Weissman AM: RING finger proteins: mediators of ubiquitin ligase activity. *Cell* 2000, 102:549-552.

[53] Qiu L, Joazeiro C, Fang N, Wang HY, Elly C, Altman Y, Fang D, Hunter T, Liu YC: Recognition and ubiquitination of Notch by Itch, a hect-type E3 ubiquitin ligase. *J Biol Chem* 2000, 275:35734-35737.

[54] Mazaleyrat SL, Fostier M, Wilkin MB, Aslam H, Evans DA, Cornell M, Baron M: Down-regulation of Notch target gene expression by Suppressor of deltex. *Dev Biol* 2003, 255:363-372.

[55] Ramain P, Khechumian R, Khechumian K, Arbogast N, Ackermann C, Heitzler P: Interactions between chip and the achaete/scute-daughterless heterodimers are required for pannier-driven proneural patterning. *Mol Cell* 2000, 6:781-790.

[56] Matsuno K, Diederich RJ, Go MJ, Blaumueller CM, Artavanis-Tsakonas S: Deltex acts as a positive regulator of Notch signaling through interactions with the Notch ankyrin repeats. *Development* 1995, 121:2633-2644.

[57] Diederich RJ, Matsuno K, Hing H, Artavanis-Tsakonas S: Cytosolic interaction between deltex and Notch ankyrin repeats implicates deltex in the Notch signaling pathway. *Development* 1994, 120:473-481.

[58] Sundaram M, Greenwald I: Suppressors of a lin-12 hypomorph define genes that interact with both lin-12 and glp-1 in Caenorhabditis elegans. *Genetics* 1993, 135:765-783.

[59] Hubbard EJ, Wu G, Kitajewski J, Greenwald I: sel-10, a negative regulator of lin-12 activity in Caenorhabditis elegans, encodes a member of the CDC4 family of proteins. *Genes Dev* 1997, 11:3182-3193.

[60] Wu L, Aster JC, Blacklow SC, Lake R, Artavanis-Tsakonas S, Griffin JD: MAML1, a human homologue of Drosophila mastermind, is a transcriptional co-activator for NOTCH receptors. *Nat Genet* 2000, 26:484-489.

[61] Portin P, Rantanen M: Interaction of master mind, big brain, neuralized and Notch genes of Drosophila melanogaster as expressed in adult morphology. *Hereditas* 1991, 114:197-200.

[62] Boulianne GL, de la Concha A, Campos-Ortega JA, Jan LY, Jan YN: The Drosophila neurogenic gene neuralized encodes a novel protein and is expressed in precursors of larval and adult neurons. *EMBO J* 1991, 10:2975-2983.

[63] Price BD, Chang Z, Smith R, Bockheim S, Laughon A: The Drosophila neuralized gene encodes a C3HC4 zinc finger. *EMBO J* 1993, 12:2411-2418.

[64] Yeh E, Zhou L, Rudzik N, Boulianne GL: Neuralized functions cell autonomously to regulate Drosophila sense organ development. *EMBO J* 2000, 19:4827-4837.

[65] Hicke L: Protein regulation by monoubiquitin. *Nat Rev Mol Cell Biol* 2001, 2:195-201.

[66] Pavlopoulos E, Pitsouli C, Klueg KM, Muskavitch MA, Moschonas NK, Delidakis C: neuralized Encodes a peripheral membrane protein involved in delta signaling and endocytosis. *Dev Cell* 2001, 1:807-816.

[67] Parks AL, Klueg KM, Stout JR, Muskavitch MA: Ligand endocytosis drives receptor dissociation and activation in the Notch pathway. *Development* 2000, 127:1373-1385.

[68] Vollrath B, Pudney J, Asa S, Leder P, Fitzgerald K: Isolation of a murine homologue of the Drosophila neuralized gene, a gene required for axonemal integrity in spermatozoa and terminal maturation of the mammary gland. *Mol Cell Biol* 2001, 21:7481-7494.

[69] Holley SA, Geisler R, Nusslein-Volhard C: Control of her1 expression during zebrafish somitogenesis by a delta-dependent oscillator and an independent wave-front activity. *Genes Dev* 2000, 14:1678-1690.

[70] Holley SA, Julich D, Rauch GJ, Geisler R, Nusslein-Volhard C: her1 and the notch pathway function within the oscillator mechanism that regulates zebrafish somitogenesis. *Development* 2002, 129:1175-1183.

[71] Gray M, Moens CB, Amacher SL, Eisen JS, Beattie CE: Zebrafish deadly seven functions in neurogenesis. *Dev Biol* 2001, 237:306-323.

[72] Jiang YJ, Brand M, Heisenberg CP, Beuchle D, Furutani-Seiki M, Kelsh RN, Warga RM, Granato M, Haffter P, Hammerschmidt M, Kane DA, Mullins MC, Odenthal J, van Eeden FJ, Nusslein-Volhard C: Mutations affecting neurogenesis and brain morphology in the zebrafish, Danio rerio. *Development* 1996, 123:205-216.

[73] Tessier-Lavigne M, Goodman CS: The molecular biology of axon guidance. *Science* 1996, 274:1123-1133.

[74] Dickson BJ: Molecular mechanisms of axon guidance. *Science* 2002, 298:1959-1964.

[75] Dingwell KS, Holt CE, Harris WA: The multiple decisions made by growth cones of RGCs as they navigate from the retina to the tectum in Xenopus embryos. *J Neurobiol* 2000, 44:246-259.

[76] van Horck FP, Weinl C, Holt CE: Retinal axon guidance: novel mechanisms for steering. *Curr Opin Neurobiol* 2004, 14:61-66.

[77] Tyers M, Willems AR: One ring to rule a superfamily of E3 ubiquitin ligases. *Science* 1999, 284:601, 603-604.

[78] Kitada T, Asakawa S, Hattori N, Matsumine H, Yamamura Y, Minoshima S, Yokochi M, Mizuno Y, Shimizu N: Mutations in the parkin gene cause autosomal recessive juvenile parkinsonism. *Nature* 1998, 392:605-608.

[79] Ravikumar B, Rubinsztein DC: Can autophagy protect against neurodegeneration caused by aggregate-prone proteins? *Neuroreport* 2004, 15:2443-2445.

[80] Kipreos ET, Gohel SP, Hedgecock EM: The C. elegans F-box/WD-repeat protein LIN-23 functions to limit cell division during development. *Development* 2000, 127:5071-5082.

[81] Harris WA, Holt CE, Bonhoeffer F: Retinal axons with and without their somata, growing to and arborizing in the tectum of Xenopus embryos: a time-lapse video study of single fibres in vivo. *Development*1987, 101:123-133.

[82] Lohof AM, Quillan M, Dan Y, Poo MM: Asymmetric modulation of cytosolic cAMP activity induces growth cone turning. *J Neurosci* 1992, 12:1253-1261.

[83] Luo Y, Raible D, Raper JA: Collapsin: a protein in brain that induces the collapse and paralysis of neuronal growth cones. *Cell* 1993, 75:217-227.

[84] Campbell DS, Regan AG, Lopez JS, Tannahill D, Harris WA, Holt CE: Semaphorin 3A elicits stage-dependent collapse, turning, and branching in Xenopus retinal growth cones. *J Neurosci* 2001, 21:8538-8547.

[85] Fujimuro M, Sawada H, Yokosawa H: Production and characterization of monoclonal antibodies specific to multi-ubiquitin chains of polyubiquitinated proteins. *FEBS Lett* 1994, 349:173-180.

[86] Everett RD: ICP0 induces the accumulation of colocalizing conjugated ubiquitin. *J Virol* 2000, 74:9994-10005.

[87] Brittis PA, Lu Q, Flanagan JG: Axonal protein synthesis provides a mechanism for localized regulation at an intermediate target. *Cell* 2002, 110:223-235.

[88] Ming GL, Wong ST, Henley J, Yuan XB, Song HJ, Spitzer NC, Poo MM: Adaptation in the chemotactic guidance of nerve growth cones. *Nature* 2002, 417:411-418.

[89] Campbell DS, Holt CE: Apoptotic pathway and MAPKs differentially regulate chemotropic responses of retinal growth cones. *Neuron* 2003, 37:939-952.

[90] Salvesen GS, Duckett CS: Apoptosis: IAP proteins: blocking the road to death's door. *Nat Rev Mol Cell Biol* 2002, 3:401-410.

[91] Suzuki Y, Nakabayashi Y, Takahashi R: Ubiquitin-protein ligase activity of X-linked inhibitor of apoptosis protein promotes proteasomal degradation of caspase-3 and enhances its anti-apoptotic effect in Fas-induced cell death. *Proc Natl Acad Sci U S A* 2001, 98:8662-8667.

[92] Piper M, Salih S, Weinl C, Holt CE, Harris WA: Endocytosis-dependent desensitization and protein synthesis-dependent resensitization in retinal growth cone adaptation. *Nat Neurosci* 2005, 8:179-186.

[93] Hu G, Zhang S, Vidal M, Baer JL, Xu T, Fearon ER: Mammalian homologs of seven in absentia regulate DCC via the ubiquitin- proteasome pathway. *Genes Dev* 1997, 11:2701-2714.

[94] Shirasaki R, Katsumata R, Murakami F: Change in chemoattractant responsiveness of developing axons at an intermediate target[see comments]. *Science* 1998, 279:105-107.

[95] Kidd T, Brose K, Mitchell KJ, Fetter RD, Tessier-Lavigne M, Goodman CS, Tear G: Roundabout controls axon crossing of the CNS midline and defines a novel subfamily of evolutionarily conserved guidance receptors. *Cell* 1998, 92:205-215.

[96] Kidd T, Russell C, Goodman CS, Tear G: Dosage-sensitive and complementary functions of roundabout and commissureless control axon crossing of the CNS midline. *Neuron* 1998, 20:25-33.

[97] Seeger M, Tear G, Ferres-Marco D, Goodman CS: Mutations affecting growth cone guidance in Drosophila: genes necessary for guidance toward or away from the midline. *Neuron* 1993, 10:409-426.

[98] Tear G, Harris R, Sutaria S, Kilomanski K, Goodman CS, Seeger MA: commissureless controls growth cone guidance across the CNS midline in Drosophila and encodes a novel membrane protein. *Neuron* 1996, 16:501-514.

[99] Krull CE: Comm-ing across the midline. *Nat Neurosci* 2005, 8:131-132.

[100] Harvey KF, Kumar S: Nedd4-like proteins: an emerging family of ubiquitin-protein ligases implicated in diverse cellular functions. *Trends Cell Biol* 1999, 9:166-169.

[101] Hicke L: Gettin' down with ubiquitin: turning off cell-surface receptors, transporters and channels. *Trends Cell Biol* 1999, 9:107-112.

[102] Wolf B, Seeger MA, Chiba A: Commissureless endocytosis is correlated with initiation of neuromuscular synaptogenesis. *Development* 1998, 125:3853-3863.

[103] Keleman K, Ribeiro C, Dickson BJ: Comm function in commissural axon guidance: cell-autonomous sorting of Robo in vivo. *Nat Neurosci* 2005, 8:156-163.

[104] Truman JW: Metamorphosis of the central nervous system of Drosophila. *J Neurobiol* 1990, 21:1072-1084.

[105] Lee T, Lee A, Luo L: Development of the Drosophila mushroom bodies: sequential generation of three distinct types of neurons from a neuroblast. *Development* 1999, 126:4065-4076.

[106] Lee T, Marticke S, Sung C, Robinow S, Luo L: Cell-autonomous requirement of the USP/EcR-B ecdysone receptor for mushroom body neuronal remodeling in Drosophila. *Neuron* 2000, 28:807-818.

[107] Raff MC, Whitmore AV, Finn JT: Axonal self-destruction and neurodegeneration. *Science* 2002, 296:868-871.

[108] Zhai Q, Wang J, Kim A, Liu Q, Watts R, Hoopfer E, Mitchison T, Luo L, He Z: Involvement of the ubiquitin-proteasome system in the early stages of wallerian degeneration. *Neuron* 2003, 39:217-225.

[109] Verma P, Chierzi S, Codd AM, Campbell DS, Meyer RL, Holt CE, Fawcett JW: Axonal protein synthesis and degradation are necessary for efficient growth cone regeneration. *J Neurosci* 2005, 25:331-342.

[110] Broadie KS, Richmond JE: Establishing and sculpting the synapse in Drosophila and C. elegans. *Curr Opin Neurobiol* 2002, 12:491-498.

[111] Keshishian H, Broadie K, Chiba A, Bate M: The drosophila neuromuscular junction: a model system for studying synaptic development and function. *Annu Rev Neurosci* 1996, 19:545-575.

[112] Jin Y: Synaptogenesis: insights from worm and fly. *Curr Opin Neurobiol* 2002, 12:71-79.

[113] Petersen SA, Fetter RD, Noordermeer JN, Goodman CS, DiAntonio A: Genetic analysis of glutamate receptors in Drosophila reveals a retrograde signal regulating presynaptic transmitter release. *Neuron* 1997, 19:1237-1248.

[114] Wu C, Wairkar YP, Collins CA, DiAntonio A: Highwire function at the Drosophila neuromuscular junction: spatial, structural, and temporal requirements. *J Neurosci* 2005, 25:9557-9566.

[115] Zhen M, Huang X, Bamber B, Jin Y: Regulation of presynaptic terminal organization by C. elegans RPM-1, a putative guanine nucleotide exchanger with a RING-H2 finger domain. *Neuron* 2000, 26:331-343.

[116] Nakata K, Abrams B, Grill B, Goncharov A, Huang X, Chisholm AD, Jin Y: Regulation of a DLK-1 and p38 MAP kinase pathway by the ubiquitin ligase RPM-1 is required for presynaptic development. *Cell* 2005, 120:407-420.

[117] Liao EH, Hung W, Abrams B, Zhen M: An SCF-like ubiquitin ligase complex that controls presynaptic differentiation. *Nature* 2004, 430:345-350.

[118] Huang Y, Baker RT, Fischer-Vize JA: Control of cell fate by a deubiquitinating enzyme encoded by the fat facets gene. *Science* 1995, 270:1828-1831.

[119] Wu Z, Li Q, Fortini ME, Fischer JA: Genetic analysis of the role of the drosophila fat facets gene in the ubiquitin pathway. *Dev Genet* 1999, 25:312-320.

[120] Wendland B: Epsins: adaptors in endocytosis? *Nat Rev Mol Cell Biol* 2002, 3:971-977.

[121] Chen X, Zhang B, Fischer JA: A specific protein substrate for a deubiquitinating enzyme: Liquid facets is the substrate of Fat facets. *Genes Dev* 2002, 16:289-294.

[122] McCabe BD, Hom S, Aberle H, Fetter RD, Marques G, Haerry TE, Wan H, O'Connor MB, Goodman CS, Haghighi AP: Highwire regulates presynaptic BMP signaling essential for synaptic growth. *Neuron* 2004, 41:891-905.

[123] Marques G, Bao H, Haerry TE, Shimell MJ, Duchek P, Zhang B, O'Connor MB: The Drosophila BMP type II receptor Wishful Thinking regulates neuromuscular synapse morphology and function. *Neuron* 2002, 33:529-543.

[124] McCabe BD, Marques G, Haghighi AP, Fetter RD, Crotty ML, Haerry TE, Goodman CS, O'Connor MB: The BMP homolog Gbb provides a retrograde signal that regulates synaptic growth at the Drosophila neuromuscular junction. *Neuron* 2003, 39:241-254.

[125] Aberle H, Haghighi AP, Fetter RD, McCabe BD, Magalhaes TR, Goodman CS: wishful thinking encodes a BMP type II receptor that regulates synaptic growth in Drosophila. *Neuron* 2002, 33:545-558.

[126] Keshishian H, Kim YS: Orchestrating development and function: retrograde BMP signaling in the Drosophila nervous system. *Trends Neurosci* 2004, 27:143-147.

[127] Zhu H, Kavsak P, Abdollah S, Wrana JL, Thomsen GH: A SMAD ubiquitin ligase targets the BMP pathway and affects embryonic pattern formation. *Nature* 1999, 400:687-693.

[128] Godenschwege TA, Simpson JH, Shan X, Bashaw GJ, Goodman CS, Murphey RK: Ectopic expression in the giant fiber system of Drosophila reveals distinct roles for roundabout (Robo), Robo2, and Robo3 in dendritic guidance and synaptic connectivity. *J Neurosci* 2002, 22:3117-3129.

[129] Murphey RK, Froggett SJ, Caruccio P, Shan-Crofts X, Kitamoto T, Godenschwege TA: Targeted expression of shibire ts and semaphorin 1a reveals critical periods for synapse formation in the giant fiber of Drosophila. *Development* 2003, 130:3671-3682.

[130] Zhao Y, Hegde AN, Martin KC: The ubiquitin proteasome system functions as an inhibitory constraint on synaptic strengthening. *Curr Biol* 2003, 13:887-898.

[131] Speese SD, Trotta N, Rodesch CK, Aravamudan B, Broadie K: The ubiquitin proteasome system acutely regulates presynaptic protein turnover and synaptic efficacy. *Curr Biol* 2003, 13:899-910.

[132] Patrick GN, Bingol B, Weld HA, Schuman EM: Ubiquitin-mediated proteasome activity is required for agonist-induced endocytosis of GluRs. *Curr Biol* 2003, 13:2073-2081.

[133] Bingol B, Schuman EM: A proteasome-sensitive connection between PSD-95 and GluR1 endocytosis. *Neuropharmacology* 2004, 47:755-763.

[134] Bingol B, Schuman EM: Synaptic protein degradation by the ubiquitin proteasome system. *Curr Opin Neurobiol* 2005, 15:536-541.

In: The Ubiquitin Proteasome System... ISBN 978-1-60021-749-4
Eds: Mario Di Napoli and Cezary Wojcik, pp. 373-391

Chapter 16

GENERAL ASPECTS OF UBIQUITIN PROTEASOME SYSTEM IN THE MATURE CENTRAL NERVOUS SYSTEM

Lukasz P. Bialy[1,2,*] and Izabela Mlynarczuk-Bialy[2,¥]
[1]Institute for Biochemistry, Medical Faculty Charité – Universitätsmedizin Berlin, Berlin, Germany;
[2]Department of Histology and Embryology, Center of Biostructure Research, Medical University of Warsaw, Warsaw, Poland.

ABSTRACT

The central nervous system (CNS) is the most complex structure known, and understanding its function is considered the 'final frontier' of biology. Neurons and their supporting glial cells form the cellular building blocks of the CNS. Individually these cells express a vast array of proteins that receive and transmit information; collectively integrating and processing this information allowing us to make sense of our environment and modify our behaviors. The process of protein degradation and turnover is essential for physiology of neuronal cells. The proteasome-ubiquitin system (UPS) plays central role in cytosolic proteolysis. It regulates the distribution of cell cycle phases, gene expression, transcription and antigen processing. In the CNS the proteasome pathway has additional functions which are due to physiology of the CNS. Many neuronal specific proteins interact with the 20S proteasome, moreover, brain proteasomes are thought to have a different proteolytic profile than proteasomes from

* Correspondence concerning this article should be addressed to Dr. Lukasz Bialy, MD; Institute for Biochemistry, Medical Faculty Charité – Universitätsmedizin Berlin, Monbijoustr. 2, 10-117 Berlin, Germany. Phone: +49-30-450-528-288; Fax +49-30-450-528-921; E-mail: lbialy@esculap.pl.

¥ Correspondence concerning this article should be addressed to Dr. Izabela Mlynarczuk-Bialy, MD, PhD; Department of Histology and Embryology - Centre for Biostructure Research, Warsaw Medical University, Chałubińskiego 5, 02-004 Warsaw, Poland. Phone: +48-22-629-5282; Fax +48-22-629-5282; E-mail: izamlynar@esculap.pl.

other tissues. Proteasomes are present ubiquitously, but not homogenously, in CNS cells (both in neurons and in glia). The subcellular localization of proteasomes differs in various parts of the CNS. Both neurons and glia are able to induce immunosubunits of proteasome under inflammatory conditions. UPS also shapes the function, development, and plasticity of synaptic connections and UPS elements are recruited in post-synaptic densities (PDS). The UPS is also essential for Long-Term Facilitation and is required for the establishment of long-term memory, moreover the blockade of UPS produces full retrograde amnesia in rat. The knowledge of the UPS in particular areas of the CNS and of the function of this system in every neuronal cell, enables to draft a more complex vision on protein turnover in the CNS, but details of all the processes remain to be discovered.

Keywords: proteasome, ubiquitin, central nervous system, synapse, glia, dendrites, NF-κB, memory.

Abbreviations

AMPA, α-amino-3-hydroxy-5-methyl-4-izoxazolepropionic acid; βAPP, beta-amyloid precursor protein; BrAAP, branched chain amino acid-preferring activity; CNS, central nervous system; GABA, γ-amino butyric acid; GLR1, glutamate receptor 1; GlyR, glycine receptor; IAP, inhibitor of apoptosis protein; IFN, interferon; IκB, inhibitor of NF-κB; IL-6, interleukin 6; LMP7, low molecular protein 7; LTD, long-term depression; LTF, long term facilitation; MECL-1, multicatalytic endopeptidase complex-like-1; MHC, major histocompatibility complex; MnSOD, manganese superoxide dismutase; NF-κB, Nuclear Factor κB; NGF, nerve growth factor; NMDA, n-methyl-D-aspartate; PDS, post-synaptic densities; PKA, protein kinase A; SNAAP, small neutral amino acid-preferring activity; SPRC, synapse associated polyribosome complex; TNF, Tumor Necrosis Factor; UIMs, ubiquitin-interacting motifs; UPS, protesome-ubiquitin system; USP, ubiquitin specific protease.

Introduction

The central nervous system (CNS) consists of the encephalon and the spinal cord. The CNS is the most complex structure in the human body and is formed by more than 100 million nerve cells (neurons) assisted by many more glial cells. Each neuron has multiple interconnections with other neurons, forming a very complex system for communication [1].

For right and proper function of nerve tissue and the CNS, precise regulation of protein synthesis and degradation is indispensable. In order to shape the development, function and plasticity of neuronal cells, the nervous system has adapted the power of the ubiquitin-proteasome system (UPS). The proteasome is a multicatalytic proteinase complex, which degrades most of the cytosolic proteins. Proteins tagged for degradation are associated with polyubiquitin chain in process named as ubiquitinylation. A cascade of E1 (ubiquitin-activating), E2 (ubiquitin-conjugating), and E3 (ubiquitin-ligases) enzymes operates the process of ubiquitynylation and specialized deubiquitynating enzymes detach ubiquitin from the substrate (compare with Chapter 3). However in face of the tremendous progress made in proteasome research the characterization of proteasome system in the CNS is only beginning to be elucidated. This text focuses on the role of the UPS in physiological function of neurons, nerve tissue and the CNS. In order to better understand the particular functions of the UPS in the CNS also some general data on structure and physiology of nerve tissue and the CNS are provided.

Histological Localization of Components of the Ubiquitin-Proteasome System in the Brain

Structurally, nerve tissue consists of two cell types: neurons and several types of glial cells, which support and protect neurons, and participate in neural nutrition and the defense processes of the central nervous system.

Neurons are responsible for reception, transmission, and processing of stimuli; triggering of certain cell activities; and release of neurotransmitters. The majority of neurons consist of 3 parts: the dendrites, which are processes receiving stimuli from the environment, sensory epithelial cells or other neurons, the cell body (perikaryon), which is the trophic center for the whole neuronal cell, containing the nucleus and surrounding cytoplasm with a highly developed rough cytoplasmic reticulum with numerous polyribosomes named Nissl bodies (or tigroid), and the axon which is a single, often very long process specialized in generating or conducting nerve impulses to other cells [2].

Glial cells, which are also present in the CNS represent three type of cells: oligodendrocytes, which participate in formation of myelin, astrocytes, which are dispread in the whole CNS and surround small blood vessels playing a role in forming the blood-brain barrier, and finally microglia, which participate in immunity processes in the CNS [2].

On sections, the CNS consists of white matter and gray matter. Gray matter is prevalent at the surface of the cerebrum and cerebellum forming the cerebral and cerebellal cortex respectively. Gray matter contains neuronal cell bodies, dendrites and the unmyelinated portions of axons as well as numerous glial cells. This is the region where synapses are localized. White matter is present in more central regions of the encephalon and contains myelinated axons and the myelin producing oligodendrocytes. Islands of gray matter localized in the white matter are called nuclei and contain cell bodies [2].

Taking into account that the proteasome function is essential for existence of each particular cell, there is no wonder that immunohistochemical studies have revealed the presence of 20S proteasomes in all structures of the central nervous system [3] (Figure 1). It has been shown by means of immunohistochemistry that the pyramidal neocortical neurons and the motor neurons in the ventral horn of spinal cord show the highest proteasome immunoreactivity within the whole CNS [4]. However the localization of the proteasome varies in the different neuronal structures [4]. In contrast to non-neuronal cells, many of neurons show high nuclear proteasomal immunoreactivity but also dendrites and axons, including synaptic terminals, contain proteasomes, indicating the role of localized proteolysis in their function [4], [5]. Also, the localization of 19S regulatory cap of 26S proteasome

(PA700) correlates to that showed for 20S proteasome [6], but also there are some subcellular divergences in localization of particular subunits of 19S cap [5].

Within the forebrain, the neocortex of cerebral hemispheres shows uniform proteasome immunoreactivity pattern with most intensive signal in pyramidal neurons of the 5th layer. In contrast to neocortex, in the paleocortex neurons of the 2nd layer are predominantly immunoreactive positive while the rest of the tissue is weakly stained. Also pyramidal cells of Amon's horn and granule cells of the dentate gyrus in the hippocampal formation as well as Purkinje cells in the cerebellum are strongly labeled [4] (Figure 1).

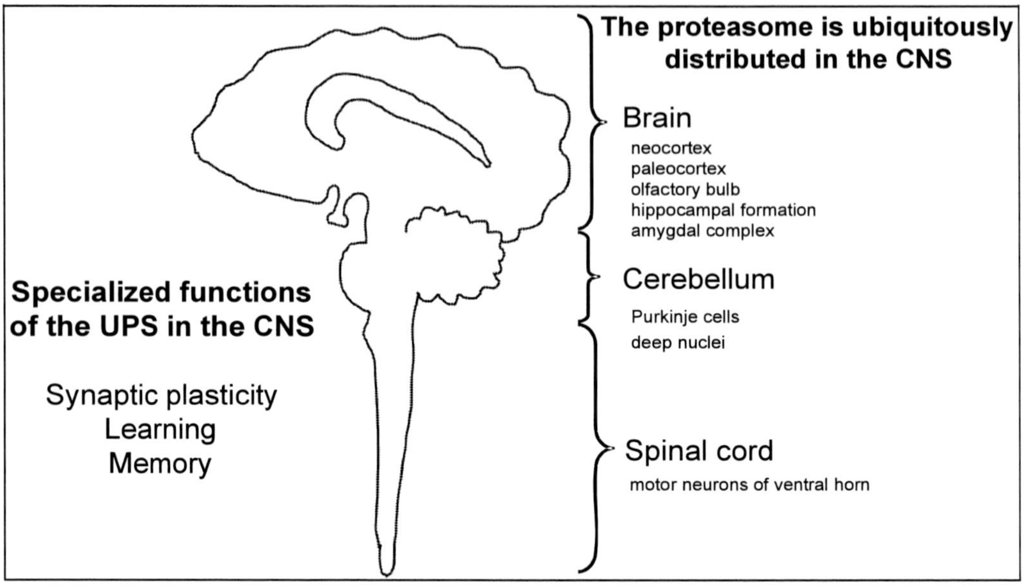

Figure 1. The CNS structures and fucnctions, in which UPS is involved. Besides of functions of the UPS common for all cells, like regulation of the cell cycle, apoptosis and gene expression in the CNS, the UPS serves also specialized functions due to its extraordinary role and physiology. This specialized functions of UPS in the CNS are regulation of synaptic plasticity, learning and memory. The proteasome system is common in all structures of the CNS (brain, cerebellum, spinal cord).

Despite the fact of so ubiquitous presence of proteasomes in the brain, the subcellular localization of this protease complex is not so obvious. However, there is no doubt that proteasomes and ubiquitin are present in all parts of neurons including nuclei, cytoplasm of perikarya, dendrites, axons as well as synaptic bulbs [3-5].

Within the cells mentioned before only in pyramidal neurons of the 5th layer of neocortex proteasomes seem to be distributed homogeneously in both cytoplasm of perikarya and in nuclei. Similarly, the motor neurons in the ventral horn of the spinal cord show the same homogenous pattern of proteasomal distribution as the neurons of the 5th layer of the neocortex. The remaining neurons of the CNS display a predominant nuclear localization of proteasomes with only slight immunoreactivity in the cytoplasm of perikarya, however cytoplasm of both perikarya and neuronal extensions are always positive immunoreactive. The examples of this group of neurons are neurons in: the 2nd to 4th and 6th layer of the neocortex, the piriform cortex and the olfactory bulb, the amygdala complex, the hippocampal formation, the paraventricular thalamic nucleus and Purkinje cells in the cerebellum. There is also a small group of neurons in the mesencephalon and the brain stem that display moderate presence of proteasomes in cytoplasm of perikarya eg.: the lateral vestibular nucleus, the intermediate gray layer of the superior colliculus, the deep nuclei of the cerebellum, the motor nuclei of the brain stem and the reticular formation [4]. But the nuclear prevalence of proteasomes within neurons in CNS is not always confirmable [3]. In various parts of the brain there can be found neurons with high cytoplasmatic staining for 20S proteasome frequently with intense perinuclear immunoreactive positive rings [5]. However, taking into account the multiplicity of possibilities of physiological states that neurons are able to reach it is conceivable that localization of proteasomes varies from one physiological stage to another e.g. as it varies in different cell cycle phases in HeLa cells [7]. Moreover, immunohistochemial data usually represent only static and momentary proteasome distribution thus it is no wonder that various studies of localization of 20S proteasomes in the brain, even made with standardized protocol using the same antibody, can provide different results of subcellular distribution, depending on different experimental conditions (Bialy LP and Wilczynski GM unpublished data).

Recently published data by the group of Laszlo gives more information about subcellular localization of 19S regulator of proteasome and ubiquitin in human (postmortem) and rat brain [5]. The authors found that ubiquitin and two ATP-ases subunits of 19S regulator (S4 and S7) are also predominantly localized in nuclei of neurons and glial cells in both human and rat brain. In case of S8 subunit of 19S regulator neurons within rat brain showed prominent nuclear immunostaining in contrast to predominantly cytoplasmic localization of S8 found in human brains. Other study of 19S subunits (S6a, S6b, S8, and S10b) revealed predominantly cytoplasmic and neuritic localization in neurons in both rat and human brain. The subcellular localization of 19S subunits and ubiquitin were rather uniform in all examined brain areas in all kinds of neuron populations. Additionaly, a subset of nuclear bodies that were strongly immunoreactive for the components of UPS was found in neurons and glial cells in rat brain. These UPS-immunopositive nuclear bodies show co-localization of 20S proteasome and ubiquitin and often are found at the periphery of the nucleolus however, the nucleoli themselves are not immunostained [5].

The Function of the UPS in the Action of the Chemical Synapse

Neurons are grouped in circuits similar to electronic circuits that are combinations of elements that form systems of various sizes and complexities. The distal part of the axon is usually branched and constitutes the terminal arborization. At distal part of each arborization, branch end bulbs are located, which interact with other neurons or non-nerve cells forming the chemical synapse [2]. Chemical synapses are specialized connections between neuronal cells by which the cells communicate to each other. Formation of synapses is a condition for neuronal survival [8]. Apart from chemical synapses in CNS also another type of synapse occurs – the electrical synapse [9]. In this chapter the function of the UPS is discussed in contest of action of the chemical synapse while the role of UPS in regulation of synaptic plasticity is reviewed in Chapter 18.

Chemical synapses transmit information to the next cell in the circuit and they are sites of functional contact between neurons or between neurons and other effector cells. At the synapse an electrical signal is converted from the presynaptic cell into a chemical signal that acts on the postsynaptic cell. The chemical synapses transmit information by releasing neurotransmitters during the signaling process. The change of electrical potential in the presynaptic cell triggers it to release neurotransmitter, which is stored in membrane-bound synaptic vesicles and is released by exocytosis. The dynamic process of chemical signal transduction requires quick protein production, degradation and turnover in the pre-synaptic and post-synaptic parts of neurons [10].

The neurotransmitter rapidly diffuses across the synaptic cleft and provokes the change in electrical potential of the postsynaptic cell by binding to transmitter gated ion channels. The main stimulating neurotransmitter of the CNS is glutamate acid that can bind to different types of glutamate receptors eg. n-Methyl-D-Aspartate type receptor (NDMA receptor) or α-amino-3-hydroxy-5-methyl-4-isoxazolepropionic acid type receptor (AMPA receptor) [10, 11]. After the neurotransmitter has been secreted, it is rapidly removed, either by specific enzymes in the synaptic cleft or by re-uptake (endocytosis) either by the terminal of releasing nerve or by surrounding glial cells. Rapid removal of neurotransmitter ensures spatial and temporal precision of signaling at a synapse: synaptic cleft is cleared before the next pulse of neurotransmitter is released and surrounding cells are prevented from the action of neurotransmitter [10]. Many molecules and mechanisms participate in the function of a synapse. The most important regulatory systems include, but are not limited to, the level of intracellular calcium, the activity of protein kinases and phosphatases, and control of gene transcription and protein translation [12]. The molecular architecture of the post synaptic terminal is highly dynamic and is composed of cytoskeletal elements, scaffolding proteins and cell surface receptors [13-15]. For example glutamate receptors at the postsynaptic membrane are linked via scaffolding molecules to a host of signaling and regulatory proteins [16]. This organization of postsynaptic proteins is termed the postsynaptic density (PSD) and it is an essential element of the synaptic plasticity (see Chapter 18).

The UPS has been shown to be involved in degradation of neuronal proteins playing a key role in signal transduction by the synapse. The UPS participates in the regulation of both short- and long- term facilitation by pre- and post-synaptic mechanisms (Figures 2 and 3).

It has been demonstrated that inhibition of the proteasome at presynaptic level results in enhanced synaptic transmission by accumulation of the essential synaptic vesicle-priming protein Dunc-13 in presynaptic buttons. The modulation of the synaptic transmission occurs within tens of minutes after the addition of proteasome inhibitor and causes approximately 50% increase in the amplitude of evoked potential [16]. Also other proteins forming the neurotransmitter releasing machinery are targets for ubiquitin dependent proteolysis. Syntaxin 1 is ubiquitinated by E3 ligase named Staring (syntaxin 1-interacting RING finger protein), that is expressed within the brain in both cytosolic and membrane fractions. Staring associates with the E2 enzyme UbcH8, which is abundant in the brain, and facilitates proteasome-dependent degradation of syntaxin1 [17]. Another pre-synaptic substrate for UPS involved in action of chemical synapse is synaptophisin, which is ubiquitinated by Siah (ubiquitin ligase of RING proteins) E3 ligases [18], (Figure 2).

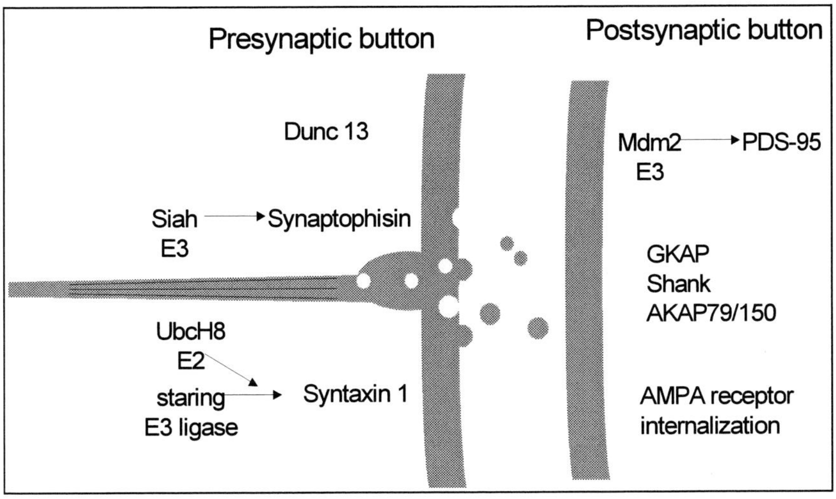

Figure 2. The UPS is involved in regulation of synaptic transmission and plasticity at both presynatic button and postsynaptic process of the given nerve cell. Some proteins are identified and known to be degraded by the proteasome. For some also E2 and E3 enzymes are known. This figure presents some examples of such proteins and their post- or pre-synaptic localization (see the text for details).

The UPS modulates also the postsynaptic density (PDS), while remodeling of the PDS is required for the plasticity of the brain and formation of long term memory. The major scaffolding protein of the postsynaptic density PDS-95 is a substrate for ubiquitinanylation by the E3 ligase Mdm2 [19]. The other scaffold proteins of the PDS shown to be ubiquitinylated are: GKAP, Shank and AKAP79/150 [20]. The ubiquitinylation of PDS-95 protein takes place as an answer to the activation of NMDA receptors. After ubiquitinylation, PDS-95 protein is rapidly removed from synaptic site by proteasome dependent degradation. This phenomenon is a part of a rapid mechanism for postsynaptic regulation of neurotransmitter receptors distribution. Mutations that prevent PDS-95 ubiquitinylation, as well as, proteasome inhibitors, block NMDA inducible internalization of AMPA-type glutamate receptors that give rise to proteasome inhibitors inducible long-term depression of the synapse [19]. Also AMPA inducible AMPA receptor internalization is stopped by proteasome inhibition [21]. It has been shown in *C. elegans* that glutamate receptor 1 (GLR1), which is a homologue of mammalian GluR1 subunit of AMPA glutamate receptor is ubiquitinylated and internalized in ubiquitin dependent mechanism [22]. There are also suggestions that AMPA receptor subunits are ubiquitinated also in mammalian brain [23].

It is known that ubiquitinylation serves not only as a signal for proteasome-dependent degradation but also plays a role in protein addressing to lysosomes [24]. In contrast to proteasome dependent proteolysis, proteins that are addressed to lysosomes are usually mono-ubiquitinylated [25]. This mechanism is involved in internalization and recycling of varieties of surface cellular receptors, by interaction with adaptor proteins of clathrin-coated pits [26]. The others neurotransmitter receptors that have been shown to be ubiquitinylated and thereafter internalized in ubiquitin dependent way are: the inhibitory glycine receptor (GlyR) [27] and the γ-aminobutyric acid A GABA(A) receptor [28]. Moreover, glycine receptor has been shown to be degraded by lysosomes [27], while in proteolysis of GABA(A) receptor both pathways (lysosomal and proteasomal) seem to be involved [29] (Figure 3).

In addition, another receptor known to be endocytosed in ubiquitin dependent manner is the Robo receptor, which regulates outgrowth of axons in *Drosophila.* The internalization of Robo receptor depends on ubiquitinylation of Comm (comisurreless) protein by Nedd4 ubiquitin ligase [30]. Ubiquitinylated Comm protein is a signal for endocytosis of the Robo receptor. Unfortunately, a homolog of Comm protein has been not found in vertebrates, so we cannot predict the possible importance of these findings for the function of the central nervous system (see Chapter 15).

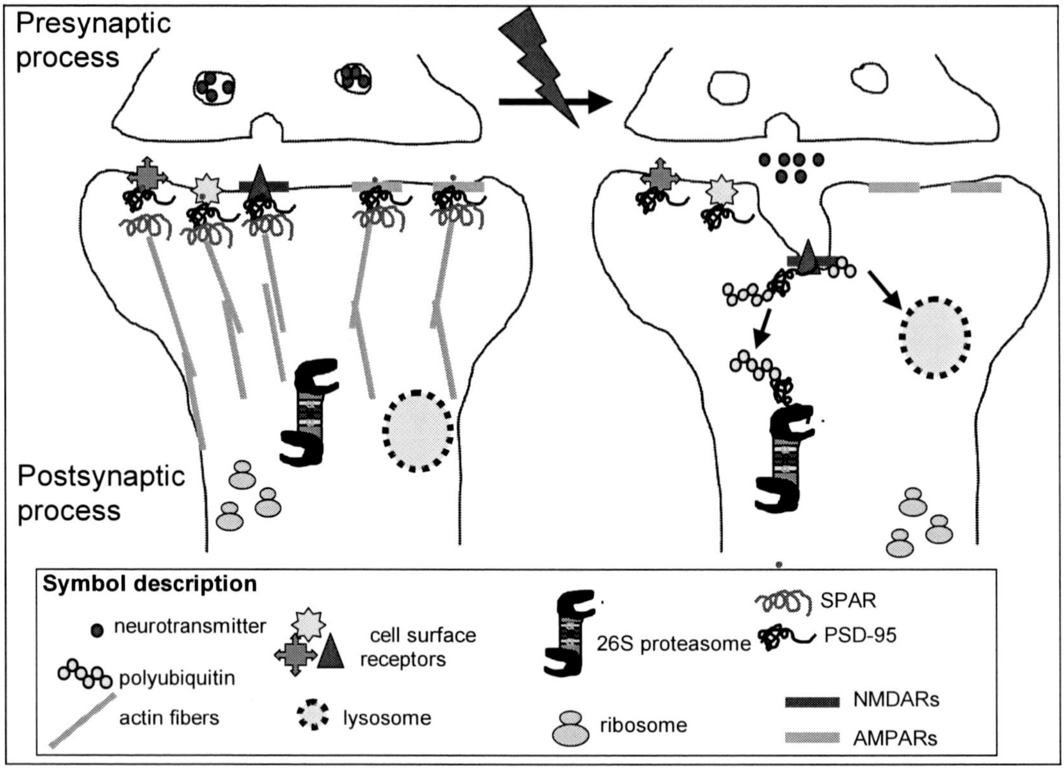

Figure 3. The involement of the UPS in the action of the synapse (For details see the text).

Deubiquitinylating Enzymes and long Term Memory – The Evidence of Compartmentalization of Ubiquitin Dependent Protein Degradation in Neurons

The majority of proteins both for structural and for transport purposes are synthesized within the perikaryon, which contains large nucleus and a highly developed rough cytoplasmic reticulum with numerous polyribosomes (Nissl substance, tigroid) [2]. However, local synthesis of proteins within dendrites and axons has been also identified [31]. The delivery of mRNA to particular subcellular domains, where translation is locally controlled plays a pivotal role in synaptic plasticity. The machinery of local translation in dendrites and axons consist mainly of synapse associated polyribosome complexes (SPRCs), mRNA and other regulatory elements. Electron microscopic analysis revealed that most of the polyribosomes in dendrites and axons are selectively positioned close to postsynaptic membrane [32]. The formation of long-term memory requires de novo protein synthesis both in the local area of synapse as well as in the perikaryon [31, 33]. The compartmentalization of protein synthesis in neurons is followed by compartmentalization of ubiquitin dependent proteolysis.

However, not only ubiquitin and proteasome but also deubiquitinating enzymes (see Chapter 4) participate in the switch from short-term to long-term memory. It has been shown that neuronal specific ubiquitin c-terminal hydrolase (Ap-uch) is involved in long-term facilitation (LTF) in *Aplysia*, by enhancing the recycling of ubiquitin in the presence of the proteasome [34]. In contrast, Ap-uch activity is not necessary for short term facilitation, since blocking of Ap-Uch completely prevented induction of long term but not short term facilitation. The target for the UPS during the formation of the long term facilitation (LTF) in *Aplysia* is the R (regulatory) subunit of cAMP-dependent protein kinase A (PKA) [35]. Ubiquitin dependent degradation of R subunits of PKA is the answer to repeated serotonin stimulation of the synapse and results in continuous activation of PKA that promotes a long-lasting increase in synaptic strength and long-term memory storage. Moreover, repeated setotonin stimulation was also found to induce Ap-uch enzyme [35].

The ubiquitin-proteasome-mediated proteolysis is required for long-term memory not only in *Aplysia* but also in mammals. Bilateral infusion of proteasome inhibitors (lactacystin or PSI) to the CA1 region of hippocampus results in total retrograde amnesia in rats for a one-trial avoidance learning when given up to 7 hours after training [36]. In addition, within 4 hours of training the total rate of ubiquitinylation increases and the levels of IκB (a substrate of the ubiquitin-proteasome cascade) decreases in hippocampus followed by stimulation of 26S activity [36].

There is also clear evidence that Vesl-1S/Homer-1a proteins, which are PDS proteins up regulated during long-term potentiation or seizure, are rapidly degraded in ubiquitin-proteasome dependent manner. After proteasome inhibition Vesl-1S/Homer-1a proteins are selectively sorted to the postsynaptic regions [37], where they interact with glutamate receptors [38], forming complexes with other scaffolding proteins of postsynaptic density e.g. with PDS-95 protein [39]. This is another example confirming the importance of function of the UPS in the formation of long term memory and action of the synapse.

Various studies reveal different effect of proteasome inhibition on formation of long term memory. This is the consequence of compartmentalization of proteasome dependent proteolysis and of the different function of the UPS in particular parts of neuron. No wonder that diverse (even opposite) physiological response to inhibition of proteasome in neuronal cell can be observed.

As it has been mentioned before, Lopez-Salon *et al.* [36] have shown that proteasome inhibitors prevent formation of long term memory in mammals. However, in some conditions, inhibition of the proteasome has been shown to increase the synaptic strength in *Aplysia* neurons and to potentate serotonin-dependent long term facilitation [40], suggesting that ubiquitin-proteasome dependent proteolysis influences synaptic strengthening by multiple mechanisms.

Also in mammals there is a strong evidence for involvement of deubiquitinating machinery in generation of long term memory. The ubiquitin-specific protease synUSP has been found to be associated with the PDS in rat. Moreover, synUSP seems to be translated locally and its mRNA has been identified in postsynaptic bulb. SynUSP protein is localized within dendrites as well as in the perikaryon but predominantly at post-synaptic density and post-synaptic lipid rafts [41]. This finding gives the basis to postulate that synUSP in mammalian might play a similar role to Ap-uch in Aplysia.

To emphasize the possible involvement of the ubiquitin-proteasome system in generation of long term memory in humans it is worth to mention that one of ubiquitin ligases (E6-AP) is a crucial protein in pathogenesis of human Angelman's syndrome and mutation in E6-AP in humans with Angelman's syndrome combines with mental retardation [42] (see Chapter 38). Moreover, mice carrying the mutation in E6-AP ubiquitin ligase show defects in long-term potentiation in their hyppocampii [43].

Glia and Immunoproteasomes

Glial cells fulfill supporting and nutrition function for neuronal cells. There are numerous glial cells in the CNS which number excess 10-50 times the number of neuronal cells. In the CNS there are three types of glial cells: microglia, which participate in defense processes in the CNS; oligodendrocytes, which participate in formation of myelin; and astrocytes, which are dispread in the whole CNS and surround blood vessels [2].

For a long time it was widely accepted that the brain is immunologically privileged because of: (*i*) the blood-brain barrier, (*ii*) lack of expression of the MHC molecules and (*iii*) lack of any immunocompetent cells [44]. Now it is known that also the CNS is able to operate immunological mechanisms, including MHC class I molecules expression (for details, see Chapter 34) [44]. The main cells sharing the function in immunological mechanisms of the CNS are microglia. Microglia derive from circulating monocytes of the monocyte-macrophage series that invade the brain during embryonic and early postnatal life [45]. Their phagocytic capacity is exercised during the brain maturation, remodeling, and injury.

In the CNS, non stimulated microglia are present as ramified cells that have small cell bodies with numerous branching processes. In response to neuronal injury and infections, ramified microglia rapidly transform into activated states [46]. Activated microglia show upregulation of the many marker antigens they share with circulating monocytes and antigen presenting cells, including the major histocompatibility (MHC) class II molecules [44, 47].

Activation of microglia is generally regarded as a double-edged response because activated microglia play cytotoxic and protective roles depending on the extent of neuronal cell injury. Microglia are the most efficient antigen presenting cells within the central nervous system [47]. Moreover, it has been suggested that microglia are involved in MHC class I-mediated antigen presentation during viral infection [48, 49].

The MHC class I antigen presentation pathway requires the 20S proteasome and two types of activator proteins – 19S or/and PA28 to generate the antigenic peptides. The antigenic peptides are transported into the endoplasmic reticulum, where they are loaded on

MHC class I molecules. Then antigenic peptides-loaded MHC class I molecules are translocated to the cell surface to activate cytotoxic T lymphocytes [50]. In response to interferon-γ (INF-γ), three constitutive β subunits of the 20S proteasome are replaced by the immunoproteasome subunits (LMP2, LMP7, MECL-1). IFN-γ-induced subunit replacement implicates increase in the epitope generation capacity of the proteasome [50, 51].

The first observations by analysis of crude extracts of the brain tissue give rise to the conclusion that immunoproteasomes are absent within the brain *in toto* [52, 53]. However, later the presence of INF-γ-inducible subunits has been confirmed in proteasomes purified from in post mortem human brain tissue [54]. Nevertheless, the expression of immunoproteasome subunits within the brain *in toto* is two (LMP7) to three (LMP2, MECL-1) times lower than in kidneys under the same condition. This non-equal expression of various immunoproteasome subunits suggests the occurrence of various subpopulations of immunproteasomes with not complete the replacement of immunosubunits [55]. Despite the fact of the lower expression of the proteasome immunosubunits within the brain, INF-γ is able to induce synthesis of immunosubunits not only in glia but also in neurons [54]. Moreover, their induction can be enhanced under pathological conditions [56].

Stohwasser *et al.* [57] have characterized the dynamics of the immunoproteasome subunit replacement in primary murine microglia and in the microglia derived cell line BV-2 upon IFN-γ and lipopolysacharide (LPS) stimulation. They have found that both IFN-γ and LPS are able to induce immunoproteasome subunits in primary microglial cells, whereas BV-2 cells responded only to IFN-γ. These observations support the idea that microglia play a pivotal role in MHC class I-mediated antigen presentation and adapts its proteasomal subunit composition during viral and/or bacterial infections to gain an optimal epitope processing in the CNS.

Microglia cells isolated from adult rats show decrease in activity of both 20S and 26S proteasomes, while the total amount of proteasones is the same in comparison to newborn microglia. This phenomenon influences mainly degradation rates of short-lived proteins that are major substrates of MHC class I restricted epitopes [58].

Neuroproteasome – Hypothesis or Reality

Neuronal cell express various proteins that are absent from other cells. Some of these neuronal specific proteins are directly engaged in the action of the UPS. The best known examples of such proteins are parkin and ataxin-3. Parkin is a RING E3 ligase, whose mutations lead to juvenile onset Parkinson's disease. Parkin binds via its ubiquitin-like domain to the 26S proteasome subunit S5a [59]. Known proteins ubiquitylated by parkin include α-synuclein and those contributing directly to the formation of Lewy bodies [60]. Another protein known to interact with the proteasome and to bind ubiquitinylated proteins is ataxin-3 [61]. Mutations within ataxin-3 are associated with polyQ disease ataxias type 3. Ataxin-3 contains three ubiquitin-interacting motifs (UIMs), two of which reside N-terminal to and C-terminal to the polyQ domain [62]. Moreover, ataxin-3 also reveals the catalytic triad of amino acids characteristic for deubiquitinating enzymes - ubiquitin-specific cysteine proteases [63]. Thus, ataxin-3 is thought to be a deubiquitinating enzyme. Recent publications show that this deubiquitinating activity regulates ubiquitination of ataxin-3 itself [64].

Studies of ataxin-3 and parkin demonstrate that proteasomes may be assembled with different polyubiquitin-binding proteins, expressing different substrate specificity. It could be assumed that such assembling proteins like ataxin-3 or parkin might influence substrate specificity of the proteasome. Taking together these data some authors postulate the presence of the 'neuro-proteasome' [60]. This speculative 'neuro-proteasome' would be formed by 20S or 26S proteasome interacting with specific neuronal proteins, conferring distinct proteasome activities in neurons (see Chapter 18).

Johnston [60] postulates that hypothetical 'neuro-proteasome' analogical from the immunoproteasome is a result of evolutionary specialization of the proteasome in the CNS. However, it has to be said that there is no evidence of any neuronal specific α and β subunits of 20S and the neuro-specialization of proteasome could be due solely to different assembling proteins and perhaps to phosphorylation of proteasome subunits [60]. Nevertheless, there are some data suggesting the altered proteolytic properties of proteasome in the brain. In crude extracts of the brain in comparison to extracts from other tissues small neutral aminoacid (SNAAP) and branched chain aminoacids (BrAAP) hydrolysing activities of the proteasome are elevated [65], but contraposition of these data onto hypotethical 'neuro-proteasome' should be carried with a big caution. Furthermore, another big protease (≈105kDa) sharing also at least two of proteasomal activities (trypsin-like and chymotrypisn-like) has been identified in the brain [66].

Taking together all these data the hypothesis of 'neuro-proteasome' is so far only the speculative construction and should be confirmed by experimental findings.

Transcription Factor NF-κB – An Example of a Particular Protein Regulating by the UPS in the CNS

One of well characterized proteins that undergo regulation by the UPS is transcription factor NF-κB. NF-κB is widely known for its ubiquitous roles in inflammation and immune responses, as well as in control of cell division and apoptosis [67]. In the nervous system apparent from these functions, NF-κB plays distinctive roles like coordination of cellular response to injury or regulation of synaptic signaling. Receptor-linked NF-κB signal transduction pathways mediated for e.g. by TNF and Fas ligand, have been also described in neurons and glias [68]. However, in the CNS NF-κB can be also activated by additional neuron-specific signals like NGF [69]. NF-κB is a dimer composed most commonly of p50 and p65 subunits. Each NF-κB monomer contains Rel region, through which it can bind to DNA and dimerize. NF-κB complex is held in the cytoplasm inactive by IκB proteins which prevent DNA binding and maintain the complex in the cytoplasm. In addition, novel NF-κB subunits may be expressed in the nervous system including neuronal κB factor [70, 71]. Upon stimulation, NF-κB inhibitors become phosphorylated and ubiquitinated as well as rapidly degraded by the 26S proteasome. Released NF-κB complex undergoes translocation to the nucleus and influences the expression of a complex array of genes in the CNS (Figure 4), those in general serve important functions in cellular response to injury and in neuronal plasticity [72]. Functional NF-κB complexes are present in essentially all cell types in the nervous system including neurons, astrocytes, oligodendrocytes and microglia [71].

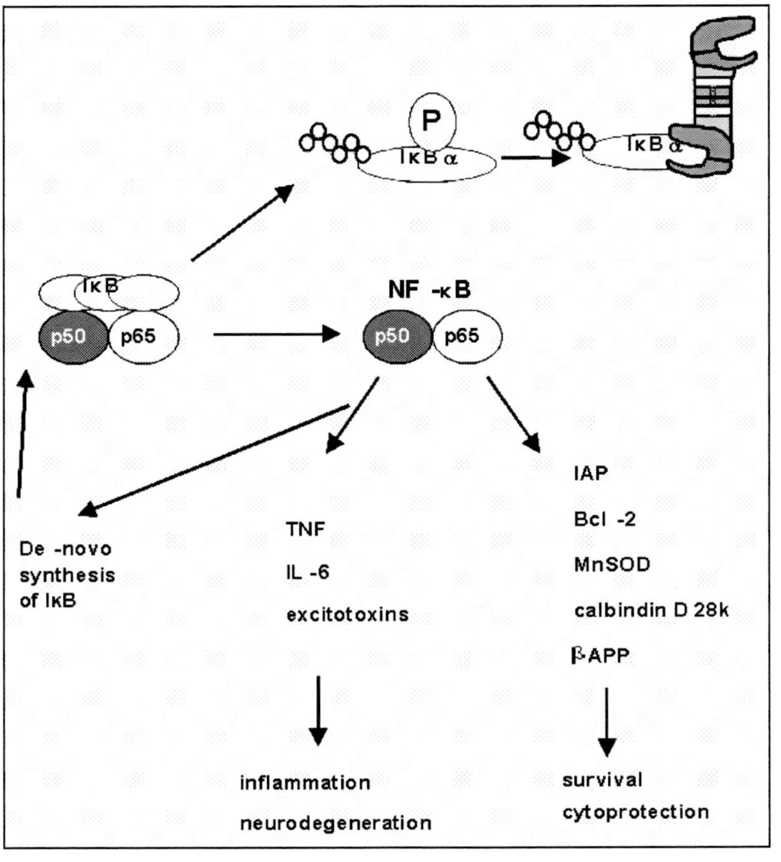

Figure 4. The UPS degrades IκB proteins thus enabling activation of NF-κB. Proteins which are trancriptionally activated by NF-κB can on one side be involved in neuronal damage, while on the other side have a cytoprotective function and promote cell survival.

Interestingly, electrical activity within neurons and synaptic transmission between neurons are potent stimuli for NF-κB activation and such neuronal activity may account for the relatively high constitutive activity of NF-kB in brain tissue compared with other tissues. Moreover, studies on neuronal excitation indicate that NF-κB might be regulated during synaptic transmission. Inducible NF-κB has been detected at presynaptic sites in the cortex, hippocampus and cerebellum [71]. This unusual localization suggests a novel synapse-to-nucleus signaling system in which transcriptional factors incited by local synaptic signals are able to activate transcriptional factors that are transmitted to the nucleus after retrograde transport [71]. To this end, experiments with p65 knockouts revealed that mice lacking p65 demonstrate spatial learning deficit [73] that is an evidence that NF-κB can regulate long-term memory. It has been shown that in mice lacking both TNF receptors after stimulation of axons of CA3 pyramidal neurons at a frequency of 1Hz there is no induction of long-term depression (LTD) [72]. When slices from wild type mice are pretreated with κB decoy DNA, LTD cannot be induced and the amplitude of LTP is significantly decreased [72]. These results provide direct evidence for involvement of NF-κB activation in synaptic plasticity. Moreover, the finding that NF-κB is activated by basal synaptic stimulation and that deletion of the p65 subunit leads to learning deficits, establishes a physiological role for this transcription factor in normal adult brain function [73]. However, some forms of learning such as those depending on the striatum are intact in p65 knockout mouse. Analysis of NF-κB binding and distribution in mouse hippocampus suggest that this transcription factor may serve both as a signaling molecule after its activation at the synapse and then as a transcription initiator upon reaching the nucleus [74].

Also glial cells undergo regulation by NF-κB, particularly in modulation of inflammatory response. Genes induced by NF-κB in glia and neurons encode among others receptor proteins, cell death and antitapoptotic agents, injury responsive factors as well as antioxidant enzymes. In particular NF-κB induces expression of TNF and IL-6 which play an important role in the onset, regulation, and propagation of immune and inflammatory responses within the CNS and that are produced in high amounts in microglia and astrocytes [75]. Proteins that protect cells against apoptosis like Bcl-2 and inhibitor of apoptosis proteins (IAPs) allow survival of neuronal cells in experimental models of stroke and seizures [75]. Finally, NF-κB regulates neuroprotective proteins like ßAPP (an injury responsive cytokine/neurotrophic factor) manganese superoxide dismutase (Mn-SOD) and calcium binding protein, calbindin-D 28k, which may play roles in modulatory calcium mediated neuronal signalling and cell death [75].

Transcription factor NF-κB is an example of very complex assembly exhibiting various effects in the CNS that is regulated and controlled among others by the UPS (see also Chapter 19). Inhibition of the proteasome can exhibit a modulatory effect on the function of CNS and memory formation mediated by NF-kB. It could cause also a positive effect in inflammatory and neurodegenerative processes by inhibiting production of proinflammatory cytokines, neurotoxic reactive oxygen species and excitotoxins by microglia.

CONCLUSIONS

To conclude it must be said that the UPS plays a crucial role in the function of the central nervous system. Due to the physiological differences of neurons the UPS serves also in specialized functions of neurons. Apart from functions known from all others cells, like regulation of the cell cycle [76], apoptosis [77] and gene expression [78], which are fulfilled by the UPS also in the neuronal cells and in the whole CNS, the UPS is involved in many neuronal-specific processes. These specialized functions of UPS in the CNS are regulation of synaptic transmission, synaptic plasticity, learning and memory. Moreover, some proteins that are expressed in non-neuronal cells (eg. NF-κB) in neurons and their neighboring cells are employed also for distinctive functions like synaptic signaling, learning and memory [75]. The UPS acts ubiquitously in all structures of the CNS as well as the proteasomal complexes were found in all neuronal cells. However, despite the fact of so ubiquitous presence of the UPS in the CNS, the subcellular localization of proteasome complexes is not so obvious and large gaps remain in our understanding of the mechanisms and consequences of UPS activity in the CNS. Many studies should be done to better characterize the involvement of the UPS in particular processes in the CNS. This knowledge is indispensable among others for prediction of the results of therapeutic intervention in the UPS in the CNS.

REFERENCES

[1] Laughlin SB, Sejnowski TJ: Communication in neuronal networks. *Science* 2003, 301:1870-1874.

[2] Stevens A, Lowe J: Nerve tissue, In Human Histology. Edited by Stevens A, Lowe J. London, New York: Gower Medical Publishing; 1992:206-225.

[3] Kamakura K, Ishiura S, Nonaka I, Sugita H: Localization of ingensin in rat central nervous system and skeletal muscle. *J Neurosci Res* 1988, 20:473-478.

[4] Mengual E, Arizti P, Rodrigo J, Gimenez-Amaya JM, Castano JG: Immunohistochemical distribution and electron microscopic subcellular localization of the proteasome in the rat CNS. *J Neurosci* 1996, 16:6331-6341.

[5] Adori C, Low P, Georgij M, Bagdy G, Laszlo L, Kovacs GG: Subcellular distribution of components of the ubiquitin-proteasome-system in non-diseased human and rat brain. *J Histochem Cytochem* 2006, 54:263-267.

[6] Nakamura T, Tanaka T, Nagano T, Yoneda T, Takagi H, Sato M: Distribution of mRNA encoding Tat-binding protein-1 (TBP-1), a component of 26S proteasome, in the rat brain. *Brain Res Mol Brain Res* 1998, 53:321-327.

[7] Wojcik C, Paweletz N, Schroeter D: Localization of proteasomal antigens during different phases of the cell cycle in HeLa cells. *Eur J Cell Biol* 1995, 68:191-198.

[8] Waites CL, Craig AM, Garner CC: Mechanisms of vertebrate synaptogenesis. *Annu Rev Neurosci* 2005, 28:251-274.

[9] Rozental R, Giaume C, Spray DC: Gap junctions in the nervous system. *Brain Res Brain Res Rev* 2000, 32:11-15.

[10] Castellucci VF, Robitaille R: The chemical synapse: mechanisms of transmission and modulation. *Can J Physiol Pharmacol* 1999, 77:631-633.

[11] Dingledine R, Borges K, Bowie D, Traynelis SF: The glutamate receptor ion channels. *Pharmacol Rev* 1999, 51:7-61.

[12] Wiersma-Meems R, Van Minnen J, Syed NI: Synapse formation and plasticity: the roles of local protein synthesis. *Neuroscientist* 2005, 11:228-237.

[13] Park SJ, Kim IB, Choi KR, Moon JI, Oh SJ, Chung JW, Chun MH: Reorganization of horizontal cell processes in the developing FVB/N mouse retina. *Cell Tissue Res* 2001, 306:341-346.

[14] Yamagata M, Sanes JR, Weiner JA: Synaptic adhesion molecules. *Curr Opin Cell Biol* 2003, 15:621-632.

[15] Luscher C, Nicoll RA, Malenka RC, Muller D: Synaptic plasticity and dynamic modulation of the postsynaptic membrane. *Nat Neurosci* 2000, 3:545-550.

[16] Sheng M: Molecular organization of the postsynaptic specialization. *Proc Natl Acad Sci U S A* 2001, 98:7058-7061.

[17] Chin LS, Vavalle JP, Li L: Staring, a novel E3 ubiquitin-protein ligase that targets syntaxin 1 for degradation. *J Biol Chem* 2002, 277:35071-35079.
[18] Wheeler TC, Chin LS, Li Y, Roudabush FL, Li L: Regulation of synaptophysin degradation by mammalian homologues of seven in absentia. *J Biol Chem* 2002, 277:10273-10282.
[19] Colledge M, Snyder EM, Crozier RA, Soderling JA, Jin Y, Langeberg LK, Lu H, Bear MF, Scott JD: Ubiquitination regulates PSD-95 degradation and AMPA receptor surface expression. *Neuron* 2003, 40:595-607.
[20] Ehlers MD: Activity level controls postsynaptic composition and signaling via the `ubiquitin-proteasome system. *Nat Neurosci* 2003, 6:231-242.
[21] Patrick GN, Bingol B, Weld HA, Schuman EM: Ubiquitin-mediated proteasome activity is required for agonist-induced endocytosis of GluRs. *Curr Biol* 2003, 13:2073-2081.
[22] Burbea M, Dreier L, Dittman JS, Grunwald ME, Kaplan JM: Ubiquitin and AP180 regulate the abundance of GLR-1 glutamate receptors at postsynaptic elements in C. elegans. *Neuron* 2002, 35:107-120.
[23] Hegde AN: Ubiquitin-proteasome-mediated local protein degradation and synaptic plasticity. *Prog Neurobiol* 2004, 73:311-357.
[24] Wojcik C: Ubiquitin-more than just a signal for protein degradation. *Trends Cell Biol* 2001, 11:397-399.
[25] Hicke L, Dunn R: Regulation of membrane protein transport by ubiquitin and ubiquitin-binding proteins. *Annu Rev Cell Dev Biol* 2003, 19:141-172.
[26] Rappoport JZ, Simon SM, Benmerah A. Understanding living clathin-coated pits. *Traffic* 2004, 5:327-337
[27] Buttner C, Sadtler S, Leyendecker A, Laube B, Griffon N, Betz H, Schmalzing G: Ubiquitination precedes internalization and proteolytic cleavage of plasma membrane-bound glycine receptors. *J Biol Chem* 2001, 276:42978-42985.
[28] Bedford FK, Kittler JT, Muller E, Thomas P, Uren JM, Merlo D, Wisden W, Triller A, Smart TG, Moss SJ: GABA(A) receptor cell surface number and subunit stability are regulated by the ubiquitin-like protein Plic-1. *Nat Neurosci* 2001, 4:908-916.
[29] Luscher B, Keller CA: Ubiquitination, proteasomes and GABA(A) receptors. *Nat Cell Biol* 2001, 3:E232-233.
[30] Myat A, Henry P, McCabe V, Flintoft L, Rotin D, Tear G: Drosophila Nedd4, a ubiquitin ligase, is recruited by Commissureless to control cell surface levels of the roundabout receptor. *Neuron* 2002, 35:447-459.
[31] Steward O, Schuman EM: Compartmentalized synthesis and degradation of proteins in neurons. *Neuron* 2003, 40:347-359.
[32] Steward O, Falk PM, Torre ER: Ultrastructural basis for gene expression at the synapse: synapse-associated polyribosome complexes. *J Neurocytol* 1996, 25:717-734.
[33] Konietzko U, Kauselmann G, Scafidi J, Staubli U, Mikkers H, Berns A, Schweizer M, Waltereit R, Kuhl D: Pim kinase expression is induced by LTP stimulation and required for the consolidation of enduring LTP. *Embo J* 1999, 18:3359-3369.

[34] Hegde AN, Inokuchi K, Pei W, Casadio A, Ghirardi M, Chain DG, Martin KC, Kandel ER, Schwartz JH: Ubiquitin C-terminal hydrolase is an immediate-early gene essential for long-term facilitation in Aplysia. *Cell* 1997, 89:115-126.

[35] Hegde AN, Goldberg AL, Schwartz JH: Regulatory subunits of cAMP-dependent protein kinases are degraded after conjugation to ubiquitin: a molecular mechanism underlying long-term synaptic plasticity. *Proc Natl Acad Sci U S A* 1993, 90:7436-7440.

[36] Lopez-Salon M, Alonso M, Vianna MR, Viola H, Mello e Souza T, Izquierdo I, Pasquini JM, Medina JH: The ubiquitin-proteasome cascade is required for mammalian long-term memory formation. *Eur J Neurosci* 2001, 14:1820-1826.

[37] Ageta H, Kato A, Fukazawa Y, Inokuchi K, Sugiyama H: Effects of proteasome inhibitors on the synaptic localization of Vesl-1S/Homer-1a proteins. *Brain Res Mol Brain Res 2001*, 97:186-189.

[38] Brakeman PR, Lanahan AA, O'Brien R, Roche K, Barnes CA, Huganir RL, Worley PF: Homer: a protein that selectively binds metabotropic glutamate receptors. *Nature* 1997, 386:284-288.

[39] Tu JC, Xiao B, Naisbitt S, Yuan JP, Petralia RS, Brakeman P, Doan A, Aakalu VK, Lanahan AA, Sheng M, Worley PF: Coupling of mGluR/Homer and PSD-95 complexes by the Shank family of postsynaptic density proteins. *Neuron* 1999, 23:583-592.

[40] Zhao Y, Hegde AN, Martin KC: The ubiquitin proteasome system functions as an inhibitory constraint on synaptic strengthening. *Curr Biol* 2003, 13:887-898.

[41] Tian QB, Okano A, Nakayama K, Miyazawa S, Endo S, Suzuki T: A novel ubiquitin-specific protease, synUSP, is localized at the post-synaptic density and post-synaptic lipid raft. *J Neurochem* 2003, 87:665-675.

[42] Kishino T, Lalande M, Wagstaff J: UBE3A/E6-AP mutations cause Angelman syndrome. *Nat Genet* 1997, 15:70-73.

[43] Jiang XY, Abrams TW: Use-dependent decline of paired-pulse facilitation at Aplysia sensory neuron synapses suggests a distinct vesicle pool or release mechanism. *J Neurosci* 1998, 18:10310-10319.

[44] Shrikant P, Benveniste EN: The central nervous system as an immunocompetent organ: role of glial cells in antigen presentation. *J Immunol* 1996, 157:1819-1822.

[45] Kaur C, Hao AJ, Wu CH, Ling EA: Origin of microglia. *Microsc Res Tech* 2001, 54:2-9.

[46] Cross AK, Woodroofe MN: Immunoregulation of microglial functional properties. *Microsc Res Tech* 2001, 54:10-17.

[47] Kim SU, de Vellis J: Microglia in health and disease. *J Neurosci Res* 2005, 81:302-313.

[48] Shrikant P, Benos DJ, Tang LP, Benveniste EN: HIV glycoprotein 120 enhances intercellular adhesion molecule-1 gene expression in glial cells. Involvement of Janus kinase/signal transducer and activator of transcription and protein kinase C signaling pathways. *J Immunol* 1996, 156:1307-1314.

[49] Kolson DL, Lavi E, Gonzalez-Scarano F: The effects of human immunodeficiency virus in the central nervous system. *Adv Virus Res* 1998, 50:1-47.

[50] Kloetzel PM: Antigen processing by the proteasome. *Nat Rev Mol Cell Biol* 2001, 2:179-187.

[51] Kloetzel PM: The proteasome and MHC class I antigen processing. *Biochim Biophys Acta* 2004, 1695:225-233.

[52] Fruh K, Gossen M, Wang K, Bujard H, Peterson PA, Yang Y: Displacement of housekeeping proteasome subunits by MHC-encoded LMPs: a newly discovered mechanism for modulating the multicatalytic proteinase complex. *Embo J* 1994, 13:3236-3244.

[53] Noda C, Tanahashi N, Shimbara N, Hendil KB, Tanaka K: Tissue distribution of constitutive proteasomes, immunoproteasomes, and PA28 in rats. *Biochem Biophys Res Commun* 2000, 277:348-354.

[54] Diaz-Hernandez M, Hernandez F, Martin-Aparicio E, Gomez-Ramos P, Moran MA, Castano JG, Ferrer I, Avila J, Lucas JJ: Neuronal induction of the immunoproteasome in Huntington's disease. *J Neurosci* 2003, 23:11653-11661.

[55] Dahlmann B, Ruppert T, Kloetzel PM, Kuehn L: Subtypes of 20S proteasomes from skeletal muscle. *Biochimie* 2001, 83:295-299.

[56] Diaz-Hernandez M, Martin-Aparicio E, Avila J, Hernandez F, Lucas JJ: Enhanced induction of the immunoproteasome by interferon gamma in neurons expressing mutant Huntingtin. *Neurotox Res* 2004, 6:463-468.

[57] Stohwasser R, Giesebrecht J, Kraft R, Muller EC, Hausler KG, Kettenmann H, Hanisch UK, Kloetzel PM: Biochemical analysis of proteasomes from mouse microglia: induction of immunoproteasomes by interferon-gamma and lipopolysaccharide. *Glia* 2000, 29:355-365.

[58] Stolzing A, Grune T: Impairment of protein homeostasis and decline of proteasome activity in microglial cells from adult Wistar rats. *J Neurosci Res* 2003, 71:264-271.

[59] Tanaka K, Suzuki T, Chiba T, Shimura H, Hattori N, Mizuno Y: Parkin is linked to the ubiquitin pathway. *J Mol Med* 2001, 79:482-494.

[60] Johnston JA, Madura K: Rings, chains and ladders: ubiquitin goes to work in the neuron. *Prog Neurobiol* 2004, 73:227-257.

[61] Chai Y, Berke SS, Cohen RE, Paulson HL: Poly-ubiquitin binding by the polyglutamine disease protein ataxin-3 links its normal function to protein surveillance pathways. *J Biol Chem* 2004, 279:3605-3611.

[62] Donaldson KM, Li W, Ching KA, Batalov S, Tsai CC, Joazeiro CA: Ubiquitin-mediated sequestration of normal cellular proteins into polyglutamine aggregates. *Proc Natl Acad Sci U S A* 2003, 100:8892-8897.

[63] Amerik AY, Hochstrasser M: Mechanism and function of deubiquitinating enzymes. *Biochim Biophys Acta* 2004, 1695:189-207.

[64] Berke SJ, Chai Y, Marrs GL, Wen H, Paulson HL: Defining the role of ubiquitin-interacting motifs in the polyglutamine disease protein, ataxin-3. *J Biol Chem* 2005, 280:32026-32034.

[65] Farout L, Lamare MC, Cardozo C, Harrisson M, Briand Y, Briand M: Distribution of proteasomes and of the five proteolytic activities in rat tissues. *Arch Biochem Biophys* 2000, 374:207-212.

[66] Vigouroux S, Furukawa Y, Farout L, S JK, Briand M, Briand Y: Peptidase activities of the 20/26S proteasome and a novel protease in human brain. *J Neurochem* 2003, 84:392-396.

[67] Li X, Stark GR: NFkappaB-dependent signaling pathways. *Exp Hematol* 2002, 30: 285-296

[68] Bruce AJ, Boling W, Kindy MS, Peschon J, Kraemer PJ, Carpenter MK, Holtsberg FW, Mattson MP: Altered neuronal and microglial responses to excitotoxic and ischemic brain injury in mice lacking TNF receptors. *Nat Med* 1996 2:788-794

[69] Carter BD, Kaltschmidt C, Kaltschmidt B, Offenhauser N, Bohm-Matthaei R, Baeuerle PA, Barde YA: Selective activation of NF-kappa B by nerve growth factor through the neurotrophin receptor p75. *Science* 1996, 272:542-545.

[70] Moerman AM, Mao X, Lucas MM, Barger SW: Characterization of a neuronal kappaB-binding factor distinct from NF-kappaB. *Brain Res Mol Brain Res* 1999, 67:303-315.

[71] O'Neill LA, Kaltschmidt C: NF-kappa B: a crucial transcription factor for glial and neuronal cell function. *Trends Neurosci* 1997, 20:252-258.

[72] Albensi BC, Mattson MP: Evidence for the involvement of TNF and NF-kappaB in hippocampal synaptic plasticity. *Synapse* 2000, 35:151-159.

[73] Meffert MK, Chang JM, Wiltgen BJ, Fanselow MS, Baltimore D: NF-kappa B functions in synaptic signaling and behavior. *Nat Neurosci* 2003, 6:1072-1078.

[74] Freudenthal R, Romano A, Routtenberg A: Transcription factor NF-kappaB activation after in vivo perforant path LTP in mouse hippocampus. *Hippocampus* 2004, 14:677-683.

[75] Mattson MP, Camandola S: NF-kappaB in neuronal plasticity and neurodegenerative disorders. *J Clin Invest* 2001, 107:247-254.

[76] Hershko A: The ubiquitin system for protein degradation and some of its roles in the control of the cell division cycle. *Cell Death Differ* 2005, 12:1191-1197.

[77] Wojcik C: Regulation of apoptosis by the ubiquitin and proteasome pathway. *J Cell Mol Med* 2002, 6:25-48.

[78] Kinyamu HK, Chen J, Archer TK: Linking the ubiquitin-proteasome pathway to chromatin remodeling/modification by nuclear receptors. *J Mol Endocrinol* 2005, 34:281-297.

In: The Ubiquitin Proteasome System… ISBN 978-1-60021-749-4
Eds: Mario Di Napoli and Cezary Wojcik, pp. 393-409

Chapter 17

Ubiquitin-Proteasome System in the Peripheral Nervous System: Functional and Morphological Aspects

Lucia Notterpek*

Departments of Neuroscience and Anatomy and Cell Biology, College of Medicine, McKnight Brain Institute, University of Florida, Gainesville, FL 32610, USA.

Abstract

Compared to the central nervous system (CNS), the involvement of the ubiquitin-proteasome system (UPS) in peripheral nerve biology has received less attention. Nonetheless, there are several emerging areas of peripheral nerve biology, as well as pathobiology, in which the ubiquitin-proteasome pathway appears to play critical roles. A major contribution to understanding the role of the UPS in nervous system function came from studies of the *Drosophila* neuromuscular junction. Overexpression of the deubiquitinating protease *fat facets* in motor neurons led to a profound disruption of synaptic growth control, indicating that a balance between ubiquitination and deubiquitination is key for synaptic development. Recent studies in a variety of neuronal model systems revealed that ubiquitination is a critical factor in basic presynaptic and postsynaptic mechanisms, as well as synaptic plasticity throughout the nervous system. In addition to the role in synaptic communication, the UPS is involved in controlling the survival and proliferation of peripheral glial cells, called Schwann cells. Similarly, the survival of cultured sympathetic neurons is also influenced by the UPS. In response to injury, peripheral nerves degenerate by an apoptosis independent mechanism known as Wallerian degeneration. This mechanism involves the fragmentation of axonal microtubules and the destruction of the axolemma. The application of proteasome inhibitors slows this degenerative process and profoundly delays Wallerian degeneration.

* Correspondence concerning this article should be addressed to Dr. Lucia Notterpek, PhD; Department of Neuroscience, McKnight Brain Institute 100 Newell Drive, Box 100244, Gainesville, FL 32610-0244, USA. Phone: 352-294-0030; Fax: 352-846-3854; Email: notterp@mbi.ufl.edu.

Therefore, it is possible that proteasome inhibitors can help to maintain axonal integrity in neurodegenerative events where axonal degeneration is involved. The UPS has also been linked to at least two distinct hereditary disorders of the PNS, namely neurofibromatosis and demyelinating neuropathies. In neurofibromatosis, Schwann cell proliferation is deregulated leading to multiple tumors, mostly schwannomas. Mutated forms of the neurofibrinin 2 tumor suppressor gene are rapidly degraded by the proteasome and thought to contribute to a loss of function phenotype of the wild type allele. Peripheral myelin protein 22 (PMP22), also known as growth arrest specific gene 3, is a short-lived Schwann cell protein that is also degraded by the proteasome. Mutations within and duplication of the PMP22 gene are linked to demyelinating neuropathies, including Charcot-Marie-Tooth disease type IA. When PMP22 is mutated or overexpressed, the degradation of the protein is slowed leading to its accumulation in cytosolic aggregates, termed aggresomes. The formation of PMP22 aggresomes is associated with an impairment of UPS activity and the recruitment of proteasome substrates and chaperones to the aggregates. These events can disrupt the cell cycle of the Schwann cells and enhance neural dysfunction. The findings described here indicate that there has been a tremendous progress in our understanding of the role of the UPS in PNS biology. Significantly, findings from the PNS have provided ground-breaking insights into basic neural mechanisms, such as synaptic communication and axonal degeneration. The involvement of the UPS in peripheral nerve disease reveals another commonality between the CNS and PNS, and will help to speed the development of novel therapeutic strategies aimed at influencing protein turnover in the nervous system.

Keywords: neuropathy, Schwann cell, myelin, demyelination, glia, aggresome, peripheral myelin protein 22.

ABBREVIATIONS

CMT1A, Charcot-Marie-Tooth disease type 1A; CNS, central nervous system; DSS, Dejerine Sottas Syndrome; GAN, giant axonal neuropathy; HNPP, hereditary neuropathy with liabiility to pressure palsies; MAP1B-LC, light chain of microtubule associated protein 1B; NGF, nerve growth factor; NMJ, neuromuscular junction; Nmnat1, nicotinamide mononucleotide adenyltrasferase 1; PMP22, peripheral myelin protein 22; PNS, peripheral nervous system; UPS, ubiquitin-proteasome system; Wt, wild type; MBP, myelin basic protein.

INTRODUCTION

Peripheral nerves are highly efficient cables that convey information between the central nervous system (CNS) and all parts of the body. The peripheral nervous system (PNS) includes all neural structures from the nerve endings to their entry into the brain or the spinal cord. The neuronal cell bodies of peripheral nerves are found in columns or groups in the CNS, or in peripheral ganglia, such as the sensory and autonomic ganglia. Schwann cells, the glial cells of the PNS, are derived from the neural crest and are essential for the maintenance and support of peripheral axons. During development, Schwann cell precursors undergo rapid proliferation giving rise to myelin- and non-myelin-forming cells. Myelin-forming Schwann cells, the counterparts of CNS oligodendrocytes, encircle segments of individual axons and wrap them with multiple layers of myelin. In a mature peripheral nerve, each Schwann cell forms associations with single internodal segments of myelin that are separated by the nodes of Ranvier. This one-to-one arrangement of Schwann cells and axons is unique compared to the relationship of a single oligodendrocyte with numerous axons and multiple internodes. Non-myelinating, ensheathing Schwann cells associate with small diameter axons and maintain contact with several nerve fiber (Remak) bundles. Schwann cells express a variety of ion channels and neurotransmitter receptors and fulfill multiple roles in the PNS [1]. The interdependence of peripheral axons and glia on one another is demonstrated by the dedifferentiation of Schwann cells in response to post-injury axonal degeneration. Similarly, under conditions of glial abnormalities, such as in heritable demyelinating neuropathies, axons degenerate. Nonetheless, under permissive conditions peripheral axons have the ability to regenerate and remyelinate, which provide unique opportunities for investigations of neural degeneration and regeneration, as well as for the development of treatment strategies in acute and chronic disease conditions.

The relative simplicity and easy accessibility of peripheral nerves and their glial cells have allowed researchers to examine complex cellular and molecular mechanisms underlying basic neuronal functions, including axonal growth and degeneration, and regeneration. In addition, synaptic targets of peripheral nerves, such as muscle tissue, are readily amenable to exogenous perturbation and *in vitro* modeling. Investigations of the neuromuscular junction (NMJ) have provided ground-breaking insights into basic synaptic mechanisms. As described below with specific examples from the literature, findings from the PNS have had a major impact on our understanding of the role of the ubiquitin-proteasome system (UPS) in neural mechanisms, including axonal degeneration/regeneration, as well as synaptic communication. A commonality among these events is the necessity for precise control of protein availability. Therefore, it is not surprising that recent reports have found the involvement of the UPS in peripheral neurite outgrowth, axonal degeneration and regeneration, as well as synaptic function. In addition, the UPS is critical for the maintenance and regulation of basic Schwann cell biology, including cell division and differentiation. Finally, alterations in the activity of the UPS have been detected in mouse models of heritable demyelinating neuropathies.

Neuronal Survival and Death

During the development of the nervous system a large number of neurons die by programmed cell death, most likely because they fail to receive sufficient quantities of trophic factors from their target cells [2]. A widely used model of neuronal death and survival are sympathetic neurons, cultured from the superior cervical ganglia of newborn rodents. In the presence of nerve growth factor (NGF) these cells survive and extend neurites, while in the absence of NGF they die by programmed cell death [3]. The first indication for the role of the UPS in the death and survival of peripheral neurons was obtained in trophic factor-deprived sympathetic neuron cultures [4]. The inclusion of low dose proteasome inhibitors, such as lactacystin (0.5-5 μM), was able to rescue the neurons from NGF-deprived cell death and prevented the cleavage of poly(ADP-ribose) polymerase.

Since these original observations nearly ten years ago, several *in vitro* studies with pharmacologic inhibitors confirmed the role of the UPS in regulating the survival of various neurons. Depending on the specific experimental conditions used, proteasome inhibition can be pro-survival or pro-apoptotic. Long-term, high dose inhibition of the proteasome leads to cell death, while shorter and/or low dose treatment promotes survival [4,5]. For example, treatment of sympathetic neuron cultures with lactacystin for 24 or 48h is associated with increased apoptotic cell death [6]. This effect is exposure and dosage dependent, with higher concentrations leading to more rapid neuronal loss. It is hypothesized that the degradation of factors necessary for the induction of apoptosis, as well as of neuroprotective molecules, are regulated by the proteasome. In one paradigm, increased degradation of certain proteins by the proteasome may be necessary for the induction of apoptosis, therefore its inhibition would be pro-survival [7]. Similarly, elevating the steady-state levels of anti-apoptotic proteins by blocking their degradation could also promote survival [7,8]. On the other hand, blocking the degradation of pro-apoptotic factors could lead to cell death.

In addition to the pharmacologic paradigms, the involvement of the UPS in regulating neuronal survival is supported by recent findings in giant axonal neuropathy (GAN) [9] (see Chapter 26). Giant axonal neuropathy is a sensory and motor neuropathy associated with mutations in the *GAN* gene, which encodes the ubiquitously expressed protein, gigaxonin. Gigaxonin binds the ubiquitin-activating enzyme E1, while simultaneously interacting with the light chain of microtubule associated protein 1B (MAP1B-LC). Overexpression of gigaxonin leads to enhanced degradation of MAP1B-LC, which can be antagonized by inhibition of the proteasome. In agreement, ablation of gigaxonin causes a substantial accumulation of MAP1B-LC. Together, these studies indicate that gigaxonin is critical for the UPS-mediated degradation of MAP1B-LC. Consequently, elevation in the levels of MAP1B-LC is associated with neuronal death, possibly by interfering with the movement of motor proteins. While the exact mechanism by which an increase in MAP1B-LC leads to neuronal death are not known, the linkage of gigaxonin with the UPS clearly demonstrates how changes in protein degradation can underlie a neurologic disease. As gigaxonin and MAP1B-LC are expressed in both CNS and PNS neurons, this example also demonstrates a common mechanism in controlling the fate of various neuronal cell types. Indeed, recent reports in primary cortical and cerebellar granule neuron cultures indicate the participation of the UPS in regulating neuron death and survival [10,11].

Axonal Guidance and Pruning

The embryonic nervous system contains exuberant projections both to appropriate and inappropriate targets. The adult pattern of connectivity is attained by selective elimination of certain connections and the maintenance of others. Neurite extension, guided by growth cones, is a dynamic process that requires the correct concentrations of certain proteins at specific sites. Pioneering studies from *Drosophila* indicate the involvement of the UPS in growth cone guidance and axonal pruning [12]. The *Drosophila* mutant *bendless* has an altered escape response which is underlined by the disruption of single neuronal connections [13]. Cloning and characterization of the gene responsible for this behavior identified bendless as an E2 ubiquitin-conjugating enzyme [14]. Complementary studies from *Xenopus* suggest that ubiquitination and degradation of growth cone guidance molecules might be a general mechanism to regulate the concentration and potentially the localization of proteins during neurite extension [15].

Similar to growth cone guidance, the UPS is also involved in axon pruning (see also Chapter 26). Selective branch elimination, or axon pruning, is utilized throughout the nervous system and underlies the establishment of precise neural circuits [16]. The elimination of motoneuron branches at the developing vertebrate NMJ is well documented and is thought to involve activity-dependent and -independent events [12]. Once again, genetic studies from *Drosophila* indicate the requirement for ubiquitin and the UPS in axonal pruning [17]. Time-lapse microscopy studies reveal that axonal degeneration is mediated by local breakdown of axonal proteins rather than retraction of processes. Degradation of the microtubule cytoskeleton precedes axon pruning which is blocked by genetic ablation of the ubiquitin-conjugating enzyme (E1) or the 19S proteasome subunit [17]. Furthermore, expression of a yeast ubiquitin protease also inhibited axon pruning. Analogous studies in developing mammalian systems are yet to be reported. Nonetheless, based on recent findings in axonal degeneration models (see below) it is likely that the results from the *Drosophila* will have wide-ranging applications.

THE UPS IN SYNAPTIC FUNCTION

Synaptic connections are a special form of intercellular communication between two neurons, or a neuron and a nonneural postsynaptic cell, such as muscle. During development, axonal growth cones differentiate into presynaptic terminals when the appropriate target cell is reached. In coordination with the postsynaptic cell, the presynaptic cell differentiates by expressing a variety of molecules involved with neurotransmitter synthesis, packaging and release. The postsynaptic cell prepares for communication by the expression and correct positioning of neurotransmitter receptor molecules and associated signaling components.

The NMJ, likely the most studied form of synaptic communication, involves a presynaptic motor neuron and a postsynaptic muscle cell. With regards to discovering the involvement of the UPS in synaptic function, studies of the *Drosophila* neuromuscular model system have been pivotal [12,18]. A genetic screen, aimed at discovering molecules whose overexpression disrupts normal synaptic development, led to the identification of the deubiquitinating protease *fat facets* as a protein involved in synaptic growth control [19]. Neuronal overexpression of *fat facets* results in a profound increase in the number of synaptic boutons and the disruption of synaptic function. Similarly, ectopic expression of a yeast deubiquitinating enzyme is associated with synaptic overgrowth and dysfunction, indicating that a balance between ubiquitination and deubiquitination is crucial for normal synaptic development.

Since this key discovery in *Drosophila*, ubiquitination and ubiquitin-dependent degradation of proteins have been tied to numerous aspects of synaptic function, including presynaptic vesicular neurotransmitter release, postsynaptic differentiation, as well as synaptic plasticity [12]. The UPS is thought to have multiple roles in these events. As monoubiquitination of target proteins is involved in endocytosis, as well as protein trafficking [20], it is not surprising that presynaptic vesicular dynamics are profoundly influenced by changes in the levels of deubiquitinating enzymes. For example, persistent monoubiquitination of the clathrin-associated protein epsin by reduced expression of a deubiquitinating enzyme alters the dynamics of vesicular fusion and endocytosis [21]. Degradation of multi-ubiquitinated proteins by the proteasome is also important in synaptic activity and has been observed in a variety of organisms [12]. For instance, acute pharmacological inhibition, or genetic disruption, of the proteasome leads to the accumulation of the essential synaptic vesicle-priming protein DUNC-13 and is associated with increased presynaptic efficacy [22]. Based on the multifaceted involvement of the UPS in synaptic function we can speculate that future studies will uncover the linkage of ubiquitin-regulated events to diseases with synaptic malfunction.

Degeneration and Regeneration of Peripheral Nerves

In response to a cut or crush injury, the distal portions of axons undergo a degenerative process known as Wallerian degeneration [23]. The distal stumps of degenerating peripheral nerves provide a supportive environment for regeneration and can even support the growth of CNS neurites. In contrast, injured axons of the mammalian brain and spinal cord rarely regenerate, likely due to the presence of inhibitory environmental factors [24]. Therefore, understanding the molecular mechanism underlying the degeneration and regeneration of peripheral axons has wide implications for neurodegenerative events throughout the nervous system.

The degenerative changes in the distal stump of a severed axon begin with the disintegration and degeneration of the axoplasma and the axolemma. In small nerve fibers, these events are completed within 24 hours of injury, while in larger axons they may take 48 hours [24]. The first indication for the involvement of the UPS in Wallerian degeneration came from the spontaneous mouse mutant strain Wld^S [25] (see also Chapter 26). Wld mice harbor a genetic alteration in the ubiquitin regulatory enzyme UFD2a (alternatively callled Ube4b) and the nicotinamide mononucleotide adenyltransferase 1 (Nmnat1) [26]. The Wld genotype is associated with extended survival of transected axons and delayed Wallerian degeneration, the exact mechanism of which is debated [27] (see also Chapter 26). Nonetheless, pharmacological inhibition of the proteasome in transected superior cervical ganglion axons increases the survival of the distal stump [28]. Likewise, the modulation of UPS activity by overexpression of the deubiquitinating enzyme UBP2 slowed axonal degeneration. Furthermore, in the same study, inhibition of the proteasome also delayed the degeneration of neurites in response to NGF deprivation. Together, these findings suggest that the UPS is involved in injury-mediated as well as developmental axon loss paradigms. While these findings are compelling, more studies are necessary to identify the specific UPS constituents and substrates in axons. A recent report suggests that rather than functioning in generalized axonal proteolysis, the activity of the UPS regulates specific signaling pathways involved in controlling axonal survival [29]. Nonetheless, it is possible that modulation of proteasome activity can help the maintenance of axonal integrity in neurodegenerative events and delay Wallerian degeneration.

SCHWANN CELL DIVISION AND NEUROFIBROMATOSIS

Controlled proliferation and differentiation of Schwann cells is a prerequisite for myelination and normal PNS biology. During development, Schwann cells proliferate and establish a one-to-one relationship with axons, followed by the secretion of a basal lamina and myelin wrapping. Compared to oligodendrocyes, the myelin-forming glial cells of the CNS, fully-differentiated myelinated Schwann cells retain the ability to dedifferentiate and proliferate in response to a variety of signals, including nerve injury. The wave of Schwann cell proliferation during normal nerve development, as well as in response to insult, is designed to attain the correct number of glial cells for the support and myelination of axons. In a variety of organisms and cell types, the UPS is a recognized regulator of cell division [30]. Thus, it is not unexpected that Schwann cell proliferation, while in part is driven by axonal signals, involves dynamic changes in the localization of UPS constituents [31]. The proliferation of Schwann cells is well-matched to the requirement of axonal growth, yet this event is associated with apoptotic cell death [32]. The exact mechanism for the induction of programmed Schwann cell death is not known, but *in vitro* studies suggest a role for the UPS [33]. Extended, 24-36h pharmacological inhibition of the proteasome by lactacystin in cultured Schwann cells induces apoptosis [33]. Similarly, in a tellurium-induced demyelination model, excess number of proliferating Schwann cells are eliminated by apoptosis, which is associated with a pronounced increase in the levels of free ubiquitin in the perikarya [34]. Independent of the studies on the involvement of the UPS in Schwann cell survival, recent reports suggest a role for the proteasome in the pathogenesis of peripheral nerve sheath tumors. Dysregulation of Schwann cell division and differentiation is associated with two autosomal dominant disorders, namely neurofibromatoses type 1 (NF1) and type 2 (NF2) [35]. Peripheral neurofibromas, the hallmarks of NF1, arise from Schwann cells and other supportive cells of peripheral nerves. Neurofibromin, tumor suppressor gene 1 whose misexpression is associated with familial NF1 is a proteasome substrate [36]. It is uncertain how neurofibromin controls cell division, but it is hypothesized that when neurofibromin levels are high, *ras* activation and cell proliferation is halted. Therefore, by altering the degradation of the wild type (Wt) neurofibromin in NF patients one could potentially upregulate its steady-state levels and arrest cell division [36].

NF2 is also associated with deregulated Schwann cell division leading to multiple tumors, mostly schwannomas [35]. The disease is caused by mutations in the neurofibrinin 2 tumor suppressor gene, known as schwannomin or merlin. Mutated forms of merlin are rapidly degraded by the proteasome and thought to contribute to a loss of function phenotype of the Wt allele [37]. Again, by altering the activity of the UPS, the levels of merlin could be regulated, providing a potential therapeutic approach to NF2. Collectively, these studies show a pivotal role for the UPS in regulating the survival and division of Schwann cells, *in vitro* and *in vivo*. Future investigations will identify additional specific targets of ubiquitination and examine if Schwann cells utilize the UPS in a unique manner to control cell cycle progression.

MYELIN PROTEIN TURNOVER

Myelin is a highly specialized extension of the Schwann cell plasma membrane, with compact and uncompacted domains. Studies from spontaneous and genetically engineered myelin protein mutant animals indicate that the stability of myelin is greatly influenced by the stochiometry of its constituents. As ubiquitination is known to be involved in the trafficking and degradation of membrane proteins [20], one might expect to uncover a role for the UPS in myelination. So far only two myelin proteins have been identified whose turnover is regulated by the UPS, namely myelin basic protein (MBP) and peripheral myelin protein 22 (PMP22). MBP is expressed by oligodendrocytes and Schwann cells and is associated with the intracellular surface of myelin [38]. MBP is thought to play a role in myelin compaction, likely through direct interactions with protein zero [39]. The degradation of MBP in a ubiquitin-dependent manner by the 26S proteasome was described nearly ten years ago [40], yet these studies have not been followed up. Based on the involvement of the UPS in several neurodegenerative disease mechanisms, it is possible that alterations in the turnover of MBP might contribute to the pathogenesis of some CNS disorders.

Peripheral myelin protein 22 (PMP22) is an integral membrane protein of Schwann cells and is largely undetectable in CNS myelin [41]. The proposed secondary structure of PMP22 contains four transmembrane domains and a single carbohydrate modification in the first extracellular loop [42]. PMP22 was first identified as one of the growth arrest specific (*gas3*) genes in serum-deprived fibroblasts [43], and then as a Schwann cell protein involved in peripheral nerve degeneration [44]. A growth regulatory function for PMP22 in Schwann cells has also been proposed [45], as elevated expression is associated with the differentiated myelinating state. In myelinated peripheral nerves, PMP22 is localized to the compact portion of myelin and is thought be a structural component [44]. Misexpression of MBP or PMP22 is associated with myelin defects in the CNS and/or the PNS, respectively [42].

In our approach to understanding the pathogenesis of PMP22-associated demyelinating neuropathies, we first determined the processing and turnover rate of the Wt protein in healthy Schwann cells [46]. Under non-myelinating and myelinating conditions, and in *ex vivo* myelinating nerves, the majority (~80%) of the newly-synthesized, endoH-sensitive PMP22 is rapidly degraded by the proteasome, presumably due to misfolding [46]. Approximately 20% of the protein acquires complex carbohydrate modification and incorporates into the Schwann cell plasma membrane. Misfolded and hard-to-fold membrane proteins are known to be degraded by ubiquitin and proteasome-dependent ER-associated degradation [47]. Indeed, inhibition of this pathway by lactacystin for five hours is associated with the condensation of PMP22 in the perinuclear region, with distinct aggresome formation by sixteen hours [48]. Aggresomes are cytoplasmic, membrane-free inclusions that accumulate in a microtubule-dependent fashion near the centrosome and are excluded from the major organelles [48,49,50].

The formation of PMP22 aggresomes under proteasome impairment demonstrates that PMP22 is a proteasome substrate. The mechanism for proteasomal targeting of the newly-synthesized wt PMP22 is not yet known, but could involve ubiquitin-dependent or ubiquitin-independent mechanisms. We detected ubiquitinated PMP22 in Schwann cells by co-immunoprecipitation experiments [51], but the specific site for this posttranslational modification, or the enzymes responsible for the ubiquitination of PMP22, are not yet known. The wt PMP22 contains three potential sites for ubiquitination, including the amino terminal and lysines 92 and 158. Additionally, PMP22 is a tetraspan membrane protein, therefore exposure of unburied hydrophobic sequences of the misfolded protein could serve as a signal for degradation.

Although aggregates of misfolded PMP22 could become detrimental to Schwann cell function, our culture studies indicate that the formation of PMP22 aggregates, unlike alpha-synuclein [52], is not toxic and under permissive conditions the cells have the ability to clear them [53]. By using a pharmacological approach in cultured cells, we demonstrated that Schwann cells can reroute the aggregated protein for lysosomal degradation via autophagy [53]. This finding suggests a link between the two main protein degradative machineries of the cell, the UPS and lysosomes. The signaling mechanisms involved in the recruitment of autophagosomes to undegraded, aggregated proteasome substrates is unknown, but appears to be dependent on microtubules and the tubulin deacetylase, HDAC6 [54].

Table 1. PMP22-linked hereditary demyelinating neuropathies.

Disorder	Age of onset	Genetic defect	Animal model
CMT1A	2nd-3rd decade	1.5 Mb duplication on chromosome 17	PMP22 overexpressor rats and mice
		Single point mutations in PMP22	Trembler and Trembler-J mice
HNPP	2nd-3rd decade	1.5 Mb deletion on chromosome 17	PMP22-deficient mice
DSS	1st decade	Single point mutations in PMP22	Trembler and Trembler-J mice

HEREDITARY DEMYELINATING NEUROPATHIES

Inherited demyelinating peripheral neuropathies are a common, heterogeneous group of diseases that lead to progressive myelin instability and slowed nerve conduction velocities. Affected individuals present with sensory and motor disturbances, muscle weakness and atrophy. A large fraction of these diseases is linked to misexpression of PMP22, a proteasome substrate (Table 1). The majority of Charcot-Marie-Tooth type 1A (CMT1A) neuropathy patients carry three copies of the *PMP22* gene, while in a smaller group; single amino acid substitutions in PMP22 are present [44]. Single point mutations in PMP22 have also been identified in Dejerine-Sottas Syndrome (DSS) patients, a more severe, congenital hypomyelinating neuropathy [55]. Yet in another group of peripheral neuropathies, called hereditary neuropathy with liability to pressure palsies (HNPP), the *PMP22* gene is deleted. The neuropathies show that alterations in *PMP22* expression cause disease by three different mechanisms, which include overexpression, underexpression and single point mutations. These genetic diseases also indicate that for normal Schwann cell function and myelination the levels of PMP22 must be tightly regulated. As even in normal Schwann cells ~80 % of the newly synthesized PMP22 is targeted for the UPS [46], in the gene duplication and point mutation disease paradigms the degradative machinery responsible for the turnover of PMP22 might become overwhelmed and lead to the formation of protein aggregates. Indeed, nerve fibers from PMP22 overproducer and Trembler J (Leu16Pro) mice contain cytosolic aggregates that are immunorecative for PMP22 and ubiquitin (Figure 1 and 2A).

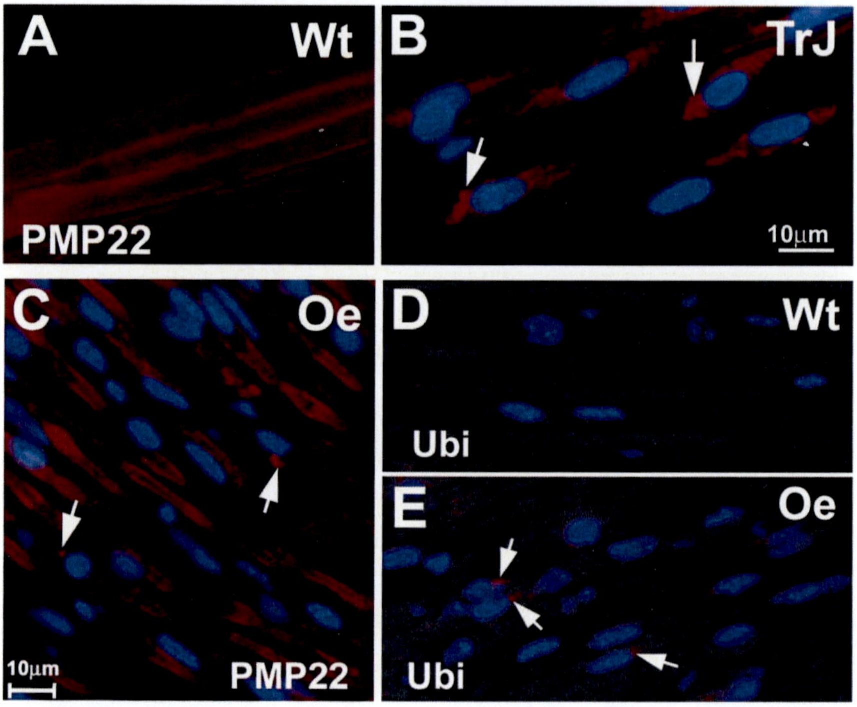

Figure 1. Protein aggregates in neuropathic nerves. (A) In nerve fibers from wild type (Wt) mice, PMP22 is detected uniformly in the Schwann cell myelin. (B) In samples from PMP22 point mutant Trembler J (TrJ) mice, the Schwann cells have rounded cell bodies and are immunoreactive for PMP22 (arrows). (C) PMP22 is similarly retained in some of the Schwann cell bodies in nerves from PMP22 overexpressor (Oe) mice (arrows). (D and E) Nerves of neuropathic mice show enhanced reactivity with anti-ubiquitin antibodies and ubiquitin-like staining is seen near nuclei (arrows). Nuclei are stained with Hoechst dye (blue), which reveals the increased numbers of Schwann cells in affected samples [68].

To examine the involvement of the UPS in PMP22 neuropathies we determined the turnover of mutated PMP22s *in vitro* [56] and in vivo (Figure 2B). Point mutant (Leu16Pro and Gly150Asp) PMP22s were epitope tagged and the trafficking and half-lives of the recombinant proteins were followed in primary Schwann cells, with or without proteasome inhibition. As we predicted, the half-lives of these mutated proteins are increased compared to wt, and inhibition of the proteasome by lactacystin or epoxomicin leads to their accumulation in cytosolic aggresomes. We also examined the trafficking and degradation of two frameshift mutated PMP22s, both of which truncate the protein after the second transmembrane domain [57]. Although these truncated proteins are missing one of the potential ubiquitination sites at lysine 158, and two of the hydrophobic transmembrane domains, they are still substrates for the proteasome. We also found that the mutant protein with higher propensity for spontaneous aggregation is associated with the less severe disease phenotype [57]. Independent studies by Isaacs and colleagues found a similar link between the formation of large PMP22 aggregates and mild disease phenotypes [58]. Therefore, based on the correlation of large PMP22 aggregates with a less severe disease phenotype [57,58] and the existence of an autophagic-dependent mechanism involved in their clearance [53], a protective role for the formation of aggregates in neuropathic Schwann cells can be proposed.

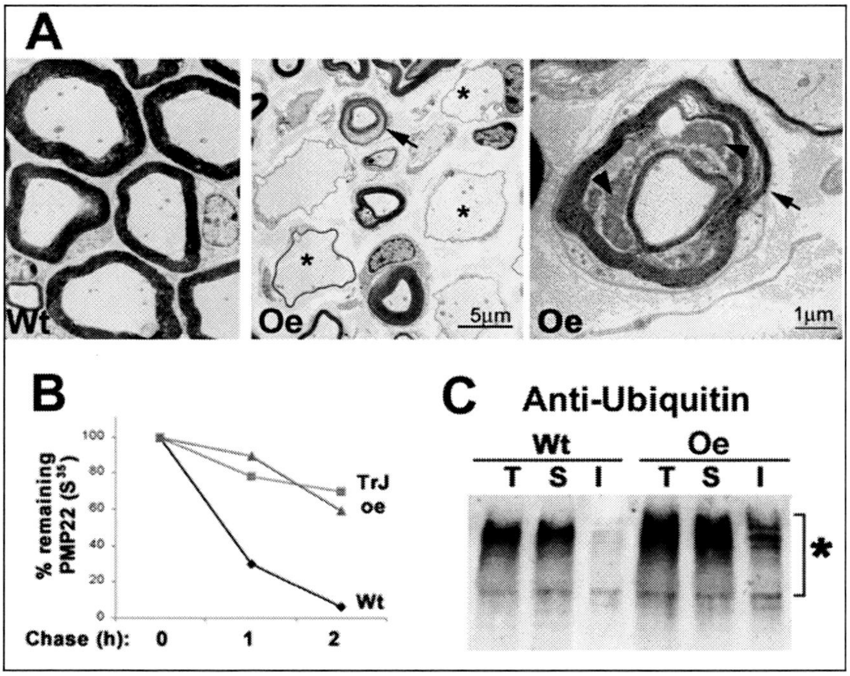

Figure 2. Ultrastructural and biochemical changes in neuropathic nerves. (A) Sciatic nerves from adult mice were processed for ultrastuctural examination. The majority of large fibers are myelinated in samples from wild type mice. In comparison, unmyelinated (asterisks) and thinly myelinated (arrow) axons are prevalent in the neuropathic sample (Oe). At higher magnification, amorphous protein aggregates (arrowheads in right panel) are seen between layers of split myelin (arrow). (B) By pulse-chase analysis, the turnover of PMP22 is reduced in TrJ and PMP22 overexpressor nerve lysates, indicating altered degradation. (C) At steady-state, the accumulation of slow-migrating ubiquitinated substrates is detected in the detergent-insoluble fraction (I) of neuropathic nerves. T: total lysate; S: detergent-soluble; I: detergent-insoluble.

Nonetheless, as aggregate-containing Schwann cells are associated with disease phenotypes it is of interest to examine the protein components of the aggregates. Cytosolic accumulation of misfolded PMP22 may have at least two consequences: (*i*) serve as a nucleation site for the aggregation of other Schwann cell molecules, including chaperones (Hsp70, αB-crystallin), myelin proteins such as MBP and constituents of the ubiquitin-proteasome pathway; and (*ii*) interfere with the activity of the UPS, which can further decrease the turnover of PMP22. The entrapment of these molecules within PMP22 aggregates could lead to a protein imbalance of the cell and contribute to the pathogenesis of the neuropathy. Chaperones transiently interact with proteins to aid their folding, trafficking and degradation, thereby contributing to the fate of newly-synthesized molecules [59]. Recent studies indicate that molecular chaperones can provide protection against neurodegenerative disease by preventing the aggregation of mutant proteins, catalyzing protein folding and multimer assembly, solubilization of aggregated proteins, and promoting the ubiquitination and degradation of abnormal proteins [60]. Therefore, the recruitment of Hsp70 and αB-crystallin to PMP22 aggregates [53,56], will decrease the available pool of these molecules and likely interfere with normal Schwann cell function. Thus, increasing the availability of molecular chaperones might be beneficial for PMP22-associated neuropathies and should be tested.

The formation of PMP22 aggregates in neuropathic nerves may also contribute to demyelination by recruitment/retention of myelin proteins. PMP22 is known to form homodimers, as well as to interact with protein zero (P0) [61,62]. In neuropathic mouse nerves, we found P0 only occasionally colocalizing with PMP22 aggregates, while MBP is detected in over a third of the aggregates [63]. MBP, an essential component of peripheral myelin, is an endogenous substrate of the proteasome [40]. The mislocalization of MBP to perinuclear aggregates and the increased detergent-insolubility of specific isoforms (18 and 21 kDa) illustrate how the formation of PMP22 aggregates can alter the trafficking of additional myelin proteins.

Another possible mechanism by which aggregates could affect Schwann cell biology and contribute to the neuropathic phenotype is through influencing proteasome activity. In cultured PMP22 overexpressor cells, the protein aggregates are immunoreactive for the 11S and 20S subunits and show the accumulation of ubiquitinated substrates in the detergent-insoluble pool [51,63] (also see Figure 2C). The formation of protein aggregates is known to impair the activity of the proteasome [64], which in turn, can further increase the accumulation of ubiquitinated substrates. Therefore, we examined whether the presence of PMP22 aggregates correlates with changes in proteasome function by measurements of chymotypsin-like and overall 20S activity [51,63]. The chymotrypsin-like activity of the proteasome in neuropathic nerve lysates, as reflected by the ability to cleave the substrate LLVY is diminished as compared to Wt. Similarly, the ability of the 26S to degrade a short-lived GFP reporter substrate (Ub^{G76V}-GFP) [65] is also impaired [51,63]. Conversely, we did not detect impairment of proteasome activity or the accumulation of ubiquitinated substrates in samples from PMP22-deficient mice.

Collectively, these studies indicate that alteration in the UPS-mediated degradation of PMP22 is a commonality among PMP22 point mutation and overexpression disease paradigms. Slowed degradation of PMP22 leads to the formation of cytosolic aggregates, which is associated with an impairment of UPS activity and the recruitment of chaperones and myelin proteins to the aggregates. Therefore, despite the initial protective nature of aggregates, they could contribute to cellular dysfunction if they are not cleared over time. Coincidently, the two mechanisms involved in the potential clearing and prevention of protein aggregation are known to change with age. Autophagic proteolysis and the ability of cells to induce chaperones, declines with age [66,67]. It is an attractive hypothesis that combinatorial effects of these changes likely contribute to the progressive subcellular pathogenesis of PMP22 neuropathies and potentially other disorders involving proteasome substrates. It is of great interest to us if upregulating the chaperone and the autophagic pathway in neuropathic mice can slow the progression of the disease and ameliorate the described subcellular changes.

Conclusions

Regulated degradation of proteins is a key mechanism in diverse biological systems. Ubiquitination and UPS-mediated degradation of neuronal and glial proteins is proving to be a central mechanism in PNS biology. The relative simplicity of the PNS allows investigators to further probe these pathways and identify specific neuronal and glial substrates. Based on the examples described here, it is anticipated that many of the findings from the PNS will be applicable to CNS as well. An interesting question will be to examine if glial cells of the CNS, astrocytes and oligodendrocytes, share the ability with Schwann cells to activate the autophagic pathway for the removal of misfolded UPS substrates. Understanding the role of the UPS in PNS biology and in pathogenesis is providing the bases for the development of new therapeutic approaches for neurodegenerative diseases.

Acknowledgements

Our work is supported by the NIH-NINDS and the National Muscular Dystrophy Association.

References

[1] Verkhratsky A, Steinhauser C: Ion channels in glial cells. *Brain Res Brain Res Rev* 2000, 32:380-412.

[2] Raff MC, Barres BA, Burne JF, Coles HS, Ishizaki Y, Jacobson MD: Programmed cell death and the control of cell survival: lessons from the nervous system. *Science* 1993, 262:695-700.

[3] Garcia I, Martinou I, Tsujimoto Y, Martinou JC: Prevention of programmed cell death of sympathetic neurons by the bcl-2 proto-oncogene. *Science* 1992, 258:302-304.

[4] Sadoul R, Fernandez PA, Quiquerez AL, Martinou I, Maki M, Schroter M, Becherer JD, Irmler M, Tschopp J, Martinou JC: Involvement of the proteasome in the programmed cell death of NGF-deprived sympathetic neurons. *EMBO J* 1996, 15:3845-3852.

[5] Lang-Rollin IC, Rideout HJ, Noticewala M, Stefanis L: Mechanisms of caspase-independent neuronal death: energy depletion and free radical generation. *J Neurosci* 2003, 23:11015-11025.

[6] Lang-Rollin I, Vekrellis K, Wang Q, Rideout HJ, Stefanis L: Application of proteasomal inhibitors to mouse sympathetic neurons activates the intrinsic apoptotic pathway. *J Neurochem* 2004, 90:1511-1520.

[7] Canu N, Barbato C, Ciotti MT, Serafino A, Dus L, Calissano P: Proteasome involvement and accumulation of ubiquitinated proteins in cerebellar granule neurons undergoing apoptosis. *J Neurosci* 2000, 20:589-599.

[8] Dimmeler S, Zeiher AM: Nitric oxide-an endothelial cell survival factor. *Cell Death Differ* 1999, 6:964-968.

[9] Allen E, Ding J, Wang W, Pramanik S, Chou J, Yau V, Yang Y: Gigaxonin-controlled degradation of MAP1B light chain is critical to neuronal survival. *Nature* 2005, 438:224-228. Epub 2005 Oct 2016.

[10] Yew EH, Cheung NS, Choy MS, Qi RZ, Lee AY, Peng ZF, Melendez AJ, Manikandan J, Koay ES, Chiu LL, Ng WL, Whiteman M, Kandiah J, Halliwell B: Proteasome inhibition by lactacystin in primary neuronal cells induces both potentially neuroprotective and pro-apoptotic transcriptional responses: a microarray analysis. *J Neurochem* 2005, 94:943-956. Epub 2005 Jun 2030.

[11] Butts BD, Hudson HR, Linseman DA, Le SS, Ryan KR, Bouchard RJ, Heidenreich KA: Proteasome inhibition elicits a biphasic effect on neuronal apoptosis via differential regulation of pro-survival and pro-apoptotic transcription factors. *Mol Cell Neurosci* 2005, 30:279-289.

[12] DiAntonio A, Hicke L: Ubiquitin-dependent regulation of the synapse. *Annu Rev Neurosci* 2004, 27:223-246

[13] Thomas JB, Wyman RJ: Mutations altering synaptic connectivity between identified neurons in Drosophila. *J Neurosci* 1984, 4:530-538.

[14] Muralidhar MG, Thomas JB: The Drosophila bendless gene encodes a neural protein related to ubiquitin-conjugating enzymes. *Neuron* 1993, 11:253-266.

[15] Campbell DS, Holt CE: Chemotropic responses of retinal growth cones mediated by rapid local protein synthesis and degradation. *Neuron* 2001, 32:1013-1026.

[16] Kantor DB, Kolodkin AL: Curbing the excesses of youth: molecular insights into axonal pruning. *Neuron* 2003, 38:849-852.

[17] Watts RJ, Hoopfer ED, Luo L: Axon pruning during Drosophila metamorphosis: evidence for local degeneration and requirement of the ubiquitin-proteasome system. *Neuron* 2003, 38:871-885.

[18] Yi JJ, Ehlers MD: Ubiquitin and protein turnover in synapse function. *Neuron* 2005, 47:629-632.

[19] DiAntonio A, Haghighi AP, Portman SL, Lee JD, Amaranto AM, Goodman CS: Ubiquitination-dependent mechanisms regulate synaptic growth and function. *Nature* 2001, 412:449-452.

[20] d'Azzo A, Bongiovanni A, Nastasi T: E3 ubiquitin ligases as regulators of membrane protein trafficking and degradation. *Traffic* 2005, 6:429-441.

[21] Chen H, Polo S, Di Fiore PP, De Camilli PV: Rapid Ca2+-dependent decrease of protein ubiquitination at synapses. *Proc Natl Acad Sci U S A* 2003, 100:14908-14913. Epub 12003 Dec 14901.

[22] Speese SD, Trotta N, Rodesch CK, Aravamudan B, Broadie K: The ubiquitin proteasome system acutely regulates presynaptic protein turnover and synaptic efficacy. *Curr Biol* 2003, 13:899-910.

[23] Waller A: Experiments on the Section of the Glossopharyngeal and Hypoglossal Nerves of the Frog, and Observations of the Alterations Produced Thereby in the Structure of Their Primitive Fibres. *Philosophical Transactions of the Royal Society of London* 1850, 140:423-429.

[24] Stoll G, Jander S, Myers RR: Degeneration and regeneration of the peripheral nervous system: from Augustus Waller's observations to neuroinflammation. *J Peripher Nerv Syst* 2002, 7:13-27.

[25] Lunn ER, Perry VH, Brown MC, Rosen H, Gordon S: Absence of Wallerian degeneration does not hinder regeneration in peripheral nerve. Eur J Neurosci 1989, 1:27-33.

[26] Mack TG, Reiner M, Beirowski B, Mi W, Emanuelli M, Wagner D, Thomson D, Gillingwater T, Court F, Conforti L, Fernando FS, Tarlton A, Andressen C, Addicks K, Magni G, Ribchester RR, Perry VH, Coleman MP: Wallerian degeneration of injured axons and synapses is delayed by a Ube4b/Nmnat chimeric gene. *Nat Neurosci* 2001, 4:1199-1206.

[27] Hall SM: Observations on the progress of Wallerian degeneration in transected peripheral nerves of C57BL/Wld mice in the presence of recruited macrophages. *J Neurocytol* 1993, 22:480-490.

[28] Zhai Q, Wang J, Kim A, Liu Q, Watts R, Hoopfer E, Mitchison T, Luo L, He Z: Involvement of the ubiquitin-proteasome system in the early stages of wallerian degeneration. *Neuron* 2003, 39:217-225.

[29] MacInnis BL, Campenot RB: Regulation of Wallerian degeneration and nerve growth factor withdrawal-induced pruning of axons of sympathetic neurons by the proteasome and the MEK/Erk pathway. *Mol Cell Neurosci* 2005, 28:430-439.

[30] Hershko A: The ubiquitin system for protein degradation and some of its roles in the control of the cell division cycle. *Cell Death Differ* 2005, 12:1191-1197.

[31] Lafarga M, Fernandez R, Mayo I, Berciano MT, Castano JG: Proteasome dynamics during cell cycle in rat Schwann cells. *Glia* 2002, 38:313-328.

[32] Nakao J, Shinoda J, Nakai Y, Murase S, Uyemura K: Apoptosis regulates the number of Schwann cells at the premyelinating stage. *J Neurochem* 1997, 68:1853-1862.

[33] Pasquini LA, Besio Moreno M, Lopez Salon M, Soto EF: Apoptosis in Schwann cell cultures is closely interrelated with the activity of the ubiquitin-proteasome proteolytic pathway. *Neurochem Res* 2002, 27:1401-1419.

[34] Calle E, Berciano MT, Fernandez R, Lafarga M: Activation of the autophagy, c-FOS and ubiquitin expression, and nucleolar alterations in Schwann cells precede demyelination in tellurium-induced neuropathy. *Acta Neuropathol (Berl)* 1999, 97:143-155.

[35] Melean G, Sestini R, Ammannati F, Papi L: Genetic insights into familial tumors of the nervous system. *Am J Med Genet C Semin Med Genet* 2004, 129:74-84.

[36] Cichowski K, Santiago S, Jardim M, Johnson BW, Jacks T: Dynamic regulation of the Ras pathway via proteolysis of the NF1 tumor suppressor. *Genes Dev* 2003, 17:449-454.

[37] Gautreau A, Manent J, Fievet B, Louvard D, Giovannini M, Arpin M: Mutant products of the NF2 tumor suppressor gene are degraded by the ubiquitin-proteasome pathway. *J Biol Chem* 2002, 277:31279-31282. Epub 32002 Jul 31218.

[38] Campagnoni AT, Macklin WB: Cellular and molecular aspects of myelin protein gene expression. *Mol Neurobiol* 1988, 2:41-89.

[39] Smith R: The basic protein of CNS myelin: its structure and ligand binding. *J Neurochem* 1992, 59:1589-1608.

[40] Akaishi T, Shiomi T, Sawada H, Yokosawa H: Purification and properties of the 26S proteasome from the rat brain: evidence for its degradation of myelin basic protein in a ubiquitin-dependent manner. *Brain Res* 1996, 722:139-144.

[41] Snipes GJ, Suter U, Welcher AA, Shooter EM: Characterization of a novel peripheral nervous system myelin protein (PMP-22/SR13). *J Cell Biol* 1992, 117:225-238.

[42] Schachner M, Martini R: Glycans and the modulation of neural-recognition molecule function. *Trends Neurosci* 1995, 18:183-191.

[43] Schneider C, King RM, Philipson L: Genes specifically expressed at growth arrest of mammalian cells. *Cell* 1988, 54:787-793.

[44] Naef R, Suter U: Many facets of the peripheral myelin protein PMP22 in myelination and disease. *Microsc Res Tech* 1998, 41:359-371.

[45] Zoidl G, Blass-Kampmann S, D'Urso D, Schmalenbach C, Muller HW: Retroviral-mediated gene transfer of the peripheral myelin protein PMP22 in Schwann cells: modulation of cell growth. *Embo J* 1995, 14:1122-1128.

[46] Pareek S, Notterpek L, Snipes GJ, Naef R, Sossin W, Laliberte J, Iacampo S, Suter U, Shooter EM, Murphy RA: Neurons promote the translocation of peripheral myelin protein 22 into myelin. *J Neurosci* 1997, 17:7754-7762.

[47] Trombetta ES, Parodi AJ: Quality control and protein folding in the secretory pathway. *Annu Rev Cell Dev Biol* 2003, 19:649-676

[48] Notterpek L, Ryan MC, Tobler AR, Shooter EM: PMP22 accumulation in aggresomes: implications for CMT1A pathology. *Neurobiol Dis* 1999, 6:450-460.

[49] Wojcik C, Schroeter D, Wilk S, Lamprecht J, Paweletz N: Ubiquitin-mediated proteolysis centers in HeLa cells: indication from studies of an inhibitor of the chymotrypsin-like activity of the proteasome. *Eur J Cell Biol* 1996, 71:311-318.

[50] Johnston JA, Ward CL, Kopito RR: Aggresomes: a cellular response to misfolded proteins. *J Cell Biol* 1998, 143:1883-1898.

[51] Fortun J, Go JC, Li J, Amici SA, Dunn WA, Jr., Notterpek L: Alterations in degradative pathways and protein aggregation in a neuropathy model based on PMP22 overexpression. *Neurobiol Dis* 2005, 30:30.

[52] Forman MS, Trojanowski JQ, Lee VM: Neurodegenerative diseases: a decade of discoveries paves the way for therapeutic breakthroughs. *Nat Med* 2004, 10:1055-1063.

[53] Fortun J, Dunn WA, Jr., Joy S, Li J, Notterpek L: Emerging role for autophagy in the removal of aggresomes in Schwann cells. *J Neurosci* 2003, 23:10672-10680.

[54] Iwata A, Riley BE, Johnston JA, Kopito RR: HDAC6 and Microtubules Are Required for Autophagic Degradation of Aggregated Huntingtin. *J Biol Chem* 2005, 280:40282-40292. Epub 42005 Sep 40228.

[55] Snipes GJ, Suter U: Molecular basis of common hereditary motor and sensory neuropathies in humans and in mouse models. *Brain Pathol* 1995, 5:233-247.

[56] Ryan MC, Shooter EM, Notterpek L: Aggresome formation in neuropathy models based on peripheral myelin protein 22 mutations. *Neurobiol Dis* 2002, 10:109-118.

[57] Johnson JS, Roux KJ, Fletcher BS, Fortun J, Notterpek L: Molecular alterations resulting from frameshift mutations in peripheral myelin protein 22: Implications for neuropathy severity. *J Neurosci Res* 2005, 4:4.

[58] Isaacs AM, Jeans A, Oliver PL, Vizor L, Brown SD, Hunter AJ, Davies KE: Identification of a new Pmp22 mouse mutant and trafficking analysis of a Pmp22 allelic series suggesting that protein aggregates may be protective in Pmp22-associated peripheral neuropathy. *Mol Cell Neurosci* 2002, 21:114-125.

[59] Frydman J: Folding of newly translated proteins in vivo: the role of molecular chaperones. *Annu Rev Biochem* 2001, 70:603-647.

[60] Sherman MY, Goldberg AL: Cellular defenses against unfolded proteins: a cell biologist thinks about neurodegenerative diseases. *Neuron* 2001, 29:15-32.

[61] Tobler AR, Liu N, Mueller L, Shooter EM: Differential aggregation of the Trembler and Trembler J mutants of peripheral myelin protein 22. *Proc Natl Acad Sci U S A* 2002, 99:483-488.

[62] D'Urso D, Ehrhardt P, Muller HW: Peripheral myelin protein 22 and protein zero: a novel association in peripheral nervous system myelin. *J Neurosci* 1999, 19:3396-3403.

[63] Fortun J, Li J, Go J, Fenstermaker A, Fletcher BS, Notterpek L: Impaired proteasome activity and accumulation of ubiquitinated substrates in a hereditary neuropathy model. *J Neurochem* 2005, 92:1531-1541.

[64] Bence NF, Sampat RM, Kopito RR: Impairment of the ubiquitin-proteasome system by protein aggregation. *Science* 2001, 292:1552-1555.

[65] Dantuma NP, Lindsten K, Glas R, Jellne M, Masucci MG: Short-lived green fluorescent proteins for quantifying ubiquitin/proteasome-dependent proteolysis in living cells. *Nat Biotechnol* 2000, 18:538-543.

[66] Bergamini E, Cavallini G, Donati A, Gori Z: The role of macroautophagy in the ageing process, anti-ageing intervention and age-associated diseases. *Int J Biochem Cell Biol* 2004, 36:2392-2404.

[67] Soti C, Csermely P: Aging and molecular chaperones. *Exp Gerontol* 2003, 38:1037-1040.

[68] Notterpek L, Shooter EM, Snipes GJ: Upregulation of the endosomal-lysosomal pathway in the Trembler-J neuropathy. *J Neurosci* 1997, 17:4190-4200.

In: The Ubiquitin Proteasome System... ISBN 978-1-60021-749-4
Eds: Mario Di Napoli and Cezary Wojcik, pp. 411-442

Chapter 18

DYNAMIC REGULATION OF SYNAPTIC PLASTICITY BY THE UBIQUITIN PROTEASOME SYSTEM

Jennifer A. Johnston[*]
Elan Pharmaceuticals, South San Francisco, CA 94080, USA.

ABSTRACT

The ubiquitin-proteasome pathway (UPS) is understood to have a major role in the regulation of the cell cycle of dividing cells. More recently, it has become clear that even non-dividing cells, like neurons, rely on the specific regulated protein degradation pathways of the UPS. Through genetic studies as well as elegant biochemical and cell biological experiments it is now clear that the UPS plays an important role in neuron specific functions, such as pathfinding during development, neurodegeneration and synaptic plasticity. The regulation of synapse formation and re-formation is central to neuronal function. Recent work has demonstrated that the regulated destruction of specific components of the synapse is a crucial determinant of the long-term status of a neuronal connection. Moreover, it is clear that neurotransmitter stimulation of receptors can begin a cascade resulting in the targeted removal of specific synapse scaffold proteins. The permanent removal of these proteins through degradation allows for the rearrangement of synapse components in a manner specific to the neurotransmitter-mediated signal.

Keywords: Proteasome, neurodegeneration, synapse, ubiquitin.

[*] Correspondence concerning this article should be addressed to Dr. Jennifer A. Johnston, PhD; Elan Pharmaceuticals 1000 Gateway Blvd., South San Francisco, CA 94080, USA.. Phone: +1-650-794-4271; Fax: +1-650-553-7196; Email: jennifer.johnston@elan.com.

Abbreviations

AchR; nicotinic acetylcholine receptor; ALS, amyotrofic lateral sclerosis; AMPA, alpha-amino3-hydroxy-5-methyl-4-isoxazole propionic acid; APC/C, anaphase-promoting complex/cyclosome; C. elegans, Caenorhabditis elegans; CREB, cyclic AMP- responsive element binding protein; E6-AP, Ebe3a gene encodes E6 associated protein; EJC, evoked junctional currents; ExJC, excitatory junctional currents; ERAD, endoplasmic reticulum associated degradation; GABA, gamma-amino-butyric acid; GFP, green fluorescent protein; GlyR, glycine receptor; GPCR, G protein-coupled receptor; HECT, homologus to E6-AP C-terminus; HTT, huntingtin; IPC, insoluble protein complexes; LGIC, ligand, gated ion channel; LTD, long-term depression; LTP, long-term potentiation; mGluR; metabotropic glutamate receptors; NMDA, N-methyl-D-aspartic acid; NMJ, neuromuscolar junction; NTAN, N-terminal amidase; PLK2, polo-like kinase 2; PKA, protein kinase A; PSD, post-synaptic density; SCA, spinocerebellar ataxia; SNARE, NEM-sensitive fusion protein attachment protein receptor; SNK, serum-induced kinase; SPAR, spine-associated Rap guanosine triphosphatase activating protein; TTX, tetrodotoxin; Ub, ubiquitin; UPS, ubiquitin-proteasome system; VAMP2, vesicle associatedmembrane protein 2.

Introduction

The brain is not a static organ. In order to create, store and retrieve information the brain has evolved as a dynamic system that can quickly respond to stimuli. The stability of neuronal synapses is an important means to connect different brain regions and to allow the transfer of biochemical and electrical signals that can be translated to us cognitively as specific sensory information or responses. Cognition, learning and memory has evolved as a system that is stable, to allow memory, but is not static, to allow learning. The stability and instability of neuronal connectivity is generally referred to a synaptic plasticity, and is currently believed to be the underlying principle for learning and memory in the brain.

Neurons are polarized cells that receive, transmit, and even process information. The cell surface of neurons is specialized, with microdomains that have localized functions. Neurons have one axon, that primarily functions to provide neurotransmitter release from terminal synapses in response to input stimuli. Neurons have many dendrites that are specialized to respond to stimuli at the post-synaptic densities (PSD), and the PSD is characterized by the presence of neurotransmitter receptor complexes. So, in one portion of the cell is a receiver of input stimuli and another portion is responsible for transmitting a signal. The specialized (and highly simplified in this chapter) functions of neuronal axons and dendrites require distinct sets of protein components. These components allow the processing of input and transmission signals to result in functional cellular responses through rearrangement of protein components at the cell surface in localized microdomains (e.g. presynapse, PSD).

A neuronal synapse, generally speaking, consists of two neurons: a presynaptic and a post-synaptic neuron. A tiny portion of each cell's surface is bound to each other through extracellular matrix molecules and cell surface proteins [1-5]. Chemical and electrical information is transferred from the pre-synaptic cell to the post-synaptic cell via regulated release of synaptic vesicle components from the pre-synaptic cell. These components can then interact with receptors on the post-synaptic cell to impart information about appropriate responses to the stimuli. While a short-term and transient stimulus can lead to post-synaptic activity in the form of an intracellular signal back to the postsynaptic cell body and nucleus, it is also possible that a continued stimulation of the post-synaptic cell can lead to a fundamental architectural change in the way the cell will respond to the stimuli in the future.

These changes in synaptic membrane components are currently thought to underlie the fundamental aspects of neuronal plasticity (review see [6]). Synapses can be made stronger or weaker depending on the type of stimuli that is received, and the type of neuron receiving the signal. Synapses that are made stronger exhibit long-term potentiation (LTP), a state in which the synapses are attuned to repeated stimuli and have altered the cohort of receptors at the cell surface to reflect this change (for review, see [7-9]). Synapses that have been made weaker due to stimuli exhibit long-term depression (LTD) and have also fundamentally changed their synaptic membrane components (for review of LTD see [10-12]). This choreography of membrane components at two apposing membrane surfaces is possible through intracellular events that facilitate the rearrangement of cell surface proteins. For example, in order for a neurotransmitter receptor to be removed from the cell surface, intracellular scaffold proteins that previously held the receptor in place, or helped the receptor get to the cell surface in the first place, must be altered in their function for the receptor. This can be achieved through post-translational modification of the protein, for example phosphorylation, which will alter its ability to interact with the receptor. Then dephosphorylation can reverse protein function to allow a return to its original interaction with the receptor. However, with the long-term cellular changes that underlie synaptic plasticity, synapses are changed in a fundamental way, requiring both protein synthesis [13,14] and protein degradation [15], whereas phosphorylation and dephosphorylation is a relatively reversible and transient change. The role for protein synthesis has been known for some years [16] and recently has a major role for protein degradation become known [15]. However, with the first suggestion that the targeted degradation of protein was involved in receptor turnover at synaptic sites, there have been numerous excellent studies that have elucidated clear roles for ubiquitination and degradation, either through the proteasome [15,17-19], or through targeting to lysosomes [20]. Ubiquitin-dependent degradation of protein can occur through the proteasome or the lysosome and is a result of substrate recognition by specific proteins that function as E3 ligases in a multi-enzyme cascade that results in the covalent addition of ubiquitin to the substrate (The general mechanisms of the enzymatic cascade of ubiquitin pathway enzymes is discussed in Chapter 3, also see [21-23]). In general, proteins that have been covalently tagged by only one ubiquitin moiety are directed to the lysosome. Proteins that have multiple ubiquitin moieties covalently attached to them are escorted to the proteasome. Monoubiquitination (Mono-ub) is largely responsible for the degradation of cell surface transmembrane proteins, because the endocytosis of the receptor in a membrane milieu is suited for fusion with the lysosomal membrane. Epsins interact with both substrate-bound Mono-ub and the endocytosing membrane and serve to aid in the direction of plasma membrane proteins to the lysosome [24-26]. Multiubiquitin (Multi-ub) is added to cytoplasmic proteins in a characteristic chain, that to serve as an effective degradation signal must equal or exceed four ubiquitin moieties [27]. In this ubiquitin chain, one ubiquitin is linked to another through the epsilon amino group of a lysine on the acceptor ubiquitin, and the C-terminal glycine of the next ubiquitin [27,28]. Each ubiquitin is added sequentially to build the chain. Multi-ub is recognized by cellular shuttle-factors to shuttle the ub-tagged protein to the proteasome for degradation [29]. One important exception to the rule that 'membrane proteins are degraded by lysosomes' can occur during the biosynthesis of a transmembrane protein. While the protein is being processed through the ER, it can be specifically targeted for removal through quality control mechanisms [30]. The quality control process will result in multiubiquitination and retrotranslocation of the membrane protein back in to the cytosol, where it is degraded by proteasomes [31]. As yet, it has not been shown that cell surface transmembrane proteins are pulled out of the membrane for degradation by cytoplasmic proteasomes. However, in the specific case of ER-associated degradation (ERAD) transmembrane proteins are recognized in the ER as degradation substrates and are retrotranslocated out of the ER by Sec61-dependent mechanisms to be degraded by cytoplasmic proteasomes after multiubiquitination [32,33]. The crucial distinction in the case of ERAD is that the transmembrane proteins never made it to the plasma membrane, and were degraded during biosynthesis as evidenced by the lack of acquisition of mature glycosylation on the proteins prior to degradation. ER quality control (ERAD), monoubiquitination and multiubiquitination are all cellular processes that happen in all cells. However, the mechanisms that govern the highly specialized regulation of synaptic proteins and information about how these events occur has only recently been revealed.

Neuronal cell biology has recently been facilitated by advances in cell culturing techniques that allow the growth and maintenance of specific types of neurons. With this technical achievement has come the ability to study the unique behavior of neurons, which are a remarkable cell type. The cells are highly polarized, and the cellular architecture is the most complex of any cell in the body. Because neurons are cells, they share many of the most basic cellular mechanisms and pathways as other cells (e.g. transcription, translation, protein secretion, etc). However, because of their highly specified and regulated functions, and their ability to transfer information along great distances requires a uniquely coordinated means of dispersing information and maintaining cellular integrity. Moreover, because the distance from the cell body to a synapse is so great, there is a cell body independent means of processing information that does not require the nucleus, although communication between cell body and synapse is tightly linked [34]. In fact, synaptosomes, which are synapses that have been separated from their respective pre and post synaptic cell bodies, can still respond to specific stimuli suggesting the synapse has some autonomy from the cell body.

Early studies showed clearly that neuronal synapses can respond to stimuli with mechanisms that require protein synthesis, and that the new proteins were synthesized from mRNA that resides close to the synapse [49,50]. However it was also shown that changes in synaptic responses to stimuli also correlated with a down-regulation of certain receptor types at the cell surface and that these changes in receptor number are regulated through endocytosis and exocytosis [51]. Recently, the focus has turned to protein degradation as an unequivocal way of altering the protein composition of a synapse. We now know that receptor number at the cell surface can be regulated biosynthetically. For example, the $GABA_A$ receptor is a multisubunit receptor that can be regulated biosynthetically by the assembly of $GABA_A$ receptor subunits in the ER. Alterations in the stability of the $GABA_A$

subunit will alter the amount of GABA receptor on the cell surface and change the cell's excitable state in a relatively long-term fashion [18]. Post-synaptically, receptor number at the cell surface can also be regulated via monoubiquitination and endocytosis to the lysosome. For example, the 5-HT receptor is internalized rapidly after serotonin administration and this internalization was shown to require a Mono-ub dependent mechanism [52]. Moreover, the internalization of a receptor through monoubiquitination and endocytosis also requires multiubiquitination and proteasome degradation of scaffold proteins that hold the receptor complexes in place at the cell surface. For example, PSD-95 is a scaffold protein that holds the AMPA receptors in place on the cell surface and specific stimuli can result in the multiubiquitination and degradation of PSD-95 by proteasomes [17]. The degradation of PSD-95 then allows the internalization of the receptor/membrane complex and delivery to the lysosome. Presynaptically, the regulation of exocytosis of neurotransmitter containing vesicles is regulated by proteasome dependent-mechanisms, and a variety of E3s have been implicated in those processes [45,46]. Finally, the sensitization of neurons and the development of LTP or LTD requires the upregulation of specific kinases, which in turn regulate their cognate targets [19]. Often, the phosphorylation of the target leads to instability and recognition for degradation by cellular proteasomes. Ubiquitin pathway enzymes known to be involved in these processes are summarized in Table I. In this chapter, we will discuss the recent information known about synaptic plasticity and the role for ubiquitin-dependent degradation.

Table I. Ubiquitin Proteasome Pathway Enzymes with a Role in Synaptic Plasticity

Neuronal Protein	Relationship to the UPS	Neuronal location	Reference
Cdh1-APC	E3 ligase	Pre, PSD	[35-37]
E6-AP	E3 ligase	PSD (likely)	[38]
Fbx2	E3 ligase	PSD	[39]
Lin-23	E3 ligase	PSD	[40]
MdM2	E3 ligase	PSD	[17]
NTAN1/UBR1	E3 ligase	Pre, PSD	[41]
Parkin	E3 ligase	Pre, PSD	[42]
Plic-1	Shuttle factor	Cytoplasmic	[18]
Siah	E3 ligase	Pre, PSD	[43, 44]
Spring	E3 ligase	Pre	[45]
Staring	E3 ligase	PSD	[46]
Uch-L1	Deubiquitinating enzyme	Pre, PSD	[47]
Usp14	Deubiquitinating enzyme	Pre, PSD	[48]

Biosynthetic Regulation of Receptor Delivery to the Cell Surface

Glutamate receptors play a central role in the neuronal excitation of the mammalian CNS. There are two types of glutamate receptors, ionotropic and metabotropic receptors.

Metabotropic receptors (mGluRs) are G protein coupled receptors (GPCR) and ionotropic glutamate receptors are ligand gated ion channels (LGIC). Generally speaking, ligand-activated metabotropic receptors can affect intracellular signaling cascades that may affect neurons excitability. Ionotropic receptors will open their ion channel in response to ligand binding, allowing for the entry of specific ions, such as Ca^{2+}, Na^{+} or Cl^{-} that will then affect intracellular processes.

MGluR1 and 5 are specific metabotropic glutamate receptors that are linked to IP3 signaling cascades, and have important roles in both long-term potentiation and depression, as well as synapse development and elimination [53-56]. MGluR1 has been shown to be a short-lived protein that can be stabilized by proteasome inhibitors, mutation of specific lysine residues or deletion mutations in the C-terminus of the protein [44]. The half-life of MGluR1 can be made even shorter when expressed in the presence of Siah-1A [44]. Siah-1A is an E3 ligase with numerous reported neuronal substrates (DCC: [57] synaptophysin: [43] and Numb: [58]). Siah-1A expression with mGluR1 resulted in an increase in the detectable ubiquitination of the mGluR1, whereas expression of a ring-finger mutant of Siah-1A did not increase ubiquitination of mGluR1, suggesting Siah-1A is responsible for degradation of a group of mGluRs [44]. The fact that the mGluRs were modified by multiubiquitin and the half-life in the presence of Siah-1A was about 10 hours shorter than in its absence suggests that Siah-1A recognizes mGluRs during biosynthesis, and probably mediate an ERAD process to regulate cellular levels of mGluRs. It is important to point out that Siah-1 is itself a short-lived protein that is regulated by the UPS pathway [59]. This suggests that there may be an extra level of control over the biosynthesis of Siah-1 substrates because there may be an as yet unidentified factor that can control the degradation of Siah-1. This unidentified factor(s) could be regulated by neuronal activity or developmental cues (in the case of DCC) to control the abundance of Siah-1 in neurons.

Ligand gated ion channels are well-known mediators of neuronal activity. Nicotinic acetylcholine receptors (AChR) are hetero-oligomeric ion channel complexes whose biogenesis and membrane insertion have been extensively studied [60]. Recently, the number of channels assembled and delivered to the cell surface was shown to increase in the presence of proteasome inhibitors [61]. Detailed analysis showed that the AChR complex subunits were regulated by multi-ubiquitination and degradation by proteasomes. The multi-ubiquitinated subunits fractionated with membranes, and displayed appropriately glycosylated forms, suggesting that the retrotranslocation of the subunits and degradation by proteasomes is a tightly coupled process [61]. The degradation of AChR subunits by ERAD appears to be a means to regulate the number of receptors on the cell surface, because the ligand binding and sedimentation profiles of the ER-associated forms of the AChR that accumulate after proteasome inhibitor treatment are preserved, suggesting appropriately folded and assembled AChR complexes [61]. Oftentimes, ERAD is involved in the removal of malfolded proteins in the ER, but this example highlights the alternative function of ERAD, which is to tightly control receptor delivery to the cell surface. In the AChR example, ERAD controls the number of cell surface receptors, but is probably regulated, perhaps like the GABAa receptors by Plic-1, to be able to quickly respond to stimuli to increase or decrease the number of receptors at the cell surface after stimuli.

$GABA_A$ receptors are responsible for fast synaptic inhibition in neurons and $GABA_A$ receptors cycle between the cell surface and endocytic compartments to regulate the number of receptors at inhibitory synapses [62]. An increase in $GABA_A$ receptor number is correlated to an increase in the inhibitory strength of the synapse. Increases in cell surface receptor could be due to inhibition of endocytosis from the cell surface, increase in recycling of endocytic pools of receptors, or an increase in the number of receptors inserted in to the membrane via biosynthetic mechanisms. The endocytosis of $GABA_A$ receptors has been shown to require clathrin and dynamin-dependent mechanisms and is as yet not linked to a monoubiquitination-dependent process [63]. In addition, PKC, GABARAP, gephyrin, growth factor tyrosine kinase receptor and Plic-1 have been implicated in regulation of $GABA_A$ at the cell surface [18,64-66]. Plic-1 protein is generally characterized as a ubiquitin-related protein because its amino terminus is 33% identical to ubiquitin, and its C-terminus contains a ubiquitin-associated (UBA) domain [67]. Plic-1 has been shown to be a putative 'shuttle factor' that binds to ubiquitin via the UBA domain and also binds to the proteasome via its ubiquitin-like domain [68,69]. Shuttle proteins have been speculated to escort ubiquitinated protein to the proteasome [29]. $GABA_A$ receptor subunits are short lived when examined by pulse chase analysis [70], and their half-life significantly increases in the presence of proteasome inhibitors, suggesting that $GABA_A$ receptor subunits are degraded by the proteasome during biosynthesis [18]. In the presence of overexpressed Plic-1, the half-life of the $GABA_A$ subunits increased approximately 30%. Overexpression of Plic-1 also correlated to an increase in $GABA_A$ receptor at the cell surface, as well as an increase in inhibitory strength of the synapses [18]. At first glance it may seem counterintuitive that Plic-1 would decrease proteasome degradation of $GABA_A$ subunits, given its function as a shuttle factor. However, a likely mechanism for the observed data is that the overexpressed Plic-1 binds to the $GABA_A$ receptor subunit prior to multiubiquitination and as such disrupts the appropriate targeting to the proteasome, increasing intracellular pools of $GABA_A$ receptor. A similar function for PLIC-1 with other substrates has been reported [68,71]. Plic-1 may also be acting to stabilize previously internalized receptors because Plic-1 can be localized to clathrin-coated vesicles and intracellular membranes [18] and has recently been linked to Eps15 [72]. Currently, while it is unknown if Plic-1 is acting on both recycling and biosynthetic pathways, or if it has specificity for one pathway, pulse-chase analysis has established the role of biosynthetic degradation [18]. Biosynthetic regulation of the assembly of cell surface protein receptors is a common mechanism in cells. Often, a particular subunit of a multimeric complex is regulated in abundance by degradation by the proteasome via ERAD. Elimination of the targeted subunit prevents receptor complex assembly and therefore regulates insertion in the plasma membrane. A rapid stabilization of the limiting subunit would allow the cell great flexibility to upregulate receptor levels without having to increase mRNA levels via a nuclear signaling mechanism and allows cells to respond rapidly to changing cellular conditions. In the example of the $GABA_A$ receptor assembly, Plic-1 may regulate the degradation of subunits as a result of specific neuronal stimuli to increase the strength of inhibitory synapses. It would be interesting to know if Plic-1 association with $GABA_A$ receptor subunits changes with signals that increase inhibitory synaptic strength.

In the three examples discussed in this section, it is clear that cell surface receptor complexes can be regulated in their delivery to the cell surface through stability of specific receptor subunits at an intermediate point in the maturation of the protein. Under certain conditions, a subunit may be rapidly degraded to decrease cell surface expression. However, under other conditions degradation is prevented and the stabilized subunit can assemble with other members of the receptor to allow increased receptor delivery to the cell surface (see summary in Figure 1B). And, it is the dynamic changes in receptors at the cell surface that defines synaptic responsiveness and plasticity for neurons.

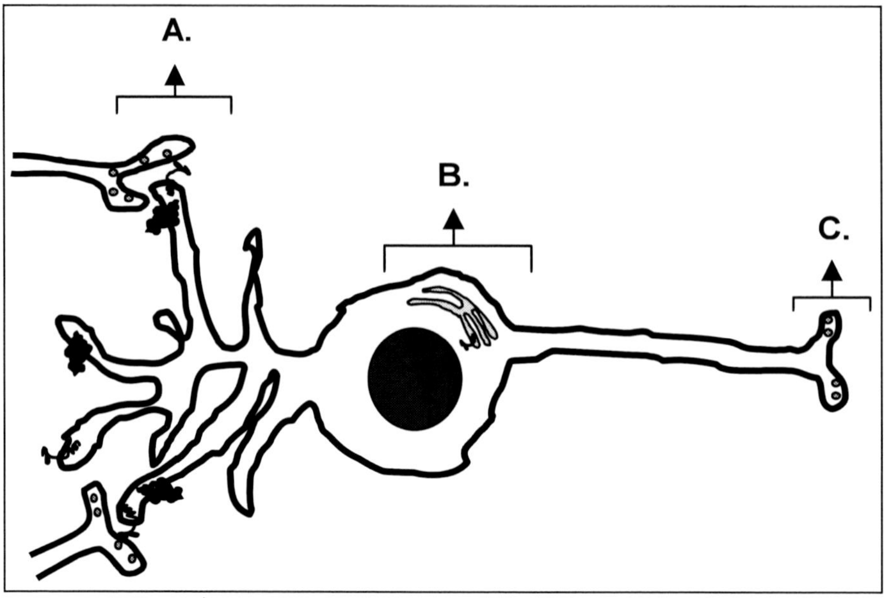

Figure 1. Schematic representation of a CNS neuron. A,B, and C represent different domains of the neuron and are discussed in relation to the UPS in detail in 1A, B and C.

Post-Synaptic Regulation of Cell Surface Receptors

Ubiquitin-dependent pathways regulate an orchestra of events at the PSD to allow for the development of LTP and LTD responses to stimuli. For the ionotropic glutamate receptors, NMDA and AMPA receptors, the ubiquitin-proteasome pathway is involved in many facets of the receptors' responsiveness to neurotransmitters at the PSD (Kainate is also a ionotropic glutamate receptor, however it is not known if the UPS is involved in its regulation). Ubiquitination events regulate the receptor subunits themselves, as well as scaffolding proteins that allow for receptor membrane insertion or removal from the plasma membrane [17,39,73]. The activation of NMDA receptors can lead to a decrease in AMPA receptors at the cell surface and stable LTD [51,74]. It now seems clear that the removal of scaffold proteins that hold cell surface receptors in place will allow cell surface remodeling in response to highly specific stimuli. The internalization of the receptor requires a monoubiquitination/lysosome event and the removal of the scaffold protein requires a multiubiquitination/proteasome event. This dynamic process is shown schematically in Figure 1A-C, and is discussed in detail below.

A *tour de force* study was reported in 2003, by Ehlers in which the author examined the fate of synaptic proteins after stimuli in both intact neurons in culture as well as synaptosomes [15]. Synaptosomes are synapses that have been isolated away from the cell body of the neuron. Because the pre and post-synaptic portions of the synapse are held together very tightly, it is possible to isolate them from their respective cell bodies. Experimentally, the presynaptic portion of the synaptosomes can be stimulated to release neurotransmitter and the post-synaptic portion will respond to the neurotransmitter by modulation of the components at the cell surface. Rearrangement of PSD components after stimulation has been suggested to involve Ca^{2+} mediated signal cascade molecules, scaffold proteins and cell surface receptor molecules.

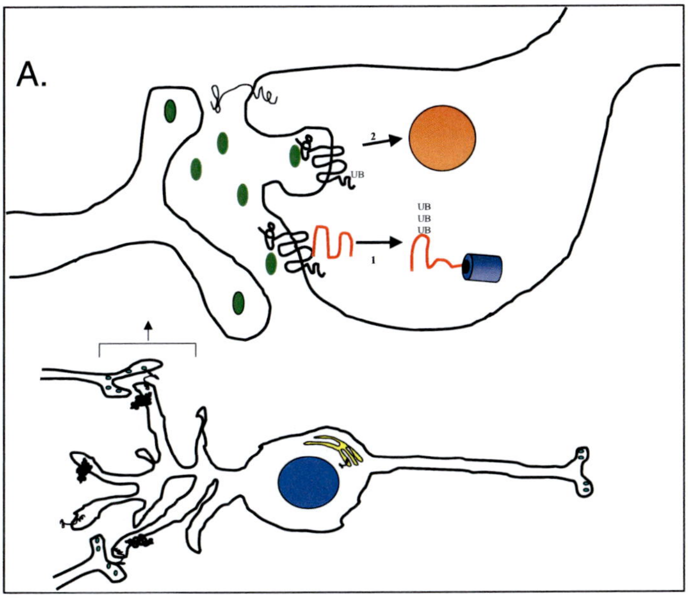

Figure 1A. PSD in a dendrite of the neuron. Presynaptic elements release neurotransmitter (green) that acts on postsynaptic receptors to generate intracellular events that regulate response to stimuli. In this diagram, a cell surface receptor is internalized after two ubiquitin dependent events: 1. Multi-ubiquitination and degradation of a receptor scaffold Protein (red) (i.e., PSD-95 or SPAR) is required for the endocytosis of a receptor; 2. Mono-ubiquitination of the cell surface receptor is required for the delivery to lysosomes (brown) for degradation.

To gain insight into the molecular identity of the specific proteins that are rearranged after stimuli in a global fashion, Ehlers used a large-scale proteomics effort complemented by a candidate approach to characterize the proteins that were short-lived after the application of stimuli [15]. Bicuculline was used to mimic activating stimuli and tetrodotoxin (TTX) was used to mimic inhibitory stimuli. A consistent pattern of proteins was seen to disappear after stimulation, and pulse-chase analysis determined that synaptic activity alters the half-life of specific synaptic proteins. In the presence of proteasome inhibitors these specific proteins were stabilized. A specific antibody to multiubiquitinated proteins was used to isolate the cohort of proteins that were multiubiquitinated after stimuli. Proteins identified included post-synaptic scaffolding proteins, Shank, GKAP and AKAP79/150. These proteins have been implicated in the delivery and maintanence of receptors at the PSD [74,75], and this recent result suggests that proteasomal degradation may remove these scaffolding proteins, allowing for dramatic alterations in the PSD. Interestingly, the cell surface receptors, NR1 and NR2b of the NMDA receptor decreased but were not multiubiquitinated. It may be that the receptors are modified with Mono-ub for lysosomal degradation, but this was not determined by Ehlers [15] as his methods were designed to only identify multiubiquitinated proteins. Importantly, this study was the first to show that synaptosomal preparations, synapses isolated for the cognate cell bodies, had active proteasomal degradation pathways. This means that the recognition proteins, E3 ligases, conjugating enzymes, E2s, and proteasomes were present in the synapses and could be activated and function autonomous from the cell body. This study and the others discussed below, opened the door to new targets of Ca+ mediated synaptic events, and a new mechanism for regulation of synaptic receptor activation and desensitization through the finality of the ubiquitin-proteasome pathway. Subsequently, it was shown that PSD-95 is ubiquitinated and degraded after specific stimuli, in contrast to the work by Ehlers [17]. PSD-95 has been implicated as a 'slot protein' that functions to maintain and deliver receptors to the cell surface in neurons [76]. The decrease in surface AMPA receptors after NMDA stimulation can be blocked by the addition of proteasome inhibitors and by a mutation in the PSD-95 that prevents ubiquitination [17]. Importantly, PSD-95 ubiquitination is dependent on the appropriate stimulus, providing further support that the ubiquitination of PSD-95 is a specific response to neuronal signaling cues. For example, in Ehlers [15] the global analysis of synaptosomal protein turnover by the proteasome did not reveal PSD-95 and this is most likely due to the difference between Bicuculline stimulation of synaptosomes versus the NMDA stimulation of neurons. These results highlight the importance of using various defined experimental paradigms to examine specific proteins of interest and their targeting to the proteasome, as it appears to be highly stimuli specific. This is not surprising, as a primary function of neuronal synapses is to create a small, localized area for specific cellular responses.

Intact cultured hippocampal neurons have also been shown to require active proteasomes for internalization of AMPA receptor subunits GluR1 and GluR2 [73]. Application of proteasome inhibitors (MG-132) prevented internalization after bicuculline stimulation, suggesting the effects were synaptic. MG-132 is a reversible proteasome inhibitor and its addition for as little as 2-5 minutes was sufficient for the observed block. GluR1,2 internalization required multiubiquitination, because a K48R mutant of ubiquitin prevented internalization. The K48R mutant would support monoubiquitination [28,77], so the fact that internalization was blocked with K48R suggests that proteasomal degradation of multiubiquitinated substrates was required for the GluR1 and 2 internalization. However, the identity of the protein(s) degraded is not yet been identified, however it is likely to be a scaffold protein similar to those identified by Ehlers (2003). Gentry *et al.*, correlated proteasome location with synaptic components and this data is consistent with a high resolution electron microscopic analysis of proteasome location in neurons that suggested a key role for the proteasome in neurons, specifically in long-term memory, over ten years ago

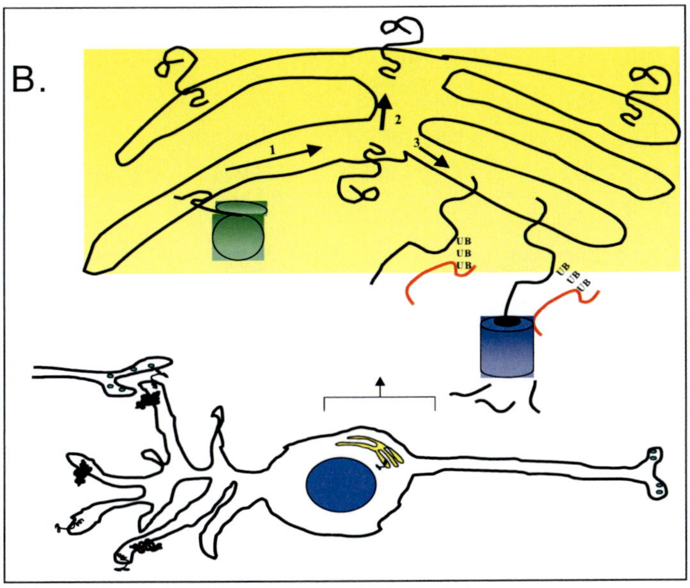

Figure 1B. Biosynthetic degradation of a cell surface receptor can regulate delivery to the cell surface. Receptor subunits are synthesized on cytoplasmic ribosomes and delivered into the endoplasmic reticulum (ER) where they are either: 1: assembled into receptor complexes and transited to the cell surface; or 2: recognized for degradation, followed by multi-ubiquitination and dislocation from the ER into cytoplasmic proteasomes. In the case for the $GABA_A$ receptor, Plic-1 (red) binds to a multi-ubiquitin ladder and to the proteasome and shuttles substrates into the mouth of the proteasome. The ER of a neuron is present throughout the neuron and is shown schematically here.

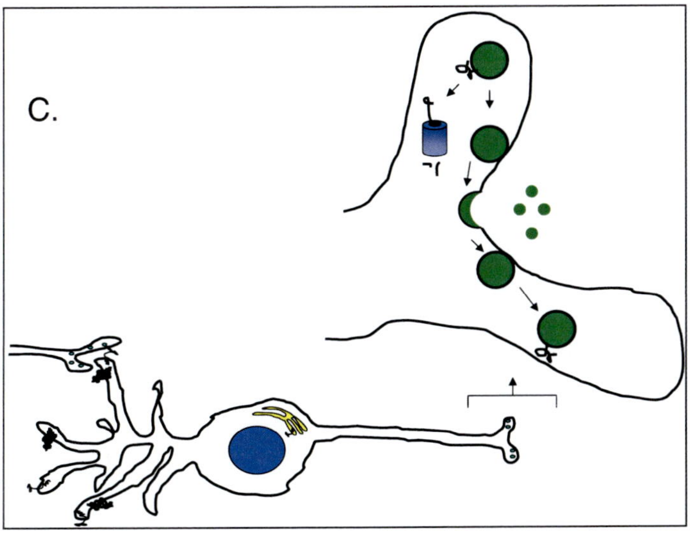

Figure 1C. Regulation of vesicle exocytosis in the presynapse. The presynapse of a neuron is responsible for the release of neurotransmitter-containing vesicles. The selective multiubiquitination and degradation of specific synaptic vesicle-associated proteins is required for vesicle fusion and exocytosis, as well as the re-endocytosis of the vesicle after release (recycling). In this example, the degradation of Dunc-13 is required for the ability of the vesicle to be primed with the plasma membrane prior to release of neurotransmitter contents.

[78]. That speculation was based on the data that demonstrated proteasomes in the nucleus and in the synapse [78], areas we now know to be the crucial cellular locations for regulation of degradation controlling synaptic plasticity. It would be highly interesting to understand if proteasomes have a dynamic spatial regulation in neurons in relation to the excitation state of the neuron. For example, in highly active neurons are proteasomes recruited to synapses, and conversely transferred to other locations in synapses that are under LTD? In general, there have not been definitive cellular studies to determine if the proteasome itself undergoes specific translocations in the cell. It is currently understood that the proteasome is found diffusely throughout the cell, facilitating degradation at almost any place in the cell. In the highly compartmentalized architecture of the neuron it may prove to be more important for the proteasome to be recruited to a particular active region. Future studies will likely address this important neuronal issue.

The inhibitory Glycine receptor (GlyR) is a ligand gated ion channel (LGIC) that is regulated in its levels at the cell surface in response to neuronal stimuli.. The GlyRs are internalized via endocytosis and this process is regulated by monoubiquitination followed by degradation in lysosomes [79]. Cell surface labeling followed by affinity isolation of GlyR complexes revealed specific ubiquitination of a fraction of the GlyRs [79]. The degradation of the GlyRs does not occur via proteasomes because lysosomotrophic agents prevented degradation of the GlyRs while proteasome inhibitors did not [79]. Because of the methods used for these experiments it was not possible to understand if the cell surface delivery of assembled subunits could be affected by ERAD mechanisms, (although previous studies suggest this may be the case [80]) but it was clear the removal of GlyR from the cell surface requires at least one monoubiquitination event for efficient targeting to lysosomes. Interestingly, because the degradation of the GlyRs was not affected by proteasome inhibitors, it suggests there is not a scaffold protein for the GlyRs that requires proteasomal degradation to allow for the remodeling of GlyR from the cell surface—this is in contrast to the AMPA receptor, which requires the degradation of PSD-95 by proteasomes to allow for those remodeling events to occur [17]. The examples presented in this section are summarized in Figure 1A.

APC/C and its Role in Synaptic Plasticity

The anaphace-promoting complex/cyclosome (APC/C) is a multisubunit protein complex that functions as an E3 ligase. APC/C is composed of catalytic and substrate recognition subunits and the particular composition of subunits will confer functional specificity for E3 ligase function (APC/C has an enormous literature, reviewed in [81]). APC can associate with cdc20 or cdh1 depending on cellular context [82,83], and each of these subunits serves to confer substrate specificity on the APC/C. While the role of APC/C in the cell cycle is well established, recent data suggests it plays an unexpected but crucial role in differentiated neurons. Cdh1 has been identified in neurons [84], and has a demonstrated role in neuronal growth [35]. Studies from *Drosophila* have recently demonstrated a role for APC/C in differentiated neurons of the neuromuscular junction (NMJ) [36]. Finally, APC/C has also been shown to be important for GluR1 ubiquitination in rodents [37]. CDH1-APC was shown to be localized to the nucleus in rat cultures, and was proposed to be a regulator of gene expression important for axonal morphogenesis [35]. However, data is now accumulating that suggests APC/C has functions to regulate synaptic plasticity and post-development neuronal functions. In the *Drosophila* model, flies mutant for the C subunit of APC/C were defective in synaptic bouton number at the NMJ, and this effect is independent of APC/C expression in the muscle target tissue [36]. Furthermore, subunits of the APC/C protein were localized to the synapses providing further evidence that the APC/C is acting in/at synapses, not via an indirect mechanism emanating from the cell body (e.g. regulation of gene expression). Interestingly, the APC/C subunits were found in the presynaptic as well as the postsynaptic target tissue [36]. Because the APC/C is a large multisubunit ligase, that contains catalytic subunits as well as substrate recognition subunits, the APC/C could function to recognize and degrade a large number of different proteins via exchange of specific substrate recognition subunits to confer situation and cell-type specificity. In this example, the APC/C may associate with different proteins pre and post-synaptically. One commonality in substrate recognition is the presence of a particular protein sequence in diverse proteins that can be recognized as a destruction signal when exposed. A strict destruction box motif (RxxLxxxxN) is recognized by cdc20-APC and cdh1-APC to a lesser extent [85] while a KEN box motif is recognized by cdh1-APC [86]. In order to gain insight into potential substrates for APC/C in *Drosophila* neurons, proteins were identified that contained destruction boxes and had characterized mutations which lead to changes in synaptic bouton number (to be consistent with the phenotype of the APC/C mutants) [36]. Liprin-α is one such protein for which mutations confer a defect in axon terminal branching, and contain three putative destruction box motifs [87]. Liprin-α is a multidomain protein that is involved in postsynaptic receptor targeting [88] and the development of presynaptic active zones [89]. Liprin-α levels were found to be higher in *Drosophila* APC/C mutants and specifically were higher in the presynaptic area. Ubiquitinated forms of Liprin-α were detectable, suggesting Liprin-α is recognized by an E3 ligase for degradation by proteasomes [36]. Highwire, another protein found to regulate synaptic bouton size and number [90] also has a destruction box motif at the C-term but its levels were not changed in APC/C mutants [36]. These studies reveal molecular information about the regulation of synaptic function that suggests there will likely be numerous E3 ligases and substrates that work in concert to regulate synaptic function, because both APC/C and Highwire are present in the synapse, but do not appear to overlap in function. APC/C regulation potentially has a role post-synaptically to regulate glutamate receptor levels, because the levels of GluRIIa are dramatically increased in the APC/C mutants in the NMJ [36]. Postsynaptically APC/C mutants change the sensitivity to neurotransmitter through an increase in the number of postsynaptic GluRIIa glutamate receptors [36]. APC/C may regulate GluRIIa levels directly, or indirectly, and examples form *Caenorhabditis elegans* (*C. elegans*) and rodents may offer some clues.

In *C. elegans*, the non-NMDA glutamate receptor is GluR1 and removal of this receptor from the postsynaptic density in response to stimuli has been shown to be dependent on ubiquitination and to require endocytosis [20]. Mutations that affect ubiquitination or endocytosis result in GluR1 accumulation at the cell surface and changes in behavior consistent with increased numbers of GluR1 at the cell surface [20]. This ubiquitination occurs indirectly through the APC/C-CDH1 complex, suggesting that endocytosis of the *C.*

elegans GluR1 may require ubiquitination and degradation of other (as yet unknown) proteins in addition to GluR1 [37]. In rodents the AMPA receptor, which is the rodent non-NMDA glutamate receptor, is removed from the plasma membrane after the multiubiquitination and degradation of PSD-95 [17]. The E3 ligase MdM2 is responsible for the recognition and ubiquitination of PSD-95 [17]. In a similar manner, the APC/C may regulate other cellular proteins that ultimately regulate the level of GluR1, or alter GluR characteristics to facilitate degradation (e.g scaffold proteins, kinases, transcription factors). Alternatively, both *C. elegans* GluR1 and *Drosophila* GluRIIa have the destruction boxes in the extracellular domain of the protein, which suggests the levels of the proteins may be regulated biosynthetically by APC/C via ERAD mechanisms. In fact, a recent study has shown that the NMDA receptor subunit NR1, is regulated in a similar manner by the FBX2/NFB42/Fbs1 E3 protein ligase and disruption of the appropriate destruction of NR1 during biosynthesis results in an increased density of NMDA receptors in response to activity stimuli [39]. A similar mechanism may regulate the GluRs in some cellular contexts, and it is likely that a combination of genetic and cell biology studies will be required to dissect the many factors that could influence AMPA levels at the surface of the postsynaptic density.

Presynaptic Protein Degradation

Clearly, protein degradation plays a role in postsynaptic responses to neurotransmitter stimulation through changes in cell surface protein abundance. The PSD is a protein rich microdomain wherein surface receptors are regulated by scaffold proteins in response to stimuli as described in the previous sections. The pre-synaptic portion of the synapse is a protein and vesicle rich domain that is highly dynamic and responsive to cellular stimuli. After activation, the pre-synaptic terminal will mobilize neurotransmitter-containing vesicles that reside just below the plasma membrane to fuse with the plasma membrane and release their contents into the synaptic space (Figure 1C). The mobilization and fusion events require exquisite coordination of intracellular signals to regulate protein function and control vesicle movement. After exocytosis, there are also recycling mechanisms to replenish the store of synaptic vesicles (for a review of the mechanisms governing synaptic vesicle assembly, release and recycling, see [91,92]). For example, the NEM-sensitive fusion protein attachment protein receptor (SNARE) complex of proteins is composed of syntaxin-1, VAMP2 and SNAP-25. This SNARE complex is a crucial regulatory component of multiple aspects of vesicle dynamics. Spring is an E3 ligase that has also been shown to be involved in pre-synaptic vesicle exocytosis [45]. Spring is localized in synapses, and is found in purified Synaptosomal preparations. Spring interacts with SNAP-25 and prevents the assembly of SNARE complex via a direct competition with syntaxin-1 for SNAP-25 binding. SNAP-25 and syntaxin-1 interaction is required for vesicle exocytosis. This suggests that Spring can act as a regulator of a critical component of vesicle exocytosis, by controlling the levels of the SNAP-25 protein. When sufficient SNAP-25 is available, it will associate with syntaxin-1 and vesicle associated membrane protein 2 (VAMP2) to form SNARE and this association will specifically allow for stimulus-evoked release of neurotransmitter via exocytosis [93]. Interestingly, because dissociation of SNAP-25 from a vesicle that has completed membrane fusion is required for endocytosis and recycling [94], there may also be a role for degradation of SNAP-25 to allow for vesicle endocytosis. Staring is an E3 ligase that has been shown to regulate syntaxin-1 levels [46]. Staring and syntaxin-1 were colocalized to synaptosomes, suggesting that at least some degradation of syntaxin-1 by Staring could occur at the nerve terminal. However, pulse-chase analysis also suggests that Staring could regulate syntaxin-1 levels via ERAD. Because syntaxin-1 is a transmembrane protein, it is likely the majority of proteasome-dependent degradation of syntaxin occurs via ERAD. There may be a role for monoubiquitination of syntaxin-1 by Staring in an independent role for regulation of endocytosis after vesicle fusion, but this was not addressed in recent publications [46]. If Staring is required for stimulated vesicle exocytosis, it should be possible to establish this using isolated synaptosomes with proteasome inhibitors, and should clarify the role of Staring (i.e., ERAD or functional regulation of vesicle dynamics through syntaxin-1) in synaptic plasticity.

Two additional E3 ligases that have been shown to act in the presynaptic area include ZNRF1 and 2 which, when mutated, can prevent Ca+-dependent exocytosis in PC12 cells [95]. The substrates of ZNRF in neurons are not yet known. Because highly similar proteins exist in *C. elegans* and *Drosophila* specific molecular mechanisms for ZNRFs in exocytosis of presynaptic vesicles may be elucidated using genetic approaches.

The N-end rule is a specific ubiquitin dependent pathway of protein degradation [96], with a well-characterized complement of E3 proteins operating in mammalian cells [97]. N-end rule dependent mechanisms also operate within the presynaptic area, although specific substrates have yet to be identified [98]. Using synthetic N-end rule substrates as sensors, pharmacological inhibition of synthetic substrate degradation was demonstrated in the presynapse, and established that local synaptic proteasomes are responsible for the degradation [98]. This result is consistent with Ehlers [15] data using synaptosomes to establish that proteasome-mediated degradation is contained entirely within the synaptic area, and does not require translocation to the cell body, or other neuronal area. One specific component of the N-end rule pathway is asparagine-specific N-terminal amidase (NTAN). Interestingly, NTAN deficient mice display a phenotype suggestive of presynaptic deficits [41] providing further evidence that the N-end rule pathway operates to regulate specific (as yet unknown) aspects of synaptic activity.

Pre-synaptic release of neurotransmitters can be regulated by substances that act directly on the pre-synaptic terminal. These substances can act to increase or decrease the amount of neurotransmitter released and their effects are measured electrophysiologically as an increase or decrease in evoked junctional currents (EJC). The changes in EJC can be effected through changes in voltage-dependent ion channels, changes in regulation of secretion of vesicles containing neurotransmitter or changes in pre-synaptic membrane potential. Recently, proteasome dependent degradation has been shown to be required for EJCs, suggesting that specific pre-synaptic events are also regulated by proteasome-dependent degradation. In general, application of proteasome inhibitors to the NMJ of *Drosophila* results in an increase in the excitatory junctional currents (ExJC) when measured using electrophysiology directly from the pre-synaptic site [98]. Protein synthesis inhibitors did not have the same increase in ExJC, suggesting protein degradation plays a key role in the strength of ExJCs. Interestingly, instead of a whole cohort of proteins degraded by the proteasome, only DUNC-13 was identified to be changed in abundance after the application of proteasome inhibitors (for less than one hour) to the NMJ [98]. DUNC-13 has been shown to be involved in controlling neurotransmission strength, through a vesicle priming mechanism [99]. Because these EJCs were not provoked by a bath application of neurotransmitter, it would be interesting to know if the cohort of short-lived proteins changes after application of GABA [100] (or other stimuli) to the NMJs. In other words, maybe the DUNC-13 is a constitutively short-lived protein, while other crucial presynaptic regulatory proteins only become proteasome substrates after a specific stimuli. An analogous case is the absence of PSD-95 degradation after Bicuculline stimulation [15], but degradation of PSD-95 after NMDA/AMPA receptor stimulation [17] (see PSD discussion above). It may be a precisely regulated mechanism of protein degradation to destabilize proteins in response to very specific neuronal stimuli, both pre- and post- synaptically.

APC/C E3 ligase has been suggested to regulate levels of Liprin-α because APC/C mutants result in an increased level of Liprin-α pre-synaptically [36]. Moreover, mutations in Liprin-α can suppress synaptic defects caused by mutations in APC/C. Liprin-α has been shown to be ubiquitinated in vivo, although it is unclear if the ubiquitination is mono-ubiquitin or multi-ub. Liprin-α interacts with DLAR receptor tyrosine phosphatase in the regulation of synaptic bouton number [87]. Liprin-α also interacts with RIM, a large scaffold protein that has been suggested to be a fundamental building block in the organization of the presynaptic architechture [89]. Therefore, while the requirement for targeted degradation of Liprin-a by APC/C in the pre-synapse is established, it is still unclear the exact molecular mechanism in which Liprin-α levels allow the appropriate development of pre-synaptic organization.

Liprin-α and Unc-13 both interact with RIM, a pre-synaptic scaffold protein [89,101]. As discussed above both Liprin-α and Unc-13 are regulated by proteasome degradation to control vesicle exocytosis, and it will be interesting to learn if there is a common regulatory role played by RIM1 in the recognition of Liprin-α and unc-13 for degradation. For example, does RIM become phosphorylated after stimuli and this phosphorylation leads to a conformational change that leads to Liprin-α and/or Unc-13 degradation? Undoubtedly as we gain more information about specific players that control vesicle exocytosis, a more complete picture of the chronology of events underlying regulation of exocytosis will be revealed. Already, the data discussed above indicate that the proteasome plays a role in both pre and post synaptic responses to stimuli, and as we become more informed about the molecular constituents of these areas, we will be able to study the role of protein degradation in regulation of these proteins. For example, an analysis of ion channel associated proteins [102] may reveal more proteins specifically regulated by ubiquitin-dependent pathways in neurotransmission.

KINASES, TRANSCRIPTION FACTORS AND E3 LIGASES

Any discussion of synaptic plasticity would be incomplete without a discussion of transcriptional changes and alterations in kinases that underlie learning and memory. From elegant, classical studies in *Aplysia* we have gained considerable insight into mechanisms of long-term facilitation. Protein Kinase A (PKA) is a cAMP-dependent protein kinase that is composed of catalytic (C) and regulatory (R) subunits. PKA has direct effect on short-term presynaptic function by modulation of ion channel activity by phosphorylation [103]. In conditions that enhance long-term facilitation (for example, training or prolonged exposure to neurotransmitter), PKA is persistently active, and this increased activity has been shown to be due to the specific degradation of the R subunits by the proteasome pathway [104,105]. In the absence of R subunits, the C subunits are highly active, and this persistant activation has been shown to be a characteristic of long-term facilitation in *Aplysia*. Protein synthesis is also required for the prolonged alteration in R-to-C subunit stoichiometry [106] and some of the induced proteins may be components of the proteasome pathway, such as a specific E3 ligase. Interestingly, UCH-L1 is also required for long-term facilitation in Aplysia [47]. UCH-L1 is a deubiquitinating enzyme that disassembles ubiquitin chains. Uch-L1 activity may be required in neurons to remove the ubiquitin chains from a substrate to stabilize it, or alternatively may be required to aid in its degradation by the proteasome because ubiquitin chains must be removed from a substrate prior to its entry into the mouth of the proteasome [47]. Yet another alternative function for Uch-L1 may be to regulate the half-life of a synaptic E3 ligase. Some ring finger E3 ligases have been shown to autoubiquitinate, and that autoubiquitination can signal turnover of the ligase [107]. In the case of MdM2 and HAUSP, HAUSP is a deubiquitinating enzyme that will prevent the autoubiquitination and degradation of MdM2, resulting in a stabilization of MdM2 and an increase in its cellular activity [108]. Uch-L1 may perform a similar function in regards to a postsynaptic E3 ligase during long-term facilitation. The requirement for both degradation of R subunits of PKA and the deubiquitination of protein demonstrates that the role of the proteasome pathway in the regulation of facilitation and potentiation pathways in neurons is complex and likely involves a multitude of opposing factors working in a kinetic balance that when shifted, results in long-term cellular (neuronal) changes.

The long-term changes that underlie learning and memory require new protein synthesis. Activation of nuclear transcription factors after neuronal stimulation is one way to regulate the levels of proteins required for synaptic changes, because transcription factors function to regulate gene expression. The CREB and C/EBP transcription factors are critically involved in learning and memory and are activated in response to calcium and cAMP (for review see [109]). cAMP, as discussed above, will activate PKA by dissociation of the R subunits from the C subunits. Activated PKA will then phosphorylate CREB (cyclic AMP- responsive element binding protein) and nuclear CREB will initiate gene transcription. Therefore the proteasomal degradation of PKA R subunits is required for the early events that lead to gene transcription events that are crucial elements for the establishment of long-term memory and learning. C/EBP is a transcription factor that is generated as a result of CREB activation, and is involved in the transcription of late response genes that are also required for learning and memory. C/EBP is degraded by the proteasome pathway when the protein is not phosphorylated, but when phosphorylated by MAP kinase is protected from degradation [110]. Probably, C/EBP degradation is a means to always have C/EBP present in cells to be able to respond rapidly to stimuli. The genes are that are induced by activity-dependent changes in transcription factors are the subject of much research since the general utility of gene microarray analysis became established. These regulated proteins are likely to be dynamically regulated in abundance and serve key roles in the neuron's response to activity. Therefore, it is also likely that many of these induced gene products will be substrates of the ubiquitin-proteasome pathway. One such regulator is the polo-like kinase 2 (PLK2) [also known as serum-induced kinase (SNK)]. PLK2 is a serine-threonine kinase that is induced by specific stimuli in the brain [111]. Recently, PLK2 has been shown to act on *Sp*ine--*a*ssociated *R*ap guanosine triphosphatase activating protein (SPAR), and that PLK2 phosphorylation of SPAR leads to SPAR degradation by proteasomes [19]. The removal of SPAR from dendritic spines leads to the remodeling of dendritic spines in response to stimuli. Because SPAR is a major dendritic scaffold protein its removal is a carefully considered decision by neurons to remodel it connectivity. PLK2 levels are also regulated by autophosphorylation and proteasomal degradation suggesting that PLK2 levels must be quickly regulated to prevent inappropriate kinase activity of PLK2 [112]. Some level of PLK2 must be in neurons tonically, even though proteins levels in unstimulated cells are low, because the application of proteasome inhibitors to unstimulated cells results in an increase in detectable PLK2 level [19]. It is likely that many of the proteins regulated by activity-dependent transcription are also degraded by the proteasome pathway in order to properly modulate synaptic changes in response to activity that result in precise remodeling events.

Ube3a gene encodes E6 associated protein (E6-AP) ubiquitin-protein ligase, best known for its ability to regulate p53 levels in the presence of the E6 viral protein [113]. E6-AP is a HECT domain (*H*omologous to *E*6-AP *C*-*T*erminus) ligase, which is a class of ligases that transfer ubiquitin to themselves prior to substrate transfer, and do not contain a RING domain [23,114,115]. In the absence of E6, E6-AP does not appear to play a major role in p53 regulation, but does have a clear role in regulating LTP in humans and mice [38]. In humans, the deletion of Ube3a (E6-AP) results in Angelman syndrome, which is characterized by a variety of abnormal neurological manifestations [116] (see Chapter 38). In order to create a mouse model of Angelman syndrome, Jiang *et al*, generated a targeted deletion in the Ube3a gene [38]. Maternal deficiency of the Ube3a gene in mouse offspring results in deficits in contextual fear conditioning (a form of learning) as well as LTP. Interestingly, baseline synaptic function was normal, suggesting the E6-AP has a function that is specific to the development of LTP. While we do not yet know the exact substrates of the E6-AP in neurons, it is clear that E6-AP is either regulating directly a signal transduction pathway or the stability of a gene product that is part of the induced genes for learning and memory, perhaps as a result of CREB or C/EBP activity.

Parkin is a RING E3 ligase that was first identified as a locus linked to autosomal recessive, early onset, Parkinson's Disease [117,118]. Loss of Parkin protein activity (either through point mutations or deletions) leads to loss of dopamine neurons and parkinsonian symptoms. Parkin has been the subject of intensive research and a complete review of all of its characteristics is beyond the scope of this review [119] (for more details see Chapter 31). However, Parkin has been linked to a variety of synaptic proteins and as such may be an important regulator of synaptic function. Parkin has been found both pre and post synaptically [42]. Presynaptically, Parkin may regulate a glycosylated form of α-synuclein [120]. Synuclein is a protein that was originally described to be involved in learning and memory and is a synaptic vesicle-associated protein [121,122]. Ubiquitinated forms of α-synuclein have been described, although those forms did not have glycosylation [123,124], so it is still unclear the exact molecular mechanisms governing Parkin and synuclein interaction, as well as synuclein ubiquitination. Parkin also regulates cdcrel-1 and 2 [125], and synaptotagmin IX [126] both synaptic vesicle associated proteins. These substrates were identified through yeast two-hybrid interactions, suggesting that Parkin directly recognizes these proteins. Unfortunately, two mouse lines that had knocked out the Parkin gene did not show any obvious presynaptic functional deficit phenotypes, although they were extensively studied [127, 128] and a third Parkin deletion mouse, showed a minor reduction in synaptic excitiability that may be a function of the background of the mouse strain and not Parkin absence [129]. Post-synaptically, Parkin interacts with CASK, a calcium dependent kinase, and a PDZ domain containing protein that has been linked to remodeling events at the PSD [130]. While Parkin does not appear to regulate levels of CASK, it may be associated with CASK as an anchor in the PSD to be proximal to potential substrates. Parkin has also been linked to ERAD pathways, specifically the degradation of a GPCR in dopamine neurons, PAEL-R [131]. Parkin appears to be upregulated by agents that induce misfolded protein stress [132] and Parkin may play dual roles to regulate specific proteins in the synapse, as well as ERAD in the cytoplasm. Parkin has been shown to be a mediator of Kainate toxicity as a part of a multi-protein complex, that serves to direct Parkin activity to specific substrates [133]. For example, the association of hsel-10 with Parkin will direct ligase activity to cyclin E and prevent neuronal apoptosis after excitotoxic stimuli [133]. Parkin activity also protects cells from α-synuclein overexpression mediated toxicity, although the exact mechanism of this protection is not known [134]. Recently, overexpression of Parkin was shown to prevent JNK activation in *Drosophila* [135], and it may be that the protective effects of Parkin in the injury models (Kainate, synuclein overexpression) may be a result of suppression of JNK activation. Therefore, it appears that Parkin is required for dopamine neuron survival, and may also be able to act as a neuroprotectant for a variety of damaging stimuli.

PATHOLOGICAL CONDITIONS AND THE UPS

As discussed in the many examples above, it is clear that proteasome activity is required for many neuronal processes, and that a defect in any aspect of proteasome pathway function could result in serious consequences for normal function. Aspects of the proteasome pathway include E3 ligases, deubiquitination, shuttle factors, unfolding enzymes resident on the proteasome or even proteasome capacity itself. A neuron that experiences a change in the ability to degrade substrates in a timely fashion may face a compounding effect resulting in cellular distress and at the extreme condition, disease.

As discussed above, a defect in Ube3a (E6-AP) or Parkin ligases can lead to neurological disease. A certain class of neurodegenerative diseases have been characterized histopathologically by immunochemistry using ubiquitin antibodies, revealing abundant ubiquitin-positive intracellular inclusions [136]. Huntington's disease, Parkinson's disease, Alzheimer's disease, Spinocerebellar Ataxia (SCA), and Amyotrophic Lateral Sclerosis (ALS) are a few of the neurological diseases that are manifested pathologically with ubiquitin positive inclusions. The accumulation of ubiquitinated protein strongly suggests that there is a defect in some aspect of the proteasome pathway, because ubiquitin conjugated protein generally does not accumulate in healthy cells [137]. These disorders can also be classified as 'misfolded protein' disorders, because some inheritable cases are caused by mutations in key genes that result in proteins which are misfolded and accumulate as insoluble ubiquitin-positive inclusions (aggresomes [138]) inside of cells [139]. Experimentally, aggresomes can be induced in cells after pharmacological inhibition of the proteasome, or by overexpression of unstable, aggregation prone proteins [138]. Interestingly, recent studies have hinted that the very earliest stages of protein aggregation, long before the formation of overt inclusions in cells can have profound effects on cell behavior. For example, recent work has suggested that ubiquitinated protein aggregates can compete for proteasome access with other normal cellular proteins [140]. The consequence of such competition in dividing cells is a delay or interruption of the cell cycle [140]. This result suggests that an excess of misfolded proteasome substrates can impact the degradation of other cellular proteins and lead to alterations in cell behavior. In the case of neurons, now that we understand some of the roles of the ubiquitin pathway in neurons, we can begin to understand what pathways may be affected in neurons that are accumulating misfolded protein as a result of decreases in proteasome degradation, that ultimately manifest as accumulations of ubiquitinated protein.

It was shown that the presence of insoluble protein complexes (IPCs) of protein existed prior to overt aggresome formation in cells [141]. IPCs are defined biochemically as SDS-resistant high molecular weight aggregates of protein [141]. Chronologically, it appears that IPCs form long before large intracellular aggresomes are manifest. It may be that IPCs create a competition for proteasome substrates that can be evidenced as early alterations in normal synaptic function, as suggested by the cellular experiments using a proteasome-green fluorescent protein (GFP) reporter system [140]. Experimental data in vivo supporting this hypothesis can be found from studies of the huntingtin (HTT) protein, that is linked to the development of Huntington's disease [142]. Transgenic animals expressing the exon-1 of HTT with an expanded CAG repeat demonstrate age-dependent degeneration accompanied by the formation of large intracellular inclusions of HTT in the brain [143]. Behavioral analysis of these animals revealed that the formation of aggregates preceded the onset of motor impairment as measured by rotorod experiments [144]. At 12 weeks of HTT expression, the animals have visible aggregates and behavior defects. While there was a clear reduction in the size of the striatum in these animals, there was no measurable cell loss, leading to the suggestion that the aggregates themselves were not toxic to cells [143]. This conclusion was supported by the use of neuronal cultures from the same animals, where the expression of the transgene did not lead to cell death after seven days [144]. While this may in fact be the case, the behavioral data clearly correlate the formation of aggregates to a loss of motor coordination, which if the transgene is continued to express, leads to greater symptomology. While it is clear there was no cell loss due to aggregate formation, these animal models may reveal that aggregate formation is adaptive and therefore protective, but the substrate competition for ubiquitin/proteasome pathway machinery eventually leads to cellular distress and dysfunction as evidenced by robust behavioral phenotypes in animal models.

There is evidence for changes in neuronal function due to expression of expanded HTT in transgenic animals. A decrease in D1 receptors in the striatum of HTT symptomatic mice has been reported [143]. D1 receptors are a GPCR, similar to the β-Adrenergic receptors

(β_2R). The levels of β_2R at the cell surface have been shown to be regulated by the proteasome pathway, in that the adapter molecule arrestin is degraded after recognition by MdM2 ligase and this allows clathrin coated pit endocytosis to remove β_2R receptors from the cell surface [145]. In a remarkable set of experiments it has been shown that the inability to deubiquitinate the arrestin associated with the β_2Rs results in the enhanced degradation of the β_2Rs in lysosomes [146]. If this mechanism is similar in the HTT animals, this suggests that the HTT, as it competes for proteasome degradation [140] and deubiquitinating enzymes, can accelerate the degradation of D1 receptors via the competition at the level of ubiquitination/degradation. HTT has been shown to be an effective competitor of the proteasome, and the formation of inclusions, as well as overexpressed HTT substrate can effectively compete proteasome function resulting in impairment [140]. D1 receptor levels may depend on arrestin similar to other GPCRs [145] although this has not yet been shown, and some of the loss of D1 receptors may be due to changes in transcription [147]. It is consistent that the loss of D1 receptors at early stages in pathology in the HTT exon 1 transgenic mice is correlated to the accumulation of HTT protein as the HTT competes for proteasome degradation with adapter proteins that may regulate D1 receptor at the cell surface. Alternatively, the biogenesis of D1 receptor and delivery to the plasma membrane may be affected by proteasome impairment, because the ER export of D1 receptors has been shown to be a crucial regulatory mechanism [148].

Pharmacological experiments may also reveal changes in other crucial cell surface molecules that are regulated by proteasome activity. For example, as discussed above, the SPAR protein is a major organizer of the dendritic spine architecture (as it can bind to actin, PSD95, Rap and contains a PDZ domain that may bind it to a number of cell surface proteins) and is degraded after phosphorylation by PLK2. However, the overexpression of SPAR leads to an enlargement of dendritic spine heads [149] suggesting that a dysregulation of SPAR degradation can lead to serious consequences in the organization of receptor proteins at the PSD. An alternative example, the number of AMPA glutamate receptors at the cell surface is suggested to be regulated by ubiquitination of PSD95 [150], and specific pharmacological agents may reveal increases or decreases in AMPA receptors as a function of proteasome competition. In addition to PSD-95 dependent removal of AMPA from the cell surface, the delivery of new GluR to the cell surface from the biosynthetic machinery can be impaired by a disruption of normal subunit assembly mechanisms—for example, the NR2B subunit is degraded by proteasomes in order to regulate the number of receptors delivered to the cell surface [151]. It may be that in conditions of proteasome impairment, the NR2B is stabilized and results in more glutamate receptor delivery to the cell surface. Either the impaired degradation of PSD-95, or the increase in GluRs could result in excitotoxicity and sensitivity to glutaminergic agonists and antagonists.

In fact, recent studies in a HTT mouse demonstrated that the onset of pathological symptoms is correlated to an increase in sensitivity to excitotoxic stimuli, as well as agonists of glutamate receptors [152]. The effect of aggregated HTT on specific subtypes of receptors has been studied in tissue culture systems [153], and it would be interesting to use the proteasome activity GFP reporter cell lines [140, 154] to determine if the pharmacological changes are coincident with an increase in GFP. Moreover, other studies with an HTT transgenic model has described resistance to quinolinic acid [155]. Both of these different transgenic HTT models are consistent with the notion that the HTT protein has an effect on the Ub/proteasome pathway that impacts the normal regulation of cell surface proteins that can manifest in changes in neuropharmacology.

As another example, a mouse model of ALS may also prove informative about early pharmacological changes that occur prior to overt symptomatic manifestation.

For example, overexpression of the SOD protein will result in an accumulation of ubiquitinated SOD in aggresomes in cells and IPCs of SOD protein occurred prior to overt SOD aggresome formation in cells. Importantly, the IPCs of SOD occur prior to early symptomology in transgenic animals for mutant SOD1 protein [141]. Interestingly, a focal loss of glutamate transporter EAAT2 was reported to coincide with early symptomology in the SOD1 mutant transgenic mice [156]. It would be important to know if EAAT2 protein biosynthesis, or turnover from the plasma membrane, is regulated by the proteasome pathway. The accumulation of SOD1 IPCs may create a competition for proteasome activity resulting in alterations in EAAT2 protein levels. Specific pharmacological agents may be able to reveal the earliest time at which changes in EAAT2 appear. It would also be of value to cross the SOD1 mutant mice to a GFP-proteasome reporter animal [157] to determine if changes in proteasome activity coincide with the earliest symptomatic behavioral and biochemical changes.

CONCLUSIONS

In summary, the complex regulation of cell surface proteins in neurons that allows for appropriate neurotransmission depends on a functional UPS, and a few basic examples have been presented here. Factors that decrease the efficiency of targeting mechanisms and degradation capacity can have profound effects on neuronal function, even leading to disease. These effects may be manifest much earlier than the appearance of large cellular ubiquitin-positive inclusions, and insightful pharmacological experiments in transgenic animals may allow for early detection of pathology, as well as reveal novel intervention points. As demonstrated by the HTT transgenic models, the inclusions and long-term proteasome inhibition does not kill cells [144], but the long term effects of dysregulated neurotransmission may lead to cell loss. A more detailed discussion of the theory of proteasome competition leading to neurological disease is presented in Johnston 2005 [158]. Undoubtedly, as we understand more of the protein players that regulate synaptic function, as well as identify specific E3 ligases involved, we will be able to understand the regulated choreography of neurotransmission and synaptic plasticity in health and disease.

REFERENCES

[1] Missler M, Sudhof TC: Neurexins: three genes and 1001 products. *Trend Gen* 1998, 14:20-26

[2] Song JY, Ichtchenko K, Sudhof TC and Brose N.: Neuroligin 1 is a postsynaptic cell-adhesion molecule of excitatory synapses. *PNAS* 1999, 96:1100-1105

[3] Uchida N, Honjo Y, Johnson KR, Wheeler MJ and Takeichi M.: The catenin/cadherin adhesion system is localized in synaptic junctions bordering transmitter release zones. *J Cell Biol* 1996, 135:767-779

[4] Inoue A, Sanes JR: Lamina-specific connectivity in the brain: regulation by N-cadherin, neurotrophins, and glycoconjugates. *Science* 1997, 276:1428-1431

[5] Tang L, Hung CP, Schuman EM: A role for the cadherin family of cell adhesion molecules in hippocampal long-term potentiation. *Neuron* 1998, 20:1165-1175

[6] Abbott LF, Nelson SB: Synaptic plasticity: taming the beast. *Nat Neurosci* 2000, 3 Suppl:1178-1183

[7] Bliss TV, Collingridge GL: A synaptic model of memory: long-term potentiation in the hippocampus. *Nature* 1993, 361:31-39

[8] Eichenbaum H: Spatial learning. The LTP-memory connection. *Natur*e 1995, 378:131-132

[9] Huang YY, Nguyen PV, Abel T and Kandel ER: Long-lasting forms of synaptic potentiation in the mammalian hippocampus. *Learn Mem* 1996, 3:74-85

[10] Linden DJ: The return of the spike: postsynaptic action potentials and the induction of LTP and LTD. *Neuron* 1999, 22:661-666

[11] Linden DJ, Connor JA: Cellular mechanisms of long-term depression in the cerebellum. *Curr Opin Neurobio* 1993, 3:401-406

[12] Linden DJ, Connor JA: Long-term synaptic depression. *Ann Rev Neurosci* 1995, 18:319-357

[13] Steward O, Schuman EM: Protein synthesis at synaptic sites on dendrites. *Ann Rev Neurosci* 2001, 24:299-325

[14] Steward O, Schuman EM: Compartmentalized synthesis and degradation of proteins in neurons. *Neuron* 2003, 40:347-359

[15] Ehlers MD: Activity level controls postsynaptic composition and signaling via the ubiquitin-proteasome system. *Nat Neurosci* 2003, 6:231-242

[16] Steward O, Levy WB: Preferential localization of polyribosomes under the base of dendritic spines in granule cells of the dentate gyrus. *J Neurosci* 1982, 2:284-291

[17] Colledge M, Snyder EM, Crozier RA, Soderling JA, Jin Y, Langeberg LK, Lu H, Bear MF, and Scott JD: Ubiquitination regulates PSD-95 degradation and AMPA receptor surface expression. *Neuron* 2003, 40:595-607

[18] Bedford FK, Kittler JT, Muller E, Thomas P, Uren JM, Merlo D, Wisden W, Triller A., Smart TG and Moss SJ: GABA(A) receptor cell surface number and subunit stability are regulated by the ubiquitin-like protein Plic-1. *Nat Neurosci* 2001, 4:908-916

[19] Pak DT, Sheng M: Targeted protein degradation and synapse remodeling by an inducible protein kinase. *Science* 2003, 302:1368-1373

[20] Burbea M, Dreier L, Dittman JS, Grunwald ME, and Kaplan JM: Ubiquitin and AP180 regulate the abundance of GLR-1 glutamate receptors at postsynaptic elements in C. elegans. *Neuron* 2002, 35:107-120

[21] Hershko A, Ciechanover A: The ubiquitin system. *Ann Rev Biochem* 1998, 67:425-479

[22] Ciechanover A: Proteolysis: from the lysosome to ubiquitin and the proteasome. *Nat Rev Mol Cell Biol* 2005, 6:79-87

[23] Ardley HR, PA: E3 ubiquitin ligases. *Essays in Biochemistry* 2005, 41:15-30

[24] Hicke L, Dunn R: Regulation of membrane protein transport by ubiquitin and ubiquitin-binding proteins. *Ann Rev Cell Dev Biol* 2003, 19:141-172

[25] Shih SC, Katzmann DJ, Schnell JD, Sutanto M, Emr SD, and Hicke L.: Epsins and Vps27p/Hrs contain ubiquitin-binding domains that function in receptor endocytosis. *Nat Cell Biol* 2002, 4:389-393

[26] Shih SC, Sloper-Mould KE, and Hicke L: Monoubiquitin carries a novel internalization signal that is appended to activated receptors. *EMBO* 2000, 19:187-198

[27] Sloper-Mould KE, Jemc JC, Pickart CM and Hicke L.: Distinct functional surface regions on ubiquitin. *J Biol Chem* 2001, 276:30483-30489

[28] Chau V, Tobias JW, Bachmair A, Mariott D, Ecker DJ, Gonda DK and Varshavsky, A.: A multiubiquitin chain is confined to specific lysine in a targeted short-lived protein. *Science* 1989, 243:1576-1583

[29] Madura K: Rad23 and Rpn10: perennial wallflowers join the melee. *Trend Biochem Sci* 2004, 29:637-640

[30] Ward CL, Omura S, Kopito RR: Degradation of CFTR by the ubiquitin-proteasome pathway. *Cell* 1995, 83:121-127

[31] Kopito RR: ER quality control: the cytoplasmic connection. *Cell* 1997, 88:427-430

[32] Wiertz EJ, Tortorella D, Bogyo M, Yu J, Mothes W, Jones TR, Rapoport TA and Ploegh HL: Sec61-mediated transfer of a membrane protein from the endoplasmic reticulum to the proteasome for destruction. *Nature* 1996, 384:432-438

[33] McCracken AA, Brodsky JL: Evolving questions and paradigm shifts in endoplasmic-reticulum-associated degradation (ERAD). *Bioessays* 2003, 25:868-877

[34] Sherff CM, Carew TJ: Coincident induction of long-term facilitation in Aplysia: cooperativity between cell bodies and remote synapses. *Science* 1999, 285:1911-1914

[35] Konishi Y, Stegmuller J, Matsuda T, Bonni S, and Bonni A.: Cdh1-APC controls axonal growth and patterning in the mammalian brain. *Science* 2004, 303:1026-1030

[36] van Roessel P, Elliott DA, Robinson IM, Prokop A., and Brand AH: Independent regulation of synaptic size and activity by the anaphase-promoting complex. *Cell* 2004, 119:707-718

[37] Juo P, Kaplan JM: The anaphase-promoting complex regulates the abundance of GLR-1 glutamate receptors in the ventral nerve cord of C. elegans. *Curr Biol* 2004, 14:2057-2062

[38] Jiang YH, Armstrong D, Albrecht U, Atkins CM, Noebels JL, Eichele G, Sweatt JD, and Beaudet AL: Mutation of the Angelman ubiquitin ligase in mice causes increased cytoplasmic p53 and deficits of contextual learning and long-term potentiation. *Neuron* 1998, 21:799-811

[39] Kato A, Rouach N, Nicoll RA, and Bredt DS: Activity-dependent NMDA receptor degradation mediated by retrotranslocation and ubiquitination. *PNAS* 2005, 102:5600-5605

[40] Dreier L, Burbea M, Kaplan JM: LIN-23-mediated degradation of beta-catenin regulates the abundance of GLR-1 glutamate receptors in the ventral nerve cord of C. elegans. *Neuron* 2005, 46:51-64

[41] Kwon YT, Balogh SA, Davydov IV, Kashina AS, Yoon JK, Xie Y, Gaur A, Hyde L, Denenberg VH and Varshavsky A.: Altered activity, social behavior, and spatial memory in mice lacking the NTAN1p amidase and the asparagine branch of the N-end rule pathway. *Mol Cell Biol* 2000, 20:4135-4148

[42] Schlossmacher MG, Frosch MP, Gai WP, Medina M, Sharma N, Forno L, Ochiishi T, Shimura H, Sharon R, Hattori N, Langston JW, Mizuno Y, Hyman BT, Selkoe DJ and Kosik KS: Parkin localizes to the Lewy bodies of Parkinson disease and dementia with Lewy bodies. *Am J Path* 2002, 160:1655-1667

[43] Wheeler TC, Chin LS, Li Y, Roundabush FL, and Li L: Regulation of synaptophysin degradation by mammalian homologues of seven in absentia. *J Biol Chem* 2002, 277:10273-10282

[44] Moriyoshi K, Iijima K, Fujii H, Ito H, Cho Y, and Nakanishi S: Seven in absentia homolog 1A mediates ubiquitination and degradation of group 1 metabotropic glutamate receptors. *PNAS* 2004, 101:8614-8619

[45] Li Y, Chin LS, Weigel C, and Li L: Spring, a novel RING finger protein that regulates synaptic vesicle exocytosis. *J Biol Chem* 2001, 276:40824-40833

[46] Chin LS, Vavalle JP, Li L: Staring, a novel E3 ubiquitin-protein ligase that targets syntaxin 1 for degradation. *J Biol Chem* 2002, 277:35071-35079

[47] Hegde AN, Inokuchi K, Pei W, Casadia A., Ghirardi M, Chain DG, Martin KC, Kandel ER, and Schwartz JH: Ubiquitin C-terminal hydrolase is an immediate-early gene essential for long-term facilitation in Aplysia. *Cell* 1997, 89:115-126

[48] Wilson SM, Bhattacharyya B, Rachel RA, Coppola V, Tessarollo L, Householder DB, Fletcher CF, Miller RJ, Copeland NG, and Jenkins NA: Synaptic defects in ataxia mice result from a mutation in Usp14, encoding a ubiquitin-specific protease. *Nat Gen* 2002, 32:420-425

[49] Chicurel ME, Terrian DM, Potter H: mRNA at the synapse: analysis of a synaptosomal preparation enriched in hippocampal dendritic spines. *J Neurosci* 1993, 13:4054-4063

[50] Martin KC, Casadio A, Zhu H, Yaping E, Rose JC, Chen M, Bailey CH, and Kandel ER: Synapse-specific, long-term facilitation of aplysia sensory to motor synapses: a function for local protein synthesis in memory storage. *Cell* 1997, 91:927-938

[51] Malinow R, Malenka RC: AMPA receptor trafficking and synaptic plasticity. *Ann Rev Neurosci* 2002, 25:103-126

[52] Zhao Y, Hegde AN, Martin KC: The ubiquitin proteasome system functions as an inhibitory constraint on synaptic strengthening. *Curr Biol* 2003, 13:887-898

[53] Aiba A, Kano M, Chen C, Stanton ME, Fox GD, Herrup K, Zwingman TA, and Tonegawa S: Deficient cerebellar long-term depression and impaired motor learning in mGluR1 mutant mice. *Cell* 1994, 79:377-388

[54] Ichise T, Kano M, Hashimoto K, Yanagihara D, Nakao K, Shigemoto R, Katsuki M, and Aiba A: mGluR1 in cerebellar Purkinje cells essential for long-term depression, synapse elimination, and motor coordination. *Science* 2000, 288:1832-1835

[55] Lu YM, Jia Z, Janus C, Henderson JT, Gerlai R, Wojtowicz JM, and Roder JC: Mice lacking metabotropic glutamate receptor 5 show impaired learning and reduced CA1 long-term potentiation (LTP) but normal CA3 LTP. *J Neurosci* 1997, 17:5196-5205

[56] Kano M, Hashimoto K, Kurihara H, Watanabe M, Inoue Y, Aiba A and Tonegawa S: Persistent multiple climbing fiber innervation of cerebellar Purkinje cells in mice lacking mGluR1. *Neuron* 1997, 18:71-79

[57] Hu G, Zhang S, Vidal M, Baer JL, Xu T and Fearon ER: Mammalian homologs of seven in absentia regulate DCC via the ubiquitin-proteasome pathway. *Gen Dev* 1997, 11:2701-2714

[58] Susini L, Passer BJ, Amzallag-Elbaz N, Juven-Gershon T, Prieur S, Privat N, Tuynder M, Gendron MC, Israel A, Amson R, Oren M and Telerman A.: Siah-1 binds and regulates the function of Numb.*PNAS* 2001, 98:15067-15072

[59] Hu G, Fearon ER: Siah-1 N-terminal RING domain is required for proteolysis function, and C-terminal sequences regulate oligomerization and binding to target proteins. *Mol Cell Biol* 1999, 19:724-732

[60] Keller SH, Taylor P: Determinants responsible for assembly of the nicotinic acetylcholine receptor. *J Gen Physiol* 1999, 113:171-176

[61] Christianson JC, Green WN: Regulation of nicotinic receptor expression by the ubiquitin-proteasome system. *EMBO* 2004, 23:4156-65

[62] Macdonald RL, Olsen RW: GABAA receptor channels. *Ann Rev Neurosci* 1994, 17:569-602

[63] Kittler JT, Delmas P, Jovanovic JN, Brown DA, Smart TG, and Moss SJ: Constitutive endocytosis of GABAA receptors by an association with the adaptin AP2 complex modulates inhibitory synaptic currents in hippocampal neurons. *J Neurosci* 2000, 20:7972-7977

[64] Essrich C, Lorez M, Benson JA, Fritschy JM, and Luscher B: Postsynaptic clustering of major GABAA receptor subtypes requires the gamma 2 subunit and gephyrin. *Nat Neurosci* 1998, 1:563-5671

[65] Wang H, Bedford FK, Brandon NJ, Moss SJ and Olsen RW: GABA(A)-receptor-associated protein links GABA(A) receptors and the cytoskeleton. *Nature* 1999, 397:69-72

[66] Connolly CN, Kittler JT, Thomas P, Uren JM, Brandon NJ, Smart TG, and Moss SJ: Cell surface stability of gamma-aminobutyric acid type A receptors. Dependence on protein kinase C activity and subunit composition. *J Biol Chem* 1999, 274:36565-36572

[67] Jentsch S, Pyrowolakis G: Ubiquitin and its kin: how close are the family ties? *Trend Cell Biol* 2000, 10:335-342

[68] Kleijnen MF, Shih AH, Zhou P, Kumar S, Soccio RE, Kedersha NL, Gill G, and Howley PM: The hPLIC proteins may provide a link between the ubiquitination machinery and the proteasome. *Mol Cell* 2000, 6:409-419

[69] Kleijnen MF, Alarcon RM, Howley PM: The ubiquitin-associated domain of hPLIC-2 interacts with the proteasome. *Mol Biol Cell* 2003, 14:3868-3875

[70] Gorrie GH, Vallis Y, Stephenson A, Whitfield J, Browning B, Smart TG, and Moss SJ: Assembly of GABAA receptors composed of alpha1 and beta2 subunits in both cultured neurons and fibroblasts. *J Neurosci* 1997, 17:6587-6596

[71] Ficklin MB, Zhao, S. and Feng, G.: Ubiquitin-1 regulates Nicotine-induced Up-regulation of Neuronal Nicotinic Acetylcholine Receptors. *J Biol Chem* 2005, 280:34088-34095

[72] Regan-Klapisz E, Sorokina I, Vorrtman J, de Keizer P, Roovers R, Verheesen P, Urbe S, Fallon L, Fon E, Verkleij A., Benmerah A., and van Bergen en Henegouwen, P: Ubiquilin recruits Eps15 into ubiquitin-rich cytoplasmic aggregates via a UIM-UBL interaction. *J Cell Sci* 2005, 118:4437-4450

[73] Patrick GN, Bingol B, Weld HA, and Schuman EM: Ubiquitin-mediated proteasome activity is required for agonist-induced endocytosis of GluRs. *Curr Biol* 2003, 13:2073-2081

[74] Snyder EM, Colledge M, Crozier RA, Chen WS, Scott JD, and Bear MF: Role for A kinase-anchoring proteins (AKAPS) in glutamate receptor trafficking and long term synaptic depression. *J Biol Chem* 2005, 280:16962-16968

[75] Naisbitt S, Kim E, Tu JC, Xiao B, Sala C, Valtschanoff J, Weinberg RJ, Worley PF, and Sheng M: Shank, a novel family of postsynaptic density proteins that binds to the NMDA receptor/PSD-95/GKAP complex and cortactin. *Neuro*n 1999, 23:569-582

[76] Kim E, Sheng M: PDZ domain proteins of synapses. *Nat Rev Neurosci* 2004, 5:771-781

[77] Pickart CM, Eddins MJ: Ubiquitin: structures, functions, mechanisms. *Biochem et Biophys Acta* 2004, 1695:55-72

[78] Mengual E, Arizti P, Rodrigo J, Gimenez-Amaya J, and Castano JG: Immunohistochemical distribution and electron microscopic subcellular localization of the proteasome in the rat CNS. *J Neurosci* 1996, 16:6331-6341

[79] Buttner C, Sadtler S, Leyendecker A, Laube B, Griffon N, Betz H, Schmalzing G: Ubiquitination precedes internalization and proteolytic cleavage of plasma membrane-bound glycine receptors. *J Biol Chem* 2001, 276:42978-42985

[80] Griffon N, Buttner C, Nicke A, Kuhse J, Schmalzing G, and Betz H.: Molecular determinants of glycine receptor subunit assembly. *EMBO* 1999, 18:4711-4721

[81] Harper JW, Burton JL, Solomon MJ: The anaphase-promoting complex: it's not just for mitosis any more. *Gen Dev* 2002, 16:2179-2206

[82] Visintin R, Prinz S, Amon A: CDC20 and CDH1: a family of substrate-specific activators of APC-dependent proteolysis. *Science* 1997, 278:460-463

[83] Schwab M, Lutum AS, Seufert W: Yeast Hct1 is a regulator of Clb2 cyclin proteolysis. *Cell* 1997, 90:683-693

[84] Gieffers C, Peters BH, Kramer ER, Dotti CG and Peters JM: Expression of the CDH1-associated form of the anaphase-promoting complex in postmitotic neurons. *PNAS* 1999, 96:11317-11322

[85] Fang G, Yu H, Kirschner MW: Direct binding of CDC20 protein family members activates the anaphase-promoting complex in mitosis and G1. *Mol Cell* 1998, 2:163-171

[86] Pfleger CM, Kirschner MW: The KEN box: an APC recognition signal distinct from the D box targeted by Cdh1. *Gen Dev* 2000, 14:655-665

[87] Kaufmann N, DeProto J, Ranjan R, Wan H, and Van Vactor D: Drosophila liprin-alpha and the receptor phosphatase Dlar control synapse morphogenesis. *Neuron* 2002, 34:27-38

[88] Wyszynski M, Kim E, Dunah AW, Passafaro M, Valtschanoff JG, Serra-Pages C, Streuli M, Weinberg RJ and Sheng M: Interaction between GRIP and liprin-alpha/SYD2 is required for AMPA receptor targeting. *Neuron* 2002, 34:39-52

[89] Schoch S, Castillo PE, Jo T, Mukherjee K, Geppert M, Wang Y, Schmitz F, Malenka RC, and Sudhof TC: RIM1alpha forms a protein scaffold for regulating neurotransmitter release at the active zone. *Nature* 2002, 415:321-326

[90] Wan HI, DiAntonio A, Fetter RD, Bergstrom K, Strauss R, and Goodman CS.: Highwire regulates synaptic growth in Drosophila. *Neuron* 2000, 26:313-329

[91] Li L, Chin LS: The molecular machinery of synaptic vesicle exocytosis. *Cell Mol Life Sci* 2003, 60:942-960

[92] Brunger AT: Structural insights into the molecular mechanism of Ca(2+)-dependent exocytosis. *Curr Op Neurobiol* 2000, 10:293-302

[93] Washbourne P, Thompson PM, Carta M, Costa ET, Mathews JR, Lopez-Benito G, Molnar Z, Becher MW, Valenzuela CF, Partridge LD and Wilsn MC: Genetic ablation of the t-SNARE SNAP-25 distinguishes mechanisms of neuroexocytosis. *Nat Neurosci* 2002, 5:19-26

[94] Bean AJ, Davanger S, Chou MF, Gerhardt B, Tsujimoto S, and Chang Y: Hrs-2 regulates receptor-mediated endocytosis via interactions with Eps15. *J Biol Chem* 2000, 275:15271-15278

[95] Araki T, Milbrandt J: ZNRF proteins constitute a family of presynaptic E3 ubiquitin ligases. *J Neurosci* 2003, 23:9385-9394

[96] Varshavsky A: The N-end rule: functions, mysteries, uses. *PNA*S 1996, 93:12142-12149

[97] Kwon YT, Reiss Y, Fried VA, Hersko A, Yoon JK, Gonda DK, Sangan P, Copeland NG, Jenkins NA and Varshavsky A: The mouse and human genes encoding the recognition component of the N-end rule pathway. *PNAS* 1998, 95:7898-7903

[98] Speese SD, Trotta N, Rodesch CK, Aravamudan B and Broadie K: The ubiquitin proteasome system acutely regulates presynaptic protein turnover and synaptic efficacy. *Curr Biol* 2003, 13:899-910

[99] Betz A, Ashery U, Rickmann M, Augustin I, Neher E, Sudhof TC, Rettig J and Brose N: Munc13-1 is a presynaptic phorbol ester receptor that enhances neurotransmitter release. Neuron 1998, 21:123-136

[100] Gutovitz S, Birmingham JT, Luther JA, Simon DJ and Marder E: GABA enhances transmission at an excitatory glutamatergic synapse. *J Neurosci* 2001, 21:5935-5943

[101] Betz A, Thakur P, Junge HJ, Ashery U, Rhee JS, Scheuss V, Rosenmund C, Rettig J and Brose N: Functional interaction of the active zone proteins Munc13-1 and RIM1 in synaptic vesicle priming. *Neuron* 2001, 30:183-196

[102] Wyszynski M, Sheng M: Analysis of ion channel associated proteins. *Meth Enzymol* 1999, 294:371-385

[103] Siegelbaum SA, Camardo JS, Kandel ER: Serotonin and cyclic AMP close single K+ channels in Aplysia sensory neurones. *Nature* 1982, 299:413-417

[104] Hegde AN, Goldberg AL, Schwartz JH: Regulatory subunits of cAMP-dependent protein kinases are degraded after conjugation to ubiquitin: a molecular mechanism underlying long-term synaptic plasticity. *PNAS* 1993, 90:7436-7440

[105] Chain DG, Hegde AN, Yamamoto N, Liu-March B and Schwartz JH: Persistent activation of cAMP-dependent protein kinase by regulated proteolysis suggests a neuron-specific function of the ubiquitin system in Aplysia. *J Neurosci* 1995, 15:7592-7603

[106] Greenberg SM, Castellucci VF, Bayley H and Schwartz JH: A molecular mechanism for long-term sensitization in Aplysia. *Nature* 1987, 329:62-65

[107] Lorick KL, Jensen JP, Fang S, Ong AM, Hatakeyama S and Weissman AM: RING fingers mediate ubiquitin-conjugating enzyme (E2)-dependent ubiquitination. *PNAS* 1999, 96:11364-11369

[108] Cummins JM, Vogelstein B: HAUSP is required for p53 destabilization. *Cell Cycle* 2004, 3:689-692

[109] Frank DA, Greenberg ME: CREB: a mediator of long-term memory from mollusks to mammals. *Cell* 1994, 79:5-8

[110] Yamamoto N, Hegde AN, Chain DG and Schwartz JH: Activation and degradation of the transcription factor C/EBP during long-term facilitation in Aplysia. *J Neurochem* 1999, 73:2415-2423

[111] Kauselmann G, Weiler M, Wulff P, Jessberger S, Konietzko U, Scafidi J, Staubli U, Bereiter-Hahn J, Strebhardt K and Kuhl D: The polo-like protein kinases Fnk and Snk associate with a Ca(2+)- and integrin-binding protein and are regulated dynamically with synaptic plasticity. *EMBO* 1999, 18:5528-5539

[112] Seeburg DP, Pak D, Sheng M: Polo-like kinases in the nervous system. *Oncogene* 2005, 24:292-298

[113] Huibregtse JM, Scheffner M, Howley PM: A cellular protein mediates association of p53 with the E6 oncoprotein of human papillomavirus types 16 or 18. *EMBO* 1991, 10:4129-135

[114] Joazeiro CA, Weissman AM: RING finger proteins: mediators of ubiquitin ligase activity. *Cell* 2000, 102:549-552

[115] Robinson PA, Ardley HC: Ubiquitin-protein ligases. *J Cell Sci* 2004, 117:5191-5194

[116] Kishino T, Lalande M, Wagstaff J: UBE3A/E6-AP mutations cause Angelman syndrome.[erratum appears in Nat Genet 1997 Apr;15(4):411]. *Nat Gen* 1997, 15:70-73

[117] Kitada T, Asakawa S, Hattori N, Matsumine H, Yamammura S, Minoshima S, Yokochi M, Mizuno Y and Shimizu N: Mutations in the parkin gene cause autosomal recessive juvenile parkinsonism. *Nature* 1998, 392:605-608

[118] Shimura H, Hattori N, Kubo S, Mizuno Y, Asakawa S, Monishima S, Shimizu N, Iwai K, Chiba T, Tanaka K and Suzuki T: Familial Parkinson disease gene product, parkin, is a ubiquitin-protein ligase. *Nat Gen* 2000, 25:302-305

[119] Kahle PJ, Haass C: How does parkin ligate ubiquitin to Parkinson's disease? *EMBO* 2004, 5:681-685

[120] Shimura H, Schlossmacher MG, Hattori N, Frosch MP, Trockenbacher A, Schneider R, Mizuno Y, Kosik KS and Selkoe DJ: Ubiquitination of a new form of alpha-synuclein by parkin from human brain: implications for Parkinson's disease. *Science* 2001, 293:263-269

[121] Clayton DF, George JM: The synucleins: a family of proteins involved in synaptic function, plasticity, neurodegeneration and disease. *Trend Neurosc* 1998, 21:249-254

[122] George JM: The synucleins. *Gen Biol* 2002, 3:REVIEWS3002

[123] Hasegawa M, Fujiwara H, Nonaka T, Wakabayashi K, Takahasi H, Lee VM, Trojanowski JQ, Mann D and Iwatsubo T.: Phosphorylated alpha-synuclein is ubiquitinated in alpha-synucleinopathy lesions. *J Biol Chem* 2002, 277:49071-49076

[124] Nonaka T, Iwatsubo T, Hasegawa M: Ubiquitination of alpha-synuclein. *Biochem* 2005, 44:361-368

[125] Zhang Y, Gao J, Chung KK, Huang H, Dawson VL and Dawson TM: Parkin functions as an E2-dependent ubiquitin- protein ligase and promotes the degradation of the synaptic vesicle-associated protein, CDCrel-1. *PNAS* 2000, 97:13354-13359

[126] Huynh DP, Scoles DR, Nguyen D and Pulst SM: The autosomal recessive juvenile Parkinson disease gene product, parkin, interacts with and ubiquitinates synaptotagmin XI. *Hum Mol Gen* 2003, 12:2587-2597

[127] Von Coelln R, Thomas B, Savitt JM, Lim KL, Sasaki M, Hess EJ, Dawson VL and Dawson TM: Loss of locus coeruleus neurons and reduced startle in parkin null mice. *PNAS* 2004, 101:10744-10749

[128] Perez FA, Palmiter RD: Parkin-deficient mice are not a robust model of parkinsonism. *PNAS* 2005, 102:2174-2179

[129] Goldberg MS, Fleming SM, Palacino JJ, Cepeda C, Lam HA, Bhatnagar A, Meloni EG, Wu N, Ackerson LC, Klapstein GJ, Gajendiran M, Roth BL, Chesselet MF, Maidment NT, Levine MS and Shen J: Parkin-deficient mice exhibit nigrostriatal deficits but not loss of dopaminergic neurons. *J Biol Chem* 2003, 278:43628-43635

[130] Fallon L, Moreau F, Croft BG, Labib N, Gu WJ and Fon EA: Parkin and CASK/LIN-2 associate via a PDZ-mediated interaction and are co-localized in lipid rafts and postsynaptic densities in brain. *J Biol Chem* 2002, 277:486-491

[131] Yang Y, Nishimura I, Imai Y, Takahasi R and Lu B: Parkin suppresses dopaminergic neuron-selective neurotoxicity induced by Pael-R in Drosophila. *Neuron* 2003, 37:911-924

[132] Imai Y, Soda M, Takahashi R: Parkin suppresses unfolded protein stress-induced cell death through its E3 ubiquitin-protein ligase activity. *J Biol Chem* 2000, 275:35661-35664

[133] Staropoli JF, McDermott C, Martinat C, Schulman B, Demireva E and Abeliovich A: Parkin is a component of an SCF-like ubiquitin ligase complex and protects postmitotic neurons from kainate excitotoxicity. *Neuron* 2003, 37:735-749

[134] Lo Bianco C, Schneider BL, Bauer M, Sajada A, Brice A, Iwatsubo T and Aebischer P: Lentiviral vector delivery of parkin prevents dopaminergic degeneration in an alpha-synuclein rat model of Parkinson's disease. *PNAS* 2004, 101:17510-17515

[135] Cha GH, Kim S, Park J, Lee E, Kim M, Lee SB, Kim JM, Chung J and Cho KS: Parkin negatively regulates JNK pathway in the dopaminergic neurons of Drosophila. *PNAS* 2005, 102:10345-10350

[136] Mayer RJ: From neurodegeneration to neurohomeostasis: the role of ubiquitin. *Drug News Persp* 2003, 16:103-108

[137] Johnston JA, Madura K: Rings, chains and ladders: ubiquitin goes to work in the neuron. *Prog Neurobiol* 2004, 73:227-257

[138] Johnston JA, Ward CL, Kopito RR: Aggresomes: a cellular response to misfolded proteins. *J Cell Biol* 1998, 143:1883-1898

[139] Carrell RW, Lomas DA: Conformational disease. *Lancet* 1997, 350:134-138

[140] Bence NF, Sampat RM, Kopito RR: Impairment of the ubiquitin-proteasome system by protein aggregation. *Science* 2001, 292:1552-1555

[141] Johnston JA, Dalton MJ, Gurney ME and Kopito RR: Formation of high molecular weight complexes of mutant Cu, Zn-superoxide dismutase in a mouse model for familial amyotrophic lateral sclerosis. *PNAS* 2000, 97:12571-12576

[142] MacDonald ME, Gines S, Gusella JF and Wheeler VC: Huntington's disease. *NeuroMol Med* 2003, 4:7-20

[143] Yamamoto A, Lucas JJ, Hen R: Reversal of neuropathology and motor dysfunction in a conditional model of Huntington's disease. *Cell* 2000, 101:57-66

[144] Martin-Aparicio E, Yamamoto A, Hernandez F, Hen R, Avila J and Lucas JJ: Proteasomal-dependent aggregate reversal and absence of cell death in a conditional mouse model of Huntington's disease. *J Neurosci* 2001, 21:8772-8781

[145] Shenoy SK, McDonald PH, Kohout TA and Lefkowitz RJ: Regulation of receptor fate by ubiquitination of activated beta 2-adrenergic receptor and beta-arrestin. *Science* 2001, 294:1307-1313

[146] Shenoy SK, Lefkowitz RJ: Trafficking patterns of beta-arrestin and G protein-coupled receptors determined by the kinetics of beta-arrestin deubiquitination. *J Biol Chem* 2003, 278:14498-14506

[147] Cha JH, Kosinski CM, Kerner JA, Alsdorf SA, Mangiarini L, Davies SW, Penney JB, Bates GP and Young AB: Altered brain neurotransmitter receptors in transgenic mice expressing a portion of an abnormal human huntington disease gene. *PNAS* 1998, 95:6480-6485

[148] Bermak JC, Zhou QY: Accessory proteins in the biogenesis of G protein-coupled receptors. *Mol Interven* 2001, 1:282-287

[149] Pak DT, Yang S, Rudolph-Correia S, Kim E and Sheng M: Regulation of dendritic spine morphology by SPAR, a PSD-95-associated RapGAP. *Neuron* 2001, 31:289-303

[150] Colledge M, Snyder EM, Crozier RA, Soderling JA, Jin Y, Langeberg LK, Lu H, Bear MF and Scott JD: Ubiquitination Regulates PSD-95 Degradation and AMPA Receptor Surface Expression. *Neuron* 2003, 40:595-607

[151] Huh KH, Wenthold RJ: Turnover analysis of glutamate receptors identifies a rapidly degraded pool of the N-methyl-D-aspartate receptor subunit, NR1, in cultured cerebellar granule cells. *J Biol Chem* 1999, 274:151-157

[152] Zeron MM, Hansson O, Chen N, Wellington CL, Leavitt BR, Brundin P, Hayden MR and Raymond LA: Increased sensitivity to N-methyl-D-aspartate receptor-mediated excitotoxicity in a mouse model of Huntington's disease. *Neuron* 2002, 33:849-860

[153] Zeron MM, Chen N, Moshaver A, Lee AT, Wellington CL, Hayden MR and Raymond LA: Mutant huntingtin enhances excitotoxic cell death. *Mol Cell Neurosci* 2001, 17:41-53

[154] Dantuma NP, Lindsten K, Glas R, Jellne M and Masucci MG: Short-lived green fluorescent proteins for quantifying ubiquitin/proteasome-dependent proteolysis in living cells. *Nat Biotech* 2000, 18:538-543

[155] Hansson O, Petersen A, Leist M, Nicotera P, Castilho RF and Brundin P: Transgenic mice expressing a Huntington's disease mutation are resistant to quinolinic acid-induced striatal excitotoxicity. *PNAS* 1999, 96:8727-8732

[156] Howland DS, Liu J, She Y, Goad B, Maragakis NJ, Kim B, Erickson J, Kulik J, DeVito L, Psaltis G, DeGennaro LJ, Cleveland DW and Rothstein JD: Focal loss of the glutamate transporter EAAT2 in a transgenic rat model of SOD1 mutant-mediated amyotrophic lateral sclerosis (ALS). *PNAS* 2002, 99:1604-1609

[157] Lindsten K, Menendez-Benito V, Masucci MG and Dantuma NP: A transgenic mouse model of the ubiquitin/proteasome system. *Nat Biotech* 2003, 21:897-902

[158] Johnston JA: Aggresomes: the proteasome, inclusion bodies and protein aggregation. In *Protein Misfolding, Aggregation and Conformational Disease*. Edited by Uversky and Fink. New York: Kluwer/Academic/Plenum Publishers; 2005

In: The Ubiquitin Proteasome System... ISBN 978-1-60021-749-4
Eds: Mario Di Napoli and Cezary Wojcik, pp. 443-493

Chapter 19

THE ROLE OF THE UBIQUITIN PROTEASOME SYSTEM IN MODIFYING CELLULAR STRESS RESPONSES

***Mario Di Napoli*[1,2,*] *and Francesca Papa*[1]**

[1]Neurological Section, SMDN—Center for Cardiovascular Medicine and Cerebrovascular Disease Prevention, I-67039 Sulmona (L'Aquila), Italy;
[2]Neurological Service, San Camillo de' Lellis General Hospital, I-02100 – Rieti, Italy.

ABSTRACT

Organisms respond to acute environmental change by orchestrating a stress response to prevent further damage. The key aspect of the stress response is a reaction to any form of macromolecular damage that exceeds a set threshold, independent of the underlying cause. It is becoming clear that higher organisms developed a complex, sensitive and maybe equally important network of regulatory pathways, relying largely on protein interactions, post-translational modifications and proteolysis. Molecular chaperones help hundreds of signalling molecules to keep their activation-competent state, and regulate various signalling processes ranging from signalling at the plasma membrane to transcription. Furthermore, molecular chaperones recognize proteins of non-native structure, prevent them from irreversible intracellular aggregation, and then act with regulatory co-chaperones in the conversion of proteins to be properly folded and in a functional state, stabilizing the phenotypes of various cells and organisms. This may be related to their low affinity for the proteins they interact with, which means that they represent weak links in protein networks. However, not every non-native protein is folded successfully. Those proteins that are not accurately folded/refolded are then directed to the ubiquitin-proteasome system (UPS) for destruction. The UPS is the predominant nonlysosomal protein degradation pathway which insures the viability,

* Correspondence concerning this article should be addressed to Dr. Mario Di Napoli, MD; Neurological Section, SMDN—Center for Cardiovascular Medicine and Cerebrovascular Disease Prevention, Via Trento, 41 67039–Sulmona (AQ) Italy Phone +39.(0)864.52716 FAX +39.1782262209 E-mail: mariodinapoli@katamail.com.

proliferation and signaling of eukaryotic organisms. In periods of stress, rapid elimination of denatured, misfolded and damaged proteins by proteasomes also becomes a critical determinant of cell fate. Stress response including its associated oxidative, nitrosylative and energetic stresses underlie neurodegeneration and evoke a discreet set of transcriptional events which have a complex and interdependent relationship with proteasomal function. Both chaperones and proteasomes act jointly together for selective removal of proteins with aberrant structure so as to keep protein homeostasis in cells. Though the precise nature of the cooperative linkage between chaperone and UPS pathways remains largely elusive so far, how slowly folding or misfolded polypeptides are targeted for proteasomal degradation, accumulating evidence from *in vivo* and *in vitro* studies shed some light on the molecular mechanisms that link proteasomes and molecular chaperones. Generally, selection of proteins for degradation is mediated by E3 ubiquitin ligases of the mechanistically distinct HECT and RING domain sub-types. Recent studies suggest that the U-box protein family represents a third class of E3 enzymes. CHIP, a U-box-containing protein, is a degradatory co-chaperone of heat-shock protein 70 (Hsp70) and Hsp90 that facilitates polyubiquitination of chaperone substrates. This mechanism affords time for a separate set of stressor-specific adaptations, designed to re-establish cellular homeostasis, to take action. Finally, the disruption of this protein folding quality control results in the accumulation of non-native protein species that can form oligomers, aggregates, and inclusions indicative of neurodegenerative disease. Many neurodegenerative diseases such as amyotrophic lateral sclerosis, Alzheimer's disease, Parkinson's disease and polyglutamine diseases, to cite the well studied, are characterized by conformational changes in proteins that result in misfolding, aggregation and intra- or extra-neuronal accumulation of amyloid fibrils. A common feature may be the formation of off-pathway folding intermediates that are unstable, self-associate, and with time lead to a chronic imbalance in protein homeostasis acting on the correct functioning of molecular chaperones and the UPS with deleterious consequences on cellular function. This has led to a hypothesis that enhancement of components of the cellular quality control machinery, specifically the levels and activities of molecular chaperones and the UPS suppress aggregation and toxicity phenotypes to allow cellular function to be restored. A detailed understanding of the molecular basis of chaperone-UPS mediated protection against neurodegeneration might lead to the development of therapies for neurodegenerative disorders that are associated with protein misfolding and aggregation.

Keywords: cellular stress response, apoptosis, molecular chaperone, environmental stress, heat shock protein.

ABBREVIATIONS

α-Syn, α-synuclein; Aβ, amyloid β; AD, Alzheimer's disease; ALS, amyotrophic lateral sclerosis; Apg-1, protein kinase essential for autophagy; APP, amyloid precursor protein; BAG, Bcl-2 binding athanogene; BiP, immunoglobulin heavy chain binding protein; CCT, Cytosolic Chaperonin containing TCP-1; CHIP, c-terminal HSC70 interacting protein; DLB, dementia with LBs; DJ1, Parkinson disease (autosomal recessive, early onset) 1; E1, Ub activating enzyme; E2s, Ub conjugating enzymes; E3s, ubiquitin ligase; ER, endoplasmic reticulum; GCIs, Glial cytoplasmic inclusions; GSH, glutathione; Grp78, glucose-regulated protein 78; HD, Huntington's Disease; HDAC6, Histone deacetylase 6; Hip, Hsc70-interacting protein; Hop, Hsc70-Hsp90-organizing protein; HS, heat shock; HSBP1, heat shock protein binding factor 1; Hsc70, heat shock cognate protein (a constitutively expressed homolog of Hsp70); HSE, heat shock elements; HSF, heat shock factors; Hsp, heat-shock protein; HSPs, heat-shock proteins; Htt, huntingtin; IAPs, inhibitors of apoptosis proteins; JNK, jun kinase; LB, Lewy bodies; LBVAD, LB variant of Alzheimer's disease; MAPK, Mitogen Activated Protein Kinase; MPTP, 1-methyl-4-phenyl-1,2,3,6-tetrahydropyridine; MSA, multiple system atrophy; NBD, N-terminal nucleotide binding domain; mtHsp70, mitochondrial Hsp70; NCBI, National Center for Biotechnology Information; NF-κB, nuclear factor-κB; NFT, neuro-fibrillary tangles; NEFs, nucleotide exchange factors; NMDA, N-methyl-D-aspartate; NO, nitric oxide; NOS, nitric oxide synthase; PACE, proteasome activation control element; PD, Parkinson's disease; PDI, protein disulfide isomerase; PINK1, phosphatase and tensin induced putative kinase 1; PolyQ, polyglutamine; PPIases, Peptidyl-prolyl cis-trans isomerases; QC, quality control; RNAi, RNA-mediated interference; SAPK, stress activated protein kinase; SBD, substrate binding domain; SBMA, spinal bulbar muscular atrophy; SCA, spinocerebellar ataxias; sHSP, small heat shock proteinheat shock protein; SN, substantia nigra; SOD, superoxide dismutase; TCP-1, tailless complex polypeptide; TRIC, TCP-1 Ring Complex; TPR, tetratricopeptide repeat motif; Ub, ubiquitin; UCHL1, Ub C-terminal hydrolase L1; UIM, ubiquitin interaction motif; UPR, unfolded protein response; UPS, ubiquitin-proteasome system, VHL, von Hippel–Lindau.

Introduction

The cellular stress response or so-called heat shock (HS) response can be defined as a reaction to the threat of macromolecular damage and it is one of the primordial intracellular defense mechanisms against stressful conditions that protect cells from sudden environmental change or frequent fluctuations in environmental factors. Prokaryotic and eukaryotic cells react to unfavorable environmental conditions by increased synthesis of stress proteins. The cellular stress response has been associated most clearly with protective effects during conditions that perturb both protein and DNA integrity following such environmental challenges in all three super-kingdoms, the archaea [1], the eubacteria [2] and eukaryotes [3,4]. Cells respond to chemical, environmental or physiological stress through a transient arrest of the cell cycle that is accompanied by widespread changes in macromolecular synthesis, degradation, trafficking, overall cellular metabolism and signal transduction pathways to cope with stressful conditions until more favorable conditions are encountered.

The main essence of the cellular stress response consists of protection of macromolecules during the initial phase of exposure to any adverse environmental condition that significantly perturbs cellular homeostasis. It is well established that one consequence of environmental or chemical stresses is the destabilization of protein conformation, leading to protein misfolding and aggregation [5]. Many types of environmental stress have been shown to cause deleterious changes in protein conformation. These factors include changes in temperature [6,7], pressure [8], pH [9], osmolarity [1], exposition to ionizing radiation [10], higher concentration of heavy metals [11], hypoxia/ischemia [12] and reactive oxygen species

[13,14]. Likewise, many of these various stresses are also known to cause DNA damage [15,16]. Exposure of cells and organisms to stresses induce the cellular stress response leading to the preferential transcription and translation of heat shock proteins (HSPs) [3,13,17,18]. Thus, HSPs enable cells to survive when there are unfavorable conditions in the outer environment [19-22]. Optimal HS response in terms of HSP synthesis and activity is essential for cell survival. In contrast, inefficient and altered HS response has been implicated in abnormal growth and development [23], ageing and apoptosis [24,25]. In the case of ageing, it has been shown that cells isolated from aged tissues and organisms, and cells undergoing replicative senescence *in vitro*, have a reduced HS response and a higher incidence of death when submitted to stress. These aspects are extensively reviewed in Chapter 22.

Under stressful conditions, when protein folding is disturbed, HSPs assist in protein refolding, in protein protection, in cellular protection from protein damage, in dissolving aggregated protein, in sequestering overloaded and damaged proteins into larger aggregates, in targeting damaged proteins for degradation and in interfering with the apoptotic program [26-28]. It seems that HSPs are able to distinguish between slightly misfolded proteins, which can be refolded, and severely misfolded proteins which should be degraded [25]. Some HSPs are known to be chaperones and are involved in the renaturation of misfolded proteins. Chaperones recognize and bind to other proteins when they are in non-native conformations and are exposing hydrophobic sequences (see Chapter 10). Their role is to minimize the aggregation of non-native proteins formed during stress. Typically, chaperones function as oligomers, if not as a complex of several different chaperones, co-chaperones and/or nucleotide exchange factors [3]. While the presence of protein chaperones facilitates the refolding of unfolded proteins, severe or prolonged stresses can lead to irreversible protein damage. Indeed, protein aggregation is associated with a number of disease states including Alzheimer's Disease (AD), Huntington's Disease (HD) Parkinson's disease (PD), amyotrophic lateral sclerosis (ALS) and prion diseases [29]. This role of HSPs has made them an obvious target of interest. Typically, HSPs function either as oligomers or in a complex with other chaperones, co-chaperones and/or nucleotide exchange factors. Moreover, the cellular stress response may play roles as yet poorly known for the stabilization of other macromolecules, such as lipid structures (membranes) and RNA.

The function of some the HSPs in ubiquitin-dependent protein degradation is also evident; ubiquitin itself also belongs to the HSP super-family [5,30-33]. The degradation of proteins by the ubiquitin-proteasome system (UPS) plays an important regulatory role in a broad range of biological processes such as cell cycle regulation, metabolic adaptations, gene regulation, development and differentiation [34-38] (see Chapter 2). Growing evidence indicates that failure to eliminate misfolded proteins can lead to the formation of potentially toxic aggregates, inactivation of functional proteins, and ultimately cell death3. The number of disease states linked to aberrant protein conformations underscores the importance of effective quality control for cell survival [39]. Importantly, the UPS degradation pathway also provides an efficient mechanism to remove damaged and aggregated proteins to prevent adverse consequences to cell maintenance and proliferation. The majority of proteasome substrates identified to date are marked for degradation by polyubiquitination. Usually, only substrates targeted to the proteasome by polyubiquitination are able to gain access to its proteolytic core although several exceptions exist [40]. Exceptions to this principle, however, are well documented and can help us understand the process proteasomes use to recognize their substrates (see Chapter 8). Examples include ornithine decarboxylase, p21/Cip1, TCRα, IκBα, c-Jun, calmodulin and thymidylate synthase [40]. Degradation of these proteins can be completely ubiquitin-independent or coexist with ubiquitin-dependent pathways. Uncoupling degradation from ubiquitin modification may reflect the evolutionary conservation of mechanisms optimized for highly specialized regulatory functions. Ubiquitination proceeds by successive attachment of a ubiquitin moieties to target substrates via a serial reaction catalyzed by three enzymes, ubiquitin activating enzyme (E1), ubiquitin conjugating enzyme (E2) and ubiquitin ligase (E3). Polyubiquitinated proteins are finally targeted to the 26S proteasome to be degraded [34-38] (see Chapter 3). Polyubiquitination of some proteins also requires so-called E4 enzymes that cooperate with E3 ligases to extend the polyubiquitin chain [41].

Because of this universal property, the cellular stress response consists of adaptations that maximize the stabilization, protection and repair of macromolecular structure and function. Such benefit carries the price of transiently decreasing the cells' capacity for most of its normal functions by draining metabolic energy and reducing the conformational flexibility of proteins and DNA. Reduced conformational flexibility decreases the efficacy of enzymes by slowing the rate at which structural changes occur in the active site during catalysis. Through similar kinetic effects, conformational flexibility is also rate- limiting for functions of other macromolecules. Despite these disadvantages, the cellular stress response shelters the ultimate cell function during adverse environmental conditions – the survival of cells.

Inducible HSPs are induced by stressful stimuli and are thought to assist in the maintenance of cellular integrity and viability, and HSPs are thought to prevent protein denaturation and incorrect polypeptide aggregation during exposure to physical and chemical insults. Although during stress the synthesis of stress proteins has increased considerably, a lot of the stress proteins are expressed as constitutive proteins, and they play the significant role even in cells which are not exposed to stress factors [42]. They are involved in antigen presentation [43], steroid receptor function [44], intracellular trafficking [45,46], nuclear receptor binding [44,47], and apoptosis [48-50], making it clear that they may also be involved in various cell signaling pathways. In the nervous system, the HSPs are induced in a variety of pathologic states, including cerebral ischemia [50], neurodegenerative diseases [51], epilepsy [52], and trauma [53]. Expression has been detected in a variety of cell populations within the nervous system, including neurons, glia, and endothelial cells [54].

This review focuses on the events evoked by cellular stress response and the regulation of molecular chaperones with their impact on the UPS function in the nervous system as well as on how misfolded, aggregation-prone proteins are handled by these two machineries in neurodegenerative diseases.

Nomenclature of Stress Proteins

Nomenclature and the Basic Division of Stress Proteins

In mammals, many HSP families comprise multiple members that differ in inducibility, intracellular localization and function (Table 1) [55]. The study of Tissieres *et al.* (1974) introduced the term '*heat shock proteins*', and it belongs to the beginning of research on the stress proteins [56]. Until now, no uniform system of naming stress proteins has been adopted but several helpful conventions are in broad use. Some designations are linked historically with the induction conditions such as Hsp and glucose-regulated protein (Grp) followed by the estimated rounded molecular mass of the protein, e.g. Hsp90. However molecular mass has become an inadequate criterion, as e.g. some HSPs include a Hsp70 domain with masses ranging from 16 to 170 kDa. Sometimes we can see another abbreviation, 'Hsc' (heat shock cognates), which has been used for the constitutive forms of HSP. Those forms of HSP are also present in non-stressed cells, and in contrast to the majority of other proteins, their intracellular concentrations do not increase during HS. In eukaryotic cells, different stress factors induce the synthesis of another, even if similar proteins or the same proteins are localized in another cellular compartment [57]. The term 'chaperone' is used very often. This term points out the function of the protein directly. It concerns stress as well as non-stress proteins, which accompany unfolded polypeptides during their cellular transport, and they promote the translocation of proteins through membranes or their integration into cellular organelles. Increased knowledge about stress proteins has lead to division of the HSP super-family into specific families based on their molecular masses and sequence homology [58]. The division of HSP into families is not yet precisely standardized. Earlier divisions into families: Hsp90, Hsp70, Hsp60 and small Hsp43 has been extended step by step to Hsp110, Hsp100, Hsp90, Hsp70, Hsp60, Hsp40, Hsp10 and small Hsp families (sHSP) [3]. Numeric indexing represents the protein molecular masses in kDa. Stress proteins are classified into appropriate families according to their approximate molecular masses, their functions in the cells and their homology in the primary structures. Currently, we have the largest amount of information about the Hsp90, Hsp70, Hsp60 and sHSP families of stress proteins. Recently a systematic nomenclature has been proposed by Sghaier, *et al.* [59] using a systematic nomenclature based on the accession number in the database of the National Center for Biotechnology Information (NCBI).

The Small Heat Shock Proteins (sHsp)

The precise functions of sHSPs including Hsp27 and the eye-lens protein αB-crystallin are incompletely understood. However, they seem to play a major role in preventing protein aggregation under conditions of cellular stress [60-64]. All members investigated so far form large oligomeric complexes of spherical or cylindrical appearance [64]. Complex formation is independent of ATP binding and hydrolysis, but appears to be regulated by temperature and phosphorylation. The structural analysis of wheat Hsp16.9 suggested that the oligomeric complex acts as a storage form rather than an enclosure for substrates, as the active chaperone

Table 1. Cellular localization and function of the major HSP families in mammals

Family	HSP	Location	Function
Hsp10	Ubiquitin	Cytosol/nucleus	Tag protein for degradation
	Hsp10	Mitochondria	Cofactor of Hsp60, tolerance of ischemia
sHsp	Hsp27 Hsp28/αB-crystallin	Cytosol/nucleus	Large oligomers, regulator of cytoskeleton, resistance to oxidative, UV and TNF stresses, tolerance of hyperthermia and ischemia, prevent aggregation, antiapoptotic
Hsp30	Hsp32 HO-1	Cytosol/nucleus	Stress inducible, regulation of heme-protein turnover, iron metabolism and oxidative stress
Hsp40	Hsp47	Endoplasmic reticulum	Heat inducible, procollagen chaperone
	Hdj 1 Hdj 2	Cytosol/nucleus	Cofactor of Hsp70, increase ATPase activity and substrate release
Hsp60	Hsp56	Cytosol	Stress inducible, bind the steroid hormone receptor complex and FK506, rotamase function
	Hsp60	Mitochondria	Constitutive/inducible, weak ATPase activity, binding and folding of imported proteins, tolerance of hyperthermia and ischemia
	Hsp65		Antitumorigenic action
	TCP-1	Cytosol/nucleus	Constitutive, weak ATPase activity, folding of actin and tubulin
Hsp70	Hsp70/Hsp72	Cytosol/nucleus	Stress inducible, ATPase activity, tolerance of hyperthermia, ischemia/hypoxia, resistance to oxidative, UV and TNF stresses, protection against protein aggregation, regulation of HS response, protection of transcription/translation, tumorigenicity, antiapoptotic
	Hsp73/HSC70	Cytosol/nucleus	Constitutive, ATPase activity, folding, trafficking, tolerance of hyperthermia, promote lysosomal degradation
	Grp75	Mitochondria	Constitutive/inducible, ATPase activity, transport and folding of polypeptides into matrix
	Grp78 (BiP)	Endoplasmic reticulum	Constitutive/inducible, ATPase activity, protein secretion and translocation and degradation into ER
Hsp90	Hsp90 α/β	Cytosol/nucleus	Partly inducible, ATPase activity, autophosphorylation, tolerance of hyperthermia, ischemia, apoptosis, role in cell cycle and proliferation and in signal transduction, prevents aggregation
	Grp94	Endoplasmic reticulum	Constitutive/inducible, ATPase activity, protein folding and secretion

ER, endoplasmic reticulum; TCP-1, tailless complex polypeptide; Grp, glucose-regulated protein; HSP, heat shock protein; BiP, immunoglobulin heavy chain binding protein; mtHsp70, mitochondrial Hsp70; HSF1, heat shock factor 1; Apg-1, protein kinase essential for autophagy.

appears to be a dimer [64]. In agreement with this notion, dissociation of the oligomeric complex formed by yeast Hsp26 was found to be a prerequisite for efficient chaperone activity [65]. Subsequent refolding may occur spontaneously or may involve cooperation with other chaperones such as Hsp70 [66]. In response to HS several different sHSP change phosphorylation status and form large (300–800 kDa) oligomers with ATP-independent chaperone activity [58]. The sHSP and Hsp90 families are known to capture unfolded proteins and create a reservoir of folding intermediates preventing further aggregation. sHSP are also able to induce an increase in cellular glutathione (GSH) levels, which works together with ascorbic acid and coenzyme Q as a kind of redox buffer capacity for the cell, and protects the mitochondrial membrane [67]. Working together with Hsp70 the complex is able to work as a kind of cytoplasmic antioxidant by covering the sensitive sites of the proteins. Hsp90 and the sHSP αB-crystallin are able to stabilize a more active form of the proteasome

[58] and overexpression of Hsp27 confers proteasome inhibitor resistance in lymphoma cells [68,69].

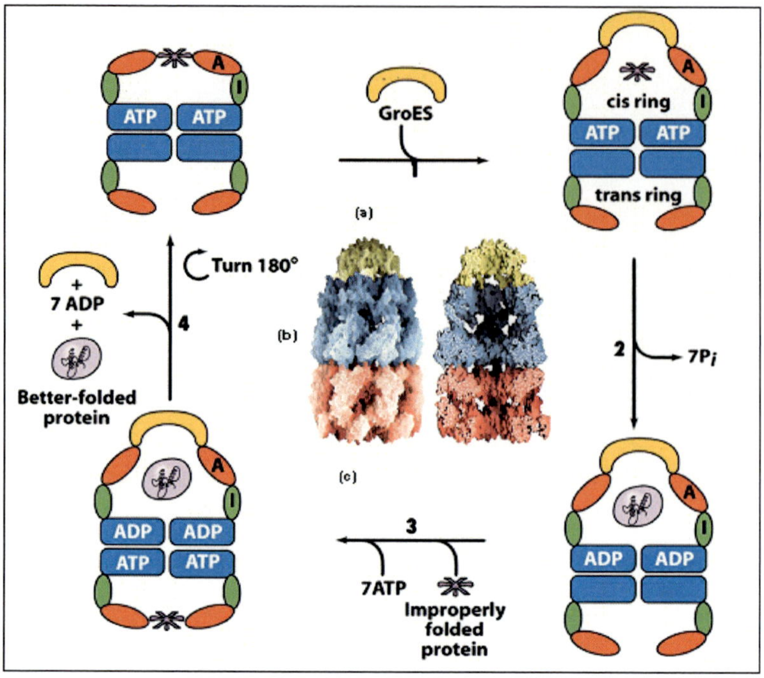

Figure 1. Mechanism of chaperonin action. The bacterial chaperonin GroEL consists of two homo-heptameric rings, which form a barrel-shaped structure: (a) GroeS; (b) domains of GroeL; (c) GroeL. (1) ATP molecules and misfolded protein binds to chaperonin; (2) ATP and GroES are bound by the same ring, which displaces the peptide into an enclosed cavity that is capped by GroES, where the peptide is folded; (3) Hydrolyses 7 ATP molecules; (4) Binding 7 ATP to trans ring and concomitant release of folded protein, ADP molecules and GroES from cis ring, binding of misfolded protein to trans ring; (5) Cis ring becomes trans ring and cycle can repeat. Some polypeptides require several cycles of binding and release to reach their native state.

Chaperonins/Hsp60

The chaperonins (the group of stress proteins belonging to the Hsp60 family) which are defined by a barrel-shaped, doublering structure [70,71], have significant roles in polypeptide folding and in protein transport in the cells as well. Members include bacterial GroEL, Hsp60 of mitochondria and chloroplasts, and the TriC–CCT complex localized in the eukaryotic cytoplasm. Based on their characteristic ring structure, a central cavity is formed, which accommodates nonnative proteins via hydrophobic interactions. The Hsp60/Hsp10 chaperonin system is localized primarily in the matrix space of mitochondria where it assists in folding, refolding and/or elimination of mitochondrial proteins [48,58]. Generally, chaperonins are able to form stable complexes with proteins, which are imported to chloroplasts and to mitochondria [58,72]. They perform their chaperone function also in co-operation with the other molecules, e.g. cnp10 and Hsp70 [73,74]. Hsp60 also has other important functions in an immune response due to its immunodominant properties [75-77].

Furthermore, bacterial Hsp60 and Hsp70 take a part in regulation of gene transcription coding for the stress proteins [78].

Conformational changes of the chaperonin subunits induced through ATP hydrolysis change the inner lining of the cavity from a hydrophobic to a hydrophilic character [79-81] (Figure 1). As a consequence the unfolded polypeptide is released into the central chamber and can proceed on its folding pathway in a protected environment [82]. The chaperonins are therefore capable of folding proteins such as actin that cannot be properly folded via other mechanisms [83].

Hsp70

The Hsp70 proteins bind to misfolded proteins promiscuously during translation or after stress-mediated protein damage [84,85]. Proteins in the Hsp70 family are known for their ability to bind peptide chains. Members of this family are highly conserved throughout evolution and are found throughout the prokaryotic and eukaryotic phylogeny.

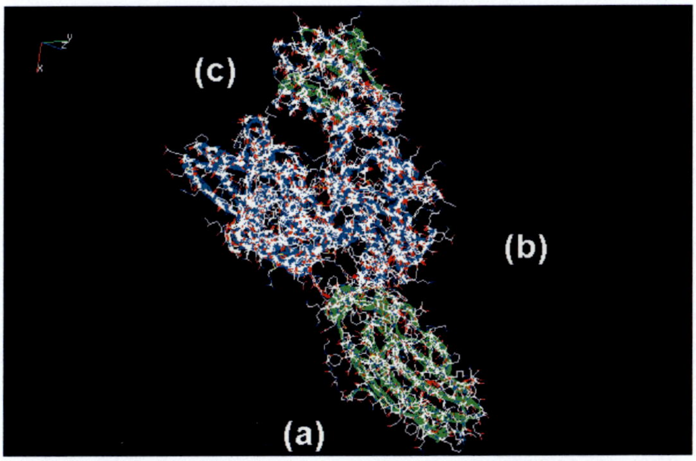

Figure 2. The structure of human Hsp70. Hsp70 proteins display a characteristic domain structure comprising an amino terminal ATPase domain (a), a peptide binding domain (b), and a carboxyl-terminal domain (c) that is supposed to form a lid over the peptide-binding domain. In cooperation with other chaperones, Hsp70s stabilize preexistent proteins against aggregation and mediate the folding of newly translated polypeptides in the cytosol as well as within organelles. These chaperones participate in all these processes through their ability to recognize nonnative conformations of other proteins. They bind extended peptide segments with a net hydrophobic character exposed by polypeptides during translation and membrane translocation, or following stress-induced damage. Drawing made with DeepView / Swiss-PdbViewer Version 3.7 (SP5) [86].

Single cells contain multiple Hsp70 homologues, even within a single cellular compartment; for example, mammalian cells express two inducible homologues (Hsp70.1 and Hsp70.3) and a constitutive homologue Hsc70 in the cytoplasm. These homologues have overlapping but not totally redundant cellular functions. They act in: (*i*) folding of newly synthesized proteins, (*ii*) transport of proteins across membranes, (*iii*) refolding of misfolded and aggregated proteins, (*iv*) control of activity of regulatory proteins and (*v*) protein degradation [86-88]. This versatility is achieved through the evolutionary amplification and diversification of *hsp70* genes, which has generated both specialized Hsp70 chaperones and more diverged Hsp110 and Hsp170 proteins. Versatility is also achieved through extensive employment of cochaperones, J proteins, and nucleotide exchange factors (NEFs), which regulate Hsp70 activity.

The structure of Hsp70 consists of three functional domains: an amino-terminal ATPase domain, a central peptide-binding cleft, and a carboxyl terminus that seems to form a lid over the peptide-binding cleft [71] (Figure 2) [89]. By X-ray analysis it was found that the ATPase domain has high structural homology with ATPase domains of hexokinases [90] and actin [91]. The chaperones recognize short segments of the protein substrate, which are composed of clusters of hydrophobic amino acids flanked by basic residues [92]. Such binding motifs occur frequently within protein sequences and are found exposed on nonnative proteins. In fact, mammalian Hsp70 binds to a wide range of nascent and newly synthesized proteins, comprising about 15–20% of total cellular proteins [82]. This percentage is most likely further increased under stress conditions. Hsp70 proteins apparently prevent protein aggregation and promote proper folding by shielding hydrophobic segments of the protein substrate via a substrate binding domain (SBD), thereby preventing aggregation and promoting proper folding. ATP binding to the N-terminal nucleotide binding domain (NBD) induces conformational changes in the adjacent SBD, which opens the substrate binding pocket and its helical lid. Conversely, substrate binding in synergy with the action of J proteins triggers ATP hydrolysis and concomitant closing of the SDB, which traps substrate proteins [93]. According to this model, ATP binding and hydrolysis by the amino-terminal ATPase domain of Hsp70 induce conformational changes of the carboxyl terminus, which lead to lid opening and closure [71] (Figure 3). ATP binding to the ATPase domain opens up the substrate binding cavity in the adjacent substrate binding domain; ATP hydrolysis closes the cavity and traps bound substrates. Stable holding of the protein substrate requires closing of the binding pocket, which is induced upon ATP hydrolysis and conversion of Hsp70 to the ADP-bound conformation. ATP binding may rearrange the interface between the NBD groove and SBD helix A, perhaps even disrupting it, thereby facilitating the opening of the SBD. Concomitantly, the linker may relocate and become more tightly associated with the connecting region, such that at no stage during the functional cycle of Hsp70 do the SBD and NBD become completely disconnected. It appears that the additional minor contact involving residues of the β sandwich of the SBD plays a critical role because a mutant with alteration in one of the involved residues has coupling defects [93]. Signal transduction between SBD and NBD thus seems to rely on integrated rearrangements of the linker and at least two contact sites within the SBD. The dynamic association of Hsp70 with nonnative polypeptide substrates thus depends on ongoing cycles of ATP binding, hydrolysis, and nucleotide exchange. Importantly, ancillary co-chaperones are employed to regulate the ATPase cycle [85,93,94]. Decreased association of certain proteins with Hsp90 and increased association with Hsp60/Hsp70 lead to their 20S proteasome-mediated degradation. Hsp70 has been shown to promote the poly-ubiquitination of damaged proteins. Ubiquitination seems also to be involved in the degradation of unfolded polypeptide by the lysosome. One major mechanism of the lysosomal degradation of proteins is dependent on Hsc73 and is responsible for the degradation of a significant amount of the cytosolic protein [95]. Co-chaperones of the Hsp40 family (also termed J proteins due to their founding member bacterial DnaJ) stimulate the ATP hydrolysis step within the Hsp70 reaction cycle and in this way promote substrate binding [96] (Figure 3). In contrast, the carboxyl terminus of Hsp70-interacting protein (CHIP) attenuates ATP hydrolysis [97].

CHIP has been implicated in the decision as to whether a target protein enters the refolding or the degradation pathway. In contrast to other E3 ligases, CHIP forms homodimers through interaction via a central coiled coil region. Dimerization is a prerequisite for the activity of CHIP in ubiquitination and thus may provide a mechanism for regulation of CHIP activity. Similarly, nucleotide exchange on Hsp70 is under the control of stimulating and inhibiting cochaperones. The Hsp70-interacting protein (Hip) slows down nucleotide exchange by stabilizing the ADP-bound conformation of the chaperone [98], whereas nucleotide exchange is stimulated by the co-chaperone BAG-1 (Bcl-2-associated athanogene 1), which assists substrate unloading from Hsp70 [99,100]. By altering the ATPase cycle, the co-chaperones directly modulate the folding activity of Hsp70. In addition to chaperone-recognition motifs, co-chaperones often possess other functional domains and therefore link chaperone activity to distinct cellular processes [101]. Indeed, the co-chaperones BAG-1 and CHIP apparently modulate Hsp70 function during protein degradation.

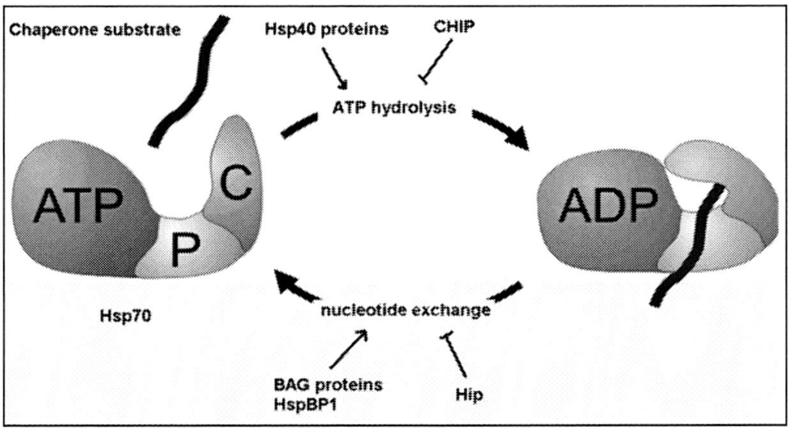

Figure 3. The Hsp70 reaction cycle. The Hsp70 structure consists of three functional domains: an amino-terminal ATPase domain (ATP), a central peptide-binding cleft (P), and a carboxyl terminus that seems to form a lid over the peptide-binding cleft (C) (see *Figure 2*). In the ATP-bound conformation, the binding pocket is open, resulting in a low affinity for the binding of a chaperone substrate. ATP hydrolysis induces stable substrate binding through a closure of the peptide-binding pocket. Substrate release is induced upon nucleotide exchange. ATP hydrolysis and nucleotide exchange are regulated by diverse co-chaperones.

The Hsp90 Family

Members of the Hsp90 family constitute 1–2% of cytosolic proteins and have stress-related as well as housekeeping functions. Hsp90 (Hsp83) is the most abundant cytosolic protein in eukaryotic cells. Its homologues were found in the endoplasmic reticulum (ER) of higher eukaryotes (Grp94 also called endoplasmin) and in prokarytic cells (HtpG). Hsp90 exists *in vivo,* as a dimer and it co-exists as a cytosolic protein in mammals. Two isoforms exist: Hsp90α and Hsp90β. Those are indicated as alpha and beta and are produced in the same quantity. In *E. coli* the HtpG (68 kDa) differs from Hsp90 by the absence of a charged area of approximately 50 amino acids, which is characteristic for the other Hsp90 homologues [102]. The Hsp70 and Hsp90 families exhibit several common features: both possess ATPase activity and are regulated by ATP binding and hydrolysis, and both are further regulated by ancillary cochaperones [103-108]. Unlike Hsp70, however, cytoplasmic Hsp90 is not generally involved in the folding of newly synthesized polypeptide chains. Instead it plays a key role in the regulation of signal transduction networks, as most of the known substrates of Hsp90 are signaling proteins, the classical examples being steroid hormone receptors and signaling kinases.

Hsp90 stabilizes damaged proteins during and after stress. Hsp90 interacts and either modulate the assembly, the stability and/or the activity of particular cellular proteins such as protein kinases, calcineurin, calmodulin, nitric oxide synthase, telomerase, steroid receptors, oncogenes and transcription factors [103-109]. Significant interest is dedicated to the pivotal role of Hsp90 into the regulation of hormonal receptors [110-115]. Receptors what are not bounded to their hormones are probably bound to Hsp90 shortly after their translation [110,116]. In case of the glucocorticotropic receptor, binding of Hsp90 leads to an enhancement capability of the receptor to bind to the steroid hormone [110,116]. The C-terminal region of the Hsp90 molecule is responsible for the receptor binding [117]. In addition, Hsp90 appears to be closely linked to the protein degradation in the cell. Hsp90 also shows direct interaction with the proteasome and might possess regulatory roles, other than determination of the fate of unfolded proteins that cooperate with co-chaperones, PA28 and CHIP [118,119]. Initially, Hsp90 was considered to inhibit the 20S proteasome [120,121] and also to protect it from oxidative stress [122] but recently it was also shown that Hsp90 interacts with the 26S proteasome and plays a principal role in the assembly and maintenance of the 26S proteasome.in yeast [123].

Hsp90 is presented as a suppressor of cryptic genetic variations by assisting mutant proteins to maintain a wild type structure and function [124]. On a molecular level, Hsp90 binds to substrates at a late stage of the folding pathway, when the substrate is poised for activation by ligand binding or associations with other factors. Consequently, Hsp90 accepts partially folded conformations from Hsp70 for further processing. In the case of the chaperone-assisted activation of the glucocorticoid hormone receptor [110,115,116] and also of the progesterone receptor [111], the sequence of events leading to attaining an active conformation is fairly well understood (Figure 4). It appears that the receptors are initially recognized by Hsp40 and are then delivered to Hsp70 [125]. Subsequent transfer onto Hsp90 requires the Hsp70/Hsp90 organizing protein (Hop), which possesses non-overlapping binding sites for Hsp70 and Hsp90 and therefore acts as a coupling factor between the two chaperones [126]. In conjunction with p23 and different cyclophilins, Hsp90 eventually mediates conformational changes that enable the receptor to reach a high-affinity state for ligand binding. Hsp90 and p23 play also a direct role in the regulation of the HS response by modulating the HSF1 activation/deactivation process. Hsp90, Hsp70, Hsp60 and p23 make heterocomplex with a variety of transcription factors and protein kinases involved in mitogenic signal transduction. The major function of this complex may be to fold the client protein and to keep it inactive until it reaches its ultimate location. There is also a potential involvement of Hsp70 and Hsp90 in DNA replication since members of these families interact with components of the eukaryotic cell cycle. Hsp70, Hsp90, Hsp27 and TCP-1 are known to bind and stabilize actin, tubulin and the microtubules/microfilament network playing a role in cellular morphology and signal transduction pathways.

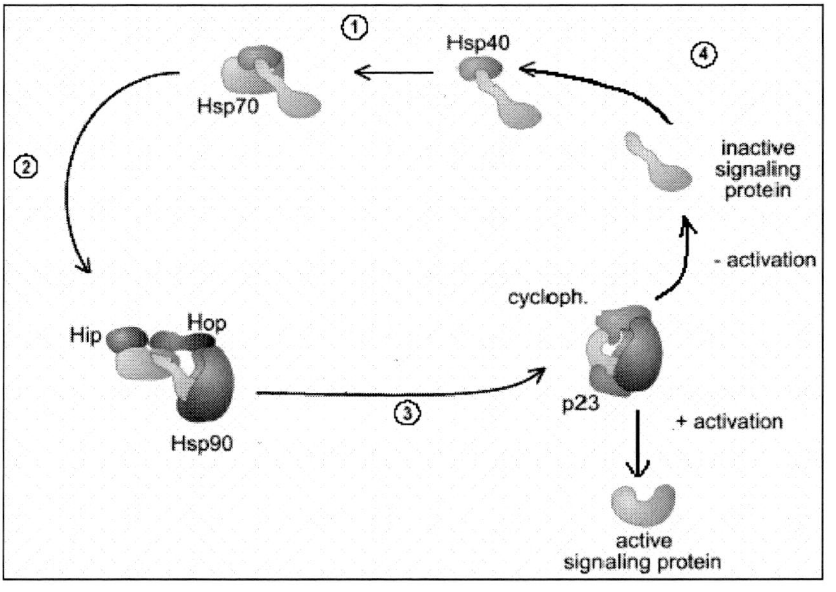

Figure 4. Hsp90 chaperone system for the regulation of signal transduction pathways. The involvement of cofactors may change depending on the target protein. The inactive signaling protein, e.g., a steroid hormone receptor, is initially recognized by Hsp40 and delivered to Hsp70 (step 1). Subsequently, a multi-chaperone complex assembles: the co-chaperone Hop is recruited to establish a physical connection between Hsp70 and Hsp90. Hop stimulates recruitment of an Hsp90 dimer that accepts the substrate from Hsp70. Presumably, ATP hydrolysis by Hsp70 releases the bound signalling protein and transfers it to the Hsp90 dimer, thus resulting in the formation of the intermediate complex (step 2). At the final stage of the chaperone pathway, Hsp90 associates with p23 and diverse cyclophilins (cycloph.) to mediate conformational changes of the signaling protein necessary to reach an activatable state (step 3). Upon activation, i.e., hormone binding in the case of the steroid receptor, the signaling protein is released from Hsp90. In the absence of an activating stimulus, the signaling protein folds back to the inactive state when released and enters a new cycle of chaperone binding and in the absence of a hormone ligand, the receptor protein may participate in another cycle (step 4).

On other signaling pathways Hsp90 serves as a scaffolding factor to permit interactions between kinases and their substrates, as is the case for Akt kinase and endothelial nitric oxide synthase (NOS) [127,128]. Since many of the Hsp90 substrate proteins are involved in regulating cell proliferation and cell death, it is not surprising that the chaperone recently emerged as a drug target in tumor therapy [106,107]. The antibiotics geldanamycin [129] and radicicol [130] specifically bind to Hsp90 in mammalian cells and inhibit the function of the chaperone by occupying its ATP-binding pocket. Drugs based on these compounds are now being developed as anticancer agents, as they potentially inactivate multiple signaling pathways that drive carcinogenesis. Remarkably, drug-induced inhibition of Hsp90 blocks the chaperone-assisted activation of signaling proteins and leads to their rapid degradation via the UPS [131-136]. Hsp90 inhibitors therefore have emerged as helpful tools to study chaperone–proteasome cooperation.

Transcription of Genes Coding for HSP

Eukaryotic stress genes are generally controlled by HSE (heat shock elements) and HSF (heat shock factors), which are bound to them closely [137]. The induction of the HS response is through the HSF working as molecular links between environmental stresses and the stress response [48,58]. HSE are present as monomers in non-stressed cells, while they form trimers in stressed cells [138]. The trimer is distinguished by a higher asymmetry compared to the monomeric form [138]. Interaction HSF-HSE can contain various number of the basic HSE units. The smallest detectable part of HSE bounded to HSF is 10 bp long, and it exists in configuration 'tail-to-tail' or 'head-to-head'. HSF covers these parts in same manner. HSF are also constitutive proteins in monomeric forms in eukaryotes (with the exception of yeasts) [139]. They are activated during the HS and form trimers [140]. HSF are redistributed from non-specific places on chromatin to separated chromosomal targets. A subpopulation of HSF molecules is transferred from the cytoplasmatic compartments into the nuclei during HS as well [138]. Thus, HSF and HSE interact if the HSF is in trimer form [141]. Binding activity of HSF is induced by various agents, which affect protein structure. These agents include: heat, slightly acidic pH, or presence of detergents [140]. Original hypothesis of HSF activation *in vitro* is based on direct formation of an oligomeric state (trimer) [142]. The impossibility of dissociation of trimers to the monomeric forms in normal conditions shows that it cannot be a reversible one-component system.

Activation of Heat Shock Factors (HSF): a Stress Regulatory Network

Vertebrates express four HSFs encoded by small multigene families and regulated by a diverse array of environmental and developmental signals [143]. The four vertebrate HSFs are expressed constitutively and co-operate functionally. HSF1 is a long-lived protein, which is present as an inactive monomer considered to be a general stress responsive factor. HSF1 is expressed ubiquitously and is activated by mild HS as well as by multiple environmental or physiological stresses [144,145]. HSF2 is a short-lived protein which is an inactive dimer refractory to typical stress stimuli except to proteasome inhibitors [146]. It is considered to be important during embryogenesis [147] and spermatogenesis [138,148-150]. HSF3 is also an inactive dimer and may exhibit complex interactions with other transcription factors governing development, growth and apoptosis, such as c-Myc and p53 [151]. HSF3 is also an important co-regulator of HSF1, activated by severe HS and chemical stress [152,153]. HSF4 constitutively binds DNA even in non-stressed cells and is preferentially expressed in muscle, brain and pancreas [154-157]. Proteasome inhibitors are the common activators of HSF1, HSF2 and HSF3 [158-160].

Regulation of HSP gene transcription is mediated by the interaction of the HSF (of which the principal one in vertebrates is HSF1) with HSEs in the HSP gene promoter regions [161]. In unstressed cells, HSF1 is located in the cytoplasm and in the nucleus. It is maintained as a non DNA-binding inactive complex both by internal coiled-coil interactions and by stoichiometric binding with Hsp90, Hsp70 and other chaperones. Under stressful conditions, cellular proteins undergo denaturation and/or polyubiquitination and sequester the chaperones capping HSF1. Inactive HSF1 is liberated and translocates into the nucleus. HSF1 is activated by trimerization and subsequent phosphorylation [48]. HSF1 is hyperphosphorylated in a *ras*-dependent manner by members of the mitogen-activated protein kinase (MAPK) subfamilies (e.g. ERK1, JNK/SAPK, p38 protein kinase) [162,163]. HSF1 is converted to phosphorylated trimers with the capacity to bind DNA, and translocates from the cytoplasm to the nucleus [164].

The precise mechanisms for stress sensing and signalling to activate HSF1, and the mechanisms by which many distinct stresses activate HSF1, are poorly understood. The signal that activates HSF1 is thought to be a flux of newly synthesized non-native proteins [164]. Other pathways of Hsp induction were discovered through a sensor of the cellular redox status which allows the preexisting Hsp33 to respond much more quickly to stress than by regulation at the transcriptional or translational level [165] or through redox-dependent activation of HSF1 by hydrogen peroxide. Recombinant mammalian HSF1 directly senses both heat and hydrogen peroxide to assemble into a homotrimer in a reversible and redox-regulated manner [166]. The sensing of both stresses requires two cysteine residues within the HSF1 DNA-binding domain that are engaged in redox-sensitive disulfide bonds [166].

Under conditions of moderate short-term stress, proteins may be partially unfolded and aggregate in a reversible manner that is dependent on the expression of a large family of protein chaperones under the transcriptional controls of HSF [167]. However, as a consequence of acute or prolonged stress, proteins may be irreversibly damaged and targeted for degradation. The downregulation of the UPS by pharmacological proteasome inhibitors causes HSF activation in yeast and mammals [168,169]. Severe HS, promote proteasome dissociation into 20S CP and 19S RP subcomplexes resulting in a rapid decline in intracellular proteolysis [123,170]. Conversely, proteasome levels increase in response to long-term mild stress. Under a multitude of mild stress conditions proteasome levels rapidly increase to participate in the removal of damaged proteins. For example in budding yeast, the transcription factor Rpn4 activates this elevated transcription. Rpn4 is a short lived protein that recognizes a proteasome activation control element (PACE) motif in their promoters, as well as in its own gene and numerous other genes [171]. Rpn4 is part of a negative feedback circuit being itself a substrate of the 26S proteasome [172-174]. Under stress conditions that require elevated levels of proteasomes to tackle the higher loads of damaged proteins, or when proteasome activity is compromised, degradation of Rpn4 is retarded. Under these circumstances transcription of *RPN4* is elevated and in turn, this leads to upregulation of proteasome levels. In this sense, direct regulation of *RPN4* by HSF provides an efficient mechanism to cope with the stress caused by unfolded proteins, through increasing the cellular protein degradation capacity [175]. HSF co-ordinates a feed-forward gene regulatory circuit for *RPN4* activation. In addition to Rpn4, other HSF target genes are also likely to function in the degradation of damaged proteins [176-178], implying a role for HSF in coordinated regulation of ubiquitin-mediated proteolysis pathways. Therefore, HSF directs a co-ordinated response for activation of the cellular ubiquitination and degradation machinery in the face of acute stress where protein unfolding is manifest. However, despite intensive searches in higher eukaryotes an Rpn4 homologue has not been identified.

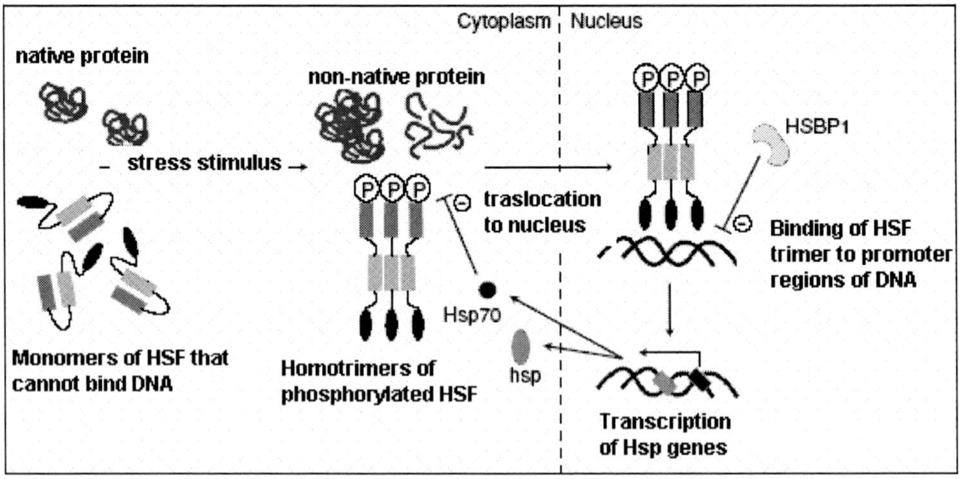

Figure 5. Regulation of transcription of heat shock protein genes by heat shock factor. Heat shock factor (HSF) is present in the cytoplasm as a latent monomeric molecule that is unable to bind to DNA. Under stressful conditions, the flux of non-native proteins (which are non-functional, prone to aggregation, protease-sensitive, and bind to chaperones) leads to phosphorylation (P) and trimerisation of HSF. The trimers translocate to the nucleus, bind the promoter regions of heat shock protein (hsp) genes and mediate hsp gene transcription. The activity of HSF trimers is downregulated by HSPs (e.g. Hsp70) and the heat shock binding protein 1 (HSBP1) that is found in the nucleus.

The generation of HSPs must be only transient, even if exposure to stress is over a prolonged period, as a continued presence of HSPs would adversely influence protein homeostasis and a variety of intracellular functions (Figure 5). One mechanism, by which the activity of HSF1 is regulated, is via the binding of Hsp70 to its transactivation domain, thereby leading to repression of heat shock gene transcription [179]. The interaction between Hsp70 and HSF1 has no effect on DNA binding or the stress induced phosphorylation state of HSF1 [179]. A second mechanism that regulates Hsp synthesis is the interaction between heat shock protein binding factor 1 (HSBP1) and the active trimeric form of HSF1 and Hsp70, thereby inhibiting the capacity of HSF1 to bind to DNA [179]. HSBP1 is predominantly localized in the nucleus and levels of HSBP1 mRNA have been shown to be increased in a variety of cell lines and animal tissues and to be unaffected by HS [179]. A third regulatory mechanism is based on CHIP (carboxy terminus of Hsp70-binding protein), a dual-function co-chaperone/ubiquitin ligase that targets a broad range of chaperone substrates for proteasomal degradation (see below) [180]. CHIP not only enhances Hsp70 induction during acute stress but also mediates its turnover during the stress recovery process. Central to this dual-phase regulation is its substrate dependence: CHIP preferentially ubiquitinates chaperone-bound substrates, whereas degradation of Hsp70 by CHIP-dependent targeting to the UPS occurs when misfolded substrates have been depleted [180].

FROM PROTEIN FOLDING TO STRESS RESPONSE

One of the most critical events in the biogenesis of a protein is the conversion of its linear amino acid sequence into the properly folded three-dimensional structure (see Chapter 12). Although the amino acid sequence of a protein contains all the information that is required to dictate proper folding into a functional, three-dimensional structure, the crowded intracellular milieu places constraints on the folding of polypeptides, thereby promoting misfolding and aggregation. As soon as a nascent polypeptide chain emerges from the ribosome, it is prone to misfolding and subsequent aggregation. As a consequence, protein folding *in vivo* is typically not spontaneous [181], and organisms have evolved a highly conserved class of proteins called molecular chaperones that prevent inappropriate interactions within and between non-native polypeptides, enhance the efficiency of *de novo* protein folding and promote the refolding of proteins that have become misfolded as a result of cellular stress [70,182-184]. Chaperones contain no specific information for polypeptide folding, but they are able to prevent production of aggregates from the nascent polypeptides [185]. It could be noticed that other heat induced proteins like peptidyl-prolyl *cis-trans* isomerases (PPIases) and protein disulphide isomerases (PDI) act directly in a formation of higher protein structures [186], and they are called 'foldases' [187]. Under certain conditions, when chaperones cannot repair misfolded proteins, chaperone-mediated targeting to the UPS or to lysosomes results in selective degradation. Thus chaperones are defined as proteins that assist other macromolecules in folding/unfolding and in assembly/disassembly of higher order structures without being components of these final structures. Several but not all stress proteins are molecular chaperones. A variety of co-chaperones are also present in the complexes with the HSPs.

Molecular Chaperones and Protein Aggregation

Protein folding occurs primarily in the cytoplasm (cytosolic protein folding for cytosolic proteins) and in the ER (oxidant protein folding for membrane associated or secreted proteins like antibodies among many others) [188]. The differing redox and ionic milieus inside these two compartments, and the different functions and destinations of the client proteins folded therein, have necessitated the existence of distinct chaperone networks [188,189]. The details on the posttranslational processes in the ER called protein quality control (QC) whereby the cell folds newly synthesized proteins and either refolds or degrades polypeptides that fail to attain or maintain a native structure are given in Chapter 13. Both networks exploit the exquisite sensitivity of cysteines to redox state, but they respond in opposite directions, reflecting the different conditions in the cytosol (reducing) and in the ER (more oxidizing) [190]. Thus, for example, the cytosolic chaperone Hsp33 forms active dimers in response to oxidation, linking the responses to thermal and oxidative stress. Folded proteins are released upon Hsp33 reduction, whereas unfolded substrates are released only in the presence of additional chaperone complexes that are able to refold them [190]. In contrast, the ER oxidoreductase PDI appears to function as a chaperone primarily when reduced. Mechanistically, cysteines are the main targets, because their thiol groups can be modified according to redox conditions. Formation of a disulfide, in particular, imposes structural constraints on the polypeptide, which can result in novel functions. Cysteines therefore provide molecular switches, which can sense and transducer redox changes in the environment by modifying protein conformation in a rapid and reversible way (reviewed in [188]). PPIases catalyse the interconversion of a peptide bond that precedes a proline residue from the *trans* (extended) to the *cis* (bent) position, which is often needed for proper protein folding. Furthermore, up-regulation of PDI is an adaptive response to accumulation of misfolded proteins in the ER that helps protect neurons from apoptosis resulting from ER stress. Uehara *et al.* showed that *S*-nitrosylated PDI was present in the brains of people who had had PD or AD [191]. *S*-nitrosylation, a reaction transferring a nitric oxide (NO) group to a critical cysteine thiol, impaired the ability of PDI to act as a chaperone and an isomerase leading to the accumulation of polyubiquitinated proteins, and to activation of the the unfolded protein response (UPR). When coexpressed with synphilin-1 in a dopaminergic neuroblastoma cell line, PDI inhibited the development of synphilin-1-dependent Lewy-body (LB)-like inclusions, a protective effect that was attenuated by NO. Exposure of cultured cortical neurons to *N*-methyl-D-aspartate (NMDA) led to NOS-sensitive accumulation of *S*-nitrosylated PDI, accumulation of polyubiquitinated proteins, and apoptosis; PDI overexpression decreased the number of polyubiquitinated and apoptotic cells and attenuated NMDA-dependent activation of the UPR [191]. In neuroblastoma cells, PDI overexpression inhibited cell death in response to ER stress, inhibition of the proteasome, or overexpression of a protein that induces the unfolded protein response, protective effects that were reversed by exposure to a NO donor. PDI has two domains that function as independent active sites with homology to the small, redox-active protein thioredoxin. During neurodegenerative disorders and cerebral ischemia, the accumulation of immature and denatured proteins results in ER dysfunction, but the upregulation of PDI represents an adaptive response to protect neuronal cells.

Hsp70 and Hsp90 are major components of the cytosolic protein quality control (QC) machinery and facilitate both protein folding and degradation [84,101]. However, is not clear how chaperone substrates are partitioned between protein folding and degradation pathways. To fold cellular proteins, Hsp70 and Hsp90 function with folding factors, termed co-chaperones (Table 2). Basic function of Hsp70 is the binding of unfolded proteins and their release. It proceeds in an ATP dependent cycle [192,193], in which: (*i*) partially unfolded protein associates with C-terminal domain Hsp70 [194,195], (*ii*) binding of co-chaperone Hsp40 started a dissociation of ATP with N-terminal domain of Hsp70 [196], which leads to a conformational change (Hsp70/ADP complex), (*iii*) Hsp40 dissociates, (*iv*) BAG-1 protein binds, (*v*) ADP/ATP exchange is initiated, (*vi*) BAG-1 dissociates, (*vii*) bounded proteins are released finally. Folding co-chaperones regulate nucleotide hydrolysis and deliver non-native proteins to the polypeptide-binding domains of Hsp70 and Hsp90. Co-chaperones that assist Hsp70 in protein folding include the Hsp40 proteins, Hip and Hop [84,101]. Hsp90 is regulated by co-chaperones that include Hop, p23 and the cyclophilins [197,198]. In some cases, Hsp70 and Hsp90 function sequentially to fold the same substrate protein and Hop acts as a coupling factor that links these chaperones to each other. To degrade proteins, Hsp70 and Hsp90 appear to interact with a set of co-chaperones that have degradatory functions, such as BAG-1, CHIP and, possibly, scythe [101]. It is proposed that domains within these co-chaperones enable them to interact both with chaperones and with components of the UPS (Table 2). The interaction of Hsp70 and Hsp90 with either folding or degradatory co-chaperones provides a mechanism to explain how they chose between the biogenesis or destruction of their substrate proteins [199,200].

Table 2. Co-chaperones of mammalian Hsp70 and Hsp90 that function as folding or degradation factors

Co-chaperone	Chaperone Hsp 70	Hsp 90	Chaperone-binding domain	Specialized sub-domains	Cellular function
Folding factors					
Hdj1/Hdj2	X		J-domain	Polipeptide binding	Helps to load Hsp70 with non-native proteins by regulating binding its ATPase activity and functioning as a chaperone.
Hop	X	X	TPR		Binds C-terminal EEVD motifs present in Hsp70 and Hsp90 and tethers these chaperones together. Facilitates substrate exchange between Hsp70 and Hsp90.
Hip	X		TPR		Binds to the ATPase domain of Hsp70. Promotes protein folding by modulating the affinity of Hsp70 for ADP.
Cyclophillin		X	TPR	Polipeptide binding	A family of co-chaperones that contains a conserved binding chaperone domain that has peptidyl-prolylisomerase activity. Facilitates late-stage Conformational maturation events for Hsp90 substrates.
P23		X		Polipeptide binding	Binds to the N-terminus of Hsp90 to stimulate ATP-binding dependent substrate release. Has a N-terminal Hsp90-binding domain and a C-terminal chaperone domain.
Degradation factors					
BAG-1	X		BAG	Ubiquitin-like	Binds to the ATPase domain of Hsp70 and functions as a nucleotide exchange factor. It utilizes its ubiquitin-like domain to target some Hsp70-bound substrates to components of the UPS.
Scythe	X		BAG	Ubiquitin-like	Has functions similar to Bag-1 as both are anti-apoptotic proteins and modulate Hsp70 chaperone activity. Interactions between Scythe and components of the UPS are proposed, but yet to be demonstrated.
CHIP	X	X	TPR	U-box	Binds Hsp70 and Hsp90 at a C-terminal TPR acceptor site. The U-box is proposed to recruit E2 enzymes to chaperones to promote ubiquitination of non-native proteins.

TPR indicates tetratricorepeat motif. Hdj-1 and Hdj-2 are members of the Hsp40 family of co-chaperones that utilize a conserved J-domain to regulate the ATPase activity of Hsp70.

Molecular Chaperones and Protein Degradation

Molecular chaperones take a part in degradation of irreversibly damaged proteins. Selective recognition of non-native proteins is the first step towards their elimination. Based on their ability to interact with non-native folding intermediates, molecular chaperones are prime candidates to aid in the triage of misfolded proteins. Once potentially damaging conformers have been identified, the cell can respond to their presence in three ways. First, cellular factors may attempt to rescue the misfolded conformations by refolding them to a functional native state. Second, the cell can sequester misfolded proteins in an attempt to prevent toxic interactions. Accordingly, chaperones alleviate the toxicity associated with aberrant protein conformations in neurodegenerative disease models [201]. Chaperones seem to alleviate toxicity by sequestering the soluble toxic oligomeric species or by modulating their conformation [201,202]: overexpressing Hsp70 suppresses the toxicity associated with various proteins including amyloid-β (Aβ) and tau in AD, α-Synuclein (α-Syn) in PD, superoxide dismutase (SOD1) in ALS, and polyglutamine (polyQ)-expanded proteins in HD, spinal bulbar muscular atrophy (SBMA) and ataxias [201]. Hsp40, induces a conformational rearrangement in mutant huntingtin [203,204] and disfavours the accumulation of specific soluble polyQ fibril intermediates [202]. Aggresome formation is apparently a specific and active cellular response when production of misfolded proteins exceeds the capacity of the UPS to tag and remove these proteins (see Chapter 12). The pathways that regulate aggresome assembly are only now being explicated. Histone deacetylase 6 (HDAC6) appears to be a key regulator of aggresome assembly [205]. HDAC6 is a microtubule-associated deacetylase that has the capacity to bind both multi-ubiquitinated proteins and dynein motors and is believed to recruit misfolded proteins to the pericentriolar region for aggresome assembly. Deletion of HDAC6 prevents aggresome formation and sensitizes cells to the toxic effects of misfolded proteins. Components of the ubiquitin–proteasome system and chaperones such as Hsp70 are abundantly present in and are actively recruited to aggresomes [206-208]. Furthermore, elevating cellular Hsp70 levels can reduce aggresome formation by stimulating proteasomal degradation [209]. It appears that these subcellular structures are major sites of chaperone–proteasome cooperation to mediate the metabolism of misfolded proteins.

The proteins, which are determined to degradation, have to be in a soluble state to be recognised by proteolytic enzymes. The precise role of chaperones in eliminating misfolded proteins is still unclear. In the simplest model, chaperones would be primarily dedicated to stabilizing and refolding nonnative polypeptides. In this case, their role in QC could be an extension of their primary role in folding; that is, to maintain the solubility of misfolded intermediates and facilitate sampling by the ubiquitination machinery. HSP preserves such state and therefore the chaperone function of HSP is essential not only for the repair of damaged proteins, but also for their degradation [210]. Not well understood are the events that lead to the cessation of efforts to fold a substrate, and the diversion of the substrate to the terminal degradative pathway. It is possible that chaperones and components of the UPS exist in a state of competition for these substrates and that repeated cycling of a substrate on and off a chaperone maintains the substrate in a soluble state and increases, in a stochastic manner, its likelihood of interactions with the ubiquitination machinery. Unfortunately, only indirect evidences, manipulating the balance between folding and degradation activities of chaperones, confirm such conclusions. In cells treated with Hsp90 inhibitors, the chaperone-assisted activation of signaling proteins is abrogated and chaperone substrates such as the protein kinases Raf-1 and ErbB2 are rapidly degraded by the UPS [131-136,211]. This appears to be due, in part, to transfer of the substrates back to Hsp70 and progression toward the UPS. Substrate interactions with chaperones and consequently their commitment, either toward the folding pathway or to their degradation via the UPS apparently serve as an essential post-translational protein QC mechanism within eukaryotic cells.

Several HSP family members as well as sHSPs, Hsp70 and Hsp90 have all been implicated to participate in protein degradation. A function for chaperones in targeting misfolded proteins for degradation has been established in various ways. Hsp27 was recently shown to stimulate the degradation of phosphorylated IκBα via the UPS, which may account for the antiapoptotic function of Hsp27 [212]. Similarly, Hsp27 facilitates the proteasomal degradation of phosphorylated tau, a microtubule-binding protein and component of protein deposits in AD [213]. Cell lines which upregulate Hsp27 in response to the treatment with proteasome inhibitors are resistant to their proapoptotic effects [68,69]. Hsp70 is required for the *in vitro* degradation of some misfolded proteins [214], whereas *in vivo* experiments implicate the yeast Hsp40 homologue Ydj1p [215]. Hsp70 participates in the degradation of apolipoprotein B100 (apoB), which is essential for the assembly and secretion of very low-density lipoproteins from the liver [216]. Hsp70 and Hsp90 are required for the degradation of the immaturely glycosylated and aberrantly folded forms of the cystic fibrosis transmembrane conductance regulator (CFTR) [200,216-219]. CFTR is an ion channel localized at the apical surface of epithelial cells. Its functional absence causes cystic fibrosis, the most common genetic disease in Caucasians [220,221]. Hsp70 also takes part in the degradation of and misfolded von Hippel-Lindau (VHL) variants [222]. In addition, overexpressing Hsp70 and Hsp40 increases the proteasome-mediated degradation of α-Syn and polyQ expanded proteins [201].

From a different point of view, recent analysis of the QC mechanisms of misfolded variants of the VHL tumour suppressor suggests that chaperones have an active role selecting proteins for degradation [222]. The observation that some chaperones specifically interact with E3s raises the possibility that, at least in some cases, chaperones could recognize misfolded proteins and subsequently mediate their polyubiquitination by directly recruiting an E3 ligase [223-225].

However, some data argue for a more direct role of chaperones in the degradation process. Although some misfolded proteins may be directly recognized by the proteasome [226], specific pathways within the UPS are probably relied on for the degradation of most misfolded and damaged proteins. Hsp70 plays an active and necessary role in the ubiquitination of some substrates [214]. This activity of Hsp70 depends on its chaperone function, indicating that conformational changes within substrates may facilitate recognition by the ubiquitination machinery. Finally, a post-ubiquitination function for chaperones has been proposed [224]. For instance, the neuronal Hsp70 cofactor HSJ1 stimulates the ubiquitination of Hsp70-bound proteins via its ubiquitin interaction motif (UIM) domains and their subsequent sorting to the proteasome [227]. A member of the Hsp70 family protein, Hsc73 is essential for the final maturation steps of the 20S proteasome from the 16S precursor complex [228]. Removal of Hsc73 results in an immediate formation of aggregates of this precursor indicating that Hsc73 keeps the assembly of intermediate complex in a soluble and probably partially unfolded state to allow subunit processing and correct folding. Considering these notions, Hsc73 might regulate the amount of 20S complex under stressful conditions. Furthermore, Hsp90 participates in the ATP-dependent assembly of the 26S proteasome [223].

Proteomics analysis of proteasome-interacting proteins revealed physical interactions between Hsp70 and Hsp90 with the 26S proteasome [223]. Evidence for a functional interplay between Hsp90 and PA28 also indicates that Hsp90 appears to compensate the loss of PA28 function in MHC class I antigen processing, suggesting that Hsp90 and PA28 operate either redundantly or specifically for generation of MHC class I ligands [229]. However, the biological relevance of these interactions is not clear at this stage. These findings may provide new mechanistic insight into the cooperative interactions between the molecular chaperone and proteolysis systems. *In vivo* and *in vitro* inactivation of Hsp90 caused dissociation of the 26S proteasomes into their constituents. Conversely, these dissociated constituents reassemble in Hsp90-dependent fashion both *in vivo* and *in vitro*. These processes are ATP-dependent and are suppressed by geldanamycin, an Hsp90 inhibitor. These results strongly suggest that the ATPase activity of Hsp90 is essential to the assembly and maintenance of the 26S proteasome and that Hsp90 plays some regulatory roles on the UPS pathway through assembly and disassembly of the 26S proteasome [123]. Thus, Hsp70- or Hsp90-bound substrates may be directed to proteasomes by virtue of direct or indirect chaperone–UPS interactions. Considering this regulatory role of Hsp90, impairment of the UPS pathway caused by protein aggregation might be partially brought about by the collapse of this regulation.

The CHIP Ubiquitin Ligase: From Chaperone to UPS Link between Folding and Degradation Systems

Major insights into molecular mechanisms that underlie the cooperation of molecular chaperones with the UPS were obtained through the functional characterization of the co-chaperone CHIP (reviewed in [101]). CHIP, originally identified as a co-chaperone of Hsp70 [97] can associate with Hsp70 and Hsp90 through the amino-terminal tetratricopeptide repeat (TPR) domain and adjacent charged domain [199,200]. Proteins containing TPR domains are involved in many protein-protein interactions; in particular, several Hsp interaction partners—including Hip, Hop, and the cyclophilins—interact with Hsc70 or Hsp90 through TPR domains [98,230]. CHIP contains three TPR domains at its amino terminus, which are used for binding to Hsp70 and Hsp90 [97,199].

Since CHIP attenuates stimulation of ATPase activity of Hsc70 by Hsp40, addition of CHIP diminishes the refolding activity of the Hsc70-Hsp40 complex for denatured substrates [97]. Besides the TPR domains, CHIP possesses a U-box domain at its carboxyl terminus [97]. The U-box domain has a tertiary structure that resembles the RING-finger domain of ubiquitin-protein ligase (E3), which covalently attaches ubiquitin to target proteins, designating them for destruction by the proteasome [231], but they lack the metal-chelating residues and instead are structured by intramolecular interactions [231].

The tissue distribution of CHIP supports the notion that it participates in protein folding and degradation decisions, as it is most highly expressed in tissues with high metabolic activity and protein turnover: skeletal muscle, heart, and brain. CHIP is also detectable in most cultured cells, and is particularly abundant in muscle and neuronal cells and in tumor-derived cell lines [97]. Intracellularly, CHIP is primarily localized to the cytoplasm under quiescent conditions [97], although a fraction of CHIP is present in the nucleus [200]. In addition, cytoplasmic CHIP traffics into the nucleus in response to environmental challenge in cultured cells, which may serve as a protective mechanism or to regulate transcriptional responses in the setting of stress [232]. As anticipated from its tertiary structure, CHIP executes E3 ubiquitin ligase activity upon specific substrates; CHIP ubiquitinates substrates of Hsp70 and Hsp90 and stimulates their degradation by the proteasome [97,199,200,233]

CHIP interacts with the terminal-terminal EEVD motifs of Hsp70 and Hsp90, similar to other TPR domain–containing co-chaperones such as Hop [126,230,234] Although the carboxy-terminal domain of the Hsp70 is the interaction domain for CHIP's amino terminal TPR domain, it is the amino terminal ATP-binding domain of Hsp70-Hsc70 that regulates substrate binding in a nucleotide-dependent fashion. The molecular cochaperones Hip and Hsp40 promote substrate binding by stabilizing the adenosine diphosphate (ADP)- bound conformation and activating ATPase activity, respectively [98], whereas BAG-1 promotes substrate release by exchanging ATP for ADP [235]. CHIP attenuates the Hsp40-stimulated ATPase and refolding activities of Hsp70 [97] and acts as an E3 ligase to facilitate the transfer of a polyubiquitin chain to misfolded substrates [233]. This activity may facilitate protein folding, perhaps by slowing the Hsc70 reaction cycle [232,236] under stressful conditions, or it may assist in loading misfolded proteins into the UPS.

Because Hsp90 also contains a TPR-acceptor site to interact with cochaperones, CHIP interacts with Hsp90 with approximately equivalent affinity to its interaction with Hsp70 [199]. This interaction results in remodeling of Hsp90 chaperone complexes, such that the co-chaperone p23, which is required for the appropriate activation of many, if not all, Hsp90 client proteins, is excluded. CHIP should inhibit the function of proteins that require Hsp90 for conformational activation. In any event, the studies are consistent in supporting a role for CHIP as a key component of the chaperone-dependent QC mechanism. CHIP efficiently targets client proteins, particularly when they are partially unfolded (as is the case for most Hsp90 clients when bound to the chaperone) or frankly misfolded as is the case for most proteins binding to Hsp70 through exposed hydrophobic residues). Consequently, CHIP appears to be chaperone-dependent E3 that ubiquitylates Hsp90-captured unfolded proteins [119]. Thus, CHIP is an ideal molecule acting as a protein QC E3 that selectively leads abnormal proteins recognized by molecular chaperones to degradation by the proteasome. For example, CHIP cooperates with Hsp70 during the degradation of immature forms of the CFTR protein at the ER membrane and during the ubiquitination of phosphorylated forms of the microtubule binding protein tau, which is of clinical importance due to its role in the pathology of AD [200,213]. These data indicate that CHIP plays a heretofore unsuspected role in coordinating the response to stress, not only by serving as a rate-limiting step in the degradation of damaged proteins but also by increasing the buffering capacity of the chaperone system to guard against stress-dependent proteotoxicity.

CHIP also mediates crosstalk between molecular chaperones and the UPS by associating with BAG-1, a protein that binds to the 26S proteasome and assists in the degradation of specific chaperone substrates [237]. Since both BAG-1 and CHIP are molecules that integrate molecular chaperones and UPS, cooperation of these two cofactors appears to reflect the fate of chaperone-captured proteins. CHIP converts Hsp70/Hsp90 chaperones into substrate recognition factors of a functional ubiquitin ligase complex, whereas BAG-1 supports binding of the Hsp70 complex to the proteasome and triggers the release of ubiquitylated substrates from HsclHsp70 for their transfer to the proteasome. Since BAG-1 and CHIP bind to different domains of Hsp70, these two co-chaperones are able to associate simultaneously with Hsp70. CHIP also regulates the association of BAG-1 with proteasome by K II-linked polyubiquitination of BAG-1 [238]. The formation of the ternary chaperone-cofactor complex might accelerate the degradation of chaperone-captured unfolded proteins by the UPS pathway [101,225,239,240].

The elucidation of CHIP's roles in the cell has helped clarify the mechanism of linkage between the cells protein folding-refolding machinery and its degradation machinery — the 2 pathways called upon to provide protein QC in the cell and thereby maintain normal cellular function, particularly in the setting of cellular stress. Characterization of CHIP function may therefore provide insights into how the cellular processes contribute to physiologic and pathologic processes at the cellular and organism level. Although a link with CHIP and pathophysiologic states is largely speculative at present, it is worth considering how the basic observations of CHIP at the level of biochemistry and cell biology can provide clues to understanding several neurological diseases.

HSPS IN THE DISEASES OF THE NERVOUS SYSTEM

Many systemic and neurodegenerative disorders, termed '*protein-misfolding disorders*', or perhaps more appropriately '*protein-conformational disorders*', are characterized by the accumulation of intracellular or extracellular protein aggregates (see Chapter 28). The HSR and Hsp have been implicated in many of these neurodegenerative diseases based on the association of chaperones with intracellular aggregates. Accumulation of protein aggregates might account for the observation that neurons, are particularly vulnerable to the detrimental effects of misfolded and/or aggregated proteins because they cannot dilute potentially toxic species through cell division and the late onset of neurodegenerative diseases that are linked to protein aggregation [241]. The difficulties of protein degradation together with an impaired protease activity and chaperone action in aging neurons, lead to a massive accumulation of these proteins and cause neurodegeneration [242]. The regulation of intracellular balance between refolding and degradation is a critical issue for cells. The collective activities of the molecular chaperones, the UPS and lysosome-mediated autophagy are normally sufficient to prevent the accumulation of misfolded proteins. Not only mutations of each protein, but also the deficiency of the chaperone or the UPS system may cause protein misfolding or aggregation. AD, PD, ALS and the polyQ diseases are all characterized by the accumulation of distinct aggregated proteins, mutations of which cause severe, inherited forms of disease. It is noteworthy to point out that proteasome inhibitors increase the frequency of ubiquitin-positive intracellular inclusions that that are characteristic of many neurodegenerative disorders [243]. Therefore, one could assume that a critical aspect of various neuronal degenerative diseases is failure of protein QC mediated by molecular chaperones and/or UPS. In addition, to sharing common morphological and biochemical features, the plaques/inclusion bodies that are characteristic of AD, PD, ALS and the polyQ diseases all co-localize with several of the same proteins, including various molecular chaperones and components of the UPS [244]. Co-localization of the protein QC machinery with inclusion bodies might reflect an irreversible sequestration and subsequent loss of function, and/or a failed attempt to refold or degrade aggregated proteins. There is a great interest in the interaction between some putative protein folding diseases and the chaperone system or the UPS pathway. Collectively, these observations have led to the hypothesis that the elevated levels of Hsp reduce or dampen aggregate formation and cellular degeneration [245,246]. Neurodegenerative diseases often occur later in life when HS genes seem to be induced poorly [245,247]. The HSR has recently been implicated in the regulation of longevity in *Caenorhabditis elegans* in a pathway that overlaps with the insulin signaling pathway [205,248]. Reduction of heat shock transcription factor HSF1 levels cause a decreased life span in *C. elegans*, similar to life span effects observed in mutants of Daf-16, a FOXO transcription factor in the insulin signaling pathway. Daf-16 and HSF1 share a subset of downstream target genes, including sHSPs. RNA interference (RNAi) experiments show that a decrease in sHSPs and other HSPs leads to a decrease in longevity [205,248]. Therefore, in addition to the prevention of diseases of aging, increased levels of HSPs may lead to increases in life span.

Alzheimer's Disease (AD) and Related Tauopathies

AD is characterized by amyloid-β peptide (Aβ) in extracellular senile plaques and tau in neuro-fibrillary tangles (NFT), aggregates that are major morphological indicators of the disease [249]. AD is the most common tauopathy, a group of familial neurodegenerative conditions distinguished by intracellular filamentous bodies composed of tau, a low molecular weight microtubule-associated protein [250] Neurons are the predominant location of tau pathology in Alzheimer's, but glial pathology manifests in corticobasal degeneration and progressive supranuclear palsy. Several studies showed the induction of sHsp proteins (Hsp27, αB-crystallin), Hsp70 and ubiquitin in neurons affected by AD and in surrounding astrocytes [251]. Immunohistochemical and immunoblotting analysis of brain tissue of AD patients showed highly induced expression of the Hsp27 in proliferating astrocytes establishing a pattern of astrocytic gliosis (klazmatodendrosis) into the areas rich in senile plaques, neurofibrillary tangles and Hirano bodies [252,253]. Increased expression of αB-crystallin is found in glial inclusions of both sporadic and familial tauopathies [250]. αB-crystallin is usually observed in a subset of astrocytic and oligodendrocytic tau inclusions as well as the neuropil thread pathology in cellular processes, but the co-expression of αB-crystallin with tau inclusions is relatively specific to tauopathies with extensive glial pathology representing a response by glia to the accumulation of misfolded or aggregated tau protein that is linked to the pathogenesis of the glial pathology and distinct from mechanisms underlying neuronal tau pathology in neurodegenerative disease [250].

Accumulated chaperones are participating in the attempts of the affected neuron to sequester the Aβ and other damaged proteins in AD [254]. Hsp27 preferentially binds pathological hyperphosphorylated tau and paired helical filaments tau in human brain samples directly but not non-phosphorylated tau reducing its concentration by facilitating its proteasomal degradation and dephosphorylation [213]. Mass spectrometry reveals that three Hsp16 family members, in addition to other molecular chaperones, coimmunoprecipitate with human Aβ in transgenic *C. elegans* [255]. Moreover, Hsp27 rescues pathological hyperphosphorylated tau-mediated cell death in cell culture [213]. Human recombinant αB-crystallin also interacts with Aβ *in vitro* and promotes β-sheet formation by Aβ [256]. Thioflavine T fluorescence assays and electron microscopy demonstrated that human Hsp27 inhibits Aβ amyloidogenesis *in vitro* much more effectively than α-crystallin, which is almost without effect [257]. Nonetheless, study of Hsp27 suggests aging-related reduction in chaperone activity contributes to AD pathogenesis. Cytoplasmic Hsp60, a specific chaperone for actin and tubulin is decreased in AD-affected neurons leaving the cytoskeletal proteins deficient and aggregated [258].

The accumulation of molecular chaperones in extracellular senile plaques found in the brains of patients with AD is difficult to explain, as most cytosolic HSPs, including Hsp70, do not contain a secretory signal sequence, and release through the classic exocytic pathway is not likely. In cultured cells, the extracellular accumulation of Hsp70 might be facilitated by a calcium induced interaction with lipid rafts [259]. It is plausible that a small fraction of cytosolic chaperones could be targeted to lipid rafts and become associated with Aβ oligomers before their eventual secretion into the extracellular space. Experimental data from cell cultures have shown that Grp78, an Hsp70 that is found in the ER, binds amyloid precursor protein (APP) and decreases the secretion of amyloid-β40 (Aβ40) and Aβ42, indicating that Grp78 might retain APP in the ER and/or shield APP from β/γ-secretase cleavage [260]. RNAi of Hsp70 and Hsp90 in cell culture increases the accumulation of insoluble, aggregated tau and impairs the association of tau with microtubules showing a functional interaction between tau, Hsp70 and Hsp90 while Hsp70 / Hsp90 expression induction has an opposite effect indicating that the activity of these chaperones maintains the native fold and function of tau [261]. Overexpression of cytosolic Hsp70 rescues neurons from intracellular Aβ42-mediated toxicity [262], probably involving chaperone-stimulated ER-associated degradation of Aβ, a process by which misfolded secretory and/or membrane proteins are recognized by a QC mechanism in the ER, retro-translocated into the cytosol and degraded by the 26S proteasome [263] (see Chapter 13). Indeed, calreticulin, an abundant ER chaperone was shown to participate in the quality control of the amyloid precursor protein [264] and the ER-homologue of Hsp70, Grp78 had an increased expression in successfully surviving neurons [265]. There are reports to show that mutant presenilin-1, an ER transmembrane protein being the most prevalent cause of early-onset familial AD, impairs the ER chaperone response and thus sensitizes the affected neuron to apoptosis. However, this latter finding could not be confirmed in other systems [266]. Hsp60, Hsp70, and Hsp90 both alone and in combination provide differential protection against intracellular Aβ stress through the maintenance of mitochondrial oxidative phosphorylation and functionality of tricarboxylic acid cycle enzymes. Notably, Aβ is found to selectively inhibit complex IV activity, an effect selectively neutralized by Hsp60. The combined effect of HSPs is to reduce the free radical burden, preserve ATP generation, decrease cytochrome c release, and prevent caspase-9 activation, all important mediators of Aβ induced neuronal dysfunction and death [267]. However, αB-crystallin inhibits Aβ fibril formation *in vitro*, although β-sheet content and neuronal toxicity of Aβ preparations increase. Possibly, αB-crystallin / Aβ complexes maintain Aβ -40 in a nonfibrillar, highly toxic form and Aβ toxicity is independent of fibril formation. In this scenario, sHSPs exacerbate rather than diminish, AD symptoms [268].

Finally, CHIP cooperates with Hsp70 during the ubiquitination of phosphorylated forms of the microtubule binding protein tau [200,213,269] and suppresses toxicity in cell culture [213], but paradoxically enhances the levels of insoluble, ubiquitylated tau [200,213] indicating that under certain pathological conditions CHIP-mediated ubiquitination might partition soluble, toxic forms of phosphorylated tau into an insoluble compartment. Similar results are also recently confirmend in AD patients and in a JNPL3 mouse brain tauopathy model [270]: CHIP was inversely proportional to sarcosyl-insoluble tau accumulation in both human and in mouse suggesting that increases in CHIP may protect against NFT formation in the early stages of AD. If confirmed, this would indicate that the quality-control machinery in a neuron might play an important role in retarding the pathogenesis of tauopathies [270]

Parkinson's Disease (PD) and Related α-Synucleinopathies

PD is an age-related disorder characterized by a progressive degeneration of dopaminergic neurons in the substantia nigra and showing a corresponding motor deficit. An increasing number of evidence shows that besides oxidative stress and mitochondrial dysfunction protein folding defects are also key elements of PD etiology. Similarly to AD, glial and astroglial cells of PD patients showed the expression of αB-crystallin and similarly to neurofibrillary tangles aggregated proteins in LB had a large content of various HSP [268]. Hsp27 and αB-crystallin appear in PD with severe dementia [271].

Familial PD is characterized by accumulation of the α-Syn protein in LB, and several studies have investigated the effects of molecular chaperones on α-Syn aggregation and toxicity. Overexpression of HDJ1 (an Hsp40) or Hsp70 in an α-Syn/synphilin 1 cell model markedly decreases the number of cells that contain inclusion bodies without effects on cell viability [272], the detergent-insoluble, high molecular mass α-Syn species and total α-Syn protein together with toxicity of transfected α-Syn [273].

Hsp70 might enhance refolding and/or promote degradation of α-Syn or mediate a biochemical change in α-Syn that suppresses its toxicity [273]. Hsp90 is the predominant Hsp examined that co-localized with α-Syn in LB, Lewy neurites, and glial cell inclusions and Hsp90 co-localizes with α-Syn filaments of LB in PD (Figure 6). Hsp90 levels are most predominantly increased in PD brains, but are also evident in other α-synucleinopathies such as dementia with LB (DLB), and multiple system atrophy (MSA) characterized by filamentous α-Syn inclusions (Figure 7) [274]. Cell culture studies also revealed that α-Syn co-immunoprecipitated preferentially with Hsp90 and Hsc70 relative to other HSPs, and exposure of cells to proteasome inhibitors resulted in increased levels of Hsp90 suggesting a role of Hsp90 in the formation of α-Syn inclusions in PD and related α- synucleinopathies [274]. Understanding the molecular interactions among α-Syn, Hsp90, and ubiquitin in α-Syn pathogenesis is a question relevant to both biological and pathological processes. Iit is plausible that when α-Syn initially becomes denatured/aggregated by stress caused by genetic

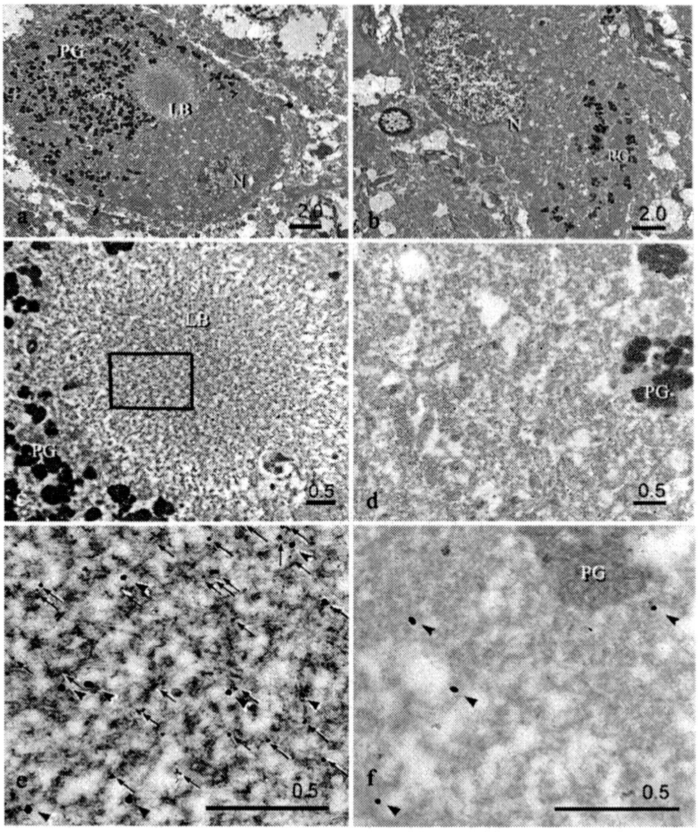

Figure 6. Immunoelectron Microscopy of Hsp90 and α-syn in Substantia Nigra LBs of PD. Double-immunoelectron microscopy reveals a close association of Hsp90 and α-syn in filamentous LBs in dopaminergic neurons of the SN in PD brain (a, c, e). α-syn-positive (10 nm gold, arrow) filaments in the LB are also Hsp90-positive (18 nm gold, arrowhead), whereas unaffected dopaminergic neurons in the same SN section (f) reveal infrequent Hsp90 labeling (arrowhead). Image e is a high-power view of the inset in c, from the center of a LB. PG, neuronal pigment; N, nucleus. Courtesy of Dr. John Q. Trojanowski, Center for Neurodegenerative Disease Research and Marian S. Ware Alzheimer Drug Discovery Program, Department of Pathology and Laboratory Medicine University of Pennsylvania School of Medicine, Philadelphia, PA 19104-4283, USA [274]. Reprinted from *Am J Pathol* 2006 168: 947-961 with permission from the American Society for Investigative Pathology. © 2006 American Society for Investigative Pathology.

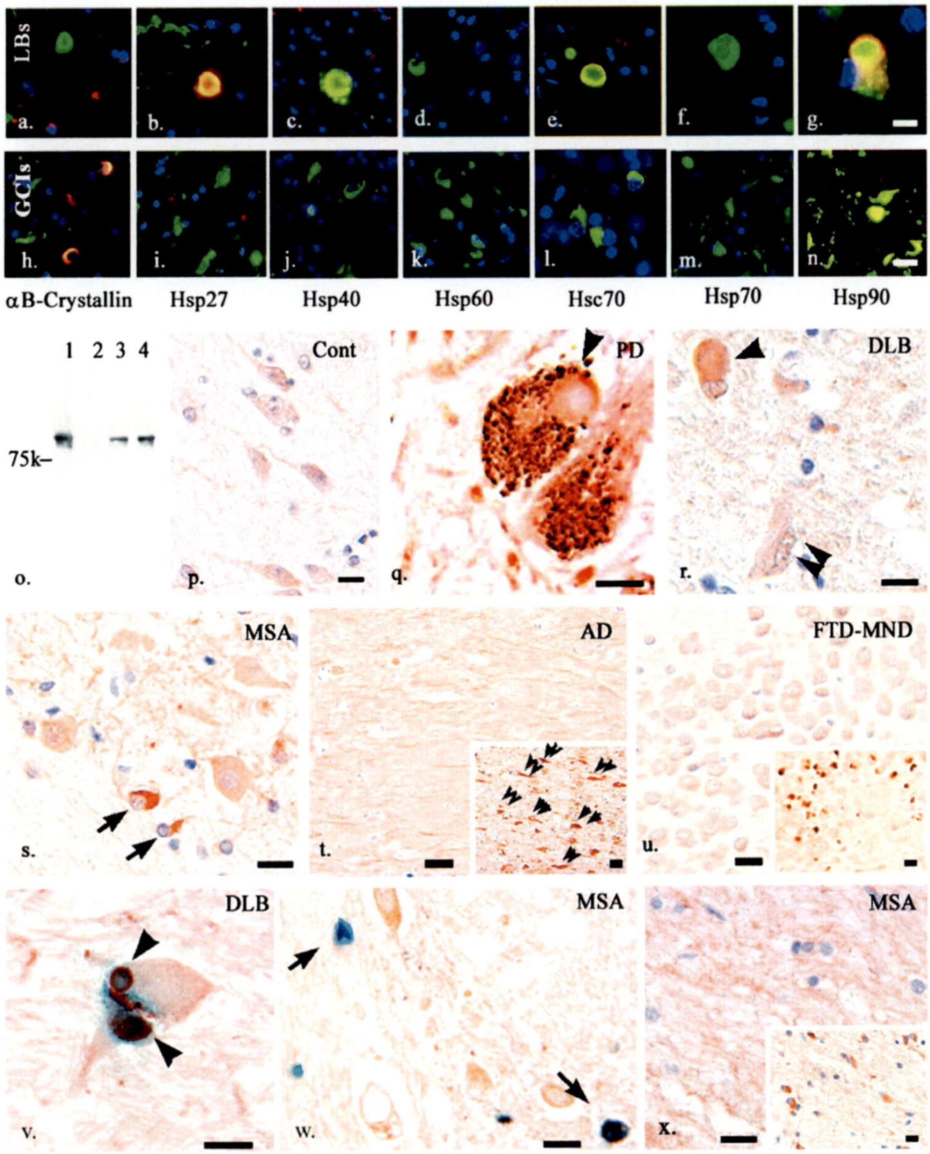

Figure 7. Screening of HSPs using double-label fluorescent immunohistochemistry (FIHC) in α-synucleinopathies. Screening of HSPs using double-label FIHC with different anti-HSP and anti-α-syn antibodies visualized by Alexa Fluor 594 or Texas Red and Alexa Fluor 488, respectively, shows co-localization of a subset of HSPs examined in LBs (a–g) and GCIs (h–n). Yellow color represents co-localization of Hsp and α-Syn. Hsp27 (b), Hsp40 (c), Hsc70 (e), and Hsp90 (g) are located in LBs, whereas αB-crystallin (h) and Hsp90 (n) are located in GCIs and threads. Specificity of Hsp90 antibodies (o), 9D2 (lanes 1 and 2), and AC88 (lanes 3 and 4) is shown in HS fraction of human (lanes 1 and 3) and mouse (lanes 2 and 4) brain homogenates. Rat 9D2 does not recognize mouse Hsp90 (lane 2). Photomicrographs in p–x show modest Hsp90 IR in neurons of normal brain (p), which also is seen in α-synucleinopathy brains, in addition to more intense Hsp90 IR in LBs (arrowhead) of the PD SN (q), DLB amygdala (r), and GCIs (arrow) in the MSA pons (s). Little or no iHsp90 IR is seen in neurofibrillary tangles of AD (t, double arrowheads) and DLB (r). t: IHC profile with an antibody against Hsp90 (9D2) or tau (AT8) (inset) from adjacent sections in the CA1 region of AD hippocampus are presented for comparison. It is apparent that Hsp90 IR is reduced in neurofibrillary tangles. u:

Modest Hsp90 IR is noted in ubiquitin inclusions in the hippocampus of the FTD-MND brain. Inset shows ubiquitin IR in inclusions on the adjacent section. v and w show double-label IHC with horseradish peroxidase-DAB and β-galactosidase-X-gal to illustrate co-localization of iHsp90 IR (brown) with α-Syn IR (blue) in LBs (arrowhead) in the midbrain of PD (v) as well as GCIs (arrow) in the pons of MSA (w). x: No CHIP IR was found on GCI in the pons of MSA, whereas the adjacent section indicates descent α-Syn IR inclusions (inset). Scale bars, 10 μm. Courtesy of Dr. John Q. Trojanowski, Center for Neurodegenerative Disease Research and Marian S. Ware Alzheimer Drug Discovery Program, Department of Pathology and Laboratory Medicine University of Pennsylvania School of Medicine, Philadelphia, PA 19104-4283, USA [274]. Reprinted from *Am J Pathol* 2006 168: 947-961 with permission from the American Society for Investigative Pathology. © 2006 American Society for Investigative Pathology.

or environmental factors, Hsp90 may engage α-Syn and successfully rescue α-Syn from further denaturing processes at this step. However, if stress persists chronically, α-Syn may eventually attain a firmly aggregated stage, wherein Hsp90 fails to rescue α-Syn from misfolding. Subsequently Hsp90 may redirect the protein to the proteasome system by facilitating its ubiquitination for degradation. Alternatively, Hsp90 may engage α-Syn preferably when it becomes denatured and/or aggregated, to promote its degradation by facilitating ubiquitination. However, the presence of HSPs and ubiquitin in inclusions points to an unsuccessful attempt to remove aggregated proteins by the proteasomal machinery. Proteasomal inhibition by MG-132 or lactacystin causes up-regulation of Hsp90, and leads to the accumulation of ubiquitinated proteins [274]. Thus, a chronic impairment of the proteasomal system, which might occur insidiously during the onset or progression of neurodegenerative diseases, may not be counteracted by the induction of HSPs, and the increase and association of Hsp90 with α-Syn aggregates might contribute to cell death.

Recent studies in *Drosophila melanogaster* indicate that Hsp70 might have a protective role in PD [275,276]. Co-expression of human Hsp70 prevents α-Syn-mediated toxicity, but, paradoxically, has no visible effect on the inclusion body phenotype at the level of light microscopy. The protective effect of Hsp70 might be attributed to the destabilization of toxic, misfolded α-Syn monomers and/or small micro-aggregates that are not visible using light microscopy. Co-expression of a dominant-negative form of *D. melanogaster* Hsp70 with α-Syn accelerates the loss of dopaminergic neurons, indicating that endogenous chaperones modestly suppress α-Syn-mediated neurodegeneration [275]. Consistent with these results, administration of geldanamycin, that specifically binds to and interferes with the activity of the molecular chaperone Hsp90 [277], a negative regulator of HSF1 protects against α-Syn toxicity in this fly model [276]. The effect of Hsp70 on neurodegeneration has been investigated also in the 1-methyl-4-phenyl-1,2,3,6-tetrahydropyridine (MPTP) mouse model of idiopathic PD [278]. Hsp70 gene transfer to dopamine neurons by a recombinant adeno-associated virus significantly protects the mouse dopaminergic system against MPTP-induced dopamine neuron loss, the associated decline in striatal dopamine levels and tyrosine hydroxylase-positive fibers and reduces MPTP-induced apoptosis in the substantia nigra. Overexpression of Hsp70 in mice that are transgenic for α-Syn significantly reduces the formation of high molecular mass, detergent-insoluble material by α-Syn [273]. It was also recently suggested that mutant α-Syn might antagonize the receptor that is involved in chaperone-mediated autophagy, thereby enhancing the accumulation of toxic misfolded proteins and resulting in cellular dysfunction [279]. Similar protective results were found in a lactacystin-induced proteasomal dysfunction model of PD [280] where Hsp70 shows to reduce lactacystin-induced dopaminergic neuronal death in the substantia nigra in part by fostering aggresome formation and lysosome-mediated autophagy [281]. Finally, overexpression of wild type DJ-1 gene, which is mutated in early onset PD with autosomal recessive inheritance, in dopaminergic cells protects cells from death induced by hydrogen peroxide and 6-hydroxydopamine through increased expression of Hsp70 [282].

Familial Amyotrophic Lateral Sclerosis

For reasons that remain unclear, motor neurons have a high threshold for induction of the HSR, which might contribute to the selective degeneration of motor neurons observed in familial-ALS probably for an impaired ability to activate the HSF1 [283]. A mouse model of familial ALS displays down-regulation of sHSPs in motor neurons and up-regulation in astrocytes [283]. Remarkably, primary spinal cord cultures fail to upregulate Hsp70 in response to heat shock, glutamate excitotoxicity or expression of mutant SOD1 with a glycine→alanine substitution at residue 93 (G93A), nor was Hsp70 increased in spinal cords of G93A SOD1 transgenic mice or sporadic or familial ALS patients [283], whereas cerebellar, cortical and pyramidal neurons, as well as astrocytes, efficiently upregulate Hsp70 in response to HS. Mouse Hsp25 colocalizes with mutant SOD1 [284], similar to results obtained with a cultured neuronal cell line. Interaction with mutant, but not wild-type SOD1 may limit antiapoptotic potential and decrease cell protection by Hsp25. αB-crystalline binds mutated SOD1 characteristic of familial ALS [285]. Over-expression of either Hsp27 or Hsp70 has a protective effect against SOD1 disease associated mutant-induced cell death. However, over-expression of Hsp27 and Hsp70 together has a greater potent anti-apoptotic effect, than when expressed singly, against the damaging effects of mutant SOD1. Familial ALS-associated SOD1 disease mutants possess enhanced death-inducing properties and lead to increased apoptosis which can be prevented by either the use of specific caspase inhibitors or Hsp27 and/or Hsp70 over-expression [286]. Increasing the level of the Hsp70 by gene transfer reduced formation of mutant SOD-containing proteinaceous aggregates in cultured primary motor neurons expressing G93A SOD1 and prolonged their survival [287].

A physical interaction between HSP70 and mutant SOD1 probably mediates neuroprototection: however, the precise mechanism remains unclear. Hsp70, Hsp40 and αB-crystallin co-immunoprecipitate with SOD1 in cell lines that express mutant, but not wild-type, SOD1 [285,288]. The interaction between Hsp70 and mutant SOD1 is evident in total cell extracts, but barely discernible in an isolated supernatant fraction of soluble proteins, indicating that Hsp70 interacts specifically with detergent insoluble SOD1 complexes suggesting that mutation-induced alteration of protein conformation may not in itself be sufficient for direct interaction with HSPs [288]. Hsp70 or Hsc70, and CHIP are also involved in proteasomal degradation of mutant SOD1. Only mutant SOD1 interacts with Hsp/Hsc70 *in vivo*, and *in vitro* experiments revealing that Hsp/Hsc70 preferentially interacts with apo-SOD1 or dithiothreitol-treated holo-SOD1, compare with metallated or oxidized forms. CHIP interacts only with mutant SOD1 and promotes its degradation. Both Hsp70 and CHIP promotes polyubiquitination of mutant SOD1-associated molecules, but not of mutant

SOD1, indicating that mutant SOD1 is not a substrate of CHIP. Moreover, mutant SOD1-associated Hsp/Hsc70, a known substrate of CHIP, is polyubiquitinated *in vivo*, and polyubiquitinated Hsc70 by CHIP interacts with the S5a subunit of the 26S proteasome *in vitro* [289].

Polyglutamine (polyQ) Expansion Diseases

Polyglutamine diseases designate a group of neurodegenerative disorders characterized by the presence of a toxic polyglutamine expansion in specific target proteins. Those diseases include HD, spinocerebellar ataxias types 1 and 3 (SCA1, SCA3), and SBMA. A feature of these diseases is the presence of ubiquitinated intraneuronal inclusions derived from the mutant proteins, which colocalize with HSPs in SCA1 and SBMA and proteasomal components in SCA1, SCA3, and SBMA [208,290]. The effect of chaperones on the aggregation and toxicity of proteins with polyQ expansions has been intensely investigated in a diverse range of models, including *in vitro* systems, yeast, worms, flies and mice. Many studies have analyzed the effect that chaperone overexpression has on inclusion body formation and toxicity of pathogenic polyQ fragments in cell culture.

Overexpression of Hsp40 consistently suppresses the formation of polyQ inclusion bodies and their toxicity [290-292]. Overexpression of TriC–CCT (Hsp60) can prevent the formation of mutant huntingtin (Htt) aggregates when it is expressed in yeast cells, mammalian cell lines and neuronal cells [293-295]. In all cases, this was associated with reduced cell death. Mutant Htt still oligomerizes in the presence of TriC–CCT, but it forms aggregates with different properties that do not seem detrimental to cell survival.

Tam *et al.* [295] investigated the effect of overexpressing each of the eight subunits of TRiC. Whereas most subunits did not prevent the formation of cellular inclusions, subunit 1 strongly inhibited toxic Htt aggregation and increased neuron viability. This protective activity was found to reside in the apical domain of the protein, which has been recently shown to contain the protein's polypeptide binding site. However, RNA knockdown of just one of the other eight subunits was enough to stimulate Htt aggregation and neuronal toxicity, which, instead, indicates that only the fully assembled TriC–CCT chaperonin complex can provide neuroprotection against mutant Htt. So, it seems that mutant Htt can oligomerize by mechanisms that can lead to the formation of either toxic or benign aggregates [295].

Finally, the protective function of TriC–CCT depends on the presence of Hsp70, and that TriC–CCT could only act on Htt after it had been processed by Hsp70 [294]. This fits with the well-known role of these proteins in normal protein regulation: Hsp70 interacts first at the point of translation to prevent premature folding events, whereas TriC–CCT functions downstream to regulate the correct folding and aggregation of proteins. TriC–CCT specifically prevents the aggregation of newly synthesized proteins by recognizing hydrophobic β-strands. Interestingly, toxic conformations of mutant Htt adopt a β-sheet structure, thereby providing a glimpse of how TriC–CCT might recognize and regulate the conformation of Htt.

Overexpression of Hsp70 also suppresses polyglutamine toxicity, and typically correlates with a decrease in inclusion body formation [296], although the ability of Hsp70 to suppress apoptosis depends on inhibition of the pro-apoptotic factors caspase 3 and caspase 9 [297]. Hsp70, which is known to inhibit JNK during the HSR, suppresses polyQ-mediated aggregation of JNK phosphatase M3/6 and activation of JNK and AP-1. Interestingly, levels of Hsp70 are down-regulated by polyQ expansion [298]. The Hsp70 chaperone increases CHIP-mediated ubiquitination of ataxin-1 *in vitro*, and the TPR domain, which mediates CHIP interactions with chaperones, is required for ataxin-1 ubiquitination in cell culture [299]. Interestingly, CHIP also interacts with and ubiquitinates unexpanded ataxin-1. Overexpression of CHIP in a *Drosophila* model of SCA1 decreases the protein steady-state levels of both expanded and unexpanded ataxin-1 and suppresses their toxicity. Transient overexpression of CHIP increases ubiquitination and the rate of degradation of polyQ-expanded huntingtin or ataxin-3 [300]. CHIP is not effective in suppressing the toxicity caused by a bare 127Q tract with only a short hemagglutinin tag, but it is very efficient in suppressing toxicity caused by a 128Q tract in the context of an N-terminal huntingtin backbone [299] and the suppressive effect is more prominent when CHIP is overexpressed along with Hsc70 [300]. Overexpression of Hsp27 was seen to decrease polyQ toxicity without altering inclusion body formation, but in a manner that correlates with a decrease in free radical production [301]. These studies highlight the possibility that chaperones facilitate neuroprotection through several distinct mechanisms, but because small, diffusible, potentially toxic polyQ assemblies could not be evaluated, an essential role for the refolding activity of chaperones cannot be ruled out. Furthermore, the stoichiometry of chaperones relative to the polyQ proteins might determine their effect on aggregation, equally important is the identification of other cellular factors that may be induced as part of the polyQ stress response.

Conclusions

HSPs and neurological diseases are evidently linked, but consequences are uncertain. There is overwhelming experimental evidence that molecular chaperones are crucial modulators of protein aggregation and neurodegeneration. Chaperoning can prevent or promote aggregate formation, and either outcome may be favorable or unfavorable, depending on the disease. The mechanism by which chaperone activity facilitates neuroprotection remains poorly understood. Prevention of abnormal protein aggregation obviously does not always benefit cells, an observation with important implications when choosing therapeutic approaches to neurological diseases. Chaperones might also help to prevent neuronal dysfunction by assisting intracellular trafficking, including synaptic transmission. The ability of molecular chaperones to interfere with oxidative stress and block apoptotic signalling pathways should also not be overlooked. A proximal step in the pathogenesis of neurodegenerative diseases linked to protein aggregation probably stems from aberrant protein interactions due to altered conformations in disease-causing proteins, effectively triggering a cascade of pathogenic events that culminates in neuronal dysfunction before the appearance of inclusion bodies. Effective therapies will probably require the simultaneous modulation of several components of the protein QC apparatus, and molecular chaperones will have a key role in these types of approach.

REFERENCES

[1] Macario AJ, Lange M, Ahring BK, De Macario EC: Stress genes and proteins in the archaea. *Microbiol Mol Biol Rev* 1999, 63:923-67.

[2] Hecker M, Volker U: General stress response of Bacillus subtilis and other bacteria. *Adv Microb Physiol* 2001, 44:35-91.

[3] Feder ME, Hofmann GE: Heat-shock proteins, molecular chaperones, and the stress response: evolutionary and ecological physiology. *Annu Rev Physiol* 1999, 61:243-282.

[4] Pearce AK, Humphrey TC: Integrating stress-response and cell-cycle checkpoint pathways. *Trends Cell Biol* 2001, 11:426-433.

[5] Goldberg AL: Protein degradation and protection against misfolded or damaged proteins. *Nature* 2003, 426:895-899.

[6] Al Fageeh MB, Smales CM: Control and regulation of the cellular responses to cold shock: the responses in yeast and mammalian systems. *Biochem J* 2006, 397:247-259.

[7] Lepock JR: How do cells respond to their thermal environment? *Int J Hyperthermia* 2005, 21:681-687.

[8] Somero GN: Adaptations to high hydrostatic pressure. *Annu Rev Physiol* 1992, 54:557-577.

[9] Rowbury RJ: Extracellular sensing components and extracellular induction component alarmones give early warning against stress in Escherichia coli. *Adv Microb Physiol* 2001, 44:215-257.

[10] Kempner ES: Damage to proteins due to the direct action of ionizing radiation. *Q Rev Biophys* 1993, 26:27-48.

[11] Farrer BT, Pecoraro VL: Heavy-metal complexation by de novo peptide design. *Curr Opin Drug Discov Devel* 2002, 5:937-943.

[12] Borkan SC, Gullans SR: Molecular chaperones in the kidney. *Annu Rev Physiol* 2002, 64:503-527.

[13] Fehrenbach E, Northoff H: Free radicals, exercise, apoptosis, and heat shock proteins. *Exerc Immunol Rev* 2001, 7:66-89.

[14] Martindale JL, Holbrook NJ: Cellular response to oxidative stress: signaling for suicide and survival. *J Cell Physiol* 2002, 192:1-15.

[15] Kasprzak KS: Oxidative DNA and protein damage in metal-induced toxicity and carcinogenesis. *Free Radic Biol Med* 2002, 32:958-967.

[16] Rydberg B: Radiation-induced DNA damage and chromatin structure. *Acta Oncol* 2001, 40:682-685.

[17] Niess AM, Dickhuth HH, Northoff H, Fehrenbach E: Free radicals and oxidative stress in exercise--immunological aspects. *Exerc Immunol Rev* 1999, 5:22-56.

[18] Jolly C, Morimoto RI: Role of the heat shock response and molecular chaperones in oncogenesis and cell death. *J Natl Cancer Inst* 2000, 92:1564-1572.

[19] Boreham DR, Mitchel RE: Regulation of heat and radiation stress responses in yeast by hsp-104. *Radiat Res* 1994, 137:190-195.

[20] Donati YR, Slosman DO, Polla BS: Oxidative injury and the heat shock response. *Biochem Pharmacol* 1990, 40:2571-2577.

[21] Polla BS: A role for heat shock proteins in inflammation? *Immunol Today* 1988, 9:134-137.

[22] Sheikh-Hamad D, Garcia-Perez A, Ferraris JD, Peters EM, Burg MB: Induction of gene expression by heat shock versus osmotic stress. *Am J Physiol* 1994, 267:F28-F34.

[23] Neuer A, Spandorfer SD, Giraldo P, Dieterle S, Rosenwaks Z, Witkin SS: The role of heat shock proteins in reproduction. *Hum Reprod Update* 2000, 6:149-159.

[24] Gabai VL, Meriin AB, Yaglom JA, Volloch VZ, Sherman MY: Role of Hsp70 in regulation of stress-kinase JNK: implications in apoptosis and aging. *FEBS Lett* 1998, 438:1-4.

[25] Soti C, Csermely P: Molecular chaperones and the aging process. *Biogerontology* 2000, 1:225-233.

[26] Ellis RJ: Protein folding: inside the cage. *Nature* 2006, 442:360-362.

[27] Ellis RJ, Minton AP: Protein aggregation in crowded environments. *Biol Chem* 2006, 387:485-497.

[28] Ellis RJ: Molecular chaperones: assisting assembly in addition to folding. *Trends Biochem Sci* 2006, 31:395-401.

[29] Forman MS, Trojanowski JQ, Lee VM: Neurodegenerative diseases: a decade of discoveries paves the way for therapeutic breakthroughs. *Nat Med* 2004, 10:1055-1063.

[30] Harrison PJ, Procter AW, Exworthy T, Roberts GW, Najlerahim A, Barton AJ, Pearson RC: Heat shock protein (hsx70) mRNA expression in human brain: effects of neurodegenerative disease and agonal state. *Neuropathol Appl Neurobiol* 1993, 19:10-21.

[31] Mayer RJ, Lowe J, Landon M, McDermott H, Tuckwell J, Doherty F, Laszlo L: Ubiquitin and the lysosome system: molecular immunopathology reveals the connection. *Biomed Biochim Acta* 1991, 50:333-341.

[32] Mayer RJ, Arnold J, Laszlo L, Landon M, Lowe J: Ubiquitin in health and disease. *Biochim Biophys Acta* 1991, 1089:141-157.

[33] Tan Z, Tu W, Schreiber SS: Downregulation of free ubiquitin: a novel mechanism of p53 stabilization and neuronal cell death. *Brain Res Mol Brain Res* 2001, 91:179-188.

[34] Varshavsky A: The ubiquitin system. *Trends Biochem Sci* 1997, 22:383-387.

[35] Ciechanover A, Brundin P: The ubiquitin proteasome system in neurodegenerative diseases: sometimes the chicken, sometimes the egg. *Neuron* 2003, 40:427-446.

[36] Ciechanover A, Schwartz AL: The ubiquitin system: pathogenesis of human diseases and drug targeting. *Biochim Biophys Acta* 2004, 1695:3-17.

[37] Ciechanover A: Proteolysis: from the lysosome to ubiquitin and the proteasome. *Nat Rev Mol Cell Biol* 2005, 6:79-87.

[38] Glickman MH, Ciechanover A: The ubiquitin-proteasome proteolytic pathway: destruction for the sake of construction. *Physiol Rev* 2002, 82:373-428.

[39] Dobson CM: Principles of protein folding, misfolding and aggregation. *Semin Cell Dev Biol* 2004, 15:3-16.

[40] Hoyt MA, Coffino P: Ubiquitin-free routes into the proteasome. *Cell Mol Life Sci* 2004, 61:1596-1600.

[41] Hoppe T: Multiubiquitylation by E4 enzymes: 'one size' doesn't fit all. *Trends Biochem Sci* 2005, 30:183-187.

[42] Harboe M, Quayle AJ: Heat shock proteins: friend and foe? *Clin Exp Immunol* 1991, 86:2-5.

[43] Tsan MF, Gao B: Heat shock protein and innate immunity. *Cell Mol Immunol* 2004, 1:274-279.

[44] Pratt WB, Morishima Y, Murphy M, Harrell M: Chaperoning of glucocorticoid receptors. *Handb Exp Pharmacol* 2006, 111-138.

[45] Ficker E, Dennis A, Kuryshev Y, Wible BA, Brown AM: HERG channel trafficking. *Novartis Found Symp* 2005, 266:57-69.

[46] Wadhwa R, Taira K, Kaul SC: An Hsp70 family chaperone, mortalin/mthsp70/PBP74/Grp75: what, when, and where? *Cell Stress Chaperones* 2002, 7:309-316.

[47] Gaestel M: Molecular chaperones in signal transduction. *Handb Exp Pharmacol* 2006, 93-109.

[48] Kiang JG, Tsokos GC: Heat shock protein 70 kDa: molecular biology, biochemistry, and physiology. *Pharmacol Ther* 1998, 80:183-201.

[49] Beere HM: Death versus survival: functional interaction between the apoptotic and stress-inducible heat shock protein pathways. *J Clin Invest* 2005, 115:2633-2639.

[50] Giffard RG, Yenari MA: Many mechanisms for hsp70 protection from cerebral ischemia. *J Neurosurg Anesthesiol* 2004, 16:53-61.

[51] Scheibel T, Buchner J: Protein aggregation as a cause for disease. *Handb Exp Pharmacol* 2006, 199-219.

[52] Lowenstein DH: The stress protein response and its potential relationship to prolonged seizure activity. *Clin Neuropharmacol* 1995, 18:148-158.

[53] Ahmed RG: Heat stress induced histopathology and pathophysiology of the central nervous system. *Int J Dev Neurosci* 2005, 23:549-557.

[54] Foster JA, Brown IR: Differential induction of heat shock mRNA in oligodendrocytes, microglia, and astrocytes following hyperthermia. *Brain Res Mol Brain Res* 1997, 45:207-218.

[55] Minowada G, Welch WJ: Clinical implications of the stress response. *J Clin Invest* 1995, 95:3-12.

[56] Tissieres A, Mitchell HK, Tracy UM: Protein synthesis in salivary glands of Drosophila melanogaster: relation to chromosome puffs. *J Mol Biol* 1974, 84:389-398.

[57] Schlesinger MJ: How the cell copes with stress and the function of heat shock proteins. *Pediatr Res* 1994, 36:1-6.

[58] Verbeke P, Fonager J, Clark BF, Rattan SI: Heat shock response and ageing: mechanisms and applications. *Cell Biol Int* 2001, 25:845-857.

[59] Sghaier H, Le Ai TH, Horiike T, Shinozawa T: Molecular chaperones: proposal of a systematic computer-oriented nomenclature and construction of a centralized database. *In Silico Biol* 2004, 4:311-322.

[60] Basha E, Lee GJ, Breci LA, Hausrath AC, Buan NR, Giese KC, Vierling E: The identity of proteins associated with a small heat shock protein during heat stress in vivo indicates that these chaperones protect a wide range of cellular functions. *J Biol Chem* 2004, 279:7566-7575.

[61] Haslbeck M, Braun N, Stromer T, Richter B, Model N, Weinkauf S, Buchner J: Hsp42 is the general small heat shock protein in the cytosol of Saccharomyces cerevisiae. *EMBO J* 2004, 23:638-649.

[62] Stromer T, Fischer E, Richter K, Haslbeck M, Buchner J: Analysis of the regulation of the molecular chaperone Hsp26 by temperature-induced dissociation: the N-terminal domail is important for oligomer assembly and the binding of unfolding proteins. *J Biol Chem* 2004, 279:11222-11228.

[63] Van Montfort R, Slingsby C, Vierling E: Structure and function of the small heat shock protein/alpha-crystallin family of molecular chaperones. *Adv Protein Chem* 2001, 59:105-156.

[64] van Montfort RL, Basha E, Friedrich KL, Slingsby C, Vierling E: Crystal structure and assembly of a eukaryotic small heat shock protein. *Nat Struct Biol* 2001, 8:1025-1030.

[65] Haslbeck M, Walke S, Stromer T, Ehrnsperger M, White HE, Chen S, Saibil HR, Buchner J: Hsp26: a temperature-regulated chaperone. *EMBO J* 1999, 18:6744-6751.

[66] Lee GJ, Roseman AM, Saibil HR, Vierling E: A small heat shock protein stably binds heat-denatured model substrates and can maintain a substrate in a folding-competent state. *EMBO J* 1997, 16:659-671.

[67] Preville X, Salvemini F, Giraud S, Chaufour S, Paul C, Stepien G, Ursini MV, Arrigo AP: Mammalian small stress proteins protect against oxidative stress through their ability to increase glucose-6-phosphate dehydrogenase activity and by maintaining optimal cellular detoxifying machinery. *Exp Cell Res* 1999, 247:61-78.

[68] Chauhan D, Li G, Shringarpure R, Podar K, Ohtake Y, Hideshima T, Anderson KC: Blockade of Hsp27 overcomes Bortezomib/proteasome inhibitor PS-341 resistance in lymphoma cells. *Cancer Res* 2003, 63:6174-6177.

[69] Chauhan D, Li G, Auclair D, Hideshima T, Podar K, Mitsiades N, Mitsiades C, Chen LB, Munshi N, Saxena S, Anderson KC: 2-Methoxyestardiol and bortezomib/proteasome-inhibitor overcome dexamethasone-resistance in multiple myeloma cells by modulating Heat Shock Protein-27. *Apoptosis* 2004, 9:149-155.

[70] Hartl FU, Hayer-Hartl M: Molecular chaperones in the cytosol: from nascent chain to folded protein. *Science* 2002, 295:1852-1858.

[71] Bukau B, Horwich AL: The Hsp70 and Hsp60 chaperone machines. *Cell* 1998, 92:351-366.

[72] van der Vies SM, Gatenby AA, Georgopoulos C: Bacteriophage T4 encodes a co-chaperonin that can substitute for Escherichia coli GroES in protein folding. *Nature* 1994, 368:654-656.

[73] Straus D, Walter W, Gross CA: DnaK, DnaJ, and GrpE heat shock proteins negatively regulate heat shock gene expression by controlling the synthesis and stability of sigma 32. *Genes Dev* 1990, 4:2202-2209.

[74] Azem A, Kessel M, Goloubinoff P: Characterization of a functional GroEL14(GroES7)2 chaperonin hetero-oligomer. *Science* 1994, 265:653-656.

[75] Luft BJ, Gorevic PD, Jiang W, Munoz P, Dattwyler RJ: Immunologic and structural characterization of the dominant 66- to 73-kDa antigens of Borrelia burgdorferi. *J Immunol* 1991, 146:2776-2782.

[76] Dieterle S, Wollenhaupt J: Humoral immune response to the chlamydial heat shock proteins hsp60 and hsp70 in Chlamydia-associated chronic salpingitis with tubal occlusion. *Hum Reprod* 1996, 11:1352-1356.

[77] Kakeya H, Udono H, Ikuno N, Yamamoto Y, Mitsutake K, Miyazaki T, Tomono K, Koga H, Tashiro T, Nakayama E, Kohno S: A 77-kilodalton protein of Cryptococcus neoformans, a member of the heat shock protein 70 family, is a major antigen detected in the sera of mice with pulmonary cryptococcosis. *Infect Immun* 1997, 65:1653-1658.

[78] Roy H, Bloom M, Milos P, Monroe M: Studies on the assembly of large subunits of ribulose bisphosphate carboxylase in isolated pea chloroplasts. *J Cell Biol* 1982, 94:20-27.

[79] Xu Z, Horwich AL, Sigler PB: The crystal structure of the asymmetric GroEL-GroES-(ADP)7 chaperonin complex. *Nature* 1997, 388:741-750.

[80] Saibil HR, Ranson NA: The chaperonin folding machine. *Trends Biochem Sci* 2002, 27:627-632.

[81] Meyer AS, Gillespie JR, Walther D, Millet IS, Doniach S, Frydman J: Closing the folding chamber of the eukaryotic chaperonin requires the transition state of ATP hydrolysis. *Cell* 2003, 113:369-381.

[82] Thulasiraman V, Yang CF, Frydman J: In vivo newly translated polypeptides are sequestered in a protected folding environment. *EMBO J* 1999, 18:85-95.

[83] Lewis SA, Tian G, Vainberg IE, Cowan NJ: Chaperonin-mediated folding of actin and tubulin. *J Cell Biol* 1996, 132:1-4.

[84] Frydman J: Folding of newly translated proteins in vivo: the role of molecular chaperones. *Annu Rev Biochem* 2001, 70:603-647.

[85] Mayer MP, Brehmer D, Gassler CS, Bukau B: Hsp70 chaperone machines. *Adv Protein Chem* 2001, 59:1-44.

[86] Craig EA: Chaperones: helpers along the pathways to protein folding. *Science* 1993, 260:1902-1903.

[87] Craig EA, Gambill BD, Nelson RJ: Heat shock proteins: molecular chaperones of protein biogenesis. *Microbiol Rev* 1993, 57:402-414.

[88] Fink AL: Chaperone-mediated protein folding. *Physiol Rev* 1999, 79:425-449.

[89] Schwede T, Kopp J, Guex N, Peitsch MC: SWISS-MODEL: An automated protein homology-modeling server. *Nucleic Acids Res* 2003, 31:3381-3385.

[90] Flaherty KM, DeLuca-Flaherty C, McKay DB: Three-dimensional structure of the ATPase fragment of a 70K heat-shock cognate protein. *Nature* 1990, 346:623-628.

[91] Flaherty KM, McKay DB, Kabsch W, Holmes KC: Similarity of the three-dimensional structures of actin and the ATPase fragment of a 70-kDa heat shock cognate protein. *Proc Natl Acad Sci U S A* 1991, 88:5041-5045.

[92] Rudiger S, Germeroth L, Schneider-Mergener J, Bukau B: Substrate specificity of the DnaK chaperone determined by screening cellulose-bound peptide libraries. *EMBO J* 1997, 16:1501-1507.

[93] Bukau B, Weissman J, Horwich A: Molecular chaperones and protein quality control. *Cell* 2006, 125:443-451.
[94] Frydman J, Hohfeld J: Chaperones get in touch: the Hip-Hop connection. *Trends Biochem Sci* 1997, 22:87-92.
[95] Cuervo AM, Dice JF: Regulation of lamp2a levels in the lysosomal membrane. *Traffic* 2000, 1:570-583.
[96] Minami Y, Hohfeld J, Ohtsuka K, Hartl FU: Regulation of the heat-shock protein 70 reaction cycle by the mammalian DnaJ homolog, Hsp40. *J Biol Chem* 1996, 271:19617-19624.
[97] Ballinger CA, Connell P, Wu Y, Hu Z, Thompson LJ, Yin LY, Patterson C: Identification of CHIP, a novel tetratricopeptide repeat-containing protein that interacts with heat shock proteins and negatively regulates chaperone functions. *Mol Cell Biol* 1999, 19:4535-4545.
[98] Hohfeld J, Minami Y, Hartl FU: Hip, a novel cochaperone involved in the eukaryotic Hsc70/Hsp40 reaction cycle. *Cell* 1995, 83:589-598.
[99] Gassler CS, Wiederkehr T, Brehmer D, Bukau B, Mayer MP: Bag-1M accelerates nucleotide release for human Hsc70 and Hsp70 and can act concentration-dependent as positive and negative cofactor. *J Biol Chem* 2001, 276:32538-32544.
[100] Sondermann H, Scheufler C, Schneider C, Hohfeld J, Hartl FU, Moarefi I: Structure of a Bag/Hsc70 complex: convergent functional evolution of Hsp70 nucleotide exchange factors. *Science* 2001, 291:1553-1557.
[101] Hohfeld J, Cyr DM, Patterson C: From the cradle to the grave: molecular chaperones that may choose between folding and degradation. *EMBO Rep* 2001, 2:885-890.
[102] Parsell DA, Taulien J, Lindquist S: The role of heat-shock proteins in thermotolerance. *Philos Trans R Soc Lond B Biol Sci* 1993, 339:279-285.
[103] Didelot C, Schmitt E, Brunet M, Maingret L, Parcellier A, Garrido C: Heat shock proteins: endogenous modulators of apoptotic cell death. *Handb Exp Pharmacol* 2006, 171-198.
[104] Zhao R, Houry WA: Hsp90: a chaperone for protein folding and gene regulation. *Biochem Cell Biol* 2005, 83:703-710.
[105] Mayer MP, Bukau B: Hsp70 chaperones: cellular functions and molecular mechanism. *Cell Mol Life Sci* 2005, 62:670-684.
[106] Burrows F, Zhang H, Kamal A: Hsp90 activation and cell cycle regulation. *Cell Cycle* 2004, 3:1530-1536.
[107] Chiosis G, Vilenchik M, Kim J, Solit D: Hsp90: the vulnerable chaperone. *Drug Discov Today* 2004, 9:881-888.
[108] Sreedhar AS, Kalmar E, Csermely P, Shen YF: Hsp90 isoforms: functions, expression and clinical importance. *FEBS Lett* 2004, 562:11-15.
[109] Obermann WM, Sondermann H, Russo AA, Pavletich NP, Hartl FU: In vivo function of Hsp90 is dependent on ATP binding and ATP hydrolysis. *J Cell Biol* 1998, 143:901-910.
[110] DeFranco DB, Csermely P: Steroid receptor and molecular chaperone encounters in the nucleus. *Sci STKE* 2000, 2000:E1.

[111] Haverinen M, Passinen S, Syvala H, Pasanen S, Manninen T, Tuohimaa P, Ylikomi T: Heat shock protein 90 and the nuclear transport of progesterone receptor. *Cell Stress Chaperones* 2001, 6:256-262.

[112] Howard KJ, Holley SJ, Yamamoto KR, Distelhorst CW: Mapping the HSP90 binding region of the glucocorticoid receptor. *J Biol Chem* 1990, 265:11928-11935.

[113] McLaughlin SH, Jackson SE: Folding and stability of the ligand-binding domain of the glucocorticoid receptor. *Protein Sci* 2002, 11:1926-1936.

[114] Polanowska-Grabowska R, Gear AR: Heat-shock proteins and platelet function. *Platelets* 2000, 11:6-22.

[115] Rajapandi T, Greene LE, Eisenberg E: The molecular chaperones Hsp90 and Hsc70 are both necessary and sufficient to activate hormone binding by glucocorticoid receptor. *J Biol Chem* 2000, 275:22597-22604.

[116] DeFranco DB, Ramakrishnan C, Tang Y: Molecular chaperones and subcellular trafficking of steroid receptors. *J Steroid Biochem Mol Biol* 1998, 65:51-58.

[117] Shaknovich R, Shue G, Kohtz DS: Conformational activation of a basic helix-loop-helix protein (MyoD1) by the C-terminal region of murine HSP90 (HSP84). *Mol Cell Biol* 1992, 12:5059-5068.

[118] Minami Y, Kawasaki H, Minami M, Tanahashi N, Tanaka K, Yahara I: A critical role for the proteasome activator PA28 in the Hsp90-dependent protein refolding. *J Biol Chem* 2000, 275:9055-9061.

[119] Murata S, Minami Y, Minami M, Chiba T, Tanaka K: CHIP is a chaperone-dependent E3 ligase that ubiquitylates unfolded protein. *EMBO Rep* 2001, 2:1133-1138.

[120] Tsubuki S, Saito Y, Kawashima S: Purification and characterization of an endogenous inhibitor specific to the Z-Leu-Leu-Leu-MCA degrading activity in proteasome and its identification as heat-shock protein 90. *FEBS Lett* 1994, 344:229-233.

[121] Wagner BJ, Margolis JW: Age-dependent association of isolated bovine lens multicatalytic proteinase complex (proteasome) with heat-shock protein 90, an endogenous inhibitor. *Arch Biochem Biophys* 1995, 323:455-462.

[122] Conconi M, Petropoulos I, Emod I, Turlin E, Biville F, Friguet B: Protection from oxidative inactivation of the 20S proteasome by heat-shock protein 90. *Biochem J* 1998, 333 (Pt 2):407-415.

[123] Imai J, Maruya M, Yashiroda H, Yahara I, Tanaka K: The molecular chaperone Hsp90 plays a role in the assembly and maintenance of the 26S proteasome. *EMBO J* 2003, 22:3557-3567.

[124] Rutherford SL, Lindquist S: Hsp90 as a capacitor for morphological evolution. *Nature* 1998, 396:336-342.

[125] Hernandez MP, Chadli A, Toft DO: HSP40 binding is the first step in the HSP90 chaperoning pathway for the progesterone receptor. *J Biol Chem* 2002, 277:11873-11881.

[126] Scheufler C, Brinker A, Bourenkov G, Pegoraro S, Moroder L, Bartunik H, Hartl FU, Moarefi I: Structure of TPR domain-peptide complexes: critical elements in the assembly of the Hsp70-Hsp90 multichaperone machine. *Cell* 2000, 101:199-210.

[127] Fontana J, Fulton D, Chen Y, Fairchild TA, McCabe TJ, Fujita N, Tsuruo T, Sessa WC: Domain mapping studies reveal that the M domain of hsp90 serves as a molecular

scaffold to regulate Akt-dependent phosphorylation of endothelial nitric oxide synthase and NO release. *Circ Res* 2002, 90:866-873.

[128] Fulton D, Fontana J, Sowa G, Gratton JP, Lin M, Li KX, Michell B, Kemp BE, Rodman D, Sessa WC: Localization of endothelial nitric-oxide synthase phosphorylated on serine 1179 and nitric oxide in Golgi and plasma membrane defines the existence of two pools of active enzyme. *J Biol Chem* 2002, 277:4277-4284.

[129] Stebbins CE, Russo AA, Schneider C, Rosen N, Hartl FU, Pavletich NP: Crystal structure of an Hsp90-geldanamycin complex: targeting of a protein chaperone by an antitumor agent. *Cell* 1997, 89:239-250.

[130] Sharma SV, Agatsuma T, Nakano H: Targeting of the protein chaperone, HSP90, by the transformation suppressing agent, radicicol. *Oncogene* 1998, 16:2639-2645.

[131] Citri A, Alroy I, Lavi S, Rubin C, Xu W, Grammatikakis N, Patterson C, Neckers L, Fry DW, Yarden Y: Drug-induced ubiquitylation and degradation of ErbB receptor tyrosine kinases: implications for cancer therapy. *EMBO J* 2002, 21:2407-2417.

[132] Marcu MG, Schulte TW, Neckers L: Novobiocin and related coumarins and depletion of heat shock protein 90-dependent signaling proteins. *J Natl Cancer Inst* 2000, 92:242-248.

[133] Mimnaugh EG, Chavany C, Neckers L: Polyubiquitination and proteasomal degradation of the p185c-erbB-2 receptor protein-tyrosine kinase induced by geldanamycin. *J Biol Chem* 1996, 271:22796-22801.

[134] Schneider C, Sepp-Lorenzino L, Nimmesgern E, Ouerfelli O, Danishefsky S, Rosen N, Hartl FU: Pharmacologic shifting of a balance between protein refolding and degradation mediated by Hsp90. *Proc Natl Acad Sci U S A* 1996, 93:14536-14541.

[135] Solit DB, Zheng FF, Drobnjak M, Munster PN, Higgins B, Verbel D, Heller G, Tong W, Cordon-Cardo C, Agus DB, Scher HI, Rosen N: 17-Allylamino-17-demethoxygeldanamycin induces the degradation of androgen receptor and HER-2/neu and inhibits the growth of prostate cancer xenografts. *Clin Cancer Res* 2002, 8:986-993.

[136] Xu W, Mimnaugh E, Rosser MF, Nicchitta C, Marcu M, Yarden Y, Neckers L: Sensitivity of mature Erbb2 to geldanamycin is conferred by its kinase domain and is mediated by the chaperone protein Hsp90. *J Biol Chem* 2001, 276:3702-3708.

[137] Swamynathan SK, Revathi CJ, Srinivas UK: Identification and characterization of promoter elements responsible for the induction of the albumin gene by heat shock in early embryonic rat liver. *DNA Cell Biol* 1996, 15:897-905.

[138] Morange M: HSFs in development. *Handb Exp Pharmacol* 2006, 153-169.

[139] Nover L, Scharf KD, Gagliardi D, Vergne P, Czarnecka-Verner E, Gurley WB: The Hsf world: classification and properties of plant heat stress transcription factors. *Cell Stress Chaperones* 1996, 1:215-223.

[140] Sorger PK: Yeast heat shock factor contains separable transient and sustained response transcriptional activators. *Cell* 1990, 62:793-805.

[141] Perisic O, Xiao H, Lis JT: Stable binding of Drosophila heat shock factor to head-to-head and tail-to-tail repeats of a conserved 5 bp recognition unit. *Cell* 1989, 59:797-806.

[142] Larson JS, Schuetz TJ, Kingston RE: Activation in vitro of sequence-specific DNA binding by a human regulatory factor. *Nature* 1988, 335:372-375.

[143] Wu C: Heat shock transcription factors: structure and regulation. *Annu Rev Cell Dev Biol* 1995, 11:441-469.

[144] Lee YK, Manalo D, Liu AY: Heat shock response, heat shock transcription factor and cell aging. *Biol Signals* 1996, 5:180-191.

[145] Liu AY, Lee YK, Manalo D, Huang LE: Attenuated heat shock transcriptional response in aging: molecular mechanism and implication in the biology of aging. *EXS* 1996, 77:393-408.

[146] Mathew A, Morimoto RI: Role of the heat-shock response in the life and death of proteins. *Ann N Y Acad Sci* 1998, 851:99-111.

[147] Rallu M, Loones M, Lallemand Y, Morimoto R, Morange M, Mezger V: Function and regulation of heat shock factor 2 during mouse embryogenesis. *Proc Natl Acad Sci U S A* 1997, 94:2392-2397.

[148] McMillan DR, Christians E, Forster M, Xiao X, Connell P, Plumier JC, Zuo X, Richardson J, Morgan S, Benjamin IJ: Heat shock transcription factor 2 is not essential for embryonic development, fertility, or adult cognitive and psychomotor function in mice. *Mol Cell Biol* 2002, 22:8005-8014.

[149] Sarge KD, Cullen KE: Regulation of hsp expression during rodent spermatogenesis. *Cell Mol Life Sci* 1997, 53:191-197.

[150] Sarge KD, Park-Sarge OK, Kirby JD, Mayo KE, Morimoto RI: Expression of heat shock factor 2 in mouse testis: potential role as a regulator of heat-shock protein gene expression during spermatogenesis. *Biol Reprod* 1994, 50:1334-1343.

[151] Tanikawa J, Ichikawa-Iwata E, Kanei-Ishii C, Nakai A, Matsuzawa S, Reed JC, Ishii S: p53 suppresses the c-Myb-induced activation of heat shock transcription factor 3. *J Biol Chem* 2000, 275:15578-15585.

[152] Shabtay A, Arad Z: Reciprocal activation of HSF1 and HSF3 in brain and blood tissues: is redundancy developmentally related? *Am J Physiol Regul Integr Comp Physiol* 2006, 291:R566-R572.

[153] Tanabe M, Kawazoe Y, Takeda S, Morimoto RI, Nagata K, Nakai A: Disruption of the HSF3 gene results in the severe reduction of heat shock gene expression and loss of thermotolerance. *EMBO J* 1998, 17:1750-1758.

[154] Fujimoto M, Izu H, Seki K, Fukuda K, Nishida T, Yamada S, Kato K, Yonemura S, Inouye S, Nakai A: HSF4 is required for normal cell growth and differentiation during mouse lens development. *EMBO J* 2004, 23:4297-4306.

[155] Nakai A, Tanabe M, Kawazoe Y, Inazawa J, Morimoto RI, Nagata K: HSF4, a new member of the human heat shock factor family which lacks properties of a transcriptional activator. *Mol Cell Biol* 1997, 17:469-481.

[156] Nakai A: New aspects in the vertebrate heat shock factor system: Hsf3 and Hsf4. *Cell Stress Chaperones* 1999, 4:86-93.

[157] Zhang Y, Frejtag W, Dai R, Mivechi NF: Heat shock factor-4 (HSF-4a) is a repressor of HSF-1 mediated transcription. *J Cell Biochem* 2001, 82:692-703.

[158] Pritts TA, Hungness ES, Hershko DD, Robb BW, Sun X, Luo GJ, Fischer JE, Wong HR, Hasselgren PO: Proteasome inhibitors induce heat shock response and increase IL-

6 expression in human intestinal epithelial cells. *Am J Physiol Regul Integr Comp Physiol* 2002, 282:R1016-R1026.
[159] Tacchini L, Dansi P, Matteucci E, Bernelli-Zazzera A, Desiderio MA: Influence of proteasome and redox state on heat shock-induced activation of stress kinases, AP-1 and HSF. *Biochim Biophys Acta* 2001, 1538:76-89.
[160] Kawazoe Y, Nakai A, Tanabe M, Nagata K: Proteasome inhibition leads to the activation of all members of the heat-shock-factor family. *Eur J Biochem* 1998, 255:356-362.
[161] Voellmy R: Transduction of the stress signal and mechanisms of transcriptional regulation of heat shock/stress protein gene expression in higher eukaryotes. *Crit Rev Eukaryot Gene Expr* 1994, 4:357-401.
[162] Kim J, Nueda A, Meng YH, Dynan WS, Mivechi NF: Analysis of the phosphorylation of human heat shock transcription factor-1 by MAP kinase family members. *J Cell Biochem* 1997, 67:43-54.
[163] Knauf U, Newton EM, Kyriakis J, Kingston RE: Repression of human heat shock factor 1 activity at control temperature by phosphorylation. *Genes Dev* 1996, 10:2782-2793.
[164] Morimoto RI: Regulation of the heat shock transcriptional response: cross talk between a family of heat shock factors, molecular chaperones, and negative regulators. *Genes Dev* 1998, 12:3788-3796.
[165] Jakob U, Muse W, Eser M, Bardwell JC: Chaperone activity with a redox switch. *Cell* 1999, 96:341-352.
[166] Ahn SG, Thiele DJ: Redox regulation of mammalian heat shock factor 1 is essential for Hsp gene activation and protection from stress. *Genes Dev* 2003, 17:516-528.
[167] Pirkkala L, Nykanen P, Sistonen L: Roles of the heat shock transcription factors in regulation of the heat shock response and beyond. *FASEB J* 2001, 15:1118-1131.
[168] Pirkkala L, Alastalo TP, Zuo X, Benjamin IJ, Sistonen L: Disruption of heat shock factor 1 reveals an essential role in the ubiquitin proteolytic pathway. *Mol Cell Biol* 2000, 20:2670-2675.
[169] Lee DH, Goldberg AL: Proteasome inhibitors: valuable new tools for cell biologists. *Trends Cell Biol* 1998, 8:397-403.
[170] Bajorek M, Finley D, Glickman MH: Proteasome disassembly and downregulation is correlated with viability during stationary phase. *Curr Biol* 2003, 13:1140-1144.
[171] Mannhaupt G, Schnall R, Karpov V, Vetter I, Feldmann H: Rpn4p acts as a transcription factor by binding to PACE, a nonamer box found upstream of 26S proteasomal and other genes in yeast. *FEBS Lett* 1999, 450:27-34.
[172] Ju D, Xie Y: Proteasomal degradation of RPN4 via two distinct mechanisms, ubiquitin-dependent and -independent. *J Biol Chem* 2004, 279:23851-23854.
[173] London MK, Keck BI, Ramos PC, Dohmen RJ: Regulatory mechanisms controlling biogenesis of ubiquitin and the proteasome. *FEBS Lett* 2004, 567:259-264.
[174] Ju D, Wang L, Mao X, Xie Y: Homeostatic regulation of the proteasome via an Rpn4-dependent feedback circuit. *Biochem Biophys Res Commun* 2004, 321:51-57.

[175] Hahn JS, Neef DW, Thiele DJ: A stress regulatory network for co-ordinated activation of proteasome expression mediated by yeast heat shock transcription factor. *Mol Microbiol* 2006, 60:240-251.

[176] Seufert W, Jentsch S: Ubiquitin-conjugating enzymes UBC4 and UBC5 mediate selective degradation of short-lived and abnormal proteins. *EMBO J* 1990, 9:543-550.

[177] Simon JR, Treger JM, McEntee K: Multiple independent regulatory pathways control UBI4 expression after heat shock in Saccharomyces cerevisiae. *Mol Microbiol* 1999, 31:823-832.

[178] Hahn JS, Hu Z, Thiele DJ, Iyer VR: Genome-wide analysis of the biology of stress responses through heat shock transcription factor. *Mol Cell Biol* 2004, 24:5249-5256.

[179] Shi Y, Mosser DD, Morimoto RI: Molecular chaperones as HSF1-specific transcriptional repressors. *Genes Dev* 1998, 12:654-666.

[180] Qian SB, McDonough H, Boellmann F, Cyr DM, Patterson C: CHIP-mediated stress recovery by sequential ubiquitination of substrates and Hsp70. *Nature* 2006, 440:551-555.

[181] Ellis RJ: Macromolecular crowding: an important but neglected aspect of the intracellular environment. *Curr Opin Struct Biol* 2001, 11:114-119.

[182] Mathew A, Mathur SK, Morimoto RI: Heat shock response and protein degradation: regulation of HSF2 by the ubiquitin-proteasome pathway. *Mol Cell Biol* 1998, 18:5091-5098.

[183] Hartl FU: Molecular chaperones in cellular protein folding. *Nature* 1996, 381:571-579.

[184] Martin J, Hartl FU: Chaperone-assisted protein folding. *Curr Opin Struct Biol* 1997, 7:41-52.

[185] Landry SJ, Gierasch LM: Polypeptide interactions with molecular chaperones and their relationship to in vivo protein folding. *Annu Rev Biophys Biomol Struct* 1994, 23:645-669.

[186] Schmid FX, Mayr LM, Mucke M, Schonbrunner ER: Prolyl isomerases: role in protein folding. *Adv Protein Chem* 1993, 44:25-66.

[187] Wang CC: Protein disulfide isomerase assists protein folding as both an isomerase and a chaperone. *Ann N Y Acad Sci* 1998, 864:9-13.

[188] Sitia R, Molteni SN: Stress, protein (mis)folding, and signaling: the redox connection. *Sci STKE* 2004, 2004:e27.

[189] Hoffmann JH, Linke K, Graf PC, Lilie H, Jakob U: Identification of a redox-regulated chaperone network. *EMBO J* 2004, 23:160-168.

[190] Graf PC, Martinez-Yamout M, VanHaerents S, Lilie H, Dyson HJ, Jakob U: Activation of the redox-regulated chaperone Hsp33 by domain unfolding. *J Biol Chem* 2004, 279:20529-20538.

[191] Uehara T, Nakamura T, Yao D, Shi ZQ, Gu Z, Ma Y, Masliah E, Nomura Y, Lipton SA: S-nitrosylated protein-disulphide isomerase links protein misfolding to neurodegeneration. *Nature* 2006, 441:513-517.

[192] Kabakov AE, Gabai VL: Heat-shock proteins maintain the viability of ATP-deprived cells: what is the mechanism? *Trends Cell Biol* 1994, 4:193-196.

[193] Palleros DR, Shi L, Reid KL, Fink AL: hsp70-protein complexes. Complex stability and conformation of bound substrate protein. *J Biol Chem* 1994, 269:13107-13114.

[194] James P, Pfund C, Craig EA: Functional specificity among Hsp70 molecular chaperones. *Science* 1997, 275:387-389.
[195] Zhu X, Zhao X, Burkholder WF, Gragerov A, Ogata CM, Gottesman ME, Hendrickson WA: Structural analysis of substrate binding by the molecular chaperone DnaK. *Science* 1996, 272:1606-1614.
[196] Osipiuk J, Walsh MA, Freeman BC, Morimoto RI, Joachimiak A: Structure of a new crystal form of human Hsp70 ATPase domain. *Acta Crystallogr D Biol Crystallogr* 1999, 55:1105-1107.
[197] Caplan AJ: Hsp90's secrets unfold: new insights from structural and functional studies. *Trends Cell Biol* 1999, 9:262-268.
[198] Young JC, Moarefi I, Hartl FU: Hsp90: a specialized but essential protein-folding tool. *J Cell Biol* 2001, 154:267-273.
[199] Connell P, Ballinger CA, Jiang J, Wu Y, Thompson LJ, Hohfeld J, Patterson C: The co-chaperone CHIP regulates protein triage decisions mediated by heat-shock proteins. *Nat Cell Biol* 2001, 3:93-96.
[200] Meacham GC, Patterson C, Zhang W, Younger JM, Cyr DM: The Hsc70 co-chaperone CHIP targets immature CFTR for proteasomal degradation. *Nat Cell Biol* 2001, 3:100-105.
[201] Muchowski PJ, Wacker JL: Modulation of neurodegeneration by molecular chaperones. *Nat Rev Neurosci* 2005, 6:11-22.
[202] Wacker JL, Zareie MH, Fong H, Sarikaya M, Muchowski PJ: Hsp70 and Hsp40 attenuate formation of spherical and annular polyglutamine oligomers by partitioning monomer. *Nat Struct Mol Biol* 2004, 11:1215-1222.
[203] Schaffar G, Breuer P, Boteva R, Behrends C, Tzvetkov N, Strippel N, Sakahira H, Siegers K, Hayer-Hartl M, Hartl FU: Cellular toxicity of polyglutamine expansion proteins: mechanism of transcription factor deactivation. *Mol Cell* 2004, 15:95-105.
[204] Barral JM, Broadley SA, Schaffar G, Hartl FU: Roles of molecular chaperones in protein misfolding diseases. *Semin Cell Dev Biol* 2004, 15:17-29.
[205] Hsu AL, Murphy CT, Kenyon C: Regulation of aging and age-related disease by DAF-16 and heat-shock factor. *Science* 2003, 300:1142-1145.
[206] Garcia-Mata R, Bebok Z, Sorscher EJ, Sztul ES: Characterization and dynamics of aggresome formation by a cytosolic GFP-chimera. *J Cell Biol* 1999, 146:1239-1254.
[207] Wigley WC, Fabunmi RP, Lee MG, Marino CR, Muallem S, DeMartino GN, Thomas PJ: Dynamic association of proteasomal machinery with the centrosome. *J Cell Biol* 1999, 145:481-490.
[208] Wyttenbach A, Carmichael J, Swartz J, Furlong RA, Narain Y, Rankin J, Rubinsztein DC: Effects of heat shock, heat shock protein 40 (HDJ-2), and proteasome inhibition on protein aggregation in cellular models of Huntington's disease. *Proc Natl Acad Sci U S A* 2000, 97:2898-2903.
[209] Dul JL, Davis DP, Williamson EK, Stevens FJ, Argon Y: Hsp70 and antifibrillogenic peptides promote degradation and inhibit intracellular aggregation of amyloidogenic light chains. *J Cell Biol* 2001, 152:705-716.

[210] Stuart RA, Cyr DM, Neupert W: Hsp70 in mitochondrial biogenesis: from chaperoning nascent polypeptide chains to facilitation of protein degradation. *Experientia* 1994, 50:1002-1011.

[211] Xu W, Marcu M, Yuan X, Mimnaugh E, Patterson C, Neckers L: Chaperone-dependent E3 ubiquitin ligase CHIP mediates a degradative pathway for c-ErbB2/Neu. *Proc Natl Acad Sci U S A* 2002, 99:12847-12852.

[212] Parcellier A, Schmitt E, Gurbuxani S, Seigneurin-Berny D, Pance A, Chantome A, Plenchette S, Khochbin S, Solary E, Garrido C: HSP27 is a ubiquitin-binding protein involved in I-kappaBalpha proteasomal degradation. *Mol Cell Biol* 2003, 23:5790-5802.

[213] Shimura H, Miura-Shimura Y, Kosik KS: Binding of tau to heat shock protein 27 leads to decreased concentration of hyperphosphorylated tau and enhanced cell survival. *J Biol Chem* 2004, 279:17957-17962.

[214] Bercovich B, Stancovski I, Mayer A, Blumenfeld N, Laszlo A, Schwartz AL, Ciechanover A: Ubiquitin-dependent degradation of certain protein substrates in vitro requires the molecular chaperone Hsc70. *J Biol Chem* 1997, 272:9002-9010.

[215] Lee DH, Sherman MY, Goldberg AL: Involvement of the molecular chaperone Ydj1 in the ubiquitin-dependent degradation of short-lived and abnormal proteins in Saccharomyces cerevisiae. *Mol Cell Biol* 1996, 16:4773-4781.

[216] Gusarova V, Caplan AJ, Brodsky JL, Fisher EA: Apoprotein B degradation is promoted by the molecular chaperones hsp90 and hsp70. *J Biol Chem* 2001, 276:24891-24900.

[217] Yang Y, Janich S, Cohn JA, Wilson JM: The common variant of cystic fibrosis transmembrane conductance regulator is recognized by hsp70 and degraded in a pre-Golgi nonlysosomal compartment. *Proc Natl Acad Sci U S A* 1993, 90:9480-9484.

[218] Meacham GC, Lu Z, King S, Sorscher E, Tousson A, Cyr DM: The Hdj-2/Hsc70 chaperone pair facilitates early steps in CFTR biogenesis. *EMBO J* 1999, 18:1492-1505.

[219] Amaral MD: Processing of CFTR: traversing the cellular maze--how much CFTR needs to go through to avoid cystic fibrosis? *Pediatr Pulmonol* 2005, 39:479-491.

[220] Kopito RR: Biosynthesis and degradation of CFTR. *Physiol Rev* 1999, 79:S167-S173.

[221] Riordan JR: Cystic fibrosis as a disease of misprocessing of the cystic fibrosis transmembrane conductance regulator glycoprotein. *Am J Hum Genet* 1999, 64:1499-1504.

[222] McClellan AJ, Scott MD, Frydman J: Folding and quality control of the VHL tumor suppressor proceed through distinct chaperone pathways. *Cell* 2005, 121:739-748.

[223] Verma R, Chen S, Feldman R, Schieltz D, Yates J, Dohmen J, Deshaies RJ: Proteasomal proteomics: identification of nucleotide-sensitive proteasome-interacting proteins by mass spectrometric analysis of affinity-purified proteasomes. *Mol Biol Cell* 2000, 11:3425-3439.

[224] Esser C, Alberti S, Hohfeld J: Cooperation of molecular chaperones with the ubiquitin/proteasome system. *Biochim Biophys Acta* 2004, 1695:171-188.

[225] Cyr DM, Hohfeld J, Patterson C: Protein quality control: U-box-containing E3 ubiquitin ligases join the fold. *Trends Biochem Sci* 2002, 27:368-375.

[226] Liu CW, Corboy MJ, DeMartino GN, Thomas PJ: Endoproteolytic activity of the proteasome. *Science* 2003, 299:408-411.

[227] Westhoff B, Chapple JP, van der SJ, Hohfeld J, Cheetham ME: HSJ1 is a neuronal shuttling factor for the sorting of chaperone clients to the proteasome. *Curr Biol* 2005, 15:1058-1064.

[228] Kloetzel PM: Antigen processing by the proteasome. *Nat Rev Mol Cell Biol* 2001, 2:179-187.

[229] Yamano T, Murata S, Shimbara N, Tanaka N, Chiba T, Tanaka K, Yui K, Udono H: Two distinct pathways mediated by PA28 and hsp90 in major histocompatibility complex class I antigen processing. *J Exp Med* 2002, 196:185-196.

[230] Demand J, Luders J, Hohfeld J: The carboxy-terminal domain of Hsc70 provides binding sites for a distinct set of chaperone cofactors. *Mol Cell Biol* 1998, 18:2023-2028.

[231] Aravind L, Koonin EV: The U box is a modified RING finger - a common domain in ubiquitination. *Curr Biol* 2000, 10:R132-R134.

[232] Dai Q, Zhang C, Wu Y, McDonough H, Whaley RA, Godfrey V, Li HH, Madamanchi N, Xu W, Neckers L, Cyr D, Patterson C: CHIP activates HSF1 and confers protection against apoptosis and cellular stress. *EMBO J* 2003, 22:5446-5458.

[233] Jiang J, Ballinger CA, Wu Y, Dai Q, Cyr DM, Hohfeld J, Patterson C: CHIP is a U-box-dependent E3 ubiquitin ligase: identification of Hsc70 as a target for ubiquitylation. *J Biol Chem* 2001, 276:42938-42944.

[234] Brinker A, Scheufler C, Von Der MF, Fleckenstein B, Herrmann C, Jung G, Moarefi I, Hartl FU: Ligand discrimination by TPR domains. Relevance and selectivity of EEVD-recognition in Hsp70 x Hop x Hsp90 complexes. *J Biol Chem* 2002, 277:19265-19275.

[235] Hohfeld J, Jentsch S: GrpE-like regulation of the hsc70 chaperone by the anti-apoptotic protein BAG-1. *EMBO J* 1997, 16:6209-6216.

[236] Kampinga HH, Kanon B, Salomons FA, Kabakov AE, Patterson C: Overexpression of the cochaperone CHIP enhances Hsp70-dependent folding activity in mammalian cells. *Mol Cell Biol* 2003, 23:4948-4958.

[237] Luders J, Demand J, Hohfeld J: The ubiquitin-related BAG-1 provides a link between the molecular chaperones Hsc70/Hsp70 and the proteasome. *J Biol Chem* 2000, 275:4613-4617.

[238] Alberti S, Demand J, Esser C, Emmerich N, Schild H, Hohfeld J: Ubiquitylation of BAG-1 suggests a novel regulatory mechanism during the sorting of chaperone substrates to the proteasome. *J Biol Chem* 2002, 277:45920-45927.

[239] Demand J, Alberti S, Patterson C, Hohfeld J: Cooperation of a ubiquitin domain protein and an E3 ubiquitin ligase during chaperone/proteasome coupling. *Curr Biol* 2001, 11:1569-1577.

[240] Wiederkehr T, Bukau B, Buchberger A: Protein turnover: a CHIP programmed for proteolysis. *Curr Biol* 2002, 12:R26-R28.

[241] Soti C, Csermely P: Chaperones come of age. *Cell Stress Chaperones* 2002, 7:186-190.

[242] Macario AJ, Conway dM: The molecular chaperone system and other anti-stress mechanisms in archaea. *Front Biosci* 2001, 6:D262-D283.

[243] Floyd JA, Hamilton BA: Intranuclear inclusions and the ubiquitin-proteasome pathway: digestion of a red herring? *Neuron* 1999, 24:765-766.

[244] Sherman MY, Goldberg AL: Cellular defenses against unfolded proteins: a cell biologist thinks about neurodegenerative diseases. *Neuron* 2001, 29:15-32.

[245] Krobitsch S, Lindquist S: Aggregation of huntingtin in yeast varies with the length of the polyglutamine expansion and the expression of chaperone proteins. *Proc Natl Acad Sci U S A* 2000, 97:1589-1594.

[246] Warrick JM, Chan HY, Gray-Board, Chai Y, Paulson HL, Bonini NM: Suppression of polyglutamine-mediated neurodegeneration in Drosophila by the molecular chaperone HSP70. *Nat Genet* 1999, 23:425-428.

[247] Soti C, Csermely P: Chaperones and aging: role in neurodegeneration and in other civilizational diseases. *Neurochem Int* 2002, 41:383-389.

[248] Morley JF, Morimoto RI: Regulation of longevity in Caenorhabditis elegans by heat shock factor and molecular chaperones. *Mol Biol Cell* 2004, 15:657-664.

[249] Citron M: Beta-secretase inhibition for the treatment of Alzheimer's disease--promise and challenge. *Trends Pharmacol Sci* 2004, 25:92-97.

[250] Dabir DV, Trojanowski JQ, Richter-Landsberg C, Lee VM, Forman MS: Expression of the small heat-shock protein alphaB-crystallin in tauopathies with glial pathology. *Am J Pathol* 2004, 164:155-166.

[251] Wilhelmus MM, Otte-Holler I, Wesseling P, de Waal RM, Boelens WC, Verbeek MM: Specific association of small heat shock proteins with the pathological hallmarks of Alzheimer's disease brains. *Neuropathol Appl Neurobiol* 2006, 32:119-130.

[252] Renkawek K, Bosman GJ, de Jong WW: Expression of small heat-shock protein hsp 27 in reactive gliosis in Alzheimer disease and other types of dementia. *Acta Neuropathol (Berl)* 1994, 87:511-519.

[253] Renkawek K, Bosman GJ, Gaestel M: Increased expression of heat-shock protein 27 kDa in Alzheimer disease: a preliminary study. *Neuroreport* 1993, 5:14-16.

[254] Kouchi Z, Sorimachi H, Suzuki K, Ishiura S: Proteasome inhibitors induce the association of Alzheimer's amyloid precursor protein with Hsc73. *Biochem Biophys Res Commun* 1999, 254:804-810.

[255] Fonte V, Kapulkin V, Taft A, Fluet A, Friedman D, Link CD: Interaction of intracellular beta amyloid peptide with chaperone proteins. *Proc Natl Acad Sci U S A* 2002, 99:9439-9444.

[256] Liang JJ: Interaction between beta-amyloid and lens alphaB-crystallin. *FEBS Lett* 2000, 484:98-101.

[257] Kudva YC, Hiddinga HJ, Butler PC, Mueske CS, Eberhardt NL: Small heat shock proteins inhibit in vitro A beta(1-42) amyloidogenesis. *FEBS Lett* 1997, 416:117-121.

[258] Schuller E, Gulesserian T, Seidl R, Cairns N, Lube G: Brain t-complex polypeptide 1 (TCP- 1) related to its natural substrate beta1 tubulin is decreased in Alzheimer's disease. *Life Sci* 2001, 69:263-270.

[259] Broquet AH, Thomas G, Masliah J, Trugnan G, Bachelet M: Expression of the molecular chaperone Hsp70 in detergent-resistant microdomains correlates with its membrane delivery and release. *J Biol Chem* 2003, 278:21601-21606.

[260] Yang Y, Turner RS, Gaut JR: The chaperone BiP/GRP78 binds to amyloid precursor protein and decreases Abeta40 and Abeta42 secretion. *J Biol Chem* 1998, 273:25552-25555.

[261] Dou F, Netzer WJ, Tanemura K, Li F, Hartl FU, Takashima A, Gouras GK, Greengard P, Xu H: Chaperones increase association of tau protein with microtubules. *Proc Natl Acad Sci U S A* 2003, 100:721-726.

[262] Magrane J, Smith RC, Walsh K, Querfurth HW: Heat shock protein 70 participates in the neuroprotective response to intracellularly expressed beta-amyloid in neurons. *J Neurosci* 2004, 24:1700-1706.

[263] McCracken AA, Brodsky JL: Evolving questions and paradigm shifts in endoplasmic-reticulum-associated degradation (ERAD). *Bioessays* 2003, 25:868-877.

[264] Johnson RJ, Xiao G, Shanmugaratnam J, Fine RE: Calreticulin functions as a molecular chaperone for the beta-amyloid precursor protein. *Neurobiol Aging* 2001, 22:387-395.

[265] Hamos JE, Oblas B, Pulaski-Salo D, Welch WJ, Bole DG, Drachman DA: Expression of heat shock proteins in Alzheimer's disease. *Neurology* 1991, 41:345-350.

[266] Lee AS: The glucose-regulated proteins: stress induction and clinical applications. *Trends Biochem Sci* 2001, 26:504-510.

[267] Veereshwarayya V, Kumar P, Rosen KM, Mestril R, Querfurth HW: Differential Effects of Mitochondrial Heat Shock Protein 60 and Related Molecular Chaperones to Prevent Intracellular beta-Amyloid-induced Inhibition of Complex IV and Limit Apoptosis. *J Biol Chem* 2006, 281:29468-29478.

[268] Stege GJ, Renkawek K, Overkamp PS, Verschuure P, Van Rijk AF, Reijnen-Aalbers A, Boelens WC, Bosman GJ, de Jong WW: The molecular chaperone alphaB-crystallin enhances amyloid beta neurotoxicity. *Biochem Biophys Res Commun* 1999, 262:152-156.

[269] Petrucelli L, Dickson D, Kehoe K, Taylor J, Snyder H, Grover A, De Lucia M, McGowan E, Lewis J, Prihar G, Kim J, Dillmann WH, Browne SE, Hall A, Voellmy R, Tsuboi Y, Dawson TM, Wolozin B, Hardy J, Hutton M: CHIP and Hsp70 regulate tau ubiquitination, degradation and aggregation. *Hum Mol Genet* 2004, 13:703-714.

[270] Sahara N, Murayama M, Mizoroki T, Urushitani M, Imai Y, Takahashi R, Murata S, Tanaka K, Takashima A: In vivo evidence of CHIP up-regulation attenuating tau aggregation. *J Neurochem* 2005, 94:1254-1263.

[271] Renkawek K, Stege GJ, Bosman GJ: Dementia, gliosis and expression of the small heat shock proteins hsp27 and alpha B-crystallin in Parkinson's disease. *Neuroreport* 1999, 10:2273-2276.

[272] McLean PJ, Kawamata H, Shariff S, Hewett J, Sharma N, Ueda K, Breakefield XO, Hyman BT: TorsinA and heat shock proteins act as molecular chaperones: suppression of alpha-synuclein aggregation. *J Neurochem* 2002, 83:846-854.

[273] Klucken J, Shin Y, Masliah E, Hyman BT, McLean PJ: Hsp70 Reduces alpha-Synuclein Aggregation and Toxicity. *J Biol Chem* 2004, 279:25497-25502.

[274] Uryu K, Richter-Landsberg C, Welch W, Sun E, Goldbaum O, Norris EH, Pham CT, Yazawa I, Hilburger K, Micsenyi M, Giasson BI, Bonini NM, Lee VM, Trojanowski JQ: Convergence of heat shock protein 90 with ubiquitin in filamentous alpha-synuclein inclusions of alpha-synucleinopathies. *Am J Pathol* 2006, 168:947-961.

[275] Auluck PK, Chan HY, Trojanowski JQ, Lee VM, Bonini NM: Chaperone suppression of alpha-synuclein toxicity in a Drosophila model for Parkinson's disease. *Science* 2002, 295:865-868.

[276] Auluck PK, Bonini NM: Pharmacological prevention of Parkinson disease in Drosophila. *Nat Med* 2002, 8:1185-1186.

[277] Whitesell L, Mimnaugh EG, De Costa B, Myers CE, Neckers LM: Inhibition of heat shock protein HSP90-pp60v-src heteroprotein complex formation by benzoquinone ansamycins: essential role for stress proteins in oncogenic transformation. *Proc Natl Acad Sci U S A* 1994, 91:8324-8328.

[278] Grunblatt E, Mandel S, Jacob-Hirsch J, Zeligson S, Amariglo N, Rechavi G, Li J, Ravid R, Roggendorf W, Riederer P, Youdim MB: Gene expression profiling of parkinsonian substantia nigra pars compacta; alterations in ubiquitin-proteasome, heat shock protein, iron and oxidative stress regulated proteins, cell adhesion/cellular matrix and vesicle trafficking genes. *J Neural Transm* 2004, 111:1543-1573.

[279] Cuervo AM, Stefanis L, Fredenburg R, Lansbury PT, Sulzer D: Impaired degradation of mutant alpha-synuclein by chaperone-mediated autophagy. *Science* 2004, 305:1292-1295.

[280] Ahn TB, Jeon BS: Protective role of heat shock and heat shock protein 70 in lactacystin-induced cell death both in the rat substantia nigra and PC12 cells. *Brain Res* 2006, 1087:159-167.

[281] Ahn TB, Jeon BS: Protective role of heat shock and heat shock protein 70 in lactacystin-induced cell death both in the rat substantia nigra and PC12 cells. *Brain Res* 2006, 1087:159-167.

[282] Zhou W, Freed CR: DJ-1 up-regulates glutathione synthesis during oxidative stress and inhibits A53T alpha-synuclein toxicity. *J Biol Chem* 2005, 280:43150-43158.

[283] Batulan Z, Shinder GA, Minotti S, He BP, Doroudchi MM, Nalbantoglu J, Strong MJ, Durham HD: High threshold for induction of the stress response in motor neurons is associated with failure to activate HSF1. *J Neurosci* 2003, 23:5789-5798.

[284] Maatkamp A, Vlug A, Haasdijk E, Troost D, French PJ, Jaarsma D: Decrease of Hsp25 protein expression precedes degeneration of motoneurons in ALS-SOD1 mice. *Eur J Neurosci* 2004, 20:14-28.

[285] Wang J, Xu G, Li H, Gonzales V, Fromholt D, Karch C, Copeland NG, Jenkins NA, Borchelt DR: Somatodendritic accumulation of misfolded SOD1-L126Z in motor neurons mediates degeneration: alphaB-crystallin modulates aggregation. *Hum Mol Genet* 2005, 14:2335-2347.

[286] Patel YJ, Payne S, de Belleroche J, Latchman DS: Hsp27 and Hsp70 administered in combination have a potent protective effect against FALS-associated SOD1-mutant-induced cell death in mammalian neuronal cells. *Brain Res Mol Brain Res* 2005, 134:256-274.

[287] Bruening W, Roy J, Giasson B, Figlewicz DA, Mushynski WE, Durham HD: Up-regulation of protein chaperones preserves viability of cells expressing toxic Cu/Zn-superoxide dismutase mutants associated with amyotrophic lateral sclerosis. *J Neurochem* 1999, 72:693-699.

[288] Shinder GA, Lacourse MC, Minotti S, Durham HD: Mutant Cu/Zn-superoxide dismutase proteins have altered solubility and interact with heat shock/stress proteins in models of amyotrophic lateral sclerosis. *J Biol Chem* 2001, 276:12791-12796.

[289] Urushitani M, Kurisu J, Tateno M, Hatakeyama S, Nakayama K, Kato S, Takahashi R: CHIP promotes proteasomal degradation of familial ALS-linked mutant SOD1 by ubiquitinating Hsp/Hsc70. *J Neurochem* 2004, 90:231-244.

[290] Chai Y, Koppenhafer SL, Bonini NM, Paulson HL: Analysis of the role of heat shock protein (Hsp) molecular chaperones in polyglutamine disease. *J Neurosci* 1999, 19:10338-10347.

[291] Kazemi-Esfarjani P, Benzer S: Genetic suppression of polyglutamine toxicity in Drosophila. *Science* 2000, 287:1837-1840.

[292] Muchowski PJ, Schaffar G, Sittler A, Wanker EE, Hayer-Hartl MK, Hartl FU: Hsp70 and hsp40 chaperones can inhibit self-assembly of polyglutamine proteins into amyloid-like fibrils. *Proc Natl Acad Sci U S A* 2000, 97:7841-7846.

[293] Kitamura A, Kubota H, Pack CG, Matsumoto G, Hirayama S, Takahashi Y, Kimura H, Kinjo M, Morimoto RI, Nagata K: Cytosolic chaperonin prevents polyglutamine toxicity with altering the aggregation state. *Nat Cell Biol* 2006, 8:1163-1170.

[294] Behrends C, Langer CA, Boteva R, Bottcher UM, Stemp MJ, Schaffar G, Rao BV, Giese A, Kretzschmar H, Siegers K, Hartl FU: Chaperonin TRiC promotes the assembly of polyQ expansion proteins into nontoxic oligomers. *Mol Cell* 2006, 23:887-897.

[295] Tam S, Geller R, Spiess C, Frydman J: The chaperonin TRiC controls polyglutamine aggregation and toxicity through subunit-specific interactions. *Nat Cell Biol* 2006, 8:1155-1162.

[296] Fujimoto M, Takaki E, Hayashi T, Kitaura Y, Tanaka Y, Inouye S, Nakai A: Active HSF1 significantly suppresses polyglutamine aggregate formation in cellular and mouse models. *J Biol Chem* 2005, 280:34908-34916.

[297] Zhou H, Li SH, Li XJ: Chaperone suppression of cellular toxicity of huntingtin is independent of polyglutamine aggregation. *J Biol Chem* 2001, 276:48417-48424.

[298] Merienne K, Helmlinger D, Perkin GR, Devys D, Trottier Y: Polyglutamine expansion induces a protein-damaging stress connecting heat shock protein 70 to the JNK pathway. *J Biol Chem* 2003, 278:16957-16967.

[299] Al Ramahi I, Lam YC, Chen HK, de Gouyon B, Zhang M, Perez AM, Branco J, de Haro M, Patterson C, Zoghbi HY, Botas J: CHIP protects from the neurotoxicity of expanded and wild-type ataxin-1 and promotes their ubiquitination and degradation. *J Biol Chem* 2006, 281:26714-26724.

[300] Jana NR, Dikshit P, Goswami A, Kotliarova S, Murata S, Tanaka K, Nukina N: Co-chaperone CHIP associates with expanded polyglutamine protein and promotes their degradation by proteasomes. *J Biol Chem* 2005, 280:11635-11640.

[301] Wyttenbach A, Sauvageot O, Carmichael J, Diaz-latoud C, Arrigo AP, Rubinsztein DC: Heat shock protein 27 prevents cellular polyglutamine toxicity and suppresses the increase of reactive oxygen species caused by huntingtin. *Hum Mol Genet* 2002, 11:1137-1151.

In: The Ubiquitin Proteasome System... ISBN 978-1-60021-749-4
Eds: Mario Di Napoli and Cezary Wojcik, pp. 495-511

Chapter 20

THE HOMEOSTATIC CONTROL OF GLUCOCORTICOID RECEPTOR IN THE CENTRAL NERVOUS SYSTEM BY THE UBIQUITIN-PROTEASOME SYSTEM

Theo Rein[1,*] and Donald B. DeFranco[2,¥]

[1]Chaperone Research Group, Max Planck Institute of Psychiatry, D-80804 Muenchen, Germany.

[2]Department of Pharmacology and Neuroscience, University of Pittsburgh School of Medicine, Pittsburgh, PA 15261 USA.

ABSTRACT

Glucocorticoid hormones exert a variety of effects on the brain and impact memory, anxiety, and CNS responses to stress. The action of these hormones is mediated primarily by soluble receptors, the corticosteroid or glucocorticoid receptors, which primarily act directly in the nucleus to regulate select networks of target genes. Multiple mechanisms account for the diversity of glucocorticoid action including cell and tissue-specific response of target genes, the differential action of receptor variants as well as complex interactions between a plethora of accessory factors that directly or indirectly modulate glucocorticoid receptor activity. Furthermore, many *in vitro* and *in vivo* model studies have revealed a relationship between expression levels of glucocorticoid receptors and cellular responsiveness to glucocorticoid hormone. Various intrinsic and extrinsic factors influence the expression of the glucocorticoid receptors and thereby impact cellular

* Correspondence concerning this article should be addressed to : Dr. Theo Rein, PhD; Chaperone Research Group, Max Planck Institute of Psychiatry, Kraepelinstrasse 2-10, D-80804 Muenchen, Germany. Phone: +49-89-30622-531; Fax: +49-89-30622-610 ; E-mail: theorein@mpipsykl.mpg.de.

¥ Correspondence concerning this article should be addressed to Dr. Donald B. DeFranco, PhD; Department of Pharmacology and Neuroscience, University of Pittsburgh School of Medicine, Pittsburgh, PA 15261 USA, Phone: +1-412-624-4259; Fax: +1-412-648-1945; E-mail: dod1@pitt.edu.

output to glucocorticoid hormone exposure. This review will focus on the current state of knowledge regarding the regulation of glucocorticoid receptor protein expression and highlight a number of recent studies that illustrate the critical importance of cellular maintenance of appropriate receptor levels in complex neuronal responses.

Keywords: glucocorticoid receptor, ubiquitin, proteasome, protein degradation, stress.

Abbreviations

AS, Angelman's Syndrome; CHIP, carboxyl-terminus of hsp70 interacting protein; CNS , central nervous system; DBD, DNA-binding domain; DUB, deubiquitinating enzymes; E6-AP, E6-Associated protein; GR, glucocorticoid receptor; Hdm2, human Mdm2 ortholog; HPA, hypothalamic pituitary adrenal axus; LBD, ligand-binding domain; Mdm2, Murine double minute 2; MR, mineralocorticoid receptor; NR, nuclear receptor; RPF1, receptor potentiation factor; RSP5, reverse Spt phenotype 5; UPS, ubiquitin-proteasome system.

Introduction

The efficient and appropriate turnover of proteins is an important mechanism that maintains dynamic cellular function and operates in many subcellular compartments. In highly specialized cells such as neurons, protein degradation is particularly critical as discrete changes in cellular morphology and response to external signals ensures the coordinated development of complex neuronal circuits and plasticity of individual cells that comprise a response network. An analysis of many components of complex intracellular signaling pathways have revealed the importance of regulated protein turnover in the termination of resulting intracellular responses. Furthermore, in many cases, the termination of signaling pathways that results from the degradation of specific protein components of the pathway occurs following an acute stimulus. Neuronal responses to steroid hormones such as glucocorticoids can be elicited following both acute and chronic exposures to hormone that occurs within normal physiological contexts. In fact, the duration and magnitude of glucocorticoid exposure has an impact on neuronal activity in various regions of the brain. Since the glucocorticoid hormone signal within cells is principally mediated by the two corticosteroid receptors, the mineralocorticoid and glucocorticoid receptor (MR and GR), this signaling pathway is particularly amenable to molecular and genetic dissection of the relationship between receptor expression and/or turnover and physiological outcome. This review will focus on the current state of knowledge regarding the regulation of GR turnover and highlight a number of recent studies that illustrate the critical importance of cellular maintenance of appropriate receptor levels in complex neuronal responses.

THE GR GENE AND PROTEIN

The corticosteroid receptors are involved in a number of processes in the brain, e.g. neuronal cell birth and cell death in the dentate gyrus [1], transmitter-evoked excitability [2], dendritic morphology [3] and long-term potentiation as well as long term depression [4-6]. These effects are mediated by the actions of glucocorticoids on select gene networks in the brain [7] and are thought to be the underpinnings of the plethora of physiological effects of these hormones, including the regulation of stress, memory, and anxiety among others. The physiological importance of GR expression becomes obvious from the widespread observation of the precipitation of pathologies through chronically elevated glucocorticoids [8].

Historically, two proteins were designated as corticosteroid receptors, the type I and type II corticosteroid or glucocorticoid receptors. These receptors were distinguished primarily on differences in their ligand-binding affinity [9]. However, the type II receptor is now considered the bona fide GR. The type I GR, although having the capacity to bind natural corticosteroids, is the mineralocorticoid receptor (MR) [10] and functions primarily, but not exclusively (see below), to mediate aldosterone effects in target tissue. The designation of type I and type II GRs is rarely used now.

GR is expressed ubiquitously and MR expression is more restricted. It is found in both sodium transporting epithelia and non epithelial tissues including the brain [11]. Since circulating corticosteroids reach every organ this allows for coordinated actions of these hormones in the CNS and peripheral tissues. MR, which in the brain lacks aldosterone specificity, has high-affinity for cortisol/corticosterone sufficient for hormone binding at basal levels and is thus occupied with glucocorticoid under conditions where circulating glucocorticoid levels are not elevated [12]. In contrast, GR has an about 10 fold lower affinity and is thus occupied by hormone mainly at elevated levels of glucocorticoid reached under stressful conditions or at the circadian peaks of hormone levels in the circulation (i.e. early morning in humans). However, differences in glucocorticoid binding affinity alone are not the sole determinant of the specificity of MR versus GR action in the brain as regional differences in the expression of these receptors occurs in brain. In particular, MR expression in the brain is more restricted as compared to the ubiquitous GR. Limbic neurons express MR at high levels, which can impact neuronal activity in cases of low-level glucocorticoid exposure [13]. Furthermore, the ability of MR to form a heterodimer with GR gives rise to further fine-tuning of hormone responsiveness [14,15].

Though very similar in structure, MR and GR are likely to regulate distinct gene networks [16]. Thus the relative occupancy of the various potential corticosteroid receptor species (i.e. MR/MR homodimers, GR/GR homodimers, MR/GR heterodimers) would be expected to elicit distinct physiological responses to hormone in cells that express both receptors.

Diversity of glucocorticoid action is also generated by the presence of multiple distinct forms of the GR. The products of the single GR gene, which is located on chromosome 5q31-32 in humans [17], mediate glucocorticoid effects in nearly all target tissues. The major protein generated by the GR gene is called GRα [18]. Alternative splicing of the major transcript from this gene generates an mRNA species that when translated generates a variant receptor isoform designated GRβ [19-21]. GRβ differs from GRα it its C-terminal region and does not have the capacity to bind glucocorticoid hormone. In addition GRβ is not as widely expressed as GRα. In cells that co-express both isoforms of the receptor, GRβ has been found to act as a dominant negative protein to limit the action of GRα [19]. GRβ may be involved in glucocorticoid resistance in peripheral tissues but its role in influencing glucocorticoid responsiveness in the brain appears to be limited [22].

In addition to distinct α and β isoforms of GR protein, various amino-terminal truncated forms of the GRα protein exist in human tissue [23]. In model cell culture experiments, these receptor variants generate unique profiles of glucocorticoid-regulated gene expression [23]. The relevance of these GRα isoforms, which are generated by usage of alternative AUG start codons, to human physiology and pathophysiology particular in the brain is not well established. The remainder of this review will focus exclusively on GRα, which will hereafter be referred to simply as GR.

GR: A Member of the Nuclear Receptor Superfamily

The GR is a member of a large super-family of nuclear receptors (NRs) that number approximately 48 in the human genome [24]. The NRs share a common structural organization with a highly conserved, centrally localized zinc-finger DNA binding domain (DBD), a less-well conserved carboxyl-terminal ligand binding domain (LBD), and a divergent amino terminal domain. The nucleotide sequence specificity for DNA binding by the NRs is dictated by the precise amino acid sequence of the NR DBDs [25] although these receptors also have the capacity to activate or repress transcription of target genes in the absence of direct DNA-binding [26-29]. NRs interact with a wide variety of DNA-binding transcription factors, and coregulator proteins that do not directly associate with DNA. The association of the receptors with these various gene regulatory proteins influences the wide-ranging transcriptional effects of NRs [30,31]. A recent review [32] summarizes studies that analyzed the expression and potential impact of various GR coregulatory proteins in the brain.

Regulation of GR Protein Turnover

In most glucocorticoid responsive cells and tissues, chronic hormone treatment leads to downregulation of GR protein [33,34]. Human leukemic cells represent an exception to this property as GR levels in these cells are increased in response to prolonged hormone treatment [35]. Interestingly, glucocorticoid induced apoptosis in leukemic cells appears to require this auto-induction of GR. While both transcriptional and posttranscriptional mechanisms contribute to hormone-dependent GR downregulation [37], this review will focus on the enhanced protein degradation for discussions of hormone-dependent GR downregulation.

The degradation of GR [37,38], like MR [39] and other steroid hormone receptors, occurs mainly via the ubiquitin-proteasome system (UPS) [40]. Proteins that are targeted to the proteasome for degradation are covalently tagged with multiple ubiquitin moieties (see

Chapter 3). The addition of ubiquitin, a 76 amino acid peptide, to target proteins is regulated by the sequential action of an ubiquitin activating enzyme (E1), ubiquitin conjugating enzyme (E2) and ubiquitin ligase (E3) [40]. E3 ligases, which are the most diverse components of the ubiquitin modification pathway are primarily responsible for generating specificity in protein ubiquitylation. The human genome contains approximately 62 known E3 ligase genes (including subunits of the multiprotein complexes) while ubiquitin conjugating enzymes and ubiquitin activating enzymes are only encoded by 25 and 2 genes, respectively. Nearly all E1-E3 genes are also expressed the brain. In general, E3 ligases fall into two broad categories: the RING (really interesting new gene)-type and the HECT (homologous to the E6-associated protein C-terminus)-type E3 ligases. The RING-type E3 ligases can be further categorized into single-protein ligases such as GRAIL and multi-subunit ligases such as SCF complexes.

A number of E3 ligases have been characterized that target GR to the proteasome. For example, the E3 ligase Hdm2 is responsible for hormone-induced downregulation of GR in cultured human umbilical vein endothelial cells exposed to DNA damaging agents or hypoxia [41]. Furthermore, in human breast cancer cell lines, the enhancement of GR degradation by estrogens is due to the estrogen-dependent induction of Hdm2 expression [42]. This represents a novel form of steroid hormone cross talk that operates at the level of regulated GR degradation. It is not known whether hormone regulation of other E3 ligases or UPS components contributes to cross talk with glucocorticoid hormone signaling pathways. Furthermore, while it seems likely that different E3 ligases will be responsible for promoting UPS-driven degradation of GR (e.g. see Table 1), it is unclear whether physiological regulation of specific E3 ligases contributes to cell or tissue specific differences in steady state GR levels either under basal or hormone stimulated conditions.

Table 1. Components of the UPS that Influence GR Function

UPS Component	Notes	References
E6-Associated Protein (E6-AP)	E3 ubiquitin protein-ligase: Ubiquitin ligase activity not required for GR coactivation	[56]
Receptor potentiation factor 1/Reverse Spt phenotype 5 (RPF1/RSP5)	E3 ubiquitin protein-ligase: Ubiquitin ligase activity not required for GR coactivation	[101]
UBCH7	Ubiquitin-conjugating enzyme: Ubiquitin conjugating activity required for steroid receptor coactivation	[102]
Murine Double Minute 2 (Mdm2) and human Mdm2 ortholog (Hdm2)	E3 ubiquitin protein-ligase: Ubiquitylation of glucocorticoid receptor requires p53	[41, 42]
Carboxy terminus of Hsp70 Interacting Protein (CHIP)	E3/E4 ubiquitin protein-ligase: Promotes hormone-independent and hormone-dependent degradation of glucocorticoid receptor	[43, 44]

The carboxyl-terminal hsp70 interacting protein (CHIP) has also been identified as an E3 ligase for GR that promotes GR degradation [43]. In model studies in fibroblast cell lines, CHIP appears to be function in maintaining basal levels of GR expression operating to enhance GR degradation in the absence of a hormonal signal [43]. However, in the HT22 hippocampal cell line, CHIP overexpression selectively triggered GR degradation in response to glucocorticoid treatment [44]. In fact, in both HT22 cells and primary rat embryonic hippocampal neuron cultures, chronic glucocorticoid treatment did not lead to GR downregulation [45]. CHIP overexpression restored hormone-dependent downregulation of GR in the HT22 cells [44]. These model *in vitro* experiments suggest that the apparent lack of GR downregulation observed in rat fetal neurons *in vivo* [46] and in primary culture could be due to reduced expression or activity of specific E3 ligases (e.g. CHIP) that act on the receptor in neurons. As mentioned above, many E3 ligases are expressed in the brain and may be subjected to unique cell-specific and developmental stage-specific profiles of expression. It remains to be tested whether such diversity in E3 ligase expression in the brain impacts the level or GR expression or impacts glucocorticoid responsiveness.

It should be noted that, analogous to most other posttranslational modifications of proteins, ubiquitylation is a reversible process. There are different classes of ubiquitin proteases (also called deubiquitinating enzymes, DUBs, see Chapter 4). A recent survey found approximately 79 putative DUBs in humans that are functional [47]. Although DUBs may be physiologically as important as the ubiquitylation machinery, they are much less well understood. At least some of the DUBs act in the brain with an impact on important brain activities, e.g., synapse function [48, 49]. We are not aware of a study explicitly linking DUBs to regulation of GR. Nevertheless, it appears that the physiological function of DUBs is just beginning to be elucidated. DUBs may provide a proofreading mechanism that enhances the fidelity of the UPS.

Many components of the UPS and accessory factors are differentially expressed in developing mouse hippocampus [50]. In addition some proteasome subunit mRNAs are also differentially regulated during development in the rat midbrain [51]. Alternative splicing of specific proteasome subunits has been found to generate distinct proteasome forms during development [52]. Thus, the UPS machinery may exhibit some plasticity during development. These changes in composition and function of the UPS may therefore affect the expression levels and turnover of specific substrates throughout development. Differential expression or function of specific UPS components is well established in the brain. For example, mutations in the E6-AP ubiquitin-protein ligase are associated with Angelman's syndrome (AS), a human disease that is characterized by severe mental retardation and motor dysfunction [53,54; and Chapter 38]. Furthermore, increased abundance of one E6-AP substrate, the p53 oncoprotein, results from an E6-AP deficiency and could underlie deficits in contextual learning and long term potentiation that occur in mouse models of AS [55]. While E6-AP has been found to interact with the progesterone receptor, its impact on nuclear receptor function, including GR, appears to be more related to its transcriptional coactivator activity [56].

RELATIONSHIP BETWEEN GR PTOTEIN LEVELS AND HORMONE RESPONSIVENESS

In principle, the crucial task of limiting responsiveness to glucocorticoid can be achieved at different levels and by different modes of action. Access of hormone to cells or organs may be limited by the action of steroid transporters in the membrane like the multidrug resistance p-glycoprotein [57]. GR function may be reduced by the expression of GR-inhibitory proteins like FKBP51 [58]. Finally, glucocorticoid action may be limited by regulation of GR protein levels. Many *in vitro* and *in vivo* model studies have examined the relationship between GR expression and cellular responsiveness to glucocorticoid hormone. In the first model study to examine this relationship, Yamamoto and co-workers engineered hepatoma cell lines by transfection with cloned GR to create sublines with varying levels of GR protein expression [59]. These studies though informative, have been superseded by recent analyses in transgenic mice, which have exquisitely illustrated the importance of GR expression levels to tissue-specific physiological effects of glucocorticoids (see Figure 1 for summary). GR was overexpressed to varying extents (i.e. approximately 20-60%) by Schütz and co-workers in various tissues of transgenic mice. Overexpression of GR in these animals led to various alterations in stress responses, glucocorticoid-induced apoptosis in thymocytes, reduced inflammatory responses and an increase resistance to endotoxic shock [60]. Furthermore, transgenic mouse models with multi-tissue increases or decreases in GR expression exhibit many alterations in behavioral responses to selective stresses [61].

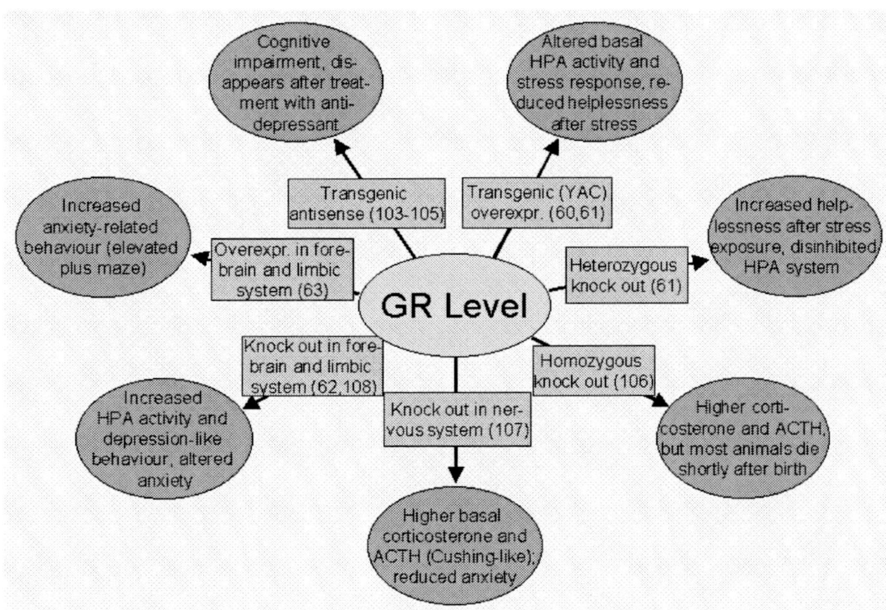

Figure 1. Summary of CNS Effects in Transgenic Mice Engineered with Alterations in GR Expression Levels. Genetic manipulation of GR gene and citation are indicated in rectangle linked to GR expression. Summary of consequences in the CNS of individual genetic manipulation of GR expression is shown in connected ovals.

The impact of altered GR expression in select brain regions has also been analyzed with transgenic mice models. Specifically, transgenic mice with reduced postnatal expression of GR in forebrain exhibit an increase in depression-like behavior as assayed by forced swim and tail suspension tests [62]. Transgenic mice with increased GR expression in forebrain showed increased anxiety-like behavior as assessed by the elevated plus maze test [63]. Forebrain overexpression and underexpression of GR also alters behavioral and biochemical responses to anti-depressants [62,63]. If these results are relevant to human behavior, genetic or environmental influences on GR expression, particularly in the brain, could have important consequences for individual responses to stress or the efficacy of anti-depressive drugs.

There is also significant evidence establishing that modulation of GR protein levels represents an important mode of action for some drugs that alter brain function. For example, since it has been recognized that the function of GR is crucial in the etiology and cure of major depression [64], a potential effect of antidepressants on GR has been investigated in numerous studies. Chronic treatment with the tricyclic antidepressant amitriptyline or with moclobemide indeed altered GR levels in rats in a dose-, time-, and brain region-specific manner [65,66]. Similarly, administration of amitriptyline or desipramine for 14 days in rats led to a significant increase of GR in hippocampus, but not in parietal cortex [67]. In contrast, exposure for only 9 days with the antidepressants fluoxetine or venlafaxine decreased GR levels in the CA3 subfield of the hippocampus [68]. Interestingly, the effect of chronic fluoxetine treatment in rats on GR levels was dependent on the age of the animals [69]. In cellular model systems, the mood stabilizer valproate pronouncedly decreased GR protein levels [70]. Similarly, a modulatory effect on GR expression by diverse antidepressants was shown in numerous other cellular model systems including primary neurons [71-74]. Since the function of the HPA axis, and thus also GR, is considered important in a number of other brain-related pathologies [8,75-77] effects of other brain-targeting drugs on GR expression were also explored. For example, the psychostimulants methamphetamine and cocaine were found to alter GR expression levels [78-82].

The plethora of findings on drug-modulated GR expression levels contrasts with the lack of mechanistic understanding of these effects in most cases. Some of the effects apparently are exerted at the transcriptional level, including GR promoter choice [83]. It is intriguing, however, that treatment with the antidepressants imipramine or citalopram affects the expression of a number of UPS genes [84]. Thus, it is tempting to speculate that the effects of antidepressant treatment on GR levels are, at least in part, due to their action on UPS genes. Further support for this hypothesis comes from the discovery that the RING finger protein kf-1 is induced in rat frontal cortex after chronic treatment with the selective serotonin reuptake inhibitor antidepressant sertraline or the tricyclic antidepressant imipramine, as well as after acute and chronic electroconvulsive treatment [85,86], a method widely used for therapy of depression. Moreover, kf-1 is induced in rat brain also by repeated transcranial stimulation, a non-invasive method used for treatment of depression [87].

Finally, a study on 24 control cases and 22 patients with schizophrenia, which is one of the brain diseases often associated with hypercortisolemia and altered GR expression [88,89], found impressive changes in proteasome and ubiquitin genes [90]. In this study, laser capture dissection was used to isolate dentate granule neurons from postmortem brain tissues for microarray expression profiling. Nine proteasome and six ubiquitin genes were downregulated in patients, one ubiquitin gene was upregulated. There is a debate about whether or not UPS dysfunction is causally related to Alzheimer's disease pathogenesis [91,92]. In any case, since the HPA axis abnormalities in Alzheimer's disease have therapeutic implications [93], it will be worthwhile exploring a possible relation between HPA axis activity, GR levels and UPS proteins, even more so since GR mutation analyses so far have not discovered alterations in Alzheimer's disease [94]. Similarly, in multiple sclerosis HPA axis activity has been found associated with disease progression [95,96] and changes in the UPS has also been detected with this disease [97], but a specific connection of GR and UPS has not yet been investigated in multiple sclerosis.

Conclusions

Hormone-dependent downregulation of GR is an important feedback mechanism that limits potential negative effects of chronic glucocorticoid stimulation as hormone responsiveness should be limited as GR levels decline. Thus, cells or tissues that are unable to downregulate GR could be particularly vulnerable to prolonged glucocorticoid exposure. This could be particularly problematic for patients undergoing chronic or high dose glucocorticoid therapy. For example, glucocorticoid therapy is widely used to promote lung maturation and cardiovascular function in premature infants [98]. However potential risks to neuronal development and metabolism may result from perinatal or neonatal glucocorticoid treatment [99]. The mechanisms responsible for these potentially adverse effects of glucocorticoids are unknown. However, since the absolute levels of GR can have an impact on complex behavioral responses (see above), the inability of fetal or neonatal neurons to downregulate GR levels [44,45] could be one result of non-physiologic, prolonged hormone exposure that alters neurodevelopment and generates subsequent behavioral or cognitive deficiencies. Selective disruptions of the GR downregulation pathway will be required in model *in vitro* and *in vivo* systems to reveal the importance of this feedback mechanism to physiological effects of prolonged glucocorticoid exposure. In human leukemic cells, glucocorticoid induced apoptosis depends not on the absolute steady state levels of GR, but on the auto-induction of receptor levels following hormone treatment [100]. Thus, the adaptation of GR protein expression to chronic glucocorticoid treatment may be more relevant to biological response to hormone than the absolute levels of the receptor expressed within that given cell type. Importantly, the dramatic results establishing behavioral consequences of altered GR expression provides compelling rationale for more detailed mechanistic analysis of the regulation of GR expression and turnover in the CNS.

References

[1] Hassan AH, von Rosenstiel P, Patchev VK, Holsboer F, Almeida OF: Exacerbation of apoptosis in the dentate gyrus of the aged rat by dexamethasone and the protective role of corticosterone. *Exp Neurol* 1996, 140:43-52.

[2] Joels M, de Kloet ER: Effects of glucocorticoids and norepinephrine on the excitability in the hippocampus. *Science* 1989, 245:1502-1505

[3] Woolley CS, Gould E, McEwen BS: Exposure to excess glucocorticoids alters dendritic morphology of adult hippocampal pyramidal neurons. *Brain Res* 1990, 531:225-231

[4] Diamond DM, Bennett MC, Fleshner M, Rose GM: Inverted-U relationship between the level of peripheral corticosterone and the magnitude of hippocampal primed burst potentiation. *Hippocampus* 1992, 2:421-430

[5] Kim JJ, Diamond DM: The stressed hippocampus, synaptic plasticity and lost memories. *Nat Rev Neurosci* 2002, 3:453-462

[6] Xu L, Anwyl R, Rowan MJ: Spatial exploration induces a persistent reversal of long-term potentiation in rat hippocampus. *Nature* 1998, 394:891-894

[7] Datson NA, van der Perk J, de Kloet ER, Vreugdenhil E: Identification of corticosteroid-responsive genes in rat hippocampus using serial analysis of gene expression. *Eur J Neurosci* 2001, 14:675-689

[8] de Kloet ER, Joels M, Holsboer F: Stress and the brain: from adaptation to disease. *Nat Rev Neurosci* 2005, 6:463-475

[9] Funder JW, Feldman D, Edelman IS: The roles of plasma binding and receptor specificity in the mineralocorticoid action of aldosterone. *Endocrinology* 1973, 92:994-1004.

[10] Arriza JL, Weinberger C, Cerelli G, Glaser TM, Handelin BL, Housman DE, Evans RM: Cloning of human mineralocorticoid receptor complementary DNA: structural and functional kinship with the glucocorticoid receptor. *Science* 1987, 237:268-275.

[11] Funder JW: Mineralocorticoid receptors: distribution and activation. *Heart Fail Rev* 2005, 10:15-22Arriza JL, Simerly RB, Swanson LW, Evans RM: The neuronal mineralocorticoid receptor as a mediator of glucocorticoid response. *Neuron* 1988, 1:887-900

[12] Arriza JL, Simerly RB, Swanson LW, Evans RM: The neuronal mineralocorticoid receptor as a mediator of glucocorticoid response. *Neuron* 1988, 1:887-900

[13] Karst H, Berger S, Turiault M, Tronche F, Schütz G, Joels M: Mineralocorticoid receptors are indispensable for nongenomic modulation of hippocampal glutamate transmission by corticosterone. *Proc Natl Acad Sci U S A* 2005, 102:19204-19207

[14] Trapp T, Rupprecht R, Castren M, Reul JM, Holsboer F: Heterodimerization between mineralocorticoid and glucocorticoid receptor: a new principle of glucocorticoid action in the CNS. *Neuron* 1994, 13:1457-1462

[15] Nishi M, Ogawa H, Ito T, Matsuda KI, Kawata M: Dynamic changes in subcellular localization of mineralocorticoid receptor in living cells: in comparison with glucocorticoid receptor using dual-color labeling with green fluorescent protein spectral variants. *Mol Endocrinol* 2001, 15:1077-1092

[16] Pearce D: A mechanistic basis for distinct mineralocorticoid and glucocorticoid receptor transcriptional specificities. *Steroids* 1994, 59:153-159

[17] Theriault A, Boyd E, Harrap SB, Hollenberg SM, Connor SM: Regional chromosomal assignment of the human glucocorticoid receptor gene to 5q31. *Hum Genet* 1989, 83: 289–291

[18] Hollenberg SM, Weinberger C, Ong ES, Cerelli G, Oro A, Lebo R, Thompson EB, Rosenfeld MG, Evans RM: Primary structure and expression of a functional human glucocorticoid receptor cDNA. *Nature* 1985, 318: 635-641

[19] Bamberger CM, Bamberger AM, de Castro M, Chrousos GP: Glucocorticoid receptor beta, a potential endogenous inhibitor of glucocorticoid action in humans. *J Clin Invest* 1995, 95: 2435–2441

[20] Oakley RH, Sar M, Cidlowski JA: The human glucocorticoid receptor beta isoform. Expression, biochemical properties, and putative function *J Biol Chem* 1996, 271: 9550–9559

[21] Zhou J, Cidlowski JA: The human glucocorticoid receptor: one gene, multiple proteins and diverse responses. *Steroids* 2005, 70: 407-417

[22] DeRijk RH, Schaaf M, Stam FJ, de Jong IE, Swaab DF, Ravid R, Vreugdenhil E, Cidlowski JA, de Kloet ER, Lucassen PJ: Very low levels of the glucocorticoid receptor beta isoform in the human hippocampus as shown by Taqman RT-PCR and immunocytochemistry. *Brain Res Mol Brain Res* 2003, 116:17-26

[23] Lu NZ, Cidlowski JA: Translational regulatory mechanisms generate N-terminal glucocorticoid receptor isoforms with unique transcriptional target genes. *Mol Cell* 2005, 18: 331-342

[24] Mangelsdorf DJ, Thummel C, Beato M, Herrlich P, Schütz G, Umesono K, Blumberg B, Kastner P, Mark M, Chambon P, Evans RM: The nuclear receptor superfamily: the second decade. *Cell* 1995, 83: 835-839

[25] Claessens F, Gewirth DT: DNA recognition by nuclear receptors. *Essays Biochem* 2004, 40: 59-72

[26] Jonat C, Rahmsdorf HJ, Park KK, Cato AC, Gebel S, Ponta H, Herrlich P: Antitumor promotion and antiinflammation: down-modulation of AP-1 (Fos/Jun) activity by glucocorticoid hormone. *Cell* 1990, 62:1189-1204

[27] Schüle R, Rangarajan P, Kliewer S, Ransone LJ, Bolado J, Yang N, Verma IM, Evans RM: Functional antagonism between oncoprotein c-Jun and the glucocorticoid receptor. *Cell* 1990, 62: 1217-1226

[28] Yang-Yen HF, Chambard JC, Sun YL, Smeal T, Schmidt TJ, Drouin J, Karin M: Transcriptional interference between c-Jun and the glucocorticoid receptor: mutual inhibition of DNA binding due to direct protein-protein interaction. *Cell* 1990, 62: 1205-1215

[29] Diamond MI, Miner JN, Yoshinaga SK, Yamamoto KR: Transcription factor interactions: selectors of positive or negative regulation from a single DNA element. *Science* 1990, 249: 1266-1272

[30] Glass CK, Rosenfeld MG: The coregulator exchange in transcriptional functions of nuclear receptors. *Genes Dev* 2000, 4: 121-141

[31] McKenna NJ, O'Malley BW: Combinatorial control of gene expression by nuclear receptors and coregulators. *Cell* 2002, 108: 465-474

[32] Meijer OC, van der Laan S, Lachize S, Steenbergen PJ, de Kloet ER: Steroid receptor coregulator diversity: What can it mean for the stressed brain? *Neuroscience* 2006, 138:891-899

[33] Bellingham DL, Sar M, Cidlowski JA: Ligand-dependent down-regulation of stably transfected human glucocorticoid receptors is associated with the loss of functional glucocorticoid responsiveness. *Mol Endocrinol* 1992, 6: 2090-2102
[34] Silva CM, Powell-Oliver FE, Jewell CM, Sar M, Allgood VE, Cidlowski JA: Regulation of the human glucocorticoid receptor by long-term and chronic treatment with glucocorticoid. *Steroids* 1994, 59: 436-442
[35] Ramdas J, Liu W, Harmon JM: Glucocorticoid-induced cell death requires autoinduction of glucocorticoid receptor expression in human leukemic T cells. *Cancer Res* 1999, 59: 1378-1385
[36] Oakley RH, Cidlowski JA: Homologous down regulation of the glucocorticoid receptor: the molecular machinery. *Crit Rev Eukaryot Gene Expr* 1993, 3: 63-88
[37] Wallace AD, Cidlowski JA: Proteasome-mediated glucocorticoid receptor degradation restricts transcriptional signaling by glucocorticoids. *J Biol Chem* 2001, 276: 42714-42721
[38] DeRoo BJ, Rentsch C, Sampath S, Young J, DeFranco DB, Archer TK: Proteasomal inhibition enhances glucocorticoid receptor transactivation and alters its subnuclear trafficking. *Mol Cell Biol* 2002, 22: 4113-4123
[39] Yokota K, Shibata H, Kobayashi S, Suda N, Murai A, Kurihara I, Saito I, Saruta T: Proteasome-mediated mineralocorticoid receptor degradation attenuates transcriptional response to aldosterone. *Endocr Res* 2004, 30:611-616
[40] Ciechanover A, Orian A, Schwartz AL: Ubiquitin-mediated proteolysis: biological regulation via destruction. *Bioessays* 2000, 22: 442-451
[41] Sengupta S, Wasylyk B: Ligand-dependent interaction of the glucocorticoid receptor with p53 enhances their degradation by Hdm2. *Genes Dev* 2001, 15: 2367-2380
[42] Kinyamu HK, Archer TK: Estrogen receptor-dependent proteasomal degradation of the glucocorticoid receptor is coupled to an increase in mdm2 protein expression. *Mol Cell Biol* 2003, 23: 5867-5881
[43] Connell P, Ballinger CA, Jiang J, Wu Y, Thompson LJ, Hohfeld J, Patterson C: The co-chaperone CHIP regulates protein triage decisions mediated by heat-shock proteins. *Nat Cell Biol* 2001, 3: 93-96
[44] Wang X, DeFranco DB: Alternative effects of the ubiquitin-proteasome pathway on glucocorticoid receptor down-regulation and transactivation are mediated by CHIP, an E3 ligase. *Mol Endocrinol* 2005, 19: 1474-1482
[45] Wang X, Pongrac J, DeFranco DB: Glucocorticoid receptors in hippocampal neurons that do not engage proteasomes escape from hormone-dependent down-regulation but maintain transactivation activity. *Mol Endocrinol* 2002, 16: 1987-1998
[46] Ghosh B, Wood CR, Held GA, Abbott BD, Lau C: Glucocorticoid receptor regulation in the rat embryo: a potential site for developmental toxicity? *Toxicol Appl Pharmacol* 2000, 164: 221-229
[47] Nijman SM, Luna-Vargas MP, Velds A, Brummelkamp TR, Dirac AM, Sixma TK, Bernards R: A genomic and functional inventory of deubiquitinating enzymes. *Cell* 2005, 123:773-786
[48] Wilson SM, Bhattacharyya B, Rachel RA, Coppola V, Tessarollo L, Householder DB, Fletcher CF, Miller RJ, Copeland NG, Jenkins NA: Synaptic defects in ataxia mice

result from a mutation in Usp14, encoding a ubiquitin-specific protease. *Nat Genet* 2002, 32:420-425

[49] Anderson C, Crimmins S, Wilson JA, Korbel GA, Ploegh HL, Wilson SM: Loss of Usp14 results in reduced levels of ubiquitin in ataxia mice. *J Neurochem* 2005, 95:724-731

[50] Mody M, Cao Y, Cui Z, Tay KY, Shyong A, Shimizu E, Pham K, Schultz P, Welsh D, Tsien JZ: Genome-wide gene expression profiles of the developing mouse hippocampus. *Proc Natl Acad Sci USA* 2001, 98:8862-8867

[51] El-Khodor BF, Kholodilov NG, Yarygina O, Burke RE: The expression of mRNAs for the proteasome complex is developmentally regulated in the rat mesencephalon. *Brain Res Dev Brain Res* 2001, 129:47-56

[52] Kawahara H, Kasahara M, Nishiyama A, Ohsumi K, Goto T, Kishimoto T, Saeki Y, Yokosawa H, Shimbara N, Murata S, Chiba T, Suzuki K, Tanaka K: Developmentally regulated, alternative splicing of the Rpn10 gene generates multiple forms of 26S proteasomes. *EMBO J* 2000, 19:4144-4153

[53] Kishino T, Lalande M, Wagstaff J: UBE3A/E6-AP mutations cause Angelman syndrome. *Nat Genet* 1997, 15:70-73

[54] Matsuura T, Sutcliffe JS, Fang P, Galjaard RJ, Jiang YH, Benton CS, Rommens JM, Beaudet AL: De novo truncating mutations in E6-AP ubiquitin-protein ligase gene (UBE3A) in Angelman syndrome. *Nat Genet* 1997, 15:74-77

[55] Jiang YH, Armstrong D, Albrecht U, Atkins CM, Noebels JL, Eichele G, Sweatt JD, Beaudet AL: Mutation of the Angelman ubiquitin ligase in mice causes increased cytoplasmic p53 and deficits of contextual learning and long-term potentiation. *Neuron* 1998, 21:799-811

[56] Nawaz Z, Lonard DM, Smith CL, Lev-Lehman E, Tsai SY, Tsai MJ, O'Malley BW: The Angelman syndrome-associated protein, E6-AP, is a coactivator for the nuclear hormone receptor superfamily. *Mol Cell Biol* 1999, 19:1182-1199

[57] Uhr M, Holsboer F, Müller MB: Penetration of endogenous steroid hormones corticosterone, cortisol, aldosterone and progesterone into the brain is enhanced in mice deficient for both mdr1a and mdr1b P-glycoproteins. *J Neuroendocrinol* 2002, 14:753-759

[58] Wochnik GM, Rüegg J, Abel GA, Schmidt U, Holsboer F, Rein T: FK506-binding proteins 51 and 52 differentially regulate dynein interaction and nuclear translocation of the glucocorticoid receptor in mammalian cells. *J Biol Chem* 2005, 280:4609-4016

[59] Vanderbilt JN, Miesfeld R, Maler BA, Yamamoto KR: Intracellular receptor concentration limits glucocorticoid-dependent enhancer activity. *Mol Endocrinol* 1987, 1: 68-74

[60] Reichardt HM, Umland T, Bauer A, Kretz O, Schütz G: Mice with an increased glucocorticoid receptor gene dosage show enhanced resistance to stress and endotoxic shock. *Mol Cell Biol* 2000, 20: 9009-901

[61] Ridder S, Chourbaji S, Hellweg R, Urani A, Zacher C, Schmid W, Zink M, Hortnagl H, Flor H, Henn FA, Schütz G, Gass P: Mice with genetically altered glucocorticoid receptor expression show altered sensitivity for stress-induced depressive reactions. *J Neurosci* 2005, 25: 6243-6250

[62] Boyle MP, Brewer JA, Funatsu M, Wozniak DF, Tsien JZ, Izumi Y, Muglia LJ: Acquired deficit of forebrain glucocorticoid receptor produces depression-like changes in adrenal axis regulation and behavior. *Proc Natl Acad Sci U S A* 2005, 102: 473-478

[63] Wei Q, Lu XY, Liu L, Schafer G, Shieh KR, Burke S, Robinson TE, Watson SJ, Seasholtz AF, Akil H: Glucocorticoid receptor overexpression in forebrain: a mouse model of increased emotional lability. *Proc Natl Acad Sci U S A* 2004, 101: 11851-11856

[64] Holsboer F: The corticosteroid receptor hypothesis of depression. *Neuropsychopharmacology* 2000, 25:477-501

[65] Reul JM, Stec I, Söder M, Holsboer F: Chronic treatment of rats with the antidepressant amitriptyline attenuates the activity of the hypothalamic-pituitary-adrenocortical system. *Endocrinology* 1993; 133:312-320

[66] Reul JM, Labeur MS, Grigoriadis DE, De Souza EB, Holsboer F: Hypothalamic-pituitary-adrenocortical axis changes in the rat after long-term treatment with the reversible monoamine oxidase-A inhibitor moclobemide. *Neuroendocrinology* 1994, 60:509-519

[67] Seckl JR, Fink G: Antidepressants increase glucocorticoid and mineralocorticoid receptor mRNA expression in rat hippocampus in vivo. *Neuroendocrinology* 1992, 55:621-626

[68] Yau JL, Noble J, Hibberd C, Seckl JR: Short-term administration of fluoxetine and venlafaxine decreases corticosteroid receptor mRNA expression in the rat hippocampus *Neurosci Lett* 2001, 306:161-164

[69] Yau JL, Hibberd C, Noble J, Seckl JR: The effect of chronic fluoxetine treatment on brain corticosteroid receptor mRNA expression and spatial memory in young and aged rats. *Brain Res Mol Brain Res* 2002, 106:117-123

[70] Basta-Kaim A, Budziszewska B, Jaworska-Feil L, Tetich M, Kubera M, Leskiewicz M, Lason W: Mood stabilizers inhibit glucocorticoid receptor function in LMCAT cells. *Eur J Pharmacol* 2004, 495:103-110

[71] Lai M, McCormick JA, Chapman KE, Kelly PA, Seckl JR, Yau JL: Differential regulation of corticosteroid receptors by monoamine neurotransmitters and antidepressant drugs in primary hippocampal culture. *Neuroscience* 2003, 118:975-984

[72] Pariante CM, Hye A, Williamson R, Makoff A, Lovestone S, Kerwin RW: The antidepressant clomipramine regulates cortisol intracellular concentrations and glucocorticoid receptor expression in fibroblasts and rat primary neurones. *Neuropsychopharmacology* 2003, 28:1553-1561

[73] Budziszewska B, Siwanowicz J, Przegalinski E: The effect of chronic treatment with antidepressant drugs on the corticosteroid receptor levels in the rat hippocampus. *Pol J Pharmacol* 1994, 46:147-152

[74] Okugawa G, Omori K, Suzukawa J, Fujiseki Y, Kinoshita T, Inagaki C: Long-term treatment with antidepressants increases glucocorticoid receptor binding and gene expression in cultured rat hippocampal neurones. *J Neuroendocrinol* 1999, 11:887-895.

[75] de Kloet ER, Joels M, Holsboer F. Stress and the brain: from adaptation to disease. *Nat Rev Neurosci* 2005, 6:463-475

[76] Knable MB, Barci BM, Webster MJ, Meador-Woodruff J, Torrey EF: Stanley Neuropathology Consortium. Molecular abnormalities of the hippocampus in severe psychiatric illness: postmortem findings from the Stanley Neuropathology Consortium. *Mol Psychiatry* 2004, 9:609-620

[77] Perlman WR, Webster MJ, Kleinman JE, Weickert CS: Reduced glucocorticoid and estrogen receptor alpha messenger ribonucleic acid levels in the amygdala of patients with major mental illness. *Biol Psychiatry* 2004, 56:844-852

[78] Webster MJ, Knable MB, O'Grady J, Orthmann J, Weickert CS: Regional specificity of brain glucocorticoid receptor mRNA alterations in subjects with schizophrenia and mood disorders. *Mol Psychiatry* 2002, 7:985-994

[79] Malaplate-Armand C, Ferrari L, Masson C, Visvikis-Siest S, Lambert H, Batt AM: Down-regulation of astroglial CYP2C, glucocorticoid receptor and constitutive androstane receptor genes in response to cocaine in human U373 MG astrocytoma cells. *Toxicol Lett* 2005, 59:203-211

[80] Kabbaj M, Yoshida S, Numachi Y, Matsuoka H, Devine DP, Sato M: Methamphetamine differentially regulates hippocampal glucocorticoid and mineralocorticoid receptor mRNAs in Fischer and Lewis rats. *Brain Res Mol Brain Res* 2003, 117:8-14

[81] Budziszewska B, Leskiewicz M, Jaworska-Feil L, Lason W: Repeated cocaine administration down-regulates glucocorticoid receptors in the rat brain cortex and hippocampus. *Pol J Pharmacol* 1996, 48:575-581

[82] Mantsch JR, Yuferov V, Mathieu-Kia AM, Ho A, Kreek MJ: Neuroendocrine alterations in a high-dose, extended-access rat self-administration model of escalating cocaine use. *Psychoneuroendocrinology* 2003, 28:836-862

[83] Yau JL, Noble J, Chapman KE, Seckl JR: Differential regulation of variant glucocorticoid receptor mRNAs in the rat hippocampus by the antidepressant fluoxetine. *Brain Res Mol Brain Res* 2004, 129:189-192

[84] Palotas M, Palotas A, Puskas LG, Kitajka K, Pakaski M, Janka Z, Molnar J, Penke B, Kalman J: Gene expression profile analysis of the rat cortex following treatment with imipramine and citalopram. *Int J Neuropsychopharmacol* 2004, 7:401-413

[85] Nishioka G, Yamada M, Kudo K, Takahashi K, Kiuchi Y, Higuchi T, Momose K, Kamijima K, Yamada M: Induction of kf-1 after repeated electroconvulsive treatment and chronic antidepressant treatment in rat frontal cortex and hippocampus. *J Neural Transm* 2003, 110:277-285

[86] Yamada M, Yamada M, Yamazaki S, Takahashi K, Nishioka G, Kudo K, Ozawa H, Yamada S, Kiuchi Y, Kamijima K, Higuchi T, Momose K: Identification of a novel gene with RING-H2 finger motif induced after chronic antidepressant treatment in rat brain. *Biochem Biophys Res Commun.* 2000, 278:150-157

[87] Kudo K, Yamada M, Takahashi K, Nishioka G, Tanaka S, Hashiguchi T, Fukuzako H, Takigawa M, Higuchi T, Momose K, Kamijima K, Yamada M: Repetitive transcranial magnetic stimulation induces kf-1 expression in the rat brain. *Life Sci* 2005, 76:2421-2429

[88] Muck-Seler D, Pivac N, Mustapic M, Crncevic Z, Jakovljevic M, Sagud M: Platelet serotonin and plasma prolactin and cortisol in healthy, depressed and schizophrenic women. *Psychiatry Res* 2004, 127:217-226

[89] Knable MB, Barci BM, Webster MJ, Meador-Woodruff J, Torrey EF: Stanley Neuropathology Consortium. Molecular abnormalities of the hippocampus in severe psychiatric illness: postmortem findings from the Stanley Neuropathology Consortium. *Mol Psychiatry* 2004, 9:609-620

[90] Altar CA, Jurata LW, Charles V, Lemire A, Liu P, Bukhman Y, Young TA, Bullard J, Yokoe H, Webster MJ, Knable MB, Brockman JA: Deficient hippocampal neuron expression of proteasome, ubiquitin, and mitochondrial genes in multiple schizophrenia cohorts. *Biol Psychiatry* 2005, 58:85-96

[91] Layfield R, Lowe J, Bedford L: The ubiquitin-proteasome system and neurodegenerative disorders. *Essays Biochem.* 2005, 41: 157-171

[92] Schmitt HP: Protein ubiquitination, degradation and the proteasome in neuro-degenerative disorders: No clear evidence for a significant pathogenetic role of proteasome failure in Alzheimer disease and related disorders. *Med Hypotheses*. 2006, 67:311-317

[93] Pomara N, Greenberg WM, Branford MD, Doraiswamy PM: Therapeutic implications of HPA axis abnormalities in Alzheimer's disease: review and update. *Psychopharmacol Bull.* 2003, 37:120-34

[94] Tsolakidou AF, Coulocheri SA, Trikkas G, Moutsatsou P: Gene analysis of the glucocorticoid receptor alpha in Alzheimer's disease. *Clin Chim Acta* 2004, 349:167-72

[95] Gold SM, Raji A, Huitinga I, Wiedemann K, Schulz KH, Heesen C: Hypothalamo-pituitary-adrenal axis activity predicts disease progression in multiple sclerosis. *J Neuroimmunol* 2005 165: 186-91

[96] Huitinga I, Erkut ZA, van Beurden D, Swaab DF: The hypothalamo-pituitary-adrenal axis in multiple sclerosis. *Ann N Y Acad Sci* 2003, 992:118-128

[97] Giordana MT, Richiardi P, Trevisan E, Boghi A, Palmucci L: Abnormal ubiquitination of axons in normally myelinated white matter in multiple sclerosis brain. *Neuropathol Appl Neurobiol* 2002, 28: 35-41

[98] Liggins GC, Howie RN: A controlled trial of antepartum glucocorticoid treatment for prevention of the respiratory distress syndrome in premature infants. *Pediatrics* 1972, 50: 515-525

[99] Yeh TF, Lin YJ, Lin Hc, Huang CC, Hsieh WS, Lin CH, Tsai SH: Outcomes at school age after postnatal dexamethasone therapy for lung disease of prematurity. *N Engl J Med* 2004, 350: 1304-1313

[100] Riml S, Schmidt S, Ausserlechner MJ, Geley S, Kofler R: Glucocorticoid receptor heterozygosity combined with lack of receptor auto-induction causes glucocorticoid resistance in Jurkat acute lymphoblastic leukemia cells. *Cell Death Differ* 2004, 11: S65-S72

[101] Imhof MO, McDonnell DP: Yeast RSP5 and its human homolog hRPF1 potentiate hormone-dependent activation of transcription by human progesterone and glucocorticoid receptors. *Mol Cell Biol* 1996, 16:2594-2605

[102] Verma S, Ismail A, Gao X, Fu G, Li X, O'Malley BW, Nawaz Z: The ubiquitin-conjugating enzyme UBCH7 acts as a coactivator for steroid hormone receptors. *Mol Cell Biol* 2004, 24:8716-8726

[103] Pepin MC, Pothier F, Barden N: Impaired type II glucocorticoid-receptor function in mice bearing antisense RNA transgene. *Nature* 1992, 355: 725-728

[104] Pepin MC, Pothier F, Barden N: Antidepressant drug action in a transgenic mouse model of the endocrine changes seen in depression. *Mol Pharmacol.* 1992, 42: 991-995

[105] Montkowski A, Barden N, Wotjak C, Stec I, Ganster J, Meaney M, Engelmann M, Reul JM, Landgraf R, Holsboer F: Long-term antidepressant treatment reduces behavioural deficits in transgenic mice with impaired glucocorticoid receptor function. *J Neuroendocrinol* 1995 7: 841-845

[106] Cole TJ, Blendy JA, Monaghan AP, Krieglstein K, Schmid W, Aguzzi A, Fantuzzi G, Hummler E, Unsicker K, Schütz G: Targeted disruption of the glucocorticoid receptor gene blocks adrenergic chromaffin cell development and severely retards lung maturation. *Genes Dev.* 1995, 9: 1608-1621

[107] Tronche F, Kellendonk C, Kretz O, Gass P, Anlag K, Orban PC, Bock R, Klein R, Schütz G: Disruption of the glucocorticoid receptor gene in the nervous system results in reduced anxiety. *Nat Genet* 1999, 23: 99-103

[108] Boyle MP, Kolber BJ, Vogt SK, Wozniak DF, Muglia LJ: Forebrain glucocorticoid receptors modulate anxiety-associated locomotor activation and adrenal responsiveness. *J Neurosci.* 2006, 26: 1971-1978.

In: The Ubiquitin Proteasome System… ISBN 978-1-60021-749-4
Eds: Mario Di Napoli and Cezary Wojcik, pp. 513-536

Chapter 21

Role of the Ubiquitin- and Proteasome System in Neuronal Apoptosis

Cezary Wójcik*

Department of Anatomy and Cell Biology, Indiana University School of Medicine, Evansville, Indiana, USA.

Abstract

The apoptotic pathway and the ubiquitin- and proteasome system (UPS) are mutually dependent and interconnected. Active caspases are able to cleave proteasomal subunits leading to a decrease in proteasome activity, while proteasomes are able to degrade active caspases. Proteasome inhibitors easily induce apoptosis in rapidly cycling cells and in particular in malignant cells, however they are relatively well tolerated by differentiated, postmitotic cells such as neurons. Moreover, they confer neuroprotection likely through the induction of heat shock proteins and inhibit apoptotic and excitotoxic neuronal death triggered by different mechanisms. Nevertheless, at higher doses they eventually induce apoptosis. Some populations of neurons are more susceptible to the proapoptotic effects of proteasome inhibitors than other, in particular the dopaminergic neurons in the striatum. Proteasome dependent steps are located both in the induction and the execution stages of apoptosis, upstream and downstream from caspase activation. Multiple mechanisms of action may be involved such as inhibition of the cell cycle, oxidative stress, protein aggregation and inhibition of NFκB activation, just to name a few. The clinical use of proteasome inhibitors raises the question of their possible neurotoxic effect, which may surface when this class of drugs will be used in the treatment of chronic disorders such as rheumatoid arthritis. The relationship between the UPS and apoptosis in neuronal cells is therefore highly complex and far from being fully understood.

* Correspondence concerning this article should be addressed to Dr. Cezary Wójcik, MD, PhD; Department of Anatomy and Cell Biology, Indiana University School of Medicine, 8600 University Blvd., Evansville, IN, USA. Phone: (812) 461 5437; Fax: (812) 465 1184; E-mail: cwojcik@iupui.edu.

Keywords: apoptosis cell death, caspases, IAPs, Bcl-2, proteasome, ubiquitin.

Abbreviations

BIR, baculovirus inhibitor of apoptosis; Cdks, cyclin-dependent kinases; DIAP, IAP from *Drosophila*; IAP, inhibitor of apoptosis; nNOS, neuronal isoform NO synthase; ODC, ornithin decarboxylase; ROS, reactive oxygen species; UBL, ubiquitin-like; UPS, ubiquitin and proteasome dependent proteolytic system.

Introduction

Apoptosis is one of the processes which shape our central nervous systems during development as well as in the post-natal life. While apoptosis of multiple neuronal and glial precursor cells is a prerequisite for a normal development of the nervous system [1], apoptosis in the adult nervous system often is a result of different kinds of injury and leads to irreversible neuronal loss. Strong injuries, such as ischemia at the core of a cerebral infarct, lead to cell death by necrosis, while apoptosis is triggered by more discrete kind of insults, such as hypoxia in the penumbra of an ischemic lesion, the action of different kinds of excitotoxins, environmental poisons and pathogens as well as through intracellular accumulation of undegraded proteins [2-10]. Ageing of the central nervous system (CNS) is characterized by a progressive apoptotic loss of neurons and there is growing evidence of apoptosis in neurodegenerative disease. However, it is still unclear whether the pathological manifestations observed in slow neurodegenerative diseases are due to neuronal loss or whether they are related to independent degenerative events in the axodendritic network [11]. Moreover, due to the plasticity and adaptability of the nervous system the loss of neuronal cells through apoptosis can proceed without clinical manifestations until the damage is considerable. It is therefore clear that therapeutic intervention aimed at the prevention and inhibition of neuronal apoptosis is of pristine importance [10,12-15].

Since the discovery of apoptosis over 30 years ago our understanding of this process has increased dramatically [16]. Last years have seen great advances in the molecular dissection of the apoptotic pathways, which are beyond the scope of this chapter [17]. While basic apoptotic machinery is similar in all cell types, neuronal cells have some peculiarities resulting from their postmitotic state and unique structure and function [8,9,18-20].

The apoptotic process can be usually divided into an induction and an execution phase. Despite the plethora of diverse stimuli leading to apoptotic induction, the execution of apoptosis involves a common step consisting usually on the activation of caspases. The latter are a family of cytoplasmic cystein proteases which cleave after aspartate residues [21]. Caspase-dependent cleavages selectively inhibit some proteins, while activate another, which in turn leads to the phenotype characteristic of apoptotic cells. Apoptosis of neuronal cells have various distinguishing features. Classic Bcl-2-dependent and caspase-mediated events account only partially for neurodegenerative changes in injured neurons. Blockage of the caspase execution machinery only temporarily rescues damaged neurons and classical apoptotic features can still appear in caspase-inhibited neurons which are characterized by cytochrome c release, chromatin condensation to irregularly shaped clumps, DNA-fragmentation, and exposure of phosphatidylserine [11].

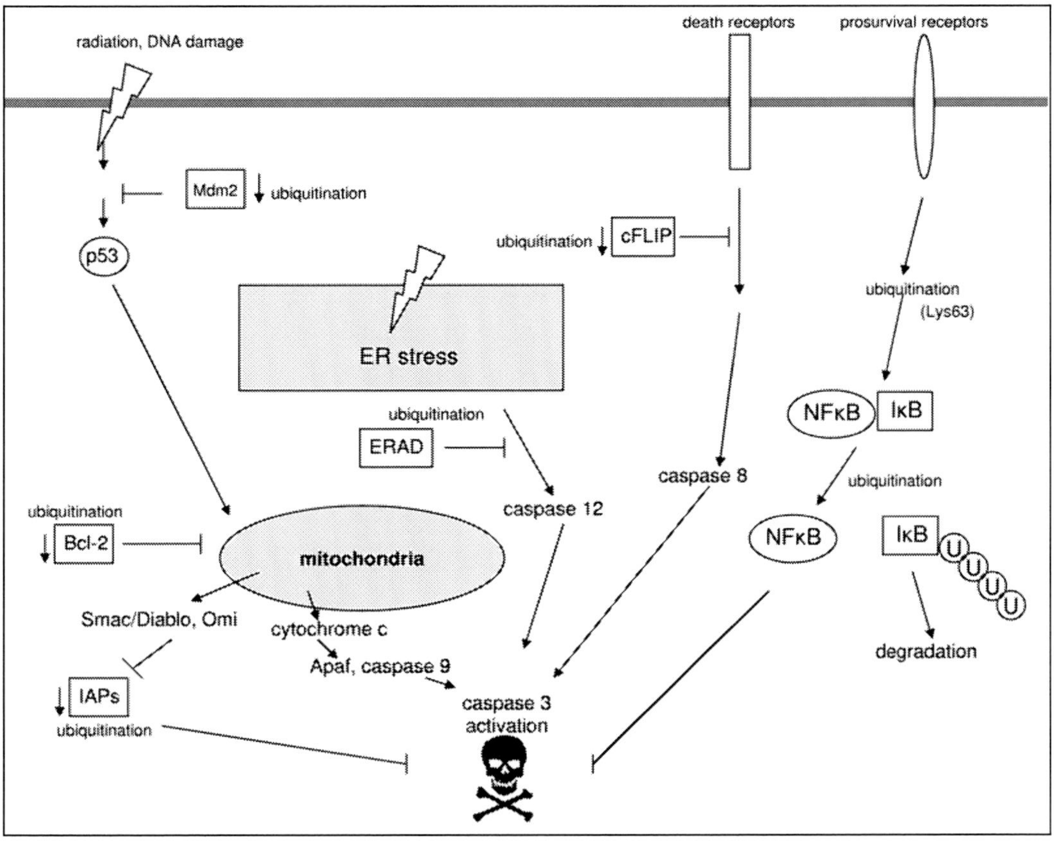

Figure 1. Schematic representation of the involvement of the UPS in apoptosis. Ubiquitination and subsequent substrate degradation by 26S proteasomes control different steps in the apoptotic pathways performing both prosurvival and proapoptotic functions (see text for details). Therefore, inhibition of the proteasome produces either prosurvival or proapoptotic effects depending on the cell type as well as depending on the duration and extent of proteasome inhibition. Complete and prolonged proteasome inhibition inevitably leads to apoptosis since it mimicks caspase-dependent inhibition of proteasome function, which is downstream of caspase-3 activation (not shown on the figure).

Most cytoplasmic proteolysis in eukaryotic cells is mediated by a different class of proteolytic enzymes, called proteasomes [22]. Proteasomes mediate not only the bulk of intracellular proteolysis but are able of selectively degrading multiple regulatory proteins. Most of the proteins degraded by the proteasomes are first tagged by the attachment of multiple ubiquitin moieties (ubiquitination) and therefore the term ubiquitin-proteasome system (UPS) is used to describe this proteolytic pathway [23]. The mechanisms of ubiquitination as well as structure and function of the proteasomes have been described in detail in the preceding chapters and therefore will not be discussed further here. Taking into account the importance of the UPS in physiology and pathology, the discovery that it is also involved in apoptosis came without a major surprise. The link between the UPS and apoptosis have been established during the study of programmed cell death in the intersegmental muscles of the hawkworm *Maduca sexta* [24]. The intimate intermingling and crosstalk of those pathways has been the subject of several reviews, however there are constantly new data which are added to this puzzle [25-31]. Figure 1 offers an overview of the multiple events on the apoptotic pathways which are controlled by ubiquitination.

EFFECTS OF PROTEASOME INHIBITORS ON NEURONAL APOPTOSIS

The UPS has recently become the subject of therapeutic intervention through the clinical use of Velcade® (bortezomib), a boronate proteasome inhibitor known formerly as PS-341. Velcade® has been approved for the treatment of patients with chemotherapy-resistant multiple myeloma [32]. A different proteasome inhibitor known as MLN-519 is evaluated for clinical trials in the early treatment of stroke [33,34]. Other proteasome inhibitors as well as drugs affecting the different enzymes of the ubiquitination cascade are down the drug discovery pipeline [35]. A detailed description of the different proteasome inhibitors is given in Chapter 40. Interestingly, the current and proposed clinical use of proteasome inhibitors is based either on their property to induce apoptosis (in cancer cells) or to prevent apoptosis (in neuronal cells), showing that proteasome inhibitors are a double-edged sword, which must be used with caution.

The proapoptotic effects of proteasome inhibitors have been studied first in different cancer cell lines [30,31]. It has been soon discovered, that in contrast to cancer cells, primary and terminally differentiated cells are relatively resistant to the action of proteasome inhibitors, requiring much higher concentrations and longer incubation times [36,37]. Moreover, in terminally differentiated resting cells such as neuronal cells, low doses of proteasome inhibitors were preventing apoptosis, for example induced by NGF deprivation in sympathetic neurons [38] or by serum deprivation of cerebellar granule cells [39]. This phenomenon was not limited to neuronal cells, since proteasome inhibitors also prevented dehamethasone-induced apoptosis of thymocytes [40] and apoptosis of lens-derived αTN4-1 cells following INF-γ treatment [41]. The cytoprotective effects of proteasome inhibitors may go beyond apoptosis, since they protects neuronal cells from glutamate toxicity, which does not involve activation of caspases and has mixed morphological features with necrosis [42]. Proteasome inhibitors are also able to delay Wallerian degeneration of axons [43,44]. Inhibition of proteasome however does not prevent neuronal necrosis [45]. Despite the lack of proapoptotic action at lower concentrations, higher doses of proteasome inhibitors and prolonged treatment inevitably induced cell death [38,40]. Caspase-9 is likely the initiator caspase in proteasome inhibitor-induced apoptosis [46].

Injection of the proteasome inhibitor PSI into the lateral ventricle of rats results in widespread neuronal apoptosis in various regions of the brain, as detected by the appearance of the ladder of fragmented apoptotic DNA [47]. Proteasomal inhibition leads to apoptotic death of mouse sympathetic neurons through the intrinsic pathway. Inhibition of caspases as well as deletion of Bax or overexpression of Bcl-2 provides complete protection against proteasomal inhibition-induced death. This proves that proteasome inhibition activates the intrinsic apoptotic pathway involving bcl-2 family members and the mitochondria [48]. The mechanism of action of proteasome inhibitors in primary neurons does not therefore seem much different from its mechanism in other cell types, such as myeloid leukemia cell lines where Bcl-2 antagonizes apoptosis induced by proteasome inhibitors [49]. Treatment with Velcade® decreases Bcl-2 expression on a transcriptional level, without effect on Bax or Bak [50]. Overexpression of Bcl-2 induces an increase in proteasome activity [51]. These results imply that Bcl-2 may be the main regulator of the effects of proteasome inhibitors.

Proteasome inhibition in cultures of primary neurons activates at first an initial neuroprotective pathway involving heat shock proteins, antioxidants and cell cycle inhibitors, which likely offers protection against damage caused by hypoxia such as found at the penumbra of the ischemic stroke. However, a continuous proteasome inhibition leads finally to the activation of the apoptotic pathway [52]. Pharmacological suppression of proteasome function induces apoptosis of primary neurons via the release of cytochrome c from mitochondria and activation of caspase-3-like proteases [53]. Moreover, proteasome inhibitors block the protective effects against apoptosis and glutamate toxicity of pigment epithelium-derived factor (PEDF) on immature cerebellar granule cells [54]. They induce apoptosis in cerebellar granule cells, which can be reverted through inhibitors of transcription and translation, as well as by specific caspase inhibitors, demonstrating that proteasomes act upstream from caspase activation preventing apoptosis [55,56]. However, proteasome inhibitors block apoptosis of cerebellar granule cells induced by reduction of extracellular potassium, preventing caspase activity and calpain-caspase-3-mediated processing of tau protein, suggesting that proteasomes are indeed involved upstream of the caspase activation, however by actually augmenting the apoptotic signal [57]. The role of the UPS in neuronal apoptosis is even more complex. Inhibitors of the proteasome reduce apoptotic-like features of the neuronal soma induced by different agents even in the presence of caspase inhibitors, suggesting that some proteasome-dependent steps act downstream of caspase activation in the execution phase of apoptosis [11].

Therefore, it looks like proteasome-mediated events take place both early and late in apoptosis, i.e. both at the initiation and execution phase, before and after activation of caspases. The final outcome of the treatment with proteasome inhibitors strongly depends on the timeframe of their action, their concentration and also on the cell type used. It looks like even among the cells found in the central nervous system, there are big differences in the involvement of the UPS in apoptosis. For example, in neurons of the cerebral cortex, thyroid hormone produces apoptosis associated with an increase of the levels of polyubiquitinated proteins, while the opposite results are obtained in cerebellar granular cells. In oligodendrocytes thyroid hormone increases apoptosis but does not produce changes in the UPS [58]. Moreover, cultured striatal neurons were more vulnerable than cortical neurons to the proapoptotic action of proteasome inhibitors, while astrocytes have shown resistance when compared to neurons [59]. Dopaminergic neurons of the substantia nigra are particularly susceptible to induction of apoptosis through the action of proteasome inhibitors, as evidenced in primary cultures of embryonic rat ventral midbrain, where proteasome inhibitors lead to apoptotic death specifically within phenotypically defined tyrosine hydroxylase (TH)-positive dopaminergic neurons, with little or no apoptotic death induced in GABAergic neurons [60].

Targets of Proteasome-Dependent Degradation During Apoptosis

Since basically any cytosolic, nuclear or ER protein can be degraded by the UPS, the fact that numerous proapoptotic and antiapoptotic proteins are indeed substrates of this system was not surprising. Inhibition of the degradation of those proteins was used and abused as an explanation for the mechanism of action of proteasome inhibitors. The proapoptotic proteins reported to be degraded by the UPS, are for example p53, Bax, Bak, Bid, Bik, Smac, Grim, Reaper and active caspases [61-67]. The antiapoptotic proteins which are substrates of the UPS include Mcl-1 and IAPs [68,69]. Moreover, degradation of both positive and negative cell cycle regulators such as cyclins [70], and cdk inhibitors [71] is controlled by the UPS.

Multiple targets of proteasome-mediated degradation during corticosteroid-induced apoptosis have been identified. They include transcription factors that regulate genes necessary for cell proliferation (e.g. c-Fos, NFκB, AP-1 [72,73]), enzymes whose activity is essential for cell proliferation (e.g. ornithine decarboxylase [74]), cell cycle regulatory proteins (e.g. $p27^{Kip1}$ [75]), and proteins that normally repress caspases (e.g. IAPs, inhibitors of apoptosis [76]). Since Bcl-2 have been reported to inhibit the increase in proteasome activity associated with corticosteroid-induced apoptosis, as well as degradation of c-Fos and $p27^{Kip1}$, the proteasome-mediated degradation of pro-survival factors may be an important control point by Bcl-2 [77]. Additional substrates have been identified in endothelial cells exposed to pro-apoptotic stimuli, such as lipopolysaccharide, TNFα and interleukin 1β in the presence of cycloheximide, where proteasome inhibitors prevent apoptosis by blocking the degradation of antiapoptotic proteins FLIP and Mcl-1 [78].

Cytochrome *c* release from mitochondria is a key step in the apoptotic process, since cytochrome *c* associates with Apaf1 and procaspases to form the apoptosome, where initiator caspases are activated, leading to the activation of executor caspases. The UPS rapidly degrades cytochrome *c* released from the mitochondria, as evidenced by a study in mouse knocked out for Apaf1. Apaf1(-/-) neuronal cells are resistant to common apoptotic stimuli and neurodegenerative inducers such as amyloid-beta peptide (typical of Alzheimer's disease) and mutant G93A superoxide dismutase 1 (typical of familial amyotrophic lateral sclerosis) [79]. Degradation of cytochrome *c* by the UPS may therefore prevent apoptosome formation unless the leak of this mitochondrial protein is massive and the UPS can not degrade it in time, before caspase activation.

The most important of all factors degraded by proteasomes during apoptosis may be the IAPs (inhibitors of apoptosis). IAPs are a family of proteins containing one or more characteristic BIR (baculovirus inhibitor of apoptosis) domains, which have multiple biological activities that include binding and inhibiting caspases, regulating cell cycle progression, and modulating receptor-mediated signal transduction. Several IAPs including XIAP and c-IAP1 in addition to three BIR domains contain a RING finger domain, which confers ubiquitin protease ligase (E3) activity. In response to an apoptotic stimulus IAPs undergo auto-ubiquitination and degradation. IAPs can bind a variety of proteins, such as caspases and TRAFs. They inactivate caspases by binding them and by targeting for degradation [80]. They also can ubiquitinate and target for degradation the IAP inhibitor, mitochondria-derived Smac/Diablo [67]. Proteasome inhibitors promote cell survival by stabilizing the IAPs [76,81]. Indeed, in sympathetic neurons a full length XIAP is rapidly degraded by the UPS upon the action of different proapoptotic stimuli, however when the RING finger is deleted preventing the E3 activity XIAP is stable and protects neurons from apoptosis, despite the fact it can not ubiquitinate caspases, Smac or other substrates. The protective effect is caused by its association with those proapoptotic proteins [69]. In an IAP from *Drosophila* (DIAP), active caspases clip the N-terminal peptide producing a functional E3 molecule, which is however very unstable within cells due to the presence of a destabilizing N-terminal residue. Caspase-cleaved DIAP is therefore quickly removed from the cells through the so called 'N-end rule' pathway [82,83].

Components of the Proteasome System Degraded by Caspases

As shown above by the degradation of E3s from the IAP family, the crosstalk between UPS and apoptotic pathway does not go only into one direction. During the execution phase of apoptosis of cerebellar granular cells a perturbation in normal UPS function occurs, and high levels of ubiquitinated proteins accumulate in the cytoplasm. Such accumulation correlates with a progressive decline of proteasome chymotrypsin and trypsin-like activities and, to a lower extent, of peptidyl-glutamyl peptide hydrolyzing activity. Both intracytoplasmic accumulation of ubiquitinated proteins and decline of proteasome function are reversed by caspase inhibitors [84], suggesting that components of the proteasome may be targets of caspase activity. Indeed, caspases cleave different UPS components too, including three subunits of the 26S proteasome itself. Capsase-mediated cleavage of the Rpt5, Rpn2 and Rpn10 subunits of the 26S proteasome, all forming part of the base of the PA700 (see Chapter 7) severe the proteolytic activity of the holoenzyme [85]. This impairment of proteasome activity can account for the increased levels of ubiquitin conjugates associated with early stages of apoptosis. It also can explain the decreased ability of the proteasome to degrade antiapoptotic proteins such as as Smac and Omi and an ubiquitin-independent substrate ornithin decarboxylase (ODC). *In vitro* induced cleavage of 26S proteasome by active caspase 3 clearly impaired its ability to degrade ODC. Inactivation of 26S proteasome activity early in apoptosis prevents the degradation of active caspases, Smac and other proapoptotic molecules providing a positive feedback loop for caspase activation and apoptotic commitment of the cell [85,86]. Moreover, 26S proteasome may be inactivated in neuronal cells through the modification of selected subunits of the PA700 with the O-linked N-acetylglucosamine moiety [87]. Drug-induced increment in this selective modification favored accumulation of polyubiquitin conjugates and p53 on the hippocampus, followed by apoptosis of some hippocampal neurons [88]. Additional evidence points out that O-linked N-acetylglucosamine modification of many neuronal proteins plays a role in neurodegeneration and therefore in neuronal apoptosis [89].

Not only components of the PA700 proteasome activator are targeted for cleavage through caspases, but also subunits of a different proteasome activator highly expressed in the brain called PA28γ/REG γ (see Chapter 7). It was degraded by caspases-3 and -7 during FAS-induced apoptosis of HeLa cells and cisplatin-induced apoptosis of MCF7 cells [90]. It is difficult to judge at this point about the possible significance of this finding, since not much is known about the possible functions of PA28γ [91,92]. Knockout mice without PA28γ do not show detectable neurological abnormalities [93]. It is worth mentioning, that it has been proposed that PA28γ contributes to the pathology of poly-glutamine tract expansion diseases, since it suppresses the proteasome active sites principally responsible for cleaving after glutamine residues [94].

There are at least three groups of E3s being inactivated during apoptosis by caspase-mediated cleavage, IAPs, parkin and Ufd2. IAPs have been discussed in the previous section. Parkin in neuronal cells is inactivated by caspases, what may directly lead to the accumulation of toxic parkin substrates and triggering cell death of dopaminergic cells in the course of Parkinson's disease [95]. Loss-of-function mutations in the parkin gene are known to result in autosomal recessive juvenile parkinsonism, which causes selective degeneration of nigrostriatal dopaminergic neurons in the absence of Lewy bodies. Overexpression of parkin protects cells of the PC12 neuronal line from the neurotoxicity of the proteasome inhibitor lactacystin, increasing at the same time the accumulation of ubiquitin-protein conjugates in the form of defined aggresomes [96]. During apoptosis induced by various stimuli caspase 6 specifically cleaves UFD2 producing an inactive 110 kDa fragment from the whole length 130 kDa protein [97]. UFD2 is the human homologue of the yeast polyubiquitination factor (E4) Ufd2p [98]. Its function is lost in congenital DiGeorge and velo-cardio-facial syndromes [99]. Recombinant UFD2 has an *in vitro* ubiquitin ligase activity, which is lost after the caspase mediated cleavage [97]. It is unknown what are the substrates of UFD2 in human cells and whether it performs a role similar to yeast E4. UFD2 tightly binds to cytoplasmic chaperone/ATP-ase called p97 or VCP (valosin-containing protein). The latter is very similar to the ATP-ases forming the base of the 19S cap (PA700) of the 26S proteasome and have a role in the degradation of selected ubiquitinated proteins [100], specially those exported from the ER [101]. Taking into account the fact, that it is induced by various stimuli in several cell lines, the inactivation of UFD2 during apoptosis may be an important event. It can be speculated, that UFD2 inactivation may prevent the ubiquitination and therefore proteasome-mediated degradation of a specific subset of substrates. It is tempting to speculate, that IAPs may be among UFD2 substrates.

Accumulation of Ubiquitinated Proteins as a Possible Trigger of Apoptosis

After considering a series of proteins whose accumulation can trigger apoptosis one must also consider the non-specific accumulation of bulk ubiquitinated proteins within the cells as a possible apoptosis trigger. Accumulation of high molecular weight ubiquitin conjugates (usually in the form of an organized perinuclear aggregate or 'aggresome') and depletion of monomeric ubiquitin is a characteristic feature of the treatment with proteasome inhibitors of diverse cell types [102-106]. Details of this process have been described in Chapter 12. It has been suggested that aggresomes represent an *in vitro* model of the inclusion bodies found in the different types of neurodegenerative disorders [107,108].

Whether aggresomes and corresponding cytoplasmic inclusions found *in vivo* contribute to neuronal death or protect cells from the toxic effects of misfolded proteins remains controversial. After treatment with proteasome inhibitors aggresomes were observed in both viable as well as apoptotic neurons [60]. Moreover, quantitative analysis revealed aggresomes in 60% of nonapoptotic cells but only in 10% of apoptotic cells. Apoptosis induced by overexpression of α-synuclein was not coupled with increased prevalence of aggresome-bearing cells [109]. Upon treatment with proteasome inhibitors mouse neurons isolated from p53-deficient animals show delayed apoptosis associated with an increased numbers of inclusions, which likely represent enhanced survival [110]. In a different model, inhibition of neuronal Cdks prevents apoptosis induced by proteasome inhibition without affecting the formation of aggresomes [111]. Therefore, rather than a trigger of apoptosis, the formation of intracellular aggregates recruits otherwise toxic and/or apoptogenic proteins to a well delimited place, where they can be stored with less harm to the cells than in a dispersed form [112].

However, accumulation of misfolded proteins induces a decrease in proteasome activity, probably by recruiting ubiquitin and proteasomes [113]. Formation of macroscopically visible aggresomes is not a prerequisite for this effect [114]. Expression of aggregation-prone proteins, such as mutant α-synuclein in neuronal cell lines results in increased sensitivity to proteasome inhibitors, leading to mitochondrial abnormalities and apoptosis [115]. Therefore, formation of aggresomes likely has a dual role: at the beginning it rescues neurons from the cytotoxic effects of dispersed aggregated proteins, however when it persists, recruitment of UPS components to the inclusion body makes the cell more vulnerable to the action of other nocive insults.

Nuclear and cytoplasmic inclusions found in polyglutamine disorders such as Huntington's disease and several ataxias are also related to aggresomes. Indeed, polyglutamine containing proteins are resistant to proteasome degradation. The longer the poly-Q repeat, the slower the degradation by the proteasome [116]. The shift of the proteasomal components from the total cellular environment to the nuclear aggregates, as well as the comparatively slower degradation of proteins with longer polyglutamines, decrease the proteasome's availability for degrading other key target proteins [114]. Expression of polyglutamine repeats is associated with altered proteasomal function and apoptosis. Impaired proteasomal function plays an important role in polyglutamine protein-induced cell death [116]. Moreover, UBB+1, a mutant ubiquitin that accumulates in the neurons of patients with Alzheimer's disease and Huntington's disease induces formation of aggresome-like inclusion bodies and neuronal apoptosis [117,118].

Heat Shock Protein in Proteasome Inhibitor-Induced Apoptosis

When misfolded proteins appear in the cell as a result of an external stimulus such as heat shock or as the result of the expression of mutated proteins such as the polyglutamine repeat-bearing proteins, heat shock proteins or molecular chaperones are induced. The interplay between the UPS and the chaperone system is described in Chapter 10 and 19.

Molecular chaperones associate with the misfolded protein trying to refold it properly. If this fails, the misfolded protein is ubiquitinated and targeted for degradation by the proteasome mediated by its association with specialized adaptors such as CHIP and Bag-1 (see below). Degradation and refolding are two complementary mechanisms for the rescue of extensive damage to cellular proteins [119]. When there is an excess of unfolded proteins and not enough chaperones available, or when the proteasome activity is impaired, the unproperly folded proteins tend to aggregate recruiting the chaperones, proteasomes and ubiquitin into the aggresomes or inclusion bodies [107,108]. Proteasome inhibitors are known to induce heat shock proteins in the cytosol and the ER lumen [120,121]. Interestingly, dopaminergic neurons, unlike other neurons within these cultures or cultured cortical neurons, fail to induce the chaperone Hsp70 in response to proteasomal inhibition. This failure may explain in part the increased sensitivity of these neurons to proteasomal inhibitors [122]. Overexpression of Hsp70 in dopaminergic neurons *in vivo* protects them from cell death induced by pharmacologic agents used to induce animal models of Parkinson's disease [123].

Induction of heat shock proteins after low level proteasome inhibition offers a neuroprotective mechanism, as it was found to protect cardiomyocytes from ischemia-related injury [124]. A recent study has shown that the stress inducible Hsp70 protects hippocampal CA1 neuronal cells from both global and focal ischemia *in vivo* and in cell culture models of ischemia/reperfusion injury *in vitro*. Hsp70 also reduced the number of protein aggregates in those neurons. Induction of Hsp70 protects also astrocytes from cell death by apoptosis and necrosis [125]. Proteasome inhibition in cultures of primary neurons activates at first an initial neuroprotective pathway involving heat shock proteins [52].

Bag-1 has been first identified as a supressor of apoptosis, which binds Bcl-2 [126]. Mammals possess several paralogues of this protein. All the Bag family members have a C-terminal BAG domain, which interacts with the ATP-ase domain of Hsp70 accelerating the ADP-ATP exchange and therefore stimulating substrate protein release [127]. Bag-1 and Bag-6 (Scythe, BAt3) have in addition to the BAG domain also an amino-terminal ubiquitin-like (UBL) domain. The UBL domain of Bag-1 mediates its association with the 26S proteasome, and since it also binds Hsp70 it recruit this major Hsp to the proteasome [128]l. Bag-1 can act as an unloading factor promoting direct delivery of unfolded proteins, substrates of Hsp70 activity, for degradation. Bag-1 interacts with CHIP, an ubiquitin ligase interacting with Hsp70 and Hsp90, which mediates ubiquitination of unfolded proteins [129,130]. CHIP co-immunoprecipitates with the polyglutamine-expanded huntingtin or ataxin-3 and associates with their aggregates. Overexpression of CHIP suppresses the aggregation and cell death mediated by expanded polyglutamine proteins and the suppressive effect is more prominent when CHIP is overexpressed along with Hsc70 [131].

Role of NFKB in Neuronal Apoptosis

NFκB is a dimeric factor recruited normally in the cytoplasm by the IkBα protein, which is a specific inhibitor of its activity. Upon stimulation by various cytokines (IL-1, TNFα, TRAIL etc.), bacterial lipopolysaccharide, UV radiation, ionizing radiation or oxidative stress a signal transduction cascade is activated which leads to the phosphorylation of IkBα, its ubiquitination, degradation by the 26S proteasome, release of NFκB its entry into the nucleus and transcription of NFκB dependent genes [132,133]. However, the very same cytokines and other stimuli may also trigger apoptosis [134]. The ultimate outcome depends on which one of the branches of this dual reponse prevails. Proteasome inhibitors greatly enhance the proapoptotic action of those stimuli by abolishing the NFκB response. This effects have been observed *in vitro* [135-138] and *in vivo* [139] in different cell types.

NFκB plays important roles in the regulation of many activities of neuronal cells, such as synaptic transmission, inflammation, neuroprotection, and neurotoxicity. In resting neurons, NFκB is present both in the cytoplasm, as an inducible-inactive complex, and in the nucleus, as a constitutive form. Regulation of its inducible activity relies on degradation of IκB(s) through the UPS [140]. A decrease in the nuclear NFκB is associated with the induction of apoptosis in cerebellar granular cells whether induced by potassium withdrawal or proteasome inhibitors [56,141]. Prosurvival signaling by neurotrophin or NGF in the central nervous system is mediated by NFκB and it becomes impaired during the ageing process [142]. Induction of NFκB protects neurons from apoptosis resulting from different insults and therefore is beneficial to the cell, however on the other side induction of NFκB pathway is associated with inflammation, which in turn may elicit more damage to the central nervous system. Therefore, while NFκB is beneficial for single cells, it is not beneficial for the nervous tissue as a whole. Inhibition of NFκB activation through proteasome inhibition elicits a strong antiinflammatory effect, which is the base of the planned use of those agents in the aftermath of stroke, to decrease the secondary tissue damage [34]. However, the use of proteasome inhibitors in stroke must be evaluated very carefully since experimental data have shown that intraventricular injection of a proteasome inhibitor induces a significant inhibition of NFκB activity *in vivo* which is associated with neuronal apoptosis [47,143]. Indeed, systemic administration of PSI, a lipid soluble proteasome inhibitor of old generation, has induced apoptosis of dopaminergic neurons associated with parkinsonism [144]. On the other hand, clinically available proteasome inhibitors do not penetrate the blood-brain barrier and their action in stroke is therefore limited to cerebral vasculature and limited regions of nervous tissue where the blood-brain barrier is impaired as a consequence of stroke.

OXIDATIVE STRESS IN NEURONAL APOPTOSIS

Oxidative stress is pivotal for the modulation of critical cellular functions of neurons and glial cells, including activation of apoptosis. Neurons are particularly susceptible to oxidative stress due to the high rate of oxidative metabolism in the brain and the low level of antioxidant enzymes compared to other somatic tissues. This is supported by fact that oxidative stress is the intracellular end point of many neurotoxic stimuli. It also represents a major cause of the neuropathology underlying a variety of neurodegenerative diseases. Mitochondrial dysfunction, involving impairment in the energetic supply of the cells, apoptosis and overproduction of reactive oxygen species (ROS) is a final common pathogenic mechanism in aging and in neurodegenerative disease [145,146]. Oxidative stress causes extensive damage to proteins, lipids and nucleic acids, either triggering apoptosis or prompting different repair mechanisms. Oxidatively damaged proteins are known substrates of the proteasome, in particular the 20S proteasome, which for its activity does not require ATP or ubiquitination [147]. Therefore, degradation of oxidized proteins in hypoxic conditions in the penumbra seems to be an important mechanism promoting neuronal survival by prevention of excessive accumulation of damaged proteins. On the other hand oxidized proteins are removed from the cytosol and delivered to the proteolytic center of the cell, where they form an organized inclusion or aggresome [148]. Oxidative stress can however damage the proteasome itself leading to a decrease in its activity [149]. Treatment of neuronal cells with agents known to promote oxidative stress, such as hydrogen peroxide or heavy metal ions induce an increase in the levels of polyubiquitinated proteins and associated apoptosis [150]. Enzymes of the antioxidant system, which exhibits a huge activity increase up to 3 h after apoptosis induction in neuronal cells, are subject to UPS-dependent proteolysis. Prevention of their proteolysis by proteasome inhibitors blocks the ROS-dependent release of cytochrome *c* and caspase-3 activation [151]. Neuronal apoptosis induced by proteasome inhibitors is associated with increased nitric oxide production mediated by nNOS (neuronal isoform of the NO synthase), whose inhibition decreases levels of apoptosis, indicating that cell death induced by proteasome inhibitors it is at least in part mediated by nitric oxide [152].

DNA is another target of oxidative damage, which is countered by a complex defense mechanism, the DNA damage response, which involves activation of cell cycle genes in postmitotic cells [146]. Increased levels of nucleic acid oxidation have been described as part of normal brain aging and have been demonstrated to occur in multiple neurological disorders. Low level proteasome inhibition increases the level of nucleic acid oxidation in primary neuron and astrocyte cultures [153].

Inhibition of the Cell Cycle

The transitions of the cell cycle are regulated by UPS-dependent proteolysis of either cyclins or cdk inhibitors [154]. It is therefore logical, that another mechanism of proapoptotic action of proteasome inhibitors involves the induction of a block in the cell cycle, which is usually generalized and involves all phases [155,156]. At first glance, this mechanism of action seems inapplicable to post-mitotic neurons, however neuronal apoptosis in general is often associated with an abortive activation of cell cycle genes [146]. Therefore the inhibitory effect of proteasome inhibitors on cell cycle progression may be actually beneficial to neuronal cells. Apoptosis induced in neuronal cells by proteasome inhibitors can be prevented by flavopiridol, a specific inhibitor of cyclin-dependent kinases (Cdks), and by overexpression of natural Cdk inhibitors. Flavopiridol blocks the phosphorylation of retinoblastoma protein (pRb) that normally occurs after proteasome inhibition. Moreover, expression of a mutant pRb that lacked phosphorylation sites was neuroprotective. It therefore seems that in cortical neurons, proteasomal inhibition leads to an apoptotic pathway dependent on Cdk activation and pRb inactivation [111].

Another protein whose stabilization by proteasome inhibitors may be resposible for induction of apoptosis is p53. p53 is the 'guardian of genome integrity' since it elicits block in the cell cycle, DNA repair and eventually apoptotic response after stress insults compromising genomic inegrity, oncogene activation and hypoxia. p53 is acting as a transcriptional regulator inducing expression of several crucial genes mediating those effects [157-159]. p53 levels are normally low in the cells, since it is constantly ubiquitinated by Mdm2 RING-finger ubiquitin ligase and then degraded by the proteasome [160]. Mdm2 ubiquitinates also itself, what targets its for degradation by proteasomes during apoptosis even in absence of active p53, while it is also cleaved by caspase-3 in p53-mediated apoptosis [161].

In situations of cellular stress p53 degradation is stopped and its levels rapidly rise. Cultured cortical neurons treated with proteasome inhibitors show an early increase of p53 levels, accompanied by nuclear translocation, while at later time points p53 is found sequestered within ubiquitinated inclusions. Neurons isolated from p53-deficient mouse neurons show delayed apoptosis upon proteasome inhibition when compared to controls, indicating the role of p53 [162].

Proteasome inhibitors can also cause accumulation of c-Myc oncoprotein, which is a critical transcriptional regulator of cellular proliferation involved in the induction of apoptosis. For example, in human glioma cells proteasome inhibitors cause elevation of c-Myc protein levels, which induces transiently FasL message, and expression of FasL protein in turn stimulates the Fas receptor-ligand apoptotic signaling pathway [163].

Conclusions

The ubiquitin- and proteasome-dependent system of protein degradation is an important player in neuronal apoptosis. Since the UPS is involved in multiple cellular functions, its mechanism of action is multimodal and involves accumulation of ubiquitinated proteins, inhibition of NFκB activation, oxidative stress, block in the degradation of cell cycle proteins, p53 and oncogene products, just to name a few. Proteasome-dependent steps on the apoptotic pathway are both upstream and downstream of caspase activation. While proteasomes can cleave and degrade caspases, active caspases on the other hand can cleave proteasomal subunits.

Proteasome inhibitors used at low doses confer neuroprotection by induction of heat shock proteins and prevention of apoptosis induced by other stimuli, however at longer expositions and higher doses they themselves induce apoptotic cell death. Certain populations of neurons, in particular dopaminergic neurons of the substantia nigra are more susceptible to proteasomal inhibition than others. This observations raises a red flag, since proteasome inhibitors have already entered the clinical realm for the treatment of multiple myeloma and other cancers. Moreover, they are evaluated for the treatment of stroke, since in animal models of cerebral ischemia they were shown neuroprotective. This fact may be due to the inpenetrability of the brain-blood barrier to proteasome inhibitors currently used in clinic. There seem to be no acute toxicity of the proteasome inhibitor Velcade® (bortezomib) towards the dopaminergic neurons, since there have been no reports of neurotoxicity of this compound other than peripheral neuropathies [164,165]. However, since this process may take years it is unknown if parkinsonism is not going to appear as a very late effect after adminsitration of Velcade® if patients enter a complete remission. The beneficial effects of proteasome inhibitors in stroke are on the other hand related to the effect of those drugs on the endothelium of brain vessels and the cells of inflammatory infiltrate in the ischemic area, where proteasome inhibitors suppress NFκB signaling resulting in a proinflammatory effect. The use of proteasome inhibitors will likely expand to other areas, including immunosupression or the treatment of severe inflammatory diseases such as rheumatoid arthritis [35,166]. When their administration will become chronic, adverse neurological effects are likely to be unmasked.

References

[1] Yeo W, Gautier J: Early neural cell death: dying to become neurons. *Dev Biol* 2004, 274:233-244.

[2] Lossi L, Merighi A: In vivo cellular and molecular mechanisms of neuronal apoptosis in the mammalian CNS. *Prog Neurobiol* 2003, 69:287-312.

[3] Mayer RJ: From neurodegeneration to neurohomeostasis: the role of ubiquitin. *Drug News Perspect* 2003, 16:103-108.

[4] Davies AM: Regulation of neuronal survival and death by extracellular signals during development. *EMBO J* 2003, 22:2537-2545.

[5] Nijhawan D, Honarpour N, Wang X: Apoptosis in neural development and disease. *Annu Rev Neurosci* 2000, 23:73-87.

[6] Yakovlev AG, Faden AI: Caspase-dependent apoptotic pathways in CNS injury. *Mol Neurobiol* 2001, 24:131-144.

[7] Love S: Apoptosis and brain ischaemia. *Prog Neuropsychopharmacol Biol Psychiatry* 2003, 27:267-282.

[8] Zhang X, Chen Y, Jenkins LW, Kochanek PM, Clark RS: Bench-to-bedside review: Apoptosis/programmed cell death triggered by traumatic brain injury. *Crit Care* 2005, 9:66-75.

[9] Prunell GF, Troy CM: Balancing neuronal death. *J Neurosci Res* 2004, 78:1-6.

[10] Ekshyyan O, Aw TY: Apoptosis in acute and chronic neurological disorders. *Front Biosci* 2004, 9:1567-1576.

[11] Volbracht C, Leist M, Kolb SA, Nicotera P: Apoptosis in caspase-inhibited neurons. *Mol Med* 2001, 7:36-48.

[12] Charriaut-Marlangue C: Apoptosis: a target for neuroprotection. *Therapie* 2004, 59:185-190.

[13] Alam JJ: Apoptosis: target for novel drugs. *Trends Biotechnol* 2003, 21:479-483.

[14] Mattson MP: Excitotoxic and excitoprotective mechanisms: abundant targets for the prevention and treatment of neurodegenerative disorders. *Neuromolecular Med* 2003, 3:65-94.

[15] Shacka JJ, Roth KA: Regulation of neuronal cell death and neurodegeneration by members of the Bcl-2 family: therapeutic implications. *Curr Drug Targets CNS Neurol Disord* 2005, 4:25-39.

[16] Kerr JF, Wyllie AH, Currie AR: Apoptosis: a basic biological phenomenon with wide-ranging implications in tissue kinetics. *Br J Cancer* 1972, 26:239-257.

[17] Yan N, Shi Y: Mechanisms of apoptosis through structural biology. *Annu Rev Cell Dev Biol* 2005, 21:35-56.
[18] Prunell GF, Arboleda VA, Troy CM: Caspase function in neuronal death: delineation of the role of caspases in ischemia. *Curr Drug Targets CNS Neurol Disord* 2005, 4:51-61.
[19] D'Mello SR, Chin PC: Treating neurodegenerative conditions through the understanding of neuronal apoptosis. *Curr Drug Targets CNS Neurol Disord* 2005, 4:3-23.
[20] Yakovlev AG, Faden AI: Caspase-dependent apoptotic pathways in CNS injury. *Mol Neurobiol* 2001, 24:131-144.
[21] Zhivotovsky B: Caspases: the enzymes of death. *Essays Biochem* 2003, 39:25-40.
[22] Rock KL, Gramm C, Rothstein L, Clark K, Stein R, Dick L, Hwang D, Goldberg AL: Inhibitors of the proteasome block the degradation of most cell proteins and the generation of peptides presented on MHC class I molecules. *Cell* 1994, 78:761-771.
[23] Glickman MH, Ciechanover A: The ubiquitin-proteasome proteolytic pathway: destruction for the sake of construction. *Physiol Rev* 2002, 82:373-428.
[24] Schwartz LM, Myer A, Kosz L, Engelstein M, Maier C: Activation of polyubiquitin gene expression during developmentally programmed cell death. *Neuron* 1990, 5:411-419.
[25] Drexler HCA: Programmed cell death and the proteasome. *Apoptosis* 1998, 3:1-7.
[26] Orlowski RZ: The role of the ubiquitin-proteasome pathway in apoptosis. *Cell Death Differ* 1999, 6:303-313.
[27] Grimm LM, Osborne BA: Apoptosis and the proteasome. *Results Probl Cell Differ* 1999, 23:209-228.
[28] Jesenberger V, Jentsch S: Deadly encounter: ubiquitin meets apoptosis. *Nat Rev Mol Cell Biol* 2002, 3:112-121.
[29] Vaux DL, Silke J: IAPs, RINGs and ubiquitylation. *Nat Rev Mol Cell Biol* 2005, 6:287-297.
[30] Wojcik C: Regulation of apoptosis by the ubiquitin and proteasome pathway. *J Cell Mol Med* 2002, 6:25-48.
[31] Wojcik C: Proteasomes in apoptosis: villains or guardians? *Cell Mol Life Sci* 1999, 56:908-917.
[32] Adams J: The proteasome: a suitable antineoplastic target. *Nat Rev Cancer* 2004, 4:349-360.
[33] Di Napoli M, Papa F: MLN-519. Millennium/PAION. *Curr Opin Investig Drugs* 2003, 4:333-341.
[34] Wojcik C, Di Napoli M: Ubiquitin-proteasome system and proteasome inhibition: new strategies in stroke therapy. *Stroke* 2004, 35:1506-1518.
[35] Wojcik C: Inhibition of the proteasome as a therapeutic approach. *Drug Discov Today* 1999, 4:188-192.
[36] Orlowski RZ, Eswara JR, Lafond-Walker A, Grever MR, Orlowski M, Dang CV: Tumor growth inhibition induced in a murine model of human Burkitt's lymphoma by a proteasome inhibitor. *Cancer Res* 1998, 58:4342-4348.

[37] Pleban E, Bury M, Mlynarczuk I, Wojcik C: Effects of proteasome inhibitor PSI on neoplastic and non-transformed cell lines. *Folia Histochem Cytobiol* 2001, 39:133-134.

[38] Sadoul R, Fernandez PA, Quiquerez AL, Martinou I, Maki M, Schroter M, Becherer JD, Irmler M, Tschopp J, Martinou JC: Involvement of the proteasome in the programmed cell death of NGF-deprived sympathetic neurons. *EMBO J* 1996, 15:3845-3852.

[39] Atlante A, Bobba A, Calissano P, Passarella S, Marra E: The apoptosis/necrosis transition in cerebellar granule cells depends on the mutual relationship of the antioxidant and the proteolytic systems which regulate ROS production and cytochrome c release en route to death. *J Neurochem* 2003, 84:960-971.

[40] Grimm LM, Goldberg AL, Poirier GG, Schwartz LM, Osborne BA: Proteasomes play an essential role in thymocyte apoptosis. *EMBO J* 1996, 15:3835-3844.

[41] Awasthi N, Wagner BJ: Interferon-gamma induces apoptosis of lens alphaTN4-1 cells and proteasome inhibition has an antiapoptotic effect. *Invest Ophthalmol Vis Sci* 2004, 45:222-229.

[42] van Leyen K, Siddiq A, Ratan RR, Lo EH: Proteasome inhibition protects HT22 neuronal cells from oxidative glutamate toxicity. *J Neurochem* 2005, 92:824-830.

[43] Coleman MP, Ribchester RR: Programmed axon death, synaptic dysfunction and the ubiquitin proteasome system. *Curr Drug Targets CNS Neurol Disord* 2004, 3:227-238.

[44] Zhai Q, Wang J, Kim A, Liu Q, Watts R, Hoopfer E, Mitchison T, Luo L, He Z: Involvement of the ubiquitin-proteasome system in the early stages of wallerian degeneration. *Neuron* 2003, 39:217-225.

[45] Bobba A, Canu N, Atlante A, Petragallo V, Calissano P, Marra E: Proteasome inhibitors prevent cytochrome c release during apoptosis but not in excitotoxic death of cerebellar granule neurons. *FEBS Lett* 2002, 515:8-12.

[46] Almond JB, Snowden RT, Hunter A, Dinsdale D, Cain K, Cohen GM: Proteasome inhibitor-induced apoptosis of B-chronic lymphocytic leukaemia cells involves cytochrome c release and caspase activation, accompanied by formation of an approximately 700 kDa Apaf-1 containing apoptosome complex. *Leukemia* 2001, 15:1388-1397.

[47] Taglialatela G, Kaufmann JA, Trevino A, Perez-Polo JR: Central nervous system DNA fragmentation induced by the inhibition of nuclear factor kappa B. *Neuroreport* 1998, 9:489-493.

[48] Lang-Rollin I, Vekrellis K, Wang Q, Rideout HJ, Stefanis L: Application of proteasomal inhibitors to mouse sympathetic neurons activates the intrinsic apoptotic pathway. *J Neurochem* 2004, 90:1511-1520.

[49] Soligo D, Servida F, Delia D, Fontanella E, Lamorte G, Caneva L, Fumiatti R, Lambertenghi DG: The apoptogenic response of human myeloid leukaemia cell lines and of normal and malignant haematopoietic progenitor cells to the proteasome inhibitor PSI. *Br J Haematol* 2001, 113:126-135.

[50] Bold RJ, Virudachalam S, McConkey DJ: Chemosensitization of pancreatic cancer by inhibition of the 26S proteasome. *J Surg Res* 2001, 100:11-17.

[51] Lee M, Hyun DH, Marshall KA, Ellerby LM, Bredesen DE, Jenner P, Halliwell B: Effect of overexpression of Bcl-2 on cellular oxidative damage, nitric oxide production, antioxidant defenses, and the proteasome. *Free Radic Biol Med* 2001, 31:1550-1559.

[52] Yew EH, Cheung NS, Choy MS, Qi RZ, Lee AY, Peng ZF, Melendez AJ, Manikandan J, Koay ES, Chiu LL, Ng WL, Whiteman M, Kandiah J, Halliwell B: Proteasome inhibition by lactacystin in primary neuronal cells induces both potentially neuroprotective and pro-apoptotic transcriptional responses: a microarray analysis. *J Neurochem* 2005, 94:943-956.

[53] Qiu JH, Asai A, Chi S, Saito N, Hamada H, Kirino T: Proteasome inhibitors induce cytochrome c-caspase-3-like protease-mediated apoptosis in cultured cortical neurons. *J Neurosci* 2000, 20:259-265.

[54] Yabe T, Wilson D, Schwartz JP: NFkappaB activation is required for the neuroprotective effects of pigment epithelium-derived factor (PEDF) on cerebellar granule neurons. *J Biol Chem* 2001, 276:43313-43319.

[55] Pasquini LA, Besio MM, Adamo AM, Pasquini JM, Soto EF: Lactacystin, a specific inhibitor of the proteasome, induces apoptosis and activates caspase-3 in cultured cerebellar granule cells. *J Neurosci Res* 2000, 59:601-611.

[56] Porcile C, Piccioli P, Stanzione S, Bajetto A, Bonavia R, Barbero S, Florio T, Schettinia G: Proteasome inhibitors induce cerebellar granule cell death: inhibition of nuclear factor-kB activation. *Ann N Y Acad Sci* 2002, 973:402-413.

[57] Canu N, Barbato C, Ciotti MT, Serafino A, Dus L, Calissano P: Proteasome involvement and accumulation of ubiquitinated proteins in cerebellar granule neurons undergoing apoptosis. *J Neurosci* 2000, 20:589-599.

[58] Pasquini LA, Marta CB, Adamo AM, Pasquini JM, Soto EF: Relationship between the ubiquitin-dependent pathway and apoptosis in different cells of the central nervous system: effect of thyroid hormones. *Neurochem Res* 2000, 25:627-635.

[59] Snider BJ, Tee LY, Canzoniero LM, Babcock DJ, Choi DW: NMDA antagonists exacerbate neuronal death caused by proteasome inhibition in cultured cortical and striatal neurons. *Eur J Neurosci* 2002, 15:419-428.

[60] Rideout HJ, Lang-Rollin IC, Savalle M, Stefanis L: Dopaminergic neurons in rat ventral midbrain cultures undergo selective apoptosis and form inclusions, but do not up-regulate iHSP70, following proteasomal inhibition. *J Neurochem* 2005, 93:1304-1313.

[61] Chen L, Madura K: Rad23 promotes the targeting of proteolytic substrates to the proteasome. *Mol Cell Biol* 2002, 22:4902-4913.

[62] Kim S, Choi K, Kwon D, Benveniste EN, Choi C: Ubiquitin-proteasome pathway as a primary defender against TRAIL-mediated cell death. *Cell Mol Life Sci* 2004, 61:1075-1081.

[63] Luciano F, Jacquel A, Colosetti P, Herrant M, Cagnol S, Pages G, Auberger P: Phosphorylation of Bim-EL by Erk1/2 on serine 69 promotes its degradation via the proteasome pathway and regulates its proapoptotic function. *Oncogene* 2003, 22:6785-6793.

[64] Olson MR, Holley CL, Yoo SJ, Huh JR, Hay BA, Kornbluth S: Reaper is regulated by IAP-mediated ubiquitination. *J Biol Chem* 2003, 278:4028-4034.

[65] Maki CG, Huibregtse JM, Howley PM: In vivo ubiquitination and proteasome-mediated degradation of p53(1). *Cancer Res* 1996, 56:2649-2654.

[66] Chang YC, Lee YS, Tejima T, Tanaka K, Omura S, Heintz NH, Mitsui Y, Magae J: mdm2 and bax, downstream mediators of the p53 response, are degraded by the ubiquitin-proteasome pathway. *Cell Growth Differ* 1998, 9:79-84.

[67] Hu S, Yang X: Cellular inhibitor of apoptosis 1 and 2 are ubiquitin ligases for the apoptosis inducer Smac/DIABLO. *J Biol Chem* 2003, 278:10055-10060.

[68] Nijhawan D, Fang M, Traer E, Zhong Q, Gao W, Du F, Wang X: Elimination of Mcl-1 is required for the initiation of apoptosis following ultraviolet irradiation. *Genes Dev* 2003, 17:1475-1486.

[69] Yu LY, Korhonen L, Martinez R, Jokitalo E, Chen Y, Arumae U, Lindholm D: Regulation of sympathetic neuron and neuroblastoma cell death by XIAP and its association with proteasomes in neural cells. *Mol Cell Neurosci* 2003, 22:308-318.

[70] Glotzer M, Murray AW, Kirschner MW: Cyclin is degraded by the ubiquitin pathway. *Nature* 1991, 349:132-138.

[71] Pagano M, Tam SW, Theodoras AM, Beer-Romero P, Del Sal G, Chau V, Yew PR, Draetta GF, Rolfe M: Role of the ubiquitin-proteasome pathway in regulating abundance of the cyclin-dependent kinase inhibitor p27. *Science* 1995, 269:682-685.

[72] He H, Qi XM, Grossmann J, Distelhorst CW: c-Fos degradation by the proteasome. an early, Bcl-2-regulated step in apoptosis. *J Biol Chem* 1998, 273:25015-25019.

[73] Ivanov VN, Nikolic-Zugic J: Biochemical and kinetic characterization of the glucocorticoid-induced apoptosis of immature CD4+CD8+ thymocytes. *Int Immunol* 1998, 10:1807-1817.

[74] Grassilli E, Benatti F, Dansi P, Giammarioli AM, Malorni W, Franceschi C, Desiderio MA: Inhibition of proteasome function prevents thymocyte apoptosis: involvement of ornithine decarboxylase. *Biochem Biophys Res Comm* 1998, 250:293-297.

[75] Gil-Gomez G, Berns A, Brady HJ: A link between cell cycle and cell death: Bax and Bcl-2 modulate Cdk2 activation during thymocyte apoptosis. *EMBO J* 1998, 17:7209-7218.

[76] Yang Y, Fang S, Jensen JP, Weissman AM, Ashwell JD: Ubiquitin protein ligase activity of IAPs and their degradation in proteasomes in response to apoptotic stimuli. *Science* 2000, 288:874-877.

[77] Distelhorst CW: Recent insights into the mechanism of glucocorticosteroid-induced apoptosis. *Cell Death Differ* 2002, 9:6-19.

[78] Bannerman DD, Tupper JC, Ricketts WA, Bennett CF, Winn RK, Harlan JM: A constitutive cytoprotective pathway protects endothelial cells from lipopolysaccharide-induced apoptosis. *J Biol Chem* 2001, 276:14924-14932.

[79] Cozzolino M, Ferraro E, Ferri A, Rigamonti D, Quondamatteo F, Ding H, Xu ZS, Ferrari F, Angelini DF, Rotilio G, Cattaneo E, Carri MT, Cecconi F: Apoptosome inactivation rescues proneural and neural cells from neurodegeneration. *Cell Death Differ* 2004, 11:1179-1191.

[80] Yang YL, Li XM: The IAP family: endogenous caspase inhibitors with multiple biological activities. *Cell Res* 2000, 10:169-177.

[81] Richter BW, Duckett CS: The IAP proteins: caspase inhibitors and beyond. *Sci STKE* 2000, 2000:E1.

[82] Ditzel M, Wilson R, Tenev T, Zachariou A, Paul A, Deas E, Meier P: Degradation of DIAP1 by the N-end rule pathway is essential for regulating apoptosis. *Nat Cell Biol* 2003, 5:467-473.

[83] Varshavsky A: The N-end rule and regulation of apoptosis. *Nat Cell Biol* 2003, 5:373-376.

[84] Canu N, Barbato C, Ciotti MT, Serafino A, Dus L, Calissano P: Proteasome involvement and accumulation of ubiquitinated proteins in cerebellar granule neurons undergoing apoptosis. *J Neurosci* 2000, 20:589-599.

[85] Sun XM, Butterworth M, MacFarlane M, Dubiel W, Ciechanover A, Cohen GM: Caspase activation inhibits proteasome function during apoptosis. *Mol Cell* 2004, 14:81-93.

[86] Friedman J, Xue D: To live or die by the sword; the regulation of apoptosis by the proteasome. *Dev Cell* 2004, 6:460-461.

[87] Zhang F, Su K, Yang X, Bowe DB, Paterson AJ, Kudlow JE: O-GlcNAc modification is an endogenous inhibitor of the proteasome. *Cell* 2003, 115:715-725.

[88] Liu K, Paterson AJ, Zhang F, McAndrew J, Fukuchi K, Wyss JM, Peng L, Hu Y, Kudlow JE: Accumulation of protein O-GlcNAc modification inhibits proteasomes in the brain and coincides with neuronal apoptosis in brain areas with high O-GlcNAc metabolism. *J Neurochem* 2004, 89:1044-1055.

[89] Lefebvre T, Guinez C, Dehennaut V, Beseme-Dekeyser O, Morelle W, Michalski JC: Does O-GlcNAc play a role in neurodegenerative diseases? *Expert Rev Proteomics* 2005, 2:265-275.

[90] Araya R, Takahashi R, Nomura Y: Yeast two-hybrid screening using constitutive-active caspase-7 as bait in the identification of PA28gamma as an effector caspase substrate. *Cell Death Differ* 2002, 9:322-328.

[91] Tanahashi N, Yokota K, Ahn JY, Chung CH, Fujiwara T, Takahashi E, Demartino GN, Slaughter CA, Toyonaga T, Yamamura K, Shimbara N, Tanaka K: Molecular properties of the proteasome activator PA28 family proteins and gamma-interferon regulation. *Genes Cells* 1997, 2:195-211.

[92] Li J, Gao X, Ortega J, Nazif T, Joss L, Bogyo M, Steven AC, Rechsteiner M: Lysine 188 substitutions convert the pattern of proteasome activation by REGgamma to that of REGs alpha and beta. *EMBO J* 2001, 20:3359-3369.

[93] Murata S, Kawahara H, Tohma S, Yamamoto K, Kasahara M, Nabeshima Y, Tanaka K, Chiba T: Growth retardation in mice lacking the proteasome activator PA28gamma. *J Biol Chem* 1999, 274:38211-38215.

[94] Goellner GM, Rechsteiner M: Are Huntington's and polyglutamine-based ataxias proteasome storage diseases? *Int J Biochem Cell Biol* 2003, 35:562-571.

[95] Kahns S, Kalai M, Jakobsen LD, Clark BF, Vandenabeele P, Jensen PH: Caspase-1 and caspase-8 cleave and inactivate cellular parkin. *J Biol Chem* 2003, 278:23376-23380.

[96] Yang H, Zhou HY, Li B, Chen SD: Neuroprotection of Parkin against apoptosis is independent of inclusion body formation. *Neuroreport* 2005, 16:1117-1121.

[97] Mahoney JA, Odin JA, White SM, Shaffer D, Koff A, Casciola-Rosen L, Rosen A: The human homologue of the yeast polyubiquitination factor Ufd2p is cleaved by caspase 6 and granzyme B during apoptosis. *Biochem J* 2002, 361:587-595.

[98] Koegl M, Hoppe T, Schlenker S, Ulrich HD, Mayer TU, Jentsch S: A novel ubiquitination factor, E4, is involved in multiubiquitin chain assembly. *Cell* 1999, 96:635-644.

[99] Pizzuti A, Novelli G, Ratti A, Amati F, Mari A, Calabrese G, Nicolis S, Silani V, Marino B, Scarlato G, Ottolenghi S, Dallapiccola B: UFD1L, a developmentally expressed ubiquitination gene, is deleted in CATCH 22 syndrome. *Hum Mol Genet* 1997, 6:259-265.

[100] Dai RM, Li CC: Valosin-containing protein is a multi-ubiquitin chain-targeting factor required in ubiquitin-proteasome degradation. *Nat Cell Biol* 2001, 3:740-744.

[101] Ye Y, Meyer HH, Rapoport TA: The AAA ATPase Cdc48/p97 and its partners transport proteins from the ER into the cytosol. *Nature* 2001, 414:652-656.

[102] Wójcik C, Schroeter D, Wilk S, Lamprecht J, Paweletz N: Ubiquitin-mediated proteolysis centers in HeLa cells: indication from studies of an inhibitor of the chymotrypsin-like activity of the proteasome. *Eur J Cell Biol* 1996, 71:311-318.

[103] Johnston JA, Ward CL, Kopito RR: Aggresomes: a cellular response to misfolded proteins. *J Cell Biol* 1998, 143:1883-1898.

[104] Wojcik C: An inhibitor of the chymotrypsin-like activity of the proteasome (PSI) induces similar morphological changes in various cell lines. *Folia Histochem Cytobiol* 1997, 35:211-214.

[105] Wojcik C: On the spatial organization of ubiquitin-dependent proteolysis in HeLa cells. *Folia Histochem Cytobiol* 1997, 35:117-118.

[106] Wigley WC, Fabunmi RP, Lee MG, Marino CR, Muallem S, Demartino GN, Thomas PJ: Dynamic association of proteasomal machinery with the centrosome. *J Cell Biol* 1999, 145:481-490.

[107] Sherman MY, Goldberg AL: Cellular defenses against unfolded proteins: a cell biologist thinks about neurodegenerative diseases. *Neuron* 2001, 29:15-32.

[108] Kopito RR, Sitia R: Aggresomes and Russell bodies. Symptoms of cellular indigestion? *EMBO Rep* 2000, 1:225-231.

[109] Tanaka M, Kim YM, Lee G, Junn E, Iwatsubo T, Mouradian MM: Aggresomes formed by alpha-synuclein and synphilin-1 are cytoprotective. *J Biol Chem* 2004, 279:4625-4631.

[110] Dietrich P, Rideout HJ, Wang Q, Stefanis L: Lack of p53 delays apoptosis, but increases ubiquitinated inclusions, in proteasomal inhibitor-treated cultured cortical neurons. *Mol Cell Neurosci* 2003, 24:430-441.

[111] Rideout HJ, Wang Q, Park DS, Stefanis L: Cyclin-dependent kinase activity is required for apoptotic death but not inclusion formation in cortical neurons after proteasomal inhibition. *J Neurosci* 2003, 23:1237-1245.

[112] Rideout HJ, Larsen KE, Sulzer D, Stefanis L: Proteasomal inhibition leads to formation of ubiquitin/alpha-synuclein- immunoreactive inclusions in PC12 cells. *J Neurochem* 2001, 78:899-908.

[113] Bence NF, Sampat RM, Kopito RR: Impairment of the ubiquitin-proteasome system by protein aggregation. *Science* 2001, 292:1552-1555.

[114] Bennett EJ, Bence NF, Jayakumar R, Kopito RR: Global impairment of the ubiquitin-proteasome system by nuclear or cytoplasmic protein aggregates precedes inclusion body formation. *Mol Cell* 2005, 17:351-365.

[115] Tanaka Y, Engelender S, Igarashi S, Rao RK, Wanner T, Tanzi RE, Sawa A, Dawson L, Dawson TM, Ross CA: Inducible expression of mutant alpha-synuclein decreases proteasome activity and increases sensitivity to mitochondria-dependent apoptosis. *Hum Mol Genet* 2001, 10:919-926.

[116] Jana NR, Zemskov EA, Wang G, Nukina N: Altered proteasomal function due to the expression of polyglutamine-expanded truncated N-terminal huntingtin induces apoptosis by caspase activation through mitochondrial cytochrome c release. *Hum Mol Genet* 2001, 10:1049-1059.

[117] De Vrij FM, Sluijs JA, Gregori L, Fischer DF, Hermens WT, Goldgaber D, Verhaagen J, Van Leeuwen FW, Hol EM: Mutant ubiquitin expressed in Alzheimer's disease causes neuronal death. *FASEB J* 2001, 15:2680-2688.

[118] de PR, Fischer DF, Maat-Schieman ML, Hobo B, de Vos RA, Brunt ER, Hol EM, Roos RA, Van Leeuwen FW: Accumulation of aberrant ubiquitin induces aggregate formation and cell death in polyglutamine diseases. *Hum Mol Genet* 2004, 13:1803-1813.

[119] Friant S, Meier KD, Riezman H: Increased ubiquitin-dependent degradation can replace the essential requirement for heat shock protein induction. *EMBO J* 2003, 22:3783-3791.

[120] Lee DH, Goldberg AL: Proteasome inhibitors cause induction of heat shock proteins and trehalose, which together confer thermotolerance in Saccharomyces cerevisiae. *Mol Cell Biol* 1998, 18:30-38.

[121] Pritts TA, Hungness ES, Hershko DD, Robb BW, Sun X, Luo GJ, Fischer JE, Wong HR, Hasselgren PO: Proteasome inhibitors induce heat shock response and increase IL-6 expression in human intestinal epithelial cells. *Am J Physiol Regul Integr Comp Physiol* 2002, 282:R1016-R1026.

[122] Rideout HJ, Lang-Rollin IC, Savalle M, Stefanis L: Dopaminergic neurons in rat ventral midbrain cultures undergo selective apoptosis and form inclusions, but do not up-regulate iHSP70, following proteasomal inhibition. *J Neurochem* 2005, 93:1304-1313.

[123] Dong Z, Wolfer DP, Lipp HP, Bueler H: Hsp70 gene transfer by adeno-associated virus inhibits MPTP-induced nigrostriatal degeneration in the mouse model of Parkinson disease. *Mol Ther* 2005, 11:80-88.

[124] Luss H, Schmitz W, Neumann J: A proteasome inhibitor confers cardioprotection. *Cardiovasc Res* 2002, 54:140-151.

[125] Giffard RG, Xu L, Zhao H, Carrico W, Ouyang Y, Qiao Y, Sapolsky R, Steinberg G, Hu B, Yenari MA: Chaperones, protein aggregation, and brain protection from hypoxic/ischemic injury. *J Exp Biol* 2004, 207:3213-3220.

[126] Matsuzawa S, Takayama S, Froesch BA, Zapata JM, Reed JC: p53-inducible human homologue of Drosophila seven in absentia (Siah) inhibits cell growth: suppression by Bag-1. *EMBO J* 1998, 17:2736-2747.

[127] Sondermann H, Scheufler C, Schneider C, Hohfeld J, Hartl FU, Moarefi I: Structure of a Bag/Hsc70 complex: convergent functional evolution of Hsp70 nucleotide exchange factors. *Science* 2001, 291:1553-1557.

[128] Luders J, Demand J, Hohfeld J: The ubiquitin-related BAG-1 provides a link between the molecular chaperones Hsc70/Hsp70 and the proteasome. *J Biol Chem* 2000, 275:4613-4617.

[129] Murata S, Minami Y, Minami M, Chiba T, Tanaka K: CHIP is a chaperone-dependent E3 ligase that ubiquitylates unfolded protein. *EMBO Rep* 2001, 2:1133-1138.

[130] Demand J, Alberti S, Patterson C, Hohfeld J: Cooperation of a ubiquitin domain protein and an E3 ubiquitin ligase during chaperone/proteasome coupling. *Curr Biol* 2001, 11:1569-1577.

[131] Jana NR, Dikshit P, Goswami A, Kotliarova S, Murata S, Tanaka K, Nukina N: Co-chaperone CHIP associates with expanded polyglutamine protein and promotes their degradation by proteasomes. *J Biol Chem* 2005, 280:11635-11640.

[132] Turco MC, Romano MF, Petrella A, Bisogni R, Tassone P, Venuta S: NF-kappaB/Rel-mediated regulation of apoptosis in hematologic malignancies and normal hematopoietic progenitors. *Leukemia* 2004, 18:11-17.

[133] Li X, Stark GR: NFkappaB-dependent signaling pathways. *Exp Hematol* 2002, 30:285-296.

[134] Chen G, Goeddel DV: TNF-R1 signaling: a beautiful pathway. *Science* 2002, 296:1634-1635.

[135] Johnson TR, Stone K, Nikrad M, Yeh T, Zong WX, Thompson CB, Nesterov A, Kraft AS: The proteasome inhibitor PS-341 overcomes TRAIL resistance in Bax and caspase 9-negative or Bcl-xL overexpressing cells. *Oncogene* 2003, 22:4953-4963.

[136] Mlynarczuk I, Hoser G, Grzela T, Stoklosa T, Wojcik C, Malejczyk J, Jakobisiak M: Augmented pro-apoptotic effects of TRAIL and proteasome inhibitor in human promonocytic leukemic U937 cells. *Anticancer Res* 2001, 21:1237-1240.

[137] Wang CY, Mayo MW, Baldwin AS, Jr.: TNF- and cancer therapy-induced apoptosis: potentiation by inhibition of NF-kappaB. *Science* 1996, 274:784-787.

[138] Van Antwerp DJ, Martin SJ, Kafri T, Green DR, Verma IM: Suppression of TNF-alpha-induced apoptosis by NF-kappaB. *Science* 1996, 274:787-789.

[139] Golab J, Stoklosa T, Czajka A, Dabrowska A, Jakobisiak M, Zagozdzon R, Wojcik C, Marczak M, Wilk S: Synergistic antitumor effects of a selective proteasome inhibitor and TNF in mice. *Anticancer Res* 2000, 20:1717-1721.

[140] Traenckner EB, Wilk S, Baeuerle PA: A proteasome inhibitor prevents activation of NF-kappa B and stabilizes a newly phosphorylated form of I kappa B-alpha that is still bound to NF-kappa B. *EMBO J* 1994, 13:5433-5441.

[141] Piccioli P, Porcile C, Stanzione S, Bisaglia M, Bajetto A, Bonavia R, Florio T, Schettini G: Inhibition of nuclear factor-kappaB activation induces apoptosis in cerebellar granule cells. *J Neurosci Res* 2001, 66:1064-1073.

[142] Carter BD, Kaltschmidt C, Kaltschmidt B, Offenhauser N, Bohm-Matthaei R, Baeuerle PA, Barde YA: Selective activation of NF-kappa B by nerve growth factor through the neurotrophin receptor p75. *Science* 1996, 272:542-545.

[143] Maggirwar SB, Sarmiere PD, Dewhurst S, Freeman RS: Nerve growth factor-dependent activation of NF-kappaB contributes to survival of sympathetic neurons. *J Neurosci* 1998, 18:10356-10365.

[144] McNaught KS, Perl DP, Brownell AL, Olanow CW: Systemic exposure to proteasome inhibitors causes a progressive model of Parkinson's disease. *Ann Neurol* 2004, 56:149-162.

[145] Emerit J, Edeas M, Bricaire F: Neurodegenerative diseases and oxidative stress. *Biomed Pharmacother* 2004, 58:39-46.

[146] Kruman II: Why do neurons enter the cell cycle? *Cell Cycle* 2004, 3:769-773.

[147] Pacifici RE, Salo DC, Davies KJ: Macroxyproteinase (M.O.P.): a 670 kDa proteinase complex that degrades oxidatively denatured proteins in red blood cells. *Free Radic Biol Med* 1989, 7:521-536.

[148] Shimizu K, Kiuchi Y, Ando K, Hayakawa M, Kikugawa K: Coordination of oxidized protein hydrolase and the proteasome in the clearance of cytotoxic denatured proteins. *Biochem Biophys Res Commun* 2004, 324:140-146.

[149] Elkon H, Melamed E, Offen D: Oxidative stress, induced by 6-hydroxydopamine, reduces proteasome activities in PC12 cells: implications for the pathogenesis of Parkinson's disease. *J Mol Neurosci* 2004, 24:387-400.

[150] Figueiredo-Pereira ME, Cohen G: The ubiquitin/proteasome pathway: friend or foe in zinc-, cadmium-, and H2O2-induced neuronal oxidative stress. *Mol Biol Rep* 1999, 26:65-69.

[151] Atlante A, Bobba A, Calissano P, Passarella S, Marra E: The apoptosis/necrosis transition in cerebellar granule cells depends on the mutual relationship of the antioxidant and the proteolytic systems which regulate ROS production and cytochrome c release en route to death. *J Neurochem* 2003, 84:960-971.

[152] Lee MH, Hyun DH, Jenner P, Halliwell B: Effect of proteasome inhibition on cellular oxidative damage, antioxidant defences and nitric oxide production. *J Neurochem* 2001, 78:32-41.

[153] Ding Q, Dimayuga E, Markesbery WR, Keller JN: Proteasome inhibition increases DNA and RNA oxidation in astrocyte and neuron cultures. *J Neurochem* 2004, 91:1211-1218.

[154] Hershko A: Roles of ubiquitin-mediated proteolysis in cell cycle control. *Curr Opin Cell Biol* 1997, 9:788-799.

[155] Wojcik C, Schroeter D, Stoehr M, Wilk S, Paweletz N: An inhibitor of the chymotrypsin-like activity of the multicatalytic proteinase complex (20S proteasome) induces arrest in G2-phase and metaphase in HeLa cells. *Europ J Cell Biol* 1996, 70:172-178.

[156] Machiels BM, Henfling ME, Gerards WL, Broers JL, Bloemendal H, Ramaekers FC, Schutte B: Detailed analysis of cell cycle kinetics upon proteasome inhibition. *Cytometry* 1997, 28:243-252.

[157] Shen Y, White E: p53-dependent apoptosis pathways. *Adv Cancer Res* 2001, 82:55-84.

[158] Wang XW: Role of p53 and apoptosis in carcinogenesis. *Anticancer Res* 1999, 19:4759-4771.

[159] Burns TF, El Deiry WS: The p53 pathway and apoptosis. *J Cell Physiol* 1999, 181:231-239.

[160] Fang S, Jensen JP, Ludwig RL, Vousden KH, Weissman AM: Mdm2 is a RING finger-dependent ubiquitin protein ligase for itself and p53. *J Biol Chem* 2000, 275:8945-8951.

[161] Cho JW, Park JC, Lee JC, Kwon TK, Park JW, Baek WK, Suh SI, Suh MH: The levels of MDM2 protein are decreased by a proteasome-mediated proteolysis prior to caspase-3-dependent pRb and PARP cleavages. *J Korean Med Sci* 2001, 16:135-139.

[162] Dietrich P, Rideout HJ, Wang Q, Stefanis L: Lack of p53 delays apoptosis, but increases ubiquitinated inclusions, in proteasomal inhibitor-treated cultured cortical neurons. *Mol Cell Neurosci* 2003, 24:430-441.

[163] Tani E, Kitagawa H, Ikemoto H, Matsumoto T: Proteasome inhibitors induce Fas-mediated apoptosis by c-Myc accumulation and subsequent induction of FasL message in human glioma cells. *FEBS Lett* 2001, 504:53-58.

[164] Adams J, Palombella VJ, Sausville EA, Johnson J, Destree A, Lazarus DD, Maas J, Pien CS, Prakash S, Elliott PJ: Proteasome inhibitors: a novel class of potent and effective antitumor agents. *Cancer Res* 1999, 59:2615-2622.

[165] Adams J: Proteasome inhibition in cancer: development of PS-341. *Semin Oncol* 2001, 28:613-619.

[166] Wojcik C: Ubiquitin- and proteasome-dependent proteolytic pathway as an emerging theapeutic target. *Emerging Therapeutic Targets* 2000, 4:1-23.

In: The Ubiquitin Proteasome System... ISBN 978-1-60021-749-4
Eds: Mario Di Napoli and Cezary Wojcik, pp. 537-552

Chapter 22

MOLECULAR CHAPERONES AND THE UBIQUITIN PROTEASOME SYSTEM IN AGING

Stuart K. Calderwood*
Beth Israel Deaconess Medical Center,
Harvard Medical School, Boston, MA 02215, USA.

ABSTRACT

Unrepaired protein damage leads to the formation of lethal protein aggregates in cells and ultimately causes cell death. Protein damage accumulates in cells due to oxidative stress, transduction with prion particles with dominant conformations or due to genetic alterations in proteins that lead to formation of insoluble aggregates. All prokaryotic and eukaryotic cells possess two main strategies to counteract these changes and avoid the accumulation of protein aggregates. These pathways for protein quality control include: (1) the protein chaperone and refolding systems and (2) targeted proteolysis of the malfolding protein. In mammalian cell the molecular chaperones heat shock proteins 70 and 90 (Hsp70 and 90) appear to play key regulatory roles in protein triage after damage. These molecular chaperones can bind to malfolded proteins, deter the aggregation cascade and then target the protein substrates towards either: (1) the pathways of refolding by chaperonin- containing folding structures or (2) can promote ubiquination of its target through mechanisms involving the ubiquitin ligase CHIP and deliver the ubiquinated protein to the proteasome for degradation. Dysregulation of this system occurs during aging and is amplified during a range of degenerative disease states. Failure of this defense system may occur at many levels and decreased expression of proteins that mediate pathways 1 and 2 appears to be involved in aging, particularly of neuronal cells.

Keywords: molecular chaperone, CHIP, proteasome, heat shock factor protein.

Abbreviations

BAG, BCl-2-associated athanogene; *C. elegans*, Caernorhabtitis elegans; CHIP, carboxy-terminus of Hsp70 interacting protein; CNS, central nervous system; Cyp40, cyclosporin A-binding immunophilin; *E. coli, Escherichia coli*; E3 ligase, ubiquitin E3 ligase; FKBP, FK506-binding protein; GSK3, glycogen synthase kinase 3; GrpE, glucose-regulated protein E; Hip, hsc70-interacting protein; Hop, Hsp70-Hsp90-organizing protein; HSF, heat shock transcription factor, Hsp, heat shock protein; HSPBP1, Hsp70-interacting protein; PCD, programmed cell death; sHsp, small heat shock proteins; TPR, tetratrico peptide repeat.

Introduction: Heat Shock Proteins

The heat shock proteins (Hsp) are products of a number of distinct gene families required for cell survival during stress, named for the approximate molecular mass of their products and include Hsp10, 27, 40, 60, 70, 90, and 110 (Table 1) [1-4]. The cytoprotective properties of the Hsps are closely linked to their primary functions as molecular chaperones [1,5,6]. Molecular chaperones are proteins with the primary function of interacting with sequences in target proteins, an interaction that results in the stabilization or enhanced folding of the target polypeptide. The intracellular reactions catalyzed by the Hsps, which led to their designation as molecular chaperones, are divided into two main categories which have been described as (1) 'protein holding' and (2) 'protein folding' [7,8]. The principal holding proteins belong to the Hsp70 and Hsp90 families which bind to unfolded sequences in polypeptide substrates, showing preference for hydrophobic regions [8,9]. Such holding interactions occur during: (1) mRNA translation when Hsp70 binds to the elongating polypeptide chain in order to prevent premature self-associations within the nascent protein, (2) during heat shock when proteins partially unfold and expose hydrophobic sequences which are bound by Hsp and (3) constitutively when Hsp90 binds to proteins with unstable tertiary structures [8,10]. Hsp70 and Hsp90 function in large complexes or chaperone machines which also contain a number of accessory proteins or co-chaperones, that bind the primary chaperone in order to mediate substrate selection, and cycles of association with and disassociation from the substrate (Table 1) [9,10]. After completion of their molecular chaperone function, Hsp 70 and 90 are actively released from protein substrates using intrinsic ATPase domains [2]. Protein folding (2), involves the Hsp60 'chaperonin' family as well as a number of other related chaperonin proteins [11,12]. Chaperonins have the primary function of folding previously unfolded polypeptide sequences. For this purpose, they self-associate in order to form large folding chambers in which the substrate protein can undergo the appropriate intramolecular interactions required to attain its correct tertiary structure in an ATP-dependent process [11,12]. These distinctions are however not absolute and Hsp70 can mediate protein folding [9]. Molecular chaperone substrates have been carefully evaluated in *Escherichia coli* (*E.*

* Correspondence concerning this article should be addressed to Dr Stuart K. Calderwood, PhD; BIDMC, 21-27 Burlington Avenue, 553B, Boston MA, 02215, USA. Phone: +1-617 632 0628; Fax: +1-617 632 0635; Email: scalderw@bidmc.harvard.edu.

coli), in which at least 340 cytosolic proteins have been shown to require the Hsp70 homolog DnaK for folding, while a distinct group of proteins is folded by the Hsp60 homolog GroEL [13]. Protein folding also involves the Hsp27 family; the small Hsp assemble into large aggregates that mediate holding and folding in an ATP independent manner [14]. In addition to this humble molecular chaperone role the Hsps also play key parts in the control of cellular metabolism [10]. Cell regulation by largely Hsp70 and Hsp90, each of which can bind stably to a number of regulatory molecules [10]. Hsp90 plays a major role in regulation of mitogenesis and cell cycle progression through such association and Hsp70 is closely involved in protection from programmed cell death (PCD) each through interaction with a number of key regulatory proteins.

Table 1. Heat Shock Proteins and co-chaperones

Protein	Function	Co-chaperones	References
Hsp27	Molecular Chaperone	none	[14]
Hsp60	Chaperonin	Hsp10	[11, 12]
Hsp70	Molecular Chaperone	Hsp40, GrpE, BAG, HspBP1, Hip, Hop, CHIP	[8, 9, 10]
Hsp90	Molecular Chaperone	p23, HOP, FKBP51, FKBP52 Cyp40, cdc37	[7, 8, 10]
Hsp110	Molecular Chaperone	none	[4]

Of central importance to the regulation of the Hsp70 and Hsp90 co-chaperone systems is the presence of acceptor sites for proteins that contain the tetratricopeptide repeat (TPR) domain at the extreme carboxy-terminus [9,10,15]. The TPR domain is formed af a number of helical structures arrayed in such a way that it forms binding region for TPR acceptor sites in interacting proteins [15]. TPR domain proteins involved in molecular chaperone function include the scaffold protein Hop with at least 2 TPR domains [15]. Hop is thus able to bind simultaneously to both Hsp70 and Hsp90 and by stabilizing their interactions, permit their coordinated activity in protein folding [9,10]. In addition, another TPR domain protein Hip aids to the ATPase cycle of Hsp70 while the immunophilins Cyp40, FKBP51 and FKBP52 bind to Hsp90 through its C-terminal TPR acceptor site and catalyze further steps in protein folding [9,10,16]. Association of the primary molecular chaperones Hsp70 and Hsp90 with these TPR domain proteins is thus essential for many of the properties of Hsp70 and Hsp90 required for protein folding.

The molecular chaperone properties of Hsps are harnessed during heat shock when a large number of cellular proteins undergo synchronous unfolding due to the chaotropic effects of heat and threaten with the cellular catastrophe of protein aggregation [17]. A similar process appears to occur in aging when proteins damaged by oxidative stress or undergoing aggregation due to dominant conformations that tend towards aggregation begin to accumulate [18,19]. Such protein aggregation is deterred by engagement of the heat shock response and the accompanying abundant expression of the Hsp cohort which recognizes denatured proteins through the holding properties of Hsp27, 70 and 90 and refold such denatured proteins with the aid of the chaperonins [20]. Interestingly studies in *E. coli* have shown that aggregation during heat shock largely involves a group of unstable proteins whose aggregation can be inhibited by upregulation of the *E. coli* Hsp70 DnaK [13]. In addition, as protein denaturation and aggregation are powerful triggers of PCD, Hsp have developed powerful anti-apoptotic properties that deter PCD and thus permit a time window for subsequent repair of the proteome [6,21]. The massive Hsp expression that occurs during the heat shock response involves facilitation of expression at each level, including the activation of the potent heat shock transcription factor 1 (HSF1), Hsp mRNA stabilization, selective Hsp translation and Hsp stabilization at the protein level [1,22].

Molecular Chaperones and Proteases Combine to Mediate Protein Quality Control

In all organisms there are proteins of differing degrees of stability which tend to become denatured and aggregated and this process occurs progressively during aging. Two major strategies exist for dealing with protein aggregates and these include (1) protein folding by molecular chaperones and (2) protein degradation by proteases. In *E. coli*, the processes are tightly coupled and the genes encoding the major cytosolic chaperones and proteases are under common control by the transcription factor Sigma^{32} which controls the heat shock regulon [13] At the functional level however, the molecular chaperones and proteases appear to operate individually to resolve protein aggregates and each can protect *E.coli* against heat shock.

Unlike in *E.coli*, in mammalian cells, proteases are not prominent members of the heat shock family [1,13]. However numerous functional links between molecular chaperones and proteases are emerging [23]. It was shown a number of years ago that HSP70 family member HSC70 plays a role in recruiting substrates to the lysosome and recent studies indicate that HSC70 mediates the entry of such substrates into the lysosomal lumen prior to degradation [23]. However, the most compelling connection between these two systems involves the protein CHIP (carboxyl terminus of HSP70 interacting protein) which bridges the molecular chaperone and ubiquitin-proteasome systems [24]. CHIP contains a U box domain that permits it to conjugate ubiquitin to its substrates and a TPR domain that permits it to bind to molecular chaperones and other TPR domain proteins [25,26].

Functional and Structural Coupling of Molecular Chaperones and the Ubiquitin Proteasome System through Chip

The detailed features of the ubiquitin-proteasome system are reviewed in other chapters of this volume and will not be repeated here (see Chapter 3). In brief, the ubiquitin-proteasome system permits the specific covalent tagging of proteins that are selected for degradation through a protein degradation machine called the proteasome [27,28]. The proteasome contains multiple proteins including a number of proteases arranged in the form of a degradation chamber which permits rapid degradation of substrates to small polypeptides [27,28]. Proteins selected for degradation by the proteasome are first tagged by the addition of a chain of low molecular proteins called ubiquitin [27,28]. Ubiquitin residues are added sequentially to specific lysine residues in the protein to be degraded [27,28]. This process, known as 'ubiquilation' is catalyzed by a three enzyme cascade: Enzymes in the cascade include E1 which recruits ubiquitin, E2 which receives ubiquitin from E1 and E3 the ubiquitin ligase that receives ubiquitin from E2 and couples it to the substrate [27,28]. While there is only one E1 enzyme and a limited number of E2s, there are many E3 ligases which reflect the varied number of substrates in the cell. The E3 ligases which are members of a number of families (the single subunit RING finger type, the multi-subunit RING-finger type and the HECT domain type, UFD2 homology (U box) proteins) permit the selection of a wide variety of motifs in the substrate in order to select the target for ubiquilation; In some cases for instance relative degrees of phosphorylation differentially regulate E3 binding permitting the regulation of protein degradation through protein kinase cascades [29]. This is exemplified by the widely studied the multi-subunit RING-finger type Fbw1 containing E3 ligases that catalyze the ubiquilation of phosphorylated IκBα, β-catenin and Cdc25 and thus control key regulatory pathways [29]. CHIP is a member of the U-box domain family [26,30]. Significantly, its target specificity is provided by its TPR domain which can bind to the molecular chaperones Hsp70, HSC70 and Hsp90 [26,30]. Thus by binding to molecular chaperones, CHIP can target denatured proteins bound to Hsp70, HSC70 or Hsp90 for degradation by catalyzing the poly-ubiquilation of the bound denatured protein [25].

Molecular Chaperones and Proteasomal Degradation

When denatured proteins and protein aggregates accumulate within the aging cell, at least three scenarios can be envisaged based on the known interactions between molecular chaperones and CHIP, as depicted in (Figure 1). Aggregates can continue to form as in *pathway (1)* and lead to cell inactivation and death. However, denatured / aggregated proteins are recognized by the denatured protein binding domains of Hsp70 (and Hsp90) and form complexes with such denatured proteins. Such complexes can next be assembled into protein refolding chaperone machines as in *pathway (2)*. The Hsp use the TPR acceptor sites in their C-terminal domains to bind to the TPR domain-containing scaffold protein HOP as well as co-chaperones Hip, Cyp40, FKB51 and FKB52 in order to initiate an ATP dependent cycle of reactions that leads to refolding of proteins towards native, functional conformations. However, the proteins can take an alternative pathway (3) when the TPR acceptor site of Hsp70 is bound to the TPR domain of CHIP instead of the other co-chaperones; in this case, the substrate protein is brought close to the U box domain of CHIP, becomes polyubiquilated in cooperation with adjacent E1 and E2 enzymes, and finally targeted to the proteasome and is rapidly degraded. Thus the cell has at least two potentially competing pathways for the resolution of protein aggregates. By competing with co-chaperones such as HOP for the TPR acceptor sites on Hsp70 and Hsp90, CHIP can inhibit the refolding pathway while other TPR domain proteins can, in turn inhibit the CHIP pathway [30].

There are in addition, further levels of regulation of the pathways which are mediated through additional co-chaperones that bind to Hsp70. For example, when Hsp70 binds to the co-chaperone BAG-1, there is enhanced interaction of the Hsp70-CHIP1 complex with the proteasome and increased degradation of Hsp70-bound substrates such as glucocorticoid receptor [25]. By contrast, another member of the BAG family, BAG-2 which also binds to Hsp70 inhibits ubiquilation by CHIP and enhances the activity of the refolding pathway [31]. Likewise another Hsp70 co-chaperone, HSPBP1 inhibits substrate ubiquilation and enhances the refolding pathway [32]. These proteins, each of which interact with the ATPase domain of Hsp70 are thus plausible regulatory molecules for the competing pathways of protein aggregate resolution by refolding or degradation and provide further layers of regulation.

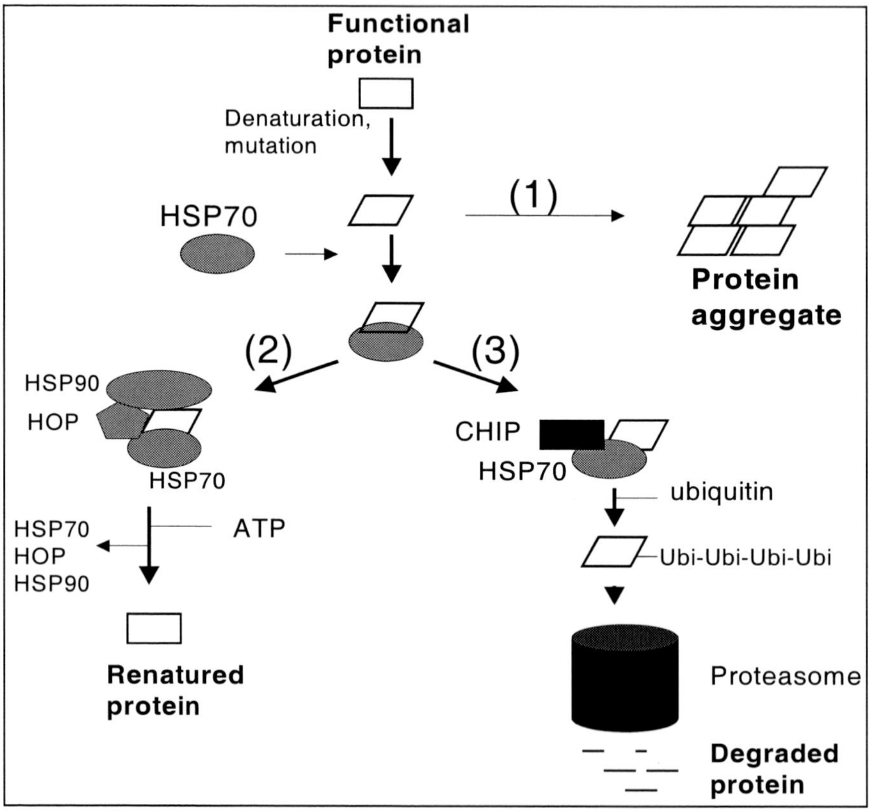

Figure 1. Quality control and salvage pathways for aggregated proteins. Functional proteins may become denatured and aggregated through damage, mutation or interaction with proteins with dominant conformations that lead to aggregation. If not arrested such aggregation [(*pathway (1)*] may lead to cell death. Denatured and aggregated proteins are however recognized by molecular chaperones such as Hsp70 and Hsp90, which form complexes with them. Binding to the Hsp may lead to salvage of the denatured protein through *pathway (2)* in which the denatured protein is refolded through a pathway involving sequential interactions with molecular chaperones Hsp70 and Hsp90 which are bridged through binding to the TPR domains of HOP. However, the proteins may enter a third pathway (3) through binding of the Hsp70 to the TPR domain of the ubiquitin E3 ligase CHIP. Recruitment of CHIP leads to polyubiquilation of the denatured protein and its degradation through the proteasome pathway. These two pathways (2 and 3) thus compete for the denatured proteins through the binding of key TPR domain proteins in each pathway to the TPR acceptor motifs on Hsp70 and Hsp90.

CHANGES IN HSP EXPRESSION AND PROTEIN QUALITY CONTROL WITH AGING

Aging is associated with the degeneration of Hsp expression with time and the loss of resistance to cellular oxidants: elevated HSF1 leads to significant increase in lifespan in *Caernorhabtitis elegans* (*C. elegans*) and *Drosophila* [18,19,33-36]. The effects of HSF1 and Hsp on longevity appear to be particularly mediated through their ability to protect motor neurones and for instance Hsp knockdown in the motor neurones of *Drosophila* causes an equivalent effect on longevity compared to a whole body knockdown [37,38]. The already fragile heat shock response of neuronal cells becomes additionally burdened during exposure to a range of neurodegenerative diseases that lead to the increasing accumulation of insoluble, aggregated proteins and inclusion bodies and the inactivation of components of the proteasomal degradation pathway [39-41]. We will discuss here the mechanisms underlying the loss of heat shock response in aging with particular emphasis on neuronal cells.

Many of the processes in neurodegeneration are accompanied by decreased expression of molecular chaperones with time and the accumulation of tangled and aggregated proteins which are toxic to neurones [42-44]. Studies on mechanisms of aging have implicated HSF1 as an important factor in longevity [33,45]. As mentioned, inhibition of HSF1 expression or function decreases lifespan while extra copies of *HSF1* increase lifespan in *C. elegans* [33,45]. This increase appears to be due to the transcriptional activation of small heat shock proteins (sHsp: *Hsp16.1, Hsp16.49; Hsp12.6*) in *C. elegans* [33]. This increase in longevity due to elevated Hsp expression appears to be related to protection of neuronal function as RNAi antagonism of HSF1 expression led to the accelerated onset of polyglutamine aggregates in *C. elegans* [33]. The sHsp (as well as Hsp70) also play a key role in longevity in *Drosophila* [46,47]. Inactivation of the sHsp family Hsp22 gene in *Drosophila* markedly decreases lifespan and, more significantly, a similar decrease in lifespan is seen if only motor neurones are targeted, strongly implicating a critical role for these tissues in aging and dependence on Hsp [37,38]. The effects of the sHsp may be related to ability to resist the toxicity associated with oxidative stress. A similar role for sHsp in human disease is provided by recent findings showing protective effects for Hsp27 in Huntington's disease [48]. Naturally occurring polymorphisms in Hsp22 and Hsp27 are also associated with motor neurone neuropathy [49]. Thus the decreased expression of sHsp or the occurrence of inactivating mutations in sHsp are associated with neuronal cell death particularly in motor neurones [49]. Heat shock transcription factor 1 (HSF1) is the mammalian regulator of the response to protein stress (including heat, oxidative, ischemic stress) and activates the transcription of heat shock protein (*HSP*) genes [50,51]. Aggregated, denatured and damaged proteins are the common proximal inducers of HSF1 activity which can be observed acutely during the response to heat shock, a potent protein denaturant [52-54]. Disruption of the *hsf1* gene in mouse embryonic fibroblasts leads to a profound loss of thermotolerance in such (*hsf1-/-*) cells and markedly increased susceptibility to heat-induced apoptosis [55]. The aging process is also associated with degeneration of the heat shock response and the thermostability of DNA binding activity of HSF1 was significantly reduced with age in a cell-free system as well as in isolated hepatocytes [56]. Two additional members of the mammalian HSF family HSF2 and HSF4 have also been discovered (reviewed by [51]).

Although their physiological function is not entirely clear, one isoform of HSF2 (HSF2A) cooperates with HSF1 in *HSP* gene transcription [57]. HSF4 functions as an HSF1 repressor [58]. One consistent finding in neuronal cells and tissues is that while glial cells express abundant HSF1 and HSF2, and mount a sturdy heat shock response, the heat shock response is deficient in aging neurones [59-61]. Similar findings were made by Batalan *et al.* who showed that HSF1 fails to be activated in motor neurones even when microinjected with plasmids encoding HSF1 expression vector, suggesting a block to the signal transduction pathways leading to HSF1 expression in these cells [62].

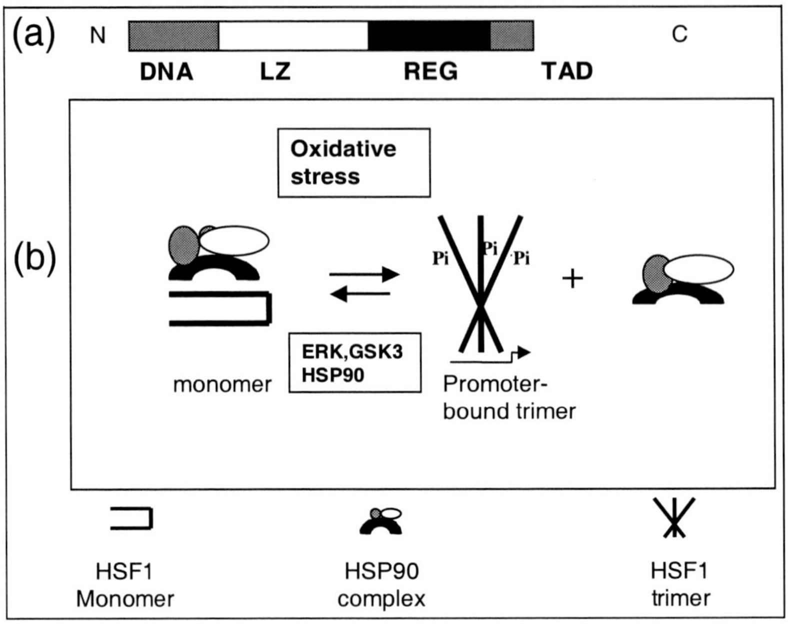

Figure 2. Heat shock factor 1 and HSP gene transcription. (A) Functional domains of HSF1: HSF1 contains an N-terminal DNA binding domain (DNA) adjacent to a trimerization domain containing arrays of leucine zipper regions (LZ). The C-terminus of HSF1 contains two trans-activation domains (TAD). The TAD is under control of the regulatory domain (REG), which converts stress signals into trans activation of Hsp promoters. (B) HSF1 response to stresses that lead to protein aggregation. Under resting conditions, intracellular HSF1 is maintained in an inactive form in high Mr complexes containing the molecular chaperone Hsp90, co-chaperones and other molecules. HSF1 is also repressed by phosphorylation in the regulatory domain by protein kinases ERK1 and GSK3 that mediate nuclear export and cytoplasmic sequestration. Exposure of cells to oxidative stress and damaged proteins results in HSF1 activation by a multi-step reaction, including escape from repressor complexes, nuclear localization, trimerization, hyperphosphorylation and activation of Hsp promoters and requires the expression of CHIP. HSF1 activation is downregulated under conditions in which GSK3 is activated as in aging neuronal cells and when CHIP levels are decreased.

In understanding the deficit in expression of Hsp during aging, it is important to consider what we know about the regulation of HSF1. Under non-stress conditions, HSF1 is transcriptionally repressed and Hsp transcription is minimal [63,64]. In the inactive complex, HSF1 is monomer that is constitutively phosphorylated and lacks the ability to bind the *cis*-acting heat shock elements (HSE) located in the promoters of *HSP* genes (Figure 2) [51,65,66]. Increased levels of denatured and aggregated proteins result in the conversion of

HSF1 from inactive monomer to DNA-binding trimer [65-68]. Activation of HSF1 is a multi-step process, involving trimerization, acquisition of HSE-binding activity and inducible phosphorylation, which results in the transcription of Hsp genes [50,51,63,64]. Hsp90 is the principal cellular repressor of HSF1 in unstressed cells and plays a major role in retaining HSF1 in an inactive state; HSF1 trimerization is accompanied by the sequestration of Hsp90 in protein aggregates and escape from Hsp90-containing HSF1 complexes in response to stress [69,70]. However, little is known regarding the potential role of Hsp90 in the decline of HSF1 during aging. Other important levels of regulation also control HSF1 activity and it was shown recently that 14-3-3 mediated nuclear exclusion represses HSF1 after activation by the protein kinases ERK1 and GSK3 and subsequent recruitment of 14-3-3 [71-73]. HSF1 thus resembles a number of factors in being subject to regulation by phosphorylation and 14-3-3 dependent alterations in nucleocytoplasmic shuttling (Figure 2) [74].

Potential Mechanisms for Decreased HSF Activity and HSP Expression in Aging

The increased levels of protein aggregation with increasing age of the organism may be due to (*i*) progressive accumulation of insoluble protein products with time, (*ii*) the decreased ability of the molecular chaperone / protein degradation system to deal with increased concentrations of aggregated proteins or, more likely (*iii*) a combination of both processes. Most studies show a decrease in *hsp* gene transcription and HSF1 activity in neurones and degeneration of HSF1 activity during aging seems a feature of most tissues [75,76]. The decrease in stress-induced Hsp expression with age was first ascribed to a decrease in HSF1 expression at the protein level and a reduced ability of HSF1 to form DNA binding trimers [59,60]. HSF1 concentration is at critical levels in most cells and activation requires trimerization, a kinetically rare event favored by higher HSF concentrations – reviewed in [51]. Decreases in HSF1 levels thus place the response beneath a key threshold for activation. Subsequent studies also indicated defect in HSF regulation in cultured motor neurones and Batalan *et al* showed that, while HSF1 was not activated in cultured neuronal cells even under conditions of HSF1 overexpression, a construct lacking the regulatory domain could be activated (Figure 1) [62]. The regulatory domain of HSF1 is a major protein interaction region and phosphorylation of this domain by the protein kinases ERK1 and GSK3 leads to HSF1 repression through recruitment of the adaptor protein 14-3-3 and nuclear exclusion (Figure 2) [72,77-79]. Nuclear exclusion leads to loss of HSF1 from *HSP* gene promoters and repression of *HSP* transcription [71,72]. In addition, our recent studies indicate that this domain may also represent a phosphodegron site for ubiquitin E3 ligase binding and modulation of GSK3 may up-and down-regulate HSF1 levels [80]. Interestingly abnormal increases in GSK3 levels occur in some neuronal disorders such as Alzheimer's disease, an effect which may lead to HSF1 repression [81]. 14-3-3 levels also become altered in areas of the brain subjected to prion diseases and the appearance of 14-3-3 proteins in the cerebrospinal fluid is characteristic of some neurodegenerative diseases [82]. It thus seems likely that age-related changes in GSK3 and 14-3-3 levels and activities are involved in the progressive loss of capacity in the heat shock response with time particularly in the CNS. The studies of Batalan also showed that increased expression of HSF2, as opposed to HSF1 in cells exposed to inhibitors of the proteosome leads to activation of Hsp expression [62]. HSF2 can cooperate with HSF1 in the activation of *HSP* promoters, and our studies showed that increased expression of the active isoform of HSF2, (HSF2A) causes a marked increase in stress-induced *HSP* transcription [57]. Altered regulation of both HSF1 and HSF2 may therefore mediate the high threshold for induction of the stress response in motor neurones. HSF1 has also been shown to be regulated by CHIP which mediates the ubiquination of denatured proteins, targeting them for degradation through the proteasomal pathway [83-85]. CHIP is essential for the transcriptional activation of HSF1 [83]. It may be significant that CHIP and HSC70 cooperate with another ubiquitin E3 ligase, Parkin in the degradation of the receptor Pael-R and that defects in this system mediate the changes involved in Parkinson's disease [86]. It would thus be instructive to examine age related CHIP activity in relation to Parkin inactivation in neuronal cells [83,86]. In addition, it has been shown that when CHIP associates with Hsp70 and Hsp90 through its TPR domains it can lead to polyubiquilation of the molecular chaperones themselves in addition to their protein cargo [26]. Dysregulation of Hsp-CHIP interactions and altered Hsp degradation may thus also underlie some of the changes in Hsp levels that accompany aging [26]. It may also be significant that CHIP associates with expanded polyglutamine repeats that accumulate in cells over time and sequestration of CHIP by high concentrations of proteins bearing polyglutamine repeats may mediate the inhibition of HSF1 in neurodegeneration and aging [87]. Finally changes in the transcription of Hsp genes could occur on chromatin, at steps after the signaling stages described above, and alterations in DNA methylation and histone modification may mediate some of the changes in activity of HSF1 and heat shock promoters during aging.

As discussed earlier, our understanding both of HSF1 regulation and *HSP* promoter function is increasing. The reduced levels of Hsp during aging, a decrease which makes neuronal tissues particularly vulnerable to protein damage and cell death, is due to age-related alterations in expression at a number of regulatory levels. Alterations in HSF1 levels, modulations in the functional regulation of HSF1 and changes in the properties of *HSP* promoters have been observed. Although it is known that *HSP* promoters are coordinately regulated by HSF, stress induction of such promoters also involves other independent effects that influence RNA polymerase II activity and chromatin structure [88,89]. In addition other transcription factors target Hsp genes including the FOXO factor DAF 16 and STAT1 and changes in the activity of these factors are likely to make an impact on age-related gene expression [33,90].

Conclusions

Mammalian cells possess powerful mechanisms for protein quality control which permit (1) the resolution of aggregated proteins by molecular chaperones and (2) the degradation of proteins that fail to be salvaged by the molecular chaperone pathway through the ubiquitin-proteasome system. Cross talk between these pathways is regulated by TPR domain proteins such as the scaffold protein HOP that permits assembly of molecular chaperone complexes and CHIP a ubiquitin E3 ligase that mediates the ubiquination and degradation of denatured proteins. These TPR domain proteins compete for Hsp-denatured protein complexes and the outcome of this competition determines the fate of the damaged protein. During aging, the rate of transcription of the Hsp genes decreases and the Hsp as well the co-chaperones become sequestered in protein aggregates. This leads to a decline in activity of the protein quality control pathways, loss of vigor and the symptoms of aging.

REFERENCES

[1] Lindquist S, Craig EA: The heat shock proteins. *Ann Rev Genet* 1988, 22:631-637.

[2] Bukau B, Horwich AL: The Hsp70 and Hsp60 chaperone machines. *Cell* 1998, 92:351-366.

[3] Tang D, Khaleque AA, Jones ER, Theriault JR, Li C, Wong WH, Stevenson MA, Calderwood SK: Expression of heat shock proteins and Hsp messenger ribonucleic acid in human prostate carcinoma *in vitro* and in tumors *in vivo*. *Cell Stress Chaperones* 2005, 10:46-59.

[4] Easton DP Kaneko Y, Subjeck JR: The hsp110 and Grp1 70 stress proteins: newly recognized relatives of the Hsp70s. *Cell Stress Chaperones* 2000, 5:276-290.

[5] Nylandsted J, Brand K, Jaattela M: Heat shock protein 70 is required for the survival of cancer cells. *Ann N Y Acad Sci.* 2000, 926:122-125.

[6] Beere HM: 'The stress of dying': the role of heat shock proteins in the regulation of apoptosis. *J Cell Sci* 2004, 117: 2641-2651.

[7] Buchner J: Hsp 90& Co.- a holding for folding. *Trends Biochem Sci* 1999, 24:136-142.

[8] Wegele H, Muller L, Buchner J: Hsp70 and Hsp90--a relay team for protein folding. *Rev Physiol Biochem Pharmacol* 2004, 151:1-44.

[9] Mayer MP Bukau B: Hsp70 chaperones: cellular functions and molecular mechanism. *Cell Mol Life Sci* 2005, 62:670-684.

[10] Pratt WB Toft DO: Regulation of signaling protein function and trafficking by the Hsp90/Hsp70-based chaperone machinery. *Exp Biol Med (Maywood)* 2003, 228:111-133.

[11] Spiess C, Meyer AS, Reissmann S, Frydman J: Mechanism of the eukaryotic chaperonin: protein folding in the chamber of secrets. *Trends Cell Biol* 2004, 14:598-604.

[12] Young JC, Agashe VR, Siegers K, Hartl FU: Pathways of chaperone-mediated protein folding in the cytosol. *Nat Rev Mol Cell Biol* 2004, 5:781-791.

[13] Tomoyasu T, Mogk A, Langen H, Goloubinoff P, Bukau B: Genetic dissection of the roles of chaperones and proteases in protein folding and degradation in the Escherichia coli cytosol. *Mol Microbiol* 2001, 40:397-413.

[14] Arrigo AP: [Heat shock proteins as molecular chaperones]. *Med Sci (Paris)* 2005, 21:619-625.

[15] D'Andrea LD, Regan L: TPR proteins: the versatile helix. *Trends Biochem Sci* 2003, 28:55-62.

[16] Cheung-Flynn J, Roberts PJ, Riggs DL, Smith DF: C-terminal sequences outside the tetratricopeptide repeat domain of FKBP51 and FKBP52 cause differential binding to Hsp90. *J Biol Chem* 2003, 278:17388-17394.
[17] Schlessinger MJ: How the cell copes with stress and the function of heat shock proteins. *Pediatric Research* 1994, 36:1-6.
[18] Lee YK, Manalo D, Liu AY: Heat shock response, heat shock transcription factor and cell aging. *Biol Signals* 1996, 5:180-191.
[19] Liu AY, Lee YK, Manalo D, Huang LE: Attenuated heat shock transcriptional response in aging: molecular mechanism and implication in the biology of aging. *Exs* 1996, 77:393-408.
[20] Hut HM, Kampinga HH, Sibon OC: Hsp70 protects mitotic cells against heat-induced centrosome damage and division abnormalities. *Mol Biol Cell* 2005, 16:3776-3785.
[21] Nylandsted J, Gyrd-Hansen M, Danielewicz A, Fehrenbacher N, Lademann U, Hoyer-Hansen M, Weber E, Multhoff G, Rohde M, Jaattela M: Heat shock protein 70 promotes cell survival by inhibiting lysosomal membrane permeabilization. *J Exp Med* 2004, 200:425-435.
[22] Calderwood S: Regulatory interfaces between the stress protein response and other gene expression programs in the cell. *Methods* 2005, 23:139-148.
[23] Agarraberes FA, Dice JF: A molecular chaperone complex at the lysosomal membrane is required for protein translocation. *J Cell Sci* 2001, 114:2491-2499.
[24] Ballinger CA, Connell P, Wu Y, Hu Z, Thompson LJ, Yin LY, Patterson C: Identification of CHIP, a novel tetratricopeptide repeat-containing protein that interacts with heat shock proteins and negatively regulates chaperone functions. *Mol Cell Biol* 1999, 19:4535-4545.
[25] Demand J, Alberti S, Patterson C, Hohfeld J: Cooperation of a ubiquitin domain protein and an E3 ubiquitin ligase during chaperone/proteasome coupling. *Curr Biol* 2001, 11:1569-1577.
[26] Jiang J, Ballinger CA, Wu Y, Dai Q, Cyr DM, Hohfeld J, Patterson C: CHIP is a U-box-dependent E3 ubiquitin ligase: identification of Hsc70 as a target for ubiquitylation. *J Biol Chem* 2001, 276:42938-42944.
[27] Hershko A: The ubiquitin system for protein degradation and some of its roles in the control of the cell division cycle. *Cell Death Differ* 2005, 12:1191-1197.
[28] Ciechanover A: Intracellular protein degradation: from a vague idea thru the lysosome and the ubiquitin-proteasome system and onto human diseases and drug targeting. *Cell Death Differ* 2005, 12:1178-1190.
[29] Cardozo T Pagano M: The SCF ubiquitin ligase: insights into a molecular machine. *Nat Rev Mol Cell Biol* 2004, 5:739-751.
[30] Connell P, Ballinger CA, Jiang J, Wu Y, Thompson LJ, Hohfeld J, Patterson C: The co-chaperone CHIP regulates protein triage decisions mediated by heat-shock proteins. *Nat Cell Biol* 2001, 3:93-96.
[31] Arndt V, Daniel C, Nastainczyk W, Alberti S, Hohfeld J: BAG-2 Acts as an Inhibitor of the Chaperone-associated Ubiquitin Ligase CHIP. *Mol Biol Cell* 2005, 16:5891-5900.

[32] Alberti S, Bohse K, Arndt V, Schmitz A, Hohfeld J: The cochaperone HspBP1 inhibits the CHIP ubiquitin ligase and stimulates the maturation of the cystic fibrosis transmembrane conductance regulator. *Mol Biol Cell* 2004, 15:4003-4010.

[33] Hsu AL, Murphy CT, Kenyon C: Regulation of aging and age-related disease by DAF-16 and heat-shock factor. *Science* 2003, 300:1142-1145.

[34] Verbeke P, Fonager J, Clark BF, Rattan SI: Heat shock response and ageing: mechanisms and applications. *Cell Biol Int* 2001, 25:845-857.

[35] Bonelli MA, Alfieri RR, Poli M, Petronini PG, Borghetti AF: Heat-induced proteasomic degradation of HSF1 in serum-starved human fibroblasts aging in vitro. *Exp Cell Res* 2001, 267:165-172.

[36] Gutsmann-Conrad A, Heydari AR, You S, Richardson A: The expression of heat shock protein 70 decreases with cellular senescence in vitro and in cells derived from young and old human subjects. *Exp Cell Res* 1998, 241:404-413.

[37] Morrow G, Battistini S, Zhang P, Tanguay RM: Decreased lifespan in absence of expression of the mitochondrial small heat shock protein Hsp22 in Drosophila. *J Biol Chem* 2004, 279:43382-43385.

[38] Morrow G, Samson M, Michaud S, Tanguay RM: Overexpression of the small mitochondrial Hsp22 extends Drosophila life span and increases resistance to oxidative stress. *FASEB J* 2004, 18:598-599.

[39] Helfand S L: Neurobiology. Chaperones take flight. *Science* 2002, 295:809-810.

[40] Cowan KJ, Diamond MI, Welch WJ: Polyglutamine protein aggregation and toxicity are linked to the cellular stress response. *Hum Mol Genet* 2003, 12:1377-1391.

[41] Berke SJ, Paulson HL: Protein aggregation and the ubiquitin proteasome pathway: gaining the UPPer hand on neurodegeneration. *Curr Opin Genet Dev* 2003, 13:253-261.

[42] Harrison PJ, Procter AW Exworthy T, Roberts GW, Najlerahim A, Barton AJ, Pearson RC: Heat shock protein (hsx70) mRNA expression in human brain: effects of neurodegenerative disease and agonal state. *Neuropathol Appl Neurobiol* 1993, 19:10-21.

[43] Muchowski PJ, Schaffar G, Sittler A, Wanker EE, Hayer-Hartl MK, and Hartl FU: Hsp70 and hsp40 chaperones can inhibit self-assembly of polyglutamine proteins into amyloid-like fibrils. *Proc Natl Acad Sci U S A* 2000, 97:7841-7846.

[44] Zatloukal K, Stumptner C, Fuchsbichler A, Heid H, Schnoelzer M, Kenner L, Kleinert R, Prinz M, Aguzzi A, Denk H: p62 Is a common component of cytoplasmic inclusions in protein aggregation diseases. *Am J Pathol* 2002, 160:255-263.

[45] Garigan D, Hsu AL, Fraser AG, Kamath RS, Ahringer J, Kenyon, C: Genetic analysis of tissue aging in Caenorhabditis elegans: a role for heat-shock factor and bacterial proliferation. *Genetics* 2002, 161:1101-1112.

[46] Kurapati R, Passananti HB, Rose MR, Tower J: Increased hsp22 RNA levels in Drosophila lines genetically selected for increased longevity. *J Gerontol A Biol Sci Med Sci* 2000, 55: 552-559.

[47] Tatar M, Khazaeli AA, Curtsinger J W: Chaperoning extended life. *Nature* 1997, 390: 30.

[48] Wyttenbach A, Sauvageot O, Carmichael J, Diaz-Latoud C, Arrigo AP, Rubinsztein DC: Heat shock protein 27 prevents cellular polyglutamine toxicity and suppresses the increase of reactive oxygen species caused by huntingtin. *Hum Mol Genet* 2002, 11:1137-1151.

[49] Benndorf R, Welsh, MJ: Shocking degeneration. *Nat Genet* 2004, 36:547-548.

[50] Voellmy R: Transduction of the stress signal and mechanisms of transcriptional regulation of heat shock / stress protein expression in higher eukaryotes. *Crit Rev Eukaryotic Gene Expr* 1994, 4:357-401.

[51] Wu C: Heat shock transcription factors: structure and regulation. *Ann Rev Cell Dev Biol* 1995, 11:441-469.

[52] Kampinga HH, Brunsting JF, Stege GJ, Burgman PW, Konings AW: Thermal protein denaturation and protein aggregation in cells made thermotolerant by various chemicals: role of heat shock proteins. *Exp Cell Res* 1995, 219:536-546.

[53] Kampinga HH: Hyperthermia, thermotolerance and topoisomerase II inhibitors. *Br J Cancer* 1995, 72:333-338.

[54] Ananthan J, Goldberg AL, Voellmy R: Abnormal proteins serve as eukaryotic stress signals and trigger the activation of heat shock genes. *Science* 1986, 232: 522-524.

[55] McMillan DR, Xiao X, Shao L, Graves K, Benjamin IJ: Targeted disruption of heat shock transcription factor 1 abolishes thermotolerance and protection against heat-inducible apoptosis. *J Biol Chem* 1998, 273:7523-7528.

[56] Heydari AR, You S, Takahashi R, Gutsmann-Conrad A, Sarge KD, Richardson A: Age-related changes in activation of heat shock transcription factor 1 in rat hepatocytes. *Exp Cell Res*. 2000, 256:83-93.

[57] He H, Soncin F, Grammatikakis N, Li Y, Siganou A, Gong J, Brown SA, Kingston RE, Calderwood SK: Elevated expression of heat shock factor 2a stimulates HSF1-induced transcription during stress. *J Biol Chem* 2003, 278:35465-35475.

[58] Nakai A, Tanabe M, Kawazoe Y, Inazawa J, Morimoto RI, Nagata K: HSF4, a new member of the human heat shock factor family which lacks properties of a transcriptional activator. *Mol Cell Biol* 1997, 17:469-481.

[59] Marcuccilli CJ, Mathur SK, Morimoto RI, Miller RJ: Regulatory differences in the stress response of hippocampal neurons and glial cells after heat shock. *J Neurosci* 1996, 16:478-485.

[60] Brown I R, Rush SJ: Cellular localization of the heat shock transcription factors HSF1 and HSF2 in the rat brain during postnatal development and following hyperthermia. *Brain Res* 1999, 821:333-340.

[61] Kaarniranta K, Oksala N, Karjalainen HM, Suuronen T, Sistonen L, Helminen HJ, Salminen A, Lammi MJ: Neuronal cells show regulatory differences in the Hsp70 gene response. *Brain Res Mol Brain Res* 2002, 101:136-140.

[62] Batulan Z, Shinder GA, Minotti S, He BP, Doroudchi MM, Nalbantoglu J, Strong MJ, Durham HD: High threshold for induction of the stress response in motor neurons is associated with failure to activate HSF1. *J Neurosci* 2003, 23:5789-5798.

[63] Price BD, Calderwood SK: Calcium is essential for multistep activation of the heat shock factor in permeabilized cells. *Mol Cell Biol* 1991, 11:3365-3368.

[64] Hensold JO, Hunt CR, Calderwood SK, Houseman DE, Kingston RE: DNA binding of heat shock factor to the heat shock element is insufficient for transcriptional activation in murine erythroleukemia cells. *Mol Cell Biol* 1990, 10: 1600-1608.

[65] Rabindran SK, Haroun RI, Clos J, Wisniewski J, Wu C: Regulation of heat shock factor trimer formation: role of a conserved leucine zipper. *Science* 1993, 259:230-234.

[66] Baler R, Dahl G, Voellmy R: Activation of human heat shock transcription is accompanied by oligomerization, modification and rapid translocation of heat shock transcription facor HSF-1. *Mol Cell Biol* 1993, 13:2486-2496.

[67] Westwood T, Wu C: Activation of *drosophila* heat shock factor: conformational changes associated with monomer-to-trimer transition. *Mol Cell Biol.* 1993, 13:3481-3486.

[68] Sarge KD, Murphy SP, Morimoto RI: Activation of heat shock gene transcription by heat shock factor 1 involves oligomerization, acquisition of DNA-binding activity, and nuclear localization and can occur in the absence of stress. *Mol Cell Biol* 1993, 13:1392-1407.

[69] Zou J, Guo Y, Guettouche T, Smith DF, Voellmy R: Repression of heat shock transcription factor HSF1 activation by Hsp90 (Hsp90 complex) that forms a stress-sensitive complex with HSF1. *Cell* 1998, 94:471-480.

[70] Guo Y, Guettouche T, Fenna M, Boellmann F, Pratt WB, Toft DO, Smith DF, Voellmy R: Evidence for a mechanism of repression of heat shock factor 1 transcriptional activity by a multichaperone complex. *J Biol Chem* 2001, 276:45791-45799.

[71] Wang X, Grammatikakis N, Siganou A, Stevenson MA, Calderwood SK: Interactions between extracellular signal regulated protein kinase 1 (ERK1), 14-3-3 epsilon and heat shock factor 1 during stress. *J Biol Chem* 2004, 279:49460-49469.

[72] Wang X, Grammatikakis N, Siganou A, Calderwood SK: Regulation of molecular chaperone gene transcription involves the serine phosphorylation, 14-3-3 epsilon binding, and cytoplasmic sequestration of heat shock factor 1. *Mol Cell Biol* 2003, 23:6013-6026.

[73] He B, Meng YH, Mivechi NH: Glycogen synthase kinase 3b and extracellular signal-regulated protein kinase inactivate heat shock transcription factor 1 by facilitating the disappearance of transcriptionally active granules after heat shock. *Mol Cell Biol* 1998, 18:6624-6632.

[74] Brunet A, Kanai F, Stehn J, Xu J, Sarbassova D, Frangioni JV, Dalal SN, DeCaprio JA, Greenberg, ME, Yaffe MB: 14-3-3 transits to the nucleus and participates in dynamic nucleocytoplasmic transport. *J Cell Biol* 2002, 156:817-828.

[75] Shamovsky I, Gershon, D: Novel regulatory factors of HSF-1 activation: facts and perspectives regarding their involvement in the age-associated attenuation of the heat shock response. *Mech Ageing Dev* 2004, 125:767-775.

[76] Walker GA, Thompson FJ, Brawley A, Scanlon T, Devaney E: Heat shock factor functions at the convergence of the stress response and developmental pathways in Caenorhabditis elegans. *FASEB J* 2003, 17:1960-1962.

[77] Chu B, Soncin F, Price BD, Stevenson MA, Calderwood SK: Sequential phosphorylation by mitogen-activated protein kinase and glycogen synthase kinase 3

represses transcriptional activation by heat shock factor-1. *J Biol Chem* 1996, 271:30847-30857.

[78] Chu B, Zhong R, Soncin F, Stevenson MA, Calderwood SK: Transcriptional activity of heat shock factor 1 at 37°C is repressed through phosphorylation on two distinct serine residues by glycogen synthase kinase 3 and protein kinase Cα and Cζ. *J Biol Chem* 1998, 273: 18640-18646.

[79] Holmberg C I, Hietakangas V, Mikhailov A, Rantanen JO, Kallio M, Meinander A, Hellman J, Miorrice N, Mackintosh C, Morimoto RI, Ericksson JE, Sistonen L: Phosphorylation of serine 230 promotes inducible transcriptional activity of heat shock factor 1. *EMBO J.* 2001, 20:3800-3810.

[80] Khaleque MA, Bharti A, Sawyer D, Gong J, Benjamin IJ, Stevenson MA, Calderwood SK: Induction of heat shock proteins by heregulin beta1 leads to protection from apoptosis and anchorage-independent growth. *Oncogene* 2005, 24:6564-6573.

[81] Bhat RV, Budd Haeberlein SL, Avila J: Glycogen synthase kinase 3: a drug target for CNS therapies. *J Neurochem* 2004, 89:1313-1317.

[82] Van Everbroeck BR, Boons J, Cras P: 14-3-3 {gamma}-isoform detection distinguishes sporadic Creutzfeldt-Jakob disease from other dementias. *J Neurol Neurosurg Psychiatry* 2005, 76:100-102.

[83] Dai Q, Zhang C, Wu Y, McDonough H, Whaley RA, Godfrey V, Li HH, Madamanchi N, Xu W, Neckers L, Cyr D, Patterson C: CHIP activates HSF1 and confers protection against apoptosis and cellular stress. *EMBO J* 2003, 22:5446-5458.

[84] Zhang C, Xu Z, He XR, Michael LH, Patterson C: CHIP, a co-chaperone/ubiquitin ligase that regulates protein quality control, is required for maximal cardioprotection after myocardial infarction in mice. *Am J Physiol Heart Circ Physiol 2005, 288: H2836-H2842.*

[85] Sahara N, Murayama M, Mizoroki T, Urushitani M, Imai Y, Takahashi R, Murata S, Tanaka K, Takashima A: In vivo evidence of CHIP up-regulation attenuating tau aggregation. *J Neurochem* 2005, 94:1254-1263.

[86] Takahashi R, Imai Y: Pael receptor, endoplasmic reticulum stress, and Parkinson's disease. *J Neurol* 2003, 250 Suppl 3: III25-29.

[87] Jana NR, Dikshit P, Goswami A, Kotliarova S, Murata S, Tanaka K, Nukina N: Co-chaperone CHIP associates with expanded polyglutamine protein and promotes their degradation by proteasomes. *J Biol Chem* 2005, 280:11635-11640.

[88] Thomson S, Hollis A, Hazzalin CA, Mahadevan LC: Distinct stimulus-specific histone modifications at Hsp70 chromatin targeted by the transcription factor heat shock factor-1. *Mol Cell* 2004, 15:585-594.

[89] Smith ST, Petruk S, Sedkov Y, Cho E, Tillib S, Canaani E, Mazo A: Modulation of heat shock gene expression by the TAC1 chromatin-modifying complex. *Nat Cell Biol* 2004, 6:162-167.

[90] Stephanou A, Latchman DS: Transcriptional regulation of the heat shock protein genes by STAT family transcription factors. *Gene Expr* 1999, 7:311-319.

In: The Ubiquitin Proteasome System… ISBN 978-1-60021-749-4
Eds: Mario Di Napoli and Cezary Wojcik, pp. 553-577

Chapter 23

IMPAIRMENT OF THE UBIQUITIN-PROTEASOME SYSTEM: A COMMON PATHOGENIC MECHANISM IN NEURODEGENERATIVE DISORDERS

***Lian Li*[*] *and Lih-Shen Chin*[¥]**

Department of Pharmacology, Center for Neurodegenerative Disease, Emory University School of Medicine, Atlanta, GA 30322-3090, USA.

ABSTRACT

The causes of various neurodegenerative diseases, particularly sporadic cases, remain unknown, but increasing evidence suggests that these diseases may share similar molecular and cellular mechanisms of pathogenesis. One prominent feature common to most neurodegenerative diseases is the accumulation of misfolded proteins in the form of insoluble protein aggregates or inclusion bodies. Although these aggregates have different protein compositions, they all contain ubiquitin and proteasome subunits, implying a failure of the ubiquitin-proteasome system (UPS) in the removal of misfolded proteins. A direct link between UPS dysfunction and neurodegeneration has been provided by recent findings that genetic mutations in UPS components cause several rare, familial forms of neurodegenerative diseases. Furthermore, it is becoming increasingly clear that oxidative stress, which results from aging or exposure to environmental toxins, can directly damage UPS components, thereby contributing to the pathogenesis of sporadic forms of neurodegenerative diseases. Aberrations in the UPS

[*] Correspondence concerning this article should be addressed to Dr. Lian Li, PhD; Department of Pharmacology, Emory University School of Medicine, 1510 Clifton Road, Atlanta, GA 30322-3090, USA. Phone: 404-727-5987; Fax: 404-727-0365; E-mail: lianli@pharm.emory.edu.

[¥] Correspondence concerning this article should be addressed to Dr. Lih-Shen Chin, PhD; Department of Pharmacology, Emory University School of Medicine, 1510 Clifton Road, Atlanta, GA, 30322-3090, USA. Phone 404-727-0361; Fax. 404-727-0365; E-mail: chinl@pharm.emory.edu.

often result in defective proteasome-mediated protein degradation, leading to accumulation of toxic proteins and eventually to neuronal cell death. Interestingly, emerging evidence has begun to suggest that impairment in substrate-specific components of the UPS, such as E3 ubiquitin-protein ligases, may cause aberrant ubiquitination and neurodegeneration in a proteasome-independent manner. This chapter provides an overview of the molecular components of the UPS and their impairment in familial and sporadic forms of neurodegenerative diseases, and summarizes present knowledge about the pathogenic mechanisms of UPS dysfunction in neurodegeneration.

Keywords: Neurodegenerative disorders, Aggregation, E3 ubiquitin-protein ligase, Deubiquitinating enzyme, Oxidative stress, Proteasome, Ubiquitin.

Abbreviations

Aβ, β-amyloid; AD, Alzheimer's disease; ALS, amyotrophic lateral sclerosis; APP, amyloid precursor protein; AR-JP, autosomal recessive juvenile Parkinsonism; CHIP, C terminus of Hsc70-interacting protein; CMT, Charcot-Marie-Tooth disease; DUB, deubiquitinating enzyme; E1, ubiquitin-activating enzyme; E2, ubiquitin-conjugating enzyme; E3, ubiquitin-protein ligase; ER, endoplasmic reticulum; ERAD, ER-associated degradation; HD, Huntington's disease; HECT, Homologous to E6AP C terminus; HOIL-1, Haem-oxidized iron regulatory protein 2 ubiquitin ligase-1; IRP2, iron regulatory protein 2; MPTP, 1-methyl-4-phenyl-1,2,3,6-tetrahydropyridine; NO, nitric oxide; PD, Parkinson's disease; PGJ2, 15-deoxy-delta (12, 14)-prostaglandin J2; PrP, Prion protein; RING, Really Interesting New Gene; RNS, reactive nitrogen species; ROS, reactive oxygen species; SCA, spinocerebellar ataxia; SCA-3, spinocerebellar ataxia type-3; SIMPLE, small integral membrane protein of the lysosome/late endosome; Ub, ubiquitin; UBA, ubiquitin-associated; UBL, ubiquitin-like; UBP, ubiquitin processing protease; UCH, ubiquitin carboxy-terminal hydrolase; UCH-L1, ubiquitin carboxy-terminal hydrolase L1; UIM, ubiquitin-interacting motif; UPS, ubiquitin-proteasome system.

Introduction

Neurodegenerative diseases are characterized by the selective loss of neurons in specific brain regions and the deposition of misfolded proteins into aggregates or inclusions, such as neurofibrillary tangles and neuritic plaques in Alzheimer's disease (AD), Lewy bodies in Parkinson's disease (PD), Bunina bodies in amyotrophic lateral sclerosis (ALS), and nuclear and cytoplasmic inclusions in polyglutamine expansion disorders such as Huntington's disease (HD) and spinocerebellar ataxias (SCAs) [1,2]. The accumulation of protein aggregates in these diseases is likely due to a chronic imbalance between the generation and clearance of misfolded proteins. Misfolded proteins can be generated by genetic mutations or chemical modifications such as oxidation and glycation. Pathogenic mutations, such as PD-linked missense mutations in α-synuclein and HD-associated polyglutamine expansions in huntingtin, have been shown to cause neurodegeneration directly by inducing abnormal protein conformations and aggregation [3]. Oxidative stress, which results from aging or exposure to pesticides and other environmental toxins [4], is a major cause of protein misfolding responsible for the progressive buildup of damaged proteins in common sporadic forms of neurodegenerative diseases. Since the ubiquitin-proteasome system (UPS) plays a major role in selective degradation of misfolded and damaged proteins, the accumulation of protein aggregates enriched with ubiquitin and components of the UPS in various neurodegenerative diseases [2,5] suggests a potential involvement of dysfunctional UPS in the formation of these aggregates.

Remarkable progress has been made over the past several years in our understanding of the UPS and its diverse roles in regulation of numerous cellular processes, including neuronal function and dysfunction. We now know that the UPS is not just a constitutive degradation machine for garbage disposal, but rather, it is a complex and tightly regulated system for controlling ubiquitination and degradation of abnormal (misfolded or damaged) as well as normal proteins in cells [6,7]. Recent identification of mutations in UPS components as the genetic defects responsible for several monogenic familial forms of neurodegenerative diseases points to a causative role of UPS dysfunction in neurodegeneration. Studies of the mutant gene products have begun to suggest that abnormal protein ubiquitination could cause neurodegeneration in a proteasome-dependent and/or a proteasome-independent manner. Moreover, emerging evidence indicates that oxidative stress directly damages the UPS, and thereby contributes to the pathogenesis of sporadic forms of neurodegenerative diseases. In this chapter, we review recent advances in our understanding of the UPS and its regulation, and discuss the pathogenic mechanisms by which impaired UPS components cause neurodegeneration in familial and sporadic forms of neurodegenerative diseases.

Molecular Components of the Ubiquitin-Proteasome System

Protein degradation via the UPS is a major intracellular proteolytic pathway that not only eliminates misfolded and damaged proteins, but also selectively degrades normal cellular proteins, and thereby regulates diverse biological processes, including differentiation, neurotransmission, and apoptosis [7,8]. In the UPS, substrates are first tagged by covalent linkage to multiple molecules of ubiquitin, a 76-amino-acid polypeptide. The ubiquitinated substrate proteins are subsequently recognized and degraded by the 26S proteasome (Figure 1). Conjugation of ubiquitin to a substrate is a multi-step process that requires sequential action of three enzymes. First, ubiquitin is activated by the ubiquitin-activating enzyme (E1) at the expense of ATP (see Chapter 3). The activated ubiquitin is then transferred to an ubiquitin-conjugating enzyme (E2). The ubiquitin-protein ligase (E3) specifically recognizes the substrate, which can be either a normal or an abnormal protein, and catalyzes the last step of the ubiquitination process, i.e., the transfer of the activated ubiquitin from the E2 to the substrate. In most cases, ubiquitin is covalently conjugated to the substrate through formation of an isopeptide bond between the carboxyl group of the C-terminal glycine residue on ubiquitin and the ε-amino group of a lysine residue on the substrate. Successive conjugation of ubiquitin moieties to a lysine residue of the previously conjugated ubiquitin results in the formation of a polyubiquitin chain.

A polyubiquitin chain containing at least four ubiquitin moieties linked through K48 serves as a signal to target substrates for degradation by the 26S proteasome [6]. The proteasome is composed of a barrel-shaped 20S catalytic core, capped on either end by a 19S regulatory complex [9]. The 19S complex recognizes polyubiquitinated substrates and assists in unfolding and translocation of the substrate into the proteolytic chamber of the 20S core for degradation into small peptides. The polyubiquitin chain is removed from the substrate prior to entering the proteolytic core, and is recycled to free ubiquitin by the action of a deubiquitinating enzyme (DUB) (see Chapter 4). Recent evidence indicates that additional factors, such as proteins containing ubiquitin-like (UBL) and ubiquitin-associated (UBA) domains, are involved in the recognition and delivery of polyubiquitinated substrates to the 26S proteasome for unfolding and degradation [10; and Chapter 5].

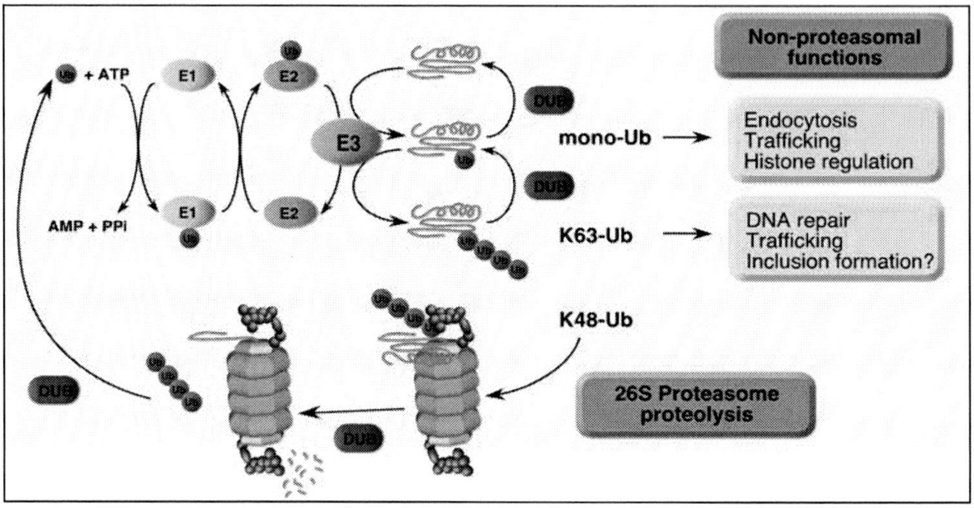

Figure 1. Molecular mechanisms of protein ubiquitination and degradation by the UPS. Ubiquitination involves a highly specific enzyme cascade in which ubiquitin (Ub) is first activated by the ubiquitin-activating enzyme (E1), then transferred to an ubiquitin-conjugating enzyme (E2), and finally covalently attached to the substrate by an ubiquitin-protein ligase (E3). Ubiquitination is a reversible posttranslational modification in which the removal of Ub is mediated by a deubiquitinating enzyme (DUB). Substrate proteins can be either monoubiquitinated or polyubiquitinated through successive conjugation of Ub moieties to an internal lysine residue in Ub. K48-linked poly-Ub chains are recognized by the 26S proteasome, resulting in degradation of the substrate and recycling of Ub. Monoubiquitination or K63-linked polyubiquitination plays a number of regulatory roles in cells that are proteasome-independent.

Although proteasome-mediated degradation is the best-known role of ubiquitination, it is becoming increasingly clear that ubiquitination also serves non-proteasomal functions to modulate protein activity, location, and interactions in a manner analogous to phosphorylation [11]. Monoubiquitination at single or multiple lysine residues of a substrate plays a signaling role in various biological processes, including endocytosis, endosomal sorting, histone modification, and viral budding [12]. Polyubiquitination linked through K63 has been shown to modulate a number of cellular functions, such as DNA repair, translation, kinase activation, and protein trafficking [6,11]. Interestingly, a recent study suggests that K63-linked polyubiquitination might have a role in the formation of protein inclusions in Parkinson's disease [13]

Protein ubiquitination and degradation via the UPS are highly specific and tightly regulated processes. Among the molecular components of the UPS, the E3 ligases are perhaps the most important players because they recruit the substrates for ubiquitination and determine the timing and specificity of protein degradation. E3 ligases are either single proteins or multi-subunit protein complexes. They are classified into two major groups: HECT (Homologous to E6AP C terminus) domain- and RING (Really Interesting New Gene) finger-containing E3s [7,14]. Another class of E3s that has been recently described is the U-box-containing E3s, such as CHIP (C terminus of Hsc70-interacting protein), which may operate as an elongation factor (also known as E4) for the assembly of polyubiquitin chains on a substrate [15]. Furthermore, E3-mediated substrate recognition and ubiquitination can be modulated by phosphorylation, oxidation, and interactions with other proteins. This modulation is the primary regulated step in ubiquitin-mediated proteolysis [7]. Consistent with a crucial role for E3 ligases in selective protein ubiquitination, it is estimated that the human genome contains more than six hundred E3 ligases, in contrast to a single E1 ubiquitin-activating enzyme and about two dozen E2 ubiquitin-conjugating enzymes [7,14,16]. Another group of key regulators of protein ubiquitination and degradation are the DUBs or deubquitinating enzymes, which catalyze the ubiquitin deconjugation reaction [17]. Although the functions of this group of enzymes remain largely unknown, the predicted presence of more than a hundred different DUBs in the human genome [17,18] suggests that DUBs play specific, diverse roles in regulation of cellular processes beyond simply recycling ubiquitin. In support of this notion, emerging evidence points to a role for DUB-mediated substrate-specific deubiquitination in modulation of gene silencing, protein trafficking, and NF-κB signaling [17].

A question important to understanding the pathogenesis of neurodegenerative diseases is how misfolded and damaged proteins are selectively recognized and degraded by the UPS. Although the ability of cells to selectively degrade abnormal proteins has been known for more than three decades, the molecular mechanisms underlying such selective degradation remain poorly understood. Recently, the U-box type E3 ubiquitin-protein ligase CHIP has been shown to ubiquitinate certain misfolded proteins in a process that requires molecular chaperone Hsp70 for recognition of misfolded substrates [19; and Chapter 19]. Furthermore, a novel E3 ligase, HOIL-1 [Haem-oxidized iron regulatory protein 2 (IRP2) ubiquitin ligase-1], can specifically recognize and ubiquitinate oxidized IRP2 but not non-oxidized IRP2, and target oxidized IRP2 for degradation by the proteasome [20]. Thus, selective ubiquitination and degradation of misfolded and damaged proteins could be mediated through the binding of an E3 ligase or its co-factor(s) to an oxidation-induced motif or exposed hydrophobic regions due to misfolding. On the other hand, there is evidence suggesting that ubiquitination is not essential for proteasomal degradation of all abnormal proteins in cells [21]. For example, several oxidized proteins such as calmodulin and ovalbumin or 'natively unfolded' proteins, such as α-synuclein, have been shown to undergo ubiquitin-independent degradation by the proteasome [22,23]. At present, it is unclear how these unfolded proteins are selectively recognized by the proteasome.

UPS AND PROTEIN AGGREGATION

Accumulation of protein aggregates containing misfolded proteins is a common pathological feature of many neurodegenerative diseases [1,2]. These protein aggregates exhibit strong immunoreactivity to antibodies against ubiquitin or ubiquitin-protein conjugates [24,25], providing the first clue that the UPS may play a role in aggregate formation and disease progression. The protein composition of these aggregates is disease-specific. For example, α-synuclein is the major component of Lewy bodies in PD [26]; tau and β-amyloid (Aβ) peptides, which are cleavage products of the the amyloid precursor protein (APP), are the main constituents of intracellular neurofibrillary tangles and extracellular amyloid plaques in AD, respectively [27]. In addition to the core components, the aggregates also contain a variety of other proteins, many of which are posttranslationally modified by oxidation and nitration [28]. The core components of protein aggregates are usually natively unfolded, aggregation-prone proteins. A critical role for the accumulation of these core abnormal proteins in disease pathogenesis is underscored by the identification of mutations in the genes encoding these proteins. For instance, missense mutations in α-synuclein and APP cause familial PD and AD, respectively [1]. Furthermore, an increase in the gene dosage, such as triplication or duplication of the α-synuclein locus [29,30], can also lead to disease, suggesting that excess levels of wild-type α-synuclein protein is sufficient to cause PD.

Whether protein aggregates are cytotoxic or cytoprotective remains a hotly debated issue. The presence of protein aggregates in nearly every known neurodegenerative disease suggests that the protein aggregates per se, or some event associated with the protein aggregation process, is toxic to neurons. In support of this notion, many of the mutations that cause dominantly inherited neurodegenerative diseases have been shown to promote protein misfolding and aggregation [3]. For example, familial PD-linked missense mutations in α-synuclein dramatically increase the propensity of α-synuclein to form aggregates both *in vitro* and *in vivo* [31,32]. Increasing evidence indicates that protein aggregation is a complex multi-step process that results in several different kinds of intermediates and products, including small, soluble oligomers; large, amorphous aggregates; and highly ordered, β-sheet-rich fibrils [33]. Recent studies suggest that small oligomers or protofibrils may be the principal toxic species responsible for neuronal cell death [3]. In contrast, the microscopically visible fibrillar aggregates or inclusions may be inert or even be neuroprotective [2,34]. In cultured cells, it has been shown that one way the cell handles excess misfolded proteins, which could result from UPS impairment or increased oxidative stress, is to collect and compartmentalize misfolded proteins in specialized inclusions called aggresomes [35; and Chapter 12]. Aggresomes are thought to be cytoprotective because they sequester toxic, aggregated proteins and may facilitate their elimination by autophagy and lysosomal degradation [35,36]. Inclusion bodies found in neurodegenerative diseases, particularly Lewy bodies in PD, seem to share some similarities with aggresomes [37]. However, it remains unresolved whether these inclusion bodies are indeed aggresomes.

The UPS plays a crucial role in protecting cells against the toxic effect of protein aggregation by degrading soluble, monomeric misfolded aggregation-prone proteins. Impairment in the UPS would increase the levels of aggregation-prone proteins and promote the formation of toxic oligomers or protofibrils. Consistent with this notion, the proteasome has been shown to degrade α-synuclein [38], tau [39,40], and Aβ [41]. It is controversial whether the degradation of these proteins requires prior ubiquitination or not, because they are natively unfolded proteins, which can undergo ubiquitin-independent degradation by the 20S proteasome *in vitro* [39,42]. In cultured cells, inhibition of the proteasome by treatment with the proteasome inhibitor MG132 or lactacystin results in increased aggregation and cytotoxicity of α-synuclein [43], tau [44], and Aβ [41]. Moreover, genetic screens in *Drosophila* have identified several loss-of-function mutants of UPS components as enhancers that augment the cytotoxicity induced by protein aggregation associated with polyglutamine expansion [45]. Recently, it was reported that systemic exposure of rats to proteasome inhibitors causes a PD-like phenotype, including the formation of ubiquitin/α-synuclein-positive, Lewy body-like inclusions [46].

In contrast to its ability to degrade soluble, monomeric misfolded proteins, the proteasome is ineffective in degrading oligomeric protofibrils and large aggregates [9]. In fact, recent studies using GFP-based reporters of UPS activity in cultured cells suggest that the function of the UPS is severely impaired by accumulation of protein aggregates [47-49]. It has been proposed that misfolded proteins or aggregates may block the 26S proteasome due to their inability to fully enter the 20S catalytic pore or to exit from the proteasome (see Chapter 14). For example, expanded polyglutamine regions have been shown to be intrinsically resistant to degradation by purified proteasomes [50] and polyglutamine-containing proteins are kinetically trapped within proteasomes [51]. Another proposed mechanism is that protein aggregates may indirectly impair UPS function by sequestering components of the UPS. Proteasome subunits and other UPS components are often found in inclusion bodies from human patients and animal models of neurodegenerative diseases [5]. This sequestration is thought to cause UPS impairment by depleting UPS components from their cellular sites of action [52]. Interestingly, both of the above models were challenged by a recent study [49], which reports that production of protein aggregates specifically targeted to either the nucleus or cytosol leads to global impairment of UPS function in both cellular compartments. Furthermore, UPS impairment can be observed in the absence of any detectable inclusion bodies, suggesting that intermediate forms of protein aggregates, such as small oligomers or protofibrils, may be the toxic species responsible for causing global UPS impairment [49].

Although impairment of the UPS by protein aggregates is an attractive hypothesis, there is no *in vivo* data to support this theory. A recent study using a GFP-based reporter of UPS activity in a mouse model of the polyglutamine disease SCA7 has revealed no evidence for UPS impairment in the vulnerable neurons even at the terminal stages of pathogenesis [53], arguing against this hypothesis. More such experiments are needed to determine whether UPS is impaired by protein aggregates in other animal models of neurodegenerative diseases.

GENETIC MUTATIONS IN UPS COMPONENTS CAUSE FAMILIAL NEURODEGENERATIVE DISEASES

Compelling evidence for a causative role of UPS dysfunction in neurodegeneration comes from identification of mutations in UPS components as the genetic defects responsible for several hereditary forms of neurodegenerative disorders (Table 1). The identified mutant genes encode either E3 or DUB enzymes, highlighting the importance of these two classes of key regulators of ubiquitination in the control of neuronal function and survival. To date, no pathogenic mutations have been found in the components of the 26S proteasome (see Chapter 28).

Table 1. UPS components mutated in familial neurodegenerative diseases

Disease	Inheritance	Protein	Function	References
Human				
Parkinson's disease	Recessive	Parkin	E3	[54]
	Dominant	UCH-L1	DUB/E3	[79]
Charcot-Marie-Tooth disease	Dominant	SIMPLE	E3	[74,75]
Spinocerebellar ataxia type-3	Dominant	Ataxin-3	DUB	[86]
Mouse				
Gracile axonal dystrophy	Recessive	UCH-L1	DUB/E3	[81]
Spongiform neurodegeneration	Recessive	Mahogunin	E3	[77,78]

What is the pathogenic mechanism by which the mutations in each identified E3 and DUB enzyme causes neurodegeneration? This is an important question because the answer will provide novel insights for understanding and treating neurodegenerative diseases. Mutations in E3 or DUB enzymes are expected to result in abnormal ubiquitination. Depending on the type of ubiquitination affected, the mutations could cause neurodegeneration through two different mechanisms (Figure 2). In the first model, aberrant K48-linked polyubiquitination resulting from mutated E3 or DUB alters protein degradation by the proteasome, leading to accumulation of toxic proteins and subsequent neurodegeneration. In the second model, aberrant monoubiquitination or K63-linked polyubiquitination resulting from mutated E3 or DUB alters crucial non-proteasomal functions, such as gene transcription and protein trafficking, and thereby causes neurodegeneration without protein aggregation. Below, we summarize current information regarding the pathogenic mechanisms of the identified mutations in each associated neurodegenerative disease.

GENETIC MUTATIONS IN E3 LIGASES

Parkin

Loss-of-function mutations in parkin, a 465-amino-acid RING-type E3 ligase, were first identified as the cause for autosomal recessive juvenile Parkinsonism (AR-JP) and subsequently found to account for ~50% of all recessively transmitted early-onset PD cases [54-56]. Interestingly, patients with parkin mutations do not exhibit Lewy body pathology.

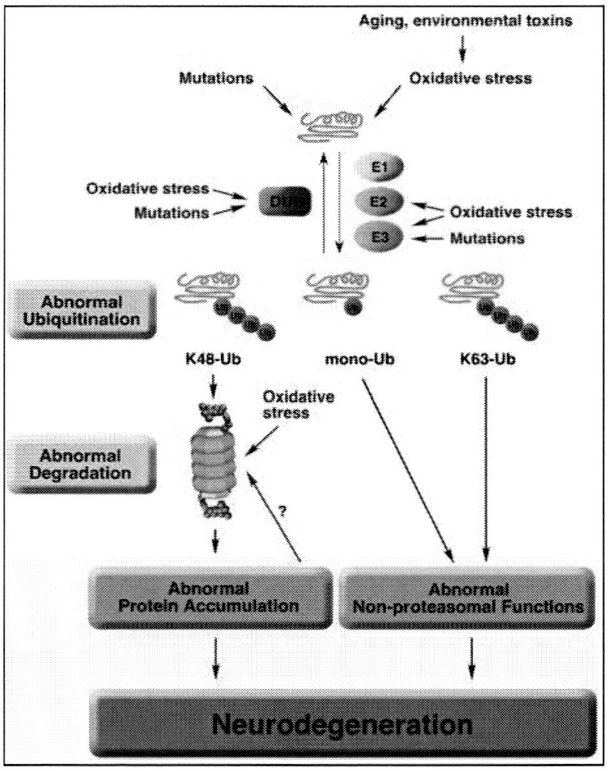

Figure 2. Possible pathogenic mechanisms by which impaired UPS components cause neurodegeneration. Genetic mutations or oxidative stress from aging and/or exposure to environmental toxins have been shown to impair the ubiquitination machinery (particularly E3 ubiquitin-protein ligases) and deubiquitinating enzymes (DUBs), resulting in abnormal ubiquitination. Depending on the type of ubiquitination affected, the impairment could cause neurodegeneration through two different mechanisms. In the first model, aberrant K48-linked polyubiquitination resulting from impaired E3s or DUBs alters protein degradation by the proteasome, leading to accumulation of toxic proteins and subsequent neurodegeneration. The proteasomes could be directly damaged by oxidative stress or might be inhibited by protein aggregation, which exacerbates the neurotoxicity. In the second model, aberrant monoubiquitination or K63-linked polyubiquitination resulting from impaired E3s or DUBs alters crucial non-proteasomal functions, such as gene transcription and protein trafficking, thereby causing neurodegeneration without protein aggregation. These two models are not mutually exclusive because a single E3 or DUB enzyme, such as parkin or UCH-L1, could regulate more than one type of ubiquitination. In addition, abnormal ubiquitination and neurodegeneration could also result from mutation or oxidative stress-induced structural changes in the protein substrates that alter their recognition and degradation by the UPS.

However, these patients display selective loss of nigral dopaminergic neurons, suggesting that Lewy body formation is not necessary for causing neurodegeneration. The lack of Lewy bodies in AR-JP patients could imply that the normal function of parkin is required for the formation of Lewy bodies, or alternatively, parkin-mediated ubiquitination may have non-proteasomal functions so that parkin mutations would cause neurodegeneration without protein aggregation (the second model in Figure 2).

The molecular mechanisms by which loss of parkin function causes neurodegeneration remain unclear. It has been widely hypothesized that loss of parkin E3 ligase function would result in accumulation of its potentially toxic substrates, which eventually leads to neurodegeneration [54-56]. *In vitro* and cell culture experiments reveal that parkin binds several E2 enzymes (UbcH7, UbcH8, Ubc6, and Ubc7), and regulates ubiquitination and degradation of a number of putative substrates, including CDCrel-1, synphilin-1, synaptotagmin XI, cyclin E, α/β tubulin, O-glycosylated α-synuclein, and the p38 subunit of the aminoacyl-tRNA synthetase complex [56-62]. In addition, parkin interacts with molecular chaperone Hsp70 and E3 ligase CHIP to facilitate ubiquitination and degradation of misfolded Pael-R [63,64] and polyglutamine-containing proteins [65]. Pael-R is a G-protein coupled receptor (GPR37) that, when overexpressed, accumulates in an unfolded form in the endoplasmic reticulum (ER) and induces ER stress, which ultimately results in cell death [63]. The Pael-R-induced cell death can be suppressed by co-overexpression of parkin [63], supporting a role for parkin in the ER-associated degradation (ERAD). Pael-R was found to accumulate in AR-JP patients [66] as well as in Lewy bodies of sporadic PD brains [67], suggesting that impairment in ERAD-related function of parkin may contribute to PD pathogenesis.

An unexpected complexity in modeling human parkin mutations in animals is that parkin knockout mice exhibit very mild deficits and do not develop PD-like phenotypes [68-71]. Surprisingly, none of the above-mentioned parkin substrates was found to accumulate in the brains of parkin knockout mice, bringing doubt to the validity of these proteins as the physiological substrates of parkin [68,69]. However, if parkin were involved in regulation of monoubiquitination or non-K48-linked polyubiquitination of its substrates, then loss of parkin function would not lead to accumulation of the substrate proteins. Indeed, parkin has recently been shown to bind the dimeric E2 enzyme UbcH13/Uev1a and promote K63-linked polyubiquitination of synphilin-1 [13,72]. These results support the possibility that loss of parkin function could cause neurodegeneration in a proteasome-independent manner by altering K63-linked polyubiquitination of its substrates.

Simple

Charcot-Marie-Tooth disease (CMT) is a heterogeneous group of inherited peripheral neuropathies that affect motor and sensory nerves of the peripheral nervous system [73]. Recently, mutations in the gene encoding a putative E3 ligase, SIMPLE (small integral membrane protein of the lysosome/late endosome), have been identified as the genetic defects responsible for an autosomal dominant form of type 1 or demyelinating CMT [74,75]. Mutations in SIMPLE have also been linked to a type 2 form of CMT, which is characterized by axonal degeneration [75]. SIMPLE contains a predicted RING finger [75] and binds the HECT-type E3 ligase Nedd4 [76], suggesting a potential function of SIMPLE as either a single subunit E3 ligase or a multi-subunit E3 ligase component. Although the E3 ligase activity of SIMPLE has not yet been determined, SIMPLE has been shown to bind Tsg101, a component of the ESCRT-I (endosomal sorting complex required for transport-I) complex that sorts monoubiquitinated membrane cargo proteins to the lysosomal pathway for degradation [76]. It is thus possible that SIMPLE may regulate monoubiquitination and subsequent trafficking of cargo proteins to the lysosomes. Mutations in SIMPLE may result in aberrant monoubiquitination and abnormal lysosomal trafficking, leading to peripheral nerve demyelination and degeneration.

Mahogunin

Spongiform neurodegeneration is a relatively rare type of pathology consisting mainly of vacuolation in neuronal cell bodies and processes, neuronal cell death, and astrocytosis. Prion diseases, also known as transmissible spongiform encephalopathies, are a group of human and animal disorders characterized by spongiform neurodegeneration and accumulation of the protease-resistant prion protein PrP-Sc in neurons (see Chapter 34). Interestingly, recent genetic studies reveal that a null mutation in the gene encoding a novel RING finger protein called mahogunin causes a recessively transmitted form of spongiform neurodegeneration in mice that includes many features of prion disease but without accumulation of protease-resistant prion protein [77,78]. *In vitro* ubiquitination assays show that recombinant mahogunin protein exhibits E2 (Ubc5)-dependent auto-ubiquitination activity, suggesting that mahogunin functions as an E3 ligase [78]. The substrates of mahogunin remain to be identified. Prion protein (PrP) does not seem to be a substrate of mahogunin because mahogunin is unable to ubiquitinate PrP *in vitro* and there is no accumulation of PrP-Sc in the mahogunin mutant mice [78]. The lack of protein aggregates in the mahogunin mutant mice provides yet another example that mutation in an E3 could cause neurodegeneration without protein aggregation (the second model in Figure 2). It would be of interest to determine whether the mahogunin protein protects from prion-induced neurodegeneration and whether the mahogunin substrates have a role in regulation of PrP metabolism and/or other cellular pathways crucial for neuronal survivial. Further characterization of the E3 ligase activity of mahogunin and its substrates should lead to novel insights into the molecular mechanisms underlying spongiform neurodegeneration and prion disease pathogenesis.

Genetic Mutations in DUB Enzymes

UCH-L1

In addition to mutations in the E3 ligase parkin, genetic evidence supporting the involvement of UPS dysfunction in PD pathogenesis comes from the identification of an

I93M missense mutation in the gene encoding ubiquitin carboxy-terminal hydrolase L1 (UCH-L1) in two siblings of a German family with autosomal dominant familial PD [79]. Although it is controversial whether the I93M variant is a pathogenic mutation or a rare polymorphism [80], a direct role for UCH-L1 in neurodegeneration has been clearly demonstrated by genetic studies in mice. Deletion of exons 7 and 8 containing the hydrolase catalytic residues of murine UCH-L1 causes gracile axonal dystrophy (*gad*), a recessively transmitted neurodegenerative disease characterized by progressive axonal degeneration in sensory and motor neurons in the gracile tracts of the spinal cord. Surprisingly, there is no neurodegeneration in the substantia nigra of the *gad* mice [81].

UCH-L1 is a highly abundant neuronal protein that possesses a well-characterized deubiquitinating activity for hydrolyzing C-terminal amides of ubiquitin to generate monomeric ubiquitin [82]. Such a hydrolase activity is believed to facilitate UPS-mediated protein degradation by recycling ubiquitin monomers [83]. In addition, UCH-L1 might also have a role in stabilizing ubiquitin monomers *in vivo* [84]. Thus, loss of UCH-L1 function would result in decreased cellular level of ubiquitin monomers and a general impairment of the UPS function, leading to toxic buildup of misfolded proteins. In support of this possibility, the *gad* mice display progressive accumulation of ubiquitin-positive protein aggregates in sensory and motor neurons [81,84].

Intriguingly, UCH-L1 was reported to possess a second, dimerization-dependent E3 ligase activity that adds ubiquitin to α-synuclein-ubiquitin conjugates via a K63 linkage [85]. This ligase activity is thought to be at least partly pathogenic because K63-linked polyubiquitination may inhibit K48 polyubiquitination-mediated α-synuclein degradation, leading to accumulation and aggregation of α-synuclein. Consistent with this idea, a S18Y polymorphic variant of UCH-L1 associated with decreased PD risk has been shown to exhibit reduced ligase activity but normal hydrolase activity *in vitro* [85]. Further characterization of UCH-L1 enzymatic activities and their substrate specificity *in vivo* is essential for understanding the role of UCH-L1 in the pathogenesis of PD and other neurodegenerative diseases.

Ataxin-3

Polyglutamine expansion in the coding region of ataxin-3 causes spinocerebellar ataxia type-3 (SCA-3; also known as Machado-Joseph disease), an autosomal dominant form of neurodegenerative polyglutamine disorder [86]. Like in other polyglutamine disease proteins such as huntingtin, the expanded polyglutamine stretch (> 50 glutamines) in ataxin-3 is thought to confer gain-of-function toxicity by inducing protein misfolding and aggregation, resulting in formation of nuclear and cytoplasmic inclusions [87]. Interestingly, homozygous SCA-3 patients with two mutant alleles exhibit earlier disease onset and more severe phenotypes than heterozygous individuals with one mutant allele [87], suggesting a role for the normal function of ataxin-3 in modulation of SCA-3 disease pathogenesis and progression.

Recent bioinformatic analysis reveals that ataxin-3 contains two ubiquitin-interacting motifs (UIMs) and a Josephin domain that shares homology with the catalytic sites of UCH (ubiquitin carboxy-terminal hydrolase) and UBP (ubiquitin processing protease) classes of DUB enzymes [88]. The solution structure of ataxin-3 Josephin domain has been solved by NMR, which confirms that this domain indeed assumes the papain-like cysteine protease fold characteristic of other DUBs [89]. Moreover, biochemical studies have demonstrated that ataxin-3 binds polyubiquitin chains through its UIMs and exhibits deubiquitinating activity [89,90]. Further analysis of ataxin-3 enzymatic activity suggests that ataxin-3 functions as a polyubiquitin chain-editing enzyme that shortens K48-linked polyubiquitin chains [89,91]. Ataxin-3 also associates with the proteasome and has been implicated in regulation of UPS-mediated protein degradation [90,92]. A very recent study has shown that, while polyglutamine expanded mutant ataxin-3 protein induces neurodegeneration in *Drosophila*, wild-type human ataxin-3 suppresses neurotoxicity induced by polyglutamine disease proteins, including mutant ataxin-3 protein itself as well as mutant huntingtin protein [93]. These data suggest that the normal function of ataxin-3 is neuroprotective.

Impairment of UPS Components by Oxidative Stress in Sporadic Neurodegenerative Diseases

Despite recent progress in identification of the genetic defects responsible for rare monogenic familial forms of neurodegenerative diseases, the causes of other forms of neurodegenerative diseases, particularly sporadic cases, remain largely unknown. Oxidative stress has been strongly implicated in the pathogenesis of many age-related neurodegenerative diseases, including AD, PD, and ALS [4,94,95]. For example, these diseases have been associated with increased production of reactive oxygen species (ROS) and/or impaired antioxidant defense systems, which could result from aging, genetic predisposition, and environmental factors [4]. Epidemiological studies suggest that exposure to pesticides, herbicides, and other environmental toxins that inhibit mitochondrial complex I, can lead to excess production of ROS and increased incidence of sporadic PD [96]. In addition to the mitochondria, the ER is also a major source of ROS [97,98]. ER stress caused by the accumulation of misfolded proteins, such as Pael-R in PD, leads to increased production of ROS which is damaging to neurons [99]. Dopaminergic neurons of the substantia nigra are thought to be particularly vulnerable to increased oxidative stress because of the intrinsic ability of dopamine to promote oxidative damage [100]. Consistent with this notion, oxidative stress induced by rotenone, paraquat, and 1-methyl-4-phenyl-1,2,3,6-tetrahydropyridine (MPTP), has been shown to produce PD-like phenotypes in rodents [101].

In spite of the overwhelming evidence linking oxidative stress to the pathogenesis of PD and other neurodegenerative diseases, relatively little is presently known of the biochemical pathways by which increased oxidative stress leads to neuronal dysfunction and, ultimately, neuronal cell death. Although it was initially thought that targets of oxidative damage by reactive oxygen species were random and indiscriminate, it has become increasingly clear that the susceptibility of proteins to oxidative damage is highly dependent on specific properties of individual proteins, such as unique sequence motifs, surface accessibility, and subcellular localization [102,103]. Emerging evidence indicates that oxidative stress can directly damage UPS components (Table 2). The oxidative damage to the UPS may contribute to neurodegeneration in sporadic neurodegenerative diseases in a manner similar to the genetic mutations of UPS components in causing familial neurodegenerative diseases (Figure 2).

Oxidative Damage to the Ubiquitination Machinery

As mentioned earlier, a majority of E3 ubiquitin-protein ligases in cells are RING finger-containing E3s. The RING finger motif is a cysteine/histidine-rich (C_3HC_4), Zn^{2+}-binding domain that serves as the E3 catalytic core with the binding site for the E2 enzyme [7,14]. The Zn^{2+}-bound cysteine thiolate anion (Cys-S^-) is more reactive than the sulfhydryl group (Cys-SH), and can be readily modified by a variety of ROS and reactive nitrogen species (RNS) [104]. For example, the cysteine residues in the RING finger of APC11, a component of the multi-subunit E3 called anaphase-promoting complex, are oxidized in response to oxidative stress induced by H_2O_2. The cysteine oxidation induces dissociation of Zn^{2+} from the RING finger and disrupts the E2-binding site, leading to the loss of E3 activity [105]. Such oxidative stress-induced inactivation mechanism may also apply to other RING-type E3 ligases, including parkin. In fact, parkin has been reported to undergo misfolding and aggregation in response to H_2O_2 [106]. Furthermore, the cysteine residues in the RING fingers of parkin have been shown to be *S*-nitrosylated by nitric oxide (NO), resulting in a dramatic reduction in parkin's E3 activity and neuroprotective function [107,108]. A very recent study has demonstrated that dopamine quinone, a reactive metabolite of dopamine oxidation, can covalently modify the cysteine residues in parkin RING fingers and functionally inactivate parkin, providing a mechanism linking the loss of parkin function with selective degeneration of dopaminergic neurons [109]. The dopamine-derived parkin adducts as well as *S*-nitrosylated parkin have been detected in brain samples from patients with sporadic PD [107-109], suggesting the involvement of oxidative and nitrosative stress-induced damage to parkin in the pathogenesis of sporadic PD.

In addition to E3s, E1 and E2 enzymes also contain reactive cysteine residues that have the potential to serve as the targets for oxidative and nitrosative stress-induced modifications. In support of this possibility, the E2 enzyme UbcH7 is robustly modified by dopamine quinone *in vitro* [109]. Since these cysteine residues participate in the formation of high-energy thioester intermediates that are crucial for the catalysis of the ubiquitination reaction [7], oxidative and nitrosative stress-induced modifications of these residues would be expected to inactivate these enzymes, leading to a general inhibition of the UPS function which ultimately results in neuronal cell death in sporadic neurodegenerative diseases.

Oxidative Damage to DUB Enzymes

The most widely used marker for oxidative damage to proteins is the presence of carbonyl groups, which can be introduced into proteins by direct oxidation of Pro, Arg, Lys, and Thr side chains, or by Michael addition reactions with products of lipid peroxidation or glycooxidation [94,102,103]. Postmortem analyses reveal that the total levels of protein carbonyls are elevated in brains from patients with AD, PD, or other neurodegenerative diseases [28,110]. However, the identities of the oxidized proteins that have been altered by carbonylation or other types of oxidation remain largely unknown. As a first step towards a molecular understanding of the pathogenic mechanism of oxidative stress in neurodegenerative diseases, we performed a search for specific protein targets of oxidative damage in sporadic AD and PD brains by using a proteomic approach that combined two-dimensional gel electrophoresis, immunological detection of protein carbonylation, and mass spectrometry [111]. Interestingly, a major target of oxidative damage in AD and PD that we identified is UCH-L1. As described earlier, UCH-L1 is a DUB/E3 dual function enzyme whose mutations have been linked to early-onset familial PD in human and to gracile axonal dystrophy in mouse.

Table 2. UPS components damaged by oxidative stress in sporadic neurodegenerative diseases

Protein	Function	Modification	In vitro	Diseases	References
Ubiquitination/deubiquitination machinery					
Parkin	E3	S-nitrosylation	Yes	PD	[107,108]
		Dopamine adduct	Yes	PD	[109]
APC11	E3 subunit	Cys oxidation	Yes	n.d.	[105]
UbcH7	E2	Dopamine adduct	Yes	n.d.	[109]
UCH-L1	DUB/E3	Carbonylation	n.d.	PD, AD	[111]
		Met oxidation	n.d.	PD, AD	[111]
		Cys oxidation	n.d.	PD, AD	[111]
		HNE adduct	Yes	n.d.	[112]
26S proteasome					
S6 ATPase	19S cap subunit	Carbonylation	Yes	n.d.	[115]
α6	20S core subunit	HNE adduct	Yes	n.d.	[116]
α2, α6, α7	20S core subunits	HNE adduct	Yes	IRI	[117]
β subunits	20S catalytic subunits	Acrolein adduct	Yes	n.d.	[118]

HNE, 4-hydroxy-2-nonenal; IRI, ischemia/reperfusion injury; n.d., not determined.

In addition to carbonylation, we found that UCH-L1 is also oxidatively modified by methionine oxidation and cysteine oxidation in sporadic AD and PD brains [111]. Oxidative damage to UCH-L1 by the identified modifications may result in irreversible alteration in the conformation and/or DUB/E3 enzymatic activities of UCH-L1, and thus has deleterious effects on neuronal function and survival similar to the pathogenic effects caused by the UCH-L1 genetic mutations as described earlier. Consistent with this notion, a recent *in vitro* study showed that the DUB activity of recombinant UCH-L1 was decreased upon oxidation of UCH-L1 by 4-hydroxy-2-nonenal, a lipid peroxidation product that generates carbonyl groups in proteins via Michael addition reactions [112]. Oxidative modifications may also render UCH-L1 itself more resistant to proteolysis and promote its aggregation into hallmark lesions of AD and PD brains. In support of this possibility, we and other groups have found the presence of abundant UCH-L1 protein in neurofibrillary tangles in AD and in Lewy bodies in PD brains [111,113].

Although UCH-L1 is the only identified DUB that is oxidatively damaged in sporadic neurodegenerative diseases, it is possible that other DUBs might also be the targets for oxidative and nitrosative stress-induced modifications. Out of the five known classes of DUB enzymes, four classes are cysteine proteases [17]. The active site cysteine residues usually have high propensity for being modified by a variety of ROS and RNS [104]. Modifications of the active site cysteine residues would inactivate DUB enzymes and result in abnormal protein ubiquitination and degradation, thereby contributing to the pathogenesis of sporadic neurodegenerative diseases (Figure 2).

Oxidative Damage to the Proteasome

Accumulating evidence indicates that oxidative stress not only impairs the ubiquitination/deubiquitination machinery, but also causes direct damage to the 26S proteasome (Table 2). The endogenous product of inflammation 15-deoxy-delta (12, 14)-prostaglandin J2 (PGJ2) is a potent inducer of intracellular oxidative stress implicated in the pathogenesis of a number of neurodegenerative diseases, including AD, PD, and ALS [114]. A recent proteomic study has shown that in human neuroblastoma SH-SY5Y cells, one of the subunits in the 19S regulatory complex of the 26S proteasome, S6 ATPase, is oxidatively damaged by carbonylation in response to oxidative stress induced by PGJ2 or H_2O_2 [115]. The oxidative damage to S6 ATPase is accompanied by a significant reduction in the S6 ATPase activity and in the ability of the 26S proteasome to degrade substrate proteins. In addition, the lipid peroxidation product 4-hydroxy-2-nonenal, a putative endogenous mediator of oxidative stress, has been shown to modify several α subunits (α2, α6, α7) of the 20S proteasome and inhibit the proteasome activity *in vitro* [116] as well as in a rat model of ischemia/reperfusion injury [117]. Furthermore, in SH-SY5Y cells, PD-associated environmental toxin rotenone has been shown to inhibit the proteasomal proteolytic activity by inducing oxidative modification of the catalytic β subunits of the 20S proteasome with the lipid peroxidation product acrolein [118].

The susceptibility of the proteasome components to oxidative stress-induced modifications as described above raises the possibility that the 26S proteasome is oxidatively damaged in brains of patients with sporadic neurodegenerative diseases. However, it remains to be determined whether this indeed is the case. Recently, the levels of 20S proteasome α (but not β) subunits and 20S proteasomal enzymatic activities have been reported to be reduced selectively in the substantia nigra of sporadic PD patients compared to age-matched controls [119,120]. Furthermore, an animal model study has shown that systemic exposure to proteasome inhibitors causes rats to develop PD-like phenotypes, including dopaminergic neurodegeneration, motor behavioral deficits, and accumulation of Lewy body-like protein aggregates [46]. These findings suggest that proteasome impairment plays a crucial role in the pathogenesis of sporadic PD.

In addition to being damaged via direct oxidation of its subunits, the proteasome may be blocked or inhibited by oxidative stress-induced misfolded proteins and aggregates

[47,48,121]. However, as pointed out earlier, the evidence supporting this view has come from studies using purified proteasome or cultured cells. It remains to be resolved whether oxidized proteins or aggregates can directly inhibit the proteasome *in vivo*.

Conclusions

The UPS is an elaborate system that not only controls protein degradation via proteasome-mediated proteolysis, but also regulates protein function via multiple types of ubiquitination. Recent genetic studies of familial forms of neurodegenerative diseases have provided direct evidence linking dysregulation of ubiquitination to neurodegeneration. The list of disease-causing mutations in E3 ubiquitin-protein ligases and deubiquitinating enzymes is growing. Identification of the physiological substrates and the cellular processes that are regulated by each of these enzymes is crucial for understanding the role of the UPS in neuronal function and survival. It is important to investigate the proteasome-dependent as well as the proteasome-independent mechanisms of aberrant ubiquitination in the pathogenesis of neurodegenerative diseases. Moreover, future studies are needed to better understand the interplay between UPS dysfunction, oxidative stress, and protein aggregation. A mechanistic understanding of the UPS and its malfunction in various neurodegenerative diseases will undoubtedly facilitate the development of novel rational therapies for treating these devastating disorders.

Acknowledgements

This work was supported in part by US National Institutes of Health Grants (AG021489, NS047199, NS050650, and NS047575) and Emory Center for Neurodegenerative Disease-Merck Scholar Award.

References

[1] Taylor JP, Hardy J, Fischbeck KH: Toxic proteins in neurodegenerative disease. *Science* 2002, 296:1991-1995.

[2] Ross CA, Poirier MA: Protein aggregation and neurodegenerative disease. *Nat Med* 2004, 10 Suppl:S10-17.

[3] Caughey B, Lansbury PT: Protofibrils, pores, fibrils, and neurodegeneration: separating the responsible protein aggregates from the innocent bystanders. *Annu Rev Neurosci* 2003, 26:267-298.

[4] Beal MF: Oxidatively modified proteins in aging and disease. *Free Radic Biol Med* 2002, 32:797-803.

[5] Schwartz AL, Ciechanover A: The ubiquitin-proteasome pathway and pathogenesis of human diseases. *Annu Rev Med* 1999, 50:57-74.

[6] Pickart CM, Fushman D: Polyubiquitin chains: polymeric protein signals. *Curr Opin Chem Biol* 2004, 8:610-616.

[7] Glickman MH, Ciechanover A: The ubiquitin-proteasome proteolytic pathway: destruction for the sake of construction. *Physiol Rev* 2002, 82:373-428.

[8] DiAntonio A, Hicke L: Ubiquitin-dependent regulation of the synapse. *Annu Rev Neurosci* 2004, 27:223-246.

[9] Pickart CM, Cohen RE: Proteasomes and their kin: proteases in the machine age. *Nat Rev Mol Cell Biol* 2004, 5:177-187.

[10] Elsasser S, Finley D: Delivery of ubiquitinated substrates to protein-unfolding machines. *Nat Cell Biol* 2005, 7:742-749.

[11] Welchman RL, Gordon C, Mayer RJ: Ubiquitin and ubiquitin-like proteins as multifunctional signals. *Nat Rev Mol Cell Biol* 2005, 6:599-609.

[12] Hicke L: Protein regulation by monoubiquitin. *Nat Rev Mol Cell Biol* 2001, 2:195-201.

[13] Lim KL, Chew KC, Tan JM, Wang C, Chung KK, Zhang Y, Tanaka Y, Smith W, Engelender S, Ross CA, Dawson VL, Dawson TM: Parkin mediates nonclassical, proteasomal-independent ubiquitination of synphilin-1: implications for Lewy body formation. *J Neurosci* 2005, 25:2002-2009.

[14] Weissman AM: Themes and variations on ubiquitylation. *Nat Rev Mol Cell Biol* 2001, 2:169-178.

[15] Hoppe T: Multiubiquitylation by E4 enzymes: 'one size' doesn't fit all. *Trends Biochem Sci* 2005, 30:183-187.

[16] Wilkinson KD, Ventii KH, Friedrich KL, Mullally JE: The ubiquitin signal: assembly, recognition and termination. Symposium on ubiquitin and signaling. *EMBO Rep* 2005, 6:815-820.

[17] Amerik AY, Hochstrasser M: Mechanism and function of deubiquitinating enzymes. *Biochim Biophys Acta* 2004, 1695:189-207.

[18] Wong BR, Parlati F, Qu K, Demo S, Pray T, Huang J, Payan DG, Bennett MK: Drug discovery in the ubiquitin regulatory pathway. *Drug Discov Today* 2003, 8:746-754.

[19] Murata S, Chiba T, Tanaka K: CHIP: a quality-control E3 ligase collaborating with molecular chaperones. *Int J Biochem Cell Biol* 2003, 35:572-578.

[20] Yamanaka K, Ishikawa H, Megumi Y, Tokunaga F, Kanie M, Rouault TA, Morishima I, Minato N, Ishimori K, Iwai K: Identification of the ubiquitin-protein ligase that recognizes oxidized IRP2. *Nat Cell Biol* 2003, 5:336-340.

[21] Grune T, Merker K, Sandig G, Davies KJ: Selective degradation of oxidatively modified protein substrates by the proteasome. *Biochem Biophys Res Commun* 2003, 305:709-718.

[22] Benaroudj N, Tarcsa E, Cascio P, Goldberg AL: The unfolding of substrates and ubiquitin-independent protein degradation by proteasomes. *Biochimie* 2001, 83:311-318.

[23] Liu CW, Corboy MJ, DeMartino GN, Thomas PJ: Endoproteolytic activity of the proteasome. *Science* 2003, 299:408-411.

[24] Mori H, Kondo J, Ihara Y: Ubiquitin is a component of paired helical filaments in Alzheimer's disease. *Science* 1987, 235:1641-1644.

[25] Lowe J, Blanchard A, Morrell K, Lennox G, Reynolds L, Billett M, Landon M, Mayer RJ: Ubiquitin is a common factor in intermediate filament inclusion bodies of diverse type in man, including those of Parkinson's disease, Pick's disease, and Alzheimer's disease, as well as Rosenthal fibres in cerebellar astrocytomas, cytoplasmic bodies in muscle, and mallory bodies in alcoholic liver disease. *J Pathol* 1988, 155:9-15.

[26] Spillantini MG, Schmidt ML, Lee VM, Trojanowski JQ, Jakes R, Goedert M: Alpha-synuclein in Lewy bodies. *Nature* 1997, 388:839-840.

[27] Selkoe DJ: Alzheimer's disease: genes, proteins, and therapy. *Physiol Rev* 2001, 81:741-766.

[28] Giasson BI, Ischiropoulos H, Lee VM, Trojanowski JQ: The relationship between oxidative/nitrative stress and pathological inclusions in Alzheimer's and Parkinson's diseases. *Free Radic Biol Med* 2002, 32:1264-1275.

[29] Singleton AB, Farrer M, Johnson J, Singleton A, Hague S, Kachergus J, Hulihan M, Peuralinna T, Dutra A, Nussbaum R, Lincoln S, Crawley A, Hanson M, Maraganore D, Adler C, Cookson MR, Muenter M, Baptista M, Miller D, Blancato J, Hardy J, Gwinn-Hardy K: alpha-Synuclein locus triplication causes Parkinson's disease. *Science* 2003, 302:841.

[30] Chartier-Harlin MC, Kachergus J, Roumier C, Mouroux V, Douay X, Lincoln S, Levecque C, Larvor L, Andrieux J, Hulihan M, Waucquier N, Defebvre L, Amouyel P, Farrer M, Destee A: Alpha-synuclein locus duplication as a cause of familial Parkinson's disease. *Lancet* 2004, 364:1167-1169.

[31] Conway KA, Harper JD, Lansbury PT: Accelerated in vitro fibril formation by a mutant alpha-synuclein linked to early-onset Parkinson disease. *Nat Med* 1998, 4:1318-1320.

[32] Giasson BI, Duda JE, Quinn SM, Zhang B, Trojanowski JQ, Lee VM: Neuronal alpha-synucleinopathy with severe movement disorder in mice expressing A53T human alpha-synuclein. *Neuron* 2002, 34:521-533.

[33] Temussi PA, Masino L, Pastore A: From Alzheimer to Huntington: why is a structural understanding so difficult? *EMBO J* 2003, 22:355-361.

[34] Arrasate M, Mitra S, Schweitzer ES, Segal MR, Finkbeiner S: Inclusion body formation reduces levels of mutant huntingtin and the risk of neuronal death. *Nature* 2004, 431:805-810.

[35] Kopito RR: Aggresomes, inclusion bodies and protein aggregation. *Trends Cell Biol* 2000, 10:524-530.

[36] Garcia-Mata R, Gao YS, Sztul E: Hassles with taking out the garbage: aggravating aggresomes. *Traffic* 2002, 3:388-396.

[37] Olanow CW, Perl DP, DeMartino GN, McNaught KS: Lewy-body formation is an aggresome-related process: a hypothesis. *Lancet Neurol* 2004, 3:496-503.

[38] Bennett MC, Bishop JF, Leng Y, Chock PB, Chase TN, Mouradian MM: Degradation of alpha-synuclein by proteasome. *J Biol Chem* 1999, 274:33855-33858.

[39] David DC, Layfield R, Serpell L, Narain Y, Goedert M, Spillantini MG: Proteasomal degradation of tau protein. *J Neurochem* 2002, 83:176-185.

[40] Shimura H, Schwartz D, Gygi SP, Kosik KS: CHIP-Hsc70 complex ubiquitinates phosphorylated Tau and enhances cell survival. *J Biol Chem* 2003, 10:1121-1159.

[41] Lopez Salon M, Pasquini L, Besio Moreno M, Pasquini JM, Soto E: Relationship between beta-amyloid degradation and the 26S proteasome in neural cells. *Exp Neurol* 2003, 180:131-143.
[42] Tofaris GK, Layfield R, Spillantini MG: alpha-synuclein metabolism and aggregation is linked to ubiquitin-independent degradation by the proteasome. *FEBS Lett* 2001, 509:22-26.
[43] Tanaka Y, Engelender S, Igarashi S, Rao RK, Wanner T, Tanzi RE, Sawa A, V LD, Dawson TM, Ross CA: Inducible expression of mutant alpha-synuclein decreases proteasome activity and increases sensitivity to mitochondria-dependent apoptosis. *Hum Mol Genet* 2001, 10:919-926.
[44] Babu JR, Geetha T, Wooten MW: Sequestosome 1/p62 shuttles polyubiquitinated tau for proteasomal degradation. *J Neurochem* 2005, 94:192-203.
[45] Fernandez-Funez P, Nino-Rosales ML, de Gouyon B, She WC, Luchak JM, Martinez P, Turiegano E, Benito J, Capovilla M, Skinner PJ, McCall A, Canal I, Orr HT, Zoghbi HY, Botas J: Identification of genes that modify ataxin-1-induced neurodegeneration. *Nature* 2000, 408:101-106.
[46] McNaught KS, Perl DP, Brownell AL, Olanow CW: Systemic exposure to proteasome inhibitors causes a progressive model of Parkinson's disease. *Ann Neurol* 2004, 56:149-162.
[47] Bence NF, Sampat RM, Kopito RR: Impairment of the ubiquitin-proteasome system by protein aggregation. *Science* 2001, 292:1552-1555.
[48] Jana NR, Zemskov EA, Wang G, Nukina N: Altered proteasomal function due to the expression of polyglutamine-expanded truncated N-terminal huntingtin induces apoptosis by caspase activation through mitochondrial cytochrome c release. *Hum Mol Genet* 2001, 10:1049-1059.
[49] Bennett EJ, Bence NF, Jayakumar R, Kopito RR: Global impairment of the ubiquitin-proteasome system by nuclear or cytoplasmic protein aggregates precedes inclusion body formation. *Mol Cell* 2005, 17:351-365.
[50] Venkatraman P, Wetzel R, Tanaka M, Nukina N, Goldberg AL: Eukaryotic proteasomes cannot digest polyglutamine sequences and release them during degradation of polyglutamine-containing proteins. *Mol Cell* 2004, 14:95-104.
[51] Holmberg CI, Staniszewski KE, Mensah KN, Matouschek A, Morimoto RI: Inefficient degradation of truncated polyglutamine proteins by the proteasome. *EMBO J* 2004, 23:4307-4318.
[52] Donaldson KM, Li W, Ching KA, Batalov S, Tsai CC, Joazeiro CA: Ubiquitin-mediated sequestration of normal cellular proteins into polyglutamine aggregates. *Proc Natl Acad Sci U S A* 2003, 100:8892-8897.
[53] Bowman AB, Yoo SY, Dantuma NP, Zoghbi HY: Neuronal dysfunction in a polyglutamine disease model occurs in the absence of ubiquitin-proteasome system impairment and inversely correlates with the degree of nuclear inclusion formation. *Hum Mol Genet* 2005, 14:679-691.
[54] Kitada T, Asakawa S, Hattori N, Matsumine H, Yamamura Y, Minoshima S, Yokochi M, Mizuno Y, Shimizu N: Mutations in the parkin gene cause autosomal recessive juvenile parkinsonism. *Nature* 1998, 392:605-608.

[55] Lucking CB, Durr A, Bonifati V, Vaughan J, De Michele G, Gasser T, Harhangi BS, Meco G, Denefle P, Wood NW, Agid Y, Brice A: Association between early-onset Parkinson's disease and mutations in the parkin gene. French Parkinson's Disease Genetics Study Group. *N Engl J Med* 2000, 342:1560-1567.

[56] Hattori N, Mizuno Y: Pathogenetic mechanisms of parkin in Parkinson's disease. *Lancet* 2004, 364:722-724.

[57] Shimura H, Hattori N, Kubo S, Mizuno Y, Asakawa S, Minoshima S, Shimizu N, Iwai K, Chiba T, Tanaka K, Suzuki T: Familial Parkinson disease gene product, parkin, is a ubiquitin-protein ligase. *Nat Genet* 2000, 25:302-305.

[58] Zhang Y, Gao J, Chung KK, Huang H, Dawson VL, Dawson TM: Parkin functions as an E2-dependent ubiquitin- protein ligase and promotes the degradation of the synaptic vesicle-associated protein, CDCrel-1. *Proc Natl Acad Sci U S A* 2000, 97:13354-13359.

[59] Chung KK, Zhang Y, Lim KL, Tanaka Y, Huang H, Gao J, Ross CA, Dawson VL, Dawson TM: Parkin ubiquitinates the alpha-synuclein-interacting protein, synphilin- 1: implications for Lewy-body formation in Parkinson disease. *Nat Med* 2001, 7:1144-1150.

[60] Huynh DP, Scoles DR, Nguyen D, Pulst SM: The autosomal recessive juvenile Parkinson disease gene product, parkin, interacts with and ubiquitinates synaptotagmin XI. *Hum Mol Genet* 2003, 12:2587-2597.

[61] Ren Y, Zhao J, Feng J: Parkin binds to alpha/beta tubulin and increases their ubiquitination and degradation. *J Neurosci* 2003, 23:3316-3324.

[62] Corti O, Hampe C, Koutnikova H, Darios F, Jacquier S, Prigent A, Robinson JC, Pradier L, Ruberg M, Mirande M, Hirsch E, Rooney T, Fournier A, Brice A: The p38 subunit of the aminoacyl-tRNA synthetase complex is a Parkin substrate: linking protein biosynthesis and neurodegeneration. *Hum Mol Genet* 2003, 12:1427-1437.

[63] Imai Y, Soda M, Inoue H, Hattori N, Mizuno Y, Takahashi R: An unfolded putative transmembrane polypeptide, which can lead to endoplasmic reticulum stress, is a substrate of Parkin. *Cell* 2001, 105:891-902.

[64] Imai Y, Soda M, Hatakeyama S, Akagi T, Hashikawa T, Nakayama KI, Takahashi R: CHIP is associated with Parkin, a gene responsible for familial Parkinson's disease, and enhances its ubiquitin ligase activity. *Mol Cell* 2002, 10:55-67.

[65] Tsai YC, Fishman PS, Thakor NV, Oyler GA: Parkin facilitates the elimination of expanded polyglutamine proteins and leads to preservation of proteasome function. *J Biol Chem* 2003, 278:22044-22055.

[66] Takahashi R, Imai Y, Hattori N, Mizuno Y: Parkin and endoplasmic reticulum stress. *Ann N Y Acad Sci* 2003, 991:101-106.

[67] Murakami T, Shoji M, Imai Y, Inoue H, Kawarabayashi T, Matsubara E, Harigaya Y, Sasaki A, Takahashi R, Abe K: Pael-R is accumulated in Lewy bodies of Parkinson's disease. *Ann Neurol* 2004, 55:439-442.

[68] Goldberg MS, Fleming SM, Palacino JJ, Cepeda C, Lam HA, Bhatnagar A, Meloni EG, Wu N, Ackerson LC, Klapstein GJ, Gajendiran M, Roth BL, Chesselet MF, Maidment NT, Levine MS, Shen J: Parkin-deficient mice exhibit nigrostriatal deficits but not loss of dopaminergic neurons. *J Biol Chem* 2003, 278:43628-43635.

[69] Itier JM, Ibanez P, Mena MA, Abbas N, Cohen-Salmon C, Bohme GA, Laville M, Pratt J, Corti O, Pradier L, Ret G, Joubert C, Periquet M, Araujo F, Negroni J, Casarejos MJ, Canals S, Solano R, Serrano A, Gallego E, Sanchez M, Denefle P, Benavides J, Tremp G, Rooney TA, Brice A, Garcia de Yebenes J: Parkin gene inactivation alters behaviour and dopamine neurotransmission in the mouse. *Hum Mol Genet* 2003, 12:2277-2291.

[70] Von Coelln R, Thomas B, Savitt JM, Lim KL, Sasaki M, Hess EJ, Dawson VL, Dawson TM: Loss of locus coeruleus neurons and reduced startle in parkin null mice. *Proc Natl Acad Sci U S A* 2004, 101:10744-10749.

[71] Perez FA, Palmiter RD: Parkin-deficient mice are not a robust model of parkinsonism. *Proc Natl Acad Sci U S A* 2005, 102:2174-2179.

[72] Doss-Pepe EW, Chen L, Madura K: Alpha-synuclein and parkin contribute to the assembly of ubiquitin lysine 63-linked multiubiquitin chains. *J Biol Chem* 2005, 280:16619-16624.

[73] Shy ME: Charcot-Marie-Tooth disease: an update. *Curr Opin Neurol* 2004, 17:579-585.

[74] Street VA, Bennett CL, Goldy JD, Shirk AJ, Kleopa KA, Tempel BL, Lipe HP, Scherer SS, Bird TD, Chance PF: Mutation of a putative protein degradation gene LITAF/SIMPLE in Charcot-Marie-Tooth disease 1C. *Neurology* 2003, 60:22-26.

[75] Saifi GM, Szigeti K, Wiszniewski W, Shy ME, Krajewski K, Hausmanowa-Petrusewicz I, Kochanski A, Reeser S, Mancias P, Butler I, Lupski JR: SIMPLE mutations in Charcot-Marie-Tooth disease and the potential role of its protein product in protein degradation. *Hum Mutat* 2005, 25:372-383.

[76] Shirk AJ, Anderson SK, Hashemi SH, Chance PF, Bennett CL: SIMPLE interacts with NEDD4 and TSG101: Evidence for a role in lysosomal sorting and implications for Charcot-Marie-Tooth disease. *J Neurosci Res* 2005, 82:43-50.

[77] Phan LK, Lin F, LeDuc CA, Chung WK, Leibel RL: The mouse mahoganoid coat color mutation disrupts a novel C3HC4 RING domain protein. *J Clin Invest* 2002, 110:1449-1459.

[78] He L, Lu XY, Jolly AF, Eldridge AG, Watson SJ, Jackson PK, Barsh GS, Gunn TM: Spongiform degeneration in mahoganoid mutant mice. *Science* 2003, 299:710-712.

[79] Leroy E, Boyer R, Auburger G, Leube B, Ulm G, Mezey E, Harta G, Brownstein MJ, Jonnalagada S, Chernova T, Dehejia A, Lavedan C, Gasser T, Steinbach PJ, Wilkinson KD, Polymeropoulos MH: The ubiquitin pathway in Parkinson's disease. *Nature* 1998, 395:451-452.

[80] Healy DG, Abou-Sleiman PM, Wood NW: Genetic causes of Parkinson's disease: UCHL-1. *Cell Tissue Res* 2004, 318:189-194.

[81] Saigoh K, Wang YL, Suh JG, Yamanishi T, Sakai Y, Kiyosawa H, Harada T, Ichihara N, Wakana S, Kikuchi T, Wada K: Intragenic deletion in the gene encoding ubiquitin carboxy-terminal hydrolase in gad mice. *Nat Genet* 1999, 23:47-51.

[82] Wilkinson KD, Lee KM, Deshpande S, Duerksen-Hughes P, Boss JM, Pohl J: The neuron-specific protein PGP 9.5 is a ubiquitin carboxyl-terminal hydrolase. *Science* 1989, 246:670-673.

[83] Wilkinson KD: Ubiquitination and deubiquitination: targeting of proteins for degradation by the proteasome. *Semin Cell Dev Biol* 2000, 11:141-148.

[84] Osaka H, Wang YL, Takada K, Takizawa S, Setsuie R, Li H, Sato Y, Nishikawa K, Sun YJ, Sakurai M, Harada T, Hara Y, Kimura I, Chiba S, Namikawa K, Kiyama H, Noda M, Aoki S, Wada K: Ubiquitin carboxy-terminal hydrolase L1 binds to and stabilizes monoubiquitin in neuron. *Hum Mol Genet* 2003, 12:1945-1958.

[85] Liu Y, Fallon L, Lashuel HA, Liu Z, Lansbury PT, Jr.: The UCH-L1 gene encodes two opposing enzymatic activities that affect alpha-synuclein degradation and Parkinson's disease susceptibility. *Cell* 2002, 111:209-218.

[86] Kawaguchi Y, Okamoto T, Taniwaki M, Aizawa M, Inoue M, Katayama S, Kawakami H, Nakamura S, Nishimura M, Akiguchi I, et al.: CAG expansions in a novel gene for Machado-Joseph disease at chromosome 14q32.1. *Nat Genet* 1994, 8:221-228.

[87] Gusella JF, MacDonald ME: Molecular genetics: unmasking polyglutamine triggers in neurodegenerative disease. *Nat Rev Neurosci* 2000, 1:109-115.

[88] Scheel H, Tomiuk S, Hofmann K: Elucidation of ataxin-3 and ataxin-7 function by integrative bioinformatics. *Hum Mol Genet* 2003, 12:2845-2852.

[89] Mao Y, Senic-Matuglia F, Di Fiore PP, Polo S, Hodsdon ME, De Camilli P: Deubiquitinating function of ataxin-3: insights from the solution structure of the Josephin domain. *Proc Natl Acad Sci U S A* 2005, 102:12700-12705.

[90] Burnett B, Li F, Pittman RN: The polyglutamine neurodegenerative protein ataxin-3 binds polyubiquitylated proteins and has ubiquitin protease activity. *Hum Mol Genet* 2003, 12:3195-3205.

[91] Burnett BG, Pittman RN: The polyglutamine neurodegenerative protein ataxin 3 regulates aggresome formation. *Proc Natl Acad Sci U S A* 2005, 102:4330-4335.

[92] Doss-Pepe EW, Stenroos ES, Johnson WG, Madura K: Ataxin-3 interactions with rad23 and valosin-containing protein and its associations with ubiquitin chains and the proteasome are consistent with a role in ubiquitin-mediated proteolysis. *Mol Cell Biol* 2003, 23:6469-6483.

[93] Warrick JM, Morabito LM, Bilen J, Gordesky-Gold B, Faust LZ, Paulson HL, Bonini NM: Ataxin-3 suppresses polyglutamine neurodegeneration in Drosophila by a ubiquitin-associated mechanism. *Mol Cell* 2005, 18:37-48.

[94] Levine RL, Stadtman ER: Oxidative modification of proteins during aging. *Exp Gerontol* 2001, 36:1495-1502.

[95] Ischiropoulos H, Beckman JS: Oxidative stress and nitration in neurodegeneration: cause, effect, or association? *J Clin Invest* 2003, 111:163-169.

[96] Jenner P: Parkinson's disease, pesticides and mitochondrial dysfunction. *Trends Neurosci* 2001, 24:245-247.

[97] Tu BP, Weissman JS: Oxidative protein folding in eukaryotes: mechanisms and consequences. *J Cell Biol* 2004, 164:341-346.

[98] Zorov DB, Bannikova SY, Belousov VV, Vyssokikh MY, Zorova LD, Isaev NK, Krasnikov BF, Plotnikov EY: Reactive oxygen and nitrogen species: friends or foes? *Biochemistry-Russia* 2005, 70:215-221.

[99] Lindholm D, Wootz H, Korhonen L: ER stress and neurodegenerative diseases. *Cell Death Differ* 2006, 13:385-392.

[100] Lotharius J, Brundin P: Pathogenesis of Parkinson's disease: dopamine, vesicles and alpha-synuclein. *Nat Rev Neurosci* 2002, 3:932-942.

[101] Dauer W, Przedborski S: Parkinson's disease: mechanisms and models. *Neuron* 2003, 39:889-909.

[102] Gracy RW, Talent JM, Kong Y, Conrad CC: Reactive oxygen species: the unavoidable environmental insult? *Mutat Res* 1999, 428:17-22.

[103] Nystrom T: Role of oxidative carbonylation in protein quality control and senescence. *EMBO J* 2005, 24:1311-1317.

[104] Claiborne A, Yeh JI, Mallett TC, Luba J, Crane EJ, 3rd, Charrier V, Parsonage D: Protein-sulfenic acids: diverse roles for an unlikely player in enzyme catalysis and redox regulation. *Biochemistry* 1999, 38:15407-15416.

[105] Chang TS, Jeong W, Lee DY, Cho CS, Rhee SG: The RING-H2-finger protein APC11 as a target of hydrogen peroxide. *Free Radic Biol Med* 2004, 37:521-530.

[106] Winklhofer KF, Henn IH, Kay-Jackson PC, Heller U, Tatzelt J: Inactivation of parkin by oxidative stress and C-terminal truncations: a protective role of molecular chaperones. *J Biol Chem* 2003, 278:47199-47208.

[107] Chung KK, Thomas B, Li X, Pletnikova O, Troncoso JC, Marsh L, Dawson VL, Dawson TM: S-nitrosylation of parkin regulates ubiquitination and compromises parkin's protective function. *Science* 2004, 304:1328-1331.

[108] Yao D, Gu Z, Nakamura T, Shi ZQ, Ma Y, Gaston B, Palmer LA, Rockenstein EM, Zhang Z, Masliah E, Uehara T, Lipton SA: Nitrosative stress linked to sporadic Parkinson's disease: S-nitrosylation of parkin regulates its E3 ubiquitin ligase activity. *Proc Natl Acad Sci U S A* 2004, 101:10810-10814.

[109] Lavoie MJ, Ostaszewski BL, Weihofen A, Schlossmacher MG, Selkoe DJ: Dopamine covalently modifies and functionally inactivates parkin. *Nat Med* 2005, 11:1214-1221.

[110] Jenner P: Oxidative stress in Parkinson's disease. *Ann Neurol* 2003, 53 Suppl 3:S26-36; discussion S36-28.

[111] Choi J, Levey AI, Weintraub ST, Rees HD, Gearing M, Chin LS, Li L: Oxidative modifications and down-regulation of ubiquitin carboxyl-terminal hydrolase L1 associated with idiopathic Parkinson's and Alzheimer's diseases. *J Biol Chem* 2004, 279:13256-13264.

[112] Nishikawa K, Li H, Kawamura R, Osaka H, Wang YL, Hara Y, Hirokawa T, Manago Y, Amano T, Noda M, Aoki S, Wada K: Alterations of structure and hydrolase activity of parkinsonism-associated human ubiquitin carboxyl-terminal hydrolase L1 variants. *Biochem Biophys Res Commun* 2003, 304:176-183.

[113] Lowe J, McDermott H, Landon M, Mayer RJ, Wilkinson KD: Ubiquitin carboxyl-terminal hydrolase (PGP 9.5) is selectively present in ubiquitinated inclusion bodies characteristic of human neurodegenerative diseases. *J Pathol* 1990, 161:153-160.

[114] Musiek ES, Milne GL, McLaughlin B, Morrow JD: Cyclopentenone eicosanoids as mediators of neurodegeneration: a pathogenic mechanism of oxidative stress-mediated and cyclooxygenase-mediated neurotoxicity. *Brain Pathol* 2005, 15:149-158.

[115] Ishii T, Sakurai T, Usami H, Uchida K: Oxidative modification of proteasome: identification of an oxidation-sensitive subunit in 26 s proteasome. *Biochemistry* 2005, 44:13893-13901.

[116] Ferrington DA, Kapphahn RJ: Catalytic site-specific inhibition of the 20S proteasome by 4-hydroxynonenal. *FEBS Lett* 2004, 578:217-223.

[117] Bulteau AL, Lundberg KC, Humphries KM, Sadek HA, Szweda PA, Friguet B, Szweda LI: Oxidative modification and inactivation of the proteasome during coronary occlusion/reperfusion. *J Biol Chem* 2001, 276:30057-30063.

[118] Shamoto-Nagai M, Maruyama W, Kato Y, Isobe K, Tanaka M, Naoi M, Osawa T: An inhibitor of mitochondrial complex I, rotenone, inactivates proteasome by oxidative modification and induces aggregation of oxidized proteins in SH-SY5Y cells. *J Neurosci Res* 2003, 74:589-597.

[119] McNaught KS, Jenner P: Proteasomal function is impaired in substantia nigra in Parkinson's disease. *Neurosci Lett* 2001, 297:191-194.

[120] McNaught KS, Belizaire R, Isacson O, Jenner P, Olanow CW: Altered proteasomal function in sporadic Parkinson's disease. *Exp Neurol* 2003, 179:38-46.

[121] Le Pecheur M, Bourdon E, Paly E, Farout L, Friguet B, London J: Oxidized SOD1 alters proteasome activities in vitro and in the cortex of SOD1 overexpressing mice. *FEBS Lett* 2005, 579:3613-3618.

In: The Ubiquitin Proteasome System… ISBN 978-1-60021-749-4
Eds: Mario Di Napoli and Cezary Wojcik, pp. 579-598

Chapter 24

NEUROPROTECTIVE AND PRO-APOPTOTIC RESPONSES OF IBIQUITIN PROTEASOME SYSTEM

Meng Shyan Choy and Nam Sang Cheung*
Department of Biochemistry, Yong Loo Lin School of Medicine, National University of Singapore, 8 Medical Drive, Singapore 117597.

ABSTRACT

The accumulation of unfolded, misfolded or damaged proteins in cells is a threat to cell survival. The ubiquitin-proteasome system (UPS) is responsible for the degradation of these abnormal proteins. UPS dysfunction has been postulated to play a key role in the pathogenesis of neurodegenerative diseases such as Alzheimer's and Parkinson's diseases. Both normal and misfolded proteins can undergo highly specific degradation by the UPS. Selective degradation of correctly folded proteins underlies many cellular regulations. Examples include degradation of cyclins or their inhibitors in the regulation of the cell cycle and the degradation of IκB in the activation of immunity responses. The endoplasmic reticulum (ER) is the site of synthesis of membrane proteins and secretory proteins. In the ER, defective or unfolded proteins are degraded [a process known as ER-associated proteins degradation (ERAD)], whereas correctly folded proteins are spared. In the familial form of Alzheimer's disease, transcriptional misreading of the stress-induced polyubiquitin gene produces ubiquitin with aberrant C-terminal extensions that competitively inhibit proteasomal function. This inhibition of UPS may impair ERAD, thereby causing the accumulation of misfolded proteins in the ER, resulting in ER stress and induction of cell death through the activation of calpain and caspase-3. Proteasome inhibitors such as lactacystin have been reported to activate the pro-apoptotic transcription factor C/EBP-homologous protein (CHOP) and to cause cell death in

* Correspondence concerning this article should be addressed to Dr. Nam Sang Cheung, PhD; Department of Biochemistry, Yong Loo Lin School of Medicine, National University of Singapore, 8 Medical Drive, Singapore 117597. Phone: 65-65164780; Fax: 65-67791453 ; E-mail: bchcns@nus.edu.sg.

cultured cortical neurons. Although the inhibition of proteasomes has been linked to cell death, recent studies have shown that, below a threshold level, proteasome inhibition can activate neuroprotective responses — proteasome inhibition has been shown to induce various molecular chaperones such as heat shock proteins (HSPs) that increase cell tolerance to the accumulation of unfolded and damaged proteins, stimulate the expression of UPS components through a feedback mechanism, and suppress inflammatory responses by inhibiting IκB degradation. Although the inhibition of proteasomes may stimulate neuroprotective responses, prolonged ER stress ultimately leads to apoptosis. Further studies to elucidate the impact of proteasomal inhibition on other cellular signaling pathways may provide insights on the interplay between the UPS and cell physiology. A better understanding of the function and activation of the neuroprotective or pro-apoptotic responses would provide a means to manipulate this pathway in order to cure diseases associated with unfolded proteins. Much remains to be discovered about the inducibility and functioning of chaperones and neuroprotective ubiquitin-proteasome pathways in neurons. Such studies would be useful, since genetic polymorphism in these protective systems and changes in their expression with ageing may play critical roles in the accumulation of unfolded or damaged proteins, and in the pathogenesis of disease. Moreover, pharmacological induction or activation of these protein repair-and-degradative systems could in future be developed into innovative therapies for neurodegenerative diseases.

Keywords: Neuroprotective, pro-apoptotic, neuron, microarray, lactacystin.

Abbreviations

ARJP, autosomal recessive juvenile parkinsonism; Atf4, activating transcription factor 4; ATF6, activating transcription factor 6; C/EBP, CCAAT/enhancer binding protein; Cebpb, C/EBP beta; CHOP, C/EBP-homologous protein; COX-2, cyclooxygenase-2; Ddit3, DNA-damaged inducible transcripts 3; EGCG, (-)-epigallocatechin-3-gallate; ER, endoplasmic reticulum; ERAD, ER associated proteins degradation; EST, expressed sequence tag; GSH, reduced glutathione; HSP, heat shock protein; ROS, reactive oxygen species; SREBP, sterol regulatory element binding proteins; UBB, ubiquitin B; UPS, ubiquitin-proteasome system.

Introduction

Aggregation of Misfolded Proteins and Neurodegeneration

The abnormal accumulation of misfolded proteins as protein aggregates in neurons is a hallmark of many neurodegenerative diseases [1] and can lead to stress and cell death [2]. Misfolded proteins tend to aggregate because the normally buried hydrophobic domains of these proteins associate with one another. Eukaryotic cells have two main strategies to counteract the formation of such protein aggregates. First, the induction of heat shock proteins (HSPs) and molecular chaperones that are involved in the protein refolding systems and second, targeted degradation of damaged and misfolded proteins by the UPS [3,4,5]. In conditions where UPS activity is affected, the homeostasis between protein synthesis and protein degradation is disrupted, and protein aggregation occurs [6]. The dysfunction of UPS has been proposed as one of the main causes of neurodegeneration [1,7,8]. The identification of a number of genes responsible for rare familial forms of neurodegenerative diseases has provided insights into the underlying mechanisms of these diseases [9,10]. For example, loss-of-function mutations in the gene encoding ubiquitin protein ligase (E3), Parkin, are linked to autosomal recessive juvenile parkinsonism (ARJP) [9,11,12], while over-expression of Parkin could counter misfolded protein stress-induced cell death [2]. Recent studies also revealed a frame-shift mutation of the ubiquitin B gene (*UBB*) that produces a variant form of ubiquitin, UBB^{+1}, a major component in intracellular protein inclusions in Alzheimer's disease and progressive supranuclear palsy. This mutation results in the absence of the C-terminal G76 of ubiquitin, preventing the ligation of UBB^{+1} to target protein substrates or poly-ubiquitin chains. Instead, UBB^{+1} is itself readily poly-ubiquitinated, so that it acts as a potent competitive inhibitor of 26S proteasome [13,14,15].

Familial types of neurodegenerative diseases, such as certain forms of Parkinson's disease, are linked to loss-of-function mutations of ubiquitin ligase. The sporadic forms of neurodegeneration are, however, often associated with ageing. In ageing cells, the capacity to handle the accumulation of misfolded or damaged proteins is insufficient to prevent their accumulation and resultant toxicity [16]. Furthermore, recent studies reported that expression of neuroprotective antioxidants and molecular chaperones decreases in ageing cells [16; and Chapter 22]. The resultant accumulation of misfolded protein aggregates leads to inhibition of the UPS.

The expression, functions and regulation of heat shock proteins (HSPs), molecular chaperones and the UPS components in mammalian cells are largely not well understood (see Chapter 19). Investigations in these areas could therefore provide valuable insights into the molecular mechanisms behind the pathogenesis of neurodegenerative diseases. Pharmacological induction or activation of these protein repair-and-degradative systems could lead to the development of innovative therapies for neurodegenerative diseases [1].

Proteasome inhibitors have been used widely in the study of the UPS system in mammalian cells [17,18]. In addition to being useful research tools for dissecting the roles of the proteasome, these inhibitors have potential applications in biotechnology and medicine. Proteasome inhibitors are known to induce cell death [19]. On the other hand, they can also protect cells against other insults. For example, the proteasome inhibitor MLN-519 has been shown to be very effective in suppressing the production of inflammatory mediators in stroke models [20]. Furthermore, proteasome inhibition can also induce the synthesis of various HSPs, which increase tolerance of cells to stressful conditions [21]. It appears that proteasome inhibition induces both protective and apoptotic effects in mammalian cells, depending on the presence or absence of a number of factors (see Chapter 21 and [19,22]). The factors that seem to be involved in these contradictory outcomes of proteasome inhibition are (*i*) the duration of exposure, (*ii*) the concentration of proteasome inhibitor used, (*iii*) the type of cells used and (*iv*) the type of proteasome inhibitor used [19,23,24].

The fundamental objectives in any neurodegeneration and neuroprotection research are to determine the factors constituting the primary event, the sequence in which these events occur, and whether they act in concurrence in the pathogenic process [25]. Our study of genes differentially expressed during lactacystin-induced neuronal apoptosis using microarray technique has shown the induction of potentially neuroprotective and pro-apoptotic transcriptional responses [22]. However, the sequence of events leading to the process of neuronal apoptosis is not clear. This paper focuses on our research findings from the time course microarray study of the genes differentially expressed during lactacystin-induced neuronal apoptosis, and on other recent findings in this area of research. Our microarray study revealed that treatment of cultured cortical neurons with 1 μM lactacystin resulted in apoptosis and a large number of genes being differentially expressed: out of a total of 12488 genes and expressed sequence tags (ESTs) in the murine genome GeneChip® U74Av2, the expressions of 1168 genes were enhanced more than two-fold, according to the one-way ANOVA, $p<0.01$. Some of these genes were potentially neuroprotective, while others were potentially pro-apoptotic. These genes were grouped according to their biological functions and the significance of the findings is described in the following section.

UBIQUITIN-PROTEASOME SYSTEM

It is now evident that the UPS plays an important role in the degradation of misfolded and unfolded proteins in neurons [1,26]. Alterations in proteasomal functions have been found in ageing (review in [27]) and neurodegenerative diseases [28]. For example, a recent study reported that the expression of genes encoding proteasome subunits were down-regulated in the *substantia nigra pars compacta* area of brains of Parkinson's disease patients, suggesting that robust transcription of proteasome subunits might play an important role in the survival of neuronal cells under stress conditions [28].

An inspection of our microarray data reveals that the genes encoding components of the UPS system are the earliest to be up-regulated during proteasomal inhibition by lactacystin (Table 1). We attribute the changes of UPS gene expression to the response to the accumulation of unfolded proteins in neurons. How these genes are regulated in mamalian cells is not clear. In yeast (*Saccharomyces cerevisiae*), however, genes encoding proteasomal subunits are preceeded by a common upstream activating cis-element called the proteasome-associated control element. This proteasome-associated control element serves as a target sequence for the transcription factor Rpn4 that activates proteasomal gene expression in a concerted manner [29]. Interestingly, Rpn4 is also a substrate of the 26S proteasome. Thus, the same protein that regulates and induces proteasome synthesis is also degraded by the proteasome, constituting a negative feedback circuit. More recently, Rnp4 was found to be responsible for the elevation of cellular levels of proteasome subunit mRNAs, in response to various stress conditions such as the abnormal accumulation of misfolded proteins, suggesting that Rpn4, in yeast, is indeed a master regulator responsible for the ability of the cell to compensate for proteasome inhibition [30,31].

The regulation of the expression of proteasomal subunit genes in mamals is not well understood. Meiners *et al.* reported that proteasomal inhibition resulted in a transient and concerted up-regulation of all 26S proteasome subunit mRNAs, as well as an enhanced synthesis of all proteasomal subunits, and an increase in the number of proteasomes. This is the first report that the number of proteasomes in mammalian cells is regulated at the transcriptional level, and that there exists an autoregulatory feedback mechanism that allows for the compensation of reduced proteasomal activity [32,33].

Table 1. Genes differentially expressed during lactacystin treatment: UPS

Ubiquitin-proteasome system			Time point (h)				
Probe id	Symbol	Gene title	4.5	7.5	24	48	Genbank
100955_at	2700084L22Rik	RIKEN cDNA 2700084L22 gene	2.24 ± 0.26	1.53 ± 0.11	1.06 ± 0.11	1.38 ± 0.16	AA989957
95600_at	Arih2	ariadne homolog 2 (Drosophila)	2.61 ± 0.14	4.11 ± 0.14	1.24 ± 0.15	-1.05 ± 0.19	AJ130975
96892_at	Psma1	proteasome subunit, alpha type 1	2.79 ± 0.09	3.37 ± 0.10	2.49 ± 0.20	-1.06 ± 0.14	AI836804
92544_f_at	Psma3	proteasome subunit, alpha type 3	2.45 ± 0.09	2.83 ± 0.06	1.74 ± 0.14	-1.04 ± 0.15	AF055983
94841_at	Psma5	proteasome subunit, alpha type 5	2.16 ± 0.08	2.69 ± 0.10	2.48 ± 0.30	1.01 ± 0.21	AW048997
93988_at	Psma7	proteasome subunit, alpha type 7	2.60 ± 0.08	3.53 ± 0.07	2.35 ± 0.13	-1.09 ± 0.19	AI836676
98113_at	Psmb1	proteasome subunit, beta type 1	1.52 ± 0.07	1.80 ± 0.06	2.00 ± 0.10	1.04 ± 0.17	U60824
94219_at	Psmb2	proteasome subunit, beta type 2	1.57 ± 0.06	2.28 ± 0.09	2.63 ± 0.39	-1.20 ± 0.22	AI853269
94025_at	Psmb3	proteasome subunit, beta type 3	2.78 ± 0.06	4.50 ± 0.20	3.22 ± 0.21	1.10 ± 0.24	AW045339
98557_f_at	Psmb4	proteasome subunit, beta type 4	2.21 ± 0.06	2.26 ± 0.07	1.95 ± 0.11	1.03 ± 0.19	U65636
101992_at	Psmb6	proteasome subunit, beta type 6	1.99 ± 0.10	2.60 ± 0.07	2.40 ± 0.16	-1.11 ± 0.18	U13393
160152_at	Psmc1	proteasome 26S subunit, ATPase 1	2.28 ± 0.12	3.79 ± 0.09	2.29 ± 0.16	-1.06 ± 0.13	U39302
95448_at	Psmc2	proteasome 26S subunit, ATPase 2	2.58 ± 0.09	2.53 ± 0.07	2.49 ± 0.26	-1.16 ± 0.13	AI839371
93734_i_at	Psmc3	proteasome 26S subunit, ATPase 3	2.71 ± 0.07	3.13 ± 0.11	2.27 ± 0.12	-1.05 ± 0.13	D49686
103319_at	Psmd10	proteasome 26S subunit, non-ATPase, 10	2.21 ± 0.10	1.19 ± 0.10	-1.21 ± 0.10	1.49 ± 0.18	AB022022
160305_at	Psmd11	proteasome 26S subunit, non-ATPase, 11	5.34 ± 0.07	5.03 ± 0.27	2.75 ± 0.17	1.02 ± 0.12	AW121693
93971_f_at	Psmd12	proteasome 26S subunit, non-ATPase, 12	3.34 ± 0.06	3.62 ± 0.20	2.56 ± 0.31	-1.09 ± 0.17	AI838669
95742_at	Psmd13	proteasome 26S subunit, non-ATPase, 13	1.59 ± 0.07	2.30 ± 0.07	1.92 ± 0.11	1.02 ± 0.15	AW045451
97274_at	Psmd14	proteasome 26S subunit, non-ATPase, 14	2.26 ± 0.06	2.72 ± 0.09	2.22 ± 0.18	-1.18 ± 0.14	Y13071
92769_at	Psmd3	proteasome 26S subunit, non-ATPase, 3	1.12 ± 0.07	2.95 ± 0.07	1.95 ± 0.12	-1.12 ± 0.11	M25149
94302_at	Psmd4	proteasome 26S subunit, non-ATPase, 4	2.85 ± 0.07	5.29 ± 0.12	3.07 ± 0.35	-1.05 ± 0.21	AF013099
96698_at	Psmd5	proteasome 26S subunit, non-ATPase, 5	1.94 ± 0.07	2.79 ± 0.13	1.46 ± 0.19	-1.06 ± 0.14	AI835520
103350_at	Psmd7	proteasome 26S subunit, non-ATPase, 7	2.38 ± 0.08	3.03 ± 0.08	1.88 ± 0.30	-1.31 ± 0.15	M64641
98522_at	Psmd8	proteasome 26S subunit, non-ATPase, 8	2.30 ± 0.13	2.38 ± 0.10	1.38 ± 0.10	-1.11 ± 0.15	AI839158
95124_i_at	Rbx1	ring-box 1	3.18 ± 0.11	2.00 ± 0.19	1.13 ± 0.15	-1.21 ± 0.17	AW122337
160205_f_at	Rnf11	ring finger protein 11	2.04 ± 0.06	1.74 ± 0.09	1.01 ± 0.11	-1.29 ± 0.11	AB024427
101966_s_at	Rnf13	ring finger protein 13	2.09 ± 0.09	3.13 ± 0.12	1.00 ± 0.10	1.13 ± 0.12	AF037206
93164_at	Rnf2	ring finger protein 2	1.79 ± 0.10	3.25 ± 0.17	-1.57 ± 0.10	-1.12 ± 0.14	Y12783
96961_at	Rnf110	ring finger protein 110	3.08 ± 0.15	1.55 ± 0.15	-1.08 ± 0.17	1.14 ± 0.17	AI503821
101069_g_at	Mkrn1	makorin, ring finger protein, 1	1.96 ± 0.09	3.10 ± 0.08	-1.25 ± 0.12	-1.32 ± 0.12	AA656621
100985_at	Siah1a	seven in absentia 1A	1.86 ± 0.07	2.33 ± 0.09	1.16 ± 0.11	-1.40 ± 0.12	Z19579
101255_at	Ubb	ubiquitin B	3.10 ± 0.07	3.62 ± 0.06	1.08 ± 0.15	-1.19 ± 0.14	X51703
95215_f_at	Ubc	ubiquitin C	4.04 ± 0.09	4.79 ± 0.07	1.39 ± 0.16	-1.25 ± 0.12	D50527
102812_i_at	Ube1dc1	ubiquitin-activating enzyme E1-domain containing 1	1.82 ± 0.11	2.30 ± 0.06	1.66 ± 0.14	1.06 ± 0.13	AW210346
93069_at	Ube2d2	ubiquitin-conjugating enzyme E2D 2	2.46 ± 0.10	2.13 ± 0.12	1.08 ± 0.14	-1.30 ± 0.13	U62483
101581_at	Ube3a	ubiquitin protein ligase E3A	3.65 ± 0.08	1.93 ± 0.09	1.24 ± 0.11	-1.20 ± 0.10	U82122
94018_at	Ubl3	ubiquitin-like 3	2.17 ± 0.06	1.57 ± 0.07	1.41 ± 0.13	1.13 ± 0.15	AW120725
95601_at	Ubqln1	ubiquilin 1	1.31 ± 0.06	2.30 ± 0.10	1.08 ± 0.10	-1.29 ± 0.17	AW125420
93303_at	Ufd1l	ubiquitin fusion degradation 1 like	2.49 ± 0.07	4.45 ± 0.07	2.05 ± 0.34	-1.43 ± 0.16	U64445
161870_at	Usp15	ubiquitin specific protease 15	3.77 ± 0.09	-3.28 ± 0.12	-1.57 ± 0.12	1.50 ± 0.24	AV359471
99085_at	Usp3	ubiquitin specific protease 3	1.42 ± 0.14	2.55 ± 0.07	1.26 ± 0.12	1.03 ± 0.13	AI021421
99086_g_at	Usp3	ubiquitin specific protease 3	1.73 ± 0.10	2.74 ± 0.11	1.14 ± 0.12	-1.23 ± 0.15	AI021421
160724_at	Usp49	ubiquitin specific protease 49	2.06 ± 0.24	2.51 ± 0.08	-1.06 ± 0.10	-1.06 ± 0.11	AJ245617

Gene expression indicated as FC±SE where FC=fold change; SE=Standard Error.

Our microarray data showed that besides those of proteasome subunits, the genes encoding ubiquitin (*Ubb*), ubiquitin-conjugating enzymes E2 (*Ubc, Ube2d2*) and ubiquitin protein ligase E3A (*Ube3a*) were also significantly up-regulated during the early phase of lactacystin treatment (Table 1). Since these proteins play important roles in the degradation of proteins, it is perhaps not surprising that these proteins, too, were up-regulated during proteasome inhibition. Both *Ubb* and *Ubc* genes were reported to be up-regulated following ischemia injury in animal models [34]. They might therefore play a role in degrading denatured proteins resulting from oxidative stress during injury [34]. It was reported recently that the overexpression of wild-type or mutant ubiquitin transgenes resulted in a small but significant delay in the onset of clinical symptoms, and mild acceleration of familial amyotrophic lateral sclerosis, respectively, in an animal model, although the neuroprotective mechanism at present is not clear [35].

The other group of proteins that shows early up-regulation is the ring finger proteins. Ring finger proteins have the consensus sequence CX2CX(9-39)CX(1-3)HX(2-3_C/HX2CX(4-48)CX2C, with Cys and His representing zinc-binding residues (review in [36]). Recent studies demonstrated that many ring finger-containing proteins are ubiquitin protein ligases (E3s) (review in [36,37]). Ring finger-containing E3s play pivotal roles in diverse cellular processes, and mutations in genes encoding these ubiquitin ligases are implicated in diseases such as juvenile parkinsonism [36]. Ring finger E3s also play key roles in the quality control of protein synthesis. For example, Hrd1p is a yeast ER membrane ring finger protein that regulates the degradation of abnormal ER protein via the UPS. Therefore, ring finger proteins might have a neuroprotective role against the accumulation of abnormally folded proteins in neurons.

In a recent study, Lee *et al.* showed that pretreatment of cultured neocortical neurons with proteasome inhibitors increased proteasome activity and reduced the vulnerability of these neurons to oxidative injury [38]. In another study, the over-expression of one proteasome subunit (Psmb5) alone was able to increase the amount of assembled proteasomes, and conferred protection against oxidative stress in primary IMR90 human fibroblasts [39]. Taken together, these observations and our microarray data (Table 1) suggest that genes that encode components of the UPS can be induced by proteasome inhibitors, and the up-regulation of these genes has a neuroprotective role in neurons.

Heat Shock Proteins and Molecular Chaperones

HSPs with a chaperoning function work together with the UPS to prevent the accumulation of misfolded, potentially toxic proteins, as well as to control the degradation of the bulk of cytoplasmic proteins. The levels of these proteins are often increased in response to stress, and they have been shown to enhance cell resistance to various insults [21,40].

HSPs and molecular chaperones facilitate the refolding of misfolded proteins to prevent them from aggregating in the cell [41]. Mammalian cells possess a number of HSPs, which are induced in response to stresses and display protective chaperone activity [42].

HSPs can act via two mechanisms to confer cellular protection. First, as molecular chaperones, HSPs are active in the formation and maintenance of the native conformation of cytosolic proteins and the stabilization of actin filaments. Second, both Hsp70 and Hsp27 inhibit the release of cytochrome c by suppressing Bid, a pro-apoptotic member of the Bcl-2 family (review by [43]). Much of the research to date has focused on the action of Hsp27 and

Hsp70 individually, but the full therapeutic benefits of these molecules may depend on a better understanding of their combined neuroprotective action.

Table 2. Genes differentially expressed during lactacystin treatment: heat shock proteins and molecular chaperones

Heat shock proteins and molecular chaperones			Time point (h)				
Probe id	Symbol	Gene title	4.5	7.5	24	48	Genbank
104589_at	C80913	expressed sequence C80913	2.04 ± 0.10	-1.11 ± 0.06	1.38 ± 0.12	1.16 ± 0.11	AF091096
98153_at	Cct3	chaperonin subunit 3 (gamma)	2.24 ± 0.10	2.54 ± 0.08	1.79 ± 0.22	-1.17 ± 0.14	L20509
160562_at	Cct7	chaperonin subunit 7 (eta)	2.00 ± 0.10	2.09 ± 0.06	1.53 ± 0.25	-1.34 ± 0.12	Z31399
96254_at	Dnajb1	DnaJ (Hsp40) homolog, subfamily B, member 1	2.11 ± 0.20	3.52 ± 0.15	1.42 ± 0.26	-1.12 ± 0.12	AB028272
98572_at	Dnajb11	DnaJ (Hsp40) homolog, subfamily B, member 11	1.84 ± 0.10	2.21 ± 0.08	1.41 ± 0.13	-1.12 ± 0.12	AW12255
93853_at	Dnajb4	DnaJ (Hsp40) homolog, subfamily B, member 4	2.19 ± 0.16	2.69 ± 0.13	1.46 ± 0.21	-1.24 ± 0.12	AA763918
104625_at	Dnajb6	DnaJ (Hsp40) homolog, subfamily B, member 6	1.61 ± 0.21	2.16 ± 0.15	-1.30 ± 0.15	-1.03 ± 0.11	AA874130
103344_at	Dnajc1	DnaJ (Hsp40) homolog, subfamily C, member 1	2.31 ± 0.30	1.86 ± 0.09	1.42 ± 0.10	1.32 ± 0.11	L16953
94422_at	Dnajc13	DnaJ (Hsp40) homolog, subfamily C, member 13	2.12 ± 0.08	1.37 ± 0.08	1.07 ± 0.15	-1.17 ± 0.11	AI842938
102414_i_at	Dnajc3	DnaJ (Hsp40) homolog, subfamily C, member 3	2.05 ± 0.14	2.25 ± 0.15	1.31 ± 0.10	1.42 ± 0.12	U28423
93211_at	Dnajc5	DnaJ (Hsp40) homolog, subfamily C, member 5	1.41 ± 0.10	3.16 ± 0.17	-1.48 ± 0.17	-1.56 ± 0.16	AF032115
102761_at	Grpel2	GrpE-like 2, mitochondrial	2.03 ± 0.12	2.07 ± 0.17	1.15 ± 0.13	-1.20 ± 0.10	AF041060
98111_at	Hsp105	heat shock protein 105	1.57 ± 0.09	2.38 ± 0.08	2.54 ± 0.53	-1.20 ± 0.13	L40406
93875_at	Hspa1a	heat shock protein 1A	-1.27 ± 0.13	4.70 ± 0.53	13.82 ± 6.14	1.78 ± 0.53	M12571
101955_at	Hspa5	heat shock 70kD protein 5 (glucose-regulated protein)	2.20 ± 0.07	2.39 ± 0.06	1.87 ± 0.17	-1.13 ± 0.11	AJ002387
96564_at	Hspa8	heat shock protein 8	2.05 ± 0.08	1.96 ± 0.10	1.14 ± 0.14	-1.18 ± 0.10	X54401
97914_at	Hspa9a	heat shock protein, A	1.68 ± 0.07	3.68 ± 0.07	2.18 ± 0.27	1.07 ± 0.16	D17666
160139_at	Hspb8	heat shock 27kDa protein 8	-1.37 ± 0.12	1.67 ± 0.16	8.08 ± 0.47	2.22 ± 0.48	AI848798
95359_at	Hspcb	heat shock protein 1, beta	2.06 ± 0.06	2.82 ± 0.11	1.46 ± 0.20	-1.08 ± 0.17	M18186
92829_at	Hspe1	heat shock protein 1 (chaperonin 10)	1.60 ± 0.07	1.38 ± 0.06	2.34 ± 0.15	1.17 ± 0.15	U09659
101207_at	Ppia	peptidylprolyl isomerase A	2.43 ± 0.07	1.98 ± 0.06	-1.08 ± 0.10	-1.18 ± 0.13	X52803
100089_at	Ppic	peptidylprolyl isomerase C	1.20 ± 0.11	-1.27 ± 0.07	1.77 ± 0.16	2.10 ± 0.16	M74227
99350_at	Sec63	SEC63-like (S. cerevisiae)	1.70 ± 0.36	2.32 ± 0.12	1.28 ± 0.10	-1.06 ± 0.13	C76102
94817_at	Serpinh1	serine (or cysteine) proteinase inhibitor, clade H, member 1	1.14 ± 0.08	3.10 ± 0.06	6.16 ± 0.15	3.30 ± 0.43	X60676

Gene expression indicated as FC±SE where FC=fold change; SE=Standard Error.

Our microarray data shows that genes which encode HSPs were up-regulated during the early phase of proteasome inhibition by lactacystin (Table 2). Among the HSPs, Hsp70 is probably the best characterized. Hsp70 can suppress both necrosis and apoptosis induced by various injuries in vivo and in vitro [44,45,46]. Hsp70 is an ATPase that can bind newly exposed hydrophobic sequences on denatured proteins. In this role, ATP hydrolysis provides the energy for cycles of association and disassociation, a process which allows the denatured protein repeated opportunities to refold to its native conformation. When correct refolding is not possible, Hsp70, in collaboration with other Hsp70-interacting proteins such as Hsp40,

CHIP (carboxyl terminus of Hsc70-interacting protein) and ubiquitin E3 ligase (see Chapter 10 and [47]), is able to target bound substrates to proteasomal degradation.

Another group of HSPs, called the small heat shock proteins (sHSPs), consists of 10 members in human and mouse (HSPB1–10), of which Hsp27 (Hspb1) and Hsp22 (Hspb8) are the best-known [48]. The genes encoding these sHSPs were observed to be up-regulated at an early time point of lactacystin treatment (Table 2). Unlike Hsp70, sHSPs have no ATPase activity and their binding to substrates appeared to be modulated by their de-oligomerization [49]. Recent findings have demonstrated that mutation of either Hsp27 or the related protein Hsp22 can be observed in specific families of patients with hereditary motor neuropathy arising from premature axonal loss, possibly due to cell death and subsequent neuronal degeneration [50]. Although the majority of studies on the protective effects of individual HSPs have concentrated on the major inducible Hsp70, a variety of evidence suggests that the sHSPs such as Hsp27 may have a more potent protective effect in the nervous system [50]. We recently demonstrated that over-expression of Hsp22 in rat PC12 cell lines could protect the cells against proteasome inhibition by MG-132 [22]. The actual neuroprotective effect of Hsp22 against MG-132 is likely to be more than the observed 25%, given the fact that the transfection efficiency was only about 30% [22].

Endoplasmic Reticulum Stress

The endoplasmic reticulum (ER) is the site of synthesis of proteins destined for secretion or targeting to the plasma membrane. Damaged or misfolded proteins can be generated during the process of protein synthesis. These misfolded proteins are retrotranslocated across the ER membrane to be degraded by cytosolic proteasomes. This process is called ER-associated degradation (ERAD) and is important in preventing the accumulation of misfolded proteins (see Chapter 13 and [51]). ERAD functions in protein quality control, where damaged or unfolded proteins are selectively targeted for degradation, while correctly folded proteins are spared [3].

When the ER becomes overwhelmed by the accumulation of large amounts of misfolded proteins, for example in the case of proteasomal inhibition, the cell triggers a specific ER stress response. This ER stress may in turn signal a change in cell status, such as the regulation of protein synthesis, or even the activation of apoptosis. Many protein-folding diseases arise from mutations which result in the impaired function of protein quality control and ER stress response, as has been observed in neurodegenerative diseases such as Alzheimer's disease [52,53].

Genes encoding ER stress-associated transcription factors such as DNA-damaged inducible transcripts 3 (Ddit3), also known as C/EBP-homologous protein (CHOP), CCAAT/enhancer binding protein (C/EBP) beta (Cebpb) and activating transcription factor 4 (Atf4), have been reported to be up-regulated during proteasomal inhibition and ER stress [22,54]. In our microarray data, genes encoding Ddit3/CHOP, Cebpb and Atf4 were all observed to be up-regulated during the early time points of lactacystin-induced neuronal apoptosis (Table 3). Ddit3/CHOP is a small nuclear protein transcription factor of the C/EBP family that is normally undetectable, but is expressed at high levels during the ER stress response in cells exposed to conditions that perturb protein folding in the ER [55]. The expression of Ddit3/CHOP in stressed cells is linked to the development of programmed cell death [56,57,58,59]. The pro-apoptotic role of Ddit3/CHOP protein is believed to suppress the expression of neuroprotective Bcl-2 in cells, causing such cells to be more susceptible to apoptosis [59,60].

Table 3. Genes differentially expressed during lactacystin treatment:ER stress

ER stress			Time point (h)				
Probe id	Symbol	Gene title	4.5	7.5	24	48	Genbank
104155_f_at	Atf3	activating transcription factor 3	3.77 ± 0.13	11.13 ± 0.38	6.68 ± 1.76	2.06 ± 0.27	U19118
100599_at	Atf4	activating transcription factor 4	3.52 ± 0.06	3.73 ± 0.09	-1.03 ± 0.25	-2.13 ± 0.12	M94087
101429_at	Ddit3	DNA-damage inducible transcript 3	2.90 ± 0.10	5.01 ± 0.12	2.65 ± 0.42	1.02 ± 0.15	X67083
95057_at	Herpud1	homocysteine-inducible, endoplasmic reticulum stress-inducible, ubiquitin-like domain member 1	1.93 ± 0.08	3.18 ± 0.16	1.17 ± 0.22	-1.32 ± 0.11	AI846938
94821_at	Xbp1	X-box binding protein 1	1.99 ± 0.08	2.30 ± 0.09	-1.21 ± 0.12	-1.74 ± 0.10	AW123880
92925_at	Cebpb	CCAAT/enhancer binding protein (C/EBP), beta	-1.13 ± 0.18	4.26 ± 0.14	3.44 ± 0.34	1.32 ± 0.13	M61007

Gene expression indicated as FC±SE where FC=fold change; SE=Standard Error; ER= endoplasmic reticulum.

INFLAMMATORY RESPONSE

Neuroinflammation plays a key role in the pathophysiology of cerebral ischemia. Proteasomal inhibition is able to attenuate the inflammatory cascade in cerebral ischemia and reduce ischemic damage [61]. The mechanism of this neuroprotection involves the deactivation of NF-κB by stabilizing its inhibitor IκB, preventing the translocation of NF-κB to the nucleus and thereby suppressing the expression of pro-inflammatory genes. The effect of proteasome inhibitors on animal models has been evaluated recently. Intravenous infusion of MLN-519 (an analog of clasto-lactacystin β-lactone) effectively attenuated the expression of cell adhesion proteins, reduced the invasion of leukocytes and hence limited brain tissue damage [62]. Our microarray analysis has shown that proteasomal inhibition by lactacystin caused an early down-regulation of genes associated with the inflammatory response (Table 4). This is consistent with the anti-inflammatory effects of proteasome inhibitors.

Other studies, however, reported that proteasome inhibitors could induce cyclooxygenase-2 (COX-2) activation in neuronal cell cultures [22,63]. Inflammatory pathways involving COX-2 and subsequent generation of prostaglandins are potential causes of neurodegeneration associated with amyotrophic lateral sclerosis [64] and Alzheimer's disease. Several regulatory elements on the murine COX-2 promoter, including a cyclic AMP response element, two C/EBP sites and a single NFκB site, have been shown to be involved in COX-2 promoter activation [65]. The up-regulation of COX-2 expression during proteasome inhibition resulting from proteasome inhibitor treatment is unlikely to be mediated by NF-κB [63]. Instead, COX-2 may be activated by Cebpb binding to the promoter region of COX-2 [65]. Since Cebpb expression is mediated by ER stress, ER stress may be the trigger for the activation of COX-2 in neurons during proteasome inhibition.

Table 4. Genes differentially expressed during lactacystin treatment: inflammatory responses

Inflammatory responses			Time point (h)				
Probe id	Symbol	Gene title	4.5	7.5	24	48	Genbank
98088_at	Cd14	CD14 antigen	-2.35 ± 0.13	-3.62 ± 0.07	-1.13 ± 0.11	1.33 ± 0.13	X13333
160511_at	Cxcl12	chemokine (C-X-C motif) ligand 12	2.37 ± 0.17	1.77 ± 0.22	-1.32 ± 0.17	-1.51 ± 0.14	L12029
103202_at	Gbp3	guanylate nucleotide binding protein 3	1.14 ± 0.18	-1.19 ± 0.15	2.87 ± 0.18	2.67 ± 0.21	AW047476
101341_at	H2-M9	histocompatibility 2, M region locus 9	-5.34 ± 0.14	-5.30 ± 0.06	-1.52 ± 0.10	-1.01 ± 0.13	AF016308
102250_at	Il27ra	interleukin 27 receptor, alpha	-2.61 ± 0.08	-2.23 ± 0.07	-1.23 ± 0.12	-1.16 ± 0.15	AF053005
93077_s_at	Ly6c	lymphocyte antigen 6 complex, locus C	-2.71 ± 0.07	-2.44 ± 0.08	-1.20 ± 0.10	-1.09 ± 0.13	D86232
96939_at	Myl9	myosin, light polypeptide 9, regulatory	1.01 ± 0.17	-1.09 ± 0.11	1.62 ± 0.16	2.18 ± 0.25	AI842649
101923_at	Pla2g7	phospholipase A2, group VII (platelet-activating factor acetylhydrolase, plasma)	-1.20 ± 0.07	-1.66 ± 0.06	-1.61 ± 0.15	2.14 ± 0.18	U34277
97944_f_at	Tcra	T-cell receptor alpha chain	-2.45 ± 0.07	-1.71 ± 0.07	-1.14 ± 0.13	1.06 ± 0.13	AF099808

Gene expression is indicated as FC±SE where FC=fold change; SE= Standard Error.

ANTIOXIDANTS

ER stress can also lead to the production of reactive oxygen species (ROS) [60,66]. Cells respond to oxidative stress by increasing the expression of genes associated with the antioxidant GSH, such as *Gsta4*, *Mgst1* and *Gclm* [67,68,69]. In our microarray study, the group of genes associated with antioxidant response, such as glutathione S-transferase, alpha 4 (*Gsta4*), microsomal glutathione S-transferase 1 (*Mgst1*) and metallothionein 1 (*Mt1*), were up-regulated at the later time points (Table 5). Gsta4 is known for its high catalytic efficiency in the conjugation of 4-hydroxinonenal and other genotoxic products of lipid peroxidation during oxidative stress [70]. Mgst1, an ER-bound enzyme known for its oxidative stress protection, was also up-regulated. Cellular GSH is an important antioxidant in the cell. The decrease in the level of cellular GSH was reported to be an early event in the pathogenesis of Parkinson's disease [71].

Besides their role in sequestration and distribution of metal ions such as copper and zinc, metallothioneins are known to provide cryoprotection from ROS. The down-regulation of metallothioneins is implicated in redox status and increased susceptibility to oxidative stress and metal-induced neurotoxicity [72] (Table 5).

Table 5. Genes differentially expressed during lactacystin treatment: Oxidative stress

Oxidative stress			Time point (h)												
Probe id	Symbol	Gene title	4.5			7.5			24			48			Genbank
94132_at	Gpx1	glutathione peroxidase 1	-2.11	±	0.07	-2.43	±	0.06	-1.10	±	0.10	-1.02	±	0.10	X03920
160335_at	Gclm	glutamate-cysteine ligase , modifier subunit	2.97	±	0.10	2.63	±	0.14	3.25	±	0.48	1.51	±	0.25	U95053
96085_at	Gsta4*	glutathione S-transferase, alpha 4	1.17	±	0.10	1.03	±	0.08	5.07	±	0.12	4.56	±	1.18	L06047
97681_f_at	Gstm3	glutathione S-transferase, mu 3	-5.19	±	0.15	-3.77	±	0.07	-1.11	±	0.11	1.06	±	0.11	J03953
93026_at	Mgst1	microsomal glutathione S-transferase 1	1.63	±	0.14	1.07	±	0.12	4.13	±	0.61	2.97	±	0.36	AW124337
93573_at	Mt1	metallothionein 1	-1.24	±	0.07	-2.35	±	0.06	2.98	±	0.10	4.14	±	1.04	V00835
101561_at	Mt2	metallothionein 2	-1.51	±	0.06	-2.20	±	0.06	2.22	±	0.11	2.77	±	0.45	K02236
100606_at	Prnp	prion protein	1.46	±	0.08	2.09	±	0.07	1.43	±	0.12	-1.02	±	0.12	M18070
100538_at	Sod1	superoxide dismutase 1, soluble	-3.26	±	0.06	-1.50	±	0.07	2.19	±	0.43	1.05	±	0.33	M35725

Gene expression is indicated as FC±SE where FC=fold change; SE= Standard Error.

Cholesterol Biosynthesis

Cholesterol, an essential component of cellular membranes, is synthesized on the ER surface [66,73]. Cholesterol is an amphipathic molecule which serves as an essential membrane component (enriched in lipid rafts), as well as a precursor of steroid hormones. In neurons, cholesterol is important for the stability of the synapse and the maintenance of synaptic plasticity [74].

Disruption of cholesterol homeostasis in neurons has been associated with neurodegenerative diseases such as Alzheimer's disease. Furthermore, down-regulation of the rate-limiting enzyme 3-hydroxy-3-methylglutaryl-Coenzyme A reductase (Hmgcr), a protein encoded by the *Hmgcr* gene, was reported to cause cell death. In cultured neurons, the inhibition of Hmgcr using inhibitors such as pravastatin was sufficient to reduce neurite growth and to induce cell death [75,76]. Our microarray data reveals that all genes involved in cholesterol biosynthesis were down-regulated 24 h upon lactacystin treatment (Table 6).

The regulation of fatty acid and cholesterol synthesis in neurons is not well understood [74,77]. How lactacystin treatment induces the down-regulation of these genes is not clear, but it is not unexpected that the synthesis of lipid and cholesterol would be affected during ER stress, since cholesterol is synthesized in the ER.

A recent report suggests that cholesterol and fatty acid biosynthesis are controlled by a common family of transcription factors, known as the sterol regulatory element binding proteins (SREBPs) [78]. Upon activation (e.g. when the ER is deficient in lipid and sterol), the ER-anchored SREBP precursor transits to the Golgi, where it undergoes a sequential two-step cleavage process to release the NH2-terminal active domain, designated "nuclear form of SREBP" or "SREBP(N)", which is transported to the nucleus. In the nucleus, it promotes the expression of many genes involved in cholesterol and fatty acid synthesis. Activating transcription factor 6 (ATF6), an ER membrane-bound transcription factor, can also undergo a similar two-step cleavage process to form ATF6(N) during ER stress. ATF6(N) translocates to the nucleus, where it directs the transcriptional activation of chaperone molecules and enzymes essential for protein folding. Zeng *et al.* showed that over-expression of ATF6(N) in HepG2 (liver cell line) could suppress the SREBP(N)-mediated transcription of *HMGCR* and 3-hydroxy-3-methylglytaryl-Coenzyme A synthase (*HMGCS*) [79]. The authors suggest that ATF6(N) can bind directly to SREBP(N) and attenuate the SREBP(N)-mediated transcription. This report suggests a direct link between ER stress and the down-regulation of lipid and cholesterol biosynthesis.

Cholesterol biosynthesis is a complex synthetic pathway that requires dozens of enzymes and large amounts of energy [74]. The down-regulation of cholesterol synthesis genes might be a step taken by cells to conserve energy, especially when they are under stress.

Table 6. Genes differentially expressed during lactacystin treatment: Cholesterol biosynthesis

Cholesterol biosynthesis			Time point (h)				
Probe id	Symbol	Gene title	4.5	7.5	24	48	Genbank
94325_at	Hmgcs1	3-hydroxy-3-methylglutaryl-Coenzyme A synthase 1	1.39 ± 0.07	-1.22 ± 0.06	-2.85 ± 0.09	-2.07 ± 0.10	AW124932
94916_at	Cyp51	cytochrome P450, 51	-1.41 ± 0.06	-1.47 ± 0.10	-3.21 ± 0.10	-3.12 ± 0.10	AW122260
160770_at	Mvd	mevalonate (diphospho) decarboxylase	-3.06 ± 0.07	-2.61 ± 0.06	-6.96 ± 0.08	-4.98 ± 0.09	AW049778
95632_f_at	Mvk	mevalonate kinase	1.51 ± 0.15	1.17 ± 0.09	-8.26 ± 0.09	-7.52 ± 0.09	AW122653
98970_at	Ggps1	geranylgeranyl diphosphate synthase 1	2.67 ± 0.13	2.79 ± 0.18	1.06 ± 0.13	-1.30 ± 0.11	AB016044
98630_at	Nsdhl	NAD(P) dependent steroid dehydrogenase-like	1.62 ± 0.09	1.01 ± 0.06	-3.50 ± 0.09	-2.91 ± 0.10	AW106745
96269_at	Idi1	isopentenyl-diphosphate delta isomerase	-1.53 ± 0.06	-2.39 ± 0.06	-5.82 ± 0.09	-4.78 ± 0.09	AA716963
104285_at	Hmgcr	3-hydroxy-3-methylglutaryl-Coenzyme A reductase	-1.77 ± 0.06	-1.53 ± 0.06	-2.43 ± 0.10	-3.38 ± 0.09	M62766
160737_at	Lss	lanosterol synthase	-1.23 ± 0.12	-1.11 ± 0.13	-2.65 ± 0.11	-3.17 ± 0.11	AW060927
100418_at	Gng2	guanine nucleotide binding protein (G protein), gamma 2 subunit	1.39 ± 0.07	2.36 ± 0.13	-1.96 ± 0.11	-2.00 ± 0.11	AW123750

Gene expression is indicated as FC±SE where FC=fold change; SE= standard Error.

Concluding Remarks

The UPS is involved in the regulation of many important biological processes in neurons. Our study using microarray GeneChip® reveals that the expression of many potentially neuroprotective and pro-apoptotic genes were enhanced during proteasome inhibition by lactacystin. When these genes were grouped according to their biological functions, it was discovered that some genes were differentially expressed during the early phase of lactacystin-induced neuronal apoptosis (UPS, HSPs and molecular chaperones, ER stress and inflammatory responses), while others were differentially expressed during the late phase of lactacystin-induced neuronal apoptosis (oxidative stress and cholesterol biosynthesis). Based on this microarray data and other published data, a hypothetical mechanism of neuronal apoptosis induced by proteasomal inhibition can be derived (Figure 1). In the course of proteasome inhibition, there is an early up-regulation of genes encoding the proteasome subunits and HSPs. The expression of these genes, probably before the onset of apoptosis, seems to be a cellular response to the abnormal aggregation of unfolded proteins in cells arising from the inhibition of proteasomal activity. Subsequently, a number of genes encoding neuroprotective antioxidants were up-regulated at a later stage, probably after the onset of apoptosis, as revealed by our microarray data, suggesting that neurons were under oxidative stress at later stages of proteasome inhibition. Our microarray data also reveals the early up-regulation of the pro-apoptotic genes associated with ER stress. The early up-regulation of these ER stress-associated genes suggests that ER stress might be the main cause of cell death in lactacystin-induced cultured cortical neurons.

Since lactacystin is an irreversible proteasome inhibitor, cells exposed to lactacystin will eventually undergo apoptosis. Our microarray data shows an early ER stress response upon lactacystin treatment, in spite of the up-regulation of neuroprotective genes such as those that encode proteasomal subunits and HSPs. As demonstrated in Figure 1, there are a few possible pathways by which ER stress can cause neuronal apoptosis. Firstly, ER stress can induce the up-regulation of pro-apoptotic transcription factors such as Ddit3/CHOP [56,57,58,59]. Secondly, ER stress can generate ROS and thus cause oxidative stress [60,66]. As discussed in the previous section, the up-regulation of genes encoding antioxidants in our microarray data could have been a response to an increase of ROS in the neurons. Thirdly, ER stress can cause the disruption of calcium homeostasis, which can result in the activation of calpain (a cytosolic calcium-activated protease). Calpain is known to cleave p35, the neuronal-specific activator of cyclin-dependent kinase 5 (cdk5), generating proteolytic fragments, one of which is p25, which is known to accumulate in the brains of patients with Alzheimer's disease [80]. p25 is known to hyper-phosphorylate tau to disrupt the cytoskeleton and promote neuronal death [80,81]. In addition, calpain can also trigger apoptosis through the activation of caspase-3 [52]. Fourthly, our microarray data show that the genes associated with cholesterol biosynthesis were down-regulated. Whether this is involved directly in the induction of neuronal apoptosis remains unclear. Lastly, ER stress might have caused the activation of inflammatory responses such as the activation of COX-2 at the later stage of neuronal apoptosis [22,63]. Although proteasome inhibitors are known to suppress inflammatory responses [61], the inflammatory pathway involving COX-2 could be activated by ER stress [65].

We recently explored the effects of the tea catechin (-)-epigallocatechin-3-gallate (EGCG) as a potential proteasome inhibitor for the induction of HSPs and UPS gene expression. EGCG is the major constituent of green tea catechins, accounting for more than 10% of the green tea extract dry weight [25]. Initially, the beneficial biological effects of

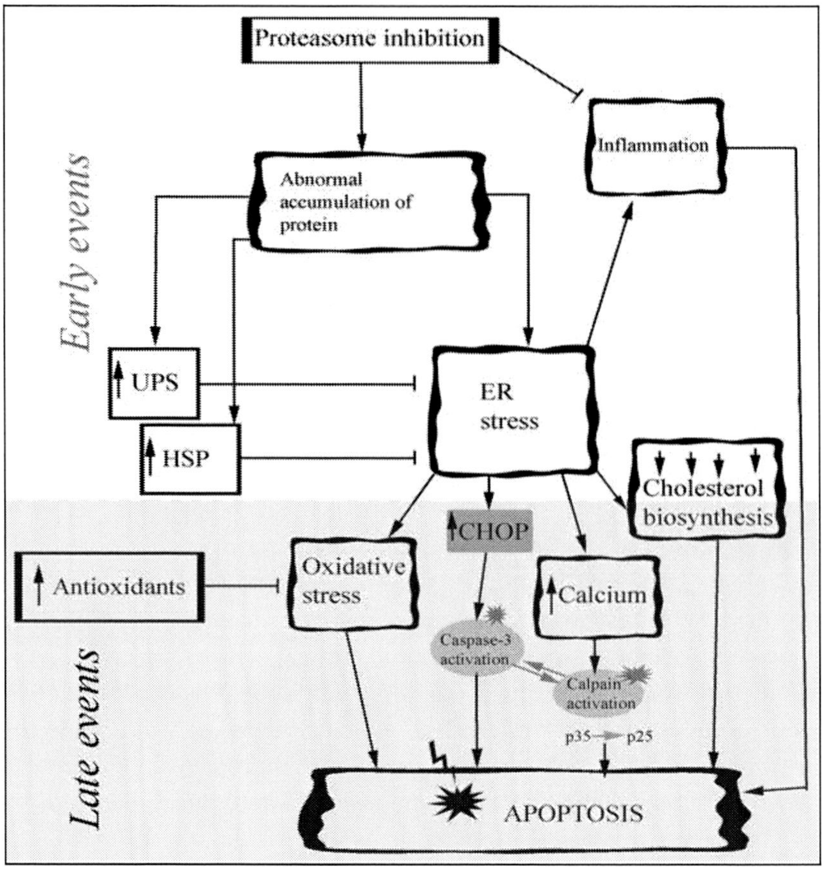

Figure 1. The hypothetical mechanism of lactacystin-induced neuronal apoptosis. The inhibition of proteasome by lactacystin induces the abnormal accumulation of proteins in neurons. To deal with this stress, the neurons activate the early neuroprotective responses through the up-regulation of UPS and HSPs. Exposure to lactacystin can also trigger ER stress responses, which lead to the activation of other pro-apoptotic responses such as 1) oxidative stress, 2) up-regulation of ER stress associated pro-apoptotic transcription factor CHOP, 3) calcium disruption, 4) decrease in cholesterol biosynthesis and 5) activation of the inflammatory responses at the late phase of lactacystin treatment.

these polyphenolic catechins were attributed to the antioxidant and iron-chelating actions of their polyphenol constituents, and to the modulation of endogenous metabolizing and antioxidant enzymes (review in [25]). However, the results of several recent studies suggest that the biological activity of the polyphenols present in green tea, especially EGCG, could be due to the inhibition of proteasome activity [82,83]. Like lactacystin, EGCG is found to specifically inhibit the chymotrypsin-like activity of proteasomes [82,84]. The experiments carried out by Nam *et al.* demonstrated that EGCG, at concentrations typically observed in the serum of green tea drinkers, could inhibit proteasomal activity, and cause arrest of cell division in the G1 phase of the cell cycle of several tumor cell lines and transformed cell lines [84]. Our microarray analysis reveals that exposure of cultured cortical neurons to a non-lethal dose of EGCG significantly up-regulated genes encoding the UPS components, such as proteasome subunits, ubiquitin and ubiquitin protein ligases. This suggests that EGCG has the potential to protect neurons against stress arising from the accumulation of unfolded proteins, since up-regulation of UPS components has been reported to be neuroprotective [32,33].

Acknowledgement

This work is supported by the National Medical Research Council research grant (R-183-000-075-213) and the Academic Research Fund of the National University of Singapore (R-183-000-060-214). We thank Dr. Teo Tian Seng and Ms Joyce Koh for their critical reading and input in this paper and Ms Seet Sze Jee for all the technical support.

References

[1] Sherman MY, Goldberg AL: Cellular defenses against unfolded proteins: a cell biologist thinks about neurodegenerative diseases. *Neurons* 2001, 29:15–32.

[2] Imai Y, Soda M, Inoue H, Hattori N, Mizuno Y, Takahashi R: An unfolded putative transmembrane polypeptide, which can lead to endoplasmic reticulum stress, is a substrate of Parkin. *Cell* 2001, 105:891–902.

[3] Hampton RY: ER-associated degradation in protein quality control and cellular regulation. *Curr Opin Cell Biol* 2002, 14:476–482.

[4] Goldberg AL: Protein degradation and protection against misfolded or damaged proteins. *Nature* 2003, 426:895–899.

[5] Taylor DM, Kabashi E, Agar JN, Minotti S, Durham HD: Proteasome activity or expression is not altered by activation of the heat shock transcription factor Hsf1 in cultured fibroblasts or myoblasts. *Cell Stress Chaperones* 2005, 10:230–241.

[6] Grune T, Jung T, Merker K, Davies KJA: Decrease proteolysis casued by protein aggregates, inclusion bodies, plaques, lipofuscin, ceroid, and 'aggresomes' during oxidative stress, aging, and disease. *Int J Biochem Cell Biol* 2004, 36:2519–2530.

[7] Halliwell B: Hypothesis: proteasomal dysfunction: a primary event in neurogeneration that leads to nitrative and oxidative stress and subsequent cell death. *Ann NY Acad Sci* 2002, 962:182–194

[8] McNaught KSP, CW Olanow: Protein aggregation in the pathogenesis of familial and sporadic Parkinson's disease. *Neurobiol Aging* 2006, 2006, 27:530-545.

[9] Bossy-Wetzel E, Schwarzenbacher R, Lipton SA: Molecular pathways to neurodegeneration. *Nature Medicine* 2004, 10:S2–S9.

[10] Moore DJ, West AB, Dawson VL, Dawson TM: Molecular pathophysiology of Parkinson's disease. *Annu Rev Neurosci* 2005, 28:57–87.

[11] Shimura H, Hattori N, Kubo S, Mizuno Y, Askawa S, Minoshima S, Shimizu N, Iwai K, Chiba T, Tanaka K, Suzuki T: Familial Parkinson disease gene product, parkin, is a ubiquitin-protein ligase. *Nature Genetics* 2000, 25:302–305.

[12] Mizuno Y, Hattori N, Mori H, Suzuki T, Tanaka K: Parkin and Parkinson's disease. *Curr Opin Neurol* 2001, 14:477–482.

[13] Lam YA, Pickart CM, Alban A, London M, Jamieson C, Ramage R, Mayer RJ, Layfield R: Inhibition of the ubiquitin-proteasome system in Alzheimer's disease. *Proc Acad Natl Sci USA* 2000, 97:9902–9906.

[14] Hope AD, de Silva R, Fischer DF, Hol EM, van Leeuwen FW, Lees AJ: Alzheimer's associated variant ubiquitin causes inhibition of the 26S proteasome and chaperone expression. *J Neurochem* 2003, 86:394–404.

[15] Song S, Jung Y: Alzheimer's disease meets the ubiquitin-proteasome system. *Trends in Mol Med* 2004, 10:565–570.

[16] Keller JN, Dimayuga E, Chen Q, Thorpe J, Gee J, Ding Q: Autophagy, proteasomes, lipofuscin, and oxidative stress in the aging brain. *Int J Biochem Cell Biol* 2004, 36:2376–2391.

[17] Adams J: The development of proteasome inhibitors as anticancer drugs. *Cancer Cell* 2003, 5:417–421.

[18] Melino G: Discovery of the ubiquitin proteasome system and its involvement in apoptosis. *Cell Death Differ* 2005, 12:1155–1157.

[19] Meriin AB, Gabai VL, Yaglom J, Shifrin VI, Sherman MY: Proteasome inhibitors activate stress kinases and induce Hsp72. *J Biol Chem* 1998, 273:6373–6379.

[20] Di Napoli M, Papa F: The proteasome system and proteasome inhibitors in stroke:controlling the inflammatory response. *Curr Opin Investig Drugs* 2003, 4:1333–1342.

[21] Bush KT, Goldberg AL, Nigam SK: Proteasome inhibition leads to a heat-shock response, induction of endoplasmic reticulum chaperones, and thermotolerance. *J Biol Chem* 1997, 272:9086–9092.

[22] Yew EHJ, Cheung NS, Choy MS, Qi RZ, Lee AY, Peng ZF, Melendez AJ, Manikandan J, Koay ES, Chiu L, Ng WL, Whiteman M, Kandiah J, Halliwell B: Proteasome inhibition by lactacystin in primary neuronal cells induces both potentially neuroprotective and pro-apoptotic transcriptional responses: a microarray analysis. *J Neurochem* 2005, 94:943–956.

[23] Butts BD, Hudson HR, Linseman DA, Le SS, Ryan KR, Bouchard RJ, Heidenreich KA: Proteasome inhibition elicits a biphasic effect on neuronal apoptosis via differential regulation of pro-survival and pro-apoptotic transcription factors. *Mol Cell Neurosci* 2005, 30:279–289.

[24] Suh J, Lee YA, Gwag BJ: Induction and attenuation of neuronal apoptosis by proteasome inhibitors in murine cortical cell cultures. *J Neurochem* 2005, 95:684–694.

[25] Mandel S, Youdim MBH: Catechin polyphenols: neurodegeneration and neuroprotection in neurodegenerative diseases. *Free Rad Biol Med* 2004, 37:304–317.

[26] Schroder M, Kaufman RJ: ER stress and the unfolded protein response. *Mutat Res* 2005, 569:29–63.

[27] Gaczynska M, Osmulski PA, Ward WF: Caretaker or undertaker? The role of the proteasome in aging. *Mech Ageing Dev* 2001, 122:235–254.

[28] Mandel S, Grunblatt E, Riederer P, Amariglio N, Hirsch JJ, Rechavi G, Youdim MBH: Gene expression profiling of sporadic Parkinson's disease substantia nigra pars compacta reveals impairment of ubiquitin-proteasome subunits, SKP1A, Aldehyde dehydrogenase, and chaperone HSC-70. *Ann NY Acad Sci* 2005, 1053:356–375.

[29] Mannhaupt G, Schnall R, Karpov V, Vetter I, Feldmann: Rpn4p acts as a transcription factor by binding to PACE, a nonamer box found upstream of 26S proteasomal and other genes in yeast. *FEBS Lett* 1999, 450:27–34.
[30] London MK, Keck BI, Ramos PC, Dohmen RJ: Regulatory mechanisms controlling biogenesis of ubiquitin and the proteasome. *FEBS Lett* 2004, 567:259–264.
[31] Glickman MH, Raveh D: Proteasome plasticity. *FEBS Lett* 2005, 579:3214–3223.
[32] Meiners S, Heyken D, Weller A, Ludwig A, Stangl K, Kloetzel, Kruger E: Inhibition of proteasome activity induces concerted expression of proteasome genes and de novo formation of mammalian proteasomes. *J Biol Chem* 2003, 278:21517–21525.
[33] Goldbaum O, Richter-Landsberg: Proteolytic stress causes heat shock protein induction, tau ubiquitination, and the recruitment of ubiquitin to tau-positive aggregates in oligodendrocytes in culture. *J Neurosci* 2004, 24:5748–5757.
[34] Noga M, Hayashi T: Ubiquitin gene expression following transient forebrain ischemia. *Brain Res Mol Brain Res* 1996, 36:261–267.
[35] Gilchrist CA, Grayt DA, Stieber A, Gonatas NK, Kopito RR: Effect of ubiquitin expression on neuropathogenesis in a mouse model of gamilial amyotrophic lateral sclerosis. *Neuropathol Appl Neurobiol* 2005, 31:20–33.
[36] Joazeiro CA, Weissman: RING finger proteins: mediators of ubiquitin ligase activity. *Cell* 2000, 102:549–552.
[37] Yang Y, Yu X: Regulation of apoptosis: the ubiquitous way. *FASEB J* 2003, 17:790–799.
[38] Lee CS, Tee LY, Warmke T, Vinjamoori A, Cai A, Fagan AM, Snider BJ: A proteasomal stress response: pre-treatment with proteasome inhibitors increases proteasome activity and reduces neuronal vulnerability to oxidative injury. *J Neurochem* 2004, 91:996–1006.
[39] Chondrogianni N, Tzavelas C, Pemberton AJ, Nezis IP, Rivett AJ, Gonos ES: Overexpression of proteasome β5 subunit increases the amount of assembled proteasome and confers ameliorated response to oxidative stress and higher survival rates. *J Biol Chem* 2005, 280:11840–11850.
[40] Ding Q, Keller JN: Proteasome and proteasome inhibition in the central nervous system. *Free Radic Biol Med* 2001, 31:574–584.
[41] Meriin AB, Sherman MY: Role of molecular chaperones in neurodegenerative disorders. *Int J Hyperhermia* 2005, 21:403–419.
[42] Ohtsuka K, Suzuki T: Roles of molecular chaperones in the nervous system. *Brain Res Bull* 2000, 53:141–146.
[43] Franklin TB, Krueger-Naug AM, Clarke DB, Arrigo A-P, Currie RW: The role of heat shock proteins Hsp70 and Hsp27 in cellular protection of the central nervous system. *Int J Hyperthermia* 2005, 21:379–392.
[44] Moser DD, Caron AW, Bourget L, Meriin AB, Sherman MY, Morimoto RI, Massie B: The chaperone function of hsp70 is required for protection against stress-induced apoptosis. *Mol Cell Biol* 2000, 20:7146–7159.
[45] Sun Y, Ouwang Y-B, Xu L, Chow AM-Y, Anderson R, Hecker JG, Giffard RG: The carbozyl-terminal domain of inducible Hsp70 protects from ischemic injury in vivo and in vitro. *J Cereb Blood Flow Metab* 2005, 1–4, in press.

[46] Lai Y, Du L, Dunsmore KE, Jenkins LW, Wong HR, Clark RSB: Selectively increasing inducible heat shock protein 70 via TAT-protein transduction protects neurons from nitrosative stress and excitotoxicity. *J Neurochem* 2005, 94:360–366.

[47] McDonough H, Patterson C: CHIP: a link between the chaperone and proteasome systems. *Cell Stress Chaperones* 2003, 8: 303–308.

[48] Carra S, Sivilotti M, Chavez Zobel AT, Lambert H, Landry J: HspB8, a small heat shock protein mutated in human neuromuscular disorders, has in vivo chaperone activity in cultured cells. *Hum Mol Genet* 2005, 14:1659–1669.

[49] Giese KC, Vierling E: Changes in oligomerization are essential for the chaperone activity of a small heat shock protein in vivo and in vitro. *J Biol Chem* 2002, 277: 46310–46318.

[50] Latchman DS: HSP27 and cell survival in neurons. *Int J Hyperthermia* 2005, 21:393–402.

[51] Sitia R, Braakman I: Quality control in the endoplasmic reticulum protein factory. *Nature* 2003, 426:891–894.

[52] Rutkowski DT, Kaufman RJ: A trip to the ER: coping with stress. *Trends in Cell Biol* 2004, 14:20–28.

[53] Groenendyk J, Michalak M: Endoplasmic reticulum quality control and apoptosis. *Acta Biochimica Polonica* 2005, paper in press.

[54] Reimertz C, Kögel D, Rami A, Chittenden T, Prehn JHM: Gene expression during ER stress-induced apoptosis in neurons: induction of the BH3-only protein Bbc3/PUMA and activation of the mitochondrial apoptosis pathway. *J Cell Biol* 2003, 162:587–597.

[55] Wang XZ, Lawson B, Brewer JW, Zinszner H, Sanjay A, Mi LJ, Boorstein R, Kreibich, Hendershot LM, Ron D: Signals from the stressed endoplasmic reticulum induce C/EBP-homologous protein (CHOP/GADD153). *Mol Cell Biol* 1996, 16:4273–4280.

[56] Zimmermann J, Erdmann D, Lalande I, Grossenbacher R, Noorani M, Fürst P: Proteasome inhibitor induced gene expression profiles reveal overexpression of transcriptional regulators ATF3, GADD153 and MAD1. *Oncogene* 2000, 19:2913–2920.

[57] Wang X-Z, Kuroda M, Sok J, Batchvarova N, Kimmel R, Chung P, Zinszner H, Ron D: Identification of novel stress-induced genes down stream of CHOP. *EMBO J* 1998, 17:3619–3630.

[58] Fribley A, Zeng Q, Wang C-Y: Proteasome inhibitor PS-341 induces apoptosis through induction of endoplasmic reticulum stress-reactive oxygen species in head and neck squamous cell carcinoma cells. *Mol Cell Biol* 2004, 24:9695–9704.

[59] Rao RV, Ellerby HM, Bredesen DE: Coupling endoplasmic reticulum stress to the cell death program. *Cell Death Differ* 2004, 11:372–380.

[60] McCullough, KD, Martindale JL, Klotz L-O, Aw T-K, Holbrook NJ: Gadd153 sensitizes cells to endoplasmic reticulum stress by down-regulating Bcl2 and perturbing the cellular redox State. *Mol Cell Biol* 2001, 21:1249–1259.

[61] Wojcik C, Di Napoli: Ubiquitin-proteasome system and proteasome inhibition: new strategies in stroke therapy. *Stroke* 2004, 35:1506–1518.

[62] Phillips JB, Williams AJ, Adams J, Elliott PJ, Tortella FC: Proteasome inhibitor PS-519 reduces infarction and attenuates leukocyte infiltration in a rat model of focal cerebral ischemia. *Stroke* 2000, 31:1686–1693.

[63] Rockwell P, Yuan H, Magnusson R, Figueiredo-Pereira ME: Proteasome inhibition in neuronal cells induces a proinflammatory response manifested by upregulation of cyclooxygenase-2, its accumulation as ubiquitin conjugates, and production of the prostaglandin PGE_2. *Arch Biochem Biophys* 2000, 374:325–333.

[64] Consilvio C, Vincent AM, Feldman EL: Neuroinflammation, COX-2, and ALS–a dual role? *Expr Neurol* 2004, 187:1–10.

[65] Cieslik K, Zhu Y, Wu KK: Salicylate suppression macrophage nitric-oxide synthase-2 and cyclo-oxygenase-2 expression by inhibiting CCAAT/enhancer-binding protein-β binding via a common signaling pathway. *J Biol Chem* 2002, 277:49304–49310.

[66] Pahl H L: Signal transduction from the endoplasmic reticulum to the cell nucleus. *Physiol Rev* 1999, 79:683–701.

[67] Coppedè F, Armani C, Della Bidia D, Petrozzi L, Bonuccelli U, Migliore L: Molecular implications of the human glutathione transferase A-4 gene (hGSTA4) polymorphisms in neurodegenerative diseases. *Mutat Res* 2005, 579:107–114.

[68] Maeda A, Crabb JW, Palczewski K: Microsomal glutathione S-transferase 1 in the retinal pigment epithelium: protection against oxidative stress and a potential role in aging. *Biochem* 2005, 44:480–489.

[69] Chen Y, Shertzer HG, Schneider SN, Nebert DW, Dalton TP: Glutamate cysteine ligase catalysis: dependence on ATP and modifier subunit for regulation of tissue glutathione levels. *J Biol Chem* 2005, in press.

[70] Hubatsch I, Ridderström M, Mannervik B: Human glutathione transferase A4-4: an Alpha class enzyme with high catalytic efficiency in the conjugation of 4-hydroxinonenal and other genotoxic products of lipid peroxidation. *J Biochem* 1998, 330:175–179.

[71] Owen AD, Schapira AH, Jenner P, Marsden CD: Oxidative stress and Parkinson's disease. *Ann N Y Acad Sci* 1996, 786:217–223.

[72] Aschner M, Cherian MG, Klaassen CD, Palmiter RD, Erickson JC, Bush AI: Metallothioneins in brain–the role in physiology and pathology. *Toxicol Appl Pharmacol* 1997, 142: 229–242.

[73] Chapman R, Sidrauski C, Walter P: Intracellular signaling from the endoplasmic reticulum to the nucleus. *Annu Rev Cell Dev Biol* 1998, 14:459–485.

[74] Pfrieger FW: Cholesterol homeostasis and function in neurons of the central nervous system. *Cell Mol Life Sci* 2003, 60:1158–1171

[75] Schulz JG, Bösel J, Stoeckel M, Megow D, Dirnagl U, Endres M: HMG-CoA reductase inhibition causes neurite loss by interfering with geranylgeranylpyrophosphate synthesis. *J Neurochem* 2004, 89:24–32.

[76] Tanaka T, Tatsuno I, Uchida D, Moroo I, Morio H, Nakamura S, Noguchi Y, Yasuda T, Kitagawa M, Saito Y, Hirai A: Geranylgeranyl-pyrophosphate, and isoprenoid of mevalonate cascade, is a critical compound for rat primary cultured cortical neurons to protect the cell death induced by 3-hydroxy-3-methylglutarul-CoA reductase inhibition. *J Neuronsci* 2000, 28:2852–2859.

[77] Koudinova NV, Kontush A, Berezov TT, Koudinov AR: Amyloid beta, neural lipids, cholesterol & Alzheimer's disease. *Neurobiol Lipids* 2003, 1:27–33. Available at: http://neurobiologyoflipids.org/content/1/6/

[78] Eberlé D, Hegarty B, Bossard P, Ferré P, Foufelle F: SREBP transcription factors: master regulators of lipid homeostasis. *Biochimie* 2004, 86:839–848.

[79] Zeng L, Lu M, Mori K, Luo S, Lee AS, Zhu Y, Shyy J Y-J: ATF6 modulates SREBP2-mediated lipogenesis. *EMBO J* 2004, 23:950–958.

[80] Lee MS, Kwon YT, Li M, Peng J, Friedlander RM, Tsai LH: Neurotoxicity induces cleavage of p35 and p25 by calpain. *Nature* 2000, 405:360–364.

[81] Kusakawa G, Saito T, Onuki R, Ishiguro K, Kishimoto T, Hisanaga S: Calpain-dependent proteolytic cleavage of the p35 cyclin-dependent kinase 5 activator to p25. *J Biol Chem* 2000, 275:17166–17172.

[82] Wan SB, Chen D, Dou QP, Chan TH: Study of the green tea polyphenols catechin-3-gallate (CG) and epicatechin-3-gallate (EGC) as proteasome inhibitors. *Bioorg Med Chem* 2004, 12:3521–3527.

[83] Shay NF, Banz WJ: Regulation of gene transcription by botanicals: novel regulatory mechanisms. *Annu Rev Nutr* 2005, 25:14.1–14.19.

[84] Nam S, Smith DM, Dou QP: Ester bond-containing tea polyphenols potently inhibit proteasome activity in vitro and in vivo. *J Biol Chem* 2001, 276:13322–13330.

In: The Ubiquitin Proteasome System… ISBN 978-1-60021-749-4
Eds: Mario Di Napoli and Cezary Wojcik, pp. 599-625

Chapter 25

UBIQUITIN PROTEASOME SYSTEM PATHWAY IN DOPAMINERGIC NEURODEGENERATION

Hideto Miwa*

Department of Neurology, Wakayama Medical University, Wakayama-city, Japan.

ABSTRACT

The ubiquitin-proteasome system (UPS) has a central role in the selective degradation of many intracellular proteins. Functional failure of the UPS may result in an abnormal accumulation of ubiquitinated, misfolded, aggregated, or oxidated proteins that should be removed from cells, finally resulting in cell death. Recent advances in genetic studies in familiar Parkinson's disease (PD) have provided important insight into the molecular pathways involved in disease pathogenesis. Proteins coded by the causal genes of familial parkinsonism, such as α-synuclein, parkin, ubiquitin carboxy-terminal hydrolase L1 (UCH-L1), and DJ-1, are possibly related to the ubiquitin-proteasome protein degradation system. Mutations in these proteins may interfere with normal protein degradation by UPS caused by 'loss of function' or 'gain of function'. A major constituent of Lewy bodies is aggregated α-synuclein. Mutant α-synuclein aggregates and resists degradation by the UPS, eventually disturbing normal cellular functions. Both parkin and UCH-L1 are components of the UPS that contribute to normal ubiquitination and de-ubiquitination mechanisms, respectively. Loss-of-function of parkin or UCH-L1 can interrupt normal protein degradation by the UPS. In addition, DJ-1 may also function to alleviate protein misfolding. The accumulated findings suggest the hypothesis that UPS failure and the subsequent proteolytic stress contribute to the etiopathogenesis that underlies dopaminergic neurodegeneration in both hereditary and sporadic PD. Experimental studies have revealed that dopaminergic neurons may be particularly vulnerable to proteasome inhibition *in vitro* and *in vivo*. Local administration of a proteasome inhibitor into the nigrostriatal system (the substantia nigra, striatum, or

* Correspondence concerning this article should be addressed to Dr. Hideto Miwa, MD; Department of Neurology, Wakayama Medical University, 811-1 Kimiidera, Wakayama-city, Wakayama 641-8510, Japan. Phone: +81-73-441-0655; Fax: +81-73-441-0655 E-mail: h-miwa@wakayama-med.ac.jp.

medial forebrain bundle) in rodents is sufficiently to induce dopaminergic neuron degeneration in the substantia nigra and the formation of α-synuclein-immunopositive intracytoplasmic inclusions similar to Lewy bodies. In addition, it has recently been reported that dopaminergic neurons progressively degenerate with Lewy-body-like inclusion formation following the systemic administration of proteasome inhibitors in rats. These experimental findings suggest that the inhibition of the UPS may be a common pathway for dopaminergic neuron death in PD, although further studies are required to establish the proteasome-inhibitor-induced PD model. If UPS failure is a key mechanism underlying dopaminergic neuron death, the next question, which is probably a more essential one, is why UPS failure occurs in dopaminergic neurons in PD. There must be primary, upstream events that affect UPS function. To date, several factors have been stressed in the pathogenetic mechanisms underlying dopaminergic neuron degeneration in PD, such as deficits in mitochondrial function, oxidative stress, neuroinflammation, and the accumulation of aberrant or misfolded proteins. Determining the principal molecular pathways that exaggerate UPS dysfunction will provide relevant clues to the understanding of the pathogenesis of sporadic and familial forms of PD. This chapter reviews the most recent advances in our knowledge on the relationships between UPS failure and dopaminergic neuron degeneration.

Keywords: Parkinson's disease, Dopaminergic neurons, α-synuclein, Lewy bodies, Oxidative stress, Ubiquitin proteasome system.

Abbreviations

ARJP, autosomal recessive juvenile parkinsonism; HO-1, heme oxygenase-1; 6-OHDA, 6-hydroxydopoamine; HNE, hydroxynonenal; LRRK2, leucine-rich repeat kinase 2; MFB, medial forebrain bundle; MPP+, 1-methyl-4-phenylpiridium ions; MPTP, 1-methyl-4-phenyl-1,2,3,6-tetrahydropyridine; NMDA, N-methyl-D-aspartate; Pael-R, parkin-associated endothelin receptor-like receptor; Paraquat, 1,1'-dimethyl-4,4'-bipyridium; PD, Parkinson's disease; PINK1, PTEN-induced kinase 1; ROS, reactive oxygen species; SN, substantia nigra; SNpc, substantia nigra pars compacta; SNpr, substantia nigra pars reticulate; UCH-L1, ubiquitin carboxy-terminal hydrolase L1; UPS, ubiquitin-proteasome system.

Introduction

Parkinson's disease (PD), first described by James Parkinson in 1817, is a neurodegenerative disorder characterized by impairment of motor functions; for example, tremors, rigidity, akinesia, and postural reflex disturbance. It is not uncommon in the elderly, and about 1 to 2% of individuals older than 60 years of age are affected. The neuropathological hallmark of PD is the preferential degeneration of dopaminergic neurons in the substantia nigra pars compacta, and the presence of intracytoplasmic inclusions termed Lewy bodies [1,2]. Lewy bodies are observed in dopaminergic neurons remaining in the substantia nigra, and in other cortical and subcortical structures in the PD brain. Since the dopaminergic neurons in the substantia nigra supply dopamine to the striatum via the nigrostriatal pathway, the degeneration of dopaminergic neurons results in a prominent decrease in the dopamine content of the striatum. By the time symptoms appear, the substantia nigra has already lost about 60% of its dopaminergic neurons and the dopamine content of the striatum is less than about 80% of normal.

Pharmacological therapy is generally effective in the early stage of PD with dopaminomimetics such as L-DOPA and dopamine receptor agonists used as first-line drugs. Despite treatment, however, PD symptoms deteriorate progressively because current therapeutics are not able to slow disease progression. In the advanced stage of PD, patients therefore face various problems that threaten their quality of life, such as motor complications due to long-term therapy, freezing and falls, psychiatric disturbances, autonomic disturbances, and dementia [3,4]. Therapeutic efficacies are therefore limited in long-term treated PD patients. Development of a disease-modifying therapy, particularly one that can slow or stop disease progression, is thus strongly desired. In line with this goal, a detailed understanding of the cause and pathogenetic mechanisms of the disease is necessary.

Although the actual etiopathogenesis of PD is undetermined, recent advances in genetic research on familial PD have provided clues. Gene mutations related to the ubiquitin proteasome system (UPS) have been identified as the cause, suggesting a possible pathogenetic link between the UPS and mechanisms underlying dopaminergic neuron degeneration [5,6]. Since the UPS pathway contributes to the detoxification and targeting of damaged or misfolded proteins for degradation, it is possible that an impairment of UPS function induces the accumulation of damaged or toxic proteins, and subsequent deterioration of intracellular circumstances, resulting in neurodegeneration or neuronal death [6-9]. Indeed, the accumulation of modified proteins and inclusion body formation have been suggested as common cellular mechanisms in neurodegenerative diseases such as PD, Alzheimer's disease, Huntington's disease, and amyotrophic lateral sclerosis [10]. Moreover, as mentioned above, the intracytoplasmic inclusions, Lewy bodies, are an important pathological feature of PD. The following section reviews recent advances in our understanding of the relationship between UPS failure and dopaminergic neuron degeneration, focusing on UPS regulation of dopaminergic neuron degeneration in experimental models.

Parkinson's Disease and the Ubiquitin Proteasome System

Lewy Bodies: a Mechanism for Intracellular Segregation of Poorly Degraded Proteins?

Lewy bodies are an important pathological feature of PD, present not only in dopaminergic neurons in the substantia nigra but also in neurons in the locus coeruleus, the dorsal motor nucleus of the vagus, and the nucleus basalis of Meynert [1]. Lewy bodies are characteristically round eosinophilic inclusions composed of a halo of radiating fibrils and a less dense core [2]. They contain many different proteins and lipids, the most abundant of which is α-synuclein [11]. α-Synuclein is found in Lewy bodies of not only sporadic but also familial PD patients with α-synuclein mutations, suggesting the significance of α-synuclein in the pathogenesis of PD. Among the other proteins found in the Lewy bodies are ubiquitin, synphilin-1, neurofilaments, parkin, UCH-L1, proteasomal elements, heat shock proteins, and torsin A [13,14].

Lewy bodies are considered the key to understanding the pathogenetic mechanisms underlying neurodegeneration in PD; however, their significance in the neurodegenerative process has not been fully elucidated. It remains unknown whether Lewy bodies are toxic or protective. It was recently suggested that the formation of Lewy bodies might be a neuroprotective event in which dopaminergic neurons attempt to sequester and compartmentalize poorly degraded proteins into insoluble aggregates [14,15]. The formation of inclusions perhaps enables neurons to survive deleterious changes in intracellular homeostasis, particularly enabling them to clear up modified, unwanted proteins in the cytoplasm. This is discussed again in detail later in this chapter.

Lessons from Gene Mutations Identified in Familial Parkinson's Disease

Most patients with PD have the sporadic form of the disease; however, some cases are familial. The gene mutations identified as the cause of familial PD are listed (Table 1). Of these, PARK1, PARK 2, PARK 4, PARK 5 and PARK 7 are thought to be closely related to the UPS, capable of interfering with normal protein degradation. In the following sections, the relationship between gene mutations and their mutant proteins are briefly reviewed.

Table 1. Loci and genes linked to familial Parkinson's disease

Locus	Chromosomal region	Gene	Inheritance	Lewy bodies
PARK1	4q21	α-synuclein	AD	+
PARK2	6q25.2-27	Parkin	AR	-
PARK3	2p13	unknown	AD	+
PARK4	4q21-22	α-synuclein triplication	AD	+
PARK5	4p14	UCHL1	AR	?
PARK6	1p35-36	PINK1	AR	?
PARK7	1p36	DJ-1	AR	?
PARK8	12p11.2-q13.1	LRRK2/dardarin	+/-	
PARK9	1p36	Unknown	AR	?
PARK10	1p32	Unknown	susceptible gene	?
NR4A2	2q22-23	Nurr1	AD	?
PARK11	2q36-37	Unknown	AD	?

α-Synuclein

α-Synuclein is a 140 amino acid protein with unknown function. It is abundant in neurons, particularly in presynaptic nerve terminals, and is a major component of the Lewy bodies. The first mutation identified was a missense mutation, A53T, in exon 4 of the gene encoding α-synuclein [16]. Two further mutations (A30P and E46K) were reported later [17,18], and importantly, α-synuclein gene triplication was identified recently [19,20], suggesting that not only gene mutations but also an increase in the amount of cellular α-synuclein can cause dopaminergic neuron degeneration. The likelihood of aggregation could be increased by increasing protein concentrations [10]. Wild-type α-synuclein is monomelic; however, at a high concentration, it polymerizes into filaments [21]. Genetic variability in the α-synuclein gene is also a risk factor for the development of sporadic PD [22].

Although α-synuclein has been considered the most important protein related to the etiopathogenesis of PD, it remains unknown whether or not α-synuclein is degraded by the UPS. In neuroblastoma cell lines, the UPS contributes to degradation of α-synuclein, with the degradation of A53T mutant α-synuclein being slower than that of wild-type [23]. On the other hand, in PC12 cells and primary mesencephalic neurons, proteasome inhibitors did not affect levels of α-synuclein [24], suggesting that α-synuclein is degraded not only by proteasomes but also autophagy, another system of protein degradation [25-27]. Therefore, the relationship between a loss of UPS function and accumulation of α-synuclein is perhaps not so simple. Moreover, the speculation that UPS failure directly causes α-synuclein accumulation should be abandoned; rather the increase in α-synuclein itself is perhaps important in inducing the impairment of UPS function. Overexpression of α-synuclein inhibits proteasome function and sensitizes cells to the toxic effects of proteasome inhibitors [28,29]. Moreover, aggregated α-synuclein is thought to bind directly to proteasomes, possibly resulting in UPS failure [30]. It was reported that A53T mutant α-synuclein causes endoplasmic reticulum stress, probably as a consequence of decreasing UPS activity [120]. Understanding in detail how α-synuclein contributes to dopaminergic neuron degeneration is currently one of the most important themes for research studies on pathogenesis of PD.

Parkin

Autosomal recessive juvenile parkinsonism (ARJP) has been described in a series of Japanese kindreds [31,32]. Neuropathologically, Lewy bodies are not present; however, various deletions and mutations have been identified in the gene for parkin [33-35], a 465 amino acid protein that functions as a ubiquitin ligase (E3) [36,37]. Parkin mutations are the most frequent cause of early-onset PD. The ubiquitin ligase activity is markedly decreased in the substantia nigra and striatum of patients with ARJP [36,37], suggesting that a loss of ubiquitin ligase function causes dopaminergic neurodegeneration in ARJP. Therefore, identification of the protein substrates of parkin is important for understanding the exact pathophysiological mechanism underlying cell death.

Parkin does not interact with or ubiquitinate native α-synuclein, indicating that native α-synuclein is not a parkin substrate, but rather it interacts with a glycosylated form of α-synuclein [37]. Parkin substrates identified so far are as follows: the α-synuclein interacting protein synphilin-1 [38], CDCrel-1 [39], and parkin-associated endothelin receptor-like receptor (Pael-R) [40]. CDCrel-1, a protein involved in cytokinesis, is thought to influence synaptic vesicle function, while Pael-R is related to endoplasmic reticulum (ER) stress and cell death. When Paer-R is overexpressed in cells, it tends to become unfolded, insoluble, and ubiquitinated, possibly resulting in ER stress-induced cell death [40] (for details on ER stress response, see Chapter 13). Parkin is known to specifically ubiquitinate this receptor [41], and is also thought to have a neuroprotective action [42,43].

UCH-L1

Ubiquitin C-terminal hydrolase-L1 (UCH-L1) is a deubiquitinating enzyme and important component of the UPS. A UCH-L1 mutation has been identified in a German family with PD [44], but is thought to be very rare [45,46]. It is suggested that UCH-L1 also has ligase as well as deubiquitinating functions [47].

DJ-1

To date, few reports have documented DJ-1 mutations. DJ-1 is thought to function as a signaling molecule, alerting the mitochondria of oxidative stress [48]. At present, the functional relationship between DJ-1 and the UPS is unclear; however, structurally, it is thought to alleviate protein misfolding, since it has similarities to bacterial HSP31 homologs [49]. A recent study revealed that DJ-1 acts as a redox-sensitive chaperone protein, increasing the solubility of α-synuclein by preventing its aggregation [50]. It is therefore suggested that mutations leading to misfolded DJ-1 might overload the UPS [51]. DJ-1 is also a sensor of oxidative damage, which also increases the aggregation of α-synuclein. Moreover, it was also reported that DJ-1 can interact with α-synuclein [52], while pathogenic mutants of DJ-1 specifically interact with parkin [53].

PINK1

Mutations in PINK1 (PTEN-induced kinase 1) have been identified as the cause of familial parkinsonism (PARK6) [54,55]. The detailed functions of PINK1 have been not elucidated, but it is known to be located in mitochondria where it is thought to protect against mitochondrial damage [54]. Experimental findings *in vitro* suggest that PINK1 might have an anti-apoptotic action [56].

LRRK2

Recently, mutations in leucine-rich repeat kinase 2 (LRRK2) [57], which encodes dardarin protein [58], were identified as the cause of autosomal dominant PD (PARK8) [59]. However, the function of dardarin remains unclear.

THE UBIQUITIN PROTEASOME SYSTEM IN SPORADIC PARKINSON'S DISEASE

A few studies have reported the protein degradation activity of UPS in autopsied brains of patients with PD. For example, impairments in 20/26S proteasome hydrolyzing activities, such as chymotrypsin- (39%), trypsin- (42%) and postacidic-like (33%) activities (see Chapter 6), are impaired in the substantia nigra but not the striatum or cortex in PD [60]. In extranigral brain areas, where neuronal loss is not observed in PD, proteasome activities are preserved, suggesting that systemic global disturbance in the catalytic and degradation ability of the proteasome itself does not occur in PD [61]. In addition, α- but not β-subunits of 26/20S proteasomes are known to be lost within dopaminergic neurons [62,63]. Thus, proteasome activity seems to decrease selectively in the substantia nigra of PD. Whether or not UPS failure is a primary or upstream event of dopaminergic neurodegeneration cascades in PD remains to be elucidated.

PROTEASOME INHIBITION OF DOPAMINERGIC NEURONS

Are Dopaminergic Neurons Particularly Vulnerable to Proteasome Inhibition?

One of the problems in determining the pathogenetic mechanisms underlying PD is understanding why dopaminergic neurons in the substantia nigra are selectively damaged. It is thus important to determine whether or not dopaminergic neurons are differentially vulnerable to proteasome inhibition. In primary ventral mesencephalic cultures, administration of lactacystin, a proteasome inhibitor, induced preferential degeneration of dopaminergic neurons with the formation of α-synuclein-immunopositive inclusions [64]. McNaught *et al.* examined uptake of [^{3}H]-dopamine and [^{14}C]-GABA, and found that only the former decreased, suggesting differential vulnerability in dopaminergic neurons. Similarly, overexpression of mutant α-synuclein also decreased proteasome function, and caused selective toxicity to catecholaminergic neurons *in vitro* [42]. Moreover, in ventral mesencephalic and striatal co-culture, tyrosine hydroxylase (TH) -immunopositive neurons are more vulnerable than TH-immunonegative neurons [65]. Dopaminergic neurons are also thought to be particularly vulnerable to proteasome inhibition *in vivo*. Infusion of proteasome inhibitor was shown to selectively induce dopaminergic neuron degeneration in the rat substantia nigra [66], as was systemic administration of PSI or epoxomycin, both of which are lipophilic proteasome inhibitors [67].

Dopaminergic Neuron Degeneration by Proteasome Inhibition In Vivo

Intranigral Administration of Proteasome Inhibitor

To date, only a limited number of studies have documented the effects of proteasome inhibition on dopaminergic neurons (Table 2). McNaught *et al.* [66] reported the effects of injection of lactacystin, a water-soluble selective proteasome inhibitor, into unilateral substantia nigra in rats, revealing that the treated animals displayed bradykinesia, a stooped posture, and contralateral head tilting. Apomorphine-induced contralateral circlings, a pharmacobehavioral characteristic of unilateral damage of the nigrostriatal dopamine pathway, were also noted in their observations. Histological investigation also revealed a prominent loss of dopaminergic neurons in the substantia nigra, and α-synuclein-immunopositive intracytoplasmic inclusions were observed in the dopaminergic neurons of the substantia nigra on the affected side. We also studied the effects of intranigral administration of proteasome inhibitor in rats, and confirmed the above findings (Figure 1). In our study, local damage of the tissue near the injection site was unavoidable to a varying degree. However, dopaminergic neurodegeneration in the substantia nigra was comparatively far more extensive, suggesting that dopaminergic neurons are highly vulnerable to proteasome inhibition.

Table 2. Animal model of dopaminergic neurodegeneration induced by proteasome inhibitor treatment

Reference	Treatment	Administration site	Animal	Results
McNaught *et al.* (2002) [66]	Lactacystin	Substantia nigra	Rat	Loss of DA neurons in the SN.α-Synuclein-immunopositive inclusionsApomorphine-induced circling (+)
Fornai *et al.* (2003) [68]	Lactacystin Epoxomycin	Striatum	Rat	Loss of DA neurons in the SNα-Synuclein-immunopositive inclusions
McNaught *et al.* (2004) [67]	Epoxomycin PSI	Systemic	Rat	layed loss of DA neurons in the I. α-Synuclein-immunopositive :lusions. Locus coeruleus, dorsal ga nucleus, nucleus basalis of eynert also involved.
Miwa *et al.* (2005) [69]	Lactacystin	Striatum	Rat	ss of DA neurons in the SN Synuclein-immunopositive :lusions. Increased oxidative ess markers in the SN
Zhang *et al.* (2005) [71]	Lactacystin	MFB	Mouse (C57BL/6)	Loss of DA neurons in the SN. Ubiquitin-immunopositive inclusions

MFB indicates medial forebrain bundle.

Intrastriatal Administration of Proteasome Inhibitor

Two studies have reported the effects of intrastriatal proteasome inhibition. Fornai *et al.* [68] reported a significant loss of dopaminergic neurons with apoptosis following intrastriatal proteasome inhibition, and suppressed proteasome inhibitor toxicity with reductions in endogenous dopamine. In the surviving neurons in the substantia nigra, intracytoplasmic inclusions immunopositive to ubiquitin, α-synuclein, and parkin were observed. They also reported that proteasome inhibitor-induced inclusions have identical ultrastructural features to inclusions obtained in *in vitro* proteasome inhibition.

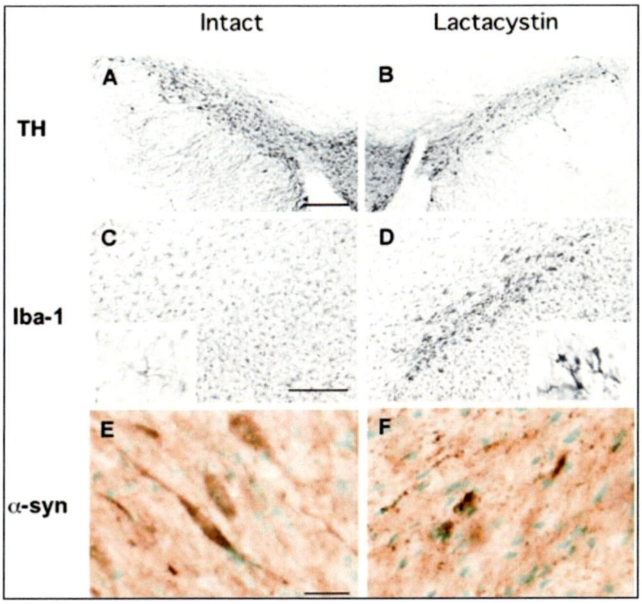

Figure 1. Injection of lactacystin (5 μg/2 μl), a proteasome inhibitor, into substantia nigra (SN) induced dopaminergic neuron degeneration, microglial activation, and the formation of inclusions in rats at 14 days postlesion. (A-B) Tyrosine hydroxylase (TH) immunostaining of the SN following unilateral intranigral injection of lactacystin. An asymmetrical loss of TH immunopositivity in the SN is demonstrated in (B) compared to the control (A). (C-D) Iba-1 immunostaining of the SN showing activated microglia following intranigral injection of lactacystin (D). Inserts show Iba-1-immunopositive microglia at a higher magnification. Iba-1-immunopositive microglia in the intact SN (C, insert) exhibited characteristics of inactive resting microglia such as small cell bodies with long thin lamellipodia. On the other hand, microglia in the lactacystin-injected SN exhibited characteristic morphological changes typical of activation, such as large cell bodies with multiple, thick and short lamellipodia (D, insert). (E-F) α-Synuclein-immunostaining of the SN. In the control SN (E), α-synuclein immunopositivity was homogeneously present in the dopaminergic neurons. On the other hand, in the lactacystin-injected SN, α-synuclein-immunopositive granules were observed in the dopaminergic neurons (F). Scale bars: 100 μm in (A-D) and 25 μm in (E, F).

We also confirmed that intrastriatal injection of lactacystin induces progressive degeneration of dopaminergic neurons with apoptosis in the rat substantia nigra [69], and found α-synuclein-immunopositive intracytoplasmic or intraneuritic inclusions in the neurons remaining in the substantia nigra. A marked increase in heme oxygenase-1 (HO-1) was also observed during the progressive loss of dopaminergic neurons. HO-1 is thought to be one of the most useful key markers of oxidative stress in nigrostriatal dopaminergic neurons, and to play a protective role against oxidative stress. Thus, the increase in HO-1 suggests that oxidative stress is extensively induced in dopaminergic neurons following proteasome inhibition in intrastriatal nerve terminals. UPS failure caused by proteasome inhibitor in the nerve terminals is therefore sufficient to induce prominent oxidative stress in dopaminergic neurons *in vivo*. We also reported an abundant accumulation of α-synuclein-immunopositive granules in the substantia nigra pars reticulata (SNpr) following intrastriatal injection of lactacystin [69]. Speculatively, the accumulation of α-synuclein-immunopositive granules in the SNpr is thought to result from the accumulation of α-synuclein in nerve terminals of the striatonigral neurons. Interestingly, proteasome inhibition of cell bodies of striatal neurons induced α-synuclein accumulation in the striatonigral nerve terminals but no prominent loss of the striatal neurons themselves.

On the other hand, proteasome inhibition of the nigrostriatal nerve terminals induced α-synuclein accumulation in their cell bodies with prominent loss of nigral neurons. This discrepancy also supports the suggestion that nigrostriatal dopaminergic neurons are more vulnerable than striatonigral GABAergic fibers. Microglial activation was also noted in the substantia nigra following intrastrial injection of proteasome inhibitor (our unpublished data), suggesting that neuroinflammation is extensively induced during intrastriatal proteasome inhibition-induced dopaminergic neuron degeneration. Speculatively, certain substances such as reactive oxygen species, which are released from damaged dopaminergic neurons, are thought to activate microglia, similar to the case of retrograde dopaminergic neuron degeneration following intrastriatal injection of MPP^+ [70].

Intra-Medial Forebrain Bundle Administration of Proteasome Inhibitor

Injection of proteasome inhibitor into the medial forebrain bundle (MFB) might also damage the nigrostriatal dopaminergic neurons. It was previously reported that intra-MFB injection of lactacystin induced dopaminergic neuron degeneration accompanied by the formation of ubiquitin-immunopositive inclusions in C57BL/6 mice [71]. In the case of mice, since the injection site of MFB is anatomically approximate to the substantia nigra, it might be possible that the injected proteasome inhibitor spread directly to the substantia nigra. However, in a preliminary experiment in our laboratory, we confirmed that intra-MFB injection of lactacystin could also induce dopaminergic neuron degeneration in the substantia nigra of rats; however, further studies are required to asses the validity of the intra-MFB proteasome inhibitor model.

Systemic Administration of Proteasome Inhibitor

One study has reported the effects of systemic administration of proteasome inhibitors in rats. McNaught *et al.* [67] systemically injected naturally occurring (epoxomycin) and synthetic (PSI, Z-Ile-Glu(O*t*Bu)-Ala-Leu-al) inhibitors into rats repeatedly for two weeks; both inhibitors are lipophilic and thus can enter the brain across the blood-brain-barrier. The animals became progressively akinetic after a latency of 1 to 2 weeks, and in their brains, dopaminergic neuron death with apoptosis and inflammation in the substantia nigra was observed. Moreover, neurodegeneration occurred in the locus coeruleus, dorsal motor nucleus of the vagus, and the nucleus basalis of Meynert, all of which are known to be involved in PD. In addition, α-synuclein-immunopositive inclusions resembling Lewy bodies were reported in neurodegenerative sites. These findings show that systemic proteasome inhibition can closely recapitulate key features of PD, suggesting that UPS failure plays a central role in cell death cascades in PD.

THE UBIQUITIN PROTEASOME SYSTEM IN ANIMAL MODELS OF PARKINSON'S DISEASE

To date, several neurotoxins have been shown to have selective toxicity to dopaminergic neurons in the substantia nigra *in vivo*; for example, 6-hydroxydopoamine (6-OHDA), MPTP and its metabolite MPP+, rotenone, and paraquat. The mechanisms of toxicity differ between each neurotoxin, and in addition, species-specific differences exist. The method or mode of toxin-delivery is also closely related to the toxic insults. The alteration in UPS function in these models is reviewed in the following.

6-Hydroxydopamine

The neurotoxin 6-hydroxydopamine (6-OHDA) is one of the most widely used toxins in animal models of PD [72]. Particularly, it is used to destroy dopaminergic neurons in rats, since rats don't show toxicity to MPTP. Unilateral stereotaxic injection of 6-OHDA into the nigrostriatal dopaminergic pathway causes dopaminergic neuron degeneration of the ipsilateral substantia nigra with severe dopamine depletion in the ipsilateral striatum. 6-OHDA-induced unilaterally dopamine-depleted animals show a characteristic circling behavior as a result of dopaminergic agonists, and have been used to assess dopaminergic actions of various drugs [72]. 6-OHDA is taken up by dopamine and noradrenaline transporters, and is selectively toxic to monoaminergic neurons (Luthman 1989). It has also been suggested that accumulation of 6-OHDA in the cytosol produces reactive oxygen species (ROS), finally resulting in neuron death [73].

A limited number of studies have documented the relationship between 6-OHDA toxicity and the UPS. Experiments *in vitro* have shown that 6-OHDA increases levels of free ubiquitin and ubiquitin-conjugated proteins, and markedly increases protein degradation [74]. Moreover, in cultured cells, proteasome activity increases following exposure to low doses of 6-OHDA, but higher doses cause UPS failure and cell death [75], suggesting that mild oxidative stress elevates proteasome activity in response to increases in protein damage. Proteasome inhibition *in vitro* might attenuate 6-OHDA-induced protein degradation and potentiate its toxicity [74]. However, in contrast, a recent report showed that low doses of proteasome inhibitor prevented 6-OHDA-induced dopaminergic neuron death in the rat substantia nigra [76].

MPTP

The neurotoxin 1-methyl-4-phenyl-1,2,3,6-tetrahydropyridine (MPTP) has been widely used for experimental analysis of dopaminergic neuron degeneration in various animal species including primates, cats, and mice [77-79]. Systemic administration of MPTP induces selective degeneration of dopaminergic neurons in the substantia nigra and dopamine deficiency-related behavioral disturbances resembling those of PD. Currently, the MPTP-model is thus regarded as a standard animal model of PD. After enzymatic conversion to 1-

methyl-4-phenylpiridium ions (MPP^+) in glial cells, systemically administered MPTP inhibits the activity of mitochondrial complex I, leading to the production of ROS or decreasing ATP production, finally inducing dopaminergic neuron death [80]. MPP^+ also binds to vesicular monoamine trasporter-2 and interacts with synaptic vesicles [81], leading to the release of dopamine and resulting in the excess of cytosolic dopamine [82]. Consequently, autooxidation of dopamine occurs, facilitating oxidative stress.

α-Synuclein in MPTP-treated animals has been documented in a number of studies. For example, α-synuclein levels were found to increase following MPTP treatment in mice and monkeys [83,84]. Moreover, α-synuclein null mice display striking resistance to MPTP, possibly due to the inability of the toxin to inhibit mitochondrial complex I [85]. Generally, in MPTP-induced models, the formation of Lewy body-like inclusion bodies is lacking in the dopaminergic neurons in the substantia nigra [86]. However, the neurotoxic effects of MPTP seem to vary and depend on the treatment protocol. A recent report showed that if MPTP is administered continuously with an osmotic minipump, not only dopaminergic neuron death but also the formation of α-synuclein-immunopositive inclusions can be induced [87]. Moreover, in this model, only continuous, not sporadic, MPTP infusions were able to inhibit UPS function. This continuous MPTP infusion model might therefore recreate a disease state that mimics PD better than acute MPTP injections [87].

Rotenone

Systemic administration of rotenone, when performed chronically, induces selective loss of dopaminergic neurons and the formation of α-synuclein-immunopositive inclusions in rats [88-90], suggesting that an impairment of mitochondrial function, particularly that of complex I, is sufficiently capable of inducing dopaminergic neuron degeneration. However, following administration of higher doses of rotenone, dopaminergic neuron degeneration was not observed. Instead, extensive basal ganglia necrosis was induced [91], similar to the pathological characteristics of mitochondrial encephalopathies such as Leigh encephalopathy [92]. The difference in these two experimental models suggests that the neurotoxic effects of rotenone are dependent on the dose or concentration of rotenone. When higher doses of rotenone were administered, severe bioenergy crisis due to severe ATP depletion was acutely induced, possibly resulting in basal ganglia necrosis with sparing dopaminergic neurons [91-92].

Chronic, long-lasting oxidative stress, but not acute energy crisis, might be closely related to selective dopaminergic neuron death. Experimental results *in vitro* suggest that selective dopaminergic neuron death and alterations in UPS function occur only with rotenone exposure that partially maintains ATP/ADP [93]. Moreover, the decreased activity of mitochondrial complex I was shown to reduce proteasome activity through oxidative modification of proteasome and aggregation with other oxidized proteins [94]. The pathophysiological relationship between decreased activity of mitochondrial complex I and the accumulation of modified proteins with/without UPS function is currently unclear; however, elucidation of this relationship will give us a clue to understanding dopaminergic neuron degeneration.

Paraquat

Herbicides are thought to be one of the environmental risk factors of PD. Systemic treatment of mice with the herbicide 1,1'-dimethyl-4,4'-bipyridium (paraquat) caused a loss of nigrostriatal dopaminergic neurons and the formation of aggregates containing α-synuclein in mice [95]. Following exposure to paraquat, brain levels of α-synuclein were also significantly increased [96]. Interestingly, in α-synuclein overexpressing mice, systemic administration of paraquat induced protein aggregates immunopositive to α-synuclein but no dopaminergic neurodegeneration, suggesting that dissociation exists between toxicant-induced α-synuclein deposition and neurodegeneration [95]. The mechanism of paraquat neurotoxicity is most likely mediated via oxidative stress, since paraquat generates superoxide both by electron transfer reactions with NADH-dependent oxireductases and by redox cycling via reactions with molecular oxygen [97,98]. Whether or not paraquat interacts with UPS function remains unknown, but it is thought to interfere via oxidative stress.

OXIDATIVE STRESS AND UPS FAILURE: A SYNERGISTIC SPIRAL OF DOPAMINERGIC NEURON DEGENERATION

Oxidative stress, as described above, likely plays one of the most important roles in the pathophysiological mechanisms underlying dopaminergic neuron degeneration in animal models of PD. The toxicity of 6-OHDA is attributed to production of ROS, and the neurotoxicity of paraquat is caused by the production of superoxide. Further, of particular importance is that both MPTP and rotenone inhibit mitochondrial complex I. In neurons, free radicals are constantly generated, particularly in the mitochondria, where electrons leak from the electron transport chain. This electron leak also occurs within mitochondrial complex I, particularly at the area proximal to the rotenone-binding site. Inhibition of mitochondrial complex I therefore results in an increased electron leak, and consequently, excessive free radicals. If mitochondrial functions are severely inhibited, mitochondrial respiration will be globally broken down, resulting in an intracellular bioenergetic crisis. On the other hand, if mitochondrial complex I is only partially inhibited, mitochondrial respiration will not be so prominently affected, but the production of free radicals might still markedly increase. Subsequently, oxidative damage will extensively involve intracellular substances such as proteins, lipids, and DNA. Indeed, in PD brains, accumulated findings suggest oxidative damage of proteins, lipids, and DNA [99-104].

Increased levels of lipid hydroperoxides [99] and basal protein oxidation [101] have been observed in the substantia nigra in PD. Levels of 8-hydroxy-2-deoxyguanosine and 8-hydroxyguanosine, markers suggestive of DNA damage, are also increased in the PD brain [103,104]. Mitochondrial DNA is also affected by oxidative damage, presumably leading to further decline in mitochondria function. In addition, levels of hydroxynonenal (HNE), a marker of lipid peroxidation, also increase in the substantia nigra in PD [102]. Under conditions of oxidative stress, not only superoxide itself but also more toxic reactive molecules such as peroxynitrite are generated, leading to post-translational modification of pathogenetically important proteins such as α-synuclein and parkin [105-106].

Dopaminergic neurons are essentially destined to suffer from oxidative stress, since free radicals are inevitably produced during dopamine auto-oxidation and dopamine metabolism [107]. It is therefore possible to speculate that oxidative stress is one of the most relevant key mechanisms underlying dopaminergic neurodegeneration in both experimental animal and human PD. The oxidized proteins, lipids, and DNA, have an unfavorable influence on intracellular homeostasis, and are occasionally directly toxic to dopaminergic neurons. On the other hand, the accumulation of such substrates, particularly oxidized proteins, might suggest that clearance of the modified proteins is inadequate. If the oxidized modified proteins, which should be degraded by the UPS and cleared from the cytoplasm, are excessively present in the cells, the formation of aggregates or inclusions might be facilitated. Indeed, characteristic intracellular inclusions or aggregates can be seen in dopaminergic neurons remaining in the substantia nigra in PD patients as well as animals with experimental parkinsonism [66-69]. Moreover, there is a recently emerging hypothesis that conformational change, mishandling, and aggregation of intracellular protein might contribute to various neurodegenerative diseases [10]. This suggests a functional deficiency of the protein degradation pathway, particularly for degradation of damaged proteins. Indeed, genetic studies suggest that a functional deficiency of the UPS might be related to dopaminergic neurodegeneration. In line with this, mishandling of abnormal, unnecessary or unfavorable proteins, some of which are modified by free radicals, must contribute to the deterioration of intracellular homeostasis.

In animal models of dopaminergic neuron degeneration (induced by rotenone, paraquat, chronically administered MPTP, and proteasome inhibitors), the formation of α-synuclein-immunopositive aggregates or inclusions resembling Lewy bodies is observed [66-69,87-90,96]; however, the significance of the formation of inclusion bodies has yet to be determined. α-Synuclein-immunopositive inclusions might also form if oxidative stress is chronic and continuous. For example, as described above, intracytoplasmic inclusions can be induced in mice if MPTP is administered chronically and continuously but not sporadically [87]. Also, in rotenone-induced dopaminergic neuron degeneration, inhibition of mitochondrial complex I is not so severe as to induce mitochondrial respiration failure [88]. It therefore appears likely that oxidative stress increases the rate of α-synuclein aggregation, as with copper and other heavy metals [108]. The aggregation of α-synuclein itself can cause oxidative stress, which, in turn, can cause conformational changes in α-synuclein, resulting in proteolytic stress [109]. Therefore, oxidative stress and α-synuclein aggregation seem to facilitate each other, and this interaction might be particularly toxic in dopaminergic neurons. Only in dopaminergic neurons does the overexpression of α-synuclein result in apoptotic cell death. Moreover, suppression of endogeneous dopamine in these cells can inhibit α-synuclein-induced apoptosis [110], implicating that an increase in α-synuclein and interaction with dopamine or related metabolic compounds might play a relevant role in dopaminergic neuron death. Experimental results showing that administration of proteasome inhibitor differentially induces the formation of α-synuclein-immunopositive inclusions and dopaminergic neuron death *in vivo* support this speculation [66-69].

The relationship between oxidative stress and UPS function therefore needs to be addressed, since there is evidence suggesting that oxidative stress contributes to impaired UPS function. It was previously reported that α-subunits of 26/20S proteasomes are particularly vulnerable to free radical- and HNE-induced damage [111]. Moreover, the ATP depletion induced by oxidative damage of mitochondria might worsen proteasome function, since ATP maintains the 26S proteasome complex [112]. In the mouse model of PD induced by continuous MPTP infusion, dopaminergic neurodegeneration and the formation of α-synuclein-immunopositive inclusions are observed, and UPS function is inhibited [87]. In turn, an impairment of UPS function might deteriorate oxidative stress. Proteasome inhibition potentiates the neurotoxicity of mitochondrial complex I inhibitors [113], and in addition, low levels of chronic proteasome inhibition *in vitro* can dramatically alter mitochondrial homeostasis by increasing free radical production and reducing mitochondrial complex I and II activity [114]. During dopaminergic neurodegeneration following intrastriatal proteasome inhibition, we also revealed a prominent increase in oxidative stress markers prior to cell death [69]. It is therefore possible that oxidative stress inhibits proteasome function while impairment of UPS function causes and worsens oxidative stress. Both oxidative stress and UPS failure therefore influence each other, synergically contributing to the deterioration of cellular functions.

A Novel Implication: Proteasome Inhibition is Neuroprotective?

The complex effects of proteasome inhibition on dopaminergic neurons have been reported, suggesting that proteasome inhibition, if its toxicity is not sufficient to induce cell death, might not only be toxic to but also have a neuroprotective effect on dopaminergic neurons. Recently, Sawada *et al.* [115] reported that sub-lethal doses of proteasome inhibitors could block MPP^+-induced dopaminergic neuronal death *in vitro* and *in vivo*. Similarly, dopaminergic neurodegeneration of rat substantia nigra following intranigral injection of 6-hydroxydopamine (6-OHDA) can be similarly prevented by co-administration of proteasome inhibitors [76]. Of course, MPP^+ and rotenone neurotoxicity are enhanced by proteasome inhibitors if the dose is sufficient to suppress proteasome activity to less than 10% [113]. It therefore appears likely that overall inhibition of proteasome activity is certainly toxic, but that partial suppression of activity might exert a neuroprotective effect. Although the neuroprotective mechanisms of proteasome inhibition are unclear, it is suggested that proteasome inhibitors might facilitate the formation of inclusions and block the toxicity of ROS production and/or toxic substances such as small oligomers (protofibrils), both of which are induced during neurotoxin-induced cell death. This speculation is not inconsistent with the finding that aggresome-associated α-synuclein is cytoprotective [15]. Speculatively, it is possible that neurons survive by avoiding unfavorable, toxic substances by sequestering them into inclusions, and that during this process, proteasome inhibition might help and facilitate inclusion formation. Further studies are herefore required to reveal the function of proteasomes in dopaminergic neurodegeneration (see Chapter 24).

UPS and Dopaminergic Neuron Degeneration in PD: A Hypothesis

Both genetic and environmental factors are important in the pathogenesis of cell death in sporadic PD. In the cascade of dopaminergic neurodegeneration, various factors involving oxidative stress, mitochondrial dysfunction, excitotoxicity, and neuroinflammation, have thought to play a relevant role [100,116-119] (Figure 2). However, these factors are not thought to be present to the same extent in all patients, and therefore, the pathogenesis likely differs between individuals. There is, however, one emerging hypothesis whereby neurodegenerative disorders such as PD have a common mechanism underlying cell death: impaired protein clearance or protein aggregation. The hypothetical speculation that PD is associated with protein mishandling is based on the following findings. First, neuropathologically, intracytoplasmic inclusions (Lewy bodies) are the hallmark of PD. Second, genetically, the proteins encoded by causal genes of familial parkinsonism are closely associated with the UPS, the protein degradation machinery [5-6,9,14,21,41]. Third, proteasome activity declines in the substantia nigra in PD. In addition, experimentally, dopaminergic neurons are differentially vulnerable to proteasome inhibition both *in vivo* and *in vitro* [42,65-69].

The hypothesis that protein mishandling and accumulation of abnormal proteins such as mutant or misfolded proteins or proteins modified by ROS might be the key mechanism underlying dopaminergic neuron degeneration in familial as well as sporadic PD is worth noting. These unwanted or potentially toxic proteins are thought to be prone to aggregation, leading to polymerization [10]. For example, in the case of mutations of α-synuclein, altered proteins are thought to be prone to aggregation. Moreover, in the case of gene-driven increases in α-synuclein, for example, α-synuclein triplication, α-synuclein aggregation is thought to be facilitated [19,20]. In the case of mutations of proteins associated with UPS machinery, such as parkin and UCH-L1, loss-of-function of protein handling increases strain on the UPS to degrade α-synuclein, particularly its modified form, and substances of parkin, possibly resulting in the formation of aggregates [6,41]. In sporadic PD, both environmental and genetic factors are thought to contribute to the generation of oxidative stress, particular by inhibiting mitochondrial complex I [100,116]. Inhibition of mitochondrial complex I induces both production of free radicals and a bioenergy crisis, leading to extensive oxidative stress. Oxidative stress involves not only proteins, but also lipids and DNA [99-104], chronically and progressively worsening intracellular homeostasis, and particularly affecting normal protein handling [7,9,14]. Oxidative stress also facilitates α-synuclein aggregation. Dopamine autooxidation also plays an important role in facilitating oxidative stress in dopaminergic neurons [107]. Further, oxidative stress might induce impairment of UPS function, in turn causing further oxidative stress. Both oxidative stress and declining UPS function might therefore be trapped in a synergic spiral that finally results in cell death. Finally, in dopaminergic neuron degeneration, activation of NMDA receptors occurs [118] and activated microglia enhance neuroinflammation [117], exerting deleterious effects on the dopaminergic neurons.

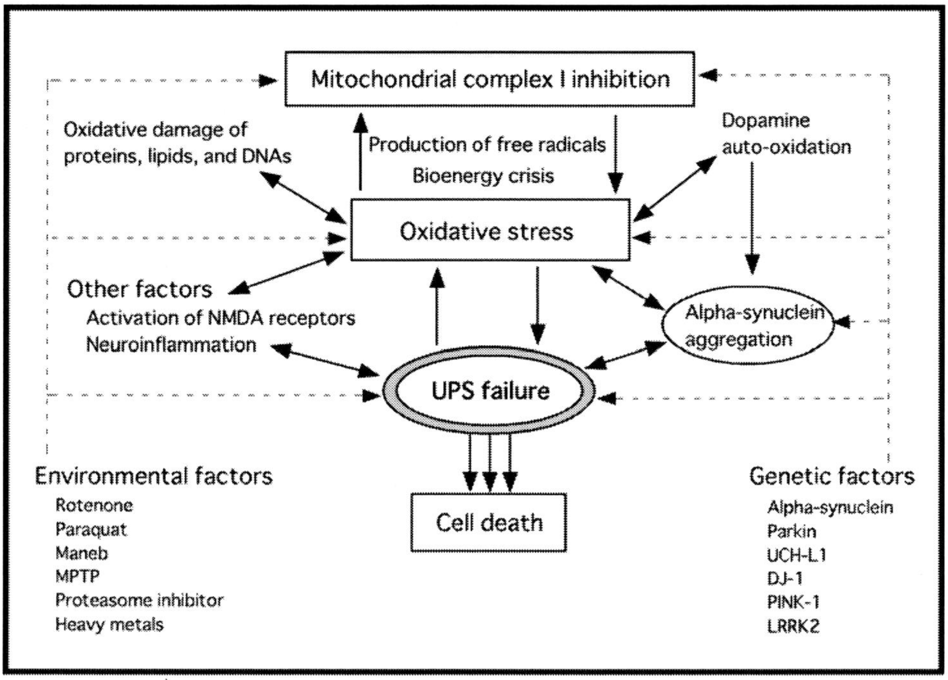

Figure 2. Schematic representation of the possible inter-relationship between factors known to contribute to the etiopathogenesis underlying dopaminergic neurodegeneration in Parkinson's disease (PD). Genetic as well as environmental factors are considered important in the pathogenesis of PD. Speculatively, mitochondrial dysfunction, particularly of complex I, is thought to result in oxidative stress and a bioenergy crisis. Oxidative damage involves proteins, lipids, and DNAs, further facilitating oxidative stress. Proteins modified by oxidative stress, such as α-synuclein, possibly accumulate and eventually aggregate, leading to polymerization. Oxidative dopamine metabolism also progresses oxidative stress. Oxidative stress might therefore induce an impairment of ubiquitin proteasome system (UPS) function, perhaps directly involving proteasome subunits. Oxidative stress and the decline in UPS function are thus perhaps trapped in a synergic spiral that finally results in cell death.

Speculatively, it is suggested that protein aggregation or the process of polymerization itself is toxic to neurons. The detailed mechanisms underlying the aggregation of α-synuclein remain uncertain; however, it is hypothesized that toxicity might closely depend on the degree of aggregation (Figure 3). Moreover, intermediates such as oligomers or protofibrils are more toxic than both the precursor protein of the aggresomes and the inclusions [10] (about aggresome; see Chapter 12).

The accumulation of α-synuclein may not be the direct result of the primary inhibition of protein degradation of UPS, because it remains unclear whether α-synuclein is actually degraded by the UPS. Secondary oxidative stress resulting from the proteasome inhibition also may be important in the accumulation of α-synuclein in neurons. Both oxidative stress and UPS failure synergistically contribute to the deterioration of cellular functions, resulting in an accumulation of α-synuclein in dopaminergic neurons. For the survival of these neurons, it may be desirable to eliminate the toxic intermediates, such as oligomers or protofibrils of α-synuclein, by successfully sequestering them into inclusions (Lewy bodies).

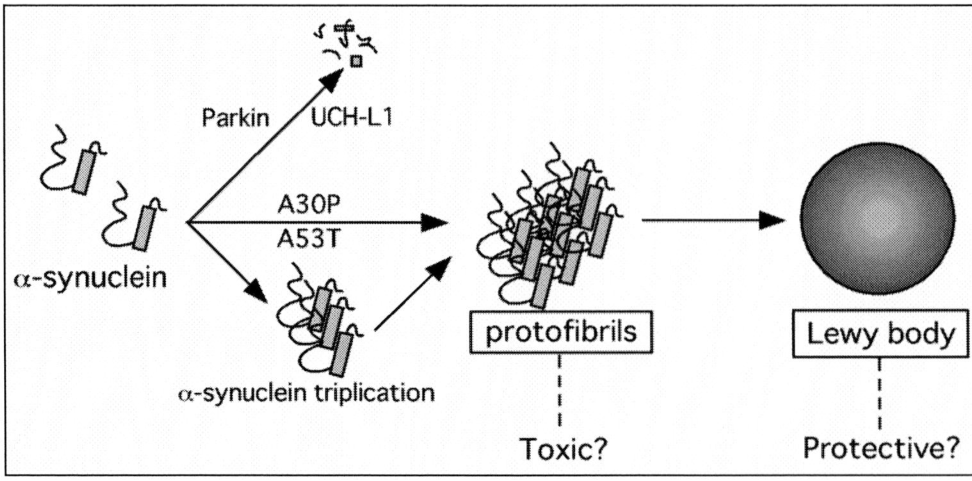

Figure 3. α-Synuclein aggregation and Lewy body formation. Mutant α-synuclein resulting from gene mutations (A30P, A53T) and abnormally excessive α-synuclein as a result of gene triplication are prone to aggregation. On the other hand, in the case of parkin mutations, α-synuclein can be degraded by the UPS, since α-synuclein is not a substrate of parkin. However, the intermediates (protofibrils) might be more toxic than the aggresome or inclusions, and the Levy body itself might be protective rather than toxic.

Conclusion

How the UPS contributes to the pathogenetic mechanisms underlying dopaminergic neurodegeneration in PD is currently a hot topic of research. There has been substantial progress in our understanding of UPS function in both normal and abnormal conditions, raising the hypothesis that UPS failure might be central in the cell death cascades involved in neurodegeneration. However, at present, there is much more to be learnt about the etiopathogenesis of PD. For example, it is necessary for us to understand more about the exact relationship between protein aggregation and neurodegeneration: why certain proteins aggregate and others don't and whether the aggregates or the process of aggregation itself is toxic. We also need to know more about the mutual relationship between UPS function and other pathogenetically important factors, such as oxidative stress, mitochondrial dysfunction, neuroinflammation, excitotoxicity mediated via NMDA receptors, and heavy metals. The mechanisms underlying dopaminergic neurodegeneration in PD appear complex; however, further accumulation of research findings will continue to provide clues to the etiopathogenesis of PD as well as aiding the development of new therapeutic strategies, particularly those addressing the modification of disease progression.

Acknowledgements

This work was supported by a Grant-in-Aid for Scientific Research from the Ministry of Education, Culture, Sports, Science and Technology, Japan (14570613, 17590889). I thank

Prof. Tomoyoshi Kondo for his valuable comments, and Tomomi Kubo and Ai Suzuki for technical assistance.

REFERENCES

[1] Braak H, Braak E, Yilmazer D, Schultz C, de Vos RA, Jansen EN: Nigral and extranigral pathology in Parkinson's disease. *Neural Transm* 1995, 46 (Suppl):15-31.

[2] Forno LS: Neuropathology of Parkinson's disease. *J Neuropathol Exp Neurol* 1996, 55(3):259-272.

[3] Olanow CW, Jankovic J: Neuroprotective therapy in Parkinson's disease and motor complications: a search for a pathogenesis-targeted, disease-modifying strategy. *Mov Disord* 2005, 20 (Suppl 11):S3-10.

[4] Mizuno Y, Kondo T, Mori H: Various aspects of motor fluctuations and their management in Parkinson's disease. *Neurology* 1994, 44 (7 Suppl 6):S29-34.

[5] Huang Y, Cheung L, Rowe D, Halliday G: Genetic contributions to Parkinson's disease. *Brain Res Brain Res Rev* 2004, 46:44-70.

[6] Hattori N, Mizuno Y: Pathogenetic mechanisms of parkin in Parkinson's disease. *Lancet* 2004, 364:722-724.

[7] McNaught KS, Olanow CW: Proteolytic stress: a unifying concept for the etiopathogenesis of Parkinson's disease. *Ann Neurol* 2003, 53 (Suppl 3):S73-84

[8] Tanaka K, Suzuki T, Hattori N, Mizuno Y: Ubiquitin, proteasome and parkin. *Biochim Biophys Acta* 2004, 1695:235-s47.

[9] Betarbet R, Sherer TB, Greenamyre JT: Ubiquitin-proteasome system and Parkinson's diseases. *Exp Neurol* 2005, 191 Suppl 1:S17-27.

[10] Ross CA, Poirier MA: Protein aggregation and neurodegenerative disease. *Nat Med* 2004, 10 Suppl:S10-17.

[11] Spillantini MG, Schmidt ML, Lee VM, Trojanowski JQ, Jakes R, Goedert M: Alpha-synuclein in Lewy bodies. *Nature* 1997, 388:839-840.

[12] Mezey E, Dehejia AM, Harta G, Tresser N, Suchy SF, Nussbaum RL, Brownstein MJ, Polymeropoulos MH: Alpha synuclein is present in Lewy bodies in sporadic Parkinson's disease. *Mol Psychiatry* 1998, 3:493-499.

[13] Kuzuhara S, Mori H, Izumiyama N, Yoshimura M, Ihara Y: Lewy bodies are ubiquitinated. A light and electron microscopic immunocytochemical study. *Acta Neuropathol (Berl)* 1988, 75:345-353.

[14] McNaught KS, Olanow CW, Halliwell B, Isacson O, Jenner P: Failure of the ubiquitin-proteasome system in Parkinson's disease. *Nat Rev Neurosci* 2001, 2:589-594.

[15] Tanaka M, Kim YM, Lee G, Junn E, Iwatsubo T, Mouradian MM: Aggresomes formed by alpha-synuclein and synphilin-1 are cytoprotective. *J Biol Chem* 2004, 279:4625-4631.

[16] Polymeropoulos MH, Lavedan C, Leroy E, Ide SE, Dehejia A, Dutra A, Pike B, Root H, Rubenstein J, Boyer R, Stenroos ES, Chandrasekharappa S, Athanassiadou A, Papapetropoulos T, Johnson WG, Lazzarini AM, Duvoisin RC, Di Iorio G, Golbe LI,

Nussbaum RL: Mutation in the alpha-synuclein gene identified in families with Parkinson's disease. *Science* 1997, 276:2045-2047.

[17] Kruger R, Kuhn W, Leenders KL, Sprengelmeyer R, Muller T, Woitalla D, Portman AT, Maguire RP, Veenma L, Schroder U, Schols L, Epplen JT, Riess O, Przuntek H: Familial parkinsonism with synuclein pathology: clinical and PET studies of A30P mutation carriers. *Neurology* 2001, 56:1355-1362.

[18] Zarranz JJ, Alegre J, Gomez-Esteban JC, Lezcano E, Ros R, Ampuero I, Vidal L, Hoenicka J, Rodriguez O, Atares B, Llorens V, Gomez Tortosa E, del Ser T, Munoz DG, de Yebenes JG: The new mutation, E46K, of alpha-synuclein causes Parkinson and Lewy body dementia. *Ann Neurol* 2004, 55:164-173.

[19] Singleton AB, Farrer M, Johnson J, Singleton A, Hague S, Kachergus J, Hulihan M, Peuralinna T, Dutra A, Nussbaum R, Lincoln S, Crawley A, Hanson M, Maraganore D, Adler C, Cookson MR, Muenter M, Baptista M, Miller D, Blancato J, Hardy J, Gwinn-Hardy K: alpha-Synuclein locus triplication causes Parkinson's disease. *Science* 2003, 302:841.

[20] Farrer M, Kachergus J, Forno L, Lincoln S, Wang DS, Hulihan M, Maraganore D, Gwinn-Hardy K, Wszolek Z, Dickson D, Langston JW: Comparison of kindreds with parkinsonism and alpha-synuclein genomic multiplications. *Ann Neurol* 2004, 55:174-179.

[21] Giasson BI, Lee VM: Are ubiquitination pathways central to Parkinson's disease? *Cell* 2003, 114:1-8.

[22] Farrer M, Maraganore DM, Lockhart P, Singleton A, Lesnick TG, de Andrade M, West A, de Silva R, Hardy J, Hernandez D: alpha-Synuclein gene haplotypes are associated with Parkinson's disease. *Hum Mol Genet* 2001, 10:1847-1851.

[23] Bennett MC, Bishop JF, Leng Y, Chock PB, Chase TN, Mouradian MM. Degradation of alpha-synuclein by proteasome: *J Biol Chem* 1999, 274:33855-33858.

[24] Biasini E, Fioriti L, Ceglia I, Invernizzi R, Bertoli A, Chiesa R, Forloni G: Proteasome inhibition and aggregation in Parkinson's disease: a comparative study in untransfected and transfected cells. *J Neurochem* 2004, 88:545-553.

[25] Zabrocki P, Pellens K, Vanhelmont T, Vandebroek T, Griffioen G, Wera S, Van Leuven F, Winderickx J: Characterization of alpha-synuclein aggregation and synergistic toxicity with protein tau in yeast. *FEBS J* 2005, 272:1386-1400.

[26] Cuervo AM, Stefanis L, Fredenburg R, Lansbury PT, Sulzer D: Impaired degradation of mutant alpha-synuclein by chaperone-mediated autophagy. *Science* 2004, 305:1292-1295.

[27] Webb JL, Ravikumar B, Atkins J, Skepper JN, Rubinsztein DC: Alpha-Synuclein is degraded by both autophagy and the proteasome. *J Biol Chem* 2003, 278:25009-25013.

[28] Tanaka Y, Engelender S, Igarashi S, Rao RK, Wanner T, Tanzi RE, Sawa A, L Dawson V, Dawson TM, Ross CA: Inducible expression of mutant alpha-synuclein decreases proteasome activity and increases sensitivity to mitochondria-dependent apoptosis. *Hum Mol Genet* 2001, 10:919-926.

[29] Stefanis L, Larsen KE, Rideout HJ, Sulzer D, Greene LA: Expression of A53T mutant but not wild-type alpha-synuclein in PC12 cells induces alterations of the ubiquitin-

dependent degradation system, loss of dopamine release, and autophagic cell death. *J Neurosci* 2001, 21:9549-9560.

[30] Snyder H, Mensah K, Theisler C, Lee J, Matouschek A, Wolozin B: Aggregated and monomeric alpha-synuclein bind to the S6' proteasomal protein and inhibit proteasomal function. *J Biol Chem* 2003, 278:11753-11759.

[31] Yokochi M, Narabayashi H, Iizuka R, Nagatsu T: Juvenile parkinsonism--some clinical, pharmacological, and neuropathological aspects. *Adv Neurol* 1984, 40:407-413.

[32] Ishikawa A, Takahashi H: Clinical and neuropathological aspects of autosomal recessive juvenile parkinsonism. *J Neurol* 1998, 245(Suppl 3):4-9.

[33] Kitada T, Asakawa S, Hattori N, Matsumine H, Yamamura Y, Minoshima S, Yokochi M, Mizuno Y, Shimizu N: Mutations in the parkin gene cause autosomal recessive juvenile parkinsonism. *Nature* 1998, 392:605-608.

[34] Hattori N, Kitada T, Matsumine H, Asakawa S, Yamamura Y, Yoshino H, Kobayashi T, Yokochi M, Wang M, Yoritaka A, Kondo T, Kuzuhara S, Nakamura S, Shimizu N, Mizuno Y: Molecular genetic analysis of a novel Parkin gene in Japanese families with autosomal recessive juvenile parkinsonism: evidence for variable homozygous deletions in the Parkin gene in affected individuals. *Ann Neurol* 1998, 44:935-941.

[35] Lucking CB, Durr A, Bonifati V, Vaughan J, De Michele G, Gasser T, Harhangi BS, Meco G, Denefle P, Wood NW, Agid Y, Brice A: Association between early-onset Parkinson's disease and mutations in the parkin gene. French Parkinson's Disease Genetics Study Group. *N Engl J Med* 2000, 342:1560-1567.

[36] Shimura H, Hattori N, Kubo S, Mizuno Y, Asakawa S, Minoshima S, Shimizu N, Iwai K, Chiba T, Tanaka K, Suzuki T: Familial Parkinson disease gene product, parkin, is a ubiquitin-protein ligase. *Nat Genet* 2000, 25:302-305.

[37] Shimura H, Schlossmacher MG, Hattori N, Frosch MP, Trockenbacher A, Schneider R, Mizuno Y, Kosik KS, Selkoe DJ: Ubiquitination of a new form of alpha-synuclein by parkin from human brain: implications for Parkinson's disease. *Science* 2001, 293:263-269.

[38] Chung KK, Zhang Y, Lim KL, Tanaka Y, Huang H, Gao J, Ross CA, Dawson VL, Dawson TM: Parkin ubiquitinates the alpha-synuclein-interacting protein, synphilin-1: implications for Lewy-body formation in Parkinson disease. *Nat Med* 2001, 7:1144-1150.

[39] Zhang Y, Gao J, Chung KK, Huang H, Dawson VL, Dawson TM: Parkin functions as an E2-dependent ubiquitin- protein ligase and promotes the degradation of the synaptic vesicle-associated protein, CDCrel-1. *Proc Natl Acad Sci U S A* 2000, 97:13354-13359.

[40] Imai Y, Soda M, Inoue H, Hattori N, Mizuno Y, Takahashi R: An unfolded putative transmembrane polypeptide, which can lead to endoplasmic reticulum stress, is a substrate of Parkin. *Cell* 2001, 105:891-902.

[41] Takahashi R, Imai Y, Hattori N, Mizuno Y: Parkin and endoplasmic reticulum stress. *Ann N Y Acad Sci* 2003, 991:101-106.

[42] Petrucelli L, O'Farrell C, Lockhart PJ, Baptista M, Kehoe K, Vink L, Choi P,Wolozin B, Farrer M, Hardy J, Cookson MR: Parkin protects against the toxicity associated with

mutant alpha-synuclein: proteasome dysfunction selectively affects catecholaminergic neurons. *Neuron* 2002, 36:1007-1019.

[43] Yang Y, Nishimura I, Imai Y, Takahashi R, Lu B: Parkin suppresses dopaminergic neuron-selective neurotoxicity induced by Pael-R in Drosophila. *Neuron* 2003, 37:911-294.

[44] Leroy E, Boyer R, Auburger G, Leube B, Ulm G, Mezey E, Harta G, Brownstein MJ, Jonnalagada S, Chernova T, Dehejia A, Lavedan C, Gasser T, Steinbach PJ, Wilkinson KD, Polymeropoulos MH: The ubiquitin pathway in Parkinson's disease. *Nature* 1998, 395:451-452.

[45] Maraganore DM, Farrer MJ, Hardy JA, Lincoln SJ, McDonnell SK, Rocca WA: Case-control study of the ubiquitin carboxy-terminal hydrolase L1 gene in Parkinson's disease. *Neurology* 1999, 53:1858-1860.

[46] Harhangi BS, Farrer MJ, Lincoln S, Bonifati V, Meco G, De Michele G, Brice A, Durr A, Martinez M, Gasser T, Bereznai B, Vaughan JR, Wood NW, Hardy J, Oostra BA, Breteler MM: The Ile93Met mutation in the ubiquitin carboxy-terminal-hydrolase-L1 gene is not observed in European cases with familial Parkinson's disease. *Neurosci Lett* 1999, 270:1-4.

[47] Liu Y, Fallon L, Lashuel HA, Liu Z, Lansbury PT Jr: The UCH-L1 gene encodes two opposing enzymatic activities that affect alpha-synuclein degradation and Parkinson's disease susceptibility. *Cell* 2002, 111:209-218.

[48] Mitsumoto A, Nakagawa Y: DJ-1 is an indicator for endogenous reactive oxygen species elicited by endotoxin. *Free Radic Res* 2001, 35:885-893.

[49] Quigley PM, Korotkov K, Baneyx F, Hol WG: The 1.6-A crystal structure of the class of chaperones represented by Escherichia coli Hsp31 reveals a putative catalytic triad. *Proc Natl Acad Sci U S A* 2003, 100:3137-3142.

[50] Shendelman S, Jonason A, Martinat C, Leete T, Abeliovich A: DJ-1 is a redox-dependent molecular chaperone that inhibits alpha-synuclein aggregate formation. *PLoS Biol* 2004, 2:e362.

[51] Miller DW, Ahmad R, Hague S, Baptista MJ, Canet-Aviles R, McLendon C, Carter DM, Zhu PP, Stadler J, Chandran J, Klinefelter GR, Blackstone C, Cookson MR: L166P mutant DJ-1, causative for recessive Parkinson's disease, is degraded through the ubiquitin-proteasome system. *J Biol Chem* 2003, 278:36588-36595.

[52] Meulener MC, Graves CL, Sampathu DM, Armstrong-Gold CE, Bonini NM, Giasson BI: DJ-1 is present in a large molecular complex in human brain tissue and interacts with alpha-synuclein. *J Neurochem* 2005, 93:1524-1532.

[53] Moore DJ, Zhang L, Troncoso J, Lee MK, Hattori N, Mizuno Y, Dawson TM, Dawson VL: Association of DJ-1 and parkin mediated by pathogenic DJ-1 mutations and oxidative stress. *Hum Mol Genet* 2005, 14:71-84.

[54] Valente EM, Abou-Sleiman PM, Caputo V, Muqit MM, Harvey K, Gispert S, Ali Z, Del Turco D, Bentivoglio AR, Healy DG, Albanese A, Nussbaum R, Gonzalez-Maldonado R, Deller T, Salvi S, Cortelli P, Gilks WP, Latchman DS, Harvey RJ, Dallapiccola B, Auburger G, Wood NW: Hereditary early-onset Parkinson's disease caused by mutations in PINK1. *Science* 2004, 304:1158-1160.

[55] Hatano Y, Li Y, Sato K, Asakawa S, Yamamura Y, Tomiyama H, Yoshino H, Asahina M, Kobayashi S, Hassin-Baer S, Lu CS, Ng AR, Rosales RL, Shimizu N, Toda T, Mizuno Y, Hattori N. Novel PINK1 mutations in early-onset parkinsonism. *Ann Neurol* 2004, 56:424-427.

[56] Petit A, Kawarai T, Paitel E, Sanjo N, Maj M, Scheid M, Chen F, Gu Y, Hasegawa H, Salehi-Rad S, Wang L, Rogaeva E, Fraser P, Robinson B, St George-Hyslop P, Tandon A: Wild-type PINK1 Prevents Basal and Induced Neuronal Apoptosis, a Protective Effect Abrogated by Parkinson Disease-related Mutations. *J Biol Chem* 2005, 280:34025-34032.

[57] Zimprich A, Biskup S, Leitner P, Lichtner P, Farrer M, Lincoln S, Kachergus J, Hulihan M, Uitti RJ, Calne DB, Stoessl AJ, Pfeiffer RF, Patenge N, Carbajal IC, Vieregge P, Asmus F, Muller-Myhsok B, Dickson DW, Meitinger T, Strom TM, Wszolek ZK, Gasser T: Mutations in LRRK2 cause autosomal-dominant parkinsonism with pleomorphic pathology. *Neuron* 2004, 44:601-607.

[58] Paisan-Ruiz C, Jain S, Evans EW, Gilks WP, Simon J, van der Brug M, Lopez de Munain A, Aparicio S, Gil AM, Khan N, Johnson J, Martinez JR, Nicholl D, Carrera IM, Pena AS, de Silva R, Lees A, Marti-Masso JF, Perez-Tur J, Wood NW, Singleton AB: Cloning of the gene containing mutations that cause PARK8-linked Parkinson's disease. *Neuron* 2004, 44:595-600.

[59] Funayama M, Hasegawa K, Ohta E, Kawashima N, Komiyama M, Kowa H, Tsuji S, Obata F: An LRRK2 mutation as a cause for the parkinsonism in the original PARK8 family. *Ann Neurol* 2005, 57:918-921.

[60] McNaught KS, Jenner P: Proteasomal function is impaired in substantia nigra in Parkinson's disease. *Neurosci Lett* 2001, 297:191-194.

[61] Furukawa Y, Vigouroux S, Wong H, Guttman M, Rajput AH, Ang L, Briand M, Kish SJ, Briand Y: Brain proteasomal function in sporadic Parkinson's disease and related disorders. *Ann Neurol* 2002, 51:779-782.

[62] McNaught KS, Belizaire R, Jenner P, Olanow CW, Isacson O: Selective loss of 20S proteasome alpha-subunits in the substantia nigra pars compacta in Parkinson's disease. *Neurosci Lett* 2002, 326:155-158.

[63] McNaught KS, Belizaire R, Isacson O, Jenner P, Olanow CW: Altered proteasomal function in sporadic Parkinson's disease. *Exp Neurol* 2003, 179:38-46.

[64] McNaught KS, Mytilineou C, Jnobaptiste R, Yabut J, Shashidharan P, Jenner P, Olanow CW: Impairment of the ubiquitin-proteasome system causes dopaminergic cell death and inclusion body formation in ventral mesencephalic cultures. *J Neurochem* 2002, 81:301-306.

[65] Nishi K: Proteasome inhibition induces remarkable and selective degeneration of dopaminergic neuron in ventral mesencephalic-striatal co-culture of rat. *Neurosci Res* 2005, 52 (suppl 1):S113.

[66] McNaught KS, Bjorklund LM, Belizaire R, Isacson O, Jenner P, Olanow CW: Proteasome inhibition causes nigral degeneration with inclusion bodies in rats. *Neuroreport* 2002, 13:1437-1441.

[67] McNaught KS, Perl DP, Brownell AL, Olanow CW: Systemic exposure to proteasome inhibitors causes a progressive model of Parkinson's disease. *Ann Neurol* 2004, 56:149-162.
[68] Fornai F, Lenzi P, Gesi M, Ferrucci M, Lazzeri G, Busceti CL, Ruffoli R, Soldani P, Ruggieri S, Alessandri MG, Paparelli A: Fine structure and biochemical mechanisms underlying nigrostriatal inclusions and cell death after proteasome inhibition. *J Neurosci* 2003, 23:8955-8966.
[69] Miwa H, Kubo T, Suzuki A, Nishi K, Kondo T: Retrograde dopaminergic neuron degeneration following intrastriatal proteasome inhibition. *Neurosci Lett* 2005, 380:93-98.
[70] Miwa H, Kubo T, Morita S, Nakanishi I, Kondo T: Oxidative stress and microglial activation in substantia nigra following striatal MPP+. *Neuroreport* 2004, 15:1039-1044.
[71] Zhang X, Xie W, Qu S, Pan T, Wang X, Le W: Neuroprotection by iron chelator against proteasome inhibitor-induced nigral degeneration. *Biochem Biophys Res Commun* 2005, 333:544-549.
[72] Marshall JF, Ungerstedt U: Supersensitivity to apomorphine following destruction of the ascending dopamine neurons: quantification using the rotational model. *Eur J Pharmacol* 1977, 41:361-367.
[73] Lotharius J, Dugan LL, O'Malley KL: Distinct mechanisms underlie neurotoxin-mediated cell death in cultured dopaminergic neurons. *J Neurosci* 1999, 19:1284-1293.
[74] Elkon H, Melamed E, Offen D: 6-Hydroxydopamine increases ubiquitin-conjugates and protein degradation: implications for the pathogenesis of Parkinson's disease. *Cell Mol Neurobiol* 2001, 21:771-781.
[75] Elkon H, Melamed E, Offen D: Oxidative stress, induced by 6-hydroxydopamine, reduces proteasome activities in PC12 cells: implications for the pathogenesis of Parkinson's disease. *J Mol Neurosci* 2004, 24:387-400.
[76] Inden M, Kondo J, Kitamura Y, Takata K, Nishimura K, Taniguchi T, Sawada H, Shimohama S: Proteasome inhibitors protect against degeneration of nigral dopaminergic neurons in hemiparkinsonian rats. *J Pharmacol Sci* 2005, 97:203-211.
[77] Langston JW, Langston EB, Irwin I: MPTP-induced parkinsonism in human and non-human primates--clinical and experimental aspects. *Acta Neurol Scand* 1984, 100 (Suppl):49-54.
[78] Schneider JS, Yuwiler A, Markham CH: Production of a Parkinson-like syndrome in the cat with N-methyl-4-phenyl-1,2,3,6-tetrahydropyridine (MPTP): behavior, histology, and biochemistry. *Exp Neurol* 1986, 91:293-307.
[79] Heikkila RE, Hess A, Duvoisin RC: Dopaminergic neurotoxicity of 1-methyl-4-phenyl-1,2,5,6-tetrahydropyridine in mice. *Science* 1984, 224:1451-1453.
[80] Przedborski S, Tieu K, Perier C, Vila M: MPTP as a mitochondrial neurotoxic model of Parkinson's disease. *J Bioenerg Biomembr* 2004, 36:375-379.
[81] Speciale SG, Liang CL, Sonsalla PK, Edwards RH, German DC: The neurotoxin 1-methyl-4-phenylpyridinium is sequestered within neurons that contain the vesicular monoamine transporter. *Neuroscience* 1998, 84:1177-1185.

[82] Rollema H, Damsma G, Horn AS, De Vries JB, Westerink BH: Brain dialysis in conscious rats reveals an instantaneous massive release of striatal dopamine in response to MPP+. *Eur J Pharmacol* 1986, 126:345-346.

[83] Vila M, Vukosavic S, Jackson-Lewis V, Neystat M, Jakowec M, Przedborski S.J Neurochem: Alpha-synuclein up-regulation in substantia nigra dopaminergic neurons following administration of the parkinsonian toxin MPTP. *J Neurochem.* 2000, 74:721-729.

[84] Purisai MG, McCormack AL, Langston WJ, Johnston LC, Di Monte DA: alpha-Synuclein expression in the substantia nigra of MPTP-lesioned non-human primates. *Neurobiol Dis* 2005, 20:898-906

[85] Dauer W, Kholodilov N, Vila M, Trillat AC, Goodchild R, Larsen KE, Staal R, Tieu K, Schmitz Y, Yuan CA, Rocha M, Jackson-Lewis V, Hersch S, Sulzer D, Przedborski S, Burke R, Hen R: Resistance of alpha -synuclein null mice to the parkinsonian neurotoxin MPTP. *Proc Natl Acad Sci* U S A 2002, 99:14524-14529.

[86] Shimoji M, Zhang L, Mandir AS, Dawson VL, Dawson TM: Absence of inclusion body formation in the MPTP mouse model of Parkinson's disease. *Brain Res Mol Brain Res* 2005, 134:103-108.

[87] Fornai F, Schluter OM, Lenzi P, Gesi M, Ruffoli R, Ferrucci M, Lazzeri G, Busceti CL, Pontarelli F, Battaglia G, Pellegrini A, Nicoletti F, Ruggieri S, Paparelli A, Sudhof TC: Parkinson-like syndrome induced by continuous MPTP infusion: convergent roles of the ubiquitin-proteasome system and alpha-synuclein. *Proc Natl Acad Sci U S A* 2005, 102:3413-3418.

[88] Betarbet R, Sherer TB, MacKenzie G, Garcia-Osuna M, Panov AV, Greenamyre JT: Chronic systemic pesticide exposure reproduces features of Parkinson's disease. *Nat Neurosci* 2000, 3:1301-1306.

[89] Alam M, Schmidt WJ: Rotenone destroys dopaminergic neurons and induces parkinsonian symptoms in rats. *Behav Brain Res* 2002, 136:317-324.

[90] Sherer TB, Kim JH, Betarbet R, Greenamyre JT: Subcutaneous rotenone exposure causes highly selective dopaminergic degeneration and alpha-synuclein aggregation. *Exp Neurol* 2003, 179:9-16.

[91] Ferrante RJ, Schulz JB, Kowall NW, Beal MF: Systemic administration of rotenone produces selective damage in the striatum and globus pallidus, but not in the substantia nigra. *Brain Res* 1997,753:157-162.

[92] Medina L, Chi TL, DeVivo DC, Hilal SK: MR findings in patients with subacute necrotizing encephalomyelopathy (Leigh syndrome): correlation with biochemical defect. *AJNR Am J Neuroradiol* 1990, 11:379-384.

[93] Zeevalk GD, Bernard LP: Energy status, ubiquitin proteasomal function, and oxidative stress during chronic and acute complex I inhibition with rotenone in mesencephalic cultures. *Antioxid Redox Signal* 2005, 7:662-672.

[94] Shamoto-Nagai M, Maruyama W, Kato Y, Isobe K, Tanaka M, Naoi M, Osawa T: An inhibitor of mitochondrial complex I, rotenone, inactivates proteasome by oxidative modification and induces aggregation of oxidized proteins in SH-SY5Y cells. *J Neurosci Res* 2003, 74:589-597.

[95] Manning-Bog AB, McCormack AL, Purisai MG, Bolin LM, Di Monte DA. Alpha-synuclein overexpression protects against paraquat-induced neurodegeneration. *J Neurosci* 2003, 23:3095-3099.

[96] Manning-Bog AB, McCormack AL, Li J, Uversky VN, Fink AL, Di Monte DA: The herbicide paraquat causes up-regulation and aggregation of alpha-synuclein in mice: paraquat and alpha-synuclein. *J Biol Chem* 2002, 277:1641-1644.

[97] Burkitt MJ, Kadiiska MB, Hanna PM, Jordan SJ, Mason RP: Electron spin resonance spin-trapping investigation into the effects of paraquat and desferrioxamine on hydroxyl radical generation during acute iron poisoning. *Mol Pharmacol* 1993, 43:257-263.

[98] Clejan L, Cederbaum AI: Synergistic interactions between NADPH-cytochrome P-450 reductase, paraquat, and iron in the generation of active oxygen radicals. *Biochem Pharmacol* 1989, 38:1779-1786.

[99] Dexter DT, Carter CJ, Wells FR, Javoy-Agid F, Agid Y, Lees A, Jenner P, Marsden CD: Basal lipid peroxidation in substantia nigra is increased in Parkinson's disease. *J Neurochem* 1989, 52:381-389.

[100] Jenner P: Oxidative mechanisms in nigral cell death in Parkinson's disease. *Mov Disord* 1998, 13 (Suppl 1):24-34.

[101] Floor E, Wetzel MG: Increased protein oxidation in human substantia nigra pars compacta in comparison with basal ganglia and prefrontal cortex measured with an improved dinitrophenylhydrazine assay. *J Neurochem* 1998, 70:268-275.

[102] Yoritaka A, Hattori N, Uchida K, Tanaka M, Stadtman ER, Mizuno Y: Immunohistochemical detection of 4-hydroxynonenal protein adducts in Parkinson disease. *Proc Natl Acad Sci U S A* 1996, 93:2696-2701.

[103] Alam ZI, Daniel SE, Lees AJ, Marsden DC, Jenner P, Halliwell B: A generalised increase in protein carbonyls in the brain in Parkinson's but not incidental Lewy body disease. *J Neurochem* 1997, 69:1326-1329.

[104] Zhang J, Perry G, Smith MA, Robertson D, Olson SJ, Graham DG, Montine TJ: Parkinson's disease is associated with oxidative damage to cytoplasmic DNA and RNA in substantia nigra neurons. *Am J Pathol* 1999, 154:1423-1429.

[105] Przedborski S, Chen Q, Vila M, Giasson BI, Djaldatti R, Vukosavic S, Souza JM, Jackson-Lewis V, Lee VM, Ischiropoulos H: Oxidative post-translational modifications of alpha-synuclein in the 1-methyl-4-phenyl-1,2,3,6-tetrahydropyridine (MPTP) mouse model of Parkinson's disease. *J Neurochem* 2001, 76:637-640.

[106] Chung KK, Thomas B, Li X, Pletnikova O, Troncoso JC, Marsh L, Dawson VL, Dawson TM: S-nitrosylation of parkin regulates ubiquitination and compromises parkin's protective function. *Science* 2004, 304:1328-1331.

[107] Lotharius J, Brundin P: Pathogenesis of Parkinson's disease: dopamine, vesicles and alpha-synuclein. *Nat Rev Neurosci* 2002, 3:932-942.

[108] Uversky VN, Li J, Fink AL: Metal-triggered structural transformations, aggregation, and fibrillation of human alpha-synuclein. A possible molecular NK between Parkinson's disease and heavy metal exposure. *J Biol Chem* 2001, 276:44284-44296.

[109] Turnbull S, Tabner BJ, El-Agnaf OM, Moore S, Davies Y, Allsop D: alpha-Synuclein implicated in Parkinson's disease catalyses the formation of hydrogen peroxide in vitro. *Free Radic Biol Med* 2001, 30:1163-1170.

[110] Xu J, Kao SY, Lee FJ, Song W, Jin LW, Yankner BA: Dopamine-dependent neurotoxicity of alpha-synuclein: a mechanism for selective neurodegeneration in Parkinson disease. *Nat Med* 2002, 8:600-606.

[111] Bulteau AL, Lundberg KC, Humphries KM, Sadek HA, Szweda PA, Friguet B, Szweda LI: Oxidative modification and inactivation of the proteasome during coronary occlusion/reperfusion. *J Biol Chem* 2001, 276:30057-30063.

[112] Voges D, Zwickl P, Baumeister W: The 26S proteasome: a molecular machine designed for controlled proteolysis. *Annu Rev Biochem* 1999, 68:1015-1068.

[113] Hoglinger GU, Carrard G, Michel PP, Medja F, Lombes A, Ruberg M, Friguet B, Hirsch EC: Dysfunction of mitochondrial complex I and the proteasome: interactions between two biochemical deficits in a cellular model of Parkinson's disease. *J Neurochem* 2003, 86:1297-1307.

[114] Sullivan PG, Dragicevic NB, Deng JH, Bai Y, Dimayuga E, Ding Q, Chen Q, Bruce-Keller AJ, Keller JN: Proteasome inhibition alters neural mitochondrial homeostasis and mitochondria turnover. *J Biol Chem* 2004, 279:20699-20707.

[115] Sawada H, Kohno R, Kihara T, Izumi Y, Sakka N, Ibi M, Nakanishi M, Nakamizo T, Yamakawa K, Shibasaki H, Yamamoto N, Akaike A, Inden M, Kitamura Y, Taniguchi T, Shimohama S: Proteasome mediates dopaminergic neuronal degeneration, and its inhibition causes alpha-synuclein inclusions. *J Biol Chem* 2004, 279:10710-10719.

[116] Dawson TM, Dawson VL: Molecular pathways of neurodegeneration in Parkinson's disease. *Science* 2003, 302(5646):819-822.

[117] Hald A, Lotharius J: Oxidative stress and inflammation in Parkinson's disease: is there a causal link? *Exp Neurol* 2005, 193:279-290.

[118] Beal MF: Excitotoxicity and nitric oxide in Parkinson's disease pathogenesis. *Ann Neurol* 1998, 44(3 Suppl 1):S110-114.

[119] Riedlerer PF: Views on neurodegeneration as a basis for neuroprotective strategies. *Med Sci Monit* 2004, 10:RA287-290.

[120] Smith WW, Jiang H, Pei Z, Tanaka Y, Morita H, Sawa A, Dawson VL, Dawson TM, Ross CA. *Hum Mol Genet* 2005; 14:3801-3811.

In: The Ubiquitin Proteasome System… ISBN 978-1-60021-749-4
Eds: Mario Di Napoli and Cezary Wojcik, pp. 627-650

Chapter 26

The Ubiquitin-Proteasome System in Axon Degeneration

Michael P. Coleman[*]
The Babraham Institute, Babraham, Cambridge, CB2 4AT, United Kingdom.

Abstract

The ubiquitin-proteasome system (UPS) has multiple roles in axon degeneration. An efficiently functioning UPS is essential for maintenance of healthy axons, but the UPS is also required to activate pathways of programmed axon death designed to remove axons in injury, disease and development. Thus, as in the cell body (Chapter 24) the UPS is a double-edged sword for axons. Genetic defects in the UPS often cause progressive degeneration of synapses and distal axons. These structures appear more vulnerable than neuronal cell bodies to failure of normal protein turnover. Similarly, pharmacological blockade with proteasome inhibitors causes neurite death *in vitro* and peripheral neuropathy *in vivo*, suggesting long axons are critically dependent on a fully functioning UPS. Thus, axon degeneration could make an important contribution to neurodegenerative disorders where UPS defects are reported, such as Alzheimer's disease, Parkinson's disease and Huntington's disease. Prominent and early axon and synapse pathology has been reported in each of these disorders. The UPS also controls axon survival by regulating nuclear and axonal events. A chimeric nuclear protein containing an N-terminal region of multiubiquitination factor Ube4b delays the degeneration of injured axons in the slow Wallerian degeneration mutant mouse (*Wld*^{S}) for several weeks. The Ube4b domain of Wld^{S} protein is required for its neuroprotective effect *in vivo* and interacts with valosin containing protein (VCP/p97) within the nucleus, an event that may influence downstream axonal mediators of this phenotype. Proteasome inhibition in axons also delays Wallerian degeneration, possibly by preventing downregulation of the MEK/ERK pathway. Rapid Wallerian degeneration in wild-type

[*] Correspondence concerning this article should be addressed to Dr. Michael Coleman, PhD; B501, The Babraham Institute, Babraham Research Campus, Babraham, Cambridge, CB2 4AT, United Kingdom. Phone: +44 1223 496315, Fax: +44 1223 496348, E-mail: michael.coleman@bbsrc.ac.uk.

axons seems to be a proactive, regulated process, similar in principle to apoptosis, albeit different in molecular details. Physical axon injury is not the only way to trigger this process because Wld^S delays also axon degeneration caused by several genetic and toxic insults. The common factor may be a blockade of anterograde axonal transport. Transport from the cell body may deliver an inhibitor of Wallerian degeneration that stops it from being triggered in healthy axons. Axonal pruning, the large-scale elimination of excess axon branches formed during development, also requires cell-autonomous action of the UPS. Cell specific deletion of genes encoding E1 and proteasome subunits in *Drosophila* blocks this process suggesting that the UPS degrades key regulators of the pruning process, or participates in the execution phase. There are interesting parallels with Wallerian degeneration, which also requires UPS activity in *Drosophila.* Both are proactive, cell-autonomous axon death programmes regulated in part by the UPS. However, at least some of the molecular details are distinct. Thus, several specific actions of the UPS participate in programmed degeneration of axons, whereas non-specific failure of the UPS can cause axon pathology. The apparent contradiction reflects the many roles of the UPS in the normal biochemistry of axons, roles that will be important to consider for effective targeting of therapeutic strategies.

Keywords: Wallerian degeneration, axonopathy, ubiquitin proteasome system, valosin containing protein, axonal pruning, Wld^S.

Abbreviations

ALS, Amyotrophic lateral sclerosis; atJ, Ataxia allele 'J'; CNS, Central nervous system; E1, Ubiquitin activating enzyme E1; E3, Ubiquitin E3 ligase; ER, Endoplasmic reticulum; ERK, Extracellular-signal related kinase; gad, Gracile axonal dystrophy (allele of Uch-l1) ; GAN: Giant axonal neuropathy; MAP1B-LC, Microtubule-associated protein 1B light chain; MEK, MAP/ERK kinase ; N70, N-terminal 70 amino acids shared by WldS and Ube4b; NMJ, Neuromuscular junction; Nmnat1, Nicotinamide mononucleotide adenylyltransferase 1; SCA7, Spinocerebellar ataxia type 7; SOD1, Superoxide dismutase 1; TBCB, Tubulin folding cofactor B; TBCE, Tubulin folding cofactor E; Ube4b, Ubiquitination factor E4b; Uch-l1, Ubiquitin carboxy-terminal hydrolase 1; UPS, Ubiquitin proteasome system ; VCP, Valosin containing protein, or p97; WldS, Slow Wallerian degeneration protein ; WldS, Slow Wallerian degeneration mouse, gene, or phenotype.

Introduction

Control of protein turnover in axons presents a unique challenge. These structures, up to 1 metre long in humans, contain up to 99% of the total protein content of the neuron. Proteins synthesized in the cell body must be delivered to their sites of action without being degraded, a delivery process that takes many days or months depending on the specific protein and its speed of transport. Thus, many axonal proteins need to have a long half-life. However, the UPS must also prevent the build-up of axonal proteins to levels that endanger the function or even survival of the axon, and it must degrade any misfolded proteins or axonal proteins damaged by free radicals. Thus, the UPS has to perform a delicate balancing act in axons even more than elsewhere. The difficulty of achieving this is reflected in the axon degeneration that results when things go wrong.

For example, ubiquitin itself moves within slow component b of axonal transport, moving at approximately 3 mm/day [1]. It takes around one month to reach the ends of the longest human axons and must be protected from degradation during this time. In the mouse axonal disease gracile axonal dystrophy (*gad*), axonal ubiquitin levels are depleted because of the absence of ubiquitin carboxy-terminal hydrolase 11 (Uch-11), a protein that stabilizes ubiquitin during its long journey down the axon [2,3]. The resulting degradation of ubiquitin leads to axon degeneration. In contrast, in the hereditary sensory and motor disorder giant axonal neuropathy it is the failure of the UPS to degrade proteins that leads to axon degeneration, in this case the light chain of microtubule associated protein 1B (MAP1B-LC) [4] and tubulin folding cofactor B (TBCB) [5].

Clearly the UPS is one contributory factor to the delicate balance between oversupply and undersupply of axonal proteins (Figure 1). The arrival of new material via axonal transport must be continuously balanced by either active retrograde axonal flow or local degradation. However, anterograde transport can significantly exceed retrograde transport for some species [6], suggesting there is an important role for distal degradative processes. The accumulation of ubiquitinated proteins in degenerating axons or neurites in Alzheimer's disease [7], Parkinson's disease [8] and Huntington's disease [9,10] and other disorders probably reflects the failure of the UPS in distal axons (see Chapters 30 to 34). Taken together with reports of early axon and synapse loss in each of these disorders [11-14], this suggests an early contributory role of axonal UPS failure to pathogenesis rather than a late, consequential role.

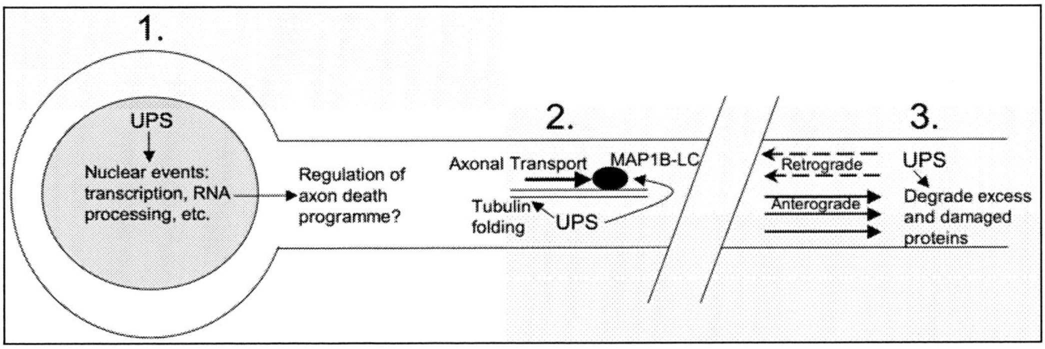

Figure 1. The UPS and axon degeneration. The UPS plays several roles in determining whether axons survive or degenerate. This diagram shows several of the known or proposed events. See text for further details. 1. Programmed axon death (Wallerian degeneration and axonal pruning) may each be to some extent determined by nuclear events that alter the expression of survival factors entering the axon. 2. The E3 gigaxonin degrades proteins that regulate tubulin folding and axonal transport, so that its absence leads to transport defects and axonal swelling. 3. An imbalance of anterograde to retrograde transport means that many of the proteins delivered to the axon must be degraded there, especially if they become damaged by free radicals or other factors. However, these proteins must be long-lived in the first place to reach sites so far distant from the cell body. Thus, UPS defects may lead to a dangerous build-up of excess normal or damaged proteins.

However, an apparently contradictory picture is also emerging. The UPS is required not only to maintain normal healthy axons, but also to bring about their programmed destruction in injury, disease or development [15-17]. These regulatory actions of the UPS are often nuclear, with downstream axonal mediators, but some are axonal (Figure 1). This chapter explores how these two conflicting roles of the UPS can co-exist, focusing on what is known at present about the specific UPS steps involved and how they are compartmentalized to different parts of the neuron. The focus here is on the degeneration of axons. For the many roles of the UPS in axonal pathfinding and synaptic growth, the reader is referred to Chapters 15 and 17.

The UPS has to Function Normally to Maintain Healthy Axons and Synapses

There is abundant circumstantial evidence that defects in the axonal UPS are important in human neuropathology (see below). However, it is important to put this in perspective by considering first an increasing number of genetic and pharmacological studies that demonstrate a causative role in axon degeneration. In a range of genetic disorders in mouse and one hereditary human disease it is now known that axon degeneration results from impairment of the UPS, and some of the molecular pathways are beginning to emerge. Moreover, axon degeneration often results when proteasomes are inhibited, both *in vitro* and *in vivo* in mice or humans.

Gracile Axonal Dystrophy (*grad*) Mice

Gracile axonal dystrophy (*gad*) is an autosomal recessive disease in mice, characterized by progressive 'dying back' of axons in the gracile tract, and loss of the extreme distal ends of both sensory and motor axons in the peripheral nervous system [18-20]. A progressive sensory ataxia and motor paresis results. The CNS pathology manifests as eosinophilic axonal swellings, or spheroids, some reaching more than 20 μm in diameter, that appear first in the medulla oblongata and then progressively appear at more caudal positions in the gracile tract. By fluorescently labeling a subset of CNS neurons, we find that this axonal dystrophy is a mixture of multiple small swellings on continuous axons, larger endbulbs on broken axons, and fragments of axoplasm disconnected from their parent axon at both ends (Figure 2). While the continuous axons in the early stages of this pathology may remain functional, and potentially even recoverable, those in which continuity has been lost are clearly damaged beyond repair.

gad is caused by a small intragenic deletion in Uch-l1, which is effectively a null mutation because the truncated protein is not detectable in homozygotes [2]. Uch-l1 appears to have several functions. As the name implies, it has carboxy-terminal hydrolase activity, capable of releasing ubiquitin from ubiquitination intermediates before they are degraded by the proteasome (see Chapter 4). The purpose of this may be to edit and regulate growing ubiquitin chains, or to recycle ubiquitin molecules for further use. The axons of *gad* mice show a decreased level of monomeric ubiquitin [3], which could be interpreted as indicating a failure to recycle ubiquitin, as without this hydrolase activity ubiquitin risks being degraded along with its target proteins. However, Uch-l1 also binds monomeric ubiquitin, suggesting that it helps maintain monomeric ubiquitin levels in other ways, akin to a chaperone or carrier protein, or preventing its targeting to lysosomal degradation [3]. Indeed a C90S mutant Uch-l1 that lacks hydrolase activity is able to increase monomeric ubiquitin concentration in cultured cells. The ubiquitin-binding model is also supported by the facts that carboxy-terminal hydrolase activity is low even in wild-type Uch-l1, and it is unclear why such an enzyme would need to be so abundant that it constitutes 1-2% of brain protein. A third function of Uch-l1 is as a ubiquitin E3 ligase, an activity present only when it dimerizes [21]. Hence, there are several ways in which the UPS of *gad* mice could be deficient.

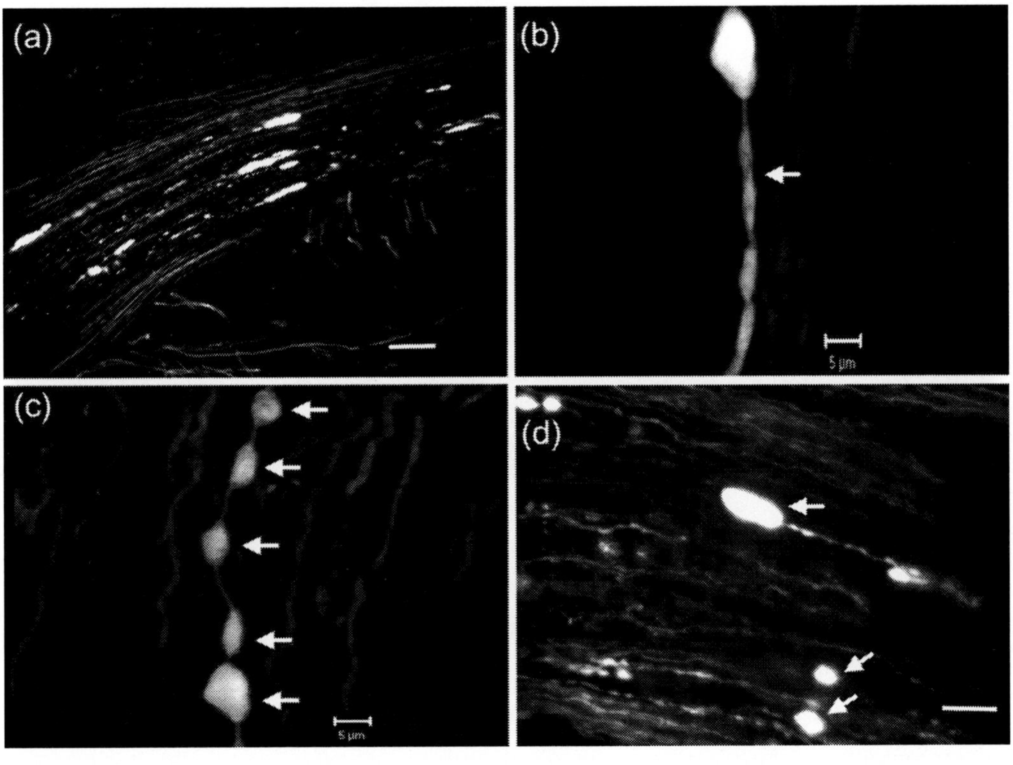

Figure 2. Axonal spheroid pathology in gad/YFP-H mice. *gad* mice, deficient in Uch-l1 [2] were crossed to YFP-H mice, in which a subset of axons are fluorescent due to restricted expression of a YFP transgene [109]. This improves longitudinal imaging of individual axons in the CNS, making it possible to determine whether spheroids arise on continuous or broken axons. The figure shows swollen axons in the cervical gracile tract at 19 weeks. Smaller swellings were often found on continuous axons and took the form of varicosities (b, arrow) and multiple 'en passant' spheroids clearly connected by strands of axoplasm (c, arrows). Larger spheroids often took the form of endbulbs (d, horizontal arrow) or isolated fragments of axoplasm (d, diagonal arrows). Scale bars: 50 µm, except where otherwise indicated. (Courtesy of Dr. Robert Adalbert, The Babraham Institute, Cambridge, UK)

It is unclear whether the null mutation in *gad* mice directly models any human disease. However, there are several genetic and biological links between Uch-l1 and Parkinson's disease. First, an I93M missense mutation, which reduces hydrolase activity, was found in two patients in a hereditary Parkinson's disease family [22], although the causative status of this change remains unconfirmed. Second, after some controversy, an association between the polymorphic variant S18Y and a reduced risk of Parkinson's has been confirmed in a large collaborative study [23]. Third, Uch-l1 shows some association with Parkinson's disease pathology such as Lewy bodies [24]. However, as Uch-l1 is such an abundant protein the significance of this immunostaining is unclear. Uch-l1 variants linked to Parkinson's disease susceptibility also cause accumulation of the hereditary Parkinsonism protein α-synuclein in cultured cells [21], and *gad* mice show accumulation of γ- and β-synuclein (though not α-synuclein) in axonal spheroids [25]. Finally, some enzyme activity studies suggest an inverse correlation in these polymorphic forms between hydrolase activity and the risk of Parkinson's disease [26], suggesting that the null mutation in *gad* mice just might represent an extreme case. However, *gad* mice die without classical Parkinson's symptoms at around 8 months of age and we cannot tell whether this early mortality hides Parkinson-like symptoms that may have appeared later. However, it is also important not to overlook the possibility that the early pathology in *gad* mice, even if not targeted to the nigro-striatal pathway, tells us something about the role of UPS-mediated distal axon and synapse loss in Parkinson's disease.

Ataxia (ax^J) Mice

Another mouse strain with a UPS defect, ataxia (ax^J), shows several similarities to *gad*. First, they share a synaptic or axonal defect, with little or no involvement of neuronal cell bodies, causing progressive ataxia and hindlimb paralysis [27,28]. Ataxia mice show defects in synaptic transmission in both CNS and peripheral nervous system [27]. Second, ataxia mice also have a recessive mutation in a deubiquitinating enzyme, ubiquitin-specific protease 14 (Usp14) [27]. The importance of deubiquitinating enzymes such as *fat facets* in regulating synapse development is discussed in Chapters 15 and 17, and here we see that similar enzymes are critical for synapse maintenance. Third, the level of monomeric ubiquitin in mouse brain and other tissues is decreased as a result of this mutation, not unlike the change in *gad* axons [28]. Earlier reports focused on the ability of Usp14 to hydrolyze monoubiquitin adducts [27] but its association with proteasomes suggests it has also a 'recycling' role, removing ubiquitin before it is degraded along with target proteins [28]. Studies of the yeast homolog also support this model, as proteasome inhibition rescues the decrease in free ubiquitin.

UBE4B Deficient Mice

The phenotype of mice that are haploinsufficient for ubiquitination factor E4b (Ube4b, or Ufd2a) is also related to that of *gad* mice. Homozygous deletion of Ube4b is embryonic lethal due to a cardiac defect, but heterozygotes show axonal dystrophy characterized by eosinophilic spheroids in the gracile tract from seven months of age [29]. As in *gad* mice, spheroids are particularly prominent in the distal regions of axons and are filled with disorganized neurofilaments and vesicular structures. A late-onset hind limb weakness results. Purkinje cells, which also degenerate in Ube4b$^{+/-}$ mice, express an Hsp70 family member and show numerous electron-dense bodies in the endoplasmic reticulum (ER), suggesting that ER stress could be the cause of axon degeneration in these mice. It may be that a cell body undergoing ER stress is unable to support the distal ends of its axon. However, a direct role for Ube4b in the axon also appears likely, especially as Ube4b is known to be present in axons [30]. Another possibility is that Ube4b has a nuclear role, as overexpression suppresses neurodegeneration in an expanded polyglutamine disorder [31] and Ube4b has also been detected in nuclei [30,32] (See also section on Wallerian degeneration, below). As the substrates of Ube4b emerge [33-36], the pathogenic mechanism will be better understood.

GIANT AXONAL NEUROPATHY

Giant axonal neuropathy (GAN) is a severe autosomal recessive sensory and motor disorder in humans characterized by abnormal accumulation of intermediate filaments in many cell types. In peripheral nerves, disorganization of the neurofilament network leads to huge axonal swellings, many times the normal axon diameter. The defective gene, gigaxonin, has been identified [37] and shown to be a ubiquitin scaffolding protein that links E1 protein to at least two substrates, either or both of which could mediate the axon pathology.

One gigaxonin substrate is microtubule associated protein 1B light chain (MAP1B-LC) [4]. Gigaxonin is required for the UPS mediated turnover of MAP1B-LC and levels of MAP1B-LC are substantially increased in gigaxonin null mice. Overexpression of MAP1B-LC in transfected primary neuronal cultures causes neurite fragmentation and eventual cell death, showing that the role of gigaxonin in facilitating MAP1B-LC degradation is critical to axonal survival, and ultimately that of the neuron too [4]. According to one model, MAP1B-LC is a negative regulator of axonal transport, and its downregulation after birth is required to allow efficient transport along longer, mature adult axons. When this downregulation fails in GAN patients, axonal death results. There is an interesting parallel with observations that overexpression of another microtubule associated protein, tau, also slows especially anterograde axonal transport [6], although tau might not be degraded by the UPS [38].

Another gigaxonin substrate, which is equally likely to mediate axon degeneration, is tubulin folding cofactor B (TBCB) [5]. Like MAP1B-LC, TBCB increases both *in vitro* and *in vivo* when gigaxonin levels decrease, leading to a fall in microtubule density in axons that can be mimicked by directly overexpressing TBCB. TBCB is one of five tubulin folding cofactors that coordinate folding of α/β-tubulin heterodimers to form the microtubule network. Loss-of-function mutation in another member of this family, TBCE, is the cause of the severe mouse axonal disorder progressive motor neuronopathy (*pmn*), in which microtubule density is also strongly decreased [39]. Thus, increasing TBCB and decreasing TBCE have similar effects on microtubule density and both are likely to cause axon degeneration. Unlike MAP1B-LC, however, TBCB is widely expressed and may mediate also the non-neuronal phenotype in GAN patients [5].

Thus, within the last seven years the mutation of four different UPS proteins has been shown to cause primary axon degeneration without significant loss of neuronal cell bodies. Some patterns are beginning to emerge. So far, we know that recessive mutation of both deubiquitinating enzymes and ubiquitin ligases can cause axon degeneration. We know that the distal ends of axons are affected first, and that the pathological manifestation is often axonal swelling, particularly within the CNS. Clearly, distal axons and their synapses are highly susceptible to UPS defects. It is not yet clear whether this is usually due to a general build up of an excess of misfolded proteins, or to an increase in one specific toxic component, as in GAN. As more examples are discovered, mechanisms should become better understood and consistent themes may emerge.

Proteasome Inhibition and Axon Pathology

Pharmacological studies also demonstrate the sensitivity of axons to UPS impairment, raising the important issue of neurological side effects from therapeutic use of proteasome inhibitors [40]. *In vitro* studies show that the proteasome inhibitor lactacystin is toxic for neurites in primary neuronal cultures at concentrations that do not affect the cell bodies [41,42]. Interestingly, lactacystin and other proteasome inhibitors also induce neurite outgrowth in neuronal-like cell lines, and proteasome subunits are upregulated during neuronal differentiation, indicating that the UPS has separate roles in axon development and maintenance [42-45]. As may be expected from these *in vitro* data, proteasome inhibitors also cause axon degeneration *in vivo*. Bortezomib (Velcade®), a dipeptide boronic acid proteasome inhibitor currently undergoing clinical trials in multiple myeloma and some solid tumors, causes peripheral neuropathy in some patients [40,46]. Long sensory fibres are particularly badly affected, resulting in neuropathic pain that is dose limiting in some patients (for more details see Chapters 44 and 45). In rats, stereotaxic injection of lactacystin or epoxomycin into the striatum causes retrograde degeneration of the nigrostriatal pathway, with initial loss of nerve terminals and later death of cell bodies [47]. Systemic administration of epoxomycin or synthetic proteasome inhibitors also causes degeneration of the nigro-striatal pathway [48]. In the systemic model it is not clear whether axons or cell bodies are the primary targets, but the susceptibility of axons to proteasome inhibition suggests a strong likelihood that they are directly affected.

The Wider Relevance of Axonal UPS Defects to Neurodegenerative Disease

The genetic and pharmacological studies described above help in several ways to understand how the UPS may contribute to axon degeneration in a wider range of neurodegenerative disorders. First, they establish the principle that axons are particularly vulnerable when the UPS is impaired. Second, they reveal molecular events that could contribute also to more widespread human disorders, highlighting the importance of studying these events in other contexts. Third, they establish patterns of pathology (e.g., distal axon degeneration, axonal swelling) that result when specific aspects of the UPS are disrupted, indicating one potential pathway that could underlie similar pathology in other disorders.

For example, axonal swelling results when the UPS is impaired in *gad* and Ube4b$^{+/-}$ mice, and in GAN patients (the phenotype of the GAN mouse model is still being characterized). Axons also swell in a wide range of human chronic or acute neurodegenerative conditions, such as Alzheimer's disease [49], amyotrophic lateral sclerosis (ALS) [50,51], traumatic brain injury [52], stroke [53], and many others [54,55]. There are of course many varied causes for these disorders, but as it has now been established that a defective UPS can lead to axonal swelling, it is important to ask whether it is a necessary step in the pathogenesis of some of these disorders.

Further support for an important, widespread role of the axonal UPS in neuropathology comes from observations that many of the swollen or dystrophic axons in a range of neurodegenerative disorders contain accumulations of ubiquitin epitopes (Table 1). Sometimes the ubiquitin signal is axon-specific. For example, in one ALS model, SOD1 and ubiquitin accumulate in vacuolated mitochondria specifically within axons, suggesting that the toxicity of mutant SOD1 specifically affects axonal mitochondria and that ubiquitination is an important early response to it [56]. The molecular status of the accumulated ubiquitin epitopes is often unclear, but in many cases it probably represents multi-ubiquitinated proteins that somehow fail to undergo degradation by proteasomes (see Chapter 23). In contrast, ubiquitin staining decreases in axons of *gad* mice and brain of *ax*^{J} mice, where loss of a deubiquitinating enzyme causes monoubiquitin levels to decrease [3,28,54]. Thus, defects in deubiquitinating enzymes might not underlie disorders in which ubiquitin epitopes accumulate in axons.

Table 1. Neurodegenerative disorders and animal models showing ubiquitin accumulation in axons

Disorder/model	Species	Site	Reference
Alzheimer's disease	Human	Cerebellum	[7]
Amyotrophic lateral sclerosis	Human, mouse and rat (SOD1^{G93A} transgenics)	Ventral horn, ventral root, corticospinal tract	[56] [110, 111]
Parabromophenylacetylurea toxicity	Rat	Brain	[112]
Cervical compressive myelopathy	Horse (race horses)	Spinal cord white matter	[113]
HIV encephalitis	Human	Brain	[114]
Huntington's disease	Human	Cerebral cortex, striatum	[9, 10]
Infantile neuroaxonal dystrophy	Human	Brain	[112]
Multiple sclerosis	Human	Around chronic plaques	[115]
Normal ageing	Rhesus monkey	Globus pallidus, substantia nigra pars reticulata	[116]
Parkinson's Disease	Human	Midbrain, brainstem	[8, 117]
Progressive supranuclear palsy	Human	Hippocampus	[118]
Spinal cord compression	Rat	Thoracic spinal cord	[119]
Tauopathy	Mouse	Brain, spinal cord	[120]
Trigeminal disorders	Cow	Trigeminal ganglia	[121]

For many of the disorders listed in Table 1 there is independent evidence that the UPS is impaired. These include Alzheimer's disease, where molecular misreading results in a mutant ubiquitin, UBB^{+1}, an aberrant protein that interferes with the growth of ubiquitin chains on targeted proteins [57]. In Parkinson's disease, many genetic causes have been traced to defects in the UPS [58,59]. As it is clear that axons are particularly vulnerable when the UPS is impaired, it is important to investigate whether axons and their synapses are the first structures to degenerate in these disorders.

In polyglutamine expansion diseases, characterized by early axon and synapse disturbance [60,61], the role of the UPS is currently controversial. Expanded polyglutamine impairs proteasome mediated degradation *in vitro* [62] but it is unclear whether similar events occur *in vivo*. The advent of ubiquitin reporter transgenic mice [63] allows UPS activity to be assessed *in vivo*. G76V mutant ubiquitin is fused to C-terminal green fluorescent protein. The mutation prevents hydrolysis by C-terminal hydrolases, leading to efficient ubiquitination of this reporter and its degradation by the proteasome. However, when proteasome degradation is blocked pharmacologically or genetically, green fluorescence accumulates. In order to characterize the expected UPS defect, this reporter system was applied to a mouse model of the polyglutamine expansion disorder spinal cerebellar ataxia 7 (SCA7). Unexpectedly, fluorescence increased only very late in the disease and even this increase could be explained by an increase in reporter transcription [64]. Nevertheless, further evidence of the importance of the UPS in polyglutamine disorders comes from the observations that both Ube4b overexpression, and loss-of-function of valosin containing protein (VCP) (see below) suppresses expanded polyglutamine disease in *Drosophila* [31,65].

Thus, circumstantial evidence strongly implicates the UPS in the pathogenesis of many neurodegenerative disorder, and the monogenic defects and pharmacological experiments described previously demonstrate the potential for these changes to be contributory and not mere consequences of pathology. Although its role remains controversial in some disorders, axonal UPS defects could be a very common trigger, or key contributory event, in many human neurodegenerative disorders.

The UPS and Axon Death Programmes

It is not always advantageous for an organism to maintain axons. Just as apoptosis removes cells in injury, disease, and development, programmes of axonal death are designed to remove axons that are no longer useful. The mechanisms are distinct from apoptosis, probably to prevent the death programme spreading to neuronal cell bodies and killing them in most cases [66]. In this way, other axonal branches may survive, and in the peripheral nervous system the proximal axonal stump has the chance to regenerate. Programmed axon death mechanisms are just beginning to emerge and it has become clear that Wallerian degeneration, the pathway that operates in injury and disease, is distinct from developmental pathways, such as axonal pruning and synapse elimination. However, the UPS is involved in both. Importantly, although the mechanisms are executed locally in the axon, they appear to be controlled to some extent from the cell nucleus [67,68]. This may help keep separate the axon maintenance actions of the UPS (see above) from the UPS components that mediate axon destruction.

WALLERIAN DEGENERATION

Transection or other severe injury to mammalian peripheral nerves leaves a long distal axon stump, disconnected from the cell body. Until this stump is removed, it blocks axon regeneration [69,70]. Wallerian degeneration is the process by which the distal stump dies, and is one of the largest scale cytoplasmic destruction processes the body has to undertake (Figure 3). For example, a motor neuron is one of the largest cells in the body and more than 99% of its cytoplasmic volume resides in the axon. A similar process is often triggered by genetic or toxic blockade of axonal transport in disorders where there is no physical axon injury [55]. It appears that once a peripheral axon is damaged beyond repair, whether by injury or other defects, it is better to destroy it and regenerate. This scenario is reminiscent of apoptosis, where sick cells are actively destroyed and replaced by division of neighboring cells. However, unlike apoptosis, Wallerian degeneration does not appear to involve caspase activation and is not blocked by caspase inhibitors, Bcl-2 overexpression, or Bax and Bak deletion [71-73]. Thus, it is distinct from apoptosis in molecular terms. Nevertheless, the destruction process is a proactive pathway and several lines of evidence now point to an important controlling role for the UPS.

As in apoptosis research, the discovery of a regulatory gene has opened up the field of Wallerian degeneration. A spontaneous mutation in C57BL/*Wld^S* mice delays Wallerian degeneration tenfold, providing a route towards understanding the degenerative mechanism [74]. An unusual mutation [75] results in the *de novo* expression of a chimeric gene [76], which confers the cell-autonomous, neuroprotective slow Wallerian degeneration (*Wld^S*) phenotype in transgenic mice [68]. An in-frame fusion protein is produced, consisting of the N-terminal 70 amino acids (N70) of multiubiquitination factor E4b (Ube4b, also called Ufd2a), a short linker region of 18 amino acids, and the entire coding sequence of the NAD^+-synthesizing enzyme nicotinamide mononucleotide adenylyltransferase 1 (Nmnat1) (Figure 3a). Despite some conflicting evidence *in vitro* [77,78], transgenic mouse experiments have now established that the Nmnat1 domain is unable to confer the *Wld^S* phenotype *in vivo*, indicating that the N70 Ube4b-derived domain contributes to the phenotype [123]. It is now important to determine whether the action of both parts of Wld^S protein are required to protect axons robustly in mammalian nerves. Surprisingly, the Wld^S protein is undetectable in axons *in vivo* [30,68,79,80], although its presence at very low levels cannot be ruled out. Instead, Wld^S appears to be a nuclear protein, whose protective effect on axons is likely to be mediated by downstream axonal factors.

Valosin-containing protein (VCP/p97), which binds directly to N70, is a candidate for one of these downstream factors [81]. VCP is an extremely abundant protein with many diverse roles throughout the cell, including a key step of the UPS during the degradation of misfolded nascent protein in the ER (See Chapter 13 and [82,83]). Wld^S partially redistributes VCP, and what appears to be associated multi-ubiquitinated proteins, into a pattern of discrete intranuclear foci [81]. This could alter the function of VCP or some of its other binding partners. Whether VCP mediates the *Wld*^{S} phenotype is still under investigation, but these changes are at least consistent with a nuclear role for Wld^S involving its Ube4b-derived N70 domain. It also fits with reports that Wld^S is able to alter gene expression, but most effectively when both N70 and Nmnat1 are expressed [84], as VCP and the UPS can contribute to control of transcription factor stability and RNA processing [36,85]. It also remains unclear whether the intranuclear foci in which Wld^S and VCP accumulate are PML bodies, as these are enriched for components of the proteasome (Chapter 10).

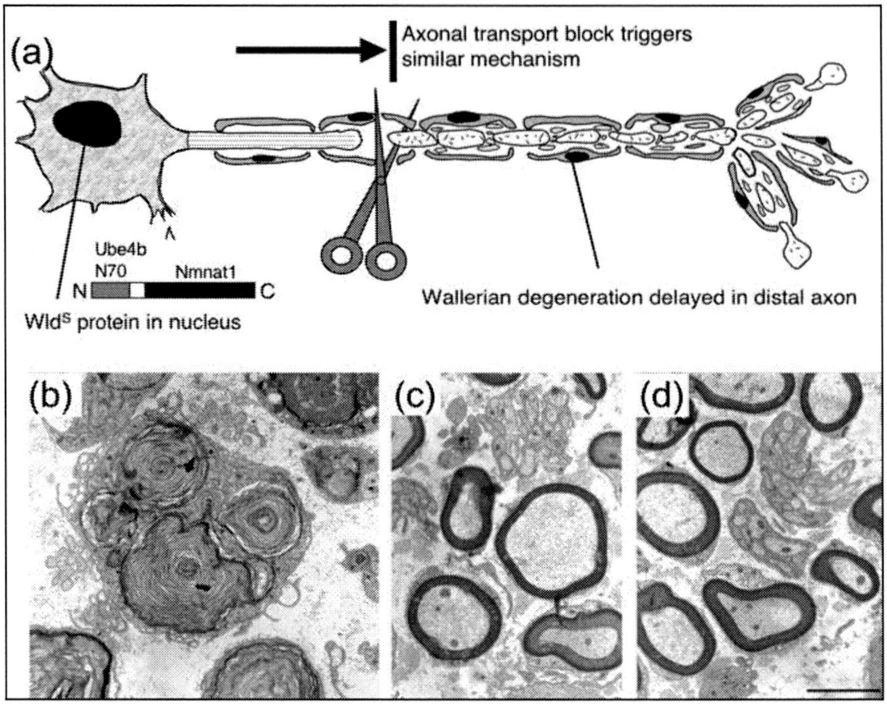

Figure 3. Wallerian degeneration and its delay in *Wld*^{S}, and *Wld*^{S} transgenic mice. (a) Wallerian degeneration is the degeneration of the distal part of an injured axon that occurs rapidly after a latent phase of approximately 36 h in wild-type mouse sciatic nerve. The Wld^S protein delays degeneration of the distal axon for several weeks and has a similar effect in several neurodegenerative disorders where axonal transport is blocked. The chimeric protein, whose N-terminal 70 amino acids are derived from Ube4b, is only detectable in the nucleus in vivo. (b-d) Prolonged survival of the distal stumps of injured axons in Wld^S (c) and transgenic Wld^S (d) mice compared to wild-type (b) 14 days after a sciatic nerve lesion. a: reprinted with permission from Coleman, M.P. and Ribchester R.R. Programmed axon death, synaptic dysfunction and the ubiquitin proteasome system. Current Drug Targets: CNS and Neurological Disorders 3: 153-60 Copyright (2004), with permission from Bentham Science Publishers Ltd.; b-d: reprinted from Coleman, M.P. and Perry, V.H. Axon pathology in neurological disease: a neglected therapeutic target; Trends in Neurosciences 25: 532-7 Copyright (2002), with permission from Elsevier. Electron micrographs courtesy of Dr. Bogdan Beirowski.

Interestingly, N70 is part of a 126 amino acid sequence that is a relatively recent development in evolution and is also necessary for ubiquitination activity in mammalian Ube4b enzyme [35,81,86]. How this sequence alters activity is unclear because it does not contain the U-box. However, one clue is that it becomes phosphorylated during mitosis and phosphorylation reduces enzyme activity [35]. As the N-terminal region of Ube4b is so influential in determining ubiquitination activity, it is reasonable to expect that the identical N70 sequence of Wld^S protein may interfere with this domain, perhaps competing with it to bind VCP, and alter ubiquitination activity.

These studies suggest that UPS-mediated events within the nucleus of a neuron predetermine what the rate of Wallerian degeneration will be if its axon should become injured, probably by altering the expression level of a downstream axonal protein. The mechanism is probably distinct from that which triggers axon degeneration when the axonal UPS fails in *gad*, ax^J, $Ube4b^{+/-}$ and GAN (see above), as different subcellular compartments appear to be involved. However, as described below, there is also evidence that the UPS has to function in axons to allow programmed axon death.

When primary neuronal cultures are grown in compartmentalized culture, and the proteasome inhibitor MG132 is added specifically to the neuritic compartment, the survival of transected neurites is delayed from around 10 hours to at least 24 hours [87]. Addition of proteasome inhibitor immediately after transection is sufficient for this effect, further indicating that it is a local mechanism within axons. As Wld^S has not been detected in axons *in vivo*, the protective effect of proteasome inhibitor in axons may be distinct from that of Wld^S. However, the possibility that Wld^S acts at very low levels in the axon has not been eliminated, and some *in vitro* studies do suggest that it can enter short neurites when expressed at high levels [78,88]. Further studies showed that axon protection by proteasome inhibition requires the extracellular-signal related kinase (ERK) signaling pathway, as simultaneous inhibition of ERK phosphorylation using the drug U0126 restores rapid neurite degeneration [87]. Another study also found a small delay in degeneration of transected neurites after pretreatment of both neurites and cell bodies with proteasome inhibitor [89], although neither study reported a delay in neurite degeneration that matches *Wld*^{S} neurons, whose neurites begin to degenerate only 3-4 days after transection in primary culture [90,91]. Finally, new data in a *Drosophila* model of slow Wallerian degeneration also indicate that an efficient UPS is required for Wallerian degeneration, as expression of the yeast ubiquitin protease UBP2 delays Wallerian degeneration in this model [122].

Programmed Axon death in Disease

Wallerian degeneration has been used as a model of axon death for over 150 years [92], but only since the discovery of the *Wld*^{S} mouse has it been possible to test the underlying hypothesis that injury-induced axon degeneration is related to axon pathology. A series of recent studies has now shown that Wld^S can delay axon degeneration in several genetic and toxic disorders in the absence of any physical injury to the axon [54,79,80,93,94]. The common factor may be a blockade of anterograde axonal transport. Transport from the cell body may deliver an inhibitor of Wallerian degeneration that stops it from being triggered in healthy axons. Thus, like Wallerian degeneration, axon degeneration in some types of motor neuron disease, Charcot-Marie-Tooth disease, toxic neuropathies and other disorders may also be controlled through nuclear UPS events with downstream axonal mediators. Wld^S also reduces neuropathic pain, consistent with partial Wallerian degeneration of a nerve being the trigger for neuropathic pain [95-97; and Chapter 27]. However, axon degeneration is not blocked in all disorders, the most notable exceptions so far being some mouse models of ALS [55,98,99]. Aβ-induced axon degeneration is also not blocked by proteasome inhibition [100] in the way that Wallerian degeneration is. Thus, a programmed axon death mechanism operates in some, but not all, neurodegenerative disorders through a mechanism related to Wallerian degeneration.

Developmental Axon Loss

Among the many roles of apoptosis is to kill cells that arise during development but have no function in the adult. The equivalent processes in axon biology are axonal pruning, where large axon branches are removed, and synapse elimination, where the number of presynaptic inputs to a target cell is reduced. Both occur without cell death, indicating that the actual mechanism is non-apoptotic [15,101].

Axonal pruning in the *Drosophila* mushroom body appears morphologically similar to Wallerian degeneration, with near simultaneous degeneration of large axonal regions [67,102]. It also shares the features of caspase independence and cell-autonomous action of the UPS, as it is blocked by cell-specific deletion of the E1 or proteasome subunits [67]. The UPS may degrade key regulators of the pruning process, or participate in the execution stage. These similarities raised the possibility that the mechanism is related to that of Wallerian degeneration [15] but new data now distinguish between the two mechanisms (Liqun Luo, personal communication). The *Wld^S* gene delays injury-induced axon degeneration in *Drosophila* but does not delay axonal pruning in either *Drosophila* or mice. Nevertheless, inhibiting the UPS does delay both Wallerian degeneration and axonal pruning in flies [67,122], suggesting there may be some overlap further downstream in the mechanisms. Furthermore, in primary neuronal cultures proteasome inhibition delays both injury-induced neurite degeneration and degeneration triggered by limiting the supply of nerve growth factor to the distal neurite, an experiment that could be regarded as an *in vitro* model of pruning [87]. Thus, axonal pruning shares some similarity to Wallerian degeneration, which may extend to some UPS mediated steps, but the available *in vivo* evidence suggests that the two processes are distinct at the molecular level.

Synapse elimination is also distinct from Wallerian degeneration of wild-type axons, as it is unaltered in *Wld^S* mice [103] and because the cellular events are sharply different from those in Wallerian degeneration. Synapse elimination proceeds by an unusual cellular mechanism known as 'axosome shedding', in which fragments of axoplasm become isolated from a retracting axon and are ultimately engulfed by Schwann cells [104]. In contrast, Wallerian degeneration in wild-type nerves progresses as a wave of fragmentation spreading rapidly over the whole axon [102]. However, motor nerve terminals at the neuromuscular junctions (NMJ) of *Wld^S* mice after experimental axotomy appear remarkably similar to NMJ undergoing synapse elimination. Terminal swellings can be seen connected by very thin axon collaterals to the main part of the distal axon, reminiscent of the immediate precursor of an axosome [105,106]. Synapse loss is one of the first events after axotomy, both in wild-type and *Wld^S* axons [107,108]. It is delayed by Wld^S, but not as robustly as Wallerian degeneration of the main axon trunk and, curiously, the neuroprotective effect of *Wld^S* is lost with age specifically at synapses [105]. Thus, while classical synapse elimination appears distinct from Wallerian degeneration, the degeneration of axotomized synapses shows similarities with both.

Conclusions

In summary, this chapter distinguishes four separate ways in which the UPS influences axon survival. First, multiple lines of genetic and toxic evidence indicate that axons require the UPS for normal maintenance and survival. Compared to other parts of the neuron axons are highly sensitive to UPS blockade. These findings have important implications for the use of proteasome inhibitors as drugs, and indeed the first clinical trials indicate that peripheral neuropathy can result. They also highlight the importance of testing for an early, contributory role of axon degeneration in disorders where UPS impairment has been reported. Second, axon destruction programmes are also regulated by the UPS, but at least partly though nuclear events. Wld^S protein may delay axon degeneration in injury and disease though its binding partner VCP, and possibly also though NAD$^+$ metabolism. These events require the N-terminal Ube4b-derived domain, suggesting that the UPS is involved in this process. Third, direct application of proteasome inhibitors to neurites blocks injury-induced degeneration through a mechanism that is probably distinct from that of the Wld^S protein. Fourth, axonal pruning, like Wallerian degeneration requires the UPS to eliminate axonal branches.

The UPS achieves these, often contradictory, effects on axon survival by temporal separation, cellular compartmentalization, and molecular specificity [16]. For example, nuclear UPS steps may control the expression of a downstream axonal inhibitor of axon degeneration in *Wld*^{S} mice, but it appears to be failure of the UPS within the axon itself that causes distal axon degeneration in *gad* mice and in GAN patients, and axonal pruning differs from Wallerian degeneration both in developmental timing and in the controlling proteins. A fuller understanding may allow us to mitigate the damage to axons caused by some proteasome inhibitor drugs such as Velcade®, to block axonal disorders that result from a defective UPS, and perhaps to mimic the *Wld*^{S} gene or its downstream mediators to target neuroprotection to axons.

References

[1] Bizzi A, Schaetzle B, Patton A, Gambetti P, Autilio-Gambetti L: Axonal transport of two major components of the ubiquitin system: free ubiquitin and ubiquitin carboxyl-terminal hydrolase PGP 9.5. *Brain Res* 1991, 548:292-299.

[2] Saigoh K, Wang YL, Suh JG, Yamanishi T, Sakai Y, Kiyosawa H, Harada T, Ichihara N, Wakana S, Kikuchi T, Wada K: Intragenic deletion in the gene encoding ubiquitin carboxy-terminal hydrolase in gad mice. *Nat Genet* 1999, 23:47-51.

[3] Osaka H, Wang YL, Takada K, Takizawa S, Setsuie R, Li H, Sato Y, Nishikawa K, Sun YJ, Sakurai M, Harada T, Hara Y, Kimura I, Chiba S, Namikawa K, Kiyama H,

Noda M, Aoki S, Wada K: Ubiquitin carboxy-terminal hydrolase L1 binds to and stabilizes monoubiquitin in neuron. *Hum Mol Genet* 2003, 12:1945-1958.

[4] Allen E, Ding J, Wang W, Pramanik S, Chou J, Yau V, Yang Y: Gigaxonin-controlled degradation of MAP1B light chain is critical to neuronal survival. *Nature* 2005, 438:224-228.

[5] Wang W, Ding J, Allen E, Zhu P, Zhang L, Vogel H, Yang Y: Gigaxonin interacts with tubulin folding cofactor B and controls its degradation through the ubiquitin-proteasome pathway. *Curr Biol* 2005, 15:2050-2055.

[6] Stamer K, Vogel R, Thies E, Mandelkow E, Mandelkow EM: Tau blocks traffic of organelles, neurofilaments, and APP vesicles in neurons and enhances oxidative stress. *J Cell Biol* 2002, 156:1051-1063.

[7] Dickson DW, Wertkin A, Mattiace LA, Fier E, Kress Y, Davies P, Yen SH: Ubiquitin immunoelectron microscopy of dystrophic neurites in cerebellar senile plaques of Alzheimer's disease. *Acta Neuropathol* 1990, 79:486-493.

[8] Gai WP, Blessing WW, Blumbergs PC: Ubiquitin-positive degenerating neurites in the brainstem in Parkinson's disease. *Brain* 1995, 118:1447-1459.

[9] Cammarata S, Caponnetto C, Tabaton M: Ubiquitin-reactive neurites in cerebral cortex of subjects with Huntington's chorea: a pathological correlate of dementia? *Neurosci Lett* 1993, 156:96-98.

[10] DiFiglia M, Sapp E, Chase KO, Davies SW, Bates GP, Vonsattel JP, Aronin N: Aggregation of huntingtin in neuronal intranuclear inclusions and dystrophic neurites in brain. *Science* 1997, 277:1990-1993.

[11] Li JY, Plomann M, Brundin P: Huntington's disease: a synaptopathy? *Trends Mol Med* 2003, 9:414-420.

[12] Scheff SW, Price DA: Synaptic pathology in Alzheimer's disease: a review of ultrastructural studies. *Neurobiol Aging* 2003, 24:1029-1046.

[13] Lotharius J, Brundin P: Pathogenesis of Parkinson's disease: dopamine, vesicles and alpha-synuclein. *Nat Rev Neurosci* 2002, 3:932-942.

[14] Martin LJ, Pan Y, Price AC, Sterling W, Copeland NG, Jenkins NA, Price DL, Lee MK: Parkinson's disease alpha-synuclein transgenic mice develop neuronal mitochondrial degeneration and cell death. *J Neurosci* 2006, 26:41-50.

[15] Luo L, O'Leary DD: Axon retraction and degeneration in development and disease. *Annu Rev Neurosci* 2005, 28:127-156.

[16] Coleman MP, Ribchester RR: Programmed axon death, synaptic dysfunction and the ubiquitin proteasome system. *Curr Drug Targets CNS Neurol Disord* 2004, 3:227-238.

[17] Korhonen L, Lindholm D: The ubiquitin proteasome system in synaptic and axonal degeneration: a new twist to an old cycle. *J Cell Biol* 2004, 165:27-30.

[18] Yamazaki K, Wakasugi N, Tomita T, Kikuchi T, Mukoyama M, Ando K: Gracile axonal dystrophy (GAD), a new neurological mutant in the mouse. *Proc Soc Exp Biol Med* 1988, 187:209-215.

[19] Kikuchi T, Mukoyama M, Yamazaki K, Moriya H: Axonal degeneration of ascending sensory neurons in gracile axonal dystrophy mutant mouse. *Acta Neuropathol (Berl)* 1990, 80:145-151.

[20] Miura H, Oda K, Endo C, Yamazaki K, Shibasaki H, Kikuchi T: Progressive degeneration of motor nerve terminals in GAD mutant mouse with hereditary sensory axonopathy. *Neuropathol Appl Neurobiol* 1993, 19:41-51.

[21] Liu Y, Fallon L, Lashuel HA, Liu Z, Lansbury PT, Jr.: The UCH-L1 gene encodes two opposing enzymatic activities that affect alpha-synuclein degradation and Parkinson's disease susceptibility. *Cell* 2002, 111:209-218.

[22] Leroy E, Boyer R, Auburger G, Leube B, Ulm G, Mezey E, Harta G, Brownstein MJ, Jonnalagada S, Chernova T, Dehejia A, Lavedan C, Gasser T, Steinbach PJ, Wilkinson KD, Polymeropoulos MH: The ubiquitin pathway in Parkinson's disease. *Nature* 1998, 395:451-452.

[23] Maraganore DM, Lesnick TG, Elbaz A, Chartier-Harlin MC, Gasser T, Kruger R, Hattori N, Mellick GD, Quattrone A, Satoh J, Toda T, Wang J, Ioannidis JP, de Andrade M, Rocca WA: UCHL1 is a Parkinson's disease susceptibility gene. *Ann Neurol* 2004, 55:512-521.

[24] Lowe J, McDermott H, Landon M, Mayer RJ, Wilkinson KD: Ubiquitin carboxyl-terminal hydrolase (PGP 9.5) is selectively present in ubiquitinated inclusion bodies characteristic of human neurodegenerative diseases. *J Pathol* 1990, 161:153-160.

[25] Wang YL, Takeda A, Osaka H, Hara Y, Furuta A, Setsuie R, Sun YJ, Kwon J, Sato Y, Sakurai M, Noda M, Yoshikawa Y, Wada K: Accumulation of beta- and gamma-synucleins in the ubiquitin carboxyl-terminal hydrolase L1-deficient gad mouse. *Brain Res* 2004, 1019:1-9.

[26] Nishikawa K, Li H, Kawamura R, Osaka H, Wang YL, Hara Y, Hirokawa T, Manago Y, Amano T, Noda M, Aoki S, Wada K: Alterations of structure and hydrolase activity of parkinsonism-associated human ubiquitin carboxyl-terminal hydrolase L1 variants. *Biochem Biophys Res Commun* 2003, 304:176-183.

[27] Wilson SM, Bhattacharyya B, Rachel RA, Coppola V, Tessarollo L, Householder DB, Fletcher CF, Miller RJ, Copeland NG, Jenkins NA: Synaptic defects in ataxia mice result from a mutation in Usp14, encoding a ubiquitin-specific protease. *Nat Genet* 2002, 32:420-425.

[28] Anderson C, Crimmins S, Wilson JA, Korbel GA, Ploegh HL, Wilson SM: Loss of Usp14 results in reduced levels of ubiquitin in ataxia mice. *J Neurochem* 2005, 95:724-731.

[29] Kaneko-Oshikawa C, Nakagawa T, Yamada M, Yoshikawa H, Matsumoto M, Yada M, Hatakeyama S, Nakayama K, Nakayama KI: Mammalian E4 is required for cardiac development and maintenance of the nervous system. *Mol Cell Biol* 2005, 25:10953-10964.

[30] Fang C, Bernades-Silva M, Coleman MP, Perry VH: The cellular distribution of the WldS chimeric protein and its constituent proteins in the central nervous system. *Neuroscience* 2005, 135:1107-1118.

[31] Matsumoto M, Yada M, Hatakeyama S, Ishimoto H, Tanimura T, Tsuji S, Kakizuka A, Kitagawa M, Nakayama KI: Molecular clearance of ataxin-3 is regulated by a mammalian E4. *EMBO J* 2004, 23:659-669.

[32] Hatakeyama S, Yada M, Matsumoto M, Ishida N, Nakayama KI: U-Box proteins as a new family of ubiquitin-protein ligases. *J Biol Chem* 2001, 276:32111-33120.

[33] Hoppe T, Cassata G, Barral JM, Springer W, Hutagalung AH, Epstein HF, Baumeister R: Regulation of the myosin-directed chaperone UNC-45 by a novel E3/E4-multiubiquitylation complex in C. elegans. *Cell* 2004, 118:337-349.

[34] Okumura F, Hatakeyama S, Matsumoto M, Kamura T, Nakayama KI: Functional regulation of FEZ1 by the U-box-type ubiquitin ligase E4B contributes to neuritogenesis. *J Biol Chem* 2004, 279:53533-53543.

[35] Spinette S, Lengauer C, Mahoney JA, Jallepalli PV, Wang Z, Casciola-Rosen L, Rosen A: Ufd2, a novel autoantigen in scleroderma, regulates sister chromatid separation. *Cell Cycle* 2004, 3:1638-1644.

[36] Richly H, Rape M, Braun S, Rumpf S, Hoege C, Jentsch S: A series of ubiquitin binding factors connects CDC48/p97 to substrate multiubiquitylation and proteasomal targeting. *Cell* 2005, 120:73-84.

[37] Bomont P, Cavalier L, Blondeau F, Ben Hamida C, Belal S, Tazir M, Demir E, Topaloglu H, Korinthenberg R, Tuysuz B, Landrieu P, Hentati F, Koenig M: The gene encoding gigaxonin, a new member of the cytoskeletal BTB/kelch repeat family, is mutated in giant axonal neuropathy. *Nat Genet* 2000, 26:370-374.

[38] Feuillette S, Blard O, Lecourtois M, Frebourg T, Campion D, Dumanchin C: Tau is not normally degraded by the proteasome. *J Neurosci Res* 2005, 80:400-405.

[39] Martin N, Jaubert J, Gounon P, Salido E, Haase G, Szatanik M, Guenet JL: A missense mutation in Tbce causes progressive motor neuronopathy in mice. *Nat Genet* 2002, 32:443-447.

[40] Kane RC, Bross PF, Farrell AT, Pazdur R: Velcade: U.S. FDA approval for the treatment of multiple myeloma progressing on prior therapy. *Oncologist* 2003, 8:508-513.

[41] Laser H, Mack TG, Wagner D, Coleman MP: Proteasome inhibition arrests neurite outgrowth and causes "dying-back" degeneration in primary culture. *J Neurosci Res* 2003, 74:906-916.

[42] Klimaschewski L, Hausott B, Ingorokva S, Pfaller K: Constitutively expressed catalytic proteasomal subunits are up-regulated during neuronal differentiation and required for axon initiation, elongation and maintenance. *J Neurochem* 2006, 96:1708-1717.

[43] Ohtani-Kaneko R, Takada K, Ligo M, Hara M, Yokosawa H, Kawashima S, Ohkawa K, Hirata K: Proteasome inhibitors which induce neurite outgrowth from PC12 cells cause different subcellular accumulations of multi-ubiquitin chains. *Neurochem Res* 1998, 23:1435-1443.

[44] Obin M, Mesco E, Gong X, Haas AL, Joseph J, Taylor A: Neurite outgrowth in PC12 cells. Distinguishing the roles of ubiquitylation and ubiquitin-dependent proteolysis. *J Biol Chem* 1999, 274:11789-11795.

[45] Fenteany G, Standaert RF, Reichard GA, Corey EJ, Schreiber SL: A beta-lactone related to lactacystin induces neurite outgrowth in a neuroblastoma cell line and inhibits cell cycle progression in an osteosarcoma cell line. *Proc Natl Acad Sci U S A* 1994, 91:3358-3362.

[46] Markovic SN, Geyer SM, Dawkins F, Sharfman W, Albertini M, Maples W, Fracasso PM, Fitch T, Lorusso P, Adjei AA, Erlichman C: A phase II study of bortezomib in the treatment of metastatic malignant melanoma. *Cancer* 2005, 103:2584-2589.

[47] Fornai F, Lenzi P, Gesi M, Ferrucci M, Lazzeri G, Busceti CL, Ruffoli R, Soldani P, Ruggieri S, Alessandri MG, Paparelli A: Fine structure and biochemical mechanisms underlying nigrostriatal inclusions and cell death after proteasome inhibition. *J Neurosci* 2003, 23:8955-8966.

[48] McNaught KS, Perl DP, Brownell AL, Olanow CW: Systemic exposure to proteasome inhibitors causes a progressive model of Parkinson's disease. *Ann Neurol* 2004, 56:149-162.

[49] Stokin GB, Lillo C, Falzone TL, Brusch RG, Rockenstein E, Mount SL, Raman R, Davies P, Masliah E, Williams DS, Goldstein LS: Axonopathy and transport deficits early in the pathogenesis of Alzheimer's disease. *Science* 2005, 307:1282-1288.

[50] Tu PH, Raju P, Robinson KA, Gurney ME, Trojanowski JQ, Lee VM: Transgenic mice carrying a human mutant superoxide dismutase transgene develop neuronal cytoskeletal pathology resembling human amyotrophic lateral sclerosis lesions. *Proc Natl Acad Sci U S A* 1996, 93:3155-3160.

[51] Oosthuyse B, Moons L, Storkebaum E, Beck H, Nuyens D, Brusselmans K, Van Dorpe J, Hellings P, Gorselink M, Heymans S, Theilmeier G, Dewerchin M, Laudenbach V, Vermylen P, Raat H, Acker T, Vleminckx V, Van Den Bosch L, Cashman N, Fujisawa H, Drost MR, Sciot R, Bruyninckx F, Hicklin DJ, Ince C, Gressens P, Lupu F, Plate KH, Robberecht W, Herbert JM, Collen D, Carmeliet P: Deletion of the hypoxia-response element in the vascular endothelial growth factor promoter causes motor neuron degeneration. *Nat Genet* 2001, 28:131-138.

[52] Cheng CL, Povlishock JT: The effect of traumatic brain injury on the visual system: a morphologic characterization of reactive axonal change. *J Neurotrauma* 1988, 5:47-60.

[53] Dewar D, Yam P, McCulloch J: Drug development for stroke: importance of protecting cerebral white matter. *Eur J Pharmacol* 1999, 375:41-50.

[54] Mi W, Beirowski B, Gillingwater TH, Adalbert R, Wagner D, Grumme D, Osaka H, Conforti L, Arnhold S, Addicks K, Wada K, Ribchester RR, Coleman MP: The slow Wallerian degeneration gene, WldS, inhibits axonal spheroid pathology in gracile axonal dystrophy mice. *Brain* 2005, 128:405-416.

[55] Coleman M: Axon degeneration mechanisms: commonality amid diversity. *Nat Rev Neurosci* 2005, 6:889-898.

[56] Sasaki S, Warita H, Murakami T, Abe K, Iwata M: Ultrastructural study of mitochondria in the spinal cord of transgenic mice with a G93A mutant SOD1 gene. *Acta Neuropathol (Berl)* 2004, 107:461-474.

[57] De Vrij FM, Sluijs JA, Gregori L, Fischer DF, Hermens WT, Goldgaber D, Verhaagen J, Van Leeuwen FW, Hol EM: Mutant ubiquitin expressed in Alzheimer's disease causes neuronal death. *FASEB J* 2001, 15:2680-2688.

[58] Chung KK, Dawson VL, Dawson TM: The role of the ubiquitin-proteasomal pathway in Parkinson's disease and other neurodegenerative disorders. *Trends Neurosci* 2001, 24(Suppl):S7-14.

[59] Moore DJ, West AB, Dawson VL, Dawson TM: Molecular pathophysiology of Parkinson's disease. *Annu Rev Neurosci* 2005, 28:57-87.

[60] Li H, Li SH, Yu ZX, Shelbourne P, Li XJ: Huntingtin aggregate-associated axonal degeneration is an early pathological event in Huntington's disease mice. *J Neurosci* 2001, 21:8473-8481.

[61] Murphy KP, Carter RJ, Lione LA, Mangiarini L, Mahal A, Bates GP, Dunnett SB, Morton AJ: Abnormal synaptic plasticity and impaired spatial cognition in mice transgenic for exon 1 of the human Huntington's disease mutation. *J Neurosci* 2000, 20:5115-5123.

[62] Bence NF, Sampat RM, Kopito RR: Impairment of the ubiquitin-proteasome system by protein aggregation. *Science* 2001, 292:1552-1555.

[63] Lindsten K, Menendez-Benito V, Masucci MG, Dantuma NP: A transgenic mouse model of the ubiquitin/proteasome system. *Nat Biotechnol* 2003, 21:897-902.

[64] Bowman AB, Yoo SY, Dantuma NP, Zoghbi HY: Neuronal dysfunction in a polyglutamine disease model occurs in the absence of ubiquitin-proteasome system impairment and inversely correlates with the degree of nuclear inclusion formation. *Hum Mol Genet* 2005, 14:679-691.

[65] Higashiyama H, Hirose F, Yamaguchi M, Inoue YH, Fujikake N, Matsukage A, Kakizuka A: Identification of ter94, Drosophila VCP, as a modulator of polyglutamine-induced neurodegeneration. *Cell Death Differ* 2002, 9:264-273.

[66] Raff MC, Whitmore AV, Finn JT: Axonal self-destruction and neurodegeneration. *Science* 2002, 296:868-871.

[67] Watts RJ, Hoopfer ED, Luo L: Axon pruning during Drosophila metamorphosis: evidence for local degeneration and requirement of the ubiquitin-proteasome system. *Neuron* 2003, 38:871-885.

[68] Mack TG, Reiner M, Beirowski B, Mi W, Emanuelli M, Wagner D, Thomson D, Gillingwater T, Court F, Conforti L, Fernando FS, Tarlton A, Andressen C, Addicks K, Magni G, Ribchester RR, Perry VH, Coleman MP: Wallerian degeneration of injured axons and synapses is delayed by a Ube4b/Nmnat chimeric gene. *Nat Neurosci* 2001, 4:1199-1206.

[69] Brown MC, Perry VH, Hunt SP, Lapper SR: Further studies on motor and sensory nerve regeneration in mice with delayed Wallerian degeneration. *Eur J Neurosci* 1994, 6:420-428.

[70] Levy D, Kubes P, Zochodne DW: Delayed peripheral nerve degeneration, regeneration, and pain in mice lacking inducible nitric oxide synthase. *J Neuropathol Exp Neurol* 2001, 60:411-421.

[71] Finn JT, Weil M, Archer F, Siman R, Srinivasan A, Raff MC: Evidence that wallerian degeneration and localized axon degeneration induced by local neurotrophin deprivation Do not involve caspases. *J Neurosci* 2000, 20:1333-1341.

[72] Burne JF, Staple JK, Raff MC: Glial cells are increased proportionally in transgenic optic nerves with increased numbers of axons. *J Neurosci* 1996, 16:2064-2073.

[73] Whitmore AV, Lindsten T, Raff MC, Thompson CB: The proapoptotic proteins Bax and Bak are not involved in Wallerian degeneration. *Cell Death Differ* 2003, 10:260-261.

[74] Lunn ER, Perry VH, Brown MC, Rosen H, Gordon S: Absence of Wallerian degeneration does not hinder regeneration in peripheral nerve. *Eur J Neurosci* 1989, 1:27-33.

[75] Coleman MP, Conforti L, Buckmaster EA, Tarlton A, Ewing RM, Brown MC, Lyon MF, Perry VH: An 85-kb tandem triplication in the slow Wallerian degeneration (Wlds) mouse. *Proc Natl Acad Sci U S A* 1998, 95:9985-9990.

[76] Conforti L, Tarlton A, Mack TG, Mi W, Buckmaster EA, Wagner D, Perry VH, Coleman MP: A Ufd2/D4Cole1e chimeric protein and overexpression of rbp7 in the slow wallerian degeneration (WldS) mouse. *Proc Natl Acad Sci U S A* 2000, 97:11377-11382.

[77] Araki T, Sasaki Y, Milbrandt J: Increased nuclear NAD biosynthesis and SIRT1 activation prevent axonal degeneration. *Science* 2004, 305:1010-1013.

[78] Wang J, Zhai Q, Chen Y, Lin E, Gu W, McBurney MW, He Z: A local mechanism mediates NAD-dependent protection of axon degeneration. *J Cell Biol* 2005, 170:349-355.

[79] Samsam M, Mi W, Wessig C, Zielasek J, Toyka KV, Coleman MP, Martini R: The Wlds mutation delays robust loss of motor and sensory axons in a genetic model for myelin-related axonopathy. *J Neurosci* 2003, 23:2833-2839.

[80] Sajadi A, Schneider BL, Aebischer P: Wld(s)-Mediated Protection of Dopaminergic Fibers in an Animal Model of Parkinson Disease. *Curr Biol* 2004, 14:326-330.

[81] Laser H, Conforti L, Morreale G, Mack TG, Heyer M, Haley JE, Wishart TM, Beirowski B, Walker SA, Haase G, Celik A, Adalbert R, Wagner D, Grumme D, Ribchester RR, Plomann M, Coleman MP: The Slow Wallerian Degeneration Protein, WldS, Binds Directly to VCP/p97 and Partially Redistributes It within the Nucleus. *Mol Biol Cell* 2005, 17:1075-84.

[82] Rabinovich E, Kerem A, Frohlich KU, Diamant N, Bar-Nun S: AAA-ATPase p97/Cdc48p, a cytosolic chaperone required for endoplasmic reticulum-associated protein degradation. *Mol Cell Biol* 2002, 22:626-634.

[83] Wang Q, Song C, Li CC: Molecular perspectives on p97-VCP: progress in understanding its structure and diverse biological functions. *J Struct Biol* 2004, 146:44-57.

[84] Gillingwater T, Wishart TM, Chen PE, Haley JE, Robertson K, MacDonald SH, Middleton S, Wawrowski K, Shipston MJ, Melmed S, Wyllie DJ, Skehel PA, Coleman MP, Ribchester RR: The neuroprotective WldS gene regulates expression of Pttg1 and erythroid differentiation regulator 1-like gene in mice and human cells. *Hum Mol Genet* 2006, 15:625-35.

[85] Welchman RL, Gordon C, Mayer RJ: Ubiquitin and ubiquitin-like proteins as multifunctional signals. *Nat Rev Mol Cell Biol* 2005, 6:599-609.

[86] Mahoney JA, Odin JA, White SM, Shaffer D, Koff A, Casciola-Rosen L, Rosen A: The human homologue of the yeast polyubiquitination factor Ufd2p is cleaved by caspase 6 and granzyme B during apoptosis. *Biochem J* 2002, 361:587-595.

[87] Macinnis BL, Campenot RB: Regulation of Wallerian degeneration and nerve growth factor withdrawal-induced pruning of axons of sympathetic neurons by the proteasome and the MEK/Erk pathway. *Mol Cell Neurosci* 2005, 28:430-439.

[88] Wang MS, Fang G, Culver DG, Davis AA, Rich MM, Glass JD: The WldS protein protects against axonal degeneration: a model of gene therapy for peripheral neuropathy. *Ann Neurol* 2001, 50:773-779.

[89] Zhai Q, Wang J, Kim A, Liu Q, Watts R, Hoopfer E, Mitchison T, Luo L, He Z: Involvement of the ubiquitin-proteasome system in the early stages of wallerian degeneration. *Neuron* 2003, 39:217-225.

[90] Buckmaster EA, Perry VH, Brown MC: The rate of Wallerian degeneration in cultured neurons from wild type and C57BL/WldS mice depends on time in culture and may be extended in the presence of elevated K+ levels. *Eur J Neurosci* 1995, 7:1596-1602.

[91] Glass JD, Brushart TM, George EB, Griffin JW: Prolonged survival of transected nerve fibres in C57BL/Ola mice is an intrinsic characteristic of the axon. *J Neurocytol* 1993, 22:311-321.

[92] Waller A: Experiments on the section of glossopharyngeal and hypoglossal nerves of the frog and observations uf the alternatives produced thereby in the structure of their primitive fibres. *Philos Trans R Soc Lond B Biol Sci* 1850, 140:423-429.

[93] Ferri A, Sanes JR, Coleman MP, Cunningham JM, Kato AC: Inhibiting axon degeneration and synapse loss attenuates apoptosis and disease progression in a mouse model of motoneuron disease. *Curr Biol* 2003, 13:669-673.

[94] Wang MS, Davis AA, Culver DG, Glass JD: WldS mice are resistant to paclitaxel (taxol) neuropathy. *Ann Neurol* 2002, 52:442-447.

[95] Myers RR, Heckman HM, Rodriguez M: Reduced hyperalgesia in nerve-injured WLD mice: relationship to nerve fiber phagocytosis, axonal degeneration, and regeneration in normal mice. *Exp Neurol* 1996, 141:94-101.

[96] Ramer MS, French GD, Bisby MA: Wallerian degeneration is required for both neuropathic pain and sympathetic sprouting into the DRG. *Pain* 1997, 72:71-78.

[97] Sommer C, Schafers M: Painful mononeuropathy in C57BL/Wld mice with delayed wallerian degeneration: differential effects of cytokine production and nerve regeneration on thermal and mechanical hypersensitivity. *Brain Res* 1998, 784:154-162.

[98] Fischer LR, Culver DG, Davis AA, Tennant P, Wang M, Coleman M, Asress S, Adalbert R, Alexander GM, Glass JD: The Wld(S) gene modestly prolongs survival in the SOD1(G93A) fALS mouse. *Neurobiol Dis* 2005, 19:293-300.

[99] Velde CV, Garcia ML, Yin X, Trapp BD, Cleveland DW: The Neuroprotective Factor Wlds Does Not Attenuate Mutant SOD1-Mediated Motor Neuron Disease. *Neuromolecular Med* 2004, 5:193-204.

[100] Song MS, Saavedra L, de Chaves EI: Apoptosis is secondary to non-apoptotic axonal degeneration in neurons exposed to Abeta in distal axons. *Neurobiol Aging*. 2006, 27:1224-1238.

[101] DiAntonio A, Hicke L: Ubiquitin-dependent regulation of the synapse. *Annu Rev Neurosci* 2004, 27:223-246.

[102] Beirowski B, Adalbert R, Wagner D, Grumme D, Addicks K, Ribchester RR, Coleman MP: The progressive nature of Wallerian degeneration in wild-type and slow Wallerian degeneration (WldS) nerves. *BMC Neurosci* 2005, 6:6.

[103] Parson SH, Mackintosh CL, Ribchester RR: Elimination of motor nerve terminals in neonatal mice expressing a gene for slow wallerian degeneration (C57Bl/Wlds). *Eur J Neurosci* 1997, 9:1586-1592.

[104] Bishop DL, Misgeld T, Walsh MK, Gan WB, Lichtman JW: Axon branch removal at developing synapses by axosome shedding. *Neuron* 2004, 44:651-661.

[105] Gillingwater TH, Thomson D, Mack TG, Soffin EM, Mattison RJ, Coleman MP, Ribchester RR: Age-dependent synapse withdrawal at axotomised neuromuscular junctions in Wld(s) mutant and Ube4b/Nmnat transgenic mice. *J Physiol* 2002, 543:739-755.

[106] Gillingwater TH, Ribchester RR: The relationship of neuromuscular synapse elimination to synaptic degeneration and pathology: Insights from Wld(S) and other mutant mice. *J Neurocytol* 2003, 32:863-881.

[107] Miledi R, Slater CR: On the degeneration of rat neuromuscular junctions after nerve section. *J Physiol* 1970, 207:507-528.

[108] Ribchester RR, Tsao JW, Barry JA, Asgari-Jirhandeh N, Perry VH, Brown MC: Persistence of neuromuscular junctions after axotomy in mice with slow Wallerian degeneration (C57BL/WldS). *Eur J Neurosci* 1995, 7:1641-1650.

[109] Feng G, Mellor RH, Bernstein M, Keller-Peck C, Nguyen QT, Wallace M, Nerbonne JM, Lichtman JW, Sanes JR: Imaging neuronal subsets in transgenic mice expressing multiple spectral variants of GFP. *Neuron* 2000, 28:41-51.

[110] Howland DS, Liu J, She Y, Goad B, Maragakis NJ, Kim B, Erickson J, Kulik J, DeVito L, Psaltis G, DeGennaro LJ, Cleveland DW, Rothstein JD: Focal loss of the glutamate transporter EAAT2 in a transgenic rat model of SOD1 mutant-mediated amyotrophic lateral sclerosis (ALS). *Proc Natl Acad Sci U S A* 2002, 99:1604-1609.

[111] Takahashi T, Yagishita S, Amano N, Yamaoka K, Kamei T: Amyotrophic lateral sclerosis with numerous axonal spheroids in the corticospinal tract and massive degeneration of the cortex. *Acta Neuropathol (Berl)* 1997, 94:294-299.

[112] Bacci B, Cochran E, Nunzi MG, Izeki E, Mizutani T, Patton A, Hite S, Sayre LM, Autilio-Gambetti L, Gambetti P: Amyloid beta precursor protein and ubiquitin epitopes in human and experimental dystrophic axons. Ultrastructural localization. *Am J Pathol* 1994, 144:702-710.

[113] Jortner BS, Scarratt WK, Modransky PD, Walton A, Perkins SK: Ubiquitin expression in degenerating axons of equine cervical compressive myelopathy. *Vet Pathol* 1996, 33:356-359.

[114] Giometto B, An SF, Groves M, Scaravilli T, Geddes JF, Miller R, Tavolato B, Beckett AA, Scaravilli F: Accumulation of beta-amyloid precursor protein in HIV encephalitis: relationship with neuropsychological abnormalities. *Ann Neurol* 1997, 42:34-40.

[115] Giordana MT, Richiardi P, Trevisan E, Boghi A, Palmucci L: Abnormal ubiquitination of axons in normally myelinated white matter in multiple sclerosis brain. *Neuropathol Appl Neurobiol* 2002, 28:35-41.

[116] Fukuda T, Shimizu J, Furuhata H, Abe T, Shimizu K, Oishi T, Ogihara M, Kubota J, Sasaki A, Sasaki K, Azuma T, Umemura S: Overexpression of heat shock proteins in pallido-nigral axonal spheroids of nonhuman aged primates. *Acta Neuropathol (Berl)* 2005, 110:145-150.

[117] Murakami T, Shoji M, Imai Y, Inoue H, Kawarabayashi T, Matsubara E, Harigaya Y, Sasaki A, Takahashi R, Abe K: Pael-R is accumulated in Lewy bodies of Parkinson's disease. *Ann Neurol* 2004, 55:439-442.

[118] Arima K, Nakamura M, Sunohara N, Nishio T, Ogawa M, Hirai S, Kawai M, Ikeda K: Immunohistochemical and ultrastructural characterization of neuritic clusters around ghost tangles in the hippocampal formation in progressive supranuclear palsy brains. *Acta Neuropathol (Berl)* 1999, 97:565-576.

[119] Li GL, Farooque M: Expression of ubiquitin-like immunoreactivity in axons after compression trauma to rat spinal cord. *Acta Neuropathol* 1996, 91:155-160.

[120] Spittaels K, Van den Haute C, Van Dorpe J, Bruynseels K, Vandezande K, Laenen I, Geerts H, Mercken M, Sciot R, Van Lommel A, Loos R, Van Leuven F: Prominent axonopathy in the brain and spinal cord of transgenic mice overexpressing four-repeat human tau protein. *Am J Pathol* 1999, 155:2153-2165.

[121] Furuoka H, Watanabe T, Matsui T, Narama I: Degenerative axonal swellings in the trigeminal ganglia of cattle. *Acta Neuropathol (Berl)* 1996, 92:409-414.

[122] Hoopfer ED, McLaughlin T, Watts RJ, Schuldiner O, O'Leary DDM, Luo L: Wld^S protection distinguishes axon degeneration following injury from naturally occurring developmental pruning. *Neuron* 2006, 50:883-895.

[123] Conforti L, Fang G, Beirowski B, Wang MS, Sorci L, Asress S, Adalbert R, Silva A, Bridge K, Huang XP, Magni G, Glass JD, Coleman MP: NAD^+ and axon degeneration revisited:Nmnat1 cannot substitute for Wld^S to delay Wallerian degeneration. *Cell Death Diff* 2006, Apr 28: [Epub ahead of print].

In: The Ubiquitin Proteasome System... ISBN 978-1-60021-749-4
Eds: Mario Di Napoli and Cezary Wojcik, pp. 651-670 © 2007 Nova Science Publishers, Inc.

Chapter 27

The Ubiquitin Proteasome System in Pain Transmission and Neuropathic Pain

Clare W. J. Proudfoot, Darren C. Robertson, Emer M. Garry and Susan M. Fleetwood-Walker*

Centre for Neuroscience Research, Division of Veterinary Biomedical Sciences, University of Edinburgh, EH9 1QH.

Abstract

Chronic pain states involve long-term biochemical and anatomical changes including plasticity at the first synapse in the spinal cord, which is crucial to the development of hyperalgesia (increased sensitivity to noxious stimuli) and allodynia (perception of innocuous stimuli as painful). Spinal dorsal horn neurons become hyperexcitable in the process of "central sensitisation", which shows partial similarity to other forms of synaptic plasticity, such as hippocampal long-term potentiation (LTP). Both processes involve pre- and post-synaptic changes, and rely on the NMDA receptor and associated proteins. The ubiquitin-proteasome system (UPS) has been implicated in central sensitisation and the development of neuropathic pain. In an animal model of neuropathic pain, proteasome inhibitors have been shown to rapidly attenuate behavioural hyperalgesia and allodynia, inhibited firing of dorsal horn neurons evoked by noxious and innocuous stimuli in neuropathic animals, or by mustard oil in normal animals. Expression of the enzyme UCH-L1 (Ubiquitin C-terminal hydrolase) was further increased in the spinal cord dorsal horn ipsilateral to neuropathy, supporting a central role for the UPS in neuropathic pain. Studies of other CNS areas have emphasised the importance of the UPS in regulation of synapse structure and neurotransmitter release and its role of such changes in plasticity. Postsynaptically, the UPS mediates changes in composition of the postsynaptic density (PSD) since activity-dependent ubiquitination

* Correspondence concerning this article should be addressed to Prof. Susan M Fleetwood-Walker, Centre for Neuroscience Research, Division of Veterinary Biomedical Sciences, University of Edinburgh, Edinburgh EH9 1QH, UK. Phone: +44-131-650-6091; Fax:+ 44-131-650-6576; E-mail: s.m.fleetwood-walker@ed.ac.uk.

regulates PSD composition and several key scaffolding molecules which are involved in pain sensitisation, undergo activity-dependent ubiquitination. PSD-95 in particular plays a key role in neuropathic pain. PSD-95 links to the NMDA receptor which is essential for central sensitisation and may regulate AMPA receptor synaptic insertion. NMDA receptor activation causes PSD-95 ubiquitination and degradation and blockade of this process prevents NMDA receptor induced AMPA-receptor recycling and long-term depression. Since alteration in levels of surface glutamate receptor expression is a key means by which synaptic strength is altered, these observations further support the idea of acute regulation of synapse function by the ubiquitin-proteasome pathway. In addition, proteasome inhibitors reduce NMDA receptor-dependent activation of the CREB and ERK/MAPK signalling cascades, indicating a further mechanism by which the UPS may influence long-term plasticity during chronic pain. The UPS is also involved in synapse development and in morphological changes in dendritic spines. Thus, a wide array of changes in protein: protein interactions, signalling events and cytostructural changes depend on UPS function during the synaptic plasticity that underlies chronic pain states. These processes may represent promising targets for the development of novel analgesic strategies.

Keywords: Pain, Neuropathy, Ubiquitin, Proteasome, Neuronal plasticity, Posterior horn, Postsynaptic density.

Abbreviations

CCI; chronic constriction injury model; CFA; Complete Freund's adjuvant; CNS; central nervous system; COX-2; cyclooxygenase-2; CREB; cyclic AMP response element binding protein; DRG; dorsal root ganglia; E1; ubiquitin-activating enzyme; E2; ubiquitin conjugating enzyme; E3; ubiquitin ligases, ERK; extracellular signal-regulated protein kinase; GluR; glutamate receptor (AMPA receptor subtype); GRIP; glutamate receptor - interacting protein; IKK; IκB kinase; IL; interleukin; LI/II; laminae I and II in superficial dorsal horn of spinal cord; LTP; long-term potentiation; MAGUK; membrane associated guanylate kinase; MAP; mitogen activated protein; mGluR; metabotropic glutamate receptor; NF-L; neurofilament-light; NR; NMDA receptor; NSAIDs; Non-steroidal anti-inflammatory drugs; NSF; N-ethylmaleimide-sensitive fusion protein; PICK1; protein interacting with C kinase 1; PKA; protein kinase A; PWL; paw withdrawal latency; PWT; paw withdrawal threshold; PSD; postsynaptic density; RVM; rostroventral medulla; Ser; serine reside; Siah1A; Seven in absentia homolog 1A; SPAR; spine-associated Rap GTPase activating protein, SPET; suspended paw elevation time; TNF-α; tumour necrosis factor-α; UCH-L1, Ubiquitin C-terminal hydrolase; UPS; ubiquitin-proteasome system.

Introduction

Normal, physiological pain serves a useful purpose, alerting us to damaging stimuli and triggering withdrawal reflexes, or forcing a state of rest to promote recovery from injury. On the contrary, chronic pain, which persists long after the original cause has resolved, serves no beneficial purpose and presents a major clinical problem. Neuropathic pain is a form of chronic pain which is caused by damage to the nervous system. It can occur as a result of trauma, diabetes and demyelinating diseases, viral infections (such as in postherpetic neuralgia), certain chemotherapeutic agents (all of which damage peripheral nerves) and also from central nervous system damage. Treatment options are limited [1]; conventional first-line treatments such as NSAIDs (Non-steroidal anti-inflammatory drugs) and opioids are ineffective [2]. Current prescribed medications are anti-convulsant drugs (e.g. gabapentin), sodium channel blockers and tricyclic antidepressants, all of which show variable efficacy and cause deleterious side effects. Under normal circumstances, a sensation of pain is evoked when noxious stimuli to the periphery evoke action potentials in the terminals of nociceptive-specific primary afferent neurons. These primary afferents terminate in the dorsal horn of the spinal cord, synapsing onto central spinal neurons, which form several pathways connecting to the thalamus and cortex. The synapse in the dorsal horn between primary afferent and second order neurons plays a key role in pain transmission and processing. Plasticity at this synapse is a major means by which behavioural sensitisation can develop. This sensitisation is reflected in hyperalgesia—heightened sensitivity to noxious stimuli, allodynia—perception of innocuous stimuli as noxious, and spontaneous pain. Various animal models have been developed in an effort to reproduce the sensory disorders accompanying human peripheral neuropathies. Partial nerve injury models mimic the most common type of injury seen in humans. Such models are highly reproducible and result in the hallmark behavioural signs of neuropathic pain; hyperalgesia and allodynia. Sensitisation as a result of plasticity in pain pathways occurs normally following injury, but is short-term and reverses as the injury heals. However, in neuropathic pain, continuing modification of the pain pathways, particularly at the early synapses in the dorsal horn, results in long-term sensitisation, which far outlasts the injury. Damage to the peripheral nerves causes some axons to degenerate, and remaining neurons display altered properties including spontaneous and ectopic firing. Phenotypic changes occur in damaged and neighbouring afferents with alterations in expression of ion channels, receptors, and levels of neurotransmitters, and *de novo* expression of certain proteins. In postsynaptic neurons, levels of receptors and cellular proteins are also altered. These changes result in central sensitisation, a form of synaptic plasticity with some analogies to hippocampal long-term potentiation (LTP) and to other forms of central nervous system plasticity resulting in a facilitation of neurotransmission due to repetitive conditioning stimuli (see Chapter 14-17; Figure 1). Activity-dependent sensitisation has been reported in rodent, cat and primate dorsal horn neurons [3,4]. A central sensitisation-like phenomenon can also be generated in other supraspinal pain processing areas: the rostroventral medulla (RVM), anterior cingulate cortex, and amygdala [5-7].

LTP and central sensitisation are both dependent on excitatory glutamatergic transmission involving AMPA and NMDA subtypes of ionotropic glutamate receptor [8-12]. AMPA receptors are composed of four subunits (GluR1-4), while NR1, NR2 (NR2A-D) and NR3 (NR3A/B) subunits all constitute the NMDA receptor family. All these subunits are differentially localised in the spinal cord with GluR1 and GluR2 subunits of AMPA receptor and NR1 and NR2B subunits of NMDA receptor being most concentrated in the major nociceptive processing area in the superficial dorsal horn (see later). Under basal conditions, acute nociceptive transmission is mediated by AMPA receptors, while NMDA receptors are inactive, due to blockade by Mg^{2+} ions in the NMDA channel pore. Under conditions of nerve injury, high-frequency stimulation produces a cumulative membrane depolarisation which leads to removal of the voltage-dependent Mg^{2+} ion block in the channel pore, allowing entry of calcium ions and resultant calcium-dependent intracellular signalling. Metabotropic glutamate receptors (mGluR1-7) have also been implicated in synaptic plasticity in hippocampus and cerebellum as well as spinal neuronal sensitisation associated with chronic pain.

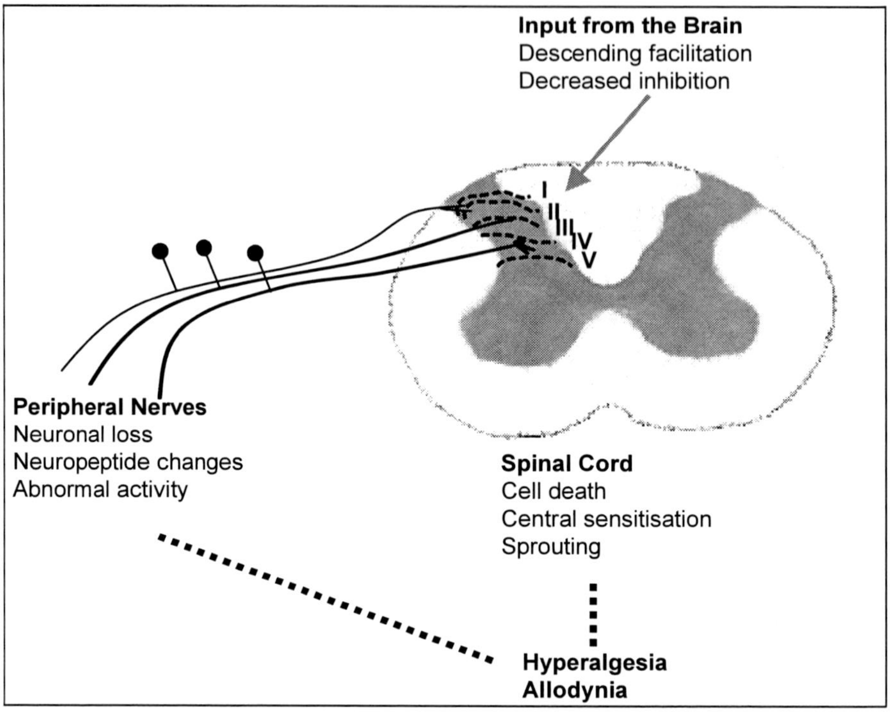

Figure 1. Peripheral and central mechanisms of neuropathic pain. Changes at the level of peripheral afferents, the dorsal horn of the spinal cord, and descending input from brain areas. All lead to the development of hyperalgesia (heightened sensitivity to noxious stimuli) and allodynia (perception of innocuous stimuli as painful) in neuropathic pain.

THE UBIQUITIN-PTOTEASOME SYSTEM (UPS)

A major intracellular pathway that is now known to act as an important regulator of such synaptic function is the ubiquitin-proteasome system [13-16]. The UPS is a highly conserved multi-enzyme mechanism that targets cytosolic proteins for degradation [17]. A sequential series of reactions is catalysed by ubiquitin-activating enzyme (E1), followed by ubiquitin conjugating enzymes (E2) and a number of ubiquitin ligases (E3), resulting in ubiquitin binding to lysine residues in the target protein [18]. Activity-dependent ubiquitination and degradation by the proteasome proteolytic complex is a mechanism by which synaptic proteins and thus synaptic strength may be regulated. Levels of ubiquitin-conjugated proteins are strongly linked to synaptic activity level, and this is particularly pronounced for postsynaptic density (PSD) proteins [19]. This pathway is involved in the regulation of synaptic plasticity in the hippocampus, whereby the proteasome inhibitors MG-132 and lactacystin can block long-term depression, a further form of lasting alteration in synaptic function [16].

Localisation of the UPS Near Synapses

The UPS has been localised intracellularly in the vicinity of the synapse, making it a plausible regulator of synaptic function. Confocal microscopy demonstrates that ubiquitin and the proteasome are abundant in dendrites of cultured hippocampal neurons, in or near putative spines [15]. Ubiquitin conjugation has been demonstrated in synaptosomal and PSD preparations, indicating that ubiquitination can occur in or near synapses [19]. It is possible that dynamic recruitment of UPS components to synaptic sites could act as a regulator of function [15]. The UPS is directly involved in affecting growth of dendritic spines by a mechanism involving degradation of the protein SPAR (spine-associated Rap GTPase activating protein) which is a postsynaptic scaffolding molecule that binds both filamentous actin and the MAGUK (membrane-associated guanylate kinase) family adapter protein prominent in postsynaptic densities, PSD-95. Degradation of SPAR by the UPS is activity-dependent, and results in depletion of both SPAR and PSD-95 from spines, leading to dissipation of spines [20].

The UPS and Neuropathic Pain

The UPS is specifically implicated in the sensitisation that occurs in the spinal cord in chronic pain [21]. Extracellular recordings of multireceptive dorsal horn neurons show heightened neuronal activity following the topical application of the chemical algogen mustard oil (allyl isothiocyanate), a selective activator of nociceptive C-fibers [22]. Ionophoresis of the ubiquitin-proteasome inhibitors lactacystin or MG-132 *in vivo* partially inhibits the mustard oil-induced increase in firing but had no effects on responses to an innocuous brush stimulus [21]. This indicates that the UPS is preferentially involved in the processing of noxious (mustard oil) but not innocuous (brush) sensory stimuli in the spinal cord under normal conditions (Table 1).

The sciatic nerve chronic constriction injury model (CCI) produces sensitisation in the neurons of the spinal dorsal horn and the behavioural correlates of hyperalgesia and allodynia indicative of neuropathic pain [23]. Spinal dorsal horn neurons from neuropathic animals show elevated responses to brush as well as mustard oil, which is representative of mechanical allodynia, but there is also a *de novo* response to noxious (4°C) cold stimulation, known as 'cold allodynia'. In extracellular recordings from the spinal neurons of neuropathic animals, the same proteasome inhibitors lactacystin and MG-132 could significantly inhibit elevated neuronal responsiveness evoked by noxious stimulation with mustard oil or cold stimulation as well as that evoked by innocuous brush stimulation [21]. This further indicates that ubiquitin is involved in the processing of noxious sensory stimuli, which, following nerve injury, includes not only mustard oil, but now also brush and cold stimulation (Table 1).

Table 1. Summary of effects of proteasome inhibitors on sensory stimulation-induced dorsal horn neuronal activity

	Normal			Nerve injury		
Stimulus type	Innocuous	Noxious		Innocuous	Noxious	
	Brush	Mustard Oil	Cold	Brush	Mustard Oil	Cold
Lactacystin	-	↓	n/a	↓	↓	↓
MG-132	-	↓	n/a	↓	↓	↓

Summary of the effects of the ionophoretic application of proteasome inhibitors lactacystin and MG-132 on dorsal horn neuron responses to innocuous (brush) or noxious (mustard oil or cold) sensory stimulation. In naïve animals, neither drug had any effect (-) on non-nociceptive brush-evoked firing, while both drugs were capable of reducing (↓) neuronal firing induced by the noxious mustard oil stimulus. The cold (4°C) stimulus was not tested (n/a) in normal animals, as it elicits behavioural reflex withdrawal responses only in neuropathic animals. Thus normally, the UPS appears to contribute only to the processing of nociceptive sensory stimuli. In neuropathic animals, proteasome inhibitors are now capable of reducing dorsal horn responses evoked not only by nociceptive mustard oil and newly developed nociceptive responses to cold stimulation but also responses evoked by non-nociceptive brush stimulation. Thus, in the sensitised state following nerve injury, the UPS is now involved in processing noxious and innocuous sensory stimuli.

In agreement with the electrophysiological results, administration of the ubiquitin-proteasome inhibitors MG-132 and epoxomicin into the intrathecal space reversed whole animal behavioural thermal hyperalgesia and mechanical and cold allodynia which develop ipsilateral to nerve injury, but had no effect in naïve animals or on uninjured contralateral paw values [Figure 2; 21]. In both behavioural and electrophysiological experiments, proteasome inhibitors exerted an effect within 20 minutes of application. Such rapidity of action suggests that UPS activity is essential for the maintenance of established neuropathic sensitisation.

Biochemical changes were also observed consistent with an involvement of the UPS in the neuropathic pain state. RT-PCR, in situ hybridisation, and Western blot analysis of the rat homologue of ubiquitin C-terminal hydrolase, UCH-L1, (also referred to as PGP 9.5), showed a significant elevation of the expression of mRNA and protein in the spinal cord ipsilateral to nerve injury with a time course paralleling the development of sensitised behavioural reflex withdrawal responses [21]. Thus, there is a clear involvement of the UPS in chronic pain states. But what are the key targets that have implications for spinal cord sensitisation?

POTENTIAL TARGETS OF THE UPS IN NEUROPATHIC PAIN

Protein Kinase A (PKA)

One possible target is protein kinase A (PKA). PKA is composed of regulatory and catalytic subunits, whereby the regulatory subunits normally maintain the kinase in an inactive state. Upon activation, the regulatory subunits of PKA dissociate and there is a translocation to the cell nucleus of the now constitutively active catalytic subunit [24]. Ubiquitin-directed proteolysis is known to degrade the regulatory subunits, leading to persistent kinase activation [13,24]. PKA is involved in pain transmission, as activation of PKA causes increased dorsal horn neuron responsiveness and leads to hyperalgesia [25-27] and mice with a knockout of specific regulatory subunits of PKA have reduced responses in the formalin model of inflammation [28]. Furthermore, enzymatic activity of PKA in the spinal cord is enhanced following nerve injury [21]. This increase is blocked by local spinal application of the ubiquitin proteasome inhibitors epoxomicin and MG-132 [21]. Thus, UPS regulation of PKA is important for nociceptive transmission.

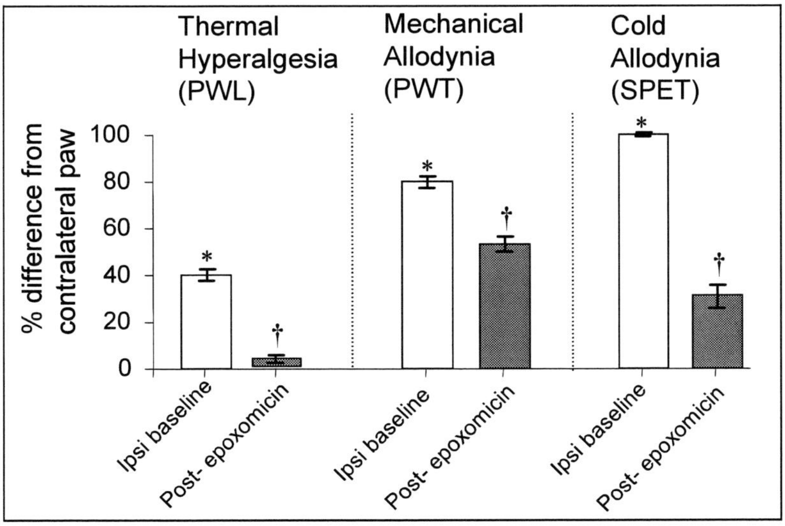

Figure 2. Reversal of neuropathic pain behaviour following epoxomicin treatment. The bar chart shows mean percentage difference in nerve-injured ipsilateral (ipsi) paw from uninjured contralateral paw, at pre-drug baseline (unshaded columns), and 10-25 minutes following intrathecal administration of the proteasome inhibitor epoxomicin (shaded columns). Data illustrate the percentage change ipsilateral to nerve injury (compared to the contralateral side) in thermal sensitivity (PWL – paw withdrawal latency to noxious thermal stimuli: approximately 40% reduction), mechanical sensitivity (PWT – paw withdrawal threshold to calibrated mechanical stimuli; approximately 80% reduction) and cold allodynia (SPET – Suspended Paw Elevation Test to 4°C water; response seen only on the ipsilateral side). Data are shown as mean ± standard error of the mean. * represents significant ipsilateral-contralateral difference ($p<0.05$, t-test for thermal, Wilcoxon test for mechanical/cold), † represents significant effect of drug ($p<0.05$, One-Way repeated measures ANOVA followed by Dunnett's post-hoc test for thermal, Friedman ANOVA on ranks followed by Dunn's post-hoc test for mechanical/cold). Nerve-injury induced sensitisation is substantially reduced by epoxomicin administration, as indicated by thermal, mechanical and cold sensory tests.

A number of downstream targets of PKA have been identified that are likely to be critical to spinal cord sensitisation and neuropathic pain states and these notably include NMDA and AMPA type glutamate receptors. The C-terminal domains of the NR1 and NR2 subunits of NMDA receptors contain sites of PKA phosphorylation that may increase the degree of Ca^{2+} entry evoked by NMDA receptor stimulation [29,30,31,32]. Electrophysiological responses of spinal neurons to NMDA are known to be facilitated by PKA [12,25,26]. NR1 phosphorylation at serine residue 897 (Ser^{897}) is increased in spinal cord following CCI and in spinal dorsal horn spinothalamic tract cells following capsaicin injection [33,34]. However, it has not yet been established whether there are any PKA regulatory affects on NMDA receptors involving the UPS.

PKA can also phosphorylate serine residue 845 (Ser^{845}) located in the intracellular C-terminal domain of the GluR1 AMPA receptor subunit [35,36]. Phosphorylation at Ser^{845} increases AMPA receptor responsiveness through a modulation of channel gating and channel open probability without affecting channel conductance [37]. GluR1 subunit phosphorylation at Ser^{845} is known to occur in the superficial (LI/II) spinal dorsal horn in response to noxious stimulation with capsaicin [38,39] and following peripheral nerve injury [Garry *et al.*, unpublished observations]. This modulation by PKA is required for the insertion of GluR1-containing receptors into synapses [40]. Correspondingly, the adenylyl cyclase activator forskolin can also elicit recruitment of functional AMPA receptors [16].

NMDA Receptors

NMDA receptors are critical to the spinal sensitisation and hyperexcitability that occurs following injury [41-43]. In the spinal cord, NMDA receptor blockade reduces frequency-dependent potentiation (wind-up) of cells to repeated C-fiber stimulation and following mustard oil application [44,45]. The NR1 subunit of the NMDA receptor is ubiquitinated by Fbx2, and binding of Fbx2 to the subunit is activity-dependent [46], suggesting that the UPS regulates the NMDA receptor in an activity-dependent manner. Ubiquitination of NR1 in HEK293 cells is blocked by co-expression with the neuronal filament protein, neurofilament-light (NF-L), [47] which has implications for NMDA receptor stability in the PSD [48].

NMDA receptor activation triggers intracellular signalling cascades, which are involved in long-term plasticity. These include two prominent pathways through the nuclear transcription factor cyclic AMP response element binding protein (CREB) and the mitogen-activated ERK-MAP kinase pathways. The CREB pathway activates transcriptional changes, leading to long-term modification of neuronal biochemistry and function. As well as its own profile of transcriptional changes, the MAP kinase pathway causes short-term modification of proteins via phosphorylation, and also activation of CREB [49]. Both pathways are associated with development of chronic pain states. Phosphorylation of p38 MAP kinase and p42/44(ERK1/2) MAP kinase is induced in the ipsilateral dorsal horn of the spinal cord and dorsal root ganglia following peripheral nerve injury (or inflammation) and inhibition of these pathways prevents the behavioural sensitisation seen following nerve injury [50-54]. Increases in phosphorylation of CREB in the spinal cord are observed following partial sciatic nerve ligation [55] and following chronic constriction injury, with a time-course paralleling development of thermal hyperalgesia [56]. Both pre-treatment with proteasome inhibitors and synaptic blockade augmented NMDA receptor dependent activation of ERK-MAP kinase, and inhibited activation of CREB [57]. CREB is also a prominent target of PKA, which may be activated by a variety of other non-NMDA receptor inputs and which is UPS-regulated (see above). Thus the UPS influences a variety of long-term transcriptional changes through modification of signalling cascades.

AMPA Receptors

AMPA receptors are involved in basal nociceptive responses but they are also critical in mediating sensitised sensory responses [58]. Agonists of the AMPA receptor result in sensitisation of behavioural nociceptive responses and increased activity of dorsal horn neurons [59,60]. AMPA receptor antagonists alter acute nociceptive withdrawal responses and reverse nerve-injury induced pain [9,58,61,62]. Virtually all dorsal horn neurons express GluR2 predominantly in the superficial (LI/II) dorsal horn, where it is known to increase following injury [38,58,63]. GluR1 expression is more restricted in the superficial dorsal horn LI/II where it has been localised at primary afferent synapses due to association with markers of unmyelinated afferents [38,63]. GluR3 and GluR4 are expressed in ventral and deep dorsal (LIII-VI) horn with weak expression in the superficial dorsal horn and are thus less likely to play a role in pain plasticity [38,63-68].

Unlike NMDA receptors, AMPA receptors are rapidly turned over [69,70], and agonist binding may cause receptor internalisation and breakdown [14,71]. Insertion and removal of AMPA receptor subunits is now thought to be a major component of synaptic plasticity such that AMPA receptor abundance determines synaptic strength [69,72,73]. AMPA receptors undergo rapid endocytosis followed by recycling or degradation in response to ligand binding, NMDA receptor activation and during long-term depression [14]. Excitatory transmission is depressed following AMPA receptor internalisation [71] and AMPA receptor trafficking from the cytosol to the membrane is thought to be important in hippocampal LTP in culture [73]. It is generally believed that GluR2/3 containing channels are constitutively inserted into the synapse while GluR1/4 containing receptor insertion is inducible (eg. by LTP) [74]. Painful chemical stimulation of the colon or somatosensory nerve injury can cause delivery of GluR1 but not GluR2/3 from cytosol to the plasma membrane in the spinal cord [76, Garry unpublished observations]. However, there is a nerve injury- and AMPA receptor stimulation-induced internalisation of spinal GluR2 that can be prevented by inhibitors of clathrin-coated vesicle endocytosis [58], a mechanism known to regulate AMPA receptor trafficking [75].

AMPA receptor surface expression is subject to regulation by the UPS [15,16,77,79]. There is ubiquitination of GLR-1, the *C. Elegans* form of mammalian GluR1, and this has been proposed as a signal for subunit endocytosis [77]. Also, AMPA- (or NMDA) receptor activation stimulated internalisation of GluR1 and GluR2 in hippocampal dendrites and spinal cord [14,15,58,78] is prevented by pre-treatment with the proteasome inhibitors MG-132 and lactacystin [15]. These compounds inhibit behavioural reflex sensitisation and suppress dorsal horn neuron hypersensitivity that occurs in response to peripheral nerve injury and noxious stimulation, as mentioned above [21]. This suggests that AMPA receptor trafficking requires UPS-dependent degradation, although no direct ubiquitination of AMPA receptors has been observed [16]. Nevertheless, ubiquitination may occur via a number of intracellular proteins in the PSD that interact with AMPA receptor subunits and are subject to degradation by the UPS [57].

GLUTAMATE RECEPTOR-INTERACTING PROTEINS

While AMPA receptor plasticity is thought to involve changes in the number of receptors at the synapse [72], NMDA receptor plasticity involves a complex of adapter and signalling proteins in the PSD [33,80].

The PSD matrix is responsible for the organisation of neurotransmitter receptors, scaffold proteins and signalling enzymes and is dynamically regulated in response to activity. At least some of these changes are mediated by the UPS. Profound changes in the PSD occur in response to prolonged activity, and these are prevented by proteasome inhibition [19]. Indeed, a diverse range of PSD proteins are controlled by UPS-dependent changes.

One such protein is PSD-95, an NMDA receptor-interacting protein that forms intermolecular complexes with signalling and other molecules to regulate receptor function. PSD-95 has a critical role in the development of neuropathic reflex behaviours following nerve injury [33]. Mice with a mutation in PSD-95 that disrupts connections to intracellular signalling cascades but retains NMDA receptor interacting abilities fail to develop the expected sensitised responses following CCI-induced nerve injury but display normal responses to the formalin model of chronic inflammation [33]. This represented the identification of one of the first proteins underlying a differential processing capacity for different types of chronic pain [81].

PSD-95 can also interact with AMPA receptors to promote their surface expression whereby over-expression of PSD-95 increases both AMPA receptor currents and the number of AMPA receptors at the synapse by inhibition of AMPA receptor internalisation [82-87]. Stimulation of AMPA receptors leads to a reduction in the number and intensity of PSD-95 puncta in hippocampal dendrites that is sensitive to the UPS inhibitor, MG-132 [82]. This could provide a mechanism for UPS-dependent modulation of AMPA receptor synaptic accumulation. Furthermore, activation of NMDA receptors causes ubiquitination and subsequent degradation of PSD-95 that is prevented in the presence of MG-132 and lactacystin [16]. However, there is some controversy as to the conditions under which ubiquitination of PSD-95 might occur, i.e. basally or in response to NMDA receptor stimulation [16,57,82]. NMDA receptor stimulation also causes a reduction in the number of synaptic AMPA receptors that is blocked when PSD-95 ubiquitination is prevented [14,16,75,88,89]; MG-132 can block the loss in synaptic GluR1 subunits that occurs following NMDA receptor activation [15,16,78].

PSD-95 interacts indirectly with AMPA receptors through the surface protein stargazin [84,86]. This trimeric interaction complex is thought to mediate surface expression and synaptic targeting of AMPA receptors [84,90]. The PDZ binding site of stargazin is regulated by PKA and mutations that mimic this phosphorylation site (T321E and T321D) can downregulate AMPA receptor function [90,91].

The GluR2 subunit interacts with a variety of intracellular proteins via sequences in its C-terminal tail. GRIP (glutamate receptor -interacting protein) and PICK1 (protein interacting with C kinase 1) have been implicated in AMPA receptor endocytosis [71,92] and receptor clustering is decreased in cultured spinal neurons with blockade of these interactions [93]. Consistent with this, there is translocation of both GRIP and PICK1 in the spinal cord following AMPA receptor stimulation [58]. Local spinal administration of interfering peptides to block interactions with PICK1 can reverse CCI-induced behavioural reflex sensitisation [58]. The related protein GRIP is also translocated from the membrane in spinal cord following AMPA receptor stimulation [58]. The translocation of both of these intracellular proteins is prevented in the presence of inhibitors of the 'clathrin-coated vesicle' pathway / clathrin mediated endocytosis [58]. However, any involvement of the UPS in this process has not yet been established.

Another GluR2 interacting protein, NSF (N-ethylmaleimide-sensitive fusion protein) is involved in receptor exocytosis [89,94] and blockade of this interaction in the spinal cord reverses hyperalgesia and allodynia that results from peripheral nerve injury [58]. This interaction is thought to be necessary for normal synaptic expression of AMPA receptors as blocking this interaction in culture results in a rundown of synaptic currents in hippocampal neurons in association with reduced synaptic expression of AMPA receptors [94,95]. Activity-dependent insertion of AMPA receptors into functionally 'silent' synapses that only express functional NMDA (but not AMPA) receptors may be responsible for the expression of LTP [96,97]. Again, it is not known if the UPS is involved in the regulation of AMPA receptor interactions with NSF.

METABOTROPIC GLUTAMATE RECEPTORS

Metabotropic glutamate receptors also play an important role in pain processing, particularly at the level of the spinal cord. Group I mGluRs (mGluR1 and mGluR5) are implicated in the development of sensitised pain states [98], and Group II (mGluR 2 and 3) and Group III (mGluR4 and 7) receptors are implicated in inhibition of pain signalling [99,100-102]. Of these, the Group I mGluRs have been shown to be ubiquinated and therefore targeted for degradation by the E3 protein Siah1A (Seven in absentia homolog 1A), which binds to long splice forms of mGluR1a and mGluR5 [103]. Other mGluRs may also be potential targets for ubiquitination and thereby regulation by the UPS.

THE UPS AND NF-κB SIGNALLING

The UPS regulates gene expression via its interaction with the transcription factor NF-κB [104]. NF-κB normally exists in a dimeric form complexed with the inhibitory regulator, IκB [105]. The IκB complex prevents the nuclear import of NF-κB, holding it in the cytoplasm and thereby inhibiting its transcriptional activity [105]. This interaction is mediated by ankyrin repeats and the ubiquitination and subsequent degradation of IκB by the proteasome is key to NF-κB regulation. The phosphorylation of IκB by IκB kinase (IKK) at two N-terminal serine residues is the event which precipitates the polyubiquitination of IκB and therefore its degradation [106]. Genes over which NF-κB exerts control and are relevant to neuropathic pain include, proinflammatory enzymes [inducible nitric oxide synthase, cyclooxygenase-2 (COX-2)], cytokines such as tumour necrosis factor (TNF)-α, interleukin (IL)-1β, IL-6, chemokines (RANTES, macrophage inflammatory protein, IL-8), adhesion factors (intracellular adhesion molecule, vascular cell adhesion molecule) and proteins involved in controlling apoptosis (Fas, FasL) [107]. Recent evidence has demonstrated that the regulation of NF-κB by the UPS has direct relevance to neuropathic pain as in the CCI model of nerve injury NF-κB activity is increased in rat lumbar DRG ipsilateral to injury [108]. Furthermore, intraneural introduction of competitive inhibitors for NF-κB has been shown to reduce hyperalgesia following spinal nerve ligation [109]. In a recent study we have shown that glial release of cytokines including TNF-α plays a key role in the sensitisation of spinal cord neurons in neuropathic pain [50]. Subcutaneous injection of Complete Freund's adjuvant (CFA), a model of inflammatory pain, has also been shown to cause the immediate activation of spinal NF-κB [110].

In the CFA model, pre-treatment with intrathecally administered inhibitors of NF-κB significantly reduces mechanical allodynia and thermal hyperalgesia [111]. Similarly the IκB kinase inhibitor, S1627 has been shown to reduce or reverse allodynia and hyperalgesia in both the zymosan-induced model of inflammatory pain and the CCI model, by preventing NF-κB dependant gene expression [107]. These changes in UPS function are likely to have marked effects on NF-κB-mediating signals that bring about central changes in protein expression, which support the development of chronic pain states.

Conclusions

It is clear that the UPS plays an important part in the sensitization of spinal cord neurons in the neuropathic pain state. Intrathecally administered UPS inhibitors display striking reversal of neuropathic sensitization even when it has become previously established, so may have potential as a novel form of analgesic for this currently intractable pain state. We and others have identified a number of key proteins that are implicated in synaptic plasticity in either spinal cord or forebrain and whose degradation or function is regulated by the UPS. It is unlikely that the role of the UPS in neuropathic pain comprises multiple components with the function of NMDA, AMPA and metabotropic glutamate receptors and major intracellular signalling pathways such as PKA and NF-κB being modulated by UPS activity. Important questions for the future will be unravelling which of these potential contributions are function-limiting in neuropathic pain, whether specific proteins or targeting mechanisms underlie dynamic actions of the UPS in synaptic plasticity and whether agents perhaps targeting such specific actions of the UPS could be developed as useful analgesics. In principle, ubiquitin-proteasome inhibitors would be expected to be beneficial when acutely applied in chronic neuropathic pain states, however, some clinical studies indicate that the first clinically available ubiquitin-proteasome inhibitor, the boronic acid derivative, bortezomib actually caused peripheral neuropathy de novo or perhaps exacerbated a pre-existing peripheral neuropathy, such as Charcot-Marie-Tooth disease [112]. It is not clear if this neuropathy develops because the agent is blocking the UPS chronically rather than acutely or because it is a side effect of this particular drug, as is found for example, with some other chemotherapeutic drugs such as the vinca alkaloid, vincristine [113,114]. Only trials with other ubiquitin-proteasome inhibitors of different chemical structures, but the same effect, will reveal this.

REFERENCES

[1] Hansson PT, Dickenson AH: Pharmacological treatment of peripheral neuropathic pain conditions based on shared commonalities despite multiple etiologies. *Pain* 2005, 113:251-254.

[2] Arner S, BA Meyerson: Lack of analgesic effect of opioids on neuropathic and idiopathic forms of pain. *Pain* 1988, 33:11-23.

[3] Kenshalo DR, Jr., Leonard RB, Chung JM, Willis WD: Facilitation of the response of primate spinothalamic cells to cold and to tactile stimuli by noxious heating of the skin. *Pain* 1982, 12:141-152.

[4] Dougherty PM, Willis WD: Enhanced responses of spinothalamic tract neurons to excitatory amino acids accompany capsaicin-induced sensitization in the monkey. *J Neurosci* 1992, 12:883-894.

[5] Neugebauer V, Li W, Bird GC, Bhave G, Gereau RWt: Synaptic plasticity in the amygdala in a model of arthritic pain: differential roles of metabotropic glutamate receptors 1 and 5. *J Neurosci* 2003, 23:52-63.

[6] Porreca F, Ossipov MH, Gebhart GF: Chronic pain and medullary descending facilitation. *Trends Neurosci* 2002, 25:319-325.

[7] Wei F, Zhuo M: Potentiation of sensory responses in the anterior cingulate cortex following digit amputation in the anaesthetised rat. *J Physiol* 2001, 532:823-833.

[8] Bliss TV, Collingridge GL: A synaptic model of memory: long-term potentiation in the hippocampus. *Nature* 1993, 361:31-39.

[9] Mao J, Price DD, Hayes RL, Lu J, Mayer DJ: Differential roles of NMDA and non-NMDA receptor activation in induction and maintenance of thermal hyperalgesia in rats with painful peripheral mononeuropathy. *Brain Res* 1992, 598:271-278.

[10] Mao J, Price DD, Hayes RL, Lu J, Mayer DJ, Frenk H: Intrathecal treatment with dextrorphan or ketamine potently reduces pain-related behaviors in a rat model of peripheral mononeuropathy. *Brain Res* 1993, 605:164-168.

[11] Tal M, Bennett GJ: Dextrorphan relieves neuropathic heat-evoked hyperalgesia in the rat. *Neurosci Lett* 1993, 151:107-110.

[12] Heppenstall PA, Fleetwood-Walker SM: The glycine site of the NMDA receptor contributes to neurokinin1 receptor agonist facilitation of NMDA receptor agonist-evoked activity in rat dorsal horn neurons. *Brain Res* 1997, 744:235-245.

[13] Hegde AN, Inokuchi K, Pei W, Casadio A, Ghirardi M, Chain DG, Martin KC, Kandel ER, Schwartz JH: Ubiquitin C-terminal hydrolase is an immediate-early gene essential for long-term facilitation in Aplysia. *Cell* 1997, *89:*115-126.

[14] Ehlers, MD: Reinsertion or degradation of AMPA receptors determined by activity-dependent endocytic sorting. *Neuron* 2000, 28:511-525.

[15] Patrick GN, Bingol B, Weld HA, Schuman EM: Ubiquitin-mediated proteasome activity is required for agonist-induced endocytosis of GluRs. *Curr Biol* 2003, 13:2073-2081.

[16] Colledge M, Snyder EM, Crozier RA, Soderling JA, Jin Y, Langeberg LK, Lu H, Bear MF, Scott JD: Ubiquitination regulates PSD-95 degradation and AMPA receptor surface expression. *Neuron* 2003, 40:595-607.

[17] Glickman MH, Ciechanover A. The ubiquitin-proteasome proteolytic pathway: destruction for the sake of construction. *Physiol Rev* 2002, 82:373-428.

[18] Hochstrasser M: Biochemistry. All in the ubiquitin family. Science 2000, 289:563-564.

[19] Ehlers MD: Eppendorf 2003 prize-winning essay. Ubiquitin and the deconstruction of synapses. *Science* 2003, 302:800-801.

[20] Pak DT, Sheng M: Targeted protein degradation and synapse remodeling by an inducible protein kinase. *Science* 2003, 302:1368-1373.

[21] Moss A, Blackburn-Munro G, Garry EM, Blakemore JA, Dickinson T, Rosie R, Mitchell R, Fleetwood-Walker SM: A role of the ubiquitin-proteasome system in neuropathic pain. *J Neurosci* 2002, 22:1363-1372.

[22] Woolf CJ, Wall PD: Relative effectiveness of C primary afferent fibers of different origins in evoking a prolonged facilitation of the flexor reflex in the rat. *J. Neurosci.* 1986, 6:1433-1442.

[23] Bennett GJ, Xie YK: A peripheral mononeuropathy in rat that produces disorders of pain sensation like those seen in man. *Pain* 1988, 33:87-107.

[24] Chain DG, Schwartz JH, Hedge AN: Ubiquitin-mediated proteolysis in learning and memory. *Mol Neurobiol* 1999, 20:125-142.

[25] Cerne R, Jiang M, Randić M: Cyclic adenosine 3'5'-monophosphate potentiates excitatory amino acid and synaptic responses of rat spinal dorsal horn neurons. *Brain Res* 1992, 596:111-123.

[26] Cerne R, Rusin KI, Randić M: Enhancement of the *N*-methyl-D-aspartate response in spinal dorsal horn neurons by cAMP-dependent protein kinase. *Neurosci Lett* 1993, 161:124-128.

[27] Sluka KA: Activation of the cAMP transduction cascade contributes to the mechanical hyperalgesia and allodynia induced by intradermal injection of capsaicin. *Br J Pharmacol* 1997, 122:1165-1173.

[28] Malmberg AB, Brandon EP, Idzerda RL, Liu H, McKnight GS, Basbaum AI: Diminished inflammation and nociceptive pain with preservation of neuropathic pain in mice with a targeted mutation of the type I regulatory subunit of cAMP-dependent protein kinase. *J Neurosci* 1997, 17:7462-7470.

[29] Tingley WG, Ehlers MD, Kameyama K, Doherty C, Ptak JB, Riley CT, Huganir RL: Characterization of protein kinase A and protein kinase C phosphorylation of the *N*-

methyl-D-aspartate receptor NR1 subunit using phosphorylation site-specific antibodies. *J Biol Chem* 1997, 272:5157-5166.

[30] Leonard AS, Hell JW: Cyclic AMP-dependent protein kinase and protein kinase C phosphorylate *N*-methyl-D-aspartate receptors at different sites. *J Biol Chem* 1997, 272:12107-12115.

[31] Lau LF, Huganir RL: Differential tyrosine phosphorylation of *N*-methyl-D-aspartate receptor subunits. *J Biol Chem* 1995, 270:20036-20041.

[32] Westphal RS, Tavalin SJ, Lin JW, Alto NM, Fraser IDC, Langeberg LK, Sheng M, Scott JD: Regulation of NMDA receptors by an associated phosphatase-kinase signaling complex. *Science* 1999, 285:93-96.

[33] Garry EM, Moss A, Delaney A, O'Neill F, Blakemore J, Husi H, Mitchell R, Grant SG, Fleetwood-Walker SM: Neuropathic sensitization of behavioural reflexes and spinal NMDA receptor/CaM kinase II interactions are disrupted in PSD-95 mutant mice. *Curr Biol* 2003, 13:321-328.

[34] Zou X, Lin Q, Willis WD: Enhanced phosphorylation of NMDA receptor 1 subunits in spinal cord dorsal horn and spinothalamic tract neurons after intradermal injection of capsaicin in rats. *J Neurosci* 2000, 20: 6989-6997.

[35] Roche KW, O'Brien RJ, Mammen AL, Bernhardt J, Huganir RL: Characterization of multiple phosphorylation sites on the AMPA receptor GluR1 subunit. *Neuron* 1996, 16:1179-1188.

[36] Mammen AL, Kameyama K, Roche KW, Huganir RL: Phosphorylation of the α-3-hydroxy-5-methylisoxazole4-propionic acid receptor GluR1 subunit by calcium/ calmodulin-dependent kinase II. *J Biol Chem* 1997, 272: 32528-32533.

[37] Banke TG, Bowie D, Lee H, Huganir RL, Schousboe A, Traynelis SF: Control of GluR1 AMPA receptor function by cAMP-dependent protein kinase. *J Neurosci* 2000, 20:89-102.

[38] Nagy GG, Al-Ayyan M, Andrew D, Fukaya M, Watanabe M, Todd AJ: Widespread expression of the AMPA receptor GluR2 subunit at glutamatergic synapses in the rat spinal cord and phosphorylation of GluR1 in response to noxious stimulation revealed with an antigen-unmasking method. *J Neurosci*. 2004, 24:5766-5777.

[39] Fang L, Wu J, Zhang X, Lin Q, Willis WD: Increased phosphorylation of the GluR1 subunit of spinal cord alpha-amino-3-hydroxy-5-methyl-4-isoxazole propionate receptor in rats following intradermal injection of capsaicin. *Neuroscience* 2003, 122:237-245.

[40] Esteban JA, Shi SH, Wilson C, Nuriya M, Huganir RL, Malinow R: PKA phosphorylation of AMPA receptor subunits controls synaptic trafficking underlying plasticity. *Nat Neurosci* 2003, 6:136 – 143.

[41] Ji RR, Woolf CJ: Neuronal plasticity and signal transduction in nociceptive neurons: implications for the initiation and maintenance of pathological pain. *Neurobiol Dis* 2001, 8:1-10.

[42] Davies SN, Lodge D: Evidence for involvement of N-methylaspartate receptors in 'wind-up' of class 2 neurons in the dorsal horn of the rat. *Brain Res* 1987, 424:402-406.

[43] Dickenson AH, and Sullivan AF: Evidence for a role of the NMDA receptor in the frequency dependent potentiation of deep rat dorsal horn nociceptive neurons following C fiber stimulation. *Neuropharmacology* 1987, 26:1235-1238.

[44] Dickenson AH, Aydar E: Antagonism at the glycine site on the NMDA receptor reduces spinal nociception in the rat. *Neurosci Lett* 1991, 121:263-266.

[45] Woolf CJ, Thompson SWN: The induction and maintenance of central sensitization is dependent on *N*-methyl-D-aspartic acid receptor activation; Implications for the treatment of post-injury pain hypersensitivity states. *Pain* 1991, 44:293-299.

[46] Kato A, Rouach N, Nicoll RA, Bredt DS: Activity-dependent NMDA receptor degradation mediated by retrotranslocation and ubiquitination. Proc Natl Acad Sci U S A 2005, 102:5600-5605.

[47] Ratnam J, Teichberg VI.: Neurofilament-light increases the cell surface expression of the N-methyl-D-aspartate receptor and prevents its ubiquitination. *J Neurochem*. 2005, 92:878-885.

[48] Ehlers MD, Fung ET, O'Brien RJ and Huganir RL: Splice variant-specific interaction of the NMDA receptor subunit NR1 with neuronal intermediate filaments. *J. Neurosci* 1998, 18:720–730.

[49] Xing J, Ginty DD, Greenberg ME: Coupling of the RAS-MAPK pathway to gene activation by RSK2, a growth factor-regulated CREB kinase. *Science* 1996, 273:959-963.

[50] Garry EM, Delaney A, Blackburn-Munro G, Dickinson T, Moss A, Nakalembe I, Robertson DC, Rosie R, Robberecht P, Mitchell R, Fleetwood-Walker SM: Activation of p38 and p42/44 MAP kinase in neuropathic pain: Involvement of VPAC(2) and NK(2) receptors and mediation by spinal glia. *Mol Cell Neurosci* 2005, 30:523-537.

[51] Kim SY, Bae JC, Kim JY, Lee HL, Lee KM, Kim DS, Cho HJ: Activation of p38 MAP kinase in the rat dorsal root ganglia and spinal cord following peripheral inflammation and nerve injury. *Neuroreport* 2002, 13:2483-2486.

[52] Svensson CI, Marsala M, Westerlund A, Calcutt NA, Campana WM, Freshwater JD, Catalano R, Feng Y, Protter AA, Scott B, Yaksh TL: Activation of p38 mitogen-activated protein kinase in spinal microglia is a critical link in inflammation-induced spinal pain processing. *J Neurochem* 2003, 86:1534-1544.

[53] Jin SX, Zhuang ZY, Woolf CJ, Ji RR: p38 mitogen-activated protein kinase is activated after a spinal nerve ligation in spinal cord microglia and dorsal root ganglion neurons and contributes to the generation of neuropathic pain. *J Neurosci* 2003, 23:4017-4022.

[54] Ji RR, Befort K, Brenner GJ, Woolf CJ: ERK MAP kinase activation in superficial spinal cord neurons induces prodynorphin and NK-1 upregulation and contributes to persistent inflammatory pain hypersensitivity. *J Neurosci* 2002, 22:478-485.

[55] Ma W, Quirion R: Increased phosphorylation of cyclic AMP response element-binding protein (CREB) in the superficial dorsal horn neurons following partial sciatic nerve ligation. *Pain* 2001, 93:295-301.

[56] Miletic G, Miyabe T, Gebhardt KJ, Miletic V: Increased levels of Homer1b/c and Shank1a in the post-synaptic density of spinal dorsal horn neurons are associated with neuropathic pain in rats. *Neurosci Lett* 2005, 386:189-193.

[57] Ehlers MD: Activity level controls postsynaptic composition and signaling via the ubiquitin-proteasome system. *Nat Neurosci* 2003, 6:231-242.

[58] Garry EM, Moss A, Rosie R, Delaney A, Mitchell R, Fleetwood-Walker SM: Specific involvement in neuropathic pain of AMPA receptors and adapter proteins for the GluR2 subunit. *Mol Cell Neurosci* 2003, 24:10-22.

[59] Aanonsen LM, Lei S, Wilcox GL: Excitatory amino acid receptors and nociceptive neurotransmission in rat spinal cord. *Pain* 1990, 41:309-321.

[60] Cumberbatch MJ, Chizh BA, Headley PM: AMPA receptors have an equal role in spinal nociceptive and non-nociceptive transmission. *Neuroreport* 1994, 5:877-880.

[61] Sorkin LS, Yaksh TL, Doom CM: Pain models display differential sensitivity to Ca^{2+}-permeable non-NMDA glutamate receptor antagonists. *Anesthesiology* 2001, 95:965-973.

[62] Nishiyama T, Yaksh TL, Weber E: Effects of intrathecal NMDA and non-NMDA antagonists on acute thermal nociception and their interaction with morphine. *Anesthesiology* 1998, 89:715-722.

[63] Hartmann B, Ahmadi S, Heppenstall PA, Lewin GR, Schott C, Borchardt T, Seeburg PH, Zeilhofer HU, Sprengel R, Kuner R: The AMPA receptor subunits GluR-A and GluR-B reciprocally modulate spinal synaptic plasticity and inflammatory pain. *Neuron* 2004, 44:637-650.

[64] Engelman HS, Allen TB, MacDermott AB: The distribution of neurons expressing calcium-permeable AMPA receptors in the superficial laminae of the spinal cord. *J Neurosci* 1999, 19:2081-2089.

[65] Shibata T, Watanabe M, Ichikawa R, Inoue Y, Koyanagi T: Differential expressions of α-amino-3-hydroxy-5-methyl-4-isoxazolepropionic acid and *N*-methyl-D-aspartate receptor subunit mRNAs between visceromotor and somatomotor neurons of the rat lumbosacral spinal cord. *J Comp Neurol* 1999, 404:172-182.

[66] Spike RC, Kerr R, Maxwell DJ, Todd AJ: GluR1 and GluR2/3 subunits of the AMPA-type glutamate receptor are associated with particular types of neuron in laminae I-III of the spinal dorsal horn of the rat. *Eur J Neurosci* 1998, 10:324-333.

[67] Furuyama T, Kiyama H, Sato K, Park HT, Maeno H, Takagi H, Tohyama M: Region-specific expression of subunits of ionotropic glutamate receptors (AMPA-type, KA-type and NMDA receptors) in the rat spinal cord with special reference to nociception. *Brain Res Mol Brain Res* 1993, 1-2:141-51.

[68] Tölle TR, Berthele A, Zieglgänsberger W, Seeburg PH, Wisden W: The differential expression of 16 NMDA and non-NMDA receptor subunits in the rat spinal-cord and in periaqueductal gray. *J Neurosci* 1993, 13:5009-5028.

[69] Sheng M, Lee SH: AMPA receptor trafficking and the control of synaptic transmission. *Cell* 2001, 105:825-828.

[70] Ziff EB: Recent excitement in the ionotropic glutamate receptor field. *Ann N Y Acad Sci* 1999, 868:465-473.

[71] Lin JW, Ju W, Foster K, Lee SH, Ahmadian G, Wyszynski M, Wang YT, Sheng M: Distinct molecular mechanisms and divergent endocytotic pathways of AMPA receptor internalisation. *Nat Neurosci* 2000, 3:1282-1290.

[72] Malenka RC, Nicoll RA: Long-term potentiation-a decade of progress? *Science* 1999, 285:1870-1874.

[73] Malinow R, Malenka RC: AMPA receptor trafficking and synaptic plasticity. *Annu Rev Neurosci* 2002, 25:103-126.

[74] Bredt D and Nicoll R: AMPA Receptor Trafficking at Excitatory Synapses. *Neuron* 2003, 40:361-379

[75] Man YH, Lin JW, Ju WH, Ahmadian G, Liu L, Becker LE, Sheng M and Wang YT: Regulation of AMPA receptor-mediated synaptic transmission by clathrin-dependent receptor internalization. *Neuron* 2000, 25:649–662.

[76] Galan A, Laird JM, Cervero F: In vivo recruitment by painful stimuli of AMPA receptor subunits to the plasma membrane of spinal cord neurons. *Pain* 2004, 112:315-323.

[77] Burbea M, Dreier L, Dittman JS, Grunwald ME, Kaplan JM: Ubiquitin and AP180 regulate the abundance of GLR-1 glutamate receptors at postsynaptic elements in *C. elegans*. *Neuron* 2002, 35:107-120.

[78] Beattie EC, Carroll RC, Yu X, Morishita W, Yasuda H, von Zastrow M, Malenka RC: Regulation of AMPA receptor endocytosis by a signaling mechanism shared with LTD. *Nat Neurosci* 2000, 3:1291-1300.

[79] Turrigiano GG: A recipe for ridding synapses of the ubiquitous AMPA receptor. *Trends Neurosci* 2002, 25:597-598.

[80] Husi H, Ward MA, Choudhary JS, Blackstock WP, Grant SG: Proteomic analysis of NMDA receptor-adhesion protein signaling complexes. *Nat Neurosci* 2000, 3:661-669.

[81] Garry EM, Fleetwood-Walker SM: Organizing pains. *Trends Neurosci* 2004, 27:292-294.

[82] Bingol B, Schuman EM: A proteasome-sensitive connection between PSD-95 and GluR1 endocytosis. *Neuropharmacology* 2004, 47:755-763.

[83] Chen L, El-Husseini A, Tomita S, Bredt DS, Nicoll RA: Stargazin differentially controls the trafficking of alpha-amino-3-hydroxyl-5-methyl-4-isoxazolepropionate and kainate receptors. *Mol Pharmacol* 2003, 64:703-706.

[84] Chen L, Chetkovich DM, Petralia RS, Sweeney NT, Kawasaki Y, Wenthold RJ, Bredt DS, Nicoll RA: Stargazin regulates synaptic targeting of AMPA receptors by two distinct mechanisms. *Nature* 2000, 408:936-943.

[85] Dakoji S, Tomita S, Karimzadegan S, Nicoll RA, Bredt DS: Interaction of transmembrane AMPA receptor regulatory proteins with multiple membrane associated guanylate kinases. *Neuropharmacology* 2003, 45:849-856.

[86] Schnell E, Sizemore M, Karimzadegan S, Chen L, Bredt DS, Nicoll RA: Direct interactions between PSD-95 and stargazin control synaptic AMPA receptor number. *Proc Natl Acad Sci U S A*. 2002, 99:13902-13907.

[87] El-Husseini AE, Schnell E, Chetkovich DM, Nicoll RA and Bredt DS: PSD-95 involvement in maturation of excitatory synapses. *Science* 2000, 290:1364-1368.

[88] Carroll RC, Beattie EC, Xia H, Luscher C, Altschuler Y, Nicoll RA, Malenka RC, von Zastrow M: Dynamin-dependent endocytosis of ionotropic glutamate receptors. *Proc Natl Acad Sci U S A* 1999, 96:14112-14117

[89] Luscher C, Xia H, Beattie EC, Carroll RC, von Zastrow M, Malenka RC, Nicoll RA: Role of AMPA receptor cycling in synaptic transmission and plasticity. *Neuron* 1999, 24:649-658

[90] Chetkovich DM, Chen L, Stocker TJ, Nicoll RA, Bredt DS: Phosphorylation of the postsynaptic density-95 (PSD-95)/discs large/zona occludens-1 binding site of stargazin regulates binding to PSD-95 and synaptic targeting of AMPA receptors. *J Neurosci* 2002, 22:5791-5796

[91] Choi J, Ko J, Park E, Lee JR, Yoon J, Lim S, Kim E: Phosphorylation of stargazin by protein kinase A regulates its interaction with PSD-95. *J Biol Chem* 2002, 277:12359-12363

[92] Perez JL, Khatri L, Chang C, Srivastava S, Osten P, Ziff EB: PICK1 targets activated protein kinase Calpha to AMPA receptor clusters in spines of hippocampal neurons and reduces surface levels of the AMPA-type glutamate receptor subunit 2. *J Neurosci* 2001, 21:5417-5428

[93] Dong H, O'Brien RJ, Fung ET, Lanahan AA, Worley PF, Huganir RL: GRIP: a synaptic PDZ domain-containing protein that interacts with AMPA receptors. *Nature* 1997, 386:279-84.

[94] Noel, J, Ralph, GS, Pickard, L, Williams, J, Molnar, E, Uney, JB, Collingridge, GL, Henley, JM: Surface expression of AMPA receptors in hippocampal neurons is regulated by an NSF-dependent mechanism. *Neuron* 1999, 23:365-376.

[95] Nishimune A, Isaac JTR, Molnar E, Noel J, Nash SR, Tagaya M, Collingridge GL, Nakanishi S, Henley JM: NSF binding to GluR2 regulates synaptic transmission. *Neuron* 1998, 21:87-97.

[96] Malenka RC, Nicoll RA: Silent synapses speak up. *Neuron* 1997, 19:473-476.

[97] Carroll RC, Beattie EC, von Zastrow M, Malenka RC: Role of AMPA receptor endocytosis in synaptic plasticity. *Nat Rev Neurosci* 2001, 2:315-324.

[98] Young MR, Blackburn-Munro G, Dickinson T, Johnson MJ, Anderson H, Nakalembe I, Fleetwood-Walker SM: Antisense ablation of type I metabotropic glutamate receptor mGluR1 inhibits spinal nociceptive transmission. *J Neurosci* 1998, 18:10180-10188.

[99] Fisher K, Lefebvre C, Coderre TJ: Antinociceptive effects following intrathecal pretreatment with selective metabotropic glutamate receptor compounds in a rat model of neuropathic pain. *Pharmacol Biochem Behav* 2002, 73:411-418.

[100] Simmons RM, Webster AA, Kalra AB, Iyengar S: Group II mGluR receptor agonists are effective in persistent and neuropathic pain models in rats. *Pharmacol Biochem Behav* 2002, 73:419-427.

[101] Chen SR, Pan HL: Distinct roles of group III metabotropic glutamate receptors in control of nociception and dorsal horn neurons in normal and nerve-injured rats. *J Pharmacol Exp Ther* 2005, 312:120-126.

[102] Neugebauer V, Chen PS, Willis WD: Groups II and III metabotropic glutamate receptors differentially modulate brief and prolonged nociception in primate STT cells. *J Neurophysiol* 2000, 84:2998-3009

[103] Moriyoshi K, Iijima K, Fujii H, Ito H, Cho Y, Nakanishi S: Seven in absentia homolog 1A mediates ubiquitination and degradation of group 1 metabotropic glutamate receptors. *Proc Natl Acad Sci U S A* 2004, 101:8614-8619.

[104] Palombella VJ, Rando OJ, Goldberg AL, Maniatis T: The ubiquitin-proteasome pathway is required for processing the NF-kappa B1 precursor protein and the activation of NF-kappa B. *Cell* 1994, 78:773-785

[105] Baeuerle PA, Baltimore D: I kappa B: a specific inhibitor of the NF-kappa B transcription factor. *Science* 1988, 242:540-546.

[106] Chen Z, Hagler J, Palombella VJ, Melandri F, Scherer D, Ballard D, Maniatis T: Signal-induced site-specific phosphorylation targets IkBα to the ubiquitin-proteasome pathway. *Genes Dev* 1995, 9:1586-1597.

[107] Tegeder I, Niederberger E, Schmidt R, Kunz S, Guhring H, Ritzeler O, Michaelis M, Geisslinger G: Specific Inhibition of IkappaB kinase reduces hyperalgesia in inflammatory and neuropathic pain models in rats. *J Neurosci* 2004, 24:1637-1645.

[108] Ma W, Bisby MA: Increase of preprotachykinin mRNA and substance P immunoreactivity in spared dorsal root ganglion neurons following partial sciatic nerve injury. *Eur J Neurosci* 1998, 10:2388-2399.

[109] Sakaue G, Shimaoka M, Fukuoka T, Hiroi T, Inoue T, Hashimoto N, Sakaguchi T, Sawa Y, Morishita R, Kiyono H, Noguchi K, Mashimo T: NF-kappa B decoy suppresses cytokine expression and thermal hyperalgesia in a rat neuropathic pain model. *Neuroreport* 2001, 12:2079-2084

[110] Chan CF, Sun WZ, Lin JK, Lin-Shiau SY: Activation of transcription factors of nuclear factor kappa B, activator protein-1 and octamer factors in hyperalgesia. *Eur J Pharmacol* 2000, 402:61-68.

[111] Lee KM, Kang BS, Lee HL, Son SJ, Hwang SH, Kim DS, Park JS, Cho HJ: Spinal NF-kB activation induces COX-2 upregulation and contributes to inflammatory pain hypersensitivity. *Eur J Neurosci* 2004, 19:3375-3381.

[112] Hamilton AL, Eder JP, Pavlick AC, Clark JW, Liebes L, Garcia-Carbonero R, Chachoua A, Ryan DP, Soma V, Farrell K, Kinchla N, Boyden J, Yee H, Zeleniuch-Jacquotte A, Wright J, Elliott P, Adams J, Muggia FM: Proteasome inhibition with bortezomib (PS-341): a phase I study with pharmacodynamic end points using a day 1 and day 4 schedule in a 14-day cycle. *J Clin Oncology* 2005, 23: 6107-6116

[113] Topp KS, Tanner KD, Levine JD: Damage to the cytoskeleton of large diameter sensory neurons and myelinated axons in vincristine-induced painful peripheral neuropathy in the rat. *J Comp Neurol* 2000, 424:563-576

[114] Quasthoff S, Hartung HP: Chemotherapy-induced peripheral neuropathy. *J Neurol* 2002, 249: 9-17.

In: The Ubiquitin Proteasome System... ISBN 978-1-60021-749-4
Eds: Mario Di Napoli and Cezary Wojcik, pp. 671-713

Chapter 28

THE UBIQUITIN PROTEASOME SYSTEM IN NEUROLOGICAL DISORDERS: FROM CONFIRMATIONAL DISEASES TO PROTEASOMEPATHIES

Mario Di Napoli[1,2,*] *and Francesca Papa*[1]

[1]Neurological Section, SMDN – Center for Cardiovascular Medicine and Cerebrovascular Disease Prevention, Sulmona (L'Aquila) I-67039, Italy;

[2]Neurological Service, San Camillo de'Lellis General Hospital, Rieti I-02100, Italy.

ABSTRACT

Protein misfolding and aggregation are common to most neurodegenerative diseases, suggesting that abnormalities of protein homeostasis contribute to pathogenesis. Protein folding inside cells is assisted by various chaperones and folding factors, and misfolded proteins are eliminated by the ubiquitin-proteasome system (UPS) and macroautophagy to ensure high fidelity of protein expression. Under certain circumstances, misfolded proteins escape the degradation process, yielding to deposit of protein aggregates such as loop-sheet polymer and amyloid fibril. Dysfunction of the UPS or macroautophagy pathways might contribute in a wide variety of neurodegenerative diseases. Some proteins, when not properly degraded through the UPS, tend to form aggregates by binding to one another to form an insoluble structure that is very difficult to disassemble. Many of the components of neurodegenerative disease aggregates have been studied for their ability to form independent aggregates *in vitro* and *in vivo* and their biological activity described. Consistent with this view, protein aggregates have been regarded in a pathogenic connotation, with most aspects of neurologic pathogenesis being largely attributed to their presence in nerve tissues. However, the neurotoxicity of protein

* Correspondence concerning this article should be addressed to Dr. Mario Di Napoli, MD; SMDN – Center for Cardiovascular Medicine and Cerebrovascular Disease Prevention, Neurological Section, Via Trento, 41 I-67039 – Sulmona (AQ) Italy. Phone: +39 (0)864 52716; Fax: +39 (0)864 52716; Email: mariodinapoli@katamail.com.

aggregates remains ambiguous as direct evidence substantiating it have long remained elusive. Primary UPS involvement in neurodegenerative diseases seems even more probable when the UPS is viewed not simply as an isolated degradation machine but rather as a complex cascade linked both to other degradation processes and to chaperone systems. In neurodegenerative diseases, perturbations of proteasome function may occur through its recruitment to or sequestration into protein aggregates, through chronic overloading of proteasome capacity by misfolded protein, or through still undetermined effects on other activities of the UPS. Collectively, these dysfunctions in proteasome activity could also be called *proteasomepathies* and differentiated from other specific protein degradation system disturbance. The identification of several degradation system disturbances will assist in choosing therapies when protein-specific disease-modifying treatments are available.

Keywords: Amyloid fibril, Conformational disease, Protein folding, Protein misfolding, Protein aggregation, ubiquitin, proteasome, neurodegenerative disease.

Introduction

Many late-onset neurodegenerative diseases are caused by aggregate prone proteins [1]. The *Conformational Diseases* [2], known also as *proteinopathies*, represent a group of conditions characterized by protein misfolding, followed by self-association and subsequent deposition of the aggregated protein in the affected tissues. These diseases include conditions in which the proteins are predominantly cytosolic, such as Parkinson's disease (PD) and adult-onset Huntington's disease (HD), predominantly intranuclear (for example, spinocerebellar ataxia type 1), aggregated in the endoplasmic reticulum (ER) as seen with neuroserpin mutations that cause familial encephalopathy with neuroserpin inclusion bodies, and secreted extracellularly such as β-amyloid (Aβ) in Alzheimer's disease (AD).

The hallmark of these otherwise unrelated diseases is the formation of aggregates containing misfolded proteins, i.e. proteins in a non-native folding state, with a concomitant gain of function which eventually leads to neuronal death [3,4]. The aggregates have various supramolecular organizations and, in most cases, form structurally well-defined insoluble fibrillar deposits [5]. In one case (prion disease), aggregates are even believed to be responsible for disease transmission [6].

The realization that abnormal protein accumulation characterizes particular disease sets has led to a classification based on the composition of the abnormal protein inclusions [7]. These deposits are usually fibrillar and have characteristic structural and histological morphologies while the biological effects of these differing protein deposits are distinctive and depend upon such features as the tissue involved and whether the deposits are intra- or extracellular. In the case of neurodegenerative diseases, the gross histopathological consequences of protein misfolding are features such as senile plaques and neurofibrillary tangles in AD, Lewy bodies (LBs) in PD and Lewy body dementia (LBD), other nuclear inclusions in the polyglutamine (poly-Q) repeat diseases such as HD and the spinocerebellar ataxias (SCAs). Whether soluble monomers, oligomers or larger aggregates are the most toxic species causing many of these previous cited conditions has been the subject of considerable debate and uncertainty [8,9]. Nevertheless, in general, it seems that the capacity to aggregate (although not necessarily the aggregates themselves) is correlated with toxicity. Thus, it is important to understand the cellular processes regulating their levels.

Folding of the protein chain to give the functional structure, and maintenance of the functional conformation are complex and critical processes. All proteins, at some point, complete their given task and must be degraded and recycled by the cell. In order to be biologically active, proteins must fold into their well-defined three-dimensional structures. The properties of the peptide bond and the amino acid side chains confer a high degree of conformational flexibility to the protein conformation, resulting in tremendous possible conformations from a single polypeptide chain [10]. Nonetheless, only one conformation that is thermodynamically the most stable state generally corresponds to the native state, which is determined by the primary sequences. Protein folding has recently been described in terms of energy landscape of a 'folding funnel' [11] (see Chapter 12). The bottom of the funnel represents the native state of the protein. At top of the funnel, the protein exists in random states with large number of conformations of high free energy. Progress down the funnel is accompanied by an increase in native-like structure as folding proceeds [12]. The surface of this folding funnel is unique for a specific polipeptide sequence under a particular set of conditions and is determined by both thermodynamic and kinetic properties of the folding polypeptide chain. Based on the realization that the unfolded and partially folded states are conformationally heterogeneous, and that there may not be a single route to the native state, models of folding have now evolved into the landscape view of protein folding [11-13], in which the unfolded polypeptide chain searches for the native conformation on a usually rugged energy surface or 'landscape', until the unique native structure is formed. Random fluctuations in the unfolded or partially folded states drive this reaction, as different native as well as nonnative contacts are sampled. In general, native interactions between residues are assumed to be more stable than non-native contacts, and as such contacts form; the number of available conformations is reduced, driving the polypeptide chain towards the native structure. Partially folded states on this landscape may be intrinsically prone to aggregation, and favorable intermolecular contacts may lead to their association and ultimately to protein misfolding diseases.

Protein folding has been studied in detail by both experimental and theoretical procedures and this has expanded the impact of studies of protein folding from a key fundamental question to a central issue in the understanding of several human diseases. Biophysical methods have been relatively successful at probing the structural consequences of some disease-associated mutations. For example, three-dimensional structures of variant proteins show the impact of particular mutations, analysis of altered folding and unfolding kinetics may illustrate the affect of a mutation on protein stability *in vitro*, and fiber diffraction and electron microscopy studies have illuminated the structure of amyloid fibrils. *In vitro* folding of some simple proteins has been investigated in detail by various optical techniques including NMR, circular dichroism and fluorescence spectroscopy, with the aid of rapid-mixing devices [14,15] (Table 1). However, *in vivo* folding inside living cells is substantially different from *in vitro* folding in many aspects even though the underlying principles are the same: (*i*) *in vivo* protein folding occurs in a crowded environment with a number of different proteins or different type of macromolecules; (*ii*) protein folding sometimes occurs before completion of its synthesis. Our understanding of the impact of alternative conformers on the cell, and how alternative conformations may compromise cellular activity or induce toxicity, is steadily growing. However, the rate of synthesis of the full-length protein may not be a major determinant of the production rate of the toxic species— the main rate limiting step under conditions of synthesis varying between 50% and 100% of physiological rates may be at the level of cleavage (if the cleavage step is saturated). The best-known example of cleavage generating a toxic product occurs during the production of Aβ peptide from full-length amyloid precursor protein (APP) [16]. However, related toxic-fragment models also seem to be relevant to certain poly-Q diseases, such as HD [17,18] and SCA3 [19].

The three main categories of nonnative conformations— namely, stable misfolded forms, unstable misfolded forms, and aggregation-prone forms— can have three different consequences: functional deficiency, dominant-negative effects, or toxic cellular effects. The first studies of protein misfolding pathologies tended to focus most on intra- or extracellular aggregation of proteins in diseases that exhibit a gain-of-function pathology. There is increasing recognition that early species may be toxic in these processes, and there is now a shift toward investigating the cellular response to aggregation, in addition to studying the impact of the aggregation itself. Another large group of diseases involves the rapid degradation of mutant protein, resulting in a loss-of-function pathology. Increasingly, studies of different conformational diseases are highlighting the fact that the cellular responses observed with various pathologies have common features. Examination of these defective folding disorders has also highlighted the normal cellular mechanisms for dealing with protein quality control (QC). Generally, extracellular proteins are taken into the cell and processed by lysosomes, while proteins in the cytosol and the nucleus are degraded by proteasomes [20]. This fact has led to the accepted view that protein QC plays a critical role in neuronal function and survival. Its importance is underscored by studies showing that manipulating QC pathways alters pathogenesis of neurodegenerative diseases [21,22]. Various components of the QC machinery likely contribute to disease mechanisms, most notably the heat shock protein (HSP) system [19,21-23] (see Chapter 19), the ubiquitin–proteasome system (UPS) [19,24-28] (see Chapter 3) and the authophagy system. This chapter focuses on the study of the many cellular consequences of protein misfolding on the pathways that clear aggregate-prone and toxic proteins and their relationship with the UPS.

Cellular Protein Folding

Many details of the folding process depend on the environment in which folding takes place. When polypeptides are synthesized in the cells, they fold in the cytoplasm after release from the ribosome or in other subcellular compartments such as ER or mitochondria after they are translocated through membranes [14,29,30]. Within the cells, proteins in the process of synthesis encounter particular challenges imposed by the crowded macromolecules before completion of folding [31]. As incompletely folded chains expose particular regions that are destined to be buried in the native state, they are prone to aggregate with other molecules because they have exposed hydrophobic surfaces. Situations are more problematic because aggregation process follows second order kinetics and therefore surpasses the first order folding process as their concentration increases [10], which condition is normally satisfied within the concentrated milieu of the cells [31]. Consequently, elaborate systems have evolved to prevent proteins from being aggregated prior to folding. The first one is molecular chaperones, and the second one is UPS, each of which is not independent of the other, and in some way they cooperate in living cells. The details on the role of chaperones as folding assistants are given in Chapters 19 and 22 while the folding process in ER is discussed in Chapter 13. For further details, the reader is remander to the corresponding chapters. Cellular machinery regulating synthesis, translocation, folding and degradation of proteins seems to operate in a very stringent manner to ensure protein aggregation is minimized (Figure 1).

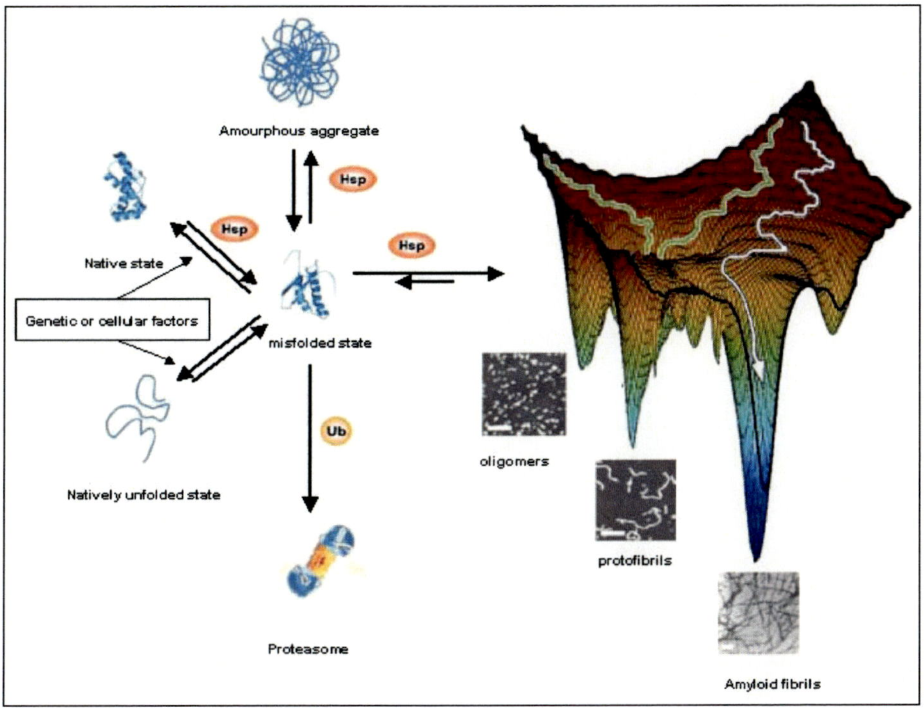

Figure 1.

Nevertheless, in a certain disease condition, misfolded protein will escape such elaborate system forming protein aggregates. The folding of newly synthesized proteins to their native conformations involves the sequential action of multiple molecular chaperones (see Chapter 19). Two major chaperone classes, Hsp70 and Hsp60, act in a tightly controlled ATP-dependent manner to bind and release unfolded or misfolded substrates, thereby enhancing substrate refolding and preventing aggregation. Furthermore, recognition of abnormal proteins by the cellular machinery leads to their ubiquitination and subsequent degradation by the 26S proteasome (see Chapters 3 and 7) (Figure 1). However, even for proteins that fold successfully to their native state and hence escape the cellular QC machinery, random conformational fluctuations can lead to the transient formation of aggregation-prone intermediate states (Figure 1). In the crowded environment of the cell, and also influenced by environmental factors, such species may then start to aggregate, forming small oligomers or larger particles that initiate the formation of intracellular aggregates. This may lead to the accumulation of large quantities of partially folded proteins and the saturation of the capacity of the QC machinery, exacerbating the formation of intracellular aggregates before refolding or degradation is possible [29] (Figure 1). Several intermediate structures have been identified using electron and atomic force microscopy in *in vitro* studies during fibril formation, including small oligomers, membrane embedded pores and protofibrils (Figure 1). Whether these structures form on-pathway or are an off-pathway product of fibril formation, and which of these structures are actually the toxic ones, are probably the most debated questions today [32].

Misfolded Proteins as Key Effectors in Neurodegenerative Diseases

A broad array of human neurodegenerative diseases are characterized by the accumulation of extracellular and/or intracellular protein aggregates sharing strikingly similar histopathological features that may hold the key to their molecular pathogenesis [33-36] suggesting a role of the biophysical properties of protein stability, aggregation and degradation in the pathology of nervous system (Table 2) [9,37,38]. It is accepted by many that misfolding and deposition imparts a gain of function that ultimately leads to neuronal death [4] although there are opponents to this point of view [39]. In these diseases, specific peptides or proteins misfold, often because of mutations, and give rise to protein aggregates. Although misfolded proteins are also considered to accumulate in neurons and glia as a result of physiological processes associated with ageing, protein overexpression and conformational rigidity has been shown to be particularly toxic to neurons under disease conditions. Neurodegenerative diseases are typically manifested at an advanced age and it is completely unclear, why the corresponding protein depositions do not occur during the first decades of life. It is suspected that molecular and/or cellular clearance mechanisms are of significance and lose their efficiency increasingly with rising age probably for an efficiency reduction of antioxidative systems [40].

In most cases, the critical event characterizing these disorders is the conversion of a native protein conformation typified by α-helix and or random structure, into β-pleated sheet aggregates. Aβ eptides (proteins), α-synuclein (α-syn), huntingtin (Htt) and prion protein (PrP) may undergo misfolding processes. During these conformation changes β-pleated sheet structures will be formed. Peptides possessing β-structure have a high tendency to form suprachemical structures: oligomers, protofibrils and fibrils. The general rule for misfolded proteins is that they are rich in β-sheet structures, formed of alternating peptide-pleated sheets. The natural, active conformation of most proteins usually comprises a mixture of α-helix and some unordered structure. In the pathological condition, the proportion of α-helical structure diminishes with a concomitant appearance of highly β-pleated sheet aggregates, characterized by a higher order fibrillar structure. In many cases, molecules of β-sheet misfolded protein will polymerize into the fibril-type structures, which characterize amyloid. The common property of the polymerized protein aggregates is proposed to be the induction of tissue damage, either by gaining a toxic activity or by losing the intrinsic biological function of the native protein.

At the cellular level, these diseases are characterized by the accumulation of aberrant proteins either intracellularly or extracellularly in specific groups of cells that subsequently undergo death [41]. The precise association between protein accumulation and cell death remains incompletely understood and may vary from disease to disease. Moreover, what is also puzzling is the observation that these inclusions are found almost exclusively within certain subgroups of neurons in specific regions of the brain regardless of the widespread tissue distribution of the core proteins [42]: aggresome like structures in prionopathies [43,44], neurofibrillary tangles and plaques in AD and tauopathies, LBs in PD and related synucleinopathies [45-47], cytoplasmic and nuclear inclusion bodies in poly-Q expansion diseases such as HD and SCA [21,48-50]. In serpinopathies, neuroserpin molecules form cytosplasmic neuronal inclusions of loopsheet polymers (Collins' bodies) in the deeper layers of the cerebral cortex and the substantia nigra [51,52]. Aggresome like structures that clumped protein aggregates surrounding nuclei, along dendrites and ubiquitin (Ub)-immunoreactive nuclear inclusions are also formed after brain ischemia [53-57]. The presence of insoluble proteinaceous deposits differ in their protein content but invariably contain components of the UPS [33]. This has led to the suggestion that a chronic imbalance between their generation and processing may be the primary cause for the formation of protein deposits [34,58]. The classic causes for protein misfolding, which lead to a loss of function, are missense mutations, protein modifications or post-translational damage, or expansion of amino acid repeats as is observed in poly-Q disorders. Misfolded proteins escape the protective mechanisms and form intractable aggregates within cells or in the extracellular space. It is thought that either the aggregates of the disease protein themselves or the work done by the aggregates, or the process of their formation confers cellular toxicity. This idea supports the notion that misfolded disease proteins act through a gain-of-function [4,59].

The pathological hallmark is the presence of insoluble intra- or extracellular inclusion bodies (IB) in affected regions of the brain. Significantly, neurons cannot dilute toxic, misfolded proteins through cell division, which coincides with a decreased proteasome and molecular chaperone activity (for a review, see [36,60,61]). The major constituents of these inclusions are misfolded proteins. IB contain aberrantly polymerized or aggregated disease protein, either full-length protein or proteolytic fragments including cytoskeletal elements such as tau, tubulin-associated proteins and neurofilaments [62-66], display amyloid-like properties, suggesting a similar fibrillar structure [66].

The disease proteins involved in conformational disorders have no obvious sequence similarities. Moreover, they display amyloid-like properties, suggesting a fibrillar structure. Almost all aggregations result in β-linkages formed by hydrogen bonding between peptide loops and sheets. Domain swapping is another mechanism of intermolecular linkage in cystatin polymer [67]. In these polymers, the individual molecules substantially retain their ordered structure. The linkage is formed by the remarkable realignment of peptide segments to give the sequential layering of β-structures known as β-amyloid fibril. Molecular models of Aβ fragments and α-syn suggest that amyloid fibrils are composed of a number of protofilaments exhibiting cross-β conformation i.e. hydrogen bonding β-sheet structure where the β-strands run perpendicular to the fibril axis [68]. Results from the experiments by X-ray fiber diffraction, cryoelectron microscopy and solid-state NMR have shown that amyloid fibrils are long, unbranched and often twisted structures a few nanometres in diameter and the organized core structure is composed of β-sheets whose strands run perpendicular to the fibril axis [69-71]. The conversion or misfolding of the protein is a product of one or many factors acting independently or in tandem. Under appropriate circumstances, misfolded monomers may oligomerize into pre-fibrillar assemblies. Such circumstances can be satisfied by mutations in the native protein, ionic strength, pH, the presence of metal ions and the concentration of the protein itself. *In vitro* studies with purified disease proteins are most consistent with nucleation- and concentration-dependent self-aggregation— that is, a self-driven reaction [72,73]. Given this, the formation of oligomers and aggregates will tend to occur as long as a critical concentration of misfolded protein is reached in the cell. Any factor that increases the intracellular concentration of misfolded disease protein would favor aggregation and inclusion formation.

Much of our information on the regulation of amyloid formation is based upon studies of the Aβ protein. It has been proposed that oligomeric seeds of Aβ facilitate further protein misfolding, thus promoting polymerization of the protein and eventual fibril formation [74,75]. The intrinsic effect of specific mutations on the rate of aggregation can be correlated to a remarkable extent with the changes in simple physicochemical properties such as hydrophobicity, secondary structure propensity and charge [76]. Unfolding of substrate proteins is the prerequisite for degradation by the proteasome, and β-structure is more difficult to unravel than α-helix or surface loop. This explains why amyloid deposits, having high content of β-structure, are not easily cleared away by the proteasome system [77].

Effects of Protein Aggregates on Neuron Physiology

A lot of mechanisms have been proposed and experimentally verified for the observed neuronal loss in the affected brain regions, most of which involve either the generation of reactive oxygen species (ROS) and induction of oxidative stress or the perturbation of Ca^{+2} homeostasis. The undegradable deposits may sequestrate components of the chaperone and degradation systems: Ub chains, parkin, proteasome subunits [78-81], HSP chaperones such as Hsp70 and Hsp27 [82,83], reducing the activity of assisted folding and proteolysis. Finally, IB often contain other proteins that, although not themselves intrinsically aggregation-prone, such as intermediate filaments, protein kinases, and transcription factors [1,41], become recruited, concentrated, or trapped within the inclusions. The molecular forces driving their co-localization to inclusions and the potential role this plays in pathogenesis are still largely unknown.

Generally, aggregation occurs in two steps. The first step characterized by a slow lag phase seems to involve the formation of soluble oligomers because of relatively nonspecific interactions. The oligomeric nucleus then rapidly grows. The lag phase can be minimized or eliminated by seeding preformed nucleus [84,85]. The whole process will form a positive feedback cycle leading to cellular toxicity. Aberrant interactions of misfolded proteins with other proteins will also cause cellular toxicity because the interacting proteins are then no more functional. Sequestration of one or more of these cellular proteins into IBs— and consequent depletion from the cytosol or nucleus— has been proposed as one possible mechanism through which protein aggregates can damage cells [86]. For example, poly-Q-expanded Htt can recruit other important cellular proteins with normal poly-Q stretches into the aggregates [87]. Misfolded and aggregated peptides appear to owe their toxicity to protein regions that become exposed on their surfaces while being buried in the interior of correctly folded native state structures. Surface exposure of large hydrophobic groups favors interactions of misfolded proteins with cell membranes with a subsequent loss of the regulation of intracellular ion balance and redox status [46,84,88,89]. Aggregated protein forms a ring-like structure and inserts into the cellular membrane as a pore, which induces cytotoxicity including calcium dysregulation, membrane depolarization, mitochondrial dysfunction, and inhibition of long-term potentiation [89].

Induction of Oxidative Stress

Numerous studies have implicated Aβ in the induction of oxidative stress as an underlying cause of pathogenesis in AD. The contributory role of both soluble and aggregated Aβ in the generation of oxidative radicals is well documented and has been demonstrated both *in vivo* and *in vitro* [90,91]. The mechanisms of Aβ-dependent ROS formation include (*i*) the participation of transition metal ions (Fe^{+2}, Cu^{+2}), ROS are generated from peroxides through Fenton chemistry reactions by these ions bound to Aβ moieties [92]; (*ii*) the activation of the receptor for advanced glycation end-products (RAGE) [93]. Aggregated Aβ has been found to bind and activate RAGE or type 2-scavenger receptors, leading to the downstream activation of NADH oxidases [94] that catalyze formation of superoxide (O_2•-) and subsequent generation of hydrogen peroxide (H_2O_2) and related species [95]; (*iii*) prolonged excitatory state of neurons by perturbation of the glutamate cycle [96]. Increased excitability of neurons by glutamate induces oxidative processes either by binding to the NMDA receptors [97] or by competing for the cysteine-antiporter system and thereby depleting intracellular glutathione (GSSH) pools from neurons and glia [98]. Elevation of intracellular Ca^{+2} levels via ionotropic glutamate receptor stimulation leads to calmodulin-dependent activation of nitric oxide synthase (NOS) and the generation of reactive nitrogen species (RNS), such as the highly neurotoxic peroxynitrite ($ONOO^-$) [99]; (*iv*) induction of inflammatory mediators (e.g. IL-1β, TNF-α) from astrocytic and reactive microglial cells [100]. The activation of NOS is also known to be mediated by the inflammatory factors released from activated astrocytes and microglia, which, besides causing upregulation of NOS expression, can increase oxidative load by induction of cyclooxygenase 2 (COX-2) [101], an enzyme with a proven contributory role to oxidative cell damage in neurodegeneration [102]. Microglial cells are also activated in AD, the process involves Aβ binding and activation of cell surface molecules such as CD45, CD40, CD36 and integrins [103]. α-Syn may interfere with ERK-signaling [104]. Htt distrupts receptor signaling mediated by EGF and NGF [105]. PrP plays important role in signal transduction pathways of neurons, the abnormal prion protein can act on MAPK signalling pathway or on the JNK-c-Jun pathway causing neurodegeneration [106].

Perturbation of Ca^{+2} Homeostasis

Regulation of Ca^{+2} dynamics in neurons is complex [107]. Activation of metabotropic glutamate receptor results in release of Ca^{+2} from ER. Two types of Ca^{+2} channels in the ER membrane regulate efflux, the inositol (1,4,5) triphosphate receptor (IP3R) and the ryanodine receptor (RYR). Ca^{+2}-release plays important role in modulating synaptic plasticity [108]. Oxidative stress is also presumed to mediate subsequent perturbation of Ca^{+2} homeostasis via additional pathways besides ionotropic glutamate receptor stimulation, through the deregulation of some equally critical ion transport systems, including ion channels (Ca^{+2}, K^{+} channels), ion pumps (Ca^{+2} pump, Na^{+}/ K^{+} ATPase), ion exchangers and cotransporters (Na^{+}/ Ca^{+2} exchanger, K^{+}/Cl^{-} cotransporter), which are vital for neurotransmission and synaptic function [109].

Altered Ca^{+2} homeostasis is believed to be of fundamental importance in the development of AD pathology since it represents a vital element of synaptic transmission. The exact sequence of events is supposedly instigated by raised levels of ROS, which in turn cause alteration in Ca^{+2} homeostasis and tau phosphorylation, a step also believed do be Aβ induced [110]. Overactivation of glutamate receptors results in a substained elevation of intracellular Ca^{2+} levels and induces alterations in the neuronal cytoskeleton similar to those seen in neurofibrillary tangles [111].

Studies performed on cultured cortical neurons and SH-SY- 5Y neuroblastoma cell lines, however, provide evidence for a different sequence of events for Aβ-mediated neurotoxicity, in which Ca^{+2} influx represents the initial step that precedes presentation of an oxidatively stressed environment and tau phosphorylation [112]. Whatever the initiating event, both oxidative stress and alteration of Ca^{+2} homeostasis are known to culminate in neurological lesions in the AD brain that either manifest as neuronal loss due to apoptosis or necrosis, or synaptic loss [92]. Apoptosis through oxidative stress has been extensively studied and it is known to involve one of many distinct pathways that include mitochondrial dysfunction and cytochrome c release, caspase and calpain non-specific activation, lipid peroxidation, DNA damage and protein oxidation, some of which are also replicated in necrosis and in processes leading to loss of Ca^{+2} homeostasis [92,109].

Perturbation of other Intracellular Pathways

Deleterious effects of misfolded proteins also include the pathological interplay with synaptosomal proteins [79,113] and the membranes of intracellular organelles, in particular those of the mitochondria, leading to the generation of ROS with subsequent oxidative stress and cellular demise through classical apoptotic pathways. Aβ and α-syn have dual effects: in monomer form these peptides have neurotrophic effect, in aggregated form they can disturb signaling pathways causing dysfunction and death of neurons. Aβ may act on ionotropic glutamate receptors and cause Ca^{2+}-influx. Wnt/β-catenin signaling is also involved in Aβ-dependent neurodegeneration [114-117].

In addition, misfolded, intracellularly stable protein aggregates can trigger transcriptional dysregulation [118,119], disruption of microtubule-dependent axonal transport [120], perturbation of membrane permeability [121], and impaired function of the UPS [27,122]. It is important to note that those mechanisms described above are not mutually exclusive but complementary to each other and may act in combination. Finally, major deficits in autophagic pathways have also been linked to the overexpression of mutant α-syn given that α-syn antagonized the receptor involved in chaperone-mediated autophagy (CMA) and thereby significantly enhanced the accumulation of the toxic, misfolded protein [123].

Experimental advances highlighted common pathogenic mechanisms among the most common neurodegenerative disorders and designated aggregated proteins, such as Aβ, α-syn, Htt and PrP, as pivotal effectors triggering neuronal demise. Though evidence correlating inclusions with neurotoxicity is substantial [124], several studies suggest compellingly that inclusions are neither necessary nor sufficient to cause neurodegeneration. The Poly-Q diseases illustrate this particularly well. In a mouse model of the Poly-Q disease SCA1, two separate molecular manipulations— mutating an E3 ligase and deleting the ataxin-1 self-association domain— reduced inclusion formation but actually worsened the disease phenotype [125,126]. A knock-in mouse model of SCA1 also showed that brain regions most susceptible to neurodegeneration had fewer inclusions than did relatively resistant regions [127]. In a *Drosophila* model of Poly-Q disease, overexpression of HSP chaperones suppressed toxicity markedly without reducing inclusions [21]. These findings, as well as, similar findings in PD models [23,28], argue that inclusion formation cannot be the principal pathogenic element in neurodegenerative diseases.

Many of the mutations that cause dominantly inherited neurodegenerative diseases have been demonstrated to dramatically increase the propensity of the mutant gene products to self-associate into protein aggregates *in vitro* and *in vivo* [128,129], supporting the widely considered hypothesis that aggregation underlies the molecular pathogenesis of many neurodegenerative disorders [1]. Amyloid diseases, the most significant class of neurodegenerative disorders associated with protein misfolding, are underscored by the aggregation of a specific protein together with a range of other components, such as additional proteins and carbohydrates, which become incorporated into amyloid deposits. Although the major protein constituents that landmark amyloidose disorders are unrelated in size or primary amino acid sequence, their aggregates exhibit common characteristics, including morphology, detergent insolubility, high β-sheet content, protease resistance, and a specific optic behavior as revealed by dye binding [130,131]. Mutations resulting in neurodegenerative diseases fall into two broad classes. The first class comprises mutations that affect proteins, irrespective of their native function, and cause them to misfold. The classic example of this one is HD. The protein encoded by the Htt gene contains a stretch of glutamine residues (or polyglutamine repeat), and the genomic DNA sequence that codes for this polyglutamine repeat is subject to misreading and expansion. When the length of the polyglutamine repeat in Htt reaches a critical threshold of approximately 35 residues, the protein becomes prone to misfolding and aggregation [21]. This appears to be the proximate cause of neurotoxicity in this invariably fatal disease. A number of other neurodegenerative diseases are caused by poly-Q expansions. For example, SCA-1 is caused by poly-Q expansions in the protein ataxin-1 [50,132]. In other diseases, protein misfolding occurs due to other mutations that induce misfolding and aggregation; for example, mutations in superoxide dismutase-1 (SOD1) lead to aggregation and neurotoxicity in amyotrophic lateral sclerosis (ALS) [133-135]. Other mutations that result in neurodegenerative diseases are instructive in that they directly implicate the UPS in the pathogenesis of these diseases. For example, mutations in the gene encoding the protein parkin are associated with juvenile-onset PD [136]. Parkin is a RING finger-containing Ub ligase, and mutations in this Ub ligase cause accumulation of target proteins that ultimately result in the neurotoxicity and motor dysfunction associated with PD [136]. Repressor screens of neurodegeneration phenotypes in animal models have also linked the molecular chaperone machinery to neurodegeneration [42,137,138]. Taken together, the pathophysiology of neurodegenerative diseases provides a compelling demonstration of the importance of the regulated metabolism of misfolded proteins and provides direct evidence of the role of both molecular chaperones and the UPS in guarding against protein misfolding and its consequent toxicity [42,137,138].

As protein misfolding poses a major threat to cell function and viability, molecular mechanisms must have evolved to prevent the accumulation of misfolded proteins and thus aggregate formation. Two protective strategies appear to be followed (see Chapter 12): (*i*) molecular chaperones are employed to stabilize nonnative protein conformations and to promote folding to the native state whenever possible [60]. Chaperones are essential for guiding folding and protecting against self-association of misfolded species into protein oligomers and aggregates. Misfolded proteins may be ‘rescued’ via chaperones and co-chaperones as has been observed for ataxin-1, parkin associated endothelin-receptor-like receptor (Pael-R) and α-Syn [139-141]. The sequestering activity of chaperones serves to prevent misfolding and aggregation [41,42,137]. (*ii*) Misfolded proteins are removed by degradation, involving, for example, the UPS and autophagy-lysosome pathways [42]. Protein fate thus appears to be determined by a tight interplay of cellular protein-folding and protein-degradation systems.

Pathways that Clear Aggregate-Prone Proteins

The autophagy-lysosome pathways and UPS are the two main routes of protein and organelle clearance in eukaryotic cells (Figure 2). Autophagy, is the mechanism by which long-lived, stable proteins are degraded, and is the only mechanism by which entire organelles such as mitochondria and peroxisomes are recycled. Autophagy was originally described as a cellular response to starvation, and one of the primary functions of autophagy is to produce amino acids from degraded proteins for the survival of the cell when nutrients are scarce. Thus, autophagy is stimulated by the decrease in amino acid content [142] and is hormonally controlled; glucagon promotes, while insulin inhibits, autophagy [143]. The lysosomal pathway of intracellular protein turnover can be further divided into three distinct pathways in higher eukaryotes: macroautophagy, CMA, and microautophagy [144].

Macroautophagy has been described as the main route for bulk protein degradation under conditions of nutrient starvation or stress. It is generally considered to be a non-specific process in organisms from yeast to humans, but a specific uptake process is conceivable, as there are precedents for cargo selection in the macroautophagy-related cytoplasm-to-vacuole trafficking (Cvt) pathway in yeast [145]. Macroautophagy is a multi-step process. First, the autophagosome, a vesicular structure possessing a double membrane presumably derived from the ER, forms in the cytoplasm, engulfing various proteins, lipids, and damaged or dysfunctional organelles. The autophagosome later fuses with primary lysosomes (or vacuoles in yeast), which are comprised of a single membrane-bound compartment harboring a host of hydrolytic enzymes. The external membrane of the autophagosome becomes part of the lysosomal membrane upon fusion. Following fusion, the complex acidifies and matures into an autophagolysosome. The autophagosome and autophagolysosome are collectively referred to as autophagic vacuoles (AVs). Finally, the inner membrane structure within the autophagolysosome disintegrates while its contents are digested, and the vacuolar contents are recycled to provide amino acids and energy as needed by the cell (Figure 2; for review, see [146]). The proteins involved in macroautophagy have been well defined in yeast, and a subset of the genes involved are known as *apg* or *aut* genes (for review, see [147,148]). The proteins involved in mammalian macroautophagy are less well known, although a number of mammalian homologues to the yeast *apg* proteins have been recently discovered. Very little information has been garnered for proteins involved in neuronal macroautophagy, a role for the lysosomal enzyme cathepsin D has been suggested by the recent introduction of cathepsin D knockout mice. The central nervous system (CNS) tissues of these mice are filled with autophagosomes and autophagolysosomes and contain ceroid lipofuscin [149], suggesting that cathepsin D is necessary for complete autophagic proteolysis.

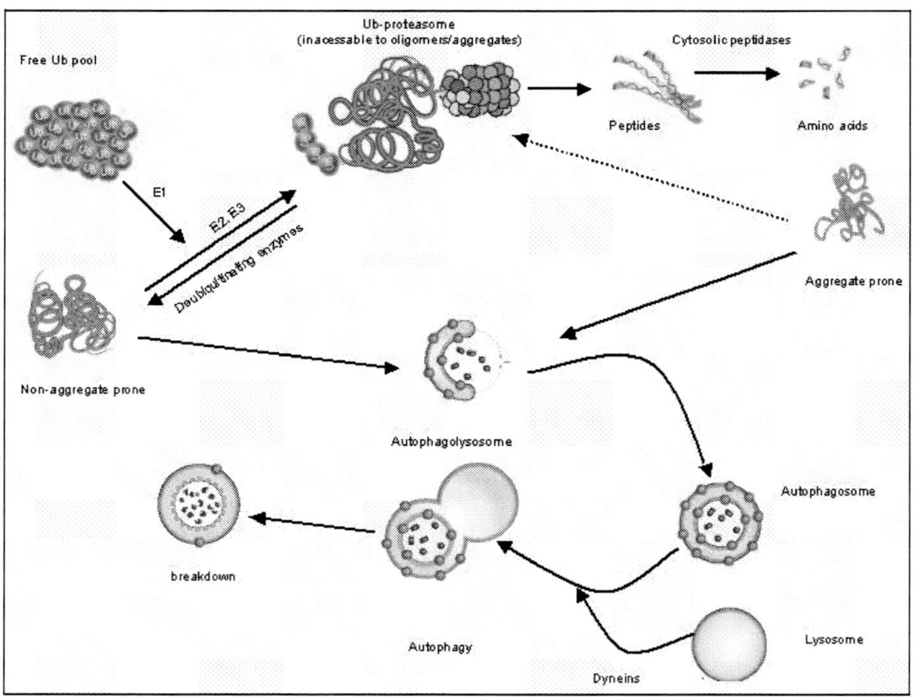

Figure 2.

Another pathway for cytosolic protein clearance through lysosomes is CMA, a pathway distinct from macroautophagy that was first identified in confluent, serum-deprived cultured fibroblasts and its defining feature is selectivity for distinct proteins [150]. This process is restricted to the elimination of proteins that possess an amino acid sequence biochemically related to the pentapeptide Lys-Phe-Glu-Arg-Gln (KFERQ) during conditions of prolonged starvation [151]. Proteins containing this motif, which is present in ~30% of cytosolic proteins [152], are recognized and bound by the cytosolic form of heat shock cognate protein of 70 kDa (HSC70), a molecular chaperone [153,154]. The list of known substrates for chaperone-mediated autophagy include annexins, transcription factors, glycolytic enzymes, and cytosolic protease subunits [155] and it is likely that a diverse array of proteins are degraded by this pathway. The selectivity of this pathway is further mediated by lgp-96 (or LAMP2A), a lysosomal integral membrane receptor of 96 kDa [156]. This transfers protein substrates to the lysosomal membrane, where, through binding to the LAMP2A, they are translocated into the lysosomal lumen and degraded. Lysosomes can also directly engulf cytoplasm by invagination, protrusion and/or septation of the lysosomal limiting membrane, a process known as microautophagy [157]. Microautophagy appeared to be responsible for the gradual, continuous turnover of cytosolic proteins that is not activated by nutritional deprivation or stress. However, this process is poorly understood in mammalian cells.

The UPS is the other and more efficient pathway in degrading wild-type proteins (Figure 2; see Chapter 3). The UPS is responsible for the regulated degradation of aged, damaged, and misfolded proteins. Proper function of the UPS is vital for cell survival, and the loss-of-function of the UPS has been found in a wide variety of neurodegenerative diseases, including AD, PD, and HD, and many others [20,158,159]. In neurodegenerative diseases, perturbations of proteasome function may occur through its recruitment to or sequestration into protein aggregates, through chronic overloading of proteasome capacity by misfolded protein, or through still undetermined effects on other activities of the UPS. Coupled with UPS dysfunction is protein aggregation, in which proteins self associate to form one large, tangled complex. Whether aggregation triggers UPS dysfunction or vice-versa is widely debated, as well as the effect of protein aggregates on cell toxicity [88,158]. The UPS is also important for the degradation of misfolded proteins in the ER. Misfolded ER proteins are retrotranslocated back into the cytosol, where they are degraded by proteasomes (a process known as endoplasmic reticulum-associated degradation, or ERAD; see Chapter 13). Many of the proteins that cause proteinopathies (including those with poly-Q expansions and α-syn) are partly dependent on the UPS for their clearance [160,161]. The inhibition of macroautophagy has much smaller effects (if any effect at all) on the clearance of wild-type Htt exon 1 fragments or wild-type α-syn than on the clearance of the mutant aggregate-prone species [160-164]. For the proteins that have access to both pathways, proteasomes are the favoured and dominating clearance route. This preference for UPS may also relate to tagging and targeting mechanisms that contribute to the selectivity of this pathway. When a cytosolic protein is aggregate-prone and a poor proteasome substrate, macroautophagy becomes a major clearance route by default and becomes more effective than the proteasome (Figure 2). PolyQ-expansion mutations, such as those seen in mutant Htt and ataxin 3 mutant forms of α-syn, and different forms of tau are strongly dependent on the macroautophagy pathway for their clearance [161-167]. Inhibition of macroautophagy by gene knockout in otherwise normal mice leads to the formation of ubiquitinated aggregates in various tissues [168] and the dependence of proteins on the macroautophagy pathway for their clearance correlates with their propensity to aggregate [160-164]. Furthermore, when normal cultured neurons are treated with proteasome inhibitors, the proteins that form aggregates seem to be degraded by macroautophagy [169]. Probably the macroautophagy is a default pathway that becomes increasingly important when an aggregate-prone substrate cannot be efficiently cleared by the proteasome, it is possible that there are also signals that preferentially target such proteins to autophagosomes.

Neurodegeneration and Autophagy Dysfunction

One of the salient pathological features of chronic neurodegenerative disease is the slow, continual loss of cells within specific neuronal populations. Apoptotic cell death has been reported in AD, PD, HD, ALS, and various prion diseases, including Creutzfeldt-Jakob disease (CJD) (for review, see [170,171]). Autophagy may play a protective role in the early stages of programmed cell death [172,173]. A number of studies have demonstrated the presence of autophagy in HD brain [174] and in animal models of HD [175,176]. The increased autophagic activity in these contexts appeared to be due to the sequestration and functional inactivation of mTOR (a negative regulator of autophagy) by the poly-Q aggregates. This may be a mechanism accounting for enhanced autophagy in diseases other than HD as mTOR sequestration was seen in the aggregates in brains of patients with a number of different poly-Q expansion diseases [167]. Thus, increased autophagy in poly-Q diseases may be due to increased autophagosome production, due to inhibition of mTOR, which negatively regulates an early stage in this pathway. While the proteasome can process soluble monomeric forms of Htt, its narrow pore precludes proteasome clearance of oligomeric and aggregated species. Furthermore, the proteasome cannot cleave within the poly-Q tract [177]. Thus, the proteasomes effectively remove the sequences flanking the poly-Q stretch and leave an expanded poly-Q stretch with almost no flanking sequences.

Likewise, autophagic degeneration was observed in AD [178,179] and alterations in the lysosomal system were observed in both human disease and animal models of AD [178,180,181]. Because proteasome activity declines in aging and is further impaired in AD [182,183], a loss of autophagy function, if it develops in AD, leaves neurons with no competent mechanism to remove abnormal and potentially toxic proteins.

Autophagic degeneration is prevalent in melanized neurons of PD brain [184]. Recent evidence provides further support for this autoptic data. PC12 cells expressing mutant α-syn are deficient in the degradation of protein substrates through the CMA pathway [185]. The upregulation of macroautophagy that occurs in PC12 cells expressing mutant α-syn appears therefore to be a compensatory response, following the primary defect at the level of CMA [185]. The primary aberrant effect appears to be an inappropriately tight binding of mutant α-syn to LAMP2a, the receptor component of CMA. In this fashion, mutant α-syn do not permit the binding and uptake of the normal endogenous substrates of CMA [185]. Thus, in this cellular model, macroautophagy is a compensatory response of the cell attempting to cope with constraints imposed on its ability to degrade proteins, and not a primary mechanism of cell death induction. It has also been hypothesized that PD may be caused, at least in part, by dysfunction of the UPS [186]. It has recently been demonstrated that neuronal cell lines treated with low doses of proteasome inhibitors activate macroautophagy as a compensatory response [187]. Similar findings have been reported earlier in HeLa cells acutely treated with proteasome inhibitors [188]. In the first such study in primary neurons, Rideout *et al.* observed marked induction of macroautophagy and activation of the lysosomal pathway in cultured cortical neurons treated with proteasome inhibitors [169]. Furthermore, pharmacological modulation of macroautophagy led to alterations in the number of LB-like ubiquitinated inclusions formed in this model, suggesting that macroautophagy may be responsible for inclusion dissolution [169]. A related issue that is quite controversial is the degradation of α-syn itself. Initial studies showed that α-syn, in some cases in a polyubiquitinated form, accumulated in cells upon proteasome inhibition, suggesting that the proteasome was responsible for α-syn degradation through the UPS [189-191]. A subsequent study, based on a purified system, showed that α-syn, as would be predicted from its natively unfolded state, does not need to be ubiquitinated to be degraded by the proteasome [192]. To complicate matters, other studies have failed to replicate the finding of accumulation of endogenous or overexpressed α-syn with proteasome inhibition [193-196]. Some of these studies in particular were performed with endogenous α-syn in PC12 cells or cultured cortical neurons [194,195] and thus may reflect more closely the normal turnover of the protein. While some of the earlier studies may have suffered from limitations due to the artificial transient nature of α-syn overexpression, the presence of epitope tags, and the expression in a non-neuronal environment, other studies have observed that proteasome inhibition impairs α-syn clearance in both differentiated and undifferentiated PC12 cells [161]. One reason for these apparent discrepancies is that α-syn may be processed via a diversity of pathways and that a small change in turnover resulting from proteasome inhibition (which may not be a major degradation route) may not rapidly or overtly translate into alterations in steadystate levels. There is now considerable evidence that α-syn levels can accumulate with lysosomal inhibition, although the exact circumstances (cellular milieu, species of α-syn, class of lysosomal inhibitor) differ across studies [161,185,196-198]. In the most physiological system examined so far, that of cultured post-natal rat dopaminergic neurons, endogenous α-syn turnover was significantly increased only by the general lysosomal inhibitor ammonium chloride, but not by the macroautophagy inhibitor 3-methyladenine. Consistent with a minor role of the proteasomes in α-syn degradation, application of epoxomicin, the selective proteasome inhibitor, provided only a very small increase in the half-life of α-syn in this setting. In fact, α-syn contains a pentapeptide motif that targets it to the pathway of CMA. Analysis in an *in vitro* system of purified lysosomes has confirmed that α-syn can be degraded by CMA. CMA appears to be the major rate-limiting pathway used for wild-type α-syn degradation in a neuronal cell context [185]. It would also be interesting to study the role of macroautophagy in α-syn clearance. In conclusion, macroautophagy appears to occur under physiologic conditions in *substantia nigra pars compacta*, and to be further induced in PD. It is likely to represent a compensatory response to the dysfunction of other intracellular degradation pathways, mitochondrial dysfunction or the accumulation of cytosolic dopamine. The attempt of the neurons to activate macroautophagy may eventually backfire, leading to further cellular dysfunction and autophagic death. Degradation of α-syn in lysosomes may be closely related to its aberrant effects. In particular, disturbance of CMA may be a primary pathogenic effect of mutant α-syn. While mutant forms of α-syn are very rare, it may be important to consider if there is dysfunction of CMA in sporadic PD.

Experimental scrapie and CJD, both prion-related diseases, promoted the activation of neuronal autophagy and formation of AVs [199,200]. Finally, lipofuscin that accumulates in epithelia and neurons of senescent organisms is associated with autophagocytosis [201].

Interestingly, mice with neuronally confined autophagy-gene knockouts develop intraneuronal aggregates and neurodegeneration [202,203]. Although defects in autophagy-restricted genes have not been described in humans, autophagy may be compromised by other means. For example, dynein function is important for autophagosome-lysosome fusion, and autophagic clearance is impaired even by hemizygous loss of dynein activity [204]. Dyneins are motor proteins that typically move cargoes towards the cell centre along microtubules, and components of this machinery are genetically compromised in some forms of motor neuron disease, either directly [for example, by a dynactin mutation in one form of human motor neuron disease [205,206], and dynein heavy-chain mutations in the legs at odd angles (*Loa*) and cramping (*Cra1*) mice] [207], or indirectly, if dynein functions are compromised by other mutations (for instance, SOD1 mutations causing familial ALS [208,209]). In some of these conditions (for example, SOD1 mutations, *Loa* mice and human dynactin mutations), protein aggregation is a feature, and this may be due in part to compromised autophagy. This possibility is supported by studies showing increased autophagosome numbers (compatible with decreased autophagosome-lysosome fusion) in the *Loa* mice, and increased aggregation and toxicity of mutant huntingtin fragments when HD mice are crossed with the *Loa* mice (or when the HD mutation is expressed in flies with hemizygous loss of either dynein heavy-chain or light-chain genes) [204]. Although one cannot claim that defective autophagy is responsible for all of the pathology resulting from such mutations, by slowing the clearance of the aggregate-prone proteins and possibly increasing susceptibility to apoptosis, its contribution of dyneins is likely to be significant [210,211]. The chaperone-mediated autophagy pathway may also be relevant to neurodegeneration. Wild-type α-syn has recently been shown to be selectively translocated into lysosomes for degradation by the CMA pathway as well as being a substrate for the proteasome. In assays *in vitro* using isolated lysosomes, the pathogenic A53T and A30P α-syn mutant proteins were shown to bind to the CMA-pathway receptor on the lysosomal membrane, but seem to act as uptake blockers, inhibiting both their own degradation and that of other substrates [185]. It will be interesting to investigate whether similar effects are mediated by these mutations *in vivo*, because such a mechanism might contribute to their toxic gain of function.

Neurodegeneration and Proteasome Dysfunction: The Proteasomepathies

The UPS, recognizing and selectively degrading misfolded and damaged proteins, protects cells against the potentially toxic effects of protein aggregation. A number of proteinopathies have been associated with decreased proteasome activity in humans or in cell models. Analyses of AD and PD human disease brain have shown modest decreases in the proteolytic activities of the proteasome [183,186]. General levels of proteasome activity are also reduced in affected neurons during the symptomatic phase in a transgenic mouse model of familial ALS [212,213]. In transfected cells overexpressing disease proteins, the formation of aggresomes is associated with decreased proteasome activity. For example, using a 'destabilized' green fluorescent protein (GFP) reporter cell line that provides a physiological readout of proteasome activity, Kopito and co-workers observed failed UPS activity as suggested by increasing in GFP fluorescence when mutant proteins became aggregated in the cell [27]. Similar results have been observed in cellular models of PD [24,191,194,214], prion diseases [215-217], poly-Q diseases [27,122,218], and ALS disease [219]. A reduction in proteasome activity, however, does not necessarily require aggregation formation. In PC12 neural cells, for example, Tanaka *et al.* showed that inducibly expressed mutant α-syn did not form aggregates yet significantly decreased peptidase activities of the proteasome, resulting in increased sensitivity to sublethal doses of proteasome inhibitors [28]. Additionally, whereas a nontoxic reduction in proteasome activity was observed in neuronal cells expressing mutant α-syn, subsequent stress with normally sub-lethal concentrations of a proteasome inhibitor resulted in apoptotic cell death [28]. These findings of proteasome inhibition have been observed in cell-culture models where the mutant protein is overexpressed. Whether this accurately mirrors the *in vivo* state, where expression levels of mutant protein are lower and the pathophysiological insult occurs over years rather than days, is unknown. Probably, under most circumstances cells are capable of handling misfolded proteins sufficiently to prevent them from exerting toxicity and/or being sequestered into inclusions. However, under circumstances of increased physiological or environmental stress or compromise of protective mechanisms with aging, the UPS may become overloaded and impaired. Indeed, despite the existence of many mouse models with intraneuronal inclusions, no one has yet reported inhibition of proteasome activity by inclusions *in vivo*. There is even some negative evidence: in some poly-Q disease mouse models, proteasome impairment is not seen at early stages [220,221]. Proteasome subunits can be cleaved by caspases in cells undergoing apoptosis [222], and it is possible that this process may explain some of the cell-model data. However, further studies of proteasome function in animal models, including inducible models, and the necessity of conducting accurate biochemical measurements are required to answer whether direct compromise of proteasome activity plays any role in pathogenesis.

Proteasome inhibition, pharmacologically or genetically, increases aggregation or inclusion formation in cells and invertebrate models [21,122,194,214,215,218,223-226]. If during the course of disease the UPS becomes compromised for any reason, this would further reduce the global rate of protein degradation and foster a cellular environment that increasingly favors aggregation and inclusion formation. Because a build-up of aggregated protein in the cell may itself directly compromise the UPS [27], a vicious cycle of protein aggregation and proteasome perturbation might ensue. In addition, at least some disease proteins appear to be degraded less efficiently than their normal counterparts — a property that also should increase their steady-state levels and favor aggregation. For example, mutant α-syn was degraded 50% slower in a cellular PD model [189], degradation rates of Htt and ataxin-1 slowed proportionally with increasing glutamine repeat length [122,125], and ataxin-7 protein levels increased over time in a SCA7 mouse model [227]. Once disease proteins have aggregated, they are probably even less efficiently degraded. Still, the UPS has the capacity to eliminate proteins stuck in inclusions, as shown in an inducible HD mouse model where inclusions were cleared once the Htt transgene was repressed [228]. The significance of this latter role is underscored by the discovery that loss-of-function mutations in genes encoding UPS components can cause neurodegenerative diseases in humans [229] and rodents [230,231] and enhance the cytotoxicity of aggregation-prone proteins linked to dominantly inherited neurodegenerative diseases [125,232].

In general, the inclusion bodies consist of insoluble, unfolded proteins that are commonly tagged with the small protein, Ub. Covalent tagging of proteins with chains of Ub generally targets them for degradation. Indeed, the UPS is the major route through which intracellular proteolysis is regulated. This strongly implicates the UPS in these disease-associated inclusions, either due to malfunction (of specific UPS components) or overload of the system (due to aggregation of unfolded/mutant proteins), resulting in subsequent cellular toxicity. The production or accumulation of intracellular protein aggregates in cells profoundly impairs the functional capacity of the UPS [27,122,233]. Although defective UPS function is a robust and reproducible response to protein aggregation, the mechanism by which protein aggregation is linked to UPS impairment remains an open and compelling question.

Apoptosis is promoted by accumulation of abnormal proteins and through depletion of HSPs that have direct inhibitory effects on apoptotic pathways in addition to their protein chaperoning activity [34]. The proteasome is the major pathway for degradation of transcription factors and other short-lived regulatory proteins, thus proteasome inhibition could alter transcription of multiple gene families includine those that promote cell death [234]. Another important function of proteasomes is degrading misfolded proteins shuttled into the cytoplasm from the ER. ER stress results from disturbance of ER calcium homeostasis or an imbalance in the amount of misfolded protein and ER chaperoning capacity; both ER and Golgi networks are disrupted for example in ALS and experimental models of mutant SOD1, implicating multiple compartments in failure of protein QC [235]. Proteasome inhibitors also disrupt mitochondrial homeostasis [236,237].

Inclusion formation is also, at least in part, a cellular response that serves to concentrate misfolded proteins and perhaps facilitate their degradation— in other words, a cell-driven process. The best example is the aggresome: an inclusion body described in many cell models that overexpress mutant proteins [41,215,238,239]. Aggresomes are cytoplasmic deposits of aggregated protein that form in a microtubule-dependent manner and localize to the microtubule organizing center. Aggresomes arise when the rate of abnormal protein production exceeds the cell's capacity to handle it, including when the proteasome is compromised [41,215,238,239]. An unresolved issue is whether the cytoplasmic inclusions seen in human diseases— such as LBs, glial cytoplasmic inclusions and perinuclear poly-Q inclusions— are in fact aggresomes. In addition, it is entirely unknown what nuclear forces contribute to the formation of the nuclear inclusions seen in poly-Q diseases and in some neurons during normal aging (e.g. the Marinesco bodies found in neurons of the substantia nigra). However, it is not clear whether these phenomena are a cause or late effects of disease. There is no doubt that proteasome inhibition enhances aggregation of mutant proteins in cells. It is less clear whether aggregates directly inhibit proteolytic activities of the proteasome in the disease state. Evidently, substrates need to be unfolded to pass through the narrow pore of the proteasome barrel, which precludes the clearance of oligomeric and aggregated proteins [233]. Two possible models could account for the impairment of UPS function by protein aggregation. One model suggests that aggregated or aggregation-prone proteins directly inhibit or 'choke' the 26S proteasome— a situation that might result from their engagement by degradation-resistant [240,241] or hard-to-unfold proteins: perturbations of proteasome function may occur through its recruitment to or sequestration into protein aggregates, through chronic overloading of proteasome capacity by misfolded protein, or through still undetermined effects on other activities of the UPS. In this manner protein aggregates directly inhibit or sequester 26S proteasomes. Because proteasome proteolysis is highly processive [242], aggregates could be inhibitors of proteasome degradation, as they are undegradable and slowly released. Recent data suggest that the proteasome's enzymatic machinery might not be able to cleave between successive glutamine residues [240,241], despite the presence of endogenous ubiquitination. This has important implications for poly-Q expansion mutations, as it suggests that the proteasome removes the flanking sequences around the expansions. On exiting from the proteasome, such isolated poly-Q stretches may be more toxic than the pre-proteasome species with the flanking sequences. However, it is likely that the isolated poly-Q peptides exiting the proteasome are rapidly degraded by as yet unidentified cytosolic peptidases, because there are a number of normal proteins with wildtype poly-Q stretches of 20-35 repeats.

This 'proteasome choking' model predicts that in cells exhibiting near complete loss of UPS function, a substantial fraction of total cellular proteasomes should be associated with protein aggregates. Thus, aggregates could simply sequester proteasomes away from cellular sites where they are required, as suggested from the observation of proteasome subunits in IBs in brains from human [243] and animal models [58] of neurodegenerative disease. Inspection of immunofluorescent images of proteasome subunit distribution in aggregate-containing cells indicates a clear enrichment of proteasome subunits in IBs. However, this enrichment is not accompanied by significant depletion of proteasomes from nuclear or cytoplasmic pools [244] and protein aggregates do not inhibit 26S proteasome-mediated degradation of Ub-dependent and Ub-independent substrates, even when present in vast molar excess argueing strongly against a requirement for direct physical interaction between proteasomes and aggregates for UPS impairment [244]. These data do not exclude the possibility that ubiquitinated aggregates might interact more tightly with proteasomes. In AD and several other neurodegenerative diseases, molecular misreading of a Ub gene leads to accumulation of an aberrant, frame-shifted Ub, UBB^{+1} [245]. UBB^{+1} accumulates in degenerating neurons of tauopathies, such as AD, progressive supranuclear palsy (PSP) and Pick's disease (PID), and of Poly-Q diseases. By contrast, synucleinopathies, such as PD, LBD and multiple system atrophy (MSA), are negative for UBB^{+1} [246]. The UBB gene codes for a polyubiquitin protein, which is post-translationally cleaved into three Ub monomers by Ub specific protease. In the UBB mRNA, a GU dinucleotide deletion occurs adjacent to a GAGAG motif in the first Ub repeat, and a CU deletion occurs in a CUCU motif in the third Ub repeat. The first deletion results in the extended Ub molecule, UBB^{+1}, which comprises 95 amino acids instead of 76, of which the last 20 amino acids are out of frame. The CU deletion results in a truncated Ub of 19 amino acids, of which the last five are out of frame. UBB^{+1} is ubiquitinated, targeted to the proteasome and then degraded by the proteasome; a low expression of the protein is thus not detectable [247]. A high expression of UBB^{+1} can block the proteasome and subsequently will result in cell death by apoptosis [248]. The inhibitory effect of UBB^{+1} has been shown in a cell-free system [249] and in living cells by monitoring the accumulation of GFP-based proteasome substrates [247,250]. This inhibition is completely dependent on the ubiquitination of UBB^{+1}. A mutant UBB^{+1} in which both lysine residues at positions 29 and 48 were replaced with arginine residues, is not targeted to the proteasome and is consequently unable to inhibit UPS activity [247]. Conditional expression of UBB^{+1} in neuroblastoma cells results in elevated expression of HSPs, which might be explained by the accumulation of misfolded proteins, following inhibition of UPS activity [250]. An induction of the heat-shock system chaperones misfolded proteins and protects the cell against a possible toxic action of these proteins. Unexpectedly, these cells were found to be less vulnerable to oxidative stress [250]. This last finding is an indication that proteasome inhibition and aggregate formation might protect neurons against oxidative stress, perhaps by compensatory mechanisms.

A second model, not mutually exclusive with the first, is that protein aggregates indirectly interfere with UPS function by sequestering or directly clogging proteasomes, it is possible that they could impair UPS function by influencing the proteasome activity or distribution of UPS modulators, inactivating or depleting a UPS activator. For example, it is has been suggested that depletion of proteasomes [251] or other UPS components [252] by sequestration into IB could account for the observed impairment of UPS function by protein aggregates. However, measurement of free Ub levels in cells with large aggregate burdens does not support this hypothesis [27]. Proteins containing expanded poly-Q have been shown to interact with and inactivate poly-Q containing transcription factors [253,254]; recent studies have suggested that this interaction can occur with an early, oligomeric form of poly-Q repeat and may depend more on the poly-Q conformation than on its aggregation state [255]. Smaller intermediate forms of protein aggregates are more toxic to cells than fibrillar forms [88,131]. Consistent with this, it has been argued that the formation of higher order aggregates and their subsequent coalescence into IBs may be cytoprotective [256,257]. Most disease-linked aggregated proteins accumulate in IBs that are characteristically restricted to either the nucleus or cytoplasm, and the effective toxicity associated with the presence of these IBs appears to be strongly influenced by the cellular compartment in which the aggregates accumulate [126,258]. Finally, it should be noted that nonnative pathogenic conformers of poly-glutamine proteins could interact with and activate UPS inhibitors. For example, the presence of nonnative undegraded protein aggregates could initiate caspase activation; one recent study reported that caspase activation results in irreversible inhibition of proteasomes via cleavage of 19S regulatory particle subunits S5a, S6', and S1 [222]. A related issue is whether a natural age-related decline in proteasome activity contributes to disease. Most inherited and acquired neurodegenerative diseases occur later in life, and some studies suggest that aged brains have lower proteasome activity [259,260]. Evidence supporting a link between aging and decreased protein surveillance comes from a *Caenorhabditis elegans* model of poly-Q disease in which a mutation in the longevity gene age-1 exacerbated aggregation and toxicity [261].

The above findings demonstrate compromise of proteasome catabolic capacity in several models of neruodegenerative diseases. However, how those relate to disease pathogenesis requires further work.

The strongest support for the possibility that the UPS has a primary role in the pathology of a wide range of proteinopathies (or in sporadic forms of diseases such as PD) comes from human/mammalian mutations that involve components of this pathway and cause neurodegeneration. Regardless of whether there is any direct inhibition of the 20S core, misfolded and aggregated disease proteins may perturb upstream elements of the UPS and, by extension, related Ub-dependent cellular processes. Just how important such upstream Ub pathways are for neurons is illustrated by the fact that early onset parkinsonism can be caused by the failure of a single Ub ligase, parkin [262] in which autosomal recessive loss-of-function mutations cause PD [190,263,264]. A number of parkin substrates have been identified, but at present it is not clear which is the major contributor to pathology in parkin deficiency. Interestingly, parkin mutations usually cause nigral degeneration without LB formation, arguing that the compromise in Ub pathways caused by parkin deficiency occurs independently of aggregate formation [262]. Heterozygous mutations in the gene encoding ubiquitin carboxy-terminal hydrolase L1 (UCHL1) have been causally implicated in families with PD [265], although the data are contentious [266]. Although the S18Y variant has been associated with sporadic PD in genetic-association studies, this association has not withstood robust confirmatory analyses [267]. UCHL1 might function not only as a ubiquitin hydrolase (hydrolysing ubiquitin chains to free ubiquitin monomers) but also as a ubiquitin ligase [191]. The role of *UCHL1* in human PD is still unclear, but loss of function is known to be deleterious, as recessive deletion of part of this gene causes gracile axonal dystrophy in mice [231]. Although this phenotype is not obviously parkinsonian, it is associated with the formation of ubiquitinated intraneuronal aggregates [231]. Mutations in CDC48/p97 (also known as valosin-containing protein, VCP) cause a dominantly inherited disease known as inclusion-body myopathy with Paget's disease of bone and frontotemporal dementia, which is also characterized by cytoplasmic and nuclear aggregates in muscle and brain tissues [268]. This pathology is consistent with the role of VCP as a component of the machinery shuttling ubiquitinated ERAD substrates to the proteasome [269,270] (although VCP has a number of other roles that may also affect pathology). In addition, mutations in the E3 ligases E6-AP and hCdc34p have been shown to enhance disease progression in a mouse model of SCA1 and toxicity in a cell model of HD, respectively [125,258]. It is important to recognize that perturbations of the UPS could be subtle and still compromise neurons by rendering them more susceptible to other cellular stresses. This was evident in a cell model in which expanded poly-Q protein caused a decrease in proteasome activity only when the cells were treated with heat shock [271].

The above genetic data, which suggest that primary genetic deficiencies of components of the ubiquitin-proteasome system are sufficient to cause neurodegeneration, are complemented by pharmacological data showing that the injection of proteasome inhibitors into adult rats causes parkinsonian features, including LB-like aggregates [272]. However, it is still not clear whether proteasome dysfunction is an important early causal factor *in vivo* in proteinopathies in which there are no primary defects in the UPS.

Although studies of human disease tissue suggest a role for proteasome perturbation in disease, proving that there is a causal link between proteasome inhibition and neurodegeneration is next to impossible in postmortem human tissue. Several groups have reported inhibition of the proteasome postmortem brain tissue. These experiments all involved homogenization of tissue and addition of fluorogenic peptides with protease cleavage sites. These substrates have been developed for the different proteolytic activities of the proteasome and they have subsequently been used to study the activity of the purified proteasome complex [273,274]. To elucidate the role of the UPS in the pathogenesis of neurodegenerative diseases, methods for measuring and manipulating proteasome activity, preferably in the brain of the living mouse, are needed. Recently, such methods have become available, and the first *in vivo* applications have been reported [220]. Based on irreversible inhibitors of the proteasome, another probe for activity has recently been developed which modifies the enzymatic moieties in the 20S core. With this activity probe, it is possible to determine the subunit-specific inhibition of the proteasome in living cells [275].

GFP-based constructs have been developed, enabling the possibility of monitoring proteasome activity *in vivo*. These constructs are based on the fact that destabilizing amino acid sequences (degrons) can confer proteasome-targeting signals to otherwise stable proteins. Independently, researchers have made destabilized versions of GFP, by fusion of CL1, a short degron from yeast, to GFP [27], using N-end rule modification [276] of the amino-terminus of GFP, by the destabilizing the amino acid arginine (R) (Ub–R–GFP) [277] or by synthesis of a Ub fusion degradation–GFP substrate (Ub^{G76V}–GFP) [277]. The Ub^{G76V}–GFP construct has been engineered into transgenic mice and has been shown to report proteasome inhibition *in vivo* [278]. However, a caveat of the GFP reporters of proteasomal function is that significant loss of activity may be required before GFP accumulates to readily visible levels [279,280] and how cells live with this level of functional impairment is not known.

An alternative mouse model to monitor inhibition of the UPS has been engineered by fusing four copies of noncleavable Ub^{G76V} to luciferase [281]. *In vivo* bioluminescence imaging of live Ub–luciferase mice thus enables repetitive testing of proteasome inhibitors. These reporter systems for the proteasome all depend on ubiquitination of the GFP- or luciferase fusion proteins, and in all likelihood the E2 and E3 ubiquitination enzymes are different for all three reporters: the Ub–R–GFP uses N-end rule Ub ligases UBR1 or UBR2 [282], the Ub^{G76V}–GFP substrate is recognized by an unidentified Ub ligase and it is suggested that the CL1 degron is recognized by doa10 [283]. Because ubiquitination is an essential step for these proteasome activity reporters to be degraded, conclusions based on these reporters should always consider this caveat but the use of two different reporters [247] might rule out potential artifacts induced by the Ub ligases. In addition, recently, ubiquitination-independent GFP-based reporter has been established based on mouse ornithine decarboxylase [284], which is known to be degraded by the 20S proteasome independently of ubiquitination. Monitoring the ubiquitination of specific substrates in living cells has recently also been facilitated by Ub mediated fluorescence complementation [285]. In this system, split GFP is reconstituted *in vivo* when a substrate is ligated to Ub molecules (ubiquitinated) or to small Ub-like modifier molecules (sumoylated). One attractive target for this system is tau, which is ubiquitinated in tangles [286] and thought to be regulated by the proteasome in AD [287,288].

Conclusion

Undoubtedly, a large number of human diseases are associated with the accumulation of a whole myriad of misfolded proteins. This accumulation can be in tissues, intra- or extracellularly, in the CNS or in the periphery and contained within inclusions, aggresomes or filaments typically of fibrillar structure and accompanying extensive neuronal cell loss, displaying a selective brain distribution. The recently discovered similarities of a number of these aggregates with a novel type of experimentally induced protein deposit, formed as a general response to discrepancies in protein turnover and designated the 'aggresome', has prompted speculations about the involvement of degradation pathways (UPS and macroautophagy) acting on such aggregate-prone cytosolic proteins. Consistent with this view, protein aggregates have been regarded in a pathogenic connotation, with the most aspects of neurologic pathogenesis being largely attributed to their presence in nerve tissues. However, the neurotoxicity of protein aggregates remains ambiguous as direct evidence substantiating it have long remained elusive. A convergence of evidence now support the notion that the actual culprits might comprise the oligomeric, non-fibrillar intermediates that arise early during the aggregation process, termed protofibrils and that the fibrillar end-stage aggregates themselves might actually serve a neuroprotective function. The UPS is more efficient in degrading wild-type proteins than basal levels of macroautophagy, however the proteasomes can show a reduced catalytic activity towards the substrate molecule with the consequent filling of the proteolytic chamber or a a very low transport to the cavity of the 20S, mostly because of an impaired or highly costly interaction between the ubiquitinated substrate and the 19S cap with a consequent accumulation of misfolding protein suggesting that the specificity of the substrate and the proteasome structure influence dynamically the degradation rate of proteasome. When a cytosolic protein is aggregate-prone and a poor proteasome substrate, macroautophagy becomes a major clearance route by default and becomes more effective than the proteasome. The above ones attest to the fact that protein aggregation remains a complex issue with a role far more enigmatic than originally thought but nonetheless important for the understanding of the pathological basis of neurodegenerative disorders. The identification of a specific protein degradation system disturbance and their corresponding clinical syndromes will assist in choosing therapies when protein-specific disease-modifying treatments are available.

References

[1] Taylor JP, Hardy J, Fischbeck KH: Toxic proteins in neurodegenerative disease. *Science* 2002, 296:1991-1995.

[2] Carrell RW, Lomas DA: Conformational disease. *Lancet* 1997, 350:134-138.

[3] Spillantini MG, Bird TD, Ghetti B: Frontotemporal dementia and Parkinsonism linked to chromosome 17: a new group of tauopathies. *Brain Pathol* 1998, 8:387-402.

[4] Hardy J, Selkoe DJ: The amyloid hypothesis of Alzheimer's disease: progress and problems on the road to therapeutics. *Science* 2002, 297:353-356.

[5] Glenner GG, Wong CW: Alzheimer's disease and Down's syndrome: sharing of a unique cerebrovascular amyloid fibril protein. *Biochem Biophys Res Commun* 1984, 122:1131-1135.

[6] Rudd PM, Wormald MR, Wing DR, Prusiner SB, Dwek RA: Prion glycoprotein: structure, dynamics, and roles for the sugars. *Biochemistry* 2001, 40:3759-3766.

[7] Galpern WR, Lang AE: Interface between tauopathies and synucleinopathies: a tale of two proteins. *Ann Neurol* 2006, 59:449-458.

[8] Ross CA, Poirier MA: Protein aggregation and neurodegenerative disease. *Nat Med* 2004, 10 Suppl:S10-S17.

[9] Ross CA, Poirier MA: Opinion: What is the role of protein aggregation in neurodegeneration? *Nat Rev Mol Cell Biol* 2005, 6:891-898.

[10] Fersht AR, Daggett V: Protein folding and unfolding at atomic resolution. *Cell* 2002, 108:573-582.

[11] Dill KA, Chan HS: From Levinthal to pathways to funnels. *Nat Struct Biol* 1997, 4:10-19.

[12] Thirumalai D, Hyeon C: RNA and protein folding: common themes and variations. *Biochemistry* 2005, 44:4957-4970.

[13] Yon JM: Protein folding: a perspective for biology, medicine and biotechnology. *Braz J Med Biol Res* 2001, 34:419-435.

[14] Dobson CM: Principles of protein folding, misfolding and aggregation. *Semin Cell Dev Biol* 2004, 15:3-16.

[15] Dobson CM: Protein aggregation and its consequences for human disease. *Protein Pept Lett* 2006, 13:219-227.

[16] Pangalos MN, Jacobsen SJ, Reinhart PH: Disease modifying strategies for the treatment of Alzheimer's disease targeted at modulating levels of the beta-amyloid peptide. *Biochem Soc Trans* 2005, 33:553-558.

[17] Wellington CL, Ellerby LM, Gutekunst CA, Rogers D, Warby S, Graham RK, Loubser O, van Raamsdonk J, Singaraja R, Yang YZ, Gafni J, Bredesen D, Hersch SM, Leavitt BR, Roy S, Nicholson DW, Hayden MR: Caspase cleavage of mutant huntingtin precedes neurodegeneration in Huntington's disease. *J Neurosci* 2002, 22:7862-7872.

[18] Luo S, Vacher C, Davies JE, Rubinsztein DC: Cdk5 phosphorylation of huntingtin reduces its cleavage by caspases: implications for mutant huntingtin toxicity. *J Cell Biol* 2005, 169:647-656.

[19] Wellington CL, Ellerby LM, Hackam AS, Margolis RL, Trifiro MA, Singaraja R, McCutcheon K, Salvesen GS, Propp SS, Bromm M, Rowland KJ, Zhang T, Rasper D, Roy S, Thornberry N, Pinsky L, Kakizuka A, Ross CA, Nicholson DW, Bredesen DE, Hayden MR: Caspase cleavage of gene products associated with triplet expansion disorders generates truncated fragments containing the polyglutamine tract. *J Biol Chem* 1998, 273:9158-9167.

[20] Glickman MH, Ciechanover A: The ubiquitin-proteasome proteolytic pathway: destruction for the sake of construction. *Physiol Rev* 2002, 82:373-428.

[21] Chan HY, Warrick JM, Gray-Board, Paulson HL, Bonini NM: Mechanisms of chaperone suppression of polyglutamine disease: selectivity, synergy and modulation of protein solubility in Drosophila. *Hum Mol Genet* 2000, 9:2811-2820.

[22] Cummings CJ, Sun Y, Opal P, Antalffy B, Mestril R, Orr HT, Dillmann WH, Zoghbi HY: Over-expression of inducible HSP70 chaperone suppresses neuropathology and improves motor function in SCA1 mice. *Hum Mol Genet* 2001, 10:1511-1518.

[23] Auluck PK, Chan HY, Trojanowski JQ, Lee VM, Bonini NM: Chaperone suppression of alpha-synuclein toxicity in a Drosophila model for Parkinson's disease. *Science* 2002, 295:865-868.

[24] Petrucelli L, O'Farrell C, Lockhart PJ, Baptista M, Kehoe K, Vink L, Choi P, Wolozin B, Farrer M, Hardy J, Cookson MR: Parkin protects against the toxicity associated with mutant alpha-synuclein: proteasome dysfunction selectively affects catecholaminergic neurons. *Neuron* 2002, 36:1007-1019.

[25] Verhoef LG, Lindsten K, Masucci MG, Dantuma NP: Aggregate formation inhibits proteasomal degradation of polyglutamine proteins. *Hum Mol Genet* 2002, 11:2689-2700.

[26] Lam YA, Lawson TG, Velayutham M, Zweier JL, Pickart CM: A proteasomal ATPase subunit recognizes the polyubiquitin degradation signal. *Nature* 2002, 416:763-767.

[27] Bence NF, Sampat RM, Kopito RR: Impairment of the ubiquitin-proteasome system by protein aggregation. *Science* 2001, 292:1552-1555.

[28] Tanaka Y, Engelender S, Igarashi S, Rao RK, Wanner T, Tanzi RE, Sawa A, Dawson L, Dawson TM, Ross CA: Inducible expression of mutant alpha-synuclein decreases proteasome activity and increases sensitivity to mitochondria-dependent apoptosis. *Hum Mol Genet* 2001, 10:919-926.

[29] Barral JM, Broadley SA, Schaffar G, Hartl FU: Roles of molecular chaperones in protein misfolding diseases. *Semin Cell Dev Biol* 2004, 15:17-29.

[30] Dobson CM: Experimental investigation of protein folding and misfolding. *Methods* 2004, 34:4-14.

[31] Ellis RJ: Macromolecular crowding: an important but neglected aspect of the intracellular environment. *Curr Opin Struct Biol* 2001, 11:114-119.

[32] Lashuel HA: Membrane permeabilization: a common mechanism in protein-misfolding diseases. *Sci Aging Knowledge Environ* 2005, 2005:e28.

[33] Ciechanover A, Orian A, Schwartz AL: Ubiquitin-mediated proteolysis: biological regulation via destruction. *Bioessays* 2000, 22:442-451.

[34] Sherman MY, Goldberg AL: Cellular defenses against unfolded proteins: a cell biologist thinks about neurodegenerative diseases. *Neuron* 2001, 29:15-32.

[35] Stefani M, Dobson CM: Protein aggregation and aggregate toxicity: new insights into protein folding, misfolding diseases and biological evolution. *J Mol Med* 2003, 81:678-699.

[36] Stefani M: Protein misfolding and aggregation: new examples in medicine and biology of the dark side of the protein world. *Biochim Biophys Acta* 2004, 1739:5-25.

[37] Trzesniewska K, Brzyska M, Elbaum D: Neurodegenerative aspects of protein aggregation. *Acta Neurobiol Exp (Wars)* 2004, 64:41-52.

[38] Dimakopoulos AC: Protein aggregation in Alzheimer's disease and other neoropathological disorders. *Curr Alzheimer Res* 2005, 2:19-28.

[39] Lee HG, Casadesus G, Zhu X, Takeda A, Perry G, Smith MA: Challenging the amyloid cascade hypothesis: senile plaques and amyloid-beta as protective adaptations to Alzheimer disease. *Ann N Y Acad Sci* 2004, 1019:1-4.

[40] Tabner BJ, Turnbull S, Fullwood NJ, German M, Allsop D: The production of hydrogen peroxide during early-stage protein aggregation: a common pathological mechanism in different neurodegenerative diseases? *Biochem Soc Trans* 2005, 33:548-550.

[41] Kopito RR: Aggresomes, inclusion bodies and protein aggregation. *Trends Cell Biol* 2000, 10:524-530.

[42] Berke SJ, Paulson HL: Protein aggregation and the ubiquitin proteasome pathway: gaining the UPPer hand on neurodegeneration. *Curr Opin Genet Dev* 2003, 13:253-261.

[43] Dimcheff DE, Portis JL, Caughey B: Prion proteins meet protein quality control. *Trends Cell Biol* 2003, 13:337-340.

[44] Chiesa R, Harris DA: Prion diseases: what is the neurotoxic molecule? *Neurobiol Dis* 2001, 8:743-763.

[45] Thompson AJ, Barrow CJ: Protein conformational misfolding and amyloid formation: characteristics of a new class of disorders that include Alzheimer's and Prion diseases. *Curr Med Chem* 2002, 9:1751-1762.

[46] Hashimoto M, Rockenstein E, Crews L, Masliah E: Role of protein aggregation in mitochondrial dysfunction and neurodegeneration in Alzheimer's and Parkinson's diseases. *Neuromolecular Med* 2003, 4:21-36.

[47] Driscoll M, Gerstbrein B: Dying for a cause: invertebrate genetics takes on human neurodegeneration. *Nat Rev Genet* 2003, 4:181-194.

[48] Wyttenbach A: Role of heat shock proteins during polyglutamine neurodegeneration: mechanisms and hypothesis. *J Mol Neurosci* 2004, 23:69-96.

[49] Jana NR, Nukina N: Recent advances in understanding the pathogenesis of polyglutamine diseases: involvement of molecular chaperones and ubiquitin-proteasome pathway. *J Chem Neuroanat* 2003, 26:95-101.

[50] Chai Y, Berke SS, Cohen RE, Paulson HL: Poly-ubiquitin binding by the polyglutamine disease protein ataxin-3 links its normal function to protein surveillance pathways. *J Biol Chem* 2004, 279:3605-3611.

[51] Briand C, Kozlov SV, Sonderegger P, Grutter MG: Crystal structure of neuroserpin: a neuronal serpin involved in a conformational disease. *FEBS Lett* 2001, 505:18-22.

[52] Carrell RW: Cell toxicity and conformational disease. *Trends Cell Biol* 2005, 15:574-580.

[53] Hu BR, Janelidze S, Ginsberg MD, Busto R, Perez-Pinzon M, Sick TJ, Siesjo BK, Liu CL: Protein aggregation after focal brain ischemia and reperfusion. *J Cereb Blood Flow Metab* 2001, 21:865-875.

[54] Liu CL, Martone ME, Hu BR: Protein ubiquitination in postsynaptic densities after transient cerebral ischemia. *J Cereb Blood Flow Metab* 2004, 24:1219-1225.

[55] Wojcik C, Di Napoli M: Ubiquitin-Proteasome System and Proteasome Inhibition: New Strategies in Stroke Therapy. *Stroke* 2004, 35:1506-1518.

[56] Di Napoli M, Papa F: The proteasome system and proteasome inhibitors in stroke: controlling the inflammatory response. *Curr Opin Investig Drugs* 2003, 4:1333-1342.

[57] Gubellini P, Bisso GM, Ciofi-Luzzatto A, Fortuna S, Lorenzini P, Michalek H, Scarsella G: Ubiquitin-mediated stress response in a rat model of brain transient ischemia/hypoxia. *Neurochem Res* 1997, 22:93-100.

[58] Cummings CJ, Mancini MA, Antalffy B, DeFranco DB, Orr HT, Zoghbi HY: Chaperone suppression of aggregation and altered subcellular proteasome localization imply protein misfolding in SCA1. *Nat Genet* 1998, 19:148-154.

[59] Selkoe DJ: Cell biology of protein misfolding: the examples of Alzheimer's and Parkinson's diseases. *Nat Cell Biol* 2004, 6:1054-1061.

[60] Muchowski PJ, Wacker JL: Modulation of neurodegeneration by molecular chaperones. *Nat Rev Neurosci* 2005, 6:11-22.

[61] Muchowski PJ: Protein misfolding, amyloid formation, and neurodegeneration: a critical role for molecular chaperones? *Neuron* 2002, 35:9-12.

[62] Lee DH, Wang HY: Differential physiologic responses of alpha7 nicotinic acetylcholine receptors to beta-amyloid1-40 and beta-amyloid1-42. *J Neurobiol* 2003, 55:25-30.

[63] Lee FJ, Liu F, Pristupa ZB, Niznik HB: Direct binding and functional coupling of alpha-synuclein to the dopamine transporters accelerate dopamine-induced apoptosis. *FASEB J* 2001, 15:916-926.

[64] Lee KS, Linden R, Prado MA, Brentani RR, Martins VR: Towards cellular receptors for prions. *Rev Med Virol* 2003, 13:399-408.

[65] Lee SJ, Liyanage U, Bickel PE, Xia W, Lansbury PT, Jr., Kosik KS: A detergent-insoluble membrane compartment contains A beta in vivo. *Nat Med* 1998, 4:730-734.

[66] Temussi PA, Masino L, Pastore A: From Alzheimer to Huntington: why is a structural understanding so difficult? *EMBO J* 2003, 22:355-361.

[67] Staniforth RA, Giannini S, Higgins LD, Conroy MJ, Hounslow AM, Jerala R, Craven CJ, Waltho JP: Three-dimensional domain swapping in the folded and molten-globule states of cystatins, an amyloid-forming structural superfamily. *EMBO J* 2001, 20:4774-4781.

[68] Serpell LC, Berriman J, Jakes R, Goedert M, Crowther RA: Fiber diffraction of synthetic alpha-synuclein filaments shows amyloid-like cross-beta conformation. *Proc Natl Acad Sci U S A* 2000, 97:4897-4902.

[69] Jimenez JL, Guijarro JI, Orlova E, Zurdo J, Dobson CM, Sunde M, Saibil HR: Cryo-electron microscopy structure of an SH3 amyloid fibril and model of the molecular packing. *EMBO J* 1999, 18:815-821.

[70] Sunde M, Blake C: The structure of amyloid fibrils by electron microscopy and X-ray diffraction. *Adv Protein Chem* 1997, 50:123-159.

[71] Petkova AT, Ishii Y, Balbach JJ, Antzutkin ON, Leapman RD, Delaglio F, Tycko R: A structural model for Alzheimer's beta -amyloid fibrils based on experimental constraints from solid state NMR. *Proc Natl Acad Sci U S A* 2002, 99:16742-16747.

[72] Chen S, Ferrone FA, Wetzel R: Huntington's disease age-of-onset linked to polyglutamine aggregation nucleation. *Proc Natl Acad Sci U S A* 2002, 99:11884-11889.

[73] Ross CA, Poirier MA, Wanker EE, Amzel M: Polyglutamine fibrillogenesis: the pathway unfolds. *Proc Natl Acad Sci U S A* 2003, 100:1-3.

[74] Jarrett JT, Lansbury PT, Jr.: Seeding "one-dimensional crystallization" of amyloid: a pathogenic mechanism in Alzheimer's disease and scrapie? *Cell* 1993, 73:1055-1058.

[75] Jarrett JT, Lansbury PT, Jr.: Amyloid fibril formation requires a chemically discriminating nucleation event: studies of an amyloidogenic sequence from the bacterial protein OsmB. *Biochemistry* 1992, 31:12345-12352.

[76] Chiti F, Stefani M, Taddei N, Ramponi G, Dobson CM: Rationalization of the effects of mutations on peptide and protein aggregation rates. *Nature* 2003, 424:805-808.

[77] Lee C, Schwartz MP, Prakash S, Iwakura M, Matouschek A: ATP-dependent proteases degrade their substrates by processively unraveling them from the degradation signal. *Mol Cell* 2001, 7:627-637.

[78] Abramov AY, Canevari L, Duchen MR: Calcium signals induced by amyloid beta peptide and their consequences in neurons and astrocytes in culture. *Biochim Biophys Acta* 2004, 1742:81-87.

[79] Lacor PN, Buniel MC, Chang L, Fernandez SJ, Gong Y, Viola KL, Lambert MP, Velasco PT, Bigio EH, Finch CE, Krafft GA, Klein WL: Synaptic targeting by Alzheimer's-related amyloid beta oligomers. *J Neurosci* 2004, 24:10191-10200.

[80] Lain E, Penke B, Delacourte A, Gundisch D, Schroder H, Witter B: Effects of Abeta1-42 fibrils and of the tetrapeptide Pr-IIGL on the phosphorylation state of the tau-protein and on the alpha7 nicotinic acetylcholine receptor in vitro. *Eur J Neurosci* 2005, 21:879-888.

[81] Le Y, Gong W, Tiffany HL, Tumanov A, Nedospasov S, Shen W, Dunlop NM, Gao JL, Murphy PM, Oppenheim JJ, Wang JM: Amyloid (beta)42 activates a G-protein-coupled chemoattractant receptor, FPR-like-1. *J Neurosci* 2001, 21:RC123.

[82] Koenigsknecht J, Landreth G: Microglial phagocytosis of fibrillar beta-amyloid through a beta1 integrin-dependent mechanism. *J Neurosci* 2004, 24:9838-9846.

[83] Lambert MP, Viola KL, Chromy BA, Chang L, Morgan TE, Yu J, Venton DL, Krafft GA, Finch CE, Klein WL: Vaccination with soluble Abeta oligomers generates toxicity-neutralizing antibodies. *J Neurochem* 2001, 79:595-605.

[84] Caughey B, Lansbury PT: Protofibrils, pores, fibrils, and neurodegeneration: separating the responsible protein aggregates from the innocent bystanders. *Annu Rev Neurosci* 2003, 26:267-298.

[85] Hetz C, Soto C: Protein misfolding and disease: the case of prion disorders. *Cell Mol Life Sci* 2003, 60:133-143.

[86] Perutz MF, Johnson T, Suzuki M, Finch JT: Glutamine repeats as polar zippers: their possible role in inherited neurodegenerative diseases. *Proc Natl Acad Sci U S A* 1994, 91:5355-5358.

[87] Schaffar G, Breuer P, Boteva R, Behrends C, Tzvetkov N, Strippel N, Sakahira H, Siegers K, Hayer-Hartl M, Hartl FU: Cellular toxicity of polyglutamine expansion proteins: mechanism of transcription factor deactivation. *Mol Cell* 2004, 15:95-105.
[88] Bucciantini M, Giannoni E, Chiti F, Baroni F, Formigli L, Zurdo J, Taddei N, Ramponi G, Dobson CM, Stefani M: Inherent toxicity of aggregates implies a common mechanism for protein misfolding diseases. *Nature* 2002, 416:507-511.
[89] Quist A, Doudevski I, Lin H, Azimova R, Ng D, Frangione B, Kagan B, Ghiso J, Lal R: Amyloid ion channels: a common structural link for protein-misfolding disease. *Proc Natl Acad Sci U S A* 2005, 102:10427-10432.
[90] Smith MA, Hirai K, Hsiao K, Pappolla MA, Harris PL, Siedlak SL, Tabaton M, Perry G: Amyloid-beta deposition in Alzheimer transgenic mice is associated with oxidative stress. *J Neurochem* 1998, 70:2212-2215.
[91] Mattson MP, Mattson EP: Amyloid peptide enhances nail rusting: novel insight into mechanisms of aging and Alzheimer's disease. *Ageing Res Rev* 2002, 1:327-330.
[92] Behl C, Moosmann B: Oxidative nerve cell death in Alzheimer's disease and stroke: antioxidants as neuroprotective compounds. *Biol Chem* 2002, 383:521-536.
[93] Yan SD, Stern D, Kane MD, Kuo YM, Lampert HC, Roher AE: RAGE-Abeta interactions in the pathophysiology of Alzheimer's disease. *Restor Neurol Neurosci* 1998, 12:167-173.
[94] Yan SD, Chen X, Fu J, Chen M, Zhu H, Roher A, Slattery T, Zhao L, Nagashima M, Morser J, Migheli A, Nawroth P, Stern D, Schmidt AM: RAGE and amyloid-beta peptide neurotoxicity in Alzheimer's disease. *Nature* 1996, 382:685-691.
[95] Behl C, Davis JB, Lesley R, Schubert D: Hydrogen peroxide mediates amyloid beta protein toxicity. *Cell* 1994, 77:817-827.
[96] Doraiswamy PM: Alzheimer's disease and the glutamate NMDA receptor. *Psychopharmacol Bull* 2003, 37:41-49.
[97] Coyle JT, Puttfarcken P: Oxidative stress, glutamate, and neurodegenerative disorders. *Science* 1993, 262:689-695.
[98] Schubert D, Piasecki D: Oxidative glutamate toxicity can be a component of the excitotoxicity cascade. *J Neurosci* 2001, 21:7455-7462.
[99] Smith MA, Richey Harris PL, Sayre LM, Beckman JS, Perry G: Widespread peroxynitrite-mediated damage in Alzheimer's disease. *J Neurosci* 1997, 17:2653-2657.
[100] Rogers J, Strohmeyer R, Kovelowski CJ, Li R: Microglia and inflammatory mechanisms in the clearance of amyloid beta peptide. *Glia* 2002, 40:260-269.
[101] Molina-Holgado E, Ortiz S, Molina-Holgado F, Guaza C: Induction of COX-2 and PGE(2) biosynthesis by IL-1beta is mediated by PKC and mitogen-activated protein kinases in murine astrocytes. *Br J Pharmacol* 2000, 131:152-159.
[102] Candelario-Jalil E, Gonzalez-Falcon A, Garcia-Cabrera M, Alvarez D, Al Dalain S, Martinez G, Leon OS, Springer JE: Assessment of the relative contribution of COX-1 and COX-2 isoforms to ischemia-induced oxidative damage and neurodegeneration following transient global cerebral ischemia. *J Neurochem* 2003, 86:545-555.
[103] Ho GJ, Drego R, Hakimian E, Masliah E: Mechanisms of cell signaling and inflammation in Alzheimer's disease. *Curr Drug Targets Inflamm Allergy* 2005, 4:247-256.

[104] Hashimoto M, Takenouchi T, Rockenstein E, Masliah E: Alpha-synuclein up-regulates expression of caveolin-1 and down-regulates extracellular signal-regulated kinase activity in B103 neuroblastoma cells: role in the pathogenesis of Parkinson's disease. *J Neurochem* 2003, 85:1468-1479.

[105] Lievens JC, Rival T, Iche M, Chneiweiss H, Birman S: Expanded polyglutamine peptides disrupt EGF receptor signaling and glutamate transporter expression in Drosophila. *Hum Mol Genet* 2005, 14:713-724.

[106] Carimalo J, Cronier S, Petit G, Peyrin JM, Boukhtouche F, Arbez N, Lemaigre-Dubreuil Y, Brugg B, Miquel MC: Activation of the JNK-c-Jun pathway during the early phase of neuronal apoptosis induced by PrP106-126 and prion infection. *Eur J Neurosci* 2005, 21:2311-2319.

[107] Berridge MJ, Lipp P, Bootman MD: The versatility and universality of calcium signalling. *Nat Rev Mol Cell Biol* 2000, 1:11-21.

[108] Mattson MP, Chan SL, Camandola S: Presenilin mutations and calcium signaling defects in the nervous and immune systems. *Bioessays* 2001, 23:733-744.

[109] Kourie JI: Mechanisms of amyloid beta protein-induced modification in ion transport systems: implications for neurodegenerative diseases. *Cell Mol Neurobiol* 2001, 21:173-213.

[110] Zheng WH, Bastianetto S, Mennicken F, Ma W, Kar S: Amyloid beta peptide induces tau phosphorylation and loss of cholinergic neurons in rat primary septal cultures. *Neuroscience* 2002, 115:201-211.

[111] Mattson MP: Antigenic changes similar to those seen in neurofibrillary tangles are elicited by glutamate and Ca2+ influx in cultured hippocampal neurons. *Neuron* 1990, 4:105-117.

[112] Ekinci FJ, Linsley MD, Shea TB: Beta-amyloid-induced calcium influx induces apoptosis in culture by oxidative stress rather than tau phosphorylation. *Brain Res Mol Brain Res* 2000, 76:389-395.

[113] Verdier Y, Huszar E, Penke B, Penke Z, Woffendin G, Scigelova M, Fulop L, Szucs M, Medzihradszky K, Janaky T: Identification of synaptic plasma membrane proteins co-precipitated with fibrillar beta-amyloid peptide. *J Neurochem* 2005, 94:617-628.

[114] Caricasole A, Copani A, Caraci F, Aronica E, Rozemuller AJ, Caruso A, Storto M, Gaviraghi G, Terstappen GC, Nicoletti F: Induction of Dickkopf-1, a negative modulator of the Wnt pathway, is associated with neuronal degeneration in Alzheimer's brain. *J Neurosci* 2004, 24:6021-6027.

[115] Caricasole A, Copani A, Caruso A, Caraci F, Iacovelli L, Sortino MA, Terstappen GC, Nicoletti F: The Wnt pathway, cell-cycle activation and beta-amyloid: novel therapeutic strategies in Alzheimer's disease? *Trends Pharmacol Sci* 2003, 24:233-238.

[116] De Ferrari GV, Inestrosa NC: Wnt signaling function in Alzheimer's disease. *Brain Res Brain Res Rev* 2000, 33:1-12.

[117] Inestrosa N, De Ferrari GV, Garrido JL, Alvarez A, Olivares GH, Barria MI, Bronfman M, Chacon MA: Wnt signaling involvement in beta-amyloid-dependent neurodegeneration. *Neurochem Int* 2002, 41:341-344.

[118] Fu L, Gao YS, Sztul E: Transcriptional repression and cell death induced by nuclear aggregates of non-polyglutamine protein. *Neurobiol Dis* 2005, 20:656-665.

[119] Robakis NK: An Alzheimer's disease hypothesis based on transcriptional dysregulation. *Amyloid* 2003, 10:80-85.

[120] Gunawardena S, Her LS, Brusch RG, Laymon RA, Niesman IR, Gordesky-Gold B, Sintasath L, Bonini NM, Goldstein LS: Disruption of axonal transport by loss of huntingtin or expression of pathogenic polyQ proteins in Drosophila. *Neuron* 2003, 40:25-40.

[121] Lashuel HA, Hartley D, Petre BM, Walz T, Lansbury PT, Jr.: Neurodegenerative disease: amyloid pores from pathogenic mutations. *Nature* 2002, 418:291.

[122] Jana NR, Zemskov EA, Wang G, Nukina N: Altered proteasomal function due to the expression of polyglutamine-expanded truncated N-terminal huntingtin induces apoptosis by caspase activation through mitochondrial cytochrome c release. *Hum Mol Genet* 2001, 10:1049-1059.

[123] Rajagopalan S, Andersen JK: Alpha synuclein aggregation: is it the toxic gain of function responsible for neurodegeneration in Parkinson's disease? *Mech Ageing Dev* 2001, 122:1499-1510.

[124] Bossy-Wetzel E, Schwarzenbacher R, Lipton SA: Molecular pathways to neurodegeneration. *Nat Med* 2004, 10 Suppl:S2-S9.

[125] Cummings CJ, Reinstein E, Sun Y, Antalffy B, Jiang Y, Ciechanover A, Orr HT, Beaudet AL, Zoghbi HY: Mutation of the E6-AP ubiquitin ligase reduces nuclear inclusion frequency while accelerating polyglutamine-induced pathology in SCA1 mice. *Neuron* 1999, 24:879-892.

[126] Klement IA, Skinner PJ, Kaytor MD, Yi H, Hersch SM, Clark HB, Zoghbi HY, Orr HT: Ataxin-1 nuclear localization and aggregation: role in polyglutamine-induced disease in SCA1 transgenic mice. *Cell* 1998, 95:41-53.

[127] Watase K, Weeber EJ, Xu B, Antalffy B, Yuva-Paylor L, Hashimoto K, Kano M, Atkinson R, Sun Y, Armstrong DL, Sweatt JD, Orr HT, Paylor R, Zoghbi HY: A long CAG repeat in the mouse Sca1 locus replicates SCA1 features and reveals the impact of protein solubility on selective neurodegeneration. *Neuron* 2002, 34:905-919.

[128] Fink AL: Protein aggregation: folding aggregates, inclusion bodies and amyloid. *Fold Des* 1998, 3:R9-23.

[129] Wanker EE: Protein aggregation and pathogenesis of Huntington's disease: mechanisms and correlations. *Biol Chem* 2000, 381:937-942.

[130] Ghiso J, Frangione B: Amyloidosis and Alzheimer's disease. *Adv Drug Deliv Rev* 2002, 54:1539-1551.

[131] Kayed R, Head E, Thompson JL, McIntire TM, Milton SC, Cotman CW, Glabe CG: Common structure of soluble amyloid oligomers implies common mechanism of pathogenesis. *Science* 2003, 300:486-489.

[132] Emamian ES, Kaytor MD, Duvick LA, Zu T, Tousey SK, Zoghbi HY, Clark HB, Orr HT: Serine 776 of ataxin-1 is critical for polyglutamine-induced disease in SCA1 transgenic mice. *Neuron* 2003, 38:375-387.

[133] Cluskey S, Ramsden DB: Mechanisms of neurodegeneration in amyotrophic lateral sclerosis. *Mol Pathol* 2001, 54:386-392.

[134] Liu J, Lillo C, Jonsson PA, Vande VC, Ward CM, Miller TM, Subramaniam JR, Rothstein JD, Marklund S, Andersen PM, Brannstrom T, Gredal O, Wong PC,

Williams DS, Cleveland DW: Toxicity of familial ALS-linked SOD1 mutants from selective recruitment to spinal mitochondria. *Neuron* 2004, 43:5-17.

[135] Wood JD, Beaujeux TP, Shaw PJ: Protein aggregation in motor neurone disorders. *Neuropathol Appl Neurobiol* 2003, 29:529-545.

[136] Imai Y, Soda M, Hatakeyama S, Akagi T, Hashikawa T, Nakayama KI, Takahashi R: CHIP is associated with Parkin, a gene responsible for familial Parkinson's disease, and enhances its ubiquitin ligase activity. *Mol Cell* 2002, 10:55-67.

[137] Rao RV, Bredesen DE: Misfolded proteins, endoplasmic reticulum stress and neurodegeneration. *Curr Opin Cell Biol* 2004, 16:653-662.

[138] Soti C, Csermely P: Chaperones and aging: role in neurodegeneration and in other civilizational diseases. *Neurochem Int* 2002, 41:383-389.

[139] Kagan BL, Azimov R, Azimova R: Amyloid peptide channels. *J Membr Biol* 2004, 202:1-10.

[140] Kajkowski EM, Lo CF, Ning X, Walker S, Sofia HJ, Wang W, Edris W, Chanda P, Wagner E, Vile S, Ryan K, McHendry-Rinde B, Smith SC, Wood A, Rhodes KJ, Kennedy JD, Bard J, Jacobsen JS, Ozenberger BA: beta -Amyloid peptide-induced apoptosis regulated by a novel protein containing a g protein activation module. *J Biol Chem* 2001, 276:18748-18756.

[141] Kar S, Quirion R: Amyloid beta peptides and central cholinergic neurons: functional interrelationship and relevance to Alzheimer's disease pathology. *Prog Brain Res* 2004, 145:261-274.

[142] Mortimore GE, Poso AR: The lysosomal pathway of intracellular proteolysis in liver: regulation by amino acids. *Adv Enzyme Regul* 1986, 25:257-276.

[143] Blommaart EF, Luiken JJ, Meijer AJ: Autophagic proteolysis: control and specificity. *Histochem J* 1997, 29:365-385.

[144] Klionsky DJ, Ohsumi Y: Vacuolar import of proteins and organelles from the cytoplasm. *Annu Rev Cell Dev Biol* 1999, 15:1-32.

[145] Yorimitsu T, Klionsky DJ: Autophagy: molecular machinery for self-eating. *Cell Death Differ* 2005, 12 Suppl 2:1542-1552.

[146] Stromhaug PE, Klionsky DJ: Approaching the molecular mechanism of autophagy. *Traffic* 2001, 2:524-531.

[147] Klionsky DJ, Emr SD: Autophagy as a regulated pathway of cellular degradation. *Science* 2000, 290:1717-1721.

[148] Ohsumi Y: Molecular dissection of autophagy: two ubiquitin-like systems. *Nat Rev Mol Cell Biol* 2001, 2:211-216.

[149] Koike M, Nakanishi H, Saftig P, Ezaki J, Isahara K, Ohsawa Y, Schulz-Schaeffer W, Watanabe T, Waguri S, Kametaka S, Shibata M, Yamamoto K, Kominami E, Peters C, von Figura K, Uchiyama Y: Cathepsin D deficiency induces lysosomal storage with ceroid lipofuscin in mouse CNS neurons. *J Neurosci* 2000, 20:6898-6906.

[150] Dice JF, Terlecky SR, Chiang HL, Olson TS, Isenman LD, Short-Russell SR, Freundlieb S, Terlecky LJ: A selective pathway for degradation of cytosolic proteins by lysosomes. *Semin Cell Biol* 1990, 1:449-455.

[151] Massey A, Kiffin R, Cuervo AM: Pathophysiology of chaperone-mediated autophagy. *Int J Biochem Cell Biol* 2004, 36:2420-2434.

[152] Dice JF: Selective degradation of cytosolic proteins by lysosomes. *Ann N Y Acad Sci* 1992, 674:58-64.

[153] Terlecky SR, Dice JF: Polypeptide import and degradation by isolated lysosomes. *J Biol Chem* 1993, 268:23490-23495.

[154] Cuervo AM, Terlecky SR, Dice JF, Knecht E: Selective binding and uptake of ribonuclease A and glyceraldehyde-3-phosphate dehydrogenase by isolated rat liver lysosomes. *J Biol Chem* 1994, 269:26374-26380.

[155] Cuervo AM, Gomes AV, Barnes JA, Dice JF: Selective degradation of annexins by chaperone-mediated autophagy. *J Biol Chem* 2000, 275:33329-33335.

[156] Cuervo AM, Dice JF: A receptor for the selective uptake and degradation of proteins by lysosomes. *Science* 1996, 273:501-503.

[157] Dice JF: Molecular determinants of protein half-lives in eukaryotic cells. *FASEB J* 1987, 1:349-357.

[158] Miller RJ, Wilson SM: Neurological disease: UPS stops delivering! *Trends Pharmacol Sci* 2003, 24:18-23.

[159] Johnston JA, Madura K: Rings, chains and ladders: ubiquitin goes to work in the neuron. *Prog Neurobiol* 2004, 73:227-257.

[160] Ravikumar B, Duden R, Rubinsztein DC: Aggregate-prone proteins with polyglutamine and polyalanine expansions are degraded by autophagy. *Hum Mol Genet* 2002, 11:1107-1117.

[161] Webb JL, Ravikumar B, Atkins J, Skepper JN, Rubinsztein DC: Alpha-Synuclein is degraded by both autophagy and the proteasome. *J Biol Chem* 2003, 278:25009-25013.

[162] Shibata M, Lu T, Furuya T, Degterev A, Mizushima N, Yoshimori T, MacDonald M, Yankner B, Yuan J: Regulation of intracellular accumulation of mutant Huntingtin by Beclin 1. *J Biol Chem* 2006, 281:14474-14485.

[163] Berger Z, Ravikumar B, Menzies FM, Oroz LG, Underwood BR, Pangalos MN, Schmitt I, Wullner U, Evert BO, O'Kane CJ, Rubinsztein DC: Rapamycin alleviates toxicity of different aggregate-prone proteins. *Hum Mol Genet* 2006, 15:433-442.

[164] Ravikumar B, Stewart A, Kita H, Kato K, Duden R, Rubinsztein DC: Raised intracellular glucose concentrations reduce aggregation and cell death caused by mutant huntingtin exon 1 by decreasing mTOR phosphorylation and inducing autophagy. *Hum Mol Genet* 2003, 12:985-994.

[165] Iwata A, Christianson JC, Bucci M, Ellerby LM, Nukina N, Forno LS, Kopito RR: Increased susceptibility of cytoplasmic over nuclear polyglutamine aggregates to autophagic degradation. *Proc Natl Acad Sci U S A* 2005, 102:13135-13140.

[166] Qin ZH, Wang Y, Kegel KB, Kazantsev A, Apostol BL, Thompson LM, Yoder J, Aronin N, DiFiglia M: Autophagy regulates the processing of amino terminal huntingtin fragments. *Hum Mol Genet* 2003, 12:3231-3244.

[167] Ravikumar B, Vacher C, Berger Z, Davies JE, Luo S, Oroz LG, Scaravilli F, Easton DF, Duden R, O'Kane CJ, Rubinsztein DC: Inhibition of mTOR induces autophagy and reduces toxicity of polyglutamine expansions in fly and mouse models of Huntington disease. *Nat Genet* 2004, 36:585-595.

[168] Komatsu M, Waguri S, Ueno T, Iwata J, Murata S, Tanida I, Ezaki J, Mizushima N, Ohsumi Y, Uchiyama Y, Kominami E, Tanaka K, Chiba T: Impairment of starvation-

induced and constitutive autophagy in Atg7-deficient mice. *J Cell Biol* 2005, 169:425-434.

[169] Rideout HJ, Lang-Rollin I, Stefanis L: Involvement of macroautophagy in the dissolution of neuronal inclusions. *Int J Biochem Cell Biol* 2004, 36:2551-2562.

[170] Offen D, Elkon H, Melamed E: Apoptosis as a general cell death pathway in neurodegenerative diseases. *J Neural Transm Suppl* 2000, 153-166.

[171] Martin LJ: Neuronal cell death in nervous system development, disease, and injury (Review). *Int J Mol Med* 2001, 7:455-478.

[172] Jellinger KA, Stadelmann C: Mechanisms of cell death in neurodegenerative disorders. *J Neural Transm Suppl* 2000, 59:95-114.

[173] Jellinger KA, Stadelmann CH: The enigma of cell death in neurodegenerative disorders. *J Neural Transm Suppl* 2000, 60:21-36.

[174] Bursch W, Ellinger A: Autophagy--a basic mechanism and a potential role for neurodegeneration. *Folia Neuropathol* 2005, 43:297-310.

[175] Kegel KB, Kim M, Sapp E, McIntyre C, Castano JG, Aronin N, DiFiglia M: Huntingtin expression stimulates endosomal-lysosomal activity, endosome tubulation, and autophagy. *J Neurosci* 2000, 20:7268-7278.

[176] Petersen A, Larsen KE, Behr GG, Romero N, Przedborski S, Brundin P, Sulzer D: Expanded CAG repeats in exon 1 of the Huntington's disease gene stimulate dopamine-mediated striatal neuron autophagy and degeneration. *Hum Mol Genet* 2001, 10:1243-1254.

[177] Venkatraman P, Wetzel R, Tanaka M, Nukina N, Goldberg AL: Eukaryotic proteasomes cannot digest polyglutamine sequences and release them during degradation of polyglutamine-containing proteins. *Mol Cell* 2004, 14:95-104.

[178] Cataldo AM, Barnett JL, Berman SA, Li J, Quarless S, Bursztajn S, Lippa C, Nixon RA: Gene expression and cellular content of cathepsin D in Alzheimer's disease brain: evidence for early up-regulation of the endosomal-lysosomal system. *Neuron* 1995, 14:671-680.

[179] Stadelmann C, Deckwerth TL, Srinivasan A, Bancher C, Bruck W, Jellinger K, Lassmann H: Activation of caspase-3 in single neurons and autophagic granules of granulovacuolar degeneration in Alzheimer's disease. Evidence for apoptotic cell death. *Am J Pathol* 1999, 155:1459-1466.

[180] Cataldo AM, Hamilton DJ, Barnett JL, Paskevich PA, Nixon RA: Properties of the endosomal-lysosomal system in the human central nervous system: disturbances mark most neurons in populations at risk to degenerate in Alzheimer's disease. *J Neurosci* 1996, 16:186-199.

[181] Yang AJ, Chandswangbhuvana D, Margol L, Glabe CG: Loss of endosomal/lysosomal membrane impermeability is an early event in amyloid Abeta1-42 pathogenesis. *J Neurosci Res* 1998, 52:691-698.

[182] Keller JN, Huang FF, Markesbery WR: Decreased levels of proteasome activity and proteasome expression in aging spinal cord. *Neuroscience* 2000, 98:149-156.

[183] Keller JN, Hanni KB, Markesbery WR: Impaired proteasome function in Alzheimer's disease. *J Neurochem* 2000, 75:436-439.

[184] Anglade P, Vyas S, Javoy-Agid F, Herrero MT, Michel PP, Marquez J, Mouatt-Prigent A, Ruberg M, Hirsch EC, Agid Y: Apoptosis and autophagy in nigral neurons of patients with Parkinson's disease. *Histol Histopathol* 1997, 12:25-31.

[185] Cuervo AM, Stefanis L, Fredenburg R, Lansbury PT, Sulzer D: Impaired degradation of mutant alpha-synuclein by chaperone-mediated autophagy. *Science* 2004, 305:1292-1295.

[186] McNaught KS, Olanow CW, Halliwell B, Isacson O, Jenner P: Failure of the ubiquitin-proteasome system in Parkinson's disease. *Nat Rev Neurosci* 2001, 2:589-594.

[187] Ding Q, Dimayuga E, Martin S, Bruce-Keller AJ, Nukala V, Cuervo AM, Keller JN: Characterization of chronic low-level proteasome inhibition on neural homeostasis. *J Neurochem* 2003, 86:489-497.

[188] Wojcik C, Schroeter D, Wilk S, Lamprecht J, Paweletz N: Ubiquitin-mediated proteolysis centers in HeLa cells: indication from studies of an inhibitor of the chymotrypsin-like activity of the proteasome. *Eur J Cell Biol* 1996, 71:311-318.

[189] Bennett MC, Bishop JF, Leng Y, Chock PB, Chase TN, Mouradian MM: Degradation of alpha-synuclein by proteasome. *J Biol Chem* 1999, 274:33855-33858.

[190] Imai Y, Soda M, Takahashi R: Parkin suppresses unfolded protein stress-induced cell death through its E3 ubiquitin-protein ligase activity. *J Biol Chem* 2000, 275:35661-35664.

[191] Liu Y, Fallon L, Lashuel HA, Liu Z, Lansbury PT, Jr.: The UCH-L1 gene encodes two opposing enzymatic activities that affect alpha-synuclein degradation and Parkinson's disease susceptibility. *Cell* 2002, 111:209-218.

[192] Tofaris GK, Layfield R, Spillantini MG: alpha-synuclein metabolism and aggregation is linked to ubiquitin-independent degradation by the proteasome. *FEBS Lett* 2001, 509:22-26.

[193] Ancolio K, Alves dC, Ueda K, Checler F: Alpha-synuclein and the Parkinson's disease-related mutant Ala53Thr-alpha-synuclein do not undergo proteasomal degradation in HEK293 and neuronal cells. *Neurosci Lett* 2000, 285:79-82.

[194] Rideout HJ, Larsen KE, Sulzer D, Stefanis L: Proteasomal inhibition leads to formation of ubiquitin/alpha-synuclein-immunoreactive inclusions in PC12 cells. *J Neurochem* 2001, 78:899-908.

[195] Rideout HJ, Stefanis L: Proteasomal inhibition-induced inclusion formation and death in cortical neurons require transcription and ubiquitination. *Mol Cell Neurosci* 2002, 21:223-238.

[196] Paxinou E, Chen Q, Weisse M, Giasson BI, Norris EH, Rueter SM, Trojanowski JQ, Lee VM, Ischiropoulos H: Induction of alpha-synuclein aggregation by intracellular nitrative insult. *J Neurosci* 2001, 21:8053-8061.

[197] Lee HJ, Khoshaghideh F, Patel S, Lee SJ: Clearance of alpha-synuclein oligomeric intermediates via the lysosomal degradation pathway. *J Neurosci* 2004, 24:1888-1896.

[198] Li W, Lesuisse C, Xu Y, Troncoso JC, Price DL, Lee MK: Stabilization of alpha-synuclein protein with aging and familial parkinson's disease-linked A53T mutation. *J Neurosci* 2004, 24:7400-7409.

[199] Boellaard JW, Kao M, Schlote W, Diringer H: Neuronal autophagy in experimental scrapie. *Acta Neuropathol (Berl)* 1991, 82:225-228.

[200] Boellaard JW, Schlote W, Tateishi J: Neuronal autophagy in experimental Creutzfeldt-Jakob's disease. *Acta Neuropathol (Berl)* 1989, 78:410-418.

[201] Terman A, Brunk UT: Lipofuscin: mechanisms of formation and increase with age. *APMIS* 1998, 106:265-276.

[202] Komatsu M, Waguri S, Chiba T, Murata S, Iwata J, Tanida I, Ueno T, Koike M, Uchiyama Y, Kominami E, Tanaka K: Loss of autophagy in the central nervous system causes neurodegeneration in mice. *Nature* 2006, 441:880-884.

[203] Hara T, Nakamura K, Matsui M, Yamamoto A, Nakahara Y, Suzuki-Migishima R, Yokoyama M, Mishima K, Saito I, Okano H, Mizushima N: Suppression of basal autophagy in neural cells causes neurodegenerative disease in mice. *Nature* 2006, 441:885-889.

[204] Ravikumar B, Acevedo-Arozena A, Imarisio S, Berger Z, Vacher C, O'Kane CJ, Brown SD, Rubinsztein DC: Dynein mutations impair autophagic clearance of aggregate-prone proteins. *Nat Genet* 2005, 37:771-776.

[205] Levy JR, Sumner CJ, Caviston JP, Tokito MK, Ranganathan S, Ligon LA, Wallace KE, LaMonte BH, Harmison GG, Puls I, Fischbeck KH, Holzbaur EL: A motor neuron disease-associated mutation in p150Glued perturbs dynactin function and induces protein aggregation. *J Cell Biol* 2006, 172:733-745.

[206] Puls I, Jonnakuty C, LaMonte BH, Holzbaur EL, Tokito M, Mann E, Floeter MK, Bidus K, Drayna D, Oh SJ, Brown RH, Jr., Ludlow CL, Fischbeck KH: Mutant dynactin in motor neuron disease. *Nat Genet* 2003, 33:455-456.

[207] Hafezparast M, Klocke R, Ruhrberg C, Marquardt A, Ahmad-Annuar A, Bowen S, Lalli G, Witherden AS, Hummerich H, Nicholson S, Morgan PJ, Oozageer R, Priestley JV, Averill S, King VR, Ball S, Peters J, Toda T, Yamamoto A, Hiraoka Y, Augustin M, Korthaus D, Wattler S, Wabnitz P, Dickneite C, Lampel S, Boehme F, Peraus G, Popp A, Rudelius M, Schlegel J, Fuchs H, Hrabe dA, Schiavo G, Shima DT, Russ AP, Stumm G, Martin JE, Fisher EM: Mutations in dynein link motor neuron degeneration to defects in retrograde transport. *Science* 2003, 300:808-812.

[208] Kieran D, Hafezparast M, Bohnert S, Dick JR, Martin J, Schiavo G, Fisher EM, Greensmith L: A mutation in dynein rescues axonal transport defects and extends the life span of ALS mice. *J Cell Biol* 2005, 169:561-567.

[209] Ligon LA, LaMonte BH, Wallace KE, Weber N, Kalb RG, Holzbaur EL: Mutant superoxide dismutase disrupts cytoplasmic dynein in motor neurons. *Neuroreport* 2005, 16:533-536.

[210] Boya P, Gonzalez-Polo RA, Casares N, Perfettini JL, Dessen P, Larochette N, Metivier D, Meley D, Souquere S, Yoshimori T, Pierron G, Codogno P, Kroemer G: Inhibition of macroautophagy triggers apoptosis. *Mol Cell Biol* 2005, 25:1025-1040.

[211] Ravikumar B, Berger Z, Vacher C, O'Kane CJ, Rubinsztein DC: Rapamycin pre-treatment protects against apoptosis. *Hum Mol Genet* 2006, 15:1209-1216.

[212] Kabashi E, Agar JN, Taylor DM, Minotti S, Durham HD: Focal dysfunction of the proteasome: a pathogenic factor in a mouse model of amyotrophic lateral sclerosis. *J Neurochem* 2004, 89:1325-1335.

[213] Cheroni C, Peviani M, Cascio P, Debiasi S, Monti C, Bendotti C: Accumulation of human SOD1 and ubiquitinated deposits in the spinal cord of SOD1G93A mice during

motor neuron disease progression correlates with a decrease of proteasome. *Neurobiol Dis* 2005, 18:509-522.

[214] Tofaris GK, Layfield R, Spillantini MG: alpha-synuclein metabolism and aggregation is linked to ubiquitin-independent degradation by the proteasome. *FEBS Lett* 2001, 509:22-26.

[215] Ma J, Lindquist S: Wild-type PrP and a mutant associated with prion disease are subject to retrograde transport and proteasome degradation. *Proc Natl Acad Sci U S A* 2001, 98:14955-14960.

[216] Ma J, Wollmann R, Lindquist S: Neurotoxicity and neurodegeneration when PrP accumulates in the cytosol. *Science* 2002, 298:1781-1785.

[217] Yedidia Y, Horonchik L, Tzaban S, Yanai A, Taraboulos A: Proteasomes and ubiquitin are involved in the turnover of the wild-type prion protein. *EMBO J* 2001, 20:5383-5391.

[218] Chan HY, Warrick JM, Andriola I, Merry D, Bonini NM: Genetic modulation of polyglutamine toxicity by protein conjugation pathways in Drosophila. *Hum Mol Genet* 2002, 11:2895-2904.

[219] Kabashi E, Durham HD: Failure of protein quality control in amyotrophic lateral sclerosis. *Biochim Biophys Acta* 2006.

[220] Bowman AB, Yoo SY, Dantuma NP, Zoghbi HY: Neuronal dysfunction in a polyglutamine disease model occurs in the absence of ubiquitin-proteasome system impairment and inversely correlates with the degree of nuclear inclusion formation. *Hum Mol Genet* 2005, 14:679-691.

[221] Diaz-Hernandez M, Hernandez F, Martin-Aparicio E, Gomez-Ramos P, Moran MA, Castano JG, Ferrer I, Avila J, Lucas JJ: Neuronal induction of the immunoproteasome in Huntington's disease. *J Neurosci* 2003, 23:11653-11661.

[222] Sun XM, Butterworth M, MacFarlane M, Dubiel W, Ciechanover A, Cohen GM: Caspase activation inhibits proteasome function during apoptosis. *Mol Cell* 2004, 14:81-93.

[223] Chai Y, Koppenhafer SL, Shoesmith SJ, Perez MK, Paulson HL: Evidence for proteasome involvement in polyglutamine disease: localization to nuclear inclusions in SCA3/MJD and suppression of polyglutamine aggregation in vitro. *Hum Mol Genet* 1999, 8:673-682.

[224] Junn E, Lee SS, Suhr UT, Mouradian MM: Parkin accumulation in aggresomes due to proteasome impairment. *J Biol Chem* 2002, 277:47870-47877.

[225] Kim S, Nollen EA, Kitagawa K, Bindokas VP, Morimoto RI: Polyglutamine protein aggregates are dynamic. *Nat Cell Biol* 2002, 4:826-831.

[226] Peters PJ, Ning K, Palacios F, Boshans RL, Kazantsev A, Thompson LM, Woodman B, Bates GP, D'Souza-Schorey C: Arfaptin 2 regulates the aggregation of mutant huntingtin protein. *Nat Cell Biol* 2002, 4:240-245.

[227] Yoo SY, Pennesi ME, Weeber EJ, Xu B, Atkinson R, Chen S, Armstrong DL, Wu SM, Sweatt JD, Zoghbi HY: SCA7 knockin mice model human SCA7 and reveal gradual accumulation of mutant ataxin-7 in neurons and abnormalities in short-term plasticity. *Neuron* 2003, 37:383-401.

[228] Yamamoto A, Lucas JJ, Hen R: Reversal of neuropathology and motor dysfunction in a conditional model of Huntington's disease. *Cell* 2000, 101:57-66.

[229] Mizuno Y, Hattori N, Mori H, Suzuki T, Tanaka K: Parkin and Parkinson's disease. *Curr Opin Neurol* 2001, 14:477-482.

[230] He L, Lu XY, Jolly AF, Eldridge AG, Watson SJ, Jackson PK, Barsh GS, Gunn TM: Spongiform degeneration in mahoganoid mutant mice. *Science* 2003, 299:710-712.

[231] Saigoh K, Wang YL, Suh JG, Yamanishi T, Sakai Y, Kiyosawa H, Harada T, Ichihara N, Wakana S, Kikuchi T, Wada K: Intragenic deletion in the gene encoding ubiquitin carboxy-terminal hydrolase in gad mice. *Nat Genet* 1999, 23:47-51.

[232] Fernandez-Funez P, Nino-Rosales ML, de Gouyon B, She WC, Luchak JM, Martinez P, Turiegano E, Benito J, Capovilla M, Skinner PJ, McCall A, Canal I, Orr HT, Zoghbi HY, Botas J: Identification of genes that modify ataxin-1-induced neurodegeneration. *Nature* 2000, 408:101-106.

[233] Verhoef LG, Lindsten K, Masucci MG, Dantuma NP: Aggregate formation inhibits proteasomal degradation of polyglutamine proteins. *Hum Mol Genet* 2002, 11:2689-2700.

[234] Almond JB, Cohen GM: The proteasome: a novel target for cancer chemotherapy. *Leukemia* 2002, 16:433-443.

[235] Turner BJ, Atkin JD: ER stress and UPR in familial amyotrophic lateral sclerosis. *Curr Mol Med* 2006, 6:79-86.

[236] Ling YH, Liebes L, Zou Y, Perez-Soler R: Reactive oxygen species generation and mitochondrial dysfunction in the apoptotic response to Bortezomib, a novel proteasome inhibitor, in human H460 non-small cell lung cancer cells. *J Biol Chem* 2003, 278:33714-33723.

[237] Sullivan PG, Dragicevic NB, Deng JH, Bai Y, Dimayuga E, Ding Q, Chen Q, Bruce-Keller AJ, Keller JN: Proteasome inhibition alters neural mitochondrial homeostasis and mitochondria turnover. *J Biol Chem* 2004, 279:20699-20707.

[238] Johnston JA, Ward CL, Kopito RR: Aggresomes: a cellular response to misfolded proteins. *J Cell Biol* 1998, 143:1883-1898.

[239] Waelter S, Boeddrich A, Lurz R, Scherzinger E, Lueder G, Lehrach H, Wanker EE: Accumulation of mutant huntingtin fragments in aggresome-like inclusion bodies as a result of insufficient protein degradation. *Mol Biol Cell* 2001, 12:1393-1407.

[240] Holmberg CI, Staniszewski KE, Mensah KN, Matouschek A, Morimoto RI: Inefficient degradation of truncated polyglutamine proteins by the proteasome. *EMBO J* 2004, 23:4307-4318.

[241] Venkatraman P, Wetzel R, Tanaka M, Nukina N, Goldberg AL: Eukaryotic proteasomes cannot digest polyglutamine sequences and release them during degradation of polyglutamine-containing proteins. *Mol Cell* 2004, 14:95-104.

[242] Goldberg AL, Akopian TN, Kisselev AF, Lee DH: Protein degradation by the proteasome and dissection of its in vivo importance with synthetic inhibitors. *Mol Biol Rep* 1997, 24:69-75.

[243] Schmidt T, Lindenberg KS, Krebs A, Schols L, Laccone F, Herms J, Rechsteiner M, Riess O, Landwehrmeyer GB: Protein surveillance machinery in brains with spinocerebellar ataxia type 3: redistribution and differential recruitment of 26S

proteasome subunits and chaperones to neuronal intranuclear inclusions. *Ann Neurol* 2002, 51:302-310.
[244] Bennett EJ, Bence NF, Jayakumar R, Kopito RR: Global impairment of the ubiquitin-proteasome system by nuclear or cytoplasmic protein aggregates precedes inclusion body formation. *Mol Cell* 2005, 17:351-365.
[245] van Leeuwen FW, de Kleijn DP, van den Hurk HH, Neubauer A, Sonnemans MA, Sluijs JA, Koycu S, Ramdjielal RD, Salehi A, Martens GJ, Grosveld FG, Peter J, Burbach H, Hol EM: Frameshift mutants of beta amyloid precursor protein and ubiquitin-B in Alzheimer's and Down patients. *Science* 1998, 279:242-247.
[246] Fischer DF, De Vos RA, Van Dijk R, De Vrij FM, Proper EA, Sonnemans MA, Verhage MC, Sluijs JA, Hobo B, Zouambia M, Steur EN, Kamphorst W, Hol EM, van Leeuwen FW: Disease-specific accumulation of mutant ubiquitin as a marker for proteasomal dysfunction in the brain. *FASEB J* 2003, 17:2014-2024.
[247] Lindsten K, De Vrij FM, Verhoef LG, Fischer DF, van Leeuwen FW, Hol EM, Masucci MG, Dantuma NP: Mutant ubiquitin found in neurodegenerative disorders is a ubiquitin fusion degradation substrate that blocks proteasomal degradation. *J Cell Biol* 2002, 157:417-427.
[248] De Vrij FM, Sluijs JA, Gregori L, Fischer DF, Hermens WT, Goldgaber D, Verhaagen J, van Leeuwen FW, Hol EM: Mutant ubiquitin expressed in Alzheimer's disease causes neuronal death. *FASEB J* 2001, 15:2680-2688.
[249] Lam YA, Pickart CM, Alban A, Landon M, Jamieson C, Ramage R, Mayer RJ, Layfield R: Inhibition of the ubiquitin-proteasome system in Alzheimer's disease. *Proc Natl Acad Sci U S A* 2000, 97:9902-9906.
[250] Hope AD, de Silva R, Fischer DF, Hol EM, van Leeuwen FW, Lees AJ: Alzheimer's associated variant ubiquitin causes inhibition of the 26S proteasome and chaperone expression. *J Neurochem* 2003, 86:394-404.
[251] Cummings CJ, Orr HT, Zoghbi HY: Progress in pathogenesis studies of spinocerebellar ataxia type 1. *Philos Trans R Soc Lond B Biol Sci* 1999, 354:1079-1081.
[252] Donaldson KM, Li W, Ching KA, Batalov S, Tsai CC, Joazeiro CA: Ubiquitin-mediated sequestration of normal cellular proteins into polyglutamine aggregates. *Proc Natl Acad Sci U S A* 2003, 100:8892-8897.
[253] Kazantsev A, Preisinger E, Dranovsky A, Goldgaber D, Housman D: Insoluble detergent-resistant aggregates form between pathological and nonpathological lengths of polyglutamine in mammalian cells. *Proc Natl Acad Sci U S A* 1999, 96:11404-11409.
[254] Nucifora FC, Jr., Sasaki M, Peters MF, Huang H, Cooper JK, Yamada M, Takahashi H, Tsuji S, Troncoso J, Dawson VL, Dawson TM, Ross CA: Interference by huntingtin and atrophin-1 with cbp-mediated transcription leading to cellular toxicity. *Science* 2001, 291:2423-2428.
[255] Schaffar G, Breuer P, Boteva R, Behrends C, Tzvetkov N, Strippel N, Sakahira H, Siegers K, Hayer-Hartl M, Hartl FU: Cellular toxicity of polyglutamine expansion proteins: mechanism of transcription factor deactivation. *Mol Cell* 2004, 15:95-105.

[256] Arrasate M, Mitra S, Schweitzer ES, Segal MR, Finkbeiner S: Inclusion body formation reduces levels of mutant huntingtin and the risk of neuronal death. *Nature* 2004, 431:805-810.

[257] Taylor JP, Tanaka F, Robitschek J, Sandoval CM, Taye A, Markovic-Plese S, Fischbeck KH: Aggresomes protect cells by enhancing the degradation of toxic polyglutamine-containing protein. *Hum Mol Genet* 2003, 12:749-757.

[258] Saudou F, Finkbeiner S, Devys D, Greenberg ME: Huntingtin acts in the nucleus to induce apoptosis but death does not correlate with the formation of intranuclear inclusions. *Cell* 1998, 95:55-66.

[259] Keller JN, Gee J, Ding Q: The proteasome in brain aging. *Ageing Res Rev* 2002, 1:279-293.

[260] Carrard G, Bulteau AL, Petropoulos I, Friguet B: Impairment of proteasome structure and function in aging. *Int J Biochem Cell Biol* 2002, 34:1461-1474.

[261] Morley JF, Brignull HR, Weyers JJ, Morimoto RI: The threshold for polyglutamine-expansion protein aggregation and cellular toxicity is dynamic and influenced by aging in Caenorhabditis elegans. *Proc Natl Acad Sci U S A* 2002, 99:10417-10422.

[262] Kitada T, Asakawa S, Hattori N, Matsumine H, Yamamura Y, Minoshima S, Yokochi M, Mizuno Y, Shimizu N: Mutations in the parkin gene cause autosomal recessive juvenile parkinsonism. *Nature* 1998, 392:605-608.

[263] Zhang Y, Gao J, Chung KK, Huang H, Dawson VL, Dawson TM: Parkin functions as an E2-dependent ubiquitin- protein ligase and promotes the degradation of the synaptic vesicle-associated protein, CDCrel-1. *Proc Natl Acad Sci U S A* 2000, 97:13354-13359.

[264] Shimura H, Hattori N, Kubo S, Mizuno Y, Asakawa S, Minoshima S, Shimizu N, Iwai K, Chiba T, Tanaka K, Suzuki T: Familial Parkinson disease gene product, parkin, is a ubiquitin-protein ligase. *Nat Genet* 2000, 25:302-305.

[265] Leroy E, Boyer R, Auburger G, Leube B, Ulm G, Mezey E, Harta G, Brownstein MJ, Jonnalagada S, Chernova T, Dehejia A, Lavedan C, Gasser T, Steinbach PJ, Wilkinson KD, Polymeropoulos MH: The ubiquitin pathway in Parkinson's disease. *Nature* 1998, 395:451-452.

[266] Moore DJ, West AB, Dawson VL, Dawson TM: Molecular pathophysiology of Parkinson's disease. *Annu Rev Neurosci* 2005, 28:57-87.

[267] Healy DG, Abou-Sleiman PM, Casas JP, Ahmadi KR, Lynch T, Gandhi S, Muqit MM, Foltynie T, Barker R, Bhatia KP, Quinn NP, Lees AJ, Gibson JM, Holton JL, Revesz T, Goldstein DB, Wood NW: UCHL-1 is not a Parkinson's disease susceptibility gene. *Ann Neurol* 2006, 59:627-633.

[268] Watts GD, Wymer J, Kovach MJ, Mehta SG, Mumm S, Darvish D, Pestronk A, Whyte MP, Kimonis VE: Inclusion body myopathy associated with Paget disease of bone and frontotemporal dementia is caused by mutant valosin-containing protein. *Nat Genet* 2004, 36:377-381.

[269] Richly H, Rape M, Braun S, Rumpf S, Hoege C, Jentsch S: A series of ubiquitin binding factors connects CDC48/p97 to substrate multiubiquitylation and proteasomal targeting. *Cell* 2005, 120:73-84.

[270] Weihl CC, Dalal S, Pestronk A, Hanson PI: Inclusion body myopathy-associated mutations in p97/VCP impair endoplasmic reticulum-associated degradation. *Hum Mol Genet* 2006, 15:189-199.

[271] Ding Q, Lewis JJ, Strum KM, Dimayuga E, Bruce-Keller AJ, Dunn JC, Keller JN: Polyglutamine expansion, protein aggregation, proteasome activity, and neural survival. *J Biol Chem* 2002, 277:13935-13942.

[272] Mytilineou C, McNaught KS, Shashidharan P, Yabut J, Baptiste RJ, Parnandi A, Olanow CW: Inhibition of proteasome activity sensitizes dopamine neurons to protein alterations and oxidative stress. *J Neural Transm* 2004, 111:1237-1251.

[273] Gaczynska M, Osmulski PA: Characterization of noncompetitive regulators of proteasome activity. *Methods Enzymol* 2005, 398:425-438.

[274] Gaczynska M, Osmulski PA: Small-molecule inhibitors of proteasome activity. *Methods Mol Biol* 2005, 301:3-22.

[275] Berkers CR, Verdoes M, Lichtman E, Fiebiger E, Kessler BM, Anderson KC, Ploegh HL, Ovaa H, Galardy PJ: Activity probe for in vivo profiling of the specificity of proteasome inhibitor bortezomib. *Nat Methods* 2005, 2:357-362.

[276] Varshavsky A: The N-end rule and regulation of apoptosis. *Nat Cell Biol* 2003, 5:373-376.

[277] Dantuma NP, Lindsten K, Glas R, Jellne M, Masucci MG: Short-lived green fluorescent proteins for quantifying ubiquitin/proteasome-dependent proteolysis in living cells. *Nat Biotechnol* 2000, 18:538-543.

[278] Lindsten K, Menendez-Benito V, Masucci MG, Dantuma NP: A transgenic mouse model of the ubiquitin/proteasome system. *Nat Biotechnol* 2003, 21:897-902.

[279] Dantuma NP, Lindsten K, Glas R, Jellne M, Masucci MG: Short-lived green fluorescent proteins for quantifying ubiquitin/proteasome-dependent proteolysis in living cells. *Nat Biotechnol* 2000, 18:538-543.

[280] Bennett EJ, Bence NF, Jayakumar R, Kopito RR: Global impairment of the ubiquitin-proteasome system by nuclear or cytoplasmic protein aggregates precedes inclusion body formation. *Mol Cell* 2005, 17:351-365.

[281] Luker GD, Pica CM, Song J, Luker KE, Piwnica-Worms D: Imaging 26S proteasome activity and inhibition in living mice. *Nat Med* 2003, 9:969-973.

[282] Kwon YT, Xia Z, An JY, Tasaki T, Davydov IV, Seo JW, Sheng J, Xie Y, Varshavsky A: Female lethality and apoptosis of spermatocytes in mice lacking the UBR2 ubiquitin ligase of the N-end rule pathway. *Mol Cell Biol* 2003, 23:8255-8271.

[283] Swanson R, Locher M, Hochstrasser M: A conserved ubiquitin ligase of the nuclear envelope/endoplasmic reticulum that functions in both ER-associated and Matalpha2 repressor degradation. *Genes Dev* 2001, 15:2660-2674.

[284] Hoyt MA, Zhang M, Coffino P: Ubiquitin-independent mechanisms of mouse ornithine decarboxylase degradation are conserved between mammalian and fungal cells. *J Biol Chem* 2003, 278:12135-12143.

[285] Fang D, Kerppola TK: Ubiquitin-mediated fluorescence complementation reveals that Jun ubiquitinated by Itch/AIP4 is localized to lysosomes. *Proc Natl Acad Sci U S A* 2004, 101:14782-14787.

[286] De Vrij FM, Fischer DF, van Leeuwen FW, Hol EM: Protein quality control in Alzheimer's disease by the ubiquitin proteasome system. *Prog Neurobiol* 2004, 74:249-270.

[287] Oddo S, Billings L, Kesslak JP, Cribbs DH, LaFerla FM: Abeta immunotherapy leads to clearance of early, but not late, hyperphosphorylated tau aggregates via the proteasome. *Neuron* 2004, 43:321-332.

[288] David DC, Layfield R, Serpell L, Narain Y, Goedert M, Spillantini MG: Proteasomal degradation of tau protein. *J Neurochem* 2002, 83:176-185.

In: The Ubiquitin Proteasome System… ISBN 978-1-60021-749-4
Eds: Mario Di Napoli and Cezary Wojcik, pp. 715-734

Chapter 29

MECHANISMS INVOLVED IN THE AGGREGATION OF UBIQUITINATED PROTEINS IN NEURODEGENERATIVE DISORDERS

Kenyon D. Ogburn and Maria E. Figueiredo-Pereira*

Department of Biological Sciences, Hunter College of City, University of New York, New York, New York 10021, USA.

ABSTRACT

The ubiquitin-proteasome system (UPS) is a highly regulated and fundamental pathway for protein degradation. It controls many key cellular mechanisms critical for cell viability and function and also removes abnormal and toxic proteins generated by a lifetime of environmental damage. Most proteins are tagged by ubiquitin prior to degradation by the UPS. Notably, abnormal protein deposits containing ubiquitinated proteins are detected in a variety of neurodegenerative diseases. Such disorders include Alzheimer's disease, Parkinson's disease, amyotrophic lateral sclerosis and Huntington's disease, to name a few. Whether these protein deposits are pathogenic or represent a coping mechanism to prolong survival of the affected cells is a hotly debated issue. Lately, tremendous strides have been made to elucidate the mechanisms regulating the accumulation and aggregation of ubiquitinated proteins associated with neurodegeneration. This chapter provides a critical overview of the latest studies addressing these mechanisms. First, we will focus on oxidative stress. The brain is considered to be unusually sensitive to oxidative damage. Moreover, many age-related neurodegenerative disorders exhibit abnormal accumulation of oxidatively damaged proteins. Although there are many tantalizing clues indicating that the proteasome is crucial for the degradation of oxidatively modified proteins, we still lack a clear understanding of how these proteins are targeted for proteasomal degradation. The

* Correspondence concerning this article should be addressed to Dr. Maria E. Figueiredo-Pereira, PhD; Department of Biological Sciences, Hunter College of City University of New York, 695 Park Avenue, New York, New York 10021, Tel. 212 650-3565; Fax. 212 772-5227; E-Mail: pereira@genectr.hunter.cuny.edu.

controversy is over the requirement of ubiquitination for the degradation of oxidatively modified proteins. Secondly, we will address the relationship between inflammation and UPS impairment. The brain was long considered to be an immunologically privileged site, particularly because of the blood brain barrier and the lack of a lymphatic system. However, more recently it has been shown that the brain mounts an inflammatory response, as noted from the occurrence of edema, microglia and astrocyte activation, local invasion of circulating immune cells and production of cytokines and other immune factors. There is abundant evidence supporting that an inflammatory reaction is mounted within the CNS following trauma, stroke, infection and seizures, all of which can augment brain damage. We will discuss how products of inflammation induce oxidative stress and the accumulation and aggregation of ubiquitinated proteins. A better understanding of the mechanisms that regulate the aggregation of ubiquitinated proteins is of clinical importance for developing therapeutic strategies to prevent and treat neurodegenerative diseases. One of the major challenges that we are faced with is to single out the UPS as a therapeutic target for preventing neurodegeneration. This challenge rests on developing therapeutic strategies that will enhance degradation of abnormal and toxic proteins without compromising the normal function of the UPS.

Keywords: protein aggregation, ubiquitin, oxidative stress, inflammation, neurodegeneration, J2 prostaglandins, dopamine toxicity, aggresome, proteasome, sequestosome.

Abbreviations

AD, Alzheimer's disease; AID, acidic interaction domain; ALS, amyotrophic lateral sclerosis; AAA, ATPases associated with diverse cellular activities; ATP, adenosine triphosphate; CNS, central nervous system; COMT, catechol-*O*-methyltransferase; CP, core particle; COX-1 and COX-2, cyclooxygenase 1 and 2; DOPA, 3,4-dihydroxyphenylalanine; Erk, extracellular signal regulated kinase; GTP, guanosine triphosphate; HEK, human embryonic kidney; HNE, 4-hydroxy-2-nonenal; IL, interleukin; iNOS, inducible nitric oxide synthase; JNK, c-Jun NH_2-terminal kinase; LBs, Lewy bodies; Lys, lysine; MAPK, Mitogen-activated protein kinase; NFκB, nuclear factor κB; PD, Parkinson's disease; PEST, proline, glutamic acid, serine and threonine; PKC, protein kinase C; PPARγ, peroxisome proliferator-activated receptor gamma; 15d-PGJ2, 15-Deoxy-Δ-12,14-prostaglandin J2; PG, prostaglandin; RING, really interesting new gene; RIP, receptor interactive protein; RNAi, RNA interference; ROS, reactive oxygen species; RP, regulatory particle; Rpn, Regulatory Particle, Non-ATPase-like; Rpt, Regulatory Particle, ATPase-like; TNF, tumor necrosis factor; TRAF6, tumor necrosis factor receptor-associated factor 6; TxA2, thromboxane A2; UBA, ubiquitin-associated domain; UBL, ubiquitin-like domain; UCH, ubiquitin-carboxyl terminal; UPS, ubiquitin-proteasome system.

Introduction

Protein deposits containing ubiquitinated proteins are detected in non-pathologic aging [1] as well as in a variety of age-related neurodegenerative disorders including Alzheimer's disease (AD), Parkinson's disease (PD) and amyotrophic lateral sclerosis (ALS), to name a few [2]. All of these protein deposits contain ubiquitinated proteins but their major structural components vary from cell type to cell type. For example, microtubule associated *tau* proteins are found in cortical neurofibrillary tangles and α-synuclein in dopaminergic Lewy bodies [3]. The presence of aggregated ubiquitinated proteins in inclusion bodies indicates impairment of the ubiquitin-proteasome system (UPS) or structural changes in protein substrates impeding their degradation. The mechanisms causing the aggregation of ubiquitinated proteins have to be addressed in order to understand the overall molecular disease processes of these neurological conditions.

In this chapter, we will discuss evidence supporting the view that oxidative stress and inflammation are critical contributors to the intracellular aggregation of ubiquitinated proteins that evade degradation by the UPS. The involvement of oxidative stress in neurodegeneration has gained support from increasing evidence of its role in neuronal death in disorders such as AD and PD. Studies with autopsied brains of AD patients show a co-localization of high levels of oxidative stress products with neurofibrillary tangles and senile plaques [4]. Signs of oxidative stress, such as lipid peroxidation, increased protein carbonyls and a decline in reduced glutathione, were also detected in the *substantia nigra* in brains of PD patients [5]. Oxidative stress, especially the production of free radicals, promotes partial unfolding of cellular proteins resulting in the exposure of previously buried hydrophobic domains to ubiquitin-conjugating [6] as well as to proteolytic enzymes [7,8]. One important cellular response to oxidative stress is an increase in intracellular proteolysis by the proteasome.

Chronic inflammation of the CNS has also been implicated in a variety of neurodegenerative disorders. Notably, the spatial and temporal distribution of pro-inflammatory cyclooxygenase-2 (COX-2) correlates with neuropathological changes in a wide variety of disorders including AD, PD and ALS [9]. These disorders, which exhibit signs of inflammation, are also associated with accumulation of ubiquitinated proteins in neuronal inclusions [10]. Notably, inflammatory processes occurring in the CNS differ from systemic inflammation. The order of events occurring in the CNS following a noxious insult includes: *(i)* immune cell proliferation, *(ii)* microglia activation, *(iii)* cytokine release and *(iv)* induction of tissue repair enzymes. Together, these responses are initiated as a defense mechanism to help limit cellular damage and repair the CNS [11]. Ironically, these same pro-inflammatory agents can incite tissue damage in both acute and chronic CNS disorders.

The overall aim of this review is to provide a challenging and sometimes provocative survey of studies supporting the view that oxidative stress and CNS inflammation are critical contributors to the formation of inclusion bodies detected in most neurodegenerative disorders. In an effort to provide a general overview of the current information available on the relationship among oxidative stress, neuroinflammation and the aggregation of ubiquitinated proteins in neurodegeneration, we will address the following topics: (*i*) Inducers of oxidative stress; (*ii*) Degradation of oxidatively modified proteins; and (*iii*) Inflammation and protein aggregation.

INDUCERS OF OXIDATIVE STRESS

Aerobic Respiration

Reactive oxygen species (ROS), such as superoxide and hydrogen peroxide, are produced by mitochondria during respiration rendering all aerobic organisms susceptible to oxidative stress. Of the total oxygen consumed during respiration, about 2% ends up as ROS, but this amount may vary depending on exposure to various stress conditions, such as environmental pollutants (metals and xenobiotics) and inflammatory cytokines. The brain is considered to be abnormally sensitive to oxidative damage because (*i*) it is enriched in the more easily peroxidizable fatty acids (20:4 and 22:6), (*ii*) relative to its small weight it consumes an excessive fraction of the body total oxygen consumption and (*iii*) it is not particularly enriched in antioxidant defenses [12]. For example, the brain is low in catalase activity containing about 10% of liver catalase. In addition, certain regions of the human brain are enriched in iron/ascorbate. If tissue organizational disruption occurs, the iron/ascorbate mixture becomes an abnormally potent pro-oxidant for brain membranes [12].

Agents that impair mitochondrial activity decrease ATP production and promote ROS formation. These cytotoxic agents change the redox equilibrium and may, for example, increase the levels of mitochondrial quinones such as Coenzyme Q_{12}. Coenzyme Q_{12} is essential for maintaining the proton gradient across the mitochondrial membrane. However, high levels of quinones elevate ROS production. Indeed, higher concentrations of quinones are found in the brains of AD patients compared to normal controls, supporting a role for these compounds in the etiology of neurodegenerative diseases [13,14].

Inflammation

In the context of neurodegenerative disorders, neuroinflammation refers to any set of responses culminating in a COX-2-induced, pro-inflammatory response in CNS tissues. Cyclooxygenases (COX-1 or COX-2) are the rate-limiting enzymes in the synthesis of prostaglandins. In the first step, cyclooxygenases convert arachidonic acid to prostaglandin G2 (PGG2). Their peroxidase activity then converts PGG2 to prostaglandin H2 (PGH2), the parental prostanoid. PGH2 is subsequently converted to a variety of products including PGD2, PGE2, PGF2α, PGI2, and TxA2 by cell-specific isomerases and synthases. Under physiological conditions, the concentrations of prostaglandins in body fluids are in the pico-nanomolar range [15]. However, under inflammatory conditions, their concentrations may reach the micromolar range at the site of damage [16,17] and as such, they may act as pro-inflammatory mediators of oxidative stress.

Prostaglandins are primarily synthesized from arachidonic acid which comes from membrane phospholipids and dietary sources [18,19]. In general, the majority of prostaglandins are considered pro-inflammatory however, they may also be anti-inflammatory [16]. In a rat model of inflammation, COX-2 was initially pro-inflammatory by way of PGE2 synthesis. Conversely, during the resolution phase of inflammation, COX-2 was shown to be anti-inflammatory by way of PGD2 and 15d-PGJ2 production. Hence, the dual role of COX-2 is quite complex. PGD2 is the major prostanoid synthesized in the mammalian CNS. This prostanoid is unstable and readily undergoes both *in vivo* and *in vitro* non-enzymatic dehydration to generate biologically active cyclopentenone J2 prostaglandins, such as PGJ2, 12d-PGJ2 and 15d-PGJ2.

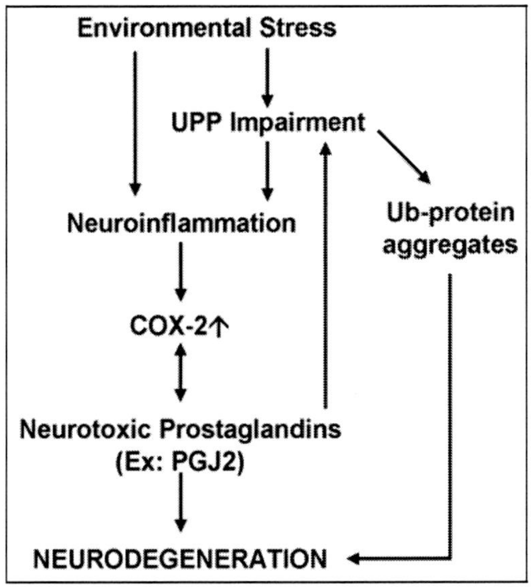

Figure 1. Environmental stress evoked by physical, chemical or microbial stimuli impairs the UPS causing aggregation of ubiquitinated (Ub) proteins. UPS impairment and environmental stress may also trigger inflammatory responses manifested by up-regulation of COX-2. Neurotoxic products of COX-2, such as PGJ2, may then accelerate the pathophysiological processes that underlie neurodegeneration. The resulting neuronal cell death may have devastating effects as, in the vast majority of cases, neurons lost to disease processes cannot be replaced.

Some of the effects of J2 prostaglandins appear to be mediated by their interaction with an intranuclear target, the peroxisome proliferator-activated receptor (PPARγ). Several studies suggest that 15d-PGJ2 is a ligand for the PPARγ receptor [20]. PPARγ is a transcription factor found in many cell types, including neurons/microglia [21], macrophages/monocytes, myocytes, fibroblast, breast cells, and human bone marrow [22,23]. During a noxious insult, the stereotypic result of macrophage/microglia activation is the production of various pro-inflammatory mediators. However, in the presence of 15d-PGJ2, the production of pro-inflammatory mediators such as IL-6, TNF-α, IL-1β and inducible nitric oxide synthase (iNOS) seems to be inhibited, thus favouring an anti-inflammatory response. In further support of this view, 15d-PGJ2 was found to inhibit IL-10 and IL-12 production by macrophages [24]. Other studies reported that PPARγ ligands induce monocytes/macrophages to respond in a pro-inflammatory manner by stimulating the expression of pro-inflammatory receptors, such as CD14 and CD11b/CD18 [25-28]. Clearly, the effects of PPARγ ligands, including 15d-PGJ2, are variable and may depend on factors such as intracellular concentrations, cell types and timing of activation of downstream targets that participate in the inflammatory response. In addition, J2 prostaglandins act through PPARγ-independent mechanisms including activation of Erk (MAPK and JNK) pathways

[29,30] and inhibition of the NFκB pathway [31,32]. This may account for the different effects of J2 prostaglandins and other PPARγ ligands [33].

J2 prostaglandins are unique among the prostaglandin family in that they have α,β-unsaturated carbonyl groups promoting Michael addition reactions with free sulfhydryl groups of cysteines in glutathione and cellular proteins [34]. These cyclopentenone prostaglandins were shown to covalently modify several proteins, including the p50 subunit of NFκB, which may explain their anti-inflammatory effects [35]. They also modify thioredoxin reductase, an enzyme that protects against oxidative damage [36] and activate Ras, a small GTPase oncogene known to activate Erk signaling pathways [37].

Recent studies suggest that J2 prostaglandins play a role in the etiology of neurodegeneration (Figure 1). Specifically, the levels of 15d-PGJ2 were found to be elevated in spinal cord motor neurons of ALS patients [38]. Additionally, J2 prostaglandins were shown to be neurotoxic and pro-oxidant agents [39], up-regulate the expression of COX-2 [30], inhibit ubiquitin isopeptidase activity [40] and Ubiquitin-carboxyl terminal hydrolase (UCH)-L1 and UCH-L3 [41] and induce the accumulation [39] and aggregation [41] of ubiquitinated proteins as well as neuronal apoptosis [42]. That UCH-L1 inhibition might be relevant to neurodegeneration is supported by the identification of a missense mutation in the gene encoding UCH-L1 in two siblings of a German family with autosomal-dominant familial PD [43]. As the UPS is a complex and tightly-regulated system, there are undoubtedly other mechanisms by which J2 prostaglandins negatively affect the UPS.

Dopamine

Parkinson's disease (PD) is characterized by the selective degeneration of dopaminergic neurons of the nigrostriatal pathway. In PD, dopaminergic neuronal loss is accompanied by Lewy bodies (LBs), which are neuronal proteinaceous cytoplasmic inclusions [44,45]. The neuronal loss is also associated with the accumulation of highly-oxidized protein aggregates, often surrounded by microglia. Upon activation, microglia release a large amount of oxidizing species, thus initiating a potentially disastrous, inflammatory cytotoxic cycle. A compensatory mechanism initiated by the cell is to increase protein degradation [46].

Among the catecholamines, dopamine is the most susceptible to autooxidation that leads to ROS formation. This is because of its high rate of oxidation to an electron-deficient quinone and a slower rate of internal cyclization of the *o*-quinone structure [47]. Conditions that promote the intracellular accumulation of dopamine might thus lead to increased dopaminergic neuronal loss [48]. Susceptibility of dopaminergic neurons to oxidative stress was shown in studies with rotenone, an environmental toxin linked specifically to PD [49].

Cyclooxygenases are also able to oxidize dopamine to dopamine quinone via their peroxidase activity. These enzymes will readily utilize dopamine as an electron donor to support their peroxidase activity generating an electron-deficient dopamine quinone as a byproduct. Dopamine quinone can then covalently bind to the sulfhydryl groups of cysteine residues on proteins. If the covalently modified cysteine is located at or near the protein active site, the binding of dopamine quinone will cause inactivation of protein function. If these protein functions are essential for cell viability, their inactivation may account for quinone-induced cytotoxicity [50].

PD pathogenesis is thus closely linked to oxidative stress [51]. Under such cytotoxic conditions, direct or indirect proteasome impairment might occur. This impairment could lead to the accumulation of cytotoxic proteins, disruption of neuronal function and ultimately cell death.

Proteotoxicity of Oxidative Stress

Understanding the sources of oxidative stress and how these conditions affect UPS activity and its substrates is relevant to neurodegeneration. Strong oxidants like the various ROS resulting from oxidative stress damage the structure of cellular proteins [52] which, if not repaired, must be removed by proteolysis to prevent their accumulation and aggregation. One of the major roles of the proteasome is to remove oxidatively modified proteins. However, we still lack a clear understanding of how oxidatively modified proteins are targeted for proteasomal degradation. Some investigators support the notion that oxidatively modified proteins in cells are removed by the 20S proteasome independently of ubiquitination [53]. Nonetheless, it is not clear how these proteins would be recognized by the 20S proteasome. Others demonstrated that there is an accumulation of ubiquitin-protein conjugates as well as increases in ubiquitin-activating and ubiquitin-conjugating enzyme activities following episodes of oxidative stress. This suggests that the ubiquitination machinery is recruited to target oxidatively modified proteins for proteasomal degradation [54]. Understanding how oxidatively modified proteins are degraded is an important issue. Not only because the brain is considered to be unusually sensitive to oxidative damage, but also because many age-related neurodegenerative disorders exhibit abnormal accumulation of oxidatively damaged proteins.

Degradation of Oxidatively Modified Proteins

It is likely that some or the majority of protein components of the cytoplasmic inclusion bodies detected in neurodegenerative disorders are aggregates of oxidatively modified proteins [55]. Although somewhat phenotypically-characterized, the biochemical mechanism leading to the formation of these protein aggregates remains poorly defined. It is well established that the ability of cells to degrade abnormal, mutated, or oxidized proteins is exceeded when the UPS is inhibited [56]. Thus, UPS impairment is likely to play an essential role in protein aggregation. These protein aggregates could themselves impair the UPS, generating a disastrous positive feedback loop. Initially, protein aggregates may not affect the UPS as the proteasome does attempt to rescue the cell from oxidative insults by degrading oxidatively modified proteins. The increased surface hydrophobicity of oxidatively modified proteins is postulated to make them more susceptible to proteasome mediated degradation [57-64]. This degradation is thought to result from recognition of the hydrophobic moieties spanning the core of the protein to its surface. Exposure of these hydrophobic patches to the intracellular milieu is dependent on protein unfolding induced by its oxidation.

Modifications of Lysine Residues on Ubiquitin

Modification of Lys residues, such as acetylation and biotinylation, occurs through nonenzymatic means. Certain ubiquitin Lys residues, such as Lys6, are targets for oxidative-modification. Like Lys48-linked polyubiquitin chains, nontraditional Lys6-linked polyubiquitin chains target proteins for degradation by the 26S proteasome. To better understand the structure-function relationship of ubiquitin and its susceptibility to oxidative stress, the modification of the Lys6 residue of ubiquitin and its physiological consequences were investigated [65]. Through mass spectrometry-based peptide mapping and N-terminal sequencing, it was demonstrated that of the seven ubiquitin lysine residues, Lys6 was the most readily labelled by sulfosuccinimidobiotin [66]. Lys6 is also susceptible to modification by *p*-nitrophenyl acetate [67], aspirin [68], acetic anhydride [69] and Oregon green succinimidyl ester [70]. It is likely that Lys6 of ubiquitin is also susceptible to modification by the highly-reactive product of lipid peroxidation, 4-hydroxy-2-nonenal (HNE) [65]. Certain modifications of ubiquitin Lys residues may render cells more vulnerable to oxidative stress because they may impair polyubiquitin chain assembly. This is supported by the finding that expression of a dominant-negative mutant form of ubiquitin (K48R) in a human teratocarcinoma NT-2 cell line and a human neuroblastoma SK-N-MC cell line increased oxidative damage and rendered cells more susceptible to HNE [71].

Lys6-biotinylated ubiquitin was found to form high molecular mass ubiquitin conjugates as readily as unmodified ubiquitin, thus it is an ideal model to study ubiquitin modification and its consequences. Ubiquitin-dependent degradation was shown to be inhibited by Lys6-biotinylated ubiquitin, as protein conjugates formed with Lys6-biotinylated ubiquitin were less susceptible to 26S proteasome degradation [65]. Expression of K6W mutant ubiquitin in lens epithelial cells caused the stabilization of UPS substrates and also rendered the transfected cells more susceptible to oxidative stress, thus mimicking the effect of Lys6-biotinylated ubiquitin. Based on these studies it is tempting to speculate that other compounds, such as J2 prostaglandins or HNE, might also interfere with the ubiquitination of proteins or their recognition by the 26S proteasome thus increasing the vulnerability of cells exposed to oxidative stress or inflammation.

20S *versus* 26S Proteasomes

Currently, it is uncertain whether it is the 26S or 20S proteasome that carries-out the *in vivo* degradation of oxidatively modified proteins. Several researchers support the notion that the 20S proteasome is sufficient for degradation of oxidatively modified proteins. It seems that degradation of the latter is ATP-independent [64] and that there is a decline in 26S proteasome activity following oxidative stress [72,73]. Furthermore, degradation of oxidatively modified proteins was not impaired in cell lines harbouring an E1-Ub-activating enzyme deficient mutant [74]. Several concerns tamper this hypothesis. In the first place, these experiments involve *in vitro* assays that assess the accumulation of ubiquitinated proteins in cell-free lysates and under conditions that may not occur in an *in vivo* setting. This paradigm does not take into account possible factors that may be absent upon cell lysate preparation. Such factors might be required for the *in vivo* ubiquitin-mediated degradation of oxidatively modified proteins. Secondly, lysozyme and ferritin were the only two substrates tested in these studies and they were exposed to high concentrations of H_2O_2, as the source of oxidative stress. These high H_2O_2 concentrations are unlikely to be ever attained intracellularly. Oxidative stress induced by H_2O_2 also affects the UPS, either by directly decreasing the activity of the 20S or 26S proteasome [75-77], or by increasing the expression and activity of at least two members of the ubiquitination machinery, namely E1 and E2 enzymes [78]. The role of ubiquitination and ultimately, the involvement of the 26S proteasome in the degradation of oxidatively modified proteins should not be discounted.

Substrate Recognition by the 19S Regulatory Particle of the 26S Proteasome

The 26S proteasome consists of a 20S core particle (CP) and one or two 19S regulatory particles (RP) that cap the entry into the catalytic core [79]. The 19S RP facilitates recruitment of polyubiquitinated proteins to the 26S proteasome and their translocation into the 20S CP. The 19S RP is composed of two sub-complexes: the lid that is distal to the 20S CP and contains polyubiquitin binding as well as deubiquitinating activities, and the base that is in contact with the pore of the 20S CP and consists of six ATPase and two non-ATPase subunits. The two latter subunits [(Rpn1(S2) and Rpn2(S1)] interact with proteins that contain ubiquitin-like domains (UBL) and may participate in proteasomal substrate targeting [79]. In addition, at least two subunits of the 19S RP bind polyubiquitin chains directly, i.e. Rpn10 (S5a) and Rpt5 (S6'), but their role in substrate targeting remains undefined [79]. It is clear that recognition of substrates by the proteasome is facilitated by its 19S RP. This form of proteasome-substrate recognition might be compromised under stress conditions, such as oxidative stress, thus impairing the degradation of proteasomal substrates and leading to their aggregation in neuronal cells [80]. More details on proteasome structures and functions are given in Chapters 6 and 7.

Modifications of the 19S Regulatory Particle of the 26S Proteasome

To address the mechanism underlying cell death induced by oxidative stress, human neuroblastoma SH-SY5Y cells were treated with an endogenous electrophile, i.e. 15d-PGJ2 that is a neurotoxic product of inflammation and a potent inducer of oxidative stress [80]. Through proteomic analysis of oxidation-sensitive proteins, the 19S RP Rpt3(S6) subunit that has ATPase activity was shown to be one of the major targets of the 15d-PGJ2 induced protein carbonylation. Furthermore, Cu^{2+}/H_2O_2-treatment of SH-Y5Y cells also resulted in the oxidative modification of Rpt3(S6) [80]. As a mechanism to explain the metal-catalyzed oxidation of Rpt3(S6), it was postulated that Cu^{2+} binds to a specific metal-binding site on Rpt3(S6) and that Cu^{2+} may react with H_2O_2 to generate ROS. These ROS would then have the potential to oxidize neighbouring amino acid residues, thereby generating carbonyl groups on the 19S RP subunit. These findings suggest that the Rpt3(S6) subunit may be one of various oxidation-sensitive subunits of the proteasome.

The Rpt3(S6) subunit of the 19S regulatory particle is one of six non-redundant ATPases all belonging to the AAA (ATPases associated with diverse cellular activities) superfamily [81]. Previous studies demonstrated that Rpt3(S6) may interact with additional 19S regulatory subunits, such as the Rpt5(S6') and Rpt1(S7) ATPases [82-85], and that these hexameric ATPase complexes are associated with substrate unfolding and translocation into the 20S CP. It is clear that 19S RP subunits are targets of oxidative insults in cells under conditions of oxidative stress mediated by ROS or products of inflammation. Oxidation of 19S RP subunits could compromise the ability of the 26S proteasome to recognize and degrade polyubiquitinated proteins leading to the accumulation of cytotoxic, oxidatively-modified proteins.

In conclusion, oxidatively modified proteins may become polyubiquitinated and targeted for degradation by the 26S proteasome. However, under conditions of oxidative stress, their degradation might be impaired because of the inability of the 19S RP to bind and unfold substrates before they enter the 20S CP. This hypothesis is supported by the finding that Rpt3(S6) down-regulation by RNAi resulted in enhanced accumulation of ubiquitinated proteins [80]. Depending on the severity of the oxidative insult, the ability of the 26S proteasome to degrade oxidatively modified proteins might be only partially compromised. However, if the oxidative insult is chronic, the overall ability of the 26S proteasome to degrade oxidatively modified proteins may become totally compromised leading to accumulation of ubiquitinated proteins to toxic levels.

Inflammation and Protein Aggregation

Oxidative Modifications

The levels of products of inflammation, such as J2 prostaglandins, may be increased rapidly in response to tissue injury or some other noxious insults. These high levels of J2 prostaglandins could exert their effect by generating ROS and thus promoting oxidative stress. Cellular proteins are susceptible to the oxidative modification of their amino acid side chains. This oxidation may radically alter the native conformation of proteins and, if not repaired, is likely to trigger their loss of function. Methionine and cysteine residues may be particularly susceptible to damage by ROS. While methionine is readily oxidized to methionine sulfoxide [86,87] cysteine sulfhydryl group oxidation may result in the formation of intra- or intermolecular disulfide bridges [88]. In addition, the prominent α,β-unsaturated ketones of J2 prostaglandins may undergo nucleophilic addition reactions with cysteine thiol groups. All of these chemical modifications can promote the aggregation of proteins or peptides, which is a hallmark of many neurodegenerative disorders. Increases in ROS in conjunction with a decline in proteasome activity, both of which may be associated with oxidative stress and/or inflammation, may thus lead to the progressive accumulation of oxidatively damaged and/or ubiquitinated proteins. Collectively, these could eventually lead to cellular dysfunction and neurodegeneration [88].

Sequestosome 1/p62

Sequestosomes, also known as p62, were first isolated in human tissues [89]. Due to their high affinity for polyubiquitin chains, sequestosomes were suggested to serve as receptors for binding and storing ubiquitinated proteins [90]. Sequestosomes contain seven structural motifs: an SH2 domain that binds the tyrosine kinase $p56^{lck}$ in a phosphotyrosine-independent manner [91], an acidic interaction domain (AID) that binds the atypical PKC ζ [92], a ZZ type ZINC finger that binds the receptor interactive protein (RIP) involved in TNFα-induced apoptosis [93], a binding site for the RING-finger protein tumor necrosis factor receptor-associated factor 6 (TRAF6) that is an E3 ubiquitin ligase [94,95], two PEST sequences and a UBA domain [96]. Sequestosomes might bind polyubiquitin chains through the C-terminal UBA domain and the proteasome through their AID domain, which is closer to their N-terminus [97]. The AID domain is proposed to be structurally similar to ubiquitin-like (UBL) domains, known to interact with the proteasome [97]. Studies with HEK cells transfected with full-length or truncated sequestosome forms indicate that sequestosomes may act as shuttles that deliver polyubiquitinated proteins to the proteasome [97] and protect the cells from apoptosis [98]. Sequestosome1/p62 was also shown to interact with K63-polyubiquitinated *tau* through its UBA domain and might regulate *tau* proteasomal degradation [99].

Sequestosome 1/p62 was detected in ubiquitin-containing intraneuronal and intraglial inclusions in a number of neurodegenerative disorders [100-103]. The expression of sequestosome 1/p62 in neuronal cells is increased by serum withdrawal conditions that trigger apoptosis and by proteasome inhibitors [104,105] as well as by expression of expanded pathologic polyglutamine repeats [106]. PGJ2, a neurotoxic product of inflammation, increased the levels of sequestosome 1/p62 and ubiquitinated proteins in a parallel manner in human neuroblastoma SK-N-SH cells [107]. Prevention of sequestosome 1/p62 up-regulation by RNA interference (RNAi) precluded the cytoplasmic aggregation of ubiquitinated proteins induced by PGJ2 but not its cytotoxicity. These data support the notion that sequestosome 1/p62 up-regulation under stress conditions contributes to the 'sequestration' of polyubiquitinated proteins into aggregates, perhaps to ensure that the proteasome is not overwhelmed. Excessive accumulation of ubiquitinated proteins, rather than their aggregation, is likely to be an important contributor to the cytotoxicity induced by the different stress agents.

Inactivation and Sequestration of Metabolic Enzymes

Catechol-*O*-methyltransferase (COMT) is an enzyme responsible for inactivating catecholamine neurotransmitters, such as dopamine, as well as catechol hormones and catechol drugs like L-DOPA, α-methyl DOPA and isoprenaline [108,109]. Treatment of human neuroblastoma SK-N-SH cells with PGJ2, a product of inflammation, reduced the expression and activity of COMT and potentiated dopamine toxicity [110]. PGJ2 also triggered the formation of large perinuclear aggregates containing COMT, thus affecting its subcellular distribution. The latter was duplicated in rat primary cortical neurons. These results suggest that COMT-impairment induced by PGJ2-treatment, may increase the concentration of dopamine (or its metabolites) to neurotoxic levels and may thus be a potential risk factor in neurodegeneration. Furthermore, aggregation of metabolic enzymes, such as COMT as well as other proteins that may be essential for cell viability, has the potential to seriously jeopardize cellular homeostasis. Under stress conditions, such as those induced by inflammation, sequestration of these proteins into aggregates may impair their activities and prevent them from promoting cell survival.

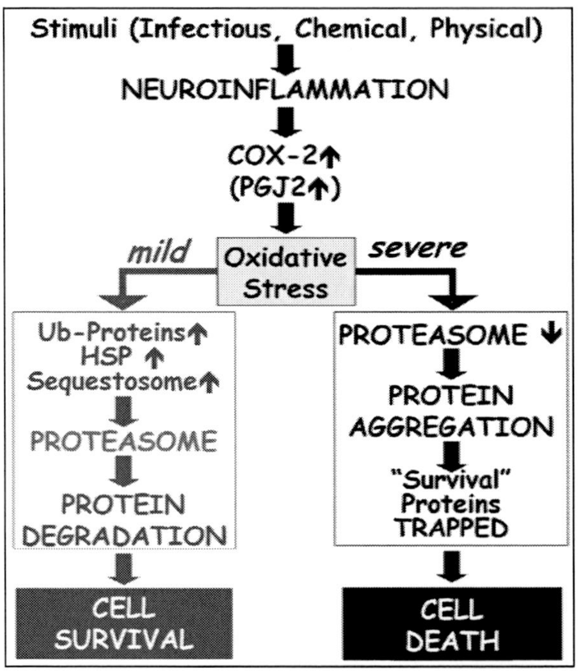

Figure 2. Scheme depicting some of the neurotoxic effects of neuroinflammation. Pro-inflammatory insults in the CNS may initiate a cascade of events that leads to increased oxidative stress and UPS impairment culminating in neurodegeneration (see text for a more detailed explanation).

CONCLUSIONS

Inflammatory processes, in particular those attributed to chronic neuroinflammation are with no doubt implicated in neurodegenerative disorders such as AD, PD and ALS. However, there is an obvious gap in the knowledge of how endogenous products of inflammation alter underlying cellular mechanisms fundamental to neurodegeneration. Our overall hypothesis (Figure 2) is that pro-inflammatory conditions may affect important cellular pathways involved in neuronal homeostasis through the production of J2 prostaglandins (PGJ2), which are derived from PGD2, the major prostaglandin synthesized in the mammalian CNS. Pro-inflammatory stimuli in the CNS induce COX-2 up-regulation and prostaglandin production. Neurotoxic prostaglandins, such as PGJ2, may trigger oxidative stress. Under mild (non-lethal) conditions the cell initiates a 'pro-survival/repair' response, which includes increased expression of heat shock proteins, ubiquitin and sequestosomes, to name a few. Increased expression of these factors indicates a cellular attempt to rescue and/or remove abnormal proteins generated by the proteotoxic effects of PGJ2, such as oxidative stress and/or protein misfolding. If the damaging effects of PGJ2 cannot be reversed by these and other repair mechanisms and the proteasome is impaired as well, this will lead to protein aggregation. These protein aggregates may sequester 'survival' proteins, such as heat shock proteins and metabolic enzymes including COMT. Once trapped in the aggregates these 'survival' factors may no longer be able to reverse the damage induced by oxidative as well as other cellular stresses. Then pro-death pathways, including apoptosis, may be activated most likely to remove damaged cells. The resulting neuronal cell death may have devastating effects as, in the vast majority of cases, neurons lost to disease processes cannot be replaced. In conclusion, pro-inflammatory insults in the CNS may initiate a cascade of events that leads to increased oxidative stress and proteasome impairment resulting in protein aggregation and culminating in neurodegeneration. Pro-inflammatory events, oxidative stress and proteasome impairment may collectively contribute to the development and exacerbation of neurodegenerative disorders.

Acknowledgements

This work was supported by NIH (RR-03037 to Hunter College from NIGMS/RCMI, GM60654 to M.E.F.-P. and T.S-G from NIGMS and GM-060665 to Hunter College from NIGMS). We apologize to those authors whose work we have been unable to cite or discuss in detail due to space constraints.

References

[1] Gray DA, Tsirigotis M, Woulfe J: Ubiquitin, proteasomes, and the aging brain. *Sci Aging Knowledge Environ* 2003, 2003:RE6.

[2] Lowe J, Blanchard A, Morrell K, Lennox G, Reynolds L, Billett M, Landon M, Mayer RJ: Ubiquitin is a common factor in intermediate filament inclusion bodies of diverse type in man, including those of Parkinson's disease, Pick's disease, and Alzheimer's disease, as well as Rosenthal fibres in cerebellar astrocytomas, cytoplasmic bodies in muscle, and mallory bodies in alcoholic liver disease. *J Pathol* 1988, 155:9-15.

[3] Alves-Rodrigues A, Gregori L, Figueiredo-Pereira ME: Ubiquitin, cellular inclusions and their role in neurodegeneration. *Trends Neurosci* 1998, 21:516-520.

[4] Markesbery WR, Carney JM: Oxidative alterations in Alzheimer's disease. *Brain Pathol* 1999, 9:133-146.

[5] Fahn S, Cohen G: The oxidant stress hypothesis in Parkinson's disease: evidence supporting it. *Ann Neurol* 1992, 32:804-812.

[6] Sadis S, Atienza CJ, Finley D: Synthetic signals for ubiquitin-dependent proteolysis. *Mol Cell Biol* 1995, 15:4086-4094.

[7] Grune T, Reinheckel T, Davies KJ: Degradation of oxidized proteins in mammalian cells. *FASEB J* 1997, 11:526-534.

[8] Chao CC, Ma YS, Stadtman ER: Modification of protein surface hydrophobicity and methionine oxidation by oxidative systems. *Proc Natl Acad Sci U S A* 1997, 94:2969-2974.

[9] Wyss-Coray T, Mucke L: Inflammation in neurodegenerative disease--a double-edged sword. *Neuron* 2002, 35:419-432.

[10] Li Z, Jansen M, Pierre S-R, Figueiredo-Pereira ME: Neurodegeneration: linking ubiquitin/proteasome pathway impairment with inflammation. *Int J Biochem Cell Biol* 2003, 35:547-552.

[11] Consilvio C, Vincent AM, Feldman EL: Neuroinflammation, COX-2, and ALS--a dual role? *Exp Neurol* 2004, 187:1-10.

[12] Floyd RA: Antioxidants, oxidative stress, and degenerative neurological disorders. *Proc Soc Exp Biol Med* 1999, 222:236-245.

[13] Dallner G, Sindelar PJ: Regulation of ubiquinone metabolism. *Free Radic Biol Med* 2000, 29:285-294.

[14] Santa-Maria I, Smith MA, Perry G, Hernandez F, Avila J, Moreno FJ: Effect of quinones on microtubule polymerization: a link between oxidative stress and cytoskeletal alterations in Alzheimer's disease. *Biochim Biophys Acta* 2005, 1740:472-480.

[15] Fukushima M: Prostaglandin J2--anti-tumour and anti-viral activities and the mechanisms involved. *Eicosanoids* 1990, 3:189-199.

[16] Gilroy DW, Colville-Nash PR, Willis D, Chivers J, Paul-Clark MJ, Willoughby DA: Inducible cyclooxygenase may have anti-inflammatory properties. *Nat Med* 1999, 5:698-701.

[17] Herschman HR, Reddy ST, Xie W: Function and regulation of prostaglandin synthase-2. *Adv Exp Med Biol* 1997, 407:61-66.

[18] Smith WL: The eicosanoids and their biochemical mechanisms of action. *Biochem J* 1989, 259:315-324.

[19] Scher JU, Pillinger MH: 15d-PGJ2: the anti-inflammatory prostaglandin? *Clin Immunol* 2005, 114:100-109.

[20] Ide T, Egan K, Bell-Parikh LC, FitzGerald GA: Activation of nuclear receptors by prostaglandins. *Thromb Res* 2003, 110:311-315.

[21] Kielian T, Drew PD: Effects of peroxisome proliferator-activated receptor-gamma agonists on central nervous system inflammation. *J Neurosci Res* 2003, 71:315-325.

[22] Greene ME, Blumberg B, McBride OW, Yi HF, Kronquist K, Kwan K, Hsieh L, Greene G, Nimer SD: Isolation of the human peroxisome proliferator activated receptor gamma cDNA: expression in hematopoietic cells and chromosomal mapping. *Gene Expr* 1995, 4:281-299.

[23] Braissant O, Foufelle F, Scotto C, Dauca M, Wahli W: Differential expression of peroxisome proliferator-activated receptors (PPARs): tissue distribution of PPAR-alpha, -beta, and -gamma in the adult rat. *Endocrinology* 1996, 137:354-366.

[24] Azuma Y, Shinohara M, Wang PL, Ohura K: 15-Deoxy-delta(12,14)-prostaglandin J(2) inhibits IL-10 and IL-12 production by macrophages. *Biochem Biophys Res Commun* 2001, 283:344-346.

[25] Nagy L, Tontonoz P, Alvarez JG, Chen H, Evans RM: Oxidized LDL regulates macrophage gene expression through ligand activation of PPARgamma. *Cell* 1998, 93:229-240.

[26] Tontonoz P, Nagy L, Alvarez JG, Thomazy VA, Evans RM: PPARgamma promotes monocyte/macrophage differentiation and uptake of oxidized LDL. *Cell* 1998, 93:241-252.

[27] Chinetti G, Fruchart JC, Staels B: Peroxisome proliferator-activated receptors (PPARs): nuclear receptors at the crossroads between lipid metabolism and inflammation. *Inflamm Res* 2000, 49:497-505.

[28] Chinetti G, Gbaguidi FG, Griglio S, Mallat Z, Antonucci M, Poulain P, Chapman J, Fruchart JC, Tedgui A, Najib-Fruchart J, Staels B: CLA-1/SR-BI is expressed in atherosclerotic lesion macrophages and regulated by activators of peroxisome proliferator-activated receptors. *Circulation* 2000, 101:2411-2417.

[29] Wilmer WA, Dixon C, Lu L, Hilbelink T, Rovin BH: A cyclopentenone prostaglandin activates mesangial MAP kinase independently of PPARgamma. *Biochem Biophys Res Commun* 2001, 281:57-62.

[30] Li Z, Jansen M, Ogburn K, Salvatierra L, Hunter L, Mathew S, Figueiredo-Pereira ME: Neurotoxic prostaglandin J2 enhances cyclooxygenase-2 expression in neuronal cells through the p38MAPK pathway: a death wish? *J Neurosci Res* 2004, 78:824-836.

[31] Rossi A, Kapahi P, Natoli G, Takahashi T, Chen Y, Karin M, Santoro MG: Anti-inflammatory cyclopentenone prostaglandins are direct inhibitors of IkappaB kinase. *Nature* 2000, 403:103-108.

[32] Straus DS, Pascual G, Li M, Welch JS, Ricote M, Hsiang CH, Sengchanthalangsy LL, Ghosh G, Glass CK: 15-deoxy-delta 12,14-prostaglandin J2 inhibits multiple steps in the NF-kappa B signaling pathway. *Proc Natl Acad Sci U S A* 2000, 97:4844-4849.

[33] Scher JU, Pillinger MH: 15d-PGJ2: the anti-inflammatory prostaglandin? *Clin Immunol* 2005, 114:100-109.

[34] Straus DS, Glass CK: Cyclopentenone prostaglandins: new insights on biological activities and cellular targets. *Med Res Rev* 2001, 21:185-210.

[35] Cernuda-Morollon E, Pineda-Molina E, Canada FJ, Perez-Sala D: 15-Deoxy-Delta 12,14-prostaglandin J2 inhibition of NF-kappaB-DNA binding through covalent modification of the p50 subunit. *J Biol Chem* 2001, 276:35530-35536.

[36] Moos PJ, Edes K, Cassidy P, Massuda E, Fitzpatrick FA: Electrophilic prostaglandins and lipid aldehydes repress redox-sensitive transcription factors p53 and hypoxia-inducible factor by impairing the selenoprotein thioredoxin reductase. *J Biol Chem* 2003, 278:745-750.

[37] Oliva JL, Perez-Sala D, Castrillo A, Martinez N, Canada FJ, Bosca L, Rojas JM: The cyclopentenone 15-deoxy-delta 12,14-prostaglandin J2 binds to and activates H-Ras. *Proc Natl Acad Sci U S A* 2003, 100:4772-4777.

[38] Kondo M, Shibata T, Kumagai T, Osawa T, Shibata N, Kobayashi M, Sasaki S, Iwata M, Noguchi N, Uchida K: 15-Deoxy-Delta (12,14)-prostaglandin J(2): the endogenous electrophile that induces neuronal apoptosis. *Proc Natl Acad Sci U S A* 2002, 99:7367-7372.

[39] Kondo M, Oya-Ito T, Kumagai T, Osawa T, Uchida K: Cyclopentenone prostaglandins as potential inducers of intracellular oxidative stress. *J Biol Chem* 2001, 276:12076-12083.

[40] Mullally JE, Moos PJ, Edes K, Fitzpatrick FA: Cyclopentenone prostaglandins of the J series inhibit the ubiquitin isopeptidase activity of the proteasome pathway. *J Biol Chem* 2001, 276:30366-30373.

[41] Li Z, Melandri F, Berdo I, Jansen M, Hunter L, Wright S, Valbrun D, Figueiredo-Pereira ME: Delta12-Prostaglandin J2 inhibits the ubiquitin hydrolase UCH-L1 and elicits ubiquitin–protein aggregation without proteasome inhibition. *Biochem Biophys Res Commun* 2004, 319:1171-1180.

[42] Kondo M, Shibata T, Kumagai T, Osawa T, Shibata N, Kobayashi M, Sasaki S, Iwata M, Noguchi N, Uchida K: 15-Deoxy-Delta (12,14)-prostaglandin J(2): the endogenous electrophile that induces neuronal apoptosis. *Proc Natl Acad Sci U S A* 2002, 99:7367-7372.

[43] Harhangi BS, Farrer MJ, Lincoln S, Bonifati V, Meco G, De Michele G, Brice A, Durr A, Martinez M, Gasser T, Bereznai B, Vaughan JR, Wood NW, Hardy J, Oostra BA, Breteler MM: The Ile93Met mutation in the ubiquitin carboxy-terminal-hydrolase-L1 gene is not observed in European cases with familial Parkinson's disease. *Neurosci Lett* 1999, 270:1-4.

[44] Forno LS: Neuropathology of Parkinson's disease. *J Neuropathol Exp Neurol* 1996, 55:259-272.

[45] Ghee M, Melki R, Michot N, Mallet J: PA700, the regulatory complex of the 26S proteasome, interferes with alpha-synuclein assembly. *FEBS J* 2005, 272:4023-4033.

[46] Shang F, Gong X, Taylor A: Activity of ubiquitin-dependent pathway in response to oxidative stress. Ubiquitin-activating enzyme is transiently up-regulated. *J Biol Chem* 1997, 272:23086-23093.

[47] Stokes AH, Hastings TG, Vrana KE: Cytotoxic and genotoxic potential of dopamine. *J Neurosci Res* 1999, 55:659-665.

[48] Gao HM, Liu B, Hong JS: Critical role for microglial NADPH oxidase in rotenone-induced degeneration of dopaminergic neurons. *J Neurosci* 2003, 23:6181-6187.

[49] Ren Y, Liu W, Jiang H, Jiang Q, Feng J: Selective vulnerability of dopaminergic neurons to microtubule depolymerization. *J Biol Chem* 2005, 280:34105-34112.

[50] Stokes AH, Hastings TG, Vrana KE: Cytotoxic and genotoxic potential of dopamine. *J Neurosci Res* 1999, 55:659-665.

[51] Mizuno Y, Ikebe S, Hattori N, Nakagawa-Hattori Y, Mochizuki H, Tanaka M, Ozawa T: Role of mitochondria in the etiology and pathogenesis of Parkinson's disease. *Biochim Biophys Acta* 1995, 1271:265-274.

[52] Stadtman ER, Berlett BS: Reactive oxygen-mediated protein oxidation in aging and disease. *Drug Metab Rev* 1998, 30:225-243.
[53] Grune T, Merker K, Sandig G, Davies KJ: Selective degradation of oxidatively modified protein substrates by the proteasome. *Biochem Biophys Res Commun* 2003, 305:709-718.
[54] Dudek EJ, Shang F, Valverde P, Liu Q, Hobbs M, Taylor A: Selectivity of the ubiquitin pathway for oxidatively modified proteins: relevance to protein precipitation diseases. *FASEB J* 2005, 19:1707-1709.
[55] Betarbet R, Sherer TB, Greenamyre JT: Ubiquitin-proteasome system and Parkinson's diseases. *Exp Neurol* 2005, 191 Suppl 1:S17-S27.
[56] Sherman MY, Goldberg AL: Cellular defenses against unfolded proteins: a cell biologist thinks about neurodegenerative diseases. *Neuron* 2001, 29:15-32.
[57] Cervera J, Levine RL: Modulation of the hydrophobicity of glutamine synthetase by mixed-function oxidation. *FASEB J* 1988, 2:2591-2595.
[58] Davies KJ: Intracellular proteolytic systems may function as secondary antioxidant defenses: an hypothesis. *J Free Radic Biol Med* 1986, 2:155-173.
[59] Davies KJ: Protein damage and degradation by oxygen radicals. I. general aspects. *J Biol Chem* 1987, 262:9895-9901.
[60] Davies KJ: Degradation of oxidized proteins by the 20S proteasome. *Biochimie* 2001, 83:301-310.
[61] Davies KJ, Delsignore ME: Protein damage and degradation by oxygen radicals. III. Modification of secondary and tertiary structure. *J Biol Chem* 1987, 262:9908-9913.
[62] Davies KJ, Delsignore ME, Lin SW: Protein damage and degradation by oxygen radicals. II. Modification of amino acids. *J Biol Chem* 1987, 262:9902-9907.
[63] Davies KJ, Lin SW, Pacifici RE: Protein damage and degradation by oxygen radicals. IV. Degradation of denatured protein. *J Biol Chem* 1987, 262:9914-9920.
[64] Grune T, Davies KJ: The proteasomal system and HNE-modified proteins. *Mol Aspects Med* 2003, 24:195-204.
[65] Shang F, Deng G, Liu Q, Guo W, Haas AL, Crosas B, Finley D, Taylor A: Lys6-modified ubiquitin inhibits ubiquitin-dependent protein degradation. *J Biol Chem* 2005, 280:20365-20374.
[66] Corsi D, Galluzzi L, Crinelli R, Magnani M: Ubiquitin is conjugated to the cytoskeletal protein alpha-spectrin in mature erythrocytes. *J Biol Chem* 1995, 270:8928-8935.
[67] Jabusch JR, Deutsch HF: Localization of lysines acetylated in ubiquitin reacted with p-nitrophenyl acetate. *Arch Biochem Biophys* 1985, 238:170-177.
[68] Macdonald JM, LeBlanc DA, Haas AL, London RE: An NMR analysis of the reaction of ubiquitin with [acetyl-1-13C]aspirin. *Biochem Pharmacol* 1999, 57:1233-1244.
[69] Macdonald JM, Haas AL, London RE: Novel mechanism of surface catalysis of protein adduct formation. NMR studies of the acetylation of ubiquitin. *J Biol Chem* 2000, 275:31908-31913.
[70] Wee KE, Lai Z, Auger KR, Ma J, Horiuchi KY, Dowling RL, Dougherty CS, Corman JI, Wynn R, Copeland RA: Steady-state kinetic analysis of human ubiquitin-activating enzyme (E1) using a fluorescently labeled ubiquitin substrate. *J Protein Chem* 2000, 19:489-498.

[71] Hyun DH, Gray DA, Halliwell B, Jenner P: Interference with ubiquitination causes oxidative damage and increased protein nitration: implications for neurodegenerative diseases. *J Neurochem* 2004, 90:422-430.

[72] Reinheckel T, Sitte N, Ullrich O, Kuckelkorn U, Davies KJ, Grune T: Comparative resistance of the 20S and 26S proteasome to oxidative stress. *Biochem J* 1998, 335 (Pt 3):637-642.

[73] Reinheckel T, Ullrich O, Sitte N, Grune T: Differential impairment of 20S and 26S proteasome activities in human hematopoietic K562 cells during oxidative stress. *Arch Biochem Biophys* 2000, 377:65-68.

[74] Shringarpure R, Grune T, Mehlhase J, Davies KJ: Ubiquitin conjugation is not required for the degradation of oxidized proteins by proteasome. *J Biol Chem* 2003, 278:311-318.

[75] Strack PR, Waxman L, Fagan JM: Activation of the multicatalytic endopeptidase by oxidants. Effects on enzyme structure. *Biochemistry* 1996, 35:7142-7149.

[76] Reinheckel T, Sitte N, Ullrich O, Kuckelkorn U, Davies KJ, Grune T: Comparative resistance of the 20S and 26S proteasome to oxidative stress. *Biochem J* 1998, 335 (Pt 3):637-642.

[77] Andersson M, Sjostrand J, Karlsson JO: Differential inhibition of three peptidase activities of the proteasome in human lens epithelium by heat and oxidation. *Exp Eye Res* 1999, 69:129-138.

[78] Obin M, Shang F, Gong X, Handelman G, Blumberg J, Taylor A: Redox regulation of ubiquitin-conjugating enzymes: mechanistic insights using the thiol-specific oxidant diamide. *FASEB J* 1998, 12:561-569.

[79] DeMartino GN: Purification of PA700, the 19S regulatory complex of the 26S proteasome. *Methods Enzymol* 2005, 398:295-306.

[80] Ishii T, Sakurai T, Usami H, Uchida K: Oxidative modification of proteasome: identification of an oxidation-sensitive subunit in 26 s proteasome. *Biochemistry* 2005, 44:13893-13901.

[81] Ogura T, Wilkinson AJ: AAA+ superfamily ATPases: common structure--diverse function. *Genes Cells* 2001, 6:575-597.

[82] Choi HS, Seol W, Moore DD: A component of the 26S proteasome binds on orphan member of the nuclear hormone receptor superfamily. *J Steroid Biochem Mol Biol* 1996, 56:23-30.

[83] Hartmann-Petersen R, Tanaka K, Hendil KB: Quaternary structure of the ATPase complex of human 26S proteasomes determined by chemical cross-linking. *Arch Biochem Biophys* 2001, 386:89-94.

[84] Ohana B, Moore PA, Ruben SM, Southgate CD, Green MR, Rosen CA: The type 1 human immunodeficiency virus Tat binding protein is a transcriptional activator belonging to an additional family of evolutionarily conserved genes. *Proc Natl Acad Sci U S A* 1993, 90:138-142.

[85] Uetz P, Giot L, Cagney G, Mansfield TA, Judson RS, Knight JR, Lockhon D, Narayan V, Srinivasan M, Pochart P, Qureshi-Emili A, Li Y, Godwin B, Conover D, Kalbfleisch T, Vijayadamodar G, Yang M, Johnston M, Fields S, Rothberg JM: A comprehensive

analysis of protein-protein interactions in Saccharomyces cerevisiae. *Nature* 2000, 403:623-627.

[86] Chao CC, Ma YS, Stadtman ER: Modification of protein surface hydrophobicity and methionine oxidation by oxidative systems. *Proc Natl Acad Sci U S A* 1997, 94:2969-2974.

[87] Levine RL, Mosoni L, Berlett BS, Stadtman ER: Methionine residues as endogenous antioxidants in proteins. *Proc Natl Acad Sci U S A* 1996, 93:15036-15040.

[88] Davies KJ, Shringarpure R: Preferential degradation of oxidized proteins by the 20S proteasome may be inhibited in aging and in inflammatory neuromuscular diseases. *Neurology* 2006, 66:S93-S96.

[89] Park I, Chung J, Walsh CT, Yun Y, Strominger JL, Shin J: Phosphotyrosine-independent binding of a 62-kDa protein to the src homology 2 (SH2) domain of p56lck and its regulation by phosphorylation of Ser-59 in the lck unique N-terminal region. *Proc Natl Acad Sci U S A* 1995, 92:12338-12342.

[90] Shin J: P62 and the sequestosome, a novel mechanism for protein metabolism. *Arch Pharm Res* 1998, 21:629-633.

[91] Vadlamudi RK, Joung I, Strominger JL, Shin J: p62, a phosphotyrosine-independent ligand of the SH2 domain of p56lck, belongs to a new class of ubiquitin-binding proteins. *J Biol Chem* 1996, 271:20235-20237.

[92] Puls A, Schmidt S, Grawe F, Stabel S: Interaction of protein kinase C zeta with ZIP, a novel protein kinase C-binding protein. *Proc Natl Acad Sci U S A* 1997, 94:6191-6196.

[93] Sanz L, Sanchez P, Lallena MJ, Diaz-Meco MT, Moscat J: The interaction of p62 with RIP links the atypical PKCs to NF-kappaB activation. *EMBO J* 1999, 18:3044-3053.

[94] Sanz L, Diaz-Meco MT, Nakano H, Moscat J: The atypical PKC-interacting protein p62 channels NF-kappaB activation by the IL-1-TRAF6 pathway. *EMBO J* 2000, 19:1576-1586.

[95] Wooten MW, Seibenhener ML, Mamidipudi V, Diaz-Meco MT, Barker PA, Moscat J: The atypical protein kinase C-interacting protein p62 is a scaffold for NF-kappaB activation by nerve growth factor. *J Biol Chem* 2001, 276:7709-7712.

[96] Geetha T, Wooten MW: Structure and functional properties of the ubiquitin binding protein p62. *FEBS Lett* 2002, 512:19-24.

[97] Seibenhener ML, Babu JR, Geetha T, Wong HC, Krishna NR, Wooten MW: Sequestosome 1/p62 is a polyubiquitin chain binding protein involved in ubiquitin proteasome degradation. *Mol Cell Biol* 2004, 24:8055-8068.

[98] Paine MG, Babu JR, Seibenhener ML, Wooten MW: Evidence for p62 aggregate formation: role in cell survival. *FEBS Lett* 2005, 579:5029-5034.

[99] Babu JR, Geetha T, Wooten MW: Sequestosome 1/p62 shuttles polyubiquitinated tau for proteasomal degradation. *J Neurochem* 2005, 94:192-203.

[100] Furukawa Y, Iseki E, Hino H, Kanai A, Odawara T, Kosaka K: Ubiquitin and ubiquitin-related proteins in neurons and dendrites of brains of atypical Pick's disease without Pick bodies. *Neuropathology* 2004, 24:38-45.

[101] Kuusisto E, Salminen A, Alafuzoff I: Ubiquitin-binding protein p62 is present in neuronal and glial inclusions in human tauopathies and synucleinopathies. *Neuroreport* 2001, 12:2085-2090.

[102] Nakano T, Nakaso K, Nakashima K, Ohama E: Expression of ubiquitin-binding protein p62 in ubiquitin-immunoreactive intraneuronal inclusions in amyotrophic lateral sclerosis with dementia: analysis of five autopsy cases with broad clinicopathological spectrum. *Acta Neuropathol (Berl)* 2004, 107:359-364.

[103] Zatloukal K, Stumptner C, Fuchsbichler A, Heid H, Schnoelzer M, Kenner L, Kleinert R, Prinz M, Aguzzi A, Denk H: p62 Is a common component of cytoplasmic inclusions in protein aggregation diseases. *Am J Pathol* 2002, 160:255-263.

[104] Kuusisto E, Suuronen T, Salminen A: Ubiquitin-binding protein p62 expression is induced during apoptosis and proteasomal inhibition in neuronal cells. *Biochem Biophys Res Commun* 2001, 280:223-228.

[105] Nakaso K, Yoshimoto Y, Nakano T, Takeshima T, Fukuhara Y, Yasui K, Araga S, Yanagawa T, Ishii T, Nakashima K: Transcriptional activation of p62/A170/ZIP during the formation of the aggregates: possible mechanisms and the role in Lewy body formation in Parkinson's disease. *Brain Res* 2004, 1012:42-51.

[106] Nagaoka U, Kim K, Jana NR, Doi H, Maruyama M, Mitsui K, Oyama F, Nukina N: Increased expression of p62 in expanded polyglutamine-expressing cells and its association with polyglutamine inclusions. *J Neurochem* 2004, 91:57-68.

[107] Wang Z, Figueiredo-Pereira ME: Inhibition of sequestosome 1/p62 up-regulation prevents aggregation of ubiquitinated proteins induced by prostaglandin J2 without reducing its neurotoxicity. *Mol Cell Neurosci* 2005, 29:222-231.

[108] Guldberg HC, Marsden CA: Catechol-O-methyl transferase: pharmacological aspects and physiological role. *Pharmacol Rev* 1975, 27:135-206.

[109] Mannisto PT, Kaakkola S: Catechol-O-methyltransferase (COMT): biochemistry, molecular biology, pharmacology, and clinical efficacy of the new selective COMT inhibitors. *Pharmacol Rev* 1999, 51:593-628.

[110] Ogburn KD, Bottiglieri T, Wang Z, Figueiredo-Pereira ME: Prostaglandin J2 reduces catechol-O-methyltransferase activity and enhances dopamine toxicity in neuronal cells. *Neurobiol Dis* 2006, 22:294-301.

In: The Ubiquitin Proteasome System... ISBN 978-1-60021-749-4
Eds: Mario Di Napoli and Cezary Wojcik, pp. 735-760

Chapter 30

The Regulation of TAU Protein Proteolysis by the Ubiquitin Proteasome System in Neurodegenerative Foldopathies

Patrice Delobel* and Isabelle Lavenir
MRC-Laboratory of Molecular Biology, Cambridge CB2 2QH, Cambridge, United Kingdom.

Abstract

Tau proteins belong to the family of Microtubule-Associated Proteins (MAPs). They are mostly expressed in neurons where they act on the assembly and the stability of tubulin. Depending on their phosphorylation state, tau proteins will modulate the polymerisation and the stability of microtubules within the axon. The primary function of tau and its differential phosphorylation enable tau to be involved in neurite outgrowth and in axonal transport. Hence, these proteins appear important for the physiology and the normal function of a neuron. However, tau proteins are also the major constituents of intraneuronal and glial inclusions described in Alzheimer's disease (AD) and many other related disorders called tauopathies. They are thought to be directly linked to the progression of neurodegeneration. For instance, the gravity of the symptoms observed in AD was shown to be closely related to the progression of the '*tau pathology*'. The recent discovery of mutations within the *tau* gene has strengthened the role of tau in the neurodegenerative processes observed in all these disorders. Indeed, the presence of certain mutations within tau can lead to its intraneuronal and intraglial aggregation and the death of the cells affected. In all these diseases tau proteins are abnormally modified when aggregated. In particular, molecular analysis revealed that this protein is

* Correspondence concerning this article should be addressed to Dr. Patrice Delobel, PhD; MRC-Laboratory of Molecular Biology, Hills Road, Cambridge CB2 2QH, Cambridge, United Kingdom. Phone: +44 (0) 1223 402320; Fax: +44 (0) 1223 402310; Email: delobel@mrc-lmb.cam.ac.uk.

hyperphosphorylated in its filamentous state. This implies that some kinases or phosphatases are involved in the abnormal processing of tau and may be responsible for its aggregation and the subsequent neurodegeneration. The intracellular accumulation of tau may also induce its hyperphosphorylation and fibrillogenesis. Inhibition of tau protein proteolysis may then be one of the mechanisms involved in its intracytoplasmic accumulation. In AD or Parkinson's disease (PD), it was shown that the activity of proteasome was inhibited. In its aggregated form, tau is also ubiquitinated suggesting that the proteasome can be involved in the degradation of tau protein. A major question concerning the proteasome impairment in AD and its involvement in the degradation of tau is its place in the cascade of events leading to tau aggregation and neurodegeneration. On one hand, a proteasome defect could contribute to the failure of the clearance of tau inclusions noticed in the disease as observed by the presence of ubiquitinated tau in the neurofibrillary tangles. On the other hand, the deficiency of this proteolytic system can also lead to tau protein accumulation, hyperphosphorylation, ubiquitination and finally to its intraneuronal aggregation.

This review focuses on the recent advances made in the understanding of the relationships between tau protein malfunction, its hyperphosphorylation and the ubiquitin-dependent proteasomal degradation of tau in AD and other tauopathies. The challenge is to pinpoint the role of the proteasome in the cascade of events leading to neurodegeneration.

Keywords: Neurodegenerative Disorders, tau, phosphorylation, aggregation, ubiquitin, proteasome.

ABBREVIATIONS

AA, Amino Acid; Aβ1-42, Amyloid beta peptide 1-42; AD, Alzheimer's Disease, ALS, Amyotrophic Lateral Sclerosis; ATP, Adenosine 5'-triphosphate; CamKII, Calcium/Calmodulin protein kinase II; CBD, Corticobasal Degeneration; Cdk, Cyclin dependent kinase; CHIP, Carboxy terminus of the Hsc70-Iteracting Protein; CKI/II, Casein kinase I and II; Ct, carboxy-terminal; DUBs, Deubiquitination Enzymes; E1, Ubiquitin-activating Enzyme; E2, Ubiquitin-conjugating Enzyme; E3, Ubiquitin-protein Ligases; FTDP-17, Frontotemporal Dementia with Parkinsonism linked to chromosome 17; GSK3α/β, Glycogen Synthase Kinase 3 alpha/beta; Gly, Glycine; Hsc: Heat Shock cognate protein; Hsp, Heat Shock protein; IR R1/R2, Inter-repeat region between the repeat R1/R2 in tau protein; MAPs, Microtubule associated proteins; MAPK, Mitogen-Activated Protein Kinase; MARK, Microtubule-Affinity Regulating Kinase; Mts, Microtubules; Nt, amino-terminal; NFD, Neurofibrillary degeneration; PD, Parkinson's disease; PDC, Parkinsonism-Dementia Complex of Guam; PDPK, Proline Directed Protein Kinases; PiD, Pick's Disease; PKA, B, C, N, Protein Kinase A, B, C and N; PP1/2A/2B, Protein Phosphatases 1/2A and 2B; Pro, Proline; PS1, Presiniline 1; PSP, Progressive Supranuclear Palsy; R1-R4, Repeat 1-4; PHFs, Paired helical filaments; SAPK, Stress-Activated Protein Kinase; Ser, serine; SFs, Straight filaments; Thr, Threonine; TTK, Tau Tubulin Kinase; Ub, Ubiquitin; UBB^{+1}, mutant Ubiquitin; UCH-L1, Ubiquitin Carboxy-Terminal Hydrolase isoenzyme L1; UPS, Ubiquitin-Proteasome System; Y, Tyrosine.

INTRODUCTION

Most neurodegenerative disorders can be viewed as diseases presenting intraneuronal accumulations of fibrillary materials and are now recognised to involve abnormalities of protein metabolism. Abnormalities of three proteins, β-amyloid, α-synuclein and tau are most frequently observed in these disorders, and represent a hallmark for more than 90% of all neurodegenerative dementias. Growing evidence also suggests that the proteasome may play a role in the accumulation of these three proteins. For example, the inhibition of the proteasome is now thought to be involved in the intraneuronal aggregation of α-synuclein and tau protein in Parkinson (PD) and Alzheimer's (AD) diseases respectively.

Neurofibrillary degeneration (NFD) corresponds to the intracellular accumulation of pathological fibrils in the cytosol of neurons [1]. In AD, NFD may be caused by the abnormal aggregation of tau proteins [2-6]. In this disorder, tau deposits form paired helical filaments (PHFs) and straight filaments (SFs). In AD, this pathological accumulation of tau fibrils was shown to be directly correlated to the progression of the NFD and the disease [7,8]. Indeed, previous work has demonstrated that the development of the NFD is progressive, hierarchical and that its progression is directly related to the severity of the symptoms observed [9]. Moreover, the intraneuronal and/or intraglial aggregation of tau protein is not restricted to AD but is a common event occurring in diseases grouped as tauopathies. This pathological hallmark was found in Progressive Supranuclear Palsy (PSP), Corticobasal Degeneration (CBD), Pick's Disease (PiD), Parkinsonism with dementia, the Amyotrophic Lateral Sclerosis/Parkinsonism-Dementia Complex of Guam (ALS/PDC) and the Frontotemporal dementia with Parkinsonism linked to chromosome 17 (FTDP-17) [10]. The latter has strengthened the evidence of the role played by tau protein in the neurodegenerative process, since FTDP-17 pathology has been linked to mutations in the *tau* gene. In June 1998, the first mutations in the *tau* gene were reported in FTDP-17 [11-13], with the current number standing at 38. Known mutations are either missense, deletion or silent mutations in the coding region or intronic mutations located close to the splice donor site of the intron after exon 10 of the *tau* gene. Functionally, most coding region mutations lead to a reduced ability of tau protein to interact with microtubules. Some coding region mutations also promote the assembly of tau protein into filaments and affect the capacity of tau to bind tau protein kinases or phosphates [14-17]. Studies on FTDP-17 have established that dysfunction of tau is sufficient to cause neurodegeneration and dementia. This is of particular interest not only for AD but also for PSP, CBD and PiD, since some mutations give rise to clinical and neuropathological phenotypes which resemble those found in these diseases. Taken together, these studies demonstrated that tau proteins have a key role in the degenerative processes observed in all these diseases.

Tau protein is abnormally modified when aggregated. The mechanism for this abnormal fibrillogenesis is not completely understood despite much evidence suggesting a hyperphosphorylation of tau before its aggregation [10]. Several studies have also suggested that an imbalance between the activity of tau protein kinases and/or phosphatases may be involved in this pathological modification of tau. In neurons, the accumulation of tau proteins in the cytoplasm may induce their hyperphosphorylation and aggregation [18,19]. Tau proteins are substrates for many intracellular proteases, including caspases, calpain 1 and the proteasome. Inhibition of one of these proteolytic systems could then lead to the intraneuronal or intraglial accumulation of tau protein. Among all proteolytic systems acting on tau, the least documented is the proteasome. In fact, it has been shown that the activity of the proteasome is impaired in AD and PD [20-22]. In tauopathies, tau proteins are also ubiquitinated when aggregated, arguing that proteasome activity may be implicated in tau degradation. Finally, recent experiments in animal models showed that an inhibition of the proteasome activity may contribute to the stabilisation of aggregates made of hyperphosphorylated tau in oligodendroglial cells [23]. All these experimental data suggest that the inhibition of the proteasome may be directly involved in the intracellular accumulation and/or aggregation of tau protein and may therefore be associated with the neurodegenerative process observed in tauopathies.

This review will focus on the specific factors, which may lead to the aggregation of tau proteins in AD and tauopathies. The dysfunction of the proteasome in these diseases and the recent advances in the understanding of the ubiquitin-dependent proteasomal degradation of tau will be also reviewed in detail. Finally, the relationships between hyperphosphorylated tau, aggregated tau and proteasome activity will be discussed.

TAU PTOTEIN: FROM ITS PHYSIOLOGICAL ROLE TO ITS ABNORMAL BEHAVIOUR

Structure and Primary Function of TAU Proteins

Tau protein belongs to the family of Microtubule-associated proteins (MAPs). In humans, these proteins are mainly expressed in neurons. Tau proteins are expressed from a unique gene located on the long arm of chromosome 17 at band position 17q21 [24,25]. In the Central Nervous System, an alternative splicing of the *tau* gene gives rise to six isoforms of tau protein ranging from 352 to 441 amino acids (Figure 1A) [26]. These variants differ from each other by the presence or the absence of exons 2 and 3 in the amino-terminal (Nt) part of the protein and exon 10 in the carboxy-terminal (Ct) part (Figure 1A). The Nt part of the protein is called the projection domain and the Ct represents the microtubule-binding domain of tau. The Ct part of the protein is responsible for the primary function of tau, i.e. the polymerisation and stability of microtubules (MTs). This side of the molecule may also be involved in the polymerisation of tau protein into intracellular fibrils. The splicing of tau is under complex temporal (developmental) and spatial (anatomical) regulation. For instance, only one isoform, lacking exons 2, 3 and 10 is present during embryogenesis. In the adult central nervous system, the six isoforms are synthesised. This spatial and differential expression of tau isovariants suggests that each of these isoforms could have a specific function within neurons, especially during brain development [10].

Originally, tau proteins were discovered as proteins binding tubulin and leading to the polymerisation and stability of MTs in cells [27-29]. This property of tau is related to the carboxy-terminal part of the protein that contains the microtubule-binding domain. Indeed, tau protein binds MTs through repetitive regions called repeat domains (R1-R4), which are encoded by exons 9-12. Each of these repeats contains a motif of 18 highly conserved amino

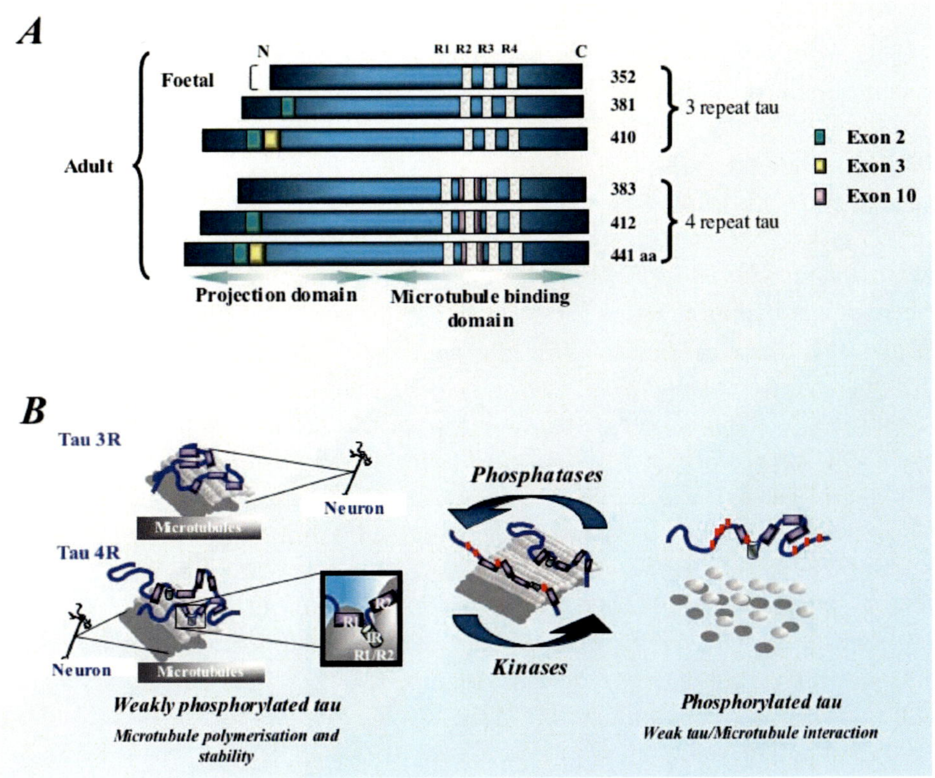

Figure 1. Tau protein: from its structure to its function. A: Schematic representation of the different tau isoforms present in the central nervous system. These variants differ by the presence or the absence of the sequences encoded by exon 2 (green), 3 (yellow) `and 10 (purple). The N-terminal part of the protein, the projection domain can interact with the plasma membrane, be involved in several signal transduction pathways or bind other molecules. The C-terminal part, the microtubule-binding domain is composed either by 3 or 4 repeats (3 or 4R). This side of the molecule is involved in the polymerisation and the stability of the microtubules. It can also bind other proteins as the presenilin1 (PS1) and the tau phosphatase PP2A [10]. B: The primary function of tau. In a weakly phosphorylated state, tau proteins (3R and 4R) will bind the microtubules and stabilise these structures within the axon. It is of note that the 4R tau binds more efficiently to microtubules than the 3R. This difference is due to the presence of the additional repeat but also to the inter-repeat region between the two first repeats (IR R1/R2). The stabilisation of the microtubules by tau will be modulated by its phosphorylation. When tau is phosphorylated by certain kinases, its interaction with the microtubules will decrease and the microtubules will depolymerise. The regulation of the phosphorylation of tau will thus change the dynamism of the microtubules, allowing the processes of neurogenesis, differentiation of neuronal cells, the axonal transport and neuronal survival.

acids indicating their importance for tau function [25,30]. The composition of the microtubule-binding domain varies with the incorporation of exon 10. Hence, the presence of exon 10 within tau will confer four microtubule-binding sequences (4R) to three tau isovariants. The three other tau isoforms only have three of these sequences and are called 3 repeat tau (3R). The insertion of this supplementary repeat is of importance for the functionality of tau, since it was demonstrated that 4R tau binds and stabilises MTs more efficiently than their 3R counterparts [31-33]. It is of note that this primary function of tau protein is regulated by its phosphorylation state since the interaction between tau and MTs involves electrostatic interactions (tau protein being basic and the microtubules containing acidic charges). The phosphorylation of tau at specific sites will then create a repulsion between tau and MTs, thus decreasing their interaction (Figure 1B) [34].

The Phosphorylation of TAU Protein: a Factor Regulating its Normal and Pathological Behaviour

The Normal phosphorylation of TAU

The phosphorylation of tau is the most important posttranslational modification of the protein and also the most studied [45]. Phosphorylation sites have been identified on tau (Table 1) thus creating a huge number of possible tau-phosphoisoforms. Some of these sites are located outside the microtubule-binding domain and are on Ser-Pro or Thr-Pro motifs. The remainder are Ser or Thr residues but do not involve a following Pro. The different phosphorylation states of tau result from the activity of specific kinases and/or phosphatases (Table 1). Two groups of kinases can be distinguished: the proline directed protein kinases (PDPK) and kinases directed against non-Ser/Thr-Pro motives (non-PDPK). The PDPK include the Mitogen-Activated Protein Kinases (MAPK), the Stress-Activated protein Kinases (SAPK), the Glycogen Synthase Kinase 3 alpha and beta (GSK3α and β), and the proteins of the Cyclin-Dependent Kinases family (Cdk1, 2 and 5). The non-PDPK are represented by the tau Tubulin Kinase (TTK), the Microtubule-Affinity Regulating Kinase (MARK), the Calcium/Calmodulin protein Kinase II (CamKII), the Casein Kinase I and II (CKI and CKII), and finally, the Protein Kinase A, B, C and N (PKA, PKB, PKC and PKN). Tau proteins are also substrates for Protein Phosphatases 1, 2A and 2B (calcineurin) [10, http://www.alzheimer-adna.com/Gb/Tau/TauPhosphoSeq.htm]. The balance between the activity of these Protein Kinases and/or Protein Phosphatases will regulate the different phosphorylation states of tau and will thus act directly on its function. It is also important to note that the activity of some kinases will be regulated by their own phosphorylation state. The tau Protein Phosphatases could thus change the phosphorylation of tau directly or by an inhibitory mechanism of certain tau protein kinases.

Little is known about the function of these phosphorylation states except that some alter the binding kinetics of tau to MTs. Indeed, *in vitro* experiments have shown that phosphorylated tau is less able to bind MTs compared to unphosphorylated tau [35,36]. However, all the phosphorylation sites on tau are not involved in this MTs binding. In fact, some motifs are more important than others for this regulation: they include the Thr205 and the Ser202, 214, 262, 324 and 356. The phosphorylation on these sites leads to a huge decrease of the affinity of tau for the MTs and the depolymerisation of the latter (Figure 1B). The kinases able to phosphorylate these sites (Table 1) and their antagonists, the phosphatases, will then regulate the interaction of tau with the MTs by changing the phosphorylation on these specific motifs. The fragile balance in the phosphorylation state of tau at these sites is crucial for the cell since the dynamics of the MTs is indispensable for the viability and the physiology of neurons not only during neurogenesis, but also for the axonal transport and the process of neuronal differentiation.

Table 1. the phosphorylation of tau proteins, associated kinases and phosphatases

	Y	T	S	T	T	T	T	T	S	S	S	S	S	S	T	S	S	T	S
			P	P	P	P	P	P					P	P	P			P	
	29	39	46	50	69	153	175	181	184	185	195	198	199	202	205	208	210	212	214
PHFs	+						+	+	+	+		+	+	+	+	+	+	+	+
CamKII																		X	X
CKI																			
CKII		X																	
Cdc2				X		X					X			X	X				
Cdk2														X	X				
Cdk5						X		X			X			X	X			X	X
GSK3α									X				X	X				X	
GSK3β			X	X	X	X	X	X	X		X	X	X	X	X	X		X	X
MAPK			X	X	X	X	X	X					X		X				
MARK																			
Phosphorylase k																			
PKA											X	X		X	X			X	X
PKB																			X
PKC																			
PKN													-	-				X	
PK35/41																			
SAPK			X				X	X		X				X	X			X	
TTK																			
PP1												X	X		X				
PP2A											X		X	X					
PP2B			X					X					X	X					

	T	T	S	S	S	T	S	S	S	S	S	S	S	S	S	T	T	S	S	T	S	S	S	S	S	S
	P	P	P															P			P					P
	217	231	235	237	238	245	258	262	285	293	305	320	324	352	356	361	377	396	400	403	404	409	412	413	416	422
PHFs	+	+	+	+	+			+							+	+		+	+	+	+	+	+	+		+
CamKII								X	X						X							X			X	
CKI																		X			X					
CKII																		X			X					
Cdc2																		X			X					
Cdk2																		X			X					
Cdk5		X	X					X										X			X					
GSK3α		X	X					X					X					X			X					
GSK3β	X	X	X															X	X	X	X			X		
MAPK	X	X	X															X			X				X	
MARK								X		X	X	X	X		X											
Phosphorylase k			X	X				X	X		X			X												
PKA						X		X		X	X	X	X		X	X						X	X		X	
PKB																										
PKC							X			X	X		X				X									
PKN	-						X					X		X												
PK35/41								X		X			X									X				
SAPK	X	X	X			X					X				X			X			X					X
TTK																										
PP1																		X								
PP2A								X										X			X					X
PP2B		X	X									X						X			X					

Summary of the phosphorylation sites on the longest human tau (441 aa) that can be phosphorylated *in vitro* or *in vivo* by tau kinases (listed on the right) and dephosphorylated by tau phosphatases PP1, PP2A and PP2B (crosses). Positive signs show the phosphorylation sites on tau that are present in tau aggregates in tauopathies. Grey colour indicates the abnormal phosphorylation sites on tau (Thr212/Ser214: AT100; Thr231: TG3; Ser422: P-Tau 422 epitopes respectively). Minus signs represent the inhibitory effects of certain phosphorylation sites on subsequent tau phosphorylation. Abbreviations used: P: Proline; S: Serine; T: Threonine and Y: Tyrosine. For references, see the website http://www.alzheimer-adna.com/Gb/Tau/TauPhosphoSeq.htm.

The Aggregation of TAU: a Process Influenced by its Hyperphosphorylation

In solution, tau protein is natively unfolded. However, this protein has itself the ability to form filamentous deposits. Indeed, native and soluble tau is able to aggregate spontaneously in solution [37]. The process of tau aggregation is complex and involves different steps which include: (*i*) A change in the conformation of the molecule, (*ii*) A nucleation of tau monomers leading to (*iii*) The aggregation into filaments (Figure 2). The first two steps of this process are relative to the primary sequence of tau. Thus, the microtubule-binding domain of tau, and notably the third repeat seems to be responsible for the aggregation [38,39]. In contrast, the Nt part of the molecule has an inhibitory effect on tau fibrillogenesis [40]. The aggregation of tau may also be driven by other sequences within the molecule able to form β-sheets structures. These structures may lead to a change in the conformation of the protein and the dimerisation of tau monomers [41]. These features of the primary sequence of tau were strengthened by the study of FTDP-17 mutations on *tau*. Indeed, certain mutations, which decrease the affinity of tau for the MTs, also affect its fibrillogenesis in presence of polyanions such as heparin [39,15]. These mutations will increase the aggregation of tau by changing the conformation of the molecule or by increasing the stability of the β-sheets formed [42,43]. Thus, tau is a soluble protein in solution but able to self-aggregate under certain experimental and physiopathological conditions.

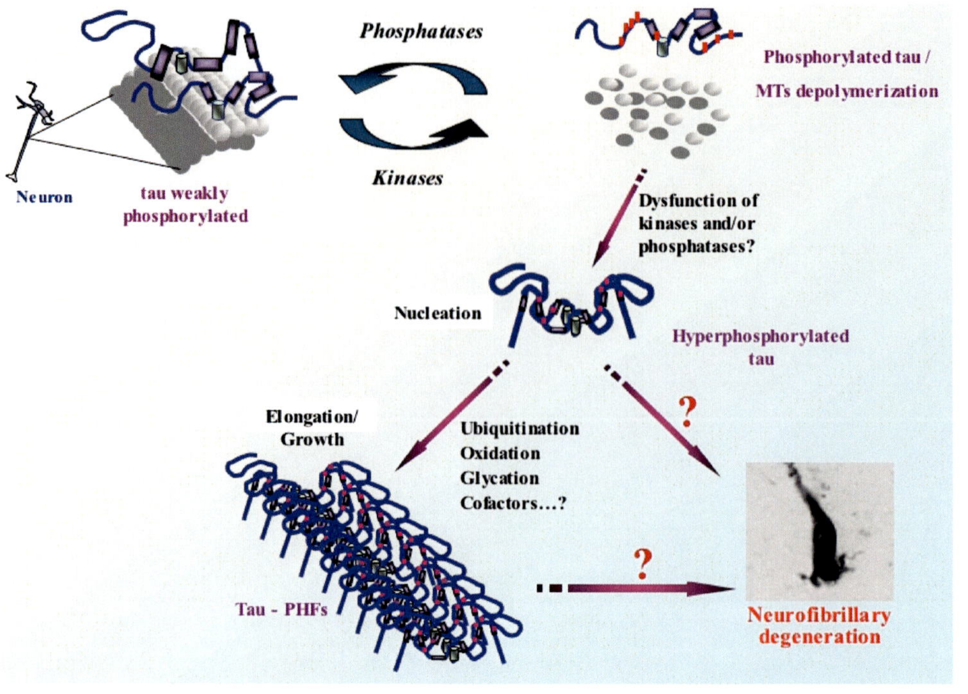

Figure 2. The pathological phosphorylation of tau, a process driving its aggregation. In Alzheimer's disease and other tauopathies, certain kinases and/or phosphatases may be deregulated. Tau proteins will then be hyperphosphorylated (increased phosphorylation on normal sites) and abnormally phosphorylated (appearance of new phosphorylation sites). This abnormal phosphorylation will induce the process of nucleation, e.g., the dimerisation of tau monomers into anti-parallels dimers. Finally, after other post-translational modifications (ubiquitination, oxidation, glycation, glycosylation and the binding of other co-factors), the hyperphosphorylated dimers will assemble into aggregates (Tau-PHFs).

This property of tau may also be influenced by different posttranslational modifications such as its hyperphosphorylation, oxidation, ubiquitination, glycation, glycosylation and proteolysis [10]. Because tau in PHFs is highly phosphorylated, it has been postulated that this modification may also be involved in the aggregation of the protein. In fact, studies have shown that tau presents an 'abnormal' phosphorylation in AD and other tauopathies. Indeed, despite the fact that many phosphorylation sites are identical between the native and the PHF-tau, there are some biochemical features that differentiate them from each other and support the concept of a pathological phosphorylation of tau in tauopathies. First, the progression of the NFD is directly correlated with the presence of a hyperphosphorylation of tau at specific sites [44]. Secondly, only the abnormally phosphorylated tau proteins observed in AD, and never the ones from control samples, can be visualised by a few phosphorylation-dependent antibodies such as AT100, AP422 or the TG antibodies [10]. The use of these antibodies also suggested that tau adopts a different conformation when phosphorylated, since with the exception of Ser422, these sites found in PHF-tau are conformation-dependent epitopes. Finally, it was suggested that the pathological phosphorylation of tau protein could act directly on its fibrillogenesis or stabilize the aggregates already formed [18]. At the molecular level, the phosphorylation of tau protein will change its conformation, increasing its fibrillogenesis [45]. It was proposed that the phosphorylation of tau would act on the aggregation process by two mechanisms: (*i*) It will promote the nucleation, a process which has been demonstrated after the phosphorylation of tau by Cdk5 [46]. The hyperphosphorylation of tau on its Ct part will then induce a change of conformation presumably starting the nucleation process [47,48]. (*ii*) The phosphorylation will also neutralise the basic charges on tau in flanking regions surrounding the microtubule-binding domain, a neutralisation which is fundamental for the self-aggregation of tau into PHF or SF [49]. Taken together, all these studies suggest that the phosphorylation of tau at specific sites is important for the formation of intraneuronal and intraglial inclusions observed in AD and other tauopathies. Another experimental approach, which supports this hypothesis is the overexpression of tau kinases or the inhibition of tau phosphatases in cellular and animal models. For instance, the constitutive expression of p25, the pathological activator of Cdk5, in a transgenic model, which already expressed the human mutant P301L tau, led to an increased phosphorylation of tau [50]. This double transgenic mouse also showed a significant deposition of tau in the cortex and the brainstem as compared to the single transgenic mouse expressing only the human P301L mutant tau. In the double transgenic model, another tau kinase, GSK3β, was associated with the NFD. The authors suggested that the phosphorylation of tau by p25/ Cdk5 released it from MTs and increased its deposition. These data and others demonstrated that the phosphorylation of tau protein could act as a catalyst for the aggregation of tau *in vivo*.

However, even if the abnormal phosphorylation of tau seems to be a key event in its pathological behaviour, this posttranslational modification is unable to fully explain the process of aggregation. Indeed, as previously shown, unphosphorylated tau is able to self-aggregate in solution. Several studies also suggested that the phosphorylation of tau at certain sites could have a protective effect on its aggregation [51]. Moreover, it was also demonstrated in different cellular models that the abnormal phosphorylation of tau, on sites found only in tauopathies is not sufficient to cause the aggregation of tau [52]. In these models, even when tau is phosphorylated on sites recognised by the antibodies AT100 and AP422/988, no tau aggregates were found, suggesting that the abnormal phosphorylation of tau at these epitopes does not necessarily lead to the aggregation of the protein. In fact, it seems that the phosphorylation of tau is an important event in the aggregation process of the protein but that this modification alone is not sufficient to induce tau fibrillogenesis *in vivo.*

Moreover, an important concern about the role of the hyperphosphorylation and/or aggregation of tau is which form of tau is really involved in cell death and neurodegeneration. Indeed, direct toxicity of the aggregates themselves has become a less attractive hypothesis to explain neurodegeneration in AD and other tauopathies. In reality, despite the fact that the aggregated tau is a general feature observed in all tauopathies and most animal models expressing mutant tau, it could be that the presence of these filamentous deposits is a compensatory mechanism employed by the cells to protect themselves against toxic tau species. In this way, it has been demonstrated in *Drosophila melanogaster* that the expression of wild type or mutant tau (carrying the R406W mutation) leads to neuronal cell death related to the accumulation of hyperphosphorylated tau but without any neurofibrillary tangle formation [53]. More recently, another study in a transgenic rodent model showed that the accumulation of NFTs is not sufficient to cause neurodegeneration [54]. In this work, the authors proved that in the mouse, the expression of human tau under the control of a repressible promoter led to neuronal loss, behaviour impairment and the development of NFTs. Surprisingly, after reduction of human tau expression, they also noted that the memory function was recovered and further neuronal loss prevented despite the accumulation of filamentous tau. They then concluded that the aggregation of tau is not sufficient to cause the neurodegeneration and cognitive decline observed in the mouse model. These studies also suggest that neuronal death is dependent on the expression of tau but that NFD is not directly caused by the intracellular aggregation of tau into PHFs. Finally, a recent study clarified these observations. Shimura and colleagues demonstrated that the accumulation of soluble hyperphosphorylated tau induced cell death. In contrast, when hyperphosphorylated tau was ubiquitinated, it formed intracellular aggregates that were no longer toxic [55]. In this case, the formation of hyperphosphorylated tau aggregates has a protective rather than a toxic effect on cells. This work also suggests that in the model used (Cos7 cells), the ubiquitination of tau is required for its aggregation. All in all, these results need to be investigated further to confirm which tau species are really involved in neurodegeneration observed in all tauopathies.

The Involvement of the Ubiquitin Proteasome System in Alzheimer's Disease and other Tauopathies

Ubiquitin (Ub) was first described in 1975 [56] as a highly conserved ubiquitous protein (see Chapter 2). Through a complex enzymatic machinery, multiple Ubs can be attached to the target protein. The polyubiquitin-tagged proteins will then be degraded by the Ubiquitin-Proteasome-System (UPS), the major system for intracellular protein degradation in eukaryotes [57]. The cascade responsible for the polyubiquitination of intracellular proteins involves distinct classes of enzymes: E1 (Ub-activating enzyme), E2 (Ub-conjugating enzyme) and E3 (Ub-protein ligases). The E1 enzyme activates Ub in an ATP-dependent mechanism. The activated Ub is then transferred to the E2 enzyme, which catalyses the covalent attachment of Ub to the target proteins. E2 enzymes may also transfer the activated Ub to an E3-Ub intermediate. Finally, E3 enzymes will be responsible for the specific recognition of the polyubiquitinated protein by the UPS [58] (see Chapter 3). The ubiquitination of proteins is a reversible process, which also involves deubiquitination enzymes (DUBs), cysteine proteases important for the control of the stability of Ub-conjugated proteins (see Chapter 4). Polyubiquitinated proteins are directed to the 26S proteasomes for degradation [59,60]. The 26S proteasome is a large enzymatic complex, ubiquitously found in all eukaryotic organisms (see Chapter 7). In fact, the 26S proteasome is formed by a proteolytic core called '20S proteasome', able to degrade unfolded proteins by an Ub-independent mechanism (see Chapter 6). The 20S core can also bind several multimeric components like the 19S regulatory units, which will allow the degradation of polyubiquitinated proteins. Together, the 20S proteasome and the 19S particles form the 26S proteasome, which is, as a whole, involved in the proteolysis of polyubiquitinated proteins [61]. The 26S proteasome may also be responsible for the degradation of unfolded proteins by an Ub-independent mechanism [62].

The degradation of certain intracytosolic unfolded or damaged proteins by the UPS sometimes involves a prior interaction of the damaged polypeptide with chaperone proteins. One major group of these molecular chaperones is represented by the Heat-shock-proteins (Hsp70, Hsc70, Hsp90 and others), which are cell constituents under normal conditions (see Chapters 10 and 19). They are essential for the folding of newly synthetised proteins and to ensure their proper intracellular localisation. They can also rescue the activity of unfolded or aggregated proteins. For instance, Hsp40 and Hsp70 can catalyse the refolding of denatured or partially denatured proteins into active forms in an ATP-dependent mechanism [63,64]. After decades of research, it was shown that these Hsps are essential for the viability of cells, where they can adopt different functions under normal or pathological conditions. These molecular chaperones can catalyse the protein folding and multimer assembly, prevent the aggregation of mutant or damaged proteins, solubilize aggregated proteins, suppress apoptotic programs and promote the ubiquitination and degradation of abnormal proteins [65]. The protective effects of molecular chaperones, the repair of unfolded proteins or their degradation are complementary but not independent mechanisms and appear to be linked at certain levels. In this way, despite their role in refolding and preventing aberrant aggregation of damaged proteins, Hsp70 and other chaperones are also required for the ubiquitination and the rapid degradation of many abnormal proteins [66]. The capacity of certain chaperones to bind unfolded polypeptides will help their future recognition by E3 enzymes, a mechanism which may facilitate the further degradation of the proteins by the UPS. The duality of the chaperone activity is important for the control of the refolding or the degradation of altered intracellular proteins and is highly advantageous for the physiology of the cells. If an altered protein cannot be adequately refolded, it will be degraded by the UPS.

As described previously, the aberrant accumulation of filamentous tau protein in NFD constitutes the neuropathological hallmark of AD and other tauopathies. In 1987, two studies described for the first time the presence of ubiquitin in PHF and so in NDF [67,68]. Since then, evidence which supports the hypothesis that the UPS may play a role in AD and other neurodegenerative disorders has grown [58]. In fact, the involvement of the UPS system in AD was suggested by an inhibition of the proteasome activity in AD brain, the accumulation of ubiquitinated proteins in NFD and the presence of proteasome subunit in disease related areas. First of all, the enzymes involved in the polyubiquitination of unfolded proteins are deregulated in AD, as demonstrated by the reduced levels of certain ubiquitinating (E1), Ub-conjugated (E2) and deubiquitinating enzymes such as UCH-L1. The activity of the UPS is also altered [20]. Indeed, the activity of the 20S proteolytic core of the proteasome (20S α 5 or the three β subunits, which are involved in confining the proteolytic chamber or in protein degradation respectively) and the S1 units (19S non-ATPase) are down-regulated in AD. Finally, another 19S ATPase (S6b) was shown to be present in NFTs [58]. The inhibition of the UPS activity in AD may be due to the presence of aggregates in NFD. Indeed, once small aggregates form within cells, they may negatively affect the activity and the function of the proteasome [69]. In AD, the inhibition of the UPS may result from a direct or indirect effect of the neurotoxicity of the amyloid peptide (Aβ1-42), which could bind and block the proteasome activity [21,70]. The inhibition of the UPS could also be caused by the aggregated tau [71]. This is in line with the studies on immunotherapeutic clearance of amyloid and tau aggregates in a mouse model of AD [72]. In this study, Aβ immunotherapy reduced not only the extracellular amyloid plaques, but also the intracellular Aβ accumulation and even led to the clearance of early tau pathology, a phenomenon mediated by the UPS. This study showed a close relationship between Aβ and tau pathologies and suggested that intracellular Aβ species could inhibit the proteasomal activity and then increase the formation of hyperphosphorylated tau aggregates, which in turn may also inhibit the UPS. Recently, another piece of evidence strengthened the role which the UPS may play in AD pathology: a new form of ubiquitin, the UBB^{+1} was shown to accumulate in the neuritic plaques and tangles in AD brains [73]. UBB^{+1} is a mutant form of ubiquitin that lacks the C-terminal Gly of wild type Ub and instead has a 19-amino acid extension. This mutant ubiquitin can itself be ubiquitinated but is unable to bind other proteins. UBB^{+1} is a powerful inhibitor of the proteasome activity when its intracellular concentration reaches a certain threshold [21,74]. This mutant ubiquitin may therefore be an important component of the UPS inhibition in AD pathology. Taken together, all these observations suggest that the activity of the UPS complex and the enzymes required for the ubiquitination of unfolded proteins are affected in AD and may participate in the development of NFD as demonstrated by the accumulation of ubiquitinated proteins in PHFs.

The major constituents of these PHFs, the tau proteins, were also shown to be ubiquitinated arguing that the UPS may be directly involved in their degradation or aggregation [75]. This modification of tau was reported later in CBD, another tauopathy [76]. Several (but not all) transgenic mouse models of human tauopathy, which over-express different mutants of human tau, have demonstrated the presence of ubiquitin in degenerating cells [77,78]. Finally, the inhibition of the proteasomal activity led to the stabilisation of aggregates made of hyperphosphorylated tau in oligodendroglial cells. In these experiments, the inhibition of the proteasome was followed by an increase of the levels of heat shock proteins αB-crystalin and Hsp70 [23]. As mentioned before, some proteins need to be recognised by molecular chaperones to be targeted to the UPS. tau proteins are such proteins. In AD, the levels of Hsp70, Hsp27, αB-crystalin and other chaperons are altered [79- 83] suggesting that they may participate in the progression of the disease. All in all, these studies suggest a close relationship between tau pathology, the failure of the UPS and the presence of molecular chaperones during neurodegeneration (see Chapter 19).

The Relationship between TAU and the UPS: A Mechanism Leading to the Degradation or the Aggregation of TAU?

As shown previously, tau is an unfolded protein in solution. As such, it can be proteolysed by the 20S proteasome by a bidirectional mechanism *in vitro* without any phosphorylation [84]. Hyperphosphorylated tau could also be degraded by an Ub-independent mechanism in cells. This mechanism involves the chaperone protein Hsp27, necessary for the recognition of phospho-tau, its targeting and its degradation by the 20S proteasome (Figure 3) and/or its dephosphorylation [85,86]. Hsp27 would then suppress the toxicity of hyperphosphorylation tau species within cells. Tau can thus be degraded by the 20S proteasome by an Ub-independent mechanism.

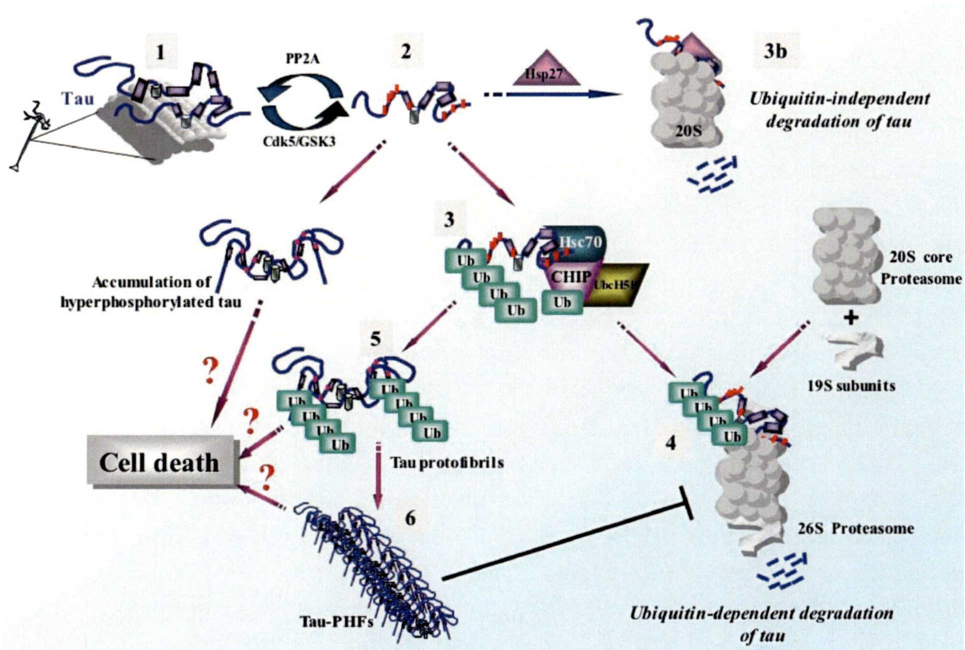

Figure 3. The degradation or aggregation of tau: process regulated by the proteasome. Tau protein can be phosphorylated by either Cdk5 or GSK3. This will lead to the detachment of tau from the microtubules (transition 1 to 2). This phosphorylation is also regulated by PP2A. Once phosphorylated, tau can be recognised by the Heat shock protein Hsp27 and then be degraded by the 20S proteasome by a ubiquitin independent mechanism (3b). Phosphorylated tau can also be recognised by the Hsp70 or the Hsc70 (3) which will allow its binding by the complex CHIP/UbcH5B. This complex will polyubiquitinate hyperphosphorylated tau, a process that could lead to tau degradation (4) or to its aggregation into tau-PHFs (5 and 6). Both mechanisms seem to prevent the cell death caused by the accumulation of soluble but hyperphosphorylated tau. However, the tau-PHFs may also inhibit the proteasome activity, which in turn will render the tau-PHFs cytotoxic for the cells. It remains to be shown what tau species may be responsible for cell death, i.e hyperphosphorylated soluble or insoluble tau (question marks).

However, tau can also be proteolyzed by the 26S proteasome following ubiquitination. Recently, three studies reported the degradation of tau by a ubiquitin-dependent pathway [55,87,88]. A common link between those three studies is the ubiquitination of tau by CHIP (carboxy terminus of the Hsc70-interacting protein), in a native [87] or hyperphosphorylated form [55]. CHIP can interact with tau through its microtubule-binding domain (amino acids 187-311), a binding which will lead to the ubiquitination of tau. It is of note that CHIP was found to colocalise with NFTs in diseases presenting hyperphosphorylated and aggregated tau, e.g., AD, PSP, CBD, PiD and FTDP-17, but to different extents [87,88]. In this way, Hatakeyama and colleagues found that CHIP is preferentially involved in the degradation of 4R tau and thus colocalised preferentially with hyperphosphorylated tau in PSP, a tauopathy which showed a specific aggregation of 4R tau isoforms [10]. In contrast, Petrucelli and colleagues argued that CHIP colocalises with aggregated tau to a greater extent in 3R (PiD=

50-70%) than in 4R + 3R (AD= 5-10%) or 4R (CBD, PSP and FTDP-17= 1-5%) tauopathies. In this study, only the tangles, and not the pre-tangle neurons with non-fibrillar phosphorylated tau, were immunoreactive for CHIP. Despite these differences, these data indicate that CHIP could act differently on tau protein depending on the isoforms considered (3R or 4R). The effect of the ubiquitination of tau by CHIP will also be different and can lead to either the degradation of the protein or its aggregation. These discrepancies may be related to the presence of different chaperone proteins, e.g., Hsp70 or Hsc70.

In the first model, CHIP leads to the ubiquitination of tau in the presence of the E2 conjugase UbcH5B, a process which increases the deposition of hyperphosphorylated tau [88]. In these experiments, tau was found to co-immunoprecipitate with CHIP but also with the Heat-shock protein Hsp70, which attenuates or antagonises the activity of CHIP on tau. Indeed, after co-transfection of COS7 cells with the mutant tau P301L and with CHIP, tau was found to be ubiquitinated in the detergent-insoluble fraction. In contrast, expression of Hsp70 selectively reduced the levels of detergent insoluble tau to undetectable amounts. These data indicate that CHIP would promote tau aggregation (following its ubiquitination) whereas Hsp70 would suppress it. In this model, Hsp70 would then protect the cells against tau aggregation and neurotoxicity. These results are consistent with another study, which showed that increased levels of Hsp70 or Hsp90 reduce tau aggregation or at least increase the association of tau for the MTs [89].

In the second model, CHIP could also ubiquitinate tau (phosphorylated or not), which can lead either to tau aggregation or to its degradation [55,87]. Shimura and colleagues demonstrated that hyperphosphorylated tau is recognised by the E2 enzyme UbcH5B and the complex formed by the E3 Ub-ligase CHIP and the chaperone protein Hsc70 (Heat-shock cognate 70 protein) rather than Hsp70. In fact, tau must be phosphorylated at specific sites to be ubiquitinated. These motifs include Ser^{199}, Ser^{202}, Thr^{205}, Ser^{396} and Ser^{404}. The kinases involved in the phosphorylation of these sites, e.g., GSK3β and Cdk5 could create these recognition sites. Their antagonist, the phosphatase PP2A eliminates the phosphorylation and prevents the ubiquitination of tau *in vitro*. In fact, the authors demonstrated that the phosphorylation of tau at these sites is necessary for the binding to Hsc70. Once phosphorylated by GSK3β or Cdk5, tau will be released from the MTs and interact with Hsc70, which will create a 'bridge' between tau and the E3 ligase CHIP. Together with the E2 enzyme, UbcH5B, CHIP will ubiquitinate the hyperphosphorylated tau (Figure 3). Then, hyperphosphorylated and ubiquitinated tau can either be degraded by the UPS or aggregate within the cell. The authors also demonstrated that both mechanisms (degradation or aggregation of tau protein) lead to cell survival and not to cell death, a process, which is apparently dependent on the accumulation of soluble but hyperphosphorylated tau species. This interesting finding is consistent with the one from Hatakeyama *et al.* [87] Indeed, in this study, 4R tau is ubiquitinated by CHIP, apparently without the requirement of any chaperone proteins. This ubiquitination leads to the degradation of tau and an increase in cell survival. The authors have also shown that even if after inhibition of the proteasome, some tau aggregates remain within the cell, they fail to induce the cell death.

Despite the differences observed between these three studies, these findings are very interesting and show that (*i*) The ubiquitination of hyperphosphorylated tau protein can lead to its degradation by the UPS. This process will lead to the clearance of toxic tau species and then to an increase in cell survival. (*ii*) Alternatively, if the ubiquitinated tau is not degraded, it will be aggregated. (*iii*) The ubiquitination of tau happens prior to the fibril formation in this model and (*iv*) The aggregation of ubiquitinated tau is also beneficial for the viability of cells since it will decrease the accumulation of soluble but hyperphosphorylated tau. These new discoveries are in line with the idea that the presence of aggregated tau is a cellular response of neurons to intracellular toxic species and that tau fibrils are not toxic for the cells.

More recently, another study discovered a new protein, p62 that may be involved in the targeting of tau to the UPS [90]. Sequestosome p62 is a newly identified cellular protein, which binds polyubiquitin chains and targets polyubiquitinated proteins to the UPS [91]. p62 was found to colocalise with hyperphosphorylated tau in AD. p62 is able to recognise polyubiquitinated tau (after ubiquitination by the E2 and E3 enzymes UbcH7 and TRAF6) and to shuttle it for its degradation by the proteasome. However, in contrast to the three previous studies, this work has also shown that if the proteasome function is impaired, polyubiquitinated-tau aggregates within cells and in this case tau fibrils are cytotoxic [90].

The Inhibition of the Proteasome: A Mechanism that Indirectly May Change the Metabolism of TAU Proteins

As shown previously, the UPS could be involved in the degradation and/or the aggregation of tau, a phenomenon which involves apparently different chaperone proteins. However, these data must be assessed with caution. In fact, the UPS is a general proteolytic system that could act on proteins directly implicated in tau proteolysis or even phosphorylation. Concerning the tau proteases, the proteasome can act on their intracellular level or activity depending on the cell type and the cell metabolism. For instance, proteasome inhibition may lead to apoptosis or prevent cell death [92-94] (see Chapter 21). In this way, the proteasome may act directly or indirectly on the activity of tau proteases, i.e., caspases, calpain or some chaperone proteins as Hsp70/90, which as described before may be involved in the degradation and/or the aggregation of tau. In some cases, when proteasome inhibition leads to caspase activation, apoptosis occurs and tau proteins are degraded by a monodirectional mechanism starting from the Ct part of tau (Figure 4) [95]. This caspase activation may also be a prerequisite for tau aggregation since it was suggested that the caspase cleavage of tau is an early event which may precede the hyperphosphorylation and aggregation of tau in AD and other tauopathies [96,97]. It is of note that again, the activation of caspases after proteasome inhibition is dependent on the cell type since in sympathetic neurons, inhibition of the proteasome leads to neuronal protection by inhibition of caspases [92]. Moreover, if Hsps are synthesised after proteasome inhibition, these chaperones will have a protective effect on cell death [98,99].

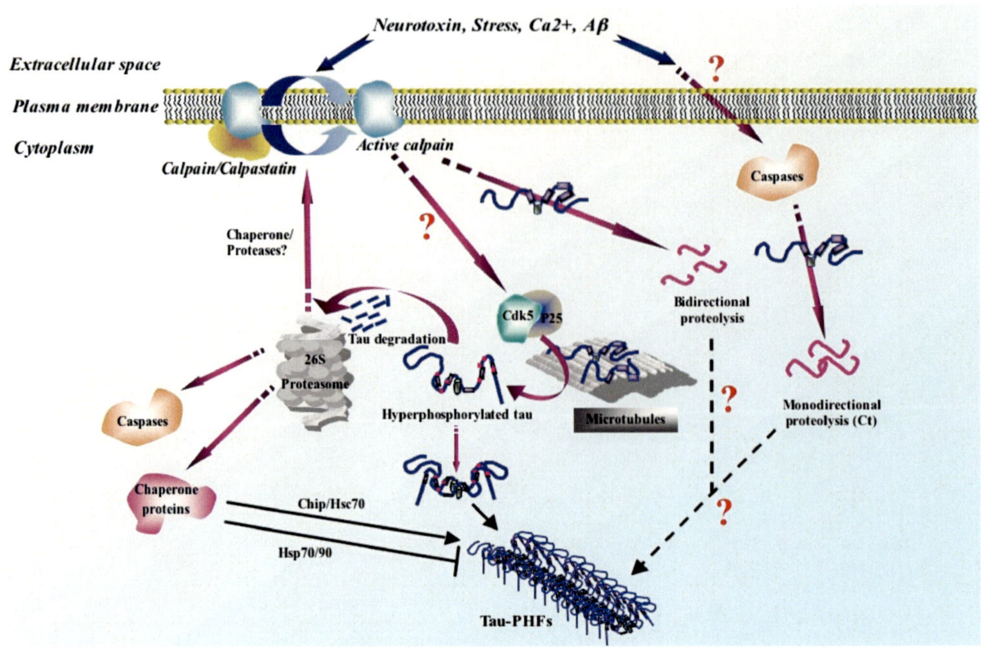

Figure 4. The UPS: a crossroad for the metabolism of tau? Hypothetical scheme representing how an inhibition of the proteasome could act on tau protein metabolism. The proteasome may act on different proteases within the cell. Depending on the cell type, the proteasome will be involved directly in the degradation and/or the aggregation of tau proteins (see *Figure 3* for more details). However, an inhibition of the proteasome can also lead to the activation of calpain1 by decreasing the levels of the endogenous calpain1 inhibitor, calpastatin. This process is dependent on newly synthesised proteins such as chaperone or other proteases. Once calpain1 is activated (a phenomenon which could also be dependent on different extracellular stimuli, i.e., neurotoxins, stress, calcium and amyloid deposits), it could lead to a complete or partial degradation of tau. The partial cleavage of tau could participate in its intraneuronal aggregation. Calpain could also over-activate Cdk5, by cleavage of its normal activator p35 into a cleaved form p25. It is of note that the p25 may be associated with Alzheimer's disease [114,115]. In this case, calpain activation will lead to the hyperphosphorylation of tau protein, a process which may enhance neurodegeneration. However, this hypothesis must be treated with caution since other studies showed no increase of p25 levels in AD and other tauopathies as compared to control samples [116,117]. The proteasome could also act on the activation of the caspases, which can degrade tau from its C-terminal extremity. This caspase-related cleavage of tau may be a prerequisite for the formation of intraneuronal tau aggregates. Finally, the proteasome can act on the intracellular chaperone proteins levels. In this case, these chaperone proteins can restore the function of tau (Hsp70/Hsp90), participate in its degradation (Hsp27, Hsp70/90 and Hsc70), or lead to the formation of aggregates (Hsc70), which are apparently not any longer neurotoxic to cells. Black arrows represent the different potential aggregation pathways of tau protein. The pink arrows show the different effects of the inhibition of the proteasome on several tau protein proteases and/or kinases. Question marks indicate the hypotheses in the literature, which are still controversial.

Other work in neuronal cell lines has also shown that the proteasome is neither directly involved in the degradation nor in the aggregation of tau [100-102]. In all these studies, the administration of proteasome inhibitors did not cause either hyperphosphorylation or aggregation of tau. In one study, inhibition of the proteasome was even shown to induce the bidirectional degradation of tau, a mechanism presumably dependent on the activation of calpain1, another tau protease [101]. In fact, in neuroblastoma cells, inhibition of the proteasome was shown to induce an activation of calpain1 by decreasing the intracellular level of its endogenous inhibitor, calpastatin. Hence, after inhibition of the proteasome, tau proteins were degraded rapidly instead of being aggregated. In this model, if the proteasome is not efficient, calpain1 is activated and tau may be proteolysed by a bidirectional process (Figure 4).

This effect is not direct and presumably requires protein synthesis since tau was not degraded if the translation was inhibited by cycloheximide. In fact, chaperone proteins or other proteases may be implicated in the degradation of calpastatin since the combination of translation and proteasome inhibitors had no effect on calpastatin levels. It was recently demonstrated that calpain is over-activated in AD [103]. This protease may also be involved in tau misfolding and aggregation. Indeed, calpain seems to be implicated in conformational changes of tau proteins after tau phosphorylation, a mechanism that could lead to tau fibrillogenesis [104]. However, it is of note that in AD, even if calpain is over-activated, this protease may not be able to degrade tau. Indeed, hyperphosphorylated or aggregated tau are less susceptible to calpain digestion [105]. Taken together, these data suggested that despite the fact that the proteasome can degrade tau directly, it could act indirectly on tau catabolism in different cells by affecting tau protease activity or the levels of different chaperone proteins. A last point, which needs to be considered is the hyperphosphorylation of tau and how the proteasome can modulate it. The proteasome can regulate the hyperphosphorylation of tau by changing the activity of several kinases and/or phosphatases.

For instance, the proteasome could directly regulate the expression and activity of cell cycle proteins [106-108]. Tau proteins were also shown to be hyperphosphorylated and abnormally phosphorylated during cell cycle progression especially during mitosis [52,109,110]. The UPS may be then involved in this type of tau phosphorylation if its inhibition leads to the accumulation of cell cycle related proteins. This is of particular interest since it was suggested that these cell cycle proteins accumulate in NFD in AD and other tauopathies and may be a prerequisite for the hyperphosphorylation and/or aggregation of tau [111-113]. In conclusion, the inhibition of the proteasome in neuronal cells can change the metabolism of tau protein in two ways: first by increasing the activity of tau proteases and secondly by leading to the accumulation or the activation of tau kinases. In both cases, the inhibition of the proteasome may then participate indirectly in the aggregation process of tau proteins in neurons.

Conclusion

The UPS is inhibited in AD and probably in other tauopathies. This dysfunction may be related to the presence of intracellular aggregates within degenerating neurons as demonstrated with cytotoxic forms of amyloid peptides or tau-PHFs. The inhibition of the proteasome could also be the result of the new mutant ubiquitin UBB^{+1}. As the UPS could be either directly or indirectly involved in tau proteolysis and/or aggregation, it seems that it is a new potential target to thwart the intraneuronal accumulation of hyperphosphorylated tau and thus the neurodegeneration observed in all tauopathies. However, the effect of the proteasome on tau degradation is dependent on the cell system considered and appears to be at least in part related to the presence of several chaperone proteins. Moreover, as the proteasome inhibition could also lead to the activation of other systems that could act on tau metabolism (and shown to be deregulated in AD and other tauopathies), further studies will be needed to really understand what are the consequences of the UPS inhibition on the NFD. Importantly, it is still necessary to determine the role of the inhibition of the UPS in the cascade of events leading to neurodegeneration and especially if the inhibition of the proteasome is a primary cause leading to neurodegeneration or whether it is a consequence of the accumulation of toxic species that may inhibit its activity. Finally, with the recent discovery that the tau-PHFs are probably not responsible for the neurodegeneration observed in tauopathies, future work must focus on how the UPS inhibition contributes to the disease and how to counter the toxic effect of tau species whatever they are. All in all, it appears that the UPS may play a central role in the development of the 'tau pathology' in tauopathies and that a better understanding of its inhibition will lead in the next decade to possible new treatments.

ACKNOWLEDGEMENTS

We thank Dr K. Virdee for helpful discussions. PD is supported by a European Molecular Biology Organization (EMBO, Heidelberg, Germany) fellowship, reference ALTF 80-2004.

REFERENCES

[1] Brion JP, Couck AM, Flament-Durand J: Ultrastructural study of enriched fractions of 'tangles' from human patients with senile dementia of the Alzheimer type. *Acta Neuropathol (Berl)* 1984, 64:148-152

[2] Brion JP, Couck AM, Passareiro E, Flament-Durand: Neurofibrillary tangles of Alzheimer's disease: an immunohistochemical study. *J. Submicrosc. Cytol.* 1985, 17:89-96

[3] Delacourte A, Defossez A: Alzheimer's disease: Tau proteins, the promoting factors of microtubule assembly, are major components of paired helical filaments. *J Neurol Sci* 1986, 76:173-186

[4] Goedert M, Wischik CM, Crowther RA, Walker JE, Klug A: Cloning and sequencing of the cDNA encoding a core protein of the paired helical filament of Alzheimer disease: identification as the microtubule-associated protein tau. *Proc Natl Acad Sci U S A* 1988, 85:4051-4055

[5] Wischik CM, Novak M, Edwards PC, Klug A, Tichelaar W, Crowther RA: Structural characterization of the core of the paired helical filament of Alzheimer disease. *Proc Natl Acad Sci U S A* 1988, 85:4884-4888

[6] Wischik CM, Novak M, Thogersen HC, Edwards PC, Runswick MJ, Jakes R, Walker JE, Milstein C, Roth M, Klug A: Isolation of a fragment of tau derived from the core of

the paired helical filament of Alzheimer disease. *Proc Natl Acad Sci U S A* 1988, 85:4506-4510

[7] Braak H, Braak E: Neuropathological stageing of Alzheimer-related changes. *Acta Neuropathol.*1991, 82: 239-59

[8] Delacourte A, Sergeant N, Champain D, Wattez A, Maurage CA, Lebert F, Pasquier F, David JP: Nonoverlapping but synergetic tau and APP pathologies in sporadic Alzheimer's disease. *Neurology* 2002, 59:398-407

[9] Delacourte A, David JP, Sergeant N, Buee L, Wattez A, Vermersch P, Ghozali F, Fallet-Bianco C, Pasquier F, Lebert F, Petit H, Di Menza C: The biochemical pathway of neurofibrillary degeneration in aging and Alzheimer's disease. *Neurology* 1999, 52:1158-1165

[10] Buee L, Bussiere T, Buee-Scherrer V, Delacourte A, Hof PR: Tau protein isoforms, phosphorylation and role in neurodegenerative disorders. *Brain Res Brain Res Rev* 2000, 33:95-130

[11] Hutton M, Lendon CL, Rizzu P, Baker M, Froelich S, Houlden H, Pickering-Brown S, Chakraverty S, Isaacs A, Grover A, Hackett J, Adamson J, Lincoln S, Dickson D, Davies P, Petersen RC, Stevens M, de Graaff E, Wauters E, van Baren J, Hillebrand M, Joosse M, Kwon JM, Nowotny P, Che LK, Norton J, Morris JC, Reed LA, Trojanowski J, Basun H, Lannfelt L, Neystat M, Fahn S, Dark F, Tannenberg T, Dodd PR, Hayward N, Kwok JB, Schofield PR, Andreadis A, Snowden J, Craufurd D, Neary D, Owen F, Oostra BA, Hardy J, Goate A, van Swieten J, Mann D, Lynch T, Heutink P: Association of missense and 5'-splice-site mutations in tau with the inherited dementia FTDP-17. *Nature* 1998, 393:702-705

[12] Poorkaj P, Bird TD, Wijsman E, Nemens E, Garruto RM, Anderson L, Andreadis A, Wiederholt WC, Raskind M, Schellenberg GD: Tau is a candidate gene for chromosome 17 frontotemporal dementia. *Ann Neurol* 1998, 43:815-825

[13] Spillantini MG, Murrell JR, Goedert M, Farlow MR, Klug A, Ghetti B: Mutation in the tau gene in familial multiple system tauopathy with presenile dementia. *Proc Natl Acad Sci U S A* 1998, 95:7737-7741

[14] Bugiani O, Murrell JR, Giaccone G, Hasegawa M, Ghigo G, Tabaton M, Morbin M, Primavera A, Carella F, Solaro C, Grisoli M, Savoiardo M, Spillantini MG, Tagliavini F, Goedert M, Ghetti B: Frontotemporal dementia and corticobasal degeneration in a family with a P301S mutation in tau. *J Neuropathol Exp Neurol* 1999, 58:667-677

[15] Goedert M, Jakes R, Crowther RA: Effects of frontotemporal dementia FTDP-17 mutations on heparin-induced assembly of tau filaments. *FEBS Lett* 1999, 450:306-311

[16] Goedert M, Satumtira S, Jakes R, Smith MJ, Kamibayashi C, White CL, 3rd, Sontag E: Reduced binding of protein phosphatase 2A to tau protein with frontotemporal dementia and parkinsonism linked to chromosome 17 mutations. *J Neurochem* 2000, 75:2155-2162

[17] Delobel P, Flament S, Hamdane M, Jakes R, Rousseau A, Delacourte A, Vilain JP, Goedert M, Buee L: Functional characterization of FTDP-17 tau gene mutations through their effects on Xenopus oocyte maturation. *J Biol Chem* 2002, 277:9199-9205

[18] Sahara N, Lewis J, DeTure M, McGowan E, Dickson DW, Hutton M, Yen SH: Assembly of tau in transgenic animals expressing P301L tau: alteration of phosphorylation and solubility. *J Neurochem* 2002, 83:1498-1508
[19] Delobel P, Mailliot C, Hamdane M, Sambo AV, Begard S, Violleau A, Delacourte A, Buee L: Stable-tau overexpression in human neuroblastoma cells: an open door for explaining neuronal death in tauopathies. *Ann N Y Acad Sci* 2003, 1010:623-634
[20] Keller JN, Hanni KB, Markesbery WR: Impaired proteasome function in Alzheimer's disease. *J Neurochem* 2000, 75:436-439
[21] Lam YA, Pickart CM, Alban A, Landon M, Jamieson C, Ramage R, Mayer RJ, Layfield R: Inhibition of the ubiquitin-proteasome system in Alzheimer's disease. *Proc Natl Acad Sci U S A* 2000, 97:9902-9906
[22] McNaught KS, Olanow CW, Halliwell B, Isacson O, Jenner P: Failure of the ubiquitin-proteasome system in Parkinson's disease. *Nat Rev Neurosci* 2001, 2:589-594
[23] Goldbaum O, Oppermann M, Handschuh M, Dabir D, Zhang B, Forman MS, Trojanowski JQ, Lee VM, Richter-Landsberg C: Proteasome inhibition stabilizes tau inclusions in oligodendroglial cells that occur after treatment with okadaic acid. *J Neurosci* 2003, 23:8872-8880
[24] Neve RL, Harris P, Kosik KS, Kurnit DM, Donlon TA: Identification of cDNA clones for the human microtubule-associated protein tau and chromosomal localization of the genes for tau and microtubule-associated protein 2. *Brain Res* 1986, 387:271-280
[25] Goedert M, Spillantini MG, Potier MC, Ulrich J, Crowther RA: Cloning and sequencing of the cDNA encoding an isoform of microtubule-associated protein tau containing four tandem repeats: differential expression of tau protein mRNAs in human brain. *Embo J* 1989, 8:393-399
[26] Goedert M, Spillantini MG, Jakes R, Rutherford D, Crowther RA: Multiple isoforms of human microtubule-associated protein tau: sequences and localization in neurofibrillary tangles of Alzheimer's disease. *Neuron* 1989, 3:519-526
[27] Weingarten MD, Lockwood AH, Hwo SY, Kirschner MW: A protein factor essential for microtubule assembly. *Proc Natl Acad Sci U S A* 1975, 72:1858-1862
[28] Witman GB, Cleveland DW, Weingarten MD, Kirschner MW: Tubulin requires tau for growth onto microtubule initiating sites. *Proc Natl Acad Sci U S A* 1976, 73:4070-4074
[29] Drubin DG, Kirschner MW: Tau protein function in living cells. *J Cell Biol* 1986, 103:2739-2746
[30] Himmler A, Drechsel D, Kirschner MW, Martin DW, Jr.: Tau consists of a set of proteins with repeated C-terminal microtubule-binding domains and variable N-terminal domains. *Mol Cell Biol* 1989, 9:1381-1388
[31] Goedert M, Jakes R: Expression of separate isoforms of human tau protein: correlation with the tau pattern in brain and effects on tubulin polymerization, *EMBO J* 1990, 9:4225-4230
[32] Goode BL, Feinstein SC: Identification of a novel microtubule binding and assembly domain in the developmentally regulated inter-repeat region of tau. *J Cell Biol* 1994, 124:769-782

[33] Goode BL, Chau M, Denis PE, Feinstein SC: Structural and functional differences between 3-repeat and 4-repeat tau isoforms. Implications for normal tau function and the onset of neurodegenetative disease. *J Biol Chem* 2000, 275:38182-38189

[34] Lee G, Neve RL, Kosik KS: The microtubule binding domain of tau protein. *Neuron* 1989, 2:1615-1624

[35] Brandt R, Lee G, Teplow DB, Shalloway D, Abdel-Ghany M: Differential effect of phosphorylation and substrate modulation on tau's ability to promote microtubule growth and nucleation. *J Biol Chem* 1994, 269:11776-11782

[36] Xie H, Litersky JM, Hartigan JA, Jope RS, Johnson GV: The interrelationship between selective tau phosphorylation and microtubule association. *Brain Res* 1998, 798:173-183

[37] Crowther RA, Olesen OF, Smith MJ, Jakes R, Goedert M: Assembly of Alzheimer-like filaments from full-length tau protein. *FEBS Lett* 1994, 337:135-138

[38] Perez M, Valpuesta JM, Medina M, Montejo de Garcini E, Avila J: Polymerization of tau into filaments in the presence of heparin: the minimal sequence required for tau-tau interaction. *J Neurochem* 1996, 67:1183-1190

[39] Arrasate M, Perez M, Armas-Portela R, Avila J: Polymerization of tau peptides into fibrillar structures. The effect of FTDP-17 mutations. *FEBS Lett* 1999, 446:199-202

[40] Perez M, Arrasate M, Montejo De Garcini E, Munoz V, Avila J: In vitro assembly of tau protein: mapping the regions involved in filament formation. *Biochemistry* 2001, 40:5983-5991

[41] Carmel G, Mager EM, Binder LI, Kuret J: The structural basis of monoclonal antibody Alz50's selectivity for Alzheimer's disease pathology. *J Biol Chem* 1996, 271:32789-32795

[42] Jicha GA, Rockwood JM, Berenfeld B, Hutton M, Davies P: Altered conformation of recombinant frontotemporal dementia-17 mutant tau proteins. *Neurosci Lett* 1999, 260:153-156

[43] von Bergen M, Friedhoff P, Biernat J, Heberle J, Mandelkow EM, Mandelkow E: Assembly of tau protein into Alzheimer paired helical filaments depends on a local sequence motif ((306)VQIVYK(311)) forming beta structure. *Proc Natl Acad Sci U S A* 2000, 97:5129-5134

[44] Augustinack JC, Schneider A, Mandelkow EM, Hyman BT: Specific tau phosphorylation sites correlate with severity of neuronal cytopathology in Alzheimer's disease. *Acta Neuropathol (Berl)* 2002, 103:26-35

[45] Daly NL, Hoffmann R, Otvos L, Jr., Craik DJ: Role of phosphorylation in the conformation of tau peptides implicated in Alzheimer's disease. *Biochemistry* 2000, 39:9039-9046

[46] Paudel HK: Phosphorylation by neuronal cdc2-like protein kinase promotes dimerization of Tau protein in vitro. *J Biol Chem* 1997, 272:28328-28334

[47] Leger J, Kempf M, Lee G, Brandt R: Conversion of serine to aspartate imitates phosphorylation-induced changes in the structure and function of microtubule-associated protein tau. *J Biol Chem* 1997, 272:8441-8446

[48] Uversky VN, Winter S, Galzitskaya OV, Kittler L, Lober G: Hyperphosphorylation induces structural modification of tau-protein. *FEBS Lett* 1998, 439:21-25

[49] Alonso A, Zaidi T, Novak M, Grundke-Iqbal I, Iqbal K: Hyperphosphorylation induces self-assembly of tau into tangles of paired helical filaments/straight filaments. *Proc Natl Acad Sci U S A* 2001, 98:6923-6928

[50] Noble W, Olm V, Takata K, Casey E, Mary O, Meyerson J, Gaynor K, LaFrancois J, Wang L, Kondo T, Davies P, Burns M, Veeranna, Nixon R, Dickson D, Matsuoka Y, Ahlijanian M, Lau LF, Duff K: Cdk5 is a key factor in tau aggregation and tangle formation in vivo. *Neuron* 2003, 38:555-565

[51] Schneider A, Biernat J, von Bergen M, Mandelkow E, Mandelkow EM: Phosphorylation that detaches tau protein from microtubules (Ser262, Ser214) also protects it against aggregation into Alzheimer paired helical filaments. *Biochemistry* 1999, 38:3549-3558

[52] Delobel P, Flament S, Hamdane M, Mailliot C, Sambo AV, Begard S, Sergeant N, Delacourte A, Vilain JP, Buee L: Abnormal Tau phosphorylation of the Alzheimer-type also occurs during mitosis. *J Neurochem* 2002, 83:412-420

[53] Wittmann CW, Wszolek MF, Shulman JM, Salvaterra PM, Lewis J, Hutton M, Feany MB: Tauopathy in Drosophila: neurodegeneration without neurofibrillary tangles. *Science* 2001, 293:711-714

[54] Santacruz K, Lewis J, Spires T, Paulson J, Kotilinek L, Ingelsson M, Guimaraes A, DeTure M, Ramsden M, McGowan E, Forster C, Yue M, Orne J, Janus C, Mariash A, Kuskowski M, Hyman B, Hutton M, Ashe KH: Tau suppression in a neurodegenerative mouse model improves memory function. *Science* 2005, 309:476-481

[55] Shimura H, Schwartz D, Gygi SP, Kosik KS: CHIP-Hsc70 complex ubiquitinates phosphorylated tau and enhances cell survival. *J Biol Chem* 2004, 279:4869-4876

[56] Goldstein G, Scheid M, Hammerling U, Schlesinger DH, Niall HD, Boyse EA: Isolation of a polypeptide that has lymphocyte-differentiating properties and is probably represented universally in living cells. *Proc Natl Acad Sci U S A* 1975, 72:11-15

[57] Glickman MH, Ciechanover A: The ubiquitin-proteasome proteolytic pathway: destruction for the sake of construction. *Physiol Rev* 2002, 82:373-428

[58] de Vrij FM, Fischer DF, van Leeuwen FW, Hol EM: Protein quality control in Alzheimer's disease by the ubiquitin proteasome system. *Prog Neurobiol* 2004, 74:249-270

[59] Rock KL, Gramm C, Rothstein L, Clark K, Stein R, Dick L, Hwang D, Goldberg AL: Inhibitors of the proteasome block the degradation of most cell proteins and the generation of peptides presented on MHC class I molecules. *Cell* 1994, 78:761-771

[60] Coux O, Tanaka K, Goldberg AL: Structure and functions of the 20S and 26S proteasomes. *Annu Rev Biochem* 1996, 65:801-847

[61] Ciechanover A: The ubiquitin-proteasome proteolytic pathway. *Cell* 1994, 79:13-21

[62] Murakami Y, Matsufuji S, Kameji T, Hayashi S, Igarashi K, Tamura T, Tanaka K, Ichihara A: Ornithine decarboxylase is degraded by the 26S proteasome without ubiquitination. *Nature* 1992, 360:597-599

[63] Bukau B, Horwich AL: The Hsp70 and Hsp60 chaperone machines. *Cell* 1998, 92:351-366

[64] Mayer MP, Bukau B: Hsp70 chaperones: cellular functions and molecular mechanism. *Cell Mol Life Sci* 2005, 62:670-684

[65] Sherman MY, Goldberg AL: Cellular defenses against unfolded proteins: a cell biologist thinks about neurodegenerative diseases. *Neuron* 2001, 29:15-32

[66] Lee DH, Sherman MY, Goldberg AL: Involvement of the molecular chaperone Ydj1 in the ubiquitin-dependent degradation of short-lived and abnormal proteins in Saccharomyces cerevisiae. *Mol Cell Biol* 1996, 16:4773-4781

[67] Mori H, Kondo J, Ihara Y: Ubiquitin is a component of paired helical filaments in Alzheimer's disease. *Science* 1987, 235:1641-1644

[68] Perry G, Friedman R, Shaw G, Chau V: Ubiquitin is detected in neurofibrillary tangles and senile plaque neurites of Alzheimer disease brains. *Proc Natl Acad Sci U S A* 1987, 84:3033-3036

[69] Bence NF, Sampat RM, Kopito RR: Impairment of the ubiquitin-proteasome system by protein aggregation. *Science* 2001, 292:1552-1555

[70] Gregori L, Fuchs C, Figueiredo-Pereira ME, Van Nostrand WE, Goldgaber D: Amyloid beta-protein inhibits ubiquitin-dependent protein degradation in vitro. *J Biol Chem* 1995, 270:19702-19708

[71] Keck S, Nitsch R, Grune T, Ullrich O: Proteasome inhibition by paired helical filament-tau in brains of patients with Alzheimer's disease. *J Neurochem* 2003, 85:115-122

[72] Oddo S, Billings L, Kesslak JP, Cribbs DH, LaFerla FM: Abeta immunotherapy leads to clearance of early, but not late, hyperphosphorylated tau aggregates via the proteasome. *Neuron* 2004, 43:321-332

[73] van Leeuwen FW, de Kleijn DP, van den Hurk HH, Neubauer A, Sonnemans MA, Sluijs JA, Koycu S, Ramdjielal RD, Salehi A, Martens GJ, Grosveld FG, Peter J, Burbach H, Hol EM: Frameshift mutants of beta amyloid precursor protein and ubiquitin-B in Alzheimer's and Down patients. *Science* 1998, 279:242-247

[74] Lindsten K, de Vrij FM, Verhoef LG, Fischer DF, van Leeuwen FW, Hol EM, Masucci MG, Dantuma NP: Mutant ubiquitin found in neurodegenerative disorders is a ubiquitin fusion degradation substrate that blocks proteasomal degradation. *J Cell Biol* 2002, 157:417-427

[75] Morishima-Kawashima M, Hasegawa M, Takio K, Suzuki M, Titani K, Ihara Y: Ubiquitin is conjugated with amino-terminally processed tau in paired helical filaments. *Neuron* 1993, 10:1151-1160

[76] Yang L, Ksiezak-Reding H: Ubiquitin immunoreactivity of paired helical filaments differs in Alzheimer's disease and corticobasal degeneration. *Acta Neuropathol (Berl)* 1998, 96:520-526

[77] Lin WL, Lewis J, Yen SH, Hutton M, Dickson DW: Filamentous tau in oligodendrocytes and astrocytes of transgenic mice expressing the human tau isoform with the P301L mutation. *Am J Pathol* 2003, 162:213-218

[78] Ikeda M, Shoji M, Kawarai T, Kawarabayashi T, Matsubara E, Murakami T, Sasaki A, Tomidokoro Y, Ikarashi Y, Kuribara H, Ishiguro K, Hasegawa M, Yen SH, Chishti MA, Harigaya Y, Abe K, Okamoto K, St George-Hyslop P, Westaway D:

Accumulation of filamentous tau in the cerebral cortex of human tau R406W transgenic mice. *Am J Pathol* 2005, 166:521-531

[79] Perez N, Sugar J, Charya S, Johnson G, Merril C, Bierer L, Perl D, Haroutunian V, Wallace W: Increased synthesis and accumulation of heat shock 70 proteins in Alzheimer's disease. *Brain Res Mol Brain Res* 1991, 11:249-254

[80] Renkawek K, Voorter CE, Bosman GJ, van Workum FP, de Jong WW: Expression of alpha B-crystallin in Alzheimer's disease. *Acta Neuropathol (Berl)* 1994, 87:155-160

[81] Renkawek K, Bosman GJ, de Jong WW: Expression of small heat-shock protein hsp 27 in reactive gliosis in Alzheimer disease and other types of dementia. *Acta Neuropathol (Berl)* 1994, 87:511-519

[82] Yoo BC, Kim SH, Cairns N, Fountoulakis M, Lubec G: Deranged expression of molecular chaperones in brains of patients with Alzheimer's disease. *Biochem Biophys Res Commun* 2001, 280:249-258

[83] Fonte V, Kapulkin V, Taft A, Fluet A, Friedman D, Link CD: Interaction of intracellular beta amyloid peptide with chaperone proteins. *Proc Natl Acad Sci U S A* 2002, 99:9439-9444

[84] David DC, Layfield R, Serpell L, Narain Y, Goedert M, Spillantini MG: Proteasomal degradation of tau protein. *J Neurochem* 2002, 83:176-185

[85] Shimura H, Miura-Shimura Y, Kosik KS: Binding of tau to heat shock protein 27 leads to decreased concentration of hyperphosphorylated tau and enhanced cell survival. *J Biol Chem* 2004, 279:17957-17962

[86] Kosik KS, Shimura H: Phosphorylated tau and the neurodegenerative foldopathies. *Biochim Biophys Acta* 2005, 1739:298-310

[87] Hatakeyama S, Matsumoto M, Kamura T, Murayama M, Chui DH, Planel E, Takahashi R, Nakayama KI, Takashima A: U-box protein carboxyl terminus of Hsc70-interacting protein (CHIP) mediates poly-ubiquitylation preferentially on four-repeat Tau and is involved in neurodegeneration of tauopathy. *J Neurochem* 2004, 91:299-307

[88] Petrucelli L, Dickson D, Kehoe K, Taylor J, Snyder H, Grover A, De Lucia M, McGowan E, Lewis J, Prihar G, Kim J, Dillmann WH, Browne SE, Hall A, Voellmy R, Tsuboi Y, Dawson TM, Wolozin B, Hardy J, Hutton M: CHIP and Hsp70 regulate tau ubiquitination, degradation and aggregation. *Hum Mol Genet* 2004, 13:703-714

[89] Dou F, Netzer WJ, Tanemura K, Li F, Hartl FU, Takashima A, Gouras GK, Greengard P, Xu H: Chaperones increase association of tau protein with microtubules. *Proc Natl Acad Sci U S A* 2003, 100:721-726

[90] Babu JR, Geetha T, Wooten MW: Sequestosome 1/p62 shuttles polyubiquitinated tau for proteasomal degradation. *J Neurochem* 2005, 94:192-203

[91] Seibenhener ML, Babu JR, Geetha T, Wong HC, Krishna NR, Wooten MW: Sequestosome 1/p62 is a polyubiquitin chain binding protein involved in ubiquitin proteasome degradation. *Mol Cell Biol* 2004, 24:8055-8068

[92] Sadoul R, Fernandez PA, Quiquerez AL, Martinou I, Maki M, Schroter M, Becherer JD, Irmler M, Tschopp J, Martinou JC: Involvement of the proteasome in the programmed cell death of NGF-deprived sympathetic neurons. *EMBO J* 1996, 15:3845-3852

[93] Qiu JH, Asai A, Chi S, Saito N, Hamada H, Kirino T: Proteasome inhibitors induce cytochrome c-caspase-3-like protease-mediated apoptosis in cultured cortical neurons. *J Neurosci* 2000, 20:259-265

[94] Wojcik C: Regulation of apoptosis by the ubiquitin and proteasome pathway. *J Cell Mol Med* 2002, 6:25-48

[95] Canu N, Barbato C, Ciotti MT, Serafino A, Dus L, Calissano P: Proteasome involvement and accumulation of ubiquitinated proteins in cerebellar granule neurons undergoing apoptosis. *J Neurosci* 2000, 20:589-599

[96] Gamblin TC, Chen F, Zambrano A, Abraha A, Lagalwar S, Guillozet AL, Lu M, Fu Y, Garcia-Sierra F, LaPointe N, Miller R, Berry RW, Binder LI, Cryns VL: Caspase cleavage of tau: linking amyloid and neurofibrillary tangles in Alzheimer's disease. *Proc Natl Acad Sci U S A* 2003, 100:10032-10037

[97] Rissman RA, Poon WW, Blurton-Jones M, Oddo S, Torp R, Vitek MP, LaFerla FM, Rohn TT, Cotman CW: Caspase-cleavage of tau is an early event in Alzheimer disease tangle pathology. *J Clin Invest* 2004, 114:121-130

[98] Bush KT, Goldberg AL, Nigam SK: Proteasome inhibition leads to a heat-shock response, induction of endoplasmic reticulum chaperones, and thermotolerance. *J Biol Chem* 1997, 272:9086-9092

[99] Kawazoe Y, Nakai A, Tanabe M, Nagata K: Proteasome inhibition leads to the activation of all members of the heat-shock-factor family. *Eur J Biochem* 1998, 255:356-362

[100] Brown MR, Bondada V, Keller JN, Thorpe J, Geddes JW: Proteasome or calpain inhibition does not alter cellular tau levels in neuroblastoma cells or primary neurons. *J Alzheimers Dis* 2005, 7:15-24

[101] Delobel P, Leroy O, Hamdane M, Sambo AV, Delacourte A, Buee L: Proteasome inhibition and Tau proteolysis: an unexpected regulation. *FEBS Lett* 2005, 579:1-5

[102] Feuillette S, Blard O, Lecourtois M, Frebourg T, Campion D, Dumanchin C: Tau is not normally degraded by the proteasome. *J Neurosci Res* 2005, 80:400-405

[103] Adamec E, Mohan P, Vonsattel JP, Nixon RA: Calpain activation in neurodegenerative diseases: confocal immunofluorescence study with antibodies specifically recognizing the active form of calpain 2. *Acta Neuropathol (Berl)* 2002, 104:92-104

[104] Shea TB, Spencer MJ, Beermann ML, Cressman CM, Nixon RA: Calcium influx into human neuroblastoma cells induces ALZ-50 immunoreactivity: involvement of calpain-mediated hydrolysis of protein kinase C. *J Neurochem* 1996, 66:1539-1549

[105] Litersky JM, Johnson GV: Phosphorylation by cAMP-dependent protein kinase inhibits the degradation of tau by calpain. *J Biol Chem* 1992, 267:1563-1568

[106] Baldin V, Cans C, Knibiehler M, Ducommun B: Phosphorylation of human CDC25B phosphatase by CDK1-cyclin A triggers its proteasome-dependent degradation. *J Biol Chem* 1997, 272:32731-32734

[107] Wu Y, Luo H, Kanaan N, Wu J: The proteasome controls the expression of a proliferation-associated nuclear antigen Ki-67. *J Cell Biochem* 2000, 76:596-604

[108] Tao GZ, Rott LS, Lowe AW, Omary MB: Hyposmotic stress induces cell growth arrest via proteasome activation and cyclin/cyclin-dependent kinase degradation. *J Biol Chem* 2002, 277:19295-19303

[109] Pope WB, Lambert MP, Leypold B, Seupaul R, Sletten L, Krafft G, Klein WL: Microtubule-associated protein tau is hyperphosphorylated during mitosis in the human neuroblastoma cell line SH-SY5Y. *Exp Neurol* 1994, 126:185-194

[110] Preuss U, Doring F, Illenberger S, Mandelkow EM: Cell cycle-dependent phosphorylation and microtubule binding of tau protein stably transfected into Chinese hamster ovary cells. *Mol Biol Cell* 1995, 6:1397-1410

[111] Vincent I, Jicha G, Rosado M, Dickson DW: Aberrant expression of mitotic cdc2/cyclin B1 kinase in degenerating neurons of Alzheimer's disease brain. *J Neurosci* 1997, 17:3588-3598

[112] Ding XL, Husseman J, Tomashevski A, Nochlin D, Jin LW, Vincent I: The cell cycle Cdc25A tyrosine phosphatase is activated in degenerating postmitotic neurons in Alzheimer's disease. *Am J Pathol* 2000, 157:1983-1990

[113] Husseman JW, Nochlin D, Vincent I: Mitotic activation: a convergent mechanism for a cohort of neurodegenerative diseases. *Neurobiol Aging* 2000, 21:815-828

[114] Patrick GN, Zukerberg L, Nikolic M, de la Monte S, Dikkes P, Tsai L-H: Conversion of p35 to p25 deregulates Cdk5 activity and promotes neurodegeneration. *Nature* 1999, 402: 615-622.

[115] Lee MS, Kwon YT, Li M, Peng J, Friedlander RM, Tsai L-H: Neurotoxicity induces cleavage of p35 to p25 by calpain. *Nature*, 2000, 405: 306-404

[116] Yoo BC and Lubec G: Neurobiology: p25 protein in neurodegeneration. *Nature*, 2001, 411: 763-764.

[117] Tandon A, Yu H, Wang L, Rogaeva E, Sato C, Chisti MA, Kawarai T, Hasegawa H, Chen F, Davies P, Fraser PE, Westaway D, St George-Hyslop PH: Brain levels of Cdk5 activator p25 are not increased in Alzheimer's or other neurodegenerative diseases with neurofibrillary tangles. *J. Neurochem*, 2003, 86(3): 572-581.

In: The Ubiquitin Proteasome System… ISBN 978-1-60021-749-4
Eds: Mario Di Napoli and Cezary Wojcik, pp. 761-791

Chapter 31

UBIQUITIN-DEPENDENT PROTEOLYSIS IN PARKINSON'S DISEASE

***Jian Feng*[1,*], *Thomas Schmidt-Glenewinkel*[2] *and Mario Di Napoli*[3,4]**
[1]Department of Physiology and Biophysics, State University of New York at Buffalo, Buffalo, NY 14214, USA;
[2]Department of Biological Sciences, Hunter College of CUNY, New York, NY10021, USA;
[3]Neurological Section, SMDN – Center for Cardiovascular Medicine and Cerebrovascular Disease Prevention, Sulmona (L'Aquila) I-67039, Italy;
[4]Neurological Service, San Camillo de'Lellis General Hospital, Rieti I-02100, Italy.

ABSTRACT

Parkinson's disease (PD) is a movement disorder characterized by the selective loss of dopaminergic (DA) neurons in substantia nigra. It is generally believed that a combination of environmental and genetic factors underlie the selective death of these DA neurons and ensuing locomotor symptoms. Significant breakthroughs in human genetic studies have recently led to the identification of several genes linked to PD. Among these genes, α-synuclein, parkin, UCH-L1 and DJ-1 play diverse roles in ubiquitin-dependent proteolysis by the 26 S proteasome, while the functions of PINK1 and LRRK2 are still largely unknown. Pathogenic mutations of α-synuclein enhance its propensity to misfold and aggregate. Parkin has a protein-ubiquitin E3 ligase activity towards a variety of substrates. When parkin is mutated, accumulation of its substrates may significantly contribute to the demise of dopaminergic neurons. Ubiquitin Carboxyl-terminal Hydroxylase L1 (UCH-L1) is a brain-specific deubiquitinating enzyme, whose catalytic activity is significantly reduced by its mutations found in a few PD cases. Among the many functions of DJ-1, its ability to counteract reactive oxygen species

* Correspondence concerning this article should be addressed to Dr. Jian Feng, PhD; Department of Physiology and Biophysics, State University of New York at Buffalo, 124 Sherman Hall, Buffalo, NY 14214, USA. Phone: 1-716-829-2345; Fax: 1-716-829-2699 ; E-mail: jianfeng@buffalo.edu.

appears to be critically involved in PD. Mutations of DJ-1 greatly affect its stability and dimerization, as well as interactions with a variety of proteins including parkin. The disparate functions of these PD-linked proteins all fall within the framework of how the cell handles misfolded and aggregated proteins. An emerging common theme is that unfolded or misfolded proteins, if not promptly removed through ubiquitin-dependent proteolysis by the 26 S proteasome, may induce Unfolded Protein Response (UPR). Cells activate the UPR program to increase the production of proteins that help to handle misfolded or unfolded proteins. Another cellular program that is operating in parallel and perhaps connected with UPR is the ability of the cell to accumulate misfolded proteins in the form of aggresomes. The formation of a single large inclusion at the centrosome area greatly minimizes the impact of dispersed aggregates of misfolded proteins. Increasing evidence suggests that the cell may activate autophagy to degrade proteins in the aggresome. Thus, a unifying theme may connect the diverse functions of several PD-linked genes to the three sets of overlapping, interdependent cellular programs: UPR, Aggresome Formation and Autophagy. Cell death could be triggered when these protective mechanisms fail. The greatest challenge is to understand the selectivity of cell death. The general functions of the ubiquitin-dependent proteolysis system (UPS), as well as those of the three protective cellular programs, must be considered within the unique cellular and physiological context of dopaminergic neurons to ultimately answer the question. Superimposed on this complexity is the impact of various toxins implicated in PD. Although this multilayered view may be an oversimplification of the involvement of UPS in PD, it allows us to summarize many pieces of seemly unrelated information into a coherent model that can be tested experimentally. Future explorations under this framework would provide us a more comprehensive view on the molecular and cellular basis of PD.

Keywords: Parkinson's disease, parkin, autophagy, α-Synuclein, UCH-L1 (ubiquitin carboxy-terminal hydrolase L1), DJ-1, Pink1 (PTEN-induced kinase 1), Leucine-rich repeat kinase 2 (LRRK2).

Abbreviations

Aβ, β-amyloid protein; DA, dopaminergic; ER, endoplasmic reticulum; CSPα, cysteine-string protein α; IBR, In-between RING fingers; LAMP2A, lysosome-associated membrane protein 2A; LB, Lewy bodies; LRRK-2:, Leucine-rich related kinase 2; MAO, Monoamine oxidase; MTOC, microtubule organizing centers; NAC, non-amyloid component; Nurr1, Nuclear – receptor related 1; PD, Parkinson disease; PINK1, PTEN – induced kinase 1; RING, Really Interesting New Gene; ROS, reactive oxygen species; SNpc substantia nigra pars compacta; UCH-L1: Ubiquitin-carboxy terminal hydrolase L1; UPR; unfolded protrein response; UPS, ubiquitin-dependent proteolysis system.

Introduction

Parkinson Disease (PD) is one of the most common neurodegenerative disorders, afflicting 1~2% of population over 65 years of age [1]. Despite the wide range of clinical manifestations associated with various forms of PD, a core set of motor symptoms, including resting tremor, muscle rigidity, bradykinesia, gait disturbance and postural instability, distinguish this disease from other movement disorders. The most prevalent and unifying pathological feature of PD is the degeneration of dopaminergic neurons in substantia nigra pars compacta (SNpc), which leads to reduced dopamine release to the striatum, a brain region critical for voluntary locomotor activities [2]. Although neurodegeneration also occurs in other brain areas, the loss of dopaminergic neurons in SNpc is far more pervasive and appears to contribute most significantly to the symptoms of PD [8]. Dopamine replacement therapy with L-DOPA/carbidopa significantly alleviates clinical symptoms in many PD patients, especially at the early stage. The obligatory role of nigral dopaminergic neurons in PD is further substantiated by lesion studies using neurotoxins such as MPTP (1,2,3,6-methyl-phenyl-tetrahydropyridine) [9], 6-OHDA (6-hydroxydopamine) [10] and rotenone [11]. These toxins selectively destroy dopaminergic neurons in SNpc, and produce symptoms remarkably similar to those of PD. Together these lines of evidence strongly suggest that the pathological hallmark and direct cause of PD is the degeneration of dopaminergic neurons in substantia nigra.

A variety of genetic and environmental factors underlie the selective demise of dopaminergic neurons in PD. The majority of PD cases occur sporadically, with no obvious inheritance pattern. A large number of epidemiological studies [11,17], especially those performed on twins [18], reveal strong environmental connections to PD. The concordance rates are virtually identical in monozygotic and dizygotic twins with age at onset older than 50. However, in PD twin pairs with age at onset younger than 50, the concordance rate in monozygotic twins (100%) is significantly higher than that in dizygotic twins (17%) [18]. This study provides very clear evidence that the common, sporadic forms of late-onset PD are highly influenced by environmental factors, while the early-onset forms of PD have a strong genetic basis. Among the environmental factors studied, the use of pesticides and herbicides has consistently been found to be a significant risk factor [19]. Recent studies in rats, using pesticides such as rotenone [11] or paraquat (*N,N'*-dimethyl-4,4'-bipyridylium) and maneb (manganese ethylenebisdithiocarbamate) [24], have demonstrated that PD-like symptoms and degeneration of DA neurons do occur in animals submitted to long-term treatment with these pesticides.

In terms of genetic factors, mutations of five genes have been definitively linked to familial forms of PD [25]. Gain-of-function mutations of α-synuclein or triplication of the wild-type allele are linked to a rare, early-onset form of PD [26]. Dominant mutations of the LRRK2 gene appear to be a frequent cause of familial PD. On the other hand, mutations of parkin, DJ-1 and PINK1 cause PD largely in a recessively manner. Among recessively-inherited PD cases, mutations of parkin seem to be most prevalent [25]. In addition to these genes, mutations of UCH-L1, NR4A2 (Nurr1) and synphilin-1 have been implicated in rare cases of PD. Furthermore, several loci have been linked to familial forms of PD, although the responsible genes have not been identified.

Many lines of evidence have suggested that both environmental and genetic factors associated with PD impinge on several common pathways that are critical to the survival and death of dopaminergic neurons. These include impairment of the Ubiquitin-Proteasome System (UPS) and ensuing stress caused by misfolded proteins, mitochondrial dysfunction, and oxidative stress. Among the genes linked to familial forms of PD, many are directly involved in the UPS and UPR misfolded protein stress. For example, parkin encodes for a protein-ubiquitin E3 ligase, and is responsible for the ubiquitination and degradation of a variety of substrates [30]. UCH-L1 is an abundant neuronal deubiquitinating enzyme that catalyze the hydrolysis of polyubiquitin chain into free ubiquitin monomer [31], which however also shows a ubiquitin ligase activity [32]. Mutations of α-synuclein or overexpression of the wild-type protein lead to fibrils and aggregates, which induce UPR [37]. On the other hand, environmental PD toxins such as rotenone disrupt mitochondrial functions by inhibiting complex I of the mitochondrial respiratory chain. Very interestingly, loss of parkin in *Drosophila* causes defects in mitochondrial morphology [38], although the underlying mechanism is unclear. Both complex I inhibition and dopamine metabolism produce reactive oxygen species that would cause oxidative stress in dopaminergic neurons. Overexpression of parkin significantly reduces dopamine-induced cell death by attenuating the production of reactive oxygen species, while DJ-1 functions as an anti-oxidant in response to oxidative stress. This chapter will focus on the involvements of PD-linked genes on ubiquitin-dependent proteolysis. By examining this particular pathway, we hope to offer some clues on the molecular and cellular mechanisms of PD.

Genetic Mutations and Misfolded Proteins Involved in Parkinson's Disease

One of the pathological hallmarks of PD is the formation of intracellular inclusions called Lewy bodies (LB), eosinophilic hyaline structures frequently found in degenerating neurons including dopaminergic neurons in substantia nigra. LB are immunoreactive with anti- ubiquitin antibodies and contain many misfolded proteins including protein products of genes linked to PD. Evidence accumulated in the past few years has increasingly suggested the critical involvement of misfolded proteins in the molecular etiology of PD, particularly various familial forms that are very often associated with early age at onset. We will discuss the roles of genetic mutations and the corresponding misfolded proteins in the molecular pathogenesis of PD.

α-Synuclein

The first discovered gene causatively linked to familial PD is α-synuclein (Figure 1), which encodes a small protein of 140 amino acid residues [39]. So far, three point mutations of this gene (A30P, A53T and E46K) have been linked to a rare form of early-onset autosomal dominant PD [40]. In addition, triplication of a genomic region containing the wild-type α-synuclein gene has been shown to segregate with PD phenotypes in a kindred from Iowa [26]. Similarly, duplication of the α-synuclein locus has also been reported in French and Italian families with autosomal-dominant PD [43]. Subsequent studies using much large number of PD families have failed to identify multiplication of the α-synuclein gene (SCNA), suggesting that such genomic changes are rare phenomena [44]. Nevertheless, this remarkable discovery mirrors the association of trisomy 21 (Down syndrome) with Alzheimer's disease [45]. Triplication of chromosome 21 has long been linked to the occurrence of Alzheimer's disease by the fifth decades of these patients [46]. This association prompted the landmark discovery of β-amyloid protein (Aβ) by Glenner and Wong from cerebrovascular amyloid deposits in patients with Down syndrome [50]. The similar link between α-synuclein triplication and PD suggest that overexpression of the wild-type α-synuclein causes PD and may be particularly toxic to nigral dopaminergic neurons. This notion is corroborated by the recent finding that genetic variations in the promoter region of the α-synuclein gene are associated with sporadic PD, which suggests that expression levels of wild-type α-synuclein affect the chance of getting PD [51].

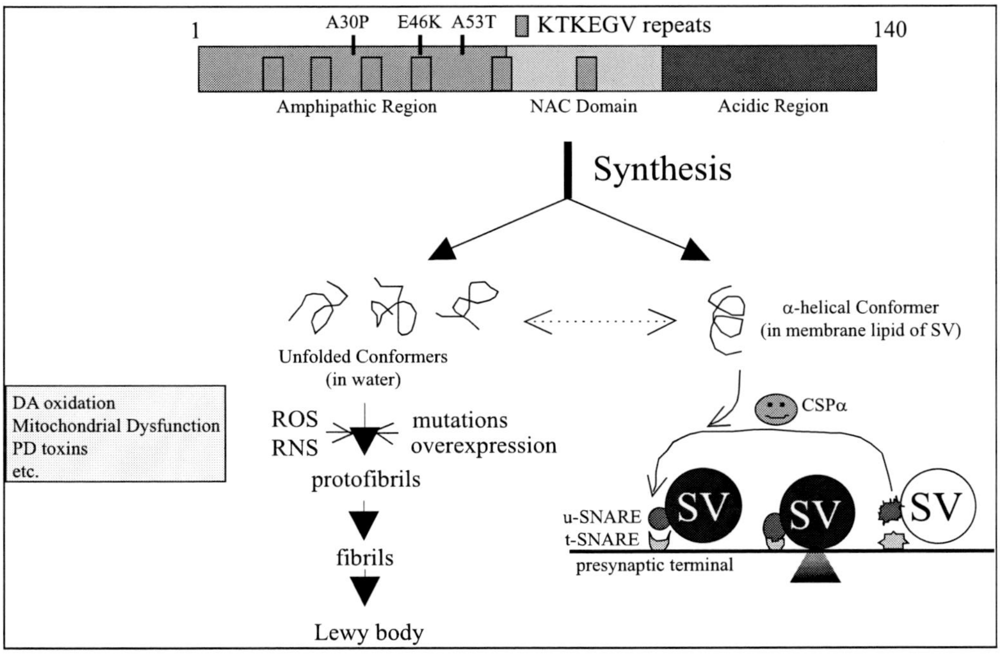

Figure 1. Biochemical properties and in vivo functions of α-synuclein. Domains of α-synuclein and positions of point mutations linked to PD are shown in the diagram. When α-synuclein is produced, it may assume different conformations depending on its environment. In aqueous solution, α-synuclein is disordered and prone to aggregate into protofibrils, fibrils, and eventually a prominent intracellular inclusion called Lewy body. Reactive oxygen species (ROS) and reactive nitrogen species (RNS) produced in dopaminergic neurons, from dopamine oxidation, mitochondrial dysfunction or the actions of PD toxins may increase the propensity of α-synuclein aggregation by chemically modifying the protein. On the other hand, α-synuclein avidly binds to membrane phospholipids and exhibits α-helical conformation in such an environment. Through an unclear mechanism, α-synuclein cooperates with cysteine string protein α (CSPα), a chaperone responsible for the folding and refolding of SNARE proteins involved in synaptic vesicle recycling. This function of α-synuclein allows it to genetically complement the obligatory roles of CSPα in synaptic vesicle recycling, neuronal survival and viability of the mice.

Together with β-synuclein, γ-synuclein and synoretin, α-synuclein belong to a family of small proteins whose functions still remain unclear despite extensive studies [52,53]. α-synuclein has an amphipathic N-terminal region that contains six imperfect repeats of the consensus sequence KTKEGV. This is followed by a hydrophobic middle region containing the NAC (non-amyloid component) domain. The C-terminal region of α-synuclein is quite acidic. Numerous studies on α-synuclein have reveal many different aspects of its physiological functions. A series of recent papers using α-synuclein knockout mice have generated a surprising yet converging theme of the *in vivo* function of this enigmatic small protein.

α-synuclein is ubiquitously expressed in the brain and is associated with synaptic vesicles and membraneous structures in the presynaptic terminals. Its genetic ablation in mice resulted in mild defects in the recycling of dopamine vesicles and dopaminergic transmission [54]. No gross neuroanatomical or neurochemical change was seen in α-synuclein knockout mice [54,55]. However, transgenic overexpression of α-synuclein abolishes the severe degeneration of nerve terminals seen in cysteine-string protein α (CSPα) knockout mice, whereas genetic ablation of α-synuclein exacerbates the phenotypes of CSPα knockouts [56]. CSPα acts as a chaperone to facilitate the folding and refolding of synaptic SNARE proteins that are required for fusion of synaptic vesicles with presynaptic terminal [60] . In the absence of CSPα, defects in synaptic vesicle recycling leads to progressive neurodegeneration at 2-3 weeks postnatal and death of the animal at 1-4 months [61]. When α-synuclein is overexpressed in these animals, it rescues the deficiencies caused by the lack of CSPα apparently through a downstream mechanism that requires the binding of α-synuclein to phospholipids [56]. Like other synucleins, α-synuclein is a soluble protein with unfolded conformations at the native state. It strongly binds to negatively-charged phospholipids and assumes α-helical conformation upon such binding [62]. α-synuclein binds to and inhibit phospholipase D [66], prevents lipid droplets from hydrolysis [67] and regulates vesicle trafficking in yeast [70]. Together, these converging lines of evidence suggest that a key physiological function of α-synuclein is to facilitate synaptic vesicle recycling at the presynaptic terminal through interaction with membrane phospholipids.

The pathogenic properties of α-synuclein seems to be linked to its flexible conformations in aqueous environment. Wild-type α-synuclein is prone to aggregate due to its disordered structure. Overexpression of the wild-type gene or its PD-linked mutations increase this propensity [71].

In dopaminergic neurons, reactive oxygen species and dopamine quinones produced during dopamine metabolism may modify α-synuclein and exacerbate its misfolding and aggregation. Thus, genetic mutations of α-synuclein (point mutations, multiplication of the locus, or variations in its promoter sequence) may cause selective degeneration of DA neurons through accelerated misfolding and aggregation due to oxidative stress induced by dopamine oxidation. This condition may be exacerbated by the defects in synaptic vesicle recycling in the absence of functional α-synuclein. It should be noted that overexpression of wild-type α-synuclein may drive the proteins into aggregated forms, away from phospholipid environment on the membrane, where the α-helical conformers of the protein normally cooperate with CSPα to facilitate the folding and refolding of SNARE proteins required for vesicle recycling at the presynaptic terminal. Consistent with this model, the A30P mutant α-synuclein, which does not bind to phospholipids, also fails to rescue CSPα knockout mice, whereas the A53T mutant, which retain the ability to bind phospholipids, rescues CSPα knockout mice as well as the wild-type α-synuclein does [56].

Parkin

Among the PD-linked genes identified so far, mutations of parkin are by far the most frequent cause of recessive PD [40,72]. In addition, parkin mutations, presumably at the heterozygous state, play a significant role in idiopathic PD, especially in cases with an early age at onset [75,76]. These lines of evidence demonstrate that parkin is one of the most prevalent genetic factors in PD.

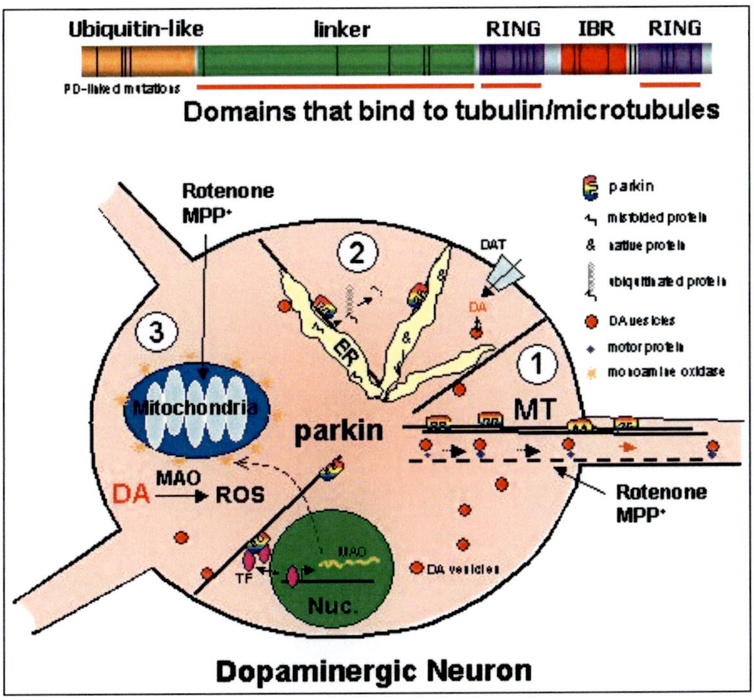

Figure 2. Cellular functions of parkin. Functional domains and locations of PD-linked point mutations are shown in the top diagram. Parkin has three domains (linker, RING1 and RING2) that provide strong, independent and redundant binding to tubulin and microtubules. This feature suggests that parkin exerts its cellular functions in the context of microtubules. Minimally, three distinct functions of parkin are linked to its interactions with microtubules. First, the binding between parkin and microtubules stabilizes the microtubule network against depolymerizing agents including PD toxins rotenone and MPP^+. Second, the location of parkin on microtubules and the attachment of ER to microtubules suggest that parkin could efficiently ubiquitinate misfolded transmembrane protein substrates (e.g. DAT), which have to be retrotranslocated to the cytosol for ubiquitination and degradation. Third, the strong binding between parkin and microtubules enable parkin to serve as a cytosolic anchor for transcription factors that regulate the expression of certain mitochondrial proteins (e.g. MAO). All three functions of parkin are directly linked to PD through stabilizing microtubule network for vesicular transport (in 1), enhancing the precision of dopaminergic transmission (in 2), and limiting dopamine oxidation by suppressing the expression of MAO (in 3).

Parkin is a protein-ubiquitin E3 ligase [77], which ubiquitinates proteins for their subsequent degradation by the 26S proteasome [78,79] (Figure 2). Recent studies from many groups, including ours, have identified a number of parkin substrates such as CDCRel-1 [82], Pael receptor [83], O-glycosylated α-synuclein [86], synphilin-1 [88], cyclin E [89], α- and β-tubulin [90], expanded polyglutamine [93], p38 [96], etc. The diversity of these substrates suggests that parkin may ubiquitinate many targets in different types of cells, as parkin and most of these substrates are widely expressed [97].

Our previous studies have shown that parkin ubiquitinates α- and β-tubulin and accelerates their degradation by the 26S proteasome; these effects are abolished by PD-linked mutations of parkin [90]. We have also found that parkin tightly binds to microtubules and tubulin through redundant strong interactions mediated by three independent domains of parkin (linker, RING1 and RING2) [101]. In taxol-mediated microtubule co-assembly assays, almost all parkin exists as a complex with microtubules in the pellet fraction. The binding between parkin and microtubules is so strong that they cannot be separated even with 3.8M of NaCl; it co-purifies with tubulin and is found in >99% pure tubulin preparation [101]. Consistent with this tight interaction, parkin exhibits punctuate subcellular localization along microtubules [90].

The ability of parkin to ubiquitinate α- and β-tubulin and to facilitate their degradation may be very important for cell survival. The synthesis of α- and β-tubulin polypeptides are very tightly regulated at the transcriptional and translational levels to ensure the equimolar production of both tubulins [104], as overexpression of either tubulin gene is toxic to the cell [105,106]. The formation of polymerization-competent α/β heterodimers requires a series of folding reactions that are perhaps the most complicated for all known proteins (reviewed in [109]. First, α or β polypeptide is folded into quasi-native conformers with the help of cytosolic chaperonins, which are ribosome-sized multisubunit complexes that facilitate protein folding in an ATP-dependent manner. Second, α and β monomers are captured by tubulin-specific folding cofactors. In a sequential, coordinated and reversible folding process catalyzed by cofactors A through E, α/β heterodimers are formed with the hydrolysis of GTP. Misfolded tubulins produced during this complicated process are quickly degraded through an unknown mechanism. As an E3 ligase for tubulins, parkin may well ubiquitinate misfolded tubulins to facilitate their degradation. Because the tubulin folding process is dependent on ATP and GTP hydrolysis, mitochondrial complex I-inhibiting PD toxins that can reduce ATP production (e.g. rotenone and MPP^+) may adversely affect the folding reaction and lead to increased production of misfolded tubulin. In addition, the ability of these PD toxins to depolymerize microtubules [113-116] would further increase the amount of tubulin that needs to be degraded. It has been known for a long time that microtubule depolymerization leads to rapid degradation of tubulin, a protein that normally has a very long half-life [117]. By ubiquitinating and degrading tubulin, parkin could prevent the same kind of toxicity caused by overexpression of tubulin [105,106]. Because nigral DA neurons have very long axons projecting to striatum, a very high percentage of total cell volume is in the axon, which contains large quantities of microtubules. Thus, exposure of these PD toxins may result in much higher demand to ubiquitinate and degrade tubulin in nigral DA neurons than that in other types of cells with smaller volume and shorter processes. Our studies have shown the TH^+ neurons in midbrain neuronal cultures are much more vulnerable than TH^-

neurons to microtubule depolymerizing agents such as rotenone, colchicine or nocodazole [118].

Many parkin substrates, such as Pael-R [83], DAT [119], synaptotagmin XI [120], CDCrel-1 [82], are transmembrane proteins or membrane-associated proteins. At least some of these proteins are prone to misfold in the endoplasmic reticulum (ER) [83,119], which causes UPR if left unchecked [121] (see Chapter 13). Previous studies have demonstrated that misfolded membrane proteins are retroversely translocated from the ER to the cytosol, where they must be immediately ubiquitinated to avoid aggregation due to the abundance of hydrophobic residues left exposed by the disordered polypeptide chain [122]. Under normal situations, the ER is attached to microtubules to maintain its morphology and stability [123-125]. The proximity of the ER to parkin, which is anchored on microtubules [90,126], gives parkin ideal accssess to misfolded substrates as they are retrotranslocated from the ER.

Many misfolded proteins are transported along microtubules to the proteolytic center of the cell, a perinuclear area around the centrosome, where under conditions of impaired proteolysis area to form a single large inclusion is formed [127], termed the 'aggresome' [128] (see Chapter 12). The localization of parkin along microtubules [90] would greatly facilitate the ubiquitination of its substrates and their subsequent transport on microtubules to the proteolytic centeraggresome. Previous studies have shown that parkin and its substrates, such as CDCrel-1, Pael-R and DAT, are accumulated in the perinuclear aggresomesarea when protein degradation is compromised by proteasome inhibitors or an excess of misfolded proteins themselves [119,129,130]. Accumulation of these proteins around the centrosome is a microtubule-dependent process, since disruption or overt stabilization of microtubules abolishes this phenomenon [130]. It appears that parkin anchored on microtubules may serve as sentinels to efficiently ubiquitinate misfolded proteins for their destruction by the 26S proteasome or transportation along microtubules to the aggresome, proteasomes should be overwhelmed by misfolded proteins. Thus, a unique combination of several features – parkin as an E3 ligase anchored on microtubules, the attachment of ER to microtubules, and the need to retrotranslocate misfolded transmembrane proteins from the ER for ubiquitination and degradation in the cytosol – makes parkin ideally suited to ubiquitinate transmembrane or membrane-associated substrates. The ubiquitination of DAT by parkin appears to be a novel mechanism to regulate the functions of DAT, which are critical to dopaminergic transmission. In addition to these functions that are dependent on the E3 ligase activity of parkin, our recent studies have shown that parkin suppresses the transcription of monoamine oxidases through an E3-independent mechanism [131]. Monoamine oxidase has two isoforms, MAO-A and MAO-B, which are encoded by two distinct genes [132]. Both MAOs are located on the cytoplasmic side of the mitochondrial outer membrane through a C-terminal tail anchored in the membrane [133,134]. MAO-catalyzed oxidative deamination of monoamines produces H_2O_2 that has been shown to damage mitochondrial DNA [135]. Thus, ROS produced during dopamine oxidation can affect mitochondrial functions and may lead to reduced ATP production and increased generation of ROS, if the electron transfer chain is disrupted. Several lines of evidence have indicated a strong involvement of MAO in PD. It has been found that PD patients have elevated MAO-B activity in the substantia nigra [136]. Consistent with these, MAO inhibitors such as deprenyl have been widely used to delay the progression of PD symptoms [137]. On the other hand, MAO-B knockout mice are resistant to the PD toxin MPTP [138], because MAO-B is responsible for the conversion of MPTP to MPP^+ [139], the active toxin that is selectively taken up by DA neurons through the dopamine transporter [140]. Thus, results from both basic research and clinical studies point to MAO as an important factor in PD pathogenesis. Our studies have shown that overexpression of parkin in the human dopaminergic neuroblastoma cell line SH-SY5Y or mouse fibroblast NIH3T3 cells significantly reduces the mRNA levels of both MAO-A and MAO-B. In B-lymphocyte cells obtained from a PD family with deletion of parkin exon 4, levels of both MAOs are significantly increased in the homozygous, but not heterozygous, carrier of this mutation [131].

This study complements several studies that have indicated the involvement of parkin in mitochondrial functions. In parkin-deficient *Drosophila*, the flight and climbing problems are caused by apoptotic degeneration of indirect flight muscles, which is accompanied by abnormal mitochondrial morphology [38]. Mitochondrial defects are also found in spermatids of male mutant flies, which are sterile due to the lack of mature sperm [38]. Consistent with this, overexpression of parkin in PC12 cells delays ceramide-induced mitochondrial swelling and subsequent release of cytochrome C and apoptosis [141]. Parkin is shown to be associated with mitochondria and may interact with an unknown target that is critical for mitochondrial functions [141]. A recent report shows that parkin enhances mitochondrial biogenesis in proliferating cells by regulating the expression of many mitochondrial proteins [142].

How does parkin, a cytosolic protein, regulate the gene expression of mitochondrial proteins such as MAOs? Parkin has a RING-IBR-RING motif in the C-terminus [143]. RING (Really Interesting New Gene) finger is a variation of zinc finger, while IBR (In-between RING fingers) domain is a zinc finger-like, putative metal-binding motif found between two RING fingers in a group of proteins implicated in transcription regulation. One member in the family, RBCK1 binds to DNA and activates transcription in *in vitro* assays [144,145]. More recently, parc (p53-associated, parkin-like cytoplasmic protein) has been found to contain the RING-IBR-RING motif in the C-terminus. Although parc also has E3 ligase activity and can ubiquitinate itself, it does not ubiquitinate p53, to which it binds strongly. Parc suppresses p53-dependent gene transcription and apoptosis by tethering it in the cytosol [146]. Previous studies have also shown that p53 is anchored in the cytoplasm through interaction with microtubules [147]. It seems likely that parc and microtubules may act in a concerted manner to sequester a portion of p53 in the cytoplasmic pool, away from its transcriptional activities in the nucleus [148]. Our previous studies have found that parkin binds to microtubules strongly and appears to be absent in the nucleus [90]. In a similar manner to parc, parkin might tether a transcription factor for mitochondrial proteins in the cytosol to regulate their transcription.

Thus, a general picture about cellular functions of parkin appears to center on its unusually strong interaction with tubulin and microtubules. Three different but connected functions of parkin need to be considered within the context of its interactions with tubulin and microtubules. First, the strong binding between parkin and microtubules stabilizes the microtubule network against depolymerizing agents [126], which exert much higher toxicity on dopaminergic neurons than non-dopaminergic neurons [118]. The unique combination of cell morphology (long axons enriched with microtubules) and neurochemistry [production of reactive oxygen species (ROS) during dopamine metabolism] of dopaminergic neurons renders these cells particularly vulnerable to PD toxins such as rotenone and MPP+, which depolymerize microtubules and inhibit mitochondrial respiratory chain. Since microtubule depolymerization induces rapid degradation of tubulin [117], the ability of parkin to ubiquitinate and degrade tubulin would also be critical for the survival of the cell in the presence of these PD toxins. Second, the effectiveness of parkin to handle misfolded transmembrane proteins such as DAT appears to be related to several important facts: (*i*) misfolded transmembrane proteins need to be retrotranslocated from the ER to the cytosol to be ubiquitinated and degraded; (*ii*) ER is attached to microtubules; (*iii*) parkin binds strongly to microtubules and exhibits punctated localization along microtubules. It seems plausible that parkin, which is anchored on microtubules, may act as sentinels to efficiently ubiquitinate misfolded transmembrane proteins as they are reverse translocated from the ER to the cytosol. If this process goes awry (e.g. when parkin is mutated and loses its E3 ligase activity), these misfolded transmembrane substrates of parkin would easily aggregate and trigger UPR. In the case of dopamine transporter, parkin increases its surface expression by ubiquitinating and degrading misfolded DAT molecules [119]. This allows native DAT conformers to oligomerize with each other into functional transporters destined for the plasma membrane. This function of parkin increases dopamine uptake and thereby enhances the temporal and spatial precision of dopaminergic transmission. Third, the ability of parkin to regulate expression of mitochondrial proteins such as MAO may be mediated by sequestration of key transcription factor(s) in the cytosol. It is quite clear at least in the case for MAO that, parkin affects gene expression through a mechanism independent of its E3 ligase activity [131]. Instead, parkin may serve as a cytosolic anchor to tether transcription factors away from DNA and transcriptional machinery in the nucleus. By suppressing the expression level of MAO, parkin may limit the production of reactive oxygen species generated during dopamine oxidation by MAO.

UCH-L1

UCH-L1 (ubiquitin carboxy-terminal hydrolase L1) was originally identified in neurons and testis of mammals [149] but is known now to occur as widespread as *Aplysia* [150] and *Drosophila* [151]. The enzyme is a member of a family of ubiquitin hydrolases which in mammals consists of UCH-L1, UCH-L3, UCH-5 and BAP1. UCH-L1 is one of the most abundant proteins in the mammalian brain (1-2% of total brain protein content) [149,152]. The primary function of the enzyme is the deubiquitination of C-terminal esters and amides of ubiquitin resulting in monomeric ubiquitin, an essential step prior to degradation of polyubiquitinated proteins by the proteasome.

UCH-L1 has been implicated in the pathogenesis of PD because of a misense mutation (I93M) in the fourth exons of the uch-l1 gene which was identified in a German family, leading to an increased susceptibility for PD. This mutation reduces enzyme activity by about 50% [153]. It has also been identified as a constituent of LB [154]. Curiously enough, another polymorphism in the uch-l1 gene (S18Y) in humans results in a decreased susceptibility for PD [32,155-157]. The S18Y mutation in the enzyme increases the hydrolase activity when compared with the wild-type enzyme [158].

However, it has been observed that UCH-L1 in its dimeric form has ubiquityl ligase activity with a potential for polyubiquitinating K63 of α-synuclein resulting in an elevated cytoplasmatic concentration of this protein [32]. The effect of the S18Y mutation is a reduction in dimerization while maintaining almost wild-type or higher hydrolase activity which would explain the reduced risk for PD.

The issue of involvement of UCH-L1 in PD is further complicated by the fact that complete lack of UCH-L1 activity as found in the *gad* mouse does not result in a Parkinson phenotype in the mouse. The gad mouse carries a spontaneous, autosomal recessive mutation in the Uchl1 gene because of an intragenic deletion of exons 7 and 8 resulting in an Uchl1 null mutant [159]. Analysis of the clinical phenotype demonstrates severe sensory ataxia at the early stage of the disease followed by motor paresis [160]. The underlying defect in the *gad* mouse is axonal degeneration in the gracile tract, the formation of spheroid bodies in nerve terminals and progressive accumulation of ubiquitinated proteins [161]. The loss of UCH-L1 function might lead to a decreased pool of available ubiquitin which in turn would interfere with protein degradation via the UPS pathway.

DJ-1

A possible role for the gene DJ-1 in PD was established when mutations in DJ-1 were identified in the PARK7 locus, causing an early onset form of PD [162]. The recessive mutation is quite rare, being responsible for an estimated 1-2% of early onset cases [163]. The DJ-1 gene has a size of about 24kb containing 8 exons and giving rise to a major mRNA of about 1kb with an open reading frame of 570bp encoding a protein of 189 amino acids [162]. The protein is widely expressed in the brain and peripheral tissues but a clear physiological role for the protein has yet to be established [164,165]. Based on its structure, the protein is related to ThiJ/PfpI/DJ1 superfamily [162]. The wild-type protein can form homodimers and some of the known mutations disrupt dimer formation –e.g. L166P but not M26I and E64D which have also been identified in patients with early onset parkinsonism. It has been suggested that the protein modulates gene expression under conditions of cellular stress, so providing a protective effect against oxidative stress [113,162]. The possible stress sensor might be the sulfhydryl group of cysteine C106 which can be oxidized to sulfonic acid, resulting in a shift of the pI of the DJ-1 protein [166,167].

There is some experimental evidence indicating that some of the oxidized form of DJ-1 are localized to mitochondria but the physiological significance of this translocation is unclear. There is currently no evidence that this step interferes with mitochondrial function and especially ATP generation which is important for proteasome assembly and function (for a review of these issues see [39]). In summary – there is currently no evidence that DJ-1 and its known mutations are linked to UPS malfunction.

PINK 1

The Pink1 (PTEN-induced kinase 1) gene was identified as part of the PARK 6 locus [168]. Homozygous mutations in this gene are causing PD of the early onset type. Although disease transmission is recessive, it should be noted that about 5% of the sporadic early-onset type of PD have single heterozygous PINK1 mutations. The gene spans 18 kb, containing 8 exons and encodes a 581 amino acids protein [169]. Based on a highly conserved protein kinase domain (amino acids 156 to 509) it has been suggested that the protein encodes a putative serine/threonine kinase of the Ca^{2+} -calmodulin type with a mitochondrial targeting sequence. Since the identified mutations are in and around the kinase domain it is assumed that loss of the kinase activity is responsible for the disease but this as well as the physiological function of the enzyme remains to be clarified.

Recent studies in *Drosophila* have shown that PINK 1 mutants show apoptotic muscle degeneration, defects in mitochondrial morphology, degeneration of dopaminergic neurons and increased sensitivity to oxidative stress [170-172]. Expression of human PINK1 in the *Drosophila* mutants rescues the defects observed in the mutants, indicating evolutionary conservation of PINK1 function. It was further noted that loss of *Drosophila parkin* showed phenotypes similar to loss of pink1 function. Overexpression of *parkin* rescues the mitochondrial morphological defects in *pink1* mutants. These and other observations indicate that *parkin* and *pink1* appear to function in the same pathway with *pink1* functioning upstream of *parkin.*

More recently it was shown that inactivation of PINK1 in *Drosophila* results in progressive loss of dopaminergic neurons and in degeneration of the ommatidia in the compound eye. The degeneration was significantly inhibited by expression of human SOD1 or by Vitamin E indicating that PINK1 contributes to neuronal survival by preventing oxidative stress [173].

A recent report indicates that PINK1 and other mitochondrial proteins can localize to the aggresome (see below) upon proteasome inhibition with MG132 as part of a process which also localizes mitochondria to the aggresome [170]. This appears to happen only under proteasome stress when either proteasome function is diminished or when there is an excessive build-up of damaged or misfolded proteins. It has been suggested that the recruitment of mitochondria to the aggresome might facilitate protein degradation by providing additional ATP.

LRRK2

Leucine-rich repeat kinase 2 (LRRK2, dardarin) recently identified as part of the PARK8 locus, is associated with late-onset PD [174]. The protein is expressed in the brain, predominantly in the basal ganglia and hippocampus but is also found in other tissues [175,176]. The LRRK2 gene is very complex, containing 51 exons and encodes a very large protein of 2527 amino acids. Sequence analysis reveals 6 independent major domains – an ankyrin repeat region, a leucine-rich repeat domain, a ROC GTPase domain, a COR (C terminal of ROC)-domain, a tyrosine kinase-like domain and a C-terminal WD40 domain

[177,178]. The region between amino acids 180 to 660 is predicted to fold as armadillo repeats. In addition to the unusual feature of containing potentially two enzymatic centers – the predicted protein kinase and the GTPase activities - LRRK2 contains four protein interaction domains, making it potentially the scaffold of a multiprotein signaling complex. This idea is supported by the fact that the 22 identified mutations in the gene, known to cause PD, are distributed across all the catalytic and protein-protein interaction domains of LRRK2.

Relevant to the discussion here is the observation that LRRK2 can interact with the E3-ligase parkin through the COR domain although there is currently no evidence that this interaction results in polyubiquitination or degradation of the protein [179]. Clearly, this issue requires more experimental clarification.

AUTOPHAGY

An alternative to the UPS for the degradation of intracellular proteins is autophagy which is carried out in the lysosomal compartment. It has been observed that perinuclear aggregates (later named 'aggresomes') of misfolded proteins formed upon proteasome inhibition are removed from the cells by formation of autophagosomes [127], a finding which was recently extended [180,181]. Autophagy is responsible for removal of bulk cytoplasmatic constituents including cell organells. This might represent a functional survival strategy of eukaryotic cells in response to sublethal damage. Based on the mechanism of autophagic vacuole formation and the delivery of material to the autophagic vacuole, three types of autophagy can be distinguished - macroautophagy, microautophagy and chaperone- mediated autophagy. Macroautophagy involves the sequestration of parts of the cytosol, including not only soluble proteins but also cell organelles into a double membrane vesicle called the autophagosome which is subsequently fused with the lysosome to provide the proteolytic and acidic pH environment required for degradation of the sequestered proteins. Microautophagy involves the wrapping of and degradation of cytosolic regions as well as cell organelles by engulfing directly into the lysosome. In chaperone- mediated autophagy the substrate proteins degraded thru this pathway share the pentapeptide motiv KFERQ. This motive is recognized in the cytosol by the chaperone hsc73, a heatshock protein of 73 kDa, leading ultimately to the binding of this complex to the LAMP2a receptor (lysome-associated membrane protein 2a) and translocation into the lumen of the lysosome. (for reviews see.[182-185]).

Autophagy is mediated by the Atg family of proteins, which consist of 29 different members in yeast (see Table 1) [186]. So far, 18 of these genes have also been shown to occur in other eukaryotic systems. Two of the Atg proteins are conjugated to acceptor protein and phospholidpid in a manner similar to ubiquitin [187-189]).

A distinct and important role of autophagy in the nervous system has been known for a long time in cases of injury and regeration of peripheral axons [190]. Only in the last few years experimental evidence also clearly revealed an important role of autophagy in neurodegenerative diseases (for a review see [185] and Chapter 28). Specifically, in PD the occurence of autophagic vacuoles engulfing mitochondria in the *substantia nigra* point to an activation of macroautophagy [191]. In cellular models of of PD – e.g. the overexpression of mutant A53T α-synuclein in PC12 cells produced a marked increase in autophagy and cell death different from apoptosis [192]. A marked decrease in chaperone-mediated autophagy was noticed in PC12 cells under these conditions because the mutant α-synuclein blocks the LAMP2a receptor, leading to an activation of macroautophagy [193,194]. It has further been observed that in neuronal cell lines inhibition of the UPS also causes activation of macroautophagy presumably as compensatory response. The molecular linkage between the UPS system, macroautopagy and chaperone-mediated autophagy is currently unknown.

Because α-synuclein constitutes a major portion of LB a large number of studies have focused on its degradation. The protein carries the KFERQ motif which would suggest chaperone-mediated autophagy as an important degradation pathway [185,193]. There exist conflicting data if α-synuclein is mainly degraded by the UPS or by autophagy *in vivo* [195]. Interestingly, addition of rapamycin which inhibits the nutrient sensor kinase mTor, leads to an increased formation of autopaghosomes resulting in an increased clearance of α-synuclein [195]. Macroautophagy might be an compensatory response if chaperone-mediated autophagy and/or the UPS is blocked or at least reduced in activity.

Table 1. Autophagy-Related Genes

Gene	Protein Function	References
Atg1	Protein Kinase	[3-7]
Atg2	Peripheral membrane interacts with Atg9	[12-16]
Atg3	E_2-like enzyme, conjugates Atg8 to phosphatidylethanoleamine (PE)	[20-23]
Atg4	Cysteine protease	[15,27-29]
Atg5	Conjugated to Atg12 via internal lysine	[33-36]
Atg6	Component of Ptdlns 3-kinase complexes I and II	[7,33,41,42]
Atg7	E_1-like enzyme, activates Atg8 and Atg12	[7,15,47,48,48,49]
Atg8	Ubiquitine-like protein, conjugated to PE via C-terminal glycine	[6,7,27,28,35,35,57-59]
Atg9	Integral membrane protein	[6,16,63-65]
Atg10	E_2-like enzyme, conjugates Atg12 to Atg5	[68,69]
Atg11	Involved in cargo recognition	[6,47]
Atg12	Ubiquitine-like protein, conjugated to Atg5 via C-terminal glycine	[34-36,73,74]
Atg13	Modifier of Atg1 activity by phosphorylation	
Atg14	Component of Ptdlns 3-kinase complex1	[33,41]
Atg15	Lipase required for breakdown of intravacuolar vesicles	[80,81]
Atg16	Part of Atg12-Atg5 complex	[6,84,85]
Atg17	Modifier of Atg1 activity	[87]
Atg18	Peripheral membrane protein required for localization of Atg2	[7,91,92]
Atg19	Cargo receptor for Cvt pathway	[94,95]
Atg20	PX domain protein needed for the Cvt pathway	[98]
ATG21	Specific to Cvt pathway	[99]
ATG22	Integral membrane protein involved in autophagic breakdown	[100]
Atg23	Needed for Cvt vesicle completion	[102]
Atg24	Sorting of nexin	[98]
Atg25	Coiled-coil protein involved in macropexphagy	[103]
Atg26	UDP-glucose:sterol glucosyltransferase-containing GRAM domain	[6,107,108]
Atg27	Ptdlns(3)P binding protein required for Cvt pathway	[110]
ATG28	Degradation of peroxisomes (pexophagy)	[111]
ATG29	Autophagosome formation	[112]

Summary of the currently known 'autophagy-related' (Atg) genes and their functions. Genes which have been demonstrated so far to occur only in yeast are *italicized.* Adapted from Klionsky *et al.*, 2003 [186].

The essential role of autophagy in neurodegeneration was recently demonstrated using conditional mutations of Atg5 and Atg7 in mice [196,197]. Atg5 $^{flox/flox}$ and Atg7 $^{flox/flox}$ mice were crossed with mice expressing *Cre-* recombinase under the nestin promoter which is expressed in neuronal precursor cells after embryonic day 10.5. Both mice developed progressive deficits in motor function which was accompanied by the accumulation of polyubiquitinated cytoplasmatic inclusion bodies. These results suggest that autophagy is essential for survival of neuronal cells.

Finally, it should be emphasized that an emerging body of data indicates that autophagy plays also an important role in a variety of other neurodegenerativer diseases as Alzheimer's disease and Polyglutamine disease (for a review see [185]) presumably because of the occurence of protein deposits are a main histopathological finding of these neurodegenrative diseases.

Conclusions

For our current understanding of PD two observations appear to be important: first, α-synuclein is the major component of filamentous LB [131] and secondly, certain point mutations in α-synuclein are pathogenic resulting in various forms of PD. However, while the case for an involvment of α-synuclein in PD is very strong, it remains somehow mysterious why wild-type α-synuclein in the majority of PD patients accumulates in LB. In addition, although PD is a major neurodegenerative illness affecting 5% of individuals older than 85 years [198,199], it is unknown why the remaining 95% of individuals are disease-free. Although α-synuclein has a propensity for misfolding and fibrillization into filamentous LB due to its conformation in an aqueous environment, other factors appear to be neccessary to trigger PD.

Genetic screens to search for the causative factor(s) of PD have resulted in the identification of 11 genetic loci which cause a disease which variably resembles PD (see Chapter 25 of this volume). At least five of these genes (PARK1, PARK2, PARK4, PARK5 and PARK7) are linked directly or indirectly to the UPS. Recent studies of the proteasome indicate that its activity declines with age (for a review, see [200]. Because of its propensity for misfolding and denaturation, the maintenance of functional levels of α-synuclein might be under the best of circumstances a race against time to eliminate misfolded forms of this protein by either refolding by molecular chaperones or by degradation via the protein degradation machinery of the cell. It has been suggested that the failure of the UPS is the main cause of PD [201], but the causes for the decline are currently unknown. Systemic administration of a lipid-soluble proteasome inhibitor able to penetrate the blood-brain bareer induces parkinsonism in rats [202], raising the question of a possible induction fo parkinsonism in patients receiving Velcade and other proteasome inhibitors. One critical factor for proteasome function is the assembly of the 26S proteasome from the 19S and 20S precursor complexes which is ATP dependent. In *Drosophila*, proteasome activity is reduced by half in old vs. young flies with a concomitant decrease in ATP levels by 50% [203]. While in rats no or only a very mild decline in ATP concentration is found with increasing age a pronounced reduction in ATP turnover has been observed with age in the parietotemporal cortex and hippocampus of rats which also performed poorly in a memory test [204]. This effect might be attributable to a decline in ATP/ADP translocase which would diminish the release of ATP from the mitochondria into the cytoplasma [205]. It is however well established that mitochondrial function in the nervous system declines with age [206,207]. A reduction in ATP levels might also impair the ability of the cell to collect protein aggregates into perinuclear the aggrosomes. It has been reported that in *Drosophila* reduced ATP levels selectively affect molecular motors [208] which might reduce the ability of the cell to form aggresomes and so to diminish the cellular burden of protein aggregates.

In the sporadic form of PD a reduction in mitochondrial complex I activity has been observed in the *substantia nigra* [209,210] indicating a reduction in ATP which would not only affect proteasome activity but also other steps of the UPS system. It is further known that treatment with the pesticide rotenone, a known inhibitor of mitochondria, results in degeneration of the nigrostriatal dopaminergic system [11].

While no detailed studies of the ATP requirements of autophagy have been carried out it is clear that this complex system requires ATP in a variety of steps. It is also known that chaperone-mediated autophagy declines with age. So, a unifying theme – the decline in the ability of cells to maintain protein structure and to degrade proteins possibly because of a decrease in ATP, might be linking unfolded protein stress, aggresome formation, UPS and autophagy. Obviously, the mechanistic details for the decline in protein degradation need to be investigated. If indeed mitochondrial malfunction is at the core of the problem appropriate scientific and clinical studies should be initiated to investigate if this will lead to new drug discoveries for the treatment of PD.

ACKNOWLEDGEMENTS

We wish to apologize to the many colleagues whose work for reasons of space could not be included in this chapter. The work of J.F. is supported by NIH grant NS41722. Work in the laboratory of T.S-G is supported by National Institute of Health Grants NIGMS GM60654 (T.S-G. and M.F.-P.), NIGMS/RCMI RR03037 core facility grant to Hunter College and a PSC-CUNY Grant 68753. Dr.T.S-G. wishes to thank Dr. Maria Figueiredo-Pereira and Dr.Yong Zhang for critical comments on the manuscript.

REFERENCES

[1] de Rijk MC, Tzourio C, Breteler MM, Dartigues JF, Amaducci L, Lopez-Pousa S, Manubens-Bertran JM, Alperovitch A, Rocca WA: Prevalence of parkinsonism and Parkinson's disease in Europe: the EUROPARKINSON Collaborative Study. European Community Concerted Action on the Epidemiology of Parkinson's disease. *J Neurol Neurosurg Psychiatry* 1997, 62:10-15.

[2] Mink JW. Basal Ganglia. In: Zigmond MJ, Bloom FE, Landis SC, Roberts JL, Squire LR, editors. Fundamental Neuroscience .San Diego, CA.: Academic Press; 1999. p. 951-72.

[3] Matsuura A, Tsukada M, Wada Y, Ohsumi Y: Apg1p, a novel protein kinase required for the autophagic process in Saccharomyces cerevisiae. *Gene* 1997, 192:245-250.

[4] Straub M, Bredschneider M, Thumm M: AUT3, a serine/threonine kinase gene, is essential for autophagocytosis in Saccharomyces cerevisiae. *J Bacteriol* 1997, 179:3875-3883.

[5] Komduur JA, Veenhuis M, Kiel JA: The Hansenula polymorpha PDD7 gene is essential for macropexophagy and microautophagy. *FEMS Yeast Res* 2003, 3:27-34.

[6] Mukaiyama H, Oku M, Baba M, Samizo T, Hammond AT, Glick BS, Kato N, Sakai Y: Paz2 and 13 other PAZ gene products regulate vacuolar engulfment of peroxisomes during micropexophagy. *Genes Cells* 2002, 7:75-90.

[7] Melendez A, Talloczy Z, Seaman M, Eskelinen EL, Hall DH, Levine B: Autophagy genes are essential for dauer development and life-span extension in C. elegans. *Science* 2003, 301:1387-1391.

[8] Poewe WH, Wenning GK: The natural history of Parkinson's disease. *Ann Neurol* 1998, 44:S1-S9.

[9] Langston JW, Ballard P, Tetrud JW, Irwin I: Chronic Parkinsonism in humans due to a product of meperidine-analog synthesis. *Science* 1983, 219:979-980.

[10] Jeon BS, Jackson-Lewis V, Burke RE: 6-Hydroxydopamine lesion of the rat substantia nigra: time course and morphology of cell death. *Neurodegeneration* 1995, 4:131-137.

[11] Betarbet R, Sherer TB, MacKenzie G, Garcia-Osuna M, Panov AV, Greenamyre JT: Chronic systemic pesticide exposure reproduces features of Parkinson's disease. *Nat Neurosci* 2000, 3:1301-1306.

[12] Shintani T, Suzuki K, Kamada Y, Noda T, Ohsumi Y: Apg2p functions in autophagosome formation on the perivacuolar structure. *J Biol Chem* 2001, 276:30452-30460.

[13] Wang CW, Kim J, Huang WP, Abeliovich H, Stromhaug PE, Dunn WA, Jr., Klionsky DJ: Apg2 is a novel protein required for the cytoplasm to vacuole targeting, autophagy, and pexophagy pathways. *J Biol Chem* 2001, 276:30442-30451.

[14] Barth H, Meiling-Wesse K, Epple UD, Thumm M: Autophagy and the cytoplasm to vacuole targeting pathway both require Aut10p. *FEBS Lett* 2001, 508:23-28.

[15] Mukaiyama H, Baba M, Osumi M, Aoyagi S, Kato N, Ohsumi Y, Sakai Y: Modification of a ubiquitin-like protein Paz2 conducted micropexophagy through formation of a novel membrane structure. *Mol Biol Cell* 2004, 15:58-70.

[16] Stromhaug PE, Bevan A, Dunn WA, Jr.: GSA11 encodes a unique 208-kDa protein required for pexophagy and autophagy in Pichia pastoris. *J Biol Chem* 2001, 276:42422-42435.

[17] Langston JW: Parkinson's disease: current and future challenges. *Neurotoxicology* 2002, 23:443-450.

[18] Tanner CM, Ottman R, Goldman SM, Ellenberg J, Chan P, Mayeux R, Langston JW: Parkinson disease in twins: an etiologic study. *JAMA* 1999, 281:341-346.

[19] Di Monte DA: The environment and Parkinson's disease: is the nigrostriatal system preferentially targeted by neurotoxins? *Lancet Neurol* 2003, 2:531-538.

[20] Schlumpberger M, Schaeffeler E, Straub M, Bredschneider M, Wolf DH, Thumm M: AUT1, a gene essential for autophagocytosis in the yeast Saccharomyces cerevisiae. *J Bacteriol* 1997, 179:1068-1076.

[21] Ichimura Y, Kirisako T, Takao T, Satomi Y, Shimonishi Y, Ishihara N, Mizushima N, Tanida I, Kominami E, Ohsumi M, Noda T, Ohsumi Y: A ubiquitin-like system mediates protein lipidation. *Nature* 2000, 408:488-492.

[22] Juhasz G, Csikos G, Sinka R, Erdelyi M, Sass M: The Drosophila homolog of Aut1 is essential for autophagy and development. *FEBS Lett* 2003, 543:154-158.

[23] Tanida I, Tanida-Miyake E, Komatsu M, Ueno T, Kominami E: Human Apg3p/Aut1p homologue is an authentic E2 enzyme for multiple substrates, GATE-16, GABARAP, and MAP-LC3, and facilitates the conjugation of hApg12p to hApg5p. *J Biol Chem* 2002, 277:13739-13744.

[24] Thiruchelvam M, Richfield EK, Baggs RB, Tank AW, Cory-Slechta DA: The nigrostriatal dopaminergic system as a preferential target of repeated exposures to combined paraquat and maneb: implications for Parkinson's disease. *J Neurosci* 2000, 20:9207-9214.

[25] Moore DJ, West AB, Dawson VL, Dawson TM: Molecular pathophysiology of Parkinson's disease. *Annu Rev Neurosci* 2005, 28:57-87.

[26] Singleton AB, Farrer M, Johnson J, Singleton A, Hague S, Kachergus J, Hulihan M, Peuralinna T, Dutra A, Nussbaum R, Lincoln S, Crawley A, Hanson M, Maraganore D, Adler C, Cookson MR, Muenter M, Baptista M, Miller D, Blancato J, Hardy J, Gwinn-Hardy K: alpha-Synuclein locus triplication causes Parkinson's disease. *Science* 2003, 302:841.

[27] Lang T, Schaeffeler E, Bernreuther D, Bredschneider M, Wolf DH, Thumm M: Aut2p and Aut7p, two novel microtubule-associated proteins are essential for delivery of autophagic vesicles to the vacuole. *EMBO J* 1998, 17:3597-3607.

[28] Kirisako T, Ichimura Y, Okada H, Kabeya Y, Mizushima N, Yoshimori T, Ohsumi M, Takao T, Noda T, Ohsumi Y: The reversible modification regulates the membrane-binding state of Apg8/Aut7 essential for autophagy and the cytoplasm to vacuole targeting pathway. *J Cell Biol* 2000, 151:263-276.

[29] Thumm M, Kadowaki T: The loss of Drosophila APG4/AUT2 function modifies the phenotypes of cut and Notch signaling pathway mutants. *Mol Genet Genomics* 2001, 266:657-663.

[30] Moore DJ: Parkin: a multifaceted ubiquitin ligase. *Biochem Soc Trans* 2006, 34:749-753.

[31] Vigouroux S, Briand M, Briand Y: Linkage between the proteasome pathway and neurodegenerative diseases and aging. *Mol Neurobiol* 2004, 30:201-221.

[32] Liu Y, Fallon L, Lashuel HA, Liu Z, Lansbury PT, Jr.: The UCH-L1 gene encodes two opposing enzymatic activities that affect alpha-synuclein degradation and Parkinson's disease susceptibility. *Cell* 2002, 111:209-218.

[33] Kametaka S, Okano T, Ohsumi M, Ohsumi Y: Apg14p and Apg6/Vps30p form a protein complex essential for autophagy in the yeast, Saccharomyces cerevisiae. *J Biol Chem* 1998, 273:22284-22291.

[34] Mizushima N, Yamamoto A, Hatano M, Kobayashi Y, Kabeya Y, Suzuki K, Tokuhisa T, Ohsumi Y, Yoshimori T: Dissection of autophagosome formation using Apg5-deficient mouse embryonic stem cells. *J Cell Biol* 2001, 152:657-668.
[35] Otto GP, Wu MY, Kazgan N, Anderson OR, Kessin RH: Macroautophagy is required for multicellular development of the social amoeba Dictyostelium discoideum. *J Biol Chem* 2003, 278:17636-17645.
[36] Mizushima N, Sugita H, Yoshimori T, Ohsumi Y: A new protein conjugation system in human. The counterpart of the yeast Apg12p conjugation system essential for autophagy. *J Biol Chem* 1998, 273:33889-33892.
[37] Mukaetova-Ladinska EB, McKeith IG: Pathophysiology of synuclein aggregation in Lewy body disease. *Mech Ageing Dev* 2006, 127:188-202.
[38] Greene JC, Whitworth AJ, Kuo I, Andrews LA, Feany MB, Pallanck LJ: Mitochondrial pathology and apoptotic muscle degeneration in Drosophila parkin mutants. *Proc Natl Acad Sci U S A* 2003, 100:4078-4083.
[39] Cookson MR: The biochemistry of Parkinson's disease. *Annu Rev Biochem* 2005, 74:29-52.
[40] Dawson TM, Dawson VL: Rare genetic mutations shed light on the pathogenesis of Parkinson disease. *J Clin Invest* 2003, 111:145-151.
[41] Kihara A, Noda T, Ishihara N, Ohsumi Y: Two distinct Vps34 phosphatidylinositol 3-kinase complexes function in autophagy and carboxypeptidase Y sorting in Saccharomyces cerevisiae. *J Cell Biol* 2001, 152:519-530.
[42] Liang XH, Jackson S, Seaman M, Brown K, Kempkes B, Hibshoosh H, Levine B: Induction of autophagy and inhibition of tumorigenesis by beclin 1. *Nature* 1999, 402:672-676.
[43] Chartier-Harlin MC, Kachergus J, Roumier C, Mouroux V, Douay X, Lincoln S, Levecque C, Larvor L, Andrieux J, Hulihan M, Waucquier N, Defebvre L, Amouyel P, Farrer M, Destee A: Alpha-synuclein locus duplication as a cause of familial Parkinson's disease. *Lancet* 2004, 364:1167-1169.
[44] Johnson J, Hague SM, Hanson M, Gibson A, Wilson KE, Evans EW, Singleton AA, Inerney-Leo A, Nussbaum RL, Hernandez DG, Gallardo M, McKeith IG, Burn DJ, Ryu M, Hellstrom O, Ravina B, Eerola J, Perry RH, Jaros E, Tienari P, Weiser R, Gwinn-Hardy K, Morris CM, Hardy J, Singleton AB: SNCA multiplication is not a common cause of Parkinson disease or dementia with Lewy bodies. *Neurology* 2004, 63:554-556.
[45] Singleton A, Myers A, Hardy J: The law of mass action applied to neurodegenerative disease: a hypothesis concerning the etiology and pathogenesis of complex diseases. *Hum Mol Genet* 2004, 13 Spec No 1:R123-R126.
[46] Lott IT, Head E: Alzheimer disease and Down syndrome: factors in pathogenesis. *Neurobiol Aging* 2005, 26:383-389.
[47] Kim J, Kamada Y, Stromhaug PE, Guan J, Hefner-Gravink A, Baba M, Scott SV, Ohsumi Y, Dunn WA, Jr., Klionsky DJ: Cvt9/Gsa9 functions in sequestering selective cytosolic cargo destined for the vacuole. *J Cell Biol* 2001, 153:381-396.
[48] Yuan W, Stromhaug PE, Dunn WA, Jr.: Glucose-induced autophagy of peroxisomes in Pichia pastoris requires a unique E1-like protein. *Mol Biol Cell* 1999, 10:1353-1366.

[49] Tanida I, Mizushima N, Kiyooka M, Ohsumi M, Ueno T, Ohsumi Y, Kominami E: Apg7p/Cvt2p: A novel protein-activating enzyme essential for autophagy. *Mol Biol Cell* 1999, 10:1367-1379.

[50] Glenner GG, Wong CW, Quaranta V, Eanes ED: The amyloid deposits in Alzheimer's disease: their nature and pathogenesis. *Appl Pathol* 1984, 2:357-369.

[51] Pals P, Lincoln S, Manning J, Heckman M, Skipper L, Hulihan M, Van den BM, De PT, Cras P, Crook J, Van BC, Farrer MJ: alpha-Synuclein promoter confers susceptibility to Parkinson's disease. *Ann Neurol* 2004, 56:591-595.

[52] Maroteaux L, Scheller RH: The rat brain synucleins; family of proteins transiently associated with neuronal membrane. *Brain Res Mol Brain Res* 1991, 11:335-343.

[53] Ji H, Liu YE, Jia T, Wang M, Liu J, Xiao G, Joseph BK, Rosen C, Shi YE: Identification of a breast cancer-specific gene, BCSG1, by direct differential cDNA sequencing. *Cancer Res* 1997, 57:759-764.

[54] Abeliovich A, Schmitz Y, Farinas I, Choi-Lundberg D, Ho WH, Castillo PE, Shinsky N, Verdugo JM, Armanini M, Ryan A, Hynes M, Phillips H, Sulzer D, Rosenthal A: Mice lacking alpha-synuclein display functional deficits in the nigrostriatal dopamine system. *Neuron* 2000, 25:239-252.

[55] Chandra S, Fornai F, Kwon HB, Yazdani U, Atasoy D, Liu X, Hammer RE, Battaglia G, German DC, Castillo PE, Sudhof TC: Double-knockout mice for alpha- and beta-synucleins: effect on synaptic functions. *Proc Natl Acad Sci U S A* 2004, 101:14966-14971.

[56] Chandra S, Gallardo G, Fernandez-Chacon R, Schluter OM, Sudhof TC: Alpha-synuclein cooperates with CSPalpha in preventing neurodegeneration. *Cell* 2005, 123:383-396.

[57] Harding TM, Morano KA, Scott SV, Klionsky DJ: Isolation and characterization of yeast mutants in the cytoplasm to vacuole protein targeting pathway. *J Cell Biol* 1995, 131:591-602.

[58] He H, Dang Y, Dai F, Guo Z, Wu J, She X, Pei Y, Chen Y, Ling W, Wu C, Zhao S, Liu JO, Yu L: Post-translational modifications of three members of the human MAP1LC3 family and detection of a novel type of modification for MAP1LC3B. *J Biol Chem* 2003, 278:29278-29287.

[59] Tanida I, Tanida-Miyake E, Nishitani T, Komatsu M, Yamazaki H, Ueno T, Kominami E: Murine Apg12p has a substrate preference for murine Apg7p over three Apg8p homologs. *Biochem Biophys Res Commun* 2002, 292:256-262.

[60] Bonini NM, Giasson BI: Snaring the function of alpha-synuclein. *Cell* 2005, 123:359-361.

[61] Fernandez-Chacon R, Wolfel M, Nishimune H, Tabares L, Schmitz F, Castellano-Munoz M, Rosenmund C, Montesinos ML, Sanes JR, Schneggenburger R, Sudhof TC: The synaptic vesicle protein CSP alpha prevents presynaptic degeneration. *Neuron* 2004, 42:237-251.

[62] Davidson WS, Jonas A, Clayton DF, George JM: Stabilization of alpha-synuclein secondary structure upon binding to synthetic membranes. *J Biol Chem* 1998, 273:9443-9449.

[63] Noda T, Kim J, Huang WP, Baba M, Tokunaga C, Ohsumi Y, Klionsky DJ: Apg9p/Cvt7p is an integral membrane protein required for transport vesicle formation in the Cvt and autophagy pathways. *J Cell Biol* 2000, 148:465-480.

[64] Lang T, Reiche S, Straub M, Bredschneider M, Thumm M: Autophagy and the cvt pathway both depend on AUT9. *J Bacteriol* 2000, 182:2125-2133.

[65] Hanaoka H, Noda T, Shirano Y, Kato T, Hayashi H, Shibata D, Tabata S, Ohsumi Y: Leaf senescence and starvation-induced chlorosis are accelerated by the disruption of an Arabidopsis autophagy gene. *Plant Physiol* 2002, 129:1181-1193.

[66] Jenco JM, Rawlingson A, Daniels B, Morris AJ: Regulation of phospholipase D2: selective inhibition of mammalian phospholipase D isoenzymes by alpha- and beta-synucleins. *Biochemistry* 1998, 37:4901-4909.

[67] Cole NB, Murphy DD, Grider T, Rueter S, Brasaemle D, Nussbaum RL: Lipid droplet binding and oligomerization properties of the Parkinson's disease protein alpha-synuclein. *J Biol Chem* 2002, 277:6344-6352.

[68] Shintani T, Mizushima N, Ogawa Y, Matsuura A, Noda T, Ohsumi Y: Apg10p, a novel protein-conjugating enzyme essential for autophagy in yeast. *EMBO J* 1999, 18:5234-5241.

[69] Mizushima N, Yoshimori T, Ohsumi Y: Mouse Apg10 as an Apg12-conjugating enzyme: analysis by the conjugation-mediated yeast two-hybrid method. *FEBS Lett* 2002, 532:450-454.

[70] Outeiro TF, Lindquist S: Yeast cells provide insight into alpha-synuclein biology and pathobiology. *Science* 2003, 302:1772-1775.

[71] Conway KA, Harper JD, Lansbury PT: Accelerated in vitro fibril formation by a mutant alpha-synuclein linked to early-onset Parkinson disease. *Nat Med* 1998, 4:1318-1320.

[72] Dekker MC, Bonifati V, van Duijn CM: Parkinson's disease: piecing together a genetic jigsaw. *Brain* 2003, 126:1722-1733.

[73] Tanida I, Nishitani T, Nemoto T, Ueno T, Kominami E: Mammalian Apg12p, but not the Apg12p.Apg5p conjugate, facilitates LC3 processing. *Biochem Biophys Res Commun* 2002, 296:1164-1170.

[74] Mizushima N, Noda T, Yoshimori T, Tanaka Y, Ishii T, George MD, Klionsky DJ, Ohsumi M, Ohsumi Y: A protein conjugation system essential for autophagy. *Nature* 1998, 395:395-398.

[75] Oliveira SA, Scott WK, Martin ER, Nance MA, Watts RL, Hubble JP, Koller WC, Pahwa R, Stern MB, Hiner BC, Ondo WG, Allen FH, Jr., Scott BL, Goetz CG, Small GW, Mastaglia F, Stajich JM, Zhang F, Booze MW, Winn MP, Middleton LT, Haines JL, Pericak-Vance MA, Vance JM: Parkin mutations and susceptibility alleles in late-onset Parkinson's disease. *Ann Neurol* 2003, 53:624-629.

[76] Scott WK, Nance MA, Watts RL, Hubble JP, Koller WC, Lyons K, Pahwa R, Stern MB, Colcher A, Hiner BC, Jankovic J, Ondo WG, Allen FH, Jr., Goetz CG, Small GW, Masterman D, Mastaglia F, Laing NG, Stajich JM, Slotterbeck B, Booze MW, Ribble RC, Rampersaud E, West SG, Gibson RA, Middleton LT, Roses AD, Haines JL, Scott BL, Vance JM, Pericak-Vance MA: Complete genomic screen in Parkinson disease: evidence for multiple genes. *JAMA* 2001, 286:2239-2244.

[77] Shimura H, Hattori N, Kubo S, Mizuno Y, Asakawa S, Minoshima S, Shimizu N, Iwai K, Chiba T, Tanaka K, Suzuki T: Familial Parkinson disease gene product, parkin, is a ubiquitin-protein ligase. *Nat Genet* 2000, 25:302-305.

[78] Hershko A, Ciechanover A: The ubiquitin system. *Annu Rev Biochem* 1998, 67:425-479.

[79] Hochstrasser M: Ubiquitin-dependent protein degradation. *Annu Rev Genet* 1996, 30:405-439.

[80] Epple UD, Suriapranata I, Eskelinen EL, Thumm M: Aut5/Cvt17p, a putative lipase essential for disintegration of autophagic bodies inside the vacuole. *J Bacteriol* 2001, 183:5942-5955.

[81] Teter SA, Eggerton KP, Scott SV, Kim J, Fischer AM, Klionsky DJ: Degradation of lipid vesicles in the yeast vacuole requires function of Cvt17, a putative lipase. *J Biol Chem* 2001, 276:2083-2087.

[82] Zhang Y, Gao J, Chung KK, Huang H, Dawson VL, Dawson TM: Parkin functions as an E2-dependent ubiquitin- protein ligase and promotes the degradation of the synaptic vesicle-associated protein, CDCrel-1. *Proc Natl Acad Sci U S A* 2000, 97:13354-13359.

[83] Imai Y, Soda M, Inoue H, Hattori N, Mizuno Y, Takahashi R: An unfolded putative transmembrane polypeptide, which can lead to endoplasmic reticulum stress, is a substrate of Parkin. *Cell* 2001, 105:891-902.

[84] Mizushima N, Noda T, Ohsumi Y: Apg16p is required for the function of the Apg12p-Apg5p conjugate in the yeast autophagy pathway. *EMBO J* 1999, 18:3888-3896.

[85] Mizushima N, Kuma A, Kobayashi Y, Yamamoto A, Matsubae M, Takao T, Natsume T, Ohsumi Y, Yoshimori T: Mouse Apg16L, a novel WD-repeat protein, targets to the autophagic isolation membrane with the Apg12-Apg5 conjugate. *J Cell Sci* 2003, 116:1679-1688.

[86] Shimura H, Schlossmacher MG, Hattori N, Frosch MP, Trockenbacher A, Schneider R, Mizuno Y, Kosik KS, Selkoe DJ: Ubiquitination of a new form of alpha-synuclein by parkin from human brain: implications for Parkinson's disease. *Science* 2001, 293:263-269.

[87] Kamada Y, Funakoshi T, Shintani T, Nagano K, Ohsumi M, Ohsumi Y: Tor-mediated induction of autophagy via an Apg1 protein kinase complex. *J Cell Biol* 2000, 150:1507-1513.

[88] Chung KK, Zhang Y, Lim KL, Tanaka Y, Huang H, Gao J, Ross CA, Dawson VL, Dawson TM: Parkin ubiquitinates the alpha-synuclein-interacting protein, synphilin-1: implications for Lewy-body formation in Parkinson disease. *Nat Med* 2001, 7:1144-1150.

[89] Staropoli JF, McDermott C, Martinat C, Schulman B, Demireva E, Abeliovich A: Parkin is a component of an SCF-like ubiquitin ligase complex and protects postmitotic neurons from kainate excitotoxicity. *Neuron* 2003, 37:735-749.

[90] Ren Y, Zhao J, Feng J: Parkin binds to alpha/beta tubulin and increases their ubiquitination and degradation. *J Neurosci* 2003, 23:3316-3324.

[91] Barth H, Thumm M: A genomic screen identifies AUT8 as a novel gene essential for autophagy in the yeast Saccharomyces cerevisiae. *Gene* 2001, 274:151-156.

[92] Guan J, Stromhaug PE, George MD, Habibzadegah-Tari P, Bevan A, Dunn WA, Jr., Klionsky DJ: Cvt18/Gsa12 is required for cytoplasm-to-vacuole transport, pexophagy, and autophagy in Saccharomyces cerevisiae and Pichia pastoris. *Mol Biol Cell* 2001, 12:3821-3838.

[93] Tsai YC, Fishman PS, Thakor NV, Oyler GA: Parkin facilitates the elimination of expanded polyglutamine proteins and leads to preservation of proteasome function. *J Biol Chem* 2003, 278:22044-22055.

[94] Scott SV, Guan J, Hutchins MU, Kim J, Klionsky DJ: Cvt19 is a receptor for the cytoplasm-to-vacuole targeting pathway. *Mol Cell* 2001, 7:1131-1141.

[95] Leber R, Silles E, Sandoval IV, Mazon MJ: Yol082p, a novel CVT protein involved in the selective targeting of aminopeptidase I to the yeast vacuole. *J Biol Chem* 2001, 276:29210-29217.

[96] Corti O, Hampe C, Koutnikova H, Darios F, Jacquier S, Prigent A, Robinson JC, Pradier L, Ruberg M, Mirande M, Hirsch E, Rooney T, Fournier A, Brice A: The p38 subunit of the aminoacyl-tRNA synthetase complex is a Parkin substrate: linking protein biosynthesis and neurodegeneration. *Hum Mol Genet* 2003, 12:1427-1437.

[97] Feng J: Genetic Factors in Parkinson's Disease and Potential Therapeutic Targets. *Curr Neuropharm* 2003, 1:301-313.

[98] Nice DC, Sato TK, Stromhaug PE, Emr SD, Klionsky DJ: Cooperative binding of the cytoplasm to vacuole targeting pathway proteins, Cvt13 and Cvt20, to phosphatidylinositol 3-phosphate at the pre-autophagosomal structure is required for selective autophagy. *J Biol Chem* 2002, 277:30198-30207.

[99] Barth H, Meiling-Wesse K, Epple UD, Thumm M: Mai1p is essential for maturation of proaminopeptidase I but not for autophagy. *FEBS Lett* 2002, 512:173-179.

[100] Suriapranata I, Epple UD, Bernreuther D, Bredschneider M, Sovarasteanu K, Thumm M: The breakdown of autophagic vesicles inside the vacuole depends on Aut4p. *J Cell Sci* 2000, 113 (Pt 22):4025-4033.

[101] Zhang K, Wong HN, Song B, Miller CN, Scheuner D, Kaufman RJ: The unfolded protein response sensor IRE1alpha is required at 2 distinct steps in B cell lymphopoiesis. *J Clin Invest* 2005, 115:268-281.

[102] Tucker KA, Reggiori F, Dunn WA, Jr., Klionsky DJ: Atg23 is essential for the cytoplasm to vacuole targeting pathway and efficient autophagy but not pexophagy. *J Biol Chem* 2003, 278:48445-48452.

[103] Monastyrska I, Kiel JA, Krikken AM, Komduur JA, Veenhuis M, van dK, I: The Hansenula polymorpha ATG25 gene encodes a novel coiled-coil protein that is required for macropexophagy. *Autophagy* 2005, 1:92-100.

[104] Cleveland DW, Pittenger MF, Feramisco JR: Elevation of tubulin levels by microinjection suppresses new tubulin synthesis. *Nature* 1983, 305:738-740.

[105] Burke D, Gasdaska P, Hartwell L: Dominant effects of tubulin overexpression in Saccharomyces cerevisiae. *Mol Cell Biol* 1989, 9:1049-1059.

[106] Weinstein B, Solomon F: Phenotypic consequences of tubulin overproduction in Saccharomyces cerevisiae: differences between alpha-tubulin and beta-tubulin. *Mol Cell Biol* 1990, 10:5295-5304.

[107] Oku M, Warnecke D, Noda T, Muller F, Heinz E, Mukaiyama H, Kato N, Sakai Y: Peroxisome degradation requires catalytically active sterol glucosyltransferase with a GRAM domain. *EMBO J* 2003, 22:3231-3241.

[108] Stasyk OV, Nazarko TY, Stasyk OG, Krasovska OS, Warnecke D, Nicaud JM, Cregg JM, Sibirny AA: Sterol glucosyltransferases have different functional roles in Pichia pastoris and Yarrowia lipolytica. *Cell Biol Int* 2003, 27:947-952.

[109] Lewis SA, Tian GL, Cowan NJ: The alpha- and beta-tubulin folding pathways. *Trends in Cell Biology* 1997, 7:479-484.

[110] Wurmser AE, Emr SD: Novel PtdIns(3)P-binding protein Etf1 functions as an effector of the Vps34 PtdIns 3-kinase in autophagy. *J Cell Biol* 2002, 158:761-772.

[111] Stasyk OV, Stasyk OG, Mathewson RD, Farre JC, Nazarko VY, Krasovska OS, Subramani S, Cregg JM, Sibirny AA: Atg28, a novel coiled-coil protein involved in autophagic degradation of peroxisomes in the methylotrophic yeast Pichia pastoris. *Autophagy* 2006, 2:30-38.

[112] Kawamata T, Kamada Y, Suzuki K, Kuboshima N, Akimatsu H, Ota S, Ohsumi M, Ohsumi Y: Characterization of a novel autophagy-specific gene, ATG29. *Biochem Biophys Res Commun* 2005, 338:1884-1889.

[113] Brinkley BR, Barham SS, Barranco SC, Fuller GM: Rotenone inhibition of spindle microtubule assembly in mammalian cells. *Exp Cell Res* 1974, 85:41-46.

[114] Cappelletti G, Maggioni MG, Maci R: Influence of MPP+ on the state of tubulin polymerisation in NGF-differentiated PC12 cells. *J Neurosci Res* 1999, 56:28-35.

[115] Cappelletti G, Pedrotti B, Maggioni MG, Maci R: Microtubule assembly is directly affected by MPP(+)in vitro. *Cell Biol Int* 2001, 25:981-984.

[116] Marshall LE, Himes RH: Rotenone inhibition of tubulin self-assembly. *Biochim Biophys Acta* 1978, 543:590-594.

[117] Cleveland DW: Autoregulated control of tubulin synthesis in animal cells. *Curr Opin Cell Biol* 1989, 1:10-14.

[118] Ren Y, Liu W, Jiang H, Jiang Q, Feng J: Selective vulnerability of dopaminergic neurons to microtubule depolymerization. *J Biol Chem* 2005, 280:34105-34112.

[119] Jiang H, Jiang Q, Feng J: Parkin increases dopamine uptake by enhancing the cell surface expression of dopamine transporter. *J Biol Chem* 2004, 279:54380-54386.

[120] Huynh DP, Scoles DR, Nguyen D, Pulst SM: The autosomal recessive juvenile Parkinson disease gene product, parkin, interacts with and ubiquitinates synaptotagmin XI. *Hum Mol Genet* 2003, 12:2587-2597.

[121] Hampton RY: ER-associated degradation in protein quality control and cellular regulation. *Curr Opin Cell Biol* 2002, 14:476-482.

[122] Tsai B, Ye Y, Rapoport TA: Retro-translocation of proteins from the endoplasmic reticulum into the cytosol. *Nat Rev Mol Cell Biol* 2002, 3:246-255.

[123] Cole NB, Lippincott-Schwartz J: Organization of organelles and membrane traffic by microtubules. *Curr Opin Cell Biol* 1995, 7:55-64.

[124] Terasaki M, Chen LB, Fujiwara K: Microtubules and the endoplasmic reticulum are highly interdependent structures. *J Cell Biol* 1986, 103:1557-1568.

[125] Waterman-Storer CM, Salmon ED: Endoplasmic reticulum membrane tubules are distributed by microtubules in living cells using three distinct mechanisms. *Curr Biol* 1998, 8:798-806.

[126] Yang F, Jiang Q, Zhao J, Ren Y, Sutton MD, Feng J: Parkin stabilizes microtubules through strong binding mediated by three independent domains. *J Biol Chem* 2005, 280:17154-17162.

[127] Wojcik C, Schroeter D, Stoehr M, Wilk S, Paweletz N: An inhibitor of the chymotrypsin-like activity of the multicatalytic proteinase complex (20S proteasome) induces arrest in G2-phase and metaphase in HeLa cells. *Eur J Cell Biol* 1996, 70:172-178.

[128] Kopito RR: Aggresomes, inclusion bodies and protein aggregation. *Trends Cell Biol* 2000, 10:524-530.

[129] Imai Y, Soda M, Murakami T, Shoji M, Abe K, Takahashi R: A product of the human gene adjacent to parkin is a component of Lewy bodies and suppresses Pael receptor-induced cell death. *J Biol Chem* 2003, 278:51901-51910.

[130] Zhao J, Ren Y, Jiang Q, Feng J: Parkin is recruited to the centrosome in response to inhibition of proteasomes. *J Cell Sci* 2003, 116:4011-4019.

[131] Jiang H, Jiang Q, Liu W, Feng J: Parkin suppresses the expression of monoamine oxidases. *J Biol Chem* 2006, 281:8591-8599.

[132] Shih JC, Chen K, Ridd MJ: Monoamine oxidase: from genes to behavior. *Annu Rev Neurosci* 1999, 22:197-217.

[133] Binda C, Newton-Vinson P, Hubalek F, Edmondson DE, Mattevi A: Structure of human monoamine oxidase B, a drug target for the treatment of neurological disorders. *Nat Struct Biol* 2002, 9:22-26.

[134] Rebrin I, Geha RM, Chen K, Shih JC: Effects of carboxyl-terminal truncations on the activity and solubility of human monoamine oxidase B. *J Biol Chem* 2001, 276:29499-29506.

[135] Hauptmann N, Grimsby J, Shih JC, Cadenas E: The metabolism of tyramine by monoamine oxidase A/B causes oxidative damage to mitochondrial DNA. *Arch Biochem Biophys* 1996, 335:295-304.

[136] Riederer P, Jellinger K: Neurochemical insights into monoamine oxidase inhibitors, with special reference to deprenyl (selegiline). *Acta Neurol Scand Suppl* 1983, 95:43-55.

[137] Youdim MB, Riederer PF: A review of the mechanisms and role of monoamine oxidase inhibitors in Parkinson's disease. *Neurology* 2004, 63:S32-S35.

[138] Grimsby J, Toth M, Chen K, Kumazawa T, Klaidman L, Adams JD, Karoum F, Gal J, Shih JC: Increased stress response and beta-phenylethylamine in MAOB-deficient mice. *Nat Genet* 1997, 17:206-210.

[139] Langston JW, Irwin I, Langston EB, Forno LS: Pargyline prevents MPTP-induced parkinsonism in primates. *Science* 1984, 225:1480-1482.

[140] Javitch JA, D'Amato RJ, Strittmatter SM, Snyder SH: Parkinsonism-inducing neurotoxin, N-methyl-4-phenyl-1,2,3,6 -tetrahydropyridine: uptake of the metabolite N-methyl-4-phenylpyridine by dopamine neurons explains selective toxicity. *Proc Natl Acad Sci U S A* 1985, 82:2173-2177.

[141] Darios F, Corti O, Lucking CB, Hampe C, Muriel MP, Abbas N, Gu WJ, Hirsch EC, Rooney T, Ruberg M, Brice A: Parkin prevents mitochondrial swelling and cytochrome c release in mitochondria-dependent cell death. *Hum Mol Genet* 2003, 12:517-526.

[142] Kuroda Y, Mitsui T, Kunishige M, Matsumoto T: Parkin affects mitochondrial function and apoptosis in neuronal and myogenic cells. *Biochem Biophys Res Commun* 2006, 348:787-793.

[143] Morett E, Bork P: A novel transactivation domain in parkin. *Trends Biochem Sci* 1999, 24:229-231.

[144] Tatematsu K, Tokunaga C, Nakagawa N, Tanizawa K, Kuroda S, Kikkawa U: Transcriptional activity of RBCK1 protein (RBCC protein interacting with PKC 1): requirement of RING-finger and B-Box motifs and regulation by protein kinases. *Biochem Biophys Res Commun* 1998, 247:392-396.

[145] Tokunaga C, Kuroda S, Tatematsu K, Nakagawa N, Ono Y, Kikkawa U: Molecular cloning and characterization of a novel protein kinase C-interacting protein with structural motifs related to RBCC family proteins. *Biochem Biophys Res Commun* 1998, 244:353-359.

[146] Nikolaev AY, Li M, Puskas N, Qin J, Gu W: Parc: a cytoplasmic anchor for p53. *Cell* 2003, 112:29-40.

[147] Giannakakou P, Sackett DL, Ward Y, Webster KR, Blagosklonny MV, Fojo T: p53 is associated with cellular microtubules and is transported to the nucleus by dynein. *Nat Cell Biol* 2000, 2:709-717.

[148] O'Brate A, Giannakakou P: The importance of p53 location: nuclear or cytoplasmic zip code? *Drug Resist Updat* 2003, 6:313-322.

[149] Wilkinson KD, Lee KM, Deshpande S, Duerksen-Hughes P, Boss JM, Pohl J: The neuron-specific protein PGP 9.5 is a ubiquitin carboxyl-terminal hydrolase. *Science* 1989, 246:670-673.

[150] Hegde AN, Inokuchi K, Pei W, Casadio A, Ghirardi M, Chain DG, Martin KC, Kandel ER, Schwartz JH: Ubiquitin C-terminal hydrolase is an immediate-early gene essential for long-term facilitation in Aplysia. *Cell* 1997, 89:115-126.

[151] Zhang N, Wilkinson K, Bownes M: Cloning and analysis of expression of a ubiquitin carboxyl terminal hydrolase expressed during oogenesis in Drosophila melanogaster. *Dev Biol* 1993, 157:214-223.

[152] Wilkinson KD, Deshpande S, Larsen CN: Comparisons of neuronal (PGP 9.5) and non-neuronal ubiquitin C-terminal hydrolases. *Biochem Soc Trans* 1992, 20:631-637.

[153] Leroy E, Boyer R, Auburger G, Leube B, Ulm G, Mezey E, Harta G, Brownstein MJ, Jonnalagada S, Chernova T, Dehejia A, Lavedan C, Gasser T, Steinbach PJ, Wilkinson KD, Polymeropoulos MH: The ubiquitin pathway in Parkinson's disease. *Nature* 1998, 395:451-452.

[154] Lowe J, McDermott H, Landon M, Mayer RJ, Wilkinson KD: Ubiquitin carboxyl-terminal hydrolase (PGP 9.5) is selectively present in ubiquitinated inclusion bodies characteristic of human neurodegenerative diseases. *J Pathol* 1990, 161:153-160.

[155] Levecque C, Destee A, Mouroux V, Becquet E, Defebvre L, Amouyel P, Chartier-Harlin MC: No genetic association of the ubiquitin carboxy-terminal hydrolase-L1

gene S18Y polymorphism with familial Parkinson's disease. *J Neural Transm* 2001, 108:979-984.

[156] Maraganore DM, Farrer MJ, Hardy JA, Lincoln SJ, McDonnell SK, Rocca WA: Case-control study of the ubiquitin carboxy-terminal hydrolase L1 gene in Parkinson's disease. *Neurology* 1999, 53:1858-1860.

[157] Satoh J, Kuroda Y: A polymorphic variation of serine to tyrosine at codon 18 in the ubiquitin C-terminal hydrolase-L1 gene is associated with a reduced risk of sporadic Parkinson's disease in a Japanese population. *J Neurol Sci* 2001, 189:113-117.

[158] Nishikawa K, Li H, Kawamura R, Osaka H, Wang YL, Hara Y, Hirokawa T, Manago Y, Amano T, Noda M, Aoki S, Wada K: Alterations of structure and hydrolase activity of parkinsonism-associated human ubiquitin carboxyl-terminal hydrolase L1 variants. *Biochem Biophys Res Commun* 2003, 304:176-183.

[159] Saigoh K, Wang YL, Suh JG, Yamanishi T, Sakai Y, Kiyosawa H, Harada T, Ichihara N, Wakana S, Kikuchi T, Wada K: Intragenic deletion in the gene encoding ubiquitin carboxy-terminal hydrolase in gad mice. *Nat Genet* 1999, 23:47-51.

[160] Yamazaki K, Wakasugi N, Tomita T, Kikuchi T, Mukoyama M, Ando K: Gracile axonal dystrophy (GAD), a new neurological mutant in the mouse. *Proc Soc Exp Biol Med* 1988, 187:209-215.

[161] Kikuchi T, Mukoyama M, Yamazaki K, Moriya H: Axonal degeneration of ascending sensory neurons in gracile axonal dystrophy mutant mouse. *Acta Neuropathol (Berl)* 1990, 80:145-151.

[162] Bonifati V, Rizzu P, van Baren MJ, Schaap O, Breedveld GJ, Krieger E, Dekker MC, Squitieri F, Ibanez P, Joosse M, van Dongen JW, Vanacore N, van Swieten JC, Brice A, Meco G, van Duijn CM, Oostra BA, Heutink P: Mutations in the DJ-1 gene associated with autosomal recessive early-onset parkinsonism. *Science* 2003, 299:256-259.

[163] Kubo S, Hattori N, Mizuno Y: Recessive Parkinson's disease. *Mov Disord* 2006, 21:885-893.

[164] Bandopadhyay R, Kingsbury AE, Cookson MR, Reid AR, Evans IM, Hope AD, Pittman AM, Lashley T, Canet-Aviles R, Miller DW, McLendon C, Strand C, Leonard AJ, bou-Sleiman PM, Healy DG, Ariga H, Wood NW, de SR, Revesz T, Hardy JA, Lees AJ: The expression of DJ-1 (PARK7) in normal human CNS and idiopathic Parkinson's disease. *Brain* 2004, 127:420-430.

[165] Olzmann JA, Brown K, Wilkinson KD, Rees HD, Huai Q, Ke H, Levey AI, Li L, Chin LS: Familial Parkinson's disease-associated L166P mutation disrupts DJ-1 protein folding and function. *J Biol Chem* 2004, 279:8506-8515.

[166] Canet-Aviles RM, Wilson MA, Miller DW, Ahmad R, McLendon C, Bandyopadhyay S, Baptista MJ, Ringe D, Petsko GA, Cookson MR: The Parkinson's disease protein DJ-1 is neuroprotective due to cysteine-sulfinic acid-driven mitochondrial localization. *Proc Natl Acad Sci U S A* 2004, 101:9103-9108.

[167] Kinumi T, Kimata J, Taira T, Ariga H, Niki E: Cysteine-106 of DJ-1 is the most sensitive cysteine residue to hydrogen peroxide-mediated oxidation in vivo in human umbilical vein endothelial cells. *Biochem Biophys Res Commun* 2004, 317:722-728.

[168] Valente EM, Bentivoglio AR, Dixon PH, Ferraris A, Ialongo T, Frontali M, Albanese A, Wood NW: Localization of a novel locus for autosomal recessive early-onset parkinsonism, PARK6, on human chromosome 1p35-p36. *Am J Hum Genet* 2001, 68:895-900.

[169] Valente EM, bou-Sleiman PM, Caputo V, Muqit MM, Harvey K, Gispert S, Ali Z, Del TD, Bentivoglio AR, Healy DG, Albanese A, Nussbaum R, Gonzalez-Maldonado R, Deller T, Salvi S, Cortelli P, Gilks WP, Latchman DS, Harvey RJ, Dallapiccola B, Auburger G, Wood NW: Hereditary early-onset Parkinson's disease caused by mutations in PINK1. *Science* 2004, 304:1158-1160.

[170] Muqit MM, bou-Sleiman PM, Saurin AT, Harvey K, Gandhi S, Deas E, Eaton S, Payne S, Venner K, Matilla A, Healy DG, Gilks WP, Lees AJ, Holton J, Revesz T, Parker PJ, Harvey RJ, Wood NW, Latchman DS: Altered cleavage and localization of PINK1 to aggresomes in the presence of proteasomal stress. *J Neurochem* 2006, 98:156-169.

[171] Park J, Lee SB, Lee S, Kim Y, Song S, Kim S, Bae E, Kim J, Shong M, Kim JM, Chung J: Mitochondrial dysfunction in Drosophila PINK1 mutants is complemented by parkin. *Nature* 2006, 441:1157-1161.

[172] Yang Y, Gehrke S, Imai Y, Huang Z, Ouyang Y, Wang JW, Yang L, Beal MF, Vogel H, Lu B: Mitochondrial pathology and muscle and dopaminergic neuron degeneration caused by inactivation of Drosophila Pink1 is rescued by Parkin. *Proc Natl Acad Sci U S A* 2006, 103:10793-10798.

[173] Wang D, Qian L, Xiong H, Liu J, Neckameyer WS, Oldham S, Xia K, Wang J, Bodmer R, Zhang Z: Antioxidants protect PINK1-dependent dopaminergic neurons in Drosophila. *Proc Natl Acad Sci U S A* 2006, 103:13520-13525.

[174] Funayama M, Hasegawa K, Ohta E, Kawashima N, Komiyama M, Kowa H, Tsuji S, Obata F: An LRRK2 mutation as a cause for the parkinsonism in the original PARK8 family. *Ann Neurol* 2005, 57:918-921.

[175] Galter D, Westerlund M, Carmine A, Lindqvist E, Sydow O, Olson L: LRRK2 expression linked to dopamine-innervated areas. *Ann Neurol* 2006, 59:714-719.

[176] Taymans JM, Van den HC, Baekelandt V: Distribution of PINK1 and LRRK2 in rat and mouse brain. *J Neurochem* 2006, 98:951-961.

[177] Mata IF, Wedemeyer WJ, Farrer MJ, Taylor JP, Gallo KA: LRRK2 in Parkinson's disease: protein domains and functional insights. *Trends Neurosci* 2006, 29:286-293.

[178] Paisan-Ruiz C, Jain S, Evans EW, Gilks WP, Simon J, van der BM, Lopez de MA, Aparicio S, Gil AM, Khan N, Johnson J, Martinez JR, Nicholl D, Carrera IM, Pena AS, de SR, Lees A, Marti-Masso JF, Perez-Tur J, Wood NW, Singleton AB: Cloning of the gene containing mutations that cause PARK8-linked Parkinson's disease. *Neuron* 2004, 44:595-600.

[179] Smith WW, Pei Z, Jiang H, Moore DJ, Liang Y, West AB, Dawson VL, Dawson TM, Ross CA: Leucine-rich repeat kinase 2 (LRRK2) interacts with parkin, and mutant LRRK2 induces neuronal degeneration. *Proc Natl Acad Sci U S A* 2005, 102:18676-18681.

[180] Webb JL, Ravikumar B, Rubinsztein DC: Microtubule disruption inhibits autophagosome-lysosome fusion: implications for studying the roles of aggresomes in polyglutamine diseases. *Int J Biochem Cell Biol* 2004, 36:2541-2550.

[181] Fortun J, Dunn WA, Jr., Joy S, Li J, Notterpek L: Emerging role for autophagy in the removal of aggresomes in Schwann cells. *J Neurosci* 2003, 23:10672-10680.

[182] Cuervo AM: Autophagy: many paths to the same end. *Mol Cell Biochem* 2004, 263:55-72.

[183] Massey AC, Zhang C, Cuervo AM: Chaperone-mediated autophagy in aging and disease. *Curr Top Dev Biol* 2006, 73:205-235.

[184] Menzies FM, Ravikumar B, Rubinsztein DC: Protective roles for induction of autophagy in multiple proteinopathies. *Autophagy* 2006, 2:224-225.

[185] Rubinsztein DC, Difiglia M, Heintz N, Nixon RA, Qin ZH, Ravikumar B, Stefanis L, Tolkovsky A: Autophagy and its possible roles in nervous system diseases, damage and repair. *Autophagy* 2005, 1:11-22.

[186] Klionsky DJ, Cregg JM, Dunn WA, Jr., Emr SD, Sakai Y, Sandoval IV, Sibirny A, Subramani S, Thumm M, Veenhuis M, Ohsumi Y: A unified nomenclature for yeast autophagy-related genes. *Dev Cell* 2003, 5:539-545.

[187] Marino G, Lopez-Otin C: Autophagy: molecular mechanisms, physiological functions and relevance in human pathology. *Cell Mol Life Sci* 2004, 61:1439-1454.

[188] Bergamini E: Autophagy: a cell repair mechanism that retards ageing and age-associated diseases and can be intensified pharmacologically. *Mol Aspects Med* 2006, 27:403-410.

[189] Komatsu M, Kominami E, Tanaka K: Autophagy and neurodegeneration. *Autophagy* 2006, 2:315-317.

[190] Matthews MR, Raisman G: A light and electron microscopic study of the cellular response to axonal injury in the superior cervical ganglion of the rat. *Proc R Soc Lond B Biol Sci* 1972, 181:43-79.

[191] Anglade P, Vyas S, Javoy-Agid F, Herrero MT, Michel PP, Marquez J, Mouatt-Prigent A, Ruberg M, Hirsch EC, Agid Y: Apoptosis and autophagy in nigral neurons of patients with Parkinson's disease. *Histol Histopathol* 1997, 12:25-31.

[192] Stefanis L, Larsen KE, Rideout HJ, Sulzer D, Greene LA: Expression of A53T mutant but not wild-type alpha-synuclein in PC12 cells induces alterations of the ubiquitin-dependent degradation system, loss of dopamine release, and autophagic cell death. *J Neurosci* 2001, 21:9549-9560.

[193] Bandhyopadhyay U, Cuervo AM: Chaperone-mediated autophagy in aging and neurodegeneration: Lessons from alpha-synuclein. *Exp Gerontol* 2006.

[194] Cuervo AM, Stefanis L, Fredenburg R, Lansbury PT, Sulzer D: Impaired degradation of mutant alpha-synuclein by chaperone-mediated autophagy. *Science* 2004, 305:1292-1295.

[195] Webb JL, Ravikumar B, Atkins J, Skepper JN, Rubinsztein DC: Alpha-Synuclein is degraded by both autophagy and the proteasome. *J Biol Chem* 2003, 278:25009-25013.

[196] Hara T, Nakamura K, Matsui M, Yamamoto A, Nakahara Y, Suzuki-Migishima R, Yokoyama M, Mishima K, Saito I, Okano H, Mizushima N: Suppression of basal autophagy in neural cells causes neurodegenerative disease in mice. *Nature* 2006, 441:885-889.

[197] Komatsu M, Waguri S, Chiba T, Murata S, Iwata J, Tanida I, Ueno T, Koike M, Uchiyama Y, Kominami E, Tanaka K: Loss of autophagy in the central nervous system causes neurodegeneration in mice. *Nature* 2006, 441:880-884.

[198] Lee VM, Trojanowski JQ: Mechanisms of Parkinson's disease linked to pathological alpha-synuclein: new targets for drug discovery. *Neuron* 2006, 52:33-38.

[199] Polymeropoulos MH, Lavedan C, Leroy E, Ide SE, Dehejia A, Dutra A, Pike B, Root H, Rubenstein J, Boyer R, Stenroos ES, Chandrasekharappa S, Athanassiadou A, Papapetropoulos T, Johnson WG, Lazzarini AM, Duvoisin RC, Di IG, Golbe LI, Nussbaum RL: Mutation in the alpha-synuclein gene identified in families with Parkinson's disease. *Science* 1997, 276:2045-2047.

[200] Farout L, Friguet B: Proteasome function in aging and oxidative stress: implications in protein maintenance failure. *Antioxid Redox Signal* 2006, 8:205-216.

[201] McNaught KS, Olanow CW, Halliwell B, Isacson O, Jenner P: Failure of the ubiquitin-proteasome system in Parkinson's disease. *Nat Rev Neurosci* 2001, 2:589-594.

[202] McNaught KS, Perl DP, Brownell AL, Olanow CW: Systemic exposure to proteasome inhibitors causes a progressive model of Parkinson's disease. *Ann Neurol* 2004, 56:149-162.

[203] Vernace VA, Arnaud L, Schmidt-Glenewinkel T, Figueiredo-Pereira ME: Aging perturbs 26S proteasome assembly in Drosophila melanogaster. *FASEB J* 2006, in press.

[204] Hoyer S, Lannert H, Latteier E, Meisel T: Relationship between cerebral energy metabolism in parietotemporal cortex and hippocampus and mental activity during aging in rats. *J Neural Transm* 2004, 111:575-589.

[205] Nohl H, Kramer R: Molecular basis of age-dependent changes in the activity of adenine nucleotide translocase. *Mech Ageing Dev* 1980, 14:137-144.

[206] Navarro A: Mitochondrial enzyme activities as biochemical markers of aging. *Mol Aspects Med* 2004, 25:37-48.

[207] Navarro A, Boveris AA: The mitochondrial energy transduction system and the aging process. *Am J Physiol Cell Physiol* 2006. Oct 4; [Epub ahead of print].

[208] Kidd T, bu-Shumays R, Katzen A, Sisson JC, Jimenez G, Pinchin S, Sullivan W, Ish-Horowicz D: The epsilon-subunit of mitochondrial ATP synthase is required for normal spindle orientation during the Drosophila embryonic divisions. *Genetics* 2005, 170:697-708.

[209] Schapira AH, Mann VM, Cooper JM, Dexter D, Daniel SE, Jenner P, Clark JB, Marsden CD: Anatomic and disease specificity of NADH CoQ1 reductase (complex I) deficiency in Parkinson's disease. *J Neurochem* 1990, 55:2142-2145.

[210] Schapira AH, Cooper JM, Dexter D, Clark JB, Jenner P, Marsden CD: Mitochondrial complex I deficiency in Parkinson's disease. *J Neurochem* 1990, 54:823-827.

In: The Ubiquitin Proteasome System… ISBN 978-1-60021-749-4
Eds: Mario Di Napoli and Cezary Wojcik, pp. 793-812

Chapter 32

THE UBIQUITIN-PROTEASOME SYSTEM IN HUNTINGTON'S DISEASE AND OTHER POLYGLUTAMINE DISEASES

Louise Kelly* and Elsdon Storey¥

Department of Neurosciences, Monash University, Alfred Hospital, Melbourne, Victoria 3001, Australia.

ABSTRACT

Huntington's disease (HD) belongs to a group of nine polyglutamine (polyQ) tract disorders, which also includes spinocerebellar ataxias (SCA's) types -1, 2, 3, 6, 7, and 17, spinobulbar muscular atrophy (SMBA) and dentatorubral-pallidoluysian atrophy (DRPLA). The proteins involved in each of these disorders show no homology to one another except for an expanded polyQ tract. Although each protein is ubiquitously expressed throughout the central nervous system (and most non-neural tissues), only a distinct subset of neurons is affected in each disease, with only partial overlap between each. A common feature of these diseases is the formation of polyQ-containing intraneuronal inclusions, which are typically also immunoreactive for ubiquitin. However, the pathogenesis of these diseases is unknown, and there is much debate as to whether it is the inclusions themselves that are pathogenic or whether they are merely markers of disease. One suggestion, on the basis of numerous studies showing co-localisation of various other proteins with the inclusions, has been that the inclusions contribute to pathogenesis, interfering with normal cellular functioning by trapping components such as transcription factors, molecular chaperones, and components of the

* Correspondence concerning this article should be addressed to Dr. Louise Kelly, Monash University, Department of Neurosciences, 4th Floor Centre Block, Alfred Hospital, Commercial Road, Melbourne, Victoria 3001, Australia. Phone: + 61 3 9276 2697; Fax:+ 61 3 9207 1043; Email: louise.kelly@med.monash.edu.au.

¥ Correspondence concerning this article should be addressed to Prof. Elsdon Storey, Monash University, Department of Neurosciences, 4th Floor Centre Block, Alfred Hospital, Commercial Road, Melbourne, Victoria 3001, Australia. Phone:+ 61 3 9276 2552; Fax:+ 61 3 9276 2458; Email: elsdon.storey@med.monash.edu.au.

ubiquitin-proteasome system (UPS), and thus preventing them from carrying out their normal functions. However, this theory is disputed, with other studies suggesting that the inclusions may in fact be a form of cellular defense. Other evidence against a toxic role for aggregates includes the short-stop HD animal model, where increased inclusion formation accompanies decreased neuronal death, and SCA-1 models in which the protein is mutated so as not to form aggregates but toxicity is still seen. As the inclusions in these polyQ diseases are ubiquitinated, the role of the UPS in their pathogenesis has come under scrutiny. While some studies show that the function of the UPS is impaired in these disorders, other studies report no loss of function. Any impairment of the UPS may relate to difficulty with degradation of expanded or perhaps just aggregated polyQ proteins, although the evidence for such difficulty is also conflicting. It has been reported that UPS components are sequestered irreversibly into aggregates of polyQ-containing huntingtin (Htt) fragments. These Htt fragments were incompletely degraded, and a stable interaction between the polyQ-bearing proteins and the proteasome was seen. Such a stable interaction with non-degradable, aggregated polyQ proteins might result in depletion of proteasomal activity. In support of this suggestion, other studies have reported that UPS impairment is seen in the presence of aggregated polyQ-bearing proteins, and that this is evident in both the cytoplasmic and the nuclear compartments, even when the aggregated protein sequences are targeted to either the nucleus or cytoplasm alone. However, this impairment is also seen in cells where there are no detectable aggregates or toxicity, suggesting that UPS overload may not be a factor in neurotoxicity. Furthermore, an animal model of SCA-7 has shown that while there is neuronal damage in susceptible cells, the UPS remains functional in these neurons. Here we review the conflicting evidence from previous studies of the UPS in polyQ disorders, and discuss both the possible roles of the UPS in the pathogenesis of these diseases and the effect of inclusion formation on the UPS.

Keywords: Huntington's disease, Polyglutamine, Spinocerebellar ataxia, ubiquitin, pathogenesis, proteasome, nuclear inclusion.

ABBREVIATIONS

AD, Alzheimer's disease; CBP, CREB-binding protein; DRPLA, dentatorubral-pallidoluysian atrophy; ER, endoplasmic reticulum; Htt, huntingtin; HD, Huntington's disease; HRPT, hypoxanthine phosphoribosyltransferase; polyQ, polyglutamine; SMBA, spinobulbar muscular atrophy; SCA, spinocerebellar ataxia; SCA-1, spinocerebellar ataxia type 1; SCA-2, spinocerebellar ataxia type 2; SCA-3, spinocerebellar ataxia type 3; SCA-7, spinocerebellar ataxia type 7, VCP, valosin-containing protein.

INTRODUCTION

The dominantly inherited polyglutamine diseases include Huntington's disease (HD), spinocerebellar ataxias (SCA's) types -1, 2, 3, 6, 7, and 17, spinobulbar muscular atrophy (SMBA) and dentatorubral-pallidoluysian atrophy (DRPLA). The various proteins involved in each of these diseases all contain a polyglutamine tract (polyQ) that is expanded beyond a specific threshold, resulting in neuronal cell toxicity [1]. The expanded polyQ tract is the only homology shared by these proteins. Despite their ubiquitous expression within (and beyond) the CNS, only certain subsets of neurons are affected, with incomplete overlap between the diseases. These expanded polyQ proteins tend to misfold and aggregate, giving rise to one of the characteristic features of these disorders: intranuclear inclusions (Figure 1). These inclusions consist of the full-length expanded polyQ protein, and/or a fragment of the mutant protein bearing the polyQ tract. Other proteins associated with these nuclear inclusions include components of the ubiquitin-proteasome system (UPS), various chaperones, and transcription factors [2, 3]. Although intranuclear inclusions are the characteristic feature of these diseases cytoplasmic inclusions have been reported in SMBA [4], HD [5], and SCA1 [5]. These cytoplasmic inclusions (aggresomes, see Chapter 12) are similar in composition to the nuclear inclusions but do not bind transcription factors, they are removed from the cell via autophagy and appear to be less toxic to the cell [5]. Whether these inclusions are pathogenic or merely markers of disease is subject to debate, with a number of studies supporting each premise.

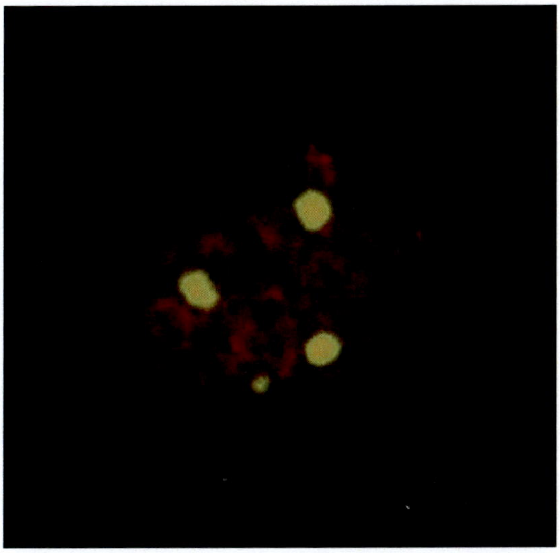

Figure 1. Intranuclear inclusions in a PC12 cell. The nucleus (red) of a PC12 cell (rat pheochromocytoma cell line) showing ataxin-1 intranuclear inclusions (green).

There is only limited information on the mechanisms of pathogenesis of the polyQ diseases, but much discussion revolves around whether it is the inclusions that are themselves toxic, or whether they are merely markers of protein misaggregation and reflect a protective mechanism of the cell trying to rid itself of misfolded proteins via the UPS. The polyQ proteins have been shown to have a propensity to misfold and aggregate [7]. This was confirmed in a mouse model with an expanded polyQ tract out of context, [inserted into the hypoxanthine phosphoribosyltransferase (HRPT) protein], which led to both inclusion formation and neurological abnormalities [8]. Further to this, an *in vitro* study showed inclusion formation and cell death following transfection of primary rat neurons with an expanded polyQ tract devoid of host proteins [9]. While these findings suggest that the polyQ tract is important in the formation of the inclusions, they do not address the question of whether it is the inclusions themselves that are pathogenic. Some studies have noted that neurotoxicity can occur without the formation of visible inclusions: for example in SCA-1 transgenic mice [10], and in a transfected cell model of HD [11]. In both studies it was shown that nuclear localization of the mutant protein rather than inclusion formation was critical for the development of neurotoxicity. A SCA-7 mouse model also showed dysfunction in neurons weeks before inclusions became visible [12], and in SCA-2 patients the most vulnerable neurons — Purkinje cells — do not appear to develop inclusions, which supports the hypothesis that inclusions may be protective [13]. Moreover, numerous different neurons are seen to have inclusions, including both those that are susceptible and those that are non-susceptible to the disease in SCA-7 [14]. A dissociation between inclusion formation (increased) and neurotoxicity (decreased) is also seen in the short-stop mouse model of HD [15]. Conversely, and despite the evidence from human SCA-2, there is evidence that neurotoxicity does not occur in cells that do not develop nuclear inclusions in a SCA-1 transgenic mouse model [10] (Figure 2).

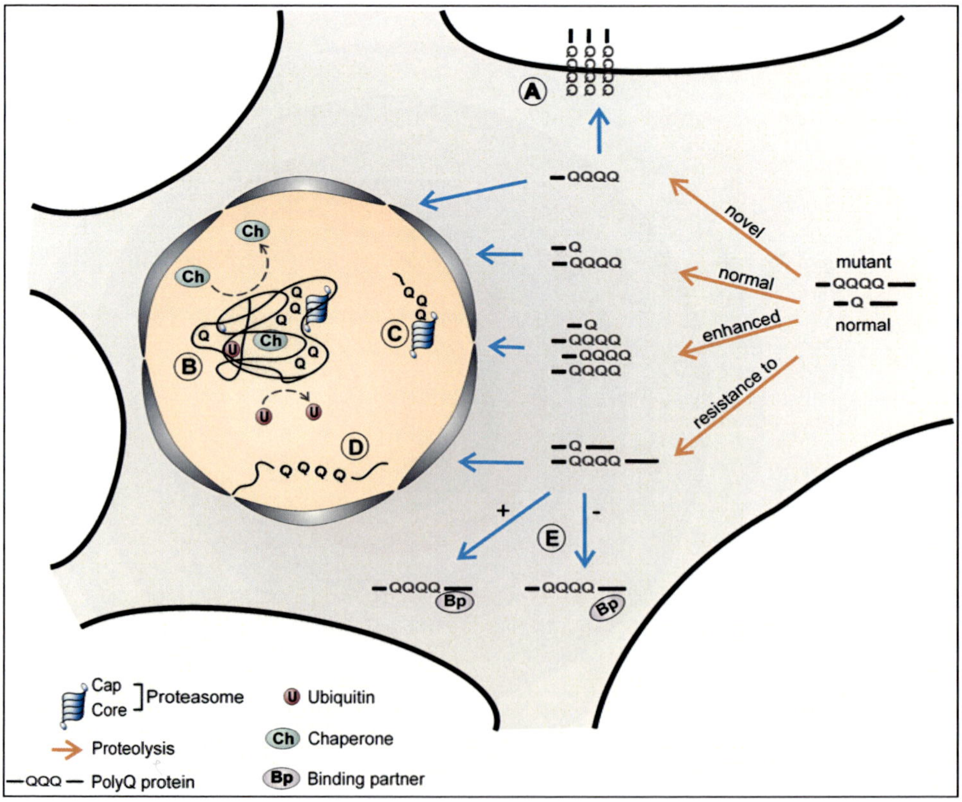

Figure 2. Possible outcomes of proteolysis of expanded polyQ proteins. Whether proteolysis of expanded polyQ proteins is unaffected, enhanced, reduced or the result of novel proteolysis there are several possible toxic outcomes for the cell. Proteolysis can lead to the formation of truncated expanded polyQ proteins which may have deleterious effects on the cell, including; (A) formation of cationic channels in the cell membrane, (B) formation of intranuclear inclusions which recruits components of the UPS, (C) resistance to proteolysis by the proteasome, (D) disruption of the nuclear matrix, (E) alteration of normal protein binding partner interactions. (Adapted from Tarlac & Storey (2003) [6])

It is apparent from these studies that the expanded polyQ proteins tend to form aggregates, but it is possible that toxicity precedes or is independent of aggregate formation. Needless to say, extensive research efforts to elucidate the mechanism(s) of polyQ disease pathogenesis continue. Given that proteins from the UPS and chaperones have been shown to interact with these inclusions, it is hardly surprising that the role of the UPS in the pathology of these diseases has come under increasing scrutiny over recent years. Much of the resulting evidence is contradictory, and consensus on the role of the UPS in these disorders is yet to be reached. The following two sections summarize the evidence for (Table 1) and against (Table 2) UPS involvement in the pathogenesis of the polyQ diseases, respectively.

Table 1. Summary of evidence *FOR* UPS involvement/impairment in the polyQ diseases

1.	UPS components found within inclusions and re-distributed in inclusion-bearing cells
	- ubiquitin
	- chaperones (HDJ2/HSDJ, HSP90□, HSP70, HSP90)
	- proteosome components (26S, 20S, 19S, 11S, LMP2,VCP)
2.	Degradation of polyQ-containing proteins by the UPS is impaired
	- polyQ cannot be degraded by the UPS, or
	- polyQ degradation by the UPS is is polyQ tract length dependent , or
	- poly Q cannot be degraded by the UPS once the polyQ protein has formed inclusions
3.	Inhibition of UPS leads to increased polyQ aggregate formation
4.	Mutations in UPS and over-expression of chaperones have an opposite effect on polyQ neurotoxicity
	- loss-of-function mutations in UPS lead to neurological disease
	- loss-of-function mutations in UPS lead to increased aggregate formation
	- over expression of chaperones leads to decreased inclusion formation
	- over-expression of chaperones leads to improved motor function in polyQ animals
5.	UPS is inhibited in cells transfect with polyQ proteins
6.	ER stress is seen cells transfected with polyQ proteins
7.	UPS involvement has been reported in other neurological diseases
	AD, PD, Prion diseases

Evidence for the Involvement of the UPS in the PolyQ Dusease

Association of Ubiquitin Proteasome System Proteins and Chaperones with PolyQ Inclusions

The inclusions seen in the polyQ diseases not only contain the polyQ proteins, but also numerous other proteins including those of the UPS, chaperones and transcription factors. Ubiquitinated polyQ inclusions have been seen in SCA-1 transgenic mice, ataxin-1 and ataxin-3 transfected cells, and affected brain regions of SCA-1 patients [16], SCA-3 patients

[17, 18], and HD patients [19], although not all inclusions seen in the brains of HD patients were ubiquitinated [20]. Interestingly, while SCA-7 patients have numerous inclusions in neurons unaffected by disease, only those inclusions in susceptible cells are ubiquitinated [21]. This finding is strengthened by the fact that ataxin-7-containing inclusions in COS cells are not ubiquitinated, suggesting that there is some cell specificity to ubiquitination, which may be necessary for — or at least a marker of — neurotoxicity. The 26S proteasome is comprised of a number of subunits which include the barrel-shaped 20S catalytic core, and the 19S and 11S regulatory caps (see Chapters 5, 6 and 7). Various components of the proteasome are also reported to be associated with polyQ inclusions, including the immunoproteasome subunit (LMP2) (a subunit that is induced in antigen-presenting cells and results in generation of peptides that are optimal for MHC-I antigen presentation) in a transgenic HD mouse [22], the 20S proteasome subunit in HD [23], and the 26S proteasome in SCA-3 [24]. SCA-1 inclusions in human and transgenic mice stain immunopositive for the 20S proteasome subunit and the heat shock protein, HDJ2/HSDJ, and these findings have been confirmed in transfected cell models [25]. Another molecule associated with the polyQ proteins and inclusions is VCP [26], a poly-ubiquitin binding protein with a function in the UPS (see Chapter 13). VCP has been shown to bind ataxin-3 in a manner dependent on the length of the polyQ tract [27, 28]. A loss-of-function in the *ter94* gene (a homologue of VCP) in a *Drosophilia* model of polyQ disease led to the suppression of polyQ- mediated neurodegeneration, though not inclusion formation, and co-expression of *ter94* and polyQ protein led to severely enhanced degeneration [29]. Over-expression of *ter94* also led to degeneration, suggesting a role as a cell death effector molecule.

While these findings demonstrate that there is interaction between the polyQ proteins and those of the UPS, and hint at impaired UPS function as a possible mechanism of neurotoxicity, they still do not confirm this suggestion. However, from these initial findings it would at least seem reasonable to hypothesise that the UPS has a role to play in the development of, or perhaps the defence against, neurotoxicity. Various experiments have therefore examined whether there is direct inhibition or overload of the UPS, preventing it from carrying out its other cellular functions.

Table 2. Summary of evidence AGAINST UPS involvement/impairment in the polyQ diseases

1.	UPS components are found within inclusions, BUT
	This is expected as it is the job of the UPS to remove misfolded proteins inclusions are dynamic and UPS components are not irreversibly bound to inclusions
2.	Degradation of polyQ-containing proteins by the UPS is not impaired
3.	Inhibition of UPS leads to increased aggregate formation, BUT
	This is expected as its the job of the UPS to remove misfolded proteins there is evidence that the aggregates are protective rather than toxic
4.	UPS function is not impaired in the polyQ diseases
5.	UPS components are not re-distributed in polyQ diseases

Degradation of PolyQ Proteins by the Ubiquitin Proteasome System

There is disagreement as to the resistance of polyQ tracts to proteasomal degradation. There are various contradictory studies that show: polyQ tracts cannot be degraded, they can be degraded, they can be degraded but only slowly, and that once they have formed inclusions they cannot be degraded.

PolyQ Tracts Cannot be Degraded by the UPS

Normally, proteins labelled for degradation with ubiquitin enter the 26S proteasome where they are digested to give rise to peptides ranging from 2 to 26 residues long. These residues are then released from the proteasome and hydrolysed by cytoplasmic and nuclear peptidases. Some recent work has suggested that eukaryotic proteasomes cannot digest the polyQ sequences within polyQ proteins, even those of non-pathogenic lengths. PolyQ-bearing proteins entering the 20S proteasome core were only partially degraded. The polyQ sequence could not be digested, but was released back into the cell [30]. Interestingly, this appears to be an evolutionary phenomenon, as proteasomes from the archaebacteria *Thermoplasma acidophilum* can digest polyQ-bearing proteins, including the polyQ sequence. Accumulation of these polyQ proteins, and the continual but unavailing attempts of the UPS to degrade them, may lead to a reduction in the cell's ability to degrade other proteins, and thus exert a toxic effect on the cell.

PolyQ Tracts can be Degraded by the UPS

In contradistinction to the results discussed in the previous section, a number of studies have shown that the UPS *can* degrade the polyQ proteins, but that this degradation is polyQ-length dependent, such that the longer the polyQ tract the slower is the degradation. Both mouse and *in vitro* models [31] have been used to show that ataxin-1 can be degraded by the UPS, with both wild-type and mutant ataxin-1 being polyubiquitinated. However, although mutant ataxin-1 is three times more resistant to degradation than the wild-type protein, inhibition of the UPS resulted in accelerated aggregate formation, suggesting that UPS-mediated proteolysis of at least some species of expanded polyQ proteins may be important. Others have confirmed that both normal and expanded polyQ-containing proteins are degraded by the UPS, with the rate of degradation being inversely proportional to the length of the polyQ tract [23]. These findings further suggest that the slow degradation of the expanded polyQ proteins and consequent re-distribution of the proteasome may lead to decreased availability and function of the UPS components in other parts of the cell. Concomitantly, it may be that the synthesis of expanded polyQ proteins is faster than the rate of their degradation by the UPS, and that this leads to the accumulation of the polyQ proteins in inclusions. Holmberg *et al.* (2004) [32] have shown that the proteasome is irreversibly bound to inclusions in cells that over-express either mutant huntingtin (Htt) or other simple expanded polyQ proteins, and that targeting the polyQ proteins for degradation by the proteasome led to only incomplete degradation both *in vitro* and *in vivo*. (These researchers used the N-end rule pathway where a ubiquitin moiety is linked to the N-terminal of a protein using a 40 amino-acid linker. The ubiquitin is readily cleaved off by ubiquitin hydrolases, leaving an unstable and readily ubiquitinated N-terminus on the protein [33]).

PolyQ Proteins are not Degraded by the UPS once they have Formed Inclusions

Contrary to the finding that mutant Htt and simple polyQ proteins could be targeted for partial degradation using the N-end rule pathway [32], in an experiment where ataxin-1 protein containing an expanded polyQ tract was targeted for degradation using an N-end rule degradation signal, soluble polyQ proteins were degraded by proteasomes, but once in aggregates they resisted degradation. Introduction of the degradation signal into mutant ataxin-1 not only reduced inclusion formation but decreased cellular toxicity [34], showing that increasing the efficiency of expanded polyQ protein degradation by the UPS is protective, and suggesting that decreased UPS activity may increase neurotoxicity.

Effects of Inhibiting Proteosomal Function on PolyQ Protein Aggregation

Inhibition of the proteasome with lactacystin in a transfected SCA-3 cell model has been shown to lead to an increase in aggregate formation which is dependent on the length of the polyQ repeat, suggesting that the proteasome plays a role in suppressing aggregate formation [24]. Although the proteasome was inhibited, its 20S catalytic core subunit was seen to associate with the aggregates [24]. In keeping with this, lactacystin inhibition of proteosomal degradation in ataxin-1 transfected cells leads to increased aggregate formation [31]. These researchers also showed that in a SCA-1 mouse model lacking the ubiquitin ligase E6-AP there was a reduction in the number of inclusions formed in Purkinje cells and that these animals had otherwise severe SCA-1 pathology, suggesting that nuclear inclusions are not necessary for neurotoxicity but that impaired degradation by the UPS may contribute to SCA-1 pathology. Moreover, addition of either of two different proteasome inhibitors (ALLN or lactacystin) dramatically increased the rate of aggregate formation by moderately expanded Htt (Q60), but had little influence on that of grossly expanded Htt (Q150) in a cellular model of HD [23].

Opposite Effects of Proteosomal Loss-of-Function Mutations and Over-Expression of chaperones

Loss of Function Mutations Affecting the UPS

Evidence for impairment of the UPS in the polyQ diseases also comes from observations in experiments in which the function of the UPS is disrupted by mutation of specific genes. Loss-of-function mutation in genes coding for various components of the UPS leads to neurodegenerative diseases in humans [Parkinson's disease (PD)] [35], rodents [36, 37] and *Drosophila* [38]. Loss-of-function mutations in cell and mouse models of SCA-1 result in enhanced cytotoxicity of the protein [31], as do mutations in the UPS in a *Drosophila* SBMA model [39]. Supporting information for the importance of the UPS in preventing toxicity comes from the report that an aberrant form of ubiquitin (resulting from a dinucleotide deletion) that inhibits the proteasome, has been found in inclusions of HD and SCA-3 patients. When this form of ubiquitin was co-transfected into cells with polyQ proteins there was an increase in inclusion formation and cytotoxicity [40]. However, in contrast, in a recent study a SCA-1 mouse model was crossed with a mouse model in which a mutant form of ubiquitin, which interferes with ubiquitin chain assembly, is expressed. No exacerbation of SCA-1 pathology was observed in the double transgenic animals, and a protective effect with regards to Purkinje cell neurotoxicity was reported, suggesting that inhibiting proteolysis of the ataxin-1 protein was protective to the cells [41].

Over-Expression of Chaperones

Conversely, reduction of polyQ protein misfolding should lead to a decrease in aggregation, and therefore perhaps in neurotoxicity. On co-expression of an expanded polyQ tract-bearing androgen receptor (the mutant protein resulting in SMBA) with the chaperone HDJ2/HSDJ, there was significant repression of inclusion formation [42]. Over-expression of the chaperone DnaJ (Hsp40 family) has also been shown to promote recognition of misfolded ataxin-1 and suppress inclusion formation [25]. Furthermore, over-expression of Hsp70 in a SCA-1 mouse model not only suppressed neuropathology but also improved the motor function in the animals [43].

Collectively, these studies suggest that aggregate formation and neurotoxicity may result from decreased or insufficient levels of function of UPS components, and that increased chaperone activity, by reducing aggregation, reduces toxicity.

Depletion and Redistribution of Ubiquitin Proteasome System Proteins and Chaperones

A number of studies have investigated whether the 26S proteasome complex, its subunits, or chaperones associated with the UPS are depleted or re-distributed within the cell in the polyQ disorders.

11S and 19S Proteasome Subunits Show a Different Distribution Pattern to the 20S Core Subunit

Various studies have described different distribution patterns of proteasomal subunits in cells containing inclusions. While most of the intranuclear inclusions in pontine nuclei examined in SCA-3 patients were immunopositive for the 11S and 19S proteasomal protein cap subunits, only a few were immunopositive for the 20S core proteasome subunit [44]. Furthermore, inclusions in a cell model of SMBA were seen to associate with the 19S proteasome cap but not the 20S core proteasome subunit [32]. On comparing the distribution of a 20S subunit in pontine nuclei in SCA-3 brains with control samples it was noted that this 20S subunit was predominantly cytoplasmic in the SCA-3 brains and predominantly nuclear in the control brains. Where there was nuclear staining for the 20S subunit in the SCA-3 brains, this was seen in the nuclear inclusions [44] and the levels of some of the subunits that make up the 20S core were markedly increased in the cytoplasm of SCA-3 pontine nuclei compare to controls, suggesting that there is either a re-distribution or an up-regulation of a 20S core subunit in SCA-3.

20S, 11S, and 19S Proteasome Subunits are Found together in Inclusions, Possibly Resulting in Functional Proteasomal Impairment

Other studies, however, have demonstrated that the 20S, 11S and 19S subcomplexes all associate with the inclusions and do not have different distribution patterns. A study of the intranuclear inclusions in human SCA-3 brain samples revealed that all the inclusions were positive for ubiquitin and a subset were positive for subunits of the 20S core, 19S cap, and 11S activator of the 26S proteasome, suggesting that the inclusions are heterogeneous at the molecular level [24]. These findings were also reflected in several transfected cell models, with the subunits of both the 20S core and 19S cap are seen to localise with the nuclear inclusions. Gel electrophoresis of these inclusions showed that the proteasomes could be separated from the polyQ proteins; thus the proteasome does not appear to be irreversibly bound to the inclusions [24] (see below).

When truncated N-terminal Htt, containing different polyQ lengths, was expressed in cell and animal models the subunits of the 20S proteasome core and 19S cap were redistributed to aggregates [23]. This, along with the slower degradation of truncated N-terminal Htt with longer length polyQ tracts, decreased the proteasomes' ability to degrade other proteins such as p53, which led to a disruption in the membrane potential of mitochondria and release of cytochrome c into the cytosol with activation of caspase-9 and caspase-3-like proteases [23].

Impairment of UPS Function by Aggregating Proteins

Further evidence for impairment of UPS function by aggregated proteins in the polyQ diseases comes from Bence *et al.* (2005) [45]. Using a degron-GFP reporter (i.e. a GFP protein with a degron label, which targets the protein for degradation by the UPS), these researchers showed that the UPS was almost completely inhibited in cells expressing two unrelated aggregating proteins, Htt and cystic fibrosis transmembrane conductance regulator protein. Another study looking at cells expressing ataxin-1 or Htt showed that there was global (cysosolic and nuclear) impairment of the UPS, even when the production of the inclusions was targeted specifically to either the cytosol or the nucleus [46]. This study therefore demonstrated that there was impairment of the UPS in cellular compartments that lack detectable inclusions.

Depletion and Re-Distribution of the Chaperones in the PolyQ Diseases

Immunostaining of intranuclear inclusions in SCA-3 patients showed that certain inclusions (about 20%) were immunoreactive for the chaperone Hsp90α (an isoform of Hsp90), fewer were immunoreactive for the chaperone HDJ-2 (a member of the Hsp40 family), and that most of the inclusions were ubiquitin positive. While the inclusions did not stain for Hsp27, Hsp60, Hsp70, or Hsc70, these chaperones were shown to be present in the cytoplasm of cells in control tissues, showing that they are present in the cells but are not recruited to the inclusions. Aside from the intense labelling of some of the intranuclear inclusions there was no obvious redistribution of the Hsp on comparison to control tissues [44]. In a cell model of SMBA, cells transfected with the androgen receptor bearing an expanded polyQ tract were seen to form inclusions that were immunoreactive for Hsp90 and HDJ-2/HSDJ but unlike the SCA-3 model the inclusions were also positive for Hsp70. The

Hsp70 staining appeared to be increased, suggesting induction of the heat shock response [32].

The Role of Endoplasmic Reticulum Stress

Endoplasmic reticulum (ER) stress occurs when initial mediators—ER-resident type I transmembrane serine/threonine protein kinases (PERK and IRE1)—are autophosphorylated, leading to cytoplasmic signal transduction, and ultimately activation of the apoptotic pathway (see Chapter 13 for more information). ER stress has been implicated as another possible pathogenic pathway in the polyQ diseases. It has been proposed that the polyQ proteins interfere with the UPS, and that this in turn induces ER stress. This hypothesis is supported by a rat phaeochromocytoma (PC12) cell model of SCA-3, in which ER stress was seen in transfected cells expressing the mutant protein. As ER stress is induced by accumulation of unfolded protein in the ER, it seemed unlikely that the stress was as a direct effect of the polyQ protein, and other factors were therefore investigated. Expression of the mutant ataxin-3 protein in mouse primary neurons led to a significant inhibition of proteasome activity. Linking these two findings are the observations that inhibition of proteasomal function in primary neurons leads to ER stress [47]. These findings suggest that poly-Q mediated depletion of the UPS led to deficient ER-associated degradation (ERAD), with accumulation of misfolded proteins within the ER and therefore ER stress [47].

Evidence for UPS Involvement in other Neurodegenerative Diseases Characterised by Protein Aggregation

It should be noted that it is not only the polyQ diseases in which impairment of the UPS has been implicated in pathogenesis. Other neurodegenerative disorders, such as Alzheimer's disease (AD), prion diseases and PD have also been shown to involve the UPS (see *Chapters 22* and *27*). Human brain sections from prion disease and AD patients were shown to have stronger staining for ubiquitin, a proteasomal subunit and Hsp72 in those brain regions areas not affected by these diseases, and weaker staining in those areas that tend to be affected [48], with the nuclei being strongly positive for proteasomal subunits: 20S, S4, S6, and S7, and ubiquitin. Other studies have also noted the presence of ubiquitin in the various types of intracellular inclusions seen in these diseases, with 20S and S6b proteasome subunits and Hsp70 all being found in neurofibrillary tangles in AD and in Lewy bodies in dementia with Lewy bodies [49, 50, 51, 52].

In another study, Ardley *et al.* (2003) [53] have shown that inhibition of proteasomal function results in accumulation of parkin (the protein mutated in the commonest form of recessively-inherited PD), with formation of non-toxic cytoplasmic inclusions. This inclusion formation was not reversed by subsequent removal of the proteasome inhibitor. This would suggest that the UPS is unable to degrade inclusions once they are formed, but can prevent formation occurring.

These findings, along with the fact that other neurodegenerative disorders such as prion diseases, AD and PD appear to show impairment of the UPS [48], would seem to suggest that the UPS has a role to play in the pathogenesis of a number of neurodegenerative diseases characterised by protein misfolding/aggregation.

Evidence Against the Involvement of the Ubiquitin-Proteasome System in the PolyQ Diseases

Given that it is the role of chaperones to ensure the proper folding of proteins, and the UPS to remove misfolded proteins from the cell, it is hardly surprising that components of this system have been found to be associated with inclusions composed of misfolded proteins. Is it therefore possible that the UPS does not actually play a part in the pathogenesis of these diseases, but rather that it forms part of an ultimately inadequate cellular response to protein misfolding and aggregation.

PolyQ Protein Inclusions are Dynamic Structures, not Necessarily Irreversibly Binding their Associated Proteins

There have been several studies looking at the nature of the intranuclear inclusions seen in the polyQ diseases. These have led to the discovery that the inclusions are not all alike, and that they are dynamic rather than static structures. A study looking at ataxin-1 inclusions noted that there were two different types of inclusions; fast and slow exchanging. Ataxin-1 was seen to move readily between the inclusion and the nucleoplasmic pool in the fast-exchanging inclusions, whereas in the slow exchanging inclusions it did not [54]. Inclusions containing relatively high levels of ubiquitin and low levels of proteasomal components also contained both fast and slow exchanging ataxin-1. However, those inclusions with low levels of ubiquitin and high levels of proteasomal components contained mainly fast exchanging ataxin-1, implying that the proteasomal components are not irreversibly trapped in inclusions, but that there is a dynamic exchange with the surrounding environment. This concept of proteins moving in and out of inclusions is also supported by the fact that Hsp70 is only transiently bound to Htt, in a dynamic interaction [55, 56]. However, this might not be the case for all the proteins associated with the polyQ inclusions. Some proteins appear to become trapped irreversibly in the inclusions, while others have a dynamic relationship with them, and this may vary between the different polyQ proteins. For example, the CREB-binding protein (CBP) has been shown to be immobile when associated with Htt and ataxin-3 inclusions, but is mobile in relation to ataxin-1 inclusions [57].

PolyQ Proteins can be Degraded by the Ubiquitin Proteasome System

As previously discussed (section '*Degradation of polyQ proteins by the ubiquitin proteasome system*') the ability of the UPS to degrade polyQ-bearing proteins has been investigated, with conflicting results. Various studies have noted that the UPS cannot, or can only slowly, degrade polyQ proteins, or that once in inclusions these proteins become more resistant to degradation. Here we will look at the evidence showing that the UPS can readily degrade the polyQ proteins.

In an *in vitro* model of HD, cells expressing mutant Htt labelled for degradation through the N-rule pathway showed that the half-life of the protein depended on both the degradation signal and the length of the polyQ tract [58]. Mutant (expanded) Htt with a shorter half life, as a result of N-rule labelling, showed delayed formation of aggregates, but appeared to be *more* toxic to the cells, suggesting that the proteins may be readily degraded by the proteasome regardless of polyQ tract length, that UPS function is not impaired, and that the inclusions are protective, or at least inert. Furthermore, evidence for polyQ degradation by the UPS comes from Michalik *et al.* (2003) [59] who used a cell model transfected with polyQ proteins of lengths Q103 or Q25 and found that both were degraded efficiently and completely. Contrary to the evidence discussed in section *'PolyQ tracts cannot be degraded by the UPS'* and section '*PolyQ proteins are not degraded by the UPS once they have formed inclusions'*, soluble polyQ proteins do appear to be degraded by proteasomes, but once the polyQ proteins form aggregates they tend to resist degradation, as shown with ataxin-1 [34] and Htt [58]. This, too, has been challenged, however: two studies using conditional transgenic mouse models of HD and SCA-1 have noted that if the gene encoding the mutant polyQ protein is silenced after inclusion formation has occurred, there is a reversal of pathology and clearance of inclusions from the affected cells [60, 61]. Of course this clearance may have been achieved by mechanisms other than UPS degradation, such as dynamic exchange with a progressively depleted soluble pool of polyQ protein. There is evidence that autophagy plays a part in removal of cytoplasmic inclusions in HD and SCA1 [5], and SMBA [4], and that lysosomes and autophagic bodies proliferate in HD and AD [62, 63]. However, the predominantly nuclear inclusions that are seen in the polyQ diseases are not able to be removed by autophagy.

There is no Impairment of the Ubiquitin Proteasome System or Redistribution of its Components in some Models of PolyQ Diseases

Using a SCA-7 knock-in mouse model with a fluorescent ubiquitin reporter, Bowman *et al.* (2005) [14] showed that early on in the disease there was neuronal dysfunction without any impairment of the UPS, and that in the later stages of the disease there was an increase in both ubiquitin mRNA and protein levels. An *in vitro* assay confirmed that proteasomal activity in the vulnerable neurons remained normal. These researchers also confirmed the findings from human cases that intranuclear inclusions were found in all cells, not just those vulnerable in SCA-7, thus showing that even if UPS components are sequestered into inclusions, this by itself is unlikely to determine cytotoxicity. Despite the findings of Schmidt *et al.* (2002) [44] (see section *'Depletion and Redistribution of ubiquitin proteasome system proteins and chaperones'*) it has also been reported that there is no depletion of UPS components in the cytosol or nucleus of cells transfected with polyQ proteins targeted to either of these compartments [46]. There may, however, be depletion of other proteins via sequestration into the inclusions, as shown in a range of neurodegenerative diseases [64, 65], and this in turn may affect other processes in the cell.

Studies on HD transgenic and age-matched control mice showed that there is no difference in proteasome activity, but that there is an age-dependant decrease in proteasome activity that could explain the formation of inclusions later on in life [66]. This is further supported by the observations that components of the UPS are suppressed at the level of transcription in aged brains [67], and there is an age-related decrease in proteasome activity seen in rat brains [68]. An *in vivo* study measuring enzymatic activity in HD mouse brains did not detect inhibition of any of the ubiquitin-proteasomal enzymes, suggesting that if impairment does occur it is not at the level of the catalytic core; in fact two of the enzymes showed *increased* activity in brain areas affected in HD [18]. Western blot analysis showed no real difference in proteasomal content, although there was an increase in some of the immunoproteasome subunits (LMP2 and LMP7), both of which were found in the neurons of control mice but at a much lower level compared to those levels seen in the HD mice. Some of the aggregates in the HD mice were also positive for LMP2 [22].

Conclusion, Methodological Problems Particularly Affecting Contemporary Studies, and Future Directions

Numerous studies have shown that the proteins of the UPS and chaperones are associated with the inclusions seen in the polyQ diseases, and indeed with protein aggregates in other neurodegenerative disorders. Both the role of the inclusions, and of the UPS in the pathogenesis of these diseases, are still unclear, however. The critical question that remains to be answered is whether the association between the UPS and polyQ-containing proteins is merely an appropriate, albeit ultimately ineffectual, response to such aggregation, or whether the association itself contributes to neurotoxicity, perhaps by interference with normal UPS function. Considerable evidence can be marshalled to support either view, as summarised in sections '*Evidence for the involvement of the ubiquitin-proteasome system in the polyQ diseases*' and '*Evidence against the involvement of the ubiquitin-proteasome system in the polyQ diseases*', respectively. There are studies showing that the inclusions irreversibly bind the UPS components and impair their functions and/or redistribute them within the various cellular compartments. Some have shown that the polyQ proteins cannot be degraded by the ubiquitin proteasome pathway and that inhibition of the UPS as a result of attempted degradation leads to increased aggregate formation and cellular toxicity. For each of these findings there are other, contradictory studies: inclusions have been shown to be dynamic, allowing movement of UPS and other proteins in and out of the inclusions; some have failed to demonstrate any impairment or redistribution of UPS proteins; and there is evidence that the polyQ proteins can be degraded by the ubiquitin proteasome system, perhaps even once they have formed aggregates.

One point that has to be considered when assessing at these studies is that many of the animal models utilize extremely long polyQ tracts, very rarely encountered in human disease, and/or promoters that lead to higher than normal expression levels of the polyQ-containing proteins. Whilst these manipulations enable a disease phenotype, and possibly inclusions, to develop during the lifespan of a mouse, these models are not truly reflective of the (typically adult-onset) polyQ diseases in humans. For example, extremely long polyQ tracts are very rare except perhaps in SCA-7, and in this disorder they result in an infantile and lethal form of the disease that is multi-systemic, not just neuronal [69]. It may be that UPS involvement in these animal models occurs because of the increased length of polyQ and/or due to the higher than endogenous expression levels of the protein - the UPS being swamped as it tries to rid the cell of the misfolded, aggregated proteins.

Another consideration is that many of the *in vivo* studies look at the brain as a whole, rather than just the affected cells. In studies suggesting that there is no UPS involvement in polyQ disease, it may be that involvement of the UPS in the small subset of neurons that is susceptible in a given disorder is masked by the lack of change in other neurons and in glia. This problem has been specifically addressed in a recent study in which a SCA-7 mouse model was used to show that there was no impairment of the UPS. The study looked specifically at the vulnerable neurons and not just at the brain as a whole to determine whether there was specific UPS impairment [14].

Certainly more studies need to be carried out to investigate what role, if any, the UPS has to play in cytotoxicity. The fact that there seems to be a difference in the dynamics of the inclusions and composition of associated proteins in different polyQ diseases might suggest that there are different mechanisms involved in each, and that the role/involvement of the UPS could be different in each disease. In addition, the age of onset and the length of the polyQ tract could determine the relative importance of that role.

A number of studies have noted that the UPS appears to be affected in old age, with reported decreased levels of 26S proteasome activity with increasing age [70] (see Chapter 22). There is suppression of the UPS at the transcription levels in aged brains [67], and age-dependant decreased proteasome activity in rodent brains [66, 68]. The chaperone system has also been reported to deteriorate with age [71]. There also appears to be an accumulation of ubiquitinated polyQ proteins in inclusions over time, which might suggest an age-related failure of the UPS [72]. This would appear to provide a possible explanation for the age-related nature of these disorders, with onset of these diseases typically being seen at around 30-50 years old, followed by steady degeneration thereafter.

References

[1] Storey E: Dominantly inherited ataxias (part I). J Clin Neurosci 1998, 5:257-264.

[2] Everett CM, Wood NW: Trinucleotide repeats and neurodegenerative disease (review). *Brain* 2004, 127:2385-2405.

[3] Ross CA, Poirier MA: Protein aggregation and neurodegenerative disease (review). *Nat Med* 2004, 10:S10-S17.

[4] Taylor JP, Tanaka F, Robitschek J, Sandoval CM, Taye A, Markovic-Plese S, Fischbeck KH: Aggresomes protect cells by enhancing the degradation of toxic polyglutmine-containing protein. *Hum Mol Gen* 2003, 12:794-757.

[5] Iwata A, Christianson JC, Bucci M, Ellerby LM, Nukina N, Forno LS, Kopito RR: Increased susceptibility of cytoplasmic over nuclear plyglutamine aggregates at autophagic degradation. *PNAS* 2005, 102:13135-13140.

[6] Tarlac V, Storey E: Role of proteolysis in polyglutamine disorders (review). *J Neuro Res* 2003, 74:406-416

[7] Zoghbi HY, Orr HT: Glutamine repeats and neurodegeneration. *Ann Rev Neurosci* 2000, 23:217-247.

[8] Ordway JM. Cearley JA. Detloff PJ: CAG-polyglutamine-repeat mutations: independence from gene context. *Phil Trans Royal Soc - Series B: Biological Sciences* 1999, 354:1083-1088.

[9] Moulder KL, Onodera O, Burke JR, Strittmatter WJ, Jonson EM: Generation of neuronal intranuclear inclusions by polyglutamine-GFP: Analysis of inclusion clearance and toxicity as a function of polyglutamine length. *J Neurosci* 1999, 19:705-715.

[10] Klement IA, Skinner PJ, Kaytor MD, Yi H, Hersch SM, Clark HB, Zoghbi HY, Orr HT: Ataxin-1 nuclear localization and aggregation: Role in polyglutamine-induced disease in *SCA1* transgenic mice. *Cell* 1998, 95:41-53.

[11] Saudou F, Finkbeiner S, Devys D, Greenberg ME: Huntingtin acts in the nucleus to induce apoptosis but death does not correlate with the formation of intranuclear inclusions. *Cell* 1998, 95:55-66.

[12] Yoo S-Y, Pennesi ME, Weeber EJ, Xu B, Atkinson R, Chen S, Armstrong DL, Wu SM, Sweatt JD, Zoghbi HY: SCA7 knockin mice model human SCA7 and reveal gradual accumulation of mutant ataxin-7 in neurons and abnormalities in short-term plasticity. *Neuron* 2003, 37:383-401.

[13] Huyhn DP, Del Bigio MR, Ho DH, Pulst S-M: Expression of ataxin-2 in brains from normal individuals and patients with Alzheimer's disease and Spinocerebellar ataxia 2. *Ann Neurol* 1999, 45:232-241.

[14] Bowman AB, Yoo S-Y, Dantuma NP, Zoghbi HY: Neuronal dysfunction in a polyglutamine disease model occurs in the absence of unbiquitin-proteasome impairment and inversely correlates with the degree of nuclear inclusion formation. *Hum Mol Gen* 2005, 14:679-691.

[15] Slow EJ, Graham RK, Osmand AP, Devon RS, Lu G, Deng Y, Pearson J, Vaid K, Bissada N, Wetzel R, Leavitt BR, Hayden MR: Absence of behavioural abnormalities and neurodegeneration *in vivo* despite widespread neuronal huntingtin inclusions. *PNAS* 2005, 102:11402-11407.

[16] Skinner PJ, Koshy BT, Cummings CJ, Klement IA, Helin K, Servadio A, Zoghbi HY, Orr HT: Ataxin-1 with an expanded glutamine tract alters nuclear matrix-associated structures. *Nature* 1997, 389:971-974.

[17] Paulson HL, Perez MK, Trittier Y, Trojanowski JQ, Subramony SH, Das SS, Vig P, Mandel J-L, Fischbeck KH, Pittman RN: Intranuclear inclusions of expanded polyglutamine protein in spinocerebellar ataxia type 3. *Neuron* 1997, 19:333-344.

[18] Perez MK, Paulson HL, Pendse SJ, Saionz SJ, Bonini NM, Pittman RN: Recruitment and the role of nuclearlocalization in polyglutamine-mediated aggregation. *J Cell Biol* 1998, 143:1457-1470.

[19] Sieradzan KA, Mechan AO, Jones L, Wanker EE, Nukina N, Mann DMA: Huntington's disease intranuclear inclusions contain truncated, ubiquitinated huntingtin protein. *Exp Neurol* 1999, 156:92-99.

[20] Kuemmerle S, Gutekunst CA, Klein AM, Li XJ, Li SH, Beal MF, Hersch SM, Ferrante RJ: Huntington aggregates may not predict neuronal death in Huntington's disease. *Ann Neurol* 1999, 46:842-849.

[21] Holmberg M, Duyckaerts C, Durr A, Cancel G, Gourfinkel-An I, Damier P, Faucheux B, Trottier Y, Hirsch EC, Agid Y, Brice A: Spinocerebellar ataxia type 7 (SCA7): a neurodegenerative disorder with neuronal intranuclear inclusions. *Hum Mol Genet* 1998, 7:913-918.

[22] Diaz-Hernandez M, Hernandez F, Martin-Aparicio E, Gomez-Ramos P, Moran MA, Castano JG, Ferrer I, Avila J, Lucas JJ: Neuronal induction of the immunoproteasome in Huntington's disease. *J Neurosci* 2003, 23:11653-11661.

[23] Jana NR, Zemskov EA, Wang G-H, Nukine N: Altered proteasomal function due to the expression of polyglutamine-expanded truncated N-terminal huntingtin induces apoptosis by caspase activation through mitochondrial cytochrome c release. *Hum Mol Gen* 2001, 10:1049-1059.

[24] Chai Y, Koppenhafer SL, Shoesmith SJ, Perez MK, Paulson HL: Evidence for proteasome involvement in polyglutamine disease: localization to nuclear inclusions in SCA3/MJD and suppression of polyglutamine aggregation *in vitro*. *Hum Mol Gen* 1999, 4:673-682.

[25] Cummings CJ, Mancini MA, Antalffy B, DeFranco D, Orr Ht, Zoghbi HY: Chaperone suppression of aggregation and altered subcellular proteasome localization imply protein misfolding in SCA1. *Nat Genet* 1998, 19:148-154.

[26] Hirabayashi M, Inoue K, Tanaka K, Nakadate K, Ohsawa Y, Kamei Y, Popiel AH, Sinohara A, Iwamatsu A, Kimura Y, Uchiyama Y, Hori S, Kakizuka A: VCP/p97 in abnormal protein aggregates, cytoplasmic vacuoles, and cell death, phenotypes relevant to neurodegeneration. *Cell Death Diff* 2001, 8:977-984.

[27] Matsumoto M, Yada M, Hatakeyama S, Ishimoto H, Tanimura T, Tsuji S, Kakizuka A, Kitagawa M, Nakayama KI: Molecular clearance of ataxin-3 is regulated by a mammalian E4. *EMBO J* 2004, 23:659-669.

[28] Burnett B, Li F, Pittman RN: The polyglutamine neurodegenerative protein ataxin-3 binds polyubiquitylated proteins and has ubiquitin protease activity. *Hum Mol Gen* 2003, 12:3195-3205.

[29] Higashiyama H, Hirose F, Yamaguchi M, Inoue YH, Fujikake N, Matsukage A, Kakizuka A: Identification of *ter94*, Drosophila VCP, as a modulator of polyglutamine-induced neurodegeneration. *Cell Death Diff* 2002, 9:264-273.

[30] Venkatraman P, Wetzel R, tanaka M, Nukina N, Goldberg AL: Eukaryotic proteasomes cannot digest polyglutamine sequences and release them during degradation of polyglutamine-containing proteins. *Mol Cell* 2004, 14:95-104.

[31] Cummings CJ, Reinstein E, Sun Y, Antalffy B, Jiang Y-H, Ciechanover A, Orr HT, Beaudet AL, Zoghbi HY: Mutation of the E6-AP ubiquitin ligase reduces nuclear frequency while accelerating polyglutamine-induced pathology in *SCA1* mice. *Neuron* 1999, 24:879-892.

[32] Holmberg CI, Staniszewski KE, Mensah KN, Matouschek A, Morimoto RI: Inefficient degradation of truncated polyglutamine proteins by the proteasome. *EMBO J* 2004, 23:4307-4318.

[33] Varshavsky A: The N-end rule: functions, mysteries, uses. *Proc Natl Acad Sci USA* 1996, 93:12142-12149.

[34] Verhoef LGGC, Lindsten K, Masucci MG, Dantuma NP: Agregate formation inhibits proteasomal degradation of polyglutamine proteins. *Hum Mol Gen* 2002, 11:2689-2700.

[35] Mizuno Y, Hattori N, Mori H, Suzuki T, Tanaka K: Parkin and Parkinson's disease. *Cur Opin Neurol* 2001, 14:477-82.

[36] He L, Lu XY, Jolly AF, Eldridge AG, Watson SJ, Jackson PK, Barsh GS, Gunn TM: Spongiform degeneration in manoganoid mutant mice. *Science* 2003, 299:710-712.

[37] Saigoh k, Wang YL, Suh JG, Yamanishi T, Sakai Y, Kiyosawa H, Harada T, Ichihara N, Wakana S, Kikuchi S, Wada K: Intragenic deletion in the gene encoding ubiquitin carboxy-terminal hydrolase in gad mice. *Nat Genet* 1999, 23:47-51.

[38] Fernandez-Funez P, Nino-Rosales ML, de Gouyon B, She W-C, Luchak JM, Martinez P, Turiegano E, Benito J, Capovilla M, Skinner PJ, McCall A, Canal I, Orr HT, Zoghbi HY, Botas J: Identification of genes that modify ataxin-1-induced neurodegeneration. *Nature* 2000, 408:101-106.

[39] Chan HYE, Warrick JM, Andriola I, Merry D, Bonini NM: Genetic modulation of polyglutamine toxicity by protein conjugation pathways in *Drosophila*. *Hum Mol Gene* 2002, 11:2895-2904.

[40] de Pril R, Fischer DF, Maat-Schieman MLC, Hobo B, de Vos RAI, Brunt ER, Hol EM, Roos RAC, van Leeuwen FW: Accumulation of aberrant ubiquitin induces aggregate formation and cell death in polyglutamine diseases. *Hum Mol Gen* 2004, 13:1803-1813.

[41] Tsirigotis M, Tang MY, Beyers M, Zhang M, Woulfe J, Gray DA: Delayed spinocerebellar ataxia in transgenic mice expressing mutant ubiquitin. *Neuropath Appl Neurobiol* 2006, 32:26-39.

[42] Stenoien DL, Cummings CJ, Adams HP, Mancini MG, Patel K, DeMartino GN, Marcelli M, Weigel NL, Mancini MA: Polyglutamine-expanded androgen receptors form aggregates that sequester heat shock proteins, proteasome components and SRC-1, and are suppressed by the HDJ-2 chaperone. *Hum Mol Gen* 1999, 8:731-741.

[43] Cummings CJ, Sun Y, Opal P, Antalffy B, Mestril R, Orr HT, Dillman WH, Zoghbi HY: Over-expression of inducible HSP70 chaperone suppresses neuropathology and improves motor function in *SCA1* mice. *Hum Mol Gen* 2001, 10:1511-1518.

[44] Schmidt T, Lindenberg KS, Krebs A, Schols L, Laccone F, Herms J, Rechsteiner M, Riess O, Landwehrmeyer GB: Protin surveillance machinery in brains with

spinocerebellar ataxia type 3: Redistribution and differential recruitment of 26S proteasome subunits and chaperones to neuronal intranuclear inclusions. *Ann Neurol* 2002, 51:302-310.

[45] Bence NF, Sampat RM, Kopito RR: Impairment of the ubiquitin-proteasome system by protein aggregation. *Science* 2001, 292:1552-1555.

[46] Bennett EJ, Bence NF, Jayakumar R, Kopito RR: Global impairment of the ubiquitin-proteasome system by nuclear or cytoplasmic protein aggregates precedes inclusion body formation. *Mol Cell* 2005, 17:351-365.

[47] Nishitoh H, Matsuzawa A, Tobiume K, Saegusa K, Takeda K, Inoue K, Hori S, Kakizuka A, Ichijo H: ASK1 is essential for endoplasmic reticulum stress-induced neuronal cell death triggered by expanded polyglutamine repeats. *Genes Devel* 2002, 16:1345-1355.

[48] Adori C, Kovacs GG, Low P, Molnar K, Gorbea C, Fellinger E, Budka H Mayer RJ, Laszlo L: The ubiquitin-proteasome system in Creutzfeld-Jacob and Alzheimer's disease: Intracellular redistribution of components correlates with neuronal vulnerability. *Neurobiol Dis* 2005, 19:427-435.

[49] Fergusson J, Landon M, Lowe J, Dawson SP, Layfield R, Hanger DP, Mayer RJ: Pathological lesions of Alzheimer's disease and dementia with Lewy bodies brains exhibit immunoreactivity to an ATPase that is a regulatory subunit of the 26S proteasme. *Neurosci Lett* 1996, 219:167-170.

[50] Hamos JE, Oblas B, Pulaski-Salo D, Welch WJ, Bole DG, Drachman DA: Expression of heat shock proteins in Alzheimer's disease. *Neurology* 1991, 41:345-350.

[51] Ii K, Ito H, Tanaka K, Hirano A: Immunocytochemical co-localization of the proteasome in ubiquitinated structures in neurodegenerative diseases and the elderly. *J Neuropathol Exp Neurol* 1997, 56:125-131.

[52] Kwak S, Masaki T, Ishiura S, Sugita H: Multicatalytic proteinase is present in Lewy bodies and neurofibrillary tangles in diffuse Lewy body disease brains. *Neurosci Lett* 1991, 128:21-24.

[53] Ardley HC, Scott GB, Rose SA, Tan NGS, Markham AF, Robinson PA: Inhibition of proteasomal activity causes inclusions formation in neuronal and non-neuronal cells overexpressing parkin. *Mol Biol Cell* 2003, 14:4541-4556.

[54] Stenoien DL, Mielke M, Mancini MA: Intranuclear ataxin-1 inclusions contain both fast and slow-exchanging components. *Nature Cell Biol* 2002, 4:806-810.

[55] Kim S, Nollen EAA, Kitagawa K, Bindokas VP, Morimoto RI: Polyglutamine protein aggregates are dynamic. *Nature* 2002, 4:826-831.

[56] Suhr ST, Senut M-C, Whitelegge JP, Faull KF, Cuizon DB, Gage FH: Identities of sequestered proteins in aggregates from cells with induced polyglutamine expression. *J Cell Biol* 2001, 153:283-294.

[57] Chai, Y, Shao J, Miller VM, Williams A, Paulson HL: Live-cell imaging revealsdivergent intracellular dynamics of polyglutamine disease proteins and supports a sequestration model of pathogenesis. *PNAS* 2002, 99:9310-9315.

[58] Kaytor MD, Wilkinson KD, Warren ST: Modulating huntingtin half-life alters polyglutamine-dependent aggregate formation and cell toxicity. *J Neurochem* 2004, 89:962-973.

[59] Michalik A, Van Broeckhoven C: Proteasome degrades soluble expanded polyglutamine completely and efficiently. *Neurobiol Dis* 2004, 16:202-211.

[60] Yamamoto A, Lucas J, Hen R: Reversal of neuropathology and motor dysfunction in a conditional model of Huntington's disease. *Cell* 2000, 101:57-66.

[61] Zu T, Duvick LA, Kaytor MD, Berlinger MS, Zoghbi HY, Clark HB, Orr HT: Recovery from polyglutamine-induced neurodegeneration in conditional *SCA1* transgenic mice. *J Neurosci* 2004, 24:8853-8861.

[62] Sapp E, Schwarz C, Chase K, Bhide PG, Young AB, Penney J, Vonsattel JP, Aronin N, DiFiglia M: Huntingtin localization in brains of normal and Huntington's disease patients. *Ann Neurol* 1997, 42:604-612.

[63] Nixon RA, Cataldo AM, Mathews PM: The endosomal-lysosomal system of neurons in Alzheimer's disease pathogenesis: A review. *Neurochem Res* 2000, 25:1161-1172.

[64] Kopito RR: Aggresomes, inclusion bodies and protein aggregation. *Trends Cell Biol* 2000, 10:524-530.

[65] Taylor JP, Hardy J, Fischbeck KH: Toxic proteins in neurodegenerative disease. *Science* 2002, 296:1991-1995.

[66] Zhou H, Cao F, Wang Z, Yu Z-X, Nguyen H-P, Evans J, Li S-H, Li X-J: Huntingtin forms toxic NH_2-terminal fragment complexes that are promoted by the age-dependent decrease in proteasome activity. *J Cell Biol* 2003, 163:109-118.

[67] Lee CK, Weindruch R, Prolla TA: Gene-expression profile of the aging brain in mice. *Nat Genet* 2000, 25:294-297.

[68] Keller JN, Huang FF, Markesbery WR: Decreased levels of proteasome activity and proteasome expression in aging spinal cord. *Neurosci* 2000, 98:149-156.

[69] Ansorge O, Giunti P, Michalik A, Van Broeckhoven C, Harding B, Wood N, Scarvilli F: Ataxin-7 aggregation and ubiquitination in infantile SCA7 with 180 CAG repeats. *Ann Neurol* 2004, 56:448-452.

[70] Goto S, Takahashi R, Kumiyama A, Radak Z, Hayashi T, Takenouchi M, Abe R: Implications of protein degradation in aging. *Ann NY Acad Sci* 2001, 928:54-64.

[71] Soti C, Csermely P: Aging and molecular chaperones. *Exp Gerontol* 2003, 38:1037-1040.

[72] Sherman MY, Goldberg AL: Cellular defences against unfolded proteins: a cell biologist thinks about neurodegenerative diseases. *Neuron* 2001, 29:15-32.

In: The Ubiquitin Proteasome System…	ISBN 978-1-60021-749-4
Eds: Mario Di Napoli and Cezary Wojcik, pp. 813-832

Chapter 33

THE ROLE OF THE UBIQUITIN PROTEASOME SYSTEM IN THE PATHOGENESIS OF PRION DISEASES

Birkir Thor Bragason and Astridur Palsdottir*

Institute for Experimental Pathology, Keldur, University of Iceland, Vesturlandsvegur, Reykjavik 112, Iceland.

ABSTRACT

Prion diseases are a group of neurodegenerative diseases that affect humans and animals. They are distinct from other neurodegenerative disorders in that they can be infectious as well as familial or sporadic. Prion diseases are characterized by long incubation periods prior to onset of symptoms, and the pathology is limited to the central nervous system consisting mainly of vacuolation in neuronal cell bodies, neuronal cell death, deposition of protein aggregates, and astrocytosis. Prion diseases were originally classified as slow viral infections; however, there is mounting evidence to support the claim that the infectious unit is a protein. A small endogenous protein, the prion protein (PrP^C; c: cellular), is a key factor in these diseases. The expression of PrP^C is highest in neurons and its precise physiological role is not clear. It is processed through the secretory system to the plasma membrane where it is predominantly an extracellular glycosyl-phosphatidyl-inositol anchored protein that contains one disulfide bond and it is di-glycosylated. The protein aggregates detected in diseased individuals contain a structurally altered protease resistant form of PrP^C, called PrP^{Sc} (Sc: scrapie). PrP^{Sc} is though to be the major part of the infectious unit. The neurotoxic mechanisms behind neuronal death in prion diseases are not clear. Loss of functional PrP^C and/or PrP^{Sc} toxicity have been suggested; however, PrP^C knockout mice are apparently normal, suggesting that loss of PrP^C is not the major cause, and toxic effects of PrP^{Sc} are limited

* Correspondence concerning this article should be addressed to Dr. Astridur Palsdottir, BSc, PhD; Institute for Experimental Pathology, Keldur, University of Iceland, Vesturlandsvegur, Reykjavik 112, Iceland. Phone: +354 585 5100; Fax: +354 567 4714; E-mail: astripal@hi.is.

to PrP^C expressing tissue. Therefore, alternate pathways of neurotoxicity have been proposed, e.g., transmembrane forms of PrP^C, and an interplay between the ubiquitin-proteasome system (UPS) and cytosolic PrP^C or cytosolic PrP^{Sc}. As with other neurodegenerative diseases the UPS has been linked with prion diseases. There are, for example, reports of ubiquitinated PrP^{Sc}, increased polyubiquitin expression, and impaired proteasome activity in prion disease and disease models. The majority of PrP^C is topologically located in the secretory system and the extracellular space. However, there are reports describing a small subset of PrP^C in the cytosol, cytosolic PrP^C, where it is subject to efficient ubiquitin-proteasome degradation. This subset of PrP^C can either arise from inefficient translocation into the endoplasmic reticulum (ER) or retrotranslocation from the ER via the ER-associated-degradation (ERAD) pathway. Cytosolic PrP^C has a tendency to aggregate if proteasome activity is inhibited. Initial reports of toxic effects of cytosolic PrP^C on cells prompted speculation whether impairment of cytosolic PrP^C degradation by the UPS due to, e.g., PrP^C mutations or perturbed ubiquitin-proteasome activity, with a resulting rise in cytosolic PrP^C concentration, could explain some of the neurotoxicity in prion diseases. However, the effects of cytosolic PrP^C between studies are in conflict, with some studies reporting toxic effects and others reporting neuroprotective effects. A recent study of the effect of mild proteasome inhibition on viability in scrapie cell-culture models has shown that cytosolic aggresome formation of PrP^{Sc}, rather than PrP^C, caused apoptosis, suggesting that accumulation of cytosolic PrP^{Sc} due to impairment of the UPS could be an important factor in the neurotoxic mechanisms at work in prion diseases.

Keywords: PrPC proteins, PrPSc proteins, Prion diseases, Protein transport, Post translational protein processing.

ABBREVIATIONS

aa, amino acid; Bax, Bcl-2 associated X protein; Bcl-2, B-cell leukemia/lymphoma 2; BSE, bovine spongiform encephalopathy; CMW, cytomegalovirus; CNS, central nervous system; CJD, Creutzfeldt-Jakob disease; CtmPrP, transmembrane PrPC with the C-terminus in the extracellular space; ER, endoplasmic reticulum; ERAD, endoplasmic reticulum associated degradation; GPI, glycosyl phosphatidyl inositol; Grb2, Growth factor receptor bound protein 2; GSS, Gerstmann-Straussler-Scheinker disease; Hsc, heat-shock cognates; Hsp, heat-shock proteins; NRAGE, neurotrophin receptor interacting MAGE homolog; PK, protein kinase; *Prnp*, prion protein gene; PrPC, normal cellular form of the prion protein; PrPSc, disease associated form of the prion protein; UPS, ubiquitin-proteasome system.

INTRODUCTION

Prion Diseases

Prion diseases, also known as transmissible spongiform encephalopathies, are a group of fatal neurodegenerative disorders that affect humans and animals. They are unique among neurodegenerative disorders in that they are infectious as well as sporadic and hereditary. The human forms of prion diseases are categorized as sporadic [sporadic Creutzfeldt-Jakob disease (sCJD)], familial [familial CJD (fCJD), Gerstmann-Straussler-Scheinker disease (GSS), and fatal familial insomnia (FFI)], or acquired/iatrogenic [Kuru, iatrogenic CJD (iCJD), and variant CJD (vCJD)]. The familial forms represent 10-15% of human prion diseases, whereas sporadic CJD is the most common form and represents about 85% of all diagnosed CJD cases with an average incidence of 1 per million per annum [1].

The most prominent prion diseases in animals are scrapie in sheep and goats, bovine spongiform encephalopathy (BSE) in cattle, chronic wasting disease in deer and elk, and transmissible mink encephalopathy. In animals, there is no conclusive evidence of genetic, or sporadic, prion diseases; the animal diseases are therefore acquired by infection. However, due to the existence of sporadic/hereditary forms of disease in humans, the same phenomena in animals cannot be excluded.

Prion diseases are characterized by long incubation periods prior to the onset of clinical illness ending fatally. The phenotype in humans consists of various neurological symptoms. The symptoms of sporadic CJD are, e.g., rapidly progressive dementia, pyramidal/extrapyramidal signs with myoclonus, and triphasic electroencephalogram discharges [1]. The underlying pathology is restricted to the central nervous system (CNS) and consists of: vacuolation of the neuropil within neuronal cell bodies and neurites, neuronal cell death, deposition of protein aggregates, and activation of astrocytes (astrocytosis) and microglia. The pathological changes result in characteristic spongiform changes due to vacuolar degeneration of grey matter in the CNS, hence the term spongiform.

The infectious agent of prion diseases differs from other pathogens. The agent is resistant to treatment that abolishes the infectivity of many conventional agents such as treatment with heat or formalin (reviewed by Prusiner [2]). In addition, the agent is resistant to ionizing radiation and UV-radiation [3]. Several hypotheses about the nature of the prion disease agent have been proposed (reviewed by Prusiner [2]). The prevalent view, introduced in 1982 by Dr. Stanley Prusiner and his colleagues at the University of California, San Francisco, is that the infectious agent is a proteinaceous particle devoid of nucleic acids [2].

The Prion Protein

The protein aggregates detected in prion diseases contain a modified form of an endogenous protein called the prion protein. This modified form of the prion protein was originally identified and characterized from infectious brain fractions of hamsters experimentally infected with scrapie by Prusiner and colleagues [4]. They coined the term 'prion', which is an abbreviation for proteinaceous infectious particle. To distinguish the endogenous, normal, form of the prion protein it is called PrP^C (C: cell), whereas the modified, disease associated form, is called PrP^{Sc} (Sc: scrapie). The difference between these two forms of the prion protein lies in a post-translational change in structure from the predominantly alpha-helical form of PrP^C to PrP^{Sc} which is mainly β-sheet [5]. In addition, the N-terminus of PrP^{Sc}, up to amino acid (aa) ~90, has been cleaved off. It is possible to distinguish between these two forms of the prion protein due to their different biochemical characteristics; PrP^C is detergent soluble and protease sensitive, whereas PrP^{Sc} is detergent insoluble and protease resistant to a degree.

PrP^C is a small protein (253 aa in humans) encoded by the prion protein gene, *PRNP*. The prion protein gene in mammals consists of two to three exons. For example, in humans and hamsters it has two exons, whereas in sheep, cattle, mice, and rats, *Prnp* has three exons. In both cases the open reading frame is completely contained within the latter/last exon and encodes PrP^C. The crucial role of this protein in prion diseases was made clear when it was demonstrated that *Prnp*$^{0/0}$ knockout mice are 'immune' to infection with prion disease (scrapie) [6].

The primary PrP^C expressing cell type in the CNS is the neuron; *Prnp* expression and protein are detected in glial cells, they are, however, in relatively low quantities compared to neurons [7]. In the periphery *Prnp* expression is detected in a wide range of cell types; the expression levels, however, are lower than in the CNS [8,9].

PrP^C is targeted for translocation into the endoplasmic reticulum (ER) by an NH_2-terminal (N-terminal) signal peptide. In the ER a COOH-terminal (C-terminal) signal sequence facilitates glycosyl-phosphatidyl-inositol (GPI)-anchor membrane attachment [10]. Structural studies of recombinant PrP^C show that the protein has a structured C-terminal part (aa 125-231) consisting of three alpha-helices and two anti-parallel β-sheets [11,12]. In contrast, the N-terminal region (aa 23-124) is flexible [13]. A characteristic feature of the N-terminus are repeats (5 in humans) consisting of 8-9 amino acids, notably histidine (P(Q/H)GGG(G/-)WGQ).

After translocation into the endoplasmic reticulum, PrP^C is processed through the secretory system during which it is glycosylated at two Asn-glycosylation sites and a single disulfide bond is formed. Fully processed, GPI anchored, PrP^C on the plasma membrane is mainly located in detergent resistant microdomains, or 'rafts' (Figure 1) [14]. There are reports that indicate that in neurons, PrP^C is located at the neuronal synapse [15]. PrP^C cycles between the plasma membrane and endosomes [16]. As is the case for other GPI-anchored proteins, PrP^C endocytosis by caveolae has been reported; however, there are also reports of endocytosis by clathrin-coated pits for chicken PrP^C [17,18]. Like other membrane proteins the ultimate fate of PrP^C seems to be degradation in the endo-lysosome system. Overall, the estimated time for PrP^C synthesis is less than 2 hours, and its half-life has been estimated at 4.5-5 hours [19].

Despite its small size, the translocation of PrP^C can yield several topological forms. In addition to the predominant GPI-anchored PrP^C form, transmembrane forms have been detected [20]. In the transmembrane forms a conserved hydrophobic region of PrP^C (aa ~110-135 in human PrP^C) serves as a transmembrane region spanning the lipid bilayer. Two topological orientations have been described, termed ^{Ntm}PrP (C-terminus in the cytosol) and ^{Ctm}PrP (N-terminus in the cytosol).

The precise physiological function of PrP^C is not clear. There are several theories. For example, PrP^C can bind Cu^{2+} via histidines in the N-terminal repeats [21]. This binding affects endocytosis [22] and a Cu^{2+} receptor function of PrP^C has been proposed. It has also been suggested that PrP^C may modulate Cu/Zn superoxide dismutase activity [23]. Reports of impaired synaptic inhibition in neurons from *Prnp*$^{0/0}$ mice [25] suggest that PrP^C may have a function in neuronal synapses. However, because other reports have not detected synaptic abnormalities in *Prnp*$^{0/0}$ mice [26,27], the functional importance of PrP^C at neuronal synapses is still unclear. An anti-apoptotic role of PrP^C has been suggested based on observations that PrP^C expression protects human primary neurons against Bax-mediated apoptosis [28], and that expression of either PrP^C or Bcl-2 rescues *Prnp* knockout neurons from apoptosis induced by serum deprivation [29]. In addition, a signalling role has been proposed, for example by interaction with caveolin-1 and stress-inducible-protein-1 resulting in Fyn activation and signalling via the cAMP/protein kinase (PKA) pathway, respectively [30,31].

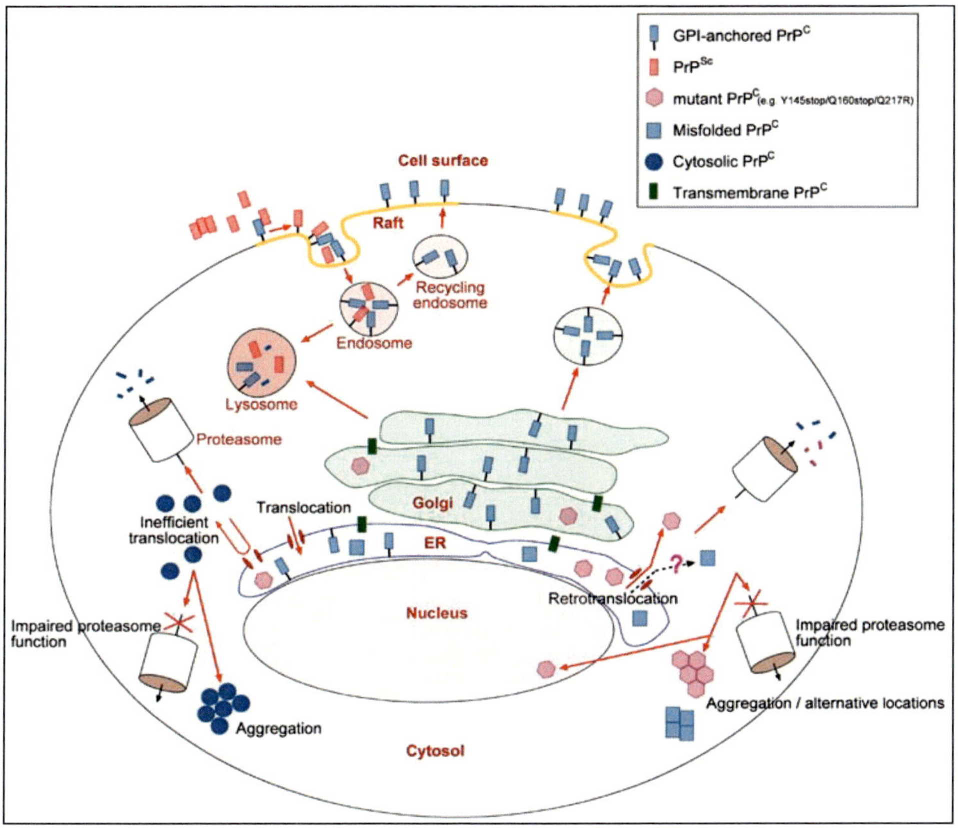

Figure 1. A schematic overview of the cellular processing of PrP^C and PrP^{Sc}. This figure is a simplified schematic overview of various PrP^C processing pathways within the cell which are addressed in the text. It shows the processing of normal GPI-anchored PrP^C through the secretory system and its localization in rafts on the cell surface. From the surface, PrP^C is internalized by endocytosis and some of it is recycled to the membrane. The interaction of PrP^C and PrP^{Sc} during infection is shown on the cell surface as well as during endocytosis. The figure shows retrotranslocation of the mutant PrP^C forms PrP^{145}, PrP^{160}, and PrP^{217}, from the endoplasmic reticulum (ER) for proteasome degradation. In light of the debate described in the text a question mark is set on the retrotranslocation of non-mutant PrP^C from the ER. The generation of a cytosolic subset of PrP^C by inefficient translocation into the ER is depicted. The consequences of an impaired ubiquitin-proteasome system, i.e., the accumulation of PrP forms in the cytosol with subsequent aggregation or alternative locations are shown. Finally, the location of transmembrane PrP^C species in the ER and Golgi is indicated.

Studies on scrapie infected cell cultures indicate that the $PrP^{C} \rightarrow PrP^{Sc}$ transformation occurs on the cell membrane and/or after endocytosis in the endo-lysosome system [32,33]. Other locations of PrP^{Sc} formation have also been suggested such as the endoplasmic reticulum [34] and the cytosol [35]. The mechanism of transformation is not precisely defined. Reports suggest that the transformation requires an interaction between PrP^{C} and PrP^{Sc}, e.g., antibodies specific for certain PrP^{C} epitopes inhibit prion propagation, perhaps by impairing PrP^{C} -PrP^{Sc} interaction [36,37]. Alternatively, the antibodies may interfere with interactions with other factors required for the transformation. Genetic studies suggest that a hitherto unidentified protein co-factor, called protein-X, is involved in the conversion [38]. During infection, PrP^{C}-PrP^{Sc} interactions result in a structural transition of endogenous host PrP^{C} into the PrP^{Sc} form in a process which imparts the biochemical characteristics of PrP^{Sc} on to the host PrP^{C} [39].

The Association of the Ubiquitin-Proteasome System and Prion Diseases

As the overview of PrP^{C} processing above details, PrP^{C} is predominantly topologically located in the secretory system or in the extracellular space from which it is internalized by endocytosis and degraded in the endo-lysosome system. Therefore, it would seem that the cytosolic ubiquitin-proteasome system (UPS) is not significantly involved in the processing of PrP^{C}. Indeed, in comparison to other neurodegenerative diseases, such as Parkinson's disease, there is a relative paucity of data on the association of the UPS and prion diseases. However, there is some evidence that indicates the involvement of the UPS, to some degree, in prion diseases and in the processing of PrP^{C} and PrP^{Sc}. First of all, there are several reports of increased ubiquitin reactivity in diseased tissue (discussed below). Second of all, there are reports that a cytosolic subset of PrP^{C} and/or PrP^{Sc} are substrates of the UPS and that impairment of the proteasome resulting in an increased concentration of cytosolic PrP species affects the viability of certain neuronal cell types (discussed below). These observations have added yet another possible mechanism of neurotoxicity in prion diseases, which at present is not completely defined.

Ubiquitin Reactivity in Prion Diseases

An increase in ubiquitin reactivity, compared to controls, has been described in several prion diseases, such as CJD [40-45], GSS [42,46], BSE [47], as well as in experimental models of scrapie in mice [48,49]. Some studies report an increase in this reactivity in association with a longer duration of disease, e.g., in CJD [41], and in the terminal stages of disease in scrapie infected mice [49]. This increased reactivity correlates with reports of an increased expression of polyubiquitin and Hsp70 genes, and a decline in proteasome function, in the terminal stages of disease in scrapie infected mice [49,50]. Furthermore, a systematic immunohistochemical analysis of the distribution of Hsp72, the 20S proteasome, and the ATPases of its 19S regulatory complex in the brains of CJD affected individuals vs.

controls [44] revealed that neuronal populations with low expression levels of these proteins were more vulnerable in disease indicating the importance of the UPS. Taken together, this data suggests that the UPS is involved in 'responding' to the disease state.

The distribution of ubiquitin reactivity seen in CJD has been reported to fall mainly into two categories as described by Ironside *et al.* [42], i.e., 'punctate' reactivity in and around PrP^{Sc} amyloid plaques and a granular pattern in the neuropil often associated with spongiform changes. In addition ubiquitin reactive inclusions are seen within neurons, and also in 'thread-like structures' in the neuropil [42]. Similar ubiquitin reactivity has also been described in mouse models of scrapie [48], in association with PrP^{Sc} deposits in GSS [46], and dot-like ubiquitin reactive structures have been described in the neuropil in BSE [47].

In addition to the ubiquitinated intraneuronal inclusions described by Ironside *et al.* [42], intracytoplasmic 'inclusion-like bodies' that react with PrP antibodies, and are somewhat morphologically reminiscent of inclusion bodies described in other neurodegenerative diseases, have, for example, been documented in CJD [51,52] and BSE [47]. However, in these studies the precise intracellular location of the inclusions was not demonstrated, i.e., whether they were in the cytosol or in a cellular compartment such as the lysosome, in which PrP^{Sc} has been detected (see below). A recent paper described the co-localization of ubiquitin with PrP aggregates in neurons by con-focal microscopy [43] and another recent report has described ubiquitinated PrP^{Sc} in the terminal stages of infection in scrapie infected mice [49]. This is interesting because ubiquitination, and the degradation of ubiquitinated proteins by the proteasome machinery, are cytosolic processes (or nuclear, as proteasomes reside in both these locations, see Chapter 11) [53]. Therefore, the existence of ubiquitinated PrP^{Sc} implies that some part of PrP^{Sc}, at least, has been in contact with the cytosol, and subsequently been targeted for degradation in that compartment [54]. In light of the results by Kristiansen *et al.* [55], who detected cytosolic aggresomes of PrP^{Sc} in scrapie infected neuronal cell lines (discussed below), it would be interesting to further characterize the precise nature of these 'inclusion-like bodies' detected in CJD.

The distribution of ubiquitin reactivity in the vicinity of spongiform changes is similar to that of PrP^{Sc} and the lysosomal proteinase cathepsin-D [42]. PrP^{Sc} has been detected in lysosomes [56] and, as mentioned, this compartment is considered to be important for the PrP^{C} to PrP^{Sc} transformation in prion diseases. PrP^{Sc} has been detected in 'late-endosome-like-organelles' along with ubiquitin [57]. Although ubiquitination of cytosolic regions of transmembrane proteins can target them for endocytosis and lysosomal degradation [58], lysosomes are, in general, not responsible for the degradation of ubiquitinated proteins. However, ubiquitinated proteins from cytosolic aggregates can end up in lysosomes due to autophagic mechanisms in response to impairment of the UPS. This has, for example, recently been described in the case of aggregated huntingtin [59]. Again, in light of the results of Kristiansen *et al.* [55], this raises the question whether a similar process might sometimes take place in prion diseases as well, which could be one explanation for the observed co-existence of PrP^{Sc} and ubiquitin in endosomes-lysosomes. Indeed, autophagic vacuoles have been described in neuronal synapses in CJD [60].

Neurotoxic Mechanisms in Prion Diseases

For the past few years there has been an ongoing debate [61] regarding the intracellular processing of PrP^C. This debate revolves around reports that a subset of PrP^C is subject to retrotranslocation from the ER for proteasome degradation by the ERAD pathway (see Chapter 13). Accumulation of this cytosolic subset of PrP^C, termed cytosolic PrP^C, following proteasome impairment can affect the viability of some neuronal cells, and has been suggested as a possible factor in the neurotoxic mechanisms in prion diseases which have not been completely defined. Before the data surrounding this debate are described (see below) it is appropriate to briefly review some of the data regarding the neurotoxic mechanisms in prion diseases.

The cell type that is lost in prion diseases is the neuron; degeneration of astrocytes is not seen [62]. Some experiments indicate that there is a selective loss of a subset of GABAergic inhibitory neurons that are immunoreactive for the calcium binding protein parvalbumin [63-66]. Other reports suggest a selective loss of glutamatergic neurons [67]. Finally, several observations in human patients, as well as in mouse models of scrapie, suggest that this neuronal cell death is by apoptosis [68-73].

As described above, the disease process in prion diseases involves conversion of host PrP^C into PrP^{Sc} by a post-translational process, supported by observations that the accumulation of PrP^{Sc} occurs without an increase in *Prnp* mRNA [74]. Inherent in this process is that for each produced PrP^{Sc} protein one PrP^C is lost. Therefore two possible reasons for the neurodegeneration are loss of functional PrP^C or toxic effects of produced PrP^{Sc}. However the changes in prion disease tissue have not been sufficiently explained with either of these possibilities (reviewed by Weissmann [75]). For example, there is a lack of phenotype in $Prnp^{0/0}$ knockout mice [76] and transgenic mice that are rendered $Prnp^{0/0}$ post-natally do not develop disease [77] suggesting that loss of PrP^C function is not the major cause of disease. However, although not the major factor in pathogenesis and in light of the ideas about PrP^C function mentioned above, loss of PrP^C might increase susceptibility to, e.g., oxidative stress, growth factor deprivation, and apoptotic cascades.

Regarding PrP^{Sc}, deposition of PrP^{Sc} aggregates is often highly associated with neuropathological changes [78], but this is not always the case, and in some instances little or no PrP^{Sc} is detected in diseased tissue [20,79,80]. Furthermore, experiments with mice show that $Prnp^{0/0}$ brain tissue is 'immune' to PrP^{Sc} produced by infected $Prnp^{+/+}$ tissue grafts in the same animal [81] and $Prnp^{0/+}$ mice have a delayed onset of symptoms after infection despite accumulating high amounts of PrP^{Sc} [82,83]. Taken together, these results show that in order for PrP^{Sc} to have an effect, the target tissue must express PrP^C, and that the disease process is correlated with PrP^C expression. A recent study has demonstrated that in order for PrP^{Sc} aggregates to cause clinical disease the target tissue must express GPI-anchored PrP^C, tissue that expresses non-GPI-anchored, secreted, PrP^C is not affected [84], suggesting that perhaps toxicity is to some extent due to abrogation of a signalling role of PrP^C.

In order to explain the pathology in prion diseases alternative pathological forms of PrP have been suggested (reviewed by Chiesa and Harris [85]). As already mentioned, PrP^C can take on two different transmembrane orientations. The ${}^{Ctm}PrP$ form of PrP^C has been associated with neurodegeneration in a case of GSS and in transgenic mice expressing *Prnp* constructs with a propensity to form ^{Ctm}PrP [20,86]. Furthermore, experiments indicate that PrP^{Sc} induces ^{Ctm}PrP formation in such mice [86]. However, most PrP^{C} mutations do not cause an increase in transmembrane forms [87], arguing against ^{Ctm}PrP as a general cause of neurotoxicity in prion diseases. The mechanism by which ^{Ctm}PrP causes neurodegeneration is unknown.

Finally, localization of PrP^{C} to other cellular compartments than the secretory system/plasma membrane, such as the cytosol, and aggregation of such forms due to impairment of the UPS has been linked to neurotoxicity in certain cell types (see below). These reports have sparked the debate mentioned above and resulted in some very interesting reports regarding the cell biology of PrP^{C}, and the association of PrP^{C} with the UPS, as discussed in detail below.

PrP^{C} and PrP^{Sc} as Substrates of the Ubiquitin-Proteasome System

There are several well documented polymorphisms/mutations in the open reading frame of the prion protein gene in mice, sheep, and humans. Most of the known mutations in the human gene are directly associated with familial forms of prion disease, whereas polymorphisms in mouse and sheep *Prnp* affect susceptibility to infection with prion diseases. The variation in the human prion protein gene falls into three categories (reviewed by Kovacs *et al.* [88]): (*a*) point mutations (some silent), most of which are located within, or near, the secondary structural elements in the C-terminus, (*b*) amber mutations (Y145stop and Q160stop), and (*c*) insertions of additional repeats in the flexible N-terminus.

How the mutations in human *PRNP* cause disease is not precisely clear. They may affect the thermodynamic stability of PrP^{C} facilitating the PrP^{C} to PrP^{Sc} conversion; however, studies suggest that this is not a general mechanism [89] although it may apply to some mutations [90]. As already mentioned, mutations in the 'transmembrane' and its vicinity have been shown to cause an increase in ^{Ctm}PrP [87]. Finally, some mutations can influence the intracellular processing of PrP^{C} as has been demonstrated for the GSS associated mutations Y145stop [91] and Q217R [92].

In contrast to the 'normal' processing of PrP^{C} through the secretory pathway described above, the majority of PrP^{C} with the Y145stop mutation (PrP^{145}) is retrotranslocated for degradation by the UPS and accumulates in the nucleus if the UPS is inhibited [91]. In addition to a nuclear localization, a mitochondrial localization of PrP^{145} has been reported, affecting mitochondrial membrane potential resulting in apoptosis [93]. Subsequently, it has been demonstrated that another truncated PrP mutant, Q160stop, is processed in a manner similar to PrP^{145}, also showing a nuclear localization following proteasome inhibition [94]. Two nuclear localization signals in PrP^{C} have been defined [95] which are, however, only active if the protein is non-glycosylated and non-GPI anchored. In contrast to PrP^{145}, the majority of PrP^{C} with the Q217R mutation (PrP^{217}) is processed through to a site distal of the cis-Golgi, whereas a small non-GPI anchored subset of PrP^{217} is retained in the ER in association with the chaperone BiP followed by retrotranslocation for proteasomal degradation [92]. These reports of the processing of mutant PrP's suggested that in some cases of prion diseases the neurotoxicity might be explained by the aberrant localization of

PrP^C to other locations than the secretory system and plasma membrane in the case of UPS impairment.

Observations from the expression of recombinant PrP^C in the yeast cytosol revealed that when PrP^C is expressed in this compartment it displays characteristics reminiscent of PrP^{Sc}, notably detergent insolubility and proteinase-K resistance in a concentration dependant manner, i.e., these attributes are enhanced with higher expression [96]. Similar observations were made for PrP^{145} in the cytosol of mammalian cells [91].

This prompted studies into whether normal PrP^C, like PrP^{145} and PrP^{217}, is to some degree, generally processed for degradation by ERAD in which case such processing, coupled with the PrP^{Sc} like characteristics of PrP^C when in this compartment, might be relevant for pathogenesis in the special case of impaired protein degradation.

The question of PrP^C retrotranslocation was addressed in two studies [97,98]. Their results showed that when mammalian cell lines expressing PrP^C are treated with proteasome inhibitors, an unglycosylated PrP^C form accumulated that was detergent insoluble and proteinase-K resistant to a degree. One of the studies showed that this PrP^C form was ubiquitinated [98]. Furthermore, fluorescence microscopy of these cells revealed that the PrP form that accumulated in response to the proteasome treatment had an altered localization [97], i.e., it was localized in Hsc70 positive cytosolic aggregates, compared to that of PrP^C in untreated cells, which resided in the secretory system and on the plasma membrane. The lack of glycosylation, and the size of the PrP species detected in these studies, indicated that the N-terminal and C-terminal signal peptides had been cleaved, and therefore that it had been subject to signal peptide cleavage within the ER and that its location within the cytosol was due to its processing by ERAD. Yedidia *et al.* [98] estimated that ~ 9% of normal PrP^C produced is degraded in the cytosol by the UPS following retrotranslocation from the ER.

In addition to these studies it has been reported that PrP^C constructs with the CJD associated mutations V203I and E211Q, and the GSS associated mutation Q212P, aggregate in vimentin-positive aggresomes in cells treated with proteasome inhibitors [99]. Finally, treatment of PrP^C expressing cells with cyclosporin-A, an inhibitor of the cyclophilin family of peptidyl-prolyl cis-trans isomerases, results in the formation of PrP^C aggresomes [100]. Cumulatively, these studies suggested that PrP^C was processed by ERAD under normal circumstances as well as in response to impaired PrP^C folding due to mutations or inhibition of the folding machinery, in which cases it formed cytosolic aggresomes.

The question whether retrotranslocated PrP^C, and its PrP^{Sc} characteristics, could be associated with pathogenesis was addressed in two studies [35,101]. They reported that (*a*) the PrP^{Sc} like characteristics of cytosolic PrP^C were 'self-perpetuating' in mammalian cells expressing PrP^C following a brief period of proteasome-inhibitor treatment [35], and (*b*) that cytosolic PrP^C was toxic to mouse neuroblastoma cells (N2a) and caused cerebellar degeneration characterized by the loss of cerebellar granular neurons in transgenic mice expressing a cytosolic PrP^C construct on a normal background [101]. These results, along with those previously discussed [97,98], prompted the hypothesis that retrotranslocation of PrP^C to the cytosol, accompanied by proteasome impairment, plays a role in generating PrP^{Sc} and thus in the pathogenesis of prion diseases.

However, reports of PrP^C retrotranslocation from the ER have been challenged [102]. In short, Drisaldi *et al.* [102] concluded from their experimental results that PrP^C is not retrotranslocated and that the observations described above were due to the experimental conditions used. They based their conclusions, among other things, on observations that (*a*) proteasome inhibition results in increased transcription from the cytomegalovirus (CMV) promoter, a common promoter in expression plasmids used in mammalian cells, and (*b*) that the PrP^C form that accumulates in response to proteasome inhibition in cells expressing PrP^C under the control of such a promoter contains an uncleaved N-terminal signal-peptide and is located at the cytosolic side of the ER, therefore indicating that it has not entered the ER and that the accumulation of this cytosolic form is due to inefficient translocation, possibly due to saturation of the translocation machinery due to elevated expression levels [102]. Therefore, they suggested that although some mutant PrP's are retrotranslocated for degradation by the proteasome, such as described above [91,92], this mechanism does not apply to normal PrP^C.

Subsequent to the report by Drisaldi *et al.* [102], evidence has been presented for endogenous cytosolic PrP^C following proteasome-inhibitor treatment of primary human neurons and cell lines, without overexpression of PrP^C from vector constructs [103-105], suggesting that this subset is not merely due to overexpression from a strong promoter. In addition, cytosolic PrP^C, in the absence of proteasome inhibition, has been described in hippocampal neurons of normal mice [106] and in transgenic mice that express a PrP^C-EGFP fusion protein from a *Prnp* promoter [107].

A recent report [105] has presented evidence for an origin of endogenous cytosolic PrP^C other than retrotranslocation, i.e., inefficient translocation of PrP^C into the ER due to inherent characteristics of the PrP^C N-terminal signal sequence resulting in a cytosolic subset of PrP^C with an uncleaved N-terminal signal peptide. These results are in line with those of Drisaldi *et al.* [102] regarding the uncleaved N-terminal signal peptide; however, Rane *et al.* [105] show that the generation of this cytosolic subset is not due to saturation of the translocation machinery, but rather these characteristics of the N-terminal signal. This might explain the reports of endogenous cytosolic PrP^C mentioned above. Rane and colleagues [105] do not rule out retrotranslocation of PrP^C; however, their data indicate that the impact of retrotranslocation on the amount of cytosolic PrP^C under normal circumstances is minor. Their estimates regarding the amount of cytosolic PrP^C indicate that PrP^C subject to proteasome degradation due to this inefficiency is approximately 20% of the total protein produced. They point out that the existence of inefficient PrP^C translocation implies that it has been evolutionary conserved, and therefore, that it is of possible physiological importance. This conservation was demonstrated by experiments showing that the N-terminal signal sequences of PrP^C from four species (human, cattle, hamster, and mouse) are equally 'inefficient' in terms of translocation [108]. Subsequent studies by this same group have revealed a marked heterogeneity in the efficiency of translocation by N-terminal signal sequences of various proteins destined for the secretory system [109], indicating that this may be a general mechanism to expand protein use to more than one compartment, as they have demonstrated for calreticulin [110]. This leads to the speculative question, raised by Rane and colleagues [105], whether the inefficiency of PrP^C translocation could be due to a physiological role for cytosolic PrP^C.

The reports detailed above agree that PrP^C in the cytosol is rapidly degraded and does not accumulate under normal circumstances. Interestingly, the reported effects of cytosolic PrP^C vary, and seem to depend on the neuronal cell type. The stabilization of cytosolic PrP^C by proteasome inhibition has been reported to cause apoptosis in PrP^C expressing mouse neuroblastoma cells (N2a) [101,103,105]. On the contrary, such cytosolic PrP^C has also been reported to protect against apoptosis under similar conditions in this cell type [55,111]. As mentioned above, transgenic mice expressing cytosolic PrP^C on a normal background had neuropathological changes consisting of gliosis and selective neurodegeneration of cerebellar granular neurons, even though the promoter used resulted in expression of cytosolic PrP^C in other neuronal cell types as well [101]. Finally, cytosolic PrP^C does not seem to be toxic in human primary neuron cultures or human neuroblastoma cell lines, in fact it counteracts Bax-mediated apoptosis in these cells [103].Therefore, not only are there conflicting results regarding the origin, and importance, of a proteasome degraded cytosolic subset of PrP^C, but also regarding the effect of cytosolic PrP^C on neuronal cells. The discrepancies regarding the effects of cytosolic PrP^C suggest that its effect may be determined by the neuronal cell type, its state, and context.

The studies described above paved the way for an interesting recent report regarding the processing of PrP^{Sc} in prion infected cell lines. Kristiansen *et al.* [55] utilized cell lines that can be infected with, and propagate, prions to study the intracellular processing of PrP^{Sc} formed in these cell lines. They found that when prion infected cells were subject to mild proteasome inhibition, chosen to represent levels of proteasome inhibition seen *in vivo* in disease or ageing, PrP^{Sc} formed cytosolic aggresomes in association with Hsc70, ubiquitin, proteasome subunits, and vimentin. Furthermore, the aggresome formation correlated with apoptosis in these cells characterized by activation of caspases-3 and -8. They also present evidence that PrP^{Sc} is associated with vimentin in scrapie infected mice, suggesting that aggresome formation also occurs *in vivo*. Of interest to the papers discussed above, the conditions of mild proteasome inhibition they used did not cause apoptosis in uninfected cells expressing PrP^C.

CONCLUSIONS

Despite its small size, the processing of PrP^C seems to be fairly complex. This is, for example, demonstrated by its ability to take on several topological forms in respect to the plasma membrane. Furthermore, the precise function of PrP^C has remained elusive. In addition to the functional question, there are several unanswered questions in the field of prion science (reviewed by Aguzzi and Polymenidou [112]), for example regarding the basis of neurotoxicity.

Reports of the retrotranslocation of mutant PrP's, and subsequently of normal PrP^C, from the ER into the cytosol, and concomitantly of the toxic effects of the accumulation of cytosolic PrP^C when the UPS is impaired, promised an explanation for neurotoxicity in prion diseases. However, due to the conflicting reports described above, the relevance of this scenario remains a matter of debate. Overall, retrotranslocation of PrP^C may be a special case for certain PrP mutants, whereas inefficient translocation of PrP^C into the ER may explain the majority of PrP^C in the cytosol at any given time. Therefore, the cytosolic location and aggregation of mutant PrP^C when the UPS is impaired could be relevant to neurotoxicity in certain cases of disease but not necessarily a general mechanism. In fact, as mentioned, a non-translocated PrP^C subset could have a physiological function rather than a toxic function.

The recent description of cytosolic PrP^{Sc} aggresomes in scrapie infected cell lines has added yet another dimension to the debate, suggesting that cytosolic PrP^{Sc}, rather than PrP^C, could be an important factor in prion disease neurotoxic mechanisms. This report raises some interesting questions and opens avenues of investigation, such as, how does PrP^{Sc} enter the cytosol? And, can these observations be connected to the increased ubiquitin-reactivity detected in prion diseases described above? For example, can such PrP^{Sc} aggresomes be detected by immunohistochemical analysis of prion disease tissue?

Elucidations of the molecular pathways that mediate the effects of cytosolic PrP^C, whether toxic or protective, and the effects of PrP^{Sc}, offer a challenge. The effects may be due to specific interactions with cytosolic proteins, or disruption of such interactions in the case of toxic effects. Therefore the identification of cytosolic PrP^C binding proteins, and the characterization of their functional significance, may aid in the understanding of the pathological/protective effects. Several cytosolic PrP^C interacting proteins have already been identified with various methods, such as Bcl-2 [113], Grb2 [114], and the neuronally expressed proteins Synapsin Ib [114] and NRAGE (neurotrophin receptor interacting MAGE homolog) [115]. Furthermore, pathogenic mechanisms elucidated in other neurodegenerative diseases characterized by aggregated proteins may, by inference, give valuable clues. For example, aggregated huntingtin has been demonstrated to disrupt functional pathways in cells by interfering with nuclear transcription factors, axonal transport, mitochondria, or the function of the UPS [59]. Considering the nuclear [91], or mitochondrial [93], localization of PrP^{145} mentioned above and reports of granular PrP deposits in the axons of neurons in CJD patients [43], the question arises whether aggregated PrP^{Sc} may have similar effects.

REFERENCES

[1] Ironside JW, Head MW: Neuropathology and molecular biology of variant Creutzfeldt-Jakob disease. *Curr Top Microbiol Immunol* 2004, 284:133-159.

[2] Prusiner SB: Novel proteinaceous infectious particles cause scrapie. *Science* 1982, 216:136-144.

[3] Alper T, Haig DA, Clarke MC: The scrapie agent: evidence against its dependence for replication on intrinsic nucleic acid. *J Gen Virol* 1978, 41:503-516.

[4] Bolton DC, McKinley MP, Prusiner SB: Identification of a protein that purifies with the scrapie prion. *Science* 1982, 218:1309-1311.

[5] Pan KM, Baldwin M, Nguyen J, Gasset M, Serban A, Groth D, Mehlhorn I, Huang Z, Fletterick RJ, Cohen FE, et al.: Conversion of alpha-helices into beta-sheets features in the formation of the scrapie prion proteins. *Proc Natl Acad Sci U S A* 1993, 90:10962-10966.

[6] Bueler H, Aguzzi A, Sailer A, Greiner RA, Autenried P, Aguet M, Weissmann C: Mice devoid of PrP are resistant to scrapie. *Cell* 1993, 73:1339-1347.

[7] Ford MJ, Burton LJ, Li H, Graham CH, Frobert Y, Grassi J, Hall SM, Morris RJ: A marked disparity between the expression of prion protein and its message by neurones of the CNS. *Neuroscience* 2002, 111:533-551.

[8] Ford MJ, Burton LJ, Morris RJ, Hall SM: Selective expression of prion protein in peripheral tissues of the adult mouse. *Neuroscience* 2002, 113:177-192.

[9] Moudjou M, Frobert Y, Grassi J, La Bonnardiere C: Cellular prion protein status in sheep: tissue-specific biochemical signatures. *J Gen Virol* 2001, 82:2017-2024.

[10] Stahl N, Borchelt DR, Hsiao K, Prusiner SB: Scrapie prion protein contains a phosphatidylinositol glycolipid. *Cell* 1987, 51:229-240.

[11] Riek R, Hornemann S, Wider G, Glockshuber R, Wuthrich K: NMR characterization of the full-length recombinant murine prion protein, mPrP(23-231). *FEBS Lett* 1997, 413:282-288.

[12] Zahn R, Liu A, Luhrs T, Riek R, von Schroetter C, Lopez Garcia F, Billeter M, Calzolai L, Wider G, Wuthrich K: NMR solution structure of the human prion protein. *Proc Natl Acad Sci U S A* 2000, 97:145-150.

[13] Donne DG, Viles JH, Groth D, Mehlhorn I, James TL, Cohen FE, Prusiner SB, Wright PE, Dyson HJ: Structure of the recombinant full-length hamster prion protein PrP(29-231): the N terminus is highly flexible. *Proc Natl Acad Sci U S A* 1997, 94:13452-13457.

[14] Naslavsky N, Stein R, Yanai A, Friedlander G, Taraboulos A: Characterization of detergent-insoluble complexes containing the cellular prion protein and its scrapie isoform. *J Biol Chem* 1997, 272:6324-6331.

[15] Herms J, Tings T, Gall S, Madlung A, Giese A, Siebert H, Schurmann P, Windl O, Brose N, Kretzschmar H: Evidence of presynaptic location and function of the prion protein. *J Neurosci* 1999, 19:8866-8875.

[16] Harris DA: Cellular biology of prion diseases. *Clin Microbiol Rev* 1999, 12:429-444.

[17] Shyng SL, Moulder KL, Lesko A, Harris DA: The N-terminal domain of a glycolipid-anchored prion protein is essential for its endocytosis via clathrin-coated pits. *J Biol Chem* 1995, 270:14793-14800.

[18] Peters PJ, Mironov A, Jr., Peretz D, van Donselaar E, Leclerc E, Erpel S, DeArmond SJ, Burton DR, Williamson RA, Vey M, Prusiner SB: Trafficking of prion proteins through a caveolae-mediated endosomal pathway. *J Cell Biol* 2003, 162:703-717.

[19] Borchelt DR, Scott M, Taraboulos A, Stahl N, Prusiner SB: Scrapie and cellular prion proteins differ in their kinetics of synthesis and topology in cultured cells. *J Cell Biol* 1990, 110:743-752.

[20] Hegde RS, Mastrianni JA, Scott MR, DeFea KA, Tremblay P, Torchia M, DeArmond SJ, Prusiner SB, Lingappa VR: A transmembrane form of the prion protein in neurodegenerative disease. *Science* 1998, 279:827-834.

[21] Brown DR, Qin K, Herms JW, Madlung A, Manson J, Strome R, Fraser PE, Kruck T, von Bohlen A, Schulz-Schaeffer W, Giese A, Westaway D, Kretzschmar H: The cellular prion protein binds copper in vivo. *Nature* 1997, 390:684-687.

[22] Pauly PC, Harris DA: Copper stimulates endocytosis of the prion protein. *J Biol Chem* 1998, 273:33107-33110.

[23] Brown DR, Besinger A: Prion protein expression and superoxide dismutase activity. *Biochem J* 1998, 334:423-429.

[24] Brown DR, Wong BS, Hafiz F, Clive C, Haswell SJ, Jones IM: Normal prion protein has an activity like that of superoxide dismutase. *Biochem J* 1999, 344 Pt 1:1-5.

[25] Collinge J, Whittington MA, Sidle KC, Smith CJ, Palmer MS, Clarke AR, Jefferys JG: Prion protein is necessary for normal synaptic function. *Nature* 1994, 370:295-297.

[26] Herms JW, Kretzchmar HA, Titz S, Keller BU: Patch-clamp analysis of synaptic transmission to cerebellar purkinje cells of prion protein knockout mice. *Eur J Neurosci* 1995, 7:2508-2512.

[27] Lledo PM, Tremblay P, DeArmond SJ, Prusiner SB, Nicoll RA: Mice deficient for prion protein exhibit normal neuronal excitability and synaptic transmission in the hippocampus. *Proc Natl Acad Sci U S A* 1996, 93:2403-2407.

[28] Bounhar Y, Zhang Y, Goodyer CG, LeBlanc A: Prion protein protects human neurons against Bax-mediated apoptosis. *J Biol Chem* 2001, 276:39145-39149.

[29] Kuwahara C, Takeuchi AM, Nishimura T, Haraguchi K, Kubosaki A, Matsumoto Y, Saeki K, Yokoyama T, Itohara S, Onodera T: Prions prevent neuronal cell-line death. *Nature* 1999, 400:225-226.

[30] Mouillet-Richard S, Ermonval M, Chebassier C, Laplanche JL, Lehmann S, Launay JM, Kellermann O: Signal transduction through prion protein. *Science* 2000, 289:1925-1928.

[31] Chiarini LB, Freitas AR, Zanata SM, Brentani RR, Martins VR, Linden R: Cellular prion protein transduces neuroprotective signals. *EMBO J* 2002, 21:3317-3326.

[32] Caughey B, Raymond GJ, Ernst D, Race RE: N-terminal truncation of the scrapie-associated form of PrP by lysosomal protease(s): implications regarding the site of conversion of PrP to the protease-resistant state. *J Virol* 1991, 65:6597-6603.

[33] Borchelt DR, Taraboulos A, Prusiner SB: Evidence for synthesis of scrapie prion proteins in the endocytic pathway. *J Biol Chem* 1992, 267:16188-16199.

[34] Beranger F, Mange A, Goud B, Lehmann S: Stimulation of PrP^{C} Retrograde Transport toward the Endoplasmic Reticulum Increases Accumulation of PrP^{Sc} in Prion-infected Cells. *J Biol Chem* 2002, 277:38972-38977.

[35] Ma J, Lindquist S: Conversion of PrP to a self-perpetuating PrP^{Sc}-like conformation in the cytosol. *Science* 2002, 298:1785-1788.

[36] Peretz D, Williamson RA, Kaneko K, Vergara J, Leclerc E, Schmitt-Ulms G, Mehlhorn IR, Legname G, Wormald MR, Rudd PM, Dwek RA, Burton DR, Prusiner SB: Antibodies inhibit prion propagation and clear cell cultures of prion infectivity. *Nature* 2001, 412:739-743.

[37] Moroncini G, Kanu N, Solforosi L, Abalos G, Telling GC, Head M, Ironside J, Brockes JP, Burton DR, Williamson RA: Motif-grafted antibodies containing the replicative interface of cellular PrP are specific for PrP^{Sc}. *Proc Natl Acad Sci U S A* 2004, 101:10404-10409.

[38] Telling GC, Scott M, Mastrianni J, Gabizon R, Torchia M, Cohen FE, DeArmond SJ, Prusiner SB: Prion propagation in mice expressing human and chimeric PrP transgenes implicates the interaction of cellular PrP with another protein. *Cell* 1995, 83:79-90.

[39] Telling GC, Parchi P, DeArmond SJ, Cortelli P, Montagna P, Gabizon R, Mastrianni J, Lugaresi E, Gambetti P, Prusiner SB: Evidence for the conformation of the pathologic isoform of the prion protein enciphering and propagating prion diversity. *Science* 1996, 274:2079-2082.

[40] Suenaga T, Hirano A, Llena JF, Ksiezak-Reding H, Yen SH, Dickson DW: Ubiquitin immunoreactivity in kuru plaques in Creutzfeldt-Jakob disease. *Ann Neurol* 1990, 28:174-177.

[41] Cammarata S, Tabaton M: Ubiquitin-reactive axons have a widespread distribution and are unrelated to prion protein plaques in Creutzfeldt-Jakob disease. *J Neurol Sci* 1992, 110:32-36.

[42] Ironside JW, McCardle L, Hayward PA, Bell JE: Ubiquitin immunocytochemistry in human spongiform encephalopathies. *Neuropathol Appl Neurobiol* 1993, 19:134-140.

[43] Kovacs GG, Preusser M, Strohschneider M, Budka H: Subcellular localization of disease-associated prion protein in the human brain. *Am J Pathol* 2005, 166:287-294.

[44] Adori C, Kovacs GG, Low P, Molnar K, Gorbea C, Fellinger E, Budka H, Mayer RJ, Laszlo L: The ubiquitin-proteasome system in Creutzfeldt-Jakob and Alzheimer disease: intracellular redistribution of components correlates with neuronal vulnerability. *Neurobiol Dis* 2005, 19:427-435.

[45] Kato S, Hirano A, Umahara T, Llena JF, Herz F, Ohama E: Ultrastructural and immunohistochemical studies on ballooned cortical neurons in Creutzfeldt-Jakob disease: expression of alpha B-crystallin, ubiquitin and stress-response protein 27. *Acta Neuropathol (Berl)* 1992, 84:443-448.

[46] Bugiani O, Giaccone G, Verga L, Pollo B, Frangione B, Farlow MR, Tagliavini F, Ghetti B: Beta PP participates in PrP-amyloid plaques of Gerstmann-Straussler-Scheinker disease, Indiana kindred. *J Neuropathol Exp Neurol* 1993, 52:64-70.

[47] Debeer S, Baron T, Bencsik A: Neuropathological characterisation of French bovine spongiform encephalopathy cases. *Histochem Cell Biol* 2003, 120:513-521.

[48] Lowe J, Fergusson J, Kenward N, Laszlo L, Landon M, Farquhar C, Brown J, Hope J, Mayer RJ: Immunoreactivity to ubiquitin-protein conjugates is present early in the disease process in the brains of scrapie-infected mice. *J Pathol* 1992, 168:169-177.

[49] Kang SC, Brown DR, Whiteman M, Li R, Pan T, Perry G, Wisniewski T, Sy MS, Wong BS: Prion protein is ubiquitinated after developing protease resistance in the brains of scrapie-infected mice. *J Pathol* 2004, 203:603-608.

[50] Kenward N, Hope J, Landon M, Mayer RJ: Expression of polyubiquitin and heat-shock protein 70 genes increases in the later stages of disease progression in scrapie-infected mouse brain. *J Neurochem* 1994, 62:1870-1877.

[51] Kovacs GG, Voigtlander T, Hainfellner JA, Budka H: Distribution of intraneuronal immunoreactivity for the prion protein in human prion diseases. *Acta Neuropathol (Berl)* 2002, 104:320-326.

[52] Tanaka S, Saito M, Morimatsu M, Ohama E: Immunohistochemical studies of the PrP^{CJD} deposition in Creutzfeldt- Jakob disease. *Neuropathology* 2000, 20:124-133.

[53] Wojcik C, DeMartino GN: Intracellular localization of proteasomes. *Int J Biochem Cell Biol* 2003, 35:579-589.

[54] Pines J, Lindon C: Proteolysis: anytime, any place, anywhere? *Nat Cell Biol* 2005, 7:731-735.

[55] Kristiansen M, Messenger MJ, Klohn PC, Brandner S, Wadsworth JD, Collinge J, Tabrizi SJ: Disease-related prion protein forms aggresomes in neuronal cells leading to caspase-activation and apoptosis. *J Biol Chem* 2005, 280:38851-38861.

[56] Grigoriev V, Escaig-Haye F, Streichenberger N, Kopp N, Langeveld J, Brown P, Fournier JG: Submicroscopic immunodetection of PrP in the brain of a patient with a new-variant of Creutzfeldt-Jakob disease. *Neurosci Lett* 1999, 264:57-60.

[57] Arnold JE, Tipler C, Laszlo L, Hope J, Landon M, Mayer RJ: The abnormal isoform of the prion protein accumulates in late-endosome- like organelles in scrapie-infected mouse brain. *J Pathol* 1995, 176:403-411.

[58] Jacob C, Cottrell GS, Gehringer D, Schmidlin F, Grady EF, Bunnett NW: c-Cbl mediates ubiquitination, degradation, and down-regulation of human protease-activated receptor 2. *J Biol Chem* 2005, 280:16076-16087.

[59] Iwata A, Riley BE, Johnston JA, Kopito RR: HDAC6 and Microtubules Are Required for Autophagic Degradation of Aggregated Huntingtin. *J Biol Chem* 2005, 280:40282-40292.

[60] Sikorska B, Liberski PP, Giraud P, Kopp N, Brown P: Autophagy is a part of ultrastructural synaptic pathology in Creutzfeldt-Jakob disease: a brain biopsy study. *Int J Biochem Cell Biol* 2004, 36:2563-2573.

[61] Priola SA, Chesebro B, Caughey B: Biomedicine. A view from the top--prion diseases from 10,000 feet. *Science* 2003, 300:917-919.

[62] Brandner S, Isenmann S, Kuhne G, Aguzzi A: Identification of the end stage of scrapie using infected neural grafts. *Brain Pathol* 1998, 8:19-27.

[63] Guentchev M, Wanschitz J, Voigtlander T, Flicker H, Budka H: Selective neuronal vulnerability in human prion diseases. Fatal familial insomnia differs from other types of prion diseases. *Am J Pathol 1*999, 155:1453-1457.

[64] Tschampa HJ, Herms JW, Schulz-Schaeffer WJ, Maruschak B, Windl O, Jastrow U, Zerr I, Steinhoff BJ, Poser S, Kretzschmar HA: Clinical findings in sporadic Creutzfeldt-Jakob disease correlate with thalamic pathology. *Brain* 2002, 125:2558-2566.

[65] Belichenko PV, Miklossy J, Belser B, Budka H, Celio MR: Early destruction of the extracellular matrix around parvalbumin- immunoreactive interneurons in Creutzfeldt-Jakob disease. *Neurobiol Dis* 1999, 6:269-279.

[66] Guentchev M, Groschup MH, Kordek R, Liberski PP, Budka H: Severe, early and selective loss of a subpopulation of GABAergic inhibitory neurons in experimental transmissible spongiform encephalopathies. *Brain Pathol* 1998, 8:615-623.

[67] Diez M, DeArmond SJ, Groth D, Prusiner SB, Hokfelt T: Decreased mk-801 binding in discrete hippocampal regions of prion- infected mice. *Neurobiol Dis* 2001, 8:692-699.

[68] Williams A, Lucassen PJ, Ritchie D, Bruce M: PrP deposition, microglial activation, and neuronal apoptosis in murine scrapie. *Exp Neurol* 1997, 144:433-438.

[69] Gray F, Chretien F, Adle-Biassette H, Dorandeu A, Ereau T, Delisle MB, Kopp N, Ironside JW, Vital C: Neuronal apoptosis in Creutzfeldt-Jakob disease. *J Neuropathol Exp Neurol* 1999, 58:321-328.

[70] Giese A, Groschup MH, Hess B, Kretzschmar HA: Neuronal cell death in scrapie-infected mice is due to apoptosis. *Brain Pathol* 1995, 5:213-221.

[71] Lucassen PJ, Williams A, Chung WC, Fraser H: Detection of apoptosis in murine scrapie. *Neurosci Lett* 1995, 198:185-188.

[72] Schatzl HM, Laszlo L, Holtzman DM, Tatzelt J, DeArmond SJ, Weiner RI, Mobley WC, Prusiner SB: A hypothalamic neuronal cell line persistently infected with scrapie prions exhibits apoptosis. *J Virol* 1997, 71:8821-8831.

[73] Dorandeu A, Wingertsmann L, Chretien F, Delisle MB, Vital C, Parchi P, Montagna P, Lugaresi E, Ironside JW, Budka H, Gambetti P, Gray F: Neuronal apoptosis in fatal familial insomnia. *Brain Pathol* 1998, 8:531-537.

[74] Oesch B, Westaway D, Walchli M, McKinley MP, Kent SB, Aebersold R, Barry RA, Tempst P, Teplow DB, Hood LE, et al.: A cellular gene encodes scrapie PrP 27-30 protein. *Cell* 1985, 40:735-746.

[75] Weissmann C: The state of the prion. *Nat Rev Microbiol* 2004, 2:861-871.

[76] Bueler H, Fischer M, Lang Y, Bluethmann H, Lipp HP, DeArmond SJ, Prusiner SB, Aguet M, Weissmann C: Normal development and behaviour of mice lacking the neuronal cell- surface PrP protein. *Nature* 1992, 356:577-582.

[77] Mallucci GR, Ratte S, Asante EA, Linehan J, Gowland I, Jefferys JG, Collinge J: Post-natal knockout of prion protein alters hippocampal CA1 properties, but does not result in neurodegeneration. *EMBO J* 2002, 21:202-210.

[78] DeArmond SJ, Mobley WC, DeMott DL, Barry RA, Beckstead JH, Prusiner SB: Changes in the localization of brain prion proteins during scrapie infection. *Neurology* 1987, 37:1271-1280.

[79] Hsiao KK, Groth D, Scott M, Yang SL, Serban H, Rapp D, Foster D, Torchia M, Dearmond SJ, Prusiner SB: Serial transmission in rodents of neurodegeneration from transgenic mice expressing mutant prion protein. *Proc Natl Acad Sci U S A* 1994, 91:9126-9130.

[80] Manson JC, Jamieson E, Baybutt H, Tuzi NL, Barron R, McConnell I, Somerville R, Ironside J, Will R, Sy MS, Melton DW, Hope J, Bostock C: A single amino acid alteration (101L) introduced into murine PrP dramatically alters incubation time of transmissible spongiform encephalopathy. *EMBO J* 1999, 18:6855-6864.

[81] Brandner S, Isenmann S, Raeber A, Fischer M, Sailer A, Kobayashi Y, Marino S, Weissmann C, Aguzzi A: Normal host prion protein necessary for scrapie-induced neurotoxicity. *Nature* 1996, 379:339-343.

[82] Bueler H, Raeber A, Sailer A, Fischer M, Aguzzi A, Weissmann C: High prion and PrP^{Sc} levels but delayed onset of disease in scrapie- inoculated mice heterozygous for a disrupted PrP gene. *Mol Med* 1994, 1:19-30.

[83] Manson JC, Clarke AR, McBride PA, McConnell I, Hope J: PrP gene dosage determines the timing but not the final intensity or distribution of lesions in scrapie pathology. *Neurodegeneration* 1994, 3:331-340.

[84] Chesebro B, Trifilo M, Race R, Meade-White K, Teng C, LaCasse R, Raymond L, Favara C, Baron G, Priola S, Caughey B, Masliah E, Oldstone M: Anchorless prion protein results in infectious amyloid disease without clinical scrapie. *Science* 2005, 308:1435-1439.

[85] Chiesa R, Harris DA: Prion diseases: what is the neurotoxic molecule? *Neurobiol Dis* 2001, 8:743-763.

[86] Hegde RS, Tremblay P, Groth D, DeArmond SJ, Prusiner SB, Lingappa VR: Transmissible and genetic prion diseases share a common pathway of neurodegeneration. *Nature* 1999, 402:822-826.

[87] Stewart RS, Harris DA: Most pathogenic mutations do not alter the membrane topology of the prion protein. *J Biol Chem* 2001, 276:2212-2220.

[88] Kovacs GG, Trabattoni G, Hainfellner JA, Ironside JW, Knight RS, Budka H: Mutations of the prion protein gene phenotypic spectrum. *J Neurol* 2002, 249:1567-1582.

[89] Liemann S, Glockshuber R: Influence of amino acid substitutions related to inherited human prion diseases on the thermodynamic stability of the cellular prion protein. *Biochemistry* 1999, 38:3258-3267.

[90] Riek R, Wider G, Billeter M, Hornemann S, Glockshuber R, Wuthrich K: Prion protein NMR structure and familial human spongiform encephalopathies. *Proc Natl Acad Sci U S A* 1998, 95:11667-11672.

[91] Zanusso G, Petersen RB, Jin T, Jing Y, Kanoush R, Ferrari S, Gambetti P, Singh N: Proteasomal degradation and N-terminal protease resistance of the codon 145 mutant prion protein. *J Biol Chem* 1999, 274:23396-23404.

[92] Jin T, Gu Y, Zanusso G, Sy M, Kumar A, Cohen M, Gambetti P, Singh N: The chaperone protein BiP binds to a mutant prion protein and mediates its degradation by the proteasome. *J Biol Chem* 2000, 275:38699-38704.

[93] Hachiya NS, Watanabe K, Kawabata MY, Jozuka A, Kozuka Y, Sakasegawa Y, Kaneko K: Prion protein with Y145STOP mutation induces mitochondria-mediated apoptosis and PrP-containing deposits in vitro. *Biochem Biophys Res Commun* 2005, 327:894-899.

[94] Lorenz H, Windl O, Kretzschmar HA: Cellular phenotyping of secretory and nuclear prion proteins associated with inherited prion diseases. *J Biol Chem* 2002, 277:8508-8516.

[95] Gu Y, Hinnerwisch J, Fredricks R, Kalepu S, Mishra RS, Singh N: Identification of cryptic nuclear localization signals in the prion protein. *Neurobiol Dis* 2003, 12:133-149.

[96] Ma J, Lindquist S: De novo generation of a PrP^{Sc}-like conformation in living cells. *Nat Cell Biol* 1999, 1:358-361.

[97] Ma J, Lindquist S: Wild-type PrP and a mutant associated with prion disease are subject to retrograde transport and proteasome degradation. *Proc Natl Acad Sci U S A* 2001, 98:14955-14960.

[98] Yedidia Y, Horonchik L, Tzaban S, Yanai A, Taraboulos A: Proteasomes and ubiquitin are involved in the turnover of the wild-type prion protein. *EMBO J* 2001, 20:5383-5391.

[99] Mishra RS, Bose S, Gu Y, Li R, Singh N: Aggresome formation by mutant prion proteins: the unfolding role of proteasomes in familial prion disorders. *J Alzheimers Dis* 2003, 5:15-23.

[100] Cohen E, Taraboulos A: Scrapie-like prion protein accumulates in aggresomes of cyclosporin A-treated cells. *EMBO J* 2003, 22:404-417.

[101] Ma J, Wollmann R, Lindquist S: Neurotoxicity and neurodegeneration when PrP accumulates in the cytosol. *Science* 2002, 298:1781-1785.

[102] Drisaldi B, Stewart RS, Adles C, Stewart LR, Quaglio E, Biasini E, Fioriti L, Chiesa R, Harris DA: Mutant PrP Is Delayed in Its Exit from the Endoplasmic Reticulum, but Neither Wild-type nor Mutant PrP Undergoes Retrotranslocation Prior to Proteasomal Degradation. *J Biol Chem* 2003, 278:21732-21743.

[103] Roucou X, Guo Q, Zhang Y, Goodyer CG, LeBlanc AC: Cytosolic prion protein is not toxic and protects against Bax-mediated cell death in human primary neurons. *J Biol Chem* 2003, 278:40877-40881.

[104] Wang X, Wang F, Sy MS, Ma J: Calpain and other cytosolic proteases can contribute to the degradation of retro-translocated prion protein in the cytosol. *J Biol Chem* 2005, 280:317-325.

[105] Rane NS, Yonkovich JL, Hegde RS: Protection from cytosolic prion protein toxicity by modulation of protein translocation. *EMBO J* 2004, 23:4550-4559.

[106] Mironov A, Jr., Latawiec D, Wille H, Bouzamondo-Bernstein E, Legname G, Williamson RA, Burton D, DeArmond SJ, Prusiner SB, Peters PJ: Cytosolic prion protein in neurons. *J Neurosci* 2003, 23:7183-7193.

[107] Barmada S, Piccardo P, Yamaguchi K, Ghetti B, Harris DA: GFP-tagged prion protein is correctly localized and functionally active in the brains of transgenic mice. *Neurobiol Dis* 2004, 16:527-537.

[108] Kim SJ, Mitra D, Salerno JR, Hegde RS: Signal sequences control gating of the protein translocation channel in a substrate-specific manner. *Dev Cell* 2002, 2:207-217.

[109] Levine CG, Mitra D, Sharma A, Smith CL, Hegde RS: The efficiency of protein compartmentalization into the secretory pathway. *Mol Biol Cell* 2005, 16:279-291.

[110] Shaffer KL, Sharma A, Snapp EL, Hegde RS: Regulation of protein compartmentalization expands the diversity of protein function. *Dev Cell* 2005, 9:545-554.

[111] Fioriti L, Dossena S, Stewart LR, Stewart RS, Harris DA, Forloni G, Chiesa R: Cytosolic prion protein (PrP) is not toxic in N2a cells and primary neurons expressing pathogenic PrP mutations. *J Biol Chem* 2005, 280: 11320-11328.

[112] Aguzzi A, Polymenidou M: Mammalian prion biology: one century of evolving concepts. *Cell* 2004, 116:313-327.

[113] Kurschner C, Morgan JI: The cellular prion protein (PrP) selectively binds to Bcl-2 in the yeast two-hybrid system. *Brain Res Mol Brain Res* 1995, 30:165-168.

[114] Spielhaupter C, Schatzl HM: PrP^C directly interacts with proteins involved in signaling pathways. *J Biol Chem* 2001, 276:44604-44612.

[115] Bragason BT, Palsdottir A: Interaction of PrP with NRAGE, a protein involved in neuronal apoptosis. *Mol Cell Neurosci* 2005, 29:232-244.

In: The Ubiquitin Proteasome System... ISBN 978-1-60021-749-4
Eds: Mario Di Napoli and Cezary Wojcik, pp. 833-847 © 2007 Nova Science Publishers, Inc.

Chapter 34

ROLE OF THE UBIQUITIN PROTEASOME SYSTEM IN ANTIGEN PRESENTATION, AUTOIMMUNE DISORDERS AND INFLAMMATION IN THE CENTRAL NERVOUS SYSTEM

Jakub Golab [*] ***and Dominika Nowis*** [¥]
Department of Immunology, The Medical University of Warsaw, Warsaw, Poland.

ABSTRACT

The proteasome plays a pivotal role during proteolytic processing of cellular proteins required for the generation of antigenic peptides presented to cytotoxic T cells by major histocompatibility complex class I molecules. The process of peptide generation is greatly improved by formation of immunoproteasomes through replacement of three β subunits with β1i (also called LMP2), β5i (LMP7) and β2i (MECL-1) and expression of PA28, a heptameric activator complex. The assembly of immunoproteasomes is stimulated by interferon-γ, a cytokine that is produced shortly after viral infection and is one of the mediators that link innate and adaptive immune responses. Numerous infectious microorganisms developed sophisticated strategies to avoid presentation of their antigenic peptides including production of proteasome-modulating molecules. Recent studies indicate a unique mechanism of epitope generation by proteasomes referred to as peptide splicing. In the brain, microglial cells are the major antigen presenting cells. However, during inflammation virtually all cells in the central nervous system can be induced to express immunoproteasomes and to present antigens in association with MHC class I molecules. The proteasome-mediated generation of peptide

* Correspondence concerning this article should be addressed to Dr. Jakub Golab, MD, PhD; Department of Immunology, The Medical University of Warsaw, Ul. Banacha 1a, F building, 02-097 Warsaw, Poland. Phone: *48-22-5992199; Fax: *48-22-5992194 E-mail: jgolab@ib.amwaw.edu.pl.

¥ Correspondence concerning this article should be addressed to Dr. Dominika Nowis, MD, PhD; Department of Immunology, The Medical University of Warsaw, Ul. Banacha 1a, F building, 02-097 Warsaw, Poland; Phone: *48-22-5992199; Fax: *48-22-5992194 E-mail: dnowis@ib.amwaw.edu.pl.

epitopes evolutionarily serves to present antigens derived from intracellular infectious microorganisms. However, proteolytic processing of intracellular proteins is by no means selective and includes processing of all proteins including self molecules that can become targets for cytotoxic T cells during several inflammatory and degenerative central nervous system diseases, such as multiple sclerosis and paraneoplastic neurological disorders. The role of ubiquitin-proteasome pathway in neuroinflammatory disorders extends beyond antigen processing for MHC class I presentation. Activation of NF-κB a key modulator of inflammatory reaction results from proteasomal degradation of its inhibitor – IκB. Proteasomes are also involved in the regulation of the activity of other transcription factors involved in the inflammatory responses including STAT proteins and Egr-1. Understanding of the underlying mechanisms involved in proteasome-mediated inflammatory processes is important for the development of novel, mechanism-based drugs.

Keywords: Antigen presentation, Autoimmune diseases, Major histocompatibility complex, Proteasome, Ubiquitin.

ABBREVIATIONS

AD, Alzheimer's disease; ALS, amyotrophic lateral sclerosis; APP, amyloid precursor protein; BBB, blood-brain barrier; bFGF, basic fibroblast growth factor; β2m, β2 microglobulin; BiP, bding protein; CD, cluster of differentiation; CNS, central nervous system; COX, cyclooxygenase; CTL, cytotoxic T lymphocyte; DRips, defective ribosomal proteins; EAE, experimental autoimmune encephalomyelitis; ER, endoplasmic reticulum; ERAP, endoplasmic reticulum-resident aminopeptidases; IKK, IκB kinase; IL, interleukin; LMP, low molecular mass protein; LPS, lipopolisaccharide; MECL, multicatalytic endopeptidase complex; MHC, major histocompatibility complex; MS, multiple sclerosis; NEMO, NF-κB essential modulator; NF-κB, nuclear factor κB; NK, natural killer; NES, nuclear-export signal; NGF, neuron growth factor; NIK, NF-κB kinase; NLS, nuclear-localization sequence; NO, nitric oxide; PD, Parkinson's disease; RHD, Rel homology domain; ROS, reactive oxygen species; SAPPα, Secreted amyloid precursor protein; SOCS, suppressor of cytokine signaling; STAT, signal transducer and activator of transcription; TAP, transporters associated with antigen presentation; TCR, T cell receptor; TNF, tumor necrosis factor.

INTRODUCTION

Central nervous system (CNS) has been regarded as an immune privileged site. This concept has been supported by both anatomic and functional observations that include descriptions of a unique blood-brain barrier (BBB), lack or low expression of MHC class I and class II molecules and lack of lymphatic drainage. However, CNS should not be completely devoid of immune protection. There must exist mechanisms to control infections and tumor development. These mechanisms clearly operate in the CNS and it seems that the unique immunity within the brain results from an active tolerance rather than passive barrier mechanisms. Accumulating evidence indicates that MHC class I molecules are indeed expressed in different brain regions by both neuronal cells and supporting glia [1]. Inability to detect MHC molecules was rather an artifact caused by using aldehyde-fixed brain sections and antibodies developed for fluorescence-activated cell sorting. Moreover, neurons can be induced to express high levels of MHC molecules following various experimental procedures including axotomy [2], exposure to cytokines [3] or manipulation of electrical activity [4]. Moreover, viral, parasitic or bacterial infections [5], neuronal transplantation [6] or even seizures [4] can also induce expression of MHC class I molecules in the brain. Intriguingly, despite the expression of MHC molecules neurons are usually resistant to lysis by cytotoxic T lymphocytes (CTL) and soluble factors released by neurons or glia can interfere with CTL activity thereby creating an environment suitable for immunomodulation [7].

Interestingly, it seems that the role of MHC class I molecules in the CNS expands beyond the mere antigen presentation for the induction of protective immunity. These molecules seem to be involved in both structural and functional plasticity and are thought to participate in the weakening and strengthening of synapses [8,9]. At least in rodents MHC class I molecules are also possibly involved in olfaction [10].

The functional significance of the immune response in the brain can be underscored by the existence of various neuroinflammatory disorders. Some of these result from the failure of self tolerance and lead to development of autoimmune disorders such as multiple sclerosis or various paraneoplastic autoimmune reactions and primarily involve adaptive immunity. Other pathologies that involve the innate immune system include degenerative disorders of the central nervous system (CNS) and include Alzheimer's disease (AD), Parkinson's disease (PD) or stroke and can be triggered secondarily for example by local deposition of amyloid β-protein [11]. The importance of the immune reactions in the CNS will possibly be even greater in the nearest future due to the development of novel therapeutic approaches to treat these diseases. For example, the unsuccessful results of the recent double-blind placebo-controlled trials in patients with PD receiving allografts of fetal dopaminergic neurons might result from the development of a classical transplantation immunity and rejection of grafted tissue [12].

The Role of the Ubiquitin-Proteasome System in Antigen Presentation

Major histocompatibility antigens (MHC) act as platforms presenting antigenic peptides to T cell receptors (TCR) on T lymphocytes. There are two groups of classical MHC molecules referred to as class I and class II molecules [13]. Class I MHC molecules are responsible for antigen presentation to 'killer' or cytotoxic CD8+ T cells (CTL), while MHC class II molecules present antigens to helper CD4+ T cells. Additionally, MHC class I molecules can be recognized by natural killer (NK) cells that depending on additional signals can become triggered or suppressed. MHC class I molecules present so called endogenous antigens i.e. peptides derived from proteins synthesized within the MHC-bearing cell. On the other hand MHC class II molecules present exogenous antigens derived from proteins endocytosed by antigen presenting cells. The genes encoding human MHC molecules (human leukocyte antigens – HLA) are the most polymorphic genes ever known and currently account to 2435 different alleles (IMTG/HLA Sequence Database: http://www.ebi.ac.uk/imgt/hla/). Since proteasomes are involved in the generation of antigenic peptides bound and presented by MHC class I molecules expressed by all nucleated cells, including neurons and other cells in the brain, the discussion in this chapter will be limited to this pathway of antigen presentation only.

General Description of MHC Class I Peptide Loading

MHC class I molecules are constitutively expressed by most nucleated cells and present endogenous peptides of 8-11 amino acids [14]. All are composed of a heavy chain (composed of three immunoglobulin-like domains) and a constant β2-microglobulin (β2m). During assembly and folding heavy chains are extensively chaperoned by cytosolic BiP proteins and finally by endoplasmic reticulum (ER) lectin calnexin [15]. Final complexing with β2m results in an exchange of calnexin into a related calreticulin followed by a formation of a large complex of proteins that resides in the ER. This protein complex is frequently referred to as a loading complex as it promotes peptide loading into MHC class I peptide-binding groove [16]. The loading complex includes MHC molecule, calreticulin as well as tapasin, TAP and Erp57 molecules [17]. Tapasin functions as a transmembrane protein that interacts with TAP and MHC molecules thus stabilizing the loading complex – it is required to retain these molecules in the ER until optimal peptides have been loaded [18,19]. Erp57 is a thiol oxidoreductase that associates with MHC class I molecules and facilitates disulfide bond formation thus influencing the conformation of these molecules [20]. Additionally Erp57 is a cysteine protease that facilitates trimming of peptides that are being loaded onto MHC class I molecules (see below). Peptide transporters (TAP1 and TAP2 proteins) shuffle peptides from the cytosol to the ER thereby supplying antigens for presentation [21]. TAP proteins transport peptides of 7 to more than 20 amino acids so some of the longer peptides can be trimmed within ER by a number of aminopeptidases (see below). The major source of peptides for antigen presentation is the ubiquitin-proteasome system but some peptides are generated in the cytosol by TPPII [22].

Where do the Peptide Antigens Come from?

This is certainly not a trivial question. Unfortunately, very little is known about the peptide processing in the CNS. An average cell contains approximately 2.6 x 10^9 proteins and produces some 4 millions new proteins every minute [23]. To avoid chaos there must exist sophisticated mechanisms that control targeting as well as turnover of cellular proteins. Proteolytic degradation is responsible for the elimination of damaged, unfolded or incomplete proteins. It is also engaged in the regulation of their function. Therefore, proteolysis might serve as an abundant source of peptide libraries derived from the majority of cellular proteins that would be sampled by MHC class I molecules (Figure 1). Normal cells infected by viruses or intracellular parasites would contain both self and foreign proteins with the latter presumed to be processed and presented in the most immunogenic manner. For many years it was thought that most peptides destined for presentation by MHC molecules would derive from old or used proteins targeted for degradation as a part of protein turnover. It was therefore unexpected to observe that the majority of antigenic peptides are derived from proteins degraded immediately after synthesis [24]. A rich source of antigens are peptide ligands derived from cryptic transcription products, such as open reading frames contained within 5' and 3' untranslated regions, alternative open reading frames, introns, or intron-exon junctions [25-27]. It is known that protein synthesis is error prone. These erroneous proteins are referred to as defective ribosomal products (DRips) and they represent between 30% to as many as 80% of newly synthesized proteins [23,28]. Since accumulation of these proteins could have disastrous consequences for the cell they are rapidly ubiquitinated and degraded by proteasomes. Therefore, MHC molecules have an immediate access to peptides derived from current profile of protein expression pattern. This strategy makes sense from the perspective of immunity. CTL are part of the surveillance system that look for MHC molecules presenting peptide antigens derived from viral or mutated proteins and exert cytotoxic activity towards the cells having such abnormal antigens. This surveillance mechanism is exquisitely sensitive as CTL can respond to as little as one MHC class I peptide complex at the cell surface [29]. This system also needs to be very rapid. In extreme conditions a viral cycle (a time necessary to enter the cell, replicate and release new progeny) is just about 4 hours. Early detection of pathogen-derived peptides is of paramount importance especially that viruses use a cohort of genes that interfere with antigen presentation [30]. Despite these pressing demands, the frequency of peptides available for MHC class I binding is extremely low. Out of 8×10^5 proteasomes that degrade 2.5 substrates per minute only one for each 500-3000 viral translation products degraded is suitable for presentation [23]. We still do not understand this apparent insufficiency in the generation of antigenic peptides.

The Role of Proteasomes in the Generation of Antigenic Peptides

Individual MHC class I molecules can bind to a limited set of peptides restricted predominantly by particular anchor residues of which one is always located at the carboxy (C) terminus [31,32]. While proteasomes form a central proteolytic system it is clearly insufficient to generate final peptides for MHC class I binding. Nonetheless, proteasomes are the only cellular proteases that generate correct C-terminus of peptides for presentation and there is no further need for C-terminal processing [33,34]. Final trimming of oligopeptides requires additional proteases that include TPPII, and other aminopeptidases such as leucine aminopeptidase, bleomycin hydrolase, puromycin-sensitive aminopeptidase, thimet oligopeptidase as well as endoplasmic reticulum-resident aminopeptidases ERAP1 and L-RAP [31].

There is a number of unresolved and sometimes paradoxical observations regarding the mechanism of peptide generation by proteasomes. They seem to cut proteins blindly irrespective of their potential further utility for the cell. Purified proteasomes can even

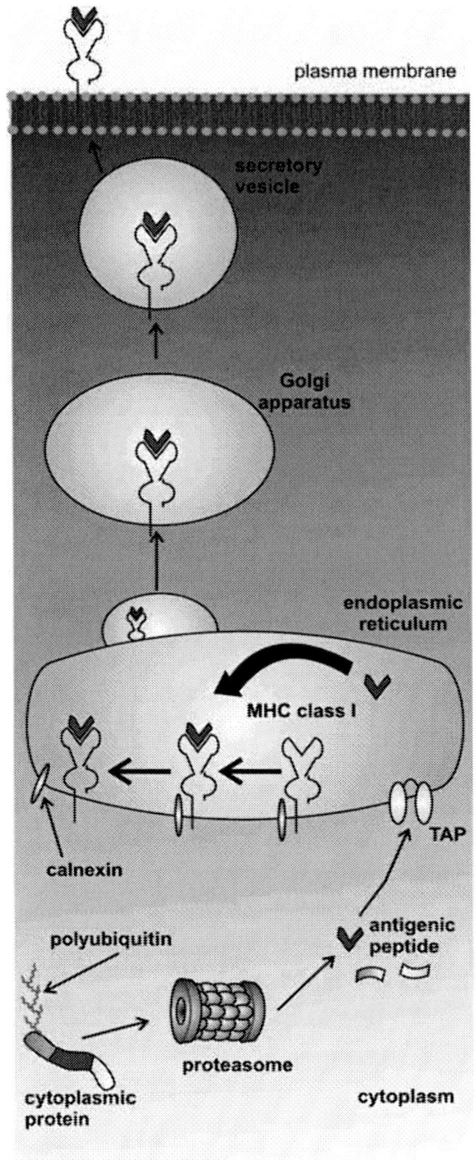

Figure 1. The pathway for presentation of antigenic peptides in association with MHC class I molecules. The majority of peptides presented by MHC class I molecules derive from defective ribosomal products (DRips). These incorrectly produced polypeptides undergo ubiquitination followed by proteasomal degradation. The peptides frequently undergo further proteolytic trimming with cytoplasmic and endoplamic reticulum-associated proteases. Transporters associated with antigen processing (TAP) translocate peptides from the cytosol to the lumen of endoplasmic reticulum where they are loaded into grooves of MHC class I pockets by the chaperone complex. Then the peptide antigen-loaded MHC molecules are transported to the surface of the cells where they can be sampled by CD8+ T cells.

destroy immunodominant epitopes [35]. Although they are required for the generation of antigenic peptides they in fact destroy many more epitopes than they generate [36,37]. Purified proteasomes degrade proteins mainly into 2-25 amino acid-long peptides. Only a small fraction of these (10%) are of correct size to fit into MHC class I molecules, with the majority (65%) being too short for this. The remaining 25% of peptides are too long to fit into MHC class I grooves [31,36,37]. This latter pool of peptides undergoes N-terminal processing to generate antigens suitable for MHC loading.

Until recently it was thought that peptides generated by proteasomes are continuous sequences derived from cellular proteins generated by simple proteolysis. However, it now seems that antigenic peptides can be the products of cutting and pasting reactions analogous to exon-intron splicing events during mRNA maturation. It was shown that antigenic peptide derived from basic fibroblast growth factor (bFGF) is in fact a nonamer consisting of residues 172 to 176 fused to residues 217 to 220 of the protein [38]. Similarly, melanocyte glycoprotein gp100-derived nonamer corresponds to residues 40 to 52 with amino acids 43 to 46 excised from the original sequence [39]. It seems that proteasomes can degrade proteins in such a way that the terminal amino group of one peptide can attack the acyl-enzyme intermediate (a complex between the remaining peptide end and a proteasomal β subunit) thus regenerating a peptide bond between two previously separated amino acids.

Fine-Tuning of Antigen Presentation with Proteasomes

An effective immune response results from a series of positive feedback loops that include improved antigen presentation. Under the influence of certain cytokines such as IFN-γ and TNF the cells improve their antigen-presenting capabilities [40]. They not only express more MHC and co-stimulatory molecules but have more dexterous machinery to process and load peptides into MHC class I peptide-binding grooves. IFN-γ induces the synthesis of proteasomal β subunits: β1i (LMP2), β2i (MECL1) and β5i (LMP7) that are incorporated into newly assembled proteasomes instead of their counterparts: β1, β2 and β5 [41,42]. These so called immunoproteasomes have different cleavage site preferences as well as a different cleavage rate. Additionally, IFN-γ induces the synthesis of a proteasomal activator PA28, a heptameric complex composed of three PA28α and four PA28β subunits that assemble with the outer α ring of the immunoproteasomes [43,44]. IFN-γ is also inducing phosphorylation of the α7 subunit of the 20S proteasome thereby facilitating disassembly of 26S proteasomes [45]. The significance of this unexpected modification is unknown but it may either suggest that free 20S proteasomes are somehow engaged in antigenic peptide generation or alternatively this modification is part of a negative regulation of antigen presentation. The significance of these modifications in the CNS is unknown. So far PA28αβ activator has not been detected in the brain [46].

Immunoproteasomes are not any better in peptide generation than regular proteasomes [40]. However, it was shown that immunosubunits are necessary for the generation of influenza-derived antigens or in the development of an effective response against hepatitis B virus [47,48]. Interestingly, it was recently shown that immunoproteasomes are incapable of generating several immunogenic antigens from self proteins [49]. This mechanism might be extremely important in negative regulation of potentially devastating autoimmune disorders that could develop during protective antiviral response.

The overall improved antigen presentation by immunoproteasomes seems to result from the capacity to generate longer polypeptides, with a correct C terminus but an extended N terminus that might facilitate TAP transport to endoplasmic reticulum where the peptides would undergo final trimming by aminopeptidases [35]. This process is greatly facilitated by PA28 that induces conformational changes in the α subunits thereby leading to the opening of the central gates of the proteasome and a faster exit of the cleavage products from the catalytic cavity [40,50].

The Role of Proteasomes in the Regulation of Inflammatory Responses in the Brain

The role of the ubiquitin-proteasome system in the development of immune response expands beyond just antigen presentation. In fact defective degradation of ubiquitinated proteins may result in up-regulation of cyclooxygenase 2 (COX-2) expression and commencement of neuroinflammation [51]. Several transcription factors involved in the regulation of inflammatory response are regulated by the ubiquitin-proteasome system. These primarily include NF-κB, which is activated by proteasome-mediated degradation of its inhibitor IκB. However, other transcription factors that induce expression of inflammation-associated molecules are degraded by this proteolytic system indicating that the regulation of inflammation is more complex than originally thought. SOCS-1, potent inhibitor of Jak kinase activity and of signaling initiated by several cytokines targets Jak1 to a perinuclear distribution resembling the microtubule organizing complex where it is degraded by proteasomes [52]. Egr-1 and STAT1 are transcription factors that induce the expression of proinflammatory genes are negatively regulated by proteasomal degradation [53,54].

Regulation of NF-κB Activity

The term NF-κB (nuclear factor kappa B) covers a family of inducible transcription factors that regulate the host immune and inflammatory responses and cellular growth properties [55]. NF-κB was first identified in the nuclei of mature B lymphocytes as a transcription factor binding an 11-bp DNA sequence in the κ-light chain enhancer [56]. The NF-κB family mediates the transcription of over 180 target genes, including genes for cell adhesion molecules, cytokines, chemokines and antiapoptotic factors [57]. NF-κB consists of homodimers or heterodimers of a family of proteins sharing a 300-acid common Rel homology domain (RHD). RHD allows DNA-binding, dimerization and nuclear localization of NF-κB [57,58]. The Rel family includes the following members: p105/50 (NF-κB1), p100/52 (NF-κB2), p65 (RelA), RelB and c-Rel [58]. Each member of the NF-κB family, except for RelB, can form homodimers, as well as heterodimers with one another [59]. The major activated form of NF-κB consists of the p65 subunit associated with either a p50 or p52 subunit [59]. In a stable state, NF-κB binds IκB. IκB is an inhibitory molecule that sequesters NF-κB in cellular cytoplasm in an inactive state, covering its RHD [57]. There are at least six IκB proteins: IκBα, IκBβ, IκBε, IκBδ, IκBγ and Bcl-3. The first three of these are stimulus-dependent regulators, while the others have distinct functions [60]. These proteins have several ankyrin repeats, a 33-amino-acid motif that mediates protein-protein interactions [59]. IκB masks dual nuclear-localization sequences (NLSs) on NF-κB subunits. There is a dynamic balance between the nuclear and cytosolic amount of inactive NF-κB-IκB complexes. IκBα covers only one of the two NLSs in the NF-κB dimer. At the same time, the nuclear-export signal (NES), a part of IκB, functions to expel the NF-κB-IκB complex from the nucleus. IκBβ, by contrast, covers both NLSs on the NF-κB dimers and preserves them in the cytoplasm [59,61]. It is well established that IκBα regulates transient, while IκBβ regulates persistent NF-κB activation. In response to a stimulus, IκBα is quickly degraded and resynthesized. The newly formed IκBα subunit has its own NLS that allows it to enter the nucleus, displace the NF-κB form its DNA binding site and transport it back to the cytoplasm [59]. The proinflammatory cytokines such as TNF or IL-1, signal to NF-κB by activating the IKK complex (IκB kinase). IKK is composed of three subunits. The subunit γ of IKK called NEMO (NF-κB essential modulator) is a regulatory component. Subunits α and β, serving as kinases, phosphorylate IκB on Ser32 and Ser36 in IκBα and Ser33 and Ser37 in IκBβ [62]. Phosphorylated IκB is recognized by the β-TrCP component of the SCF ubiquitination complex that in consequence leads to the IκB ubiquitination and degradation by the 26S proteasome. β-TrCP is critical for the preserving IκB phosphorylation prior to ubiquitination. Degradation of IκB by IKK frees NF-κB to stably translocate to the nucleus where it induces expression of target genes [61,63]. The role of proteasomes in the regulation of NF-κB activity is more complex. 26S proteasome not only degrades IκB proteins but also mediates the proteolytic cleavage of the p105 precursor to the p50 subunit of NF-κB1. Moreover, the activity of the IKK kinase depends on the formation of unusual polyubiquitynated chains linked by Lys63 [64].

Cellular response to a wide range of different stimuli leads to NF-κB activation. Among them, there are cytokines (TNF superfamily, IL-1, IL-18], inducers of the reactive oxygen species (ROS) such as hydrogen peroxide, infectious agents (bacterial as well as viral), inducers of apoptosis, carcinogens, tumor promoters and diverse kinds of stress (change in the cellular pH, hypoxia, presence of heavy metal ions) [56]. These stimuli reveal the role of NF-κB in cellular adaptation to stress. There are two ways of NF-κB activation. The classical signaling pathway is mediated by IKKβ and leads to the phosphorylation and degradation of IκB. The stimuli that result in this pathway include predominantly proinflammatory agents (TNF, IL-1, LPS and double-stranded RNA). The nonclassical pathway of NF-κB activation involves IKKα and results in p100 phosphorylation and its cleavage to the p52 subunit. IKKα is activated by the upstream kinase NIK. This signaling pathway is activated by the lymphotoxin β receptor (LTβR) in the stromal cells to produce the B-lymphocyte chemoattractant required for the proper lymphoid organs development. [61,63,65]. NF-κB is also modified posttransationaly. For its transcriptional activity, NF-κB requires the phosphorylation of p65 by the MAP kinases [61].

The Proteasome-Mediated Regulation of NF-κB Activity in Neuro-Inflammation and Neurodegenerative Disorders

NF-κB is a crucial transcription factor for glial and neuronal cell functions. It is involved in the processes of neuronal plasticity, neurodegeneration and neuronal development [66]. Interestingly, in the CNS neurons, a high level of constitutive NF-κB has been detected, probably due to the electrical activity within neurons and synaptic transmission, which are the potent stimuli for the NF-κB activation [67]. Inducible activity of NF-κB is present in presynaptic as well as postsynaptic sites and in the neuronal cytoplasm [66]. As stated above, NF-κB is a critical regulator of neuronal apoptosis and plays an extremely important role in survival of neurons exposed to cell injury, mostly by upregulating a wide spectrum of antiapoptotic and antioxidant genes [68]. On the other hand, by promotion of production of cytotoxic agents (such as NO) NF-κB may lead to the apoptosis of surrounding cells. In the microglia, activation of NF-κB stimulates the production of ROS and excitotoxins, which are toxic to the environment but at the same time activated microglia produce neurotrophic factors (e.g. NGF, bFGF, TNF), which are essential for the proper neuronal function and prevention of their apoptosis [67].

The role of NF-κB in the initiation and progression of the neurodegenerative disorders is complex. In Alzheimer's disease (AD), increased NF-κB activity is detected in the immediate vicinity of amyloid plaques. Amyloid β-peptide (Aβ) as well as glycated tau proteins can activate NF-κB [62,67]. This activation may, actually, be neuroprotective, while TNF preserves neurons from Aβ-induced cell death in the NF-κB-dependent manner [67]. Secreted amyloid precursor protein (sAPPα) can also activate NF-κB and is regarded as a neuroprotectant [66]. Moreover, decreased NF-κB activity is linked to the early-onset inherited form of AD. Cells expressing mutation in the *Presenilin-1* gene have an insufficient level of activated NF-κB, that leads to their death. Based on the data from in vitro and in vivo studies, the activation of NF-κB in the amyloid deposits seems to be a cytoprotective response [67]. NF-κB is involved in the pathogenesis of Parkinson's and Huntington diseases and amyotrophic lateral sclerosis (ALS). The increased NF-κB activity in the affected neurons may represent, similarly to AD, an early protective response [67]. NF-κB is highly active at sites of inflammation. The NF-κB-mediated production of excessive amount of proinflammatory cytokines, chemokines, adhesion molecules, matrix metalloproteinases, COX-2 and iNOS play a crucial role in the exacerbation of the pathologic processes [69].

In multiple sclerosis (MS), a neuronal disorder with strong inflammatory background, NF-κB activity is detected at high levels in microglia of active plaques [67]. In the experimental autoimmune encephalomyelitis (EAE), an animal MS model, the positive role of NF-κB inhibition (predominantly by the use of proteasome inhibitors) has been reported [70]. Moreover, some predisposing alleles in the inhibitors of NF-κB genes leading to the excessive activation of NF-κB increase MS risk [71]. Activation of NF-κB is also crucial for the progression of brain infections caused by viruses, including, for example, human polyoma JC, EBV and measles viruses [66,68]. NF-κB alone is required for HIV replication in astrocytes and is involved in the pathogenesis of HIV-induced encephalitis [66]. The use of agents inhibiting NF-κB activity in the therapy of the neurological disorders is complicated.

They may attenuate the inflammatory reactions but on the other hand, even exacerbate the neurodegenerative processes and inhibit memory and learning ability [67].

CONCLUSIONS

The role of the ubiquitin-proteasome system in antigen presentation is relatively well known. However, most of the identified details of the protein processing machinery derive from studies on fibroblasts and professional antigen presenting cells such as dendritic cells and macrophages. Very little is known about potential similarities or differences in antigen presentation between peripheral tissues and brain. Elucidation of these differences should prove useful in designing effective therapeutic strategies aimed at thwarting the pathological immune responses in the course of autoimmune and neurodegenerative diseases. The role of the ubiquitin-proteasome system in the regulation of innate or inflammatory disorders is much more complex. Proteasomes are engaged in the degradation of inhibitors (IκB) of transcription factors (NF-κB) thereby augmenting the inflammatory response. Therefore, the use of proteasome inhibitors might seem justified in the treatment of several neuroinflammatory diseases. However, proteasomes are also engaged in the degradation of egr-1, STAT proteins which also drive inflammatory responses. Moreover, inhibition of NF-κB pathway might result in impaired neuroprotection conferred by this transcription factor.

REFERENCES

[1] Boulanger LM, Shatz CJ: Immune signalling in neural development, synaptic plasticity and disease. *Nat Rev Neurosci* 2004, 5: 781-788.

[2] Maehlen J, Schroder HD, Klareskog L, Olsson T, Kristensson K: Axotomy induces MHC class I antigen expression on rat nerve cells. *Neurosci Lett* 1988, 92: 8-13.

[3] Neumann H, Cavalie A, Jenne DE, Wekerle H: Induction of MHC class I genes in neurons. *Science* 1995, 269: 549-552.

[4] Corriveau RA, Huh GS, Shatz CJ: Regulation of class I MHC gene expression in the developing and mature CNS by neural activity. *Neuron* 1998, 21: 505-520.

[5] Binder GK, Griffin DE: Interferon-gamma-mediated site-specific clearance of alphavirus from CNS neurons. *Science* 2001, 293: 303-306.

[6] Pakzaban P, Deacon TW, Burns LH, Dinsmore J, Isacson O: A novel mode of immunoprotection of neural xenotransplants: masking of donor major histocompatibility complex class I enhances transplant survival in the central nervous system. *Neuroscience* 1995, 65: 983-996.

[7] Rall GF, Mucke L, Oldstone MB: Consequences of cytotoxic T lymphocyte interaction with major histocompatibility complex class I-expressing neurons in vivo. *J Exp Med* 1995, 182: 1201-1212.

[8] Huh GS, Boulanger LM, Du H, Riquelme PA, Brotz TM, Shatz CJ: Functional requirement for class I MHC in CNS development and plasticity. *Science* 2000, 290: 2155-2159.

[9] Boulanger LM, Huh GS, Shatz CJ: Neuronal plasticity and cellular immunity: shared molecular mechanisms. *Curr Opin Neurobiol* 2001, 11: 568-578.

[10] Leinders-Zufall T, Brennan P, Widmayer P, Chandramani PS, Maul-Pavicic A, Jager M, Li XH, Breer H, Zufall F, Boehm T: MHC class I peptides as chemosensory signals in the vomeronasal organ. *Science* 306: 2004, 1033-1037.

[11] Martini R, Toyka KV: Immune-mediated components of hereditary demyelinating neuropathies: lessons from animal models and patients. *Lancet Neurol* 2004, 3: 457-465.

[12] Winkler C, Kirik D, Bjorklund A: Cell transplantation in Parkinson's disease: how can we make it work? *Trends Neurosci* 2005, 28: 86-92.

[13] Trombetta ES, Mellman I: Cell biology of antigen processing in vitro and in vivo. *Annu Rev Immunol* 2005, 23: 975-1028.

[14] York IA, Rock KL: Antigen processing and presentation by the class I major histocompatibility complex. *Annu Rev Immunol* 1996, 14: 369-396.

[15] Paulsson K, Wang P: Chaperones and folding of MHC class I molecules in the endoplasmic reticulum. *Biochim Biophys Acta* 2003, 1641: 1-12.

[16] Cresswell P, Bangia N, Dick T, Diedrich G: The nature of the MHC class I peptide loading complex. *Immunol Rev* 1999, 172: 21-28.

[17] Paulsson KM, Wang P: Quality control of MHC class I maturation. *FASEB J* 2004, 18: 31-38.

[18] Grandea AG 3rd, Golovina TN, Hamilton SE, Sriram V, Spies T, Brutkiewicz RR, Harty JT, Eisenlohr LC, Van Kaer L: Impaired assembly yet normal trafficking of MHC class I molecules in Tapasin mutant mice. *Immunity* 2000, 13: 213-222.

[19] Schoenhals GJ, Krishna RM, Grandea AG 3rd, Spies T, Peterson PA, Yang Y, Fruh K: Retention of empty MHC class I molecules by tapasin is essential to reconstitute antigen presentation in invertebrate cells. *EMBO J* 1999, 18: 743-753.

[20] Urade R, Takenaka Y, Kito M: Protein degradation by ERp72 from rat and mouse liver endoplasmic reticulum. *J Biol Chem* 1993, 268: 22004-22009.

[21] Androlewicz MJ, Ortmann B, van Endert PM, Spies T, Cresswell P: Characteristics of peptide and major histocompatibility complex class I/beta 2-microglobulin binding to the transporters associated with antigen processing (TAP1 and TAP2). *Proc Natl Acad Sci U S A* 1994, 91: 12716-12720.

[22] Seifert U, Maranon C, Shmueli A, Desoutter JF, Wesoloski L, Janek K, Henklein P, Diescher S, Andrieu M, de la Salle H, Weinschenk T, Schild H, Laderach D, Galy A, Haas G, Kloetzel PM, Reiss Y, Hosmalin A: An essential role for tripeptidyl peptidase in the generation of an MHC class I epitope. *Nat Immunol* 2003, 4: 375-379.

[23] Princiotta MF, Finzi D, Qian SB, Gibbs J, Schuchmann S, Buttgereit F, Bennink JR, Yewdell JW: Quantitating protein synthesis, degradation, and endogenous antigen processing. *Immunity* 2003, 18: 343-354.

[24] Reits EA, Vos JC, Gromme M, Neefjes J: The major substrates for TAP in vivo are derived from newly synthesized proteins. *Nature* 2000, 404: 774-778.

[25] Boon T, Van Pel A: T cell-recognized antigenic peptides derived from the cellular genome are not protein degradation products but can be generated directly by

transcription and translation of short subgenic regions. A hypothesis. *Immunogenetics* 1989, 29: 75-79.

[26] Mayrand SM, Green WR: Non-traditionally derived CTL epitopes: exceptions that prove the rules? *Immunol Today* 1998, 19: 551-556.

[27] Saulquin X, Scotet E, Trautmann L, Peyrat MA, Halary F, Bonneville M, Houssaint E: +1 Frameshifting as a novel mechanism to generate a cryptic cytotoxic T lymphocyte epitope derived from human interleukin 10. *J Exp Med* 2002, 195: 353-358.

[28] Schubert U, Anton LC, Gibbs J, Norbury CC, Yewdell JW, Bennink JR: Rapid degradation of a large fraction of newly synthesized proteins by proteasomes. *Nature* 2000, 404: 770-774.

[29] Lehner PJ: The calculus of immunity: quantitating antigen processing. *Immunity* 2003, 18: 315-317.

[30] Petersen JL, Morris CR, Solheim JC: Virus evasion of MHC class I molecule presentation. *J Immunol* 2003, 171: 4473-4478.

[31] Rock KL, York IA, Goldberg AL: Post-proteasomal antigen processing for major histocompatibility complex class I presentation. *Nat Immunol* 2004, 5: 670-677.

[32] Rock KL, York IA, Saric T, Goldberg AL: Protein degradation and the generation of MHC class I-presented peptides. *Adv Immunol* 2002, 80: 1-70.

[33] Rock KL, Goldberg AL: Degradation of cell proteins and the generation of MHC class I-presented peptides. *Annu Rev Immunol* 1999, 17: 739-779.

[34] Mo XY, Cascio P, Lemerise K, Goldberg AL, Rock K: Distinct proteolytic processes generate the C and N termini of MHC class I-binding peptides. *J Immunol* 1999, 163: 5851-5859.

[35] Cascio P, Hilton C, Kisselev AF, Rock KL, Goldberg AL: 26S proteasomes and immunoproteasomes produce mainly N-extended versions of an antigenic peptide. *EMBO J* 2001, 20: 2357-2366.

[36] Kisselev AF, Akopian TN, Woo KM, Goldberg AL: The sizes of peptides generated from protein by mammalian 26 and 20 S proteasomes. Implications for understanding the degradative mechanism and antigen presentation. *J Biol Chem* 1999, 274: 3363-3371.

[37] Toes RE, Nussbaum AK, Degermann S, Schirle M, Emmerich NP, Kraft M, Laplace C, Zwinderman A, Dick TP, Muller J, Schonfisch B, Schmid C, Fehling HJ, Stevanovic S, Rammensee HG, Schild H: Discrete cleavage motifs of constitutive and immunoproteasomes revealed by quantitative analysis of cleavage products. *J Exp Med* 2001, 194: 1-12.

[38] Hanada K, Yewdell JW, Yang JC: Immune recognition of a human renal cancer antigen through post-translational protein splicing. *Nature* 2004, 427: 252-256.

[39] Vigneron N, Stroobant V, Chapiro J, Ooms A, Degiovanni G, Morel S, van der Bruggen P, Boon T, Van den Eynde BJ: An antigenic peptide produced by peptide splicing in the proteasome. *Science* 2004, 304: 587-590.

[40] Kloetzel PM: Generation of major histocompatibility complex class I antigens: functional interplay between proteasomes and TPPII. *Nat Immunol* 2004, 5: 661-669.

[41] Driscoll J, Brown MG, Finley D, Monaco JJ: MHC-linked LMP gene products specifically alter peptidase activities of the proteasome. *Nature* 1993, 365: 262-264.

[42] Ortiz-Navarrete V, Seelig A, Gernold M, Frentzel S, Kloetzel PM, Hammerling GJ: Subunit of the '20S' proteasome (multicatalytic proteinase) encoded by the major histocompatibility complex. *Nature* 1991, 353: 662-664.

[43] Dick TP, Ruppert T, Groettrup M, Kloetzel PM, Kuehn L, Koszinowski UH, Stevanovic S, Schild H, Rammensee HG: Coordinated dual cleavages induced by the proteasome regulator PA28 lead to dominant MHC ligands. *Cell* 1996, 86: 253-262.

[44] Groettrup M, Soza A, Eggers M, Kuehn L, Dick TP, Schild H, Rammensee HG, Koszinowski UH, Kloetzel PM: A role for the proteasome regulator PA28alpha in antigen presentation. *Nature* 1996, 381: 166-168.

[45] Bose S, Stratford FL, Broadfoot KI, Mason GG, Rivett AJ: Phosphorylation of 20S proteasome alpha subunit C8 (alpha7) stabilizes the 26S proteasome and plays a role in the regulation of proteasome complexes by gamma-interferon. *Biochem J* 2004, 378: 177-184.

[46] Rechsteiner M, Hill CP: Mobilizing the proteolytic machine: cell biological roles of proteasome activators and inhibitors. *Trends Cell Biol* 2005, 15: 27-33.

[47] Sibille C, Gould KG, Willard-Gallo K, Thomson S, Rivett AJ, Powis S, Butcher GW, De Baetselier P: LMP2+ proteasomes are required for the presentation of specific antigens to cytotoxic T lymphocytes. *Curr Biol* 1995, 5: 923-930.

[48] Sijts AJ, Ruppert T, Rehermann B, Schmidt M, Koszinowski U, Kloetzel PM: Efficient generation of a hepatitis B virus cytotoxic T lymphocyte epitope requires the structural features of immunoproteasomes. *J Exp Med* 2000, 191: 503-514.

[49] Morel S, Levy F, Burlet-Schiltz O, Brasseur F, Probst-Kepper M, Peitrequin AL, Monsarrat B, Van Velthoven R, Cerottini JC, Boon T, Gairin JE, Van den Eynde BJ: Processing of some antigens by the standard proteasome but not by the immunoproteasome results in poor presentation by dendritic cells. *Immunity* 2000, 12: 107-117.

[50] Stohwasser R, Salzmann U, Giesebrecht J, Kloetzel PM, Holzhutter HG: Kinetic evidences for facilitation of peptide channelling by the proteasome activator PA28. *Eur J Biochem* 2000, 267: 6221-6230.

[51] Li Z, Jansen M, Pierre SR, Figueiredo-Pereira ME: Neurodegeneration: linking ubiquitin/proteasome pathway impairment with inflammation. *Int J Biochem Cell Biol* 2003, 35: 547-552.

[52] Vuong BQ, Arenzana TL, Showalter BM, Losman J, Chen XP, Mostecki J, Banks AS, Limnander A, Fernandez N, Rothman PB: SOCS-1 localizes to the microtubule organizing complex-associated 20S proteasome. *Mol Cell Biol* 2004, 24: 9092-9101.

[53] Bae MH, Jeong CH Kim SH, Bae MK, Jeong JW, Ahn MY, Bae SK, Kim ND, Kim CW, Kim KR, Kim K: W. Regulation of Egr-1 by association with the proteasome component C8. *Biochim Biophys Acta* 2002, 1592: 163-167.

[54] Yokosawa N, Yokota S, Kubota T, Fujii N: C-terminal region of STAT-1alpha is not necessary for its ubiquitination and degradation caused by mumps virus V protein. *J Virol* 2002, 76: 12683-12690.

[55] Yamamoto Y, Gaynor RB: Therapeutic potential of inhibition of the NF-kappaB pathway in the treatment of inflammation and cancer. *J Clin Invest* 2001, 107: 135-142.

[56] Garg A, Aggarwal BB: Nuclear transcription factor-kappaB as a target for cancer drug development. *Leukemia* 2002, 16: 1053-1068.

[57] Aradhya S, Nelson DL: NF-kappaB signaling and human disease. *Curr Opin Genet Dev* 2001, 11: 300-306.

[58] Delhalle S, Blasius R, Dicato M, Diederich MA: Beginner's Guide to NF-κB Signaling Pathways. *Ann N Y Acad Sci* 2004, 1030: 1-13.

[59] Li Q, Verma IM: NF-kappaB regulation in the immune system. *Nat Rev Immunol* 2002, 2: 725-734.

[60] Baeuerle PA: IkappaB-NF-kappaB structures: at the interface of inflammation control. *Cell* 1998, 95: 729-731.

[61] Dixit V, Mak TW: NF-kappaB signaling. Many roads lead to madrid. *Cell* 2002, 111: 615-619.

[62] Makarov SS: NF-kappaB as a therapeutic target in chronic inflammation: recent advances. *Mol Med Today* 2000, 6: 441-448.

[63] Pomerantz JL, Baltimore D: Two pathways to NF-kappaB. *Mol Cell,* 2002 10: 693-695.

[64] Wojcik C, Di Napoli M: Ubiquitin-proteasome system and proteasome inhibition: new strategies in stroke therapy. *Stroke* 2004, 35: 1506-1518.

[65] Mercurio F, Manning AM: NF-kappaB as a primary regulator of the stress response. *Oncogene* 1999, 18: 6163-6171.

[66] O'Neill LA, Kaltschmidt C: NF-kappa B: a crucial transcription factor for glial and neuronal cell function. *Trends Neurosci* 1997, 20: 252-258.

[67] Mattson MP, Camandola S: NF-kappaB in neuronal plasticity and neurodegenerative disorders. *J Clin Invest* 2001, 107: 247-254.

[68] Baldwin AS Jr: Series introduction: the transcription factor NF-kappaB and human disease. *J Clin Invest* 2001, 107: 3-6.

[69] Tak PP, Firestein GS: NF-kappaB: a key role in inflammatory diseases. *J Clin Invest,* 107: 7-11, 2001.

[70] Elliott PJ, Zollner TM, Boehncke WH: Proteasome inhibition: a new anti-inflammatory strategy. *J Mol Med* 2003, 81: 235-245.

[71] Miterski B, Bohringer S, Klein W, Sindern E, Haupts M, Schimrigk S, Epplen JT: Inhibitors in the NFkappaB cascade comprise prime candidate genes predisposing to multiple sclerosis, especially in selected combinations. *Genes Immun* 2002, 3: 211-219.

In: The Ubiquitin Proteasome System… ISBN 978-1-60021-749-4
Eds: Mario Di Napoli and Cezary Wojcik, pp. 849-861

Chapter 35

The Functional Role of Proteasome Antibodies in Neurological Disorders

Anette Storstein[1,*] *and Christian Vedeler*[1,2]
[1]Department of Neurology, Haukeland University Hospital, and
[2]Institure of Clinical Medicine, University of Bergen, N-5021 Bergen, Norway.

Abstract

The proteasome is involved in a number of critical intracellular processes. A major function of the proteasome is non-lysosomal degradation of intracellular proteins, in particular defect or damaged proteins. Targeted proteins are attached to ubiquitin, a process which is catalysed by three enzymes, E1 – E3. In human brain, the activity of proteasome is varying in different regions and with age. The normal function of proteasome in the nervous system is as essential as in other tissues, and perhaps even more, due to the limited capability of renewal of neurons and glial cells. Indeed, inhibition of proteasome alone has been shown to induce neuron death in vitro. There is considerable interest concerning the role of the ubiquitin-proteasome pathway in pathological neurodegeneration. This is partly due to the observation that proteasome activity decreases with normal aging of the brain. The main reason, however, is the accumulation of disease-related proteins in aggregates within neurons or glial cells that is a major feature of many neurodegenerative diseases, as Parkinson's disease, Huntington's disease, amyotrophic lateral sclerosis and Alzheimer's disease. Besides proteolysis, another vital task of the proteasome is the processing of intracellular proteins to be presented by the MHC class I molecules to cytotoxic T lymphocytes on the surface of the cell. Antibodies to proteasome have been identified in patients with autoimmune diseases as systemic lupus erythematosus and Sjögren syndrome. Proteasome antibodies have recently also been identified in serum from patients with immune-mediated neurological diseases, as multiple sclerosis (MS) and paraneoplastic cerebellar

* Correspondence concerning this article should be addressed to Dr. Anette Storstein, Department of Neurology, Haukeland University Hospital, N-5021 Bergen, Norway. Phone: + 47-559-75044; FAX +47-559-75165 E-mail: anette.storstein@helse-bergen.no.

degeneration (PCD). In addition, autoreactive T cells to proteasome have been identified in MS patients. The presence of circulating proteasome antibodies suggests a more widespread affection of the immune system than indicated by the organ-specific nature of MS and PCD. A humoral response to proteasome can be triggered by the elevated proteasome levels that are found in some autoimmune diseases secondary to tissue damage. Primary damage of cells is a plausible explanation for an immune response targeted to an intracellular organelle. PCD is associated with a systemic tumour, and apoptosis of dying tumour cells could result in cross-presentation of intracellular antigens to the immune system and evoke immune responses, whereas the mechanism of initiation of systemic immune responses to proteasome in MS is unknown. The functional role of antibodies to proteasome in chronic inflammatory neurological disease remains to be determined. If the antibodies are indeed of pathogenic importance, as has been shown for antibodies to intracellular targets in SLE, the action of these antibodies might mimic the action of synthetic proteasome inhibitors. Such inhibitors exert their action through disturbance of protein breakdown, inhibition of antigen presentation and inhibition of proliferation, as well as the induction of apoptosis. In animal models, proteasome inhibitors also have potent anti-inflammatory effects. Thus, immune responses to proteasome in chronic inflammatory disease can potentially be both harmful, through interference with normal proteasome function, and beneficial, by suppressing inflammation. This review article aims to evaluate the current literature on antibodies to proteasome in neurological diseases, and to discuss the potential importance of these responses.

Keywords: Neurodegenerative disease, autoimmune diseases, antibody formation, proteasome complex, paraneoplastic neurological syndromes.

Abbreviations

ANA, Anti-nuclear antibodies; dsDNA, double-stranded DNA; CNS, Central nervous system; CSF, Cerebrospinal fluid; ELISA, Enzyme-linked immunosorbent assay; MHC, Major histocompatibility complex; MS, Multiple sclerosis; PCD, Paraneoplastic cerebellar degeneration; PEM, Paraneoplastic encephalomyelitis; SLE, Systemic lupus erythematosus; UPS, Ubiquitin-proteasome system.

Introduction

The proteasome and its unique role in the maintenance of normal function in all eukaryotic cells has received a lot of scientific attention during the later years. The ubiquitin-proteasome system (UPS) is crucial in the degradation of intracellular proteins and the processing of antigenic peptides for presentation by MHC class I molecules. The integrity of UPS function in the central nervous system (CNS) is particularly important due to the essential function and vulnerability of neurons, and their limited capability for renewal. The role of the UPS in the CNS in normal state seems to go far beyond proteolysis; for instance, UPS is crucial in the development of the nervous system [Chapter 15]. The role of the proteasome in neurodegeneration is of specific interest to neurologists and neuropathologists, as these diseases are associated with the accumulation of misfolded proteins, forming intracellular aggregates [Chapter 23]. Disturbance of the UPS is now believed to participate in the pathogenesis of neurodegenerative disorders as Parkinson's disease, prion diseases and Huntington's chorea. Another central topic of research interest is the use of proteasome inhibitors, which is emerging as a major treatment strategy in some types of cancer [1], but which also seems to have therapeutic potentials in the field of cerebrovascular disease [2, and Chapters 39 and 41].

The aim of this chapter is to discuss the current literature on naturally occurring antibodies to proteasome. The number of reports on such antibodies in neurological disease is limited, however, the research on humoral responses to proteasome in connective tissue disorders has been more extensive. Some features are shared by neurological and rheumatological autoimmune disorders, in particular the presence of antibodies to intracellular antigens, and we will draw some parallels between these groups of diseases. This review aims to evaluate the potential functional importance of proteasome antibodies and their role in autoimmune diseases as systemic lupus erythematosus (SLE) and multiple sclerosis (MS).

Circulating Proteasome

The levels of circulating proteasome in plasma are particularly high in autoimmune diseases as Sjögren syndrome, myositis, autoimmune hepatitis and SLE [3]. These levels correlate with disease activity and cellular damage, and are higher in patients with systemic disease than in patients with milder and more limited affection [4]. Thus, the level of circulating proteasome has been suggested as a marker of disease severity and activity in autoimmune disease [4], and humoral immune responses to proteasome could simply reflect that cellular damage results in the release of proteasome and exposal of normally hidden antigenic epitopes, resulting in the formation of antibodies.

High levels of circulating proteasome have also been found in patients with metastatic malignant melanoma [5]. As malignant melanoma is a tumour type particularly prone to trigger immune responses and often contains inflammatory infiltrates, circulating proteasome in these patients could reflect immune activation secondary to tumour growth. Elevated circulating proteasome levels are also found in patients with solid tumours, leukemia and myeloproliferative syndromes [6,7], and similar to what has been found in Sjögren syndrome and myositis, the level of plasma proteasome correlates with disease activity. This is an interesting observation, which could explain why proteasome antibodies are more common in paraneoplastic cerebellar degeneration (PCD) than in paraneoplastic encephalomyelitis (PEM). Patients with PCD often have ovarian cancer which is disseminated at the time of diagnosis [8]. On the other hand, patients with PEM usually harbour very small lung tumours that can be initially undetectable [9].

Finally, a marked increase of circulating proteasome levels are detected in patients with septic states and in trauma patients, and these levels were significantly higher than in patients who had undergone abdominal surgery [10].

Diseases Associated with Proteasome Antibodies

There are several reports on naturally occurring antibodies to the ubiquitously expressed proteasome. The research has mainly focused on antibody prevalence in different autoimmune diseases. Although there are a few reports on antibodies to proteasome in neurological diseases, most of the reports concern systemic autoimmune diseases (Table 1). One disorder that has consistently been associated with proteasome antibodies is SLE. SLE is of interest to neuroimmunologists and neurologists for several reasons. Not only is SLE regarded as a prototype of systemic autoimmune disease, but in addition, affection of the nervous system is very common in SLE [11,12]. Antibodies to proteasome have been detected in a high proportion of patients with SLE, varying from to 35% [13] to 58% [14]. Circulating antibodies to proteasome are also found in patients with polymyositis and dermatomyositis [13,14] and primary Sjögren syndrome [15,16]. Proteasome antibodies are also reported to be found in patients with diabetes mellitus type I, which, in contrast to many of the other diseases with proteasome antibodies, is an organ-specific autoimmune disease. However, in this group of diabetes patients, the proteasome antibodies are associated with a higher risk of developing other autoimmune diseases [17]. Proteasome antibodies have not been identified in patients with rheumatoid arthritis, systemic scleroderma, primary sclerosing cholangitis and autoimmune thyroid diseases [18]. Thus, proteasome antibodies certainly seem to be most prevalent in systemic autoimmune diseases, of which, in particular, SLE and Sjögren syndrome are associated with B cell hyperreactivity [19,20].

Table 1. The reported frequencies of proteasome antibodies in different diseases, detected by immunoblotting or ELISA. The frequency of proteasome antibodies in healthy controls is reported to be 0% [15, 16, 22] to 2% [14]

Disease	Percentage of positive patients [ref]		
Systemic lupus erythematosus	35% [18]		58% [14]
Primary Sjögren syndrome	16% [16]		39% [15]
Sarcoidosis		7% [16]	
Polymyositis/dermatomyositis	19% [16]		62% [14]
Behcet's disease		19% [16]	
Rheumatoid arthritis	0% [15]		5% [14]
Vasculitis		0% [16]	
Multiple sclerosis	13% [22]		66% [16]
Paraneoplastic cerebellar degeneration		43% [22]	
Paraneoplastic encephalomyelitis		11% [22]	
Cancer without paraneoplastic disease	0% [22]		15% [22]

The research of anti-proteasome immune responses in inflammatory disease primarily affecting the nervous system is so far quite limited. However, in a recent study, antibodies to proteasome were detected in the serum of more than 60% of patients with MS [16]. The prevalence of proteasome antibodies was about the same in patients with relapsing-remitting or and primary progressive MS, but higher in patients with secondary progressive disease

(58% and 50% *versus* 80%). However, the numbers of patients in the two latter groups are quite small, and the authors concluded that the presence of antibodies to proteasome was not restricted to a particular subgroup of MS patients. Proteasome antibodies were found in the cerebrospinal fluid (CSF) of about 80% of the patients with serum antibodies. Interestingly, of the patients who were tested at the time of their first attack, about 67% harboured proteasome antibodies [16]. These patients had not yet been subjected to immunomodulating therapy as interferon, which could have influenced the prevalence of proteasome antibodies. The early detection of antibodies in MS patients are in line with studies of patients with SLE, in whom the majority have detectable anti-nuclear antibodies several years prior to the onset of clinical disease [21].

The presence of proteasome antibodies has also been investigated in paraneoplastic neurological syndromes. In paraneoplastic cerebellar degeneration (PCD), patients with a malignant tumour of the breast, lung or ovary mount a cellular and humoral immune response to antigens shared by malignant cells and normal Purkinje cells of the cerebellum. Antibodies to proteasome were detected in the serum of 45% of PCD patients with breast or ovarian cancer [22]. In contrast to MS patients, however, the humoral response to proteasome in PCD seems to take place mainly in the systemic compartment, as proteasome antibodies were not found in the CSF, even in seropositive patients. In PEM, which is usually associated with small cell lung cancer, the prevalence of serum proteasome antibodies was significantly lower than in PCD [22].

Low levels of antibodies like anti-nuclear antibodies (ANA) are detected in a percentage of healthy individuals [23]. However, antibodies to proteasome is consistently not found in normal control sera [15,16,22]. In addition, antibodies to proteasome have not been detected in patients with cancer of the gastrointestinal tract [15], ovary or lung [22] without concomitant autoimmune disease. Thus, the anti-proteasome response seems so far to be limited to patient groups with manifest autoimmune disease. The prevalence of serum antibodies to proteasome in patients in preclinical stages of autoimmune disease has not been investigated.

The eukaryotic proteasome contains a number of specific subunits [24]. The humoral responses to proteasome have been found to be heterogenous, and when antisera are tested by ELISA or immunoblotting, they recognize different subunits of both α and β subtype. Many patients have antibodies to several subunits, indicating a polyclonal activation [15,16,22]. One possible explanation for this polyclonal response is inter- or intramolecular epitope spreading within the proteasome, resulting in antibodies directed to different proteasome subunits [25]. In addition, crossreactivity between the two proteasome subtypes, may account for some of the variation in reactivity patterns [15]. The polyclonality can be of relevance for the possible functional importance of the antibodies, as they can bind to different parts of the proteasome. Relative levels of individual components of the UPS vary between different regions of the brain [26], and polyclonal antibodies could therefore also have varying local effects depending on the expression of the antigens.

Interestingly, antibodies to components of the proteasome activator complex PA28 (anti-PA28α and anti-Ki antibodies) have been detected in patients with SLE (23%) and Sjögren syndrome (24%) [27]. The anti-Ki antibodies may associate with particular clinical subsets [28]. The functional importance of the anti-PA28α and anti-Ki antibodies is uncertain. The high prevalence of such antibodies in these patient groups indicates that the proteasome can be the target of different antibodies, which may interact with the function of the proteasome on several levels.

The onset of production of antibodies to a ubiquitously expressed target antigen as the proteasome could be initiated by defects in the clearing of apoptotic cells, as suggested in SLE [29], where dying cells are thought to be a major source of antigenic material to trigger autoantibody formation [30]. It is likely that the proteasome is exposed to the immune system by the same mechanism. The anti-proteasome immune response found in many of these patients indicates an antigen-driven immune response. However, the correlation of high levels of circulating proteasome and of antibodies to proteasome in the same individuals is an issue that has not been investigated. Such an association would strengthen the hypothesis that proteasome antibody production is brought on by and is a secondary marker for tissue damage.

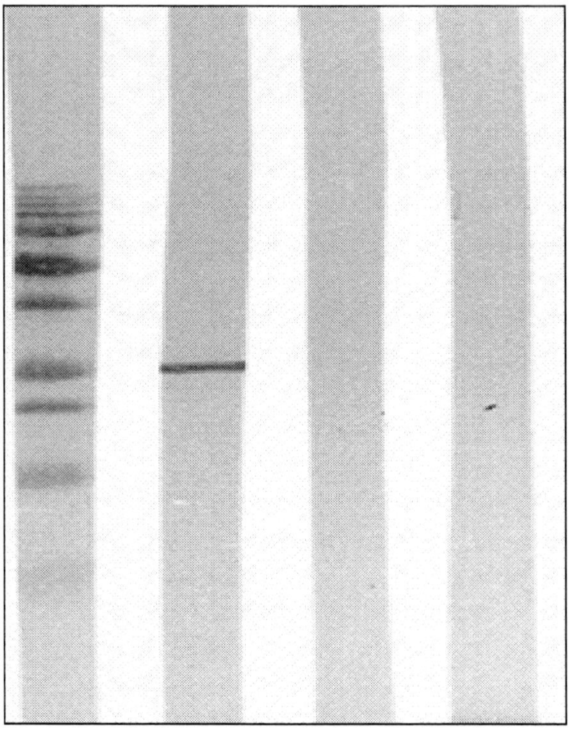

Figure 1. Representative immunoblot of 20S proteasome, purified from human red blood cells, separated by 14% SDS-PAGE. The left lane represents standard molecular weights. In second lane from the left, the blot is probed with serum from a patient with paraneoplastic cerebellar degeneration, showing that the serum (diluted 1:100) reacts with a proteasomal protein with a molecular weight of 25kDa. In lane 3, there is no reaction with the cerebrospinal fluid (diluted 1:20) of the same patient; lane 4 shows negative reaction with serum from a healthy individual. Reprinted from: Journal of Neuroimmunology, volume 165; Storstein A, Knudsen A, Vedeler CA: "Proteasome antibodies in paraneoplastic cerebellar degeneration" , pages 172-178. Copyright (2005), with permission from Elsevier.

Detection of Proteasome Antibodies

The most common method by which antibodies to proteasome are detected is Western blotting, or immunoblotting. In this assay, purified proteasome complex or recombinant proteasome subunits is separated by SDS-PAGE electrophoresis, and probed with human serum or cerebrospinal fluid [13,14,16,22] (Figure 1). The sera often react with several proteasome proteins of different molecular weights, thus indicating a polyclonal response directed at various proteasome subunits [16]. Immunoblotting of recombinant proteasome subunits can identify the different antibody specificities in positive sera. Recombinant antigens can also be used for epitope mapping to detect the antigenic epitope in the different proteasome subunits [16].

Some authors have used enzyme-linked immunosorbent assays (ELISA) to detect proteasome antibodies. This assay also employs purified proteasome to coat multititer plates; sera are then added and allowed to react, and bound antibodies are detected enzymatically by a microplate reader. The specificity of the ELISA assay can be increased by using proteasome-specific monoclonal antibodies (sandwich ELISA) [14]. Sandwich ELISA is also the preferred technique for detection of circulating proteasome [4].

Finally, proteasome antibodies can also be detected by the use of immunohistochemistry, allowing human sera to react with preparations of cell cultures [22] (Figure 2).

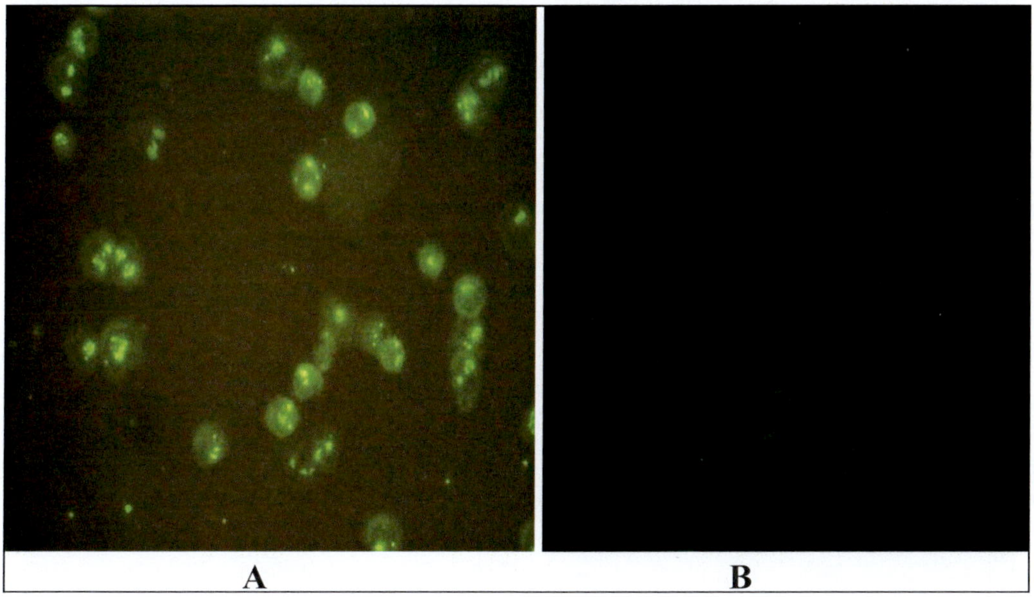

Figure 2. Cancer cells, prepared as slides and stained by serum from a patient with paraneoplastic cerebellar degeneration and high levels of proteasome antibodies. (A) Multilocular intracellular staining produced by serum diluted 1:2000. (B) When the serum has been pre-absorbed with 20S protein, the staining is abolished, showing that antibodies in the serum react with 20S to form antigen-antibody complexes that do not stain the cancer cells. Reprinted from: *Journal of Neuroimmunology*, volume 165; Storstein A, Knudsen A, Vedeler CA: "Proteasome antibodies in paraneoplastic cerebellar degeneration", pages 172-178. Copyright (2005), with permission from Elsevier.

Pathogenic Role of Proteasome Antibodies

The proteasome antibodies are targeted to an intracellular antigen that is located in the nucleus as well as the cytoplasm, and which is ubiquitously expressed in all eukaryotic cells. The pathogenic role of antibodies to intracellular antigens has been a controversial issue for a long time. Many systemic autoimmune diseases are associated with antibodies to intracellular targets. It is generally believed that nuclear and cytoplasmic antigens are inaccessible to circulating antibodies and that such antibodies are highly unlikely to be of pathogenic importance. However, some antibodies are able to penetrate the cell membrane, among them anti-dsDNA antibodies, anti-RNP antibodies and anti-SSB/La antibodies [31].

Even if the antibodies are successful in traversing the cell membrane, it is still debated whether antibodies, once internalized, can initiate cell damage, and the evidence for this is mainly experimental. However, recent research has shown that humoral responses directed to intracellular targets may indeed participate actively in the pathogenesis of autoimmune disease. In SLE, anti-dsDNA antibodies antigens seem to be both of pathogenic importance as well as important markers of disease [21,32]. It has also been hypothesized that antibodies to intracellular antigens may exert functional effects in systemic sclerosis and in rheumatoid arthritis [33]. Thus, although the pathogenicity of antibodies to intracellular targets is still a debated issue, future research may change the present impression of these antibodies as pure diagnostic markers.

Whether antibodies to proteasome participate in the pathogenesis of autoimmune diseases, neurological or otherwise, has not been clarified. In general, there are certain criteria that need to be fulfilled for an antibody to be considered as pathogenic. First, the antibodies should be associated with a high degree of disease specificity, which is not the case for the proteasome antibodies. Even if autoimmunity is a common feature in the diseases in which proteasome antibodies are found, the clinical spectrum is still very wide. Second, the proteasome antibodies do not seem to cluster with distinct clinical phenotypes. Third, there is no certain correlation between the level of proteasome antibodies and disease activity, although this has not been systematically investigated. Finally, the pathogenic effects of antibody-antigen interactions *in vivo* should be reproducible in experimental systems. However, no such models have been established for proteasome antibodies.

Which arguments are in favour of functional effects of proteasome antibodies? The antibodies have been detected at a very early stage of disease, i.e. in MS, suggesting that they are a contributing factor, and not only a secondary response to the inflammation and cell damage [16]. Additionally, the proteasome antibodies are of the IgG and IgM subclass [16]; and the IgG subclasses of proteasome antibodies in PCD are IgG1 and IgG2 [22]. These results indicate an immune response that is antigen-driven and T cell-dependent, as is often the case of pathogenic antibodies.

If antibodies to proteasome exert functional effects inside the blood-brain barrier, intrathecal antibody synthesis is to be expected. In the study by Mayo and coworkers, antibodies to proteasome were also detected in the CSF of about 80% of the seropositive MS patients [16]. Although it is noted that the concentration of proteasome antibodies, based on the total IgG content, was enriched in the CSF of one single patient, the report does not comment on intrathecal antibody production in the seropositive patients. Intrathecal IgG production is very common in MS, shown by multiple oligoclonal IgG bands detected by isoelectric focusing of the CSF [34]. The specificity of the IgG represented by these bands is not known, but it remains to be seen whether some are directed against proteasomes.

The lack of intrathecal proteasome antibodies in PCD strongly suggests that they do not participate in the pathogenesis of Purkinje cell death in this disorder. On the other side, the anti-proteasome response in PCD may be a part of the immune responses directed at tumoural antigens, such as antigens presented by ovarian tumours. In ovarian cancer, the majority of patients have advanced malignancy at the time of diagnosis [35]. Thus, the antibodies to proteasome found in PCD patients could reflect tumour burden and cell death. In patients with PEM, the associated tumour is usually very small and often occult, perhaps resulting in a less extensive presentation of antigenic tumour material to the immune system. This may partly explain why proteasome antibodies are less common in PEM than in PCD [22].

Coexisting cellular responses to proteasome have only been investigated in MS. Autoreactive T cells to proteasome were detected in 40% of analysed MS sera, but proliferation of peripheral blood mononuclear cells was not detected in patients without proteasome antibodies [16]. The presence of autoreactive T cells is an argument in favour of a pathogenic role of the anti-proteasome immune response in MS, but must be regarded with some caution, as these results have not been confirmed by others.

FUNCTIONAL EFFECTS OF PROTEASOME ANTIBODIES

The functional effects of proteasome antibodies are uncertain. This is partly due to the low serum levels of proteasome antibodies in autoimmune disease, which is usually in the range of 1/100 – 1/2000 [22]. Furthermore, the antibodies are directed to an intracellular target, leaving doubt as to their pathogenicity. The polyclonality of the proteasome antibodies and the complexity of proteasome function still suggest that there are multiple potential molecular pathways for the antibodies to exert their effects. Also, conformational diversities could allow the antibodies to recognize more than one target epitope. The UPS has a complex and central role in the maintenance of neurons, their excitability and outgrowth, in neuroprotection and in neurometabolism [2]. The UPS is also important in synaptic function and synaptic plasticity [36], perhaps also by its close association with MHC class I function and antigen expression [37]. Thus, interaction with the proteasome can have disastrous effects in neurons, which the role of proteasome dysfunction in neurodegenerative disorders has clearly demonstrated [38]. In these diseases, proteasome inhibition seems to participate in neuronal death, probably through multiple effects, including elevated intracellular levels of protein oxidation. Whether antibodies could inhibit proteasome function by similar mechanisms as found in neurodegenerative diseases, is not known. However, such effects could work in orchestra with the multitude of other humoral and cellular immune responses in immune-mediated diseases like MS, to increase neuronal vulnerability. The potential effects of proteasome antibodies in a pro-inflammatory environment are likely to differ from effects under normal conditions. In systemic autoimmune inflammation, there is usually an upregulation of interferon-inducible proteasome subunits, whereas in Sjögren syndrome, deficiency of a specific proteasome subunit has been identified [39]. To relate this finding to the specificity of proteasome antibodies in Sjögren syndrome would thus be of great interest.

We have shown that proteasome antibodies stain living cancer cells, indicating IgG internalisation, but viability studies did not reveal any direct cytotoxic effects of such antibodies on cancer cells *in vitro* [22]. Whether natural antibodies could bind and inhibit the pathways of the UPS remains uncertain. However, even if the lack of cytotoxicity suggests that the antibodies do not interfere with vital cellular mechanisms concerning growth and differentiation, the antibodies could still affect other functions as degradation and presentation of antigens on MHC class I molecules, potentially influencing the presentation of intracellular antigens to the immune system. The UPS has particular functions within the nervous system, in particular concerning synaptic transmission, and antibody effects inside the CNS may thus differ from the effects observed in cell cultures of neuronal as well as non-neuronal origin. By suppression of the activation of nuclear factor-kappa B (NF-κB), synthetic inhibitors of the UPS can mediate anti-inflammatory effects, which is of potential great therapeutic interest in CNS inflammation [40]. If antibodies to proteasome can act in the same inhibitory fashion, their presence in inflammatory CNS diseases as MS could be favourable.

An interesting observation is that the prevalence of proteasome antibodies is even higher in MS than in systemic inflammatory diseases like SLE. There are immunological similarities between MS and SLE, and there could possibly be shared pathways and mechanisms in the pathogenesis of these diseases. Nevertheless, the proteasome antibodies may not reflect a specific immune response. The spectrum of autoimmune diseases in which proteasome antibodies have been detected indicates that these antibodies reflects a global abnormality in the B cell regulation and function in certain patient groups that have a predilection for autoimmune disease and a lowered threshold for the production of autoantibodies [16,41]. It is thus possible that the proteasome antibodies represent a bystander immune response without direct pathogenic effects.

Conclusion

Proteasome antibodies are detected in patients with systemic and organ-specific autoimmune diseases. The significance of these antibodies is uncertain. It is likely that they reflect cellular damage and release of intracellular antigens and represent bystander immune responses. A primary pathogenic function seems less probable, however, in an inflammatory micro-environment, the antibodies may still exert functional effects. This is an issue that needs to be clarified in experimental models.

An interesting aspect of proteasome antibodies is why they are present in some patients and not in others, and whether proteasome antibodies are related to the severity and prognosis of disease. The limited number of reports on proteasome antibodies does not allow for conclusions in this matter, but further studies should aim to investigate possible correlations of the presence and level of proteasome antibodies and the long term prognosis of MS and other immune-mediated diseases.

REFERENCES

[1] Adams J: The development of proteasome inhibitors as anticancer drugs. *Cancer Cell* 2004; 5:417-421.

[2] Di Napoli M, McLaughlin B: The ubiquitin-proteasome system as a drug target in cerebrovascular disease: therapeutic potential of proteasome inhibitors. *Curr Opin Investig Drugs* 2005; 6:686-699.

[3] Mountz JD: Significance of increased circulating proteasome in autoimmune disease. *J Rheumatol* 2002; 29:2027-2030.

[4] Egerer K, Kuckelkorn U, Rudolph PE, Ruckert JC, Dorner T, Burmester GR, Kloetzel PM, Feist E: Circulating proteasomes are markers of cell damage and immunologic activity in autoimmune diseases. *J Rheumatol* 2002; 29:2045-2052.

[5] Stoebner PE, Lavabre-Bertrand T, Henry L, Guiraud I, Carillo S, Dandurand M, Joujoux JM, Bureau JP, Meunier L: High plasma proteasome levels are detected in patients with metastatic malignant melanoma. *Br J Dermatol* 2005; 152:948-953.

[6] Wada M, Kosaka M, Saito S, Sano T, Tanaka K, Ichihara A: Serum concentration and localization in tumor cells of proteasomes in patients with hematologic malignancy and their pathophysiologic significance. *J Lab Clin Med* 1993; 121:215-223.

[7] Lavabre-Bertrand T, Henry L, Carillo S, Guiraud I, Ouali A, Dutaud D, Aubry L, Rossi JF, Bureau JP: Plasma proteasome level is a potential marker in patients with solid tumors and hemopoietic malignancies. *Cancer* 2001; 92:2493-2500.

[8] Shams'ili S, Grefkens J, de Leeuw B, van den Bent M, Hooijkaas H, van der Holt B, Vecht C, Sillevis Smitt P: Paraneoplastic cerebellar degeneration associated with antineuronal antibodies: analysis of 50 patients. *Brain* 2003; 126:1409-1418.

[9] Darnell RB, Posner JB. Observing the invisible: successful tumor immunity in humans. *Nat Immunol* 2003; 4:201.

[10] Roth GA, Moser B, Krenn C, Roth-Walter F, Hetz H, Richter S, Brunner M, Jensen-Jarolim E, Wolner E, Hoetzenecker K, Boltz-Niculescu G, Ankersmit KJ: Heightened levels of circulating 20S proteasome in critically ill patients. *Eur J Clin Invest* 2005; 35:399-403.

[11] Abbott NJ, Mendonca LL, Dolman DE: The blood-brain barrier in systemic lupus erythematosus. *Lupus* 2003; 12:908-915.

[12] Ferreira S, D'Cruz DP, Hughes GR: Multiple sclerosis, neuropsychiatric lupus and antiphospholipid syndrome: where do we stand? *Rheumatology* (Oxford) 2005; 44:434-442.

[13] Arribas J, Luz Rodriguez M, Alvarez-Do Forno R, Castano JG: Autoantibodies against the multicatalytic proteinase in patients with systemic lupus erythematosus. *J Exp Med* 1991; 173(2):423-427.

[14] Feist E, Dorner T, Kuckelkorn U, Schmidtke G, Micheel B, Hiepe F, Burmester GR, Kloetzel PM: Proteasome alpha-type subunit C9 is a primary target of autoantibodies in sera of patients with myositis and systemic lupus erythematosus. *J Exp Med* 1996; 184(4):1313-1318.

[15] Feist E, Kuckelkorn U, Dorner T, Donitz H, Scheffler S, Hiepe F, Kloetzel PM, Burmester GR: Autoantibodies in primary Sjogren's syndrome are directed against proteasomal subunits of the alpha and beta type. *Arthritis Rheum* 1999; 42:697-702.
[16] Mayo I, Arribas J, Villoslada P, Alvarez DoForno R, Rodriguez-Vilarino S, Montalban X, de Sagarra MR, Castano JG: The proteasome is a major autoantigen in multiple sclerosis. *Brain* 2002; 125:2658-2667.
[17] Kordonouri O, Meyer K, Egerer K, Hartmann R, Scheffler S, Burmester GR, Kuckelkorn U, Danne T, Feist E: Prevalence of 20S proteasome, anti-nuclear and thyroid antibodies in young patients at onset of type 1 diabetes mellitus and the risk of autoimmune thyroiditis. *J Pediatr Endocrinol Metab* 2004; 17:975-981.
[18] Mayo I, Arizti P, Pares A, Oliva J, Doforno RA, de Sagarra MR, Rodes J, CAstano JG: Antibodies against the COOH-terminal region of E. coli ClpP protease in patients with primary biliary cirrhosis. *J Hepatol* 2000; 33:528-536.
[19] Hansen A, Lipsky PE, Dorner T: Immunopathogenesis of primary Sjogren's syndrome: implications for disease management and therapy. *Curr Opin Rheumatol* 2005; 17:558-565.
[20] Tsubata T: B cell abnormality and autoimmune disorders. *Autoimmunity* 2005; 38:331-337.
[21] Arbuckle MR, McClain MT, Rubertone MV, Scofield RH, Dennis GJ, James JA, Harley JB: Development of autoantibodies before the clinical onset of systemic lupus erythematosus. *N Engl J Med* 2003; 349:1526-1533.
[22] Storstein A, Knudsen A, Vedeler CA: Proteasome antibodies in paraneoplastic cerebellar degeneration. *J Neuroimmunol* 2005; 165:172-178.
[23] Pisetsky DS: Anti-DNA and autoantibodies. *Curr Opin Rheumatol* 2000; 12:364-368.
[24] Brooks P, Fuertes G, Murray RZ, Bose S, Knecht E, Rechsteiner MC, Hendil KP, Tanaka K, Dyson J, Rivett J: Subcellular localization of proteasomes and their regulatory complexes in mammalian cells. *Biochem J* 2000; 346:155-161.
[25] Vanderlugt CL, Miller SD: Epitope spreading in immune-mediated diseases: implications for immunotherapy. *Nat Rev Immunol* 2002; 2:85-95.
[26] Klimaschewski L: Ubiquitin-dependent proteolysis in neurons. *News Physiol Sci* 2003; 18:29-33.
[27] Matsushita M, Takasaki Y, Takeuchi K, Yamada H, Matsudaira R, Hashimoto H: Autoimmune response to proteasome activator 28alpha in patients with connective tissue diseases. *J Rheumatol* 2004; 31:252-259.
[28] Cavazzana I, Franceschini F, Vassalini C, Danieli E, Quinzanini M, Airo P, Cattaneo R: Clinical and serological features of 35 patients with anti-Ki autoantibodies. *Lupus* 2005; 14:837-841.
[29] Nagy G, Koncz A, Perl A: T- and B-cell abnormalities in systemic lupus erythematosus. *Crit Rev Immunol* 2005; 25:123-140.
[30] Gaipl US, Voll RE, Sheriff A, Franz S, Kalden JR, Herrmann M: Impaired clearance of dying cells in systemic lupus erythematosus. *Autoimmun Rev* 2005; 4:189-194.
[31] Lim PL, Zouali M: Pathogenic autoantibodies: Emerging insights into tissue injury. *Immunol Lett* 2005; 103:17-26.

[32] Arbuckle MR, James JA, Kohlhase KF, Rubertone MV, Dennis GJ, Harley JB: Development of anti-dsDNA autoantibodies prior to clinical diagnosis of systemic lupus erythematosus. *Scand J Immunol* 2001; 54:211-219.

[33] Benoist C, Mathis D: A revival of the B cell paradigm for rheumatoid arthritis pathogenesis? *Arthritis Res* 2000; 2:90-94.

[34] Lunding J, Midgard R, Vedeler CA: Oligoclonal bands in cerebrospinal fluid: a comparative study of isoelectric focusing, agarose gel electrophoresis and IgG index. *Acta Neurol Scand* 2000; 102:322-325.

[35] Bjorge T, Engeland A, Sundfor K, Trope CG: Prognosis of 2,800 patients with epithelial ovarian cancer diagnosed during 1975-94 and treated at the Norwegian Radium Hospital. *Acta Obstet Gynecol Scand* 1998; 77:777-781.

[36] Upadhya SC, Hegde AN: Ubiquitin-proteasome pathway components as therapeutic targets for CNS maladies. *Curr Pharm Des* 2005; 11:3807-3828.

[37] Darnell RB: Immunologic complexity in neurons. *Neuron* 1998; 21:947-950.

[38] Hattori N, Mizuno Y: Pathogenetic mechanisms of parkin in Parkinson's disease. *Lancet* 2004; 364:722-724.

[39] Krause S, Kuckelkorn U, Dorner T, Burmester GR, Feist E, Kloetzel PM: Immunoproteasome subunit LMP2 expression is deregulated in Sjogren's syndrome but not in other autoimmune disorders. *Ann Rheum Dis* 2006; 65:1021-1027.

[40] Wojcik C, Di Napoli M: Ubiquitin-proteasome system and proteasome inhibition: new strategies in stroke therapy. *Stroke* 2004; 35:1506-1518.

[41] Davidson A, Diamond B: Autoimmune diseases. *N Engl J Med* 2001; 345:340-350.

In: The Ubiquitin Proteasome System… ISBN 978-1-60021-749-4
Eds: Mario Di Napoli and Cezary Wojcik, pp. 863-874 © 2007 Nova Science Publishers, Inc.

Chapter 36

THE UBIQUITIN PROTEASOME SYSTEM IN THE PATHOBIOLOGY OF HUMAN GLIOMAS

Marco Piccinini[1], Maria Teresa Rinaudo[1], and Davide Schiffer[2,*]
[1]Department of Medicine and Experimental Oncology, Section of Biochemistry, University of Turin, Turin, Italy;
[2]Centre of Neuro-bio-oncology, Foundation Policlinico di Monza/University of Turin, Vercelli, Italy.

ABSTRACT

In higher eukaryotic cells, the 26S proteasome is the central component of the ubiquitin-proteasome system (UPS), in which it provides for the degradation of cytoplasmic and nuclear proteins, usually tagged with ubiquitin oligomers, and their resolution into short peptides. This pathway is involved in the control of a large array of cellular processes including protein turnover, digestion of damaged, mutant and viral proteins, cell cycle regulation, cell division, differentiation and development. Furthermore, it is also implicated in DNA repair, stress, immune and inflammatory responses, apoptosis, cell surface receptor modulation, transcription factor processing and activation, etc. Proteins belonging to different molecular pathways playing an important role in glioma progression or regression may undergo degradation or processing via the UPS, and consequently be inactivated, or conversely activated following proteasome inhibition. In GBM there is a striking shift of the balance constitutive/immunoproteasome towards the latter; paralleled by depression of the chymotrypsin-like activity. This is in opposition to its expected enhancement, being this activity higher in the immunoproteasome with respect to the standard proteasome. A better understanding of this discrepancy as well as of the enhanced apoptosis associated

* Correspondence concerning this article should be addressed to Dr. Davide Schiffer, MD; Centro Ricerche di Neuro-bio-oncologia, Fondazione Policlinico di Monza, Via Pietro Micca, 29, 13100 Vercelli, Italy. Phone: +39-0161-3691; Fax: +39-0161-369109 ; E-mail: davide.schiffer@cnbo.it.

with proteasome inhibition, as observed in GBM, could be helpful in designing novel therapeutic strategies.

Keywords: nervous tissue neoplasms, glioblastoma enhanced pathways, ubiquitin-proteasome system, apoptosis, proteasome inhibitors, standard proteasome, immunoproteasome.

ABBREVIATIONS

AFX, transcription factor also known as FOXO4; Akt, protein kinase B PKB; BAD, proapoptotic tumor suppressor protein; Bcl-2, bcl-Associated Death Protein; CDK, cyclin dependent kinase; c-Myc, proto-oncogene protein c-myc-transcription factor product of the oncogene c-myc; E2F, transcription Factor of E2 gene; E2F4, transcription repressor E2F4; EGFR, epidermal growth factor receptor; Fas,. transmembrane fas receptor also defined CD95 or Apo1 antigen; FKHR, forkhead transcription factor; FRAP, FKBP12 rapamycin-associated protein; GBM, glioblastoma; LLnL, Leu-Leu-norLeu-al; L+R, ligand+receptor; MAPK, mitogen activated protein kinase; Mdm2, E3-like ubiquitin-protein ligase product of mdm2 gene; MG132, Z-Leu-Leu-Leu-al ; mTOR, mammalian target of rapamycin protein kinase; NFκB, transcription nuclear factor κB; p14, cell cycle regulator or tumor suppressor gene; p21WAF1, CDK inhibitor p21; p27, CDK inhibitor p27Kip1; PDGFR, platelet derived growth factor receptor; PIP3, phosphatidylinositol-3-phosphate; pRb-P, retinoblastoma-associated protein-phosphorylated; PS341, Z-Ile-Glu(octaBut)-Ala-Leu-al; PTEN, phosphatase and tensin homolog deleted on chromosome 10; Ras, product of Ras proto-oncogene, a monomeric G-Protein; SV40, tumor virus SV40; TNFα, tumor necrosis factor α; TP53, transcription protein 53; TRAIL, TNF-related apoptosis-inducing ligand; UPS, ubiquitin-proteasome system.

THE UBIQUITIN SYSTEM AND GLIOBLASTOMA

In higher eukaryotic cells, the proteasome constitutes the central protease of the ubiquitin-proteasome system (UPS), playing a role in the cytoplasmic and nuclear degradation at neutral pH of most intracellular proteins, preferentially if tagged with ubiquitin oligomers (see Chapters 3, 6 and 7). Originally, the UPS was described as a way to provide for the digestion of misfolded, damaged, mutant, or viral proteins and their transformation into short peptides further degraded to single amino acids by specific peptidases. However, evidence is growing that this pathway is also involved in the degradation or processing of a variety of proteins, frequently short-lived regulatory proteins, implicated in different vital cellular processes including cell cycle regulation, cell division, differentiation and development, DNA transcription and repair, apoptosis, modulation of cell surface receptors, ion channels and secretory pathways, response to stress and extracellular challenges, immune surveillance, inflammation, etc [1]. Therefore, direct or indirect aberrations of the UPS result in a variety of pathologies, including malignancies.

Proteins involved in different molecular pathways, playing an important role in glioma progression or regression, may undergo degradation or processing by the UPS and consequently be inactivated or conversely activated following proteasome inhibition. In this regard, there are many examples. The key regulation of the cell cycle check-point G1-S is exerted by p53 together with Mdm2 and $p14^{ARF}$. The N-terminus of Mdm2 binds to the transactivation domain of p53 and inhibits its transcriptional activity; furthermore, Mdm2 regulates p53 protein level, because p53 is targeted for nuclear export and cytoplasmic degradation following ubiquitination by Mdm2 which in the UPS functions as an E3 ubiquitin ligase. Interestingly, Mdm2 is frequently amplified in glioblastomas. Another example of UPS intervention is given by p27/Kip.1 which regulates G1-S transition inhibiting cyclin E- and A-CDK2 complexes; p27/Kip.1 is expressed in G0 and G1 and decreases in cells entering into S-phase. It is post-translationally regulated by degradation into the UPS. SCF complexes (Skip1, Cul-1, F-box protein) are a class of ubiquitin ligases ensuring the specific recognition and ubiquitination of different substrates through different F-box proteins. Skp2 belongs to this group of proteins and is required for G1-S transition targeting p27/Kip.1 for ubiquitination and degradation. In glioblastoma, an inverse relationship is appreciable between Skp2 and p27/Kip.1. Among other proteins degraded into UPS, IκBα must be mentioned preventing NFκB translocation to the nucleus.

Pathways of Astrocytic Glioma Progression and Alterations at the Gene Level

In astrocytic gliomas, a progression is realized through anaplasia, which indicates loss of the phenotypic features of a certain stage of differentiation, with regression to a more immature stage. A diffuse astrocytoma grade II, through anaplastic astrocytoma grade III, may transform into a IV grade glioblastoma (GBM). In the course of anaplasia, pathologic aspects typical of the different stages are associated with genetic alterations, as depicted in Table 1.

Table 1. Progressing anaplasia and genetic alterations

Tumor stage	Associated pathology	Genetics
Astrocytoma	Proliferation	TP53 mutations
	Apoptosis	PDGFR over-expression
		7p, 22q losses
Anaplastic Astrocytoma	Cell cycle deregulation	CDKN2A/p16 deletion
		RB mutation
		CDK4 amplification
		9q, 13q, 19q, 11p losses
Glioblastoma	Necroses	EGFR amplification/truncation
	Angiogenesis	PTEN mutations
	Clonal selection	pRb pathway alterations

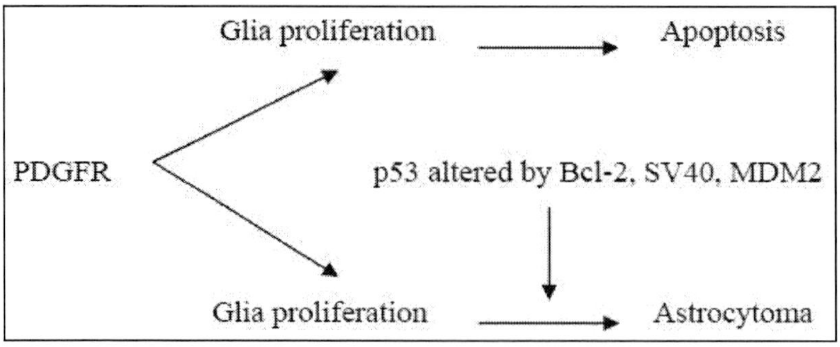

Figure 1. PDGFR stimulation leads to apoptosis if p53 is wild type and to astrocytoma if it is inactivated.

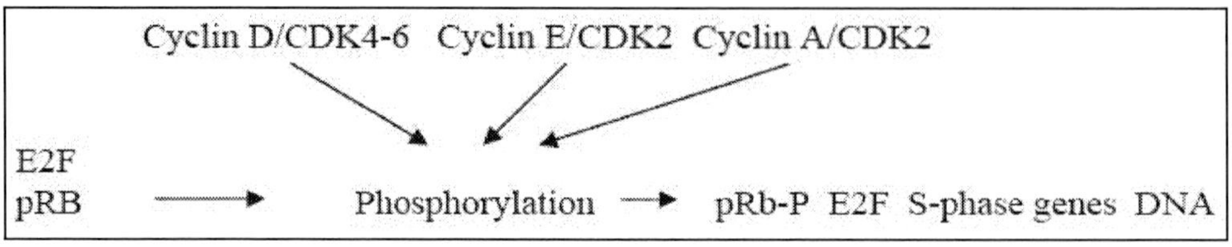

Figure 2. pRb pathway. Complexes of different cyclins and kinases regulate the phosphorylation of pRb and activation of the transcription factor E2F leading to DNA gene expression.

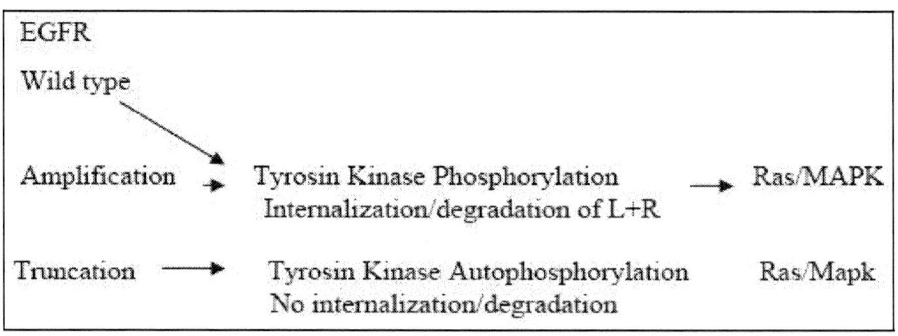

Figure 3. Amplified or truncated EGFR keep active Ras pathway and proliferation.

Two types of GBM have been identified: primary GBM arising as such from the beginning and secondary GBM originating from the transformation of a pre-existing astrocytoma. It is worth mentioning that the two GBM types show different molecular assets, with TP53 mutations prevailing in the secondary type and EGFR amplification, PTEN mutations and other genetic alterations in the primary type [2]. Tumour progression is due to genotypic instability followed by genotypic and then phenotypic heterogeneity, giving rise to new clones characterized by increasing proliferation rate and mutability, which substitute the predecessors in a process of selection by competition. Inactivation of tumour suppressor genes and accumulation of mutations are the basis of this event. The genetic alterations are distributed across several molecular pathways [3,4], which include proteins susceptible to degradation by UPS. These pathways are: PDGFR stimulation of glia (Figure 1) which may lead to apoptosis with wild type p53 and to tumour development if p53 is inactivated by different ways; cell cycle regulation through cyclins and the relevant kinases with the consequent phosphorylation of pRb and activation of the transcription factor E2F (Figure 2); EGFR signalling keeping the Ras pathway active for proliferation with amplification or truncation of the receptor (Figure 3); p14-Mdm2-p53 loop which can inactivate p53 (Figure 4); PTEN–Akt circuit which can make a switch between cell proliferation and apoptosis (Figure 5).

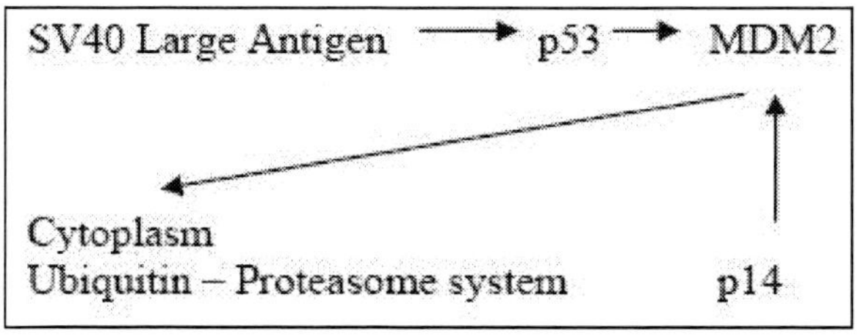

Figure 4. p53 can be inactivated by Mdm2 and exported to the cytoplasm where it is degraded by the proteasome; Mdm2 is regulated by p14.

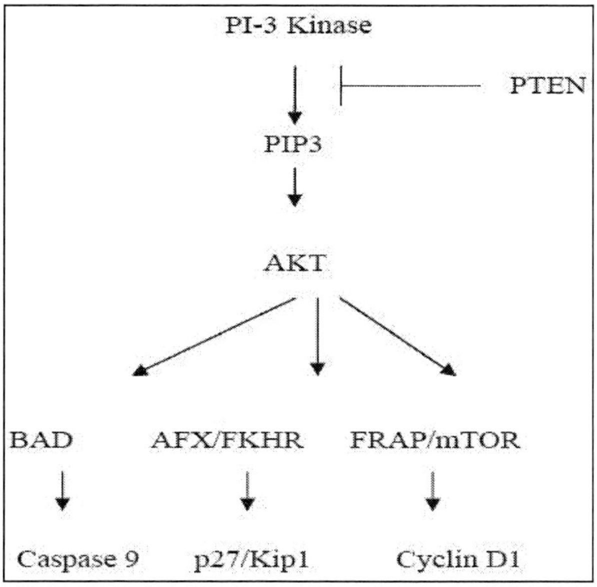

Figure 5. Mutated PTEN frees Akt which is a switch between apoptosis and proliferation.

Proteasome Inhibitors in Glioblastomacell Lines and Explants

Proteasomes, as the central proteolytic machinery of the UPS, play a pivotal role in controlling cell proliferation and differentiation as well as programmed cell death (apoptosis) in a variety of normal and tumour cells [5-10]. Their implication in the regulation of apoptosis largely relies on the activation of NFκB (see Chapter 21). Constitutively, NFκB is prevented from nuclear translocation because sequestered in the cytoplasm following its binding with the inhibitory protein IκB. NFκB as an active transcription factor, is a dimer made up by p50 and p65/Rel subunits; p50 arises through processing of the inactive precursor p105. Both p105 and IκB enter UPS provided they have been phosphorylated and tagged with ubiquitin oligomers. The former will be processed into mature p50, whereas the latter will be fully degraded. Once processed and IkB-free, p50 binds with p65 and the heterodimer translocates to the nucleus, where it transactivates a series of genes encoding proteins implicated in promoting cell proliferation and immune and inflammatory responses, as well as in awakening the antiapoptotic surveillance system [11-13]. The latter event partly relies on caspase-3 and -8 degradation by proteasomes [14]; however, caspase-mediated degradation of some proteasome subunits occurs in early stages of apoptosis [15,16], revealing that the execution of the apoptotic progression is finely tuned by a network of signals that upregulate or slow down its progression.

The involvement of proteasomes in the development of gliomas is supported by: (*i*) the status of NFκB, which is strongly activated in glioblastoma specimens and accumulates in the nucleus in parallel with tumour progression [17,18]; (*ii*) a number of findings carried out in conditions of proteasome function blockade by cell permeable inhibitors (see Chapter 40). For example, glioma cells exposed to proteasome inhibitors, whether peptide aldehydes, such as LLnL and MG132, or non-peptide molecules, such as lactacystin and epooxomicin [19,20], undergo apoptosis. This event is associated with activation of caspase-3 [21-23].

Another selective and strong proteasome inhibitor is PS-341, also called bortezomib, a boronic acid dipeptide, which exhibits prominent effects *in vitro* and *in vivo* against several solid tumours [24], and is the first approved drug in the proteasome inhibitor class of anticancer agents [25]. In human GBM cell lines and in primary GBM explants, PS-341 arrests cells in G2/M with a concomitant decrease in the percentage of cells in S phase. These events are associated with increased expression of $p21^{WAF1}$, $p27^{Kip.1}$, and cyclin B1 proteins, decreased levels of CDK2, CDK4 and the transcription factor E2F4. All these events take place along with reduced transcriptional activity of NFκB [17,18,26]. Furthermore, in GBM cells, PS-341 enhances TRAIL- and TNFα-induced cell death and apoptosis, suggesting that it may be considered for an effective therapy in patients with gliomas [26].

It is debated how TNFα and TRAIL signaling cascades, on the one hand, lead to NFκB activation, which is known to promote the expression of antiapoptotic genes, while on the other hand, they favour apoptosis. TRAIL- and TNFα-induced apoptosis implies activation of the caspase cascade as well as of NFκB [27-29]. Since caspases are proteasome substrates, it is possible that the proteasome inhibition, such as that determined by PSI-341, together with TRAIL- and TNFα -induced caspase activation gives rise to accumulation of caspases, because no longer degraded by the proteasome. However, in murine cortical cell lines, activation or conversely attenuation of the apoptotic program occurs depending on whether MG132-induced blockade of proteasome function was partial or complete [30].

Furthermore, in glioma cells, proteasome inhibitor-induced mitochondrial-independent caspase-3-dependent apoptosis relies on c-Myc protein stabilization, in turn responsible for a transient increase of Fas ligand (FasL) message to stimulate the apoptotic signaling pathway. In fact, in these cells, c-Myc protein accumulation is associated with a markedly increased expression of FasL mRNA along with slightly increased Fas-CD95 receptor mRNA; these events precede in time the activation of caspase-3. Indirectly, what above described strengthens the involvement of proteasomes as anti-apoptotic factors in malignant glioma cells [23].

Lastly, in human and rat GBM cell lines, the UPS has been implicated in the stability of the eukaryotic Elongation Factor-2 (eEF-2) kinase, a highly conserved calcium/calmodulin-dependent enzyme involved in the regulation of protein translation and cell proliferation. Rapid changes in the activity and quantity of the kinase are observed upon cell growth stimulation and up-regulasted kinase activity appears as a feature of malignant cell growth. In GBM cells, eEF-2 kinase behaves as a relatively short-lived protein, with a half-life of less than 6 hours. Treatment of these cells with MG132 results in accumulation of the kinase in ubiquitin-tagged forms, with a consequent prolonged protein half-life. In this regard, MG132 is considered as an effective agent in the treatment of certain forms of cancer, including GBM [31].

PROTEASOMES IN HUMAN GBM

In contrast with the large array of findings that indirectly strengthen the involvement of proteasomes in tumorigenesis, there is little known evidence that proteasomes are functionally and structurally modified in malignant cells, including glioma cells. This gap of knowledge likely results from the lack of an adequate control tissue.

In human cortical tissue, as well as in unstimulated mouse microglial cells, the 20S proteasome is expressed as a constitutive (standard) proteasome and as an immunoproteasome. The former comprises the coordinated assembly of 14 α and 14 β subunits, arranged in an $\alpha_7\beta_7\beta_7\alpha_7$ stoichiometry. In the latter, the active constitutive subunits β_1 β_2 and β_5 are replaced by their interferon (INF)-γ inducible homologue counterparts, defined as iβ_1/LMP2, iβ_2/MECL-1 and iβ_5/LMP7 [32-34]. These changes give rise to modifications of the peptidase activity of the proteasome (see Chapter 34). In fact, replacement of β_1 with LMP2 implies suppression of the peptidyl-glutamyl peptide hydrolysing (PGPH) activity, a feature of β_1, and substitution with chymotrypsin-like activity which is shared by LMP7; replacement of β_2 with MECL-1 implies enhancement of the trypsin-like activity [35-37]. As a consequence, the proteolytic activity of the proteasome shifts towards the generation of peptides with a hydrophobic or basic amino-acidic residue at their C-terminus; these peptides are preferentially taken up by MHC-class I proteins for presentation to cytotoxic T lymphocytes, with a consequent immune response [35]. However, in contrast to proteasomes from other sources, in particular from professional antigen presenting cells, 20S proteasomes from brain cortical regions and unstimulated microglial cells are characterized by low levels of the three inducible subunits [33] and by low levels of LMP2 and LMP7, but undetectable MECL-1 [32], respectively. Neurons and resting microglial cells are poor or destitute of immunoreactivity, particularly the former [32,38-41]; however, in brains under appropriate stimuli, microglial cells become the major antigen presenting cells and respond to pathologic events [32,41].

In the human brain, the 20S proteasome PGPH activity is higher than in the kidney, a tissue largely involved in the immune response, whereas the opposite is true for the chymotrypsin-like and trypsin-like activities, both involved in the generation of antigenic peptides [33]. In primary microglia cultures lengthily treated with INF-γ, MECL-1 and LMP7 subunits accumulate and in parallel of the constitutive counterpart β2 disappears [32]. Lastly, in human brain, the 20S proteasome, besides as a single independent particle, also exists as a complex in association with the multimeric protein 11S regulator (11SReg), also defined as proteasome activator 28 (PA28) [34]. The regulator comprises two types of subunits, 7 α and 6 β, arranged in two caps, one at the top and the other at the bottom of the proteasome unit. Expression of the regulator is under the control of INF-γ and its regulatory role consists of enhancing the proteasome peptidase activity, namely the chymotrypsin-like activity [42], thus favouring the generation of peptides with increased affinity for MHC class I molecules.

The 20S proteasome in various tissues is mainly expressed as a high molecular mass protein complex, derived from its ATP-dependent association with the regulatory component, the 19S complex or PA700; this assembly gives rise to the 26S proteasome, the central protease of UPS. The 26S proteasome has been purified and widely characterized in bovine, but not human brain [43], although reasons exist to believe that in human brain it has properties comparable to those of other tissues (Piccinini *et al.* unpublished observations). The 26S proteasome is likely to be expressed also in mouse glial cells [32], although a final confirmation is still lacking.

The characterization of the proteasome catalytic core in human GBM has recently been attained by a comparison between the structural and functional properties of the 20S proteasome isolated and purified from fresh surgery specimens of tumour tissue and those of the 20S proteasome from fresh peritumoral, histologically normal tissue.

In GBM, the 20S proteasome has properties largely similar to those of control tissue, as well as of human and bovine brain cortical regions and glial cells [32-34,43]. In human GBM, proteasome structural properties are preserved, since in all GBM surgical specimens examined, the 20S proteasome was constituted by the three active subunits β1, β2, and β5, as well as by their counterpart IFN-γ inducible subunits, LMP2, MECL-1 and LMP7. However, in the proteasome from 70% GBM specimens, the three inducible subunits are much more expressed than in controls (Figure 6); surprisingly, this feature is associated with a strongly depressed chymotrypsin-like activity, as opposed to the unvaried trypsin-like one which is. This discrepancy between proteasome functional and structural properties is also a feature of tissues in conditions of stress, such as those induced by oxygen radicals [44], and thus it can be taken as a marker of the metabolic disorders underlying the tumour malignancy.

The incorporation of the 11SReg in the 20S proteasome is instead unaltered in all GBM specimens [34]; however, because of the influence of the regulator on proteasome chymotrypsin-like activity, it can be concluded that the aforementioned role of 11SReg does not take place in GBMs, particularly in those where the INF-γ inducible subunits accumulate, and again, this may be a mark of malignancy.

CONCLUSION

Collectively, the reported observations indicate that in GBM, the UPS plays a great role as an anti-apoptotic system by eliminating controlled cell death and promoting factors favouring cell proliferation. This role largely relies on the efficiency of the proteasome. As for the UPS, gliomas do not behave differently from other tumours. Therefore, throwing light on the discrepancy between proteasome functional and structural properties in gliomas and ascertaining whether this discrepancy also occurs in other tumours might uncover a common dysfunction of a huge number of cell processes. This could be a crucial step in designing multifunctional therapeutic drugs common to all tumours. Interestingly, proteasomes have been revealed to be quite sensitive to cytostatic drugs [45,46].

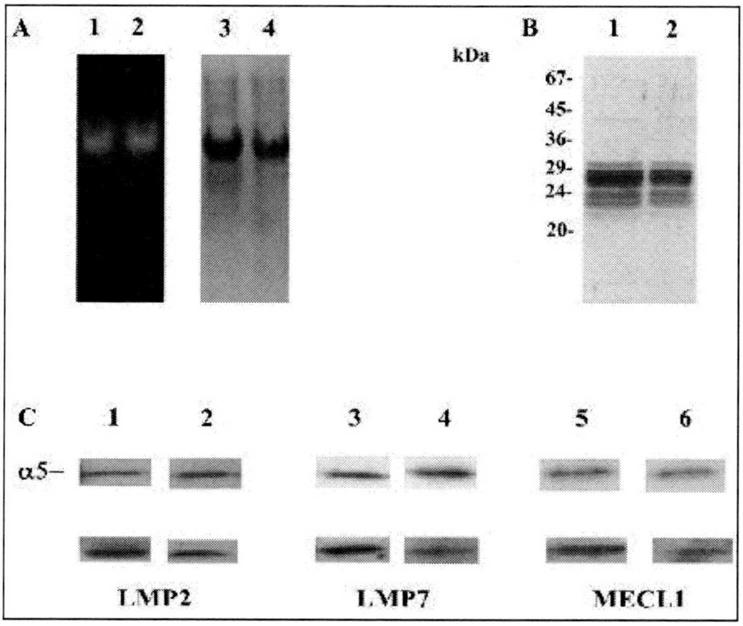

Figure 6. The 20S proteasome from human glioblastomas (G-20S) and control specimens (C-20S). Panel A: Resolution of G-20S (lanes 1, 3) and C-20S (lanes 2, 4) by non-denaturing gel electrophoresis and visualization by gel overlying with the fluorogenic peptide Suc-L-L-V-Y-AMC, substrate of the proteasome chymotrypsin-like activity (lanes 1, 2) or Coomassie Blue staining (lanes 3, 4). Panel B: Resolution of G-20S (lane 1) and C-20S (lane 2) by denaturing gel electrophoresis (SDS-PAGE) and protein band visualization by silver staining. Panel C: G-20S (lanes 1, 3, 5) and C-20S (lanes 2, 4, 6) inducible subunits LMP2, LMP7 and MECL-1 resolved by SDS-PAGE, electroblotted on PVDF membranes and identified by immunodecoration by selective antibodies. Blots were re-probed by a monoclonal antibody to the 20S proteasome constitutive α5 subunit for equal protein loading. (From Piccinini *et al.* 2005; [34])

References

[1] Ciechanover A, Iwai K: The ubiquitin system: from basic mechanisms to the patient bed. *IUBMB Life* 2004; 56:193-201.

[2] Kleihues P, Ohgaki H: Primary and secondary glioblastomaa, from concept to clinical diagnosis. *Brain Pathol* 1999; 1:44-51.

[3] Maher E, Furnari FB, Bachoo RM, Rowitch DH, Mouis DN, Cavenee WK, De Pinho RA: Malignant gliomas: genetic and biology of a grave matter. *Genes & Develop* 2001; 15:1311-1333.

[4] Collins VP: Brain tumours: classification and genes. *J. Neurol Neurosurg Psychiatry* 2004; 75:(Suppl II) ii2-ii11.

[5] Lecker SH, Solomon V, Mitch WE, Goldberg AL: Muscle protein breakdown and the critical role of the ubiquitin-proteasome pathway in normal and disease states. *J Nutr* 1999; 12:227S-237S.

[6] Kisselev AF, Goldberg AL: Proteasome inhibitors: from research tools to drug candidates. *Chem Biol* 2001; 8:739-758.

[7] Glickman MA, Ciechanover A: The ubiquitin-proteasome proteolytic pathway: destruction for the sake of construction. *Physiol Rev* 2002; 82:373-428.

[8] Almond JB, Cohen GM: The proteasome: a novel target for cancer chemotherapy. *Leukemia* 2002; 16:433-443.

[9] Husom AD, Peters EA, Kolling EA, Fugere NA, Thompson LV, Ferrington DA: Altered proteasome function and subunit composition in aged muscle. *Arch Biochem Biophys* 2004; 421:67-76.

[10] Ciechanover A: The ubiquitin proteolytic system: from a vague idea, through basic mechanisms, and onto human diseases and drug targeting. *Neurology* 2006; 66 (2 Suppl 1):S7-S19.

[11] Baldwin AS Jr: The NF-kB and IkB proteins: new discoveries and insights. *Annu Rew Immunol* 1996; 14:649-681.

[12] Karin M, Ben-Neriah Y: Phosphorylation meets ubiquitination: the control of NF-[kappa]B activity. *Annu Rev Immunol* 2000; 18:621-663.

[13] Baud V, Karin M: Signal transduction by tumor necrosis factor and its relatives. *Trends Cell Biol* 2001; 11:372-377.

[14] Kim S, Choi K, Kwon D, Benveniste EN, Choi C: Ubiquitin-proteasome pathway as a primary defender against TRAIL-mediated cell death. *Cell Mol Life Sci* 2004; 61:1075-1081.

[15] Friedman J, Xue D: To live or die by the sword: the regulation of apoptosis by the proteasome. *Dev Cell* 2004; 6:460-461.

[16] Sun XM, Butterworth M, MacFarlane M, Dubiel W, Ciechanover A, Cohen GM: Caspase activation inhibits proteasome function during apoptosis. *Mol Cell* 2004; 14:81-93.

[17] Hayashi S, Yamamoto M, Ueno Y, Ikeda K, Ohshima K, Soma G, Fukushima T: Expression of nuclear factor-kappa B, tumor necrosis factor receptor type 1, and c-Myc in human astrocytomas. *Neurol Med Chir (Tokyo)* 2001; 41:187-195.

[18] Nagai S, Washiyama K, Kurimoto M, Takaku A, Endo S, Kumanishi T: Aberrant nuclear factor-kappaB activity and its participation in the growth of human malignant astrocytoma. *J Neurosurg* 2002; 96:909-917.

[19] Fenteany G, Staendaert RF, Lane WS, Choi S, Corey EJ, Schreiber SL: Inhibition of the proteasome activities and subunit-specific amino-terminal threonine modification by lactacystin. *Science* 1995; 268:726-731.

[20] Meng L, Mohan R, Kwok BH, Elofsson M, Sin N, Crews CM: Epoxomicin, a potent and selective proteasome inhibitor, exhibits in vivo antiinflammatory activity. *Proc Natl Acad Sci USA* 1999; 96:10403-10408.

[21] Wagenknecht B, Hermisson M, Eitel K, Weller M: Proteasome inhibitors induce p53/p21-independent apoptosis in human glioma cells. *Cell Physiol Biochem* 1999; 9:117-125.

[22] Wagenknecht B, Hermisson M, Groscurth P, Liston P, Krammer PH, Weller M: Proteasome inhibitor-induced apoptosis of glioma cells involves the processing of multiple caspases and cytochrome c release. *J Neurochem* 2000; 75:2288-2297.

[23] Tani E, Kitagawa H, Ikemoto H, Matsumoto T: Proteasome inhibitors induce Fas-mediated apoptosis by c-Myc accumulation and subsequent induction of FasL message in human glioma cells. *FEBS Lett* 2001; 504:53-58.

[24] An J, Sun YP, Adams J, Fischer M, Belldegrun A, Rettig MB: Drug interactions between proteasome inhibitor bortezomib and cytotoxic chemotherapy, tumor necrosis factor (TNF) alpha, and TNF-related apoptosis-inducing ligand in prostate cancer. *Clin Cancer Res* 2003; 39:4537-4545.

[25] Mani A, Gelmann EP: The ubiquitin-proteasome pathway and its role in cancer. *J Clin Oncol* 2005; 23:4776-4789.

[26] Yin D, Zhou H, Kumagai T, Liu G, Ong JM, Black KL, Koeffler HP: Proteasome inhibitor PS-341 causes cell growth arrest and apoptosis in human glioblastoma multiforme (GBM). *Oncogene* 2005; 24:344–354.

[27] Franco AV, Zhang XD, Van Berkel E, Sanders JE, Zhang XY, Thomas WD, Nguyen T, Hersey P: The role of NF-kappa B in TNF-related apoptosis-inducing ligand (TRAIL)-induced apoptosis of melanoma cells. *J Immunol* 2001; 166:5337-5345.

[28] Dai Y, Rahmani M, Grant S: An intact NF-kappaB pathway is required for histone deacetylase inhibitor-induced G1 arrest and maturation in U937 human myeloid leukemia cells. *Cell Cycle* 2003; 2:467-472.

[29] Dai Y, Rahmani M, Grant S: Proteasome inhibitors potentiate leukemic cell apoptosis induced by the cyclin-dependent kinase inhibitor flavopiridol through a SAPK/JNK- and NF-kappaB-dependent process. *Oncogene* 2003; 22:7108-7122.

[30] Suh J, Lee YA, Gwag BJ: Induction and attenuation of neuronal apoptosis by proteasome inhibitors in murine cortical cell cultures. *J Neurochem* 2005; 9:684-694.

[31] Arora S, Yang JM, Hait WN: Identification of the ubiquitin-proteasome pathway in the regulation of the stability of eukaryotic elongation factor-2 kinase. *Cancer Res* 2005; 65:3806-3810.

[32] Stohwasser R, Giesebrecht J, Kraft R, Muller EC, Hausler KG, Kettenmann H, Hanisch UK, Kloetzel PM: Biochemical analysis of proteasomes from mouse microglia: induction of immunoproteasomes by interferon-gamma and lipopolysaccharide. *Glia* 2000; 29:355-365.

[33] Piccinini M, Mostert M, Croce S, Baldovino S, Papotti M, Rinaudo MT: Interferon-gamma-inducible subunits are incorporated in human brain 20S proteasome. *J Neuroimmunol* 2003; 135:135-140.

[34] Piccinini M, Rinaudo MT, Anselmino A, Ramondetti C, Buccinna B, Fiano V, Ghimenti C, Schiffer D: Characterization of the 20S proteasome in human glioblastomas. *Anticancer Res* 2005; 25:3203-3210.

[35] Tanaka K, Kasahara M: The MHC class I ligand-generating system: roles of immunoproteasomes and the interferon-γ-inducible proteasome activator PA28. *Immunol Rev* 1998; 163:161-176.

[36] Rock KL, Goldberg AL: Degradation of cell proteins and the generation of MHC class I-presented peptides. *Annu Rev Immunol* 1999; 17:739-779.

[37] Noda C, Tanahashi N, Shimbara N, Hendil KB, Tanaka K: Tissue distribution of constitutive proteasomes, immunoproteasomes and PA28 in rats. *Biochem Biophys Res Commun* 2000; 277:348-354.

[38] Früh K, Yang Y, Arnold D, Chambers J, Wu L, Waters JB, Spies T, Peterson Per A: Alternative exon usage and processing of the major histocompatibility complex-encoded proteasome subunits. *J Biol Chem* 1992; 267:22131-22140.

[39] Neumann H, Cavlie A, Jenne DE, Wekerle H: Induction of MHC class I genes in neurons. *Science* 1995; 269:549-552.

[40] Mavria G, Hall KT, Jones RA, Blair GE: Transcriptional regulation of MHC class I gene expression in rat oligodendrocytes. *Biochem J* 1998; 330:155-161.

[41] Xiao B G, Link H: Immune regulation within the central nervous system. *J Neurol Sci* 1998; 157:1-12.

[42] Rechsteiner M, Realini C, Ustrell V: The proteasome activator 11S REG (PA28) and class I antigen presentation. *Biochem J* 2000; 345:1-15.

[43] Piccinini M, Tazartes O, Mostert M, Musso A, DeMarchi M, Rinaudo MT: Structural and functional characterization of 20S and 26S proteasomes from bovine brain. *Mol Brain Res* 2000; 76:103-114.

[44] Carrard G, Dieu M, Raes M, Toussaint O, Friguet B: Impact of ageing on proteasome structure and function in human lymphocytes. *Int J Biochem Cell Biol* 2003; 35:728-739.

[45] Piccinini M, Tazartes O, Mezzatesta C, Ricotti E, Bedino S, Grosso F, Dianzani U, Tovo PA, Mostert M, Musso A, Rinaudo MT: Proteasomes are a target of the anti-tumour drug vinblastine. *Biochem J* 2001; 356:835-841.

[46] Piccinini M, Mostert M, Rinaudo MT: Proteasomes as drug targets. *Curr Drug Targets* 2003; 4:657-671.

In: The Ubiquitin Proteasome System… ISBN 978-1-60021-749-4
Eds: Mario Di Napoli and Cezary Wojcik, pp. 875-896

Chapter 37

THE UBIQUITIN PROTEASOME SYSTEM IN THE PATHOBIOLOGY OF HUMAN PITUITARY TUMORS

Mădălina Muşat[2], Márta Korbonits[1] and Ashley B. Grossman[1,*]
[1]Department of Endocrinology, St. Bartholomew's Hospital, London, UK;
[2]Department of Endocrinology, Carol Davila University of Medicine and Pharmacy, Bucharest, Romania.

ABSTRACT

Pituitary tumors are usually benign lesions, but their tumorigenic process may constitute a model of the initial stages of carcinogenesis. Two major theories have been subject to most investigation: hormonal (usually hypothalamic factors) and/or growth factor over-stimulation, or a molecular defect within the pituitary itself. Oncogenes and tumor suppressor genes involved in other types of tumor do not appear to play a major role in the pathogenesis of pituitary tumors. In addition, germline genetic disorders, in which pituitary tumors are a common feature, have not shed much light on the pathogenesis of the more common sporadic tumors. An increasing number of reports point to deregulation of the cell cycle in these tumors, while transgenic disruption of the cell cycle machinery frequently leads to pituitary tumors in animal models. Cell cycle progression during G1, S and G2 phases is normally regulated by the fluctuation in the concentration of cyclins, cyclin-dependent kinases (CDKs) and their inhibitors, while securin, separin and cohesin regulate progression through M phase. This is mainly achieved through the programmed degradation of these proteins within the ubiquitin-proteasome system (UPS), but also by transcriptional regulation and subcellular compartmentalization. Alterations of these processes result in uncontrolled proliferation, aneuploidy and tumorigenesis. Aberrations of one or more components of the

* Correspondence concerning this article should be addressed to Prof. Dr. Ashley B. Grossman, Centre for Endocrinology, 5th Floor King George V Building, St.Bartholomew's Hospital, West Smithfield, London, EC1A 7BE,UK. Phone: +44-207-6018343; Fax: +44-207-6018505; E-mail: A.B.Grossman@qmul.ac.uk.

pRb/p16/cyclinD1/CDK4 pathway have been shown in 80% of pituitary tumors. We have shown that low levels of nuclear p27 in human pituitary tumors associate with increased degradation of the protein through the UPS. Human securin, identified as the product of pituitary tumor transforming gene (PTTG), is over-expressed in human pituitary tumors. This can cause aneuploidy and inhibition of p53 actions towards cell cycle arrest, DNA repair and apoptosis. PTTG also contributes to pituitary tumorigenesis by modulation of angiogenesis. Degradation of PTTG is ubiquitin-dependent and promotes the initiation of anaphase and exit from mitosis. Incomplete PTTG degradation through the anaphase-promoting complex/cyclosome (APC/C) secondary to PTTG over-expression results in doubling of chromosome numbers. Whether the cell cycle changes reported in pituitary tumors are truly causal remains uncertain and it is more likely that alteration in signaling pathways feed into the cell cycle which then executes an aberrant set of instructions that result in cell proliferation. Excessive regulatory hormone stimulation can lead to an increased number of cells in the pituitary in various physiological or pathological states. Animal models also provide data that in the presence of excessive hypothalamic hormone stimulation, adenoma formation can occur. Hormonal (usually hypothalamic factors) and/or growth factor over-stimulation of the pituitary is dependent on signaling through membrane and/or nuclear receptors. A number of these receptors such as protein G- coupled receptors, tyrosine-kinase receptors, growth hormone, glucocorticoid and estrogen receptors are down-regulated via degradation through the ubiquitin proteasome system. Various anomalies of receptor expression observed in pituitary tumors may be explained through excessive or incomplete degradation, which may then cause aberrant signaling in different proliferative pathways to result in tumor formation. Increasing research in the field of ubiquitin-proteasome degradation of various proteins involved in pituitary proliferation is likely to provide new insights into pituitary tumorigenesis.

Keywords: human pituitary tumors, ubiquitin-proteasome system, cell cycle.

Abbreviations

Akt, Protein kinase B; APC/C, Anaphase-Promoting Complex/Cyclosome; CKS1, CDC2-associated protein; CREB, cAMP Response Element-Binding; D2R, dopaminergic receptor type 2; ER, estrogen receptor; GH, Growth hormone; GHR, Growth hormone receptor; GHRH, GH-releasing hormone; GR, glucocorticoid receptor; IGF-I, insulin-like growth factor I; ICER, inducible cAMP early repressor; Jab1, Jun Activation Domain-Binding Protein; MAPK, mitogen-activated protein kinase; MEN-1, multiple endocrine neoplasia type 1; MIF, Macrophage inhibitory factor; PI3K, phosphatidylinositol-3-kinase; PKA, protein kinase A; PRKAR1A, protein kinase A regulatory subunit 1α; PRLR, prolactin receptor; PTEN, protein and tensin homolog deleted on chromosome 10; PTTG , pituitary tumor transforming gene; SKP2, S-phase kinase interacting protein 2; TRK, tyrosine kinase receptor; UPS, ubiquitin-proteasome system.

INTRODUCTION

Pituitary tumors account for an average of 10% of intracranial tumors. Usually small, benign lesions, they may be clinically important in that they can affect the whole endocrine system as well as being locally invasive into the cavernous sinuses, optic chiasm or brain. Despite extensive research in the past 30 years, the molecular basis of pituitary tumorigenesis remains controversial. This review will concentrate on molecular changes in these tumors, and particularly alterations in the ubiquitin-proteasome system (UPS).

There are two major theories which have been subject to most investigation: hormonal (usually hypothalamic factors) and/or growth factor over-stimulation, or a molecular defect within the pituitary itself. In the presence of excessive hypothalamic hormone stimulation, as in longstanding untreated end-organ failure, reactive pituitary adenomas can occur. This is also suggested by transgenic animal models that develop pituitary tumors when excessively exposed to hypothalamic releasing factors [1]. However, there is substantial evidence in favor of the monoclonal nature of pituitary tumors [1]. This argues for an intrinsic molecular defect as the primary initiating event in tumor formation, although there may be hypothalamic factors that accelerate or modulate the process of tumorigenesis. Thus, the question arises as to the nature of the precise molecular pathology underlying such tumors.

Human genetic disease can in some instances be associated with endocrine neoplasia, and it was therefore thought likely that the identification of the genetic cause of multiple endocrine neoplasia type 1 (MEN-1), in which some 40% of patients develop pituitary tumors, would explain the somatic mutation in the majority of sporadic pituitary tumors. This has proved not to be the case, with somatic mutations of the MEN-1 gene, *menin*, accounting for no more than 1–2% of sporadic tumors [2]. We have also shown that the expression of *menin* mRNA is not different in pituitary adenomas compared with normal tissue [3]. Similarly, Carney syndrome is an autosomal dominant genetic disorder in which there is an approximately 10% prevalence of somatotroph adenomas. The genetic basis for one type of this disease is now known and it involves a mutation of the protein kinase A regulatory subunit on chromosome 17. Several studies have, however, been unable to demonstrate somatic mutations of this gene in sporadic pituitary tumors, nor alterations in the level of its mRNA expression [4]. At present, therefore, human germline disorders have shed little light on the pathogenesis of the much more common sporadic tumors, although a locus at 2p16, positionally identified as the probable locus for both the second type of Carney syndrome, or the gene for familial acromegaly at 11q13 [5], may be more enlightening when finally identified.

CELL CYCLE ALTERATIONS IN PITUITARY TUMORS

An alternative approach to the problem of pituitary tumorigenesis is through data derived from animal models, particularly those involving gene additions and knockouts. Transgenic disruption of the cell cycle machinery frequently leads to pituitary adenomas in animal models. The cell cycle is the process by which cells grow, replicate their genome and divide. Its control system operates through cyclical interaction of proteins that induce and coordinate proper progression through the cycle. Driving the cell cycle is mainly dependent on the fluctuations in the concentration of cyclins, cyclin-dependent kinases (CDK) and their inhibitors (CDKI) achieved through the programmed transcription and degradation of these proteins by proteolysis within the UPS (Figure 1).

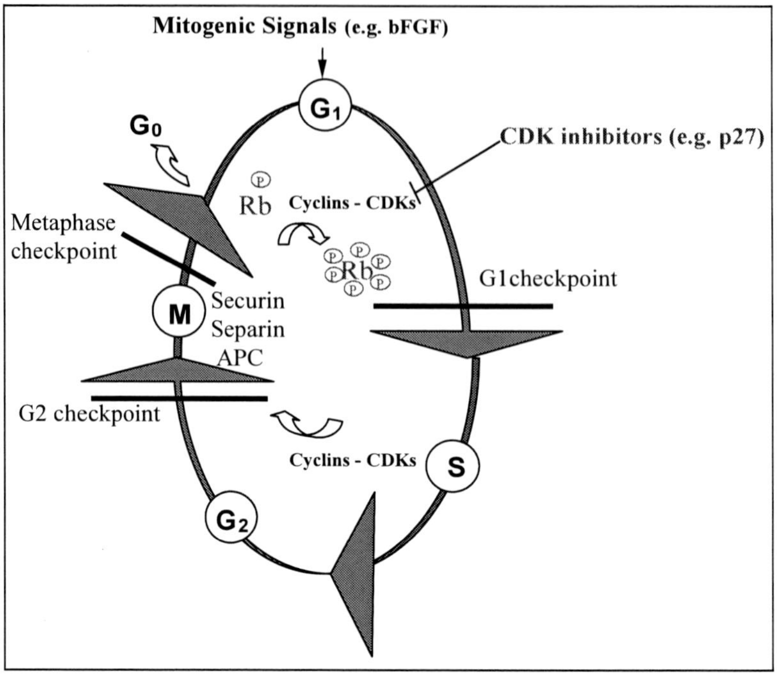

Figure 1. Cell cycle control. Cell cycle progression through G1, S and G2 is mainly dependent on the fluctuation in concentrations of cyclins, CDKs and CDKI. In early G1, in response to mitogenic signals, cyclin-CDKs complexes phosphorylate Rb, resulting in activation of transcription factors which participate in the generation of molecules required for G1/S transition. Beyond G1 checkpoint the presence of mitogens is no longer required for the cell to enter a new round of division. At G2 checkpoint the quality of the new synthetized DNA is checked before entering mitosis. Progression through M phase is ensured by the interaction between separin, securin and cohesin. At metaphase/anaphase transition, when all chromosomes are attached to the spindle, APC is activated to trigger separin degradation allowing progression through anaphase.

To ensure proper progression through the cycle, cells have developed a series of checkpoints where feedback signals conveying information about the downstream processes can delay progress into a new phase until they have successfully completed the previous one, and also provide regulation by signals from the environment, such as mitogens, growth factors, etc. The major checkpoint in mammalian cells is in G1, known as the restriction point. As mammalian cells undergo a period of mitogen dependence before entering the cell cycle, the transition beyond the restriction point represents a commitment to a new round of division, regardless of the presence of mitogens (Figure 1). The three retinoblastoma family members pRb/p105 [6,7] p107 [8,9] and Rb2/p130 [10], negatively control cell cycle progression between G1 and S phases. Before G1 phase progression is initiated, Rb is underphosphorylated and thus able to repress cell cycle progression. Cyclin-CDK mediated phosphorylation of Rb is the most likely mechanism that turns off the anti-proliferative actions of Rb at G1/S transition. In early G1, in response to mitogenic signals, CDK4 & 6 in cyclin D complexes partially phosphorylate Rb, resulting in partial activation of E2F/DP transcription factors which participate in the generation

of molecules required for G1/S transition, including cyclin E. CDK2 sequentially activates E-type cyclins (cyclin E1-E2). The cyclin E-CDK2 complex completes Rb phosphorylation which now releases the transcription factors allowing them to carry out specific tasks in cell-cycle progression, such as cyclin A synthesis. CDKI counteract CDK actions, either by blocking their activation, or by impairing substrate/ATP access [11]. There are two types of CDKI. The INK family (INK4a/p16 [12]; INK 4b/p15 [13]; INK4c/p18 and INK4d/p19 [14] exert inhibitory activity by binding to CDK4 and CDK6. These proteins exert their actions by competing with D-type cyclins for CDK subunits and thus preventing phosphorylation of pRb and inhibiting progress through G1/S. Members of the WAF/KIP family (WAF1/p21; KIP1/p27; KIP2/p57) form heterodimeric complexes with G1/S CDKs and inhibit kinase activity of CDK2-cyclin E complexes [15,16].

The UPS plays a central role in the regulation of cell growth and proliferation by controlling the abundance of key cell cycle proteins. Increasing evidence indicates that unscheduled proteolysis of many cell cycle regulators contributes significantly to tumorigenesis and is indeed found in many types of human cancers. Aberrant proteolysis with oncogenic potential is elicited by two major mechanisms: defective degradation of positive cell cycle regulators (i.e., proto-oncoproteins) and enhanced degradation of negative cell cycle regulators (i.e., tumor suppressor proteins) [17]. In many cases, increased protein stability is the result of mutations in the substrate that prevent the recognition of the protein by the ubiquitin-mediated degradation machinery. Alternatively, the specific recognition proteins mediating ubiquitination (ubiquitin ligases) are not expressed or harbor mutations rendering them inactive. In contrast, the over-expression of a ubiquitin ligase may result in the enhanced degradation of a negative cell cycle regulator [17].

The regulation of the G1/S transition appears to be one site of particular sensitivity in the provenance of pituitary tumors. In particular, aberrations of one or more components of the pRb/p16/cyclin D1/CDK4 pathway seem to be a frequent event (80%) in pituitary tumor formation [18,19]. Loss of pRb and p16 protein expression in these tumors was suggested to be mostly due to methylation in their gene-promoter region [20-22]. Cyclin D1 is over-expressed in aggressive and non-functioning pituitary tumors [23,24], and this occurs in the absence of the cyclin D1 gene CCND1 allelic imbalance (i.e. gene amplification) suggesting that there are additional mechanisms responsible for deregulating cyclin D1 expression in human pituitary tumorigenesis [23]. Abnormal cyclin D degradation through the UPS could theoretically be involved [17,25], but this has not yet been shown to be aberrant in pituitary tumors.

More data have been gathered on the ubiquitination of cyclin E and p27 in pituitary tumors. $p27^{-/-}$ mice show an increased growth rate due to increased cellularity, testicular and ovarian cell hyperplasia and infertility, and hyperplasia of the pituitary intermediate lobe, with nearly 100% mortality caused by such a 'benign' pituitary tumor. Although the p27 gene was not found to be mutated in human pituitary tumors [26], and its mRNA expression was similar in tumor samples in comparison with normal pituitaries, the load of p27 protein expression in pituitary adenomas, especially in corticotroph adenomas and pituitary carcinomas, was shown to be much lower than that in normal pituitary tissue [27-29]. On the contrary, cyclin E expression was increased in corticotroph adenomas [24], which may be related to the particularly low levels of nuclear p27 in these tumors. Both p27 and cyclin E are degraded through the UPS. The E3 enzyme, known as ubiquitin ligase/SCF complex, catalyses the transfer of ubiquitin (Ub) groups to a lysine residue in the target protein and also controls the specificity of the ubiquitination (Figure 2). The SCF complex consist of 3 core subunits (Cul, Skp1, Roc/Rbx1) that couple to one of several F-box proteins, named after their F-box motif - a highly conserved sequence of amino acids.. While the F-box is the SCF-binding domain, the F-box protein also has a substrate-binding domain to ensure specificity. In the case of p27, the specific F-box protein is Skp2 (S-phase kinase interacting protein 2, named as it was discovered through its interaction with cyclin A-CDK2 complex). Skp2 cooperates with Cks1 (CDC2-associated protein) to undergo allosteric alterations allowing it to bind phosphorylated p27 (Figure 2). The highest level of p27 ubiquitination occurs at the G1/S transition, targeting lysine residues 134, 135 and 165 on the p27 molecule [30].

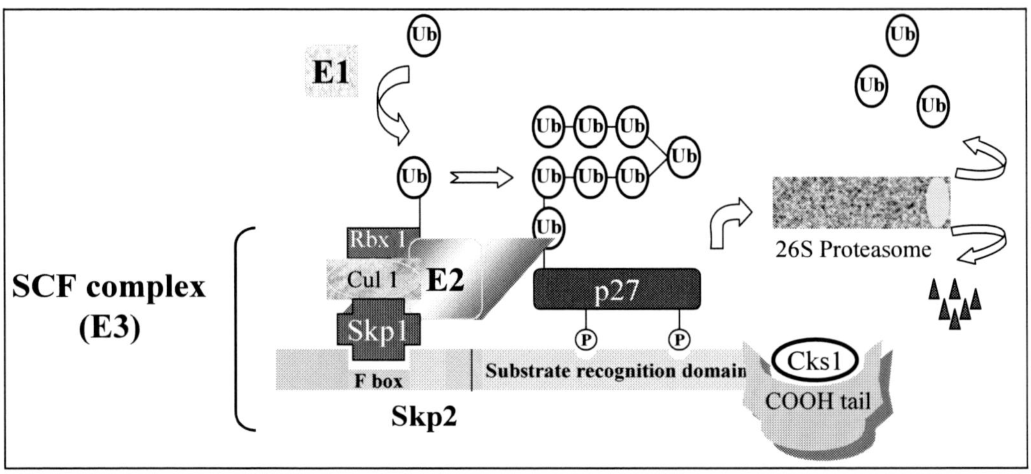

Figure 2. Degradation of p27 by the UPS. Ubiquitination is a specific process that is signalled by a degradation signal – degron – in the substrate protein. In response, a cascade of enzymes generically termed E1 (ubiquitin activating enzyme), E2 (ubiquitin conjugating enzyme) and E3 (ubiquitin ligase), catalyse the addition of ubiquitin polymers to the protein substrates. The ring finger type E3 that contains a SCF complex consists of 3 core subunits (Cul, Skp1, Roc/Rbx1) that couple to one of several F-box or D-box proteins. While the F-box is the SCF-binding domain, the F-box protein has also a substrate-binding domain to ensure specificity. In the case of p27, the specific F-box protein is Skp2 (S-phase kinase interacting protein 2,). Skp2 cooperates with Cks1 to undergo allosteric alterations allowing it to bind phosphorylated p27 (adapted from Nakayama *et al.* [92]).

At the G1/S transition, the increasing cyclin E-CDK2 activity is responsible for nuclear phosphorylation of p27 on Thr 187 [31]; p27 therefore can bind to cyclin E-CDK2 in two conformations: in a tight state, in the presence of high ATP concentrations under which the kinase activity is inhibited [32], and secondly in a loose state, at low concentrations of ATP, under which CDK2 phosphorylates p27. Thus, once cyclin E-CDK2 is activated, it can trigger p27 degradation accounting for the irreversibility of the subsequent entry to S phase

[33]. p27 mutants (Thr187/Ala) which are resistant to phosphorylation by cyclin E-CDK2 are resistant to ubiquitination [32]. p27 mutants that can be phosphorylated, but cannot bind the cyclin E-CDK2 complex, have also been claimed to be refractory to ubiquitination. In most pituitary adenomas phosphorylation at Thr187 occur in a similar manner to that seen in the normal pituitary [34]. However, this appears to be greatly increased in corticotroph adenomas.

Jab1, which enables p27 to be exported from the nucleus and would thus enhance its cytoplasmic degradation, was not obviously over-expressed in adenomas sufficient to account for diminished nuclear p27 [34]. Macrophage inhibitory factor (MIF), which has been reported to bind and hence inactivate Jab1, was also not changed in a direction that would explain the loss of nuclear p27 [35]. Increased Skp2 expression could play at least a part in p27 depletion, but overall levels of Skp2 mRNA and protein were not significantly different between normal pituitary tissue and pituitary adenomas [36]. However, tumors with low p27 protein expression did show significantly higher Skp2 expression than samples with normal p27 protein expression, suggesting that Skp2 may play a role in at least part of this process [36]. No difference was observed in Cks1 mRNA levels between normal pituitaries and pituitary adenomas; Cks1 protein expression was not assessed [37].

PTEN (protein and tensin homolog deleted on chromosome 10), the tumour suppressor function of PTEN as shown by the analysis of hereditary cancer in Cowden syndrome, controls PI3K (phosphatidylinositol-3-kinase) action and also collaborates with the transcription factor p53 in DNA repair, apoptosis, senescence and inhibition of angiogenesis It has been suggested that PTEN-deficiency in mouse embryonic stem cells causes a decrease of p27 levels with a concomitant increase in Skp2. Conversely, in human glioblastoma cells, ectopic PTEN expression leads to p27 accumulation, which is accompanied by a reduction in Skp2 [38]. The study of PTEN expression in normal and tumorous human pituitary revealed a direct correlation between nuclear PTEN and p27 levels (Figure 3) [39]. The degradation of PTEN is most likely mediated by a proteasome-dependent pathway, as there is evidence that PTEN is polyubiquitinated [40]. Post-translational regulation of PTEN has also been reported in different cell types [41,42]. Under-expression of *nuclear* PTEN in the pituitary tumors compared to normal tissue may possibly relate to enhanced protein degradation leading to the increased proliferation of the tumors.

Although the nuclear ubiquitin ligase Skp2 is implicated in p27 degradation, proteolysis of p27 at the G0-G1 transition proceeds normally in Skp2(-/-) cells [43]. These data suggest the existence of a Skp2-independent pathway for the degradation of p27 at G1 phase. Kamura *et al.* have described a previously unidentified E3 complex, KPC (Kip1/p27 ubiquitination-promoting complex), consisting of KPC1 and KPC2 [43]. KPC1 contains a RING-finger domain, and KPC2 contains a ubiquitin-like domain and two ubiquitin-associated domains. KPC interacts with and ubiquitinates p27 and is localized to the cytoplasm. Over-expression of KPC promoted the degradation of p27, whereas a dominant-negative mutant of KPC1 delayed p27 degradation [43]. The nuclear export of p27 seems to be necessary for KPC-mediated proteolysis. Depletion of KPC1 by RNA interference also inhibited p27 degradation. KPC thus probably controls degradation of p27 in G1 phase after export of the latter from the nucleus [43]. There are no data with respect to KPC function in pituitary tumors.

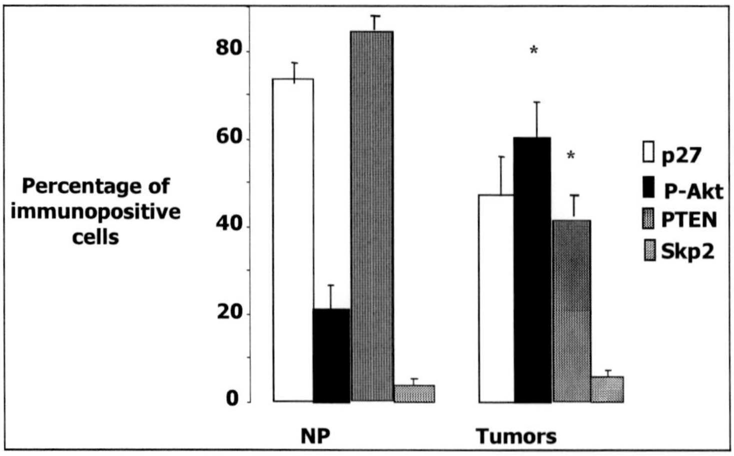

Figure 3. Immunohistochemical expression of p27, phospho-(Ser 473) Akt, PTEN and Skp2 in human pituitary. Normal pituitary (NP) have a higher expression of p27 than pituitary tumors, which positively correlates with PTEN expression. On the contrary, phospho-Akt is overexpressed in pituitary tumors compared to normals (*$p<0.05$)[39]. Skp2 overall expression was not found significantly different between the two categories [36].

Cyclin E is also degraded by the UPS, and this degradation is regulated by both CDK2 binding and Cdk2 catalytic activity [44]. Free cyclin E is readily ubiquitinated and degraded by the proteasome. Binding to CDK2 protects cyclin E from ubiquitination, and this protection is reversed by CDK2 activity in a process that involves phosphorylation of cyclin E itself. Clurman *et al.* proposed a model in which CDK2 activity initiates cyclin E degradation by promoting the disassembly of cyclin E-CDK2 complexes, followed by the ubiquitination and proteolysis of free cyclin E [44]. The specific F-box protein that recognizes cyclin E for degradation is hCDC4/Fbw/Ago [45-47], an equivalent of Skp2 for p27. Morris *et al.* explored hCDC4/Fbw/Archipelago/Ago expression in order to assess cyclin E degradation. These studies indicate that Ago mRNA, contrary to expectation, is *over-expressed* in all subtypes of pituitary adenomas compared with normal pituitary [48].

In M phase the negative regulation of cell cycle progression is dependent on a protease named separin/separase/ESP1 [49]. The critical target of separin is cohesin Scc1p, which tethers sister chromatids together at metaphase. The cleavage of cohesin by separin triggers anaphase. Negative regulation of separin is achieved by another protein known as securin [50]. Securin blocks the separin protein until activation of APC/C. Securin-separin interaction is thus essentially for the maintenance of euploidy. Human securin is identical to the product of PTTG [50,51]. In situ hybridization revealed securin/PTTG expression in non-functioning and in GH-secreting adenomas but not in normal pituitary tissue. Using a more sensitive detection method, reverse transcription polymerase chain reaction (RT-PCR), low-level securin expression was detected in normal pituitary. Expression levels in normal pituitary tissue were compared with those in 54 pituitary tumors using comparative RT-PCR to reveal that most tumor samples expressed higher levels of securin [52]. Securin is localized to both the nucleus and cytoplasm and interacts with several protein partners. Several tumorigenic mechanisms are proposed for human securin: securin and fibroblast growth factor (FGF) form a positive feedback loop and stimulate tumor angiogenesis; securin transactivates c-myc or other pro-proliferation genes; securin over-expression causes aneuploidy [53] and securin expression inhibits p53 DNA binding, transactivation and p53-dependent apoptosis [54]. p53 also mediates DNA damage-induced inhibition of securin transcription [54]. The oncogenic nature of increased expression of securin may result from chromosome gain or loss, produced by errors in chromatid separation [50]. Securin is targeted to degradation through the UPS at the beginning of anaphase.

APC/C is a multisubunit ubiquitin ligase that mediates the proteolysis of cell cycle proteins in mitosis and G1. APC is activated at the beginning of anaphase to target securin for proteolysis, thus releasing separin and allowing chromatid separation in M phase. APC activity is controlled by Mad2, a component of the mitotic checkpoint, which ensures that all kinetochores become attached to microtubules [55]. In addition, to display full ubiquitin ligase activity, APC must bind to the Cdc20 and Cdh1 accessory factors, which are responsible for securin and cyclin B degradation [55]. Cyclin B and securin ubiquitination depends on a destruction box (D box) sequence in these proteins. Kraft *et al.* showed that Cdh1 specifically crosslinks to the APC subunit Cdc27 and that Cdh1 binding to APC depends on the presence of Cdc27, implying that APC is activated by the association of Cdh1 with Cdc27, enabling APC to recognize the D box of substrates via Cdh1's propeller domain [56]. Cdh1 maintains APC activated beyond anaphase, through G1, until it is inactivated by CDKs. It is not known whether securin degradation is also altered in pituitary tumors, but drugs able to modify this process along with antisense strategies [57] may be of benefit in tumors over-expressing securin.

Hormone and Receptor Dysregulation in Pituitary Tumors

Dysregulation of normal signaling pathways in pituitary adenomas have been sought to explain the specific alterations of secretory tumors. In acromegaly, there is evidence of deranged feedback regulation in individual somatotrophs. Regulation of the somatotroph cell involves hypothalamic stimulating (GH-releasing hormone, GHRH) and inhibiting factors (somatostatin), as well as feedback regulation by GH and insulin-like growth factor (IGF-I). The first mutation ever identified in human pituitary tumors was the activating mutation in the α-subunit of the G_s protein (gsp) linked to GHRH receptor [58]. The mutation, present in about 40% of human somatotroph adenomas, results in elevated camp levels that activates protein kinase A, which phoshorylates (i.e. activates) the camp response element binding protein (CREB) and leads to sustained GH hypersecretion and cell proliferation. High levels of phosphorylated CREB are present in somatotroph adenomas, yet only 25% of high CREB expressing tumors harbor gsp^+ mutations. This could be explained by an additional activation of CREB signaling as constitutive activation [59] or impaired control over its activation. The inducible cAMP early repressor (ICER) is a powerful transcriptional inhibitor that plays an important role in the regulation of the CREB-dependent transcriptional response, including CREB itself. ICER activity is primarily determined by its intracellular concentration and its degradation occurs by phosphorylation and ubiquitination. Two elegant studies have addressed ICER ubiquitination in pituitary cell lines [60,61]. In a GH3 rat somatotroph tumor cell line, lactacystin, a specific proteasome inhibitor, decreased the rate of ICER degradation associated with the accumulation of ICER-ubiquitin conjugates [60]. Activation of the mitogen-activated protein kinase (MAPK) pathway in mouse pituitary AtT20 cells increases ICER phosphorylation. ICER phosphorylation was abrogated by inhibition of the MAPK pathway either by cAMP or directly by the MAPK inhibitor PD098059. The MAPKs extracellular signal-regulated kinases 1 and 2 physically interact with ICER and mediate the phosphorylation of ICER on a critical serine residue (Ser-41) [61]. ICER degradation might be an important mechanism to limit the negative effect of ICER on the cAMP-inducible transcriptional response. Thus, the identification of the proteolytic system(s) involved in ICER degradation would be essential to understand somatotroph tumor cells with high expression of CREB.

Feedback regulation of GH release at the pituitary level is exerted by GH and IGF-I via their receptors. Study of mRNA and protein expression of the GH receptor and the type 1 IGF receptor genes in a range of pituitary tumors revealed decreased expression of both these genes in somatotroph adenomas, suggesting that decreased negative feedback via GH and IGF-I might play a role in the uncontrolled GH release in somatotroph adenomas [62]. The sequencing of the coding region of the GH receptor gene in 15 somatotroph adenomas, identified no other alteration apart from known polymorphisms [62]. No mutations had previously been found in the IGF receptor β-subunit in 19 somatotroph adenomas [2]. Little is known about GH receptor downregulation in the pituitary. Current knowledge is that an active ubiquitination system is required for both uptake (endocytosis) and degradation of the receptor in the lysosomes [63] and only the cell surface expression of dimerized GH receptors is controlled by the ubiquitin system; the uptake of the receptor is a continuous process, independent of both GH-binding and Jak2 signal transduction [64].

Somatostatin is produced in the periventricular and arcuate nuclei of the hypothalamus and has a profound inhibitory effect on GH release. It has been suggested that down-regulation of somatostatin receptors could play a role in the increase of GH release from the pituitary. However, mutations of somatostatin receptors are only very rarely described in pituitary tumors [65] and with the ubiquitination of a few G-protein coupled receptors reported [66] this remains uncertain with respect to the somatostatin receptor.

Prolactinomas are tumors arising from lactotroph cells of the pituitary. Dopamine is the hypothalamic inhibitor for lactotroph cells, while estrogens have a powerful stimulatory effect during pregnancy, resulting in pituitary enlargement. Knocking out the dopaminergic receptor D2R in mice gives rise to lactotroph tumors [65]. In humans, mutations of D2R are infrequent in prolactinomas, but low expression of D2R is associated with unresponsiveness to dopaminergic therapy [65]. D2R belongs to the G-protein coupled receptor family and further studies on its degradation should be addressed. Prolactinomas contain the highest concentration of estrogen receptors of all the pituitary tumor types. Estrogen-mediated effects in normal and neoplastic pituitary appear to be highly dependent on the expression of estrogen receptor-alpha (Erα) and beta (ERβ), which have varying transcriptional activities. In animal studies, estrogen treatment is associated with an increase in many factors shown to promote tumorigenesis including securin, vascular endothelial growth factor, galanin and estrogens can cause true adenomas in rodents [65]. Regulation of estrogen receptor concentration is a key component in limiting estrogen responsiveness in target cells.

Nevertheless, the mechanisms governing ER concentration in the lactotroph cells of the anterior pituitary, a major site of estrogen action, are undetermined. Alarid *et al.* used a lactotroph cell line, PR1, to explore the regulation of ER protein by estrogen [67]. Estrogen treatment resulted in an approximate 60% decrease in ER steady state protein levels. The estrogen-induced degradation of ER protein could be prevented by pretreatment with peptide aldehyde inhibitors of proteasome protease, whereas inhibitors of calpain and lysosomal proteases were ineffective [67]. More recent studies from the same group showed that thyroid hormone could prevent estrogen-induced proteolysis of ERα protein in lactotroph cells of the pituitary [68]. The stabilization of ERα protein by thyroid hormone represents a selective blockade against estradiol-stimulated degradation, because thyroid hormone (but not glucocorticoid) can protect estrogen-activated ERα [68]. However, thyroid hormone did not prevent estrogen-induced induction of prolactin gene expression or the ability of ERα to stimulate proliferation. These results demonstrate that estrogen-induced proteolysis of ERα is not a general requirement for the transcriptional effects of the receptor, and they demonstrate that proteolytic regulation is a pathway by which other endocrine factors can indirectly modulate ERα activity [68]. In GH3 cells, forskolin stimulates ERα transcription through the protein kinase A (PKA) pathway and prevents estradiol-induced ERα degradation [69]. These data suggest a mechanism of ERα transcriptional activation by PKA that is distinct from estradiol activation and that may contribute to the synergistic transcriptional activation of ERα by ligand-dependent and PKA-dependent pathways [69]. The impact of ER modulation through proteolysis by the 26S proteasome on pituitary tumor formation has not been addressed to date.

Prolactin has a negative feedback effect on its own secretion via the prolactin receptor, both at the level of the hypothalamus and the pituitary. Normal lactotrophs express high levels of prolactin receptor and prolactinomas have a higher mRNA expression of prolactin receptor [70]. At the post-translational level, there is a negative regulation of prolactin receptor stability by its ligand. Prolactin promotes interaction between its receptor and the F-box protein ß-TrCP2, which functions as a substrate recognition subunit of the SCF (ß-TrCP) E3 ubiquitin ligase conveying the receptor to proteolysis by the ubiquitin proteasome system [71]. While this may be part of a normal ligand-dependent down-regulation of receptor in target tissues, at the pituitary level high prolactin may trigger prolactin receptor excessive degradation, abrogating the normal negative feedback on the lactotrophs that could become more susceptible to hypersecretion and proliferation.

The UPS regulates the turnover of many receptors including members of the nuclear receptor superfamily, such as receptors for thyroid hormone, androgen, glucocorticoid (GR), progesterone, retinoic acid, 9-cis retinoic acid and vitamin D [68]. In mouse corticotroph AtT-20 cells, it has been shown that chronic glucocorticoid treatment causes a down-regulation of GR levels [72]. Vedeckis *et al.* showed that chronic glucocorticoid treatment reduces the amount of GR mRNA to about 50% of that in untreated cells [73]. These studies point to possible mechanisms whereby the responsiveness of the cell to steroid hormones is altered by the regulation of the steroid receptor protein and mRNA levels. For these receptors, proteasome inhibition interferes with steroid-mediated transcription by increased accumulation of the GR, confirming that it is likewise a substrate for the UPS [74,75]. It is not know whether the relative resistance to glucocorticoid feedback encountered in corticotroph adenomas causing Cushing's disease is related to GR-dependent down-regulation of GR levels. In general, there is little evidence that GR is abnormal in structure or function in the majority of corticotrophinomas [65].

Kovacs *et al.* studied ubiquitin expression in 31 non-tumorous pituitary glands and 133 pituitary adenomas by immunocytochemical techniques [76]. Normal non-tumorous hypophyses were immunonegative for ubiquitin. Ubiquitin immunoreactivity was present in 3% to 30% of corticotrophs containing Crooke's hyaline in 10 of 12 glucocorticoid-treated patients: 58 adenomas showed ubiquitin-immunoreactive cells. Ubiquitin immunoreactivity was found in cytokeratin immunopositive filamentous inclusions of Crooke's cell adenomas and in fibrous bodies of somatotroph adenomas. Forty-five adenomas showed diffuse cytoplasmic immunopositivity. No correlation was revealed between ubiquitin immunoreactivity, hormone content, and bromocriptine or octreotide treatments. The results are consistent with the interpretation that ubiquitin immunoreactivity in non-tumorous corticotrophs containing Crooke's hyaline and in various adenomas is secondary to glucocorticoid excess or to altered metabolic activity. Whether ubiquitin expression reflects increased ubiquitin synthesis or increased activity of the UPS remains to be elucidated [76].

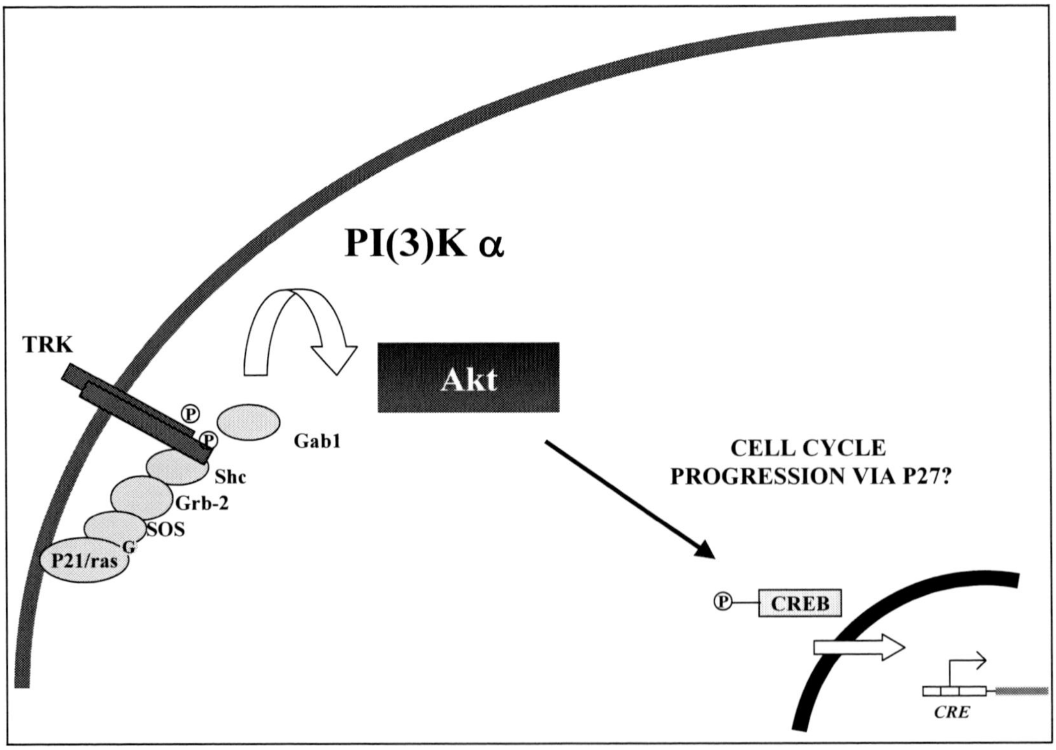

Figure 4. Enhanced Akt signaling in pituitary tumors. Enhanced Akt signaling may be triggered by constitutively active, over-expressed or ligand-activated tyrosine kinase receptors.In response, phosphatidyl-inositol 3 kinase (PI3K) recruits Akt to the cell membrane where it becomes active. Secondary to its activation, Akt triggers a cascade of responses: cell cycle progression, inhibition of apoptosis, increased cell growth and motility that may be responsible for tumor progression.

SIGNAL TRANSDUCTION: PROTEIN KINASES

Mitogenic signaling by receptor tyrosine kinases which involves increased activity of PI3K and over-activation of Akt triggers a cascade of responses – cell growth, proliferation, survival and increased motility – which drive tumor progression in breast, ovarian, prostate, pancreatic and thyroid cancers [77]. The role of Akt in tumorigenesis is due to phosphorylation and relocalization of key regulatory molecules involved in apoptosis, cell growth and proliferation. With respect to cell cycle progression, Akt has been shown to be involved in preventing cyclin D1 degradation, and to negatively influence the expression and localization of cell cycle inhibitors such as p21 and p27 [78]. Over-expression and activation of the Akt pathway has been shown in pituitary tumors (Figure 3), and we have speculated that cell-cycle changes observed in such tumors are secondary to these more proximate alterations [39]. Since Akt is a major downstream signaling molecule of tyrosine kinase receptors with growth factor ligands, our data are most compatible with an abnormality at this level as the primary driver of pituitary tumorigenesis (Figure 4).

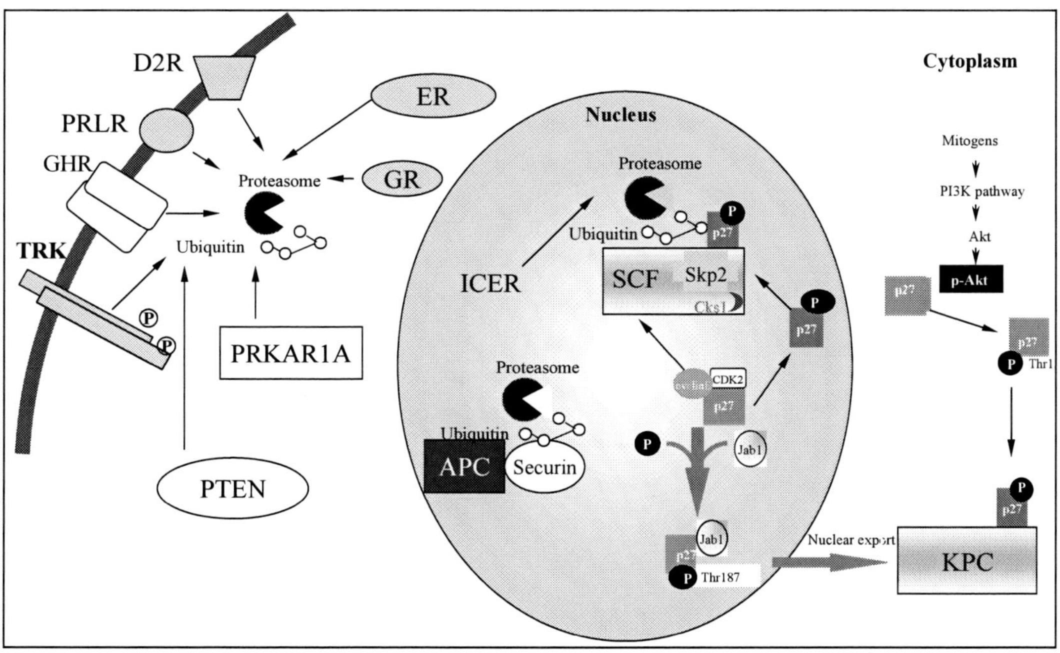

Figure 5. Potential sites of action of UPS in the pituitary. Within the nucleus, p27 is phosphorylated by Cdk2 and is either recognized by Skp2, (part of E3/SCF) and destroyed via nuclear proteasomes or exported to cytoplasm where a parallel system (E3/KPC) triggers its ubiqutination and proteolysis. Mitogens acting through tyrosine kinase receptors (TRK) activate PI3K-Akt pathway that phosphorylates p27 on Thr157, thus sequestrating p27 in the cytoplasm, away from its nuclear targets. Securin/PTTG degradation occurs via E3/APC that triggers its nuclear ubiquitination at the beginning of anaphase. Other proteins involved in pituitary regulation and tumorigenesis may be down-regulated by the ubiquitin proteasome system: inducible cAMP early repressor (ICER), PTEN phosphatase, protein kinase A regulatory subunit1 (PRKAR1A), growth hormone receptor (GHR), prolactin receptor (PRLR), dopamine receptor (D2R), glucocorticoid receptor (GR), estrogen receptor (ER).

Many growth factors and their receptors have been studied as candidates in pituitary tumor formation. These include transforming growth factor α and β (TGF-α, TGF β), epidermal growth factor (EGF) and epidermal growth factor receptor (EGFR), fibroblast growth factors (FGFs) and their receptors (FGFR 1-4) [79]. While there is circumstantial evidence that growth factor signaling can cause transformation, there are few data regarding this crucial aspect of ligand-mediated receptor internalization and degradation through the ubiquitin proteasome pathway in human pituitary tumors (Figure 5).

The protein kinase A regulatory subunit 1α (PRKAR1A) has been identified as the gene responsible for the Carney complex type I [80]. Carney complex is a multiple neoplasia syndrome characterized by spotty skin pigmentation, cardiac and soft tissue (skin, mucous membrane) myxomas, psammomatous melanotic schwannomas and endocrine tumors including Cushing's syndrome from nodular adrenocortical hyperplasia, pituitary adenomas (acromegaly or prolactinoma), Sertoli cell tumors and Leydig cell tumors [80]. The mechanism of tumorigenesis in patients with Carney syndrome involves constitutive activation of camp-dependent protein kinase A.

Several studies have been unable to demonstrate somatic mutations of PRKAR1A in sporadic pituitary tumors, or alterations in the level of its mRNA expression [4,81]. The two regulatory subunits (R1 and R2) of protein kinase A are differentially expressed in cancer cell lines and exert diverse roles in growth control. Lania *et al.* have addressed the expression of the PKA regulatory subunits R1A, R2A, and R2B in a series of 30 pituitary adenomas and the effects of subunit activation on cell proliferation [82]. Immunohistochemistry documented low or absent R1A levels in all tumors, whereas R2A and R2B were highly expressed, thus resulting in an unbalanced R1/R2 ratio. The low levels of R1A were, at least in part, due to proteasome-mediated degradation [82]. The effect of the R1/R2 ratio on proliferation was assessed in GH3 cells, which showed a similar unbalanced pattern of R-subunit expression, and in growth hormone-secreting adenomas. The R2-selective cAMP analog 8-Cl cAMP, and R1A RNA silencing, both stimulated cell proliferation and increased cyclin D1 expression, respectively, in human and rat adenomatous somatotrophs. These data show that a low R1/R2 ratio promotes proliferation of transformed somatotrophs and are consistent with the Carney complex model in which R1A inactivating mutations further unbalance this ratio in favor of R2 subunits. These results suggest that low expression of R1A protein may favor cAMP-dependent proliferation of transformed somatotrophs [82]. In all, these data show that the UPS is active in anterior pituitary cells and more studies are needed to clarify to what extent this pathway is altered in various pituitary tumors subtypes.

The Ubiquitin Proteasome System as Potential Target in Pituitary Tumors Therapy

The proteasome is a multi-catalytic proteinase complex responsible for the degradation of most intracellular proteins, including proteins crucial to cell cycle regulation and programmed cell death, or apoptosis. In preclinical cancer models, proteasome inhibitors induce apoptosis, have in vivo anti-tumor effect, and sensitize malignant cells and tumors to the proapoptotic effects of conventional chemotherapeutics and radiation therapy [83,84].

Transformed cells display greater susceptibility to proteasome inhibition than non-malignant cells, making proteasome inhibition promising as a novel approach to the treatment of cancer. First-generation proteasome inhibitors lacked usefulness because of broad specificity and irreversible binding to the proteasome. However, the later synthesis of the peptide boronic acid proteasome inhibitor bortezomib allowed for selective, reversible binding. Basic investigations have reported the anti-tumor activity of bortezomib in a variety of hematological and solid tumor models and have demonstrated the ability of bortezomib to enhance chemosensitivity and overcome cellular mechanisms of drug resistance [85]. Bortezomib (Velcade™; formerly PS-341), the first such inhibitor to undergo clinical testing, has demonstrated impressive anti-tumor activity and manageable toxicities in Phase I and II trials both as a single agent, and in combination with other drugs. In the recent APEX trial bortezomib proved to be superior to high-dose dexamethasone for the treatment of patients with multiple myeloma who have had a relapse after one to three previous therapies [86].

Previous studies showed that proteasome inhibitors induced apoptosis in growth hormone- and prolactin-secreting rat pituitary tumor cells *in vitro*, but not in normal pituitary cells. Three proteasome inhibitors, PSI, MG-132 and lactacystin, caused apoptosis in these cells [87]. In the light of clinical trials with new proteasome inhibitors, further studies to address their use in pituitary tumors are needed.

Other unconventional anti-tumor therapies have been shown to influence ubiquitination and degradation by the proteasome complex. Liu *et al.* showed that vitamin D and its analog EB1089 could selectively arrest pituitary corticotroph growth and induce p27 accumulation by reducing p27 association with Skp2 and with CDK2 [88]. These findings highlight vitamin D analogs as targets for drug development in the treatment of inoperable corticotroph adenomas [88].

CONCLUSIONS

While the UPS has been discovered as a machinery of protein degradation, its role has more recently been widened including several other processes such as membrane receptor internalization [89], control of ribosomal function [90] and DNA repair [91]. In addition to transcriptional and translational regulatory processes, targeting of cellular proteins for proteasome-mediated degradation provides an important regulatory step by which cell function can be modulated. This model system suggests the need for endocrine studies to consider non-genomic and non-receptor-mediated processes that alter cell function in general and that might contribute to the appearance and progression of pituitary tumors, and new data might lead to development of new drugs for these tumors.

References

[1] Clayton RN, Farrell WE: Pituitary tumour clonality revisited. *Front Horm Res* 2004, 32:186-204.

[2] Grossman AB, Korbonits M: Akting and cycling: a tale of the pituitary. *Horm Res* 2004, 62 Suppl 3:117-123.

[3] Satta MA, Korbonits M, Jacobs RA, Bolden-Dwinfour DA, Kaltsas GA, Vangeli V, Adams E, Fahlbusch R, Grossman AB: Expression of menin gene mRNA in pituitary tumours. *Eur J Endocrinol* 1999, 140:358-361.

[4] Kaltsas GA, Kola B, Borboli N, Morris DG, Gueorguiev M, Swords FM, Czirjak S, Kirschner LS, Stratakis CA, Korbonits M, Grossman AB: Sequence analysis of the PRKAR1A gene in sporadic somatotroph and other pituitary tumours. *Clin Endocrinol (Oxf)* 2002, 57:443-448.

[5] Gadelha MR, Une KN, Rohde K, Vaisman M, Kineman RD, Frohman LA: Isolated familial somatotropinomas: establishment of linkage to chromosome 11q13.1-11q13.3 and evidence for a potential second locus at chromosome 2p16-12. *J Clin Endocrinol Metab* 2000, 85:707-714.

[6] Lee WH, Bookstein R, Hong F, Young LJ, Shew JY, Lee EY: Human retinoblastoma susceptibility gene: cloning, identification, and sequence. *Science* 1987, 235:1394-1399.

[7] Lee WH, Shew JY, Hong FD, Sery TW, Donoso LA, Young LJ, Shew JY, Lee EY: The retinoblastoma susceptibility gene encodes a nuclear phosphoprotein associated with DNA binding activity. *Nature* 1987, 329:642-645.

[8] Yeung RS, Bell DW, Testa JR, Mayol X, Baldi A, Grana X, Klinga-Levan K, Knudson AG, Giordano A: The retinoblastoma-related gene, RB2, maps to human chromosome 16q12 and rat chromosome 19. *Oncogene* 1993, 8:3465-3468.

[9] Zhu L, van den HS, Helin K, Fattaey A, Ewen M, Livingston D, Dyson N, Harlow E: Inhibition of cell proliferation by p107, a relative of the retinoblastoma protein. *Genes Dev* 1993, 7:1111-1125.

[10] Mayol X, Grana X, Baldi A, Sang N, Hu Q, Giordano A: Cloning of a new member of the retinoblastoma gene family (pRb2) which binds to the E1A transforming domain. *Oncogene* 1993, 8:2561-2566.

[11] Pavletich NP: Mechanisms of cyclin-dependent kinase regulation: structures of Cdks, their cyclin activators, and Cip and INK4 inhibitors. *J Mol Biol* 1999, 287:821-828.

[12] Serrano M, Hannon GJ, Beach D: A new regulatory motif in cell-cycle control causing specific inhibition of cyclin D/CDK4. *Nature* 1993, 366:704-707.

[13] Hannon GJ, Beach D: p15INK4B is a potential effector of TGF-beta-induced cell cycle arrest. *Nature* 1994, 371:257-261.

[14] Chan FK, Zhang J, Cheng L, Shapiro DN, Winoto A: Identification of human and mouse p19, a novel CDK4 and CDK6 inhibitor with homology to p16ink4. *Mol Cell Biol* 1995, 15:2682-2688.

[15] Harper JW: Cyclin dependent kinase inhibitors. *Cancer Surv* 1997, 29:91-107.

[16] Nakayama K, Nakayama K: Cip/Kip cyclin-dependent kinase inhibitors: brakes of the cell cycle engine during development. *Bioessays* 1998, 20:1020-1029.

[17] Bashir T, Pagano M: Aberrant ubiquitin-mediated proteolysis of cell cycle regulatory proteins and oncogenesis. *Adv Cancer Res* 2003, 88:101-144.

[18] Ogino A, Yoshino A, Katayama Y, Watanabe T, Ota T, Komine C Yokoyama T, Fukushima T: The p15(INK4b)/p16(INK4a)/RB1 pathway is frequently deregulated in human pituitary adenomas. *J Neuropathol Exp Neurol* 2005, 64:398-403.

[19] Simpson DJ, Frost SJ, Bicknell JE, Broome JC, McNicol AM, Clayton RN, Farrell WE: Aberrant expression of G(1)/S regulators is a frequent event in sporadic pituitary adenomas. *Carcinogenesis* 2001, 22:1149-1154.

[20] Woloschak M, Yu A, Post KD: Frequent inactivation of the p16 gene in human pituitary tumors by gene methylation. *Mol Carcinog* 1997, 19:221-224.

[21] Simpson DJ, Bicknell JE, McNicol AM, Clayton RN, Farrell WE: Hypermethylation of the p16/CDKN2A/MTSI gene and loss of protein expression is associated with nonfunctional pituitary adenomas but not somatotrophinomas. *Genes Chromosomes Cancer* 1999, 24:328-336.

[22] Simpson DJ, Hibberts NA, McNicol AM, Clayton RN, Farrell WE: Loss of pRb expression in pituitary adenomas is associated with methylation of the RB1 CpG island. *Cancer Res* 2000, 60:1211-1216.

[23] Hibberts NA, Simpson DJ, Bicknell JE, Broome JC, Hoban PR, Clayton RN, Farrell WE: Analysis of cyclin D1 (CCND1) allelic imbalance and overexpression in sporadic human pituitary tumors. *Clin Cancer Res* 1999, 5:2133-2139.

[24] Jordan S, Lidhar K, Korbonits M, Lowe DG, Grossman AB: Cyclin D and cyclin E expression in normal and adenomatous pituitary. *Eur J Endocrinol* 2000, 143:R1-R6.

[25] Diehl JA, Zindy F, Sherr CJ: Inhibition of cyclin D1 phosphorylation on threonine-286 prevents its rapid degradation via the ubiquitin-proteasome pathway. *Genes Dev* 1997, 11:957-972.

[26] Dahia PL, Aguiar RC, Honegger J, Fahlbush R, Jordan S, Lowe DG, Lu X, Clayton RN, Besser GM, Grossman AB: Mutation and expression analysis of the p27/kip1 gene in corticotrophin-secreting tumours. *Oncogene* 1998, 16:69-76.

[27] Bamberger CM, Fehn M, Bamberger AM, Ludecke DK, Beil FU, Saeger W, Schulte HM: Reduced expression levels of the cell-cycle inhibitor p27Kip1 in human pituitary adenomas. *Eur J Endocrinol* 1999, 140:250-255.

[28] Lidhar K, Korbonits M, Jordan S, Khalimova Z, Kaltsas G, Lu X, Clayton RN, Jenkins PJ, Monson JP, Besser GM, Lowe DG, Grossman AB: Low expression of the cell cycle inhibitor p27Kip1 in normal corticotroph cells, corticotroph tumors, and malignant pituitary tumors. *J Clin Endocrinol Metab* 1999, 84:3823-3830.

[29] Qian X, Jin L, Grande JP, Lloyd RV: Transforming growth factor-beta and p27 expression in pituitary cells. *Endocrinology* 1996, 137:3051-3060.

[30] Shirane M, Harumiya Y, Ishida N, Hirai A, Miyamoto C, Hatakeyama S Nakayama K, Kitagawa M: Down-regulation of p27(Kip1) by two mechanisms, ubiquitin-mediated degradation and proteolytic processing. *J Biol Chem* 1999, 274:13886-13893.

[31] Nguyen H, Gitig DM, Koff A: Cell-free degradation of p27(kip1), a G1 cyclin-dependent kinase inhibitor, is dependent on CDK2 activity and the proteasome. *Mol Cell Biol* 1999, 19:1190-1201.

[32] Sheaff RJ, Groudine M, Gordon M, Roberts JM, Clurman BE: Cyclin E-CDK2 is a regulator of p27Kip1. *Genes Dev* 1997, 11:1464-1478.

[33] Vlach J, Hennecke S, Amati B: Phosphorylation-dependent degradation of the cyclin-dependent kinase inhibitor p27. *EMBO J* 1997, 16:5334-5344.

[34] Korbonits M, Chahal HS, Kaltsas G, Jordan S, Urmanova Y, Khalimova Z, Harris PE, Farrell WE, Claret FX, Grossman AB: Expression of phosphorylated p27(Kip1) protein and Jun activation domain-binding protein 1 in human pituitary tumors. *J Clin Endocrinol Metab* 2002, 87:2635-2643.

[35] Pyle ME, Korbonits M, Gueorguiev M, Jordan S, Kola B, Morris DG, Meinhardt A, Powell MP, Claret FX, Zhang Q, Metz C, Bucala R, Grossman AB: Macrophage migration inhibitory factor expression is increased in pituitary adenoma cell nuclei. *J Endocrinol* 2003, 176:103-110.

[36] Musat M., Korbonits M, Pyle ME, Gueorguiev M, Kola B, Morris DG, Powell MP, Dumitrache C, Poiana C, Grossman AB: The expression of the F-box protein Skp2 is negatively associated with p27 expression in human pituitary tumors. *Pituitary* 2002, 5:235-242.

[37] Musat M, Korbonits M, Grossman AB: Cell Cycle Dysregulation in Pituitary Tumors. In *Molecular Pathology of the Pituitary*. Edited by Kontogeorgos G., Kovacs K, eds. Karger. 2004: 34-62.

[38] Mamillapalli R, Gavrilova N, Mihaylova VT, Tsvetkov LM, Wu H, Zhang H, Sun H: PTEN regulates the ubiquitin-dependent degradation of the CDK inhibitor p27(KIP1) through the ubiquitin E3 ligase SCF(SKP2). *Curr Biol* 2001, 11:263-267.

[39] Musat M, Korbonits M, Kola B, Borboli N, Hanson MR, Nanzer AM, Grigson J, Jordan S, Morris DG, Gueorguiev M, Coculescu M, Basu S, Grossman AB: Enhanced protein kinase B/Akt signalling in pituitary tumours. *Endocr Relat Cancer* 2005, 12:423-433.

[40] Tolkacheva T, Boddapati M, Sanfiz A, Tsuchida K, Kimmelman AC, Chan AM: Regulation of PTEN binding to MAGI-2 by two putative phosphorylation sites at threonine 382 and 383. *Cancer Res* 2001, 61(13):4985-4989.

[41] Waite KA, Eng C: BMP2 exposure results in decreased PTEN protein degradation and increased PTEN levels. *Hum Mol Genet* 2003, 12(6):679-684.

[42] Wu W, Wang X, Zhang W, Reed W, Samet JM, Whang YE, Ghio AJ: Zinc-induced PTEN protein degradation through the proteasome pathway in human airway epithelial cells. *J Biol Chem* 2003, 278:28258-28263.

[43] Kamura T, Hara T, Matsumoto M, Ishida N, Okumura F, Hatakeyama S, Yoshida M, Nakayama K, Nakayama KI: Cytoplasmic ubiquitin ligase KPC regulates proteolysis of p27(Kip1) at G1 phase. *Nat Cell Biol* 2004, 6:1229-1235.

[44] Clurman BE, Sheaff RJ, Thress K, Groudine M, Roberts JM: Turnover of cyclin E by the ubiquitin-proteasome pathway is regulated by cdk2 binding and cyclin phosphorylation. *Genes Dev* 1996, 10:1979-1990.

[45] Moberg KH, Bell DW, Wahrer DC, Haber DA, Hariharan IK: Archipelago regulates Cyclin E levels in Drosophila and is mutated in human cancer cell lines. *Nature* 2001, 413:311-316.

[46] Strohmaier H, Spruck CH, Kaiser P, Won KA, Sangfelt O, Reed SI: Human F-box protein hCdc4 targets cyclin E for proteolysis and is mutated in a breast cancer cell line. *Nature* 2001, 413:316-322.

[47] Koepp DM, Schaefer LK, Ye X, Keyomarsi K, Chu C, Harper JW, Elledge SJ: Phosphorylation-dependent ubiquitination of cyclin E by the SCFFbw7 ubiquitin ligase. *Science* 2001, 294:173-177.

[48] Morris D.G., Bonner S, Korbonits M, Grossman A. The cyclin E specific F-box protein Archipelago is overexpressed in pituitary tumours. Endocrine Abstracts Spring 2005, 7.

[49] Uhlmann F, Lottspeich F, Nasmyth K: Sister-chromatid separation at anaphase onset is promoted by cleavage of the cohesin subunit Scc1. *Nature* 1999, 400:37-42.

[50] Zou H, McGarry TJ, Bernal T, Kirschner MW: Identification of a vertebrate sister-chromatid separation inhibitor involved in transformation and tumorigenesis. *Science* 1999, 285:418-422.

[51] Zhang X, Horwitz GA, Prezant TR, Valentini A, Nakashima M, Bronstein MD, Melmed S: Structure, expression, and function of human pituitary tumor-transforming gene (PTTG). *Mol Endocrinol* 1999, 13:156-166.

[52] Zhang X, Horwitz GA, Heaney AP, Nakashima M, Prezant TR, Bronstein MD, Melmed S: Pituitary tumor transforming gene (PTTG) expression in pituitary adenomas. *J Clin Endocrinol Metab* 1999, 84:761-767.

[53] Yu R, Melmed S: Oncogene activation in pituitary tumors. *Brain Pathol* 2001, 11:328-341.

[54] Yu R, Melmed S: Pituitary tumor transforming gene: an update. *Front Horm Res* 2004, 32:175-185.

[55] Romero F, Gil-Bernabe AM, Saez C, Japon MA, Pintor-Toro JA, Tortolero M: Securin is a target of the UV response pathway in mammalian cells. *Mol Cell Biol* 2004, 24:2720-2733.

[56] Kraft C, Vodermaier HC, Maurer-Stroh S, Eisenhaber F, Peters JM: The WD40 propeller domain of Cdh1 functions as a destruction box receptor for APC/C substrates. *Mol Cell* 2005, 18:543-553.

[57] Solbach C, Roller M, Peters S, Nicoletti M, Kaufmann M, Knecht R: Pituitary tumor-transforming gene (PTTG): a novel target for anti-tumor therapy. *Anticancer Res* 2005, 25:121-125.

[58] Landis CA, Masters SB, Spada A, Pace AM, Bourne HR, Vallar L: GTPase inhibiting mutations activate the alpha chain of Gs and stimulate adenylyl cyclase in human pituitary tumours. *Nature* 1989, 340:692-696.

[59] Bertherat J, Chanson P, Montminy M: The cyclic adenosine 3',5'-monophosphate-responsive factor CREB is constitutively activated in human somatotroph adenomas. *Mol Endocrinol* 1995, 9:777-783.

[60] Folco EJ, Koren G: Degradation of the inducible cAMP early repressor (ICER) by the ubiquitin-proteasome pathway. *Biochem J* 1997, 328 (Pt 1):37-43.

[61] Yehia G, Schlotter F, Razavi R, Alessandrini A, Molina CA: Mitogen-activated protein kinase phosphorylates and targets inducible cAMP early repressor to ubiquitin-mediated destruction. *J Biol Chem* 2001, 276:35272-35279.

[62] Kola B, Korbonits M, Diaz-Cano S, Kaltsas G, Morris DG, Jordan S, Metherell L, Powell M, Czirjak S, Arnaldi G, Bustin S, Boscaro M, Mantero F, Grossman AB: Reduced expression of the growth hormone and type 1 insulin-like growth factor receptors in human somatotroph tumours and an analysis of possible mutations of the growth hormone receptor. *Clin Endocrinol (Oxf)* 2003, 59:328-338.

[63] Strous GJ, van Kerkhof P, Govers R, Ciechanover A, Schwartz AL: The ubiquitin conjugation system is required for ligand-induced endocytosis and degradation of the growth hormone receptor. *EMBO J* 1996, 15:3806-3812.

[64] Strous GJ, dos Santos CA, Gent J, Govers R, Sachse M, Schantl J, van Kerkhof P: Ubiquitin system-dependent regulation of growth hormone receptor signal transduction. *Curr Top Microbiol Immunol* 2004, 286:81-118.

[65] Korbonits M, Morris DG, Nanzer A, Kola B, Grossman AB: Role of regulatory factors in pituitary tumour formation. *Front Horm Res* 2004, 32:63-95.

[66] Shenoy SK, McDonald PH, Kohout TA, Lefkowitz RJ: Regulation of receptor fate by ubiquitination of activated beta 2-adrenergic receptor and beta-arrestin. *Science* 2001, 294:1307-1313.

[67] Alarid ET, Bakopoulos N, Solodin N: Proteasome-mediated proteolysis of estrogen receptor: a novel component in autologous down-regulation. *Mol Endocrinol* 1999, 13:1522-1534.

[68] Alarid ET, Preisler-Mashek MT, Solodin NM: Thyroid hormone is an inhibitor of estrogen-induced degradation of estrogen receptor-alpha protein: estrogen-dependent proteolysis is not essential for receptor transactivation function in the pituitary. *Endocrinology* 2003, 144:3469-3476.

[69] Tsai HW, Katzenellenbogen JA, Katzenellenbogen BS, Shupnik MA: Protein kinase A activation of estrogen receptor alpha transcription does not require proteasome activity and protects the receptor from ligand-mediated degradation. *Endocrinology* 2004, 145:2730-2738.

[70] Jin L, Qian X, Kulig E, Scheithauer BW, Calle-Rodrigue R, Abboud C, Davis DH, Kovacs K, Lloyd RV: Prolactin receptor messenger ribonucleic acid in normal and neoplastic human pituitary tissues. *J Clin Endocrinol Metab* 1997, 82:963-968.

[71] Li Y, Kumar KG, Tang W, Spiegelman VS, Fuchs SY: Negative regulation of prolactin receptor stability and signaling mediated by SCF(beta-TrCP) E3 ubiquitin ligase. *Mol Cell Biol* 2004, 24:4038-4048.

[72] Svec F, Rudis M: Glucocorticoids regulate the glucocorticoid receptor in the AtT-20 cell. *J Biol Chem* 1981, 256:5984-5987.

[73] Vedeckis WV, Ali M, Allen HR: Regulation of glucocorticoid receptor protein and mRNA levels. *Cancer Res* 1989, 49(8 Suppl):2295s-2302s.

[74] Deroo BJ, Rentsch C, Sampath S, Young J, DeFranco DB, Archer TK: Proteasomal inhibition enhances glucocorticoid receptor transactivation and alters its subnuclear trafficking. *Mol Cell Biol* 2002, 22:4113-4123.

[75] Wallace AD, Cidlowski JA: Proteasome-mediated glucocorticoid receptor degradation restricts transcriptional signaling by glucocorticoids. *J Biol Chem* 2001, 276:42714-42721.

[76] Marin F, Cheng Z, Kovacs K: Ubiquitin immunoreactivity in corticotrophs following glucocorticoid treatment and in pituitary adenomas. *Arch Pathol Lab Med* 1993, 117:254-258.

[77] Ringel MD, Hayre N, Saito J, Saunier B, Schuppert F, Burch H, Bernet V, Burman KD, Kohn LD, Saji M: Overexpression and overactivation of Akt in thyroid carcinoma. *Cancer Res* 2001, 61:6105-6111.

[78] Vivanco I, Sawyers CL: The phosphatidylinositol 3-Kinase AKT pathway in human cancer. *Nat Rev Cancer* 2002, 2:489-501.

[79] Asa SL, Ezzat S: Molecular basis of pituitary development and cytogenesis. *Front Horm Res* 2004, 32:1-19.

[80] Casey M, Vaughan CJ, He J, Hatcher CJ, Winter JM, Weremowicz S, Montgomery K, Kucherlapati R, Morton CC, Basson CT: Mutations in the protein kinase A R1alpha regulatory subunit cause familial cardiac myxomas and Carney complex. *J Clin Invest* 2000, 106:R31-R38.

[81] Yamasaki H, Mizusawa N, Nagahiro S, Yamada S, Sano T, Itakura M, Yoshimoto K: GH-secreting pituitary adenomas infrequently contain inactivating mutations of PRKAR1A and LOH of 17q23-24. *Clin Endocrinol (Oxf)* 2003, 58:464-470.

[82] Lania AG, Mantovani G, Ferrero S, Pellegrini C, Bondioni S, Peverelli E, Braidotti P, Locatelli M, Zavanone ML, Ferrante E, Bosari S, Beck-Peccoz P, Spada A.: Proliferation of transformed somatotroph cells related to low or absent expression of protein kinase a regulatory subunit 1A protein. *Cancer Res* 2004, 64:9193-9198.

[83] Fujita E, Mukasa T, Tsukahara T, Arahata K, Omura S, Momoi T: Enhancement of CPP32-like activity in the TNF-treated U937 cells by the proteasome inhibitors. *Biochem Biophys Res Commun* 1996, 224:74-79.

[84] Voorhees PM, Dees EC, O'Neil B, Orlowski RZ: The proteasome as a target for cancer therapy. *Clin Cancer Res* 2003, 9:6316-6325.

[85] Richardson PG, Mitsiades C, Hideshima T, Anderson KC: Proteasome inhibition in the treatment of cancer. *Cell Cycle* 2005, 4:290-296.

[86] Richardson PG, Sonneveld P, Schuster MW, Irwin D, Stadtmauer EA, Facon T, Harousseau JL, Ben Yehuda D, Lonial S, Goldschmidt H, Reece D, San Miguel J.F, Blade J, Boccadoro M, Cavenagh J, Dalton WS, Boral AL, Esseltine DL, Porter J B, Schenkein D, Anderson KC: Bortezomib or high-dose dexamethasone for relapsed multiple myeloma. *N Engl J Med* 2005, 352:2487-2498.

[87] Yu R, Ren SG, Melmed S: Proteasome inhibitors induce apoptosis in growth hormone- and prolactin-secreting rat pituitary tumor cells. *J Endocrinol* 2002, 174:379-386.

[88] Liu W, Asa SL, Ezzat S: Vitamin D and its analog EB1089 induce p27 accumulation and diminish association of p27 with Skp2 independent of PTEN in pituitary corticotroph cells. *Brain Pathol* 2002, 12:412-419.

[89] Terrell J, Shih S, Dunn R, Hicke L: A function for monoubiquitination in the internalization of a G protein-coupled receptor. *Mol Cell* 1998, 1:193-202.

[90] Spence J, Gali RR, Dittmar G, Sherman F, Karin M, Finley D: Cell cycle-regulated modification of the ribosome by a variant multiubiquitin chain. *Cell* 2000, 102:67-76.

[91] Russell SJ, Reed SH, Huang W, Friedberg EC, Johnston SA: The 19S regulatory complex of the proteasome functions independently of proteolysis in nucleotide excision repair. *Mol Cell* 1999, 3:687-695.

[92] Nakayama KI, Hatakeyama S, Nakayama K: Regulation of the cell cycle at the G1-S transition by proteolysis of cyclin E and p27Kip1. *Biochem Biophys Res Commun* 2001, 282:853-860.

In: The Ubiquitin Proteasome System… ISBN 978-1-60021-749-4
Eds: Mario Di Napoli and Cezary Wojcik, pp. 897-921

Chapter 38

The Role of the Ubiquitin-Proteasome System in Epilepsy and Seizure Susceptibility

Edward Glasscock*
Department of Neurology, Baylor College of Medicine, Houston, TX 77030, USA.

Abstract

One of the most common neurological disorders today is epilepsy. Epilepsy is a chronic brain disease characterized by recurrent, spontaneous seizures resulting from abnormal synchronization of neurons in the central nervous system. Seizures can stem from a variety of brain insults including head trauma, fever, illness, and electroconvulsive shock. However, one of the most important factors governing seizure susceptibility appears to be genetic predisposition. Epilepsy often results from inheritance of one or a combination of several predisposing genetic factors that disturb the balance of excitatory and inhibitory neural networks in the brain. More than 70 genes have been linked to epilepsy from work done on inherited disorders in humans, mice, and fruit flies. These genes encode a wide variety of products ranging from ion channel proteins to tRNAs. Recently, a relationship has begun to emerge between epilepsy and genes of the ubiquitin-proteasome system (UPS). The UPS is the molecular machinery responsible for the degradation of cytoplasmic proteins in the cell. A protein is marked for proteolytic processing in the UPS by the addition of ubiquitin molecules that target the protein to the proteasome, a multisubunit complex that reduces it to small peptides and amino acids. Defects in UPS genes have been linked to epilepsy and altered seizure susceptibility in humans, mice, dogs, and most recently in fruit flies. The two human UPS genes linked to epilepsy are *UBE3A* and *EPM2B*. These genes both encode E3 ubiquitin ligase proteins, the enzymes directly responsible for mediating the transfer of

* Correspondence concerning this article should be addressed to Dr. Edward Glasscock, PhD; Department of Neurology, Baylor College of Medicine, One Baylor Plaza, Houston, TX 77030, USA. Phone: +1-713-798-5862; Fax: +1-713-798-7528; E-mail: glasscoc@bcm.tmc.edu.

ubiquitin to substrates to mark them for proteasomal degradation. Defects in the human *UBE3A* gene lead to Angelman syndrome, a complex genetic disease marked by epilepsy in conjunction with other neurological manifestations such as mental retardation and ataxia. Mutation of the human *EPM2B* gene causes a severe and ultimately fatal form of progressive myoclonus epilepsy known as Lafora disease, which is characterized by the occurrence of starchy inclusion bodies within cells. Similar to humans, disruption of the homologs of *UBE3A* and *EPM2B* in animal models leads to altered seizure susceptibility and epilepsy. Finally, the most recent addition to the list of epilepsy-related UPS genes is *mei-P26* in the fruit fly *Drosophila melanogaster*. Mutation of the *mei-P26* gene has recently been shown to drastically decrease seizure susceptibility, essentially curing epilepsy in *Drosophila* models of the disease. The protein encoded by *mei-P26* resembles E3 ubiquitin ligases but this function has not yet been tested biochemically. Although the mechanisms by which these genes regulate seizure susceptibility are still under investigation, it is becoming increasingly clear that UPS-related genes play a critical role in the etiology of epilepsy and human seizure disorders.

Keywords: epilepsy, seizure, E3 ubiquitin ligase, RING domain, HECT domain, Lafora disease, Angelman syndrome, bang sensitive.

Abbreviations

AS, Angelman syndrome; BS, bang sensitive; ECS, electroconvulsive shock; EEG, electroencephalogram; ERAD, endoplasmic reticulum-associated degradation; GS, glycogen synthase; GSK3, glycogen synthase kinase 3; LB, Lafora body; LD, Lafora disease; LTP, long-term potentiation; nAChRs, nicotinic acetylcholine receptors; TRIM, tripartite motif; UPS, ubiquitin-proteasome system.

Introduction

Epilepsy is a prevalent brain disease characterized by the presence of recurrent and spontaneous, unprovoked seizures [1]. A seizure is a transient change in behavior resulting from abnormal, involuntary, and rhythmic firing of cortical neurons in the brain [1,2]. One in ten people will experience a seizure in their lifetime [1]. Seizures may result from a provoking factor in an otherwise healthy brain, such as a metabolic abnormality, alcohol withdrawal, acute neurological insult, narcotics intoxication, or high fever in childhood [1]. Eliminating exposure to the provoking factor is usually sufficient to alleviate the occurrence of seizures [1]. However, spontaneous seizures persist without provocation in up to 30% of people who experience an initial seizure due to a chronic brain disorder known as epilepsy [3]. The prevailing paradigm is that epilepsy results from disruption of the balance of excitatory and inhibitory neural networks in the brain leading to hyperexcitability [4]. Although this imbalance can be accomplished by brain damaging insults, such as head traumas or tumors, the predominant factor in the etiology of epilepsy and seizure susceptibility appears to be genetic predisposition [4].

Epilepsy is governed by a strong genetic component. Of the more than 40 million people worldwide with epilepsy, genetics is predicted to play a role in up to 60% of cases [4]. The first evidence for a genetic contribution in epilepsy came from numerous studies of familial incidence and twins. Parents, siblings, and children of epileptics have a two- to threefold increased risk of developing epilepsy due to shared genetic susceptibility [5]. Twin studies reveal identical (monozygotic) twins who are genetically indistinguishable exhibit higher concordance rates for the disease than fraternal (dizygotic) twins who are less genetically similar [6].

Although a clear genetic component exists for epilepsy, elucidation of the genes responsible for causing the disease has proved challenging for a variety of reasons. First, epilepsy usually conforms to a non-mendelian, complex inheritance pattern, in which manifestation of the disease depends on the inheritance of multiple interacting susceptibility genes [7]. Second, the disease shows a high degree of genetic heterogeneity: multiple genotypes can produce identical phenotypes [7]. Third, epilepsy frequently exhibits incomplete penetrance and phenotypic variability, likely due to the presence of genetic modifiers that mask or enhance phenotypes [7].

Despite these genetic complexities, the past few decades have seen the identification of more than 70 different epilepsy genes from work done in humans, mice, and fruit flies [8]. Single gene defects responsible for both idiopathic and symptomatic forms of epilepsy have been identified, providing insight into the mechanisms underlying seizures. In idiopathic epilepsy, no obvious structural or metabolic defect is present so genetics is presumed to be the primary factor [4]. Genes associated with idiopathic epilepsy almost invariably code for ion channels [7]. In symptomatic epilepsy, seizures occur as a byproduct of an obvious structural or metabolic defect. Currently, over 160 mendelian genetic disorders have been characterized that cause symptomatic epilepsy as one component of a neurologically complex phenotype [9]. Genes responsible for symptomatic epilepsy fall into three different categories including mutations associated with brain developmental abnormalities, progressive neurodegeneration, and disturbed energy metabolism [4]. The gene products represented across these categories are diverse, often encoding molecules with no obvious connection to neuronal excitability, including tRNAs, actin-binding proteins, and RNA-binding proteins [4].

Over the last several years, a relationship has begun to emerge between components of the ubiquitin-proteasome system (UPS) and epilepsy. The UPS is the molecular machinery responsible for the control of cellular protein levels by mediating the degradation of cytoplasmic proteins [10]. A protein is marked for proteolytic processing in the UPS by the addition of at least four ubiquitin molecules to a lysine residue on the substrate, a process called polyubiquitination, which targets the protein to the proteasome (see Chapter 3 and [10]). The proteasome is a multisubunit complex that reduces polyubiquitinated proteins to small peptides and amino acids (see Chapters 5 and 6 and [10]). The marking of proteins for degradation by the proteasome takes place in a three step multi-enzyme process [10]. In the first step, the E1 ubiquitin-activating enzyme activates ubiquitin, a small 76 amino acid polypeptide, in an ATP-dependent manner with the formation of a thioester linkage between the E1 and ubiquitin. In the second step, the E1 transfers the activated ubiquitin to an E2 ubiquitin conjugase. In the third step, the E2 cooperates with an E3 ubiquitin ligase to attach the ubiquitin to a protein substrate or to a ubiquitin already attached to the protein. While the human genome encodes only one major E1 enzyme, it contains approximately 50 E2s and hundreds of E3s [11]. The E3 ligases are of critical importance in the UPS since they confer substrate specificity through their numerous protein-protein interactions [12]. Details are given in Chapter 3.

At least two main classes of E3s exist based on the catalytic domain that mediates ubiquitin transfer: (*i*) those with a HECT domain and (*ii*) those with a RING or related domain [12]. Both classes of E3s bind E2s [12]. Members of the HECT (*h*omologous to *E*6AP *c*arboxy-*t*erminus) class directly participate in the ubiquitination of substrates by forming a covalent thioester bond with ubiquitin, which is received from an E2 and then transferred to the substrate [10]. HECT E3s are usually single, large proteins that contain the HECT domain towards the C-terminus and some kind of substrate recognition motif towards the N-terminus [12]. The HECT domain assumes a bilobal structure with a broad cleft where the two lobes meet. At the base of this cleft lies a critical cysteine residue that forms the necessary thioester intermediate required for the transfer of ubiquitin to substrate [13]. Typical N-terminal substrate recognition motifs found in HECT E3s include C2 modules for interaction with phospholipids and proteins and WW domains for binding proline-rich PPxY motifs [12].

The RING (*R*eally *I*nteresting *N*ew *G*ene) E3s differ from the HECT E3s in structure and catalytic function. In contrast to the action of HECT E3s, members of the RING class do not form a bond with ubiquitin [10]. Instead they indirectly participate in ubiquitination by approximating the substrate to the E2 which transfers the ubiquitin [12]. The canonical RING E3s are defined by the consensus sequence [$CX_2CX_{(9\text{-}39)}CX_{(1\text{-}3)}HX_{(2\text{-}3)}C/HX_2CX_{(4\text{-}48)}CX_2C$] in which eight cysteines and histidines form a cross-brace structure that coordinates two zinc atoms [14,15]. A few non-canonical but structurally related RING domains have also been characterized including the PHD (*p*lant *h*omeo*d*omain) and the U-box (*U*FD2-homology domain) [12]. The PHD domain closely resembles the RING consensus sequence except that it contains a cysteine rather than a histidine in the fourth coordination position and a conserved tryptophan before the seventh conserved cysteine residue [16]. The U-box domain is only distantly related to the RING by sequence analysis and has no conserved zinc coordinating residues, but its structure is predicted to be similar to the RING since they share the same conserved charged and polar residues [16]. Whereas HECT E3s are usually single protein ligases, RING E3s function as either single proteins or multiprotein complexes [12]. Single protein RING E3s contain a RING domain as well as a substrate recognition motif, such as SH2 (*S*rc-*h*omology *2*), SH3 (*S*rc-*h*omology *3*), or NHL (*N*CL-1, *H*T2A, and *L*IN-41) domains, in the same protein [16]. Multiprotein RING E3s contain a catalytic RING domain but lack the necessary motifs for substrate recognition [12]. Therefore, these RING E3s mediate degradation by forming multiprotein complexes with other proteins that contain substrate recognition elements [12].

Recent discoveries have revealed that E3 ubiquitin ligases are important regulators of seizure susceptibility. Over the last decade, defects in three different E3 ligase genes have been linked to epilepsy, as summarized in Table 1. In 1997, *UBE3A* became the first UPS gene associated with a human disease when it was found that maternal loss of function of *UBE3A* causes Angelman syndrome, a complex genetic disorder characterized by epilepsy, mental retardation, and a happy disposition [17,18]. The *UBE3A* gene encodes an E3 ubiquitin ligase protein belonging to the HECT class. Similar *UBE3A* mutations in the mouse also cause increased seizure susceptibility [19]. In 2003, *EPM2B* was added to the list of UPS mutations associated with epilepsy when it was found that disruption of the gene causes Lafora disease, a progressive myoclonus epilepsy [20]. The *EPM2B* gene encodes an E3 ubiquitin ligase protein belonging to the single protein RING class. Subsequently, *EPM2B* became the first gene associated with canine epilepsy when it was found that a repeat expansion in the coding region of the gene causes seizures in miniature wirehaired dachshunds [21]. Finally, in 2005, the *mei-P26* gene in the fruit fly was found to cure epilepsy in *Drosophila* models of the disease [22]. The *mei-P26* gene is predicted to encode a RING class E3 ubiquitin ligase enzyme. Therefore, multiple connections between the UPS and seizure susceptibility are now known to exist.

This review focuses on the relationship between defects in the UPS and epilepsy and seizure susceptibility. A section is devoted to each of the E3 ligase genes associated with epilepsy, including *UBE3A*, *EPM2B*, and *mei-P26*, along with discussion of the diseases, phenotypes, and mechanisms associated with each gene. The last section summarizes key points and takes a glimpse at the future of epilepsy research as it relates to the UPS.

Table 1. E3 ubiquitin ligases associated with epilepsy and seizure susceptibility

Gene	Protein	E3 catalytic domain	E3 activity	Target substrates	Disease/phenotype (organism)
UBE3A	E6AP	HECT	Yes	p53, HHR23A, McM7, E6AP	epilepsy assoc. with Angelman syndrome (humans), audiogenic seizures (mice)
EPM2B (*NHLRC1*)	malin	RING	Yes	laforin, malin, glycogen synthase?	epilepsy assoc. with Lafora disease (humans), epilepsy (dogs)
mei-P26	MEI-P26	RING	?	?	seizure suppressor (fruit flies)

THE *UBE3A* GENE AND ANGELMAN SYNDROME

Angelman syndrome (AS) was the first human disease to be associated with a genetic defect in the UPS. The salient features of the disease, which was first described by Harry Angelman in 1965, include epilepsy, mental retardation, absent speech, ataxia, dysmorphic facial features, frequent laughter, and a happy disposition [23,24]. Angelman first observed the syndrome in three children which he termed 'puppet children' because of their puppet-like jerky movements and propensity for laughter [23]. The prevalence of AS is estimated to be about one in 10,000 people [25]. The disease accounts for up to 6% of all children who experience mental retardation coupled with spontaneous seizures [26]. Symptoms of AS typically become apparent within the first four years of life [27]. Although treatment for AS is limited to management of symptoms, normal life spans are possible when the patients are otherwise healthy [28].

Spontaneous seizures are a prominent clinical feature of AS. It is estimated that about 90% of AS patients have epilepsy [29,30]. Seizure onset often begins in infancy or early childhood, usually before the age of five and with an average age of two years old [29]. As the patient ages, their seizure manifestation often changes [31]. About 40% of patients experience febrile seizures during infancy [29]. In childhood, a variety of seizure types are observed including tonic-clonic, atypical absence, myoclonic, tonic, and status epilepticus [29]. In adulthood, atypical absence seizures, myoclonic seizures, or a combination of the two predominate [29]. The most effective treatment for epilepsy in AS patients appears to be valproic acid and benzodiazepines, such as clonazepam [29,30].

In addition to seizures, the electroencephalogram (EEG) of AS patients shows a highly abnormal, rhythmic interictal profile that is characteristic of the disease. EEGs are a useful and non-invasive tool for detecting and localizing abnormal brain activity by means of scalp electrodes. Three different EEG patterns are typically seen in AS, either in isolation or in combination. The most common pattern consists of prolonged runs of high-voltage 2-3 Hz activity, mixed with spikes and sharp waves, in the frontal lobe regions [29]. The second most common pattern is generalized, high-voltage 4-6 Hz activity, with or without spikes [29]. Finally, the third and least common pattern is spikes and sharp waves mixed with high-voltage 3-4 Hz posterior activity, which is facilitated by eye closure [29]. These EEG patterns are characteristic of AS and can facilitate diagnosis, especially since they can occur before the onset of seizures and in patients without seizures [29,32,33].

AS arises from loss of function of the maternal copy of the *UBE3A* gene [17,18]. The *UBE3A* gene encodes the E6-associated protein (E6AP), a HECT class E3 ubiquitin ligase enzyme. The gene was first discovered by virtue of its ability to mediate the degradation of the tumor suppressor protein p53 in the presence of human papillomavirus type 16 and 18 E6 protein [34,35]. Only maternal loss of function of *UBE3A* causes AS because *UBE3A* expression is regulated by genomic imprinting [36,37]. Genomic imprinting is a form of epigenetic inheritance that refers to the silencing of genes based on their parental origin [38]. Unlike most genes, which are equally expressed from both chromosomes, imprinted genes are only expressed from either the maternal or paternal chromosome. Gene silencing by genomic imprinting usually involves the marking, or imprinting, of DNA by methylation at CpG islands. This DNA methylation silences genes by inhibiting their accessibility to the transcriptional machinery. Genomic imprinting occurs during gametogenesis when the inherited marks or imprints are erased and reset to conform to the sex of the individual who inherited them. In the case of *UBE3A*, genomic imprinting leads to the silencing of the gene on the paternally inherited chromosome and expression of the gene from only the maternal chromosome. However, *UBE3A* only appears to be regulated by imprinting in the brain, such that most E6AP in the brain is maternally derived [39,40]. Therefore, in AS when there is a maternal loss of function of *UBE3A*, the brain is devoid of functional E6AP. The mechanism by which genomic imprinting controls *UBE3A* expression is not fully understood, but it appears to involve regulation by a paternal *UBE3A* antisense transcript rather than differential methylation of the gene itself [41].

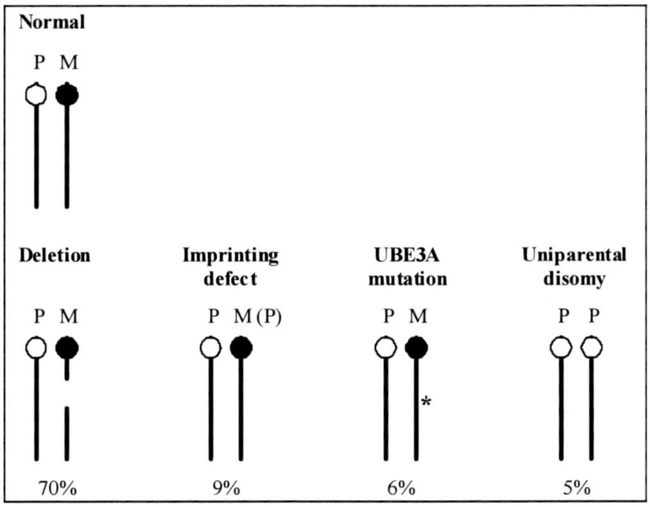

Figure 1. Summary of the genetic mechanisms responsible for abnormal expression of *UBE3A* from chromosome 15 in Angelman syndrome. Normally, an individual inherits one intact copy of chromosome 15 from each parent. In the brain, genomic imprinting silences *UBE3A* on the paternal chromosome by a complex mechanism that is not fully understood. Therefore, most E6AP in the brain is maternally derived. At least four known classes of genetic mutations can cause improper maternal *UBE3A* expression leading to AS. Deletion of the entire maternal *UBE3A* genomic region is the most common genetic abnormality occurring in 70% of patients. The second most common genetic cause of AS is imprinting defects, which occurs in 9% of patients. When imprinting defects occur, the maternal chromosome resembles the paternal chromosome in its imprint status, leading to decreased *UBE3A* expression. In 6% of patients, AS is caused by intragenic *UBE3A* mutation on the maternal chromosome. Finally, 5% of AS patients exhibit a rare genetic condition known as uniparental disomy in which both copies of chromosome 15 are inherited from the father.

Disruption of the maternal *UBE3A* gene occurs by a variety of complex genetic mechanisms which can be categorized into five main groups (Figure 1). The first group which accounts for about 70% of AS patients is deletion of the chromosome region 15q11-13 which removes *UBE3A* [42]. This deletion also includes at least 10 other genes which may modify the symptoms of AS patients. One such gene is *GABRB3* which encodes the β3 subunit of the γ-aminobutyric acid $(GABA)_A$ receptor. The deletion of *GABRB3* is thought to exacerbate seizures in AS patients by dysregulating inhibitory GABAergic neurotransmission [32,43]. The second group which accounts for about 9% of AS patients is imprinting defects [42]. This category includes mutations which disrupt a nearby (≥100 kb) imprinting center involved in the switching of the *UBE3A* imprint status between maternal and paternal. The third group which accounts for about 6% of AS patients is intragenic *UBE3A* mutations [42]. Several different *UBE3A* mutations have been found that cause AS including small deletions (2-4 bp), small duplications (5 bp), missense mutations, nonsense mutations, and frameshift mutations [17,18]. Most of these molecular lesions result in truncated E6AP, while some affect the protein catalytic domain. The fourth group which accounts for about 5% of AS patients is uniparental disomy [42]. In these cases, two paternal *UBE3A* chromosomes are inherited instead of one from each parent. The final group which accounts for the remaining 10% of AS patients is those which have an unknown etiology that does not fit any of the above groups [42]. These individuals may be misdiagnosed or may suffer AS due to a previously unknown mechanism. However, they may also have uncommon mutations affecting *UBE3A* expression that have gone undetected.

A mouse model of AS has been generated by gene targeted disruption of *UBE3A* to yield a null mutation [19]. *UBE3A* deficient mice faithfully recapitulate many of the important aspects of the disease and provide some insight into the mechanisms underlying AS. Like humans, in mice only maternal loss of function of *UBE3A* causes pathology. Mice lacking maternal *UBE3A* exhibit many characteristics observed in the human condition including seizures, abnormal EEG, ataxia, and learning deficits. Seizures can be induced in maternal loss of function *UBE3A* mice by audiogenic stimulation. The EEGs of maternally deficient *UBE3A* mice are analogous to humans in that they exhibit continuous, bilateral 3 sec spike waves that are intermixed with polyspike and slow wave discharges. Mice without maternal *UBE3A* show learning deficits related to hippocampal dysfunction including defective contextual associative learning and impaired long-term potentiation (LTP). These learning deficits appear to be linked to aberrant autophosphorylation and activity of calcium/calmodulin-dependent protein kinase II (CaMKII) in the hippocampus without an overall increase in CaMKII concentration [44]. Interestingly, the mouse model of AS also exhibits tissue dependent genomic imprinting like humans. In mice, the maternal chromosome appears to be the predominant source of *UBE3A* protein in the hippocampus and cerebellar Purkinje cells. It has not been determined yet whether this kind of regional imprinting takes place in human brain. Finally, increased cytoplasmic p53 protein levels were observed in the hippocampus and cerebellar Purkinje cells of maternally deficient *UBE3A* mice, corresponding to the regions where *UBE3A* is absent in these mice. This abnormality correlates with the original observation that E6AP mediates p53 degradation [34]. Interestingly, in light of the p53 abnormality in *UBE3A* mice, the brain of one AS patient was subsequently examined and also shown to exhibit increased p53 levels in the cerebellar Purkinje cells [19]. Therefore, the mouse model of AS provides insight into the human disease and may be a valuable tool for understanding how mutation of an E3 ubiquitin ligase can lead to seizures and epilepsy.

Although several possibilities exist, the simplest explanation for *UBE3A* disruption causing epilepsy involves the improper regulation of E6AP substrates. The E6AP protein is composed of a non-catalytic N-terminal portion and a catalytic C-terminal HECT domain [45]. The non-catalytic N-terminus is likely involved with conferring substrate recognition and specificity, while the C-terminal HECT domain directly catalyzes the attachment of ubiquitin to substrate by means of a ubiquitin-thioester intermediate [35]. Most of the non-truncating AS-associated *UBE3A* mutations map to the catalytic cleft of the HECT domain, implicating the E3 ligase activity of E6AP in the etiology of AS and epilepsy [13,46]. These mutations affect the ability of E6AP to ubiquitinate its target substrates for degradation by impairing the formation of ubiquitin-thioester intermediate, the transfer of ubiquitin to substrate, and the stability of the E6AP protein [47]. The lack of E3 ligase activity likely leads to dysregulation of its target substrates. So far four targets of E6AP-mediated degradation have been identified including a tumor suppressor (p53), a human homolog of the yeast DNA repair protein Rad23 (HHR23A), a protein involved in the initiation of DNA replication (Mcm7), and E6AP itself [34,48-50]. None of these target substrates appear to have an obvious connection to epilepsy. However, there is some evidence that *UBE3A*-mediated degradation of p53 may be important. The p53 protein promotes apoptosis in the central nervous system (CNS) following seizure-induced brain injury [51]. In addition, as previously mentioned, the absence of E6AP is correlated with increased levels of p53 in the brains of mice and humans [19]. It is not clear yet whether p53 is important in the pathogenesis of AS, but the possibility still exists that an increase in p53-mediated cell death in certain parts of the brain could increase seizure susceptibility and lead to epilepsy. However, no *p53* mutations have yet been found that are associated with the occurrence of spontaneous seizures. In addition, evidence is conflicting about whether E6AP can promote p53 degradation in the absence of viral E6 protein. Whatever the mechanism by which a maternal *UBE3A* deficiency affects seizure susceptibility, the reality will likely be complex, possibly involving the coordinate dysregulation of multiple E6AP target substrates most of which have probably not been identified.

The *EPM2B* Gene and Lafora Disease

Lafora disease (LD) is one of five progressive myoclonus epilepsy (PME) syndromes and the only one caused by a known genetic defect in the UPS [52]. The PMEs all share the common characteristics of myoclonic seizures, tonic-clonic seizures, and progressive neurological deterioration, including cerebellar ataxia and dementia [53]. In the case of LD, patients also exhibit pathognomonic starchlike inclusions called Lafora bodies (LBs) which are thought to contribute to the observed spontaneous seizures and neurological decline [52,54]. LBs were first described in 1911 by Gonzalo R. Lafora for whom the disease is named [54]. LBs are composed of periodic acid-Schiff positive glycogen molecules with irregular branching, termed polyglucosans [52]. These polyglucosans form dense aggregates in the cytoplasm of neurons in the brain, as well as most other tissues including muscle, liver, and skin [52]. In the brain, LBs tend to accumulate in neuronal perikarya and in dendrites, including at synapses, but not in axons [52,55]. The vast quantity of LBs and their near complete occupation of the postsynaptic space suggest they almost certainly contribute to the neurological manifestations of LD [56]. In support of a causative role for LBs in the disease, polyglucosan inclusions predate behavioral abnormalities such as epilepsy in a mouse model of LD [57]. In addition, the lack of dendritic LBs in another mouse model of the disease appears to protect against epilepsy [55]. However, some debate still exists as to the role of LBs in the pathogenesis of LD since neuronal degeneration does not always correlate with the presence of LBs [57].

LD usually affects people in their teenage years and is ultimately fatal. Although patients exhibit seemingly normal development during their first decade of life, between the ages of 12 and 17 most begin to experience seizures which progressively worsen [52]. The first sign of trouble is usually a generalized tonic-clonic seizure [52]. However, during the course of the disease LD patients also experience myoclonic seizures and occipital seizures with transient blindness, visual hallucinations or photoconvulsion, and atypical absence, atonic, and complex partial seizures [52]. Valproic acid aids in the control of seizures but with the passage of time, they begin to exhibit increasing occurrence and increasing intractability [52].

Usually following seizure onset, degeneration of the central nervous system soon becomes evident in the form of dysarthria, ataxia, emotional disturbance, confusion, and dementia [52]. Eventually, complications from neurodegeneration and status epilepticus lead to death within 2-10 years of disease onset, usually by the age of 20 [52,58]. Currently, no cure exists for LD and treatment is palliative [53]. Diagnosis is usually performed by a skin biopsy to look for the presence of LBs in patients with epilepsy [58]. LD is a rare disorder with an unknown prevalence [58]. Most cases originate in the Mediterranean countries of Europe, the Middle East, India, Pakistan, and Northern Africa [57,59]. Cultures that practice consanguineous marriages exhibit a higher frequency of LD [59].

The EEG profile of LD patients is highly abnormal and progressively worsens. At disease onset, generalized spike-wave discharges with occipital predominance occur with a frequency of 3 Hz [52,60,61]. Other types of epileptiform discharges recorded include generalized single-spike, polyspike, sharp-wave, spike-wave, and slow-wave bursting in various combinations [60]. Aside from epileptic waveforms, EEG background activity and sleep features appear relatively normal in the beginning [60]. As the disease progresses, epileptic activity increases. Spike-wave seizures occur at faster frequencies of 6-12 Hz [61]. Patients begin to exhibit sensitivity to light-induced spike-wave discharges and occipital spikes [59,60]. The EEG background slows from the normal alpha range (8-13 Hz) to the theta/delta range (≤8 Hz) [59,60]. Eventually, normal sleep features, such as sleep spindles, are lost [60]. In the latter stages of the disease, multifocal epileptiform discharges become prevalent, especially in the occipital region [53]. Notably, the background slowing and epileptiform discharges that are typical of LD can occur before the onset of symptoms [59]. Therefore, as is the case with AS, the EEGs of LD patients show abnormalities characteristic of the disease that can aid diagnosis.

LD is an autosomal recessive disorder caused by disruption of the *EPM2B* gene or the *EPM2A* gene [20,62]. *EPM2B*, also known as *NHLRC1*, encodes a single protein E3 ubiquitin ligase called malin that contains a RING domain and an NHL domain [20]. The NHL domain is composed of six NHL repeats that form a six-bladed β-propeller structure that mediates protein-protein interactions [63]. It is likely that the malin NHL domain acts as a substrate-interacting motif specifying targets for the E3 ubiquitin ligase activity of the malin RING domain [64]. LD patients exhibit a wide spectrum of epilepsy mutations affecting either the RING or NHL domains, including insertions and deletions leading to frameshifts, missense mutations, and nonsense mutations [20,65]. The only animal model of epilepsy caused by *EPM2B* disruption exists in dogs where expansion of a 12 nucleotide repeat in the gene's coding region leads to canine LD and epilepsy [21].

The other gene implicated in LD is *EPM2A*, which encodes a protein called laforin [62]. Laforin is a dual-specificity phosphatase with a carbohydrate-binding domain that participates in glycogen binding [66,67]. Laforin is proposed to play a role in the regulation of glycogen metabolism by promoting proper glycogen production and/or inhibiting aberrant glycogen accumulation in the form of LBs [55,64]. A role for laforin in the latter process is supported by its colocalization with LBs and by its binding preference for LBs over glycogen both *in vitro* and *in vivo* [55]. About 48% of LD cases result from *EPM2A* mutation, while 40% stem from disruption of the E3 ligase *EPM2B* [68]. The remaining 12% of cases appear linked to a distinct third locus that has not yet been identified at the molecular level [68].

The current model for how malin loss-of-function leads to LD and epilepsy is complex and involves its interaction with laforin and another as yet unidentified protein. *In vitro* and *in vivo* experiments have verified that the malin RING domain confers E3 ubiquitin ligase activity to the protein [56,64]. Furthermore, malin is known to polyubiquitinate laforin thereby promoting its degradation by the proteasome [56,64]. Mutations found in LD patients that affect either the RING or NHL domains of malin abolish this malin-dependent degradation of laforin *in vitro* [64]. These observations correlate with clinical data showing that polyclonal α-laforin antibodies only detect laforin in the tissues of LD patients with defective malin and not in the tissues of non-LD patients with intact malin [55]. However, these biochemical and clinical findings appear to contradict our genetic knowledge that either malin or laforin loss of function causes epilepsy.

To resolve this apparent conflict, the existence of another protein in the pathway, protein X, has been proposed [64]. In this hypothesis, ubiquitination by malin and dephosphorylation by laforin work together to extinguish the activity of protein X and allow proper glycogen metabolism; however, when protein X is dysregulated by malin or laforin deficiency, LBs accumulate leading to epilepsy [64]. While the specifics of the malin-laforin interaction still need to be determined, at least four possibilities can be envisioned for a pathway involving malin, laforin, and protein X. First, laforin could activate malin by dephosphorylation and activated malin could then mediate proteasomal degradation of protein X (Figure 2A). Second, malin could inactivate protein X by polyubiquitination leading to proteasomal degradation and laforin could inactivate protein X by dephosphorylation (Figure 2B). Third, laforin could dephosphorylate protein X, thereby triggering its proteasomal degradation mediated by malin (Figure 2C). Fourth, malin and laforin could act together in a multiprotein complex to dephosphorylate and ubiquitinate protein X leading to its degradation (Figure 2D). Another question that remains to be answered is whether malin-dependent degradation of laforin is secondary, only occurring after protein X has been destroyed. It is possible that destruction of laforin by malin is required for degradation of protein X to proceed [64]. Although the identity of predicted protein X is unknown, one attractive candidate is glycogen synthase (GS) [56]. GS elongates glycogen strands by catalyzing the addition of glucose units, thereby promoting LB formation [55]. Malin binds GS *in vitro* but whether it mediates the degradation of GS is still not clear [56]. Laforin also appears to interact with GS via glycogen synthase kinase 3 (GSK3), the primary inhibitor of GS [56]. Laforin dephosphorylates GSK3, which activates it to inhibit GS [56]. In addition, laforin may also physically interfere with GS activity by competing with it for a binding site on R5, a molecular scaffold protein that forms a complex with GS and glycogen particles to further elongate the glycogen strands [55]. Therefore, if GS is indeed protein X, a double-pronged negative feedback pathway may exist whereby polyglucosans are detected by laforin which inhibits GS via GSK3 and via malin-mediated degradation, as depicted in Figure 3 [56]. Although much is already known about LD, further investigation is required to fully understand the relationship between malin, laforin, and epilepsy.

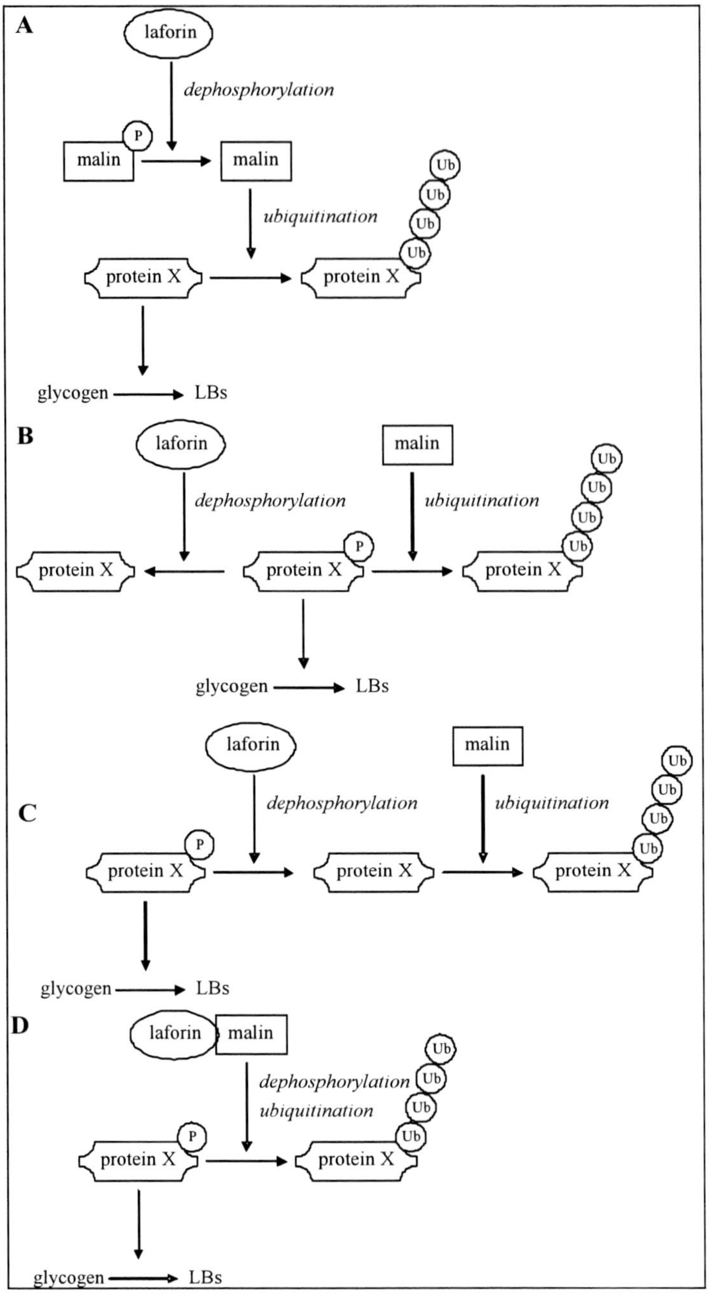

Figure 2. Possible models for interactions between malin, laforin, and the predicted protein X. In all of the models, it is assumed that protein X is involved in proper glycogen metabolism and that overactivity of protein X leads to LB accumulation. (A) Laforin activates malin by dephosphorylation and then activated malin mediates degradation of protein X. (B) Malin inactivates protein X by mediating its proteasomal degradation, while laforin inactivates protein X by dephosphorylation. (C) Laforin dephosphorylates protein X, which triggers its malin-dependent proteasomal degradation. (D) Laforin and malin sequentially dephosphorylate and ubiquitinate protein X as part of a multiprotein complex that may include other unknown proteins.

THE *MEI-P26* GENE AND SEIZURE SUSCEPTIBILITY

The *mei-P26* gene in *Drosophila melanogaster* represents a unique locus with respect to the UPS and epilepsy. The *mei-P26* gene, which putatively encodes an E3 ubiquitin ligase, is the first example of a UPS gene that decreases seizure susceptibility. Furthermore, it is the only *Drosophila* model of a UPS gene mutation connected to seizure disorders. The ability of *mei-P26* to regulate seizure susceptibility was discovered in a screen for modifier mutations that could suppress seizures in bang-sensitive (BS) paralytic mutants [22]. BS paralytics are a well-characterized class of *Drosophila* behavioral mutant used to model epilepsy in the fly [69-71]. They derive their name from their unique behavioral response to mechanical stimulation, such as a tap of the culture vial, which elicits a hyperactive seizure episode followed by temporary paralysis [69]. BS paralytics also exhibit extreme sensitivity to seizures induced by electroconvulsive shock (ECS) displaying seizure thresholds that are 5-10 times lower than wild-type [70,72,73]. Multiple strains of BS mutants have been identified including among others *bangsenseless* (*bss*; gene unknown), *easily shocked* (*eas;* ethanolamine kinase gene), and *slamdance* (*sda*; aminopeptidase gene) [74-76].

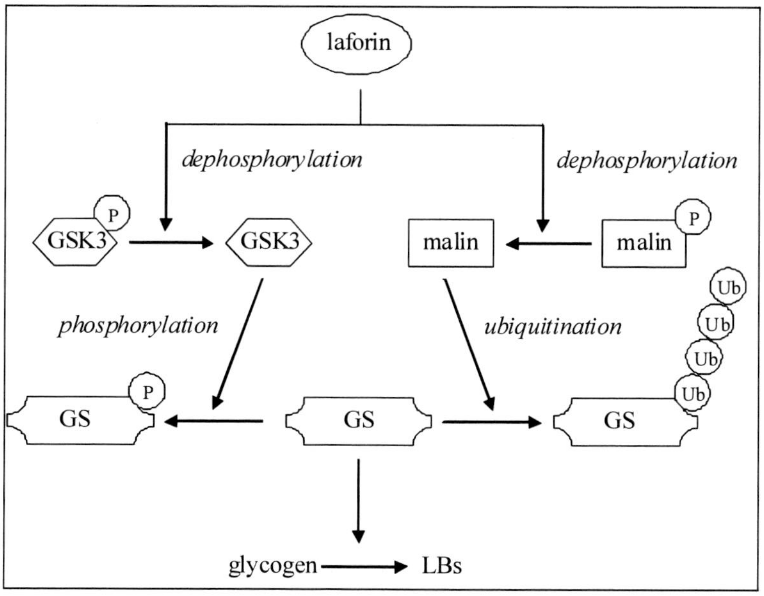

Figure 3. Possible model for a double-pronged negative feedback pathway involving malin, laforin, and GS. If GS is protein X, then a double-pronged negative feedback pathway could exist in which laforin mediates inactivation of GS via GSK3 and via malin. Inactivation of GS would protect against LBs. This model is a combination of A and B in *Figure 2* and it represents only one possibility for how malin and laforin could interact with GS. Other scenarios could also be imagined.

Loss of function of *mei-P26* reduces seizure susceptibility behaviorally and electrophysiologically, essentially curing epilepsy in some *Drosophila* models of the disease. Behaviorally, disruption of *mei-P26* almost completely eliminates seizures and paralysis in *eas* and *sda* 'epileptic' flies following mechanical stimulation [22]. In addition, *mei-P26* mutation shortens the recovery time from BS paralysis by as much as 19% in the few flies that do still seize [22]. This reduction in recovery time is reminiscent of the effects in BS flies of human anticonvulsant drugs, which can attenuate paralysis times by up to 50% [77]. Electrophysiologically, mutation of *mei-P26* drastically increases the seizure threshold of BS flies. The seizure threshold here is defined as the minimum voltage required to induce seizure activity in the indirect flight muscles by administration of a high-frequency ECS to the brain of the fly. Disruption of *mei-P26* raises the seizure threshold of *eas* flies by almost three times and restores the *sda* seizure threshold to the wild-type range [22]. When analyzed outside of a BS genetic background, *mei-P26* mutants exhibit extremely high seizure thresholds that are at least three times higher than wild-type values and even exceed those measured for Na^+ channel and gap junction mutants [22]. When taken together, the behavioral and electrophysiological seizure suppression shown by *mei-P26* mutants strongly suggests that the *mei-P26* gene plays a fundamental and critical role in the regulation of seizure susceptibility.

In addition to exerting effects on seizure susceptibility, the *mei-P26* gene also plays a role in meiosis and nervous system development. Mutants of *mei-P26* were originally discovered in a *P* element transposon mutagenesis screen for mutations that affect segregation of chromosomes during meiosis [78]. Mutation of the gene was found to cause numerous germline defects, including reduced meiotic recombination, nondisjunction, female sterility, and tumorous ovaries [79]. In another screen for genes required for *Drosophila* nervous system development, RNA interference of *mei-P26* was found to cause breaks and disorganization of the ventral nerve cord in embryos, as well as disorganization of the peripheral nervous system [80]. Thus, the *mei-P26* gene appears to be a crucial factor for several fundamental biological processes.

The *mei-P26* gene is predicted to encode a RING-type E3 ubiquitin ligase with multiple conserved protein domains. The N-terminal portion of MEI-P26 contains the RING as part of an RBCC domain, or tripartite motif (TRIM) [79]. RBCC domains are composed of a *R*ING finger followed by two *B*-box motifs (type 1 and type 2) and a *c*oiled-*c*oil region [81]. In general, RBCC domains are thought to homo-oligomerize through their coiled-coil motifs resulting in formation of multiprotein complexes, which specify different subcellular compartments [81]. The C-terminal end of MEI-P26 contains an NHL protein-protein interaction domain consisting of six NHL repeats [79]. The NHL domain appears to be particularly important for proper MEI-P26 function since mutation of a critical, conserved residue in the NHL domain contributes to the observed seizure-related and meiotic phenotypes [22]. The presence of a RING finger and an NHL domain in MEI-P26 is reminiscent of the composition of malin, which also contains these domains. Although E3 ligase activity has not yet been verified for MEI-P26, ubiquitin ligase activity has been determined for at least nine other RBCC proteins including TRIM5, Efp, and Mid1 [81].

The closest mammalian homologs of MEI-P26 are Tripartite motif protein 3 (TRIM3) in *Homo sapiens*, *Mus musculus*, and *Rattus norvegicus* and Tripartite motif protein 2 (TRIM2) in *Homo sapiens* and *Mus musculus* [22]. Each of these proteins possesses a RBCC-NHL protein domain configuration like MEI-P26. However, they only contain a single B-box motif whereas MEI-P26 has two. Little is known about the function of TRIM3 and TRIM2 proteins in mammals but they do appear to be important for nervous system function and seizure susceptibility. Mouse TRIM2, which is also known as NARF (*N*eural *a*ctivity-related *R*ING *f*inger protein), is expressed in the hippocampus and upregulated following seizure-related neural activity [82]. Rat TRIM3, which is also known as BERP (*B*rain *e*xpressed *R*ING *f*inger protein), shows expression in neurites and growth cones and disruption of its NHL domain causes neurites to be unresponsive to nerve growth factor [83]. Whether TRIM3 and TRIM2 proteins behave as E3 ubiquitin ligases is still unknown.

The mechanism by which *mei-P26* loss-of-function reduces seizure susceptibility has not been determined and is still under investigation. It is tempting to speculate that MEI-P26 acts as an E3 ligase that mediates degradation of substrates related to nervous system excitability and development. Although *mei-P26* disruption reduces neuronal excitability as measured by seizure threshold, the identity of the individual neurons or populations of neurons that are affected is unclear. However, the neurons of the giant fiber neural circuit, which are monitored for seizure activity during electrophysiology experiments, can probably be ruled out as they exhibit normal firing thresholds and performance characteristics [22]. Since the *mei-P26* gene acts as a tumor suppressor in ovarian tissue, it is feasible that it could have a similar role in nervous tissue. Therefore, one strong possibility is that MEI-P26 acts on substrates to negatively regulate synaptic development and growth. In that case, loss of function of *mei-P26* could lead to overgrowth of synapses, disrupting the balance of excitatory and inhibitory networks in favor of inhibition. Since our knowledge about MEI-P26 is still somewhat limited, research needs to be done on the biochemical function of MEI-P26 to verify that it indeed acts as an E3 ligase as sequence analysis predicts. Assuming that it does, the next step is to identify the substrates of MEI-P26. Identification of these substrates may yield a wealth of information related to seizure susceptibility since these interacting proteins are potential seizure suppressors themselves and possible targets for therapeutic intervention in humans.

A good example of the type of substrate that could interact with an E3 ligase like MEI-P26 is found in the Homer-1a protein. This protein, also called Vesl-1S, was first isolated as an immediate-early gene product that is upregulated in rat hippocampus following pharmacologically induced seizure [84,85]. Overexpression of Homer-1a in mice reduces susceptibility to seizures following corneal stimulation and slows the progression of kindling-induced seizures [86]. Experiments in cell lines and cultured neurons have shown that application of proteasome inhibitors selectively promotes the expression and ubiquitination of Homer-1a proteins, while expression of other related Homer proteins is unaffected [87]. Thus, Homer-1a appears to have a role as a seizure suppressor that is specifically regulated by the UPS.

Homer-1a belongs to a family of postsynaptic scaffolding proteins that is involved in the localization and function of group I metabotropic glutamate receptors (mGluRs) [88]. Homer-1a is one of at least 17 related Homer proteins that are encoded by the genes *homer-1*, *homer-2*, and *homer-3* in humans [89]. Homer proteins interact with numerous binding partners including group I mGluRs, inositol trisphosphate (IP_3) receptors, ryanodine receptors, C-type transient receptor potential (TRPC) channels, and other postsynaptic scaffolding molecules such as Shank [90]. Because most Homer proteins are capable of multimerizing, they can form physical links between mGluRs or TRPC channels and the IP_3 or ryanodine receptors of intracellular Ca^{2+} stores [90]. In addition, the interaction between Homer and Shank proteins allows the coupling of ionotrophic *N*-methyl-D-aspartate (NMDA) glutamate receptors to mGluRs through an interaction involving another postsynaptic scaffolding protein called PSD-95 [91]. Interestingly, PSD-95 has also been shown to affect seizure susceptibility in addition to being regulated by the UPS [92,93]. Although most Homer proteins can bind one another to form multiprotein complexes, Homer-1a lacks the motif necessary for those interactions [94]. Instead, it acts as an endogenous dominant negative, competitively interfering with the binding of multimerizing Homers to form these complexes [94]. The dominant negative and anti-epileptogenic properties of Homer-1a suggest that compounds that mimic the natural action of Homer-1a by disrupting Homer complex formation may have potential as anticonvulsant agents [86].

Conclusion

A role for the UPS in regulating seizure susceptibility has become increasingly clear with the discovery of three different E3 ubiquitin ligases related to epilepsy. Defects in two of the E3 genes, *UBE3A* and *EPM2B*, cause epilepsy in humans and in animal models when disrupted. Abnormal maternal *UBE3A* expression is responsible for AS, a neurological disease marked by epilepsy, mental retardation, and ataxia. Disruption of *EPM2B* causes LD, which is characterized by progressive myoclonus epilepsy, neurodegeneration, and polyglucosan inclusion bodies. The other E3 gene related to epilepsy, *mei-P26*, has the opposite effect of *UBE3A* and *EPM2B* mutations, curing epilepsy in *Drosophila* models when its function is lost. Although the exact mechanism by which these genes alter seizure susceptibility is unknown, the answer almost certainly lies in the identity of their target substrates.

A relationship between the UPS and epilepsy is perhaps not surprising in light of the numerous roles the UPS plays in the nervous system [95]. The UPS is important for the wiring and neuronal connectivity of the nervous system. This function first became apparent with the identification of the *bendless* mutant in *Drosophila*. In *bendless* flies, disruption of an E2 ubiquitin conjugase gene causes the giant fiber neuron to fail to make the appropriate synaptic connections [96]. The UPS also helps regulate synaptic transmission. Expression of synaptic proteins, such as synaptophysin, are affected by proteasomal degradation, as is the abundance of ion channels proteins, such as $GABA_A$ receptors, NMDA-type glutamate receptors, and nicotinic acetylcholine receptors (nAChRs) [97-100]. Membrane proteins, such as ion channels, are subject to degradation by the UPS via a molecular pathway involving endoplasmic reticulum-associated degradation (ERAD) [101]. ERAD mediates the degradation of misfolded and unassembled proteins by dislocating them from the ER to the cytoplasm where they are polyubiquitinated, targeting them to the proteasome (see Chapter 13 and [102]). Finally, the UPS is important for synaptic plasticity (see Chapter 18). In *Aplysia*, proteasomal degradation of regulatory subunits of cyclic-AMP-dependent protein kinase (PKA) is required for long-term facilitation (LTF) [103]. As mentioned earlier, *UBE3A* defects are associated with impaired LTP and learning in mice [19]. Thus, regulation of seizure susceptibility can be added to the already numerous roles that the UPS performs in the nervous system.

In the future, it will be interesting to see if a relationship exists between epilepsy and abnormal ion channel expression caused by UPS defects. Ion channels are the predominant type of proteins implicated in human idiopathic epilepsy. Of the approximately 13 human genes associated with idiopathic epilepsy, 12 encode ion channels [7]. In contrast, symptomatic epilepsies, such as AS and LD, result from defects in a wider variety of genes, which often do not have obvious connections to neuronal excitability. However, it would not be surprising to find secondary effects on ion channel expression in some symptomatic epilepsies, especially in AS and LD where the potential exists for myriad substrate interactions. Transcriptional profiling of a mouse knockout of the *EPM2A* gene that encodes laforin provides some indication that secondary ion channel defects may contribute to epilepsy in LD. In *EPM2A* knockout mice, five different genes related to ion transport show significantly altered expression, including the potassium channel β2 subunit gene, *KCNAB2*, which has been associated with epilepsy [104,105]. Whether ion channel expression is altered in *EPM2B* deficient animals remains to be seen.

In addition to regulation by the UPS via ERAD, ion channel expression can also be regulated by an alternative mode of ubiquitin-mediated degradation that exists via the endocytic pathway [12]. In this pathway, instead of being polyubiquitinated, membrane proteins are monoubiquitinated, which signals them to be internalized by endocytosis. Following endocytosis, proteins are either recycled back to the membrane or targeted to the lysosome for degradation. This monoubiquitin-dependent internalization has been shown to regulate expression of glutamate and glycine receptors in the nervous system [106,107]. Thus, E3 ligases may participate in regulating ion channel expression by polyubiquitination leading to proteasomal degradation or by monoubiquitination leading to protein internalization and/or lysosomal degradation.

The *UBE3A*, *EPM2B*, and *mei-P26* genes likely represent only the tip of the iceberg with respect to the number of UPS genes related to epilepsy. It seems safe to assume that many more epilepsy genes remain to be identified when one considers that genetics is presumed to play a role in 60% of cases and that the approximately 20 known human epilepsy genes represent only a small fraction of patients [4,108]. In light of the hundreds of known E3 ligase genes alone, most of which remain uncharacterized, it seems plausible that many of the outstanding epilepsy genes may encode components of the UPS. While the most likely candidates are nervous system-specific E3 ligases, it will be interesting to see if any other types of UPS genes are found to play a role in seizure susceptibility. Possibilities include E2 enzymes or de-ubiquitinating enzymes, which oppose the action of E3 ligases by removing ubiquitin molecules. Identification of the target substrates for the E3 ligases will also be crucial in the future. Whether any of the E3s associated with epilepsy target ion channel proteins will be interesting to see. In the coming years, epilepsy research related to the UPS should yield some exciting discoveries that will provide insight into the fundamental processes that regulate excitability in the nervous system.

ACKNOWLEDGEMENTS

This work was supported by an NIH postdoctoral training grant, NS43124.

References

[1] Shneker BF, Fountain NB: Epilepsy. *Dis Mon* 2003, 49:426-478.

[2] McNamara JO: Cellular and molecular basis of epilepsy. *J Neurosci* 1994, 14:3413-3425.

[3] Hauser WA, Annegers JF, Rocca WA: Descriptive epidemiology of epilepsy: contributions of population-based studies from Rochester, Minnesota. *Mayo Clin Proc* 1996, 71:576-586.

[4] Bate L, Gardiner M: Molecular genetics of human epilepsies. *Expert Rev Mol Med* 1999, 1:1-22.

[5] Ottman R, Annegers JF, Risch N, Hauser WA, Susser M: Relations of genetic and environmental factors in the etiology of epilepsy. *Ann Neurol* 1996, 39:442-449

[6] Berkovic SF, Howell RA, Hay DA, Hopper JL: Epilepsies in twins: genetics of the major epilepsy syndromes. *Ann Neurol* 1998, 43:435-445.

[7] Kullman DM: Genetics of epilepsy. *J Neurol Neurosurg Psychiatry* 2002, 73(Suppl II):ii32-ii35.

[8] Noebels JL: The biology of epilepsy genes. *Annu Rev Neurosci* 2003, 26:599-625.

[9] Robinson R, Gardiner M: Genetics of childhood epilepsy. *Arch Dis Child* 2000, 82:121-125.

[10] Johnston JA, Madura K: Rings, chains, and ladders: ubiquitin goes to work in the neuron. *Prog Neurobiol* 2004, 73:227-257.

[11] Jiang Y-h, Beaudet AL: Human disorders of ubiquitination and proteasomal degradation. *Curr Opin Pediatr* 2004, 16:419-426.

[12] d'Azzo A, Bongiovanni A, Nastasi T: E3 ubiquitin ligases as regulators of membrane protein trafficking and degradation. *Traffic* 2005, 6:429-441.

[13] Huang L, Kinnuca E, Wang G, Beaudenon S, Howley PM, Huibregtse JM, Pavletich NP: Structure of an E6AP-UbcH7 complex: insights into ubiquitination by the E2-E3 enzyme cascade. *Science* 1999, 286:1321-1326.

[14] Joazeiro CAP, Weissman AM: RING finger proteins: mediators of ubiquitin ligase activity. *Cell* 2000, 102:549-552.

[15] Jackson PK, Eldridge AG, Freed E, Furstenthal L, Hsu JY, Kaiser BK, Reimann DR: The lore of the RINGs: substrate recognition and catalysis by ubiquitin ligases. *Trends Cell Biol* 2000, 10:429-439.

[16] Fang S, Weissman AM: A field guide to ubiquitylation. *Cell Mol Life Sci* 2004, 61:1546-1561.

[17] Kishino T, Lalande M, Wagstaff J: *UBE3A*/E6-AP mutations cause Angelman syndrome. *Nat Genet* 1997, 15:70-73.

[18] Matsuura T, Sutcliff JS, Fang P, Galjaard R-J Jiang Y-h, Benton CS, Rommens JM, Beaudet AL: *De novo* truncating mutations in E6-AP ubiquitin-protein ligase gene (*UBE3A*) in Angelman syndrome. *Nat Genet* 1997, 15:74-77.

[19] Jiang, Y-h, Armstrong D, Albrecht U, Atkins CM, Noebels JL, Eichele G, Sweatt JD, Beaudet AL: Mutation of the Angelman ubiquitin ligase in mice causes increased cytoplasmic p53 and deficits of contextual learning and long-term potentiation. *Neuron* 1998, 21:799-811.

[20] Chan EM, Young EJ, Ianzano L, Munteanu I, Zhao X, Christopoulos CC, Avanzini G, Elia M, Ackerley CA, Jovic NJ, Bohlega S, Andermann E, Rouleau GA, Delgado-Escueta AV, Minassian BA, Scherer SW: Mutations in *NHLRC1* cause progressive myoclonus epilepsy. *Nat Genet* 2003, 35:125-127.

[21] Lohi H, Young EJ, Fitzmaurice SN, Rusbridge C, Chan EM, Vervoot M, Turnbull J, Zhao X-C, Ianzano L, Paterson AD, Sutter NB, Ostrander EA, Andre C, Shelton GD, Ackerley CA, Scherer SW, Minassian BA: Expanded repeat in canine epilepsy. *Science* 2005, 307:81.

[22] Glasscock E, Singhania A, Tanouye MA: The mei-P26 gene encodes a RING finger B-box coiled-coil-NHL protein that regulates seizure susceptibility in Drosophila. *Genetics* 2005, 170:1677-1689.

[23] Angelman H: "Puppet" children. A report on three cases. *Dev Med Child Neurol* 1965, 7:681-688.

[24] Williams CA, Angleman H, Clayton-Smith J, Driscoll DJ, Hendrickson JE, Knoll JHM, Magenis RE, Schinzel A, Wagstaff J, Whidden EM, Zori RT: Angelman Syndrome consensus for diagnostic criteria. *Am J Med Genet* 1995, 56:237-238.

[25] Petersen MB, Brondum-Nielsen K, Hansen LK, Wulff K: Clinical, cytogenetic, and molecular diagnosis of Angelman syndrome: estimated prevalence rate in a Danish county. *Am J Med Genet* 1995, 60:261-262.

[26] Kyllerman M: On the prevalence of Angelman syndrome. *Am J Med Genet* 1995, 59: 405.

[27] Laan LAEM, v Haeringen A, Brouwer OF: Angelman syndrome: a review of clinical and genetic aspects. *Clin Neurol Neurosurg* 1999, 101:161-170.

[28] Guerrini R, Carrozzo R, Rinaldi R, Bonanni P: Angelman syndrome: etiology, clinical features, diagnosis, and management of symptoms. *Pediatr Drugs* 2003, 5:647-661.

[29] Laan LAEM, Renier WO, Arts WFM, Buntinx, IM, vd Burgt IJAM, Stroink H, Beuten J, Zwinderman, KH, van Dijk JG, Brouwer OF: Evolution of epilepsy and EEG findings in Angelman syndrome. *Epilepsia* 1997, 38:195-199.

[30] Galvan-Manso M, Campistol J, Conill J, Sanmarti FX: Analysis of the characteristics of epilepsy in 37 patients with the molecular diagnosis of Angelman syndrome. *Epileptic Disord* 2005, 7:19-25.

[31] Buoni S, Grosso S, Pucci L, Fois A: Diagnosis of Angelman syndrome: clinical and EEG criteria. *Brain & Dev* 1999, 21:296-302.

[32] Dan B, Boyd SG: Angelman syndrome reviewed from a neurophysiological perspective. The UBE3A-GABRB3 hypothesis. *Neuropediatrics* 2003, 34:169-176.

[33] Boyd SG, Harden A, Patton MA: The EEG in the early diagnosis of the Angelman (happy puppet) syndrome. *Eur J Pediatr* 1988, 147:508–513.

[34] Huibregtse JM, Scheffner M, Howley PM: A cellular protein mediates association of p53 with the E6 oncoprotein of human papillomavirus types 16 or 18. *EMBO J* 1991, 10:4129-4135.

[35] Scheffner M, Nuber U, Huibretse JM: Protein ubiquitination involving a E1-E2-E3 enzyme ubiquitin thioester cascade. *Nature* 1995, 373:81-83.

[36] Malcolm S, Clayton-Smith J, Nichols M, Robb S, Webb T, Armour JA, Jeffreys AJ, Pembrey ME: Uniparental paternal disomy in Angelman's syndrome. *Lancet* 1991,337:694-697.

[37] Cassidy SB, Dykens E, Williams CA: Prader-Willi and Angelman syndromes: sister imprinted disorders. *Am J Med Genet (Semin Med Genet)* 2000, 97:136-146.

[38] Bartolomei MS, Tilghman SM: Genomic imprinting in mammals. *Annu Rev Genet* 1997, 31:493-525.

[39] Vu TH, Hoffman AR: Imprinting of the Angelman syndrome gene, *UBE3A*, is restricted to brain. *Nat Genet* 1997, 17:12-13.

[40] Rougeulle C, Glatt H, Lalande M: The Angelman syndrome candidate gene, *UBE3A*/E6-AP, is imprinted in brain. *Nat Genet* 1997, 17:14-15.

[41] Chamberlain SJ, Brannan CI: The Prader-Willi syndrome imprinting center activates the paternally expressed murine Ube3a antisense transcript but represses paternal Ube3a. *Genomics* 2001, 73:316-322.

[42] Jiang Y-h, Lev-Lehman E, Bressler J, Tsai T-F, Beaudet AL: Genetics of Angelman syndrome. *Am J Hum Genet* 1999, 65:1-6.

[43] Minassian BA, DeLorey TM, Olsen RW, Philippart M, Bronstein Y, Zhang Q, Guerrini R, Van Ness P, Livet MO, Delgado-Escueta AV: Angelman syndrome: correlations between phenotypes and genotypes. *Ann Neurol* 1998, 43:485-493.

[44] Weeber EJ, Jiang Y-h, Elgersma Y, Varga AW, Carrasquillo Y, Brown SE, Christian JM, Mirnikjoo B, Silva A, Beaudet AL, Sweatt JD: Derangement of hippocampal calcium/calmodulin-dependent protein kinase II in a mouse model for Angelman mental retardation syndrome. *J Neurosci* 2003, 23:2634-2644.

[45] Schwarz SE, Rosa JL, Scheffner M: Characterization of human hect domain family members and their interaction with UbcH5 and UbcH7. *J Biol Chem* 1998, 273:12148-12154.

[46] Nawaz Z, Lonard DM, Smith CL, Lev-Lehman E, Tsai SY, Tsai M-j, O'Malley BW: The Angelman syndrome-associated protein, E6-AP, is a coactivator for the nuclear hormone receptor superfamily. *Mol Cell Biol* 1999, 19:1182-1189.

[47] Cooper EM, Hudson AW, Amos J, Wagstaff J, Howley PM: Biochemical analysis of Angelman syndrome-associated mutations in the E3 ubiquitin ligase E6-associated protein. *J Biol Chem* 2004, 279:41208-41217.

[48] Kumar S, Talis AL, Howley PM: Identification of HHR23A as a substrate for E6-associated protein-mediated ubiquitination. *J Biol Chem* 1999, 274:18785-18792.

[49] Kuhne C, Banks L: E3-ubiquitin ligase/E6-AP links multicopy maintenance protein 7 to the ubiquitination pathway by a novel motif, the L2G box. *J Biol Chem* 1998, 273:34302-34309.

[50] Nuber U, Schwarz SE, Scheffner M: The ubiquitin-protein ligase E6-associated protein (E6-AP) serves as its own substrate. *Eur J Biochem* 1998, 254:643-649.

[51] Xiang H, Hochman DW, Saya H, Fujiwara T, Schwartzkroin PA, Morrison RS: Evidence for p53-mediated modulation of neuronal viability. *J Neurosci* 1996, 16:6753-6765.

[52] Minassian BA: Lafora's disease: towards a clinical, pathologic, and molecular synthesis. *Pediatr Neurol* 2001, 25:21-29.

[53] Shahwan A, Farrell M, Delanty N: Progressive myoclonic epilepsies: a review of genetic and therapeutic aspects. *Lancet Neurol* 2005, 4:239-248.

[54] Lafora GR, Gluck B: Beitrag zur histopathologie der myoklonischen epilepsie. *Z Ges Neurol Psychiat* 1911, 6:1-14.

[55] Chan EM, Ackerley CA, Lohi H, Ianzano L, Cortez MA, Shannan P, Scherer SW, Minassian BA: Laforin preferentially binds the neurotoxic starch-like polyglucosans, which form in its absence in progressive myoclonus epilepsy. *Hum Mol Genet* 2004, 13:1117-1129.

[56] Lohi H, Ianzano L, Zhao X-C, Chan EM, Turnbull J, Scherer SW, Ackerley CA, Minassian BA: Novel glycogen synthase kinase 3 and ubiquitination pathways in progressive myoclonus epilepsy. *Hum Mol Genet* 2005, 14:2727-2736.

[57] Ganesh S, Delgad-Escueta AV, Sakamoto T, Avila MR, Machado-Salas J, Hoshii Y, Akagi T, Gomi H, Suzuki T, Amano K, Agarwala KL, Hasegawa Y, Bai D-S, Ishihara T, Hashikawa T, Itohara S, Cornford EM, Niki H, Yamakawa K: Targeted disruption of the *Epm2a* gene causes formation of Lafora inclusion bodies, neurodegeneration, ataxia, myoclonus epilepsy and impaired behavioral responses in mice. *Hum Mol Genet* 2002, 11:1251-1262.

[58] Zupanc ML, Legros B: Progressive myoclonic epilepsy. *Cerebellum* 2004, 3:156-171.

[59] Acharya JN, Satishchandra P, Asha T, Shankar SK: Lafora's disease in South India: a clinical, electrophysiologic, and pathologic study. *Epilepsia* 1993, 34:476-487.

[60] Kobayashi K, Iyoda K, Ohtsuka Y, Ohtahara S, Yamada M: Longitudinal clinicoelectrophysiologic study of a case of Lafora disease proven by skin biopsy. *Epilepsia* 1990, 31:194-201.

[61] Yen C, Beydoun A, Drury I: Longitudinal EEG studies in a kindred with Lafora disease. *Epilepsia* 1991, 32:895-899.

[62] Minassian BA, Lee JR, Herbrick JA, Huizenga J, Soder S, Mungall AJ, Dunham I, Gardner R, Fong CY, Carpenter S, Jardim L, Satishchandra P, Andermann E, Snead OC 3rd, Lopes-Cendes I, Tsui LC, Delgado-Escueta AV, Rouleau GA, Scherer SW: Mutations in a gene encoding a novel protein tyrosine phosphatase cause progressive myoclonus epilepsy. *Nat Genet* 1998, 20:171-174.

[63] Slack FJ, Ruvkun G: A novel repeat domain that is often associated with RING finger and B-box motifs. *Trends Biochem Sci* 1998, 23:474-475.

[64] Gentry MS, Worby CA, Dixon JE: Insights into Lafora disease: malin is an E3 ubiquitin ligase that ubiquitinates and promotes the degradation of laforin. *Proc Natl Acad Sci USA* 2005, 102:8501-8506.

[65] Ianzano L, Zhang J, Chan EM, Zhao X-C, Lohi H, Scherer SW, Minassian BA: Lafora progressive myoclonus epilepsy mutation database-EPM2A and NHLRC1 (EPM2B) genes. *Hum Mutat* 2005, 26:397.

[66] Ganesh S, Agarwala KL, Ueda K, Akagi T, Shoda K, Usui T, Hashikawa T, Osada H, Delgado-Escueta AV, Yamakawa K: Laforin, defective in the progressive myoclonus epilepsy of Lafora type, is a dual-specificity phosphatase associated with polyribosomes. *Hum Mol Genet* 2000, 9:2251-2261.

[67] Wang J, Stuckey JA, Wishart MJ, Dixon JE: A unique carbohydrate binding domain targets the lafora disease phosphatase to glycogen. *J Biol Chem* 2002, 277:2377-2380.

[68] Chan EM, Omer S, Ahmed M, Bridges LR, Bennett C, Scherer SW, Minassian BA: Progressive myoclonus epilepsy with polyglucosans (Lafora disease). Evidence for a third locus. *Neurology* 2004, 63:565-567.

[69] Benzer S: From the gene to behavior. *J Am Med Assoc* 1971, 218:1015-1022.

[70] Kuebler D, Tanouye MA: Modifications of seizure susceptibility in *Drosophila. J Neurophysiol* 83:998-1009.

[71] Hekmat-Scafe DS, Dang KN, Tanouye MA: Seizure suppression by gain-of-function *escargot* mutations. *Genetics* 2005, 169:1477-1493.

[72] Pavlidis P, Tanouye MA: Seizures and failures in the giant fiber pathway of *Drosophila* bang-sensitive paralytic mutants. *J Neurosci* 1995, 15:5810-5819.

[73] Kuebler D, Zhang H, Ren X, Tanouye MA: Genetic suppression of seizure-susceptibility in *Drosophila. J Neurophysiol* 2001, 86:1211-1225.

[74] Ganetzky B, Wu C-F: Indirect suppression involving behavioral mutants with altered nerve excitability in *Drosophila melanogaster*. *Genetics* 1982, 100:597-614.

[75] Pavlidis P, Ramaswami M, Tanouye MA: The Drosophila *easily shocked* gene: a mutation in a phospholipids pathway causes seizure, neuronal failure, and paralysis. *Cell* 1994, 79:23-33.

[76] Zhang H, Tan J, Reynolds E, Kuebler D, Faulhaber S, Tanouye M: The Drosophila *slamdance* gene: a mutation in an aminopeptidase can cause seizure, paralysis and neuronal failure. *Genetics* 2002, 162:1283-1299.

[77] Tan JS, Lin F, Tanouye MA: Potassium bromide, an anticonvulsant, is effective at alleviating seizures in the *Drosophila* bang-sensitive mutant *bang senseless*. *Brain Res* 2004, 1020:45-52.

[78] Sekelsky JL, McKim KS, Messina L, French RL, Hurley WD, Arbel T, Chin GM, Deneen B, Force SJ, Hari KL, Jang JK, Laurencon AC, Madden LD, Matthies HJ, Milliken DB, Page SL, Ring AD, Wayson SM, Zimmerman CC, Hawley RS: Identification of novel Drosophila meiotic genes recovered in a *P*-element screen. *Genetics* 1999, 152:529-542.

[79] Page SL, McKim KS, Deneen B, Van Hook TL, Hawley RS: Genetic studies of *mei-P26* reveal a link between the processes that control germ cell proliferation in both sexes and those that control meiotic exchange in Drosophila. *Genetics* 2000, 155:1757-1772.

[80] Ivanov A I, Rovescalli AC, Pozzi P, Yoo S, Mozer B, Li HP, Yu SH, Higashida H, Guo V, Spencer M, Nirenberg M: Genes required for *Drosophila* nervous system development identified by RNA interference. *Proc Natl Acad Sci USA* 2004, 101:16216-16221.

[81] Meroni G, Diez-Rouz G: TRIM/RBCC, a novel class of 'single protein RING finger' E3 ubiquitin ligases. *BioEssays* 2005, 27:1147-1157.

[82] Ohkawa N, Kokura K, Matsu-Ura T, Obinata T, Konishi Y, Tamura TA: Molecular cloning and characterization of neural activity-related RING finger protein (NARF): a new member of the RBCC family is a candidate for the partner of myosin V. *J Neurochem* 2001, 78:75-87.

[83] El-Husseini AE, Vincent SR: Cloning and characterization of a novel RING finger protein that interacts with class V myosins. *J Biol Chem* 1999, 274:19771-19777.

[84] Kato A, Ozawa F, Saitoh Y, Hirai K, Inokuchi K: *vesl*, a gene encoding VASP/Ena family related protein, is upregulated during seizure, long-term potentiation and synaptogenesis. *FEBS Letters* 1997, 412:193-189.

[85] Brakeman PR, Lanahan AA, O'Brien R, Roche K, Barnes CA, Huganir RL, Worley PF: Homer: a protein that selectively binds metabotropic glutamate receptors. *Nature* 1997, 386:284-288.

[86] Potschka H, Krupp E, Ebert U, Gumbel C, Leichtlein C, Lorch B, Pickert A, Kramps S, Young, K, Grune U, Keller A, Welschof M, Vogt G, Xiao B, Worley PF, Loscher W, Hiemisch H: Kindling-induced overexpression of *Homer 1A* and its functional implications for epileptogenesis. *Eur J Neurosci* 2002, 16:2157-2165.

[87] Ageta H, Kato A, Hatakeyama S, Nakayama K, Isojima Y, Sugiyama H: Regulation of the level of Vesl-1S/Homer-1a proteins by ubiquitin-proteasome proteolytic systems. *J Biol Chem* 2001, 276:15893-15897.

[88] Fagni L, Worley PF, Ango F: Homer as both a scaffold and Transduction molecule. *Sci STKE* 2002, 137:RE8

[89] Soloviev MM, Ciruela F, Chan W, McIlhinney RAJ: Molecular characterisation of two structurally distinct groups of human homers, generated by extensive alternative splicing. *J Mol Biol* 2000, 295:1184-1200.

[90] Ehrenbgruber MU, Kato A, Inokuchi K, Hennou S: Homer/Vesl proteins and their roles in CNS neurons. *Mol Neurobiol* 2004, 29:213-227.

[91] Tu JC, Xiao B, Naisbitt S, Yuan JP, Petralia RS, Brakeman P, Doan A, Aakalu VK, Lanahan AA, Sheng M, Worley PF: Coupling of mGluR/Homer and PSD-95 complexes by the Shank family of postsynaptic density proteins. *Neuron* 1999, 23:583-592.

[92] Chen WF, Chang H, Huang LT, Lai MC, Yang CH, Wan TH, Yang SN: Alterations in long-term seizure susceptibility and the complex of PSD-95 with NMDA receptor from animals previously exposed to perinatal hypoxia. *Epilepsia* 2006, 47:288-296.

[93] Colledge M, Snyder EM, Crozier RA, Soderling JA, Jin Y, Langeberg LK, Lu H, Bear MF, Scott JD: Ubiquitination regulates PSD-95 degradation and AMPA receptor surface expression. *Neuron* 2003, 40:595-607.

[94] Xiao B, Tu JC, Petralia RS, Yuan JP, Doan A, Breder CD, Ruggiero A, Lanahan AA, Wenthold RJ, Worley PF: Homer regulates the association of group 1 metabotropic glutamate receptors with multivalent complexes of Homer-related, synaptic proteins. *Neuron* 1998, 21:707-716.

[95] Hedge AN, DiAntonio A: Ubiquitin and the synapse. *Nat Rev Neurosci* 2002, 3:854-861.

[96] Oh CE, McMahon R, Benzer S, Tanouye MA: *bendless*, a *Drosophila* gene affecting neuronal connectivity, encodes a ubiquitin-conjugating enzyme homolog. *J Neurosci* 14:3166-3179.

[97] Wheeler TC, Chin L-S, Li Y, Roudabush FL, Li L: Regulation of synaptophysin degradation by mammalian homologues of *Seven in Absentia*. *J Biol Chem* 2002, 277:10273-10282.

[98] Bedford FK, Kittler JT, Muller E, Thomas P, Uren JM, Merlo D, Wisden W, Triller A, Smart TG, Moss SJ: $GABA_A$ receptor cell surface number and subunit stability are regulated by the ubiquitin-like protein Plic-1. *Nat Neurosci* 2001, 4:908-916.

[99] Kato A, Rouach N, Nicoll RA, Bredt DS: Activity-dependent NMDA receptor degradation mediated retrotranslocation and ubiquitination. *Proc Natl Acad Sci* USA 2005, 102:5600-5605.

[100] Christianson JC, Green WN: Regulation of nicotinic receptor expression by the ubiquitin-proteasome system. *EMBO J* 2004, 23:4156-4165.

[101] Abriel H, Staub O: Ubiquitylation of ion channels. *Physiology (Bethesda)* 2005, 20:398-407.

[102] Meusser B, Hirsch C, Jarosch E, Sommer T: ERAD: the long road to destruction. *Nat Cell Biol* 2005, 7:766-772.

[103] Hedge AN, Goldberg AL, Schwartz JH: Regulatory subunits of the cAMP-dependent protein kinases are degraded after conjugation to ubiquitin: a molecular mechanism underlying long-term synaptic plasticity. *Proc Natl Acad Sci USA* 1993, 90:7436-7440.

[104] Ganesh S, Tsurutani N, Amano K, Mittal S, Uchikawa C, Delgado-Escueta AV, Yamakawa K: Transcriptional profiling of a mouse model for Lafora disease reveals dysregulation of genes involved in the expression and modification of proteins. *Neurosci Lett* 2005, 387:62-67.

[105] Heilstedt HA, Burgess DL, Anderson AE, Chedrawi A, Tharp B, Lee O, Kashork CD, Starkey DE, Wu YQ, Noebels JL, Shaffer LG, Shapira SK: Loss of the potassium channel beta-subunit gene, KCNAB2, is associated with epilepsy in patients with 1p36 deletion syndrome. *Epilepsia* 2001, 42:1103-1111.

[106] Burbea M, Dreier L, Dittman JS, Grunwald ME, Kaplan JM: Ubiquitin and AP180 regulate the abundance of GLR-1 glutamate receptors at postsynaptic elements in *C. elegans*. *Neuron* 2002, 35:107-20.

[107] Buttner C, Sadtler S, Leyendecker A, Laube B, Griffon N, Betz H, Schmalzing G: Ubiquitination precedes internalization and proteolytic cleavage of plasma membrane-bound glycine receptors. *J Biol Chem* 2001, 276:42978-85.

[108] Puranam RS, McNamara JO: Seizure disorders in mutant mice: relevance to human epilepsies. *Curr Opin Neurobiol* 1999, 9:281-287.

In: The Ubiquitin Proteasome System... ISBN 978-1-60021-749-4
Eds: Mario Di Napoli and Cezary Wojcik, pp. 921-960 © 2007 Nova Science Publishers, Inc.

Chapter 39

The Ubiquitin Proteasome System in Cerebral Ischemia

Mario Di Napoli[1,2,*] ***and Francesca Papa***[1]

[1]Neurological Section, SMDN—Center for Cardiovascular Medicine and Cerebrovascular Disease Prevention, I-67039 – Sulmona (L'Aquila), Italy;
[2]Neurological Service, San Camillo de' Lellis General Hospital, I-02100 – Rieti, Italy.

Abstract

Ischemic stroke is caused by obstruction of blood flow to the brain, resulting in energy failure that initiates a complex series of metabolic events, ultimately causing neuronal death. Cell death occurs by a necrotic pathway characterized by either ischemic/homogenizing cell change or edematous cell change. Death also occurs via an apoptotic-like pathway that is characterized, minimally, by DNA laddering and a dependence on caspase activity and, optimally, by those properties, additional characteristic protein and phospholipid changes, and morphological attributes of apotosis. Death may also occur by autophagocytosis. This review is directed at understanding how the ubiquitin-proteasome system (UPS) participates in global and focal cerebral ischemia. These are the two principal rodent models for human disease. Proteasomes are large multicatalytic protease complexes that are found in the cytosol and in the nucleus of eukaryotic cells with a central role in cellular protein turnover. The UPS is the predominant nonlysosomal protein degradation pathway which insures the viability, proliferation and signaling of eukaryotic organisms. Overwhelming data exists implicating a critical role of the UPS in cerebral ischemic injury. Ischemic and hypoxic trauma and their associated oxidative, nitrosylative and energetic stress underlie neurodegeneration following stroke and evoke a discreet set of transcriptional events which have a complex and interdependent relationship with proteasomal function. Rapid elimination of denatured, misfolded and damaged proteins by the proteasome becomes a

* Correspondence concerning this article should be addressed to Dr. Mario Di Napoli, MD; Neurological Section, SMDN—Center for Cardiovascular Medicine and Cerebrovascular Disease Prevention, Via Trento, 41 67039–Sulmona (AQ) Italy. Phone +39.(0)864.52716 Fax +39.1782262209 E-mail: mariodinapoli@katamail.com.

critical determinant of cell fate. Proof of principle has been obtained from animal models of cerebral ischemia in which proteasome inhibitors reduce neuronal and astrocytic degeneration, cortical infarct volume, infarct neutrophil infiltration, and nuclear factor-κB (NF-κB) immunoreactivity. This neuroprotective efficacy has been observed when proteasome inhibitors have been used 6 hours after ischemic insult. Strategies aimed at effecting long lasting changes in proteasomal function are not recommended given the growing body of evidence implicated long term proteasomal dysfunction in chronic neurodegenerative disease. These effects are likely due to the fact that the UPS is also essential for cellular growth, metabolism and repair. These effects of proteasomal inhibition make development of short lived proteasome inhibitors or compounds which can spatially and temporally regulated the UPS desirable clinical targets. Preclinical studies in animal models indicate that the use of specific proteasome inhibitors may be valuable in treating a host of acute neurological disorders including ischemic stroke. Proteasome inhibition could be a potential treatment option for stroke.

Keywords: proteasome, ubiquitin, stroke, NFκB, HIF-1α, glutamate, NMDA, excitotoxicity, oxidative stress, neurons, neurotoxicity, inflammation.

Abbreviations

AMP, adenosine monophosphate; ARNT, aryl hydrocarbon receptor nuclear translocator; ATP, adenosine triphosphate; BAG, Bcl-2 binding athanogene; BBB, blood-brain barrier; Bcl2, B-cell lymphoma protein 2; CHIP, c-terminal HSC70 interacting protein; ChTL, chymotrypsin-like; CyDK, Cyclin-Dependent Kinases; CNS, central nervous system; CP, core particle; DUBs, deubiquitinating enzymes; E1, Ub activating enzyme; E2s, Ub conjugating enzymes; E3s, ubiquitin ligase; EGCG, (-)-epigallocatechin-3-gallate; EPO, erytropoietin; ER, endoplasmic reticulum; HECT, homologous to E6-associated protein C-terminus; HIF-1, hypoxia-inducible transcription factor 1; HIF PHDs, 3-4 HIF prolyl hydroxylases; HIP, HSC70-interacting protein; HOP, HSC70-Hsp90-organizing protein; HSC70, heat shock cognate protein 70 (a constitutively expressed homolog of Hsp70); Hsp, heat-shock protein; IAPs, inhibitors of apoptosis proteins; IDD, iron-dependent degradation; IKK, IκB kinase; IRP, iron regulatory protein; MCAo, middle cerebral artery occlusion; MAPK, Mitogen Activated Protein Kinase; NF-κB, nuclear factor-κB; NO, nitric oxide; ODD, oxygen-dependent domain; Pael-R, Pael receptor; PARP, poly-ADPribose polymerase; PD, Parkinson's disease; PDI, protein disulfide isomerase; PGJ2, Prostaglandin J2; PGPH, peptidylglutamyl peptide hydrolase; PKC, Protein Kinase C; PS64, peptide substrate 64; PSD-95, Postsynaptic Density 95; RKT, Receptor tyrosine kinases; RING, really interesting new gene; ROS, reactive oxygen species; RP, regulatory particle; Rpn, Regulatory particle non-ATPase; Rpt, Regulatory particle triple ATPase; r-tPA, tissue plasminogen activator; Siah, Seven in absentia homologs; TL, trypsin like; Ub, ubiquitin; UBPs, Ub-specific proteases; UCHL1, Ub C-terminal hydrolase L1; UPR, unfolded protein response; UPS, ubiquitin-proteasome system; VEGF, vascular endothelial growth factor; VHL, von Hippel–Lindau.

INTRODUCTION

Cerebral ischemia or stroke is characterized by a vessel obstruction of blood flow to the brain. There are essentially two types of stroke: (*i*) global or total loss of blood flow to the brain caused by events such as cardiac arrest, or (*ii*) focal (regional) arising from local interruption of blood flow to the brain due to an artery blockage. Animal models have been developed to mimic the human condition [1]. The global model involves the occlusion of bilateral common carotid and vertebral arteries, while focal ischemia is induced by either transient or permanent middle cerebral artery occlusion (MCAo). Stroke results in disruption of glucose and oxygen supply that ultimately leads to apoptotic and necrotic cell death, and development of an infarction. Focal ischemia is characterized by an ischemic core surrounded by a 'penumbra' region that has partial reduction in blood flow due to the presence of collateral arteries. If left untreated the infarct can propagate into the penumbra. The ischemic core is generally considered unsalvageable, whereas the penumbra may be rescued by timely intervention and poses a target for the development of therapeutic treatment.

Ischemic neuronal cell death is a principal neuropathological feature of stroke and constitutes a major source of morbidity and mortality, especially in the elderly. The widespread use of antihypertensive and antiplatelet drugs has reduced the age-adjusted incidence of stroke, but little progress has been made in stroke treatment [2]. The efficacy of treatments directed at revascularization, anticoagulation, inhibition of platelet function and thrombolysis is marginal or disputed. Moreover, neuroprotective strategies have predominately been aimed at inhibiting ischemia-induced over activation of glutamate receptors (*excitotoxicity*), but have not been successful [3]. The therapeutic potential for developing small molecule inhibitors for the treatment of stroke is significant given that only a small area of the affected brain tissue, the ischemic core, is irreversibly damaged at the onset of stroke. A much larger volume of the brain tissue surrounding the ischemic core, known as penumbra, has the potential to recover much of its function [4].

The pathophysiological mechanisms of brain injury following stroke include essential roles for inflammation, energetic dysfunction and protein destabilization [5]. Targeting individual inflammatory mediators has, however, proven difficult due to redundancy in function [6]. The ischemic cell death proceeds through several stages that include several changes initiated by ischemia and reperfusion that are very likely to play major roles in cell death. These include inhibition (and subsequent reactivation) of electron transport, decreased ATP, decreased pH, increased cell Ca^{2+}, release of glutamate, increased arachidonic acid, and also gene activation leading to cytokine synthesis, synthesis of enzymes involved in free radical production, and accumulation of leukocytes. Energy failure is the initial metabolic event in stroke The energy needs of the brain are supplied by metabolism of glucose and oxygen for the phosphorylation of ADP to ATP. Most of the ATP generated is utilized in the brain in maintaining intracellular homeostasis and transmembrane ion gradients of sodium, potassium, and calcium. Energy failure results in rapid loss of ATP and uncontrolled leakage of ions across the cell membrane that results in membrane depolarization and release of the neurotransmitters such as glutamate and dopamine [7,8]. Excess glutamate release and stimulation of its receptors result in activation of phospholipases [9-11], phospholipid hydrolysis, and arachidonic acid release [11]. These changes lead to the activation of damaging events of free radicals and their product peroxynitrite, the actions of the Ca^{2+}-dependent protease calpain, the activity of phospholipases, the activity of poly-ADPribose polymerase (PARP), and the activation of the apoptotic pathway producing the long-term changes in macromolecules or key metabolites. The long-term damaging effects of these macromolecular and metabolite changes on critical cell functions and structures lead to the defined end stages of cell damage. These targeted functions and structures include the plasmalemma, the mitochondria, the cytoskeleton, protein synthesis, and kinase activities. Ultimately these processes lead to apoptotic or necrotic cell death [12]. The major driving force for the necrotic cell death process appears to be the generation of free radicals and peroxynitrite. Effects of a large number of damaging changes can be explained on the basis of their ability to generate free radicals in early or late stages of damage.

According to this description, the key step in ischemic cell death is a consideration of how the changes occurring during and after ischemia, including gene activation and synthesis of new proteins, conspire to produce damaging levels of free radicals and peroxynitrite, to activate calpain and other Ca^{2+}-driven processes that are damaging, and to initiate the apoptotic process. Cerebral ischemia results in selective increased mRNA levels of genes involved in stress, inflammation, transcription and plasticity, and decreased mRNA levels of genes which control neurotransmitter function and ionic balance [13]. An underlying pattern, to which there are important exceptions, is that the protective proteins are synthesized in relatively nonvulnerable cells and are not synthesized in vulnerable cells, or cells that will die. This raises the possibility that the absence of these proteins may be important in development of cell death. In addition to proteins that promote cell death—such as the proapoptotic protein Bax and the caspases—survival-promoting proteins like Bcl2 family members, heat-shock protein (Hsp), inhibitors of apoptosis proteins (IAPs), hypoxia-inducible transcription factor 1α (HIF-1α), nuclear factor-κB (NF-κB) and a host of other transcription factors are also induced [14-17]. Vulnerable cells, in global or focal ischemia, suffer from the absence of synthesis of these protective proteins, or in some cases from their downregulation. This hypothesis is supported by the phenomenon of tolerance, mild ischemic injury increases resistance to subsequent stressors in a protein synthesis dependent manner [18]. Evidence suggests that tolerance results from synthesis of proteins that are normally not synthesized in the vulnerable cells but that is synthesized in cells that do not go on to die and survive the initial stress [18]. Why these proteins, as well as immediate early genes, are not synthesized in vulnerable cells is unknown. It could result from the general downregulation of synthesis, although this is not a compelling argument, but a critical determinant of cell fate also becomes a rapid elimination of denatured, misfolded and damaged proteins by the ubiquitin (Ub)-proteasome system (UPS) (see Chapter 2). The UPS is an ATP-dependent protein degradation system degrading short-lived proteins under normal metabolic conditions, as well as, bulk degradation of long-lived proteins, partial digestion/processing of some regulatory proteins, and antigen presentation. Among the key regulatory proteins degraded by the 26S proteasome are the transcription factor c-Fos, M-, S-, and G1-phase specific cyclins, cyclin dependent kinase inhibitors, p53, and a host of oncoproteins [19-21]. The degradation of cellular proteins is a highly complex, temporally controlled and tightly regulated process essential for mitosis, energy and ion homeostasis, revascularization and repair. With the multitude of protein substrates targeted and the myriad of processes involved, it is not surprising that aberrations in the UPS are implicated in the pathogenesis of cerebral ischemia [22-24]. However, despite intensive research in this area, key questions remain unanswered. Among these are the modes of specific and timed recognition for the degradation of the many substrates, and the mechanisms that underlie aberrations in the system that lead to pathogenesis of ischemic cell death. The advent of powerful new genomic and proteomic technologies has substantially enhanced our understanding of the evolution of the ischemic injury. Understanding the identity and interactions of proteins as a part of complex networks is the central problem in the post-genome era. In this review, we will focus on the recent advances in our understanding of UPS in cerebral ischemia, as well as, the novel approaches for stroke intervention which these studies have revealed (see Chapter 41).

THE UPS IN CEREBRAL ISCHEMIA

Ub-Protein and Ubiquitinated Inclusions in Cerebral Ischemia

While the chronic loss of proteasomal function has been implicated in the pathogenesis of protein aggregations observed in a host of neurological diseases [25,26], the role of aberrant proteasomal degradation and ubiquination in acute conditions including stroke is just beginning to be addressed [23,24]. Aberrant proteasomal function has been observed in experimental models of stroke [23,24,27,28] and loss of 26S proteasomal activity reportedly occurs rapidly (i.e. within 10 minutes) following global ischemia [29]. After global ischemia the expression of Rpn2 subunit of the 26S proteasome was elevated at 12 hours in the dentate gyrus [30]. After 24 hours, Rpn2 increased its expression significantly in both the CA1 and dentate gyrus compared with control animals. This alteration in proteasome expression was also associated with the change of transcriptional factor SEF-2 [30].

Poly-Ub-conjugated proteins are accumulated in hippocampal cells following transient global ischemia [31,32], following focal ischemia [33,34] and *in vitro* models of neuronal hypoxia [35]. However, biochemical characterization of the Ub-immunoreactive material is still lacking. Ub immunoreactivity in inclusions is based solely on the demonstrated 'conjugate' specificity of the antibodies utilized in the studies. In cerebral ischemia, cell inclusions containing Ub-proteins are found predominantly in the neuronal soma, dendrites, and axons, and they were associated with intracellular membranous structures during the postischemic phase [36,37]. Ub-proteins are localized in lysosomal vesicles and in late endosome-like organelles of the neurons in the ischemic area [36,37]. Some of these proteins might include cell surface receptors that require ubiquitination for internalization [38-41]. The endosomal/lysosomal deposits might also derive from an unsuccessful attempt by the cell to eliminate the Ub-proteins by autophagy. After ischemia, cytosolic proteins can be rapidly sequestered into autophagic vesicles assembled from pre-existing endoplasmic reticulum (ER) or Golgi membranes.

High-resolution confocal microscopy showed also that clumped protein aggregates surrounding nuclei and along dendrites were formed after brain ischemia [37]. Irreversible aggregation of translational components, i.e. the ribosomes and their associated nascent polypeptides, initiation factors, translational chaperones and degradation enzymes, has been reported after brain ischemia [42]. Translational complex components consisting of small ribosomal subunit protein 6, large subunit protein 28, eukaryotic initiation factor-3η, co-translational chaperone heat shock cognate protein 70 (HSC70) and co-chaperone Hsp40-Hdj1, as well as co-translational ubiquitin ligase c-terminus of Hsp70-interacting protein (CHIP) are all irreversibly clumped into large abnormal protein aggregates after ischemia [42]. Translational components were also highly ubiquitinated. Ub-immunoreactive nuclear inclusions are also detected in specific hyppocampal neurons [33]. These aggregates contain ubiquinated proteins which can also be found in the neuronal soma, dendrites, and axons. In the post-ischemic phase, ubiquinated proteins are associated with intracellular membranous structures [36,37]. Ub-proteins in neuronal lysosomal vesicles and in late endosome-like organelles in the ischemic area [36,37] may be result from an attempt to eliminate accumulating Ub-proteins by autophagy.

The abnormal Ub immunoreactivity is primarily associated with the tau-related protein and intermediate filament network [43,44]. These observations, together with the idea that the intermediate filament system is the most stress-sensitive element of the cytoskeleton [45], suggest a coupling of the stress response with the formation of inclusions at the site of cellular damage.

Short episodes of ischemia induce a transient decrease in the activity of the 26S proteasome, followed by accumulation of Ub-proteins in the gerbil cortex [28]. Similarly, expression of Ub and other heat shock genes is rapidly increased following the exposure of hippocampal neurons to hypoxia [35]. Free Ub is dramatically decreased in both gerbil and rat hippocampus within hours after ischemia and then recovers over the next 1–3 days in all cell types except CA1 cells that are destined for death [35,46,47]. This depletion may be caused by impaired conversion from conjugated to free Ub and/or failure of *de novo* Ub synthesis but unfortunately, there is no further evidence to this effect. This phenomenon may be involved in damage, possibly by allowing buildup of partially denatured proteins [35]. This leads to speculate that cerebral ischemia is associated with an inability of the neuron to degrade protein aggregates. In general, Ub-proteins do not accumulate in healthy cells and are rapidly degraded by the Ub/ATP-dependent pathway. Failure to eliminate the Ub-protein deposits might result from a malfunction of the Ub/ATP-dependent pathway or from structural changes in the protein substrates, rendering them inaccessible to proteolysis.

It has been reported that proteasome inhibition is sufficient to induce both the formation of cellular aggresomes composed of aggregation-prone proteins and neuronal death [48,49], but it remains to be determined whether protein aggregation causes neuronal degeneration or is a rescuing mechanism in the cell (see Chapter 12).

UPS Response to Hypoxia

While brief and acute hypoxia does not impair proteasome function [34], a clear inhibitory effect of hypoxia on proteasome function is evident only after prolonged hypoxic periods [50] and in the presence of a concomitant inflammatory reaction [51,52]. Repeated and intermittent episodes of hypoxia decrease also markedly proteasomal activity in aged brains of Sprague-Dawley rats [53]. Recently, a specific defense system against hypoxia has been described also in the brain, which uses the hypoxia-inducible factor-1 (HIF-1) as transcription factor [50]. One of the most rapid changes induced by hypoxia is an increase in the expression of the HIF-1. The HIF-1 complex is composed of two protein subunits: HIF-1β/ARNT (aryl hydrocarbon receptor nuclear translocator), which is constitutively expressed, and HIF-1α, which is rapidly ubiquitinated and degraded to near negligible levels in normoxic cells. HIF-1α coordinates the response to prolonged hypoxia which pertains to glycolysis, glucose transport, vasodilation, and angiogenesis [50]. The rapid turnover of HIF-1α in normal cells is mediated by hydroxylation on two proline residues within a conserved oxygen dependent degradation domain. In the presence of oxygen, these prolyl residues are enzymatically hydroxylated which promotes interaction with the von Hippel–Lindau (VHL) tumour suppressor protein, leading to ubiquitylation by the VHL associated E3 ligase, and subsequent proteasomal destruction. As a consequence of gene mutations, individuals with VHL disease are susceptible to retinal angiomas, hemangioblastomas of brain and spinal cord, and tumors in other areas including pancreas, kidney and adrenal glands [54]. These tissues are generally highly vascularized tissues and overproduce peptides and hormones such as erytropoietin (EPO) and vascular endothelial growth factor (VEGF) involved in angiogensis and hypoxia sensing. The hydroxylation of a proline residue (HIF1α P564) is mediated by 3-4 HIF prolyl hydroxylases (HIF PHDs) (Figure 1) which themselves are substantially altered by stress and subject to proteasomal degradation. Seven in absentia homologs (Siah) E3 Ub ligases are induced by hypoxia and their binding to HIF PHDs removes the impetus to degrade HIF resulting in HIF complex stabilization, nuclear translocation and transcription. The importance of this feedback loop has recently been demonstrated in Siah knockout cells where HIF-1α expression does not increase as oxygen levels drop [55]. Thus, the function and stability of HIF are intimately linked to proteasomal function. Not surprisingly, inhibitors of proteasome activity protect the degradation of the HIF-1 complex [56], and produce a dramatic accumulation of HIF-1α protein [57], resulting in accelerated formation of vascular structures *in vitro* [58]. This mechanism could contribute to the rescue of the penumbra of an ischemic lesion. However, the mechanism by which hypoxia inhibits VHL-mediated ubiquitination of HIF-1α is currently unknown. As for hypoxic signaling, Tanaka *et al.* [59] have recently shown that protein disulfide isomerase (PDI), an enzyme which also catalyzes protein folding reactions, is also upregulated in response to CNS hypoxia. Like Hsp70, PDI is intimately linked with the proteasome. It shuttles proteins for proteasomal degradation via interactions with the Ub like protein ubiquilin. Moreover, enhancing PDI function confers cytoprotection following stroke. Given the linkage between energetic status and protein aggregation, it is tempting to speculate that the loss of E3 Ub ligases, such as Siah, into proteinacious aggregates formed during ischemia and chronic neurodegenerative diseases compromises the ability of neurons to mount an appropriate stress response through HIF and other transcriptional elements [60].

Finally, another oxygen-sensing mechanism has also been shown to link intracellular responses to hypoxia with the proteasome. The iron regulatory protein (IRP) -1 and IRP-2 are post-transcriptional regulators of cellular iron metabolism that bind to an iron-responsive element in the untranslated region of mRNAs involved in iron homeostasis and can serve as a link between iron homeostasis and oxidative stress sensing [61,62]. IRP-2 is regulated by iron in a heme-dependent manner via its iron-dependent degradation (IDD) domain, a specific 73

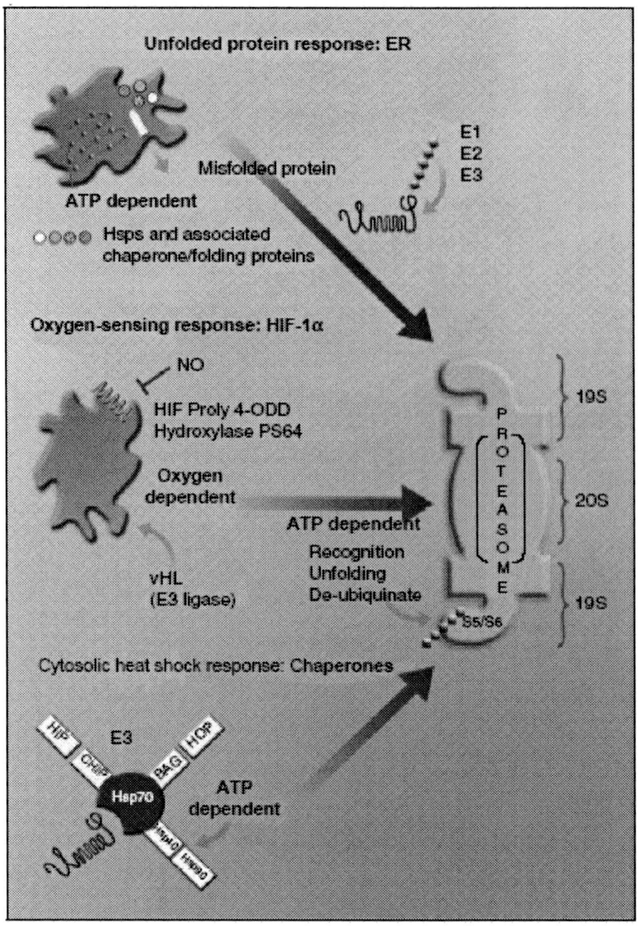

Figure 1. Cellular response to ischemic stress. Multiple energy and oxygen-dependent pathways converge on the proteasome in order to remove damaged proteins and activate transcription factors with roles in increasing oxygenation, vascularization and survival. Examples of these pathways are shown for the unfolded protein response in the endoplasmic reticulum (ER), the stabilization of hypoxin-reducible transcription factor (HIF)-1α and the chaperone triage system. Protein trafficking and assembly in the ER is responsive to multiple energetic and ionic perturbations, including calcium buffering, redox regulation, caspase activation and post-translational modification of proteins. These functions are dependent upon ATP content and redox-sensitive proteins, including the Sec ATPases (SEC). Misfolded proteins are targeted to the proteasome through ubiquitination (E1, E2 and E3 enzymes). Oxygen sensitivity is demonstrated in the HIF-1α subunit of the HIF transcription factor. Under normoxia, HIF hydroxylases modify proline residues within the oxygen-dependent domain (ODD) of HIF allowing for rapid ubiquination by the von Hippel–Lindau (vHL) E3 ligase. Anaerobic conditions block these modifications leading to enhanced HIF-1a expression and transcriptional activity. Finally, chaperones play an essential role in the stress response to ischemia. Some of the interactions of the stressinducible chaperone Hsp70 are illustrated. Competition for Hsp70 binding leads to altered chaperone activity, in which the protein moves from a primary role in refolding, to one of targeting proteins to the proteasome to be degraded. Again, the function of this protein is dependent upon ATPase activity of interacting chaperones. BAG, Bcl-2 binding athanogene; CHIP, c-terminal Hsc70 (a constitutively expressed homolog of Hsp70)-interacting protein; HIP, Hsc70-interacting protein; HOP, Hsc70-Hsp90-organizing protein; Hsp, heat shock protein; NO, nitric oxide; PS 64, peptide substrate 64. Reprinted with permission from Di Napoli M and McLaughlin BA: The proteasome ubiquitin system as a drug target in cerebrovascular disease: The therapeutic potential of proteasome inhibitors Current Opinion in Investigational Drugs 2005 6(7): 686-699, 2005 © The Thomson Corporation.

amino acid domain that can target heterologous proteins for ubiquitin mediated degradation after iron-mediated oxidative modification [63,64]. Yamanaka *et al.* [65] recently reported the identification of the Ub ligase that recognizes oxidized IRP-2. IDD is believed to bind heme and, in the presence of oxygen, produce a superoxide radical that oxidizes the IDD domain, thus generating a binding site for the IRP-2–Ub ligase HOIL-1 [65].

UPS and Hypoxic Preconditioning

Hypoxic preconditioning, or hypoxia-induced tolerance, refers to a brief period of hypoxia that protects against an otherwise subsequent lethal insult. Hypoxic preconditioning protects the brain, heart and retina, and indeed all other organs that have been tested, against several types of injury, including ischemia [66,67]. Hypoxia protects against both focal and global ischemia in the neonatal and adult brain [66,67]. Ischemic preconditioning reduce significantly protein aggregation and virtually eliminate neuronal death in CA1 neurons. Biochemical analyses reveal that ischemic preconditioning decrease accumulation of ubiquitin-conjugated proteins (Ub-proteins) and reduced free Ub depletion after brain ischemia. Furthermore, ischemic preconditioning also reduce redistribution of HSC70 and Hdj1 from cytosolic fraction to protein aggregate-containing fraction after brain ischemia [68]. Two types of ischemic tolerance have been reported for the brain: delayed and rapid ischemic tolerance. Delayed ischemic tolerance develops over 1–3 days *in vivo* or 24 hours *in vitro*, is mediated by a gene-based mechanism and requires new protein synthesis [18,69-71]. In contrast, rapid ischemic tolerance is protein synthesis-independent and occurs within 1 hour of the preconditioning ischemia [72]. Rapid ischemic tolerance induced by hypoxic preconditioning has been hypothesized to be due, at least in part, to rapid degradation of the pro-apoptotic Bcl-2 family member Bim by UPS [72]. Proteasome inhibitors prevent Bim degradation and block rapid ischemic tolerance [72]. Paradoxically, however, it was established that proteasome inhibitors prevent the necrotic and apoptotic cell death, but in the same concentration abolish the effects of preconditioning and postconditioning [73,74]. The data available show that the specific protective effect of proteasome inhibitors could be caused by autophagy activation although the mechanisms of this phenomenon are still in need of thorough investigation [73,74].

UPS Response to Ischemia

In the postischemic hippocampus, conjugated Ub accumulates and free Ub is depleted, suggesting impaired proteasome function [33,34,75]. The accumulation of conjugated Ub may reflect hypofunction of downstream proteasome activity that normally degrades ubiquitinated proteins. Moreover, direct injection of a proteasome inhibitor into the lateral ventricles of the rat induced DNA fragmentation in various CNS areas, suggesting that suppression of proteasome is able to induce neuronal apoptosis [76] (see Chapter 21). Therefore, it is reasonable to speculate that proteasome malfunction may in part underlie the molecular events of the ischemia induced neuronal death. Decreased proteasome activity at the ischemic core and the surrounding tissues allows accumulation of oxidized proteins, resulting in formation of protein aggregates, ER stress (see below), impairment of cell function and eventually cell death. Indeed, in an experimental ischemia of rat brains a 60% elevation of Ub conjugate levels in the ischemic compared with the non-ischemic animals was observed within 1 hour of recovery. The conjugate immunoreactivity remained at this level for 6 hours but eventually declined to control levels by 24 hours of recovery [33,34]. Increased formation of poly-Ub conjugates was accompanied with a significant increase in the transcription levels of poly-Ub genes [77].

UPS Response to ATP Depletion

The ubiquitination and degradation processes are heavily dependent upon ATP. Loss of ATP resulting from oxygen deprivation has profound and immediate consequences on a host of cellular functions including proteasomal function [27,78,79].

One hour transient focal cerebral ischemia induces marked depletion of the E3 ligase parkin but does not affect the levels of E2s. Parkin up-regulation has been shown to protect cells from injury induced by ER stress, which suggests that parkin depletion may increase the sensitivity of neurons to stimuli managed by the ER including the unfolded protein response (UPR) and calcium dysregulation which are induced by loss of cellular ATP levels [80]. In the ischemic core, ATP- and Ub-dependent degradation mediated by the 26S proteasome is impaired, while the ATP- and Ub-independent degradation mediated by 20S proteasome is unimpeded. This is likely due to the dissociation of 26S proteasomes, which occurs under stress, into 20S proteasomes and PA700 caps. Following ischemia in the gebril cortex, the 26S proteasome chymotrypsin-like (ChTL) activity decreases, whereas the 20S proteasome ChTL activity increases [28]. While the 26S proteasome activity recovers in many regions following reperfusion, particularly vulnerable areas including the CA1 region of the hippocampus, PA700 and 20S proteasomes do not fully reassociate, an aspect which may explain the delayed neuronal cell death in these regions [27].

UPS Response to Intracellular pH Levels

Another consequence of oxygen deprivation is the increase in cellular acidosis. The cellular site of action of low pH is not completely resolved. Effects of cellular acidosis on the UPS which appear to exist but have not yet been extensively investigated, are quite reasonable. While the molecular site of action of changes in pH has not been localized within the proteasome, it is clear that loss of proteasome activity occur as pH levels decrease. The 20S proteasome has three different proteolytic activities including trypsin-like (TL), chymotrypsin-like (ChTL) and caspase-like, all of which are less effective with decreasing pH [81]. Proteasomes may also be altered by direct transient denaturation induced by pH, indirectly by enhanced formation of free radical formation (via iron delocalization and the Fenton reaction) [82] or more specifically by altering catalytic activity of the proteasomal complex as well as by altering the action of Ub-protein-ligase complexes [83,84].

UPS Response to Intracellular Ca^{2+} Levels

Loss of Ca^{2+} homeostasis, often in the form of cytoplasmic increases, leads to cell injury in cerebral ischemia [9]. The effects of proteasome inhibitors on intracellular Ca^{2+} levels was tested in murine neocortical cultures. Although multiple destructive processes are activated by Ca^{2+}, lethal outcomes are determined largely by Ca^{2+}-induced mitochondrial permeability transition. Exposure of murine neocortical cultures to proteasome inhibitors (0.1-10 μmol clasto-lactacystin β-lactone or MG-132) for 48 h resulted in widespread neuronal death associated with a reduction in intracellular free calcium associated with intracellular calcium starvation [85].

UPS and Mitochondrial Function

Retention of the mitochondrial membrane potential during increases in $[Ca^{2+}]_i$ favors mitochondrial Ca^{2+} uptake and overload, resulting in mitochondrial permeability transition and cell death. In contrast, dissipation of mitochondrial membrane potential reduces mitochondrial Ca^{2+} uptake, retards mitochondrial permeability transition, and delays death, even in cells with large $[Ca^{2+}]_i$ increases. The rates of mitochondrial membrane potential dissipation and mitochondrial Ca^{2+} uptake may determine cellular sensitivity to Ca^{2+} toxicity under pathological conditions, including ischemic injury. Chronic proteasome inhibition is deleterious to cells as it severely impedes mitochondrial calcium buffering, results in loss of mitochondrial membrane potential ($\Delta\psi_m$), formation of dense mitochondrial deposition, cytochrome-c release into the cytosol with secondary dilation of the rough ER, formation of cytoplasmic vacuoles and activation of caspase-3 [25,86-88].

UPS and Reactive Oxygen Species (ROS)

After cerebral ischemia-reperfusion injury there is a time-dependent decrease in proteasome activity in the affected area, which is associated with oxidative stress, and not with decreased expression of proteasome subunits [25,89]. ROS are known to modify several proteasome subunits (α1, α2 and α4) and impair proteasome activity [90,91]. 20S proteasomes can degrade mildly oxidized proteins without previous ubiquitination, however they are unable to degrade extensively oxidized proteins [92,93]. Moreover, oxidative damage enhances the effects of proteasome inhibition, leading to protein aggregation and cell death [94]. Because, metal ions appear to be delocalized from the proteins in the ischemic and postischemic phase, the effect of oxidative stress induced by neurotoxic metal ions on the properties of the brain 20S proteasome has also been studied showing that metal-catalyzed oxidation strongly affects the functions of the brain 20S proteasome although the catalytic subunits seem to be differently influenced by oxidative phenomena: TL activity showed gradual activation while ChTL and peptidylglutamyl peptide hydrolase (PGPH) activities were substantially inhibited [82]. At the same time, the intracellular redox status, probably through the level of oxidized proteins, is an important element that can either activate or down-regulate the 20S proteasome ChTL activity in living cells [95] acting by a feedback mechanism because the antioxidant system is also subjected to the proteasome-dependent proteolysis [96].

UPS and Protein Synthesis

There is very little known about rates of protein synthesis and degradation after ischemia. The average turnover time of brain proteins are 3–5 days [97]. Protein synthesis does indeed seem to be a crucial parameter in the pathological process of ischemic cell death, because there is a close relationship between shutdown of translation and extent of cell death. Translation is severely suppressed after transient ischemia in all cells in which the energy state is disturbed during vascular occlusion. It recovers in all resistant areas but never recovers in cells vulnerable even to a short interruption of blood supply [98,99]. Protein synthesis is a complex process that can be subdivided into the initiation step where various initiation factors, including the initiator Met-tRNA, mRNA and ribosomal subunits, are needed to form the initiation complex, the elongation step where the new polypeptide chain is synthesized, and the termination step.

If normal protein degradation is blocked after ischemia, it would reduce the impact of synthesis inhibition. There is a marked decrease in Ub immunoreactivity in both gerbil and rat hippocampus within hours after global ischemia and then recovers over the next 1–3 days in all cell types except CA1 cells that are destined for death [33,100]. During focal ischemia, blocking proteasome activity was extremely protective to the core of the lesion [101]. One explanation for this is that generalized protein degradation makes a major contribution to ischemic core damage. Indeed, protein content within the core of a focal lesion is severely reduced. However, another possible explanation is that proteasome activity is damaging because it allows NF-κB activation [102]. Cell death might not result from a functional defect in one or more of key processes; rather, it may result from continued activation of perpetrators set in motion by the ischemia, with ultimate breakdown of the cell as a unit. Besides the UPS, during ischemia there may be an activation of Ca^{2+}-dependent calpains and lysosomal cathepsins, which degrade material delivered by autophagy. Thus, at the doses used in vivo, it is possible that proteasome inhibitors are also blocking activity of other proteolytic systems, either directly or indirectly. The current generation of cell-permeable intracellular proteasome inhibitors also inhibit calpains. Indeed two specific calpain inhibitors Val-Phe (MDL-28170) and Leu-aminobutyrate (AK-275), have been found to provide strong protection in focal ischemia.[103]

UPS and the Damage to the Cytoskeleton

The dissolution of the microtubular network contributes significantly to apoptotic or necrotic cell death [9]. Known mechanisms of microtubule dissociation include MAP2 phosphorylation or proteolysis, dissociation of the putative microtubule-stabilizing protein STOP, decreased GTP/GDP, or proteolysis of tubulin [9]. Prostaglandin J2 (PGJ2), an endogenous product of inflammation, disrupts also the cytoskeleton in neuronal cells [104,105]. Furthermore, PGJ2 perturbe microtubule polymerization in vitro and decrease the number of free sulfhydryl groups on tubulin cysteines together with 26S proteasome assembly and activity, which preced the accumulation of ubiquitinated proteins as detergent/salt-insoluble aggregates [104,105] (see Chapter 29).

Proteasome inhibitors prevent Wallerian degeneration *in vitro* and *in vivo*, stabilizing microtubular cytoskeleton in the axons [45,106] (see Chapter 26). Since increased proteolysis of different cytoskeletal elements is one of the early events in the penumbra of an ischemic lesion, it is likely, that such a mechanism also contributes to the neuroprotection in stroke. At the same time, ischemia disrupts the neuronal cytoskeleton and increases intracellular Ub levels, presumably to promote the removal of abnormal proteins by the Ub/ATP-dependent pathway [107]. As a consequence of the massive increase in intracellular calcium during ischemic episodes, tau proteins are hyperphosphorylated and are no longer able to stimulate tubulin polymerization [107]. This leads to destabilization of the microtubule system and could activate a cascade of events that compromise the proteolytic machinery.

UPS and Protein Kinases

Permanent, or long-term, inactivation of protein kinases or phosphatases could lead to initiation of apoptosis or could lead to the permanent alteration of proteins involved in cell membrane or mitochondrial function, the cytoskeleton or protein synthesis. Such effects could thus make a major contribution to ischemic cell damage [9]. The UPS has been implicated in regulating the levels of many cellular proteins of the signal transduction pathways [108]. There is a direct relationship between the phosphorylation/dephosphorylation cascade of the signal transduction pathways and the targeting of the regulatory proteins for ubiquitination. These interacting systems are seen for Protein Kinase C (PKC) system [109-112], chaperone system [113-115], Mitogen Activated Protein Kinase (MAPK) system [116,117], Cyclin-Dependent Kinases (CyDK) system [118-121], and calcineurin (calmodulin-dependent phosphatase) [122]. Proteasome inhibitors demonstrate that many proteins of the signal transduction pathways are regulated by degradation via the UPS and their use is associated with multiple perturbations in expression/activation of signaling- and survival-related proteins.

UPS and Heat Shock Protein (Hsp)

The pattern of Hsp expression after ischemia is very similar to that of the immediate early genes. Messenger RNA for Hsp70 and Hsp90 begin to rise within a few minutes of the insult in all regions, and persist [9,123]. As in many chronic neurodegenerative diseases, Hsps co-localize with Ub-proteins in the inclusion bodies [44,77,124]. The Hsps are highly conserved, abundantly expressed proteins with diverse functions including the assembly of multiprotein complexes, transportation of nascent polypeptides and regulation of protein folding [125]. Hsp70 is the major inducible Hsp found in cells [126] and Hsp70 and its constitutively expressed homologue HSC70 interact with many of the same binding partners and client proteins. In addition to aiding in protein refolding, the Hsp70 family can sequester activated caspases and other cell death proteins [127-129]. Expression of the Hsp70 gene is increased in cerebral ischemia in both global and focal models of cerebral ischemia [77,124,130] and in human ischemic stroke [44], indicating activation of the stress response (see Chapter 19). The induction of focal cerebral ischemia in rats by MCAo has been shown to increase, over time, the mRNA levels of the Hsp27 and Hsp70. However, the levels of Hsp90 mRNA remain constant. In contrast, during global ischemia, Hsp70 and Hsp90 mRNA levels are both raised, particularly in the CA1 neurons in the hippocampus [9,123].

Hsps may confer resistance to ischemia, by preserving proteasome function and attenuating the toxicity of proteasome inhibition reducing oxidative stress by alleviating the load of misfolded proteins destined for degradation by the UPS [25]. It is suggested that Hsp90 induces conformational changes that affect the ChTL and PGPH activities expressed by the X and Y subunits, respectively [131], depending on the activation state of the proteasome [132]. Hsp90 also plays a role in the assembly and maintenance of 26S proteasomes since functional loss of Hsp90 results in 26S proteasome dissociation [133]. An intriguing scenario is that a Hsp90 dissociation–association cycle might be envisaged for the 26S proteasome and it might be regulated in responding to ATP depletion during cerebral ischemia. It is interesting to note that the involvement of an Hsp90 chaperone in an ATP dependent manner assembly of the 26S proteasome indicates that changes in the physiological state of Hsp90 after cerebral ischemia may alter the amounts of the 26S proteasome [133]. In this regard, previous studies indicating that reduced assembly of the 26S proteasome is due to reduced availability of ATP after cerebral ischemia [23,24,27,28] are conceivable with a reduction of an ATP dependent manner assembly of the 26S proteasome by Hsp90 or with the disassembly of the 26S proteasome as a stress response regulated by Hsp90. While Hsp90 is required for the 26S proteasome assembly, it is also responsible for refolding stress-damaged proteins and thereby might be sequestered to those damaged proteins after an ischemic insult.

Conversely, Ub dramatically decreased in both gerbil and rat hippocampus within hours after ischemia and then recovers over the next 1–3 days in all cell types except CA1 cells that are destined for death [33,100]. This phenomenon may be involved in damage, possibly by allowing buildup of partially denatured proteins, but unfortunately, there is no further evidence to this effect. The accumulation of misfolded proteins in the cytosol leads to increased expression of Hsp, while accumulation of such proteins in the ER stimulates the expression of many ER resident proteins, most of which function as molecular chaperones. Chaperones are connected to proteasomes in at least four ways. First, chaperones can deliver substrates to the proteasome [134]. In a similar fashion, the chaperone VCP/Cdc48 is required for the degradation of several Ub pathway substrates [135,136]. VCP is a member of the AAA family of ATPases (see Chapter 13). The large hexameric ATPase appears to function as a protein separase able to remove ubiquitylated monomers from multisubunit complexes. In some cases, the liberated proteins are degraded by the 26S proteasome; in other cases, the separated proteins may change their intracellular location. The proteasome also degrades ER membrane proteins [137-140]. If these ER membrane proteins possess a large cytoplasmic domain, their proteasomal degradation can require Hsps 40, 70, and 90 as well as VCP. Hsp90 is required to assemble and stabilize the 26S proteasome, providing a third connection between chaperones and proteasomes [133]. Hsp90 is also able to bind and suppress peptide hydrolysis by the 20S proteasome [141]. Finally, both chaperones and proteasomes are induced by the accumulation of denatured proteins within eukaryotic cells. Recently, inhibitors of the proteasome have been identified that can block the rapid degradation of abnormal cytosolic and ER-associated proteins [16]. Proteasome inhibitors induce the expression of various Hsps and ER chaperones due to the accumulation of sufficient amounts of abnormal proteins and/or the inhibition of degradation of a key regulatory factor (e.g. heat-shock factor).

Ischemic injury, ROS generation and injuries that induce protein denaturation increase Hsp70 protein expression [126]. Overexpression of Hsp70 protects against glutamate toxicity, ischemia and oxidative injury [142,143]. Hsp70 functions as a part of a multiprotein complex and association with different binding partners can dramatically alter Hsp70 function. For instance, the E3 Ub chain formation protein CHIP competes for c-terminal binding to Hsp70 with HOP. Similarly, BAG-1 competes with CHIP for N-terminal binding. Formation of BAG-1/Hsp70/CHIP complexes is thought to redirect Hsp activity away from protein refolding and towards ubiquination and proteasomal degradation (Figure 1) [144].

CHIP's Ub ligase activity and link to proteasomal function has been shown to be critical in mediating protein refolding and degradation in other degenerative conditions including familial Parkinson's disease (PD) caused by mutations in the Parkin gene, the cystic-fibrosis transmembrane-conductance regulator which controls chloride-ion channel function and protein tau which is altered in individuals with a number of neurodegenerative diseases including Alzheimer's disease [145-147]. Proteasome activity seems to be correlated with the stability of the neurodegenerative disease-associated proteins and their fragments, which are responsible for the generation of disease pathology. For PD, it is well known how the UPS malfunction occurs: (*i*) mutations of an E3 Ub ligase, Parkin, in PD abrogates its enzymatic activity for ubiquitinating substances, including unfolded Pael receptor (Pael-R), and is toxic for neurons [145,148], and (*ii*) the loss of Ub C-terminal hydrolase L1 (UCHL1) activity by mutations increases susceptibility to PD [149]. Furthermore, the accumulation of Ub coniugates seems to be uncommon in neuronal death: the Ub coniugates are accumulated by protein amyloid-β in primary cortical neuron cell cultures and proteasome inhibition potentiates amyloid-β induced neuronal death [117], although treatment with proteasome inhibitors effectively reduces neuronal and astrocytic degeneration during the ischemic stress produced by stroke [24]. The temporal window of proteasomal inhibition is essential to determining fate as long term inhibition of the proteasome is a potent neurotoxic stimuli [150]. However, mild proteasomal perturbation is a highly effective mechanism to induce neuroprotective protein expression, block deleterious effects associated with inflammation and enhance energetic status. These observations speak of the therapeutic potential of delivering small molecule therapies which can spatially and or temporally restrict proteasomal degradation and enhance cell survival.

The non-lethal activation of ROS and even caspases is also essential for neuroprotection [151]. The molecular chaperone Hsp70 is upregulated by the initial stressor and it subsequently performs essential functions in protein refolding as well as targeting denatured proteins to the proteasome for degradation. Inhibitors of proteasomal function rapidly increase Hsp70 expression and like many other chaperones and chaperone binding proteins, Hsp70 is capable of blocking a host of acute and chronic conditions including metabolic diseases, cancer, autoimmune disorders and neurodegeneration [152].

UPS and Ischemic Core Damage

It does seem to be prevented by a proteasome inhibitor. This is a somewhat surprising result, the inhibitor actually extended penumbral damage somewhat (possibly due to the prevention of normally anti-apoptotic effect of NF-κB). It seems quite unlikely that the effect at the core is due to NF-κB blockade because damage there seems unlikely to require synthesis of new protein(s). Thus the result suggests that the activation of proteasome 20S contributes to cell death by causing breakdown of specific protein(s) [101,153].

UPS and Excitotoxic Cell Death

The early phase of necrosis induced due to glutamate neurotoxicity apparently does not require proteasome activation in cerebellar granule cultured cells [154]. The loss of ATP which occurs in ischemic injury results in an inability to remove glutamate from the synaptic cleft by ATP dependent transport, reversal of the glutamate transporters and excessive stimulation of NMDA and other glutamate receptors [155,156]. The overstimulation of NMDA receptors leads to calcium influx [157] which is associated with enhanced cytotoxic activation of nitric oxide synthase (NOS) [158], calpain [159], phospholipase A_2 [160], and MAPK [161]. NMDA-induced calcium entry can also uncouple respiration from ATP synthesis and result in production of free radicals enhancing this cytotoxic cascade [155,156,162]. The proteasome inhibitor MG-132 (1-4 μg) does not affect the NF-κB activation in rat striatal neurons by NMDA receptor stimulation involving IκB-α degradation by a caspase-3-like cysteine protease dependent mechanism [163]. Proteasome inhibition can prevent cytochrome-c release in cerebellar granule cells undergoing apoptosis, thus improving cell survival, but not necrosis [154]. However, glutamate receptor antagonists might also exacerbate proteasome inhibition-induced neuronal death [85].

A complex interaction feedback loop exists between glutamate mediated excitation and proteasomal degradation in that stimulation of glutamate receptors by NMDA application leads to ubiquitination of scaffolding protein Postsynaptic Density 95 (PSD-95) via the E3 Ub ligase Mdm2 and its proteasomal degradation [38,164-166]. This process requires calcium-dependent dephosphorylation of PKA substrates [167,168]. PSD-95 degradation likely has important ramifications on synaptic transmission, stability and plasticity particularly given that PSD-95 also associated with glutamatergic AMPA receptors which are endocytosed in an NMDA receptor dependent manner [38,164,165] (see Chapter 18). Moreover, Arundine and Tymianski [169] have shown that a HIV-Tat fusion protein containing C' terminal NMDA residues designed to interfer with PSD-95/receptor interactions blocks excitotoxic cell death *in vitro* as well as ischemia *in vivo*. This suggests that proteasomal degradation of PSD-95 is essential to dampen excitotoxic insults and that global proteasome inhibition would have long lasting effects on synaptic efficacy and possibly lead to sustained glutamatergic transmission and enhanced neurotoxicity under pathophysiological conditions. The observation that NMDA receptor stimulation upregulates expression of the p112 proteasome subunit suggests that glutamate receptor antagonists may exacerbate cell death induced by exposure to proteasome inhibitors [85,85]. In sum, these observations suggest that an important reciprical relationship exists between the proteasome and synaptic activity (see Chapter 18).

Proteasome Interaction with Apoptogenic Proteins

Caspases are a family of highly conserved cysteine proteases which are required for programmed cell death [170]. Caspases cleave substrates including structural proteins, DNA repair and cleavage enzymes, kinases, as well as proteins which inhibit apoptosis [171]. Activation of caspases 1, 3, 8, 9, and 11 and release of cytochrome *c* have all been observed in cerebral ischemia [172]. Moreover, inhibition of caspases reduces tissue damage and improves neurological outcome [173,174], even when delivered following damage [175,176]. During apoptosis, caspases cleave subunits of the 19S regulatory complex of the proteasome including: S6' (Rpt5) and S5a (Rpn10), whose role is to recognize polyubiquitinated substrates, as well as S1 (Rpn2), cooperates with other subunits to hold together the lid and base. These cleavage events inhibit the proteasomal degradation of Ub-dependent and -independent cellular substrates, including proapoptotic molecules such as Smac, caspases themselves, p53 and other proapoptotic molecules. Caspase cleavage of the proteasome is thought to enhance the execution of the apoptotic program by providing a feed-forward amplification loop [177]. The existence of this positive feedback loop in caspase-mediated proteasome inactivation suggests that blocking 19S proteasome is vital for the proper function of the apoptotic machinery. Another consequence of this proteolysis is the partial dissociation of the 19S regulatory complex from the 20S proteasome. Uncoupling of the RP from the CP may diminish the proteasome- mediated ATP consumption and preserve cellular energy for certain ATP-sensitive steps of apoptosis. More details on proteasome interaction with apoptogenic proteins are presented in Chapter 21.

UPS and Gene-Mediated Effects Acting on NF-κB

Inflammatory pathways are regulated by a limited number of transcription factors, the most important being NF-κB. NF-κB is a collective name for dimeric transcription factors of the Rel family of proteins. Its most abundant form is the cytoplasmic p65/p50 dimer, bound to IκBα [178]. Free radical damage following ischemia is probably partially mediated by NF-κB: a signal transduction cascade is activated leading to the phosphorylation of IκBα on Ser 32 and 36 by the multimeric IKK (IκB kinase) complex. IKK-mediated phosphorylation triggers the ubiquitination of IκBα by the E3 ligase $SCF^{\beta TRCP}$. Ubiquitinated IκBα is targeted to the 26S proteasome [179]. Once IκBα is degraded, the nuclear localization signal of NF-κB is unmasked allowing its translocation to the nucleus, where it binds to promoter regions of several proinflammatory genes inducing their expression and thus amplifying the inflammatory response. The p50 subunit of NF-κB is generated from the p105 precursor by limited proteolytic cleavage mediated by the 26S proteasome [180,181]. The intensity of NF-κB activation depends upon various factors, including the variable E3 activity of the $SCF^{\beta TRCP}$ complex, which is regulated by a reversible covalent modification with the Ub-like protein NEDD8 [182]. Finally, the activity of the IKK kinase depends on the formation of unusual poly-Ub chains linked by Lys63. It is unclear yet what is the role of those chains, however they are not targeting the ubiquitinated protein for degradation by the proteasome, unlike the chains formed by Lys48 linkages [183].

Preclinical trials aimed at blocking NF-κB activity using gene therapy, peptides, small molecules, and proteasome inhibitors while other neuroprotective compounds (e.g. antioxidants, natural products, salycilates, and NSAIDs) have also been shown to block NF-κB [184]. Although the NF-κB pathway is antiapoptotic for individual cells [185,186], it is damaging the neurons after ischemia, when it is activated in endothelial and microglial cells as well as in infiltrating leukocytes, leading to a massive production of inflammatory mediators, which in turn lead to neuronal damage. NF-κB is activated in the core and penumbra 1 day after 90-min temporary focal ischemia of the cortex, and the proteasome inhibitor MLN-519 attenuates damage measured 24 h after 2-h ischemia [101] as well as induction of inflammatory cytokines and cell adhesion molecules [101,153,187] which may reflect prevention of damage via increased global proteolysis or it may reflect the prevention of NF-κB activation. However, because the proteasome pathway is required for NF-κB activation, the result may reflect the importance of NF-κB in focal damage. If so, it shows that the ischemic core is most susceptible to damage via this system. NF-κB drives the transcription of many pro-inflammatory cytokines (IL-1β and TNF-α), enzymes (COX-2, iNOS), which are damaging in both focal and global ischemia, and also of cell adhesion molecules, such as ICAM-1 and selectins of endotelial cells, fibronectin and laminin of the extracellular matrix, and integrins and L-selectins of neutrophiles that are damaging in focal ischemia [186,188]. Both these responses were blunted by the proteasome inhibitors (n-tosyl-Phe-chloromethylketone and MLN-519) [101,153,187].

UPS and Endoplasmic Reticulum (ER) Disfunction

In the cytoplasm and mitochondria, where calcium activity is normally low, a prolonged excessive rise in free calcium levels is believed to be toxic; in the ER, in contrast, Ca^{2+} activity is relatively high and severe stress is caused by a depletion of ER Ca^{2+} stores. These Ca^{2+}-dependent processes are fundamental to normal cell function. Under conditions of ER dysfunction unfolded proteins accumulate in the ER lumen, a signal responsible for activation of the UPR and the ER-associated degradation (ERAD) [135,136] (see also Chapter 13). In acute cerebral ischemia, the ER Ca^{2+} pool is a primary target of toxic metabolites or intermediates, such as prostaglandins and ROS, produced during the pathological process. Affected neurons need to activate the entire UPR to cope with the severe form of stress induced by ER dysfunction. This stress response is however hindered under conditions where protein synthesis is suppressed to such an extent that processed xbp1 mRNA is not translated into the processed XBP1 protein [XBP1(proc)]. ER dysfunction observed after cerebral ischemia may also be triggered by impairment of the UPS. Newly synthesized peptides are processed and folded in the lumen of the ER before being secreted or incorporated into membranes. Processing and folding is a complex process termed ER quality control that is guided and monitored by a number of chaperons and folding enzymes [189]. Blocking proteasome activity triggers activation of the ER stress response [137-140], indicating that even in the physiological state proteins are created that cannot be properly processed and folded and therefore need to be degraded at the proteasome. The observation that transient cerebral ischemia impairs the UPS [27,37,139,190] suggests that these dysfunctions may contribute to ischemia-induced ER stress. Detergent-insoluble polyubiquitinated protein aggregates and a marked decrease in free ubiquitin levels have indeed been found in vulnerable neurons subjected to transient cerebral ischemia implying that the ER/UPS is impaired by ischemia [36,37]. Transient cerebral ischemia thus triggers a pathological process that is in many respects similar to those underlying degenerative diseases of the brain where the ER/UPS is also thought to be involved [191]. Activation of ERAD is important for the degradation of unfolded proteins through the UPS resulting in further ER stress. ER functioning is thus impaired in two different ways: first by the direct action of toxic intermediates, produced in the course of the pathological process, hindering vital ER reactions, and second, by the inability of cells to fully activate UPR and ERAD, leaving them unable to withstand the severe form of stress induced by ER dysfunction [192].

The Hypothesis of the Dual Role of the UPS in Stroke

The role of the UPS in the CNS is only just beginning to be elucidated. At the present, it is difficult to outline a single and clearly defined role of the UPS in cerebral ischemia and to establish what are the exact reasons why proteasome inhibitors have an apparently well-defined neuroprotective effect in stroke models [101,153,193-199].

Several natural and synthetic compounds that act as proteasome inhibitors have been reported with different chemical characteristics. Their pharmacological classes and mechanism of action are summarized in Table 1 [22-24]. These compounds either block the active sites of 20S proteasome core particle non-covalently, by specific hydrogen bonds with the main-chain atoms of the protein, without modifying the nucleophilic Thr1 residue or targeting all individual active subunits with comparable affinity (peptide vinyl sulfones) (see Chapter 40).

It is clear that there are numerous possible modes of action for proteasome inhibitors in protecting neurones and glia from ischemic damage (see Chapters 41 and 42). While the specificity of all proteasome inhibitors is not yet known, several of these compounds have been tested in stroke models (Table 2), although not all their possible mechanisms of action are clearly established. One of the best understood drugs is MLN-519 which consistently reduces cerebral infarct volume after MCAo in a dose dependent manner with a therapeutic window of up to six hours after onset of ischemia [200,201].

Table 1. Representative classes of proteasome inhibitors

Proteasome inhibitors	Class of compound	Mechanism of action
Naturally occurring		
Lactacystin	Prodrug for the β-lactone structure	It binds covalently to subunit β5 of mammalian proteasomes It also inhibits cathepsin A and tripeptidyl peptidase II.
Aclacinomycin	Aklavinone	It inhibits the ChTL activity of the proteasome without effects on cathepsin B, stimulated trypsin, and inhibited chymotrypsin and, to a lesser extent, calpain.
Eponemycin	α'-β'-epoxyketone	The compound binds covalently the β5, β5i and β1i catalytic subunits of the 20S proteasome and selectively inhibits the three major proteasome proteolytic activities at different rates.
Epoxomycin	α'-β'-epoxyketone	The compound binds covalently to the β5i, β5, β2i and β2 catalytic subunits of the 20S proteasome and inhibits primarily the ChTL activity. It does not inhibit other proteases (i.e. calpain, cathepsin B, papain, trypsin, chymotrypsin) at concentrations up to 50 M.
PR-39	Cathelicidin	The compound is a highly basic arginine/proline-rich peptide and reversibly binds to the α7 subunit of the proteasome.
Synthetic		
Peptide mimetic	No specific proteasome inhibitors	No specific proteasome inhibitors.
MLN-519	Synthetic agent similar to lactacystin	The compound reacts with the proposed catalytic nucleophile Oγ of Thr^1 on the β5 subunit.
MG-132	Tripeptidyl aldehyde	Commonly used reversible inhibitor of the ChTL activity of the proteasome, also inhibits cathepsins and calpains.
CEP-1612	Dipeptidyl aldehyde	The compound forms reversible covalent adducts with the proposed catalytic nucleophile, Oγ of Thr^1 on the β-subunit. It also inhibit lysosomal and Ca^{+2}-activated proteases.
CVT-634	Dipeptide benzamide	Inhibitor of the ChTL activity.
Bortezomib Velcade® (formely PS-341)	Dipeptidyl boronic acid inhibitors	Inhibitor of the ChTL activity, approved by FDA for the treatment of multiple myeloma and other malignancies.
2-aminobenzylstatine, formely NVP-AFB340 and NVP-AFD314	Boronic acid derivatives	A structure-based optimization approach improved the potency of this series with the most potent compound achieving an IC_{50} value of 7 nM against the ChTL activity. In addition, these compounds demonstrated good selectivity against the PGPH and TL proteasomal activities(all IC_{50} values > 20μM).
Vinyl sulfone tripeptides	Vinyl sulfone moieties of tripeptides	Competitive inhibitors that are largely specific for individual β subunits of the 20S proteasome but also inhibit intracellular cysteine proteases.
Ritonavir	HIV-1 protease inhibitor.	The compound is also a weak, low micromolar inhibitor of the chymotryptic activity of the 20S proteasome by binding the proteasome subunit β5 and β5i.

Table 2. Animal models of cerebral ischemia and proteasome inhibitor treatment

Study	Year	Treatment	Model of ischemia (Strain)	Results
Buchan AM *et al.* [195]	2000	CVT-634 (50 mg/kg, i.p.)	90 min MCAo 1 & 7 day recovery (inbred SH, rats)	Smaller infarct of 13±2% (P<0.01) and 2±2% (P<0.001) of hemispheric volume at 1 day and 7 days.
Phillips JB *et al.* [101]	2000	MLN-519 (0.003-0.1 mg/kg, i..v.)	Temporary MCAo 24 & 72 h recovery (Sprague-Dawley, rats)	Neuroprotection approached 60%. Neutrophil infiltration at 24h was significantly decreased (63 to 70%, P<0.05). The neuroprotective effect is in part caused by a reduction in the leukocyte inflammatory response.
Zhang L *et al.* [202]	2001	MLN-519 (1.0 mg/kg, i.v.)	Embolic stroke model 7 day recovery (Wistar, rats)	Combination treatment with r-tPA even at 6h significantly (P<0.05) reduced infarct volume, improved neurological recovery, and did not increase the incidence of hemorrhagic transformation and extends the neuroprotective effect to at least 6 hours after embolization.
Williams AJ *et al.* [153]	2003	MLN-519 (1.0 mg/kg, i.v.)	Temporary MCAo 24 h recovery (Sprague-Dawley, rats)	Treatment up to 6h after MCAo (4h after reperfusion) reduced neuronal and astrocytic degeneration, decreased cortical infarct volume (-48%), and increased neurologic recovery (+51%), with a reduced neutrophil infiltration (-38%), and a decrease in activated NF-κB immunoreactivity (-45%).
Berti R *et al.* [193]	2003	MLN-519 (1.0 mg/kg, i.v.)	Temporary MCAo 3-72 h recovery (Sprague-Dawley, rats)	The most striking effects of i.v. treatment were associated with reductions in ICAM-1 expression at 3 h followed by reductions in E-selectin (12-72 h). Less dramatic reductions were observed in IL-1β (3-24 h) and TNF-α (24 h) with no apparent effects on IL-6 and VCAM-1 mRNA levels

Table 2 (continued)

Study	Year	Treatment	Model of ischemia (Strain)	Results
Williams AJ *et al.* [194]	2004	MLN-519 (1.0 mg/kg, i.v.)	Temporary MCAo 72 h recovery (Sprague-Dawley, rats)	Core infarct volume in MLN519-treated rats was reduced with delayed treatments of 6, 8, or 10 hours after injury (P<0.05) and was associated with reductions in neuronal and axonal degeneration. Neuroprotection treatment provides an extended treatment window of up to 10 hours.
Williams AJ *et al.* [196]	2005	MLN-519 (1.0 mg/kg, i.v.)	Temporary MCAo 14 day recovery (Sprague-Dawley, rats)	Not significant difference in 2-week post-injury survival following MLN519 delayed treatment (at 10, 24, and 48 h post-occlusion). Significant reduction of the percent loss of tissue in the ipsilateral brain hemisphere (-17%), significant increase of body weight (+39%) and significant improvement in overall neurological function across the 2-week recovery period.
Stagliano NE *et al.* [197]	2006	Bortezomib (0.2 mg/kg i.v.)	Permanent Temporary MCAo 24h recovery (Wistar, rats)	A single dose of VELCADE® given 1 h post-MCAo, resulted in a 40% decrease in infarct volume (P<0.05) and a 38% decrease in neurologic deficits (P<0.01).
Henninger N *et al.* [199]	2006	Bortezomib (0.2 mg/kg i.v.)	*Model 1*: permanent MCAo 24h recovery (Wistar-Kyoto, rats) *Model 2*: embolic MCAo 24h recovery (Sprague-Dawley, rats)	A significant reduction of infarct volume in permanent (-42.7%, –32.8% and –4.4%; 1, 2 and 3 h treated group, respectively; P<0.05 for 1 and 2 h, treatment; at 3 hours not neuroprotective) and embolic (-58%; P=0.002) stroke models with a inhibited whole blood proteasome activity (-77%).

Table 2 (continued)

Study	Year	Treatment	Model of ischemia (Strain)	Results
Zhang L *et al.* [198]	2006	Bortezomib (0.2 mg/kg i.v.)	Embolic stroke model 24h recovery 7 day ischemic lesion volume (Wistar, rats)	A significant ($P<0.05$) reduction of secondary thrombosis, inflammatory responses, blood brain barrier disruption, infarct volume and neurological functional deficit when administrated within 4 h after stroke onset. Combination of bortezomib and tPA extends the thrombolytic window for stroke to 6 h.

MLN-519 exhibits a neuroprotective effect following focal brain ischemia with a 50 to 60% reduction of infarct volume, decreased leukocyte infiltration as well as prevention of NF-κB activation following reperfusion [101,153]. When MLN-519 was used in combination with tissue plasminogen activator (r-tPA) in a embolic stroke model, it decreased infarct volume and improved neurological outcome 1 week after the ischemic episode, as well as eliminated hemorrhage associated with t-PA treatment even when given 6 h after vessel occlusion [202]. These neuroprotective effects was also replicated in models of cerebral hemorrhage [23].

Recently preliminary experimental studies have also suggested neuroprotective effects of bortezomib in ischemic stroke models and cerebral ischemia [197-199]. Bortezomib (Velcade®, PS-341), a potent and selective inhibitor of the proteasome, was recently approved by the Food and Drug Administration for use in patients with refractory and relapsed multiple myeloma. Bortezomib is also effective preclinically in the treatment of inflammatory related diseases, such as chronic polyarthritis and liver inflammation (see Chapter 42). Within the limits of detection, bortezomib is a reversible inhibitor, which does not penetrate the blood-brain bareer (BBB)[203]. A single dose of bortezomib given 1 hr post- MCAo, resulted in a 40% decrease in infarct volume and a 38% decrease in neurological deficit in a rat permanent MCAo model [197]. The functional and histopathologic protection was accompanied by a 67% inhibition of whole blood proteasome activity, a level of inhibition which is commonly achieved in cancer patients in the clinic [197]. The potential neuroprotective effects of bortezomib were also evident in an embolic model of MCAo [199] and in combination with delayed thrombolytic therapy on a rat model of embolic focal cerebral ischemia [198]. Treatment with bortezomib reduced adverse cerebrovascular events including secondary thrombosis, inflammatory responses, and BBB disruption, and hence reduces infarct volume and neurological functional deficit when administrated within 4 h after stroke onset [198]. Combination of bortezomib and rtPA extended the thrombolytic window for stroke to 6 h, which is associated with the improvement of vascular patency and integrity. Real time RT-PCR of endothelial cells isolated by laser-capture microdissection from brain tissue and Western blot analysis show that bortezomib upregulated endothelial NOS expression and blocked NF-κB activation.

Reduction of cerebral infarct volume by proteasome inhibitors may depend on a combination of previously described effects and from the net balance of their positive and negative effects in modulating the cellular metabolic pathways. The proteasome clearly represents a central target for the processing and metabolism of proteins with critical roles in excitability, outgrowth, neuroprotection, metabolism and repair and its role is just beginning to be understood. Given this complexity, no single role of the UPS is likely to emerge in cerebral ischemia. Indeed, considerable controversy exists over the role of the UPS in neuronal ischemic cell death which may result from differences in experimental approaches and endpoints. However, in many animal models proteasome inhibitors exert a surprisingly consistent neuroprotective effect. While time- and dose-dependent proteasome inhibition will likely promote neuronal survival following stroke by helping neurons evoke chaperones, dampen inflammation and promote revascularization and energetic repletion. However, neurons subjected to prolonged and/or complete proteasome inhibition are likely to be severely damaged resulting from inability to resume requisite communication, protein turnover and protein trafficking. This hypothesis is supported by the observation that the temporal window of decreased proteasome activity is coincident with the therapeutic window during which proteasome inhibitors promote neuronal survival following stroke. The evidence presented in this review suggest that UPS may be involved in the acute neurodegenerative phase that occurs immediately after stroke, but that after this period, it assumes its normal physiological functions, which include promotion of neuronal survival (*the yin-yang effect*). Proteasome inhibition prevents the death of neurons immediately after cerebral ischemia, but may start to kill them thereafter. The problem of defining the temporal window during which proteasome inhibitors are efficacious is particularly important given that protracted use of these compounds results in neuronal cell death [88,154,190,204]. This phenomenon is observed in different systems. Prolonged exposure of murine neocortical or spinal cultures to proteasome inhibitors resulted in widespread neuronal death [204]. Proteasome inhibition also induced a time- and dose-dependent increase in poly-ADP-ribosylation in the neural PC6 cell line and in primary hippocampal neuron cultures [190] and dopamine neurotoxicity increased in the presence of proteasome inhibitors in a neural PC12 cell line [205]. By contrast, repair mediated by UPS appears to be long lasting. This is in agreement with the physiological function of proteasome in the nervous system during development [206-209]. In addition to the temporal constraints, important spatial considerations must also be addressed in terms of the means to deliver a proteasome inhibitory cocktail given that the neurotransmitter profile and metabolic activity of cells alter the efficacy of proteasome inhibitors [16]. Hypoxic endothelia showed a >10-fold increase in sensitivity to proteasome inhibitors [52].

Proteasome inhibition occurs during cerebral ischemia-reperfusion injury and is mediated, at least in part, by oxidative stress, which also directly activates NF-κB [25,85,210]. Proteasome inhibition may be the means by which oxidative stress mediates neuronal cell death. After cerebral ischemia-reperfusion injury, there is a time-dependent decrease in proteasome activity that is not associated with decreased expression of proteasome subunits. At the same time, a time- and dose-dependent proteasome inhibition promotes neuronal survival after stroke and help neurons to maintain their physiological functions. Probably, proteasome activity plays a double role in ischemic damage, on one hand, post-ischemic impairment of proteasome activity leads to accumulation of Ub-conjugates, contributing to loss of neuronal function, on the other hand, proteasome activity is associated with a developing inflammatory response by activation of NF-κB-mediated transcription in both neuronal and non-neuronal cells [211]. While neurons can withstand relatively long periods with intracellular accumulations of ubiquitinated proteins such as found in neurodegenerative disorders [212-215], they are very sensitive to damage elicited by an inflammatory response. Therefore proteasome inhibitors are considered to be of interest in stroke medicine, since they are able to prevent NF-κB activation [216] and therefore reduce the ischemic damage following stroke [195]. However, the role of NF-κB in the brain is unclear. *In vitro*, NF-κB activation can be either protective or deleterious. Cell culture studies have clearly shown that activation of NF-κB in neurons protects them against excitotoxic and metabolic insults relevant to the pathogenesis of stroke [19]. Data from studies of mice lacking the p50 subunit of NF-κB suggest that, overall, NF-κB activation enhances ischemic neuronal death, but its effects differ between cell types such that, whereas activation of NF-κB in microglia promotes ischemic neuronal degeneration, activation of NF-κB in neurons may increase their survival after a stroke [19]. The neuroprotective effects of proteasome inhibitors *in vivo* probably involve non-neuronal mechanisms, primarily in the vasculature within the ischemic area by the downregulated expression of genes in microvascular endothelial cells that encode for inflammatory cytokines and adhesion molecules [23,201,217,218]. Radiolabeled proteasome inhibitors did not show any evidence of brain penetration when administered at times when blood-brain barrier integrity was weakest (at 2 and 24 h after injury) in an ischemic stroke model [101]. At the same time, proteasome inhibitors prevent the disruption of the integrity of the microvascular beds partially based on their inhibitory action on matrix metalloproteinases [219].

An increasing number of patients have been receiving treatment with proteasome inhibitors without serious adverse reactions. The most serious possible complication of inhibiting proteasomes may involve the CNS since 'aggresome' formation have been evoked as a model for neurodegenerative disorders [220,221]. The degradation of most cytoplasmic and nuclear proteins depends on the activity of the UPS, while the lysosomal enzymes degrade mostly extracellular proteins and a fraction of cytoplasmic proteins engulfed by autophagy [222]. Short treatment with proteasome inhibitors has important therapeutic potential for increasing survival following stroke. As the role of the ubiquination machinery, the factors which dictate substrate specificity and the interactions of ischemia induced biochemical changes on discreet regions of the proteasome become known, we will likely be able to design better therapeutic agents to more selectively target components of the UPS.

Researchers and pharmaceutical companies are now poised to develop highly relevant biochemical and cell-based high-throughput screening assays to identify small molecule inhibitors of the proteasome. Cell-based screens using noninvasive bioluminescence imaging of 26S proteasome function is particularly intriguing strategy to meet these goals and transgenic mice have been developed which have green fluorescent protein tagged proteasome substrates which can be followed overtime to determine the regional and temporal effectiveness of proteasome inhibitors [223]. There are currently more than 20 clinical trials in the US and Canada targeting the proteasome, non-invasive methods to image protein degradation in humans is not yet possible making it difficult to assess if proteasome inhibitors reached their targets and their stability in vivo.

The preclinical profile of even the current group of non-selective proteasome inhibitors is superior to many previously investigated compounds, and the observed protection afforded by proteasome inhibitors has been replicated in a variety of model systems. While these compounds are promising agents for the treatment of acute ischemic injury, there is an immediate need for basic research on the pharmacokinetics, safety profile and toxicity of these compounds prior to entering more rigorous tests of clinical efficacy.

References

[1] Traystman RJ: Animal models of focal and global cerebral ischemia. *ILAR J* 2003, 44:85-95.

[2] Caso V, Hacke W: The very acute stroke treatment: fibrinolysis and after. *Clin Exp Hypertens* 2002, 24:595-602.

[3] Ikonomidou C, Turski L: Why did NMDA receptor antagonists fail clinical trials for stroke and traumatic brain injury? *Lancet Neurol* 2002, 1:383-386.

[4] Dirnagl U, Iadecola C, Moskowitz MA: Pathobiology of ischaemic stroke: an integrated view. *Trends Neurosci* 1999, 22:397.

[5] del Zoppo G, Ginis I, Hallenbeck JM, Iadecola C, Wang X, Feuerstein GZ: Inflammation and stroke: putative role for cytokines, adhesion molecules and iNOS in brain response to ischemia. *Brain Pathol* 2000, 10:95-112.

[6] Stanimirovic D, Satoh K: Inflammatory mediators of cerebral endothelium: a role in ischemic brain inflammation. *Brain Pathol* 2000, 10:113-126.

[7] Adibhatla RM, Hatcher JF, Larsen EC, Chen X, Sun D, Tsao FH: CDP-choline significantly restores phosphatidylcholine levels by differentially affecting phospholipase A2 and CTP: phosphocholine cytidylyltransferase after stroke. *J Biol Chem* 2006, 281:6718-6725.

[8] Adibhatla RM, Hatcher JF: Cytidine 5'-diphosphocholine (CDP-choline) in stroke and other CNS disorders. *Neurochem Res* 2005, 30:15-23.

[9] Lipton P: Ischemic cell death in brain neurons. *Physiol Review* 1999, 79:1431-1568.

[10] Phillis JW, O'Regan MH: The role of phospholipases, cyclooxygenases, and lipoxygenases in cerebral ischemic/traumatic injuries. *Crit Rev Neurobiol* 2003, 15:61-90.

[11] Rao AM, Hatcher JF, Kindy MS, Dempsey RJ: Arachidonic acid and leukotriene C4: role in transient cerebral ischemia of gerbils. *Neurochem Res* 1999, 24:1225-1232.

[12] Mattson MP, Culmsee C, Yu ZF: Apoptotic and antiapoptotic mechanisms in stroke. *Cell Tissue Res* 2000, 301:173-187.

[13] Raghavendra Rao VL, Bowen KK, Dhodda VK, Song G, Franklin JLGNR, Dempsey RJ: Gene expression analysis of spontaneously hypertensive rat cerebral cortex following transient focal cerebral ischemia. *J Neurochem* 2002, 83:1072-1086.

[14] Onteniente B, Couriaud C, Braudeau J, Benchoua A, Guegan C: The mechanisms of cell death in focal cerebral ischemia highlight neuroprotective perspectives by anti-caspase therapy. *Biochem Pharmacol* 2003, 66:1643-1649.

[15] Eldadah BA, Faden AI: Caspase pathways, neuronal apoptosis, and CNS injury. *J Neurotrauma* 2000, 17:811-829.

[16] Wojcik C: Regulation of apoptosis by the ubiquitin and proteasome pathway. *J Cell Mol Med* 2002, 6:25-48.

[17] Endres M, Dirnagl U: Ischemia and stroke. *Adv Exp Med Biol* 2002, 513:455-73:455-473.

[18] Stenzel-Poore MP, Stevens SL, Xiong Z, Lessov NS, Harrington CA, Mori M, Meller R, Rosenzweig HL, Tobar E, Shaw TE, Chu X, Simon RP: Effect of ischaemic preconditioning on genomic response to cerebral ischaemia: similarity to neuroprotective strategies in hibernation and hypoxia-tolerant states. *Lancet* 2003, 362:1028-1037.

[19] Mattson MP, Camandola S: NF-kB in neuronal plasticity and neurodegenerative disorders. *J Clin Invest* 2001, 107:243-254.

[20] Yang Y, Fang S, Jensen JP, Weissman AM, Ashwell JD: Ubiquitin protein ligase activity of IAPs and their degradation in proteasomes in response to apoptotic stimuli. *Science* 2000, 288:874-7.

[21] Suzuki Y, Nakabayashi Y, Takahashi R: Ubiquitin-protein ligase activity of X-linked inhibitor of apoptosis protein promotes proteasomal degradation of caspase-3 and enhances its anti-apoptotic effect in Fas-induced cell death. *Proc Natl Acad Sci U S A* 2001, 98:8662-8667.

[22] Di Napoli M, McLaughlin B: The ubiquitin-proteasome system as a drug target in cerebrovascular disease: therapeutic potential of proteasome inhibitors. *Curr Opin Investig Drugs* 2005, 6:686-699.

[23] Di Napoli M, Papa F: The proteasome system and proteasome inhibitors in stroke: controlling the inflammatory response. *Curr Opin Investig Drugs* 2003, 4:1333-1342.

[24] Wojcik C, Di Napoli M: Ubiquitin-Proteasome System and Proteasome Inhibition: New Strategies in Stroke Therapy. *Stroke* 2004, 35:1506-1518.

[25] Ding Q, Keller JN: Proteasomes and proteasome inhibition in the central nervous system. *Free Radic Biol Med* 2001, 31:574-584.

[26] Gray DA, Tsirigotis M, Woulfe J: Ubiquitin, proteasomes, and the aging brain. *Sci Aging Knowledge Environ* 2003, 2003:RE6.

[27] Asai A, Tanahashi N, Qiu JH, Saito N, Chi S, Kawahara N, Tanaka K, Kirino T: Selective proteasomal dysfunction in the hippocampal CA1 region after transient forebrain ischemia. *J Cereb Blood Flow Metab* 2002, 22:705-710.

[28] Kamikubo T, Hayashi T: Changes in proteasome activity following transient ischemia. *Neurochem Int* 1996, 28:209-212.

[29] Kim CH, Kim JH, Hsu CY, Ahn YS: Zinc is required in pyrrolidine dithiocarbamate inhibition of NF-[kappa]B activation. *FEBS Letters* 1999, 449:28-32.

[30] Wigle D, Ho W, Lo D, Francis J, Eubanks JH, Wallace MC: Altered expression levels of SEF-2 and p112 in the rat hippocampus after transient cerebral ischemia:

identification by mRNA differential display. *J Cereb Blood Flow Metab* 1999, 19:435-442.

[31] Kahns S, Kalai M, Jakobsen LD, Clark BF, Vandenabeele P, Jensen PH: Caspase-1 and caspase-8 cleave and inactivate cellular parkin. *J Biol Chem* 2003, 278:23376-80.

[32] Kahns S, Lykkebo S, Jakobsen LD, Nielsen MS, Jensen PH: Caspase-mediated parkin cleavage in apoptotic cell death. *J Biol Chem* 2002, 277:15303-8.

[33] Gubellini P, Bisso GM, Ciofi-Luzzatto A, Fortuna S, Lorenzini P, Michalek H, Scarsella G: Ubiquitin-mediated stress response in a rat model of brain transient ischemia/hypoxia. *Neurochem Res* 1997, 22:93-100.

[34] Vannucci SJ, Mummery R, Hawkes RB, Rider CC, Beesley PW: Hypoxia-ischemia induces a rapid elevation of ubiquitin conjugate levels and ubiquitin immunoreactivity in the immature rat brain. *J Cereb Blood Flow Metab* 1998, 18:376-385.

[35] Magnusson K, Wieloch T: Impairment of protein ubiquination may cause delayed neuronal death. *Neurosci Lett* 1989, 96:264-270.

[36] Hu BR, Martone ME, Jones YZ, Liu CL: Protein aggregation after transient cerebral ischemia. *J Neurosci* 2000, 20:3191-3199.

[37] Hu BR, Janelidze S, Ginsberg MD, Busto R, Perez-Pinzon M, Sick TJ, Siesjo BK, Liu CL: Protein aggregation after focal brain ischemia and reperfusion. *J Cereb Blood Flow Metab* 2001, 21:865-875.

[38] Ehlers MD: Activity level controls postsynaptic composition and signaling via the ubiquitin-proteasome system. *Nat Neurosci* 2003, 6:231-242.

[39] Rocca A, Lamaze C, Subtil A, Dautry-Varsat A: Involvement of the ubiquitin/proteasome system in sorting of the interleukin 2 receptor beta chain to late endocytic compartments. *Mol Biol Cell* 2001, 12:1293-1301.

[40] van Kerkhof P, Alves dos Santos CM, Sachse M, Klumperman J, Bu G, Strous GJ: Proteasome inhibitors block a late step in lysosomal transport of selected membrane but not soluble proteins. *Mol Biol Cell* 2001, 12:2556-2566.

[41] van Kerkhof P, Strous GJ: The ubiquitin-proteasome pathway regulates lysosomal degradation of the growth hormone receptor and its ligand. *Biochem Soc Trans* 2001, 29:488-493.

[42] Liu CL, Ge P, Zhang F, Hu BR: Co-translational protein aggregation after transient cerebral ischemia. *Neuroscience* 2005, 134:1273-1284.

[43] Uchihara T, Tsuchiya K, Kondo H, Hayama T, Ikeda K: Widespread appearance of Alz-50 immunoreactive neurons in the human brain with cerebral infarction. *Stroke* 1995, 26:2145-2148.

[44] Tytell M, Brown WR, Moody DM, Challa VR: Immunohistochemical assessment of constitutive and inducible heat-shock protein 70 and ubiquitin in human cerebellum and caudate nucleus. *Mol Chem Neuropathol* 1998, 35:97-117.

[45] Zhai Q, Wang J, Kim A, Liu Q, Watts R, Hoopfer E, Mitchison T, Luo L, He Z: Involvement of the ubiquitin-proteasome system in the early stages of wallerian degeneration. *Neuron* 2003, 39:217-225.

[46] Deshpande JK, Bergstedt K, Linden T, Kalimo H, Wieloch T: Ultrastructural changes in the hippocampal CA1 region following transient cerebral ischemia: evidence against programmed cell death. *Exp Brain Res* 1992, 88:91-105.

[47] Morimoto T, Ide T, Ihara Y, Tamura A, Kirino T: Transient ischemia depletes free ubiquitin in the gerbil hippocampal CA1 neurons. *Am J Pathol* 1996, 148:249-257.

[48] McNaught KS, Mytilineou C, Jnobaptiste R, Yabut J, Shashidharan P, Jennert P, Olanow CW: Impairment of the ubiquitin-proteasome system causes dopaminergic cell death and inclusion body formation in ventral mesencephalic cultures. *J Neurochem* 2002, 81:301-306.

[49] Zhang HG, Wang J, Yang X, Hsu HC, Mountz JD: Regulation of apoptosis proteins in cancer cells by ubiquitin. *Oncogene* 2004, 23:2009-2015.

[50] Schmidt-Kastner R, Zhang BT, Webster K, Kietzmann T, Zhao W, Busto R, Ginsberg MD: Hypoxia-inducible factor-1 (hif-1) in experimental brain ischemia. *Scientific World Journal* 2002, 2:123-124.

[51] Stanimirovic D, Zhang W, Howlett C, Lemieux P, Smith C: Inflammatory gene transcription in human astrocytes exposed to hypoxia: roles of the nuclear factor-kappaB and autocrine stimulation. *J Neuroimmunol* 2001, 119:365-376.

[52] Zund G, Uezono S, Stahl GL, Dzus AL, McGowan FX, Hickey PR, Colgan SP: Hypoxia enhances induction of endothelial ICAM-1: role for metabolic acidosis and proteasomes. *Am J Physiol* 1997, 273:C1571-C1580.

[53] Gozal D, Row BW, Kheirandish L, Liu R, Guo SZ, Qiang F, Brittian KR: Increased susceptibility to intermittent hypoxia in aging rats: changes in proteasomal activity, neuronal apoptosis and spatial function. *J Neurochem* 2003, 86:1545-1552.

[54] Richard S, Campello C, Taillandier L, Parker F, Resche F: Haemangioblastoma of the central nervous system in von Hippel-Lindau disease. French VHL Study Group. *J Intern Med* 1998, 243:547-553.

[55] Nakayama K, Frew IJ, Hagensen M, Skals M, Habelhah H, Bhoumik A, Kadoya T, Erdjument-Bromage H, Tempst P, Frappell PB: Siah2 Regulates Stability of Prolyl-Hydroxylases, Controls HIF1[alpha] Abundance, and Modulates Physiological Responses to Hypoxia. *Cell* 2004, 117:941-952.

[56] Salceda S, Caro J: Hypoxia-inducible factor 1alpha (HIF-1alpha) protein is rapidly degraded by the ubiquitin-proteasome system under normoxic conditions. Its stabilization by hypoxia depends on redox-induced changes. *J Biol Chem* 1997, 272:22642-22647.

[57] Kallio PJ, Wilson WJ, O'Brien S, Makino Y, Poellinger L: Regulation of the hypoxia-inducible transcription factor 1alpha by the ubiquitin-proteasome pathway. *J Biol Chem* 1999, 274:6519-6525.

[58] Li J, Post M, Volk R, Gao Y, Li M, Metais C, Sato K, Tsai J, Aird W, Rosenberg RD, Hampton TG, Sellke F, Carmeliet P, Simons M: PR39, a peptide regulator of angiogenesis. *Nat Med* 2000, 6:49-55.

[59] Tanaka S, Uehara T, Nomura Y: Up-regulation of protein-disulfide isomerase in response to hypoxia/brain ischemia and its protective effect against apoptotic cell death. *J Biol Chem* 2000, 275:10383-10393.

[60] Soucek T, Cumming R, Dargusch R, Maher P, Schubert D: The regulation of glucose metabolism by HIF-1 mediates a neuroprotective response to amyloid beta peptide. *Neuron* 2003, 39:43-56.

[61] Fillebeen C, Pantopoulos K: Redox control of iron regulatory proteins. *Redox Rep* 2002, 7:15-22.

[62] Kim S, Ponka P: Role of nitric oxide in cellular iron metabolism. *Biometals* 2003, 16:125-135.

[63] Iwai K, Klausner RD, Rouault TA: Requirements for iron-regulated degradation of the RNA binding protein, iron regulatory protein 2. *EMBO J* 1995, 14:5350-5357.

[64] Iwai K, Drake SK, Wehr NB, Weissman AM, LaVaute T, Minato N, Klausner RD, Levine RL, Rouault TA: Iron-dependent oxidation, ubiquitination, and degradation of iron regulatory protein 2: implications for degradation of oxidized proteins. *Proc Natl Acad Sci U S A* 1998, 95:4924-4928.

[65] Yamanaka K, Ishikawa H, Megumi Y, Tokunaga F, Kanie M, Rouault TA, Morishima I, Minato N, Ishimori K, Iwai K: Identification of the ubiquitin-protein ligase that recognizes oxidized IRP2. *Nat Cell Biol* 2003, 5:336-340.

[66] Lu GW, Yu S, Li RH, Cui XY, Gao CY: Hypoxic preconditioning: a novel intrinsic cytoprotective strategy. *Mol Neurobiol* 2005, 31:255-271.

[67] Sharp FR, Ran R, Lu A, Tang Y, Strauss KI, Glass T, Ardizzone T, Bernaudin M: Hypoxic preconditioning protects against ischemic brain injury. *NeuroRx* 2004, 1:26-35.

[68] Liu C, Chen S, Kamme F, Hu BR: Ischemic preconditioning prevents protein aggregation after transient cerebral ischemia. *Neuroscience* 2005, 134:69-80.

[69] Chen J, Graham SH, Zhu RL, Simon RP: Stress proteins and tolerance to focal cerebral ischemia. *J Cereb Blood Flow Metab* 1996, 16:566-577.

[70] Dirnagl U, Simon RP, Hallenbeck JM: Ischemic tolerance and endogenous neuroprotection. *Trends Neurosci* 2003, 26:248-254.

[71] Kirino T: Ischemic tolerance. *J Cereb Blood Flow Metab* 2002, 22:1283-1296.

[72] Meller R, Cameron JA, Torrey DJ, Clayton CE, Ordonez AN, Henshall DC, Minami M, Schindler CK, Saugstad JA, Simon RP: Rapid degradation of Bim by the ubiquitin-proteasome pathway mediates short-term ischemic tolerance in cultured neurons. *J Biol Chem* 2006, 281:7429-7436.

[73] Dosenko VE, Nagibin VS, Tumanovska LV, Moibenko AA: Protective Effect of Autophagy in Anoxia-reoxygenation of Isolated Cardiomyocyte? *Autophagy* 2006, 2.

[74] Dosenko VE, Nagibin VS, Tumanovskaya LV, Zagoriy VY, Moibenko AA, Vaage J: Proteasomal proteolysis in anoxia-reoxygenation, preconditioning and postconditioning of isolated cardiomyocytes. *Pathophysiology* 2006, 13:119-125.

[75] Ide T, Takada K, Qiu JH, Saito N, Kawahara N, Asai A, Kirino T: Ubiquitin stress response in postischemic hippocampus under nontolerant and tolerant conditions. *J Cereb Blood Flow Metab* 1999, 19:750-756.

[76] Taglialatela G, Kaufmann JA, Trevino A, Perez-Polo JR: Central nervous system DNA fragmentation induced by the inhibition of nuclear factor kappa B. *Neuroreport* 1998, 9:489-493.

[77] Noga M, Hayashi T, Tanaka J: Gene expressions of ubiquitin and hsp70 following focal ischaemia in rat brain. *Neuroreport* 1997, 8:1239-1241.

[78] Lam YA, Lawson TG, Velayutham M, Zweier JL, Pickart CM: A proteasomal ATPase subunit recognizes the polyubiquitin degradation signal. *Nature* 2002, 416:763-767.

[79] Lee C, Schwartz MP, Prakash S, Iwakura M, Matouschek A: ATP-dependent proteases degrade their substrates by processively unraveling them from the degradation signal. *Mol Cell* 2001, 7:627-637.

[80] Mengesdorf T, Jensen PH, Mies G, Aufenberg C, Paschen W: Down-regulation of parkin protein in transient focal cerebral ischemia: A link between stroke and degenerative disease? *PNAS* 2002, 99:15042-15047.

[81] Hampton MB, Fadeel B, Orrenius S: Redox regulation of the caspases during apoptosis. *Ann N Y Acad Sci* 1998, 854:328-335.

[82] Amici M, Forti K, Nobili C, Lupidi G, Angeletti M, Fioretti E, Eleuteri AM: Effect of neurotoxic metal ions on the proteolytic activities of the 20S proteasome from bovine brain. *J Biol Inorg Chem* 2002, 7:750-756.

[83] Fruh K, Gossen M, Wang K, Bujard H, Peterson PA, Yang Y: Displacement of housekeeping proteasome subunits by MHC-encoded LMPs: a newly discovered mechanism for modulating the multicatalytic proteinase complex. *EMBO J* 1994, 13:3236-3244.

[84] Hershko A: Roles of ubiquitin-mediated proteolysis in cell cycle control. *Curr Opin Cell Biol* 1997, 9:788-799.

[85] Snider BJ, Tee LY, Canzoniero LM, Babcock DJ, Choi DW: NMDA antagonists exacerbate neuronal death caused by proteasome inhibition in cultured cortical and striatal neurons. *Eur J Neurosci* 2002, 15:419-428.

[86] Wagenknecht B, Hermisson M, Groscurth P, Liston P, Krammer PH, Weller M: Proteasome inhibitor-induced apoptosis of glioma cells involves the processing of multiple caspases and cytochrome c release. *J Neurochem* 2000, 75:2288-2297.

[87] Ling YH, Liebes L, Zou Y, Perez-Soler R: Reactive oxygen species generation and mitochondrial dysfunction in the apoptotic response to Bortezomib, a novel proteasome inhibitor, in human H460 non-small cell lung cancer cells. *J Biol Chem* 2003, 278:33714-33723.

[88] Qiu JH, Asai A, Chi S, Saito N, Hamada H, Kirino T: Proteasome inhibitors induce cytochrome c-caspase-3-like protease-mediated apoptosis in cultured cortical neurons. *J Neurosci* 2000, 20:259-265.

[89] Keller JN, Gee J, Ding Q: The proteasome in brain aging. *Ageing Res Rev* 2002, 1:279-293.

[90] Bulteau AL, Lundberg KC, Humphries KM, Sadek HA, Szweda PA, Friguet B, Szweda LI: Oxidative modification and inactivation of the proteasome during coronary occlusion/reperfusion. *J Biol Chem* 2001, 276:30057-30063.

[91] Carrard G, Bulteau AL, Petropoulos I, Friguet B: Impairment of proteasome structure and function in aging. *Int J Biochem Cell Biol* 2002, 34:1461-1474.

[92] Grune T, Merker K, Sandig G, Davies KJ: Selective degradation of oxidatively modified protein substrates by the proteasome. *Biochem Biophys Res Commun* 2003, 305:709-718.

[93] Shringarpure R, Grune T, Mehlhase J, Davies KJ: Ubiquitin conjugation is not required for the degradation of oxidized proteins by proteasome. *J Biol Chem* 2003, 278:311-318.

[94] Demasi M, Davies KJ: Proteasome inhibitors induce intracellular protein aggregation and cell death by an oxygen-dependent mechanism. *FEBS Lett* 2003, 542:89-94.

[95] Kretz-Remy C, Arrigo AP: Modulation of the chymotrypsin-like activity of the 20S proteasome by intracellular redox status: effects of glutathione peroxidase-1 overexpression and antioxidant drugs. *Biol Chem* 2003, 384:589-595.

[96] Atlante A, Bobba A, Calissano P, Passarella S, Marra E: The apoptosis/necrosis transition in cerebellar granule cells depends on the mutual relationship of the antioxidant and the proteolytic systems which regulate ROS production and cytochrome c release en route to death. *J Neurochem* 2003, 84:960-971.

[97] Grange E, Gharib A, Lepetiti P, Guilaud J, Sarda N, Bobillier P: Brain protein synthesis in the conscious rat using L-[35S]methionine: relationship of methionine specific activity between plasma and precursor compartment and evaluation of methionine metabolic pathways. *J Neurochem* 1992, 59:1437-1443.

[98] Bodsch W, Takahashi K, Barbier A, Ophoff BG, Hossmann KA: Cerebral protein synthesis and ischemia. *Prog Brain Res* 1985, 63:197-210.

[99] Thilmann R, Xie Y, Kleihues P, Kiessling M: Persistent inhibition of protein synthesis precedes delayed neuronal death in postischemic gerbil hippocampus. *Acta Neuropathol (Berl)* 1986, 71:88-93.

[100] Franklin JL, Johnson EMJ: Control of neuronal size homeostasis by trophic factor-mediated coupling of protein degradation to protein synthesis. *J Cell Biol* 1998, 142:1313-1324.

[101] Phillips JB, Williams AJ, Adams J, Elliott PJ, Tortella FC: Proteasome inhibitor PS519 reduces infarction and attenuates leukocyte infiltration in a rat model of focal cerebral ischemia. *Stroke* 2000, 31:1686-1693.

[102] Baldwin ASJ: The NF-kB and IkB proteins: new discoveries and insights. *Annu Rev Immunol* 1996, 14:649-681.

[103] Markgraf CG, Velayo NL, Johnson MP: Six-hour window of opportunity for calpain inhibition in focal cerebral ischemia in rats. *Stroke* 1998, 29:152-158.

[104] Ogburn KD, Figueiredo-Pereira ME: Cytoskeleton/Endoplasmic reticulum collapse induced by prostaglandin j2 parallels centrosomal deposition of ubiquitinated protein aggregates. *J Biol Chem* 2006, 281:23274-23284.

[105] Wang Z, Aris VM, Ogburn KD, Soteropoulos P, Figueiredo-Pereira ME: Prostaglandin J2 Alters Pro-survival and Pro-death Gene Expression Patterns and 26 S Proteasome Assembly in Human Neuroblastoma Cells. *J Biol Chem* 2006, 281:21377-21386.

[106] Conforti L, Tarlton A, Mack TG, Mi W, Buckmaster EA, Wagner D, Perry VH, Coleman MP: A Ufd2/D4Cole1e chimeric protein and overexpression of Rbp7 in the slow Wallerian degeneration (WldS) mouse. *Proc Natl Acad Sci U S A* 2000, 97:11377-11382.

[107] Dewar D, Graham DI, Teasdale GM, McCulloch J: Cerebral ischemia induces alterations in tau and ubiquitin proteins. *Dementia* 1994, 5:168-173.

[108] Fuchs SY, Fried VA, Ronai Z: Stress-activated kinases regulate protein stability. *Oncogene* 1998, 17:1483-1490.

[109] Lu Z, Liu D, Hornia A, Devonish W, Pagano M, Foster DA: Activation of protein kinase C triggers its ubiquitination and degradation. *Mol Cell Biol* 1998, 18:839-845.

[110] Okuda H, Saitoh K, Hirai S, Iwai K, Takaki Y, Baba M, Minato N, Ohno S, Shuin T: The von Hippel-Lindau tumor suppressor protein mediates ubiquitination of activated atypical protein kinase C. *J Biol Chem* 2001, 276:43611-43617.

[111] Skinner PJ, Vierra-Green CA, Clark HB, Zoghbi HY, Orr HT: Altered trafficking of membrane proteins in purkinje cells of SCA1 transgenic mice. *Am J Pathol* 2001, 159:905-913.

[112] Youdim MB, Amit T, Falach-Yogev M, Am OB, Maruyama W, Naoi M: The essentiality of Bcl-2, PKC and proteasome-ubiquitin complex activations in the neuroprotective-antiapoptotic action of the anti-Parkinson drug, rasagiline. *Biochem Pharmacol* 2003, 66:1635-1641.

[113] Dimmeler S, Breitschopf K, Haendeler J, Zeiher AM: Dephosphorylation targets Bcl-2 for ubiquitin-dependent degradation: a link between the apoptosome and the proteasome pathway. *J Exp Med* 1999, 189:1815-1822.

[114] Meriin AB, Gabai VL, Yaglom J, Shifrin VI, Sherman MY: Proteasome inhibitors activate stress kinases and induce Hsp72. Diverse effects on apoptosis. *J Biol Chem* 1998, 273:6373-6379.

[115] Schulte TW, An WG, Neckers LM: Geldanamycin-induced destabilization of Raf-1 involves the proteasome. *Biochem Biophys Res Commun* 1997, 239:655-659.

[116] Wu HM, Wen HC, Lin WW: Proteasome inhibitors stimulate interleukin-8 expression via Ras and apoptosis signal-regulating kinase-dependent extracellular signal-related kinase and c-Jun N-terminal kinase activation. *Am J Respir Cell Mol Biol* 2002, 27:234-243.

[117] Song S, Kim SY, Hong YM, Jo DG, Lee JY, Shim SM, Chung CW, Seo SJ, Yoo YJ, Koh JY, Lee MC, Yates AJ, Ichijo H, Jung YK: Essential role of E2-25K/Hip-2 in mediating amyloid-beta neurotoxicity. *Mol Cell* 2003, 12:553-563.

[118] Dai Y, Rahmani M, Grant S: Proteasome inhibitors potentiate leukemic cell apoptosis induced by the cyclin-dependent kinase inhibitor flavopiridol through a SAPK/JNK- and NF-kappaB-dependent process. *Oncogene* 2003, 22:7108-7122.

[119] Kamura T, Hara T, Kotoshiba S, Yada M, Ishida N, Imaki H, Hatakeyama S, Nakayama K, Nakayama KI: Degradation of p57Kip2 mediated by SCFSkp2-dependent ubiquitylation. *Proc Natl Acad Sci U S A* 2003, 100:10231-10236.

[120] Patrick GN, Zhou P, Kwon YT, Howley PM, Tsai LH: p35, the neuronal-specific activator of cyclin-dependent kinase 5 (Cdk5) is degraded by the ubiquitin-proteasome pathway. *J Biol Chem* 1998, 273:24057-24064.

[121] Rideout HC, Wang Q, Park DS, Stefanis L: Cyclin-dependent kinase activity is required for apoptotic death but not inclusion formation in cortical neurons after proteasomal inhibition. *J Neurosci* 2003, 23:1237-1245.

[122] Casciati A, Ferri A, Cozzolino M, Celsi F, Nencini M, Rotilio G, Carri MT: Oxidative modulation of nuclear factor-kappaB in human cells expressing mutant fALS-typical superoxide dismutases. *J Neurochem* 2002, 83:1019-1029.

[123] Wagstaff MJ, Collaco-Moraes Y, Aspey BS, Coffin RS, Harrison MJ, Latchman DS, de Belleroche JS: Focal cerebral ischaemia increases the levels of several classes of heat shock proteins and their corresponding mRNAs. *Brain Res Mol Brain Res* 1996, 42:236-244.

[124] Nowak TS, Jr., Bond U, Schlesinger MJ: Heat shock RNA levels in brain and other tissues after hyperthermia and transient ischemia. *J Neurochem* 1990, 54:451-458.

[125] Welch WJ: Heat shock proteins functioning as molecular chaperones: their roles in normal and stressed cells. *Philos Trans R Soc Lond B Biol Sci* 1993, 339:327-33.

[126] Yenari MA, Giffard RG, Sapolsky RM, Steinberg GK: The neuroprotective potential of heat shock protein 70 (HSP70). *Mol Med Today* 1999, 5:525-531.

[127] Beere HM, Wolf BB, Cain K, Mosser DD, Mahboubi A, Kuwana T, Tailor P, Morimoto RI, Cohen GM, Green DR: Heat-shock protein 70 inhibits apoptosis by preventing recruitment of procaspase-9 to the Apaf-1 apoptosome. *Nat Cell Biol* 2000, 2:469-75.

[128] Saleh A, Srinivasula SM, Balkir L, Robbins PD, Alnemri ES: Negative regulation of the Apaf-1 apoptosome by Hsp70. *Nat Cell Biol* 2000, 2:476-83.

[129] Lee JE, Yenari MA, Sun GH, Xu L, Emond MR, Cheng D, Steinberg GK, Giffard RG: Differential neuroprotection from human heat shock protein 70 overexpression in in vitro and in vivo models of ischemia and ischemia-like conditions. *Exp Neurol* 2001, 170:129-39.

[130] Hara T, Mies G, Hata R, Hossmann KA: Gene expressions after thrombolytic treatment of middle cerebral artery clot embolism in mice. *Stroke* 2001, 32:1912-1919.

[131] Lu X, Michaud C, Orlowski M: Heat shock protein-90 and the catalytic activities of the 20 S proteasome (multicatalytic proteinase complex). *Arch Biochem Biophys* 2001, 387:163-171.

[132] Conconi M, Petropoulos I, Emod I, Turlin E, Biville F, Friguet B: Protection from oxidative inactivation of the 20S proteasome by heat-shock protein 90. *Biochem J* 1998, 333 (Pt 2):407-415.

[133] Imai J, Maruya M, Yashiroda H, Yahara I, Tanaka K: The molecular chaperone Hsp90 plays a role in the assembly and maintenance of the 26S proteasome. *EMBO J* 2003, 22:3557-3567.

[134] Takayama S, Reed JC: Molecular chaperone targeting and regulation by BAG family proteins. *Nat Cell Biol* 2001, 3:E237.

[135] Hampton RY: ER-associated degradation in protein quality control and cellular regulation. *Curr Opin Cell Biol* 2002, 14:476-482.

[136] Nishikawa SI, Fewell SW, Kato Y, Brodsky JL, Endo T: Molecular chaperones in the yeast endoplasmic reticulum maintain the solubility of proteins for retrotranslocation and degradation. *J Cell Biol* 2001, 153:1061-1070.

[137] Bush KT, Goldberg AL, Nigam SK: Proteasome inhibition leads to a heat-shock response, induction of endoplasmic reticulum chaperones, and thermotolerance. *J Biol Chem* 1997, 272:9086-9092.

[138] Hampton RY: ER stress response: getting the UPR hand on misfolded proteins. *Curr Biol* 2000, 10:R518-R521.

[139] Mengesdorf T, Proud CG, Mies G, Paschen W: Mechanisms underlying suppression of protein synthesis induced by transient focal cerebral ischemia in mouse brain. *Exp Neurol* 2002, 177:538-546.

[140] Paschen W: Endoplasmic reticulum: a primary target in various acute disorders and degenerative diseases of the brain. *Cell Calcium* 2003, 34:365-383.

[141] Lu X, Michaud C, Orlowski M: Heat shock protein-90 and the catalytic activities of the 20 S proteasome (multicatalytic proteinase complex). *Arch Biochem Biophys* 2001, 387:163-171.

[142] Xu L, Lee JE, Giffard RG: Overexpression of bcl-2, bcl-XL or hsp70 in murine cortical astrocytes reduces injury of co-cultured neurons. *Neurosci Lett* 1999, 277:193-7.

[143] Hoehn B, Ringer TM, Xu L, Giffard RG, Sapolsky RM, Steinberg GK, Yenari MA: Overexpression of HSP72 after induction of experimental stroke protects neurons from ischemic damage. *J Cereb Blood Flow Metab* 2001, 21:1303-9.

[144] Cyr DM, Hohfeld J, Patterson C: Protein quality control: U-box-containing E3 ubiquitin ligases join the fold. *Trends Biochem Sci* 2002, 27:368-375.

[145] Imai Y, Soda M, Hatakeyama S, Akagi T, Hashikawa T, Nakayama KI, Takahashi R: CHIP is associated with Parkin, a gene responsible for familial Parkinson's disease, and enhances its ubiquitin ligase activity. *Mol Cell* 2002, 10:55-67.

[146] Shimura H, Schwartz D, Gygi SP, Kosik KS: CHIP-Hsc70 complex ubiquitinates phosphorylated tau and enhances cell survival. *J Biol Chem* 2004, 279:4869-4876.

[147] Meacham GC, Patterson C, Zhang W, Younger JM, Cyr DM: The Hsc70 co-chaperone CHIP targets immature CFTR for proteasomal degradation. *Nat Cell Biol* 2001, 3:100-105.

[148] Imai Y, Soda M, Murakami T, Shoji M, Abe K, Takahashi R: A product of the human gene adjacent to parkin is a component of Lewy bodies and suppresses Pael receptor-induced cell death. *J Biol Chem* 2003, 278:51901-51910.

[149] Maraganore DM, de Andrade M, Lesnick TG, Farrer MJ, Bower JH, Hardy JA, Rocca WA: Complex interactions in Parkinson's disease: a two-phased approach. *Mov Disord* 2003, 18:631-636.

[150] Ding Q, Bruce-Keller AJ, Chen Q, Keller JN: Analysis of gene expression in neural cells subject to chronic proteasome inhibition. *Free Radic Biol Med* 2004, 36:445-455.

[151] McLaughlin BA, Erhardt JA, Legos JJ, White R, Barone FC, Parsons AA, Aizenman E: Caspase activation is required for neuroprotection in ischemic preconditioning. *Proc Natl Acad Sci USA* 2003, 100:715-720.

[152] Golab J, Bauer TM, Volker D, Cord N: Role of the ubiquitin-proteasome pathway in the diagnosis of human diseases. *Clin Chim Acta* 2004, 340:27-40.

[153] Williams AJ, Hale SL, Moffett JR, Dave JR, Elliott PJ, Adams J, Tortella FC: Delayed treatment with MLN519 reduces infarction and associated neurologic deficit caused by focal ischemic brain injury in rats via antiinflammatory mechanisms involving nuclear factor-kappaB activation, gliosis, and leukocyte infiltration. *J Cereb Blood Flow Metab* 2003, 23:75-87.

[154] Bobba A, Canu N, Atlante A, Petragallo V, Calissano P, Sarra E: Proteasome inhibitors prevent cytochrome c release during apoptosis but not in excitotoxic death of cerebellar granule neurons. *FEBS Lett* 2002, 515:8-12.

[155] Peng TI, Jou MJ, Sheu SS, Greenamyre JT: Visualization of NMDA receptor-induced mitochondrial calcium accumulation in striatal neurons. *Exp Neurol* 1998, 149:1-12.

[156] Dugan LL, Sensi SL, Canzoniero LM, Handran SD, Rothman SM, Lin TS, Goldberg MP, Choi DW: Mitochondrial production of reactive oxygen species in cortical neurons following exposure to N-methyl-D-aspartate. *J Neurosci* 1995, 15:6377-6388.

[157] Mattson MP: Excitotoxic and excitoprotective mechanisms: abundant targets for the prevention and treatment of neurodegenerative disorders. *Neuromolecular Med* 2003, 3:65-94.

[158] Dawson VL, Dawson TM, London ED, Bredt DS, Snyder SH: Nitric oxide mediates glutamate neurotoxicity in primary cortical cultures. *Proc Natl Acad Sci U S A* 1991, 88:6368-6371.

[159] Siman R, Noszek JC, Kegerise C: Calpain I activation is specifically related to excitatory amino acid induction of hippocampal damage. *J Neurosci* 1989, 9:1579-1590.

[160] Dumuis A, Sebben M, Haynes L, Pin JP, Bockaert J: NMDA receptors activate the arachidonic acid cascade system in striatal neurons. *Nature* 1988, 336:68-70.

[161] Stanciu M, Wang Y, Kentor R, Burke N, Watkins S, Kress G, Reynolds I, Klann E, Angiolieri MR, Johnson et al: Persistent activation of ERK contributes to glutamate-induced oxidative toxicity in a neuronal cell line and primary cortical neuron cultures. *J Biol Chem* 2000, 275:12200-12206.

[162] Reynolds IJ, Hastings TG: Glutamate induces the production of reactive oxygen species in cultured forebrain neurons following NMDA receptor activation. *J Neurosci* 1995, 15:3318-3327.

[163] Qin Z, Wang Y, Chasea TN: A caspase-3-like protease is involved in NF-kappaB activation induced by stimulation of N-methyl-D-aspartate receptors in rat striatum. *Brain Res Mol Brain Res* 2000, 80:111-122.

[164] Colledge M, Snyder EM, Crozier RA, Soderling JA, Jin Y, Langeberg LK, Lu H, Bear MF, Scott JD: Ubiquitination regulates PSD-95 degradation and AMPA receptor surface expression. *Neuron* 2003, 40:595-607.

[165] Ehrlich I, Malinow R: Postsynaptic Density 95 controls AMPA Receptor Incorporation during Long-Term Potentiation and Experience-Driven Synaptic Plasticity. *J Neurosci* 2004, 24:916-927.

[166] Liu CL, Martone ME, Hu BR: Protein ubiquitination in postsynaptic densities after transient cerebral ischemia. *J Cereb Blood Flow Metab* 2004, 24:1219-1225.

[167] Cohen NA, Brenman JE, Snyder SH, Bredt DS: Binding of the inward rectifier K+ channel Kir 2.3 to PSD-95 is regulated by protein kinase A phosphorylation. *Neuron* 1996, 17:759-767.

[168] Suen PC, Wu K, Xu JL, Lin SY, Levine ES, Black IB: NMDA receptor subunits in the postsynaptic density of rat brain: expression and phosphorylation by endogenous protein kinases. *Brain Res Mol Brain Res* 1998, 59:215-228.

[169] Arundine M, Tymianski M: Molecular mechanisms of calcium-dependent neurodegeneration in excitotoxicity. *Cell Calcium* 2003, 34:325-337.

[170] Thornberry NA, Lazebnik Y: Caspases: Enemies Within. *Science* 1998, 281:1312-1316.

[171] Fischer U, Janicke RU, Schulze-Osthoff K: Many cuts to ruin: a comprehensive update of caspase substrates. *Cell Death Differ* 2003, 10:76-100.

[172] Benchoua A, Guegan C, Couriaud C, Hosseini H, Sampaio N, Morin D, Onteniente B: Specific Caspase Pathways Are Activated in the Two Stages of Cerebral Infarction. *J Neurosci* 2001, 21:7127-7134.

[173] Friedlander RM, Gagliardini V, Hara H, Fink KB, Li W, MacDonald G, Fishman MC, Greenberg AH, Moskowitz MA, Yuan J: Expression of a dominant negative mutant of interleukin-1 beta converting enzyme in transgenic mice prevents neuronal cell death induced by trophic factor withdrawal and ischemic brain injury. *J Exp Med* 1997, 185:933-40.

[174] Trapp T, Korhonen L, Besselmann M, Martinez R, Mercer EA, Lindholm D: Transgenic mice overexpressing XIAP in neurons show better outcome after transient cerebral ischemia. *Mol Cell Neurosci* 2003, 23:302-313.

[175] Hara H, Fink K, Endres M, Friedlander RM, Gagliardini V, Yuan J, Moskowitz MA: Attenuation of transient focal cerebral ischemic injury in transgenic mice expressing a mutant ICE inhibitory protein. *J Cereb Blood Flow Metab* 1997, 17:370-5.

[176] Fink K, Zhu J, Namura S, Shimizu-Sasamata M, Endres M, Ma J, Dalkara T, Yuan J, Moskowitz MA: Prolonged therapeutic window for ischemic brain damage caused by delayed caspase activation. *J Cereb Blood Flow Metab* 1998, 18:1071-6.

[177] Sun XM, Butterworth M, MacFarlane M, Dubiel W, Ciechanover A, Cohen GM: Caspase activation inhibits proteasome function during apoptosis. *Mol Cell* 2004, 14:81-93.

[178] Perkins ND: The Rel/NF-kappa B family: friend and foe. *Trends Biochem Sci* 2000, 25:434-440.

[179] Chen Z, Hagler J, Palombella VJ, Melandri F, Scherer D, Ballard D, Maniatis T: Signal-induced site-specific phosphorylation targets I kappa B alpha to the ubiquitin-proteasome pathway. *Genes Dev* 1995, 9:1586-1597.

[180] Lin L, DeMartino GN, Greene WC: Cotranslational biogenesis of NF-kappaB p50 by the 26S proteasome. *Cell* 1998, 92:819-828.

[181] Rape M, Jentsch S: Taking a bite: proteasomal protein processing. *Nat Cell Biol* 2002, 4:E113-E116.

[182] Amir RE, Iwai K, Ciechanover A: The NEDD8 pathway is essential for SCF(beta - TrCP)-mediated ubiquitination and processing of the NF-kappa B precursor p105. *J Biol Chem* 2002, 277:23253-23259.

[183] Wang M, Suzuki T, Kitada T, Asakawa S, Minoshima S, Shimizu N, Tanaka K, Mizuno Y, Hattori N: Developmental changes in the expression of parkin and UbcR7, a parkin- interacting and ubiquitin-conjugating enzyme, in rat brain. *J Neurochem* 2001, 77:1561-1568.

[184] Epinat J-C, Gilmore TD: Diverse agents act at multiple levels to inhibit the Rel/NF-[kappa]B signal transduction pathway. *Oncogene* 1999, 18:6896-6909.

[185] Schulze-Osthoff K, Ferrari D, Riehemann K, Wesselborg S: Regulation of NF-kB activation by MAP kinase cascades. *Immunobiology* 1997, 198:35-49.

[186] Wallach D: Cell death induction by TNF: a matter of self control. *Trends Biochem Sci* 1997, 22:107-109.

[187] Salminen A, Liu PK, Hsu Y: Alteration of transcription factor binding activities in the ischemic brain. *Biochem Biophys Res Commun* 1994, 212:939-944.

[188] Rothlein R, Czajkowski M, O'Neill MM, Marlin SD, Mainlofi E, Merluzzi VJ: Induction of intracellular adhesion molecule 1 on primary and continuous cell lines by proinflammatory cytokines. *J Immunol* 1988, 141:1665-1669.

[189] Hammond C, Helenius A: Quality control in the secretory pathway: retention of a misfolded viral membrane glycoprotein involves cycling between the ER, intermediate compartment, and Golgi apparatus. *J Cell Biol* 1994, 126:41-52.

[190] Keller JN, Markesbery WR: Proteasome inhibition results in increased poly-ADP-ribosylation: implications for neuron death. *J Neurosci Res* 2000, 61:436-442.

[191] Mattson MP, Chan SL: Neuronal and glial calcium signaling in Alzheimer's disease. *Cell Calcium* 2003, 34:385-397.

[192] Dong Z, Wang JZ, Yu F, Venkatachalam MA: Apoptosis-resistance of hypoxic cells: multiple factors involved and a role for IAP-2. *Am J Pathol* 2003, 163:663-671.

[193] Berti R, Williams AJ, Velarde LC, Moffett JR, Elliott PJ, Adams J, Yao C, Dave JR, Tortella FC: Effect of the proteasome inhibitor MLN519 on the expression of inflammatory molecules following middle cerebral artery occlusion and reperfusion in the rat. *Neurotox Res* 2003, 5:505-514.

[194] Williams AJ, Berti R, Dave JR, Elliot PJ, Adams J, Tortella FC: Delayed treatment of ischemia/reperfusion brain injury: extended therapeutic window with the proteosome inhibitor MLN519. *Stroke* 2004, 35:1186-1191.

[195] Buchan AM, Li H, Blackburn B: Neuroprotection achieved with a novel proteasome inhibitor which blocks NF-κB activation. *Neuroreport* 2000, 11:427-430.

[196] Williams AJ, Myers TM, Cohn SI, Sharrow KM, Lu XC, Tortella FC: Recovery from ischemic brain injury in the rat following a 10 h delayed injection with MLN519. *Pharmacol Biochem Behav* 2005, 81:182-189.

[197] Stagliano NE, Ren JM, Riodan W, Finklestein SP: Velcade®, a first-in-class proteasome inhibitor, is efficacious in rat permanent middle cerebral artery occlusion. *J Cereb Blood Flow Metab* 2005, 25:S38.

[198] Zhang L, Zhang ZG, Liu X, Hozeska A, Stagliano N, Riordan W, Lu M, Chopp M: Treatment of embolic stroke in rats with bortezomib and recombinant human tissue plasminogen activator. *Thromb Haemost* 2006, 95:166-173.

[199] Henninger N, Sicard KM, Bouley J, Fisher M, Stagliano NE: The proteasome inhibitor VELCADE reduces infarction in rat models of focal cerebral ischemia. *Neurosci Lett* 2006, 398:300-305.

[200] Williams AJ, Dave JR, Tortella FC: Neuroprotection with the proteasome inhibitor MLN519 in focal ischemic brain injury: relation to nuclear factor kappaB (NF-kappaB), inflammatory gene expression, and leukocyte infiltration. *Neurochem Int* 2006, 49:106-112.

[201] Di Napoli M, Papa F: MLN-519 Millenium/PAION. *Curr Opin Invest Drug* 2003, 4:333-341.

[202] Zhang L, Zhang ZG, Zhang RL, Lu M, Adams J, Elliott PJ, Chopp M: Postischemic (6-Hour) treatment with recombinant human tissue plasminogen activator and proteasome inhibitor PS-519 reduces infarction in a rat model of embolic focal cerebral ischemia. *Stroke* 2001, 32:2926-2931.

[203] Adams J: The proteasome: a suitable antineoplastic target. *Nat Rev Cancer* 2004, 4:349-360.

[204] Kikuchi S, Shinpo K, Takeuchi M, Tsuji S, Yabe I, Niino M, Tashiro K: Effect of geranylgeranylaceton on cellular damage induced by proteasome inhibition in cultured spinal neurons. *J Neurosci Res* 2002, 69:373-381.

[205] Keller JN, Huang FF, Dimayuga ER, Maragos WF: Dopamine induces proteasome inhibition in neural PC12 cell line. *Free Radic Biol Med* 2000, 29:1037-1042.

[206] Takada K, Kanda T, Ohkawa K, Matsuda M: Ubiquitin and ubiquitin-protein conjugates in PC12h cells: changes during neuronal differentiation. *Neurochem Res* 1994, 19:391-398.

[207] Flann S, Hawkes RB, Riederer BM, Rider CC, Beesley PW: Changes in ubiquitin immunoreactivity in developing rat brain: a putative role for ubiquitin and ubiquitin conjugates in dendrite outgrowth and differentiation. *Neuroscience* 1997, 81:173-187.

[208] El-Khodor BF, Kholodilov NG, Yarygina O, Burke RE: The expression of mRNAs for the proteasome complex is developmentally regulated in the rat mesencephalon. *Brain Res Dev Brain Res* 2001, 129:47-56.

[209] Wojcik C, Wilk S: Changes in proteasome expression and activity during differentiation of neuronal precursor NTera 2 clone D1 cells. *Neurochem Int* 1999, 34:131-136.

[210] Keller JN, Huang FF, Zhu H, Yu J, Ho YS, Kindy TS: Oxidative stress-associated impairment of proteasome activity during ischemia-reperfusion injury. *J Cereb Blood Flow Metab* 2000, 20:1467-1473.

[211] Stephenson D, Yin T, Smalstig EB, Hsu MA, Panetta J, Little S, Clemens J: Transcription factor nuclear factor-kappa B is activated in neurons after focal cerebral ischemia. *J Cereb Blood Flow Metab* 2000, 20:592-603.

[212] Bailey CK, Andriola IF, Kampinga HH, Merry DE: Molecular chaperones enhance the degradation of expanded polyglutamine repeat androgen receptor in a cellular model of spinal and bulbar muscular atrophy. *Hum Mol Genet* 2002, 11:515-523.

[213] Halliwell B: Hypothesis: proteasomal dysfunction: a primary event in neurogeneration that leads to nitrative and oxidative stress and subsequent cell death. *Ann N Y Acad Sci* 2002, 962:182-194.

[214] Jenner P: Oxidative stress in Parkinson's disease. *Ann Neurol* 2003, 53 Suppl 3:S26-S36.

[215] Mayer RJ: From neurodegeneration to neurohomeostasis: the role of ubiquitin. *Drug News Perspect* 2003, 16:103-108.

[216] Elliott PJ, Zollner TM, Boehncke WH: Proteasome inhibition: a new anti-inflammatory strategy. *J Mol Med* 2003, 81:235-245.

[217] Howard EF, Chen Q, Cheng C, Carroll JE, Hess D: NF-kappa B is activated and ICAM-1 gene expression is upregulated during reoxygenation of human brain endothelial cells. *Neurosci Lett* 1998, 248:199-203.

[218] Schmedtje JF, Ji YS, Liu WL, Dubois RN, Runge MS: Hypoxia induces cyclooxygenase-2 via the NF-kappa-B p65 transcription factor in human vascular endothelial cells. *J Biol Chem* 1997, 272:601-608.

[219] Kolev K, Skopal J, Simon L, Csonka E, Machovich R, Nagy Z: Matrix metalloproteinase-9 expression in post-hypoxic human brain capillary endothelial cells:

H2O2 as a trigger and NF-kappaB as a signal transducer. *Thromb Haemost* 2003, 90:528-537.

[220] Kopito RR: Aggresomes, inclusion bodies and protein aggregation. *Trends Cell Biol* 2000, 10:524-530.

[221] Sherman MY, Goldberg AL: Cellular defenses against unfolded proteins: a cell biologist thinks about neurodegenerative diseases. *Neuron* 2001, 29:15-32.

[222] Glickman MH, Ciechanover A: The ubiquitin-proteasome proteolytic pathway: destruction for the sake of construction. *Physiol Rev* 2002, 82:373-428.

[223] Lindsten K, Menendez-Benito V: A transgenic mouse model of the ubiquitin/proteasome system. *Nat Biotechnol* 2003, 21:897-902.

In: The Ubiquitin Proteasome System… ISBN 978-1-60021-749-4
Eds: Mario Di Napoli and Cezary Wojcik, pp. 961-987 © 2007 Nova Science Publishers, Inc.

Chapter 40

Pharmacology of the Ubiquitin Proteasome System: Proteasome Inhibitors and Modulators

***Halina Ostrowska*[1,*] *and Marek Z. Wojtukiewicz*[2,¥]**

[1]Department of Biology, Faculty of Pharmacy, [2]Department of Oncology, Faculty of Medicine, Medical University, Bialystok, Poland.

Abstract

The ubiquitin-proteasome system (UPS) is of key importance in the degradation of misfolded/abnormal proteins, viral proteins and many short-lived proteins that play vital roles in cell proliferation, differentiation, apoptosis and inflammatory processes. Therefore, both the ubiquitin - protein conjugation system and the 26S proteasomes constitute important target for pharmacological intervention. A number of small molecule inhibitors that target the 20S proteasome are described in the literature as possible anti-cancer and anti-inflammatory agents. Among them, a dipeptidyl boronic acid bortezomib (Velcade®, PS-341) is currently used in clinical practice for the treatment of relapsed or refractory myeloma, while MLN-519, a synthetic analog of microbial lactacystin, is under clinical evaluation for the treatment of ischemic cerebral stroke. Novel highly selective inhibitors of the 20S proteasome that bind noncovalently to the substrate binding-sites only and that exhibit less cytotoxic effects against normal cells, have recently been generated by combinatorial chemistry or identified by high-throughput screening of the pharmaceutical company's compound archives. Furthermore, a number of agents used in conventional therapies, dietary chemopreventive compounds and toxins

* Correspondence concerning this article should be addressed to Dr. Halina Ostrowska, Department of Biology Medical University of Bialystok, Poland; 1 Kilinskiego st. 15-089 Bialystok, Poland. Phone: +48 (85)-748-54-60; Fax: +48 (85)-748-54-61; E-mail: halost@amb.edu.pl.

¥ Correspondence concerning this article should be addressed to Dr. Marek Z. Wojtukiewicz, Department of Oncology, Medical University of Bialystok, Poland. 12 Ogrodowa st. 15-027 Bialystok, Poland. Phone (Fax): +48 (85)-7320481; E-mail: mwojtuk@amb.edu.pl.

block 26S proteasome-dependent protein degradation, either by inhibiting the 20S proteolytically active subunits or by regulating the expression or function of non-proteolytic subunits of this complex. Moreover, selected small molecules function as 20S proteasome activators and may be considered as potential therapeutics in pathological states resulting from the loss of proteasome activity, such as neurodegenerative disorders. Finally, more recently particular attention has been focused on small molecules that could block protein degradation at the level of their ubiquitination or could direct disease- promoting proteins for ubiquitination and degradation. These groups of modulators are currently being tested in preclinical settings for their therapeutic potential in cancer and inflammatory diseases.

The chapter summarizes the current knowledge on the chemistry of synthetic and natural inhibitors and modulators of the UPS, their mode of action and potential therapeutic relevance in the therapy of various human diseases with special respect to the central nervous system (CNS) pathology.

Keywords: Inhibitors, Modulators, Chemical structure, Pharmacology, Proteasome, Ubiquitin.

Abbreviations

ALLN,acetyl-Leu-Leu-norleucinal; BMSCs, bone marrow stromal cells; CNS, central nervous system; CVT- 634; 5-methoxy-1-indanone-3-acetyl-Leu-D-Leu-1-indanylamide; Cbz, N-benzyloxcarbonyl; ChT-L, chymotrypsin-like; E3, ubiquitin protein ligase; EGCG, epigallocatechin-3 gallate; Fu, fumagillol; HIF-1α hypoxia-inducible factor-1α; HIV, human immunodeficiency virus; IκB, nuclear factor κB inhibitor; LPS, lipopolysaccharide; MG-132, Cbz-Leu-Leu-leucinal; MHC, major histocompatibility complex; MM, multiple myeloma; NF-κB, nuclear factor-κB; NIP, nitrophenol derivative; Ntn-hydrolases; N-terminal nucleophile hydrolase; PGPH, peptidylglutamylpeptide hydrolyzing; PR, proline and arginine rich peptide; protacs, proteolysis targeting chimeric molecules; PS-341, pyrazinyl(Pyz)carbonyl-Phe-Leu boronic acid; MLN-519 C-7 n-propyl analogue of clasto-lactacystin/β-lactone; PSI, Cbz-Ile-Glu (O-t-Bu)-Ala-leucinal; SCF, Skp1p-Cullin-F-box protein complex; SMPI, small molecule proteolysis inducers ; T-L, trypsin-like; TNF, tumor necrosis factor; UPS, ubiquitin-proteasome system; VHL, von Hippel-Lindau; vs, vinyl sulfone; Z, N-benzyloxycarbonyl.

Introduction

During the last decade a remarkable progress has been made in understanding the role of the ubiquitin-proteasome system (UPS) in physiological and pathological processes in humans [1], including its significance in the central nervous system [2]. The UPS constitutes the major non-lysosomal proteolytic pathway responsible for the elimination of abnormal or damaged proteins, antigen proteins destined for presentation on the major histocompatibility complex (MHC) class I molecules as well as for targeted destruction of many short-lived

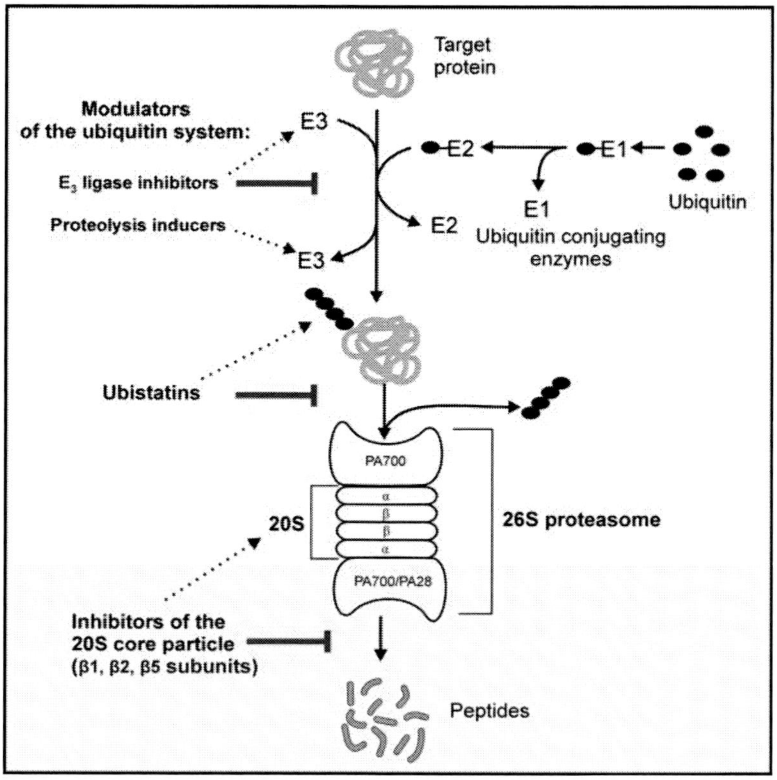

Figure 1. The ubiquitin proteasome system as a potential therapeutic target. Known targets are indicated by arrows with broken lines. Protein ubiquitination occurs through the activity of the ubiquitin-activating enzyme (E1), ubiquitin-conjugating enzymes (E2) and ubiquitin-protein ligases (E3). Polyubiquitinated protein is then degraded by the 26S proteasome (PA700-20S-PA700 or PA700-20S-PA28 complexes). Inhibitors of substrate specific E3 prevent (red) protein ubiquitination and subsequent degradation by the 26S proteasome [15].Ubistatins block the binding of ubiquitinated proteins to the 26S proteasome by targeting lysine 48-linked multi-ubiquitin chains [16]. Small molecule proteolysis inducers can direct a disease-promoting protein for ubiquitination and degradation [17]. Proteasome inhibitors block (red) degradation of ubiquitinated proteins by binding to the proteolytically active subunits located in two inner β-rings of the 20S core particle [9].

proteins, including those that control cell cycle progression (e.g. cyclins, cyclin kinase inhibitor $p27^{Kip1}$, $p21^{Waf1/Cip1}$), apoptosis (e.g. p53, Bax, c-IAP) and nuclear factor-kappaB (NF-κB) transcriptional pathway (i.e. IκBα) [reviewed in 1,3]. Therefore, abnormal activity of the UPS or failure of this system contribute to the development of a number of human disorders (e.g. cancer, inflammatory and neurodegenerative disorders), suggesting that both the ubiquitin-protein conjugating system and the 26S proteasome proteolytic pathway constitute attractive targets for pharmacological intervention [reviewed in 4-8] (Figure 1). Since an important role was first ascribed to the 26S proteasome in the pathogenesis of cancer and inflammatory diseases, specific and effective inhibitors of the 20S catalytic core particle have become of great interest as possible therapeutic agents [reviewed in 9]. First inhibitors of the 20S proteasome were either chemically synthesized on the basis of the structure of synthetic peptide substrates specific for proteasomal chymotrypsin-like (ChT-L), trypsin-like (T-L) and peptidylglutamylpeptide (PGPH, caspase-like) activities or isolated from biological extracts on the ground of their initial anti-tumor and anti-inflammatory activities. Those, specific for the ChT-L activity of the 20S proteasome have attracted great interest, because the inhibition of the activity was sufficient to arrest cell cycle, to induce apoptosis and to prevent the expression of many NF-κB -dependent genes both *in vitro* and *in vivo* [reviewed in 9-12]. Two different proteasome inhibitors are now in clinical trials for the treatment of cancer and stroke (see below).

A new avenue for the investigation of the 20S proteasome as a therapeutic target has been opened thanks to the implementation of a high-throughput screening of the pharmaceutical company's small molecule libraries and the utility of the 20S proteasome crystal structure into the design of novel inhibitors with low levels of cytotoxicity against normal cells [13,14]. This is of particular relevance in the light of the findings that proteasome function is of paramount importance for the maintenance of cellular homeostasis and its complete blockage is lethal for the individual cell and the organism as a whole.

Because of this, much attention is now being paid to the development of small molecules that could block protein degradation at the level of the ubiquitination (i.e. inhibitors of ubiquitin-protein E3 ligases, ubistatins) [15,16] or could direct a disease-promoting protein to the ubiquitination and degradation pathway [17] (Figure 1). It is now considered that such modulators may offer greater therapeutic promise in certain malignancies and inflammatory processes than classical inhibitors of the 20S proteasome.

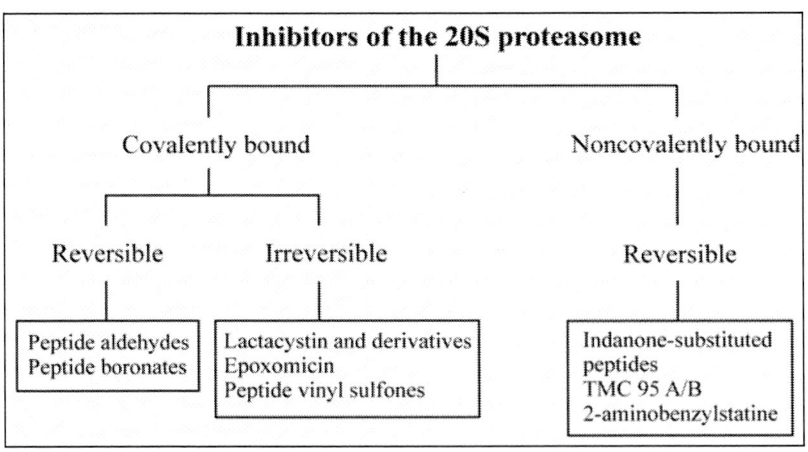

Figure 2. Classification of the 20S proteasome inhibitors. Proposed classification of the 20S proteasome inhibitors based on their mode of action probed by using the X-ray crystallographic analysis of the 20S core particle [13,14,19].

PHARMACOLOGICAL INHIBITORS OF THE 20S/26S PROTEASOMES

The 20S proteasome, the first member of a new threonine class of proteases (N-terminal nucleophile hydrolase, Ntn-hydrolase), exhibits three distinct well-characterized activities (i.e. ChT-L, T-L and PGPH) [18] that can individually be targeted by small molecule inhibitors [13]. Based on the inhibitory mechanism probed by using the X-ray crystallographic analysis of the 20S proteasome, the small molecule inhibitors can be classified into two major groups: 1/those which bind covalently to the catalytic active N-terminal threonine (Thr1Oγ) located on the subunits β1 (PGPH), β2 (T-L) and β5 (ChT-L) and 2/ those which react noncovalently with the substrate binding sites, essential for the catalytic activity (Figure 2).

Covalently bound inhibitors are either reversible analogues of the transition state (i.e. peptide aldehydes, boronates) or irreversible inhibitors (i.e. microbial lactacystin and epoxomicin, and synthetic peptide vinyl sulfones). They have been commonly used to explore the role of the UPS in physiological and pathological processes, indicating that specific inhibition of the 20S proteasome activity can offer a new approach to the development of anticancer and anti-inflammatory drugs [9-12].

Noncovalently bound inhibitors form a new structural class of reversible substrate-binding site modifiers (i.e. indanone-substituted peptides, cyclic tripeptide TMC-95, 2-aminobenzylstatine derivative) that are now being explored for their therapeutic usefulness (see below).

Covalently Bound Inhibitors of the 20S Core Particle

A number of peptide-derived and naturally occurring small molecules that directly target the 20S proteasome have been synthesized and evaluated [reviewed in 9, recently in 20]. Those, which are effective therapeutically in animal models of human diseases and currently tested in clinical trials are presented in Figure 3.

Peptide Aldehydes

Peptide aldehydes were the first generation of the 20S proteasome inhibitors [21]. Among them, the most important are ALLN (N-acetyl-Leu-Leu-norleucinal, calpain inhibitor I), MG-132 (Cbz-Leu-Leu-leucinal) and PSI (Z-Ile-Glu (O-t-Bu)-Ala-leucinal) (Figure 3). They consist of a highly hydrophobic peptide portion that binds to the substrate binding sites and the pharmacophore aldehyde moiety that reacts with the hydroxyl group of the N-terminal catalytic Thr1Oγ to form reversible adduct [19].

Peptide aldehyde ALLN binds covalently to the N- terminal Thr1Oγ located on all three proteolytically active β subunits (β1, β2, β5) [13]; it potently and reversibly blocks the ChT-L activity (IC_{50} of 2.1μM), while less effectively inhibits the T-L and PGPH activities of the 20S proteasome [21]. Much higher potency against other cellular proteases (e.g. calpain, cathepsin B) limited its usefulness in cell- based studies and in animal models. However, the structure of this inhibitor served as a model for synthesis of more potent and selective noncovalent inhibitor of the ChT-L activity of the 20S proteasome (see below), as well as for the design of homo- or heterobivalent inhibitors linked via polyethyleneglycol (PEGx) inhibiting the ChT-L activity or both the ChT-L and T-L activities at nanomolar concentrations [13].

Peptide aldehydes, MG 132 and PSI inhibit predominantly the ChT-L activity of the 20S proteasome [9,22]. They enter cells and effectively block the 26S proteasome-mediated protein degradation at micromolar concentrations [23,24]. However, the major disadvantages of these inhibitors as drug candidates include their poor stability within the cells and the lack of specificity, since at higher concentrations that can also inhibit cathepsin B and calpains. Despite this, both MG-132 and PSI have proved useful in uncovering the role of the 26S proteasomes in a number of cellular processes, which was later confirmed with the use of more specific inhibitors lactacystin and epoxomicin [reviewed in 9,25]. In particular, PSI and MG-132 were the first inhibitors that helped demonstrate the importance of the 26S proteasome in IκBα degradation and NF-κB activation in stimulated cells [26], in the generation of antigen peptides presented on MHC class I molecules [24] as well as in the apoptotic pathway in cancer cell lines [27]. They were also effective therapeutically in various animal models of human diseases [reviewed in 8,12], including cardiovascular disorders [28]. In relation to the latter pathologies, PSI effectively prevented hypertension and vascular hypertrophy in experimental model of deoxycortecosterone (DOCA)-salt - induced hypertension and inhibited ischemic acute renal failure (ARF) in rats, through the suppression of endothelin-1 (ET-1) production in the aorta and kidney *via* inhibition of NF-κB transcriptional pathway [28]. More recently we have demonstrated that PSI prevents experimental arterial thrombosis and suppresses indirectly platelet aggregation *ex vivo* in renovascular hypertensive rats [29]. The exact mechanism, through which it exerts its antithrombotic activity, is not yet known, however there exists evidence that the inhibition of 26S proteasome-mediated NF-κB activation is sufficient to prevent the expression of the tissue factor (TF) in monocytes during extracorporeal circulation [30].

It is of interest that rats subjected to several systemically injected doses of PSI developed progressive parkinsonism with bradykinesia, rigidity, tremor and an abnormal posture confirmed by neurodegeneration observed in postmortem examinations what is suggestive that PSI is a candidate Parkinson disease (PD)-inducing toxin [31]. The effects of PSI were also mimicked by more highly selective proteasome inhibitor epoxomicin.

Peptide Boronates

Much more potent and selective inhibitors of the 20S proteasome are peptides possessing the boronic acid pharmacophore group instead of the aldehyde group [32] (Figure 3). Peptide boronates form a stable tetrahedral adduct and dissociate from the proteasome at a slower rate than the peptide aldehydes [19], thus the inhibition is practically irreversible.

Among various peptide boronates tested, the most important is bortezomib (Velcade®, previously known as PS-341) (Figure 3), because it is the first proteasome inhibitor that has progressed to clinical trials for the treatment of cancer (see below). The structure of this inhibitor combines both the N-blocked high affinity peptide portion (pyrazinyl(Pyz)carbonyl-

Phe-Leu), which binds selectively to the substrate binding site of the ChT-L activity, and the C-terminal boronic acid group that binds to the N-terminal Thr1Oγ on β5 subunit to form slowly-reversible pseudo-covalent adduct [33].

In cell-based studies, bortezomib fulfilled all the criteria for the treatment of cancer: (*i*) it crossed cell membranes, (*ii*) selectively and reversibly inhibited the 20S and 26S proteasome-dependent protein degradation (i.e. IκBα cyclin E, p53, p27) in a number of tumor cell lines, (*iii*) showed cytotoxity against 60 cancer cell lines derived from multiple human tumors in the National Cancer Institute *in vitro* screen, (*iv*) exhibited relatively few toxic effects on normal cells, (*v*) induced apoptosis in tumor cells resistant to chemotherapy or radiation, (*vi*) down-regulated cytokine-induced expression of tumor necrosis factor-α (TNF-α), vascular cell adhesion molecule-1 (VCAM-1), vascular endothelial growth factor (VEGF), and (*vii*) unlike peptide aldehydes, it was refractory to removal from cancer cells by the multidrug resistance carrier system [33-35]. Afterwards, bortezomib was demonstrated to be therapeutically active in human tumor xenograft models of a wide range of hematological malignancies and solid tumors, both as a single agent and in combination with standard chemotherapeutics [reviewed in 36]. On the basis of the preclinical studies bortezomib has recently received the US Food and Drug Administration (FDA) approval for the treatment of patients with relapsed and refractory multiple myeloma (MM) [reviewed in 37]. Bortezomib is also under investigation for its therapeutic efficacy in other hematological malignancies [38] and solid tumors [39]. The state of the art information concerning pharmacology, pharmacokinetics and practical application of bortezomib was recently reviewed by Schwartz and Davidson [40]. In general, bortezomib appears to be well tolerated with mild/moderate and manageable side effects. In MM patients it produced 35% overall response rate and 10% complete responses. The major molecular mechanism, through which bortezomib mediates anti-MM activity involves the induction of p53-dependent and p53 - independent apoptotic pathway in the MM cells resistant to conventional therapy and down-regulation of the expression of tumor-promoting proteins in bone marrow stromal cells (BMSCs), including IL-6, the key growth and survival factor produced by these cells [41]. The mode of action of bortezomib depends largely on the inhibition of anti-apoptotic and pro-inflammatory NF-κB both in the MM cells and in BMSCs [42]. Gene expression profiling and proteomic analysis of the MM cells have recently demonstrated that subtoxic concentration of bortezomib down-regulates the expression of several other proteins involved in the cellular response to genotoxic stress [43].

Recent screening of the pharmaceutical company's compound collection has identified a novel potent but rapidly reversible lactam boronic acid based inhibitor of the 20S proteasome [44] (Figure 3).

Lactacystin and Derivatives

Lactacystin (Figure 3), a non-peptide compound produced by *Streptomyces lactacysteinaeus*, was first found to be a neurite outgrowth-inducer in a murine neuroblastoma cell line [45]. Further extensive studies of Fenteany and co-workers [46] have demonstrated that the radiolabeled lactacystin targets the 20S proteasome subunit X (β5) in the cell and binds irreversibly to the N-terminal catalytic threonine residues of the purified 20S proteasome. X-ray crystal structure of the yeast 20S proteasome-lactacystin complex has confirmed that it binds irreversibly to the N-terminal Thr1Oγ on β5 subunit harboring the ChT-L activity [13]. Lactacystin is, in fact, a precursor for cell-permeable inhibitory active intermediate *clasto*-lactacystin β-lactone (Figure 3), which is formed following the removal of N-acetylcysteine in aqueous solution [47]. On binding, the β-lactone ring undergoes nucleophilic attack by the hydroxyl group of the N-terminal threonine residue to form a stable acyl – enzyme complex [19]. This adduct is however slowly ($t_{½}$ about 20h) hydrolyzed by water, thus the inhibition of proteasome activity is reversed within the cell [47]. Lactacystin/β-lactone inhibits the ubiquitin-mediated degradation of short and long lived proteins in a wide range of cell lines, without affecting other known cellular proteases [48]. However, it is capable of inhibiting lysosomal serine carboxypeptidase A (cathepsin A) in cell lysates [49] and in tumor cell lines [50] and cytosolic tripeptidyl peptidase II [51].

Lactacystin has been extensively used to dissect the function of the 26S proteasome in physiological and pathological processes in a broad range of human cell lines [48]. An important discovery that was a consequence of the use of the lactacystin/β-lactone in cell-based studies was that the inhibition of proteasomes could either induce or prevent apoptosis; the phenomenon depends on the cell type, model system, the length of time of cell exposure and the concentration of the inhibitor [reviewed in 25]. This allowed better understanding of the role played by 26S proteasomes in the apoptotic pathway in a number of cell types, including normal and transformed neuronal and glial cells [52-55, and for review see in 2]. Curiously enough, sub-toxic concentrations of lactacystin in combination with a cAMP-stimulating agent are able to induce extensive apoptosis of neuroblastoma cells in an *in vitro* study [56].

In animal models, lactacystin helped to uncover the role of proteasomes in angiogenesis and in the production of plasminogen activator [57]. It also confirmed the significance of the 26S proteasome-dependent NF-κB activation pathway in the early phase of renal injury in animal model of ischemia/reperfussion [58]. Furthermore, lactacystin helped to better understand the significance of proteasomes in Parkinson's disease (PD), both in the early phase of dopaminergic neuronal death and in inclusion body formation in an experimental model of hemiparkinsonian rats [59].

Importantly, lactacystin is a good example of a synergy between basic research and drug development. After lactacystin/β-lactone had been identified as an inhibitory active constituent exhibiting poor stability within the cells [47], this compound (named omuralide), served as a model to the synthesis of the *clasto*-lactacystin β-lactone-based compounds that were endowed with increased stability and potency [60]. One of them, MLN-519 (C-7 n-propyl analogue of *clasto*-lactacystin *β*-lactone, Figure 3) has been shown in an *in vitro* study to inhibit the 26S proteasome-dependent NF-κB activation pathway with lower IC_{50} values (1-5 μM) than the natural product. Then, MLN-519 was successfully used in animal models of myocardial reperfusion injury [61,62]; focal cerebral ischemia as a single agent [63] and in combination with tissue plasminogen activator (t-PA) [64]; middle cerebral artery occlusion and reperfusion [65]; and in other inflammatory diseases, including experimental autoimmune encephalomyelitis, which is an animal model of multiple sclerosis [reviewed in 11]. The preclinical studies have indicated that MLN-519 exerts cytoprotective effects by the inhibition of 26S proteasome-mediated NF-κB activation in pro-inflammatory cells infiltrating the ischemic tissue. MLN-519 is currently under clinical evaluation for the treatment of acute stroke and myocardial infarction [66,67]. Detailed information concerning the role of the UPS and MLN-519 therapy in stroke is described in the respective chapters of this book (see Chapters 41 and 42) and reviewed elsewhere [68,69].

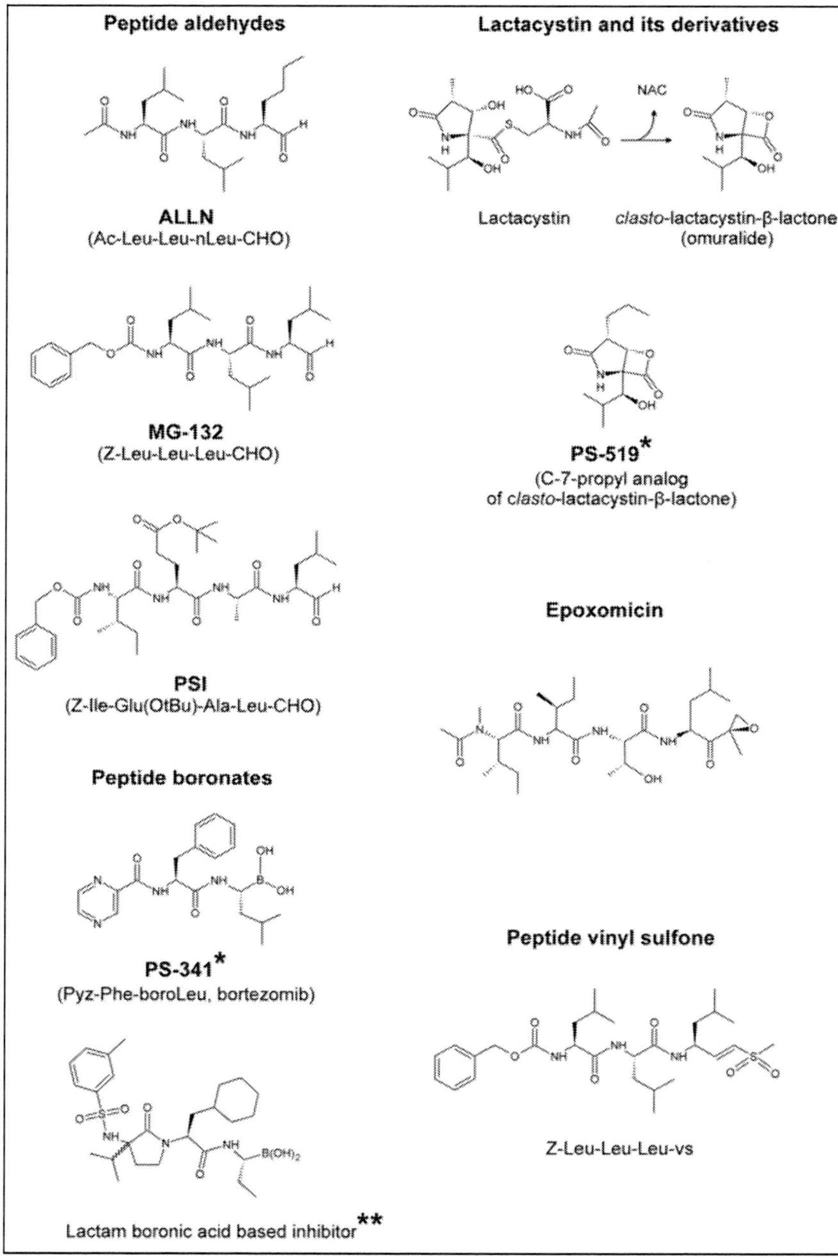

Figure 3. Chemical structure of the 20S proteasome inhibitors explored for therapeutic usefulness in animal models and (*) currently in clinical trials. (**) Lactam boronic acid based inhibitor is a newly synthesized PS-341 analog; its cellular effects are currently under investigation [44].

Other Covalently Bound Inhibitors

Epoxomicin, an active microbial α', β'-epoxyketone peptide, and synthetic peptide vinyl sulfones (Figure 3) are strong irreversible inhibitors of the 20S proteasome.

Epoxomicin has been identified as a potent and highly selective inhibitor of IκB degradation and NF-κB activation by the 26S proteasomes in cytokine-stimulated cells and as an anti-inflammatory agent in the murine ear edema assay [70]. X -ray analysis of the crystals of 20S proteasome complexed with epoxomicin revealed that the epoxy group binds covalently to the N-terminal catalytic Thr1Oγ located on the three active β subunits to form a stable morpholino adduct, specific for the small class of Ntn-hydrolases [13]. Thus, epoxomicin, unlike most other proteasome inhibitors, is highly specific for the 20S proteasome and does not inhibit other cellular proteases examined so far [70]. However, since epoxomicin inhibits irreversibly all three proteolytic activities of the 20S proteasome, it provides a powerful tool for the investigation of the UPS function in living cells, while is useless as a drug candidate.

Similarly, the synthetic peptide vinyl sulfones (vs), such as Z-Leu-Leu-Leu-vs (Figure 3), Ac-Arg-Leu-Leu-N-vs, synthesized by Bogyo and co-workers [71], bind covalently to the N - terminal Thr1Oγ located on all three catalytically active β-subunits [19] and inhibit the ChT-L, T-L, and PGPH activities of the 20S proteasome in an irreversible manner. Therefore, as epoxomicin, they have limited application as drug candidates. Peptide vinylsulfone, Z-Leu-Leu-Leu-vs tagged with biotin is applied for the purpose of affinity chromatography, while its radiolabeled nitrophenol derivative, ^{125}I-NIP-Leu-Leu-Leu-vs is a useful probe for active site labeling of the proteasome in the living cells [71].

Interestingly, more recent studies have demonstrated that synthetic trileucine methyl vinyl sulfonates, inhibiting the T-L activity of the 20S proteasome, have anti-trypanosomal activity in the sub-nanomolar concentrations [72].

Noncovalent Inhibitors of the 20S Core Particle

Promising results with classical covalently bound inhibitors of the 20S proteasome as anticancer and anti-inflammatory agents led to widespread interest into the design and synthesis of novel noncovalent inhibitors of the 20S proteasome that could exhibit lower levels of cytotoxicity against normal cells.

One of the first noncovalent inhibitors of the 20S proteasome, named CVT-634 (Figure 4), was synthesized on the basis of the structure of a calpain inhibitor I (Ac-Leu-Leu-norleucinal) [73]. CVT-634 consists of a high affinity dipeptide Ac-Leu-Leu that binds to the substrate binding site of the β5 subunit, and a less reactive, but potentially hydratable indanone group in the C-terminus, instead of norleucinal moiety. CVT-634 inhibits selectively the ChT-L activity of the purified 20S proteasome (IC_{50} 0.2μM), without modifying the N-terminal catalytic Thr1Oγ of β5 subunit. It also inhibits the 20S proteasome in human tumor cell lines with IC_{50} 10-20μM and does not affect other cellular proteases. In cell-based studies, CVT-634 inhibits growth of many cancer cell lines and blocks the cytokine-induced NF-κB activation pathway [74]. Of interest, CVT-634, similarly to MLN-519, is able to reduce infarct volume in a focal model of cerebral ischemia [75].

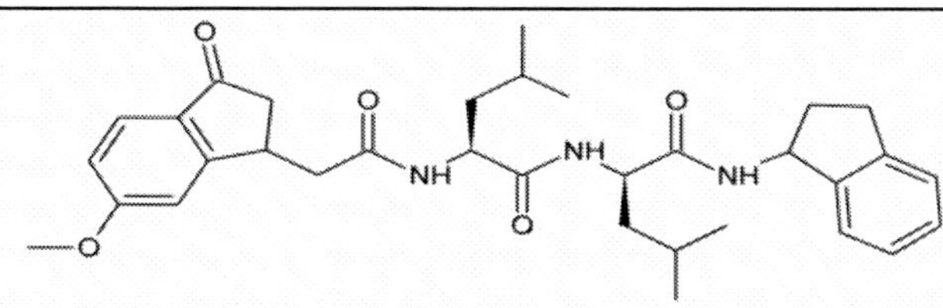

CVT-634 (5-methoxy-1-indanone-3-acetyl-Leu-D-Leu-1-indanylamide)

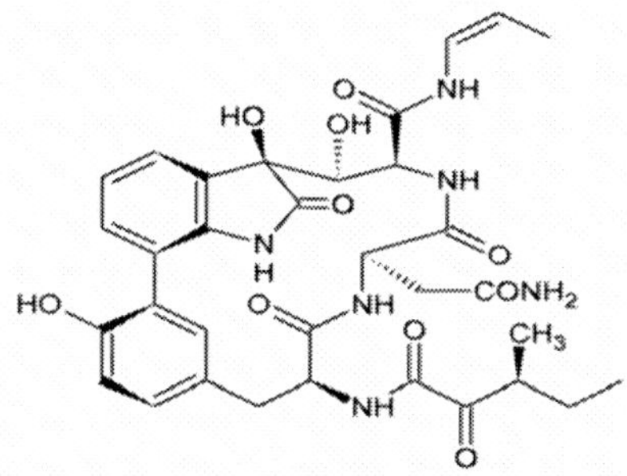

TMC-95A

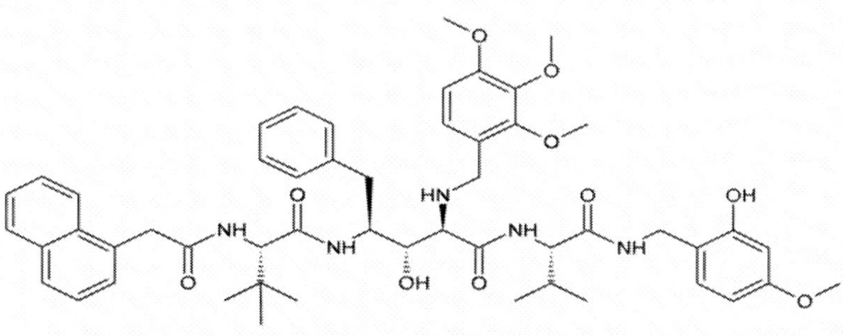

2-aminobenzylstatine derivative

Figure 4. Noncovalent inhibitors of the 20S proteasome core particle.

Koguchi and co-workers [76] isolated from the fermentation broth of *Apiospora montagnei*, a series of cyclic tripeptides, named TMC-95 A-D, which were able to inhibit reversibly the 20S proteasome without affecting other known proteases. TMC-95A (Figure 4) that selectively inhibited the 20S proteasome in the low nanomolar range, was then synthesized by Albrecht and Williams [77]. X-ray crystallographic analysis of the yeast 20S proteasome- TMC 95A complex has revealed that it does not modify the N-terminal catalytic Thr1Oγ, but it forms characteristic hydrogen bonds with the substrate-binding sites of all three proteolytically active β-subunits (β1, β1, β5) [13,78]. This compound has been found to be cytotoxic against HCT-116 human colon carcinoma cells (IC_{50} 4.4 μM) and HL-60 human promyelocytic leukemia cells (IC_{50} 9.8 μM) [76]. It also induces neurite outgrowth in PC12 rat pheochromocytoma cells (1 - 20μM) with less toxic effects than covalently bound proteasome inhibitors [79].

In a search for novel cytotoxic and antiproliferative agents that act by noncovalent inhibition of the 20S proteasome, Garcia-Echeverria and co-workers [80] applied a high-throughput screening of a series of the 2-aminobenzylstatine derivatives from a Novartis library archives. Modification of the 2-aminobenzylstatine derivatives, originally synthesized to target the human immunodeficiency virus -1 (HIV-1) aspartyl protease, led to the discovery of a potent and highly selective inhibitor of the 20S proteasome (Figure 4) that bound exclusively to the substrate-binding site of the subunit β5 harboring the ChT-L activity [80]. Recent structure-based optimization approach has allowed improving the potency of the 2-aminobenzylstatine derivative from micromolar to nanomolar concentrations [14].

Conventional Pharmacological Agents Targeting the 20S/26S Proteasomes

The proteasome-dependent protein degradation is also influenced by a number of agents used in conventional therapies, dietary chemopreventive compounds and toxins [reviewed in 8, 20]. Herein we present the examples of the drugs that directly target the 20S proteolytic subunits or regulate the expression and function of non-proteolytic proteasome subunits necessary for proteasomal activity.

Ritonavir (Figure 5), an inhibitor of the HIV-1 protease, which is used for the treatment of HIV-infected patients and AIDS disease, inhibits noncovalently and reversibly the ChT-L activity of the purified 20S proteasome (IC_{50} of 3μM) and moderately enhances the T-L activity of this complex ('two-site modifier') [81]. Ritonavir has been shown to markedly reduce the presentation of lymphocytic choriomeningitis virus (LCMV) epitopes on MHC class I molecules of infected cells at therapeutically relevant concentrations (IC_{50} of 5μM), suggesting that its anti-proteasome activity may be beneficial in the treatment of autoimmune diseases and organ transplantation [82,83]. At higher concentrations (i.e. 50μM) Ritonavir causes accumulation of intracellular ubiquitinated proteins and prevents lipopolysaccharide (LPS)-induced degradation of the IκBα [82]. Recent studies indicate that Ritonavir, through a selective inhibition of the ChT-L activity of the 20S proteasome, is able to reduce the rate of proliferation and to induce apoptosis in several tumor cell lines [84]. It also exerts cytostatic and cytotoxic effects on glioma cells *in vitro*, while is unable to control tumor growth *in vivo* most likely due to the blood-tumor 'barrier' [85]. Interestingly, Ritonavir, similarly to proteasome inhibitor MLN-519, prevents clinical symptoms of autoimmune encephalomyelitis in rats by inhibiting the activation and infiltration of T-cells into the CNS [86].

Two immunosuppressive agents that also influence the proteasome-dependent protein degradation are cyclosporin A and rapamicin (Figure 5). Cyclosporin A acts as an uncompetitive inhibitor of the ChT-L activity of the 20S proteasome [87], while the structurally unrelated rapamicin inhibits the expression of the proteasome activator, PA28α/β both at mRNA and protein levels in activated T cells [88]. PA28 is known to stimulate the LMP-containing 20S proteasomes (so called immuno-proteasomes) and their complexes with

PA700 regulator in order to generate short peptides for antigen presentation on MHC class I molecules [89]. It seems that immunosuppressive activity of rapamicin and cyclosporin A depends, at least in part, on the inhibition of antigen production by proteasomes. In addition, cyclosporin A inhibits the 26S proteasome-dependent IκB degradation and NFκBp105 processing in lipopolysaccharide (LPS) stimulated cells [87], suggesting that the inhibition of 20S proteasome is the mechanism responsible for its anti-inflammatory activity.

Aclacinomycin (Aclarubicin, Figure 5), an antitumor drug isolated from *Streptomyces,* acts as an uncompetitive and reversible inhibitor specific for the ChT-L activity of the purified 20S proteasome [90]. Both aglycone and sugar moieties of the aclacinomycin A are required for inhibitory activity. Another anticancer agent vinblastine (a *Vinca* alcaloid) inhibits reversibly and noncompetitively all three proteolytic activities of the purified 20S proteasome [91]. In cell-based studies aclacinomycin and vinblastine have been demonstrated to inhibit degradation of polyubiquitinated proteins [90], including signal-induced degradation of IκBα [91].

Similarly, a toxic fungal epipolythiodioxopiperazine metabolite, gliotoxin (Figure 5), which is known to inhibit NF-κB activation pathway and to induce apoptosis, has been recognized as a highly selective noncompetitive inhibitor of the ChT-L activity of the purified 20S proteasome [92]. Inhibition of 26S proteasome-dependent degradation of IκB, β-catenin and p53 protein in the cells treated with gliotoxin is reversed by dithiothreitol (DTT), which reduces the disulfide bridge in gliotoxin molecule that is essential for its biological effects.

More recently, it has been demonstrated that a green tea ester-bond-containing polyphenol (-) -epigallocatechin-3 gallate [(-)-EGCG] (Figure 5) of known anticancer properties, inhibits the 20S proteasome by a mechanism similar to that of lactacystin/β-lactone, namely through irreversible binding to the catalytic active N-terminal Thr1Oγ on β5 subunit harboring the ChT-L activity [93]. Furthermore, both (-)- EGCG and its (+)enantiomeric analogs target proteasomes in cancer cell lines, leading to the inhibition of proliferation, induction of apoptosis and suppression of colony formation. Particularly interesting is that the peracetate ester of (-)-EGCG, synthesized in an attempt to enhance a low bioavailability of (-)-EGCG, inhibits proteasome activity in intact leukemic cells, but exhibits no inhibitory action on purified 20S proteasome, indicating that it may function as a pro-drug of (-)-EGCG [94].

A natural antibacterial peptide, PR-39 (proline-and arginine-rich peptide), unlike other proteasome inhibitors, blocks 26S proteasome-mediated protein degradation by reversible binding to the catalytically inactive subunit α7 of the 20S core particle [95,96]. Importantly, such interaction is sufficient to inhibit IκB degradation and NF-κB activation, without affecting total proteasome-dependent proteolysis in the cell culture assays. PR39 has been shown to suppress VCAM-1 and ICAM-1 gene expression in TNF-α - activated human endothelial cells and to reduce myocardial infarct size in a mouse model of myocardial infarction [95].

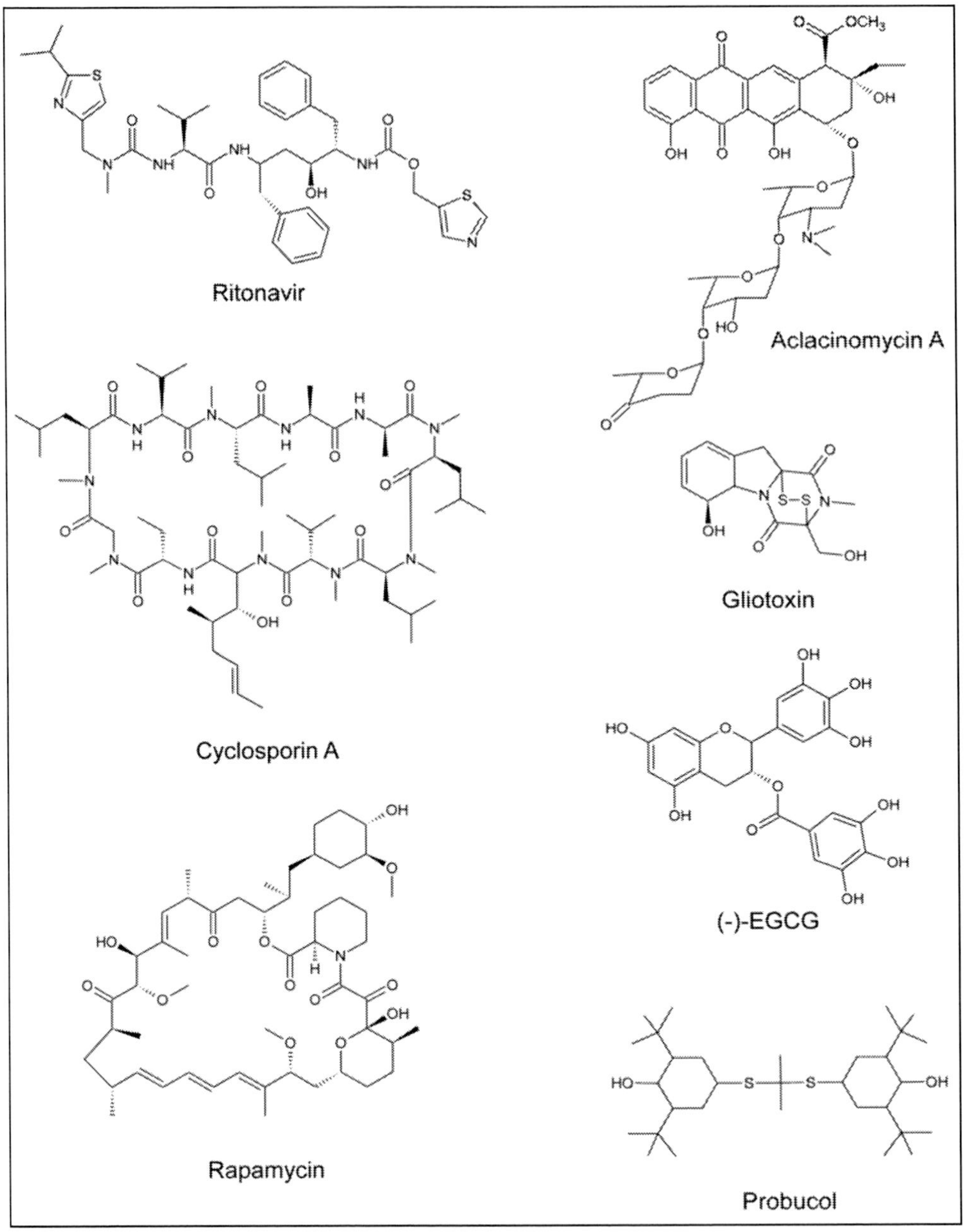

Figure 5. Conventional pharmacological agents targeting the 20S/26S proteasomes.

Furthermore, an anti-atherogenic phenolic antioxidant, probucol (Figure 5) and its analogue BO-653, have been both shown to suppress 26S proteasome-dependent IκBα degradation in activated human endothelial cells via specific inhibition of gene expression for the α-type 20S proteasome subunits PMSA1 (HC2), PMSA2 (HC3), PMSA3 (HC8) and PMSA4 (HC9) [97]. The exact mechanisms by which these phenolic compounds regulate gene expression of the α subunits of the 20S proteasome are still obscure.

Finally, some natural and synthetic small molecules function as proteasomal activators. These include cardiolipin (diphosphatidylglycerol) that stimulates the ChT-L and PGPH activities of the 20S proteasome [98] and the synthetic peptidyl alcohol Z-Ile-Glu (O-t-Bu)-Ala -leucinol that enhances the ChT-L activity of the purified 20S proteasome [99]. Interestingly, Z-Ile-Glu (O-t-Bu)-Ala -leucinol alters the profile of peptides generated from ovalbumin immunodominant epitopes by the PA28α-20S proteasome, and thus behaves as a proteasome modulator. It cannot be excluded that synthetic small molecule activators, like Z-Ile-Glu (O-t-Bu)-Ala -leucinol, will be tested as potential therapeutics for the regulation of immune response or in some pathologies resulting from the loss of proteasome activity, such as neurodegenerative disorders.

Pharmacological Modulators of the Ubiquitin Pathway

A novel approach to drug development is to design small molecule compounds that instead of inhibiting the 20S proteasome, could block protein degradation at the step of the ubiquitination or that could target a disease-promoting protein for the ubiquitination and degradation by 26S proteasomes. Several small molecule inhibitors of the ubiquitin-protein-conjugating system have been synthesized and evaluated [for recent review see in 5, 15, 20], however many of them have a limited therapeutic usefulness due to the lack of specificity. It may turn out that ubistatins, compounds with a novel mechanism of action [16], may be among them. Ubistatins (sulfonated stilbenes) were discovered by chemical genetic screens for inhibitors of cyclin B degradation in *Xenopus* extracts and were found to block the degradation of ubiquitinated Sic1 by the 26S proteasome without affecting proteasomal activity assayed with the use of synthetic peptide substrates [16]. Out of the ubistatins, the most potent compound is ubistatin A (Figure 6A). Nuclear magnetic resonance (NMR) titration experiments have indicated that ubistatin A interacts specifically with the Lys 48-linked ubiquitin chains, thus preventing binding of ubiquitynated Sic1 to the specific receptors (Rad23 and Rpn10) in 19S regulator particle and its translocation into the 20S proteasome (see in Figure 1). Novel highly selective small molecules targeting the ubiquitination system include certain E3 ligase inhibitors and proteolysis inducers (Figure 6 B-D).

Inhibitors of Substrate Specific E3 Ligase

The ubiquitin-protein E3 ligases transfer activated ubiquitin from one of the ubiquitin-conjugating E2s directly to a lysine residue in the target protein to generate polyubiquitin chain [reviewed in 1, 100] (see in Figure 1). Such proteins are then recognized by the 19S regulatory particle and directed to the 20S core particle for destruction, following the removal of the polyubiquitin chains by deubiquitinating enzymes (DUBs). There are hundreds of various E3 ligases, but only few of them are well characterized with the respect to the substrate specificity [reviewed recently in 101-103, and also in Ref. 1]. Their overexpression and low levels of the target proteins have been demonstrated in many human tumors.

The most extensively studied E3 ligases as potential candidates for drug discovery include the ubiquitin E3 ligase, Mdm2 that ubiquitinates the tumor suppressor protein p53 [104] and the E3 ligases that specifically ubiquitinate the phosphorylated NF-κB inhibitor pIκBα [105,106].

The structural analysis of the Mdm2 ligase prompted the investigators to search for small molecule inhibitors that could bind to the p53 binding site, thus preventing its ubiquitination and degradation by the 26S proteasome. Recent screening of diverse libraries of synthetic chemicals has identified a series of cis-imidazoline analogs, named nutlins (1-3) that bind to the hydrophobic p-53-binding pocket in the Mdm-2 molecule with IC_{50} values ranging from 100 to 300nM [104]. Nutlin-3 (Figure 6B) exhibits the most potent binding activity. In cell-based assay nutlin-3 prevented p53 ubiquitination, resulting in the activation of the p53 pathway in certain cancer cells. It also inhibited tumor growth in an experimental model in mice. More studies are however required to find out whether *in vivo* activity of the nutlins is only limited to tumor tissues.

The IκBα is ubiqutinated on certain lysine residues by the $SCF^{\beta TRCP}$ (Skp1-Cullin-F-box) ubiquitin ligase complex that recognizes exclusively the phosphorylated form of IκBα (pIκBα) in cytokine - activated cells [reviewed in 107]. The identification of the putitative pIκBα-E3 ligase recognition motif allowed synthesis of a short pIκBα peptide antagonist that after microinjection, specifically inhibited IκBα ubiquitination and its subsequent degradation by the 26S proteasome in TNF-α-stimulated HeLa cells and abolished the expression of inducible cell adhesion molecule, E-selectin [105].

Recently, a small molecule compound, Ro106-9920 (Figure 6C) was identified using the screening paradigm as a highly selective inhibitor of IκBα ubiquitination and NF-κB activation in cell-free assay and in TNF-α - or LPS- stimulated cells [106]. It also prevents cytokine production *in vivo* in rats. Ro106-9920 does not however, block the $SCF^{\beta TRCP}$ but it specifically targets another, yet unidentified, IκBα E3 ligase.

Proteolysis Inducers

Another interesting approach is a selective ubiquitination of the disease-promoting proteins, including those that are not known to be ubiquitinated ('targeted proteolysis'). Several chimeric F-box proteins have recently been designed to recruit a selected cellular protein (e.g. β-catenin, retinoblastoma, cyclin A, cdk2) for ubiquitination and degradation by the 26S proteasomes [reviewed in 17]. These large protein-based compounds require however, special delivery system, eg. viral or liposomal vehicles, what makes them poor drug candidates.

On the other hand, two types of small molecule proteolysis inducers seem to be potentially useful therapeutics in the future. These include the proteolysis-targeting chimeric molecules, named protacs, and the small molecule proteolysis inducers (SMPI) (Figure 6D).

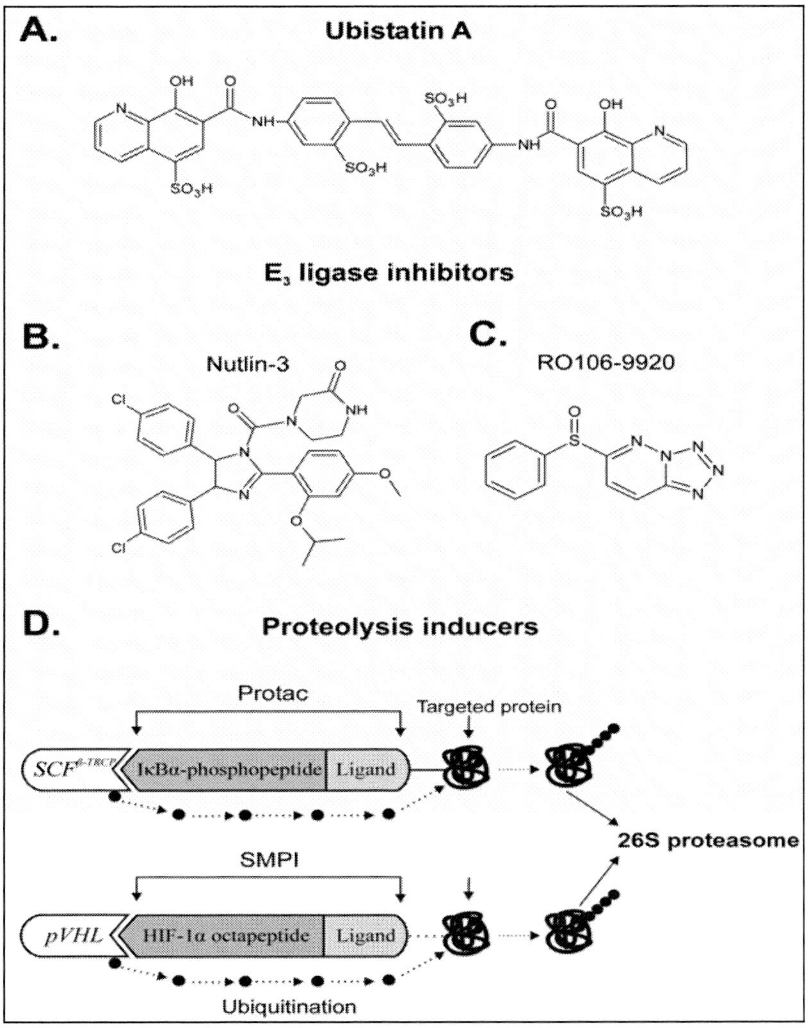

Figure 6. Small molecule modulators of the ubiquitin system. Chemical structures of: (A) ubistatin A [16]; (B) nutlin-3 [104]; (C) RO106-9920 [106]. (D) schematic presentation of proteolysis inducers consisting of IκBα-phosphopeptide (protacs) or HIF-1α (hypoxia-inducible factor-1α) octapeptide (SMPI, small molecule proteolysis inducers) linked to the ligand that specifically recognize and bind a disease-promoting protein [108-110]. SCFβ-TRCP ubiquitin ligase or pVHL (von Hippel-Lindau) ligase ubiquitinate target protein (shown as black dots), which is then proteolyzed by the 26S proteasome. For details, see text.

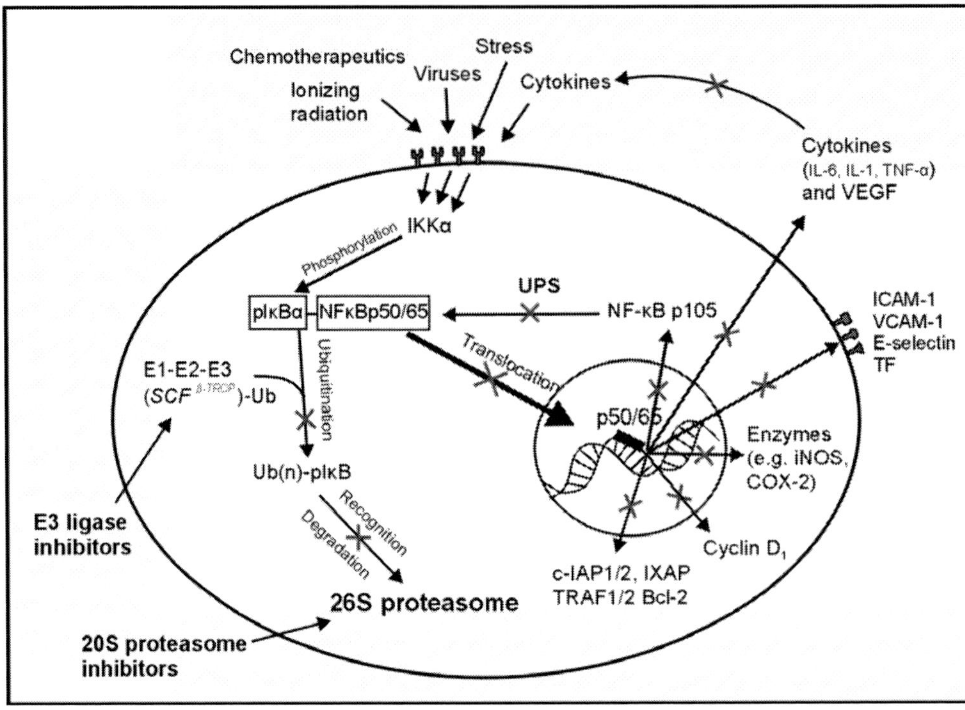

Figure 7. Diagram representing the inhibition of the NF-κB pathway by UPS inhibitors as a possible common mechanism of their anti-tumor and anti-inflammatory activities in certain malignancies and inflammatory diseases. Inflammatory signals, viruses, oxidative stress, certain chemotherapeutics and ionizing radiation activate the IκB kinases (IKKs) that phosphorylate the IκBα, which is then ubiquitinated by specific SCF^{TRCP} ligase and subsequently degraded by the 26S proteasome. The active NFκBp50/65 translocates into the nucleus where it activates the expression of a number of genes that suppress apoptosis and differentiation, induce cell proliferation, promote angiogenesis and metastasis and potentate inflammation [107, 111]. A cell-permeable pIκBα phosphopeptide antagonist of $SCF^{\beta\text{-}TRCP}$ blocks pIκBα ubiquitination resulting in the inhibition of IκBα degradation by the 26S proteasome. Cell-permeable inhibitors of the 20S proteasome inhibit degradation of ubiquitinated IκBα resulting in the blockade of p50/65 nuclear translocation.

The protacs were synthesized on the basis of well-characterized internal recognition motif triggering the binding of IκBα to $SCF^{\beta\text{-}TRCP}$ ligase (10-amino acid IκBα phosphopeptide domain), which was linked covalently with the following ligands: 1/ ovalicin, an anti-angiogenic inhibitor that specifically recognizes the metionine aminopeptidase-2 (Met-AP-2) (protac-1), 2/ estradiol, which binds noncovalently to the estrogen receptor implicated in the progression of breast cancer (protac-2) or 3/ dihydroxytestosterone (DHT) that recognizes the androgen receptor, a known promotor of prostate cancer growth (protac-3) [108,109]. Thus, following binding, the $SCF^{\beta\text{-}TRCP}$ ligase ubiquitinates target protein, which is then proteolyzed by the 26S proteasome complex. In a cell-based assay Protac-3 (1μM) has been shown to recruit the androgen receptor to ubiquitination by the $SCF^{\beta\text{-}TRCP}$ ligase followed by subsequent degradation by the 26S proteasome [109]. However, a major disadvantage of the phosphopeptide-based protacs is their poor membrane permeability and bioavailability.

More recently, Zhang and co-workers [110] synthesized small molecule proteolysis inducers (SMPIs) that are able to penetrate cells. They consist of a cell-permeable synthetic octapeptide derived from hypoxia-inducible factor-1α (HIF-1α), which is specifically recognized by the phosphorylated form of von Hippel-Lindau tumor suppressor (pVHL) ligase E3 and the covalently bound ligand domain (estradiol or anti-angiogenic fumagillol (Fu) instead of ovalicin). The Fu-SMPI selectively induces the ubiquitination of Met-AP-2 by pVHL E3 ligase and its degradation by the 26S proteasome in lung cancer cells, while the estradiol-based SMPI recruits the estrogen receptor for ubiquitnation and degradation by the 26S proteasome in MCF-7 breast cancer cells.

CONCLUSIONS

The UPS is essential for intracellular homeostasis, since the degradation of 80-90% of cytosolic, nuclear and membrane proteins is kept under its tight control. The UPS degrading proteins are those involved in cell proliferation, apoptosis and inflammation as well as misfolded, inappropriately stabilized or damaged proteins, etc. It is not astonishing that the abberation in the UPS function contributes to various debilitating diseases, in particular cancer, cardiovascular disorders and neurodegenerative diseases. Therefore, the attention of researches, pharmaceutical companies and clinicians has focused on this vital system. Over the last decade a number of small molecule inhibitors of the UPS have been tested in experimental settings. However, pleiotropic functions of the UPS implicated toxic adverse effects of its inhibitors in preclinical studies, which limited their clinical use. As a consequence, only two small molecule inhibitors of the 20S proteolytic core particle were developed and entered clinical trials in cancer and stroke patients. Rapid progress in combinatorial chemistry, structure-based design and high-throughput screening of pharmaceutical companies' small molecule libraries have accelerated developments in the identification of new classes of proteasome inhibitors that are cautiously awaited to become therapeutic agents. Furthermore, better understanding of the function of the UPS has opened another new avenue of pharmacological intervention into the ubiquitin protein-conjugating system that allows (*i*) enhanced ubiquitination and degradation of selected tumor-promoting proteins [17], and (*ii*) prevention of the ubiquitination and subsequent degradation of tumor suppressor proteins [15], as well as IκBα [105], an inhibitor of NF-κB transcriptional factor, stimulating the expression of specific cellular genes, whose products contribute to the pathogenesis of some human cancers and inflammatory diseases [reviewed in 107 and 111] (Figure 7). In fact, the inhibition of NF-κB pathway by the 20S proteasome inhibitors (i.e. PS-341 and MLN-519 is though to be a major mechanism of their anti-neoplastic and anti-inflammatory effects, especially in tumors resistant to conventional therapy and in stroke or myocardial infarction, respectively.

All in all, it is truly fascinating to witness how rapidly the advances in basic research are translated into innovative and promising therapies designed to combat diseases that have so far been incurable.

Acknowledgments

We thank Justyna Ostrowska-Roszczenko and Kamil Kruszewski for assistance in the preparation of the manuscript and the figures.

References

[1] Glickman MH, Ciechanover A: The ubiquitin-proteasome proteolytic pathway: destruction for the sake of construction. *Physiol Rev* 2001, 82:373-428.

[2] Ding Q, Keller JN: Proteasomes and proteasome inhibition in the central nervous system. *Free Radical Biol Med* 2001, 31:574-584.

[3] Coux O, Tanaka K, Goldberg AL: Structure and functions of the 20S and 26S proteasomes. *Annu Rev Biochem* 1996, 65:801-847.

[4] Ciechanover A, Schwartz AL: The ubiquitin system: pathogenesis of human diseases and drug targeting. *Biochim Biophys Acta* 2004, 1695:3-7.

[5] Burger A, Seth AK: The ubiquitin-mediated protein degradation pathway in cancer: therapeutic implications. *Eur J Cancer* 2004, 40:2217-2229.

[6] Adams J: The proteasome: structure, function, and role in the cell. *Canc Treat Rev* 2003, 29 (Suppl 1):3-9.

[7] Herrmann J, Ciechanover A, Lerman LO, Lerman A: The ubiquitin-proteasome system in cardiovascular diseases- a hypothesis extended. *Cardiovasc Res* 2004, 61:11-21.

[8] Wójcik C: Ubiquitin-and proteasome-dependent pathway of protein degradation as an emerging therapeutic target. *Emerging Therapeutic Targets* 2000, 4:89-111.

[9] Kisselev AF, Goldberg AL: Proteasome inhibitors: from research tools to drug candidates. *Chem Biol* 2001, 8:739-758.

[10] Adams J: Proteasome inhibitors as new anticancer drugs. *Curr Opin Oncol* 2002, 14:628-634.

[11] Elliott PJ, Zollner TM, Boehncke WH: Proteasome inhibition: a new anti-inflammatory strategy. *J Mol Med* 2003, 81:235-245.

[12] Rivett AJ, Gardner RC: Proteasome inhibitors: from in vitro uses to clinical trials. *J Pept Sci* 2000, 6:478-488.

[13] Groll M, Huber R: Inhibitors of eukaryotic 20S proteasome core particle: a structural approach. *Biochim Biophys Acta* 2004, 1695:33-44.

[14] Furet A, Imbach P, Furst P, Lang M, Noorani M, Zimmermann J, Garcia-Echeverria C: Modeling of the binding mode of a non-covalent inhibitor of the 20S proteasome. Application to structure-based analogue design. *Bioorg Med Chem Lett* 2001, 11:1321-1324.

[15] Robinson PA, Ardley HC: Ubiquitin-protein ligases-novel therapeutic targets? *Curr Protein Pept Sci* 2004, 5:163-76.

[16] Verma R, Peters NR, D'Onofrio M, Tochtrop GP, Sakamoto KM, Varadan R, Zhang M, Coffino P, Fushman D, Deshaiesn RJ, King RW: Ubistatins inhibit proteasome-dependent degradation by binding the ubiquitin chain. *Science* 2004, 306:117-120.

[17] Zhou P: Targeted protein degradation. *Curr Opin Chem Biol* 2005, 9:51-55.

[18] Orlowski M, Wilk S: Catalytic activities of the 20S proteasome, a multicatalytic proteinase complex. *Archiv Biochem Biophys* 2000, 383:1-16.

[19] McCormack TA, Baumeister W, Grenier L, Moomaw C, Plamondon L, Pramanik B, Slaughter C, Soucy R, Stein R, Zühl F, Dick L: Active site-directed inhibitors of *Rhodococcus* 20S proteasome. Kinetics and mechanism. *J Biol Chem* 1997, 272:26103-26109.

[20] Delcros JG, Floc'h MB, Prigent C, Arlot-Bonnemains Y: Proteasome inhibitors as therapeutic agents: current and future strategies. *Curr Med Chem* 2003, 10:479-503.

[21] Vinitsky A, Cardozo C, Lorenzimo L, Michaud C, Orlowski M: Inhibition of the proteolytic activity of the multicatalytic proteinase complex (proteasome) by substrate-related peptidyl aldehydes. *J Biol Chem* 1994, 269:29860-29866.

[22] Wilk S, Figueiredo-Pereira ME: Synthetic inhibitors of the multicatalytic proteinase complex (proteasome). *Enzyme Protein* 1993, 47:306-313.

[23] Figueiredo-Pereira ME, Berg KA, Wilk S: A new inhibitor of the multicatalytic proteinase complex (20S proteasome) induces accumulation of ubiquitin-protein conjugates in a neuronal cell. *J Neurochem* 1994, 63:1578-1581.

[24] Rock KL, Gramm C, Rothstein L, Clark K, Stein R, Dick L, Hwang D, Goldberg AL: Inhibitors of the proteasome block the degradation of most cell proteins and the generation of peptides presented on MHC class I molecules. *Cell* 1994, 78:761-771.

[25] Wojcik C: Regulation of apoptosis by the ubiquitin and proteasome pathway. *J Cell Mol Med* 2002, 6:25-48.

[26] Traenckner EB, Wilk S, Baeurele PA: A proteasome inhibitor prevents activation of NF-kappaB and stabilizes a newly phosphorylated form of I kappa B-alpha that is still bound to NF-kappa B. *EMBO J* 1994, 13:5433-5441.

[27] Lopes UG, Erhardt P, Yao R, Cooper GM: p53-dependent induction of apoptosis by proteasome inhibitors. *J Biol Chem* 1997; 272:12893-12896.

[28] Takaoka M, Ohkita M, Matsumura Y: Pathophysiological role of proteasome-dependent proteolytic pathway in endothelin-1-related cardiovascular diseases. *Curr Vasc Pharmacol* 2003, 1:19-26.

[29] Ostrowska JK, Wojtukiewicz MZ, Chabielska E, Buczko W, Ostrowska H: Proteasome inhibitor prevents experimental arterial thrombosis in renovascular hypertensive rats. *Thromb Haemost* 2004, 92:171-177.

[30] Morgan EN, Pohlman TH, Vocelka C: Nuclear factor kappa B mediates a procoagulant response in monocytes during extracorporeal circulation. *J Thorac Cardiovasc Surg* 2003, 125:165-171.

[31] McNaught KS, Perl DP, Brownell AL, Olanow CW: Systemic exposure to proteasome inhibitors causes a progressive model of Parkinson disease. *Ann Neurol* 2004, 56:149-162.

[32] Adams J, Behnke M, Chen S, Cruishank AA, Dick LR, Grenier L, Klunder JM, Ma YT, Plamondon L, Stein RL: Potent and selective inhibitors of the proteasome: dipeptidyl boronic acids. *Bioorg Med Chem Lett* 1998, 8:333-338.

[33] Gardner RC, Assinder SJ, Christie G, Mason GG, Markwell R, Wadsworth H, McLaughlin M, King R, Chabot-Flechter MC, Breton JJ, Allsop D, Rivett AJ: Characterization of peptidyl boronic acid inhibitors of mammalian 20S and 26S

proteasome and their inhibition of proteasomes in cultured cells. *Biochem J* 2000, 346:447-454.

[34] Adams J, Palombella V, Sausville EA, Johnson J, Destree A, Lazarus DD, Maas J, Pien CS, Prakash S, Elliott PJ: Proteasome inhibitors: a novel class of potent and effective antitumor agents. *Cancer Res* 1999, 59:2615-2622.

[35] Hideshima T, Richardson P, Chauhan D, Palombella VJ, Elliot PJ, Adams J, Anderson KC: The proteasome inhibitor PS-341 inhibits growth, induces apoptosis, and overcomes drug resistance in human multiple myeloma cells. *Cancer Res* 2001, 61:3071-3076.

[36] Adams J: Development of the proteasome inhibitor PS-341. *The Oncologist* 2002, 7:9-17.

[37] Adams J, Kauffman M: Development of the proteasome inhibitor Velcade (Bortezomib). *Cancer Invest* 2004, 22:304-11.

[38] Orlowski RZ, Stinchcombe TE, Mitchell BS, Shea TC, Baldwin AS, Stahl S, Adams J, Esseltine DL, Elliott PJ, Pien CS, Guercidini R, Anderson JK, Depcik-Smith ND, Bhagat R, Lehman MJ, Novick SC, O'Connor OA, Soignet SL: Phase I trial of the proteasome inhibitor PS-341 on patients with refractory hematologic malignancies. *J Clin Oncol* 2002, 20:4420-4427.

[39] Lenz HJ: Clinical update: proteasome inhibitors in solid tumors. *Canc Treat Rev* 2003, 29:41-48.

[40] Schwartz R, Davidson T: Pharmacology, pharmacokinetics, and practical applications of bortezomib. *Oncology* 2004, 18:14-21.

[41] Hideshima T, Mitsiades C, Akiyama M, Hayashi T, Chauhan D, Richardson P, Schlossman R, Podar K, Munshi NC, Mitsiades N, Anderson KC: Molecular mechanisms mediating antimyeloma activity of protesome inhibitor PS-341. *Blood* 2003, 101:1530-1534.

[42] Hideshima T, Chauhan D, Richardson P, Mitsiades C, Mitsides N, Hayashi T, Munshi N, Dang L, Castro A, Palombella V, Adams J, Anderson KC : NF-κB as a therapeutic target in multiple myeloma. *J Biol Chem* 2002, 277:16639-16647.

[43] Mitsiades N, Mitsiades CS, Richardson PG, Poulaki V, Tai Y-Y, Chauhan D, Fanourakis G, Gu X, Bailey C, Joseph M, Libermann TA, Schlossman R, Munshi NC, Hideshima T, Anderson KC: The proteasome inhibitor PS-341 potentates sensitivity of multiple myeloma cells to conventional chemotherapeutic agents: therapeutic applications. *Blood* 2003, 101:2377-2380.

[44] Purandare AV, Wan H, Laing N, Benbatoul K, Vaccaro W, Poss MA: Identification of a potent and rapidly reversible inhibitor of the 20S-proteasome. *Bioorg Med Chem Lett* 2004, 14:4701-4704.

[45] Omura S, Fujimoto T, Otoguro K, Matsuzaki K, Moriguchi R, Tanaka H, Sasaki Y: Lactacystin, a novel microbial metabolite, induces neuritogenesis of neuroblastoma cells. *J Antibiot (Tokyo)* 1991, 44:113-116.

[46] Fenteany G, Standaert RF, Lane WS, Choi S, Corey EJ, Schreiber SL: Inhibition of proteasome activities and subunit-specific amino-terminal threonine modification by lactacystin. *Science* 1995, 268:726-731.

[47] Dick LR, Cruikshank AA, Destree AT, Grenier L, McCormack TA, Melandri FD, Nunes SL, Palombella VJ, Parent LA, Plamondon L, Stein RS: Mechanistic studies on the inactivation of the proteasome by lactacystin in cultured cells. *J Biol Chem* 1997, 272:182-188.

[48] Fenteany G, Schreiber SL: Lactacystin, proteasome function, and cell fate. *J Biol Chem* 1998, 273:8545-8548.

[49] Ostrowska H, Wojcik C, Omura S, Worowski K: Lactacystin, a specific inhibitor of the proteasome, inhibits human platelet lysosomal cathepsin A-like enzyme. *Biochem Biophys Res Commun* 1997, 234:729-732.

[50] Ostrowska H, Wójcik C, Wilk S, Omura S, Kozlowski L, Stoklosa T, Worowski K, Radziwon P: Separation of cathepsin A-like enzyme and the proteasome: evidence that lactacystin/β-lactone is not a specific inhibitor of proteasome. *Int J Biochem Cell Biol* 2000, 32:747-757.

[51] Geier E, Pfeifer G, Wilm M, Lucchiari-Hartz M, Baumeister W, Eichmann K, Niedermann G: A giant protease with potential to substitute for some functions of the proteasome. Science 1999, 283:978-981.

[52] Porcile C, Piccioli P, Stanzione S, BajettoA, Bonavia R, Barbero S, Florio T, Schettinia G: Proteasome inhibitors induce cerebellar granule cell death: inhibition of nuclear factor-κB activation. *Ann NY Acad Sci* 2002, 973:402-413.

[53] Rideout HJ, Wang Q, Park DS, Stefanis L: Cyclin-dependent kinase activity is required for apoptotic death but not inclusion formation in cortical neurones after proteasomal inhibition. *J Neurosci* 2003, 23:1237-1245.

[54] Lang-Rollin I, Vekrellis K, Wang Q, Rideout HJ, Stefanis L: Application of proteasomal inhibitors to mouse sympathetic neurons activates the intrinsic apoptotic pathway. *J Neurochem* 2004, 90:1511-1520.

[55] Lee CS, Tee LY, Warmke T, Vinjamoori A, Cai A, Fagan AM, Snider BJ: A proteasomal stress response: pre-treatment with proteasome inhibitors increases proteasome activity and reduces neuronal vulnerability to oxidative injury. *J Neurochem* 2004, 91:996-1006.

[56] Kumar B, Hanson AJ, Prasad KN: Sensitivity of proteasome to its inhibitors increases during cAMP-induced differentiation of neuroblastoma cells in culture and causes decreased viability. *Cancer Lett* 2004, 204:53-59.

[57] Oikawa T, Sasaki T, Nakamura M, Shimamura M, Tanahashi N, Omura S, Tanaka K: The proteasome is involved in angiogenesis. *Biochem Biophys Res Commun* 1998, 8:243-248.

[58] Itoh M, Takaoka M, Shibata A, Okhita M, Matsumura Y: Preventive effect of lactacystin, a selective proteasome inhibitor, on ischemic acute renal failure in rats. *J Pharmacol Exp Ther* 2001, 298:501-507.

[59] Inden M, Kondo JI, Kitamura Y, Takata K, Nishimura K, Taniguchi T, Sawada H, Shimohama S: Proteasome inhibitors protect against degeneration of nigral dopaminergic neurons in hemiparkinsonian rats. *J Pharmacol Sci* 2005, 97:203-211.

[60] Soucy F, Plamondon L, Behnke M, Roush W: Synthesis of *clasto*-lactacystin β-lactone and analogs thereof. *J Am Chem Soc* 1999, 121:9967-9976.

[61] Campbell B, Adams J, Shin YK, Lefer AM: Cardioprotective effects of a novel proteasome inhibitor following ischemia and reperfusion in the isolated perfused rat heart. *J Mol Cell Cardiol* 1999, 31:467-476.

[62] Pye J, Ardeshirpour F, McCain A, Bellinger DA, Merricks E, Adams J, Elliott PJ, Pien C, Fisher TH, Baldwin AS Jr, Nichols TC: Proteasome inhibition ablates activation of NF-κB in myocardial reperfusion and reduces reperfusion injury. *Am J Physiol Heart Circ Physiol* 2003, 264:H919-H926.

[63] Phillips JB, Williams AJ, Adams J, Elliott PJ, Tortella FC: Proteasome inhibitor PS519 reduced infarction and attenuates leukocyte infiltration in a rat model of focal cerebral ischemia. *Stroke* 2000, 31:1686-1693.

[64] Zhang L, Zhang ZG, Zhang RL, Lu M, Adams J, Elliott PJ, Chopp M: Postischemic (6-hour) treatment with recombinant human tissue plasminogen activator and proteasome inhibitor PS-519 reduces infarction in a rat model of embolic focal cerebral ischemia. *Stroke* 2001, 32:2926-2931.

[65] Berti R, Williams AJ, Velarde LC, Moffett JR, Elliott PJ, Adams J, Yao C, Dave JR, Tortella FC: Effect of the proteasome inhibitor MLN519 on the expression of inflammatory molecules following middle cerebral artery occlusion and reperfusion in the rat. *Neurotox Res* 2003, 5:505-514.

[66] Shah IM, Lees KR, Pien CP, Elliott PJ: Early clinical experience with the novel proteasome inhibitor PS-519. *Br J Clin Pharmacol* 2002, 54:269-276.

[67] Di Napoli M, Papa F: MLN-519: Milenium/PAION. *Curr Opin Invest Drugs*.2003, 4:333-341.

[68] Di Napoli M, Papa F: The proteasome system and proteasome inhibitors in stroke: controlling the inflammatory response. *Curr Opin Invest Drugs* 2003, 4:1333-1342.

[69] Wojcik C, Di Napoli M: Ubiquitin-proteasome system and proteasome inhibition: new strategies in stroke therapy. *Stroke* 2004, 35:1506-1518.

[70] Meng L, Mohan R, Kwok BHB, Elofsson M, Sin N, Crews CM: Epoxomicin, a potent and selective proteasome inhibitor, exhibits *in vivo* anti-inflammatory activity. *Proc Natl Acad Sci USA* 1999, 96:10406-10408.

[71] Bogyo M, McMaster JS, Gaczynska M, Tortorella D, Goldberg AL, Ploegh H: Covalent modification of the active site threonine of proteasomal β subunits and the *Escherichia coli* homolog HIV by a new class of inhibitors. *Proc Natl Acad Sci USA* 1997, 94:6629-6634.

[72] Steverding D, Spackman RW, Royle HJ, Glenn RJ: Trypanocidal activities of trileucine methyl vinyl sulfone proteasome inhibitors. *Parasitol Res* 2005, 95:73-76.

[73] Lum RT, Nelson MG, Joly A, Horsma AG, Lee G, Meyer SM, Wick MM, Schow SR: Selective inhibition of the chymotrypsin-like activity of the 20S proteasome by 5-methoxy-1-indanone dipeptide benzamides. *Bioorg Med Chem Lett* 1998, 8:209-214.

[74] Lum RT, Kerwar SS, Meyer SM, Nelson MG, Schow SR, Shiffman D, Wick MM, Joly A: A new structural class of proteasome inhibitors that prevent NF-kappa B activation. *Biochem Pharmacol* 1998, 55:1391-1397.

[75] Buchan AM, Li H, Blackburn B: Neuroprotection achieved with a novel proteasome inhibitor which blocks NF-kappaB activation. *Neuroreport* 2000, 11:427-430.

[76] Koguchi Y, Kohno J, Nishio M, Takahashi K, Okuda T, Ohnuki T, Komatsubara S: TMC-95A, B, C, and D, novel proteasome inhibitors produced by Apiospora montagnei Sacc. TC 1093. *J Antibiot* 2000, 53:105-109.

[77] Albrecht BK, Williams RM: A concise, total synthesis of the TMC-95A/B proteasome inhibitor. *Proc Natl Acad Sci 2004*, 101:11949-11954.

[78] Kaiser M, Groll M, Siciliano C, Assfalg-Mechleidt I, Weyher E, Kohno J, Milbradt AG, Renner C, Moroder L: Binding mode of TMC-95A analogues to eukaryotic 20S proteasome. *Chem Biochem* 2004, 5:1256-1266.

[79] Inoue M, Zhai H, Sakazaki H, Furuyama H, Fukuyama Y, Hirama M: TMC-95A, a reversible proteasome inhibitor, induces neurite outgrowth in PC12 cells. *Bioorg Med Chem Lett* 2004, 14:663-665.

[80] García-Echeverría C, Imbach P, France D, Fűrst P, Lang M, Noorani M, Scholz D, Zimmermann J, Furet P: A new structural class of selective and non-covalent inhibitors inhibitors of the chymotrypsin-like activity of the 20S proteasome. *Bioorg Med Chem Lett* 2001, 11:1417-1319.

[81] Schmidtke G, Holzhütter H-G, Bogyo M, Kairies N, Groll M, Giuli R, Emch S, Groettrup M: How an inhibitor of the HIV-I protease modulates proteasome activity. *J Biol Chem* 1999, 10:35734-35740.

[82] Andre P, Groettrup M, Kleneman P, de Giuli R, Booth BL, Cerundolo V, Bonneville M, Joterau F, Zinkernagel RM, Lotteau V: An inhibitor of HIV-1 protease modulates proteasome activity, antigen presentation and T cell responses. *Proc Natl Acad Sci USA* 1998, 95:13120-13124.

[83] Groettrup M, Schmidtke G: Selective proteasome inhibitors: modulators of antigen presentation?. *Drug Discov Today* 1999, 4:63-71.

[84] Gaedicke S, Firat-Geier E, Constantiniu O, Lucchiari-Hartz M, Freudenberg M, Galanos C, Niedermann G: Antitumor effect of the human immunodeficiency virus protease inhibitor ritonavir: induction of tumor-cell apoptosis associated with perturbation of proteasomal proteolysis. *Cancer Research* 2002, 62:6901-6908.

[85] Laurent N, de Bouard S, Guillamo JS, Christov C, Zini R, Jouault H, Andre P, Lotteau V, Peschanski M: Effects of the proteasome inhibitor ritonavir on glioma growth in vitro and in vivo. *Mol Cancer Ther* 2004, 3:129-136.

[86] Hosseini H, Andre P, Lefevre N, Viala L, Walzre T, Peschanski M, Lotteau V: Protection against experimental autoimmuno encephalomyelitis by a proteasome modulator. *J Neuroimmunol* 2001, 30:233-244.

[87] Meyer S, Kohler NG, Joly A: Cyclosporine A is an uncompetitive inhibitor of proteasome activity and prevents NF-κB activation. *FEBS Letters* 1997, 413:354-358.

[88] Wang X, Omura S, Szweda LI, Yang Y, Bérard J, Seminaro J, Wu J: Rapamycin inhibits proteasome activator expression and proteasome activity. *Eur J Immunol* 1997, 27:2781-2786.

[89] Sijts A, Sun Y, Janek K, Kral S, Paschen A, Schadendorf D, Kloatzel PM: The role of the proteasome activator PA28 in MHC class I antigen processing. *Mol Immunol* 2002, 39:165-169.

[90] Figueiredo-Pereira ME, Chen WE, Li J, Johdo O: The antitumor drug aclacinomycin A, which inhibits the degradation of ubiquitinated proteins, shows selectivity for the

chymotrypsin-like activity of the bovine pituitary 20S proteasome. *J Biol Chem* 1996, 271:16455-16459.

[91] Piccininin M, Tazartes O, Mezztesta C, Ricotti E, Bedino S, Grosso F, Dianzani U, Tovo PA, Mostert M, Musso A, Rinaudo MT: Proteasomes are a target of the anti-tumor drug vinblastine. *Biochem J* 2001, 356:835-841.

[92] Kroll M, Arenzana-Seisdedos F, Bachelerie F, Thomas D, Friguet B, Conconi M: The secondary fungal metabolite gliotoxin targets proteolytic activities of the proteasome. *Chem Biol* 1999, 6:689-698.

[93] Kazi A, Wang Z, Kumar N, Falsetti SC, Chan TH, Dou QP: Structure-activity relationships of synthetic analogs of (-)-epigallocatechin-3-gallate as proteasome inhibitors. *Anticancer Res* 2004, 24:943-954.

[94] Lam WH, Kazi A, Kuhn DJ, Chow LM, Chan AS, Dou QP, Chan TH: A potential prodrug for a green tea polyphenol proteasome inhibitor: evaluation of the peracetate ester of (-)-epigallocatechin gallate [(-)-EGCG]. *Bioorg Med Chem* 2004, 12:5587-5593.

[95] Gao Y, Lecker S, Post MJ, Hietaranta AJ, Li J, Volk R, Lim, Sato K, Saluja AK, Steer M, Goldberg AL, Simons M: Inhibition of ubiquitin-proteasome pathway-mediated IκBα degradation by a naturally occurring antibacterial peptide. *J Clin Invest* 2000, 106:439-448.

[96] Gaczynska M, Osmulski PA, Gao Y, Post MJ, Simons M: Proline-and arginine-rich peptides constitute a novel class of allosteric inhibitors of proteasome activity. *Biochemistry* 2003, 42:8663-8670.

[97] Takabe W, Kodama T, Hamakubo T, Tanaka K, Suzuki T, Aburatani H, Matsukawa N, Noguchi N: Anti-atherogenic antioxidants regulate the expression and function of proteasome α- type subunits in human endothelial cells. *J Biol Chem* 2001, 276:40497-40501.

[98] Ruiz de Mena I, Mahillo E, Arribas J, Castano JG: Kinetic mechanism of activation by cardiolipin (diphosphatidylglycerol) of the rat liver multicatalytic proteinase. *Biochem J* 1993, 296:93-97.

[99] Wilk S, Chen W-E, Magnusson RP: Modulation of the PA28α-20S proteasome interaction by a peptidyl alcohol. *Arch Biochem Biophys* 1999, 362:283-290.

[100] Ciechanover A, Orian A, Schwartz AL: Ubiquitin-mediated proteolysis: biological regulation via destruction. *BioEssays* 2000, 22:442-451.

[101] Scheffner M, Whitaker NJ: Human papillomavirus-induced carcinogenesis and the ubiquitin-proteasome system. *Semin Cancer Biol* 2003, 13:59-67.

[102] Michael D, Oren M: The p53-Mdm2 module and the ubiquitin system. *Semin Cancer Biol* 2003, 13:49-58.

[103] Deshaies RJ: SCF and Cullin/Ring H2 - based ubiquitin ligases. *Annu Rev Cell Dev Biol* 1999, 15:435-467.

[104] Vassilev LT, Vu BT, Graves B, Carvajal D, Podlaski F, Filipovic Z, Kong N, Kammlott U, Lukacs C, Klein C, Fotouhi N, Liu EA: *In vivo* activation of the p53 pathway by small-molecule antagonists of MDM2. *Science* 2004, 303:844-848.

[105] Yaron A, Gonen H, Alkalay I, Hatzubai A, Jung S, Beyth S, Mercurio F, Manning AM, Ciechanover A, Ben-Neriah Y: Inhibition of NF-κB cellular function via specific targeting of the IκB-ubiquitin ligase. *EMBO J* 1997, 16:6486-6494.

[106] Swinney DC, Xu YZ, Scarafia LE, Lee I, Mak AY, Gan QF, Ramesha CS, Mulkins MA, Dunn J, So OY, Biegel T, Dinh M, Volkel P, Barnett J, Dalrymple SA, Lee S, Huber M: A small molecule ubiquitination inhibitor blocks NF-κB-dependent cytokine expression in cells and rats. *J Biol Chem* 2002, 277:2357-23581.

[107] Baldwin AS: The NF-κB and IκB proteins: new discoveries and insights. *Annu Rev Immunol* 1996, 14:649-651.

[108] Sakamoto KM, Kim KB, Kumagai A, Mercurio F, Crews CM, Deshaies RJ: Protacs: chimeric molecules that target proteins to the Skp1-cullin-F box complex for ubiquitination and degradation. *Proc Natl Acad Sci* 2001, 98:8554-8559.

[109] Sakamoto KM, Kim KB, Verma R, Ransick A, Stein B, Crews CM, Deshaies RJ: Development of protacs to target cancer-promoting proteins for ubiquitination and degradation. *Mol Cell Proteomics* 2003, 2:1350-1358.

[110] Zhang D, Baek S-H, Ho A, Kim K: Degradation of target protein in living cells by small-molecule proteolysis inducers. *Bioorg Med Chem Lett* 2004, 14:645-648.

[111] Yamamoto Y, Gaynor RB: Therapeutic potential of inhibition of the NF-κB pathway in the treatment of inflammation and cancer. *J Clin Invest* 2001, 107:135-142.

In: The Ubiquitin Proteasome System... ISBN 978-1-60021-749-4
Eds: Mario Di Napoli and Cezary Wojcik, pp. 989-1010© 2007 Nova Science Publishers, Inc.

Chapter 41

PRECLINICAL DATA ON THE USE OF PROTEASOME INHIBITORS: A NEW APPROACH TO TREATMENT OF NERVOUS SYSTEM DISEASES

***Anthony J. Williams*[1,*], *Jitendra R. Dave*[1], *Peter Elliott*[2] *and Frank C. Tortella*[1,¥]**

[1]Walter Reed Army Institute of Research, Department of Applied Neurobiology, Silver Spring, MD 20910 USA;

[2]Sirtris Pharmaceuticals Inc., Boston, MA 02139 USA.

ABSTRACT

The ubiquitin-proteasome system (UPS), aside from a major degradation pathway of intracellular proteins, involves the modulation of key proteins that control cellular physiology through cell cycle regulation, immune response, and activation of gene expression. The core enzymatic molecule of the UPS is the 20S proteasome. Alterations in the proteasome proteolytic pathway have been contributed to protein alterations associated with aging and, in fact, dysregulation of the UPS has been linked to several disease states including neurodegenerative diseases (i.e. Parkinson's, Alzheimer's, and Huntington's disease), malignancies, and inflammatory-related diseases. As such, strong preclinical data now exist supporting the use of reversible proteasome inhibitors to treat a variety of disease states including cancer, rheumatoid arthritis, asthma, psoriasis, autoimmune encephalomyelitis, myocardial infarction, and ischemic brain injury.

---

* Correspondence concerning this article should be addressed to Dr. Anthony J. Williams, PhD; Department of Applied Neurobiology, Division of Psychiatry and Neuroscience, Walter Reed Army Institute of Research, 503 Robert Grant Ave, Room 2A40 Silver Spring, MD 20910 USA; Phone: +1-301-319-9461; Fax: +1-301-319-9905; E-mail: anthony.williams@na.amedd.army.mil.

¥ Opinions, interpretations, conclusions, and recommendations are those of the authors and are not necessarily endorsed by the U.S. Army (para 4-3, AR 360-5).

Currently, the proteasome inhibitor Velcade® is approved for treatment of multiple myeloma. Phase I safety trials have also been completed with the proteasome inhibitor MLN-519, at doses capable of reducing blood proteasome activity by 80%. Experimental studies with MLN-519 have indicated significant neuroprotective treatment effects in animal models of ischemia/reperfusion injury at doses that reduce blood 20S proteasome activity by 40-80%. Following focal ischemic brain injury in rats, treatment with MLN-519 has been associated with a significant reduction of brain infarction along with improved neurological outcome and electrophysiological brain activity as evaluated up to two weeks post-injury. Importantly, MLN-519 exhibited a wide therapeutic treatment window with a delayed initial treatment of up to 6-10 h post-injury. The therapeutic efficacy has been linked to an attenuation of aberrant gene expression; in particular, studies with MLN-519 have indicated that treatment of ischemic brain injury in rats is associated with a reduction of the nuclear factor (NF)-κB mediated neuro-inflammatory response, where, following injury, MLN-519 treatment has been shown to reduce activated NF-κB immunoreactivity and attenuate the increase in both cytokine (TNF-α, IL-1β, and IL-6) and cellular adhesion molecule (ICAM-1 and E-selectin) expression. MLN-519 also provided dramatic reductions of both neutrophil and macrophage infiltration into the injured rat brain. Similar anti-inflammatory effects of proteasome inhibition have been observed in other experimental inflammatory disease models as well. The aim of this chapter is to review the experimental and clinical data relating to the role of the proteasome in CNS disorders and to evaluate the potential use of proteasome inhibitors to treat CNS disease.

Keywords: MLN-519, brain injury, ischemia, NF-κB, neuroprotection, proteasome inhibitor.

Abbreviations

AD, Alzheimer's Disease; ALS, Amyotrophic Lateral Sclerosis; CNS, Central Nervous System; GFAP glial fibrillary acidic protein; HD, Huntington's Disease; ICAM intercellular adhesion molecole; IL, interleukin; I-κB, Inhibitory κB; MCAo, Middle Cerebral Artery Occlusion; NF-κB, Nuclear Factor κB; PD, Parkinson's Disease; UPS, Ubiquitin-Proteasome System.

Introduction

The proteasome is an enzymatic complex present in all central nervous system (CNS) cells. The barrel shaped core of the proteasome, i.e. the 20S proteasome, and its numerous β subunits are responsible for the proteolytic activity of the UPS. Proteasome function plays an essential role in maintaining cellular homeostasis through influencing the balance of intracellular protein turnover in order to address specific proteolytic needs of a cell's given condition [3-7]. For the many roles of the proteasome, the reader is referred to Chapters 6 and 7. Although a major role of the UPS is the proteolysis of misfolded or damaged proteins, one of the most important physiological roles of the proteasome involves the degradation of several short-lived regulatory proteins that control the expression of multiple genes. In disease, a shift from the normal homeostatic role of the proteasome to a neurodegenerative state induced by a loss of normal proteasomal function or through activation of aberrant gene expression cascades has been proposed to aggravate the underlying condition. In fact, recent reports have implicated altered proteasome function in the progression of neurodegenerative diseases, malignancies, and inflammatory-related disorders [3,8-16]. A flurry of active research is now focused on evaluation of novel treatment strategies targeting the proteasome in order to help improve recovery from these types of disorders. It is the focus of this chapter to review the preclinical evidence supporting the use of proteasome inhibitors to treat acute CNS disorders with a focus on cerebral ischemia.

THE PROTEASOME AND CNS DISEASE

It is well established that physiological and pathological stressors such as inflammatory stimuli can influence proteasome activity within cells. While relatively little is known about the expression of the various proteasome and proteasome-related moieties in the brain (see Chapter 16), immunohistochemical studies have confirmed that the 20S proteasome is localized throughout the CNS [17] in neurons, glia, and synapses [18] with the majority of CNS proteasome activity residing in the cytosol [13]. Alteration of proteasomal function in any of these cellular compartments can lead to eventual cellular dysfunction. Two distinct and contrasting roles of the proteasome in disease progression have been described: (*i*) 'loss of function' due to decreased proteasomal activity and intracellular protein aggregation within the cell or (*ii*) 'gain of function' due to proteasome-mediated activation of the aberrant gene expression associated with disease. In the case of loss of proteasomal function, disease progression may benefit from strategies to re-activate homeostatic levels of proteasome function within the cell, while in the latter case, transient inhibition of the proteasome aimed at reducing pathological gene expression may be beneficial.

Loss of normal levels of proteasome activity can occur with aging but has also been linked to a variety of neurodegenerative diseases including Alzheimer's disease (AD) [14], Parkinson's disease (PD) [16], Huntington's disease (HD) [15], and amyotrophic lateral sclerosis (ALS) [19,20]. Most post-mitotic cells continually produce new proteins and a delicate balance in overall protein level is maintained through active degradation of old or damaged proteins. As such, loss of function of the proteasome or UPS appears to be involved in the eventual accumulation of unwanted protein aggregates. The underlying causes of proteasomal dysfunction within the cell can be multi-factorial involving alteration of one or more specific enzymes of the UPS or other entities that affect the normal function of the proteasome. In particular, mutations in proteins associated with neurodegenerative disease progression have the ability to directly inhibit proteasome activity leading to a feed-forward process of proteasome inactivation and increased intracellular protein aggregation linked to the promotion of cell death [14]. A common theme among several neurodegenerative disorders is the presence of protein aggregates including the presence of ubiquitin and ubiquitin-conjugated proteins, suggesting an overall loss of proteasomal activity within these cells. Several types and classifications of protein aggregates have been characterized in a variety of disease states (i.e. AD, PD, HD, and ALS) and are implicated in disease progression. In addition, loss of proteasomal activity has been linked to the activation of programmed axon death, loss of synaptic function [21] and induction of oxidative or nitrative stress on the brain [15]. For the many roles of the UPS in aging and neurodegenerative diseases, the reader is referred to Chapters 22 through and 33.

Aside from a loss of function of proteasome activity, the proteasome can also play a direct role in disease progression through active participation in the expression of disease-related genes and promotion of cellular disability/death. As such, modulation of proteasome activity with the use of proteasome inhibitors has been studied as a therapeutic target due to the role of the proteasome in regulating gene expression. One such regulatory system under control of the UPS, that has received considerable recent attention, is the NF-κB pathway. NF-κB activation can promote the expression of a variety of target genes including those involved in apoptosis and inflammation. In particular, an increase in the activity of NF-κB has been associated with refractory tumorigenesis in certain types of cancer cells [8,22] and in the upregulation of pro-inflammatory genes involved in the several inflammatory conditions including the neuronal degeneration associated with ischemic brain injury [11].

Proteasome Inhibition: A Therapeutic Perspective

Proteasome inhibitors are currently being evaluated for the treatment of a variety of disease states including cancer, asthma, psoriasis, multiple sclerosis, myocardial infarction, and stroke [8-11]. Several distinct classes of natural and synthetic compounds have been discovered that exhibit proteasome inhibitory activity including aldehyde derivatives (CEP-1612, MG-132) [9,11], dipeptide benzamide derivatives (CVT-634) [9], dipeptide boronic acid substitutes (PS-341, MLN-273, NVP-AFB340, NVP-AFD314) [23], vinyl sulfone tripeptides [24], HIV-1 protease inhibitors (ritonavir) [25], lactacystin derivatives (MLN-519) [26,27], epoxyketones (epoxomicin) [28], 2-aminobenzylstatines [11], and the porcine polypeptide PR-39 [29] (see Chapter 40). These compounds have provided a valuable research tool for studying the role of the proteasome in cellular function and disease. Among CNS related disorders, proteasome inhibition has also shown promise for the treatment of ischemia-induced brain injury.

Extensive basic research studies within the past decade have provided a wealth of information concerning the complex interaction of the cell death mechanisms that underlie the evolution of neurodegenerative brain injury [30,31]. The effects of proteasome inhibition on CNS cells can be multiple; including neuroprotection against injury and inflammation, stimulation of neurite outgrowth, enhancement of synaptic transmission, as well as neurotoxicity at higher concentrations and long exposure periods [9]. Proteasome inhibition has also been shown to prevent degradation of heat shock protein inducible factor 1α and can stimulate angiogenesis in vitro [32]. For acute CNS disorders such as stroke or traumatic brain injury, research has focused on defining the delayed cell death pathways involved in the progression of injury including the contributions of inflammation and apoptotic-like cell death [33]. In fact, treatment with neuroprotective agents that target these delayed injury cascades have shown promise in preclinical studies. In particular, anti-inflammatory compounds such as proteasome inhibitors, that target UPS-mediated gene expression, have shown excellent pre-clinical efficacy for the treatment of acute ischemic brain injury with a wide therapeutic window of opportunity for initiation of treatment following injury onset [9]. Of these compounds, the lactacystin derivative MLN-519 has been most extensively evaluated in *in vivo* animal models as a potential treatment of cerebral ischemia.

Cerebral Ischemia

Ischemic brain injury due to insults such as stroke is a leading cause of death in the United States, as well as being a crucial player in brain damage due to blunt head trauma. A host of therapeutic interventions against ischemic brain injury have been tested clinically, most of which involved targeting ionic homeostasis with the use of voltage- or ligand-gated ion channel blockers [34]. Despite aggressive research into developing neuroprotection treatments for brain injury, clinical trials have met with limited success. Administration of tissue plasminogen activator to relieve thrombo-embolic vascular occlusions is the only approved therapy but treatment is limited to the first 3 hours post-injury [33,35]. Unfortunately, most stroke victims do not receive medical care within 3 hours indicating the continued need for development of novel compounds for the treatment of ischemic brain injury.

Several concomitant factors have likely contributed to the failure of clinical neuroprotection trials in acute brain injury including poor pre-clinical pharmacodynamic evaluation. One critical limitation of several experimental neuroprotection agents has been their narrow therapeutic treatment window allowing only limited time from injury to offer any therapeutic relief. Inclusion of a surrogate marker of drug activity, evaluation of long-term recovery in injury models more closely resembling the clinical situation, improved patient selection criteria and trial design based on the pre-clinical modeling, and targeting an adequate clinical dose have also been suggested to enhance the probability of success in future clinical trials [36]. Clearly, the development of pre-clinical (i.e. animal) testing must focus on these factors in order to improve opportunities for successful transitions from pre-clinical to clinical studies where secondary cell death mechanisms are the necessary target of delayed treatment effects. Thus, a search for novel pharmacological agents has been initiated targeting downstream or delayed injury processes [37]. Such targets include inhibition of the acute inflammatory response induced following injury to the brain [38,39].

Cerebral ischemia is a pathological process caused by a loss of blood flow to the brain that involves a complex interaction between various molecular perpetrators of cell death. A sustained decrease in cerebral blood flow can lead to inhibition of oxidative phosphorylation and result in a maelstrom of cytotoxic activity. At the cellular level, once ionic homeostasis is lost, proteases, phospholipases, and free radical formation all participate in the eventual demise of the cell. The end result is cell death defined by either abrupt necrotic cell damage due to excitotoxic overload or activation of delayed apoptotic cascades [30,40]. Initially only a small area, the core infarct, is irreversibly damaged following an ischemic attack. The surrounding tissue, known as the peri-infarct or penumbral region, is subject to further damage through secondary cell death cascades. As such, these secondary mechanisms of injury represent a viable opportunity for pharmacological treatment strategies.

Similar to other neurodegenerative disorders, loss of function of the UPS following ischemic injury has been shown to occur and may be related to a variety of factors including changes in intracellular pH, Ca^{2+}, or presence of reactive oxygen species (see Chapter 39). In ischemic brain tissues the 26S proteasome appears to be more sensitive to ischemia, most likely due to disassociation into PA700 end caps and 20S proteasomes, whereas the ATP independent and ubiquitin-independent degradation mediated by 20S proteasome proceeds without obstacles [9]. The proteasome also plays a critical role in the activation of transcription factors such as NF-κB that promote a neuro-inflammatory response following ischemic injury. In fact, several recent experimental studies have indicated that the acute neuro-inflammatory response may be a principal mediator of secondary cell death responses [41]. Thus, proteasome inhibitors offer the potential for neuroprotective mediation of ischemic brain injury through anti-inflammatory effects. A general reduction of injury-induced protein degradation may also occur with proteasome inhibition, including the preservation of pro-survival molecules such as heat shock proteins [42]. Proteasome inhibitors have also been shown to prevent Wallerian degeneration induced by axonal transection through stabilization of axonal microtubules [43]. The preclinical studies supporting the use of proteasome inhibitors to treat cerebral ischemia and the accumulating evidence of an anti-inflammatory mechanism of action are addressed below.

Treatment of Cerebral Ischemia with the Proteasome Inhibitor MLN-519

MLN-519 (Millennium Pharmaceuticals Inc., Cambridge, MA, USA) is a synthetic analog of *clasto*-lactacystin β-lactone, derived from the bacterial metabolite lactacystin and developed as a small molecule therapeutic agent and potent inhibitor of 20S proteasome activity [26]. Several pre-clinical reports have indicated that proteasome inhibitors are protective against ischemia/reperfusion-type injuries to organ systems including the heart and brain [11,27]. Specifically, recent studies have indicated that MLN-519 is effective in the treatment of myocardial reperfusion injury in pigs [44]. The protective effects of proteasome inhibition have also been reported in models of focal ischemic brain injury in rodents (Table 1). Two standard models of experimental stroke are described in Table 1 induced either by advancing an intraluminal filament into the brain via the internal carotid artery to induce either a transient or permanent middle cerebral artery occlusion (MCAo) [45] or by using an embolic occlusion model [46]. Following 90 min of focal cerebral ischemia the proteasome inhibitor CVT-634 resulted in significantly smaller brain infarct volumes as assessed 1 and 7 days post-injury [45] while studies with MLN-519 have indicated the effects of proteasome inhibition to extend the treatment window of tissue plasminogen activator in an embolic clot model of ischemia/reperfusion brain injury in rats. Several recent studies have also indicated the potent neuroprotective effect of MLN-519 in the intraluminal filament model of focal ischemic brain injury in rats with a wide therapeutic treatment window of up to 10 hours post-injury [1,45,48,49]. Overall, the preclinical neuroprotection profile for treatment of ischemic brain injury has been well established with similar results across several laboratories.

Table 1. Summary of published reports evaluating the neuroprotection efficacy of treatment with proteasome inhibitors in rodent models of focal cerebral ischemia [middle cerebral artery occlusion (MCAo)

Study	Year	Drug (mean dose)	Model	Summary of Treatment Effects
Buchan *et al.* [45]	2000	CVT-634 (50 mg/kg, i.p.)	90 min MCAo 1 & 7 day recovery	Reduction of brain infarct volume
Phillips *et al.* [46]	2000	MLN-519 (0.003-0.1 mg/kg, i.v.)	2h MCAo 24 & 72 h recovery	Reduced brain infarction, improved neurological outcome, improved electroencephalographic activity, and reduction in brain leukocyte infiltration.
Zhang *et al.* [49]	2001	MLN-519 (1.0 mg/kg, i.v.)	embolic occlusion 7 day recovery	Reduced brain infarction, improved neurological outcome, 6 h therapeutic window, reduced hemorrhagic transformation associated with recombinant human tissue plasminogen activator treatment, and reduction in brain leukocyte infiltration.
Berti *et al.* [71]	2003	MLN-519 (1.0 mg/kg, i.v.)	2h MCAo 3-72 h recovery	Reduction in NF-κB reactivity. Reduction of ICAM-1 (3h), IL-1β (6-12h), E-selectin (12-24h), and TNF-α (24h) gene expression.
Williams *et al.* [47]	2003	MLN-519 (1.0 mg/kg, i.v.)	2h MCAo 24 h recovery	Reduced brain infarction, improved neurological outcome, 6 h therapeutic window, 80-90% transient reduction of proteasome activity, reduction of NF-κB reactivity particularly in endothelial cells and leukocytes, reduced brain GFAP reactivity, and reduction in brain leukocyte infiltration.
Williams *et al.* [1]	2004	MLN-519 (1.0 mg/kg, i.v.)	2h MCAo 72 h recovery	Reduced brain infarction, improved neurological outcome, 10 h therapeutic window, reduction in E-selectin, ICAM-1, TNF-α, and IL-6 gene expression, and reduction in brain leukocyte infiltration.
Williams *et al.* [48]	2005	MLN-519 (1.0 mg/kg, i.v.)	2h MCAo 14 day recovery	Reduced mortality, reduced brain tissue loss, improved neurological outcome, and improved weight gain.

ICAM indicates intercellular adhesion molecole; IL, interleukin; NF-κB, Nuclear Factor κB; GFAP glial fibrillary acidic protein.

MLN-519: Potency/Efficacy

Dose response studies with MLN-519 (0.01-0.3 mg/kg, delivered i.v. at 2 h post-injury) indicated a significant treatment effect to reduce core brain infarction by up to 60% as evaluated 24 h following transient MCAo in rats [45]. The neuroprotective reduction in brain infarct volume was also associated with significant improvements in neurological outcome and increase in cortical electroencephalographic activity in the injured brain hemisphere [45]. In a subsequent study a higher dose of MLN-519 (1.0 mg/kg), which transiently reduced blood proteasome activity by 80-90%, was shown to significantly reduce cortical brain infarction and improve neurological outcome as evaluated out to 24 h post-injury even when the initial injection was delayed to 6 h post-occlusion [48]. In a study by Zhang *et al.* [46] MLN-519 treatment of embolic stoke in rats was shown to significantly reduce infarct volume and improve neurological recovery alone and in combination with the 'clot-buster' recombinant human tissue plasminogen factor with a therapeutic window also out to at least 6 h post-injury without inducing hemorrhagic transformation of the infarct (a primary concern with the use of clot-dissolving drugs). Based on these pre-clinical reports, the neuroprotective effects of MLN-519 were shown to be both dose and time dependent with transient reductions in blood proteasome activity to levels shown to be safe in phase I clinical safety trials [50]. Additionally, the ability to monitor a patient's blood proteasome activity level provides a surrogate measure of drug activity allowing direct correlation of treatment to outcome, which will be a distinct advantage in clinical brain injury trials [36].

MLN-519: Therapeutic Treatment Window

One critical facet of preclinical neuroprotection drug development is the assessment of the full therapeutic treatment window of a given neuroprotection compound. Specifically, it is important to evaluate the amount of time that can elapse post-injury, in which treatment can be delayed, and still significantly improve outcome measures [36]. The optimal therapeutic treatment window of MLN-519 (1.0 mg/kg) has been shown to occur with delayed injections up to 6-10 h post-injury to improve outcome in brain injured rats out to 3 days or even up to 2 weeks following MCAo [1,49]. Specifically, core brain infarct volume following MCAo was significantly reduced with delayed treatments of MLN-519 beginning 6, 8, or 10 h after onset of injury and was associated with reductions in neuronal and axonal degeneration in the injured brain hemisphere [1]. Following 2 weeks recovery, MLN-519 treatment (1.0 mg/kg, delivered i.v. at 10, 24, and 48 h post-occlusion) reduced both mortality and the volume of brain tissue loss, improved body weight gain, and was associated with a superior overall neurological outcome compared to vehicle control animals [49]. Overall, MLN-519 has been shown to possess potent neuroprotective properties related to a reduction of histopathological brain injury and improved neurological and physiological outcome from ischemia/reperfusion brain injury with a wide therapeutic treatment window of up to 10 h post-injury. Critically, the 6-10 h therapeutic treatment window of MLN-519 corresponded to an anti-inflammatory reduction of NF-κB mediated gene expression of cytokines and cellular adhesion molecules, most of which showed peak increases in mRNA transcript levels during this acute period (Figure 1).

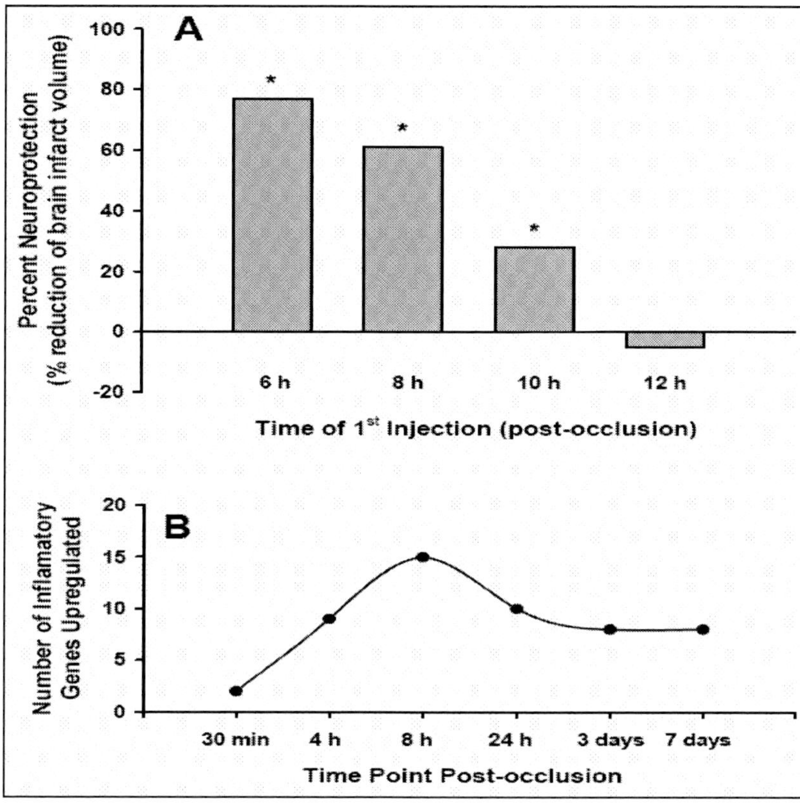

Figure 1. Comparison of the neuroprotective efficacy of MLN-519 to time of peak inflammatory gene expression following focal ischemic brain injury in rats. The 6-10 h therapeutic window of MLN-519 (A, data derived from [1]) corresponds to time period of peak inflammatory gene expression as determined by microarray analysis (B) following 2 h of transient middle cerebral artery occlusion in the rat (Data derived from [2]).

MLN-519: Safety Profile

The availability of clinical assays to measure blood proteasome activity offer the advantage of being able to monitor patients to within safe therapeutic levels of proteasome inhibition shown to be neuroprotective in animal models (i.e 40-80% reduction of blood proteasome activity) [50] while avoiding potential cellular toxicity associated with complete or extended periods of proteasome inhibition [9]. In MCAo injured rats, a single bolus dose of MLN-519 (1.0 mg/kg, i.v.) transiently reduces blood 20S proteasome activity by 87-89% (measured 1 h post-injection), which returned to control pre-injury levels by 24 h [48]. However, it has been established that high concentrations of proteasome inhibitors are toxic to cells including neurons [13]. A study by Taglialatela et al. [51] indicated that intracerebroventricular injection of a proteasome inhibitor directly into normal rat brain can induce DNA fragmentation. Safe therapeutic treatment with MLN-519 may be related to the route of administration studied (i.v.) along with the poor blood brain barrier permeability of MLN-519 in normal or brain injured animals [11,45]. In animals, toxicity with proteasome inhibition generally results from an extended period of excessive proteasome inhibition (i.e. >95% for 24 h or more) [50]. Toxicity following MLN-519 administration does not occur until doses reach 8 mg/kg or higher in rats.

Specific side effects at this dose include gastrointestinal problems and lowering of blood pressure [50]. In brain-injured animals, the neuroprotective effects of MLN-519 (1.0 mg/kg) did not significantly alter physiological parameters (i.e. blood pressure, heart rate, blood gases, or body temperature) following MCAo injury. Overall, the timing of drug administration, cell-type affected and dose used may all be important factors in assessing effective therapy with proteasome inhibitors. Based on neuroprotective efficacy of MLN-519, with effective doses ranging from 0.03-1.0 mg/kg, a safety index (toxic/neuroprotective dose) can be estimated to at least 8 or higher. Overall, patients may benefit from transient proteasome inhibition protocols shown to be safe and effective in clinical trials. However, care providers may want to consider the presence of an active neurodegenerative disease in a patient as preclusion to the therapeutic use of proteasome inhibitors due to the potential exacerbation of an already dysfunctional UPS-related disease.

MLN-519: Clinical Trials

Phase I clinical trials with MLN-519 in normal human volunteers have indicated that 20S proteasome activity can be safely reduced by 70-80% (1.6 mg/m^2) following a single bolus injection [50]. The effects of proteasome inhibition with MLN-519 on blood proteasome activity were transient and up to 2 additional doses could be administered at 24 h intervals following initial dosing without treatment-emergent symptoms or abnormality of laboratory tests [50]. Minor adverse events included a transient altered taste sensation, discomfort in injection arm, headache, or flu-like symptoms but were not dose-related and were observed in both MLN-519 (28%) and placebo (41%) treated subjects [50]. These clinical data indicate that proteasome inhibition is achievable to levels that have been associated with significant neuroprotective effects in animal models of brain injury. Given the safety profile of MLN-519, treatment could potentially be delivered as early as possible during the acute post-injury period (i.e. in the home or ambulance by trained medical staff). Thereafter, 'clot-busters' such as tissue plasminogen activator could be administered, if appropriate, in the hospital. This pharmaceutical 'cocktail' approach to the treatment of acute brain injury has been proposed as a strategy to optimize therapy due to the multitude of physiological factors and molecular perpetrators that act in concert towards progression of injury to the brain. In addition, therapies that target the more advanced phase of the injury could be used to induce regeneration of damaged neural tissue (e.g. growth factors or stem cell therapy) to help promote the overall functional recovery of the brain-injured patient.

THE PROTEASOME AND NF-κB MEDIATED INFLAMMATION

The proteasome plays a key role in the NF-κB-mediated inflammatory response (Figure 2). The transcription factor NF-κB is constitutively expressed within the cytosol but is normally bound to an inhibiting molecule IκB. A variety of inflammatory signals can induce the phosphorylation of IκB, targeting it for degradation through the UPS. The release and translocation of NF-κB into the nucleus can then stimulate inflammatory gene expression. As such, proteasome inhibition represents one method of inhibiting the activation of the NF-κB-meditated inflammatory response [8,9].

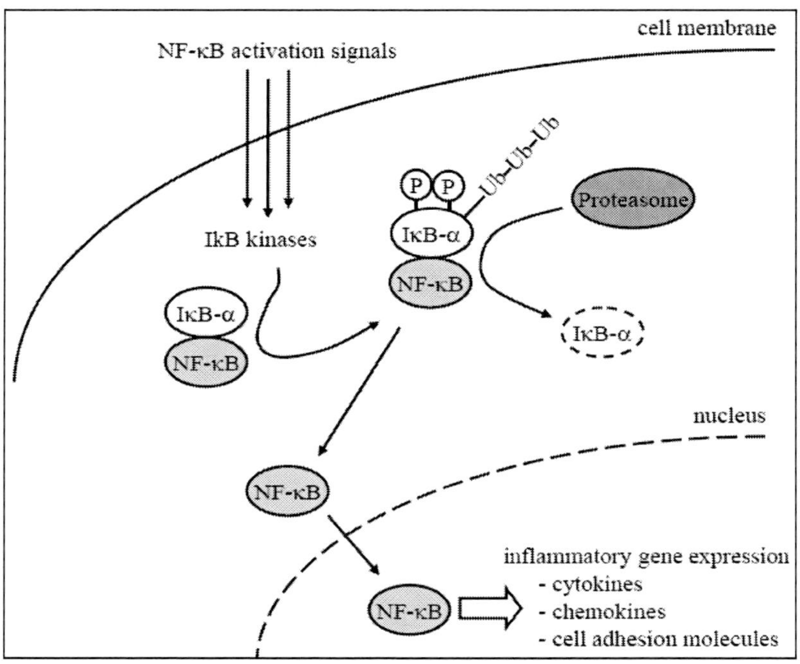

Figure 2. Proteasome and the NF-□B pathway. Stimulation of the NF-κB pathway by activation signals, such as inflammatory cytokines and reactive oxygen species, activate intracellular IκB kinases. IκB kinases induce the phosphorylation and subsequent ubiquitination of the IκB molecule. Ubiquinated IκB is then targeted for degradation by the proteasome. Release of constitutively expressed NF-κB from its inhibitory molecule, IκB, allows activation and translocation of NF-κB to the nucleus. Nuclear NF-κB is responsible for expression of numerous pro-inflammatory genes.

The activation of NF-κB within the brain has been reported to occur in a variety of CNS injury models including spinal cord injury [52], traumatic brain injury [53], and both global [54] and focal brain ischemia [55-58]. Following MCAo injury in rats, the activation of NF-κB has been reported as early as 3 h post-injury and was sustained out to at least 72 h in the ipsilateral cerebral cortex and striatum [59]. Based on cellular morphology, early activation of NF-κB appears in large multipolar neurons and astrocytes (3 h) while at 12-24 h activated NF-κB appears in small round or rod-shaped cells suggestive of infiltrating leukocytes or microglia [59]. The activation, nuclear translocation, and DNA binding of NF-κB following experimental ischemic brain injury has been reported in several other studies as well and has been implicated in the promotion of brain injury [56,60].

The inhibition of NF-κB activity has shown therapeutic efficacy in several experimental inflammatory disease models including neuronal injury [61]. Mice lacking the p50 subunit of NF-κB (p50 knockout) have been reported to have reduced brain injury following focal brain ischemia as compared to wild type controls [55]. Inhibition of NF-κB expression with pyrrolidine dithiocarbamate reduces the injury associated with focal or global ischemia with a post-injury therapeutic window of up to 6 h [58,60]. However, depending on the drug and dose used, inhibition of NF-κB can be toxic to the normal [51] or ischemic brain [57]. The inhibition of NF-κB in specific cell types may have contrasting effects on outcome due to the complex array of molecular events involved in NF-κB signaling and the role of NF-κB activation in different cell types [62]. NF-κB activation has also been linked to cerebral remodelling and may promote neuroplasticity during the advanced stages of recovery from brain injury [63]. A critical factor in assessing the inhibition of NF-κB as a therapeutic treatment is to evaluate the toxic versus therapeutic dose range and assessment of cell-specific activity. Progress towards understanding the role of NF-κB regulation on cell physiology is an active area of research and is elegantly reviewed in a series of recent review articles [62,64-68].

Treatment of Cerebral Ischemia with MLN-519: Mechanism of Action

In ischemia/reperfusion injury, such as that encountered during stroke, there are delayed biochemical events leading to 'abnormal' *de novo* gene expression and protein synthesis. These changes in cellular pathology are marked by increases in activated microglia and leukocyte infiltration of neutrophils and macrophages initiated by intra-luminal adhesion molecule attachment and diapedesis across the blood brain barrier. As a neuro-inflammation causality promoting secondary brain damage in peri-infarct zones this process, which has been relatively well defined in both experimental and clinical studies of ischemic brain injury, represents a 'delayed cellular target' for achieving neuroprotection [69]. Inflammatory mediators, including cytokines, adhesion molecules and infiltrating leukocytes all play a major role coordinating the pro-inflammatory response to an ischemic event. As such, site-specific targeting of anti-inflammatory mechanisms through regulation of these inflammatory proteins represents an exciting and promising research platform for explorations in neuroprotection strategies against cerebral ischemic episodes. In this regard, specific, non-toxic inhibitors of the 20S proteasome and proteasome-related systems (i.e. NF-κB) with compounds such as MLN-519 indeed represent viable opportunities as a stroke therapy [9]. Based on pre-clinical evidence, the neuroprotective effect of MLN-519 is strongly associated with an anti-inflammatory mechanism of action although other potential mechanisms of action have not been ruled out including the direct inhibition of injury-induced protein degradation [9].

Inhibition of NF-κB Activity

Proteasome inhibition has been associated with a concomitant reduction in activation of the transcription factor NF-κB. For example, MLN-519 has been reported to inhibit the formation of NF-κB-DNA complexes in activated T-cells *in vitro* and a significant dose-dependent reduction of super antigen mediated T-cell proliferation [70]. Treatment of focal cerebral ischemia with MLN-519 has also been directly associated with an attenuation of activated NF-κB. A single injection of MLN-519 (1.0 mg/kg, i.v.) given 2 h following a MCAo in rats was associated with a significant reduction in immunoreactivity for the activated form of NF-κB from 3-72 h following ischemic brain injury as compared to vehicle treatment, particularly in endothelial cells [71]. Similarly, in a subsequent study, a 6 h delayed injection of MLN-519 reduced optical density of activated NF-κB immunoreactivity predominately noted in endothelial cells (47% reduction) and leukocytes (45% reduction) as compared to neurons and glial cells (34% reduction) [48]. Two interesting aspects of these studies were apparent: (*i*) NF-κB activation was not completely inhibited with MLN-519 and (*ii*) the inhibitory effect favored vascular cells and peripheral leukocytes over resident brain cells (i.e. neurons/glia). This cell-specific effect of MLN-519 is most likely related to the poor brain penetrability of MLN-519 [11,46].

Inflammatory Gene Expression

Microarray gene studies of the injured rat brain following transient focal brain ischemia have mapped the post-injury expression profile of multiple inflammatory genes post-injury. In particular, a host of pro-inflammatory cytokines, chemokines and cellular adhesion molecules are upregulated with the peak expression levels between 4-24 h post-injury [2,59]. Direct evidence has indicated that several of these pro-inflammatory markers are toxic to cells and can promote cell death although the role of inflammation in brain injury is still controversial [41,72]. However, several preclinical studies have indicated the therapeutic efficacy of treatment of brain injury with anti-inflammatory agents [31,41].

A wide selection of anti-inflammatory drugs are currently available that target one or more mediators of the inflammatory cascade. However, treatment of brain-injured subjects with anti-inflammatory agents aimed at reducing injury severity may not be optimal. One potential complication is related to the redundancy in function of the multitude of mediators involved in inflammation [41]. In response, recent approaches have focused on modulation of transcriptional factors, such as NF-κB, that control the gene expression of multiple pro-inflammatory molecules [55]. The use of compounds such as proteasome inhibitors is one method to block the activation of NF-κB and reduce expression of a host of inflammatory genes, representing a novel approach to treatment of ischemic brain injury [8-10,22].

MLN-519 treatment of MCAo injury in rats has been associated with an attenuation of the increase in both cytokine and cell adhesion molecule expression (Figure 3). MLN-519, delivered 2 h post-MCAo, reduced IL-1β expression from 6-24 h (33-43% decrease) but appeared to have little effect on the expression of TNF-α and IL-6 although a slight but significant reduction of TNF-α was measured at 24 h [71]. Immunohistochemical staining for each of these cytokines revealed a strong presence in perivascular cells [71]. Interestingly, by delaying the initial injection of MLN-519 to 6 h post-injury, significant reductions were measured in both TNF-α (46%) and IL-6 (58%) at 24 h post-injury [1]. It appears that delaying the initial post-injury injection of MLN-519 corresponds to a differential effect on cytokine expression as compared to injections initiated at the time of reperfusion (2 h). Importantly, however, a 6 h delayed injection of MLN-519, which corresponds to the time of peak cytokine expression, was still able to significantly reduce inflammatory gene mRNA levels and was associated with a neuroprotective outcome. Other models of inflammatory disorders have also indicated the therapeutic efficacy of proteasome inhibitors to reduce inflammatory gene expression [11]. In particular, in a model of streptococcal induced polyarthiritis in rats, significantly lower serum levels of the pro-inflammatory factors IL-1 and IL-6 were measured in PS-341 treated rats as compared to the increase in these factors measured following vehicle treatment [73].

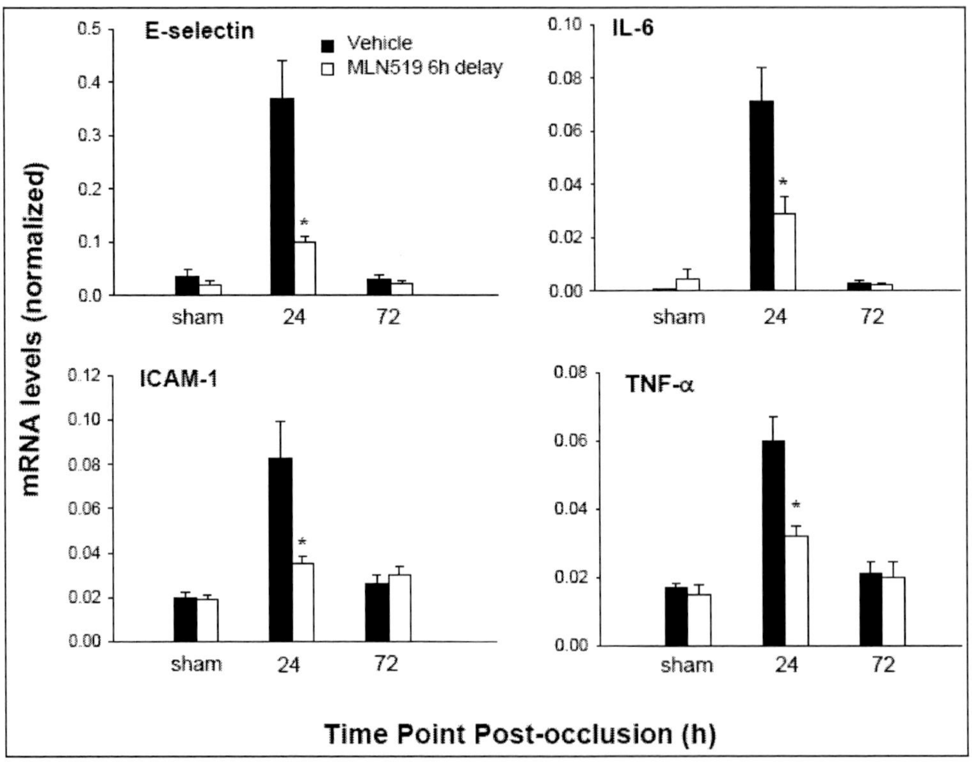

Figure 3. Reduction of inflammatory gene expression with MLN-519. Effect of MLN-519 treatment (1.0 mg/kg, delivered i.v. starting 6 h post-occlusion) on cytokine and cell adhesion molecule mRNA expression after 2 h of transient middle cerebral artery occlusion in the rat. (Data derived from report by [1]).

Similar to the effects on cytokine expression, a 2 h delayed injection of MLN-519 was associated with dramatic decreases in the cellular adhesion molecule mRNA expression of ICAM-1 at 3 h post-injury, while E-selectin was decreased at 12 and 24 h as compared to vehicle treated animals [71]. Reductions in ICAM-1 (58%) and E-selectin (72%) were also verified at 24 h post-injury with a 6 h delayed injection of MLN-519 [1].

Immunohistochemical staining of these cellular adhesion molecules indicated expression was predominately within infiltrating leukocytes and endothelial cells [71].

Inflammatory Cell Infiltration

The upregulation of pro-inflammatory cytokines, chemokines, and cellular adhesion molecules following brain injury has been linked to formation of cerebral edema and infiltration of inflammatory cells into the injured brain [74-76]. The resulting diapedesis of inflammatory cells involves the interaction of several cell surface molecules under control of NF-κB including integrins, selectins, and members of the immunoglobulin superfamily (IgCAM). Cell permeable inflammatory messengers such as chemokines are released from the injured brain and attract circulating peripheral leukocytes to the injured endothelium where they bind newly expressed P- and E-selectin. The interaction between cell surface integrins and IgCAMs then promote the diapedesis of peripheral leukocytes across the blood brain barrier and into injured brain regions where they participate in the neuro-inflammatory process [74,75].

Neuronal injury following an ischemic attack can evolve over several hours to days post-injury [77,78]. Strong evidence now exists indicating that active, delayed injury processes are involved in, and ultimately determine, the eventual degree of cell survival following injury [39]. The injury-induced inflammatory response is a complex and multi-step process involving inflammatory gene upregulation, release of chemotaxic agents into the blood stream, and eventual activation/recruitment of peripheral leukocytes to the site of injury. Diapedesis of inflammatory cells into the injured brain begins with an initial phase of neutrophil infiltration, which peaks at 24 h post-injury, followed by macrophage infiltration at 72 h post-injury [78]. Inflammatory cells not only dispose of cellular debris but also are a major source of post-injury toxins including reactive oxygen species and pro-inflammatory cytokines [41]. Microvascular occlusion may also occur due to the intravascular collection of inflammatory cells around the site of injury [75]. Although inflammatory cell-mediated phagocytosis is inherent to the natural healing process, pro-inflammatory molecules can promote further injury to the vulnerable penumbral regions that surround an ischemic lesion in a 'feed-forward' process of inflammatory cell-mediated injury [41,54].

Proteasome inhibitors have shown efficacy in reducing inflammatory cell infiltration in a variety of inflammatory-related disorders. Reduction of inflammatory infiltrate has been observed following streptococcal induced polyarthiritis in rats when the proteasome inhibitor PS-341 was administered after onset of the disease and was associated with minimal degradation of articular cartilage of the joints [73]. In an animal model of pulmonary eosinophilia induced by allergen exposure, intratracheal administration of MLN-519 induced a significant dose-dependent reduction of leukocytes in lung lavage fluid [79]. MLN-519 has also been shown to decrease CNS infiltration of T-cells in a mouse model of autoimmune encephalomylelitis (an experimental model of multiple sclerosis) and was associated with an improvement in clinical disease score with fewer relapses [80]. Additionally, MLN-519 has been reported to significantly reduce the inflammatory infiltrate of lesional human plaque-stage psoriasis skin grafts in mice [70].

Treatment of ischemic brain injury with proteasome inhibitors has also been shown to affect the diapedesis of cells into the injured brain. An interesting aspect of treatment with proteasome inhibitors is that different cell types may react differently to proteasome inhibition particularly following injury [81]. For example, the sensitivity of human endothelial cells to proteasome inhibition increases 10-fold following exposure to hypoxic conditions [82]. Studies evaluating the treatment effects of MLN-519 following rat MCAo injury have repeatedly indicated the neuroprotective reduction in brain infarction is associated with a reduction in both neutrophil and macrophage infiltration in both cortical and subcortical brain regions with up to a 10 h delayed injection of MLN-519 [1,45,83]. Overall, MLN-519 treatment reduced neutrophil infiltration by 36-60% as evaluated 24 h post-injury [45,48] and reduced both neutrophil (32-79%) and macrophage (up to 32-80%) infiltration as evaluated at 72 h post-injury [1,45]. Similarly, following embolic stroke in rats, significant reductions in inflammatory cells were also observed with delayed injections of MLN-519 (initiated 2, 4, or 6 h post-injury) [46]. This dramatic effect of MLN-519 treatment on inflammatory cell infiltration may be directly related to reducing endothelial cell-induced attraction of peripheral leukocytes, providing a key anti-inflammatory mechanism of action related to the improved outcome following experimental cerebral ischemia.

Conclusion

Based on experimental data, the proteasome appears to play a crucial role in the progression of several different disease states including CNS disorders. In many cases, neurodegenerative disorders are associated with a loss of function of proteasome activity or the UPS in general. These types of disorders may benefit from strategies to reestablish normal levels of proteasomal activity. However, the proteasome is also a therapeutic target in several types of disease due to the role it plays in promotion of the aberrant gene expression associated with disease progression. Pre-clinical data supports the use of proteasome inhibitors to treat inflammatory related disorders such as injury to the CNS due to cerebral ischemia. In fact, the proteasome inhibitor MLN-519 appears to provide an excellent therapeutic window for the treatment of ischemia/reperfusion brain injury in rats. The neuroprotective effects of MLN-519 have been associated with an improvement in histopathological and functional recovery with a 6-10 h therapeutic treatment window with no significant detrimental changes observed in physiological parameters between vehicle and drug-treated animals. MLN-519 treatment also correlated to a reduction in blood proteasome activity by 80-90% with a dosing schedule shown to be safe in normal, healthy human volunteers.

The mechanism of action of proteasome inhibition to treat ischemic brain injury appears to involve an anti-inflammatory effect via reduction of NF-κB mediated inflammatory gene expression. MLN-519 treatment has been shown to reduce activated NF-κB immunoreactivity, attenuate inflammatory gene expression, and reduce infiltration of inflammatory neutrophils and macrophages in the injured rat brain. Additionally, the 6-10 h therapeutic window of MLN-519 correlates well to peak increases in cytokine and cellular adhesion molecule mRNA upregulation. As such, one of the primary mechanisms of action of

MLN-519 in ischemic brain injury may involve the inhibition of cellular adhesion molecule-mediated infiltration of peripheral inflammatory cells into the brain, which represents a potentially 'non-neuronal' mechanism of neuroprotective activity. Overall, pre-clinical brain injury studies support the therapeutic use of proteasome inhibitors, which represent a new approach for the treatment of CNS disorders such as clinical stroke through attenuation of the inflammatory-mediated neuropathology associated with ischemia/reperfusion injury to the brain. The superior neuroprotective treatment profile of novel compounds such as MLN-519 to treat brain injury may ultimately play a key role, alone or in conjunction with other drug therapies, as part of an overall therapeutic regiment to promote recovery of the brain-injured patient.

References

[1] Williams AJ, Berti R, Dave JR, Elliot PJ, Adams J, Tortella FC: Delayed treatment of ischemia/reperfusion brain injury: extended therapeutic window with the proteosome inhibitor MLN519. *Stroke* 2004, 35:1186-1191.

[2] Lu XC, Williams AJ, Yao C, Berti R, Hartings JA, Whipple R, Vahey MT, Polavarapu RG, Woller KL, Tortella FC, Dave JR: Microarray analysis of acute and delayed gene expression profile in rats after focal ischemic brain injury and reperfusion. *J Neurosci Res* 2004, 77:843-857.

[3] Glickman MH, Ciechanover A: The ubiquitin-proteasome proteolytic pathway: destruction for the sake of construction. *Physiol Rev* 2002, 82:373-428.

[4] Bochtler M, Ditzel L, Groll M, Hartmann C, Huber R: The proteasome. *Annu Rev Biophys Biomol Struct* 1999, 28:295-317.

[5] Hershko A, Ciechanover A: The ubiquitin system. *Annu Rev Biochem* 1998, 67:425-479.

[6] Davies KJ: Degradation of oxidized proteins by the 20S proteasome. *Biochimie* 2001, 83:301-310.

[7] Benaroudj N, Tarcsa E, Cascio P, Goldberg AL: The unfolding of substrates and ubiquitin-independent protein degradation by proteasomes. *Biochimie* 2001, 83:311-318.

[8] Elliott P, Ross J: The proteasome: a new target for novel drug therapies. *Am J Clin Pathol* 2001, 116:637-646.

[9] Wojcik C, Di Napoli M: Ubiquitin-proteasome system and proteasome inhibition: new strategies in stroke therapy. *Stroke* 2004, 35:1506-1518.

[10] Di Napoli M, Papa F: The proteasome system and proteasome inhibitors in stroke: controlling the inflammatory response. *Curr Opin Investig Drugs* 2003, 4:1333-1342.

[11] Elliott PJ, Zollner TM, Boehncke WH: Proteasome inhibition: a new anti-inflammatory strategy. *J Mol Med* 2003, 81:235-245.

[12] Kukan M: Emerging roles of proteasomes in ischemia-reperfusion injury of organs. *J Physiol Pharmacol* 2004, 55:3-15.

[13] Keller JN, Gee J, Ding Q: The proteasome in brain aging. *Ageing Res Rev* 2002, 1:279-293.

[14] Song S, Jung YK: Alzheimer's disease meets the ubiquitin-proteasome system. *Trends Mol Med* 2004, 10:565-570.

[15] Halliwell B: Hypothesis: proteasomal dysfunction: a primary event in neurogeneration that leads to nitrative and oxidative stress and subsequent cell death. *Ann N Y Acad Sci* 2002, 962:182-194.

[16] McNaught KS, Perl DP, Brownell AL, Olanow CW: Systemic exposure to proteasome inhibitors causes a progressive model of Parkinson's disease. *Ann Neurol* 2004, 56:149-162.

[17] Mengual E, Arizti P, Rodrigo J, Gimenez-Amaya JM, Castano JG: Immunohistochemical distribution and electron microscopic subcellular localization of the proteasome in the rat CNS. *J Neurosci* 1996, 16:6331-6341.

[18] Chain DG, Hegde AN, Yamamoto N, Liu-Marsh B, Schwartz JH: Persistent activation of cAMP-dependent protein kinase by regulated proteolysis suggests a neuron-specific function of the ubiquitin system in Aplysia. *J Neurosci* 1995, 15:7592-7603.

[19] Ciechanover A, Brundin P: The ubiquitin proteasome system in neurodegenerative diseases: sometimes the chicken, sometimes the egg. *Neuron* 2003, 40:427-446.

[20] Berke SJ, Paulson HL: Protein aggregation and the ubiquitin proteasome pathway: gaining the UPPer hand on neurodegeneration. *Curr Opin Genet Dev* 2003, 13:253-261.

[21] Coleman MP, Ribchester RR: Programmed axon death, synaptic dysfunction and the ubiquitin proteasome system. *Curr Drug Targets CNS Neurol Disord* 2004, 3:227-238.

[22] Elliott P, Shah I, Lees K, Pien C, Tortella F, Adams J: Proteasome Inhibition - A novel target for anti-stroke therapy. *Restorative Neurology and Neuroscience* 2000, 16:156.

[23] Adams J, Behnke M, Chen S, Cruickshank AA, Dick LR, Grenier L, Klunder JM, Ma YT, Plamondon L, Stein RL: Potent and selective inhibitors of the proteasome: dipeptidyl boronic acids. *Bioorg Med Chem Lett* 1998, 8:333-338.

[24] Bogyo M, McMaster JS, Gaczynska M, Tortorella D, Goldberg AL, Ploegh H: Covalent modification of the active site threonine of proteasomal beta subunits and the Escherichia coli homolog HslV by a new class of inhibitors. *Proc Natl Acad Sci U S A* 1997, 94:6629-6634.

[25] Schmidtke G, Holzhutter HG, Bogyo M, Kairies N, Groll M, de Giuli R, Emch S, Groettrup M: How an inhibitor of the HIV-I protease modulates proteasome activity. *J Biol Chem* 1999, 274:35734-35740.

[26] Soucy F, Grenier L, Behnke M, Destree A, McCormack T, Adams J, Plamondon L: A novel and efficient synthesis of a highly active analogue of *clasto*-lactacystin b-lactone. *J Am Chem Soc* 1999, 121:9967-9976.

[27] Di Napoli M, Papa F: MLN-519. Millennium/PAION. *Curr Opin Investig Drugs* 2003, 4:333-341.

[28] Elofsson M, Splittgerber U, Myung J, Mohan R, Crews CM: Towards subunit-specific proteasome inhibitors: synthesis and evaluation of peptide alpha',beta'-epoxyketones. *Chem Biol* 1999, 6:811-822.

[29] Gao Y, Lecker S, Post MJ, Hietaranta AJ, Li J, Volk R, Li M, Sato K, Saluja AK, Steer ML, Goldberg AL, Simons M: Inhibition of ubiquitin-proteasome pathway-mediated I

kappa B alpha degradation by a naturally occurring antibacterial peptide. *J Clin Invest* 2000, 106:439-448.

[30] Lipton P: Ischemic cell death in brain neurons. *Physiol Rev* 1999, 79: 1431-1568.

[31] Bramlett HM, Dietrich WD: Pathophysiology of cerebral ischemia and brain trauma: similarities and differences. *J Cereb Blood Flow Metab* 2004, 24:133-150.

[32] Li J, Post M, Volk R, Gao Y, Li M, Metais C, Sato K, Tsai J, Aird W, Rosenberg RD, Hampton TG, Sellke F, Carmeliet P, Simons M: PR39, a peptide regulator of angiogenesis. *Nat Med* 2000, 6:49-55.

[33] Janardhan V, Qureshi AI: Mechanisms of ischemic brain injury. *Curr Cardiol Rep* 2004, 6:117-123.

[34] Muir K, Lees K: Clinical experience with excitatory amino acid antagonist drugs. *Stroke* 1995, 26:503-513.

[35] DeGraba TJ, Pettigrew LC: Why do neuroprotective drugs work in animals but not humans? *Neurol Clin* 2000, 18:475-493.

[36] Fisher M, for S.T.A.I.R.: Recommendations for advancing development of acute stroke therapies: stroke therapy academic industry roundtable. *Stroke* 2003, 34:1539-1546.

[37] Heiss W, Fink G, Huber M, Herholz K: Positron emission tomography imaging and the therapeutic window. *Stroke* 1993, 24:I50-53; discussion I48-59, I54-56.

[38] Lucchesi B: Complement activation, neutrophils, and oxygen radicals in reperfusion injury. *Stroke* 1993, 24:I41-47; discussion I38-40.

[39] Barinaga M: Stroke-damaged neurons may commit cellular suicide. *Science* 1998, 281:1302-1303.

[40] Martin L, Al-Abdulla N, Brambrink A, Kirsch J, Sieber F, Portera-Cailliau C: Neurodegeneration in excitotoxicity, global cerebral ischemia, and target deprivation: a perspective on the contributions of apoptosis and necrosis. *Brain Res Bull* 1998, 46:281-309.

[41] Stanimirovic D, Satoh K: Inflammatory mediators of cerebral endothelium: a role in ischemic brain inflammation. *Brain Pathol* 2000, 10:113-126.

[42] Ding Q, Keller JN: Proteasome inhibition in oxidative stress neurotoxicity: implications for heat shock proteins. *J Neurochem* 2001, 77:1010-1017.

[43] Zhai Q, Wang J, Kim A, Liu Q, Watts R, Hoopfer E, Mitchison T, Luo L, He Z: Involvement of the ubiquitin-proteasome system in the early stages of wallerian degeneration. *Neuron* 2003, 39:217-225.

[44] Pye J, Ardeshirpour F, McCain B, DA, Merricks E, Adams J, Elliott P, Pien C, Fischer T, Baldwin A, Nichols T: Proteasome inhibition ablates activation of NF-kB in myocardial reperfusion and reduces reperfusion injury. *Am J Physiol Heart Circ Physiol* 2003, 284:H919-H926.

[45] Buchan AM, Li H, Blackburn B: Neuroprotection achieved with a novel proteasome inhibitor which blocks NF-kappaB activation. *Neuroreport* 2000, 11:427-430.

[46] Phillips J, Williams A, Adams J, Elliott P, Tortella F: Proteasome inhibitor PS519 reduces infarction and attenuates leukocyte infiltration in a rat model of focal cerebral ischemia. *Stroke* 2000, 31:1686-1693.

[47] Williams AJ, Hale SL, Moffett JR, Dave JR, Elliott PJ, Adams J, Tortella FC: Delayed Treatment With MLN519 Reduces Infarction and Associated Neurologic Deficit

Caused by Focal Ischemic Brain Injury in Rats via Antiinflammatory Mechanisms Involving Nuclear Factor-kappaB Activation, Gliosis, and Leukocyte Infiltration. *J Cereb Blood Flow Metab* 2003, 23:75-87.

[48] Williams AJ, Myers TM, Cohn SI, Sharrow KM, Lu XC, Tortella FC: Recovery from ischemic brain injury in the rat following a 10 h delayed injection with MLN519. *Pharmacol Biochem Behav* 2005, 81:182-189.

[49] Zhang L, Zhang ZG, Zhang RL, Lu M, Adams J, Elliott PJ, Chopp M: Postischemic (6-Hour) treatment with recombinant human tissue plasminogen activator and proteasome inhibitor PS-519 reduces infarction in a rat model of embolic focal cerebral ischemia. *Stroke* 2001, 32:2926-2931.

[50] Shah I, Lees K, Pien C, Elliott P: Early clinical experience with the novel proteasome inhibitor PS-519. *Br J Clin Pharmacol* 2002, 54:269-276.

[51] Taglialatela G, Kaufmann J, Trevino A, Perez-Polo J: Central nervous system DNA fragmentation induced by the inhibition of nuclear factor kappa B. *Neuroreport* 1998, 9:489-493.

[52] Bethea J, Castro M, Keane R, Lee T, Dietrich W, Yezierski R: Traumatic spinal cord injury induces nuclear factor-kappaB activation. *J Neurosci* 1998, 18:3251-3260.

[53] Nonaka M, Chen X, Pierce J, Leoni M, McIntosh T, Wolf J, Smith D: Prolonged activation of NF-kappaB following traumatic brain injury in rats. *J Neurotrauma* 1999, 16:1023-1034.

[54] Clemens J, Stephenson D, Yin T, Smalstig E, Panetta J, Little S: Drug-induced neuroprotection from global ischemia is associated with prevention of persistent but not transient activation of nuclear factor-kappaB in rats. *Stroke* 1998, 29:677-682.

[55] Schneider A, Martin-Villalba A, Weih F, Vogel J, Wirth T, Schwaninger M: NF-kB is activated and promotes cell death in focal cerebral ischemia. *Nat Med* 1999, 5:554-559.

[56] Stephenson D, Yin T, Smalstig E, Hsu M, Panetta J, Little S, Clemens J: Transcription factor nuclear factor-kappa B is activated in neurons after focal ischemia. *J Cereb Blood Flow Metab* 2000, 20:592-603.

[57] Hill W, Hess D, Carroll J, Chandramohan G, Howard E, Chen Q, Cheng C, Martin-Studdard A, Waller J, Beswick R: The NF-kB inhibitor diethyldithiocarbamate (DDTC) increases brain cell death in a transient middle cerebral artery occlusion model of ischemia. *Brain Res Bull* 2001, 55:375-386.

[58] Nurmi A, Vartiainen N, Pihlaja R, Goldsteins G, Yrjanheikki J, Koistinaho J: Pyrrolidine dithiocarbamate inhibits translocation of nuclear factor kappa-B in neurons and protects against brain ischaemia with a wide therapeutic time window. *J Neurochem* 2004, 91:755-765.

[59] Berti R, Williams A, Moffett J, Hale S, Velarde L, Elliott P, Yao C, Dave J, Tortella F: Real time PCR mRNA analysis of the inflammatory cascade associated with ischemia reperfusion brain injury. *J Cereb Blood Flow Metab* 2002, 22:1068-1079.

[60] Nurmi A, Lindsberg PJ, Koistinaho M, Zhang W, Juettler E, Karjalainen-Lindsberg ML, Weih F, Frank N, Schwaninger M, Koistinaho J: Nuclear factor-kappaB contributes to infarction after permanent focal ischemia. *Stroke* 2004, 35:987-991.

[61] Verma IM: Nuclear factor (NF)-kappaB proteins: therapeutic targets. *Ann Rheum Dis* 2004, 63 Suppl 2:ii57-ii61.

[62] Hayden MS, Ghosh S: Signaling to NF-kappaB. *Genes Dev* 2004, 18:2195-2224.

[63] Barger SW, Moerman AM, Mao X: Molecular mechanisms of cytokine-induced neuroprotection: NFkappaB and neuroplasticity. *Curr Pharm Des* 2005, 11:985-998.

[64] Bonizzi G, Karin M: The two NF-kappaB activation pathways and their role in innate and adaptive immunity. *Trends Immunol* 2004, 25:280-288.

[65] Chen LF, Greene WC: Shaping the nuclear action of NF-kappaB. *Nat Rev Mol Cell Biol* 2004, 5:392-401.

[66] Karin M, Yamamoto Y, Wang QM: The IKK NF-kappa B system: a treasure trove for drug development. *Nat Rev Drug Discov* 2004, 3:17-26.

[67] Kucharczak J, Simmons MJ, Fan Y, Gelinas C: To be, or not to be: NF-kappaB is the answer--role of Rel/NF-kappaB in the regulation of apoptosis. *Oncogene* 2003, 22:8961-8982.

[68] Ruland J, Mak TW: Transducing signals from antigen receptors to nuclear factor kappaB. *Immunol Rev* 2003, 193:93-100.

[69] Wang X: Investigational anti-inflammatory agents for the treatment of ischaemic brain injury. *Expert Opin Investig Drugs* 2005, 14:393-409.

[70] Zollner TM, Podda M, Pien C, Elliott PJ, Kaufmann R, Boehncke WH: Proteasome inhibition reduces superantigen-mediated T cell activation and the severity of psoriasis in a SCID-hu model. *J Clin Invest* 2002, 109:671-679.

[71] Berti R, Williams A, Velarde L, Elliott P, Adams J, Yao C, Dave J, Tortella F: Effect of the proteasome inhibitor MLN519 on expression of inflammatory molecules following middle cerebral artery occlusion and reperfusion in the rat. *Neurotox Res* 2003, 5:505-514.

[72] Emerich D, Dean R, Bartus R: The role of leukocytes following cerebral ischemia: pathogenic variable or bystander reaction to emerging infarct? *Exp Neurol* 2002, 173:168-181.

[73] Palombella VJ, Conner EM, Fuseler JW, Destree A, Davis JM, Laroux FS, Wolf RE, Huang J, Brand S, Elliott PJ, Lazarus D, McCormack T, Parent L, Stein R, Adams J, Grisham MB: Role of the proteasome and NF-kappaB in streptococcal cell wall-induced polyarthritis. *Proc Natl Acad Sci U S A* 1998, 95:15671-15676.

[74] Yoshimoto T, Houkin K, Tada M, Abe H: Induction of cytokines, chemokines and adhesion molecule mRNA in a rat forebrain reperfusion model. *Acta Neuropathol (Berl)* 1997, 93:154-158.

[75] Danton GH, Dietrich WD: Inflammatory mechanisms after ischemia and stroke. *J Neuropathol Exp Neurol* 2003, 62:127-136.

[76] Frijns CJ, Kappelle LJ: Inflammatory cell adhesion molecules in ischemic cerebrovascular disease. *Stroke* 2002, 33:2115-2122.

[77] Clark R, Lee E, Fish C, White R, Price W, Jonak Z, Feuerstein G, Barone F: Development of tissue damage, inflammation and resolution following stroke: an immunohistochemical and quantitative planimetric study. *Brain Res Bull* 1993, 31:565-572.

[78] Clark R, Lee E, White R, Jonak Z, Feuerstein G, Barone F: Reperfusion following focal stroke hastens inflammation and resolution of ischemic injured tissue. *Brain Res Bull* 1994, 35:387-392.

[79] Elliott PJ, Pien CS, McCormack TA, Chapman ID, Adams J: Proteasome inhibition: A novel mechanism to combat asthma. *J Allergy Clin Immunol* 1999, 104:294-300.

[80] Vanderlugt CL, Rahbe SM, Elliott PJ, Dal Canto MC, Miller SD: Treatment of established relapsing experimental autoimmune encephalomyelitis with the proteasome inhibitor PS-519. *J Autoimmun* 2000, 14:205-211.

[81] Wojcik C: Regulation of apoptosis by the ubiquitin and proteasome pathway. *J Cell Mol Med* 2002, 6:25-48.

[82] Zund G, Uezono S, Stahl GL, Dzus AL, McGowan FX, Hickey PR, Colgan SP: Hypoxia enhances induction of endothelial ICAM-1: role for metabolic acidosis and proteasomes. *Am J Physiol* 1997, 273:C1571-1580.

[83] Williams A, Lu X-C, Hartings J, Tortella F: Neuroprotection assessment by topographic electroencephalographic analysis: effects of a sodium channel blocker to reduce polymorphic delta activity following ischaemic brain injury in rats. *Fund Clin Pharm* 2003, 17:581-593.

In: The Ubiquitin Proteasome System... ISBN 978-1-60021-749-4
Eds: Mario Di Napoli and Cezary Wojcik, pp. 1011-1035© 2007 Nova Science Publishers, Inc.

Chapter 42

Clinical Experience of Proteasome Inhibitors in Central Nervous System Diseases

Imtiaz M. Shah[1,*] and Mario Di Napoli[2,3]
[1]Mansionhouse Unit, Victoria Infirmary, Glasgow, Scotland. G41 3DX;
[2]Neurological Service, San Camillo de'Lellis General Hospital, Rieti I-02100, Italy;
[3]Neurological Section, SMDN – Center for Cardiovascular Medicine and Cerebrovascular Disease Prevention, Sulmona (L'Aquila) I-67039, Italy.

Abstract

The proteasome is an enzyme, which is present within all cells, from yeast to humans. It has a central role in the proteolytic degradation of the vast majority of intracellular proteins. Among the key proteins modulated by the proteasome are those involved in controlling inflammatory processes, cell cycle regulation, and gene expression. Agents that inhibit the proteasome have been shown to be active in numerous animal models of inflammation and cancer. Two proteasome inhibitors are under clinical evaluation. MLN-519 is being studied for the treatment of reperfusion injury that occurs following cerebral ischemia and myocardial infarction. The other, Bortezomib (Velcade®), has recently been licensed for the clinical treatment of multiple myeloma. It is also undergoing further evaluation for the treatment of chronic lymphocytic leukemia and a variety of solid tumors. The proteasome may also have an important role in the evolution of HIV-related disorders including AIDS and inflammatory disorders. Therapeutic strategies using proteasome inhibitors for the treatment of these conditions have now entered preclinical development. MLN-519 is a small-molecular-weight lactacystin analogue developed by Millenium (LeukoSite) for the potential treatment of inflammatory disease and stroke using a novel ubiquitin proteasome enzyme inhibitor

* Correspondence concerning this article should be addressed to Dr. Imtiaz M. Shah, MSc MRCP; Mansionhouse Unit, Victoria Infirmary, Glasgow. Scotland. G41 3DX. Phone: +44 141 434 0902; Fax: +44 141 201 6305; Email: I.M.Shah@clinmed.gla.ac.uk.

approach. The reperfusion that follows an ischemic event provides both positive and negative factors that affect the overall outcome of the cerebral tissue. The ischemic endothelium upregulates the expression of cell adhesion molecules, which then attract the circulating leukocytes. Once bound to the endothelium, these cells diapedese into the tissue and are responsible for the destruction and much of the subsequent tissue damage. MLN-519 attenuates the expression of these cellular proteins, reduces the invasion of leukocytes and hence limits tissue damage. MLN-519 has demonstrated a neuroprotective effect in rat models of middle cerebral artery temporary occlusion. MLN-519 reduces infarct volume, brain edema and increases neurological recovery with a reported therapeutic window of at least 6-hours. These effects are associated with a temporary reduction of circulanting 20S proteasome activity (70-80%), with a reduced leukocyte infiltration and decreased nuclear factor-κB activation. Similar protective results have also been reported in experimental myocardial infarction models in rats and pigs: MLN-519 protects cardiac tissue from ischemia and maintains its functionality as demonstrated by preserved left ventricular developed pressure and contractile function. These data demonstrate substantial clinical value, since many patients are admitted to the hospital hours after the stroke or heart attack has occurred and reperfusion has begun. That inhibition of the proteasome can be of benefit under these clinically-relevant conditions demonstrates their potential in these common life-threatening diseases. An explorative phase I trial has demonstrated that MLN-519 is well tolerated by healthy subjects at levels that are maximally neuroprotective in experimental conditions. It is currently undergoing further evaluation for clinical trials in acute stroke and myocardial infarction.

Keywords: Proteasome inhibitors, NF-κB, Neuroinflammation, Stroke.

Abbreviations

6-OHDA, 6-hydroxyl dopamine; AD, Alzheimer's disease; AMC, 7-amino-4-methyl-coumarin; BBB, blood-brain barrier; ChTL, chymotrypsin-like; EEG, electroencephalogram; eNOS, endothelial nitric oxide synthase; ER, endoplasmic reticulum, ERAD, ER-associate degradation; hsp, heat shock proteins; HD, Huntington´s disease; HIV, human immunodeficiency virus; IAP, inhibitor of apoptosis proteins; ICAM, intracellular adhesion molecule; IκB, inhibitory-κB; IKK, IκB kinase; IL, interleukin; MCAo, middle cerebral artery occlusion; MM, multiple myeloma; NF-κB, nuclear factor κB; PD, Parkinson's disease; PGPH, peptidylglutamyl peptide-hydrolyzing; PSI, Cbz-Ile-Glu (O-t-Bu)-Ala-leucinal; RT-PCR, real time polymerase chain reaction; TAT, tyrosine amino transferase; TL, trypsin-like; TNF –α, Tumour Necrosis Factor–α, UPS, ubiquitin proteasome system; VCAM-1, Vascular Cell Adhesion Molecole-1.

Introduction

Protein homeostasis is critical to several biological processes and the ubiquitin proteasome system (UPS), which processes more than 80% of all cellular proteins, is the principal mechanism for degradation of proteins. It plays an essential role in regulating the intracellular concentration of specific proteins, thereby maintaining homeostasis within the cells. The UPS controls global protein degradation rates and the precisely timed degradation of regulatory proteins, such as cyclins, transcription factors and oncogene products (which are important in cancer and inflammatory disorders) [1-3]. Therefore, targeting the regulation of protein production and destruction has been a major focus of drug research [4-6]. Accordingly, the proteasome has emerged as an attractive target for drug discovery. The 26S proteasome is a large multi-catalytic protease, which has a major role in intracellular proteolysis [7]. It is present within all eukaryotic cells and its function has now been recognised as being crucial for cellular function. The proteasome protects the cell against oxidative stress, and prevents the accumulation of redundant and damaged proteins. Dysfunction of the proteasome has been implicated in the pathogenesis of neurodegenerative conditions like Parkinson's (PD) and Alzheimer's Disease (AD) [8]. However, inhibition of the proteasome has also been shown to be beneficial in the treatment of certain cancers and inflammatory conditions [9,10]. Inhibition of the 26S proteasome prevents this targeted proteolysis, which can affect multiple signalling cascades within the cell. Manipulating both the global proteolysis rates and the degradation of particular substrates are both attractive goals for drug design [4,5]. This chapter will review the expanding role of the presently available proteasome inhibitors and their potential clinical uses in the treatment of nervous system diseases.

THE UBIQUITIN-PROTEASOME SYSTEM

Current research has focused on the UPS and its role in the pathophysiology of disease processes [11]. Significant advances have been made in the scientific understanding of the fundamental importance of the UPS in processes beyond the proteolytic degradation of damaged, oxidised, or misfolded proteins. In nonclinical studies, targeted degradation of key regulatory proteins by the 26S proteasome has been shown to be involved in controlling the cell cycle, transcription, apoptosis, angiogenesis, cell migration, and metastasis. Several in vitro studies have shown that inhibition of the proteasome affects the temporal stability of various cell-cycle regulatory proteins, especially those that are short lived. Cyclins, cyclin-dependent kinase inhibitors, and tumour suppressors have all been shown to be substrates for the UPS (Table 1). Inhibiting the degradation of regulators such as p27, p21, and p53 has been implicated as a mechanism by which proteasome inhibition impairs tumour cell growth and survival. The role of the proteasome in cell cycle regulation has provided the opportunity for exploring the therapeutic potential of proteasome inhibition.

The UPS regulates the action of transcription factor Nuclear Factor–κB (NF-κB) and this has been an important target for therapeutic drug research [10]. NF-κB is an inducible transcription factor of the *Rel* family, sequestered in the cytoplasm by the IκB family of proteins. NF-κB exists in several dimeric forms, but the p50/p65 heterodimer is the predominant one [12]. NF-κB, in its inactive form, is normally complexed to the inhibitory protein, inhibitory-κB (IκB) and the complex is sequestered in the cytoplasm, preventing NF-κB mediated gene transcription. Phosphorylation of IκB by IκB kinase (IKK) leads to activation of NF-κB [13]. Phosphorylated IκB is then ubiquitinated, and subsequently undergoes degradation by the proteasome [14]. In response to stimuli, IκB is phosphorylated, resulting in its ubiquitination and subsequent degradation by the proteasome. NF-κB is thereby released and is rapidly translocated to the nucleus, where it promotes transcription of genes encoding cytokines [e.g. tumour necrosis factor (TNF-α), interleukin-1 (IL-1), IL-6], stress response factors (e.g. cyclooxygenase-2, nitric oxide, 5-lipoxygenase), cell cycle regulators, and cell adhesion molecules (e.g. ICAM-l, VCAM-l, E-selectin). Activation of NF-κB also increases expression of anti-apoptotic proteins such as IAP-l and IAP-2 and the Bcl-2 family [15]. Proteasome inhibition has therefore been an attractive target for drug research in an attempt to reduce NF-κB activation by preventing IκB degradation, thus resulting in a dampening of the inflammatory response [16,17]. Activation of the transcription factor NF-κB can occur when cells are stimulated by cytokines, antigens, oxidative triggers, and cell-cell contact. The role of NF-κB in stimulated cells is to promote transcription of proinflamamtory genes, including cytokines, cell adhesion molecules, and pro-angiogenic molecules. It also suppresses apoptosis in favour of cell growth and migration.

Table 1. Proteins targeted by the ubiquitin-proteasome pathway in non-clinical studies

Class of Proteins	Protein	Protein function
Cyclins and related proteins	Cyclins A, B, D, E	Cell-cycle progression
	Cyclin-dependent (CDK) inhibitors (p27, p21)	Regulation of cyclin kinase activity
Tumour suppressor	p53	Transcription factor
Oncogenes	c-fos c-jun	These are two different transcription factors which form a complex; they are also proto-oncogenes
	c-myc	Transcription factor
	N-myc	Transcription factor
Inhibitory proteins	IκB	Inhibitor of NF-κB
	p130	Inhibitor of E2F-l
Enzymes	cdc25 phosphatase	Phosphatase
	Tyrosine amino transferase (TAT)	Tyrosine metabolism

Aberrant NF-κB signalling is an important feature of several neurological disorders like stroke, epilepsy, amyotrophic lateral sclerosis, AD, PD, and Huntington´s diseases (HD). Traumatic brain and spinal cord injuries can also be classified as neurodegenerative. Insult to the brain or spinal cord in these disorders induces a cascade of signalling events that stimulate NF-κB activation in neurons, and injury responsive glial cells [12]. In ischemic brain injury (lack of oxygen in brain leading to neuronal death), which can result in a stroke, the activity of NF-κB is found to be very high. For example, NF-κB (p50) activation has been reported to enhance ischemic neuronal death [17]. In the case of epileptic seizures, NF-κB activity has been found to be rapidly increased in hippocampal neurons within 4-16 hours following kainite-induced seizures [18,19]. In rat models of traumatic injury, levels of NF-κB activity are increased in cerebral cortex within hours of the insult [20]. NF-κB activity has been shown to be increased in certain chronic neurodegenerative disorders like AD, PD, HD, and multiple sclerosis [20]. By inducing the production and release of inflammatory cytokines, reactive oxygen molecules and excitotoxins, activation of NF-κB in microglia and astrocytes may contribute to neuronal degeneration. However, activation of NF-κB in neurons can also promote their survival by inducing the expression of genes encoding anti-apoptotic proteins such as Bcl-2 and the antioxidant enzyme Mn-superoxide dismutase.

From the above information, it is clear that NF-κB is an important transcription factor involved in a range of neurological disorders. Its activation and regulation in different disease states is variable, in a sense that it may be important in the generation of pathogenesis, or it may itself be activated by disease specific proteins. Hence, if the regulation of NF-κB can be selectively controlled, most of these disease processes can be dealt with at a very early stage, when they are more dependent upon cellular factors induced by NF-κB. The degradation of IκB after its phosphorylation is an important event in the activation of NF-κB. Thus, with the knowledge of the molecular mechanisms involving ubiquitination and proteasome dependent degradation, mentioned above, some proteasome inhibitors have been developed in an attempt to modify the NF-κB – proteasome pathway.

The Proteasome as a Therapeutic Drug Target

The proteasome is an abundant multi-enzyme complex that is involved in the degradation of intracellular proteins (see Chapters 6 and 7). Intracellular levels of a vast number of different proteins are regulated by polyubiquitination and subsequent turnover mediated by the proteasome. Interference with proteasome function therefore leads to disturbances in a variety of cellular activities. The 26S proteasome is formed by a core 20S barrel-shaped structure [21]. This is made up of 7 different α and 7 different β sub-units, which are arranged into four stacked heptameric rings. The two outer α rings sandwich the inner core composed of two β rings. The α-subunits provide stability to the proteasome, while the proteolytic sites of the proteasome are located within the individual β-subunits. Each enzymatic site is directed towards the centre of the 20S complex. The regulatory 19S complex (or PA700) caps each end of the 20S core and controls substrate entry into the proteasome. Substrates enter through narrow openings of the outer rings of the 20S proteasome, and must be inserted within the central chamber formed by the β inner core in order to be degraded [7]. This view may be oversimplified, since there are now emerging examples of regulation at the level of proteasome and specific substrate recognition mediated by ubiquitin- and proteasome-binding adaptor proteins such as Rad23 [22].

The substrate protein is initially targeted by the enzyme ubiquitin ligase E3, which ubiquitinates the protein (see Chapter 3 and [23]). Further ubiquitin molecules are subsequently added to create a polyubiquitin chain. The 19S complex contains the ubiquitin recognition sites and this regulates protein entry into the 20S core. The polyubiquitinated protein binds to the 19S sub-unit, which cleaves the ubiqutin chain using ATPases, and sends the unfolded protein into the proteasome core for processing. The β-rings have three active sites having chymotrypsin-like (ChTL), trypsin-like (TL), and peptidylglutamyl peptide-hydrolyzing (PGPH like or caspase–like) activities. These proteolytic sites degrade the substrate protein into smaller peptide chains. Cleavage of the protein occurs by nucleophilic attack by an N-terminal threonine residue from the β sub-units [24].

Proteasome inhibitors have been designed to target these enzymatic sites, and most interfere with the ChTL activity of the proteasome [25,26]. The proteasome inhibitors comprise five important classes of chemical agents: peptide aldehydes, peptide vinyl sulfones, peptide boronates, peptide epoxyketones, and β-lactones. The role of these drugs have been investigated for potential clinical treatments [10,27]. Several drugs targeting the proteasome are currently in preclinical and clinical testing (Table 2) (see Chapter 40).

Table 2. Drugs targeting the proteasome currently in preclinical and clinical testing

Reagent	Principle	Company/Reference	Experimental Effects	Clinical Trial/Status
Bortezomib (Velcade®, PS-341)	Dipeptide boronic acid	Millennium	Stabilization of cell-cycle and proapoptotic proteins, inhibition of antiapoptotic proteins, effects on tumor microenvironment	FDA approved for relapsed and refractory multiple myeloma, ongoing trials for several tumors
MLN519 (MLN-519)	Lactacystin derivative	Millennium	Potent anti-inflammatory and neuroprotective effects	Phase 1, intended for acute stroke and myocardial infection
Epoxomicin, Eponemycin	Streptomyces epoxyketones	Meng *et al.*, 1999 [28,29]	Cytotoxic effects in various tumor cells	Preclinical
Nitrophenyl trileucine vinyl sulfone	Trileucine vinyl sulfone	Bogyo *et al.*, 1997 [30]	Irreversible inhibition of trypsin- and chymotrypsin-like proteasome activities	Preclinical
Ritonavir	Peptidomimeti c protease inhibitor	Abbott	HIV protease inhibitor, also inhibits chymotrypsin-like activity of proteasome	Approved for AIDS, phase 2 studies in tumor patients to be started

Companies: Abbott Laboratories (www.abbott.com), Millennium Pharmaceuticals, Inc- (www.mlnm.com).

Due to the importance of the UPS one can foresee that a proteasomal inhibitor would be a toxic agent not suitable for drug development; however different in vitro cell lineages display an astounding variation in the degree of sensitivity to proteasome inhibitors. Moreover, the potent proteasome inhibitor bortezomib (Velcade®) manufactured by Millennium Pharmaceuticals is relatively well tolerated by human patients while it is an effective killer of different cancer cells [31,32] The main focus of research in neurological disease has been on the treatment of neuroinflammation associated with cerebrovascular disease [33]. However, targeting the proteasome may not be sufficient to inactivate the NF-κB pathway and more specific inhibitors targeting the ubiquitin ligases [E3] are currently being researched [34]. The great advantage of such drugs is their enormous specificity intrinsically related to the ubiquitination mechanism, which relies on the enormous variety of different E3s, many of them specialized only to ubiquitinate just a very limited subset of substrates. For example, MDM2 is the ubiquitin ligase which ubiquitinates itself and the crucial tumor suppressor p53 and therefore may be a target for cancer drug design [35] while inhibition of $SCF^{\beta TRCP}$ or other E3s which ubiquitinate IkBα may therefore be a suitable target for anti-inflammatory drugs [36]. The design of E3-specific inhibitors pose a great challenge and has not yet gone beyond early laboratory work [34].

NF-κB and Cerebrovascular Disease

Neuroinflammation has been recognised as an important component in the pathophysiology of acute cerebrovascular disease [37]. Acute stroke results in the activation of an inflammatory cascade, which can lead to increased cerebral infarct size and worsen clinical outcome [38,39]. Various inflammatory mediators have been implicated in this inflammatory process and therapeutic agents are being researched in an attempt to modify this damaging response. Animal models of acute ischemic stroke have shown increased NF-κB activity and this has been identified as an important common pathway in neuroinflammation [17,40]. NF-κB triggers the release of interleukins (IL-1, IL-6), TNF-α and cell adhesion molecules (CAM's) [41]. The formation of reactive oxygen species and TNF-α itself results in further activation of NF-κB [42]. This leads to neutrophil infiltration and these inflammatory mediators are important in the progression of acute stroke [43].

Levels of IL-1 rapidly rise during the acute phase of stroke, and can remain elevated for several days [44]. Studies using IL-1 receptor antagonists have shown a reduction in infarct size in rat models of stroke [45]. TNF-α levels also rise after stroke and can persist for up to 5 days [46]. Overexpression of TNF-α receptors also occurs during acute ischemic stroke [47]. Intracerebral administration of TNF-α prior to middle cerebral artery occlusion (MCAo) significantly enlarges infarct size [48]. However, TNF-α may have beneficial effects during the recovery phase of stroke in cerebral re-modelling. IL-6 levels also rise during the acute period of stroke and again correlate with larger infarct size and poor clinical outcome [39]. IL-6 and TNF-α further trigger the activation of iNOS and COX-2 enzymes which are also implicated in neuronal damage [49]. Leucocyte infiltration of the cerebral tissue also occurs as part of the neuroinflammatory response [50]. This requires the interaction of CAM's and chemokines. This allows the leucocytes to roll along the endothelium which then undergo diapedesis [51]. Animal models of stroke have shown a strong correlation between CAM's and cerebral infarct size [52]. Anti-ICAM treatment, so far, has been unsuccessful in patients with acute ischemic stroke. The Enlimomab study used a monoclonal antibody against ICAM-1, which was administered within 6 hours of ischemic stroke onset [53]. The 3-month outcome mortality data and adverse events were worse in the enlimomab group and it appeared that there may have been a pro-inflammatory response.

In an attempt to modify release of all these inflammatory mediators a central upstream target has been extensively researched. NF-κB inactivation is an attractive therapeutic option and proteasome inhibitors are being tested in the treatment of acute stroke. However, NF-κB action may be beneficial during the recovery phase of stroke and in cerebral re-modelling [12]. Therefore detailed evaluation of any drug would be required.

PRE-CLINICAL STUDIES

Most pre-clinical stroke studies have been performed on MLN-519, an analogue of lactacystin [54]. Lactacystin is a bacterial metabolite, which was the initial lead compound found to have proteasomal inhibition activity (a) (Figure 1). Lactacystin acts via a β-lactone ring intermediate called clasto-lactacystin-β-lactone (b). The β-lactone ring reacts with the active site of the proteasome and it mainly inhibits the chymotrypsin-like activity [55]. The β-lactone intermediate irreversibly reacts with the threonine hydroxyl group of the active site [56]. The proteasome inhibitor MLN-519 is also based around the clasto-lactacystin-β-lactone strucure; (1R – [1S, 4R, 5S] – 1 – (1 – hydroxy – 2 – methylpropyl) – 4 – propyl – 6 – oxa – 2 – azabicyclo [3.2.1] heptane – 3, 7 – dione) (c). Again, the β-lactone ring is the active part of MLN-519, which forms a covalent bond with the threonine hydroxyl group of the proteasome enzymatic site [26].

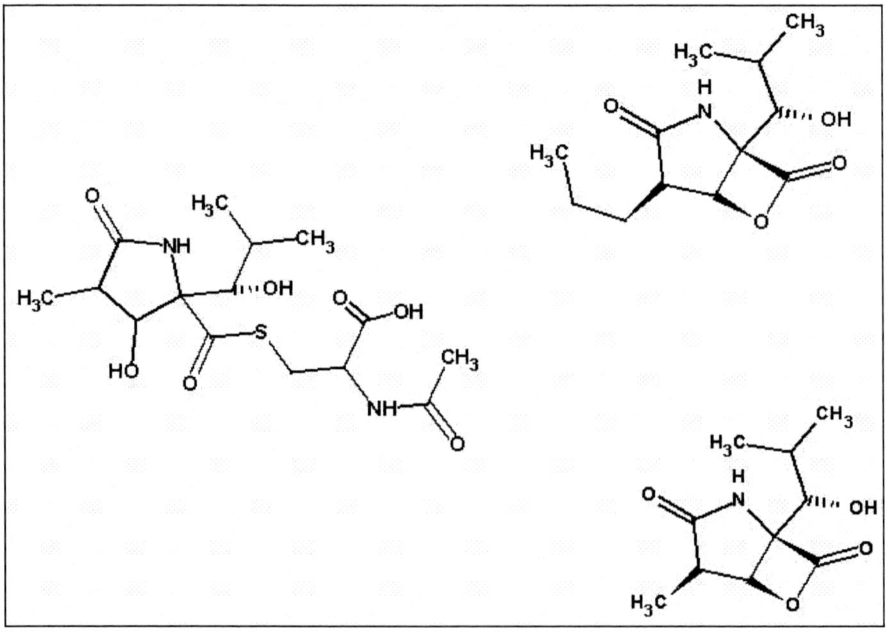

Figure 1. Structures of (a)-lactacystin, (b) clasto-lactacystin-β-lactone , and (c) MLN-519.

Pre-clinical studies in rat models of acute ischemic stroke have shown a reduction in cerebral infarct volume after treatment with MLN-519. MCAo rat models were treated with an intravenous infusion of MLN-519, 2 hours after the onset of acute ischemia. MLN-519 doses ranging from 0.01mg/kg to 0.3 mg/kg were shown to significantly reduce cerebral infarct volumes from 183±42 mm^3 to 138±30mm^3, respectively. Dose response analysis of infarct volume at 24hrs showed that neuroprotection approached 50-60% with the highest doses (0.1mg/kg to 0.3mg/kg). Evaluation of the rats using electroencephalogram (EEG) and assessment of neurological function showed significant improvements after treatment with MLN-519. Neutrophil infiltration was also significantly decreased in infarcted tissue by up to 70%. Further studies have shown that MLN-519 significantly reduces neutrophil infiltration via inhibition of the NF-κB – proteasome pathway [54]. Analysis of inflammatory mediators showed significant reductions in ICAM and E-selectin levels. Immunohistochemical analysis showed that the drug was most active within the endothelial cells and leucocytes [57].

The use of thrombolytic therapy in the treatment of acute ischemic stroke is being increasingly used in acute stroke units [58]. Patients presenting within three hours of symptom onset, and with no contraindications to treatment, can be treated with recombinant tissue-plasminogen activator (rt-PA). The use of neuroprotective agents in combination with rt-PA is an attractive option in an attempt to minimise cerebral tissue damage. MLN-519 in combination with rt-PA showed significant reductions in cerebral infarct volumes in rat models of MCAo [59]. Combination treatment given up to 6 hours post-ischemia significantly improved neurological recovery and reduced infarct volume. There was no increased incidence of hemorrhagic transformation in the combined treatment groups, compared with controls and MLN-519 alone.

This may therefore be an attractive treatment combination to consider for future clinical trials [60]. Rat models of stroke have also shown that NF-κB plays an important role in peri-lesional inflammation associated with intracerebral hemorrhage [61]. Therefore proteasome inhibition treatment may be beneficial in both acute ischemic and hemorrhagic stroke. The promising results of MLN-519 in pre-clinical stroke animal models has led to clinical trials of this drug.

Clinical Studies

The Phase I study of MLN-519 was performed on healthy male volunteers [62]. This study was a randomised double-blind placebo-controlled trial. It was designed to provide initial safety, tolerability and pharmacodynamic data on MLN-519 in humans. The initial study looked at ascending single bolus doses of MLN-519, in order to establish a therapeutic dose range. This was followed by a multi-dose study, when single bolus doses of MLN-519 were administered over three consecutive days at 24-hour intervals. The primary end-points were the effects of MLN-519 on blood 20S proteasome inhibition and drug toxicity. The aim was to achieve maximal dosing at 80% proteasome inhibition.

In pre-clinical drug safety studies MLN-519 toxicity was seen in doses above 8 mg/kg. Gastro-intestinal side effects, weight loss and possible blood pressure lowering was observed in rat studies. These effects were noted when proteasome inhibition was > 95% for extended periods of time (>24hrs). In vitro studies of MLN-519 showed the development of neurotoxicity in spinal neurons after 24hr incubation [63]. Little toxicity was observed at

80% proteasome inhibition and therefore this level was the target for maximal dosing with MLN-519 [62].

The maximal pharmacodynamic effect of MLN-519 in this phase I study was observed at approximately 1 hour post-dosing. The single dose study investigated doses ranging from 0.012 mg/m^2 to 1.6 mg/m^2. The maximal dose tested (1.6 mg/m^2, ~0.267 mg/kg) reduced the proteasome activity in circulating whole blood cells from 1.31 to 0.60 nmol/AMC/s/mg protein (~54%) [62,64]. The triple-dosing study used doses between 0.5 mg/m^2 to 1.6 mg/m^2. Male subjects in 17 groups of four were randomly assigned to receive a bolus dose of MLN-519 or placebo. Within each group three subjects received the drug and one placebo. Up to 80% blood 20S proteasome inhibition was achieved with the maximal dose of MLN-519, at 1.6 mg/m^2 (Figure 2). There were no significant changes in biochemical or hematological parameters observed. ECG and hemodynamic monitoring were normal. There were no drug dose related side effects. Minor side effects were transient and noted only on drug administration – altered test sensation and discomfort in the injection arm. There were no significant differences noted between the treatment and placebo groups, suggesting the symptoms were associated with the diluent and not the drug.

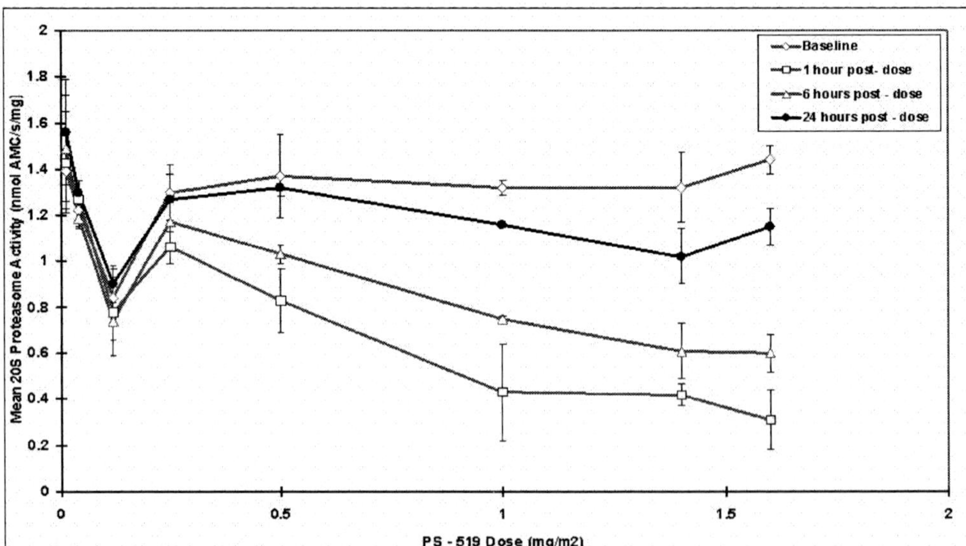

Figure 2. Single dose MLN-519 and 20S proteasome activity with incremental dosing. Maximum proteasome inhibition was achieved at 1-hour post-dosing. Approximately 80% proteasome inhibition was achieved with the highest dosing of 1.6 mg/m2. Proteasome activity returned to within normal limits at 24 hours post-dosing. (Copyright permission: Shah IM *et al.*: *Br J Clin Pharmacol* 2002, 54,269-276) [62].

It was not possible to obtain pharmacokinetic data on MLN-519, due to its rapid plasma clearance. Pre-clinical radio-labelled studies of MLN-519 have shown rapid clearance of the drug, within the first 10 minutes of administration. It is mainly taken up by endothelial cells and leukocytes and does not cross the blood brain barrier [54]. The pharmacodynamic data was therefore more important as this provides us with details of proteasomal and drug activity [65].

Maximal proteasome inhibition was seen at 1-hour post-dosing and activity returned to basal levels within 24 hours. This was also observed in the triple dosing study (Figure 3). Neuroinflammation associated with acute ischemic stroke has been shown to persist for up to 72 hours post-stroke [39]. Hence, the rationale for the triple dosing study in an attempt to mimic a potential neuroprotective treatment regime in acute stroke patients. There was no evidence of accumulation of the pharmacodynamic effect of MLN-519 with repeated dosing. The short action of MLN-519 would be beneficial in acute stroke patients, as NF-κB activity has been associated with cerebral re-modelling [12]. Prolonged treatment may be detrimental in stroke recovery and rehabilitation. Proteasome function has also been shown to decline with ageing [66]. As elderly patients will be the main treatment population, then safety and tolerability data must also be studied in this group. More clinical trials are being planned to further evaluate proteasome inhibitors in the treatment of acute stroke [33,64,67].

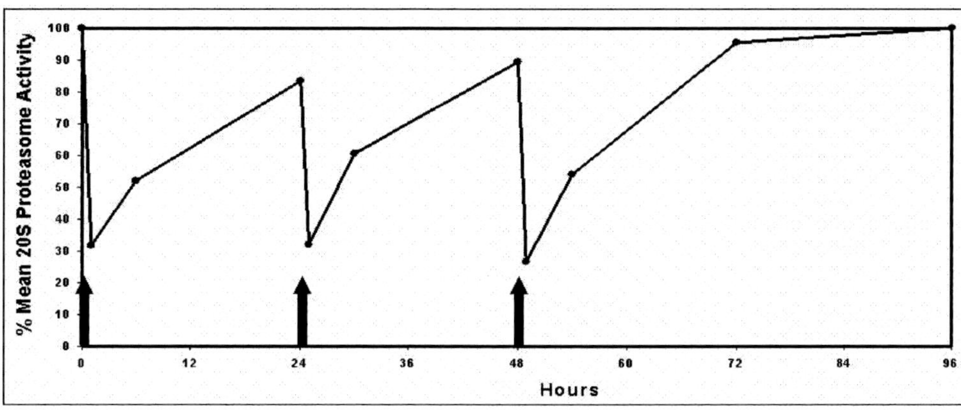

Figure 3. Triple dose MLN-519 and 20S proteasome activity (Dose: 1.6 mg/m^2). Three consecutive doses of MLN-519 were administered 24 hours apart (arrows). Proteasome activity returned to within normal limits at 24 hours post-dosing and there was no cumulative effect of consecutive dosing.(Copyright permission: Shah IM *et al.*: *Br J Clin Pharmacol* 2002, 54:269-76) [62].

Bortezomib

The ubiquitin-proteasome pathway plays a significant role in the degradation of regulatory proteins required for cell cycle progression and mitosis [68]. A disruption in the regulation of these cell cycle proteins results in abnormal cell division and tumorogenesis [69]. Proteasome inhibitors selectively target and induce apoptosis in proliferating cancer cells. Therefore these drugs have been an important area of therapeutic research in cancer treatment [70].

Bortezomib (Velcade®; PS-341), a dipeptidyl boronic acid proteasome blocker developed by Millennium Pharmaceuticals (Cambridge, MA) in cooperation with Johnson and Johnson (New Brunswick, NJ), was tested against a panel of 60 tumor cell lines and displayed promising anticancer properties. Bortezomib is the first proteasome inhibitor to be licensed for clinical use [71,72]. It was approved by the FDA in May 2003 for the treatment of patients who have received at least two prior therapies for multiple myeloma (MM), and have demonstrated disease progression on the last therapy. Phase 2 studies demonstrated a response to bortezomib in 27% of patients irrespective of prior treatments [72]. It is currently being used in the treatment of the hematological malignancy, MM. It is given as an intravenous injection. Inhibition of NF–κB and reduction in IL-6 levels are postulated mechanisms of action [73]. Bortezomib is cleared rapidly from the circulation and only pharmacodynamic data has been available. Maximal 20S proteasome inhibition with bortezomib has been observed at 1-hour post dosing but its duration of action is longer compared to MLN-519 [32]. Proteasome inhibition returned to baseline at 72 hours post-dosing. Early pre-clinical studies showed toxicity with bortezomib was dose dependent and reversible. Bortezomib has also been shown to be effective in drug resistant MM cells and this important property is being further investigated [74]. This is consistent with preclinical findings that the response of MM cells to the drug is independent of their sensitivity to other chemotherapeutics. Nontoxic doses of bortezomib also sensitize several other cancer cells to chemotherapy in vitro and in animal models. It was suggested that the proapoptotic effect of proteasome inhibitors might be due to the observed stabilization of IκB resulting in NF-κB inhibition and the down-regulation of antiapoptotic NF-κB target genes [75]. Another observation was the down-regulation of Bcl-2 in pancreatic cancer cells treated with bortezomib, which would in itself probably be sufficient to induce apoptosis by misbalancing pro- and antiapoptotic proteins [76]. Accumulation of cell cycle inhibitors such as p21 and concomitant cell cycle arrest and apoptosis might also play a role [77]. Bortezomib like other proteasome inhibitors, e.g., lactacystin, also induce p53 accumulation, even though p53 does not seem to be crucial for sensitization of tumor cells [78]. Further clinical trials of bortezomib are ongoing for treatment of other cancers and in combination with other chemotherapeutic agents [79,80]. Peripheral neuropathy, thrombocytopenia and gastro-intestinal disturbance have been the most common side effects associated with treatment (see Chapter 43). However, proteasome inhibition of up to 80% was well tolerated and maximal clinical dosing has been aimed to achieve this level of inhibition.

There are only few recent preliminary experimental studies on the utility of bortezomib in ischemic stroke and cerebral ischemia (see Chapter 39). A single dose of bortezomib given 1 hr post-MCAo, resulted in a 40% decrease in infarct volume and a 38% decrease in neurological deficit in a rat permanent MCAo model [81]. The functional and histopathologic protection was accompanied by a 67% inhibition of whole blood proteasome activity, a level of inhibition which is commonly achieved in cancer patients in the clinic [81]. The potential neuroprotective effects of bortezomib were also evident in an embolic model of MCAo [82] and in combination with delayed thrombolytic therapy on a rat model of embolic focal cerebral ischemia [83]. Treatment with bortezomib reduced adverse cerebrovascular events including secondary thrombosis, inflammatory responses, and blood-brain barrier (BBB) disruption, and hence reduces infarct volume and neurological functional deficit when administrated within 4 h after stroke onset [83] Combination of bortezomib and rtPA extended the thrombolytic window for stroke to 6 h, which is associated with the improvement of vascular patency and integrity. Real time RT-PCR of endothelial cells isolated by laser-capture microdissection from brain tissue and Western blot analysis showed that bortezomib upregulated endothelial nitric oxide synthase (eNOS) expression and blocks NF-κB activation [83].

The NF-κB-Proteasome Pathway and other Inflammatory-Mediated Conditions

The effects of proteasome inhibition have also been investigated in other inflammatory conditions [84]. MLN-519 has been tested in an experimental mouse model of autoimmune encephalomyelitis [85]. This was shown to reduce relapses of the disease and a reduction in T-cell activation. This suggests that proteasome inhibitors may be beneficial in other neuroinflammatory conditions like multiple sclerosis. MLN-519 has also shown beneficial effects in a delayed-type hypersensitivity model of asthma [86]. Other common inflammatory conditions, in animal models of psoriasis and arthritis, have also shown significant improvement with proteasome inhibition [87,88].

MLN-519 has also been studied in animal models of myocardial injury and reperfusion [89]. Proteasome inhibition was shown to reduce polymorphonuclear leukocyte infiltration in a ischemia/reperfusion rat heart model. The study showed a reduction in the expression of CAM, P-selectin, and preservation of cardiac function. At 1 mg/kg, MLN-519 treated rats showed a final left ventricular developed pressure of 98±3% in an ischemia reperfusion model, compared to 52±8% in the control group. In a porcine model of myocardial reperfusion injury, MLN-519 was shown to inhibit the activation of NF-κB [90]. There was also a reduction in the release of the myocardial enzymes creatinine kinase MB and troponin I, and a significant reduction in myocardial infarct size. Myocardial function was also preserved. Proteasome inhibitors may therefore have a role in the treatment of vascular reperfusion injury and further research is ongoing [64,91]. The role of NF-κB in the pathophysiology of atherosclerosis is also being further researched [92].

Therapeutic Aspects of Proteasome Inhibition in Nervous System Diseases

It is clear that the proteasome represents a central target for the processing and metabolism of multiple proteins whose critical roles in cellular function are being elucidated through the use of selective inhibitors. To avoid eliciting the significant side effects associated with complete inhibition of the proteasome and the possible immunosuppressive effects (with increased risk of infection and cancer) from persistent suppression of NF-κB activation, it is critical that we understand how to partially and temporally attenuate proteasome function to elicit the desired therapeutic effect. Taking into account the central role of the UPS in the biology of eukaryotic cells [2], proteasome inhibitors can be thought of as deadly toxins without any therapeutic value. Unexpectedly, they have been relatively well tolerated drugs because apparently most normal cells tolerate high levels of proteasome inhibition [5,93,94]. The availability of selective proteasome inhibitors such as MG-132 [95] and PSI [96] have boosted research of cellular effects of UPS inhibition. These effects included accumulation of ubiquitinated proteins within the cells [95,96], often in the form of an organized pericentrosomal aggregate called 'aggresome' [97,98], as well as blocking different stages of the cell cycle [98,99] and apoptosis [100,101]. While proteasome inhibition easily induced apoptosis in various cancer cell lines, primary cells such as diploid human fibroblasts are relatively resistant to the action of proteasome inhibitors, requiring much higher concentrations and longer incubation times [102,103]. Moreover, in terminally differentiated, resting cells low doses of proteasome inhibitors were cytoprotective, e.g. preventing apoptosis induced by nerve growth factor (NGF) deprivation in sympathetic neurons [104], serum deprivation of cerebellar granule cells [105], dehamethasone treatment of thymocytes [106] and IFNγ treatment of lens-derived αTN4-1 cells [107]. Even in some cancer cell lines proteasome inhibition was shown to be cytoprotective [108]. Nevertheless, higher doses of proteasome inhibitors and prolonged treatment inevitably induced cell death [104,106] (see Chapter 21).

The effects of proteasome inhibitors can be divided into general, substrate-independent effects and specific, substrate-dependent effects. Specific effects of proteasome inhibition depend on the inhibition of degradation of specific substrates such as cyclins, IκBα, p53 etc. and have been discussed in the previous chapters of this book (see Chapters 7 and 19). Specific effects may be very different depending on the cell type involved: they may be either neutral, beneficial or detrimental to the organism as a whole. General effects of proteasome inhibition do not depend on the cell type and the kind of substrate involved. They result from the accumulation of ubiquitinated proteins, which occurs even after moderate inhibition of the proteasome (see Chapter 12). However, the cell has developed mechanisms to defend against misfolded and aggregated proteins. The first line of defense involves the many molecular chaperones that aid in the normal folding and also refolding of abnormal conformations back into the native state. If this fails, abnormal proteins can be targeted for degradation by covalent attachment of polyubiquitin followed by targeting to the proteasome and degradation [109,110]. The presence of ubiquitin, chaperones and proteasome components in inclusions presumably represents cellular defenses overwhelmed by the excessive aggregation within cells. Even the inclusions themselves are the outcome of an active process by which the cell collects irreversibly aggregated protein, translocates it to an ‘aggresome’ near the nucleus by active transport and attempts to eliminate it, probably by autophagic or other lysosomal-like processes [97,111]. Accumulation of unfolded proteins in the cytosol induces heat shock response characterized by overexpression of cytosolic chaperones such as the heat shock proteins (Hsp), Hsp70 or Hsp90, while accumulation of unfolded proteins in the endoplasmic reticulum secondary to the inhibition of ERAD (ER-associate degradation) induces unfolded protein response [112,113] (see Chapters 11 and 13).

Ubiquitinated proteins are not dispersed randomly in the cytosol and nucleus but instead tend to accumulate into discrete subcellular domains, which may correspond to regions of increased protein turnover or ‘proteolytic centers’ [98,115]. In the cytoplasm ubiquitinated proteins coalesce into a single aggregate around the centrosome by a centripete microtubule-dependent transport [98]. This central aggregate, sometimes refered to as the ‘aggresome’ recruits proteasomes and other components of the UPS further impairing its function [97,116]. At the same time ubiquitinated proteins in the nuclei coalesce into discrete subnuclear domains called PML bodies [117,118]. It is a matter of debate whether the formation of these cytoplasmic and nuclear inclusions are deleterious to the cells or not [97,119]. However, it is possible to dissociate the proapoptotic effects of proteasome inhibitors from their effects on ‘aggresome’ formation suggesting that those aggregates are cytoprotective, recruiting unfolded proteins which may have otherwise wreaked havoc within the cells [120,121]. Moreover, 'aggresomes' are often viewed as models of aggregates occurring in human neurons in several neurodegenerative disorders including AD and PD. It is often argued, that neurons which survive do have those aggregates which protected them from cell death, while cells which died probably failed to do so [97,119].

Potential adverse effects of proteasome inhibitors reported from the bedside may stem from either the specific or general effects. Increasing number of patients are receiving treatment with proteasome inhibitors without serious adverse reactions being reported. Most experience is coming from the use of Velcade®. The most frequent adverse effects occurring in ~30% of patients involved weakness, nausea, diarrhoea, decreased appetite, constipation, thrombocytopenia, peripheral neuropathy, pyrexia, vomiting and anemia. These effects were mostly reversible after discontinuation of therapy and/or manageable with secondary treatments. Two cases of death may have been possibly related to the use of proteasome inhibitors [32,122]. The Phase I trial of MLN-519, a different proteasome inhibitor also showed no major adverse effects at all at doses which corresponded to the desired therapeutic values. Minor effects included irritation at the site of injection, altered taste sensation, headache and flu-like syndrome [62]. Another set of clinical data comes from the use of the HIV protease inhibitor Ritonavir in AIDS patients, because Ritonavir is also a bona fide proteasome inhibitor [123]. Compared to Velcade® and MLN-519, Ritonavir is administered continuously over prolonged periods of time eliciting therefore effects of a chronic impairment of proteasome function. The most common adverse effects observed in almost 30% of patients involved nausea and vomiting [124]. Hyperlipidemia induced by chronic Ritonavir administration has been linked to increased hepatic lipoprotein production, caused by the prevention of proteasome-mediated degradation of apoB and activated sterol regulatory element binding proteins in the liver [125]. It is likely that prolonged administration of either Velcade® or MLN-519 will lead to similar effects. All reported adverse effects seem to stem from the specific inhibition of degradation of particular proteins. Prolonged exposure of neural cells in culture to very low concentrations of proteasome inhibitors clearly affects the profile of gene expression therefore a similar situation is very likely to happen in vivo [126].

It is clear, that the current experience with proteasome inhibitors in clinical settings is very encouraging, since they seem to be well tolerated drugs with manageable side effects [127]. Does the accumulation of ubiquitinated proteins around the centrosomes and in the nuclei of cells throughout the entire body somehow affect the function of our cells? The most serious possible complication of inhibiting proteasomes may involve the central nervous system since 'aggresome' formation has been postulated as a model for neurodegenerative disorders [97,119]. Neither MLN-519 nor Velcade® efficiently penetrate the brain-blood bareer (BBB) [32,62].

Penetration of Ritonavir through the BBB is considerable, however its levels in the central nervous system are too low to efficiently control HIV infection in AIDS-related dementia [128,129]. Since patients treated with Ritonavir do often develop AIDS-related dementia it is difficult to assess whether the reported adverse neurological effects of Ritonavir are directly related to the drug or to the underlaying disease. However, Ritonavir induces significantly more adverse reactions than other HIV protease inhibitors which do not inhibit the proteasome [124].

Other proteasome inhibitors that penetrate more efficiently the BBB – such as PSI or epoxomicin – cause the key features of PD after prolonged exposure [130] although showed protective effects on DA cell death in the rat PD model using 6-hydroxyl dopamine (6-OHDA) [131]. This implies that the relationship between proteasome inhibitors, UPS and the pathogenesis of neurodegenerative disorders may be more complex than we thought. (For details, see Chapters 23, 25 and 28).

Short courses of treatment with proteasome inhibitors such as the one pursued in stroke patients bears relatively few and mild adverse reactions versus a high possible benefit. Even prolonged treatments such as treatment of HIV and cancer patients is relatively safe, taking into account the seriousness of the primary disease and the life expectancy. The problems of long-term adverse reactions will certainly surface when a population of patients treated with proteasome inhibitors survive for several years. They will require monitoring to detect any possible sign of increased incidence of neurodegenerative disorders or any other effects which may have been caused by the widespread accumulation of ubiquitinated proteins within cells of different tissues. Perhaps focusing on specific aspects of the UPS may provide more promise for neuroprotective efficacy rather than the simpler and less specific proteasome inhibition. The interaction of these events is complex, and the outcome of therapeutic interventions aimed at these elements of cellular injury is uncertain without more rational and specific targeting of these mechanisms and knowledge of the underlying state of the organism with respect to these factors.

CONCLUSION

With the recent licensing of Bortezomib, the first proteasome inhibitor in clinical use, the ubiquitin–proteasome pathway has become an important target for therapeutic drug research. Important research areas include inflammatory conditions, vascular disease, HIV and cancer treatments.

The main focus of therapeutic targeting in neurological disease has been in the neuroinflammation associated with acute stroke [8,33]. Both, Bortezomib and MLN-519 have shown encouraging results in rat models of acute ischemic stroke, with reduced infarct size and improved neurological outcome. Clinical evaluation of these proteasome inhibitors is ongoing and further safety data are required before these drugs can be licensed for the treatment of acute stroke.

In vitro studies of proteasome inhibitors have shown that disruption in the balance of this important regulatory pathway can lead to different pathophysiological disease processes. Chronic dysfunction of the UPS has been associated with neurodegenerative conditions like PD and AD. The UPS also plays an important role in regulation of cell growth and gene expression [26]. The action of NF-κB must also be taken into account as disruption of its function could be detrimental to the body's immune response [132]. This broad function of the UPS must be thoroughly assessed before the licensing of proteasome inhibitors for clinical use.

The development of proteasome inhibitors has been an exciting new area of molecular and translational medicine. The elucidation of these molecular pathways has provided us with many new therapeutic drug targets, which will hopefully aid us in the development of specific drugs for the treatment of neurological disease associated with the UPS.

REFERENCES

[1] Pickart CM: Mechanisms underlying ubiquitination. *Annu Rev Biochem* 2001, 70:503-533.

[2] Glickman MH, Ciechanover A: The ubiquitin-proteasome proteolytic pathway: destruction for the sake of construction. *Physiol Rev* 2002, 82:373-428.

[3] DeMartino GN, Slaughter CA: The proteasome, a novel protease regulated by multiple mechanisms. *J Biol Chem* 1999, 274:22123-22126.

[4] Wójcik C: Ubiquitin- and proteasome-dependent proteolytic pathway as an emerging theapeutic target. *Emerging Therapeutic Targets* 2000, 4:1-23.

[5] Wojcik C: Inhibition of the proteasome as a therapeutic approach. *Drug Discov Today* 1999, 4:188-189.

[6] Magill L, Walker B, Irvine AE: The proteasome: a novel therapeutic target in haematopoietic malignancy. *Hematology* 2003, 8:275-283.

[7] Wolf DH, Hilt W: The proteasome: a proteolytic nanomachine of cell regulation and waste disposal. *Biochim Biophys Acta* 2004, 1695:19-31.

[8] Moore DJ, Dawson VL, Dawson TM: Role for the ubiquitin-proteasome system in Parkinson's disease and other neurodegenerative brain amyloidoses. *Neuromolecular Med* 2003, 4:95-108.

[9] Adams J: Proteasome inhibition: a novel approach to cancer therapy. *Trends Mol Med* 2002, 8:S49-S54.

[10] Elliott PJ, Zollner TM, Boehncke WH: Proteasome inhibition: a new anti-inflammatory strategy. *J Mol Med* 2003, 81:235-245.

[11] Schwartz AL, Ciechanover A: The ubiquitin-proteasome pathway and pathogenesis of human diseases. *Annu Rev Med* 1999, 50:57-74.

[12] Mattson MP, Camandola S: NF-kappaB in neuronal plasticity and neurodegenerative disorders. *J Clin Invest* 2001, 107:247-254.

[13] Viatour P, Merville MP, Bours V, Chariot A: Phosphorylation of NF-[kappa]B and I[kappa]B proteins: implications in cancer and inflammation. *Trends in Biochemical Sciences* 2005, 30:43-52.

[14] Chen ZJ: Ubiquitin signalling in the NF-kappaB pathway. *Nat Cell Biol* 2005, 7:758-765.

[15] Delhalle Sylv, Blasius Roma, Dicato Mari, Diederich Marc: A Beginner's Guide to NF-{kappa}B Signaling Pathways. *Ann NY Acad Sci* 2004, 1030:1-13.

[16] Wojcik C, Di Napoli M: Ubiquitin-Proteasome System and Proteasome Inhibition: New Strategies in Stroke Therapy. *Stroke* 2004, 35:1506-1518.

[17] Schneider A, Martin-Villalba A, Weih F, Vogel J, Wirth T, Schwaninger M: NF-kappaB is activated and promotes cell death in focal cerebral ischemia. *Nat Med* 1999, 5:554-559.

[18] Won SJ, Ko HW, Kim EY, Park EC, Huh K, Jung NP, Choi I, Oh YK, Shin HC, Gwag BJ: Nuclear factor kappa B-mediated kainate neurotoxicity in the rat and hamster hippocampus. *Neuroscience* 1999, 94:83-91.

[19] Prasad AV, Pilcher WH, Joseph SA: Nuclear factor-kappa B in rat brain: enhanced DNA-binding activity following convulsant-induced seizures. *Neurosci Lett* 1994, 170:145-148.

[20] Mattson MP: NF-kappaB in the survival and plasticity of neurons. *Neurochem Res* 2005, 30:883-893.

[21] Adams J: The proteasome: structure, function, and role in the cell. *Cancer Treat Rev* 2003, 29 Suppl 1:3-9.

[22] Chen L, Madura K: Rad23 promotes the targeting of proteolytic substrates to the proteasome. *Mol Cell Biol* 2002, 22:4902-4913.

[23] Pickart CM, Eddins MJ: Ubiquitin: structures, functions, mechanisms. *Biochimica et Biophysica Acta (BBA) - Molecular Cell Research* 2004, 1695:55-72.

[24] Kisselev AF, Songyang Z, Goldberg AL: Why does threonine, and not serine, function as the active site nucleophile in proteasomes? *J Biol Chem* 2000, 275:14831-14837.

[25] Groll M, Huber R: Inhibitors of the eukaryotic 20S proteasome core particle: a structural approach. *Biochim Biophys Acta* 2004, 1695:33-44.

[26] Kisselev AF, Goldberg AL: Proteasome inhibitors: from research tools to drug candidates. *Chem Biol* 2001, 8:739-758.

[27] Voorhees PM, Dees EC, O'Neil B, Orlowski RZ: The proteasome as a target for cancer therapy. *Clin Cancer Res* 2003, 9:6316-6325.

[28] Meng L, Kwok BH, Sin N, Crews CM: Eponemycin exerts its antitumor effect through the inhibition of proteasome function. *Cancer Res* 1999, 59:2798-2801.

[29] Meng L, Mohan R, Kwok BH, Elofsson M, Sin N, Crews CM: Epoxomicin, a potent and selective proteasome inhibitor, exhibits in vivo antiinflammatory activity. *Proc Natl Acad Sci U S A* 1999, 96:10403-10408.

[30] Bogyo M, McMaster JS, Gaczynska M, Tortorella D, Goldberg AL, Ploegh H: Covalent modification of the active site threonine of proteasomal beta subunits and the Escherichia coli homolog HslV by a new class of inhibitors. *Proc Natl Acad Sci U S A* 1997, 94:6629-6634.

[31] Aghajanian C, Soignet S, Dizon DS, Pien CS, Adams J, Elliott PJ, Sabbatini P, Miller V, Hensley ML, Pezzulli S, Canales C, Daud A, Spriggs DR: A phase I trial of the novel proteasome inhibitor PS341 in advanced solid tumor malignancies. *Clin Cancer Res* 2002, 8:2505-2511.

[32] Orlowski RZ, Stinchcombe TE, Mitchell BS, Shea TC, Baldwin AS, Stahl S, Adams J, Esseltine DL, Elliott PJ, Pien CS, Guerciolini R, Anderson JK, Depcik-Smith ND, Bhagat R, Lehman MJ, Novick SC, O'Connor OA, Soignet SL: Phase I trial of the proteasome inhibitor PS-341 in patients with refractory hematologic malignancies. *J Clin Oncol* 2002, 20:4420-4427.

[33] Di Napoli M, McLaughlin B: The ubiquitin-proteasome system as a drug target in cerebrovascular disease: therapeutic potential of proteasome inhibitors. *Curr Opin Investig Drugs* 2005, 6:686-699.

[34] Sun Y: Targeting E3 ubiquitin ligases for cancer therapy. *Cancer Biol Ther* 2003, 2:623-629.

[35] Chene P: Inhibiting the p53-MDM2 interaction: an important target for cancer therapy. *Nat Rev Cancer* 2003, 3:102-109.

[36] Swinney DC, Xu YZ, Scarafia LE, Lee I, Mak AY, Gan QF, Ramesha CS, Mulkins MA, Dunn J, So OY, Biegel T, Dinh M, Volkel P, Barnett J, Dalrymple SA, Lee S, Huber M: A small molecule ubiquitination inhibitor blocks NF-kappa B-dependent cytokine expression in cells and rats. *J Biol Chem* 2002, 277:23573-23581.

[37] Zheng Z, Yenari MA: Post-ischemic inflammation: molecular mechanisms and therapeutic implications. *Neurol Res* 2004, 26:884-892.

[38] Montaner J, Rovira A, Molina CA, Arenillas JF, Ribo M, Chacon P, Monasterio J, Alvarez-Sabin J: Plasmatic level of neuroinflammatory markers predict the extent of diffusion-weighted image lesions in hyperacute stroke. *J Cereb Blood Flow Metab* 2003, 23:1403-1407.

[39] Smith CJ, Emsley HC, Gavin CM, Georgiou RF, Vail A, Barberan EM, del Zoppo GJ, Hallenbeck JM, Rothwell NJ, Hopkins SJ, Tyrrell PJ: Peak plasma interleukin-6 and other peripheral markers of inflammation in the first week of ischaemic stroke correlate with brain infarct volume, stroke severity and long-term outcome. *BMC Neurol* 2004, 4:2.

[40] Stephenson D, Yin T, Smalstig EB, Hsu MA, Panetta J, Little S, Clemens J: Transcription factor nuclear factor-kappa B is activated in neurons after focal cerebral ischemia. *J Cereb Blood Flow Metab* 2000, 20:592-603.

[41] Zhang W, Potrovita I, Tarabin V, Herrmann O, Beer V, Weih F, Schneider A, Schwaninger M: Neuronal activation of NF-kappaB contributes to cell death in cerebral ischemia. *J Cereb Blood Flow Metab* 2005, 25:30-40.

[42] Xie H, Ray PE, Short BL: NF-{kappa}B Activation Plays a Role in Superoxide-Mediated Cerebral Endothelial Dysfunction After Hypoxia/Reoxygenation. *Stroke* 2005, 36:1047-1052.

[43] Dirnagl U, Iadecola C, Moskowitz MA: Pathobiology of ischaemic stroke: an integrated view. *Trends Neurosci* 1999, 22:391-397.

[44] Zhang Z, Chopp M, Goussev A, Powers C: Cerebral vessels express interleukin 1beta after focal cerebral ischemia. *Brain Res* 1998, 784:210-217.

[45] Garcia JH, Liu KF, Relton JK: Interleukin-1 receptor antagonist decreases the number of necrotic neurons in rats with middle cerebral artery occlusion. *Am J Pathol* 1995, 147:1477-1486.

[46] Intiso D, Zarrelli MM, Lagioia G, Di Rienzo F, Checchia DA, Simone P, Tonali P, Cioffi Dagger RP: Tumor necrosis factor alpha serum levels and inflammatory response in acute ischemic stroke patients. *Neurol Sci* 2004, 24:390-396.

[47] Dziewulska D, Mossakowski MJ: Cellular expression of tumor necrosis factor a and its receptors in human ischemic stroke. *Clin Neuropathol* 2003, 22:35-40.

[48] Barone FC, Arvin B, White RF, Miller A, Webb CL, Willette RN, Lysko PG, Feuerstein GZ: Tumor Necrosis Factor-{alpha} : A Mediator of Focal Ischemic Brain Injury. *Stroke* 1997, 28:1233-1244.

[49] Nogawa S, Forster C, Zhang F, Nagayama M, Ross ME, Iadecola C: Interaction between inducible nitric oxide synthase and cyclooxygenase-2 after cerebral ischemia. *Proc Natl Acad Sci U S A* 1998, 95:10966-10971.

[50] Price CJ, Menon DK, Peters AM, Ballinger JR, Barber RW, Balan KK, Lynch A, Xuereb JH, Fryer T, Guadagno JV, Warburton EA: Cerebral neutrophil recruitment, histology, and outcome in acute ischemic stroke: an imaging-based study. *Stroke* 2004, 35:1659-1664.

[51] Frijns CJ, Kappelle LJ: Inflammatory cell adhesion molecules in ischemic cerebrovascular disease. *Stroke* 2002, 33:2115-2122.

[52] Zhang RL, Chopp M, Zaloga C, Zhang ZG, Jiang N, Gautam SC, Tang WX, Tsang W, Anderson DC, Manning AM: The temporal profiles of ICAM-1 protein and mRNA expression after transient MCA occlusion in the rat. *Brain Res* 1995, 682:182-188.

[53] Use of anti-ICAM-1 therapy in ischemic stroke: results of the Enlimomab Acute Stroke Trial: *Neurology* 2001, 57:1428-1434.

[54] Williams AJ, Hale SL, Moffett JR, Dave JR, Elliott PJ, Adams J, Tortella FC: Delayed treatment with MLN519 reduces infarction and associated neurologic deficit caused by focal ischemic brain injury in rats via antiinflammatory mechanisms involving nuclear factor-kappaB activation, gliosis, and leukocyte infiltration. *J Cereb Blood Flow Metab* 2003, 23:75-87.

[55] Fenteany G, Schreiber SL: Lactacystin, proteasome function, and cell fate. *J Biol Chem* 1998, 273:8545-8548.

[56] Fenteany G, Standaert RF, Lane WS, Choi S, Corey EJ, Schreiber SL: Inhibition of proteasome activities and subunit-specific amino-terminal threonine modification by lactacystin. *Science* 1995, 268:726-731.

[57] Berti R, Williams AJ, Velarde LC, Moffett JR, Elliott PJ, Adams J, Yao C, Dave JR, Tortella FC: Effect of the proteasome inhibitor MLN519 on the expression of inflammatory molecules following middle cerebral artery occlusion and reperfusion in the rat. *Neurotox Res* 2003, 5:505-514.

[58] Bravata DM: Intravenous thrombolysis in acute ischaemic stroke: optimising its use in routine clinical practice. *CNS Drugs* 2005, 19:295-302.

[59] Zhang L, Zhang ZG, Zhang RL, Lu M, Adams J, Elliott PJ, Chopp M: Postischemic (6-Hour) treatment with recombinant human tissue plasminogen activator and proteasome inhibitor PS-519 reduces infarction in a rat model of embolic focal cerebral ischemia. *Stroke* 2001, 32:2926-2931.

[60] Gladstone DJ, Black SE, Hakim AM: Toward wisdom from failure: lessons from neuroprotective stroke trials and new therapeutic directions. *Stroke* 2002, 33:2123-2136.

[61] Hickenbottom SL, Grotta JC, Strong R, Denner LA, Aronowski J, Macdonald RL: Nuclear Factor-{kappa}B and Cell Death After Experimental Intracerebral Hemorrhage in Rats ò Editorial Comment. *Stroke* 1999, 30:2472-247a.

[62] Shah IM, Lees KR, Pien CP, Elliott PJ: Early clinical experience with the novel proteasome inhibitor PS-519. *Br J Clin Pharmacol* 2002, 54:269-276.

[63] Kikuchi S, Shinpo K, Takeuchi M, Tsuji S, Yabe I, Niino M, Tashiro K: Effect of geranylgeranylaceton on cellular damage induced by proteasome inhibition in cultured spinal neurons. *J Neurosci Res* 2002, 69:373-381.

[64] Di Napoli M, Papa F: MLN-519. Millennium/PAION. *Curr Opin Investig Drugs* 2003, 4:333-341.

[65] Lightcap ES, McCormack TA, Pien CS, Chau V, Adams J, Elliott PJ: Proteasome inhibition measurements: clinical application. *Clin Chem* 2000, 46:673-683.

[66] Keller JN, Gee J, Ding Q: The proteasome in brain aging. *Ageing Res Rev* 2002, 1:279-293.

[67] Aronowski J, Hall CE: New horizons for primary intracerebral hemorrhage treatment: experience from preclinical studies. *Neurol Res* 2005, 27:268-279.

[68] King RW, Deshaies RJ, Peters JM, Kirschner MW: How proteolysis drives the cell cycle. *Science* 1996, 274:1652-1659.

[69] Sherr CJ: Cancer cell cycles. *Science* 1996, 274:1672-1677.

[70] Adams J: Proteasome inhibitors as new anticancer drugs. *Curr Opin Oncol* 2002, 14:628-634.

[71] Adams J, Kauffman M: Development of the proteasome inhibitor Velcade (Bortezomib). *Cancer Invest* 2004, 22:304-311.

[72] Richardson PG, Sonneveld P, Schuster MW, Irwin D, Stadtmauer EA, Facon T, Harousseau JL, Ben Yehuda D, Lonial S, Goldschmidt H, Reece D, San Miguel JF, Blade J, Boccadoro M, Cavenagh J, Dalton WS, Boral AL, Esseltine DL, Porter JB, Schenkein D, Anderson KC: Bortezomib or high-dose dexamethasone for relapsed multiple myeloma. *N Engl J Med* 2005, 352:2487-2498.

[73] Chauhan D, Uchiyama H, Akbarali Y, Urashima M, Yamamoto K, Libermann TA, Anderson KC: Multiple myeloma cell adhesion-induced interleukin-6 expression in bone marrow stromal cells involves activation of NF-kappa B. *Blood* 1996, 87:1104-1112.

[74] Hideshima T, Richardson P, Chauhan D, Palombella VJ, Elliott PJ, Adams J, Anderson KC: The proteasome inhibitor PS-341 inhibits growth, induces apoptosis, and overcomes drug resistance in human multiple myeloma cells. *Cancer Res* 2001, 61:3071-3076.

[75] Berenson JR, Ma HM, Vescio R: The role of nuclear factor-kappaB in the biology and treatment of multiple myeloma. *Semin Oncol* 2001, 28:626-633.

[76] Bold RJ, Virudachalam S, McConkey DJ: Chemosensitization of pancreatic cancer by inhibition of the 26S proteasome. *J Surg Res* 2001, 100:11-17.

[77] Adams J, Palombella VJ, Sausville EA, Johnson J, Destree A, Lazarus DD, Maas J, Pien CS, Prakash S, Elliott PJ: Proteasome inhibitors: a novel class of potent and effective antitumor agents. *Cancer Res* 1999, 59:2615-2622.

[78] An WG, Hwang SG, Trepel JB, Blagosklonny MV: Protease inhibitor-induced apoptosis: accumulation of wt p53, p21WAF1/CIP1, and induction of apoptosis are independent markers of proteasome inhibition. *Leukemia* 2000, 14:1276-1283.

[79] Hamilton AL, Eder JP, Pavlick AC, Clark JW, Liebes L, Garcia-Carbonero R, Chachoua A, Ryan DP, Soma V, Farrell K, Kinchla N, Boyden J, Yee H, Zeleniuch-Jacquotte A, Wright J, Elliott P, Adams J, Muggia FM: Proteasome Inhibition With Bortezomib (PS-341): A Phase I Study With Pharmacodynamic End Points Using a Day 1 and Day 4 Schedule in a 14-Day Cycle. *J Clin Oncol* 2005, 23:6107-6116.

[80] Ma MH, Yang HH, Parker K, Manyak S, Friedman JM, Altamirano C, Wu ZQ, Borad MJ, Frantzen M, Roussos E, Neeser J, Mikail A, Adams J, Sjak-Shie N, Vescio RA, Berenson JR: The proteasome inhibitor PS-341 markedly enhances sensitivity of multiple myeloma tumor cells to chemotherapeutic agents. *Clin Cancer Res* 2003, 9:1136-1144.

[81] Stagliano NE, Ren JM, Riodan W, Finklestein SP: Velcade®, a first-in-class proteasome inhibitor, is efficacious in rat permanent middle cerebral artery occlusion. *J Cereb Blood Flow Metab* 2005, 25:S38.

[82] Henninger N, Sicard KM, Bouley J, Fisher M, Stagliano NE: The proteasome inhibitor VELCADE® reduces infarction in rat models of focal cerebral ischemia. *Neurosci Lett* 2006, 398:300-305.

[83] Zhang L, Zhang ZG, Liu X, Hozeska A, Stagliano N, Riordan W, Lu M, Chopp M: Treatment of embolic stroke in rats with bortezomib and recombinant human tissue plasminogen activator. *Thromb Haemost* 2006, 95:166-173.

[84] Elliott PJ, Ross JS: The proteasome: a new target for novel drug therapies. *Am J Clin Pathol* 2001, 116:637-646.

[85] Vanderlugt CL, Rahbe SM, Elliott PJ, Dal Canto MC, Miller SD: Treatment of established relapsing experimental autoimmune encephalomyelitis with the proteasome inhibitor PS-519. *J Autoimmun* 2000, 14:205-211.

[86] Elliott PJ, Pien CS, McCormack TA, Chapman ID, Adams J: Proteasome inhibition: A novel mechanism to combat asthma. *J Allergy Clin Immunol* 1999, 104:294-300.

[87] Palombella VJ, Conner EM, Fuseler JW, Destree A, Davis JM, Laroux FS, Wolf RE, Huang J, Brand S, Elliott PJ, Lazarus D, McCormack T, Parent L, Stein R, Adams J, Grisham MB: Role of the proteasome and NF-kappaB in streptococcal cell wall-induced polyarthritis. *Proc Natl Acad Sci U S A* 1998, 95:15671-15676.

[88] Zollner TM, Podda M, Pien C, Elliott PJ, Kaufmann R, Boehncke WH: Proteasome inhibition reduces superantigen-mediated T cell activation and the severity of psoriasis in a SCID-hu model. *J Clin Invest* 2002, 109:671-679.

[89] Campbell B, Adams J, Shin YK, Lefer AM: Cardioprotective effects of a novel proteasome inhibitor following ischemia and reperfusion in the isolated perfused rat heart. *J Mol Cell Cardiol* 1999, 31:467-476.

[90] Pye J, Ardeshirpour F, McCain A, Bellinger DA, Merricks E, Adams J, Elliott PJ, Pien C, Fischer TH, Baldwin AS, Jr., Nichols TC: Proteasome inhibition ablates activation of NF-kappa B in myocardial reperfusion and reduces reperfusion injury. *Am J Physiol Heart Circ Physiol* 2003, 284:H919-H926.

[91] Meiners S, Laule M, Rother W, Guenther C, Prauka I, Muschick P, Baumann G, Kloetzel PM, Stangl K: Ubiquitin-Proteasome Pathway as a New Target for the Prevention of Restenosis. *Circulation* 2002, 105:483-489.

[92] de Winther MP, Kanters E, Kraal G, Hofker MH: Nuclear factor kappaB signaling in atherogenesis. *Arterioscler Thromb Vasc Biol* 2005, 25:904-914.

[93] Golab J: Proteasome inhibitors in the treatment of cancer. *Drug Discov Today* 2003, 8:575.

[94] Adams J: Potential for proteasome inhibition in the treatment of cancer. *Drug Discov Today* 2003, 8:307-315.

[95] Rock KL, Gramm C, Rothstein L, Clark K, Stein R, Dick L, Hwang D, Goldberg AL: Inhibitors of the proteasome block the degradation of most cell proteins and the generation of peptides presented on MHC class I molecules. *Cell* 1994, 78:761-71.

[96] Figueiredo-Pereira ME, Berg KA, Wilk S: A new inhibitor of the chymotrypsin-like activity of the multicatalytic proteinase complex (20S proteasome) induces accumulation of ubiquitin-protein conjugates in a neuronal cell. *J Neurochem* 1994, 63:1578-1581.

[97] Kopito RR: Aggresomes, inclusion bodies and protein aggregation. *Trends Cell Biol* 2000, 10:524-530.

[98] Wojcik C, Schroeter D, Wilk S, Lamprecht J, Paweletz N: Ubiquitin-mediated proteolysis centers in HeLa cells: indication from studies of an inhibitor of the chymotrypsin-like activity of the proteasome. *Eur J Cell Biol* 1996, 71:311-318.

[99] Machiels BM, Henfling ME, Gerards WL, Broers JL, Bloemendal H, Ramaekers FC, Schutte B: Detailed analysis of cell cycle kinetics upon proteasome inhibition. *Cytometry* 1997, 28:243-252.

[100] Wojcik C, Stoklosa T, Giermasz A, Golab J, Zagozdzon R, Kawiak J, Wilk S, Komar A, Kaca A, Malejczyk J, Jakobisiak M: Apoptosis induced in L1210 leukaemia cells by an inhibitor of the chymotrypsin-like activity of the proteasome. *Apoptosis* 1997, 2:455-462.

[101] Drexler HC: Activation of the cell death program by inhibition of proteasome function. *Proc Natl Acad Sci* 1997, 94:855-60.

[102] Orlowski RZ, Eswara JR, Lafond-Walker A, Grever MR, Orlowski M, Dang CV: Tumor growth inhibition induced in a murine model of human Burkitt's lymphoma by a proteasome inhibitor. *Cancer Res* 1998, 58:4342-8.

[103] Pleban E, Bury M, Mlynarczuk I, Wojcik C: Effects of proteasome inhibitor PSI on neoplastic and non-transformed cell lines. *Folia Histochem Cytobiol* 2001, 39:133-134.

[104] Sadoul R, Fernandez PA, Quiquerez AL, Martinou I, Maki M, Schroter M, Becherer JD, Irmler M, Tschopp J, Martinou JC: Involvement of the proteasome in the programmed cell death of NGF-deprived sympathetic neurons. *EMBO J* 1996, 15:3845-52.

[105] Atlante A, Bobba A, Calissano P, Passarella S, Marra E: The apoptosis/necrosis transition in cerebellar granule cells depends on the mutual relationship of the antioxidant and the proteolytic systems which regulate ROS production and cytochrome c release en route to death. *J Neurochem* 2003, 84:960-971.

[106] Grimm LM, Goldberg AL, Poirier GG, Schwartz LM, Osborne BA: Proteasomes play an essential role in thymocyte apoptosis. *EMBO J* 1996, 15:3835-3844.

[107] Awasthi N, Wagner BJ: Interferon-gamma induces apoptosis of lens alphaTN4-1 cells and proteasome inhibition has an antiapoptotic effect. *Invest Ophthalmol Vis Sci* 2004, 45:222-229.

[108] Nawrocki ST, Sweeney-Gotsch B, Takamori R, McConkey DJ: The proteasome inhibitor bortezomib enhances the activity of docetaxel in orthotopic human pancreatic tumor xenografts. *Mol Cancer Ther* 2004, 3:59-70.

[109] Ciechanover A, Brundin P: The ubiquitin proteasome system in neurodegenerative diseases: sometimes the chicken, sometimes the egg. *Neuron* 2003, 40:427-446.

[110] Goldberg AL: Protein degradation and protection against misfolded or damaged proteins. *Nature* 2003, 426:895-899.

[111] Ravikumar B, Rubinsztein DC: Can autophagy protect against neurodegeneration caused by aggregate-prone proteins? *Neuroreport* 2004, 15:2443-2445.

[112] Lee DH, Goldberg AL: Proteasome inhibitors cause induction of heat shock proteins and trehalose, which together confer thermotolerance in Saccharomyces cerevisiae. *Mol Cell Biol* 1998, 18:30-38.

[113] Kawazoe Y, Nakai A, Tanabe M, Nagata K: Proteasome inhibition leads to the activation of all members of the heat-shock-factor family. *Eur J Biochem* 1998, 255:356-362.

[114] McCracken AA, Brodsky JL: A molecular portrait of the response to unfolded proteins. *Genome Biol* 2000, 1:REVIEWS1013.

[115] Wojcik C: On the spatial organization of ubiquitin-dependent proteolysis in HeLa cells. *Folia Histochem Cytobiol* 1997, 35:117-118.

[116] Bence NF, Sampat RM, Kopito RR: Impairment of the ubiquitin-proteasome system by protein aggregation. *Science* 2001, 292:1552-1555.

[117] Baumann CT, Ma H, Wolford R, Reyes JC, Maruvada P, Lim C, Yen PM, Stallcup MR, Hager GL: The glucocorticoid receptor interacting protein 1 (GRIP1) localizes in discrete nuclear foci that associate with ND10 bodies and are enriched in components of the 26S proteasome. *Mol Endocrinol* 2001, 15:485-500.

[118] Wojcik C, DeMartino GN: Intracellular localization of proteasomes. *Int J Biochem Cell Biol* 2003, 35:579-589.

[119] Sherman MY, Goldberg AL: Cellular defenses against unfolded proteins: a cell biologist thinks about neurodegenerative diseases. *Neuron* 2001, 29:15-32.

[120] Rideout HJ, Wang Q, Park DS, Stefanis L: Cyclin-dependent kinase activity is required for apoptotic death but not inclusion formation in cortical neurons after proteasomal inhibition. *J Neurosci* 2003, 23:1237-1245.

[121] Tanaka M, Kim YM, Lee G, Junn E, Iwatsubo T, Mouradian MM: Aggresomes Formed by {alpha}-Synuclein and Synphilin-1 Are Cytoprotective. *J Biol Chem* 2004, 279:4625-4631.

[122] Richardson P: Clinical update: proteasome inhibitors in hematologic malignancies. *Cancer Treat Rev* 2003, 29 Suppl 1:33-39.

[123] Andre P, Groettrup M, Klenerman P, de Giuli R, Booth BL, Jr., Cerundolo V, Bonneville M, Jotereau F, Zinkernagel RM, Lotteau V: An inhibitor of HIV-1 protease modulates proteasome activity, antigen presentation, and T cell responses. *Proc Natl Acad Sci U S A* 1998, 95:13120-13124.

[124] Bonfanti P, Valsecchi L, Parazzini F, Carradori S, Pusterla L, Fortuna P, Timillero L, Alessi F, Ghiselli G, Gabbuti A, Di Cintio E, Martinelli C, Faggion I, Landonio S, Quirino T: Incidence of adverse reactions in HIV patients treated with protease inhibitors: a cohort study. Coordinamento Italiano Studio Allergia e Infezione da HIV (CISAI) Group. *J Acquir Immune Defic Syndr* 2000, 23:236-245.

[125] Riddle TM, Schildmeyer NM, Phan C, Fichtenbaum CJ, Hui DY: The HIV protease inhibitor ritonavir increases lipoprotein production and has no effect on lipoprotein clearance in mice. *J Lipid Res* 2002, 43:1458-1463.

[126] Ding Q, Bruce-Keller AJ, Chen Q, Keller JN: Analysis of gene expression in neural cells subject to chronic proteasome inhibition. *Free Radic Biol Med* 2004, 36:445-455.

[127] Adams J: The proteasome: a suitable antineoplastic target. *Nat Rev Cancer* 2004, 4:349-360.

[128] Anthonypillai C, Sanderson RN, Gibbs JE, Thomas SA: The distribution of the HIV protease inhibitor, ritonavir, to the brain, cerebrospinal fluid, and choroid plexuses of the guinea pig. *J Pharmacol Exp Ther* 2004, 308:912-920.

[129] van dS, I, Vos CM, Nabulsi L, Blom-Roosemalen MC, Voorwinden HH, de Boer AG, Breimer DD: Assessment of active transport of HIV protease inhibitors in various cell lines and the in vitro blood--brain barrier. *AIDS* 2001, 15:483-491.

[130] McNaught KS, Perl DP, Brownell AL, Olanow CW: Systemic exposure to proteasome inhibitors causes a progressive model of Parkinson's disease. *Ann Neurol* 2004, 56:149-162.

[131] Setsuie R, Kabuta T, Wada K: Does proteosome inhibition decrease or accelerate toxin-induced dopaminergic neurodegeneration? *J Pharmacol Sci* 2005, 97:457-460.

[132] Karin M, Lin A: NF-kappaB at the crossroads of life and death. *Nat Immunol* 2002, 3:221-227.

In: The Ubiquitin Proteasome System… ISBN 978-1-60021-749-4
Eds: Mario Di Napoli and Cezary Wojcik, pp. 1037-1054© 2007 Nova Science Publishers, Inc.

Chapter 43

Clinical Application of Proteasome Inhibitor Bortezomib: Characterization of Neurotoxicity

Jens Voortman*[*] and Giuseppe Giaccone
Department of Medical Oncology, VU University Medical Center, Amsterdam, The Netherlands.

Abstract

Bortezomib (Velcade®) is a dipeptide boronic acid proteasome inhibitor that specifically targets the chymotryptic-like proteolytic activity of the 20S proteasome. It has shown great potential as a novel anti-cancer agent and has been approved for the treatment of multiple myeloma. Therapeutic development as a single agent or in combination with other agents is ongoing in hematological malignancies as well as in various solid tumor types. Furthermore, application of proteasome inhibition therapy in other areas of disease is being explored, such as prevention of reperfusion injury following acute ischemic stroke and management of chronic inflammatory diseases. Bortezomib is generally well tolerated. However, one of the most serious overall as well as dose limiting toxicities has been peripheral neuropathy. Bortezomib-induced peripheral neuropathy constitutes a length-dependent, sensory rather than motor, axonal, small rather than large fiber, polyneuropathy. In agreement with its small fiber neuropathy characteristics, neuropathic pain and symptoms of autonomic dysfunction have also been frequently reported upon bortezomib treatment. Risk factors for bortezomib neurotoxicity include pre-existent neuropathy and prior treatment with neurotoxic (anti-cancer) agents. Additionally, individual susceptibility, and not only cumulative dose, is of great importance. After discontinuing bortezomib therapy, neuropathy resolves in approximately half of the patients. Nevertheless, in severe cases,

* Correspondence concerning this article should be addressed to Dr. Jens Voortman, MD; Department of Medical Oncology, VU University Medical Center, De Boelelaan 1117, 1081 HV Amsterdam, The Netherlands. Phone: +31 (0)20 4444 300; Fax: +31 (0)20 4444 355; E-mail: j.voortman@vumc.nl.

pharmacologic management of (autonomic) neuropathy and neuropathic pain is required up to several months after discontinuation of bortezomib. The exact biological mechanism of peripheral neuropathy induced by systemic proteasome inhibition therapy still has to be elucidated. Detrimental effects of proteasome inhibition on nerve terminal protein homeostasis as well as myelin production by Schwann cells might explain the high incidence of neurotoxicity in bortezomib-treated patients.

Keywords: bortezomib, neurotoxicity syndromes, peripheral nervous system diseases, proteasome endopeptidase complex, adverse effects, antineoplastic agents,

Abbreviations

APEX, assessment of proteasome inhibition for extending remissions; AUC, area under the curve; C_{max}, maximum plasma concentration; CMAP, compound muscle action potential; CMTX1A, X-linked Charcot-Marie-Tooth disease 1A; CNS, central nervous system; CREST, clinical response and efficacy study of bortezomib in the treatment of relapsing multiple myeloma; CTCAE, common terminology criteria for adverse events; d, day; DLT, dose limiting toxicity; FDA, food and drug administration; GR, grade; IENF, intra-epidermal nerve fiber density; MM, multiple myeloma; MTD, maximum tolerated dose; n, number (of patients); NCS, nerve conduction studies; NF-κB, nuclear factor kappa B; PBMC, peripheral blood mononuclear cell; pegLD, pegylated liposomal doxorubicin; 20S PI, 20S proteasome inhibition; PK, pharmacokinetic; PMP22, peripheral myelin protein 22; PN, peripheral neuropathy; SNAP, sensory nerve action potential; q, every; SUMMIT, study of uncontrolled multiple myeloma managed with proteasome inhibition therapy; TNS, total neuropathy score; UPS, ubiquitin-proteasome system.

Introduction

Growing consciousness of the pivotal role of the proteasome in normal cell physiology as well as in human disease propelled the development of proteasome inhibitors for research and therapeutic applications [1-3].

So far two proteasome inhibitors, MLN-519 and bortezomib, have been evaluated clinically [4]. Bortezomib (Velcade®, formerly known as PS-341, LDP-341 and MLN-341) has been dominating the field of proteasome inhibition therapy, with over a hundred clinical studies conducted or ongoing and currently approved for the treatment of multiple myeloma (MM) patients. So far, only one clinical study in healthy volunteers has been published with the lactacystin β-lactone-derived proteasome inhibitor MLN-519 [5]. Boronate proteasome inhibitors, such as bortezomib, are characterized by a pharmacophore containing a functional boronic acid group. They display remarkable selectivity towards the proteasome. Reversible binding to the chymotrypsin-like proteolytic activity, localized within the β5 subunit of the 20S core of the proteasome, results in a temporary inhibition of intracellular proteasome activity [6]. In *in vitro* studies, bortezomib emerged as a promising novel anti-cancer agent, demonstrating unique anti-tumor properties [7]. Bortezomib, and proteasome inhibitors in general, show a striking selectivity in cytotoxicity towards malignant cells compared to normal cells [8-11]. In early clinical experience, exceptional results were obtained in the treatment of patients with relapsing MM, who were refractory to conventional therapies [12]. The Clinical Response and Efficacy Study of Bortezomib in the Treatment of Relapsing Multiple Myeloma (CREST) study, a phase 2 study in heavily pretreated MM patients, showed response rates of 30% and 38% in patients treated with twice-weekly bortezomib dosed at 1.0 and 1.3 mg/m^2 respectively [13]. In 2003, on the basis of this and another comparable, larger phase 2 study, the Study of Uncontrolled Multiple Myeloma Managed with Proteasome Inhibition Therapy (SUMMIT), bortezomib received fast-track FDA approval for treatment of patients with relapsed refractory MM [14]. Additional indications for treatment with bortezomib are likely to follow considering the observed efficacy in other hematological malignancies such as mantle cell lymphoma and Waldenström macroglobulinemia as well as in several solid tumor types [15-17].

Preclinical studies attribute a critical part of bortezomib's activity in hematological malignancies to the inhibitory effect of proteasome inhibition on activation of transcription factor nuclear factor kappa B (NF-κB) [8,18,19]. It is therefore not a surprising development that systemic proteasome inhibition therapy is expanding to other areas of disease characterized by pathologic activation of NF-κB mediated signaling, such as chronic inflammatory diseases and reperfusion injury following acute ischemic stroke [20-22].

Bortezomib Pharmacology

Pharmacokinetic Profile

Bortezomib is administered as a short intravenous injection. The bortezomib plasma profile is best described by a two-compartment pharmacokinetic (PK) model with a rapid initial distribution half-life ($t_{1/2\alpha}$: 0.22 to 0.46 hours), followed by a longer terminal elimination phase ($t_{1/2\beta}$: > 10 days) [23].

Though in general plasma PK profiles are consistent among patients, maximum plasma concentration (C_{max}) values did not show an apparent relationship to dose and varied significantly among individuals [23]. Animal studies, using radioactively labeled bortezomib, showed rapid distribution into most tissues with the exception of brain, testis and some parts of eye and optic nerve [24]. In humans, over 30 metabolites, none pharmacologically active, were identified [25].

Pharmacodynamic Profile

As detection of bortezomib in serum proved to be difficult by its rapid removal from the vascular compartment, early-on in the clinical development of bortezomib a pharmacodynamic assay was developed to monitor the degree of 20S proteasome inhibition (20S PI) in peripheral blood mononuclear cells (PBMCs) [26]. It was demonstrated that 20S

PI in peripheral PBMCs was related both to the dose of bortezomib in milligrams per square meter, and to the total dose of bortezomib in milligrams. However, at higher doses of bortezomib, 20S PI is not an ideal surrogate for the bortezomib serum level as it reaches a plateau at around 60-75% [23]. Peak proteasome inhibition is reached one hour post injection, and proteasome activity in PBMCs typically restores to baseline levels 72 hours after injection of bortezomib. Administration of bortezomib is therefore bound to a maximum frequency of twice per week, allowing proteasome activity to restore before the next dose is administered [12].

It must be noted that proteasome expression levels and subtype distribution differ among cell types, and even within single cells the relative proportion of proteasome-subtypes, such as immunoproteasomes, is subject to change [27,28]. Furthermore, proteasome activity is different in normal cells compared to malignant cells [11]. Thus far comparative analysis of the level of proteasome inhibition in PBMCs and simultaneously obtained tumor samples were conducted in only a handful of patients. A comparable outcome for the degree of 20S PI in PBMCs and tumor samples was found [23,29]. However, 20S PI in the bone marrow was shown to be around half of that observed in the matched whole blood sample. There are no indications that tolerance or tachyphylaxis develop upon repeated doses of bortezomib [23,29].

Incidence of Bortezomib-Induced Neuropathy

Phase 1 Studies

Four different dosing schedules of bortezomib were assessed in seven phase 1 studies, which served to determine safety, and the maximum tolerated dose of bortezomib in each schedule. Six studies were conducted in adult patients and one in pediatric patients (see Table 1).

The first study in humans evaluated a weekly schedule of bortezomib on days 1, 8, 15, and 22 of a 35-day cycle. In following phase 1 studies, more dose-intense twice-weekly schedules were assessed: days 1 and 4 of a 14 day cycle; days 1, 4, 8 and 11 of a 21 day cycle; days 1, 4, 8, 11, 15, 18, 22 and 25 of a 42 day cycle [12,30-33]. Patients receiving twice-weekly bortezomib had their therapy most commonly interrupted during the third week of treatment, the major toxicity being malaise and fatigue [12,32]. Therefore twice-weekly bortezomib for two weeks followed by the third week off, days 1, 4, 8 and 11 of a 21-day cycle, became the standard regimen.

Sensory neuropathy was among the most commonly observed dose limiting toxicities, next to thrombocytopenia, diarrhea, fatigue and (orthostatic) hypotension. Occurrence of severe peripheral neuropathy did not allow further dose-escalation of bortezomib in three out of seven dose finding studies [29,30,32].

Table 1. Incidence of peripheral neuropathy in phase 1 dose finding studies with bortezomib

Studies	No.	Malignancy	Schema	MTD (mg/m²)	DLT (mg/m²)	Neurotoxicity	
Papandreou (2004)	53	advanced solid tumors	d1, 8, 15, 22 q35d	1.60	2.0 diarrhea hypotension	Overall, any grade GR3: 0.13-1.8 mg/m² 2.0 mg/m²	8% 0% 20%
Hamilton (2005)	40	advanced solid tumors/ lymphomas	d1, 4 q14d	1.75	1.75 + 1.90 *sensory neuropathy* fatigue diarrhea	Overall, any grade: GR3: 0.25-1.45 mg/m² 1.75 mg/m² 1.90 mg/m²	28% 0% 8% 29%
Aghajanian (2002)	42	advanced solid tumors	d1, 4, 8, 11 q21d	1.56	1.56 *sensory neuropathy* diarrhea	Overall, any grade: GR3: 0.13-1.08 mg/m² 1.30 mg/m² 1.56 mg/m²	12% 0% 33% 17%
Dy (2005)	I:28	advanced solid tumors/ lymphomas/ multiple myeloma	I: d1, 4, 8, 11, 15, 18, 22, 25 q42d	I: 1.50	I: 1.50 + 1.70 trombocytopenia *sensory neuropathy*	I: Overall, any grade GR3: 0.50-1.25 mg/m² 1.50-1.70 mg/m²	18% 0% 6%
	II:16		II: d1, 4, 8, 11 q21d	II:1.50	I: 1.60 trombocytopenia	II: Overall, any grade: GR3:	19% 0%
Orlowski (2002)	27	refractory hematologic malignancies	d1, 4, 8, 11, 15, 18, 22, 25 q42d	1.04	1.20 + 1.38 trombocytopenia fatigue and malaise hyponatriemia hypokalemia	Overall, any grade: GR3:	19% 0%
Cortes (2004)	15	refractory/ relapsed acute leukemias	d1, 4, 8, 11, 15, 18, 22, 25 q42d	1.25	1.50 orth. hypotension nausea diarrhea hypokalemia fluid retention	Overall, any grade: GR3:	13% 0%
Blaney (2004)	15	recurrent/ refractory solid tumors *(pediatric patients)*	d1, 4, 8, 11 q21d	1.20	1.60 trombocytopenia	Grade 1 >GR1	7% 0%

n indicates number of patients; d, day; q, every; MTD, maximum tolerated dose; DLT, dose limiting toxicity; GR, grade.

In clinical studies, sensory or motor neuropathy is graded according to the Common Terminology Criteria for Adverse Events (CTCAE) (see Table 2) [34]. Grade 3 neuropathy, which is generally regarded as dose limiting, implies function is affected, e.g. of the hand or foot, however there is no interference with 'activities of daily living' (self-care). Overall incidence of drug-related neuropathy, all grades, in phase 1 studies in adult patients varied from 8% to 28% with a clear rise in incidence and severity at higher dose and higher dose-intensity of bortezomib [12,23,29,30,32,33]. In the weekly schedule overall incidence of peripheral neuropathy was relatively low at 8% compared to 12 to 28% in the more dose-intense twice-weekly schedules of bortezomib. It was furthermore observed pre-existing neuropathy and prior neurotoxic treatment predisposes to bortezomib-induced neurotoxicity [30]. In a phase 1 study in pediatric patients a twice-weekly schedule of bortezomib was evaluated, at 1.2 or 1.6 mg/m^2 per dose [31]. Levels of 20S PI were comparable to those found in adults. No dose-limiting peripheral neuropathy was observed in pediatric patients and hardly any neurotoxicity was observed in this small study: only 1 out of 15 patients experienced a grade 1 neuropathy. An explanation for this could be that more than 80% of patients received just one cycle of therapy. Furthermore 10 out of 15 patients were treated at 1.2 mg/m^2, which is lower than the adult MTD. Age-related differences in proteasome activity in children compared to adults might also play a role [35,36].

Table 2. Grading of neuropathy and neuropathic pain according to the Common Terminology Criteria for Adverse Events (CTCAE)

	Grade				
Adverse Event	**0**	**1**	**2**	**3**	**4**
			NEUROLOGY		
neuropathy-motor	normal	asymptomatic, weakness on exam/ testing only	symptomatic weakness interfering with function, but not interfering with activities of daily living	weakness interfering with activities of daily living; bracing or assistance to walk	life-threatening; disabling e.g. paralysis
neuropathy-sensory	normal	asymptomatic; loss of deep tendon reflexes or paresthesia (including tingling) but not interfering with function	sensory loss alteration or paresthesia (including tingling), interfering with function, but not interfering with activities of daily living	sensory alteration or paresthesia interfering with activities of daily living	disabling
			PAIN		
neuropathic pain	none	mild pain not interfering with function	moderate pain: pain or analgesics interfering with function, but not interfering with activities of daily living	severe pain: pain or analgesics severely interfering with activities of daily living	disabling

Phase 2 and 3 Studies

In phase 2 and 3 studies, which served to assess efficacy of bortezomib treatment in MM patients as well as several other hematological malignancies and solid tumor types, peripheral neuropathy was among the most frequently observed toxicities necessitating dose reductions or discontinuation of treatment.

In the previously mentioned SUMMIT and CREST phase 2 studies, a total number of 256 patients with relapsed and/ or refractory MM were treated with 1.0 or 1.3 mg/m^2 bortezomib twice-weekly. Peripheral neuropathy was reported in 35% of patients whereby the number of events per 100 patient doses steadily increased through the first 5 cycles, peaking at 5.3% at cycle 5, and steadily decreased thereafter. In the SUMMIT study (202 patients), cumulative dose-related peripheral neuropathy was considered ‘the most clinically significant adverse event’. Dose reductions for peripheral neuropathy were necessary in 12% of patients and 5% of patients discontinued treatment due to peripheral neuropathy.

Especially patients with pre-existent neuropathy and/ or prior exposure to neurotoxic anti-cancer agents, such as taxanes and platinum compounds, are more prone to develop high-grade neuropathy [35]. The incidence of treatment-emergent grade 3 neuropathy was 16% in patients who had peripheral neuropathy prior to bortezomib treatment compared to 3% in patients without prior peripheral neuropathy [13,14,37].

In the largest study with bortezomib conducted to date, the Assessment of Proteasome Inhibition for Extending Remissions (APEX) phase 3 study, 669 patients with relapsed MM were randomized to receive either high-dose dexamethasone (336 patients) or twice-weekly bortezomib (333 patients) at 1.3 mg/m^2 [38]. Regarding neurotoxicity, incidence was comparable to that observed in the SUMMIT and CREST studies. In the bortezomib arm, overall incidence of peripheral neuropathy was 36%. Peripheral neuropathy necessitated early discontinuation of treatment in 8% of patients, making it the predominant reason for treatment discontinuation. Incidence of grade 3 and 4 neuropathy was 7% and 1% respectively. In the dexamethasone arm overall incidence of peripheral neuropathy was 9%, grade 3 and 4 incidence was below 1%. Furthermore, in the bortezomib arm paresthesias were reported in 21% of patients, grade 3 in 2% of patients, compared to 8% and 0% respectively in the dexamethasone arm.

In an extension study of the SUMMIT and CREST studies, allowing patients who benefited from the treatment to continue bortezomib treatment beyond the initial eight three-week cycles, only in 14% of patients new or worsening of peripheral neuropathy was observed, compared to 30% when these same patients participated in the original SUMMIT or CREST studies [39]. An important consequence of this observation could be that maintenance therapy with bortezomib might be possible in certain patients not susceptible for its neurotoxic side effects. It is important to realize in this regard that bortezomib is not only being developed for treatment of malignant neoplastic diseases but also in chronic inflammatory diseases [20].

In phase 1 studies, a higher maximum tolerated dose was established in solid tumor patients compared to patients with hematological malignancies, 1.5 mg/m^2 and 1.3 mg/m^2 respectively. In part this was due to a greater bone marrow reserve in solid tumor patients, resulting in less pronounced bortezomib-induced thrombocytopenia. So far, the sizes of published studies in solid tumor patients were relatively small, varying from 16 to 37 patients.

Striking is the great variability in incidence of peripheral neuropathy between different studies in solid tumor patients. A study in patients with metastatic neuro-endocrine tumors who did not have any prior chemotherapy, showed the highest overall and grade 3 incidence at 63% and 37% respectively. Other studies with a similar or higher dose-intensity report overall incidences of 21% to 48% and grade 3 incidences of 3% to 15% [40-46].

Larger studies in solid tumor patients will have to show whether or not overall and treatment-emergent neurotoxicity is comparable to what has been observed in the large studies conducted in MM patients. A comparison will remain difficult, as most hematological patients included in published studies with bortezomib were generally more exposed to prior neurotoxic treatments than solid tumor patients.

Combination Studies

As inhibition of proteasome activity is increasingly considered as a rational target for chemosensitization, bortezomib is being combined with other chemotherapeutic agents in many studies [47]. However, an undesirable effect of combining proteasome inhibition and chemotherapy could be potentiation of neurotoxicity, especially when bortezomib is combined with other neurotoxic agents.

A randomized phase 2 study in advanced non-small cell lung cancer patients, who had received one prior chemotherapy regimen, compared bortezomib alone to the combination of docetaxel and bortezomib. More grade 3 neuropathy was observed in the bortezomib alone arm than in the combination arm, 15% vs. 5% respectively. It must be noted that bortezomib was dosed higher at 1.5 mg/m^2 in the bortezomib-alone arm vs. 1.3 mg/m^2 in the combination arm [48].

In a dose finding study combining bortezomib with pegylated liposomal doxorubicin (pegLD) in 42 patients with hematological malignancies, the overall incidence of neuropathy was 55% [49]. Grade 3 incidence was 6%, 17% and 33% at doses of 1.30 mg/m^2, 1.40 mg/m^2 and 1.50 mg/m^2, respectively. The dose of pegLD was kept constant at 30 mg/m^2 in all cohorts. Neuropathy as observed in this study was related only to bortezomib dose and comparable to, or slightly higher than incidences in single agent bortezomib studies at similar dose-intensity. It must be noted that patients in this study were heavily pretreated with a median number of five prior, often neurotoxic, chemotherapy regimens, which is a predisposing factor for bortezomib-induced neuropathy.

In another small dose finding study bortezomib was combined with a fixed dose of carboplatin (area under the curve [AUC] 5). Fifteen patients with recurrent ovarian or primary peritoneal cancer refractory to at least one platinum-based chemotherapeutic regimen, were treated. Even though patients had received prior neurotoxic treatment and bortezomib was combined with a relatively neurotoxic chemotherapy, overall incidence of sensory neuropathy remained rather low at 27%. Grade 3 dose limiting sensory neuropathy was observed only in one patient treated at the highest dose level of 1.5 mg/m^2 bortezomib [50].

Preliminary results of a dose finding study combining bortezomib with gemcitabine and cisplatin in 34 chemonaive patients with advanced solid tumors indicate there was no potentiation of neurotoxicity [51]. Furthermore, a patient has been described who developed grade 3 peripheral neuropathy following bortezomib treatment, which improved to grade 1 despite subsequent treatment with cisplatin-based chemotherapy [43].

Overall, preliminary results indicate that when bortezomib is administered concurrently with other chemotherapeutic drugs such as pegylated liposomal doxorubicin, somewhat neurotoxic drugs such as docetaxel and carboplatin and a known neurotoxic chemotherapeutic agent such as cisplatin, no potentiation of neurotoxicity occurs. Future and currently running combination studies will have to show whether or not neurotoxicity is manageable, especially in combination with other reputedly neurotoxic agents such as thalidomide, paclitaxel and vincristine. It might be that prior damage to peripheral nerves due to prior neurotoxic treatment or other illnesses predisposes more to bortezomib-induced neurotoxicity than concurrent combinations of bortezomib and other (neurotoxic) agents in chemonaive or second line treated patients.

Clinical Manifestations

Peripheral Neuropathy

In animal studies with bortezomib, the nervous system had already been determined as a target organ of toxicity. Neurotoxicity of repeat dose bortezomib in animals included axonal swelling and degeneration in peripheral nerves, dorsal spinal roots, and tracts of the spinal cord as well as multifocal hemorrhage and necrosis in the brain [24,52].

Clinical signs of bortezomib-induced neuropathy in patients included tingling, pain, diminished pinprick, vibratory and temperature sense. Motor symptoms are uncommon. Typically paresthesias were more intense in the distal lower limbs compared to the hands, increasing in severity with every dose of bortezomib. Neuropathic pain was also mainly confined to the legs and feet [30,43,50,53]. Nerve conduction studies (NCS) showed reduced or absent sensory nerve action potential (SNAP) amplitude and reduced peroneal compound muscle action potential (CMAP) amplitude and which is consistent with an axonal polyneuropathy. Skin-biopsies showed decreased intra-epidermal nerve fiber (IENF) density, which is typical for 'dying back' axonopathy [54]. A strong correlation was found between Total Neuropathy Score (TNS), a validated measure of peripheral nerve function, SNAP amplitude values and cumulative bortezomib dose [53,55]. Bortezomib-induced neuropathy has been characterized by NCS and quantitative sensory testing as a length-dependent, sensory, axonal polyneuropathy with predominantly small fiber (A-δ myelinated afferent; nociceptive C unmyelinated afferent) involvement [43,53,37].

Small fiber neuropathies are generally characterized by peripheral pain [57]. Concordantly, bortezomib-induced neuropathy can also be extremely painful, occasionally requiring narcotic analgesia and other drugs such as gabapentin and amitriptyline for pain management. This can be necessary for many months following discontinuation of bortezomib [23,30,50]. Apart from several described cases of severe neuropathic pain associated with bortezomib treatment, overall incidence of neuropathic pain is somewhat unclear when reviewing the published studies. In one study, all patients with neuropathy were reported to experience neuropathic pain and this was clearly specified in the adverse events listing, in other studies, a pain component has not been specified [43]. Though not clearly designated as neuropathic pain, 'pain in limb' was reported in the SUMMIT study in 13% of patients, grade 3 in 7% of patients. In the APEX study pain in limb was reported in 15% of patients in the bortezomib arm vs. 7% in the dexamethasone treated arm. Grade 3 pain in limb was solely observed in the bortezomib arm with an incidence of 2% [38]. In the phase 1 dose finding study evaluating weekly bortezomib, pain in limb was seen in 3% of patients at the lower seven dose levels, compared to 30% of patients at the highest three dose levels of bortezomib (1.6 - 2.0 mg/m^2) [23]. It is therefore likely the frequently reported 'pain in limb' reported in studies with bortezomib, is related to a small fiber like neuropathy induced by bortezomib.

In the SUMMIT study complete resolution or improvement of PN was observed 'in the majority of patients' [13,14,37,39]. In the APEX study patients treated with bortezomib who developed grade 2 or higher peripheral neuropathy had in 46% complete resolution (return to baseline) and in 5% improvement of symptoms at last assessment. Median time to resolution was 3.5 months [38]. De facto in 49% of patients with grade 2 or higher peripheral neuropathy, symptoms did not improve after discontinuation of bortezomib. Furthermore, worsening or onset of neuropathy after discontinuation of bortezomib has been described in a few patients [43].

Generally speaking, bortezomib-induced neuropathy is dose-dependent. However, it can manifest itself after a single dose of bortezomib while some patients will not develop neuropathy, even upon prolonged exposure. The median cumulative dose at the onset of neuropathy was determined in one 16-patient study at 13.3 mg/m^2 ranging from 1.5 to 28.5 mg/m^2 [43]. This suggests individual susceptibility varies greatly.

Autonomic Instability

Proposed as a small fiber neuropathy based on clinical findings and diagnostic testing, it is expected that autonomic functions be affected as well. Small A-δ fibers and C fibers carry autonomic functions such as bowel movements and blood pressure in addition to temperature and pain sensation [56].

In phase 1 studies, a dose-dependent increase in incidence of hypotension was observed and in two studies (orthostatic) hypotension was one of the dose-limiting toxicities [23,33]. Hypotension did not appear to be related to cardiac failure nor adrenal dysfunction. At higher dose levels, 'autonomic instability', characterized by postural hypotension and syncope were partly attributed to bortezomib therapy and reported to resolve upon discontinuation of bortezomib treatment [12,23].

In a phase 2 study evaluating bortezomib treatment in sixteen patients, six out of ten patients with grade 2 to 3 peripheral neuropathy also showed grade 2 to 3 symptoms 'possibly related to autonomic neuropathy' such as orthostatic hypotension, syncope, dizziness, ileus and abdominal cramps. With the exception of abdominal cramps, symptoms indicative for autonomic dysfunction occurred at cumulative doses equal or greater to those at onset of peripheral sensory neuropathy. Only in one patient symptoms 'possibly related to autonomic neuropathy' (grade 3 ileus) were observed *without* symptoms of peripheral sensory neuropathy. In contrast, in four out of ten patients with peripheral sensory neuropathy, there were no symptoms indicating autonomic neuropathy [43]. In the APEX phase 3 study, adverse events implicated in autonomic dysfunction such as constipation, abdominal pain and headache were reported more also frequently in the bortezomib arm compared to the dexamethasone arm with overall incidences of 42% vs. 15%, 16% vs. 4% and 26% vs. 13% respectively [38].

It is likely autonomic neuropathy represents a later event in bortezomib-induced neurotoxicity, following or coinciding with, but generally not preceding peripheral sensory neuropathy.

Ototoxicity

In one case report severe irreversible bilateral hearing loss after bortezomib therapy has been described in a MM patient with a minor prior hearing impairment [57]. Typical about this case was that not only medium and high frequencies were affected, as in cisplatin-induced ototoxicity, but all frequencies. Furthermore deterioration of hearing loss in the low frequency range continued for twelve months. This may have been an idiosyncratic effect in this one patient, as no other reports of bortezomib-induced ototoxicity have been published, even though many patients treated with bortezomib had been exposed to known ototoxic agents such as cisplatin. Nevertheless, for the time being, caution is warranted for ototoxic effects of bortezomib treatment, especially in patients with prior hearing loss, or when combining bortezomib with ototoxic drugs such as cisplatin.

Central Nervous System Effects

So far there are no indications that bortezomib therapy is associated with effects on the CNS. In 6% of patients of the extended phase 2 CREST/ SUMMIT studies, with treatment durations up to eleven months, several CNS events, such as memory impairment and mental state changes, were reported. However, they were not clearly attributed to bortezomib, but rather to disease progression and confounding illnesses [39].

Bortezomib-Induced Peripheral Neuropathy

Pathogenesis

Pathologic states associated with the ubiquitin-proteasome system (UPS) can be the result of either a *loss of function* or a *gain of function* of this system. In the first scenario, proteins are stabilized, in the latter proteins are degraded in an abnormal or accelerated fashion. In the case of blunt and abrupt inhibition of proteasome function by a chemical inhibitor such as bortezomib, *loss of function* of the UPS occurs resulting in stabilization of proteasome-degraded proteins.

Ubiquitination plays a role among others in neuronal survival, synaptogenesis and axon function [58-60; and Chapters 15-18]. The marked neurotoxicity occurring upon systemic proteasome inhibition is therefore not surprising. Furthermore, neurodegenerative disorders are strongly associated with aberrations in the UPS, either as a primary cause or as a secondary consequence [61].

Nerve conduction studies and quantitative sensory testing indicate that bortezomib causes a length dependent axonal sensory neuropathy with predominantly small fiber involvement [37].

Considering the role of the UPS in the peripheral nerve system, it is very important to distinguish differences in effects of proteasome inhibition on the neuron cell body compared to the distal axon. Neuronal differentiation and neurite outgrowth is known to be promoted by proteasome inhibition [62-64]. In fact the first proteasome inhibitor, lactacystin, was discovered using a neurite outgrowth assay [65].

In contrast to these neuritogenic effects on the neuron cell body, proteasome inhibition has been shown to have detrimental effects on growth and long-term maintenance of mature axons. In sympathetic and sensory explant cultures from mice, proteasome inhibition led to 'dying-back' degeneration of nerve terminals [66]. A proposed explanation for this finding was the intense and continuous anterograde and retrograde protein transport occurring in the axon terminal. Proteins delivered to the axon terminals by anterograde axonal transport must either be transported back to the cell body by retrograde axonal transport, or be degraded in situ. Inhibition of proteasomal degradation could simply disturb the balance between delivery and degradation of axonal proteins, causing them to accumulate to toxic levels in nerve terminals resulting in degenerative axonopathy [66].

In addition to generally reported degenerative axonopathy, demyelinating neuropathy has also been associated with bortezomib treatment [23,67]. The role of the UPS in demyelinating neuropathies such as X-linked Charcot-Marie-Tooth disease 1A (CMTX1A) has been studied quite extensively (see Chapters 17 and 26). Peripheral myelin protein 22 (PMP22) is a 22-kDa glycoprotein mainly expressed by Schwann cells. Correct expression levels of PMP22 are essential for normal peripheral nerve function. Duplication of the gene and concomitant high expression levels are associated with peripheral demyelinating and axonal neuropathies [68,69]. Levels of PMP22, a protein with a very short half-life, are regulated by the UPS. Preclinical studies have shown that, when the proteasome is inhibited, PMP22 accumulates in perinuclear aggresomes. The exact mechanism by which PMP22 accumulation and aggresome formation might contribute to cellular alterations and demyelination needs yet to be elucidated. A possible mechanism is that intracellular retention will reduce the amount of protein that is incorporated into myelin resulting in peripheral nerve demyelination and dysfunction of Schwann cells [70].

CNS side effects have not been clearly associated with bortezomib treatment. Aside from headache, no marked increase in incidence of memory loss or altered motor function have been related to bortezomib treatment. This could be a result of a limited crossing of the blood brain barrier by bortezomib. Potentially, for CNS effects to occur, they might require a more sustained level of inhibition over a longer period of time.

Conclusions

Peripheral sensory neuropathy has been reported as an important and common toxicity of bortezomib treatment, occurring in about one third of patients. Neuropathy induced by bortezomib is predominantly cumulative though individual susceptibility varies greatly. Risk factors for bortezomib-induced neurotoxicity are prior exposure to neurotoxic agents as well as pre-existent neuropathy. Features of bortezomib neuropathy are characteristic for a small fiber neuropathy, characterized by a more sensory than motor neuropathy, neuropathic pain and autonomic dysfunction. Treatment is symptomatic with analgesics in case of neuropathic pain. Symptoms resolve in over half of patients after discontinuation of bortezomib therapy.

Detrimental effects of proteasome inhibition on nerve terminal protein homeostasis as well as myelin production by Schwann cells might explain the high incidence of neurotoxicity in bortezomib-treated patients.

References

[1] King RW, Deshaies RJ, Peters J-M, Kirschner MW: How proteolysis drives the cell cycle. *Science* 1996, 274:1652-1659.

[2] Adams J: The proteasome: a suitable antineoplastic target. *Nat Rev Cancer* 2004, 4:349-360.

[3] Schwartz P, Alan L, Ciechanover M: The ubiquitin-proteasome pathway and pathogenesis of human diseases. *Ann Rev Med* 1999, 50:57-74.

[4] Chauhan D, Catley L, Li G, Podar K, Hideshima T, Velankar M, Mitsiades C, Mitsiades N, Yasui H, Letai A, Ovaa H, Berkers C, Nicholson B, Chao TH, Neuteboom ST, Richardson P, Palladino MA, Anderson KC: A novel orally active proteasome inhibitor induces apoptosis in multiple myeloma cells with mechanisms distinct from Bortezomib. *Cancer Cell* 2005, 8:407- 19.

[5] Shah IM, Lees KR, Pien CP, Elliott PJ: Early clinical experience with the novel proteasome inhibitor PS-519. *Br J Clin Pharmacol* 2002, 54:269-76.

[6] Voorhees PM, Orlowski RZ: The proteasome and proteasome inhibitors in cancer therapy. *Annu Rev Pharmacol Toxicol* 2006, 46:189-213.

[7] Adams J, Palombella VJ, Sausville EA, Johnson J, Destree A, Lazarus DD, Maas J, Pien CS, Prakash S, Elliott PJ: Proteasome Inhibitors: A novel class of potent and effective antitumor agents. *Cancer Res* 1999, 59:2615-2622.

[8] Hideshima T, Richardson P, Chauhan D, Palombella VJ, Elliott PJ, Adams J, Anderson KC: The proteasome inhibitor PS-341 inhibits growth, induces apoptosis, and overcomes drug resistance in human multiple myeloma cells. *Cancer Res* 2001, 61:3071-3076.

[9] An B, Goldfarb RH, Siman R, Dou QP: Novel dipeptidyl proteasome inhibitors overcome Bcl-2 protective function and selectively accumulate the cyclin-dependent kinase inhibitor p27 and induce apoptosis in transformed, but not normal, human fibroblasts. *Cell Death Differ* 1998, 5:1062-1075.

[10] Orlowski RZ, Eswara JR, Lafond-Walker A, Grever MR, Orlowski M, Dang CV: Tumor growth inhibition induced in a murine model of human Burkitt's lymphoma by a proteasome inhibitor. *Cancer Res* 1998, 58:4342-4348.

[11] Masdehors P, Merle-Beral H, Maloum K, Omura S, Magdelenat H, Delic J: Deregulation of the ubiquitin system and p53 proteolysis modify the apoptotic response in B-CLL lymphocytes. *Blood* 2000, 96:269-274.

[12] Orlowski RZ, Stinchcombe TE, Mitchell BS, Shea TC, Baldwin AS, Stahl S, Adams J, Esseltine DL, Elliott PJ, Pien CS, Guerciolini R, Anderson JK, Depcik-Smith ND, Bhagat R, Lehman MJ, Novick SC, O'Connor OA, Soignet SL: Phase I trial of the proteasome inhibitor PS-341 in patients with refractory hematologic malignancies. *J Clin Oncol* 2002, 20:4420-4427.

[13] Jagannath S, Barlogie B, Berenson J, Siegel D, Irwin D, Richardson PG, Niesvizky R, Alexanian R, Limentani SA, Alsina M, Adams J, Kauffman M, Esseltine DL, Schenkein DP, Anderson KC: A phase 2 study of two doses of bortezomib in relapsed or refractory myeloma. *Br J Haematol* 2004, 127:165-172.

[14] Richardson PG, Barlogie B, Berenson J, Singhal S, Jagannath S, Irwin D, Rajkumar SV, Srkalovic G, Alsina M, Alexanian R, Siegel D, Orlowski RZ, Kuter D, Limentani SA, Lee S, Hideshima T, Esseltine DL, Kauffman M, Adams J, Schenkein DP, Anderson KC: A phase 2 study of bortezomib in relapsed, refractory myeloma. *N Engl J Med* 2003, 348:2609-2617.

[15] O'Connor OA, Wright J, Moskowitz C, Muzzy J, MacGregor-Cortelli B, Stubblefield M, Straus D, Portlock C, Hamlin P, Choi E, Dumetrescu O, Esseltine D, Trehu E, Adams J, Schenkein D, Zelenetz AD: Phase II clinical experience with the novel proteasome inhibitor bortezomib in patients with indolent non-Hodgkin's lymphoma and mantle cell lymphoma. *J Clin Oncol* 2005, 23:676-684.

[16] Dimopoulos MA, Anagnostopoulos A, Kyrtsonis MC, Castritis E, Bitsaktsis A, Pangalis GA: Treatment of relapsed or refractory Waldenstrom's macroglobulinemia with bortezomib. *Haematologica* 2005, 90:1655-1658.

[17] Ludwig H, Khayat D, Giaccone G, Facon T: Proteasome inhibition and its clinical prospects in the treatment of hematologic and solid malignancies. *Cancer* 2005, 104:1794-1807.

[18] Berenson JR, Ma HM, Vescio R: The role of nuclear factor-kappaB in the biology and treatment of multiple myeloma. *Semin Oncol* 2001, 28:626-633.

[19] Pham LV, Tamayo AT, Yoshimura LC, Lo P, Ford RJ: Inhibition of constitutive NF-{kappa}B activation in mantle cell lymphoma B cells leads to induction of cell cycle arrest and apoptosis. *J Immunol* 2003, 171:88-95.

[20] Elliott PJ, Zollner TM, Boehncke W-H: Proteasome inhibition: a new anti-inflammatory strategy. *J Mol Med* 2003, 81:235-245.

[21] Phillips JB, Williams AJ, Adams J, Elliott PJ, Tortella FC, Clemens JA: Proteasome inhibitor PS519 reduces infarction and attenuates leukocyte infiltration in a rat model of focal cerebral ischemia. *Stroke* 2000, 31:1686-1693.

[22] Williams AJ, Berti R, Dave JR, Elliot PJ, Adams J, Tortella FC: Delayed treatment of ischemia/reperfusion brain injury: extended therapeutic window with the proteosome inhibitor MLN519. *Stroke* 2004, 35:1186-1191.

[23] Papandreou CN, Daliani DD, Nix D, Yang H, Madden T, Wang X, Pien CS, Millikan RE, Tu SM, Pagliaro L, Kim J, Adams J, Elliott P, Esseltine D, Petrusich A, Dieringer P, Perez C, Logothetis CJ: Phase I trial of the proteasome inhibitor bortezomib in patients with advanced solid tumors with observations in androgen-independent prostate cancer. *J Clin Oncol* 2004, 22:2108-2121.

[24] European Medicines Agency: Scientific discussion for the approval of Velcade. Upd. until 1 May 2004. http://www.emea.eu.int/humandocs/PDFs/EPAR/velcade/166104en6.pdf. Accessed December 1, 2005.

[25] Pekol T, Daniels JS, Labutti J, Parsons I, Nix D, Baronas E, Hsieh F, Gan LS, Miwa G: Human metabolism of the proteasome inhibitor bortezomib: identification of circulating metabolites. *Drug Metab Dispos* 2005, 33:771-777.

[26] Lightcap ES, McCormack TA, Pien CS, Chau V, Adams J, Elliott PJ: Proteasome inhibition measurements: clinical application. *Clin Chem* 2000, 46:673-83.

[27] Bandoh N, Ogino T, Cho HS, Hur SY, Shen J, Wang X, Kato S, Miyokawa N, Harabuchi Y, Ferrone S: Development and characterization of human constitutive proteasome and immunoproteasome subunit-specific monoclonal antibodies. *Tissue Antigens* 2005, 66:185-194.

[28] Altun M, Galardy PJ, Shringarpure R, Hideshima T, LeBlanc R, Anderson KC, Ploegh HL, Kessler BM: Effects of PS-341 on the Activity and Composition of Proteasomes in Multiple Myeloma Cells. *Cancer Res* 2005, 65:7896-7901.

[29] Hamilton AL, Eder JP, Pavlick AC, Clark JW, Liebes L, Garcia-Carbonero R, Chachoua A, Ryan DP, Soma V, Farrell K, Kinchla N, Boyden J, Yee H, Zeleniuch-Jacquotte A, Wright J, Elliott P, Adams J, Muggia FM: Proteasome inhibition with bortezomib (PS-341): a phase I study with pharmacodynamic end points using a day 1 and day 4 schedule in a 14-day cycle. *J Clin Oncol* 2005, 23:6107-6116.

[30] Aghajanian C, Soignet S, Dizon DS, Pien CS, Adams J, Elliott PJ, Sabbatini P, Miller V, Hensley ML, Pezzulli S, Canales C, Daud A, Spriggs DR: A phase I trial of the novel proteasome inhibitor PS341 in advanced solid tumor malignancies. *Clin Cancer Res* 2002, 8:2505-2511.

[31] Blaney SM, Bernstein M, Neville K, Ginsberg J, Kitchen B, Horton T, Berg SL, Krailo M, Adamson PC: Phase I study of the proteasome inhibitor bortezomib in pediatric patients with refractory solid tumors: a Children's Oncology Group study (ADVL0015). *J Clin Oncol* 2004, 22:4804-4809.

[32] Dy GK, Thomas JP, Wilding G, Bruzek L, Mandrekar S, Erlichman C, Alberti D, Binger K, Pitot HC, Alberts SR, Hanson LJ, Marnocha R, Tutsch K, Kaufmann SH, Adjei AA: A phase I and pharmacologic trial of two schedules of the proteasome inhibitor, PS-341 (bortezomib, Velcade), in patients with advanced cancer. *Clin Cancer Res* 2005, 11:3410-3416.

[33] Cortes J, Thomas D, Koller C, Giles F, Estey E, Faderl S, Garcia-Manero G, McConkey D, Ruiz SL, Guerciolini R, Wright J, Kantarjian H: Phase I study of bortezomib in refractory or relapsed acute leukemias. *Clin Cancer Res* 2004, 10:3371-3376.

[34] National Cancer Institute: Cancer Therapy Evaluation Program (CTEP). Common Terminology Criteria for Adverse Events v3.0 (CTCAE). http://ctep.cancer.gov/forms/CTCAEv3.pdf. Accessed December 1, 2005.

[35] Rivett AJ, Bose S, Pemberton AJ, Brooks P, Onion D, Shirley D, Stratford FL, Forti K: Assays of proteasome activity in relation to aging. *Exp Gerontol* 2002, 37:1217-1222.

[36] Keller JN, Hanni KB, Markesbery WR: Possible involvement of proteasome inhibition in aging: implications for oxidative stress. *Mech Ageing Dev* 2000, 113:61-70.

[37] Richardson PG, Briemberg H, Jagannath S, Barlogie B, Berenson J, Singhal S, Traynor A, Siegel D, Irwin D, Schuster M, Srkalovic G, Alexanian R, Rajkumar SV, Limentani S, Alsina M, Orlowski R, Kuter D, Esseltine D, Adams J, Schenkein DP, Wen P, Amato A, Anderson KC: Peripheral neuropathy following bortezomib (Velcade[tm], formerly PS-341) therapy in patients with advanced multiple myeloma (MM): characterization and reversibility. *Proceedings of the American Society of Hematology* 2003. San Diego. *Blood* 2003, 102:149a, abstract 512

[38] Richardson PG, Sonneveld P, Schuster MW, Irwin D, Stadtmauer EA, Facon T, Harousseau JL, Ben-Yehuda D, Lonial S, Goldschmidt H, Reece D, San-Miguel JF, Blade J, Boccadoro M, Cavenagh J, Dalton WS, Boral AL, Esseltine DL, Porter JB, Schenkein D, Anderson KC, Assessment of Proteasome Inhibition for Extending Remissions (APEX) Investigators: Bortezomib or high-dose dexamethasone for relapsed multiple myeloma. *N Engl J Med* 2005, 352:2487-2498.

[39] Berenson JR, Jagannath S, Barlogie B, Siegel DT, Alexanian R, Richardson PG, Irwin D, Alsina M, Rajkumar SV, Srkalovic G, Singhal S, Limentani S, Niesvizky R, Esseltine DL, Trehu E, Schenkein DP, Anderson K: Safety of prolonged therapy with bortezomib in relapsed or refractory multiple myeloma. *Cancer* 2005, 104: 2141-2148.

[40] Davis NB, Taber DA, Ansari RH, Ryan CW, George C, Vokes EE, Vogelzang NJ, Stadler WM: Phase II trial of PS-341 in patients with renal cell cancer: a University of Chicago phase II consortium study. *J Clin Oncol* 2004, 22:115-119.

[41] Kondagunta GV, Drucker B, Schwartz L, Bacik J, Marion S, Russo P, Mazumdar M, Motzer RJ: Phase II trial of bortezomib for patients with advanced renal cell carcinoma. *J Clin Oncol* 2004, 22:3720-3725.

[42] Mackay H, Hedley D, Major P, Townsley C, Mackenzie M, Vincent M, Degendorfer P, Tsao MS, Nicklee T, Birle D, Wright J, Siu L, Moore M, Oza A: A phase II trial with pharmacodynamic endpoints of the proteasome inhibitor bortezomib in patients with metastatic colorectal cancer. *Clin Cancer Res* 2005, 11:5526-33.

[43] Shah MH, Young D, Kindler HL, Webb I, Kleiber B, Wright J, Grever M: Phase II study of the proteasome inhibitor bortezomib (PS-341) in patients with metastatic neuroendocrine tumors. *Clin Cancer Res* 2004, 10:6111-6118.

[44] Robert G, Maki ASK, Scheu K, Yamada J, Wadler S, . Antonescu CR, Wright JJ, Schwartz GK: A multicenter phase II study of bortezomib in recurrent or metastatic sarcomas. *Cancer* 2005, 103:1431-1438.

[45] Markovic SN, Geyer SM, Dawkins F, Sharfman W, Albertini M, Maples W, Fracasso PM, Fitch T, Lorusso P, Adjei AA, Erlichman C: A phase II study of bortezomib in the treatment of metastatic malignant melanoma. *Cancer* 2005, 103:2584-2589.

[46] Stevenson JP, Nho CW, Johnson SW: Effects of bortezomib (PS-341) on NF-KappaB activation in peripheral blood mononuclear cells (PBMCs) of advanced non-small cell lung cancer (NSCLC) patients: a phase II/ pharmacodynamic trial. *Proceedings of the 39th Annual Meeting of the American Society of Clinical Oncology* 2003. Chicago. *J Clin Oncol* 2003, abstract 7145.

[47] Yang HH, Ma MH, Vescio RA, Berenson JR: Overcoming drug resistance in multiple myeloma: the emergence of therapeutic approaches to induce apoptosis. *J Clin Oncol* 2003, 21:4239-4247.

[48] Fanucchi MP, Fossella F, Fidias P, Natale RB, Belt RJ, Carbone DP, Govindan R, Raez L, Robert F, Schiller J: Bortezomib ± docetaxel in previously treated patients with advanced non-small cell lung cancer (NSCLC): a phase II study. *Proceedings of the 41st Annual Meeting of the American Society of Clinical Oncology* 2005. Orlando. *J Clin Oncol* 2005, abstract 7034.

[49] Orlowski RZ, Voorhees PM, Garcia RA, Hall MD, Kudrik FJ, Allred T, Johri AR, Jones PE, Ivanova A, Van Deventer HW, Gabriel DA, Shea TC, Mitchell BS, Adams J, Esseltine DL, Trehu EG, Green M, Lehman MJ, Natoli S, Collins JM, Lindley CM, Dees EC: Phase 1 trial of the proteasome inhibitor bortezomib and pegylated liposomal doxorubicin in patients with advanced hematologic malignancies. *Blood* 2005, 105:3058-3065.

[50] Aghajanian C, Dizon DS, Sabbatini P, Raizer JJ, Dupont J, Spriggs DR: Phase I trial of bortezomib and carboplatin in recurrent ovarian or primary peritoneal cancer. *J Clin Oncol* 2005, 23:5943-5949.

[51] Voortman J, Smit EF, Kuenen BC, Peters GJ, van de Velde H, Giaccone G: A phase 1B, open-label, dose-escalation study of bortezomib in combination with gemcitabine and cisplatin in the first-line treatment of patients with advanced solid tumors: preliminary results: *Proceedings of the 13th European Cancer Conference* 2005. Paris. *Eur J Cancer* 2005, abstract 1468.

[52] Millennium Pharmaceuticals: Velcade® (bortezomib) for Injection Prescribing Information for the United States. Issued June 2005.

[53] Chaudhry V, Cornblath DR, George B, Borrello I: Characteristics of Velcade-induced peripheral neuropathy. *Meeting of the Peripheral Nerve Society* 2005. Tuscany, Italy. Unnumbered abstract.

[54] Lauria G, Cornblath DR, Johansson O, McArthur JC, Mellgren SI, Nolano M, Rosenberg N, Sommer C, European Federation of Neurological Societies: EFNS guidelines on the use of skin biopsy in the diagnosis of peripheral neuropathy. *Eur J Neurol* 2005, 12:747-758.

[55] Cornblath DR, Chaudhry V, Carter K, Lee D, Seysedadr M, Miernicki M, Joh T: Total neuropathy score: Validation and reliability study. *Neurology* 1999, 53:1660-1664.

[56] Hoitsma E, Reulen JPH, de Baets M, Drent M, Spaans F, Faber CG: Small fiber neuropathy: a common and important clinical disorder. *J Neurol Sci* 2004, 227:119-130.

[57] Engelhardt M, Muller AM, Maier W, Wasch R: Severe irreversible bilateral hearing loss after bortezomib (VELCADE) therapy in a multiple myeloma (MM) patient. *Leukemia* 2005, 19:869-70.

[58] Yi JJ, Ehlers MD: Ubiquitin and protein turnover in synapse function. *Neuron* 2005, 47:629-632.

[59] Allen E, Ding J, Wang W, Pramanik S, Chou J, Yau V, Yang Y: Gigaxonin-controlled degradation of MAP1B light chain is critical to neuronal survival. *Nature* 2005, 438:224-228.

[60] Korhonen L, Lindholm D: The ubiquitin proteasome system in synaptic and axonal degeneration: a new twist to an old cycle. *J Cell Biol* 2004, 165:27-30.

[61] Ciechanover A, Brundin P: The ubiquitin proteasome system in neurodegenerative diseases: sometimes the chicken, sometimes the egg. *Neuron* 2003, 40:427-446.

[62] Fenteany G, Standaert R, Reichard G, Corey E, Schreiber S: A {beta}-lactone related to lactacystin induces neurite outgrowth in a neuroblastoma cell line and inhibits cell cycle progression in an osteosarcoma cell line. *PNAS* 1994, 91:3358-3362.

[63] Ohtani-Kaneko R, Takada K, Iigo M, Hara M, Yokosawa H, Kawashima S, Ohkawa K, Hirata K: Proteasome inhibitors which induce neurite outgrowth from PC12h cells cause different subcellular accumulations of multi-ubiquitin chains. *Neurochem Res* 1998, 23:1435-1443.

[64] Obin M, Mesco E, Gong X, Haas AL, Joseph J, Taylor A: Neurite outgrowth in PC12 Cells. Distinguishing the roles of ubiquitylation and ubiquitin-dependent proteolysis. *J Biol Chem* 1999, 274:11789-11795.

[65] Omura S, Fujimoto T, Otoguro K, Matsuzaki K, Moriguchi R, Tanaka H, Sasaki Y: Lactacystin, a novel microbial metabolite, induces neuritogenesis of neuroblastoma cells. *J Antibiot (Tokyo)* 1991, 44:113-116.

[66] Laser H, Mack TG, Wagner D, Coleman MP: Proteasome inhibition arrests neurite outgrowth and causes "dying-back" degeneration in primary culture. *J Neurosci Res* 2003, 74:906-16.

[67] Umapathi T, Chaudhry V: Toxic neuropathy. *Curr Opin Neurol* 2005, 18:574-80.

[68] Chance PF, Abbas N, Lensch MW, Pentao L, Roa BB, Patel PI, Lupski JR: Two autosomal dominant neuropathies result from reciprocal DNA duplication/deletion of a region on chromosome 17. *Hum Mol Genet* 1994, 3:223-228.

[69] Ryan MC, Shooter EM, Notterpek L: Aggresome Formation in Neuropathy Models Based on Peripheral Myelin Protein 22 Mutations. *Neurobiol Dis* 2002, 10:109-118.

[70] Notterpek L, Ryan MC, Tobler AR, Shooter EM: PMP22 accumulation in aggresomes: implications for CMT1A pathology. *Neurobiol Dis* 1999, 6:450-60.

Book Glossary

11 S cap	See: PA28.
19S cap	Also known as PA700 (proteasome activator of 700 kDa). A subcomplex of the 26S proteasome that acts as a receptor of ubiquitinated proteins, deubiquitinates proteins, unfolds them, opens the entrance to the 20S proteasome and translocates the polypeptide chains into it. Subdivided into lid and base complexes linked by Rpn10 (S5a). Different nomenclatures of the subunits exist, but the most widely accepted is the Rpn/Rpt nomenclature, where Rpt designate proteasomal ATP-ases (Rpt1-6) and Rpn designate non-proteasomal subunits (Rpn1-3 and Rpn5-12). Few additional components lack the Rpn names (e.g. UCH37).
20S proteasome	Multisubunit proteolytic core of the 26S proteasome; formerly known as multicatalytic proteinase complex (MPC); is a member of the N-terminal nucleophile hydrolases family; is organized in four stacked 7-membered rings composed of 2 outer α and 2 inner β rings arranged in a cylinder-like shape, with a central catalytic chamber and two antechambers. The entrance to the antechambers is occluded in inactive (latent) 20S proteasomes by N-terminal extensions of several α subunits. Activation of the 20S proteasome can be achieved by physical or chemical means or by binding to activators such as PA28 and PA700. It is characterized by the presence of three main hydrolytic activities, referred to as chymotrypsin-like activity (ChTL), trypsin-like activity (TL) and post-glutamyl peptide-hydrolyzing (PGPH) activity (also known as caspase-like). 20S proteasomes can degrade only short peptides and some denatured proteins, they can not recognize or bind polyubiquitinated proteins. Different nomenclature of the 20S subunits exist, but the most widely accepted is the α/β nomenclature, which distinguishes 7 α subunits (α1-α7) and 10 β subunits (β1-β7 and β1i, β2i, β5i).
26S proteasome	Composed of the core 20S proteasome associated with one or two 19S caps (PA700). The 26S proteasome are the active form of proteasomes, engaged in the degradation of ubiquitinated proteins. The 19S cap provides the capacity to bind polyubiquitinated proteins, deubiquitinate them, unfold them and translocate them to the central 20S proteasome, which provides the proteolytic sites.

AAA (atpases associated with various cellular activities) proteins	A superfamily of enzymatic machines that posses a structurally conserved ATPase domain and diverse cellular functions. Includes the proteasomal ATP-ases (Rpt1-6) and VCP.
Adhesion molecule	Specialized cell surface molecules that are involved in interactions between different cells. Include different families of proteins such as cadherins, integrins, selectins, ICAM, etc.
After eight	A zebrafish homologue of the Drosophila delta gene.
Aggresome	Inclusion body, which is assembled around centrosomes in the area of the proteolytic center of the cell through centripete microtubule-mediated transport; it is enriched in ubiquinated proteins, chaperones and proteasomes, and surrounded by a cage of intermediate filaments. Arise as a consequence of proteasome inhibition or/and massive overexpression of misfolded proteins. Aggresomes are often considers in vitro models of inclusion bodies found in neurodegenerative disorders.
Akt/Protein kinase B	A serine/threonine kinase that has a wide range of substrates; it acts downstream of PI3K to regulate many biological processes, such as proliferation, apoptosis and growth and is involved in tumorigenesis.
Alfa(α)-synuclein	One in a family of structurally related proteins that are prominently expressed in the central nervous system. Aggregated α-synuclein proteins form brain lesions that are hallmarks of some neurodegenerative diseases (synucleinopathies). The gene for α-synuclein, which is called SNCA, is on chromosome 4q21. One form of hereditary Parkinson's disease is due to mutations in SNCA. Another form of hereditary Parkinson disease is due to a triplication of SNCA. See also: Parkinson's disease.
Algogen:	A substance that produces pain e.g. capsaicin, mustard oil.
Allodynia:	The perception of normally innocuous stimuli such as light touch as painful. This is a frequent symptom of neuropathic pain.
Allosteric effect	Coupling of conformational change between two widely separated binding sites.
Alzheimer's disease (AD)	A degenerative disease of the brain associated with the development of protein deposits in the cerebral cortex and characterized by confusion, disorientation, memory failure, speech disturbances.
Ampa	Alpha-amino-3-hydroxy-5-methyl-4-isoxazole propionic acic. Agonist for AMPA receptor, mimics the effects of glutamate.
Amygdale	(Latin, *corpus amygdaloideum*) an almond-shaped set of neurons located deep in the brain's medial temporal lobe. Shown to play a key role in the processing of emotions, the amygdala is a part of the limbic system.
Amyloid fibrils	Structures formed by many disease-causing proteins when they

	aggregate.Amyloid fibrils share common biochemical characteristics such as detergent insolubility, high β-sheet content and a cross β structure, protease resistance and the ability to bind lipophilic dyes, such as congo red.
Amyotrophic lateral sclerosis / parkinsonism dementia complex of guam (als/pdc)	ALS is a progressive neurodegenerative disorder involving primarily the motor neurons of the cerebral cortex (upper motor neurons, umn), brain stem and spinal cord (lower motor neurons, lmn). Guamanian als/pdc amyotrophic lateral sclerosis/parkinsonism dementia complex of Guam (als/pdc) is a chronic neurodegenerative disorder highly prevalent in the native chamorro population of guam island in the western pacific. The etiopathogenesis of this disorder is not yet elucidated, although it has been hypothesized that environmental factors such as aluminium or neurotoxins might be involved. Neuropathologically, guamanian als/pdc shows a severe cortical atrophy and neuronal loss. The neuropathological hallmark of als/pdc is the widespread nft formation, especially in the isocortex and hippocampal formation.
Anaphase promoting complex (apc)	A large multi-protein complex consisting of at least 11 subunits with E3 ubiquitin ligase activity, implicated in mediating proteolysis during cell-cycle progression. Recently, roles for the APC in neural development have been described. Also known as the cyclosome.
Angiogenesis	The formation of blood vessels, such as occurs during embryogenesis, tissue repair or tumorigenesis.
Antigen	A substance that stimulates an immune response, especially the production of antibodies.
Anti-nuclear antibodies	Autoantibodies to nuclear antigens found in different systemic autoimmune diseases, in particular systemic lupus erythematosus and diseases affecting the connective tissue.
Antisense	Oligonucleotides with a sequence that is complementary to the mRNA of a given molecule can be used to block its translation. The subsequent temporary elimination of the protein of interest often provides useful information on its biological function.
Aplysia	Sea slug that belongs to the family Aplysiidae and is a genus of sea hares.
Apoptosis	A process of programmed cell death which can be initiated (initiation phase) through an extrinsic (TNF, FAS ligand) or intrinsic pathway (DNA damage, oxidative stress), leading to activation of caspases, in particular caspase 3 (execution phase). Activation of caspases brings upon cleavage of multiple cellular proteins, including some proteasomal subunits. Subsequent activation of endonucleases induces internucleosomal DNA cleavage. Morphological features of apoptosis include loss of cell

	adhesion, peripheral chromatin condensation, and cell blebbing, ultimately leading to the division of the cell into multiple apoptotic bodies, which ar ephagocytosed by surrounding cells. In distinction to necrosis, apoptosis does not elicit an inflammatory response.
Arborisation	The process by which an axonal or dendritic growth cone changes its morphology upon reaching its target by the extension and retraction of branch tips to form an elaborate stucture resembling a tree or arbor.
Ariadne	A Drosophila gene encoding a RING finger domain-containing protein. Null mutants are lethal and exibit a large number of abnormalities such as in neural development including axon pathfinding.
Armadillo superfamily	The armadillo repeat is an approximately 40 amino acids long tandemly repeated sequence motif first identified in the Drosophila segment polarity gene product armadillo, a protein that mediates cell adhesion. Similar repeats were later found in the mammalian armadillo homolog β-catenin, the junctional plaque protein plakoglobin, the adenomatous polyposis coli (APC) tumor suppressor protein, and a number of other proteins. These proteins exert several functions through interactions of their tandem armadillo repeats domain with diverse binding partners. The proteins combine structural roles as cell-contact and cytoskeleton-associated proteins and signaling functions by generating and transducing signals affecting gene expression.
Ataxia	Impairment of the ability to perform smoothly coordinated voluntary movements.
Ataxin	Protein associated with spinocerebellar ataxia, e.g. ataxin-1 is the protein associated with SCA-1. The ataxin proteins contain a polyglutamine tract.
Autophagy	A process by which areas of cytoplasm including entire organelles are surrounded by membranes, likely derived from the endoplasmic reticulum. Once the entire area of cytoplasm is secluded it fuses with primary lysosomes, leading to the degradation of enclosed cytoplasmic structures. Autophagy is controlled by a complex cascade of enzymatic reactions including the covalent conjugation of two different ubiquitin-like proteins of the Atg family. Autophagy is triggered in situations of starvation providing to the cell a steady source of nutrients, however excessive autophagy leads to autophagic cell death.
Axon	Also called nerve fiber, is a long slender projection of a nerve cell, or neuron, that conducts electrical impulses away from the neuron's cell body or soma.
Axonal dystrophy	Generic term for mis-shapen axons in pathology, encompassing both larger spheroids and smaller varicosities.

Axonal pruning	The cell-autonomous, programmed removal of superfluous axonal branches that arise during development
Axonal spheroid	Focal swelling of an axon, usually in the CNS, to many times its usual diameter, typically 10-50 μm. Spheroids are often filled with disorganised cytoskeleton and organelles, and many stain positively for APP.
Axonal transport	The flow of proteins, organelles and other axonal components along the axon. Axonal transport is bidirectional, with anterograde transport moving away from the cell body and retrograde transport moving towards it. Components also move with different speeds, classified as slow and fast axonal transport.
baculovirus inhibitor of apoptosis protein repeat (BIR)	BIR is a domain of tandem repeats separated by a variable length linker that seems to confer cell death-preventing activity. The BIR domains characterize the Inhibitor of Apoptosis (IAP) family of proteins (MEROPS proteinase inhibitor family I32, clan IV) that suppress apoptosis by interacting with and inhibiting the enzymatic activity of both initiator and effector caspases (MEROPS peptidase family C14). Several distinct mammalian IAPs including XIAP, c-IAP1, c-IAP2, and ML-IAP, have been identified, and they all exhibit antiapoptotic activity in cell culture. The functional unit in each IAP protein is the baculoviral IAP repeat (BIR), which contains approximately 80 amino acids folded around a zinc atom. Most mammalian IAPs have more than one BIR domain, with the different BIR domains performing distinct functions. For example, in XIAP, the third BIR domain (BIR3) potently inhibits the catalytic activity of caspase-9, whereas the linker sequences immediately preceding the second BIR domain (BIR2) selectively targets caspase-3 or -7.
Bag-1	BAG-1 (also known as RAP46; BCL2-associated athanogene) is an anti-apoptotic protein, which has been shown previously to interact with molecular chaperones of the Hsp70/Hsc70 family, characterized by an N-terminal ubiquitin-like domain, which binds to the 26S proteasome, providing a link between molecular chaperones and the UPS.
Bang-sensitive paralytic	A class of fly mutant that exhibits hyperactive seizure behavior followed by temporary paralysis in response to mechanical stimulation, such as a "bang" of the culture vial.
Base of the 19s cap (pa700)	A subdivision of the 19S cap (PA700) formed by a hexameric ring of proteasomal ATP-ases (Rpt1-6) as well as two largest, non-ATP-ase subunits of the PA700 (Rpn1 and 2).The base attaches to the α ring of the 20S proteasome on one side and to the lid of the PA700 on the other, through the Rpn10 subunit, which forms the hinge region.
Basic helix-loop-	A family of positive and negative regulators of transcription i.e.

helix (bhlh)	transcription factors characterised by a basic (positively charged) helix-loop-helix motif mediating sequence specific DNA recognition.
Basic helix–loop–helix (bhlh)	A structural motif present in many transcription factors that is characterized by two α-helices separated by a loop.The helices mediate dimerization, and the adjacent basic region is required for DNA binding.
Bcl2	A protein that promotes the survival of neurons by stabilizing mitochondrial membranes and decreasing oxidative stress.
Bcl-2 family (b-cell lymphoma-2 family).	These are proteins with a structural similarity to Bcl-2, the prototypical inhibitor of apoptosis. The Bcl-2 family comprises proteins that both block and enhance apoptosis.
Bendless	A Drosophila gene which when mutated leads to defects in the giant fibre axon pathfiding and escape-jump response. Bendless encodes a ubiquitin-conjugating enzyme and was the first mutation in a component of the UPS which exhibited axon pathfinding defects.
Beta (β)-Amyloid	An amyloid derived from a larger precursor protein (APP: Amyloid precursor protein) and is a component of the neurofibrillary tangles and plaques characteristic of Alzheimer's disease.
Beta (β)-Sheet structures	The β sheet (also β-pleated sheet or β strand) is a commonly occurring form of regular secondary structure in proteins. It consists of a stretch of amino acids whose peptide backbones are almost fully extended, resulting in an elongated pleatlike structure in which the peptide carbonyls point in alternating directions relative to the plane of the sheet. A typical strand is about five to ten amino acids long. In the most common usage, β strand refers to a single continuous stretch of amino acids adopting an extended conformation and involved in hydrogen bonds; by contrast, a β sheet refers to an assembly of such strands that are hydrogen-bonded to each other. However, the term 'β sheet' is also sometimes used as a synonym of 'βstrand', i.e., for a single segment of extended, hydrogen-bonded amino acids.
Bir motif	A ~70 amino-acid zinc-finger motif called the baculoviral inhibitor of apoptosis repeat. The number of BIR domains in a given IAP varies from one to three, but they are invariably present at the amino-terminus of the protein, and mediate the interaction with caspases.
Blm3p	Initially identified as an extragenic suppressor of the blm3-1 mutation in a genetic screen to detect genes controlling sensitivity to bleomycin, a drug that induces DNA double strand breaks. Yeast ortholog of the mammalian 20S proteasome activator PA200. It is now designated as Blm10p.
Bone morphogenetic	Members of the transforming growth factor βfamily of molecules

protein (bmp)	having multiple roles in development, including synaptogenesis.
Bortezomib (a dipeptidyl boronic acid, ps-341)	The first proteasome inhibitor that has progressed to clinical trails for the treatment of multiple myeloma (MM) and other cancers. Received the US Food and Drug Administration approval (Velcade^{TM}, Millenium Pharmaceuticals, Cambridge, MA) for the treatment of patients with relapsed and refractory MM. The mode of action depends largely on the inhibition of anti-apoptotic and anti-inflammatory NF-κB pathway both in the MM cells and in bone marrow stromal cells.
Braap	Proteasomal activity cleaving after branched-chain amino acids.
Brain derived neurotrophic factor	A neurotrophin playing roles in proliferation, differentiation and survival of neurons during development, as well as in the synaptic activity and plasticity in many groups of mature neurones.
Brain stem	The lower part of the brain, adjoining and structurally continuous with the spinal cord.
Calcium/calmodulin protein kinase ii (camkii)	Ca^{2+}/Calmodulin-Dependent Protein Kinase II (CaMKII) is a serine/threonine kinase. It is a Ca^{2+} /calmodulin-dependent, truncated monomer (1-325 amino acid residues) of the subunit. Autophosphorylation of threonine 286 in the presence of Ca^{2+} and calmodulin activates CaMKII and produces substantial Ca^{2+} /calmodulin-independent activity.
Calpain/calpastatin	Calpain is a calcium-activated proteinase of eukaryotic cells that itself activates several cellular enzymes. In erythrocytes, it affects several proteins important for the determination of cellular shape and deformability. This protease is also involves in apoptosis, cytoskeletal reorganization and muscle protein degradation. Calpain exists as a heterodimer composed of a small regulatory subunit and one of three large catalytic subunits, designated calpain1, 2 and 3. Calpastatin regulates calpain by inhibiting both the proteolytic activity of calpain and its binding to membranes. Calpastatin exists in two types, tissue type (100-120 kDa) and erythrocyte type (70 kDa), resulting from both alternative splicing and proteolytic processing.
Capsaicin	Irritant chemical responsible for the burning sensation of chilli peppers, activates the ion channel TRPV1 (transient receptor potential V1).
Carney syndrome	Multiple Neoplasia Syndrome characterized by spotty skin pigmentation, cardiac and soft tissue (skin, mucous membrane) myxomas, psammomatous melanotic schwannomas and endocrine tumors including Cushing's syndrome from nodular adrenocortical hyperplasia, pituitary adenomas (acromegaly or prolactinoma), Sertoli cell tumors and Leydig cell tumors.
Casein kinase i and ii (chi/ii)	Casein Kinases (CKI and II) are serine/threonine protein kinases. Numerous isoforms have been described, most with monomeric

	structure.
Caspases	Caspases are members of the cysteine-aspartic acid protease (caspase) family and are generated by a unique gene. They exist as inactive proenzymes which undergo proteolytic processing at conserved aspartic residues to produce two subunits, large and small, that dimerize to form the active enzyme Their sequential activation plays a central role in the execution-phase of cell apoptosis. Caspase 3 cleaves and activates caspases 6, 7 and 9 (executor caspases), and the protein itself is processed by caspases 4, 8, 9 and 10 (initiator caspases).
Cbc^{vhl} complex	An SCF-related complex of elongin B, elongin C, cullin-2 and the RING-finger protein Rbx1/Roc-1.The substrate recognizing subunit pVHL binds to the elongin B/C complex through a motif known as the Socs box. It is believed that the von Hippel–Lindau cancer syndrome is a direct consequence of a loss of cellular CBC^{VHL}-mediated ubiquitylation activity.
Cdk inhibitor p21	It binds to and inhibits the activity of cyclin-CDK2 or -CDK4 complexes, and thus functions as a regulator of cell cycle progression at G1. The expression of the coding gene is tightly controlled by the tumor suppressor protein p53, through which this protein mediates the p53-dependent cell cycle G1 phase arrest in response to a variety of stress stimuli. This protein can interact with proliferating cell nuclear antigen (PCNA), a DNA polymerase accessory factor, and plays a regulatory role in S phase DNA replication and DNA damage repair.
Cdk inhibitor $p27^{kip1}$	It binds to and prevents the activation of cyclin E-CDK2 or cyclin D-CDK4 complexes, and thus controls the cell cycle progression at G1. The degradation of this protein, which is triggered by its CDK dependent phosphorylation and subsequent ubiquitination by SCF complexes, is required for the cellular transition from quiescence to the proliferative state.
Central sensitisation	The phenomenon of synaptic facilitation in the dorsal horn of the spinal cord, which occurs following tissue damage or nerve injury, whereby activation of spinal cord neurons can be elicited by a weaker sensory signal than in normal animals. This sensitisation has some characteristics of synaptic plasticity associated with learning and memory (for example, hippocampal long-term potentiation).
Centrosome	A structure adjacent to the nucleus formed by a pair of centrioles surrounded by amorphous pericentriolar material, serving as a microtubule-organizing center.
CFTR	See: Cystic fibrosis transmembrane conductance regulator.
Chaperones	Proteins whose function is to assist other proteins in achieving proper folding. Many chaperones are heat shock proteins, that is,

	proteins expressed in response to elevated temperatures or other cellular stresses. Other chaperones are involved in folding newly made proteins as they are extruded from the ribosome. Although most newly synthesized proteins can fold in absence of chaperones, a minority strictly requires them.
Chaperonin	A subclass of chaperones, forming cylindrical stacked complexes with an inner cavity, which assist the folding of nascent, non-native polypeptides into their native, functional state.
Charcot-Marie-Tooth disease	A group of peripheral neuropathies either affecting primarily the axons (type II) or the Schwann cells (type I).
Chemical synapse	Specialized junction through which neurons signal to one another and to non-neuronal cells such as muscles or glands.
Chemokines	Class of pro-inflammatory cytokines that have the ability to attract and activate leukocytes.
Chip (carboxyl terminus of hsp70-interacting protein):	A co-chaperone that negatively regulates the ATPase and chaperone activities of Hsc70. It has U box-dependent ubiquitin ligase activity that targets chaperone substrates for proteasome-dependent degradation.
Cht-l	Chymotrypsin-like activity; proteasomal activity cleaving after hydrophobic residues and associated to the β5 subunit.
Cks1 (cdc2-associated protein)	Cofactor for Skp2; induces allosteric alterations in Skp2 molecule, allowing it to bind phosphorylated substrate
Clpap (clpxp)	A major multicomponent protease of Escherichia coli, consists of a proteolytic component, ClpP, in association with an ATP-hydrolyzing, chaperone component, ClpA(X).
Cluster of differentiation (cd)	Molecules are cell surface molecules recognized by specific sets of antibodies. The CD nomenclature is established during International Workshops and Conferences on Human Leukocyte Differentiation Antigens (HLDA). This system was meant to classify many monoclonal antibodies, generated by different laboratories, against various cell surface molecules on leukocytes. Around 300 CD molecules have been identified.
Combinatorial chemistry	A technology for synthesizing and characterizing collections of chemical compounds (libraries) and the screening of those libraries for compounds with useful properties; the modern approach to drug discovery.
Conformational diseases	Conformational diseases are diseases where cellular functions are compromised because of misfolded proteins. The conceptional framework of conformational diseases is found in the cellular protein quality control systems which in the normal and young cell eliminate misfolded proteins. Misfolding may occur in proteins with an intrinsic ability to aggregate and in oxidatively damaged proteins, which accumulate by ageing. If the protein quality control systems are not sufficiently efficient cell toxic protein complexes

	may accumulate. This pathogenesis is a major contributing factor in the development of late onset neurodegenerative disorders.
Consanguineous	Related by blood, usually referring to marriage between close relatives.
Constitutive proteasome	20S proteasome whose active subunits are β1, β5 and β2, in contrast to the immunoproteasome.
Corticobasal degeneration (cbd)	A slowly progressive disorder characterized by neurodegenerative changes of certain brain regions, including the cerebral cortex (particularly the frontal and parietal lobes) and parts of the basal ganglia. Most patients initially develop symptoms in their 60s or 70s. Primary findings may include stiffness (rigidity); slowness of movement (bradykinesia); loss of the ability to coordinate and execute certain purposeful movements of the arms or legs (limb apraxia); the sensation that a limb is not one's own ('alien limb phenomenon'); and other sensory abnormalities. Affected individuals may also develop slurred, labored speech (dysarthria); dystonia; and irregular, involuntary, 'shock-like' contractions of certain muscle groups, particularly of the hands and forearms, that may be provoked or aggravated by voluntary movement and certain external stimuli (action and reflex myoclonus).
Covalently bound inhibitors of the 20s proteasome	Peptide or non-peptide – based small molecules bearing diverse functional groups that bind covalently to the hydroxyl group of the N-terminal threonine-1 residue of each catalytic β-subunit; inhibit the activity either reversibly or irreversibly; commonly used to explore the role for the proteasome in physiology and pathology; exhibit anti-cancer and anti-inflammatory activities.
Cowden syndrome	A hereditary predisposition to tumors: hamartomas of the skin, mucous membranes, breast and thyroid that is caused by PTEN mutations
Creb (camp response element-binding protein)	CREB binds to cAMP-responsive gene promoters that have in common an 8-base enhancer known as the cAMP-response element (CRE). Cyclic AMP (cAMP) second messenger pathways provide a chief means by which cellular growth, differentiation, and function can be influenced by extracellular signals.
Cyclins	These function as positive regulatory subunits of cyclindependent kinases (CDKs). Cyclin–CDK complexes are usually activated at specific points during the cell cycle and have a specific set of substrates.
Cystic fibrosis transmembrane conductance regulator (Cftr)	A multispanning transmembrane chloride ion channel and regulator of other transporters; mutations cause cystic fibrosis.
Cytokine	Small proteins or biological factors that are released by cells and have specific effects on cell-cell interaction, communication, and

	behavior of other cells.
Deadly seven	A zebrafish homologue of the Drosophila notch gene.
Degradasome	A coupled multiprotein system that physically links the 26S proteasome and one or more ubiquitin receptors with components of the ubiquitin conjugation system and other ancillary factors.
Degron	Amino acid sequence, conformational determinant or chemically modified protein structure that confers metabolic instability to proteins and acts as a degradation signal. Examples of degrons include the cyclin destruction box and destabilizing N-terminal amino acids.
Deleted in colorectal carcinoma	One of the receptors for the axon guidance cue Netrin-1 identified for its potential role as a tumour suppressor gene.
Delta	A neurogenic gene first indentified in Drosophila which when mutated leads to an excess of neurones differentiating. Delta is a transmembrane protein which interacts with the Notch protein medeating the process of lateral inhibition.
Deltex	Identified in Drosophila as a positive regulator of the Notch signalling pathway. Deltex contains a RING-H2 domain at the c-terminus and two copies of a WWE protein-protein interaction domain indicating it may be an E3 ubiquitin ligase.
Demyelinating neuropathy	Peripheral nerve disease associated with the loss or destruction of myelin from Schwann cells.
Demyelination	Loss of the myelin sheath surrounding myelinated axons, which occurs as a result of disease or damage.
Dendrites	Projections of a neuron (usually branched) that act to conduct the electrical stimulation received from other cells to and from the cell body (from Greek dendron - tree).
Dentatorubral Pallidoluysian Atrophy	This is a rare neurodegenerative disease reported mostly in Japan. It is characterised by epilepsy, chorea and ataxia. It is caused by the expansion of a CAG nucleotide repeat in a gene on chromosome 12. Like Huntington's disease the onset of the symptoms are earlier, and the disorder more severe, if the defective gene is inherited paternally. Early onset and severe symptoms are more marked the longer the CAG repeat.
Deubiquitinating enzymes (dubs)	Multiple cellular enzymes able to cleave the peptide and isopeptide (isopeptidases) bonds formed between ubiquitin molecules or between ubiquitin and the substrate protein. DUBs perform main three functions: 1) they are necessary components of the 26S proteasome, removing polyubiquitin chain before the substrate is degraded; 2) they are required for the generation of free ubiquitin from the products of the ubiquitin-fusion genes; 3) they edit and rescue ubiquitinated substrates, opposing the action of the E1-E2-E3 cascade.
Dislocation	See: retrotranslocation.

Dlk-1	A mitogen activated kinase kinase kinase which in *C. elegans* is regulated via ubiquitination and regulates synaptogenesis downstream of RPM-1, the C elegans homologue of the Drosophila highwire gene.
Dopamine:	An endogenous catecholamine and major transmitter in the extrapyramidal system of the brain important in regulating movement. In the synthesis of catecholamines from tyrosine, it is the immediate precursor to norepinephrine and epinephrine.
Dorsal horn	The area of the spinal cord where the majority of sensory afferents terminate, comprising the laminae I-VI.
Dysarthria	Speech disorder resulting from the inability to properly control the muscles of the mouth.
E1	Ubiquitin-activating enzyme (UBA); an enzyme that activates ubiquitin by forming a ubiquitin–E1 thiol ester bond, first step in the ubiquitination cascade. There are two isoforms of the E1 in humans, E1A and E1B.
E2	Ubiquitin conjugating enzyme (UBC); an enzyme that conjugates ubiquitin by transferring the activated ubiquitin from an E1 and forming an ubiquitin–E2 thiol ester bond; it interacts with specific E3 enzymes. There are ~20 known E2s in the human genome.
E3	Ubiquitin ligases; enzymes whcih bring upon specificity to the UPS, recognizing the substrate to be ubiquitinated. There are over 700 different E3s in the human genome, which contain either the HECT (Homologous to the E6-AP Carboxyl Terminus) domain or the RING (really interesting new gene) domain (or the closely related U-box domain). Ubiquitination can occur in two ways: Directly from E2, catalysed by RING domain E3s and via a thiol linkage to the E3 enzyme, catalysed by HECT domain E3s. E3s containing the HECT domain are monomeric, while enzymes containing the RING domain form multisubunit complexes, such as the APC or the SCF.
E4	Ubiquitin chain elongation factor; A specialized ubiquitin ligase that is capable of elongating oligoubiquitinated substartes. Known human E4s include Ufd2A and B.
Ecm29	~200-kDa HEAT-repeat protein that associates with the 26S proteasome. Many species are present in mouse brain ranging from 55 kDa to greater than 250 kDa and are likely to arise by alternative splicing and/or proteolytic processing. Ecm29 has been proposed to function as an adaptor to link the 26S proteasome to endocytic, secretory, transport and protein quality control pathways.
Ectopic neurite outgrowth	A class of mutations identified in C. elegans with defects in the axon outgrowth of specific neurone types.
Electroencephalogra	A recording of the summated electrical potentials of neurons in the

m (eeg)	cerebral cortex using scalp electrodes.
Embolus	A clot formed by platelets or leukocytes that blocks a blood vessel.
Endbulbs	Large swellings, up to 50 μm diameter, that develop terminally on both proximal and distal axon stumps after transection.
Endocytosis	The process by which eukaryotic cells take up material from the outside by invagination of the plasma membrane.
Endothelium	The layer of epithelial cells that lines the blood and lymph vessels of the body.
Ephb2	A member of the erythropoietin producing hepatocellular family of tyrosine kinase receptors playing roles in tunourigenesis and axon guidance.
Epigenetic inheritance	Transmission of phenotypic changes without alteration of the genetic code; includes traits trasmited through the pattern of DNA methylation and through the pattern of posttranslational modification of histones.
Epilepsy	A brain disease characterized by the presence of recurrent and spontaneous, unprovoked seizures
Erad (endoplasmic reticulum-associated degradation)	The process by which lumenal and transmembrane proteins present in the endoplasmic reticulum are degraded; not all ERAD depends on the UPS, and not all ERAD involves retrotranslocation from the ER to the cytosol, however the term ERAD is often used to describe the retrotranslocation of substrates from the ER followed by their ubiquitination and UPS-dependent degradation. ERAD is a quality control mechanism, which prevents the accumulation of misfolded proteins in the ER or their secretion to the extracellular space, it therefore counteracts ER stress and complements the unfolded protein response. Depending on whether the misfolded domain is localized within the lumen, the membrane or the cytosol, the ERAD pathway can be subdivided into ERAD-L, ERAD-M and ERAD-C, with different sets of factors required for each pathway. All subdivisions of ERAD converge on the cytosolic site of the ER membrane, where substrates are delivered to the 26S proteasome, often with the assistance of the VCP-Ufd1-Npl4 complex.
Erythropoietin	A renal hormone that is induced by anaemia and that activates haemoglobin synthesis by bone marrow red-cell precursors.
Esrom	The zebrafish homologue of the Drosophila highwire gene. Esrom functions as an E3 ubiquitin ligase and is required for the topographic mapping of zebrafish RGCs in the optic tectum.
Etiology	The cause or origin of disease.
Familial cylindromatosis	See. Turban tumor syndrome.
familial	An autosomal dominantly inherited dementia, histologically

encephalopathy with neuroserpin inclusion bodies (fenib),	characterized by unique neuronal inclusion bodies, and biochemically by polymers of the neuron-specific serpin, neuroserpin.
Fat facets	A member of the Drosophila deubiquitinating enzyme family which may antagonize ubiquitin-dependent mechanisms.
Frazzled	The Drosophila homologue of the netrin receptor DCC.
Frontotemporal dementia with parkinsonism linked to chromosome 17 (ftdp-17)	Hereditary Frontotemporal Dementia with Parkinsonism-17 (FTDP-17) is a progressive dementia that can present with a variety of clinical features, including behavioral and cognitive changes, psychiatric symptoms, language disturbances, and/or motor dysfunction. Onset of these symptoms typically occurs between 40 and 60 years of age. In all these diseases, the symptoms observed were related to mutation in the tau gene.
Gamma γ-Interferon	Immunomodulatory cytokine.
Ganglion mother cell	During neurogenesis in insects neuroblats undergo eight waves of mitosis giving rise to progeny known as the ganglion mother cell. Each ganglion mother cell performs one equal cell division yielding two neurones.
Genomic imprinting	A phenomenon of epigenetic inheritance in which a gene's expression pattern is dependent on the parent-of-origin.
Giant axonal neuropathy (GAN)	GAN is a rare hereditary motor and sensory neuropathy (HSMN) that severely affects the central nervous system. The first symptoms appear in early childhood. This disorder is characterized by abnormalities in the peripheral and central nervous systems including low muscle tone (hypotonia), muscle weakness, decreased reflexes, impaired muscle coordination (ataxia), seizures and mental retardation. Pale, tightly curled hair is frequently seen in those affected. Giant axonal neuropathy follows autosomal recessive genetic inheritance.
Gigaxonin	Gigaxonin controls protein degradation, and is essential for neuronal function and survival. Gigaxonin is now known as a ubiquitin scaffolding protein that controls MAP1B-LC degradation. Mutations in the GAN gene, which encodes the ubiquitously expressed protein gigaxonin, results in a sensory and motor neuropathy called Giant Axonal Neuropathy (GAN). Features of GAN include axonal degeneration. See Giant axonal neuropathy.
Glass bottomed boat	The Drosophila homologue of bone morphogenetic protein.
Glioblastoma	It is the most frequent astrocytic gliomas and the most malignant of neuroepithelial tumors. It can be primary, arising as such since the beginning, or secondary by malignant transformation of a previous astrocytoma.
Glucocorticoid	A steroid hormone synthesized and secreted by specialized cells in the adrenal cortex that exerts wide-ranging effects on nearly every

	tissue in the body including the brain.
Glucocorticoid receptor	The receptor for glucocorticoid hormones (e.g. cortisol) that is a member of the nuclear receptor superfamily.
Growth cone	The tip of a developing axon or dendrite with long thin filopodia and lamellipodia responsible for sensing the environment and guiding the axon or dendrite to its targeted during development or regeneration.
Gsk3 α and β glycogen synthase kinase3 alpha and beta)	GSK-3 is a serine/threonine kinase, which is involved in many cell functions, including, insulin pathway, growth factor and nutrient signaling, cell division, apoptosis, modulation of transcription factors AP-1 and CREB, and specification of cell fate1. Phosphorylation of GSK-3β on serine 9 results in its inactivation. GSK-3β has also been shown to phosphorylate tau, the major component of neurofibrillary lesions of Alzheimer's disease
Guidance cue	A signal which guides the growth cones of axons or denrites to their correct target during neural development. Examples of guidance cues include the netrin and semaphorin families.
Hcdc4/fbw/archipelago/ago.	F-box protein with 7 tandem tryptophan-aspartic acid repeats. It binds directly to cyclin E and is thought to target it for ubiquitin-mediated degradation.
Hdj-1/Hdj-2	Hdj-1 and Hdj-2 are members of the Hsp40 family of co-chaperones that utilize a conserved J-domain to regulate the ATPase activity of Hsp70.
HEAT repeat	Named after huntingtin, eukaryotic elongation factor 3, the PR65/A subunit of protein phosphatase 2A, and the target of rapamycin (TOR) lipid kinase, HEAT repeats are α-helical domains composed of roughly 50 amino acid residues which pack together to form elongated superhelices or solenoids. Canonical HEAT repeats consist of two helices, which form helical hairpins that stack upon one another into a single domain with a continuous hydrophobic core. These domains are found in a wide variety of proteins of differing activities that function as scaffold, anchoring or adaptor proteins.
Heat shock protein	Heat shock proteins (Hsps) are a group of molecular chaperones, which are normally up-regulated when a cell undergoes various type of environmental stresses such as heat, cold or oxygen deprivation.
Hect	A protein domain homologous to the E6-assopciated protein (E6AP) C terminus characteristic of this family of ubiquitin ligases.
Hereditary inclusion body myopathy (h-IBM) associated with Paget disease of bone (PDB) and	IBMPFD is a rare, complex and ultimately lethal, autosomal dominant disorder (MIM 605382). IBMPFD features adult-onset proximal and distal muscle weakness (clinically resembling limb girdle muscular dystrophy), early-onset PDB in most cases, and premature FTD. Mutations in the valosin-containing protein (VCP)

frontotemporal dementia (FTD) (IBMPFD)	on chromosome 9p13-p12 were recently found to be associated with IBMPFD.
High throughput screening (hts)	The process in which thousands of compounds are screened against a known or unknown target and the ones exhibiting the biggest positive effect are taken on for more detailed analysis; can be performed using microwell-or cell array-based assays or using pooled libraries.
Highwire	A potential Drosophila RING-H2 E3 ubiquitin ligase which negatively regulates synaptogenesis.
Hippocampus	A part of the brain located inside the temporal lobe, forming a part of the limbic system and plays a part in memory and spatial navigation. The name derives from its curved shape in coronal sections of the brain, which to some resembles a seahorse (Greek: hippokampos).
Hrs (hepatocyte growth factor-regulated tyrosine kinase substrate)	Hrs is localised to early endosomes in a manner that requires phosphoinositide 3-kinase (PI 3-kinase) activity.
Hsp100	Heat shock protein 100 (Hsp100) chaperones are members of the AAA+ protein family (adenosine triphosphatases with diverse activities) that share a common ATPase domain and form large ring-shaped structures. In yeast, Hsp104, the best-characterized Hsp100, regulates protein aggregation, disaggregation and thermotolerance, but no mammalian homologue has been identified so far.
Hsp40	Hsp40 co-chaperones bind Hsp70 through a conserved J-domain and stimulate ATP hydrolysis, resulting in a conformational switch that closes the substrate-binding pocket of Hsp70 and facilitates the capture of non-native protein substrates. Hsp40s also bind protein substrates and target these substrates to Hsp70, enhancing the efficiency of the Hsp70/Hsp40 refolding cycle. Higher eukaryotes have many Hsp40 family members,whose differential expression or localization might regulate the substrate specificity of conserved Hsp70 family members.
Hsp60	Hsp60 chaperones are heptameric complexes of identical subunits stacked back to back in a double-ring structure that contains a large central cavity in which protein folding is thought to occur. In eukaryotes, Hsp60 family members (also called Group I chaperonins) are found in the mitochondria, and cooperate with a cofactor of the Hsp10 family. A second class of chaperonins (Group II chaperonins) is found in the eukaryotic cytosol but has no HSP10 cofactor. The best-characterized Group II chaperonin is TRiC, which comprises eight subunits per ring encoded by

	different genes. TRiC is thought to be crucial for the folding of actin and tubulin in the eukaryotic cytosol.
Hsp70	Hsp70 chaperones (with HSP40s, their co-chaperones) assist in the stabilization and folding of many substrates and are found in most cellular compartments. In humans, 11 genes that encode Hsp70 family members have been identified, including the constitutive cytosolic member heat shock cognate 70 (HSC70), the stress-induced cytosolic HSP70, the endoplasmic reticulum-localized glucose-regulated protein 78 (GRP78) and the mitochondrial GRP75. All Hsp70 proteins have a conserved amino-terminal ATPase domain that binds and hydrolyses ATP, and a carboxy (C)-terminal substrate-binding domain.
Hsp90	Hsp90 chaperones are an essential component of the eukaryotic cytosol, where they stabilize misfolded proteins and regulate the activity of various signalling proteins, including steroid hormone receptors, tyrosine kinases, nitric oxide synthase and calcineurin.
Huntingtin (htt)	The polyglutamine containing protein associated with Huntington's disease.
Huntington's disease (hd)	An autosomal, dominantly inherited disorder characterized by the onset of progressive chorea (involuntary, forcible, rapid, jerky movements), dementia, and ataxia. Huntington's disease is a polyglutamine tract disorder.
Hybrid proteasomes	Proteasomes composed of a 20S proteasome capped on one end by a 19S cap (PA700) and on the other end by 11S cap (PA28).
Hyperalgesia:	Heightened sensitivity to noxious stimuli which can be a short-term effect following tissue damage, or can be chronic, as occurs in neuropathic pain.
Hypothalamic-Pituitary-Adrenal (HPA) axis	The hierarchy of stress hormones that serve to ultimately regulate the synthesis and secretion of glucocorticoids from cells in the adrenal cortex. For example, neuropeptides (e.g. corticotrophin releasing factor or CRF) secreted from specialized neuroendocrine cells of the hypothalamus stimulate the secretion of hormones from specialized cells of the anterior pituitary (e.g. adrenocorticotropic hormone or ACTH), which ultimate regulate hormone (e.g. cortisol) synthesis and secretion from cells in the adrenal cortex. A negative feedback loop operates at all levels of this axis driven by the secreted hormones.
Iap (inhibitor of apoptosis proteins)	All contain one or more baculoviral IAP repeat motifs involved in mediating protein-protein interactions. Many IAPs also possess a RING domain which enables recruitment of ubiquuitin-conjugating enzymes and catalyses the transfer of ubiquitin onto target proteins. IAP protein levels can themselves be regulated by ubiquitin-mediated proteolysis.
Icer (inducible camp	A splice variant of CREM (multiexonic gene that encodes both

early repressor)	activators and antagonists of cAMP-inducible transcription) that is induced by activation of the adenylyl cyclase signal transduction pathway. ICER serves as a dominant-negative repressor of cAMP-induced transcription.
Idiopathic epilepsy	Epilepsy with no known cause.
Immediate-early gene	A gene that is transcribed rapidly and transiently in response to cellular stimulation.
Immunoproteasome	20S proteasome assembling the three catalytically-active subunits β1i, β5i and β2i which replace their constitutive homologues under the influence of γ-interferon.
Inclusion bodies	Cellular structures found inside neurons that are composed of aggregated proteins, including amyloid fibrils, molecular chaperones and components of the UPS. Recent studies indicate that the formation of inclusion bodies correlates with neuronal survival and is a protective response.
Infarct	An area of tissue death due to a local lack of oxygen.
Inflammation	A localized protective reaction of tissues to irritation, injury or infection haracterized by heat, pain, redness, swelling and sometimes loss of function.
Inflammatory pain:	Pain arising from tissue damage, or release of inflammatory agents, which can be short-term or chronic.
Interictal	The interval between seizures.
Intrathecal	The fluid-filled space between the spinal cord and surrounding dural membrane.
Ionophoresis, ionophoretic	In electrophysiological studies, a method of drug application close to the recording site in nervous tissue. The process involves ejection of an ionised pharmacological agent from the tip of a glass electrode using a small electric current arranged in opposite polarity.
Ischemia	A low oxygen state usually due to obstruction of the arterial blood supply or inadequate blood flow leading to hypoxia in the tissue.
Isoelectric focusing	An immunoblotting technique used for the detection of immunoglobuline synthesis inside the central nervous system (intrathecal antibody synthesis).
isopeptidases	See: deubiquitinating enzymes.
Iκb-kinase (ikk)	The 700–900-kDa IκB-kinase (IKK) complex includes the catalytic subunits IKKα and IKKβ and the regulatory subunit IKKγNEMO. Both catalytic substrates are involved in the activation of NF-κB transcription factors, but they do so by distinct mechanisms and substrates.As shown by genetic studies, IKKβ is essential for inducible IκB phosphorylation and degradation.
J2 prostaglandins	A group of potent hormone-like lipid compounds that are derived from arachidonic acid, contain 20 carbon atoms including a five-

	carbon ring, and modulate inflammation.
Jamm	Jab1/MPN/Mov34 proteases are metallo-enzymes that have JAMM or MPN+ metal-binding domain for deubiquitinating activity.
Jnk	The c-Jun amino-terminal kinase belongs to the group of mitogen-activated protein kinases (MAPKs) and is activated in mammalian cells by environmental stress, pro-inflammatory cytokines and mitogenic stimuli. JNK regulates the activities of many transcription factors, and is required for the regulation of inflammatory responses, cell proliferation and apoptosis.
Josephin domain	The Josephin domain is an eukaryotic protein module of about 180 residues, which occurs in stand-alone form in Josephin-like proteins, and as an amino-terminal domain associated with two or three copies of the ubiquitin-interacting motif (UIM) in ataxin 3-like proteins. It is a mainly α helical cysteine-protease domain predicted to be active against ubiquitin chains or related substrates. The Josephin domain contains two conserved histidines and one cysteine that is required for the ubiquitin protease activity.
Kennedy Disease	See: (X-linked) Spinal and Bulbar Muscular Atrophy.
Ki antigen	Old name for PA28γ
Lactacystin	Lactacystin is a microbial metabolite originally isolated from Streptomyces that is now widely used as a selective inhibitor of the 20S proteasome
Lateral inhibition	The process during neurogenesis by which neuroblasts inhibit their neighbouring cells from becoming neurones.
Leukocyte	white corpuscles in the blood involved with host defenses.
Leukodystrophy	A disorder of the *white matter* of the brain, the part of the brain that contains myelinated *nerve* fibers. The white matter is white because it is the color of *myelin*, the insulation covering the nerve fibers. (The white matter is as opposed to the *gray matter*, the *cortex* of the brain which contains the nerve cell bodies). The white matter is involved in the conduction of nerve impulses in the brain
Lewy bodies:	intracytoplasmic, eosinophilic, round α Synuclein (αS)-positive inclusions found in neurons. The presence of Lewy bodies is a histological hallmark of Parkinson's disease (PD). They are found typically in the substantia nigra and locus coeruleus but are also seen in the neocortex.
Lid of the 19S cap (PA700)	The lid is a subdivision of the 19S cap (PA700) distal to the 20S proteasome, which attaches to the base through a hinge formed by Rpn10 and is formed by all the subunits of the PA700 with the exception of proteasomal ATP-ases (Rpt1-6), Rpn1 and Rpn2.
Ligand gated ion channel	Transmembrane ion channel that will open and close to allow transport of ions in response to binding of neurotransmitter (or other chemical signal). For example, in neurons, calcium channels open in response to specific stimuli and this entry of ionic calcium

	is important for regulation of many events in neurons.
Ligand of numb-protein X	A Drosophila RING finger containing E3 ubiqitin ligase which ubiqutinates and targets numb for ubiqitin-mediated proteolysis regulating Notch signalling.
Lipid peroxidation	An autocatalytic process in which free radicals attack double bonds in membrane lipids, resulting in structural damage to membranes and the liberation of toxic aldehydes such as 4-hydroxynonenal.
lipid rafts	Membrane microdomains, formed by high concentrations of sphingolipids and cholesterol immersed in a phospholipidrich environment, that are involved in specialized pathways of protein/lipid transport and signalling
Liquid facets	The Drosophila homologue of the vertebrate endocytic protein epsin and is a target of the deubiquitinating enzeme fat facets
Long-term depression	The theory that down-regulation of a post-synaptic receptor will lead to loss of responsiveness to neurotransmitter stimulation, resulting in a depressed response to stimuli in the future
Long-term potentiation (ltp).	An enduring increase in the amplitude of excitatory postsynaptic potentials as a result of high-frequencyb (tetanic) stimulation of afferent pathways. It is measured both as the amplitude of excitatory postsynaptic potentials and as the magnitude of the postsynaptic cell population spike. LTP is most frequently studied in the hippocampus and is often considered to be the cellular basis of learning and memory in vertebrates.
Lysosome	A membrane-bound organelle characterized by a low pH that contains high concentrations of hydrolytic enzymes including multiple proteases, mainly of the cathepsin family.
Machado-Joseph Disease	also called spinocerebellar ataxia type III, is a rare, inherited, ataxia (lack of muscular control) affecting the central nervous system and characterized by the slow degeneration of particular areas of the brain called the hindbrain. Patients with MJD may eventually become crippled and/or paralyzed but their intellect remains intact. The onset of symptoms of MJD varies from early teens to late adulthood. Three forms of Machado-Joseph Disease are recognized: Types MJD-I, MJD-II, and MJD-III. The differences in the types of MJD relate to the age of onset and severity. Earlier onset usually produces more severe symptoms.
Macrophage	phagocytic cell of mammalian tissues that become activated in response to foreign materials or tissue injury and play an important role in killing foreign cells, release of pro-inflammatory substances, and antigen presentation.
Major histocompatibility molecules (MHC)	are cell surface proteins found on most cells of the body. There are three classes of MHC molecules. Class I and II molecules participate in presentation of antigens to T cells. MHC class I molecules typically interact with the cell surface receptor of a type

	of lymphocytes known as killer or cytotoxic T cells, while MHC class II molecules present antigens to helper T cells.
Maps	Microtubule associated proteins. Proteins involved in the polymerisation and the stability of microtubules.
Mdm2 gene	encodes for E3-like ubiquitin-protein ligase, a nuclear phosphoprotein that binds and inhibits transactivation by tumor protein p53, as part of an autoregulatory negative feedback loop. Overexpression of this gene can result in excessive inactivation of tumor protein p53, diminishing its tumor suppressor function. This protein has E3 ubiquitin ligase activity, which targets tumor protein p53 for proteasomal degradation.
Medea	A Drosophila SMAD mediating intracellular signalling downstream of BMP receptors.
MEN-1 (multiple endocrine neoplasia type 1)	autosomal dominant condition that describes the association of the occurrence of tumors involving two or more endocrine glands: parathyroid hyperplasia, pancreatic endocrine tumors, pituitary adenomas
Mendelian disorder	a disease that adheres to single gene inheritance patterns, such as autosomal dominant, autosomal recessive, X-linked dominant, and X-linked recessive
Menin	gene located on 11q13 encodes menin, a tumor suppressor gene that is mutated in MEN-1.
Mesencephalon	(or midbrain) is the middle of three vesicles that arise from the neural tube of developing brain.
Metabotropic glutamate receptors	A group of cell surface receptors that bind the excitatory neurotransmitter glutamate (while glutamate is also an amino acid used for making proteins it can also act as neurotransmitter in the nervous system. Generally, these receptors are seven-pass transmembrane G-protein coupled receptors, and the binding of ligand to the receptor results in activation of a biochemical signalling pathway inside the neuron.
Microarray	DNA Microarrays, commonly known as gene chips, are small, solid supports (glass slides or silicon chips) onto which the sequences from thousands of genes are immobilized at fixed locations. Microarrays may be used to assay the gene expression of thousand of genes simultaneously within a single sample or to compare gene expression in two different tissue samples or cell types.
Microtubule binding domain of tau protein	The Microtubule binding domain was given to the C-terminal part of tau protein. This side of the molecule is involved in the binding and the stability of microtubules. It will differ by the incorporation or not of the exon 10 of the tau gene.
Microtubule-affinity regulating kinase	kinase with an apparent molecular mass of 110 kDa that phosphorylates the neuronal MAPs tau and MAP2 and the

(mark)	ubiquitous MAP4 on their homologous KXGS motifs. The kinase caused rapid detachment of all three MAPs from microtubules, resulting in high dynamic instability, and was therefore termed MARK (MAP/microtubule affinity-regulating kinase).
Mind bomb	A zebrafish RNG E3 ubiqutin ligase which regulates Notch-Delta signalling via promoting the ubiquitination and internalization of Delta.
Mineralocorticoid receptor	The receptor for mineralocorticoid hormones (e.g. aldosterone) that is a member of the nuclear receptor superfamily
Mjd (Machado Joseph disease proteases)	Machado Joseph disease proteases are cysteine proteases that have Josephin domain for deubiquitinating activity.
Molecular chaperones	See: chaperones.
Morula	The Drosophila orthologoue of the anaphase promoting complex subunit 2 gene.
Mtoc (microtubule organizing center)	Cellular structure from which the microtubular cytoskeleton radiates towards cell periphery; usually a synonym for the centrosome plus the associated pericentrosomal material, however e.g. in human oocytes MTOC are actually acentrosomal, made only of the pericentrosomal material.
Mts (microtubules)	A type of filamentous protein polymer found in the cytoplasm of eukaryotic cells, polymer of α and β tubulin arranged into a 13 protofilaments forming an empty tubule of ~24 nm of diamteter and varying length from several micrometers to possible millimeters in axons of nerve cells. MTs occurs singly or in pairs, triplets, or bundles. Microtubules is one of the main components of the cytoskeleton, they emanate from the MTOC located close to the nucleus and project to the periphery. In axons, MTs are involved in retrograde and anterograde transport.MTs also form the centrosomes, basal bodies, cilia and flagella as well as the spindle during mitosis and meiosis.
Multivesicular bodies (mbvs)	Late endosomal organelles that form by invagination of the endosomal membrane to form intraluminal vesicles with subsequent fusion of the MVBs with lysosomes.
Mushroom body	The olfactory learning and memory centre in insects.
Myelin:	Multilayed, lipid-rich membrane that wraps nerves to increase the efficiency of signal propagation along axons.
Nedd8 (neddylation)	Nedd8 is a ubiquitin-like small protein modifier. The Nedd8 conjugation process, called neddylation, is similar to ubiquitination. Neddylation utilizes the E1 activating-enzyme complex composed of two subunits, APP-BP1 and UBA3, and the E2 conjugating-enzyme, UBC12. The only known substrates of neddylation are Cullin family proteins (Cul1, Cul2, Cul3, Cul4A,

	Cul4B, and Cul5) which have been shown to be modified by Nedd8 in mammalian cells.
N-end rule	The rule which determines that protein stability depends on specific amino acids present at the N-terminus. While for example Met is characteristic of long lived proteins, N-terminal Arg induces quick ubiquitination and proteasomal degradation of proteins. Destabilizing amino acids at the N-terminus are the "degrons".
Neocortex	a part of the brain of mammals (also birds and reptiles); the top layer of the cerebral hemispheres. Other names neopallium and isocortex.
Nerve growth factor: (NGF)	A naturally occurring molecule in the body which stimulates the growth and differentiation of the sympathetic and certain sensory nerves. NGF is a protein that consists of 3 types of polypeptide chains -- alpha, beta and gamma -- that interact to form the protein. The NGF β chain (NGF β) is solely responsible for the nerve growth stimulating activity of NGF. The NGF β gene is in chromosome band 1p22.
Netrin-1	A secreted protein with homology to the extracellular matrix molecule laminin with attractive and repulsive effects on growing axons.
Neuralized	A Drosohila E3 ubiquitin ligase which regulates the Notch-Delta signalling pathway via the ubiqutination of Delta.
Neuralized homology repeat	A novel protein domain of unknown function identified in the neuralized gene.
Neurodegeneration	Progressive damage or death of neurons causing a gradual decline of bodily functions regulated by the affected parts of the nervous system
Neurofibrillary degeneration (NFD)	neurofibrillary degeneration is the formation of coarse, argentophilic, intracytoplasmic fibres, often in complex tangles within intracranial nerve cells that are undergoing aging. NFD corresponds to the intracellular accumulation of pathological fibrils in the cytosol of neurons. In Alzheimer's disease, NFD may be caused by the abnormal aggregation of tau proteins.
Neurofibrillary tangles (nfts)	Accumulation of twisted protein fragments inside neurons. Neurofibrillary tangles are one of the characteristics structural abnormalities found in the brains of patients with Alzheimer's disease patients. Upon autopsy, the presence of amyloid plaques and neurofibrillary tangles is used to positively diagnose Alzheimer's disease. In Alzheimer's disease and other tauopathies, tangles are mainly formed by abnormally modified tau protein.
Neurofibromatosis	Autosomal dominant disorders associated with deregulated Schwann cell proliferation and are classified as type 1 or type 2. The hallmark of NF1 is neurofibromas, while NF2 is associated with schwannomas.

Neurogenesis	The stage of development during which neuronal precursors cells proliferate to generate neurones.
Neuropathic pain	Chronic pain arising from damage to the nervous system, either to the peripheral nerves or to the central nervous system.
Neuropathy	Any disorder affecting any segment of the peripheral nervous system.
Neuroserpin	In the central nervous system, *neuroserpin* (NSP) is a serpin thought to *regulate t-PA* enzymatic activity.
Neutrophil	a white blood cell with conspicuous cytoplasmic granules (granulocyte) involved in host defense.
NF-κB	nuclear factor-κB; a ubiquitous prosurvival and proinflammatory transcription factor composed of two different subunits of the Rel family. Inactive NF-κB is bound to an inhibitory protein IκB in the cytoplasm, which masks its nuclear localization signal. Upon activation of specific receptors (for example TNF R1), IκBα is phosphorylated, ubiquitinated and degraded by the proteasome releasing active NF-κB, which translocates to the nucleus and induces the expression of specific genes.
NF-κb essential modulator (NEMO)	is a regulatory subunit of the inhibitor of IκB kinase (IKK) complex. It contains multiple coiled-coil motifs and a zinc finger at the COOH-terminal end. NEMO is required for the assembly and activation of the IKK complex in response to a wide range of NF-κB-stimulating signals, including tumor necrosis factor (TNF).
Nissl's bodies	Chromophil substance in the form of granules found in the cell bodies and dendrites of neurons, but is absent from axons. They consist principally of the ribose type of nucleic acid and nucleoprotein and stain strongly with basic aniline dyes. They are concerned with protein synthesis and metabolism; their condition varies with physiological and pathological conditions.
Nmda	N-methyl-D-aspartic acid. Agonist for the NMDA receptor, mimics the effects of glutamate. A synthetic amino acid derivative that is useful in neurochemical research to distinguish between different glutamate receptor subtypes.
Nob1p (Nin One Binding Protein)	Nin One Binding Protein involved in the maturation of 20S proteasome by Ump1
Nociception, nociceptive:	The sensory detection of noxious stimuli, which may result in a sensation of pain.
Noncovalent inhibitors of the 20S proteasome	small molecules that bind reversibly to the substrate binding sites in the active sites located on the β-subunits without modifying the catalytic active N-terminal threonine; generated by combinatorial chemistry or identified by high throughput screening ; exhibit low cytotoxicity against normal cells.
Nondisjunction	failure of chromosomes to segregate during meiosis

Non-dopaminergic neurons:	neurons that use neurotransmitters other than dopamine. Non-dopaminergic neurons affected by PD include the noradrenergic neurons in the locus coeruleus, serotonergic cells in the dorsal raphe, cholinergic cells in the nucleus basalis of Meynert and pyramidal neurons in parts of the hippocampal formation
Normoxic	At or containing a normal level of oxygen.
N-terminal nucleophile hydrolase (Ntn hydrolases)	superfamily of three known enzymes that use the side chain of the amino-terminal residue as the nucleophile in the catalytic attack at the carbonyl carbon. The nucleophile (protor donor) is threonine in the 20S proteasome, serine in penicillin acylase and cysteine in glutamine PRPP amidotransferase.
Ntn-hydrolases	class of enzymes which perform their catalytic activities relaying on the N-terminal aminoacid residue as nucleophile
Nuclear export signal (NE)	is an amino acid sequence used to localize the protein to the cell nucleus through the nuclear pore complex. Usually, this signal consists of a few short sequences of positively charged lysines or arginines.
nuclear factor-κb (nf-κb)	A family of transcription factors important for pro-inflammatory and antiapoptotic responses. They are activated by the phosphorylation and subsequent ubiquitindependent proteolytic degradation of their respective inhibitors, known as inhibitor of κB (IκB). Phosphorylation of IκB occurs through tissuespecific kinases, IκB kinase 1 (IKK1) and IKK2.
Nuclear receptor	A specific family of hormone, vitamin or small metabolite receptors that share a common structural and functional organization including a zinc-finger DNA-binding domain and carboxyl-terminal ligand-binding domain. These receptors are generally localized predominantly in the nucleus but can also be found in the cytoplasm.
Nucleation	A process by which the addition of a small amount of preaggregated protein to a monomeric preparation of the same protein robustly accelerates the assembly of amyloid fibrils.
NZB domain	Also known as NZF. Putative zinc finger domain found at the C-terminus of Npl4, a cofactor of valosin-containing protein, that binds polyubiquitylated substrates. Similar domains are also found in TAK1-binding protein (TAB2), Vps36 and RBCK2. NZB domains appear to bind both Lys-48- and Lys-63-linked ubiquitin chains.
Oligoclonal bands	Immunoglobulins visualized as discrete bands by isoelectric focusing. If the bands are present in the cerebrospinal fluid only and not in the corresponding serum, this is interpreted as a sign of intrathecal immunoglobuline synthesis. Oligoclonal bands in the cerebrospinal fluid are found in different inflammatory diseases affecting the central nervous system, in particular in multiple

	sclerosis
Omuralide	the name for the highly selective proteasome inhibitor β-clasto-lactacystin proposed by Corey group in honor of discovery of the lactacystin by Omura research group; its synthetic analog, MLN-519, is currently under clinical evaluation for the treatment of acute stroke and myocardial infarction.
Open gate proteasome	The 20S core particle of the proteasome is characterized by a central axial channel, which is gated at both ends. Regulatory subunits, such as the 19S particle, or the presence of substrates can modify the size of the gates. The proteasome therefore can assume dynamically both an open and close gate conformation. Mutant proteasomes with deletion of the α3 N-terminal chain are open gate proteasomes, which are able to degrade substrates at a faster rate than the wilde type.
Out (Ovarian tumor proteases)	A Drosophila protein involved in oocyte morphogenesis. OUT are deubiquitinating enzymes that have homology to viral cysteine proteases in its catalytic domain sequence. As the putative catalytic cysteine is replaced by a serine it is not clear whether OTU is an active protease or in inactivated protease homologue.
Oxidative stress	A build-up of free radicals and H_2O_2 resulting in cell dmage and disease.
P element	a small segment of DNA called a transposable element that is capable of moving from one genomic location to another in Drosophila
Pa200	The mammalian ortholog of blm3p (now called blm10p), a ~200-kDa protein composed of numerous HEAT repeats, that binds either one or both ends of the 20S proteasome and activates peptide hydrolysis by the 20S particle in vitro. PA200 is a nuclear protein proposed to link proteasomes to repair mechanisms at DNA double strand breaks.
Pa28	proteasome activator formed by different members of the family of small 28 kDa proteins termed α, β or γ that share significant sequence homology with one another and function as ATP-independent homo- (α/α, γ/γ) or hetero- (α/β) heptameric rings that cap one or both ends of the 20S catalytic core particle. Binding of PA28 oligomers to the 20S proteasome causes gating and activation of the proteasome's peptidase activity. Also known as 11S cap. PA28 γ known formerly as Ki antigen.
PAC1-PAC2 complex	associates with the α-subunits before α-rings are complete; functios as a scaffold for α-ring assembly
Pael-R	A putative G protein-coupled transmembrane polypeptide identified as an intracellular substrate of Parkin.
Pam	Protein associated with Myc, originally identified from a human cDNA library by its interaction with the transcriptional activating

	domain of the c-terminus of Myc. PAM is the human homolgoue of the Drosophila highwire and C.elegans RPN-1 genes.
PAN (proteasome-activating nucleotidase)	The ATPase complex from archaebacteria that is highly homologous to the ATPases of the eukaryotic 19S proteasome-regulatory complex. Quality control: A system for 'proof-reading' that distinguishes native from non-native protein conformations.
Paraneoplastic cerebellar degeneration	Cerebellar dysfunction characterised by loss of balance and coordination, with a subacute onset, that occurs in association with cancer
Paraneoplastic encephalomyelitis	Encephalomyelitis, in particular affecting the brainstem and the limbic system, that is associated with different malignancies
Parkin	Parkin is an E3 ligase in the ubiquitin-proteasome system. Many mutations in parkin have been associated with a familial form of Parkinson's disease termed autosomal recessive juvenile parkinsonism. How loss of function of parkin leads to dopaminergic cell death in this disease is unclear. The prevailing hypothesis is that parkin helps degrade one or more proteins toxic to dopaminergic neurons. Putative substrates of parkin include synphilin-1, CDC-rel1, CDC-rel2, cyclinE, p38 tRNA synthase, Pael-R, synaptotagmin XI, synphilin-1, sp22 and parkin itself. See also Ubiquitin ligase.
Parkinson's disease (PD)	is a neurodegenerative movement disorder, clinically characterized by a resting tremor, rigidity, hypokinesia and postural instability. The neuropathological hallmarks are intraneuronal Lewy bodies and dystrophic neurites (Lewy neurites), which both contain aggregated proteins, such as a-synuclein, ubiquitinated proteins, parkin and Pael-R (a parkin substrate). The loss of dopaminergic neurons in the substantia nigra pars compacta is the major cause of the clinical movement problems but it has been shown that more widespread neuropathology is present in the brains of PD patients, including degeneration of noradrenergic, serotonergic, peptidergic and cholinergic systems, before degeneration of the substantia nigra occurs. The autosomal dominant and recessive forms of PD are caused by mutations in the genes encoding a-synuclein, two ubiquitination enzymes (parkin and UCH-L1), the molecular chaperone DJ-1 and the signaling molecule leucine-rich repeat kinase 2. Moreover, overexpression of a-synuclein by duplication or triplication of the gene can also lead to PD. Similar to other neurodegenerative diseases, PD is primarily a sporadic disorder with a complex etiology.
Parvalbumin immunoreactive neurons	Parvalbumin is a calcium binding protein that belongs to the so called 'EF-hand' family of calcium binding proteins. Other notable members are calbindin-D28K and calretinin.These three proteins have been used as markers for distinct subpopulations of cortical

	interneurons.
Pathognomonic	distinctively characteristic for a particular disease to the point of aiding diagnosis
Paw withdrawal latency:	Behavioural measurement – recording of the latency for an animal to withdraw its paw from a noxious thermal stimulus.
Paw withdrawal threshold: :	Behavioural measurement – recording of the mechanical force or pressure required for an animal to withdraw its paw.
PC domain	Repeat of 35-40 residues found in subunits S1 and S2 of the 26S proteasome and subunit Apc1 of the anaphase-promoting complex or cyclosome. The most highly conserved feature in these repeats is an alternating pattern of large aliphatic residues and glycine or alanine. The variable part of the repeats contains a pattern of hydrophobic and hydrophilic residues with a periodicity of 3.6 typical of amphipathic helices. These repeats are proposed to fold into structures resembling α-helical toroids.
Pelizaeus-Merzbacher disease:	A disorder of the central nervous system (CNS) in which there is loss of myelin, the sheath around the nerves. The disease is clinically characterized by nystagmus (rhythmical oscillation of the eyes), impaired motor development, tremor, progressive spasticity (increased muscle tone), ataxia (wobbliness), choreoathetotic movements, and dysartria (difficulty speaking). Pelizaeus-Merzbacher disease (PMD) in its classical form manifests in infancy or early childhood and progresses to severe spasticity and ataxia. The lifespan may be shortened.PMD is due to mutation in the gene PLP1. This gene is located on the X chromosome in band Xq22. The disease describes an X-linked pattern of inheritance with boys who have the mutation affected with the disease while females with the mutation are carriers. The PLP1 gene encodes proteolipid protein (PLP), the most abundant protein of the myelin sheath in the CNS. The mutation in PLP1 in PMD results in loss of myelin and that, in turn, causes the neurological abnormalities. The severity of myelin loss is dependent on the particular PLP1 mutation and can range from early lethal forms of PMD to a mild disorder known as spastic paraplegia type 2 (SPG2). Among the mutations in the PLP1 gene locus that can cause PMD is a duplication of PLP1 in which the duplicated region may be far away from the original PLP locus in chromosome region Xq22. The PLP1 duplication is almost always present in the mothers of affected boys and usually can be traced to the maternal grandfather.
Penetrance	the frequency of expression of a phenotype for a given genotype
Pgph	Peptidylglutamyl-peptide hydrolysing activity; proteasomal activity cleaving after acidic residues and associated to the β1 subunit Also called ‘caspase-like’ activity or “post-acidic” activity.
Pharmacology	the study of the drugs with respect to their origin, nature,

	properties, and mechanisms of actions and their effects on living tissues and organisms.
Phfs: Paired helical filaments	Filaments found in degenerating neurons in Alzheimer's disease. These filaments are mostly composed of abnormal tau proteins. These filaments are also characteristic of other neuropathological disorders called "tauopathies".
Phosphorylation / Hyperphosphorylation of tau protein	The phosphorylation is the addition of a phosphate group to a compound by an enzyme (e.g., thymidine kinase, tyrosine kinase…). Phosphorylation is an essential step in many cellular processes. Phosphorylation may cause conformational changes in proteins or activate particular enzymes. Concerning tau proteins, its phosphorylation will change its activity in neurons and mostly its capacity to bind and polymerase Microtubules. The hyperphosphorylation of tau is an increase of its normal phosphorylation and an appearance of new phosphorylation sites on tau molecule. The hyperphosphorylation of tau is thought to be involved in its intraneuronal aggregation as described in Alzheimer's disease and other tauopathies.
Phr-1	The mouse homologue of the Drosophila highwire gene, a potential E3 ubiquitin ligase.
PI3K (phosphatidylinositol-3-kinase)	large family of enzymes that catalyse the phosphorylation of inositol-containing lipids, thus transmitting signals from tyrosine kinases and G-protein coupled receptors. PI3K pathway regulates proliferation, growth, apoptosis and cytoskeletal rearrangement.
Pick's Disease (pid)	Pick's disease is a dementing illness associated with deterioration of the frontal and temporal lobes of the brain. Symptoms may include a decline in social behavior (including disinhibition, tactlessness, and breaches of interpersonal etiquette), emotional blunting, apathy, changes in eating habits (including increased appetite, weight gain, and increased preference for sweets), attention problems, decreased insight, speech and language problems (including reduced speech ability, repetition of phrases heard, reduced use of nouns, difficulty naming objects, loss of word meaning, diminished writing ability, and mutism), and difficulty recognizing faces. Though Alzheimer's disease and other forms of dementia can sometimes cause similar symptoms, Pick's disease is more likely to cause certain deficits in behavior and speech (such as disinhibition or loss of nouns), while memory and visuospatial function (which are frequently affected by Alzheimer's Disease) tend to be relatively spared. Also, the onset of Pick's Disease (usually between the ages of 45 and 65) is earlier than is normally seen in Alzheimer's disease.
Piriform cortex	part of paleopallium that together with olfactory cortex relates to the perception of smells; present in amphibians, reptiles, birds and

	mammals
Polyglutamine (polyq) diseases	Group of neurodegenerative diseases that includes Huntington's, Spinal Bulbar Muscular Atrophy, Dentatorubral Pallidoluysian Atrophy and a number of Spinocerebellar Ataxias (1, 2, 3, 6, 7, 17). These diseases are caused by mutation within the coding regions of several unrelated proteins resulting in the expansion of polyglutamine tracts within these proteins. A primary hallmark of polyQ diseases is the presence of intracellular, often nuclear, polyQ inclusion bodies deposited within the diseased brains of polyQ patients. PolyQ inclusions contained the polyQ-expanded protein along with chaperones and components of the ubiquitin-proteasome system.
Polyglutamine tract	A repeated sequence of glutamine residues within a protein.
Postsynaptic density	The intracellular area immediately beneath the postsynaptic membrane of a synapse, which contains a high density of specialised proteins, particularly receptor binding proteins associated with synaptic function.
Preselenins	Presenilins are essential components of γ-secretase, a protease complex catalyzing intramembrane proteolysis of various type I membrane proteins, including the amyloid precursor protein and the Notch receptor. Important in the pathogenesis of Alzheimer's disease and in normal development and expressed in many tissues, presenilins (PS1 and PS2) are proteins with multiple transmembrane domains and are processed into N-terminal and C-terminal fragments (NTF and CTF). Both proteins are predominantly located within the endoplasmic reticulum (ER) and early Golgi apparatus. The exact functions associated with PS proteins have not been fully characterized yet.
Prevalence	the total number of cases of a disease in a given population at a given time.
Primary afferents:	Peripheral sensory neurons, which are activated by stimuli to the periphery and synapse in the dorsal horn of the spinal cord.
Prion	This term is an abbreviation for "proteinaceous infectious particle", the putative infectious agent of prion diseases according to the prion hypothesis.
Prion diseases	A group of neurodegenerative diseases that affect humans and animals, also known as transmissible spongiform encephalopathies. An example of a human prion disease is Creutzfeldt-Jakob disease, examples of animal prion diseases are scrapie in sheep and bovine spongiform encephalopathy in cattle.
Prion protein	A protein that plays a key role in prion diseases. The normal form of the prion protein is called PrP^C (C: cellular). In diseased tissue PrP^C has been post-translationally modified into a disease associated form, often called PrP^{Sc} (Sc: scrapie), which is thought

	to be part of the infectious unit of prion diseases, the prion.
PRKAR1A (protein kinase A regulatory subunit 1(α)	381-amino acid protein. The holoenzyme of PKA, a tetramer consisting of 2 regulatory and 2 catalytic subunits, is inactive in the absence of cAMP. Activation occurs when 2 cAMP molecules bind to each regulatory subunit, eliciting a reversible conformational change that releases active catalytic subunits. Germline mutations in PRKAR1A, an apparent tumor-suppressor gene, are responsible for the Carney complex phenotype.
Progressive supranuclear palsy (PSP)	Progressive supranuclear palsy (PSP) (or the Steele-Richardson-Olszewski syndrome, after the Canadian physicians who described it in 1963) is a rare degenerative disorder involving the gradual deterioration and death of selected neurons in the brain. Typical effects are problems with control of gait and balance, and an inability to aim the eyes properly, especially in the vertical directions (downward gaze palsy). Other symptoms may be alterations of mood and behavior, depression and apathy as well as mild dementia. There is currently no effective treatment for the disease.
Projection domain of tau protein	The projection domain was given to the N-terminal part of Tau proteins.
Proline directed protein kinases (PDPK)	The PDPK are kinases, which phosphorylate serine and threonine only if these amino acids are followed by a praline.
Proteasome activator	Group of molecules that include the 19S subunit of the 26S proteasome, PA28 and PA200 (blm10p) which bind one or both ends of the 20S proteasome and activate its catalytic activity in an ATP-dependent (19S subunit) or ATP-independent manner (PA28 and PA200).
Proteasome cleavage prediction algorithm	Bioinformatics tools based on experimental data on proteasome degradation and on mathematical models. These algorithms are useful to predict the position of proteasome cleavage (also known as cleavage site) in a given amino acid sequence. These algorithms are used in immunology to predict intracellularly generated antigenic peptides presented on MHC class I molecules in the context of antigen presentation.
Proteasome inhibitors	These are classified into four groups: lactacystin and β-lactone derivates, vinyl sulfones, peptide aldehydes and peptide boronates. The aldehyde and boronate inhibitors are reversible and more amenable to clinical use.
Proteasomes	A term used to describe either 20S proteasomes (constitutive or immunoproteasomes) or 26S proteasomes.
Protein activator	a binding protein that regulates positively another protein
Protein aggregate	An abnormal protein assembly that results from the cohesion of two or more misfolded monomeric proteins. Protein aggregates that

	form amyloid fibrils are often resistant to solubilization with ionic detergents after boiling.
Protein aggregation	the clumping of proteins, in particular proteins which are misfolded or which contain multiple β-sheets. Protein aggregation depends mostly on hydrophobic interactions.
Protein Kinase A,B,C and N (PKA/B/C and N)	PKA: In cell biology, cAMP-dependent protein kinase (cAPK), also known as protein kinase A (PKA, EC 2.7.1.37), refers to a family of enzymes whose activity is dependent on the level of cyclic AMP (cAMP) in the cell. Each PKA is a holoenzyme that consists of two regulatory and two catalytic subunits. Under low levels of cAMP, the holoenzyme remains intact and is catalytically inactive. When the concentration of cAMP rises (e.g. activation of adenylate cyclases by certain G protein-coupled receptors, inhibition of phosphodiesterases which degrade cAMP), cAMP binds to the two binding sites on the regulatory subunits, which then undergo a conformational change that releases the catalytic subunits. The free catalytic subunits can then catalyze the transfer of ATP terminal phosphates to protein substrates at serine, or threonine residues. This phosphorylation usually results in a change in activity of the substrate. Since PKAs are present in a variety of cells and act on different substrates, PKA and cAMP regulation are involved in many different pathways. In addition, the effects of PKA phosphylation are generally transient because protein phosphatases quickly dephosphorylate PKA targets. PKB: Akt1, also known as "Akt" or protein kinase B (PKB) is an important molecule in mammalian cellular signaling. There are three genes in the "Akt family": Akt1, Akt2, and Akt3. Akt1 is involved in cellular survival pathways, by inhibiting apoptotic processes. Akt1 is also able to induce protein synthesis pathways, and is therefore a key signaling protein in the cellular pathways that lead to skeletal muscle hypertrophy, and general tissue growth. Since it can block apoptosis, and thereby promote cell survival, Akt1 has been implicated as a major factor in many types of cancer. Akt2 is an important signaling molecule in the Insulin signaling pathway. It's required to induce glucose transport. The role of Akt3 is less clear, though it appears to be predominantly expressed in brain. PKC: Protein kinase C ('PKC', EC 2.7.1.37) is actually a family of protein kinases consisting of ~10 isozymes. They are divided into three subfamilies: conventional (or classical), novel, and atypical based on their second messenger requirements. Conventional (c) PKCs contain the isoforms I, II, and III. These require Ca2+, diacylglycerol (DAG), and a phospholipid such as phosphatidylcholine for activation. Novel (n)PKCs isoforms require DAG, but do not require Ca2+ for activation. Thus,

	conventional and novel PKCs are activated through the same signal transduction pathway as phospholipase C. On the other hand, Atypical (a)PKCs require neither Ca2+ nor diacylglycerol for activation.
Protein Phosphatase 1, 2A and 2B (PP1/2A and 2B)	Protein phosphatases are enzymes that remove phosphate groups that have been attached to amino acid residues of proteins by protein kinases. Whereas a kinase enzymatically adds a phosphate to a protein, a phosphatase's purpose is phosphate removal. It should be noted that phosphate addition and removal do not necessarily correspond to enzyme activation or inhibition, and that several enzymes have separate phosphorylation sites for activating or inhibiting functional regulation. CDK, for example can be either activated or deactivated depending on the speicific amino acid residue being phosphorylated. The phosphates are important in signal transduction by regulating the proteins they are attached to. To reverse the regulatory effect, the phosphate has to be removed. This occurs on its own by hydrolysis or is mediated by protein phosphatases. Serine and threonine phosphates are stable under physiological conditions, so a phosphatase has to remove the phosphate to reverse the regulation. There are four known groups: PP1, PP2A, PP2B (AKA calcineurin) and PP2C. The first three have sequence homology in the catalytic domain, but differ in substrate specifity. Ser/Thr-specific protein phosphatases are regulated by their location within the cell and by specific inhibitor proteins.
Proteolysis inducers	small molecules designated to recruit a disease-promoting protein for ubiquitination and degradation by the 20S proteasome; include proteolysis-targeting chimeric molecules (protacs), and the small molecule proteolysis inducers (SMPI); currently being tested in preclinical settings for therapeutic potential in selected types of cancer.
Proteolytic center of the cell	An area around the centrosome, enriched in proteasomes and other components of the UPS; multiple substrates of the UPS arrive to the p.c. by microtubule mediated transport, where they are ubiquitinated and degraded. When the UPS is overwhelmed, its degradative capacity is diminished leading to the accumulation of proteins forming a structured aggregate, or aggresome.
Proto-oncogene protein c-myc-transcription factor	is the product of v-myc myelocytomatosis viral oncogene homolog, and is a multifunctional, nuclear phosphoprotein that plays a role in cell cycle progression, apoptosis and cellular transformation. It functions as a transcription factor that regulates transcription of specific target genes.
PTEN (protein and tensin homolog	encodes a protein-and lipid phosphatase that controls PI3K (phosphatidylinositol-3-kinase) cascade and also intervenes in cell

deleted on chromosome 10)/ MMAC1 (mutated in multiple advanced cancers-1)	cycle regulation, DNA repair, apoptosis, senescence, and inhibition of angiogenesis
PTTG (pituitary tumor transforming gene)	located on chromosome 5q33, encodes human securin that blocks sister chromatid separation until the beginning of anaphase. Its over-expression can cause aneuploidy.
Purkinje cells	Inhibitory neurons in the cerebellum that use GABA γ-aminobutyric acid) as their neurotransmitter.Their cell bodies are situated beneath the molecular layer, and their dendrites branch extensively in this layer. Their axons project into the underlying white matter, and they provide the only output from the cerebellar cortex.
Reactive oxygen species (ROS).	Highly reactive oxygenbased molecules with an unpaired electron in their outer orbital that are capable of damaging proteins, lipids and nucleic acids. Examples include hydrogen peroxide and hydroxyl radicals.
Reperfusion	the restoration of blood flow to an organ or tissue.
Retrotranslocation	also called dislocation. A process by which proteins destined for degradation by cytoplasmic 26S proteasomes (ERAD) are extracted from the ER in a direction opposite to their insertion during protein synthesis (translocation). Retrotranslocation occurs either through the same Sec61 channel used for translocation, or through a specialized channel composed of derlins, Doa10 and other proteins.
RING-FINGER	The RING (really interesting new gene) consensus sequence is: CX2CX(9–39)CX(1–3)HX(2–3)C/HX2CX(4–8)CX2C. The cysteines and histidines represent metal binding sites. The first, second, fifth and sixth of these bind one zinc ion and the third, fourth, seventh and eighth bind the second zinc ion.
Ring-finger proteins	A family of proteins that are structurally defined by the presence of the zinc-binding RING-finger motif. Many RING-finger proteins are ubiquitin ligases or subunits thereof.
RNA interference	A process by which small interfering RNAs induce cleavage of specific mRNAs within cells inducing a functional konockdown of a specific gene product. While RNAi is an ancient mechanism which evolved as a response against certain viruses and retrotransposomes, it can be exploited for experimental purposes or for therapy.
Roundabout	A transmembrane receptor for the slit family of ligands involved in mediating axon guidance.
Scf ubiquitin-ligase complex	A multisubunit E3 ubiquitin ligase, which is composed of Skp1, cullin-1 protein, F-box protein, and Rbx1/Roc-1 RINGfinger protein. The F-box protein is the substrate recruiting factor.

Schwann cell:	Glial cells of the peripheral nervous system, responsible for myelin formation and other supportive functions.
Semaphorin 1A	A transmembrane member of the semaphorin family of molecules which may have an attractive effect on growing axons during development.
Semaphorin 3A	A secreted member of the semaphorin family of molecules implicated in having attractive and repulsive effect on growing axons during development.
Sequestosome 1	a protein with seven structural motifs: an SH2 domain that binds the tyrosine kinase $p56^{lck}$, an acidic interaction domain (AID) that binds the atypical PKC ζ, a ZZ type ZINC finger that binds the receptor interactive protein (RIP) involved in TNFα-induced apoptosis, a binding site for the RING-finger protein tumor necrosis factor receptor-associated factor 6 (TRAF6), two PEST sequences and a UBA domain that binds polyubiquitin chains
SKP2 (S-phase kinase interacting protein 2 / CDK2/Cyclin A-associated protein p45)	F box protein, gene located on chromosome 5p13, recognizes protein substrates (e.g.p27) for ubiquitination by SCF complex/ ubiquitin ligase.
SMAD Small mothers against decapentaplegic	Intracellular mediators of the BMP signalling pathway.
Smad ubiquitin regulatory factor	Regulate BMP singalling pathway by targeting SMADs for ubiquitin-mediated proteolysis.
Small heat shock proteins	sHsps have a molecular mass of less than 40 kDa and assemble into large, oligomeric structures that resemble a hollow ball.All sHSPs contain a conserved, C-terminal α-crystallin domain of about 100 residues that mediates oligomeric assembly. Similar to HSP90 chaperones, sHSPs transiently interact with and stabilize misfolded substrates, conceivably until the HSP70/HSP40 system can actively refold them.
Small inhibitory rna	A small RNA molecule that interferes with normal RNA processing, causing rapid degradation of the endogenous RNA and thereby precluding translation. This provides a simple way of studying the effects of the absence of a gene product in simple organisms and in cells.
Snaap	proteasomal activity cleaving after small neutral amino acids
Somatosensory:	The perception of sensory (such as mechanical, thermal and chemical) stimuli.
Spherical and annular oligomers	Metastable structures observed in many amyloid-forming proteins that might be on a pathway to fibril formation. These structures have been proposed to be the principal toxic entities that mediate

	neuronal dysfunction.
Spinobulbar muscular atrophy (SMBA):	A polyglutamine tract disorder that manifests as a neuromuscular disease due to an expanded polyglutamine tract in the androgen receptor.
spinocerebellar ataxia, type III	See: Machado-Joseph Disease.
Spinocerebellar Ataxias (scas):	A group of dominantly inherited diseases linked by the presence of a polyglutamine repeat in the relevant protein. They are of predominantly late-onset and may be subdivided based on clinical features and genetic mapping.
STAM (signal-transducing adaptor molecule)	STAM interacts with Hrs and may therefore be involved in endocytosis/vesicular transport.
Straight filaments (sfs)	straight filaments are found in Alzheimer's disease and other tauopathies such as the corticobasal degeneration and the progressive supranuclear palsy. They correspond to a specific aggregation of tau protein (4R tau).
Stress activated Protein Kinases (SAPK)	Protein kinases are enzymes that modify other proteins by adding phosphate groups to them (phosphorylation), changing their function radically. About thirty percent of proteins can be modified by kinases. Disregulated kinase activity is the root cause of many diseases, especially cancers, as kinases regulate many aspects of cell growth, movement, and apoptosis. Kinase-inhibiting drugs are being developed to treat several diseases. Kinases are stress-activated when specific events such as DNA damage or an overload of Ca2+ ions is detected. There are also kinases that are activated only by stress, referred to as c-Jun N-terminal kinases, or JNKs. These stress activated protein kinases respond to stress stimuli like cytokines, ultrafiolet radiation, heat schock, and osmotic shock. They're also involved in cell differentiation and apoptosis.
Substantia nigra:	a dark band of gray matter deep within the brain where pigmented cells manufacture the neurotransmitter dopamine for movement control. Degeneration of cells in this region lead to the neurologic movement disorder PD.
Sumo (Sumoylation)	Small Ubiquitin-related Modifier or SUMO proteins are a family of small proteins, most are around 100 amino acids in length and 12 kDa in mass, that are covalently attached to and detached from other proteins in cells to modify their function, using a mechanism analogous to, but distinct from, ubiquitin. The exact length and mass varies between SUMO family members and depends on which organism the protein comes from. In contrast to ubiquitin, SUMO is not used to tag proteins for degradation. Mature SUMO

	is produced when the last four amino acids of the C-terminus have been cleaved off. Sumoylation is a post-translational modification involved in various cellular processes, such as nuclear-cytosolic transport, transcriptional regulation, apoptosis, protein stability, response to stress, and progression through the cell cycle.
Suppressor of deltex	A Drosophila E3 ubiquitin ligase that regulates the Notch signalling pathway.
Suppressors of cytokine signaling (SOCS)	proteins comprise a family of cytoplasmic negative feedback regulators of cytokine signaling. These proteins inhibit JAK kinases activated by numerous cytokine receptors.
Suspended paw elevation time:	Behavioural measurement – test for cold allodynia in neuropathic animals – recording suspension of the injured paw from 4°C water over several seconds following immersion of the paws.
Symptomatic epilepsy	epilepsy with an identifiable cause, such as a brain tumor, brain trauma, or some other neurological disorder.
Synapse	Region of contact between the axon of the pre-synaptic neuron and the dendrite of the post-synaptic neuron. Synapses are specialized structures that create a microdomain between cells. This allows for the localized release of neurotransmitter from the pre-synaptic axon, resulting in the interaction of neurotransmitter with the neurotransmitter receptors present on the surface of the dendrite.
Tau tubulin kinase (ttk)	TTK can phosphorylate serine, threonine and tyrosine hydroxyamino acids. This kinase is associated with cell proliferation and was shown to phosphorylate tau protein.
Tauopathies	Neurodegenerative disorders involving deposition of abnormal tau protein isoforms in neurons and glial cells in the brain. Pathological aggregations of tau proteins are sometimes associated with mutation of the tau gene on chromosome 17 in patients with FTDP-17.
T-l (trypsin-like activity)	Proteasomal activity cleaving after basic residues and associated to the β2 subunit
Tnf-receptor family	Members of this family function as trimers and multimers of trimers, and can trigger proliferation, survival, differentiation or death. A subfamily that comprises the death receptors Fas/CD95 and TNF-R1, as well as some other members of this family, contains a cytoplasmic region — the death domain — which is essential for inducing apoptosis. However, at the same tiem TNF-R1 can also elicit a prosurvival and proinflammatory response, activating NF-κB.
Tnf-receptor-associated factors (trafs)	These are adaptor proteins for various cell-surface receptors.Most TRAFs encode a RING-finger motif at their amino-terminus; in the case of TRAF2 and TRAF5, the RING-finger is required for NF-κB activation.

TPR (tetratricorepeat motif)	The TPR domain is a 34-residue helix-turn-helix motif that facilitates protein–protein interactions. The TPRs of HOP, the cyclophillins and CHIP bind their cognate chaperones at the same C-terminal EEVD motif. Hip's TPRs enable it to bind the ATPase domain of Hsp70, but not the EEVD site.
Trail (tnf-related apoptosis inducing ligand)	This induces apoptosis preferentially in transformed cells. In contrast to other death-inducing ligands, TRAIL is expressed in a wide range of tissues.
Transport mechanism of substrate	Mechanisms responsible for the transport of substrate molecules within the 20S proteolytic chamber of the proteasome. Substrates enter the proteasome core particle in a partially unfolded or ubiquitinated state. The substrate composition influences the interaction with the external alpha ring of the 20S and/or with the regulatory caps subunits (19S, PA28). The rate at which substrates are degraded within the 20S can strongly be influenced by the transport mechanism. The forces defining the transport mechanism are complex and still largely unknown.
Transporters associated with antigen processing (TAP proteins)	are proteins of the endoplasmic reticulum responsible for the transport of cytosolic peptides to the lumen of ER. This process uses energy from ATP degradations and is necessary for the loading of peptide antigens into grooves of MHC class I molecules.
Tric/CCT	Group II chaperonins, such as TCP-1 ring complex in the eukaryote
Tubulin	is the protein which makes up microtubules. Microtubules are assembled from dimers of α- and β-tubulin. Each of these subunits has three domains. γ-tubulin is important in the nucleation and polar orientation of microtubule. Tubulin binds GTP and assembles onto the (+) ends of microtubules in the GTP-bound state. Once assembled into microtubules, it hydrolyzes GTP into GDP. The GDP-bound form of tubulin will disassemble from the tip of a microtubule, though it will not spontaneously fall out of the middle. This GTP cycle is essential for the dynamic instability of the microtubule. Tubulin was long thought to be specific to eukaryotes. Recently, however, the prokaryotic cell division protein FtsZ was shown to be evolutionarily related to tubulin. Delta and epsilon tubulin have been found to localize at centrioles and may play a role in forming the mitotic spindle during mitosis. Alpha and Beta Tubulins are proteins that have a molecular weight of approximately 55 kiloDaltons (kDa) each.
Tubulin associated unit (tau)	Tau proteins are mainly expressed in neurons (6 isoforms in the central nervous system) where they act on the polymerisation and stability of Microtubules. These proteins belong to the family of Microtubule associated proteins (MAPs).
Tumor suppressor	regulates cell cycle, specifically the transition from G0 to G1. It

protein p53	has low levels in normal cells and high levels in tumor cells. It contains DNA-binding, oligomerization and transcription activation domains. It binds as a tetramer to a p53-binding site and activates downstream genes inhibiting growth and/or invasion, functioning as a tumor suppressor. Mutants of p53, frequent in many human cancers, fail to bind the consensus DNA binding site, and hence cause loss of tumor suppressor activity. Alterations of the TP53 gene occur not only as somatic mutations in human malignancies, but also as germline mutations.
Tumour-necrosis factor-α (tnf-α)	A prototypic member of a family of cytokines that interact with several receptors, among them receptors that are responsible for eliciting apoptosis.
Turban tumor syndrome.	Also called Familial cylindromatosis. A genetic syndrome in which numerous benign tumors of skin adnexa (such as the sweat glands) develop, principally on the head and neck. This disorder is inherited in an autosomal manner and is caused by mutation of the CYLD gene on chromosome 16q12-q13. Mutation of CYLD has been likened to having faulty brakes on a car. Instead of a pileup of cars, a pileup of cells results. Topical application of aspirin, another type of brake on cell proliferation, may possibly be useful. the turban tumor syndrome.
Ubb+1	Mutant ubiquitin. UBB+1 is a mutant form of ubiquitin that lacks the C-terminal Gly of wild type ubiquitin and instead has a 19 amino-acids extension. This mutant can itself be ubiquitinated but is unable to bind other proteins. UBB+1 is a powerful inhibitor of the proteasome activity when its intracellular concentration reaches a certain threshold.
Ubiquitin	a heat stable low molecular mass (~8 kDa) protein formed by 76 amino acids, largely preserved during the phylogenesis, which is covalently attached through isopeptide bonds to substrate proteins (ubiquitination), often forming multiple adducts in form of polyubiquitin chains, serving in form of oligomers and in an ATP dependent manner binds the protein destined to destruction in the ubiquitin proteolytic system.
Ubiquitin activating enzyme (UBA)	See: E1.
Ubiquitin and proteasome-dependent proteolytic system (UPS)	is the major nuclear and cytoplasmic proteolytic system which involves the degradation of proteins by 20S and 26S proteasomes, usually after previous ubiquitination. It requires energy in form of ATP for both ubiquitination and degradation by the 26S proteasomes.
Ubiquitin conjugating enzyme (UBC)	See: E2.

Ubiquitin C-terminal hydrolases Uch	Ubiquitin C-terminal hydrolases are cysteine proteases that generate free ubiquitin mainly from ubiquitin adducts and ubiquitin precursors.
Ubiquitin fusion degradation (UFD)	a subdivision of the UPS in the cytosol, where uncleavable ubiquitin fused to other proteins destabilizes these proteins and induces their ubiquitination and proteasomal degradation.
Ubiquitin ligase	See: E3.
Ubiquitin receptors	Class of proteins that contain ubiquitin-binding domains that are known to associate directly with either mono- or polyubiquitin signals in partner proteins or proteasomal substrates rather than participate in ubiquitylation reactions. Known ubiquitin-binding domains include the ubiquitin-interacting motif (UIM), the ubiquitin-associated (UBA) domain, the ubiquitin-conjugating enzyme variant (UEV), VHS (Vps 27, Hrs, STAM), NZF (Npl4 zinc finger), and the polyubiquitin-associated zinc finger (PAZ) among others.
Ubiquitin-associated (UBA) domain	Small domain of about 40-55 residues whose three-dimensional structure is a compact three-helix bundle of low sequence conservation. UBA domains have a relatively high affinity for Lys-48-linked polyubiquitin chains and UBA domain proteins have been implicated in the ubiquitin fusion degradation pathway involving Lys-29-linked ubiquitin chains. Affinity of UBA domain proteins for monoubiquitin is ~10-500 μM whereas affinity towards polyubiquitin chains is between 2 and 4 orders of magnitude higher.
Ubiquitination (ubiquitylation)	The process of covalent attachement of ubiquitin to other proteins, which is achieved either through an isopeptide bond between an ε-amino group of Lys of the substrate and the C-terminal Gly of ubiquitin or through a peptide bond between N-terminal amino group of the substrate and the C-terminal Gly of ubiquitin. Monoubiquitination is the attachement of a single ubiquitin moiety to the substrate, multiubiquitination is the attachement of multiple ubiquitin moieties to different sites on the substrate, while polyubiquitination is the assembly of polyubiquitin chains, where one ubiquitin is conjugated to another through isopeptide linkages involving one of the 6 Lys present in the ubiquitin molecule. Ubiquitination is carried on by the E1-E2-E3 cascade of enzymes.
Ubiquitin-conjugating domain (ubc)	The ~16-kDa ubiquitin conjugating domain of E2s harbours the active-site cysteine residue that is required for the formation of a thioester-linked E2-ubiquitin complex.
Ubiquitin-interacting motif (UIM)	Composed of approximately 20 amino acids, UIMs are characterized by "LALAL" motifs within the sequence of the ubiquitin-binding sites of subunit S5a of the 26S proteasome. UIMs are present in diverse protein families, including proteins

	involved in ubiquitylation and ubiquitin metabolism, proteolysis, and endocytosis. They bind either mono- or polyubiquitin with affinities ranging between 100 and 400 μM.
Ubiquitin-like (UBL) domain	This domain is found at or near the N-terminus of proteins and is defined by a stretch of 45-80 residues with significant sequence homology and very similar three-dimensional structure to ubiquitin. UBL domains bind to the 26S proteasome, where they may dock on the S1, S2 or S5a subunits of the 19S regulatory complex. Most UBL-containing proteins have functions related to the ubiquitin-proteasome system and may promote the assembly of proteasomal supercomplexes or degradasomes.
Ubiquitin-specific proteases (usp)	See: deubiquitinating enzymes.
UBX domain	The UBX domain comprises ~80-residue C-terminal modules structurally related to ubiquitin in spite of low sequence conservation with the latter. UBX domain proteins can be grouped into five evolutionarily conserved families represented by the human cofactor p47, and the Y33K, FAF-1, UBXD1 and Rep-8 proteins. The UBX domain has been proposed to function as a general binding module for valosin-containing protein/cdc48, an hexameric segregase that dissociates protein complexes.
Ump1	short-lived protein identified as a chaperone necessary for a correct proteasome assembly and maturation
Unfolded protein response (UPR)	A concerted cellular reaction to the presence of misfolded proteins in the ER. It involves three main branches, depending on the activity of three ER transmembrane proteins: PEK/PERK, IRE1 and ATF6. The initial response involves activation of PEK/PERK, which phosphorylates cytsosolic eIF2α, leading to translational attenuation of most gene products accompanied by specific translational activation of a specialized array of gene products characterized by the presence of IRES or 5' upstream alternative ORFs. Subsequently activation of the IRE1 endonuclease leads to the alternative cytosolic splicing of the transcription factor XBP1, while transit to the Golgi of ATF6 leads to the proteolytic cleavage of its transactivation domain. XBP1 and ATF6 induce the transcription of an array of genes which include ER chaperones, proteins involved in ERAD, and other aspects of tehs ecretory pathway. UPR has therefore a cytoprotective function, however when overactive, it can lead to apoptosis through a pthway involving activation of caspase 12 in mice or its functional homolog in humans, caspase 4.
UT3 domain	Region corresponding to amino acids 1-211 of the sequence of Ufd1. Structurally, it contains a double-ψ β-barrel fold and a αβ roll that resemble the N-terminal region of valosin-containing

	protein (VCP). The UT3 domain of Ufd1 contains two non-overlapping ubiquitin-binding regions located in the N-terminal double-ψ β-barrel domain: One that binds monoubiquitin with low affinity (K_d of 1-2 mM), and a region that binds polyubiquitin in Lys-48 linkage.
VCP (valosin-containing protein)	An essential, ubiquitous and abundant ATPase of the AAA family of 97 kDa, forming a ring-shaped homohexamer of a 97 kDa . VCP is able to bind polyubiquitin chains and misfolded proteins. It associates with over 30 different proteins forming complexes involved in diverse cellular activities, including membrane fusion, mitosis, apoptosis, nucleotide repair and ubiquitin-dependent degradation of proteins. VCP functions in the UPS often in association with the Ufd1-Npl4 heterodimer. In particular, VCP-Ufd1-Npl4 is believed to participate in the retrotranslocation of proteins from the ER and their delivery to the 26S proteasome (ERAD). Also known as p97, Ter94 (in Drosophila) and Cdc48 (in yeast).
VHS domain	The VHS (Vps-27, Hrs and STAM) domain is a ~150-residue long domain that contains eight α-helices (α1 through α8) and a C-terminal extension, and are often found at the N-terminus of proteins involved in membrane targeting/cargo recognition along the endocytic and secretory pathways. The eight α-helices fold into a curved double-layer superhelical structure with concave and convex surfaces. The first two helical hairpins (α1-α2 and α3-α4) within the VHS domain resemble HEAT repeats, whereas the third repeat, consisting of helices α5, α6 and α7, is reminiscent of the three-helix ARM repeat. VHS domains interact with sorting receptors and sorting signals within the cargo such as ubiquitin.
Wallerian degeneration	The degeneration of an axon distal to a site of injury, which begins to occur at about 1.5 days after a lesion. Wallerian degeneration is delayed approximately tenfold in rats or mice that carry the dominantly acting slow Wallerian degeneration (Wld^S) gene.
WD40 repeat	Minimally conserved domain of approximately 40-60 amino acid residues characterized by a glycine-histidine (GH) dipeptide 11 to 24 residues from its N-terminus, and separated approximately 40 amino acids from the end tryptophan-aspartic acid (WD) dipeptide. WD40 proteins are speculated to form a circularized β propeller structure that functions as a scaffold to which proteins bind through coordination with residues on the top and bottom surfaces of the propeller to constitute multiprotein complex assemblies. Most WD40-repeat proteins contain a cluster of at least 4 repeats and participate in many essential biological functions ranging from signal transduction, RNA synthesis and processing, vesicular trafficking, and cytoskeletal assembly to cell cycle regulation and

	apoptosis.
Wishful thinking	A Drosophila type II BMP receptor involved in retrograde signalling during synaptic development.
X-linked Spinal Bulbar Muscular Atrophy	Also called Kennedy Disease is a rare, slowly progressive muscular disorder that affects males only and is inherited as an X-linked genetic trait. Uncontrollable twitching (fasciculations) followed by weakness and wasting of the muscles becomes apparent some time after the age of fifteen. The muscles of the face, lips, tongue, mouth, throat, vocal chords, trunk and limbs may be affected. Very large calves may also be found in some patients with this disorder. Kennedy disease is caused by a mutation in the androgen receptor (AR) gene. Androgen insensitivity leads to abnormal swelling of the breasts (gynecomastia), small testes and infertility.
XPC binding domain	Region in Rad23 that binds to Rad4, the yeast homolog of Xeroderma pigmentosum group C complementing protein (XPC) promoting the assembly of a multiprotein nucleotide excision repair (NER) complex at the site of DNA lesions. The XPC domain of Rad23 also binds cytosolic peptide:N-glycanase allowing the formation of a degradasome for the turning over of N-linked glycoproteins dislocated from the endoplasmic reticulum.

Appendix I. Nomenclature of proteasome subunits
[Reprinted with the permission of BIOMOL International LP (www.biomol.com)]

Nomenclature						Gene		1° Acc. #	Seq. length	MW
Baumeister, *et al.* [1]	'Old' human	Coux *et al.* [2]	Groll *et al.* [3]	Miscellaneous	UniProtKB [4]	Human	Yeast (S.c.)	(human)	(amino acids)	(Da)
20S α-type subunits										
α1	iota	Pro-α6	α1_sc	Pros27, p27k, C7, Prs2, Y8, Prc2, Scl1	α6	PSMA6	PRS2	P60900	246	27399
α2	C3	Pro-α2	α2_sc	Pre8, Prs4, Y7	α2	PSMA2	PRS4	P25787	233	25767
α3	C9	Pro-α4	α3_sc	Pre9, Prs5, Y13	α4	PSMA4	PRS5	P25789	261	29484
α4	C6	Pro-α3	α4_sc	XAPC-7, Pre6	α7	PSMA7	PRE6	O14818	248	27887
α5	zeta	Pro-α1	α5_sc	Pup2, Doa5	α5	PSMA5	PUP2	P28066	241	26411
α6	C2	Pro-α5	α6_sc	nu, Pros30, p30k, Pre5	α1	PSMA1	PRE5	P25786	263	29556
α7	C8	Pro-α7	α7_sc	Pre10, Prs1, C1, Prc1	α3	PSMA3	PRS1	P25788	254	28302
20S β-type subunits										
β1	Y	Pro-β3	β1_sc	delta, Lmp9, Pre3	β6	PSMB6	PRE3	P28072	239/205	25358/21904
β1i	Lmp2	Pro-β3		Ring12	β9	PSMB9	-	P28065	219/199	23264/21276
β2	Z	Pro-β2	β2_sc	Lmp19, MC14, Pup1	β7	PSMB7	PUP1	Q99436	277/234	29965/25218
β2i	MECL-1	Pro-β2		Lmp10	β10	PSMB10	-	P40306	273/234	28936/24648
β3	C10	Pro-β6	β3_sc	theta, Pup3	β3	PSMB3	PUP3	P49720	205	22949
β4	C7	Pro-β4	β4_sc	Pre1, C11	β2	PSMB2	PRE1	P49721	201	22836
β5	X	Pro-β1	β5_sc	epsilon, Lmp17, MB1, Pre2, Doa3, Prg1	β5	PSMB5	PRE2	P28074	208/204	22897/22458
β5i	Lmp7	Pro-β1		Ring10, Y2, C13	β8	PSMB8	-	P28062	276/204	30354/22660
β6	C5	Pro-β5	β6_sc	gamma, Pre7, Prs3, C5, Pts1	β1	PSMB1	PRS3	P20618	241	26489
β7	N3	Pro-β7	β7_sc	beta, Pros26, Pre4	β4	PSMB4	PRE4	P28070	264/219	29192/24380

Appendix I. Nomenclature of proteasome subunits (cont.d)

19S Regulator (19S cap, PA700)								
Nomenclature				Gene		1° Acc #	Seq. length	MW
Finley, *et al.* [5]	Dubiel *et al.* [6]	Miscellaneous	UniProtKB [4]	Human	Yeast (S.c.)	(human)	(amino acids)	(Da)
19S (PA700) regulator ATPase subunits								
Rpt1	S7	p48, Mss1, Yta3, Cim5	Subunit 7	PSMC2	CIM5	P35998	432	48503
Rpt2	S4	p56, Yhs4, Yta5, Mts2	Subunit 4	PSMC1	YTA5	P62191	440	49185
Rpt3	S6b	p48, Tbp7, Yta2, Ynt1, MS73	Subunit 6b	PSMC4	YTA2	P43686	418	47336
Rpt4	S10b	p42, Sug2, Pcs1, Crl13, CADp44	Subunit 10b	PSMC6	SUG2	P62333	389	44173
Rpt5	S6a	p50, Tbp1, Yta1	Subunit 6a	PSMC3	YTA1	P17980	439	49204
Rpt6	S8	p45, Trip1, Sug1, Cim3, Crl3, Tby1, Tbp10, m56	Subunit 8	PSMC5	SUG1	P62195	406	45626
19S (PA700) regulator non-ATPase subunits								
Rpn1	S2	p97, Trap2, Nas1, Hrd2, Rpd1, Mts4	Subunit 2	PSMD2	HRD2	Q13200	908	100200
Rpn2	S1	p112, Sen3	Subunit 1	PSMD1	SEN3	Q99460	953	105836
Rpn3	S3	p58, Sun2	Subunit 3	PSMD3	SUN2	O43242	534	60978
Rpn4		Son1, Ufd5			RPN4	Q03465(Sc)	531	60153
Rpn5		p55, Nas5	Subunit 12	PSMD12	YDL147W	O00232	455	52773
Rpn6	S9	p44.5, Nas4/6?	Subunit 11	PSMD11	YDL097C	O00231	421	47333
Rpn7	S10a	p44, HUMORF07	Subunit 6	PSMD6		Q15008	389	45531
Rpn8	S12	p40, Mov-34h, Nas3	Subunit 7	PSMD7	YOR261C	P51665	324	37025
Rpn9	S11	p40.5, Les1, Nas7	Subunit 13	PSMD13		Q9UNM6	376	42918
Rpn10	S5a	p54, ASF1, Sun1, Mcb1, Mbp1	Subunit 4	PSMD4	SUN1	P55036	377	40736
Rpn11	S13	Poh1, Mpr1, Pad1h	Subunit 14	PSMD14	MPR1	O00487	310	34577
Rpn12	S14	p31, Nin1, Mts3	Subunit 8	PSMD8	NIN1	P48556	257	30005
Rpn13		YLR421C			RPN13	O13563(Sc)	156	17902
	S5b	p50.5	Subunit 5	PSMD5		Q16401	503	56065
	S15	p27-L	Subunit 9	PSMD9	NAS2	O00233	223	24654
		p28, Gankyrin, Nas6	Subunit 10	PSMD10		O75832	226	24428

11S Activator (11S cap, PA28)								
Nomenclature					Gene	1° Acc #	Seq. length	MW
Dubiel, *et al.* [7]	Ma, *et al.* [8]	Realini, *et al.* [9]	Kandil, *et al.* [10]	UniProtKB [4]	(human)	(human)	(amino acids)	(Da)
11Sα	PA28α	REGα		Subunit 1	PSME1	Q06323	249	28723
11Sβ	PA28β	REGβ		Subunit 2	PSME2	Q9UL46	238	27230
11Sγ	PA28γ	REGγ	Ki antigen	Subunit 3	PSME3	P61289	254	29506

REFERENCES

[1] Baumeister W, Walz J, Zuhl F, Seemuller E: The proteasome: paradigm of a self-compartmentalizing protease. *Cell* 1998, 92:367-380.

[2] Coux O, Nothwang HG, Silva P, I, Recillas TF, Bey F, Scherrer K: Phylogenic relationships of the amino acid sequences of prosome (proteasome, MCP) subunits. *Mol Gen Genet* 1994, 245:769-780.

[3] Groll M, Ditzel L, Lowe J, Stock D, Bochtler M, Bartunik HD, Huber R: Structure of 20S proteasome from yeast at 2.4 A resolution. *Nature* 1997, 386:463-471.

[4] Apweiler R, Bairoch A, Wu C. The Universal Protein Resource. UniProtKB 2006 September 15 [cited 2006 Sep 15];Available from: URL: http://www.ebi.uniprot.org/index.shtml

[5] Finley D, Tanaka K, Mann C, Feldmann H, Hochstrasser M, Vierstra R, Johnston S, Hampton R, Haber J, Mccusker J, Silver P, Frontali L, Thorsness P, Varshavsky A, Byers B, Madura K, Reed SI, Wolf D, Jentsch S, Sommer T, Baumeister W, Goldberg A, Fried V, Rubin DM, Toh-e A,.: Unified nomenclature for subunits of the Saccharomyces cerevisiae proteasome regulatory particle. *Trends Biochem Sci* 1998, 23:244-245.

[6] Dubiel W, Ferrell K, Rechsteiner M: Subunits of the regulatory complex of the 26S protease. *Mol Biol Rep* 1995, 21:27-34.

[7] Dubiel W, Pratt G, Ferrell K, Rechsteiner M: Purification of an 11 S regulator of the multicatalytic protease. *J Biol Chem* 1992, 267:22369-22377.

[8] Ma CP, Slaughter CA, DeMartino GN: Identification, purification, and characterization of a protein activator (PA28) of the 20 S proteasome (macropain). *J Biol Chem* 1992, 267:10515-10523.

[9] Realini C, Jensen CC, Zhang Z, Johnston SC, Knowlton JR, Hill CP, Rechsteiner M: Characterization of recombinant REGalpha, REGbeta, and REGgamma proteasome activators. *J Biol Chem* 1997, 272:25483-25492.

[10]Kandil E, Kohda K, Ishibashi T, Tanaka K, Kasahara M: PA28 subunits of the mouse proteasome: primary structures and chromosomal localization of the genes. *Immunogenetics* 1997, 46:337-344.

INDEX

6

A

B

C

D

E

F

G

H

J

K

L

M

N

O

P

Q

R

S

T

U

V

W

X

Y

Z